MW00837584

BIOMARKERS IN TOXICOLOGY

BIOMARKERS IN TOXICOLOGY

SECOND EDITION

Edited by

RAMESH C. GUPTA, DVM, MVSc, PhD, DABT, FACT, FACN, FATS
Professor and Head, Toxicology Department
Breathitt Veterinary Centre
Murray State University
Hopkinsville, Kentucky
United States

Academic Press is an imprint of Elsevier
125 London Wall, London EC2Y 5AS, United Kingdom
525 B Street, Suite 1650, San Diego, CA 92101, United States
50 Hampshire Street, 5th Floor, Cambridge, MA 02139, United States
The Boulevard, Langford Lane, Kidlington, Oxford OX5 1GB, United Kingdom

Library of Congress Cataloging-in-Publication Data
A catalog record for this book is available from the Library of Congress

British Library Cataloguing-in-Publication Data
A catalogue record for this book is available from the British Library

ISBN: 978-0-12-814655-2

For information on all Academic Press publications visit our website at https://www.elsevier.com/books-and-journals

Publisher: Mica H. Haley
Acquisition Editor: Erin Hill-Parks
Editorial Project Manager: Kristi Anderson
Production Project Manager: Mohanapriyan Rajendran
Cover Designer: Victoria Pearson

Typeset by TNQ Technologies

Printed in the United States of America
Last digit is the print number: 9 8 7 6 5 4 3 2 1

This book is dedicated to my daughter Rekha, wife Denise,
and parents the late Chandra and Triveni Gupta.

Contents

6. Mechanistic Toxicology Biomarkers in *Caenorhabditis elegans*

VANESSA A. FITSANAKIS, REKEK NEGGA, AND HOLLY E. HATFIELD

7. Potential of Small Animals in Toxicity Testing: Hope From Small World

SHIWANGI DWIVEDI, SUMIT SINGH VERMA, CLINTON D'SOUZA, NIKEE AWASTHEE, ANURAG SHARMA, AND SUBASH CHANDRA GUPTA

8. Alternative Animal Toxicity Testing and Biomarkers

GOPALA KRISHNA, PRIYA A. KRISHNA, SARYU GOEL, AND KAVYA A. KRISHNA

9. Adverse Outcome Pathways and Their Role in Revealing Biomarkers

MAGDALINI SACHANA

II SYSTEMS TOXICITY BIOMARKERS

10. Central Nervous System Toxicity Biomarkers

LUCIO G. COSTA

11. Peripheral Nervous System Toxicity Biomarkers

TIRUPAPULIYUR V. DAMODARAN

12. Cardiovascular Toxicity Biomarkers

CSABA K. ZOLTANI

33. Biomarkers of Chemical Mixture Toxicity

ANTONIO F. HERNÁNDEZ, FERNANDO GIL, AND ARISTIDIS M. TSATSAKIS

34. Biomarkers of Toxic Solvents and Gases

SZABINA A. STICE

IV
BIOTOXINS BIOMARKERS

35. Freshwater Cyanotoxins

GURJOT KAUR

36. Mycotoxins

ROBERT W. COPPOCK AND MARGITTA M. DZIWENKA

37. Poisonous Plants: Biomarkers for Diagnosis

KIP E. PANTER, KEVIN D. WELCH, AND DALE R. GARDNER

V
PHARMACEUTICALS AND NUTRACEUTICALS BIOMARKERS

38. Biomarkers of Drug Toxicity and Safety Evaluation

ARTURO ANADÓN, MARÍA ROSA MARTÍNEZ-LARRAÑAGA, IRMA ARES, AND MARÍA ARÁNZAZU MARTÍNEZ

39. Risk Assessment, Regulation, and the Role of Biomarkers for the Evaluation of Dietary Ingredients Present in Dietary Supplements

SANDRA A. JAMES-YI, COREY J. HILMAS, AND DANIEL S. FABRICANT

40. Nutriphenomics in Rodent Models: Impact of Dietary Choices on Toxicological Biomarkers

MICHAEL A. PELLIZZON AND MATTHEW R. RICCI

VI
NANOMATERIALS AND RADIATION

41. Engineered Nanomaterials: Biomarkers of Exposure and Effect

ENRICO BERGAMASCHI, MARY GULUMIAN, JUN KANNO, AND KAI SAVOLAINEN

42. Biomarkers of Exposure and Responses to Ionizing Radiation

ROGER O. McCLELLAN

VII
CARCINOGENS BIOMONITORING AND CANCER BIOMARKERS

43. Biomonitoring Exposures to Carcinogens

SOFIA PAVANELLO AND MARCELLO LOTTI

44. Genotoxicity Biomarkers: Molecular Basis of Genetic Variability and Susceptibility

SZABINA A. STICE, SUDHEER R. BEEDANAGARI, SURYANARAYANA V. VULIMIRI, SNEHA P. BHATIA, AND BRINDA MAHADEVAN

45. Epigenetic Biomarkers in Toxicology

ANIRUDH J. CHINTALAPATI AND FRANK A. BARILE

46. Risk Factors as Biomarkers of Susceptibility in Breast Cancer

CAROLINA NEGREI AND BIANCA GALATEANU

47. Pancreatic and Ovarian Cancer Biomarkers

GEORGE GEORGIADIS, CHARALAMPOS BELANTIS, CHARALAMPOS MAMOULAKIS, JOHN TSIAOUSSIS, WALLACE A. HAYES, AND ARISTIDIS M. TSATSAKIS

48. Prostate Cancer Biomarkers

CHARALAMPOS MAMOULAKIS, CHARALAMPOS MAVRIDIS, GEORGE GEORGIADIS, CHARALAMPOS BELANTIS, IOANNIS E. ZISIS, IORDANIS SKAMAGKAS, IOANNIS HERETIS, WALLACE A. HAYES, AND ARISTIDIS M. TSATSAKIS

VIII

DISEASE BIOMARKERS

49. Biomarkers of Alzheimer's Disease

JASON PITT

50. Biomarkers of Parkinson's Disease

HUAJUN JIN, ARTHI KANTHASAMY, VELLAREDDY ANANTHARAM, AND ANUMANTHA G. KANTHASAMY

51. Biomarkers for Drugs of Abuse and Neuropsychiatric Disorders: Models and Mechanisms

PUSHPINDER KAUR MULTANI, NITIN SAINI, RAVNEET KAUR, AND VANDANA SAINI

52. Osteoarthritis Biomarkers

RAMESH C. GUPTA, AJAY SRIVASTAVA, RAJIV LALL, AND ANITA SINHA

53. Pathological Biomarkers in Toxicology

MELITON N. NOVILLA, VINCENT P. MEADOR, STEWART B. JACOBSON, AND JESSICA S. FORTIN

54. Oral Pathology Biomarkers

ANUPAMA MUKHERJEE

IX
SPECIAL TOPICS

55. Biomarkers of Mitochondrial Dysfunction and Toxicity

CARLOS M. PALMEIRA, JOÃO S. TEODORO, RUI SILVA, AND ANABELA P. ROLO

56. Biomarkers of Blood–Brain Barrier Dysfunction

REKHA K. GUPTA AND RAMESH C. GUPTA

57. Biomarkers of Oxidative/Nitrosative Stress and Neurotoxicity

DEJAN MILATOVIC, SNJEZANA ZAJA-MILATOVIC, RAMESH C. GUPTA

58. Cytoskeletal Disruption as a Biomarker of Developmental Neurotoxicity

ALAN J. HARGREAVES, MAGDALINI SACHANA, AND JOHN FLASKOS

59. MicroRNA Expression as an Indicator of Tissue Toxicity and a Biomarker in Disease and Drug-Induced Toxicological Evaluation

GOPALA KRISHNA, SAURABH CHATTERJEE, PRIYA A. KRISHNA, AND RATANESH KUMAR SETH

60. Citrulline: Pharmacological Perspectives and Role as a Biomarker in Diseases and Toxicities

SHILPA N. KAORE AND NAVINCHANDRA M. KAORE

X

APPLICATIONS OF BIOMARKERS IN TOXICOLOGY

61. Analysis of Toxin- and Toxicant-Induced Biomarker Signatures Using Microarrays

SADIKSHYA BHANDARI AND MICHAEL A. LYNES

62. Biomarkers Detection for Toxicity Testing Using Metabolomics

DAVID J. BORTS

63. Transcriptomic Biomarkers in Safety and Risk Assessment of Chemicals 1125

DAVID T. SZABO AND AMY A. DEVLIN

64. Percellome Toxicogenomics Project as a Source of Biomarkers of Chemical Toxicity

JUN KANNO

65. Proteomics in Biomarkers of Chemical Toxicity

CHRISTINA WILSON-FRANK

66. Biomarkers for Testing Toxicity and Monitoring Exposure to Xenobiotics

JORGE ESTÉVEZ, EUGENIO VILANOVA, AND MIGUEL A. SOGORB

67. Biomarkers in Epidemiology, Risk Assessment and Regulatory Toxicology

A.M. FAN, P. COHN, S.H. YOU, AND P. LIN

Contributors

Arturo Anadón Department of Pharmacology and Toxicology, Faculty of Veterinary Medicine, Universidad Complutense de Madrid, Madrid, Spain

Vellareddy Anantharam Parkinson's Disorder Research Laboratory, Iowa Center for Advanced Neurotoxicology, Department of Biomedical Sciences, Iowa State University, Ames, IA, United States

Anthony E. Archibong Department of Microbiology, Immunology & Physiology, Meharry Medical College, Nashville, TN, United States

Irma Ares Department of Pharmacology and Toxicology, Faculty of Veterinary Medicine, Universidad Complutense de Madrid, Madrid, Spain

Adam D. Aulbach Charles River Laboratories, Mattawan, MI, United States

Nikee Awasthee Department of Biochemistry, Institute of Science, Banaras Hindu University, Varanasi, India

Aryamitra Banerjee Study Director, Toxicology Research Laboratory, University of Illinois at Chicago, Chicago, IL, United States

Leah D. Banks Department of Biochemistry, Cancer Biology, Neuroscience & Pharmacology, Meharry Medical College, Nashville, TN, United States

Frank A. Barile St. John's University, College of Pharmacy and Health Sciences, Department of Pharmaceutical Sciences, Toxicology Division, New York, United States

Sudheer R. Beedanagari Bristol Myers Squibb, New Brunswick, NJ, United States

Charalampos Belantis Department of Urology, University General Hospital of Heraklion, University of Crete, Medical School, Heraklion, Crete, Greece

Enrico Bergamaschi Laboratory of Toxicology and Occupational Epidemiology, Department of Public Health Science and Pediatrics, University of Turin, Italy

Sadikshya Bhandari Molecular and Cell Biology, University of Connecticut, Storrs, CT, United States

Sneha P. Bhatia Research Institute for Fragrance Materials, NJ, United States

Karyn Bischoff Cornell University, Diagnostic Toxicologist, New York State Animal Health Diagnostic Center, Ithaca NY, United States

David J. Borts Department of Veterinary Diagnostic and Production Animal Medicine, College of Veterinary Medicine, Iowa State University, Ames, IA, United States

Emily Brehm Department of Comparative Biosciences, University of Illinois at Urbana-Champaign, Urbana, IL, United States

Subash Chandra Gupta Department of Biochemistry, Institute of Science, Banaras Hindu University, Varanasi, India

Saurabh Chatterjee Department of Environmental Health Sciences, Arnold School of Public Health, University of South Carolina, Columbia, SC, United States

Catheryne Chiang Department of Comparative Biosciences, University of Illinois at Urbana-Champaign, Urbana, IL, United States

Anirudh J. Chintalapati St. John's University, College of Pharmacy and Health Sciences, Department of Pharmaceutical Sciences, Toxicology Division, New York, United States

P. Cohn New Jersey Department of Health (retired), Trenton, NJ, United States

Robert W. Coppock Toxicologist and Associates, Ltd, Vegreville, AB, Canada

Lucio G. Costa Department of Environmental and Occupational Health Sciences, University of Washington, Seattle, WA, United States; Department of Medicine & Surgery, University of Parma, Parma, Italy

Tirupapuliyur V. Damodaran Department of Biological and Biomedical Sciences, North Carolina Central University, Durham, NC, United States

Clinton D'Souza Division of Environmental Health and Toxicology, Nitte University Center for Science Education and Research (NUCSER), Deralakatte, India

Wolf-D. Dettbarn Vanderbilt University, Nashville, TN, United States

Amy A. Devlin US Food and Drug Administration, Silver Spring, MD, United States

Robin B. Doss Toxicology Department, Breathitt Veterinary Center, Murray State University, Hopkinsville, Kentucky, United States

Shiwangi Dwivedi Division of Environmental Health and Toxicology, Nitte University Center for Science Education and Research (NUCSER), Deralakatte, India

Margitta M. Dziwenka Faculty of Medicine & Dentistry, Division of Health Sciences Laboratory Animal Services, Edmonton, AB, Canada

Jorge Estévez Instituto de Bioingeniería, Universidad Miguel Hernández de Elche, Spain

Daniel S. Fabricant President and CEO, Natural Products Association, Washington D.C., United States

A.M. Fan California Environmental Protection Agency (retired), Oakland/Sacramento, CA, United States

Vanessa A. Fitsanakis Department of Pharmaceutical Sciences, Northeast Ohio Medical University, Rootstown, OH, United States

John Flaskos Laboratory of Biochemistry and Toxicology, Faculty of Veterinary Medicine, Aristotle University of Thessaloniki, Thessaloniki, Greece

Jodi A. Flaws Department of Comparative Biosciences, University of Illinois at Urbana-Champaign, Urbana, IL, United States

Swaran J.S. Flora National Institute of Pharmaceutical Education and Research, Raebareli, U.P., India

Sue M. Ford Department of Pharmaceutical Sciences, College of Pharmacy & Health Sciences, St. John's University, Jamaica, NY, United States

Jessica S. Fortin College of Veterinary Medicine, Michigan State University, East Lansing, MI, United States

Domniki Fragou Laboratory of Forensic Medicine & Toxicology, School of Medicine, Aristotle University of Thessaloniki, Greece

Shayne C. Gad Principal of Gad Consulting Services, Raleigh, NC, United States

Bianca Galateanu Department of Biochemistry and Molecular Biology, University of Bucharest, Bucharest, Romania

Dale R. Gardner United States Department of Agriculture, Agricultural Research Service, Poisonous Plant Research Laboratory, Logan, UT, United States

George Georgiadis Department of Urology, University General Hospital of Heraklion, University of Crete, Medical School, Heraklion, Crete, Greece

Fernando Gil Department of Legal Medicine and Toxicology, University of Granada School of Medicine, Granada, Spain

Saryu Goel Nonclinical Expert, Leesburg, VA, United States

Mary Gulumian Toxicology Research Projects NIOH, School of Pathology, University of the Witwatersrand South Africa

P.K. Gupta Director Toxicology Consulting Group, Former Principal Scientist and Chief Division of Pharmacology and Toxicology, IVRI, Advisor, World Health Organization (Geneva), Bareilly, India

Ramesh C. Gupta Toxicology Department, Breathitt Veterinary Center, Murray State University, Hopkinsville, Kentucky, United States

Rekha K. Gupta School of Medicine, University of Louisville, Louisville, KY, United States

Sharon Gwaltney-Brant Veterinary Information Network, Mahomet, IL, United States

Alan J. Hargreaves School of Science and Technology, Nottingham Trent University, Nottingham, United Kingdom

Kelly L. Harris Department of Biochemistry, Cancer Biology, Neuroscience & Pharmacology, Meharry Medical College, Nashville, TN, United States

Kenneth J. Harris Department of Biochemistry, Cancer Biology, Neuroscience & Pharmacology, Meharry Medical College, Nashville, TN, United States

Holly E. Hatfield Department of Biochemistry, University of Cincinnati Medical School, Cincinnati, OH, United States

Wallace A. Hayes University of South Florida College of Public Health, Tampa, FL, United States; Michigan State University, East Lansing, MI, United States

Ioannis Heretis Department of Urology, University General Hospital of Heraklion, University of Crete, Medical School, Heraklion, Crete, Greece

Antonio F. Hernández Department of Legal Medicine and Toxicology, University of Granada School of Medicine, Granada, Spain

Corey J. Hilmas Senior Vice President of Scientific and Regulatory Affairs, Natural Products Association, Washington D.C., United States

Darryl B. Hood College of Public Health, Ohio State University, Columbus, OH, United States

Pasi Huuskonen School of Pharmacy/Toxicology, Faculty of Health Sciences, University of Eastern Finland, Kuopio, Finland

Stewart B. Jacobson Shin Nippon Biomedical Laboratories USA, Ltd, Everett, WA, United States

Sandra A. James-Yi Associate Principle Scientist, Product Safety/Nutritional Toxicology, Mary Kay Inc., Addison, Texas, United States

Huajun Jin Parkinson's Disorder Research Laboratory, Iowa Center for Advanced Neurotoxicology, Department of Biomedical Sciences, Iowa State University, Ames, IA, United States

Jun Kanno Japan Bioassay Research Center, Japan Organization of Occupational Health and Safety, Hadano, Japan

Arthi Kanthasamy Parkinson's Disorder Research Laboratory, Iowa Center for Advanced Neurotoxicology, Department of Biomedical Sciences, Iowa State University, Ames, IA, United States

Anumantha G. Kanthasamy Parkinson's Disorder Research Laboratory, Iowa Center for Advanced Neurotoxicology, Department of Biomedical Sciences, Iowa State University, Ames, IA, United States

Shilpa N. Kaore Raipur Institute of Medical Sciences, Raipur, India

Navinchandra M. Kaore Raipur Institute of Medical Sciences, Raipur, India

Bhupendra S. Kaphalia Department of Pathology, University of Texas Medical Branch, Galveston, TX, United States

Vesa Karttunen School of Pharmacy/Toxicology, Faculty of Health Sciences, University of Eastern Finland, Kuopio, Finland

Gurjot Kaur Human and Environmental Toxicology, Department of Biology, University of Konstanz, Konstanz, Germany

Ravneet Kaur Aveley, Western Australia

Prasada Rao S. Kodavanti Neurotoxicology Branch, Toxicity Assessment Division, National Health and Environmental Effects Research Laboratory, Office of Research and Development, U.S. Environmental Protection Agency, Research Triangle Park, NC, United States

Urmila P. Kodavanti Environmental Public Health Division, National Health and Environmental Effects Research Laboratory, Office of Research and Development, United States Environmental Protection Agency, Durham, North Carolina, United States

George A. Kontadakis Laboratory of Optics and Vision and Ophthalmology Department, University of Crete, Heraklion, Greece

Gopala Krishna Nonclinical Consultants, Ellicott City, MD, United States

Priya A. Krishna Nonclinical Consultants, Ellicott City, MD, United States

Kavya A. Krishna Nonclinical Consultants, Ellicott City, MD, United States

Maria Kummu Research Unit of Biomedicine, Pharmacology and Toxicology, Faculty of Medicine, University of Oulu, Oulu, Finland

George D. Kymionis Jules Gonin Eye Hospital, Faculty of Biology and Medicine, University of Lausanne, Lausanne, Switzerland

Rajiv Lall Vets Plus Inc., Menomonie, WI, United States

P. Lin National Institute of Environmental Health Sciences, National Health Research Institutes, Taiwan

Bommanna G. Loganathan Department of Chemistry and Watershed Studies Institute, Murray State University, Murray, KY, United States

Jarkko Loikkanen European Chemicals Agency, Helsinki, Finland

Marcello Lotti Dipartimento di Scienze Cardio-Toraco-Vascolari e Sanità Pubblica, Università degli Studi Padova, Padova, Italy

Michael A. Lynes Molecular and Cell Biology, University of Connecticut, Storrs, CT, United States

Brinda Mahadevan Medical Safety & Surveillance, Abbott Laboratories, Columbus, OH, United States

Jitendra K. Malik Division of Pharmacology and Toxicology, Indian Veterinary Research Institute, Izatnagar, Bareilly, Uttar Pradesh, India

Charalampos Mamoulakis Department of Urology, University General Hospital of Heraklion, University of Crete, Medical School, Heraklion, Crete, Greece

Jane A. Mantey Department of Biochemistry, Cancer Biology, Neuroscience & Pharmacology, Meharry Medical College, Nashville, TN, United States

María Rosa Martínez-Larrañaga Department of Pharmacology and Toxicology, Faculty of Veterinary Medicine, Universidad Complutense de Madrid, Madrid, Spain

María Aránzazu Martínez Department of Pharmacology and Toxicology, Faculty of Veterinary Medicine, Universidad Complutense de Madrid, Madrid, Spain

Charalampos Mavridis Department of Urology, University General Hospital of Heraklion, University of Crete, Medical School, Heraklion, Crete, Greece

Roger O. McClellan Independent Advisor, Toxicology and Risk Analysis, Albuquerque, New Mexico, United States

Vincent P. Meador Covance Laboratories, Inc., Global Pathology, Madison, WI, United States

Lars Friis Mikkelsen Ellegaard Göttingen Minipigs A/S, Dalmose, Denmark

Dejan Milatovic Charlottesville, VA, United States

Ida R. Miller Mukherjee Institute of Psychiatry and Human Behavior, Bambolim, Goa, India

Anupama Mukherjee Oral Pathology, Microbiology and Forensic Odontology, Goa Dental College and Hospital, Panaji, India

Pushpinder Kaur Multani Johnson & Johnson, Janssen Research & Development, Malvern, PA, United States

Päivi Myllynen NordLab, Oulu, Finland

Kirsi Myöhänen European Chemicals Agency, Helsinki, Finland

Rekek Negga Department of Biology, King University, Bristol, TN, United States

Carolina Negrei Departament of Toxicology, Faculty of Pharmacy, "Carol Davila" University of Medicine and Pharmacy, Bucharest, Romania

Meliton N. Novilla Shin Nippon Biomedical Laboratories USA, Ltd, Everett, WA, United States; School of Veterinary Medicine, Purdue University, West Lafayette, IN, United States

Stephanie Padilla Integrated Systems Toxicology Division, Office of Research and Development, U.S. Environmental Protection Agency, Research Triangle Park, NC, United States

Carlos M. Palmeira Center for Neurosciences and Cell Biology of the University of Coimbra and Department of Life Sciences of the University of Coimbra Largo Marquês de Pombal, Coimbra, Portugal

Kip E. Panter United States Department of Agriculture, Agricultural Research Service, Poisonous Plant Research Laboratory, Logan, UT, United States

Markku Pasanen School of Pharmacy/Toxicology, Faculty of Health Sciences, University of Eastern Finland, Kuopio, Finland

Daniel J. Patrick Charles River Laboratories, Mattawan, MI, United States

Sofia Pavanello Dipartimento di Scienze Cardio-Toraco-Vascolari e Sanità Pubblica, Università degli Studi Padova, Padova, Italy

Henrik Duelund Pedersen Ellegaard Göttingen Minipigs A/S, Dalmose, Denmark

Olavi Pelkonen Research Unit of Biomedicine, Pharmacology and Toxicology, Faculty of Medicine, University of Oulu, Oulu, Finland

Michael A. Pellizzon Research Diets, Inc., New Brunswick, NJ, United States

Jason Pitt University of Evansville, Evansville, IN

Argyro Plaka Laboratory of Optics and Vision and Ophthalmology Department, University of Crete, Heraklion, Greece

Aramandla Ramesh Department of Biochemistry, Cancer Biology, Neuroscience & Pharmacology, Meharry Medical College, Nashville, TN, United States

Saniya Rattan Department of Comparative Biosciences, University of Illinois at Urbana-Champaign, Urbana, IL, United States

Jenni Repo School of Pharmacy/Toxicology, Faculty of Health Sciences, University of Eastern Finland, Kuopio, Finland

Matthew R. Ricci Research Diets, Inc., New Brunswick, NJ, United States

Anabela P. Rolo Center for Neurosciences and Cell Biology of the University of Coimbra and Department of Life Sciences of the University of Coimbra Largo Marquês de Pombal, Coimbra, Portugal

Magdalini Sachana Organization for Economic Cooperation and Development (OECD), Paris, France

Heidi Sahlman School of Pharmacy/Toxicology, Faculty of Health Sciences, University of Eastern Finland, Kuopio, Finland

Nitin Saini Johnson & Johnson, Janssen Research & Development, Malvern, PA, United States

Vandana Saini Washington, United States

Kai Savolainen Nanosafety Research Centre, Finnish Institute of Occupational Health, Helsinki, Finland

Ratanesh Kumar Seth Department of Environmental Health Sciences, Arnold School of Public Health, University of South Carolina, Columbia, SC, United States

Abha Sharma National Institute of Pharmaceutical Education and Research, Raebareli, U.P., India

Anurag Sharma Division of Environmental Health and Toxicology, Nitte University Center for Science Education and Research (NUCSER), Deralakatte, India

Elina Sieppi Research Unit of Biomedicine, Pharmacology and Toxicology, Faculty of Medicine, University of Oulu, Oulu, Finland

Rui Silva Center for Neurosciences and Cell Biology of the University of Coimbra and Department of Life Sciences of the University of Coimbra Largo Marquês de Pombal, Coimbra, Portugal

Anita Sinha Vets Plus Inc., Menomonie, WI, United States

Iordanis Skamagkas Department of Urology, University General Hospital of Heraklion, University of Crete, Medical School, Heraklion, Crete, Greece

Samantha J. Snow Environmental Public Health Division, National Health and Environmental Effects Research Laboratory, Office of Research and Development, United States Environmental Protection Agency, Durham, North Carolina, United States

Miguel A. Sogorb Instituto de Bioingeniería, Universidad Miguel Hernández de Elche, Spain

Ajay Srivastava Vets Plus Inc., Menomonie, WI, United States

Szabina A. Stice Division of Biotechnology and GRAS Notice Review, Center for Food Safety and Applied Nutrition, US Food and Drug Administration, College Park, MD, United States

Markus Storvik School of Pharmacy/Toxicology, Faculty of Health Sciences, University of Eastern Finland, Kuopio, Finland

David T. Szabo PPG Industries Incorporated, Pittsburgh, PA, United States

João S. Teodoro Center for Neurosciences and Cell Biology of the University of Coimbra and Department of Life Sciences of the University of Coimbra Largo Marquês de Pombal, Coimbra, Portugal

Aristidis M. Tsatsakis Center of Toxicology Science & Research, Medical School, University of Crete, Heraklion, Greece

John Tsiaoussis Laboratory of Anatomy-Histology-Embryology, Medical School, University of Crete, Heraklion, Crete, Greece

Kirsi Vähäkangas School of Pharmacy/Toxicology, Faculty of Health Sciences, University of Eastern Finland, Kuopio, Finland

Sumit Singh Verma Department of Biochemistry, Institute of Science, Banaras Hindu University, Varanasi, India

Eugenio Vilanova Instituto de Bioingeniería, Universidad Miguel Hernández de Elche, Spain

Suryanarayana V. Vulimiri National Center for Environmental Assessment, Environmental Protection Agency (EPA), Washington DC, United States

Genoa R. Warner Department of Comparative Biosciences, University of Illinois at Urbana-Champaign, Urbana, IL, United States

Kevin D. Welch United States Department of Agriculture, Agricultural Research Service, Poisonous Plant Research Laboratory, Logan, UT, United States

Christina Wilson-Frank Purdue University, College of Veterinary Medicine, Animal Disease Diagnostic Laboratory, Department of Comparative Pathobiology, West Lafayette, IN, United States

S.H. You Institute of Food Safety and Risk Management, National Taiwan Ocean University, Keelung, Taiwan

Snjezana Zaja-Milatovic PAREXEL International, Alexandria, VA, United States

Ioannis E. Zisis Department of Urology, University General Hospital of Heraklion, University of Crete, Medical School, Heraklion, Crete, Greece

Csaba K. Zoltani Emeritus US Army Research Lab, Aberdeen Proving Ground, MD, United States

Foreword

The first edition of *Biomarkers in Toxicology*, edited by Ramesh Gupta, was published in 2014. The whole area of biomarkers, not only in toxicology, is rapidly developing, partly because of the availability of highly sophisticated analytical equipment, and so the second edition of this book is greatly welcomed. The second edition contains 12 new chapters, and most of the rest have been updated.

Merriam Webster defines a biomarker as a distinctive biological or biologically derived indicator (as a metabolite) of a process, event, or condition (as, for example, aging, disease, or oil formation). There are other definitions, for example, in Environmental Health Criteria 222 Biomarkers. In Risk Assessment http://www.inchem.org/documents/ehc/ehc/ehc222.htm#1.0 biomarkers are defined thus "A biomarker is any substance, structure or process that can be measured in the body or its products and influence or predict the incidence of outcome or disease." The subject of the present book is biomarkers in toxicology, but it should be remembered that biomarkers include substances used in the detection of numerous diseases, including the autoimmune diseases, which are not generally thought to be toxicological in origin. However, biomarkers are crucial to toxicology and allied disciplines such as epidemiology and risk assessment.

The earliest toxicological biomarkers of exposure date from before precise analytical techniques were available and include the Kayser–Fleischer ring (described 1902/3), usually indicative of copper accumulation in the cornea in cases of Wilson's disease, as well as lead lines in the gums associated with lead toxicity. The cherry red color noted as a (rather unreliable) clinical sign in carbon monoxide poisoning may also be described as a biomarker.

One of the earliest biomarkers relying on biochemical analytical techniques was measurement of cholinesterase activity, initially whole blood cholinesterase and later plasma pseudocholinesterase (butyrylcholinesterase) and red blood cell acetylcholinesterase. Cholinesterase measurements were introduced at defense laboratories after World War II as a screening test for excessive exposure to organophosphate compounds: a 20% depression in activity was considered to mandate cessation of exposure of individuals to organophosphate nerve agents (the basis of the 20% figure is obscure, but seemed protective). Pseudocholinesterase and red blood cell acetylcholinesterase measurements now have numerous uses in worker protection, clinical diagnosis of poisoning, and human and experimental animal studies (including regulatory ones) in relation to the use of organophosphate and other anticholinesterase pesticides. Since World War II biomarkers of toxicity have ballooned in importance and number in worker and consumer protection and clinical and experimental toxicology and are also widely used in regulation of chemicals in animal and, less commonly, human experimental studies. Toxicological biomarkers are also used in allied disciplines, for example, epidemiology, and may be used to estimate loads of exposure in populations being investigated.

In toxicology, biomarkers are often divided into biomarkers of exposure, of effect, and of susceptibility, and all of these are dealt with in this book, which is extremely wide-ranging. The book has an initial introductory part, including discussion of rodent, nonhuman primate, and zebrafish and *Caenorhabditis elegans* models for toxicological testing. There are two new chapters in this part: firstly, *Drosophila melanogaster, Eisenia fetida,* and *Daphnia magna* for toxicity testing and biomarkers and secondly, adverse outcome pathways and biomarkers.

Part II, systems toxicity biomarkers, comprises chapters on biomarkers of toxicity in relation to all important organs and organ systems. There is an additional chapter on reproductive and developmental toxicity biomarkers, and another on ototoxicity biomarkers. Part III, renamed chemical agents, solvents, and gases toxicity biomarkers, deals with biomarkers in relation to the toxicity of specific groups of compounds and comprises chapters on pesticides, as well as polychlorinated biphenyls (PCBs), polybrominated biphenyls (PBBs), brominated flame retardants, polycyclic aromatic hydrocarbons (PAHs), bisphenol A, melamine and cyanuric acid, and metals; there is a very useful chapter on biomarkers of chemical mixture toxicity. Part IV (biotoxins biomarkers) has three chapters on, respectively, freshwater cyanotoxins, mycotoxins, and poisonous plants: biomarkers for diagnosis (of poisoning). Part V covers pharmaceuticals and nutraceuticals, with chapters on drug toxicity biomarkers and nutriphenomic biomarkers together with a new chapter on biomarkers of toxicity for dietary ingredients contained in dietary supplements. Part VI covers nanomaterials and radiation, with two chapters, one on biomarkers of exposure and effect of engineered nanomaterials and the other on biomarkers of exposure and effects of radiation.

Part VII is entitled carcinogens biomonitoring and cancer biomarkers and contains six chapters. These are on biomonitoring exposures to carcinogens, genotoxicity biomarkers, epigenetic biomarkers in toxicology, breast cancer biomarkers, pancreatic and ovarian cancer biomarkers, and prostate cancer biomarkers. Part VIII is called disease biomarkers and deals with biomarkers of Alzheimer's disease, biomarkers of Parkinson's disease, biomarkers for drugs of abuse and neuropsychiatric disorders: models and mechanisms, osteoarthritis biomarkers, pathological biomarkers in toxicology and oral pathology biomarkers. Of these, the chapters on osteoarthritis and oral pathology are new. Part IX is called special topics. This part of the book contains chapters on biomarkers of mitochondrial dysfunction and toxicity, biomarkers of blood-brain barrier dysfunction, biomarkers of oxidative/nitrosative stress and neurotoxicity, cytoskeletal disruption as a biomarker of developmental neurotoxicity, membrane transporters and transporter substrates as biomarkers for drug pharmacokinetics, pharmacodynamics, and toxicity/adverse events, and citrulline: pharmacological perspectives and role as a biomarker in diseases and toxicity. Of these chapters, that on the blood-brain barrier dysfunction is new and is particularly welcome as the blood-brain barrier is very important in protecting the central nervous system against toxicants. The last part of the book is on applications of biomarkers. It contains three new chapters: biomarkers detection for toxicity testing using microarray technology, metabolomics, and proteomics. Also there are chapters on transcriptomic biomarkers, percellome toxicogenomics, biomarkers in computational toxicology, biomarkers in biomonitoring of xenobiotics and biomarkers in toxicology, risk assessment, and environmental chemical regulations.

The 67-chapter book has an outstanding array of authors from the United States, Canada, Denmark, Finland, Greece, India, Italy, Japan, Portugal, Romania, and Spain. Professor Gupta deserves our gratitude for assembling such a distinguished group of experts to produce so comprehensive a book on this rapidly growing and very important field.

Timothy C Marrs
Edenbridge
United Kingdom

P A R T I

TOXICITY TESTING MODELS AND BIOMARKERS

CHAPTER

1

Introduction

Ramesh C. Gupta

Toxicology Department, Breathitt Veterinary Center, Murray State University, Hopkinsville, Kentucky, United States

Biomarkers can broadly be defined as indicators or signaling events in biological systems or samples of measurable changes at the molecular, biochemical, cellular, physiological, pathological, or behavioral levels in response to xenobiotics. The Biomarkers Definitions Working Group of the National Institutes of Health (NIH) has defined the biomarker as "a characteristic that is objectively measured and evaluated as an indicator of normal biological processes or pharmacological responses to a therapeutic agent." In the field of toxicology, biomarkers have been classified as markers of exposure, effect, and susceptibility. Measurement of biomarkers reflects the time course of an injury and provides information on the molecular mechanisms of toxicity. These biomarkers provide us the confidence of accurate diagnosis, prognosis, and treatment. The biomarkers of early chemical exposure can occur in concert with biomarkers of early disease detection, and that information aids in avoiding further chemical exposure and in strategic development of a novel treatment, including personalized medicine (i.e., treating the patient, and not the disease). In essence, with the utilization of specific biomarkers, an ounce of prevention can be worth a pound of treatment.

Biomarkers are used in drug development, during preclinical and clinical trials, for efficacy and safety assessment. Biomarkers can reveal valuable information regarding diagnosis, prognosis, and predict treatment efficacy or toxicity; serve as markers of disease progression; and serve as auxiliary endpoints for clinical trials (Stern et al., 2018), with the ultimate goal of delivering safe and effective medicines to patients (Lavezzari and Womack, 2016: Gerlach et al., 2018). In addition, a biomarker in drug development should be ethically acceptable (Hey, 2017). Safety biomarkers can be used to predict, detect, and monitor drug-induced toxicity during both preclinical studies and human clinical trials. Developing and validating highly sensitive methods for measurement of biomarkers and understanding the resultant data are complex processes that require a great deal of time, effort, and intellectual input. Furthermore, understanding drug metabolism seems essential in some cases, as the metabolite of a drug can be used as a biomarker, and the drug and/or its metabolite has to be patented by the United States Patent Office and by a similar governmental office/agency in other countries. In the past, many drugs were developed with biomarker assays that guided their use, and this trend is likely to continue in the future for drug discovery and development. With the judicious use of biomarkers, as in evidence-based medicine, patients are most likely to benefit from select treatments and least likely to suffer from their adverse effects. On the contrary, utilization of a bad biomarker can be as harmful to a patient as a bad drug. Therefore, biomarkers need to be validated and evaluated by an accredited laboratory, which participates in a proficiency testing program, to provide a high level of confidence to both clinicians and patients.

In the toxicology field, biomarkers should be specific, accurate, sensitive, validated, biologically or clinically relevant, and easy and fast to perform to be useful as predictive tools for toxicity testing and surveillance and for improving quantitative estimates of exposure and dose. Therefore, biomarkers are utilized for biomonitoring data that are useful in a variety of applications, from exposure assessment to risk assessment, management, and regulations (Ganzleben et al., 2017).

In the early 1990s, Dr. Maria Cristina Fossi from the University of Siena, Italy, emphasized the approach for the development and validation of nondestructive biomarkers over destructive biomarkers in the field of toxicology. She described the ideal biomarker as being measurable in readily available tissues or biological products and obtainable in a noninvasive way; related

Biomarkers in Toxicology, Second Edition
https://doi.org/10.1016/B978-0-12-814655-2.00001-3

to exposure and/or degree of harm to the organism; directly related to the mechanism of action of the contaminants; highly sensitive with techniques that require minimal quantities of sample and are easy to perform and cost-effective; and suitable for different species.

The development and validation of new techniques in the laboratory may provide the basis for a valuable field method. But, before a new biomarker's application, some basic information is required, such as dose–response relationships, and biological and environmental factors, which can influence the baseline values of responses. It is important to mention that, when dealing with a biochemical or metabolic biomarker, species differences can be the biggest challenge for any toxicologist.

Biomarkers have applications in all areas of toxicology, especially in the fields of pesticides, metals, mycotoxins, and drugs. In the case of veterinary toxicology, biomarkers of plant toxins deserve equal attention. Farmers, pesticide application workers, and greenhouse workers are exposed to pesticides by direct contact and their family members can be exposed via secondhand exposure. Measurement of residues of pesticides, and their metabolites and metals in urine, serves as the most accurate and reliable biomarkers of exposure in agriculture, industrial, and occupational safety and health settings. Recent evidence suggests that in utero or early life exposure to certain pesticides, metals, and other environmental contaminants may cause neurodegenerative (Alzheimer's, Parkinson's, schizophrenia, Huntington's, ALS, and others) and cardiovascular diseases, diabetes, and cancer later in life. In these diseases and many others, specific and sensitive biomarkers play important roles in early diagnosis, and this can serve as the cornerstone for timely therapeutic intervention.

Mycotoxin-related toxicity, carcinogenesis, and other health ailments are encountered in man and animals around the world. In developing countries, where regulatory guidelines are not strictly followed, adverse health effects (especially reproductive and developmental effects) are devastating. In these scenarios, early biomarkers of exposure play a pivotal role in avoiding further exposure to the contaminated food/feed and thus safeguard human and animal health.

With the current knowledge of system biology, proteomics, metabonomics, toxicogenomics, and various mathematical and computational/chemometric modelings, undetectable biomarkers can be discovered and these biomarkers can predict how tissues respond to toxicants and drugs and/or their metabolites, and how the tissue damage and repair processes compromise the tissue's function. Imaging and chemometric biomarkers are of greater sensitivity and carry more information than conventional biomarkers, as they detect (1) low levels of chemical exposure (exposure biomarker) and (2) an early tissue response (endogenous response biomarker). The priority will always be for the development of a noninvasive approach over an invasive approach, and nondestructive biomarkers over destructive biomarkers, but this may not be possible in all cases.

In 2011, the Joint SOT/EUROTOX Debate proposed that "biomarkers from blood and urine will replace traditional histopathological evaluation to determine adverse responses," identifying and comparing the strengths and limitations of histopathology with serum and urine biomarkers. Unlike histopathological techniques, blood and urine biomarkers are noninvasive, quantifiable, and of translational value. Of course, the complete replacement of histopathological biomarkers with blood and urine may not be possible in the near future, as in some instances histopathological biomarkers will still be used because of recent developments in invaluable molecular pathology techniques.

For the quest of developing the most sensitive and reliable biomarkers, integration of novel and existing biomarkers with a multidisciplinary approach appears fruitful. Furthermore, a multibiomarker approach seems more informative and accurate than a single biomarker approach. By now, microRNAs (miRNAs) have been well recognized as reliable and robust biomarkers for early detection of diseases, birth defects, pathological changes, cancer, and toxicities (Quiat and Olson, 2013; Wang et al., 2013; Bailey and Glaab, 2018). Because they are stable in biofluids, such as blood, there is rapidly growing interest in using miRNAs as diagnostic, prognostic, and predictive biomarkers, and the outlook for the clinical application of miRNA discoveries is promising, especially in molecular medicine. Soon, incorporating pharmacological and toxicological targeting of miRNAs into the development of innovative therapeutic strategies will be routine. Still, more innovative biomarkers need to be developed that will be highly sensitive (biotechnology-based techniques), require minimum quantities of sample, and will promise high-throughput screening.

At the recent meetings of the Society of Toxicology, the EUROTOX, and International Congress of Toxicology, a large number of toxicologists emphasized the importance of biomarkers in health, disease, and toxicity. Accordingly, *Biomarkers in Toxicology*, second edition has been prepared to meet the challenges of today's toxicologists, pharmacologists, environmentalists, and physicians in academia, industry, and government. This reference book is of particular interest to those in governmental agencies, such as NIH, USEPA, USFDA, USDA, NIOSH, OSHA, CDC, REACH, EFSA, etc. This is the most comprehensive biomarkers book to date as it covers every possible aspect of exposure, effects, and susceptibility to chemicals. There are many novel topics in this volume that are not covered in any previous book. This edition identifies and establishes the most sensitive, accurate, unique, and validated biomarkers that can be used as indicators of exposure and effect(s) of chemicals, and chemical-related long-term diseases, such as cardiovascular, metabolic and neurodegenerative diseases, and cancer. Sixty-seven chapters are organized under eight sections with a user-friendly format, and each chapter is enriched with current literature and references for further reading. This book begins with general concepts of toxicity and safety testing and biomarker development using various animal and animal alternative models, adverse outcome pathways, followed by biomarkers of system/organ toxicity, chemicals, solvents, gases, and biotoxins. There are several chapters on biomarkers of pharmaceuticals, nutraceuticals, petroleum products, chemical mixtures, radiation, engineered nanomaterials, epigenetics, genotoxicity, and carcinogens. In the disease section, chapters cover the biomarkers of Alzheimer's, Parkinson's, neuropsychiatric disorders, osteoarthritis, and some other pathological conditions. Under special topics, chapters are included on mitochondrial dysfunction and toxicity, the blood–brain barrier, oxidative/nitrosative stress, developmental neurotoxicity, miRNAs as indicators of tissue injury, and citrulline in diseases and toxicity. Lastly, a large number of chapters are dedicated to the application of biomarkers in toxicology, including the latest strategies and technologies in the development of biomarkers, biomarkers in drug development, safety evaluation, and toxicity testing and integration of biomarkers in biomonitoring of chemical exposure and risk assessment, especially in the context of industrial, environmental, and occupational medicine and toxicology.

The editor remains indebted to the contributors of this book for their hard work and dedication. These contributors are highly qualified and considered authorities in the fields of toxicology, pharmacology, pathology, biochemistry, and human and veterinary medicine. He expresses his gratitude to Ms. Denise Gupta and Ms. Robin B. Doss for their untiring support in technical assistance and text and reference checking. Finally, the editor would like to thank Ms. Kristi Anderson, Ms. Kattie Washington, Ms. Kathy Padilla, and Mr. Mohana Priyan Rajendran (the editorial staff at Academic Press/Elsevier) for their immense support at every stage of the production of this book.

References

Bailey, W.J., Glaab, W.E., 2018. Accessible miRNA as novel toxicity biomarkers. Int. J. Toxicol. 37 (2), 116–120.

Ganzleben, C., Antignac, J.-P., Barouki, R., et al., 2017. Human biomonitoring as a tool to support chemicals regulation in the European Union. Int. J. Hyg. Env. Health 220, 94–97.

Gerlach, C.V., Derzi, M., Ramaiah, S.K., et al., 2018. Industry perspective on biomarker development and qualification. Clin. Pharmacol. Ther. 103 (1), 27–31.

Hey, S.P., 2017. Ethical challenges in biomarker-driven drug development. Clin. Pharmacol. Ther. 103 (1), 23–25.

Lavezzari, G., Womack, A.W., 2016. Industry perspective on biomarker qualification. Clin. Pharmacol. Ther. 99 (2), 208–213.

Quiat, D., Olson, E.N., 2013. MicroRNA in cardiovascular disease: from pathogenesis to prevention and treatment. J. Clin. Invest. 123 (1), 11–18.

Stern, A.D., Alexander, B.M., Chandra, A., 2018. Innovation incentives and biomarkers. Clin. Pharmacol. Ther. 103 (1), 34–36.

Wang, K., Yuan, Y., Li, H., et al., 2013. The spectrum of circulating RNA: a window into systems toxicology. Toxicol. Sci. 132 (2), 478–492.

CHAPTER

2

Rodent Models for Toxicity Testing and Biomarkers

Shayne C. Gad

Principal of Gad Consulting Services, Raleigh, NC, United States

INTRODUCTION

Three rodent species are widely used in toxicology: the rat, the mouse, and the hamster. Two of these, the rat and mouse, are the most widely used in experimental biology and medicine. These have formed the basis for exploring the efficacy of drugs and for the identification and evaluation of toxicities associated with exposure to drugs, industrial and agricultural chemicals, and understanding the mechanisms of their toxicity since toxicology became an identified discipline.

A large (and growing) set of biomarkers are known for use in identifying and determining the relative (and relevant) risks to humans or other target species. These include:

- Body weights
- Clinical pathology (hematology)
- Clinical chemistry
- Organ weights
- Gross histologic changes at necropsy
- Immunogenicity
- Microscopic evaluation of tissues
- Changes in physiologic functions and electrophysiology
- Effects on specific genomic markers

Biomarkers are measurements of test model (animal) parameters that can provide important quantitative data about the biological state of the test model, which are predictive of effects in humans. These biomarkers in toxicology are preferably shared by both test animals and humans and in a manner that the relationship of findings in one species to another is known.

Accordingly, we will proceed to understand the current uses of these three rodent species as predictive models for effects in humans, as well as how they are measured and what their normal ranges are.

As these potential pieces of data are overviewed and considered, it is important to remember that each of these biomarkers is a part of the overall picture as to what the model is predicting as per potential adverse effects in humans. Meaningful safety assessment requires that all the data be incorporated in an integrated safety assessment. Because dose/toxicodynamic relationships will vary with level of exposure of test animals, it is also necessary that multiple (traditionally at least three) "dose" levels be evaluated.

The picture becomes both more complex but also clearer as to relevance as new biomarkers are identified and became understandable. These include proteomics (Amacher, 2010), new clinical chemistry parameters, immune system responses, and real time functional physiologic system measurements by telemetrized instrumentation (Gad, 2013).

Table 2.1 (Gad, 2013) presents the current most relevant associations of biomarkers with renal and liver toxicity, whereas Table 2.2 (adopted from Gad, 2013) presents an overview of the association between classical clinical chemistry parameters and specific target organ toxicities (Table 2.3). Table 2.4 summarizes causes associated with hematological findings in the rat.

Over the last 10 years, diligent efforts under the rubric of the Critical Path Initiative have led to the identification of a more specific set of clinical chemistry biomarkers for key potential target organs.

Biomarkers in Toxicology, Second Edition
https://doi.org/10.1016/B978-0-12-814655-2.00002-5

Heart	Troponins	Zethelius et al. (2008)
Kidney	KIM-1, Albumin, Beta-2-Microglobulin	Hoffmann et al. (2010), Ozer et al. (2010), and Vaidya et al. (2010)
Liver	DILI (Drug Induced Liver Injury)l ALT, BUN, coagulation factor	Shi et al. (2010)

Other recently identified biomarkers for specific targets include:

1. *Mitochondrial Dysfunction*. Increased uptake of calcium (because ATP depletion) by mitochondria activates phospholipases, resulting in accumulation of free fatty acids. These cause changes in the permeability of mitochondrial membranes, such as the *mitochondrial permeability transition*.
2. *Progressive Loss of Phospholipids*. Increased degradation by endogenous phospholipases and inability of the cell to keep up with synthesis of new phospholipids (reacylation, an ATP-dependent process).
3. *Cytoskeletal Abnormalities*. Activated proteases lyse cytoskeletal elements and cell swelling causes detachment of cell membrane from cytoskeleton; stretching of the cell membrane results in increased membrane damage.

TABLE 2.1 Classic Associations in Toxicology

Liver Toxicity	Renal Toxicity
Increased plasma activity of liver marker enzymes, e.g., alanine and aspartate aminotransferases	Increased water consumption and urine volume. Urine parameters may change, e.g., enzymes and cellular debris.
Decreased plasma total protein concentration	Increased plasma concentrations of urea and creatinine. Proteinuria.
Increased coagulation times due to decreased synthesis of coagulation factors	Severe renal toxicity may lead to decreased erythrocyte parameters due to effects on erythrocyte synthesis
Increased liver weight due to enzyme induction or accumulation of lipid or glycogen	Increased kidney weight
Change in color or size at necropsy	Change in color or size at necropsy
Histological findings such as necrosis or centrilobular hypertrophy due to enzyme induction	Histological change, e.g., basophilic tubules or necrosis, papillary necrosis, or glomerular changes.

4. *Reactive Oxygen Species*. Produced within the cell by infiltrating neutrophils and macrophages, especially after restoration of blood flow to an area (reperfusion injury). Cell injury triggers release of a number of inflammatory cytokines and chemokines that amplify the host immune response and attract neutrophils to the site.
5. *Lipid Breakdown Products*. Unesterified free fatty acids, acyl carnitine, and lysophospholipids. These have a detergent effect on membranes and may exchange with membrane phospholipids, causing permeability changes.

THE RAT

Use in Toxicological Research

Ideally, safety testing of products intended for use in humans, or to which humans could be exposed, should be done in humans. The data from humans would apply without reservation to complex human physiology and cellular/biochemical mechanisms and human risk assessment. Unfortunately, humans cannot be used for this purpose. Therefore, the choice of an appropriate species for toxicology studies should be based on a comparison of the pharmacokinetics, target pharmacodynamics, and metabolism of the test compound in different laboratory species and man. In the absence of this data, this choice is often based on practicality and economics. The rat has become a species of choice because of the metabolic similarities, as well as their small size, relatively docile nature, short life span, and short gestation period. The extensive use of the rat in research has led to the development of a large historical database of their nutrition, diseases, and general biology.

Characteristics

Although the rat is a species of choice in toxicology research because of the many physiological similarities and anatomical characteristics, differences exist that must be considered when designing and conducting studies with this animal. Rats are obligate nose breathers; as such an inhaled test material is subject to nasal filtration and absorption. The placenta is considerably more porous in the rat. This difference may increase the chance of fetal exposure to an administered test material or increase the overall level of fetal exposure to an administered test material. The overall distribution of intestinal microflora is different in the rat, which may lead to differences in the metabolism of an orally administered test material. These and other differences in the rat may lead to positive signs of toxicity to a test material that may not be present in a different species. There are

TABLE 2.2 Association of Changes in Biochemical Parameters With Actions at Particular Target Organs (Gad, 2013)

Parameter	Blood	Heart	Lung	Kidney	Liver	Bone	Intestine	Pancreas	Notes
Albumin				↓	↓				Produced by the liver; very significant reductions indicate extensive liver damage
ALP					↑	↑	↑		Elevations usually associated with cholestasis; bone alkaline phosphatase tends to be higher in young animals
ALT (formerly SGPT)					↑				Elevations usually associated with hepatic damage or disease
AST (formerly SGOT)		↑		↑	↑			↑	Present in skeletal muscle and heart and most commonly associated with damage to these
Beta-2-Microglobulin				↑					
Bilirubin (total)	↑				↑				Usually elevated due to cholestasis, due to either obstruction or hepatopathy
BUN				↑	↓				Estimates blood filtering capacity of the kidneys; does not become significantly elevated until the kidney function is reduced 60%–75%
Calcium				↑					Can be life threatening and result in acute death
Cholinesterase				↑	↓				Found in plasma, brain, and RBC
CPK		↑							Most often elevated due to skeletal muscle damage but can also be produced by cardiac muscle damage; can be more sensitive than histopathology
Creatinine				↑					Also estimates blood filtering capacity of kidney as BUN does
Glucose								↑	Alterations other than those associated with stress uncommon and reflect an effect on the pancreatic islets or anorexia
GGT					↑				Elevated in cholestasis; this is a microsomal enzyme, and levels often increase in response to microsomal enzyme induction
HBDH		↑			↑				–
KIM-1				↑					
LDH		↑	↑	↑	↑				Increase usually due to skeletal muscle, cardiac muscle, or liver damage; not very specific
Protein (total)				↓	↓				Absolute alterations usually associated with decreased production (liver) or increased loss (kidney); can see increase in case of muscle wasting (catabolism)
SDH					↑↓				Liver enzyme that can be quite sensitive but is fairly unstable; samples should be processed as soon as possible
Trophonin		↑							

ALP, alkaline phosphatase; *BUN*, blood urea nitrogen; *CPK*, creatinine phosphokinase; *GGT*, gamma glutamyl transferase; *HBDH*, hydroxybutyric dehydrogenase; *LDH*, lactic dehydrogenase; *RBCs*, red blood cells; *SDH*, sorbitol dehydrogenase; *SGOT*, serum glutamic oxaloacetic transaminase (also called AST [aspartate amino transferase]); *SGPT*, serum glutamicpyruvic transaminase (also called ALT [alanine amino transferase]); ↑, increase in chemistry values; ↓, decrease in chemistry values.

TABLE 2.3 Liver Enzymes

"Liver Enzyme"	Nomenclature	Plasma-Tissue Sources	Cellular Location
AST (SGOT)	Aspartate Aminotransferase	Liver, Heart, Skeletal Muscle, Kidney, Brain, RBCs	Mitochondria, Cytoplasm
ALT (SGPT)	Alanine Aminotransferase	Mostly liver, Heart, Skeletal Muscle	Cytoplasm
Alk Phos (AP)	Alkaline Phosphatase	Bile ducts, GI tract, Bone, Placenta	Membranes
GGT	Gamma-glutamyl transferase	Liver, Kidney, Heart	Membranes
GDH	Glutamate Dehydrogenase	Liver, Kidney Skeletal Muscle	Mitochondria
SDH	Sorbitol Dehydrogenase	Mostly liver	Cytoplasm
LDH	Lactate Dehydrogenase	Heart, Skeletal Muscle, RBCs, Lung, Liver, All tissues	Cytoplasm

RBC, red blood cell.

TABLE 2.4 Some Probable Conditions Affecting Hematological Changes (Gad, 2013)

Parameter	Elevation	Depression	Parameter	Elevation	Depression
Red blood cells (RBCs)	1. Vascular shock 2. Excessive diuresis 3. Chronic hypoxia 4. Hyperadreno corticism	1. Anemias a. Blood Loss b. Hemolysis c. Low RBC production	Platelets		1. Bone marrow depression 2. Immune disorder
Hematocrit	1. Increased RBC 2. Stress 3. Shock a. Trauma b. Surgery 4. Polycythemia	1. Anemias 2. Pregnancy 3. Excessive hydration	Neutrophils	1. Acute bacterial infections 2. Tissue necrosis 3. Strenuous exercise 4. Convulsions 5. Tachycardia 6. Acute hemorrhage	
Hemoglobin	1. Polycythemia (increased in production of RBC)	1. Anemias 2. Lead Poisonings	Lymphocytes	1. Leukemia 2. Malnutrition 3. Viral infections	
Mean cell volume	1. Anemias 2. B-12 deficiency	1. Iron deficiency	Monocytes	1. Protozoal infections	
Mean corpuscular hemoglobin	1. Reticulocytosis	1. Iron deficiency	Eosinophils	1. Allergy 2. Irradiation 3. Pernicious anemia 4. Parasitism	
White blood cells	1. Bacterial infections 2. Bone marrow stimulation	1. Bone marrow depression 2. Cancer chemotherapy 3. Chemical intoxication 4. Splenic disorders	Basophils	1. Lead poisoning	

also differences (though generally less striking) between different strains of rats and sometimes even between the animals supplied by difference sources.

Strain Differences

Breeding rats for specific characteristics has produced some physiological differences between strains of rats. Some of these differences are known to affect how the various strains react to toxicants. Among others, strain specific differences have been found in sensitivity to thiourea (Dieke and Richter, 1945), sensitivity to acetaminophen nephrotoxicity (Newton et al., 1985a,b), the incidence of spontaneous glomerular sclerosis (Bolton et al., 1976), sensitivity to the carcinogenic actions of 7,12-dimethylbenz(a)anthracene (Boyland and Sydnor, 1962), the effects of trimethyltin on operant behavior and hippocampal glial fibrillary acidic protein (GFAP) (MacPhail et al., 2003), differences in renal carcinogenesis (Hino et al., 2003), differences in cytochrome P4501A1 gene expression caused by 2,3,7,8-tetrachlorodibenzo-p-dioxin in the liver (Jana et al., 1998), susceptibility to 4-nitroquinoline 1-oxide induce carcinoma (Kitano et al., 1992), and differences in the levels of drug-metabolizing enzymes (Page and Vesell, 1969). In recent years, research and breeding programs have been focused on producing inbred and outbred strains focused on specific disease models and susceptibility to the development of certain carcinomas. When choosing a strain for use, it is important to consider these differences.

Of importance for carcinogenicity studies, strain differences have been found in the incidence of spontaneous tumors. Table 2.5 gives the incidence of spontaneous tumors found in commonly used strains in carcinogenicity studies. The historical incidence is important to the analysis of a study in that a high spontaneous rate may mask a small test material–related increase in tumor incidence.

Because of lower spontaneous tumor rates, the Wistar has become the most popular strain in toxicological research.

Normal Physiological Values

General values for selected physiological parameters are given in Tables 2.6 and 2.7. Normal values will vary based on the strain of animal, supplier, feed, and housing conditions. These tables should be used as a point of reference only.

STUDY DESIGNS

The length and design of toxicology studies used to predict human risk are governed by guidelines issued by regulatory bodies such as the US Food and Drug Administration (FDA), the International Conference on Harmonization (ICH), the Environmental Protection Agency (EPA), and their counterparts worldwide. Toxicology studies are divided into a series of three sets of studies that are required for each phase of clinical trials. For initial approval to begin clinical trials, the following studies are required. The length of dosing in the toxicology studies varies depending on the intended length in clinical trials. A test compound intended to be a repeat dose study for up to 28 days in duration initially requires a two phase study in which a maximum tolerated dose (MTD) following a single administration is determined followed by a second phase during which the test compound is administered daily at dose levels based on the MTD for 5–7 days (Table 2.8). Following the completion of the MTD study a 14 or 28 Day Repeat Dose study should be conducted (Table 2.9). These studies assess the effects of a test compound at dosages that do not cause immediate toxic effects.

In support of Phase 2 clinical trials, longer-term subchronic and chronic toxicity studies (Table 2.10) should be conducted. Subchronic and chronic toxicity studies are designed to assess the test compound effects following prolonged periods of exposure. The highest dosage level in each of these studies should produce a toxic effect such that target organs may be identified. The lowest dosage level should provide a margin of safety that exceeds the human clinical dose and ideally allows for the definition of no observable effect level. Alternatively, when effects related to the pharmacological mechanism of the test compound or when observed effects may be related to treatment with the test compound but may not be of toxicologic significance, a no observable adverse effect level (NOAEL) may be determined.

In addition to the subchronic and chronic toxicity studies in support of Phase 2 clinical trials, reproductive

TABLE 2.5 Incidence of Common Spontaneous Tumors in Fischer 344 and CD (SD)IGS Rats

		% Tumors in Untreated Rats							
		CD(SD)IGS		CD (SD)		Fisher		Wistar	
Organ	Tumor Type	Male	Female	Male	Female	Male	Female	Male	Female
Adrenal gland	Pheochromocytoma	10.0	2.3	11.3	2.3	11.9	3.2	3.2	1.3
Mammary gland	Fibroadenoma	1.4	44.5	1.3	16.7	0.8	7.1	1.2	30.2
Pancreas	Islet cell adenoma	3.6	1.4	4.0	0.3	1.5	0.2	5.3	1.9
Pituitary gland	Adenoma pars distalis	33.6	56.8	35.7	50.3	12.4	28.2	41.1	65.8
Testis	Interstitial cell tumor	1.8		7.0		74.6		4.3	
Thyroid gland	C-Cell Adenoma	10.5	5.0	5.0	5.7	12.5	8.2	10.1	10.7

Adapted from Charles, R., 2011. Spontaneous Neoplastic Lesions in the Crl:CD BR Rat. Charles River Laboratories, Inc., Massachusetts; Mitsumori, K., Watanabe, T., Kashida, Y., 2001. Variability in the Incidence of Spontaneous Tumors in CD (SD) IGS, CD (SD), F244 and Wistar Hannover Rats in Biological Reference Data on CD(SD) IGS Rats, Yokohama, CD(SD) IGS Study Group.

TABLE 2.6 Selected Normative Data

HUSBANDRY	
Room temperature (°C)	18–26
Relative humidity (%)	30–70
Ventilation (air change/h)	10
Light/dark cycle (h)	12–14/12–10
Minimum cage floor size	
Housed individually (cm^2)	350
Breeding with pup (cm^2)	800
Group housed (cm^2 adult)	250
GENERAL	
Life span (years)	2.5–3.0
Surface area (cm^2)	0.03–0.06
Chromosome number (diploid)	42
Water consumption (mL/100 g/day)	10–12
Food consumption (g/day)	20–40
Average body temperature (°C)	37.5
REPRODUCTION	
Puberty (males and females)	50 ± 10 days
Breeding season	All year
Type of estrous cycle	Polyestrous
Length of estrous cycle	4–5 days
Duration of estrous	10–20 h
Mechanism of ovulation	Spontaneous
Time of ovulation	7–10 h after onset of estrous
Time of implantation	Late day 4 or 5[a]
Length of gestation	21–23 days
Litter size	8–16 pups
Birth weight	5–6 g
Eyes open	10–12 days
Weaning age/weight	21 days/40–50 g
CARDIOVASCULAR	
Arterial blood pressure	
Systolic (mmHg)	116–145
Diastolic (mmHg)	76–97
Heart rate (beats/min)	296–388
Cardiac output (mL/min)	10–80
Blood volume (mL/kg)	64
PULMONARY	
Respiration (breaths/min)	100–140
Tidal volume (mL)	1.1–2.5
Compliance (mL/cm H_2O)	0.3–0.9
Resistance (cm H_2O/mL s)	0.1–0.55
Pattern	Obligate nasal
RENAL	
Urine volume	15–30 mL/24 h
Na^+ excretion	200 mmol/L/24 h
K^+ excretion	150 mmol/L/24 h
Urine osmolarity	2000 mOsm/kg H_2O
Urine pH	7.3–8.5
Urine specific gravity	1.01–1.07
Urine creatinine	6 μmol/L/24 h
Glomerular filtration rate	1.0 mL/min/100 g body weight

[a]*The estrous cycle length may vary from 4 to 5 days between strains. Time of implantation may vary based upon the length of the estrous cycle and is dependent upon Day 0 or the first day sperm is found in the vagina.*

Data from Baker, H.J., Lindsey, J.R., Weisbroth, S.H., 1979. Housing to control research variables. In: Baker, H.J., Lindsey, J.R., Weisbroth, S.H. (Eds.), The Laboratory Rat, vol. 1. Academic Press, New York, pp. 169–192; Hofsteller, J., Svekow, M.A., Hickman, D.L., 2006. Morphophysiology. In: Svekow, M.A., Weisbroth, S.H., Franlin, C.L. (Eds.), The Laboratory Rat, second ed. vol. I. Academic Press, New York, pp. 93–125; Peplow, A., Peplow, P., Hafez, E., 1974. Parameters of reproduction. In: Vo, I., Melby, E., Altmon, N. (Eds.), Handbook of Laboratory Animal Science. CRC Press, Boca Raton, pp. 107–116; Waynforth, H., Flecknell, P., 1980. Experimental and Surgical Technique in the Rat, second ed. Elsevier Academic Press; Sharp, P., LaRegina, M., 1998. The Laboratory Rat. Academic Press, Philadelphia; Van Zutphen, L.F.M., Baumans, V., Beynen, A.C., 1993. Principles of Laboratory Animal Science. Elsevier, Amsterdam.

safety studies may also be required. Reproductive toxicity studies are typically required for a test compound intended to be administered to women of childbearing age or may affect male reproduction. These studies include an assessment of the potential effects of the test compound on general fertility and reproductive performance (Segment I), developmental toxicity (Segment II), or affect perinatal and postnatal development (Segment III). The highest dose in reproductive studies should be chosen so that administration causes some minimal toxicity. Typically, a dose range finding pilot study in a small number of animals should be conducted prior to initiating the definitive reproductive toxicology studies. Examples of protocols designed to meet the ICH guidelines are presented in Tables 2.9, 2.9A, and 2.10.

In support of Phase 3 clinical trials, two carcinogenicity studies may be required (Table 2.8), one in rats and one in mice. Typically 18 months to 2 years in duration, this type of study is designed to assess the potential of the test compound to induce neoplastic lesions. The highest dosage in a carcinogenicity study should cause minimal toxicity when administered via the intended route for clinical use. The preclinical studies required

TABLE 2.7 Growth Rates in Selected Rat Strains

	Age (days)							
	Crl:CD (SD)IGSBR		Crl:(WI)BR		Crl:(LE)BR		CDF(F-344)/CrlBR	
Weight (g)	M	F	M	F	M	F	M	F
Up to 50	Up to 23	Up to 23	Up to 23	Up to 25	Up to 21	Up to 21	Up to 23	Up to 23
51–75	24–28	24–29	24–28	26–30	22–25	22–26	24–29	24–29
76–100	29–34	30–35	29–32	31–34	26–29	27–31	30–34	30–35
101–125	35–37	36–39	33–35	35–40	30–34	32–36	35–39	36–42
126–150	38–42	40–44	36–40	41–47	35–37	37–43	40–45	43–55
151–175	43–45	45–50	41–44	48–56	38–42	44–50	46–50	56–72
176–200	46–49	51–56	45–48	57–64	43–46	51–55	51–57	73–105
201–225	50–52	57–70	49–52	65–81	47–49	56–69	58–63	105+
226–250	53–56	71–84	53–56	82–105	50–55	70–86	64+	
251–275	57–59	84–105	57–61	106+	56–58	87–102		
276–300	60–65	106+	62–67		59–64	103+		
301–325	66–71		68–73		65–70			
326–350	72–77		74–79		71–80			
351–375	78–87		80–87		81–90			
376+	88+		88+		91+			

Adapted from Charles, R., 2004. Growth Rates in Selected Rat Strains. Charles River Laboratories, Inc., Massachusetts.

in support of the clinical trials are dependent on the intended route and frequency of administration of the test compound and the intended age group to be treated (Tables 2.11–2.14).

ROUTES OF TEST ARTICLE ADMINISTRATION

Oral Routes

Rodents have several unique characteristics to be considered regarding the oral administration of test compounds. One of the most important characteristics is the lack of an emetic response. The lack of this response allows for a higher dose of a potential emetic compound to be administered and evaluated. Many compound and excipients may cause emesis in dogs or other large animal species and may lead to a low level of exposure and erratic blood levels. A second factor to consider is that rodents are nocturnal and eat most of their food at night. When maintained on a 12 h light–dark cycle, rats have been found to consume 75% of their daily food intake during the dark cycle (Wong and Oace, 1981). This should be taken into consideration when designing an oral gavage study and determining when the animal may be dosed. Early in the light cycle, animals are more likely to have a full stomach and complications associated with dosing may occur if large volumes of test article are administered. In addition, a full stomach may affect gastric emptying and the rate of absorption of an orally administered test compound.

Techniques for oral administration of test compounds include mixing in the diet, via gavage or stomach tube, via capsule, or in drinking water. The most widely used methods of oral administration are the dietary and gavage techniques.

Dietary Versus Gavage Methods

The choice between dietary and gavage dosing techniques is typically based on several factors. A scientific decision can only be made with a knowledge of the pharmacokinetics of the test compound administered by both methods. Other considerations that may be used in making this decision are as follows.

The dietary method can be used if a compound can be mixed with the diet, is stable under storage conditions in the diet, and is palatable to the animal. A major advantage of the dietary method is that it requires less manpower to perform the study. The diet mixing process can be performed weekly or, if stability allows, less often. The mixing and feeding process is less labor-intensive than gavaging rats on a daily basis.

TABLE 2.8 Maximum Tolerated Dose Study in Rats

Phase A	Oral MTD Study	
	Males	Females
Dose Level 1	3	3
Dose Level 2	3	3
Dose Level 3	3	3
Dose Level 4	3	3

Phase B	7-Day Oral Range Finding Study			
	Main Study		Toxicokinetics	
	Males	Females	Males	Females
Control	5	5	–	–
Low dose	5	5	9	9
Mid dose	5	5	9	9
High dose	5	5	9	9

Experimental Design:
In Phase A, the dose level will be increased until the maximum tolerated dose (MTD) is determined. The MTD is a dose that produces neither mortality nor more than a 10% decrement in body weight nor clinical signs of toxicity. In Phase B, animals will be dosed daily for 7 days at fractions of the single dose MTD to estimate a repeat dose MTD.
Dose Route/Frequency:
As requested.
Phase A: Once.
Phase B: Once per day for 7 consecutive days.
Observations: *Twice daily in both phases (mortality/moribundity).*
Detailed Clinical Observations: *Daily in both phases.*
Body Weights: *Daily in both phases.*
Food Consumption: *Daily.*
Clinical Pathology (Phase B only): *Hematology, clinical chemistry, and urinalysis evaluations on all surviving main study animals at termination.*
Necropsy (Phase B only): *Tissues saved for possible future histopathological evaluation.*
Organ Weights (Phase B only): *Adrenals, brain, heart, kidneys, liver, lungs, ovaries with oviducts, pituitary, prostate, salivary glands, seminal vesicles, spleen, thyroid with parathyroid, thymus, testes, uterus.*
Toxicokinetics: *Blood collected on days 1 and 7 (three cohorts consisting of three animals/sex/treatment group bled twice to equal six time points), calculation of* C_{max}, T_{max}, $AUC_{0\text{-}24}$, *and* $T_{1/2}$.

Several disadvantages also exist in using the dietary method. Methods must be developed and validated to prove homogeneity and stability. This is not as easy a process as with a suspension or solution. The dietary method is also less exact than the gavage method, in that the concentration of compound mixed in the feed is based on predicted feed consumption and body weights. In addition, if the feed is not palatable to the animal, or the test compound makes the animal ill, feed consumption may be reduced thereby reducing exposure to the test compound. In addition, the facility and control animals may be exposed to the test compound through dust or vapors.

TABLE 2.9 14 or 28 Day Repeat Dose Toxicity Study in Rats

	Main Study*		Toxicokinetics	
	Males	Females	Males	Females
Vehicle control	10	10	–	–
Low dose	10	10	$9 + 3^a$	$9 + 3^a$
Mid dose	10	10	$9 + 3^a$	$9 + 3^a$
High dose	10	10	$9 + 3^a$	$9 + 3^a$

Observations: Twice daily (mortality/moribundity).
Detailed Clinical Observation: *Weekly.*
Functional Observational Battery: *Pretest and Day 14 or 25.*
Body Weights: *Weekly.*
Food Consumption: *Weekly.*
Ophthalmology: *All animals prior to test article administration; all surviving main study animals at study termination.*
Clinical Pathology: *Hematology, clinical chemistry, and urinalysis evaluations on all surviving main study animals at termination.*
Toxicokinetics: *Blood collected on Days 1 and 14 or 27 (three cohorts consisting of three animals/sex/treatment group bled twice to equal six time points); TK modeling. The use of subsets of all the animals in a test group is called "spare sampling," intended to avoid the need for additional ("Satellite") groups of animals. Note that although six time points are commonly collected, more may be required or taken to adequately characterize a drug's pharmacokinetics.*
Necropsy: *All main study animals; toxicokinetic animals euthanized and discarded.*
Organ Weights: *Adrenals, brain, heart, kidneys, liver, lungs, ovaries with oviducts, pituitary, prostate, salivary glands, seminal vesicles, spleen, thyroid with parathyroid, thymus, testes, uterus.*
Slide Preparation/Microscopic Pathology: *All animals in the vehicle control and high dose groups and all found dead animals: full set of standard tissues; low and mid dose group target organs (to be determined); gross lesions from all animals.*
[a]Three additional animals/sex/treatment group included as replacement animals.
** Should also refer to table note "a".*

The gavage method may be used when the test compound is not stable in the diet or may not be palatable to the animals. In addition, the gavage method is preferable when evaluating toxicokinetics or pharmacokinetics. As with dietary mixtures, test compound administered via gavage as a solution or suspension should be analyzed for homogeneity, stability, and concentration. Methods for solution or suspension may be easier to develop than those required for dietary mixtures. For Good Laboratory Practices (GLP) studies, evaluation of homogeneity, stability, and concentration should be conducted for every study. If the same methodology and batch size are used for multiple studies, homogeneity may be established once. Stability of the test compound in solution or suspension should be determined under the testing conditions in the proposed vehicle. Typically, stability for toxicology studies is established for between 7 and 14 days. If the test compound is not found to be stable, stability of shorter duration may be established. Lastly, concentration analysis should be established for each dose level and should be periodically evaluated during longer-term studies.

With the gavage method of dosing, a more precise amount of the test compound can be delivered and

TABLE 2.9A 28 Day Repeat Dose Toxicity Study With Immunophenotyping in Rats

	Main Study		Toxicokinetics	
	Males	**Females**	**Males**	**Females**
Vehicle control	10	10		
Low dose	10	10	9 + 3[a]	9 + 3[a]
Mid dose	10	10	9 + 3[a]	9 + 3[a]
High dose	10	10	9 + 3[a]	9 + 3[a]

Observations: Twice daily (mortality/moribundity).
Detailed Clinical Observation: *Weekly.*
Functional Observational Battery: *Pretest and Day 25.*
Body Weights: *Weekly.*
Food consumption: *Weekly.*
Ophthalmology: *All animals prior to test article administration; all surviving main study animals at study termination.*
Clinical Pathology: *Hematology, clinical chemistry, and urinalysis evaluations on all surviving main study animals at termination.*
Immunotoxicology: *Immunophenotyping of blood leukocytes by flow cytometry on all surviving main study animals at termination. NK cell assay on blood leukocytes of all surviving main study animals at termination. May include identification of any antidrug antibodies (ADAs).*
Toxicokinetics: *Blood collected on Days 1 and 27 (three cohorts consisting of three animals/sex/treatment group bled twice to equal six time points).*
Necropsy: *All main study animals; toxicokinetic animals euthanized and discarded.*
Organ Weights: *Adrenals, brain, heart, kidneys, liver, lungs, ovaries with oviducts, pituitary, prostate, **salivary** glands, seminal vesicles, spleen, thyroid with parathyroid, thymus, testes, uterus, two lymph nodes (e.g., mesenteric, axillary, popliteal, etc.) including the lymph node draining the route of administration.*
Slide Preparation/Microscopic Pathology: *All animals in the vehicle control and high dose groups and all found dead animals: full set of standard tissues (add Peyer's patch, extra lymph node); low and mid dose group target organs; gross lesions from all animals.*
[a]Three additional animals/sex/treatment groups included as replacement animals; the control animals will not be evaluated for toxicokinetics.

may reduce the amount of test compound required to complete the study. This becomes important when evaluating the effects of a pharmaceutical, as the required dose levels and exposure levels to show safety may be lower than that required for a pesticide or chemical. A disadvantage of the gavage method is that it involves handling of the rat for each dosing. Handling of the rat has been shown to increase corticosterone levels (Barrett and Stockham, 1963) and may affect study results. Additionally, daily intubation may lead to death due to esophageal puncture or inhalation pneumonia.

Dietary Method

When utilizing the dietary method, the test compound is mixed with the diet and administered to the animals either ad libitum or the diet is presented to the animals for a fixed amount of time each day. The dosage received by an animal is regulated by varying the concentration of test compound in the diet based on the predicted food consumption and body weight.

TABLE 2.10 Subchronic and Chronic Toxicity Study in Rats

	Main Study		Toxicokinetics	
	Males	**Females**	**Males**	**Females**
Vehicle control	15	15	–	–
Low dose	15	15	9+3[a]	9+3[a]
Mid dose	15	15	9+3[a]	9+3[a]
High dose	15	15	9+3[a]	9+3[a]

Observations: Twice daily (mortality/moribundity).
Detailed Clinical Observation: *Weekly.*
Body Weights: *Weekly.*
Food Consumption: *Weekly.*
Ophthalmology: *All animals prior to test article administration; all surviving main study animals at study termination.*
Clinical Pathology: *Hematology, clinical chemistry, and urinalysis evaluations on all surviving main study animals at termination.*
Toxicokinetics: *Blood collected on Days 1 and 90 (three cohorts consisting of three animals/sex/treatment group bled twice to equal six time points); TK modeling.*
Necropsy: *All main study animals; toxicokinetic animals euthanized and discarded.*
Organ Weights: *Adrenals, brain, heart, kidneys, liver, lungs, ovaries with oviducts, pituitary, prostate, salivary glands, seminal vesicles, spleen, thyroid with parathyroid, thymus, testes, uterus.*
Slide Preparation/Microscopic Pathology: *All animals in the vehicle control and high dose groups and all found dead animals: full set of standard tissues; low and mid dose group target organs; gross lesions from all animals.*
[a]Three additional animals/sex/treatment group included as replacement animals.

Food consumption and body weight predictions are based on historical laboratory data for early time points in a study. As the study progresses, growth and food consumption curves can be established for each group

TABLE 2.11 Study of Fertility and Early Embryonic Development to Implantation in Rats

	Males	**Females**
Vehicle control	25	25
Low dose	25	25
Mid dose	25	25
High dose	25	25

Dose Route/Frequency: Males dosed began 28 days before mating and continued until euthanasia. Females dosed began 14 days before mating and continued through Day 7 of gestation (implantation).
Observations: *Twice daily (mortality/moribundity).*
Clinical Examinations: *Observations for clinical signs, body weights, and food consumption measurements recorded during the study period. Beginning at initiation of test article administration, females examined daily to establish estrous cycle.*
Uterine Examinations: *Performed on dams on Day 13 of gestation. Gravid uterine weight and the weight of the ovaries recorded. Total number of corpora lutea and implantations, location of resorptions, and embryos recorded. Females subjected to necropsy, and reproductive organs and gross lesions fixed for possible microscopic evaluation.*
Evaluation of Males: *Following disposition of females, the males were euthanized and subjected to a necropsy. The testes and epididymides weighed, and analysis of sperm parameters (concentration, motility, and morphology) performed. Reproductive organs and gross lesions fixed for possible microscopic evaluation.*
Statistical Analysis: *Standard.*

TABLE 2.12 Embryo-Fetal Development in Rats

	Time Mated Females
Vehicle control	25
Low dose	25
Mid dose	25
High dose	25

Dose Route/Frequency: Dosing initiated on Day 6 of gestation and continued to include Day 17 of gestation.
Observations: *Twice daily (mortality/moribundity).*
Clinical Examinations: *Daily Gestation Days 6 through 20.*
Body weights/Food Consumption: *Gestation Days 0, 6, 9, 12, 15, 18, and 20.*
Cesarean Section/Necropsy: *Litters will be delivered by cesarean section on Day 20 of gestation. Gravid uterine weight will be recorded. Total number of corpora lutea, implantations, early and late resorptions, live and dead fetuses, and sex and individual body weights of fetuses will be recorded. External abnormalities of fetuses will be recorded. Approximately one-half of the fetuses will be processed for visceral abnormalities, and the remaining fetuses will be processed for skeletal abnormalities. Dams will be subjected to a necropsy and gross lesions and target organs (if known) will be saved.*

TABLE 2.14 Carcinogenicity Study in Rats

	Main Study		6-Month Satellite	
	Males	Females	Males	Females
Vehicle control	60	60	20	20
Low dose	60	60	20	20
Mid dose	60	60	20	20
High dose	60	60	20	20

Study Desgin: Group as per Table 2.14.
Observations: *Twice daily (mortality/moribundity).*
Detailed Clinical Observations: *Once weekly.*
Body Weights: *Weekly for first 13 weeks, monthly thereafter.*
Food Consumption: *Weekly for first 13 weeks, monthly thereafter.*
Ophthalmology: *All animals pretest and all survivors prior to terminal sacrifice.*
Clinical Pathology:
Main Study: Hematology at termination.
6-Month Satellite: Hematology, clinical chemistry, and urinalysis evaluations on all surviving satellite animals at termination.
Necropsy: *All animals.*
Slide Preparation/Microscopic Pathology: *All animals, full set of standard tissues, all masses, and all lesions.*
Statistical Analysis: *Standard.*

and group mean data can be used to predict future food consumption. Different concentrations of the test compound and diet should be made for each sex.

Test compounds and diets are mixed in two steps: (1) the compound and about 10% of the total amount of diet are blended in a premix, then (2) the premix and the remainders of the diet are mixed. The total amount of diet to be mixed is first weighed out, the 10% is separated into the premix. To make the premix, the entire test compound and an aliquot of the diet (from the 10%) are put into a mortar. These ingredients are ground with a pestle

TABLE 2.13 Pre- and Postnatal Development, Including Maternal Function in Rats

	P Generation (F0)		F1 Generation	
	Males	Females	Males	Females
Vehicle control	NA	25	25	25
Low dose	NA	25	25	25
Mid dose	NA	25	25	25
High dose	NA	25	25	25

Number in Study: P Generation—100 females, F1 Generation—100 males, 100 females.
Dose Route/frequency: *Once daily to P animals from Gestation Day (GD) 6 to Postnatal Day (PND) 21. F1 animals not dosed.*
Observations: *Twice daily (mortality/moribundity).*
Clinical Observations: *P females—daily during treatment/F1 adults—weekly.*
Body Weights:
P females—GD 0, 6, 10, 14, 17, and 20, PND 0, 7, 10, 14, and 21.
F1 males—Weekly through termination.
F1 females—Weekly until evidence of copulation detected, then GD 0, 7, 10, and 13.
Food Consumption: *P females—On corresponding body weight days during gestation/lactation.*
Vaginal Smears: *All F1 females during a 21-day cohabitation period until evidence of copulation is detected.*
Litter Evaluations: *All F1 offspring, count, body weight, sex, clinical observations on PND 0, 4, 7, 14, 21; behavioral and developmental evaluation of four males and four females from each litter for static righting, pinna detachment, cliff aversion, eye opening, air drop righting reflex, neuropharmacological evaluation, auditory response. One male and one female (selected for the next generation) tested for sexual maturation (vaginal opening, preputial separation), motor activity/emotionality, and passive avoidance.*
Cesarean Section: *On GD 13, F1 females for location of viable and nonviable embryos, early and late resorptions, number of total implantations, and corpora lutea.*
Sperm Evaluation: *May be conducted on F1 males if evidence of reduced fertility is noted (additional cost).*
Necropsy: *Gross lesions/target organs fixed for possible microscopic evaluation (additional cost).*
All P females at PND 22 as well as all F1 weanlings not selected for F1 generation.
All F1 females at GD 13.
All F1 males after termination of F1 cesarean sections.

until the mixture appears homogeneous. The mixture and the remainder of the premix are then layered in a small capacity mixer and mixed for 5–10 min. The time for this mixing process can be varied if analysis shows the total mixture is not homogeneous. For the final mix, the premix and the remainder of the diet are layered in a large capacity mixer. The mixing time will vary with the type of blender and can be varied if the analysis shows the total mixture is not homogeneous.

Several types of blenders are available for the mixing process; these include open-bowl "kitchen" mixers, V or PK blenders, and Turbula mixers. Metal parts should be ground to eliminate electrostatic forces. In addition, alternative methods of dietary administration such as microencapsulation may be used for volatile, reactive, or unpalatable chemicals.

Gavage Method

In the gavage procedure, the test compound is administered by passing a feeding tube or gavage needle attached to a syringe down the esophagus into the stomach.

Test Article Preparation

If not already a liquid, the test compound is prepared for administration by adding it to the appropriate vehicle. The choice of vehicle will depend on the characteristics of the compound and whether it is to be administered as a suspension or a solution. In addition, consideration must be given to the effects of the vehicle on the rat (Gad and Chengelis, 1998). Common vehicles used include water and food grade oils such as corn oil. Suspensions are made when aqueous vehicles are desired and the test compound is not soluble. Suspending agents such as methylcellulose are added to increase the viscosity and hold the compound in suspension. Other agents such as Tween 80, ethanol, polyethyleneglycol 400 (PEG 400), and others may be used as wetting or stabilizing agents.

Equipment

Soft catheters made of silastic or polyethylene (e.g., infant feeding tubes), stainless steel gavage needles with smooth ball-shaped tips, or polyethylene gavage needles with ball-shaped tips are commonly used. All are commercially available and are relatively inexpensive. Although the soft catheter minimizes the chance of esophageal trauma, liquid can leak past the catheter and back up the esophagus and be aspirated. The ball-shaped tips of the stainless steel gavage needles reduce the chances of tracheal injections; however, if an animal struggles while the needle is in the esophagus, the rigid needle increases the chances of perforating the esophagus. The polyethylene gavage needle incorporates the best of both the soft catheter and the stainless steel needle, but because of the flexible nature of the needle, the risk for tracheal injection is increased.

Conybeare and Leslie (1980) found that deaths in gavage studies were a result of aspiration of small amounts of irritant solutions or acidic, hypertonic solutions. They also found that the use of a ball-tip 4 mm in diameter helped to eliminate deaths related to dosing. With gentle handling, the animals will be acclimated to the techniques used and dosing will become easier.

Aspiration and tracheal administration of test compound as well as esophageal trauma have been associated with gavage dosing and may lead to difficulty in interpretation of the study. The catheter and the needles all have risks inherent in their use; therefore, care should be taken when using these tools and animal technicians should be properly trained. The choice of the appropriate catheter or needle should be left up to the technician and should be whatever the technician has been trained and is most comfortable with.

Technique

The description below is appropriate for either a gavage needle or catheter; for simplicity, only the needle will be mentioned in the description. Prior to picking up the animal, the syringe should be attached to the needle and filled with the appropriate amount of test compound to be delivered. Any air bubbles should be eliminated and the needle wiped clean of residual test compound. This is done so that the animal does not taste the test compound and residual test compound is not aspirated as the needle is passed down the esophagus. If the dosing liquid is distasteful, the animal may struggle after repeated dosing and increase the chances of being injured.

To position the animals for gavage, it should be grasped by the skin of the back and neck ensuring that the head, neck, and back are in a straight line. Alternatively, the animals can be grasped about the shoulders, with the index finger and thumb on either side of the head. The objective is to firmly hold the animals to be able to control any struggling if it occurs and to also prevent the animal from being able to bite the technician. For even more control, the animal may be placed on a table or brought up against the operator's chest.

Once the animal is in position, the needle can be inserted into the mouth of the animal, moved over the tongue, and down into the esophagus. The length of the needle should be inserted into the animal. A slight rotation of the needle may help with insertion into the esophagus. If the needle is inserted into the trachea, the animal may struggle. The syringe should be grasped lightly such that, if the animal does struggle, the chances of an esophageal tear are minimized. If the animal continues to struggle, the needle should be withdrawn to allow the animal to calm down, and then dosing should be attempted again. Alternatively, if a catheter is used, as the tube is placed into the mouth, it should be placed to the side between the molars. This is done because the tube may by bitten or transected if passed too close to the front teeth.

With the needle in place, the test compound should be slowly expelled into the animal. If administered to rapidly, reflux may occur and the test compound may back up into the esophagus, resulting in an inaccurate dose being given and possible aspiration of the test compound. Once the dose has been delivered, the needle should be withdrawn and the animal observed for any signs of distress or respiratory difficulty. An experienced technician should be able to dose between five and seven animals per minute without causing discomfort to the animals and with minimal dosing-related deaths.

Gavage liquids are commonly administered at a volume of 5–10 mL/kg body weight. The volume should be enough to be delivered accurately, but not so much that it will adversely affect the animal. The maximum volume should be no more than 20 mL/kg. If using volumes greater than 10 mL/kg, it may be advisable to fast the animals for several hours prior to dosing. This will ensure that the stomach is empty prior to dosing and able to handle the larger volume. This option should be considered carefully, as fasting can affect the rate of absorption and clearance from the stomach. In addition, the choice of housing and bedding should be considered when dosing with large volume as rats have the tendency to eat the bedding, which may hinder gavage dosing. In addition, the volume chosen can have an effect on the results of the study and volumes greater the 10 mL/kg should only be used when issues of solubility and exposure exist. Ferguson (1962) found that a change in dose volume of from 5% to 1% of body weight could reduce mortality rate from approximately 95%–5%, respectively, at equivalent doses.

Neonatal Administration

Neonatal intragastric injections can be made orally with thin silicone tubing (Gibson and Becker, 1967; Smith and Kelleher, 1973) or by intragastric injection with a 27-gauge needle through the abdominal wall (Worth et al., 1963; Bader and Klinger, 1974). The oral method using silicone tubing is performed in a similar manner to the previously described method in adult rats. The intragastric injection through the abdominal wall is performed by first locating the stomach in the upper left quadrant of the abdomen and then carefully inserting the needle through the abdominal wall into the stomach taking care that the animal does not move. The syringe should be gently aspirated to ensure proper placement and then the injection completed and the needle withdrawn.

Capsule

To eliminate the possibility of dosing errors and to deal with compounds that cannot be delivered through conventional means, methods have been developed for the administration of capsules into the esophagus of the rat. The test compound may be prefabricated into a small capsule or the test article may be weighed and placed into commercially available capsules. An individual capsule is then placed into a specially designed cup in the end of a gavage needle, and the needle is then inserted into the esophagus of the rat. The capsule is then pushed out of the cup into the esophagus using either air or a rod inside the needle. The needle is then withdrawn and the capsule moves down into the stomach by peristaltic action. Only a small amount of test compound can be administered as a single dose using this method, but multiple capsules can be administered sequentially in the same dosing session.

Water

As an alternative to dietary administration, compounds that are water soluble, palatable to the rat, and stable in water may be administered via the drinking water. This method offers similar advantages as adding a test compound to the diet. Additionally, compounds will be more easily mixed and analyses will be more easily developed than when a compound is in the diet. However, spillage of water makes measurement of the actual dose received difficult.

Intravenous Route

One of the most common methods of administration of test compound is via intravenous (iv) injection or infusion. The iv route is often the route of choice for compounds that have poor bioavailability via the oral route or have a short half-life. Several issues must be considered when administering a test compound intravenously. The compound must be soluble in an acceptable iv vehicle or excipient, must be able to be administered as a solution, and should be sterile or sterile filtered prior to administration. In addition, when designing a study, the pharmacokinetic profile of the test compound administered intravenously should be considered. Study activities such as clinical observations and functional observational battery should be planned around the expected time of greatest plasma concentration.

A variety of veins may be used for iv injections (Diehl et al., 2001). These include the lateral tail (caudal), jugular, femoral, saphenous, lateral marginal, dorsal metatarsal, sublingual, and dorsal penile vein. Although most of these are superficial, and easily available for injection, several require the use of anesthesia or more than one technician and may be of limited use in repeat dose studies. Although anesthesia may be acceptable for acute studies or surgical model, its repeated use may have an effect on the toxicity of a test compound.

Lateral Tail Vein

The lateral tail veins are currently the most widely used for iv injections in the rat. The veins are easily visible, especially in young animals and injections can be performed by one person without the use of anesthesia. The technician performing the function should be well trained and care should be taken to ensure that the lateral veins are being accessed and not the dorsal or ventral artery of the vein.

Bolus Injection

The animal should be placed in an appropriate restrainer. This typically consists of a solid tube in which the animal is placed into headfirst and has a stop that is placed behind the animal with a hole that allows the tail to hang out the back. The restraint tube is designed to be secure enough that the animal cannot move, back out, or turn, but can still breathe comfortably. Once secure, the tail should be cleaned and the vein may be dilated with heat. This may be accomplished by placing the tail in warm water (40–45°C), placed under a heat lamp, or wrapped with warm gauze. Care must be taken to avoid using excessive heat as tissue damage may result. Minasian (1980) describes a tourniquet made from a plastic syringe and thread. If used, this should not be left on for an extended period of time.

When performing an injection, the end of the tail should be held firmly and taut with the thumb and index finger of one hand. A 23-gauge needle attached to an appropriately sized syringe should be held with the bevel up at a shallow angle parallel to the vein. The skin of the tail is then pierced and the needle advanced until resistance is no longer felt. The plunger of the syringe should then be aspirated to ensure proper placement of the needle. The use of a needle with a clear or transparent hub will facilitate confirmation of correct placement. Blood backflow into the needle confirms entry into the vein. Alternatively, a butterfly needle with an extension line may be used. The butterfly needle with an extension set precludes the need to hold the tail, needle, and syringe. When using this type of setup, the butterfly needle is attached to an extension set and syringe that is filled with the test compound. The tail may be taped to the table, and the butterfly needle is then inserted into the vein and placement is verified by aspiration on the syringe. Once confirmed, the butterfly needle may also be taped in place. This prevents the needle from pulling out of the vein during dosing. This type of setup can be very useful when administering large volumes of test article as a slow bolus over several minutes or when the test compound may be irritating or mildly caustic. Taping the animal's tail in place prevents the animal from pulling the tail out of the fingers of the technician.

If repeated dosing is to be performed, the initial venipunctures should be performed as close to the tip of the tail as possible. During the injection, if the needle comes out of the vein, a bleb will form under the skin. The needle should be repositioned immediately to prevent infiltration of the solution around the vein. Infiltration of an irritating solution can cause necrosis and make future injections difficult or impossible. Injection of 2 mL/100 g body weight can be accomplished without stress to the rat. Barrow (1968) found that injections of volumes over this amount produced respiratory difficulty and pulmonary edema.

Tail Vein Infusions

Tail vein infusions are convenient because catheter placement can be accomplished without anesthesia. A 23-gauge or smaller needle connected to an extension set is inserted into the tail. The needle and extension set is then secured to the tail with tape. The extension set is attached to a syringe that is placed on a pump and the test compound can be infused. The tail may be taped to a wooden stick or tongue depressor to further protect the needle from being dislodged. Over the needle, catheters are also commercially available and offer the advantage that the needle is removed once the catheter is placed in the vein and may help to prevent further penetration of the vein wall and subsequent perivascular dosing (Rhodes and Patterson, 1979). Advantages that this technique has over permanent indwelling catheters are that the catheter is removed following dosing and will not become occluded and the animal doses not have to undergo anesthesia and a surgical procedure to place the catheter. Permanent catheters have a tendency over time to develop a fibrin flap or become clotted, thus losing patency. A major disadvantage is that the animals have to be restrained during the infusion, which may causes stress and alter the results of the study. When using this technique, the duration of the infusion should be limited so that the length of time the animal is restrained is limited.

An alternative technique using the lateral tail vein involves placing a catheter in the vein and wrapping the tail in a similar manner as previously described, then a lightweight protective cover attached to a tether system is placed around the tail to hold the catheter or needle in place.

Jugular Vein

Although this route has been used for bolus injections, it is most widely used as a site for cannulation from indwelling catheters. The indwelling catheter requires surgical implantation under anesthesia.

Bolus Injection

Although injections can be made by exposing the jugular vein by incision, this method is not acceptable for repeated dosing. The jugular vein can be accessed for test compound administration without exposing the vein. The animal can either be anesthetized or restrained on the back. The head is positioned to either the left or the right for access to the respective jugular vein. A 23-gauge needle fitted to a syringe with the bevel up is inserted in a cephalocaudal direction into the angle made by the neck and shoulder. The needle should enter the vein anterior to the point at which it passes between the pectoralis muscle and the clavicle. When about one-half the length of the needle has penetrated the skin, the bevel should be in the lumen of the vessel. Insertion of the needle through the muscle stabilizes the needle and minimizes bleeding. Caution should be used when using this technique as it is considered to be a "blind stick" into the vessel, and damage to the vessel may occur. Repeated access of the vessel is not recommended.

Infusion

For the purpose of continuous infusion of the test compound over extended periods of time or for repeated short-term infusions, implanted catheters in the jugular vein may be used. For implantation of a jugular catheter, the animal is first anesthetized and placed in dorsal recumbency, and the surgical site is prepared. A midline incision is then made in the neck, and a section of the jugular vein is dissected free. Manipulation of the vein should be limited to prevent vasospasm. A cephalic ligature is then tied and the vein elevated. A small incision is then made in the vein, and the catheter is passed into the vein and tied in place. The other end of the catheter is then tunneled subcutaneously (sc) to between the scapula where the catheter is exteriorized. The catheter should be filled with an anticoagulant solution such as heparin when not in use. When correctly positioned, the tip of the cannula will be at the junction of both vena cava. If placing catheters into young animals, enough of the catheter should be inserted to allow for growth of the animal. Care should be taken that the catheter is not inserted too far as the tip may be pushed into the right ventricle of the heart. Improper placement of the catheter may lead to administration of the test compound directly into the heart, which can cause complications.

Similar to the jugular vein, administration of test article via the femoral vein requires an implanted catheter. For implantation of a femoral catheter, the animal is first anesthetized and placed in dorsal recumbency, and the surgical site is prepared. A midline incision is then made in the inguinal area and a section of the femoral vein is dissected free. Manipulation of the vein should be limited to prevent vasospasm. A ligature is then tied and the vein elevated. A small incision is then made in the vein and the catheter is passed into the vein and tied in place. The other end of the catheter is then tunneled sc to between the scapula where the catheter is exteriorized. The catheter should be filled with an anticoagulant solution such as heparin when not in use. When correctly positioned, the tip of the cannula will be position in the vena cava. For longer-term infusions, the femoral vein catheter may be preferable as patency is easier to maintain and the risk of damage to the heart from the catheter is avoided.

Several commercial vendors offer surgical support services and for an additional fee will implant either jugular or femoral catheters. These vendors will typically have a specific methodology for implant, but will accept requests for modifications such as catheter type, exteriorization site, etc. The typical catheter implanted may be manufactured from polyethylene, polypropylene, or silastic. In recent years, manufacturers have developed catheters impregnated or ionically bound with heparin. These materials may help to prolong the life of the catheter.

The useful lifetime for jugular and femoral catheters is quite variable; the lumen of the cannula may eventually become obstructed by a blood clot or fibrous mass. The position of the tip of the catheter is important. Clot formation is less likely to occur if the tip of the catheter is placed in the venous stream rather than in the jugular vein (Popovic and Popovic, 1960). It is recommended for repeated short-term infusions, and when test article is not being infused, a slow infusion of saline will help to prolong the life of the catheter.

Prior to use, the patency of the catheter should be checked by removing the anticoagulant lock, check for blood draw back, and then flushing with saline or Lactated Ringers solution. Alternatively, patency can be checked by injecting 3–6 mg of pentobarbital solution (0.05–0.10 mL of a 60 mg/mL solution) into the catheter (Weeks, 1972). If the catheter is patent, the rat will lose its righting reflex and become ataxic within 10–15 s of injection. The rat will recover in 10–15 min.

Rats will destroy the catheter if it is left unprotected or in easy reach of the forepaws. By exteriorizing the catheter between the scapula, the rat will not be able to chew on the catheter. For the purpose of continuous infusion, several manufacturers have developed tether systems and catheter sheaths made of metal that prevent the animal from chewing on the catheter. These systems typically consist of a jacket with an attached tether through which the catheter is passed. The catheter then attaches to a swivel that prevents the catheter from becoming kinked. The swivel then attaches to a second catheter that can be attached to a syringe or pump for administration of the test compound (Guo and Zhou, 2003). When performing long-term infusion studies, the effects of the catheter and harness should be considered. Infections, septicemia, a variety of visceral lesions, endothelial lesions, and increased platelet consumption have been observed in cannulated animals (Hysell and Abrams, 1967; Meuleman et al., 1980; Vilageliu et al., 1981). Decreased or erratic weight gains and decreased liver and thymus weights have been observed in tethered animals. These changes may be attributed to the stress involved in chronic tethering of the animals.

An alternative to an exteriorized catheter is to attach a sc port to the catheter that can be accessed via a transcutaneous needle stick. This type of setup helps to prevent infections that can occur with transcutaneous catheters. One of the pitfalls of this sc port is that the port may only be accessed a finite number of time. In addition, care has to be taken to ensure the port and catheter are properly flushed of all test compound and blood as clots can easily form. Administration of small volumes of test compound may be accomplished using a sc implanted osmotic pump. This type of pump is connected to the catheter after being filled with the test compound and implanted in a sc pocket. This allows for continuous administration of small amounts of compound without the need for a jacket and tether system.

Saphenous, Lateral Marginal, and Metatarsal Veins

These veins in the leg and foot are easily visualized and can be injected without anesthesia; however, assistance is required. Shaving the area over the saphenous or lateral marginal vein makes visualization easier. During injection it is necessary for one technician to restrain the animal and occlude the vessel to cause it to dilate. Wiping the skin over the vein with 70% alcohol or with gauze soaked in hot water will help to dilate the vessel and increase the possibility of success. The second technician then performs the injection, in which a 26–27-gauge needle should be used.

Dorsal Penis Vein

When administering test article via the dorsal penis vein, it is preferable to use anesthesia. Lightly anesthetizing the animals with an inhaled anesthetic such as isoflurane or CO_2/O_2 will prevent the animal from struggling and increase the possibility of a successful injection. This procedure requires two technicians to perform the injection. One technician holds the animal by the skin on the back and the feet and tail. The vertebral column is then hyperextended. The second technician then grasps the tip of the penis between the thumb and forefinger, and injects the test solution into the dorsal vein using a 26–30-gauge needle.

Sublingual Vein

Although the method of sublingual vein injection has the disadvantage of requiring anesthesia, it only requires one technician. Ideally, the animal should be anesthetized with an inhaled anesthetic such as isoflurane or CO_2/O_2, but injectable anesthesia may also be used. The animal should be placed in dorsal recumbency with the head toward the operator. The test compound may be administered by holding the tongue between the thumb and forefinger; using a 26–30-gauge needle, the vein is entered at a very shallow angle and the injection is performed. After completion of the injection, the bleeding can be stopped using direct pressure. Once the bleeding has stopped, a small cotton-wool pledget should be placed over the vein and the tongue placed back in the mouth. The animal will spit the cotton out on regaining consciousness.

Intraperitoneal Route

Test compounds injected into the peritoneal cavity will be absorbed into the portal circulation and transported to the liver. As a result, the compound will be subjected to the metabolic activity of the liver prior to being circulated to the remainder of the animal. Based on the level of blood flow and circulatory surface area in the peritoneal cavity, compounds injected intraperitoneally (ip) will be absorbed quickly.

Intraperitoneal administration of test compounds in the rat can be performed by one person. The animal should be picked up by the scruff of the neck and back and held firmly in dorsal recumbency. This position will allow for proper access to the peritoneal cavity. The belly of the animal should be visually divided into quadrants and a needle (<21 gauge) should be inserted anteriorly into one of the lower quadrants just lateral to the midline. Aspiration of the syringe prior to injection will help the investigator to determine if the needle is positioned appropriately.

Intramuscular Route

Intramuscular (im) injection of compounds will result in the rapid absorption into general circulation due to the abundant supply of blood vessels. However, the speed of absorption will not be as fast as with an ip injection. Acceptable sites in the rat are the quadriceps, the thigh, and the triceps. This procedure can be done with one or two technicians. The selected muscle mass should be stabilized with the thumb and forefinger of one hand while restraining the animal and guiding the needle into the muscle with the other hand. A 21-gauge or smaller needle should be used. The needle should be lightly aspirated to ensure the tip of the needle is not in a blood vessel and then the compound is slowly injected. A slow injection with a minimal volume will help to minimize pain. Approximately 1 mL/kg of solution can be injected per site. If larger volumes are required, multiple injection sites should be used.

Subcutaneous Route

Absorption following sc injection is typically slower than following im injection. This may be advantageous if a relatively sustained period of absorption is desired. Another advantage of the sc route versus the im route is that a much large volume of test compound can be administered. Five to ten milliliter can be easily injected with little to no discomfort to the animals. This can be beneficial for test compounds that have limited solubility. Suitable sites for injection are the ventral body, the flank, and shoulders. To perform the injection, the skin is grasped between the thumb and forefinger and raised to make a tent. The needle (<20 gauge) is inserted through the skin to make the injection. Injection sites can be varied for multiple-dose studies where the solution is a potential irritant.

To minimize the stress of manipulation and to provide a means for continuous infusion, a perforated cannula or catheter can be implanted transcutaneously. The cannula can then be secured using a tether system similar to that used for iv infusions. Using this system, the test compound can be infused continuously. Mucha (1980) injected sodium pentobarbital directly into the cannula and showed it was absorbed much more rapidly than following injection. An alternate method of infusion of small volumes over an extended period of time is through the use of an osmotic minipump. The pump is filled with the test compound and implanted sc under anesthesia. These pumps are commercially available and offer the advantage of a continuous infusion at a constant rate without the animals being encumbered with the infusion apparatus. This method would work only when the solutions are stable at body temperature for the duration of the infusion period. In recent years several types of absorbable microspheres have been developed that can act as carriers for the test compound.

Topical Route

The rat has not traditionally been used as a model in skin irritation or sensitization studies. However, the rat has been used in systemic toxicity studies where the skin is used as a portal of entry for whole body exposure or in skin painting studies where the carcinogenic potential is being assessed. In a comparison of absolute absorption rates of several compounds, Bartek et al. (1972) found dermal absorption rates in the rat tended to be slightly lower than in the rabbit, and higher than in the monkey, swine, and human.

Exposure in dermal studies is usually to the anterior dorsal portion of the back. The skin should be shaved weekly or 24 h prior to skin painting. Care should be taken to ensure that the skin is not damaged during shaving, as this can increase the rate of absorption of the test compound. The test area for application should be clearly marked; for repeat dose studies, the area of this site is often 10% of the body area. Usually 0.25–1.0 mL of the test solution is applied in skin painting studies. The amount of a cream or ointment applied will vary with the test compound and desired total dose administered. Dosing is typically performed every day.

The actual dose in a dermal toxicity study is determined by the amount of compound absorbed, therefore factors that influence absorption should be considered. Several design features in a topical study may affect absorption; abraded skin will tend to absorb faster than intact skin; test compound may adhere to or build up at the site of exposure and may impede absorption (and test compound may be chemically changed owing to exposure to air or light); the test compound may be licked or scratched from the site; and the test compound may be ingested by the animal.

Several techniques have been developed to avoid removal or ingestion of the test compound. For acute studies, Rice and Ketterer (1977) described a cable-type restrainer attached to a stainless steel plate. Loops just behind the front and just in front of the hind legs hold the animal immobile. Other methods may be used to reduce stress to the animals and allow the animal mobility. One method is the use of an "Elizabethan" collar. This is a 4–5-cm wide strip of plastic or metal that fits around the neck of the animal. This prevents the animal from being able to turn its head to gain access to its back. Consideration should be given to the use of this method as the collar also may prevent the animal from being able to properly eat from a feed jar. A second method is to wrap the animal with gauze and then with plastic wrap. Care should be taken when wrapping the animal to ensure that the wrap is not too tight. This type of covered exposure may affect the absorption of the test compound. Other types of harness, collars, and acrylic chambers may be used, and the appropriate technique should be chosen based on the intended length of exposure and the efficacy with which the technique can be performed by the technicians.

Rectal Route

The rectal route is not a routinely used method of administration in toxicology. However, administration by this route is sometimes required to support drugs given rectally by suppository. For dosing, the animal is held by the base of the tail and a stainless steel, and ball-tipped gavage needle (5 cm) or vinyl tube (6 cm) attached to a syringe is inserted into the rectum. Care must be taken not to damage the rectum when inserting the needle. The syringe should be held lightly: the weight of the needle and syringe propel the needle. The animal can either be awake or anesthetized. If animals are awake, excretion of the unabsorbed test compound may occur. Methods to control excretion have included ligation of the rectum (Nishihata et al., 1984) or various types of septums that are tied or glued in place (DeBoer et al., 1980; Iwamoto and Watanabe, 1985). Anesthetized animals can be placed on an inclined board to retard expulsion.

An important factor in rectally dosing the rat is that the depth of deposition of the test compound will affect the rate of absorption and should be standardized. Drugs subject to extensive first-pass metabolism, such as propranolol and lidocaine, have been found to be much more bioavailable when injected close to the anus rather than in the upper areas of the rectum (DeBoer et al., 1980; Iwamoto and Watanabe, 1985). The reason for this difference in bioavailability appears to be that the venous return in the upper rectal area is through the upper rectal vein that feeds back into the portal circulation and then into the liver. The venous return in the lower rectum is through the lower hemorrhoidal veins and is not connected to the portal system, but goes directly to the inferior vena cava (Iwamoto and Watanabe, 1985).

Intranasal Route

With the increasing number of drugs being delivered nasally, methods have been developed to support this route. For administration in unanesthetized animals, the appropriate volume of test material is drawn into the tip of a pipette or other appropriate dosing implement. The tip of the dosing implement is placed directly over but not into the nostril to be dosed, and the test material is instilled into the nostril. The animal will aspirate the test material into the nasal passage. This can be repeated for the opposite nostril, or the opposite nostril can be used as a control treatment. In the event that the animal sneezes or the test compound is otherwise expelled, the nostril should be retreated.

Inhalation Route

Owing to the complexities and equipment involved in generating, maintaining, and measuring appropriate atmospheres, the inhalation study is one of the most technically difficult to perform. This section is not written to provide a complete discussion of the skills necessary to perform an inhalation study, but will deal with the general considerations about the three major steps in exposing rats by the inhalation route: generation of the test atmosphere, exposure of the test animals, and measurement and characterization of the test atmosphere.

Intratracheal Administration

An alternative to inhalation administration is to instill the test compound into the trachea of the animal. Techniques have been developed to instill the test compound into the trachea safely and repeatedly. To perform the procedure, the animal must first be anesthetized, preferably with a gas anesthesia such as isoflurane. Once anesthetized, a speculum is inserted into the mouth and passed into the trachea. A syringe and needle with a 5-cm piece of tubing is used to instill the test compound. The tubing is then passed over the speculum and the test compound is administered. Volumes should be limited to 2 mL or less of test compound.

REGIMEN

There are a number of factors that can be manipulated to maximize both administered and absorbed levels of a drug, as well as increasing both tolerance and target tissue specificity. Most of these are beyond the scope of this volume (Gad and Spainhour, 2017), but regimens (how often and at what intervals a test article is administered) are fundamental tools of the experimental animal researcher. Once, twice (bid), or three (tid) times a day are readily available approaches. Likewise, for parenteral routes, an administration can be done quickly (bolus) or over a period of time (infusion).

END POINT MEASUREMENT TECHNIQUES

Observations and Physical Examinations

Rats are routinely monitored during toxicology studies as an assessment of their general health and to define the effects of the test article. In acute and subchronic studies, animals may be observed frequently in an effort to define short-term pharmacologic changes induced by the test compound, which may become apparent at peak blood levels. Specifically, in acute toxicology studies clinical observations will help to establish a MTD. In chronic studies, these observations are critical in tracking tumor development and for determining animals in extremis, which should be euthanized for humane reasons and to prevent autolysis and tissue loss. Arnold et al. (1977) provide a useful description of a clinical assessment program for chronic studies.

Daily observations are performed first thing in the morning and last thing before leaving in the afternoon to assess the health of the animals and identify animals that may be in extremis. In this observation, behavioral status, respiratory signs, skin, eyes, and excretory products are noted. Care should be taken to disturb the animal as little as possible, as this may induce stress and affect the animal's behavior. The animal should be picked up and examined more closely if abnormalities are detected. Special attention should be paid to the amount of feces present, because a decrease in fecal output may be the first signs of a watering system malfunction. In acute studies or where pharmacological effects are expected, animals may be examined continuously or at peak plasma levels.

A more thorough physical examination should be done weekly. Each animal is taken from its cage and placed on an examination table where respiration, behavior, general appearance, and locomotion are observed. The technician should then pick up the animal, examine its body orifices, skin, and coat, and perform a palpation of the trunk and limbs to check for tumors. The detection and tracking of the size and fate of masses (potential tumors) is essential for carcinogenicity studies.

Animals that have experienced severe weight loss over the previous week, a progressive decline in weight over several weeks, or other severe clinical signs should be observed more frequently and marked for possible euthanasia. Body and feeder weights can be measured as part of the physical examination or as a separate function. Performance of these operations as a part of physical examinations will minimize animal handling and potential stress. If the operations are combined, it is important that the physical examinations be done completely and not rushed.

Neurobehavioral Examination

Neurobehavioral examinations are included in toxicology studies to assess the behavioral and neurological effects of test compounds. These examinations, which may be done as part of acute or repeat dose toxicity studies or studies specifically designed to assess only neurobehavioral effects, typically involve screens consisting of an abbreviated functional observational battery and some measure of locomotor activity (Annau, 1986). The EPA has written guidelines on the design and conduct of these studies (OECD, 1997, 2008). To meet EPA guidelines, the screen is performed prior to the start of treatment, then periodically during the course of treatment. It may be performed as a separate study, on satellite groups of animals in conjunction with the main study, or on animals in the main portion of the toxicology study. The ICH in 2002 adapted guidelines requiring neurobehavioral assessment of all new pharmaceuticals prior to initiation of Phase I clinical trials. To meet ICH guidelines, a more formal functional observational battery should be conducted and may be conducted on main study animals or may be conducted as a separate study. Where initial screens indicate the possibility of a test compound–related change, a more specialized series of tests may be performed to assess the nature of the effect and the extent of the central nervous system involvement. These secondary tests evaluate motor and sensory function as well as cognitive ability. Examples of secondary tests include sensory-evoked potential experiments and schedule-controlled behavior studies. Descriptions of these secondary tests can be found in Annau (1986).

Functional Observational Battery

The typical observational battery (FOB) includes observation of home-cage and open-field activity as well as measurements of reflexive, physiological, and neuromuscular function (Gad, 1982; Moser et al., 1988; Moser and Ross, 1996) as outlined in Table 2.15. Observations and measurements that have become standard for an FOB evaluation are in Table 2.16. The order of measurement should be consistent, progressing from the least interactive to the most interactive measurements. Home-cage observations are made first. Assessments of posture, clonic movements, tonic movements, and palpebral closure (if the animal's eyes can be seen) are taken prior to removing the animal from its cage. However, it may be necessary to pull the cage from the cage bank or remove the cage cover to see the animal's eyes. The animals are then transferred to the open field. During the transfer, certain physical observations are made. The technician removes the animal from the cage. The technician holds the animal and notes increased or decreased body tone as well as such observations as bite marks, soiled fur appearance, missing toe nails, emaciation (shallow stomach, prominent spinal vertebrae), or death. In addition, observations of lacrimation, palpebral closure, piloerection, exophthalmus, and salivation are also made. The animal is then placed in the open-field apparatus for a set period of time. Measurements of rearing, urination, and defecation are made immediately at the end of the assessment period. Assessment of clonic movements, tonic movements, gait, mobility score, arousal, respiration, stereotypic behavior, and bizarre behavior may be made

TABLE 2.15 Functional Observational Battery in Rats

	Number of Animals	
	Males	**Females**
Control	10	10
Low dose	10	10
Mid dose	10	10
High dose	10	10

Dosing: The test article will be administered by the required route.
Functional Observational Battery: *FOB evaluations will be conducted prior to dosing, at the estimated time of peak effect, and 24 h postdose. The evaluations are as follows:*
Home Cage Observations:
Assessments of posture, clonic movements, tonic movements, and palpebral closure (if the animal's eyes can be seen) are taken prior to removing the animal from its cage.
Handling Observations:
Observations of ease of removal, handling reactivity, lacrimation, palpebral closure, piloerection, and salivation are made on removal of the animal from the home cage.
Open Field Observations:
The animal is placed in the center of an open-field testing box (measuring 20″ × 20″ × 8″). Clean absorbent paper may be used to cover the bottom of the box if required by protocol. Using a stopwatch, the animal's stay in the box is timed for 3 min. Measurements of rearing, urination, and defecation are made immediately at the end of the 3 min. Assessment of clonic movements, tonic movements, gait, mobility score, arousal, vocalization, respiration, stereotypic behavior, and bizarre behavior may be made immediately after the 3 min have ended or the technician may continue to observe the animal for a longer period of time to allow for more accurate assessment.
Sensorimotor Observations:
The approach response, touch response, click response, and tail-pinch response (stimulus reactivity tests) are performed while the animal is in the open-field apparatus, after the 3 min time period is over and all other measurements have been recorded. The animal is removed from the open field apparatus for the pupil response, righting reflex, thermal response, hind limb splay, and grip strength measurements.
Physiological Evaluations:
The animal's body weight and rectal temperature is measured and recorded.
Clinical Examination: *Following each FOB assessment (additional observations will be conducted prior to dosing for locomotor activity, if requested).*

immediately after the time period has ended or the technician may continue to observe the animal for a longer period to allow for more accurate assessment. When the open-field assessment has been completed, the animal is removed from the open-field apparatus for the approach response, touch response, click response, tail pinch response, pupil response, righting reflex, thermal response, hind limb splay, and grip strength measurements. After completion of the manipulative assessments, the physiological evaluations are completed.

Procedures used during the performance of an FOB must be standardized because some observations made have a subjective component. If at all possible, a single observer should be used throughout a single study. If not possible, a single observer should conduct all assessments of an animal. In addition, technicians should be blinded to the treatment conditions for each animal.

Locomotor Activity

Methods used for recording motor activity include direct observation and automated techniques such as photocell devices and mechanical measurements (MacPhail et al., 1989). In direct observations, the observer can make quantitative measurements of the frequency, duration, or sequencing of various motor components of behavior or qualitative records on the presence or absence of certain components of activity. Photocell devices record the number of times an animal interrupts a beam in specially designed chambers. In mechanical chambers, the animal's movements result in a vertical or horizontal displacement of the chamber; records are kept of the chamber's movements. There are advantages and disadvantages of each technique. In direct observation, record can be made of behavior, such as convulsions, which may not be observed when using the photocell or mechanical methods. A disadvantage of the direct observation method is that the animal may be influenced by the presence of the observer. Advantages to the photocell and mechanical methods are that the data are captured electronically, the observer does not have to be present, and the computer system can graphically present the data in the form of lines crossed or a map of the activity.

To make activity determinations, an animal or group of animals is put into an observation or recording chamber and activity is recorded for a specific period of time. Because activity will normally decline over the course of the session, the length of the observation period is important. The EPA guideline specifies that activity should approach asymptotic levels by the last 20% of the session. Haggerty (1989) used a 15-min recording session, accumulating data over three 5-min intervals. Because a large number of environmental conditions can affect motor activity, e.g., sound level, cage design, lighting, temperature, and humidity, or odors, it is important to minimize variations in the test environment.

Cardiovascular Parameters

Examinations of the cardiovascular system may be scheduled into toxicology studies or performed when the cardiovascular system is a suspected target of the test compound. The ICH has adapted guidelines requiring cardiovascular safety assessment of all new pharmaceuticals prior to initiation of Phase I clinical trials. This guideline recommends that this assessment be conducted in a nonrodent species, but cardiovascular assessment can be performed for screening purposes or as additional support data.

TABLE 2.16 Functional Observation Battery

Home Cage	Open Field	Manipulative	Physiological and Neuromuscular
Posture	Rearing	Ease of Removal From Cage	Body Weight
Clonic Movements	Urination	Handling Reactivity	Body Temperature
Tonic Movements	Defecation	Lacrimation	Hind limb extensor strength
Palpebral Closure	Clonic Movements	Palpebral Closure	Grip Strength
	Tonic Movements	Piloerection	Hind limb Splay
	Gait	Exophthalmus	
	Mobility	Salivation	
	Ataxia	Approach Response	
	Arousal	Touch Response	
	Vocalizations	Click Response	
	Respiration	Tail Pinch Response	
	Stereotypical	Pupil Response	
	Bizarre Behavior	Eye blink response	
		Forelimb extension	
		Hind limb extension	
		Righting reflex	
		Thermal Response	

From Gad, S.C., 1982. A neuromuscular screen for use in industrial toxicology. J. Toxicol. Environ. Health 9, 691–704; Moser, V.C., McCormick, J.P., Creason, J.P., et al., 1988. Comparison of chlordimeform and carbaryl using a functional observational battery. Fundam. Appl. Toxicol. 11, 189–206; Haggerty, G.C., 1989. Development of tier I neurobehavioral testing capabilities for incorporation into pivotal rodent safety assessment studies. J. Am. Coll. Toxicol. 8, 53–69.

Electrocardiography

Although the dog has traditionally been the species of choice in toxicology studies of effects of ECG (also called EKG), research with the rat has progressed and increased over the years. Detweiler (1981) and Detweiler et al. (1981) provide an excellent review of the use of electrocardiography in toxicology studies in the rat. This section will present a general discussion on the aspects of the recording methods and interpretation of the ECG in the rat.

Recording Methods

Restraint

One of the disadvantages of traditional methods of studying ECGs in rats is that it is difficult to keep the animals still while recording. It is important that the animal remain in a constant position during the procedure using skin leads to avoid muscular artifact in the ECG. Various forms of restraint have been tested, each of which requires acclimation to the procedure prior to evaluation. These methods included restraining the rat in a supine position using rubber gloves (Hundley et al., 1945), pinning the animal to a board, boards with clamps, and plastic tubes with slits on either side, which allow for placement of the electrodes (Zbinden et al., 1980; Spear, 1982). Various forms of anesthesia have also been evaluated.

No conclusion has been reached about the best method of restraint. The basic concerns are that manual methods and physical restraint require acclimation to allow the animal to become accustomed to the procedure, and tracings can be reasonably free of muscular artifact. Also, varying pressures of clamps or handling during the restraint may affect the results. The use of anesthesia has been shown to produce changes in the ECG and the possibility of drug interactions between the anesthetic and the test compound may occur.

Position

The most common positions are the prone or ventral recumbency position when animals are awake and the supine or dorsal recumbency position when anesthetized. Beinfield and Lehr (1956) compared the positions and concluded that the prone position produced an increased R wave and it avoided unfavorable cardiac rotation and an undesirable variation in the projection of the special QRS loop.

Tethered

Robineau (1988) developed an electrode system that can be implanted sc a few days prior to recording. The device has a disconnect that is exteriorized between the scapula. The advantage of this system is that a cable can be connected to the plug and the ECG can be taken in unrestrained rats. The disadvantage of this method is that it should only be used for short-term studies as the implant provides a source for infection in the animal.

Leads

Most investigators use Einthoven's bipolar limb lead system, lead I (right and left foreleg), lead II (right foreleg and left hind leg), and lead III (left foreleg and left hind leg), with and without the augmented unipolar limb leads aVR (right foreleg), aVL (left foreleg), and aVF (one of the hind legs). Because foreleg position can alter the scalar ECG wave amplitudes, investigators must standardize foreleg positions during recording. When implanting telemetry leads, the leads are placed on the right clavical and the most caudal rib on the left side in a modified lead II configuration.

In the past, various types of leads have been used to connect the ECG wires. These included the use of hypodermic needles inserted under the skin, small gauge insulated copper wires wrapped around the shaved distal portion of the limbs, or alligator clips. In addition, platinum-tipped pin electrodes are commercially available, which provide a good quality signal with limited discomfort to the animal.

Telemetry

For the last 10 years, techniques have been developed for monitoring cardiovascular parameters via telemetry in the rat (Kuwahara et al., 1994; Kramer et al., 1995; Ichimaru and Kuwaki, 1998); they have been developed, improved, and used increasingly. Totally implantable battery-operated systems that can monitor several physiological parameters including ECG have been developed. The implants are available with ECG leads that can be positioned in the Lead II configuration for monitoring ECG continuously for extended periods of time in a freely moving animal. As the leads are attached sc to the musculature, the signal is of a higher quality than skin leads. Also, the animal is not affected by the observer, as the animal can be monitored remotely. One of the disadvantages is that these implants are only designed to monitor a single lead.

ECG Waveform

The major points to notice about the rat ECG are that the conventional waves of the mammalian ECG (P, QRS, and T) are all identifiable in the rat ECG, there is no isoelectric line during the electrocardiographic complex, and there is no ST segment. The duration of the standard intervals evaluated in the ECG of the rat are as follows: P, 10–20; PR Interval 35–50; QRS complex 12–25, and QT interval 38–80 ms (Detweiler, 1981). The duration of the intervals is related to the heart rate; as the heart rate increases, the intervals shorten and as the heart rate slows, the intervals prolong. These intervals can also be affected by administration of test compounds and are the basis for the requirement to assess the effects of a test compound on cardiac function. Specifically, prolongation of the QT interval has been correlated with a phenomenon call Torsade de Pointe or sudden cardiac death. Several classes of compounds, such as antihistamine and Ca^{2+} channel blockers have been shown to prolong the QT interval (Gras and Llenas, 1999). Each wave of the ECG represents either a depolarization or repolarization of the atria and ventricles of the heart. For example, the P wave is an electrical representation of the depolarization of the right atria and the T wave is an electrical representation of the repolarization of the left ventricle. Spear (1982) provides a more in-depth discussion of the waveforms and the electrophysiology of the heart. Several computerized systems that are capable of recognizing the independent waveforms and measuring the intervals have been developed. Caution should be taken when using these systems as they are typically programmed to recognize a normal ECG, and the presence of arrhythmias may be missed or interval may not be measured correctly.

Heart Rate

The heart rate can be calculated from standard limb lead ECGs by measuring the distance between the two peaks of the R wave. This distance is then divided by the chart speed (i.e., 50 mm/s) to calculate the RR Interval in seconds. This is then divided into 60 s to calculate the heart rate in beats per minute. Using the formulas below, the heart rate of an RR Interval of 10 mm is measured on a chart printed at 50 mm/s = 300 beats/min.

RR Interval (mm/beat)/Chart Speed (mm/s) = RR interval (s/beat)

(60 s/min)/RR interval (s/beat) = Heart Rate (Beats/min)

Many environmental factors can affect the heart rate of the rat, such as excessive manipulation, technicians the animal is not familiar with, new environments, etc. Therefore, when evaluating ECG using a restrained method, it is important to acclimate the animal to the test procedures prior to starting. Detweiler (1981), in a review of the literature, found that published heart rates for rats varied between 250 and 750 beats/min. Awake, restrained adult rats had heart rates from 330 to

600 beats/min and well-acclimated restrained adult animals had heart rates from 250 to 350 beats/min. Heart rate in unrestrained telemetrized animals has been reported to be between 225 and 350 beats/min (Guiol et al., 1992; Zbinden, 1981).

Blood Pressure

During the conduct of a toxicology study it may be necessary to monitor blood pressure. This can be done using either indirect or direct methods. Caution should be taken when using indirect methods as the values obtained may be variable. The direct method involves the implantation of an arterial cannula for measurement of blood pressure. This method is more reliable, but it has a limited period during which the cannula may remain patent. It is recommended that for definitive assessment of the hemodynamic effects of a test compound, a nonrodent species such as a dog or a nonhuman primate be used.

Indirect Measurement

Indirect methods of blood pressure measurement detect systolic blood pressures by the occlusion of arterial inflow of blood and the subsequent detection of the pressure at which the first arterial pulsation occurs. The two places where indirect measurements can be made on the rat are the tail and the hindpaw.

Tail Cuff Method

The tail cuff method monitors pressures in the ventral caudal artery. In this method, the animal is put into a restrainer that allows for free access to the tail. An inflatable cuff is then placed around the base of the tail and the pressure is increased until flow stops. The pressure is then slowly released until flow resumes. The cuff pressure at the time when flow resumes is the systolic blood pressure. Various methods have been used to determine when this occurs.

Because the caudal pulse is rather weak, preheating of the animals in boxes at temperatures of 30–42°C for periods up to 10 min may be necessary to dilate the caudal artery. This technique should be used with caution, as previously discussed changes in body temperature can have widespread effects on the animal and may produce unexpected test compound effects.

The placement and width of the tail cuff is important. There is a gradient in pressure along the caudal artery, which amounts to 4.5 mmHg/cm. For this reason, the cuff should be placed close to the base of the tail and this should be standardized. If multiple reading will be done over time, marking the placement of the cuff with an indelible marker will help to standardize placement. In addition, variation in the width of the rubber tubing can be a source of error. Bunag (1973) found that the most accurate readings were given by a 15-mm cuff; shorter cuffs gave falsely elevated readings and longer cuffs gave low readings.

Hindpaw Method

Measurement of blood pressure in the hindpaw does not measure the pressure in a specific vessel. In this method, the animal is placed in a restrainer, as in the tail cuff method. A pressure cuff is placed around the ankle to occlude blood flow and blood pressure is measured as the cuff pressure is released and blood flow returns. As in the tail cuff method, several techniques have been used to determine the return of blood flow. These include visual observation (Griffith, 1934), photoelectric cell (Kersten et al., 1947), and oximeter (Korol and McShane, 1963). The advantage of the photoelectric and oximeter methods is that they do not require preheating to dilate the vessels of the hindpaw. The oximeter method measures mean arterial pressure rather than systolic pressure. None of the indirect measurements are able to evaluate the complete hemodynamic cycle. They evaluate only systolic or mean arterial pressure, but not systolic, diastolic and mean arterial pressure together.

Direct Measurement

The direct measurement techniques involve the cannulation of an artery with the blood pressure being determined with a manometer or transducer connected to the free end of the cannula. This is true for types of direct measurement including telemetry. Surgery and cannula placement utilize similar techniques as those previously described for placement of a venous catheter. In short, the artery is isolated and a small incision is made in the femoral artery. The catheter is then inserted into the hole and passed into the abdominal aorta. The carotid artery may also be used for this procedure, but care should be taken not to insert too far as the catheter may be passed into the left ventricle of the heart. If this occurs, the blood pressure waveform will change in appearance. The left ventricular waveform has a similar systolic pressure as an arterial blood pressure waveform. But the diastolic pressure is much different. If the catheter has been placed in the left ventricle, the diastolic pressure will be 0 mmHg or slightly negative. If this occurs the catheter should be backed out into the aortic arch.

Where chronic use is desired, cannulas are typically run sc and exteriorized between the scapula or at the back of the head. A carotid artery catheter can be expected to remain patent for 3–5 weeks (Ross, 1977; Andrews et al., 1978), whereas an abdominal aorta catheter may remain patent for several months. Care should be taken when using arterial catheters for long periods as fibrin deposits can build up on the catheter or clots may form in the catheters. The risk exists that these deposits or clots could be expelled during the flushing of the catheter or during normal movement of the animal. These clots or deposit once free may occlude other vessels downstream and in the case of a carotid catheter may cause a stroke to occur.

Blood Collection Techniques

Blood samples are routinely collected in safety studies to determine: (1) direct test compound effects on the blood or bone marrow, (2) effects on other organs as indicated by the contents of the blood, for instance, leakage enzymes such as aspartate aminotransferase (AST), and (3) blood levels of the test compound or its metabolites. A variety of techniques have been described for the collection of blood from the rat. The choice of a specific technique may depend on factors such as: (1) the volume to be collected, (2) if the animal is to survive the procedure, (3) the frequency with which samples will need to be collected, (4) whether anesthetics can be used, (5) likelihood of the animal surviving the procedure, and (6) the impact of organ damage resulting from the procedure. An adult rat has a blood volume of about 50 mL/kg; approximately 10% of the total blood volume can be collected from the rat in a single draw without adversely affecting hematology parameters. For longer-term studies, 10 mL/kg per 2 weeks is a reliable guideline for volume of blood drawn. If these volumes are exceeded, hematocrit and red cell mass may be reduced on evaluation (Diehl et al., 2001).

Technique, anesthetic used for blood draws, and treatment of the animals (i.e., fasted or not fasted) should be standardized throughout a study if repeated samples are being taken. The technique, anesthetic, and handling of the animal may produce effects on the hematology or clinical chemistry parameters to be evaluated.

Retro-Orbital Plexus

The retro-orbital plexus is a commonly used site for periodic sampling during the course of a study. This method has been shown to be a reliable method for the repeated collection of blood samples. Collection from this site should always be conducted under anesthesia to reduce pain and stress to the animal. Light anesthesia with a mixture of carbon dioxide and oxygen will minimize struggling of the animal and will help to ensure a quick collection with little injury to the animal.

Blood is collected using a microcapillary tube or the fine end of a Pasteur pipet. The tube is inserted into the orbit of the eye at an anterior angle formed by the lids and the nictitating membrane. A slight thrust past the eyeball will make the tube enter the slightly resistant horny membrane of the sinus. The tube may be rotated slightly as it is inserted. Once the sinus has been punctured, blood will fill the tube. Once the tube is filled, the blood may be allowed to drip out of the end of the tube into an appropriate collection tube. If the flow stops, the tube may be pulled out or advanced slightly to reestablish flow.

In the hands of an experienced technician, there is minimal risk to the animal. Studies have shown that repeated collection of blood from the retro-orbital sinus can produce histological and behavioral changes that may require an animal to be removed from study (McGee and Maronpot, 1979; van Heck, 1998). Several serious side effects of this collection method have been documented. These include retro-orbital hemorrhage, corneal ulceration, keratitis, pannus formation, damage to the optic nerve, and fracture of the orbital bones. When designing the study, the method of blood collection should be taken into consideration. If an important end point of the study is ophthalmological examinations, this method should not be used. In addition, anesthetization of the animals may affect other study end points.

Tail

A tail vein bleed offers a visible target and is of minimal risk to the animal. Blood will flow faster if the tail has been warmed causing vasodilation. This may be accomplished by dipping the tail in warm water (40–45°C), placing the animal in a warming cabinet for 5–10 min, or warming the tail with a heat lamp. This method does not require that the animal to be anesthetized, but the animal should be restrained such that the tail is held immobile. Several methods may be employed to collect blood from the tail. These include clipping the end of the tail off, vein puncture, or artery puncture.

Tail Clip

For this method, the animal may be lightly anesthetized or placed in a restraint tube. To collect blood, 2–3 mm of the distal part of the tail is amputated with sharp scissors or a scalpel blade. Blood of 3–4 mL can be collected in 20–30 s from a 200–250 g rat. When collection has been completed, the cut surface may be cauterized with a hot spatula or glass rod. This method produces a reliable volume of blood, but it should not be used for repeated sampling over extended periods of time.

Venipuncture

Animals should be restrained in a holder that allows complete access to the tail. The tail should then be cleaned and pressure may be applied to the base of the tail causing the vein to dilate. The vein should be punctured using a 21-gauge needle, and the blood can either be slowly withdrawn into a syringe through the needle or the needle may be removed and the tail allowed to bleed freely. The use of a butterfly needle may help facilitate the collection of blood as it is less likely to be dislodged if the tail moves. Collection of blood from the tail vein typically yields samples of between 0.5 and 1.0 mL. On completion of sample collection the needle should be withdrawn and pressure is applied to the tail to stop the bleeding (Frank et al., 1991).

Arterial Puncture

This method is conducted in a similar manner to a venipuncture in that the animal is placed in an appropriate restraint tube and the tail is cleaned. A 21-gauge needle is then inserted into the artery in the midventral surface of the tail close to the distal end. The blood is then withdrawn into a syringe or allowed to flow freely into a collection tube. The animal's blood pressure will ensure whether the blood continues to flow until pressure is applied to the wound.

Cardiac Puncture

This method should always be performed under general anesthesia and should not be used as a survival collection technique. This technique offers a rapid method for collection of a large volume of blood from the rat. It is possible to collect between 5 and 7 mL from a 300–350 g rat using this technique, and it is possible to exsanguinate the animal using this method.

To collect blood using this method the animal should be anesthetized with a combination of carbon dioxide and oxygen or carbon dioxide alone. The animal is then placed in dorsal recumbency and the heart is located. The heart may be located by placing the index finger over the fourth and fifth left ribs and the thumb on the right side of the thorax. The collection needle (25–26 gauge, 1–2-cm long) should be inserted at a 45-degree angle into the heart. Once the needle is introduced, the syringe should be aspirated slightly to produce a vacuum. The needle is then advanced until blood is obtained.

Abdominal Aorta and Vena Cava

Collection from the abdominal aorta offers a convenient way to exsanguinate an animal and obtain a maximal amount of blood (Popovic and Popovic, 1960). To perform the procedure, the animal is anesthetized and the aorta exposed by dissection. A section of the aorta distal to the diaphragm is then exposed and the proximal end is clamped. The aorta is then cut and the distal end placed in a collection tube and the clamp is released. This method will allow for collection of a maximal amount of blood in a short period of time. This method should only be used as a terminal procedure. Alternatively, the aorta may be accessed using a needle or a butterfly needle and samples may then be collected into a syringe.

Winsett et al. (1985) described a method of repeated sampling from the vena cava of conscious rats. An assistant holds that animal while the operator grasps the animal just below the last rib. The needle is inserted 1 cm to the right of the spinous process of the first lumbar vertebra at a 45 degrees angle until the needle touches the bone. The needle is slightly withdrawn and then advanced at a slightly shallower angle to miss the bone and access the vena cava. The maneuver of first identifying the bone is essential to the procedure. This procedure may be used for repeated collection, but care should be taken to ensure the animal does not struggle during the collection. If the animal struggles, the needle may lacerate the vena cava causing the death of the animal.

Jugular Vein

The jugular vein provides a means for chronic blood sampling that is of low risk to the animals' health. This method does not require the use of anesthesia and can be accomplished through the proper restraint and positioning of the animal.

The unanesthetized method requires two technicians to perform the collection. The animal is placed on a restraint board in dorsal recumbency. The forelimbs are tied down to the board and one technician holds the hind limbs of the animal. The second technician grasps the animals head and turns it down and away from the desired collection site, right or left jugular. The needle (21 gauge) is then inserted into the middle of the triangle formed by the neck, shoulder, and clavicle, parallel to the body. As the needle is inserted, the syringe is aspirated until blood is observed in the syringe. The collection is then completed and the site is held off for approximately 30 s. It is important that the technician holding the hind limbs of the animal continually observe the respiratory rate of the animal. If the head is turned too far or is held in the wrong position for too long, the animal may go into respiratory distress. This method may be used for repeated collections with a high level of success.

Alternatively, the animal may be anesthetized and the jugular vein may be surgically exposed. The animal should be prepared using standard aseptic technique and the ventral neck should be shaved. The jugular vein may be exposed by incision of the skin and dissection of the sc tissues. A needle (20 gauge) may then be inserted into the vein through the pectoral muscle, directing the needle toward the head. Inserting the needle through the muscle will help to stabilize the needle during the collection. Once the needle is in place, the blood should flow freely with little to no aspiration. Care should be taken when aspirating the syringe as too much pressure will cause the vein to collapse. Once the collection is complete, the incision can be closed with a wound clip. This method may be used as an alternative if the animal does not require serial bleeds or as a method to replace a terminal bleed. If the procedure is used for a terminal bleed, strict aseptic technique is not necessary.

Proximal Saphenous and Metatarsal Vein

A small amount of blood (0.1–0.2 mL) can be collected from an animal at minimal risk to their health utilizing the proximal saphenous and metatarsal veins. No anesthesia is required.

Proximal Saphenous Vein

The inner aspect of the thigh of the hind limb should be shaved free of hair. While one technician holds the animal and compresses the inguinal area to dilate the vein, a second technician creates a longitudinal nick in the vein with a 20-gauge needle or a hematocrit lancet. The blood can then be collected into heparinized capillary tubes. This method works well for repeated sampling of small amounts of blood but would not be appropriate for blood volumes required for evaluation of clinical pathology parameters.

Metatarsal Vein

This procedure can be conducted with or without an assistant. The animal is restrained and a nick is made in the vessel with a needle. The blood can then be collected into a capillary tube or through the needle into a syringe.

Sublingual Vein

When using the sublingual vein for blood sampling, the animal should be anesthetized and then cradled in the palm of the hand. By holding the animal's head between the thumb and index fingers, the head can be stretched back and the skin of the face pulled backward. This will force the mouth open and the tongue against the palate. The right or left vein should be cut with iris scissors and the animal held such that blood drips into the collection tube. The bleeding can then be stopped by applying pressure with a gauze pad or cotton-tipped applicator.

Decapitation

Decapitation should only be performed by trained technicians with the appropriate equipment. There are several commercially available small animal guillotines that should be used to perform this technique. This technique is appropriate when a maximal blood volume is desired and contamination of the sample is not considered to be an issue. To perform this method, the head is first removed and the animal is held over the collection vessel and arterial and venous blood is allowed to drain from the body.

Cannulation

Although the blood collection methods described above will provide sufficient volumes and quality of sample for the majority of toxicology studies, specific protocols may require blood to be sampled from animals that have been subjected to a minimum of handling or from specific sites within the body of the animal. Cannulation of a specific artery or vein will typically meet this requirement, though in rodents, the usable life span of the cannula is limited by the length of time the cannula remains patent. Yoburn et al. (1984) compared jugular, carotid, and femoral cannulas for long-term sampling of blood. They found the femoral artery cannula was preferable in terms of patency and postsurgical weight loss. Collection from a cannula is the same regardless of implant site. The cannula is typically exteriorized between the animal's scapula and a stylet is inserted into the end of the cannula. For the purpose of collection, the stylet is removed and a needle is inserted into the cannula with a syringe attached. The heparin lock is then drawn out of the cannula until blood is observed in the syringe. The syringe is then removed from the needle and a new syringe is used to collect the sample. The collection syringe is then removed from the needle and replaced with a syringe filled with the desired solution for locking the cannula, typically a heparin dextrose solution. Once the cannula is flushed and locked with the heparin solution, the stylet is replaced. Using the appropriate technique for flushing and locking the cannula is the key to maintaining the patency of the cannula.

Jugular Vein

The cannulation procedure is the same as previously described for infusion techniques. The cannula has been found to remain patent for a variable period of time, and the length of time the cannula remains patent is directly related to the skill of the technician collecting the samples. It is important to remember that a cannula placed in the venous system can easily develop clots in the cannula if not flushed and locked properly. Various methods of anchoring the cannula to the rats back or head have been developed for ease of sampling or as a connection point for continuous infusion. Each of these methods has been developed such that the exteriorized cannula is positioned so that the animal cannot gain access to it.

Inferior Vena Cava

The inferior vena cava appears to provide a site for long-lived cannulas. The cannula can be surgically placed either directly into the vena cava through an abdominal surgery or can be advanced into the vena cava from the femoral vein. Either way, the cannula has been shown to remain patent for months (Kaufman, 1980). In either implantation, the cannula is then tunneled sc and exteriorized between the scapula in the same manner as the jugular vein cannula.

Abdominal Aorta

The most common method for placing a cannula in the abdominal aorta is via the femoral artery. The cannula is inserted into the femoral artery and advanced to the level of the just above the kidneys. The opposite end of the cannula is then tunneled to and exteriorized between the scapula. This is a minor surgical procedure in which the animal recovers easily. Similar methods are employed when collecting blood from the arterial catheter as those used for the venous catheter. Caution should be taken when removing the stylet from the cannula, as a cannula in the arterial system is under pressure. If not properly prepared, the animal could quickly lose a large amount of blood. It is recommended that the technician place a clamp on the cannula prior to removal of the stylet. Once the blood collection is complete, the cannula should be thoroughly flushed with saline prior to locking the cannula. The tip of an arterial cannula is positioned such that it is against the flow of blood. Because of this, clots and fibrin deposits develop easily on the end of the cannula limiting the usable life.

Subcutaneous Ports

An exteriorized cannula is a source of contamination, infection, and is subject to destruction by the animal or other animals if group housed. Several types of sc ports have been developed for implantation along with the cannula. These ports are designed such that test materials may be injected or infused through them or blood may be collected from the port. In most cases, the port is implanted sc on the dorsal side of the animal. As with exteriorized cannula, the skill of the technician accessing, collecting samples, flushing, and locking the port directly affects the usable life of the port.

Urine Collection

Urine is generally collected in toxicology studies to assess kidney function. The most common method used for urine collection is a commercially available stainless steel cage. The cage is designed such that the urine and feces are separated by a cone-shaped device. The urine drains off the collecting walls into a tube and the feces fall into an inverted cone. Food and water are made available in such a way that the urine will not be contaminated. Although this type of cage produces urine of acceptable quality for normal urinalysis, the sample may be contaminated with hair or feces. Other methods for urine collection in the rat include cystocentesis, which involves a needle stick into the bladder, and cannulation of the bladder.

Necropsy

The necropsy is the link between antemortem findings and histological observation. It is an essential portion of the toxicology study, and because a necropsy will involve the processing of a large number of animals, it is important that the procedure is well planned (Black, 1986). At a prenecropsy meeting involving the pathologist, prosectors, and the study director, necropsy responsibilities can be discussed. Additionally, the study director can summarize clinical findings and potential target organs. The prosectors should be familiar with the protocol and amendments for the study involving the animals being necropsied. The protocol should clearly state which tissues are to be collected and weighed and how the tissues should be preserved. During the necropsy, devices such as checklists or prelabeled compartmentalized trays should be present to ensure that all required organs are taken and weighed. In recent years commercially available computer software has been developed to assist in the collection and weighing of tissues at necropsy. This is an electronic copy of the checklist, but it also provides a method for recording observations in a consistent manner. Copies of the last clinical observations should be present at the necropsy, so the prosectors are alerted to lesions that may be present and require special attention. Palpation records are particularly important at carcinogenicity study necropsies to ensure that all masses detected at the last examination are confirmed and collected.

The necropsy will involve a check of animal identification and sex, an external examination of the animal, an in situ examination of all tissues and organs (prior to dissection, and the collecting and weighing of the required tissues).

SUMMARY

In summary, there are several advantages to the use of the rat in toxicology studies. Because of its widespread use in many fields of biology, there is a large historical database of information about the anatomy and needs of the species. This knowledge, along with information about the species' metabolism and response to toxicants, has shown the rat to be a generally good model for the prediction of the human response to toxicants. Rats have a life span of 24–30 months, which is convenient for chronic toxicity and carcinogenicity studies where animals need to be exposed for the majority of their lifetime. The short gestation time and large litter size make the rat a good model for reproductive studies. The development of specific pathogen-free rats and improvement in husbandry has eliminated most of the disease outbreaks that may have introduced variability into a study. The lack of an emetic response allows for the testing of higher dosages of compound that may cause vomiting in other species. The small size of the rat is useful in that a large number can be housed economically. The size is also useful in that smaller amounts of test compound are required to gain maximal exposure.

The relatively small size of the rat is also one of its major disadvantages. The amount of blood that can be taken from the animal is limited, thus limiting the number of parameters that can be investigated or the number of toxicokinetic samples that can be collected from a single animal. This problem can be overcome by adding additional animals in interim sacrifice groups or by collecting toxicokinetic samples from cohorts of animals at different times. In most cases it is recommended that toxicokinetic samples be collected from satellite animals and not from the main study animals being used for evaluation of toxicity. However, an increased number of animals means increased work in the conduct of the study. The small size and relatively active nature of the rats makes some procedures, such as iv dosing or collection of electrocardiograms, difficult. These issues have been overcome with the use of suitable restrainers or in some cases, anesthesia. The rat has been used successfully in toxicology research for close to a century and will continue to be used for the foreseeable future.

The Mouse

Use in Toxicological Research

As discussed earlier, the choice of a species for toxicity testing is based on consideration of a range of variables. Ideally, if toxicity testing is intended to provide information on the safety of a test article in or by humans, the species chosen for testing should be most similar to humans in the way it handles the test article pharmacodynamically. Substantial differences in absorption, distribution, metabolism, or elimination between test species and the target species, e.g., humans, will reduce the predictive value of the test results. From a practical standpoint, often the pharmacokinetics is unknown in humans or the variety of available test species at the time of species selection. For this reason, testing is usually conducted in at least two species. Generally, one of those species is usually a rodent and one a nonrodent.

Mice have many advantages as test animals for toxicity testing. They are small, relatively economical to obtain, house, and care for, and they are generally easy to handle. Mice are generally more economical than rats in these respects. Although mice may attempt to escape or bite handlers, with regular, gentle handling they are easily managed. Other advantages of the species include a short gestation period and a short natural life span. These characteristics allow studies that include evaluation of reproductive performance or exposure to a test article for periods approaching the expected life span (e.g., evaluation of carcinogenic potential) to be conducted in a practical time frame. High quality, healthy mice are available from reliable commercial suppliers. Many genetically well-defined highly inbred, specifically or randomly outbred strains are available. Mice have been used in biomedical research for hundreds of years, and because of this, many technical procedures have been developed for use with the species, and a vast body of historical data is available for most strains. This historical database includes information on optimal nutritional and housing requirements in addition to data such as the expected background incidence of various diseases and types of tumors in untreated animals, and is continuously being added to (Blackwell et al., 1995). An additional advantage for mice, particularly when testing highly humanized biological products, is that transgenic mice that have the gene encoding the specific human pharmacodynamic drug receptor can now be readily developed.

There are also disadvantages to using mice, and most are related to the small size of the animal and the limits that this imposes. The smaller size and higher metabolic rate compared with the rat renders the species less hearty than rats. Deviations in environmental conditions such as an air conditioning failure or failure in an automatic watering system typically have more severe effects on the smaller species such as mice than the same deviations have on rats. Owing to their high level of natural activity, most strains of mice will not become as docile or easy to handle as rats that have received equivalent handling. Small size often precludes or renders more difficult a number of procedures that are commonly conducted in toxicity testing, such as the collection of large samples or repeated samples of blood and urine, electrocardiographic evaluation, and some necropsy evaluations. The FDA provides Human Equivalent Dose interspecies conversion factors for converting most animal model NOAELs or such to a human equivalent. The conversion factor for the mouse is 12.1, meaning an observed NOAEL in a mouse of 12.1 mg/kg/day would be converted to 1.00 mg/kg/day for conservative human administration. This relatively high value imposes a de facto penalty on using mice as one of the species in a safety assessment.

This section will provide brief summaries of some of the normal physiological values and salient features of the species and some of the specific strains that may be useful in selecting an appropriate species and strain for toxicity testing.

Normal Physiological Values

Selected normal physiological values for mice are shown in Tables 2.17 and 2.18. Median survival of a number of groups of Charles River CD-I outbred mice is shown in Table 2.19.

These normal values will vary depending on the strain of mouse, supplier, condition at arrival, type of feed, environmental and housing conditions, and, in some cases, time of year. These data should be considered as a reference but will not necessarily represent experience in any particular laboratory.

TABLE 2.17 Normal Physiological Values General and Reproductive

GENERAL	
Life span	
Average	1–3 years
Maximum reported	4 years
Adult weight	
Male	20–40 g
Female	18–40 g
Surface area	0.03–0.06 cm^2
Chromosome number (diploid)	40
Food consumption	4–5 g/day
Water consumption	5–8 mL/day ad libitum
Body temperature	36.5°C
Oxygen consumption	1.69 mL/g/h
REPRODUCTIVE	
Age, sexual maturity	
Male	50 days (20–35 g)
Female	18–40 g
Breeding season	Continuous, cyclic
Estrus cycle	4–5 days
Gestation period	
Average	19 days
Range	17–21 days
Litter size	
Average	12
Range	1–23
Birth weight	1.5 g
Age begin dry food	10 days
Age at weaning	16–21 days (10–12 g)

Data derived from Fox, J.G., Bennett, S.W., 2015. Biology and diseases of mice. In: Fox, J.G., Anderson, L., Otto, G., Prichett-Corning, K., Wary, M. (Eds.), Laboratory Animal Medicine. Academic Press, New York, pp. 31–89.

Species Differences

Mice are similar to other common laboratory animal species and to humans in many ways, yet the differences should not be underestimated and must be understood. Mice have a high metabolic rate compared to other species. This fact alone may result in increased or decreased toxicity of a test article, depending on the specific mechanism of intoxication. In many cases, high metabolic rate may be associated with rapid absorption, distribution, metabolism, and elimination of a test article. It may also lead to higher systemic C_{max} levels of toxicants such as reactive oxygen species. Mice are obligate nose breathers and have more convoluted nasal passages than humans. This may result in an excess of respirable test article deposited in the nasal passages, resulting in either increased or decreased relative toxicity, depending on the most critical site of absorption. The small size of the mouse compared to other common laboratory species offers a significant advantage if the test article is expensive or in short supply. As an approximation, a mouse weighs about 10% as much as a rat, about 5% as much as a guinea pig, about 1% as much as a rabbit, and less than 1% as much as a dog or primate. Material requirements to administer equivalent dose levels are usually proportional to body weight, so the test article savings associated with the mouse are evident. The small size of a mouse results in high surface area to body mass ratio, which in turn causes the mouse to be relatively intolerant of thermal and water balance stresses. The kidneys of a mouse have about twice the glomerular filtering surface per gram of body weight as a rat, and owing to the specific architecture of the murine kidney, they are capable of producing urine that is about four times as concentrated as the highest attainable human concentrations (Fox and Bennett, 2015). These characteristics of renal architecture and function may be important to the toxicity of some test articles. Mice differ from most species by the formation of a persistent vaginal plug after mating. The presence of a vaginal plug is easily detected, is considered evidence

TABLE 2.18 Normal Physiological Values Cardiovascular and Respiratory

CARDIOVASCULAR	
Heart rate	
Average	600/min
Range	320–800/min
Blood pressure	
Systolic	113–160 mmHg
Diastolic	102–110 mmHg
Blood volume	
Plasma	45 mL/kg
Whole	78 mL/kg
Hematocrit	41.5%
RBC life span	20–30 days
RBC diameter	6.6 microns
Plasma pH	7.2–7.4
RESPIRATORY	
Rate	
Average	163/min
Range	320–800/min
Tidal volume	
Average	0.18 mL
Range	0.09–0.38 mL
Minute volume	
Average	24 mL/min
Range	11–36 mL/min

RBC, red blood cell.
Data derived from Fox, J.G., Bennett, S.W., 2015. Biology and diseases of mice. In: Fox, J.G., Anderson, L., Otto, G., Prichett-Corning, K., Wary, M. (Eds.), Laboratory Animal Medicine. Academic Press, New York, pp. 31–89, and from the Animal Diet Reference Guide, Purina Mills, Inc. (1987).

TABLE 2.19 Median Survival of 16 Groups of Control Mice (%)

	Period of Time on Study (months)			
Sex	**6 (%)**	**12 (%)**	**18 (%)**	**21 (%)**
Male	98	91	63	46
Female	98	95	74	68

Data represent median survival of Charles River CD-I outbred albino mice enrolled in 24-month chronic toxicity studies at pharmaceutical or contract toxicology laboratories.
Adapted from Lang, P.L., February 1989a. Spontaneous Neoplastic Lesions in the B6C3F/CrIBR Mouse. Charles River Monograph. Charles River Laboratories, Wilmington, Massachusetts; Lang, P.L., Fall 1989b. Survival of Crl: CD-I BR Mice during Chronic Toxicology Studies. Charles River Laboratories Reference Paper. Wilmington, Massachusetts.

TABLE 2.20 Normal Body Weights in Grams of Selected Strains of Mice

	Outbred Strains						**Inbred Strains**						**Hybrid**	
	CD-1		**CF-I**		**CFW**		**C3H**		**C57BU6**		**BALB/c**		**B6C3FI**	
Age (days)	**M**	**F**	**M**	**F**	**M**	**F**	**M**	**F**	**M**	**F**	**M**	**F**	**M**	**F**
21	12	11	12	11	9	9	–	–	–	–	–	–	–	–
28	20	18	18	17	16	13	17	16	14	13	16	14	16	14
35	27	22	24	21	19	17	18	17	17	14	17	16	20	17
42	30	24	27	22	24	20	20	18	19	16	18	17	22	18
49	33	26	28	24	27	22	24	23	21	17	20	18	24	19
56	35	27	30	26	28	23	27	26	22	18	21	19	26	21

Data derived from Charles River Growth Charts.

of mating, and is a useful characteristic during the conduct of reproductive studies.

It is also frequently the case in pharmaceutical research and development that the nonclinical efficacy model for a new drug is in the mouse, making it the natural choice for rodent evaluation of the drug.

Strain Differences

In addition to differences between mice and other species, there are important differences among different strains of mice. The appropriate choice of a strain of mice for a particular toxicity study should consider the specific objectives of the study and the specific characteristics of candidate strains that might assist or hinder in achieving those study objectives.

One difference among strains is in the normal body weights of various strains at different ages. These differences are summarized for selected strains available from the Charles River Breeding Laboratories in Table 2.20. Outbred strains tend to be larger at maturity than inbred strains, with the CD-I strain reaching the highest mean weights at 56 days of age of those strains in Table 2.20. The CF1 strain has been reported to be highly resistant to mouse typhoid and to be relatively resistant to salmonellosis (Hill, 1981). Nude or athymic strains of mice are more sensitive to tumor development than heterozygous strains. These sensitive strains develop the same types of tumors as those seen in more conventional strains, but the incidences are higher and the latency periods shorter. There is a wide spectrum of susceptibility to spontaneous lung tumors in various strains of mice, and evidence suggests that there is a high correlation between spontaneous incidence and chemical inducibility in those various strains (Shimkin and Stoner, 1975).

The inbred strain A mouse appears to be the most susceptible to lung tumors and forms the basis of a

TABLE 2.21 Incidence of Spontaneously Occurring Neoplastic Lesions

	CD-I		B6C3Fl	
Location and Lesion	M	F	M	F
LYMPHORETICULAR TUMORS				
Lymphosarcoma			6.0	12.0
Lymphocytic leukemia			1.3	1.4
Lymphoma	3.7	9.9		
Histiocytic lymphoma			0.5	1.4
Histiocytic sarcoma			0.1	1.4
Lymphoblastic lymphoma	1.1	1.7	0.2	0.1
Lymphocytic lymphoma	2.6	1.7	0.6	1.7
Reticulum cell sarcoma	2.6	5.1	0.4	0.2
SKIN/SUBCUTIS				
Fibrosarcoma	0.2	0.5	1.0	0.5
Mammary gland				
Adenocarcinoma		1.7		0.9
LUNG				
Bronchiolar/alveolar adenoma	4.0	2.9	8.3	3.3
Bronchiolar/alveolar carcinoma	3.5	3.1	1.9	0.6
Alveolar type II carcinoma	11.7	13.9	0.2	0.1
Alveolar type II adenoma			2.5	1.2
Adenoma			1.2	0.7
LIVER				
Nodular hepatocellular proliferation	5.4	1.7	0.5	0.1
Hepatocellular adenoma	5.6	0.8	17.2	7.1
Hepatocellular carcinoma	7.3	1.0	13.2	2.4
Hemangioma	1.0	1.2	0.7	0.3
Hemangiosarcoma	1.0	0.2	0.5	0.1
REPRODUCTIVE SYSTEM				
Ovary				
Cystadenoma		1.1		0.3
UTERUS				
Endometrial stromal proliferation		3.3		2.9
Endometrial sarcoma		1.9		0.6
Leiomyoma		1.0		
Leiomyosarcoma		1.0		0.3
Hemangioma		1.0		0.9
PITUITARY				
Adenoma		3.4	0.3	7.9
THYROID GLAND				
Follicular cell adenoma	0.2		0.8	2.4
ADRENAL				
Cortical adenoma	8.6	1.0	0.4	0.3
HARDERIAN GLAND				
Cystadenoma			1.9	1.6
Adenoma			1.5	0.6

Lesions occurring at spontaneous incidence of ~:1% in either sex of Charles River CD-1 or 136C3171 mice strain.
Data from Charles River Breeding Laboratories, compiled from control animals on 24-month studies completed between 1978 and 1986.

lung tumor bioassay, with tumors inducible within 8 weeks or less of treatment. Susceptibility of various strains to the initiation and/or promotion of skin tumors has also been shown to differ greatly (Chouroulinkov et al., 1988; Steinel and Baker, 1988). The incidences of selected spontaneously occurring neoplastic lesions in CD-I (outbred) and B60171 (hybrid) strains are compared in Table 2.21.

The number of strain-related differences in susceptibility to various test articles and environmental conditions exceeds the scope of this chapter, but additional information is available (Nebert et al., 1982).

STUDY DESIGNS

Most toxicity and teratology studies conducted in mice are designed to provide information on potential human toxicity. Test substances are typically administered by the expected route of human exposure. A pharmaceutical product that is intended for oral administration (tablet, capsule, solution, or suspension) or a food additive would generally be administered by the oral route. Oral administration to mice is usually accomplished by administration of a solution or suspension by oral gavage, by mixture of the test substance with the diet, or less commonly added to the drinking water.

The specific design of toxicity studies should be tailored to the objective to be achieved and to any specific characteristics of the test substance. Many features of study design will be predicated on guidelines and practices of regulatory agencies such as the FDA or EPA in the United States or their counterparts in other countries to which the results of the study are submitted in support of a safety claim. Recommendations for study length (duration of dosing) fall in this category.

Toxicity studies are usually conducted in order of increasing duration of dosing, beginning with acute toxicity studies. When this regimen is followed, each study provides progressively more useful information for the selection of doses for the next, longer study.

Acute Toxicity Studies

Acute toxicity studies are conducted to evaluate the effects of a single substance. Usually each animal receives a single dose of the test substance in this study design. On rare occasion, repeated doses may be administered, but in any event, all doses are administered within 24 h or less. Historically, a primary objective of acute toxicity testing was to determine an LD_{50} dose, or that dose which would be lethal to 50% of the animals treated. To achieve this objective, groups of mice, often numbering 10 or more per sex, are treated with a single dose of the test substance. Depending on the rate of survival in the initial group(s), additional groups are added to the study at higher and/or lower doses such that most animals that receive the highest doses die and most that receive the lowest doses survive. Survival is assessed at some predetermined interval after dosing, usually 7 or 14 days, but occasionally as early as 24 h. The resultant dose–response data can be analyzed by a statistical method such as probit analysis (Finney, 1971) to provide an estimate of the median lethal dose (LD_{50}) and some measure of the precision of that estimate, such as the 95% fiducial limits. There are very few scientifically valid reasons to include determination of the LD_{50} as a significant objective of acute toxicity testing, and most regulatory agencies have dropped their requirements for a specific value for the LD_{50}, and animal welfare considerations preclude the use of the large numbers of animals previously required.

A more contemporary design for acute toxicity testing attempts to derive a maximum amount of information from a minimum number of animals. Study objectives include determination of the most important clinical signs attributable to high doses of the test substance, time of onset and remission of those signs, possible determination of a minimum lethal dosage, and in the event of lethality, the sequence and timing of effects leading to death or recovery. These objectives are achieved by means of a comprehensive schedule of animal observations following dosing. These objectives can usually be achieved by treating from one to three groups of three to five mice/sex/group at different doses.

Traditionally, acute toxicity testing of potential new pharmaceutical products is conducted in at least three species, with one being a nonrodent, and by at least two routes of administration, one of which is the intended clinical route. Mice are the most frequently selected rodent species for acute toxicity testing. The choice of routes of administration depends on the intended clinical route and on how much is already known about the oral bioavailability of the test substance. If the intended clinical route is oral, acute testing by oral gavage with a solution or suspension is of primary importance. If other clinical routes are anticipated (e.g., iv or dermal), they represent good secondary routes for acute testing. Ordinarily, at least one parenteral route is used for acute testing, and that route may be intravenous (IV) if the product is soluble in a fairly innocuous vehicle (e.g., water or saline) or ip as a suspension if the product is insoluble in an aqueous (or other innocuous) vehicle. If the intended clinical route is not oral, the oral route is usually selected as a secondary route for acute toxicity testing to provide information relevant to accidental oral ingestion. A rough estimate of oral bioavailability can be based on a comparison of the acute toxicity associated with various doses administered by the oral and parenteral routes. Acute toxicity testing conducted for other purposes is usually more limited in scope. Most regulatory agencies no longer require a full complement of species and routes of administration to render decisions on acute toxicity.

There are a few characteristics of acute toxicity testing that are not common in other toxicity protocols. In a typical repeated dose toxicity study, several groups of animals are treated concurrently with predetermined doses of test substance and a control substance. To reduce animal use in acute toxicity testing, studies that include more than one dose group are usually dosed sequentially, with an interval of at least 24 h between dosing of subsequent groups. This allows the effects of the previous dose to be fully manifested and allows selection of the subsequent dose to provide the highest probability of contributing more useful information. Another unusual aspect of acute toxicity studies is the nutritional status of the animals at dosing. Because some schools believe that the results of acute toxicity testing are more reliable if all animals are in a uniform nutritional state, mice to be dosed orally are often fasted overnight prior to dosing. Fasting allows dose volumes to be higher than in repeated dose studies, and because dosing only occurs on 1 day, dietary stress is considered tolerable. The scientific merits of this practice are debatable, but fasting is "traditional" in oral acute toxicity studies (and in initial human clinical studies). Although the practice of conducting gross necropsies at the end of acute toxicity studies is growing in popularity, this practice rarely yields useful information. The toxicity resulting from acute exposure is usually associated with a biochemical or functional imbalance rather than with a change in the gross or microscopic architecture of an organ system. Changes observable at gross necropsy are more often associated with repeated dosing at sublethal levels. For similar reasons, microscopic examination of tissue is rarely conducted in acute toxicity studies unless there is some scientific reason to expect it would be useful.

The results of a well-designed acute toxicity study can help to predict likely target organ systems and possible outcome in the event of massive human overexposure, can help in establishing risk categories for EPA or Department of Transportation classification, and can help in dose selection for the initial repeated dose toxicity tests to be conducted. An example of an acute toxicity study design is in Table 2.22.

Short-Term Toxicity Studies

The objective of short-term or subchronic toxicity studies is to describe and define the toxicity associated with repeated administration of high, but generally survivable doses of a test substance. This may include identification of target organs and systems, definition of the maximum survivable repeated dose, and the highest "clean" or no effect dose. Short-term repeated dose studies also serve as dose range-finding studies for longer-term repeated dose studies.

TABLE 2.22 Typical Acute Toxicity Study Design for Mice

Number of mice/sex/dose group	3–5
Number of dose groups	1–3
Number of control groups	None
Dosing frequency	Single dose
Dosing days	I day
Survival checks	Not done (part of Clin. Obs.)
Clinical observations	4 or more on day of treatment, then 1–2 daily
Physical examinations	Not done
Body weights	Prior to dosing
Feed consumption	Not done
Number of reversal mice	None
Duration of reversal period	Not applicable
Blood collection	Not done
Hematology parameters	Not done
Clinical chemistry parameters	Not done
Urine collection	Not done
Necropsy	Gross (increasingly, but rarely useful)
Tissue collection	Rarely (specific cause only)

Short-term toxicity studies range in duration of dosing from about 7 to 90 days. Mice typically receive a single, daily dose of the test substance, 7 days/week by the expected clinical route of administration. If the test substance is administered in the diet (or rarely, the drinking water), that admixture is available continuously. Short-term studies usually include three to four groups of mice exposed to different dose levels of the test substance, and an additional group exposed to the carrier to serve as a control for the effects of treatment. Group sizes for these studies are on the order of 5–10 mice/sex/dose. Ideally, dose levels should be selected for these studies such that a few animals die at the highest dose prior to the completion of dosing (to assure exposure to the maximum survivable dose), and all survive at the lowest dose with minimal evidence of toxic effects. The middle dose or doses should be set at approximately equal log increments between the high and low doses. It is important to begin to identify the highest dose level that is free of serious toxic effects to determine whether the test substance is likely to be toxic to humans at the expected therapeutic dose or exposure level.

Parameters monitored in a typical short-term repeated dose study may include daily observations for clinical signs of toxicity and mortality, weekly physical examinations, body weight and feed consumption, and terminal measurement of serum glucose and urea concentrations, serum aspartate aminotransferase, alanine aminotransferase (ALT), and alkaline phosphatase activity. Animals found dead or killed by design are typically submitted for gross necropsy, and selected tissues, such as adrenal gland, bone (sternum, including marrow), brain, heart, kidney, liver, lung, testis, and thymus are collected, weighed (except for bone and lung), and processed for routine microscopic examination by a qualified veterinary pathologist. An example of a short-term toxicity study design is in Table 2.23.

Chronic Toxicity Studies (26 weeks–2 years)

The objective of chronic, or long-term, toxicity studies is to refine the description of the toxicity associated with long-term administration of high, survivable doses of a test substance. Chronic toxicity studies are more commonly conducted in rats than in mice, but such studies can be conducted in mice, and this discussion describes objectives and practices for conducting such studies. Target organs and systems have usually been identified prior to the conduct of chronic studies, but it is the chronic study that provides the best opportunities to understand the subtle changes associated with

TABLE 2.23 Typical Short-Term Toxicity Study Design for Mice

Number of mice/sex/dose group	5–40
Number of dose groups	3–4
Number of control groups	1
Dosing frequency	Once, daily
Dosing days	Daily for 7–90 days
Survival checks	1–2 daily
Clinical observations	Daily
Physical examinations	Weekly
Body weights	Weekly
Feed consumption	Weekly
Number of reversal mice	None
Duration of reversal period	None
Blood collection	Terminal, all animals
Hematology parameters	None
Clinical chemistry parameters	Limited
Urine collection	Not done
Necropsy	Gross, all animals
Tissue collection	Limited list, all animals

long-term administration of high doses and to focus more closely on the highest "clean" or no effect dose. Chronic toxicity studies also serve to refine the doses to be administered in the carcinogenicity studies that typically follow them.

Chronic, or long-term, toxicity studies range in duration of dosing from about 26 weeks to as long as 2 years, but now it is, most commonly, only 6-months in rodents. Single, daily doses of the test substance are administered by the expected clinical route of administration. If the substance is intended for oral administration, the convenience and economy of administration in the diet (or rarely the water) becomes important. Diet admixtures are made available ad libitum unless they must be removed for a specific procedure during the study. Chronic studies usually include three groups of mice exposed to different dose levels of the test substance and an additional group exposed to the carrier to serve as a control for the effects of treatment. Chronic toxicity studies often include "reversal groups," or subsets of each dose group that are not sacrificed immediately on the completion of treatment. The purpose of the reversal groups is to determine whether any toxic effects associated with treatment are permanent or subject to recovery or reversal. Mice in the reversal groups may be allowed from 2 to 4 weeks of recovery time from the end of treatment until necropsy. Group sizes for chronic studies are on the order of 20–50 mice/sex/dose. Sizes of the reversal groups, if they are included, may be about 25%–35% of the original dose groups. Dose levels should be selected for these studies such that there is substantial toxicity at the highest dose, but few if any treatment-related deaths. The low dose in chronic studies should confirm, or if necessary, refine previous estimates of the highest dose level that is free of serious toxic effects, and thereby reinforce previous estimates of the relative safety of the test substance at the expected human dose or exposure level. The middle dose should be the approximate geometric mean of the high and low doses.

Parameters monitored in a typical chronic toxicity study may include daily observation for moribundity and mortality, weekly physical examinations, body weight and feed consumption, and terminal measurement of serum glucose and urea concentrations, serum aspartate aminotransferase, ALT, and alkaline phosphatase activity. Hematological parameters for mice are often limited to differential smear evaluations. Although red blood cell (RBC) counts, white blood cell counts, and hemoglobin concentrations can be determined, the values for mice are somewhat variable from many of the commonly used laboratory instruments, so these parameters are not always evaluated. Animals found dead or killed by design are typically submitted for gross necropsy, and a list of 30–50 selected tissues are collected, weighed (except for bone and lung), and processed for routine microscopic examination by a qualified veterinary pathologist. An example of a chronic toxicity study design is in Table 2.24.

Carcinogenicity Studies (18–24 months)

The objective of carcinogenicity studies is to determine whether the test substance is a carcinogen when administered at maximum tolerable doses for a period approaching the life expectancy of the mouse.

This objective is simpler in many respects than the objective of the longer toxicity studies. It is assumed that by the time carcinogenicity studies are undertaken, the chronic toxicity studies have been essentially completed, and the actual toxicity of the substance is about as well understood as it can be based on animal studies. Carcinogenicity studies are not usually encumbered by tests to further the understanding of toxicity, but rather are focused to maximize the ability to answer the single question of carcinogenicity.

Carcinogenicity studies in mice are generally designed to expose the animals for a period of 18 months to 2 years. Improvements in animal husbandry during the past decade have increased the life expectancy of most strains of mice, so there has been a

TABLE 2.24 Typical Chronic Toxicity Study Design for Mice

Number of mice/sex/dose group	20–50
Number of dose groups	3
Number of control groups	1
Dosing frequency	Once, daily
Dosing days	Daily 26–52 weeks
Survival checks	Daily
Clinical observations	Not done
Physical examinations	Weekly
Body weights	Weekly
Feed consumption	Weekly
Number of reversal mice	25%–35% of main groups
Duration of reversal period	2–4 weeks
Blood collection	Terminal, all animals
Hematology parameters	Dif. smear, RBC, WBC
Clinical chemistry parameters	Limited list
Urine collection	Not done
Necropsy	Gross, all animals
Tissue collection	Comprehensive list, all animals

RBC, red blood cell.

tendency to extend carcinogenicity studies in mice to 2 years. The study design normally contains provisions to allow for termination of the study prior to the intended end point if excessive mortality is encountered. This provision is intended to ensure that an adequate number of survivors are sacrificed with successful collection of all necessary tissues for a meaningful, statistical analysis of tumor incidence. Carcinogenicity studies usually include three groups of mice exposed to different dose levels of the test substance and an additional group exposed to the carrier to serve as a control for the effects of treatment. Group sizes for these studies are on the order of 50–70 mice/sex/dose.

The high dose in carcinogenicity studies should be the MTD. This dose should produce evident toxicity by the end of dosing. A commonly held minimum criterion for evident toxicity is a decrement in body weight or body weight gain of at least 10% from the control to the high-dose group. If the test substance is not very toxic, most regulatory agencies will accept a carcinogenicity study in which the high dose is the highest dose that can be practically administered, even though that dose does not produce evident toxicity. In this example, the animal has, in effect, "tolerated" the highest dose that could be administered.

The selection of lower doses is not as critical as in chronic studies because the concept of a "clean dose" of a known carcinogen is not widely accepted, and it will be of little value in attempting to commercialize a product.

The middle dose may become important if the high dose has been inadvertently set too high, resulting in excessive toxicity and early mortality. In that case a middle dose that elicits evident toxicity without excessive mortality may become an acceptable MTD and may effectively "salvage" the study. As in chronic studies, the middle dose should be at the approximate geometric mean of the high and low doses.

The most important data generated in a carcinogenicity study are the histopathology data. Of particular importance are the data on the incidence of various types of malignancies in the different treatment groups. Control or even untreated mice normally will have some "background" incidence of various types of malignancies over the course of a carcinogenicity study. The key question then is whether the treated groups have a significantly higher incidence of "normally expected" tumor types than the control group, whether they have occurrences of "nonnormal" tumor types that are not seen (concurrently or historically) in control animals, and, most importantly, whether such incidence is attributable to the test substance.

Other parameters monitored in a typical carcinogenicity study may include daily observation for survival and moribundity, periodic physical examinations, periodic examinations for palpable masses, body weight, and feed consumption (especially important in dietary admix studies), and for some studies periodic peripheral blood smears and terminal red and white blood cell counts. Animals found dead or killed by design are submitted for gross necropsy, and a comprehensive list of 40–50 prescribed tissues plus any tissue masses, suspected tumors, and an identifiable regional lymph tissue are collected from each animal to be processed and examined histologically by a qualified veterinary pathologist for evidence of carcinogenicity. An example of a carcinogenicity study design is in Table 2.25.

Teratology Studies

Mice are occasionally used in teratology studies to assess the effects of test substances on congenital defects in the young when administered to pregnant females. Mice have a regular, short estrus cycle, a short gestation period, high fertility, and typically produce relatively large litters of young. Mice rank behind rats and rabbits as the species of choice for assessing teratology, but there may be good scientific reasons to use mice in some instances. The conduct of Segment II teratology studies in mice and rabbits with Segment I and III studies in rats provides an opportunity to evaluate teratogenicity

TABLE 2.25 Typical Carcinogenicity Study Design for Mice

Number of control groups	2
Dosing frequency	Once, daily
Dosing days	Daily for 18–24 months
Survival checks	Daily
Clinical observations	Not done
Physical examinations	Monthly during first year, 2x/month thereafter
Body weights	Weekly during first 2 months, 2x/month thereafter
Feed consumption	Weekly during first 2 months, 2x/month thereafter
Blood collection	Optional periodic, terminal
Hematology parameters	Periodic peripheral smears, terminal smears, RBC, WBC
Clinical chemistry parameters	Not done
Necropsy	Gross, all animals (including those found dead during study.)

RBC, red blood cell.

in three species. Although there is an advantage in conducting teratology studies in at least two species, preferably one of which is a nonrodent, some substances (e.g., certain antibiotics) are especially toxic to rabbits, making teratology testing impractical in that species. In that case, mice become the second best species available (behind rats) for teratology testing and are usually the choice as the second species. A significant disadvantage in using mice for teratology studies is that they are much more cannibalistic than rats or rabbits. This characteristic renders the species unusable for Segment I and III studies, in which pregnant females are allowed to deliver their young, and limits them to use in Segment II studies, in which the young are delivered by cesarean section on gestation day 18, prior to delivery. Mating can be confirmed in mice by daily inspection of cohabitating females for the presence of a vaginal plug. The copulatory plug in mice is much more persistent than in rats, in which mating must be confirmed by vaginal lavage and microscopic examination.

Segment II Teratology Studies

A Segment II teratology (embryo-fetal development) study is conducted to assess the effects of a test substance on fetal survival and congenital malformations (teratology). Females are mated and monitored daily for the presence of copulatory plugs. The presence of a plug confirms that mating has occurred, and the day of discovery is defined as gestation day 1. Mated females are dosed with the test substance from gestation days 6 through 15, a period that begins soon after implantation (day 5), and continues through completion of organogenesis (day 13). This dose period exposes the young throughout the period of organogenesis, but it tends to minimize preimplantation embryotoxicity and postorganogenesis maternal and fetal toxicity. The young are delivered by cesarean section on gestation day 18, prior to normal parturition on day 19, to avoid cannibalism. The maternal reproductive organs are examined for numbers of corpora lutea, implantations, resorptions, and live and dead fetuses. The fetuses are weighed, sexed, and examined for gross, visceral, and skeletal abnormalities. An example of a Segment II teratology study design is in Table 2.26.

Genetic Toxicity Studies

The objective of genetic toxicity testing is to identify and describe the effects of agents that specifically produce genetic alterations at subtoxic doses. Mice are used in a variety of genetic toxicity study designs in an effort to achieve this objective. Neither a comprehensive listing of genetic toxicity procedures using mice nor a comprehensive description of any number of those procedures is within the scope of this chapter. Rather, a few of the most commonly employed procedures will be summarized. The reader is referred to other sources such as Brusick and Fields (2014), DeAngelis et al.

TABLE 2.26 Typical Segment II Teratology Study Design for Mice

Number of female mice/dose group	20
Number of dose groups	3
Number of control groups	1
Dosing frequency	Once, daily
Dosing days	Days 6–15 of gestation
Survival checks	Daily
Clinical observations	Daily
Physical examinations	Not done
Body weights	1–3X/week
Feed consumption	Not done
Number of reversal mice	None
Duration of reversal period	Not done
Blood collection	Not done
Caesarean section	Day 18 of gestation
Necropsy	All dams gross, all fetuses external and 113 of fetuses visceral exam at caesarean section

x, should be 3 times/week

(2007), and Dean (1983, 1984) for more detailed discussions of genetic toxicity, and more comprehensive descriptions of some of the specific tests used in that field. Kaput (2005) reviews the interaction between nutrition and genomics.

Mouse Micronucleus Assay

The objective of the mouse micronucleus assay is to determine whether a test article causes disruption and separation or breakage of chromosomes. The mouse micronucleus assay is one of the most commonly conducted in vivo tests for genetic toxicity. Comparison of the incidence of micronuclei in proliferating cells from treated versus control mice provides an indirect measurement of chromosome damage in somatic cells. Micronuclei can only be formed as a result of disruption and separation or breakage of chromosomes, followed by cell division. The preferred cells for evaluation are newly formed erythrocytes in mouse bone marrow because the micronuclei formed in these cells are not expelled during the last division in which the nucleus is extruded from the normoblast.

One or more dose levels of test article and a control treatment are administered to separate groups of at least five mice per sex. The highest dose should be the MTD, or one that produces some evidence of cytotoxicity. It is important that high quality (e.g., specific pathogen-free) mice of known genetic stability and consistent species, strain, source, age, weight, and clinical condition be used to assure comparability with historical controls. Each animal typically receives a single dose of test or control article. Bone marrow samples are collected at a minimum of three different intervals after dosing, ranging from 12 to 72 h. At least 1000 polychromatic erythrocytes are evaluated for the presence of micronuclei for each test and control mouse. An example of a mouse micronucleus assay study design is in Table 2.27.

TABLE 2.27 Typical Mouse Micronucleus Assay Study Design

Number of mice/sex/dose group	5 or more
Number of dose groups	1 or more
Number of control groups	2, one positive, one negative
Dosing frequency	Single dose
Dosing days	One day
Survival checks	Daily
Clinical observations	Not done
Body weights	Prior to dosing
Feed consumption	Not done
Bone marrow collection	3 or more intervals, from 12 to 72 h after dosing
Polychromatic erythrocytes evaluated/mouse	1000 minimum

Heritable Translocation Assay

The objective of the heritable translocation assay (HTA) is to assess the potential of a test article to induce reciprocal translocations between chromosomes in germ cells of treated male mice. This assay has the advantage of detecting transmissible genetic alterations, which are potentially more damaging to the gene pool than nontransmissible or lethal changes. Induced translocations can be detected by mating the F1 progeny of treated males with untreated, unrelated females. Translocations will be evidenced by a reduction in the number of viable fetuses sired by affected males. The presence of translocation figures in meiotic metaphase serves as cytogenetic verification of the presence of reciprocal translocations.

The HTA typically consists of three groups of 30 male mice each, treated with different dose levels of the test article, and a negative control group treated with the dosing vehicle. A positive control group is optional. As the period of spermatogenesis in the mouse is about 7 weeks, all treated and control animals are dosed on a daily basis for 7 weeks. The dosing route is usually oral gavage, and the dose levels are selected on the basis of the oral LD_{50}. The high dose is typically one-eighth of the LD_{50}, and the medium and low doses are about one-third and one-tenth of the high dose, respectively. On completion of dosing, each male is mated to two females. Two hundred healthy males are selected from the offspring of each of these groups of mating and allowed to reach sexual maturity, whereupon each male is mated to three virgin females. The females from this second mating are sacrificed about 3 weeks after cohabitation with the males was initiated, and the number of living fetuses and resorbing embryos present in the uteri are counted. Any male that produces 10 or more living fetuses in any one female is considered fertile, and no further mating is needed. Males falling below those criteria are mated to an additional set of three females, and the evaluation is repeated. Failing males may be remated up to three times. Males that never succeed in producing at least one litter of 10 or more living fetuses are considered sterile (or semisterile) and are sacrificed, and their gonadal cells are examined for cytogenetic evidence of translocations. An example of an HTA study design is in Table 2.28.

Microbial Host-Mediated Assay

The objective of the microbial host-mediated assay is to determine the ability of a mammalian system (e.g., the mouse) to metabolically activate or detoxify a test article with respect to its mutagenic potential. The mutagenic potential is measured by means of one of a variety of

TABLE 2.28 Typical Mouse Heritable Translocation Assay Study Design

Number of male/mice/dose group	30
Number of dose groups	3
Number of control groups	1–2 (negative; pos. optional)
Dosing frequency	Once daily
Dosing days	Daily, 7 weeks
Survival checks	Daily
Clinical observations	Not done
Mating of dosed males	On completion of dosing
Number of F, males mated/group	200
Number of F examined cytogenetically for translocations	All sterile and semisterile
Tissue collection	Testes

microorganisms, depending on the specific types of mutations being investigated (e.g., base-pair substitution, frame shift). Some of the microorganisms used for this type of testing include various strains of *Salmonella typhimurium*, *Escherichia coli*, and *Neurospora crassa*. The results of the host-mediated assay can be compared with the direct effect of the test article on the same test strain of microorganisms to determine whether the host (e.g., mouse) is metabolically activating, detoxifying, or having no effect on the mutagenic potential of the test article. The microbial host-mediated assay, then, is an attempt to combine the convenience of microorganisms for detecting hereditary changes with the metabolism of the test article gained by administration to a whole animal.

The microbial host-mediated assay consists of three groups of about 10 mice each, treated with different dose levels of the test article, a negative control group treated with the dosing vehicle, and a positive control group. Doses are administered daily for up to about 5 days. The high dose is usually about one-half of the LD_{50}, with the medium and low doses about one-third and one-tenth of the high dose, respectively. The preferred route of administration is oral, but im or ip injection may be used if necessary. At the end of the period of test article administration, the test strain of microorganism is administered, usually intravenously. Following appropriate incubation periods (e.g., 1, 2, and 4 h), mice are sacrificed and the microorganisms are collected, frequently from liver tissue. The collected microorganisms are grown on minimal agar plates to assess mutation rate. An example of a microbial host-mediated assay study design is in Table 2.29.

Special Studies

The diversity of toxicity study designs using the mouse as a test system to examine a specific hypothesis defies description. Many of these designs do not strictly fit within the major section headings chosen for this chapter and might be referred to as "special studies." This chapter is intended to focus on the more commonly conducted types of toxicity studies in mice, but it might be useful to describe two special study designs here as examples of some of the interesting end points that can be evaluated in this species.

Mouse Ear Swelling Test

The concept of a mouse ear swelling test (MEST) is that sensitization can be detected by measuring edema in the ear of a mouse that results from topical application of a test article to an animal that has been previously sensitized by means of dermal application to the abdomen (Gad et al., 1986).

The objective of the MEST is to provide an alternative test for dermal sensitization potential that makes more efficient use of animals, labor, and other resources than traditional study designs conducted in guinea pigs. A MEST test is often preceded by a dose-finding activity to identify the highest concentration of test article that is no more than minimally irritating to the abdominal skin, and the highest concentration that is nonirritating to the ear. These concentrations are then employed in the sensitization assay.

The sensitization assay is carried out in three phases: an induction phase, a challenge phase, and if necessary, a rechallenge phase. The sensitization assay requires 15 mice treated with the test article and 10 treated with a control substance, typically the vehicle used for the test article. During the induction phase (study day 1), the abdomen is shaved using a small animal clipper, and 20 μL of 1:1 emulsion of Freund's complete adjuvant in distilled water is injected intradermally (id) at each of two abdominal sites on opposite sides of the ventral midline. Once daily on study days 1, 2, 3, and 4, the stratum corneum is stripped from the abdominal skin using adhesive tape, and the appropriate concentration of test article is applied to the abdomen at a volume of 100 μL.

On study day 11, the challenge phase is initiated. The appropriate concentration of test article is applied to the skin of the left ears of all treated and 5 of the 10 control mice at a volume of 10 μL, and an equal volume of the control (vehicle) is applied to the right ears of those same animals. Thicknesses of both ears of all treated and the five selected control mice are measured on days 12 and 13. Any mouse with a left ear thickness ~120% of its right ear thickness is considered to be a positive responder. If one or more mice are judged positive in the absence of primary irritation (any control

TABLE 2.29 Typical Mouse Microbial Host-Mediated Assay

Number of mice/dose group	10
Number of dose groups	3
Number of control groups	2 (one positive, one negative)
Dosing frequency	Once, daily
Dosing days	Daily for 5 days
Survival checks	Daily
Clinical observations	Not done
Body weights	Prior to dosing
Test organism administered	After last dose
Test organism recovered	e.g., 1, 2, and 4 h after administration
Necropsy	Not done
Tissue collection	Liver or other appropriate tissue for organism recovery

TABLE 2.30 Typical Mouse Ear Swelling Test Study Design

Number of male mice/dose group	15/treated group, 5/control group
Number of dose groups	1
Number of control groups	2
Dosing frequency	Once daily
Dosing days	Days 1, 2, 3, and 4 (induction)
	Day 11 (challenge)
	Day 18 (optional rechallenge)
Survival checks	Daily
Clinical observations	Not done
Body weights	Not done
Ear thickness measured	Day 10, 12, and 13 (challenge)
	Day 17, 19, and 20 (rechallenge)
Necropsy	Not done
Tissue collection	None

mice with left ear thickness >110% of right ear thickness) on study day 12 or 13, the test article is considered to be a sensitizer. Evidence of primary irritation requires that the test be repeated using a lower concentration for the challenge dose.

If results are negative (all mice with a left ear thickness increases of <10%) or equivocal (some mice with increases of 10%–19%, but none >20%), a rechallenge is conducted on study days 17 through 20. Baseline ear thicknesses are measured on study day 17. The test article is applied to the right ears of all test mice and the control mice that were not used during the initial challenge on study day 18. Ear thicknesses of the right ears of all test mice and the new control mice are measured on study days 19 and 20, and they are compared to the baseline thicknesses taken on day 32. The criteria for positive response are the same as for the initial challenge. An example of a MEST study design is in Table 2.30.

Dermal Carcinogenicity (Skin Painting) Study

The concept of the dermal carcinogenicity (or skin painting) study is that carcinogens, or of more recent interest, cocarcinogens and tumor promoters can be evaluated or their potencies compared in as little as a few months of testing.

The carcinogenicity of some chemicals (e.g., polycyclic aromatic hydrocarbons) can be detected easily by the production of papillomas or carcinomas. Mouse skin apparently functions in this system because it contains enzymes necessary to produce the active intermediates that lead to initiation. Tars from tobacco, coal, and various petroleum products show active carcinogenic potential in this system, although many of the same products are not carcinogenic when administered systemically. Hepatic detoxification of systemically administered doses may account for this difference.

This study design has been especially useful in recent years in the study of cocarcinogens and tumor promoters. In a typical study design, groups of 25–50 mice might receive from one to a few systemic doses of a known tumor initiator. Following receipt of the initiator, the fur over the anterior portion of the back is shaved, and the suspected tumor initiators are applied to the skin of the back at a frequency of 2–3 times per week. Shaving will need to be repeated approximately weekly. Development of papillomas or carcinomas of the skin is readily visible in the shaved area. Active chemicals are often detected within a few months' treatment. This study is generally intended to continue treatment for a period of about 30–40 weeks, but it may be continued for up to 2 years if necessary.

Obvious advantages of this study design include its relative efficiency in terms of animal numbers and labor and its relative brevity compared to a conventional 18- to 24-month carcinogenicity bioassay. It is reasonable to conclude that positive findings of carcinogenicity in this test would make a conventional carcinogenicity bioassay unnecessary. Negative findings in a dermal carcinogenicity test, however, would not assure the absence of carcinogenic potential, and a conventional bioassay would still be necessary.

Disadvantages of this procedure include difficulty in accurate quantification of dose, as the topically applied dose can run off the animal, be scratched or licked off, or can accumulate as a crust, effectively reducing absorption. Another criticism centers around the fact that if treatment with a suspected promoter is interrupted after a period of 60 days or so, evident papillomas often regress, raising the question of whether they represented sites of true carcinogenicity.

ROUTES OF TEST SUBSTANCE ADMINISTRATION

Techniques are available to administer test substances to mice by most routes of potential human exposure. The choice of a route of administration for a toxicity study should consider the expected route of human exposure and any other scientific objectives that need to be achieved to facilitate safe use of the test substance. Of the various routes available, most test articles will have the most rapid onset of effects and the greatest potency when administered by the iv route, followed in approximately descending order by the inhalation, ip, sc, im, id, oral, and topical routes (Eaton and Gilbert, 2013). The expected route of human exposure is probably the most important single determinant of route for toxicity testing. In the discussion that follows, the most commonly employed routes for toxicity testing in mice, oral dosing, will be discussed first, followed by the less commonly employed routes.

Oral Administration

Oral administration is probably the most frequently used route of exposure for toxicity testing in mice. Many products are intended for oral administration to humans, and many others are subject to accidental ingestion. Oral administration subjects the test substance to limitations of absorption and metabolism that are similar but not necessarily identical to those in humans. Mice, like rats, differ from many other species in that they do not have an emetic response. For this reason, large doses of substances that would cause emesis in dogs or primates will be retained in the stomachs of mice. Although this characteristic facilitates testing at high doses and maximizes potential exposure to toxic effects, it may lead to an overestimate of potential human toxicity because the animal lacks the protective aspect of the emetic response. Another area in which mice differ from dogs and primates is that mice are nocturnal. This characteristic adds some pharmacokinetic variables to the equation for extrapolating toxicity findings from the mouse to the human. Doses that are administered during the day are administered to animals that are in the lower phase of their circadian metabolic cycles. This may mean slower absorption, slower metabolism to either more or less toxic metabolites, and/or slower elimination of the test substance. Conversely, test substances administered in the diet or drinking water will be largely consumed at night, as that is when mice consume most of their daily intake of food and water. Although this regimen more closely approximates human consumption during the active part of the day, it makes observation of the animals during the period of peak exposure and metabolism difficult. The three means of oral administration are oral gavage, dietary admixtures, and mixture with the drinking water.

Gavage

Oral gavage offers the advantages of precisely measured doses that can be administered at precise times. Doses can be administered during the day so animals can be conveniently observed for toxic effects during the first few hours after dosing. Volatile substances and those that lack stability over longer periods in the presence of diet, air, or water can be effectively administered by this method. Gavage allows administration of unpalatable substances that might not be accepted in the diet or water. There are disadvantages associated with gavage administration. The test substance must either be a liquid or be soluble or suspendible in a liquid vehicle system. The method is relatively labor-intensive compared to diet admix. The processes of daily handling and intubation of all animals engenders the risk of injury during the intubation process, including esophageal puncture and aspiration of test article. In addition, the process of frequent handling causes stress to the animals. Although it is convenient to administer doses during the day, daytime is the period of lowest metabolic activity for nocturnal species such as mice. This circadian effect may not be most representative of diurnal species such as humans.

Description of Technique

Gavage administration entails intubation of the mouse with an intubation needle or soft plastic catheter attached to a graduated hypodermic syringe. The dose is administered into the esophagus. Intubation needles for mice are typically constructed of stainless steel tubing with a stainless steel ball tip to reduce the probability of esophageal perforation and reflux and aspiration of the dose. Acceptable tubing sizes range from 22- to 18-gauge, with the larger bore reserved for older mice (e.g., $\geq$25g). Tubing length is not critical but may range from 1 to 3 in. The ball tip is typically 1.25–2.25 mm in diameter. Intubation needles are available commercially (e.g., Popper & Sons, Inc., New Hyde Park, New York) in straight and curved configurations. The choice of shape is a question of personal preference on the part of the dosing technician. Prior to dosing, the test substance must be prepared in a liquid form at an appropriate concentration. Liquid test substances may require dilution. Solid substances will require either dissolution or suspension in an innocuous vehicle. The preferred vehicle is water. If the substance is insoluble in water, various agents may be added to improve wetting (e.g., 0.1% v/v polysorbate 80) and to reduce settling (e.g., 0.5% w/v methylcellulose). Suspensions should be analyzed prior to administration to assure proper concentration, homogeneity, and stability of the substance in the suspending vehicle. Appropriate dose volumes for gavage administration are in the range of 5–10 mL/kg of body weight, but volumes as high as 20 mL/kg can be administered carefully, particularly in acute studies in which the mice have been fasted prior to dosing.

For the actual process of dose administration, the mouse should be weighed, and the individual dose calculated. The appropriate dose volume should be drawn into the dosing syringe and any air bubbles should be expelled. The mouse is then picked up by the skin of the back and neck, and the head is tipped back to form a straight line from the nose through the back of the throat and to the stomach.

The intubation needle is inserted to the back of the mouth, then gently tipped back, if necessary, to enter the esophagus. The mouse will usually facilitate entry into the esophagus by swallowing the ball of the needle. One successful approach is to envision the tip of the sternum as a "target" for the tip of the intubation needle. When properly positioned, the tube can easily be inserted to a reasonable depth, but it need not reach the stomach. When in position, the dose should be administered slowly to avoid reflux, but promptly enough to reduce the likelihood that the mouse will struggle and injure itself.

Dietary Admixtures

Oral administration by dietary admixtures offers several advantages, including ease of administration, minimal handling of animals for dosing, and elimination of the risk of injury associated with intubation. The method offers relatively precise dose administration for the group (better than water mix; not as good as gavage), as both mean food consumption and mean body weight for periods of a week or longer are easily measured. Dry, insoluble substances can be administered easily, and administration of test substances to mice during the "awake" phase of their circadian cycle is an advantage. There are disadvantages associated with dietary admix. Accuracy of individual doses is lower than with gavage, and there is not a single identifiable time of dosing. Volatile substances and those that lack stability over periods of at least 4–7 days in the presence of diet, air, or water are precluded from this method. Diet admixtures must be sampled and the samples be analyzed periodically to assure proper concentration and homogeneity of the mixture during the course of the study. Unpalatable test substances typically result in reduced dietary intake, which leads to an increase in the concentration of test substance in the diet during subsequent weeks in an attempt to reachieve the desired doses. The increased concentrations may be even less palatable, leading to further reduction in dietary intake, and in some cases, eventually to malnutrition. In any dietary study, results should be evaluated carefully to discriminate between changes associated with altered nutritional status and true test substance toxicity.

Description of Technique

Oral administration of dietary admix entails presentation of a mixture of the test substance in the diet in place of the normal diet received by the animals. The concentration of test article in the diet is adjusted, based on the most recently collected data on mean body weight and food consumption for each sex and dose group, to provide the desired doses of test article during the period in question. Early in a study, when body weight and food consumption are changing due to rapid growth of the animals, projections of the mean body weight and food consumption for the coming period should be based on both the most recent measurements and the rate of change (slope of the plot) of those parameters over several recent periods. If test article stability permits, a convenient period for measurement of body weight and food consumption is about 1 week.

Test substances are usually mixed with the meal form of rodent diet in one or more of a variety of mechanical blenders. Common types of blenders include the Turbula, paddle style, and twin shell, or "Y" blenders. When large quantities of diet admix are required, it may be advantageous to blend the test substance with the diet in two steps, preparing a premix of 1–2 kg in a blender such as a Turbula, and then adding the premix to a larger-scale blender, such as a paddle blender. This procedure often produces a more homogeneous mix in a shorter blending period.

Following mixing, the appropriate concentration of blend is dispensed into animal feeders for presentation to the mice. Feeders are weighed when they are placed in the cages and when they are removed to determine the average amount of feed consumed per day. Mice are weighed at the same time, both to determine body weight gain over the period and to calculate the food consumption in grams per kilogram of body weight per day. Based on this calculation, the concentration of test substance in the diet may be varied up or down to more closely approach the intended doses of test substance in milligrams per kilograms per day.

Among potential problems that may be encountered in conducting dietary admix studies, two may lead to inappropriate calculations of feed consumption, leading to incorrect calculation of concentrations for future periods. These problems are excessive feed spillage, which may result from mice digging or playing in their feeders, and contamination of the feeder with urine and feces, which may result from mice living in their feeders. Feeders should be checked daily for excessive spillage at the time survival checks and/or observations are conducted, and excessive spillage should be documented. Contamination with urine and feces will lead to an incorrectly high feeder weight at the end of the feed consumption period and an underestimate of true feed consumption.

Excessive contamination should be documented. Animals with excessive spillage and those with excessive contamination of feeders should be excluded from the calculation of mean body weight and food consumption used to prepare future concentrations of diet admix.

Drinking Water

Administration of a water-soluble test substance in the drinking water has many of the same advantages as administrations of a dietary admix. This method is rarely used for toxicity studies, however, because of the difficulty in accurately measuring the quantity of water actually consumed. Although graduated water bottles can be used, spillage due to mice rubbing against sipper tubes, inefficient drinking, evaporation, vibration, and leaking bottles makes these measurements imprecise. If that practical problem could be solved, administration in the drinking water would bear little conceptual difference from dietary admix. Stability of environmental temperature and humidity is essential to the conduct of a study in which the test article is mixed in the drinking water, as increased temperature and/or lowered humidity lead to increased water consumption. Water consumption must remain relatively stable to allow calculation of appropriate concentrations of test article in water to achieve study objectives.

Description of Technique

From a practical standpoint, administration of test article in the drinking water has many similarities to administration by dietary admix. It is essential that the test article be both soluble and chemically stable in water for the period of presentation. That period should be in the range of about 2 days to a maximum of about 1 week. Water remaining in a water bottle for periods in excess of a week may become heavily contaminated with bacteria. Analogous to the situation with diet admix, the precision of dose administration is directly related to the accuracy with which average daily water consumption and body weight can be measured. Concentrations of test article in water should be adjusted, if necessary, after each measurement of average daily water consumption and body weight to assure precise dose administration.

As solubility in water is a prerequisite for this method, mixing procedures are usually simpler than for diet admixes. Homogeneity analysis should not be required for a true solution, but samples should be analyzed regularly to confirm that concentrations are what they were intended to be.

Water can be provided in graduated bottles. Contents of the bottles should be recorded at the beginning of the consumption period (but after the bottles are placed on the cages to accommodate spillage during that operation), and again at the end of the consumption period to determine average daily water consumption. Evidence should be documented. Difficulty in accurately measuring the amount of water actually consumed by the mice is the largest disadvantage to this method of administration.

Intravenous Injection

Intravenous injection offers the advantages of immediate, complete systemic availability of a precise dose at a known point in time. The process of absorption is eliminated, as it is the possibility that some or all of the test article may be metabolized by the liver prior to distribution to the systemic circulation and target organs. Most substances exhibit the biggest potency and rapidity of onset of activity of all routes of administration when administered intravenously. Intravenous administration provides a useful benchmark against which absorption and bioavailability from administration by other routes can be compared.

It is essential that test articles administered intravenously be in solution at the time of administration and remain in solution after injection. Solutions that are subject to precipitation by changes in pH, temperature, or osmolarity should be confirmed at physiological conditions to assure that they will not precipitate after injection. Introduction of insoluble particles, such as those in a suspension, introduces a high probability of embolism, particularly in the pulmonary capillary bed, which will produce severe moribundity or death. The toxicologist is left with the problem of differentiating such moribundity or mortality from the true toxicity of the test article.

Other characteristics of an iv solution that should be evaluated prior to undertaking an iv toxicity study involving repeated dosing include the potential for the solution to cause hemolysis or vascular or sc irritation. Hemolysis may be a result of the administration of solutions of inappropriate osmolarity (hypotonic solutions are particularly damaging); in which case the problem can be resolved by adjusting the osmolarity of the solution.

Ideally, solutions for iv administration should be isotonic to blood and have a pH of about 7. Usually a pH in the range of about 5–9 will be acceptable. Solutions that cause appreciable vascular or sc irritation may result in sufficient injury to the veins and surrounding tissue to preclude repeated administration.

The rate of iv injection is an important variable that must be controlled fairly precisely to achieve reproducible results within a study. If the iv toxicity of two or more test articles is to be compared, it is essential that each be administered at the same rate. Intravenous injections can be administered as a bolus over a period as short as a few seconds, as a continuous, 24-h/day infusion, or over just about any interval in between.

There is nothing particularly magical about any specific dosing period, but it is critical that it remains constant. As a practical matter, an injection period of about 2 min is a reasonable upper limit for hand-held injections into caudal veins of reasonable numbers of mice. Longer periods lead to very slow injections and increase the risk of extravasation, are time-consuming, and are difficult and tiring for the toxicologist. Injection periods much shorter than 1 min increase the likelihood that an inordinately high peak plasma concentration may precede mixing with the total blood volume and may compromise the survival of the animal. A corollary to the artificially high peak plasma concentration associated with a short injection period can occur if the injection is administered at an uneven rate. It is particularly critical that the rate of injection not be increased during the last half of the injection period, as this is a time when the animal has already received an initial "loading," and the deleterious effect of increasing the rate of administration will be amplified. In our laboratory, we have found that administration of volumes of 5–10 mL/kg body weight (0. 15–0.30 mL for a 30-g mouse) administered evenly over a period of 2 min represents a good compromise. One of the biggest disadvantages of iv administration is that it is a very labor-intensive procedure, requiring more time per animal than any other route. In addition, repeated, daily administration to caudal veins of mice for periods longer than 2–4 weeks becomes technically difficult owing to the accumulation of scar tissue and occasional trauma.

Long duration or continuous, 24-h/day infusions, though possible, are technically difficult and will not be discussed in detail here. As a practical matter, long-duration injections or continuous infusions are typically administered through a surgically implanted catheter using an infusion pump. The catheter is usually placed in a large, superficial vein such as the jugular or femoral vein, and then exteriorized at a site such as between the scapulae, which is difficult for the animal to chew or scratch. The surgical procedure is relatively simple. The difficulty lies in keeping the cannula patent and secure in the vein during recovery after surgery, then through the period of dosing. The maximum volume of infusion, even over a continuous 24-h period should not exceed about 20–30 mL/kg/day, which will typically be less than 1 mL administered over a 24-h period. This infusion rate is so slow as to be difficult to administer, even with a high-quality infusion pump. The catheter must be attached to the mouse in a way that prevents mutilation of the catheter by the animal without undue limitation of mobility.

Description of Technique

Intravenous administration entails injection of the desired dose into an appropriate vein. In a typical study, mice will receive a single injection daily over a period of about 2 min or less. Such injections are usually administered into a lateral IMM (tail) vein, using a hypodermic needle attached to a graduated syringe. Hypodermic needles used for caudal vein injections in mice are usually no more than 1" in length and in the range of about 23–25 gauge. The smallest hypodermic syringe that will contain full dose volume will provide the greatest precision in dose measurement, but 1-cc disposable syringes are often used.

The needle should be installed on the syringe such that the bevel of the needle faces the side of the syringe that will be used to measure the dose. This will allow the graduations to be read easily when the needle has been inserted into the vein in an up configuration. A stopwatch that is easy to read is useful for timing dose administration, and an electronic timer that is activated by a foot pedal is most convenient. A convenient restrainer should hold the mouse securely, but without undue risk of suffocation or injury, while allowing five accesses to the tail. A source of warm water and/or a tourniquet device are useful, as is a supply of small gauze sponges. A tourniquet can be constructed from a disposable plastic syringe and a piece of suture. The larger the syringe size, the more pressure it is able to exert on the suture loop. Relatively large (e.g., size 0) braided silk suture should be used for the loop to minimize the risk of cutting the skin of the tail. The suture loop is attached to the plunger inside the barrel of the syringe and then threaded out through the tip of the syringe.

Prior to dosing, the dosing solutions should be prepared. Solutions should be analyzed periodically to assure that proper concentrations are being attained. Each mouse should be weighed and its dose calculated. As previously noted, doses in the range of 5–10 mL/kg are acceptable for injection over periods of about 30 s to

2 min. The mouse is placed into the restrainer, and the appropriate dose is drawn into the syringe. Any air bubbles should be expelled from the syringe. This is most important for iv injections. A tourniquet device can be applied at this point but should not be applied too tightly. The objective is to block venous return, but not arterial supply, thus dilating the veins. As an alternative to a tourniquet, some toxicologists prefer to warm the tail with a gauze sponge wetted in warm (not hot) water to enhance vasodilation. The tail is now held in one hand while the needle is inserted with the other. The needle should be inserted with the bevel up to minimize the chance of puncturing through both sides of the vein. Successful venipuncture will result in the reflux of a small amount of blood into the hub of the needle. Owing to the small volume of blood that usually refluxes, this phenomenon will be most easily visualized if needles that have transparent "flashback" hubs are used. The initial attempt at venipuncture should be made toward the tip of the tail, such that if the vein is missed, a subsequent attempt can be made closer to the base of the tail without risk that the dose will leak out of the initial hole. When the needle is securely in the vein, it can be held with the "tail-holding hand" while the plunger of the tourniquet is depressed to open the vein with the "dosing hand."

Now that the needle is in the vein and the vein is open, the timer can be started, and the dose can be administered. One convenient method to assure even dose administration is to divide the dosing period and the dose volume into a convenient number of parts. A 2-min dosing period might be divided into eight 15-s intervals, and the dose volume divided by eight. The doser can then administer one-eighth of the total dose volume over each 15-s interval for 2 min to assure a relatively even rate of injection. When the full dose has been administered, a clean dry gauze sponge should be pinched over the injection site and the needle should be withdrawn. Maintaining pressure on the site of the injection for 10–30 s after withdrawal of the needle is usually adequate to prevent bleeding. The mouse can now be removed from the restrainer. As iv injections typically result in a rapid onset of activity, it is often appropriate to observe a mouse for the first few minutes after dosing for clinical signs of toxicity.

Intraperitoneal Injection

The ip (intraperitoneal) route of administration generally offers the second most rapid absorption of a test article among the parenteral routes, with systemic availability second only to iv injection. Rapid absorption is conferred by the large surface area of the lining of the peritoneal cavity, and by the rich blood supply to that area.

Intraperitoneal administration leads to absorption primarily through the portal circulation. As a result, test articles that are metabolized by the liver are subjected to extensive (or even complete) metabolism prior to reaching systemic circulation and target organs, unless, of course, the target organ is the liver, in which case toxicity may even be amplified. Test articles that are excreted in the bile are similarly subject to elimination prior to reaching the systemic circulation and target organs. Water-insoluble mixture aqueous suspensions, for example, can be administered by the ip route. This may provide the opportunity for rapid systemic absorption of lipid-soluble or certain other test articles. Solutions or suspensions for ip injection should be adjusted to a pH in the range of about 5–9 to reduce the potential for irritation. Osmolarity of the dosing formulation is not critical, as it is for iv injection. Dose volumes for ip administration are in the range of 5–10 mL/kg/day, but volumes as high as 20 mL/kg/day are acceptable, particularly if the study is of limited duration, or if it is known that the test article will be absorbed by the ip route.

One of the most significant disadvantages of ip administration is the risk of peritonitis. Peritonitis can result from any of three primary causes: physical irritation caused by accumulation of a truly insoluble or irritating test article in the peritoneal cavity, introduction of exogenous microbiological contamination, or microbiological contamination resulting from injury to the gastrointestinal tract or urinary bladder. The potential for a test article to produce physical irritation or chemical peritonitis can be assessed in studies of one to a few days in duration. Although physical or chemical peritonitis is the most frequently seen form of peritonitis in toxicity studies, it is still found with only a small percentage of test articles. Mice are relatively resistant to microbiological infection, so microbiological peritonitis is even less common than physical or chemical peritonitis. Peritonitis resulting from injury during the injection process is extremely rare when injections are administered by qualified toxicologists. There is a slight risk to the animals of physical injury to a major organ or vessel during the injection process, but again this is extremely rare in the hands of qualified dosers.

Description of Technique

Intraperitoneal injections are administered into the peritoneal cavity using a hypodermic needle attached to a graduated syringe. Each mouse receives a single daily dose, administered as a bolus, for the duration of the toxicity study. The injection should be administered into the animal's lower, right abdominal quadrant to minimize the risk of injury to the liver, spleen, and bladder. For initial training purposes in dosing by the ip route, it is useful to sacrifice a mouse, in which the abdominal cavity is opened to expose the internal organs that may be susceptible to injury during the injection procedure. This will allow a novice to hold the animal in a dosing position and clearly visualize where the lobes of the liver, the spleen, and the urinary bladder will be, and the area of less vulnerability between these organs. Hypodermic needles used for ip injections to mice need be no longer than about 5/8 in and should be the smallest diameter that will allow easy injection of the dose volume to minimize the trauma to the abdominal wall with commensurate potential for leakage. Needles in the range of 23–25 gauge are appropriate for use with solutions and suspensions of low viscosity. Suspensions of high viscosity may require the use of needles with a larger bore. Needles as large as 19–20 gauge can be used, but they require great care to avoid injury and leakage of the test article from the injection site.

Prior to initiation of a toxicity study, dosing formulations should be prepared, and samples are analyzed for concentration and homogeneity of suspensions, if appropriate. Each mouse should be weighed and its dose calculated. The appropriate dose is then drawn into syringe, and air bubbles are expelled. The mouse is picked up with one hand, and held with the ventral surface toward the doser. Movement of the animals' right hind leg should be restricted to limit interference with the needle during dosing. The needle should be inserted at an angle of about 15–30 degrees into the abdominal cavity to facilitate penetration of the abdominal wall.

The location should be to the right of the midline (to avoid the spleen) at a position about midway between the lower edge of the liver and the urinary bladder to a depth of about 1 cm (3/8 in.). Following insertion, the needle is withdrawn slightly, moved about, and the angle of insertion is reduced to assure that the tip has not penetrated or snagged any internal organs. The dose is now administered as a bolus and the needle is withdrawn. If a large-bore needle has been used, it may be necessary to apply gentle finger pressure over the injection site for a few seconds to prevent leakage of the dose.

Intramuscular Injection

The im route of administration is less commonly used in toxicity testing, but it may be appropriate if the test article is intended for im administration to humans. The im route generally results in slower absorption of a test article, with lower peak plasma levels, but more sustained effects than iv or ip injection. The rate of absorption can be influenced by the amount of vascular perfusion of the tissue surrounding the injection, the vehicle, and the injected volume, which indirectly may alter the surface area of tissue available for absorption. Coadministration of a vasodilator generally increases the rate of absorption, whereas coadministration of a vasoconstrictor generally decreases that rate. Administration of the test article as a solution or suspension in a viscous, poorly absorbed vehicle generally retards absorption. The ability to control the rate of absorption can be a significant advantage in some cases, as it allows the toxicologist to administer a dose of a test article that may be absorbed over a period of many hours or even days. This can be especially useful in the case of test articles that have short half-lives after absorption, as a result of rapid metabolism and/or elimination. Limitations to im dosing include the limited number of muscle groups in the mouse that are large enough to accept dosing, e.g., the muscles of the posterior aspect of the femoral region and the small dose volume that can be administered. If possible, the same injection sites should not be treated every day to allow time for absorption and recovery from the trauma of dosing. This means that while a single acute dose might be divided into the hind limbs, repeated daily doses should be administered into alternate limbs. The dose volume should not exceed 1.0 mL/kg per injection site, or about 0.03 mL for a 30-g mouse, and smaller volumes are preferable. An acute study in which each animal is dosed once would allow 1 mL/kg to be administered into each hind limb, for a total dose volume of 2 mL/kg. This dose volume coupled with the limit of solubility or suspendability of the test article in the vehicle selected may restrict the maximum dose of test article below toxic levels. A further limitation on toxicity testing by the im route is that the formulation to be injected must not cause significant local irritation, particularly if repeated doses will be administered. This limitation may require that a separate study be conducted to assess im irritation potential prior to initiation of a repeated dose study by this route. Intramuscular injection is more labor-intensive than most other routes with the exception of iv injection.

Description of Technique

Intramuscular injections are administered into the large muscle groups of the posterior aspect of the femoral region using a hypodermic needle attached to a graduated syringe. Each mouse receives a single daily dose, administered as a bolus into alternate hind limbs for the duration of the study. Hypodermic needles used for im injection should be the smallest diameter that will allow injection, but in the range of 27 gauge up to a maximum of about 23 gauge. Prior to dosing, the same procedures for formulation, analysis, weighing of mice, and calculation of doses should be followed as those recommended for ip dosing. The dose is drawn into a syringe, and air bubbles are expelled. The mouse may be held by an assistant, and the needle inserted to the approximate center of the muscle mass. The dose is injected as a bolus, and the needle is withdrawn. The muscle may be massaged gently to distribute the dose prior to returning the mouse to its cage.

Subcutaneous Injection

The sc route of administration is not commonly used in toxicity testing, but may be appropriate if the test article is intended for sc administration to humans, or as a more practical substitute for im testing in mice. The sc route is similar in many characteristics of absorption to the im route, and generally results in slower absorption of a test article, with lower peak plasma levels, but more sustained effects than ip administration. The rate of sc absorption can also be influenced by the amount of vascular perfusion of the tissue surrounding the injection, the vehicle, and the injected volume, which indirectly may alter the surface area of tissue available for absorption. Coadministration of a vasodilator generally increases the rate of absorption, whereas coadministration of a vasoconstrictor generally decreases that rate. Administration of the test article as a solution or suspension in a viscous, poorly absorbed vehicle generally retards absorption. The ability to control the rate of absorption is similar to that seen with im injection and can offer the same advantages, as it may allow the toxicologist to administer a dose of a test article that may be absorbed over a period of many hours or even days. Some limitations to im dosing do not apply to sc dosing. Subcutaneous doses can be injected at a wide variety of sites, if necessary. In addition, dose volumes of up to 10–20 mL/kg/day may be administered repeatedly if they are well absorbed and do not cause excessive local irritation. These large dose volumes allow administration of much higher total doses than can be administered im. It may still be necessary to conduct a separate study to assess sc irritation potential prior to initiation of a repeated dose toxicity study. Subcutaneous injections can easily be administered to mice without assistance.

Description of Technique

Subcutaneous injections are administered into the region beneath the skin using a hypodermic needle attached to a graduated syringe. Each mouse receives a single daily dose, administered as a bolus. Daily doses may be administered at the same site if absorption is complete and irritation is minimal, but sc trauma may be reduced if the injection site can be changed from day-to-day. Hypodermic needles used for sc injection should have the smallest diameter that will allow injection, but in the range of about 26 gauge up to a maximum of about 20 gauge. Although larger volumes can be administered, dose volumes of about 10 mL/kg/day are preferable.

Prior to dosing, the same procedures for formulation, analysis, weighing of mice, and calculation of doses should be followed as those recommended for ip dosing. The dose is drawn into a syringe, and air bubbles are expelled. The mouse is grasped by a fold of skin. One of the most convenient injection sites is the skin in the midscapular region, which allows the restraint and dosing of the mouse with minimal risk of being bitten. The needle is inserted through the skin into the sc region. The dose is injected as a bolus, and the needle is withdrawn. The injection site may be pinched for a few seconds to prevent leakage, and the area around the injection site may be massaged gently to distribute the dose prior to returning the mouse to its cage.

Intradermal Injection

Intradermal injection is not a route that is commonly used for toxicity studies. It may be appropriate to test products intended for id administration to humans by that same route in mice, and studies of limited duration are technically feasible. The id route offers the advantage of slow absorption owing to the poor vascular perfusion of the skin relative to tissues in other areas of potential administration. This slow absorption is typically associated with longer time to onset of effects, lower peak plasma levels, but more sustained effects than routes that result in faster absorption. To the extent that the test article may be metabolized by the skin, the id route would be expected to offer greater opportunity for such metabolism than sc injection, but less than topical administration. Injected volume for id dosing should be limited to about 1 mL/kg per injection site or less, with smaller volumes preferred if repeated doses will be administered. It is acceptable to administer id doses at multiple sites simultaneously if higher total doses are required. Irritating formulations of the test article must be avoided, especially if multiple doses will be administered, as ulceration and necrosis of the skin can result.

Description of Technique

Intradermal injections can be administered at a variety of accessible sites, but the skin of the abdomen or back is often used. The area in which the injections will be administered should be shaved with a small animal clipper to allow good visualization during and after dosing. Doses can be administered using a small hypodermic needle attached to a graduated tuberculin syringe. Needle diameter should be limited to 27 gauge or smaller, and 30 gauge is preferable. The use of a needle with an id bevel is not necessary. Prior to dosing, the same procedures for test article preparation and analysis of formulations, weighing of mice, and calculation of doses should be followed as recommended for ip dosing. The dose is drawn into the syringe and air bubbles are expelled. The mouse is held in one hand, and the needle is inserted into the skin at a shallow angle with the bevel of the needle up to avoid penetration into the sc space. With practice, the toxicologist can feel the needle penetrate into the sc space, if that happens by accident, and can relocate the needle prior to injection. A properly administered id dose will appear as a small bleb on the surface of the skin. A dose administered into the sc space will not appear as a bleb, as the dose will be distributed over a larger area.

Topical Administration

The topical route of administration is occasionally used for toxicity testing. This choice of route may be appropriate for testing the systemic and local toxic effects of substances intended for human topical administration or which are likely to come into accidental contact with human skin. Data suggest that the mouse is one of the less appropriate species for extrapolation of percutaneous toxicity to the human, as the permeability of mouse skin (as well as rat and rabbit skin) is substantially higher than the permeability of human skin (Maibach and Wester, 1989). Nevertheless, topical application to mice may be appropriate in special cases, such as the conduct of the MEST for dermal sensitization potential (Gad et al., 1986).

Historically, the skin was perceived as a barrier to absorption. It is now clear that lipophilic compounds are readily absorbed into and across the skin, and also that the skin may be a source of significant metabolism of some chemicals (Maibach and Wester, 1989). Variables, in addition to lipophilicity, that are likely to affect dermal absorption include the integrity of the skin at the treatment site, the vehicle employed for dosing, occlusion, and/or restraint of the mouse after treatment, and whether the test article is washed off after some prescribed period. Variations on the integrity of the skin include totally intact skin, skin from which the outer epidermal layers have been tape stripped using a surgical adhesive tape (e.g., Dermiclear), and skin that has been abraded. The presumption is that nonlipophilic test articles will penetrate stripped (thinned) or abraded (interrupted) skin more extensively than they would in intact skin. The proper choice of a vehicle may enhance the permeability of the skin to a nonlipophilic chemical. Occlusion of the treatment site and/or restraint of the mouse after application of a topical dose improve retention of the dose in contact with the skin and reduce the probability that the animal will orally ingest the topical dose. Washing excess test article from the treatment site after a prescribed time will limit the exposure period to a known interval. The appropriate choices for the above (and other variables) in topical toxicity study design are a function of the specific objectives of the study, and the physical and chemical characteristics of the test article.

Description of Technique

As the number of procedural variables for topical dosing is so great, the procedures described for topical dosing in the MEST (Gad et al., 1986) will be described as a representative technique. In that procedure, the hair is clipped from the treatment site (e.g., abdomen or back) on the first day of treatment and the epidermal layer is tape stripped until the site has a slightly shimmy appearance, typically 10–20 applications and removals of a surgical adhesive tape such as Dermiclear (Johnson & Johnson Co., New Brunswick, NJ). Next, a fixed volume, e.g., 100 μL, of the test article in a volatile solvent such as ethanol is applied to the treatment site. The solvent is allowed to dry using a warm air blower if necessary, and the animal is returned to its cage. On subsequent treatment days the tape-stripping operation can be reduced to about 5–10 applications of adhesive tape to achieve the shinny appearance.

Inhalation

The inhalation route of administration offers the most rapid absorption of most test articles, with systemic availability second only to iv injection. Efficiency of absorption by the inhalation route is conferred by the large surface area of the respiratory system, the close proximity of the inner alveolar surface to the blood circulating through the lungs, and the fact that the entire cardiac output passes through the lungs with each circuit of the blood through the body. Absorption of inhaled agents proceeds via one or more of the following mechanisms depending on specific characteristics of the agent: direct absorption into the blood stream, absorption from the gastrointestinal tract following deposit in the nasopharynx or transport by mucociliary escalation and swallowing, and/or lymphatic uptake following deposit in the alveoli.

Inhalation studies are particularly useful in estimating the risk of accidental or occupational exposure to a gas, vapor, dust, fume, or mist as well as in evaluating the toxicity of agents that are intended to be administered by inhalation. Administration by inhalation is the most technologically complex means of routine exposure, and a comprehensive description of the procedures is beyond the scope of this chapter. The reader is referred to other works (e.g., McClellan and

Henderson, 1989; Leikauf, 2013; Kennedy and Trochimowicz, 1982) for further description. Rather, this discussion will be limited to some of the advantages, disadvantages, and variables to be considered in inhalation testing. The primary advantage of inhalation is rapid, effective absorption. The primary disadvantage is the technological complexity of the method, with the associated risk of technical error and disregard of an important variable.

For a mouse to inhale a test article, the mouse must be placed in an environment that contains the test article in the form of a gas, vapor, dust, fume, or mist. The test article must exist in a particle size that is inspirable, generally having an aerodynamic diameter from 1 to about 10 A. Particle size dictates the location where the test article will be deposited and absorbed in the respiratory tract. Larger particles are deposited in the nasopharyngeal region, with successively smaller particles deposited in the trachea, bronchial, bronchiolar, and finally the alveolar region for particles of about 1 μ or less. The technology of particle generation and uniform distribution through the exposure apparatus is complex in itself. In addition to generating and uniformly distributing the test atmosphere, care must be exercised to capture the exhaust from the exposure apparatus, such that the test article can be contained without contamination of the laboratory or environment. Exposure periods can range from a few minutes, appropriate for test articles that may pose only an acute exposure risk to continuous exposure over a prolonged period, appropriate for test articles that may pose a risk of long-term environmental or occupational exposure. Exposure apparatus generally takes the form of a chamber that contains the whole animal or groups of animals, or a device that exposes only the head or nose of the animal(s) to the experimental atmosphere.

Chamber (Whole Body)

Inhalation chambers allow relatively large numbers of mice to be exposed simultaneously without restraint. The aerodynamic considerations are complex, but simpler than for a head-only or nose-only exposure system. The flow rate through a chamber must be adequate to provide temperature and humidity control. Disadvantages of whole body chambers include the tendency for test article to accumulate in the fur, from which it can be ingested; on the skin and eyes, which may interfere with the intended route of exposure and the difficulty in monitoring respiratory volume and rate of individual animals.

Head/Nose Exposure (Head Only/Nose Only)

Head- or nose-only exposure apparatus limits exposure of the mouse to the test article by routes, other than inhalation, as only a small amount of skin and fur are exposed to the test environment. In addition, it is possible to monitor respiratory volume and rate of individual animals with some of the head-or-nose-only equipment. Disadvantages to this equipment include the fact that only a relatively small number of animals can be simultaneously exposed, and those animals must be restrained in a position that keeps their heads or noses in close contact with the exposure apparatus. This restraint imposes stress on the animals and virtually precludes continuous exposure, as the processes of eating and drinking are not possible with most of this equipment. The restrainer may limit the animals' ability to dissipate excess body heat.

END POINT DATA COLLECTION

Types of data that are routinely collected during the conduct of toxicity studies in mice fall into three broad categories: clinical observations and physical examinations, clinical laboratory evaluations, and postmortem procedures. Cardiovascular parameters are not measured in routine toxicology studies. Heart rates in awake mice have been measured in the range of 300 to more than 800 beats/min (Hoyt et al., 2007). Reliable blood pressure measurements are best made by cannulation of a major artery, such as the carotid. Such procedures require anesthesia and surgery, neither of which is especially desirable during the course of a study that may be of long duration and involve many animals.

Clinical Observations and Physical Examinations

Clinical observations entail the recording of effects that can be detected by direct observation, such as abnormal gait and body weight. For the sake of this discussion, a variety of parameters that can be observed or measured directly will be discussed in this section. Clinical observations often provide the first indication of which physiological systems are being affected by the test article.

Mice should be observed regularly throughout the in-life portion of a toxicity study. The type and frequency of these observations should be tailored to meet the scientific objectives of the specific study. Most effects observed following administration of acute (single) doses occur within a relatively short time after dosing. As acute iv doses are often associated with almost immediate effects, it might be appropriate to observe treated mice within 5 min, at about 15, 30, and 60 min, and again at 2 and 4 h after dosing. Observations should be repeated at least once daily on all subsequent study days throughout the postdosing observation period. This schedule should provide information on the times of onset, peak activity, and remission from toxic effects as well as information on the sequence and severity of effects observed. The high intensity of data collection on the day of dosing in acute studies requires that the system for conducting and recording observations be simple and time efficient. Typically, a system of "exception reporting" is used, in which observations of exceptions from the norm are recorded, and the absence of comment on a system (e.g., respiration) implies that parameter is normal. Clinical observations in repeated dose studies should be conducted at approximately the same time each day to assure that changes in findings over the course of the study can be attributed to the accumulation of or adaptation to toxic effects rather than incidental changes attributable to circadian rhythm or time after dosing. Minimally, all animals should be observed early in the day, prior to daily dosing, and it is highly desirable to conduct at least one additional daily observation at 2–4 h after dosing (or late in the day) to be aware of effects that may be associated with higher blood levels of test article usually found from a few minutes to a few hours after dosing.

The simplest form of clinical observation is an observation for survival and moribundity. This or a higher level of observation must be conducted at least once daily in all toxicity studies. The next level of observation is an observation for clinical signs of toxicity, such as abnormal level of spontaneous motor activity, abnormal gait, abnormal respiration, and abnormal quantity or quality of fecal output.

The next level of observation is more structured, and is typically conducted about once in a week during studies of a few weeks' duration to as infrequently as about once in a month during the later phases of 26-week to 2-year studies. During the conduct of a physical examination, specific parameters are evaluated, such as quality of coat, body orifices (for excessive or unusual discharges), eyes, respiratory sounds, and in studies longer than about 26 weeks, animals are examined carefully for evidence of visible or palpable masses. Body weight and feed consumption are typically monitored in studies longer than a few days. An appropriate interval for measuring body weight and feed consumption is about a week. These two parameters should be measured concurrently, such that changes in one can be compared directly to changes in the other. In longer studies, in which the mice have reached maturity and body weight gain has approached zero, the frequency at which body weight and feed consumption are measured can be reduced to as infrequently as once per month. The interval over which they are measured would remain at about a week, however.

Clinical Laboratory Evaluations

Clinical laboratory evaluations of mice refer to evaluations of blood and urine. Blood is routinely collected at sacrifice in repeated dose studies, and small quantities (e.g., about 0.10 mL) of blood can be collected at interim periods during the course of the study for the purpose of evaluating differential smears or other limited objectives. Interim (nonterminal) blood samples can be collected by retro-orbital venous plexus puncture, cardiac puncture, and tail snip, among other techniques. Each of these techniques has certain disadvantages. Retro-orbital puncture is technically difficult and may require anesthesia or immobilization of the animal. Cardiac puncture typically requires anesthesia, and cardiac injury may compromise the histological evaluation of cardiac tissue. Tail snip often yields samples that are contaminated with extravascular, extracellular fluids. Any administration of anesthetic agents during the study of a test article that is not thoroughly understood engenders some risk to the interpretation of the study, as potential interactions of the anesthetic with the metabolism or direct effect of the test article are nearly impossible to predict. Blood collected at the time of sacrifice is typically drawn from the inferior vena cava or the abdominal aorta while the mouse is under anesthesia. In the case of terminal blood collection, potential interaction of the anesthetic agent with the test article, induction of liver enzymes, etc., is not an issue.

Parameters evaluated in blood samples drawn from mice include evaluation of differential smears for morphological abnormalities and differential white counts, measurement of serum glucose and urea concentrations, serum aspartate aminotransferase, ALT, and alkaline phosphatase activity. RBC counts, white blood cell counts, and hemoglobin concentrations can be measured, but these parameters are quite variable for mice on many of the commonly used laboratory instruments, so they are often omitted. In addition, bone marrow smears may be prepared but are usually only prepared at sacrifice in mice. Bone marrow smears may help in understanding hematological changes. Caution should be exercised in comparing experimental data with results obtained from the literature or with results obtained on different instrumentation or by different procedures. For greatest utility, a set of normal values should be compiled for the laboratory procedures and equipment used to produce the data in the toxicity study.

As a practical matter, urine is not usually collected in routine toxicity studies. The primary difficulty in conducting urinalysis is that the mouse produces a very small volume of urine during a reasonable collection period (e.g., 16–24 h), and of that volume, considerable and variable quantities are lost to evaporation and on the surfaces of the collection apparatus. As a result, attempts to evaluate urinary concentrations of practically anything can be very misleading.

Postmortem Procedures

Postmortem procedures, literally those procedures performed after the death of the animal, include confirmation of the identification number and sex of the animal, an external examination, examination of the significant internal organs in place prior to removal, and then removal, weighing of appropriate organs, and collection of tissue specimens for histological processing and microscopic examination. The microscopic examination of tissue specimens by a qualified veterinary pathologist may be the single most important source of information in understanding the toxicity of a test article. The pathologist's findings should be carefully integrated with the other study data (e.g., clinical observations, body weights, feed consumption, and clinical laboratory findings) to fully comprehend the effects of the test article on the mouse under the conditions of study.

Ordinarily, the list of tissues to be routinely weighed, collected, and processed for histological examination will be specified in the study protocol. In addition to the tissues specified in the protocol, specimens are usually collected of all lesions or target organs that have been identified during the course of the study or at gross necropsy. It is important to provide the necropsy staff with a current list of abnormal clinical observations, especially any evidence of visible or palpable masses, as this is the time when the visible and palpable lesions can be linked to the histopathological evaluation of those lesions. Every effort should be made to locate all lesions described and collect representative tissue from those sites.

A detailed description of necropsy procedures is beyond the scope of this discussion. It should be emphasized that the necropsy process, particularly when conducted on a large number of animals at the scheduled termination of a study, is a process in which a large number of tissue samples may be collected and a similarly large quantity of data may be gathered during a short period of time. As such, this process presents many opportunities for loss or misidentification of samples and data. A rigorous system of accounting for which tissues have been collected from each animal, and for tracking the samples and data collected is critical to the accurate interpretation of the toxicity study.

SUMMARY

In summary, the mouse is one of the most useful species for toxicity testing. Mice have been used in biomedical research for hundreds of years. As a result of long usage, many techniques have been developed to dose and evaluate mice, and a wealth of historical data has been accumulated in the literature. A wide variety of genetic strains have been developed for specific purposes. It is often possible to select a strain for testing that is particularly vulnerable or resistant to either the test article or a particular type of lesion that might be expected to be associated with that test article. The small size of the mouse confers economy in acquisition, husbandry, handling, and test article consumption. The relatively short gestation period and life span of the mouse are useful in conducting reproductive studies, or studies in which the test article will be administered for a high percentage of the lifetime of the animal.

The small size of the mouse is responsible for most of the disadvantages of the species as well. The species is relatively susceptible to environmental stress. Small size and blood volume makes it difficult or impossible to collect multiple samples of blood and urine over short periods of time. Assays that might require large volumes of blood or urine are precluded. Certain physiological evaluations, such as electrocardiograms, are difficult owing to the small size and high activity level of the species.

The Hamster

The hamster is the third most frequently used laboratory animal following the rat and mouse at a level of ~146,000/year (Renshaw et al., 1975; Silverman, 2012), although its use in toxicology is somewhat limited. Although historically the hamster saw extensive use in carcinogenesis testing, as will be overviewed, this has changed. It has many attractive features as a laboratory animal because of its reproduction ease, unique anatomical and physical features, rapid physiological development, short life span, low incidence of spontaneous diseases, and a high susceptibility to induced pathological agents. Hamsters historically have been used in several fields, especially in carcinogenesis because of its low incidence of spontaneous tumors, but currently most of their use is seen in testing associated with buccal delivery of drugs (Gad, 2016). Hamsters have also played an important role in blood vessel physiology because their cheek pouches with thin vascularized walls are very accessible. The hamster is also a major model in diabetes research.

SPECIES

The following is a discussion of the eight hamster species maintained in the laboratory. Table 2.31 lists these hamsters' common and scientific names and their chromosome numbers.

Syrian Hamster

The Syrian hamster is the most common laboratory hamster. Eighty percent of all hamsters used in research are Syrian. The remaining 20% are primarily Chinese, followed distantly by European, Armenian, Rumanian, Turkish, South African, and Dzungarian hamsters. The Syrian was originally native to the arid, temperate regions of Southeast Europe and Asia Minor. It was first described as a new species (*Cricetus auratus*) in 1839. For almost 100 years, no hamsters were caught in the wild. The only proof that the species existed was the preservation in alcohol of two hamsters, one in London and the other in Beirut. Specimens of the species were finally obtained from the wild starting in the 1900s and have since been bred in multitudes in captivity. It lives in deep tunnels that ensure cool temperatures and increased humidity. It is a nocturnal animal. The Syrian is virtually tailless and has smooth short hair. Normal coloration is reddish gold with a grayish white ventral portion. The dorsal side may have a black stripe. The ears are pointed with dark coloration; and the eyes are small, dark, and bright. The average life span is 2 years, but these animals can live up to 3 years. The animal is 14–19 cm in length and weighs 114–140 g at adulthood. The female is usually heavier and longer than the male.

The Syrian hamster was introduced into the laboratory in 1930 to study the Mediterranean disease kala-azar. Israel Aharoni (Hebrew University, Jerusalem, Israel) collected 11 young golden hamsters from Syria in 1930 while on a zoological expedition. The litter with their mother had been found in their burrow 2.5 m under a wheat field. Aharoni and his wife kept the hamsters in their house until one night when they all escaped. Nine hamsters were recovered and given to the animal facilities supervisor of the Weizmann & Seiff Institute, Jerusalem, Israel. Of the nine, five escaped the first night in the new facility, leaving only one female. The female was mated and gave birth to a litter of healthy pups. In a year's time, these hamsters produced more than 300 offspring and were the forbearers of today's laboratory-bred Syrian hamster. There is no record of any further captures of Syrian hamsters from the wild.

The Syrian has been involved in oncology, virology, endocrinology, physiology, parasitology, genetics, and pharmacology research. The cheek pouch of the Syrian hamster has provided the technology for studying microcirculation and the growth of human tumors.

Chinese Hamster

The Chinese hamster is native to China. It is 39–46 g in weight and 9 cm long at adulthood. Its life span is 2.5–3.0 years under laboratory conditions. Though the Chinese hamster is smaller than the Syrian, its testicles, spleen, and brain are larger.

The Chinese hamster was originally used for the first time in 1919. Mice were extremely scarce at the time so hamsters were used to determine the best therapy for the patients with pneumonia. The Chinese hamster was also used to study TB, influenza, diphtheria, and rabies. Robert B. Watson, in December 1948 (right before the Communist takeover of China), was given 10 female and 10 male hamsters from C. H. Hu of the Peking Union Medical College. Watson placed the hamsters on what he believes was one of the last Pan Am flights out of China to San Francisco. From San Francisco, the hamsters were sent to New York. V. Schwenter of the Harvard Medical School obtained the hamsters and eventually successfully bred them in the laboratory. Of the original 20 hamsters, 4 of the females and 3 of the males produced offspring that gave rise to the present Chinese hamster population. The Harvard colony has since become extinct; however, colonies were established at the Upjohn Company (which became Pharmacia and most recently was acquired by Pfizer), Kalamazoo, Michigan, and the C. H. Best Institute in Toronto, Ontario, in the 1960s.

TABLE 2.31 Common and Species Names and Chromosome Number

Common Name	Species Name	Chromosome Number
Syrian (Golden)	*Mesocricetus auratus*	44
Chinese (Striped, Black)	*Cricetus griseus* or *barabensis*	22
European (Common, Black, Field)	*Cricetus cricetus*	22
South African	*Mystromys albicaudatus*	32
Rumanian (Newtoni's)	*Mesocritceus newtoni*	38
Turkish (Kurdanti)	*Mesocricetus auratus*	42/44
Armenian (Gray, Migratory)	*Cricetulus migratorius*	22
Dzungarian (Hairy-footed)	*Phodopus sungorus*	28

The Chinese hamster has been used primarily in research for cytogenetics because of its low chromosome number (Fenner, 1986), and in diabetes mellitus because (1) some strains have very high incidences of the disease and (2) the course of the disease in this species is similar to that seen in humans.

European Hamster

The European hamster was first found in a West Germany industrial area. Its natural habitat is the lowlands of Central and Eastern Europe. The European hamster is a very aggressive animal, and in the wild each adult lives in its own burrow. It has a white face and feet, bodies are dorsally reddish brown and ventrally black with white patches laterally. They are about the size of a guinea pig, averaging 27–32 and 22–25 cm in length and weighing 450 and 350 g for males and females, respectively. Males reach sexual maturity at 60 days of age, whereas females at 80–90 days of age, and they are mainly seedeaters. In the wild, they hibernate in the winter months. In their natural habitat, European hamsters can live up to 8 years, whereas under laboratory conditions, the average life span is 5 years. This reduction is believed to be due to the lack of hibernation afforded to a laboratory-raised European hamster (Mohr and Ernst, 1987).

The European hamster has been used only in hibernation studies and in inhalation studies because its tidal volumes are the largest of any laboratory rodent species.

Armenian Hamster

The Armenian hamster is native to the Union of the Soviet Socialist Republic (USSR). Its body size, weight, care, and maintenance are similar to that of the Chinese hamster.

The Armenian hamster was first introduced as a laboratory animal in 1963. It was brought to the United States as a part of the US–USSR. Cultural Exchange Program. Scientists in the United States wanted to find more species of the dwarf hamster (like the Chinese), and the Armenian species has been the only species found. Although the Armenian hamster has been used on a limited basis, its research use has been in cytogenetics and oncology.

Turkish Hamster

The Turkish hamster is native to Iran and Turkey. It was originally trapped in 1962. As an adult, its average body weight is 150 g and its average life span is a little less than 2 years, although they have lived as long as 4 years. Some populations of the Turkish hamster have a diploid number of 42 and others a number of 44. These hamsters interbreed readily and produce offspring with a diploid number of 44. Hamsters with diploid number of 42 hibernate less than those with 44. Besides hibernation research, Turkish hamsters have been used in immunology, genetics, and reproductive behavior research (Yerganian, 1972; Cantrell and Padovan, 1987).

Rumanian Hamster

The Rumanian hamster was initially trapped and described in 1965. It is native to the Bucharest area and is used in the laboratories surrounding that area. Its care, size, and management are similar to that of the Syrian hamster, although it does not reproduce as well as the Syrian. The Rumanian hamster adult averages 100 g weight. Its face is more pointed and ratlike than the Syrian hamster, but it is similar in appearance to the Turkish hamster.

Dzungarian Hamster

The Dzungarian hamster is very timid. The males are 11 cm long and 40–50 g in weight; the females are 9 cm long and weigh 30 g at maturity. The Dzungarian hamster has a short tail about 1 cm in length, which is usually hidden by the body fur. The fur on the dorsal side is gray with a dark-brown or black stripe from the nape of the neck to the base of the tail. The ventral fur is white. The average life span has been reported to 1 year by Herberg et al. (1980) and 2 years by Heldmaier and Steleinlechner (1981).

The Dzungarian hamster, native to the USSR, was originally trapped in Siberia and provided to the United States by the USSR. The present Dzungarian hamster population is the result of the mating of one female to two males who were domesticated in 1965. The Dzungarian has been used in research involving photoperiodism, the pineal gland, and thermoregulation.

South African Hamster

The South African is the only member of its genus and the only hamster native to Africa. The first colony was established in South Africa in 1941. In its natural habitat, South African hamster is a nocturnal, solitary burrowing rodent. Unlike other hamsters, it does not have cheek pouches. The hamster has gray to brown fur on its dorsal aspect with white on the ventral surface, feet, and tail. The tail is 5–8 cm long. Its ears are erect and the eyes are dark and bright. Adult males and females weigh 145 and 95 g, respectively (Hall et al., 1967). The average life span is 2.4 years, with a maximum life span of 6.2 years (Davis, 1963). The

United States received its first South African hamster in 1962. These hamsters have not had much of an impact in biomedical research except in diabetes mellitus and infectious disease research.

SPONTANEOUS TUMORS

Table 2.32 lists the most common spontaneous tumors in hamsters by incidence and their total incidence. The most frequent tumors are seen in the adrenal cortex and intestinal tract, followed by the lymphoreticular system, the endometrium, endocrine system, and ovaries of aging females (Sher, 1982). The benign tumors found are usually adenomas of the adrenal cortex and polyps of the intestinal tract. In a study by Dontenwill et al. (1973), adenocarcinomas were age related with a rate of greater than 50% in hamsters over 100 weeks of age. The rate of small intestinal adenocarcinomas (0.8%) seen by Fabry (1985) was higher in hamsters than in rats or mice. Lymphosacromas are the most common malignant tumors of the Syrian hamster. Tumors of the liver, pituitary, lung, urinary bladder, and mammary gland are practically unknown in the hamster, but these do occur spontaneously in older rats and mice (Homburger et al., 1979a,b). Genetic drifts seen in many colonies of hamsters may influence the rate of spontaneous tumors as in the rat and mouse.

TABLE 2.32 Incidence of Spontaneous Tumors in Syrian Hamsters

Neoplasm	Males	Females	Total
Adrenal adenoma	12.7	9.4	11.0
Lymphoreticular neoplasm	3.7	2.3	3.0
Uterus endometrial polyp		3.0	
Uterus endometrial carcinoma		3.0	
Adrenal carcinoma	3.0	2.0	2.5
Pancreas islet cell adenoma	3.7	1.3	2.5
Vagina papilloma		2.0	
Stomach papilloma	1.7	1.7	1.7
Thyroid carcinoma	1.0	2.0	1.5
Uterus leiomyoma		1.0	
Small intestine adenocarcinoma	0.3	1.3	0.8
Pituitary adenoma	0.0	1.3	0.7
Pancreas islet cell carcinoma	0.7	0.7	0.7
Ovary fibroma		0.7	
Ovary theca cell tumor		0.7	

ANIMAL IDENTIFICATION

Hamsters are usually identified by tagging, punching, or coding of the ear or ear tattooing, which is done aseptically.

DOSING PROCEDURES

Oral Administration

To dose a hamster orally (p.o.), the animal is grasped by the skin of the neck and back. The gavage tube (metal 18- or 20-gauge) or a polyethylene catheter (2–3 cm in length) is passed into the mouth via the interdental space. The tube is passed gently into the esophagus and the fluid administered. The method is similar to the procedure done in the mouse and rat.

Subcutaneous Administration

The hamster is restrained for sc dosing as described for oral dosing. The needle is placed into the skin that is tented by pulling up a fold of skin on the back firmly between the thumb and index finger immediately behind the head. The injection is made into the skin parallel to the back. The hamster's loose skin enables large volumes to be injected sc in comparison to other rodent species of the same size (Collins, 1979).

Intradermal Administration

To administer an agent id, the skin over the desired injection site is shaved. Holding the animal as described above, the needle (30-gauge) is advanced just a few millimeters into the skin. If there is suddenly no resistance, then the needle can been pushed through the skin. Withdraw and advance the needle again. Following administration of the material, a small welt will be visible.

Intramuscular Administration

To administer an im dose, the muscles of the posterior and anterior thighs of the hamster are the most frequently used sites. The animal is restrained as described for oral dosing by an assistant and one leg is held by the doser. The quadriceps are held between the forefinger and the thumb of the doser. The material is injected into the muscle mass.

Intraperitoneal Administration

To dose a hamster ip, the animal is grasped as described previously. The needle is pushed parallel to the line of the leg through the abdominal wall into the peritoneal cavity. Following the leg line avoids administration into the urinary bladder or the liver. Administration may occur when there is no resistance to the needle passage.

Intravenous Administration

It is best that the hamster is anesthetized to administer materials iv. The veins that can be used are the femoral, jugular, and cephalic. The areas must be shaved, a skin incision made to expose the vein, and then a needle may be placed into the vein and the material administered.

BLOOD COLLECTION TECHNIQUES

Retro-orbital Method

The retro-orbital is the method of choice for collecting blood from the hamster. The method used is that as described for the rat. A 23-gauge needle or a microhematocrit tube may be used to obtain the blood. Three milliliters of blood may be collected retro-orbitally; however, for repeated sampling a volume of 0.5 mL is best for the animal. The use of anesthesia is preferable in the hamster.

Cardiac Puncture

Cardiac puncture in hamsters requires practice because the heart can be difficult to locate or can rotate away from the needle. A 25-gauge 3/8-in needle is the suggested equipment. A safe volume to draw from the heart with minimal damage is 1–2 mL. Repeated sampling from the heart is not advised because the mortality rate due to the blood withdrawn can be high (Wechsler, 1983). The use of anesthesia is suggested in the hamster. On exsanguination, 5 mL can be withdrawn from a 95-g hamster (Schermer, 1967).

Tail Clipping Method

Tail clipping is a method suitable only for a maximum of six samples because the tail is so short. To facilitate blood flow, place a suction bell (which is connected to a water pump) on the base of the tail. Anesthesia is not necessary if the hamster is placed in a narrow tube with the hind legs protruding so they can be held.

Femoral Vein Method

To collect blood from the femoral vein, a tourniquet is placed above the stifle and the fur over the area clipped. A skin incision is made to expose the vein and a 25-gauge 5/8-in needle is placed into the vein. The blood is collected from the needle hub with a capillary microcontainer or a microhematocrit tube.

Jugular Vein Method

To collect blood from the jugular vein, the hamster should be anesthetized, the area shaved over the jugular, a skin incision made, and a 25-gauge 5/8-in needle placed into the vein and the blood withdrawn.

Saphenous Vein

First a body tube is used to immobilize the animal. The skin over the ankle is stretched, allowing for ease in shaving the area above or around the vein. A 25-gauge, 5/8-in needle is slid into the vein and blood is withdrawn. This method is as of 2005 the recommended approach to nonterminal blood collection (Hem et al., 1998).

URINE COLLECTION

The best method is to collect the urine over 17–24 h as the animal voids. A preservative such as thymol can be added to the collection vessel before starting. Catheterization of the ureter can be done; however, there is always the chance of blood or tissue contamination. A hamster's urine is normally a thick, milky fluid.

PHYSICAL PARAMETERS

Neonatal Body Weights

At birth, Syrian hamsters weigh 2–3 g and Chinese hamsters weigh 1.5–2.5 g. Neonate hamsters are hairless with the ears and eyes closed. Table 2.33 lists information about the early development of the eight laboratory hamsters.

Body Weights and Weight Gains

The average adult body weight for each of the species is discussed in the Species section of this chapter.

In a study by Borer et al. (1977), Syrian hamsters gained 2 g/day from birth to 5 weeks of age. From days 30 (weight = 65 g) to 70 the hamsters gained 1 g/day. From day 70–88 the hamsters gained 0.3 g/day. Syrian hamsters at maturity weighed from 100 to 135 g. Normal fetal and maternal weight gain is well documented in Davis (1989).

TABLE 2.33 Neonatal Data for Hamsters

Species	Birth Weight (gm)	Day Eyes Open	Day Ears Open	Day Pups Solid Food	Day Weaned	Teeth at Birth
Syrian	2–3	15	5	7–10	21	Yes
Chinese	1.5–2.5	10–14	10–14		21–25	
Dzungarian	1.8	10	3–4	10	16–18	Yes
European						No[a]
Armenian		14	14	14	18	
Rumanian						
Turkish		12–13		12–13	20	
South African	6.5	16–25	3–5	21–25	No[a]	

[a]*Incisors erupt at 3–5 days.*

Dentition

The dentition of a Syrian hamster is monophyodont, bunodont, and brachyodont. The incisors of the Syrian species grow irregularly depending on age and sex of the hamster. The adult European hamster has one set of permanent teeth that consists of four continuously growing incisors and 12 molars.

Life Spans

The average life spans for each of the laboratory species are discussed in the Species section of this chapter.

Sexual Maturity

Sexual maturity for female hamsters begins at 4–6 weeks of age. At this time, there are 10 mature and 25 reserve follicles in each ovary. A female's first spontaneous estrus and ovulation are at 4–5 weeks and 30 days, respectively. In young immature females (4 weeks of age), ovulation may be artificially induced with 30 IU pregnant horse serum (Magalhaes, 1970). The estrus cycle is 94 h or >4 days in length with four distinct stages: proestrus, estrus, metestrus, and diestrus. Identification of the stage can be determined from differentiation of a vaginal smear through examination of cell types. During estrus (which occurs before and after ovulation) the female will show lordosis and will mate. The end of estrus is the appearance of a copious postovulatory discharge. The discharge is creamy, white, opaque, and very viscous with a pungent cheesy odor. Ovulation occurs regularly every 4 days, 9–10 h after the peak concentrations of luteinizing hormone (LH). The breeding life of a hamster is usually about 10–12 months, or after the production of six litters.

Syrian hamster males reach sexual maturity at 6–7 weeks of age and have a breeding life of about 1 year. The testicles of male Syrian and Chinese hamsters descend at day 26 and 30, respectively.

The secondary sex characteristic of the Syrian hamster is a scent organ (flank organ), which is located on the flank. The male touches the female's organ with his paws during copulation. The intensity of the pigmentation is an indication of androgen activity. The males have a darker pigmentation than females. The pigmentation is first seen at 25 days of age and is more marked at 35 days of age.

Respiratory Rate and Oxygen Consumption

The minimum respiratory rate (breaths/min) is 33, and the maximum is 127. The average respiratory rate is 74 and the mean resting respiratory rate is 30–33 breaths/min (Robinson, 1968). Hamsters are nose breathers and have a resting oxygen consumption of 2.3 mL/g/h. It is possible to implant sensors to measure these, but such is uncommon in hamsters.

BLOOD PRESSURE

The blood pressure measured by cannulation of the carotid artery of the hamster is 111 mmHg (Storia et al., 1954). The blood pressure measured on the cheek pouch by Berman et al. (1955) was 90 ± 11.3 mmHg. Another measurement of blood pressure by photoelectric tensiometry was 108 mmHg. Although there is no direct comparison of these measurements because of the techniques used, the values obtained are similar and can be used as reference values for each method described.

HEART RATE

The mean heart rate (beats/min) of the Syrian hamster is 450 with a range of 300–600.

ECG PATTERNS

The rate of contraction in the Syrian hamster is 350–500 beats/min. The PQ interval is 48 ms with a range of 40–60, the QRS interval is 15 ms with a range of 13–20, and the T- and P-wave amplitudes are 0.33 t 0.07 mV and 0.19 t 0.03 mV, respectively. For measurement of ECGs, the hamster needs to be anesthetized because of its aggressive behavior. The ECG tracings of hamsters resemble human ECG tracings (Lossnitzer et al., 1977).

CLINICAL LABORATORY

Clinical chemistry values for Syrian hamsters are listed in Table 2.34. The following will be a discussion of several interesting aspects of the hamster and its

TABLE 2.34 Clinical Chemistry Values for Syrian Hamsters

Test	Units	Male Mean Values	Female Mean Values	Range Both Sexes
Bilirubin	(mg/dL)	0.42	0.36	0.20–0.74
Cholesterol	(mg/dL)	54.8	51.5	10.0–80.0
Creatinine	(mg/dL)	1.05	0.98	0.35–1.65
Glucose	(mg/dL)	73.4	65.0	32.6–118.0
Urea nitrogen	(mg/dL)	23.4	20.8	12.5–26.0
Uric acid	(mg/dL)	4.58	4.36	1.80–5.30
Sodium	(mEq/L)	128	134	106–146
Potassium	(mEq/L)	4.66	5.30	4.0–5.9
Chloride	(mEq/L)	96.7	93.8	85.7–112.0
Bicarbonate	(mEq/L)	37.3	39.1	32.7–44.1
Phosphorus	(mg/dL)	5.29	6.04	3.4–8.24
Calcium	(mg/dL)	9.52	10.4	7.4–12.0
Magnesium	(mg/dL)	2.54	2.20	1.9–3.5
Amylase	(Somogyi units/dL)	175	196	120–250
ENZYMES				
Alkaline phosphatase	(IU/L)	17.5	15.4	3.2–30.5
Acid phosphatase	(IU/L)	7.45	6.90	3.9–10.4
Alanine transaminase	(IU/L)	26.9	20.6	11.6–35.9
Aspartate transaminase	(IU/L)	124	77.6	37.6–168
Creatinine phosphokinase	(IU/L)	101	85.0	50–190
Creatinine kinase	(IU/L)	23.1		
Lactic dehydorgenase	(IU/L)	115	110	56.0–170.0
SERUM PROTEINS				
Total protein	(g/dL)	6.94	7.25	4.3–7.7
Albumin	(g/dL)	3.23	3.50	2.63–4.10
α_1-globulin	(g/dL)	0.64	0.55	0.30–0.95
α_2-globulin	(g/dL)	1.85	1.70	0.9–2.70
β-globulin	(g/dL)	0.56	0.83	0.1–1.35
γ-globulin	(g/dL)	0.71	0.67	0.15–1.28
A/G ration		0.87	0.93	0.58–1.24

Mitruka, B.M., Rawnsley, H.M., 1981. Clinical Biochemical and Hematological Reference Values in Normal Experimental Animals and Humans, second ed. Masson, New York.

clinical chemistry values. In comparison to humans, the hamster has lower bilirubin, cholesterol, alkaline phosphatase, creatine phosphokinase, lactic dehydrogenase (LDH), and A/G ratio values and higher blood urea nitrogen, bicarbonate, phosphorus, amylase, acid phosphatase, AST, and α2-globulin values. Because hamsters are such deep day sleepers, blood collection times should be noted because during light photoperiods clinical chemistry values can be variable. The anesthesia used may also affect clinical chemistry values. These must be taken into account when analyzing clinical chemistry data.

Glucose

Thiobarbiturate anesthesia may produce glucose levels as high as 300 mg/dL in adult male Syrian hamsters. The hyperglycemia can exist for up to 5 h after anesthesia exposure, and there is no relationship to the length of hyperglycemia compared to the duration of the anesthesia. Hyperglycemia has also been reported in hibernating hamsters (Newcomer et al., 1987).

Lipids

The lipids found in the hamster are cholesterol, phospholipids, triglycerides, and fatty acids. Cholesterol concentrations in hamsters are the highest when compared to other laboratory animals, but lower than human levels. Short photoperiods (10 h or less) can cause a decrease in cholesterol, but other lipids such as plasma triglycerides are not affected. Serum lipids do increase in the hamster during hibernation. A strain of spontaneous hypercholesterolemic Syrian hamsters show increases in cholesterol when exposed to low temperatures.

Urea Nitrogen

Hamsters who develop kidney disease during the aging process have increased urea nitrogen levels as seen in other laboratory animals.

ENZYMES

Serum obtained by cardiac puncture may be contaminated by AST, LDH, ALT, and creatine phosphokinase owing to the high concentrations of these enzymes already present in the heart.

Alkaline Phosphatase

The alkaline phosphatase in the hamster is composed of isoenzymes from bone, liver, and intestine. Alkaline phosphatase is a more sensitive indicator of liver damage than bilirubin or ALT. Dramatic increases are usually indicative of bile duct obstruction. Immature hamsters have elevation of values two- to threefold compared to adults.

Alanine Aminotransferase

ALT is specific for liver damage in dogs, cats and rats. In the hamster, ALT levels are increased in both acetaminophen-induced and viral-induced hepatic necrosis.

Aspartate Aminotransferase

The activity of AST is low but increases following muscle injury. Increased AST levels have been seen in hamsters with liver neoplasms.

Creatine Kinase and Lactic Dehydrogenase

Cardiomyopathic Syrian hamsters show elevated creatine kinase and LDH activities. Normal creatine kinase levels are 23.1 IU/L, whereas in cardiomyopathic hamsters the levels are 730 IU/L.

Thyroid Hormones

Thyroid hormones are of interest because of the hamster's hibernation. Basal T_3 and T_4 decrease with age. T_4 levels in 3-month-old hamsters are 6.75 ± 0.75 μg/dL and 3.59 t 0.16 μg/dL in 20-month-old hamsters. T_3 levels in 3-month-old hamsters are 62 ± 2 ng/dL and 42 ± 3 ng/dL in 20-month-old hamsters. These changes are also seen in humans and in other rodent species. Older hamsters show less of an increase in T_3 and T_4 levels after thyroid-stimulating hormone (TSH) administration. During short photoperiods there is a decrease in TSH, T_3, and T_4. Lower temperatures also cause decreases in T_3 and T_4 levels, whereas in cold conditions there is an increase T_3 and a decrease in T_4. Pregnant hamsters may metabolize thyroid hormones differently because there is a decrease in protein-bound iodine during pregnancy.

Reproductive Hormones

During estrus there is one LH surge and the follicle-stimulating hormone (FSH) is biphasic. The first burst of FSH occurs concurrently with LH. The second burst is thought to be responsible for the initiation and/or maintenance of follicular growth for the next estrus cycle. Maintenance of functional corpora lutea is believed to be performed by a combination of prolactin, FSH, and a small amount of LH.

Progesterone is the dominant hormone during the first 2 days of estrus, decreasing on day 3 but rising again on day 4. The levels of progesterone on days 1–2 are dependent on LH. Estradiol levels are low for the first 2 days and increase and decrease on day 4.

Adrenal Hormones

In the hamster, corticosterone and cortisol are secreted by the adrenal cortex. Corticosterone levels are 3–4 times higher than cortisol during the day. Adrenocorticotropic hormone stimulation increases both hormones; however, cortisol levels are stimulated at a higher rate. Basal cortisol levels are 0.45 ± 0.04 μg/dL and 0.38 ± 0.09 μg/dL in males and females, respectively, and corticosteroid levels are 7.4 ± 1.9 mg/dL (Tomson and Wardop, 1987). Pregnant hamsters can produce large quantities of cortisol (30 μg/dL), whereas nonpregnant females have relatively low levels (0.3 μg/dL) in comparison to other species. Increased plasma cortisol levels are seen after exposure to chronic stress.

Glucocorticoid levels follow the circadian pattern seen in other rodents.

PROTEINS

Chinese hamsters with spontaneous diabetes have 10%–30% of their total proteins being α_2, whereas control hamsters have only 3%–8%. Asymptomatic hamsters with significantly elevated α_2 proteins do develop chemical or clinical diabetes later on.

HEMATOLOGY VALUES

Hematological values for Syrian hamsters are listed in Table 2.35. Hematological values for European and Chinese hamsters are listed in Table 2.36. The blood volume of a hamster is 6%–9% of its body weight. The maximum safe volume for one bleeding is 5.5 mL/kg. The practical volume from adult hamsters for diagnostic use is 1 mL. Hematological values for a hamster vary considerably because they are deep day sleepers, although values between males and females are similar. The variations seen are changes in blood volume and quantity of blood components.

Erythrocytes

Erythrocytes in the hamster have a diameter of 5–7 μm. South African hamsters have larger erythrocytes than other hamsters and laboratory rodents. A small portion of erythrocytes show polychromasia (Schermer, 1967). Erythrocytes have a life span of 50–78 days. Desai (1968) saw an increase in erythrocyte longevity during hibernation. Nucleated or basophilic cells are rare, but reticulocytes are found from 3% to 4.9%.

Leukocytes

Hamster leukograms are similar to those of other laboratory rodents. During photoperiods the total leukocyte counts range from 5000 to 10,000/μL, whereas during hibernation counts drop to 2500/μL. During sleep, the lymphocyte:neutrophil ratio is 45%:45%, whereas in awake animals the neutrophil percentage varies between 17% and 35%.

Coagulation

Hamster blood starts to coagulate at 15–20 s (Schermer, 1967) with a mean coagulation time of 142 s (Desai, 1968).

Trypanosomes

Hamsters frequently have trypanosomes in the blood. These microorganisms are parasitic but not pathogenic. They have been observed in other laboratory animals, including sheep and monkeys, but not in the large numbers as seen in the hamster. The counts of trypanosomes can be very large, sometimes equaling the leukocyte counts. Professor Enigk of the Bernhardt-Nocht Institute, Germany (personal communication) found that 50% of his hamster colony had trypanosomes. Transmission is not known; however, trypanosomes are not considered harmful and will not interfere with the outcome of a hamster study (Schermer, 1967).

BLOOD GASES AND pH

The values for blood gases are Pa_{O_2} 71.8 ± 4.9 mmHg, Pa_{CO_2} 41.1 ± 2.4 mmHg, HCO_3 29.9 ± 2.9 mEq/L, and the blood pH is 7.48 ± 0.03. Measurements of exercising hamsters show an increase of 12.9 ± 7.9 in Pa_{O_2}, a decrease of 6.6 ± 2.6 Pa_{CO_2}, and a decrease of 3.5 ± 2.3 in HCO_3 concentrations.

URINE VALUES

The range of urine volume is 5.1–8.5 mL/24 h in normal hamsters. In diabetic Chinese hamsters, the urine volume can be as high as 25 mL/day. Sodium

TABLE 2.35 Hematological Values for Syrian Hamsters

		Male		Female	
Test	**Units**	**Mean**	**Range**	**Mean**	**Range**
RBC	×10E6/mm^3	7.5	4.7–10.3	6.96	3.96–9.96
HgB	g/dL	16.8	14.4–19.2	16.0	13.1–18.9
MCV	μ^3	70.0	64.0–77.6	70.0	64.0–76.0
MCH	μμg	22.4	19.9–24.9	23.0	20.2–25.8
MCHC	%	32.0	27.5–36.5	32.6	27.8–37.4
Hct	%	52.5	47.9–57.1	49.0	39.2–58.8
Sedimentation rate	mm/h	0.64	0.32–0.96	0.50	0.30–0.70
Platelets	×10E6/mm^3	410	367–573	360	300–490
WBC	×10E6/mm^3	7.62	5.02–10.2	8.56	6.48–10.6
Neutrophils	×10E6/mm^3	1.68	1.11–2.25	2.48	1.88–3.08
Eosinophils	×10E6/mm^3	0.07	0.04–0.12	0.06	0.04–0.08
Basophils	×10E6/mm^3	0.08	0.05–0.10	0.04	0.03–0.05
Lymphocytes	×10E6/mm^3	5.6	3.69–7.51	5.81	4.41–7.20
Monocytes	×10E6/mm^3	0.19	0.12–0.26	0.20	0.16–0.25

RBC, red blood cell.
Mitruka, B.M., Rawnsley, H.M., 1981. Clinical Biochemical and Hematological Reference Values in Normal Experimental Animals and Humans, second ed. Masson, New York.

and potassium concentrations are 70 and 120 mmol/L, respectively. The pH is basic, proteins are excreted about 10 times the rate of humans, and cholesterol is the main lipid excreted.

SPECIES PECULIARITIES

The hamster cheek pouch is unique because it accepts heterologously neoplastic tissue but rejects normal human tissue. This led to the discovery that a biological difference existed between malignant and nonmalignant tissues. The cheek pouch has been used for the transplanting of neoplastic tissue for evaluation of growth. When this method was standardized, it became a screening tool for chemotherapeutic agents (Newcomer et al., 1987).

The cheek pouch is transparent and very accessible. It is ideally suited for in vivo studies of microcirculation and the behavior of formed blood elements. These features have also made the hamster the model of choice for buccal administration and evaluation of oral and mucosal irritation. The hamster cheek pouch is, indeed, the standard model for evaluating acute or cumulative (28 day) irritation/tissue tolerance, as well as providing a popular model for the induction and study of squamous cell neoplasia (Heller et al., 1996).

STRAIN-RELATED CONSIDERATIONS

In a study by Althoff and Mohr (1973) comparing the chronic respiratory response of the Chinese, Syrian, and European hamsters to diethylnitrosamine (DEN) and dibutylnitrosamine (DBN), strain-related differences were found. In the Chinese hamster, DEN did not produce neoplasms in the respiratory tract. In the Syrian species, DEN caused tumors in the trachea and lungs followed by the nasal cavity. In European hamsters, DEN produced benign and malignant tumors in the respiratory tract and caused death after 15 weeks of daily treatment.

In the Chinese hamster, DBN caused papillary tumors and malignant neoplasms in the nasal and paranasal cavities. In the Syrian species, DBN caused tumors, primarily in the trachea, and then the nasal cavities and lungs. At the high dosages in European hamsters, DBN caused carcinogenic effects in the trachea, lungs,

TABLE 2.36 Hematological Values for European and Chinese Hamsters

Parameter	Units	European Range	Chinese Range
RBC	$\times 10E6/mm^3$	6.04–9.10	4.4–9.1
HgB	g/dL	13.4–15.5	10.7–14.1
MCV	μ^3	58.7–71.4	53.6–65.2
MCH	μμg	18.6–22.5	15.5–19.1
MCHC	%	26.4–32.5	27.0–32.0
Hct	%	44.0–49.0	36.5–47.7
WBC	$\times 10E6/mm^3$	3.4–7.6	2.7–9.6
Neutrophils	$\times 10E6/mm^3$	3.5–41.6	14.8–23.6
Eosinophils	$\times 10E6/mm^3$	0–2.1	0.3–3.1
Basophils	$\times 10E6/mm^3$	0–0.2	0.0–0.5
Lymphocytes	$\times 10E6/mm^3$	50.0–95.0	68.1–84.8
Monocytes	$\times 10E6/mm^3$	0–1.0	0–2.4

RBC, red blood cell.
Mitruka, B.M., Rawnsley, H.M., 1981. Clinical Biochemical and Hematological Reference Values in Normal Experimental Animals and Humans, second ed. Masson, New York.

and nasal cavities, whereas low dosage groups had lung carcinomas.

As indicated by this study, hamsters are good models to study respiratory carcinogenesis; however, different species can have different responses to a chemical. The spontaneous rate of respiratory tumors, metabolism, and the nature of chemical need to be known before cross-species extrapolation can be done.

TYPICAL STUDY PROTOCOLS

Carcinogenicity Toxicity Testing

Hamsters are highly suitable animals for carcinogenicity testing because of a low incidence of spontaneous tumor development, but they are highly susceptible to experimentally induced carcinogenesis. The incidence of spontaneous tumors in Syrian hamsters is reported to be lower than the incidence seen in mice or rats (Homburger et al., 1979a; Althoff and Mohr, 1973). Although the hamster does have a short life span, substance-related effects and neoplasms develop rapidly, during which spontaneous diseases and tumors may not occur. However, hamsters are infrequently used in testing, and a 2004 inquiry of (the more than 90) known contract laboratories succeeded in only identifying one with current experience in performing such studies.

The carcinogenicity protocols used for rats are satisfactory for hamster studies. Changes that need to be incorporated are that blood collections should be kept to a minimum and the length of the study usually is shortened (to 96 weeks) owing to the hamster's shorter life span.

Hamsters are specifically recommended for long-term testing with aromatic amines, polycyclic hydrocarbons, and other agents suspected of being pulmonary carcinogens (Aufderheide et al., 1989). Urinary bladder carcinomas induced by aromatic amines can take up to 7 years to induce in dogs, but they can cause neoplasms in less than 1 year in hamsters (Witschi et al., 1993).

Nitrosamines caused tumors in the hamster fore stomach, liver, pancreas, nasal cavity, lung, trachea, and occasionally the esophagus. The common site for nitrosamine tumor induction in the rat is the esophagus, demonstrating species specificity for a target organ site. Hamsters are not more susceptible to nitrosamines because some nitrosamines are more toxic in the rat than the hamster and vice versa (Newcomer et al., 1987). The hamster does show a nitrosamine-induced pancreatic tumor similar to pancreatic tumors in humans.

Inhalation and Intratracheal Studies

Inhalation studies constitute a significant portion of the toxicological research using the hamster as the test species. The hamster is deemed useful because it has a lower incidence of spontaneous respiratory tumors and of respiratory diseases (Werner et al., 1979), and its respiratory epithelium is more similar to that of humans than other laboratory rodents. The hamster has similar lung absorption characteristics to those of the rat and mouse for aldehydes, ozones, and other irritant gases (Morris, 1997; Steinberg and Gleeson, 1990). Because of its more mixed breathing pattern, it compares favorably to the rat (an obligatory nose-breather) for studying fiber and particulate inhalation (King-Herbert et al., 1997; Geiser et al., 1990; Gelzleichter et al., 1999; Warheit and Harsky, 1993; Warheit et al., 1997). In studies with cigarette smoke, certain in-bred species of the hamster are the only laboratory rodents where carcinogenesis can be induced by inhalation. A laryngeal cancer in the hamster is caused by tar fractions or cigarette smoke. This cancer has been found to be histologically identical to the cancer seen in humans (Homburger et al., 1979a). The species thus continues to be popular in studying the toxicity of cigarette smoke and its mechanisms (Tafassian et al., 1993; DiCarlantonio and Talbot, 1999).

Acute and subacute inhalation toxicity studies using the hamster have studied nickel monoxide (NiO), cobaltous oxide (CoO), and chrysotile asbestos. Hamsters exposed to asbestos for 11 months developed asbestoses and those exposed to NiO developed pneumoconiosis, occupational disease states seen in humans.

In whole body exposure, inhalation study designs, the exposure chamber has to be large enough to allow an adequate number of animals to be exposed simultaneously. It should be equipped with the means to regulate temperature and humidity and have identical chambers for all treatment groups. The hamsters should be housed separately; however, if there are space limitations, animals may be housed in groups. The position of the cages should be rotated from exposure-to-exposure. The animals should be housed in an area other than the inhalation chamber when not being treated. This will reduce the contamination of the exposure chamber by bodily fluids.

Current practice makes nose only exposure with animals restrained in tables more popular, but hamsters do show extensive stress-related physiology responses to such restraint, including marked body weight loss (King-Herbert et al., 1997).

Animals are randomly distributed to test groups based on body weights as performed in other types of animal studies. The animals should have free access to water at all times. Feed should be available when animals are not being exposed unless exposure times are very lengthy. If feed is provided during exposure, then the feed is also being exposed to the test materials and may be an important aspect of the study. The number of exposures whether once or several times a day and the length of exposure time can be decided by the investigator; however, once decided, exposure time should remain consistent throughout the study. The concentration and particle size of the aerosol should be determined periodically and the aerosols should be evenly distributed in the chamber (Raabe et al., 1973). Data may be collected concerning clinical signs, body weights, pharmacokinetics, mortality, hematological and clinical chemistry functions, organ weights, and gross and microscopic observations.

For intratracheal instillation studies, the same procedures as described for inhalation studies are used except animals are exposed to the control and test articles via intratracheal administration. The common dose volume is 0.2 mL per animal, and the animal is usually anesthetized during dose administration. The number of treatments per day and the length of the study may be decided by the principal investigator.

Teratology Studies

The hamster has been a popular alternative species for teratology and reproductive toxicity studies because of its predictable estrus, short pregnancy period, rapid embryonic development, and a low incidence of spontaneous malformations (Wlodorezyk et al., 1995; Williams et al., 1991; Wolf et al., 1999; Gomez et al., 1999; DeSesso et al., 1998).

Retinoic acid (vitamin A) has been shown to be a teratogen in hamsters (Frierson et al., 1990; Willhite et al., 1996, 2000). Thalidomide was found to be a teratogen in certain in-bred strains of the Syrian hamster making it a viable alternative to rabbits. Other hamster teratogens are dinocap (Rogers et al., 1989), cyclophosphamide (Shah et al., 1996), hydrocortisone, colchicine, vincristine, vinblastine, heavy metals such as cadmium compounds, organic and inorganic mercury, 2,4,5-T (2,4,5-trichlorophenoxyacetic acid), and 2,4-D (2,4-dichlorophenoxyacetic acid) alone or contaminated with dioxin (Newcomer et al., 1987). The study design for teratology studies should take into account the embryonic development of the hamster, the strain-specific fertility seen in hamsters, and the age of the mother.

Toxicology Studies

The protocols used for rats in acute and long-term toxicity studies are satisfactory for the hamster; however, blood collection should be kept to a minimum and the length of the test may need to be adjusted (shortened) because of the shorter life span of the hamster.

The majority of toxicology work involving hamsters has been buccal inhalation and respiratory studies, although intratracheal has also been a popular route because of the possibility of precise control of doses (Biswas et al., 1993). The hamster has been found to be a useful model in the study of toxicity; however, it has some biochemical and physiological characteristics not seen in other rodent species. The hamster has a strong resistance to certain pharmacological agents such as barbiturates, morphine, and colchicine, but it is very sensitive to halothane and sevoflurane-induced changes in diaphragmatic contractility (Kagawa et al., 1998). The hamster oral LD_{50} of colchicine is 600 times the lethal dose known to humans, whereas morphine when given to hamsters at the LD_{50} does not produce a narcotic reaction. These characteristics should not be considered a barrier to the use of hamsters in toxicology studies.

Of 304 compounds evaluated by the IARC, 130 were carcinogenic in at least one rodent species. Of the 130, only 38 compounds were tested in both hamsters and rats, 35 were tested in hamsters and mice, and 78 tested in both rats and mice. Of those tested in hamsters and rats, 84% of the compounds had similar results in both species, 86% in both mice and hamsters, and in mice and rats 90% had similar results. Based on this information, hamsters are not more or less sensitive to toxicity than other rodent species used in long-term testing (Arnold and Grice, 1979).

CHINESE HAMSTER OVARY CELL CHROMOSOME ABERRATIONS

The purpose of this assay is to evaluate the ability of a compound to induce chromosomal aberrations (clastogenic responses) in the Chinese hamster ovary (CHO) cells. The realization that fully 60% of drugs listed in the *Physicians' Desk Reference* have positive CHC chromosome aberration findings associated with them have raised concerns about this assay.

The CHO cells used in this assay may be obtained from the American Tissue Culture Collection, Rockville, Maryland (Brusick and Fields, 2014). The original cells were obtained from a Chinese hamster.

The assay is divided into two parts, nonactivation and activation, with S9 rat liver as the activating agent. EMS (0.5 μL/mL) is the positive agent for the nonactivation, whereas for the activating studies it is dimethylnitrosamine at 0.5 μL/mL. The solvent used to dissolve the test article is used as the solvent vehicle for the control and positive control articles. The dosages selected for the test article are one toxic (loss of growth potential) and four lower (usually in a half-log series) concentrations. These dosages are determined in a range-finder with the cells exposed to the test article for 4 h and incubated for 24 h.

The CHO cells are grown in 10% fetal calf serum/Ham's F12 media. The cell density should be kept at $1.5 \times 10^6/75\ cm^2$ plastic flask. For the assay, approximately 0.25×106 cells per well per test article concentration are tested. The cells are exposed to the test article for 2 h at 37°C. Cells used in the activation section receive the S9 rat liver along with the test article before the 2-h activation. The cells are then washed with sterile saline and given fresh media. For each of the test article dosages, half of the plates per treatment group will receive 5-bromo-2′-deoxyuridine at a concentration of 10 μM. The cells are then incubated for 24 h with colcemid (2×10^{-7} M) added at hour 17. After incubation, the metaphase cells are collected by mitotic shake-off. These cells are swollen with 0.0075 M KCl solution, washed with a methanol: acetic acid (3:1) fixative, dropped onto glass slides, and air dried. The slides are stained with 10% Giemsa (pH 6.8) and scored for chromosomal aberrations such as chromatid and chromosome gaps, breaks, and chromatid deletions.

SYRIAN HAMSTER EMBRYO CELL TRANSFORMATION ASSAY

Another genotoxicity assay currently being promoted for use as an alternative for the CHO and a screening assay for clastogens is the Syrian hamster embryo (SHE) (LeBoeuf et al., 1996). The SHE cell transformation assay evaluates the potential of chemicals to induce morphological transformation in karyotypically normal primary cells. Induction of transformation has been shown to correlate well with the carcinogenicity of many compounds in the rodent bioassay. Historically the assay has not received widespread use because of technical difficulty (Leonard and Lauwarys, 1990). This was the subject of an entire issue of *Mutation Research* (1996) and is currently being further evaluated by regulatory agencies, using a low pH technique that is technically easier to use and reproduce (LeBoeuf et al., 1996; Kerckaert et al., 1996; Aadema et al., 1996; Isfort et al., 1996; Custer et al., 2000; Isfort et al., 1996).

MODELS OF DISEASES

Cardiomyopathy

Cardiomyopathy in hamsters originates from a genetically (recessive autosomal gene) determined metabolic defect that induces degenerative lesions in all striated muscles with particular intensity and consistency in the heart (Bajusz, 1969; Gertz, 1973). The clinical and pathological aspects of the disease resemble nonvascular myocardial disease in man. Animals appear normal, although there is cardiac muscle degeneration. The lesions appear histologically in both sexes at 350 days of age for males and 25–30 days of age for females. The first lesion is an acute myolysis with primary dissolution of the myofilaments. This lesion is healed and replaced by connective tissue by day 100 (Gertz, 1973). The disease becomes clinically apparent with a whole body sc edema; however, there are no ECG changes to foretell edema formation or to tell degree of lesion formation. The ECG changes are observed after lesion formation and consist of alterations in pathways of cardiac excitation and high-frequency alterations in the QRS interval. In the late stages of the disease, ascites, hydrothorax, and hydropericardium appear. In the terminal stage, animals are hyperpneic and cyanotic. At necropsy, the liver, spleen, kidney, heart, and other visceral organs show congestive changes such as enlargement and increases in volumes. Cardiomyopathic hamsters show an edema response to therapy with digitalis, diuretics, and salt restrictions. The average life span of hamsters with this disease is 146 days. Because this model responds to therapy in a manner similar to humans and the disease state has similar manifestations, it is considered well suited to study heart failure in humans due to cardiac muscle impairment.

Dental Caries

Caries is a disease of poorly developed and poorly calcified teeth in the hamster. It is considered transmissible and infectious. It can be induced in normal hamsters by inoculation with cariogenic microflora. Animals are infected by adding microflora to drinking water or exposing the animals to infected feces. The carious lesions are in the molars. The lesions start with changes in enamel translucency, surface depressions, and fissures. The disintegration goes into the dentine with eventual exposure of the pulp. Bacterial (putrefactive) infections, inflammation, and complete necrosis of the molar occurs. With the testing of this model, fluoride was found to be beneficial to hamsters and this prompted clinical trials in humans (Keyes, 1960).

Diabetes Mellitus

Diabetes mellitus was first described in 1959 and 1969 in the Chinese hamster and South African hamsters, respectively (Stuhlman, 1979). Only certain strains of the Chinese species are affected, and the disease is probably transmitted by a recessive gene. The disease in the Chinese hamster has a rapid onset between 1 and 3 months of age. The indications of the disease are polyuria, polydipsia, glycosuria, and ketonuria. A normal Chinese hamster has glucose levels of 110 ± 6 mg/100 mL, whereas a diabetic hamster has levels of 200–800 mg/mL. The diabetes mellitus seen in the Chinese hamster is very similar to the disease state in humans. There is a variation in the syndromes (chronic, insulin dependent), the occurrence of secondary manifestations (cataracts), discrepancies in established parameters of native insulin values, and what role heredity plays in the disease.

In the South African hamster, which has a 22% incidence of diabetes mellitus, the disease is inherited as a nonsex linked polygenic trait. With the South African hamster, the hyperglycemia varies in incidence, age of onset, degree of severity, and rate of progression, as in humans, but it is not influenced by age, sex, or state of obesity. Obesity is not a feature seen in the diabetic South African hamster, though it is seen in the Chinese hamster.

Either of these models will provide insight into diabetes mellitus. These models can be used to study the disease pathogenesis, development of secondary complications, exact genetic mechanisms, and possible therapeutic regimens.

Leprosy

The hamster was first injected with leprosy bacilli in 1937. It was the first time a laboratory animal was found to be susceptible to the agent. In recent years, leprosy bacilli has been grown in cell culture systems and in the tail and foot pads of the mouse, so the role of the hamster in leprosy research has declined (Frenkel, 1987).

Muscular Dystrophy

The muscular dystrophy (MD) syndrome can be induced in hamsters by feeding them a diet deficient in vitamin E from weaning until death. The gross appearance of the disease state does not occur before the 11th or 12th month. Strains of cardiomyopathic Syrian hamsters (BIO 14.6 and BIO 53.38) do have genetic dispositions for this disease. The clinical signs appear 60–200 days after birth, and all skeletal muscles, including the heart, are affected. The lesions are pleomorphic, characterized by focal degeneration of myofibrils, coagulation necrosis, the formation of contraction clots, and alignment of muscle nuclei in chainlike rows within the fiber (Homburger and Bajusz, 1970). In the final stage of the syndrome, the myofibrils convert to granular mass in which the nuclei have disintegrated. Earliest changes occur at 33 days of age with death usually by 220 days due to cardiac failure. These hamsters have morphological manifestations of MD.

This model is excellent to study the mechanism of MD from an in vivo system to intact animals.

Osteoarthritis and Degenerative Joint Disease

Osteoarthritis and degenerative joint disease are diseases that a hamster may develop in old age. These diseases are rare in hamsters under 2 years of age. The diseases are characterized by separation of the zone of calcification of the cartilage with sclerosis and dislocation of the bone, fibrillation of ligaments, and fibrosis of the synovial membrane. Organisms such as *Mycoplasma*, *Streptobacillus moniliformis*, *Corynebacterium kutscheri*, hormonal imbalances, and chemical and physical agents are associated with the disease. The disease in hamsters is similar to that seen in humans (Handler, 1965), which makes this an excellent model to study.

Pancreatic Cancer

Pancreatic cancer is a very difficult neoplasm to induce in laboratory animals. It can be induced in Syrian hamsters with nitrosamines. The neoplasms, histogenesis, and enzymatic patterns seen in human pathogenesis are similar to those in the hamster. Occasionally, hamsters develop diabetes during the pancreatic carcinogenesis, as is also seen in the pathogenesis in humans.

CONCLUDING REMARKS AND FUTURE DIRECTIONS

Toxicology is a continuously evolving field, borrowing its every expanding set of technologies primarily from other fields of biomedical and biological research. Although the actual integration of new technologies at the laboratory level into the field is fairly prompt, it is generally not accompanied by any of the already employed technologies (for our purpose, end points or biomarkers measured in standard studies) being deleted. Also not prompt is the inclusion of new technologies or biomarkers being reflected in changes in regulatory guidelines.

So while the last 5 years have seen proteomic more relevant clinical chemistry end points, species-specific immune functions measures, and measurement of in vivo changes of physiologic function (heart rate, blood pressure, EKG parameters, respiratory rate) and internal chemical end points (blood gases, glucose and glucagon) by implanted telemetry sensors (Gad, 2013), these have not been significantly utilized (or offered by) CRO's or integrated into guidelines, regulations, or integrated safety assessments.

Progress will entail the available new technologies being incorporated into our standard approaches to safety assessment, and hopefully a thoughtful consideration of which of the large set of measurements now taken have relevance and should be carried forward into a new synthesis of study designs.

References

Aadema, M.J., Isfort, R.J., Thompson, E.D., et al., 1996. The low pH Syrian hamster embryo (SHE) cell transformation assay: a revitalized role in carcinogen prediction. Mut. Res. 356, 5–9.

Althoff, J., Mohr, U., 1973. Comparative studies in three hamster species related to respiratory carcinogenesis. In: Spiegel, A. (Ed.), The Laboratory Animal in Drug Testing. Gustav Fischer Verlag, Stuttgart, pp. 229–232.

Amacher, D., 2010. The discovery and development of proteomic safety biomarkers for the detection of drug-induced liver toxicity. Toxicol. Appl. Pharmacol. 245, 134–142.

Andrews, D.I., Jones, D.R., Simpson, F.O., 1978. Direct recording of arterial blood pressure and heart rate in the conscious rat. J. Pharm. Pharmacol. 30, 524–525.

Annau, Z., 1986. Neurobehavioral Toxicology. The Johns Hopkins University Press, Baltimore.

Arnold, D.L., Grice, H.C., 1979. The use of the Syrian hamster in toxicology studies, with emphasis on carcinogenesis bioassay. Prog. Exp. Tumor Res. 24, 222–234.

Arnold, D., Charbonneau, S., Zawidzka, Z., Grice, H., 1977. Monitoring animal health during chronic toxicity studies. J. Environ. Pathol. Toxicol. 1, 227–239.

Aufderheide, M., Thiedemann, K.U., Riebe, M., Kohler, M., 1989. Quantification of proliferative lesions in hamster lungs after chronic exposure to cadmium aerosols. Exp. Pathol. 37, 259–263.

Bader, M., Klinger, W., 1974. Intragastric and intracardial injections in newborn rats. Methodical investigation. Z. Versuchstierk. 16, 40–42.

Bajusz, E., 1969. Hereditary cardiomyopathy: a new disease model. Am. Heart J. 77, 686–696.

Baker, H.J., Lindsey, J.R., Weisbroth, S.H., 1979. Housing to control research variables. In: Baker, H.J., Lindsey, J.R., Weisbroth, S.H. (Eds.), The Laboratory Rat, vol. 1. Academic Press, New York, pp. 169–192.

Barrett, A.M., Stockham, M.A., 1963. The effect of housing conditions and simple experimental procedures upon the corticosterone level in the plasma of rats. J. Endocrinol. 26, 97–105.

Barrow, M.V., 1968. Modified intravenous injection technique in rats. Lab. Anim. Care 18, 570–571.

Bartek, M.J., LaBudde, J.A., Maibach, H.I., 1972. Skin permeability *in vivo*: comparison in rat, rabbit, pig, and man. J. Invest. Dermatol. 58, 114–123.

Beinfield, W.H., Lehr, D., 1956. Advantages of ventral position in recording electrocardiogram of the rat. J. Appl. Physiol. 9, 153–156.

Berman, H.J., Lutz, B.R., et al., 1955. Blood pressure of golden hamsters as affected by nembutal sodium and X-irradiation (abstract). Am. J. Physiol. 183, 597.

Biswas, G., Raj, H.G., Allomeh, A., Saxena, M., et al., 1993. Comparative kinetic studies on aflatoxin B. Binding to pulmonary and hepatic DNA of rat and hamster receiving the carcinogen intratracheally. Teratog. Carcinog. Mutagen. 13, 259–268.

Black, H.E., 1986. A manager's view of the "musts" in a quality necropsy. In: Hoover, B.K., Baldwin, J.K., Uelner, A.F., Whitmire, C.E., Davies, C.L., Bristol, D.W. (Eds.), Managing Conduct and Data Quality of Toxicology Studies. Princeton Scientific, Princeton, New Jersey, pp. 249–255.

Blackwell, B., Bucci, T., Hart, R., et al., 1995. Longevity, body weight, and neoplasia in ad-libitum fed and diet- restricted C57BL6 mice fed NIH-31 open formula diet. Toxicol. Pathol. 23 (5), 570–582.

Bolton, W.K., Benton, F.R., Maclay, J.G., et al., 1976. Spontaneous glomerular sclerosis in aging Sprague-Dawley rats. Am. J. Pathol. 85, 277–302.

Borer, K.T., Kelch, R.P., White, M.P., et al., 1977. The role of the septal area in the neuroendocrine control of growth in the adult golden hamster. Neuroendocrinology 23 (3), 133–150.

Boyland, E., Sydnor, K., 1962. The induction of mammary cancer in rats. Br. J. Cancer 16, 731–739.

Brusick, D.J., Fields, W.R., 2014. Genetic toxicology. In: Hayes, A.W. (Ed.), Principles and Methods of Toxicology, sixth ed. Raven Press, New York, pp. 1173–1204.

Bunag, R., 1973. Validation in awake rats of a tail-cuff method for measuring systolic pressure. J. Appl. Physiol. 34, 279–282.

Cantrell, C.A., Padovan, D., 1987. Other hamsters: biology, care and use in research. In: Van Hoosier, G.L., McPherson, C.W. (Eds.), Laboratory Hamsters. Academic Press, Orlando, Florida, pp. 369–386.

Charles, R., 2004. Growth Rates in Selected Rat Strains. Charles River Laboratories, Inc., Massachusetts.

Charles, R., 2011. Spontaneous Neoplastic Lesions in the Crl:CD BR Rat. Charles River Laboratories, Inc., Massachusetts.

Chouroulinkov, I., Lasne, C., Phillips, D., Grover, P., 1988. Sensitivity of the skin of different mouse strains to the promoting effect of 12-0-tetradecanoyl-phorbol-13-acetate. Bull. Cancer 75, 557–565.

Collins, G.R., 1979. Hamster. In: The Manual for Laboratory Animal Technicians. American Association Laboratory Animal Science, Joliet, Illinois, pp. 121–130. Publication 67-3.

Conybeare, G., Leslie, S., 1980. Effect of quality and quantity of diet on survival and tumor incidence of outbred swiss mice. Food Cosmet. Toxicol. 18, 65–75.

Custer, L., Gibson, D.P., Aardema, M.J., et al., 2000. A refined protocol for conducting the low pH 6.7 Syrian hamster embryo (SHE) cell transformation assay. Mut. Res. 455, 129–139.

Davis, D.H.S., 1963. Wild rodents as laboratory animals and their contributions to medical research in South Africa. S. Afr. J. Med. Sci. 28, 53–69.

Davis, F.C., 1989. Daily variation in maternal and fetal weight gain in mice and hamsters. J. Exp. Zool. 250 (3), 273–282.

Dean, B.J., 1983. Report of the UKEMS Sub-committee on Guidelines for Mutagenicity Testing, Part 1, Basic Test Battery, Etc. United Kingdom Environmental Mutagen Society.

Dean, B.J. (Ed.), 1984. Report of the UKEMS Sub-committee on Guidelines for Mutagenicity Testing, Part 11, Supplemental Tests, Etc. United Kingdom Environmental Mutagen Society.

DeAngelis, M.H., Michel, D., Wagner, S., et al., 2007. Chemical mutagenesis. In: Fox, J.G., Stephen, S. (Eds.), The Mouse in Biomedical Research, second ed., vol. IV. Academic Press, New York, pp. 225–260.

DeBoer, A.G., Breimer, D.D., Pronk, J., et al., 1980. Rectal bioavailability of lidocaine in rats: absence of significant first-pass elimination. J. Pharm. Sci. 69, 804–807.

Desai, R.G., 1968. Hematology and microcirculation. In: Hoffman, R., Robinson, P.E., Magalhaes, H. (Eds.), The Golden Hamster Its Biology and Use in Medical Research. Iowa State University Press, Ames, pp. 185–191.

DeSesso, J.M., Jacobson, C.F., Scialli, A.R., et al., 1998. An assessment of the developmental toxicity of inorganic arsenic. Reprod. Toxicol. 12, 385–433.

Detweiler, D.K., 1981. The use of electrocardiography in toxicological studies with rats. In: Budden, R., Detweiler, D.K., Zbinden, G. (Eds.), The Rat Electrocardiogram in Pharmacology and Toxicology. Pergamon Press, New York, pp. 83–116.

Detweiler, D.K., Saatmon, R.A., De Baecke, P.J., 1981. Cardiac arrhythmias accompanying sialodacryoadenitis in the rat. In: Budden, R., Detweiler, D.K., Zbinden, G. (Eds.), The Rat Electrocardiogram in Pharmacology and Toxicology. Pergamon Press, New York, pp. 129–134.

DiCarlantonio, G., Talbot, P., 1999. Inhalation of mainstream and sidestream cigarette smoke retards embryo transport and slows muscle contraction in oviducts of hamsters (*Mesocricetus auratus*). Biol. Reprod. 61 (3), 651–656.

Diehl, K.H., Hull, R., Morton, D., et al., 2001. A good practice guide to the administration of substances and removal of blood, including routes and volumes. J. Appl. Toxicol. 21, 15–23.

Dieke, S.H., Richter, C.P., 1945. Acute toxicity of thiourea to rats in relation to age, diet, strain, and species variation. J. Pharmacol. Exp. Ther. 83, 195–202.

Dontenwill, W., Chevallier, H.T., Harke, H.R., et al., 1973. Investigations of the effect of chronic cigarette-smoke inhalation in the Syrian golden hamster. JNCI 51, 1781–1832.

Eaton, D.L., Gilbert, S.G., 2013. Principles of toxicology. In: Doull, J., Klaassen, C.D. (Eds.), Casarett and Doull's Toxicology: The Basic Science of Poisons, eighth ed. Macmillan, New York, pp. 13–48.

Fabry, A., 1985. The incidence of neoplasms in Syrian hamsters with particular emphasis on intestinal neoplasia. Arch. Toxicol. Suppl. 8, 124–127.

Fenner, F., 1986. In: Bhatt, P., Jacoby, R., Morse, M., New, A. (Eds.), Viral and Mycoplasm Infections of Laboratory Rodents: Effects on Biomedical Research. Academic Press, San Diego, pp. 24–25.

Ferguson, H.C., 1962. Dilution of dose and acute oral toxicity. Toxicol. Appl. Pharmacol. 4, 759–762.

Finney, D.J., 1971. Probit Analysis, third ed. Cambridge University Press. pp. 50–80, 100–124.

Fox, J.G., Bennett, S.W., 2015. Biology and diseases of mice. In: Fox, J.G., Anderson, L., Otto, G., Prichett-Corning, K., Wary, M. (Eds.), Laboratory Animal Medicine. Academic Press, New York, pp. 31–89.

Frank, P., Schoenhard, G.L., Burton, E., 1991. A method for rapid and frequent blood collection from the rat tail vein. J. Pharmacol. Methods 26, 233–238.

Frenkel, P., 1987. Experimental biology: use in infectious disease research. In: Van Hoosier, G., McPherson, C. (Eds.), Laboratory Hamsters. Academic Press, Orlando, Florida, pp. 227–249.

Frierson, M.R., Mielbach, F.A., Kocklar, D.M., 1990. Comater-automated structure evaluation (CASE) of retinoids in teratogenesis bioassays. Fundam. Appl. Toxicol. 14, 408–428.

Gad, S.C., 1982. A neuromuscular screen for use in industrial toxicology. J. Toxicol. Environ. Health 9, 691–704.

Gad, S.C., 2013. Animal Models in Toxicology, third ed. CRC Press, Boca Raton, FL.

Gad, S.C., 2016. Drug Safety Evaluation, third ed. John Wiley & Sons, Hoboken, BJ.

Gad, S.C., Chengelis, C.P., 1998. Acute Toxicology Testing Perspectives and Horizons, second ed. CRC Press, Boca Raton, Florida.

Gad, S.C., Spainhour, C.B., 2017. Routes, Regimens and Formulations: Pathways to Improved Target Receptor Delivery Levels and Specificity, and In Vivo Tolerance. CRC Press, Boca Raton, FL.

Gad, S.C., Dunn, B.J., Dobbs, D.W., et al., 1986. Development and validation of an alternative dermal sensitization test: the mouse ear swelling test (MEST). Toxicol. Appl. Pharmacol. 84, 93–114.

Geiser, M., Cruz-Orine, L.M., Hof, V.I., et al., 1990. Assessment of particle retention and clearance in the intrapulmonary conducting airways of hamster lungs with the fractionator. J. Microsc. 160, 75–88.

Gelzleichter, T.R., Bermudez, E., Mangum, J.B., et al., 1999. Comparison of pulmonary and pleural responses of rats and hamsters to inhaled refractory ceramic fibers. Toxicol. Sci. 49, 93–101.

Gertz, E., 1973. Animal model of human disease. Animal model: cardiomyopathic Syrian hamster. Am. J. Pathol. 70, 151–154.

Gibson, J.E., Becker, B.A., 1967. The administration of drugs to one day old animals. Lab. Anim. Care 17, 524–527.

Gomez, J., Macina, O.T., Mattison, D.R., et al., 1999. Structural determinants of developmental toxicity in hamsters. Teratology 60, 190–205.

Gras, J., Llenas, J., 1999. Effects of H1 antihistamines on animal models of QTc prolongation. Drug Saf. 21 (Suppl. 1), 39–44.

Griffith, J., 1934–35. Indirect method for determining blood pressure in small animals. Proc. Soc. Exp. Biol. Med. 32, 394–396.

Guiol, C., Ledoussal, C., Syrge, J.M., 1992. A radiotelemetry system for chronic measurement of blood pressure and heart rate in the unrestrained rat: validation of the method. J. Pharmacol. Toxicol. Methods 28 (2), 99–105.

Guo, Z., Zhou, L., 2003. Dual tail catheters for infusion and sampling in rats as an efficient platform for metabolic experiments. Lab. Anim. 32 (2), 45–48.

Haggerty, G.C., 1989. Development of tier I neurobehavioral testing capabilities for incorporation into pivotal rodent safety assessment studies. J. Am. Coll. Toxicol. 8, 53–69.

Hall, A., Persing, R.L., White, D.C., et al., 1967. Mystromys albicaudatus as a laboratory species. Lab. Anim. Care 17 (2), 180–188.

Handler, A., 1965. Spontaneous lesions of the hamster. In: Ribelin, W., McCoy, J. (Eds.), The Pathology of Laboratory Animals. Thomas, Springfield, Illinois, pp. 210–240.

Heldmaier, G., Steleinlechner, S., 1981. Seasonal pattern and energetics of short daily torpor in the Dzungarian hamster, *Phosupua aunfoeua*. Oceologia 48, 265–270.

Heller, B., Kluftinger, A.M., Davis, N.L., et al., 1996. A modified method of carcinogenesis induction in the DMBA hamster cheek pouch model of squamous neoplasia. Am. J. Surg. 172, 678–680.

Hem, A., Smith, A.J., Solberg, P., 1998. Saphenous vein puncture for blood sampling of the mouse, rat, hamsters, gerbil, Guinea pig, ferret and mink. Lab. Anim. 32, 364–368.

Herberg, L., Buchanan, K.D., Herbetz, L.M., et al., 1980. The Dzungarian hamster, a laboratory animal with inappropriate hyperglycemia. Comp. Biochem. Physiol. A65A, 35–60.

The CFI mouse, history and utilization. In: Hill, B.F. (Ed.), 1981. Charles River Digest, vol. 20 (2). Charles River Breeding Laboratories, Wilmington, Massachusetts.

Hino, O., Kobayashi, T., Momose, S., et al., 2003. Renal carcinogenesis: genotype, phenotype and drama type. Cancer Sci. 94, 142–147.

Hoffmann, D., Adler, M., Vaidya, V., et al., 2010. Performance of novel kidney biomarkers in preclinical toxicity studies. Toxicol. Sci. 116 (1), 8–22.

Hofsteller, J., Svekow, M.A., Hickman, D.L., 2006. Morphophysiology. In: Svekow, M.A., Weisbroth, S.H., Franlin, C.L. (Eds.), The Laboratory Rat, second ed., vol. I. Academic Press, New York, pp. 93–125.

Homburger, F., Bajusz, E., 1970. New models of human disease in Syrian hamsters. JAMA 212 (4), 604–610.

Homburger, F., Adams, R.A., Soto, E., 1979a. The special suitability of inbred Syrian hamsters for carcinogenesis testing. Arch. Toxicol. Suppl. 2, 445–450.

Homburger, F., Adams, R.A., Soto, E., et al., 1979b. Susceptibility and resistance to chemical carcinogenesis in inbred Syrian hamsters. Prog. Exp. Tumor Res. 24.

Hoyt, R.F., Hawkins, J.V., St.Claire, M.B., et al., 2007. Physiology. In: Fox, J.G., Barthold, S.W., Davisson, M.J. (Eds.), The Mouse in Biomedical Research, second ed., vol. III. Academic Press, New York, pp. 23–122.

Hundley, J.M., Ashburn, L.L., Sebrell, W.H., 1945. The electrocardiogram in chronic thiamine deficiency in rats. Am. J. Physiol. 144, 404–414.

Hysell, D.K., Abrams, G.D., 1967. Complications in the use of indwelling vascular catheters in laboratory animals. Lab. Anim. Care 17, 273–280.

Ichimaru, Y., Kuwaki, T., 1998. Development of an analysis system for 24-hour blood pressure and heart rate variability in the rat. Psychiatry Clin. Neurosci. 52 (2), 169–172.

Isfort, R.J., Kerckaert, G.A., LeBoeuf, R.A., 1996. Comparison of the standard and reduced pH Syrian hamster embryo (SHE) cell in vitro transformation assays in predicting the carcinogenic potential of chemicals. Mut. Res. 356, 11–63.

Iwamoto, K., Watanabe, J., 1985. Avoidance of first-pass metabolism of propranolol after rectal administration as a function of the absorption site. Pharm. Res. 2, 53–54.

Jana, N.R., Sarkar, S., Yonemoto, J., et al., 1998. Strain differences in cytochrome P451A1 gene expression caused by 2,3,7,8-tetrachlorodibenzo-p-dioxin in the rat liver: role of the aryl hydrocarbon receptor and its nuclear translocator. Biochem. Biophys. Res. Commun. 248 (3), 554–558.

Kagawa, T., Maekawa, N., Mikawa, K., et al., 1998. The effects of halothane and sevoflurane on fatigue-induced changes in hamster diaphragmatic contractility. Anesth. Analg. 86, 392–397.

Kaput, J., 2005. Decoding the pyramid: a systems biological approach to nutrigenomics. Ann. N.Y. Acad. Sci. 1055, 64–79.

Kaufman, S., 1980. Chronic, nonocclusive, and maintenance-free central venous cannula in the rat. Am. J. Physiol. 239, R123–R125.

Kennedy Jr., G.L., Trochimowicz, H.J., 1982. Inhalation toxicology. In: Hayes, A.W. (Ed.), Principles and Methods of Toxicology. Raven Press, New York, pp. 185–207.

Kerckaert, G.A., Isfort, R.J., Carr, G.J., et al., 1996. A comprehensive protocol for conducting the Syrian hamster embryo cell transformation assay at pH 6.70. Mut. Res. 356, 65–84.

Kersten, H., Brosene Jr., W.G., Ablondi, F., et al., 1947. A new method for the indirect measurement of blood pressure in the rat. J. Lab. Clin. Med. 32, 1090–1098.

Keyes, P.H., 1960. The infectious and transmissible nature of experimental dental caries. Arch. Oral Biol. 1, 304–320.

King-Herbert, A.P., Hesterburg, T.W., Thevenaz, P.P., et al., 1997. Effects of immobilization restraint on Syrian golden hamsters. Lab. Anim. Sci. 47, 362–366.

Kitano, M., Hatano, H., Shisa, H., 1992. Strain difference of susceptibility to 4-nitroquinoline 1-oxide-induced tongue carcinoma in rats. Jpn. J. Cancer Res. 83 (8), 843–850.

Korol, B., McShane, W., 1963. A new method for indirect recording of arterial pressure in unanesthetized rats. J. Appl. Physiol. 18, 437–439.

Kramer, K., Grimbergen, J., van der Gracht, L., et al., 1995. The use of telemetry to record electrocardiogram and heart rate in freely swimming rats. Methods Find. Exp. Clin. Pharmacol. 17 (2), 107–112.

Kuwahara, M., Yayou, K., Ishii, K., et al., 1994. Power spectral analysis of heart rate variability as a new method for assessing autonomic activity in the rat. J. Electrocardiol. 27, 333–337.

Lang, P.L., February 1989a. Spontaneous Neoplastic Lesions in the B6C3F/CrIBR Mouse. Charles River Monograph. Charles River Laboratories, Wilmington, Massachusetts.

Lang, P.L., Fall 1989b. Survival of Crl: CD-I BR Mice during Chronic Toxicology Studies. Charles River Laboratories Reference Paper, Wilmington, Massachusetts.

LeBoeuf, R.A., Kerckaert, G.A., Aardema, M.J., et al., 1996. The pH 6.7 Syrian hamster embryo cell transformation assay for assessing the carcinogenic potential of chemicals. Mut. Res. 356, 85–127.

Leikauf, G.D., 2013. Toxic responses of the respiratory system. In: Doull, J., Klaassen, C.D. (Eds.), Casarett and Doull's Toxicology: The Basic Science of Poisons, eighth ed. Macmillan, New York, pp. 691–732.

Leonard, A., Lauwarys, R., 1990. Mutagenicity, carcinogenicity, and teratogenicity of cobalt metal and cobalt compounds. Mut. Res. 239, 17–27.

Lossnitzer, K., Grewe, N., Konrad, A., et al., 1977. Electrographic changes in cardiomyopathic Syrian hamsters (strain BIO 8262). Basic Res. Cardiol. 72, 421–435.

MacPhail, R.C., Peele, D.B., Crofton, K.M., 1989. Motor activity and screening for neurotoxicity. J. Am. Coll. Toxicol. 8, 117–125.

MacPhail, R.C., O'Callaghan, J.P., Cohn, J., 2003. Acquisition, steady-state performance, and the effects of trimethyltin on the operant behavior and hippocampal GFAP of Long-Evans and Fischer 344 rats. Neurotoxicol. Teratol. 25 (4), 481–490.

Magalhaes, H., 1970. Hamsters. In: Hafez, E.S.E. (Ed.), Reproduction and Breeding Techniques for Laboratory Animals. Lea & Febiger, Philadelphia, pp. 258–272.

Maibach, H.I., Wester, R.C., 1989. Percutaneous absorption: in vivo methods in humans and animals. J. Am. Coll. Toxicol. 8, 803–813.

McClellan, R.O., Henderson, R.F., 1989. Concepts in Inhalation Toxicology. Hemisphere, New York.

McGee, M.A., Maronpot, R.R., 1979. Harderian gland dacryoadenitis in rats resulting from orbital bleeding. Lab. Anim. Sci. 29, 639–641.

Meuleman, D.G., Vogel, G.M.T., Van Delft, A.M.L., 1980. Effects of intra-arterial cannulation on blood platelet consumption in rats. Thromb. Res. 20, 45–55.

Minasian, H., 1980. A simple tourniquet to aid mouse tail venipuncture. Lab. Anim. 14, 205.

Mitruka, B.M., Rawnsley, H.M., 1981. Clinical Biochemical and Hematological Reference Values in Normal Experimental Animals and Humans, second ed. Masson, Publishing, New York.

Mitsumori, K., Watanabe, T., Kashida, Y., 2001. Variability in the Incidence of Spontaneous Tumors in CD (SD) IGS, CD (SD), F244 and Wistar Hannover Rats in Biological Reference Data on CD(SD) IGS Rats. Yokohama, CD(SD) IGS Study Group.

Mohr, U., Ernst, H., 1987. The European hamster: biology, care and use in research. In: Van Hoosier, G.L., McPherson, C.W. (Eds.), Laboratory Hamsters. Academic Press, Orlando, Florida, pp. 351–366.

Morris, J.B., 1997. Uptake of acetaldehyde vapor and aldehyde dehydrogenase levels in the upper respiratory tracts of the mouse, rat, hamster and Guinea pig. Fundam. Appl. Toxicol. 35, 91–100.

Moser, V.C., Ross, J.F., 1996. US EPA/AIHC Training Video and Reference Manual for a Functional Observational Battery. US Environmental Protection Agency, Washington, DC.

Moser, V.C., McCormick, J.P., Creason, J.P., et al., 1988. Comparison of chlordimeform and carbaryl using a functional observational battery. Fundam. Appl. Toxicol. 11, 189–206.

Mucha, R.F., 1980. Indwelling catheter for infusions into subcutaneous tissue of freely-moving rats. Physiol. Behav. 24, 425–428.

Nebert, D.W., Jensen, N.M., Shinozuka, H., et al., 1982. The *Ah* phenotype. Survey of forty-eight rat strains and twenty inbred mouse strains. Genetics 100, 79–97.

Newcomer, C.E., Fitts, D.A., Goldman, B.D., et al., 1987. Experimental biology: other research uses of Syrian hamsters. In: Van Hoosier, G.L., McPherson, C.W. (Eds.), Laboratory Hamsters. Academic Press, Orlando, Florida.

Newton, J.F., Yoshimoto, M., Bernstein, J., et al., 1985a. Acetaminophen nephrotoxicity in the rat. 1. Strain differences in nephrotoxicity and metabolism. Toxicol. Appl. Pharmacol. 69, 291–306.

Newton, J.F., Pasino, D.A., Hook, J.B., 1985b. Acetaminophen neurotoxicity in the rat: quantitation of renal metabolic activation in vivo. Toxicol. Appl. Pharmacol. 78, 3946.

Nishihata, T., Takahagi, H., Yamamoto, M., et al., 1984. Enhanced rectal absorption of cefinetazole and cefoxitin in the presence of epinephrine metabolites in rats and a high-performance liquid chromatographic assay for cephamycin antibiotics. J. Pharm. Sci. 73, 109–112.

OECD (Organization for Economic Cooperation and Development), 1997. Guideline for the Testing of Chemicals No. 424: 'Neurotoxicity Study in Rodents'. Adopted July 21, 1997.

OECD (Organization for Economic Cooperation and Development), 2008. Guideline for the Testing of Chemicals No. 407: 'Repeated Dose 28-day Oral Oxicity Study in Rodents'. Adopted October 16, 2008.

Ozer, J., Dieterle, F., Troth, S., et al., 2010. A panel of urinary biomarkers to monitor reversibility of renal injury and a serum marker with improved potential to assess renal function. Nat. Biotechnol. 28, 486–494.

Page, J.G., Vesell, E.S., 1969. Hepatic drug metabolism in ten strains of Norway rat before and after pretreatment with phenobarbital. Proc. Soc. Exp. Biol. Med. 131, 256–261.

Peplow, A., Peplow, P., Hafez, E., 1974. Parameters of reproduction. In: Vo, I., Melby, E., Altmon, N. (Eds.), Handbook of Laboratory Animal Science. CRC Press, Boca Raton, pp. 107–116.

Popovic, V., Popovic, P., 1960. Permanent cannulation of aorta and vena cava in rats and ground squirrels. J. Appl. Physiol. 15, 727–728.

Raabe, 0. G., Bennick, J.E., Light, M.E., et al., 1973. An important apparatus for acute inhalation exposure of rodents to radioactive aerosols. Toxicol. Appl. Pharmacol. 26, 264–273.

Renshaw, H.W., Van Hoosier, G.L., Amend, N.K., 1975. Survey of naturally occurring diseases of the Syrian hamster. Lab. Anim. 9, 179–191.

Rhodes, M.L., Patterson, C.E., 1979. Chronic intravenous infusion in the rat: a nonsurgical approach. Lab. Anim. Sci. 29, 82–84.

Rice, D.P., Ketterer, D.J., 1977. Restrainer and cell for dermal dosing of small laboratory animals. Lab. Anim. Sci. 27, 72–75.

Robineau, P., 1988. A simple method for recording electrocardiograms in conscious, unrestrained rats. J. Pharmacol. Methods 19, 127–133.

Robinson, P.F., 1968. General aspects of physiology. In: Hoffman, R., Robinson, P.E., Magalhaes, H. (Eds.), The Golden Hamster Its Biology and Use in Medical Research. Iowa State University Press, Ames, pp. 111–118.

Rogers, J.M., Burkhead, L.M., Barbee, B.D., 1989. Effects of dinocap on otolith development: evaluation of mouse and hamster fetuses at term. Teratology 39 (6), 515–523.

Ross, A.R., 1977. Measurement of blood pressure in unrestrained rats. Physiol. Behav. 19, 327–329.

Schermer, S., 1967. The golden hamster. In: The Blood Morphology of Laboratory Animals. F. A. Davis, Philadelphia, pp. 75–84.

Shah, R.M., Izadnegahdar, M.F., Hehn, B.M., et al., 1996. *In vivo/in vitro* studies on the effects of cyclophosphamide on growth and differentiation of hamster palate. Anti-cancer Drugs 7, 204–212.

Sharp, P., LaRegina, M., 1998. The Laboratory Rat. Academic Press, Philadelphia.

Sher, S.P., 1982. Tumors in control hamsters, rats and mice: literature tabulation. CRC Crit. Rev. Toxicol. 10 (1), 51–59.

Shi, Q., Hong, H., Senior, J., et al., 2010. Biomarkers for drug-induced liver injury. Expert Rev. Gastroenterol. Hepatol. 4 (6), 798.

Shimkin, M.B., Stoner, G.D., 1975. Lung tumors in mice: application to carcinogenesis bioassay. Adv. Cancer Res. 21, 1–58.

Silverman, J., 2012. Hamsters: biomedical research techniques. In: Suckow, M.A., Stevens, K.A., Wilson, R.P. (Eds.), The Laboratory Rabbit, Guinea Pig, Hamster, and Other Rodents. Academic Press, San Diego, CA, pp. 779–797.

Smith, C., Kelleher, P., 1973. Rat alpha-fetoprotein heterogeneity: affinity chromatography. Biochem. Biophys. Acta 317, 231–235.

Spear, J.F., 1982. Relationship between the scaler electrocardiogram and cellular electrophysiology of the rat heart. In: Budden, R., Detweiler, D.K., Zbinden, G. (Eds.), The Rat Electrocardiogram in Pharmacology and Toxicology. Pergamon Press, New York, pp. 29–40.

Steinberg, J., Gleeson, J., 1990. The pathobiology of ozone-induced damage. Environ. Health Int. J. 45 (2), 80–87.

Steinel, H.H., Baker, R.S.U., 1988. Sensitivity of HRA/Skh hairless mice to initiation/promotion of skin tumors by chemical treatment. Cancer Lett. 41, 63–68.

Storia, L., Bohr, D., Vocke, L., 1954. Experimental hypertension in the hamster. Am. J. Physiol. 94, 685–688.

Stuhlman, R.A., 1979. Animal model: spontaneous diabetes mellitus in *Mystromys albicaudatus*. Am. J. Pathol. 94, 685–688.

Tafassian, A.R., Snider, R.H., Nylen, E.S., et al., 1993. Heterogeneity studies of hamster calcitonin following acute exposure to cigarette smoke: evidence for monomeric secretion. Anat. Rec. 236 (1), 253–256.

Tomson, F.N., Wardop, K.J., 1987. Clinical chemistry and hematology. In: Van Hoosier, G.L., McPherson, C.W. (Eds.), Laboratory Hamsters. Academic Press, Orlando, FL, pp. 43–59.

Vaidya, V., Ozer, J., Dieterle, F., et al., 2010. Kidney injury molecule-1 outperforms traditional biomarkers of kidney injury in preclinical biomarker qualification studies. Nat. Biotechnol. 28, 478–485.

van Heck, H., 1998. Orbital sinus blood sampling in rats, and performed by different animal technicians. Lab. Anim. Sci. 32, 377–386.

Van Zutphen, L.F.M., Baumans, V., Beynen, A.C., 1993. Principles of Laboratory Animal Science. Elsevier, Amsterdam.

Vilageliu, J., Arano, A., Bruseghini, L., 1981. Endothelial damage induced by polyethylene catheter in the rat. Methods Find. Exp. Clin. Pharmacol. 3, 279–281.

Warheit, D.B., Harsky, M.A., 1993. Role of alveolar machrophage chemotaxis and phagocytosis in pulmonary clearance responses to inhaled particles: comparisons among rodent species. Microsc. Res. Tech. 26, 412–422.

Warheit, D.B., Snajdr, S.I., Hartsky, M.A., et al., 1997. Lung proliferative and clearance responses to inhaled para-aramid RFP in exposed hamsters and rats: comparisons with chrysotile asbestos fibers. Environ. Health Perspect. 105, 1219–1222.

Waynforth, H., Flecknell, P., 1980. Experimental and Surgical Technique in the Rat, second ed. Elsevier Academic Press.

Wechsler, R., 1983. Blood collection techniques and normal values for ferrets, rabbits, and rodents: a review. Vet. Med. Small Anim. Clin. 78 (5), 713–717.

Weeks, J.R., 1972. Long-term intravenous infusion. Methods Psychobiol. 2, 155–167.

Werner, A.P., Stuart, B.O., Sanders, C.L., 1979. Inhalation studies with Syrian golden hamsters. Prog. Exp. Tumor Res. 24, 177–198.

Willhite, C.C., Dawson, M.I., Reichart, U., 1996. Receptor-selective retinoid agonists and teratogenic activity. Drug Metab. Rev. 28, 105–119.

Willhite, C.C., Lovey, A., Eckhalf, C., 2000. Distribution, teratogenicity, and embryonic delivered dose of retinoid Ro 23-9223. Toxicol. Appl. Pharmacol. 164, 171–175.

Williams, J., Price, C.J., Sleet, R.B., et al., 1991. Codeine: developmental toxicity in hamsters and mice. Fundam. Appl. Toxicol. 16, 401–413.

Winsett, O.E., Townsend Jr., C.M., Thompson, J.C., 1985. Rapid and repeated blood sampling in the conscious laboratory rat: a new technique. Am. J. Physiol. 249, G145–G146.

Witschi, H., Wilson, D.W., Plopper, C.G., 1993. Modulation of *N*-nitrosodiethylamine-induced hamster lund tumors by ozone. Toxicology. 77, 193–202.

Wlodorezyk, B., Biernachi, B., Minta, M., et al., 1995. Male golden hamster in male reproductive toxicology testing: assessment of protective activity of selenium in acute cadmium intoxication. Bull. Environ. Contam. Toxicol. 54, 907–912.

Wolf, C.J., Ostby, J.S., Gray, L.E., 1999. Gestational exposure to 2, 3, 7, 8-tetrachlorodibenzo-p-dioxin (TCDD) severely alters reproductive function of female hamster offspring. Toxicol. Sci. 51, 259–264.

Wong, M.A., Oace, S.M., 1981. Feeing pattern and gastrointestinal transit rate of rats under different room lighting schedule. Lab. Anim. Sci. 31 (4), 362–365.

Worth, H.M., Kachmann, C., Anderson, R.C., 1963. Intragastric injection for toxicity studies with newborn rats. Toxicol. Appl. Pharmacol. 5, 719–727.

Yerganian, G., 1972. History and cytogenetics of hamsters. Prog. Exp. Tumor Res. 16, 2–41.

Yoburn, B.C., Morales, R., Inturrisi, C.E., 1984. Chronic vascular catheterization in the rat: comparison of three techniques. Physiol. Behav. 33, 89–94.

Zbinden, G., 1981. Spontaneous and induced arrhythmias in rat toxicology studies. In: Budden, R., Detweiler, D.K., Zbinden, G. (Eds.), The Rat Electrocardiogram in Pharmacology and Toxicology. Pergamon Press, New York, pp. 117–128.

Zbinden, G., Kleinert, R., Rageth, D., 1980. Assessment of emetine cardiotoxicity in a subacute toxicity experiment in rats. J. Cardiovasc. Pharmacol. 2, 155–164.

Zethelius, B., Berglund, L., Sundstrom, J., et al., 2008. Use of multiple biomarkers to improve the prediction of death from cardiovascular causes. N. Engl. J. Med. 358, 2107–2116.

CHAPTER

3

Göttingen Minipigs as Large Animal Model in Toxicology

Henrik Duelund Pedersen, Lars Friis Mikkelsen

Ellegaard Göttingen Minipigs A/S, Dalmose, Denmark

PIGS AND MINIPIGS IN TRANSLATIONAL RESEARCH

Over the last 2 decades, a significant increase in the use of pigs and especially minipigs in translational research, including nonclinical toxicology and safety assessment, has led to an enhanced understanding of human diseases and improvement in human health. The advantages of using the pig (*Sus scrofa*) for translational research include a defined genome sequence and similarities to humans in terms of anatomy, physiology and biochemistry. The minipig also offers the additional advantages of reduced size and decreased time to sexual maturity (Monticello and Haschek, 2016).

The RETHINK project (Forster et al., 2010) highlighted the use of the minipig as an alternative to the traditionally used nonrodent species, that is, the dog and the nonhuman primate. This conclusion was later confirmed by the Preclinical Safety Leadership Group (DruSafe) of the International Consortium for Innovation and Quality in Pharmaceutical Development, and the group concluded that the minipig is a viable nonrodent model for the development of small molecules and dermal products (Colleton et al., 2016). The latest developments in the use of pigs and especially minipigs in translational research, with an emphasis on toxicological pathology and nonclinical safety assessment, were addressed in a recent special issue of *Toxicologic Pathology* (Monticello and Haschek, 2016). Considering the comparative biology and practical features of toxicology testing in minipigs, it has been stated that the minipig represents a favorable profile as a nonrodent toxicology model in terms of similarity to man and applicability to different study types (Bode et al., 2010).

From information in the public domain and especially from within the European Medicines Agency and the US Food and Drug Administration, minipigs have already been fully accepted for use as a nonrodent species in toxicology testing of pharmaceutical products (Van der Laan et al., 2010). They are also recognized by the International Conference on Harmonization and the International Organization for Standardization Guidelines as a suitable animal model due to their similarities to man. Thus minipigs may be considered an acceptable choice as a nonrodent species, provided adequate scientific justification for this choice is made.

From an economic perspective, the costs of testing in minipigs are not significantly greater than the costs for a study in dogs (Van der Laan et al., 2010). The cost of testing in nonhuman primates is greater than the cost in minipigs. Furthermore, there is political and societal pressure on decreasing the use of nonhuman primates, which also is reflected in the European legislation (EU Directive 2010/63/EU).

LEGISLATION AND SPECIES SELECTION

Ensuring human safety in clinical trials and when medicines are on the market requires toxicology testing in a rodent and a nonrodent mammal. Selecting the correct nonrodent or large animal species is very important to maximize human safety, clinical benefit, and animal welfare (Smith and Hubrecht, 2001). Several factors should be considered in the large animal species selection process, including available scientific information, experience with and availability of a full range of species, and the need to balance scientific, ethical, and legal constraints. In terms of the latter, the European legislation (Directive 2010/63/EU) only allows the use of nonhuman primates if it is undertaken with a view to the avoidance, prevention, diagnosis, or treatment

Biomarkers in Toxicology, Second Edition
https://doi.org/10.1016/B978-0-12-814655-2.00003-7

of debilitating or potentially life-threatening clinical conditions in human beings and if—for scientific reasons—a species other than nonhuman primates cannot be used. Furthermore, Directive 2010/63/EU clearly states that the species to be used should be that with the lowest capacity to experience pain, suffering, distress, or lasting harm. The use of nonhuman primates in scientific procedures is still necessary in biomedical research, but due to their genetic proximity to human beings and their highly developed social skills, their use raises specific ethical and practical problems in terms of meeting their behavioral, environmental, and social needs in a laboratory environment. In contrast, the housing of Göttingen Minipigs is simple, and it is much easier to keep minipigs to a good standard of welfare under laboratory conditions (Ellegaard et al., 2010). Ethical arguments also need to be considered where the use of nonhuman primates is of the greatest concern to the public. In addition, funding bodies recognize concerns about the use of nonhuman primates in research and the difficulties associated with meeting the environmental, behavioral, and social needs of these highly intelligent animals in a laboratory (National Centre for the Replacement, Refinement and Reduction of Animals in Research (NC3Rs), London, UK).

The toxicology testing of pharmaceuticals in animals is subject to regulation by national and international bodies. Many published guidelines make recommendations regarding the scope of testing and methods to be used for each specific type of product. Some of these guidelines also recommend which animal species should be used. Overall, a variety of experimental approaches and technologies are available to assist in the selection of the relevant species for the conduct of toxicological studies. These methods for analysis, however, should be selected prudently based on the test article and its pharmacology to facilitate the minimization of risks and create better awareness of risks during clinical development of the drug candidate (Subramanyam et al., 2008). Species selection must always be made on a case-by-case basis by balancing the benefits and scientific evidence relating to the predictivity of the animal model against the harm that may accrue to the animals both from the test procedures and their lifetime experience within the laboratory environment (Webster et al., 2010). Therefore minipigs should be considered up-front as nonrodent species for legislative, scientific, economic, and ethical reasons.

GÖTTINGEN MINIPIGS AS AN ANIMAL MODEL

The most commonly used minipig species for translational research is the Göttingen Minipig (Fig. 3.1).

FIGURE 3.1 Göttingen minipig is the most commonly used minipig species for translational research.

Göttingen Minipigs have been available in the major R&D regions for more than 3 decades, and huge amounts of data and many publications are available with relevant and important background data. For example, data comparing minipig, dog, monkey, and human drug metabolism and disposition (Dalgaard, 2015), data on spontaneous background pathology (Jeppesen and Skydsgaard, 2015; Helke et al., 2016), and data on vehicle systems and excipients used in Göttingen Minipigs (Weaver et al., 2016), all highlighting and validating the use of Göttingen Minipigs in toxicological research. Furthermore, when using Göttingen Minipigs, the genetic management and standardization (Simianer and Köhn, 2010) plus the high health status (www.minipigs.dk) are advantages compared with other minipig strains along with the low body weight (Fig. 3.2).

Apart from the scientific justification, it is very easy to fulfill the natural behavioral needs of the minipig, which include social housing in harmonious groups in a challenging and spacious environment, giving them the opportunity to choose specific sites for defecation and urination while keeping their sleeping areas dry. Straw should be provided as comfortable bedding and to keep the animals occupied with rooting and chewing activities (Bollen et al., 2010). In some experimental settings, the provision of straw or hay can be a challenge and could be replaced by other types and combinations of environmental enrichment, such as balls or chains hung from the ceiling or roof of the cage (Swindle, 2007) to still fulfill their strong instinct for rooting, chewing, and exploring. It is important to provide an environment with sufficient complexity allowing expression of these key behaviors—to avoid boredom, frustration, and stereotypies (Ball, 2012). In order to facilitate stress-free handling, dosing, and sampling, the focus on acclimatizing and training minipigs should be highlighted. In order to minimize the stress related to

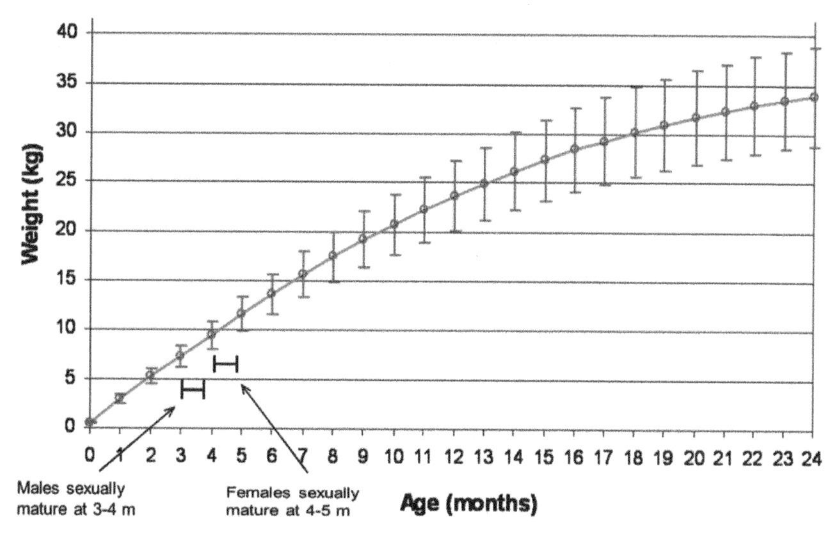

FIGURE 3.2 Growth curve illustrating how male Göttingen Minipigs increase linearly in body weight with approximately 2 kg/month for the first year, and 1 kg/month for the following year to reach their full size at approximately 2 years of age. Also, notice the early age at which Göttingen Minipigs become sexually mature.

handling and consequent impact on experimental variability, positive-reinforced socialization to human contact and training of the minipigs for procedures that they will experience are essential (Ellegaard et al., 2010), which easily can be applied through relevant training of staff in species-specific handling, dosing, and sampling (www.minipigs.dk).

DOSING ROUTES AND FORMULATIONS

All dosing routes commonly used in the development of new drugs are regularly used in Göttingen Minipigs. Guidelines have been published describing dosing routes in general and the administration volumes considered good practice in commonly used species (Diehl et al., 2001). The tolerability of the minipig to the many different excipients used in drug development studies has been reviewed, and a substantive conclusion was that minipigs tolerate several of these, such as cremophor and polysorbate 80, much better than the dog (Weaver et al., 2016). Also, the tolerability to many compounds is higher and more human-like in the minipig than in the dog. For instance, dogs show excessive vomiting in response to many compounds, including anticancer compounds (Mahl et al., 2016), and have a low tolerance to nonsteroidal anti-inflammatory drugs (NSAIDs), expressed as gastrointestinal lesions (Swindle et al., 2012).

Dermal and subcutaneous dosing is especially attractive in the Göttingen Minipig because of the similarities of the skin and subcutaneous tissue with the corresponding tissues in humans. The Göttingen Minipig has been shown to be a good model for in vitro permeation through human skin, considering that the age of the Minipig can influence the flux of certain compounds (Qvist et al., 2000; Yamamoto et al., 2017). The pale skin and the sparse and nonpigmented body hair make dermal effects easy to evaluate, and the Göttingen Minipig is, therefore, the species of choice for dermatotoxicology studies in most laboratories, and in the article by Willard-Mack et al. (2016), study designs and practical considerations of such studies are discussed.

Subcutaneous dosing is commonly used, and the minipig offers major advantages regarding this route due to similarities with human subcutaneous tissue. In humans, a subcutaneous injection is often made by injecting into a skinfold, leading to injection into subcutaneous fat. In minipigs, it is also possible to inject into subcutaneous fat by injecting behind the base of the ears. To be certain that the correct type of tissue is present at the expected depth, the injection site can be examined by ultrasound prior to injection and a suitable site marked by pen or tattoo. In other (furry) experimental animal species, a subcutaneous injection is made by lifting the skin and injecting into the loose connective tissue under the skin. Should it be desired, it is also possible in minipigs to inject into such loose connective tissue, for instance in the medial thigh region. The difference in subcutaneous tissue type can be important when assessing local toxicity, as such differences can lead to differences in compound distribution and tissue reaction toward the formulation. Also, regarding the pharmacokinetics, such tissue differences can play a key role, as suggested by findings using compounds as different as fast-acting insulins and monoclonal antibodies (Plum et al., 2000; Zheng et al., 2012).

The oral route is the most commonly used, and there are several ways of dosing orally in Göttingen Minipigs.

To ensure proper dosing and assist when dosing by gavage, a dosing chair, such as the one described in Ellegaard Göttingen Minipigs Newsletter Number 47 (www.minipigs.dk), can be very helpful. The advantages of using the minipig for oral dosing studies are the fact that the gastric pH in the fasted minipig is much less variable and more comparable to that in humans than it is in the dog (Preuße and Skaanild, 2012). Minipigs are also much less inclined to vomit as a response to various compounds and formulations than the dog. The gastric emptying of the minipig, on the other hand, is longer than that observed in humans (Suenderhauf et al., 2014; Christiansen et al., 2015). By removing the straw bedding and potentially replacing it with another enrichment type, such as cotton towels, the stomach content remaining after an overnight fast can be markedly reduced (Suenderhauf et al., 2014).

Several types of catheters can be used for intravenous access in Göttingen Minipigs; depending on the specific study needs. An educational package describing some of the many options is available on the Ellegaard Göttingen Minipig homepage (www.minipigs.dk). Recommended maximum blood sample volumes have been published (Diehl et al., 2001), and the same publication also describes different sampling sites that can be used if a catheter is not in place.

Other routes of administration might be used in more rare cases, including inhalation (Koch et al., 2001; Windt et al., 2010), intraperitoneal (Overgaard et al., 2017), and intraocular (Proksch et al., 2008).

AGE AND SEX

Göttingen Minipigs become sexually mature at an early age. Based on testicular and epididymal histology, puberty in male animals was recently found to occur at 8 weeks of age (Taberner et al., 2016). In female Göttingen Minipigs, hormonal analyses showed that puberty occurs at approximately 4–6 months (Peter et al., 2016). Based on this, it is common for pharmaceutical companies to use minipigs for toxicological studies at an age of approximately 4 months, when the animals weigh approximately 8–11 kg. As can be seen on the growth curve in Fig. 3.2, Göttingen Minipigs continue to grow until they reach a body weight of 30–40 kg at 2 years of age, if they are fed restrictively according to the recommendations provided by the breeder (www.minipigs.dk).

As a reflection of the growth curve, the levels of various bone biomarkers decrease markedly during the first 12 months of age, moderately during the next 12 months, and very little after 24 months of age (Tsutsumi, 2012).

If allowed to eat ad libitum, even feeding a regular pig diet leads to severe obesity after approximately 1–1½ years, with female minipigs weighing approximately 90 kg and having 45% body fat (Raun et al., 2007). Intact males are protected, to some degree, from diet-induced obesity and perturbed glucose metabolism and have very high circulating levels of testosterone and estradiol (Christoffersen et al., 2010, 2013).

HEMATOLOGY

Table 3.1 shows hematology reference values for Göttingen Minipigs. Reference data are also available for piglets and animals that are 3 and 6 months of age at www.minipigs.dk. It is important to acknowledge that age has a significant effect on many of the values. For instance, juvenile Göttingen Minipigs have been found to have markedly higher white blood cell counts than adult animals (Jorgensen et al., 1998). Regarding the platelet count, several studies have found that blood stabilized with standard concentrations of ethylenediamine tetraacetic acid (EDTA) can spontaneously aggregate over time, resulting in falsely low platelet counts (Olsen et al., 2001; Erkens et al., 2017). The life-span of minipig platelets (8–10 days) and red blood cells (80–100 days), as well as neutrophils, is similar to that found in humans (Moroni et al., 2011).

It has been shown that Göttingen Minipigs are a useful species for use in coagulation analyses (Petroianu et al., 1998). Again, there are important age differences. Fetal minipigs, for instance, have a lower activity of most coagulation factors (Petroianu et al., 1998). Most markedly, fetal minipigs have 10-fold lower fibrinogen concentrations compared with adult minipigs—a finding that is in accordance with findings in premature human infants (Petroianu et al., 1998). It should also be noted that not all immunoassays usually used in humans will work in pigs (Münster et al., 2002). Finally, it is important to consider that dyslipidemic animals are in a procoagulant state, likely caused in part by increased platelet activity (Kreutz et al., 2011).

BLOOD BIOCHEMISTRY

A blood panel is included in most studies, and reference values of commonly used biochemistry parameters for Göttingen Minipigs are shown in Table 3.2. Values for piglets and older animals 3–6 months of age may be found at www.minipigs.dk. Age and sex have an influence on many serum chemistry parameters (Jorgensen et al., 1998), and there can be minor differences in values from laboratory to laboratory. However, these issues rarely pose a problem because a control group is

TABLE 3.1 Hematology reference values for Göttingen Minipigs (3 + 6 Months)

Animals 3 months of age	WBC (×10³/μL)	RBC (×10⁶/μL)	HGB (g/dL)	HCT (%)	MCV (fL)	MCH (pg)	MCHC (g/dL)	RDW (%)	HDW (g/dL)
Male ($n = 49$)	11.15 ± 2.08	7.94 ± 0.68	12.2 ± 0.8	40.0 ± 2.6	50.6 ± 4.1	15.5 ± 1.5	30.6 ± 0.8	16.2 ± 1.0	1.91 ± 0.10
95% reference range	7.88–14.44	6.39–9.17	10.7–13.6	35.2–44.8	45.1–60.6	13.4–19.0	29.3–32.0	14.7–18.3	1.73–2.09
Female ($n = 49$)	11.01 ± 2.32	8.35 ± 0.56	12.6 ± 0.7	41.0 ± 2.1	49.3 ± 3.1	15.2 ± 1.1	30.7 ± 0.7	15.8 ± 1.0	1.90 ± 0.14
95% reference range	7.91–15.81	7.52–9.30	11.2–13.8	37.1–44.5	45.4–57.0	13.5–17.6	29.6–32.0	14.0–18.1	1.66–2.16

	PLT (×10³/μL)	MPV (fL)	NEUT-abs (×10³/μL)	LYMPH-abs (×10³/μL)	MONO-abs (×10³/μL)	EOS-abs (×10³/μL)	BASO-abs (×10³/μL)
Male ($n = 49$)	457 ± 135	9.2 ± 0.8	4.27 ± 1.80	6.21 ± 1.38	0.31 ± 0.10	0.26 ± 0.26	0.03 ± 0.05
95% reference range	232–688	7.8–10.9	1.95–8.93	3.70–8.70	0.15–0.51	0.07–0.88	0.01–0.07
Female ($n = 49$)	508 ± 154	9.1 ± 0.9	3.30 ± 1.40	7.05 ± 1.55	0.34 ± 0.10	0.20 ± 0.11	0.04 ± 0.02
95% reference range	273–888	7.7–10.7	1.69–7.26	4.81–9.36	0.15–0.55	0.08–0.41	0.01–0.08

	NEUT (%)	LYMPH(%)	MONO (%)	EOS (%)	BASO (%)
Male ($n = 49$)	37.4 ± 11.2	56.5 ± 10.5	2.8 ± 0.8	2.3 ± 2.2	0.3 ± 0.3
95% reference range	22.1–68.1	27.9–71.4	1.6–4.8	0.6–8.5	0.1–0.6
Female ($n = 49$)	29.4 ± 8.6	64.6 ± 8.3	3.1 ± 0.8	1.9 ± 0.9	0.3 ± 0.2
95% reference range	19.0–48.4	46.5–75.4	1.8–4.5	0.7–4.3	0.1–0.6

Animals 6 months of age	WBC (×10³/μL)	RBC (×10⁶/μL)	HGB (g/dL)	HCT (%)	MCV (fL)	MCH (pg)	MCHC (g/dL)	RDW (%)	HDW (g/dL)
Male ($n = 34$)	9.96 ± 2.13	8.20 ± 0.70	13.2 ± 0.8	41.5 ± 2.6	50.9 ± 4.0	16.1 ± 1.5	31.7 ± 0.7	16.2 ± 1.1	2.03 ± 0.16
95% reference range	7.05–13.82	6.91–9.33	11.5–14.3	36.0–46.0	45.1–58.4	14.0–19.0	30.3–32.8	14.6–18.3	1.70–2.30
Female ($n = 50$)	10.22 ± 1.88	8.46 ± 0.67	13.1 ± 1.0	41.4 ± 3.1	49.1 ± 3.8	15.6 ± 1.4	31.7 ± 0.6	16.5 ± 0.9	1.99 ± 0.16
95% reference range	6.53–13.18	7.06–9.60	11.2–14.8	35.0–48.1	44.5–58.6	13.8–18.9	30.4–32.8	15.1–18.4	1.75–2.32

	PLT (×10³/μL)	MPV (fL)	NEUT-abs (×10³/μL)	LYMPH-abs (×10³/μL)	MONO-abs (×10³/μL)	EOS-abs (×10³/μL)	BASO-abs (×10³/μL)
Male ($n = 34$)	467 ± 171	9.2 ± 1.9	5.07 ± 1.91	4.29 ± 1.01	0.38 ± 0.13	0.15 ± 0.10	0.03 ± 0.02
95% reference range	116–718	7.5–14.1	2.29–8.96	2.94–6.47	0.19–0.67	0.04–0.40	0.01–0.06
Female ($n = 50$)	448 ± 84	8.9 ± 1.0	3.1 ± 1.2	6.25 ± 1.41	0.44 ± 0.14	0.28 ± 0.12	0.03 ± 0.01
95% reference range	296–570	7.7–11.1	1.5–6.0	4.29–9.41	0.21–0.73	0.12–0.59	0.01–0.06

	NEUT (%)	LYMPH (%)	MONO (%)	EOS (%)	BASO (%)
Male ($n = 34$)	49.7 ± 11.2	44.0 ± 10.7	3.8 ± 1.1	1.6 ± 1.1	0.3 ± 0.1
95% reference range	31.3–63.9	29.6–61.0	2.2–6.1	0.4–4.5	0.1–0.5
Female ($n = 50$)	30.4 ± 8.9	61.4 ± 8.5	4.4 ± 1.4	2.9 ± 1.3	0.3 ± 0.1
95% reference range	17.7–49.7	43.5–75.8	1.9–7.4	1.0–5.6	0.1–0.6

Mean ± standard deviation. Data were obtained using an Advia 120 Hematology Analyzer. Ages are 3 months ±7 days and 6 months ±7 days.
BASO% = *basophils percentage;* BASO-abs = *basophils absolute;* EOS% = *eosinophils percentage;* EOS-abs = *eosinophils absolute;* HCT, *hematocrit;* HDW, *hemoglobin distribution width;* HGB, *hemoglobin;* LYMPH%, *lymphocytes percentage;* LYMPH-abs, *lymphocytes absolute;* MCH, *mean corpuscular hemoglobin;* MCHC, *mean corpuscular hemoglobin concentration;* MCV, *mean corpuscular volume;* MONO%, *monocytes percentage;* MONO-abs, *monocytes absolute;* MPV, *mean platelet volume;* NEUT%, *neutrophils percentage;* NEUT-abs, *neutrophils absolute;* PLT, *platelet;* RBC, *red blood cell;* RDW, *red cell distribution width;* WBC, *white blood cell.*
Data courtesy of Marshall BioResources, North Rose, New York, USA.

TABLE 3.2 Serum chemistry reference values for Göttingen Minipigs (3 + 6 Months)

Animals 3 months of age	Glucose (mg/dL)	Urea N (mg/dL)	Creatinine (mg/dL)	Sodium (mmol/L)	Potassium (mmol/L)	Na:K	Chloride (mmol/L)
Male ($n = 46$)	95 ± 20	7 ± 3	0.8 ± 0.1	143 ± 3	5.4 ± 0.7	27 ± 3	100 ± 2
95% reference range	59–132	2–14	0.7–1.1	139–149	4.4–6.5	22–33	96–103
Female ($n = 45$)	89 ± 13	6 ± 2	0.9 ± 0.1	143 ± 2	5.4 ± 0.5	27 ± 3	101 ± 2
95% reference range	68–117	3–11	0.8–1.2	140–147	4.5–6.4	23–32	97–104

	Calcium (mg/dL)	Magnesium (mg/dL)	Phosphorus (mg/dL)	Cholesterol (mg/dL)	Total Protein (g/dL)	Albumin (g/dL)	Globulin (g/dL)	A/G
Male ($n = 46$)	11.3 ± 0.4	2.6 ± 0.2	8.2 ± 1.0	79 ± 10	5.7 ± 0.3	3.6 ± 0.3	2.0 ± 0.2	1.8 ± 0.2
95% reference range	10.7–11.9	2.4–3.0	6.8–9.6	62–99	5.1–6.3	3.2–4.1	1.7–2.4	1.5–2.1
Female ($n = 45$)	11.3 ± 0.4	2.7 ± 0.2	8.0 ± 0.9	93 ± 12	6.0 ± 0.3	3.8 ± 0.2	2.3 ± 0.2	1.7 ± 0.2
95% reference range	10.5–12.1	2.4–3.0	6.6–9.7	73–119	5.4–6.5	3.4–4.1	1.9–2.7	1.4–2.0

	Amylase (U/L)	AST (U/L)	ALT (U/L)	CK (U/L)	Alk Phos (U/L)	GGT (U/L)	Tot Bili (mg/dL)
Male ($n = 46$)	681 ± 148	29 ± 9	62 ± 9	377 ± 363	198 ± 37	73 ± 11	0.3 ± 0.1
95% reference range	473–1018	17–56	47–82	145–1387	141–262	57–99	0.2–0.5
Female ($n = 45$)	706 ± 154	27 ± 7	55 ± 7	437 ± 408	193 ± 29	81 ± 14	0.3 ± 0.1
95% reference range	512–956	18–39	44–66	159–1567	140–247	60–107	0.2–0.4

Animals 6 months of age	Glucose (mg/dL)	Urea N (mg/dL)	Creatinine (mg/dL)	Sodium (mmol/L)	Potassium (mmol/L)	Na:K	Chloride (mmol/L)
Male ($n = 37$)	79 ± 17	7 ± 3	1.0 ± 0.1	143 ± 4	5.5 ± 0.6	26 ± 3	99 ± 3
95% reference range	55–111	4–13	0.8–1.3	139–148	4.6–6.7	21–32	95–104
Female ($n = 45$)	82 ± 12	9 ± 2	1.0 ± 0.1	142 ± 3	5.4 ± 0.6	26 ± 3	100 ± 2
95% reference range	64–101	6–13	0.7–1.2	137–148	4.7–6.7	21–30	96–104

	Calcium (mg/dL)	Magnesium (mg/dL)	Phosphorus (mg/dL)	Cholesterol (mg/dL)	Total Protein (g/dL)	Albumin (g/dL)	Globulin (g/dL)	A/G
Male ($n = 37$)	10.8 ± 0.7	2.5 ± 0.2	7.7 ± 0.8	53 ± 11	6.3 ± 0.4	4.2 ± 0.4	2.1 ± 0.2	2.0 ± 0.3
95% reference range	9.8–12.0	2.2–3.1	6.5–9.0	45–82	5.7–7.0	3.4–4.9	1.8–2.4	1.5–2.5
Female ($n = 45$)	11.0 ± 0.4	2.5 ± 0.2	7.2 ± 0.7	81 ± 16	6.2 ± 0.3	4.0 ± 0.3	2.2 ± 0.2	1.8 ± 0.2
95% reference range	10.5–11.9	2.2–2.8	5.8–8.3	50–109	5.3–6.7	3.5–4.3	1.8–2.5	1.5–2.3

	Amylase (U/L)	AST (U/L)	ALT (U/L)	CK (U/L)	Alk Phos (U/L)	GGT (U/L)	Tot Bili (mg/dL)
Male ($n = 37$)	548 ± 124	35 ± 15	73 ± 13	538 ± 616	160 ± 46	88 ± 15	0.3 ± 0.1
95% reference range	372–827	19–69	55–106	131–2000	95–270	65–121	0.2–0.5
Female ($n = 45$)	611 ± 111	24 ± 6	62 ± 10	333 ± 224	148 ± 34	70 ± 10	0.3 ± 0.1
95% reference range	444–852	15–35	46–83	145–864	91–212	53–86	0.2–0.4

Mean and standard deviation. Data were obtained using a Vitros 250 chemistry analyzer. Ages are 3 months ± 7 days and 6 months ±14 days.
Alk Phos, *alkaline phosphatase*; ALT, *alanine aminotransferase*; AST, *aspartate aminotransferase*; BUN, *blood urea nitrogen*; CK, *creatine kinase*; GGT, *gamma glutamyltransferase*; Na:K, *sodium:potassium*; Tot Bili, *total bilirubin*.
Data courtesy of Marshall BioResources, North Rose, New York, USA.

most often included in the study design. In young Göttingen Minipigs, values of several parameters (urea, bilirubin, globulin, alkaline phosphatase, and triglycerides) have been shown to decrease from 2 to 9 weeks of age, and albumin was found to increase over the same age range (Feyen et al., 2016).

A wide range of less common blood biochemistry markers has been used with success, and a literature search will usually help to show the types that can be used in each setting, as well as a reference to specific methods. A few guiding comments might, however, be useful for some of the parameters. For example, one should be aware that glycosylated hemoglobin (HbA1c) cannot be used in any pigs as measure of long-term plasma glucose levels because porcine erythrocytes are practically impermeable to glucose (Higgins et al., 1982). The plasma levels of fructosamine can be used instead (Christoffersen et al., 2007). Insulin is often measured in studies focused on diabetes, and C-peptide is often used as an indicator of insulin secretion due to its longer plasma half-life in humans. In pigs, however, C-peptide has practically the same half-life as insulin (Oyama et al., 1975), and therefore this parameter offers little advantage, as the curves for insulin and C-peptide will be almost identical (Christoffersen et al., 2013). Also, it should be noted that the plasma level of lactate is higher in Göttingen Minipigs than in domestic pigs, so it is, therefore, important that reference values for blood lactate concentrations in pigs reflect the breed of interest (Alstrup, 2016).

The concentrations of plasma lipids will vary according to the diet fed to the minipigs. With an atherogenic diet, a marked dyslipidemia can be obtained, leading to severe human-like atherosclerotic changes within 5–10 months (Ludvigsen et al., 2015).

ORGAN WEIGHT, BACKGROUND FINDINGS, AND COMPARATIVE ASPECTS

The pig is a monogastric omnivore, and many organs show great resemblance in anatomy and physiology to the corresponding organs in humans. For instance, the heart has a similar coronary artery distribution with negligible coronary collateral circulation, which is in contrast to the dog. The kidneys are very similar to human kidneys in function and ability to concentrate the urine (Schmidt-Nielsen and O'Dell, 1961). The skin and subcutaneous tissues are very similar to the corresponding tissues in humans, and the liver shows great resemblance to the human liver in terms of metabolism, especially regarding the cytochrome P450 enzymes (Dalgaard, 2015). Many more examples could be mentioned, and the comparative biology of the minipig and usage in regulatory toxicology studies has been reviewed in detail elsewhere (Bode et al., 2010; Swindle et al., 2012; Schomberg et al., 2016).

Reference values for organ weights for Göttingen Minipigs aged 8 weeks and 6 months are shown in Table 3.3. In the table, the organ weights are given both as absolute

TABLE 3.3 Organ weights (absolute and relative) for Göttingen Minipigs

Parameter		8 Weeks		6 Months	
		Males	Females	Males	Females
Body weight, kg	Mean	4.3	3.9	14.2	14.6
	SD	0.9	0.7	1.6	2.2
	Min	2.2	2.5	11.2	11.6
	Max	6.1	5.3	17.4	19.8
	N	23	20	20	20
Adrenals	Mean	0.461	0.444	1.133	1.130
	SD	0.064	0.082	0.185	0.183
	Min	0.350	0.322	0.892	0.774
	Max	0.572	0.647	1.574	1.402
	N	23	20	20	20
Adrenals, rel	Mean	0.0113	0.0114	0.0081	0.0079
	SD	0.0031	0.0013	0.0014	0.0016
	Min	0.0068	0.0093	0.0057	0.0052
	Max	0.0216	0.0138	0.0101	0.0115
	N	23	20	20	20
Brain	Mean	43.36	43.06	62.34	62.32
	SD	3.52	2.36	4.08	3.12
	Min	34.34	38.82	54.81	54.67
	Max	49.98	47.67	68.92	68.43
	N	23	20	20	20
Brain, rel	Mean	1.052	1.131	0.445	0.434
	SD	0.194	0.190	0.061	0.064
	Min	0.754	0.861	0.361	0.309
	Max	1.561	1.661	0.608	0.561
	N	23	20	20	20
Epididymides	Mean	4.24		20.98	
	SD	1.33		3.98	
	Min	2.06		14.26	
	Max	7.31		32.63	
	N	23		20	
Epididymides, rel	Mean	0.098		0.149	
	SD	0.017		0.027	
	Min	0.063		0.110	
	Max	0.120		0.230	
	N	23		20	

Continued

TABLE 3.3 Organ weights (absolute and relative) for Göttingen Minipigs—cont'd

Parameter		8 Weeks		6 Months	
		Males	Females	Males	Females
Eyes	Mean	5.38	5.18	9.16	8.95
	SD	0.39	0.36	0.69	0.64
	Min	4.49	4.64	8.24	7.89
	Max	5.88	5.95	11.16	10.07
	N	23	20	20	20
Eyes, rel	Mean	0.131	0.136	0.065	0.063
	SD	0.026	0.022	0.009	0.010
	Min	0.090	0.102	0.053	0.045
	Max	0.204	0.190	0.086	0.083
	N	23	20	20	20
Heart	Mean	22.60	19.63	73.72	59.76
	SD	4.12	3.51	5.85	8.98
	Min	12.38	13.58	67.05	42.22
	Max	29.63	27.10	83.57	78.19
	N	23	20	20	20
Heart, rel	Mean	.0.534	0.504	0.524	0.412
	SD	0.047	0.045	0.050	0.056
	Min	0.461	0.418	0.402	0.345
	Max	0.630	0.601	0.606	0.567
	N	23	20	20	20
Kidneys	Mean	27.26	24.08	66.44	58.66
	SD	5.55	6.04	9.34	11.16
	Min	11.93	13.27	48.02	38.85
	Max	37.26	37.41	81.75	75.87
	N	23	20	20	20
Kidneys, rel	Mean	0.640	0.611	0.470	0.401
	SD	0.052	0.064	0.060	0.048
	Min	0.542	0.456	0.399	0.318
	Max	0.735	0.720	0.599	0.483
	N	23	20	20	20
Liver	Mean	105.68	98.24	237.03	218.24
	SD	23.72	22.23	25.78	33.69
	Min	48.91	59.14	188.54	161.65
	Max	148.66	139.08	301.99	304.81
	N	23	20	20	20
Liver, rel	Mean	2.48	2.51	1.68	1.50
	SD	0.31	0.32	0.16	0.14

TABLE 3.3 Organ weights (absolute and relative) for Göttingen Minipigs—cont'd

Parameter		8 Weeks		6 Months	
		Males	Females	Males	Females
	Min	1.98	2.05	1.38	1.32
	Max	3.14	3.13	1.89	1.87
	N	23	20	20	20
Lungs	Mean	40.24	35.59	80.39	78.04.
	SD	7.54	6.88	6.23	12.49
	Min	20.46	22.83	64.92	59.14
	Max	53.37	49.77	91.67	108.35
	N	23	20	20	20
Lungs, rel	Mean	0.951	0.911	0.570	0.535
	SD	0.094	0.065	0.039	0.053
	Min	0.719	0.750	0.487	0.448
	Max	1.141	1.057	0.647	0.661
	N	23	20	20	20
Ovaries	Mean		0.05		2.08
	SD		0.02		1.08
	Min		0.02		0.85
	Max		0.10		4.81
	N		19		20
Ovaries, rel	Mean		0.00124		0.01441
	SD		0.00051		0.00745
	Min		0.00065		0.00624
	Max		0.00235		0.03204
	N		19		20
Parathyroids	Mean	0.007	0.007	0.051	0.029
	SD	0.003	0.005	0.043	0.011
	Min	0.003	0.002	0.021	0.006
	Max	0.017	0.021	0.143	0.047
	N	23	20	15	19
Parathyroids, rel	Mean	0.000177	0.000184	0.000359	0.000197
	SD	0.000100	0.000097	0.000305	0.000077
	Min	0.000070	0.000060	0.000140	0.000050
	Max	0.000450	0.000440	0.001140	0.000340
	N	23	20	15	19
Pituitary	Mean	0.040	0.043	0.095	0.100
	SD	0.007	0.006	0.018	0.017
	Min	0.028	0.031	0.062	0.072

TABLE 3.3 Organ weights (absolute and relative) for Göttingen Minipigs—cont'd

Parameter		8 Weeks		6 Months	
		Males	Females	Males	Females
	Max	0.057	0.054	0.150	0.137
	N	23	20	20	20
Pituitary, rel	Mean	0.000961	0.001120	0.000679	0.000689
	SD	0.000143	0.000205	0.000148	0.000122
	Min	0.000774	0.000824	0.000449	0.000543
	Max	0.001273	0.001520	0.001056	0.000975
	N	23	20	20	20
Prostate	Mean	2.08		11.73	
	SD	1.04		3.83	
	Min	0.68		6.85	
	Max	4.32		20.86	
	N	23		20	
Prostate, rel	Mean	0.0471		0.0836	
	SD	0.0177		0.0282	
	Min	0.0244		0.0514	
	Max	0.0942		0.1469	
	N	23		20	
Spleen	Mean	8.36	7.98	22.29	20.58
	SD	1.95	2.37	4.01	4.26
	Min	4.94	3.80	14.94	12.42
	Max	12.54	12.22	29.43	30.19
	N	23	20	20	20
Spleen, rel	Mean	0.198	0.202	0.158	0.141
	SD	0.037	0.041	0.027	0.025
	Min	0.139	0.136	0.108	0.091
	Max	0.273	0.271	0.224	0.208
	N	23	20	20	20
Testes	Mean	13.67		42.05	
	SD	4.28		8.44	
	Min	6.36		24.44	
	Max	21.53		56.62	
	N	23		20	
Testes, rel	Mean	0.318		0.298	
	SD	0.065		0.062	
	Min	0.200		0.188	
	Max	0.457		0.423	

Continued

TABLE 3.3 Organ weights (absolute and relative) for Göttingen Minipigs—cont'd

Parameter		8 Weeks		6 Months	
		Males	Females	Males	Females
	N	23		20	
Thyroid	Mean	0.294	0.265	0.982	0.971
	SD	0.091	0.069	0.479	0.202
	Min	0.183	0.142	0.578	0.678
	Max	0.544	0.386	2.802	1.422
	N	23	20	20	20
Thyroid, rel	Mean	0.00687	0.00681	0.00703	0.00667
	SD	0.00129	0.00165	0.00381	0.00123
	Min	0.00480	0.00450	0.00420	0.00499
	Max	0.00939	0.01170	0.02189	0.01027
	N	23	20	20	20
Uterus	Mean		1.63		83.04
	SD		0.61		57.45
	Min		0.90		12.76
	Max		3.82		180.16
	N		20		20
Uterus, rel	Mean		0.0421		0.5761
	SD		0.0153		0.3948
	Min		0.0247		0.0778
	Max		0.1006		12.129
	N		20		20

The 23 8-week-old male minipigs were 61 ± 4 days of age (range 55–65 days), while the 20 female minipigs were 57 ± 3 days of age (range 52–61 days). The 20 6-month-old male minipigs were 198 ± 11 days old (range 183–235 days), while the 20 females were 191 ± 10 days old (range 170–209 days).

weights and as a percentage of the body weight, as often done in toxicology studies. A normalization to body surface area (BSA) would, however, in most cases be more correct biologically. For example, using the formula $BSA(m^2) = 0.121BW(kg)^{0.575}$ (Swindle et al., 2012). The background pathological findings in minipigs have been published and excellent references are available describing both common and infrequent findings (Jeppesen and Skydsgaard, 2015; Helke et al., 2016; McInnes and McKeag, 2016).

SAFETY PHARMACOLOGY

The marked similarity between Göttingen Minipigs and humans regarding the cardiovascular system extends into aspects important for safety pharmacology

TABLE 3.4 Electrocardiographic data in Göttingen Minipigs

Year	2013	2014	2015	2016	2017
Total Number of ECGs	270	385	282	557	361
Normal ECGs	249 (92.2)	359 (93.2)	270 (95.7)	521 (93.5)	346 (95.8)
Tall R wave (mild)	0 (0.0)	0 (0.0)	0 (0.0)	0 (0.0)	1 (0.3)
Tall R wave (marked)*	0 (0.0)	0 (0.0)	0 (0.0)	0 (0.0)	0 (0.0)
Deep S wave II, III, aVF (mild)	0 (0.0)	0 (0.0)	0 (0.0)	0 (0.0)	0 (0.0)
Deep S wave II, III, aVF (moderate)*	0 (0.0)	0 (0.0)	0 (0.0)	0 (0.0)	0 (0.0)
Deep S wave II, III, aVF (marked)*	0 (0.0)	0 (0.0)	0 (0.0)	0 (0.0)	0 (0.0)
Deep S wave (single lead)	0 (0.0)	0 (0.0)	0 (0.0)	0 (0.0)	1 (0.3)
Deep Q wave	0 (0.0)	0 (0.0)	0 (0.0)	0 (0.0)	1 (0.3)
Large T wave (mild)	2 (0.7)	4 (1.0)	0 (0.0)	1 (0.2)	0 (0.0)
Large T wave questionable	0 (0.0)	0 (0.0)	0 (0.0)	0 (0.0)	0 (0.0)
Notching of R wave	0 (0.0)	0 (0.0)	0 (0.0)	0 (0.0)	0 (0.0)
Notching of QRS wave	0 (0.0)	0 (0.0)	0 (0.0)	0 (0.0)	0 (0.0)
Mild ST segment depression/ elevation	0 (0.0)	0 (0.0)	0 (0.0)	0 (0.0)	0 (0.0)
First-degree heart block*	0 (0.0)	0 (0.0)	0 (0.0)	0 (0.0)	0 (0.0)
Second-degree heart block*	0 (0.0)	0 (0.0)	0 (0.0)	1 (0.2)	0 (0.0)
Third-degree heart block*	0 (0.0)	0 (0.0)	0 (0.0)	0 (0.0)	0 (0.0)
Premature atrial contractions*	16 (5.9)	20 (5.2)	6 (2.1)	27 (4.8)	10 (2.8)
Premature ventricular contractions*	2 (0.7)	0 (0.0)	0 (0.0)	0 (0.0)	1 (0.3)
Sinus bradycardia*	0 (0.0)	0 (0.0)	0 (0.0)	0 (0.0)	0 (0.0)
Sinus tachycardia	0 (0.0)	1 (0.3)	1 (0.4)	2 (0.4)	0 (0.0)
Small complexes	0 (0.0)	0 (0.0)	1 (0.4)	0 (0.0)	0 (0.0)
Right axis shift	0 (0.0)	0 (0.0)	0 (0.0)	0 (0.0)	0 (0.0)
Left axis shift	1 (0.4)	1 (0.3)	3 (1.1)	4 (0.7)	4 (1.1)
Tall P wave	0 (0.0)	0 (0.0)	1 (0.4)	0 (0.0)	0 (0.0)
Wide P wave	0 (0.0)	0 (0.0)	0 (0.0)	0 (0.0)	0 (0.0)
Short P-R interval	0 (0.0)	1 (0.3)	1 (0.4)	0 (0.0)	0 (0.0)
Long sinus pause	NR	NR	NR	1 (0.2)	0 (0.0)

Numbers in parentheses are percentages. * indicates abnormality of concern. Tracings generated by a Mortara ELI 250 Cardiograph (speed 50 mm/s; 10 mm/mV). Minipigs were evaluated in a sling using the Nehb-Spoerri lead method.
Data courtesy of Marshall BioResources, North Rose, New York, USA.

studies. Thus detailed studies of the ion channels in the heart and the repolarization process support the use of the Göttingen Minipig as nonrodent species in cardiovascular safety pharmacology studies (Bode et al., 2010; Laursen et al., 2011; Arlock et al., 2017). The background prevalence of arrhythmias and other ECG abnormalities is low, as can be seen from Table 3.4. As stated in the table, the most common arrhythmia is premature atrial contractions, but these arrhythmias often disappear in unstressed animals. To obtain reliable data, safety pharmacology studies should be performed using telemetry. It is possible to measure cardiovascular parameters without the use of telemetry in seemingly relaxed animals restrained in a sling and, in this way, obtain values that might be used for some research purposes (Cimini and Zambraski, 1985). However, the blood pressure and heart rate obtained in such a setting will usually be higher than for truly relaxed and unrestrained animals, and the variation in the data will also often be larger. Examples of telemetry implant placement and typical data obtained can be found in Ellegaard Göttingen Minipigs Newsletter Number 50 (www.minipigs.dk). It is also possible to simultaneously measure left ventricular pressure (Markert et al., 2009) or to use very small implants and amplify the signal by means of a jacketed system.

There are a few important considerations when designing safety pharmacology studies in minipigs. First, Göttingen Minipigs are often active for a few hours after feeding (depending on the manner in which they are fed), a fact that is reflected in the heart rate and blood pressure (Stubhan et al., 2008). This should be taken into consideration when planning the time of dosing and feeding. Second, body temperature has been found to have a marked influence on the QT interval in Göttingen Minipigs (Laursen et al., 2011), a fact that is especially important to consider in explorative studies in anesthetized animals and in studies of compounds which have an influence on body temperature. Finally, it is recommended that the heart rate dependency of the QT interval should be corrected individually (Stubhan et al., 2008).

IMMUNE SYSTEM

The importance of the pig as a livestock species, together with its long-standing role as a model for human infectious diseases and vaccine development (Meurens et al., 2012; Murtaugh, 2014), explains that the immune system in the pig is extremely well characterized and that there is a large toolbox available to study it. In the last decade, knowledge of the porcine immune system and the tools to study it have increased tremendously. Much is known about the porcine immune system, which in many aspects is quite like that in humans (Bode et al., 2010; Dawson et al., 2013, 2017a; Rubic-Schneider et al., 2016; Saalmüller and Gerner, 2016). Since no major differences between the immune system in pigs and minipigs have been reported so far, minipigs are thus well-suited for studies involving the immune system.

Structurally, a few aspects should be acknowledged. For example, pig lymph nodes have a unique "inverted" structure, with reversed cortical and medullary areas and lymph flow, resulting in lymphocytes leaving the lymph nodes via the high-endothelial venules rather than the medulla. The porcine mucosa-associated lymphoid tissue shows a high degree of similarity with other mammalian species including man. However, in pigs, Peyer patches can be seen in the jejunum with 1 long Peyer patch in the terminal ileum. The structure of the porcine thymus and the composition of T-cell progenitors are very similar to that of other mammalian species, albeit much of the thymus is located in the neck (Kuper et al., 2016). The thymus increases in size after birth, reaching the largest relative weight within the first month postnatally and the largest absolute weight 3–6 months postnatally (Kuper et al., 2016).

Regarding the innate immune system, detailed gene family analyses have revealed a large overall pig–human homology of pattern recognition receptors, and transcriptome analyses using next-generation sequencing revealed human-like responses to interferon (IFN)-γ and lipopolysaccharide (LPS) (Dawson et al., 2017a). In this context, it should be mentioned that LPS has surprisingly low potency in cynomolgus macaques (Picha et al., 2004). The major acute phase proteins in pigs and minipigs include C-reactive protein (CRP), major acute-phase protein, and haptoglobin (Heegaard et al., 2011; Christoffersen et al., 2015). Regarding the complement system, an aspect of significant importance regarding immunological safety is complement activation-related pseudoallergy caused by formulations containing Cremophor EL or liposomes. The Göttingen Minipig was recently shown to be a sensitive and reproducible animal model for regulatory testing for this (Jackman et al., 2016). It should also be mentioned that pigs differ from humans in the presence of a larger number of pulmonary intravascular macrophages in addition to the alveolar and interstitial macrophages (Bode et al., 2010). The body temperature in the Göttingen Minipig is usually in the range of 37.5–39.5°C but occasionally slightly higher or lower. However, above 40°C in an unstressed animal is usually a sign of disease (Moroni et al., 2014).

The adaptive immune responses in pigs and minipigs are also human-like, and in line with that pigs have been used in studies aimed at improving human vaccines against, for instance, tuberculosis and cancer (Bruffaerts et al., 2015; Overgaard et al., 2015). A major difference from humans includes that there are no maternally derived antibodies in newborn piglets; until they acquire them via the colostrum (Bode et al., 2010). It is also characteristic of pigs that they have high proportions of γδ T-cells and $CD4^+CD8^+$ T cells (Rubic-Schneider et al., 2016).

There is a large immunology toolbox available for studies of the porcine immune system. Tools are available for studying the humoral and the cellular immune response, allowing the identification of different porcine B- and T-cell subsets and their production of relevant cytokines, such as INF-γ, tumor necrosis factor α, and interleukin 17 (Talker et al., 2013; Summerfield et al., 2015; Bøje et al., 2016; Braun et al., 2017; Käser et al., 2017). In fact, flow cytometry tools have improved immensely in recent years, resulting in an increased understanding of porcine immune cells and their function. Furthermore, the genome is known in the Göttingen Minipig (Vamathevan et al., 2013), and a very large manually curated gene database is available for a vast number of porcine immune targets (Dawson et al., 2017b), allowing for instance multiplex qPCR. Also, recent developments have been made within next-generation swine leukocyte antigen (SLA) typing (Sørensen et al., 2017), and methods for recombinant antibody expression and tetramers for antigen-specific T-cell responses in swine have been developed (Baratelli et al., 2017). Thus it is now possible to determine which immune cells are involved in response to a specific pathogen, which immune modulators these cells produce, and into which memory cells they develop.

CONCLUDING REMARKS AND FUTURE DIRECTIONS

Göttingen Minipigs are well-defined genetically and microbiologically, available worldwide, and are very similar to humans in their anatomy and physiology. For these reasons, they are widely used as nonrodent species in many types of pharmacological and toxicological studies. Significant improvements in areas such as gene-editing techniques, clinical diagnostic modalities, and analytical methods will likely render Göttingen Minipigs even more useful for such studies in the future.

Gene-editing techniques have improved significantly in recent years with vast perspectives in terms of generating novel genetically engineered porcine models (Klymiuk et al., 2016; Redel and Prather, 2016). These improvements imply that new genetically altered models can be produced much easier and faster and with less off-target effects. One area of interest is the generation of platform models, as exemplified by the generation of minipigs with severe combined immunodeficiency and minipigs that spontaneously develop cancer (Schook et al., 2015; Redel and Prather, 2016). Another area could be generation of models tailored for the development of specific novel new medicines, such as by humanization regarding a specific target. A third area is the generation of improved disease models with translation potential to humans, allowing research into the disease mechanisms and the best treatment. Finally, the area of transplantation is evolving rapidly, with improvements in both humanization, treatments to prevent rejection, and methods (using the CRISPR-Cas9 gene-editing technique) that can reduce the risk of transmission of porcine endogenous retroviruses (Niu et al., 2017).

Improvements are continuously being made in clinical diagnostic modalities, and as minipigs have a suitable size, these techniques can often be used with no or minor modifications. For instance, an optimized technique for optical coherence tomography scanning of ocular segments was recently described in Göttingen Minipigs, and in the same article, a panel of tissue-based biomarkers of the eye were established (Atzpodien et al., 2016). Another example involving imaging is the use of echocardiography, which is easy to perform in minipigs, to monitor cardiac dimensions and function (van den Dorpel et al., 2017). More advanced imaging of atherosclerosis using intravascular ultrasound and positron emission tomography can also be performed, as described in Ellegaard Göttingen Minipigs Newsletter Number 46 (www.minipigs.dk). Another example involves the use of laser Doppler flowmetry to measure local blood flow (Poort et al., 2015). Finally, telemetry techniques are continuously improving and are now being used in many more areas than cardiovascular safety pharmacology studies. Such a technique was recently used to monitor the bladder function in female Göttingen Minipigs (Huppertz et al., 2015).

Analytical methods are also continuously improving. As mentioned in the section on the immune system, the genome is known in the Göttingen Minipig (Vamathevan et al., 2013), and a very large manually curated gene database is available for a vast number of porcine immune targets (Dawson et al., 2017b), allowing multiplex qPCR and expression analyses. The number of antibodies with documented good binding in the pig is also rapidly increasing, enabling advances in methods such as immune histochemistry, flow cytometry, and enzyme-linked immunosorbent assay (ELISA). Finally, in vitro and in silico methods are improving that will aid the selection of the most optimal nonrodent species for regulatory nonclinical toxicology studies. An example of this is the project aiming at establishing a panel of recombinant cytochrome P450 enzymes from Göttingen Minipigs, as described in Ellegaard Göttingen Minipigs Newsletter Numbers 47 and 50 (www.minipig.dk).

In summary, high-quality in vivo pharmacology and toxicology studies yielding reliable results can be performed in Göttingen Minipigs, and as discussed previously, the areas and issues that might be addressed using this species are expanding. A simple PubMed search for articles using the search term "Göttingen Minipigs" or synonyms thereof suggests that this occurs at an exponential rate.

References

Alstrup, A.K., 2016. Blood lactate concentrations in Göttingen Minipigs compared with domestic pigs. J. Am. Assoc. Lab. Anim. Sci. 55, 18–20.

Arlock, P., Mow, T., Sjöberg, T., et al., 2017. Ion currents of cardiomyocytes in different regions of the Göttingen minipig heart. J Pharmacol. Toxicol. Methods 86, 12–18.

Atzpodien, E.A., Jacobsen, B., Funk, J., et al., 2016. Advanced clinical imaging and tissue-based biomarkers of the eye for toxicology studies in minipigs. Toxicol. Pathol. 44, 398–413.

Ball, R.S., 2012. Husbandry and management. In: McAnulty, P.A., Dayan, A.D., Ganderup, N.C., Hastings, K.L. (Eds.), The Minipig in Biomedical Research. CRC Press, Boca Raton, FL, pp. 17–36.

Baratelli, M., Pedersen, L.E., Trebbien, R., et al., 2017. Identification of cross-reacting T-cell epitopes in structural and non-structural proteins of swine and pandemic H1N1 influenza A virus strains in pigs. J. Gen. Virol. 98, 895–899.

Bode, G., Clausing, P., Gervais, F., et al., 2010. The utility of the minipig as an animal model in regulatory toxicology. J Pharmacol. Toxicol. Methods 62, 196–220.

Bollen, P.J.A., Hansen, A.K., Alstrup, A.K.O., 2010. The Laboratory Swine, second ed. CRC Press, Boca Raton, FL (Chapter 2).

Bøje, S., Olsen, A.W., Erneholm, K., et al., 2016. A multi subunit Chlamydia vaccine inducing neutralizing antibodies and strong IFN-γ+ CMI responses protects against a genital infection in minipigs. Immunol. Cell Biol. 94, 185–195.

Braun, R.O., Python, S., Summerfield, A., 2017. Porcine B cell subset responses to toll-like receptor ligands. Front. Immunol. 8, 1044.

Bruffaerts, N., Pedersen, L.E., Vandermeulen, G., et al., 2015. Increased B and T cell responses in *M. bovis* Bacille Calmette-Guérin vaccinated pigs co-immunized with plasmid DNA encoding a prototype tuberculosis antigen. PLoS One 10, e0132288.

Christiansen, M.L., Müllertz, A., Garmer, M., et al., 2015. Evaluation of the use of Göttingen Minipigs to predict food effects on the oral absorption of drugs in humans. J. Pharmaceut. Sci. 104, 135–143.

Christoffersen, B., Golozoubova, V., Pacini, G., et al., 2013. The young Göttingen Minipig as model of childhood and adolescent obesity: influence of diet and gender. Obesity 21, 149–158.

Christoffersen, B.O., Gade, L.P., Golozoubova, V., et al., 2010. Influence of castration-induced testosterone and estradiol deficiency on obesity and glucose metabolism in male Göttingen Minipigs. Steroids 75, 676–684.

Christoffersen, B.O., Grand, N., Golozoubova, V., et al., 2007. Gender-associated differences in metabolic syndrome-related parameters in Göttingen Minipigs. Comp. Med. 57, 493–504.

Christoffersen, B.O., Jensen, S.J., Ludvigsen, T.P., et al., 2015. Age- and sex-associated effects on acute-phase proteins in Göttingen Minipigs. Comp. Med. 65, 333–341.

Cimini, C.M., Zambraski, E.J., 1985. Non-invasive blood pressure measurement in Yucatan miniature swine using tail cuff sphygmomanometry. Lab. Anim. Sci. 35, 412–416.

Colleton, C., Brewster, D., Chester, A., et al., 2016. The use of minipigs for preclinical safety assessment by the pharmaceutical industry: results of an IQ DruSafe minipig survey. Toxicol. Pathol. 44, 458–466.

Dalgaard, L., 2015. Comparison of minipig, dog, monkey and human drug metabolism and disposition. J. Pharmacol. Toxicol. Methods 74, 80–92.

Dawson, H.D., Loveland, J.E., Pascal, G., et al., 2013. Structural and functional annotation of the porcine immunome. BMC Genomics 14, 332.

Dawson, H.D., Smith, A.D., Chen, C., Urban Jr., J.F., 2017a. An in-depth comparison of the porcine, murine and human inflammasomes; lessons from the porcine genome and transcriptome. Vet. Microbiol. 202, 2–15.

Dawson, H.D., Chen, C., Gaynor, B., et al., 2017b. The porcine translational research database: a manually curated, genomics and proteomics-based research resource. BMC Genomics 18, 643.

Diehl, K.H., Hull, R., Morton, D., et al., 2001. A good practice guide to the administration of substances and removal of blood, including routes and volumes. J. Appl. Toxicol. 21, 15–23.

Directive 2010/63/EU of the European Parliament and of the Council of 22 September 2010 on the Protection of Animals Used for Scientific Purposes, September 22, 2010.

Ellegaard, L., Cunningham, A., Edwards, S., et al., 2010. Welfare of the minipig with special reference to use in regulatory toxicology studies. J. Pharmacol. Toxicol. Methods 62, 167–183.

Erkens, T., Van den Sande, L., Witters, J., et al., 2017. Effect of time and temperature on anticoagulant-dependent pseudothrombocytopenia in Göttingen Minipigs. Vet. Clin. Pathol. 46, 416–421.

Feyen, B., Penard, L., Van Heerden, M., et al., 2016. "All pigs are equal" does the background data from juvenile Göttingen Minipigs support this? Reprod. Toxicol. 64, 105–115.

Forster, R., Bode, G., Ellegaard, L., et al., 2010. The RETHINK project on minipigs in the toxicity testing of new medicines and chemicals: conclusions and recommendations. J. Pharmacol. Toxicol. Methods 62, 236–242.

Heegaard, P.M.H., Stockmarr, A., Piñeiro, M., et al., 2011. Optimal combinations of acute phase proteins for detecting infectious disease in pigs. Vet. Res. 42, 50. https://doi.org/10.1186/1297-9716-42-50.

Helke, K.L., Nelson, K.N., Sargeant, A.M., et al., 2016. Background pathological changes in minipigs: a comparison of the incidence and nature among different breeds and populations of minipigs. Toxicol. Pathol. 44, 325–337.

Higgins, P.J., Garlick, R.L., Bunn, H.F., 1982. Glycosylated hemoglobin in human and animal red cells. Role of glucose permeability. Diabetes 31, 743–748.

Huppertz, N.D., Kirschner-Hermanns, R., Tolba, R.H., Grosse, J.O., 2015. Telemetric monitoring of bladder function in female Göttingen Minipigs. BJU Int. 116, 823–832.

Jackman, J.A., Mészáros, T., Fülöp, T., et al., 2016. Comparison of complement activation-related pseudoallergy in miniature and domestic pigs: foundation of a validatable immune toxicity model. Nanomedicine 12, 933–943.

Jeppesen, G., Skydsgaard, M., 2015. Spontaneous background pathology in Göttingen Minipigs. Toxicol. Pathol. 43, 257–266.

Jorgensen, K.D., Ellegaard, L., Klastrup, S., Svendsen, O., 1998. Haematological and clinical chemical values in pregnant and juvenile Göttingen Minipigs. In: Svendsen, O. (Ed.), The Minipig in Toxicology, pp. 181–190. Scand. J. Lab. Animal Sci. 25 (Suppl. 1).

Käser, T., Pasternak, J.A., Delgado-Ortega, M., et al., 2017. Chlamydia suis and *Chlamydia trachomatis* induce multifunctional CD4 T cells in pigs. Vaccine 35, 91–100.

Klymiuk, N., Seeliger, F., Bohlooly, Y.M., et al., 2016. Tailored pig models for preclinical efficacy and safety testing of targeted therapies. Toxicol. Pathol. 44, 346–357.

Koch, W., Windt, H., Walles, M., et al., 2001. Inhalation studies with the Göttingen Minipig. Inhal. Toxicol. 13, 249–259.

Kreutz, R.P., Alloosh, M., Mansour, K., et al., 2011. Morbid obesity and metabolic syndrome in Ossabaw miniature swine are associated with increased platelet reactivity. Diabetes Metab. Syndr. Obes. 4, 99–105.

Kuper, C.F., van Bilsen, J., Cnossen, H., et al., 2016. Development of immune organs and functioning in humans and test animals: implications for immune intervention studies. Reprod. Toxicol. 64, 180–190.

Laursen, M., Olesen, S.P., Grunnet, M., et al., 2011. Characterization of cardiac repolarization in the Göttingen minipig. J. Pharmacol. Toxicol. Methods 63, 186–195.

Ludvigsen, T.P., Kirk, R.K., Christoffersen, B.O., Pedersen, H.D., et al., 2015. Göttingen Minipig model of diet-induced atherosclerosis: influence of mild streptozotocin-induced diabetes on lesion severity and markers of inflammation evaluated in obese, obese and diabetic, and lean control animals. J. Transl. Med. 13, 312. https://doi.org/10.1186/s12967-015-0670-2.

Mahl, A., Dincer, Z., Heining, P., 2016. The potential of minipigs in the development of anticancer therapeutics. Toxicol. Pathol. 44, 391–397.

Markert, M., Stubhan, M., Mayer, K., et al., 2009. Validation of the normal, freely moving Göttingen Minipig for pharmacological safety testing. J. Pharmacol. Toxicol. Methods 60, 79–87.

McInnes, E.F., McKeag, S., 2016. A brief review of infrequent spontaneous findings, peculiar anatomical microscopic features, and potential artifacts in Göttingen Minipigs. Toxicol. Pathol. 44, 338–345.

Meurens, F., Summerfield, A., Nauwynck, H., et al., 2012. The pig: a model for human infectious diseases. Trends Microbiol. 20, 50–57.

Monticello, T.M., Haschek, W.M., 2016. Swine in translational research and drug development. Toxicol. Pathol. 44, 297–298.

Moroni, M., Lombardini, E., Salber, R., et al., 2011. Hematological changes as prognostic indicators of survival: similarities between Göttingen Minipigs, humans, and other large animal models. PLoS One 6, e25210.

Moroni, M., Port, M., Koch, A., et al., 2014. Significance of bioindicators to predict survival in irradiated minipigs. Health Phys. 106, 727–733.

Münster, A.B., Olsen, A.K., Bladbjerg, E., 2002. Usefulness of human coagulation and fibrinolysis assays in domestic pigs. Comp. Med. 52, 39–43.

Murtaugh, M.P., 2014. Advances in swine immunology help move vaccine technology forward. Vet. Immunol. Immunopathol. 159, 202–207.

National Centre for the Replacement, Refinement & Reduction of Animals in Research, London, UK. www.nc3rs.org.uk.

Niu, D., Wei, H.J., Lin, L., et al., 2017. Inactivation of porcine endogenous retrovirus in pigs using CRISPR-Cas9. Science 357, 1303–1307.

Olsen, A.K., Bladbjerg, E.M., Jensen, A.L., Hansen, A.K., 2001. Effect of pre-analytical handling on haematological variables in minipigs. Lab. Anim. 35, 147–152.

Overgaard, N.H., Frøsig, T.M., Jakobsen, J.T., et al., 2017. Low antigen dose formulated in CAF09 adjuvant favours a cytotoxic T-cell response following intraperitoneal immunization in Göttingen Minipigs. Vaccine 35, 5629–5636.

Overgaard, N.H., Frøsig, T.M., et al., 2015. Establishing the pig as a large animal model for vaccine development against human cancer. Front. Genet. 6, 286.

Oyama, H., Horino, M., Matsumura, S., et al., 1975. Immunological half-life of porcine proinsulin c-peptide. Horm. Metab. Res. 7, 520–521.

Peter, B., DeRijk, E.P.C.T., Zeltner, A., Emmen, H.H., 2016. Sexual maturation in the female Göttingen Minipig. Toxicol. Pathol. 44, 482–485.

Petroianu, G.A., Maleck, W.H., Werth, W.A., et al., 1998. Blood coagulation, platelets and haematocrit in male, pregnant, and fetal Göttingen Minipigs. In: Svendsen, O. (Ed.), The Minipig in Toxicology, pp. 211–219. Scand. J. Lab. Animal Sci. 25 (Suppl. 1).

Picha, K.M., Tam, S.H., Khandekar, V.S., et al., 2004. Effects of lipopolysaccharide on toxicologically important endpoints in *Cynomolgus macaques*. Preclinica 2, 427–434.

Plum, A., Agersø, H., Andersen, L., 2000. Pharmacokinetics of the rapid-acting insulin analog, insulin aspart, in rats, dogs, and pigs, and pharmacodynamics of insulin aspart in pigs. Drug Metab. Dispos. 28, 155–160.

Poort, L.J., Bloebaum, M.M., Böckmann, R.A., et al., 2015. Assessment of local blood flow with laser Doppler flowmetry in irradiated mandibular and frontal bone, an experiment in Göttingen Minipigs. J. Craniomaxillofac. Surg. 43, 2071–2077.

Preuße, C., Skaanild, M.T., 2012. Minipigs in absorption, distribution, metabolism, and excretion (ADME) studies. In: McAnulty, P., Dayan, A.D., Ganderup, N.C., Hastings, K.L. (Eds.), The Minipig in Biomedical Research. Taylor & Francis, Boca Raton, FL, pp. 143–158.

Proksch, J.W., Driot, J.Y., Vandeberg, P., Ward, K.W., 2008. Nonclinical safety and pharmacokinetics of intravitreally administered human-derived plasmin in rabbits and minipigs. J. Ocul. Pharmacol. Ther. 24, 320–332.

Qvist, M.H., Hoeck, U., Kreilgaard, B., et al., 2000. Evaluation of Göttingen minipig skin for transdermal in vitro permeation studies. Eur. J. Pharm. Sci. 11, 59–68.

Raun, K., VonVoss, P., Knudsen, L.B., 2007. Liraglutide, a once-daily human glucagon-like peptide-1 analog, minimizes food intake in severely obese minipigs. Obesity 15, 1710–1716.

Redel, B.K., Prather, R.S., 2016. Meganucleases revolutionize the production of genetically engineered pigs for the study of human diseases. Toxicol. Pathol. 44, 428–433.

Rubic-Schneider, T., Christen, B., Brees, D., Kammüller, M., 2016. Minipigs in translational immunosafety sciences: a perspective. Toxicol. Pathol. 44, 315–324.

Saalmüller, A., Gerner, W., 2016. The immune system of swine. In: Hein, W., Gordon, J.R., Guidos, C.J., Rolink, A., Ratcliffe, M.J.H. (Eds.), Encyclopedia of Immunobiology, vol. 1. Academic Press, Oxford, pp. 538–548.

Schmidt-Nielsen, B., O'Dell, R., 1961. Structure and concentrating mechanism in the mammalian kidney. Am. J. Physiol. 200, 1119–1124.

Schomberg, D.T., Tellez, A., Meudt, J.J., et al., 2016. Miniature swine for preclinical modeling of complexities of human disease for translational scientific discovery and accelerated development of therapies and medical devices. Toxicol. Pathol. 44, 299–314.

Schook, L.B., Collares, T.V., Darfour-Oduro, K.A., et al., 2015. Unraveling the swine genome: implications for human health. Annu. Rev. Anim. Biosci. 3, 219–244.

Simianer, H., Köhn, F., 2010. Genetic management of the Göttingen Minipig population. J. Pharmacol. Toxicol. Methods 62, 221–226.

Smith, D., Hubrecht, R., 2001. Good Practice Guidelines. The Selection of Non-rodent Species for Pharmaceutical Toxicology. Laboratory Animal Science Association (LASA), Staffordshire, UK.

Sørensen, M.R., Ilsøe, M., Strube, M.L., et al., 2017. Sequence-based genotyping of expressed swine leukocyte antigen class I alleles by next-generation sequencing reveal novel swine leukocyte antigen class I haplotypes and alleles in Belgian, Danish, and Kenyan fattening pigs and Göttingen Minipigs. Front. Immunol. 8, 701.

Stubhan, M., Markert, M., Mayer, K., et al., 2008. Evaluation of cardiovascular and ECG parameters in the normal, freely moving Göttingen Minipig. J. Pharmacol. Toxicol. Methods 57, 202–211.

Subramanyam, M., Rinaldi, N., Mertsching, E., Hutto, D., 2008. Selection of relevant species. In: Cavagnaro, J.A. (Ed.), Preclinical Safety Evaluation of Biopharmaceuticals; a Science-Based Approach to Facilitating Clinical Trials. John Wiley & Sons, Hoboken, NJ, pp. 179–205.

Suenderhauf, C., Tuffin, G., Lorentzen, H., et al., 2014. Pharmacokinetics of paracetamol in Göttingen Minipigs: in vivo studies and modelling to elucidate physiological determinants of absorption. Pharm. Res. 31, 2696–2707.

Summerfield, A., Auray, G., Ricklin, M., 2015. Comparative dendritic cell biology of veterinary mammals. Annu. Rev. Anim. Biosci. 3, 533–557.

Swindle, M.M., 2007. Biology, handling, husbandry and anatomy. In: Swindle, M.M. (Ed.), Swine in the Laboratory: Surgery, Anesthesia, Imaging, and Experimental Techniques. CRC Press, Boca Raton, FL, pp. 1–33.

Swindle, M.M., Makin, A., Herron, A.J., et al., 2012. Swine as models in biomedical research and toxicology testing. Vet. Pathol. 49, 344–356.

Taberner, E., Navratil, N., Jasmin, B., et al., 2016. Pubertal age based on testicular and epididymal histology in Göttingen Minipigs. Theriogenology 86, 2091–2095.

Talker, S.C., Kaser, T., Reutner, K., et al., 2013. Phenotypic maturation of porcine NK- and T-cell subsets. Dev. Comp. Immunol. 40, 51–68.

Tsutsumi, H., 2012. Skeletal system. In: McAnulty, P., Dayan, A.D., Ganderup, N.C., Hastings, K.L. (Eds.), The Minipig in Biomedical Research. Taylor & Francis, Boca Raton, FL, pp. 283–292.

Vamathevan, J.J., Hall, M.D., Hasan, S., et al., 2013. Minipig and beagle animal model genomes aid species selection in pharmaceutical discovery and development. Toxicol. Appl. Pharmacol. 270, 149–157.

van den Dorpel, M.M.P., Heinonen, I., Snelder, S.M., et al., 2017. Early detection of left ventricular diastolic dysfunction using conventional and speckle tracking echocardiography in a large animal model of metabolic dysfunction. Int. J. Cardiovasc. Imaging. https://doi.org/10.1007/s10554-017-1287-8 [Epub ahead of print].

Van der Laan, J.W., Brightwell, J., McAnulty, P., et al., 2010. Regulatory acceptability of the minipig in the development of pharmaceuticals, chemicals and other products. J. Pharmacol. Toxicol. Methods 62, 184–195.

Weaver, M.L., Grossi, A.B., Schützsack, J., et al., 2016. Vehicle systems and excipients used in Minipig drug development studies. Toxicol. Pathol. 44, 367–372.

Webster, J., Bollen, B., Grimm, H., Jennings, M., 2010. Ethical implications of using minipig in regulatory toxicology studies. J. Pharmacol. Toxicol. Methods 62, 160–166.

Willard-Mack, C., Ramani, T., Auletta, C., 2016. Dermatotoxicology: safety evaluation of topical products in minipigs: study designs and practical considerations. Toxicol. Pathol. 44, 382–390.

Windt, H., Kock, H., Runge, F., et al., 2010. Particle deposition in the lung of the Göttingen Minipig. Inhal. Toxicol. 22, 828–834.

Yamamoto, S., Karashima, M., Sano, N., et al., 2017. Utility of Göttingen Minipigs for prediction of human pharmacokinetic profiles after dermal drug application. Pharm. Res. 34, 2415–2424.

Zheng, Y., Tesar, D.B., Benincosa, L., et al., 2012. Minipig as a potential translatable model for monoclonal antibody pharmacokinetics after intravenous and subcutaneous administration. mAbs 4, 243–255.

CHAPTER

4

Nonhuman Primates in Preclinical Research

Adam D. Aulbach, Daniel J. Patrick

Charles River Laboratories, Mattawan, MI, United States

INTRODUCTION

Nonhuman primates (NHPs) continue to be instrumental tools in biomedical research because of their anatomic, physiologic, and immunologic similarities to humans. In recent years, there has been an increased emphasis on translational discoveries enabling researchers to provide insights directly relevant to human patient populations. It is through translational discoveries such as next-generation renal biomarkers and advanced immunology assessments that NHP animal models provide their greatest contributions. The 2 most common NHPs used in biomedical research studies are the cynomolgus (*Macaca fascicularis*) and rhesus macaques (*Macaca mulatta*). Both species play a vital role in a wide range of studies ranging from preclinical pharmaceutical development to animal model research on metabolism, obesity, behavior, and aging. More recently the squirrel monkey (*Saimiri* spp.) and the African green monkey (*Cercopithecus aethiops*) have gained popularity in research because of their low cost and accessibility (Ward et al., 2008). Because cynomolgus macaques are by far the most common NHP used in preclinical safety assessment, the remainder of this chapter will primarily focus on this species unless otherwise specified.

The preclinical safety testing of biotherapeutics poses a particular challenge in selecting a relevant animal species for use in toxicology studies. Accurate prediction of the target effects of antibody-based therapeutics requires testing in a species which shows cross-reactivity, as these compounds are highly specific to their targets. Because of the restricted reactivity of many human-specific antibody compounds, NHPs are often the only available species for efficacy and safety studies (Jonker, 1990; Chapman et al., 2007). Furthermore, there is an increasing demand for the characterization of the immunotoxicologic effects of these compounds. Routine assessment of the immune system often includes the T-cell–dependent antibody response (TDAR) in which antibody responses to standardized antigens such as keyhole limpet hemocyanin and tetanus toxoid, are measured in NHPs treated with a new drug candidate. TDAR testing is commonly performed in NHPs and required by both the Food and Drug Administration and the Committee for Medicinal Products for Human Use for new drugs (Piccotti et al., 2005; Caldwell et al., 2007; Lebrec et al., 2011). Although similar assays are also commonly performed in rodents, the lack of cross-reactivity in rodent species and dissimilarity of the rodent immune system with that of humans often dictate the need for more applicable models such as NHPs (ICH S6 addendum, 2011).

CLINICAL PATHOLOGY IN NHPS

Recommendations of standard core testing for hematology, coagulation, clinical chemistry, and urinalysis in NHPs used in biomedical research studies have been published (Weingand et al., 1996; Tomlinson et al., 2013) and generally represent the consensus opinions of the major regulatory bodies and the international scientific community. Core testing panels are listed in Table 4.1 and do not generally differ from those required in other laboratory species. Several key guidance documents (FDA, 2003; OCED, 2009) have aided in harmonization of these practices which are aimed at providing key information on organ function/injury, pathophysiologic mechanisms, and overall health of the animals. Additional parameters or adjustments to the core panels may be necessary based on the type and duration of the study (e.g., acute, subacute, and chronic), type of compound being tested, and/or anticipated pathologic effects.

Depending on the volume desired, restraint method, and size of the animal, blood can be obtained from several venipuncture sites in NHPs. For large volume

Biomarkers in Toxicology, Second Edition
https://doi.org/10.1016/B978-0-12-814655-2.00004-9

TABLE 4.1 Standard Clinical Pathology Parameters in Nonhuman Primate Toxicology Studies

Hematology	Coagulation	Clinical Chemistry	Urinalysis
Erythrocytes (RBC)	aPTT	Sodium	Volume
Hemoglobin (Hgb)	Prothrombin time	Potassium	Specific gravity
Hematocrit (Hct)	Fibrinogen	Chloride	pH
MCV		Inorganic phosphorus	
MCHC		Calcium	
MCH		Alkaline phosphatase (ALP)	
Reticulocytes		Alanine aminotransferase (ALT)	
Platelets		Aspartate aminotransferase (AST)	
Leukocytes		Total bilirubin	
Neutrophils		Gamma-glutamyl transferase (GGT)	
Lymphocytes		Urea nitrogen	
Monocytes		Creatinine	
Eosinophils		Glucose	
Basophils		Total protein	
		Albumin	
		Globulin	
		Total cholesterol	
		Triglyceride	

aPTT, activated partial thromboplastin time; *MCHC*, mean cell hemoglobin concentration; *MCV*, mean cell volume.

collections, the femoral vein is most commonly used. For small samples, a sterile lancet may be used to puncture the marginal ear vein, a finger, or a heel. Although some animals may be trained to allow blood collections without the use of restraint, pharmacologic and manual restraint methods (e.g., squeeze cage and restraining chair) are commonly used for this purpose. Guidelines for maximum blood collection volumes vary somewhat by institution but are commonly in the range of 10 mL/kg in a 14-day period or up to 15% of the animal's total blood volume in a single collection. In addition, volumes of up to 20% may be collected over a 24-h period for the purposes of toxicokinetic evaluations (Diehl et al., 2001). Average total circulating blood volumes in NHPS range from approximately 60 to 70 mL/kg with minor species differences (Bernacky et al., 2002).

Primates are prone to eliciting a significant excitement/stress response during the handling and restraint associated with blood collection. Common stress effects on white blood cells include increases in total leukocyte, neutrophil, and/or lymphocyte counts although stress-induced decreases in lymphocyte and eosinophil counts are also common. Effects of stress on serum biochemical parameters may include increases in glucose and/or cortisol concentrations. Anesthetic agents are sometimes used for chemical restraint during prestudy or on-study examinations to minimize stress and facilitate handling; however, administration of anesthetics can have direct effects on clinical pathology values which must be considered (Kim et al., 2005). Increases in muscle injury enzyme activities (e.g., aspartate aminotransaminase [AST], creatine kinase [CK], lactate dehydrogenase [LDH], and/or alanine aminotransaminase [ALT]) can be seen as a result of intramuscular dosing of anesthetic agents not only because of minor muscle trauma but also directly from certain anesthetics agents, including ketamine (Aulbach et al., 2017). The effects of stress are often greatest during initial handling and restraint in newly acquired animals but tend to diminish as animals become conditioned to the processes. These features emphasize the importance of prestudy acclimatization and performing sham blood draws before study initiation to reduce the impact of these variables.

Preanalytical Variation

Because of their longevity and varied origins, NHPs used in biomedical research often have complex histories leading to the potential for substantial preanalytical variations among a host of clinical pathology endpoints. Variations because of age, gender, species, geographic origin, duration in captivity, and estrus status have all been identified. In particular, substantial variations in lymphocyte counts, total protein, globulin, and C-reactive protein (CRP) levels are often encountered among wild-caught NHPs because of their varied background and environmental exposure to antigens and naturally occurring disease conditions (Hall and Everds, 2003). Consideration should be given to the potential effects of these variables when interpreting clinical pathology data from NHPs.

Relative to other laboratory animal species, NHPs are often older and, as a result, have had exposure to differing husbandry, nutritional, and/or environmental practices. Examples of age-related effects on biochemical endpoints reported in NHPs include increases in urea nitrogen/creatinine in aged macaques (Kapeghian et al., 1984), reduced CRP levels in younger animals (Lamperez and Thomas, 2005), and progressive decreases in bone growth markers (e.g., bone-specific alkaline phosphatase [ALP]) with stabilization of elevated calcium concentrations, as younger animals reach skeletal maturity (Perretta et al., 1991). Variations among clinical pathology endpoints generally necessitate the establishment of both gender- and species-specific reference intervals although age stratification of historical data is usually unnecessary.

Subclinical infections with *Plasmodium* spp., unicellular parasites that cause malaria in their hosts, are an additional source of preanalytical variation that is not uncommonly encountered among wild-caught NHPs used in research studies. Malaria in NHPs imported from endemic areas has been reported to be as high as 43% in some regions (Ameri, 2010). Clinical or subclinical infections may confound data interpretation, especially in cases involving immunosuppression or immunotoxicity testing. Possible sequela of subclinical infections in NHPs includes periodic reductions in red cell mass associated with hemolysis, chronically elevated lymphocyte counts, and increases in serum globulins and/or total protein. Clinical disease can result in significant morbidity and/or mortality, especially in cases where the animals are immunosuppressed, stressed, or have had experimental surgical manipulation, including splenectomy (Sasseville and Mansfield, 2009). It is becoming increasingly more common to conduct polymerase chain reaction (PCR) testing for *Plasmodium* spp. prior importing NHP into the stock colony or testing facility.

Hematology

The importance of evaluating hematology in animal toxicology studies is emphasized by the often sensitive nature of the hematologic system to a wide range of pharmacologic compounds. Factors contributing to this sensitivity include the high mitotic rate of rapidly dividing hematopoietic cells, the short half-life of most blood cell types, and the systemic exposure of many test compounds. In addition to being a requirement among most of the major global regulatory agencies for preclinical work, effects on the hematologic system observed in preclinical safety studies have a high concordance rate (91%) with those observed in human clinical trials (Olson et al., 2000).

Guidelines for core hematology parameters have remained largely unchanged over recent years and are listed in Table 4.1 (Weingand et al., 1996; Tomlinson et al., 2013). Additional parameters obtained from modern hematology instruments such as the Advia 120/2120 (e.g., red cell distribution width (RDW), platelet distribution width (PDW), mean platelet volume (MPV), etc.) have recently gained favor and may be evaluated in conjunction with traditional endpoints to add supportive or mechanistic information to studies, which have identified the hematologic system as a target organ.

Red cell mass, if used to represent the overall circulating and functional red blood cell (RBC) volume, includes erythrocyte count (RBCs), hemoglobin concentration (HgB), and the calculated hematocrit value. All 3 of these individual analytes generally behave in parallel; however, HgB alone is often preferred as a stand-alone representative for red cell mass, as it is less likely to be artifactually influenced by analytical variables and its common use in human medicine. Mean cell volume (MCV) represents the average size of erythrocytes, whereas mean cell hemoglobin concentration (MCHC) represents the average hemoglobin concentration within circulating erythrocytes. Erythropoietin is the principal growth factor governing erythropoiesis and is produced by the peritubular interstitial fibroblasts of the kidney in response to low tissue oxygen tension (i.e., hypoxia). As tissue oxygenation decreases, erythropoietin will be released stimulating the bone marrow to produce more erythrocytes. Reticulocytes are young erythrocytes that are released in increased numbers from the bone marrow in response to increased demand. In such cases, an increased MCV with concurrent decrease in MCHC will usually correlate to the increase in reticulocyte release, which represents the increased cell size and relative decreased hemoglobin concentration seen in reticulocytes.

The most common cause of increased red cell mass in NHPs in preclinical safety studies is dehydration/hemoconcentration, which is a common sequela among animals that are anorexic, have gastrointestinal illness (e.g., diarrhea and vomiting), or otherwise having toxic effects that result in poor health (Hall and Everds, 2008). Increased red cell mass in these cases will usually be accompanied by supporting clinical evidence (e.g., increased skin tenting and sunken eyes), increases in total protein, electrolytes, and/or urea nitrogen and/or creatinine.

Direct toxic effects on the hematologic system more commonly result in reductions in red cell mass (anemia). These effects usually fall into 1 of 3 general categories, which include (1) blood loss, (2) increased erythrocyte destruction, or (3) decreased erythrocyte production. Evaluation of reticulocyte counts, erythrocyte morphologic abnormalities on blood smears, histopathologic data, and clinical observations, with consideration for the study design and changes in concurrent controls, will help determine the underlying cause of red cell mass changes. When evaluating conditions involving reductions in red cell mass, it is critical to determine to what extent the bone marrow is involved. This is accomplished by evaluating reticulocyte concentration to conclude whether there is an appropriate regenerative response (increases in reticulocytes). Cynomolgus monkeys typically have resting reticulocyte percentages of <2.5%, which generally correspond to absolute concentrations of 20,000–60,000 cells/μL (Drevon-Gaillot et al., 2006). In cases where there is an increased demand for erythrocytes, as in anemia, reticulocyte concentrations will rise within 2–3 days postinsult and can reach magnitudes of 100,000–300,000 cells/μL depending on the severity, timing, and cause of the toxic insult.

Blood loss anemia occurs as a result of internal (e.g., body cavity or gastrointestinal tract) or external (may include intestinal tract) hemorrhage as related to trauma, surgery, platelet/coagulation disorders, gastrointestinal toxicity, or cardiac and vascular insults. The key feature that supports the conclusion of a blood loss anemia is a moderate to strong regenerative response (increase in reticulocyte count). Platelets are also often reflexively increased in blood loss via either splenic contraction and/or because of the cross-reactivity between erythropoietin and thrombopoietin receptors (Stockham and Scott, 2008). Concurrent decreases of serum protein (e.g., total protein and albumin) are often seen in cases of overt blood loss. Iatrogenic blood loss is a common cause of decreased red cell mass in NHP studies, and is typically the result of repeated blood collections for toxicokinetic, pharmacodynamic, and/or biomarker analyses. Concurrent controls can serve as a comparator for the expected magnitude of red cell mass decreases and reticulocyte responses to help exclude a concurrent test article–related effect. Blood loss related to menses is also commonly seen in many female primate species on toxicity studies and can result in minimally to mildly lower red cell mass and increased reticulocyte counts. The magnitude of these effects is dependent on the duration of the menstrual phase, with the most pronounced changes seen in animals with menstrual phases lasting 7 days or more (Ochi et al., 2016).

Hemolysis is the early destruction of RBCs and may occur by direct lysis within the blood vessels or heart (intravascular) or by phagocytosis by cells of the mononuclear phagocyte system (extravascular; see Table 4.2). *Intravascular hemolysis* is relatively uncommon in toxicity studies but typically occurs secondary to red cell swelling or direct damage to the red cell membrane following intravenous administration of the test article. In vitro hemolysis testing can help identify test articles that are incompatible with blood and prevent the occurrence of intravascular hemolysis on toxicity studies. Intravascular hemolysis may also occur secondary to mechanical fragmentation of erythrocytes but is relatively uncommon in toxicology studies. Fragmented erythrocytes, including schistocytes, acanthocytes, and keratocytes, may be seen on blood smears. Lesions of the cardiovascular system, including vasculitis and endothelial damage, may accompany these findings. Alterations in erythrocyte membrane lipids may increase erythrocyte fragility and result in fragmentation of erythrocytes as well (Stockham and Scott, 2008).

Hemolysis on toxicity studies more commonly results from *extravascular hemolysis*, which occurs secondary to altered erythrocyte structure or membrane alterations. Abnormal erythrocytes are phagocytized by cells of the mononuclear phagocyte system as they pass through sinuses of the spleen, liver, or bone marrow.

TABLE 4.2 Characteristics of Intravascular Versus Extravascular Hemolysis

Characteristic	Intravascular Hemolysis	Extravascular Hemolysis
Site	Vasculature or the heart	Macrophages in the spleen, liver, and bone marrow
Timing	May be immediate	Typically following ≥2 doses
Clinical pathology findings	↓ RBC, ↑ reticulocytes ± Hemoglobinemia/uria ±↑ TBIL	Decreased red cell mass Increased reticulocyte counts Increased serum bilirubin (with increased urine bilirubin)
RBC morphology findings	*Mechanical fragmentation*: Acanthocytes, schistocytes, keratocytes	*Immune-mediated*: Spherocytes or agglutination *Oxidative injury*: Heinz bodies, eccentrocytes, and/or methemoglobinemia
Other findings	Hemoglobin in the renal tubule epithelial cells Local injection site irritation	Splenomegaly Erythrophagocytosis or iron pigment in the spleen, liver, or bone marrow

RBC, red blood cells; *TBIL*, total bilirubin.

TABLE 4.3 Classic Leukocyte Patterns

Physiologic Influence	WBC	Neutrophils	Lymphocytes	Monocytes	Eosinophils
Inflammation	↑↑	↑↑	±↑	±↑	–
Epinephrine	↑↑	↑↑	↑	–	–
Cortisol	↑	↑	↓	↑	↓
Marrow suppression	↓	↓	±↓	↓	↓
Lymphoid organ depletion	↓	–	↓↓	–	–

WBC, white blood cell.

Immune-mediated injury and oxidative injury are the 2 most common causes of extravascular hemolysis in toxicology studies. Immune-mediated hemolysis typically occurs as an idiosyncratic reaction after several doses of test article. These reactions may be unpredictable, lack a dose response, and occur in only a few animals on study. Large molecule drugs, those that undergo haptenization, and those that are administered via intravenous or intramuscular routes are more likely to be immunogenic (Renard and Rosselet, 2017).

Decreased erythrocyte production, or erythroid suppression, is the most common form of hematotoxicity observed in NHPs on preclinical studies. Direct bone marrow toxicity, which can be a common toxicity with high doses of many chemotherapeutic compounds, results in injury to the erythroid precursors in the bone marrow resulting in reduced reticulocyte production and eventual decrease in mature red cell mass. In short-duration (e.g., 7 days) studies, even complete destruction of erythroid precursors in the bone marrow may not result in meaningful effects on circulating red cell mass because of the substantial half-life of mature erythrocytes in NHPs (average 85–100 days; Moore, 2000). The key features in identification of erythroid suppression are reductions in reticulocytes and MCV in the face of reduced red cell mass. Correlative histologic findings in the bone marrow (e.g., decreased hematopoietic cellularity) are common in these cases.

Leukocyte or myeloid responses observed in NHPs in toxicology studies usually fall into one of several common categories, which include stress-induced (increased leukocytes), inflammation (increased leukocytes), myeloid suppression (decreased leukocytes), and lymphoid organ toxicity (decreased lymphocytes). Stress-induced changes on myeloid parameters occur as a result of concurrent effects of catecholamine (epinephrine/norepinephrine) and glucocorticoid release on circulating blood cells. These responses can be incited by such factors as blood collection, restraint, handling, solitary housing, fasting, and blood volume depletion among others. The classic "stress response" observed in most animal species is typically characterized by increases in neutrophil concentration and concurrent decreases in the concentrations of lymphocytes and eosinophils that result in an overall increase in total leukocytes (see Table 4.3). In NHPs, this response is more varied and usually results in increased neutrophils with increased or decreased lymphocyte concentrations (Bennett et al., 1992). As stated previously, stress leukograms in NHPs are commonly observed at the first and/or second hematology collections before dosing (i.e., pretest/baseline). These effects become diminished over time as animals become more habituated to blood collections and handling procedures (Aulbach et al., 2017).

Classic inflammatory changes in NHPs are characterized by increases in neutrophil counts, with or without increases in monocytes and lymphocytes, and/or a shift toward release of immature neutrophils (band cells) into circulation. Careful consideration must be given, as these findings often overlap or occur concurrently with stress-induced effects; hence, it is important to correlate these changes to clinical observations (lethargy, fever, lesions, etc.), microscopic findings of inflammation, and/or other signals of inflammation, including changes in acute phase proteins (e.g., fibrinogen, CRP, albumin). It should be noted that primary inflammatory effects of test compounds are not uncommon and will sometimes lack direct microscopic tissue correlates.

Decreased myeloid production, or myeloid suppression, is a common form of bone marrow toxicity observed in preclinical studies in NHPs. Similar to effects on erythroid precursors as described previously, direct bone marrow injury targeting the myeloid cell line results in reduced granulocyte (mostly neutrophils) production and eventual decrease in mature circulating neutrophils. If severe reductions in neutrophil counts occur (<1000 cells/μL) in NHPs, clinical immunosuppression and secondary bacterial infection and sepsis are likely to ensue.

Lymphoid organ effects are common in toxicity studies and are generally considered independent of bone marrow toxicity in NHPs. Although NHPs harbor substantial numbers of lymphocytes within the bone marrow, circulating and functional lymphocyte counts are not typically impacted by direct insults to the bone marrow alone (Reagan et al., 2011). Circulating lymphocyte counts often correlate with lymphoid organ effects, particularly those observed in the thymus, spleen, and lymph nodes (decreased organ weights and/or decreased cellularity). Direct lymphoid organ toxicity usually results in dose-responsive decreases in circulating lymphocyte concentrations without impacting neutrophil or other granulocyte counts (see Table 4.3).

Coagulation and Hemostasis

Platelets are produced by megakaryocytes in the bone marrow and constitute the third major cell line within the hematopoietic system. They are discussed in this section as they are centrally involved in the process of hemostasis, which also involves a series of complex interactions with coagulation, fibrinolytic factors, and blood vessels. Bone marrow toxicity affecting the megakaryocytic cell line typically results in reductions in circulating platelet concentrations. Clinical observations and biologically relevant effects are generally not observed until severe reductions are reached (e.g., <15,000 platelets/μL) and include skin and mucosal petechiae and ecchymoses, mucosal bleeding, hematuria, and hyphema (bleeding into the eye), which can be life-threatening (Stockham and Scott, 2008). Thrombopoietin is the primary cytokine regulating the proliferation and differentiation of megakaryocyte progenitor cells and will increase secondary to decreased platelet mass. Platelet numbers in circulation vary considerably by species and reflect a balance of production, destruction, consumption, and redistribution to tissues like the spleen. Overall, platelets circulate for approximately 5–9 days in most species (Russell, 2010). *Decreased platelet counts* may occur from decreased or defective production by megakaryocytes or myelosuppression. Because of the circulating lifespan of platelets, low platelet counts resulting from decreased production will not be immediately apparent and are typically noted 1–2 weeks following induction of myelosuppression.

Activated partial thromboplastin time (aPTT) and prothrombin time (PT) are the standard assays used in the NHP to evaluate functional coagulation ability in preclinical safety studies (FDA, 2003; OCED, 2009). For evaluation of PT and aPTT, whole blood should be collected in trisodium citrate anticoagulant (3.2% or 3.8%) at a 9:1 blood to anticoagulant ratio. Prolongations in aPTT and PT on toxicology studies may reflect decreased production of coagulation factors by the liver (hepatic disease), increased coagulation factor inactivation or consumption (secondary to localized or disseminated coagulation), dilution of coagulation factors with excessive administration of plasma-poor fluids, or inhibition of coagulation factors by drugs, antibodies, or fibrin-degradation products. Spurious prolongations in aPTT and/or PT may occur with difficult sample collection, poor clinical condition of the animal, or with an inappropriate citrate ratio. Alterations in coagulation times should be interpreted with knowledge of the assay system and reagent set. Assay-to-assay variation can be quite substantial requiring background knowledge of historical data from the particular laboratory being used and/or the use of concurrent control animals. Even a minimally prolonged coagulation time usually represents an approximately >70% reduction in the factors involved in the assay; hence substantial effects on the biological system occurred before any effect on assay results are seen. This phenomenon is confounded by the fact that in many cases, these findings will occur in the absence of clinical observations or microscopic lesions, a reflection of the coagulation system's redundant functional capacity. Interfering substances causing false artifactual effects on coagulation tests have been documented and should be considered when effects occur in the absence of other correlative findings. Pegylated protein compounds have been shown to artifactually prolong certain aPTT reagent sets (Aulbach et al., 2012).

Fibrinogen is a positive acute-phase protein synthesized by the liver and is found in the plasma and platelet granules. Fibrinogen plays a critical role in hemostasis and is consumed during coagulation when fibrinogen is converted fibrin. Fibrinogen is also an acute phase protein, meaning its levels will increase under inflammatory conditions. For this reason, fibrinogen is a useful marker of inflammation in NHPs, and is an important link between inflammation and hemostasis. Increased fibrinogen concentration can result in shortening of PT and aPTT; however, because inflammation and coagulation occur over a continual spectrum of severity, and inflammation-induced activation of coagulation may eventually advance to hypocoagulation (Kurata et al., 2003), prolongations or shortening of coagulation times may be observed in conjunction with clinical and microscopic evidence of inflammation on toxicology studies.

Clinical Biochemistry

Key clinical chemistry analytes evaluated in NHPs are listed in Table 4.4 (Weingand et al., 1996; Tomlinson et al., 2013). This collection of assays is aimed at identifying major effects on the renal, hepatobiliary, metabolic, and fluid/electrolyte regulating systems. Additional markers of cardiac injury (e.g., cardiac troponin I [cTnI]), muscle injury, inflammation (e.g., acute phase

TABLE 4.4 Summary of Hepatocellular and Hepatobiliary Biomarkers

Hepatocellular Injury Markers	Hepatobiliary Injury Markers
Alanine aminotransferase (ALT)	Total bilirubin (TBIL)
Aspartate aminotransferase (AST)	Alkaline phosphatase (ALP)
Sorbitol dehydrogenase (SDH)	Gamma-glutamyltransferase (GGT)
Glutamate dehydrogenase (GLDH)	Total bile acids (TBA)
	5′ nucleotidase (5NT)
INDICATORS OF ALTERED HEPATIC FUNCTION	
↑ Total bilirubin (TBIL)	↓ Glucose (GLU)
↑ Total bile acids (TBA)	↑ Coagulation times (aPTT and PT)
↓ Albumin (ALB)	↓ Fibrinogen
↓ Urea nitrogen	↓ Cholesterol

proteins), hormone characterization, and immunotoxicity (e.g., immunoglobulins and cytokines) may be added to this panel as the weight-of-evidence dictates.

Electrolyte evaluations typically measure concentrations of sodium, chloride, potassium, and phosphorus and are used as indicators of fluid and electrolyte balance, hydration status, acid/base status, renal function, gastrointestinal health, and even endocrine function. It is imperative to integrate electrolyte information with other clinical pathology endpoints and clinical observations before making final conclusions, as many cases involving electrolyte alterations involve multiple interrelated processes. Increases in sodium and chloride are consistent with dehydration; however, this should be substantiated by identifying concurrent increases in protein values, red cell mass, and/or markers of reduced glomerular filtration rate such as increased urea nitrogen, creatinine, and phosphorus. Reductions in sodium and chloride are common in cases of severe gastrointestinal fluid losses (i.e., vomiting and watery feces), diuresis, and/or prolonged polyuria (Stockham and Scott, 2008). The most common cause of electrolyte losses in NHPs is generally secondary to gastrointestinal distress/toxicity.

Urea nitrogen and creatinine are widely used in NHPs to assess renal function and represent the traditional method for detecting renal injury in toxicology studies. As in humans, extensive renal injury can occur before meaningful elevations in urea nitrogen and creatinine are observed, which makes them less sensitive than some of the newer biomarkers of renal injury discussed in greater detail elsewhere in this text (e.g., kidney injury marker 1 (KIM-1), neutrophil gelatinase-associated lipocalin (NGAL), cystatin C, microglobulins). Renal biomarkers generally have been classified into several groups indicating their association with injury to a specific part of the nephron (e.g., proximal tubule, glomerulus, and collecting duct). Recent work suggests that there is a fair amount of overlap and redundancy among individual biomarkers in regards to their specific nephron segment; hence classifying biomarker effects to either the glomerulus or the tubules appears to be the most appropriate approach in most cases (Ennulat and Adler, 2015). The magnitude, timing, and sensitivity of positive signals among specific biomarkers will differ between test compounds and between assay methodologies and laboratories; hence the selection of specific biomarkers should be made with consideration of assay availability, species, anticipated pathophysiology, and analyte stability (Vlasakova et al., 2014). Any effects observed on renal injury markers should be correlated to microscopic findings in the kidneys to make definitive conclusions as effects on clinical pathology endpoints without histologic evidence of injury may indicate submicroscopic or even nonrenal causes (e.g., dehydration). Given the high sensitivity of many of the newer renal injury biomarkers, it is not uncommon to see signals among biomarker data in the absence of histologic effects. Under these circumstances, effects are generally considered submicroscopic or molecular in nature without meaningful biologic or toxicologic relevance.

Global regulatory guidelines dictate that for the detection of hepatocellular injury in nonclinical animal studies, a minimum of 2 or 3 of the following tests be included in clinical chemistry panel: ALT, AST, sorbitol dehydrogenase (SDH), glutamate dehydrogenase (alternatively known as GDH), and total bile acids (FDA, 2003; OCED, 2009). For the detection of hepatobiliary injury, at least 2 or 3 of the following should be included: ALP, total bilirubin (TBIL), gamma-glutamyltransferase (GGT), 5′ nucleotidase, and total bile acids. Many of the clinical pathology endpoints used in nonclinical toxicology studies to characterize effects on the liver are cell leakage markers that are indicators of liver tissue injury (e.g., ALT and SDH), whereas others are inducible markers of biliary stasis/injury (e.g., ALP, bilirubin, and GGT). The sensitivity of ALP as an indicator of cholestasis and biliary injury in NHPs is much poorer than in many other laboratory species (e.g., dog); hence it is important to corroborate with histologic findings and avoid being too dogmatic about ALP signals in NHPs. In macaques, GGT has been described as a more sensitive and specific marker for biliary disease than ALP (Hall and Everds, 2003). Bilirubin metabolism in NHPs, measured as TBIL, closely resembles that of humans and is an additional marker for hepatobiliary injury and function. Insults to the hepatobiliary system will often result in signals in multiple hepatobiliary markers (e.g., GGT, ALP, ALT, and TBIL). Both hepatic injury and muscle injury can cause elevations in ALT and AST because of their tissue distribution in the liver, both skeletal and heart muscle, and the kidney (Mohr et al., 1971). Mild to moderate elevations in ALT and

AST of skeletal muscle origin are commonly seen in NHP species following normal chemical restraint (Kim et al., 2005). Consequently, caution must be taken in attributing elevations of ALT and AST to hepatic injury when concurrent elevations in CK and/or LDH are present. It is important to correlate microscopic alterations in the liver or muscle to elevations in these injury markers before making final conclusions about the significance of such findings, as elevations in ALT and AST in the absence of microscopic tissue alterations are not uncommon (Boone et al., 2005).

Urinalysis

Urinalysis is typically included as part of the minimum required database for nonclinical toxicology studies in most species. Recommended core tests include determination of urine physical characteristics (color, clarity, and volume) and determination of specific gravity or osmolality, pH, total protein, glucose, blood, ketones, and bilirubin (FDA, 2003; OCED, 2009). Additional testing, including creatine clearance, protein-to-creatinine ratio, fractional excretion, or renal injury biomarkers, may be necessary if the compound is a known renal toxicant or a more thorough characterization of renal effects is desired.

Urinalysis data are heavily influenced by collection technique. The challenges associated with collecting appropriate urine samples from large numbers of animals on toxicology studies are not trivial. Most commonly, voided urine is collected into a pan system over a defined period of time (typically 16 h) because of cost and labor-intensive nature of other collection techniques. This method, however, is prone to inaccurate results because of contamination by other material collected in the pan, including food, water, feces, vomitus, hair, and residual cleaning solution. Various methods (collection on ice, fasting, gauze/mesh filters, etc.) can be used in nonclinical toxicology studies to minimize urine contaminants as much as possible. Contaminates may lead to a variety of false-positive results, including effects on urine volumes, glucose, protein, occult blood, and microscopic artifacts (Aulbach et al., 2015; Siska et al., 2017). Institutions should make operational efforts to minimize the frequency of these collection variables.

BACKGROUND MICROSCOPIC LESIONS IN NHPS

In preclinical research drug safety evaluation, one of the most important factors in the accurate determination of test article–related effects is an awareness and solid understanding of all potential spontaneous microscopic changes (changes that are not induced by the test article) in the particular animal model. These spontaneous changes are also referred to as "incidental" or "background" changes. It is important to note that these background changes may arise during the study or they can be preexisting, and they can also be exacerbated (or even sometimes diminished) by a procedure or test article. Determining the relationship of the test article to these changes often relies on the experience of the pathologist, the features and biology of the effect, how the severity and incidence of the change vary across groups, other study data, and historical control data. The importance of historical control data in NHP studies cannot be underestimated, especially because these studies typically have small numbers of animals (usually 3–4) in each group, which increases the possibility that a background change will only be present in the treated animals and absent in controls. Possibly representing one of the unwritten Murphy's-related toxicology laws, a background change will often only be present in the high-dose group, which can greatly confound the interpretation of the study and the accurate determination of spontaneous versus drug-induced findings. The following sections will provide a short description of some of the more notable background lesions in cynomolgus macaques, both from a preclinical research perspective and those involving "biomarker friendly" tissues such as the heart, liver, kidneys, and skeletal muscle. It is important to note that there is no generic NHP, and the background lesions can vary greatly between species, strain, and geographic origin.

Mononuclear Cell Infiltrates

Mononuclear cell infiltrates, also sometimes referred to as lymphocytic or lymphoplasmacytic infiltrates, are by far the most common background lesion in cynomolgus macaques. These infiltrates are typically comprised of focal aggregates of lymphocytes, plasma cells, and macrophages within a tissue, with little or no evidence of damage to the surrounding parenchyma (such as renal tubules, gastric glands, cardiac myocytes, or hepatocytes in the kidney, stomach, heart, and liver, respectively). Because of the typical lack of associated parenchymal damage, it is highly unlikely that these infiltrates would be detectable with current commonly available biomarkers. It has been demonstrated that serum cTnI concentrations are not increased in cynomolgus monkeys with myocardial mononuclear cell infiltrates, and thus serum cTnI does not appear to be a biomarker that can be used to screen cynomolgus macaques for this spontaneous cardiac finding (Dunn et al., 2012). Mononuclear cell infiltrates are of unknown cause and typically observed in many cynomolgus macaque tissues, with the most commonly affected sites being the liver (61%), kidneys (29%), heart (26%), salivary glands (21%), and stomach (12%) (Chamanza et al., 2010). There are variations in the degree of mononuclear cell infiltrates between cynomolgus monkeys from different sources. When compared with Mauritian monkeys, cynomolgus monkeys from Southeast Asia (The Philippines and Vietnam) have been reported to have increased mononuclear cell infiltrates, which correlated with an increased white blood cell count (Drevon-Gaillot et al., 2006). Mononuclear cell infiltrates are also present in the uvea of the eyes of cynomolgus macaques in approximately 25% of drug safety evaluation studies and are not associated with other ocular histopathologic or ophthalmoscopic findings (Chamanza et al., 2010; Sinha et al., 2006). The most common site of mononuclear cell infiltrates is the liver, and these infiltrates can be present diffusely throughout the parenchyma or within periportal regions. The liver infiltrates are usually minimal in magnitude and can occasionally be associated with minimal focal hepatocellular necrosis. Typically, clinical chemistry changes (e.g., increased ALT or AST) are not associated with hepatic mononuclear cell infiltrates (Chamanza et al., 2010; Sato et al., 2012).

Mononuclear cell infiltrates in the heart can be of particular concern because they can be more difficult to distinguish from inflammation, can be associated with myocyte injury, and are also one of the primary histopathologic lesions associated with *Trypanosoma cruzi* infection (Chagas disease). *T. cruzi* protozoal amastigotes are often difficult to detect histologically, precluding a definitive microscopic diagnosis. Mononuclear cell infiltrates within the stomach (sometimes referred to as chronic gastritis) are one of the most common background findings in cynomolgus macaques, with recorded incidences of 100% affected animals reported in nearly half of the studies. These lesions are not usually associated with clinical signs, and *Helicobacter pylori* and *Helicobacter heilmannii*–like organisms are frequently observed in both sections with mononuclear cell infiltrates and in normal-appearing sections (Brady and Carville, 2012; Drevon-Gaillot et al., 2006; Sato et al., 2012).

Cardiovascular

Various heart muscle (myocardial) background lesions in cynomolgus macaques, such as degeneration/necrosis, hemorrhage, and fibrosis, can mimic test-article-related cardiac toxicity. Cardiomyopathy is characterized by cardiomyocyte disarray with cytoplasmic pallor to stippling and increased nuclear size (karyomegaly), fibrosis, and mononuclear cell infiltration. The disrupted cardiomyocytes are immunoreactive to desmin and cTnI antibodies and have a normal cross-striation pattern by phosphotungstic acid-hematoxylin (PTAH), indicating the chronic cardiomyopathy is not associated with active cardiomyocyte damage (Zabka et al., 2009). Myocardial degeneration and necrosis, which in advanced cases is also associated with inflammation (myocarditis) and fibrosis or mineralization, has also been reported as an idiopathic heart background lesion in cynomolgus macaques (Chamanza et al., 2006, 2010; Drevon-Gaillot et al., 2006; Keenan and Vidal, 2006). Minimal to mild transient increases in cTnI can be detected in monkeys after oral, subcutaneous, or intravenous dosing of common vehicles, as well as serial chair restraint for venipuncture blood collection assays (Reagan et al., 2017). A higher incidence and severity of myocardial degeneration/necrosis, subendocardial hemorrhage with hemosiderin, myocardial fibrosis, and arterial medial degeneration/hemorrhage have been reported in Mauritian-source cynomolgus macaques (Vidal et al., 2010). Background vascular lesions include accumulation of mucopolysaccharides in the intima of the aorta, subendocardial areas of the base of the heart, and coronary arteries (most commonly without lipid accumulation, although occasionally atherosclerotic lesions have also been observed). Vasculocentric inflammation, which either involves the wall of vessels (vasculitis, arteritis) or tissues surrounding vessels (perivascular inflammation, perivasculitis, periarteritis), can be observed spontaneously in cynomolgus macaques and is usually only observed in one or a few tissues, with the most common sites being the heart, intestines, and epididymis (Chamanza et al., 2006, 2010; Sato et al., 2012). A spontaneous vasculitis that involves more tissues and includes vascular necrosis, resembling polyarteritis nodosa in humans, has been observed in cynomolgus macaques (Albassam et al., 1993; Porter et al., 2003). Background vasculocentric lesions can be extremely problematic in toxicology because similar-appearing drug-induced vascular lesions can be an incredibly serious toxicity and can pose a significant challenge to the advancement of the compound into humans. Periarteritis in the lungs, usually with edema, can be associated with continuous intravenous infusions.

Gastrointestinal and Hepatic

Acute inflammation of the stomach (acute gastritis) occurs at a much lower incidence than chronic mononuclear cell inflammatory infiltrates and is often associated with hemorrhage, erosions, ulceration, and/or glandular microabscesses (Chamanza et al., 2010). Serum pepsinogen I has been reported as a possible noninvasive biomarker for detection of gastric mucosal injury in monkeys (Ennulat et al., 2017). Gastric ulceration is uncommon, usually subclinical, and can be associated with *Helicobacter* infection; however, when the ulceration is severe, the animals may present with anorexia, weight loss, lethargy, and vomiting of blood. Upper gastrointestinal tract hemorrhage can be associated with vomiting, regenerative anemia, and an elevated urea nitrogen. Background lesions in the intestine are uncommon but can occasionally be associated with diarrhea and diarrhea-related clinical pathologies such as dehydration, electrolyte disturbances, hemoconcentration, and, if associated with hemorrhage, hypoproteinemia, and anemia. It is important to note that diarrhea in cynomolgus macaques is usually of unknown cause (Chamanza et al., 2010). Amyloid can be deposited within the intestines (amyloid enteropathy) or systemically throughout numerous tissues, most commonly the liver and spleen. Intestinal amyloidosis can cause protein loss into the intestinal lumen (protein losing enteropathy) and can be associated with hypoalbuminemia.

Renal

Interstitial inflammation of the kidneys (interstitial nephritis or tubulointerstitial nephritis) consists primarily of infiltrates of lymphocytes, plasma cells, and small numbers of neutrophils and can be associated with tubular regeneration, interstitial fibrosis, intraepithelial pigment, and mineralization (Chamanza et al., 2010; Sato et al., 2012). Kidney glomerular lesions can include focal membranous and proliferative glomerulonephritis and glomerulosclerosis (Drevon-Gaillot et al., 2006; Sato et al., 2012). There are usually no gross pathologic findings or systemic disease associated with these glomerular or interstitial changes (Chamanza et al., 2010).

Skeletal Muscle

Background lesions in skeletal muscle are rare but important to recognize, as they need to be separated from drug-induced skeletal muscle injury. Biomarkers of muscle injury include alanine aminotransferase (ALT), aspartate aminotransferase (AST), CK, and/or LDH. Some of the more common non–drug-related background lesions in cynomolgus macaques include focal accidental trauma, vaccination-related inflammation, and infection with *Sarcocystis* spp. coccidian parasites (Chamanza et al., 2010). Sarcocystosis is usually not associated with other lesions; however, occasionally, cases exhibit inflammation characterized by infiltrates of lymphocytes, plasma cells, and eosinophils, as well as myofiber degeneration and fibrosis.

CONCLUDING REMARKS AND FUTURE DIRECTIONS

The science of toxicology will undoubtedly continue to evolve as scientists look to improve on the current approach of assessing the potential risk to human health. There is substantial social, economic, and scientific pressure to find ways to minimize the use of animals in preclinical research, particularly the use of NHPs. Toxicogenomics, proteomics, and metabolomics are just a few of the many exciting technologies with potential to revolutionize current approaches to preclinical toxicity testing. The success of these efforts will likely drive the next generation of toxicologists away from such heavy reliance on animal test systems.

In this chapter, we largely focused our discussion on traditional markers of toxicity and pathology. Although many of the new testing modalities are promising, it is important to remember that novel biomarkers and new testing methodologies must be benchmarked over time against well-characterized gold standards before they can be relied on for routine use. What good is a new renal injury biomarker if its relation to microscopic pathology or clinical pathology assessments are not known? Moving forward, we need to pursue new technologies without losing sight of traditional toxicology principles, which will continue to play a central role in toxicology assessments for the foreseeable future.

References

Albassam, M.A., Lillie, L.E., Smith, G.S., 1993. Asymptomatic polyarteritis in a cynomolgus monkey. Lab. Anim. Sci. 43, 628–629.

Aulbach, A., Cirino, K., Holland, K., Denham, S., 2012. Interference of activated partial thromboplastin time assays by polyethylene glycol conjugated protein compounds. In: Poster Abstract, 31st Annual STP Meeting.

Aulbach, A., Schultze, E., Tripathi, N., et al., 2015. Factors affecting urine reagent strip blood results in dogs and nonhuman primates and interpretation of urinalysis in preclinical toxicology studies: a Multi-Institution Contract Research Organization and BioPharmaceutical Company Perspective. Vet. Clin. Pathol. 44 (2), 229–233.

Aulbach, A., Provencher, A., Tripathi, N., 2017. Influence of study design variables on clinical pathology data. Toxicol. Pathol. 45 (2), 288–295.

Ameri, M., 2010. Laboratory diagnosis of malaria in nonhuman primates. Vet. Clin. Pathol. 39, 5–19.

Bennett, J., Gossett, K., McCarthy, M., et al., 1992. Effects of ketamine hydrochloride on serum biochemical and hematologic variables in rhesus monkeys (*Macaca mulatta*). Vet. Clin. Pathol. 21 (1), 15–18.

Bernacky, B., Susan, V., Gibson, M., Michale, E., Keeling, E., Abee, C., 2002. Nonhuman primates. In: Fox, J.G., Anderson, L.C., Loew, F.M., Quimby, F.W. (Eds.), Laboratory Animal Medicine. Academic Press, San Diego, pp. 676–791.

Boone, L., Meyer, D., Cusick, P., et al., 2005. Selection and interpretation of clinical pathology indicators of hepatic injury in preclinical studies. Vet. Clin. Pathol. 34, 182–188.

Brady, A.G., Carville, A.A.L., 2012. Digestive system diseases of nonhuman primates. In: Abee, C.R., Mansfield, K., Tardif, S.D., Morris, T. (Eds.), Nonhuman Primates in Biomedical Research, Diseases (American College of Laboratory Animal Medicine), second ed., vol. II. Academic Press, New York, pp. 589–628.

Caldwell, R., Guirguis, M., Kornbrust, E., 2007. Evaluation of various KLH dosing regimens for the cynomolgus monkey TDAR assay. In: Abstract 1723, SOT Annual Meeting.

Chamanza, R., Parry, N.M., Rogerson, P., et al., 2006. Spontaneous lesions of the cardiovascular system in purpose-bred laboratory nonhuman primates. Toxicol. Pathol. 34, 357–363.

Chamanza, R., Marxfeld, H.A., Blanco, A.I., et al., 2010. Incidences and range of spontaneous findings in control cynomolgus monkeys (*Macaca fascicularis*) used in toxicity studies. Toxicol. Pathol. 38, 642–657.

Chapman, K., Pullen, N., Graham, M., et al., 2007. Preclinical safety testing of monoclonal antibodies: the significance of species relevance. Nat. Rev. Drug Discov. 6, 120–126.

Diehl, K., Hull, R., Morton, D., et al., 2001. A good practice guide to the administration of substances and removal of blood, including routes and volumes. J. Appl. Toxicol. 21, 15–23.

Drevon-Gaillot, E., Perron-Lepage, M., Clement, C., et al., 2006. A review of background findings in cynomolgus monkeys (*Macaca fascicularis*) from three different geographical origins. Exp. Toxicol. Pathol. 58, 77–88.

Dunn, M.E., Coluccio, D., Zabka, T.S., et al., 2012. Myocardial mononuclear cell infiltrates are not associated with increased serum cardiac troponin I in cynomolgus monkeys. Toxicol. Pathol. 40, 647–650.

Ennulat, D., Adler, S., 2015. Recent successes in the identification, development, and qualification of translational biomarkers: the next generation of kidney injury biomarkers. Toxicol. Pathol. 43, 62–69.

Ennulat, D., Lynch, K., Kimbrough, C., et al., 2017. Evaluation of pepsinogen I as a biomarker of drug-induced gastric mucosal injury in cynomolgus monkeys. Toxicol. Pathol. 45 (2), 296–301.

Food and Drug Administration (FDA), 2003. General Guidelines for Designing and Conducting Toxicity Studies. Redbook 2000: Center for Food Safety and Applied Nutrition; IV.B.1.

Hall, R.L., Everds, N.E., 2003. Factors affecting the interpretation of canine and nonhuman primate clinical pathology. Toxicol. Pathol. 31, 6–10.

Hall, R.L., Everds, N.E., 2008. Principles of clinical pathology for toxicology studies. In: Principles and Methods of Toxicology, pp. 1317–1354.

International Conference on Harmonisation (IHC), 2011. Guidance for Industry S6 Preclinical Safety Evaluation of Biotechnology-Derived Pharmaceuticals, Addendum.

Jonker, M., 1990. The importance of nonhuman primates for preclinical testing of immunosuppressive monoclonal antibodies. Semin. Immunol. 2 (427), 436.

Kapeghian, J.C., Bush, M.J., Verlangieri, A.J., 1984. Changes in selected serum biochemical and Ekg values with age in cynomolgus macaques. J. Med. Primatol. 13 (5), 283–288.

Keenan, C.M., Vidal, J.D., 2006. Standard morphologic evaluation of the heart in the laboratory dog and monkey. Toxicol. Pathol. 34, 67–74.

Kim, C., Lee, H., Han, S., et al., 2005. Hematological and serum biochemical values in cynomolgus monkeys anesthetized with ketamine hydrochloride. J. Med. Primatol. 34 (2), 96–100.

Kurata, M., Sasayama, Y., Yamasaki, N., et al., 2003. Mechanism for shortening PT and APTT in dogs and rats – effect of fibrinogen on PT and APTT. J. Toxicol. Sci. 28 (5), 439–443.

Lamperez, A.J., Thomas, J.R., 2005. Normal C-reactive protein values for captive chimpanzees (*Pan troglodytes*). Contemp. Top. 44 (5), 25–26.

Lebrec, H., Cowan, L., Lagrou, M., et al., 2011. An inter-laboratory retrospective analysis of immunotoxicological endpoints in nonhuman primates: T-cell dependent antibody responses. J. Immunotoxicol. 8 (3), 238–250.

Mohr, J., Mattenheimer, R., Holmes, A., et al., 1971. Enzymology of experimental liver disease in marmoset monkeys. Enzyme 12, 99–116.

Moore, D., 2000. Hematology of nonhuman primates. In: Schalms Veterinary Hematology, fifth ed., pp. 1133–1144 (Chapter 176).

Ochi, T., Yamada, A., Nafanuma, Y., et al., 2016. Effect of road transportation on the serum biochemical parameters of cynomolgus monkeys and beagle dogs. J. Vet. Med. Sci. 78, 889–893.

Olson, H., Betton, G., Robinson, D., et al., 2000. Concordance of the toxicity of pharmaceuticals in humans and in animals. Regul. Toxicol. Pharmacol. 32, 56–67.

Organization for Economic Co-operation and Development (OECD), 2009. OECD Guidelines for the Testing of Chemicals. Publications de l'OCDE.

Perretta, G., Violante, A., Scarpulla, M., et al., 1991. Normal serum biochemical and hematological parameters in *Macaca fascicularis*. J. Med. Primatol. 20, 345–351.

Piccotti, J., Alvey, J., Reindel, J., et al., 2005. T-cell antibody response: assay development in cynomolgus monkeys. J. Immunotoxicol. 2, 191–196.

Porter, B.F., Frost, P., Hubbard, G.B., 2003. Polyarteritis nodosa in a cynomolgus macaque (*Macaca fascicularis*). Vet. Pathol. 40, 570–573.

Reagan, W., Barnes, R., Harris, P., et al., 2017. Assessment of cardiac troponin I responses in nonhuman primates during restraint, blood collection, and dosing in preclinical safety studies. Toxicol. Pathol. 45 (2), 335–343.

Reagan, W., Irizarry-Rovira, A., Poitout-Belissent, F., , et al.Bone Marrow Working Group ASVCP/STP, 2011. Toxicol. Pathol. 39, 435–448.

Renard, D., Rosselet, A., 2017. Drug-induced hemolytic anemia: pharmacological aspects. Transfus. Clin. Biol. 24, 110–114.

Russell, K.E., 2010. Platelet kinetics and laboratory evaluation of thrombocytopenia. In: Weiss, D.J., Wardrop, K.J. (Eds.), Schalm's Veterinary Hematology, sixth ed. Wiley-Blackwell, Ames, IA.

Sasseville, V., Mansfield, K., 2009. Overview of known non-human primate pathogens with potential to affect colonies used for toxicity testing. J. Immunotoxicol. 1–14.

Sato, J., Doi, T., Kanno, T., et al., 2012. Histopathology of incidental findings in cynomolgus monkeys (*Macaca fascicularis*) used in toxicity studies. J. Toxicol. Pathol. 25, 63–101.

Sinha, D.P., Cartwright, M.E., Johnson, R.C., 2006. Incidental mononuclear cell infiltrate in the uvea of cynomolgus monkeys. Toxicol. Pathol. 34, 148–151.

Siska, W., Meyer, D., Schultze, A., Brandoff, C., 2017. Identification of contaminant interferences with cause positive urine reagent test strip reactions in a cage setting for the laboratory-housed nonhuman primate, Beagle dog, and Sprague Dawley rat. Vet. Clin. Pathol. 46 (1), 85–90.

Stockham, S., Scott, M., 2008. Fundamentals of Veterinary Clinical Pathology, second ed., pp. 53–102

Tomlinson, L., Boone, L., Ramaiah, L., et al., 2013. Best practices for veterinary toxicologic clinical pathology, with emphasis on the pharmaceutical and biotechnology industries. Vet. Clin. Pathol. 23 (3), 252–256.

Vidal, J.D., Drobatz, L.S., Holliday, D.F., et al., 2010. Spontaneous findings in the heart of Mauritian-origin cynomolgus macaques (*Macaca fascicularis*). Toxicol. Pathol. 38, 297–302.

Vlasakova, K., Erdos, Z., Troth, S., et al., 2014. Evaluation of the relative performance of 12 urinary biomarkers for renal safety across 22 rat sensitivity and specificity studies. Toxicol. Sci. 8 (1), 3–20.

Ward, K., Coon, D., Magiera, D., et al., 2008. Exploration of the African green monkey as a preclinical pharmacokinetic model: intravenous pharmacokinetic parameters. Drug Metab. Disopos. 36 (4), 715–720.

Weingand, K., Brown, G., Hall, R., et al., 1996. Harmonization of animal clinical pathology testing in toxicity and safety studies. The Joint Scientific Committee for International Harmonization of Clinical Pathology Testing. Fundam. Appl. Toxicol. 29, 198–201.

Zabka, T.S., Irwin, M., Albassam, M.A., 2009. Spontaneous cardiomyopathy in cynomolgus monkeys (*Macaca fascicularis*). Toxicol. Pathol. 37, 814–818.

CHAPTER

5

Biomarkers of Toxicity in Zebrafish

Stephanie Padilla

Integrated Systems Toxicology Division, Office of Research and Development, U.S. Environmental Protection Agency, Research Triangle Park, NC, United States

INTRODUCTION

Given that zebrafish have now become an accepted model for toxicological studies in both the human health and ecological communities, this chapter takes stock of the status of biomarkers of toxicity that have been proposed for the zebrafish model. Most of what are discussed herein are biomarkers of effect, but the line between biomarkers of effect and exposure is often not well defined. Generally the chapter strives to present examples of many different types of biomarkers for many classes of toxicological "icities," but it should not be considered a comprehensive survey of all the zebrafish biomarker literature. Rather, the chapter gives a critical appraisal of the state of the effort to develop and identify biomarkers of toxicity in zebrafish.

ZEBRAFISH BACKGROUND

The zebrafish, a small (3–5 cm), freshwater teleost fish, not native to the Western Hemisphere (Spence et al., 2008), is a vertebrate species that is easy and inexpensive to rear and maintain in the laboratory. It is a model that is especially convenient for developmental studies because the offspring mature rapidly, with organogenesis completed by 3 days, and by 6 days the larvae are free swimming and feeding. The overall body plan is much like other vertebrates' and includes toxicologically relevant organ systems such as a liver with a large complement of cytochrome P450s (Tao and Peng, 2009; Goldstone et al., 2010; Otte et al., 2010; Stegeman et al., 2010; Weigt et al., 2011), a thyroid gland (Blanton and Specker, 2007; Porazzi et al., 2009; Walpita et al., 2009), and a blood–brain barrier (Jeong et al., 2008; Eliceiri et al., 2011). The model does, however, lack some other toxicologically relevant organs such as the mammary glands, lungs, and a prostate gland. Nevertheless, there are many examples where key developmental signaling pathways and their regulation are conserved between fish and mammals. A comparison of the mammalian and zebrafish genomes revealed long-range conserved synteny (i.e., "the maintenance of gene linkage on chromosomes of different species"; Kikuta et al., 2007), attesting to the similarity in important regulatory sequences and overall similarity in the genomes (Kikuta et al., 2007; Howe et al., 2013). Additionally, zebrafish have been reported to possess orthologs for 86% of the 1318 human drug targets tested (Gunnarsson et al., 2008).

Because of the concordance between zebrafish and mammalian developmental pathways, the zebrafish model is often promoted for studying mammalian disease (reviewed in Penberthy et al., 2002; Shin and Fishman, 2002). For example, it has been suggested that zebrafish would be an excellent model for ocular disease (Gestri et al., 2012), ocular motor disorders (Maurer et al., 2011), autism (Tropepe and Sive, 2003), and, in general, for the molecular dissection of developmental pathways (Dodd et al., 2000; Fetcho, 2007; Duggan et al., 2008; Gunnarsson et al., 2008; Ito et al., 2010; Spitsbergen and Kent, 2003; Xi et al., 2010; Xu and Zon, 2010).

ECOTOXICOLOGICAL BIOMARKERS OF TOXICITY

Even though the zebrafish is not a species indigenous to either the United States or Europe, it is also often used in ecotoxicological research as a surrogate for indigenous species. This is partly because of the large literature base for zebrafish and the good correlation between zebrafish toxicity and toxicity tests in other fish species (Braunbeck et al., 2005; Lammer et al., 2009; Vaughan and van Egmond, 2010).

Zebrafish biomarkers are used in ecotoxicological studies to determine if environmental samples elicit

Biomarkers in Toxicology, Second Edition
https://doi.org/10.1016/B978-0-12-814655-2.00005-0

toxic responses in zebrafish and, in some cases, to identify the type of stressor, i.e., zebrafish toxicity profiles are used as biomarkers of aquatic pollution. In these cases, zebrafish are being employed as a sentinel species, and the information gathered is more concerned with the toxicity spectrum of environmental sample(s) rather than fish toxicity per se. One of the earliest and most commonly used ecotoxicological biomarkers is the P450 enzyme CYP1A for pollution monitoring (reviewed in Goksøyr, 1995), and it is still in use; for example, in ecosystem surveys assessing petroleum contamination (dos Anjos et al., 2011) or for monitoring PCB contamination (Hung et al., 2012). The CYP1A P450 subfamily, regulated by the aryl hydrocarbon receptor, is used as a biomarker because it is often integral to the detoxification of many exogenous chemicals (both xenobiotics and drugs) and is induced by polyaromatic hydrocarbons. There are also examples where zebrafish are used in situ to assess the toxic potential of bodies of water. Using caged zebrafish (Strömqvist et al., 2010), different bodies of water were assessed for the induction of CYP1A (biomarker of xenobiotic exposure) and vitellogenin (biomarker of endocrine disruption). In some cases, the biomarker is less diagnostic for the type of pollutant and is instead used to determine if the body of water is adverse to fish development. An example of this is a recent study (Fuentealba González et al., 2011) where the biomarker, arginase activity, was established as an indicator of growth in young zebrafish. Arginase is involved in the urea cycle catalyzing the conversion of L-arginine into L-ornithine and urea and is important in the regulation of the arginine/ornithine levels within the organism. Using the data on whole fish arginase activity, accompanied by information on hatching and mortality, the authors demonstrated adverse effects of some environmental samples on developing zebrafish. Using transgenic technology to create "reporter" fish for environmental monitoring might allow rapid assessment of pollution monitoring. The development and use of these fluorescent sentinels became quite common in the last 10–15 years (Carvan et al., 2000; Hung et al., 2012). The fish can be engineered to respond to various classes of toxicants using fluorescent probes coupled to various response elements. These include the aromatic hydrocarbon response element, electrophile response element, metal response element, estrogen response element, retinoic acid, and retinoid X response elements, among others (reviewed in Carvan et al., 2000).

EXAMPLES OF BIOMARKERS OF FISH TOXICITY

The majority of biomarker development in zebrafish is devoted to biomarkers of toxicological effect. These biomarkers indicate whether a specific chemical or class of chemicals has perturbed some aspect of fish development or fitness. There are many of these biomarkers of toxicity advanced in the literature, but, for the most part, they lack demonstrated specificity. It is also often unclear how sensitive the biomarker actually is because many of the studies are conducted at concentrations that produce overt toxicity. Indicators of overt toxicity include mortality, nonhatching of embryos, or classical morphological signs (e.g., edema, curved spine, trunk shortening). Table 5.1 shows a sampling of the many different types of biomarkers and biomarker approaches in zebrafish studies. The table includes studies in developing animals, adult animals, acute and long-term treatment scenarios, biomarkers in various organ systems, and some uses of reporter fish. Heat shock protein 70, for example, was employed as a biomarker of chemical stress in many of the studies. Heat shock protein 70, highly conserved across phyla, is a protein chaperone that, during periods of temperature or chemical stress, will help maintain protein integrity within the cell and also may inhibit apoptosis. There are six examples of studies that used heat shock protein 70 as a biomarker of toxicity in zebrafish embryos or larvae exposed to a variety of chemical stressors (Table 5.1). In some cases, the investigators assessed heat shock protein message (Kreiling et al., 2007); in others, protein expression was assessed (Blechinger et al., 2002; Hu et al., 2008; Osterauer et al., 2010); and in others, green fluorescent protein (GFP) expression tethered to human heat shock protein promoter was assessed (Seok et al., 2006, 2007). In most cases, an increase in heat shock protein biomarker was noted in the presence of the chemical stressor; however, there was one exception where a decrease was noted (Kreiling et al., 2007). In some instances, heat shock protein appears to be a sensitive biomarker of toxicity (showing a significant change at doses below those that elicit overt toxicity) (e.g., Blechinger et al., 2002; Hu et al., 2008). In other instances, changes in the heat shock protein 70 biomarker were only noted at overtly toxic doses (Seok et al., 2006; Osterauer et al., 2010). The specificity of heat shock protein 70 as a biomarker of toxic effect is unknown, but it appears to be relatively nonspecific, which might be desirable for a general biomarker of toxicity. So few chemicals, however, have been tested in zebrafish according to which usefulness of heat shock protein 70 as a general biomarker of toxicity for many types of chemical classes is unknown. In fact, many of the biomarkers identified in zebrafish suffer from this shortcoming: so few chemicals have been tested that it is impossible to determine how predictive a biomarker is for any type of toxic insult and/or how diagnostic it is for a certain class of chemicals. Thus, to ascertain either predictive or diagnostic power, a wide chemical space (many different

TABLE 5.1 A Sampling of the Different Types of Biomarkers and Biomarker Approaches in Zebrafish Studies

Biomarker	Reporter Fish?	Adults or Larvae?	Chemical(s) Tested	Duration	Results	References
ATP, ADP, or AMP hydrolysis	No	Adults	$HgCl_2$, $Pb(CH_3COO)_2$	1, 4, or 30 days	In general, with both heavy metals, decreased hydrolysis with increasing exposure duration, but with exceptions	Senger et al. (2006)
Arginase	No	Embryos	Malathion	6 days	Dose-related decrease	Fuentealba González et al. (2011)
Catalase	No	Adults	Paclobutrazol	96 h	Increased at nonlethal concentrations	Ding et al. (2009)
Catalase	No	Adults	Cadmium	21 days	Decreased in both liver and ovary	Banni et al. (2011)
Carboxylesterase	No	Embryos	Methyl paraoxon	48 h	Dose–response decrease at nonovertly toxic doses	Küster and Altenburger (2006)
Cholinesterase	No	Embryos	Methyl paraoxon	48 h	Dose–response decrease at nonovertly toxic doses	Küster and Altenburger (2006)
Cholinesterase	No	Embryos and adults	Chromium (V1)	6 days (embryos); 4 days (adults)	Decreased only in embryos, but at an overtly toxic dose	Domingues et al. (2010)
Copper transporters (message)	No	Larvae (5 dpf)	$CuCl_2$, Cu_2O nanoparticles	96 h (?)	Both chemicals increased expression at low doses	Chen et al. (2011)
Cytochrome P450 (CYP1A1)	Yes	Larvae (7 dpf)	PCBs	24 h	Dose–response change in fluorescence	Hung et al. (2012)
Glutathione peroxidase	No	Adults	Cadmium	21 days	Decreased activity in both liver and ovary	Banni et al. (2011)
Glutathione peroxidase	No	Adults	Microcystin-LR, Microcystin-RR	4, 7, 15 days	Duration-, dose-, tissue-, and chemical-dependent increases and decreases	Pavagadhi et al. (2012)
Glutathione reductase	No	Adults	Microcystin-LR, Microcystin-RR	4, 7, 15 days	Duration-, dose-, tissue-, and chemical-dependent increases and decreases	Pavagadhi et al. (2012)
Glutathione-*S*-transferase	No	Adults	Microcystin-LR, Microcystin-RR	4, 7, 15 days	Duration-, dose-, and tissue-dependent increases and decreases	Pavagadhi et al. (2012)
Glutathione-*S*-transferase	No	Embryos and adults	Chromium (V1)	6 days (embryos); 4 days (adults)	Decreased in both embryos and adults at and below overtly toxic doses	Domingues et al. (2010)

Continued

TABLE 5.1 A Sampling of the Different Types of Biomarkers and Biomarker Approaches in Zebrafish Studies—cont'd

Biomarker	Reporter Fish?	Adults or Larvae?	Chemical(s) Tested	Duration	Results	References
Glutathione-*S*-transferase (message)	No	Larvae (5 dpf)	$CuCl_2$, Cu_2O nanoparticles	96 h (?)	$CuCl_2$ decreased expression below overtly toxic doses; Cu_2O nanoparticles only decreased expression at overtly toxic doses	Chen et al. (2011)
Heat shock protein 70 expression	No	Embryos	$PtCl_2$	7 days	Increased expression, but at a dose above those which demonstrated histological changes	Osterauer et al. (2010)
Heat shock protein 70 expression	No	Embryos	Tetrabromobisphenol A; hexabromocyclododecane	96 h	Both chemicals caused increases at doses that did not produce overt toxicity	Hu et al. (2008)
Heat shock protein message (*hsc70*)	No	Embryos	PCBs (Aroclor 1254)	24 h	Decreased expression	Kreiling et al. (2007)
Heat shock protein 70-4	Yes	Larvae (72 h)	Cadmium	96 h	Increased expression at nonovertly toxic doses	Blechinger et al. (2002)
Human heat shock protein promoter	Transiently transfected; EGFP reporter	Larvae (72 hpf)	Arsenite	96 h	Dose–response increased fluorescence	Seok et al. (2007)
Human heat shock protein promoter	Transiently transfected; EGFP reporter	Larvae (72 hpf)	Copper sulfate	3 h	Increased fluorescence, but also overt toxicity	Seok et al. (2006)
Lactate dehydrogenase	No	Embryos and adults	Chromium (V1)	6 days (embryos); 4 days (adults)	Only altered in embryos: increases at low doses; decreased at overtly toxic doses	Domingues et al. (2010)
Lipid peroxidation	No	Embryos	Tetrabromobisphenol A; hexabromocyclododecane	96 h	Both chemicals caused increases at doses that did not produce overt toxicity	Hu et al. (2008)
Lipid peroxidation	No	Embryos	Diclofenac	90 min	Decreased peroxidation at lowest dose only; at higher doses, no differences from control	Feito et al. (2012)
Malondialdehyde accumulation	No	Adults	Cadmium	21 days	Increased accumulation in both liver and ovary	Banni et al. (2011)
Metallothionein (message)	No	Larvae (5 dpf)	$CuCl_2$, Cu_2O nanoparticles	96 h (?)	Both chemicals increased expression at nonovertly toxic doses	Chen et al. (2011)
MicroRNAs	No	Embryos	Perfluorooctane sulfonate	24 or 120 h	Up- and downregulation, but only an overtly toxic dose used	Zhang et al. (2011)

Superoxide dismutase (SOD)	No	Adults	Paclobutrazol	96 h	Increased at nonlethal concentrations	Ding et al. (2009)
SOD	No	Embryos	Tetrabromobisphenol A; hexabromocyclododecane	96 h	Both chemicals caused changes at doses that did not produce overt toxicity, but with dissimilar dose–response relationships	Hu et al. (2008)
SOD	No	Adults (3 months)	Methyl parathion, Cadmium chloride, Zinc sulfate	96 h	All three chemicals caused an initial increase (24 h) followed by a decrease in activity	Ling et al. (2011)
SOD	No	Adults	Microcystin-LR, Microcystin-RR	4, 7, 15 days	Duration-, dose-, and tissue- dependent increases and decreases	Pavagadhi et al. (2012)
SOD	No	Adults	Cadmium	21 days	Decreased activity in both liver and ovary	Banni et al. (2011)
SOD protein expression	No	Adults	Microcystin-LR	30 days	Dose–responsive increase	Wang et al. (2010)
SOD (message)	No	Larvae (5 dpf)	$CuCl_2$, Cu_2O nanoparticles	96 h (?)	$CuCl_2$ only increased expression at overtly toxic dose; Cu_2O nanoparticles increased expression at and below overt toxicity	Chen et al. (2011)
Thyroxine (T_4)	No	Adults (3 months)	Ammonium perchlorate	12 weeks	No effect, even though there were histological changes in thyroid follicles	Mukhi et al. (2005)

? representing uncertainity

types of chemicals) needs to be tested with a given biomarker of toxicity.

The gene expression surveys that have also been used to uncover possible biomarkers of toxicity in zebrafish are not included in Table 5.1. In general, these have been used with very few toxic compounds to instead identify a large number of possible genetic biomarkers. In addition, there is rarely any follow-up to validate the biomarkers by determining the universality and specificity of the proposed biomarkers. An example of a common approach (and the common shortcomings) is a recent study (Park et al., 2012) that assessed gene expression using the Affymetrix GeneChip Zebrafish Array. They studied responses in larval zebrafish aged 3 days (72 h) to 7 days postfertilization after exposure to two different dose levels of two serotonin reuptake inhibitors (fluoxetine hydrochloride or sertraline hydrochloride). Having more than one compound to study and including two doses and two time points are positive aspects that should allow the investigators to identify strongly associated gene expression changes. The authors, however, concluded that they had identified biomarkers of serotonin reuptake inhibitors, but without additional compounds (both predicted actives and nonactives) it is impossible to assess the specificity of their proposed biomarkers. In another very detailed study that investigated sublethal methyl mercury toxicity in embryo/larval zebrafish using many different approaches, Ho et al. (2013) not only performed DNA microarrays to identify changed genes but also confirmed those using quantitative reverse transcription polymerase chain reaction (RT-qPCR) and then followed with in situ expression analyses to delineate the tissue- and region-specific responses. In doing so they found that even within the central nervous system, the in situ expression changes varied across regions, reinforcing some of the negative aspects of performing whole-animal analysis of any biochemical change. In other words, if the entire animal is analyzed for expression changes, a very focused change in one small area may be missed because of the "dilution" effect of the rest of the animal. Instead of individual genes, another approach is to assess gene network perturbation. Using reverse transcription polymerase chain reaction (RT-PCR), Liu et al. (2013) related the toxicity of two organophosphate ester flame retardants on six gene networks that involved aryl hydrocarbon receptors, peroxisome proliferator–activated receptor alpha, estrogenic receptors, thyroid hormone receptor alpha, glucocorticoid receptor, and mineralocorticoid receptor–centered gene networks. An especially important control in the study was defining what network perturbation looked like using chemicals that would be considered positive controls for each of the networks. In one of the few papers to assess the toxicity of multiple chemicals, Weil et al. (2009) analyzed the expression of a few ($n = 7$) sentinel genes, correlating gene expression changes with the chronic toxicity of 14 different chemicals. In general, there was a very good correlation between the two endpoints and, interestingly, expressions of *cyp1a, hmox1* (heme oxygenase [decycling] 1) were two of the most sensitive biomarkers of toxicity. In this case, however, *hsp70* was not a sensitive biomarker. This approach of maximizing the number of chemicals while limiting the number of genes appears to be a viable avenue for biomarker identification.

Realizing that developing biomarkers of toxicity in zebrafish appears to still be in its infancy, it may be worthwhile to ponder exactly what would be needed to establish an accepted and useful biomarker of toxicity in zebrafish. As mentioned above, one of the most important aspects of developing a biomarker would be to test a large chemical space to determine the diagnostic and predictive values of any given biomarker. In addition, combining biomarker development to mechanism discovery is also an attractive approach. An interesting example of this type of approach is a recent investigation of the toxicity of cholinesterase inhibition in developing zebrafish (Klüver et al., 2011). In this study, the investigators assessed gene expression changes in young zebrafish larvae exposed to one acetylcholinesterase inhibitor (azinphos-methyl) and two noninhibitors (1,4-dimethoxybenzene and 2,4-nitrophenol). They found that small heat shock protein (*hspb11*) expression was the most sensitive biomarker of exposure to an acetylcholinesterase inhibitor. They followed this observation with other studies that allowed them to conclude that *hspb11* may not only be a sensitive biomarker of developmental exposure to acetylcholinesterase inhibitors but also that it is involved in muscle development in zebrafish. Although this is a powerful study because it employed both positive and negative compounds, follow-up investigations assessing many more chemicals (both inhibitors and noninhibitors) would provide more convincing evidence of the utility of *hspb11* as a biomarker of anticholinesterase toxicity. Moreover, it must be noted that, at least in this case, cholinesterase inhibition was also a sensitive biomarker of exposure to azinphos-methyl.

AN EXAMPLE OF AN ESTABLISHED BIOMARKER: VITELLOGENIN

Are there any examples of moderately well-established biomarkers of toxicity in zebrafish and biomarkers that are relatively specific and predictive? A review of the literature (e.g., Marin and Matozzo, 2004; Scholz and Mayer, 2008) indicates that vitellogenin may be a predictive and specific biomarker for endocrine disruption.

Vitellogenin(s) is a family of glycolipoproteins that are precursors to the egg yolk protein found in a mature egg. Under hormonal control, vitellogenin is normally produced in the female liver and then transported via the blood to the maturing oocytes for packaging in the yolk (Lubzens et al., 2010). Changes in vitellogenin protein expression, message, or even DNA methylation status have been associated with exposure to endocrine-disrupting chemicals (Nilsen et al., 2004; Tong et al., 2004; Hoffmann et al., 2006; Kausch et al., 2008; Strömqvist et al., 2010; Dang et al., 2011). In their extensive review (Dang et al., 2011) comparing various endpoints (including vitellogenin changes) and biomarkers of endocrine disruption, it was concluded that vitellogenin was one of the three most sensitive endpoints in zebrafish tests for endocrine disruptors; the other two were measures of fecundity and gonad histology. In most of these assays, male zebrafish are exposed to putative endocrine-disrupting chemical(s) with assessment of vitellogenin induction in the liver. Given that the levels of vitellogenin are close to nothing in the normal male liver, this assay for endocrine disruption has a wide range of sensitivity. There has also been an effort in the last 5–10 years to use induced vitellogenin expression at much younger ages to assess endocrine disruption. In this manner, endocrine disruption may be incorporated into commonly used, short-term developmental assessments of toxicity in zebrafish. An example of this type of study (Muncke and Eggen, 2006) assessed *vitellogenin 1* mRNA expression in control and 17α-ethinylestradiol (EE2; 100, 1000, or 2000 ng/L; 0.34, 3.37, or 6.75 nM)–treated embryos at 24, 48, 72, 96, and 120 hours postfertilization (hpf) normalized to the expression of a housekeeping gene (β-actin) over time. Through the use of important additional controls, the authors found that the expression of β-actin was stable from 24 to 120 hpf and also noted that the 17α-ethinylestradiol treatment did not change the levels of β-actin. 17α-Ethinylestradiol treatment, however, did change *vitellogenin 1* mRNA expression levels (Fig. 5.1) at all ages except the youngest (24 hpf). Moreover, even larvae as young as 72 h were capable of mounting significant induction of *vitellogenin 1* at the lowest nominal concentration of 17α-ethinylestradiol tested, and it appears that the 120-h-old larvae would be capable of responding to even lower concentrations of this endocrine active chemical. This pattern of response supports the use of young embryos as sentinels for some endocrine-disrupting chemicals.

In an example that illustrates the specificity of vitellogenin as a biomarker and also demonstrates how increasing the chemical space for an assay focuses the biomarker universe, Kausch et al. (2008) surveyed the gene expression overlap for three estrogenic compounds: 17β-estradiol (E2), bisphenol A, and genistein. Interestingly, although all three compounds are known to be estrogenic, there was very little overlap in their gene expression profiles (Fig. 5.2). In fact, only five genes were altered by all three compounds, and all of them encode vitellogenin in zebrafish, prompting the authors to conclude that "these genes are the only suitable markers for exposure to different estrogenic compounds," meaning that the only biomarker that was reliably changed by all three estrogenic compounds was vitellogenin. If the authors had only used two instead of three chemicals to assess the overlap and

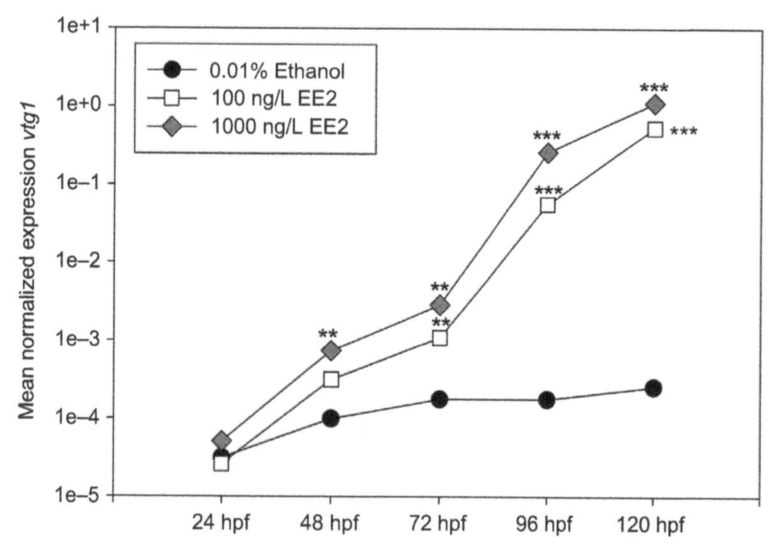

FIGURE 5.1 Expression of *vitellogenin 1* (*vtg1*) relative to β-actin in zebrafish embryos of different ages (hours postfertilization; hpf) treated with 17α-ethinylestradiol (EE2). * Indicates significant difference from the concurrent control. *Data are redrawn from Figure 4 of Muncke, J., Eggen, R.I., 2006. Vitellogenin 1 mRNA as an early molecular biomarker for endocrine disruption in developing zebrafish* (Danio rerio). *Environ. Toxicol. Chem. 25, 2734–2741.*

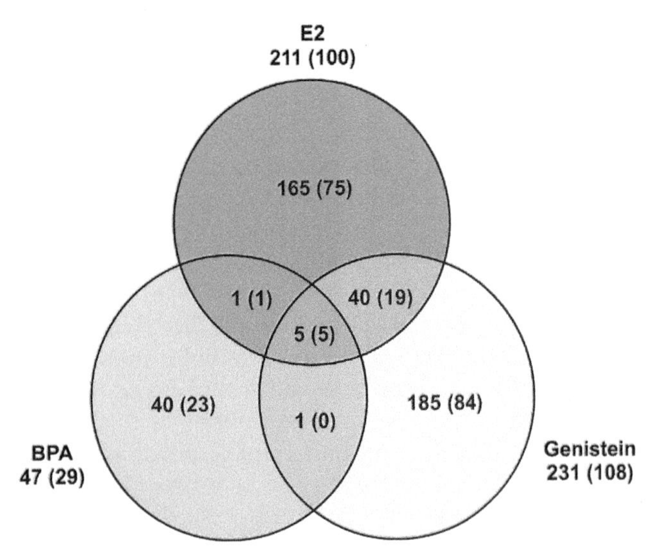

FIGURE 5.2 Venn diagram of the up- and downregulated ETs (i.e., expressed transcripts: open reading frames that are part of known genes) and TCs (i.e., tentative consensus sequences; open reading frames that show homologies to genes) in animals treated with the E2 (17β-estradiol), bisphenol A, or genistein. The first number indicates all regulated ETs and TCs. The numbers in brackets indicate the number of ETs. The overlapping areas represent the number of TCs and ETs that are regulated by two or all three chemicals. *Reproduced with permission from Figure 3 of Kausch, U., Alberti, M., Haindl, S., Budczies, J., Hock, B., 2008. Biomarkers for exposure to estrogenic compounds: gene expression analysis in zebrafish (*Danio rerio*). Environ. Toxicol. 23, 15–24. © 2008 Wiley Periodicals, Inc.*

identify possible biomarkers of estrogenicity, their conclusions would have been very different as 17β-estradiol and genistein perturbed 45 common genes.

CAUTIONS

A review of the zebrafish biomarker literature reveals that there are many variables that influence the expression of any given biomarker, and the investigator must be aware of these variables and understand how they can affect the changes in a biomarker of toxicity.

1. Time: The time after dosing or the duration of dosing can affect whether a biomarker is different from control and in some cases the direction of change. It has been demonstrated that different biomarkers show different time courses of change, so one biomarker of toxicity does not predict the time course of other biomarkers (Hoffmann et al., 2006; Senger et al., 2006; Sancho et al., 2010; Pavagadhi et al., 2012).
2. Dose (Concentration): Dose has a powerful influence on biomarker expression, and, interestingly, there are many exceptions to a linear dose–response relationship. Nonmonotonic dose–response relationships are often reported. Both inverted U- (Hoffmann et al., 2006; Chen et al., 2011; Feito et al., 2012; Gündel et al., 2012; Hung et al., 2012; Pavagadhi et al., 2012) and U (Chen et al., 2011)-shaped dose–response relationships have been noted. In many (but not all) cases, an inverted U pattern in dose and response is noted as overt toxicity is approached. Under these circumstances, an increase in the biomarker is noted at the nonovertly toxic concentrations, but as overtly toxic doses are reached, biomarker expression decreases as the animal becomes moribund.
3. Tissue: In some instances, the tissue of origin for the biomarker also determines the pattern and degree of biomarker expression. For example, the expression pattern of vitellogenin or estrogen receptor genes (Islinger et al., 2003) or the degree of DNA methylation of the *vitellogenin 1* gene (Strömqvist et al., 2010) after 17α-ethinylestradiol exposure is highly tissue-specific. In their in situ hybridization analysis of the gene expression changes of zebrafish treated with methyl mercury, Ho et al. (2013) provide an elegant visual demonstration of the marked differences in tissue biomarker responses. To complicate the issue further, the shape of the dose–response relationship, the time course, and the lowest effective dose may vary among tissues (Pavagadhi et al., 2012).
4. Sex and Age: Investigators should be cautioned not to extrapolate automatically biomarker patterns from one sex to another or from one age to another. It has been demonstrated that the protein expression patterns of adult male and female zebrafish treated with prochloraz (imidazole fungicide) are markedly different, with very little overlap between the sexes (Biales et al., 2011). In assessing the effects of chromium exposure on three biomarkers (cholinesterase, glutathione-S-transferase, and lactate dehydrogenase) in larval and adult zebrafish, it was noted that, in general, larval zebrafish showed more biomarker perturbation at lower nominal concentrations of chromium (IV) (Domingues et al., 2010) (Table 5.1) than adults. The same general pattern was noted in a study (Jin et al., 2009b) of genes coding for vitellogenin and estrogen receptors in four different ages of zebrafish (embryo, larval, juvenile, and adult) exposed to 17β-estradiol or nonylphenol: the younger animals showed increased sensitivity (i.e., lowest effective concentration), whereas the older animals tended to show a higher magnitude of induction. One very important variable that was not controlled for or even assessed in any of these studies was bioavailability of the compound—differences in internal dose due to sex or age may help explain some of the differences.
5. Animal Husbandry: Jin et al. (2009a) provide data demonstrating that photoperiod and ambient

temperature influenced the transcription levels of estrogen-responsive genes in adult male zebrafish. In general, longer photoperiods and higher temperatures produced increased levels of induction, but shorter photoperiods and lower temperatures produced more sensitive responses (i.e., lowest effect concentrations). Animal husbandry variables are generally not recognized for the power they may exert over the results of investigations into biomarkers of toxicity or exposure.

CONCLUDING REMARKS AND FUTURE DIRECTIONS

Current efforts to develop biomarkers of toxicity in zebrafish appear to be disorganized and incomplete. The only biomarker that appears to have been tested with multiple compounds in multiple laboratories is vitellogenin, and vitellogenin induction can be regarded as a biomarker for estrogenic compounds. As the zebrafish toxicological community moves forward with the identification of other biomarkers of toxicity, there is a need for rigorous testing to establish the sensitivity and specificity of all proposed biomarkers.

Acknowledgments

The author wishes to thank Dr. Robert MacPhail, Dr. William Mundy, and Dr. Aimen Farraj for reviewing earlier versions of this manuscript. The author is also grateful to Deborah Hunter and John Havel for construction of the figures. This manuscript has been reviewed by the US Environmental Protection Agency and approved for publication. Approval does not signify that the contents reflect the views of the Agency, nor does mention of trade names or commercial products constitute endorsement or recommendation for use.

References

Banni, M., Chouchene, L., Said, K., Kerkeni, A., Messaoudi, I., 2011. Mechanisms underlying the protective effect of zinc and selenium against cadmium-induced oxidative stress in zebrafish *Danio rerio*. Biometals 24, 981–992.

Biales, A.D., Bencic, D.C., Villeneuve, D.L., Ankley, G.T., Lattier, D.L., 2011. Proteomic analysis of zebrafish brain tissue following exposure to the pesticide prochloraz. Aquat. Toxicol. 105, 618–628.

Blanton, M.L., Specker, J.L., 2007. The hypothalamic-pituitary-thyroid (HPT) axis in fish and its role in fish development and reproduction. Crit. Rev. Toxicol. 37, 97–115.

Blechinger, S.R., Warren, J.T., Kuwada, J.Y., Krone, P.H., 2002. Developmental toxicology of cadmium in living embryos of a stable transgenic zebrafish line. Environ. Health Perspect. 110, 1041–1046.

Braunbeck, T., Boettcher, M., Hollert, H., 2005. Towards an alternative for the acute fish LC_{50} test in chemical assessment: the fish embryo toxicity test goes multi-species – an update. ALTEX 22, 87–102.

Carvan, M.J., Dalton, T.P., Stuart, G.W., Nebert, D.W., 2000. Transgenic zebrafish as sentinels for aquatic pollution. Ann. N.Y. Acad. Sci. 919, 133–147.

Chen, D., Zhang, D., Yu, J.C., Chan, K.M., 2011. Effects of Cu_2O nanoparticle and $CuCl_2$ on zebrafish larvae and a liver cell-line. Aquat. Toxicol. 105, 344–354.

Dang, Z., Li, K., Yin, H., Hakkert, B., Vermeire, T., 2011. Endpoint sensitivity in fish endocrine disruption assays: regulatory implications. Toxicol. Lett. 202, 36–46.

Ding, F., Song, W.H., Guo, J., Gao, M.L., Hu, W.X., 2009. Oxidative stress and structure-activity relationship in the zebrafish (*Danio rerio*) under exposure to paclobutrazol. J. Environ. Sci. Health Part B 44, 44–50.

Dodd, A., Curtis, P.M., Williams, L.C., Love, D.R., 2000. Zebrafish: bridging the gap between development and disease. Hum. Mol. Genet. 9, 2443–2449.

Domingues, I., Oliveira, R., Lourenço, J., 2010. Biomarkers as a tool to assess effects of chromium (VI): comparison of responses in zebrafish early life stages and adults. Comp. Biochem. Physiol. C Toxicol. Pharmacol. 152, 338–345.

dos Anjos, N.A., Schulze, T., Brack, W., 2011. Identification and evaluation of *cyp1a* transcript expression in fish as molecular biomarker for petroleum contamination in tropical fresh water ecosystems. Aquat. Toxicol. 103, 46–52.

Duggan, C.D., DeMaria, S., Baudhuin, A., Stafford, D., Ngai, J., 2008. Foxg1 is required for development of the vertebrate olfactory system. J. Neurosci. 28, 5229–5239.

Eliceiri, B.P., Gonzalez, A.M., Baird, A., 2011. Zebrafish model of the blood-brain barrier: morphological and permeability studies. Methods Mol. Biol. 686, 371–378.

Feito, R., Valcárcel, Y., Catalá, M., 2012. Biomarker assessment of toxicity with miniaturised bioassays: diclofenac as a case study. Ecotoxicology 21, 289–296.

Fetcho, J.R., 2007. The utility of zebrafish for studies of the comparative biology of motor systems. J. Exp. Zool. (Mol. Dev. Evol.) 308B, 550–562.

Fuentealba González, P., Llanos-Rivera, A., Carvajal Baeza, N., Uribe Pérez, E., 2011. Xenobiotic-induced changes in the arginase activity of zebrafish (*Danio rerio*) eleutheroembryo. Environ. Toxicol. Chem. 30, 2285–2291.

Gestri, G., Link, B.A., Neuhauss, S.C., 2012. The visual system of zebrafish and its use to model human ocular diseases. Dev. Neurobiol. 72, 302–327.

Goksøyr, A., 1995. Use of cytochrome P450 1A (CYP1A) in fish as a biomarker of aquatic pollution. Arch. Toxicol. Suppl. 17, 80–95.

Goldstone, J.V., McArthur, A.G., Kubota, A., 2010. Identification and developmental expression of the full complement of cytochrome P450 genes in zebrafish. BMC Genom. 11, 643.

Gündel, U., Kalkhof, S., Zitzkat, D., 2012. Concentration-response concept in ecotoxicoproteomics: effects of different phenanthrene concentrations to the zebrafish (*Danio rerio*) embryo proteome. Ecotoxicol. Environ. Saf. 76, 11–22.

Gunnarsson, L., Jauhiainen, A., Kristiansson, E., Nerman, O., Larsson, D.G., 2008. Evolutionary conservation of human drug targets in organisms used for environmental risk assessments. Environ. Sci. Technol. 42, 5807–5813.

Ho, N.Y., Yang, L., Legradi, J., 2013. Gene responses in the central nervous system of zebrafish embryos exposed to the neurotoxicant methyl mercury. Environ. Sci. Technol. 47, 3316–3325.

Hoffmann, J.L., Torontali, S.P., Thomason, R.G., 2006. Hepatic gene expression profiling using Genechips in zebrafish exposed to 17alpha-ethinylestradiol. Aquat. Toxicol. 79, 233–246.

Howe, K., Clark, M.D., Torroja, C.F., 2013. The zebrafish reference genome sequence and its relationship to the human genome. Nature 496, 498–503.

Hu, J., Liang, Y., Chen, M., Wang, X., 2008. Assessing the toxicity of TBBPA and HBCD by zebrafish embryo toxicity assay and biomarker analysis. Environ. Toxicol. 24, 334–342.

Hung, K.W., Suen, M.F., Chen, Y.F., 2012. Detection of water toxicity using cytochrome P450 transgenic zebrafish as live biosensor: for polychlorinated biphenyls toxicity. Biosens. Bioelectron. 31, 548–553.

Islinger, M., Willimski, D., Völkl, A., Braunbeck, T., 2003. Effects of 17a-ethinylestradiol on the expression of three estrogen-responsive genes and cellular ultrastructure of liver and testes in male zebrafish. Aquat. Toxicol. 62, 85–103.

Ito, T., Ando, H., Suzuki, T., 2010. Identification of a primary target of thalidomide teratogenicity. Science 327, 1345–1350.

Jeong, J.Y., Kwon, H.B., Ahn, J.C., 2008. Functional and developmental analysis of the blood-brain barrier in zebrafish. Brain Res. Bull. 75, 619–628.

Jin, Y., Chen, R., Sun, L., Liu, W., Fu, Z., 2009a. Photoperiod and temperature influence endocrine disruptive chemical-mediated effects in male adult zebrafish. Aquat. Toxicol. 92, 38–43.

Jin, Y., Chen, R., Sun, L., 2009b. Induction of estrogen-responsive gene transcription in the embryo, larval, juvenile and adult life stages of zebrafish as biomarkers of short-term exposure to endocrine disrupting chemicals. Comp. Biochem. Physiol. C Toxicol. Pharmacol. 150, 414–420.

Kausch, U., Alberti, M., Haindl, S., Budczies, J., Hock, B., 2008. Biomarkers for exposure to estrogenic compounds: gene expression analysis in zebrafish (*Danio rerio*). Environ. Toxicol. 23, 15–24.

Kikuta, H., Laplante, M., Navratilova, P., 2007. Genomic regulatory blocks encompass multiple neighboring genes and maintain conserved synteny in vertebrates. Genome Res. 17, 545–555.

Klüver, N., Yang, L., Busch, W., 2011. Transcriptional response of zebrafish embryos exposed to neurotoxic compounds reveals a muscle activity dependent *hspb11* expression. PLoS One 6, e29063.

Kreiling, J.A., Creton, R., Reinisch, C., 2007. Early embryonic exposure to polychlorinated biphenyls disrupts heat-shock protein 70 cognate expression in zebrafish. J. Toxicol. Environ. Health 70, 1005–1013.

Küster, E., Altenburger, R., 2006. Comparison of cholin- and carboxylesterase enzyme inhibition and visible effects in the zebra fish embryo bioassay under short-term paraoxon-methyl exposure. Biomarkers 11, 341–354.

Lammer, E., Carr, G.J., Wendler, K., 2009. Is the fish embryo toxicity test (FET) with the zebrafish (*Danio rerio*) a potential alternative for the fish acute toxicity test? Comp. Biochem. Physiol. C Toxicol. Pharmacol. 149, 196–209.

Ling, X., Zhang, Y., Lu, Y., Huang, H., 2011. Superoxide dismutase, catalase and acetylcholinesterase: biomarkers for the joint effects of cadmium, zinc and methyl parathion contamination in water. Environ. Technol. 32, 1463–1470.

Liu, C., Wang, Q., Liang, K., 2013. Effects of tris(1,3-dichloro-2-propyl) phosphate and triphenyl phosphate on receptor-associated mRNA expression in zebrafish embryos/larvae. Aquat. Toxicol. 128–129, 147–157.

Lubzens, E., Young, G., Bobe, J., Cerda, J., 2010. Oogenesis in teleosts: how eggs are formed. Gen. Comp. Endocrinol. 165, 367–389.

Marin, M.G., Matozzo, V., 2004. Vitellogenin induction as a biomarker of exposure to estrogenic compounds in aquatic environments. Mar. Pollut. Bull. 48, 835–839.

Maurer, C.M., Huang, Y.Y., Neuhauss, S.C., 2011. Application of zebrafish oculomotor behavior to model human disorders. Rev. Neurosci. 22, 5–16.

Mukhi, S., Carr, J.A., Anderson, T.A., Patiño, R., 2005. Novel biomarkers of perchlorate exposure in zebrafish. Environ. Toxicol. Chem. 24, 1107–1115.

Muncke, J., Eggen, R.I., 2006. Vitellogenin 1 mRNA as an early molecular biomarker for endocrine disruption in developing zebrafish (*Danio rerio*). Environ. Toxicol. Chem. 25, 2734–2741.

Nilsen, B.M., Berg, K., Eidem, J.K., 2004. Development of quantitative vitellogenin-ELISAs for fish test species used in endocrine disruptor screening. Anal. Bioanal. Chem. 378, 621–633.

Osterauer, R., Köhler, H.R., Triebskorn, R., 2010. Histopathological alterations and induction of hsp70 in ramshorn snail (*Marisa cornuarietis*) and zebrafish (*Danio rerio*) embryos after exposure to $PtCl_2$. Aquat. Toxicol. 99, 100–107.

Otte, J.C., Schmidt, A.D., Hollert, H., Braunbeck, T., 2010. Spatio-temporal development of CYP1 activity in early life-stages of zebrafish (*Danio rerio*). Aquat. Toxicol. 100, 38–50.

Park, J.W., Heah, T.P., Gouffon, J.S., Henry, T.B., Sayler, G.S., 2012. Global gene expression in larval zebrafish (*Danio rerio*) exposed to selective serotonin reuptake inhibitors (fluoxetine and sertraline) reveals unique expression profiles and potential biomarkers of exposure. Environ. Pollut. 167, 163–170.

Pavagadhi, S., Gong, Z., Hande, M.P., 2012. Biochemical response of diverse organs in adult *Danio rerio* (zebrafish) exposed to sub-lethal concentrations of microcystin-LR and microcystin-RR: a balneation study. Aquat. Toxicol. 109, 1–10.

Penberthy, W.T., Shafizadeh, E., Lin, S., 2002. The zebrafish as a model for human disease. Front. Biosci. 7, d1439–d1453.

Porazzi, P., Calebiro, D., Benato, F., Tiso, N., Persani, L., 2009. Thyroid gland development and function in the zebrafish model. Mol. Cell. Endocrinol. 312, 14–23.

Sancho, E., Villarroel, M.J., Fernandez, C., Andreu, E., Ferrando, M.D., 2010. Short-term exposure to sublethal tebuconazole induces physiological impairment in male zebrafish (*Danio rerio*). Ecotoxicol. Environ. Saf. 73, 370–376.

Scholz, S., Mayer, I., 2008. Molecular biomarkers of endocrine disruption in small model fish. Mol. Cell. Endocrinol. 293, 57–70.

Senger, M.R., Rico, E.P., de Bem Arizi, M., 2006. Exposure to Hg^{2+} and Pb^{2+} changes NTPDase and ecto-5′-nucleotidase activities in central nervous system of zebrafish (*Danio rerio*). Toxicology 226, 229–237.

Seok, S.H., Baek, M.W., Lee, H.Y., 2007. Quantitative GFP fluorescence as an indicator of arsenite developmental toxicity in mosaic heat shock protein 70 transgenic zebrafish. Toxicol. Appl. Pharmacol. 225, 154–161.

Seok, S.H., Park, J.H., Baek, M.W., 2006. Specific activation of the human HSP70 promoter by copper sulfate in mosaic transgenic zebrafish. J. Biotechnol. 126, 406–413.

Shin, J.T., Fishman, M.C., 2002. From zebrafish to human: modular medical models. Annu. Rev. Genomics Hum. Genet. 3, 311–340.

Spence, R., Gerlach, G., Lawrence, C., Smith, C., 2008. The behaviour and ecology of the zebrafish, *Danio rerio*. Biol. Rev. Camb. Philos. Soc. 83, 13–34.

Spitsbergen, J.M., Kent, M.L., 2003. The state of the art of the zebrafish model for toxicology and toxicologic pathology research—advantages and current limitations. Toxicol. Pathol. 31, 62–87.

Stegeman, J.J., Goldstone, J.V., Hahn, M.E., 2010. Perspectives on Zebrafish as a Model in Environmental Toxicology. Academic Press, pp. 367–439.

Strömqvist, M., Tooke, N., Brunström, B., 2010. DNA methylation levels in the 5′ flanking region of the *vitellogenin I* gene in liver and brain of adult zebrafish (*Danio rerio*)—sex and tissue differences and effects of 17α-ethinylestradiol exposure. Aquat. Toxicol. 98, 275–281.

Tao, T., Peng, J., 2009. Liver development in zebrafish (*Danio rerio*). J. Genet. Genomics 36, 325–334.

Tong, Y., Shan, T., Poh, Y.K., 2004. Molecular cloning of zebrafish and medaka vitellogenin genes and comparison of their expression in response to 17β-estradiol. Gene 328, 25–36.

Tropepe, V., Sive, H.L., 2003. Can zebrafish be used as a model to study the neurodevelopmental causes of autism? Genes Brain Behav. 2, 268–281.

Vaughan, M., van Egmond, R., 2010. The use of the zebrafish (*Danio rerio*) embryo for the acute toxicity testing of surfactants, as a possible alternative to the acute fish test. ATLA 38, 231–238.

Walpita, C.N., Crawford, A.D., Janssens, E.D., Van der Geyten, S., Darras, V.M., 2009. Type 2 iodothyronine deiodinase is essential for thyroid hormone-dependent embryonic development and pigmentation in zebrafish. Endocrinology 150, 530–539.
Wang, M., Chan, L.L., Si, M., Hong, H., Wang, D., 2010. Proteomic analysis of hepatic tissue of zebrafish (*Danio rerio*) experimentally exposed to chronic microcystin-LR. Toxicol. Sci. 113, 60–69.
Weigt, S., Huebler, N., Strecker, R., Braunbeck, T., Broschard, T.H., 2011. Zebrafish (*Danio rerio*) embryos as a model for testing proteratogens. Toxicology 281, 25–36.
Weil, M., Scholz, S., Zimmer, M., Sacher, F., Duis, K., 2009. Gene expression analysis in zebrafish embryos: a potential approach to predict effect concentrations in the fish early life stage test. Environ. Toxicol. Chem. 28, 1970–1978.
Xi, Y., Ryan, J., Noble, S., 2010. Impaired dopaminergic neuron development and locomotor function in zebrafish with loss of pink1 function. Eur. J. Neurosci. 31, 623–633.
Xu, C., Zon, L.I., 2010. The Zebrafish as a Model for Human Disease. Academic Press, pp. 345–365.
Zhang, L., Li, Y.Y., Zeng, H.C., 2011. MicroRNA expression changes during zebrafish development induced by perfluorooctane sulfonate. J. Appl. Toxicol. 31, 210–222.

CHAPTER

6

Mechanistic Toxicology Biomarkers in *Caenorhabditis elegans*

Vanessa A. Fitsanakis[1], *Rekek Negga*[2], *Holly E. Hatfield*[3]

[1]Department of Pharmaceutical Sciences, Northeast Ohio Medical University, Rootstown, OH, United States; [2]Department of Biology, King University, Bristol, TN, United States; [3]Department of Biochemistry, University of Cincinnati Medical School, Cincinnati, OH, United States

INTRODUCTION

Biomarkers are important clinical and diagnostic tools for determining and staging human disease or exposure to toxic substances. Monitoring changes in biomarkers also provides a way to determine if dietary or pharmacological interventions are also improving (or worsening) pathology progression. It is often challenging, however, to both develop and assess how experimentally induced changes in their levels may positively or negatively affect general human physiology. The difficulty is partially due to the complexity of human organ systems. Thus, much simpler organisms are routinely used to develop and evaluate the significance of important potential biomarkers. Although the transparent nematode *Caenorhabditis elegans (C. elegans)* has been used for decades in the field of developmental biology (Brenner, 1974), it is only within the last 10–15 years that its use has gained prominence across other disciplines (Leung et al., 2008). For example, *C. elegans* has been used to screen neurotoxic compounds (Avila et al., 2012), assess cellular respiration (Koopman et al., 2016), and investigate the efficacy of potential therapeutic drugs (Ma et al., 2018).

The growing popularity of *C. elegans* partially relates to the fact that their simple anatomical and physiological organization, as well as their genome, is closely related to that of humans (Silverman et al., 2009; Varshney et al., 2011). In fact, between 40% and 75% of human genes identified as important in disease progression or etiology (Calahorro and Ruiz-Rubio, 2011; Corsi, 2006) or those related to aging (Sutphin et al., 2017) have analogs in *C. elegans*. In addition, data suggest that even stem cell populations in these worms are similar to those found in humans and other, more evolutionarily advanced, organisms (Joshi et al., 2011; Morgan et al., 2011). Along with the ease with which their genome can be manipulated (Fay, 2006), *C. elegans* possesses several characteristics that make this worm an exceptionally versatile and useful model organism. These characteristics include the following: small size, short life span, transparent bodies, and self-fertilizations (Blaxter, 2011; Brenner, 1974).

As clinical medicine and drug development move toward genetic classification of disease subtypes, and both personalized medicine and pharmaceutical therapies, the need for a robust model organism in which potential biomarkers can be assessed is pressing. The focus of this chapter is to demonstrate how markers of oxidative stress and mitochondrial inhibition, found in a number of human diseases, can be monitored *in vivo* in *C. elegans*. While not meant to be exhaustive in nature, the goal is to provide information regarding how *C. elegans* can be and is being used to elucidate biochemical pathways involved in diseases, what the potential physiological roles for these biomarkers are, and how further insight may benefit research and clinical communities.

C. ELEGANS AND BIOMARKERS

In the 1970s, Dr. Sydney Brenner proposed to the scientific community that *C. elegans* be used as a model organism to study genetics associated with development (Brenner, 1974; Sulston and Brenner, 1974). Many reasons for proposing these worms as a new model organism were very pragmatic: huge quantities could be

Biomarkers in Toxicology, Second Edition
https://doi.org/10.1016/B978-0-12-814655-2.00006-2

grown in a short period of time; multiple generations could be produced within days or weeks; a suitable food source (*Escherichia coli*) was easy to obtain and maintain; they could be grown in Petri dishes at a variety of temperatures; the total number of cells in the organism (959 cells, of which 302 are neurons, in the hermaphrodite) made it easy to determine cellular etiology and fate; its transparency facilitated light microscopy of gross internal structures and physiological processes; transparency also rendered it ideal for labeling endogenous proteins with fluorometric tags (e.g., green fluorescent protein [GFP]); the nervous system was quickly mapped (Ward et al., 1975; White et al., 1976); and many homologs and orthologs are evolutionarily conserved through humans (Shaye and Greenwald, 2011).

As work with *C. elegans* progressed, the developmental etiology of each individual cell was mapped and published (Sulston, 1988). Later, complete neural circuitry and connectivity was determined (Chalfie and White, 1988). In more recent years, as genetic information became increasingly important for understanding development and disease risk or disease progression, it was a momentous step forward when the worm's entire genome was determined and published (Consortium, 1998). This latter achievement alone permitted the research community to begin to easily manipulate its genome (*i.e.*, formation of *gfp* constructs, development of knockout animals, and insertion of human genes). More recently, it was in *C. elegans* that microRNAs were discovered (Sharp, 1999). As such, it became plausible to generate "conditional knockout" animals by feeding them small interfering RNA (siRNA) sequences, via specifically engineered *E. coli*, to target individual genes at precise times during the life cycle. More recently, protocols have been developed that enable researchers to easily modify the genome of *C. elegans* using clustered regularly interspaced short palindromic repeats (CRISPRs)/CRISPR-associated protein-9 (Cas-9) technologies (Farboud, 2017; Prior et al., 2017).

Once the human genome was sequenced, it then became possible to align genetic sequences from both species to expand our knowledge of homologous and orthologous genes in these worms. Thus it was feasible to search for and compare gene sequences (and by extension, protein function) common in both humans, worms, and other model organisms. For example, much of the work associated with cellular apoptotic mechanisms was originally conducted and completed in *C. elegans* (Hengartner and Horvitz, 1994). It was later confirmed that similar pathways existed in higher organisms, including humans.

Perhaps not surprisingly, many laboratories began to work diligently on producing *gene::gfp* constructs, developing unique knockout strains, and engineering worms with human disease–associated genes. Remarkably, much of the genetic information associated with *C. elegans* is not only published but also readily available. Many of the mutant strains produced by laboratories around the world are archived at the University of Minnesota at the *Caenorhabditis* Genetics Center (http://www.cbs.umn.edu/cgc), a program partially funded by the National Institutes of Health—Office of Research Infrastructure Programs. Here, thousands of strains of *C. elegans* are archived and frozen so that laboratories from around the world can request them for a fee. This provides the opportunity for numerous researchers to benefit from the work of others without duplicating the production of identical strains. In addition, a public database (http://www.wormbase.org) dedicated to proteins and genes of known structure and function is also available through the collaborative work of the "Worm Community." Finally, the Worm Community was one of the first to provide "open access" material through the publication of *WormBook* (http://www.wormbook.org). Each of these valuable resources encourages open exchange of information, protocols, mutant worm and *E. coli* strains, siRNA sequences, and other materials to facilitate research in numerous fields.

As the popularity of this model organism has grown, many high-throughput assays have also been developed that allow researchers to monitor differences in gene and protein regulation and expression as well as document changes in stereotypical movements (Swierczek et al., 2011) following exposure to toxicants. High-throughput screening using complex object parametric analysis and sorting (COPAS) technology and instrumentation allows researchers to analyze and sort large numbers of worms over a short period of time (10–50 worms/s) based on size (Boyd et al., 2010). In addition, many different hybridization gene plates are available to determine whether genes associated with distinct pathways (cellular respiration, detoxication, metabolism, cell cycle, etc.) are altered following xenobiotic treatment. These technologies not only provide the opportunity to assess whether toxicants induce changes associated with a specific disease but also the ability to determine whether pharmacological intervention successfully targets and modulates molecules associated with these diseases.

Although many of the genes and proteins in *C. elegans* are analogous to those found in higher organisms, including humans, it is important to emphasize that the function and/or location of the proteins may not be identical. Furthermore, many of the organs and organ systems found in humans are absent in this model organism. On the other hand, the structural and functional similarities are such that vast insight can be gleaned from studying human biomarkers in *C. elegans* related to transcriptional and translational regulation, cellular targets, and effects of various xenobiotics on these molecules. As such, this organism can function as a unique model system for mechanistic and high-throughput studies related to biomarkers associated with human disease and toxic exposure.

MECHANISTIC BIOMARKERS

Oxidative Stress Markers

Increased oxidative stress is a nonspecific marker associated with numerous diseases and toxic insults. It is often accompanied by mitochondrial inhibition because of the increased production of reactive oxygen species (ROS) resulting from inhibition of electron transport chain enzyme complexes. Oxidative stress is often associated with metal toxicity, particularly Fe, Mn, Cu, Co, and Ni, due to the fact that they can participate in the Fenton reaction (Fig. 6.1).

Pesticide toxicity (e.g., organophosphates, rotenone, paraquat, mancozeb) and numerous diseases (Parkinson's disease [PD], amyotrophic lateral sclerosis, cardiovascular disease, stroke) are other factors associated with increased oxidative stress. It is fairly common to use fluorometric and colorimetric probes in cell culture and other *in vivo* model systems to measure the presence of ROS and/or reactive nitrogen species (RNS). In humans and other animals, however, protein adducts (Mseddi et al., 2017), DNA adducts (Mazlumoglu et al., 2017), and lipid peroxidation products (Tsikas, 2017) are frequently measured in blood, urine, cerebrospinal fluid, and other biological samples. Fortunately, all of these products can also be measured in *C. elegans*. As such, this can facilitate moving from *C. elegans* to other, more complex, organisms and then back to worms. For example, dihydroethidium (DHE) has been successfully used to assess increased oxidative stress (Chikka et al., 2016). Other laboratories have combined results from multiple assays using DHE, 10-acetyl-3,7-dihydroxyphenoxazine, and hydroxyphenyl fluorescein (Fig. 6.2) to try to delineate the principal ROS formed in *C. elegans* following exposure to various pesticides (Bailey et al., 2017). Data from these probes

$$Fe^{3+} + O_2^{\bullet -} \rightarrow Fe^{2+} + O_2$$
$$Fe^{2+} + H_2O_2 \rightarrow Fe^{3+} + OH^- + {}^{\bullet}OH$$
$$\overline{}$$
$$O_2^{\bullet -} + H_2O_2 \xrightarrow{Fe^{3+}} OH^- + {}^{\bullet}OH + O_2$$

FIGURE 6.1 Schematic of the Fenton reaction, which can also be catalyzed with transition metals other than iron (Fe). *Photo by Unknown Author is licensed under CC BY-SA.*

can also be generated quite rapidly and simply because *C. elegans* is transparent. Results, then, can be used to directly guide studies in higher organisms regarding ROS and/or RNS production.

Oxidative stress, however, is typically accompanied by increased transcription and translation of antioxidant genes. Because antioxidant and detoxication proteins have different affinities for various oxidative stress molecules, it is often important to determine the genes and proteins involved in detoxication and repair following a toxic insult. To assess this, many laboratories have created transgenic worms with a gene::fluorescent protein construct. Thus, studies designed to determine whether *superoxide dismutase* (*sod*) transcription is upregulated could treat transgenic worms with *sod::gfp* instead of wild-type (N2) worms. In the presence of increased superoxide production, requiring *sod::gfp* transcription and translation to aid in detoxication, treated worms would have greater green fluorescence than that of controls. But *C. elegans* also have other antioxidant proteins—catalase, peroxidase, glutathione-S-peroxidase, and heat shock proteins. All of these can be tagged with GFP to monitor differential expression of the various protective genes. Other laboratories have developed transgenic worms with GFP-labeled proteins that are responsible for regulating entire oxidative stress pathways. For example, the nuclear respiratory factor (NRF), designated SKN (**skin**-head) in worms, is upstream of many proteins involved in regulating and extending life span. This regulation is thought to be partially related to the ability of NRF (SKN) to increase transcription and translation pathways of other proteins involved in oxidative/general stress responses (An and Blackwell, 2003; Benedetto et al., 2010). Finally, fluorescently labeled worm strains are even available to monitor the *in vivo* production of cytochrome P450 proteins (Table 6.1).

For studies involving toxicant exposure, the aforementioned techniques can be combined to determine (1) whether oxidative stress is increased, (2) which biomolecules are the primary targets (proteins, nucleic acids, or lipids), and (3) the identity of proteins involved in detoxication. On the other hand, numerous deletion mutants, RNA interference sequences, or CRISPR/Cas 9 target sequences are also available (or can be synthesized) to assess what happens to ROS or RNS production in the absence of detoxication enzymes. Conversely, if therapeutic compounds are being screened to determine efficacy, these same endpoints can be used to assess whether pharmacological intervention was successful. Furthermore, worms mature from egg to egg-laying adults in 2 to 3 days, depending on the temperature, and live for an average of 20–24 days. Therefore, it is possible to assess how exposure to toxicants at different life stages may adversely affect oxidative stress,

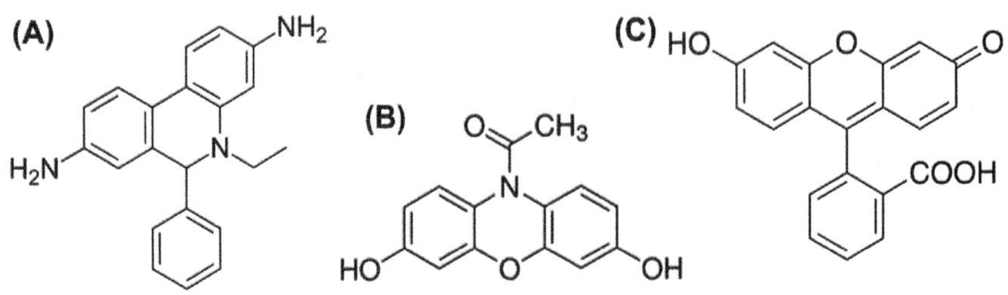

FIGURE 6.2 Structure of common fluorescent oxidative stress probes that can be used in *Caenorhabditis elegans*. (A) Dihydroethidium; (B) 10-acetyl-3,7-dihydroxyphenoxazine; and (C) hydroxyphenyl fluorescein. *(A–C) Photo by Unknown Author is licensed under CC BY-SA.*

antioxidant gene response, or detoxication (P450) gene response. While the results should be confirmed in higher model organisms, this approach permits a much faster screening, potentially over multiple generations, than is currently possible with more traditional animal models.

Mitochondrial Markers

As is the case with oxidative stress, many environmental toxicants and human diseases result in, or are marked by, mitochondrial inhibition. The target could be one of the electron transport chain complexes, as in the case of rotenone (Hoglinger et al., 2003), or enzymes in the citric acid cycle, as with Mn (Crooks et al., 2007). On the other hand, it is well-documented that mitochondrial inhibition is associated with numerous diseases, from glaucoma (He et al., 2008) and PD (Betarbet et al., 2006) to nonalcoholic fatty liver disease (Berge et al., 2016). As such, many researchers are concerned about whether exposure to their compounds of interest also promotes mitochondrial inhibition.

As with oxidative stress, many probes and markers used in cell culture can also be valuable in assessing endpoints in *C. elegans*. For example, tetramethylrhodamine ethyl ester (TMRE) and similar fluorophores have long been used to assess the integrity of the mitochondrial transmembrane proton gradient ($\Delta\Psi$). This positively charged dye crosses the mitochondrial membranes and accumulates within mitochondria. Although TMRE is predominantly used as a marker of $\Delta\Psi$, some have shown that its accumulation also correlates to mitochondrial fusion and fission events (Twig et al., 2008), cytochrome *c* release, and apoptosis (Crowley et al., 2016). In addition to using fluorescent probes, protocols exist that permit researchers to assess how well individual complexes in the electron transport chain are able to accept electrons. In these studies, colorimetric compounds that function as electron carriers and change color following oxidation or reduction can be used to determine whether complex II or complex IV is fully functional and able to participate in redox reactions (Grad et al., 2007). These protocols require that worms are permeabilized and fixed before incubation with the

TABLE 6.1 *Caenorhabditis elegans* Strains With GFP-Labeled Proteins Associated With Antioxidants and Detoxication Enzymes

Human Gene	Strain	Genotype
Superoxide dismutase	SD1746	muIs84 [sod-3p::GFP + rol-6(su1006)]. GFP should be visible at low magnification.
Superoxide dismutase	CF1580	muIs84 [(pAD76) sod-3p::GFP + rol-6(su1006)]. Green expression in head, tail, and around vulva.
Catalase	GA800	wuIs151 [ctl-1(+) + ctl-2(+) + ctl-3(+) + myo-2p::GFP].
Glutathione-*S*-transferase	CL2166	dvIs19 [(pAF15)gst-4p::GFP::NLS] III. Oxidative stress-inducible GFP.
Glutathione-*S*-transferase	CL691	dvIs19 [(pAF15) gst-4p::GFP::NLS] III. Oxidative stress-inducible GFP.
Heat shock protein	AGD926	zcIs4 [hsp-4::GFP] V. uthIs269 [sur5p::hsf-1::unc-54 3′UTR + myo-2p::tdTomato::unc-54 3′ UTR]. ER stress resistance.
Heat shock protein	CF2962	muEx420 [hsp-12.6p::RFP(NLS) + odr-1p::RFP].
Nuclear respiratory factor (nrf)	GR2198	mgTi1 [rpl-28p::skn-1a::GFP::tbb-2 3′UTR + unc-119(+)] I. Expresses SKN-1A with C-terminal GFP tag.
CYP3A4	SRU1	unc-119(ed3) III; jrIs1 [cyp-13A7p::GFP + unc-119(+)].

All strains in the table are available from the *Caenorhabditis* Genetics Center (CGC). *GFP*, green fluorescent protein.

respective reagents, which would preclude intervention with putative therapeutics and further analysis. Other methods for determining mitochondrial inhibition that easily lend themselves to real-time and longitudinal studies involve using traditional oxygen consumption studies with Clark's electrode (Bailey et al., 2017), or using the Seahorse metabolic analyzer (Luz et al., 2016). One advantage to the latter two methods is that extensive data exist in isolated mitochondria, cells, and tissues, which makes it easier to compare results in *C. elegans* and facilitate translation into other model organisms.

Closely associated with mitochondrial dysfunction is mitophagy, or the process of removing damaged mitochondria from the cytoplasm through the proteasome. Although mitophagy, as well as proteasomal elimination of other damaged organelles and proteins, is part of normal cellular "housekeeping," dysregulation can lead to increased cell death. Conversely, maintenance of mitophagy is required to extend the life span in many model organisms, including *C. elegans* (Lionaki et al., 2013). As more is understood about this important process, its regulation, and how it contributes to neurodegeneration and aging, researchers are looking for ways to assess this in vivo during both normal aging and disease states following toxicant exposure. Again, various fluorescent probes originally designed for cell culture have found their way into *C. elegans* research. For example, worms can be incubated in MitoTracker Green to facilitate the mitochondrial visualization (Dingley et al., 2010). Alternatively, fluorescent-tagged proteins associated with mitophagy have been labeled in transgenic worm strains (Table 6.2). This latter approach allows researchers to treat worms with toxicants suspected of modulating mitophagy and then visually determining whether the respective treatment had any effect on this process. On the other hand, these strains could also be used to assess whether potential therapeutic interventions reversed abnormal mitophagy.

NEURODEGENERATIVE DISEASES AND GENE–ENVIRONMENT INTERACTIONS

Recent advances in the fields of genome sequencing and bioinformatics have enabled the scientific community to more easily examine gene–environment interactions in the context of major diseases. This approach is particularly attractive to those studying chronic neurodegenerative diseases that lack obvious Mendelian inheritance, or consistent links to toxicant exposures. Thus, many researchers now consider the possibility that these diseases result from "multiple hits" (Fig. 6.3). These "hits" could be the (1) presence of a "susceptibility gene" that renders a person more vulnerable to an environmental exposure, (2) exposure to two (or more) chemicals that target the same organelle (i.e., mitochondria) or work via the same mechanism (i.e., oxidative stress), or (3) multiple genes working in concert with each other to adversely affect a specific cellular process (i.e., autophagy). The literature linking environmental toxicants to neurodegenerative diseases is rather compelling in the cases of mitochondrial inhibition

TABLE 6.2 *Caenorhabditis elegans* Strains With Fluorescence-Labeled Proteins Associated With Autophagy

Human Gene	Strain	Genotype
Microtubule-associated protein 1 light chain 3 (MAP-LC3)	DLM1	uwaEx1 [eft-3p::CERULEAN-VENUS::lgg-1 + unc-119(+)]. Labeled autophagy marker (LGG-1 in worms) is expressed with protein required for embryogenesis and the germline development.
Microtubule-associated protein 1 light chain 3 (MAP-LC3)	DLM10	uwaSi5 [myo-3p::CERULEAN-VENUS::lgg-1 + unc-119(+)] II. LGG-1 tagged with oxCerulean–oxVenus double fluorescent protein (dFP) for monitoring autophagy in body wall muscle.
Microtubule-associated protein 1 light chain 3 (MAP-LC3)	DLM11	uwaEx6 [rab-3p::CERULEAN-VENUS::lgg-1 + unc-119(+)]. Lgg-1 tagged with oxCerulean–oxVenus double fluorescent protein (dFP) for monitoring autophagy in neurons.
Microtubule-associated protein 1 light chain 3 (MAP-LC3)	MAH215	sqIs11 [lgg-1p::mCherry::GFP::lgg-1 + rol-6]. Tandem-tagged autophagy reporter strain.
Microtubule-associated protein 1 light chain 3 (MAP-LC3)	YQ243	wfIs232 [app-1p::mCherry::H2B::unc-54 + unc-119(+)]. mCherry::H2B expression in the nuclei of intestinal cells.
Phosphatase and tensin homolog (PTEN)–induced kinase 1 (PINK1)	AGD926	byEx655 [pink-1p::pink-1::GFP + myo-2p::mCherry + herring sperm DNA]. Pink-1 translational GFP fusion generated by cloning the complete pink-1 genomic fragment.

All strains in the table are available from the *Caenorhabditis* Genetics Center (CGC).

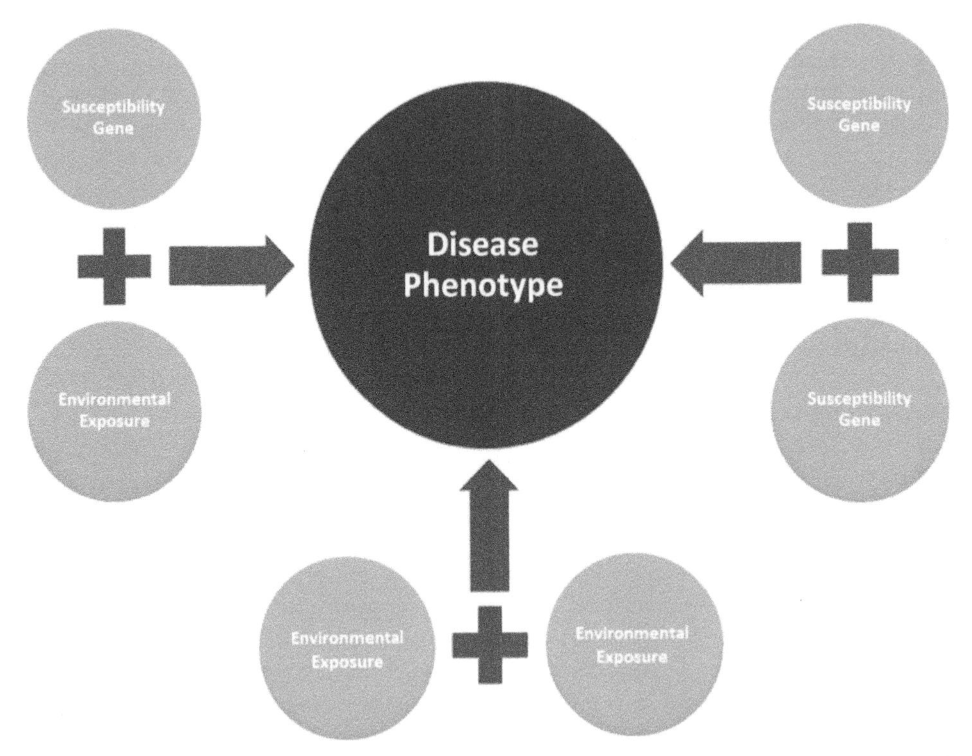

FIGURE 6.3 Multi-hit hypothesis. This figure depicts various ways, based on a multi-hit hypothesis, that a person could ultimately manifest a disease phenotype.

and increased oxidative stress, which are not unique, however, to neurodegeneration. Because *C. elegans* (1) have many genes that are homologous or orthologous to human genes, (2) are easily manipulated to create deletion mutants, (3) can have human disease–related genes inserted, and/or (4) can be assessed with methods routinely used in cell cultures and mammals, these worms provide a unique high-throughput opportunity to determine how genes, toxicants, or potential therapeutics alter biomolecules or cellular pathways.

Alzheimer's Disease

Alzheimer's disease (AD) is the most common neurodegenerative disease in the United States, affecting almost six million people. Despite being multifaceted, the disease is most commonly associated with memory loss and a diminished ability to engage in complex thought processing and analysis. Because AD is highly correlated with aging, it is likely that the number of cases of AD will continue to increase substantially as people live longer. The causes of AD are not well understood, but some studies suggest that people who engage in abundant social, physical, and mental activities are less likely to be diagnosed with or suffer from AD (or even age-related dementia) than those who do not (Qiu et al., 2009). Unfortunately, current treatments focus predominantly on slowing the disease progression. Tragically, there is no known cure.

The pathology of AD is well documented (Calderon-Garciduenas and Duyckaerts, 2017). At autopsy, neurons, particularly those in the prefrontal cortex, frontal cortex, and temporal lobe, show an increase in intracellular β-amyloid plaques (Aβ) and phosphorylated tau (p-tau). This results in intraneuronal tangles that compromise the neuron's anatomical and physiological integrity (Mattsson et al., 2012; Revett et al., 2013). The Aβ responsible for plaque formation results from abnormal cleavage of amyloid precursor protein (APP). Tangles, another pathological hallmark of AD, are formed when hyperphosphorylated tau aberrantly interacts with microtubule proteins. Both Aβ plaques and tau tangles are thought to ultimately lead to neuronal death. Although mutations in APP, presenilin 1 (PSEN1), and PSEN2 account for 6%–8% of early-onset AD diagnoses, the etiology of the vast majority of AD cases is unknown (Patterson et al., 2008). Increased plasma levels of Aβ have been observed in families with AD, people with mutations in APP, PSEN, or those who have Down's syndrome (Irizarry, 2004). Although work is rapidly progressing to bring AD-specific biomarkers to the clinic (Henriques et al., 2018; Scarano et al., 2016), longitudinal studies are required to determine how these biomarkers change over time in the presence and absence of pharmacological intervention.

Fortunately, the simplicity of the nervous system of *C. elegans* has been able to help researchers identify mechanisms and protein interactions involved in AD. Biomarkers with a demonstrated link to this disease, such as Aβ accumulation, are present in worms. In other cases, the worms can be easily manipulated to include or produce these markers. In addition, many genes associated with AD have orthologs in *C. elegans* (Table 6.3). For example, *C. elegans* express cytoplasmic $A\beta_{3-42}$, which elicits a toxic effect similar to that observed in humans who express $A\beta_{1-42}$ (McColl et al., 2012). Furthermore, incorporating endpoints mentioned in previous sections with AD-specific transgenic worms can increase the screening power available to those in the scientific community. But using *C. elegans* has also resulted in more rapid data generation time (weeks) due to both the short life span and availability of the worm's sequenced genome. Furthermore, 60%–80% of human genes associated with disease phenotypes have been identified in *C. elegans* (Keowkase, 2010). As a result of this sequence knowledge, scientists can more easily target AD genes and proteins of interest in the nematode.

Parkinson's Disease

PD is the second most common neurodegenerative disease in the United States, next to AD. The characteristic bradykinesia, resting tremor, rigidity, and postural instability typically result when 70%–80% of the dopaminergic neurons in the *substantia nigra, pars compacta*, die. Although PD is predominantly associated with aging, approximately 5%–10% of the cases, many of which are classified as "early-onset," can be attributed to genetic mutations (Rousseaux et al., 2012). On the other hand, head trauma, chronic exposure to various heavy metals (i.e., Fe, Mn), living and/or working in an agricultural environment, and acute exposure to high concentrations of various pesticides (i.e., rotenone, maneb, paraquat) also increase a person's risk of PD or parkinsonism. In general, however, most incidents of PD do not have a clear etiology and are thus referred to as "idiopathic" (Fritsch et al., 2012). As with most neurodegenerative diseases, treatment typically focuses on reducing disease symptoms rather than halting or reversing the actual disease progression. Thus, identification of biomarkers of this disease that would aid early diagnosis would be of benefit to patients and their family members.

Many biomarkers are associated with mutations in both autosomal dominant and recessive genes (Lesage and Brice, 2009). In general, Lewy bodies, predominantly composed of α-synuclein deposits, are seen in

TABLE 6.3 Genes Altered in Alzheimer's Disease and *Caenorhabditis elegans* Analog

Human Gene	Function in Humans	*C. elegans* Genes	Function in *C. elegans*	Location
Tau	Neurofibrils structure	*Ptl-1*	Microtubule-binding protein required for intracellular transportation and stabilization	Microtubules
Dynein, light chain, LC8-type 1 (DYNLL1)	Dynein light chain	*dlc-1*	Cellular transportation	Adult body wall muscle, pharynx, posterior intestine, and nervous system
Kinesin-1	Kinesin heavy chain	*unc-116, unc104*	Transportation and localization of vesicles	Neurons
Amyloid-like protein 1/2 (APLP1/2)	Memory	*Ard-1, Apl-1*	Molting, morphogenesis, pharyngeal pumping, and progression past L1 larva stage	Neurons, muscles, hypodermis, and supporting cells of larva and adults
Presenilins	Aβ cleavage	*Sel-12, Spr-1, Spr-5, Spe-4, Hop-1*	Influential in gonadal, germline, and vulval formation	Within receiving cells, fibrous body-membranous organelles and tubulin in spermatocytes
Amyloid beta (A4) precursor protein-binding, family B, member (2APBB2)	APP signal transduction	*Feh-1*	Embryonic and larval viability; pharyngeal pumping	Pharynx and neuronal subpopulations
Eukaryotic translation initiation factor 6 (Eif6)	Translation initiation factor that prevents association of the 40S and 60S ribosomal subunits	*Eif-6*	Normal growth and cell division, ribosome assembly	Ubiquitous expression in somatic larval and adult tissues

Comparison of genes associated with Alzheimer's disease in humans and their corresponding family members in *C. elegans*. Where possible, putative function is included for the *C. elegans* gene product (Bayer et al., 1999; Blanco et al., 1998; Corneveaux et al., 2010; Grundke-Iqbal et al., 1986; Lazarov et al., 2005; Putney, 2000).

multiple brain regions of PD patients at autopsy. Recessive mutations in *parkin, phosphatase and tensin homolog (PTEN)–induced putative kinase 1 (pink1)*, or *Parkinson disease 7 (PARK-7* or *DJ-1)* are thought to confer abnormalities in the ubiquitin–proteosome complex and autophagy (Coppede, 2012). Mutations identified in the *leucine-rich repeat kinase 2 (LRRK2)* gene and parkin result in defective oxidative stress sensors and mitochondrial dysfunction (Fig. 6.4). Fortunately, not only are many of these genes found in *C. elegans*, but, in most cases, they are also actually expressed in the nervous system of the worm (Table 6.4). In addition, like their human homologs and orthologs, they are associated with mitochondria, oxidative pathways, and neuronal integrity (Harrington et al., 2010; Saha et al., 2009). Identifying and characterizing these genes and their protein products is important, and in many cases knocking down or knocking out genes of interest helps determine what effect loss-of-function mutations have on the associated biochemical pathways. Again, by coupling the genetic tractability of *C. elegans* with endpoints already identified in cell culture, rodents, and humans, the ability to screen for PD-related toxicants and potential therapeutics dramatically increases relative to screening only in mammals.

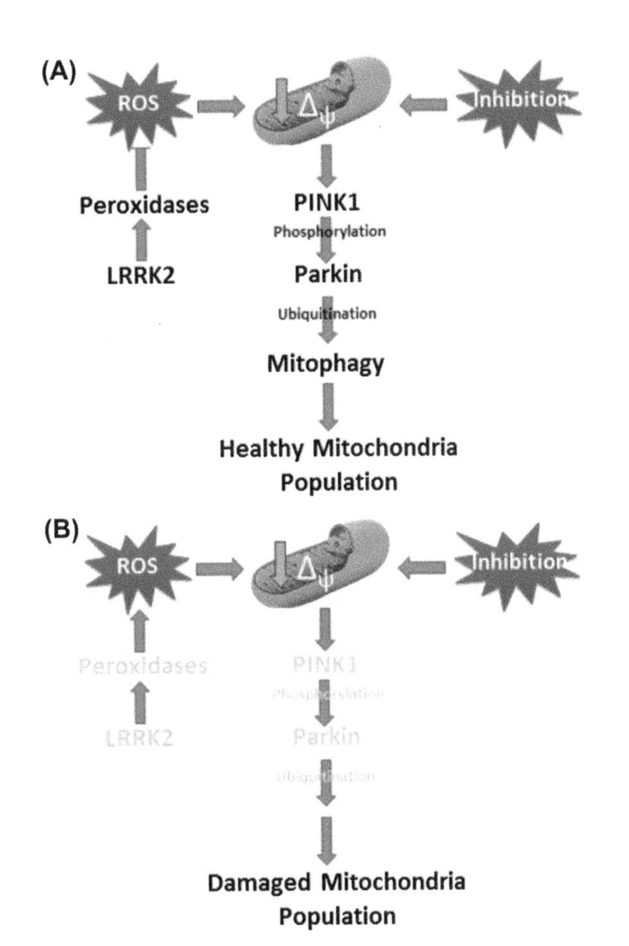

FIGURE 6.4 Genetic mutations leading to mitochondrial damage. (A) Depiction of genes relevant to PD that help protect cells from mitochondrial damage and oxidative stress. (B) When these gene products (proteins) no longer function in familial or early-onset PD, oxidative stress can increase, and mitochondria can be damaged. *PD*, Parkinson disease; *ROS*, reactive oxygen species.

CONCERNS UNIQUE TO *C. ELEGANS*

For researches who have never used *C. elegans*, there are several husbandry-related issues that are somewhat unique to this model organism. Thus, it is important to maintain good laboratory records, provide detailed protocols in manuscripts, and pay attention to the worm strains and bacterial strains used throughout each study. In this respect, using worms is partially analogous to using cell culture; conversely, worms are intact organisms with all of the complex physiology that accompanies *in vivo* work. Thus, as researchers choose *C. elegans* to complement or inform their work in more traditional organisms, it is important to keep in mind the following: (1) life stage at both the time of treatment and endpoint assessment, (2) data obtained from unique synchronizations, (3) strain characteristics, and (4) environmental control and stability.

As with almost all other organisms, the physiology of worms changes during their various life stages. Furthermore, during the early larval stages, *C. elegans* rely primarily on glycolysis and the glyoxylate cycle for energy production but shift to oxidative phosphorylation during later larval stages and early adulthood (Braeckman et al., 2008). In addition to respiration, control of transcription and translation, apoptosis regulation, and expression of antioxidant and other protective genes also change throughout development and with age. To minimize data variation, worms should be "synchronized," so they are all in the same life cycle. This is often accomplished by lysing gravid worms and isolating eggs for studies requiring large numbers. On the other hand, if relatively few worms are needed, worms in the same life cycle can by physically picked from one plate and moved to a new treatment plate. Synchronization via bleaching to lyse the worms can also minimize any fungal contamination that may be on the plates. Following synchronization, however, care must be taken to ensure adherence to the determined time point when worms will be treated, particularly if they are treated during larval stages. Because they grow and mature so rapidly, changing treatment timing by even an hour may mean that worms have progressed to a different life stage. The same is true for assessing endpoints: a strict time table must be followed and documented in published materials/methods sections to aid in data reproducibility. If longitudinal studies are required, 5′-fluorodeoxyuridine (FUdR) can be

TABLE 6.4 Genes Altered in Parkinson's Diseases and *Caenorhabditis elegans* Analog

Human Gene	Function in Humans	*C. elegans* Gene	Function in *C. elegans*	Location
Microtubule-associated protein tau (Tau)	Microtubule stabilization	*Pgrn-1*	Regulates clearance of apoptotic cells	Intestinal organelles
Parkinson protein 7 (PARK7/DJ-1)	Speculated to be a redox sensitive molecular chaperone	*Djr-1.1*	May promote stability of the transcription factors SKN-1 or SKNR-1 (homologs of mammalian NFE2L2)	Primarily in the intestine and localizes to the nucleus and the cytoplasm
ATPase type 13A2 (ATP13A2/PARK9)	Inorganic cation transport	*Catp-6, Lagr-1, Ymel-1*	Embryo development; fertility	Pharyngeal muscles
PTEN-induced putative kinase 1 (PINK1/PARK6)	Serine/threonine protein kinase localized to mitochondria	*Pink-1*	Normal response to ROS, mitochondria homeostasis, neurite outgrowth, and brood size	Neurons, musculature, and vulvar tissues in larvae and adults
Leucine-rich repeat kinase 1 (LRRK1)	Unknown, however in mice seem to aid in organogenesis and mutations	*Lrk-1*	Presynaptic vesicles protein	Golgi apparatus
NADH–Ubiquinone oxidoreductase flavoprotein 2	Reducing agent	*F53F4.10*	Facilitates electron transport in the electron transport chain	Predicted to be a mitochondrial protein
PARKIN	Component of a multiprotein E3 ubiquitin ligase complex	*Pdr-1*	Proteasome-dependent degradation of proteins in response to cytosolic and ER stress	Predicted to be cytosolic
Ubiquitin carboxyl-terminal esterase L1 (UCHL1)	Recycling polymeric chains of ubiquitin	*Ubh-1*	Cellular senescence and aging	Predicted to be cytosolic

Comparison of genes associated with Parkinson's disease in humans and their corresponding family members in *C. elegans*. Where possible, putative function is included for the *C. elegans* gene product (Birkmayer et al., 1993; Biskup et al., 2007; Maraganore et al., 2004; Mollenhauer et al., 2006; Weihofen et al., 2008; Zhang et al., 2005).

used to sterilize *C. elegans* and prevent additional egg hatching, which would introduce mixed-aged populations (Lionaki and Tavernarakis, 2013). It is also the case that one unique synchronization is usually considered an interexperimental replicate in *C. elegans* research and should be noted as such in published protocols.

As with most vertebrate model organisms, differences in strains will often lead to differences in results as well. The same is true for *C. elegans*. Although the research strains are derived from the canonical wild-type (Bristol or N2) strain, genetic mutations that make these worms so useful can also change some phenotypes. This should not be surprising, however. In addition, although the vast majority of reproduction in *C. elegans* involves hermaphrodite self-fertilization, which minimizes genetic variability, genetic drift has still been noted (Cutter et al., 2009). To minimize the effect this may have on results, it is good practice to order new stocks from the CGC on a regular basis or to pull up new stocks from those frozen in the research laboratory. This latter process is similar to what is done with cell culture stocks. This process is also important for the bacterial cultures that are used to feed the *C. elegans* because genetic variation (e.g., gene knockdown via iRNA sequences) can be introduced by feeding worms genetically modified *E. coli*. Maintaining good notes and observing changes in behavior or endpoints will help researchers identify possible changes in the genetics of a strain.

Finally, these worms are very sensitive to changes in their environment. For example, in about 50 h, worms raised at 25°C can move from eggs laid on a plate to adults capable of laying their own eggs (Fig. 6.5). If the temperature is reduced to 16°C, worms will not be mature enough to lay eggs until about 90 h after laying egg (Altun et al., 2002–2010). This means that temperatures must be kept steady throughout a study, so times for treatment and endpoint assessment will take place at the same life stage each time. In a similar manner, the tonicity of the plates, the treatment solutions, and washing solutions should also be carefully monitored. Although the protective cuticle surrounding worms can provide some protection from lysing (in a hypotonic solution) or undergoing crenation (in a hypertonic solution), these osmotic changes will still induce stress in the animal. This stress can adversely affect gene

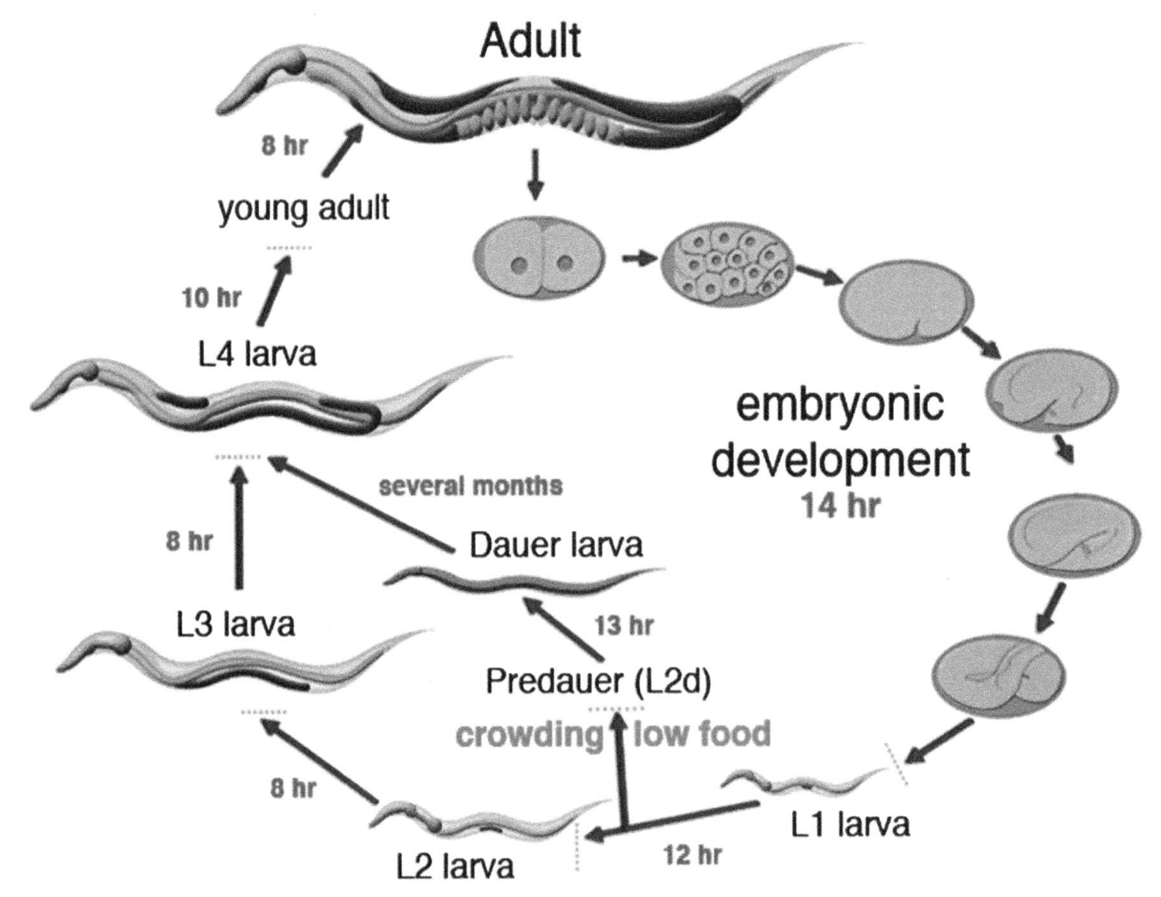

FIGURE 6.5 Life cycle of *Caenorhabditis elegans* at 25°C. The rate at which *C. elegans* proceed through its life cycle is temperature-dependent, and worms progress faster at higher temperatures. Each stage is shown, along with the corresponding size and cartoon depiction of the way the worm looks under a microscope. *Photo by Unknown Author is licensed under CC BY-SA.*

transcription and translation, which can alter experimental results. With careful monitoring, however, almost all of these confounders can be controlled or even eliminated.

CONCLUDING REMARKS AND FUTURE DIRECTIONS

Although humans and *C. elegans* are far removed from each other on the evolutionary tree, they still share remarkable genetic homology and orthology. As such, they are a valuable model organism for studying biomarkers of human disease. Not only does their transparency make it relatively easy to follow protein production using fluorescent tags, but the ease with which their genes can be manipulated through siRNA, CRISPR/Cas 9, the ability to insert relevant human genes, and the rapid production of traditional knockout animals makes them an attractive model organism. Furthermore, their ability to reproduce both hermaphroditically and through male-hermaphrodite fertilization allows for greater control over the introduction of genetic variation than what is available with more commonly used organisms.

C. elegans also has a much shorter life span than other model organisms. This makes it an ideal model for studying diseases associated with aging. On the other hand, it also means that more unique replicates of experiments can also be completed in a shorter period of time, even in studies that do not require "aged" animals. Using *C. elegans* to further characterize and explore the potential usefulness of biomarkers linked to these pathologies, one can significantly reduce the amount of time needed to study the effects of aging. It is also possible to determine if and/or how the expression of those proteins or the regulation of specific genes used as biomarkers may change over the life span of the organism. Such information is important to verify that alterations in biomarker levels or activities are not strictly age related, but predominantly disease related.

Finally, *C. elegans* offers a unique opportunity to study protein and gene regulation in an *in vivo* environment that rivals the ease of use, cost, and genetic manipulation of many *in vitro* systems. As such, these worms can be used to determine whether various pharmaceutical interventions have any positive effect on the concentration or activity of the various biomarkers of interest. This provides a unique opportunity to explore the role of treatments, to a first approximation, in an animal model that has relevance to human disease. Taken together, these examples suggest that *C. elegans* is a useful model organism that should be utilized to a greater extent in the context of biomarkers and human diseases.

Acknowledgments

The authors would like to thank Ashley Norton for her editorial assistance in the preparation of this chapter.

References

Altun, Z., Herndon, L., Crocker, C., et al., 2002–2010. Wormatlas. Retrieved from: http://www.wormatlas.org.

An, J.H., Blackwell, T.K., 2003. Skn-1 links *C. elegans* mesendodermal specification to a conserved oxidative stress response. Genes Dev. 17 (15), 1882–1893.

Avila, D., Helmcke, K., Aschner, M., 2012. The *Caenorhabiditis elegans* model as a reliable tool in neurotoxicology. Hum. Exp. Toxicol. 31 (3), 236–243.

Bailey, D.C., Todt, C.E., Burchfield, S.L., et al., 2017. Chronic exposure to a glyphosate-containing pesticide leads to mitochondrial dysfunction and increased reactive oxygen species production in *Caenorhabditis elegans*. Environ. Toxicol. Pharmacol. 57, 46–52.

Bayer, T.A., Cappai, R., Masters, C.L., et al., 1999. It all sticks together—the APP-related family of proteins and Alzheimer's disease. Mol. Psychiatry 4 (6), 524–528.

Benedetto, A., Au, C., Avila, D.S., et al., 2010. Extracellular dopamine potentiates Mn-induced oxidative stress, lifespan reduction, and dopaminergic neurodegeneration in a *bli-3*-dependent manner in *Caenorhabditis elegans*. PLoS Genet. 6 (8).

Berge, R.K., Bjorndal, B., Strand, E., et al., 2016. Tetradecylthiopropionic acid induces hepatic mitochondrial dysfunction and steatosis, accompanied by increased plasma homocysteine in mice. Lipids Health Dis. 15, 24.

Betarbet, R., Canet-Aviles, R.M., Sherer, T.B., et al., 2006. Intersecting pathways to neurodegeneration in Parkinson's disease: effects of the pesticide rotenone on DJ-1, alpha-synuclein, and the ubiquitin-proteasome system. Neurobiol. Dis. 22 (2), 404–420.

Birkmayer, J.G., Vrecko, C., Volc, D., et al., 1993. Nicotinamide adenine dinucleotide (NADH): a new therapeutic approach to Parkinson's disease. Comparison of oral and parenteral application. Acta Neurol. Scand. Suppl. 146, 32–35.

Biskup, S., Moore, D.J., Rea, A., et al., 2007. Dynamic and redundant regulation of *lrrk2* and *lrrk1* expression. BMC Neurosci. 8, 102.

Blanco, G., Irving, N.G., Brown, S.D., et al., 1998. Mapping of the human and murine X11-like genes (APBA2 and APBA2), the murine Fe65 gene (APBB1), and the human Fe65-like gene (APBB2): genes encoding phosphotyrosine-binding domain proteins that interact with the Alzheimer's disease amyloid precursor protein. Mamm. Genome 9 (6), 473–475.

Blaxter, M., 2011. Nematodes: the worm and its relatives. PLoS Biol. 9 (4), e1001050.

Boyd, W.A., McBride, S.J., Rice, J.R., et al., 2010. A high-throughput method for assessing chemical toxicity using a *Caenorhabditis elegans* reproduction assay. Toxicol. Appl. Pharmacol. 245 (2), 153–159.

Braeckman, B.P., Houthoofd, K., Vanfleteren, J.R., 2008. Intermediary Metabolism. WormBook.

Brenner, S., 1974. The genetics of *Caenorhabditis elegans*. Genetics 77, 71–94.

Calahorro, F., Ruiz-Rubio, M., 2011. *Caenorhabditis elegans* as an experimental tool for the study of complex neurological diseases: Parkinson's disease, Alzheimer's disease and autism spectrum disorder. Invert. Neurosci. 11 (2), 73–83.

Calderon-Garciduenas, A.L., Duyckaerts, C., 2017. Alzheimer disease. Handb. Clin. Neurol. 145, 325–337. https://doi.org/10.1016/B978-0-12-802395-2.00023-7.

Chalfie, M., White, J., 1988. The nervous system. In: Wood, W.B. (Ed.), The Nematode *Caeorhabditis elegans*. Cold Spring Harbor Laboratory Press, Cold Spring Harbor, pp. 337–392.

Chikka, M.R., Anbalagan, C., Dvorak, K., et al., 2016. The mitochondria-regulated immune pathway activated in the *C. elegans* intestine is neuroprotective. Cell Rep. 16 (9), 2399–2414.

Consortium, T.C. e. S., 1998. Genome sequence of the nematode *C. elegans*: a platform for investigating biology. Science 282, 2012–2018.

Coppede, F., 2012. Genetics and epigenetics of Parkinson's disease. Sci. World J. 2012, 489830.

Corneveaux, J.J., Liang, W.S., Reiman, E.M., et al., 2010. Evidence for an association between KIBRA and late-onset Alzheimer's disease. Neurobiol. Aging 31 (6), 901–909.

Corsi, A.K., 2006. A biochemist's guide to *Caenorhabditis elegans*. Anal. Biochem. 359 (1), 1–17.

Crooks, D.R., Ghosh, M.C., Braun-Sommargren, M., et al., 2007. Manganese targets m-aconitase and activates iron regulatory protein 2 in AFf5 GABAergic cells. J. Neurosci. Res. 85 (8), 1797–1809.

Crowley, L.C., Christensen, M.E., Waterhouse, N.J., 2016. Measuring mitochondrial transmembrane potential by TMRE staining. Cold Spring Harb. Protoc. 2016 (12) https://doi.org/10.1101/pdb.prot087361.

Cutter, A.D., Dey, A., Murray, R.L., 2009. Evolution of the *Caenorhabditis elegans* genome. Mol. Biol. Evol. 26 (6), 1199–1234.

Dingley, S., Polyak, E., Lightfoot, R., et al., 2010. Mitochondrial respiratory chain dysfunction variably increases oxidant stress in *Caenorhabditis elegans*. Mitochondrion 10 (2), 125–136.

Farboud, B., 2017. Targeted genome editing in *Caenorhabditis elegans* using CRISPR/Cas9. Wiley Interdiscip. Rev. Dev. Biol. 6 (6) https://doi.org/10.1002/wdev.287.

Fay, D. (Ed.), 2006. Forward Genetics and Genetics Mapping: The *C. elegans* Research Community. WormBook.

Fritsch, T., Smyth, K.A., Wallendal, M.S., et al., 2012. Parkinson disease: research update and clinical management. South. Med. J. 105 (12), 650–656.

Grad, L.I., Sayles, L.C., Lemire, B.D., 2007. Isolation and functional analysis of mitochondria from the nematode *Caenorhabditis elegans*. In: Leister, D., Herrmann, J. (Eds.), Methods in Molecular Biology: Mitochondria: Practical Protocols, vol. 372. Humana Press Inc., Totowa, NJ.

Grundke-Iqbal, I., Iqbal, K., Tung, Y.C., et al., 1986. Abnormal phosphorylation of the microtubule-associated protein tau (tau) in Alzheimer cytoskeletal pathology. Proc. Natl. Acad. Sci. U.S.A. 83 (13), 4913–4917.

Harrington, A.J., Hamamichi, S., Caldwell, G.A., et al., 2010. *C. elegans* as a model organism to investigate molecular pathways involved with Parkinson's disease. Dev. Dyn. 239 (5), 1282–1295.

He, Y., Leung, K.W., Zhang, Y.H., et al., 2008. Mitochondrial complex I defect induces ROS release and degeneration in trabecular meshwork cells of POAG patients: protection by antioxidants. Invest. Ophthalmol. Vis. Sci. 49 (4), 1447–1458. https://doi.org/10.1167/iovs.07-1361.

Hengartner, M.O., Horvitz, H.R., 1994. Programmed cell death in *Caenorhabditis elegans*. Curr. Opin. Genet. Dev. 4 (4), 581–586.

Henriques, A.D., Benedet, A.L., Camargos, E.F., et al., 2018. Fluid and imaging biomarkers for Alzheimer's disease: where we stand and where to head to. Exp. Gerontol. https://doi.org/10.1016/j.exger.2018.01.002.

Hoglinger, G.U., Feger, J., Prigent, A., et al., 2003. Chronic systemic complex I inhibition induces a hypokinetic multisystem degeneration in rats. J. Neurochem. 84 (3), 491–502.

Irizarry, M.C., 2004. Biomarkers of Alzheimer disease in plasma. NeuroRx 1 (2), 226–234.

Joshi, P.M., Riddle, M.R., Djabrayan, N.J., et al., 2011. *Caenorhabditis elegans* as a model for stem cell biology. Dev. Dyn. 239 (5), 1539–1554.

Keowkase, R., 2010. Fluoxetine protects against amyloid-beta toxicity, in part via daf-16 mediated cell signaling pathway, in *Caenorhabditis elegans*. Neuropharmacology 358–365.

Koopman, M., Michels, H., Dancy, B.M., et al., 2016. A screening-based platform for the assessment of cellular respiration in *Caenorhabditis elegans*. Nat. Protoc. 11 (10), 1798–1816.

Lazarov, O., Morfini, G.A., Lee, E.B., et al., 2005. Axonal transport, amyloid precursor protein, kinesin-1, and the processing apparatus: revisited. J. Neurosci. 25 (9), 2386–2395.

Lesage, S., Brice, A., 2009. Parkinson's disease: from monogenic forms to genetic susceptibility factors. Hum. Mol. Genet. 18 (R1), R48–R59.

Leung, M.C., Williams, P.L., Benedetto, A., et al., 2008. *Caenorhabditis elegans*: an emerging model in biomedical and environmental toxicology. Toxicol. Sci. 106 (1), 5–28.

Lionaki, E., Markaki, M., Tavernarakis, N., 2013. Autophagy and ageing: insights from invertebrate model organisms. Ageing Res. Rev. 12 (1), 413–428.

Lionaki, E., Tavernarakis, N., 2013. Assessing aging and senescent decline in *Caenorhabditis elegans*: cohort survival analysis. Methods Mol. Biol. 965, 473–484.

Luz, A.L., Lagido, C., Hirschey, M.D., et al., 2016. *In vivo* determination of mitochondrial function using luciferase-expressing *Caenorhabditis elegans*: contribution of oxidative phosphorylation, glycolysis, and fatty acid oxidation to toxicant-induced dysfunction. Curr. Protoc. Toxicol. 69, 25.28.21–25.28.22.

Ma, L., Zhao, Y., Chen, Y., et al., 2018. *Caenorhabditis elegans* as a model system for target identification and drug screening against neurodegenerative diseases. Eur. J. Pharmacol. 819, 169–180.

Maraganore, D.M., Lesnick, T.G., Elbaz, A., et al., 2004. UCHL1 is a Parkinson's disease susceptibility gene. Ann. Neurol. 55 (4), 512–521.

Mattsson, N., Rosen, E., Hansson, O., et al., 2012. Age and diagnostic performance of Alzheimer disease csf biomarkers. Neurology 78 (7), 468–476.

Mazlumoglu, M.R., Ozkan, O., Alp, H.H., et al., 2017. Measuring oxidative DNA damage with 8-hydroxy-2′-deoxyguanosine levels in patients with laryngeal cancer. Ann. Otol. Rhinol. Laryngol. 126 (2), 103–109.

McColl, G., Roberts, B.R., Pukala, T.L., et al., 2012. Utility of an improved model of amyloid-beta (abeta(1)(-)(4)(2)) toxicity in *Caenorhabditis elegans* for drug screening for Alzheimer's disease. Mol. Neurodegener. 7, 57.

Mollenhauer, B., Trenkwalder, C., von Ahsen, N., et al., 2006. Beta-amlyoid 1-42 and tau-protein in cerebrospinal fluid of patients with Parkinson's disease dementia. Dement. Geriatr. Cogn. Disord. 22 (3), 200–208.

Morgan, D.E., Crittenden, S.L., Kimble, J., 2011. The *C. elegans* adult male germline: stem cells and sexual dimorphism. Dev. Biol. 346 (2), 204–214.

Mseddi, M., Ben Mansour, R., Gargouri, B., et al., 2017. Proteins oxidation and autoantibodies' reactivity against hydrogen peroxide and malondialdehyde-oxidized thyroid antigens in patients' plasmas with Graves' disease and Hashimoto thyroiditis. Chem. Biol. Interact. 272, 145–152.

Patterson, C., Feightner, J.W., Garcia, A., et al., 2008. Diagnosis and treatment of dementia: risk assessment and primary prevention of Alzheimer disease. CMAJ 178 (5), 548–556.

Prior, H., Jawad, A.K., MacConnachie, L., et al., 2017. Highly efficient, rapid and co-CRISPR-independent genome editing in *Caenorhabditis elegans*. G3 (Bethesda) 7 (11), 3693–3698.

Putney Jr., J.W., 2000. Presenilins, Alzheimer's disease, and capacitative calcium entry. Neuron 27 (3), 411–412.

Qiu, C., Kivipelto, M., von Strauss, E., 2009. Epidemiology of Alzheimer's disease: occurrence, determinants, and strategies toward intervention. Dialogues Clin. Neurosci. 11 (2), 111–128.

Revett, T.J., Baker, G.B., Jhamandas, J., et al., 2013. Glutamate system, amyloid β peptides and tau protein: functional interrelationships and relevance to Alzheimer disease pathology. J. Psychiatry Neurosci. 38 (1), 6–23.

Rousseaux, M.W., Marcogliese, P.C., Qu, D., et al., 2012. Progressive dopaminergic cell loss with unilateral-to-bilateral progression in a genetic model of Parkinson disease. Proc. Natl. Acad. Sci. U.S.A. 109 (39), 15918–15923.

Saha, S., Guillily, M.D., Ferree, A., et al., 2009. LRRK2 modulates vulnerability to mitochondrial dysfunction in *Caenorhabditis elegans*. J. Neurosci. 29 (29), 9210–9218.

Scarano, S., Lisi, S., Ravelet, C., et al., 2016. Detecting Alzheimer's disease biomarkers: from antibodies to new bio-mimetic receptors and their application to established and emerging bioanalytical platforms-a critical review. Anal. Chim. Acta 940, 21–37.

Sharp, P.A., 1999. RNAi and double-strand RNA. Genes Dev. 13 (2), 139–141.

Shaye, D.D., Greenwald, I., 2011. Ortholist: a compendium of *C. elegans* genes with human orthologs. PLoS One 6 (5), e20085.

Silverman, G.A., Luke, C.J., Bhatia, S.R., et al., 2009. Modeling molecular and cellular aspects of human disease using the nematode *Caenorhabditis elegans*. Pediatr. Res. 65 (1), 10–18.

Sulston, J.E., 1988. Cell lineage. In: Wood, W.B. (Ed.), The Nematode *Caenorhabditis elegans*. Cold Spring Harbor Laboratory Press, Cold Spring Harbor, NY.

Sulston, J.E., Brenner, S., 1974. The DNA of *Caenorhabditis elegans*. Genetics 77 (1), 95–104.

Sutphin, G.L., Backer, G., Sheehan, S., et al., 2017. *Caenorhabditis elegans* orthologs of human genes differentially expressed with age are enriched for determinants of longevity. Aging Cell 16 (4), 672–682.

Swierczek, N.A., Giles, A.C., Rankin, C.H., et al., 2011. High-throughput behavioral analysis in *C. elegans*. Nat. Methods 8 (7), 592–598.

Tsikas, D., 2017. Assessment of lipid peroxidation by measuring malondialdehyde (MDA) and relatives in biological samples: analytical and biological challenges. Anal. Biochem. 524, 13–30.

Twig, G., Elorza, A., Molina, A.J., et al., 2008. Fission and selective fusion govern mitochondrial segregation and elimination by autophagy. EMBO J. 27 (2), 433–446.

Varshney, L.R., Chen, B.L., Paniagua, E., et al., 2011. Structural properties of the *Caenorhabditis elegans* neuronal network. PLoS Comput. Biol. 7 (2), e1001066.

Ward, S., Thomson, N., White, J.G., et al., 1975. Electron microscopical reconstruction of the anterior sensory anatomy of the nematode *Caenorhabditis elegans*. J. Comp. Neurol. 160 (3), 313–337.

Weihofen, A., Ostaszewski, B., Minami, Y., et al., 2008. PINK1 Parkinson mutations, the CDC37/HSP90 chaperones and parkin all influence the maturation or subcellular distribution of PINK1. Hum. Mol. Genet. 17 (4), 602–616.

White, J.G., Southgate, E., Thomson, J.N., et al., 1976. The structure of the ventral nerve cord of *caenorhabditis elegans*. Philos. Trans. R. Soc. Lond. B Biol. Sci. 275 (938), 327–348.

Zhang, L., Shimoji, M., Thomas, B., et al., 2005. Mitochondrial localization of the Parkinson's disease related protein DJ-1: implications for pathogenesis. Hum. Mol. Genet. 14 (14), 2063–2073.

CHAPTER

7

Potential of Small Animals in Toxicity Testing: Hope from Small World

Shiwangi Dwivedi[1], *Sumit Singh Verma*[2], *Clinton D'Souza*[1], *Nikee Awasthee*[2], *Anurag Sharma*[1], *Subash Chandra Gupta*[2]

[1]**Division of Environmental Health and Toxicology, Nitte University Center for Science Education and Research (NUCSER), Deralakatte, India;** [2]**Department of Biochemistry, Institute of Science, Banaras Hindu University, Varanasi, India**

INTRODUCTION

Rapid industrialization, unsafe use of agrochemicals, and various anthropogenic activities can introduce several chemicals into the environment. In general, the chemicals can be categorized as pesticides, polychlorinated biphenyls, metals, polycyclic aromatic hydrocarbons, volatile organic compounds, organic solvents, dioxins, complex chemical mixtures, food additives, etc (Martin et al., 2010; Nesatyy and Suter, 2007; Sutton et al., 2012). The production and use of these chemicals are part of the industrial and agricultural revolutions. These chemical–biological interactions produce serious risk to the environment and to human health (Asgari et al., 2017; Cao et al., 2016; Gupta et al., 2017; Jayaraj et al., 2016).

The major chemical hazards share three common features: (1) environmental persistence, (2) tendency of accumulation in living systems, and (3) serious health adversities. Chemicals released into the environment pose hazards to the ecosystem by interacting at population, community, cellular, and molecular levels (del Carmen and Fuiman, 2005; Houlahan et al., 2000). Therefore, assessing the effects of these toxicants on the environment and on humans is one of major biomedical concern. Research in the field of exposure assessment and hazardous identification is undergoing fundamental change due to scientific and technical advancements.

The last two decades have seen interest on the rapid detection of cellular and organismal toxicity for better predictability. Moreover, chronic exposure of environmental chemicals leads to serious health hazards, such as neurological disorders, hepatic and renal problems, and cancer. Therefore, biomarker-based exposure assessment can be used to monitor the toxic effects of chemicals. Biomarkers are the variations at the biochemical, cellular, molecular, and structural level that are measurable in biological samples (Hook et al., 2014). Apart from early and specific chemical toxic assessment, biomarkers also allow a ranking of different exposures in the laboratory and in the field. In addition, the absolute measured effects through biomarkers can be applied to regulatory environmental agencies.

The use of model organisms has always been at the core of toxicology testing to ensure environmental safety. The resultant information forms the hub of our biological knowledge to date. However, relating such laboratory assessment to the ecosystem can only be achieved by investigation with an appropriate group of organisms. Due to similar developmental pathways and gene homology, mammalian or higher models are the "gold standard" in toxicology research. However, due to regulatory, scientific, and ethical concern, an emphasis has been placed on reducing the use of higher mammals for toxicological research and testing (Badyal and Desai, 2014; Doke and Dhawale, 2015). Moreover, toxicology research needs alternative, rapid, and cost-effective methods to improve prediction of human outcomes. The focus has been given to use and develop alternative animal models and techniques in relation to the principles of the 3Rs, *Reduction*: reducing the use of animals for experiments; *Replacement*: developing alternative methods and models for experiments; and *Refinement*: improving the research and testing methods to

Biomarkers in Toxicology, Second Edition
https://doi.org/10.1016/B978-0-12-814655-2.00007-4

preclude pain and discomfort to the organisms wherever possible in biological research (Beekhuijzen, 2017; Festing et al., 1998).

Within the last two decades, various small/nonmammalian/in vitro models have been developed and designed to replace mammalian models to investigate the effects of toxicants. Due to the availability of genomic and proteomic information, short life span, high reproductive capacity, cost-effectiveness, and technical advancement, these models have become more popular. These animals have been employed for toxicity testing at the molecular, cellular, organism, and population levels (Table 7.1 and Fig. 7.1). A number of chemicals have been tested using these organisms (Table 7.2). The focus of this chapter is to discuss the contribution of three major experimental small animal models (*Drosophila melanogaster*, *Daphnia magna*, and *Eisenia fetida*) in toxicity assessment of environmental chemicals.

TABLE 7.1 Biomarkers Used for Toxicity Testing in Small Organisms

Drosophila melanogaster	*Eisenia fetida*	*Daphnia magna*
MOLECULAR AND CELLULAR BIOMARKER		
Heat shock protein 22, 23, 26, 27, 60, 70, 83	Heat shock protein 60, 70, 72, 90	Heat shock protein 60, 70, 90
Methuselah	Calreticulin	–
Reactive oxygen species	Reactive oxygen species	Reactive oxygen species
Malondialdehyde	Malondialdehyde	Malondialdehyde
Protein carbonyl	Protein carbonyl	Protein concentration
Superoxide dismutase	Superoxide dismutase	Superoxide dismutase
Catalase	Catalase	Catalase
Glutathione	Glutathione	Glutathione
Glutathione peroxidase	Guaiacol peroxidase	Selenium-dependent glutathione-peroxidase
Glutathione S-transferase	Glutathione S-transferase	Glutathione S-transferase
DNA damage check point: *Rad1*, recombination repair protein 1 (*Rrp1*)	DNA repair: poly(ADP-ribose) polymerase	DNA repair (*REV1*)
Dcp1 (caspase 3)	Caspase 8	–
Cytochrome P450: *CYP6d5*, *CYP6a8*, and *CYP12d1*, alcohol dehydrogenase, metallothionein	Calcium hemostasis, metallothionein-2	Linoleic acid, prostaglandins, thromboxanes, metallothionein
Accessory gland proteins	Ecdysone receptor, membrane-associated progesterone receptor, and adiponectin receptor	Methylfarnesoate, estrogen receptor, androgen receptor, and aryl-hydrocarbon receptor
Diptericin, drosomycin, AMP, TOLL, PGRP-LC	AMP, lectins	Lip, Ltb4dh
DEVELOPMENTAL TOXICITY PARAMETERS		
Emergence assay	Growth and mortality	Assessing motility
REPRODUCTIVE PARAMETERS		
Fecundity, fertility, and hatchability	Fertility, cocoon production, and hatching pattern	Fecundity and fertility
NEUROTOXICITY PARAMETERS		
Acetyl-cholinesterase	Acetyl-cholinesterase	Cholinesterase
Jumping and climbing assay	–	Swimming assay
IMMUNOTOXICITY PARAMETERS		
Hemocyte viability and count	Coelomocytes cell viability and count	Hemocyte viability and count
Melanization assay and phagocytosis assay	Phagocytosis assay	Melanization assay

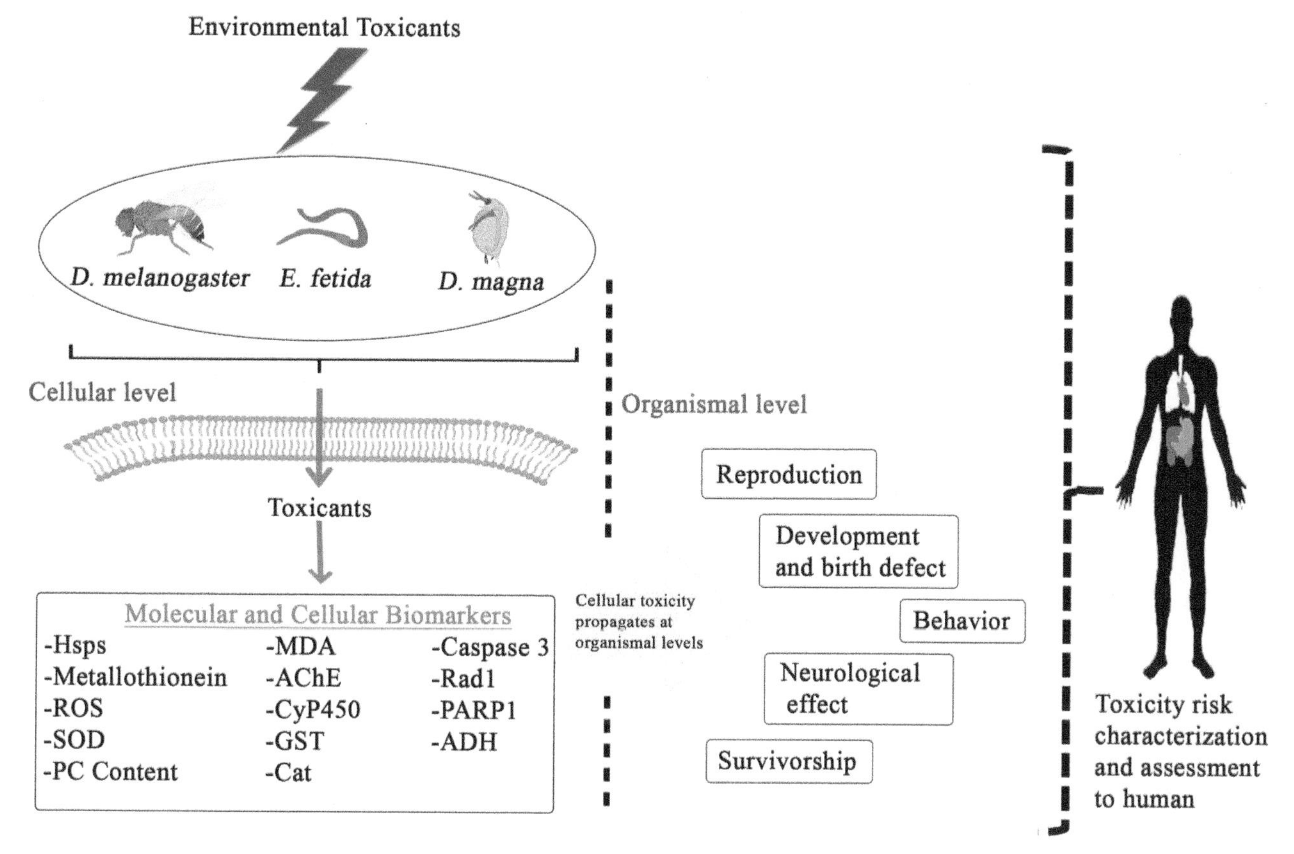

FIGURE 7.1 Schematic diagram showing utility of small organisms in toxicity testing.

TOXICITY TESTING USING SMALL ANIMALS

Drosophila melanogaster

Due to its well-defined developmental biology and genetics, *D. melanogaster* has been used as an in vivo tool in biomedical research. This model has been instrumental in deciphering the pathogenesis of human diseases including aging, cancer, immunity, and some neurodegenerative diseases (Buchon et al., 2014; He and Jasper, 2014; McGurk et al., 2015; Richardson & Kumar, 2002; Rudrapatna et al., 2012; Yadav et al., 2016). Because of its short life span, small size, low maintenance cost, high reproductive capacity, highly conserved genetic architecture, and ease of availability of mutants and transgenic lines, this model has great potential in the assessment of human relevant toxicological pathways. FlyBase (http://flybase.org) is a freely available online database for the biological information of *Drosophila*. *Drosophila* strains (mutants, transgenic, RNAi, Gal4) are also available from four major *Drosophila* stock centers (Bloomington Drosophila Stock Center, Kyoto Stock Center, NIG-FLY, and Vienna *Drosophila* RNAi Center) with nominal charges.

Drosophila ***Model in the Prediction of Cellular Toxicity***

Depicting the effect of environmental toxicants at the cellular level is crucial for human risk assessment. Environmental toxicants are different in nature; therefore cellular toxicity could be assessed at varied forms of regulation on different levels. Among the multiple cellular responses, upregulation of heat shock proteins (Hsps) is considered an early toxicity biomarker (Mahmood et al., 2014). Increased Hsps level in the cell prevents protein aggregation and other protein modification (Takalo et al., 2013). However, initially Hsps was discovered by observing puffing in the *Drosophila* salivary gland chromosome due to the elevated temperature (Ritossa, 1962). Apart from an elevated temperature, it is now clear that Hsps can be induced by other stressors including environmental chemicals (Eder et al., 2007; Kim et al., 2014; Mahmood et al., 2014). Hsps are classified into high (90, 70, and 60 kDa) and low molecular (15–30 kDa) weight families, and the Hsp70 family appears most likely to be induced following environmental exposure (Gupta et al., 2010).

Within the last two decades, transgenic *Drosophila* construct containing a lacZ reporter gene fused with a Hsp70 promoter (*hsp70-lacZ*) have been used for quantification of *hsp70* expression in *Drosophila* larvae exposed to

TABLE 7.2 Environmental Chemicals Tested Using Small Organisms

Chemicals	*Drosophila melanogaster*	*Eisenia fetida*	*Daphnia magna*
Heavy metals	Cadmium, chromium, Iron, copper, nickel, lead, zinc, manganese	Nickel, cadmium, zinc, copper, lead	Cadmium, mercury, copper, zinc, chromium
Organochlorine	Endosulfan, dieldrin, lindane, DDT, β-hexachlorocyclohexane	α-endosulfan	Endosulfan
Organophosphate	Acephate, chlorpyrifos, dichlorvos, malathion, fenitrothion, monocrotophos, methyl parathion, azamethiphos, prothiofos, diazinon	Chlorpyrifos, malathion, fenitrothion	Paraoxon, parathion, chlorpyrifos glyphosate and glyphosate-copper complexes, diazinon, dichlorvos
Triazine	Atrazine	Terbuthylazine	Atrazine
Carbamates	Methomyl, carbofuran	–	Carbendazim
Pyrethroids	Cypermethrin, permethrin, deltamethrin	Cyfluthrin, deltamethrin	Deltamethrin
Organic compound	4-vinylcyclohexene, alcohol formaldehyde, acetaldehyde benzene, toluene perchloroethylene, methyl methanesulfonate xylene, bromobenzene, n-butylbenzene, ethylbenzene, p-isopropyltoluene, naphthalene, styrene, 1,2,3- trichlorobenzene, 1,2,4-trichlorobenzene, 1,2,4-trimethylbenzene, and 1,3,5-trimethylbenzene, benzo(a) pyrene,bisphenol, polycyclic aromatic hydrocarbons	Octane, undecane, decane, 3-methyl heptane, 2,4-dimethyl heptane, 3,3-dimethyl octane, 2,2,4,6,6-pentamethyl heptane, 2,4-di tert buyl phenol, benzo(a) pyrene, bisphenol, polycyclic aromatic hydrocarbons, perfluorooctanoic acid, metaldehyde, chloroacetamide, 3,4-dichloroaniline, pentachlorophenol	3,4-dichloroaniline, chlorothalonil, nitrofen
Inorganic compound	Sodium fluoride	–	Ferric oxide
Nanoparticles	Silica nanoparticle, sliver nanoparticle, titanium dioxide nanoparticle	Silver nanoparticle	Diamond nanoparticle, titanium dioxide
Phthalates	Di-(2-ethylhexyl) phthalate	Di-(2-ethylhexyl) phthalate	–
Herbicides	Glyphosate, atrazine, rotenone	Glyphosate, butachlor, clomazone	Glyphosate, atrazine, rotenone
Fungicide	Mercurial fungicide, agallol	–	

various environmental chemicals, such as heavy metals, pesticides, volatile organic compounds, and industrial and municipal waste (Bhargav et al., 2008; Gupta et al., 2005; Kumar et al., 2011; Mukhopadhyay et al., 2003; Nazir et al., 2003; Siddique et al., 2008; Singh et al., 2009).

In addition to *Hsp70*, other members of the *Hsp* family have also been proposed for toxicity testing. Only one class of gene/protein cannot accurately assess the toxicity of a wide range of pollutants (de Pomerai, 1996; Gupta et al., 2010). Significant induction in the other classes of *hsps* such as *hsp83*, *hsp26*, *hsp23*, and *hsp22* in *Drosophila* larvae after exposure to silica nanoparticles, endosulfan, and volatile organic compounds has also been observed (Pandey et al., 2013; Sharma et al., 2012; Singh et al., 2009).

Parallel to Hsps induction, generation of reactive oxygen species (ROS) and modulation in antioxidant capacity of an organism have been used as indicators of cellular toxicity (Birben et al., 2012; Magder, 2006). Excess ROS levels due to toxicants may cause protein, lipid, and DNA damage in the cell, and eventually various disease emergencies (Fulda et al., 2010). As a protective mechanism, organisms have developed enzymatic antioxidants [e.g., catalase (CAT), glutathione peroxidase (GPx), glutathione reductase (GR), superoxide dismutase (SOD)] and nonenzymatic antioxidants [e.g., glutathione (GSH), vitamins C and D] that detoxify ROS (El Golli-Bennour and Bacha, 2011). The *Drosophila* model has been used successfully to assess ROS along with alteration in oxidative stress markers. These markers include antioxidant enzyme activity (SOD, CAT), GSH-content, lipid peroxidation (LPO), and protein carbonyl (PC) content after exposure to a range of environmental toxicants, such as silica nanoparticles, organophosphate and organochlorine insecticides, industrial and municipal waste, and monocyclic aromatic hydrocarbons (Bhargav et al., 2008; Gupta et al., 2010; Gupta et al., 2007; Pandey et al., 2013; Siddique et al., 2008; Singh et al., 2009).

The immune system is considered the first line of defense against pathogenic organisms. This is a very sensitive system which may be altered by stress factors present in the environment. To investigate Cr(VI)-induced immunotoxicity and its possible mechanism, Pragya et al. (2014) used a *Drosophila* model system. Exposure to Cr(VI) led to an extreme reduction in hemocyte population due to ROS-induced cell death, which could be rescued by the overexpression of SOD. In addition, Gao et al. (2014) also observed that paraquat-induced oxidative stress led to aberrant differentiation of hemocytes in the lymph gland (*Drosophila* hematopoietic organ), which could be rescued by overexpression of E-cadherin. Singh et al. (2011) showed the protective effects of curcumin supplementation against benzene-, toluene-, and xylene-induced oxidative stress-mediated cell death in third instar *Drosophila* larvae.

Cytochrome P450 and GST activities are associated with the detoxification process of xenobiotics and play a vital role in resistance development in the exposed organism (Omiecinski et al., 2011). In *Drosophila*, the gut, malpighian tubules, and the fat body are major sites for cytochrome P450-mediated detoxifications. Several reports have underlined the upregulation in cytochrome P450 and GST activities in *Drosophila* exposed to toxicants (Rajak et al., 2017; Sharma et al., 2012; Terhzaz et al., 2015). The metabolic enzymes (*CYP6d5, CYP6a8,* and *CYP12d1*) have been involved in the degradation of caffeine (Coelho et al. (2015).

Genotoxicity testing represents an essential component of hazardous assessment of environmental toxicants and the comet assay has been widely used as a genotoxic testing method (Martins and Costa, 2015). Recently, the comet assay has been developed in *Drosophila* midgut, brain cells, and hemocytes and used to asses genotoxicity of toxicants ranging from industrial and municipal waste, insecticides, volatile organic compounds, food additives, and heavy metals (Carmona et al., 2011; Sharma et al., 2011, 2012; Siddique et al., 2005a; b; 2008, 2009; Gupta et al., 2005; Singh et al., 2011). In addition, DNA repair pathways/enzymes are considered biomarkers for human biomonitoring (Collins and Azqueta, 2012). In this context, the comet assay has been applied to different *Drosophila* repair-deficient strains to find out affected DNA repair pathways against genotoxic agents such as industrial waste leachates, chromium, and dichlorvos (Mishra et al., 2011, 2014; Siddique et al., 2008).

Prediction of Developmental, Reproductive, and Behavioral Toxicity Using Drosophila melanogaster

Usually acute toxicity tests of short duration are used to extrapolate the safety levels of anthropogenic chemicals (Gormley and Teather, 2003). However, assessing long-term effects on organisms on the basis of these acute tests may be complex and will differ in nature (Roex et al., 2001). Therefore, systematic studies with environmentally relevant exposures of chemicals are required to provide essential information to understand their potential long-term impacts i.e., on development, behavior, and reproduction of the exposed organisms. Developmental toxicity is the study of unwanted effects on the development of an organism which may be the result of exposure to a chemical before conception, during prenatal development, or postnatally. The principal manifestations of developmental toxicity include the following: (1) embryolethality, (2) malformation, (3) growth retardation, and (4) functional deficit (O'Bryant et al., 2011; Sabra et al., 2017; Thompson and Bannigan, 2008; Ueker et al., 2016). A number of studies have shown reduced and delayed emergence pattern, reduced body size, and weight in *Drosophila* when exposed to environmental contaminants such as organophosphates, organochlorines, herbicides, industrial and municipal solid wastes, and heavy metals during early life stage (Abnoos et al., 2013; Al-Momani and Massadeh, 2005; Bhargav et al., 2008; Gupta et al., 2007; Marcus and Fiumera, 2016; Sharma et al., 2011, 2012 Siddique et al., 2009).

Developmental exposure of *Drosophila* larvae to endosulfan has been shown to cause hind leg deformities such as truncation and/or fusion of tarsal segments (tarsomeres) in newly emerged flies (Sharma et al., 2012). Interestingly, these leg deformities are reminiscent of those observed in vertebrates (Mobarak and Al-Asmari, 2011). Ranganathan et al. (1987) observed malformation in legs and wings in the ethanol-exposed *Drosophila*. In another study, methylmercury inhibited alcohol dehydrogenase activity, when co-administrated with alcohol to *Drosophila* during the developmental period (Chauhan and Chauhan, 2016). This result advocates that alcohol consumption by methylmercury-exposed pregnant mothers may have a negative effect on the developing fetus. It has been well established that exposure to environmental chemicals during critical developmental periods leads to serious reproductive, neurological, and behavioral abnormalities (Grandjean and Landrigan, 2014; Heindel et al., 1994; Tiwari et al., 2011).

Researchers have used the *Drosophila* model to evaluate environmental toxicant–induced reproductive emergency. Reproductive toxicity includes the analysis of reproductive ability (reproductive performance), courtship behavior, and molecular analysis. In flies, the male *Drosophila* initiates courtship behavior in response to various sensory inputs including auditory, visual, and olfactory signals to achieve reproductive success (Pan et al., 2012).

Accumulating evidence indicates the susceptibility of courtship behavior to exposure to environment chemicals. Chauhan et al. (2017) observed altered courtship behavior in male *Drosophila* exposed to mercury due to the reduced wing extension. Similar altered courtship was observed in flies in response to lead (Hirsch et al., 2003). Gayathri and Krishnamurthy (1983) developed a method to assess the reproductive performance. In this method, first instar *Drosophila* larvae are exposed to toxicant and allowed to grow throughout the development phase. Virgin male and female flies from exposed and normal larvae are collected and pair mating is established (control male X control female, treated male X treated female, control male X treated female, control female X treated female). These mating pairs are transferred to fresh vials every day for 10 days. To analyze reproductive performance (fecundity, fertility, and hatchability), the number of eggs laid is counted each day. Several studies explored this method to evaluate reproductive adversities induced by an environmental chemical.

Bhargav et al. (2008) and Siddique et al. (2009) showed that exposure of *Drosophila* larvae to industrial/municipal effluent posed negative effect on reproductive performance. Similarly, reduced fecundity and hatchability were recorded when first instar *Drosophila* larvae were exposed to dichlorvos, chlorpyrifos, cypermethrin, and volatile organic compounds (Gupta et al., 2007; Mukhopadhyay et al., 2006; Singh et al., 2009).

Recently, Khatun et al. (2017) observed a reduction in reproductive performance along with morphological alteration in the ovary after developmental exposure *Drosophila* to sodium fluoride. Epidemiological studies have underlined that approximately 50% of infertility cases are associated with male-related factors. Two recent studies are of particular relevance to male reproductive toxicity. In one study, exposure of male *Drosophila* to dibutyl phthalate (DBP), a known reproductive toxicant, caused reduced sperm quality, seminal protein, increased oxidative stress, and altered activity of estrogen receptors (Misra et al., 2014a; b). Sperm competition is the mating process between spermatozoa of two or more separate males to fertilize the same egg. Misra et al. (2014a,b) showed that the exposure of adult *Drosophila* to endosulfan leads to poor sperm competition and hampered reproductive output even at a low level of exposure. Expression analysis of gene encoding reproductive proteins in accessary gland (Acps) in exposed *Drosophila* provides a key link between environmental chemicals and reproductive outputs. Siddique et al. (2009) found downregulation of some seminal protein encoding genes (Acp70A and Acp36DE) in *Drosophila* exposed to tannery waste leachate. Additionally, Gupta et al. (2007) observed dysregulation in the expression of Acps in *Drosophila* males emerging from food contaminated with organophosphate compounds. Collectively, these findings prove *Drosophila* a quick, reliable genetic model to evaluate reproductive toxicity. The normal physiological activity of an organism determines its behavior. Furthermore, behavior changes are an integrated output of chemical toxicity (Pan et al., 2017; Wang and Xing, 2008). In this context, the *Drosophila* model has been used to study various environmental toxicant-induced behavioral changes that are quite similar to those of mammals (Kaun et al., 2012).

The developmental exposure of flies to endosulfan, acephate, atrazine, and silver nanoparticles showed reduced jumping and climbing behavior, along with reduced acetylcholinesterase (AChE) activity (Figueira et al., 2017; Raj et al., 2017; Rajak et al., 2017; Sharma et al., 2012). These authors concluded that the effects of these toxicants on climbing and jumping activity are mostly due to the disturbance in the dopaminergic system and induction of oxidative stress. Similarly, interrupted climbing activity was observed when adult *Drosophila* were exposed to rotenone, 4-vinylcyclohexene, and paraquat (Abolaji et al., 2015; Shukla et al., 2014; Sudati et al., 2013). To further understand the protective mechanism associated with behavioral toxicity, Pandey et al. (2016) demonstrated the protective role of Hsp27 in *Drosophila* against prolonged dichlorvos exposure at organismal level, including locomotor performance. Moreover, subsequent studies demonstrated that mutation in the *methuselah* gene (an aging-associated gene) confers resistance to locomotory defects in paraquat and dichlorvos exposure (Pandey et al., 2015; Shukla et al., 2014). St Laurent et al. (2013) observed amelioration in rotenone-induced locomotor impairment following administration of sodium butyrate (fatty acid derivative). Apart from developmental, reproductive, and behavioral toxicity, assessment of the chronic influence of environmental chemicals on life expectancy is one of the major concerns in toxicological science. The *Drosophila* model has been used in assessing life expectancy. Nazir et al. (2001) and Sharma et al. (2012) have shown that chlorpyrifos (an organophosphate compound) and endosulfan (an organochlorine compound) can reduce the survivorship of exposed adult *Drosophila* by 19% and 29%, respectively.

Toxicomics Profiling Using Drosophila

The application of an omics field to study the totality of toxic chemicals in cells is called Toxicomics. Toxicomics aim to examine the effect of toxic chemicals in cells through high-throughput experiments such as microarray and molecular interactions. The use of omics technology in toxicology has been increased in recent years. Global analysis of gene, protein, and metabolic expression changes associated with chemical exposure helped environmental toxicologists to better predict adverse response to toxicants (Joseph, 2017; Ramirez et al., 2013; Singh et al., 2010). In this context, transcriptomic data revealed the modulation of genes associated with the developmental process, stress and immune response, and metabolic process in third instar *Drosophila* larvae exposed to endosulfan (Sharma et al., 2011). Transcriptomic data derived from third instar *Drosophila* larvae exposed to Cr (VI) revealed the presence of a double strand DNA damage-triggered repair response (Mishra et al., 2013). Eom et al. (2017) studied a transcriptomic response of *Drosophila* that inhaled toluene and formaldehyde. Severe inhalation toxicity was accompanied with dysregulation of genes associated with important biological processes such as oxidative stress, immunity, and behavioral response. A similar cellular response was reported by Gu et al. (2008) in a murine model after long-term exposure of toluene and formaldehyde. Moskalev et al. (2015) studied the transcriptional response of ionizing radiation in *Drosophila* and observed misregulated genes involved in DNA damage response, aging, oxidative stress, immunity, and the metabolic process. Likewise, Xue et al. (2017) demonstrated that misregulated genes are involved in different cellular and molecular process in male *Drosophila* exposed to ethylparaben (an endocrine disruptor). Insects and mammals deploy similar mechanisms of toxicant detoxification. In this context, Mitchell et al. (2015) conducted a microarray study against α-amanitin (principal toxin of deadly poisonous mushroom) and found several genes/mechanisms that seem to provide α-amanitin resistance.

Changes in the expression profiling of miRNAs has also been explored in toxicology (Yu and Cho, 2015). Chandra et al. (2015) identified critical miRNAs in Cr(VI)-exposed *Drosophila* larvae. These miRNAs were involved in development and differentiation, DNA damage repair, and oxidation–reduction processes. Shukla et al. (2016) reported paraquat toxicity in *Drosophila* brain by metabolic profiling. Paraquat toxicity was associated with changes in energy metabolism, amino acids metabolism, lipid metabolism, and nucleotide metabolism. In conclusion, *Drosophila* helps in assessing the toxicological impact of several chemicals. Because of the conservation of the genome, observations from *Drosophila* can be extrapolated with humans.

Predication of Toxicity Using *Eisenia fetida*

E. fetida (common name: earthworm) is the major fraction of soil fauna communities in most ecosystems and represents a major biomass (>80%) in soil. Darwin's research established a keystone role of earthworms within terrestrial ecosystems and an understanding of how environmental change impacts survival and reproduction. This model has been widely used for ecotoxicological studies. The Organization for Economic Cooperation and Development (OECD) provides a number of protocols to test environmental chemical-induced cellular and organ toxicity using the earthworm. Over the years, a significant shift has been recorded in the development of end point toxicological parameters using the earthworm, i.e., from mortality analysis as the only end point to biomarker analysis at cellular, molecular, and behavioral levels (OECD, 1984). *E. fetida* is the most commonly used earthworm species in terrestrial ecotoxicology due to its (1) importance in soil structure, (2) easy breeding and rapid propagation, (3) elementary position in the food chain, and (4) its continuous exposure to pollutants present in the soil.

Biomarker research focuses on understanding the effects of environmental pollutants on different levels of biological organization, i.e., from cell to population. Studies conducted to analyze the impact of varied pollutants showed that *E. fetida* is a standard testing organism to assess toxicity at all biological organizational levels. During the last decade, changes in stress gene expression (Hsps and metallothioneine), enzyme-based parameters, and DNA damage have been used as a biomarker in *E. fetida* to assess the impact of environmental pollutants. Upregulation in the expression of *hsp70* and metallothionein was reported after exposure to imidacloprid (Dittbrenner et al., 2012), cadmium and siduron (Uwizeyimana et al., 2017), copper sulfate (Xiong et al., 2014), di-(2-ethylhexyl) phthalate (Ma et al., 2017), triclosan (Lin et al., 2014), α-endosulfan (Nam et al., 2017) zinc, copper, and lead (Homa et al., 2015).

Apart from Hsp and metallothionein, oxidative stress appears to be an important element in regulation of the response triggered by environmental stressors. A number of studies highlight the modulation in ROS, MDA levels, SOD, CAT, and lipid peroxidation activities when *E. fetida* are exposed to cyfluthrin (type II pyrethroids), silver nanoparticle, and imidacloprid (Li et al., 2017; Patricia et al., 2017; Zhang et al., 2014). Apart from oxidative stress biomarkers, changes in digestive enzyme activity directly determine the physiological status of the organism. Cellulase is one of the crucial digestive enzymes in the earthworm. Zhang et al. (2014) reported decreased activity of cellulase with prolonged exposure of *E. fetida* to imidacloprid. Similar results were obtained when *E. fetida* was exposed to the herbicide acetochlor (Xiao et al., 2006).

Recently, Ma et al. (2017) exposed *E. fetida* to sublethal concentrations of di-(2-ethylhexyl) phthalate for 14 days and observed an increase in tail DNA ratio (through Comet assay) in their coelomocytes (immune cells). Other parameters, like total cell number and viability of coelomocytes, have been also developed as biomarkers of exposure (Homa et al., 2015). Garcia-Velasco et al. (2016) used cell number and viability as a biomarker in the coelomocytes of earthworms exposed to silver nanoparticles and demonstrated its severe effect at higher biological complexity levels. Very recently, Martínez-Guitartea and his team investigated the molecular biomarker against *E. fetida* exposed to bisphenol (endocrine disruptors). The group observed changes in the expression of genes associated with endocrine function (membrane-associated progesterone receptor, ecdysone receptor), epigenetic mechanisms [DNA (cytosine-5)-methyltransferases 1], DNA damage (PARP1), and stress pathway (HSC70 4) (Novo et al., 2018). Mayilswami et al. (2016) performed differential gene expression study in perfluorooctanoic acid exposed *E. fetida*. The group observed dysregulated genes involved predominantly in the apoptotic process, calcium signaling, lipid metabolism, neuronal development, and reproduction. Similarly, a transcriptomic study by Srinithi et al. (2017) in benzo(a)pyrene-exposed *E. fetida* revealed altered mRNA expression involved in various biological processes, i.e., protein transport, calcium homeostasis, lipid metabolism, and development. Moreover, Wang et al. (2010) identified differentially regulated proteins involved in transcription, translation, the tricarboxylic acid cycle, amino acid metabolic process, protein phosphorylation, and the glucose metabolic process during cadmium exposure of *E. fetida*.

In earthworms, the body wall muscles resemble vertebrate-like cholinergic neuromuscular junctions (Rosenbluth, 1972), which contain the enzyme AChE. A number of studies have provided evidence that inhibition in AChE activity in *E. fetida* could be used as potential biomarker of behavior (Nam et al., 2015; Wang et al., 2015; Zhiqun et al., 2017). Mortality, growth, and reproduction are the most commonly used population health measurement to evaluate chemical toxicity. A number of studies have shown that growth and mortality of the earthworm could be used as a sensitive marker to assess the toxicity of environmental toxicants. Milanović et al. (2014) found galition G-5, terbis, and gardene exposures significantly increased mortality and reduced the weight and growth of the *E. fetida*. A similar decline in growth was also observed when *E. fetida* was exposed to bisphenol, chlorpyrifos, cadmium, and lead, which had long-term negative effects on the organism (Jūratė Žaltauskaitė and Sodienė, 2010; Verdu et al., 2018; Zhou et al., 2007). The growth of earthworms is more severely affected at the juvenile stage than the adult stage (Dasgupta et al., 2012; Zhou et al., 2007).

The earthworm has also been used to evaluate reproductive toxicity. The reproductive biomarkers include cocoon production and hatching pattern, viability of the worms, and sexual maturation (De Silva et al., 2009; Gestel and Dis, 1988; Robidoux et al., 1999). Earthworms produce cocoons in which their embryos develop. Accumulating evidence suggests that cocoon production can be used as a sensitive reproductive parameter for exposure to nickel (Yan et al., 2011), perchlorate (Landrum et al., 2006), zinc (Spurgeon and Hopkin, 1996), and silver nanoparticles (Makama et al., 2016). Similarly, hatchability of the cocoon has been proposed as a sensitive reproductive parameter for endosulfan carbaryl and chlorpyrifos exposure (Dasgupta et al., 2012). In addition, a single dose of malathion given to *E. fetida* was found to reduce the spermatic viability in spermatheca, sperm count, and quality, altering the cell proliferation and modifying the DNA structure of spermatogonia (Espinoza-Navarro and Bustos-Obregon, 2005). In conclusion, growth, mortality, and reproductive performance of earthworms are useful bioindicators for ecotoxicity studies.

Prediction of Toxicity Using *Daphnia magna*

The *Daphnia* species, commonly called "water flea", are common planktonic (1–5 mm long) invertebrate organisms which inhabit freshwater ecosystems and are an important food source for fish. For a long time, *Daphnia* has been considered a key member of the aquatic food chain as an intermediate link between primary and secondary productivity and its development as an alternative and sensitive model for ecotoxicological assessments (Taipale et al., 2016; Tatarazako and Oda, 2007). *D. magna* is one of the oldest models for ecotoxicology and evolutionary biology. The factors that make *D. magna* popular for experimental work include the following: 1) sensitivity toward toxic substance, 2) short life cycle, 3) small size, 4) high fecundity, 5) wide spatial distribution, and 6) availability of omics-based tools (Abe et al., 2014; Guilhermino et al., 2000). Furthermore, *D. magna* is recommended for aquatic environmental studies in OECD chemical testing guidelines (OECD, 2004).

The use of *Daphnia* for toxicological evaluation began in the early 1900s when Ernest Warren introduced *D. magna* to study the toxicity of sodium chloride (Warren, 1900). In addition, the widespread use of *D. magna* as a bioindicator is due to the fact that acute toxicity is considered a primary and sensitive test for the assessment of chemicals. During the last two decades, a number of biochemical, cellular, and molecular markers have been established in *D. magna* along with traditional ecotoxicity parameters. *Daphnia* swimming performance is considered a sensitive biomarker of toxicity. For example, when *D. magna* was exposed to copper, chromium, and titanium dioxide, the average swimming duration and speed of the organism was reduced in a concentration-dependent manner (Nikitin et al., 2018; Noss et al., 2013; Untersteiner et al., 2003). Dichlorvos, malathion, and parathion are also known to alter behavioral strength of the organism (Ren et al., 2009a,b). Travel distance is another swimming parameter for the locomotor activity of *Daphnia*. A reduction in travel distance was found in *D. manga* when the organism was exposed to a glyphosate copper mixture (Hansen and Roslev, 2016).

Assessment of fecundity and survivorship are the traditional biomarkers of ecotoxicity. Mendonca et al. (2011) reported the adverse effect of diamond nanoparticle on survival and reproduction of *D. manga*. The group also observed that the diamond nanoparticles adhere to the exoskeleton surface and accumulate within the gastrointestinal tract, leading to the blockage of food absorption by the gut cells. A similar observation (reduction in life span, growth, reproduction, and accumulation and persistence of nanoparticles) was reported by chronic exposure of *D. magna* ZnSO4 and ZnO nanoparticles (Bacchetta et al., 2016). The reproduction and survival of *D. magna* is affected by its exposure to chlorpyrifos, diazinon, cadmium, and triphenylphosphate (Geffard et al., 2008; Jemec et al., 2007; Yuan et al., 2018; Yuzhi et al., 2017). In arthropods, ecdysone signaling regulates molting and embryonic development. Palma et al. (2009) reported that endosulfan sulfate can decrease the molting process and produce embryonic toxicity, due to an anti-ecdysteroid activity of the chemical. Similarly, Toumi et al. (2013) assessed the acute and chronic toxicity of deltamethrin through analyzing longevity, molting, day of first brood, number of broods, embryotoxicity, and sex determination. The authors concluded that deltamethrin acts as an endocrine disruptor, which leads to decreasing juvenile number and growth, abnormalities in development, and interference in sex determination.

The *D. magna* model has also been used to assess the toxicity of chemicals using biomarkers at the cellular and molecular levels. For example, modulation in Hsps and metallothionine, GST, AChE activity, and ROS levels has been reported when the organism is exposed to toxicants (Dominguez et al., 2015; Ferrario et al., 2018; Haap et al., 2016; Oliveira et al., 2015; Oropesa et al., 2017; Parolini et al., 2018). Water quality has also been assessed employing the Comet assay in *D. magna* (Pellegri et al., 2014). For example, multigenerational exposure of carbendazim significantly increases DNA damage (Silva et al., 2017). In addition, coexposure of carbendazim with triclosan showed synergetic DNA damage in the exposed *D. magna* (Silva et al., 2015). Poynton et al. (2007) compared the distinct expression profiles in *D. magna* subjected to Cu, Cd, and Zn exposures and identified known biomarkers, i.e., metallothioneins and a ferritin (iron-response element). Transcriptomics analysis of nickel-exposed *D. magna* revealed an alteration in mRNA expression involved in metabolic processes, transport, and cell signaling. Moreover, in the same study mRNA profiling of a nickel and cadmium mixture pointed out the interactive molecular responses and the complex nature of mixture toxicity (Vandenbrouck et al., 2009). Collectively, *Daphnia* can be used as a simple platform to understand the impact of environmental chemicals on an ecological system.

CONCLUDING REMARKS AND FUTURE DIRECTIONS

As discussed in this chapter, *Drosophila*, the earthworm, and *Daphnia* represent an excellent alternative to a higher model organism for toxicity testing. These organisms can be used for toxicity testing at the molecular, cellular, and animal levels. The advent of advanced molecular and genetic tools has provided a basis for examining the potential of small organisms in toxicity testing by using multiple readouts. The most commonly used biomarker at the molecular level are the Hsps. Additionally, other biomarkers such as antioxidant enzymes, DNA damage response genes, metabolic enzymes, and behavioral and reproductive parameters have also been used for toxicity testing. The lower genetic redundancy, short life cycle, high conservation of signaling pathways, and ease in genetic manipulation are major advantages associated with these organisms. A number of toxicants such as heavy metals, pesticides, nanoparticles, volatile organic compounds, and additives have been tested using small organisms. In spite of several examples provided in this chapter, small organisms have not been recommended for routine toxicity testing due to the complex biological architecture of the human being. We hope that ongoing studies across the world will help in placing these animals in the forefront of biomarkers for toxicity testing.

Acknowledgments

SCG wish to acknowledge the financial support from Science and Engineering Research Board (ECR/2016/000034) and University Grants Commission [No. F.30–112/2015 (BSR)]. Sharma's laboratory is supported by Science and Engineering Research Board (ECR/2016/001863). SSV and SD are financially supported from DBT (DBT/2017/BHU/786) and NITTE (NU16PHDBS07), respectively.

References

Abe, F.R., Coleone, A.C., Machado, A.A., et al., 2014. Ecotoxicity and environmental risk assessment of larvicides used in the control of

Aedes aegypti to *Daphnia magna* (Crustacea, Cladocera). J. Toxicol. Environ. Health 77 (1–3), 37–45.

Abnoos, H., Fereidoni, M., Mahdavi-Shahri, N., et al., 2013. Developmental study of mercury effects on the fruit fly (*Drosophila melanogaster*). Interdiscipl. Toxicol. 6 (1), 34–40.

Abolaji, A.O., Kamdem, J.P., Lugokenski, T.H., et al., 2015. Ovotoxicants 4-vinylcyclohexene 1,2-monoepoxide and 4-vinylcyclohexene diepoxide disrupt redox status and modify different electrophile sensitive target enzymes and genes in *Drosophila melanogaster*. Redox. Biol. 5, 328–339.

Al-Momani, F.A., Massadeh, A.M., 2005. Effect of different heavy-metal concentrations on *Drosophila melanogaster* larval growth and development. Biol. Trace Elem. Res. 108 (1–3), 271–277.

Asgari Lajayer, B., Ghorbanpour, M., Nikabadi, S., 2017. Heavy metals in contaminated environment: ddestiny of secondary metabolite biosynthesis, oxidative status and phytoextraction in medicinal plants. Ecotoxicol. Environ. Saf. 145, 377–390.

Bacchetta, R., Maran, B., Marelli, M., et al., 2016. Role of soluble zinc in ZnO nanoparticle cytotoxicity in *Daphnia magna*: a morphological approach. Environ. Res. 148, 376–385.

Badyal, D.K., Desai, C., 2014. Animal use in pharmacology education and research: the changing scenario. Indian J. Pharmacol. 46 (3), 257–265.

Beekhuijzen, M., 2017. The era of 3Rs implementation in developmental and reproductive toxicity (DART) testing: current overview and future perspectives. Reprod. Toxicol. 72, 86–96.

Bhargav, D., Pratap Singh, M., Murthy, R.C., et al., 2008. Toxic potential of municipal solid waste leachates in transgenic *Drosophila melanogaster* (hsp70-lacZ): hsp70 as a marker of cellular damage. Ecotoxicol. Environ. Saf. 69 (2), 233–245.

Birben, E., Sahiner, U.M., Sackesen, C., et al., 2012. Oxidative stress and antioxidant defense. World. Allergy. Organ. J. 5 (1), 9–19.

Buchon, N., Silverman, N., Cherry, S., 2014. Immunity in *Drosophila melanogaster*-from microbial recognition to whole-organism physiology. Nat. Rev. Immunol. 14 (12), 796–810.

Cao, J., Xu, X., Hylkema, M.N., et al., 2016. Early-life xposure to widespread environmental toxicants and health risk: a Focus on the immune and respiratory systems. Ann. Glob. Health. 82 (1), 119–131.

Carmona, E.R., Guecheva, T.N., Creus, A., et al., 2011. Proposal of an in vivo comet assay using haemocytes of *Drosophila melanogaster*. Environ. Mol. Mutagen. 52 (2), 165–169.

Chandra, S., Pandey, A., Chowdhuri, D.K., 2015. MiRNA profiling provides insights on adverse effects of Cr(VI) in the midgut tissues of *Drosophila melanogaster*. J. Hazard Mater. 283, 558–567.

Chauhan, V., Chauhan, A., 2016. Effects of methylmercury and alcohol exposure in *Drosophila melanogaster*: potential risks in neurodevelopmental disorders. Int. J. Dev. Neurosci. 51, 36–41.

Chauhan, V., Srikumar, S., Aamer, S., et al., 2017. Methylmercury exposure induces sexual dysfunction in male and female *Drosophila Melanogaster*. Int. J. Environ. Res. Publ. Health. 14 (10).

Coelho, A., Fraichard, S., Le Goff, G., et al., 2015. Cytochrome P450-dependent metabolism of caffeine in *Drosophila melanogaster*. PLoS One 10 (2), e0117328.

Collins, A.R., Azqueta, A., 2012. DNA repair as a biomarker in human biomonitoring studies; further applications of the comet assay. Mutat. Res. 736 (1–2), 122–129.

Dasgupta, R., Chakravorty, P.P., Kaviraj, A., 2012. Effects of carbaryl, chlorpyrifos and endosulfan on growth, reproduction and respiration of tropical epigeic earthworm, *Perionyx excavatus* (Perrier). J. Environ. Sci. Health. B. 47 (2), 99–103.

de Pomerai, D., 1996. Heat-shock proteins as biomarkers of pollution. Hum. Exp. Toxicol. 15 (4), 279–285.

De Silva, P.M., Pathiratne, A., van Gestel, C.A., 2009. Influence of temperature and soil type on the toxicity of three pesticides to *Eisenia andrei*. Chemosphere 76 (10), 1410–1415.

del Carmen, A.M., Fuiman, L.A., 2005. Environmental levels of atrazine and its degradation products impair survival skills and growth of red drum larvae. Aquat. Toxicol. 74 (3), 229–241.

Dittbrenner, N., Capowiez, Y., Köhler, H.-R., et al., 2012. Stress protein response (Hsp70) and avoidance behaviour in *Eisenia fetida, Aporrectodea caliginosa* and *Lumbricus terrestris* when exposed to imidacloprid. J. Soils. Sedim. 12, 198–206.

Doke, S.K., Dhawale, S.C., 2015. Alternatives to animal testing: a review. Saudi. Pharm. J. 23 (3), 223–229.

Dominguez, G.A., Lohse, S.E., Torelli, M.D., et al., 2015. Effects of charge and surface ligand properties of nanoparticles on oxidative stress and gene expression within the gut of *Daphnia magna*. Aquat. Toxicol. 162, 1–9.

Eder, K.J., Kohler, H.R., Werner, I., 2007. Pesticide and pathogen: heat shock protein expression and acetylcholinesterase inhibition in juvenile Chinook salmon in response to multiple stressors. Environ. Toxicol. Chem. 26 (6), 1233–1242.

El Golli-Bennour, E., Bacha, H., 2011. Hsp70 expression as biomarkers of oxidative stress: mycotoxins' exploration. Toxicology 287 (1–3), 1–7.

Eom, H.J., Liu, Y., Kwak, G.S., et al., 2017. Inhalation toxicity of indoor air pollutants in *Drosophila melanogaster* using integrated transcriptomics and computational behavior analyses. Sci. Rep. 7, 46473.

Espinoza-Navarro, O., Bustos-Obregon, E., 2005. Effect of malathion on the male reproductive organs of earthworms, *Eisenia foetida*. Asian. J. Androl. 7 (1), 97–101.

Ferrario, C., Parolini, M., De Felice, B., et al., 2018. Linking sub-individual and supra-individual effects in *Daphnia magna* exposed to sub-lethal concentration of chlorpyrifos. Environ. Pollut. 235, 411–418.

Festing, M.F., Baumans, V., Combes, R.D., et al., 1998. Reducing the use of laboratory animals in biomedical research: problems and possible solutions. Altern. Lab. Anim. 26 (3), 283–301.

Figueira, F.H., de Quadros Oliveira, N., de Aguiar, L.M., et al., 2017. Exposure to atrazine alters behaviour and disrupts the dopaminergic system in *Drosophila melanogaster*. Comp. Biochem. Physiol. C Toxicol. Pharmacol. 202, 94–102.

Fulda, S., Gorman, A.M., Hori, O., et al., 2010. Cellular stress responses: cell survival and cell death. Int. J. Cell. Biol. 2010, 214074.

Gao, H., Wu, X., Simon, L., et al., 2014. Antioxidants maintain E-cadherin levels to limit *Drosophila prohemocyte* differentiation. PLoS One 9 (9), e107768.

Garcia-Velasco, N., Gandariasbeitia, M., Irizar, A., et al., 2016. Uptake route and resulting toxicity of silver nanoparticles in *Eisenia fetida* earthworm exposed through Standard OECD Tests. Ecotoxicology 25 (8), 1543–1555.

Gayathri, M.V., Krishnamurthy, N.B., 1983. Studies on the toxicity of the mercurial fungicide Agallol 3 in *Drosophila melanogaster*. Environ. Res. 24, 89–95.

Geffard, O., Geffard, A., Chaumot, A., et al., 2008. Effects of chronic dietary and waterborne cadmium exposures on the contamination level and reproduction of *Daphnia magna*. Environ. Toxicol. Chem. 27 (5), 1128–1134.

Gestel, C.A.M.V., Dis, W.A. v, 1988. The influence of soil characteristics on the toxicity of four chemicals to the earthworm *Eisenia fetida* andrei (Oligochaeta). Biol. Fertil. Soils. 6, 262–265.

Gormley, K.L., Teather, K.L., 2003. Developmental, behavioral, and reproductive effects experienced by Japanese medaka (*Oryzias latipes*) in response to short-term exposure to endosulfan. Ecotoxicol. Environ. Saf. 54 (3), 330–338.

Grandjean, P., Landrigan, P.J., 2014. Neurobehavioural effects of developmental toxicity. Lancet Neurol. 13 (3), 330–338.

Gu, Y., Fujimiya, Y., Kunugita, N., 2008. Long-term exposure to gaseous formaldehyde promotes allergen-specific IgE-mediated immune responses in a murine model. Hum. Exp. Toxicol. 27 (1), 37–43.

Guilhermino, L., Diamantino, T., Silva, M.C., et al., 2000. Acute toxicity test with *Daphnia magna*: an alternative to mammals in the prescreening of chemical toxicity? Ecotoxicol. Environ. Saf. 46 (3), 357–362.
Gupta, P., Thompson, B.L., Wahlang, B., et al., 2017. The environmental pollutant, polychlorinated biphenyls, and cardiovascular disease: a potential target for antioxidant nanotherapeutics. Drug. Deliv. Transl. Res. https://doi.org/10.1007/s13346-017-0429-9.
Gupta, S.C., Mishra, M., Sharma, A., et al., 2010a. Chlorpyrifos induces apoptosis and DNA damage in *Drosophila* through generation of reactive oxygen species. Ecotoxicol. Environ. Saf. 73 (6), 1415–1423.
Gupta, S.C., Sharma, A., Mishra, M., et al., 2010b. Heat shock proteins in toxicology: how close and how far? Life. Sci. 86 (11–12), 377–384.
Gupta, S.C., Siddique, H.R., Mathur, N., et al., 2007a. Adverse effect of organophosphate compounds, dichlorvos and chlorpyrifos in the reproductive tissues of transgenic *Drosophila melanogaster*: 70kDa heat shock protein as a marker of cellular damage. Toxicology 238 (1), 1–14.
Gupta, S.C., Siddique, H.R., Mathur, N., et al., 2007b. Induction of hsp70, alterations in oxidative stress markers and apoptosis against dichlorvos exposure in transgenic *Drosophila melanogaster*: modulation by reactive oxygen species. Biochim. Biophys. Acta 1770 (9), 1382–1394.
Gupta, S.C., Siddique, H.R., Saxena, D.K., et al., 2005. Hazardous effect of organophosphate compound, dichlorvos in transgenic *Drosophila melanogaster* (hsp70-lacZ): induction of hsp70, anti-oxidant enzymes and inhibition of acetylcholinesterase. Biochim. Biophys. Acta 1725 (1), 81–92.
Haap, T., Schwarz, S., Kohler, H.R., 2016. Metallothionein and Hsp70 trade-off against one another in *Daphnia magna* cross-tolerance to cadmium and heat stress. Aquat. Toxicol. 170, 112–119.
Hansen, L.R., Roslev, P., 2016. Behavioral responses of juvenile *Daphnia magna* after exposure to glyphosate and glyphosate-copper complexes. Aquat. Toxicol. 179, 36–43.
He, Y., Jasper, H., 2014. Studying aging in Drosophila. Methods 68 (1), 129–133.
Heindel, J.J., Chapin, R.E., Gulati, D.K., et al., 1994. Assessment of the reproductive and developmental toxicity of pesticide/fertilizer mixtures based on confirmed pesticide contamination in California and Iowa groundwater. Fund. Appl. Toxicol. 22 (4), 605–621.
Hirsch, H.V., Mercer, J., Sambaziotis, H., et al., 2003. Behavioral effects of chronic exposure to low levels of lead in *Drosophila melanogaster*. Neurotoxicology 24 (3), 435–442.
Homa, J., Rorat, A., Kruk, J., et al., 2015. Dermal exposure of *Eisenia andrei* earthworms: effects of heavy metals on metallothionein and phytochelatin synthase gene expressions in coelomocytes. Environ. Toxicol. Chem. 34 (6), 1397–1404.
Hook, S.E., Gallagher, E.P., Batley, G.E., 2014. The role of biomarkers in the assessment of aquatic ecosystem health. Integrated Environ. Assess. Manag. 10 (3), 327–341.
Houlahan, J.E., Findlay, C.S., Schmidt, B.R., et al., 2000. Quantitative evidence for global amphibian population declines. Nature 404 (6779), 752–755.
Jayaraj, R., Megha, P., Sreedev, P., 2016. Organochlorine pesticides, their toxic effects on living organisms and their fate in the environment. Interdiscipl. Toxicol. 9 (3–4), 90–100.
Jemec, A., Tisler, T., Drobne, D., et al., 2007. Comparative toxicity of imidacloprid, of its commercial liquid formulation and of diazinon to a non-target arthropod, the microcrustacean *Daphnia magna*. Chemosphere 68 (8), 1408–1418.
Joseph, P., 2017. Transcriptomics in toxicology. Food Chem. Toxicol. 109 (Pt 1), 650–662.
Jūratė Žaltauskaitė, Sodienė, I., 2010. Effects of total cadmium and lead concentrations in soil on the growth, reproduction and survival of earthworm *Eisenia fetida*. EKOLOGIJA 56, 10–16.
Kaun, K.R., Devineni, A.V., Heberlein, U., 2012. *Drosophila melanogaster* as a model to study drug addiction. Hum. Genet. 131 (6), 959–975.
Khatun, S., Rajak, P., Dutta, M., et al., 2017. Sodium fluoride adversely affects ovarian development and reproduction in *Drosophila melanogaster*. Chemosphere 186, 51–61.
Kim, B.M., Rhee, J.S., Jeong, C.B., et al., 2014. Heavy metals induce oxidative stress and trigger oxidative stress-mediated heat shock protein (hsp) modulation in the intertidal copepod *Tigriopus japonicus*. Comp. Biochem. Physiol. C Toxicol. Pharmacol. 166, 65–74.
Kumar, V., Ara, G., Afzal, M., et al., 2011. Effect of methyl methanesulfonate on hsp70 expression and tissue damage in the third instar larvae of transgenic *Drosophila melanogaster* (hsp70-lacZ). Bg. Interdiscip. Toxicol. 4 (3), 159–165.
Landrum, M., Canas, J.E., Coimbatore, G., et al., 2006. Effects of perchlorate on earthworm (*Eisenia fetida*) survival and reproductive success. Sci. Total Environ. 363 (1–3), 237–244.
Li, L., Yang, D., Song, Y., et al., 2017. The potential acute and chronic toxicity of cyfluthrin on the soil model organism, *Eisenia fetida*. Ecotoxicol. Environ. Saf. 144, 456–463.
Lin, D., Li, Y., Zhou, Q., et al., 2014. Effect of triclosan on reproduction, DNA damage and heat shock protein gene expression of the earthworm *Eisenia fetida*. Ecotoxicology 23 (10), 1826–1832.
Ma, T., Zhou, W., Chen, L., et al., 2017. Toxicity effects of di-(2-ethylhexyl) phthalate to *Eisenia fetida* at enzyme, cellular and genetic levels. PLoS One 12 (3), e0173957.
Magder, S., 2006. Reactive oxygen species: toxic molecules or spark of life? Crit. Care 10 (1), 208.
Mahmood, K., Jadoon, S., Mahmood, Q., et al., 2014. Synergistic effects of toxic elements on heat shock proteins. BioMed Res. Int. 2014, 564136.
Makama, S., Piella, J., Undas, A., et al., 2016. Properties of silver nanoparticles influencing their uptake in and toxicity to the earthworm *Lumbricus rubellus* following exposure in soil. Environ. Pollut. 218, 870–878.
Marcus, S.R., Fiumera, A.C., 2016. Atrazine exposure affects longevity, development time and body size in *Drosophila melanogaster*. J. Insect Physiol. 91–92, 18–25.
Martin, M.T., Dix, D.J., Judson, R.S., et al., 2010. Impact of environmental chemicals on key transcription regulators and correlation to toxicity end points within EPA's ToxCast program. Chem. Res. Toxicol. 23 (3), 578–590.
Martins, M., Costa, P.M., 2015. The comet assay in environmental risk assessment of marine pollutants: applications, assets and handicaps of surveying genotoxicity in non-model organisms. Mutagenesis 30 (1), 89–106.
Mayilswami, S., Krishnan, K., Megharaj, M., et al., 2016. Gene expression profile changes in *Eisenia fetida* chronically exposed to PFOA. Ecotoxicology 25 (4), 759–769.
McGurk, L., Berson, A., Bonini, N.M., 2015. *Drosophila* as an *in vivo* model for human neurodegenerative disease. Genetics 201 (2), 377–402.
Mendonca, E., Diniz, M., Silva, L., et al., 2011. Effects of diamond nanoparticle exposure on the internal structure and reproduction of *Daphnia magna*. J. Hazard Mater. 186 (1), 265–271.
Milanović, J., Milutinović, T., Mirjana, S., 2014. Effects of three pesticides on the earthworm *Eisenia fetida* (Savigny 1826) under laboratory conditions: assessment of mortality, biomass and growth inhibition. Eur. J. Soil Biol. 62, 127–131.
Mishra, M., Sharma, A., Negi, M.P., et al., 2011. Tracing the tracks of genotoxicity by trivalent and hexavalent chromium in *Drosophila melanogaster*. Mutat. Res. 722 (1), 44–51.
Mishra, M., Sharma, A., Shukla, A.K., et al., 2014. Genotoxicity of dichlorvos in strains of *Drosophila melanogaster* defective in DNA repair. Mutat. Res. Genet. Toxicol. Environ. Mutagen 766, 35–41.
Mishra, M., Sharma, A., Shukla, A.K., et al., 2013. Transcriptomic analysis provides insights on hexavalent chromium induced DNA

double strand breaks and their possible repair in midgut cells of *Drosophila melanogaster* larvae. Mutat. Res. 747–748, 28–39.

Misra, S., Kumar, A., Ratnasekhar, C., et al., 2014a. Exposure to endosulfan influences sperm competition in *Drosophila melanogaster*. Sci. Rep. 4, 7433.

Misra, S., Singh, A.C.H.R., Sharma, V., et al., 2014b. Identification of *Drosophila*-based endpoints for the assessment and understanding of xenobiotic-mediated male reproductive adversities. Toxicol. Sci. 141 (1), 278–291.

Mitchell, C.L., Yeager, R.D., Johnson, Z.J., et al., 2015. Long-Term resistance of *Drosophila melanogaster* to the mushroom toxin alpha-amanitin. PLoS One 10 (5), e0127569.

Mobarak, Y.M., Al-Asmari, M.A., 2011. Endosulfan impacts on the developing chick embryos: morphological, morphometric and skeletal changes. Int. J. Zool. Res. 7, 107–127.

Moskalev, A., Zhikrivetskaya, S., Krasnov, G., et al., 2015. A comparison of the transcriptome of *Drosophila melanogaster* in response to entomopathogenic fungus, ionizing radiation, starvation and cold shock. BMC Genom. 16 (Suppl. 13), S8.

Mukhopadhyay, I., Saxena, D.K., Chowdhuri, D.K., 2003. Hazardous effects of effluent from the chrome plating industry: 70 kDa heat shock protein expression as a marker of cellular damage in transgenic *Drosophila melanogaster* (hsp70-lacZ). Environ. Health Perspect. 111 (16), 1926–1932.

Mukhopadhyay, I., Siddique, H.R., Bajpai, V.K., et al., 2006. Synthetic pyrethroid cypermethrin induced cellular damage in reproductive tissues of *Drosophila melanogaster*: Hsp70 as a marker of cellular damage. Arch. Environ. Contam. Toxicol. 51 (4), 673–680.

Nam, T.H., Jeon, H.J., Mo, H.H., et al., 2015. Determination of biomarkers for polycyclic aromatic hydrocarbons (PAHs) toxicity to earthworm (*Eisenia fetida*). Environ. Geochem. Health 37 (6), 943–951.

Nam, T.H., Kim, L., Jeon, H.J., et al., 2017. Biomarkers indicate mixture toxicities of fluorene and phenanthrene with endosulfan toward earthworm (*Eisenia fetida*). Environ. Geochem. Health. 39 (2), 307–317.

Nazir, A., Mukhopadhyay, I., Saxena, D.K., et al., 2001. Chlorpyrifos-induced hsp70 expression and effect on reproductive performance in transgenic *Drosophila melanogaster* (hsp70-lacZ) Bg9. Arch. Environ. Contam. Toxicol. 41 (4), 443–449.

Nazir, A., Mukhopadhyay, I., Saxena, D.K., et al., 2003. Evaluation of toxic potential of captan: induction of hsp70 and tissue damage in transgenic *Drosophila melanogaster* (hsp70-lacZ) Bg9. J. Biochem. Mol. Toxicol. 17 (2), 98–107.

Nesatyy, V.J., Suter, M.J., 2007. Proteomics for the analysis of environmental stress responses in organisms. Environ. Sci. Technol. 41 (20), 6891–6900.

Nikitin, O.V., Nasyrova, E.I., Nuriakhmetova, V.R., et al., 2018. Toxicity assessment of polluted sediments using swimming behavior alteration test with *Daphnia magna*. Earth. Env. Sci. 107, 012068.

Noss, C., Dabrunz, A., Rosenfeldt, R.R., et al., 2013. Three-dimensional analysis of the swimming behavior of *Daphnia magna* exposed to nanosized titanium dioxide. PLoS One 8 (11), e80960.

Novo, M., Verdu, I., Trigo, D., et al., 2018. Endocrine disruptors in soil: effects of bisphenol A on gene expression of the earthworm *Eisenia fetida*. Ecotoxicol. Environ. Saf. 150, 159–167.

O'Bryant, S.E., Edwards, M., Menon, C.V., et al., 2011. Long-term low-level arsenic exposure is associated with poorer neuropsychological functioning: a Project FRONTIER study. Int. J. Environ. Res. Publ. Health. 8 (3), 861–874.

OECD, O. f. E.C-O.A.D., 1984. Test 207: Earthworm, acute toxicity tests. In: Organization for Economic Co-operation and Development (Ed.), OECD Guidelines for Testing of Chemicals.

OECD, O. f. E.C-O.A.D., 2004. OECD Guideline for Testing of Chemicals No. 202, Daphnia sp. Acute Immobilisation Test. OECD, Paris.

Oliveira, L.L., Antunes, S.C., Goncalves, F., et al., 2015. Evaluation of ecotoxicological effects of drugs on *Daphnia magna* using different enzymatic biomarkers. Ecotoxicol. Environ. Saf. 119, 123–131.

Omiecinski, C.J., Vanden Heuvel, J.P., Perdew, G.H., et al., 2011. Xenobiotic metabolism, disposition, and regulation by receptors: from biochemical phenomenon to predictors of major toxicities. Toxicol. Sci. 120 (Suppl 1), S49–S75.

Oropesa, A.L., Novais, S.C., Lemos, M.F., et al., 2017. Oxidative stress responses of *Daphnia magna* exposed to effluents spiked with emerging contaminants under ozonation and advanced oxidation processes. Environ. Sci. Pollut. Res. Int. 24 (2), 1735–1747.

Palma, P., Palma, V.L., Matos, C., et al., 2009. Effects of atrazine and endosulfan sulphate on the ecdysteroid system of *Daphnia magna*. Chemosphere 74 (5), 676–681. https://doi.org/10.1016/j.chemosphere.2008.10.021.

Pan, Y., Meissner, G.W., Baker, B.S., 2012. Joint control of Drosophila male courtship behavior by motion cues and activation of male-specific P1 neurons. Proc. Natl. Acad. Sci. U. S. A. 109 (25), 10065–10070.

Pan, Y., Yan, S.W., Li, R.Z., et al., 2017. Lethal/sublethal responses of *Daphnia magna* to acute norfloxacin contamination and changes in phytoplankton-zooplankton interactions induced by this antibiotic. Sci. Rep. 7, 40385.

Pandey, A., Chandra, S., Chauhan, L.K., et al., 2013. Cellular internalization and stress response of ingested amorphous silica nanoparticles in the midgut of *Drosophila melanogaster*. Biochim. Biophys. Acta 1830 (1), 2256–2266.

Pandey, A., Khatoon, R., Saini, S., et al., 2015. Efficacy of methuselah gene mutation toward tolerance of dichlorvos exposure in *Drosophila melanogaster*. Free Radic. Biol. Med. 83, 54–65.

Pandey, A., Saini, S., Khatoon, R., et al., 2016. Overexpression of hsp27 rescued neuronal cell death and reduction in life- and health-span in *Drosophila melanogaster* against prolonged exposure to dichlorvos. Mol. Neurobiol. 53 (5), 3179–3193.

Parolini, M., De Felice, B., Ferrario, C., et al., 2018. Benzoylecgonine exposure induced oxidative stress and altered swimming behavior and reproduction in *Daphnia magna*. Environ. Pollut. 232, 236–244.

Patricia, C.S., Nerea, G.V., Erik, U., et al., 2017. Responses to silver nanoparticles and silver nitrate in a battery of biomarkers measured in coelomocytes and in target tissues of *Eisenia fetida* earthworms. Ecotoxicol. Environ. Saf. 141, 57–63.

Pellegri, V., Gorbi, G., Buschini, A., 2014. Comet Assay on Daphnia magna in eco-genotoxicity testing. Aquat. Toxicol. 155, 261–268.

Poynton, H.C., Varshavsky, J.R., Chang, B., 2007. *Daphnia magna* ecotoxicogenomics provides mechanistic insights into metal toxicity. Environ. Sci. Technol. 41 (3), 1044–1050.

Pragya, P., Shukla, A.K., Murthy, R.C., et al., 2014. Over-expression of superoxide dismutase ameliorates Cr(VI) induced adverse effects via modulating cellular immune system of *Drosophila melanogaster*. PLoS One 9 (2), e88181.

Raj, A., Shah, P., Agrawal, N., 2017. Sedentary behavior and altered metabolic activity by AgNPs ingestion in *Drosophila melanogaster*. Sci. Rep. 7 (1), 15617.

Rajak, P., Dutta, M., Khatun, S., et al., 2017. Exploring hazards of acute exposure of Acephate in *Drosophila melanogaster* and search for l-ascorbic acid mediated defense in it. J. Hazard Mater. 321, 690–702.

Ramirez, T., Daneshian, M., Kamp, H., et al., 2013. Metabolomics in toxicology and preclinical research. ALTEX 30 (2), 209–225.

Ranganathan, S., Davis, D.G., Leeper, J.D., et al., 1987. Effects of differential alcohol dehydrogenase activity on the developmental toxicity of ethanol in *Drosophila melanogaster*. Teratology 36 (3), 329–334.

Ren, Z., Li, L.-Z., Zha, J.M., et al., 2009b. The avoidance responses of *Daphnia magna* to the exposure of organophosphorus pesticides in an on-line biomonitoring system. Environ. Model. Assess. 14, 405–410.

Ren, Z., Li, Z., Ma, M., et al., 2009a. Behavioral responses of *Daphnia magna* to stresses of chemicals with different toxic characteristics. Bull. Environ. Contam. Toxicol. 82 (3), 310–316.

Richardson, H., Kumar, S., 2002. Death to flies: *Drosophila* as a model system to study programmed cell death. J. Immunol. Meth. 265 (1–2), 21–38.

Ritossa, F., 1962. A new puffing pattern induced by heat shock and DNP in *Drosophila*. In: Experientia, 18, pp. 571–573.

Robidoux, P.Y., Hawari, J., Thiboutot, S., et al., 1999. Acute toxicity of 2,4,6-trinitrotoluene in earthworm (*Eisenia andrei*). Ecotoxicol. Environ. Saf. 44 (3), 311–321.

Roex, E.W., Giovannangelo, M., van Gestel, C.A., 2001. Reproductive impairment in the zebrafish, *Danio rerio*, upon chronic exposure to 1,2,3-trichlorobenzene. Ecotoxicol. Environ. Saf. 48 (2), 196–201.

Rosenbluth, J., 1972. Myoneural junctions of two ultrastructurally distinct types in earthworm body wall muscle. J. Cell Biol. 54 (3), 566–579.

Rudrapatna, V.A., Cagan, R.L., Das, T.K., 2012. Drosophila cancer models. Dev. Dynam. 241 (1), 107–118.

Sabra, S., Malmqvist, E., Saborit, A., et al., 2017. Heavy metals exposure levels and their correlation with different clinical forms of fetal growth restriction. PLoS One 12 (10), e0185645.

Sharma, A., Mishra, M., Ram, K.R., et al., 2011a. Transcriptome analysis provides insights for understanding the adverse effects of endosulfan in *Drosophila melanogaster*. Chemosphere 82 (3), 370–376.

Sharma, A., Mishra, M., Shukla, A.K., et al., 2012. Organochlorine pesticide, endosulfan induced cellular and organismal response in *Drosophila melanogaster*. J. Hazard Mater. 221–222, 275–287.

Sharma, A., Shukla, A.K., Mishra, M., et al., 2011b. Validation and application of *Drosophila melanogaster* as an in vivo model for the detection of double strand breaks by neutral Comet assay. Mutat. Res. 721 (2), 142–146.

Shukla, A.K., Pragya, P., Chaouhan, H.S., et al., 2014. Mutation in *Drosophila methuselah* resists paraquat induced Parkinson-like phenotypes. Neurobiol. Aging 35 (10), 2419 e2411–2419 e2416.

Shukla, A.K., Ratnasekhar, C., Pragya, P., et al., 2016. Metabolomic analysis provides insights on paraquat-induced Parkinson-like symptoms in *Drosophila melanogaster*. Mol. Neurobiol. 53 (1), 254–269.

Siddique, H.R., Chowdhuri, D.K., Saxena, D.K., et al., 2005a. Validation of *Drosophila melanogaster* as an *in vivo* model for genotoxicity assessment using modified alkaline Comet assay. Mutagenesis 20 (4), 285–290.

Siddique, H.R., Gupta, S.C., Dhawan, A., et al., 2005b. Genotoxicity of industrial solid waste leachates in *Drosophila melanogaster*. Environ. Mol. Mutagen. 46 (3), 189–197.

Siddique, H.R., Gupta, S.C., Mitra, K., et al., 2008a. Adverse effect of tannery waste leachates in transgenic *Drosophila melanogaster*: role of ROS in modulation of Hsp70, oxidative stress and apoptosis. J. Appl. Toxicol. 28 (6), 734–748.

Siddique, H.R., Mitra, K., Bajpai, V.K., et al., 2009. Hazardous effect of tannery solid waste leachates on development and reproduction in *Drosophila melanogaster*: 70kDa heat shock protein as a marker of cellular damage. Ecotoxicol. Environ. Saf. 72 (6), 1652–1662.

Siddique, H.R., Sharma, A., Gupta, S.C., et al., 2008b. DNA damage induced by industrial solid waste leachates in *Drosophila melanogaster*: a mechanistic approach. Environ. Mol. Mutagen. 49 (3), 206–216.

Silva, A.R., Cardoso, D.N., Cruz, A., et al., 2015. Ecotoxicity and genotoxicity of a binary combination of triclosan and carbendazim to *Daphnia magna*. Ecotoxicol. Environ. Saf. 115, 279–290.

Silva, A.R., Cardoso, D.N., Cruz, A., et al., 2017. Multigenerational effects of carbendazim in *Daphnia magna*. Environ. Toxicol. Chem. 36 (2), 383–394.

Singh, M.P., Mishra, M., Sharma, A., et al., 2011. Genotoxicity and apoptosis in *Drosophila melanogaster* exposed to benzene, toluene and xylene: attenuation by quercetin and curcumin. Toxicol. Appl. Pharmacol. 253 (1), 14–30.

Singh, M.P., Reddy, M.M., Mathur, N., et al., 2009. Induction of hsp70, hsp60, hsp83 and hsp26 and oxidative stress markers in benzene, toluene and xylene exposed *Drosophila melanogaster*: role of ROS generation. Toxicol. Appl. Pharmacol. 235 (2), 226–243.

Singh, S., Singhal, N.K., Srivastava, G., et al., 2010. Omics in mechanistic and predictive toxicology. Toxicol. Mech. Meth. 20 (7), 355–362.

Spurgeon, D.J., Hopkin, S.P., 1996. Effects of metal-contaminated soils on the growth, sexual development, and early cocoon production of the earthworm *Eisenia fetida*, with particular reference to zinc. Ecotoxicol. Environ. Saf. 35 (1), 86–95.

Srinithi, M., Krishnan, K., Naidu, R., et al., 2017. Transcriptome analysis of *Eisenia fetida* chronically exposed to benzo(a)pyrene. Env. Technol. Innov. 7, 54–62.

St Laurent, R., O'Brien, L.M., Ahmad, S.T., 2013. Sodium butyrate improves locomotor impairment and early mortality in a rotenone-induced *Drosophila* model of Parkinson's disease. Neuroscience 246, 382–390.

Sudati, J.H., Vieira, F.A., Pavin, S.S., et al., 2013. Valeriana officinalis attenuates the rotenone-induced toxicity in *Drosophila melanogaster*. Neurotoxicology 37, 118–126.

Sutton, P., Woodruff, T.J., Perron, J., et al., 2012. Toxic environmental chemicals: the role of reproductive health professionals in preventing harmful exposures. Am. J. Obstet. Gynecol. 207 (3), 164–173.

Taipale, S.J., Galloway, A.W., Aalto, S.L., et al., 2016. Terrestrial carbohydrates support freshwater zooplankton during phytoplankton deficiency. Sci. Rep. 6, 30897.

Takalo, M., Salminen, A., Soininen, H., et al., 2013. Protein aggregation and degradation mechanisms in neurodegenerative diseases. Am. J. Neurodegener. Dis. 2 (1), 1–14.

Tatarazako, N., Oda, S., 2007. The water flea *Daphnia magna* (Crustacea, Cladocera) as a test species for screening and evaluation of chemicals with endocrine disrupting effects on crustaceans. Ecotoxicology 16 (1), 197–203.

Terhzaz, S., Cabrero, P., Brinzer, R.A., et al., 2015. A novel role of *Drosophila* cytochrome P450-4e3 in permethrin insecticide tolerance. Insect Biochem. Mol. Biol. 67, 38–46.

Thompson, J., Bannigan, J., 2008. Cadmium: toxic effects on the reproductive system and the embryo. Reprod. Toxicol. 25 (3), 304–315.

Tiwari, A.K., Pragya, P., Ravi Ram, K., et al., 2011. Environmental chemical mediated male reproductive toxicity: *Drosophila melanogaster* as an alternate animal model. Theriogenology 76 (2), 197–216.

Toumi, H., Boumaiza, M., Millet, M., et al., 2013. Effects of deltamethrin (pyrethroid insecticide) on growth, reproduction, embryonic development and sex differentiation in two strains of *Daphnia magna* (Crustacea, Cladocera). Sci. Total Environ. 458–460, 47–53.

Ueker, M.E., Silva, V.M., Moi, G.P., et al., 2016. Parenteral exposure to pesticides and occurence of congenital malformations: hospital-based case-control study. BMC Pediatr. 16 (1), 125.

Untersteiner, H., Kahapka, J., Kaiser, H., 2003. Behavioural response of the cladoceran *Daphnia magna* STRAUS to sublethal Copper stress—validation by image analysis. Aquat. Toxicol. 65 (4), 435–442.

Uwizeyimana, H., Wang, M., Chen, W., 2017. Evaluation of combined noxious effects of siduron and cadmium on the earthworm *Eisenia fetida*. Environ. Sci. Pollut. Res. Int. 24 (6), 5349–5359.

Vandenbrouck, T., Soetaert, A., van der Ven, K., et al., 2009. Nickel and binary metal mixture responses in *Daphnia magna*: molecular fingerprints and (sub)organismal effects. Aquat. Toxicol. 92 (1), 18–29.

Verdu, I., Trigo, D., Martinez-Guitarte, J.L., et al., 2018. Bisphenol A in artificial soil: effects on growth, reproduction and immunity in earthworms. Chemosphere 190, 287–295.

Wang, D., Xing, X., 2008. Assessment of locomotion behavioral defects induced by acute toxicity from heavy metal exposure in nematode *Caenorhabditis elegans*. J. Environ. Sci. (China) 20 (9), 1132–1137.

Wang, K., Mu, X., Qi, S., et al., 2015. Toxicity of a neonicotinoid insecticide, guadipyr, in earthworm (*Eisenia fetida*). Ecotoxicol. Environ. Saf. 114, 17–22.

Wang, X., Chang, L., Sun, Z., et al., 2010. Analysis of earthworm *Eisenia fetida* proteomes during cadmium exposure: an ecotoxicoproteomics approach. Proteomics 10 (24), 4476–4490.

Warren, E., 1900. Memoirs: on the reaction of *Daphnia magna* (Straus) to certain changes in its environment. Quart. J. Microscop. Sci. 43, 199–224.

Xiao, N., Jing, B., Ge, F., et al., 2006. The fate of herbicide acetochlor and its toxicity to *Eisenia fetida* under laboratory conditions. Chemosphere 62 (8), 1366–1373.

Xiong, W., Ding, X., Zhang, Y., et al., 2014. Ecotoxicological effects of a veterinary food additive, copper sulphate, on antioxidant enzymes and mRNA expression in earthworms. Environ. Toxicol. Pharmacol. 37 (1), 134–140.

Xue, P., Zhao, X., Qin, M., et al., 2017. Transcriptome analysis of male *Drosophila melanogaster* exposed to ethylparaben using digital gene expression profiling. J. Insect Sci. 17 (4).

Yadav, A.K., Srikrishna, S., Gupta, S.C., 2016. Cancer drug development using *Drosophila* as an *in vivo* tool: from Bedside to Bench and Back. Trends Pharmacol. Sci. 37 (9), 789–806.

Yan, Z., Wang, B., Xie, D., et al., 2011. Uptake and toxicity of spiked nickel to earthworm *Eisenia fetida* in a range of Chinese soils. Environ. Toxicol. Chem. 30 (11), 2586–2593.

Yu, H.W., Cho, W.C., 2015. The role of microRNAs in toxicology. Arch. Toxicol. 89 (3), 319–325.

Yuan, S., Li, H., Dang, Y., et al., 2018. Effects of triphenyl phosphate on growth, reproduction and transcription of genes of *Daphnia magna*. Aquat. Toxicol. 195, 58–66.

Yuzhi, S., Mindong, C., Junying, Z., 2017. Effects of three pesticides on superoxide dismutase and glutathione-S-transferase activities and reproduction of *Daphnia magna*. Arch. Environ. Protect. 43, 80–86.

Zhang, Q., Zhang, B., Wang, C., 2014. Ecotoxicological effects on the earthworm *Eisenia fetida* following exposure to soil contaminated with imidacloprid. Environ. Sci. Pollut. Res. Int. 21 (21), 12345–12353.

Zhiqun, T., Jian, Z., Junli, Y., et al., 2017. Allelopathic effects of volatile organic compounds from Eucalyptus grandis rhizosphere soil on *Eisenia fetida* assessed using avoidance bioassays, enzyme activity, and comet assays. Chemosphere 173, 307–317.

Zhou, S.P., Duan, C.Q., Fu, H., et al., 2007. Toxicity assessment for chlorpyrifos-contaminated soil with three different earthworm test methods. J. Environ. Sci. (China) 19 (7), 854–858.

CHAPTER

8

Alternative Animal Toxicity Testing and Biomarkers

Gopala Krishna[1], *Priya A. Krishna*[1], *Saryu Goel*[2], *Kavya A. Krishna*[1]
[1]Nonclinical Consultants, Ellicott City, MD, United States; [2]Nonclinical Expert, Leesburg, VA, United States

INTRODUCTION

The subject of animal testing is highly controversial and has become an emotional issue in recent years. Therefore, it is important to recognize the growing impact on alternatives to animal testing, such as, in vitro toxicity, metabolism, and efficacy testing. It is evident that animal testing is an essential part of basic research, new product discovery, and the improvement process. In this regard, animal testing is defined as the use of nonhumans, such as vertebrate and nonvertebrate animals or organisms, in scientific experimentation in search of answers for the good of mankind and the pursuit of basic knowledge. It is estimated that up to 100 million vertebrate animals ranging from zebrafish to nonhuman primates are used annually worldwide. Likewise, invertebrates such as fruit flies (*Drosophila*) and earthworms are also commonly used in research. While the uses of vertebrates are heavily regulated, invertebrates are excluded from such regulations. Much of the animal testing around the world is conducted under the strict global regulatory guidelines and/or guidelines of respective countries, which set fundamental standards for the humane use of animals for training, experimentation, biological testing, research, or for other related purposes. These regulations and guidelines require minimizing harm to animals through 3Rs: reduction, refinement, and replacement (Russell and Burch, 1959), by searching for alternatives, including consideration for hierarchical use of species and in vitro methods.

In product discovery and development, "preclinical studies" is the term that encompasses all activities in nonhuman species. This scientific discipline includes everything that is not purely chemical or formulation-oriented nor involves humans. These studies are conducted using cells, organs, unique animal models, and multiple animal species in large numbers with an objective to obtain data on the mechanism of product's action, adverse effects (toxicity to genetic material, reproductive system, and ability to cause cancer) over a range of doses, pharmacokinetics, and metabolism of the substance that is being tested. For pharmaceuticals, these studies define safety margins for human clinical trials. The process of drug development from bench to bedside is illustrated in Fig. 8.1.

With the advancement of science and technology, particularly in biomarkers, in vitro testing, and instrumentation, tremendous progress has been made to achieve 3Rs. Biomarkers are defined as parameters such as pharmacological, physiological, pharmacokinetic, molecular markers such as genes, proteins, metabolites, and clinical chemistry, which can be used to predict a toxic event in an individual animal and/or biological system (in vitro). These markers are used to extrapolate to a similar toxic endpoint across species, including humans. An ideal biomarker is described as accessible, noninvasive, sensitive, specific, inexpensive, translational (able to cross the bridge between basic and clinical research), predictive of the extent of injury, and accurate. Generally, biomarkers are utilized in three forms: (1) biomarkers of effect—quantifiable health impairment, such as, in the form of change of liver enzymes in blood, DNA/chromosomal changes, clinical signs; (2) biomarkers of exposure—concentration over time of an agent or its metabolite that can be measured in body fluids; and (3) biomarkers of susceptibility—inherent susceptibility of an organism to toxic insult, such as polymorphisms in cytochrome P450 enzymes. Advancements in instrumentation leading to technologies such as functional/pharmacological magnetic resonance imaging (fMRI/phMRI), mass spectrometry, and flow cytometry

Biomarkers in Toxicology, Second Edition
https://doi.org/10.1016/B978-0-12-814655-2.00008-6

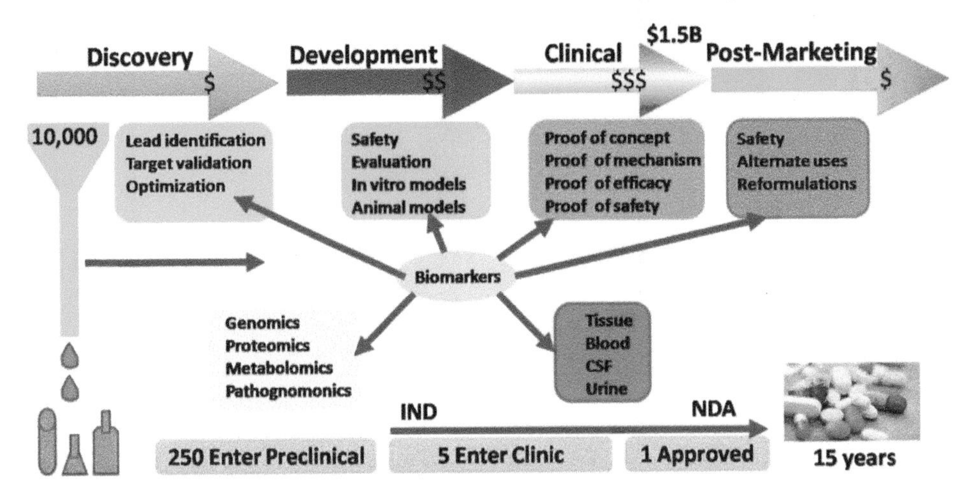

FIGURE 8.1 Use of biomarkers in product development from discovery to approval and postmarketing.

have accelerated such efforts leading to achieve the goal of 3Rs. The objective of this chapter is to provide a brief overview of the reasons for animal surrogate strategies, descriptions of select systems used for toxicity, metabolism, efficacy testing, and the role of biomarkers in the form of toxicity endpoints used in product development (Benford et al., 2000; Blaauboer et al., 2012; Bradlaw, 1989; Doke and Dhawale, 2015; Goldberg and Frazier, 1989; Hartung, 2014; McKeehan et al., 1990; Schaeffer, 1990; Tainsky, 2009; Smith, 2012; Vaidya and Mendrick, 2012).

USE OF ALTERNATIVES TO ANIMAL TESTING AND BIOMARKERS

To deal with the concerns of animal testing, alternatives have been sought. Most scientists in academia, industry, and government around the globe agree that animal testing should be ethical and cause as little suffering as possible, and that alternatives to animal testing need to be developed. The "3Rs," first described by Russell and Burch (1959), are guiding principles for the use of animals in research around the globe:

- **Reduction** refers to strategies that enable researchers to obtain comparable levels of information from fewer animals, or to obtain more information from the same number of animals.
- **Refinement** refers to strategies that alleviate or minimize potential pain, suffering or distress, and enhance animal welfare for the animals still used.
- **Replacement** refers to the preferred use of nonanimal methods over animal methods whenever it is possible to achieve the same objective(s).

These guiding principles are the fundamental reasons for the development of alternative procedures, a rationale for involving animals and appropriateness of the species, the number of animals to be used in research, and product development and training.

REASONS FOR DEVELOPING ALTERNATIVES TO ANIMAL TESTS AND BIOMARKERS

There are multiple reasons to develop and practice alternatives to animal testing of which two key reasons are described. First, economy and efficiency play a key role. In vitro tests, once developed and validated, provide toxicity and efficacy information in a cost-effective and time-saving manner. Such information can also increase the efficiency of whole-animal studies and decrease the number of animals required for toxicity, metabolism, and efficacy testing. The relative simplicity, space, and time-saving characteristics of in vitro methods are considered advantageous, albeit with some concerns that in vitro methods do not recapitulate the in vivo microenvironment and homeostasis.

Second, in vitro testing using human cell lines has been useful in securing relevant information for human risk assessment and understanding molecular mechanisms, sensitivity, and resistance to targeted therapy, thereby opening up opportunities to explore responses to existing and emerging therapies. Human cells, ethically obtained and successfully established in vitro, may provide information about a toxicant or therapeutic agent that is relevant to human risk and/or efficacy, such as mechanism of action and metabolism in human cells, and can provide the basis for selecting a suitable animal model for repeat-dose toxicity testing. This can be illustrated as an operating concept of in vitro toxicology where manifestations of changes at the molecular level can be studied at the organ level (Fig. 8.2). It is proposed that the endpoints used in the 3Rs be considered as biomarkers, where appropriate. It is also known that test substances interact with various cells, organelles, and organ systems pharmacologically and toxicologically to produce their effects (Fig. 8.3), which can be used for efficacy and safety assessment (Lee et al., 2013). Likewise, human cell lines have been used in search of effective emerging and existing therapies for certain diseases. For example, it is known that there are over 50 types of breast cell lines, which originated from about 150 human breast tumors (Neve et al., 2006). These cell lines retain most of the genetic information for autonomous growth. These cell lines have been successfully used to understand molecular mechanisms of sensitivity/resistance to inhibitors with exquisite sensitivity for the EGFR (epidermal growth factor receptor), HER2 (human epidermal growth factor receptor 2), MET (hepatocyte growth factor receptor), and BRAF kinase inhibitors (McDermott et al., 2007). These cell lines indeed paved the way for the use of trastuzumab (Herceptin) as an adjuvant therapy in the treatment of breast cancer in some individuals. Another example would be the increased BCR-ABL expression, which is an excellent biomarker of Gleevec resistance, leading to personalized medicine—meaning treating the patient with a drug that works with minimal to no adverse effect (Tainsky, 2009; Khalil et al., 2010).

EXAMPLES OF ORGANIZATIONS' RESEARCHING AND FUNDING ALTERNATIVES TO ANIMAL TESTING AND BIOMARKER DEVELOPMENT

- **Johns Hopkins University Center for Alternatives to Animal Testing (CAAT-a, 2013)**
  The Johns Hopkins University Center for Alternatives to Animal Testing (CAAT-a, 2013) has worked with scientists since 1981 to find new methods to replace the use of laboratory animals in experiments, reduce the number of animals tested, and refine necessary tests to eliminate pain and distress to test animals. The CAAT promotes humane science by supporting the creation, development, validation, and use of alternatives to animals in research, product safety testing, and education. It is not an activist group; rather, it seeks to effect change by working with scientists in industry, government, and academia.
- **Interagency Coordinating Committee on the Validation of Alternative Methods (ICCVAM, 2013)**
  Traditionally, consumer products, medical devices, chemicals, and new drugs are tested on animals to predict toxicity, metabolism, and efficacy in humans. Scientists in ICCVAM are working to promote the development and validation of alternative test methods. Alternative test methods are those that accomplish one or more of the 3Rs principles.

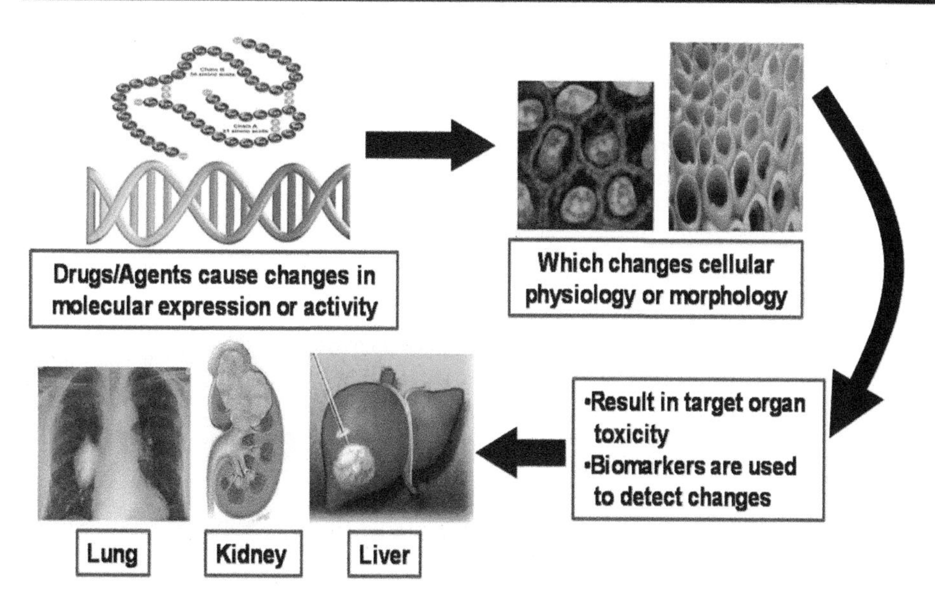

FIGURE 8.2 Toxic test substances cause changes at the molecular level (DNA or protein), which are expressed at the cellular level. This in turn results in target organ toxicity (e.g., liver, kidney), which can be studied and extrapolated to whole animals or humans based on in vitro studies.

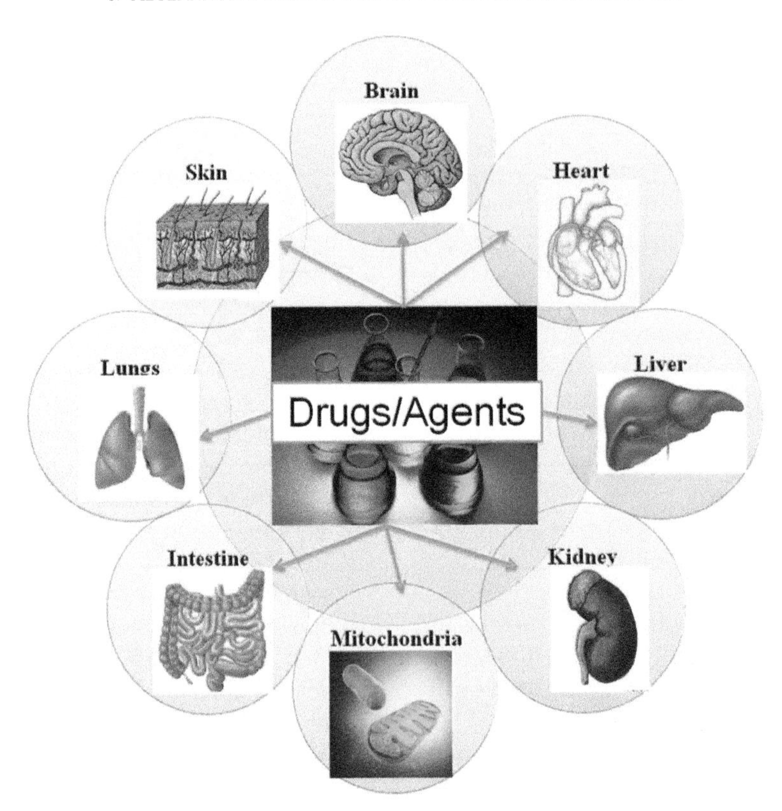

FIGURE 8.3 Drugs or agents interact with various cells, organelles, and organ systems to produce their pharmacological and/or toxicological effects. These effects can now be studied using various in silico or in vitro tools, minimizing the use of animal testing.

- **University of California–Davis, Center for Animal Alternatives (CAAT-b, 2013)**
  The UC Davis Center for Animal Alternatives disseminates information on alternatives to animal experiments with information on the most current methods for improving all aspects of animal care.
- **Dr. Hadwen Trust**
  The Dr. Hadwen Trust for Humane Research is a British biomedical research charity, founded in 1970 (Dr. Hadwen Trust, 2013). It funds peer-reviewed alternatives to animal testing in relation to health issues such as cancer, heart disease, meningitis, and Alzheimer's disease. Trust researchers have cultured human cartilage, which has helped to show how arthritis drugs work; developed a three-dimensional model of human teeth and jawbone for dental research; created a computer model of a human fetus; and used mathematical modeling to create drug treatment for non–Hodgkin's lymphoma and neuroblastoma. Over 200 papers about the Trust's work have been published in scientific journals.
- **National Centre for the Replacement, Refinement, and Reduction of Animals in Research**
  The European Centre for the Validation of Alternative Methods (ECVAM) launched an online database on toxicology of nonanimal alternative test methods in 2006 (NC3R, 2013). Categories at present include in vitro methods, quantitative structure–activity relationship (QSAR) models, and a bibliographic section, aiming to develop alternatives to animal testing. This describes general guidance to researchers and associated veterinary and animal care staff using vertebrates (live animals or animal products) in bioscience research funded by various agencies. It sets out the expectations of the funding bodies for the use of such animals in research and is therefore useful to ethics committees, referees, and members involved in reviewing research proposals. Implementation of the principles in this guidance is a condition of receiving funds from the funding bodies.
- **REACH (Registration, Evaluation, Authorization, and Restriction of Chemicals)**
  This organization is the European Community Regulation on chemicals and their safe use (EC, 1907/2006). The aim of REACH (2013) is to improve the protection of human health and the environment through better and earlier identification of the intrinsic properties of chemical substances. At the same time, REACH aims to enhance innovation and competitiveness of the EU chemicals industry, thus enhancing information sharing and minimizing unnecessary experimentation in animals. REACH is executed through a public database in which consumers and professionals can find hazard information.

ACHIEVING 3RS WITH MINIMAL TO NO ANIMAL TESTING BY USING BIOMARKERS

Over the years, several methods have been identified and practiced where possible, in achieving the 3Rs. For example,

1. In silico computer simulation
2. In vitro cell culture techniques
3. Integrated toxicity testing using fewer animals and collecting the most information by selecting different experimental endpoints, improved use of anesthetics, analgesics, and replacing open surgery with minimally invasive endoscopic surgeries, and procedures related to animals and a myriad of other things, which will minimize potential pain and distress
4. Microdosing, in which the basic behavior of the drug is assessed using human volunteers receiving doses well below those expected to produce whole-body effects (FDA, 2006)
5. Epidemiologic and clinical studies and genetic monitoring, in which studies of human populations have provided important information about the causes of many diseases, such as the relationships between cholesterol and heart disease, the mechanism of transmission of HIV, and chemical exposures and birth defects. Scientists can "see" abnormalities and track treatment progress in the brains of victims of Alzheimer's and Parkinson's diseases, schizophrenia, epilepsy, and brain injury using sophisticated scanning technologies (CT, PET, and MRI). All drugs must undergo clinical testing before being approved; carefully crafted clinical research is the best way to determine human reactions to new drugs.

Some of these strategies are described below.

In Silico Computer Simulation

There are a variety of in silico computer simulation software programs available in the public domain and on a fee-based service. Computers can often predict the toxicity of chemicals, including their potential to cause cancer or birth defects, based on their molecular structure. Computer simulations can also predict the metabolism and distribution of chemicals in human tissues.

Lhasa (2013) Limited is a not-for-profit organization that promotes the sharing of data and knowledge in chemistry and life sciences. It has developed software tools such as DEREK (**D**eductive **E**stimation of **R**isk based on **E**xisting **K**nowledge) for Windows, Meteor, and Vitic to facilitate such sharing. DEREK for Windows and Meteor are knowledge-based expert systems that predict the toxicity and metabolism of a chemical, respectively. Vitic is a chemically intelligent toxicity database. These software tools include illustrations on (1) the use of data entry and editing tools for the sharing of data and knowledge within organizations; (2) the use of proprietary data to develop nonconfidential knowledge that can be shared between organizations; (3) the use of shared expert knowledge to refine predictions; (4) the sharing of proprietary data between organizations through the formation of data-sharing groups; and (5) the use of proprietary data to validate predictions. Sharing of chemical toxicity and metabolism data and knowledge in this way offers a number of benefits, including the possibilities of rapid scientific progress and reductions in the use of animals in testing. Maximizing the accessibility of data also becomes increasingly crucial as in silico systems move toward the prediction of more complex phenomena for which limited data are available. Recently, it has been shown that the use of in silico methods is the way forward for assessing the genotoxicity and potential carcinogenicity/repeat-dose general toxicology of large numbers of chemicals (Hayashi et al., 2005). More importantly, it is expected that these methods will contribute to the reduction of animal use.

There are several other in silico methods that have been developed and validated, for example, Leadscope, MCase, Topkat, Toxtree, and Osiris software, as well as QSAR models developed by the FDA with proprietary databases (Contrera et al., 2007). It is proposed that, depending on the type of molecule being tested and the type of safety information being sought, one or more software programs may be used to understand the structural alteration alters for the endpoint desired (Krishna et al., 2014). As the data outputs are reviewed, a consensus and reproducibility of alerts in at least two or more databases may be considered as significant alerts.

Structure Activity Relationship (SAR) software, such as, DEREK, Toxtree, Ecostar, OECD QSAR Toolbox, can perform predictions for multiple toxicity endpoints and species for new drugs and agents. This software utilized either the knowledge base or rules to make determination using data from previously tested compounds and algorithms that consider a multitude of factors, including physical, chemical, and biological properties of compounds and substituent's groups. The predications are used to prioritize and determine urgency for further testing in animals, as needed (Fig. 8.4).

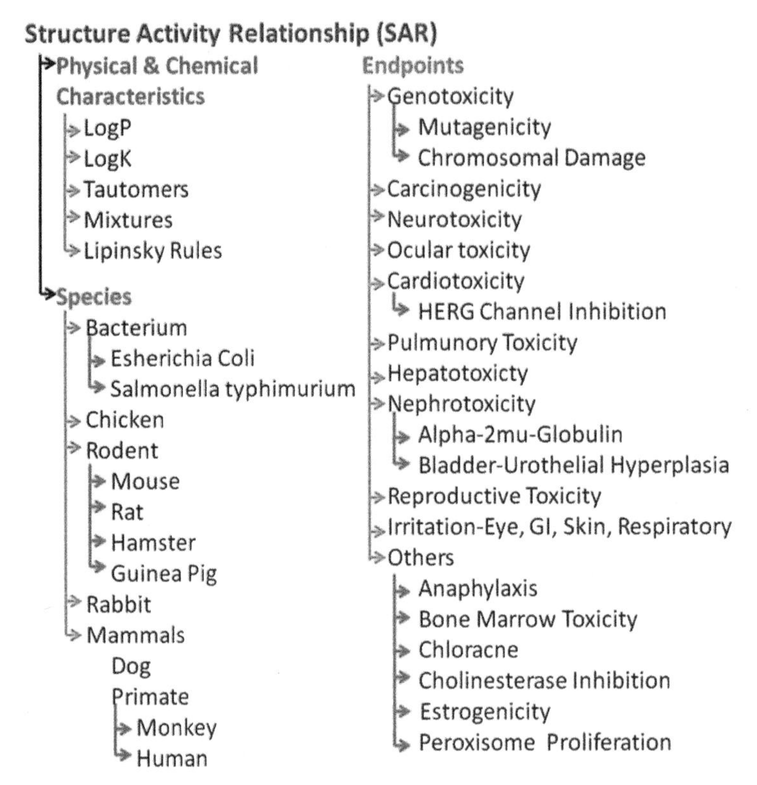

FIGURE 8.4 In silico endpoints and biomarkers commonly used to achieve 3Rs.

In Vitro Cell Culture Techniques

In vitro cell culture is currently the most successful and promising alternative to animal use. Isolated cells, tissues, and organs can be prepared and maintained in culture by methods that preserve properties and characteristics of the same cells, tissues, and organs in vivo (Krishna et al., 1984, 1985, 1986; Wilcox and Bruner, 1990; Krishna, 2010). Using such in vitro systems will permit data to be generated under controlled experimental conditions that are, in the absence of many complicating factors, characteristics of experiments in whole animals. For example, cultured cells have also been developed to create monoclonal antibodies thereby replacing the use of animals required to undergo a procedure likely to cause pain and distress. An enormous amount of valuable in vitro cell culture research is conducted today. The National Disease Research Interchange provides more than 130 types of human tissues to scientists investigating more than 50 diseases, including cancer, diabetes, and glaucoma (NDRI, 2013). Cell and tissue cultures are used to screen new therapies and to test for product safety. Genetic microarrays are being used to predict liver toxicity by measuring gene expression in human liver cells (Lipp, 2007; Tainsky, 2009).

Several in vitro tests have been validated and/or are currently being evaluated to replace the in vivo tests that cause pain and suffering to animals. These tests are used to determine, for example, phototoxicity, skin and/or eye irritation and corrosivity, immunotoxicity, bone marrow toxicity, phospholipidosis, kidney toxicity, and genotoxicity. Most frequently used in vitro tests are listed in Table 8.1.

Toxicity biomarkers in the in vitro context are defined as quantitative measurable characteristics that serve as indicators of a pathologic process or related biochemical or molecular events. There is great value in developing preclinical assays that are predictive of human clinical outcomes and in vitro methods to test for drug-induced liver injury, for example, are well recognized. High-throughput screening and high-content screening have enabled the scientific community to efficiently screen large new chemical entity (NCE) libraries. Continued advancement in the development of instrumentation and biomarkers has strengthened researchers' ability to deliver suitable and predictable assays. The combination of both methods and biomarkers may elucidate mechanisms of hepatocyte damage and significantly improve the understanding of drug safety and efficacy profile. When such methods and biomarkers are shown to have in vitro–in vivo correlation, the risk–benefit ratio for a particular drug's development can be more effectively managed, leading to safer and more successful outcomes for patients and pharmaceutical companies (Moeller, 2010).

TABLE 8.1 Most Frequently Used In Vitro Tests in Achieving 3Rs

Test	Targets	Benefits
3T3 NRU Phototoxicity Test	In vitro identification of a phototoxicity	Partial replacement of in vivo studies for the investigation of phototoxic effects.
Bovine Corneal Opacity/Permeability Assay (BOCP)	In vitro alternative to in vivo rabbit Draize test for eye irritation and corrosivity	Replacement of in vivo studies to test eye irritation or severe damage.
Human Corneal Epithelium (HCE Test)	In vitro alternative to in vivo rabbit Draize test	Same as above
3D Human Skin Model	In vitro alternative to in vivo rabbit Draize test for skin irritation and corrosivity	Same as above for skin irritation or corrosion. Can detect photoreactive chemicals if applied in UV light.
Mishell–Dutton Culture	In vitro alternative to immunotoxicity assay in the rodent (plaque assay). The immune reaction occurs in vitro	Assay done in blood cell cultures, no animal sacrifice. Blood of different species can be tested, compared, enable better prediction for humans
CFU-GM Assay	In vitro screen for bone marrow toxicity mainly for indication oncology	Early detection of bone marrow toxicity
Phospholipidosis Assay	In vitro screen for phospholipidosis-inducing drug	Early identification of drugs with unfavorable properties
Biomarkers for nephrotoxicity	Earlier and sensitive prediction of kidney toxicity	More sensitive biomarkers
Integration of genotoxicity tests into repeat-dose toxicity studies (IWGT working group)	Rat micronucleus test and comet assay included in repeat-dose studies	Better correlation to overall assessment

Limitations of Alternative In Vitro Tests

A variety of limitations of in vitro tests have been described. Briefly, the in vitro test systems are not available for all tissues and organs. Furthermore, normal systemic mechanism of absorption, distribution, metabolism, and excretion systems is absent under culture conditions. In vitro systems lack the complex, interactive effects of the immune, blood, endocrine systems, nervous system, and other integrated elements of the whole animal. Therefore, in vitro tests cannot be used to study the complex nature of systemic toxicity. Finally, validation of new methods is time consuming and expensive. Acceptance of in vitro tests as alternatives to traditional toxicity testing in whole animals is expected to be slow. However, a significant amount of progress has been made toward regulatory acceptance of such methods. The REACH program by the European Union, for example, has implemented acceptance of in vitro data for regulatory purposes.

APPLICATIONS OF IN VITRO TESTS AND BIOMARKERS

Genetic Toxicology Testing, In Vitro

A battery of in vitro genetic toxicology tests has been routinely used for over 30 years to predict the carcinogenicity of substances that are of safety concern. Such assays include, for example, bacterial assays and in vitro cell culture assays for chromosomal aberration assessment and cytotoxicity assessment. The effects on DNA and cells caused by test agents are indirectly observed and enumerated using multiple genetic and cytotoxicity endpoints as biomarkers (Sankaranarayanan et al., 2000; Gollapudi and Krishna, 2000; Hancock et al., 2004; Kirkland et al., 2005).

Bacterial Mutation Assay (Ames Assay)

This assay measures genetic damage at the single base level in DNA by using five or more test strains of bacteria. Each of the *Salmonella typhimurium* and *Escherichia coli* strains used in the assay have a unique mutation that turns off histidine biosynthesis in *Salmonella* or tryptophan biosynthesis in *E. coli*. Because of these original mutations, the bacteria require exogenous histidine or tryptophan to survive and will starve to death if grown without these essential nutrients (auxotrophy). The key to the assay is that the bacteria can undergo a reverse mutation turning the essential gene back on and permitting the cell to grow in the absence of either histidine or tryptophan (prototrophy). A specific type of mutation, either a base-pair substitution or frameshift mutation, creates each bacterial strain. The standard Ames assay designs include preliminary

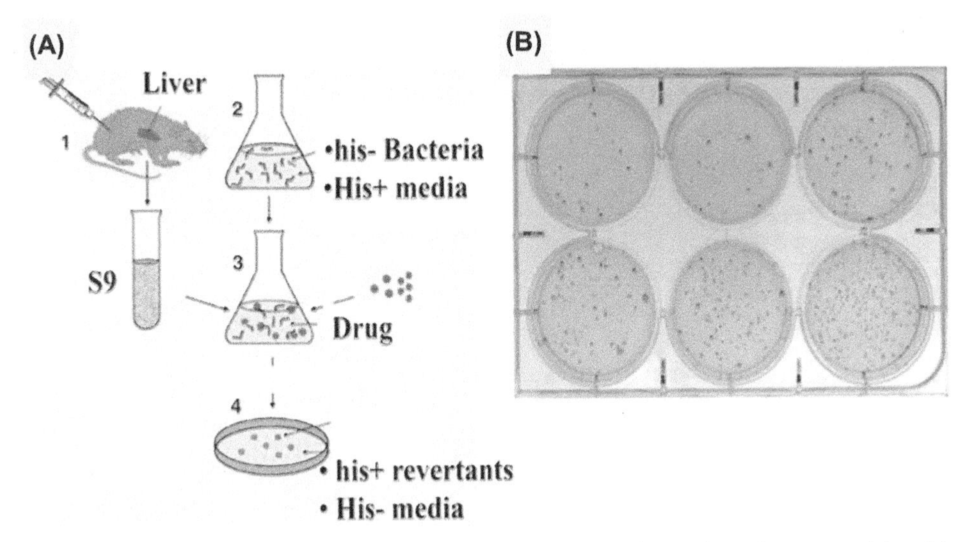

FIGURE 8.5 The in vitro Ames assay in specially constructed strains of bacteria is used to evaluate the mutagenicity of drugs or chemicals. (A) Principle: (1) S9 liver microsomal fraction is prepared from rats pretreated with a liver enzyme inducer; (2) Histidine mutant bacteria are grown in the presence of histidine media; (3) S9, bacteria, and the drug are mixed in top agar containing trace amounts of histidine to allow several DNA replications and poured on to agar plates with minimal media without the histidine and incubated; (4) If the test substance and/or its metabolite(s) causes mutations, and then it would reverse the his^- gene into his^+ gene, which can grow in minimal media and form colonies, thus, considered a mutagen. (B) The results of an Ames test are shown in a 6-well plate: from the top left each well represents control, dose 1, 2, 3, 4, and positive control, showing an increasing number of colonies with dose, suggesting the test article as a positive Ames mutagen.

toxicology tests or combined toxicity and mutation tests followed by a definitive mutation assay. In both toxicity and mutation tests, test strains are combined with S9 mix (prepared from the livers of Aroclor-treated rats) buffer, a test or control article, a trace of histidine or tryptophan, and molten agar. The bacteria use the trace of histidine or tryptophan to undergo several cell divisions but will stop growing once they have run out, leaving a characteristic "background lawn" that decreases in density with increasing toxicity. After 48 h, only those cells that have undergone a reverse mutation turning the essential gene back on have survived, producing mutant colonies. The background lawn density is scored followed by counting the number of revertant colonies. Mutation results are reported as revertants per plate. These assays are performed in the presence and absence of exogenous metabolic activation systems (S9 mix) (Fig. 8.5).

In Vitro Micronucleus and Chromosome Aberration Assays in Cultured Cells (Cytogenetic Assays)

In vitro cytogenetic analysis provides a valuable technique for evaluation of damage to chromosomes on the basis of direct observation and classification of chromosomal aberrations. Cells arrested in the metaphase are examined microscopically for both numerical and structural chromosome aberrations. An alternate cytogenetic approach is the observation of micronuclei, which represent chromosomal-breakage fragments or lagging chromosomes in the cell cytoplasm (Fig. 8.6; Krishna et al., 1989, 1992). The cytogenetic assays may be performed either in vivo or in vitro (Fig. 8.7). In vitro cytogenetic assays are performed in the presence and absence of S9 mix. These assays are also used in human biomonitoring in clinical research and/or in occupational exposures (Fig. 8.8).

Comet or Single-Cell Gel Electrophoresis Assay

The comet assay is a microgel electrophoresis technique that detects DNA damage and repair in individual cells (Fig. 8.7). The damage is represented by an increase of DNA fragments that have migrated out of the cell nucleus in the form of a characteristic streak similar to the tail of a comet. The DNA fragments are generated by DNA double strand breaks, single strand breaks, and/or strand breaks induced by alkali-labile sites in the alkaline version of the assay. The length and fragment content of the tail is directly proportional to the amount of DNA damage. Comet assay can be conducted in both in vitro and in vivo test systems and are increasingly used in genotoxic testing of industrial chemicals, agrochemicals, and pharmaceuticals. Comet assay is rapid (results are available in days), simple to perform, requires small amounts of the test substance (25–50 mg), and can be performed in almost any eukaryotic cell (different animal target organs). The comet assay serves as an important tool in early drug development as a mechanistic and genotoxic predictor (Tice et al., 2000).

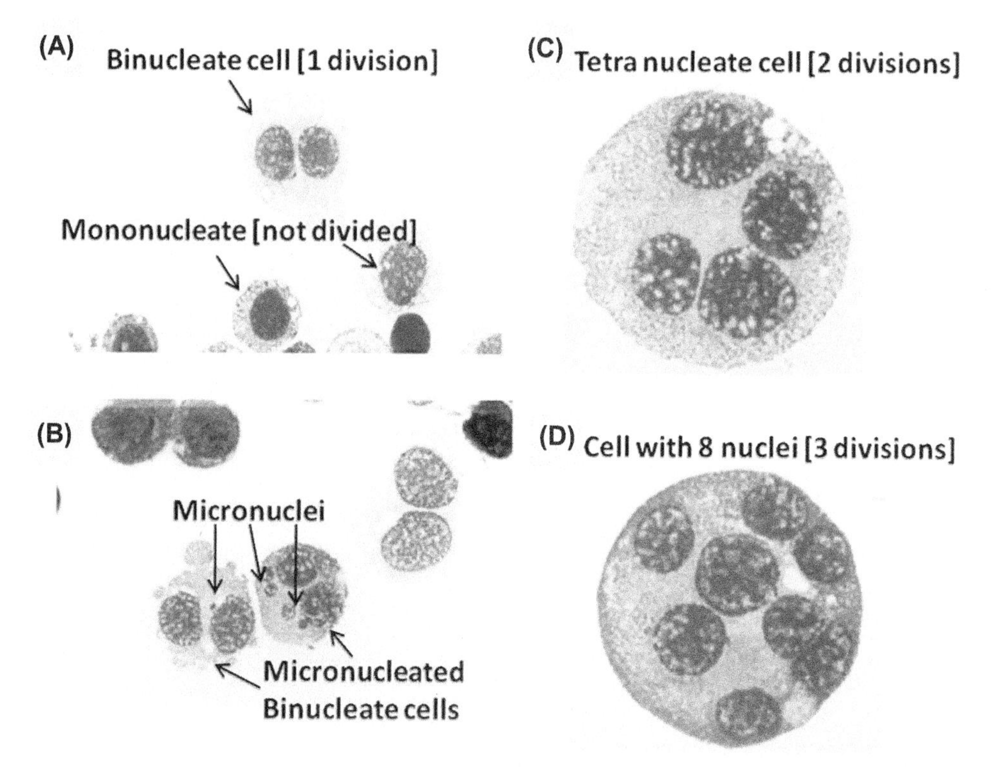

FIGURE 8.6 The in vitro and/or in vivo/in vitro micronucleus assay determines the clastogenic effects of chemicals on the chromosomes. This can be evaluated in vitro using human peripheral blood lymphocytes cultured in the presence of drug. (A) Normal mononucleate cells (not divided in culture) and binucleate cells that have undergone 1 division, (B) micronuclei in binucleate cells, (C) tetranucleate cell that has undergone two divisions, and (D) cell with 8 nuclei that have undergone 3 divisions. This assay is used to evaluate clastogenicity (chromosome breakage), aneugenicity (chromosome lagging due to spindle dysfunction), and cytotoxicity (delay in cell cycling) with appropriate staining techniques.

Integrated Genotoxicity Testing Using Fewer Animals and Collecting Most Information

The concept of integrating toxicity assessment employs fewer animals, but a variety of relevant toxicity data can be collected and evaluated (Krishna et al., 1994, 1995a,b; 1998, 2000; Fig. 8.7). As an example, an integrated in vivo genotoxicity testing philosophy and a practical approach, as applied to pharmaceuticals, is currently practiced and is recommended by the International Conference on Harmonization (ICH) guidance (ICH, 2006). In this case, a rodent (primarily rat) micronucleus assay is integrated with routine 2- to 4-week repeat-dose toxicity and toxicokinetic studies. This approach has several advantages: (1) it utilizes the general principles of toxicology that govern the overall toxicity profile of a test substance; (2) factors such as the dose and/or route of drug administration, drug metabolism, principles of toxicokinetics, and saturation of defense mechanisms are considered in evaluating genotoxicity; (3) it uses the concept of administering multiple tolerable doses aiding in achieving steady-state plasma drug levels, which is more relevant for risk assessment when compared with high acute doses; and (4) it helps minimize the amount of drug, the number of animals used, and other resources. This integration approach can be extended to other toxicology studies, and other relevant genotoxicity endpoints may be assessed. Based on experiences reported in the literature, integrating micronucleus assessment in routine toxicology testing is promising and should be utilized when practical. A number of genotoxicity endpoints can be easily evaluated in peripheral blood lymphocytes of subjects exposed by genetic monitoring during clinical trials and/or while on medication (Fig. 8.7).

Unique Zebrafish Model

For years, mouse models have been used for testing new drugs, despite their high cost. More recently, the zebrafish has been identified as an important vertebrate model for studying the development of embryos and pathogenesis of human diseases (Lieschke and Currie, 2007). When compared with mammalian models, experimental results show that zebrafish embryos exhibit similar responses to drugs for cardiovascular diseases, and antiangiogenesis and anticancer drugs (Hill et al., 2005). This suggests that the zebrafish model can be used as a bridge between the in vitro model and the in vivo model in the drug

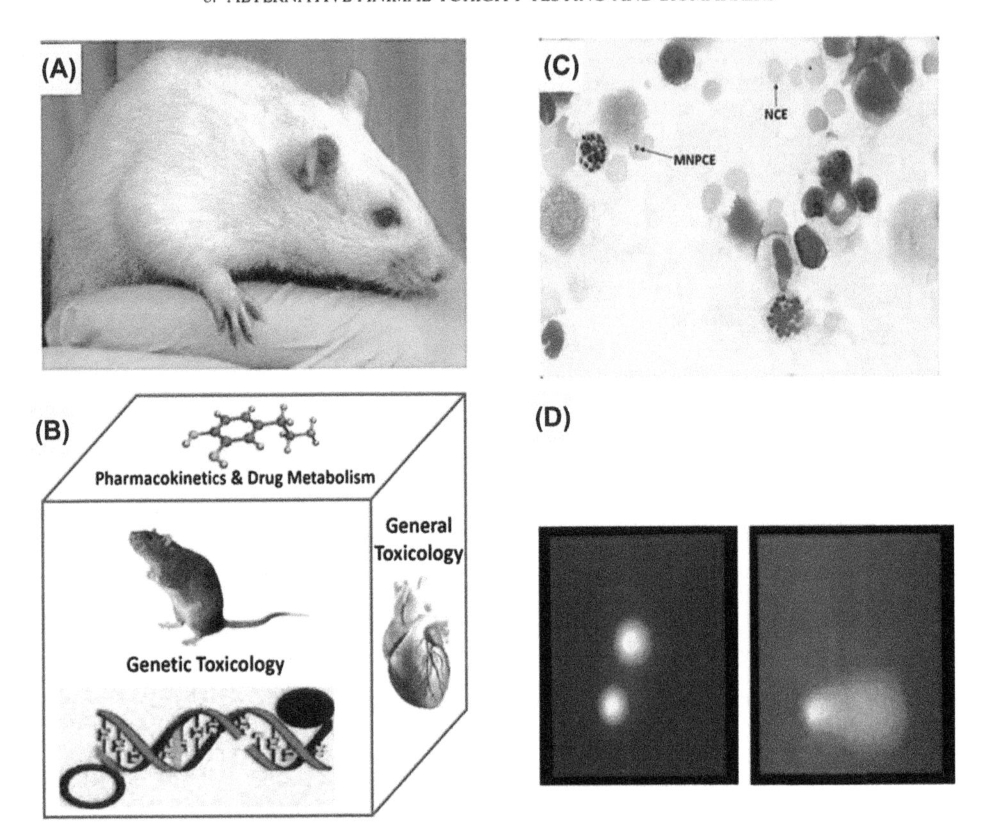

FIGURE 8.7 Integrated in vivo genotoxicity of drugs or agents can be evaluated using samples obtained during the general toxicity study, (A) The samples are obtained from an animal (e.g., rat). (B) The drug is exposed to in vivo biochemical processes, including metabolism and opportunity to interact with the DNA, thereby providing an intact animal with homeostatic assessment of the genotoxic potential of a drug; and (C) the blood sample/bone marrow is obtained and stained to detect the presence of micronuclei in polychromatic erythrocytes (PCE); if positive, then the drug is considered as clastogenic. (D) Cells, particularly from liver, are processed and tested for DNA strand breaks in the comet assay. The breaks in the DNA are observed as a comet-shaped migration of DNA from the nucleus of the cell following gel electrophoresis in treated compared with the intact round nucleus suggesting unaffected DNA (control).

discovery and screening processes. In addition, pollutants induce similar pathological changes in zebrafish and other mammalians.

In Vitro Pharmacologic Activity Testing

Pharmacology is the study of drug effects in biological systems, including therapeutic or toxic effects. The field of in vitro pharmacology takes advantage of the technological advances in studying drug effects in the laboratory settings using cell culture models and isolated tissue assays and predicts such effects to in vivo models (Irwin et al., 2008). In vitro pharmacology data provide invaluable information on the compounds' chances to succeed in the clinical phase. Multiple contract research organizations offer a variety of panels of validated in vitro pharmacological assays that cover a broad range of targets. Apart from ascertaining the potency of a NCE on target, assays are also made available for selectivity screening and profiling. These assays are available either as stand-alone or as a screening platform.

Isolated Tissues

In the study of the molecular mechanism of drug action, receptors are the most important link in the chain of events. The characteristics of the receptors are studied by the observations and quantitation of some physiological response elicited by a tissue in response to a drug. This constitutes the pharmacological approach and is an indirect study of the drug–receptor interaction. Pharmacological activity of a compound can be easily, quickly, and inexpensively obtained from isolated tissue experiments. Because of this simplicity, researchers in the early 1900s were able to define the interaction between drugs and receptors, predict drug effects, and formulate mathematical models of drug receptor theory. Isolated tissue techniques revolutionized and dominated the field until the end of the 20th century. Readers are directed to excellent reviews by Kenakin (1984, 2002), where the author discusses some very important concepts in isolated tissue methods and procedures for measuring drug activity.

A variety of isolated tissues for each receptor class have been described in the literature. Three of the

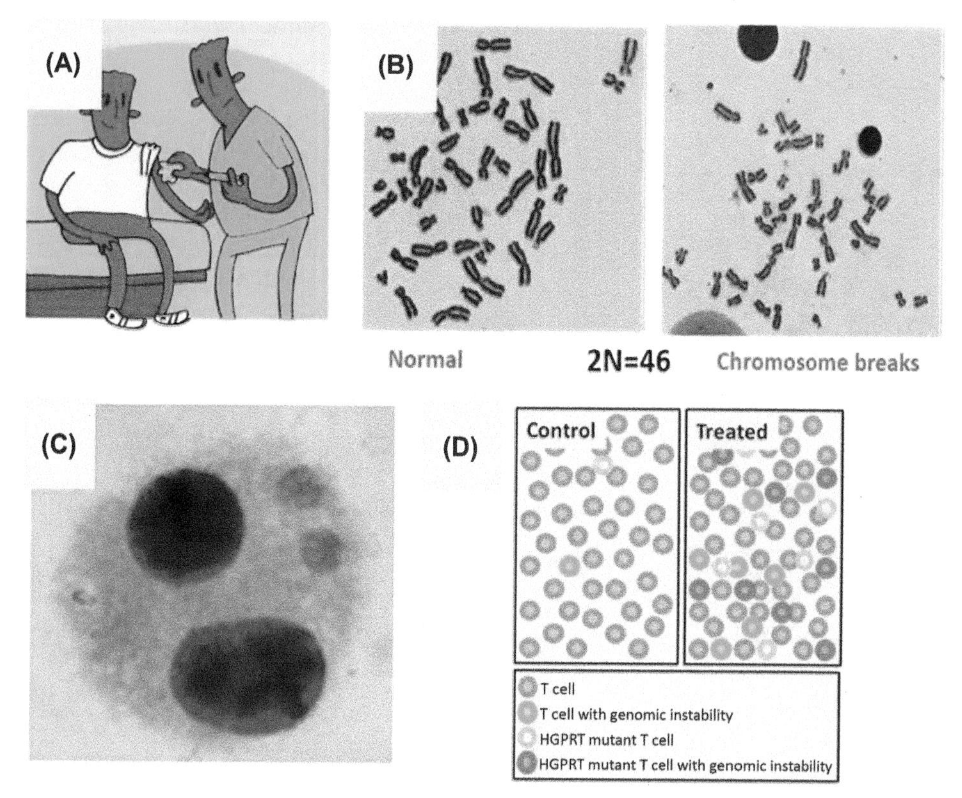

FIGURE 8.8 Monitoring humans during clinical trials using peripheral blood lymphocytes from treated subjects can be collected and evaluated for a number of genotoxicity endpoints such as, chromosomal aberrations, micronuclei, and gene mutations. (A) Small amount of blood is collected from the subject in clinical trial to isolate peripheral blood lymphocytes for further genetic testing. (B) The normal intact chromosomes are shown from untreated subjects. (C) Breaks in chromosomes are seen in cells obtained from a subject who received treatment (if the test agent is clastogenic). (D) HGPRT gene mutation is observed in cells using various staining techniques and flow cytometry.

most widely used classes of isolated tissues are from cardiac muscle, vascular smooth muscle, gastrointestinal, and other smooth muscles.

Receptor Binding Studies

Radioligand binding studies were developed to identify and characterize receptor sites in various tissues. These studies provided insights into molecular interactions between ligands (endogenous and exogenous) and receptors and clarified the modes of drug action and the general mechanism that can lead to altered receptor function. Two principal parameters that are determined with these studies are the "density (B_{max})" of a given class of receptor sites and their "affinities (K_m)" for various ligands. The values determined from these studies are compared with those derived from in vitro and in vivo functional studies. There are subtle differences between pharmacological methods and radioligand binding procedures in regard to the characterization of receptors. Binding studies lack a functional response. However, agonists' interactions with binding sites and their sensitivity to temperature, various ions, and often to guanine nucleotides can provide some indication of efficacy. Disruption of the tissue to prepare membranes or solubilization may alter the pharmacological characteristics of receptor sites. With membrane preparations, there are few diffusional barriers, and thus the actual concentration of the drug in the receptor compartment can be fairly well estimated. In addition, radioligand binding studies identify ligand-binding sites, which may or may not represent bona fide receptors. Thus, results must be treated cautiously.

In a drug discovery program, radioligand binding studies are developed and performed for several therapeutic targets such as G protein–coupled receptors (GPCRs), ion channels (including hERG), and transporters using membranes from recombinant cells as well as native tissues.

GPCR functional assays: In addition, functional assays are used as a follow-up screen on GPCR targets. Ligand binding to GPCR promotes G protein coupling initiating signal transduction pathways, which trigger a series of cellular responses. These data can be used to substantiate the binding data.

Enzyme assays: Enzymes are increasingly recognized as potential drug targets in drug discovery. Multiple contract research organizations offer assays for several enzyme classes using different readouts like fluorescence, luminescence, time-resolved fluorescence, and radiometry. The following are a few examples of enzymes that have been validated and are available for applications in pharmacology research. Hundreds of in vitro tests are available that utilize isolated or cloned animal and/or human receptors, ion channels, enzymes, transporters, and other components to determine the efficacy and safety of drugs or chemicals. Select examples of commonly used pharmacological assays are listed in Table 8.2.

In Vitro Drug Disposition Studies

Absorption, distribution, metabolism, and elimination (ADME) are the processes by which the body handles chemicals, referred to as drug disposition studies. The process of pharmacodynamics, however, describes how chemicals or test substances affect the living organisms at the molecular, cellular, organ, host, and population levels (Heinrich-Hirsch et al., 2001). As understanding of the molecular basis for ADME properties evolves, it is becoming possible to move beyond studying pharmacokinetics in intact people or animals. By combining data collected in vitro characterizing specific processes with computer models describing the interplay among the pharmacokinetic processes in the whole body, one can predict what would be expected to occur if intact organisms were exposed. Integrating data or predictions about chemical concentrations in different organs in the body with data or predictions about effects those chemicals would be expected to cause provides a long-term path toward predicting beneficial (therapeutic) or adverse (toxicological) effects. The success of this long-term path depends on continued research on the basic biological processes underlying pharmacokinetics, further development of in vitro systems both for ADME processes and for effects, and improved computer models. The major uses of in vitro ADME studies include the following:

1. To predict pharmacokinetics in vivo
2. To provide and/or characterize ADME processes and effects seen in vivo using in vitro assays
3. To predict drug–drug interactions

In vitro ADME studies help in optimization of pharmacokinetic characteristics of a drug candidate, identify the metabolic pathways, and provide valuable inputs for the design of in vivo studies. Multiple contract research organizations (the reader is suggested to visit these websites for additional information—e.g., MDS labs, CEREP) offer both standardized and customized assays as per sponsor's requirement for ADME parameters. These services include

- Physicochemical studies
  - pKa
  - LogD/LogP
  - Solubility (Thermodynamic/Kinetic)
  - Chemical stability at different pH
- Absorption/distribution Assays
  - PAMPA (**P**arallel **A**rtificial **M**embrane **p**ermeability **A**ssay)
  - Caco2 or MDCK (unidirectional/bidirectional)
  - P-gp efflux
  - Protein binding (ultrafiltration and equilibrium dialysis)
- Metabolic stability
  - Microsomal stability (rat, human, etc.)
  - Hepatocyte stability
  - Metabolite identification
  - Biological stability (plasma, serum, blood, simulating fluid)
- CYP450 Assays
  - CYP inhibition (CYP 1A2, CYP 2C9, CYP 2C19, CYP 2D6, CYP 3A4)
  - Mechanism based inhibition
  - Half-life determination/IC_{50} determination
  - CYP pathway identification
  - CYP450/metabolite identification
  - CYP induction
- Drug transporter assays
  - Identification of substrates and/or inhibitors
  - Induction of transporter (P-gp)

An illustration of drug transporters expressed in cells of various tissues and utilized to understand drug–drug interaction in vitro is shown in Fig. 8.9. Results from such studies are used in extrapolating effects on humans (FDA, 2012).

Major Challenges of In Vitro Absorption, Distribution, Metabolism, and Elimination

1. A major challenge in predicting in vivo pharmacokinetics is in characterizing the association between the in vitro conditions and the in vivo situation.
   a. For instance, measurements of glucuronyl transferase enzymatic activity using microsomes, a preparation of subcellular organelles, are highly dependent on experimental conditions, in part due to inaccessibility of the enzyme without solubilizing or pore-forming agents.
   b. Accurate quantitation of the microsomal content of liver tissue has been another concern.
   c. Similarly, there have been comparisons of the metabolic capabilities of genetically expressed enzymes, microsomes, isolated liver cells, and other systems for studying metabolism.

TABLE 8.2 List of Select Pharmacological Assays and Biomarkers Used in Research to Help Achieve Goals of 3Rs

RECEPTORS—G PROTEIN–COUPLED RECEPTORS (GPCRS)					
Adenosine	Calcitonin	Endothelin	Histamine	Neuromedin U	Prokineticin
Adrenergic	Cannabinoid	GABA	Leukotrienes	Neuropeptide Y	Relaxin
Angiotensin	Chemokines	Galanin	Melanocortin	Neurotensin	Serotonin
Apelin	Cholecystokinin	Glucagon	Muscarinic	Orexin	Urotensin-II
Bombesin	Complement 5a	Glutamate	Neurokinin	Purinegic	Vasopressin
Bradykinin	Dopamine	Neuropeptide W/B	Neuropeptide S	Protanoid	Somatostatin
Calcium sensing	Free fatty acids	Growth hormone	N-formyl peptide	Parathyroid hormone	Proteinase activate
Corticotropin-releasing factor	Glycoprotein hormone	Lysophospholipids	Orphan (Class A)	Thyrotropin-releasing hormone	Vasoactive intestinal peptide
Calcitonin gene–related peptide	Gonadotropin-releasing hormone	Melanin concentrating hormone	Opioid and opioid like	Platelet-activating factor	
ION CHANNELS—VOLTAGE-GATED CHANNELS					
Calcium channels	Potassium channels	Sodium channels	Transient receptors potential channels		
TRANSPORTERS					
Adenosine transporter	Cannabinoid transporter	Serotonin	Dopamine		
GABA	ATP-binding cassette (ABC) transporter	Choline transporter	Norepinephrine		
Solute carrier (SLC) transporters					
KINASES					
Protein-tyrosine kinases	Protein-serine/threonine kinases	Atypical kinases	Other kinases		
EPIGENETIC AND DNA RELATED					
DNA methyltransferase	Demethylases	Histone methyltransferase	Histone acetyltransferases		
DNA repair and mitotic enzymes		Topoisomerases			
OTHERS					
Phosphatases	Proteases	Cyclooxygenases	HIV proteases		

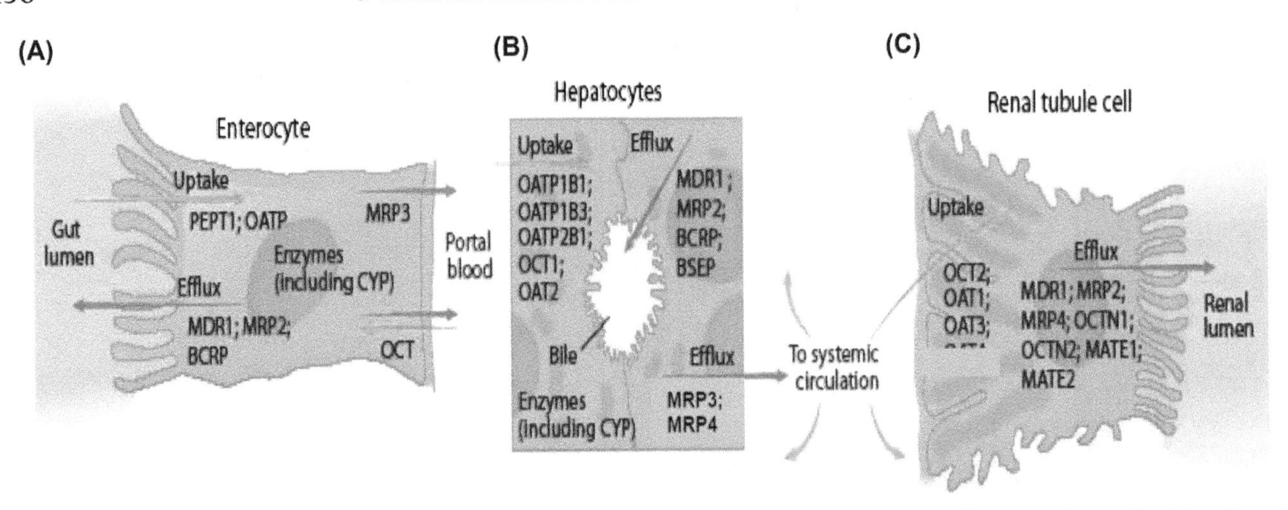

FIGURE 8.9 Illustration of efflux and uptake transporters in the gut wall (A), liver (B), and kidneys (C) that may be involved in a drug absorption, distribution, metabolism, and excretion (ADME).

d. Similar issues exist for characterizing transporter activities, which are often essential to absorption, tissue distribution, and excretion.
e. Limited in vitro methods to assess some processes, such as biliary excretion and reabsorption or urinary excretion (recent reports of sandwich cultures of hepatocytes to predict biliary excretion and expression of renal transporters in different systems like oocytes).

2. Another major challenge for predicting in vivo pharmacokinetics is developing and implementing computer simulation and modeling.
   a. A range of modeling approaches can be useful, with the long-range goal being to make predictions for the intact organism from the formula and structure of the chemicals.
   b. Considerations to link in vitro to in vivo—(1) mathematically describing selected ADME processes and (2) integrating multiple pharmacokinetic processes within physiologically based models to track chemical disposition within the "in vivo" context.

The ADME of drugs can be predicted using different in vitro assays. The data obtained from individual or a combination of assays are utilized to predict the in vivo ADME in animals or human. Select examples of ADME and PK parameters commonly used are listed in Table 8.3. For comparative purposes, an animal toxicokinetic model is shown in Fig. 8.10, and the toxicodynamics model with molecular mechanisms is shown in Fig. 8.11. These could also be considered as PK-PD modeling with allometric scaling for human risk assessment.

Advances in Technologies

With the advancement of computer science and continued improvement in instrumentation and technologies such as fMRI, phMRI, mass spectrophotometry, and flow cytometry have become useful tools with high resolution and discrimination of biological processes (Kirshna et al., 1993; Criswell et al., 1998a,b, 2003; Chin et al., 2010; Lake, 2011; Darzynkiewicz et al., 2011; Dertinger et al., 2011; Canavan, 2013). Also, with appropriate dyes as diagnostic cell markers and imaging, these tools are able to generate valuable toxicity and efficacy data, while serving as biomarkers. Thus, they have contributed to the 3Rs and continue to make a difference in accelerating the discoveries in the fields of biomarkers, toxicity, efficacy, and overall risk assessment to humans (Andersen and Krewski, 2010). The systems' applications are briefly described in Table 8.4 and a few examples illustrating these technologies are shown in Fig. 8.12.

TABLE 8.3 In Vitro ADME and PK Tests

Permeability	CYP induction
Blood stability	CYP phenotyping
Plasma stability	UGT inhibition
Blood partitioning	UGT phenotyping
Partition coefficient	Metabolite identification
Metabolic stability	Drug transporters
CYP inhibition	Protein binding

Systemic Toxicity Testing in Product Development

Keeping the concept of 3Rs and biomarkers in mind, significant progress has been made in the safety evaluation of substances, as this is the major area where a large

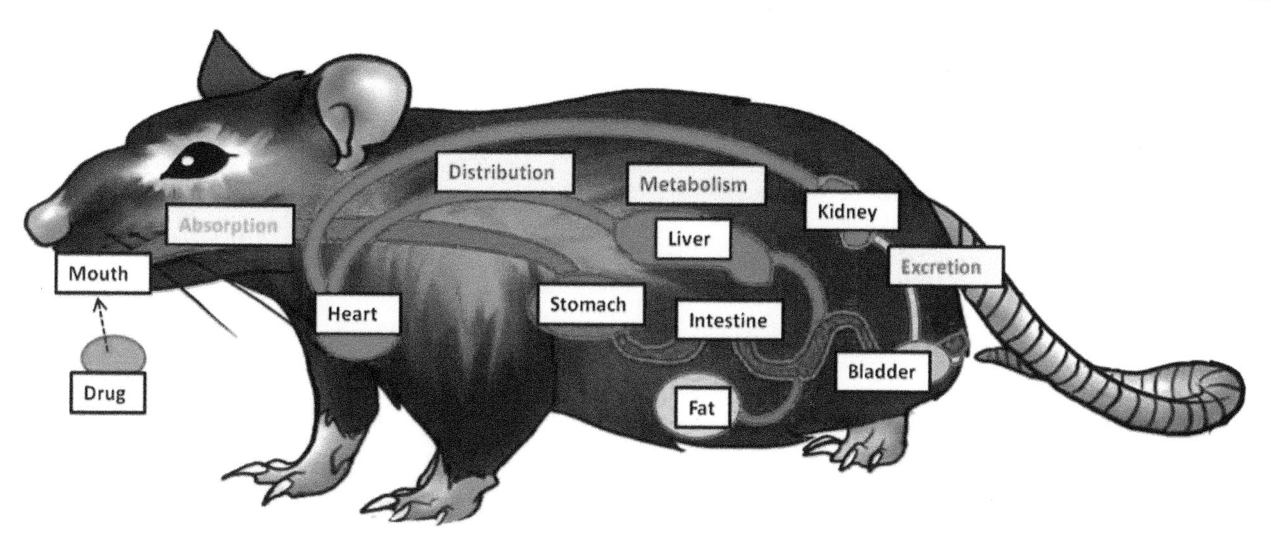

FIGURE 8.10 Toxicokinetics: an illustration of how the animal handles the drug and its metabolites.

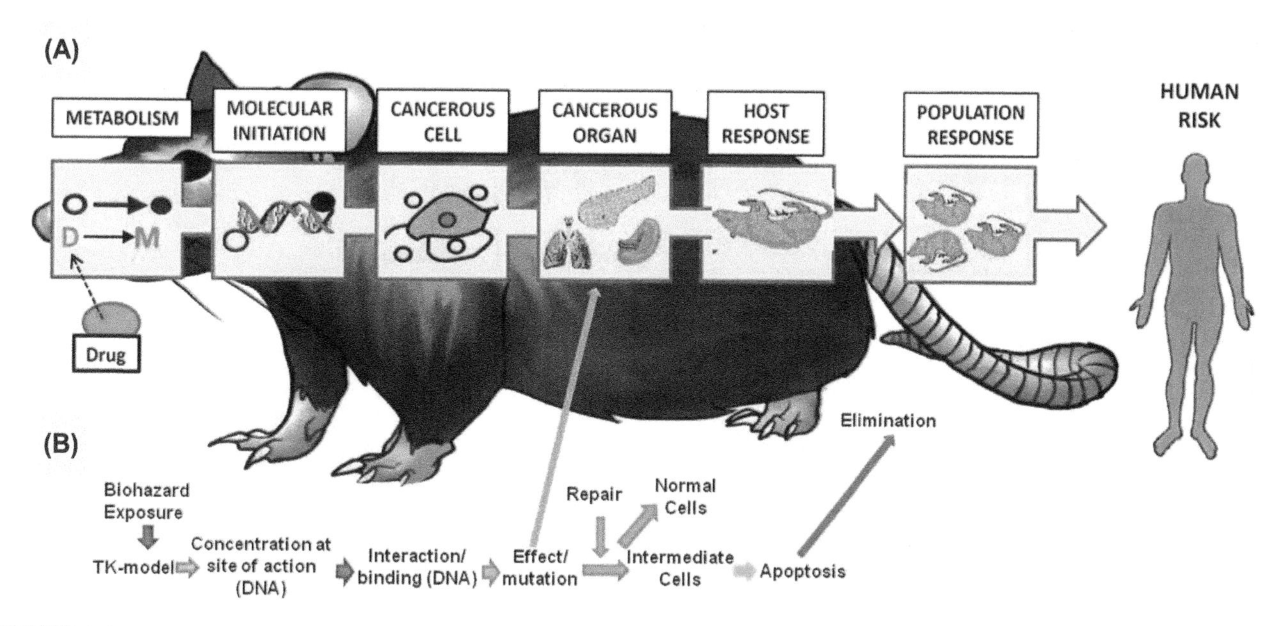

FIGURE 8.11 Toxicodynamics: (A) An illustration of how a drug handles the animal and its relationship to human risk. (B) Molecular toxicodynamics interactions.

numbers of animals are used (Fig. 8.13). Traditional animal testing has not been a solution to current day research needs (FDA, 2018; SOT, 2013a,b, 2018). For example, it requires a lot of resources and time to test all substances in the environment, their utility in testing new substances such as nanoparticles or cell therapies is questioned, the limited predictivity of traditional tests for human health effects, and animal welfare considerations. Furthermore, the scheduled 2013 marketing ban on cosmetic ingredients tested for systemic toxicity in animals has given more impetus to use new ways of safety assessment and a critical review of alternatives was sought (Adler et al., 2011; Hartung, 2011, 2014) and integrated testing strategies for safety assessments were proposed (Hartung et al., 2013). To this end, a roadmap for how to overcome the recognized scientific gaps for the full replacement of systemic toxicity testing using animals is proposed (Basketter et al., 2012). A multiprong global approach has been initiated in addressing toxicokinetics, skin sensitization, repeated-dose toxicity, carcinogenicity, and reproductive toxicity testing with deliberations among scientific experts periodically and proposed timely updates.

TABLE 8.4 Vision for Toxicity Testing for the 21st Century

Tools	Applications
High-throughput screens	Efficiently identify critical toxicity pathway perturbations across a range of doses and molecular and cellular targets
Stem cell biology	Develop in vitro toxicity pathway assays using human cells produced from directed stem cell differentiation
Functional genomics	Identify the structure of cellular circuits involved in toxicity pathway responses to assist computational dose-response modeling
Bioinformatics	Interpret complex multivariable data from HTS and genomic assays in relation to target identification and effects of sustained perturbations on organs and tissues
Systems biology	Organize information from multiple cellular response pathways to understand integrated cellular and tissue responses
Computational systems biology	Describe dose–response relationships based on perturbations of cell circuitry underlying toxicity pathway responses giving rise to thresholds, dose-dependent transitions, and other dose-related biological behaviors
PBPK models	Identify human exposure situations likely to provide tissue concentrations equivalent to in vitro activation of toxicity pathways
Structure–activity relationships	Predict toxicological responses and metabolic pathways based on the chemical properties of environmental agents and comparison to other active structures
Biomarkers	Establish biomarkers of biological change representing critical toxicity pathway perturbations

HTS, high-throughput screening.
Modified from Andersen, M.E., Krewski, D., 2010. The vision of toxicity testing in the 21st century: moving from discussion to action. Toxicol. Sci. 117, 17–24.

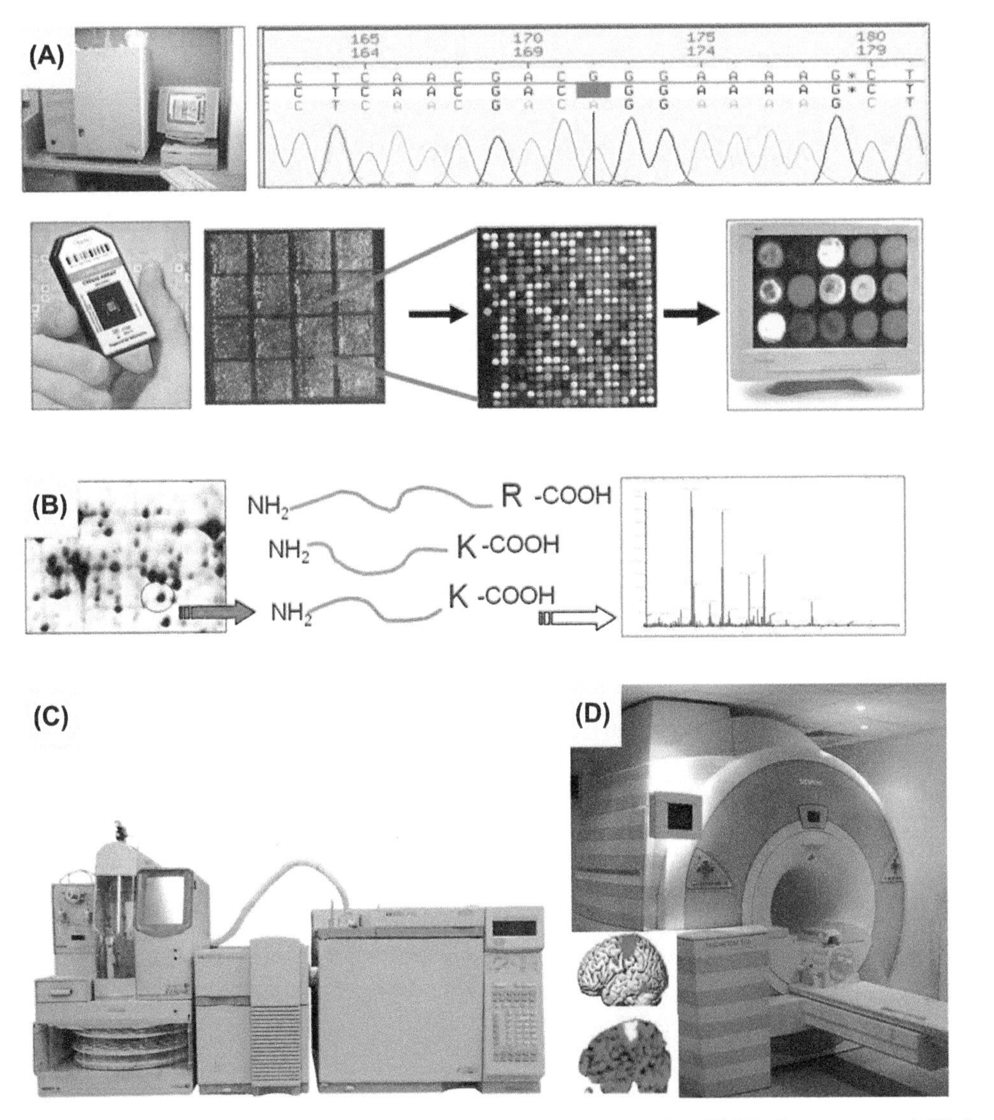

FIGURE 8.12 Select tools of modern molecular toxicology: (A) Genomics, (B) Proteomics, (C) Metabonomics, and (D) functional magnetic resonance imaging.

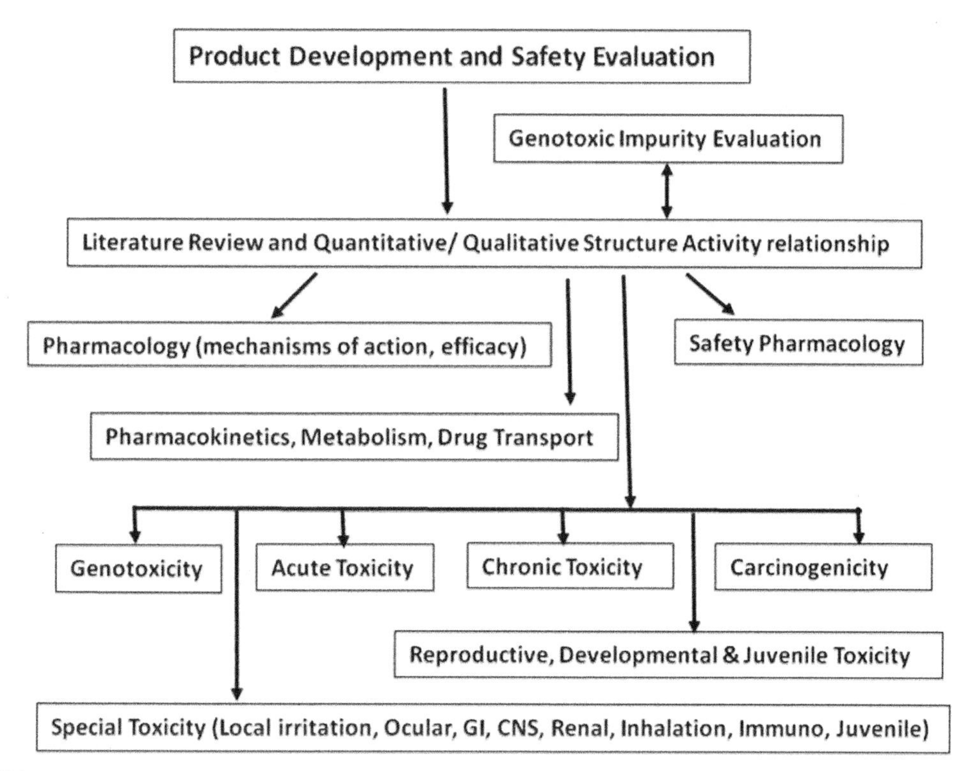

FIGURE 8.13 Multiple layers of product development and use of efficacy and toxicity endpoints as biomarkers.

CONCLUDING REMARKS AND FUTURE DIRECTIONS

Identification, development, and validation of biomarkers have played a significant role in achieving the goals of the reduction, refinement, and replacement (3Rs). Currently, a global effort is underway to recognize and put well-thought-out assays, instrumentation, and strategies in place for product discovery and development. Global regulatory agencies are working in collaboration with academia and industry, so the stakeholders, the thought, the talk, and the action are on the same page. Meaningful cooperation is sought, as needed, to attain a common goal of achieving the 3Rs while using biomarkers as we discover, develop, market, and cherish man-made advancements. This kind of collective and collaborative effort and efficient use of resources with funding from multiple bodies will continue to provide impetus in fully achieving the ultimate goal of reducing, refining, and replacing animal use, as applicable, in efficacy and toxicity testing and finding viable alternatives at all times.

Moving forward with continued innovations in biomarkers, instrumentation, and collaborative effort by the stakeholders is expected to deliver desired results in 3Rs. The ideas, vision, and initial promise in developing and utilizing technologies, such as "multi-organs a chip," an organ-by-organ, would become a reality. This is based on the premises that the "rat is not human" and animal models cannot fully mimic human patients, diseases, and toxicity. One such example is asthma, where the constriction of airways and all of the other characteristics of the disease are different. However, in many other situations, the human chip technology seems to be promising and serving as a closer surrogate to a human, at least in the drug discovery stages of drug development (Rockoff, 2013; Hartung, 2014; FDA, 2018). With these advancements, the concerns of excessive use of animals in product development would continue to decrease.

References

Adler, S., Basketter, D., Creton, S., et al., 2011. Alternative (non-animal) methods for cosmetics testing: current status and future prospects – 2010. Arch. Toxicol. 85, 367–485.

Andersen, M.E., Krewski, D., 2010. The vision of toxicity testing in the 21st century: moving from discussion to action. Toxicol. Sci. 117, 17–24.

Basketter, D., Clewell, H., Kimber, I., et al., 2012. A roadmap for the development of alternative (non-animal) methods for systemic toxicity testing – t^4 report. ALTEX 29, 3–91.

Benford, D.J., Hanley, B.A., Bottrill, K., et al., 2000. Biomarkers as predictive tools in toxicity testing: the report and recommendations of ECVAM workshop 40. ATLA 28, 119–131.

Blaauboer, B.J., Boekelheide, K., Clewell, H.J., et al., 2012. t4 Workshop report: the use of biomarkers of toxicity for integrating in vitro hazard estimates into risk assessment for humans. Transatlantic Think Tank for Toxicology. altweb.jhsph.edu/altex/29_4/t4Report29_4.pdf.

Bradlaw, J.A., 1989. Evaluation of drug and chemical toxicity with cell culture systems. Fund. Appl. Toxicol. 6, 598–606.

CAAT-a, 2013a. Center for Alternatives to Animal Testing. The Johns Hopkins University. Available at: http://caat.jhsph.edu/.

CAAT-b, 2013b. Center for Alternatives to Animal Testing. University of California. Available at: http://ora.research.ucla.edu/RSAWA/ARC/Pages/alternatives/alternatives-animal-main.aspx

Canavan, N., 2013. Mass Spec Maturing in the Age of Proteomics. Available at: http://www.dddmag.com/articles/2013/06/mass-spec-maturing-age-proteomics.

Chin, C., Upadhyay, J., Marek, G., et al., 2010. Awake rat pharmacological magnetic resonance imaging as a translational pharmacodynamic biomarker: metabotropic glutamate 2/3 agonist modulation of ketamine-induced blood oxygenation level dependence signals. Pharmacol. Exp. Ther. 336, 709–715.

Contrera, J.F., Kruhlak, N.L., Matthews, E.J., et al., 2007. Comparison of MC4PC and MDL-QSAR rodent carcinogenicity predictions and the enhancement of predictive performance by combining QSAR models. Regul. Toxicol. Pharmacol. 49, 172–182.

Criswell, K.A., Krishna, G., Zielinski, D., et al., 2003. Validation of a flow cytometric acridine orange micronuclei methodology in rats. Mutat. Res. 528, 1–18.

Criswell, K.A., Krishna, G., Zielinski, D., et al., 1998a. Use of acridine orange in: flow cytometric assessment of micronuclei induction. Mutat. Res. 414, 63–75.

Criswell, K.A., Krishna, G., Zielinski, D., et al., 1998b. Use of acridine orange in: flow cytometric evaluation of erythropoietic cytotoxicity. Mutat. Res. 414, 49–61.

Darzynkiewicz, Z., Smolewski, P., Holden, E., et al., 2011. Laser scanning cytometry for automation of the micronucleus assay. Mutagenesis 26, 153–161.

Dertinger, S.D., Torous, D.K., Hayashi, M., et al., 2011. Flow cytometric scoring of micronucleated erythrocytes: an efficient platform for assessing in vivo cytogenetic damage. Mutagenesis 26, 139–145.

Doke, S.K., Dhawale, S.C., 2015. Alternatives to animal testing: review. Saudi Pharm. J. 23, 223–229.

Dr. Hadwen Trust, 2013. Dr Hadwen Trust for Humane Research. Available at: http://www.drhadwentrust.org/.

FDA, 2006. Guidance for Industry, Investigators, and Reviewers. Exploratory IND Studies. Available at: http://www.fda.gov/downloads/Drugs/GuidanceComplianceRegulatoryInformation/Guidances/ucm078933.pdf.

FDA, 2012. Guidance for Industry, Drug Interaction Studies — Study Design, Data Analysis, Implications for Dosing, and Labeling Recommendations. Available at: http://www.fda.gov/downloads/Drugs/GuidanceComplianceRegulatoryInformation/Guidances/ucm292362.pdf.

FDA, 2018. Cosmetics Safety Q&A: Animal Testing. https://www.fda.gov/Cosmetics/ResourcesForYou/Consumers/ucm167216.htm.

Goldberg, A.M., Frazier, J.M., 1989. Alternatives to animals in toxicity testing. Sci. Am. 261, 24–30.

Gollapudi, B., Krishna, G., 2000. Practical aspects of mutagenicity testing strategy: an industrial perspective. Mutat. Res. 455, 21–28.

Hancock, A., Diehl, M., Faghih, R., et al., 2004. In vitro optimization of structure activity relationships of analogues of A-331440 combining radioligand receptor binding assays and micronucleus assays of potential antiobesity histamine H3 receptor antagonists. Basic Clin. Pharmacol. Toxicol. 95, 144–152.

Hartung, T., 2011. From alternative methods to a new toxicology. Eur. J. Pharm. Biopharm. 77, 338–349.

Hartung, T., 2014. Thomas Hartung Lecture: "Safe Drugs and Products without Animal Testing?". https://ozsheba.wordpress.com/2014/02/18/thomas-hartung-lecture-safe-drugs-and-products-without-animal-testing/.

Hartung, T., Luechtefeld, T., Maetens, A., et al., 2013. Integrated testing strategies for safety assessments. ALTEX 30, 3–18.

Hayashi, M., Kamata, E., Hirose, A., et al., 2005. In silico assessment of chemical mutagenesis in comparison with results of *Salmonella* microsome assay on 909 chemicals. Mutat. Res. 588, 129–135.

Heinrich-Hirsch, B., Madle, S., Oberemm, A., et al., 2001. The use of toxicodynamics in risk assessment. Toxicol. Lett. 120, 131–141.

Hill, A.J., Teraoka, J., Heideman, W., et al., 2005. Zebrafish as a model vertebrate for investigating chemical toxicity. Toxicol. Sci. 86, 6–19.

ICCVAM, 2013. Interagency Coordinating Committee on the Validation of Alternative Methods. Available at: http://iccvam.niehs.nih.gov/.

ICH, 2006. International Conference on Harmonization: Guidance for Industry and Review Staff Recommended Approaches to Integration of Genetic Toxicology Study Results. Available at: www.fda.gov/downloads/Drugs/.../Guidances/ucm079257.pdf.

Irwin, W., Jelic, D., Antolovic, R., 2008. Biomarkers for drug discovery: important aspects of *in vitro* assay design for hts and hcs bioassays. Croat. Chem. Acta 81, 23–30.

Kenakin, T.P., 2002. Isolated tissues. In: Enna, S.J., et al. (Eds.), Current Protocols in Pharmacology, vol. 1. John Wiley & Sons, New York. Section 4.

Kenakin, T.P., 1984. The classification of drugs and drug receptors in isolated tissues. Pharmacol. Rev. 36, 165–222.

Khalil, S.H., Abu-Amero, K.K., Al Mohareb, F., et al., 2010. Molecular monitoring of response to imatinib (Glivec) in chronic myeloid leukemia patients: experience at a tertiary care hospital in Saudi Arabia. Genet. Test. Mol. Biomark. 14, 67–74.

Kirkland, D., Aardema, M., Henderson, L., et al., 2005. Evaluation of the ability of a battery of three *in vitro* genotoxicity tests to discriminate rodent carcinogens and non-carcinogens—I. Sensitivity, specificity and relative predictivity. Mutat. Res. 584, 1–256.

Krishna, G., 2010. Alternatives to animal experimentation in research. In: Jagadeesh, et al. (Eds.), Biomedical Research – From Ideation to Publication. Wolters Kluwer/Lippincott, Williams & Wilkins, pp. 155–177.

Krishna, K., Goel, S., Krishna, G., 2014. SAR genotoxicity and tumorigenicity predictions for 2-2-MI and 4-MI using SAR software. Toxicol. Mech. Methods 24 (4), 284–293.

Kirshna, G., Brott, D., Urda, G., et al., 1993. Comparative micronucleus quantitation in pre- and post-column fractionated mouse bone marrow by manual and flow methods. Mutat. Res. 302, 119–127.

Krishna, G., Fiedler, R., Theiss, J.C., 1992. Simultaneous analysis of chromosome damage and aneuploidy in cytokinesis-blocked V79 Chinese hamster lung cells using an antikinetochore antibody. Mutat. Res. 282, 79–88.

Krishna, G., Kropko, M., Theiss, J.C., 1989. The use of cytokinesis-block method for the analysis of micronuclei in V79 Chinese hamster lung cell line: results on mitomycin C and cyclophosphamide. Mutat. Res. 222, 63–69.

Krishna, G., Nath, J., Ong, T., 1986. Murine bone marrow culture system for cytogenetic analysis. Mutat. Res. 164, 91–99.

Krishna, G., Nath, J., Ong, T., 1985. Preparation of Chinese hamster bone marrow and spleen primary cell cultures for sister chromatid exchange and chromosomal aberration studies. J. Tissue Cult. Methods 9, 199–203.

Krishna, G., Nath, J., Ong, T., 1984. Preparation of mouse bone marrow primary cultures for sister chromatid exchange and chromosomal aberration studies. J. Tissue Cult. Methods 9, 193–198.

Krishna, G., Urda, G., Lalwani, N.D., 1995a. Immunofluorescent and confocal laser cytometric analyses of centromeres in V79 cells. Mutat. Res. 328, 1–9.

Krishna, G., Urda, G., Tefera, W., et al., 1995b. Simultaneous evaluation of dexamethasone-induced apoptosis and micronuclei in rat primary spleen cell cultures. Mutat. Res. 332, 1–8.

Krishna, G., Urda, G., Theiss, J., 1998. Principles and practices of integrating genotoxicity evaluation into routine toxicology studies: a pharmaceutical industry perspective. Environ. Mol. Mutagen. 32, 115–120.

Krishna, G., Urda, G., Theiss, J.C., 1994. Comparative mouse micronucleus evaluation in bone marrow and spleen using immunofluorescence and Wright's Giemsa. Mutat. Res. 323, 11–20.

Krishna, G., Urda, G., Paulissen, J., 2000. Historical vehicle and positive control micronucleus data in mice and rats. Mutat. Res. 453, 45–50.

Lake, F., 2011. Hunting biomarkers for Huntington's disease: H2AFY. Biomark. Med. 5, 817–820.

Lee, S.L., No, D.Y., Kang, E., et al., 2013. Spheroid-based three-dimensional liver-on-a-chip to investigate hepatocyte–hepatic stellate cell interactions and flow effects. Lab Chip, Advance Article. https://doi.org/10.1039/C3LC50197C.

Lhasa, 2013. Available at: http://www.lhasalimited.org/.

Lieschke, G.J., Currie, P.D., 2007. Animal models of human disease: zebrafish swim into view. Nat. Rev. Genet. 8, 353–367.

Lipp, E., 2007. Safety/toxicity issues in drug discovery. Genet. Eng. News 27, 5. Available at: http://www.genengnews.com/gen-articles/safety-toxicity-issues-in-drug-discovery/2044/.

McDermott, U., Sharma, S.V., Dowell, L., et al., 2007. Identification of genotype-correlated sensitivity to selective kinase inhibitors by using high-throughput tumor cell line profiling. Proc. Natl. Acad. Sci. U.S.A. 104, 19936–19941.

McKeehan, W.L., Barne, S.D., Reid, L., et al., 1990. Frontiers in mammalian cell culture. In Vitro Cell. Dev. Biol. 26, 9–23.

Moeller, T., 2010. From in Vitro to in Vivo. Drug Discovery & Development. Available at: http://www.dddmag.com/articles/2010/06/vitro-vivo.

NC3R, 2013. National Centre for the Replacement, Refinement and Reduction of Animals in Research. Available at: http://www.nc3rs.org.uk.

NDRI, 2013. National Disease Research Interchange. Available at: http://www.ndriresource.org/.

Neve, R.M., Chin, K., Fridlyand, J., et al., 2006. A collection of breast cancer cell lines for the study of functionally distinct cancer subtypes. Cancer Cell 10, 515–527.

REACH, 2013. Registration, Evaluation, Authorization and Restriction of Chemicals. Available at: http://ec.europa.eu/environment/chemicals/reach/reach_intro.htm.

Rockoff, J.D., 2013. Forget lab rats: testing asthma drugs on a microchip. Wall Str. J. Next in tech-section, June 17, 2013, Dow Jones.

Russell, W., Burch, R.L. (Eds.), 1959. The Principles of Humane Experimental Technique. Methuen Press, London, UK.

Sankaranarayanan, K., Ferguson, L., Gentile, J., et al., 2000. Editorial: protocols in mutagenesis, a special issue of mutation research. Mutat. Res. 455, 1–2.

Schaeffer, W.I., 1990. Terminology associated with cell, tissue, and organ culture, molecular biology and molecular genetics. In Vitro Cell. Dev. Biol. 26, 97–101.

Smith, K., 2012. Brain imaging: fMRI 2.0. Nature 484, 24–26.

SOT, 2013a. Alternative Toxicology Test Methods: Reducing, Refining, and Replacing Animal Use for Safety Testing. Available at: http://www.toxicology.org/pr/ToxTopics/TT3_InVitro_SOT.pdf.

SOT, 2013b. Opportunities to Modify Current Regulatory Testing Guidelines and Advance the Assessment of Carcinogenicity Risk in the 21st Century. Available at: http://www.toxicology.org/pr/ToxTopics/TT6_Cancer.pdf.

SOT, 2018. The What, When and How of Using Data from Alternative Testing Methods in Chemical Safety Assessment. http://www.toxicology.org/events/am/am2018/continuing-education.asp.

Tainsky, M., 2009. Genomic and proteomic biomarkers for cancer: a multitude of opportunities. Biochim. Biophys. Acta 1796, 176–193.

Tice, R.R., Agurell, E., Anderson, D., et al., 2000. The single cell gel/comet assay: guidelines for in vitro and in vivo genetic toxicology testing. Environ. Mol. Mutagen. 35, 206–221.

Vaidya, V.S., Mendrick, D.L., 2012. Applications of biomarkers in the assessment of health and disease. In: Continuing Education Course at SOT Meeting, San Francisco, CA.

Wilcox, D.K., Bruner, L.H., 1990. In vitro alternatives for ocular safety testing: an outline of assays and possible future developments. ATLA 18, 117–128.

CHAPTER

9

Adverse Outcome Pathways and Their Role in Revealing Biomarkers

Magdalini Sachana

Organization for Economic Cooperation and Development (OECD), Paris, France

INTRODUCTION

Decades of research in the field of toxicology have led to a vast amount of scientific knowledge and a correspondingly large literature that is stored in peer-reviewed journals and text books. While the literature on this particular form is suitable and useful for researchers, it is very challenging for regulators, who assess the safety of chemicals, to use this knowledge that is spread in so many sources and is not organized in a suitable and easy accessible manner. This limits the ability to apply knowledge from toxicological studies and to take advantage of mechanistic information that could be useful for enriching regulatory decisions.

Another challenge that the regulators face today is the considerable high number of untested chemicals with significant uncertainty concerning their environmental and human health safety that need to be assessed at a short time frame. To close this data gap with the current approach that relies almost completely on animal testing is not achievable because this is far too resource-intensive concerning money, time, and number of animals. In parallel, the research community and industry have been investing significant time, effort, and resources in innovation by developing cutting-edge cellular models (e.g., three-dimensional organotypic cultures, human induced pluripotent stem cells, microfluidic organs on a chip, etc.), computational methods (e.g., in silico, modeling), and technologies (e.g., omics, high-throughput, and high-content) that could advance chemical safety assessment, complement, and potentially replace animal testing in long term. A variety of sophisticated in vitro methods are now available for a number of endpoints used for different regulatory purposes, which have the potential to close this data gap by allowing targeting testing that is based on mechanistic understanding that is captured in "toxicity pathways."

In recent years, the concept of "toxicity pathways" and its use in regulatory arena gained support from both researchers and regulators. First, the International Program on Chemical Safety (IPCS) published the framework for using mode of action information to determine human relevance of animal data (Sonich-Mullin et al., 2001) that was further refined later (Boobis et al., 2006, 2008). In 2007, the United States National Academy of Science (NAS) published the Report on Toxicity Testing in the 21st Century: A Vision and a Strategy in which emphasis to the "toxicity pathways" based on similar principles was suggested as one of the tools to help the paradigm shift in toxicity testing. This paradigm shift involves the evaluation of perturbations of biological pathways by chemicals in well-designed in vitro models preferably of human origin instead of rodents to predict adverse outcomes (AOs) in humans (NRC, 2007).

In 2012, the Organization for Economic Cooperation and Development (OECD) initiated a program that aimed to evolve the concept of adverse outcome pathways (AOPs), which was first introduced by Ankley et al. (2010). An AOP is not intending to provide a comprehensive description of every aspect of the biology and toxicology; instead it is capturing only the critical steps in a toxicity pathway. A reductionist approach to biology and toxicology is followed to reveal and establish the important events at each major level of biological organization (e.g., molecular, cellular, tissue, organ, individual) that if do not occur then the downstream sequence of events does not take place, including the AO.

Although the AOP concept is reasonable new, it has its roots to a certain extent to the biomarker concept

Biomarkers in Toxicology, Second Edition
https://doi.org/10.1016/B978-0-12-814655-2.00009-8

that was initially introduced in 1990s (WHO, 1993, 2001) and it is presented in this book by Anna M. Fan in Chapter 69 (Fan, 2018). The biomarker concept is also serving regulatory purposes and makes use of mechanistic knowledge that aligns to a certain degree with the current AOP thinking.

The chapter will focus on the AOP framework, and it will provide some principles related to AOP development and assessment. It will also explore the differences and similarities between the two regulatory frameworks (i.e., biomarkers and AOPs), and it will close by presenting some relevant examples.

ADVERSE OUTCOME PATHWAYS: PRINCIPLES OF DEVELOPMENT AND ASSESSMENT

The development of AOPs is an international effort that relies on existing biological and toxicological knowledge concerning the linkage between two anchors of a toxicity pathway—the molecular initiating event (MIE), which represents the initial interaction(s) of a stressor with a biomolecule within or on the body of an organism, and an AO (Fig. 9.1). The AO is a specialized key event (KE) that represents the final step of an AOP, which should be relevant to regulatory decision-making in chemical safety (i.e., corresponding to an apical endpoint or measurements that are done in a test guideline study such as developmental neurotoxicity, carcinogenicity, reproductive toxicity etc.). On the other hand, the MIE captures the interaction (e.g., covalent binding, hydrogen bonding, electrostatic interaction, etc.) between a chemical and the biomolecules (i.e., DNA, proteins etc.) within an organism, and it is the trigger for the following steps in the pathway.

Between the MIE and the AO, there is a series of KEs that are captured. KEs represent important steps that need to take place at different levels of biological organization that ranges from the cellular to the organ level (Fig. 9.2). The KEs need to be both measurable and essential, and so a defined biological perturbation can progress and eventually culminate with the manifestation of a specific AO. A KE can be the altered activity of an enzyme, upregulation or downregulation of a gene, increased or decreased levels of a hormone or protein, histopathological findings in a tissue, functional changes of an organ, etc.

The linkages between KEs are named key event relationships (KERs) that are described in the form of a one-way arrow relationship, linking an upstream "causing" KE to a downstream "responding" KE (Fig. 9.1). The knowledge captured in KER descriptions permits the AOP developer to collect biological and toxicological information and support the evaluation of the level of confidence in each KER and consequently in the whole AOP. Building confidence on KERs contributes to increased uptake of AOPs for regulatory application. Having confidence in data points measured at low levels of biological organization (i.e., molecular and cellular) facilitates the prediction of outcomes at higher levels of organization (i.e., organism and population level). Overall, the accurate description of KERs in an AOP provides the scientific basis for deriving the probable degree of alteration of a downstream KE from the measured state of an upstream KE.

An AOP can be seen as a simplification of the biology describing a single sequence of events that begin with an MIE that moves along a series of intermediate KEs to final reach the AO. A single AOP aims to describe how one particular molecular perturbation may cause one AO and it does not aim to cover all possible AOs that the perturbation may cause nor every possible pathway through which a particular AO may occur. Since the beginning of the OECD AOP program, it was recognized that MIEs, KEs, and AOs may be shared by

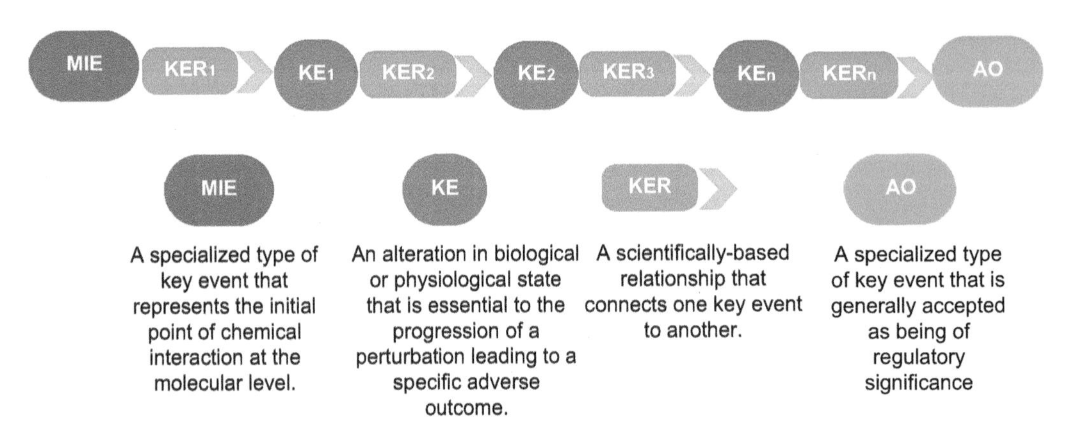

FIGURE 9.1 A schematic representation of the adverse outcome pathway (AOP) framework. An AOP is triggered by a molecular initiating event (MIE), an initial interaction with a biological target that leads to a sequential cascade of key events (KEs), linked to each other by key event relationships (KERs) to result in an adverse outcome (AO) of regulatory relevance.

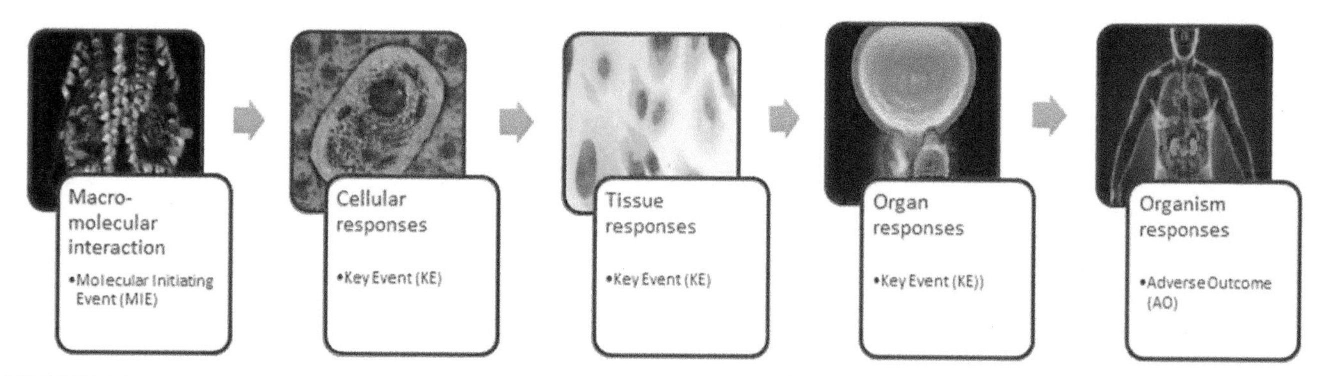

FIGURE 9.2 Graphical representation of a hypothesized adverse outcome pathway. Key events (KEs) are represented in an increasing order of biological organization starting from cellular, continuing to tissue/organ, and ending to the organism level.

more than one AOP, and that a knowledge base (KB) is required to capture this sharable information. For this reason, the AOP development process takes place making use of an IT platform named AOP-KB that permits that the described KEs and KERs are modular and independent units that can be reused by other AOP developers. By having the development of the AOPs done in an electronic platform, it also permits the building and conceptualization of AOP networks.

When two or more AOPs have one or more KEs in common, then the assembly of these AOPs in one place constructs an AOP network. The components of an AOP (KEs and KERs) are described in a modular fashion within the AOP-KB that allows the creation of AOP networks that emerge from the description of individual AOPs that share KEs. The vision is that as more AOPs are introduced to the AOP-KB, the constructed AOP networks will be able to capture gradually all the available scientific knowledge relevant to a number of possible AOs that an MIE might trigger or the range of pathways through which a particular AO may happen. AOP networks are considered the real tool that the regulators will use in the future for most of the regulatory endpoints because for the majority of AOs, a number of distinct mechanisms are involved, with the rare exception of some topical endpoints, e.g., skin sensitization. Similarly, capturing all the potential MIEs that are involved in the manifestation of an AO within a network can help regulators to assess exposure to mixtures of chemicals (Knapen et al., 2015, 2018; Villeneuve et al., 2018).

Indeed, AOPs try to partially depict what happens at the cellular level, where networks of molecular reactions occur that can be organized into higher-order interconnected pathways. Molecules are synthesized, degraded, transported from one location to another, and assembled into complexes and higher-order structures with other molecules. Intensive toxicological studies reveal the interaction and the effects of chemicals to signal transduction pathways, hormonal pathways, and other aspects of cell biology, as well as to specific genes and their proteins. By annotating all of these pathways in a single, consistent format, the AOP-KB systematically provides mechanistic knowledge that can be used for regulatory purposes and inform about existing knowledge gaps.

It is recognized that AOP descriptions reflect current knowledge and might present some knowledge gaps. For this reason, AOP descriptions are considered "living documents" that can evolve as new scientific information becomes available. AOPs are helpful to depict knowledge gaps that can trigger targeted scientific research that involves the collaboration between experts in various areas. Once new scientific evidence becomes available, an AOP can be revised and reassessed for its completeness and the relevance to regulators.

Within the OECD program on AOPs, member countries jointly develop and review AOPs and make them available to the regulators for further use. For this reason, a guidance document and a users' handbook on AOP development and assessment have been developed (OECD, 2016a, 2016b, 2016c, 2017). AOP descriptions that are developed within the OECD AOP development program and reach a mature stage undergo peer-review as per procedures outlined by the OECD, which involve two phases, the internal and the external review (Delrue, 2016). The Extended Advisory Group on Molecular Screening and Toxicogenomics (EAGMST) that oversees the AOP development process is responsible for reviewing the AOPs for completeness (internal review) and scientific rigor (external review), and for approving the final AOPs.

The AOP framework provides a transparent and scientifically based way to depict and make available current biological and toxicological knowledge that allows the prediction of subsequent KEs and AOs, once confidence has been built on KERs. This knowledge can help assay developers to invest money and effort on KE measurements that can be made efficiently using in vitro, in silico, and in chemico approaches, which are cost-effective and have a predictive value for adverse effects at higher levels of organization that are relevant to regulatory decision-making. AOPs are not only useful for assay development but also they can be applied in more regulatory contexts as it is discussed below.

ADVERSE OUTCOME PATHWAY FRAMEWORK: SIMILARITIES AND DIFFERENCES WITH BIOMARKERS

In response to some incidences related to chemical-induced human health adverse effects and environmental disasters that have happened in the past, there has been a continuous effort by regulatory agencies to improve the hazard assessment of the chemicals that are used and released in the environment. These efforts have led to ever-growing interest in developing and applying frameworks that rely on the understanding of the biological processes that can take place in an organism after chemical exposure. The concept of biomarkers was introduced in the 1990s (WHO, 1993, 2001) aiming to enhance hazard identification. Recently, the AOP framework came into play and is considered suitable for application in different regulatory contexts (OECD, 2016a). Both frameworks share lots of similarities and differences, and the AOP framework has a lot to learn and eventually build on the experiences gained from the implementation of the biomarkers framework.

There are three main types of biomarkers, namely biomarkers of exposure, effect, and susceptibility. However, in this occasion we compare only the biomarkers of effect with the AOPs, as a separate framework, named the aggregates exposure pathways (Teeguarden et al., 2016) deals with the exposure to chemicals. Regarding susceptibility, this is dealt within the applicability domain for each AOP and it is an additional consideration that needs to be taken into account during the risk assessment process. The taxa, life stage, and sex applicability of AOPs help to understand how available data can be extrapolated from in vitro to in vivo, including potential human relevance.

A biomarker of effect captures a biological response after exposure to a chemical that is linked to the manifestation of an AO or disease, and there are many similarities between this concept and the concept of AOPs (OECD, 2017; Ryan et al., 2007). One similarity is that biomarkers of effect and the KEs that are present in the AOPs both aim to capture responses at the molecular or cellular level that happens early in time after exposure to chemicals, so they can be used to predict AOs or diseases that manifest later in time. Because of this joint focus on predicting AOs, biomarkers of effect and KEs share common principles. First, both need to be measured in a relatively routine manner and preferably should be easy to collect and analyze. Second, KEs and biomarkers of effects need to be supported by toxicological and biological evidence, so they can be plausible and have a potential predictive validity by establishing dose-response relationships. Finally, temporal concordance is very important to establish KEs and biomarkers of effect as their early observation before the AO or disease is key for their predictive value. AOPs and biomarkers are also linked to global efforts to advance chemical risk assessment, and they require similar level of validation.

Despite the similarities between biomarkers of effects and AOPs, these are very close but are still separate and distinct concepts. AOPs try to capture the whole pathway that leads to an AO, providing a more holistic approach and full mechanistic understanding, whereas the biomarkers seem to be only a snapshot of the AOP that resembles a nonadjacent KER that links the biomarker with the AO or the disease, without taking into account all the mechanistic knowledge that is available. Moreover, building confidence on all the KEs present in an AOP that are described through the collection of biological and toxicological information permits the extrapolation of data measured at the molecular or cellular level to predict outcomes at the organism or population level. Biomarkers of effect tend to rely and focus on the use of molecular epidemiological data, meaning that human data are evaluated without requiring any interspecies extrapolation analysis. In contrast, AOPs tend to use mainly in vitro and in silico data to support the KERs that are present at molecular and cellular levels, and both in vitro–in vivo extrapolation and interspecies extrapolation are required when data derived from nonhuman-based cellular models. However, epidemiological data can also be used in the AOP context as it allows the support of KEs that appear closer to a higher level of biological organization. Other differences are present in the way in which AOPs and biomarkers of effect are used in the regulatory decision-making. Biomarkers are tools that can be applied directly in the process of risk assessment and tends not only to assess the presence or absence of an AO or disease but also to provide, when feasible, valuable dose-response data for risk assessment (Swenberg et al., 2008). On the other hand, AOPs use dose-response data to support the confidence and their utility in the regulatory context, and consequently they play a supportive role in the process of risk assessment that is explained fully in the last section of the present chapter. Furthermore, AOPs can be a source for establishing biomarkers of effect, and it discussed in the following section of this chapter.

Given the differences and similarities in definition, methodology, tools, and application domains for AOPs and biomarkers of effect, it can be concluded that both are very useful frameworks in the field of chemical risk assessment, and AOPs can potentially support the identification of new biomarkers that could subsequently inform chemicals and safety regulations.

ADVERSE OUTCOME PATHWAYS AS A TOOL TO RETRIEVE BIOMARKERS

The AOP-KB is an open-data resource of chemical-induced toxicity pathways. It consists of two main modules: the AOP-Wiki (https://aopwiki.org/) and the Effectopedia (https://www.effectopedia.org/), where the development of qualitative and quantitative AOPs, respectively, takes place. The eAOP portal (https://aopkb.oecd.org/) is the main entrance point for the KB, which is a search engine for AOPs and KEs that are available in both AOP-Wiki and Effectopedia.

Besides of enabling the development of AOPs by scientists, the AOP-KB can also aid regulators, who seek mechanistic information to support the chemical evaluation process. Another role for the AOP-KB would be to facilitate the identification of biomarkers that then can be further validated before their implementation for risk assessment purposes.

The KEs as they are described following the guidance provided in the OECD Users' Handbook (OECD, 2017) can be a real valuable source for retrieving biomarkers. A KE description contains information on the biological state, the biological compartment, and its general role in the biology. The description of the biological compartment in which a KE is measured can provide important information because biomarkers should be easily collected using noninvasive procedures. For example, a KE that is measured in a fluid (e.g., plasma, cerebrospinal fluid) has more potential to become a biomarker compared with a KE that can only be measured in a tissue.

In the dedicated section named "How it is Measured or Detected," more information related to KE can be obtained regarding the reliability of the methods with which a KE can be measured. These methods can be either advanced methods that underwent validation and became test guidelines or methods published in the peer-reviewed literature. Methods used to measure KEs can be found in the OECD Test Guidelines website (http://www.oecd.org/chemicalsafety/testing/oecdguidelinesforthetestingofchemicals.htm) and the EURL ECVAM Database Service on Alternative Methods to Animal Experimentation (DB-ALM) (https://ecvam-dbalm.jrc.ec.europa.eu/).

Significant information can also be obtained from a section, which is dedicated to the essentiality of the KEs within an AOP. In this section, experimental data are provided showing that by blocking a KE, the downstream KEs, including the AO, are prevented to happen. Blockage of the downstream KEs and the toxicity pathway mainly can be recorded using data from studies that are based on knockout models' specific inhibitors, induced mutants, overexpression constructs, and other single-gene perturbations. Identifying biomarkers with cause–effect relationship is highly important, and besides the essentiality of KEs, the biological plausibility of the KERs in an AOP can inform about the mechanistic (i.e., structural or functional) relationship between KEupstream and KEdownstream.

Although the biological plausibility followed by essentiality are the most influential criteria to determine the degree of confidence in an overall hypothesized AOP that will then guide its potential regulatory application (Meek et al., 2014a,b), still toxicological data using chemical stressors that define the associations between KEs within an AOP are considered important, especially for establishing biomarkers of effect. The dose-response concordance and temporal relationships between and across the series of KEs within an AOP can be retrieved from experimental studies in in vivo, in vitro, and in silico models and can inform biomarkers of effect that are applied in risk assessment.

It is recognized that a KE that occurs at lower doses and earlier time points than the following KEs of an AOP based on experimental data has better chances to become a biomarker. Furthermore, if the incidence or frequency of a KE is greater than that of the remaining KEs within an AOP, then it could satisfy most of the principles for defining biomarkers of effect.

Overall, the AOP-KB is a source for exploring KEs that have the potential to become biomarkers of effect. The AOP-KB currently contains more than one thousand KEs and is constantly increasing due to contributions from the scientific community. All the available data in the KB have not been reviewed or published yet through the OECD process. However, as the AOPs reach a maturation stage and enter the OECD review process, they can obtain OECD-approved status. This pool of KEs is worth exploring in order to identify even well-established and validated biomarkers of effect that can facilitate chemical risk assessment.

BIOMARKERS DEPICTED FROM THE ADVERSE OUTCOME PATHWAY–KNOWLEDGE BASE

Based on the AOP-KB, using the key word "biomarker," commonly recognized or emerging biomarkers that are associated with KEs of specific AOPs were identified. Information was derived mainly from the AOP-Wiki platform. Table 9.1 presents the results after conducting this search in the AOP-KB. The potential biomarkers are not presented or organized in any particular order and are

TABLE 9.1 Potential Biomarkers Revealed After Searching the Scientific Knowledge Stored in AOP-KB

Potential Biomarkers and Associated KEs	AOP
Oxidative stress (plasma protein carbonyl content)	AOP 220: Chronic Cyp2E1 activation leading to liver cancer
Increase in vascular resistance (noninvasive ultrasound-based method to evaluate flow-mediated vasodilation)	AOP 149: Peptide oxidation leading to hypertension
Decrease in tetrahydrobiopterin (tetrahydrobiopterin plasma levels)	AOP 149: Peptide oxidation leading to hypertension
Promotion of SIX-1–positive basal-type progenitor cells	AOP 167: Early-life estrogen receptor activity leading to endometrial carcinoma in the mouse
Increased induced mutations in critical genes	AOP 46: Mutagenic mode of action leading to hepatocellular carcinoma
Formation of hemoglobin adducts	AOP 31: Oxidation of iron in hemoglobin leading to hematotoxicity
Altered gene expression specific to CAR activation	AOP 107: CAR activation leading to hepatocellular adenomas and carcinomas in the mouse and the rat
Reduction in plasma vitellogenin concentrations	AOP 23: Androgen receptor agonism leading to reproductive dysfunction AOP 25: Aromatase inhibition leading to reproductive dysfunction
Inhibition of acetylcholinesterase (AChE)	AOP 16: Acetylcholinesterase inhibition leading to acute mortality
Reduced levels of brain-derived neurotrophic factor	AOP 13: Chronic binding of antagonist to NMDARs during brain development induces impairment of learning and memory abilities AOP 12: Chronic binding of antagonist to NMDARs during brain development leads to neurodegeneration with impairment in learning and memory in aging AOP 54: Inhibition of Na+/I– symporter leads to learning and memory impairment

AOP, adverse outcome pathway; *CAR*, constitutive androstane receptor; *KB*, knowledge base; *KE*, key event; *NMDAR*, N-methyl-D-aspartate receptor.

applicable to both human health and environmental risk assessment.

The list of potential biomarkers depicted after searching the AOP-KB and shown in Table 9.1 is neither exhaustive nor definitive. This is only a first attempt aiming to illustrate how AOP-KB can be used to retrieve biomarkers of effect by using available information that can be found in KEs, KERs, and AOPs. Different strategies than the one applied here can potentially be followed to retrieve biomarkers using the AOP-KB. Depending on the AO, all the relevant AOPs can be explored, and by digging in the information that is stored in KEs and KERs, new potential biomarkers of effect can emerge.

REGULATORY USE OF ADVERSE OUTCOME PATHWAYS

In the chemical industry, mainly at the stage of research and development, different in vitro and in silico models are already being used routinely. However, alternative to animal methods are hardly used for regulatory purposes in different steps of hazard and risk assessment. An important hurdle that impedes the use of data derived from nonanimal methods is the lack of mechanistic information and the association of these data with an AO that is relevant to regulatory decision. The availability of toxicological mechanistic data depends on the published scientific knowledge and the way that is stored, so it can be accessible by regulators.

Toward this goal, the OECD program on AOPs has spearheaded the development of processes to review hypothesized AOPs and built the AOP-KB, an ontology-driven database that combines existing biological knowledge with empirical data derived from toxicological experimental studies (Delrue et al., 2016). The AOPs stored in the KB are getting gradually reviewed and published in the OECD series on AOPs (http://www.oecd-ilibrary.org/environment/oecd-series-on-adverse-outcome-pathways_2415170x). The number of AOPs with OECD-approved status, although limited at the moment, is steadily increasing, and OECD encourages the uptake of AOPs by regulators in their everyday work.

It is also important to recognize that, due to its dependence on volunteer efforts from OECD member countries, the AOP-KB currently includes a very incomplete sample of all known toxicity pathways, with coverage varying across endpoints that are linked to legal data requirements and taxonomic groups. Across the toxicological endpoints, coverage is strongest in the area of neurotoxicity, carcinogenicity, and endocrine disruption based on data derived from the OECD workplan. Coverage is also good for both human health and environmental-related endpoints. Nonetheless, the absence of certain AOPs from the KB needs to be interpreted with care, as in most cases it does not mean that biological and toxicological knowledge is not available but that nobody until now has volunteered to develop certain AOPs and make them available through the AOP-KB.

The AOP-KB is a recently introduced tool that was launched in 2014 and allows not only the development of AOPs but also the query by users who search for mechanistic information. Indeed, it has been acknowledged that AOPs can help to progress in many areas of regulatory toxicology and, most importantly, to support integrated approaches for testing and assessment (IATA) (Tollefsen et al., 2014). AOPs can be applied as a framework to develop IATA as it can facilitate (1) the evaluation of existing information on chemical(s) undergoing assessment in a structured way and potentially conclusion and regulatory decision on the hazard based on existing information; (2) the identification and generation of certain type of data derived from nonanimal methods that might increase the confidence level concerning evidence of a particular hazard; and (3) determination if additional information is required to make a regulatory decision (OECD, 2016b; Sachana and Leinala, 2017; Sakuratani et al., 2018).

The AOP-KB allows users to query diseases, AOs, molecular markers, and reports on available AOPs and KEs ranked by semantic similarity to the query. The access to the AOPs and KEs after searching through the AOP-KB permits the visualization of networks of AOPs that provide a more complete picture of the number of mechanisms involved. This visualization can primarily inform in vitro assay development that can further help the selection of methods that can become candidates for OECD Test Guidelines. It is clear that identifying KEs that are essential to induce an AO will allow those who develop alternative methods to direct resources toward the development of testing methods derived from knowledge captured in AOPs or networks of AOPs. Significant progress has already been made in the area of skin sensitization, where the availability of an AOP, with OECD-approved status, related to this endpoint guided the advancement of in vitro testing (Delrue et al., 2016) and defined approaches (OECD, 2016c; Casati et al., 2018).

The MIE in each AOP, which is a specialized KE, captures the specific interaction of chemical(s) with biological systems and can be applied toward the development of mechanistically based structure–activity relationships (SARs). Establishing SARs by using AOPs can assist not only to predict whether a chemical with similar chemical structure can trigger an AOP but also facilitate the chemical grouping and subsequent read-across (OECD, 2014). In case that in vitro assays are available for a number of essential KEs for a given AOP or network of AOPs, these results can be used to develop SARs or, when quantifiable, to develop Quantitative structure–activity relationship (QSARs) that can be used as stand-alone prediction or to confirm the grouping of chemicals as it happened in the case of skin sensitization (Dimitrov et al., 2016). Using mechanistic information derived from AOPs, the robustness of the data gap–filling approach information can be improved when grouping of chemicals is applied compared with grouping of chemicals approaches that are solely based on their structural similarity.

CONCLUDING REMARKS AND FUTURE DIRECTIONS

The AOP framework is useful for many regulatory purposes, including the identification of biomarkers. Developed AOPs are stored in the AOP-KB and undergo reviewing to acquire OECD-approved status. The AOP-KB is in the process to be redesigned completely to support intuitive access to the AOP browsing and data analysis tools. The new AOP-KB will retain some current features, but the aim is to make it an attractive and user-friendly application that satisfies not only the needs of the AOP developers but also the requirements of regulators that use AOPs in their risk assessment. By improving the flexibility and performance of the AOP browser and by creating a new pathway diagram visualization tool, the development and use of AOPs from the scientific and regulatory community, respectively, is expecting to increase.

References

Ankley, G.T., Bennett, R.S., Erickson, R.J., et al., 2010. Adverse outcome pathways: a conceptual framework to support ecotoxicology research and risk assessment. Environ. Toxicol. Chem. 29, 730–741.

Boobis, A.R., Cohen, S.M., Dellarco, V., et al., 2006. IPCS framework for analyzing the relevance of a cancer mode of action for humans. Crit. Rev. Toxicol. 36 (10), 781–792.

Boobis, A.R., Doe, J.E., Heinrich-Hirsch, B., et al., 2008. IPCS framework for analyzing the relevance of a noncancer mode of action for humans. Crit. Rev. Toxicol. 38 (2), 87–96.

Casati, S., Aschberger, K., Barroso, J., Casey, W., et al., 2018. Standardisation of defined approaches for skin sensitisation testing to support regulatory use and international adoption: position of the International Cooperation on Alternative Test Methods. Arch. Toxicol. 92, 611–617.

Delrue, N., Sachana, M., Sakuratani, Y., et al., 2016. The adverse outcome pathway concept: a basis for developing regulatory decision-making tools. Altern. Lab. Anim. 44, 417–429.

Dimitrov, S.D., Diderich, R., Sobanski, T., et al., 2016. QSAR toolbox – workflow and major functionalities. SAR QSAR Environ. Res. 19, 1–17.

Fan, A.M., 2018. Biomarkers in toxicology, risk assessment, and environmental chemical regulations. In: Gupta, R.C. (Ed.), Biomarkers in Toxicology. Academic Press/Elsevier, Amsterdam (in press).

Knapen, D., Vergauwen, L., Villeneuve, D.L., et al., 2015. The potential of AOP networks for reproductive and developmental toxicity assay development. Reprod. Toxicol. 56, 52–55.

Knapen, D., Angrish, M.M., Fortin, M.C., et al., 2018. Adverse outcome pathway networks I: development and applications. Environ. Toxicol. Chem. https://doi.org/10.1002/etc.4125 (in press).

Meek, M.E., Boobis, A.R., Cote, I., et al., 2014a. New developments in the evolution and application of the WHO/IPCS framework on mode of action/species concordance analysis. J. Appl. Toxicol. 34, 1–18.

Meek, M.E., Palermo, C.M., Bachman, A.N., et al., 2014b. Mode of action human relevance (MOA/HR) framework – evolution of the Bradford Hill considerations and comparative analysis of weight of evidence. J. Appl. Toxicol. 34, 595–606.

NRC, 2007. Toxicity Testing in the 21st Century: A Vision and a Strategy. The National Academies Press, Washington.

OECD (Organisation for Economic Co-operation and Development), 2014. Guidance on Grouping of Chemicals, second ed. http://olisweb.oecd.org/vgn-ext-templating/ENV-JM-MONO(2014)4-ENG.pdf?docId=JT03356214&date=1397492303403&documentId=618451&organisationId=1&fileName=JT03356214.pdf.

OECD, 2016a. Guidance Document on Developing and Assessing Adverse Outcome Pathways. http://www.oecd.org/officialdocuments/publicdisplaydocumentpdf/?cote=env/jm/mono(2013)6&doclanguage=en.

OECD, 2016b. Guidance Document for the Use of Adverse Outcome Pathways in Developing Integrated Approaches to Testing and Assessment (IATA). http://www.oecd.org/officialdocuments/publicdisplaydocumentpdf/?cote=env/jm/mono(2016)67&doclanguage=en.

OECD, 2016c. Guidance Document on the Reporting of Defined Approaches and Individual Information Sources to Be Used within Integrated Approaches to Testing and Assessment (IATA) for Skin Sensitisation. http://www.oecd.org/officialdocuments/publicdisplaydocumentpdf/?cote=env/jm/mono(2016)29&doclanguage=en.

OECD, 2017. Users' Handbook Supplement to the Guidance Document for Developing and Assessing AOPs. https://one.oecd.org/document/ENV/JM/MONO(2016)12/en/pdf.

Ryan, P.B., Burke, T.A., Hubal, E.A., et al., 2007. Using biomarkers to inform cumulative risk assessment. Environ. Health Perspect. 115, 833–884.

Sachana, M., Leinala, E., 2017. Approaching chemical safety assessment through application of integrated approaches to testing and assessment: combining mechanistic information derived from adverse outcome pathways and alternative methods. Appl. In Vitro Toxicol. 3, 227–233.

Sakuratani, Y., Horie, M., Leinala, E., 2018. Integrated approaches to testing and assessment: OECD activities on the development and use of adverse outcome pathways and case studies. Basic Clin. Pharmacol. Toxicol. https://doi.org/10.1111/bcpt.12955 (in press).

Sonich-Mullin, C., Fielder, R., Wiltse, J., et al., 2001. International Programme on Chemical Safety. IPCS conceptual framework for evaluating a mode of action for chemical carcinogenesis. Regul. Toxicol. Pharmacol. 34, 146–152.

Swenberg, J.A., Fryar-Tita, E., Jeong, Y.C., et al., 2008. Biomarkers in toxicology and risk assessment: informing critical dose-response relationships. Chem. Res. Toxicol. 21, 253–265.

Teeguarden, J.G., Tan, Y., Edwards, S.W., et al., 2016. Completing the link between exposure science and toxicology for improved environmental health decision making: the aggregate exposure pathway framework. Environ. Sci. Technol. 50, 4579–4586.

Tollefsen, K.E., Scholz, S., Cronin, M.T., et al., 2014. Applying adverse outcome pathways (AOPs) to support integrated approaches to testing and assessment (IATA). Regul. Toxicol. Pharmacol. 70, 629–640.

Villeneuve, D.L., Angrish, M.M., Fortin, M.C., et al., 2018. Adverse outcome pathway networks II: network analytics. Environ. Toxicol. Chem. 4124 https://doi.org/10.1002/etc (in press).

WHO, 1993. Biomarkers and Risk Assessment: Concepts and Principles. International Programme on Chemical Safety (ICPS), Environmental Health Criteria 155. WHO, Geneva, Switzerland, pp. 1–35.

WHO, 2001. Biomarkers in Risk Assessment: Validity and Validation. International Programme on Chemical Safety (ICPS), Environmental Health Criteria 222. WHO, Geneva, Switzerland, pp. 1–154.

PART II

SYSTEMS TOXICITY BIOMARKERS

CHAPTER

10

Central Nervous System Toxicity Biomarkers

Lucio G. Costa[1,2]

[1]Department of Environmental and Occupational Health Sciences, University of Washington, Seattle, WA, United States; [2]Department of Medicine & Surgery, University of Parma, Parma, Italy

INTRODUCTION

The human nervous system is the most complex organ systems in terms of structure and function. It contains billions of neurons, each forming thousands of synapses leading to a very large number of connections. It also contains perhaps 10 times more glial cells (astrocytes, oligodendrocytes, microglia) than neurons, which play important roles in the overall development and functioning of the nervous system (Barres, 2008). The central nervous system (CNS) with the help of the peripheral nervous system detects and relays sensory information inside and outside the body, directs motor functions, and integrates thought processes, learning, and memory. Such functions and their complexity, together with some intrinsic characteristics (e.g., mature neurons do not divide; they are highly dependent on oxygen and glucose), make the nervous system particularly vulnerable to toxic insults.

Neurotoxicity has been defined as "any adverse effect on the chemistry, structure or function of the nervous system, during development or at maturity, induced by chemical or physical influences" (Costa, 1998a). Morphological changes such as neuronopathy (a loss of neurons), axonopathy (a degeneration of the neuronal axon), or myelinopathy (a loss of the glial cells surrounding the axon), or other gliopathies, are considered important adverse effects, even if they are mild or partially reversible. All neurochemical changes, in the absence of structural damage, are also considered adverse effects, even if they are fully reversible.

Over 200 chemicals are known to be neurotoxic to humans (Grandjean and Landrigan, 2006), and for almost twice as many there is evidence of at least animal toxicity (Spencer et al., 2000). Neurotoxic chemicals include metals, organic solvents, pesticides, various organic substances, drugs, and natural compounds (Table 10.1). In addition to chemical exposures, head concussions leading to mild traumatic brain injury (mTBI) can also cause neurobehavioral, neurological, and neuropathological alterations (Mayer et al., 2017; Dixon, 2017; Steenerson and Starling, 2017).

The above definition of neurotoxicity also indicates that there are potential differences in susceptibility to toxicants between the developing and the mature nervous system. In most cases, the developing nervous system is more sensitive to adverse effects than the adult nervous system, as indicated, for example, by the most deleterious effects of ethanol, methylmercury, or lead when exposure occurs in utero or during childhood (Giordano and Costa, 2012). Furthermore, the blood–brain barrier (BBB), which protects the mature nervous system from the entry of a number of substances, appears to be poorly developed at birth and during the first few years of life (Jensen and Catalano, 1998).

Measuring Central Nervous System Dysfunctions

Neurotoxic effects can be detected in the course of standard toxicity testing (acute, subacute, subchronic, chronic, developmental/reproductive toxicity) required by regulatory agencies worldwide. However, specific guidelines exist to further probe the potential neurotoxicity of chemicals (USEPA, 1998a; OECD, 1997). Such tests are performed in rodents and are meant to assess specific effects of the tested chemical on the nervous system. The United States Environmental Protection Agency guidelines focus on a functional observational battery, on measurements of motor activity, and on neuropathological examinations (USEPA, 1998a). The Organization for Economic Cooperation and Development guidelines, similarly, focus on clinical observations, on functional tests (e.g., motor

Biomarkers in Toxicology, Second Edition
https://doi.org/10.1016/B978-0-12-814655-2.00010-4

TABLE 10.1 Selected Examples of Compounds From Different Chemical Classes Known to Be Neurotoxic in Humans

Chemical Class	Compound	Neurotoxic Effect
Metals	Manganese	**Extrapyramidal syndrome**
	Methylmercury	**Cerebellar syndrome, visual dysfunction, encephalopathy, CNS teratogenicity**
	Lead	Peripheral neuropathy (myelinopathy), **encephalopathy**
	Thallium	Peripheral neuropathy, **optic neuropathy**
Organic solvents	Ethanol	**Acute, chronic encephalopathy, CNS teratogenicity (fetal alcohol syndrome)**
	n-Hexane	Peripheral neuropathy
	Methanol	**Optical neuropathy**
	Toluene	**Encephalopathy, CNS teratogenicity (fetal solvent syndrome)**
Pesticides	Carbamates	**Cholinergic syndrome**
	Chlorinated cyclodienes	**Seizures**
	Methyl bromide	**Acute encephalopathy**, peripheral neuropathy, **optic neuropathy**
	Organophosphates	**Cholinergic syndrome**, delayed peripheral neuropathy (some)
Other organic substances	Acrylamide	Peripheral neuropathy
	Cyanide	**Seizures**
	Hydrogen sulfide	**Acute encephalopathy**
	Polychlorinated biphenyls	**Behavioral developmental neurotoxicity**
Drugs	Cisplatin	Peripheral neuropathy
	Chlorpromazine	**Extrapyramidal disorders, seizures**
	Thalidomide	Peripheral neuropathy, teratogenicity
	Valproic acid	**Acute encephalopathy, CNS teratogenicity**
Natural compounds	Botulinum toxin	Neuromuscular transmission syndrome
	Domoic acid	**Encephalopathy, neuronopathy, seizures**
	Ricin	**Neuronopathy**
	Tetrodotoxin	Ion channel syndrome (Na^+ channels)

CNS effects are in bold. *CNS*, central nervous system.
Selected from Spencer, P.S., Schaumburg, H.H., Ludolph, A.C. (Eds), 2000. Experimental and Clinical Neurotoxicology. Oxford University Press, Oxford, pp. 1–1310 and adapted from Costa, L.G., Giordano, G., Guizzetti, M., 2011. Predictive models for neurotoxicity assessment. In: Xu, J.J., Urban, L. (Eds.), Predictive Toxicology in Drug Safety. Cambridge University Press, pp. 135–152.

activity, sensory reactivity to stimuli), and on neuropathology (OECD, 1997). These batteries are not meant to provide a complete evaluation of neurotoxicity but to act as a Tier 1 screening for potential neurotoxicity. If no effects are seen at the appropriate dose level and if the chemical structure of the substance and/or its metabolites does not suggest concern for potential neurotoxicity, the substance may be considered as not neurotoxic. On the other hand, positive findings can be followed up by further testing (Tier 2) in case of commonly existing substances with commercial value or wide exposure; for new chemical entities, development of the molecule may instead be abandoned. The decision to carry out additional studies should be thus made on a case-by-case approach and may depend on factors such as the intended use of the chemical, the possibility for human exposure, and its potential accumulation in biological systems. Such Tier 2 studies may include specialized behavioral tests, electrophysiological and neurochemical measurements, and additional morphologic studies. Examples are tests for measuring learning and memory, measurements of nerve conduction velocity, and biochemical parameters related to neurotransmission or to indices of cell integrity and functions (Costa, 1998a,b).

As said, the nervous system undergoes gradual development that continues well after birth in both animals and humans. While, on one hand, the developing nervous system may more readily adapt to, or compensate for, functional losses as a result of a toxic insult, damage to the nervous system during key periods of brain development may also result in long-term, irreversible damage (Costa, 1998a). Evidence that developmental exposure to chemicals and drugs may alter behavioral functions in young animals began to be described in the early 1970s. The field of developmental neurotoxicology thus evolved from the disciplines of neurotoxicology, developmental toxicology, and experimental psychology (Makris et al., 2009). In response to this issue, developmental neurotoxicity (DNT) testing guidelines were developed both in the United States and in Europe (USEPA, 1998b; OECD, 2007). Exposure to the test chemicals is from gestational day 6 to postnatal day 10 or 21 to the mother, thus ensuring exposure in utero and through maternal milk. Tests involve measurements of developmental landmarks and reflexes, motor activity, auditory startle test, learning and memory tests, and neuropathology (USEPA, 1998b; OECD, 2007). As for neurotoxicity testing, DNT testing has been proven to be useful and effective in identifying compounds with DNT potential (Makris et al., 2009). This is not to say that current DNT testing guidelines cannot be improved; indeed, it has been pointed out that they may be overly sensitive and produce a high rate of false positives (Claudio et al., 2000) or, in contrast, that they may be too insensitive and not enough comprehensive (Cory-Slechta et al., 2001).

In the past several years, the need to develop acceptable alternatives to conventional animal testing has been increasingly recognized by toxicologists to address problems related to the escalating costs and time required for toxicity assessments, the increasing number of chemicals being developed and commercialized, the need to respond to recent legislations (e.g., Registration Evaluation and Authorization of Chemicals and the Cosmetics Directive (76/768/EEC) in the European Union), and efforts aimed at reducing the number of animals used for toxicity testing (Gartlon et al., 2006; Costa et al., 2011). Hence, efforts have been directed toward the development of alternative models, utilizing either mammalian cells in vitro or nonmammalian model systems, which could serve as tools for neurotoxicity and DNT testing, particularly for screening purposes (Baumann et al., 2014; Nishimura et al., 2015; Vassallo et al., 2017; Sandström et al., 2017).

BIOMARKERS IN NEUROTOXICOLOGY

The term biomarker has acquired many meanings over the years. For example, the presence of chloracne may be considered a biomarker of exposure and of health effect of dioxin-like compounds, and behavioral alterations in children associated with lead exposure can be considered biomarkers of subtle neurotoxic effects of this metal. A decreased nerve conduction velocity can be considered a biomarker of a peripheral neuropathy, and the presences of specific lesions in the postmortem human brain are considered biomarkers of specific neurodegenerative diseases. Thus, biomarkers are often indicators of adverse health effects associated with a certain disease or toxic exposure. With regard to CNS biomarkers, such indicators can be represented by behavioral changes, which can be assessed in animals or in humans (e.g., memory deficits), by electrophysiological changes (e.g., in the electroencephalogram), by morphological changes (e.g., neuronal death in specific cell regions and/or gliosis), and by biochemical/molecular changes (e.g., decrease in the activity of an enzyme or in the level of a growth factor).

In a broad context, biomarkers of CNS toxicity can thus be defined as indicators of alterations of CNS function. As said, behavioral and/or electrophysiological alterations in an individual are indicative of potential changes in CNS function and can be considered "biomarkers" of CNS toxicity. In this chapter, however, the term biomarker is used to mean biological/biochemical/molecular markers, which can be measured by chemical, biochemical, or molecular biological techniques. This stricter definition still encompasses measurements that may be carried out in vivo in the target organ, the brain, in animals, and/or in cells (or other more complex systems, e.g., brain slice) in vitro. As such they may be useful in animal studies to identify specific alterations or in vitro for the same purpose or for developing alternative testing approaches (Costa et al., 2011). This chapter further focuses on biomarkers that can be measured in humans and must thus be present in rather easily and ethically obtainable body fluids (e.g., blood, urine, cerebrospinal fluid [CSF]). By providing information on the cellular substrates involved in behavioral changes and/or correlates of neuropathological lesions, biochemical and molecular approaches are relevant to the process of assessing the impact of human exposure to neurotoxicants (Costa, 1998b, 1992). Knowledge of the specific target of a neurotoxicant (Costa, 1997) or of a specific lesion in a neurodegenerative disorder or in other sources of neurotoxicity (e.g., traumatic brain injury) can be very useful to develop biomarkers of effect for use in animal toxicity studies, as well as in epidemiological investigations in humans, while adding an important component to the hazard characterization aspects of the risk assessment process. Finally, while all biomarkers indicated above would give indications of the effects of toxicants on the CNS, one has to recognize that biomarkers of exposure and of susceptibility also exist (NRC, 1987). In some cases, there may be some overlap, as certain markers of exposure may also represent biomarkers of effects (see Case study 1: The organophosphorus insecticides).

Biomarkers of Exposure

An ideal biomarker of exposure should be chemical-specific, available by minimally invasive techniques, detectable even at very low levels, easy to measure, and relatable quantitatively to certain prior exposures (Henderson et al., 1989). These concepts would also apply to biomarkers of exposure to neurotoxicants. The traditional biomonitoring of chemicals (and/or their metabolites) in biological fluids (urine, blood) and in other accessible tissues (hair, dentine pulp) has been widely used in neurotoxicology. Examples abound in the areas of solvents, metals, pesticides, and industrial chemicals. The metal lead (Pb) causes severe DNT (learning disabilities, low IQ) in children at low levels, peripheral neurotoxicity (myelinopathy) in adults at intermediate levels, and severe encephalopathy at even higher levels. Such levels are usually determined by measurements of Pb levels in blood, though other measurements (e.g., Pb levels in dentine pulp or in bone) have also been utilized. For other metals (e.g., arsenic or mercury), measurements of their concentrations in hair has also proven useful, as they reflect prior and/or cumulative exposure rather than the recent exposure provided by blood measurements (Costa, 1996). For solvents, measurements of their concentrations in blood or in breath, or of their metabolites in the urine, are very common methods to determine exposure. A limitation of such approaches lies in the fact that these measurements only reflect recent exposures. For chemicals that form covalent adducts to macromolecules, such adducts can be measured, and depending on the half-life of the chosen macromolecule, levels of these adducts would reflect exposure over weeks to months. Examples of neurotoxic chemicals for which this strategy has been used include *n*-hexane, carbon disulfide, and acrylamide (Graham et al., 1995; Valentine et al., 1993; Calleman et al., 1994).

Biomarkers of Susceptibility

Genetic factors can modulate the response to neurotoxic chemicals. Genetic polymorphisms have been identified for a number of enzymes involved in the metabolism of xenobiotics, for example, several members of the cytochrome P450 (CYP), N-acetyltransferase, and glutathione transferase (GST) families (Costa and Eaton, 2006). In addition to their major hepatic localization, such enzymatic systems are also expressed in the CNS where they can contribute to in situ activation or detoxification of neurotoxicants (Farin and Omiecinski, 1993). Organic neurotoxic compounds can be bioactivated or detoxified by these same enzymes, and genetic polymorphisms may thus play a role in differential susceptibility to CNS effects. Relevant research on the influence of genetic polymorphisms has been carried out, for example, in the context of studies on the role of environmental factors in the etiology of neurodegenerative diseases such as Parkinson's disease (PD). Investigations on familial PD have so far revealed more than a dozen autosomal dominant or autosomal recessive gene mutations responsible for variants of the disease (Houlden and Singleton, 2012; Dexter and Jenner, 2013). These include mutations of α-synuclein, parkin, DJ-1, PINK-1, and several other genes (Houlden and Singleton, 2012; Dexter and Jenner, 2013; Lill, 2016). In addition, hypotheses to explain the etiology of sporadic PD have focused on potential gene–environment interactions, namely susceptibility genes that may contribute to increased risk of PD as a result of exposure to neurotoxic agents. For example, as 1-methyl-4-phenyl-1,2,3,6-tetrahydropyridine is bioactivated to a potent dopaminergic neurotoxicant by monoamine oxidase B (MAO-B), various studies have reported of associations between MAO-B polymorphisms and PD (Costa et al., 1997). Other studies have focused on polymorphisms of enzymes involved in xenobiotic metabolism, such as CYP2D6, CYP2E1, or GSTs (BenMoyal-Segal and Soreq, 2006; Singh et al., 2008), in oxidative stress or dopaminergic neurotransmission (e.g., superoxide dismutase, dopamine receptor D2, dopamine transporter; Kelada et al., 2006; Singh et al., 2008) or in proinflammatory cytokines (e.g., interleukin 1 beta; Lee et al., 2016).

Biomarkers of Effects

Biomarkers of effects should be indicative of early modifications that precede functional or structural damage. In this context, knowledge of the mechanism(s) that lead to the ultimate neurotoxic effect is essential or at least extremely useful (Manzo et al., 1996; Costa, 1998b). Unfortunately, the exact mechanism (or mode) of action for most neurotoxicants is still unknown, and this has slowed down progress in this area. Because of the complexity of the nervous system and the diversity of manifestations of neurotoxicity, in addition to the multiplicity of cellular and biochemical targets, it is difficult to develop and validate generic markers for neurotoxicity. The scenario for neurotoxicity is indeed much more complex than that for other target organs of toxicity, e.g., the liver, where hepatotoxicity can be predicted by a few specific features (e.g., mitochondrial damage, oxidative stress, intracellular glutathione), allowing the development of potentially highly predictive screening approaches (Xu et al., 2008).

With regard to neurotoxicity, one area that has received some attention is that of neurotransmission. A number of chemicals affect various aspects of neurotransmission, including neurotransmitter synthesis, uptake and degradation, receptor interactions, and second messenger systems (Costa, 1992; Castoldi et al., 1994; Costa and Manzo, 1995). Such changes should be measurable in an easily and ethically accessible peripheral tissue and should mirror identical changes occurring in target tissue (e.g., a certain brain region). Examples of this strategy, utilized with organophosphorus (OP) insecticides, are described in the case study below. Another example is represented by measurements of MAO-B, an enzyme involved in the metabolism of dopamine, in platelets. Exposure of rats to styrene (a neurotoxic solvent used in the production of polystyrene) causes a decrease in MAO-B in the CNS (Husain et al., 1980), and platelet MAO-B activity is significantly decreased in styrene-exposed workers (Checkoway et al., 1994; Bergamaschi et al., 1996), while the enzyme is not modified in platelets of workers exposed to perchloroethylene (Checkoway et al., 1994), suggesting a certain degree of specificity. Alterations in platelet MAO-B activity may thus reflect similar alterations occurring in the CNS and serve as a biomarker of effect in case of exposure to certain neurotoxic chemicals.

A different strategy is that of measuring in accessible tissue markers of brain injury (Roberts et al., 2015). Changes in neuronal- or glial cell–specific proteins or of certain stress proteins in the brain have been proposed as in situ biomarkers of neurotoxicity (O'Callaghan and Sriram, 2005; Petzold, 2005; Rajdev and Sharp, 2000). For example, levels of glial fibrillary acidic protein (GFAP) provide indication of gliosis and are increased on neuronal toxicity (O'Callaghan and Sriram, 2005). Neuron-specific enolase (NSE), the isoform of enolase with three $\gamma\gamma$ homodimers, and S-100β, a calcium-binding protein primarily synthesized in astrocytes, are two other examples of relatively brain-specific proteins. Some studies have investigated changes in the levels of these and other proteins in different biofluids, such as CSF and blood. For example, levels of S-100β increase in blood (and also in CSF) after traumatic brain damage, stroke, cardiac operations, and a variety of neurodegenerative diseases (Rothermundt et al., 2003; Seco et al., 2012). As CNS toxicity is often associated with increased BBB permeability, it has been suggested that increases in S-100β levels indicate BBB defects (Marchi et al., 2004). Levels of GFAP and of ubiquitin C-terminal hydrolase-1 (UCH-L1; another potential marker of neuronal injury) have been found elevated in the CSF of rats on exposure to the neurotoxicant kainic acid (Glushakova et al., 2012). Interestingly, UCH-L1 was increased at early time points, whereas GFAP increased over time, in agreement with what observed in the hippocampus (Glushakova et al., 2012). Additional biomarkers have also been investigated in relationship to PD. For example, α-synuclein levels are decreased in the CSF of PD patients and so are levels of DJ-1 (Hong et al., 2010; Haas et al., 2012), though negative or inconclusive findings have been reported for the same biomarkers in blood (Haas et al., 2012; Lin et al., 2012).

MicroRNAs (miRNAs) are endogenous, small noncoding RNAs (approximately 22 nucleotides) that regulate gene expression, and over 60% of proteins are believed to be controlled posttranscriptionally by miRNAs (Friedman et al., 2009). Recently, miRNAs have been found to be potential targets or effectors of neurotoxicants (Tal and Tanguay, 2012; Roberts et al., 2015). Because some miRNAs are produced at high concentrations within cells in a tissue-specific manner and are reported to be remarkably stable in plasma, they may represent biomarkers to monitor tissue injury. For example, Laterza et al. (2009) showed that miR-124 can be used to monitor ischemia-related brain injury starting at 8 h and peaking at 24 h after occlusion (an approximately 150-fold increase in plasma relative to the sham surgery control group), whereas Ogata et al. (2015) found increases in blood levels of miR-9 and of miR-384, which correlated with acute neurotoxicity of trimethyltin.

The following sections present two case studies related to the use of biomarkers for neurotoxicity. The first one deals with OP compounds, an important class of insecticides whose mechanisms of toxicity have been for the most part elucidated. For these compounds biomarkers of exposure, susceptibility and effects are available. The second case study deals with mTBI caused by head concussion. In this case diffuse damage is present in the CNS, and biomarkers of effects attempt to capture its severity and time course of recovery.

CASE STUDY 1: THE ORGANOPHOSPHORUS INSECTICIDES

The OP insecticides are discussed in detail as examples of neurotoxic chemicals for which various types of biomarkers can be defined (Table 10.2).

Toxicity and Neurotoxicity of Organophosphorus

On acute exposure, OPs cause a number of central and peripheral effects, which are typical of an overstimulation of the cholinergic system (Costa, 2006). These are due to the ability of OPs to inhibit the activity of acetylcholinesterase (AChE), the enzyme which degrades the neurotransmitter acetylcholine and the only mean by which this neurotransmitter can be removed from the synaptic cleft. Inhibition of AChE leads to accumulation

TABLE 10.2 Biomarkers for Organophosphates

Type of Biomarker	Example
Biomarker of exposure	Plasma cholinesterase (BuChE) Red blood cell acetylcholinesterase OP in blood OP metabolites in urine OP adducts to blood proteins
Biomarker of susceptibility	Paraoxonase 1 status BuChE polymorphisms CYPs
Biomarker of effect	Red blood cell acetylcholinesterase Muscarinic receptors in lymphocytes Lymphocyte neuropathy target esterase

AChE, acetylcholinesterase; *BuChE*, butyrylcholinesterase; *CYP*, cytochrome P450; *OP*, organophosphorus.

of acetylcholine at cholinergic synapses, with an ensuing overstimulation of cholinergic receptors and the development of a cholinergic crisis; the latter includes increased sweating and salivation, profound bronchial secretion, bronchoconstriction, miosis, increased gastrointestinal motility, diarrhea, tremors, muscular twitching, and a number of CNS effects (dizziness, lethargy, fatigue, headache, mental confusion, depression of respiratory centers, convulsions, coma). AChE inhibition is due to phosphorylation of the active site of the enzyme, which can be followed, depending on the chemical structure of the OP, by the loss of an alkyl chain in a process known as "aging." When this occurs, AChE can be considered as being irreversibly inhibited. A few OPs may also cause another type of toxicity, known as organophosphate-induced delayed polyneuropathy (OPIDP). Signs and symptoms include tingling of the hands and feet, followed by sensory loss, progressive muscle weakness and flaccidity of the distal skeletal muscles of the lower and upper extremities, and ataxia (Lotti and Moretto, 2005). These may occur 2–3 weeks after a single exposure, when signs of the acute cholinergic syndrome have subsided. OPIDP, which can be classified as a distal sensorimotor axonopathy, is not related to AChE inhibition. Rather, the molecular target appears to be an esterase, present in nerve tissues and other tissues, named neuropathy target esterase (NTE). Phosphorylation of NTE by OPs is similar to that observed for AChE. Several OPs, depending on their chemical structure, can inhibit NTE and so can also some non-OPs, such as certain carbamates and sulfonyl fluorides. However, only OPs whose chemical structure leads to aging of phosphorylated NTE (by a process analogous to that described for AChE) can cause OPIDP. Other compounds that inhibit NTE but cannot undergo the aging reaction are not neuropathic, indicating that inhibition of NTE catalytic activity may not be the final mechanism of axonal degeneration.

Biomarkers of Exposure

Because of the role of acetylcholine as a neurotransmitter in both the CNS and the peripheral nervous system (somatic and autonomic), AChE is widely distributed throughout the body. Moreover, AChE is also present in erythrocytes, though its physiological role in these cells, which are devoid of synaptic contacts, has not been elucidated. Another cholinesterase, known as pseudocholinesterase or butyrylcholinesterase (BuChE), coded by a gene different from that coding for AChE and with different substrate specificity, is present in several tissues including plasma, where its physiological role has not been fully elucidated. On exposure, OPs inhibit the activity of both enzymes in tissues and in blood; hence, measurements of AChE and BuChE activities in red blood cells and in plasma have been extensively used as biomarkers of exposure OPs. A lower activity of erythrocyte AChE and of plasma BuChE would indicate that exposure to an OP (or to a similarly acting compound) has occurred. Neither measurement is specific for a specific OP. Additionally, other insecticides, such as the carbamates, also inhibit AChE and BuChE, though the enzymes' activities recover within a few hours on inhibition by a carbamate compound.

When used as biomarkers of exposure in population studies, the issue of interpersonal variability in AChE and BuChE activities should be carefully considered. When possible, baseline values should be obtained for each individual, and variations below these activity levels rather than absolute levels should be used to assess exposure. In the absence of preexposure measurements, repeated postexposure measurements should be obtained at different intervals. In this case, a significant increase in enzyme activity over time would indicate recovery from an initial exposure to an OP (Coye et al., 1987). A 30% or greater decrease of plasma BuChE from preexposure baseline usually raises a red flag and requires health and workplace surveillance and removal of the worker from the exposure; however, the toxicological significance of such decrease is still much debated (Carlock et al., 1999; USEPA, 2000). Another important issue relates to which between erythrocyte AChE and plasma BuChE is a better indicator of exposure to OPs. Many OPs appear to preferentially inhibit plasma BuChE, suggesting that this enzyme may be a more sensitive indicator of exposure; however, this is not true for all OP compounds. Furthermore, plasma BuChE activity displays a higher variability because it can be affected by other exogenous agents (e.g., drugs), physiological or pathological conditions (e.g., pregnancy or liver damage), and genetic background (Chatonnet and Lockridge, 1989; Lockridge, 1990). As AChE is the target for OP neurotoxicity, measurements of AChE activity in red blood cells are also used as a biomarker of effect for OPs (see below). NTE activity has been found in blood, namely in lymphocytes and platelets (Bertoncin et al., 1985; Maroni and Bleecker, 1986), and measurements of NTE activity in these cells may serve as indication of exposure to potentially neuropathic OPs. Such measurements may also represent markers of effects of neuropathic OPs (see below).

Several analytical methods are available to measure OPs and their metabolites in body fluids; the parent compound is measured in blood, whereas metabolites are measured in urine. These measurements are rarely carried out in the clinical setting but are extensively utilized in epidemiological studies; indeed, determination of metabolite levels in urine is the most practical method to estimate exposure to OPs (Maroni et al., 2000). Such metabolites include alkyl phosphate derivatives and chemical residues (the "leaving group") specific for each compound. The alkyl phosphates or alkyl dithiophosphates are the result of metabolism of parent compounds or their oxygen analogs by CYPs or esterases. They are not specific for a certain OP but are useful to assess exposure to (or internal dose of) several OPs. Other metabolites are specific for certain OP compounds; for example, *p*-nitrophenol in urine is an indicator of exposure to parathion or methyl parathion, while 3,5,6-trichloropyridinol is useful to assess exposure to chlorpyrifos or methyl chlorpyrifos. Although measurements of urinary metabolites of OPs have been widely used to assess exposure to OP from occupational, environmental, and dietary sources, caution should be exercised when interpreting results, as dialkyl phosphates, and also leaving groups, can be found in the environment, including food and drinks, as a result of OP degradation. This would lead to an overestimate of OP exposure (Lu et al., 2005).

More recently, there has been an increasing interest in the analysis of blood protein as biomarkers of OP exposure because the OP-adducted proteins have longer half-lives and are modified only by the active insecticide. Proteomics techniques (e.g., liquid chromatography–mass spectrometry [MS] and matrix-assisted laser desorption/ionization time-of-flight MS) have become important tools for the biomonitoring of OP-adducted proteins in blood, as they allow for the accurate and sensitive identification of OP-inhibited proteins based on the change in mass of the adducted active-site peptides following enzymatic digestion (Thompson et al., 2010). Among such proteins, BuChE has been extensively investigated. BuChE has a half-life of ~10 days in plasma, whereas red blood cell AChE has a 120 days life span; however, the concentration of BuChE in blood is 10-fold higher than AChE. The critical step in this approach is the purification of the BuChE from plasma, and the use of immunoaffinity purification should be helpful (Marsillach et al., 2013). Another protein that has been proposed as a biomarker of OP exposure is acylpeptide hydrolase (APH) present in several tissues, including the brain and the erythrocytes, where it is believed to contribute to the degradation of proteins (Fujino et al., 2000). Adducts of OPs to red blood cell APH have been recently characterized (Marsillach et al., 2011), supporting its potential use as biomarker of OP exposure (Quistad et al., 2005). Additional proteins present in blood, with which OPs form adducts, are plasma albumin and monocyte carboxylesterase (Marsillach et al., 2013).

Biomarkers of Susceptibility

Genetic polymorphisms are known to affect the toxicity of several environmental chemicals, including OPs (Costa and Eaton, 2006). One of the best characterized polymorphisms is that of the enzyme paraoxonase 1 (PON1), whose name derives from its ability to detoxify paraoxon, the active metabolite of the OP parathion. Other substrates for this enzyme, which is classified as an A-esterase, are chlorpyrifos oxon, diazinon oxon and the nerve agent sarin. PON1 is secreted from the liver, and high activity levels are found in plasma. The finding that plasma PON1 activity is multimodally distributed in human populations (Geldmacher von Mallinckrodt and Diebgen, 1988) indicated the existence of polymorphisms and suggested that individuals with low activity may be more sensitive to OP toxicity. Biochemical and molecular studies led to the purification and cloning of PON1 from different species (Furlong et al., 1991; Hassett et al., 1991) and to the identification of a Q(Gln)/R(Arg) polymorphism at position 192, with the R_{192} alloform displaying a high hydrolytic activity toward chlorpyrifos oxon and paraoxon (the latter only in vitro) (Humbert et al., 1993; Li et al., 2000). In addition to these qualitative differences, there is also a large interindividual variability in the levels of protein circulating in plasma, which are likely due to polymorphisms in the promoter region of PON1 (Costa et al., 2013). Determination of the "PON1 status" of an individual by a two-dimensional assay with two substrates (Richter et al., 2004) provides information on an individual's PON1 polymorphism and the level of PON1 expression.

Administration of purified or recombinant human PON1 to rats or mice provides protection against OP toxicity (Li et al., 1993, 2000; Stevens et al., 2008). Further studies in $PON1^{-/-}$ mice and in humanized PON1 mice (expressing either the Q192 or the R192 allele over a knockout background) (Shih et al., 1998; Cole et al., 2005) have provided additional strong evidence that PON1 is an important determinant of susceptibility for those OPs whose oxons are substrate for this enzyme. Evidence for a role of PON1 in OP toxicity in humans is still limited, though important findings have been emerging in the past few years related to exposures to OP insecticides (see Costa et al., 2013). For example, a series of studies found that low PON1 activity was associated with chronic central and/or peripheral nervous system abnormalities, at times referred to as "dippers' flu," associated with exposure of sheep dippers to diazinon (Cherry et al., 2011). In pesticide handler exposed to various OPs (most notably, chlorpyrifos), low PON1 catalytic efficiency (Q192) and low plasma PON1 activity were associated with increased degrees of plasma BuChE inhibition from baseline levels (Hoffman et al., 2009). Overall, the human studies available so far provide initial evidence that low PON1 status may increase susceptibility to adverse effects of certain OP insecticides. However, it is still uncertain whether PON1 status may play a significant role at lower dose levels of exposure (Timchalk et al., 2002; Cole et al., 2005).

CYPs are also important for the activation and detoxication of OP insecticides. Variant forms of several CYP genes have been identified, and these polymorphisms confer differences in catalytic activity or level of expression, which may result in varying rates of oxidation of xenobiotics among individuals. However, information on the role of specific CYP isozymes in the metabolism of OPs is still limited and mostly derived from in vitro studies; hence, the potential contributions of such genetic polymorphisms to OP toxicity susceptibility are still unknown, thereby limiting the usefulness of their molecular measurements (Eaton, 2000; Costa, 2001), though newer physiologically based pharmacokinetic/pharmacodynamic (PBPK/PD) modeling efforts appear promising (Foxenberg et al., 2011).

A large number of genetic polymorphisms have been also described for BuChE, with at least 39 identified genetic variants with nucleotide alterations in the coding region (Lockridge and Masson, 2000). Several of the rare variants are silent (i.e., they have <10% of normal activity), while the most common variants (e.g., atypical variants) have a reduced activity. Individuals with genetic variants of BuChE with no or low activity would be predicted to be more susceptible to OP toxicity, as suggested by a study in Brazilian farmers (Fontoura-da-Silva and Chautard-Friere-Maira, 1996), but this assumption has been challenged and appears to be valid only in the case of OP nerve agents (Lockridge et al., 2016).

Biomarkers of Effect

Changes in activity of blood cholinesterases would be reliable biomarkers of effect of OPs, if they reflect similar changes occurring in target tissues, i.e., the CNS and the muscles, particularly the diaphragm. Evidence indicates that erythrocyte AChE correlates with brain and diaphragm activity better than plasma BuChE, as observed, for example, in animals exposed to chlorpyrifos, paraoxon, or disulfoton (Padilla et al., 1994; Fitzgerald and Costa, 1993). In another study, in which rats were treated for 5 weeks with chlorpyrifos, a strong correlation between whole blood and brain cholinesterase activity was found both during dosing and also during the recovery period, though a much more reliable correlation was present during dosing, when the inhibition has reached steady state (Padilla, 1995). There is also a good correlation between the severity of signs and symptoms of poisoning and the degree of inhibition of red blood cell AChE (Lotti, 2001).

Repeated exposure to OPs has been shown to induce the development of tolerance to their toxicity, which is in part mediated by a decrease in cholinergic receptors (particularly the muscarinic ones) to compensate for the prolonged increase in acetylcholine levels (Costa et al., 1982). Such changes in muscarinic receptors may indeed serve as a protective mechanism against overstimulation by acetylcholine, but they may also result in higher brain function impairment, such as memory deficits (McDonald et al., 1988). Some subtypes of muscarinic receptors have also been identified at the protein and mRNA level in lymphocytes (Costa et al., 1988, 1994), and prolonged exposure to the OP disulfoton was shown to cause a similar decrease of muscarinic receptor density in brain areas and in lymphocytes (Costa et al., 1990; Fitzgerald and Costa, 1993).

With regard to OPIDP, there is a good correlation between lymphocytes and brain NTE (Schwab and Richardson, 1986; Makhaeva et al., 2007), suggesting that lymphocytic NTE may be used as a putative biomarker of effect of neuropathic OP compounds. A good example of its application is a case report of an attempted suicide with the insecticide chlorpyrifos where, based on the degree of lymphocyte NTE inhibition shortly after exposure, it was correctly predicted that a neuropathy would develop well after recovery from acute cholinergic poisoning had occurred (Lotti et al., 1986). However, in a study of workers exposed to the defoliant DEF, inhibition of lymphocyte NTE was observed, but no clinical or electrophysiological signs of OPIDP were detected (Lotti et al., 1983). Nevertheless, it is apparent that knowledge of the target of toxicity for organophosphates has allowed the successful use of peripheral biomarkers of effects. To date, these represent some of the few validated effect biomarkers in neurotoxicology (Costa, 1996; Manzo et al., 1996).

CASE STUDY 2: MILD TRAUMATIC BRAIN INJURY

Sport-related brain concussions have been considered as a "public health epidemic," with 1.6–3.8 million episodes occurring annually (Steenerson and Starling, 2017). These concussions, also defined as mTBI, are often seen not only in American football or ice hockey players but also in military personnel (Dixon, 2017; Mayer et al., 2017). More severe cases can evolve in chronic traumatic encephalopathy (CTE), which can only definitely be diagnosed postmortem (Ling et al., 2017). The symptomatology of mTBI is usually divided into four domains: (1) physical (headache, dizziness), (2) cognitive (concentration, memory), (3) emotional (depression, anxiety, mood lability), and (4) sleep (hypersomnia, insomnia), of which the first two are the most commonly reported (Steenerson and Starling, 2017). Routine neuroimaging is not helpful in identifying any neuropathological change. Yet, there is evidence of massive neurotransmitter dysregulation, particularly excessive glutamate release, altered mitochondrial functions, neuroinflammation, changes in cerebral blood flow, and diffuse axonal injury (Dixon, 2017; Steenerson and Starling, 2017; Mayer et al., 2017). Of concern is that mTBI, and particularly CTE, may lead to neuropsychiatric manifestation and to neurodegenerative diseases such as Alzheimer's disease. Also of relevance is the need of understanding the time course of recovery from the initial injury to assess the ability to return to play or service.

The diagnosis of mTBI is difficult and the medical field remains fragmented both in terms of the diagnostic (different criteria proffered by multiple medical organizations) and prognostic factors that influence patient care (Mayer et al., 2017). Hence, attention has been devoted in recent years to biomarkers of mTBI that may allow a more certain diagnosis and a follow-up during the recovery period to complement clinical approaches (Dambinova et al., 2016). Such biomarkers are present mostly in blood and reflect alterations of basic CNS cellular processes rather that specific molecular changes as in case of OPs. A summary of biomarkers utilized or proposed for use in mTBI is shown in Table 10.3. Blood biomarkers are usually divided to indicate injury to neurons or astroglial cells. Thus, for example, NSE would be a marker of neuronal damage, whereas S-100β would represent damage to astrocytes (Papa, 2016; Zettenberg and Blennow, 2016). Increases in the levels of these biomarkers in blood reflect the severity of the injury and the time course of recovery. In a study in professional ice hockey players, increases of blood tau (a marker of axonal injury) and of S-100β were found immediately after concussion, suggesting axonal and astroglial injury, and levels gradually decreased during rehabilitation; no changes in NSE

TABLE 10.3 Some Biofluid Biomarkers for Mild Traumatic Brain Injury

Type of Biomarker	Biomarker	Pros	Cons
Neuronal Injury	NSE (neuron-specific enolase)	Enriched in neuronal cell bodies.	Also present in oligodendrocytes and erythrocytes
	UCH-L1 (ubiquitin C-terminal hydrolase L1)	Enriched in neuronal cytoplasm	Not CNS specific. Useful in more severe cases
Axonal injury	Tau protein	Microtubule-associated protein (particularly unmyelinated axons)	Various forms of tau are present
	NFL-L and NFL-H (neurofilament light and heavy)	Associated with large myelinated axons	Extra-CNS contribution unknown
	Alpha-II spectrin breakdown products	Cytoskeletal protein abundant in nerve terminals	Only limited information available
Astroglial injury	S-100β	Calcium-binding protein in astrocytes. One of the most promising	Discrepancies linked to nonstandardized measurement methods. Lack of specificity
	GFAP (glial fibrillary acidic protein)	Expressed almost exclusively in astrocytes	Elevated in other conditions
Oligodendrocyte injury	MBP (myelin basic protein)	Specific to the myelin sheath	Scarcely studied
Other	MicroRNAs	May provide indication for several different damages	Still not validated

CNS, central nervous system.

were found (Shahim et al., 2014). In another study in 34 mTBI patients, increases in tau protein, GFAP, and amyloid β42 (a protein associated with Alzheimer's disease) were found 24 h after concussion and up to 30–90 days afterward (Bogolovsky et al., 2017). Elevated tau was also reported in a group of sport-related concussions, and higher initial tau was associated with more prolonged return to play time (Gill et al., 2017). Meier et al. (2017) found an increase in S-100β and of UCH-L1 in a group of 32 concussed athletes. Levels of serum neurofilament light (a protein expressed predominantly in myelinated axons) were found to be elevated in boxers and hockey players experiencing concussions (Shahim et al., 2017). Recent studies have also explored the use of blood miRNA levels for diagnosis and prognosis of mTBI. So far, a number of miRNA (e.g., miR-21, miR142-3, miR-335, miR423-3p) look promising (Mitra et al., 2017; Di Pietro et al., 2017), but much more work is needed in this area. In a recent systematic review, Papa et al. (2015) identified 13 studies exploring the use of biomarkers of mTBI in sport athletes; most studies had fewer than 100 patients and a total of 11 different biomarkers were analyzed. Their conclusion was that work is still needed to validate biomarkers for concussion, but there is a strong potential for biomarkers to provide "diagnostic, prognostic, and monitoring information postinjury" (Papa et al., 2015).

CONCLUDING REMARKS AND FUTURE DIRECTIONS

This overview of biomarkers for CNS toxicity has evidenced how the road for developing and validating biomarkers of neurotoxic effects is still long and difficult. A variety of tools exist to measure biomarkers of exposure to CNS toxicants; this is by far the most advanced area, in which various approaches allow for assessment of recent and/or long-term exposures to neurotoxicants. The area of biomarkers of susceptibility has not been extensively investigated with regard to CNS toxicants, but many examples exist with regard to biomarkers of susceptibility in a number of CNS diseases (e.g., PD). With regard to biomarkers of effects, it appears that new opportunities will arise from increasing knowledge of the mechanisms and target of neurotoxicants (Costa, 1998b). Clearly, CNS toxicity can be manifested in very many ways, each characterized by different cellular, biochemical, and molecular substrates. The development of one or a few biomarkers of CNS toxicity appears unlikely. Rather, biomarkers may be specific for a class of chemicals (e.g., OPs) or for certain toxicological processes (e.g., neuronal death or demyelination). In this respect, the examples of biomarkers being studied and developed for mTBI may be useful and should be further explored in relationship to exposure to neurotoxic chemicals. The ultimate goal should, however, remain that of understanding the chain of events that lead to CNS toxicity, with the hope that one or more early biochemical/molecular alterations, which would precede irreversible damage, may be mirrored in changes in accessible tissues. This would thus be defined as a new biomarker for neurotoxic effects.

Acknowledgments

Research by the author has been supported by grant from the National Institute of Environmental Health Sciences (P30ES07033, P42ES04696, R01ES022949, R01ES028273).

References

Barres, B.A., 2008. The mistery and magic of glia: a perspective on their roles in health and disease. Neuron 60, 430–440.

Baumann, J., Berenys, M., Gassmann, K., Fritsche, E., 2014. Comparative human and rat "neurosphere assay" for developmental neurotoxicity testing. Curr. Protoc. Toxicol. 59, 12.21.1–12.21.24.

BenMoyal-Segal, L., Soreq, H., 2006. Gene-environment interactions in sporadic Parkinson's disease. J. Neurochem. 97, 1740–1755.

Bergamaschi, E., Mutti, A., Cavazzini, S., et al., 1996. Peripheral markers of neurochemical effects among styrene-exposed workers. Neurotoxicology 17, 753–759.

Bertoncin, D., Russolo, A., Caroldi, S., Lotti, M., 1985. Neuropathy target esterase in human lymphocytes. Arch. Environ. Health 40, 139–144.

Bogolovsky, T., Wilson, D., Chen, Y., et al., 2017. Increases if plasma levels of glial fibrillary acidic protein, tau, and amyloid β up to 90 days after traumatic brain injury. J. Neurotrauma 34, 66–73.

Calleman, C.J., Wu, Y., He, F., et al., 1994. Relationship between biomarkers of exposure ad neurological effects in a group of workers exposed to acrylamide. Toxicol. Appl. Pharmacol. 126, 361–371.

Carlock, L.L., Chen, W.L., Gordon, E.B., et al., 1999. Regulating and assessing risks of cholinesterase-inhibiting pesticides: divergent approaches and interpretations. J. Toxicol. Environ. Health Part B, Crit. Rev. 2, 105–160.

Castoldi, A.F., Coccini, T., Rossi, A., et al., 1994. Biomarkers in environmental medicine: alterations of cell signaling as early indicators of neurotoxicity. Funct. Neurol. 9, 101–109.

Chatonnet, A., Lockridge, O., 1989. Comparison of butyrylcholinesterase and acetylcholinesterase. Biochem. J. 260, 625–634.

Checkoway, H., Echeverria, D., Moon, J.D., et al., 1994. Platelet monoamine oxidase B activity in workers exposed to styrene. Int. Arch. Occup. Environ. Health 66, 359–362.

Cherry, N., Mackness, M., Mackness, B., et al., 2011. "Dippers' flu" and its relationship to PON1 polymorphisms. Occup. Environ. Med. 68, 211–217.

Claudio, L., Kwa, W.C., Russell, A.L., Wallinga, D., 2000. Testing methods for developmental neurotoxicity of environmental chemicals. Toxicol. Appl. Pharmacol. 164, 1–14.

Cole, T.B., Walter, B.J., Shih, D.M., et al., 2005. Toxicity of chlorpyrifos and chlorpyrifos oxon in a transgenic mouse model of the human paraoxonase (PON1) Q192R polymorphism. Pharmacogenet. Genomics 15, 589–598.

Cory-Slechta, D.A., Crofton, K.M., Foran, J.A., et al., 2001. Methods to identify and characterize developmental neurotoxicity for human health risk assessment. I: behavioral effects. Environ. Health Perspect. 109 (Suppl. 1), 79–91.

Costa, L.G., 1992. Effect of neurotoxicants on brain neurochemistry. In: Tilson, H., Mitchell, C. (Eds.), Neurotoxicology. Raven Press, New York, pp. 101–123.

Costa, L.G., 1996. Biomarkers research in neurotoxicology: the role of mechanistic studies to bridge the gap between the laboratory and epidemiological investigations. Environ. Health Perspect. 104 (Suppl. 1), 55–67.

Costa, L.G., 1997. Targets and mechanisms of neurotoxicity. Adv. Occup. Med. Rehab. 3, 149–156.

Costa, L.G., 1998a. Neurotoxicity testing: a discussion of in vitro alternatives. Environ. Health Perspect. 106 (Suppl. 2), 505–510.

Costa, L.G., 1998b. Biochemical and molecular neurotoxicology: relevance to biomarker development, neurotoxicity testing and risk assessment. Toxicol. Lett. 102–103, 417–421.

Costa, L.G., 2001. Pesticide exposure. Differential risk for neurotoxic outcomes due to enzyme polymorphisms. Clin. Occup. Environ. Med. 1, 511–523.

Costa, L.G., Manzo, L., 1995. Biochemical markers of neurotoxicity: research strategies and epidemiological applications. Toxicol. Lett. 77, 137–144.

Costa, L.G., 2006. Current issues in organophosphate toxicology. Clin. Chim. Acta 366, 1–13.

Costa, L.G., Schwab, B.W., Murphy, S.D., 1982. Tolerance to anticholinesterase compounds in mammals. Toxicology 25, 79–87.

Costa, L.G., Kaylor, G., Murphy, S.D., 1988. Muscarinic cholinergic binding sites on rat lymphocytes. Immunopharmacology 16, 139–149.

Costa, L.G., Kaylor, G., Murphy, S.D., 1990. In vitro and in vivo modulation of cholinergic muscarinic receptors in rat lymphocytes and brain by cholinergic agents. Int. J. Immunopharmacol. 12, 67–76.

Costa, L.G., Eaton, D.L. (Eds.), 2006. Gene-Environment Interactions: Fundamentals of Ecogenetics. John Wiley & Sons, Hoboken, NJ, pp. 1–557.

Costa, L.G., Giordano, G., Guizzetti, M., 2011. Predictive models for neurotoxicity assessment. In: Xu, J.J., Urban, L. (Eds.), Predictive Toxicology in Drug Safety. Cambridge University Press, pp. 135–152.

Costa, L.G., Giordano, G., Cole, T.B., et al., 2013. Paraoxonase 1 (PON1) as a genetic determinant of susceptibility to organophosphate toxicity. Toxicology 307, 115–122.

Costa, P., Traver, D.J., Auger, C.B., Costa, L.G., 1994. Expression of cholinergic muscarinic receptor subtypes mRNA in rat blood mononuclear cells. Immunopharmacology 28, 113–123.

Costa, P., Checkoway, H., Levy, D., et al., 1997. Association of a polymorphism in intron 13 of the monoamine oxidase B gene with Parkinson disease. Am. J. Med. Genet. 18, 154–156.

Coye, M.J., Barnett, P.G., Midtling, J.E., et al., 1987. Clinical confirmation of organophosphate poisoning by serial cholinesterase analyses. Arch. Int. Med. 147, 438–442.

Dambinova, S.A., Maroon, J.C., Sufrinko, A.M., et al., 2016. Functional, structural, and neurotoxicity biomarkers in integrative assessment of concussions. Front. Neurol. 7, 172.

Dexter, D.T., Jenner, P., 2013. Parkinson disease: from pathology to molecular disease mechanisms. Free Rad. Biol. Med. 62, 132–144.

Di Pietro, V., Ragusa, M., Davies, D., et al., 2017. MicroRNAs as novel biomarkers for the diagnosis and prognosis of mild and sever traumatic brain injury. J. Neurotrauma 34, 1948–1954.

Dixon, K.J., 2017. Pathophysiology of traumatic brain injury. Phys. Med. Rehab. Clin. North Am. 28, 215–225.

Eaton, D.L., 2000. Biotransformation enzyme polymorphism and pesticide susceptibility. Neurotoxicology 21, 101–111.

Farin, F.M., Omiecinski, C.J., 1993. Regiospecific expression of cytochrome P450s and microsomal epoxide hydrolase in human brain tissue. J. Toxicol. Environ. Health 40, 317–335.

Fitzgerald, B.B., Costa, L.G., 1993. Modulation of muscarinic receptors and acetylcholinesterase activity in lymphocytes and in brain areas following repeated organophosphate exposure in rats. Fund. Appl. Toxicol. 20, 210–216.

Fontoura-da-Silva, S.E., Chautard-Friere-Maira, E.A., 1996. Butyrylcholinesterase variants (BChE and CHE1 loci) associated with erythrocyte acetylcholinesterase inhibition in farmers exposed to pesticides. Hum. Hered. 46, 142–147.

Foxenberg, R.J., Ellison, C.A., Knaak, J.B., et al., 2011. Cytochrome P450-specific human PBPK/PD models for the organophosphorus pesticides: chlorpyrifos and parathion. Toxicology 285, 57–66.

Friedman, R.C., Farh, K.K., Burge, C.B., Bartel, D.P., 2009. Most mammalian mRNAs are conserved targets of microRNAs. Genome Res. 19, 92–105.

Fujino, T., Watanabe, K., Beppu, M., et al., 2000. Identification of oxidized protein hydrolase of human erythrocytes as acylpeptide hydrolase. Biochim. Biophys. Acta 1478, 102–112.

Furlong, C.E., Richter, R.J., Chapline, C., Crabb, J.W., 1991. Purification of rabbit and human paraoxonase. Biochemistry 20, 10133–10140.

Gartlon, J., Kinsner, A., Bal-Price, A., et al., 2006. Evaluation of a proposed in vitro test strategy using neuronal and non-neuronal cell systems for detecting neurotoxicity. Toxicol. In Vitro 20, 1569–1581.

Geldmacher von Mallinckrodt, M., Diebgen, T.L., 1988. The human serum paraoxonase-polymorphism and specificity. Toxicol. Environ. Chem. 18, 79–96.

Gill, J., Merchant-Borna, K., Jeromin, A., et al., 2017. Acute plasma tau relates to prolonged return to play after concussion. Neurology 88, 595–602.

Giordano, G., Costa, L.G., 2012. Developmental neurotoxicity: some old and new issues. ISRN Toxicol. 12. ID 814795.

Glushakova, O.Y., Jeromin, A., Martinez, J., et al., 2012. Cerebrospinal fluid protein biomarker panel for assessment of neurotoxicity induced by kainic acid in rats. Toxicol. Sci. 130, 158–167.

Graham, D.G., Amarnath, V., Valentine, W.M., et al., 1995. Pathogenetic studies of hexane and carbon disulfide neurotoxicity. Crit. Rev. Toxicol. 25, 91–112.

Grandjean, P., Landrigan, P.J., 2006. Developmental neurotoxicity of industrial chemicals. Lancet 368, 2167–2178.

Haas, B.R., Stewart, T.H., Zhang, J., 2012. Premotor biomarkers for Parkinson's disease-a promising direction of research. Transl. Neurodegener. 1, 1–11.

Hassett, C., Richter, R.J., Humbert, R., et al., 1991. Characterization of cDNA clones encoding rabbit and human serum paraoxonase: the mature protein retains its signal sequence. Biochemistry 30, 10141–10149.

Henderson, R.F., Bechtold, W.E., et al., 1989. The use of biological markers in toxicology. Crit. Rev. Toxicol. 20, 65–82.

Hoffman, J.N., Keifer, M.C., Furlong, C.E., et al., 2009. Serum cholinesterase inhibition in relation to paraoxonase-1 (PON1) status among organophosphate-exposed agricultural pesticide handlers. Environ. Health Perspect. 117, 1402–1408.

Hong, Z., Shi, M., Chung, K.A., et al., 2010. DJ-1 and alpha-synuclein in human cerebrospinal fluid as biomarkers of Parkinson's disease. Brain 133, 713–726.

Houlden, H., Singleton, A.B., 2012. The genetics and neuropathology of Parkinson's disease. Acta Neuropathol. 124, 325–338.

Humbert, R., Adler, D.A., Disteche, C.M., et al., 1993. The molecular basis of the human serum paraoxonase polymorphisms. Nat. Genet. 3, 73–76.

Husain, R., Srivastava, S.P., Mushtaq, M., Seth, P.K., 1980. Effect of styrene on levels of serotonin, noradrenaline, dopamine and activity of acetylcholinesterase and monoamine oxidase in rat brain. Toxicol. Lett. 7, 47–50.

Jensen, K.F., Catalano, S.M., 1998. Brain morphogenesis and developmental neurotoxicology. In: Slikker, W., Chang, L.W. (Eds.), Handbook of Developmental Neurotoxicology. Academic Press, San Diego, pp. 3–41.

Kelada, S.N., Checkoway, H., Kardia, S.L., et al., 2006. 5′ and 3′ region variability in the dopamine transporter gene (SLC6A3), pesticide exposure and Parkinson's disease risk: a hypothesis-generating study. Hum. Mol. Genet. 15, 3055–3062.

Laterza, O.F., Lim, L., Garrett-Engele, P.W., et al., 2009. Plasma microRNAs as sensitive and specific biomarkers of tissue injury. Clin. Chem. 55, 1977–1983.

Lee, P.C., Raaschou-Nielsen, O., Lill, C.M., et al., 2016. Gene-environment interactions linking air pollution and inflammation in Parkinson's disease. Environ. Res. 151, 713–720.

Li, W.F., Costa, L.G., Furlong, C.E., 1993. Serum paraoxonase status: a major factor in determining resistance to organophosphates. J. Toxicol. Environ. Health 40, 337–346.

Li, W.F., Costa, L.G., Richter, R.J., et al., 2000. Catalytic efficiency determines the in vivo efficacy of PON1 for detoxifying organophosphates. Pharmacogenetics 10, 767–779.

Lill, C.M., 2016. Genetics of Parkinson's disease. Mol. Cell. Probes 30, 386–396.

Lin, X., Cook, T.J., Zabetian, C.P., et al., 2012. DJ-1 isoforms in whole blood as potential biomarkers of Parkinson disease. Sci. Rep. 2, 954.

Ling, H., Neal, J.W., Revesz, T., 2017. Evolving concepts of chronic traumatic encephalopathy as a neuropathological entity. Neuropathol. Appl. Neurobiol. 43, 467–476.

Lockridge, O., 1990. Genetic variants of human serum cholinesterase influence metabolism of the muscle relaxant succinylcholine. Pharmacol. Ther. 47, 35–60.

Lockridge, O., Masson, P., 2000. Pesticide and susceptible populations: people with butyrylcholinesterase genetic variants may be at risk. Neurotoxicology 21, 113–126.

Lockridge, O., Norgren, R.B., Johnson, R.C., Blake, T.A., 2016. Naturally occurring genetic variants of human acetylcholinesterase and butyrylcholinesterase and their potential impact on the risk of toxicity from cholinesterase inhibitors. Chem. Res. Toxicol. 29, 1381–1392.

Lotti, M., 2001. Clinical toxicology of anticholinesterases in humans. In: Krieger, R. (Ed.), Handbook of Pesticide Toxicology. Academic Press, San Diego, pp. 1043–1085.

Lotti, M., Becker, C.E., Aminoff, M.J., et al., 1983. Occupational exposure to the cotton defoliants DEF and merphos: a rational approach to monitoring organophosphorus-induced delayed neurotoxicity. J. Occup. Med. 25, 517–522.

Lotti, M., Moretto, A., Zoppellani, R., et al., 1986. Inhibition of lymphocytic neuropathy target esterase predicts the development of organophosphate-induced delayed neuropathy. Arch. Toxicol. 59, 176–179.

Lotti, M., Moretto, A., 2005. Organophosphate-induced delayed polyneuropathy. Toxicol. Rev. 24, 37–49.

Lu, C., Bravo, R., Caltabiano, L.M., et al., 2005. The presence of dialkylphosphates in fresh fruit juices: implications for organophosphorus pesticide exposure and risk assessments. J. Toxicol. Environ. Health Part A 68, 209–227.

Makhaeva, G.F., Malygin, V.V., Strakhova, N.N., et al., 2007. Biosensor assay of neuropathy target esterase in whole blood as a new approach to OPIDN risk assessment: review of progress. Hum. Exp. Toxicol. 26, 273–282.

Makris, S.L., Raffaele, K., Allen, S., et al., 2009. A retrospective performance assessment of the developmental neurotoxicity study in support of OECD test guideline 426. Environ. Health Perspect. 117, 17–25.

Manzo, L., Arigas, F., Martinez, E., et al., 1996. Biochemical markers of neurotoxicity. A review of mechanistic studies and applications. Hum. Exp. Toxicol. 15 (Suppl. 1), S20–S35.

Marchi, N., Cavaglia, M., Fazio, V., et al., 2004. Peripheral markers of blood-barrier damage. Clin. Chim. Acta 342, 1–12.

Maroni, M., Bleecker, M.L., 1986. Neuropathy target esterase in human lymphocytes and platelets. J. Appl. Toxicol. 6, 1–7.

Maroni, M., Colosio, C., Ferioli, A., Faut, A., 2000. Biological monitoring of pesticide exposure: a review. Toxicology 143, 1–118.

Marsillach, J., Richter, R.J., Kim, J.H., et al., 2011. Biomarkers of organophosphorus (OP) exposures in humans. Neurotoxicology 32, 656–660.

Marsillach, J., Costa, L.G., Furlong, C.E., 2013. Protein adducts as biomarkers of exposure to organophosphorus compounds. Toxicology 307, 46–54.

Mayer, A.R., Quinn, D.K., Master, C.L., 2017. The spectrum of mild traumatic brain injury. A review. Neurology 89, 1–10.

McDonald, B.E., Costa, L.G., Murphy, S.D., 1988. Spatial memory impairment and central muscarinic receptor loss following prolonged treatment with organophosphates. Toxicol. Lett. 40, 47–56.

Meier, T.B., Nelson, L.D., Huber, D.L., et al., 2017. Prospective assessment of acute blood markers of brain injury in sport-related concussion. J. Neurotrauma 34, 3134–3142.

Mitra, B., Rau, T.F., Surendran, N., et al., 2017. Plasma micro-RNA biomarkers for diagnosis and prognosis after traumatic brain injury: a pilot study. J. Clin. Neurosci. 38, 37–42.

Nishimura, Y., Murakami, S., Ashikawa, Y., et al., 2015. Zebrafish as a systems toxicology model for developmental neurotoxicity testing. Congenit. Anom. 55, 1–16.

NRC (National Research Council), 1987. Biological markers in environmental health research. Environ. Health Perspect. 74, 3–9.

O'Callaghan, J.P., Sriram, K., 2005. Glial fibrillary acidic protein and related glial proteins as biomarkers of neurotoxicity. Expert Opin. Drug Saf. 4, 433–442.

OECD (Organization for Economic Co-operation and Development), 1997. Test Guideline 424. OECD Guideline for Testing of Chemicals. Neurotoxicity Study in Rodents. OECD, Paris.

OECD (Organization for Economic Co-operation and Development), 2007. Test Guideline 426. OECD Guideline for Testing of Chemicals. Developmental Neurotoxicity Study. OECD, Paris.

Ogata, K., Sumida, K., Miyata, K., et al., 2015. Circulating miR-9* and miR-384-5p as potential indicators for trimethyltin-induced neurotoxicity. Toxicol. Pathol. 43, 198–208.

Padilla, S., 1995. Regulatory and research issues related to cholinesterase inhibition. Toxicology 102, 215–220.

Padilla, S., Wilson, V.Z., Bushnell, P.J., 1994. Studies on the correlation between blood cholinesterase inhibition and "target tissue" inhibition in pesticide-treated rats. Toxicology 92, 11–25.

Papa, L., 2016. Potential blood-based biomarkers for concussion. Sport Med. Arthrosc. Rev. 24, 108–115.

Papa, L., Ramia, M.M., Edwards, D., et al., 2015. Systematic review of clinical studies examining biomarkers of brain injury in athletes after sports-related concussions. J. Neurotrauma 32, 661–673.

Petzold, A., 2005. Neurofilament phosphor-forms: surrogate markers for axonal injury, degeneration and loss. J. Neurol. Sci. 233, 183–198.

Quistad, G.B., Klintenberg, R., Casida, J.E., 2005. Blood acylpeptide hydrolase activity is a sensitive marker for exposure to some organophosphate toxicants. Toxicol. Sci. 86, 291–299.

Rajdev, S., Sharp, F.R., 2000. Stress proteins as molecular markers of neurotoxicity. Toxicol. Pathol. 28, 105–112.

Richter, R.J., Jampsa, R.L., Jarvik, G.P., et al., 2004. Determination of paraoxonase 1 status and genotypes at specific polymorphic sites. In: Maines, M., Costa, L.G., Reed, D.J., Hodgson, E. (Eds.), Current Protocols in Toxicology. John Wiley & Sons, New York, pp. 4.12.1–4.12.19.

Roberts, R.A., Aschner, M., Calligaro, D., et al., 2015. Translational biomarkers of neurotoxicity: a health and environmental sciences institute perspective on the way forward. Toxicol. Sci. 148, 332–340.

Rothermundt, M., Peters, M., Prehn, J.H.M., Arolt, V., 2003. S100B in brain damage and neurodegeneration. Microsc. Res. Tech. 60, 614–632.

Sandström, J., Eggerman, E., Charvet, I., et al., 2017. Development and characterization of human embryonic stem cell-derived 3D neural tissue model for neurotoxicity testing. Toxicol. In Vitro 38, 124–135.
Schwab, B.W., Richardson, R.J., 1986. Lymphocyte and brain neurotoxic esterase: dose and time dependence of inhibition in the hen examined with three organophosphorus esters. Toxicol. Appl. Pharmacol. 83, 1–9.
Seco, M., Edelman, J.J.B., Wilson, M.K., et al., 2012. Serum biomarkers of neurologic injury in cardiac operations. Ann. Thorac. Surg. 94, 1026–1033.
Shahim, P., Tegner, Y., Wildon, D.H., et al., 2014. Blood biomarkers for brain injury in concussed professional ice hockey players. JAMA Neurol. 71, 684–692.
Shahim, P., Zettenberg, H., Tegner, Y., et al., 2017. Serum neurofilament light as a biomarker for mild traumatic brain injury in contact sports. Neurology 88, 1788–1794.
Shih, D.M., Gu, L., Xia, Y.R., et al., 1998. Mice lacking serum paraoxonase are susceptible to organophosphate toxicity and atherosclerosis. Nature 394, 284–287.
Singh, M., Khan, A.J., Shah, P.P., et al., 2008. Polymorphism in environment responsive genes and association with Parkinson disease. Mol. Cell. Biochem. 312, 131–138.
Spencer, P.S., Schaumburg, H.H., Ludolph, A.C. (Eds.), 2000. Experimental and Clinical Neurotoxicology. Oxford University Press, Oxford, pp. 1–1310.
Steenerson, K., Starling, A.J., 2017. Pathophysiology of sport-related concussion. Neurol. Clin. 35, 403–408.
Stevens, R.C., Suzuki, S.M., Cole, T.B., et al., 2008. Engineered recombinant human paraoxonase 1 (rHuPON1) purified from *Escherichia coli* protects against organophosphate poisoning. Proc. Natl. Acad. Sci. U.S.A. 105, 12780–12784.
Tal, T.L., Tanguay, R.L., 2012. Non-coding RNAs-novel targets in neurotoxicity. Neurotoxicology 33, 530–544.
Thompson, C.M., Prins, J.M., George, K.M., 2010. Mass spectrometric analyses of organophosphate insecticide oxon protein adducts. Environ. Health Perspect. 118, 11–19.
Timchalk, C., Kousba, A., Poet, T.S., 2002. Monte Carlo analysis of the human chlorpyrifos oxonase (PON1) polymorphism using a physiologically based pharmacokinetic and pharmacodynamic (PBPK/PD) model. Toxicol. Lett. 135, 51–59.
USEPA (United States Environmental Protection Agency), 1998a. Health Effects Test Guidelines. OPPTS 870.6200. Neurotoxicity Screening Battery. USEPA, Washington, DC.
USEPA (United States Environmental Protection Agency), 1998b. Health Effects Test Guidelines. OPPTS 870.6300. Developmental Neurotoxicity Study. USEPA, Washington, DC.
USEPA (U.S. Environmental Protection Agency), 2000. The Use of Data on Cholinesterase Inhibition for Risk Assessments of Organophosphorus and Carbamate Pesticides. USEPA, Washington, DC, pp. 1–50. Accessible online at: http://www.epa.gov/pesticides/trac/science/cholin.pdf.
Valentine, W.M., Graham, D.G., Anthony, D.C., 1993. Covalent cross-linking of erythrocyte spectrin by carbon disulfide in vivo. Toxicol. Appl. Pharmacol. 121, 71–77.
Vassallo, A., Chiappalone, M., de Camargo Lopes, R., et al., 2017. A multi-laboratory evaluation of microelectrode array-based measurements of neural network activity for acute neurotoxicity testing. Neurotoxicology 60, 280–292.
Xu, J.J., Henstock, P.V., Dunn, M.C., et al., 2008. Cellular imaging predictions of clinical drug-induced liver injury. Toxicol. Sci. 105, 97–105.
Zettenberg, H., Blennow, K., 2016. Fluid biomarkers for mild traumatic brain injury and related conditions. Nat. Rev. Neurol. 12, 563–574.

CHAPTER

11

Peripheral Nervous System Toxicity Biomarkers

Tirupapuliyur V. Damodaran

Department of Biological and Biomedical Sciences, North Carolina Central University, Durham, NC, United States

INTRODUCTION

The great degree of complexity involved in the structural and functional aspects of vertebrate nervous system in general and the advanced capabilities of human central nervous system (CNS) and peripheral nervous system (PNS) in particular provide huge number of potential biomarkers that may be very useful in many branches of health care industry. Both CNS and PNS activities are the basis for every response involving thoughts, actions, and emotions. Along with the endocrine system, the nervous system plays a major role in the body's homeostasis in human and other vertebrates. Because of the need to be rapid and immediate, its cells communicate by electrical and chemical signals that are highly specific. The dynamic nature of responses of various organisms to their extrinsic (external environment) and other intrinsic (homeostatic) factors to ensure their survival depends on the structural and functional integrity of the nervous system. The unique status of this organ system also makes it relatively more susceptible to toxic mechanisms arising from various day-to-day life activities. These neurotoxic mechanisms can be of short-term or long-term duration, as well as either acute or chronic. The drastic outcome from these neurotoxic mechanisms can be immediate or delayed. These mechanisms may have general effects or may affect very specific parts of this multiunit hierarchical structure. Some of these neurotoxic mechanisms may get resolved quickly by the body's repair mechanisms, whereas other types may need rigorous therapeutic protocols. Because of the nervous system's omnipresent role in many other organ systems, the toxic mechanisms that affect the nervous system may have direct or indirect, mild, or profound effects on other organ systems. Thus, accurate and prompt identification of toxic mechanisms responsible for any neurological disease conditions due to known and unknown toxic mechanisms becomes imperative for effective treatment and management protocols. Although the clinical presentation of the disease condition may be useful for preliminary diagnosis for symptomatic treatment approaches, permanent cure may be possible only after disease-specific treatments. The developing nervous system may be more or less susceptible to neurotoxic insult depending on the stage of development. Biomarkers used in adults are often applicable for use during development, but developmental stage-specific and aging-related differences may be important (Slikker and Bowyer, 2005). Although both inherited and acquired PNS disorders may have common pathways that can be identified by biomarker identification and validation, there may be unknown layers of complexity that need to be understood for better treatment protocols.

Thus, PNS biomarkers (measurable end points by cell, molecular, and clinical parameters) are very important for various applications connected with the accurate diagnosis, treatment, and management of neurological and other organ systems–related disorders. These biomarkers are also useful for assessing clinical responses, identification of risks, selection of doses, and other aspects in drug development and evaluation. Positive correlation between symptoms at all levels (molecular, cellular, and phenotypical) of abnormal and/or normal presentations (both qualitatively and quantitatively) of chemicals, metabolites, and other clinical and measurable morphological features can help define the biomarker of choice. As per the National Institute of Health (NIH) biomarker working group, a biomarker is a characteristic that can be objectively measured as an indicator of normal biological processes, pathogenic processes, or a pharmacological response to a therapeutic intervention. The five phases proposed by the NIH in the evaluation of biomarkers are in sequence: (1) discovery using genomics or proteomics, (2) developing an assay that is portable and reproducible,

Biomarkers in Toxicology, Second Edition
https://doi.org/10.1016/B978-0-12-814655-2.00011-6

(3) measuring sensitivity and specificity, (4) affirming the measurement in a large cohort, and (5) determining the risks and benefits of using the new diagnostic biomarker (Raad et al., 2014).

Hence, biomarkers can be effectively used to identify (1) whether the exposure or change has occurred, (2) the route of exposure or location of the change, (3) the pathway of exposure, (4) resulting short-term or long-term effects, and (5) outcomes of therapeutic interventions to bring the organism to its original healthy status. Biomarkers can be used to identify organ system and organ-specific changes at all levels. In this chapter, emerging new developments in the broad field of PNS-related biomarkers in health and disease status, with major emphasis on humans and related animal models, are presented.

TYPES OF BIOMARKERS OF THE PERIPHERAL NERVOUS SYSTEM

The complexity of PNS structure and function makes it a daunting task to enlist the potential targets, mechanisms of neurotoxicity, valid biomarkers, and candidate biomarkers from many biological and clinical end points. The complexity and extensive nature of the PNS makes it comparatively more susceptible and at the same time more resilient for recovery from injuries (compared to the CNS). In general, the biomarkers can be identified by carefully studying various aspects of PNS structure and function.

PERIPHERAL NERVOUS SYSTEM BIOMARKER METHODOLOGY

Applying appropriate methodology is the major factor in the rapid identification and validation of biomarkers of PNS neurotoxicity. To make provisional diagnosis of the nature of toxic mechanisms in any PNS toxicity–related clinical presentation, universal methods to collect clinical data using standard methods of neurology clinic settings are essential. Biomedical informatics with its ever-increasing branches of special categories has become an important tool for collection, storage, retrieval, and dissemination of relevant information.

The following are the commonly employed methodologies of such categories, providing valuable clinical biomarker data sets (Fig. 11.1).

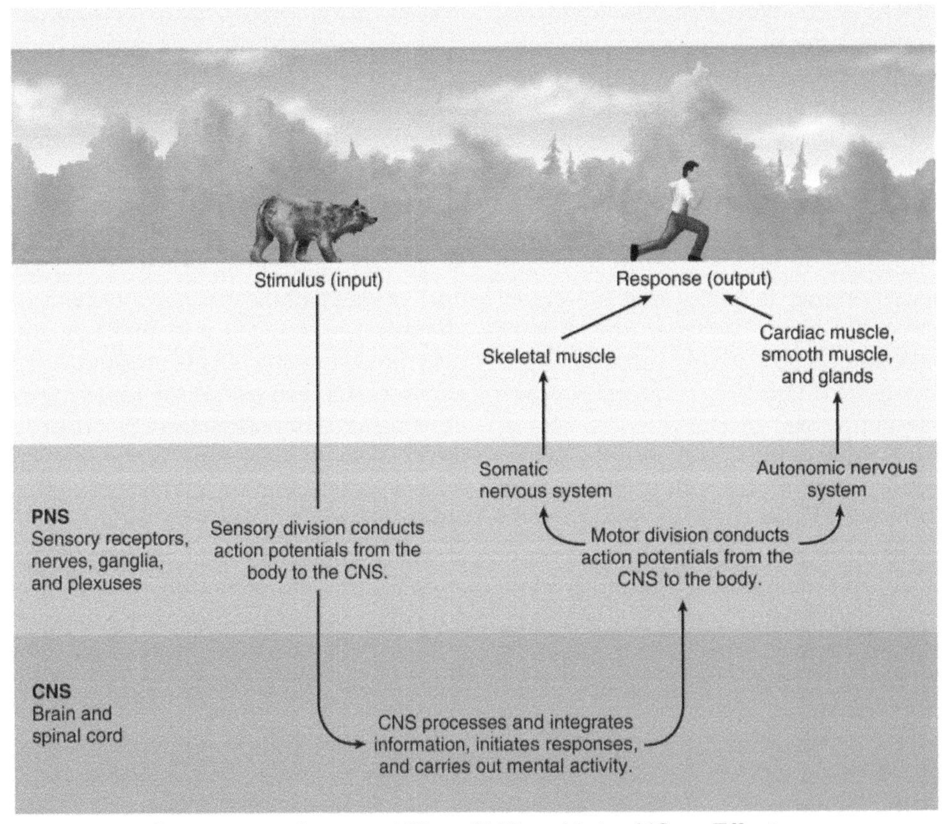

FIGURE 11.1 Organization of the nervous system: The sensory division of the peripheral nervous system (PNS) detects stimuli and conducts action potentials to the central nervous system (CNS). The CNS interprets incoming action potentials and initiates action potentials that are conducted through the motor division to produce a response. The motor division is divided into the somatic nervous system and the autonomic nervous system.

NEUROLOGICAL AND PHYSICAL EXAMINATION

Protocol

In humans, standard neurological physical examinations (NPxs) (Starks et al., 2012) include the following: Assessments of vibration perception and proprioception are generally performed on the great toes, bilaterally. Achilles deep tendon reflexes are examined bilaterally, and Romberg test performance, tandem gait, and postural tremor are routinely assessed. Clinical examination results are recorded as normal, equivocal, or abnormal. For all tests performed bilaterally (ankle reflex, toe proprioception, and toe vibration), examination results are classified as "abnormal" if ratings are abnormal or equivocal bilaterally, abnormal unilaterally and equivocal on the contralateral side, or abnormal unilaterally and missing (because of injury/amputation) on the contralateral side. For tests without laterality (postural tremor, Romberg test, and tandem gait), combined abnormal and equivocal results are used to create a dichotomized variable (normal vs. not normal) for each outcome (Starks et al., 2012). Electrophysiological measures, hand strength, sway speed, and vibrotactile threshold are also employed (Starks et al., 2012).

SAMPLE REQUIREMENTS FOR BIOMARKER TESTING

Adequate quantity of tissue (by biopsy) of investigation can be obtained for various biomarker testing procedures. The tissues include blood, cerebrospinal fluid (CSF), nerve tissue, and target (end organ) tissue. CT-guided needle biopsy and other standard blood collection procedures are used for tissue sampling. A nerve biopsy may help distinguish between demyelination (damage to the myelin sheath covering the nerve) and axon degeneration (destruction of the axonal portion of the nerve cell). It may also help to identify an inflammatory neuropathy or confirm specific diagnoses. A skin biopsy is helpful to differentiate certain disorders that might affect the small nerve fibers, as may be the case with painful sensory axonal neuropathies. A muscle or other tissue biopsy is used to diagnose and identify damage caused to muscles and organs as a result of various disorders.

Biomedical Imaging and Informatics

Various images spanning the scale from microscopic and molecular to whole body visualization, encompassing many areas of clinical medicine, are important for selecting, validating, and biomonitoring PNS disorders. Recent years have witnessed impressive advances in the use of magnetic resonance imaging (MRI) and single-photon emission CT for the assessment of patients with PNS and CNS disorders (Poewe et al., 2017). MRI is an imaging technique used to produce high-quality images of the inside of the human body. A scanner emits a strong magnetic field inside the brain (or elsewhere in the body) and produces signals that are analyzed by a computer to produce detailed images. Complementary to the clinical evaluation, conventional MRI provides crucial pieces of information for the diagnosis of acquired and hereditary PNS and CNS disorders (Filippi and Agosta, 2010). Newer quantitative magnetic resonance (MR)–based techniques, such as magnetization transfer MRI, diffusion tensor MRI, proton MR spectroscopy, and functional MRI, are contributing to elucidate the mechanisms that underlie injury, repair, and functional adaptation in patients with nervous system disorders (Filippi and Agosta, 2010). Diffusion weighted imaging (DWI) and diffusion tensor imaging (DTI) are functional MRI techniques that are able to evaluate and quantify the movement of water molecules within biological structures. In the field of trauma and peripheral nerve or plexus injury, several derived parameters from DWI and DTI studies, such as apparent diffusion coefficient or fractional anisotropy among others, can be used as potential biomarkers of neural damage, providing information about fiber organization, axonal flow, or myelin integrity (Noguerol et al., 2017).

Electrodiagnostic Tests

Electrodiagnostic tests measure the electrical activity of muscles and nerves. By measuring the electrical activity, one can determine if there is nerve damage, the extent of the damage, and potentially the cause of the damage. Frequently, noninvasive neurological evaluations such as electromyography and nerve conduction velocity (NCV) testing are used. Quantitative sensory testing and autonomic testing are also commonly used.

Blood Tests

Blood tests are commonly employed to check for various factors that affect the nervous system directly or indirectly (vitamin deficiencies, toxic elements, evidence of an abnormal immune response). Depending on the individual situation, certain laboratory tests to identify potentially treatable causes for neuropathy by the following tests are employed: (1) Antibodies to nerve components (e.g., anti-MAG antibody); (2) antibodies to cytoskeletal components (NF; glial fibrillary acidic protein (GFAP); and tubulin, etc.); (3) vitamin B12 and folate levels; (4) thyroid, liver, and kidney functions; (5) vasculitis evaluation; (6) oral glucose tolerance test; (7)

antibodies related to celiac disease; (8) Lyme disease; (9) HIV/AIDS; and (10) hepatitis C and B.

Testing for Toxins

Toxins, poisons, and chemicals can cause peripheral neuropathy. This can happen through drug or chemical abuse or through exposure to industrial chemicals in the workplace or in the environment (after either limited or long-term exposure). Common causes include exposure to lead, mercury, arsenic, and thallium. Some organic insecticides and solvents can result in neuropathies. Sniffing glue or other toxic compounds can also cause peripheral neuropathy. Certain herbal medicines, especially those that are particularly rich in mercury and arsenic, can lead to peripheral neuropathy. Hence blood profiling for the presence of parent or metabolic products or metabolites of the drug pathways is commonly used to identify the basis of toxicity (Fig. 11.2).

Molecular Diagnostics

Molecular profiling techniques such as (1) sequencing, (2) quantitative real-time polymerase chain reaction (PCR), (3) amplification of refractory mutation system (allele-specific PCR), (4) fragment length analysis (to detect any deletions within the coding portion of the gene), (5) high-performance liquid chromatography (to distinguish between wild type and mutated

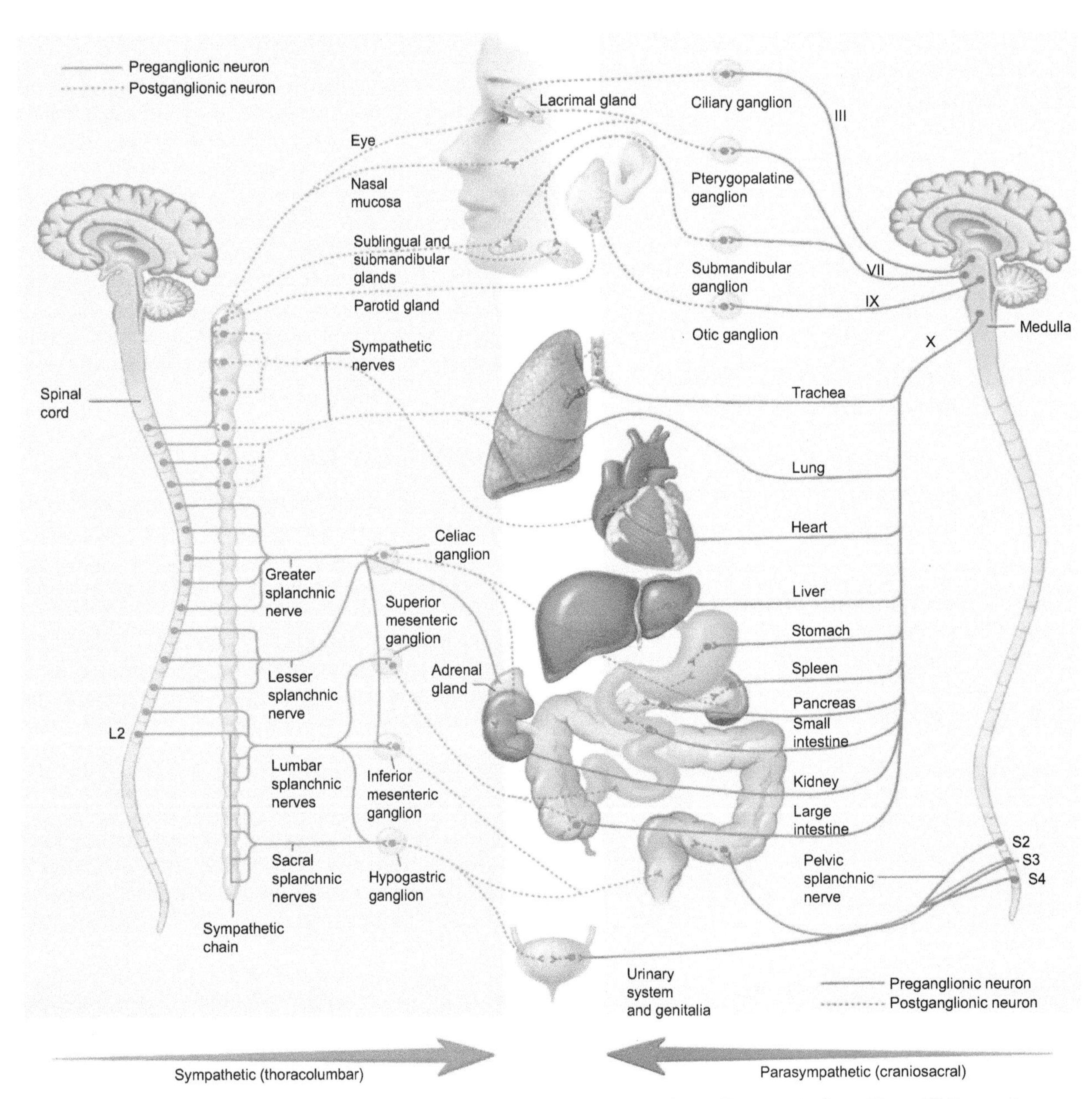

FIGURE 11.2 Innervation of organs by the autonomic nervous system. Preganglionic fibers are indicated by *solid lines* and postganglionic fibers are indicated by *dotted lines*.

DNA strands using heteroduplex formation), and (6) immunohistochemistry and fluorescence/chromogenic in situ hybridization are used to detect any defects at gene sequence level.

Molecular, Cellular, and Histological Techniques

Analysis of patient samples from various medical clinics and from animal model systems using various combinations of molecular, cellular, and histological techniques has become part of the routine laboratory protocols. Details of nucleic acid extractions (DNA and RNA), northern blotting, semiquantitative real-time PCR, and real-time PCR protocols have been described in detail in several publications (Damodaran, 2009; Damodaran et al., 2011). Similarly, methods of protein extraction and analysis using western blotting have also been described in several publications (Damodaran et al., 2009, 2011). Classical histopathological analysis and other nervous system–related structure-specific staining techniques were also described in detail in recent articles (Damodaran, 2009; Damodaran et al., 2011).

Toxicogenomics and Proteomics Approaches

RNA purity verification by Agilent analysis, microarray chip hybridization, data capturing, data analysis, Tree view analysis and clustering (Fig. 11.3), data mining for classifications, principal component analysis of gene expression (Fig. 11.4), and gene validation approaches are described in detail in many publications (Damodaran et al., 2006a,b). Functional analysis of the proteins expressed in a specific tissue and/or stage of toxic exposure is an essential step toward understanding biological processes. The proteins have been shown to undergo posttranslational modifications such as phosphorylation, glycosylation, protein processing, and other proteolytic cleavage. Both two-dimensional gels and automated mass spectrometry–based approaches for proteome and phosphorylation site analysis, as well as proteome data analysis, have been described in detail from several publications. Binding assays to unravel the mechanics of protein–protein and protein–enzyme interactions and the methodology for studying axonal transport have been described in detail in numerous articles.

Metabonomics

Neurotoxicity occurs after the neurotoxicants enter into the blood circulation, travel to sites of vulnerability, and gain access to tissues. Metabolic transformation is a key mechanism by which the toxicants become either more toxic or less toxic. These processes predominantly produce more polar metabolites that facilitate the increased excretion (Abou-Donia and Nomeir, 1986). Thus, the process called toxicokinetics encompassing absorption, tissue distribution, storage, biotransformation, and elimination plays a major role in PNS neurotoxicity. Metabolic status during these stages can be assessed by measuring the quantity and type of metabolites. Detailed information about the toxicological outcome of toxicant exposure is required to make decisions about the long-term therapy protocols. Global profiling techniques are rapid in gathering data on the impact made on transcripts, proteins, and metabolites within the cell, tissue, organ, and organ systems of the

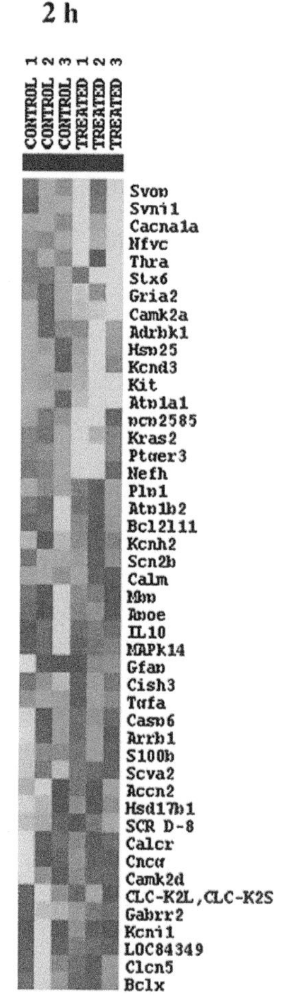

FIGURE 11.3 Tree view analysis of global gene expression at the 2 h time point is presented here. Hierarchical clustering was applied through the Cluster software program, and the results were visualized in Tree view (check http://www.microarrays.org/software for computational details). The expression level of each gene relative to the median expression level across all samples was represented by color: one representing log expression ratios of overexpressed genes and the other representing log expression ratios of underexpressed genes. The remarkable homogeneity of the coloring pattern validates the reproducibility of the replicate data from the control and treated groups (Damodaran et al., 2006b).

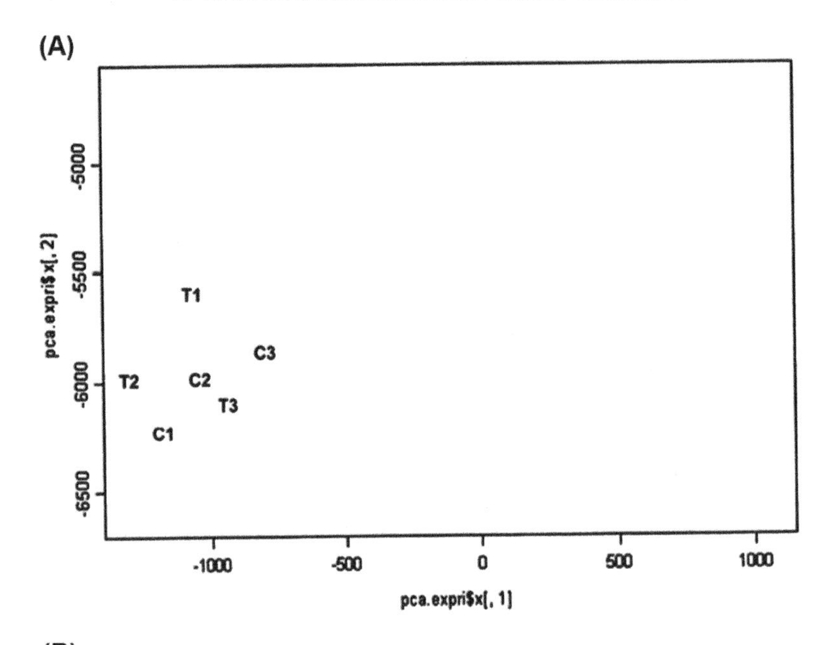

(B)

In a 3-D plot, we can see C1-3 is separating from T1-3.

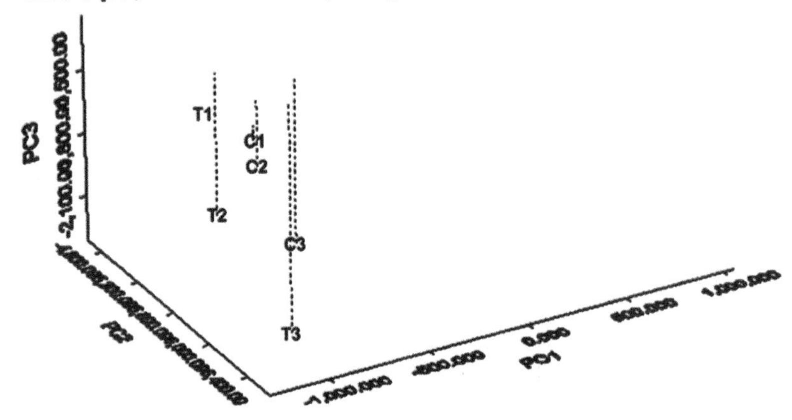

Clustering and Visual Representation of the Expression Data

FIGURE 11.4 Principal component analysis (PCA) of gene expression was prepared on a three-dimensional platform and is presented here as a graph providing a global view of the differences among the control (C1, C2, and C3) and treatment (T1, T2, and T3) groups (11.4 A). There is a clear separation among the controls and treatment groups (11.4 B), indicating differences in their total gene expression profiles (Damodaran et al., 2006b).

living organism. Metabonomics/metabolomics is one such approach that can provide a global profile of metabolites (Griffin et al., 2006). High-resolution nuclear magnetic resonance along with statistical pattern recognition analysis can provide valuable data, and detailed stepwise protocols are provided in several publications.

BIOMARKERS OF TRAUMATIC BRAIN INJURY AFFECTING PERIPHERAL NERVOUS SYSTEM

Traumatic brain injury (TBI) is a leading cause of death, disability, and resource consumption per year. There are two kinds of brain injury in TBI such as primary and secondary injuries. Secondary injury occurs over a period of hours or days following the initial trauma and results from the activation of different pathways such as inflammation, coagulation, oxidation, and apoptosis that may involve PNS (Lorente, 2015). Primary injury refers to the initial physical forces applied to the brain at the moment of impact. The secondary PNS injuries have been shown to induce the levels of some circulating biomarkers such as SP, sCD40L, TIMP-1, MSA, and CK-18 fragmented, in patients with TBI (Lorente, 2015).

Biomarkers of Nerve Tissue Injury in the Peripheral Nervous System

Because of the higher abundance of Schwann cells in the PNS, regeneration after an injury is much faster and with a higher rate of nerve repair. The proximity of the cut ends of the nerve after an injury decides the higher likelihood of repairing of the injured nerve in the PNS. Clearing up irregular segments and other pieces of debris at the site of injury by macrophages facilitates the new sprout of synthesis of essential proteins and other cytoskeletal structures at the site of injury. Besides Schwann cells, satellite glial cells also undergo mitosis and other changes in the PNS. These cell types have been implicated in the genesis and maintenance of pain in the injured nerves of the PNS (Jasmin et al., 2010). Thus the coordinated functions of several important molecules in various cell types are required for this repair pathway. Table 11.1 provides a list of the biomarker genes/proteins identified by several studies on Schwann cells and myelination.

Biomarkers of Changes in Electrophysiology

Dai et al. (2012) showed that the compound action potential durations (CAPDs), compound muscle action potential amplitudes (CAPAs), and conduction velocities of sciatic-tibial nerve (NCVs) exhibited progressively abnormal changes over time in colistin-treated female mice. These changes were in correlation with ultrastructural changes. These results from electrophysiological methods provide valuable data about the status of the PNS nerves after acute or chronic neurotoxicity.

Ion Channels as Biomarkers

Peripheral sensory neurons are adapted to recognize danger to the organism by virtue of their sensitivity to intense mechanical, thermal, and irritant chemical stimuli. Although CNS-expressed sodium channels also occur in peripheral nerves, several additional channels occur mainly in dorsal root ganglion dorsal root ganglia (DRG) cells. Differential roles for Nav1.7, Nav1.8, and Nav1.9 in sensory transduction and nociception and transient receptor potential (TRP) ion channels conducting nonselective entry of cations on activation by various noxious stimuli were reported. TRPV1 is activated by high temperatures, low pH, and capsaicin, the vallinoid irritant component of chili peppers. TRPA1 mediates the detection of reactive chemicals including environmental irritants such as tear gas and industrial isothiocyanates, but more importantly, it is also activated during tissue injury by endogenous molecular signals including 4-hydroxynonenal and prostaglandins (Chiu et al., 2013). An extensive list of nervous system–related ion channel genes used for expression profiling (SABiosciences, 2013) is provided in Table 11.2. It is noteworthy to mention here that the highest number of altered gene expression and highest level of up- or downregulation were recorded for ion channel genes in global gene expression studies on sarin exposure at early time points (Damodaran et al., 2006a,b).

TABLE 11.1 Protein Markers of Schwann Cells

Myelin-forming Schwann cell: P0, P2, myelin basic protein, CNPase, proteolipid protein
Non–myelin-forming Schwann cell: common markers: NCAM, GFAP, NGF receptor, L1, A5E3, Ran-2, S100, vimentin, laminin, galactocerebroside, seminolipid
Myelin-associated glycoprotein, proteins/enzymes involved in myelination, demyelination, and other functions of myelin: Peripheral myelin protein 22, connexin 32 (Cx32/GJB1), periaxin, GDAP1 (ganglioside-induced differentiation-associated protein 1), NF-L (neurofilament light chain), dynamin-2 (DNM2), GARS (glycyl-tRNA synthetase), myelin-specific enzyme 2′,3′-cyclic nucleotide 3′-phosphohydrolase, transcriptional regulators of myelin, EGR2 (early growth response 2), Sox10 (sry-related high-mobility group box-containing gene 10), other PNS proteins, gamma-enolase, creatine kinase-B, beta-S100 protein

NGF, nerve growth factor; *PNS*, peripheral nervous system.

TABLE 11.2 Nervous System–Related Channels as Biomarkers

CALCIUM CHANNELS
Voltage-gated: CACNA1A, CACNA1B, CACNA1C, CACNA1D, CACNA1G, CACNA1I, CACNB1, CACNB2, CACNB3, CACNG2, CACNG4
Ligand-gated: RYR3
Transient receptor potential (TRP) channels: TRPA1, TRPC1, TRPC3, TRPC6, TRPM1, TRPM2, TRPM6, TRPM8, TRPV1, TRPV2, TRPV3, TRPV4
POTASSIUM CHANNELS
Delayed rectifier: KCNA1, KCNA2, KCNA5 (KV1.5), KCNA6, KCNB1, KCNB2, KCNH1, KCNH2, KCNQ1, KCNQ2, KCNQ3
Inward rectifier: KCNH2, KCNH6, KCNH7, KCNJ1, KCNJ11, KCNJ12, KCNJ13, KCNJ14, KCNJ15, KCNJ16, KCNJ2, KCNJ3, KCNJ4, KCNJ5, KCNJ6, KCNJ9, KCNK1
Calcium-activated: KCNMA1, KCNMB4, KCNN1, KCNN2, KCNN3
Other voltage-gated potassium channels: HCN1, HCN2, KCNC1, KCNC2, KCND2, KCND3, KCNH3
Modifier subunits: KCNAB1, KCNAB2, KCNAB3, KCNS1
SODIUM CHANNELS
Amiloride sensitive: ACCN1, ACCN2, ACCN3
Voltage-gated: SCN10A, SCN11A, SCN1A, SCN1B, SCN2A, SCN2B, SCN3A, SCN8A, SCN9A
Chloride channels: BEST1, CLCN2, CLCN3, CLCN7
Sodium/chloride transport: SLC12A5

SCN, suprachiasmatic nucleus.

Biomarkers of Axonal Injury

Axon degeneration is a hallmark consequence of chemical neurotoxicant exposure (e.g., acrylamide), mechanical trauma (e.g., nerve transection, spinal cord contusion), deficient perfusion (e.g., ischemia, hypoxia), and inherited neuropathies (e.g., infantile neuroaxonal dystrophy). Regardless of the initiating event, degeneration in the PNS and CNS progresses according to a characteristic sequence of morphological changes. These shared neuropathologic features suggest that subsequent degeneration, although induced by different injury modalities, might evolve via a common mechanism. Studies indicate that Ca^{2+} accumulation in injured axons has significant neuropathic implications and is a potentially unifying mechanistic event. It was proposed that diverse injury processes can lead to axon degeneration by an increase in intraaxonal Na^{+} in conjunction with a loss of K^{+} and axolemmal depolarization. A spontaneous mutation called WldS (slow Wallerian degeneration) showed us that Wallerian degeneration is not a passive disintegration of axon, when the axon is severed from the cell body, but an active process mediated by Nmnat protein (Mack et al., 2001). More recently, using a powerful forward genetics screen, a new pathway that involves Sarm protein has been found to prevent axonal degeneration after traumatic injury (Osterloh et al., 2012).

Biomarkers of Altered Axonal Transport

Axon transport mechanisms play a major role in transporting nutrients, organelles, and other molecules toward the presynaptic terminals by a process called anterograde transport, whereas the retrograde transport is a process by which damaged organelles and recycled plasma membrane (packed in endocytotic vesicles) are transported back to the neuron cell body. There are many major human neurodegenerative pathological disorders, such as Alzheimer's disease, Parkinson's disease (PD), and amyotrophic lateral sclerosis, which display axonal pathologies including abnormal accumulations of proteins and organelles, highlighting that disruption of axonal transport is an early and perhaps causative event. Disruption to axonal transport can occur via a number of routes such as damage to (1) molecular motors, (2) microtubules, (3) cargoes (such as inhibiting their attachment to motors), and (4) mitochondria, which supply energy for molecular motors (De Vos et al., 2008). Similarly, defective axonal transport has been shown to be involved in organophosphate-induced delayed neurotoxicity (OPIDN) (Damodaran et al., 2011). At cellular and tissue level axonal degeneration accompanied by synergistic neuronal and glial cell death pathways (e.g., apoptosis, necrosis, autophagy) have been implicated as major factors in neurotoxicity (Damodaran et al., 2011).

Biomarkers of Axonal Regeneration

Successful axonal regeneration is controlled by both intrinsic factors and extrinsic factors such as support or inhibition of regeneration by glial cells, scar formation, and vascular supply (Hoke et al., 2013). The role of mTOR and STAT3 pathways in modulating the intrinsic capacity of neurons to regenerate has been proposed (Hoke et al., 2013). Deletion of the PTEN gene in retinal ganglion neurons was shown to activate the mTOR pathway and enhanced regeneration after optic nerve crush (Park et al., 2008). Manipulation of both STAT3 and mTOR pathways leads to a synergistic increase in axonal regeneration in the optic nerve crush model (Sun et al., 2011). In the adult animal, regeneration has to occur over very long distances. As the rate of axonal elongation is determined by the rate of "slow" axonal transport, this is a very slow process leading to chronic denervation changes in the Schwann cells in the distal nerve and target tissues such as muscle (Hoke et al., 2013). A potential approach to overcome this challenge is to "speed up" the rate of regeneration by overexpressing a heat shock protein (hsp27) that plays a key role in axon outgrowth in injured neurons. There are several genes (Table 11.3) identified as playing significant roles in neurogenesis and neurotrophic signaling in the PNS (SABiosciences, 2013).

Potential Biomarkers of Pain

Neuropathic pain is defined as pain initiated or caused by a primary lesion or dysfunction of the nervous system. Pain is the outcome of noxious environmental stimuli, mechanical tissue damage, and infectious or other pathological disease processes. This can be the result of (1) spinal cord injury, (2) peripheral nerve damage resulting from diabetes or other autoimmune diseases, (3) treatment with anticancer drugs that affect axon integrity, or (4) postherpetic neuralgia (Goins et al., 2012). There are two types of pain: (1) neuropathic pain and (2) inflammatory pain. Neuropathic pain often results from damage to the PNS, CNS, and peripheral tissue damage, whereas inflammation generally initiates inflammatory pain. Neuropathic and inflammatory pain both cause activation of nociceptors (damage-sensing neurons) that innervate the skin, muscle, and viscera and terminate in the laminae of the spinal cord dorsal horn. Nociceptors conduct information to the CNS via neurotransmission (action potentials) generated by a variety of ion channels and receptors (purinergic, opioid, and cannabinoid) leading to secondary activation of neuronal pathways. This stage is followed by synaptic transmission (via glutamate, serotonin, and dopamine systems). Inflammatory mediators released by immune cells and damaged neurons can modulate the transduction nociception process.

TABLE 11.3 Candidate Biomarker Genes Involved in Neurotrophic Signaling, Neuropeptides Functions, and Neurogenesis in the PNS

Neurotrophins and receptors: ADCYAP1R1, ARTN, BDNF, CD40 (TNFRSF5), CNTF, CNTFR, CRHBP, CRHR1, CRHR2, FRS2, FRS3, FUS, GDNF, GFRA1, GFRA2, GFRA3, GMFB, GMFG, MAGED1, MT3, NF1, NGF, NGFR, NGFRAP1, NR1I2, NRG1, NRG2, NTF3, NTF4, NTRK1, NTRK2, PSPN, PTGER2, TFG, TRO, VGF
NEUROPEPTIDES AND RECEPTORS
Neuropeptide hormone: CRH, NPFF, NPY, PNOC, UCN
Bombesin receptors: GRPR
Cholecystokinin receptors: CCKAR
Galanin receptors: GALR1, GALR2
Neuropeptide Y receptors: NPY1R, NPY2R, PPYR1
Neurotensin receptors: NTSR1
Tachykinin and receptors: TACR1
Other neuropeptides and receptors: NPFFR2 (GPR74), HCRT, MC2R
NEUROGENESIS
Peripheral nervous system development: GFRA3
Negative regulation of neurogenesis: MT3
Other neurogenesis related genes: ARTN, BDNF, CBLN1, CNTF, CNTFR, FGF2, GDNF, MEF2C, NELL1, NGFR, NRG1, NTF3, NTRK1, NTRK2, STAT3, TFG
Transmission of nerve impulse: CBLN1, CNTF, CRH, GALR1, GALR2, HCRT, NPFF, NPY, NTSR1, PNOC

ADC, apparent diffusion coefficient; *PNS*, peripheral nervous system.

TABLE 11.4 Potential Biomarker Genes Related to Pain and Related Disorders

CONDUCTION OF PAIN
Ion channels: TRPA1, TRPV1, TRPV3
Sodium channels: SCN10A, SCN11A, SCN3A, SCN9A, SLC6A2
Potassium channels: KCNIP3, KCNJ6, KCNQ2, KCNQ3
Purinergic receptors: ADORA1, P2RX3, P2RX4, P2RX7, P2RY1
Opioid receptors: OPRD1, OPRK1, OPRM1
Cannabinoid receptors: CNR1, CNR2
SYNAPTIC TRANSMISSION
Glutamate receptors: GRIN1, GRIN2B, GRM1, GRM5
Serotonin receptors: HTR1A, HTR2A
Calcium channel: CACNA1B
MODULATION OF PAIN RESPONSES
Eicosanoid metabolism: PLA2G1B, PTGER1, PTGER3, PTGER4, PTGES, PTGES2, PTGES3, PTGS1, PTGS2
Inflammation: ACE, ALOX5, BDKRB1, CALCA, CCK, CCKBR, CCL2, CCR2, CD200, CD4, CHRNA4, CSF1, CX3CR1, DBH, EDN1, EDNRA, FAAH, GCH1, IL-10, IL-18, IL-1α, IL-1β, IL-2, IL-6, ITGAM, ITGB2, MAPK1, MAPK14, MAPK3, MAPK8, PENK, PNOC, PROK2, TAC1, TACR1, TLR2, TLR4, TNF
Neurotransmitters: ADRB2, COMT, DBH, MAOB, PDYN, PENK, PNOC
Neurotrophins: BDNF, GDNF, NGF, NTRK1
Parkin complex: HSPA4 (HSP70), PARK7, STUB1
Parkin substrate: ATXN2, ATXN3, GPR37, SYT11
Cell adhesion: APC, APP, CDH8, FN1, NFASC, NRXN3, PTEN, TPBG
Ubiquitination: CDC27, CUL2, FBXO9, LRRK2, PAN2, PARK2, PINK1, SKP1, STUB1, UBB, UBA1, UBE2I, UBE2K, UBE2L3, UCHL1, USP34
Inflammation: FN1, PRDX2, YWHAZ
APOPTOSIS
Proapoptosis: APC, APP, CASP1 (ICE), CASP3, CASP8 (FLICE), CASP9, CUL2, MAPK9 (JNK2), PSEN2, PTEN
Antiapoptosis: APC, BDNF, CASP3, CASP9, NEFL, NR4A2 (NUR77), OPA1, PPID, PRDX2, PSEN2, PTEN, SLC25A4, SNCA, TCF7L2, UBB, YWHAZ
Mitochondria: CASP3, CASP7, CASP8 (FLICE), HSPA4 (HSP70), LRRK2, NEFL, OPA1, PARK7, PINK1, PTEN, SLC25A4, SNCA, TH, UCHL1, VAMP1, VDAC3, YWHAZ
Synaptic vesicles: LRRK2, SEPT5, SV2B, SYNGR3, SYT1, SYT11, TH
SIGNAL TRANSDUCTION
Dopaminergic: D4S234E, DDC, DRD2, HTR2A, NR4A2 (NUR77), PARK2, PARK7, PINK1, SEPT5, SLC6A3, SNCA, TH
GABAergic: DRD2, GABBR2
MAP kinase: APC, FGF13, MAPK9 (JNK2), PRDX2, RGS4
Notch: APP, PSEN2, SPEN
Cytoskeletal organization: APC, CDC42, MAPT, NEFL, PARK2
Ion transport: ATP2B2, CADPS, CXXC1, DRD2, EGLN1, GBE1, GRIA3, HTR2A, KCNJ6, NSF, PSEN2, S100B, SRSF7, SLIT1, SNCA, VDAC3
Transporters: ATP2B2, GRIA3, SLC18A2, SLC6A3, SLC25A4, SV2B, SYT1, SYT11, VDAC3
Others: ALDH1A1, BASP1, CHGB, DLK1, NCOA1, NTRK2, RTN1

Consequently, the excitability of spinal neurons is modulated via activation of resident microglia that releases growth factors (such as brain-derived neurotrophicfactor), chemokines, and cytokines. Endogenous opioid peptides and arachidonic acid metabolites acting through G protein–coupled receptors also modulate neuronal excitability. Table 11.4 provides the list of genes, identified from above-mentioned pathways (SABiosciences, 2013).

Potential Biomarkers of Psychiatric Diseases, Neurodegenerative Diseases, and Insomnia

Dysregulation of GABAergic or glutamatergic synaptic transmission results in a wide variety of nervous system disorders, including chronic pain, psychiatric diseases, neurodegenerative diseases, and insomnia. The GABA neurotransmitter system includes the GABAA and GABAC classes of ligand-gated ion channels. The glutamate neurotransmitter system includes NMDA, AMPA, and kainate ligand-gated ion channels. Key enzymes synthesize GABA or glutamate as necessary, which are then transported into synaptic vesicles. Release of GABA or glutamate from vesicles activates postsynaptic GABA-responsive or glutamate-responsive ion channels, respectively, initiating downstream G protein signaling to propagate neurotransmission. There are a number of genes essential for the synthesis and transport of GABA and glutamate, as well as responsive ion channels and downstream signaling. Studying and monitoring these genes at all levels may yield insights into the interaction of these excitatory and inhibitory neuronal systems during essential cognitive functions, thereby facilitating the process of biomarker development (Table 11.5: SABiosciences, 2013).

Potential Biomarkers of Motor Function, Emotional Behavior, Temperature Regulation, Sensory Perception, Locomotion, and Psychosis

Two of the major neurotransmitter systems such as dopamine and serotonin play a major role in the previously mentioned aspects. Dopamine affects brain processes that control both motor and emotional behavior and plays a role in the brain's reward mechanism. Serotonin is critical in temperature regulation, sensory perception, locomotion, sleep, and psychosis. Pharmacological agents targeting dopaminergic/serotoninergic neurotransmission have been clinically used to manage several neurological and psychiatric disorders, including PD, schizophrenia, bipolar disorder, depression, attention deficit and hyperactivity disorder, and addiction. The dopaminergic/serotoninergic receptors act through diverse G protein–coupled and G protein–independent mechanisms that trigger downstream intracellular signal transduction events involving the cAMP/PKA, PI-3Kinase/AKT, phospholipase A2, and phospholipase C pathways. These pathways in turn regulate various functions including synthesis, transport, and degradation of dopamine and serotonin, as well as the transcriptional regulation key genes linked to multiple neuropathological conditions (Table 11.6: SABiosciences, 2013).

TABLE 11.5 Potential Biomarker Genes Related to Glutamatergic and GABAergic Synapses

GLUTAMATERGIC SYNAPSE
Receptors: GRIA1, GRIA2, GRIA3, GRIA4, GRIK1, GRIK2, GRIK4, GRIK5, GRIN1, GRIN2A, GRIN2B, GRIN2C, GRM1, GRM2, GRM3, GRM4, GRM5, GRM6, GRM7, GRM8
Downstream signaling: ADCY7, APP, CACNA1A, CDK5R1, CLN3, DLG4 (PSD95), GNAI1, GNAQ, HOMER1, HOMER2, ITPR1, MAPK1 (ERK2), PLA2G6, PLCB1, SHANK2
Transport and secretion: ADORA1, ADORA2A, AVP, BDNF, IL1B, P2RX7, SLC17A6, SLC17A8, SLC1A1, SLC1A2, SLC1A3, SLC1A6, SLC17A6, SLC17A7, SLC17A8, SLC38A1, SLC7A11, SNCA
Metabolism: ALDH5A1, GAD1, GLS, GLUL, PRODH, SLC1A3, SRR
GABAERGIC SYNAPSE
Receptors: GABBR1, GABBR2, GABRA1, GABRA2, GABRA4, GABRA5, GABRA6, GABRB1, GABRB3, GABRD, GABRE, GABRG1, GABRG2, GABRG3, GABRQ, GABRR1, GABRR2
Downstream signaling: ADCY7, ADORA1, ADORA2A, CACNA1A, CACNA1B, GNAI1, GNAQ, GPHN, SNCA
Transport and secretion: CACNA1A, NSF, P2RX7, SLC1A3, SLC32A1, SLC38A1, SLC6A1, SLC6A11, SLC6A12, SLC6A13
Metabolism: ABAT, ALDH5A1, GAD1, GLS, GLUL, PHGDH, SLC1A3

Potential Biomarkers of Sleeping Disorders (Apnea, Insomnia, and Desynchronosis)

Synchronization of the circadian clock occurs via light stimulus of the hypothalamic suprachiasmatic nucleus

TABLE 11.6 Potential Biomarker Genes Related to Dopamine and Serotonin

RECEPTORS
Dopamine: DRD1, DRD2, DRD3, DRD4, DRD5
Serotonin: HTR1A, HTR1B, HTR1D, HTR1E, HTR1F, HTR2A, HTR2B, HTR2C, HTR3A, HTR3B, HTR4, HTR5A, HTR6, HTR7
SYNTHESIS AND DEGRADATION
Dopamine: COMT, DBH, DDC, MAOA, TH
Serotonin: MAOA, MAOB, TDO2, TPH1, TPH2
Dopamine and serotonin transporters: SLC6A3 (DAT), SLC6A4 (SERT)
SIGNAL TRANSDUCTION PATHWAYS
cAMP/PKA pathway: ADCY1, ADCY2, ADCY3, ADCY5, CASP3, CDK5, CREB1, DUSP1, FOS, PRKACA, MAPK1, PPP1R1B (DARPP32)
PI3K/AKT pathway: PIK3CA, PIK3CG, AKT1, AKT2, AKT3, GSK3A, GSK3B
PLA2 pathway: ALOX12, CYP2D6, PDE4A, PDE4B, PDE4C, PDE4D, PDE10A, PLA2G5
PLC pathway: ITPR1, PLCB1, PLCB2, PLCB3
G protein–coupled receptor regulation: ADRB1, ADRB2, ADRBK1, ADRBK2, APP, ARRB1, ARRB2, GRK4, GRK5, GRK6, SNCA, SNCAIP
Dopamine and serotonin gene targets: BDNF, EPHB1, GDNF, GFAP, NR4A1 (NUR77), NR4A3 (NOR1), PDYN, PTGS2, SLC18A1, SLC18A2 (VMAT2), SYN2

PLC, phospholipase C.

TABLE 11.7 Potential Biomarker Genes of Sleep/Wake Cycle

Circadian clock: ARNTL (BMAL1), ARNTL2, BHLHE40 (DEC1), BHLHE41, CLOCK, CSNK1E, CRY1, CRY2, NR1D1, NR1D2 (REV-ERB), PER1, PER2, PER3, RORA, TIMELESS
Casein kinases: CSNK1A1, CSNK1D, CSNK1E, CSNK2A1, CSNK2A2
CREB signaling: AANAT, CAMK2A, CAMK2B, CAMK2D, CAMK2G, CHRNB2, CREB1, CREB3, KCNMA1, HTR7, MAPK1, MAPK14, MAPK3, MAT2A, PRKACB, PRKACG, PRKAR1A, PRKAR1B, PRKAR2A, PRKAR2B, PRKCA, PRKCB, PROKR2
LIGHT-SENSING PROTEINS
Melatonin receptors: MTNR1A, MTNR1B
Opsins: OPN3, OPN4
Others: CHRNB2, CRX
Circadian-regulated transcription factors: ALAS1, EGR1, EGR3, EPO, ESRRA, HLF, IRF1, MYOD1, NFIL3, NKX2-5, PAX4, POU2F1, RORB, RORC, SMAD4, SP1, SREBF1, STAT5A, TEF, TFAP2A, TGFB1, WEE1, DBP, PPARA
Other common circadian-regulated genes: CARTPT, CCRN4L, FBXL21, FBXL3, HEBP1, NCOA3, NMS, NPAS2, NR2F6, PPARGC1A, PRF1, PTGDS, SLC9A3

(SCN) in the brain and via hormone signaling from the SCN in peripheral tissues. Interacting positive and negative circadian gene feedback loops at the transcriptional and posttranslational level set up the circadian "oscillator" and ensure tight control over transcription factors regulating expression of the appropriate genes required during circadian days or nights. Genes regulated by circadian rhythms are involved in a diverse range of biological processes that affect physiology, metabolism, and behavior. Although the circadian rhythm target genes in its "output" pathways vary widely from tissue to tissue, the transcription factors encoded by central clock and clock-controlled genes are mostly shared across all cell types. Sleeping disorders (such as apnea, insomnia, and desynchronosis) disrupt the timing of the circadian clock, requiring re-entrainment and causing fatigue. The list of genes identified to be involved in these processes is given in Table 11.7 (SABiosciences, 2013).

Biomarkers of Peripheral Nervous System Infection

Interestingly, sensory neurons share many of the same pathogens and danger molecular recognition receptor pathways as innate immune cells, which enable them also to detect pathogens. Germline-encoded pattern recognition receptors (PRRs), which recognize broadly conserved exogenous pathogen-associated molecular patterns, include members of the Toll-like receptor (TLR) family. Following PRR activation, downstream signaling pathways are turned on that induce cytokine production and activation of adaptive immunity. In addition to TLRs, innate immune cells are activated during tissue injury by endogenous-derived danger signals, also known as damage-associated molecular patterns or alarmins. These danger signals include uric acid and heat shock proteins released by dying cells during necrosis, activating immune cells during noninfectious inflammatory responses. Furthermore, activation of sensory neurons by the TLR7 ligand imiquimod leads to activation of an itch-specific sensory pathway. A key means of communication between immune cells and nociceptor neurons is through cytokines. Interleukin (IL) 1 beta and tumor necrosis factor alpha (TNF-α) are two important cytokines released by innate immune cells during inflammation, which in turn induce activation of p38 map kinases leading to increased membrane excitability. Nerve growth factor (NGF) and prostaglandin E(2) are also major inflammatory mediators released from immune cells that act directly on peripheral sensory neurons to cause sensitization (Chiu et al., 2013).

MicroRNAs as Potential Biomarkers of Peripheral Nervous System Injury

The microRNAs (miRNAs) are a recently identified class of small regulatory RNA molecules found in most organisms. They are encoded by the genomes and processed into 22 nucleotide products, which play important roles in the posttranscriptional regulation of important cellular pathways in diverse biological processes, including cell proliferation, apoptosis, tissue morphogenesis, tumorigenesis, and heart disease (Dugas and Notterpek, 2011). Recently, changes in miRNA expression profiles have been detected in different injury models, and emerging evidence strongly indicates that these changes promote neurons to survive by shifting their physiology from maintaining structure and synaptic transmission (Wu and Murashov, 2013). The list of miRNAs that showed altered expression is presented in Table 11.8.

Yu et al. (2011) noted altered miRNA expression following sciatic nerve resection in dorsal root ganglia of rats. The expression pattern of miR-206 was different from that of miR-1 and miR-133a. Notably, two genes (Hnrpu and Npy) and one gene (Ptprd) were potentially regulated both in the denervated and reinnervated muscle by miR-1 and miR-133a, respectively. There were six potential target genes (Hnrpu, Lsamp, MGC108776, Mef2, Npy, and Ppfibp2) of the upregulated miR-206 in the reinnervated muscle. Among these, three (Hnrpu, Npy, and MGC108776) were potentially regulated by both miR-1 and miR-206. Because the Mef2 transcription factor was reported to promote the transformation of type II fast glycolytic fibers into type I slow oxidative fibers, the upregulation of miR-206 with decreased expression of the Mef2 transcript in the 4-month reinnervated muscle, which presented type II fiber

TABLE 11.8 MicroRNAs Whose Expression Changed the Most in the Referenced Studies

Yun et al. (2010), Verrier et al. (2010)
miR-138, miR-138, miR-129
miR-140, miR-338, miR-145
miR-146b, miR-193
miR-195, miR-222
miR-204, miR-29a
miR-27b
miR-30a
miR-338-3p
miR-34a
Highly upregulated: miR-21, miR-221

predominance 4 months after nerve microanastomosis, might indicate the role of miR-206 in determining the fiber type after peripheral nerve regeneration. Table 11.9 provides the details of target genes for these altered miRNAs and their functional role (Jeng et al., 2009).

Chemical-Induced Peripheral Neuropathy

Because of the need for complete eradication of cancer cells, a higher dose of drugs is used in cancer chemotherapy (Table 11.10), which very often results in chemical-induced peripheral neuropathy (CIPN), a dose-limiting neurotoxic effect of chemotherapy, which can lead to early cessation of cancer treatment (Wang et al., 2012). This can result in an increased risk of relapses and decreased survival rate. Biomarkers to assess the susceptibility or tolerance level of such treatment approaches can be very useful in making decisions about the choice, dose, and duration of the drugs that can help those cancer patients. Inflammatory cascade activation, proinflammatory cytokine, upregulation, and neuroimmune communication pathways play essential roles in the initiation and progression of CIPN. Most notably, TNF-α, IL-1β, IL-6, and CCL2 are involved in neuropathic pain. These cytokines could be used as biomarkers for predicting the onset of painful peripheral neuropathy and early axonal damage (Wang et al., 2012). Reactive oxygen species (ROS) have been shown to play a causal role in the development and maintenance of paclitaxel-induced pain (Fidanboylu et al., 2011), and hence the levels of ROS can be used as a biomarker for certain drugs.

Pharmacogenomic Biomarkers for Preventing Chemical-Induced Peripheral Neuropathy

Identifying genetic or molecular biomarkers for effective diagnostics of patients with high risk of developing neurotoxicity from a given chemotherapeutic agent could significantly impact the quality of life of cancer

TABLE 11.9 Differential Expression of MicroRNAs (miRNAs) After Peripheral Denervation and Renervation

Condition	Denervation (4 months)	Reinnervation (4 months)				
Upregulated miRNAs	rim-miR-1	rno-miR-133a	rno-miR-1	rno-miR-133a	rno-miR-206	
	Ampd1	LOC49S749	Akap9	G0s2	Cit	Hnrpu
	Atp2c1	LOC680782	Elf2	Hnrpu	Gdnf	Lsamp
	Cacnb2	MGC108776	Gas6	Hspb1	Mcf2l	MGC108776
	Cenpc1	Mns1	Gp1bb	MGC108776	Ptprd	Mef2
	Chac2	Npy	I12rg	Npy	RT1-A2	Npy
	Cltc	Rasa1	Klhl24	Sgk	Stau1	Ppfibp2
Potential target genes	Ddx5	Serpina3k	LOC500110	Sox6	Sv2b	
	Echdc1	Ubc	Nexn	Wif1		
	G6pdx	Usp33	Nr3c1			
	Havcr2		Pdk4			
	Hnrpu		Plp			
	Hspd1		Ptprd			
	LOC292666		Rnf103			
	LOC299282		Zfp131			

Modified from Jeng, S.F., Rau, C.S., Liliang, P.C., 2009. Profiling muscle-specific microRNA expression after peripheral denervation and reinnervation in a rat model. J. Neurotrauma. 26, 2345–2353.

TABLE 11.10 Chemotherapeutic Agents Causing Peripheral Neuropathy

PLATINUM COMPOUNDS
Cisplatin, Oxaliplatin, Carboplatin, Taxanes, Paclitaxel, Docetaxel, Vinca alkaloids, Vincristine, Vinblastine, Vinorelbine
OTHER AGENTS
Bortezomib, Thalidomide, Lenalidomide, Colistin
ANTIALCOHOL DRUGS
Disulfiram
ANTICONVULSANTS
Phenytoin (Dilantin)
HEART OR BLOOD PRESSURE MEDICATIONS
Amiodarone, Hydralazine, Gigaxonin, Mitofusin, Perhexiline,
INFECTION FIGHTING DRUGS
Metronidazole (Flagyl), Nitrofurantoin, Thalidomide, INH (Isoniazid),
SKIN CONDITION TREATMENT DRUGS
Dapsone

TABLE 11.11 Single-Nucleotide Polymorphisms (SNPs) of Chemical-Induced Peripheral Neuropathy (CIPN) Neurotoxicity

1. Paclitaxel (Taxol) treatment related: CYPC28, CYP3A5, ABCB1, FANCD2
2. Vincristine treatment related:

   ADRB2, CAMKK1, CYP2C9, NFATC2, ID3, SLC10A2, CYP2C8
3. Cisplatin treatment related:

   GSTP1, GSTM1, GSTM3, ITGB3, AGXT, ERCC
4. FOLFOX-4 treatment related:

   ITGB3 Leu59Pro; GSTP1 Ile105Val; AGXT Pro11Leu; AGXT Ile340Met;
5. Docetaxel treatment related: ABCB1, Ser893Ala/Thr

Adapted from Miltenburg, N.C., Boogerd, W., 2014. Complications of treatment. Chemotherapy-induced neuropathy: a comprehensive survey. Cancer Treat Rev 40, 872–882bib_Miltenburg_and_Boogerd_2014.

survivors. There have been some candidate gene approaches to evaluate single-nucleotide polymorphisms (SNPs) in genes involved in the pharmacokinetic and pharmacodynamic properties of neurotoxic drugs, but no reliable biomarker has thus far been identified to detect patients at high risk of developing CIPN (Brewer et al., 2016). Variants in genes relevant to drug disposition, metabolism, and drug action can influence the patient's exposure and sensitivity to the drug. For paclitaxel, most studies focus on mutations in biologically relevant candidate genes such as CYP2C8, CYP3A4, CYP3A5, FANCD2, and ABCB1. Hertz et al. (2014) demonstrated an association between the CYP2C8*3 variant and a twofold increase in risk for paclitaxel-induced peripheral neuropathy across racial groups. Interindividual variability has also been found in the therapeutic target of paclitaxel, β-tubulin Iiα (TUBB2α) messenger RNA (mRNA) expression, and two genetic variants in the promoter region of the gene are involved in this variation. In addition, genome-wide association studies were performed to identify genetic markers predictive of peripheral neuropathy. SNPs in FGD4, FZD3, and EPHA5 were identified to be associated with a severity or dose at onset of paclitaxel-induced sensory peripheral neuropathy in patients with both European and African ancestry (Baldwin et al., 2012). In another study, a gene named RFX2 was identified to be associated with increased risk of paclitaxel-induced neurotoxicity (Wheelar et al., 2013). In a similar integration method using Asian cell lines and patient cohorts noted a decrease in AIPL1 level, resulting in decreased sensitivity of neurons to paclitaxel. In addition, breakpoint cluster region protein was also identified as protecting the neurons from paclitaxel-induced toxicity (Komatsu et al., 2015). The set of genes that drive heritability of paclitaxel-induced neuropathy was found to be implicated in axonogenesis and, more specifically, the regulation of axon outgrowth (Chhibber et al., 2014). SNPs of CIPN neurotoxicity has been listed in Table 11.11.

Epigenetic Biomarkers of Axonal Regeneration

Gadd45 alpha, one of the key regulators of epigenetic modification of gene expression, is highly upregulated after axonal injury. In parallel, many genes involved in axonal elongation and synaptic activity are demethylated on axonal injury. Further studies are required to verify whether it is possible to use epigenetic manipulation to bring the adult neurons to a more "immature" state to promote better regeneration (Hoke et al., 2013). Peripheral myelination has been shown to be disrupted in various PNS disorders. Nuc-ErbB3 has been identified to be a master transcriptional repressor involved in epigenetic regulation H3K27me3 levels and methyltransferase activity during peripheral myelination, thus making it a good biomarker (Ness et al., 2016).

Biomarkers From Studies on Inherited Peripheral Neuropathies

A number of peripheral neuropathies are inherited, of which a number of them are associated with documented defects in axonal transport (De Vos et al., 2008). In Charcot-Marie-Tooth disease (CMT) type 2E, caused by mutations in the neurofilament light gene (NF-L), mutant protein aggregates perturb the cytoskeletal network and disrupt the normal localization of mitochondria, leading to impaired axonal transport (Dequen et al., 2010). In CMT2A (mutations in mitofusin), axonal transport of mitochondria was compromised (Baloh, 2007) and in giant axonal neuropathy (mutations in gigaxonin) showed impaired retrograde axonal transport (Ding et al., 2006). X-linked demyelinating/type I CMT neuropathy, caused by mutations in connexin 32 (Vavlitou et al., 2010), showed reduced axonal transport and axonal pathology before demyelination occurs. Dynein dysfunction appears directly involved in at least a subset of these diseases. Dynein is regulated by a number of protein complexes, notably, by dynactin. Several studies have supported indirectly the involvement of dynein in neurodegeneration associated with Alzheimer's disease, PD, Huntington's disease, and motor neuron diseases (Eschbach and Dupuis, 2011). First, axonal transport disruption represents a common feature occurring in neurodegenerative diseases. Second, a number of dynein-dependent processes, including autophagy or clearance of aggregation-prone proteins, are found defective in most of these diseases. Third, a number of mutant genes in various neurodegenerative diseases are involved in the regulation of dynein transport. This includes, notably, mutations in the P150Glued subunit of dynactin, which are found in Perry syndrome and motor neuron diseases. Interestingly, gene products that are mutant in Huntington's disease, PD, motor neuron disease, or spinocerebellar ataxia are also involved in the regulation of dynein motor activity or of cargo binding (Eschbach and Dupuis, 2011).

The hereditary sensory and autonomic neuropathies encompass a number of inherited disorders that are associated with sensory dysfunction (depressed reflexes, altered pain, and temperature perception) and varying degrees of autonomic dysfunction (gastroesophageal reflux, postural hypotension, and excessive sweating).Types I–V have been shown to have mutations in the following genes: SPTLC1, HSN2, IKBKAP, NTRK1 (TRKA), and NTRK1. Neurodegenerative diseases characterized by brain, spinal cord, and PNS involvement often show widespread accumulations of tau aggregates. List of genes involved in hereditary PNS disorders is given in Table 11.12.

TABLE 11.12 Candidate Biomarker Genes Connected With Inherited Peripheral Nervous System (PNS) Disorders

THE CANDIDATES THAT WERE GENOTYPED FOR PNS DISORDERS
ACCN2, ACE, ACTB, ACTGl, ACTRlA, ACTRlB, AD0RA2A, ADRA2B, AGT, AGTRl, AKRlBl, AKTl, AKT2, APC, ARPI l, AXINl, BMF, CACNAlA, CACNAlB, CAPZAl, CAPZA2, CAPZA3, CAPZB, CD86, CMT2A,CMT2E, COMT, CTLA4, CTNNBl, CTSS, CYP3A4, CYP3A5, DCTNl, DCTN2, DCTN3, DCTN4, DCTN6, DNCLl, DNCL2A, DNM2, DVLl, DVL2, DVL3, DYNClHl, DYNClIl, DYNC1I2, DYNClLIl, DYNC1LI2, DYNC2H1, DYNC2LI1, DYNLL2, DYNLRB2, ECGFl, EGR2, FGD4, FIG4, GARS, GCHl, GDAPl, GJBl, GJB2, GJB3, GJEl, GLRA3, GLS2, GLUL, GSK3A, GSK3B, HAPl, HSN2, HSPBl, HSPB8, HTRlB, IKBKAP, IL6, KIFlA, KIFlB, KIF3A, KIF3B, KIF5A, KIF5B, KIF5C, LITAF, LMNA, MAPKl, MAP1B LC, MAPKlO, MAPKI l, MAPK 12, MAPKl 3, MAPK14, MAPK3, MAPK4, MAPK6, MAPK7, MAPK8, MAPK9, MClR, MFN2
OTHER GENES CONNECTED WITH INHERITED PNS DISORDERS
MPZ, MTMR2, MTMR13/set-binding factor 2,NDRGl, NEFL, NF-L, NFE2L2, NGFB, NPY, NR 112, NTRKl, OPRDl, OPRKl, OPRLl, OPRMl, PLPl, PMP22, PNOC, POLG, POLG2, PONl, PRPSl, PRX, PSMBl, PSMBlO, PSMB2, PSMB3, PSMB4, PSMB5, PSMB6, PSMB7, PSMB8, PSMB9, PTGERl, PTGER2, PTGER3, PTGER4, PTGSl, PTGS2,RILP, SBF2, SCN3A, SCN9A, SH3TC2, SLC12A6, SPTBNl, SPTBN2, SPTBN4, SPTBN5, SPTLCl, SURFl, TCFl, TCF4, TH, TNF, TRAK2, TRPVl, TRPV4, TTR, VIP, WNTl, WNTlOA, WNTlOB, WNTI l, WNT16, WNT2, WNT2B, WNT3, WNT3A, WNT4, WNT5A, WNT5B, WNT6, WNT7A, WNT8A, WNT8B, WNT9A, WNT9B and YARS

Multiple System Atrophy–Related Biomarkers

Multiple system atrophy (MSA) is a unique proteinopathy that differs from other α-synucleinopathies, as the pathological process resulting from accumulation of aberrant alpha synuclein (α-syn) involves the oligodendroglia rather than neurons, although both pathologies affect multiple parts of the brain, spinal cord, autonomic, and PNS (Jellinger and Wenning, 2016). Biomarkers include those based on neuroimaging (MRI, DTI, MR spectroscopy, PET), peripheral biomarkers (by skin, peripheral nerve, and other biopsies), and fluid biomarkers (from serum, plasma, and CSF). Currently, the clinically most useful markers may comprise a combination of the light chain of neurofilaments, which is consistently elevated in MSA compared with controls and PD, metabolites of the catecholamine pathway (dopamine and norepinephrine), and proteins such as α-syn, DJ-1, and total tau (Jellinger and Wenning, 2016). In this multipronged approach, using several end point–specific biomarkers has been shown to be very effective.

Biomarkers of Demyelination Injuries

Axonal degeneration contributes to persistent neurological disability, affecting PNS and CNS. Demyelination has been shown to induce ATF3 (activating transcription factor 3) in dorsal root ganglia (DRG) neurons. ATF3 expression is modulated by retrograde axonal transport of c-Jun NH2-terminal kinase (JNK) (Lindwall and Kanje, 2005). Axonally derived signaling pathways can also be initiated by neurotrophins and the myelin proteins, NOGO, MAG, and oligodendrocyte myelin glycoprotein. Similar pathways and additional transcription factors may be modulated following demyelination in both PNS and CNS. Mitochondrial function in Schwann cells has been shown to be important in maintenance of axons, and disruption of mitochondria in Schwann cells can also lead to axonal degeneration (Viader et al., 2011). Hence, gene products involved in the mitochondrial structure and function may qualify for biomarker status.

Biomarkers of Neuromuscular Junction Injuries

Neurotoxicants that target the neuromuscular junction are classified into two classes: (1) Those that alter the phenotypic synthesis, storage, or release of acetylcholine (ACh) and (2) those that disrupt the postsynaptic function of the effector cells, associated with the recognition of ACh by its target receptor, activation of conductances through the receptor-associated ion channel, or impaired hydrolysis of ACh. The neuromuscular junction has been the specific target for neurotoxins such as botulinum toxin or many of the cholinesterase inhibitors. Altered physiological splicing mechanisms of genes for agrin (AGRN), acetylcholinesterase (AChE), MuSK (MUSK), acetylcholine receptor (AChR) α1 subunit (CHRNA1), and collagen Q (COLQ) in humans and their roles in diseases involving neuromuscular junction have been shown. Molecular dissection of splicing mutations in patients with CMS (congenital myasthenic syndromes) reveals that exon P3A is alternatively skipped by hnRNP H, polypyrimidine tract-binding protein 1, and hnRNP L. Similarly, analysis of an exonic mutation in COLQ exon 16 in a CMS patient discloses that constitutive splicing of exon 16 requires binding of serine arginine-rich splicing factor 1. Intronic and exonic splicing mutations in CMS enable us to dissect molecular mechanisms underlying alternative and constitutive splicing of genes expressed at the neuromuscular junction (Ohno et al., 2017). These genes and their variants may be used as diagnostic and prognostic biomarkers.

Biomarkers From Surrogate Tissue Analysis for Peripheral Nervous System Disorders

Surrogate tissue analysis can be very useful for monitoring toxicant exposure and effect, disease development and progression, and drug efficacy testing (Tang et al., 2006). There are many accessible tissues (e.g., blood, urine, CSF, cord blood, buccal cells, or milk) that can be used as a surrogate tissue. Furthermore, recent studies from human disease conditions such as neurofibromatosis type 1, Tourette syndrome, and anticonvulsant drugs in pediatric epilepsy indicated the existence of quantifiable markers (Tang et al., 2006). Global gene expression studies using whole blood from patients with posttraumatic stress, stroke, and migraine yielded valuable data on candidate biomarkers (Segman et al., 2005). Erythrocyte AChE for organophosphates, free erythrocyte protoporphyrin for lead, lymphocyte neurotoxicity target enzyme (NTE) for organophosphates, blood aminolevulinic acid dehydratase (ALA-D) for intermittent porphyria, and carboxyhemoglobin (CO-Hb) for carbon monoxide are few examples of good surrogate markers.

Biomarkers of Cholinergic System and Its Relevance to the Peripheral Nervous System

Acetylcholine (ACh) is widely distributed in the nervous system and has been implicated as playing a critical role in cerebral cortical development, cortical activity, controlling of cerebral blood flow, and the sleep–wake cycle, as well as in modulating cognitive performances and learning and memory processes (Schliebs and Arendt, 2011).

Peripheral Nervous System Biomarkers for Exposure to Organophosphates

Organophosphate insecticides share a common mechanism of toxicity through inhibitory effects on cholinesterase enzymes in the nervous system. Hydrolysis of the organophosphates yields a dialkyl phosphate and the leaving group, which do not inhibit cholinesterase enzymes and can be measured as biomarkers of exposure to organophosphates, as the metabolites are subsequently eliminated from the body in the urine. An individual with acute symptomatic overexposure to organophosphates will usually have an abnormally low level of activity of cholinesterase enzymes measured in the serum (as butyrylcholinesterase) or in red blood cells. Paraoxonase (or PON1) enzyme is responsible for the hydrolysis and deactivation of organophosphates. Polymorphisms in the PON1 gene exist in humans adding variability in OP exposure response. In a recent study on the associations between OP pesticide use and PNS function, it was shown that long-term exposure to OP pesticides is associated with signs of impaired PNS function among pesticide applicators (Starks et al., 2012).

Biomarkers of Organophosphorus-Ester Induced Delayed Neurotoxicity

organophosphorus compound-induced delayed neurotoxicity (OPIDN) is a neurodegenerative disorder characterized by ataxia progressing to paralysis with a concomitant central and peripheral distal axonopathy (Damodaran et al., 2011). Although the exact mechanism of OPIDN development is yet to be determined, cholinergic inhibition of AChE and differential alterations of several upstream and downstream molecules and their associated biochemical machinery significantly contributes to the pathophysiology of OPIDN (Damodaran, 2009). AChE inhibitors (e.g., sarin, diisopropylfluorophosphate (DFP)) exposure may alter the regulation of the cholinergic system differentially, depending on the dose, duration, and mode of exposure as evidenced by numerous studies, thus affecting various functions such as muscle function, cognition, and sleep (Damodaran et al., 2006a,b; Damodaran, 2009). Alterations in the levels of mRNA of CaM kinase II alpha subunit, neural filament triplet proteins, GFAP, vimentin, alpha tubulin, beta tubulin subtypes, and glyceraldehyde-3-phosphate dehydrogenase (GAPDH) have been shown (Damodaran et al., 2009). Immediate early induction of c-fos and c-jun, PKA, CREB, and p-CREB in DFP-treated hen brain has been shown (Damodaran et al., 2009). The protein levels of protein kinase C, CaM kinase II, and several phosphatases (i.e., phosphatase 1 [PP1], and phosphatase 2A [PP2A], phosphatase 2B [PP2B]) were altered in the spinal cord of DFP-treated hens (Damodaran et al., 2011). Altered expressions of MAP2 mRNA (Damodaran et al., 2014), TAU mRNA (Damodaran et al., 2015), JUN-D mRNA (Damodaran and Abou-Donia, 2016), and GAP-43 mRNA (Damodaran et al., 2017) in OPIDN conditions warrant further studies to validate these genes as potential biomarkers for OP-induced OPIDN.

Sarin (molecular structural analogue of DFP) exposure initiated several validated and hitherto unreported pathways immediately (within 15 min of exposure) and a significant number of important genes (pathways) showed persistently modified levels at 2 h and 3 months posttreatment (Damodaran et al., 2006a,b). Furthermore, validation and candidate gene studies showed a similar overall pattern for both DFP (exposure to hens) and sarin (exposure to rats), in spite of the differences in species tested and structure of the analog used (Damodaran, 2009). Thus, there are common and evolutionarily conserved pathways (gene clusters) that are sensitive to AChE and NTE inhibition and probably other esterases' inhibition, irrespective of the test organism and differences in the structures of OPs tested. These organophosphorus esters-exposure related physiological genomic and nongenomic effects (such as hormonal, inflammatory, and other types) very often persist for a long time, if appropriate remedial measures are not taken (Damodaran, 2009).

AUTONOMIC NERVOUS SYSTEM

Autonomic nervous system (ANS) activity may be categorized into several measurable physiological systems or pathways, activated through sympathetic and parasympathetic mechanisms. These pathways provide a framework for a comprehensive review of stress reactivity measures. Measuring ANS activity include approximately 50 tools identified according to the ANS pathway, which are listed in the ANS Measures Table in Appendix A of the monograph published by Defense Centers of Excellence for Psychological Health and TBI (website: www.dcoe.health.mil) (Brierley-Bowers et al., 2011). Some of these tools, such as the electrocardiogram, are established and widely used. The utility of each tool is dependent on its purpose of use, target environment, and the sensitivity of measurement needed. The three ANS pathways most often used to measure the impact of mind-body skills, as seen in Table 11.2, are cardiac, vascular, and respiratory. These are the ANS pathways that physicians monitor via the vital signs (blood pressure, heart rate, temperature, and respiratory rate). Cardiac measurement, vascular measurement, respiratory measurement, galvanic skin response measurement, catecholamine measurement, cortisol measurement, pupillary response measurement, salivary amylase measurement, sweat analysis (Brierley-Bowers et al., 2011), and intracranial pressure analysis (Wostyn and De Deyn, 2017) are routinely employed.

Autoimmune Autonomic Disorders

Autoimmune autonomic disorders occur because of an immune response directed against sympathetic, parasympathetic, and enteric ganglia, autonomic nerves, or central autonomic pathways. In general, peripheral autoimmune disorders manifest with either generalized or restricted autonomic failure, e.g., isolated gastrointestinal dysmotility. Peripheral autoimmune autonomic disorders include autoimmune autonomic ganglionopathy, paraneoplastic autonomic neuropathy, and acute autonomic and sensory neuropathy (Mckeon and Benarroch, 2016). Some other disorders are characterized immunologically by paraneoplastic antibodies with a high positive predictive value for cancer, such as antineuronal nuclear antibody, type 1 (ANNA-1: anti-Hu) (Mckeon and Benarroch, 2016). Table 11.13 provides the list of biomarkers relevant to autoimmune disorders of PNS.

Autoantibodies as Biomarkers for Peripheral Nervous System Disorders

Recently, autoantibodies have been proposed as the new generation class of biomarkers because of their long-term presence in serum compared with their counterpart antigens (Raad et al., 2014). PNS axons and myelin have unique potential protein, proteolipid, and ganglioside antigenic determinants. Despite the existence of a blood–nerve barrier, both humoral and cellular immunity can be directed against peripheral axons and myelin. Among putative biomarkers, myelin-

TABLE 11.13 Biomarkers of Autoimmune Disorders Affecting Peripheral Nervous System (PNS)

Gene polymorphisms: FcγR, HLA, CD1, CD95, TNF-α, mannose-binding lectin, macrophage mediators, TCR, TLR4, killer immunoglobulin-like receptor, glucocorticoid receptor

Cytokines: IFN-γ, TNF-α, IL-17, IL-22, IL-18, IL-1β, IL-10, IL-6, IL-12, IL-16, IL-23, IL-37,TGF-β1

Complements: C3, C5b-9, C5a

Chemokines: CCL2, CCL3, CX3CL1, CXCL2, CXCL10, CCL7, CCL27, CXCL9, CXCL12

Others: Erythropoietin, heat shock protein, apolipoprotein E, C-reactive protein, neopterin, matrix metalloproteinases, reactive oxygen species, cell adhesion molecules, microRNA-155, osteopontin

BNB/B-CSF-B damage-associated biomarkers:

Brain-derived proteins: Prealbumin, transthyretin, S100B, cystatin C, prostaglandin D(2) synthase, hypocretin-1

Blood-derived proteins: Haptoglobin, fibrinogen, Apo A-IV, ApoH, vitamin D–binding protein, α-1-antitrypsin

PNS damage-associated biomarkers:

Myelin sheath–associated biomarkers: Autoantibodies to ganglioside, neurofascin, gliomedin, P0, PMP22, P214–25, connexin 32, α6β4, phospholipids

Neuron-component-associated biomarkers: Neurofilaments, tau proteins, 14-3-3 proteins, neuron-specific enolase

Other biomarkers for GBS: Creatine kinase heparin sulfate glycosaminoglycans, glial fibrillary acid protein, triglyceride, and hyponatremia

Adapted from Wang, Y., Sun, S., Zhu, J., et al., 2015. Biomarkers of Guillain-Barré syndrome: some recent progress, more still to be explored. Mediat. Inflamm. 2015, 564098.

associated glycoprotein and several antiganglioside autoantibodies have shown statistically significant associations with specific neuropathic syndromes (Fehmi et al., 2018). A detailed list of autoantibodies and corresponding clinical symptoms is given in Table 11.14.

Nonimmune Nodopathies

Peripheral nodopathies can also be induced by toxic, ischemic, nutritional, and genetic mechanisms (Fehmi et al., 2018). As many as 70% of patients who are critically ill develop neuropathies, reduced nodal voltage-gated sodium channels activity, and axonal depolarization (Dalaka et al., 2016). There is currently no recognized treatment for critical illness neuropathy, but a future ability to detect the early stages of this disease process could pave the way for disease modifying therapies aimed at preventing subsequent axonal degeneration. Acute and chronic ischemic neuropathies, often a result of systemic vasculitis presumably arising via an impact on the energy-dependent ion channel or pumps, leading to axonal degeneration (Williams et al., 2014).

TABLE 11.14 Autoantibodies to Gangliosides as Biomarkers

Clinical symptom	Antigen (potential targets)
Severe disability, mechanical ventilation	GD1a/GD1b, GD1b/GT1b
Pure motor GBS, conduction blocks at intermediate nerve	GM1/GalNac-GD1a Rb, GM1/PA, GM1/GD1a, GM1/GT1b
AMAN reversible conduction failure	LM1/GA1
Axonal sensory motor(AMSAN)	GM1, GD1a, GM1b, GalNAc-GD1a
Motor (AMAN)	GM1/GalNac-GD1a complex, GD3
Sensory/ataxic	GD1b, other disialosyl epitopes, sulfatide
Pharyngeal-cervical-brachial	GT1a, GQ1b
Miller Fisher	GQ1b (occasionally GT1a)
GBS/CMV infection	GM2
Multifocal motor neuropathy(MMN)	IgM GM1 (also asialo GM1, GM2, GD1a or NS6S)
Sensory/ataxic (CANOMAD)	IgM GD1b and other disialosyl epitopes
Gait ataxia with late-onset polyneuropathy (GALOP)	Sulfatide, Galopin (central nervous system [CNS] white matter antigen)
Distal acquired demyelinating symmetric (DADS)	IgM gammopathy, MAG, sulfatide
POEMS syndrome	IgG or IgA lambda gammopathy, ↑VEGF levels
Facial nerve involvement	GD1a
Reversible conduction failure, ataxia	GD1b
Ophthalmoparesis	GD3, *O*-Acetyl GT3
Ophthalmoplegia	GT1a
Ataxia	GT1b, *O*-Acetyl GT3
Dysfunction, distal weakness, low amplitudes for the compound muscle action potentials, facial palsy	9-*O*-Acetyl GD 1b

Adapted from Wang, Y., Sun, S., Zhu, J., et al., 2015. Biomarkers of Guillain-Barré syndrome: some recent progress, more still to be explored. Mediat. Inflamm. 2015, 564098.

Role of Peripheral Nervous System in Parkinson's Disease and Lewy Body Disease

ANS involvement occurs at early stages in both PD and incidental Lewy body disease, affecting the sympathetic, parasympathetic, and enteric nervous systems (ENSs) (Cersosimo and Benarroch, 2012). It has been proposed that α-syn pathology in PD has a distal to proximal progression along autonomic pathways. The ENS is affected before the dorsal motor nucleus of the vagus, and distal axons of cardiac sympathetic nerves degenerate before there is loss of paravertebral sympathetic ganglion neurons. Consistent with neuropathological findings, some autonomic manifestations such as constipation or impaired cardiac uptake of norepinephrine precursors occur at early stages of the disease even before the onset of motor symptoms. Biopsy of peripheral tissues may constitute a promising approach to detect α-syn neuropathology in autonomic nerves and a useful early biomarker of PD. PD is among the most frequent neurodegenerative disorders associated with aging. Several nonmotor manifestations, such as autonomic, sleep, and olfactory dysfunction, may occur at early stages of disease and precede the onset of motor symptoms. Strong evidence indicates that autonomic involvement in PD starts in the periphery. Both its early involvement and the accessibility of peripheral tissues make the peripheral autonomic system an attractive target for detection of early biomarkers of the disease (Fanar et al., 2012).

Neural Stress System of Peripheral Nervous System

The two main arms of the stress system include the ANS and the hypothalamic–pituitary–adrenal axis. These two neural stress systems coordinate the response of many other physiological systems to a stressor, including the immune and cardiovascular systems, bringing the body back to homeostasis. The nervous and immune systems communicate with each other in a bidirectional manner. Collection of sweat and saliva and measurement of heart rate variability are noninvasive methods that can be applied to evaluate neuroimmune interactions. Levels of neuropeptides, such as NPY, VIP, SP, and CGRP, were shown to have tight correlation with levels in plasma in patients with a history of depression (Marques-Deak et al., 2006; Marques et al., 2010). The sympathetic and parasympathetic (vagus nerve, mediated by ACh) branches of the ANS play an important role in the regulation of the cardiovascular system and the immune system, where the parasympathetic nervous system has been shown to have an antiinflammatory influence on the immune system.

BIOMARKERS OF ENTERIC NERVOUS SYSTEM

Glutamate is the major excitatory neurotransmitter in the ENS. Prolonged stimulation of enteric ganglia by glutamate caused necrosis and apoptosis in enteric neurons. Glutamate immunoreactive neurons were found in cultured myenteric ganglia, and a subset of enteric neurons expressed NMDA (NR1, NR2A/B), AMPA (GluR1, GluR2/3), and kainate (GluR5/6/7) receptor subunits. Glutamate receptors were clustered on enteric neurites. Excitotoxicity in the ENS suggests that overactivation of enteric glutamate receptors may contribute to the intestinal damage produced by anoxia, ischemia, and excitotoxins present in food (Kirchgessner et al., 1997).

EMERGING AVENUES OF PERIPHERAL NERVOUS SYSTEM BIOMARKER DEVELOPMENTS: EXOSOMES

Most, if not all, types of mammalian cells release small membranous vesicles known as exosomes. In addition to their protein content, these vesicles have recently been shown to contain mRNA and miRNA species. Roles for these vesicles include cell–cell signaling, removal of unwanted proteins, and transfer of pathogens, such as prions, between cells. Exosomes can be isolated from circulating fluids such as serum, urine, and CSF, and they provide a potential source of biomarkers for neurological conditions. Exosomes have been proposed as being a means of intercellular communication in the normal physiology of the nervous system; several reports have demonstrated the release of exosomes by different cell types such as astrocytes or microglial neurons (Corrado et al., 2013). Recently, exosomes may also play a key role in neuronal communication during neurodegenerative diseases, such as Alzheimer's, PD, or prion diseases. For example, Alzheimer's disease and PD are also characterized by the accumulation of misfolded proteins; recent studies have shown that the misfolded protein incorporation into exosomes protects them from degradation and also facilitates their delivery over long distances. PD is characterized by intracellular aggregates of α-syn, the Lewy bodies, in dopaminergic neurons. Alvarez-Erviti et al. (2011) showed that α-syn released from cells overexpressing the protein is efficiently transferred to recipient normal cells through exosomes. Moreover, Surgucheva et al. (2012) showed that another member of the synuclein family, γ-synuclein, secreted from neuronal cells into exosomes, can be transmitted to glial cells, thus promoting the aggregation of intracellular proteins. In Alzheimer's disease, exosomal proteins were found to accumulate in the plaques of Alzheimer's disease (AD) patient brains, thus suggesting that exosome-associated amyloid peptides may be involved in plaque formation. Another characteristic of Alzheimer's disease is the intraneuronal aggregation of abnormally modified microtubule-associated tau proteins. Saman et al. (2012) have recently shown that tau proteins are mainly secreted through exosomes in vitro and in vivo.

PERSONALIZED MEDICINE

The convergence of advancing research using the latest "Omics" technology in biomarker development and drug testing approaches has made it feasible to have an accurate, individual specific targeted therapy for PNS toxicity disorders.

CONCLUDING REMARKS AND FUTURE DIRECTIONS

In this chapter, the latest evolving concepts about PNS toxicity—related biomarkers are provided. Extensive lists of proven biomarkers and candidate biomarkers identified so far have been updated. The following future directions may deserve immediate attention and action: (1) the role of posttranslational modifications such as phosphorylation and conformational alternates (resulting from pathological mechanisms), (2) research on exosomes is still in its infancy, so there is a need to identify relevant PNS mechanisms, (3) role of miRNA expression in PNS toxicity, (4) more studies on inherited disorders of PNS, (5) epigenetics of PNS toxicity, (6) more global approaches to gather useful data on PNS toxicity biomarkers, and (7) multiplex approaches for biomarker-based PNS toxicity assessment and management.

References

Abou-Donia, M.B., Nomeir, A.A., 1986. The role of pharmacokinetics and metabolism in species Similarly, defective axonal transport has been shown to be involved in organophosphate-induced delayed neurotoxicity (OPIDN) sensitivity of neurotoxic agents. Fundam. Appl. Toxicol. 6, 190–207.

Alvarez-Erviti, Y., Seow, H., Yin, H., 2011. Delivery of sirna to the mouse brain by systemic injection of targeted exosomes. Nat. Biotechnol. 29, 341–345.

Baldwin, R.M., Owzar, K., Zembutsu, H., 2012. genome-wide association study identifies novel loci for paclitaxel-induced sensory peripheral neuropathy in CALGB 40101. Clin. Cancer Res. 18 (18), 5099–5109.

Baloh, R.H., 2007. Mitochondrial dynamics and peripheral neuropathy. Neuroscientist 14, 12–18.

Brierley-Bowers, P., Sexton, S., et al., 2011. Measures of Autonomic Nervous System Regulation. Defense Centers of Excellence for Psychological Health and Traumatic Brain Injury, Arlington, Virginia, pp. 1–24.

Brewer, J.R., Morrison, G., Dolan, M.E., et al., 2016. Chemotherapy-induced peripheral neuropathy: current status and progress. Gynecol. Oncol. 140, 176–183.

Cersosimo, M.G., Benarroch, E.E., 2012. Autonomic involvement in Parkinson's disease: pathology, pathophysiology, clinical features and possible peripheral biomarkers. J. Neurol. Sci. 313, 57–63.

Chhibber, A., Mefford, J., Stahl, E.A., et al., August 2014. Polygenic inheritance of paclitaxel-induced sensory peripheral neuropathy driven by axon outgrowth gene sets in CALGB 40101 (Alliance). Pharmacogenomics J. 14 (4), 336–342.

Chiu, I.M., Von Hehn, C.W., Woolf, C.J., 2013. Neurogenic inflammation – the peripheral nervous system's role in host defense and immunopathology. Nat. Neurosci. 15, 1063–1067.

Corrado, C., Raimondo, S., Chiesi, A., 2013. Exosomes as intercellular signaling organelles involved in health and disease: basic science and clinical applications. Int. J. Mol. Sci. 14, 5338–5366.

Dai, C., Li, J., Lin, W., 2012. Electrophysiology and ultrastructural changes in mouse sciatic nerve associated with colistin sulfate exposure. Toxicol. Mech. Meth. 22, 592–612.

Dalaka, M.C., Gooch, C., 2016. Close to the node but far enough: what nodal antibodies tell us about CIDP and its therapies. Neurology 86 (9), 796–797.

Damodaran, T.V., Patel, A.G., Greenfield, S.T., et al., 2006a. Gene expression profiles of the rat brain both immediately and 3 months following acute sarin exposure. Biochem. Pharmacol. 71, 497–520.

Damodaran, T.V., Greenfled, S.T., Patel, A.G., et al., 2006b. Toxicogenomic studies of the rat brain at an early time point following acute sarin exposure. Neurochem. Res. 31, 361–381.

Damodaran, T.V., 2009. Molecular and transcriptional responses to sarin exposure. In: Gupta, R.C. (Ed.), Handbook of Toxicology of Chemical Warfare Agents. Academic Press/Elsevier, Amsterdam, pp. 665–682.

Damodaran, T.V., Gupta, R.P., Attia, M.K., et al., 2009. DFP initiated early alterations of PKA/p-CREB pathway and differential persistence of beta-tubulin subtypes in the CNS of hens contributes to OPIDN. Toxicol. Appl. Pharmacol. 240, 132–142.

Damodaran, T.V., Attia, M.K., Abou-Donia, M.B., 2011. Early differential cell death and survival mechanisms initiate and contribute to the development of OPIDN: a study of molecular, cellular, and anatomical parameters. Toxicol. Appl. Pharmacol. 256, 348–359.

Damodaran, T.V., Abou-Donia, M.B., 2014. Differential expression of MAP-2 (microtubule associated protein-2) gene transcripts in the central nervous system of hens treated with DFP (diisopropyl phosphorfluoridate). In: Proceedings of 53rd Ann. Soc. Toxicol. Toxicol. Sci. 26.

Damodaran, T.V., Attia, M.K., Abou-Donia, M.B., 2015. Differential protein expression of CaMKinase (CaMK) II α and β subunits as well as tau mRNA expression contribute to the development of OPIDN. In: 54th Ann. Soc. Toxicol. Toxicol. Sci., vol. 27.

Damodaran, T.V., Abou-Donia, M.B., 2016. Early differential induction of Jun-D in the central nervous system of hens treated with diisopropylphosphorofluoridate (DFP) may be involved in OPIDN. In: 55th Ann Soc Toxicol. Toxicol. Sci., vol. 28.

Damodaran, T.V., Attia, M.K., Abou-Donia, M., 2017. Potential role of differential and persistent overexpression of GAP-43 mRNA in DFP-induced OPDIN related tissue specific pathways in cell survival and cell death mechanisms. In: 56th Ann Soc Toxicol. Toxicol. Sci., vol. 29.

Dequen, F., Filali, M., Larivière, R.C., 2010. Reversal of neuropathy phenotypes in conditional mouse model of Charcot-Marie-Tooth disease type 2E. Hum. Mol. Genet. 19, 2616–2629.

De Vos, K.J., Grierson, A.J., Ackerley, S., et al., 2008. Role of axonal transport in neurodegenerative diseases. Annu. Rev. Neurosci. 31, 151–173.

Ding, J., Allen, E., Wang, W., 2006. Gene targeting of GAN in mouse causes a toxic accumulation of microtubule-associated protein 8 and impaired retrograde axonal transport. Hum. Mol. Genet. 15, 1451–1463.

Dugas, J.C., Notterpek, L., 2011. MicroRNAs in oligodendrocyte and Schwann cell differentiation. Dev. Neurosci. 33, 14–20.

Eschbach, J., Dupuis, L., 2011. Cytoplasmic dynein in neurodegeneration. Pharmacol. Ther. 130, 348–363.

Fanar, P., Wong, Y.A., Husted, K.H., 2012. Cerebrospinal fluid-based kinetic biomarkers of axonal transport in monitoring neurodegeneration. J. Clin. Invest. 122, 3159–3169.

Fehmi, J., Scherer, S.S., Willison, et al., 2018. Nodes, paranodes and neuropathies. J. Neurol. Neurosurg. Psychiatry 89, 61–71.

Fidanboylu, M., Griffiths, L.A., Flatters, S.J., et al., 2011. Global inhibition of reactive oxygen species (ROS) inhibits paclitaxel-induced painful peripheral neuropathy. PLoS One 6, e25212.

Filippi, M., Agosta, F., 2010. Imaging biomarkers in multiple sclerosis. J. Magn. Reson. Imag. 31, 770–788.

Goins, W.F., Cohen, J.B., Glorioso, J.C., 2012. Gene therapy for the treatment of chronic peripheral nervous system pain. Neurobiol. Dis. 48, 255–270.

Griffin, J.L., Waters, N., Burczybski, M.E., et al., 2006. Metabonomics: metabolic profiling and pattern recognition analysis of body fluids and tissues for characteristion of drug toxicity and disease diagnosis. In: Burczybski, M.E., Rokett, J.C. (Eds.), Surrogate Tissue Analysis: Genomic, Proteomic, and Metabolomic Approaches. Taylor and Francis, Boca Raton, Fl, pp. 143–164.

Hertz, D.L., Roy, S., Jack, J., 2014. Genetic heterogeneity beyond CYP2C8*3 does not explain differential sensitivity to paclitaxel-induced neuropathy. Breast Cancer Res. Treat 145 (1), 245–254.

Hoke, A., Simpson, D.M., Freeman, R., 2013. Challenges in developing novel therapies for peripheral neuropathies: a summary of the Foundation for Peripheral Neuropathy Scientific Symposium 2012. J. Peripher. Nerv. Syst. 18, 2013.

Jasmin, L., Vit, J.P., Bhargava, A., et al., 2010. Can satellite glial cells be therapeutic targets for pain control? Neuron Glia Biol. 6, 63–71.

Jellinger, K.A., Wenning, G.K., 2016. Multiple system atrophy: pathogenic mechanisms and biomarkers. J. Neural. Transm. 123, 555–572.

Jeng, S.F., Rau, C.S., Liliang, P.C., 2009. Profiling muscle-specific microRNA expression after peripheral denervation and reinnervation in a rat model. J. Neurotrauma 26, 2345–2353.

Kirchgessner, A.L., Liu, M.T., Alcantara, F., 1997. Excitotoxicity in the enteric nervous system. J. Neurosci. 17, 8804–8816.

Komatsu, M., Wheeler, H.E., Chung, S., et al., October 1, 2015. Pharmacoethnicity in paclitaxel-induced sensory peripheral neuropathy. Clin. Cancer Res. 21 (19), 4337–4346.

Lindwall, C., Kanje, M., 2005. Retrograde axonal transport of JNK signaling molecules influence injury induced nuclear changes in p-c-Jun and ATF3 in adult rat sensory neurons. Mol. Cell. Neurosci. 29, 269–282.

Lorente, L., 2015. New prognostic biomarkers in patients with traumatic brain injury. Arch. Trauma Res. 4 (4), e30165.

Mack, T.G., Reiner, M., Beirowski, B., 2001. Wallerian degeneration of injured axons and synapses is delayed by a Ube4b/Nmnat chimeric gene. Nat. Neurosci. 4, 1199–1206.

Marques-Deak, A., Giovanni, C., Fandeh, E., 2006. Measurement of cytokines in sweat patches and plasma in healthy women: validation in a controlled study. J. Immunol. Methods 315, 99–109.

Marques, A.H., Silverman, M.N., Sternberg, E.M., 2010. Evaluation of the stress-related system applying non-invasive methodologies: salivary cortisol, heart rate variability and measurements of neuroimmune biomarkers sweat. Neuroimmunomodulation 17, 205–208.

Mckeon, A., Benarroch, E.E., 2016. Autoimmune autonomic disorders. Handb. Clin. Neurol. 133, 405–416.

Miltenburg, N.C., Boogerd, W., 2014. Complications of Treatment. Chemotherapy-induced neuropathy: a comprehensive survey. Cancer Treat Rev. 40, 872–882.

Ness, J.K., Skiles, A.A., Yap, E.H., et al., June 2016. Nuc-ErbB3 regulates H3K27me3 levels and HMT activity to establish epigenetic repression during peripheral myelination. Glia 64 (6), 977–992.

Noguerol, T.M., Barousse, R., Socolovsky, M., et al., 2017. Quantitative magnetic resonance (MR) neurography for evaluation of peripheral nerves and plexus injuries. Quant. Imag. Med. Surg. 7 (4), 398–421.

Ohno, K., Rahman, M.A., Nazim, M., et al., August 2017. Splicing regulation and dysregulation of cholinergic genes expressed at the neuromuscular junction. J. Neurochem. 142 (Suppl. 2), 64.

Osterloh, J.M., Yang, J., Rooney, T.M., 2012. dSarm/Sarm1 is required for activation of an injury-induced axon death pathway. Science 337, 481–484.

Park, K.K., Liu, K., Hu, Y., 2008. Promoting axon regeneration in the adult CNS by modulation of the PTEN/mTOR pathway. Science 322, 963–966.

Poewe, W., Seppi, K., Tanner, C.M., et al., March 2017. Parkinson disease. Nat. Rev. Dis. Primers 23 (3), 17013.

Raad, M., Nohra, E., Chams, N., et al., 2014. Autoantibodies in traumatic brain injury and central nervous system trauma. Neuroscience 281, 16–23.

SABiosciences, Qiagen Company, 2013. SABiosciences, Qiagen Company, RT Profiler PCR Arrays. Frederick, MD, USA.

Saman, S., Kim, W., Raya, M., 2012. Exosome-associated tau is secreted in tauopathy models and is selectively phosphorylated in cerebrospinal fluid in early Alzheimer disease. J. Biol. Chem. 287, 3842–3849.

Schliebs, R., Arendt, T., 2011. The cholinergic system in aging and neuronal degeneration. Behav. Brain Res. 221, 1–9.

Segman, R.H., Shefi, N., Goltser-Dubner, T., et al., 2005. Peripheral blood mononucleat cell gene expression profiles identify emergent post-traumatic stress disorders among traumatic survivors. Mol. Psychiatr. 10, 500–513.

Slikker, W.T., Bowyer, J.F., 2005. Biomarkers of adult and developmental neurotoxicity. Toxicol. Appl. Pharmacol. 206, 255–260.

Starks, S.E., Hoppin, J.A., Kamel, F., 2012. Peripheral nervous system function and organophosphate pesticide use among licensed pesticide applicators in the agricultural health study. Environ. Health Perspect. 120, 515–520.

Sun, F., Park, K.K., Belin, S., 2011. Sustained axon regeneration induced by co-deletion of PTEN and SOCS3. Nature 480, 372–375.

Surgucheva, I., Sharov, V., Surguchov, A., 2012. γ-synuclein: seeding of α-synuclein aggregation and transmission between cells. Biochemistry 51, 4743–4754.

Tang, Y., Gilbert, D.L., Glauser, T.A., et al., 2006. Blood genomic fingerprints of brain diseases. In: Burczynski, M.E., Rockett, J.C. (Eds.), Surrogate Tissue Analysis: Genomic, Proteomic, and Metabolomic Approaches. CRC Press LLC, Boca Raton, FL, pp. 31–46.

Vavlitou, V., Sargiannidou, I., Markoullis, K., 2010. Axonal pathology precedes demyelination in a mouse model of X-linked demyelinating/type I Charcot-Marie Tooth neuropathy. J. Neuropathol. Exp. Neurol. 69, 945–958.

Verrier, J.D., Semple-Rowland, S., Madorsky, I., 2010. Reduction of Dicer impairs Schwann cell differentiation and myelination. J. Neurosci. Res. 88, 2558–2568.

Viader, A., Golden, J.P., Baloh, R.H., 2011. Schwann cell mitochondrial metabolism supports long-term axonal survival and peripheral nerve function. J. Neurosci. 31, 10128–10140.

Wang, X.M., Wang, T.J., Lehky, M.J., 2012. Discovering cytokines as targets for chemotherapy-induced painful peripheral neuropathy. Cytokine 59, 3–9.

Wang, Y., Sun, S., Zhu, J., et al., 2015. Biomarkers of Guillain-Barré syndrome: some recent progress, more still to be explored. Mediat. Inflamm. 2015, 564098.

Williams, P.R., Marincu, B.N., Sorbara, C.D., et al., 2014. A recoverable state of axon injury persists for hours after spinal cord contusion in vivo. Nat. Commun. 5, 5683.

Wu, D., Murashov, A.K., 2013 Apr 1. Molecular mechanisms of peripheral nerve regeneration: emerging roles of microRNAs. Front. Physiol. 4, 55. https://doi.org/10.3389/fphys.2013.00055. eCollection 2013.

Wostyn, P., De Deyn, P.P., November 2017. Intracranial pressure-induced optic nerve sheath response as a predictive biomarker for optic disc edema in astronauts. Biomark. Med. 11 (11), 1003–1008.

Yu, B., Zhou, S., Wang, Y., Ding, G., Ding, F., Gu, X., 2011. Profile of microRNAs following rat sciatic nerve injury by deep sequencing: implication for mechanisms of nerve regeneration. PLoS One 6 (9), e24612 doi.

Yun, B., Anderegg, A., Menichella, D., Wrabetz, L., Feltri, M.L., Awatramani, R., 2010 Jun 2. MicroRNA-deficient Schwann cells display congenital hypomyelination. J. Neurosci. 30 (22), 7722–7810. https://doi.org/371/journal.pone.0024612. Epub 2011 Sep 13.

CHAPTER

12

Cardiovascular Toxicity Biomarkers

Csaba K. Zoltani

Emeritus US Army Research Lab, Aberdeen Proving Ground, MD, United States

INTRODUCTION

Cardiac biomarkers reflect the current status of the heart. A marker may be a protein, an enzyme, or a hormone whose concentration deviates from the norm. A change in electrical activity, a structural change of the organ, an alteration in a physiologic measure of a functional process, such as blood pressure (BP), may indicate pathology. Biomarkers also serve as surrogate end points, useful for evaluation of benefit or harm of clinical interventions.

The deleterious response to an externally introduced substance mirrors the analogous effects of disease-producing toxins and ensuing pathological changes that alter cardiac function. Cardiac toxicity caused by xenobiotics can be from chemicals, drugs, pesticides, and heavy metals. Vascular toxicity is indicated by changes in wall structure and behavior, by a loosening of atherosclerotic deposits, or their injection into the bloodstream. Xenobiotics primarily interfere with metabolic processes or alter ionic processes causing necrosis and inflammation of the myocardium. As a consequence, when biomarkers are released, their presence gives an indication of the suspected cause and may indicate morbidity. Biomarkers thus serve both for risk stratification and also as prognosticators.

Because of poor specificity of the traditional biomarkers, such as aspartate aminotransferase (AST), lactate dehydrogenase (LDH), and creatine kinase (CK), several new markers have been developed, such as troponins, myosin light chain 3, and fatty acid binding proteins (FABP3). Because of their tissue specificity, validation, timeliness, and moderate cost of assay, microRNAs are replacing some of these conventional markers with a new palette of cardiac biomarkers.

The primary biomarkers measured in body fluids (blood, serum, and plasma) among others are troponin T and I, which rise within hours after myocardial injury, and usually indicate necrosis of the tissue. Natriuretic peptides (ANP and BNP), an intracellular fatty acid–binding protein (FABP3) and a marker for myocardial damage, are released into plasma after myocardial injury. Measurements of an increase in myoglobin, because of its lack of specificity and CK, have fallen out of favor (Kehl et al., 2012). Cardiotoxicity is also evidenced by changes in the ionic flow in the heart vasculature, measured, and expressed by the electrocardiogram (ECG), where deviation from the norm is usually the first indicator of cardiac injury. Noteworthy change in the QRS, the complex that reflects the depolarization of the ventricles, and an increase in the amplitude of the T-wave (repolarization of the ventricles), and ST-segment elevation (may indicate MI or left bundle branch block) and R-R (time between beat changes) are initial indicators of injury.

PubMed, when queried under "cardiac biomarkers" in October 2017, offered over 51,000 references. Their current utilization has also been summarized in several recently published books by Januzzi (2011), Maisel and Jaffe (2016), Patel and Preedy (2016), and Preedy and Patel (2017), as well as by Morrow (2006). Reviews by Jaffe et al. (2006), Braunwald (2008), Martin-Ventura et al. (2009), and Wettersten and Maisel (2016) give an overview of currently used markers. Excellent and comprehensive review was given by Upadhyay (2015), for all facets of cardiovascular disorders.

At present, we are witnessing the changing of the guard in cardiac biomarkers. The protein-based markers are being supplemented and replaced by miRNAs that are more sensitive, cell- and organ-specific, and faster in delineating the source and cause of cardiac injury.

Biomarkers in Toxicology, Second Edition
https://doi.org/10.1016/B978-0-12-814655-2.00012-8

Epigenetics offers new insights into environmental influence on cardiac functioning. Heart failure (HF) can be caused by subdued energy production because of mitochondrial DNA dysfunction. Biomarkers and epigenetic biomarkers of these processes are discussed in detail.

PHYSIOLOGY OF THE CARDIOVASCULAR SYSTEM

Basic Functions

The cardiovascular system delivers oxygen, nutrients, and hormones to tissues of the body and removes waste products, carbon dioxide, and metabolic end products with the aid of several organs, including the lungs, liver, and kidneys. It consists of a four-chambered pump and two circulating systems: the pulmonary and the systemic circulation. The pulmonary system transfers blood to the lungs to eliminate waste gases and then capture oxygen. The systemic circulation channels blood with the newly acquired oxygen through arteries, arterioles, and capillaries, which facilitate the exchange of gases and nutrients for waste products, and returns it to the heart by means of the venules and veins.

The myocardial tissue is similar to skeletal muscle but differs in several essential aspects. Cardiac myocytes are composed of contractile elements, thick and thin myofibrils. Protein myosins constitute the former, whereas thin elements are made of actins. Tropomyosin and troponin interdigitate with the thick filaments. Myocytes joined by intercalated disks facilitate the electrical propagation. Cardiac muscle fibers consist of cells serially connected and laying parallel to each other. The intercalated disks fuse the cells allowing easy ion movement and thus communication and progress of the action potential. In essence, the cardiac muscle is a syncytium of cardiac cells with an interconnection allowing rapid spread of the action potential.

The syncytia of the atria and the ventriculum are separated; thus the signal (potential) goes by special fibers, allowing the atria to contract ahead of the ventricles.

The excitation–contraction coupling of cardiac muscle differs from skeletal muscle in that additional calcium ions diffuse into the sarcoplasm from the T-tubules, which have diameters much larger than those of skeletal muscle, during the action potential. The contraction of the cardiac muscle depends on the calcium ion concentration in the extracellular fluid.

Cardiac Cycle

The events of a cardiac cycle, from the beginning of one heartbeat to the next, consist of a period during which the heart fills with blood, referred to as diastole. The cycle is initiated by generation of an action potential at the sinus node, which is located at the superior lateral wall of the right atrium near the superior vena cava. The action potential proceeds through the atria to the AV node and then into the ventricles. This moves the blood into the body's vascular system. From the right ventricle, the deoxygenated blood passes to the lungs where it is oxygenated, then back to the left atrium and the left ventricle, where it passes to the body.

Nervous System

The autonomic nervous system, consisting of the sympathetic and parasympathetic portions, innervates the heart. Sympathetic stimulation is very effective in substantially doubling the output, i.e., the pumping capacity of the heart, including the ejection pressure when required. The parasympathetic (vagus) nerves can decrease the output noticeably. Inhibition of the sympathetic nerves modulates the pumping, and depresses the heart rate (HR) and the strength of the muscle contraction. Strong stimulation of the vagal system decreases the heart muscle contraction. Vagal fibers are predominantly on the atria, and thus HR, and not the strength of the contraction of the ventricles, is affected.

Sympathetic efferent nerves are located at the atria and the ventricles along the outer wall of blood vessels and the sinoatrial (SA) and atrioventricular (AV) nodes. Release of the neurotransmitter norepinephrine occurs at varicosities. The vagus nerve, constituting the parasympathetic nervous system, regulates the heartbeat and mainly innervates the SA and AV nodes. The vagus nerve exerts an inhibitory action, whereas the sympathetic nervous system exerts excitatory actions, stimulating the sinus node and the atrioventricular conduction.

Immune System

Cardiac injury activates inherent immune mechanisms to alleviate the ensuing change in the affected area. This involves the initiation of inflammatory reactions, such as the toll-like receptors (TLRs) (Shintani et al., 2013), and reactive oxygen species (ROS) generation that in turn induces nuclear factor (NF)-κB activation responsible for cytokine production. In addition, transfer growth factor TGF-β (a multifunctional cytokine), mediator of inflammation and fibrosis (Frangogiannis, 2008), plays an important role. Thus, several cardiac biomarkers reflect the response of the immune system.

Neurohormonal System

The central nervous system's afferent nerve fibers carry information from the baro-, chemo-, and volume receptors that enable the regulation of the cardiac and vascular functions. Autonomic nerves and circulating hormones are the important conveyors in this process.

In HF, caused by disease or xenobiotics, neurohormonal modulation occurs (Francis, 2011). Vasoconstrictor hormones that are antinatriuretic and antidiuretic, as well as vasodilator hormones, are activated. As the disease progresses, the former hormones are overwhelmed by the latter, resulting in an elevation of plasma noradrenaline, a sign of left ventricular dysfunction culminating in ventricular arrhythmia. Subsidiary effects include the downregulation of calcium-regulating genes and apoptosis, and thus a loss of myocytes.

With an increase of the sympathetic nervous activity, atrial and brain natriuretic peptides (ANP and BNP) are synthesized and released by the atria in response to stretch partly caused by activation of the renin–angiotensin–aldosterone system, resulting in water retention. Initially ANP and BNP are instrumental in controlling HF, but as the severity increases, their effect is attenuated. The change in the ANP/BNP ratio, in effect a cardiac biomarker, gives a measure of the LV dysfunction. Other hormones, including endothelins and growth hormone, are elevated.

Right ventricular failure induces neurohormonal activation. Plasma norepinephrine rises with pulmonary artery pressure and atrial natriuretic peptide levels (Nootens et al., 1995).

Vascular System

The vascular system consists of blood vessels whose diameters are controlled by the vascular wall autonomic nerves and metabolic and biochemical signals from outside the blood vessels; it responds to and adapts to changing conditions. Vasoactive substances are released by the endothelial cells that control homeostasis and inflammatory responses.

Vascular toxicity is characterized by degenerative and inflammatory changes in blood vessels by alteration in membrane function and structure. Modulation of the contractile proteins in the vascular cells leads to the loss of homeostasis.

Xenobiotics

Xenobiotics, chemicals found in an organism but not normally present, affect body functions in a myriad of ways. This is expressed by organelles, genetic dysfunctions, mitochondrial or sarcolemmal injury, and interference with ion homeostasis. The immune system reacts and biomarkers are independently released from the myocardium into the circulation. There are indicators of impending changes, including necrosis and inflammation, as the immune system tries to counteract the presence of xenobiotics.

Oxidative stress is also caused by xenobiotics that promote cardiac toxicity (Costa et al., 2013). Sources include catecholamines, amphetamines, and excessive alcohol use.

EPIGENETIC MODIFICATIONS

Epigenetics refers to heritable transformations that are not solely caused by the genetic code per se but processes that activate or deactivate genes. The primary activities of chromatin modifications are DNA methylation and those controlled by miRNAs and noncoding RNAs. The effect is variable expression of genetic information reflecting environmental conditions.

The primary epigenetic modification involves DNA methylation that involves covalent binding of a methyl group to the 5′-position of a cytosine in the CpG (cystosine preceding guanosine) dinucleotides. CpG-islands harbor sites of transcription initiation and transcriptional repression (Lorenzen et al., 2012) by impeding the binding of transcription factors in the promoter regions of genes.

Modification of histones around which DNA is wound is catalyzed by enzymes regulating the accessibility of the DNA for transcription. Histone methylation impacts transcriptional activation and repression and thus cardiac function. MicroRNAs play an important role in impaired protein production, epigenetic silencing. Long noncoding RNAs also impact the development of disease. Thus specific miRNA expression profiles are biomarkers for treatment prognosis.

Epigenetic biomarkers thus are useful in predicting and following the progress of cardiovascular disease (Garcia-Gimenez, 2015) and genome-wide profiling technologies (Dirks et al., 2016). Genome-wide profiling technologies enable the identification of epigenetic biomarkers. Bisulfite-based DNA methylation profiling is preferred as it is not affected by freezing or chemical fixation and requires relatively small samples. Other techniques take advantage of mutated histone-modifying enzymes directly affecting posttranslational histone alterations. ChIP-Seq, under development, is another technique likely to be adopted. It analyzes protein interactions with the DNA to identify the binding sites of associated proteins.

Circulating posttranslational-modified histones expressed in acetylation, methylation, and phosphorylation are also considered biomarkers. The modifications can modify the structure of the chromatin thereby activating or silencing of genes.

Identification of epigenetic biomarkers in cardiovascular diseases (Backs and McKinsey, 2016) has markedly contributed to the unraveling of chromatin-based environment influenced mechanisms (Webster et al., 2013; Stefanska and MacEwan, 2015; Kaikkonen et al., 2011).

MITOCHONDRIAL DYSFUNCTION

The large number of mitochondria, structures within cells, produce energy by oxidative phosphorylation. Simple sugars and oxygen are used to create adenosine triphosphate (ATP). Enzyme complexes carry out the oxidative phosphorylation.

Other cellular activities, including apoptosis and production of heme and cholesterol, are also performed by the mitochondria. Mitochondria also carry their own DNA that contains 37 genes. Thirteen of these genes take part in making enzymes involved in oxidation phosphorylation. Others produce RNAs that are instrumental in amino acids.

The primary task of the mitochondria, energy production, involves electron transport and phosphorylation apparatus as ATP is synthesized in the mitochondrion by addition of a third phosphate group to ADP by oxidative phosphorylation.

Mitochondrial dysfunction is at the root of a number of heart diseases. HF is associated with decreased mitochondrial biogenesis (Brown et al., 2017; Dorn et al., 2015). Sites of mitochondrial defects are within the electron transport and phosphorylation apparatus.

A number of biomarkers signal mitochondrial dysfunction. Changes in mitochondrial genome to nuclear genome (Mt/N) ratio in circulating cells signal dysfunction (Malik and Czajka, 2013). Increase in ROS with unpaired mitochondrial electron transport chain activity is also a sign (Frijhoff et al., 2015). In patients with mitochondrial disorders, FGF-21 is elevated and GDF15 is increased.

PROTEOMICS

Proteomics, the characterization of proteins within a biological system, has furnished insight into the behavior of cardiovascular entities and led to the discovery of biomarkers (Lam et al., 2016). Changes in proteome configurations describe the pathogenesis, protein alterations corresponding to the elicited pathological conditions.

Mass spectroscopy measurement of changes in plasma proteome due to cardiac dysfunction is available. The structure and abundance change of proteins are markers of disease. For example, in case of HF, a decrease in phosphorylation of protein kinase A located in the N-terminal extension of cardiac troponin I, as well as at other sites, is apparent. Proteomics data in cardiovascular medicine now offer new biomarkers and access to explanation of transformation processes (Anderson and Anderson, 2002).

CARDIAC TOXICITY

Toxic compounds, including drugs, can directly affect the heart. Typical effects are the formation of lesions and biochemical changes, including enzyme induction or inhibition with changes in metabolic pathways.

One of the characteristics of our environment is the acceptance of the introduction and distribution of man-made chemicals without the benefit of knowing possible side effects. Foods are part of this process and, in health care, substances that have not been a part of natural evolution are part of the treatment given to overcome undesirable genetically processed changes.

Many drugs, including the widely used antineoplastic adriamycin, are cardiotoxic. Adriamycin's toxicity is dosage-dependent and time-delayed. Although the precise mechanism of action is unknown, it is known that certain medications bind to cardiolipin on the membrane of myocytes. In addition, iron and redox reactions, resulting in the production of superoxide anions, may contribute to abnormal protein processing and hyperactivated innate immune responses, thereby causing inhibition of neuregulin-1 signaling (Shi et al., 2010).

Hydralazine, a drug used for hypertension, causes myocardial necrosis. The metabolism of allylamine, through amine oxidases, causes the formation of acrolein, a toxic compound. Likewise, heavy metals, including cobalt, cause cardiomyopathy by interfering with calcium ions in muscle tissue. There are countless other examples. These chemicals adversely affect the heart, and cardiac biomarkers give testimony to their effect.

BIOMARKER OF MYOCARDIAL IONIC DYSFUNCTION

Electrocardiogram

Among the cardiac biomarkers, the ECG, because of its ease of use and specificity in many instances, is one of the most widely used assays for detection of cardiac damage. Primarily, T-wave elevation (ventricular repolarization), ST elevation/depression (ventricle systolic depolarization) in leads I, II, III, or a limb lead of the 12-lead ECG is a hallmark of myocardial infarction (MI) and other cardiac conditions, and Q-wave evolution is an important feature that may indicate cardiac damage. The 12-lead ECG has a sensitivity of 45%–68% for MI (Menown et al., 2000) based on ST elevation. QT interval change can be considered a biomarker of sudden cardiac death, whereas the PR interval should be considered as a biomarker and indicator of atrial fibrillation. Drug-induced QT prolongation is caused by blockade of delayed rectifier potassium current, raising the possibility of arrhythmia. Overexpression of nitric oxide (NO) synthase 1 adaptor protein in cardiac myocytes leads to action potential shortening and is associated with QT interval and sudden cardiac death (Lilly, 2003).

Electrocardiographic biomarkers have been extensively used making long-term prognosis on cardiovascular events in postmenopausal women without known cardiovascular disease (Gorodeski et al., 2011). Fourteen ECG biomarkers, associated with long-term prognosis, were selected. They reflected autonomic tone, (ventricular rate and variability), atrial conduction (P-wave characteristics) QT duration, and ventricular repolarization. Random survival forests, a nonparametric decision tree–based approach, was used to identify clinical and ECG predictors of mortality.

The widely used Minnesota Code (MC) and Novacode (NC) to identify Q-wave, ST-segment, and T-wave abnormalities are also strong predictors for CHD mortality.

METABOLOMICS

The metabolite footprints left by cellular processes are suitable biomarkers for the determination of therapeutics (Pintus et al., 2017; Ussher et al., 2016). Ischemia, CVD, and HF cause measurable cellular, functional changes; and certain components, because of changes in their occurrence, play important roles.

Cardiovascular diseases (CVD) are estimated to cause approximately one-third of the deaths in the developed world. PTX3, marker of inflammation and vascular damage, plays an important role in CVD (Fornai et al., 2016). Amyloid A is a recognized biomarker for acute coronary syndrome (Yayan, 2013). Low-density lipoprotein receptor-1 is significantly elevated in ACS and is considered a biomarker for coronary artery disease and marker for potential plaque rupture.

BIOMARKERS OF WALL STRETCH

Natriuretic Peptide

Peptide hormones regulate cardiovascular, endocrine, and renal homeostasis. Wall stretch and ischemia is the predominant mechanism for synthesis and release of natriuretic peptides from cardiac atria and the ventricles (Maisel et al., 2002). The biological effect is mediated by their binding to cell surface receptors of the cardiovascular system, and this acts as an antagonist to the major vasoconstrictor neurohormonal axis. Many factors, not only cardiac conditions, can affect the presence of NPB (B-type natriuretic peptide) and NT-proBNP (N-terminal fragment of its prohormone) (Balion et al., 2008).

In cardiac damage, biomarkers such as atrial natriuretic peptide (ANP) and B-type natriuretic peptide (BNP) are cardiac-derived polypeptides encoded by a gene on chromosome 1 and produced both in the atria and the ventricles; the concentration of BNP is lower than that of ANP and C-type natriuretic peptide (CNP), a 22-amino acid polypeptide formed in the brain and in the vascular endothelium. These three polypeptides result from cleavages of pre-prohormones and prohormones. ANP and BNP are secreted in plasma with half-lives of ~1 h and 20 min. The circulating level of CNP is very low. Wall stretch is primarily responsible for the synthesis and release of BNP, and its release from cardiac myocytes is activated in HF. Norepinephrine and proinflammatory cytokines are simultaneously released. BNP's physiological actions include relaxation of vascular smooth muscle and venous and arterial dilatation. BNP affects left ventricular filling patterns by a lusitropic (myocardial relaxation) effect. BNP and NT-proBNP are markers of ventricular dysfunction and ventricular wall distension. CNPs are endogenous polypeptide mediators with natriuretic and vasodilator effects. CNP is derived from the brain and the vascular endothelium. Its circulating level in the bloodstream is low. Its effect is vasodilatory by activation of potassium channels. Contrary to ANP and BNP, CNP does not directly participate in natriuretic activity. Mainly, the cardiovascular effect of these peptides is vasodilatory, i.e., to decrease vascular reactivity (Table 12.1).

Circulating levels of ANP and BNP are elevated in congestive heart failure (CHF) (Tang et al., 2008). BNP levels above 100 pg/mL indicate the presence of HF.

TABLE 12.1 Biomarkers of Cardiac Toxicity

Biomarker	Source	Injury Type
Myoglobin	Muscle including the myocardium	Released on muscle damage
C-Reactive protein (CRP)	Necrosis causes the release of cytokines	Inflammation of arteries causes elevated level: precursor
	Triggering synthesis in the liver; binds to and clears	To heart attack
	Apoptotic cells	
Creatine kinase (CK-MB)	Myocardium enzyme; catalyzes conversion of ADP into ATP	Myocardial infarction elevated with muscle damage
Cardiac troponin	Cardiomyocytes, several types	Irreversible myocardial injury
B-type natriuretic peptide	Ventricular myocardium	Pressure overload, wall tension, heart failure
microRNA	Over 1000 types each reflecting the location and type of injury under validation	Significant volume change in plasma, serum, and blood
		On cardiac injury; the type depends on the injury
		Location and is very rapidly detectable in body fluids

The concentration reflects the severity of symptoms, such as the degree of LV (left ventricle) dysfunction.

Choline

A marker of ischemia, choline is released early into the circulation after cleaving of the phospholipids that are the constituents of cell membranes. In ACS (acute coronary syndrome) choline is increased. In conjunction with troponin, it is a marker of major adverse cardiac events. As biomarker it is used for prognostication of cardiac arrest and death (Zeisel, 2000; Danne et al., 2003). It is used to discriminate between high- and low-risk subgroups in troponin-positive patients.

Fatty Acid–Binding Protein

The cytoplasm of striated muscle cells contains fatty acid–binding proteins (FABPs) that transport long-chain unesterified fatty acids. The heart-type FABP (H-FABP) is rapidly released from the myocardium into circulation in response to myocardial injury (Azzazy et al., 2006). Because of the fact that it is also present in renal failure, its specificity for MI remains unresolved. However, it is 20 times more specific to cardiac muscle than myoglobin, and it is 10-fold lower in skeletal than cardiac muscle (Ghani et al., 2000). Levels of H-FABP exceeding 6.48 μg/L indicate a significantly higher probability of future adverse events than do low troponin levels or absent indicators of necrosis.

Mir-1, the muscle-specific microRNA, regulates FABP3, making its assessment in the plasma useful for determining cardiac pathological conditions (Catalucci et al., 2015). The inverse relationship between circulating level of FABP3 and miR-1 is a useful biomarker.

BIOMARKERS OF NECROSIS

Troponin

Troponins are regulatory proteins and part of the contractile mechanism of the cardiac muscle. Troponin is bound within the filament of the contractile apparatus. When cardiac myocytes are damaged, troponin is released into the circulation. At first the cytosolic pool is released, and then the structurally bound troponin enters the circulation. Elevated levels indicate myocardial damage. Cardiac troponin I levels of 1.0 μg/L or higher or cardiac troponin T levels of 0.1 μg/L or higher are considered elevated (Peacock et al., 2008).

Troponin is a complex of three regulatory proteins, troponin C (TnC), troponin T (TnT), and troponin I (TnI), which are integral to nonsmooth muscle contraction in cardiac muscle. They are located between actin filaments of muscle tissue. TnC binds to calcium ions and produces conformational change in TnI. TnT binds to tropomyosin and TnT binds to actin.

Troponin is difficult to detect in unaffected muscle, but troponin levels rise several hours after the onset of myocardial injury, such as MI. It is detectable up to 10 days after onset of injury. The degree of elevation of troponin also gives prognostic information on the subsequent outcome (Keller et al., 2009).

Troponin plays an important role during excitation–contraction coupling. During excitation, calcium ions bind to TnC; it interacts with tropomyosin to unblock active sites between the myosin filament and actin allowing cross-bridge cycling and thus contraction of the myofibrils that constitute the systole. Tension is reduced as the extracellular calcium entering the myocardial cells is reduced. Phosphorylation of TnI inhibits calcium binding of TnC whereby tropomyosin blocks the interaction of actin with myosin. During the subsequent diastole, one calcium ion is exchanged for three sodium ions. With FDA's recent approval of the next-generation assay, Roche's TnT Gen 5 Stat that delivers more sensitivity, the interpretation of the results is expected to lead to new standards.

Creatine Kinase

CK is an enzyme that catalyzes the conversion of phosphocreatine, the energy reservoir for regeneration of ATP, degrading ATP to ADP. CK consists of two subunits, a muscle-type M and a brain-type B. CK consists of three isoenzymes: CK-MM in skeletal muscle, CK-BB in brain, and CK-MB in heart. The myocardium has about 30% CK-MB, and thus lacks the specificity for cardiac damage. As noncardiac damage may also cause a rise, it is not used for MI diagnosis. CK-MB increases and disappears 12–36 h following the onset of symptoms. CK-MB usually returns within 2 days to baseline values. A relative index value, the ratio of CK-MB/CK, has a normal value 1. Values greater than 2.5–3.0 indicate heart damage. CK-MB has specificity not only for cardiac tissue but also rises for blunt chest trauma and cocaine abuse. It is no longer used as the main indicator of cardiac necrosis (Apple, 1999).

Myoglobin

Myoglobin is an iron- and oxygen-binding protein and it is related to hemoglobin. Myoglobin is only found in the bloodstream on injury of the muscles. It contains the heme prosthetic group that has iron-containing porphyrin in the center (Wittenberg and Wittenberg, 2003).

As one of the earliest biomarkers, it is a heme protein that is abundant in the cytoplasm of cardiac muscle cells. It appears in circulation within an hour of myocardial necrosis. Its concentration increases before that of troponin. Myoglobin is ubiquitous in skeletal muscle and therefore it lacks specificity for myocardial injury. Its concentration is elevated also in renal insufficiency but displays excellent early sensitivity, increases more rapidly than troponin and CK, but also has a short plasma half-life. Thus, elevated myoglobin has low specificity for acute myocardial infarction. Its normal value is in the range of 17.4–105.7 ng/mL.

BIOMARKERS OF INFLAMMATION

C-Reactive Protein

Among the biomarkers of inflammation and swelling of the arteries, the highly sensitive C-reactive protein (hs-CRP) is also widely used as an indicator of atherothrombosis.

CRP stimulates monocyte release of cytokines, including IL-1, IL-6, and tumor necrosis factor (TNF)-alpha and expression of intercellular adhesion molecule and vascular adhesion molecule by endothelial cells (Pasceri et al., 2000). Because CRP detects inflammation anywhere in the body, it cannot give a specific location of the inflammation, but it is considered to play a role in the initiation of cardiovascular disease and is one of the strongest predictors of risk. At 3.0 mg/L, it confers high risk for CVD (Peters et al., 2013).

Galectin-3

A sign of HF is the increase of the plasma levels of Galectin-3. Primarily found in the cytoplasm and secreted into the extracellular space, it is involved in inflammatory and fibrotic processes and associated with remodeling HF (de Boer et al., 2009). With its expression in the heart lower than in several other organs, it has strong value in predicting HF.

There are a number of other markers under consideration with the potential of becoming cardiac biomarkers, such as copeptin for acute coronary syndrome (ACS) (Nickel et al., 2012), and soluble FMS-like tyrosine kinase-1 (sFlt-1) (Stratz et al., 2012), but more work needs to be done before their clinical utility is established.

Myeloperoxidase

Myeloperoxidase (MPO) is a hemoprotein expressed by polymorphonuclear neutrophils (PMNs) and it is secreted during leukocyte degranulation (Nicholls and Hazen, 2005). This enzyme is a marker of vascular inflammation and characteristic of acute coronary syndrome. High levels of MPO are associated with endothelial dysfunction as this reduces levels of NO and increases the risk of cardiovascular diseases. It is used in diagnosing HF. This inflammatory biomarker has been approved by the US Food and Drug Administration (FDA).

Tumor Necrosis Factor

TNF signals an inflammatory process and is a marker of HF. Excessive expression is associated with cardiac hypertrophy, fibrosis, and contractile dysfunction and indicates abnormal left atrial function and left systolic and diastolic dysfunction.

Proinflammatory cytokines, such as TNF, are produced by ischemia and act as an autocrine contributor to myocardial dysfunction. TNF depresses myocardial contractile function and reduces ejection function. The heart produces TNF receptors and IL-1 receptor antagonists that enter into the circulation.

TNF disrupts calcium handling, which affects excitation–contraction coupling. In addition, apoptosis ensues a characteristic of chronic HF and arrhythmogenic right ventricular dysplasia (Meldrum, 1998; Feldman et al., 2000). Recently, levels of circulating TNF receptors were correlated with infarct size and left ventricular dysfunction (Nilsson et al., 2013) and troponin levels. The increase in the circulating levels of TNF (sCD40L) is a marker of HF and vascular disorder.

REMODELING

ST2 and sST2

ST2 is a mediator of inflammation with marked soluble ST2 (sST2) concentration rise in autoimmune disease, fibrosis, and tissue injury. ST2 concentration changes are prevalent in cardiovascular diseases, including HF and MI (Weinberg et al., 2002). sST2 is a powerful tool for prognosis, exceeding even natriuretic peptide assays (Januzzi et al., 2007).

sST2 Soluble

A biomarker of cardiomyocyte stretch, ST2 is a member of the interleukin receptor family, which signals left ventricular dysfunction and fibrosis. Elevated levels of sST2 are characteristic of HF (Mueller et al., 2008; Shah and Januzzi, 2010). sT2 levels markedly increase after myocardial infarction and positively correlate with CK (Weinberg et al., 2002). Deletion of the ST2 gene enhances cardiac hypertrophy and fibrosis and ventricular dysfunction. ST2 elevation correlates with AMI and the infarct size.

Matrix Metalloproteinase-9

In instances of cardiac remodeling, matrix metalloproteinase is significantly increased and is associated with inflammation, extracellular matrix degradation, and cardiac dysfunction. MMP-9 is a potential biomarker (Halade et al., 2013).

Growth Differentiation Factor 15

GDF15 is a cytokine that acts in regulating inflammatory and apoptotic pathways and is upregulated after cardiac injury (Ago and Sadoshima, 2006; Kempf et al., 2006, 2007). High levels of GDF15 correlate with recurrent MI.

NEUROHORMONAL ACTION

The overexpression of neurohormonal biologically active molecules, including norepinephrine that is an effector of myocardial diffraction and thrombogenesis in ACS (Blann, 1993; Spiel et al., 2008) and angiotensin II, which are synthesized in the myocardium, exerts toxic effects and contributes to HF. In vascular disease, elevated von Willebrand factor level is predictive. The TNF promotes left ventricular dysfunction (Bozkurt et al., 1998), whereas angiotensin II contributes to necrosis (Tan et al., 1991). In the failing human heart, the concentration of norepinephrine correlates with the degree of HF. Chronic catecholamine stimulation is cardiotoxic. Thus, increased activation of the sympathetic nervous system and a higher norepinephrine level go hand in hand with deteriorating HF. Antagonizing neurohormones obviate the progression of HF (Mann, 1999).

Fibrinogen

Fibrinogen is a marker for inflammation and it promotes atherothrombosis (coagulation) by altering the hemodynamic properties of blood. Elevated levels indicate an increased likelihood of coronary artery disease and vascular disease, but these remain under investigation (Danesh et al., 2005). Fibrinogen is a protein that helps blood to clot. The normal level is between 200 and 400 mg/L; higher levels enhance clot formation and increase artery wall injury.

Copeptin

A stable C-terminal fragment of arginine vasopressin (AVP), copeptin is a hormone important for cardiovascular function. The normal level of this glycosylated peptide is increased considerably in response to myocardial injury. Elevated values are associated with HF in AMI (Khan et al., 2007). This nonnecrotic biomarker is used to rule out myocardial infarction. A concentration of 0.7 pmol/L suggests a significant HF risk.

VASCULAR BIOMARKERS

Vascular damage usually is a result of endothelial cell dysfunction and increased permeability, resulting in swelling and necrosis. Multiple mediators are usually present and drugs may also play a prominent role in perturbing the endothelial cell matrix, triggering inflammatory signals, and enhancing cellular adhesion. Inflammatory cytokines are released and ROS are generated that enhance necrosis and cause vascular injury (Louden et al., 2006; Tardif et al., 2006; Zhang et al., 2010).

Endothelial cell activation results in the release of von Willebrand Factor (vWF), vascular endothelial growth factor, caveolin-1, and NO. The propeptide vWFpp in plasma is a marker of vascular injury because endothelial cells are the sole source of this protein so that even in moderate vascular injury it records significant elevation.

A potential diagnostic marker of vascular injury is the smooth muscle α actin (SMA). Vascular lesions also result from increased NO production and sustained NO synthase (NOS) activity. Excessive generation of NO leads to cell damage or cell death.

Vascular endothelial cells produce and release vWF in response to inflammatory cytokines and thrombin.

Homocysteine

Homocysteine is a protein that is created when the amino acid methionine is metabolized. Homocysteine damages arteries in high concentrations, beyond 10.8 μmol/L. It inhibits the structural components collagen, elastin, and proteoglycans of an artery (Wierzbicki, 2007). Hyperhomocysteinemia (Ganguly and Alam, 2015) changes the lipid profile, inflammation and fibrinolysis and damages endothelial cells increasing the risk of cardiovascular disease. Level of homocysteine has been suggested as a biomarker.

Lp-PLA_2

Lp-PLA_2, a vascular biomarker, also known as platelet-activating factor, correlates with other inflammatory risk factors. It is an enzyme that degrades oxidized phospholipids, resulting in proinflammatory and cytotoxic products (Uydu et al., 2013). Lp-PLA_2 levels signal the development of peripheral arterial disease.

Placental Growth Factor

The placental growth factor (PIGF) is a biomarker of vascular inflammation. The stimulated angiogenesis and the produced collagen-degrading enzymes with adhesion to vascular surfaces lead to plaque rupture

(Heeschen et al., 2003; Glaser et al., 2011). PIGF is useful in risk stratification.

Pregnancy-Associated Plasma Protein-A

PAPP-A (pregnancy-associated plasma protein A) in dimeric form is significantly elevated in unstable coronary disease (Body and Ferguson, 2006). IGFBP-4, a substrate for dPAPP-A, is a marker for plaque rupture.

OTHER MARKERS

Heart Rate Variability

Neural injury may lead to arrhythmia. Variation in the R-R interval of the ECG, often referred to as HRV (HR variability), is a marker of sympathetic and parasympathetic influences on modulation of heart rate. Reduced HRV predicts increased risk of mortality. Reduced HRV reflects an increase in sympathetic activity that predisposes to myocardial ischemia and arrhythmia (Tsuji et al., 1996).

Rho Kinase

One of the symptoms of CHF, characterized by inflammation, is vasoconstriction where Rho kinases (ROCKs) are an effector. CHF causes ROCK activity to be considerably higher than normal, as well as protein levels of ROCK and activity of RhoA. Circulating ROCK activity in leucocytes is a biomarker for CHF risk (Dong et al., 2012). Combining ROCK activity with NT-proBNP has been used to predict long-term mortality.

ROCK has been confirmed to be involved in cardiovascular disease pathologies, including vascular muscle cell hypercontraction and endothelial dysfunction (Shi and Wei, 2013). Fasudil, a potent Rho-kinase vasodilator and inhibitor is being used for the treatment of ischemic diseases. Increased ROCK activity reflects CHF.

Plasma Ceramides

Ceramides are a family of lipid molecules originating from cell membranes, are linked to inflammatory processes they contribute to the regulation of cells, and are carried by LDL and become embedded in arterial wall containing atherosclerotic plaques where activation of platelets increases the likelihood of clot formation. There are several kinds of these lipids; ratios of different kinds are now used as markers.

Elevated levels of ceramides in the plasma are associated with the risk of myocardial infarction and mortality. High levels increase the likelihood of cardiovascular events 3–4 times. Ceramide measurement is important because ceramide ratios indicate likelihood of cardiovascular events in patients with stable coronary artery disease and stable acute coronary syndrome. However, causality was not demonstrated.

MARKERS OF DRUG-INDUCED TOXICITY

Considerable time delay for the appearance of cardiac toxicity is characteristic of a number of prescribed drugs. Adriamycin (doxorubicin), an anthracycline, used for the treatment of inflammatory breast cancer, is effective as long as the total dosage over time is limited. Only years later does doxorubicin-induced cardiac toxicity become apparent (Geisberg and Sawyer, 2010). Anthracyclines cause cardiomyocyte damage (Zhang et al., 2012) by the formation of free radicals that disrupt cellular function and cause apoptosis (Sawyer, 2013). Upregulated expression of miR-146a in doxorubicin causes cardiotoxicity by targeting the ErbB4 gene (Horie et al., 2010).

At the initiation of treatment with Adriamycin, there were no positive cardiac biomarkers measured for the patient whose ECG (Fig. 12.1) was taken after the lapse of a number of years following the first use of the drug. The drug invoked biomarkers that clearly showed an inverted T-wave in the V1–V3 leads, a right bundle block manifested by a QPR considerably extended over time. Simultaneously, measurement of troponin I gave a value of 0.2 ng/mL and natriuretic peptide array of 162 pg/mL, which were all in the normal range. Interestingly enough, in some cases N-terminal pro-B-type natriuretic peptide, marker of ventricular dysfunction, increased without clinical signs of cardiotoxicity (Kilickap et al., 2005; Altena et al., 2009). Over time, the cardiac toxicity of the drug-induced bradycardia and pacemaker-induced bradycardia and pacemaker syndrome, with asynchrony of the atria and ventricles prevented proper pumping of the blood.

Changes in the LVEF (left ventricular ejection fraction) were noted as a consequence of anthracycline chemotherapy (Volkova and Russell, 2011). Cardiotoxicity is usually noted within a year of the completion of treatment, although in some cases it may be delayed.

CARDIAC BIOMARKERS OF ILLEGAL DRUGS, CHEMICALS, AND TERROR AGENTS

Refer to Table 12.2 for cardiac biomarkers of several cardiotoxicity-inducing illegal drugs, pharmaceuticals, and environmental chemicals.

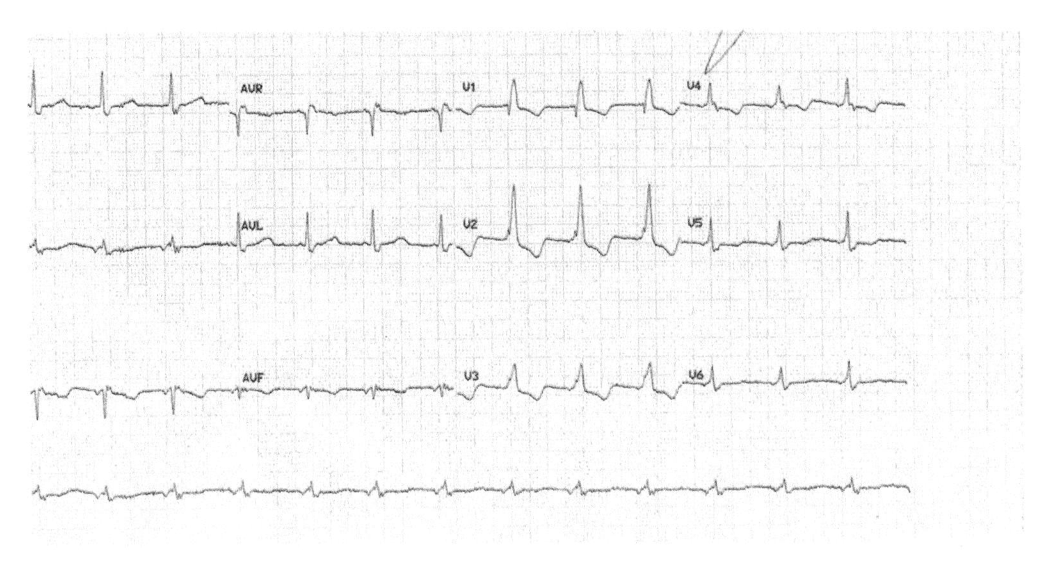

FIGURE 12.1 ECG of patient treated with Adriamycin.

Anthrax

Anthrax, caused by *Bacillus anthracis*, induces vascular insufficiency by curtailing neuronal NO synthase. Vascular function depends heavily on NOS (Casadei, 2006) and is involved in the response to injury of the myocardium. In experiments with anthrax lethal toxin–treated mice, ejection fraction and decreased ejection fraction, and contractile functions were noted (Moayeri et al., 2009). Consequently, markers of cardiac injury, myoglobin, troponin I, and heart fatty acid protein (H-FABP) levels were elevated. In addition, heart morphology changes are evident, and elevation in the permeability of blood vessel membranes is also caused by anthrax (Table 12.2).

Pharmaceuticals

Anthracyclines, extensively used for treatment of solid tumors and hematological malignancies, cause cardiomyopathy and CHF, which is refractory to treatment. Their cardiotoxicity is attributed to abnormal protein processing, immune response, and inhibition of neuregulin-1 signaling. Anthracyclines block the functioning of topoisomerase II (TOP2) (Sawyer, 2013) of tumor growth.

The side effect of TOP inhibition includes dysregulation of mitochondrial biogenesis. The innate immune system is dysregulated through miR-146 (Horie et al., 2010; Shi et al., 2010). The biomarker of drug-induced cardiac toxicity is primarily the ECG, which shows the breakdown of ionic conduction.

Cardiac glycosides elevate intracellular sodium and calcium transfer. An overload of calcium causes cardiotoxicity, arrhythmia, and premature ventricular contraction.

Slowed AV conduction results in bradycardia. An overdose may cause ventricular tachycardia and possibly ventricular fibrillation. Elements of the dysfunction of the ionic conduction system are shown by the classical biomarkers (Ma et al., 2001; Nagai et al., 2013).

Antidepressant serotonin reuptake inhibitors can cause cardiac effects, including prolonged QT_c, arrhythmia, and torsade de pointes (Leonard et al., 2011; Deshmukh et al., 2012). ECG monitoring is the best use of cardiac biomarkers. Several of the tricyclic antidepressants may cause cardiac sodium channel blockade, thereby increasing the cardiac action potential and causing a delay in atrioventricular conduction (Thanacoody and Thomas, 2005).

Antihistamines selectively bind to histamine H1 receptors. These inhibit the repolarizing potassium currents, I_K (delayed rectifier), and the inward rectifier i_{k1}. Blockade of Ca^{2+}, Na^+ is also important. Blockade of the HERG channel increases the QT_c interval. Cardiotoxic effects of antihistamines include QT_c prolongation, K^+ channel blockade, and increased action potential duration (Zhang, 1997).

Illegal Drugs

Cardiac damage due to cocaine is myriad and no single biomarker can give a definite diagnosis. Cocaine causes irreversible structural damage in a number of organs, including the heart. It stimulates the sympathetic nervous system, inhibits the reuptake of catecholamines, and increases the sensitivity of adrenergic nerve endings to norepinephrine (Turillazzi et al., 2012). Myocardial infarction due to coronary artery vasoconstriction, thrombus formation, and arrhythmia is also observed.

ECG measurements are insufficient cardiomarkers. Troponin is also elevated, though the source may not be singular (Agewall et al., 2011). The activation of α

TABLE 12.2 Cardiac Biomarkers of Cardiotoxicity-Inducing Xenobiotics

Xenobiotic	Mechanism of Cardiac Damage	Biomarker	Reference
PHARMACEUTICAL			
Anthracycline	Altered calcium homeostasis mitochondrial injury	Tachycardia	Sawyer (2013)
			Shi et al. (2010)
	Electrical conductivity damaged	Congestive heart failure	
Cardiac glycoside	Calcium overload	Congestive heart failure	Nagai et al. (2013)
		Arrhythmia, bradycardia	
		Slowed AV conduction	
Antidepressants	Anticholinergic effects	ECG abnormalities	Leonard et al. (2011)
	Depressed inward sodium, calcium		Deshmukh et al. (2012)
	Currents, outward potassium current		
Antihistamines	Block of delayed rectifier potassium	ECG: inverted T-waves, prolonged QT	Delgado et al. (1998)
	Channel, prolonged action potential	AV block,	Golightly and Greos (2005)
		fibrillation	Zhang (1997)
ILLEGAL DRUGS			
Cocaine	Blockade of norepinephrine reuptake	Chest pain	Keller and Lemberg (2003)
	Marked release of norepinephrine	ST elevation	Turillazzi et al. (2012)
	Increase of norepinephrine levels at	T-wave inversion	Phillips et al. (2009)
	Nerve terminals	Troponin elevation	
	Inhibition of nerve impulse conduction	Myocardial infarction	
Cannabis	Increased coagulation	Heart rate, blood pressure changes	Pratap and Korniyenko (2012)
	Biphasic effect on nervous system	Hypotension with prolonged use	Zuurman et al. (2009)
	THC, the active ingredient, binds to	Arrhythmias	Jarai and Kunos (2002)
	Cannabinoid receptors in cells	ST-segment elevation on the ECG	
Methamphetamine	Stimulates the central nervous system,	Prolonged QT interval,	Watts and McCollester (2006)
	Reuptake catecholamines blocked,	ST-T changes, bradycardia,	Haning and Goebert (2007)
	Hyperstimulation of postsynaptic	Hypertension, increased heart rate,	
	Neuron receptors, inhibits mitochondrial	Blood pressure,	
	Monoamine oxidase	Torsade de pointes	
ENVIRONMENTAL CHEMICALS			
Arsenic	Vascular pathologies	CRP (C-reactive protein) elevated	States et al. (2009)
	Production of reactive oxygen	Prolongation of QT on the ECG	Druwe et al. (2012)
	Disorder of heart's conduction	Arrhythmic effect	
Lead	Accumulation of ALA in blood	Level in blood exceeds. 10 μg/dL	Kakkar and Jaffery (2005)
	Increased central sympathetic	Hypertension	Glenn et al. (2003)
	Nervous system activity	QRS, ST-interval prolonged	
	Elevated norepinephrine	Increased heart rate, cardiac contractility	

Continued

TABLE 12.2 Cardiac Biomarkers of Cardiotoxicity-Inducing Xenobiotics—cont'd

Xenobiotic	Mechanism of Cardiac Damage	Biomarker	Reference
Organophosphate	Cholinergic crisis, overstimulation of	QT of ECG prolonged, ST segment	Gul et al. (2012)
	Muscarinic and nicotinic receptors	Elevation, low amplitude T-waves,	Liu et al. (2012)
	Acetylcholinesterase inhibition	Torsade de pointes	Bar-Meir et al. (2007)
		Ventricular fibrillation	
Cyanide	Cardiovascular collapse	Cyanide in whole blood, rapidly	Muncy et al. (2012)
	Inhibits multiple enzyme systems,	Metabolized, short half-life	
	Cytochrome oxidase	ST elevation	
	Disrupts ATP production	QT segment shortening	
		PR prolongation	

and β adrenergic receptor sites, leading to increased catecholamine levels (Keller and Lemberg, 2003), is also significant. Calcium handling of the myocytes and changes in the conduction system leads to QRS and QT_c changes.

Cannabis (marijuana) through tetrahydrocannabinol (THC), its active psychoactive ingredient, triggers a number of cardiovascular system responses (Jarai and Kunos, 2002). Increased heart rate (HR) and lowered BP are present. THC binds to cannabinoid receptors of several organs, including the heart, and overstimulates the receptors. In addition, THC causes overproduction of a protein, ApoC-111, which is linked to triglycerides present in blood (Zuurman et al., 2009).

There have been reports of cannabis-caused arrhythmia and recently (Pratap and Korniyenko, 2012) a case of ST-segment elevation on the ECG. Several cardiac biomarkers can respond to the presence of cannabis.

Stimulation of the central nervous system produces the euphoria of amphetamine by the blockade of catecholamine reuptake and hyperstimulation of the postsynaptic neuron receptors. In addition, various noncatecholaminergic nervous pathways are stimulated.

Cardiovascular collapse, ventricular tachyarrhythmia, as well as MI, heart failure, prolongation of QT_c, and other ECG abnormalities are common (Watts and McCollester, 2006; Haning and Goebert, 2007).

Environmental Chemicals

In many parts of the world, arsenic enters the food chain through contaminated groundwater. Arsenic causes oxidative stress by reducing the activation of eNOS and it increases the phosphorylation of myosin light-chain kinase (Singh et al., 2011). Serious side effects of arsenic include the cardiovascular system, causing altered myocardial depolarization, prolonged QT_c interval, and ST-segment changes (Quatrehomme et al., 1992; Cullen et al., 1995). Long-term exposure to arsenic also damages the vascular system, causing "black foot disease," endemic in Taiwan. Raynaud's phenomenon, a peripheral vascular disease of spasm of the digital arteries and constriction of blood vessels, besides significant increase in systolic BP, is seen after long-term exposure to arsenic (States et al., 2009; Druwe et al., 2012).

The cardiac abnormalities caused by lead are primarily hypertension and heart rate variability (Navas-Acien et al., 2007). In addition to high blood lead content (Glenn et al., 2003; Kakkar and Jaffery, 2005), there is some evidence of possible autonomic dysfunction.

Weaponized Chemicals

Cyanide shuts down the aerobic pathway by inhibiting cytochrome oxidase activity, at the mitochondrial level. Cells cannot maintain themselves on anaerobic respiration. Lactic acid buildup deprives cells of the energy source needed for the sustenance of life. In addition, the CN molecule's affinity for hemoglobin reduces the available oxygen supply (Wexler et al., 1947; Muncy et al., 2012).

Organophosphates are used as pesticides and chemical warfare agents which cause serious cardiac dysfunction. Among these, QT_c interval prolongation, and changes in QRS and right bundle band block are especially important (Vijayakumar et al., 2011). Acetylcholinesterase inhibition and delayed neuropathy play prominent roles. The hyperstimulation of receptors and the intermediate syndrome characterized by proximal muscle weakness is also present. Considerable information is contained in the work of Bar-Meir et al. (2007), Gul et al. (2012), and Liu et al. (2012).

FUTURE TRENDS IN CARDIOVASCULAR BIOMARKERS

In view of the limitations of individually used cardiac biomarkers and the need to determine the source and location of the cardiac problem on presentation, the search for biomarkers with higher sensitivity and specificity continues. Along these lines, multimarker approaches, whereby several markers are tested, have gained more promise (Sabatine et al., 2002; McCann et al., 2009). Gerszten et al. (2011) reported that troponin, C-reactive protein (CRP), and myeloperoxidase (MPO), as well as soluble CD40 ligands, were successful in predicting a cardiovascular event half a year in advance. It was noted that, individually, the markers did not yield the same prediction.

Current biomarkers reveal essentially similar pathways, thus proteomics offers an approach that may yield at present novel indicators of cardiac injury. Blood contains a number of high abundance proteins; for example, albumin contributes over 50% of the total protein mass, and 99% of the plasma protein mass is taken up by only 22 proteins, including albumin and immunoglobulin. An estimated one million different proteins exist in the blood but they range in concentration up to 11 orders of magnitude, from micro- to femtomoles per liter. Troponins are in the nanomolar range, whereas TNF is in the femtomolar range. Following myocardial injury, the leakage of proteins into the blood and subsequent posttranslational modifications open up a new avenue for biomarker identification.

A number of proteins associated with cardiovascular pathology appear to be viable biomarkers and are under close scrutiny for possible clinical use.

MicroRNAs as Cardiac Biomarkers

Currently used cardiac biomarkers in body fluids are derived from specific proteins. Their use incurs drawbacks of nonspecificity, lack of abundance, posttranslational modification, and lack of timely availability. Blood-based biomarkers are also based on antibodies, but these exhibit cross-reactivity with other proteins.

The recent realization that tissues can secrete miRNAs (microribonucleic acids), which exert an effect on other cells and tissues and play prominent roles in vascular and metabolic diseases, is now being explored for detection of cellular injury and in essence disease detection. Noncoding endogenously expressed microRNA is a short ribonucleic acid (RNA) molecule of eukaryotic cells. They posttranscriptionally regulate gene expression of their target mRNA (messenger RNA). The human genome encodes over a 1000 miRNAs, and miRNAs' expression is organ-specific, including the vasculature, with reaction to cardiac injury similar to that with cardiac enzymes. They have specific functions and regulatory roles in cellular processes (Tijsen et al., 2012). MiRNAs are endogenous, noncoding RNAs that regulate messenger RNA (mRNA) or protein levels validating their use as biomarkers (Fichtlscherer et al., 2011; Tijsen et al., 2012; van Rooij, 2012), miRNAs are endogenous, noncoding RNAs that regulate messenger RNA (mRNA) or protein levels by promoting mRNA degradation or by attenuating protein translation. Some miRNAs are located in body fluids, but most are found within cells. Distinct expression profiles are found among the different fluid types. In addition, because of the presence of ribonucleases that catalyze the degradation of RNA, miRNAs are enclosed in vesicles to protect them from RNA digestion; thus, their stability is assured. miRNAs are negative regulators of gene expression. The number of miRNA genes has been estimated to be over 1000 in the human genome, and each miRNA regulates a large number of target genes.

To protect them from degradation, miRNAs are included in lipoprotein complexes called microvesicles. Vesicles, also referred to as microparticles, are formed in successive steps that initiate ectosome formation during cell activation (Burnier et al., 2009). Particles also constitute a transcellular delivery system for biological information and intercellular communication (Hunter et al., 2008). By altering the expression of genes in the vicinity (Burnier et al., 2009; Mause and Weber, 2010) and at remote sites by transferring genetic information, this can result in the reprogramming of the phenotype of the target cell. These are actively secreted from cells in vesicles and exosomes, but the amount of RNA recovered from plasma is low.

Circulating miRNAs in body fluids are stable, their sequences are evolutionarily conserved and they are tissue-, cell-, and pathology-specific. Their presence in the fluid is detected by real-time PCR arrays but because of the low level of total RNA in the blood, it is difficult to measure their concentration. Next-generation sequencing (Creemers et al., 2012) has shown that there are over 100 different miRNAs in serum.

In addition to tissue specificity and sensitivity, presence of miRNA in cardiac injury correlates with troponin level. Change in miRNA, specific to a time, can be correlated with cardiac injury. miRNA leaks out of injured or apoptotic cells and those that lost their nucleic acid content. The serum level of miRNA increases up to a 1000-fold after MI and mirrors troponin I (Corsten et al., 2010).

For all practical purposes, and as their validation proceeds, miRNAs will be accepted as cardiac biomarkers with distinctive advantages over conventional ones. See Table 12.3, also (Sahu, 2014).

There are several microRNAs whose increase in plasma are specific and indicate different aspects of cardiac injury. MicroRNAs are named using the prefix miR and a number that reflects the order in which the gene was discovered. Sometimes the form used is hsa-mir-number, the first letters signifying the organism. Latter suffixes denote related sequences. For example, in an isoproterenol-induced myocardial injury, miR-208 increased significantly in rats (Ji et al., 2009). Because the miRNA signature is cell-specific, it greatly aids in the diagnosis. In the case of MI, six miRNAs, miR-1,

TABLE 12.3 Cardiac Circulating miRNAs

Cardio Injury	miRNA Biomarker		Upregulated		Downregulated	Function	Primary References
Myocardial infarction (MI)	miR-15	miR-1	miR-1			Effect on infarct size	Hullinger et al. (2012)
	miR-21		miR-133	miR-15	miR-29	Role in cell apoptosis	Ono et al. (2011)
	miR-214		miR-208				Aurora et al. (2012)
	miR-499		miR-21				
			miR-214				
			miR-499				
Vascular injury	miR-21	miR-222			miR-126	Disrupts vascular integrity	Urbich et al. (2008)
Epithelium	miR-155					Inflammation	
	miR-126						Harris et al. (2008)
Fibrosis	miR-221	miR-222				Controls angiogenesis	Meder et al. (2011)
							van Solingen et al. (2009)
Heart Failure	miR-21	miR-423	miR-21		miR-30	Left ventricular remodeling	Thum et al. (2008)
	miR-129	miR-30	miR-29b		miR-182	Hypertrophy	van Rooij (2012)
	miR-210	miR-182	miR-129		miR-526	Fibrosis	
	miR-211	miR-526	miR-211			Reactivation of fetal gene program	
	miR-212	miR-423-5p					Tijsen et al. (2012)
Hypertrophy	miR-208a	miR-208a				Effect on conduction system	
	miR-199	miR-199				Regulates calcineurin signaling	
	miR5133	miR-1	miR-499			Compensate for damage to tissue	Shieh et al. (2011)
	miR-21	miR-499					
Arrhythmia	miR-208a	miR-208a				Induces arrhythmia	Oliveira-Carvalho et al. (2013)
	miR-1					Regulates hypertrophy	Terentyev et al. (2009)
	miR-133	miR-133				Targets pacemaker channel	Montgomery et al. (2011)
						Signals cardiac injury	Zhang (1997)
Drug Cardiotoxicity	miR-146a	miR-146				Reduces ErbB4 expression,	Horie et al. (2010)
						Modulates immune response	

miR-133a, miR-133b, and miR-499-5p, increased, though by uneven amounts (D'Alessandra et al., 2010). Also, even though miR-208 is undetectable in healthy individuals, it significantly increases after MI (Wang et al., 2010). The observation that miR-1 and miR-133 regulate cardiac electrical properties, with the former triggering cardiac arrhythmias and both targeting the pacemaker current, has significant implications (Terentyev et al., 2009). There are indications that miR-1 may alter the calcium signaling, inducing arrhythmia.

Cardiac injury is reflected in changes of functioning and the constitution of the organ which in turn reflects the engaged miR and its concentration change. The indicated gene may be up- or downregulated. Changes in genetic involvement are markers of cardiac injury.

Cardiomyocyte hypertrophy is indicated by changes in miR-1 and miR-195. In fact, inhibition of endogenous miR-18b and miR-21 enhances hypertrophy (Tatsuguchi et al., 2007). Hypertrophic mouse hearts show an upregulation, a fourfold increase in miR-21 (Cheng et al., 2007). Inhibition of miR-21 exhibits an antigrowth effect. miR-133 also controls cardiac hypertrophy; in fact, overexpression of miR-133 is a key regulator of cardiac hypertrophy (Care et al., 2007), whereas overexpression of miR-195 is a sign of pathological growth (van Rooij et al., 2006).

Ion channels, and thus the electrical activity of the heart, are targets of miR-1 and miR-133 (Yang et al., 2007; Ono et al., 2011) that can deregulate the expression of cardiac ion channels and promote arrhythmogenesis (Terentyev et al., 2009). Overexpression of miR-208a disturbs cardiac conduction (Callis et al., 2009).

Chemotherapeutic drugs express cardiotoxicity even at moderate doses. Doxorubicin, an anthracycline, upregulates miR-146a, and ErbB2 and ErbB4, expressed in cardiomyocytes, when inhibited, causes CHF (Horie et al., 2010).

In vascular scenarios, specific miRNAs regulate gene expression on the posttranscriptional level by inhibiting the translation of protein from mRNA or the degradation of mRNA. The integrity of the endothelial monolayer is signaled by the state of the mRNAs. These have distinctive expression profiles that play crucial roles in cardiogenesis and regulate endothelial cell function. miRNA-21 is aberrantly overexpressed in vascular injury (Ji et al., 2007). Proangiogenic miRNAs are miR-37b and miR-130a, whereas miR-221 and miR-222 oppose cell proliferation and angiogenesis. Overexpression reduces endothelial NO, the key regulator for endothelial cell growth and vascular remodeling. Impaired bioavailability is a marker of atherosclerosis and ischemic cardiomyopathy. Vascular inflammation and diseases are also signaled by miR-155, miR-21, and miR-126 (Urbich et al., 2008).

Dysregulation of multiple genes occurs in myocardial infarction (Salic and De Windt, 2012). In particular, miR-21 is dysregulated in infarcted areas, whereas upregulated elsewhere. Considerably elevated levels of miR-1 and miR-133 are present in the circulation and appear significantly sooner than troponin. The level of miR-1 correlates with QRS widening on the ECG. The most upregulated miRs in MI were miR-208b and miR-499, with both favorably correlating with troponin (TnT) levels (Corsten et al., 2010). Regulation of ischemic/reperfusion cardiac injury and dysfunction via antithetical regulation of Hsp20 (heat-shock protein) is associated with miR-320, and miR-1 and miR-133 produce opposing effects on apoptosis (Ren et al., 2009).

Cardiac fibrosis is facilitated by the connective tissue growth factor (CTGF) that is characterized by miR-133 and miR-30 (Duisters et al., 2009). Active fibrotic processes play a part in diastolic heart failure (DHF), and increased myocardial fibrosis is part of hypertensive heart disease (HHD). CTGF is a secreted protein that induces ECM (extracellular matrix) synthesis. A fibrotic ECM impairs cardiomyocyte function. It plays an important role in HHD. Next-generation sequencing technologies (Creemers et al., 2012), not biased by the currently used PCR and microarray platforms, reveal that mature miRNAs have an extensive degree of variation at the terminal nucleotides that may influence miRNA target specificity. These isomiRs may also be pathology-specific. If this turns out to be so, a new set of biomarkers would be available. Sequence variants (Humphreys et al., 2012) of miRNAs and their effect on cellular processes and subsequently on pathologies (Dorn, 2012) would be significant (see Table 12.4).

RISK PREDICTION OF CARDIOVASCULAR EVENTS

The utility of biomarkers for identifying sources and causes of cardiac dysfunction is well established. Risk prediction, affecting decisions on treatment, using the information derived from biomarkers, has also advanced. Multimarker risk scores (Wang et al., 2012) based on the Framingham Heart Study and the work of Arts et al. (2015) compared the predictive ability of four established cardiovascular risk models. The established multimarker score based on the chosen biomarkers uses the multivariable Cox regression model and the Kaplan–Meier method. The C statistic was used in the classification of risks. These, based on the receiver operating characteristic curve, moderate discriminations between patients with and without cardiovascular events diagnosed with early rheumatoid arthritis.

Biomarkers of endothelial activation correlate with chemokines but less frequently with inflammation. Combining circulating biomarkers from different pathways yields improved discrimination for end points.

TABLE 12.4 Potential Cardiac Biomarkers Up- or Downregulated Cardiac miRNAs

miR-1	Ionic channels affected, arrhythmia, hypertrophy, MI
miR-15	Hypertrophy
miR-21	Increased after vessel injury
miR-23	Hypertrophy
miR-24	Myocardial infarct size, activated in endothelial cells
	After MI
miR-29	Ischemia reperfusion injury
miR-126	Vascular integrity
miR-133	Affects ionic conduction, targets connective tissue growth factor
miR-145	Downregulated in carotid artery after mechanical injury
miR-155	Increased in heart failure
miR-195	Hypertrophy
mir-206	Markedly increased in MI
miR-208	Overexpression causes hypertrophy, arrhythmia
miR-214	Ischemic injury
miR-221	Blocks neointima formation after vessel injury, regulates angiogenesis
miR-320	Targets heat-shock protein ischemia
miR-499	Stress-induced dysfunction

Imaging biomarkers, especially coronary calcium scores, improve risk discrimination, although more data are needed before it can be included. Genetic biomarkers are also considered to be potentially of less importance.

CONCLUDING REMARKS AND FUTURE DIRECTIONS

Advancement of the field of cardiac biomarkers has been substantial. Development of high-sensitivity assays now allow diagnoses to be made earlier than before on samples containing lower concentration of the suspected marker. In addition to the standard markers, now over 20 items have been assessed.

The uses of miRNAs, endogenous noncoding RNA molecules that play a regulating role in gene expression, affecting an estimated one-third of the genome, have also gained prominence. Their involvement results in the impairment of protein synthesis from target RNAs that affect a number of important biological processes, including metabolism and proliferation. Measured up-/downregulation of circulating miRNAs can now be used as a biomarker of diagnosis, prognosis, and treatment.

Biomarkers assessment has gained from the recognition of the importance of the consideration of epigenetic modifications of the source of the sample. Also, using insights from metabolomics and proteomics effect on the sample is helpful. Mitochondrial dysfunction assessment has also yielded useable biomarkers. Dysfunction assessment thus is more specific and timely than it was before.

Cardiac biomarker's promises are exceeding expectations.

References

Agewall, S., Giannitsis, E., Jernberg, T., Katus, H., 2011. Troponin elevation in coronary vs. non-coronary disease. Eur. Heart J. 32, 404–411.

Ago, T., Sadoshima, J., 2006. GDF15, a cardioprotective TGF-beta superfamily protein. Circ. Res. 98, 294–297.

Altena, R., Perik, P.J., van Veldhuisen, D.J., et al., 2009. Cardiovascular toxicity caused by cancer treatment: strategies for early detection. Lancet Oncol. 10, 391–399.

Anderson, N.L., Anderson, N.G., 2002. The human plasma proteome. Mol. Cell. Proteomics 1, 845–867.

Apple, F.S., 1999. Tissue specificity of cardiac troponin I, cardiac troponin T and creatine kinase-MB. Clin. Chim. Acta 284, 151–159.

Arts, E.E., Popa, C., Den Broeder, A.A., et al., 2015. Performance of four current risk algorithms in predicting cardiovascular events in patients with early rheumatoid arthritis. Ann. Rheum. Dis. 74, 668–674.

Aurora, A.B., Mahmoud, A.I., Luo, X., et al., 2012. MicroRNA-214 protects the mouse heart from ischemic injury by controlling Ca^2 overload and cell death. J. Clin. Invest. 122, 1222–1232.

Azzazy, H.M., Pelsers, M.M., Christenson, R.H., 2006. Unbound free fatty acids and heart-type fatty acid-binding protein: diagnostic assays and clinical applications. Clin. Chem. 52, 19–29.

Backs, J., McKinsey, T.A. (Eds.), 2016. Epigenetics in Cardiac Disease. Springer Int. Pub., Cham, Switzerland.

Balion, C.M., Santaguida, P., McKelvie, R., et al., 2008. Physiological, pathological, pharmacological, biochemical and hematological factors affecting BNP and NT-proBNP. Clin. Biochem. 41, 231–239.

Bar-Meir, E., Schein, O., Eisenkraft, A., et al., 2007. Guidelines for treating cardiac manifestations of organophosphates poisoning with special emphasis on long QT and torsades de pointes. Crit. Rev. Toxicol. 37, 279–285.

Blann, A., 1993. Von Willebrand factor and the endothelium in vascular disease. Br. J. Biomed. Sci. 50, 125–134.

Body, R., Ferguson, C., 2006. Pregnancy-associated plasma protein A: a novel cardiac marker with promise. Emerg. Med. J. 23, 875–877.

Bozkurt, B., Kribbs, S.B., Clubb Jr., F.J., et al., 1998. Pathophysiologically relevant concentrations of tumor necrosis factor-alpha promote progressive left ventricular dysfunction and remodeling in rats. Circulation 97, 1382–1391.

Braunwald, E., 2008. Biomarkers in heart failure. N. Engl. J. Med. 358, 2148–2159.

Brown, D.A., Perry, J.B., Allen, M.E., et al., 2017. Mitochondrial function as a therapeutic target in heart failure. Nat. Rev. Cardiol. 14, 238–250.

Burnier, L., Fontana, P., Kwak, B.R., Angelillo-Scherrer, A., 2009. Cell-derived microparticles in haemostasis and vascular medicine. Thromb. Haemost. 101, 439–451.

Callis, T.E., Pandya, K., Seok, H.Y., et al., 2009. MicroRNA-208a is a regulator of cardiac hypertrophy and conduction in mice. J. Clin. Invest. 119, 2772–2786.

Care, A., Catalucci, D., Felicetti, F., et al., 2007. MicroRNA-133 controls cardiac hypertrophy. Nat. Med. 13, 613–618.

Casadei, B., 2006. The emerging role of neuronal nitric oxide synthase in the regulation of myocardial function. Exp. Physiol. 91, 943–955.

Catalucci, D., Latronico, M.V.G., Condorelli, G., 2015. FABP3 as biomarker of heart pathology. In: Preedy, V.R., Patel, V.B. (Eds.), General Methods in Biomarker Research and Their Applications. Springer, Dordrecht, pp. 439–454.

Cheng, Y., Ji, R., Yue, J., et al., 2007. MicroRNAs are aberrantly expressed in hypertrophic heart: do they play a role in cardiac hypertrophy? Am. J. Pathol. 170, 1831–1840.

Corsten, M.F., Dennert, R., Jochems, S., et al., 2010. Circulating MicroRNA-208b and MicroRNA-499 reflect myocardial damage in cardiovascular disease. Circ. Cardiovasc. Genet. 3, 499–506.

Costa, V.M., Carvalho, F., Duarte, J.A., et al., 2013. The heart as a target for xenobiotic toxicity: the cardiac susceptibility to oxidative stress. Chem. Res. Toxicol. 26, 1285–1311.

Creemers, E.E., Tijsen, A.J., Pinto, Y.M., 2012. Circulating microRNAs: novel biomarkers and extracellular communicators in cardiovascular disease? Circ. Res. 110, 483–495.

Cullen, N.M., Wolf, L.R., St Clair, D., 1995. Pediatric arsenic ingestion. Am. J. Emerg. Med. 13, 432–435.

D'Alessandra, Y., Devanna, P., Limana, F., et al., 2010. Circulating microRNAs are new and sensitive biomarkers of myocardial infarction. Eur. Heart J. 31, 2765–2773.

Danesh, J., Lewington, S., Thompson, S.G., et al., 2005. Plasma fibrinogen level and the risk of major cardiovascular diseases and nonvascular mortality: an individual participant meta-analysis. J. Am. Med. Assoc. 294, 1799–1809.

Danne, O., Mockel, M., Lueders, C., et al., 2003. Prognostic implications of elevated whole blood choline levels in acute coronary syndromes. Am. J. Cardiol. 91, 1060–1067.

de Boer, R.A., Voors, A.A., Muntendam, P., et al., 2009. Galectin-3: a novel mediator of heart failure development and progression. Eur. J. Heart Fail. 11, 811–817.

Delgado, L.F., Pferferman, A., Solé, D., Naspitz, C.K., 1998. Evaluation of the potential cardiotoxicity of the antihistamines terfenadine, astemizole, loratadine, and cetirizine in atopic children. Ann. Allergy Asthma Immunol. 80, 333–337.

Deshmukh, A., Ulveling, K., Alla, V., et al., 2012. Prolonged QTc interval and torsades de pointes induced by citalopram. Tex. Heart Inst. J. 39, 68–70.

Dirks, R.A.M., Stunnenberg, H.G., Marks, H., 2016. Genome-wide epigenomic profiling for biomarker discovery. Clin. Epigenet. 8, 122–139.

Dong, M., Liao, J.K., Fang, F., et al., 2012. Increase Rho kinase activity in congestive heart failure. Eur. J. Heart Fail. 14, 965–973.

Dorn 2nd, G.W., 2012. Decoding the cardiac message: the 2011 Thomas W. Smith memorial lecture. Circ. Res. 110, 755–763.

Dorn 2nd, G.W., Vega, R.B., Kelly, D.P., 2015. Mitochondrial biogenesis and dynamics in the developing and diseased heart. Genes Dev. 29, 1981–1991.

Druwe, I.L., Sollome, J.J., Sanchez-Soria, P., et al., 2012. Arsenite activates NFkappaB through induction of C-reactive protein. Toxicol. Appl. Pharmacol. 261, 263–270.

Duisters, R.F., Tijsen, A.J., Schroen, B., et al., 2009. miR-133 and miR-30 regulate connective tissue growth factor: implications for a role of microRNAs in myocardial matrix remodeling. Circ. Res. 104, 170–178.

Feldman, A.M., Combes, A., Wagner, D., et al., 2000. The role of tumor necrosis factor in the pathophysiology of heart failure. J. Am. Coll. Cardiol. 35, 537–544.

Fichtlscherer, S., Zeiher, A.M., Dimmeler, S., 2011. Circulating microRNAs: biomarkers or mediators of cardiovascular diseases? Arterioscler. Thromb. Vasc. Biol. 31, 2383–2390.

Fornai, F., Carrizzo, A., Forte, M., et al., 2016. The inflammatory protein Pentraxin 3 in cardiovascular disease. Immun. Ageing 13, 25–34.

Francis, G.S., 2011. Neurohormonal control of heart failure. Cleve. Clin. J. Med. 78 (Suppl. 1), S75–S78.

Frangogiannis, N.G., 2008. The immune system and cardiac repair. Pharmacol. Res. 58, 88–111.

Frijhoff, J., Winyard, P.G., Zarkovic, N., et al., 2015. Clinical relevance of biomarkers of oxidative stress. Antioxid. Redox Signal. 23, 1144–1170.

Ganguly, P., Alam, S.F., 2015. Role of homocysteine in the development of cardiovascular disease. Nutr. J. 14, 6.

Garcia-Gimenez, J.L., 2015. Epigenetic Biomarkers and Diagnostics. Academic Press, Waltham, MA.

Geisberg, C., Sawyer, D.B., 2010. Mechanism of anthracycline cardiotoxicity and strategies to decrease cardiac damage. Curr. Hypertens. Rep. 12, 404–410.

Gerszten, R.E., Asnani, A., Carr, S.A., 2011. Status and prospects for discovery and verification of new biomarkers of cardiovascular disease by proteomics. Circ. Res. 109, 463–474.

Ghani, F., Wu, A.H., Graff, L., et al., 2000. Role of heart-type fatty acid-binding protein in early detection of acute myocardial infarction. Clin. Chem. 46, 718–719.

Glaser, R., Peacock, W.F., Wu, A.H., et al., 2011. Placental growth factor and B-type natriuretic peptide as independent predictors of risk from a multibiomarker panel in suspected acute coronary syndrome (Acute Risk and Related Outcomes Assessed with Cardiac Biomarkers [ARROW]) study. Am. J. Cardiol. 107, 821–826.

Glenn, B.S., Stewart, W.F., Links, J.M., et al., 2003. The longitudinal association of lead with blood pressure. Epidemiology 14, 30–36.

Golightly, L.K., Greos, L.S., 2005. Second-generation antihistamines: actions and efficacy in the management of allergic disorders. Drugs 65, 341–384.

Gorodeski, E.Z., Ishwaran, H., Kogalur, U.B., et al., 2011. Use of hundreds of electrocardiographic biomarkers for prediction of mortality in postmenopausal women. Circ. Cardiovasc. Qual. Outcomes 4, 521–532.

Gul, E.E., Can, I., Kusumoto, F.M., 2012. Case report: an unusual heart rhythm associated with organophosphate poisoning. Cardiovasc. Toxicol. 12, 263–265.

Halade, G.V., Jin, Y., Lindsey, M.L., 2013. Matrix metalloproteinase (MMP)-9: a proximal biomarker for cardiac remodeling and a distal biomarker for inflammation. Pharmacol. Ther. 139, 32–40.

Haning, W., Goebert, D., 2007. Electrocardiographic abnormalities in methamphetamine abusers. Addiction 102 (Suppl. 1), 70–75.

Harris, T.A., Yamakuchi, M., Ferlito, M., et al., 2008. MicroRNA-126 regulates endothelial expression of vascular cell adhesion molecule 1. Proc. Natl. Acad. Sci. U.S.A. 105, 1516–1521.

Heeschen, C., Aicher, A., Lehmann, R., et al., 2003. Erythropoietin is a potent physiologic stimulus for endothelial progenitor cell mobilization. Blood 102, 1340–1346.

Horie, T., Ono, K., Nishi, H., et al., 2010. Acute doxorubicin cardiotoxicity is associated with miR-146a induced inhibition of the neuregulin-ErbB pathway. Cardiovasc. Res. 87, 656–664.

Hullinger, T.G., Montgomery, R.L., Seto, A.G., et al., 2012. Inhibition of miR-15 protects against cardiac ischemic injury. Circ. Res. 110, 71–81.

Humphreys, D.T., Hynes, C.J., Patel, H.R., et al., 2012. Complexity of murine cardiomyocyte miRNA biogenesis, sequence variant expression and function. PLoS One 7, e30933.

Hunter, M.P., Ismail, N., Zhang, X., et al., 2008. Detection of microRNA expression in human peripheral blood microvesicles. PLoS One 3, e3694.

Jaffe, A.S., Babuin, L., Apple, F.S., 2006. Biomarkers in acute cardiac disease: the present and the future. J. Am. Coll. Cardiol. 48, 1–11.

Januzzi, J.L.J. (Ed.), 2011. Cardiac Biomarkers in Clinical Practice. Jones and Bartlett Publishers, Sudbury, MA.

Januzzi Jr., J.L., Peacock, W.F., Maisel, A.S., et al., 2007. Measurement of the interleukin family member ST2 in patients with acute dyspnea: results from the PRIDE (Pro-Brain Natriuretic Peptide Investigation of Dyspnea in the Emergency Department) study. J. Am. Coll. Cardiol. 50, 607–613.

Jarai, Z., Kunos, G., 2002. Cardiovascular effects of cannabinoids. Orv. Hetil. 143, 1563–1568.

Ji, R., Cheng, Y., Yue, J., et al., 2007. MicroRNA expression signature and antisense-mediated depletion reveal an essential role of microRNA in vascular neointimal lesion formation. Circ. Res. 100, 1579–1588.

Ji, X., Takahashi, R., Hiura, Y., et al., 2009. Plasma miR-208 as a biomarker of myocardial injury. Clin. Chem. 55, 1944–1949.

Kaikkonen, M.U., Lam, M.T.Y., Glass, C.K., 2011. Non-coding RNAs as regulators of gene expression and epigenetics. Cardiovasc. Res. 90, 430–440.

Kakkar, P., Jaffery, F.N., 2005. Biological markers for metal toxicity. Environ. Toxicol. Pharmacol. 19, 335–349.

Kehl, D.W., Iqbal, N., Fard, A., et al., 2012. Biomarkers in acute myocardial injury. Transl. Res. 159, 252–264.

Keller, K.B., Lemberg, L., 2003. The cocaine-abused heart. Am. J. Crit. Care 12, 562–566.

Keller, T., Zeller, T., Peetz, D., et al., 2009. Sensitive troponin I assay in early diagnosis of acute myocardial infarction. N. Engl. J. Med. 361, 868–877.

Kempf, T., Horn-Wichmann, R., Brabant, G., et al., 2007. Circulating concentrations of growth-differentiation factor 15 in apparently healthy elderly individuals and patients with chronic heart failure as assessed by a new immunoradiometric sandwich assay. Clin. Chem. 53, 284–291.

Kempf, T., Eden, M., Strelau, J., et al., 2006. The transforming growth factor-beta superfamily member growth-differentiation factor-15 protects the heart from ischemia/reperfusion injury. Circ. Res. 98, 351–360.

Khan, S.Q., Dhillon, O.S., O'Brien, R.J., et al., 2007. C-terminal provasopressin (copeptin) as a novel and prognostic marker in acute myocardial infarction: leicester Acute Myocardial Infarction Peptide (LAMP) study. Circulation 115, 2.

Kilickap, S., Barista, I., Akgul, E., et al., 2005. cTnT can be a useful marker for early detection of anthracycline cardiotoxicity. Ann. Oncol. 16, 798–804.

Lam, M.P.Y., Ping, P., Murphy, E., 2016. Proteomics research in cardiovascular medicine and biomarker discovery. J. Am. Coll. Cardiol. 68, 2819–2830.

Leonard, C.E., Bilker, W.B., Newcomb, C., et al., 2011. Antidepressants and the risk of sudden cardiac death and ventricular arrhythmia. Pharmacoepidemiol. Drug Saf. 20, 903–913.

Lilly, L. (Ed.), 2003. Pathophysiology of Heart Disease. Lippincott Williams & Wilkins, Philadelphia.

Liu, S.H., Lin, J.L., Weng, C.H., et al., 2012. Heart rate-corrected QT interval helps predict mortality after intentional organophosphate poisoning. PLoS One 7, e36576.

Lorenzen, J.M., Martino, F., Thum, T., 2012. Epigenetic modifications in cardiovascular disease. Basic Res. Cardiol. 107, 245–255.

Louden, C., Brott, D., Katein, A., et al., 2006. Biomarkers and mechanisms of drug-induced vascular injury in non-rodents. Toxicol. Pathol. 34, 19–26.

Ma, G., Brady, W.J., Pollack, M., Chan, T.C., 2001. Electrocardiographic manifestations: digitalis toxicity. J. Emerg. Med. 20, 145–152.

Malik, A.N., Czajka, A., 2013. Is mitochondrial DNA content a potential biomarker of mitochondrial dysfunction? Mitochondrion 13, 481–492.

Maisel, A., Jaffe, A.S. (Eds.), 2016. Cardiac Biomarkers: Case Studies and Clinical Correlations. Springer Int. Pub., Cham, Switzerland.

Maisel, A.S., Krishnaswamy, P., Nowak, R.M., et al., 2002. Rapid measurement of B-type natriuretic peptide in the emergency diagnosis of heart failure. N. Engl. J. Med. 347, 161–167.

Mann, D.L., 1999. Mechanisms and models in heart failure: a combinatorial approach. Circulation 100, 999–1008.

Martin-Ventura, J.L., Blanco-Colio, L.M., Tunon, J., et al., 2009. Biomarkers in cardiovascular medicine. Rev. Esp. Cardiol. 62, 677–688.

Mause, S.F., Weber, C., 2010. Microparticles: protagonists of a novel communication network for intercellular information exchange. Circ. Res. 107, 1047–1057.

McCann, C.J., Glover, B.M., Menown, I.B., et al., 2009. Prognostic value of a multimarker approach for patients presenting to hospital with acute chest pain. Am. J. Cardiol. 103, 22–28.

Meder, B., Keller, A., Vogel, B., et al., 2011. MicroRNA signatures in total peripheral blood as novel biomarkers for acute myocardial infarction. Basic Res. Cardiol. 106, 13–23.

Meldrum, D.R., 1998. Tumor necrosis factor in the heart. Am. J. Physiol. 274, R577–R595.

Menown, I.B., Mackenzie, G., Adgey, A.A., 2000. Optimizing the initial 12-lead electrocardiographic diagnosis of acute myocardial infarction. Eur. Heart J. 21, 275–283.

Moayeri, M., Crown, D., Dorward, D.W., et al., 2009. The heart is an early target of anthrax lethal toxin in mice: a protective role for neuronal nitric oxide synthase (nNOS). PLoS Pathog. 5, e1000456.

Montgomery, R.L., Hullinger, T.G., Semus, H.M., et al., 2011. Therapeutic inhibition of miR-208a improves cardiac function and survival during heart failure. Circulation 124, 1537–1547.

Morrow, D.A. (Ed.), 2006. Cardiovascular Biomarkers. Humana Press, Totowa, NJ.

Mueller, T., Dieplinger, B., Gegenhuber, A., et al., 2008. Increased plasma concentrations of soluble ST2 are predictive for 1-year mortality in patients with acute destabilized heart failure. Clin. Chem. 54, 752–756.

Muncy, T.A., Bebarta, V.S., Varney, S.M., et al., 2012. Acute electrocardiographic ST segment elevation may predict hypotension in a swine model of severe cyanide toxicity. J. Med. Toxicol. 8, 285–290.

Nagai, R., Kinugawa, K., Inoue, H., et al., 2013. Urgent management of rapid heart rate in patients with atrial fibrillation/flutter and left ventricular dysfunction. Circ. J. 77, 908–916.

Navas-Acien, A., Guallar, E., Silbergeld, E.K., Rothenberg, S.J., 2007. Lead exposure and cardiovascular disease a systematic review. Environ. Health Perspect. 115, 472–482.

Nicholls, S.J., Hazen, S.L., 2005. Myeloperoxidase and cardio-vascular disease. Arterioscler. Thromb. Vasc. Biol. 25, 1102–1111.

Nickel, C.H., Bingisser, R., Morgenthaler, N.G., 2012. The role of copeptin as a diagnostic and prognostic biomarker for risk stratification in the emergency department. BMC Med. 10, 7.

Nilsson, L., Szymanowski, A., Swahn, E., Jonasson, L., 2013. Soluble TNF receptors are associated with infarct size and ventricular dysfunction in ST-elevation myocardial infarction. PLoS One 8, e55477.

Nootens, M., Kaufmann, E., Rector, T., et al., 1995. Neurohormonal activation in patients with right ventricular failure from pulmonary hypertension: relation to hemodynamic variables and endothelin levels. J. Am. Coll. Cardiol. 26, 1581–1585.

Oliveira-Carvalho, V., Carvalho, V.O., Bocchi, E.A., 2013. The emerging role of miR-208a in the heart. DNA Cell Biol. 32, 8–12.

Ono, K., Kuwabara, Y., Han, J., 2011. MicroRNAs and cardio-vascular diseases. FEBS J. 278, 1619–1633.

Pasceri, V., Willerson, J.T., Yeh, E.T., 2000. Direct proinflammatory effect of C-reactive protein on human endothelial cells. Circulation 102, 2165–2168.

Patel, V.B., Preedy, V.R., 2016. Biomarkers in Cardiovascular Disease. Springer Science+Business Media, Dordrecht, NL.

Peacock, W.F., De Marco, T., Fonarow, G.C., et al., 2008. Cardiac troponin and outcome in acute heart failure. N. Engl. J. Med. 358, 2117–2126.

Peters, S.A., Visseren, F.L., Grobbee, D.E., 2013. Biomarkers. Screening for C-reactive protein in CVD prediction. Nat. Rev. Cardiol. 10, 1–14.

Phillips, K., Luk, A., Soor, G.S., et al., 2009. Cocaine cardiotoxicity. Am. J. Cardiovasc. Drugs 9, 177–196.

Pintus, R., Bassareo, P.P., Dessi, A., et al., 2017. Metabolomics and cardiology: toward the path of perinatal programming and personalized medicine. BioMed Res. Int. 2017, 6970631.

Pratap, B., Korniyenko, A., 2012. Toxic effects of marijuana on the cardiovascular system. Cardiovasc. Toxicol. 12, 143–148.

Preedy, V.R., Patel, V.B., 2017. Biomarkers in Disease:Methods, Discoveries and Applications. Springer Int Pub.

Quatrehomme, G., Ricq, O., Lapalus, P., et al., 1992. Acute arsenic intoxication: forensic and toxicologic aspects (an observation). J. Forensic Sci. 37, 1163–1171.

Ren, X.P., Wu, J., Wang, X., et al., 2009. MicroRNA-320 is involved in the regulation of cardiac ischemia/reperfusion injury by targeting heat-shock protein 20. Circulation 119, 2357–2366.

Sabatine, M.S., Morrow, D.A., de Lemos, J.A., et al., 2002. Multimarker approach to risk stratification in non-ST elevation acute coronary syndromes: simultaneous assessment of troponin I, C-reactive protein, and B-type natriuretic peptide. Circulation 105, 1760–1763.

Sahu, S.C., 2014. microRNAs in Toxicology and Medicine. Wiley.

Salic, K., De Windt, L.J., 2012. MicroRNAs as biomarkers for myocardial infarction. Curr. Atheroscler. Rep. 14, 193–200.

Sawyer, D.B., 2013. Anthracyclines and heart failure. N. Engl. J. Med. 368, 1154–1156.

Shah, R.V., Januzzi Jr., J.L., 2010. ST2: a novel remodeling biomarker in acute and chronic heart failure. Curr. Heart Fail. Rep. 7, 9–14.

Shi, J., Wei, L., 2013. Rho kinases in cardiovascular physiology and pathophysiology: the effect of fasudil. J. Cardiovasc. Pharmacol. 62, 341–354.

Shi, Y., Moon, M., Dawood, S., et al., 2010. Mechanisms and management of doxorubicin cardiotoxicity. Herz 36, 296–305.

Shieh, J.T., Huang, Y., Gilmore, J., Srivastava, D., 2011. Elevated miR-499 levels blunt the cardiac stress response. PLoS One 9, e19481.

Shintani, Y., Kapoor, A., Kaneko, M., et al., 2013. TLR9 mediates cellular protection by modulating energy metabolism in cardiomyocytes and neurons. Proc. Natl. Acad. Sci. U.S.A. 110, 5109–5114.

Singh, A.P., Goel, R.K., Kaur, T., 2011. Mechanisms pertaining to arsenic toxicity. Toxicol. Int. 18, 87–93.

Spiel, A.O., Gilbert, J.C., Jilma, B., 2008. Von Willebrand factor in cardiovascular disease: focus on acute coronary syndromes. Circulation 117, 1449–1459.

States, J.C., Srivastava, S., Chen, Y., Barchowsky, A., 2009. Arsenic and cardiovascular disease. Toxicol. Sci. 107, 312–323.

Stefanska, B., MacEwan, D.J., 2015. Epigenetics and pharmacology. Br. J. Pharmacol. 172, 2701–2704.

Stratz, C., Amann, M., Berg, D.D., et al., 2012. Novel biomarkers in cardiovascular disease: research tools or ready for personalized medicine? Cardiol. Rev. 20, 111–117.

Tan, L.B., Jalil, J.E., Pick, R., et al., 1991. Cardiac myocyte necrosis induced by angiotensin II. Circ. Res. 69, 1185–1195.

Tang, W.H., Francis, G.S., Morrow, D.A., et al., 2008. National Academy of Clinical Biochemistry Laboratory Medicine practice guidelines: clinical utilization of cardiac biomarker testing in heart failure. Clin. Biochem. 41, 210–221.

Tardif, J.C., Heinonen, T., Orloff, D., Libby, P., 2006. Vascular biomarkers and surrogates in cardiovascular disease. Circulation 113, 2936–2942.

Tatsuguchi, M., Seok, H.Y., Callis, T.E., et al., 2007. Expression of microRNAs is dynamically regulated during cardiomyocyte hypertrophy. J. Mol. Cell. Cardiol. 42, 1137–1141.

Terentyev, D., Belevych, A.E., Terentyeva, R., et al., 2009. miR-1 overexpression enhances Ca(2+) release and promotes cardiac arrhythmogenesis by targeting PP2A regulatory subunit B56alpha and causing CaMKII-dependent hyperphosphorylation of RyR2. Circ. Res. 104, 514–521.

Thanacoody, H.K., Thomas, S.H., 2005. Tricyclic antidepressant poisoning: cardiovascular toxicity. Toxicol. Rev. 24, 205–214.

Thum, T., Gross, C., Fiedler, J., et al., 2008. MicroRNA-21 contributes to myocardial disease by stimulating MAP kinase signalling in fibroblasts. Nature 456, 980–984.

Tijsen, A.J., Pinto, Y.M., Creemers, E.E., 2012. Circulating microRNAs as diagnostic biomarkers for cardiovascular diseases. Am. J. Physiol. Heart Circ. Physiol. 303, H1085–H1095.

Tsuji, H., Larson, M.G., Venditti Jr., F.J., et al., 1996. Impact of reduced heart rate variability on risk for cardiac events. The Framingham heart study. Circulation 94, 2850–2855.

Turillazzi, E., Bello, S., Neri, M., et al., 2012. Cardiovascular effects of cocaine: cellular, ionic and molecular mechanisms. Curr. Med. Chem. 19, 5664–5676.

Upadhyay, R.K., 2015. Emerging risk biomarkers in cardiovascular diseases and disorders. J. Lipids 1215, 971453. https://doi.org/10.1155/2015/971453.

Urbich, C., Kuehbacher, A., Dimmeler, S., 2008. Role of microRNAs in vascular diseases, inflammation, and angiogenesis. Cardiovasc. Res. 79, 581–588.

Ussher, J.R., Elmariah, S., Gerszten, R.L., Dyck, J.R.B., 2016. The emerging role of metabolomics in the diagnosis and prognosis of cardiovascular disease. J. Am. Coll. Cardiol. 68, 2850–2870.

Uydu, H.A., Bostan, M., Yilmaz, A., et al., 2013. Comparision of inflammatory biomarkers for detection of coronary stenosis in patients with stable coronary, artery disease. Eur. Rev. Med. Pharmacol. Sci. 17, 112–118.

van Rooij, E., 2012. Introduction to the series on microRNAs in the cardiovascular system. Circ. Res. 110, 481–482.

van Rooij, E., Sutherland, L.B., Liu, N., et al., 2006. A signature pattern of stress-responsive microRNAs that can evoke cardiac hypertrophy and heart failure. Proc. Natl. Acad. Sci. U.S.A. 103, 18255–18260.

Volkova, M., Russell, R., 2011. Anthracycline Cardiotoxicity: prevalence, pathogenesis and treatment. Curr. Cardiol. Rev. 7, 214–229.

van Solingen, C., Seghers, L., Bijkerk, R., et al., 2009. Antagomir-mediated silencing of endothelial cell specific microRNA-126 impairs ischemia-induced angiogenesis. J. Cell Mol. Med. 13, 1577–1585.

Vijayakumar, S., Fareedullah, M., Ashok Kumar, E., Mohan Rao, K., 2011. A prospective study on electrocardiographic findings of patients with organophosphorus poisoning. Cardiovasc. Toxicol. 11, 113–117.

Wang, G.K., Zhu, J.Q., Zhang, J.T., et al., 2010. Circulating microRNA: a novel potential biomarker for early diagnosis of acute myocardial infarction in humans. Eur. Heart J. 31, 659–666.

Wang, T.J., Wollert, K.C., Larson, M.C., et al., 2012. Prognostic utility of novel biomarkers of cardiovascular stress: the Framingham Heart Study. Circulation 126, 1596–1604.

Watts, D.J., McCollester, L., 2006. Methamphetamine-induced myocardial infarction with elevated troponin I. Am. J. Emerg. Med. 24, 132–134.

Webster, A.L.H., Shu-Ching Yan, M., Marsden, P.A., 2013. Epigenetics and cardiovascular disease. Can. J. Cardiol. 29, 46–57.

Weinberg, E.O., Shimpo, M., De Keulenaer, G.W., et al., 2002. Expression and regulation of ST2, an interleukin-1 receptor family member, in cardiomyocytes and myocardial infarction. Circulation 106, 2961–2966.

Wettersten, N., Maisel, A.S., 2016. Biomarkers for heart failure: an update for practitioners of internal medicine. Am. J. Med. 129, 560–567.

Wexler, J., Whittenberger, J.L., Dumke, P.R., 1947. The effect of cyanide on the electrocardiogram of man. Am. Heart J. 34, 163–173.

Wierzbicki, A.S., 2007. Homocysteine and cardiovascular disease: a review of the evidence. Diab. Vasc. Dis. Res. 4, 143–150.

Wittenberg, J.B., Wittenberg, B.A., 2003. Myoglobin function reassessed. J. Exp. Biol. 206, 2011–2020.

Yang, B., Lin, H., Xiao, J., et al., 2007. The muscle-specific microRNA miR-1 regulates cardiac arrhythmogenic potential by targeting GJA1 and KCNJ2. Nat. Med. 13, 486–491.
Yayan, J., 2013. Emerging families of biomarkers for coronary artery disease: inflammatory mediators. Vasc. Health Risk Manag. 9, 435–456.
Zeisel, S.H., 2000. Choline: an essential nutrient for humans. Nutrition 16, 669–671.
Zhang, J., Defelice, A.F., Hanig, J.P., Colatsky, T., 2010. Biomarkers of endothelial cell activation serve as potential surrogate markers for drug-induced vascular injury. Toxicol. Pathol. 38, 856–871.
Zhang, M., 1997. Chemistry underlying the cardiotoxicity of antihistamines. Curr. Med. Chem. 4, 171–184.
Zhang, S., Liu, X., Bawa-Khalfe, T., et al., 2012. Identification of the molecular basis of doxorubicin-induced cardiotoxicity. Nat. Med. 18, 1639–1642.
Zuurman, L., Ippel, A.E., Moin, E., van Gerven, J.M., 2009. Biomarkers for the effects of cannabis and THC in healthy volunteers. Br. J. Clin. Pharmacol. 67, 5–21.

CHAPTER

13

Respiratory Toxicity Biomarkers

Samantha J. Snow, Urmila P. Kodavanti

Environmental Public Health Division, National Health and Environmental Effects Research Laboratory, Office of Research and Development, United States Environmental Protection Agency, Durham, North Carolina, United States

INTRODUCTION

The respiration function of gas exchange is accomplished by lungs in land mammals. The respiratory system is composed of upper airways (nose and larynx), lower airways (bronchi and bronchioles), and the alveolar region (parenchyma). The structure of the respiratory system is complex and adapted to the atmospheric conditions that surround the organism. For example, because of the potential of lung exposure to large particles during burrowing in rodents, the upper airways form complex structures that enable removal and detoxification of the components inhaled, whereas, in man, the structure of the nasal airway is less complex. These habitat-related morphological adaptations are also reflected in the manner in which rats and humans breathe: rats are obligate nose breathers, whereas humans are oropharyngeal breathers. Relative to other organ systems in the body, the lung is unique in that airway and alveolar epithelial lining is directly in contact with the external atmosphere. The air–liquid interface within the alveolar region enables gas exchange to occur between the lung and vascular system for sustaining vitality. A thin layer of surfactant lining the alveolar epithelium reduces the surface tension of this air–liquid interface, thus protecting the lung from collapsing.

Because of the direct exposure of respiratory epithelium to the external air, any respirable toxic components that contaminate air can be inhaled and injure the respiratory tract. In most instances, the lung will be exposed to these materials as complex mixtures after these components are released in the air and photochemically aged. Lung injuries can also be inflicted by circulating factors, such as drugs, hormones, and other distally originating reactive substances within the body. Moreover, low atmospheric pressure, especially at high altitude, can cause lung edema acutely.

Depending on the physicochemical nature of these inhalants, the injury to the respiratory system can vary in location and lung cell type. Acute lung injury is often reversible but can also lead to longer and even permanent damage. In certain cases, some inhaled particles remain in the lung over the lifetime producing first acute inflammation, followed by a short-lived recovery, and later development of chronic lung disease, such as chronic obstructive pulmonary disease (COPD), fibrosis, or cancer. Lung injuries, thus, vary with the type and the degree of insult.

In context to the respiratory system, we can define biomarkers measured in biological samples obtained in a minimally or more invasive manner as the indicators that can reflect the nature of injury to the respiratory system. These may represent constituents released from damaged respiratory cells within the nasal passages or lung for which the degree of change will relate to the extent of injury. There is a unique opportunity available to diagnose nasal and lung injuries and diseases using a lavage technique. The ability to perform bronchoalveolar lavage (BAL) or nasal lavage (NAL) negates the need for more invasive techniques such as biopsies for disease diagnosis. This is in addition to determining biomarkers of lung toxicity in the exhaled breath, sputum, circulation, and urine samples. In this chapter, we will first provide a brief overview of the structure of the respiratory system and the types of pulmonary injuries (because the choice of biomarker selection is based on this) and then elaborate on biomarkers of lung injury. Descriptions of biomarkers determined in BAL fluid (BALF), exhaled breath, sputum, and NAL fluid (NALF); blood, serum, and plasma; and urine will be provided, with special emphasis on the use of emerging new technologies for biomarker identification. Although sputum specimens are routinely used for diagnosis in humans, they are not readily available when working

Biomarkers in Toxicology, Second Edition
https://doi.org/10.1016/B978-0-12-814655-2.00013-X

with laboratory animal models. Because the biomarkers measured in sputum are also measured in BALF and NALF, we will describe the biomarkers for all of these components together in the later discussion. Although BAL has generally been used as a terminal procedure in laboratory animals, it is often used in humans as a minimally invasive technique where a specific lung lobe is lavaged for biomarker assessment or disease diagnosis. We will include this approach for biomarker analysis because BAL can be performed repeatedly as a survival procedure on small laboratory rodents, as well as on large animals. The fluid recovered by BAL provides a unique opportunity to examine very specific biomarkers of lung toxicity. We will also briefly cover the sampling methodologies, especially for exhaled breath and BAL.

With the emergence of new biotechnologies in recent years, the field of biomarkers has seen revolutionary growth. Because high-throughput technologies enable global measures of hundreds to thousands of biomarkers in a single sample, the pattern of change in multiple functionally related biomarkers is given more emphasis than a single molecule–based assessment. This is also true for respiratory disease–specific biomarkers. We are on a steep slope of learning regarding biomarkers and their specificity. Because multiple disease processes share some common mechanisms, such as inflammation and increased oxidation, the biomarkers pertaining to these processes are similar for many illnesses or injuries, and thus, markers of these processes will not be specific to a given disease. We will also try to cover the newly emerging biomarkers with special emphasis to respiratory injuries and diseases.

THE RESPIRATORY SYSTEM

The respiratory system has evolved to absorb molecular oxygen from the air in exchange for carbon dioxide, providing energy for all cells within the organism. There are two major compartments of the respiratory system that will be important to consider with regard to biomarkers of toxicity: the airway passages used to transport air and the alveoli where gas exchange occurs.

The airway passages begin with the nose, followed by the trachea, which divides into two main bronchi, and then several small bronchioles within each lung. Different types of epithelial cells involved in various functions line the airway. The predominant cell type of the airway surface is the pseudostratified columnar ciliated cells, which function to remove particles encountered on the mucus layer from inhaled air and protect the lung from injury. Dome-shaped club cells found in small airways possess drug-metabolizing enzymes and also function as stem cells to replenish lost ciliated epithelial cells. Goblet cells produce and secrete mucus at the apical surface of airway epithelium. The mucus layer helps remove particulates and pathogens via mucociliary clearance, modulates innate immune response, and detoxifies reactive inhaled substances. The thickness of the mucus layer is anatomically related to the diameter of the airway and the number of goblet cells contributing to secretion (Voynow and Rubin, 2009; Widdicombe, 2002). The basement membrane supports airway epithelial cells, and the smooth muscle layer surrounds the interstitial space around the basement membrane. Dendritic cells project between airway epithelial cells and function as antigen-presenting cells that, on encountering antigen (particulate, microbial, or soluble), interact with T lymphocytes and direct antibody production by plasma cells to orchestrate humoral immune responses (Cook and Bottomly, 2007). The airway epithelium is innervated by sensory vagal C-fibers. When stimulated, C-fibers evoke classical reflex reaction leading to bronchoconstriction and cough that, in turn, stimulates parasympathetic tone causing bradycardia and hypotension (Pisi et al., 2009).

The bronchioles terminate into small alveolar sacs that are lined with type 1 and type 2 alveolar epithelial cells. In humans and primates, terminal bronchioles, in addition to alveolar sacs, perform the air exchange function. Type 1 epithelial cells cover most of the surface area of the alveoli. Type 2 cells function to synthesize and secrete surfactant material, in the form of lamellar bodies, as well as other proteins. Type 1 cells are in close proximity to the capillary walls, enabling diffusion of carbon dioxide from blood to air and oxygen from air to blood (Galambos and Demello, 2008; Herzog et al., 2008). Interstitial tissue supporting these structures comprises an extracellular matrix, including collagen, elastin, and myofibroblasts. The connective tissue network interlinks alveoli, and the entire unit is encapsulated by the pleural mesothelial layer, which provides anatomical structure to the lung. Injury to alveolar units can be detected using BAL and analysis of the recovered fluid for a number of biomarkers.

Alveolar type 2 epithelial cells function to secrete surfactant material that is made up of approximately 80% phospholipids, 20% neutral lipids, and proteins. The surfactant material is layered thinly over the entire alveolar surface, providing stability to the alveoli, preventing collapse, and preserving patency. It also functions as an important host-defense mechanism. Among the four surfactant proteins (SPs), SP-A and SP-D play roles in host defense, whereas hydrophobic SP-B and SP-C are involved in adsorption and spreading of the surfactant material along the alveolar lining (Enhorning, 2008; Griese, 1999). These proteins, when released in the circulation and detected in BALF, serve as important biomarkers of lung injury.

Free macrophages within the alveoli function to protect the lung from inhaled pathogens, bacteria, and particles. Macrophages also engulf excess surfactant material and neutrophils that are recruited to the lung after an injury. In addition, macrophages perform important innate immune function by expressing cytokines that are involved in mounting an inflammatory response and programming the resolution of inflammation.

CAUSES OF LUNG INJURY AND DISEASE

The respiratory system encounters a variety of toxic substances during inhalation. The soluble and gas components of air pollution can injure the epithelial lining layer, the cells underneath, and the entire respiratory tract. The inhaled substances include reactive gases, volatile organic substances, infectious agents, metals, pesticides, nanomaterials, biological substances, tobacco smoke, asbestos, and other respirable particulate matter (PM) of environmental origin. Many of these substances are removed readily by the mucociliary escalator and cleared by alveolar macrophages, but, in the process, lung injury is inflicted. In addition, some inhaled particles and fibers are unable to be cleared because of damage of the clearance system, leading to persistent lung injury. Occupational exposures involving metal fumes, silica, and asbestos have been shown to contribute to lung injury and chronic pulmonary disease burden (Clapp et al., 2008; Cohen et al., 2008). Lung injury can also be induced by decreased ambient pressure at high altitude. Sepsis-related lung injury is also common. Pulmonary exposures to circulating drugs have often been associated with lung toxicity. Examples include monocrotaline-induced pulmonary hypertension, bleomycin-induced pulmonary fibrosis, and pulmonary phospholipidosis induced by cationic amphiphilic drugs (Kodavanti and Mehendale, 1990). In addition, lung edema can occur following left heart failure (Gehlbach and Geppert, 2004). Clinical use of ventilators in infants is also associated with lung damage early in life, which affects normal lung growth and vulnerability to adult lung diseases.

Among environmental exposures, agricultural and industrial dusts, emissions from refineries, and vehicular emissions are the major causal factors of lung injury. These exposures generally occur as complex mixtures of organic and inorganic species, gases, and soluble and insoluble PM. Because each chemical and physical entity is likely to have unique lung site–specific impact, the causes of lung injury and subsequent diseases are complex and involve multiple pathologies. COPD caused by cigarette smoke is a prime example of such complexity. COPD is characterized by airway mucus hypersecretion, airway and alveolar inflammation, emphysema, and, often, overlapping asthma. Assessment of multiple biomarkers can provide more precise diagnosis in such complex diseases.

TYPES OF LUNG INJURIES AND PATHOLOGIES

Acute Lung Injuries

Acute lung injury can occur soon after exposure to noxious substances. Depending on the chemical nature and the concentration of inhalant, the injury can occur to the upper or lower respiratory tract or both. For example, exposure of rodents to high levels of sulfur dioxide produces airway inflammation and mucus hypersecretion but very little alveolar damage, as characterized by morphological analysis and determination of BALF markers (Kodavanti et al., 2006). In the case of upper airway injury, goblet cells within submucosal glands and epithelium are stimulated to produce more mucus to protect the epithelium and accelerate removal of noxious substances. The dynamic process of mucus secretion involves activation of epidermal growth factor receptor and IL-12 through FoxA2, TTF-1, SPDEF, and GABAAR signaling. These factors transcriptionally regulate MARCKS, SNAREs, and MUC proteins (Voynow and Rubin, 2009). The upper airway inflammation and mucus hypersecretion, if sustained, can lead to chronic bronchitis. In a chronic lung injury, these changes promote airway fibrosis, which leads to development of airway resistance and decreased ventilation.

Damage to the alveolar region is characterized by pulmonary edema, activation of alveolar macrophages, secretion of cytokines and immune mediators, and extravasation of inflammatory cells into the interstitial and alveolar lung compartments. The alveolar lining layer of surfactant is also altered after inhalation of respirable particles, gases, or other substances. The type of cell infiltration depends on the nature of initial damage and the animal species. For example, exposure of rats to ozone will lead to neutrophilic inflammation, whereas sensitizing substances, such as ovalbumin, will cause eosinophils to extravasate along with neutrophils (Bauer et al., 2010; Schneider et al., 2012). This inflammatory response is produced by cytokines and chemokines from epithelium and alveolar macrophages. Often, acute lung injury, depending on the degree of lung damage, type of insult, and duration, can lead to fibrosis and chronic inflammation.

During lung injury, mediators can be released basolaterally and into the circulation. SPs also have the potential to get released into the blood stream, serving as specific circulating biomarkers of alveolar damage. The mediators released apically from the damaged lung cells and infiltrating inflammatory cells can be isolated by the BAL process. The resulting BALF can be analyzed for both the injury-causing substance and biomarkers of injury/inflammation. Furthermore, pulmonary injury can trigger a neurohumoral response, resulting in release of sympathetic modulators in the circulation (Kodavanti, 2016).

Acute respiratory distress syndrome (ARDS) in children is associated with a deficiency in surfactant during early lung growth (Turell, 2008). Along with injury to the lung cells, the surfactant can also be modified chemically when encountering mechanical stress, noxious endogenous or inhaled substances, or due to abnormalities in surfactant synthesis and secretion. Under disease conditions, the hydrolysis of phosphatidylcholine by phospholipases is thought to chemically modify the surfactant, reducing its ability to maintain airway patency (Ackerman et al., 2003). As SPs are involved in lung injuries and diseases, their ability to translocate into the circulation and be detected in serum serves as a biomarker specific to lung cell injuries.

Asthma

Asthma is characterized by chronic lung inflammation and airway hyperresponsiveness (Mauad and Dolhnikoff, 2008). Multiple environmental, host genetic, and physiological factors have been linked to different degrees of pathological features of this spectrum disorder (Bloemen et al., 2007). T helper (Th) lymphocytes are involved in inflammatory processes. Inhaled particulates, cellular components of Gram-negative bacteria, and proteoglycans can activate pattern-recognition receptors, stimulate dendritic cells, and increase Th2 polarization (Hamid and Tulic, 2009). Once sensitized, the airways react to subsequent exposure with exacerbated inflammation and bronchoconstriction in asthma. The differentiation of Th cells to a Th2 phenotype is critical in the development of allergenicity and airway hyperresponsiveness associated with asthma. Th2-mediated elaboration of cytokines IL-4, IL-5, IL-9, and IL-13 leads to recruitment and activation of mast cells, eosinophils, macrophages, endothelial cells, fibroblasts, and smooth muscle cells (Hamid and Tulic, 2009). Neutrophilic inflammation is also noted with involvement of $CD4^+$ T lymphocytes and regulatory T cells. Collectively, Th2 cytokines serve as specific biomarkers of asthma (Fahy, 2009).

Bronchitis, Emphysema, and Chronic Obstructive Pulmonary Disease

Bronchitis is characterized by increased mucus production and inflammation in airways. When a mucus-produced cough lasts for more than 2 months per year, it is termed as chronic bronchitis. Emphysema is characterized by destruction of peripheral and central lung tissues, which generally occurs over a long period. Bronchitis and emphysema, when manifested together, are presented clinically by the obstruction of airways collectively known as COPD. Exposure to cigarette smoke and smoke from biomass burning in developing countries has been linked to the development of COPD (Prasad et al., 2012; Tuder and Petrache, 2012). Both airway and parenchymal injuries are inflicted by numerous toxic components found in cigarette and biomass smoke. The degree of damage to each lung compartment varies between individuals based on their genetic background. The disease is characterized by increased productive cough, resulting from chronic mucus hypersecretion, chronic inflammation (primarily neutrophilic), and destruction of alveolar walls, leading to decrement in lung function over time. A predominance of $CD8^+$ T lymphocytes and increased neutrophils, as well as macrophages, in COPD patients has been demonstrated (Barnes et al., 2006). $CD8^+$ T cells are thought to secrete interferon-γ and interferon-inducible protein 10, which are considered dominant features of the Th1 phenotype (Barnes et al., 2006). The macrophages in smokers are impaired functionally and exhibit greater elastolytic activity (Minematsu et al., 2011). Greater elastolytic activity of alveolar milieu, due to an imbalance between proteases/antiproteases and apoptosis, plays an important role in alveolar destruction. Because of the complex pathologies associated with COPD, therapeutic approaches have achieved limited success. Often, COPD patients express airway hyperresponsiveness associated with asthma. The involvement of both Th1 and Th2 phenotypes and the coexistence of these features are not well understood. Numerous biomarkers have been analyzed in BALF, sputum, and exhaled breath to determine disease characteristics.

Pulmonary Fibrosis and Granuloma

Pulmonary fibrosis is characterized by the presence of inflammation, increased fibroblast proliferation, and collagen synthesis. A number of occupational and environmental exposures to ambient PM, metals such as vanadium, reactive gases, asbestos, and silica, as well as drug treatments, can injure the lung and lead to fibrosis and/or granuloma (Khalil et al., 2007; Warheit et al., 2001). In susceptible individuals, granuloma formation occurs around the accumulated foreign materials that are not cleared by the mucociliary system and macrophages (Warheit et al., 2001). Occasionally, idiopathic lung fibrosis also occurs in susceptible individuals. Bleomycin, an antineoplastic drug, is known to induce pulmonary fibrosis in rodents and is associated with abnormal or aberrant tissue repair resulting from increased expression of growth factors and collagens (Chaudhary et al., 2007). HIF-1α and VEGF activation via PI3 kinase have been shown to contribute to collagen- and fibroproliferative effects of bleomycin (Chaudhary et al., 2007). A number of cytokines and the associated inflammatory responses have been involved in the pathogenesis of lung fibrosis. The evaluation of mediators, cytokines, and growth factors in BALF and plasma, along with lung function outcomes, can help with disease diagnosis.

Pulmonary Hypertension

Pulmonary hypertension, a complex and multifaceted disease, is characterized by an increase in the vascular pressure within the lung causing shortness of breath, fainting, and, in more severe cases, heart failure. Pulmonary hypertension can originate from either artery or vein. Hypoxia or thromboembolism can also result in pulmonary hypertension. In some cases, no specific etiologies can be identified. Before 2003, pulmonary hypertension was classified into two broad categories, primary and secondary, but has since been reclassified into five categories based on an increased understanding of the pathobiologic mechanisms (Nef et al., 2010): (1) pulmonary arterial hypertension, which can be idiopathic, familial, or caused by other diseases; (2) pulmonary hypertension associated with left heart disease; (3) hypertension associated with lung diseases such as COPD, interstitial lung disease, reduced alveolar ventilation, high altitude, and, in some cases, developmental abnormalities; (4) pulmonary arterial thromboembolism; and (5) other miscellaneous causes. Despite the causes being very different, there are some pathobiologic commonalities in pulmonary hypertension, such as vasoconstriction resulting from endothelial dysfunction and increased cell proliferation with decreased apoptosis in pulmonary microvasculature (Wilkins, 2012).

Pulmonary Phospholipidosis

Pulmonary phospholipidosis is characterized by accumulation of surfactant phospholipids in alveolar space and in macrophages. Long-term treatment with cationic amphiphilic drugs, such as amiodarone and chlorphentermine, and classes of other antiarrhythmic and antipsychotic drugs have been associated with pulmonary phospholipidosis (Kodavanti and Mehendale, 1990). These drugs selectively accumulate in the lung where they inhibit phospholipases leading to accumulation of surfactant phospholipids (Kodavanti and Mehendale, 1990). Exposure to silica has also been associated with pulmonary phospholipidosis (Kawada et al., 1989). Amiodarone, an antiarrhythmic drug, also induces pulmonary phospholipidosis while causing lung cell injury and fibrosis in patients (Bedrossian et al., 1997). A recent study shows that rats maintained on a fish oil—rich diet have increased accumulation of lipid-laden macrophages in the lung (Snow et al., 2018). BALF lipids and foamy macrophages could serve as biomarkers of pulmonary phospholipidosis.

SAMPLING METHODS FOR ANALYSIS OF BIOMARKERS OF LUNG TOXICITY

The sampling for lung toxicity biomarkers can involve noninvasive collection of exhaled breath, sputum, urine, and blood and invasive collection of BALF (Fig. 13.1). BAL is used extensively in humans for determination of toxicity and lung disease, as well as in clinical studies. In animal toxicology, BAL is generally performed as a terminal procedure; however, repeated BAL has also been used as a survival procedure (Novak et al., 2006). Because fluid recovered from BAL provides the most critical information of the lung injury, we will first discuss BAL and biomarker assessment in BALF.

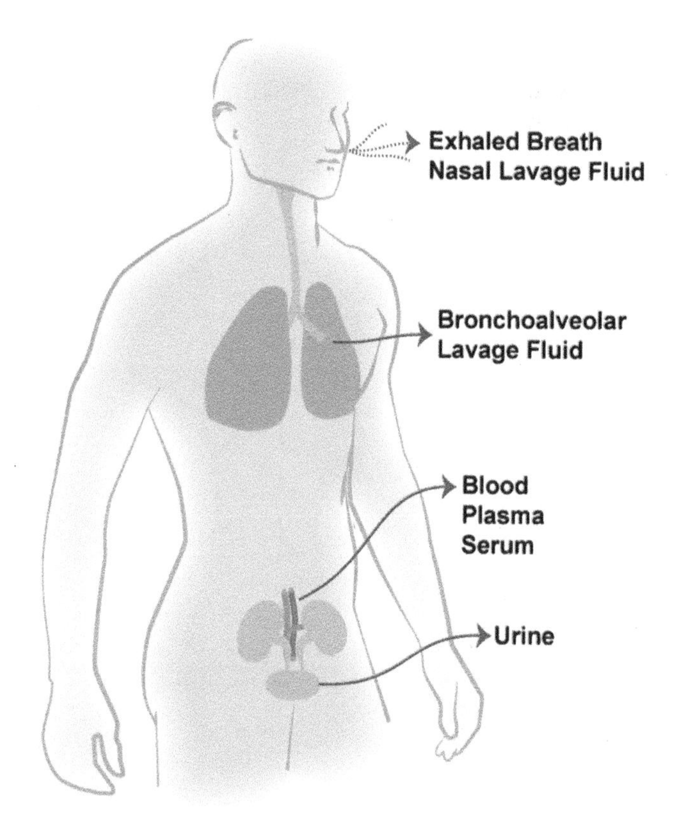

FIGURE 13.1 Schematic of sampling sites for collecting biological materials for determination of lung injury biomarkers.

Bronchoalveolar Lavage

BAL is the unique technique that enables one to sample airway and alveolar lining fluid content to determine the state of lung health. This technique is invasive but provides highly accurate and specific diagnosis of the type of lung injury and causal factors, especially in cases where diagnosis is very difficult based on blood or urine analysis. In humans, segmental lavage is performed using bronchoscopy, whereas, in animals, one lung lobe or the entire lung is lavaged as a survival or nonsurvival technique (Fig. 13.2). The technique involves slow infusion of saline or calcium- and magnesium-free phosphate-buffered saline into the trachea of anesthetized animals or in a specific bronchus of humans. This fluid is then withdrawn slowly back into the syringe in a stepwise manner. Specific treatment and recovery protocols are followed for humans and rodents (Henderson, 2005; Hoffman, 2008; Reynolds, 2011). In some cases, this protocol might not be safe for susceptible humans and is not recommended. In rodents, this procedure can be repeated several times, but several days of recovery are needed prior to performing the technique again.

The volume of BALF collected in rodents and humans is fairly consistent. Underlying injury can, however, affect the fluid recovery. The fluid recovered from BAL can be processed for analysis of cell content and a variety of biomarkers. The recovered fluid can be centrifuged to separate cells. In healthy animals and humans, these cells generally should contain 80%–95%, or even more, alveolar macrophages, whereas the remaining cells comprise neutrophils, lymphocytes, and eosinophils. There is a wide variation in the number of these nonmacrophage-type immune cells in humans and animals, depending on the animal species and strains. For example, the BALF of guinea pigs contains more eosinophils than that of rats (Kodavanti et al., 1995, 2000). The cell-free BALF can be assessed for a variety of cytokines, epithelial and endothelial injury markers, and markers of exposure, which can inform the type of injury and potential therapeutic intervention.

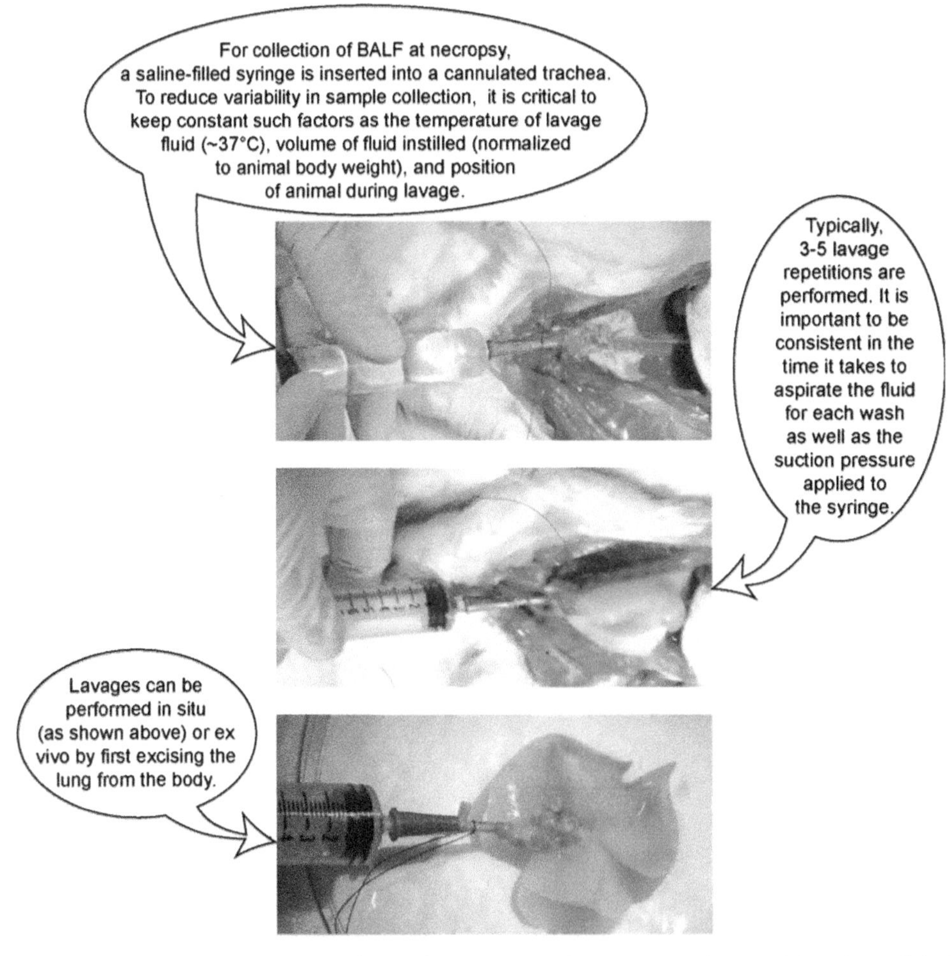

FIGURE 13.2 Photographs of bronchoalveolar lavage fluid (BALF) collection in a rodent model performed during necropsy.

Sampling Sputum, Nasal Lavage, and Exhaled Breath

The protocol for inducing and collecting sputum is employed for humans to sample upper airways. The nebulized hypertonic saline is administered in the nostrils at increasing volume after inducing bronchodilation with inhaled albuterol (Vatrella et al., 2007). The patient is encouraged to expectorate sputum through voluntary coughing. The specimen is collected in a sterile beaker and processed. The sputum can also be generated spontaneously and thus collected; however, the consistency of biomarkers in this material is often a limiting factor. The thicker mucous material can be separated from saliva in the specimen and analyzed for cellular markers (Aitken et al., 2003). Generally, the volume of sputum will be sufficient from patients having COPD and chronic bronchitis, but it is often limited in healthy individuals. The analysis of antioxidants, cytokines, and inflammatory cells can often provide mechanistic insights into the underlying disease.

NAL can be performed in rodents to simulate the procedure of induced sputum in humans. Using a similar concept to BALF collection, this terminal procedure involves a slow infusion of saline via the nasopharyngeal passage through the nasal cavity to collect the NALF from the nostrils of the animal (Cho et al., 2012). Following collection, NALF is processed in an identical manner to BALF for analysis of cellular makeup and biomarkers of injury. This technique is typically used in cases where the pollutant is known to primarily affect the nasal passages and upper airways (Snow et al., 2017).

The collection of human exhaled breath is a simple noninvasive technique. However, the methodology for collection of exhaled breath and constituting condensate (EBC) remains inconsistent and is highly influenced by temperature change. The high degree of variability hampers its utility in identifying exposure- and pathology-related biomarkers (Rosias, 2012). Mainly, the issues with methodologies involve differences in the conditions of the subjects, condenser devices, and preservation/storage of the samples. Based on the interest in identifying biomarkers in exhaled breath, the American Thoracic Society and the European Respiratory Society organized a task force to develop guidelines on collection and analysis of biomarkers in exhaled breath (Ahmadzai et al., 2013; Grob et al., 2008). However, these methodologies still vary between laboratories (Carter et al., 2012; Popov, 2011; Rosias, 2012). Recently, more sophisticated devices and controlled exhalation techniques have provided fairly consistent sampling of exhaled breath for assessment of biomarkers (Pleil et al., 2017; Winters et al., 2017). Numerous biomarkers have been identified in EBC and linked to disease conditions.

Blood and Urine Sampling

Lung injury–specific biomarkers (i.e., inhaled substances and particles, SPs, and club cell secretory proteins) have been detected in the blood of humans and animals with pulmonary injury or diseases. In humans, blood and urine are collected routinely during their clinical visits. In animals, blood is generally drawn into the syringe directly from the abdominal aorta, vein, or by cardiac puncture at necropsy. Often, orbital, tail, or other surface bleeding techniques are employed for repeated sampling of blood for biomarker analysis in animals. For the collection of urine, special metabolic cages are used to collect samples in a noninvasive manner; however, urine also can be collected directly from the bladder during necropsy.

TYPES OF LUNG TOXICITY BIOMARKERS

One can classify biomarkers broadly into two categories: (1) biomarkers of exposure and (2) biomarkers of injury or disease. Exposure biomarkers are critical in epidemiology, clinical, and toxicological studies where the substance that is inhaled is subsequently detected in BALF, NALF, blood, EBC, sputum, or urine samples, thus providing conclusive evidence of causal factor (Fig. 13.1). On the other hand, toxicity biomarkers that involve disease-specific proteins, metabolites, or nucleic acids that are detected in the collection samples will need to be further assessed to gain knowledge about the dynamic nature of the injury or disease to develop therapeutic approaches. In this section, the biomarkers of inhalation exposure, lung toxicity, and diseases will be described.

Exposure Biomarkers

A variety of inhaled parent compounds or constituents, when detected in the blood, BALF, NALF, EBC, sputum, and urine, can provide evidence of the cause of lung injury. This is particularly important in regard to injuries and diseases caused by environmental and occupational exposures. Perhaps, the most commonly studied biomarkers of exposure are the plasma levels of cotinine (a metabolite of nicotine) and carboxyhemoglobin in smokers (Liu et al., 2011). Inhaled chemicals, metals, PM components (primarily nonessential metals such as arsenic, nickel, vanadium, and organic constituents of exhausts), benzo[a]pyrenes, and nanoparticles have all been detected in the circulation. These exposure biomarkers can provide evidence of the type and degree of exposure and, often, information about the temporality of lung changes. Exposure biomarkers can also yield information about the likely pathologies in the lung.

Translocation of nonessential metals, associated with inhalation exposure to industrial combustions, has been well studied in relation to air pollution. We have shown that nonessential metals, such as nickel and vanadium, are translocated rapidly from the lung after exposure of rats to a combustion source PM containing these metals (Wallenborn et al., 2007). Translocation of an essential metal, zinc, was confirmed by exposing rats to ^{70}Zn, a stable zinc isotope (Wallenborn et al., 2009). Metals have also been detected in EBC in individuals occupationally exposed to metal-containing dusts (Felix et al., 2013). There is evidence that some manufactured nanomaterials can be detected in the circulation on pulmonary exposure; however, often the concentration of nanomaterials has been very small (Choi et al., 2010), and, in case of exposure to carbon nanomaterials, the detection methodologies are still inadequate.

Translocation of reactive polycyclic aromatic hydrocarbons of diesel exhaust into the systemic circulation has been shown in a few studies. For example, a significant portion of diesel soot–associated benzo[a]pyrene adherent to carbon core, as would be expected for components of diesel exhaust, rapidly translocated systemically after pulmonary exposure in dogs (Gerde et al., 2001). However, the detection methods for organic constituents in biological tissue samples are often inadequate and lack sensitivity. Nevertheless, there are many opportunities for identification of exposure biomarkers, especially when exposures are expected to contain metals and nanomaterials.

Biochemical Biomarkers

Lung toxicity and disease biomarkers can be measured in the exhaled breath, sputum, BALF, NALF, serum, and urine (Fig. 13.1). Depending on the sample being analyzed, the type of biomarker examined could vary somewhat. However, many of the biomarkers in the sputum, BALF, NALF, and even blood samples are common. This section covers the description of biomarkers shown to be altered in lung toxicity and pathology (Table 13.1). It also provides evidence of their relation to lung injury and their pathophysiological roles. For example, in rats exposed to high levels of sulfur dioxide, the degree of BALF neutrophilic inflammation clearly reflected the severity of airway mucus secretion in a rat strain–dependent manner (Fig. 13.3) (Kodavanti et al., 2006). Thus, the biomarkers analyzed in BALF provide highly accurate information on the state of lung pathology.

This section does not include every known biomarker of lung toxicity and its biology, but rather the most commonly used biomarkers are described, with an emphasis on their functional correlation with lung toxicity and disease.

TABLE 13.1 Commonly Used Biomarkers for Determining Various Types of Lung Injuries

Possible Indication	Biomarker
Lung cell injury (biochemical)	Protein, albumin, lactate dehydrogenase, *N*-acetylglucosaminidase, γ-glutamyltransferase, lysozyme, alkaline phosphatase
Inflammation	Neutrophils, lymphocytes, eosinophils, macrophages, monocytes
	Cytokines, chemokines
	Lymphocyte cell surface markers ($CD8^+$, $CD4^+$)
Phagocytic activity (biochemical)	β-Glucuronidase, *N*-acetylglucosaminidase, α-mannosidase, arylsulfatase, acid hydrolase
Airway injury and mucus hypersecretion	KL-6, CC16, Muc5AC, Mucin1, CK-19, SLX
Alveolar epithelial and endothelial cell injury	RAGE, vWf-Ag, alkaline phosphatase
Oxidative stress	HO-1, TBA reactive substance, isoprostane, 4-hydroxynonenal, 8-hydroxy guanosine, H_2O_2
Oxidant/antioxidant imbalance	Ascorbate, glutathione, uric acid, albumin, ferritin, transferrin, lactoferrin, ECSOD, total antioxidant capacity
Coagulation and thrombosis	TF, PAI-1, protein C, thrombomodulin
Surfactant abnormality	Surfactant proteins
Alveolar destruction	Desmosine, elastin and collagen fragments, matrix metalloproteases, antiproteases

CC16, club cell protein 16; *CK19*, cytokeratin fragment 19; *ECSOD*, extracellular superoxide dismutase; *HO-1*, hemeoxygenase-1; *KL-6*, Krebs von den Lungen-6; *Muc5AC*, mucin5AC; *PAI-1*, plasminogen activator inhibitor-1; *RAGE*, receptor for advanced glycation endproduct; *SLX*, carbohydrate antigen Sialyl Lewis (x); *TBA*, thiobarbituric acid reactive substances; *TF*, tissue factor; *vWf-Ag*, Von Willebrand Factor antigen.

Biomarkers in Exhaled Breath Condensate

EBC is used widely to identify biomarkers of exposure, lung toxicity, and systemic diseases in humans. Different breathing maneuvers (i.e., cough, forced exhalation, slow exhalation) can be utilized during collection to increase the likelihood of obtaining samples from select airway regions (Larsson et al., 2017). This technique, however, is used less extensively in laboratory animals because of the difficulties with adequate sample collection and data interpretation. EBC contains nearly 3500 components that primarily are volatile organic components, nitrogen oxide, inflammatory molecules, oxidation by-products (hydrogen peroxide and isoprostanes, nitrates and nitrites, prostanoids, and leukotrienes), and cytokines (Popov, 2011). Cytokines, such as

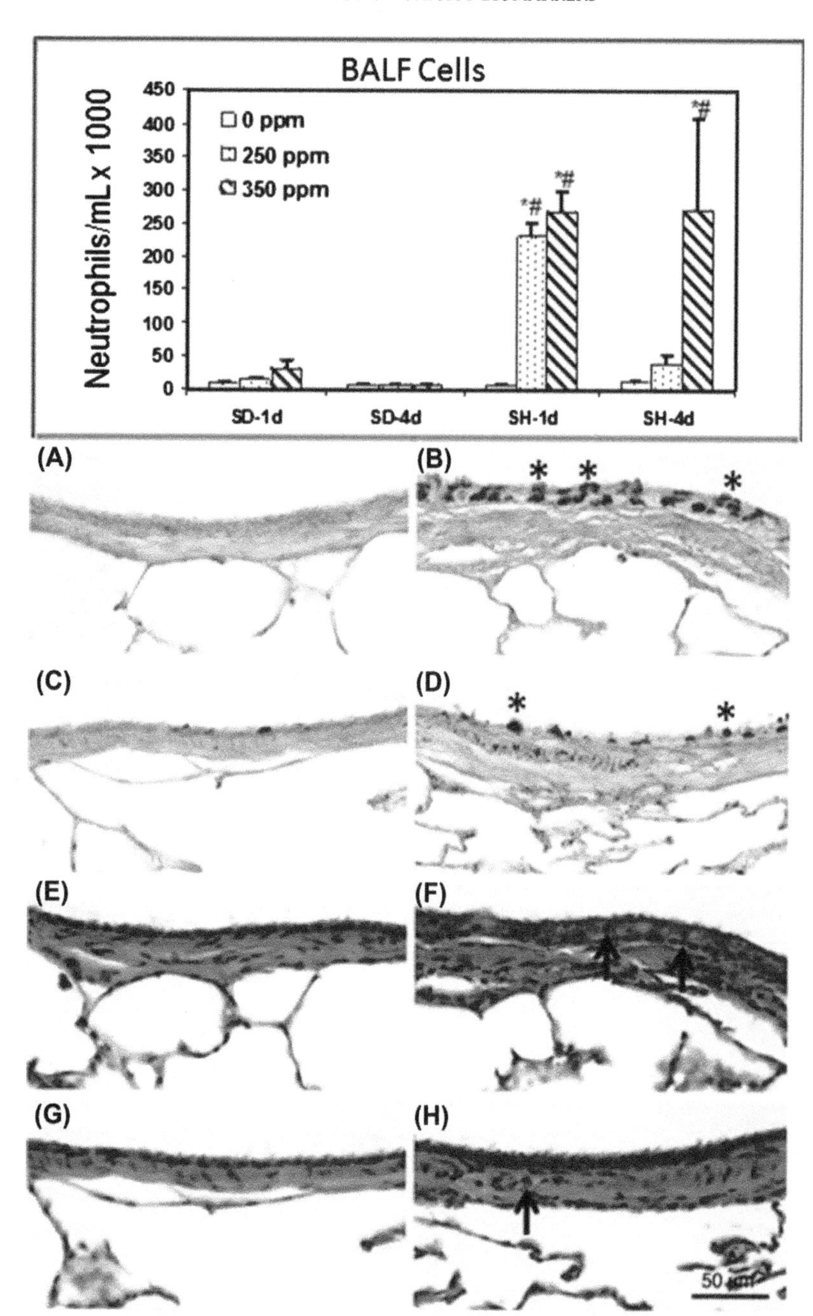

FIGURE 13.3 Bronchoalveolar lavage fluid (BALF) neutrophils correlate with airway injury and mucus hypersecretion in Sprague Dawley (SD) and spontaneously hypertensive (SH) rats exposed to high levels of sulfur dioxide (SO_2). Rats were exposed to clean air or SO_2 for 5 h/day for four consecutive days, and the necropsies were performed either 1 or 4 days following the last exposure. The upper panel shows the number of neutrophils in the BALF. The lower panel shows representative histological staining of the lung sections for mucin and inflammation in SD and SH rats exposed to filtered air or SO_2. The extent of airway mucin secretion and inflammation was evaluated in rats at 4 days following the last SO_2 exposure. Serial lung tissue sections were stained with AB-PAS (mucin) and H&E, and representative airway changes are shown in the figure. (A–D) show AB-PAS–stained lung sections (A, SH-air; B, SH-SO_2; C, SD-air; and D, SD-SO_2) and (E–H) show H&E-stained serial lung section (E, SH-air; F, SH-SO_2; G, SD-air; and H, SD-SO_2) from air- or SO_2-exposed rats at 4 days following last exposure. *Asterisks* in (B) and (D) show purple mucus staining of airway epithelial cells. *Arrows* in (F) and (H) indicate inflammatory cells in airways. *Reproduced with permission from Kodavanti, U.P., Schladweiler, M.C., Ledbetter, A.D., et al., 2006. The spontaneously hypertensive rat: an experimental model of sulfur dioxide-induced airways disease. Toxicol. Sci. 94 (1), 193–205.*

TNF-α and IL-6, have been detected in breath condensate of rats after staphylococcal pneumonia (Heidemann et al., 2011). EBC biomarkers have been especially useful in occupational scenarios (Cherot-Kornobis et al., 2012). The use of EBC for volatile organic biomarkers has provided insights into lung diseases (van de Kant et al., 2012). Exhaled nitrogen oxide measured using a handheld electrochemical detector in COPD patients

(Maniscalco et al., 2008) can also be used for other disease conditions to assess pulmonary inflammation. The levels of ferritin and superoxide dismutase, indicators of oxidative stress, are suggested as potential biomarkers of lung cancer (Carpagnano et al., 2012). Metabolomic approaches using NMR technologies have provided information on specific components of EBC in a number of pulmonary pathologies (de Laurentiis et al., 2008). Although significant advances have been made in the analysis of biomarkers in EBC, there still remain inconsistencies in sampling techniques (Ahmadzai et al., 2013; Grob et al., 2008). Recent recommendations for standardization of sampling, analysis, and reporting of data were created by a task force set up by the European Respiratory Society to address this issue (Horvath et al., 2017).

Biomarkers in Bronchoalveolar Lavage Fluid, Nasal Lavage Fluid, and Sputum

Many of the biomarkers assessed in BALF, NALF, and sputum are common, especially cytokines and inflammatory cells. However, the concentrations of other injury markers are likely to be different. Sputum and NALF content represent primarily the upper airway lining components, whereas BALF will contain both upper and deep lung components. Thus, the injury biomarkers in BALF give fairly accurate information about lower airway and deep lung injury. As discussed above, cells and the cell-free fluid recovered from BALF can provide information about exposure biomarkers because the lung lining fluid encounters all inhaled substances and pathogens. The BALF biomarkers described below can also be measured in sputum and NALF samples.

Total Protein and Albumin

In healthy individuals and animals, there is limited protein in the alveolar lining; however, injury to the airways and alveolar structures leads to leakage of vascular protein into the alveolar space. Increased synthesis of cellular protein can also leak proteins from epithelial cells into the apical side of the airways. BALF analysis of total protein provides the simplest and most accurate measure of acute lung injury and vascular leakage. Protein levels can be measured in BALF samples obtained from patients, human clinical studies, and research experiments with laboratory animals (Henderson, 2005; Hoffman, 2008). Because albumin is present at extremely low concentrations in the alveolar lining of healthy individuals and animals but can leak into the airspace on injury, the measure of albumin can provide evidence of the presence of blood proteins. Analyses of protein and albumin are made in numerous toxicological studies and in human clinical studies involving lung injuries and diseases. It should be noted that BALF albumin and proteins are the best indicators of acute lung injury, but, in some chronic lung diseases, their measurements might not reflect the degree and nature of lung pathology (Fig. 13.4). Standard colorimetric detection techniques are used to measure protein and albumin in cell-free BALF. A number of assays are available and widely used. In clinical situations, these assays are modified for automated analysis of multiple samples.

Activities of Lactate Dehydrogenase, γ-Glutamyltransferase, *N*-Acetylglucosaminidase, and Proteases/Antiproteinases

A number of enzymes can be released from injured airways and alveolar epithelial cells, which are readily detected in BALF, especially after acute lung injury. Their increases following chronic lung injuries or diseases will need to be examined with caution because not all chronic injuries are likely to be associated with active cell damage. When the activities of all these enzymes are measured, a more comprehensive assessment of lung cell injury can be made.

Lactate dehydrogenase (LDH) is a cytosolic enzyme released from cells on injury and is used widely for in vitro and in vivo toxicological studies to examine cytotoxicity (Schwartz, 1991). In many cases of lung injury, this enzyme is released into the airway lining and is detected in the BALF of animals and humans (Emad and Emad, 2008; Kodavanti et al., 2000). In occupational and acute injuries, this marker is useful; however, in chronic lung diseases, it might not be consistently increased in all types of pathologies. γ-Glutamyltransferase (GGT) is a membrane-bound enzyme involved in transfer of glutamyl moiety of glutathione (GSH) to other amino acids and dipeptides. Two isoforms of this enzyme are involved in various cellular functions. Lung epithelial cells, such as club cells and type II epithelial cells, express this enzyme that is released in the lining fluid on lung cell injury (Jean et al., 2003). Increased enzyme activity in BALF is associated with experimental lung injury and has proven to be highly sensitive in detecting a small degree of lung cell injury (Bass et al., 2015; Padilla-Carlin et al., 2011); however, its use as a biomarker of lung injury and pathology in humans is limited. Similar to GGT, *N*-acetylglucosaminidase activity assessment in the BALF of animals following experimental lung injury has proven to be useful in detecting lung cell damage (Snow et al., 2014). This lysosomal enzyme is involved in the degradation of polysaccharides and glycoconjugates containing *N*-acetyl glucosamine residues. It has been shown to be released from alveolar macrophages during phagocytosis of foreign particles (Wereszczynska-Siemiatkowska et al., 2000). The mechanism of its release in lung lining fluid is not well understood, and its increases in BALF following experimental injury are likely to be dependent on the type of insult

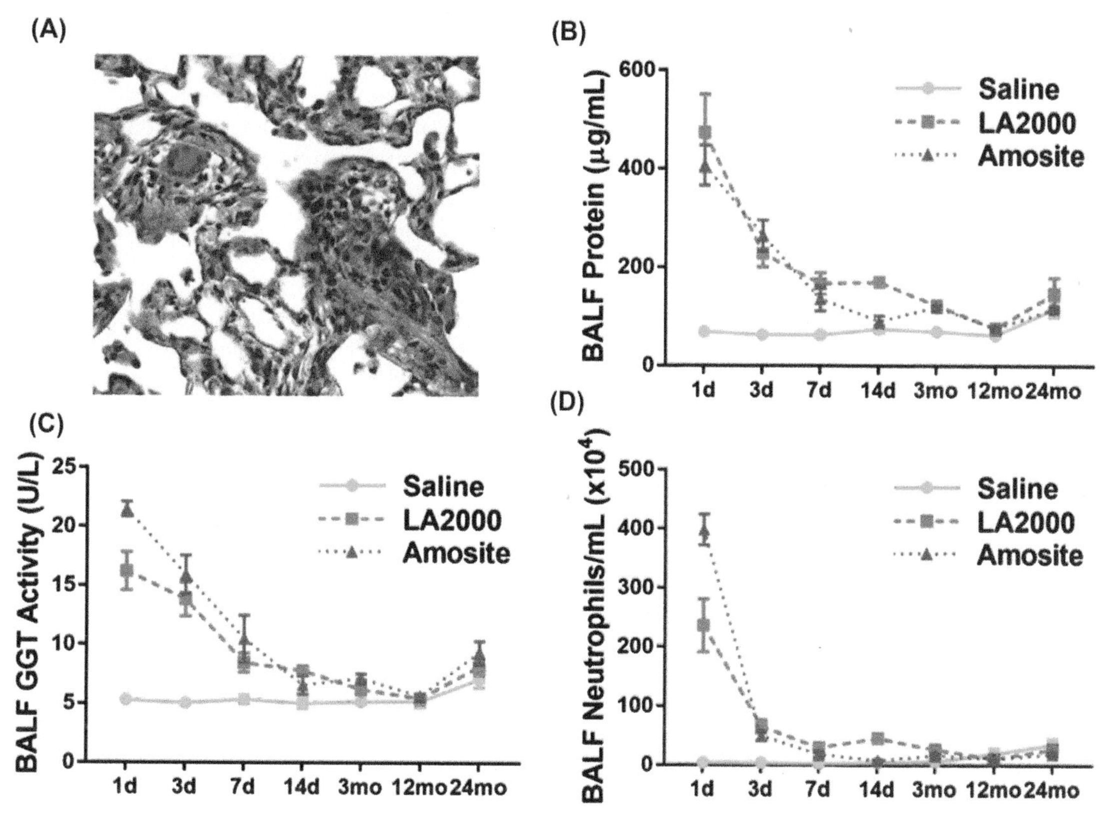

FIGURE 13.4 Temporal correlation between lung pathology and the levels of bronchoalveolar lavage fluid (BALF) injury and inflammation markers in Fischer 344 rats following exposure to asbestos. Rats were exposed once intratracheally to saline, Libby amphibole (LA2000, 5 mg/rat), or amosite (0.5 mg/rat) asbestos fibers suspended in saline. (A) Lung section stained with H&E and trichrome staining for collagen demonstrating fibrosis at 12 months after single instillation of LA2000. Measurements for (B) total protein, (C) γ-Glutamyltransferase (GGT) activity, and (D) neutrophils in BALF were measured at 1, 3, 7, and 14 days and 3, 12, and 24 months after the single instillation of asbestos fibers. Note that asbestos induces acute lung inflammation and injury (pathology not shown) that resulted in long-term fibrosis. The BALF biomarkers showed neutrophilic inflammation and increases in lung injury markers that were more pronounced shortly after exposure, which returned to baseline over time. However, lung fibrosis persisted for at least 12 months (A).

on alveolar macrophages. Alkaline phosphatase, which is present in type 2 alveolar epithelial cells, can be released in the alveolar lining and serve as a biomarker of type 2 cell damage (Aiso et al., 2010). The validity of these enzymes in BALF as biomarkers of lung cell damage will need to be verified in chronic disease conditions. The activities of these enzymes in BALF accurately reflect lung cell injury status, especially at early time points.

A variety of proteases involved in degradation of extracellular matrix proteins are shown to be increased in BALF after lung injury. In chronic diseases such as emphysema and COPD, matrix protein degradation is increased, resulting in impairment of structural integrity and contractile function of the lung. Especially in cases of cigarette smoke–induced lung disease, increases in proteases can cause protease/antiproteinase imbalance (Nyunoya et al., 2011). BALF levels of proteases are also augmented in patients with pulmonary infection (El-Solh et al., 2009) and ARDS (O'Kane et al., 2009). However, because a given protease might be involved in many different diseases and injuries, the diagnostic values for these biomarkers are limited (Vuorinen et al., 2007). Protease inhibitors are often induced as a compensatory mechanism. These proteins can be detected in BALF by antibody-based techniques. Determination of their proteolytic activity by special gel-based assays can provide insight into the signaling pathways and the type of pathology.

Oxidation By-Products and Antioxidants

Because the lung is exposed continuously to high levels of molecular oxygen, the potential for oxidation of structural macromolecules exists, especially in a disease condition where appropriate antioxidant homeostasis cannot be maintained. The lung lining fluid is enriched with a variety of chemical antioxidants (i.e., ascorbate, urate, and GSH) and proteins, such as extracellular superoxide dismutase and iron-binding proteins (i.e., lactoferrin, transferrin, and ferritin). When the balance between oxidants and antioxidants is shifted to increased oxidation, lung injury ensues.

A high concentration of ascorbate is present in the alveolar lining fluid of humans and has been shown to be the first line of defense in acute lung injury (Behndig et al., 2009). Humans, primates, and guinea pigs lack the mechanism to synthesize ascorbate in the body and, thus, require dietary supplementation, whereas laboratory rodents (rats and mice) can produce ascorbic acid. Specialized transporters exist on cells to transport ascorbate at the apical surface of the airway lining (Wilson, 2005). Determination of the level of ascorbate in BALF reflects the antioxidant capacity in experimental lung injury. The depletion of BALF ascorbate is associated with vascular leakage and inflammation following experimental lung injury (Kari et al., 1997). The levels of ascorbate in experimental cigarette smoke exposure are decreased because of increased oxidative burden (Ghio et al., 2008); however, in human COPD, asthma, or other diseases, the use of ascorbate as biomarker of oxidative stress is not well established. Urate, another enzyme with known antioxidant function, can also be measured in BALF (Vyas et al., 2001). Reduced ascorbate and urate can be measured colorimetrically or by high-performance liquid chromatography in acidified and deproteinized BALF (Molina-Diaz et al., 1998). However, accurate quantification of reduced ascorbate in biological samples poses a significant challenge because of its labile nature and the presence of chemical molecules that interact with ascorbate in BALF. Generally, biological samples can be stabilized by acidic deproteinization with perchloric and metaphosphoric acid (Roginsky et al., 1997).

GSH, a γ-glutamyl cysteinyl glycine tripeptide that is a major cellular antioxidant present in millimolar concentrations in cells, has also been shown to function as an antioxidant in alveolar lining fluid. However, the levels of GSH present in BALF are much lower, in the micromolar range, especially in animals. As a cellular antioxidant, GSH has a major role in thiol homeostasis and regulation of the activities of a variety of proteins whereas, in alveolar lining, it might function as a scavenger of reactive electrophilic species. GSH levels have been shown to be altered in a number of experimental lung injuries and in human diseases; however, depending on the pathologies, increases or decreases have been noted (Fitzpatrick et al., 2011; Kontakiotis et al., 2011). Often, the ratio of reduced and oxidized GSH has been used as a measure of cellular oxidative stress (Dalle-Donne et al., 2007). A number of colorimetric and chromatographic techniques can be used to measure GSH in BALF. Because GSH is also a labile molecule, sample preservation and storage become highly critical in accurate measurement. The use of GSH as a biomarker might be restricted to certain types of acute lung injuries where the changes in BALF GSH levels are expected.

Other airway lining antioxidants include extracellular superoxide dismutase, which is normally present within interstitial spaces and alveolar lining where it is a scavenger of superoxide radicals (Fattman et al., 2003). Following stress, the activity of extracellular superoxide dismutase has been shown to increase. This BALF protein can be detected using immunological techniques and can serve as a biomarker of oxidative stress.

Iron is the most abundant metal component of the body, is redox active, and is regulated precisely by tightly controlled transport mechanisms. Iron-binding proteins play a critical role in maintaining antioxidant homeostasis and removal of nonheme iron. High levels of ferritin, lactoferrin, and transferrin are detected in BALF, especially following lung cell injury (Ghio et al., 2008). Iron-binding proteins in BALF serve as biomarkers of acute lung injury and chronic diseases associated with increased iron overload (Ghio et al., 2008; Shannahan et al., 2010). Because these proteins are changed in a number of pathological conditions and after acute lung injury, the specificity of these proteins for a given lung disease cannot be ascertained. In addition to measuring BALF levels of iron-binding proteins, the measurement of iron-binding capacities using clinical assays can provide insight into the nature of lung pathology.

In many cases of lung injury, 4-hydroxynonenal (trans-4-hydroxy-2-nonenal; 4-HNE) is measured as a biomarker of oxidative stress. 4-HNE is produced from oxidation of polyunsaturated lipids under oxidative conditions associated with lung injury (Eder et al., 2008). Increased 4-HNE levels are observed in a number of diseases, such as ARDS and some types of cancers (Eder et al., 2008). At low concentrations, 4-HNE might induce proliferation, and high concentrations can lead to cellular apoptosis (Sunjic et al., 2005). Similarly, F(2)-isoprostanes, prostaglandin-like chemicals, have also been detected in BALF, and their increases have been associated with lung injuries (Janssen, 2008). F(2)-isoprostanes are formed in vivo by nonenzymatic oxidation of arachidonic acid, and their assessment in BALF can provide overall oxidative stress status of the lung. A number of studies have also shown increases in levels of malondialdehyde, which is produced by oxidative modification of lipids. Accurate measurement of its levels is challenging; however, increases have been shown in some lung injuries (Nemmar et al., 2012). Because oxidative stress and increased oxidation of macromolecules are associated with almost all lung pathologies and injuries, the changes in these markers will not provide specific insights into the type of disease or injury.

Cytokines, Chemokines, and Growth Factors

Cytokines and chemokines are secreted by lung epithelial, smooth muscle, interstitial, and endothelial cells and alveolar macrophages. Depending on the host genetics, nature of lung cell damage, and type of disease, the dynamics of secretion of these molecules vary. The ability to analyze the pattern of change among many functionally related cytokines, as opposed to individual ones, allows for a more accurate prediction of the type of disease and its progression and severity. For example, increased secretion of Th2 cytokines will be indicative of an asthma phenotype (Hartl et al., 2009), whereas increased interferon gamma (IFN-γ) and associated cytokines might underlie infection or COPD (Chung, 2005). The increased secretion of these cytokines is noted during an immune response and can also occur in injuries not mediated by immune mechanisms. Some of these cytokines and chemokines serve to stimulate proinflammatory reaction, whereas others are involved in the maintenance of homeostasis. On activation of cellular receptors and subsequent signaling through MAP kinase and NF-κB pathways, the expression of cytokines is induced (Chung, 2005; Hartl et al., 2009). Therefore, proteins released into the airway lining fluid can serve as biomarkers of underlying lung inflammation, infection, or chronic disease. There are numerous studies that show increases in the cytokines and chemokines in BALF in experimental and clinical settings. The release of cytokines and chemokines from cells recovered during BAL can also provide insights into the mechanisms of lung injury or disease. A variety of immunoassays, including single-protein detection and multiplex platforms, are available to determine the levels of these cytokines and chemokines in the BALF of animals and humans.

Bronchoalveolar Lavage Fluid Cells as a Determinant of Lung Inflammation

Lung injury leads to recruitment of inflammatory cells from blood to the lung compartments through highly regulated immunological processes. Inflammatory cells that are recruited in the airways and alveolar lining are recovered in BALF. These cells can be mounted on a slide using cytospin centrifugation and stained for their microscopic examination and identification. Quantification of relative cell types is used for assessment of lung inflammation in many inhalation toxicological and human exposure studies. In lung diseases, these cells provide information on the nature of pathology, whereas the surface characterization of receptors can predict molecular mechanisms.

Lung injuries and pathologies associated with inhalation exposure to particulate and fibrous materials result in macrophage phagocytosis of these inhaled materials. The inhaled particles in the 0.2–10 μm diameter range can be identified within alveolar macrophages at the light microscopy level, which can provide insight into the type of exposure. Macrophages undergo phagocytosis and pinocytosis, leading to activation (Zhang and Kaminski, 2012). Inflammatory cells are recruited into the alveolar space following lung injury through elaboration of cytokines and chemokines from activated macrophages and epithelial cells. This process brings neutrophils, eosinophils, lymphocytes, and monocytes into the lung and air spaces (Fig. 13.5). The type and the degree of cell infiltration depend on the physicochemical property of the inhalant and the degree of initial injury. The analysis of cellular content in BALF, thus, presents the most precise and accurate determination of the underlying disease process. The BALF of healthy individuals and animals primarily contains macrophages; therefore, the presence of other cell types, such as neutrophils, eosinophils, and lymphocytes, can clearly predict the nature of inflammation and the type of underlying disease (Barnes et al., 2006). The identification and quantification of inflammatory cells can provide insight into the type of pathology. For example, nonimmune lung injury involving innate response will be associated with increases in BALF neutrophils, whereas adaptive immune changes might be associated with eosinophilic and lymphocytic inflammatory response. Flow cytometry analysis for surface markers is often used to determine immunological mechanisms. In many cases, these cells recovered from healthy and diseased patients are cultured to study their potential to secrete cytokines on stimulation and their pathobiologic mechanisms.

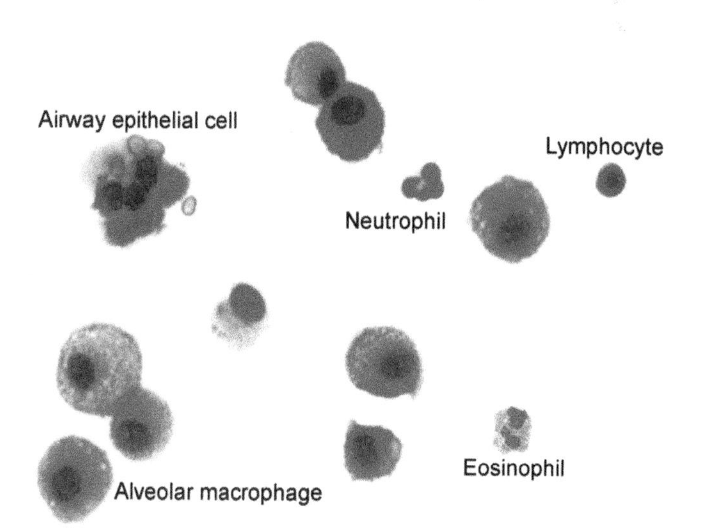

FIGURE 13.5 Inflammatory cells in lavage fluid of rats. The cells isolated via bronchoalveolar lavage (BAL) were centrifuged to prepare cell differential slides using a cytospin centrifuge and stained with the DiffQuick staining kit (Fischer Scientific). Each cell type is marked. In healthy rats, most cells (85%–95%) are alveolar macrophages; however, after acute lung injury, more neutrophils, eosinophils, and lymphocytes are found in BAL fluid (BALF). Often, sloughed-off ciliated epithelial cells are present because of procedural injury.

Circulating Biomarkers of Lung Injury and Pathology

A variety of circulating components that are changed in response to lung injury, or as a result from crossing from the lung into the bloodstream, can serve as biomarkers. Because blood can be collected in a minimally invasive manner, as opposed to the invasive collection of BALF, identifying circulating biomarkers is more feasible. The blood sample collection is performed routinely in any clinical setting and in experimental studies. However, because only a fraction of the lung content can be translocated to the circulation (depending on the characteristics of the inhalant), the sensitivity of detection and the degree of change in levels are less robust than in the BALF. Many lung injury biomarkers commonly found in BALF are also present in the circulation. New approaches and discoveries have led to significant advances in global assessment of biomarkers in the blood. Here, we will discuss several lung toxicity biomarkers individually and as a group to enable a pattern of change to be examined.

Surfactant, Other Lung-Specific Proteins, and Immunological Biomarkers

SPs are made primarily in the lung and are released into the circulation following injury and under disease conditions. Therefore, their detection in the circulation can provide highly specific information concerning lung injury and pathology. Of the four types of SPs, SP-A and SP-D have been used frequently as biomarkers of lung disease and innate immune function (Takahashi et al., 2006). Depending on the nature of injury, the type of SP released in the circulation differs. In restrictive lung diseases, the expression of SP-A and SP-D is increased, and thus, they serve as specific biomarkers. SP-A is detected in COPD patients but not in patients with fibrosis (Ohlmeier et al., 2008). The levels of circulating SP-D have been shown to correlate with alveolitis (Hant et al., 2009). SP-D has also been shown to increase in the circulation of idiopathic pulmonary fibrosis patients (Hant et al., 2009). Temporal increases in SP-A and SP-B in plasma have been shown to result from progression in alveolar capillary permeability in patients with ARDS (Schmidt et al., 2007). Detection of SP-A and SP-D in circulation is predictive of interstitial lung diseases (Cheng et al., 2003; Greene et al., 2002). Because these proteins are secreted in limited quantities, their detection sensitivity is critical in determining the correlation between the severity of lung injury and the degree by which the protein level increases in the circulation. Plasma SPs can be analyzed using standard antibody-based immunological techniques.

Krebs von den Lungen-6 (KL-6), now classified as a human MUC-1 glycoprotein, has been shown to be elevated about 70%–100% in patients with various interstitial lung diseases (Ishikawa et al., 2011). Changes in KL-6 correlate with alveolitis and scleroderma (Hant et al., 2009). Club cell secretory protein 16 (CC-16) has been detected in the sera of ARDS patients (Doyle et al., 1998). CC-16 also increased after acute ozone exposure in humans (Blomberg et al., 2003) and has been proposed as a sensitive biomarker of acute lung injury. This protein is made in club cells within airway epithelium. Napsin A, which is a protease expressed in alveolar type II cells and also renal proximal tubules, has been shown to be a highly sensitive indicator of idiopathic pulmonary fibrosis, especially in the absence of renal disease (Samukawa et al., 2012). Increased levels in the blood are indicative of lung adenocarcinoma. CC-18, derived from alveolar macrophages, has been shown to increase in the circulation during lung fibrosis (Prasse and Muller-Quernheim, 2009). Many of these biomarkers are likely to be changed with multiple lung diseases, often limiting their specificity and predictive value for a given pulmonary disease.

A number of other circulating proteins have been implicated in lung injury. Smokers who have demonstrated reduced diffusion capacity at the capillary barrier have been shown to have increased blood lysozyme, a marker of inflammatory cells. This has been shown to correlate with the degree of impairment in lung diffusion capacity (Schmekel et al., 2013). Increases in soluble Fas (sCD95)/APO-1, involved in antagonizing cellular Fas receptors and apoptotic cell signaling, have been demonstrated to be increased in coal miners (Hoffmeyer et al., 2010). Increased levels of cytokeratin-18 immune complexes were present in patients with pulmonary fibrosis (Dobashi et al., 2000). Serum levels of antineutrophil cytoplasmic antibodies (ANCAs) have been shown to increase in response to a cell-mediated immune response to antigens in the respiratory tract, which generally is associated with granuloma formation in experimental and occupational asbestos and silica exposures (Rihova et al., 2005). ANCAs have been implicated in many immune disorders and are proposed to be specific to lung immune reaction in patients with granuloma. Pentraxin-3, a receptor involved in innate immune response, has been shown to increase after acute lung injury and in ARDS (Mauri et al., 2008). Ligand CCR5, involved in trafficking inflammatory cells, has been shown to be increased in the plasma of patients with Wegener's granulomatosis. Because ANCAs were positive for all patients with CCR5 increases, ANCAs might be involved in common pathways controlling inflammation (Zhou et al., 2003).

Some immunological mediators have the potential for being considered as biomarkers for infections. In individuals infected with influenza A, high levels of circulating CD-4 T lymphocytes were noted (Zhao et al., 2012), suggesting that these lymphocytes could serve as potential biomarkers for this infection; however, their involvement in other types of inflammatory processes will need to be taken into consideration. A thymus- and activation-regulated chemokine (TARC)/CCL17, which is a functional ligand for CCR4, has been involved in a variety of pulmonary and other diseases and seems to track with eosinophilia (Miyazaki et al., 2007). This marker provides a highly sensitive indication of disease in patients with acute eosinophilic pneumonia, which closely resembles acute lung injury associated with ARDS (Miyazaki et al., 2007). These studies suggest that there are likely a number of immune response–related biomarkers that can be considered for further validation. The type of any immune biomarker change, however, can depend on the type of lung damage or infection.

Collagen and Elastin Fragments

Collagen and elastin are present at high levels in the lung and provide a unique structural framework. Injury to the structural components of the lung results in proteolytic degradation of extracellular matrix–associated collagen and elastin. The fragments of degradation by-products can then be detected in BALF, blood, and even in urine. Their presence in the plasma is indicative of lung structural impairment. Some of the fragments of collagen and elastin are highly reactive and have been shown to induce inflammatory responses within the lung (Overbeek et al., 2013). Desmosine and isodesmosine, degradation products of elastin, have been identified in circulation and are considered fairly specific markers of emphysematous changes in the lung (Viglio et al., 2007). Other elastin degradation products have also been identified and are named elastokines based on their ability to cause inflammatory reactions (Cantor and Shteyngart, 2004). Endostatin, a proteolytic fragment of the basement membrane collagen XVIII, has been shown to inhibit angiogenesis via action on endothelial cells (Kanazawa, 2007), whereas procollagen type I and type III amino terminal propeptides have been associated with ARDS (Meduri et al., 1998). These fragments can serve as biomarkers of underlying lung disease (Ostroff et al., 2010). Because other organ injuries can also fragment collagen and elastin, the analysis of by-products in plasma should accompany additional measures of lung injury. The determination of these fragments is limited to the availability of specific antibodies.

Cytokines, Metalloproteases, and Acute-Phase Proteins

It has been shown that lung injury or infection can lead to increases in circulating cytokines, such as IL-6 and TNF-α; metalloproteases, such as MMP-9; and a number of acute-phase proteins (Calfee et al., 2011). Circulating cytokines as biomarkers have been implicated in a number of pulmonary diseases (Tzortzaki et al., 2007; Tzouvelekis et al., 2005). Increased levels of circulating cytokines have also been implicated in coal miners (Zou et al., 2011). Increased MMP-9 plasma levels have been shown to occur in patients with COPD (Bolton et al., 2009). We have shown that, in rats exposed to asbestos fibers, there is a marked increase in a number of acute-phase proteins and cytokines, such as lipocalin-2 and osteopontin, but no increases in IL-6 or TNF-α could be noted in F344 rats (Shannahan et al., 2012a). A number of experimental air pollution studies have shown changes in circulating cytokines; however, the increases have been inconsistent relative to the degree of lung injury. It is likely that animal strain, temporality, and other host factors, in addition to the types of lung injuries, play important roles. Thus, the use of circulating cytokines and other proteins in experimental lung toxicity or lung diseases as biomarkers should be considered in conjunction with other types of disease measures.

Adhesion molecules, such as P-selectin, E-selectin, ICAM-1, and VCAM-1, are involved in extravasation of inflammatory cells to the lung and are likely to be changed systemically. Many of these cell surface receptors have their soluble form present in the circulation and are involved in regulatory mechanisms. As an example, increases in plasma-soluble ICAM-1 and E-selectin were detected in pediatric acute lung injury (Ballabh et al., 2003). Relative to circulating cytokines, the changes in soluble receptors after local pulmonary injury as biomarkers of diseases are less explored. Identification and validation of soluble receptors of adhesion molecules hold a great opportunity for discovery of novel biomarkers and as therapeutic targets.

Thrombosis Biomarkers

Lung injury can induce microvascular thrombosis by stimulation of the extrinsic coagulation pathway. Microvascular thrombosis plays an important role in pulmonary injuries and chronic diseases. During initiation of the coagulation cascade, the extrinsic pathway is triggered by activation of tissue factor (TF) and factor VII, but, subsequently, the intrinsic pathway is triggered, which involves factors VIII, X, and XI (Idell, 2003). It has been shown that acute lung injury induced by PM exposure can induce thrombosis in the vasculature away from the site of injury, suggesting that pulmonary-derived factors in the circulation are contributing to this systemic response (Nemmar et al., 2006). TF, which initiates the thrombosis process after acute injury, has been shown to increase in the plasma of humans after exposure to air pollutants (Emmerechts et al., 2012). It is also likely that platelet activation and secretion of reactive coagulation mediators can trigger microvascular thrombosis. The neurohumoral factors released from activated mast cells might also contribute to systemic thrombogenic effect on lung injury (Nemmar et al., 2006). We have shown that the platelet aggregation and numbers were changed in rats after exposure to asbestos fibers (Shannahan et al., 2012b). Circulating thrombomodulin levels have been increased in cancer patients undergoing radiation therapy (Hauer-Jensen et al., 1999). Moreover, circulating von Willebrand factor and P-selectin levels are also increased following acute lung injury (Ochoa et al., 2010). The consideration of thrombosis biomarkers for diagnosis of lung injuries and diseases is attractive and will be critical for acute lung injuries and experimental studies. Examination of thrombogenic processes is performed routinely in clinics. Their use as biomarkers of chronic lung disease can be of value for selected lung diseases.

Circulating Progenitor Cells

An injury or inflammation response in the lung triggers bone marrow to release progenitor cells into the circulation to induce compensatory repair at the site of injury (Yang et al., 2016). Increases in circulating progenitor cells of various lineages have been noted in multiple studies and have been proposed to serve as biomarkers; however, more research is needed with greater emphasis on the correlation between the degree of increase with the level of lung damage, disease severity, and the time course of their release. Progenitor endothelial cells have been shown to increase in pulmonary fibrosis and bronchopulmonary dysplasia (Borghesi et al., 2009). It has been demonstrated that epithelial progenitor cells are recruited at the site of lung injury in a mouse model (Gomperts et al., 2006). A number of studies have shown the increases in mesenchymal progenitor cells in interstitial lung injuries, such as fibrosis (McNulty and Janes, 2012; Smadja et al., 2013). The origin and release of progenitor cells in response to injury provides novel mechanisms involved in tissue repair, and, if validated, these biomarkers can be predictive of the precise nature of underlying lung injury and pathology.

Microparticles

Microparticles are formed by shedding or blebbing of cell membrane. These particles include microvesicles, exosomes, and apoptotic bodies as well as cellular contents. These particles also contain deoxyribonucleic acid (DNA), ribonucleic acid (RNA), microRNA (miRNA), phospholipids, and molecules derived from the parent cell (Nieri et al., 2016). They are small and range from 0.05 to 3 μm in diameter. In normal circumstances, these particles are present at a low concentration in the circulation, but, under pathological conditions and following an injury, more of these particles are released into the circulation (Piccin et al., 2007). There is a growing interest in using these microparticles as biomarkers of specific organ and cell injuries. Moreover, the components associated with these particles can be released or fused with endothelial cells and can trigger a variety of systemic responses. Thus, their role as biomarkers and also in inducing pathobiologic processes away from their site of origin is of great importance. As far as predicting their role in lung injury, it has been shown that circulating microparticles from smokers can provide insight into the ongoing lung injury (Gordon et al., 2011). Microparticles have been shown to contain TF that, on its release, can induce thrombogenic effects on the vasculature (Mackman, 2006). Acute lung injury leads to increased microparticles into the circulation (McVey et al., 2012), whereas microparticles found in the sputum have been shown to have proinflammatory effects (Porro et al., 2013). More research is needed in using microparticles and associated cell content as biomarkers. A number of flow cytometry–based techniques are available to isolate microparticles from blood plasma and characterize their properties. Examination of their properties provides important mechanistic information about the underlying disease. Understanding of their formation, function, and relevance to the underlying disease may result in new therapeutic strategies for vascular diseases secondary to lung injury.

miRNA

miRNAs are small noncoding RNAs that function in transcription and translation processes. In mammalian cells, miRNAs function by pairing with complementary messenger RNA (mRNA) sequences, causing gene silencing during the translational process. Mammalian miRNAs exhibit only a partial complementarity to target mRNA, and, thus, one miRNA can target several mRNAs, and, likewise, one mRNA can be targeted by a number of miRNA molecules. miRNAs play important roles in many cellular processes such as apoptosis, differentiation, growth, and cell signaling. The presence of miRNA in the circulation has sparked a tremendous interest in their potential use as biomarkers and therapeutic targets as well as their role in gene regulation (Paul et al., 2018). Because of their encapsulation in microvesicles, exosomes, or apoptotic bodies or when complexed with proteins, miRNAs that escape degradation by endogenous ribonucleases are released into the circulation where they can serve as biomarkers (Harrill et al., 2016). A number of techniques are available to isolate miRNA from blood plasma, and high-throughput platforms are available to profile miRNA globally.

Although few reports are available for noncancer-related lung diseases, more research has been conducted on miRNAs as biomarkers of lung cancer. It has been shown that miRNAs are expressed differentially in patients with lung cancer, tuberculosis, pneumoconiosis, and ARDS and in smokers (Abd-El-Fattah et al., 2013). Genome-wide analysis has been used to identify miRNA in coal workers with pneumoconiosis (Guo et al., 2013). Toxicology-based clinical studies have also been performed to examine miRNA expression levels in sputum and blood samples collected from individuals exposed to ozone (Fry et al., 2014), PM (Bollati et al., 2015), and diesel exhaust (Yamamoto et al., 2013). There is a great potential in using miRNAs as biomarkers of lung injuries and diseases and for therapeutic targeting; however, more studies are needed to understand their specific role in translational repression in an organ-specific manner.

Biomarker Identification Using Proteomic and Metabolomic Profiling

With the emergence of new biotechnologies, there is a growing interest in using global serum/plasma proteomic and metabolomic profiling for identifying novel biomarkers of various lung pathologies, including cancer (Cheung and Juan, 2017; Kumar et al., 2017). Proteomic profiling of blood can identify nonabundant proteins that are mediators of pathogenic and injury processes (Mehan et al., 2013). Proteomic analysis has also been applied to BALF samples to detect markers of early-stage lung cancer (Hmmier et al., 2017). Metabolomic analysis will identify small metabolites of all classes, including carbohydrates, amino acids, and lipids (Johnson et al., 2016). The characteristics of these metabolites or group of metabolites can reveal processes modified during injury to the lung. Generally, in identifying biomarkers, the patterns of change in multiple functional groups of biomarkers are better indicators of underlying disease or injury than that of an individual marker. For example, an experimental ventilator-induced rat lung injury model shows changes in serum metabolites and provides information about the pathological processes in the lung (Naz et al., 2013). Alterations in proteomic profiles provide distinct differences between cancer patients and those with asthma (Izbicka et al., 2012), suggesting that disease-specific biomarker profiles can be identified using proteomic and metabolomic approaches. COPD patients exhibit changes in amino acid metabolism based on plasma proteomic and metabolomic analyses (Ubhi et al., 2012). The metabolomic approach is also used in identifying cancer signatures (Wedge et al., 2011). Recently, metabolomic analysis has been used to investigate systemic changes following an acute ozone exposure in humans (Miller et al., 2016) and animals (Miller et al., 2015). A number of high-throughput platforms are available using combinational and mass spectrometric techniques that enable global profiling of plasma and serum samples. The use of proteomic and metabolomic techniques has provided novel insights into chronic diseases, the cross talks among pathological processes and cellular metabolism, and identification of novel biomarkers.

Biomarker Identification Using Genomic Profiling of Nucleated Blood Cells

Gene expression profiling of nucleated cells in the blood of patients with lung injury and in experimental studies can provide insights into the type of disease and injury. Circulating mononuclear cells respond to injury induced in the lung and are involved in inflammatory and immune processes. The specific mRNA expression differences between healthy and diseased individuals can lead to identification of new biomarkers and pathobiologic processes. The blood gene expression pattern of tuberculosis patients revealed humoral immunity changes (Cliff et al., 2013). Patients with ARDS show increased expression of pre-elafin, which can serve as a biomarker (Wang et al., 2008). Unique gene signatures have also been detected in blood of advanced non–small cell lung carcinoma patients (Chen et al., 2013). Use of combinational techniques including global single nucleotide polymorphism (SNP) analysis and genomic profiling has been employed for identifying new biomarkers in COPD patients (Kim et al., 2012). Transcriptional analysis of COPD patients' blood samples has revealed involvement of neutrophil proteases in the disease pathogenesis, whereas those with tuberculosis revealed inflammatory gene changes (Cliff et al., 2013) and also involvement of interferon-inducible signature (Berry et al., 2010). Thus, transcriptional profiling of blood cells during various stages in the disease process can provide mechanistic information about the disease progression and pathogenesis. Different gene expression platforms and whole genome, as well as RNA-sequencing arrays, are available for these applications; however, validation of these biomarker signatures remains a challenge.

Urine Biomarkers of Lung Injury and Disease

Unlike circulating, sputum, NALF, and BALF biomarkers, there are a limited number of injury-related molecules that can be filtered and collected in urine, and, thus, relatively few biomarkers have been identified that link to lung injury or diseases. Urine biomarkers include the components of lung structure and oxidation by-products that escape filtration across all passages. Desmosine, a structural component of elastin, and other fragments are released on lung injury. In various pathological conditions, these markers have been detected in the urine (Fill et al., 2006; Turino et al., 2011). Oxidation by-products, such as 8-hydroxydeoxyguanosine, have also been detected in urine from patients with bronchopulmonary dysplasia (Joung et al., 2011). A bombisen-like peptide has been shown to be present in individuals with bronchopulmonary dysplasia (Cullen et al., 2002). Leukotriene by-products of arachidonic acid metabolism have been found in the urine of smokers (Duffield-Lillico et al., 2009). Increased urinary mesothelin was noted in patients with mesothelioma, suggesting that this could serve as a biomarker for lung cancer patients for diagnosis of mesothelioma (Creaney et al., 2010). The sensitivity of these urinary biomarkers and their predictive value for early diagnosis, however, are limited in lung injuries and disease. Newer metabolomic techniques might enable more thorough analysis for identification of biomarkers in urine samples.

CONCLUDING REMARKS AND FUTURE DIRECTIONS

The field of biomarkers has recently exploded as a result of numerous new discoveries in biological mechanisms and also the availability of high-throughput biotechnologies. This revolution in biology has shifted the focus of research from target organ biology to systemic biology. New interactive biological processes and pathways have been discovered that provide insights into the systems interactions. This is especially true for pulmonary diseases and toxicities, which recently have been linked to cardiovascular and neuronal impairments and metabolic syndrome. The great interest in understanding circulating factors that are changed as a result of lung injury and their ability to affect distant organs has rejuvenated our interest in identifying new biomarkers. A variety of approaches have been introduced recently—blood genomic, serum proteomic, and metabolomic approaches. In this exciting time, biomarker discoveries will influence therapeutic approaches for lung diseases. In toxicology, the emerging new biomarkers will provide the most critical knowledge of systems biology.

Acknowledgments

We thank Drs. Marie Hargrove, Andres Henriquez, and Ian Gilmour for their critical reviews of the manuscript.

Disclaimer

The research described in this article has been reviewed by the US Environmental Protection Agency's National Health and Environmental Effects Research Laboratory and approved for publication. Approval does not signify that the contents necessarily reflect the views and the policies of the Agency, nor does mention of trade names or commercial products constitute endorsement or recommendation for use.

References

Abd-El-Fattah, A.A., Sadik, N.A., Shaker, O.G., et al., 2013. Differential microRNAs expression in serum of patients with lung cancer, pulmonary tuberculosis, and pneumonia. Cell Biochem. Biophys. 67 (3), 875–884.

Ackerman, S.J., Kwatia, M.A., Doyle, C.B., et al., 2003. Hydrolysis of surfactant phospholipids catalyzed by phospholipase A2 and eosinophil lysophospholipases causes surfactant dysfunction: a mechanism for small airway closure in asthma. Chest 123 (3 Suppl.), 355S.

Ahmadzai, H., Huang, S., Hettiarachchi, R., et al., 2013. Exhaled breath condensate: a comprehensive update. Clin. Chem. Lab. Med. 51 (7), 1343–1361.

Aiso, S., Yamazaki, K., Umeda, Y., et al., 2010. Pulmonary toxicity of intratracheally instilled multiwall carbon nanotubes in male Fischer 344 rats. Ind. Health 48 (6), 783–795.

Aitken, M.L., Greene, K.E., Tonelli, M.R., et al., 2003. Analysis of sequential aliquots of hypertonic saline solution-induced sputum from clinically stable patients with cystic fibrosis. Chest 123 (3), 792–799.

Ballabh, P., Kumari, J., Krauss, A.N., et al., 2003. Soluble E-selectin, soluble L-selectin and soluble ICAM-1 in bronchopulmonary dysplasia, and changes with dexamethasone. Pediatrics 111 (3), 461–468.

Barnes, P.J., Chowdhury, B., Kharitonov, S.A., et al., 2006. Pulmonary biomarkers in chronic obstructive pulmonary disease. Am. J. Respir. Crit. Care Med. 174 (1), 6–14.

Bass, V.L., Schladweiler, M.C., Nyska, A., et al., 2015. Comparative cardiopulmonary toxicity of exhausts from soy-based biofuels and diesel in healthy and hypertensive rats. Inhal. Toxicol. 27 (11), 545–556.

Bauer, A.K., Travis, E.L., Malhotra, S.S., et al., 2010. Identification of novel susceptibility genes in ozone-induced inflammation in mice. Eur. Respir. J. 36 (2), 428–437.

Bedrossian, C.W., Warren, C.J., Ohar, J., et al., 1997. Amiodarone pulmonary toxicity: cytopathology, ultrastructure, and immunocytochemistry. Ann. Diagn. Pathol. 1 (1), 47–56.

Behndig, A.F., Blomberg, A., Helleday, R., et al., 2009. Antioxidant responses to acute ozone challenge in the healthy human airway. Inhal. Toxicol. 21 (11), 933–942.

Berry, M.P., Graham, C.M., McNab, F.W., et al., 2010. An interferon-inducible neutrophil-driven blood transcriptional signature in human tuberculosis. Nature 466 (7309), 973–977.

Bloemen, K., Verstraelen, S., Van Den Heuvel, R., et al., 2007. The allergic cascade: review of the most important molecules in the asthmatic lung. Immunol. Lett. 113 (1), 6–18.

Blomberg, A., Mudway, I., Svensson, M., et al., 2003. Clara cell protein as a biomarker for ozone-induced lung injury in humans. Eur. Respir. J. 22 (6), 883–888.

Bollati, V., Angelici, L., Rizzo, G., et al., 2015. Microvesicle-associated microRNA expression is altered upon particulate matter exposure in healthy workers and in A549 cells. J. Appl. Toxicol. 35 (1), 59–67.

Bolton, C.E., Stone, M.D., Edwards, P.H., et al., 2009. Circulating matrix metalloproteinase-9 and osteoporosis in patients with chronic obstructive pulmonary disease. Chron. Respir. Dis. 6 (2), 81–87.

Borghesi, A., Massa, M., Campanelli, R., et al., 2009. Circulating endothelial progenitor cells in preterm infants with bronchopulmonary dysplasia. Am. J. Respir. Crit. Care Med. 180 (6), 540–546.

Calfee, C.S., Ware, L.B., Glidden, D.V., et al., 2011. Use of risk reclassification with multiple biomarkers improves mortality prediction in acute lung injury. Crit. Care Med. 39 (4), 711–717.

Cantor, J.O., Shteyngart, B., 2004. How a test for elastic fiber breakdown products in sputum could speed development of a treatment for pulmonary emphysema. Med. Sci. Monit. 10 (1), RA1–4.

Carpagnano, G.E., Lacedonia, D., Palladino, G.P., et al., 2012. Could exhaled ferritin and SOD be used as markers for lung cancer and prognosis prediction purposes? Eur. J. Clin. Invest. 42 (5), 478–486.

Carter, S.R., Davis, C.S., Kovacs, E.J., 2012. Exhaled breath condensate collection in the mechanically ventilated patient. Respir. Med. 106 (5), 601–613.

Chaudhary, N.I., Roth, G.J., Hilberg, F., et al., 2007. Inhibition of PDGF, VEGF and FGF signalling attenuates fibrosis. Eur. Respir. J. 29 (5), 976–985.

Chen, Y.C., Hsiao, C.C., Chen, K.D., et al., 2013. Peripheral immune cell gene expression changes in advanced non-small cell lung cancer patients treated with first line combination chemotherapy. PLoS One 8 (2), e57053.

Cheng, I.W., Ware, L.B., Greene, K.E., et al., 2003. Prognostic value of surfactant proteins A and D in patients with acute lung injury. Crit. Care Med. 31 (1), 20–27.

Cherot-Kornobis, N., Hulo, S., de Broucker, V., et al., 2012. Induced sputum, exhaled NO, and breath condensate in occupational medicine. J. Occup. Environ. Med. 54 (8), 922–927.

Cheung, C.H.Y., Juan, H.F., 2017. Quantitative proteomics in lung cancer. J. Biomed. Sci. 24 (1), 37.

Cho, S.H., Oh, S.Y., Zhu, Z., et al., 2012. Spontaneous eosinophilic nasal inflammation in a genetically-mutant mouse: comparative study with an allergic inflammation model. PLoS One 7 (4), e35114.

Choi, H.S., Ashitate, Y., Lee, J.H., et al., 2010. Rapid translocation of nanoparticles from the lung airspaces to the body. Nat. Biotechnol. 28 (12), 1300–1303.

Chung, K.F., 2005. Inflammatory mediators in chronic obstructive pulmonary disease. Curr. Drug Targets - Inflamm. Allergy 4 (6), 619–625.

Clapp, R.W., Jacobs, M.M., Loechler, E.L., 2008. Environmental and occupational causes of cancer: new evidence 2005-2007. Rev. Environ. Health 23 (1), 1–37.

Cliff, J.M., Lee, J.S., Constantinou, N., et al., 2013. Distinct phases of blood gene expression pattern through tuberculosis treatment reflect modulation of the humoral immune response. J. Infect. Dis. 207 (1), 18–29.

Cohen, R.A., Patel, A., Green, F.H., 2008. Lung disease caused by exposure to coal mine and silica dust. Semin. Respir. Crit. Care Med. 29 (6), 651–661.

Cook, D.N., Bottomly, K., 2007. Innate immune control of pulmonary dendritic cell trafficking. Proc. Am. Thorac. Soc. 4 (3), 234–239.

Creaney, J., Musk, A.W., Robinson, B.W., 2010. Sensitivity of urinary mesothelin in patients with malignant mesothelioma. J. Thorac. Oncol. 5 (9), 1461–1466.

Cullen, A., Van Marter, L.J., Allred, E.N., et al., 2002. Urine bombesin-like peptide elevation precedes clinical evidence of bronchopulmonary dysplasia. Am. J. Respir. Crit. Care Med. 165 (8), 1093–1097.

Dalle-Donne, I., Rossi, R., Giustarini, D., et al., 2007. S-glutathionylation in protein redox regulation. Free Radic. Biol. Med. 43 (6), 883–898.

de Laurentiis, G., Paris, D., Melck, D., et al., 2008. Metabonomic analysis of exhaled breath condensate in adults by nuclear magnetic resonance spectroscopy. Eur. Respir. J. 32 (5), 1175–1183.

Dobashi, N., Fujita, J., Murota, M., et al., 2000. Elevation of anti-cytokeratin 18 antibody and circulating cytokeratin 18: anti-cytokeratin 18 antibody immune complexes in sera of patients with idiopathic pulmonary fibrosis. Lung 178 (3), 171–179.

Doyle, I.R., Hermans, C., Bernard, A., et al., 1998. Clearance of Clara cell secretory protein 16 (CC16) and surfactant proteins A and B from blood in acute respiratory failure. Am. J. Respir. Crit. Care Med. 158 (5 Pt 1), 1528–1535.

Duffield-Lillico, A.J., Boyle, J.O., Zhou, X.K., et al., 2009. Levels of prostaglandin E metabolite and leukotriene E(4) are increased in the urine of smokers: evidence that celecoxib shunts arachidonic acid into the 5-lipoxygenase pathway. Canc. Prev. Res. 2 (4), 322–329.

Eder, E., Wacker, M., Wanek, P., 2008. Lipid peroxidation-related 1,N2-propanodeoxyguanosine-DNA adducts induced by endogenously formed 4-hydroxy-2-nonenal in organs of female rats fed diets supplemented with sunflower, rapeseed, olive or coconut oil. Mutat. Res. 654 (2), 101–107.

El-Solh, A.A., Amsterdam, D., Alhajhusain, A., et al., 2009. Matrix metalloproteases in bronchoalveolar lavage fluid of patients with type III *Pseudomonas aeruginosa* pneumonia. J. Infect. 59 (1), 49–55.

Emad, A., Emad, V., 2008. The value of BAL fluid LDH level in differentiating benign from malignant solitary pulmonary nodules. J. Canc. Res. Clin. Oncol. 134 (4), 489–493.

Emmerechts, J., Jacobs, L., Van Kerckhoven, S., et al., 2012. Air pollution-associated procoagulant changes: the role of circulating microvesicles. J. Thromb. Haemostasis 10 (1), 96–106.

Enhorning, G., 2008. Surfactant in airway disease. Chest 133 (4), 975–980.

Fahy, J.V., 2009. Eosinophilic and neutrophilic inflammation in asthma: insights from clinical studies. Proc. Am. Thorac. Soc. 6 (3), 256–259.

Fattman, C.L., Schaefer, L.M., Oury, T.D., 2003. Extracellular superoxide dismutase in biology and medicine. Free Radic. Biol. Med. 35 (3), 236–256.

Felix, P.M., Franco, C., Barreiros, M.A., et al., 2013. Biomarkers of exposure to metal dust in exhaled breath condensate: methodology optimization. Arch. Environ. Occup. Health 68 (2), 72–79.

Fill, J.A., Brandt, J.T., Wiedemann, H.P., et al., 2006. Urinary desmosine as a biomarker in acute lung injury. Biomarkers 11 (1), 85–96.

Fitzpatrick, A.M., Teague, W.G., Burwell, L., et al., 2011. Glutathione oxidation is associated with airway macrophage functional impairment in children with severe asthma. Pediatr. Res. 69 (2), 154–159.

Fry, R.C., Rager, J.E., Bauer, R., et al., 2014. Air toxics and epigenetic effects: ozone altered microRNAs in the sputum of human subjects. Am. J. Physiol. Lung Cell Mol. Physiol. 306 (12), L1129–L1137.

Galambos, C., Demello, D.E., 2008. Regulation of alveologenesis: clinical implications of impaired growth. Pathology 40 (2), 124–140.

Gehlbach, B.K., Geppert, E., 2004. The pulmonary manifestations of left heart failure. Chest 125 (2), 669–682.

Gerde, P., Muggenburg, B.A., Lundborg, M., et al., 2001. The rapid alveolar absorption of diesel soot-adsorbed benzo[a]pyrene: bioavailability, metabolism and dosimetry of an inhaled particle-borne carcinogen. Carcinogenesis 22 (5), 741–749.

Ghio, A.J., Hilborn, E.D., Stonehuerner, J.G., et al., 2008. Particulate matter in cigarette smoke alters iron homeostasis to produce a biological effect. Am. J. Respir. Crit. Care Med. 178 (11), 1130–1138.

Gomperts, B.N., Belperio, J.A., Rao, P.N., et al., 2006. Circulating progenitor epithelial cells traffic via CXCR4/CXCL12 in response to airway injury. J. Immunol. 176 (3), 1916–1927.

Gordon, C., Gudi, K., Krause, A., et al., 2011. Circulating endothelial microparticles as a measure of early lung destruction in cigarette smokers. Am. J. Respir. Crit. Care Med. 184 (2), 224–232.

Greene, K.E., King Jr., T.E., Kuroki, Y., et al., 2002. Serum surfactant proteins-A and -D as biomarkers in idiopathic pulmonary fibrosis. Eur. Respir. J. 19 (3), 439–446.

Griese, M., 1999. Pulmonary surfactant in health and human lung diseases: state of the art. Eur. Respir. J. 13 (6), 1455–1476.

Grob, N.M., Aytekin, M., Dweik, R.A., 2008. Biomarkers in exhaled breath condensate: a review of collection, processing and analysis. J. Breath Res. 2 (3), 037004.

Guo, L., Ji, X., Yang, S., et al., 2013. Genome-wide analysis of aberrantly expressed circulating miRNAs in patients with coal workers' pneumoconiosis. Mol. Biol. Rep. 40 (5), 3739–3747.

Hamid, Q., Tulic, M., 2009. Immunobiology of asthma. Annu. Rev. Physiol. 71, 489–507.

Hant, F.N., Ludwicka-Bradley, A., Wang, H.J., et al., 2009. Surfactant protein D and KL-6 as serum biomarkers of interstitial lung disease in patients with scleroderma. J. Rheumatol. 36 (4), 773–780.

Harrill, A.H., McCullough, S.D., Wood, C.E., et al., 2016. MicroRNA biomarkers of toxicity in biological matrices. Toxicol. Sci. 152 (2), 264–272.

Hartl, D., Lee, C.G., Da Silva, C.A., et al., 2009. Novel biomarkers in asthma: chemokines and chitinase-like proteins. Curr. Opin. Allergy Clin. Immunol. 9 (1), 60–66.

Hauer-Jensen, M., Kong, F.M., Fink, L.M., et al., 1999. Circulating thrombomodulin during radiation therapy of lung cancer. Radiat. Oncol. Invest. 7 (4), 238–242.

Heidemann, S.M., Sandhu, H., Kovacevic, N., et al., 2011. Detection of tumor necrosis factor-alpha and interleukin-6 in exhaled breath condensate of rats with pneumonia due to staphylococcal enterotoxin B. Exp. Lung Res. 37 (9), 563–567.

Henderson, R.F., 2005. Use of bronchoalveolar lavage to detect respiratory tract toxicity of inhaled material. Exp. Toxicol. Pathol. 57 (Suppl. 1), 155–159.

Herzog, E.L., Brody, A.R., Colby, T.V., et al., 2008. Knowns and unknowns of the alveolus. Proc. Am. Thorac. Soc. 5 (7), 778–782.

Hmmier, A., O'Brien, M.E., Lynch, V., et al., 2017. Proteomic analysis of bronchoalveolar lavage fluid (BALF) from lung cancer patients using label-free mass spectrometry. BBA Clin. 7, 97–104.

Hoffman, A.M., 2008. Bronchoalveolar lavage: sampling technique and guidelines for cytologic preparation and interpretation. Vet. Clin. N. Am. Equine Pract. 24 (2), 423–435 vii–viii.

Hoffmeyer, F., Henry, J., Borowitzki, G., et al., 2010. Pulmonary lesions and serum levels of soluble Fas (sCD95) in former hard coal miners. Eur. J. Med. Res. 15 (Suppl. 2), 60–63.

Horvath, I., Barnes, P.J., Loukides, S., et al., 2017. A European Respiratory Society technical standard: exhaled biomarkers in lung disease. Eur. Respir. J. 49 (4).

Idell, S., 2003. Coagulation, fibrinolysis, and fibrin deposition in acute lung injury. Crit. Care Med. 31 (4 Suppl.), S213–S220.

Ishikawa, N., Hattori, N., Tanaka, S., et al., 2011. Levels of surfactant proteins A and D and KL-6 are elevated in the induced sputum of chronic obstructive pulmonary disease patients: a sequential sputum analysis. Respiration 82 (1), 10–18.

Izbicka, E., Streeper, R.T., Michalek, J.E., et al., 2012. Plasma biomarkers distinguish non-small cell lung cancer from asthma and differ in men and women. Cancer Genomics Proteomics 9 (1), 27–35.

Janssen, L.J., 2008. Isoprostanes and lung vascular pathology. Am. J. Respir. Cell Mol. Biol. 39 (4), 383–389.

Jean, J.C., Liu, Y., Joyce-Brady, M., 2003. The importance of gamma-glutamyl transferase in lung glutathione homeostasis and antioxidant defense. Biofactors 17 (1–4), 161–173.

Johnson, C.H., Ivanisevic, J., Siuzdak, G., 2016. Metabolomics: beyond biomarkers and towards mechanisms. Nat. Rev. Mol. Cell Biol. 17 (7), 451–459.

Joung, K.E., Kim, H.S., Lee, J., et al., 2011. Correlation of urinary inflammatory and oxidative stress markers in very low birth weight infants with subsequent development of bronchopulmonary dysplasia. Free Radic. Res. 45 (9), 1024–1032.

Kanazawa, H., 2007. Role of vascular endothelial growth factor in the pathogenesis of chronic obstructive pulmonary disease. Med. Sci. Monit. 13 (11), RA189–195.

Kari, F., Hatch, G., Slade, R., et al., 1997. Dietary restriction mitigates ozone-induced lung inflammation in rats: a role for endogenous antioxidants. Am. J. Respir. Cell Mol. Biol. 17 (6), 740–747.

Kawada, H., Horiuchi, T., Shannon, J.M., et al., 1989. Alveolar type II cells, surfactant protein A (SP-A), and the phospholipid components of surfactant in acute silicosis in the rat. Am. Rev. Respir. Dis. 140 (2), 460–470.

Khalil, N., Churg, A., Muller, N., et al., 2007. Environmental, inhaled and ingested causes of pulmonary fibrosis. Toxicol. Pathol. 35 (1), 86–96.

Kim, D.K., Cho, M.H., Hersh, C.P., et al., 2012. Genome-wide association analysis of blood biomarkers in chronic obstructive pulmonary disease. Am. J. Respir. Crit. Care Med. 186 (12), 1238–1247.

Kodavanti, U.P., 2016. Stretching the stress boundary: linking air pollution health effects to a neurohormonal stress response. Biochim. Biophys. Acta 1860.

Kodavanti, U.P., Hatch, G.E., Starcher, B., et al., 1995. Ozone-induced pulmonary functional, pathological, and biochemical changes in normal and vitamin C-deficient Guinea pigs. Fund. Appl. Toxicol. 24 (2), 154–164.

Kodavanti, U.P., Mehendale, H.M., 1990. Cationic amphiphilic drugs and phospholipid storage disorder. Pharmacol. Rev. 42 (4), 327–354.

Kodavanti, U.P., Schladweiler, M.C., Ledbetter, A.D., et al., 2006. The spontaneously hypertensive rat: an experimental model of sulfur dioxide-induced airways disease. Toxicol. Sci. 94 (1), 193–205.

Kodavanti, U.P., Schladweiler, M.C., Ledbetter, A.D., et al., 2000. The spontaneously hypertensive rat as a model of human cardiovascular disease: evidence of exacerbated cardiopulmonary injury and oxidative stress from inhaled emission particulate matter. Toxicol. Appl. Pharmacol. 164 (3), 250–263.

Kontakiotis, T., Katsoulis, K., Hagizisi, O., et al., 2011. Bronchoalveolar lavage fluid alteration in antioxidant and inflammatory status in lung cancer patients. Eur. J. Intern. Med. 22 (5), 522–526.

Kumar, N., Shahjaman, M., Mollah, M.N.H., et al., 2017. Serum and plasma metabolomic biomarkers for lung cancer. Bioinformation 13 (6), 202–208.

Larsson, P., Bake, B., Wallin, A., et al., 2017. The effect of exhalation flow on endogenous particle emission and phospholipid composition. Respir. Physiol. Neurobiol. 243, 39–46.

Liu, J., Liang, Q., Frost-Pineda, K., et al., 2011. Relationship between biomarkers of cigarette smoke exposure and biomarkers of inflammation, oxidative stress, and platelet activation in adult cigarette smokers. Cancer Epidemiol. Biomark. Prev. 20 (8), 1760–1769.

Mackman, N., 2006. Role of tissue factor in hemostasis and thrombosis. Blood Cells Mol. Dis. 36 (2), 104–107.

Maniscalco, M., de Laurentiis, G., Weitzberg, E., et al., 2008. Validation study of nasal nitric oxide measurements using a hand-held electrochemical analyser. Eur. J. Clin. Invest. 38 (3), 197–200.

Mauad, T., Dolhnikoff, M., 2008. Pathologic similarities and differences between asthma and chronic obstructive pulmonary disease. Curr. Opin. Pulm. Med. 14 (1), 31–38.

Mauri, T., Coppadoro, A., Bellani, G., et al., 2008. Pentraxin 3 in acute respiratory distress syndrome: an early marker of severity. Crit. Care Med. 36 (8), 2302–2308.

McNulty, K., Janes, S.M., 2012. Stem cells and pulmonary fibrosis: cause or cure? Proc. Am. Thorac. Soc. 9 (3), 164–171.

McVey, M., Tabuchi, A., Kuebler, W.M., 2012. Microparticles and acute lung injury. Am. J. Physiol. Lung Cell Mol. Physiol. 303 (5), L364–L381.

Meduri, G.U., Tolley, E.A., Chinn, A., et al., 1998. Procollagen types I and III aminoterminal propeptide levels during acute respiratory distress syndrome and in response to methylprednisolone treatment. Am. J. Respir. Crit. Care Med. 158 (5 Pt 1), 1432–1441.

Mehan, M.R., Ostroff, R., Wilcox, S.K., et al., 2013. Highly multiplexed proteomic platform for biomarker discovery, diagnostics, and therapeutics. Adv. Exp. Med. Biol. 735, 283–300.

Miller, D.B., Ghio, A.J., Karoly, E.D., et al., 2016. Ozone exposure increases circulating stress hormones and lipid metabolites in humans. Am. J. Respir. Crit. Care Med. 193 (12), 1382–1391.

Miller, D.B., Karoly, E.D., Jones, J.C., et al., 2015. Inhaled ozone (O3)-induces changes in serum metabolomic and liver transcriptomic profiles in rats. Toxicol. Appl. Pharmacol. 286 (2), 65–79.

Minematsu, N., Blumental-Perry, A., Shapiro, S.D., 2011. Cigarette smoke inhibits engulfment of apoptotic cells by macrophages through inhibition of actin rearrangement. Am. J. Respir. Cell Mol. Biol. 44 (4), 474–482.

Miyazaki, E., Nureki, S., Ono, E., et al., 2007. Circulating thymus- and activation-regulated chemokine/CCL17 is a useful biomarker for discriminating acute eosinophilic pneumonia from other causes of acute lung injury. Chest 131 (6), 1726–1734.

Molina-Diaz, A., Ortega-Carmona, I., Pascual-Reguera, M.I., 1998. Indirect spectrophotometric determination of ascorbic acid with ferrozine by flow-injection analysis. Talanta 47 (3), 531–536.

Naz, S., Garcia, A., Rusak, M., et al., 2013. Method development and validation for rat serum fingerprinting with CE-MS: application to ventilator-induced-lung-injury study. Anal. Bioanal. Chem. 405 (14), 4849–4858.

Nef, H.M., Mollmann, H., Hamm, C., et al., 2010. Pulmonary hypertension: updated classification and management of pulmonary hypertension. Heart 96 (7), 552–559.

Nemmar, A., Hoylaerts, M.F., Nemery, B., 2006. Effects of particulate air pollution on hemostasis. Clin. Occup. Environ. Med. 5 (4), 865–881.

Nemmar, A., Subramaniyan, D., Zia, S., et al., 2012. Airway resistance, inflammation and oxidative stress following exposure to diesel exhaust particle in angiotensin II-induced hypertension in mice. Toxicology 292 (2–3), 162–168.

Nieri, D., Neri, T., Petrini, S., et al., 2016. Cell-derived microparticles and the lung. Eur. Respir. Rev. 25 (141), 266–277.

Novak, Z., Petak, F., Banfi, A., et al., 2006. An improved technique for repeated bronchoalveolar lavage and lung mechanics measurements in individual rats. Respir. Physiol. Neurobiol. 154 (3), 467–477.

Nyunoya, T., March, T.H., Tesfaigzi, Y., et al., 2011. Antioxidant diet protects against emphysema, but increases mortality in cigarette smoke-exposed mice. COPD 8 (5), 362–368.

O'Kane, C.M., McKeown, S.W., Perkins, G.D., et al., 2009. Salbutamol up-regulates matrix metalloproteinase-9 in the alveolar space in the acute respiratory distress syndrome. Crit. Care Med. 37 (7), 2242–2249.

Ochoa, C.D., Wu, S., Stevens, T., 2010. New developments in lung endothelial heterogeneity: von Willebrand factor, P-selectin, and the Weibel-Palade body. Semin. Thromb. Hemost. 36 (3), 301–308.

Ohlmeier, S., Vuolanto, M., Toljamo, T., et al., 2008. Proteomics of human lung tissue identifies surfactant protein A as a marker of chronic obstructive pulmonary disease. J. Proteome Res. 7 (12), 5125–5132.

Ostroff, R.M., Bigbee, W.L., Franklin, W., et al., 2010. Unlocking biomarker discovery: large scale application of aptamer proteomic technology for early detection of lung cancer. PLoS One 5 (12), e15003.

Overbeek, S.A., Braber, S., Koelink, P.J., et al., 2013. Cigarette smoke-induced collagen destruction; key to chronic neutrophilic airway inflammation? PLoS One 8 (1), e55612.

Padilla-Carlin, D.J., Schladweiler, M.C., Shannahan, J.H., Kodavanti, U.P., et al., 2011. Pulmonary inflammatory and fibrotic responses in Fischer 344 rats after intratracheal instillation exposure to Libby amphibole. J. Toxicol. Environ. Health 74 (17), 1111–1132.

Paul, P., Chakraborty, A., Sarkar, D., et al., 2018. Interplay between miRNAs and human diseases. J. Cell. Physiol. 233 (3), 2007–2018.

Piccin, A., Murphy, W.G., Smith, O.P., 2007. Circulating microparticles: pathophysiology and clinical implications. Blood Rev. 21 (3), 157–171.

Pisi, G., Olivieri, D., Chetta, A., 2009. The airway neurogenic inflammation: clinical and pharmacological implications. Inflamm. Allergy - Drug Targets 8 (3), 176–181.

Pleil, J.D., Wallace, A., Madden, M.C., 2017. Exhaled breath aerosol (EBA): the simplest non-invasive medium for public health and occupational exposure biomonitoring. J. Breath Res. 12.

Popov, T.A., 2011. Human exhaled breath analysis. Ann. Allergy Asthma Immunol. 106 (6), 451–456 quiz 457.

Porro, C., Di Gioia, S., Trotta, T., et al., 2013. Pro-inflammatory effect of cystic fibrosis sputum microparticles in the murine lung. J. Cyst. Fibros. 12 (6), 721–728.

Prasad, R., Singh, A., Garg, R., et al., 2012. Biomass fuel exposure and respiratory diseases in India. Biosci. Trends 6 (5), 219–228.

Prasse, A., Muller-Quernheim, J., 2009. Non-invasive biomarkers in pulmonary fibrosis. Respirology 14 (6), 788–795.

Reynolds, H.Y., 2011. Bronchoalveolar lavage and other methods to define the human respiratory tract milieu in health and disease. Lung 189 (2), 87–99.

Rihova, Z., Maixnerova, D., Jancova, E., et al., 2005. Silica and asbestos exposure in ANCA-associated vasculitis with pulmonary involvement. Ren. Fail. 27 (5), 605–608.

Roginsky, V.A., Barsukova, T.K., Bruchelt, G., et al., 1997. Iron bound to ferritin catalyzes ascorbate oxidation: effects of chelating agents. Biochim. Biophys. Acta 1335 (1–2), 33–39.

Rosias, P., 2012. Methodological aspects of exhaled breath condensate collection and analysis. J. Breath Res. 6 (2), 027102.

Samukawa, T., Hamada, T., Uto, H., et al., 2012. The elevation of serum napsin A in idiopathic pulmonary fibrosis, compared with KL-6, surfactant protein-A and surfactant protein-D. BMC Pulm. Med. 12, 55.

Schmekel, B., Blomstrand, P., Venge, P., 2013. Serum lysozyme - a surrogate marker of pulmonary microvascular injury in smokers? Clin. Physiol. Funct. Imaging 33 (4), 307–312.

Schmidt, R., Markart, P., Ruppert, C., et al., 2007. Time-dependent changes in pulmonary surfactant function and composition in acute respiratory distress syndrome due to pneumonia or aspiration. Respir. Res. 8, 55.

Schneider, B.C., Constant, S.L., Patierno, S.R., et al., 2012. Exposure to particulate hexavalent chromium exacerbates allergic asthma pathology. Toxicol. Appl. Pharmacol. 259 (1), 38–44.

Schwartz, M.K., 1991. Lactic dehydrogenase. An old enzyme reborn as a cancer marker? Am. J. Clin. Pathol. 96 (4), 441–443.

Shannahan, J.H., Alzate, O., Winnik, W.M., et al., 2012a. Acute phase response, inflammation and metabolic syndrome biomarkers of Libby asbestos exposure. Toxicol. Appl. Pharmacol. 260 (2), 105–114.

Shannahan, J.H., Schladweiler, M.C., Richards, J.H., et al., 2010. Pulmonary oxidative stress, inflammation, and dysregulated iron homeostasis in rat models of cardiovascular disease. J. Toxicol. Environ. Health 73 (10), 641–656.

Shannahan, J.H., Schladweiler, M.C., Thomas, R.F., et al., 2012b. Vascular and thrombogenic effects of pulmonary exposure to Libby amphibole. J. Toxicol. Environ. Health 75 (4), 213–231.

Smadja, D.M., Mauge, L., Nunes, H., et al., 2013. Imbalance of circulating endothelial cells and progenitors in idiopathic pulmonary fibrosis. Angiogenesis 16 (1), 147–157.

Snow, S.J., Cheng, W.Y., Henriquez, A., et al., 2018. Ozone-induced vascular contractility and pulmonary injury are differentially impacted by diets enriched with coconut oil, fish oil, and olive oil. Toxicol. Sci. 163.

Snow, S.J., McGee, J., Miller, D.B., et al., 2014. Inhaled diesel emissions generated with cerium oxide nanoparticle fuel additive induce adverse pulmonary and systemic effects. Toxicol. Sci. 142 (2), 403–417.

Snow, S.J., McGee, M.A., Henriquez, A., et al., 2017. Respiratory effects and systemic stress response following acute acrolein inhalation in rats. Toxicol. Sci. 158 (2), 454–464.

Sunjic, S.B., Cipak, A., Rabuzin, F., et al., 2005. The influence of 4-hydroxy-2-nonenal on proliferation, differentiation and apoptosis of human osteosarcoma cells. Biofactors 24 (1–4), 141–148.

Takahashi, H., Sano, H., Chiba, H., et al., 2006. Pulmonary surfactant proteins A and D: innate immune functions and biomarkers for lung diseases. Curr. Pharm. Des. 12 (5), 589–598.

Tuder, R.M., Petrache, I., 2012. Pathogenesis of chronic obstructive pulmonary disease. J. Clin. Invest. 122 (8), 2749–2755.

Turell, D.C., 2008. Advances with surfactant. Emerg. Med. Clin. N. Am. 26 (4), 921–928 viii.

Turino, G.M., Ma, S., Lin, Y.Y., et al., 2011. Matrix elastin: a promising biomarker for chronic obstructive pulmonary disease. Am. J. Respir. Crit. Care Med. 184 (6), 637–641.

Tzortzaki, E.G., Lambiri, I., Vlachaki, E., et al., 2007. Biomarkers in COPD. Curr. Med. Chem. 14 (9), 1037–1048.

Tzouvelekis, A., Kouliatsis, G., Anevlavis, S., et al., 2005. Serum biomarkers in interstitial lung diseases. Respir. Res. 6, 78.

Ubhi, B.K., Cheng, K.K., Dong, J., et al., 2012. Targeted metabolomics identifies perturbations in amino acid metabolism that subclassify patients with COPD. Mol. Biosyst. 8 (12), 3125–3133.

van de Kant, K.D., van der Sande, L.J., Jobsis, Q., et al., 2012. Clinical use of exhaled volatile organic compounds in pulmonary diseases: a systematic review. Respir. Res. 13, 117.

Vatrella, A., Bocchino, M., Perna, F., et al., 2007. Induced sputum as a tool for early detection of airway inflammation in connective diseases-related lung involvement. Respir. Med. 101 (7), 1383–1389.

Viglio, S., Annovazzi, L., Luisetti, M., et al., 2007. Progress in the methodological strategies for the detection in real samples of desmosine and isodesmosine, two biological markers of elastin degradation. J. Sep. Sci. 30 (2), 202–213.

Voynow, J.A., Rubin, B.K., 2009. Mucins, mucus, and sputum. Chest 135 (2), 505–512.

Vuorinen, K., Myllarniemi, M., Lammi, L., et al., 2007. Elevated matrilysin levels in bronchoalveolar lavage fluid do not distinguish idiopathic pulmonary fibrosis from other interstitial lung diseases. APMIS 115 (8), 969–975.

Vyas, J.R., Currie, A., Dunster, C., et al., 2001. Ascorbate acid concentration in airways lining fluid from infants who develop chronic lung disease of prematurity. Eur. J. Pediatr. 160 (3), 177–184.

Wallenborn, J.G., Kovalcik, K.D., McGee, J.K., et al., 2009. Systemic translocation of (70)zinc: kinetics following intratracheal instillation in rats. Toxicol. Appl. Pharmacol. 234 (1), 25–32.

Wallenborn, J.G., McGee, J.K., Schladweiler, M.C., et al., 2007. Systemic translocation of particulate matter-associated metals following a single intratracheal instillation in rats. Toxicol. Sci. 98 (1), 231–239.

Wang, Z., Beach, D., Su, L., et al., 2008. A genome-wide expression analysis in blood identifies pre-elafin as a biomarker in ARDS. Am. J. Respir. Cell Mol. Biol. 38 (6), 724–732.

Warheit, D.B., Hart, G.A., Hesterberg, T.W., et al., 2001. Potential pulmonary effects of man-made organic fiber (MMOF) dusts. Crit. Rev. Toxicol. 31 (6), 697–736.

Wedge, D.C., Allwood, J.W., Dunn, W., et al., 2011. Is serum or plasma more appropriate for intersubject comparisons in metabolomic studies? An assessment in patients with small-cell lung cancer. Anal. Chem. 83 (17), 6689–6697.

Wereszczynska-Siemiatkowska, U., Dlugosz, J.W., Siemiatkowski, A., et al., 2000. Lysosomal activity of pulmonary alveolar macrophages in acute experimental pancreatitis in rats with reference to positive PAF-antagonist (BN 52021) effect. Exp. Toxicol. Pathol. 52 (2), 119–125.

Widdicombe, J.H., 2002. Regulation of the depth and composition of airway surface liquid. J. Anat. 201 (4), 313–318.

Wilkins, M.R., 2012. Pulmonary hypertension: the science behind the disease spectrum. Eur. Respir. Rev. 21 (123), 19–26.

Wilson, J.X., 2005. Regulation of vitamin C transport. Annu. Rev. Nutr. 25, 105–125.

Winters, B.R., Pleil, J.D., Angrish, M.M., et al., 2017. Standardization of the collection of exhaled breath condensate and exhaled breath aerosol using a feedback regulated sampling device. J. Breath Res. 11 (4), 047107.

Yamamoto, M., Singh, A., Sava, F., et al., 2013. MicroRNA expression in response to controlled exposure to diesel exhaust: attenuation by the antioxidant N-acetylcysteine in a randomized crossover study. Environ. Health Perspect. 121 (6), 670–675.

Yang, C., Jiang, J., Yang, X., et al., 2016. Stem/progenitor cells in endogenous repairing responses: new toolbox for the treatment of acute lung injury. J. Transl. Med. 14, 47.

Zhang, Y., Kaminski, N., 2012. Biomarkers in idiopathic pulmonary fibrosis. Curr. Opin. Pulm. Med. 18 (5), 441–446.

Zhao, Y., Zhang, Y.H., Denney, L., et al., 2012. High levels of virus-specific $CD4^+$ T cells predict severe pandemic influenza A virus infection. Am. J. Respir. Crit. Care Med. 186 (12), 1292–1297.

Zhou, Y., Huang, D., Farver, C., Hoffman, G.S., 2003. Relative importance of CCR5 and antineutrophil cytoplasmic antibodies in patients with Wegener's granulomatosis. J. Rheumatol. 30 (7), 1541–1547.

Zou, J., du Prel Carroll, X., Liang, X., et al., 2011. Alterations of serum biomarkers associated with lung ventilation function impairment in coal workers: a cross-sectional study. Environ. Health 10, 83.

CHAPTER

14

Hepatic Toxicity Biomarkers

Gurjot Kaur

Human and Environmental Toxicology, Department of Biology, University of Konstanz, Konstanz, Germany

INTRODUCTION

The liver is the largest internal organ in the human body and is the main organ for metabolism and detoxification of drugs and environmental chemicals. In addition, it regulates a myriad of functions such as glucose synthesis and storage, decomposition of red blood cells, plasma protein synthesis, hormone production, and bile formation. Anatomically, the liver lies slightly below the diaphragm and anterior to the stomach, a position that facilitates maintaining metabolic homeostasis of the body. Two distinct blood supplies, the portal vein (PV) and the hepatic artery (HA), feed the liver. The PV carries blood containing digested nutrients from the gastrointestinal tract, spleen, and pancreas, whereas the HA carries oxygenated blood from the lungs. The human liver consists of four lobes and each lobe is made up of many lobules at the microscopic level. The classical lobule is a hexagonal-shaped unit centered on a central vein (CV). In each functional unit, blood enters the lobules from the PV and HA, and then flows down past the cord of hepatocytes. The lobule is divided into three zones (Fig. 14.1): (1) periportal (Zone 1) is the closest to the entering blood supply and has the highest oxygen tension; (2) centrilobular (Zone 3) abuts the CV and has the poorest oxygenation; and (3) midzonal (Zone 2) is intermediate. Because of the blood flow from the stomach and intestine, the liver is the first organ to encounter a number of insults from ingested metals, drugs, and environmental toxicants (Klaassen et al., 2013).

Because of acute or chronic exposure to significantly high toxicant levels, liver functions can be adversely affected. For example, acetaminophen (APAP) is a widely used over-the-counter analgesic and antipyretic in the United States. When used at recommended therapeutic doses, APAP is rarely associated with liver injury. Unfortunately, APAP can cause fatal acute liver failure due to the production of a highly reactive hepatotoxic metabolite at doses exceeding therapeutic doses (Ciejka et al., 2016).

Because the liver is often exposed to the highest concentrations of oral drugs, it is not surprising that the liver is often the target organ with ensuing drug-induced liver injury (DILI). DILI is a major challenge for the pharmaceutical industry and public health because DILI is a common cause of drug development termination, drug restrictions, and postmarketing drug withdrawal. Currently, more than 1000 drugs have been reported to be associated with DILI (Shi et al., 2010). In addition, the increasing popularity of dietary supplements may contribute to the high incidence of DILI because some dietary supplements alter the toxicity of concomitantly administered drugs (Abebe, 2002; Salminen et al., 2012; Gwaltney-Brant, 2016; Gupta et al., 2018).

DILI has been classified into hepatocellular injury, cholestatic injury, and mixed hepatocellular/cholestatic injury. Hepatocellular injury involves cellular damage of the hepatocytes such as the centrilobular necrosis caused by APAP. This type of damage is often associated with elevated serum alanine aminotransferase (ALT) levels due to leakage from damaged hepatocytes. Cholestatic injury involves damage to some part of the bile processing or excretion apparatus, resulting in impaired processing and/or excretion. This type of injury is often associated with elevated serum bilirubin and alkaline phosphatase (ALP), indicating alterations in bile homeostasis and/or bile duct (BD) injury. Mixed injury presents with a mixture of both types of effects. Unfortunately, drugs rarely produce a single clear clinical picture, making the diagnosis of DILI difficult. For example, amoxicillin/clavulanic acid usually causes cholestatic injury but can also produce acute hepatocellular or a mixed type injury. The histopathological analysis of liver biopsies is the most definitive way to diagnose and differentiate between different DILIs. However, the biopsies are invasive and not routinely performed.

Biomarkers in Toxicology, Second Edition
https://doi.org/10.1016/B978-0-12-814655-2.00014-1

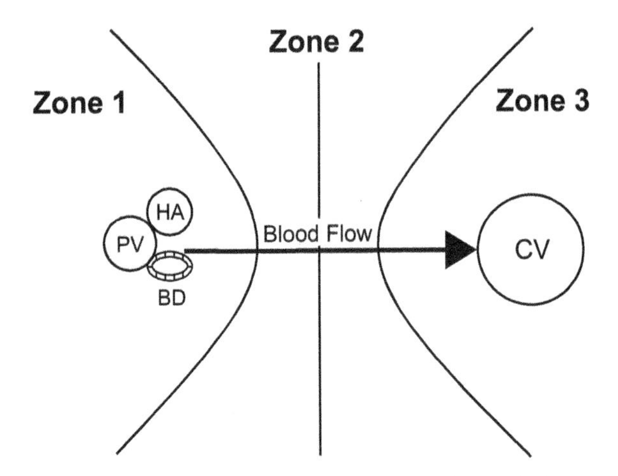

FIGURE 14.1 **Microscopic structure of the liver lobule.** Blood, supplied by portal vein (PV), or hepatic artery (HA), enters the liver lobule through the portal triad, which encompasses PV, HA, and bile duct (BD). Blood flows from Zone 1 (best oxygenation) to Zone 2 and out of the central vein (CV) in Zone 3 (poorest oxygenation).

Serum ALT activity has historically been used as a major biomarker for liver injury in humans and in preclinical studies. Damaged hepatocytes release ALT into the extracellular space with subsequent passage into the blood. However, ALT elevations can also reflect nonhepatic injury, particularly skeletal muscle injury. Additionally, the concentrations of ALT do not discriminate between different etiologies of liver injury, and ALT elevations can occur after a critical therapeutic window has passed. Another disadvantage is that ALT levels do not provide any insight into disease prognosis (Kaplowitz, 2005).

Adaptation during continued exposure to DILI drugs has been observed in preclinical and clinical studies. For example, 25% of Alzheimer's disease patients receiving tacrine experience transient, asymptomatic increases in serum ALT. The majority of these patients adapt to the drug as indicated by a return of serum ALT to baseline levels despite continued treatment. Most statins can also cause serum ALT elevations in a subset of treated patients, but often occur in the absence of histological evidence of injury. Therefore, elevated ALT levels do not necessarily indicate the severity of the liver injury and patient adaptability to the stressor or progression to total liver injury.

Another known liver toxicity biomarker is total bilirubin level (TBL), which is associated with altered bile homeostasis and/or hepatobiliary injury. The functional reserve of the liver for processing bilirubin is large; therefore, substantial hepatic injury often occurs before alterations in TBL are observed, making it an insensitive biomarker. By the time TBL is elevated, there may already be substantial loss of liver function, placing the patient in danger of liver failure. Therefore, unavailability of ideal clinical laboratory tests and inefficient prognosis necessitate development of new biomarkers that are not only sensitive but are also specific and prognostic.

Hy Zimmerman first noted that "drug-induced hepatocellular jaundice is a serious lesion. The mortality rate ranges from 10% to 50%." He realized that when drugs cause substantial hepatocyte injury that affects overall liver function and, in particular, causes jaundice as the result of impaired bilirubin processing and transport, the hepatotoxicity is likely to lead to a life-threatening event (Senior, 2006). The combination of two biomarkers, ALT and bilirubin, defined as "Hy's Law," indicates more severe injury than serum enzyme elevations alone. The significance of Hy's Law is that the combination of ALT and bilirubin, neither of which by itself is sufficiently specific, seems to be highly specific for serious liver injury. Therefore, the FDA recommends a combination of tests, including serum ALT, aspartate aminotransferase (AST), ALP activities, and TBL concentration, to identify potential DILI.

In addition to the biomarkers described above, there are symptoms commonly associated with liver injury. Abdominal pain, enlargement of the liver and spleen, distended belly full of fluid, or enlarged breasts in men are common signs of liver injury (Chopra, 2001). In combination with biomarkers of liver injury, these nonspecific and specific symptoms can help a physician identify liver injury and its seriousness. Unfortunately, there are limited tools at the physician's disposal for determining the patient's prognosis.

OVERVIEW OF LIVER PHYSIOLOGY, TOXICITY, AND PATHOLOGY

Bioactivation and Detoxification

The liver is the main metabolic organ in the body and is considered a viable defense against environmental and metabolic toxins. In general, all foreign compounds (xenobiotics) are potentially toxic. To minimize potential injury, the liver is well equipped with metabolizing enzymes termed Phase I and Phase II enzymes and Phase III transporters (Xu et al., 2005). The coordinated metabolism and transport typically make the xenobiotic less toxic and more water soluble, thereby aiding in its elimination from the body. Classically, the initial metabolic step is referred to as Phase I or the oxidation phase. This step is most frequently catalyzed by cytochrome P450 (CYP450) enzymes, which derive their name from their maximal absorbance at 450 nm when bound to carbon monoxide. They are highly expressed in liver (specifically in hepatocytes) and to a lesser extent in the epithelium lining of the gastrointestinal tract. They are typically found associated with the lipid membrane of the endoplasmic reticulum and the outer membrane of mitochondria. CYP1, CYP2, and CYP3 families are major enzymes mediating drug metabolism (Danielson, 2002). In general, Phase I CYP metabolism introduces a functional group, e.g., a hydroxyl group, to a xenobiotic. Phase I metabolism may not be a detoxifying function by itself because the metabolite may be more reactive and toxic than the parent compound, and thus it often works in conjunction with subsequent Phase II metabolism to produce a more water-soluble and less toxic compound. The functional group added by Phase I metabolism is often the target of Phase II conjugation. During this phase, the compound may undergo glucuronidation, sulfation, conjugation with glutathione, methylation, N-acetylation, or conjugation with amino acids. Typically these modifications make the compound significantly more hydrophilic, thereby enhancing excretion in the bile and urine (Klaassen et al., 2013).

Phase III is often referred to as a "metabolic" process; however, it essentially involves the transport of xenobiotics and their metabolites across biological membranes with no further metabolic alterations of the compound's structure. Phase III transporters are found in a wide array of organs/biological membranes, and they typically facilitate the removal of xenobiotics and their metabolites from the body. There are many different Phase III transporters with each one having specificity for different types of molecules. Human drug transporters can be classified under two major superfamilies: solute carrier (SLC) and ATP-binding cassette (ABC). Transporters of the SLC superfamily are the organic cation transporters (OCTs/SLC22A), the multidrug and toxin extrusion transporters (MATE transporters/SLC47A), the organic anion transporters (OATs/SLC22A), and the organic anion transporting polypeptides (OATPs/SLCO). Members of the ABC superfamily are important in drug toxicity, and these include P-glycoprotein (MDR1/ABCB1), multidrug resistance-associated protein (MRP/ABCC), and breast cancer resistance protein (BCRP/ABCG2) (DeGorter et al., 2012).

Many Phase I and Phase II enzymes show significant individual-specific variability resulting in dissimilar levels of exposure to reactive metabolites. Polymorphism and environmental exposure are the two primary causes of this variability. Polymorphisms in drug-metabolizing enzymes may cause dramatic differences in drug detoxification as demonstrated by various genome-wide association studies (GWAS) (Ahmed et al., 2016).

Protein Synthesis and Catabolism

The liver plays a vital role in protein metabolism such as deamination and transamination of amino acids, plasma protein synthesis, and removal of ammonia as urea in the urine. Several enzymes required in the amino acid metabolism pathways (for example, ALT/AST) are commonly assayed in serum to assess liver damage because they are present at much higher concentrations in the liver than in other organs (Dejong et al., 2007). Almost all blood proteins, except gamma globulins, are synthesized in the liver. Albumin, the main protein in human blood, plays a major role in maintaining plasma osmotic pressure and transportation of lipids and hormones. Some liver injury can affect the concentrations of plasma albumin and result in hypoalbuminemia and hyperglobulinemia (compensatory rise to offset the fall in albumin level) (Farrugia, 2010). Fibrinogen and blood clotting factors (e.g., Factor XIII) are another group of plasma proteins synthesized by the liver. During blood clot formation, soluble fibrinogen is converted into insoluble fibrin strands through a catalytic activation cascade of blood clotting factors. When liver function is impaired, impaired synthesis of the blood clotting factors can lead to excess bleeding. Toxic ammonia, a metabolic product of amino acid deamination, is quickly converted by liver to less toxic urea. Inefficient removal of ammonia from the circulation in humans results in central nervous system disease called hyperammonemia-related coma (Klaassen et al., 2013).

Bilirubin Processing

A very important function of the liver is to process bilirubin. Unconjugated bilirubin is the yellowish heme degradation product, derived from the breakdown of erythrocytic hemoglobin. Bilirubin when unconjugated is very hydrophobic and binds albumin during blood circulation. One of the main reasons for elevated unconjugated bilirubin levels is displacement from albumin by competing hydrophobic drugs or fatty acids. Such an increase of bilirubin may lead to increase in total bilirubin levels causing jaundice, manifested as yellowing of the skin and eye sclera. Jaundice is commonly seen in infants because higher serum levels of unconjugated bilirubin exist due to low metabolic capacity of the neonatal liver (Cohen et al., 2010). Other health problems that display jaundice include hemolysis (rupturing of erythrocytes). Eventually, bilirubin is removed from the blood by conjugation with glucuronic acid in hepatocytes in the liver. Once conjugated, bilirubin derivatives are secreted into bile for further metabolism and elimination.

Covalent Binding

A large body of evidence has shown that reactive metabolites play an important role in the pathogenesis of

DILI. Biotransformation of drugs results principally in detoxification. However, in certain cases, highly toxic reactive intermediates are generated, typically during Phase I metabolism. Such metabolites are short lived; however, if not detoxified, they can form covalent modifications of biological macromolecules and thereby damage proteins and nucleic acids. Reactive metabolites have also been identified from liver injury–inducing drugs such as APAP, tamoxifen, diclofenac, and troglitazone (Park et al., 2005).

In the case of APAP, it is predominantly metabolized by the Phase II metabolic pathways of glucuronidation and sulfation. A small portion of APAP is metabolized by Phase I enzymes to the reactive metabolite N-acetyl-p-benzoquinone imine (NAPQI) and subsequently detoxified by conjugation with glutathione (GSH). Following a toxic dose of APAP, the glucuronidation and sulfation pathways are saturated and APAP is metabolized by the Phase I pathways to NAPQI. NAPQI is detoxified by GSH conjugation; however, once GSH stores are depleted, NAPQI covalently binds cellular macromolecules such as proteins. This covalent binding occurs mainly in hepatocytes resulting in disruption of cellular function and eventual necrosis (Hinson et al., 2010). Interestingly, protein adducts can be detected in the serum of APAP-overdosed patients, most likely because of their release from necrotic hepatocytes (Muldrew et al., 2002). Thus, quantification of the protein adducts is useful in the diagnosis of APAP-overdosed patients. Although covalent binding often occurs during hepatotoxic injury, it is not necessarily sufficient to cause hepatotoxicity. For instance, 3-hydroxyacetanilide (AMAP, regioisomer of APAP) and APAP show similar levels of covalent binding in vivo, but AMAP is not hepatotoxic, even at very high doses (Roberts et al., 1990). Similarly, troglitazone is converted to reactive metabolites that generate a high level of covalent binding, but a causal relationship with hepatoxicity has not been established (Yamazaki et al., 1999). Hence, the utility of protein adducts as DILI biomarkers has been limited because of localization of protein adducts exclusively in the liver and not in blood.

Furthermore, even if covalent binding equates poorly with hepatotoxicity in many cases, the cellular stress responses activated by reactive metabolites may play a decisive role in the toxicity. GSH is a critical cellular antioxidant, important in combating cellular oxidative stress that damages cellular macromolecules. In APAP overdose, NAPQI reacts with GSH leading to depletion of this cytoprotective molecule. NAPQI covalently binds to mitochondrial proteins, increases reactive oxygen species (ROS) production, impairs mitochondrial respiration, thereby causing ATP depletion and opening the mitochondrial permeability transition (MPT) pore and finally making the mitochondrial inner membrane permeable to solutes up to 1500 Da (Jaeschke and Bajt, 2005; James et al., 2003). These events lead to the onset of MPT, which is a common pathway leading to both necrotic and apoptotic cell death. Induction of MPT causes mitochondrial depolarization, uncoupling of oxidative phosphorylation, and organelle swelling, which all result in decreased ATP synthesis and cell death (Kim et al., 2003). Many *in vitro* assays have been developed to measure GSH, ATP, ROS, mitochondrial functions, and cell viability, which might be useful as *in vitro* DILI biomarkers in predicting hepatotoxicity.

Necrosis and Apoptosis

Liver cell death is a characteristic feature of many liver diseases, such as cholestasis, hepatitis, and ischemia/reperfusion (Malhi et al., 2006). There are two patterns of cell death: necrosis (death of a large group of cells) and apoptosis (ATP-dependent individual cell death). Cell death may be restricted to certain cell types such as hepatocytes, cholangiocytes (BD cells), endothelial cells, stellate cells, or Kupffer cells (Ramachandran and Kakar, 2009). DILI is commonly observed as acute hepatocellular necrosis (Luedde et al., 2014). Hepatotoxicity may manifest pathologically in either a zonal or a diffuse pattern. Often, hepatotoxicants that cause damage primarily through a reactive metabolite will produce hepatic necrosis in a zonal pattern because even though drug concentrations are highest in Zone 1, most CYPs necessary for bioactivation are concentrated in Zone 3. For a hepatotoxicant such as APAP, CYPs in Zone 1 are at negligible levels and cannot form reactive metabolite. In contrast, CYPs in Zone 3 are at much higher levels and form the reactive metabolite NAPQI, restricting covalent binding to this zone (Bessems and Vermeulen, 2001).

Inflammation (Immune) Hepatitis

Occasionally, reactive drug or bacterial metabolites can act as "haptens" by binding to proteins and resulting in immune responses. Emerging evidence supports the hypothesis that immune cells such as phagocytes, such as Kupffer and stellate cells, and newly recruited leukocytes play critical roles in liver injury. Although the main function of phagocytes is to remove dead cells and cell debris during liver regeneration, toxic mediators generated in this process can attack stressed hepatocytes (Adams et al., 2010). Hepatic inflammation is thus a common histopathological finding in a wide range of liver diseases. DILI is also frequently associated with lymphocytic infiltration where progression and severity of liver injury is determined by the extent of the inflammation. For example, halothane, an inhalation anesthetic, is a classic example whose reactive metabolites cause immune-allergic hepatitis. Clinical hallmarks of immunological perturbation include time of presentation, general clinical features, and greatly enhanced reaction on reexposure to the drug (Martin et al., 1993).

Steatosis

Steatosis or "fatty liver" is caused by insulin resistance due to obesity (Saadeh, 2007). Histologically, it is characterized by multiple round, empty cytoplasmic vacuoles and excessive fat in the hepatocytes. Steatosis can be induced by acute exposure to many chemicals, e.g., carbon tetrachloride. Often, drug-induced steatosis is reversible and does not lead to cell death. However, it may make the liver more susceptible to other insults such as ethanol consumption (Sato et al., 1981) or it can develop into steatohepatitis (Saito et al., 2007) causing significant liver injury.

REVIEW OF EXISTING BIOMARKERS OF LIVER TOXICITY

The latest version of "FDA guidance for industry DILI: Premarketing clinical evaluation" recommends the use of a combination of four tests as DILI biomarkers (FDA, 2009). Table 14.1 lists these four biomarkers and other common serum biomarkers used in preclinical and clinical screening.

Alanine Aminotransferase

Clinical chemistry data are routinely used for noninvasive monitoring of liver disease in preclinical species and humans, and ALT is the most widely used clinical biomarker. ALT is responsible for the metabolism (transamination) of alanine and is found at much higher concentrations in the liver compared with other organs. When hepatocellular injury occurs, the liver-abundant enzyme ALT will leak into the extracellular space and enter the blood, with a slow clearance rate and a half-life of approximately 42 h (Ozer et al., 2008; Shi et al., 2010). The typical reference range is 7–35 IU/L in females and 10–40 IU/L for males (WebMD, 2017). An elevation of serum ALT activity is often reflective of liver cell damage. Unfortunately, extrahepatic injury, such as muscle injury, can also lead to elevations in ALT, making it not entirely hepatospecific. In addition, fenofibrate was found to increase ALT gene expression in the absence of apparent liver injury (false positive) (Edgar et al., 1998), whereas hepatotoxin microcystin-LR suppressed ALT gene expression (false negative) (Solter et al., 2000). Even though extrahepatic injury, such as muscle damage, can lead to an increase in ALT, serum ALT is the most widely used and universally accepted biomarker for DILI. It is deemed as the clinical chemistry gold standard for DILI detection and has been used at the FDA to facilitate regulatory decision-making. Periodic monitoring of serum ALT is also a common recommendation given to clinical practices to reduce the risks of liver injury when patients are taking drugs with known DILI potential. Present studies have suggested that measuring the ALT isozymes, ALT1 and ALT2, may aid in differentiating the source of injury. ALT1 has been noted to be localized in human hepatocytes, renal tubular epithelial cells, and salivary gland epithelial cells. ALT2, on the other hand, is localized to the human adrenal gland cortex, neuronal cells bodies, cardiac myocytes, skeletal muscle fibers, and the pancreas. Compared to ALT1, ALT2 contributes less to the total serum ALT and is probably a reflection of mitochondrial damage. A novel immunoassay has been developed to discriminate human ALT1 and ALT2 activities, and it may improve the prognosis (Shi et al., 2015).

Aspartate Aminotransferase

Based on the same rationale as ALT, AST is a standard biomarker for DILI and is well accepted by clinicians. AST is responsible for the metabolism (transamination) of aspartate. Even though sensitivity of the AST assay is lower than ALT, presumably because of its expression in extrahepatic organs such as heart and muscle, it is still a widely used liver biomarker. Apparently, the ratio between serum ALT and AST is useful in differentiating DILI from extrahepatic organ injury and diagnosing acute alcoholic hepatitis and cirrhosis when AST/ALT ratio is 2:1.

At least two isoenzymes of AST have been found: a cytosolic AST and a mitochondrial AST (mAST). The relative contributions of cytosolic AST or mAST to serum AST elevation have not been critically assessed. It remains unestablished whether AST isoenzymes are susceptible to drug-driven induction or inhibition (Ozer et al., 2008; Shi et al., 2010).

Alkaline Phosphatase

ALP is located in the liver and its concentration in serum increases when the BDs are blocked. ALP is yet another diagnostic biomarker recommended in the FDA guidelines, and it is widely adopted by clinicians (FDA, 2009; Shi et al., 2010). The Council for International Organizations of Medical Sciences consensus criteria considers a more than twofold isolated elevation of serum ALP, or ALT/ALP ratio of no more than 2, as a key biomarker of cholestatic DILI (Devarbhavi, 2012). However, it should be noted that conditions other than DILI, such as bone disease and pregnancy, also elevate

TABLE 14.1 Summary of Current Clinical Biomarkers of Liver Toxicity

Biomarker	Tissue Localization	Injury	Specific Damage Marker	Comments
ALT	Primarily localized to liver	Elevated in blood because of liver necrosis and with heart and skeletal muscle injury (necrosis)	Hepatocellular necrosis	Commonly used to assess hepatocellular injury
AST	Localized in heart, brain, skeletal muscle, and liver	Elevated in blood because of liver or extra hepatic tissue injury	Hepatocellular Necrosis	Less specific than ALT
TBL	Taken up, conjugated in liver, and secreted into bile	Marker of hepatobiliary injury and liver function; also increased because of hemolysis	Cholestasis, biliary; Liver function	Conventional biliary injury; in conjunction with ALT, better indicator of disease severity in humans
ALP	Broad tissue localization	Marker of hepatobiliary injury	Cholestasis	Conventional biliary injury; associated with drug-induced cholestasis in humans
GGT	Activity localized more to kidney than liver, pancreas	Marker of hepatobiliary injury	Cholestasis, biliary	Conventional biliary injury; high sensitivity in humans, elevation can be caused by alcohol or heart disease
Albumin	Main constituent of serum total protein	Decreased in blood with chronic liver disease	Liver function	Liver fails to synthesize enough protein, especially albumin
Ammonia	Converted to less toxic urea in liver	Elevation in blood to liver injury	Liver function	Failure of the conversion in severe liver disease
Cholesterol/ triglycerides		Increased in blood because of the failure of bile elimination	Liver function	Liver fails to remove them to bile ducts
Clotting time		Increased with severe liver injury	Liver function	Liver fails to produce coagulation factors, increased clotting time; international normalized ratio equivalent to prothrombin time
Urobilinogen		Low level in urine because of biliary obstruction	Liver function	Colorless product of bilirubin reduction, similar role to bilirubin

ALP, alkaline phosphatase; *ALT*, alanine aminotransferase; *AST*, aspartate aminotransferase; *GGT*, gamma-glutamyltransferase; *TBL*, total bilirubin level.
Adapted from Ozer, J., Ratner, M., Shaw, M., et al., 2008. The current state of serum biomarkers of hepatotoxicity. Toxicology 245, 194–205.

ALP. Thus, it should not be regarded as a specific biomarker of cholestatic DILI but rather partially predictive of biliary obstruction during liver injury (Reust and Hall, 2001).

Total Bilirubin

Total bilirubin is a composite of unconjugated (extrahepatic) and conjugated (hepatic) bilirubin. Increased Total bilirubin level or TBL causes jaundice and can indicate metabolism problems in the liver such as reduced hepatocyte uptake, impaired bilirubin conjugation, or reduced bilirubin secretion (Greer, 2014). Therefore, serum bilirubin concentration is a real-time liver function biomarker, which measures the ability of the liver to clear bilirubin from the blood as it circulates through the liver. Although serum transaminase levels indicate the rate of enzyme release from injured cells, the TBL indicates rates of enzyme degradation (Senior, 2006).

Other Existing Hepatotoxicity Biomarkers

Although some liver function tests are not sufficiently sensitive to be used as diagnostic biomarkers of hepatotoxicity, they are elevated in severe liver diseases and are indicative of liver injury. Therefore, these biomarkers are used primarily to confirm liver toxicity and indicate the extent of damage to liver function. Conventional biomarkers falling into this category are gamma-glutamyltransferase (GGT), serum total protein (albumin), ammonia, cholesterol/triglycerides, fibrinogen, prothrombin time, and urobilinogen (Greer, 2014).

Elevated serum GGT activity has a similar profile as ALP in detecting disease of the biliary system. Usually, ALP is the first test for biliary disease followed by GGT for providing a value to verify that the ALP elevations are due to biliary injury. Large quantities of alcohol intake can significantly increase the serum GGT level. Slightly elevated GGT is associated with myocardial infarction and heart failure (Betro and Edwards, 1973).

For both necrotic and obstructive liver diseases, serum total protein electrophoresis patterns are typically abnormal. In the acute stages of hepatitis, albumin is usually low and the γ-globulin fraction is elevated because of a huge increase in antibody production owing to the production of acute phase proteins. In addition, biliary cirrhosis may also cause β-globulin/antibody elevation. Similarly, blood ammonia, normally converted by the liver to less toxic urea, is elevated because of liver dysfunction. However, the liver has a high functional reserve for ammonia conversion, so it often takes a significant amount of injury to alter and detect changes in ammonia levels. Thus, blood ammonia elevations indicate end-stage liver disease and a high risk of hepatic coma.

During hepatic uptake of lipoprotein cholesterol, a portion is enzymatically converted to bile salt. Only hepatocytes have the rate-limiting enzyme, cholesterol 7α-hydroxylase, in the multistep conversion (Berkowitz et al., 1995). In acute hepatic necrosis, triglycerides may be elevated because of hepatic lipase deficiency. When the bile cannot be eliminated, cholesterol and triglycerides accumulate in the blood as low-density cholesterol. Additionally, as the liver is also responsible for production of blood coagulation factors, the clotting time increases because of impaired synthesis in the liver. However, it is not a sensitive biomarker because it only happens in the late stage of liver disease (Burtis et al., 2012).

Another insufficient biomarker is urobilinogen, a colorless product of bilirubin reduction in urine. Conversely, urobilinogen level can also indicate liver dysfunction. Low urine urobilinogen may result from biliary obstruction or complete obstructive jaundice. Because urobilinogen is formed in the intestine by bacteria, broad-spectrum antibiotic treatment can significantly decrease its level because of the damage of intestinal bacterial flora (Greer, 2014).

REVIEW OF EMERGING BIOMARKERS

There is a clear need to find additional biomarkers in serum and/or urine that can be measured in conjunction with ALT or outperform ALT with respect to specificity for liver injury. The omics methods (Table 14.2) are well suited to identify novel biomarkers of hepatotoxicity (Table 14.3) with desired specificity and sensitivity, with the potential to indicate injury earlier than the existing serum biomarkers.

Genetic Biomarkers of Hepatotoxicity

In the past 20 years, many GWAS have been conducted to test the association between genetic polymorphisms and DILI. Human leukocyte antigen (HLA) genotype has been established as a risk factor for DILI, including in drugs where immune-related toxicity was not suspected previously (Daly and Day, 2012). For example, in GWAS using DILI patients versus control subjects, HLA-B*5701 genotype was shown to be a major risk factor and ST6GAL1 as a possible cofactor of the individual vulnerability to flucloxacillin-induced DILI (Daly et al., 2009). As flucloxacillin-induced DILI occurs in 1 in 10,000 patients, it is estimated that screening for HLA-B*5701 would provide a positive predictive value of 0.12% (Alfirevic and Pirmohamed, 2012). At present, GWAS on DILI have focused either on drugs that are very widely prescribed or on those that are newly licensed; therefore, it is unclear if HLA genotype will be the strongest risk factor for DILI linked to a range of drugs. Another concern is that GWAS can reveal genetic determinants of DILI susceptibility, but they do not establish the mechanism of injury or identify individuals at imminent risk.

Genomics Biomarkers of Hepatotoxicity

Over the past decade, the use of a genomics approach to identify patterns of changes in mRNA transcripts, referred as toxicogenomics, has gained popularity for identification of novel DILI biomarkers (Yu et al., 2017). Most studies have employed microarray analysis of the rodent liver to identify unique gene expression profiles as biomarkers and elucidate the molecular mechanisms of DILI drugs. These genomics data sets, including multiple biomarkers, can be used to generate gene expression signatures associated with different drugs that cause liver injury. However, the use of liver tissue–based genomic biomarkers is not optimal because biomarkers requiring liver biopsy have limited value in a clinical setting, especially for monitoring DILI progress in patients. Thus, genomic biomarkers from a minimally invasive source, such as blood, have

TABLE 14.2 Common Platforms Used for Omics Biomarkers Discovery and Technical Challenges Associated With These Platforms

	Technological Platforms	Technical Challenges
Genetics	Genome Wide Association Study (GWAS)	• Weak associations of genetics biomarkers • Multiple polymorphic genes contribute to drug induced liver injury (DILI) • Pretreatment screening not cost-effective
Transcriptomics	Microarray	• Variability in sample storage and RNA isolation
	Quantitative real-time polymerase chain reaction (QRT-PCR)	• Lick of standard normalization method for circulating miRNAs
	Next-generation sequencing (NGS)	• Roles of miRNAs remain unclear
Proteomics	Matrix-assisted laser desorption/ionization tandem mass spectrometry (MALDI-MS/MS) Sodium dodecyl sulfate polyacrylamide gel electrophoresis mass Spectrometry (Gels-MS)	• Protein expression in constant flux • Large diversity and heterogeneity of proteome • Difficult to automate • Wide dynamic range of cellular protein expression • Limited dynamic range and detection of MS platforms • Relatively low throughput
Metabolomics	Nuclear magnetic resonance (NMR) Gas Chromatography/Mass Spectrometry (GC/MS) Liquid chromatography/mass spectrometry (LC/MS) Fourier transform infrared spectroscopy (FT-IR)	• Metabolites altered by diet and environment • Large diversity in chemical and physical properties • Wide range of chemical species requires multiple analytical platforms • Concentrations range over nine orders of magnitude

Adapted from Yang et al. (2012).

been explored extensively in preclinical animals and humans. Interestingly, applying blood transcriptomic signatures generated from rodents to human blood data enabled differentiation of APAP overdose patients from healthy controls (Bushel et al., 2007). This study indicated that alterations in genes involved in an inflammatory response were the best discriminators between subtoxic/nontoxic and toxic exposure to APAP. In another study, a downregulation of mitochondrial genes involved in complex I of the oxidative phosphorylation pathway was seen in human blood samples after APAP exposure (Fannin et al., 2010). A comprehensive approach to demonstrate that transcriptomic signatures extracted from blood can predict liver injury caused by a wide variety of hepatotoxicants has been published (Huang et al., 2010).

Recently, next-generation sequencing (NGS) technologies have become available for fast, inexpensive sequencing of whole static genomes and dynamic transcriptomes (Goodwin et al., 2016). Although the microarray platform monitors the expression levels of most annotated genes within the cell, the powerful and rapidly evolving NGS technology allows for precise quantification of gene expression including transcripts that have not been sequenced previously. Initially, there was concern regarding the comparability of NGS to microarray platforms in terms of gene expression and biological variability in a real-life toxicological study design. The Microarray Quality Control group evaluated the robustness of NGS for detecting differentially expressed genes and reported that both NGS and microarrays generate consistent biological interpretation (Su et al., 2011). However, the NGS data were fundamentally different from the microarray data, and translating these short sequences to genomic biomarkers must overcome several obstacles. One of the major challenges lies in the handling of immense volumes of NGS sequence data. Powerful bioinformatics tools are needed to assure sequence quality, conduct sequence alignment to the relevant genome and transcriptome, and provide biological interpretation from complex datasets. Over the past years, NGS has provided a more comprehensive understanding of complex transcriptomes, with turnaround time and cost comparable with those of microarrays (Goodwin et al., 2016). NGS technologies are anticipated to accelerate toxicogenomics research and play a pivotal role in identifying new DILI biomarkers.

Proteomic Biomarkers of Hepatotoxicity

Proteomics evaluates the differential protein expression between groups to determine novel protein biomarkers in toxicity and disease. Analyses involve a separation step, either gel based or gel free, followed by tandem mass spectrometry to identify proteins. To date, proteomics has successfully discovered several promising biomarkers of acute kidney injury (AKI), e.g., neutrophil gelatinase-associated lipocalin (NGAL),

TABLE 14.3 Summary of Emerging Biomarkers of Liver Toxicity

Biomarker Candidate	Biofluid Evaluated	Origin	Proposed Indication	Reference(s)
Cytokines	Plasma	Produced by all liver cells but primarily Kupffer cells	Cellular stress in the liver	Andrés et al. (2003), Ding et al. (2003a,b), Lacour et al. (2005), Steuerwald et al. (2013)
Interleukin-1	Plasma	Produced by a variety of cells	Cellular response to tissue damage	Akbay et al. (1999), Lacour et al. (2005)
Glutathione S-transferase P-form	Serum	Present in the hepatocytes	Hepatocellular injury	Fella et al. (2005), Glückmann et al. (2007)
Cytokeratin-18	Serum	Expressed by epithelial cells	Apoptosis or necrosis	Robles-Diaz et al. (2016)
High mobility group box protein I	Serum	Found in a wide range of tissues	Necrosis and inflammation	Robles-Diaz et al. (2016)
Glutamate dehydrogenase (GLDH)	Serum	Primarily found in the liver and to a lesser degree in the kidney and skeletal muscle	Hepatocellular necrosis	Robles-Diaz et al. (2016)
Malate dehydrogenase	Serum	Localized in mitochondria and extramitochondrial compartment; found primarily in liver but also in skeletal muscle, heart, and brain	Hepatocellular necrosis	Robles-Diaz et al. (2016)
Purine nucleoside phosphorylase	Serum	Primarily in the liver but also present in heart muscle and brain; mainly in cytoplasm of endothelial cells, Kupffer cells, and hepatocytes	Hepatocellular necrosis	Ozer et al. (2008), Robles-Diaz et al. (2016)
Paraoxanase 1	Serum	Produced primarily in the liver but also found in kidney, brain, and lung	Hepatocellular necrosis	Ozer et al. (2008), Robles-Diaz et al. (2016)
Glutathione S-transferase alpha	Serum	Liver specific	Hepatocellular necrosis	Ozer et al. (2008)
Apolipoprotein E	Serum	Produced in the liver but also found in brain and kidney	Hepatocellular necrosis	Robles-Diaz et al. (2016)
Bile acids	Urine, Serum	Synthesized primarily in the liver	Liver dysfunction including intrahepatic cholestasis	Xiong et al. (2014)
Ophthalmic acid	Serum	Analog of glutathione produced along a similar biosynthetic route as glutathione	Oxidative stress and glutathione depletion following hepatotoxic insult	Dello et al. (2013), Xiong et al. (2014)
5-Oxoproline	Urine, Serum	Intermediate in the synthesis of glutathione	Oxidative stress and glutathione status	Kumar et al. (2010), Xiong et al. (2014)
Steroids	Urine	Metabolites of cholesterol	Oxidative stress and liver damage	Kumar et al. (2012), Xiong et al. (2014)
Acylcarnitines	Urine, Serum	Located in heart, muscle, brain, liver, and kidney	Failure of fatty acid oxidation	Chen et al. (2009), Zhang et al. (2010)
Fatty acids	Serum	Nontissue specific	Disruption of fatty acid β-oxidation	Zhang et al. (2010)
miRNA-122	Plasma/serum	Liver-specific expression	Viral-, alcohol-, and chemical-induced liver injury; hepatocarcinoma	Ding et al. (2012), McGill and Jaeschke (2015), Robles-Diaz et al. (2016)
miRNA-192	Plasma/serum	Liver-specific expression	Chemical-induced liver injury	McGill and Jaeschke (2015)
miR-291a-5p	Urine	Unknown	Chemical-induced liver injury	McGill and Jaeschke (2015)

Adapted from Robles-Diaz, M., Medina-Caliz, I., Stephens, C., et al., 2016. Biomarkers in DILI: one more step forward. Front. Pharmacol. 7, 267 and Ozer, J., Ratner, M., Shaw, M., et al., 2008. The current state of serum biomarkers of hepatotoxicity. Toxicology 245, 194–205.

cystatin C (Cys C), kidney injury molecule-1 (KIM-1), liver-type fatty acids binding protein (L-FABP), and interleukin-18 (IL-18) (Gobe et al., 2015). These markers show promise as diagnostic markers and are being validated. The successful application of proteomics to determine more accurate biomarkers of AKI indicates the potential to also provide novel biomarkers of hepatotoxicity. There have been multiple literature reports of potential classes of protein biomarkers of hepatotoxicity.

One class of protein biomarkers includes the cytokines. The cytokines are generally associated with inflammation, immune reactivity, tissue injury or repair, and organ dysfunction and include interleukins, growth factors, interferons, tumor necrosis factors, and chemokines. All liver cells are capable of producing cytokines and the plasma levels may be indicative of cellular response (Laverty et al., 2010). Their production may be related to an initial toxic injury that activates Kupffer cells, a major source of cytokine production in the liver. The cytokine networks have been shown to mediate the hepatic response to diverse xenobiotics including APAP and PPAR ligands. However, the increase in cytokines is transient and declines rapidly, making it necessary to establish a time line over which they should be evaluated following dosing. IL-1 has been proposed to be a biomarker for liver toxicity (Lacour et al., 2005). IL-1 is a proinflammatory cytokine that can induce apoptosis, proliferation, and inflammatory processes. The activation of such proinflammatory cytokines can ultimately lead to activation of other processes, leading to generalized liver damage. The cellular stress response pathways are also activated in response to a toxic stimulus and either mount a homeostatic response or make cell fate decisions (Laverty et al., 2010; Simmons et al., 2009).

Amacher (2010) summarizes reported proteomic biomarkers associated with cellular stress response or toxicity pathways. Potential protein biomarkers of cellular stress response include proteins of the annexin family, anabolic and catabolic functions, and drug metabolism. The annexin family proteins act as calcium sensors and promote plasma membrane repair (Draeger et al., 2011). Additionally, carbonic anhydrase III, aflatoxin B1 aldehyde reductase, and GST-P, which play important roles in hepatocarcinogenicity, are reported in a study of chemically induced hepatocarcinogenesis (Fella et al., 2005).

Other potential protein markers of hepatocellular stress include the keratins and high mobility group box protein 1 (HMGB-1). Keratins are responsible for cell structure and integrity; the blood level of cytokeratin-18 (CK18) has been used to monitor apoptosis and necrosis (Cummings et al., 2008). HMGB-1 is proposed as a marker of inflammation as well as necrosis (Robles-Diaz et al., 2016). Similar to ALT and AST, these biomarkers are present in serum once hepatocytes have been damaged, thus resulting in leakage of the target protein into circulation (Ozer et al., 2008). However, all of these biomarkers are at early stages of development, and it remains to be determined whether they will be more selective than ALT and AST.

These emerging serum biomarkers include glutamate dehydrogenase (GLDH), purine nucleoside phosphorylase (PNP), malate dehydrogenase (MDH), paraoxonase 1 (PON1), sorbitol dehydrogenase, serum F protein, glutathione-S-transferase alpha (GSTα), and arginase I (Ozer et al., 2008). GLDH is a mitochondrial enzyme found primarily in the centrilobular region of the liver. GLDH plays a role in amino acid oxidation and serum activity increases with hepatocellular injury. GLDH activity is more liver specific than ALT and AST. In several recent studies, elevations of GLDH, similar to those observed with ALT, were reported in APAP-overdosed patients and patients receiving heparin. MDH catalyzes the reversible conversion of malate into oxaloacetate and is a constitutive enzyme in the citric acid cycle. MDH is a periportal enzyme whose release into the serum indicates tissue damage. Both GLDH and MDH levels are consistent in healthy populations, and they are strongly associated with elevated ALT in a broad range of liver injuries (Robles-Diaz et al., 2016).

PNP is an enzyme involved in purine metabolism and it is primarily located in the cytoplasm of endothelial cells, Kupffer cells, and hepatocytes. Several animal studies reported that PNP is released into hepatic sinusoids during liver damage in rodents. However, the value of PNP as a human liver injury biomarker has not been confirmed. PON1 is primarily produced in the liver and released into the circulation bound to high-density lipoproteins. Unlike other serum markers that indicate leakage, a decrease in PON1 is noted in serum after liver damage. Therefore, it is likely that liver damage reduces PON1 synthesis and secretion. PON1 is decreased after dosing with phenobarbital and APAP and in humans with chronic liver disease (Ozer et al., 2008). In a separate study evaluating the baseline level of emerging serum biomarkers, PON1 levels were found to be higher in African Americans compared with Caucasians (Robles-Diaz et al., 2016).

Although sorbitol dehydrogenase (SDH) is primarily located in the cytoplasm and mitochondria of liver, kidney, and seminal vesicles and is a marker of acute hepatocellular injury in rodents, serum F protein is a sensitive and specific marker of liver damage with a strong correlation to histopathology in humans. GSTα expression is restricted to liver and kidney and may serve as a region-specific marker of liver injury with a high concentration in centrilobular cells (Ozer et al., 2008).

An additional protein, serum arginase I, is highly liver specific and has been shown to have strong correlations to AST and ALT activities (Ozer et al., 2008). Finally, apolipoprotein E (APOE) expression has been linked to hepatotoxicity in two separate studies. APOE is protein synthesized and secreted by hepatocytes with its major function being transport and distribution of lipids and cholesterol to cells. Therefore, it has a major role in determining the metabolic fate of lipids and proteins.

A hepatotoxicity protein biomarker panel based on a targeted proteomics approach has been developed for use in pharmaceutical toxicology assessment (Collins et al., 2012). The selected reaction monitoring (SRM) proteomics panel was developed in part based on the hepatotoxicity biomarkers reported within the review by Amacher (2010). In total, the panel contained 48 biomarker candidates from multiple pathways including, but not limited to, xenobiotic metabolism, PPARα/PXR activation, fatty acid metabolism, and oxidative stress. This proof of principle study demonstrated the ability to develop a robust, high-throughput, customizable SRM assay to evaluate putative protein markers of hepatotoxicity.

Metabolomics Biomarkers in Hepatotoxicity

Metabolomics involves the measurement of the metabolite pool that exists within a cell or tissue under a particular set of cellular conditions. The metabolic profile is greatly influenced by both genetic and environmental factors, thereby providing phenotype-specific data that can be evaluated in a longitudinal manner. Metabolomics analyses focus on the discovery of novel, clinically relevant biomarkers in easily obtained biofluids such as urine and serum. The major analytical platforms for metabolomics include nuclear magnetic resonance (NMR) spectroscopy and liquid chromatography (LC)–mass spectrometry (MS). As hepatotoxicity is the major cause for drug-related adverse events, metabolomics has been employed in multiple preclinical studies to identify more selective markers of DILI.

Metabolites from several major pathways, e.g., bile acids (BAs), are reported in multiple studies. Synthesis, secretion, and recycling of BAs are critical functions of the liver. The BAs are important signaling molecules in liver and intestine, and accumulation of BAs due to impaired secretion or biliary obstruction can cause hepatocyte damage. Multiple studies of hepatotoxicity have noted altered profiles of BAs. Because galactosamine is a known hepatotoxin, a targeted LC-MS method was employed to investigate the metabolic effects of galactosamine by rapidly profiling BAs. Galactosamine administration significantly increased taurine- and glycine-conjugated BAs in rat serum (Want et al., 2010).

Similarly, other hepatotoxicants also alter BA profiles and have shown distinct mechanism-based patterns of changes (Coen, 2010). APAP, carbon tetrachloride (CCl4), and α-naphthylisothiocyanate (ANIT) are classical hepatotoxicants that induce altered BA profiles after a toxic dose. APAP and CCl4 increased urinary cholic acid and lithocholic acid (Kumar et al., 2012). CCl4 and ANIT induced liver failure and elevated the serum levels of BAs compared to control (Yang et al., 2008). These compounds cause distinct types of liver injury and showed different patterns of altered BAs.

Isoniazid, which is used in the prevention and treatment of tuberculosis, is also hepatotoxic, and a study of serum metabolome in wildtype and Cyp2e1-null mice indicated that accumulation of BAs and free fatty acids may play a role in isoniazid-induced hepatotoxicity (Cheng et al., 2013). In a study of APAP-induced toxicity, increases in serum BAs concurrent with a decrease in the expression of BA synthesis-related genes (Cyp7a1 and Cyp8b1) and cholesterol transporter gene (Abcg8) and increases in the expression of BA transporter genes (Mrp3 and Mrp4) suggested a mild form of intrahepatic cholestasis (Sun et al., 2013). The aforementioned results indicate that BAs may be a sensitive marker of DILI. Additionally, the specific BAs may be able to differentiate specific types of liver injury.

Hepatotoxicants that form reactive metabolites generally induce an oxidative stress response ultimately resulting in cell death. GSH is a protective molecule that is consumed as it scavenges ROS to prevent cellular damage. GSH production proceeds through the folate-dependent transmethylation and transsulfuration pathways and thus, intermediaries in these pathways may serve as biomarkers of hepatotoxicants-induced oxidative stress. One such molecule is ophthalmic acid (OA), an analog of GSH in which the cysteine residue is replaced by 2-aminobutyrate. The synthesis of OA is catalyzed by the same enzymes that synthesize GSH, γ-glutamylcysteine synthetase (Gcl), and glutathione synthetase. The depletion of GSH may activate Gcl and correspondingly increase the synthesis of OA.

In a study of APAP hepatotoxicity in a mouse model, serum OA was linked to APAP-induced oxidative stress and GSH depletion. GSH depletion occurred with a concomitant increase in OA suggesting that the OA biosynthetic pathway had been upregulated (Dello et al., 2013). However, more studies are needed to fully understand the relationship between OA and the multiple stages of liver damage. Nevertheless, the multiple biosynthetic pathways involved in the biosynthesis of GSH and OA make it difficult to clearly understand the relationship between the two molecules and the oxidative stress response. 5-Oxoproline (5-OP; pyroglutamate) is an intermediate in the GSH biosynthesis pathway, and it may be more directly related to GSH content and cell status than OA. Hepatotoxicants such as APAP (Emmett, 2014), bromobenzene (Waters et al.,

2006), and ethionine (Skordi et al., 2007) induce elevations of 5-OP in biofluids and tissues in animal studies. These drugs are known to induce oxidative stress and deplete GSH. Increased 5-OP is seen in humans with inborn errors of metabolism that affect GSH synthesis, indicating its direct relation with GSH (Tokatli et al., 2007). Chronic use of APAP can also elevate 5-OP because of metabolic acidosis (Duewall et al., 2010). 5-OP is increased in human liver epithelial cells following exposure to APAP, whereas cellular GSH content is decreased (Dello et al., 2013). GSH consumption also occurs in hydrazine-induced hepatotoxicity, and 5-OP is increased in a dose-dependent manner in urine, plasma, and liver tissue (Bando et al., 2011). Based on available literature, 5-OP appears to be directly coupled to GSH depletion and, therefore, may be more reflective of GSH status than OA.

Kumar et al. (2012) employed a global profiling method for the initial discovery of potential urinary biomarkers of APAP, CCl4, and methotrexate-induced hepatotoxicity. BAs, steroids, and amino acids were selected as liver toxicity biomarkers. In the hydrazine-induced hepatotoxicity study, multiple amino acids were elevated in urine and plasma (Bando et al., 2011). These included the amino acid precursors of GSH, cysteine, glutamine, and glycine. The amino acid changes were associated with hydrazine-induced fatty acid degeneration and glycogen depletion in the liver. Amino acid metabolism was also altered in response to bromobenzene- and galactosamine-induced hepatic necrosis (Gonzalez et al., 2012). A metabolic profiling study of changes in rat urine related to dosing with the compound atorvastatin identified estrone, cortisone, proline, cystine, 3-ureidopropionic acid, and histidine as markers of liver toxicity (Kumar et al., 2012). Metabolic profiling identified 3-hydroxy-20-deoxyguanosine and octanoylcarnitine as urinary markers of valproic acid–induced hepatotoxicity in rats (Lee et al., 2009). APAP is known to inhibit fatty acid beta-oxidation; therefore, Chen et al. (2009) evaluated metabolites specifically related to this pathway in mouse serum. The results were consistent with a disruption of fatty acid β-oxidation with an accumulation of long-chain acylcarnitines and free fatty acids in serum. The pattern of accumulation of acylcarnitines in the above study indicated that they might be useful as complementary biomarkers for monitoring APAP-induced hepatotoxicity or other compounds that disrupt fatty acid β-oxidation. Other studies of APAP hepatotoxicity have also reported altered fatty acids and acylcarnitines (Bi et al., 2013; Xiong et al., 2014). Multiple biological matrices were evaluated by NMR after exposure of rats to aflatoxin-B1 (AFB1) (Zhang et al., 2010). Significant elevations in amino acids in plasma and liver tissue indicated that AFB1 altered protein biosynthesis. Hence, BAs, 5-OP, and OA, fatty acids, amino acids, and steroids are promising potential markers of hepatotoxicity that warrant further investigation.

A subset of metabolomics referred to as metabolomics flux analysis evaluates real-time synthesis and turnover rates in specific pathways through use of a ^{13}C-labeled precursor. Stable isotope ^{13}C-labeled glucose was used to investigate the toxic effects of valproic acid (VPA) on the plasma, urine, liver, brain, and kidney metabolites (Beger et al., 2009). Results indicated that VPA disrupted the flux of acetate and its disposal via plasma cholesterol, causing liver toxicity. Usnic acid, a dietary supplement promoted for weight loss, was shown to be cytotoxic to rat primary hepatocytes in a time- and concentration-dependent manner and isotopomer distributions from flux analysis indicated a reduction in oxidative phosphorylation and gluconeogenesis at the high dose. The results from flux analysis of ^{13}C-labeled glucose were consistent with cytotoxicity and ATP depletion in the cells. Lately, Fan et al. (2012) reviewed stable isotope-resolved metabolomics and potential clinical applications.

MicroRNAs as Biomarkers of Hepatotoxicity

MicroRNAs (miRNAs) are ~22 nucleotides long, single-stranded, noncoding RNAs that have recently been recognized as novel agents exercising posttranscriptional control over most eukaryotic genomes. miRNAs are highly conserved among species, ranging from worms to humans, revealing their very ancient hierarchy. The human genome has been predicted to encode over 2000 miRNAs, which are usually produced in a cell- or tissue-specific manner and are predicted to regulate the activity of ~60% of human genes (Friedman et al., 2009). Owing to their minimally invasive nature and unique stability in different body fluids, miRNAs in biofluids hold a unique position for use as a preclinical and clinical DILI biomarker (Weber et al., 2010). Using an APAP-induced mouse model of DILI, Wang et al. (2009) reported that the level of many plasma miRNAs inversely correlated with hepatic miRNAs, indicating that hepatic injury caused the release of miRNAs into the circulation. Specifically, miRNA-122 and miRNA-192, which are predominantly expressed in the liver, increased in the plasma with concurrent decrease in the liver. The increases in both miRNAs were detected earlier than the increase of ALT. The increase of serum miRNA-122 and miRNA-192 was confirmed later in patients with APAP poisoning (Starkey Lewis et al., 2011). The change in miRNA-122 was significantly smaller in serum from patients with liver injury due to hepatitis or hepatocarcinoma (Ding et al., 2012). Although the functions of the 10 common miRNAs remain unknown, the possible target genes of these miRNAs are related to cell death, cell-to-cell signaling, and major metabolic pathways.

BIOMARKER QUALIFICATION AND VALIDATION

Applying omics-based technologies in rodents treated with a variety of hepatotoxic drugs, together with a better understanding of the hepatotoxicity mechanisms, may facilitate the identification of novel DILI biomarkers. However, it is worthwhile to point out that none of the new DILI biomarkers has been qualified for preclinical or clinical use from a regulatory perspective. Many of the potential biomarkers discussed in this chapter are not necessarily liver specific and could be a result of other tissue injury. True qualification of new biomarkers will ultimately require large numbers of samples from animals and patients treated with many different drugs.

A stringent qualification process is required to validate their specificity and sensitivity for DILI before they can supplement or replace existing biomarkers. Qualification has been described as "the process of linking a biomarker to a preclinical or clinical end point or to a biological process in a specific context" (Wagner, 2008). FDA and ICH have issued guidance on biomarker qualification and the content of data submissions. These guidance documents provide a foundation for qualifying a biomarker for a given context of use, such as preclinical versus clinical. Although specific testing and qualification plans are not provided in the guidance documents, they do highlight the need for robust data to develop a new biomarker (Food and Administration, 2011).

Nevertheless, a biomarker will require a clearly defined context of use as well as sufficient data to support a full review of its performance characteristics within that context. It is required that the biomarker be measured reliably on multiple analytical platforms. Finally, to successfully translate a biomarker from the preclinical to the clinical setting, the marker must be qualified for its intended use and should correlate with lesions observed by histopathology. As part of the qualification process, it will be necessary to quantify a marker and provide a range of normal values in a control state. Metabolomics biomarkers can be quantified by mass spectrometry based on a calibration curve or comparison of intensities after spiking a sample with an isotope-labeled standard. NMR metabolite data are quantified based on the concentration of an internal chemical shift standard. Protein biomarkers can also be quantified by mass spectrometry methods that include a labeling procedure and tandem MS analysis. A Bradford assay can be used for quantification of protein. Gene or miRNA biomarkers can be quantified using real-time polymerase chain reaction (PCR), microarray, and NGS platforms based on fluorescence signals.

CONCLUDING REMARKS AND FUTURE DIRECTIONS

DILI is a major cause for limiting the use of a drug or its removal from the market. At present, the current DILI biomarkers sometimes fail to identify a toxic compound in the preclinical development stages and even in clinical trials. Among the existing biomarkers, ALT and AST are general indicators of hepatocellular injury; ALP is reflective of the cholestatic DILI and elevated TBL is associated with increased DILI severity. The omics technologies are well suited to identify novel biomarkers of DILI that can be measured in easily obtained biofluids. Genomics, proteomics, and metabolomics methodologies have produced many candidate DILI biomarkers for future investigations. In many cases, the same markers have been noted in multiple studies with overlapping cellular pathways such as GSH depletion as an initial response to a cellular stress. The measurement of circulating miRNAs has appeared to be promising in identifying new biomarkers of liver injury. Further studies are needed to evaluate the sensitivity and specificity of the emerging biomarkers as well as to validate the omics biomarkers to develop tests that are clinically useful and cost-effective.

References

Abebe, W., 2002. Herbal medication: potential for adverse interactions with analgesic drugs. J. Clin. Pharm. Therapeut. 27, 391–401.
Adams, D.H., Ju, C., Ramaiah, S.K., et al., 2010. Mechanisms of immune-mediated liver injury. Toxicol. Sci. 115, 307–321.
Ahmed, S., Zhou, Z., Zhou, J., et al., 2016. Pharmacogenomics of drug metabolizing enzymes and transporters: relevance to precision medicine. Genomics Proteomics Bioinformatics 14, 298–313.
Akbay, A., Cinar, K., Uzunalimoglu, Ö., et al., 1999. Serum cytotoxin and oxidant stress markers in *N*-acetylcysteine treated thioacetamide hepatotoxicity of rats. Hum. Exp. Toxicol. 18, 669–676.
Alfirevic, A., Pirmohamed, M., 2012. Predictive genetic testing for drug-induced liver injury: considerations of clinical utility. Clin. Pharmacol. Ther. 92, 376–380.
Amacher, D.E., 2010. The discovery and development of proteomic safety biomarkers for the detection of drug-induced liver toxicity. Toxicol. Appl. Pharmacol. 245, 134–142.
Andrés, D., Sánchez-Reus, I., Bautista, M., Cascales, M., 2003. Depletion of Kupffer cell function by gadolinium chloride attenuates thioacetamide-induced hepatotoxicity: expression of metallothionein and HSP70. Biochem. Pharmacol. 66, 917–926.
Bando, K., Kunimatsu, T., Sakai, J., et al., 2011. GC-MS-based metabolomics reveals mechanism of action for hydrazine induced hepatotoxicity in rats. J. Appl. Toxicol. 31, 524–535.

Beger, R.D., Hansen, D.K., Schnackenberg, L.K., et al., 2009. Single valproic acid treatment inhibits glycogen and RNA ribose turnover while disrupting glucose-derived cholesterol synthesis in liver as revealed by the [U-C(6)]-d-glucose tracer in mice. Metabolomics 5, 336–345.

Berkowitz, C.M., Shen, C.S., Bilir, B.M., et al., 1995. Different hepatocytes express the cholesterol 7α-hydroxylase gene during its circadian modulation in vivo. Hepatology 21, 1658–1667.

Bessems, J.G., Vermeulen, N.P., 2001. Paracetamol (acetaminophen)-induced toxicity: molecular and biochemical mechanisms, analogues and protective approaches. Crit. Rev. Toxicol. 31, 55–138.

Betro, M., Edwards, J., 1973. Gamma-glutamyl transpeptidase and other liver function tests in myocardial infarction and heart failure. Am. J. Clin. Pathol. 60, 679–683.

Bi, H., Li, F., Krausz, K.W., et al., 2013. Targeted metabolomics of serum acylcarnitines evaluates hepatoprotective effect of Wuzhi tablet (*Schisandra sphenanthera* extract) against acute acetaminophen toxicity. Evid. Based Complement Alternat. Med. 2013, 985257.

Burtis, C.A., Ashwood, E.R., Bruns, D.E. (Eds.), 2012. Tietz Textbook of Clinical Chemistry and Molecular Diagnostics, fifth ed. Saunders, Philadelphia.

Bushel, P., Heinloth, A., Li, J., et al., 2007. Blood gene expression signatures predict exposure levels. Proc. Natl. Acad. Sci. U. S. A. 104, 18211–18216.

Chen, C., Krausz, K.W., Shah, Y.M., et al., 2009. Serum metabolomics reveals irreversible inhibition of fatty acid β-oxidation through the suppression of PPARα activation as a contributing mechanism of acetaminophen-induced hepatotoxicity. Chem. Res. Toxicol. 22, 699–707.

Cheng, J., Krausz, K.W., Li, F., et al., 2013. CYP2E1-dependent elevation of serum cholesterol, triglycerides, and hepatic bile acids by isoniazid. Toxicol. Appl. Pharmacol. 266, 245–253.

Chopra, S., 2001. The Liver Book: A Comprehensive Guide to Diagnosis, Treatment, and Recovery. Simon and Schuster, New York.

Ciejka, M., Nguyen, K., Bluth, M.H., et al., 2016. Drug toxicities of common analgesic medications in the emergency department. Clin. Lab. Med. 36, 761–776.

Coen, M., 2010. A metabonomic approach for mechanistic exploration of pre-clinical toxicology. Toxicology 278, 326–340.

Cohen, R.S., Wong, R.J., Stevenson, D.K., 2010. Understanding neonatal jaundice: a perspective on causation. Pediatr. Neonatol. 51, 143–148.

Collins, B.C., Miller, C.A., Sposny, A., et al., 2012. Development of a pharmaceutical hepatotoxicity biomarker panel using a discovery to targeted proteomics approach. Mol. Cell. Proteomics 11, 394–410.

Cummings, J., Ward, T.H., Greystoke, A., et al., 2008. Biomarker method validation in anticancer drug development. Br. J. Pharmacol. 153, 646–656.

Daly, A.K., Day, C.P., 2012. Genetic association studies in drug-induced liver injury. Drug Metab. Rev. 44, 116–126.

Daly, A.K., Donaldson, P.T., Bhatnagar, P., et al., 2009. HLA-B* 5701 genotype is a major determinant of drug-induced liver injury due to flucloxacillin. Nat. Genet. 41, 816–819.

Danielson, P.B., 2002. The cytochrome P450 superfamily: biochemistry, evolution and drug metabolism in humans. Curr. Drug Metabol. 3, 561–597.

DeGorter, M.K., Xia, C.Q., Yang, J.J., et al., 2012. Drug transporters in drug efficacy and toxicity. Annu. Rev. Pharmacol. Toxicol. 52, 249–273.

Dejong, C.H., van de Poll, M.C., Soeters, P.B., et al., 2007. Aromatic amino acid metabolism during liver failure. J. Nutr. 137, 1579S–1585S.

Dello, S.A., Neis, E.P., de Jong, M.C., et al., 2013. Systematic review of ophthalmate as a novel biomarker of hepatic glutathione depletion. Clin. Nutr. 32, 325–330.

Devarbhavi, H., 2012. An update on drug-induced liver injury. J. Clin. Exp. Hepatol. 2, 247–259.

Ding, H., Huang, J.A., Tong, J., et al., 2003a. Influence of Kupffer cells on hepatic signal transduction as demonstrated by second messengers and nuclear transcription factors. World J. Gastroenterol. 9, 2519–2522.

Ding, H., Peng, R., Reed, E., et al., 2003b. Effects of Kupffer cell inhibition on liver function and hepatocellular activity in mice. Int. J. Mol. Med. 12, 549–557.

Ding, X., Ding, J., Ning, J., et al., 2012. Circulating microRNA-122 as a potential biomarker for liver injury. Mol. Med. Rep. 5, 1428–1432.

Draeger, A., Monastyrskaya, K., Babiychuk, E.B., 2011. Plasma membrane repair and cellular damage control: the annexin survival kit. Biochem. Pharmacol. 81, 703–712.

Duewall, J.L., Fenves, A.Z., Richey, D.S., et al., 2010. 5-Oxoproline (pyroglutamic) acidosis associated with chronic acetaminophen use. Proc. (Bayl. Univ. Med. Cent.) 23, 19–20.

Edgar, A.D., Tomkiewicz, C., Costet, P., et al., 1998. Fenofibrate modifies transaminase gene expression via a peroxisome proliferator activated receptor α-dependent pathway. Toxicol. Lett. 98, 13–23.

Emmett, M., 2014. Acetaminophen toxicity and 5-oxoproline (pyroglutamic acid): a tale of two cycles, one an ATP-depleting futile cycle and the other a useful cycle. Clin. J. Am. Soc. Nephrol. 9, 191–200.

Fan, T.W.M., Lorkiewicz, P.K., Sellers, K., et al., 2012. Stable isotope-resolved metabolomics and applications for drug development. Pharmacol. Ther. 133, 366–391.

Fannin, R.D., Russo, M., O'connell, T.M., et al., 2010. Acetaminophen dosing of humans results in blood transcriptome and metabolome changes consistent with impaired oxidative phosphorylation. Hepatology 51, 227–236.

Farrugia, A., 2010. Albumin usage in clinical medicine: tradition or therapeutic? Transfus. Med. Rev. 24, 53–63.

FDA, 2009. FDA Guidance for Industry. Drug-induced Liver Injury. Premarketing Clinical Evaluation. https://www.fda.gov/downloads/Drugs/GuidanceComplianceRegulatoryInformation/Guidances/UCM174090.pdf.

Fella, K., Glückmann, M., Hellmann, J., et al., 2005. Use of two-dimensional gel electrophoresis in predictive toxicology: identification of potential early protein biomarkers in chemically induced hepatocarcinogenesis. Proteomics 5, 1914–1927.

FDA, 2011. Guidance for Industry, E16: biomarkers related to drug or biotechnology product development: context, structure, and format of qualification submissions. https://www.fda.gov/downloads/drugs/guidancecomplianceregulatoryinformation/guidances/ucm267449.pdf.

Friedman, R.C., Farh, K.K.H., Burge, C.B., et al., 2009. Most mammalian mRNAs are conserved targets of microRNAs. Genome Res. 19, 92–105.

Glückmann, M., Fella, K., Waidelich, D., et al., 2007. Prevalidation of potential protein biomarkers in toxicology using iTRAQ™ reagent technology. Proteomics 7, 1564–1574.

Gobe, G.C., Coombes, J.S., Fassett, R.G., et al., 2015. Biomarkers of drug-induced acute kidney injury in the adult. Expert Opin. Drug Metabol. Toxicol. 11, 1683–1694.

Gonzalez, E., van Liempd, S., Conde-Vancells, J., et al., 2012. Serum UPLC-MS/MS metabolic profiling in an experimental model for acute-liver injury reveals potential biomarkers for hepatotoxicity. Metabolomics 8, 997–1011.

Goodwin, S., McPherson, J.D., McCombie, W.R., 2016. Coming of age: ten years of next-generation sequencing technologies. Nat. Rev. Genet. 17, 333–351.

Greer, J.P., 2014. Wintrobe's Clinical Hematology. Wolters Kluwer Health Adis (ESP).

Gupta, R.C., Srivastava, A., Lall, R., 2018. Toxicity potential of nutraceuticals. In: Nicolotti, O. (Ed.), Computational Toxicology: Methods and Protocol. Springer, Heidelberg, pp. 367–394.

Gwaltney-Brant, S.M., 2016. Nutraceuticals in hepatic diseases. In: Gupta, R.C. (Ed.), Nutraceuticals: Efficacy, Safety and Toxicity. Academic Press/Elsevier, Amsterdam, pp. 87–99.

Hinson, J.A., Roberts, D.W., James, L.P., 2010. Mechanisms of acetaminophen-induced liver necrosis. Handb. Exp. Pharmacol. 196, 369–405.

Huang, J., Shi, W., Zhang, J., et al., 2010. Genomic indicators in the blood predict drug-induced liver injury. Pharmacogenomics J. 10, 267–277.

Jaeschke, H., Bajt, M.L., 2005. Intracellular signaling mechanisms of acetaminophen-induced liver cell death. Toxicol. Sci. 89, 31–41.

James, L.P., Mayeux, P.R., Hinson, J.A., 2003. Acetaminophen-induced hepatotoxicity. Drug Metab. Dispos. 31, 1499–1506.

Kaplowitz, N., 2005. Idiosyncratic drug hepatotoxicity. Nat. Rev. Drug Discov. 4, 489–499.

Kim, J.-S., He, L., Lemasters, J.J., 2003. Mitochondrial permeability transition: a common pathway to necrosis and apoptosis. Biochem. Biophys. Res. Commun. 304, 463–470.

Klaassen, C., Casarett, L., Doull, J., 2013. Casarett and Doull's Toxicology: The Basic Science of Poisons. McGraw Hill Professional, New York.

Kumar, B.S., Chung, B.C., Kwon, O.S., et al., 2012. Discovery of common urinary biomarkers for hepatotoxicity induced by carbon tetrachloride, acetaminophen and methotrexate by mass spectrometry-based metabolomics. J. Appl. Toxicol. 32, 505–520.

Kumar, B.S., Lee, Y.-J., Yi, H.J., et al., 2010. Discovery of safety biomarkers for atorvastatin in rat urine using mass spectrometry based metabolomics combined with global and targeted approach. Anal. Chim. Acta 661, 47–59.

Lacour, S., Gautier, J.-C., Pallardy, M., et al., 2005. Cytokines as potential biomarkers of liver toxicity. Cancer Biomark. 1, 29–39.

Laverty, H.G., Antoine, D.J., Benson, C., et al., 2010. The potential of cytokines as safety biomarkers for drug-induced liver injury. Eur. J. Clin. Pharmacol. 66, 961–976.

Lee, M.S., Jung, B.H., Chung, B.C., et al., 2009. Metabolomics study with gas chromatography–mass spectrometry for predicting valproic acid–induced hepatotoxicity and discovery of novel biomarkers in rat urine. Int. J. Toxicol. 28, 392–404.

Luedde, T., Kaplowitz, N., Schwabe, R.F., 2014. Cell death and cell death responses in liver disease: mechanisms and clinical relevance. Gastroenterology 147, 765–783.

Malhi, H., Gores, G.J., Lemasters, J.J., 2006. Apoptosis and necrosis in the liver: a tale of two deaths? Hepatology 43, S31–S44.

Martin, J.L., Kenna, J.G., Martin, B.M., et al., 1993. Halothane hepatitis patients have serum antibodies that react with protein disulfide isomerase. Hepatology 18, 858–863.

McGill, M.R., Jaeschke, H., 2015. MicroRNAs as signaling mediators and biomarkers of drug-and chemical-induced liver injury. J. Clin. Med. 4, 1063–1078.

Muldrew, K.L., James, L.P., Coop, L., et al., 2002. Determination of acetaminophen-protein adducts in mouse liver and serum and human serum after hepatotoxic doses of acetaminophen using high-performance liquid chromatography with electrochemical detection. Drug Metab. Dispos. 30, 446–451.

Ozer, J., Ratner, M., Shaw, M., et al., 2008. The current state of serum biomarkers of hepatotoxicity. Toxicology 245, 194–205.

Park, B.K., Kitteringham, N.R., Maggs, J.L., et al., 2005. The role of metabolic activation in drug-induced hepatotoxicity. Annu. Rev. Pharmacol. Toxicol. 45, 177–202.

Ramachandran, R., Kakar, S., 2009. Histological patterns in drug-induced liver disease. J. Clin. Pathol. 62, 481–492.

Reust, C.E., Hall, L., 2001. What is the differential diagnosis of an elevated alkaline phosphatase (AP) level in an otherwise asymptomatic patient? J. Fam. Pract. 50, 496.

Roberts, S.A., Price, V.F., Jollow, D.J., 1990. Acetaminophen structure-toxicity studies: in vivo covalent binding of a nonhepatotoxic analog, 3-hydroxyacetanilide. Toxicol. Appl. Pharmacol. 105, 195–208.

Robles-Diaz, M., Medina-Caliz, I., Stephens, C., et al., 2016. Biomarkers in DILI: one more step forward. Front. Pharmacol. 7, 267.

Saadeh, S., 2007. Nonalcoholic fatty liver disease and obesity. Nutr. Clin. Pract. 22, 1–10.

Saito, T., Misawa, K., Kawata, S., 2007. 1. Fatty liver and non-alcoholic steatohepatitis. Intern. Med. 46, 101–104.

Salminen, W.F., Yang, X., Shi, Q., et al., 2012. Green tea extract can potentiate acetaminophen-induced hepatotoxicity in mice. Food Chem. Toxicol. 50, 1439–1446.

Sato, C., Matsuda, Y., Lieber, C.S., 1981. Increased hepatotoxicity of acetaminophen after chronic ethanol consumption in the rat. Gastroenterology 80, 140–148.

Senior, J.R., 2006. How can 'Hy's law' help the clinician? Pharmacoepidemiol. Drug Saf. 15, 235–239.

Shi, Q., Hong, H., Senior, J., et al., 2010. Biomarkers for drug-induced liver injury. Expert Rev. Gastroenterol. Hepatol. 4, 225–234.

Shi, Q., Yang, X., Mattes, W.B., et al., 2015. Circulating mitochondrial biomarkers for drug-induced liver injury. Biomarkers Med. 9, 1215–1223.

Simmons, S.O., Fan, C.-Y., Ramabhadran, R., 2009. Cellular stress response pathway system as a sentinel ensemble in toxicological screening. Toxicol. Sci. 111, 202–225.

Skordi, E., Yap, I.K., Claus, S.P., et al., 2007. Analysis of time-related metabolic fluctuations induced by ethionine in the rat. J. Proteome Res. 6, 4572–4581.

Solter, P., Liu, Z., Guzman, R., 2000. Decreased hepatic ALT synthesis is an outcome of subchronic microcystin-LR toxicity. Toxicol. Appl. Pharmacol. 164, 216–220.

Starkey Lewis, P.J., Dear, J., Platt, V., et al., 2011. Circulating microRNAs as potential markers of human drug-induced liver injury. Hepatology 54, 1767–1776.

Steuerwald, N.M., Foureau, D.M., Norton, H.J., et al., 2013. Profiles of serum cytokines in acute drug-induced liver injury and their prognostic significance. PLoS One 8, e81974.

Su, Z., Li, Z., Chen, T., et al., 2011. Comparing next-generation sequencing and microarray technologies in a toxicological study of the effects of aristolochic acid on rat kidneys. Chem. Res. Toxicol. 24, 1486–1493.

Sun, J., Ando, Y., Ahlbory-Dieker, D., et al., 2013. Systems biology investigation to discover metabolic biomarkers of acetaminophen-induced hepatic injury using integrated transcriptomics and metabolomics. J. Mol. Biomark Diagn. S1, 002.

Tokatli, A., Kalkanoglu-Sivri, H.S., Yuce, A., et al., 2007. Acetaminophen-induced hepatotoxicity in a glutathione synthetase-deficient patient. Turk. J. Pediatr. 49, 75–76.

Wagner, J.A., 2008. Strategic approach to fit-for-purpose biomarkers in drug development. Annu. Rev. Pharmacol. Toxicol. 48, 631–651.

Wang, K., Zhang, S., Marzolf, B., et al., 2009. Circulating microRNAs: potential biomarkers for drug-induced liver injury. Proc. Natl. Acad. Sci. U. S. A. 106, 4402–4407.

Want, E.J., Coen, M., Masson, P., et al., 2010. Ultra performance liquid chromatography-mass spectrometry profiling of bile acid metabolites in biofluids: application to experimental toxicology studies. Anal. Chem. 82, 5282–5289.

Waters, N.J., Waterfield, C.J., Farrant, R.D., et al., 2006. Integrated metabonomic analysis of bromobenzene-induced hepatotoxicity: novel induction of 5-oxoprolinosis. J. Proteome Res. 5, 1448–1459.

Weber, J.A., Baxter, D.H., Zhang, S., et al., 2010. The microRNA spectrum in 12 body fluids. Clin. Chem. 56, 1733–1741.

WebMD, 2017. What Is an Alanine Aminotransferease (ALT) Test? https://www.webmd.com/a-to-z-guides/alanine-aminotransferase-test#1.

Xiong, Y.H., Xu, Y., Yang, L., et al., 2014. Gas chromatography–mass spectrometry-based profiling of serum fatty acids in acetaminophen-induced liver injured rats. J. Appl. Toxicol. 34, 149–157.

Xu, C., Li, C.Y., Kong, A.N., 2005. Introduction of phase I, II and III drug metabolism/transport by xenobiotics. Arch Pharm. Res. 28, 249–268.

Yamazaki, H., Shibata, A., Suzuki, M., et al., 1999. Oxidation of troglitazone to a quinone-type metabolite catalyzed by cytochrome P-450 2C8 and P-450 3A4 in human liver microsomes. Drug Metab. Dispos. 27, 1260–1266.

Yang, L., Xiong, A., He, Y., et al., 2008. Bile acids metabonomic study on the CCl4-and α-naphthylisothiocyanate-induced animal models: quantitative analysis of 22 bile acids by ultraperformance liquid chromatography–mass spectrometry. Chem. Res. Toxicol. 21, 2280–2288.

Yang, X., Salminen, W., Schnackenberg, L., 2012. Current and emerging biomarkers of hepatotoxicity. Curr Biomark Find 2, 43–55.

Yu, M., Zhu, Y., Cong, Q., et al., 2017. Metabonomics research progress on liver diseases. Can. J. Gastroenterol. Hepatol. 2017, 8467192.

Zhang, L., Ye, Y., An, Y., et al., 2010. Systems responses of rats to aflatoxin B1 exposure revealed with metabonomic changes in multiple biological matrices. J. Proteome Res. 10, 614–623.

CHAPTER

15

Conventional and Emerging Renal Biomarkers

Sue M. Ford

Department of Pharmaceutical Sciences, College of Pharmacy & Health Sciences, St. John's University, Jamaica, NY, United States

INTRODUCTION

Determination of renal status and prognostication of the course of disease are important in clinical practice due to the human and economic costs of kidney diseases. Approximately 30 million adults (14.8%) have chronic kidney disease (CKD), many of whom are unaware that they have the condition (Centers for Disease Control and Prevention, 2017). Tremendous variability in etiology and severity of the underlying abnormalities can confound detection and diagnosis. The traditional clinical tools for assessing renal function are often inadequate for exposing early or mild impairment and lack specificity for evaluating injury, and the search for sensitive and specific clinical biomarkers of renal injury has increased dramatically in the past 20 years. Improved biomarkers are also needed for field screening and studies of populations at risk from exposure to environmental contaminants or toxins. Drug development is another area in which biomarker exploration has been extensive. Early recognition of nephrotoxic liability in preclinical studies allows pharmaceutical companies to eliminate less-promising drug candidates at the earliest stage possible. In addition, researchers studying the underlying mechanisms of nephrotoxic agents need reliable and expedient ways to assess kidney function in laboratory animals.

The complexity of the kidneys presents numerous challenges in assessing renal status. The kidneys are sensitive to events such as exposure to toxic chemicals, low perfusion, immune system activity, and prerenal pathological conditions such as diabetes, each of which may affect renal structure and function in distinct ways. Until recently, the term acute renal failure (ARF) was used to describe a myriad of conditions with inadequate standardization of clinical definitions, making it difficult to satisfactorily compare reports in the literature (Kellum and Hoste, 2008). However, two developments have provided structure to classification of renal impairment. First, standard criteria for stages of renal dysfunction were established. In 2003, guidelines for CKD were published (Eknoyan, 2012; Levey et al., 2003) defining it as kidney damage or glomerular filtration rate (GFR) less than 60 mL/min for more than 3 months. Classification guidelines for acute kidney injury were proposed in 2004. Termed the RIFLE classification, the stages are Risk–Injury–Failure–Loss–End-stage renal disease (Cruz et al., 2009). The criteria for assigning the stages are based on change in blood creatinine or GFR and urine flow rates during a specified interval (6, 12, or 24 h). In 2007 the acute kidney injury network proposed a classification based on modification of RIFLE; this system is referred to as the AKIN classification and is based on changes in serum creatinine and the urine output during the specified time. The Kidney Disease: Improving Global Outcomes (KDIGO) working group proposed a set of criteria that are a combination of RIFLE and AKIN (Kellum et al., 2012). The recognition that newer biomarkers need to be developed and incorporated into the AKI staging framework was formalized at the 10th Acute Dialysis Quality Initiative (ADQI) Consensus Conference (Murray et al., 2014). The other significant development was a change in terminology from *acute renal failure* to *acute kidney injury.* Acute kidney injury (AKI) now encompasses the spectrum of renal impairment, from subtle pathological and/or structural changes not detectable with current functional measurements to the extreme of severe dysfunction (Kellum and Hoste, 2008).

CHARACTERISTICS OF BIOMARKERS

Renal biomarkers have been categorized into "traditional" and "novel," although several of the "novel"

Biomarkers in Toxicology, Second Edition
https://doi.org/10.1016/B978-0-12-814655-2.00015-3

TABLE 15.1 Selection Characteristics for Biomarkers

Sensitivity	Responsiveness to changes in the quantity of the parameter
Specificity	Responsive only to the desired parameter
Interference by coexisting conditions	Disease, medications
Interference by physiological variables	Age, sex, body mass
Analytical considerations	Including cost, time, validation, analytical sensitivity and specificity

biomarkers have been used for at least a decade. Generally the traditional markers are indices of function, whereas newer biomarkers have been developed to detect injury, serving as signals for specific damage such as tubular necrosis or damage to the glomerulus. Renal biomarkers may be used to evaluate current status, to monitor function over time, or to predict the risk for decline in renal function (Table 15.1).

TRADITIONAL BIOMARKERS

Traditional assessments of kidney function have relied on blood and urinary parameters that reflect functional parameters such as GFR and/or renal blood flow (blood urea nitrogen, creatinine), glomerular patency (albuminuria), tubular function (β-microglobulinuria), or concentrating ability (polyuria). However, some of the most common are affected by extrarenal events and should be interpreted with consideration of the individual patient. Blood urea nitrogen (BUN) levels are related to diet, creatinine production is altered during increased muscle turnover, and urine output is affected by the state of hydration and possible endocrine diseases. The traditional biomarkers have several disadvantages for clinical assessment, most notably that they are able to reveal changes in function or structure only after significant damage has occurred. However, they still can be useful in monitoring the status of patients with CKD (Pena et al., 2017).

Glomerular Filtration Rate

The parameter most watched clinically is GFR, which can be determined by the renal clearance of model compounds. The model compounds should be freely filtered at the glomerulus and neither reabsorbed nor secreted in the tubule. The equation for renal clearance is $Cl = (Cu \times Q)/Cp$, where Cu is the concentration of the compound in urine, Q is the urine flow rate, and Cp is the concentration in plasma. There are several compounds that can be used for measured GFR (mGFR) include inulin, ^{125}I-iothalamate, or ^{51}Cr-EDTA. Obtaining a measured GFR is the preferred approach but impractical on a routine basis, so that endogenous molecules are used. For noncritical routine estimates, serum BUN is used as an indicator of renal function. The production of urea is sufficiently constant to provide a reasonable index of renal function, with serum BUN being inversely proportional to GFR. Urea is reabsorbed in the nephron back into the bloodstream, which compromises its value as a biomarker; nonetheless, it is easy to measure and well established.

Creatinine has been the mainstay for assessing renal function and GFR clinically. Creatinine has the advantages of being produced at a constant rate, freely filtered, and not being reabsorbed. There are several problems with creatinine including that its production and plasma concentration can be increased during excessive muscle breakdown or decreased following prolonged illness and loss of muscle mass. Nonetheless, serum creatinine (sCr) is routinely measured as a renal function test and can be a valuable index when renal dysfunction is anticipated, such as monitoring kidney function when nephrotoxic drugs are used (Caires et al., 2019). A more robust value is creatinine clearance (Clcr), which can be used to provide an estimated GFR (eGFR), requiring a serum creatinine value and a timed urine sample. Various equations that correct for secretion, age, and sex have been developed, so that Clcr and eGFR are the recommended parameters for clinical evaluation of glomerular filtration. Another endogenous molecule that is increasingly being used as an indicator of GFR is cystatin C (discussed later in the chapter).

The options for evaluating GFR in patients at risk have increased over the years, and it is important to understand the advantages and limitations of the various methods (Schaeffner, 2017). BUN and sCr are easy and inexpensive methods; however, the kidneys have significant reserve capacity and measuring renal dysfunction with eGFR may be insensitive, often showing abnormalities only after considerable renal function is lost. In addition, there is variability among criteria in the standard guidelines (RIFLE, AKIN, KDIGO), and some judgment in use may be advised. For example, increases in sCr that were smaller than recommended clinical guidelines were shown to be correlated with poor renal outcomes after surgery (Machado et al., 2014, Tolpin et al., 2012).

Proteinuria

The glomerular capillaries are the barrier to distribution of large plasma proteins into urine. Large proteins such as albumin and IgM are impeded by the capillaries, whereas smaller proteins pass through the filtration barrier into the tubular fluid. In normal kidneys, only small amounts of large proteins such as albumin and IgM are filtered, most of which is degraded by the proximal tubule epithelium. Smaller proteins and peptides are filtered across the glomerular barrier depending on size, charge, and configuration and then degraded in the tubules. Consequently, there is generally little protein detectable in the urine of humans with healthy kidneys. Proteinuria may occur due to pathology of the glomerular filtration barrier, as in the case of the nephrotic syndrome, as well as damage to the proximal tubule. The type of protein in the urine can be diagnostic of the site of injury. Glomerular damage may result in large proteins appearing in the urine, which overwhelms the modest ability of the proximal tubule to remove them; in contrast, injury to the proximal tubule will impair the ability to remove smaller proteins such as β2-microglobulin or retinal binding protein from the tubular fluid. Thus, if glomerular filtration is reduced, β2-microglobulin will increase in the blood, whereas if the proximal tubule cells are damaged, β2-microglobulin will be found in the urine.

Urinary albumin is a more conventional marker of renal dysfunction, particularly CKD, which may still find utility in AKI as an adjunct to the primary classification systems (Bolisetty and Agarwal, 2011; Sugimoto et al., 2016). In fact, a recent analysis of preadmission data for patients with dialysis-requiring AKI indicated that the level of proteinuria (dipstick) was correlated with the risk of nonrecovery (Lee et al., 2018). Several specific proteins are currently being used as indices of renal dysfunction either individually (see following sections) or as part of biomarker panels (Chen et al., 2017a), and application of proteomics profiling with mass spectroscopy techniques may facilitate the identification of new biomarker candidates (Chen et al., 2017b).

Enzymuria

Enzymes released from damaged cells of the tubule have been used as markers of injury inasmuch as appearance of these enzymes in the urine is specific to kidney and they have been used in research with laboratory animals for several decades. The brush border enzymes alkaline phosphatase (AP), γ-glutamyltranspeptidase (GGT), the lysosomal enzyme N-acetyl-β-glucosaminidase (NAG), and isoforms of the cytoplasmic enzyme glutathione-S-transferase may be detected in the urine following tubular injury. Although their use for human studies has been minimal, in the future they may be valuable as part of a biomarker panel.

NEWER BIOMARKERS

Cystatin C

Cystatin C was first observed in 1961 as a "post-γ" protein detected in small amounts in the serum and urine of humans. Its presence in urine of the patients was clearly associated with renal tubular dysfunction (Butler and Flynn, 1961). The protein was sequenced and characterized in 1982 by A. Grubb, although the function was as yet unknown (Grubb and Löfberg, 1982). It was recognized as a cysteine protease inhibitor (Barrett et al., 1984; Turk and Bode, 1991), homologous with chicken cystatin (Filler et al., 2005) and renamed cystatin C (Barrett et al., 1984). Investigation of cystatin C as an indicator of GFR began as early as the mid-1980s. Cystatin C has characteristics that make it a good candidate as both an index of function (GFR) and tubular injury. It is produced by all nucleated cells at a constant rate and is cleared only by the kidneys. It is a low−molecular weight (13.3 kDa) basic protein and thereby freely filterable at the glomerulus. It is not secreted, and although it is reabsorbed by proximal tubule epithelia, it is completely catabolized within the cells and does not re-enter the plasma (Filler et al., 2005).

Because of these features, the plasma levels of cystatin C are inversely related to GFR, so that it has long been recognized that increased cystatin C plasma or serum levels reflect decreased GFR (Grubb, 1992). eGFR values calculated with serum cystatin C have been shown to correlate with 24-h creatinine clearance (Diego et al., 2016) and iohexol clearance (Schwartz et al., 2012). Whereas serum cystatin C is an index of GFR, urinary cystatin C has potential as a tubular injury marker because its presence in the urine in the absence of massive proteinuria would indicate tubular damage.

As with other biomarkers, there are some limitations that should be noted. Cystatin C has been shown to be influenced by large doses of glucocorticoids and thyroid dysfunction; such factors may influence the lack of specificity for AKI noted in a study of critically ill children (Safdar et al., 2016). Unlike creatinine, which can be measured in both serum and urine, cystatin C appears at very low levels in the urine of normal kidneys so that estimates of eGFR with cystatin use only the serum concentration. The production of cystatin C should be constant, but on a population basis, serum values have been correlated with physiological factors such as diabetes, body size, and inflammation (Grubb et al., 2011; Shlipak et al., 2006; Stevens et al., 2009; Zeng et al., 2017; Zhang et al., 2017). In addition, there are reports of large intraindividual variability compared to creatinine, which would limit its usefulness in monitoring individuals over time (Filler et al., 2005; Keevil et al., 1998). As with other clinical indices of GFR, knowledge of such factors will aid interpretation of the results and selection of the appropriate equations (Lassus and Harjola, 2012).

Cystatin C was more closely related to GFR changes in diabetics than was serum creatinine (Shlipak et al., 2006). Herget-Rosenthal et al. (2004) selected 85 patients in intensive care who, though starting with normal measures of renal function, were at risk for AKI due to various causes such as ischemia, prerenal disease (e.g., diabetes), nephrotoxicity, or sepsis. Their status was categorized according to the RIFLE criteria. Serum cystatin C in this study was found to detect the development of AKI 1 to 2 days earlier than serum creatinine. In contrast, Royakkers et al. (2011) concluded that serum or urine cystatin C values were poor indicators for the development of AKI. In a study of patients who ingested toxic chemicals, serum cystatin C responded later and underestimated the extent of AKI (Wijerathna et al., 2017). Serum cystatin C was more closely associated with development and worsening of AKI than serum creatinine in ICU patients with sepsis (Leem et al., 2017) but was not superior to sCr in a different adult ICU population (Diego et al., 2016). The utility of cystatin C to evaluate the renal function of neonates shows promise (Askenazi et al., 2012), although there are few validation studies (Kandasamy et al., 2013).

Neutrophil Gelatinase–Associated Lipocalin

Neutrophil gelatinase–associated lipocalin (NGAL) is a 25-kDa polypeptide that is produced in numerous tissues and in laboratory animals. It was discovered as a protein secreted by neutrophils and bound to gelatinase (Kjeldsen et al., 1993). Lipocalin proteins bind and transport small molecules, and it is believed that one of the important functions of NGAL is the binding of bacterial siderophores and the transport of iron into cells (Singer et al., 2013). NGAL can be induced in both neutrophils and epithelia by various stimuli including inflammation, infections, and neoplasia. It has been shown to have protective functions against infection and ischemic kidney injury (Ma et al., 2011). It may play a role in repair of damaged proximal tubule epithelia, although the nature of the NGAL involvement—contributor to injury or repair effector—is unclear (Lippi and Cervellin, 2013; Mishra et al., 2003).

In healthy individuals NGAL is expressed at low levels in various tissues, filtered at the glomerulus, and then reabsorbed by the proximal tubule (Chakraborty et al., 2012; Charlton et al., 2014; Singer et al., 2013). It is resistant to proteases (Kjeldsen et al., 1993; Mishra et al., 2003), which increases its stability before assay. NGAL is expressed by several tissues, including the nephron, in response to various pathological conditions (Singer et al., 2013). Using microarray technology, Mishra et al. (2003) identified NGAL as one of the genes upregulated in the mouse kidney within the first few hours after ischemic injury. Histochemical staining showed that NGAL protein was barely detectable in the kidneys of naïve mice but was upregulated in the proximal tubules within 3 h of ischemia. Following AKI, NGAL production is induced in the distal segments of the nephron, and it is released into the blood and lumen, resulting in elevation of both blood and urine levels (Tecson et al., 2017). In addition to increasing renal production of NGAL following injury, damage to the nephron may enhance NGAL presence in the urine by impairing proximal tubular reabsorption.

Serum NGAL has been shown to be associated with various causes of renal dysfunction including type 2 diabetes (Bacci et al., 2017), toxic injury (Luo et al., 2016), and high-salt diet in rodents (Washino et al., 2018). NGAL appeared in the urine of rats before other renal injury markers of ischemia, such as N-acetyl-β-D-glucosaminidase and β2-microglobulin (Mishra et al., 2003). They also demonstrated that NGAL was released into the media of cultured human proximal tubule cells depleted of ATP and could be detected in the urine of mice treated with the nephrotoxicant cisplatin.

NGAL can be detected in as little as 1 μL of urine. There are a number of validated clinical analytical platforms that allow results in 15–30 min (Hassanzadeh et al., 2015) Its concentration may increase 10,000-fold in response to injury, whereas in plasma the increase may be up to 100-fold (Lippi and Plebani, 2012). There are complicating factors that must be considered when interpreting NGAL data. For example, nonrenal sources such as neutrophils or urinary tract infections with leukocyturia may contribute to its concentration in some cases.

Kidney Injury Molecule-1

Kidney Injury Molecule-1 (KIM-1)[1] is a 104-kDa membrane protein resembling certain cell adhesion molecules (Ichimura et al., 1998). It is not detectable in normal kidneys but is expressed at high levels on the apical surface of proximal tubule cells following injury

[1] KIM-1 refers to the human form of the molecule, and Kim-1 refers to the rat form. Although there are amino acid sequence differences (654), this chapter will use KIM-1 for all species in discussing its use as a biomarker, except when studies cited specify Kim-1 or there is a notable species difference in human–animal performance. As a consequence of its independent discovery and naming by immunologists, KIM-1 is also referred to as TIM-1 (Rees and Kain, 2008).

or in certain types of renal carcinoma. Structurally, each molecule of KIM-1 has extracellular mucin and immunoglobulin domains (Ichimura et al., 1998), which comprise an ectodomain that is constitutively shed by metalloproteinase cleavage (Bailly et al., 2002; Lim et al., 2012). This process is increased by human serum albumin or TNF-α (Blasco et al., 2011). The dedifferentiation and proliferation of proximal tubules which are noted after injury, result in increased KIM-1 and its fragments in urine, leading to its use as an injury biomarker (Ichimura et al., 2004). A number of possible functions for KIM-1 in post–injury repair processes have been proposed, including regulation of adhesion of cells to the tubular basement membrane (Bailly et al., 2002), cell motility to facilitate covering denuded areas of the tubule (Ichimura et al., 1998), and inducing epithelial cells to take on phagocytic characteristics to remove dead cells (Ichimura et al., 2008; Rees and Kain, 2008). However, the question of whether KIM-1, such as NGAL, may actually contribute to the development of renal injury has been raised (Lim et al., 2013).

In rat kidneys subjected to reperfusion or toxicant injury, Kim-1 expression was found to be specific to proliferating cells in the S3 segment of the proximal tubule, with little response in the convoluted section (Ichimura et al., 1998, 2004). The magnitude of the increase was greater than and preceded any increase in serum creatinine, urinary NAG, or proteinuria (Sabbisetti et al., 2013). KIM-1 immunohistochemical staining of biopsy was more sensitive than histology for detecting tubular injury in transplanted kidneys (Zhang et al., 2008).

The possible beneficial role of KIM-1 in tissue repair is bolstered by the study of Zhang et al. (2008) revealing that KIM-1 staining from patients with biopsies showing acute tubular injury was correlated with recovery 18 months later. The group with higher KIM-1 staining scores had greater recovery, measured by BUN, serum creatinine, and estimated GFR. In contrast, Jin et al. (2013) observed that high serum levels of KIM-1 were correlated with lower survival rates, and a similar result was noted for urinary KIM-1 excretion (van Timmeren et al., 2007).

Urinary KIM-1 correlated well with blood lead levels in an occupational setting suggesting that it may be useful in early diagnosis of kidney injury in workers (Zhou et al., 2016a). A study of patients whose kidneys were subjected to ischemia during partial nephrectomy for renal cancer shows that urinary NGAL and KIM-1 were elevated within 1 h (Abassi et al., 2013). The levels remained high for 72 h for NGAL and 24 h for KIM-1 suggesting that NGAL can be a more reliable indicator of AKI in these patients. Similarly, whereas urinary NGAL correlated with CKD in a human population (Bhavsar et al., 2012) or tubular damage from a high-salt diet in WKY rats (Washino et al., 2018), KIM-1 did not in either case. The use of KIM-1 as an index or a predictor of renal status may depend on the conditions and calculations in which the value is used. For example, Helmersson-Karlqvist et al. (2016) studied day-to-day variability of urinary cystatin C and KIM-1 in healthy senior citizens (average age 75) and concluded that adjustment of uKIM-1 by uCr is advisable, particularly when spot urines are used. Koyner et al. (2015) observed that for cardiac surgery patients, the postoperative association between biomarkers such as KIM-1 with AKI is influenced by the preoperative eGFR.

Interleukin 18

Interleukin 18 (IL-18) is a widely distributed cytokine. It exists in a precursor form until cleaved by caspase-1 after which it has proinflammatory actions. In the normal human kidney it is found primarily in the distal segments of the nephron (Gauer et al., 2007) and is upregulated, including in the proximal tubule, following AKI and CKD (Liang et al., 2007). IL-18 appears to have a contributory role in the development of several types of renal injury, including lipopolysaccharide LPS-induced AKI (Nozaki et al., 2017) and ischemia (Melnikov et al., 2001). Numerous reports suggest that urinary IL-18 is significantly increased in various types of renal injury, including ischemia, posttransplantation acute tubular necrosis, liver cirrhosis, and cancer chemotherapy (Leslie and Meldrum, 2008; Puthumana et al., 2017; Zubowska et al., 2013). The specificity and sensitivity of IL-18 detection in these conditions suggest it has great potential as an early biomarker of injury. However, interpretation of IL-18 may be confounded by coexisting conditions such as sepsis (Li et al., 2013).

Liver-type Fatty Acid Binding Protein

Fatty acid binding proteins comprise a family of small proteins, which include the liver-type fatty acid binding protein (L-FABP; FABP-1). They are expressed in several tissues besides the liver, including the renal proximal tubules. L-FABP orchestrates the transport of fatty acids into mitochondria and peroxisomes for energy metabolism. In the kidneys it binds the breakdown products of lipid peroxidation facilitating their excretion into the urine, preventing damage by reactive oxygen species (Xu et al., 2015). L-FABP is released into the urine following different types of injury and has been extensively studied as biomarkers. The serum concentrations of these proteins increase in renal injury, which is convenient for cases where urine volume is drastically reduced. L-FABP concentrations have been shown to predict and correlate with renal status following cardiac surgery

(Moriyama et al., 2016), toxic exposure (Ferguson et al., 2010; Igarashi et al., 2013; Pelsers, 2008), or diabetic nephropathy (Colhoun and Marcovecchio, 2018).

Genomics, Proteomics, Metabolomics

The complexity of the kidney structure and function in health and disease makes it difficult to conceive of and develop a single biomarker that can give an adequate snapshot of renal status in an individual at a specific time, particularly when the need for information is critical. A panel of biomarkers is more likely to detect important changes and give a richer description of renal impairment. In fact, several companies are marketing immunology-based biomarker panels for assessment of renal function. The −omic analyses take this further in describing a pattern of gene, protein, or metabolite changes that can be analyzed and compared with fingerprints of various pathologies.

Genomic analysis has the ability to reveal increased expression of proteins and signaling molecules related to kidney injury, elucidating mechanistic information that can identify potential biomarkers and therapies (Devarajan et al., 2003; Minowa et al., 2012). Amin et al. (2004) used microarray to analyze kidney samples from rats treated with cisplatin, gentamicin, or puromycin at various doses and times ranging from 4 h to 7 days. They observed changes in gene expression in pathways related to creatinine biosynthesis, kinase signaling, cell cycle, renal transporters, renal injury, regenerative responses, drug metabolism, and resistance. The changes were related to severity of pathology, depending on time and dose. A larger-scale rodent study with 33 nephrotoxicants (Kondo et al., 2009) revealed changes in gene expression in similar pathways. An important issue with using genomic analysis for human studies is obtaining tissue samples (Ju et al., 2012). Even when biopsy material is available, frozen samples may not be available inasmuch as those taken for histological evaluation are formalin-fixed and embedded in paraffin, which fragments RNA during extraction. Some studies have obtained promising results with urine (Ju et al., 2012; Zhou et al., 2016b). The analysis of urine for nucleic acids, proteins, and transcription factors may be facilitated by their inclusion in exosomes (Gyurászová et al., 2018).

Characterization of protein expression has been productively exploited in studying renal biomarkers (Devarajan, 2007; 2008; Slocum et al., 2012; Zhang et al., 2014). Proteomic profiling has become feasible due to advances in mass spectrometry technology for proteins, specifically surface-enhanced laser desorption/ionization time-of-flight mass spectrometry (SELDI-TOF-MS), matrix-assisted laser desorption/ionization time-of-flight mass spectrometry (MALDI-TOF-MS), and Multidimensional Protein Identification Technology (MUDPIT) (Devarajan, 2007; Raj et al., 2012; Smith, 2002). Proteomic analysis is amenable to biofluid analysis, particularly urine, serum, and cell culture media. The ability to detect multiple peptides and proteins in a sample will facilitate the discovery of new biomarkers and unraveling of mechanisms. However, for clinical use the MS methods are still expensive and need validation. Commercial renal protein array kits with antibodies to cytokines and renal injury biomarkers offer a useful alternative.

Metabolomics is the outgrowth of the concept of metabolic profiling or metabolic pattern analysis of metabolites in biofluids, initiated in the 1940s and further pursued in the 70s (Gates and Sweeley, 1978). The analytical technologies, GC/MS, LC/MS, and NMR, that are used to analyze samples for small-molecule analytes are available in more facilities and are augmented by the powerful data analysis systems that make the −omics practical. Metabolomics is only recently being applied to the nephrology for the study of diseases, toxicities, and discovery of biomarkers (Ezaki et al., 2017; Mussap et al., 2014; Slocum et al., 2012; Weiss and Kim, 2012; Zhang et al., 2012). Much work needs to be done to validate the procedures and to overcome problems when working with fluids such as urine, including bacterial contamination and the influence of diet on the metabolome (Weiss and Kim, 2012). The development of more powerful bioinformatics tools for the analysis of −omics data (Breit and Weinberger, 2016) will enhance the ability to construct and interpret profiles and fingerprints of various renal pathologies for prediction, diagnosis, and intervention.

CONCLUDING REMARKS AND FUTURE DIRECTIONS

In view of the complexity of renal structure and function, and the wide variety of conditions that are of concern, it is unlikely that a single biomarker will be found that can serve all the purposes of an ideal biomarker that can detect, diagnose, and predict. The use of individual biomarkers and biomarker panels will need to be validated for factors such as age (Ferguson et al., 2010), sex (Tsuji et al., 2017), and appropriate time for evaluating (Susantitaphong et al., 2013'U.S. Food & Drug Administration). Currently the FDA has qualified several renal biomarkers for preclinical Good Laboratory Practice (GLP) studies (U.S. Food & Drug Administration) but not for clinical trials. The requirements for assessment and prediction, and the resources available, will be different for animal studies than clinical drug trials and still different for monitoring patients. It is likely that panels of serum and/or urine biomarkers can be developed for (1) general assessment of patient renal function in routine clinical tests, (2) monitoring of renal function in patients at risk for particular renal pathology such as diabetic nephropathy, ischemia, transplant rejection, drug monitoring, (3) diagnosing the causes or injury site of renal disease, and (4) investigative preclinical and clinical uses. The biomarkers selected for the panels would be those for which sufficient studies have been completed to provide evidence for their individual or combined ability to answer the specific question being asked at the requisite level of sensitivity.

References

Abassi, Z., Shalabi, A., Sohotnik, R., et al., 2013. Urinary NGAL and KIM-1: biomarkers for assessment of acute ischemic kidney injury following nephron sparing surgery. J. Urol. 189 (4), 1559–1566.

Amin, R.A., Vickers, A.E., Sistare, F., et al., 2004. Identification of putative gene-based markers of renal toxicity. Environ. Health Perspect. 112 (4), 465–479.

Askenazi, D.J., Koralkar, R., Hundley, H.E., et al., 2012. Urine biomarkers predict acute kidney injury in newborns. J. Pediatr. 161 (2), 270–275.

Bacci, M.R., Chehter, E.Z., Azzalis, L.A., et al., 2017. Serum NGAL and cystatin C comparison with urinary albumin-to-creatinine ratio and inflammatory biomarkers as early predictors of renal dysfunction in patients with type 2 diabetes. Kidney Int. Rep. 2 (2), 152–158.

Bailly, V., Zhang, Z.W., Meier, W., et al., 2002. Shedding of kidney injury molecule-1, a putative adhesion protein involved in renal regeneration. J. Biol. Chem. 277 (42), 39739–39748.

Barrett, A.J., Davies, M.E., Grubb, A., 1984. The place of human g-trace (cystatin C) amongst the cysteine proteinase inhibitors. Biochem. Biophys. Res. Commun. 120 (2), 631–636.

Bhavsar, N.A., Köttgen, A., Coresh, J., et al., 2012. Neutrophil gelatinase-associated lipocalin (NGAL) and kidney injury molecule 1 (KIM-1) as predictors of incident CKD stage 3: the Atherosclerosis Risk in Communities (ARIC) study. Am. J. Kidney Dis. 60 (2), 233–240.

Blasco, V., Wiramus, S., Textoris, J., et al., 2011. Monitoring of plasma creatinine and urinary gamma-glutamyl transpeptidase improves detection of acute kidney injury by more than 20%. Crit. Care Med. 39 (1), 52–56.

Bolisetty, S., Agarwal, A., 2011. Urine albumin as a biomarker in acute kidney injury. Am. J. Physiol. Ren. Physiol. 300 (3), F626–F627.

Breit, M., Weinberger, K.M., 2016. Metabolic biomarkers for chronic kidney disease. Arch. Biochem. Biophys. 589, 62–80.

Butler, E.A., Flynn, F.V., 1961. The occurrence of post-gamma protein in urine: a new protein abnormality. J. Clin. Pathol. 14 (2), 172–178.

Caires, R.A., da Costa e Silva, V.T., Burdmann, E.A., et al., 2019. Chapter 39-Drug-induced acute kidney injury. In: Ronco, C., Bellomo, R., Kellum, J.A., Ricci, Z. (Eds.), Critical Care Nephrology, third ed., pp. 214–221 Philadelphia.

Centers for Disease Control and Prevention, 2017. National Chronic Kidney Disease Fact Sheet. Retrieved from: https://www.cdc.gov/kidneydisease/pdf/kidney_factsheet.pdf.

Chakraborty, S., Kaur, S., Guha, S., Batra, S.K., 2012. The multifaceted roles of neutrophil gelatinase associated lipocalin (NGAL) in inflammation and cancer. Biochim. Biophys. Acta Rev. Cancer 1826 (1), 129–169.

Charlton, J.R., Portilla, D., Okusa, M.D., 2014. A basic science view of acute kidney injury biomarkers. Nephrol. Dial. Transplant. 29 (7), 1301–1311.

Chen, Y.F., Thurman, J.D., Kinter, L.B., et al., 2017a. Perspectives on using a multiplex human kidney safety biomarker panel to detect cisplatin-induced tubular toxicity in male and female Cynomolgus monkeys. Toxicol. Appl. Pharmacol. 336, 66–74.

Chen, L., Smith, J., Mikl, J., et al., 2017b. A multiplatform approach for the discovery of novel drug-induced kidney injury biomarkers. Chem. Res. Toxicol. 30 (10), 1823–1834.

Colhoun, H.M., Marcovecchio, M.L., 2018. Biomarkers of diabetic kidney disease. Diabetologia 61 (5), 996–1011.

Cruz, D.N., Ricci, Z., Ronco, C., 2009. Clinical review: RIFLE and AKIN - time for reappraisal. Crit. Care 13 (3), 211–220.

Devarajan, P., 2007. Proteomics for biomarker discovery in acute kidney injury. Semin. Nephrol. 27 (6), 637–651.

Devarajan, P., 2008. Proteomics for the investigation of acute kidney injury. In: Thongboonkerd, V. (Ed.), Proteomics in Nephrology - towards Clinical Applications. Karger, Basel, pp. 1–16.

Devarajan, P., Mishra, J., Supavekin, S., et al., 2003. Gene expression in early ischemic renal injury: clues towards pathogenesis, biomarker discovery, and novel therapeutics. Mol. Genet. Metabol. 80 (4), 365–376.

Diego, E., Castro, P., Soy, D., et al., 2016. Predictive performance of glomerular filtration rate estimation equations based on cystatin C versus serum creatinine values in critically ill patients. Am. J. Health Syst. Pharm. 73 (4), 206–215.

Eknoyan, G., 2012. A decade after the KDOQI CKD guidelines: a historical perspective. Am. J. Kidney Dis. 60 (5), 686–688.

Ezaki, T., Nishiumi, S., Azuma, T., et al., 2017. Metabolomics for the early detection of cisplatin-induced nephrotoxicity. Toxicol. Res. 6 (6), 843–853.

Ferguson, M.A., Vaidya, V.S., Waikar, S.S., et al., 2010. Urinary liver-type fatty acid-binding protein predicts adverse outcomes in acute kidney injury. Kidney Int. 77 (8), 708–714.

Filler, G., Bökenkamp, A., Hofmann, W., et al., 2005. Cystatin C as a marker of GFR – history, indications, and future research. Clin. Biochem. 38 (1), 1–8.

Gates, S.C., Sweeley, C.C., 1978. Quantitative metabolic profiling based on gas chromatography. Clin. Chem. 24 (10), 1663–1673.

Gauer, S., Sichler, O., Obermüller, N., et al., 2007. IL-18 is expressed in the intercalated cell of human kidney. Kidney Int. 72 (9), 1081–1087.

Grubb, A., 1992. Diagnostic value of analysis of cystatin C and protein HC in biological fluids. Clin. Nephrol. 38 (Suppl 1), S20–S27.

Grubb, A., Löfberg, H., 1982. Human γ-trace, a basic microprotein: amino acid sequence and presence in the adenohypophysis. Proc. Natl. Acad. Sci. U. S. A. 79 (9), 3024–3027.

Grubb, A., Björk, J., Nyman, U., et al., 2011. Cystatin C, a marker for successful aging and glomerular filtration rate, is not influenced by inflammation. Scand. J. Clin. Lab. Invest. 71 (2), 145–149.

Gyurászová, M., Kovalčíková, A., Bábíčková, J., et al., 2018. Cell-free nucleic acids in urine as potential biomarkers of kidney disease. J. Appl. Biomed. 16 (3), 157–164. https://doi.org/10.1016/j.jab.2018.01.007.

Hassanzadeh, R., Lasserson, D., Fatoba, S., et al., 2015. Point-of-care Neutrophil Gelatinase-associated Lipocalin (NGAL) Tests. Horizon Scanning Report 0042. Retrieved from NIHR Diagnostic Evidence Cooperative Oxford: https://www.community.healthcare.mic.nihr.ac.uk/reports-and-resources/horizon-scanning-reports.

Helmersson-Karlqvist, J., Ärnlöv, J., Carlsson, A.C., et al., 2016. Urinary KIM-1, but not urinary cystatin C, should be corrected for urinary creatinine. Clin. Biochem. 49 (15), 1164–1166.

Herget-Rosenthal, S., Marggraf, G., Husing, J., et al., 2004. Early detection of acute renal failure by serum cystatin C. Kidney Int. 66 (3), 1115–1122.

Ichimura, T., Bonventre, J.V., Bailly, V., et al., 1998. Kidney injury molecule-1 (KIM-1), a putative epithelial cell adhesion molecule containing a novel immunoglobulin domain, is up-regulated in renal cells after injury. J. Biol. Chem. 273 (7), 4135–4142.

Ichimura, T., Hung, C.C., Yang, S.A., et al., 2004. Kidney injury molecule-1: a tissue and urinary biomarker for nephrotoxicant-induced renal injury. Am. J. Physiol. Ren. Physiol. 286 (3), F552–F563.

Ichimura, T., Asseldonk, E.J., Humphreys, B.D., et al., 2008. Kidney injury molecule-1 is a phosphatidylserine receptor that confers a phagocytic phenotype on epithelial cells. J. Clin. Invest. 118 (5), 1657.

Igarashi, G., Iino, K., Watanabe, H., et al., 2013. Remote ischemic preconditioning alleviates contrastinduced acute kidney injury in patients with moderate chronic kidney disease. Circ. J. 77 (12), 3037–3044.

Jin, Z.K., Tian, P.X., Wang, X.Z., et al., 2013. Kidney injury molecule-1 and osteopontin: new markers for prediction of early kidney transplant rejection. Mol. Immunol. 54 (3–4), 457–464.

Ju, W., Smith, S., Kretzler, M., 2012. Genomic biomarkers for chronic kidney disease. Transl. Res. 159 (4), 290–302.

Kandasamy, Y., Smith, R., Wright, I.M.R., 2013. Measuring cystatin C to determine renal function in neonates. Pediatr. Crit. Care Med. 14 (3), 318–322.

Keevil, B.G., Kilpatrick, E.S., Nichols, S.P., et al., 1998. Biological variation of cystatin C: implications for the assessment of glomerular filtration rate. Clin. Chem. 44 (7), 1535–1539.

Kellum, J.A., Hoste, E.A.J., 2008. Acute kidney injury: epidemiology and assessment. Scand. J. Clin. Lab. Invest. 68 (S241), 6–11.

Kellum, J.A., Lameire, N., Aspelin, P., et al., 2012. KDIGO clinical practice guideline for acute kidney injury. Section 2: AKI definition. Kidney Int. Suppl. 2 (1), 19–36.

Kjeldsen, L., Johnsen, A.H., Sengeløv, H., et al., 1993. Isolation and primary structure of NGAL, a novel protein associated with human neutrophil gelatinase. J. Biol. Chem. 268 (14), 10425–10432.

Kondo, C., Minowa, Y., Uehara, T., et al., 2009. Identification of genomic biomarkers for concurrent diagnosis of drug-induced renal tubular injury using a large-scale toxicogenomics database. Toxicology 265 (1–2), 15–26.

Koyner, J.L., Coca, S.G., Thiessen-Philbrook, H., et al., 2015. Urine biomarkers and perioperative acutekidney injury: the impact of preoperative estimated GFR. Am. J. Kidney Dis. 66 (6), 1006–1014.

Lassus, J., Harjola, V.P., 2012. Cystatin C: a step forward in assessing kidney function and cardiovascular risk. Heart Fail. Rev. 17 (2), 251–261.

Lee, B.J., Go, A.S., Parikh, R., et al., 2018. Pre-admission proteinuria impacts risk of non-recovery after dialysis-requiring acute kidney injury. Kidney Int. 93 (4), 968–976.

Leem, A.Y., Park, M.S., Park, B.H., et al., 2017. Value of serum cystatin C measurement in the diagnosis of sepsis-induced kidney injury and prediction of renal function recovery. Yonsei Med. J. 58 (3), 604–612.

Leslie, J.A., Meldrum, K.K., 2008. The role of interleukin-18 in renal injury. J. Surg. Res. 145 (1), 170–175.

Levey, A.S., Coresh, J., Balk, E., et al., 2003. National Kidney Foundation practice guidelines for chronic kidney disease: evaluation, classification, and stratification. Ann. Intern. Med. 139 (2), 137–147.

Li, Y.H., Li, X.Z., Zhou, X.F., et al., 2013. Impact of sepsis on the urinary level of interleukin-18 and cystatin C in critically ill neonates. Pediatr. Nephrol. 28 (1), 135–144.

Liang, D., Liu, H.-F., Yao, C.-W., et al., 2007. Effects of interleukin 18 on injury and activation of human proximal tubular epithelial cells. Nephrology 12 (1), 53–61.

Lim, A.I., Chan, L.Y.Y., Lai, K.N., et al., 2012. Distinct role of matrix metalloproteinase-3 in kidney injury molecule-1 shedding by kidney proximal tubular epithelial cells. Int. J. Biochem. Cell Biol. 44 (6), 1040–1050.

Lim, A.I., Tang, S.C.W., Lai, K.N., et al., 2013. Kidney injury molecule-1: more than just an injury marker of tubular epithelial cells? J. Cell. Physiol. 228 (5), 917–924.

Lippi, G., Cervellin, G., 2013. Neutrophil gelatinase-associated lipocalin (NGAL), neutrophils, and CKD: which comes first? Am. J. Kidney Dis. 61 (1), 184.

Lippi, G., Plebani, M., 2012. Neutrophil gelatinase-associated lipocalin (NGAL): the laboratory perspective. Clin. Chem. Lab. Med. 50 (9), 1483–1487.

Luo, Q.H., Chen, M.L., Chen, Z.L., et al., 2016. Evaluation of KIM-1 and NGAL as early indicators for assessment of gentamycin-induced nephrotoxicity in vivo and in vitro. Kidney Blood Press. Res. 41 (6), 911–918.

Ma, Q., Devarajan, S.R., Devarajan, P., 2016. Amelioration of cisplatin-induced acute kidney injury by recombinant neutrophil gelatinase-associated lipocalin. Renal. Fail. 38 (9), 1476–1482.

Machado, M.N., Nakazone, M.A., Maia, L.N., 2014. Acute kidney injury based on KDIGO (Kidney Disease Improving Global Outcomes) criteria in patients with elevated baseline serum creatinine undergoing cardiac surgery. Rev. Bras. Cir. Cardiovasc. 29 (3), 299–307.

Melnikov, V.Y., Ecder, T., Fantuzzi, G., et al., 2001. Impaired IL-18 processing protects caspase-1–deficient mice from ischemic acute renal failure. J. Clin. Invest. 107 (9), 1145–1152.

Minowa, Y., Kondo, C., Uehara, T., et al., 2012. Toxicogenomic multigene biomarker for predicting the future onset of proximal tubular injury in rats. Toxicology 297 (1–3), 47–56.

Mishra, J., Ma, Q., Prada, A., et al., 2003. Identification of neutrophil gelatinase-associated lipocalin as a novel early urinary biomarker for ischemic renal injury. J. Am. Soc. Nephrol. 14 (10), 2534–2543.

Moriyama, T., Hagihara, S., Shiramomo, T., et al., 2016. Comparison of three early biomarkers for acute kidney injury after cardiac surgery under cardiopulmonary bypass. J. Intensive Care 4, 6.

Murray, P.T., Mehta, R.L., Shaw, A., et al., 2014. Potential use of biomarkers in acute kidney injury: report and summary of recommendations from the 10th Acute Dialysis Quality Initiative consensus conference. Kidney Int. 85 (3), 513–521.

Mussap, M., Noto, A., Fanos, V., et al., 2014. Emerging biomarkers and metabolomics for assessing toxic nephropathy and acute kidney injury (AKI) in neonatology. Biomed. Res. Int. 16, 602526.

Nozaki, Y., Hino, S., Ri, J.H., et al., 2017. Lipopolysaccharide-induced acute kidney injury is dependent on an IL-18 receptor signaling pathway. Int. J. Mol. Sci. 18 (12), 16.

Pelsers, M.M.A.L., 2008. Fatty acid-binding protein as marker for renal injury. Scand. J. Clin. Lab. Invest. 68, 73–77.

Pena, M.J., Stenvinkel, P., Kretzler, M., et al., 2017. Strategies to improve monitoring disease progression, assessing cardiovascular risk, and defining prognostic biomarkers in chronic kidney disease. Kidney Int. Suppl. 7 (2), 107–113.

Puthumana, J., Ariza, X., Belcher, J.M., et al., 2017. Urine interleukin 18 and lipocalin 2 are biomarkers of acute tubular necrosis in patients with cirrhosis: a systematic review and meta-analysis. Clin. Gastroenterol. Hepatol. 15 (7), 1003–1013.

Raj, D.A.A., Fiume, I., Capasso, G., et al., 2012. A multiplex quantitative proteomics strategy for protein biomarker studies in urinary exosomes. Kidney Int. 81 (12), 1263–1272.

Rees, A.J., Kain, R., 2008. Kim-1/Tim-1: from biomarker to therapeutic target? Nephrol. Dial. Transplant. 23 (11), 3394–3396.

Royakkers, A.A.N.M., Korevaar, J.C., van Suijlen, J.D.E., et al., 2011. Serum and urine cystatin C are poor biomarkers for acute kidney injury and renal replacement therapy. Intensive Care Med. 37 (3), 493–501.

Sabbisetti, V.S., Ito, K., Wang, C., et al., 2013. Novel assays for detection of urinary KIM-1 in mouse models of kidney injury. Toxicol. Sci. 131 (1), 13–25.

Safdar, O.Y., Shalaby, M., Khathlan, N., et al., 2016. Serum cystatin is a useful marker for the diagnosis of acute kidney injury in critically ill children: prospective cohort study. BMC Nephrol. 17, 8.

Schaeffner, E., 2017. Determining the glomerular filtration rate—an overview. J. Ren. Nutr. 27 (6), 375–380.

Schwartz, G.J., Schneider, M.F., Maier, P.S., et al., 2012. Improved equations estimating GFR in children with chronic kidney disease using an immunonephelometric determination of cystatin C. Kidney Int. 82 (4), 445–453.

Shlipak, M.G., Praught, M.L., Sarnak, M.J., 2006. Update on cystatin C: new insights into the importance of mild kidney dysfunction. Curr. Opin. Nephrol. Hypertens. 15 (3).

Singer, E., Markó, L., Paragas, N., et al., 2013. Neutrophil gelatinase-associated lipocalin: pathophysiology and clinical applications. Acta Physiol. 207 (4), 663–672.

Slocum, J.L., Heung, M., Pennathur, S., 2012. Marking renal injury: can we move beyond serum creatinine? Transl. Res. 159 (4), 277–289.

Smith, R.D., 2002. Trends in mass spectrometry instrumentation for proteomics. Trends Biotechnol. 20 (12), s3–s7.

Stevens, L.A., Schmid, C.H., Greene, T., et al., 2009. Factors other than glomerular filtration rate affect serum cystatin C levels. Kidney Int. 75 (6), 652–660.

Sugimoto, K., Toda, Y., Iwasaki, T., et al., 2016. Urinary albumin levels predict development of acute kidney injury after pediatric cardiac surgery: a prospective observational study. J. Cardiothorac. Vasc. Anesth. 30 (1), 64–68.

Susantitaphong, P., Siribamrungwong, M., Doi, K., et al., 2013. Performance of urinary liver-type fatty acid-binding protein in acute kidney injury: a meta-analysis. Am. J. Kidney Dis. 61 (3), 430–439.

Tecson, K.M., Erhardtsen, E., Eriksen, P.M., et al., 2017. Optimal cut points of plasma and urine neutrophil gelatinase-associated lipocalin for the prediction of acute kidney injury among critically ill adults: retrospective determination and clinical validation of a prospective multicentre study. BMJ Open 7 (7), e016028.

van Timmeren, M.M., van den Heuvel, M.C., Bailly, V., et al., 2007. Tubular kidney injury molecule-1 (KIM-1) in human renal disease. J. Pathol. 212 (2), 209–217.

Tolpin, D.A., Collard, C.D., Lee, V.V., et al., 2012. Subclinical changes in serum creatinine and mortality after coronary artery bypass grafting. J. Thorac. Cardiovasc. Surg. 143 (3), 682.

Tsuji, S., Sugiura, M., Tsutsumi, S., et al., 2017. Sex differences in the excretion levels of traditional and novel urinary biomarkers of nephrotoxicity in rats. J. Toxicol. Sci. 42 (5), 615–627.

Turk, V., Bode, W., 1991. The cystatins: protein inhibitors of cysteine proteinases. FEBS Lett. 285 (2), 213–219.

US Food & Drug Administration. List of qualified biomarkers. Retrieved from: https://www.fda.gov/Drugs/DevelopmentApprovalProcess/DrugDevelopmentToolsQualificationProgram/BiomarkerQualificationProgram/ucm535383.htm.

Washino, S., Hosohata, K., Jin, D.N., et al., 2018. Early urinary biomarkers of renal tubular damage by a high-salt intake independent of blood pressure in normotensive rats. Clin. Exp. Pharmacol. Physiol. 45 (3), 261–268.

Weiss, R.H., Kim, K., 2012. Metabolomics in the study of kidney diseases. Nat. Rev. Nephrol. 8 (1), 22–33.

Wijerathna, T.M., Gawarammana, I.B., Dissanayaka, D.M., et al., 2017. Serum creatinine and cystatin C provide conflicting evidence of acute kidney injury following acute ingestion of potassium permanganate and oxalic acid. Clin. Toxicol. 55 (9), 970–976.

Xu, Y., Xie, Y., Shao, X., et al., 2015. L-FABP: a novel biomarker of kidney disease. Clin. Chim. Acta 445, 85–90.

Zeng, X.-F., Lu, D.-X., Li, J.-M., et al., 2017. Performance of urinary neutrophil gelatinase-associated lipocalin, clusterin, and cystatin C in predicting diabetic kidney disease and diabetic microalbuminuria: a consecutive cohort study. BMC Nephrol. 18, 233–243.

Zhang, P.L., Rothblum, L.I., Han, W.K., et al., 2008. Kidney injury molecule-1 expression in transplant biopsies is a sensitive measure of cell injury. Kidney Int. 73 (5), 608–614.

Zhang, A., Sun, H., Wang, P., et al., 2012. Recent and potential developments of biofluid analyses in metabolomics. J. Proteomics 75 (4), 1079–1088.

Zhang, W., Zhang, L., Chen, Y.X., et al., 2014. Identification of nestin as a urinary biomarker for acute kidney injury. Am. J. Nephrol. 39 (2), 110–121.

Zhang, J.J., Wu, X.H., Gao, P.Z., et al., 2017. Correlations of serum cystatin C and glomerular filtration rate with vascular lesions and severity in acute coronary syndrome. BMC Cardiovasc. Disord. 17, 47.

Zhou, R., Xu, Y.H., Shen, J., et al., 2016a. Urinary KIM-1: a novel biomarker for evaluation of occupational exposure to lead. Sci. Rep. 6, 7.

Zhou, X.B., Qu, Z., Zhu, C., et al., 2016b. Identification of urinary microRNA biomarkers for detection of gentamicin-induced acute kidney injury in rats. Regul. Toxicol. Pharmacol. 78, 78–84.

Zubowska, M., Wyka, K., Fendler, W., et al., 2013. Interleukin 18 as a marker of chronic nephropathy in children after anticancer treatment. Dis. Markers 35 (6), 811–818.

CHAPTER

16

Gastrointestinal Toxicity Biomarkers

Aryamitra Banerjee[1], *Ramesh C. Gupta*[2]

[1]Study Director, Toxicology Research Laboratory, University of Illinois at Chicago, Chicago, IL, United States;
[2]Toxicology Department, Breathitt Veterinary Center, Murray State University, Hopkinsville, Kentucky, United States

INTRODUCTION

The gastrointestinal (GI) tract serves as the normal entry and absorption route for dietary products, which makes it prone to direct exposure to toxicants. This organ presents a high degree of structural complexity with varied tissue types serving different functions, which essentially renders it sensitive to a multitude of natural, chemical, and radiation assaults at all times.

Traditionally, the occurrence of GI toxicity has been monitored using histopathological examination, presence of occult blood in feces, evaluation of GI content, and prediction of effects on GI absorption and contractility using in vitro/ex vivo systems.

However, there has been a constant search for clinically relevant minimally invasive/noninvasive biomarkers to better achieve prediction and prognosis of GI damage.

GI damage is most commonly encountered due to (1) nonsteroidal anti-inflammatory drugs (NSAID) overuse/sensitivity, (2) chemotherapy, (3) radiation, and (4) metabolic disorders. Key biomarkers have been known and assessed in different biological samples for monitoring of disease progression and prognosis (Fig. 16.1).

NSAID-INDUCED GASTROINTESTINAL TOXICITY

The primary mode of action of NSAIDs through the inhibition of the catalyzing enzyme cyclooxygenase (d) is also known to result in GI side effects, ranging from mild dyspepsia to GI hemorrhage and perforation. The most common adverse effects of NSAIDs include upper GI tract damage with manifestation as gastric and intestinal ulcers (Cryer, 2000; Carr et al., 2017). Gastroduodenal mucosal injury due to NSAIDs includes not only a simple topical injury but also complex mechanisms with both local and systemic effects. The systemic effects principally result from the inhibition of endogenous prostaglandin synthesis (Schoen and Vender, 1989), which in turn leads to a decrease in epithelial mucus, secretion of bicarbonate, mucosal blood flow, epithelial proliferation, and mucosal resistance to injury (Whittle, 1977; Wolfe and Soll, 1988). The impairment in mucosal resistance permits further injury by endogenous factors, such as acid, pepsin, and bile salts, as well as by exogenous factors such as other NSAIDs and possibly alcohol and other toxicants.

CHEMOTHERAPY-INDUCED GASTROINTESTINAL TOXICITY

Chemotherapy is well associated with various side effects, with toxicity to the GI tract being a major clinical concern. Chemotherapy-induced GI toxicity is the key dose-limiting factor for any chemotherapeutic regimen which is not simply targeted to the upper gastroduodenal mucosa but also extends along the entire GI tract (Denham and Abbott, 1991) and hence chemotherapy-induced GI mucositis causes a major oncological challenge. The mucositis development pathway includes the activities of mitogen-activated protein kinase signaling, nuclear factor kappa B (NF-κB) signaling, Fos/Jun signaling, and Wnt signaling (Bowen et al., 2007). Cytokines, matrix metalloproteinases (MMPs), ceramide, and COX-2 are also activated as key downstream mediators of this mucosal disruption mechanism (Logan et al., 2008; Al-Dasooqi et al., 2010).

Biomarkers in Toxicology, Second Edition
https://doi.org/10.1016/B978-0-12-814655-2.00016-5

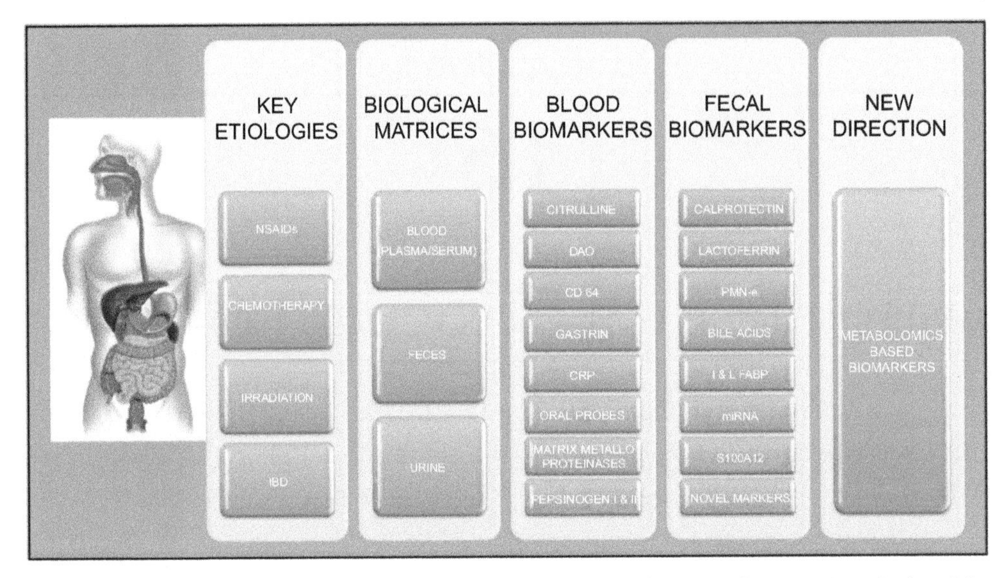

FIGURE 16.1 Summary of causes, analytical matrix, and key biomarkers currently in use for gastrointestinal toxicity monitoring across species.

RADIATION-INDUCED GASTROINTESTINAL TOXICITY

Complete manifestation of symptoms of the GI syndrome resulting as an outcome of acute radiation syndrome usually occurs with a dose greater than approximately 10 Gy (1000 rad) in humans. The pathophysiological mechanisms of gastric syndrome are complex and involve loss of clonogenic crypt cells with eventual depopulation of the intestinal villi, faulty regeneration of the intestinal stem cells postirradiation, and a systemic inflammatory response syndrome from a multitude of cytokines and growth factors released into the systemic circulation following radiation exposure and translocation of gut microbes (Potten, 1998; Marshman et al., 2002). Rapid turnover of intestinal epithelial cells causes the intestinal mucosa to be especially sensitive to high radiation exposure during radiation therapy or in any other nuclear exposure. Thus, maintenance of intestinal homeostasis is critical to combat radiation-induced GI injury (Saha et al., 2012). Organisms most commonly respond to irradiation by altering the expression and/or the posttranslational modifications of some proteins in cells, tissues, and/or organic fluids, such as serum or urine. Hence, it is perceived that protein expression profiling can be useful to evaluate a protein or a group of protein expression changes that can clearly differentiate between irradiated and nonirradiated biological systems, and between recoverable or irrecoverable radiation injuries. For further details on biomarkers of radiation-induced GI effects refer to Ray et al. (2014).

INFLAMMATORY BOWEL DISEASES

The two major forms of the chronic inflammatory bowel diseases (IBDs) are Crohn's disease (CD) and ulcerative colitis (UC). CD is characterized by discontinuous regions of intestinal inflammation normally involving the terminal ileum and colon, but can eventually affect any part of the GI tract. The inflammatory mechanism of UC is limited to the mucosa and submucosa of the colon, almost invariably affecting the rectum eventually. Biomarkers are currently used in conjunction with routine testing in the clinical care of patients with IBD (Iskandar and Ciorba, 2012).

BIOMARKERS FOR GASTROINTESTINAL DAMAGE

A number of GI toxicity biomarkers have been tested to investigate their usefulness in tracking adverse events. During this process, multiple analytical methodologies have been described for in vivo applications to track GI function in a number of species and in accessible biologic fluids such as blood and urine.

BLOOD BIOMARKERS

Citrulline

Citrulline is an intermediate metabolic amino acid produced mainly by enterocytes of the small intestine.

Levels of citrulline have been correlated with chemotherapeutically reduced enterocyte mass, independent of nutritional and inflammatory status. Plasma citrulline measurements have been used most recently in myeloablative treatment regimens as being indicative of mucosal mass (Butler, 2008; Crenn et al., 2008). Analysis of blood citrulline is challenging, and routine analysis has been limited to enzymatic assays, which are often less sensitive. However, Crenn et al. (2008) presented the utilization of high-performance liquid chromatography (HPLC) and mass spectrometry, to present citrulline as a promising GI biomarker. Alteration in diet and gut microflora is expected to alter biochemical and metabolic pathways that may lead to changes in metabolite measurements such as citrulline. However, these variables are likely to be normalized throughout the study design with the maintenance of uniform diet, age, and sex. The LC-MS/MS method is inherently highly specific for citrulline measurements and the presence of interfering metabolite components can be easily nullified.

Recently, Carr et al. (2017) reported that as a small intestinal toxicity, plasma citrulline appears to have some potential in nonclinical safety studies. However, further work is required to determine its applicability in humans as well as other preclinical species. The application of citrulline to in vitro models of intestinal disease/toxicity also requires further exploration.

Diamine Oxidase

Diamine oxidase (DAO) is an enzyme that catabolizes a variety of substrates including histamine and diamines. It is the degradative enzyme in the catabolic pathway of polyamines found in high activity in the mature upper villus cells of the rat intestinal mucosa (Luk et al., 1983; Wolvekamp and Debruin, 1994). Blood DAO activity has been well correlated with DAO activity in the villi of the small intestinal mucosa and also with the severity of small intestinal mucosal lesions induced by anticancer drugs (Luk et al., 1980; Tsujikawa et al., 1999).

However, the measurement of blood DAO has been an analytical challenge because of the extremely low measurable levels in the blood. Wolvekamp and de Bruin (1994) and Tsunooka et al. (2004) reported that DAO measurement is easily confounded by the fact that plasma levels rise markedly on heparin stimulation prior to blood draws (Wolvekamp and Debruin, 1994; Tsunooka et al., 2004). This causes blood DAO level use to be both impractical and inconvenient for routine GI toxicity screening in a preclinical setting. Thus, this biomarker measurement has not gained sufficient popularity in clinical or preclinical use.

CD64

Tillinger et al. (2009) suggested the flow cytometric measurement of neutrophilic CD64 as a useful biomarker for gastroenterologic injury in patients with inflammatory and functional diseases of the intestine. CD64, a high-affinity FcRI receptor for IgG, serves as a ligand for acute phase reactants, C-reactive protein (CRP) and amyloid P (Tillinger et al., 2009). CD64 was found to respond well to therapeutic interventions and was significantly higher in patients with IBD. Hence, it found an application as a biomarker to discriminate between gastroenterologic disorders. At this time, mouse antibodies are available for preclinical application and antibodies in other species are being researched to evaluate CD64 in other relevant preclinical in vivo toxicology studies in addition to clinical application.

Gastrins

Gastrins are a family of sequence-related carboxyamidated peptides produced by the endocrine G cells of the gastric antrum and duodenum in response to various GI stimuli. The mechanisms for gastrin release from G cells are not well understood, although antral distension, partially digested proteins, amino acids, and vagal stimulation appear to be involved in the process. Increased serum gastrin levels are generally associated with duodenal ulcers, bacterial infections, tumors, and other causative factors associated with GI damage. Serum gastrin levels were measured in rats treated with omeprazole and the effects of omeprazole on normal and ethanol-damaged gastric mucosa were observed (Fattaha and Abdel-Rahman, 2000). Serum gastrin levels were found to increase significantly in omeprazole-treated animals versus control or ethanol-treated rats. It has also been shown that chronic use of antisecretory drugs may induce hypergastrinemia by inflammatory mechanisms and the release of gastrin from antral G cells is closely regulated by neuroendocrine mechanisms possibly sensitive to local pH. The acidity of the gastric lumen has also been seen to decrease gastrin secretion through the release of somatostatin from D cells (Alvarez et al., 2007). Plasma gastrin levels were found to have increased in rats treated with acetic acid, which is known to induce gastric ulcers (Sun et al., 2002). Increased serum gastrin concentrations were also observed in dogs with chronic lymphocytic–plasmacytic enteritis compared to dogs without GI disease (Garcia-Sancho et al., 2005). Hence, a great deal of evidence exists to define gastrin as a good biomarker for GI modification, though a clear understanding of the mechanism is currently not available.

C-reactive Protein

CRP is a hepatic protein produced in response to a variety of acute and chronic inflammatory conditions and is often useful as an indirect biomarker of GI inflammation. Cytokines associated with active IBD [i.e., interleukin (IL)-6, tumor necrosis factor alpha (TNFα), and IL-1β] stimulate increased production of CRP by hepatocytes over baseline levels (Darlington et al., 1986). Such elevated CRP levels are not specific to IBD as levels are also increased in various viral and bacterial infections, other autoimmune disorders, malignancy, and other toxicities resulting in tissue necrosis (Pepys and Hirschfield, 2003). CRP serves as a robust biomarker as it is easily and reliably measured across diagnostic laboratories and has a short plasma half-life of ~19 h, which is determined by synthesis rather than degradation (Vigushin et al., 1993).

Orally Administered Probes

Fecal and urinary excretion of orally administered probes has been used to assess mucosal permeability as a marker of GI tract dysfunction and drug-induced intestinal barrier damage. Urine sucrose and ^{51}Cr-EDTA measured by radioimmunoassay correlated well with gross toxicological and pathological changes in a rat model of GI damage induced by various anticancer agents (Yanez et al., 2003). The use of ^{51}Cr-EDTA is confined to animal studies due to its radioactive nature and is most often not practical (Butler, 2008). D-xylose and 3-oxy-methyl-D-glucose (3-OMG) have been proposed as additional permeability biomarkers to reflect intestinal epithelial absorptive capacity and rhamnose/3-OMG ratio to assess intestinal transport function (Bjarnason et al., 1995). However, Butler (2008) argued that these tests measure small intestinal leakiness and additionally require HPLC for estimation and hence are also cumbersome. The ^{13}C sucrose breath test is also used to evaluate the absorptive capacity of the small intestine after ingestion of an isotope substrate and has been used in several animal models of chemotherapy-induced intestinal damage (Butler, 2008). The permeability of the small intestine can be used to indicate barrier function in a relatively noninvasive manner. In a healthy GI tract, disaccharides do not usually transfer across the mucosa. Using this concept, other methods utilizing differential absorption of monosaccharide and disaccharide in the small intestine, such as L-rhamnose and lactulose, were initiated to assess sugar permeability in celiac patients with villous atrophy (Menzies et al., 1979). The sugar permeability test measures urinary excretion of orally administered nonmetabolizable sugars and hence monitors the functioning of the small intestine (Keefe et al., 1997). The dual-permeability test using L-rhamnose and lactulose determines small bowel surface area and enterocyte tight junction concentration, respectively (Keefe et al., 1997). A correlation between high-dose chemotherapeutic effects (enteropathy and permeability of tight junctions) and altered monosaccharide and disaccharide urine concentrations has been established (Keefe et al., 1997; Daniele et al., 2001; Melichar et al., 2001).

Matrix Metalloproteinases

MMPs are a group of zinc-dependent endopeptidases that work toward extracellular matrix turnover (Al-Dasooqi et al., 2010). Recent research has identified these proteases as important mediators of chemotherapy-induced gut damage in rats (Page-McCaw et al., 2007). Hence, MMPs can be used as tissue-specific markers of chemotherapy-induced gut injury. However, the changes in MMPs have not yet been monitored sufficiently at this time for their systemic presence to be useful as toxicity biomarkers.

Pepsinogen I and II

Pepsinogen I and II, the precursors of pepsin, are produced by the gastric mucosa and released into the gastric lumen and peripheral circulation (Samloff and Liebman, 1973). Severe inflammation of gastric mucosa and its progression toward atrophic gastritis is correlated with a change in serum biomarkers pepsinogen I and II. The atrophy of the corpus mucosa leads to low synthesis of pepsinogen I and hence its low release into the serum. Advanced gastric atrophy and hypo- or achlorhydrias cause very low levels of pepsinogen I in contrast to pepsinogen II, which remains elevated in serum (Samloff, 1982; Samloff et al., 1982).

It has been previously reported that pepsinogen II is a good marker for the diagnosis of any type of gastritis as it can differentiate between subjects with gastritis from those with normal mucosa. The progression of gastritis to acute and pangastritis is associated with a continual increase in serum pepsinogen II levels, whereas the levels of pepsinogen I in the early stage of acute gastritis remain unchanged and in the normal range (Haj-Sheykholeslami et al., 2008).

FECAL BIOMARKERS

Calprotectin

Calprotectin is a calcium-binding protein found in abundance in neutrophils, where it accounts for 60% of the soluble proteins in the cytosol. Lower concentrations are found in monocytes and reactive macrophages.

Calprotectin is a sensitive, stable marker that is unaffected by medication, dietary supplements, or enzymatic degradation. This neutrophil-derived protein reflects the flux of leukocytes into the intestinal lumen. Studies indicate that increased levels of fecal calprotectin are found in IBD (Roseth et al., 1992; Bunn et al., 2001), colonic cancer (Roseth et al., 1993), and NSAID treatment (Meling et al., 1996; Tibble et al., 1999), suggesting it is a sensitive marker of GI inflammation. Accordingly, fecal calprotectin is now used for clinical assessment of GI inflammation in humans, using standard ELISA procedures.

Calprotectin, a 36.5 kDa nonglycosylated protein, plays a central role in neutrophil defense, and its fecal level correlates well with the numbers of neutrophils infiltrating the intestinal mucosa and the overall severity of GI inflammation. It is known to be stable in feces for several days after excretion (Roseth et al., 1999). However, the presence of calprotectin in feces appears to be directly related to neutrophil infiltration rather than GI damage. The release of calprotectin is most likely a consequence of cell disruption and death (Voganatsi et al., 2001), but it is also actively secreted and is hence useful as a fecal marker (Rammes et al., 1997).

Lactoferrin

Lactoferrin, an iron-binding glycoprotein, is secreted by most mucosal membranes and is a major component of the secondary granules of polymorphonuclear neutrophils, which constitute a primary component of the acute inflammatory response (Levay and Viljoen, 1995; Baveye et al., 1999). Other hematopoietic cells such as monocytes and lymphocytes do not contain lactoferrin (Naidu et al., 1997). During intestinal inflammation, leukocytes infiltrate the mucosa, resulting in an increase in the concentration of lactoferrin in the feces (Guerrant et al., 1992). Lactoferrin is resistant to proteolysis and unaffected by multiple freeze–thaws, providing a useful marker in feces as an indicator of intestinal inflammation.

Polymorphonuclear Neutrophil Elastase

Polymorphonuclear neutrophil elastase (PMN-e) is a neutral proteinase normally stored in the azurophil granules of polymorphonuclear neutrophils. It is released by activation of these cells as a mediator of inflammation (Poullis et al., 2002). Studies are ongoing to establish the utility of PMN-e as a useful GI fecal biomarker.

Bile Acids

Bile acids are minimally absorbed in the proximal small intestine, and the bile acid pool flows to the distal ileum, where the acids are reabsorbed by the enterohepatic transport system and then returned to the liver by the portal vein (Westergaard, 2007). Remaining bile acids are excreted in feces. The fecal bile acids are a complex mixture of metabolites of bile acids produced by intestinal microorganisms (Hirofuji, 1965). As a result, increases in fecal bile acids are indicative of malabsorption, which can cause diarrhea. However, the ability of fecal bile acids to accurately detect GI damage and/or decreased GI functionality is questionable based on assay reliability, though it is commonly used as a GI marker for clinical applications.

I-FABP and L-FABP

The small intestine is the initial site of dietary fatty acid uptake, and proximal intestinal enterocytes contain two fatty acid-binding proteins (FABPs), namely liver FABP (L-FABP) and intestinal FABP (I-FABP), at relatively high concentrations (Bass, 1985). I-FABP occurs in the enterocytes of the small intestine and constitutes 2% of enterocyte protein (Ockner and Manning, 1974; Sacchettini et al., 1990). Studies have described the use of I-FABP for the detection of rat intestinal injury after acute ischemic diseases (Gollin et al., 1993), rejection (Marks and Gollin, 1993; Morrissey et al., 1996), and necrotic enterocolitis (Gollin and Marks, 1993). Intestinal cells also express L-FABP, which occurs in liver and additionally in kidney (Ockner et al., 1982; Bass et al., 1989). I-FABP and L-FABP are very small proteins, which are released rapidly from GI enterocytes into the blood after cellular damage, making them good biomarker candidates.

MicroRNA

MicroRNAs (miRNAs) are small, endogenous noncoding RNAs that act as posttranscriptional regulators of gene expression. As emerging biomarkers of several organ toxicities, miRNAs are also being evaluated for their potential to track damage to the gastric and intestinal mucosa. On cell death, miRNAs are released into the surrounding environment and are stably present in the circulation (Gilad et al., 2008; Mitchell et al., 2008). Therefore, detection of tissues-specific or tissue-enriched miRNA in biological samples is expected to be a good biomarker of damage to that tissue. Serum miRNAs have been identified as biomarkers for liver toxicity (Wang et al., 2009) and myocardial injury (Adachi et al., 2010), as well as for lung, colorectal, ovarian, and prostate cancers (Mitchell et al., 2008). Urine miRNAs are being studied as indicators of renal tubular damage and a fecal miRNA assay is being developed as a screening tool for sporadic human colon cancer and active UC (Ahmed et al., 2009). The miRNA named as miR-194 was chosen for initial experiments as it is highly expressed in the small intestine and colon, although it is not entirely specific (Godwin et al., 2010). Therefore, the presence of GI-specific or GI-enriched miRNA in feces or any other relevant biological sample could be indicative of GI toxicity.

Fecal S100A12

S100A12 is similar to calprotectin in its calcium-binding properties (Kaiser et al., 2007). This protein activates NF-κB signal transduction and enhances cytokine release (Mendoza and Abreu, 2009). S1000A12 is also detectable in serum, but the fecal assay is more sensitive and specific for IBD (Manolakis et al., 2010). Further work with this biomarker is ongoing.

Other Novel Approaches

Several other fecal biomarkers are being investigated for use in GI conditions and, though promise exists, these alternatives have shown lower correlation to disease activity (van der Sluys Veer et al., 1999). These include lysozyme, leukocyte esterase, elastase, myeloperoxidase, TNFα, IL-beta, IL-4, IL-10, alpha-1 antitrypsin, and alpha-2-macroglobulin (Desai et al., 2007; Mendoza and Abreu, 2009; Lewis, 2011). M2-pyruvate kinase is possibly the most promising of these new fecal biomarkers (Judd et al., 2011). Gene expression analysis has shown that the presumed adipocyte-specific protein, adipsin, had similarly elevated mRNA levels and is a potential marker for Notch/Hes-1 signaling disruption effects in the GI tract of rats. These protein levels were found to be upregulated in the intestinal contents and feces of FGSI-treated rats, which could be monitored noninvasively (Searfoss et al., 2003). Identification of such biomarkers that may predict these irradiation or subsequent treatment-associated adverse events is of great importance. In this context, peripheral blood biomarkers or noninvasive biomarkers would be preferred, as these are less invasive than a colonic biopsy. In another study, increases in expression of neutrophil activation markers, CD177 and CEACAM1, were found to be associated with the occurrence of GI immune-related adverse events (Shahabi et al., 2013), which is also being studied further for biomarker application.

CONCLUDING REMARKS AND FUTURE DIRECTIONS

Although GI toxicity is a significant dose-limiting safety concern noted in multiple therapeutic areas, there are no GI biomarkers that can accurately track, precede, or reliably correlate with predicted and histologic evidence of GI damage. While significant efforts have been made within the pharmaceutical industry, academia, and consortia to address the biomarker gaps in other target organs such as liver, kidney, and muscle (cardiac and skeletal), there have been no concerted efforts in the area of GI biomarkers. Moreover, the markers available are not suitable to be used across species to be appropriate for clinical as well as nonclinical research applications. Attempts are currently ongoing to qualify several biomarkers as discussed, to be applicable for diagnostic and research use in humans and as laboratory animals.

One of the key new directions of novel development in GI biomarker research is the understanding of metabolome-based biomarkers.

METABOLOME-BASED BIOMARKERS

The metabolomics approach is believed to have the potential to reveal novel biochemical outcomes of toxicant administration that can lead to mechanistic insights and identification of biomarkers of cause and/or effect (Robertson et al., 2011; Suzuki et al., 2014). Metabolomic fingerprinting can be conducted on minimally prepared peripheral biological fluids, and sample differentiation is an outcome of spectroscopic or chromatographic data. These "analytical fingerprints" comprise thousands of individual data points that relate to the composition of the sample, which may indicate their molecular origin. Such a data set is used as input into multivariate statistical tools such as principal component analysis. This approach avoids the time-consuming step of data annotation and therefore may be rapidly applied for establishing differentiation between individuals or groups, with applications during in vivo screening of similar drug candidates (Robosky et al., 2002; Dieterle et al., 2006; Robertson et al., 2011), prescreening animals before extensive toxicological evaluations, and creating toxicological classification models (Lindon et al., 2005). Evaluation of effects on gut flora has been a difficult task so far but metabolomic analyses should simplify that task such that gut flora as a potential target could be routinely studied in toxicity studies (Backshall et al., 2009). In one recent study, irradiation-induced metabolic changes in GI tissue of CD2F1 mice was estimated using ultra-performance liquid chromatography coupled with electrospray time-of-flight mass spectrometry. GI injury is a critical determinant of survival after exposure to irradiation and results from this metabolomics analysis showed a clear dose- and time-dependent response to GI tissue injury. GI tissue metabolomics profiles provided an information-rich matrix that led to the identification of metabolites such as lipids and organic acids, as well as amino acids, that may impact overall GI metabolism and in part explain irradiation-induced GI toxicity (Ghosh et al., 2013).

In conclusion, it is safe to say that there still exists a lacuna in terms of a robust GI damage marker, and further focused research on identification and qualification of specific and rapid analytical methods for GI toxicity biomarkers for clinical and nonclinical application is the need of the hour.

References

Adachi, T., Nakanishi, M., Otsuka, Y., 2010. Plasma microRNA 499 as a biomarker of acute myocardial infarction. Clin. Chem. 56, 1183–1185.

Ahmed, F.E., Jeffries, C.D., Vos, P.W., 2009. Diagnostic microRNA markers for screening sporadic human colon cancer and active ulcerative colitis in stool and tissue. Cancer Genom. Proteom. 6, 281–295.

Al-Dasooqi, N., Gibson, R.J., Bowen, J.M., 2010. Matrix metalloproteinases are possible mediators for the development of alimentary tract mucositis in the dark agouti rat. Exp. Biol. Med. 235, 1244–1256.

Alvarez, A., Ibiza, M.S., Andrade, M.M., et al., 2007. Gastric antisecretory drugs induce leukocyte-endothelial cell interactions through gastrin release and activation of CCK-2 receptors. J. Pharmacol. Exp. Ther. 323, 406–413.

Backshall, A., Alferez, D., Teichert, F., 2009. Detection of metabolic alterations in non-tumor gastrointestinal tissue of the Apc(Min/+) mouse by (1)H MAS NMR spectroscopy. J. Proteome Res. 8, 1423–1430.

Bass, N.M., 1985. Function and regulation of hepatic and intestinal fatty acid binding proteins. Chem. Phys. Lipids 38, 95–114.

Bass, N.M., Barker, M.E., Manning, J.A., et al., 1989. Acinar heterogeneity of fatty acid binding protein expression in the livers of male, female and clofibrate-treated rats. Hepatology 9, 12–21.

Baveye, S., Elass, E., Mazurier, J., et al., 1999. Lactoferrin: a multifunctional glycoprotein involved in the modulation of the inflammatory process. Clin. Chem. Lab. Med. 37, 281–286.

Bjarnason, I., Macpherson, A., Hollander, D., 1995. Intestinal permeability – an overview. Gastroenterology 108, 1566–1581.

Bowen, J.M., Gibson, R.J., Cummins, A.G., et al., 2007. Irinotecan changes gene expression in the small intestine of the rat with breast cancer. Cancer Chemother. Pharm. 59, 337–348.

Bunn, S.K., Bisset, W.M., Main, J.C., et al., 2001. Fecal calprotectin as a measure of disease activity in childhood inflammatory bowel disease. J. Pediatr. Gastr. Nutr. 32, 171–177.

Butler, R.N., 2008. Measuring tools for gastrointestinal toxicity. Curr. Opin. Support. Palliat. Care 2, 35–39.

Carr, D.F., Ayehunie, S., Davies, A., et al., 2017. Towards better models and mechanistic biomarkers for drug-induced gastrointestinal injury. Pharmacol. Ther. 172, 181–194.

Crenn, P., Messing, B., Cynober, L., 2008. Citrulline as a biomarker of intestinal failure due to enterocyte mass reduction. Clin. Nutr. 27, 328–339.

Cryer, B., 2000. NSAID gastrointestinal toxicity. Curr. Opin. Gastroenterol. 16, 495–502.

Daniele, B., Secondulfo, M., De Vivo, R., 2001. Effect of chemotherapy with 5-fluorouracil on intestinal permeability and absorption in patients with advanced colorectal cancer. J. Clin. Gastroenterol. 32, 228–230.

Darlington, G.J., Wilson, D.R., Lachman, L.B., 1986. Monocyte-conditioned medium, Interleukin-1, and tumor-necrosis-factor stimulate the acute phase response in human hepatoma-cells in vitro. J. Cell Biol. 103, 787–793.

Denham, J.W., Abbott, R.L., 1991. Concurrent cisplatin, infusional fluorouracil, and conventionally fractionated radiation-therapy in head and neck cancer–dose-limiting mucosal toxicity. J. Clin. Oncol. 9, 458–463.

Desai, D., Faubion, W.A., Sandborn, W.J., 2007. Review article: biological activity markers in inflammatory bowel disease. Aliment. Pharmacol. Ther. 25, 247–255.

Dieterle, F., Schlotterbeck, G.T., Ross, A., et al., 2006. Application of metabonomics in a compound ranking study in early drug development revealing drug-induced excretion of choline into urine. Chem. Res. Toxicol. 19, 1175–1181.

Fattaha, N.A.A., Abdel-Rahman, M.S., 2000. Effects of omeprazole on ethanol lesions. Toxicol. Lett. 118, 21–30.

Garcia-Sancho, M., Rodriguez-Franco, F., Sainz, A., et al., 2005. Serum gastrin in canine chronic lymphocytic-plasmacytic enteritis. Can. Vet. J. 46, 630–634.

Ghosh, S.P., Singh, R., Chakraborty, K., 2013. Metabolomic changes in gastrointestinal tissues after whole body radiation in a murine model. Mol. Biosyst. 9, 723–731.

Gilad, S., Meiri, E., Yogev, Y., 2008. Serum microRNAs are promising novel biomarkers. PLoS One 3, e3148.

Godwin, J.G., Ge, X., Stephan, K., 2010. Identification of a microRNA signature of renal ischemia reperfusion injury. Proc. Natl. Acad. Sci. USA 107, 14339–14344.

Gollin, G., Marks, W.H., 1993. Elevation of circulating intestinal fatty acid binding protein in a luminal contents-initiated model of NEC. J. Pediatr. Surg. 28, 367–370.

Gollin, G., Marks, C., Marks, W.H., 1993. Intestinal fatty acid binding protein in serum and urine reflects early ischemic injury to the small bowel. Surgery 113, 545–551.

Guerrant, R.L., Araujo, V., Soares, E., 1992. Measurement of fecal lactoferrin as a marker of fecal leukocytes. J. Clin. Microbiol. 30, 1238–1242.

Haj-Sheykholeslami, A., Rakhshani, N., Amirzargar, A., 2008. Serum pepsinogen I, pepsinogen II, and gastrin 17 in relatives of gastric cancer patients: comparative study with type and severity of gastritis. Clin. Gastroenterol. Hepatol. 6, 174–179.

Hirofuji, S., 1965. Stero-bile acids and bile sterols. LXXII. Fecal bile acids in the dog. J. Biochem. 58, 27–33.

Iskandar, H.N., Ciorba, M.A., 2012. Biomarkers in inflammatory bowel disease: current practices and recent advances. Transl. Res. 159, 313–325.

Judd, T.A., Day, A.S., Lemberg, D.A., et al., 2011. Update of fecal markers of inflammation in inflammatory bowel disease. J. Gastroen. Hepatol. 26, 1493–1499.

Kaiser, T., Langhorst, J., Wittkowski, H., 2007. Faecal S100A12 as a non-invasive marker distinguishing inflammatory bowel disease from irritable bowel syndrome. Gut 56, 1706–1713.

Keefe, D.M., Cummins, A.G., Dale, B.M., 1997. Effect of high-dose chemotherapy on intestinal permeability in humans. Clin. Sci. (Lond.) 92, 385–389.

Levay, P.F., Viljoen, M., 1995. Lactoferrin: a general review. Haematologica 80, 252–267.

Lewis, J.D., 2011. The utility of biomarkers in the diagnosis and therapy of inflammatory bowel disease. Gastroenterology 140, 1817–1826.

Lindon, J.C., Keun, H.C., Ebbels, T.M.D., 2005. The consortium for metabonomic toxicology (COMET): aims, activities and achievements. Pharmacogenomics 6, 691–699.

Logan, R.M., Gibson, R.J., Bowen, J.M., 2008. Characterisation of mucosal changes in the alimentary tract following administration of irinotecan: implications for the pathobiology of mucositis. Cancer Chemother. Pharm. 62, 33–41.

Luk, G.D., Bayless, T.M., Baylin, S.B., 1980. Diamine oxidase (Histaminase) – a circulating marker for rat intestinal mucosal maturation and integrity. J. Clin. Invest. 66, 66–70.

Luk, G.D., Bayless, T.M., Baylin, S.B., 1983. Plasma post-heparin diamine oxidase – sensitive provocative test for quantitating length of acute intestinal mucosal injury in the rat. J. Clin. Invest. 71, 1308–1315.
Manolakis, A.C., Kapsoritakis, A.N., Georgoulias, P., 2010. Moderate performance of serum S100A12, in distinguishing inflammatory bowel disease from irritable bowel syndrome. BMC Gastroenterol. 10, 118.
Marks, W.H., Gollin, G., 1993. Biochemical detection of small intestinal allograft rejection by elevated circulating levels of serum intestinal fatty acid binding protein. Surgery 114, 206–210.
Marshman, E., Booth, C., Potten, C.S., 2002. The intestinal epithelial stem cell. Bioessays 24, 91–98.
Melichar, B., Kohout, P., Bratova, M., et al., 2001. Intestinal permeability in patients with chemotherapy-induced stomatitis. J. Cancer Res. Clin. Oncol. 127, 314–318.
Meling, T.R., Aabakken, L., Roseth, A., Osnes, M., 1996. Faecal calprotectin shedding after short-term treatment with non-steroidal anti-inflammatory drugs. Scand. J. Gastroenterol. 31, 339–344.
Mendoza, J.L., Abreu, M.T., 2009. Biological markers in inflammatory bowel disease: practical consideration for clinicians. Gastroenterol. Clin. Biol. 33 (Suppl. 3), S158–S173.
Menzies, I.S., Laker, M.F., Pounder, R., 1979. Abnormal intestinal permeability to sugars in villous atrophy. Lancet 2, 1107–1109.
Mitchell, P.S., Parkin, R.K., Kroh, E.M., 2008. Circulating microRNAs as stable blood-based markers for cancer detection. Proc. Natl. Acad. Sci. USA 105, 10513–10518.
Morrissey, P.E., Gollin, G., Marks, W.H., 1996. Small bowel allograft rejection detected by serum intestinal fatty acid-binding protein is reversible. Transplantation 61, 1451–1455.
Naidu, A.S., Arnold, R., Hutchens, T.W., Lönnerdal, B., 1997. Influence of lactoferrin on host-microbe interactionsLactoferrin. In: Hutchens, T.W., Lönnerdal, B. (Eds.), Lactoferrin. Humana Press, Totowa NJ, pp. 259–275.
Ockner, R.K., Manning, J.A., 1974. Fatty acid-binding protein in small intestine. Identification, isolation, and evidence for its role in cellular fatty acid transport. J. Clin. Invest. 54, 326–338.
Ockner, R.K., Manning, J.A., Kane, J.P., 1982. Fatty acid binding protein. Isolation from rat liver, characterization, and immunochemical quantification. J. Biol. Chem. 257, 7872–7878.
Page-McCaw, A., Ewald, A.J., Werb, Z., 2007. Matrix metalloproteinases and the regulation of tissue remodelling. Nat. Rev. Mol. Cell Biol. 8, 221–233.
Pepys, M.B., Hirschfield, G.M., 2003. C-reactive protein: a critical update. J. Clin. Invest. 112, 299 (vol. 111, pg 1805, 2003).
Potten, C.S., 1998. Stem cells in gastrointestinal epithelium: numbers, characteristics and death. Philos. T. R Soc. B. 353, 821–830.
Poullis, A., Foster, R., Northfield, T.C., Mendall, M.A., 2002. Review article: faecal markers in the assessment of activity in inflammatory bowel disease. Aliment. Pharmacol. Ther. 16, 675–681.
Rammes, A., Roth, J., Goebeler, M., 1997. Myeloid-related protein (MRP) 8 and MRP14, calcium-binding proteins of the S100 family, are secreted by activated monocytes via a novel, tubulin-dependent pathway. J. Biol. Chem. 272, 9496–9502.
Ray, K., Hudak, K., Citrin, D., et al., 2014. Biomarkers of radiation injury and response. In: Gupta, R.C. (Ed.), Biomarkers in Toxicology. Academic Press/Elsevier, Amsterdam, pp. 673–687.
Robertson, D.G., Watkins, P.B., Reily, M.D., 2011. Metabolomics in toxicology: preclinical and clinical applications. Toxicol. Sci. 120, S146–S170.
Robosky, L.C., Robertson, D.G., Baker, J.D., Rane, S., Reily, M.D., 2002. In vivo toxicity screening programs using metabonomics. Comb. Chem. High T. Scr. 5, 651–662.
Roseth, A.G., Fagerhol, M.K., Aadland, E., Schjonsby, H., 1992. Assessment of the neutrophil dominating protein calprotectin in feces – a methodologic study. Scand. J. Gastroenterol. 27, 793–798.
Roseth, A.G., Kristinsson, J., Fagerhol, M.K., 1993. Faecal calprotectin: a novel test for the diagnosis of colorectal cancer? Scand. J. Gastroenterol. 28, 1073–1076.
Roseth, A.G., Schmidt, P.N., Fagerhol, M.K., 1999. Correlation between faecal excretion of indium-111-labelled granulocytes and calprotectin, a granulocyte marker protein, in patients with inflammatory bowel disease. Scand. J. Gastroenterol. 34, 50–54.
Sacchettini, J.C., Hauft, S.M., Van Camp, S.L., et al., 1990. Developmental and structural studies of an intracellular lipid binding protein expressed in the ileal epithelium. J. Biol. Chem. 265, 19199–19207.
Saha, S., Bhanja, P., Liu, L.B., 2012. TLR9 agonist protects mice from radiation-induced gastrointestinal syndrome. PLoS One 7.
Samloff, I.M., 1982. Pepsinogens I and II: purification from gastric mucosa and radioimmunoassay in serum. Gastroenterology 82, 26–33.
Samloff, I.M., Liebman, W.M., 1973. Cellular localization of the group II pepsinogens in human stomach and duodenum by immunofluorescence. Gastroenterology 65, 36–42.
Samloff, I.M., Varis, K., Ihamaki, T., Siurala, M., Rotter, J.I., 1982. Relationships among serum pepsinogen I, serum pepsinogen II, and gastric mucosal histology. A study in relatives of patients with pernicious anemia. Gastroenterology 83, 204–209.
Schoen, R.T., Vender, R.J., 1989. Mechanisms of nonsteroidal anti-inflammatory drug-induced gastric damage. Am. J. Med. 86, 449–458.
Searfoss, G.H., Jordan, W.H., Calligaro, D.O., 2003. Adipsin, a biomarker of gastrointestinal toxicity mediated by a functional gamma-secretase inhibitor. J. Biol. Chem. 278, 46107–46116.
Shahabi, V., Berman, D., Chasalow, S.D., 2013. Gene expression profiling of whole blood in ipilimumab-treated patients for identification of potential biomarkers of immune-related gastrointestinal adverse events. J. Transl. Med. 11, 75.
Sun, F.P., Song, Y.G., Cheng, W., et al., 2002. Gastrin, somatostatin, G and D cells of gastric ulcer in rats. World J. Gastroenterol. 8, 375–378.
Suzuki, M., Nishiumi, S., Matsubara, A., et al., 2014. Metabolome analysis for discovering biomarkers of gastroenterological cancer. J. Chromatogr. B 966, 59–69.
Tibble, J.A., Sigthorsson, G., Foster, R., 1999. High prevalence of NSAID enteropathy as shown by a simple faecal test. Gut 45, 362–366.
Tillinger, W., Jilch, R., Jilma, B., 2009. Expression of the high-affinity IgG receptor FcRI (CD64) in patients with inflammatory bowel disease: a new biomarker for gastroenterologic diagnostics. Am. J. Gastroenterol. 104, 102–109.
Tsujikawa, T., Uda, K., Ihara, T., 1999. Changes in serum diamine oxidase activity during chemotherapy in patients with hematological malignancies. Cancer Lett. 147, 195–198.
Tsunooka, N., Maeyama, K., Hamada, Y., 2004. Bacterial translocation secondary to small intestinal mucosal ischemia during cardiopulmonary bypass. Measurement by diamine oxidase and peptidoglycan. Eur. J. Cardio. Thorac. Surg. 25, 275–280.
van der Sluys Veer, A., Biemond, I., Verspaget, H.W., Lamers, C.B., 1999. Faecal parameters in the assessment of activity in inflammatory bowel disease. Scand. J. Gastroenterol. (Suppl. 230), 106–110.
Vigushin, D.M., Pepys, M.B., Hawkins, P.N., 1993. Metabolic and scintigraphic studies of radioiodinated human C-reactive protein in health and disease. J. Clin. Invest. 91, 1351–1357.
Voganatsi, A., Panyutich, A., Miyasaki, K.T., Murthy, R.K., 2001. Mechanism of extracellular release of human neutrophil calprotectin complex. J. Leukocyte Biol. 70, 130–134.
Wang, K., Zhang, S., Marzolf, B., 2009. Circulating microRNAs, potential biomarkers for drug-induced liver injury. Proc. Natl. Acad. Sci. USA 106, 4402–4407.

Westergaard, H., 2007. Bile acid malabsorption. Curr. Treat. Options Gastroenterol. 10, 28–33.
Whittle, B.J.R., 1977. Mechanisms underlying gastric mucosal damage induced by indomethacin and bile-salts, and actions of prostaglandins. Br. J. Pharmacol. 60, 455–460.
Wolfe, M.M., Soll, A.H., 1988. The physiology of gastric-acid secretion. N. Engl. J. Med. 319, 1707–1715.
Wolvekamp, M.C.J., Debruin, R.W.F., 1994. Diamine oxidase – an overview of historical, biochemical and functional aspects. Digest Dis. 12, 2–14.
Yanez, J.A., Teng, X.W., Roupe, K.A., et al., 2003. Chemotherapy induced gastrointestinal toxicity in rats: involvement of mitochondrial DNA, gastrointestinal permeability and cyclooxygenase-2. J. Pharm. Pharmaceut. Sci. 6, 308–314.

CHAPTER

17

Reproductive Toxicity Biomarkers

Emily Brehm, Saniya Rattan, Catheryne Chiang, Genoa R. Warner, Jodi A. Flaws

Department of Comparative Biosciences, University of Illinois at Urbana-Champaign, Urbana, IL, United States

INTRODUCTION

Biomarkers are measureable characteristics of various biological processes in the body and are often used to reflect how chemical exposures impact normal reproduction (Strimbu and Tavel, 2010). Chemicals such as polychlorinated biphenyls, pesticides, and plasticizers are considered endocrine disruptors (Schug et al., 2011), and these chemicals have been shown to affect biomarkers (Fig. 17.1). Biomarkers can be measured throughout development and reproduction in both males and females. In males, measurable biomarkers include examining different parameters of spermatogenesis, steroidogenesis in the testes, puberty, and reproductive aging (Figs. 17.2 and 17.3). In females, biomarkers include examining the process of folliculogenesis, steroidogenesis in the ovary, puberty, the uterus, the placenta, and reproductive aging (Figs. 17.2 and 17.3). This chapter discusses these biomarkers in both males and females and describes how chemicals such as endocrine disruptors can affect these endpoints throughout the reproductive life span.

MALE REPRODUCTIVE BIOMARKERS

Spermatogenesis

Testes are the male reproductive glands. The main functions of the testes are to produce sperm and androgens, mainly testosterone. Mammalian testes are made of seminiferous tubules and interstitial compartments. Within the epithelium of the seminiferous tubules are somatic Sertoli cells and the developing germ cells (Zirkin et al., 2011). During spermatogenesis, diploid spermatogonial stem cells divide mitotically to become spermatocytes. This process is followed by two meiotic divisions, which produce haploid round spermatids, and eventually the differentiation of the spermatids into elongated, mature sperm (Zirkin et al., 2011). Besides differentiation, some spermatogonial stem cells are able to self-renew, which maintains a pool of spermatogonia that is needed for the continuous production of gametes (Zirkin et al., 2011). Spermatogenesis has been morphologically defined in stages. Specifically, Hess and de Franca described in great detail the 12 separate stages in the mouse (Hess & de Franca, 2008).

Many biomarkers have been used to evaluate the effects of endocrine disruptors and other toxicants on spermatogenesis. An important biomarker for examining spermatogenesis includes examining the morphology of the testes and epididymis. For example, Park et al. (2002) morphologically examined rat testes and found that di(2-ethylhexyl) phthalate (DEHP) caused apoptosis, necrosis, and loss of spermatogenic cells, resulting in testicular atrophy. Furthermore, adult exposure to bisphenol A (BPA) inhibited spermiation, characterized by an increase in stage VII sperm and a decrease in stage VIII sperm, in the seminiferous epithelium (Liu et al., 2013).

Other major biomarkers of spermatogenesis include examination of different semen parameters, including sperm number and concentration. When observing sperm number, the total number of sperm in an entire ejaculate is calculated by multiplying the sperm concentration by the semen volume (Cao et al., 2011). Sperm concentration refers to the number of sperm per unit volume of semen (Cao et al., 2011). Hemocytometer chambers and computer-aided sperm analysis (CASA) are used to evaluate sperm numbers and sperm concentrations (Mortimer, 2000; Mortimer et al., 2015). A hemocytometer is a thick glass slide with an indentation that defines two separate counting chambers. Sperm are

Biomarkers in Toxicology, Second Edition
https://doi.org/10.1016/B978-0-12-814655-2.00017-7

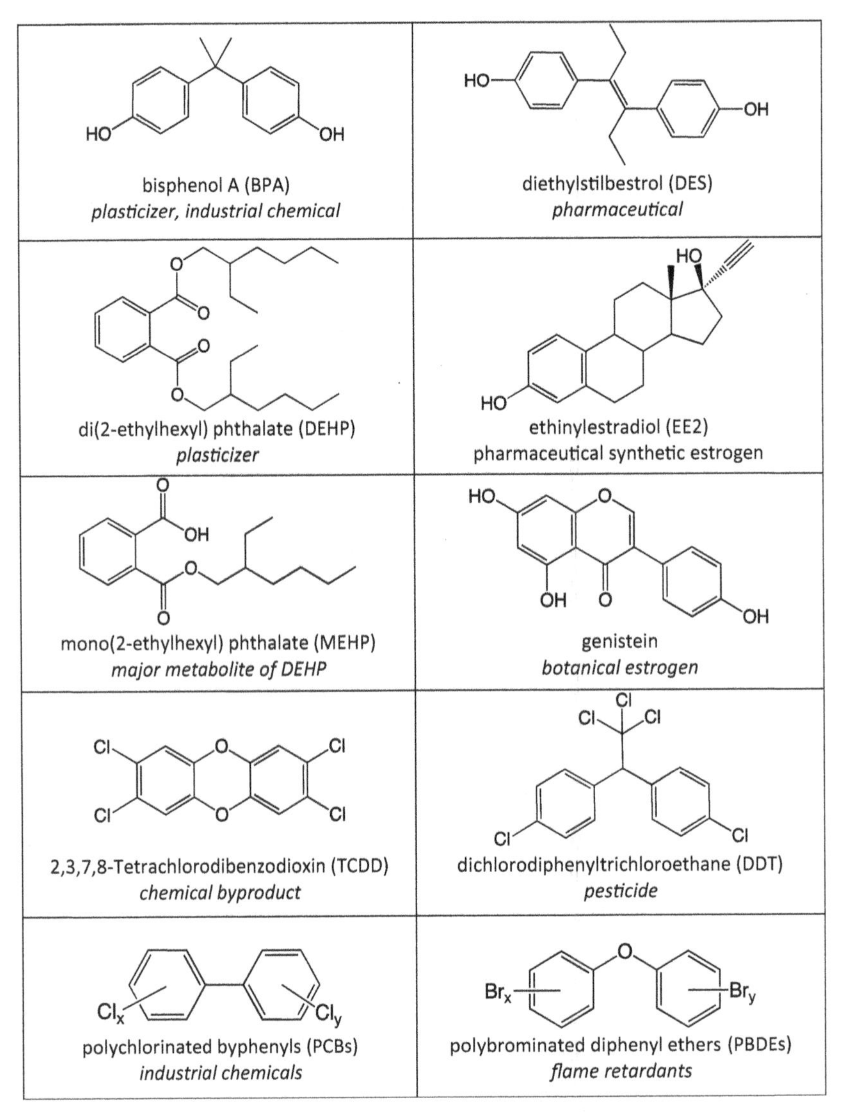

FIGURE 17.1 Select endocrine-disrupting chemicals (EDCs) discussed in this chapter.

loaded onto the slide, and a cover slip with a counting grid is placed on top. Sperm are then manually counted within a known area. Several studies have shown that endocrine-disrupting chemicals (EDCs) affect these biomarkers. For example, urinary BPA concentrations were associated with low sperm concentrations in men (Meeker et al., 2010). In another study, prenatal exposure to DEHP decreased sperm counts in the F1 generation of rats (Chen et al., 2015).

Another important biomarker of sperm quality is motility. Sperm motility is categorized as progressive motility, nonprogressive motility, or immotility (Cao et al., 2011). Besides evaluating sperm number and concentrations, CASA evaluates sperm motility and morphology (Mortimer, 2000; Mortimer et al., 2015). Studies have used CASA to show that polycyclic aromatic hydrocarbons (PAHs) and pesticides decreased sperm motility in humans (Tavares et al., 2016).

Sperm morphology is another biomarker of spermatogenesis. In humans, normal sperm morphology consists of a head, neck, midpiece, principal piece, and end piece (Cao et al., 2011). The head should be smooth and generally oval in shape with a well-defined acrosomal region (Cao et al., 2011). The midpiece should be slender and about the same length as the sperm head, and the principal piece should have a uniform diameter and be thinner than the midpiece (Cao et al., 2011). Abnormal sperm morphologies of the head include tapered, pyriform or pear-shape, round and small, amorphous, vacuolated, and/or having a small acrosomal area (Cao et al., 2011). Midpiece defects include a bent neck, asymmetry, and thick or thin insertion (Cao et al., 2011). The morphological abnormalities of the tail or principal piece include shortness, bent, coiled, and/or hooked structures (Cao et al., 2011). One technique to assess sperm morphology includes collecting

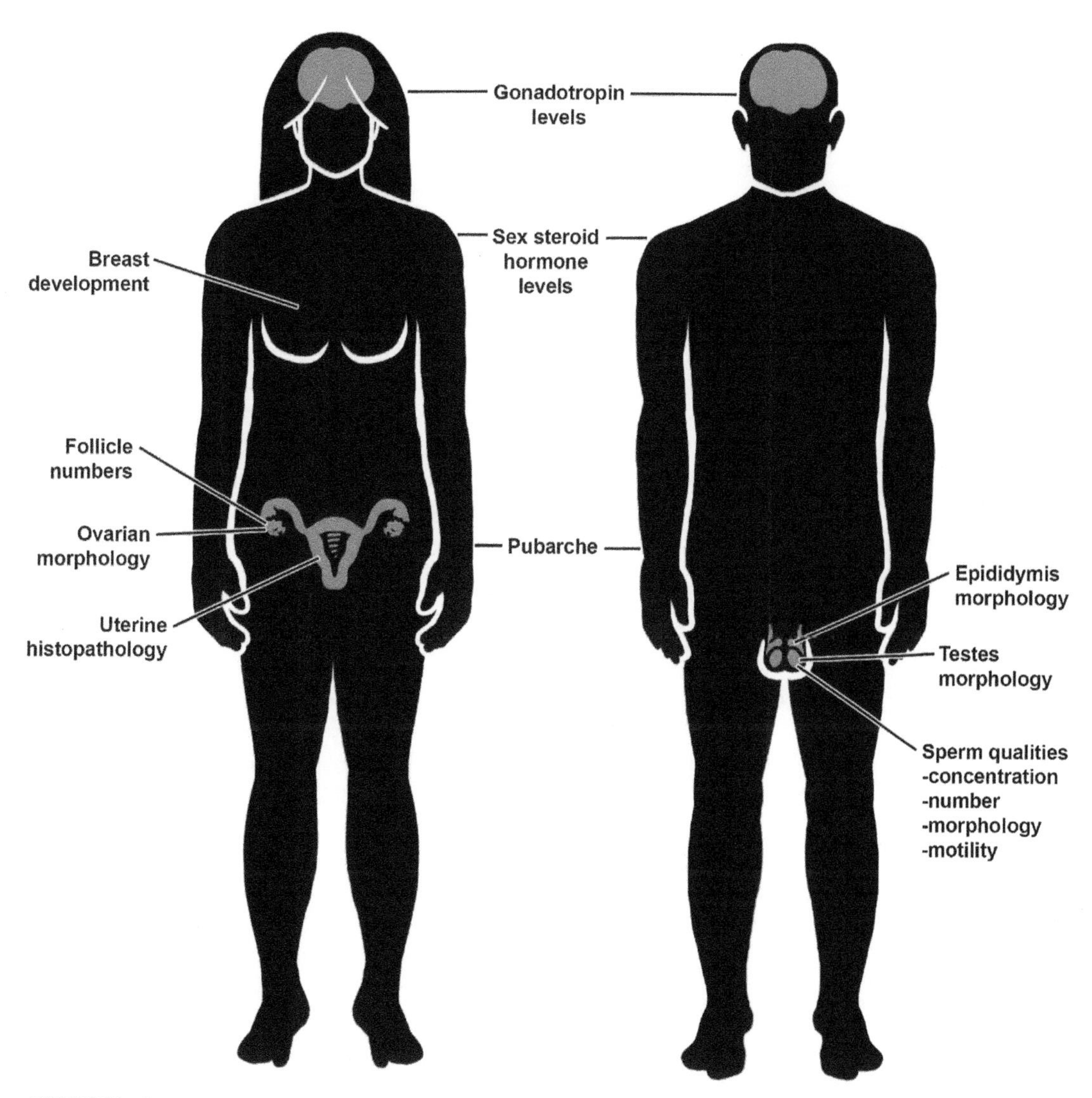

FIGURE 17.2 The diagram of biomarkers of development and reproduction measured in females and males.

wet sperm and mounting it on clean, grease-free slides that contain buffered formalin with eosin–nigrosin stain and morphologically examining 100 sperm per sample under an oil-immersion lens using a light microscope (Otubanjo and Mosuro, 2001). Agarwal et al. (1986) used this technique to determine that DEHP increased the percent of abnormal sperm in adult rats. Furthermore, in men, increases in urinary BPA concentration were associated with declines in sperm morphology (Meeker et al., 2010).

Other important biomarkers of spermatogenesis are pH and vitality of sperm. Semen pH is measured by mixing a semen sample and spreading a drop of it onto pH paper and then obtaining the pH level according to color (Cao et al., 2011). The pH of seminal fluid may play an important role in maintaining the viability and quality of sperm, but it can also ensure fertilization (Zhou et al., 2015b). Sperm vitality refers to the proportion of spermatozoa that are alive in a sample. It is measured by taking wet sperm and staining with eosin to observe whether the sperm are alive or dead (Talwar and Hayatnagarkar, 2015). If the sperm show any sign of the stain, the sperm are dead (Talwar and Hayatnagarkar, 2015). When studying viability of sperm, Tavares et al. (2013) treated human sperm with p,p′-dichlorodiphenyldichloroethylene (p,p′-DDE) and discovered that the chemical decreased the percentage of live sperm compared with control.

In vitro cultures can be used as biomarkers to examine direct effects of toxicants on sperm (Livera et al., 2000). Various in vitro techniques have been used to show that BPA increased sperm activity and accelerated capacitation-associated protein tyrosine phosphorylation in rats (Wan et al., 2016). Furthermore, mono(2-ethylhexyl) phthalate (MEHP), the principal metabolite of DEHP, has been shown to affect spermatogenic cells, leading to necrosis in rat testes *in vitro* (Andriana et al., 2004).

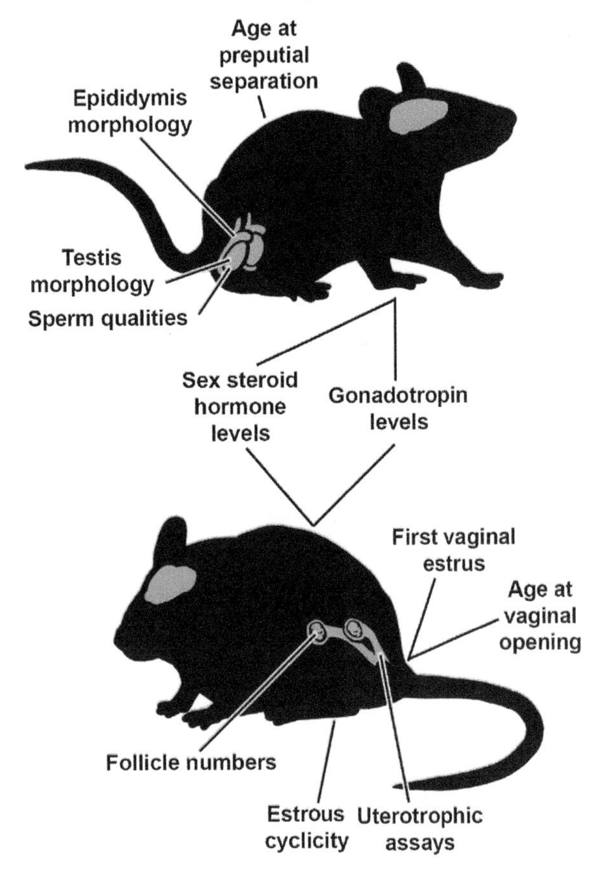

FIGURE 17.3 The diagram of biomarkers of development and reproduction measured in male and female rodents.

Synthesis of Steroids, Gonadotropins, and Peptide Hormones in Males

Reproduction is heavily controlled via hormonal interactions between the hypothalamus, pituitary, and the gonads (Yen et al., 2014). Together, these organs are known as the hypothalamic–pituitary–gonadal (HPG) axis. In males, the gonadal role in the HPG axis is filled by the testes. A variety of hormones play essential roles within the HPG axis and are required for normal male fertility and reproduction. Gonadotropin-releasing hormone (GnRH) is released in a pulsatile manner by specialized neurons in the arcuate nucleus of the mediobasal hypothalamus. GnRH travels through a series of blood vessels that facilitates rapid communication between the hypothalamus and the anterior pituitary known as the hypophyseal portal system. GnRH then stimulates gonadotropes within the anterior pituitary to secrete the gonadotropin hormones, luteinizing hormone (LH), and follicle-stimulating hormone (FSH). LH and FSH travel to the testes where they stimulate steroidogenic activity. Leydig cells and Sertoli cells are responsible for the production of the sex steroid hormone, testosterone, and the peptide hormone, inhibin B, respectively. Leydig cells, Sertoli cells, germ cells, and spermatozoa collectively produce estradiol within the testes (Hess, 2003). Testosterone, inhibin B, and estradiol all play negative feedback roles to help control the secretion of gonadotropins (Yen et al., 2014).

Chemicals can affect hormone levels through a variety of mechanisms, including interfering with feedback systems, interfering with communication between organs, altering enzymatic expression, altering enzymatic activity, directly activating or inhibiting the production of hormones, and inducing changes in hormone metabolism. Hormone levels can be used as biomarkers for normal reproductive health, or lack thereof, in males. Hormone levels are commonly measured via an enzyme-linked immunosorbent assay (ELISA). ELISAs work by utilizing conjugated antibodies that bind specifically to a particular antigen (i.e., the hormone being measured). These antibodies are conjugated to an enzyme that produces a color change when given a specific colorless substrate. The color intensity produced by the enzyme conjugated to the antibody can be used to determine the concentration of the hormone in question. A variety of ELISAs exist, and depending on the type, the color intensity may be directly or inversely correlated with the concentration of the hormone being measured (Crowther, 2008). Another type of assay that is commonly used for measuring levels of hormones is the radioimmunoassay (RIA). RIAs use radiolabeled antigens in a known concentration to compete with the unlabeled antigens present in the sample being tested. After the radiolabeled and unlabeled antigens compete for binding to the antibodies during an incubation period, any unbound antigens are washed away. Only the antibodies bound to radiolabeled or unlabeled antigens remain. Radioactivity produced by the remaining radiolabeled antigen can be measured and used to assess the concentration of the unlabeled antigen (Yalow, 1987).

Using the aforementioned assays, several studies have investigated the effects of various chemicals on the HPG axis in males. Carou et al. (2008) found that exposure to low doses of the UV filter 4-methylbenzylidene camphor (4-MBC) decreased LH, FSH, and GnRH levels in adult male rats. In addition, 4-MBC administered to pregnant rats has been shown to alter LH, FSH, and GnRH levels in male offspring (Ponzo and Silvia, 2013). Axelstad et al. (2011) found that octyl methoxycinnamate (OMC), another UV filter, reduced testosterone levels in male rats that were exposed from gestation through lactation. Faber and Hughes found that the phytoestrogen, genistein, decreased pituitary responsiveness to GnRH in postnatally exposed rats (Faber and Hughes, 1991). DEHP has been shown to reduce testosterone levels and alter LH and FSH levels in males (Ponzo and Silvia, 2013). *In vitro* procedures can also be used to assess the effects of chemicals on steroid production.

Chauvigne et al. (2011) found that treatment with MEHP significantly decreased testosterone production in cultured fetal rat testes.

Male Puberty

Puberty is the period of maturation of the HPG axis, leading to the development of the genital organs and physical changes that result in the ability to sexually reproduce (Delemarre-van de Waal, 2002). Puberty is an important process because it marks the beginning of sexual development and growth. In both humans and rodents, the onset of puberty is governed by a complex neural network in the hypothalamus that regulates the secretion of gonadotropin hormones and gonadal growth and function. Puberty consists of whole-body changes from sex organs to skeletal growth. However, early onset of puberty in males is associated with increased risk of heart attack, hypertension, and type 2 diabetes (Zhu and Chan, 2017).

A critical biomarker for puberty in males is the formation of adult Leydig cells. To measure the presence and number of adult Leydig cells, studies often use immunohistochemistry (IHC) and histological evaluation of the testes. Fetal Leydig cells are round and present as clusters, whereas adult Leydig cells are spindle-shaped and contain smooth endoplasmic reticulum and few lipid droplets. Adult Leydig cells are more heavily stained by 3β-hydroxysteroid dehydrogenase (3β-HSD) antibody than fetal Leydig cells (Ge and Hardy, 2007). Multiple studies have used adult Leydig cell formation and function as a biomarker of puberty. For instance, prenatal exposure to dibutyl phthalate (DBP) has been shown to delay the development of adult Leydig cells in the rat testes during puberty (Chen et al., 2017). Exposure to soy milk rich with glyphosate during pubertal development has been shown to increase the percentage of degenerated Leydig cells in male rats (Nardi et al., 2017). Furthermore, Ziram, an agricultural fungicide, delayed pubertal development of rat Leydig cells by inhibiting androgen production and steroidogenic enzyme activities within the Leydig cell (Guo et al., 2017).

Another important biomarker of puberty in males is testosterone levels. Specifically, testosterone levels in male humans increase from a prepubertal concentration of 0.2 to 6 ng/mL, which is typical for an adult man (Plant et al., 2015). Testosterone is important during puberty because it stimulates fertility and male secondary sexual characteristics (Ge and Hardy, 2007). Multiple studies have measured testosterone levels as a biomarker for puberty and have shown that EDCs can interfere with normal testosterone levels during puberty. For instance, exposure to DBP during prenatal development lowered serum testosterone levels during pubertal development in rats (Chen et al., 2017). Exposure to soy milk rich with glyphosate during pubertal development decreased serum testosterone levels in male rats (Nardi et al., 2017). In addition, maternal serum levels of dichlorodiphenyltrichloroethane (DDT) and dichlorodiphenyldichloroethylene (DDE) were inversely related to LH and testosterone levels in pubertal boys. Furthermore, polybrominated diphenyl ether (PBDE) concentrations in maternal serum were associated with increased levels of LH and testosterone in pubertal boys (Eskenazi et al., 2017).

In humans, the onset of puberty is assessed using the Tanner Scale. In males, the Tanner Scale assesses the growth of the gonads (gonadarche) and pubic hair (pubarche) (Rockett et al., 2004). Gonadarche is the dramatic increase in testicular volume from 2 to 20–25 mL due to the initiation of spermatogenesis (Plant et al., 2015). Gonadarche and pubarche are important biomarkers of puberty because they mark the physical changes associated with the cellular and hormonal changes in puberty. The Tanner Scale has five stages. Tanner stage 1 describes prepubertal gonad size and indicates no pigmented pubic hair. Tanner stage 5 describes adult testicular and penile size and adult pubic hair growth and pattern (Rockett et al., 2004). This scale is often used as a biomarker for endocrine disruption (Tanner and Davies, 1985). For example, one study has shown that peripubertal urinary levels of BPA are positively associated with an earlier onset of genital and pubic hair development, but delayed pubertal progression in boys (Wang et al., 2017).

In male rodents, biomarkers of puberty include preputial separation and the presence of motile sperm in the epididymis. Measurement of preputial separation includes the detection of the separation of the prepuce from the glans penis (Prevot, 2015). Preputial separation is an important biomarker because delayed preputial separation indicates anti-androgenic activity, impaired growth, and altered onset of puberty (Melching-Kollmuss et al., 2014). The presence of motile sperm in the epididymis is also an important indicator of puberty because it indicates proper testosterone levels and initiation of spermatogenesis (Prevot, 2015). Studies have shown that EDCs interfere with these biomarkers of puberty. For example, early postnatal exposure to pyrethroids, a class of insecticides, accelerated the onset of preputial separation in male mice (Ye et al., 2017). Ethinyl estradiol (EE2) exposure during prenatal development delayed the age of preputial separation in male rats (Ferguson et al., 2014). Furthermore, prepubertal silver nanoparticle exposure reduced the production of motile sperm and delayed the onset of preputial separation in male rats (Sleiman et al., 2013).

Male Reproductive Aging

Reproductive aging is defined as a loss of reproductive capabilities due to irreversible changes at the cellular and molecular levels over time. In males, the decline in reproductive capacity is less pronounced compared with females because males are able to father offspring throughout their life span. However, declining reproductive capacity occurs in males. Reduced endocrine function, altered testicular morphology, and declining sperm parameters are biomarkers of reproductive aging in male mammals (reviewed in Gunes et al., 2016). These biomarkers are commonly quantified to measure toxicity associated with reproductive aging in human and animal models.

The normal interactions within the HPG axis change with age. GnRH secretion from the hypothalamus and testosterone levels from the testes decrease with age; however, basal LH levels increase with age (reviewed in Araujo and Wittert, 2011). The most relevant hormone to measure as a biomarker for reproductive aging in males is total testosterone. Testosterone levels in aging men are often measured by liquid chromatography–tandem mass spectrometry (LC-MS/MS) because LC-MS/MS is considered to be a more sensitive, precise, and reliable test than RIAs and ELISAs (reviewed in Araujo and Wittert, 2011). LC-MS/MS measures the mass-charge ratio of charged particles. Specifically, the LC-MS/MS transitions solid or liquid samples into gaseous particles, ionizes vaporized samples, and detects and quantifies ions. LC-MS/MS utilizes selective reaction monitoring (SRM) to detect serum testosterone levels. SRM allows highly specific detection of testosterone with the mass-charge ratio of charged particles (Grebe and Singh, 2011). Although it is relatively easy to measure testosterone levels, few studies have used testosterone levels as a biomarker to investigate the effects of toxicant exposure on male reproductive aging. One study has shown that prenatal exposure to DEHP accelerates reproductive aging by decreasing testosterone levels and increasing LH levels in 19-month-old mice (Barakat et al., 2017).

Multiple morphological changes occur in the testis with age, and these changes can be used as biomarkers of reproductive aging. Specifically, mean testicular weight, Leydig cell responsiveness and numbers, and the numbers of germ cells and Sertoli cells in the aging testes decrease over time. Furthermore, the seminiferous tubules narrow in aging testes (Gunes et al., 2016). Changes in these biomarkers of reproductive aging in males have been exacerbated with toxicant exposure. For example, prenatal exposure to DEHP increased testicular germ cell apoptosis in 22-month-old mice compared to controls (Barakat et al., 2017). In utero exposure to DBP reduced Leydig cell function in an age dependent manner compared to controls in rats (Motohashi et al., 2016).

In addition to morphological changes in the testis, the prostate gland, an accessory sex organ, undergoes morphological changes with age. Benign prostatic hyperplasia (BPH), an age-related disease, is a common biomarker of reproductive aging because it is associated with hyperplasia of basal cells and stromal cells around the urethra (Gunes et al., 2016). BPH can cause lower urinary tract symptoms that negatively impact the quality of life for men by creating obstructive symptoms in voiding such as increased urgency and frequency to urinate, weak urine stream, and sensation of incomplete emptying of the bladder (Bechis et al., 2014). To measure lower urinary tract symptoms caused by BPH, experimental animals can undergo a spontaneous void spot assay. In this assay, rodents are placed in a cage lined with paper for a set period of time. After this time, the animals are removed and the paper is checked for urine spots and analyzed for voiding behavior (Ricke et al., 2016). Furthermore, the prostate gland can be removed from experimental animals and histologically examined for increased cell proliferation, weight, and volume (Ricke et al., 2016; Wu et al., 2011). In humans, BPH can be detected by a general examination, hormone tests, and a digital rectal examination (Alanazi et al., 2017).

These prostate biomarkers have been used in studies that indicate toxicant exposure disrupts the aging prostate. Specifically, prenatal and lactational exposure to 2,3,7,8-tetrachlorodibenzo-*p*-dioxin (TCDD) worsened urinary dysfunction by increasing voiding behavior and prostatic proliferation in adult mice (Ricke et al., 2016). Furthermore, low doses of BPA increased the weight, proliferation, and volume of the prostate gland in adult rats (Wu et al., 2011).

Sperm parameters such as semen volume, sperm concentration, sperm motility, sperm morphology, and total sperm count are biomarkers of male reproductive function. These biomarkers are known to decline with age (Gunes et al., 2016). CASA and hemocytometer chambers are used to evaluate these parameters. Several studies have used these biomarkers of sperm to demonstrate that toxicant exposure negatively impacts sperm parameters with aging. For example, prenatal exposure to DEHP decreased sperm concentration and reduced motile sperm in 22-month-old mice (Barakat et al., 2017).

FEMALE REPRODUCTIVE BIOMARKERS

Folliculogenesis

The ovary consists of a surface epithelium, and an ovarian cortex containing ovarian follicles, stroma, and corpora lutea. Some major functions of the ovary include maturation and ovulation of the oocyte for fertilization. Furthermore, the ovary is responsible for secreting sex steroid hormones, including estrogens, progesterone, and testosterone. Within the ovary, follicles undergo several developmental changes in a process known as folliculogenesis (Hannon and Flaws, 2015). Follicles are first formed during the later stages of fetal life in the human and during early postnatal life in the rodent (Hannon and Flaws, 2015). During early folliculogenesis, germ cells form, develop into nests, and undergo germ cell nest breakdown to form primordial follicles (Hannon and Flaws, 2015). Once the primordial follicle population is established, the follicles will become part of the follicle reserve, undergo atresia, or grow into primary follicles (Hannon and Flaws, 2015). Primary follicles consist of an oocyte surrounded by a single layer of cuboidal granulosa cells. These primary follicles then will develop into preantral follicles that contain an oocyte surrounded by multiple layers of cuboidal granulosa cells and theca cells. Preantral follicles develop into antral follicles that consist of an oocyte surrounded by numerous layers of cuboidal granulosa cells, theca cells, and a fluid-filled antrum. Changes in folliculogenesis are often used as biomarkers in female reproductive toxicology studies. Specifically, these biomarkers include examining oocyte numbers, morphological evaluation of the ovaries/oocytes/follicles, using culture systems, and measuring gene expression.

To examine oocyte numbers as a biomarker of folliculogenesis, many laboratories have adopted certain protocols to evaluate follicle numbers throughout an ovary. For example, every 10th section of the ovary is used to count all follicle types including primordial, primary, preantral, and antral follicles (Flaws et al., 1994). This procedure can be used to count raw follicle numbers or percent of each follicle type (i.e., raw number divided by the total number of follicles multiplied by 100). By calculating the percent of each follicle type, it is possible to examine shifts in the follicle pool. Another ovarian structure that can be examined as a biomarker is the corpus luteum. The presence of this biomarker is an indication that ovulation occurred in the animal. Corpora lutea are large structures containing large and small luteal cells and are counted by following the progression of the corpora lutea in each ovarian section. When examining folliculogenesis, follicle numbers have been used as a biomarker in studies that have determined that BPA, methoxychlor (MXC), TCDD, and genistein alter folliculogenesis (Patel et al., 2015). Furthermore, phthalates, particularly DEHP, have been shown to accelerate folliculogenesis (Hannon et al., 2014). Interestingly, prenatal DEHP exposure has been shown to decrease folliculogenesis in the F1 generation of mice, but accelerate folliculogenesis in the F3 generation of mice (Rattan et al., 2017). Furthermore, when examining corpora lutea as a biomarker, DEHP decreased the volume of the corpus luteum in cows (Kalo et al., 2015).

Besides examining numbers of follicles, the morphology of ovaries and their follicles can be examined as biomarkers of female reproductive toxicology. Germ cell nest breakdown, atresia, and abnormal follicles can be detected and quantified in histological sections. Follicles are considered atretic if at least 10% of the follicle contains apoptotic bodies, and abnormal follicles include follicles with multiple oocytes, multinuclei, fragmented oocytes, and/or fragmented nuclei (Hannon et al., 2016; Zhou and Flaws, 2017). When examining the biomarker of germ cell nest breakdown, *in vitro* BPA exposure has been shown to inhibit germ cell nest breakdown in female mice (Zhou et al., 2015a). Furthermore, atresia and abnormal follicles have been used as biomarkers in studies showing that a mixture of phthalates decreased atresia in antral follicles, but induced the occurrence of oocyte fragmentation in female mice (Zhou and Flaws, 2017).

Culture systems on whole ovaries or follicles can be used to observe changes in follicle and oocyte development. Cultures can also measure follicle metabolism and steroidogenesis. Follicle cultures utilize an experimental technique designed to isolate intact follicles from whole ovaries and to measure follicle growth and metabolism (Hartshorne, 1997). In addition, whole neonatal ovaries can be cultured to examine the same parameters. When culturing follicles, ovaries from young adult rodents are ideal because they contain many antral follicles. As described by Hannon et al. (2015a), the antral follicle culture technique begins by placing isolated, untreated antral follicles into a 96-well culture plate containing unsupplemented α-minimal essential medium. Chemicals for treatment are prepared in a supplemented media that is added to the wells, and follicles are cultured in medium for 24–96 h in an incubator (Hannon et al., 2015a). At specific time points during culture, follicle size is measured and follicles and culture media are collected for further analysis (Hannon et al., 2015a).

Whole neonatal ovaries can be cultured using similar methods to those used to culture antral follicles. Ovaries are removed from neonatal pups, placed in media droplets on culture plate inserts that float on supplemented media, and placed in four-well culture plates (Hannon et al., 2015a). Chemicals for treatment are prepared in a supplemented media that is added to the wells and on top of the ovary to keep it from drying out, and supplemented medium is replaced daily (Hannon et al., 2015a). When the culture is finished, ovaries are collected for gene expression analysis, IHC, or histological evaluation, and the media are collected for further evaluation of hormone levels (Hannon et al., 2015a).

These culture techniques have been used in studies showing that *in vitro* exposure to isoliquiritigenin inhibited antral follicle growth and altered estradiol, testosterone, and progesterone levels (Mahalingam et al., 2016). *In vitro* exposure to phthalates has been shown to inhibit antral follicle growth and alter steroidogenesis in mice (Hannon and Flaws, 2015). In another study, BPA exposure prevented germ cell nest breakdown by altering the expression of key ovarian apoptotic genes in neonatal ovary cultures (Zhou et al., 2015a). When examining gene expression using *in vitro* cultures, MEHP accelerated primordial follicle recruitment, possibly by overactivation of the ovarian phosphatidylinositol 3-kinase (PI3K) signaling pathway in a neonatal ovary culture (Hannon et al., 2014). Furthermore, MEHP increased the expression of the proapoptotic gene *Aifm1*, but decreased the proapoptotic gene *Bok* and the antiapoptotic gene *Bcl2l10* in the antral follicle culture (Craig et al., 2014).

Synthesis of Steroids, Gonadotropins, and Peptide Hormones in Females

In females, the HPG axis consists of the hypothalamus, pituitary, and the ovaries. Just as in males, GnRH is released from neurons within the arcuate nucleus of the mediobasal hypothalamus in a pulsatile manner. GnRH then travels to the anterior pituitary where it stimulates the release of the gonadotropins, LH, and FSH, into the bloodstream. LH and FSH stimulate the theca cells and granulosa cells, respectively, in ovarian follicles to produce a variety of hormones. LH stimulates theca cells to make progesterone. The progesterone is converted to androgens, which are taken up by the granulosa cells. FSH then stimulates granulosa cells to convert these androgens into estrogens. FSH also stimulates granulosa cells to produce the peptide hormones, inhibin A and inhibin B. During the majority of the female cycle, estradiol exerts negative feedback at the hypothalamic and pituitary levels to reduce the production of FSH and LH. Furthermore, progesterone exerts negative feedback directly on the hypothalamus to decrease GnRH secretion. Inhibin A and B exert negative feedback on the pituitary to reduce the release of FSH, specifically. However, as the dominant follicle(s) within the ovary produces rising levels of estradiol, a threshold is reached that leads to estradiol exerting a strong positive feedback on the pituitary, therefore causing the preovulatory LH surge needed for ovulation. After ovulation, the system resets and is controlled by the negative feedback loops. Thus, hormones in the female act through a complex network of negative and positive feedback loops to facilitate the cyclical nature of female fertility observed in a variety of species (Yen et al., 2014).

Just as in males, chemicals may affect hormone levels through a variety of mechanisms, and because female fertility depends on a carefully regulated cycle of fluctuating hormones, alteration of hormone levels can impede female fertility. Therefore, studies often use hormones as a biomarker of normal or abnormal female fertility. Using ELISAs and RIAs, many studies have examined the effects of a variety of chemicals on hormonal profiles in females. Using incubated hypothalamic samples from treated female rats, Rasier et al. (2007) found that treatment with the insecticide DDT increased secretion of GnRH *in vitro* in female rats. Carou et al. (2009) found that 4-MBC increased LH and FSH levels in the serum of female rats exposed from the gestational period through adulthood. Some studies have found that phytoestrogens decrease LH pulse amplitude and GnRH-induced LH secretion by the anterior pituitary (Dickerson and Gore, 2007). DEHP exposure has been shown to reduce progesterone and estradiol levels in rats, but has equivocal effects on LH (Ponzo and Silvia, 2013; Somasundaram et al., 2016). DEHP has also been shown to suppress the production of progesterone, testosterone, androstenedione, and estradiol in cultured antral follicles in mice (Hannon et al., 2015b). Interestingly, DBP decreased estradiol levels while increasing FSH levels in adult female mice (Sen et al., 2015). Polychlorinated biphenyls (PCBs) have also been shown to interfere with steroidogenesis in female rats exposed from gestation through lactation (Dickerson and Gore, 2007).

Female Puberty

Puberty is a period of development marked by the maturation of the genital organs, development of secondary sex characteristics, and the first occurrence of menstruation in women (Plant et al., 2015). In girls, puberty is assessed using the Tanner Scale. Specific for girls, the Tanner Scale assesses breast development (thelarche) and pubic hair growth. The age at which these developmental processes and the age at first menstruation (menarche) are also taken into consideration when determining the onset of puberty (Rockett et al., 2004). Late onset of puberty in girls is associated with decreased bone mineral density, increased risk of bone fractures, increased risk of osteoporosis later in life, and increased risk of coronary heart disease. However, early onset of puberty in girls is associated with an increased risk of breast cancer, endometrial cancer, and coronary heart disease (Zhu and Chan, 2017). In female rodents, puberty is determined by the presence of a vaginal opening and the first vaginal estrus (Prevot, 2015).

Key biomarkers that determine the timing of puberty involve the release of GnRH from the hypothalamus and the release of FSH and LH from the anterior pituitary, leading to the first preovulatory surge of gonadotropins. This surge is responsible for pubertal activation of the pituitary–ovarian axis. Levels of pituitary gonadotropins increase in association with puberty. GnRH forms pulsatile surges to facilitate the LH surge. GnRH peptides and blood gonadotropin concentrations can be measured using RIAs (Plant et al., 2015). Such measurements are used as biomarkers for puberty. One study has used these biomarkers to show that DEHP and MEHP levels in urine are associated with GnRH-dependent precocious puberty in young girls (Bulus et al., 2016).

The first ovulation is another important biomarker of puberty in females. The ability of the ovary to secrete high levels of estrogens over a short period of time (i.e., 24 h) during the estrous cycle is critical for the timing of the preovulatory GnRH/FSH/LH surge (Prevot, 2015). This surge is often linked to the age at vaginal opening. Vaginal opening is an external indicator of the onset of puberty in rodents. Following vaginal opening, rodents exhibit their first vaginal estrus (Prevot, 2015). To measure the first vaginal estrus, evaluation of vaginal cell cytology is required (Gal et al., 2014). A rodent is considered to be in estrus when most of the vaginal cells are cornified epithelial cells, diestrus when mostly leukocytes and few nucleated epithelial cells are present, and proestrus when mainly nucleated epithelial cells are visible in the lavage. Finally, a rodent is considered to be in metestrus if all three cell types are observed, usually with few nucleated epithelial cells. Several studies have used these biomarkers to evaluate the effects of EDCs on the onset of puberty. For example, studies show that neonatal diethylstilbestrol (DES) exposure accelerated the age at vaginal opening and induced abnormal estrous cyclicity in female rats (Ohmukai et al., 2017). Another endocrine disruptor, ethylene glycol monomethyl ether, delayed the age at vaginal opening in female rats (Taketa et al., 2017). Maternal exposure to fluoxetine, an antidepressant, delayed the onset of vaginal opening and first estrus in female rat offspring (Dos Santos et al., 2016).

Thelarche and menarche are important biomarkers used to assess pubertal development in girls. An abnormal onset of thelarche may indicate abnormal gonadotropin secretion that subsequently stimulates cyst formation and estradiol secretion by the ovary (Neely and Crossen, 2014). In addition, abnormal onset of menarche indicates abnormal HPG axis activation (Neely and Crossen, 2014). According to the Breast Cancer and Environment Research Program (BCERP), prepubertal exposure to various EDCs is associated with precocious or delayed thelarche and menarche in females. Specifically, high levels of enterolactone and mono(3-carboxypropyl) phthalate in urine are associated with delayed thelarche and menarche (Wolff et al., 2017). Increased levels of 2,5-dichlorophenol in urine are also associated with precocious thelarche and menarche in girls (Wolff et al., 2017). Furthermore, increased exposure to isoflavones is associated with delayed breast development in young girls (Cheng et al., 2010).

Uterus

The uterus is highly sensitive to EDC compounds and is itself a biomarker of exposure to estrogenic substances. Uterotrophic assays, histology, and gene expression profiling are complementary biomarkers of uterine sensitivity to hormones and endocrine disruptors. The mouse or rat uterotrophic assay has been used since the 1930s in pharmaceutical development of hormone-mimicking substances such as DES and more recently has been validated by the Organization for Economic Cooperation and Development (OECD) for identification of estrogen agonists and antagonists (Owens and Ashby, 2002; Owens and Koëter, 2003). The uterotrophic assay measures increases in uterine weight compared with body weight in response to estrogen receptor–mediated signaling in the presence of estrogenic substances. It is performed on immature or ovariectomized rodents without endogenous estrogens to minimize the baseline uterine weight before treatment (Owens and Ashby, 2002). The rodent uterotrophic assay has been used to identify the estrogenic activity of EDCs including BPA, DDT, the PCB mixture Aroclor 1221, and the organochlorine pesticides endosulfan and mirex (Gore et al., 2015). The synthetic estrogens DES and EE2 are typically used as positive controls. The immature mouse uterotrophic assay has also been used to identify estrogen antagonists such as dibenzyl phthalate (DBzP) (Zhang et al., 2011).

Histopathology of the uterus is another method of determining exposure to endocrine-active substances. Paired with uterotrophic assays, histologic analyses of the uterus reveal swelling of the stromal endometrium and luminal epithelium as uterine weight increases in the presence of estrogenic chemicals (Orphanides et al., 2004). However, in histological analysis of uterine tissues independent of complementary assays, it can be difficult to distinguish direct uterine toxicity from responses to changes in the overall hormonal environment of the reproductive tract or primary ovarian toxicity. Mitotic inhibitors and cytotoxins are two classes of compounds known to directly impact uterine cell populations (Li and Davis, 2007). In particular, epithelial cell height is commonly measured because it can increase up to five-fold in rodents in response to estrogens such as EE2, DES, and BPA (Owens and Ashby, 2002; Steinmetz et al., 1998). The antibacterial agent triclosan, which has been shown to increase uterotrophic response to EE2, enhances EE2-induced epithelial cell height growth (Louis et al., 2013).

Gene expression changes in the presence of estrogens increase protein production. These changes partially contribute to the uterine growth measured in the uterotrophic assay and to the visible changes in histologic analysis of uterine tissue. Changes in expression of protein metabolism, cell cycle, proliferation, DNA replication, RNA metabolism, mRNA transcription, and blood vessel development genes have been demonstrated in the presence of estrogens, and these changes correlate with uterine weight increase and histologic changes (Orphanides et al., 2004). In addition, EDCs have been linked to distinctive patterns of gene expression. DES, BPA, and TCDD have been shown to alter the Wnt-signaling pathway (Calhoun et al., 2014; Gore et al., 2015). PBDE 99 decreased progesterone receptor expression (Ceccatelli et al., 2006). In the presence of the MXC metabolite 2,2-bis(p-hydroxyphenyl)-1,1,1-trichloroethane, gene expression changes in the mouse uterus vary by ligand specificity (estrogen-like, antiestrogen, etc.) for estrogen and androgen nuclear receptors (Waters et al., 2001).

Placenta

The placenta is a unique and important organ that controls the transfer of nutrients and other substances from the mother to the fetus during pregnancy. Placental toxicity may indicate fetal toxicity as most xenobiotics pass through the placenta on the way to the fetus, making placental biomarkers vital for monitoring fetal exposures. However, species differences in structure and function of the placenta preclude the use of most animal models for human studies (Myllynen et al., 2005).

During pregnancy, the placenta synthesizes and regulates hormones. Imbalances in endogenous hormone levels are one biomarker of placental stress due to xenobiotics. Human chorionic gonadotropin (hCG), produced throughout pregnancy, is altered by organochlorines, BPA, phthalates, smoking, heavy metals, and other estrogenic substances (Adibi et al., 2017; Myllynen et al., 2005). Phthalates have also been shown to alter androgen steroidogenesis during pregnancy (Lyche et al., 2009).

The placenta also expresses metabolic enzymes, including several cytochrome P450s (CYPs) and both phase 1 (modification) and phase 2 (conjugation) enzymes. Metabolites measured in placental tissue and cord blood are one of the most commonly used biomarkers of placental and fetal toxicity. Ethyl glucuronide, the major metabolite of ethanol, a known fetal toxicant, can be detected in placental tissue following maternal alcohol consumption (Morini et al., 2011). Metabolites of phthalates, organochlorine pesticides, PCBs, PBDEs, and other environmental contaminants are readily detected using mass spectrometry methods (Leino et al., 2013). Toxic metabolites such as epoxides that form DNA adducts can also serve as biomarkers. PAH adducts such as benzo(a)pyrene from smoking and air pollution as well as aflatoxin adducts can be detected using methods such as immunoassays, mass spectrometry, and ^{32}P-postlabeling (Carlberg et al., 2000; Partanen et al., 2010).

Gene expression changes are another biomarker of placental toxicity. Altered expression of transporter proteins, stress proteins, signaling molecules, and metabolizing enzymes have all been linked to exposures to xenobiotics (Myllynen et al., 2005; Vahakangas and Myllynen, 2009). For example, phthalate alteration of steroidogenesis can be monitored by gene expression changes of CYP (cytochrome P450) (Adibi et al., 2010). In particular, the induction of CYP1A1, the gene for enzyme responsible for biotransformation of xenobiotics, is a sensitive biomarker of exposure to environmental contaminants (Whyatt et al., 1995). *CYP1A1* is regulated by the aryl hydrocarbon receptor (AhR), which is in turn responsive to xenobiotic ligands including polycyclic aromatic hydrocarbons, organochlorines, and dioxins (Mimura and Fujii-Kuriyama, 2003; Stejskalova and Pavek, 2011).

Female Reproductive Aging

Reproductive aging in women is an irreversible process that begins before birth and lasts through the menopausal transition. It consists of changes in sex steroid, peptide, and gonadotropin hormone levels, abnormal menstrual cyclicity, changes in ovarian histology and morphology, and decreases in fertility (Burger, 2006). Typically, sex steroid (estradiol, progesterone, and testosterone) and peptide hormone (inhibin B and anti-Müllerian hormone [AMH]) levels decrease, whereas gonadotropin (FSH and LH) levels increase with reproductive aging (de Kat et al., 2016; Santoro, 2002). In women, anovulatory cycles and abnormal hormone levels become more common as they progress through the menopausal transition and enter reproductive senescence (Santoro, 2002). In rodents, abnormal hormone levels lead to irregular estrous cyclicity, and eventually acyclicity (Scarbrough and Wise, 1990). In both women and animal models, ovarian histology significantly changes with aging and is characterized by a drastic decrease in the size of the primordial follicle pool (te Velde and Pearson, 2002). In rodents, formation of ovarian cysts is common as an animal ages (Vidal et al., 2013). These changes in hormones, cyclicity, and ovarian histology lead to subfertility/infertility with age.

Many biomarkers have been used to examine the effects of toxicants on reproductive aging. One of these biomarkers is measurement of the levels of reproductive hormones such as estradiol, testosterone, progesterone, and inhibin B by ELISAs or RIAs. Hannon et al. (2016) used both ELISAs and RIAs to determine the effects of DEHP on sex steroid, peptide, and gonadotropin hormone levels in aging female mice. They showed that adult exposure to DEHP decreased the levels of inhibin B and the ratio of estradiol to progesterone in mice at 9 months of age, indicating that DEHP exposure accelerates reproductive aging. Similarly, Shi et al. (2007) used ELISAs to determine the effects of TCDD on reproductive aging in rats, and they found that TCDD decreased estradiol levels in rats at 9 months of age.

Another biomarker of reproductive aging includes the monitoring of estrous cyclicity. In rodents, estrous cyclicity is monitored by subjecting animals to daily vaginal lavage for a set amount of time (ranging from days, to weeks, to months). As rodents age, and their hormones change, this leads to irregular cyclicity, and eventually an animal becomes acyclic (Gee et al., 1983; Gore et al., 2011; Scarbrough and Wise, 1990). This biomarker has been used to show that adult exposure to DEHP altered cyclicity in mice at 6 months of age and increased time spent in diestrus/metestrus in mice at 9 months age, indicating that the mice may be exhibiting early reproductive aging (Hannon et al., 2016). Furthermore, Aroclor 1221 increased the length of estrous cycles in rats (Walker et al., 2013).

An additional biomarker for reproductive aging includes examining the morphological appearance of the ovary. Within the ovary, follicles undergo several developmental changes in a process known as folliculogenesis (Hannon and Flaws, 2015). Follicles undergo many changes beginning as germ cells, developing into germ nests, and undergoing germ cell nest breakdown to form primordial follicles. Once the primordial follicle population is established, the follicle will become part of the follicle reserve, undergo atresia, or grow into primary follicles. With normal aging, the primordial follicle pool decreases dramatically, leading to reproductive senescence (te Velde and Pearson, 2002). Several studies have used follicle counts as a biomarker for the ovarian reserve. For example, in one study, adult DEHP exposure decreased primordial follicle numbers and total follicle numbers, leading to early reproductive aging in female mice 9 months after dosing (Hannon et al., 2016). Furthermore, prenatal exposure to DEHP altered folliculogenesis in both the F1 and F3 generations of 1-year-old mice (Brehm et al., 2018).

Besides counting follicle numbers, the general morphology of an ovary can be examined as a biomarker of ovarian function. Rodents often develop cystic follicles as they spend more time in persistent estrus and have excess levels of estrogen and progesterone with age (Vidal et al., 2013). These cysts can arise from follicles, corpora lutea, rete ovarii, surface epithelium, ovarian bursae, or embryonic remnants. The presence of cysts has been used as a biomarker in studies on phthalates. For example, Zhou et al. (2017a,b) showed that a mixture of phthalates increased the occurrence of cystic ovaries at 13 months of age in both the F1 and F2 generations of female mice.

Due to the decline in follicle numbers and quality with reproductive aging, fertility decreases as well. Fertility-related indices in rodents can be analyzed by calculating different endpoints including pregnancy rate (the number of pregnant females divided by the number of breeding pairs multiplied by 100) and fertility index (the number of pregnant females divided by the number of females with a vaginal sperm plug multiplied by 100) (Ziv-Gal et al., 2015). These rates have been used as biomarkers in several studies on EDCs. For example, *in utero* BPA exposure reduced the ability of female mice to maintain pregnancies as they became older and drastically decreased the percent of fertile females at 9 months of age (Wang et al., 2014; Ziv-Gal et al., 2015). Prenatal exposure to a phthalate mixture increased pregnancy loss in the F1 generation and reduced the pregnancy rate and fertility index in mice at 9 months of age (Zhou et al., 2017b). In women, as fertility declines during the menopausal transition, occurrence of hot flashes increases. Phthalate metabolites have been associated with an increased risk of ever experiencing hot flashes and more frequent hot flashes in perimenopausal women (Ziv-Gal et al., 2016).

CONCLUDING REMARKS AND FUTURE DIRECTIONS

Measuring biomarkers is a resourceful way to observe effects of chemical exposure on different biological processes including development and reproduction (Figs. 17.2 and 17.3). Changes in spermatogenesis, folliculogenesis, puberty, and reproductive aging are measured by examining the morphology of sperm and ovarian follicles. ELISAs and RIAs are used to measure the effects of chemicals on steroidogenesis, puberty, and reproductive aging. Furthermore, the Tanner Scale is a tool to measure puberty. Finally, gene expression is used as a biomarker for changes in the uterus, placenta, folliculogenesis, spermatogenesis, and steroidogenesis.

BPA and phthalates, particularly DEHP, are commonly associated with changes in reproductive biomarkers because of their endocrine-disrupting abilities in both males and females (Fig. 17.1). BPA has been shown to affect spermatogenesis, folliculogenesis, the uterus, the placenta, and reproductive aging in both males and females. Furthermore, DEHP affects several biomarkers including spermatogenesis, folliculogenesis, steroidogenesis, reproductive aging, and puberty in both males and females. However, the mechanisms of action are often unknown. Thus, future studies are needed to examine the mechanisms of these different EDCs and how their toxicity affects the biomarkers of reproduction. Furthermore, results from human epidemiological studies are not always consistent and vary with both population and methods. Therefore, future studies are needed to better observe the effects of EDCs on biomarkers of development and reproduction in humans.

References

Adibi, J.J., Whyatt, R.M., Hauser, R., et al., 2010. Transcriptional biomarkers of steroidogenesis and trophoblast differentiation in the placenta in relation to prenatal phthalate exposure. Environ. Health Perspect. 118 (2), 291–296. https://doi.org/10.1289/.

Adibi, J.J., Zhao, Y., Zhan, L.V., et al., 2017. An investigation of the single and combined phthalate metabolite effects on human chorionic gonadotropin expression in placental cells. Environ. Health Perspect. 125 (10), 107010.

Agarwal, D.K., Eustis, S., Lamb, J.C.T., et al., 1986. Effects of di(2-ethylhexyl) phthalate on the gonadal pathophysiology, sperm morphology, and reproductive performance of male rats. Environ. Health Perspect. 65, 343–350.

Alanazi, A.B., Alshalan, A.M., Alanazi, O.A., et al., 2017. Epidemiology of senile prostatic enlargement among elderly men in Arar, Kingdom of Saudi Arabia. Electron. Physician 9 (9), 5349–5353.

Andriana, B.B., Tay, T.W., Tachiwana, T., et al., 2004. Effects of mono(2-ethylhexyl) phthalate (MEHP) on testes in rats in vitro. Okajimas Folia Anat. Jpn. 80 (5–6), 127–136.

Araujo, A.B., Wittert, G.A., 2011. Endocrinology of the aging male. Best Pract. Res. Clin. Endocrinol. Metabol. 25 (2), 303–319.

Axelstad, M., Boberg, J., Hougaard, K.S., et al., 2011. Effects of pre- and postnatal exposure to the UV-filter octyl methoxycinnamate (OMC) on the reproductive, auditory and neurological development of rat offspring. Toxicol. Appl. Pharmacol. 250 (3), 278–290.

Barakat, R., Lin, P.P., Rattan, S., et al., 2017. Prenatal exposure to DEHP induces premature reproductive senescence in male mice. Toxicol. Sci. https://doi.org/10.1093/toxsci/kfw248.

Bechis, S.K., Otsetov, A.G., Ge, R., Olumi, A.F., 2014. Personalized medicine for the management of benign prostatic hyperplasia. J. Urol. 192 (1), 16–23.

Brehm, E., Rattan, S., Gao, L., Flaws, J.A., 2018. Prenatal exposure to di(2-ethylhexyl) phthalate causes long-term transgenerational effects on female reproduction in mice. Endocrinology 159 (2), 795–809.

Bulus, A.D., Asci, A., Erkekoglu, P., et al., 2016. The evaluation of possible role of endocrine disruptors in central and peripheral precocious puberty. Toxicol. Mech. Meth. 26 (7), 493–500.

Burger, H.G., 2006. Physiology and endocrinology of the menopause. Medicine 34 (1), 27–30.

Calhoun, K.C., Padilla-Banks, E., Jefferson, W.N., et al., 2014. Bisphenol A exposure alters developmental gene expression in the fetal rhesus macaque uterus. PLoS One 9 (1), e85894.

Cao, X.W., Lin, K., Li, C.Y., Yuan, C.W., 2011. A review of WHO laboratory manual for the examination and processing of human semen (fifth ed.). Zhonghua Nan Ke Xue 17 (12), 1059–1063.

Carlberg, C.E., Moller, L., Paakki, P., et al., 2000. DNA adducts in human placenta as biomarkers for environmental pollution, analysed by the (32)P-HPLC method. Biomarkers 5 (3), 182–191.

Carou, M.E., Deguiz, M.L., Reynoso, R., et al., 2009. Impact of the UV-B filter 4-(methylbenzylidene)-camphor (4-MBC) during prenatal development in the neuroendocrine regulation of gonadal axis in male and female adult rats. Environ. Toxicol. Pharmacol. 27 (3), 410–414.

Carou, M.E., Ponzo, O.J., Cardozo Gutierrez, R.P., et al., 2008. Low dose 4-MBC effect on neuroendocrine regulation of reproductive axis in adult male rats. Environ. Toxicol. Pharmacol. 26 (2), 222–224.

Ceccatelli, R., Faass, O., Schlumpf, M., Lichtensteiger, W., 2006. Gene expression and estrogen sensitivity in rat uterus after developmental exposure to the polybrominated diphenylether PBDE 99 and PCB. Toxicology 220 (2–3), 104–116.

Chauvigne, F., Plummer, S., Lesne, L., et al., 2011. Mono-(2-ethylhexyl) phthalate directly alters the expression of Leydig cell genes and CYP17 lyase activity in cultured rat fetal testis. PLoS One 6 (11), e27172.

Chen, J., Wu, S., Wen, S., et al., 2015. The mechanism of environmental endocrine disruptors (DEHP) induces epigenetic transgenerational inheritance of cryptorchidism. PLoS One 10 (6), e0126403.

Chen, X., Li, L., Li, H., et al., 2017. Prenatal exposure to di-n-butyl phthalate disrupts the development of adult Leydig cells in male rats during puberty. Toxicology 386 (Suppl. C), 19–27.

Cheng, G., Remer, T., Prinz-Langenohl, R., et al., 2010. Relation of isoflavones and fiber intake in childhood to the timing of puberty. Am. J. Clin. Nutr. 92 (3), 556–564.

Craig, Z.R., Singh, J., Gupta, R.K., Flaws, J.A., 2014. Co-treatment of mouse antral follicles with 17β-estradiol interferes with mono-2-ethylhexyl phthalate (MEHP)-induced atresia and altered apoptosis gene expression. Reprod. Toxicol. 45, 45–51.

Crowther, J., 2008. Enzyme linked immunosorbent assay (ELISA). In: Walker, J.M., Rapley, R. (Eds.), Molecular Biomethods Handbook. Humana Press, Totowa, NJ, pp. 657–682.

de Kat, A.C., van der Schouw, Y.T., Eijkemans, M.J., et al., 2016. Back to the basics of ovarian aging: a population-based study on longitudinal anti-Mullerian hormone decline. BMC Med. 14 (1), 151.

Delemarre-van de Waal, H.A., 2002. Regulation of puberty. Best Pract. Res. Clin. Endocrinol. Metabol. 16 (1), 1–12.

Dickerson, S.M., Gore, A.C., 2007. Estrogenic environmental endocrine-disrupting chemical effects on reproductive neuroendocrine function and dysfunction across the life cycle. Rev. Endocr. Metab. Disord. 8 (2), 143–159.

Dos Santos, A.H., Vieira, M.L., de Azevedo Camin, N., et al., 2016. In utero and lactational exposure to fluoxetine delays puberty onset in female rats offspring. Reprod. Toxicol. 62, 1–8.

Eskenazi, B., Rauch, S.A., Tenerelli, R., et al., 2017. In utero and childhood DDT, DDE, PBDE and PCBs exposure and sex hormones in adolescent boys: the CHAMACOS study. Int. J. Hyg. Environ. Health 220 (2, Part B), 364–372.

Faber, K.A., Hughes Jr., C.L., 1991. The effect of neonatal exposure to diethylstilbestrol, genistein, and zearalenone on pituitary responsiveness and sexually dimorphic nucleus volume in the castrated adult rat. Biol. Reprod. 45 (4), 649–653.

Ferguson, S.A., Law, C.D., Kissling, G.E., 2014. Developmental treatment with ethinyl estradiol, but not bisphenol A, causes alterations in sexually dimorphic behaviors in male and female Sprague Dawley rats. Toxicol. Sci. 140 (2), 374–392.

Flaws, J.A., Doerr, J.K., Sipes, I.G., Hoyer, P.B., 1994. Destruction of preantral follicles in adult rats by 4-vinyl-1-cyclohexene diepoxide. Reprod. Toxicol. 8 (6), 509–514.

Gal, A., Lin, P.C., Barger, A.M., et al., 2014. Vaginal fold histology reduces the variability introduced by vaginal exfoliative cytology in the classification of mouse estrous cycle stages. Toxicol. Pathol. 42 (8), 1212–1220.

Ge, R., Hardy, M.P., 2007. Regulation of Leydig cells during pubertal development. In: Payne, A.H., Hardy, M.P. (Eds.), The Leydig Cell in Health and Disease. Humana Press, Totowa, NJ, pp. 55–70.

Gee, D.M., Flurkey, K., Finch, C.E., 1983. Aging and the regulation of luteinizing hormone in C57BL/6J mice: impaired elevations after ovariectomy and spontaneous elevations at advanced ages. Biol. Reprod. 28 (3), 598–607.

Gore, A.C., Chappell, V.A., Fenton, S.E., et al., 2015. EDC-2: the Endocrine Society's second scientific statement on endocrine-disrupting chemicals. Endocr. Rev. 36 (6), E1–E150.

Gore, A.C., Walker, D.M., Zama, A.M., et al., 2011. Early life exposure to endocrine-disrupting chemicals causes lifelong molecular reprogramming of the hypothalamus and premature reproductive aging. Mol. Endocrinol. 25 (12), 2157–2168.

Grebe, S.K.G., Singh, R.J., 2011. LC-MS/MS in the clinical laboratory – where to from here? Clin. Biochem. Rev. 32 (1), 5–31.

Gunes, S., Hekim, G.N., Arslan, M.A., Asci, R., 2016. Effects of aging on the male reproductive system. J. Assist. Reprod. Genet. 33 (4), 441–454.

Guo, X., Zhou, S., Chen, Y., et al., 2017. Ziram delays pubertal development of rat Leydig cells. Toxicol. Sci. 160 (2), 329–340.

Hannon, P.R., Brannick, K.E., Wang, W., Flaws, J.A., 2015a. Mono(2-ethylhexyl) phthalate accelerates early folliculogenesis and inhibits steroidogenesis in cultured mouse whole ovaries and antral follicles. Biol. Reprod. 92 (5), 120.

Hannon, P.R., Brannick, K.E., Wang, W., et al., 2015b. Di(2-ethylhexyl) phthalate inhibits antral follicle growth, induces atresia, and inhibits steroid hormone production in cultured mouse antral follicles. Toxicol. Appl. Pharmacol. 284 (1), 42–53.

Hannon, P.R., Flaws, J.A., 2015. The effects of phthalates on the ovary. Front. Endocrinol. 6, 8.

Hannon, P.R., Niermann, S., Flaws, J.A., 2016. Acute exposure to di(2-Ethylhexyl) phthalate in adulthood causes adverse reproductive outcomes later in life and accelerates reproductive aging in female mice. Toxicol. Sci. 150 (1), 97–108.

Hannon, P.R., Peretz, J., Flaws, J.A., 2014. Daily exposure to di(2-ethylhexyl) phthalate alters estrous cyclicity and accelerates primordial follicle recruitment potentially via dysregulation of the phosphatidylinositol 3-kinase signaling pathway in adult mice. Biol. Reprod. 90 (6), 136.

Hartshorne, G.M., 1997. In vitro culture of ovarian follicles. Rev. Reprod. 2 (2), 94–104.

Hess, R.A., 2003. Estrogen in the adult male reproductive tract: a review. Reprod. Biol. Endocrinol. 1, 52.

Hess, R.A., de Franca, L.R., 2008. Spermatogenesis and cycle of the seminiferous epithelium. In: Cheng, C.Y. (Ed.), Molecular Mechanisms in Spermatogenesis. Springer New York, New York, NY, pp. 1–15.

Kalo, D., Hadas, R., Furman, O., et al., 2015. Carryover effects of acute DEHP exposure on ovarian function and oocyte developmental competence in lactating cows. PLoS One 10 (7), e0130896.

Leino, O., Kiviranta, H., Karjalainen, A.K., et al., 2013. Pollutant concentrations in placenta. Food Chem. Toxicol. 54, 59–69.

Li, S., Davis, B., 2007. Evaluating rodent vaginal and uterine histology in toxicity studies. Birth Defects Res. B Dev. Reprod. Toxicol. 80 (3), 246–252.

Liu, C., Duan, W., Li, R., et al., 2013. Exposure to bisphenol A disrupts meiotic progression during spermatogenesis in adult rats through estrogen-like activity. Cell Death Dis. 4, e676.

Livera, G., Rouiller-Fabre, V., Durand, P., Habert, R., 2000. Multiple effects of retinoids on the development of Sertoli, germ, and Leydig cells of fetal and neonatal rat testis in culture. Biol. Reprod. 62 (5), 1303–1314.

Louis, G.W., Hallinger, D.R., Stoker, T.E., 2013. The effect of triclosan on the uterotrophic response to extended doses of ethinyl estradiol in the weanling rat. Reprod. Toxicol. 36, 71–77.

Lyche, J.L., Gutleb, A.C., Bergman, A., et al., 2009. Reproductive and developmental toxicity of phthalates. J. Toxicol. Environ. Health B Crit. Rev. 12 (4), 225–249.

Mahalingam, S., Gao, L., Eisner, J., et al., 2016. Effects of isoliquiritigenin on ovarian antral follicle growth and steroidogenesis. Reprod. Toxicol. 66, 107–114.

Meeker, J.D., Ehrlich, S., Toth, T.L., et al., 2010. Semen quality and sperm DNA damage in relation to urinary bisphenol A among men from an infertility clinic. Reprod. Toxicol. 30 (4), 532–539.

Melching-Kollmuss, S., Fussell, K.C., Buesen, R., et al., 2014. Antiandrogenicity can only be evaluated using a weight of evidence approach. Regul. Toxicol. Pharmacol. 68 (1), 175–192.

Mimura, J., Fujii-Kuriyama, Y., 2003. Functional role of AhR in the expression of toxic effects by TCDD. Biochim. Biophys. Acta Gen. Subj. 1619 (3), 263–268.

Morini, L., Falcon, M., Pichini, S., et al., 2011. Ethyl-glucuronide and ethyl-sulfate in placental and fetal tissues by liquid chromatography coupled with tandem mass spectrometry. Anal. Biochem. 418 (1), 30–36.

Mortimer, S.T., 2000. CASA–practical aspects. J. Androl. 21 (4), 515–524.

Mortimer, S.T., van der Horst, G., Mortimer, D., 2015. The future of computer-aided sperm analysis. Asian J. Androl. 17 (4), 545–553.

Motohashi, M., Wempe, M.F., Mutou, T., et al., 2016. In utero-exposed di(n-butyl) phthalate induce dose dependent, age-related changes of morphology and testosterone-biosynthesis enzymes/associated proteins of Leydig cell mitochondria in rats. J. Toxicol. Sci. 41 (2), 195–206.

Myllynen, P., Pasanen, M., Pelkonen, O., 2005. Human placenta: a human organ for developmental toxicology research and biomonitoring. Placenta 26 (5), 361–371.

Nardi, J., Moras, P.B., Koeppe, C., et al., 2017. Prepubertal subchronic exposure to soy milk and glyphosate leads to endocrine disruption. Food Chem. Toxicol. 100 (Suppl. C), 247–252.

Neely, E.K., Crossen, S.S., 2014. Precocious puberty. Curr. Opin. Obstet. Gynecol. 26 (5), 332–338.

Ohmukai, H., Negura, T., Tachibana, S., Ohta, R., 2017. Genetic variation in low-dose effects of neonatal DES exposure in female rats. Reprod. Toxicol. 73, 322–327.

Orphanides, G., Ashby, J., Kimber, I., et al., 2004. Phenotypic anchoring of gene expression changes during estrogen-induced uterine growth. Environ. Health Perspect. 112, 1589–1606.

Otubanjo, O.A., Mosuro, A.A., 2001. An in vivo evaluation of induction of abnormal sperm morphology by some anthelmintic drugs in mice. Mutat. Res. 497 (1–2), 131–138.

Owens, J.W., Ashby, J., 2002. Critical review and evaluation of the uterotrophic bioassay for the identification of possible estrogen agonists and antagonists: in support of the validation of the OECD uterotrophic protocols for the laboratory rodent. Crit. Rev. Toxicol. 32 (6), 445–520.

Owens, W., Koëter, H.B.W.M., 2003. The OECD program to validate the rat uterotrophic bioassay: an overview. Environ. Health Perspect. 111 (12), 1527–1529.

Park, J.D., Habeebu, S.S., Klaassen, C.D., 2002. Testicular toxicity of di-(2-ethylhexyl)phthalate in young Sprague-Dawley rats. Toxicology 171 (2–3), 105–115.

Partanen, H.A., El-Nezami, H.S., Leppanen, J.M., et al., 2010. Aflatoxin B1 transfer and metabolism in human placenta. Toxicol. Sci. 113 (1), 216–225.

Patel, S., Zhou, C., Rattan, S., Flaws, J.A., 2015. Effects of endocrine-disrupting chemicals on the ovary. Biol. Reprod. 93 (1), 20.

Plant, T.M., Terasawa, E., Witchel, S.F., 2015. Chapter 32-Puberty in Non-human Primates and Man Knobil and Neill's Physiology of Reproduction, fourth ed. Academic Press, San Diego, pp. 1487–1536.

Ponzo, O.J., Silvia, C., 2013. Evidence of reproductive disruption associated with neuroendocrine changes induced by UV-B filters, phthalates and nonylphenol during sexual maturation in rats of both gender. Toxicology 311 (1–2), 41–51.

Prevot, V., 2015. Chapter 30-Puberty in Mice and Rats Knobil and Neill's Physiology of Reproduction, fourth ed. Academic Press, San Diego, pp. 1395–1439.

Rasier, G., Parent, A.S., Gerard, A., et al., 2007. Early maturation of gonadotropin-releasing hormone secretion and sexual precocity after exposure of infant female rats to estradiol or dichlorodiphenyltrichloroethane. Biol. Reprod. 77 (4), 734–742.

Rattan, S., Brehm, E., Gao, L., et al., 2017. Prenatal exposure to di(2-ethylhexyl) phthalate (DEHP) disrupts ovarian function in a transgenerational manner in female mice. Biol. Reprod. 98.

Ricke, W.A., Lee, C.W., Clapper, T.R., et al., 2016. In utero and lactational TCDD exposure increases susceptibility to lower urinary tract dysfunction in adulthood. Toxicol. Sci. 150 (2), 429–440.

Rockett, J.C., Lynch, C.D., Buck, G.M., 2004. Biomarkers for assessing reproductive development and health: Part 1–Pubertal development. Environ. Health Perspect. 112 (1), 105–112.

Santoro, N., 2002. The menopause transition: an update. Hum. Reprod. Update 8 (2), 155–160.

Scarbrough, K., Wise, P.M., 1990. Age-related changes in pulsatile luteinizing hormone release precede the transition to estrous acyclicity and depend upon estrous cycle history. Endocrinology 126 (2), 884–890.

Schug, T.T., Janesick, A., Blumberg, B., Heindel, J.J., 2011. Endocrine disrupting chemicals and disease susceptibility. J. Steroid Biochem. Mol. Biol. 127 (3–5), 204–215.

Sen, N., Liu, X., Craig, Z.R., 2015. Short term exposure to di-n-butyl phthalate (DBP) disrupts ovarian function in young CD-1 mice. Reprod. Toxicol. 53, 15–22.

Shi, Z., Valdez, K.E., Ting, A.Y., et al., 2007. Ovarian endocrine disruption underlies premature reproductive senescence following environmentally relevant chronic exposure to the aryl hydrocarbon receptor agonist 2,3,7,8-tetrachlorodibenzo-p-dioxin. Biol. Reprod. 76 (2), 198–202.

Sleiman, H.K., Romano, R.M., Oliveira, C.A., Romano, M.A., 2013. Effects of prepubertal exposure to silver nanoparticles on reproductive parameters in adult male Wistar rats. J. Toxicol. Environ. Health 76 (17), 1023–1032.

Somasundaram, D.B., Selvanesan, B.C., Ramachandran, I., Bhaskaran, R.S., 2016. Lactational exposure to di (2-ethylhexyl) phthalate impairs the ovarian and uterine function of adult offspring rat. Reprod. Sci. 23 (4), 549–559.

Steinmetz, R., Mitchner, N.A., Grant, A., et al., 1998. The xenoestrogen bisphenol A induces growth, differentiation, and c-fos gene expression in the female reproductive tract. Endocrinology 139 (6), 2741–2747.

Stejskalova, L., Pavek, P., 2011. The function of cytochrome P450 1A1 enzyme (CYP1A1) and aryl hydrocarbon receptor (AhR) in the placenta. Curr. Pharmaceut. Biotechnol. 12 (5), 715–730.

Strimbu, K., Tavel, J.A., 2010. What are biomarkers? Curr. Opin. HIV AIDS 5 (6), 463–466.

Taketa, Y., Mineshima, H., Ohta, E., Nakano-Ito, K., 2017. The effects of ethylene glycol monomethyl ether on female reproductive system in juvenile rats. J. Toxicol. Sci. 42 (6), 707–713.

Talwar, P., Hayatnagarkar, S., 2015. Sperm function test. J. Hum. Reprod. Sci. 8 (2), 61–69.

Tanner, J.M., Davies, P.S., 1985. Clinical longitudinal standards for height and height velocity for North American children. J. Pediatr. 107 (3), 317–329.

Tavares, R.S., Escada-Rebelo, S., Correia, M., et al., 2016. The non-genomic effects of endocrine-disrupting chemicals on mammalian sperm. Reproduction 151 (1), R1–R13.

Tavares, R.S., Mansell, S., Barratt, C.L., et al., 2013. p,p'-DDE activates CatSper and compromises human sperm function at environmentally relevant concentrations. Hum. Reprod. 28 (12), 3167–3177.

te Velde, E.R., Pearson, P.L., 2002. The variability of female reproductive ageing. Hum. Reprod. Update 8 (2), 141–154.

Vahakangas, K., Myllynen, P., 2009. Drug transporters in the human blood-placental barrier. Br. J. Pharmacol. 158 (3), 665–678.

Vidal, J., Mirsky, M., Colman, K., et al., 2013. Reproductive system and mammary gland. In: Toxicologic Pathology. CRC Press, pp. 717–830.

Walker, D.M., Kermath, B.A., Woller, M.J., Gore, A.C., 2013. Disruption of reproductive aging in female and male rats by gestational exposure to estrogenic endocrine disruptors. Endocrinology 154 (6), 2129–2143.

Wan, X., Ru, Y., Chu, C., et al., 2016. Bisphenol A accelerates capacitation-associated protein tyrosine phosphorylation of rat sperm by activating protein kinase A. Acta Biochim. Biophys. Sin. 48 (6), 573–580.

Wang, W., Hafner, K.S., Flaws, J.A., 2014. In utero bisphenol A exposure disrupts germ cell nest breakdown and reduces fertility with age in the mouse. Toxicol. Appl. Pharmacol. 276 (2), 157–164.

Wang, Z., Li, D., Miao, M., et al., 2017. Urine bisphenol A and pubertal development in boys. Int. J. Hyg. Environ. Health 220 (1), 43–50.

Waters, K.M., Safe, S., Gaido, K.W., 2001. Differential gene expression in response to methoxychlor and estradiol through ER, ER, and AR in reproductive tissues of female mice. Toxicol. Sci. 63 (1), 47–56.

Whyatt, R.M., Garte, S.J., Cosma, G., et al., 1995. CYP1A1 messenger RNA levels in placental tissue as a biomarker of environmental exposure. Cancer Epidemiol. Biomark. Prev. 4 (March), 147–153.

Wolff, M.S., Pajak, A., Pinney, S.M., et al., 2017. Associations of urinary phthalate and phenol biomarkers with menarche in a multiethnic cohort of young girls. Reprod. Toxicol. 67 (Suppl. C), 56–64.

Wu, J.H., Jiang, X.R., Liu, G.M., et al., 2011. Oral exposure to low-dose bisphenol A aggravates testosterone-induced benign hyperplasia prostate in rats. Toxicol. Ind. Health 27 (9), 810–819.

Yalow, R.S., 1987. Radioimmunoassay: historical aspects and general considerations. In: Patrono, C., Peskar, B.A. (Eds.), Radioimmunoassay in Basic and Clinical Pharmacology. Springer Berlin Heidelberg, Berlin, Heidelberg, pp. 1–6.

Ye, X., Li, F., Zhang, J., et al., 2017. Pyrethroid insecticide cypermethrin accelerates pubertal onset in male mice via disrupting hypothalamic-pituitary-gonadal axis. Environ. Sci. Technol. 51 (17), 10212–10221.

Yen, S.S.C., Strauss, J.F., Barbieri, R.L., 2014. Yen and Jaffe's Reproductive Endocrinology Physiology, Pathophysiology, and Clinical Management (pp. 1 online resource (xiii, 908 pp.)). Retrieved from: http://www.library.uiuc.edu/proxy/go.php?url=http://www.sciencedirect.com/science/book/9781455727582.

Zhang, Z., Hu, Y., Zhao, L., et al., 2011. Estrogen agonist/antagonist properties of dibenzyl phthalate (DBzP) based on in vitro and in vivo assays. Toxicol. Lett. 207 (1), 7–11.

Zhou, C., Flaws, J.A., 2017. Effects of an environmentally relevant phthalate mixture on cultured mouse antral follicles. Toxicol. Sci. 156 (1), 217–229.

Zhou, C., Gao, L., Flaws, J.A., 2017a. Exposure to an environmentally relevant phthalate mixture causes transgenerational effects on female reproduction in mice. Endocrinology 158 (6), 1739–1754.

Zhou, C., Gao, L., Flaws, J.A., 2017b. Prenatal exposure to an environmentally relevant phthalate mixture disrupts reproduction in F1 female mice. Toxicol. Appl. Pharmacol. 318, 49–57.

Zhou, C., Wang, W., Peretz, J., Flaws, J.A., 2015a. Bisphenol A exposure inhibits germ cell nest breakdown by reducing apoptosis in cultured neonatal mouse ovaries. Reprod. Toxicol. 57, 87–99.

Zhou, J., Chen, L., Li, J., et al., 2015b. The semen pH affects sperm motility and capacitation. PLoS One 10 (7), e0132974.

Zhu, J., Chan, Y.M., 2017. Adult consequences of self-limited delayed puberty. Pediatrics 139 (6).

Zirkin, B.R., Brown, T.R., Jarow, J.P., Wright, W.W., 2011. Chapter 3-endocrine and paracrine regulation of mammalian spermatogenesis. In: Norris, D.O., Lopez, K.H. (Eds.), Hormones and Reproduction of Vertebrates. Academic Press, London, pp. 45–57.

Ziv-Gal, A., Gallicchio, L., Chiang, C., 2016. Phthalate metabolite levels and menopausal hot flashes in midlife women. Reprod. Toxicol. 60, 76–81.

Ziv-Gal, A., Wang, W., Zhou, C., Flaws, J.A., 2015. The effects of in utero bisphenol A exposure on reproductive capacity in several generations of mice. Toxicol. Appl. Pharmacol. 284 (3), 354–362.

CHAPTER

18

Biomarkers of Toxicity in Human Placenta

Kirsi Vähäkangas[1], *Jarkko Loikkanen*[3], *Heidi Sahlman*[1], *Vesa Karttunen*[1], *Jenni Repo*[1], *Elina Sieppi*[2], *Maria Kummu*[2], *Pasi Huuskonen*[1], *Kirsi Myöhänen*[3], *Markus Storvik*[1], *Markku Pasanen*[1], *Päivi Myllynen*[4], *Olavi Pelkonen*[2]

[1]School of Pharmacy/Toxicology, Faculty of Health Sciences, University of Eastern Finland, Kuopio, Finland; [2]Research Unit of Biomedicine, Pharmacology and Toxicology, Faculty of Medicine, University of Oulu, Oulu, Finland; [3]European Chemicals Agency, Helsinki, Finland; [4]NordLab, Oulu, Finland

INTRODUCTION

Human placenta in between the mother and the fetus is a very special organ in many ways. In human reproduction, placenta is the gastrointestinal tract, lungs, liver, and kidneys of the fetus and its proper functioning is essential for fetal health and development as well as for the future health of the child (Burton et al., 2016) (Fig. 18.1). Furthermore, placental tissue is mainly of fetal origin and placental changes are directly related to fetal development. Placental biomarkers of toxicity may thus also reflect direct toxicity in the fetus. Biomarkers of placental toxicity have been searched for in studies using animals, trophoblastic cells lines, human placental tissue, and human blood samples. They would be useful in clinical medicine, e.g., when

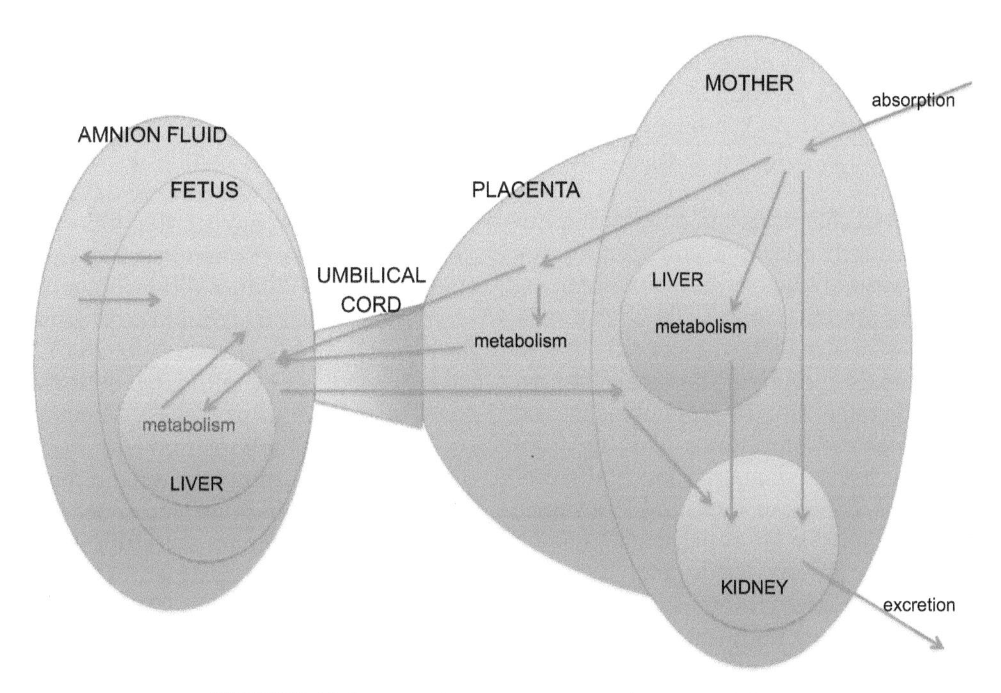

FIGURE 18.1 Toxicokinetics in mother, placenta, and fetus.

Biomarkers in Toxicology, Second Edition
https://doi.org/10.1016/B978-0-12-814655-2.00018-9

evaluating potential drug toxicity in placenta. In toxicology, biomarkers that inform about exposure and/or significant toxic effects in the placenta would be very useful in risk assessment.

The problems in extrapolating placental studies in animals to humans are the conspicuous interspecies differences in placental structure and function and in many biochemical and molecular features, e.g., expression of enzymes. Gene expression in placenta reflects its evolutionary history. Probably due to the more recent development of placenta compared to other tissues, placental transcription factors potentially regulating xenobiotic metabolism, transporters, and other placental proteins are very different from the ones present in other tissues (Pavek and Dvorak, 2008).

At term, human discoid placenta weighs about 500 g, its diameter is about 25 cm, and it is about 3 cm thick. There is thus a lot of material and human placental tissue can be easily studied after birth. However, it is important to remember that at term placenta represents tissue at the end of its lifespan. Placenta itself goes through major anatomical, biochemical, and functional changes during the 9 months of its lifespan. This is reflected in many placental structures, functions, and expression of both proteins and RNAs. Early placentas are less frequently available for study after spontaneous or induced abortions. During pregnancy, chorionic villous biopsies can be taken only by clinical indications because of the risks for the pregnancy.

From an ethical point of view, pregnant women are regarded as a special group. Requesting the placenta, especially in the emotionally sensitive situation after abortion, creates ethical issues, which have to be considered before the actual research (Halkoaho et al., 2010). The two most important ethical issues are safety of the mother and fetus, which may be endangered if samples are taken during pregnancy, and informed consent, which should be voluntarily solicited from the mother before taking the placenta.

Various environmental chemicals can be found in human placenta and can clearly affect placental physiology and molecular pathways. Drugs and other chemical compounds can induce placental toxicity through a variety of mechanisms, which may be functional also in other tissues. A genotoxic compound has the potential to induce genomic damage also in trophoblastic and other cells of the placenta. Compounds toxic to mitochondria can be expected to damage placental mitochondria as well. Thus there are measurable entities, e.g., DNA adducts, which can be anticipated to serve as biomarkers of toxicity in any tissue, including the placenta. There are other biomarkers specific to the developmental stage, reflecting both fetal and placental function, e.g., hormone levels in blood during pregnancy. New emerging levels of regulation of cellular functions, e.g., noncoding RNA, give new possibilities for biomarker development. For instance, some microRNAs seem to be expressed primarily and extensively in the placenta.

Biomarkers of placental toxicity, analyzed in placental tissue or in another matrix, would be useful for evaluation of placental and potentially also fetal toxicity in both the clinical setting and environmental studies. On the other hand, biomarkers of toxicity in placental tissue may reflect general toxicity.

PLACENTAL DEVELOPMENT AND STRUCTURE

The development of human placenta begins 6–7 days after fertilization when the fertilized egg in the form of a blastocyst implants into the uterine wall (Benirschke et al., 2006). At this point the blastocyst contains about 100–250 cells that form the outer and inner cell mass. The inner cell mass is called the embryoblast and develops into embryo, umbilical cord, and amnion, whereas the outer cell mass, or trophoblast, develops into placenta. In the trophoblast the outer layer of cells forms the syncytiotrophoblast by the fusion of neighboring trophoblastic cells. The inner layer of cells contains cytotrophoblasts, which remain nonfused and divide to create material for the syncytiotrophoblast. The nuclei of the syncytiotrophoblast lose their capacity to divide.

The formation of a full layer of the syncytiotrophoblast is completed by 12 days after the fertilization, when the uterine epithelium closes over the deeply implanted blastocyst. About 13 days after fertilization, cytotrophoblastic proliferation increases and cytotrophoblasts together with syncytiotrophoblast form the placental primary villi (Sadler, 2004; Benirschke et al., 2006). The first fetal capillaries in the villous structure can be found about 18–20 days after fertilization. Further development of the villous capillary system gives rise to the extraembryonic vascular system and contact with the maternal circulation. Placental development continues throughout the pregnancy. For example, the syncytiotrophoblastic (syncytial) knots, which are aggregates of syncytial nuclei, increase with gestational age (Benirschke et al., 2006). Their abundance can be used to evaluate villous maturity, but also can be an indicator of poor uteroplacental circulation.

For the exchange of nutrients and gases, human placenta has a complex structure with two circulations, fetal and maternal, separated by a thin layer consisting of several cell types (Fig. 18.2). The maternal uterine surface of the placenta is called the basal plate, and the fetal (amnionic) surface of the placenta is known as a chorionic plate (Benirschke et al., 2006). Between the basal

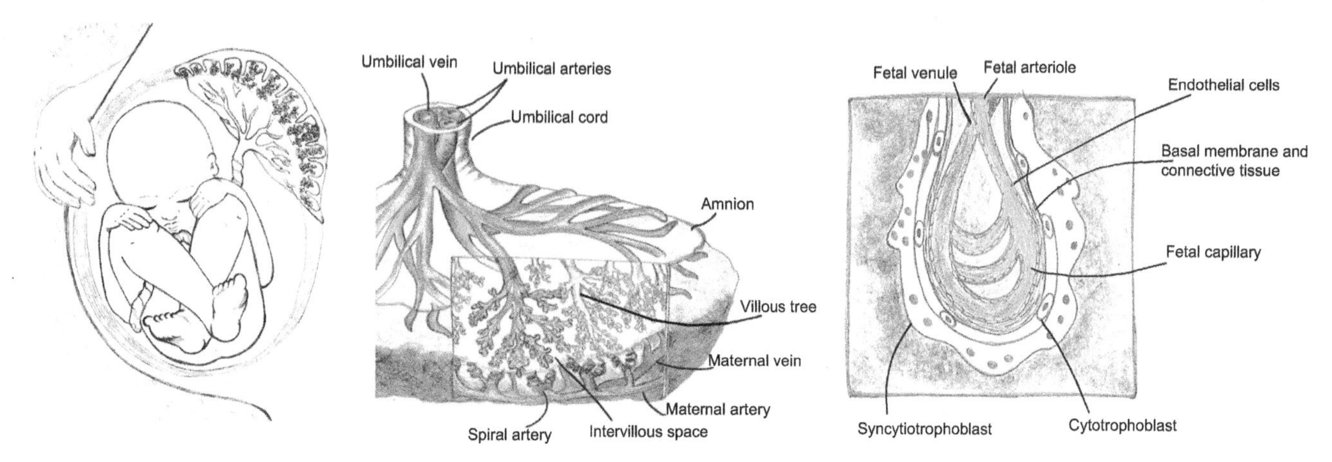

FIGURE 18.2 Anatomy of term human placenta (figure by Dr. Chiara Mannelli).

and chorionic plates there is the intervillous space. Villous trees branching from the chorionic plate are directly in contact with maternal blood that circulates through the intervillous space. The chorionic arteries and veins from villous trees gather to form bigger vessels and, finally, the three vessels of the umbilical cord. Oxygen and nutrients are delivered to the fetus by one umbilical vein, and two umbilical arteries bring the waste products back to the maternal blood circulation via the placenta.

Term human placenta has 10–40 slightly elevated areas at the basal surface, called maternal cotyledons (Benirschke et al., 2006). Each maternal cotyledon is occupied by one or several villous trees. Maternal cotyledons correspond fairly well to the positions of the villous trees. The exchange of all compounds between maternal and fetal blood occurs through the surface of villous trees. This fetomaternal distance decreases to about one-tenth of the diffusion distance in the second month toward term. At the end of pregnancy only two cell layers (syncytiotrophoblast and the endothelial lining of fetal vessels, with some connective tissue in between) separate maternal and fetal blood (Sadler, 2004; Benirschke et al., 2006).

TOXIC AND HORMONALLY ACTIVE CHEMICALS IN HUMAN PLACENTA

Most compounds cross human placenta quite easily, although the ABC (ATP binding cassette) efflux transporters may slow down the transfer of at least some compounds (Vähäkangas et al., 2011). Compounds can also accumulate in the placenta, especially lipid-soluble chemicals (Ala-Kokko et al., 2000). Polychlorinated dibenzo-*p*-dioxin (PCDD) and polychlorinated dibenzo-*p*-furan (PCDF) congeners and Co-PCBs (dioxin-like coplanar polychlorinated biphenyls) with a high toxic equivalence factor are examples of lipid-soluble environmental compounds that accumulate in human placenta (Suzuki et al., 2005). In vivo many toxic compounds have been detected in placental and fetal samples (Table 18.1). Presence of toxic chemicals in cord blood or meconium proves that the fetus is exposed and compounds measurable in placental tissue probably have reached the fetus as well. In addition, numerous compounds measured in cord blood and/or meconium implicate placental exposure. These include legal and illegal drugs, and heavy metals (Crinnion, 2009). It is obvious that all compounds that interfere with normal placental development and function can also disturb fetal development or lead to miscarriage (Burton et al., 2016). Changes in normal placental function can occur at many levels, e.g., transport of nutrients and waste products, cell signaling, production and release of hormones and enzymes, and cellular growth and maturation (Myllynen et al., 2005).

Ethanol is a good example of a toxic compound known to affect the development of placenta and its functions (Burd et al., 2007). Ethanol can be toxic for the placenta by impairing normal transfer of amino acids, zinc, and glucose (Fisher and Karl, 1988). Together with smoking and nutritional deficiencies, prenatal alcohol exposure can lead to poor or limited placental development, and consequent placental dysfunction or abruption can even threaten fetal or maternal life. Ethanol is also directly fetotoxic and possible fetal outcomes include prematurity, growth impairment, neurocognitive problems, birth defects, and mortality. In perfused human placental cotyledon, ethanol induces vasoconstriction, which lasts as long as ethanol is present (Taylor et al., 1994). Ethanol may also induce oxidative stress in placental villi (Kay et al., 2000). This has been associated with placental toxicity in the form of impaired oxygen transport and followed by fetal acidosis (Acevedo et al., 1997).

TABLE 18.1 Chemicals, Their Fetotoxicity, and Potential Biomarkers in Human Placental Tissue or Fetal or Maternal Specimens After In Vivo Exposure

Chemical Group and Examples	Possible Fetotoxicity	Potential Biomarker	Matrix Measured
AMINES			
Acrylamide	Impaired fetal growth[1]	DNA-adducts[2]	Cord blood[2]
Glycidamide	Impaired fetal growth[1]	DNA-adducts[2]	Cord blood[2]
METALS		Compounds Themselves	
Cadmium	Reduced birth weight[3,4]	Metallothioneins, lipidperoxidases[5]	Placenta[6], cord blood[6,7], maternal blood[6,7]
Lead	Immune system effects, later in life increased risk of asthma and allergy Effects on neuro development[4,8]	Metallothioneins, lipid peroxidases, ALAD[5]	Placenta[6], cord blood[6,7], maternal blood[6,7]
Mercury	Effects on neuro development[4,8]		Placenta[6], cord blood[6,7,9–11] meconium[9,10], infant's hair[9], maternal blood[6,7]
Manganese	Intrauterine growth retardation[12]	MAO-activity[13]	Cord blood[13], maternal blood[13]
MYCOTOXINS			
Aflatoxins	Low birth weight[14]; growth faltering in the first year of life[15]	DNA-adducts[16]	Placenta[16], cord blood[14,16–19], maternal blood[17,18]
Ochratoxins	NA		Cord blood[19–21], maternal blood[20], breast milk[20,21]
PAH COMPOUNDS	Gene mutations[22]	PAHs themselves[23–26]	Placenta[27], cord blood[23,26]
		DNA-adducts[27,28]	Placenta[28], cord blood[29,30]
Acenaphthylene, acenaphthene, fluorine, phenanthrene, anthracene, fluoranthene, pyrene, benz[a] anthracene, chrysene, benzo[b] fluoranthene, benzo[k] fluoranthene, benzo[a]pyrene, dibenz[a,h]anthracene, indeno [1,2,3 cd]pyrene, benzo[g,h,i] perylene,			
OTHER ENVIRONMENTAL POLLUTANTS		Compounds Themselves	
OCPs	Impaired fetal growth[31]		Placenta[32], maternal blood[23,32,33], cord blood[32,33], breast milk[23,33]
Parathion, malathion, etc.	NA		Cord blood[32,34]
PBDEs	NA		Placenta[35], maternal blood[33,36,37], cord blood[33,36,37], breast milk[33,36]
PCBs	Effects on developing neurological system and sex hormone levels; recurrent infections[8]		

TABLE 18.1 Chemicals, Their Fetotoxicity, and Potential Biomarkers in Human Placental Tissue or Fetal or Maternal Specimens After In Vivo Exposure—cont'd

Chemical Group and Examples	Possible Fetotoxicity	Potential Biomarker	Matrix Measured
PLASTICIZER			
Bisphenol A	Reduction in fetal body weight[38]		Placenta[39,40], cord blood[41–43], amniotic fluid[44], fetal liver[39], maternal blood[41–45]
SOCIAL DRUGS			
Ethanol	FASD[46]	Ethyl glucuronide[47–50], ethyl-sulfate[49,50], fatty acid ethyl esters[48]	Meconium[48,50], placenta[49], fetal tissues[49]
Nicotine	Adversely affects the development of many organ systems, especially brain and lungs[61]	Cotinine[51],	Placenta[51], cord blood[53], meconium[54]
		Trans-3′-hydroxycotinine[52]	Meconium[52]
(ILLEGAL) DRUGS			
Cocaine	Restricted fetal growth[55]	Benzoylecgonine, m-hydroxybenzoylecgonine, ecgonine methyl ester[56]	Placenta[56], umbilical cord[56], meconium[54,56,57]
Methadone	Neonatal abstinence syndrome	EDDP[56]	Placenta[56], umbilical cord[56], meconium[54,56,57]
Heroin	Impaired fetal growth; risk of neonatal abstinence syndrome[59]	6-Monoacetylmorphine[58]	Meconium[58]
Morphine	Impaired fetal growth[59]	Morphine-3-glucuronide, morphine-6-glucuronide[56]	Placenta[56], umbilical cord[56], meconium[52,54,56]
Cannabis	Adverse fetal growth trajectories[60]		Meconium[57]

ALAD, delta-aminolevulinic acid dehydratase; *EDDP*, 2-ethylidene-1,5-dimethyl-3,3-diphenylpyrrolidine; *NA*, not available; *OCPs*, organochlorine pesticides; *PAH*, polycyclic aromatic hydrocarbon; *PBDEs*, polybrominated diphenyl ethers; *PCBs*, polychlorinated biphenyls.

1 Pedersen et al., 2012; 2 von Stedingk et al., 2011; 3 Menai et al., 2012; 4 Gundacker and Hengstschläger, 2012; 5 Serafim et al., 2012; 6 Al-Saleh et al., 2011; 7 Butler Walker et al., 2006; 8 Crinnion, 2009; 9 Ramirez et al., 2000; 10 Unuvar et al., 2007; 11 Wu et al., 2013; 12 Wood, 2009; 13 Abdelouahab et al., 2010; 14 Abdulrazzaq et al., 2002; 15 Turner et al., 2007; 16 Hsieh and Hsieh, 1993; 17 De Vries et al., 1989; 18 Denning et al., 1990; 19 Jonsyn et al., 1995; 20 Postupolski et al., 2006; 21 Biasucci et al., 2011; 22 Perera et al., 2002; 23 Tsang et al., 2011; 24 Perera et al., 2005a; 25 Perera et al., 2005b; 26 Sexton et al., 2011; 27 Yu et al., 2011; 28 Manchester et al., 1988; 29 Perera et al., 1999; 30 Phillips, 2002; 31 Dewan et al., 2013; 32 Whyatt et al., 2003; 33 Jaraczewska et al., 2006; 34 Whyatt et al., 2005; 35 Ma et al., 2012; 36 Guvenius et al., 2003; 37 Park et al., 2009; 38 Ranjit et al., 2010; 39 Cao et al., 2012; 40 Jiménez-Díaz et al., 2010; 41 Lee et al., 2008; 42 Schönfelder et al., 2002; 43 Wan et al., 2010; 44 Padmanabhan et al., 2008; 45 Yamada et al., 2002; 46 Medina, 2011; 47 Matlow et al., 2013; 48 Bakdash et al., 2010; 49 Morini et al., 2011; 50 Pichini et al., 2005; 51 Vyhlidal et al., 2012; 52 Gray et al., 2010; 53 Berlin et al., 2010; 54 Ostrea et al., 1998; 55 Schempf, 2007; 56 de Castro et al., 2013; 57 Bar-Oz et al., 2003; 58 Pichini et al., 2009; 59 Wagner et al., 1998; 60 El Marroun et al., 2009; 61 England et al., 2017.

Maternal cigarette smoking during pregnancy is associated with increased risk of adverse birth outcomes, such as low birth weight and infant mortality (Vyhlidal et al., 2012). In human placenta, cotinine, one of the main metabolites of nicotine, can cause vasoconstriction by activating placental phospholipase A2 and increasing the effects of prostaglandin E2 (Rama Sastry et al., 1999). Vasoconstriction in placenta is associated with increased frequency of spontaneous abortions and preterm births. Human placenta expresses at mRNA level all and at protein level at least nine nicotine receptor subtypes (α2, α3, α4, α5, α7, α9 β1, β2, and δ; Machaalani et al., 2014). In rodent placenta nicotine exposure leads to disrupted placental morphology as well as reduced 11β-hydroxysteroid dehydrogenase activity associated with poor fetal growth and health later in life (Zhou et al., 2018). Polycyclic aromatic hydrocarbons (PAHs), ubiquitous environmental contaminants, are also among the main components of tobacco smoke. Many carcinogenic and mutagenic PAHs induce oxidative stress and are suspected to have adverse effects on reproductive outcomes (Al-Saleh et al., 2013). The fetus can be exposed to PAHs and many other toxic chemicals also via maternal food (Myöhänen and Vähäkangas, 2012). One of the most studied PAHs, carcinogenic benzo(a)pyrene (BP; Class 1; IARC, 2010), is found also in smoked, flamed, and grilled food (Miller and Ramos, 2001). BP-DNA adducts are found in human placenta (Manchester et al., 1988). Even caffeine use during pregnancy was shown to reduce birth weight of the fetus (Sengpiel et al., 2013). The doses with such effects were lower than current recommendations during pregnancy.

In human placental tissue many heavy metals, such as cadmium (Cd), copper (Cu), mercury (Hg), and lead (Pb), have been found (Kantola et al., 2000; Needham et al., 2011; Gundacker and Hengstschläger, 2012), and this may be the case with other heavy metals as well. Heavy metals affect many normal functions of human placenta (Kantola et al., 2004; Gundacker and Hengstschläger, 2012), and at least Pb and nickel (Ni) accumulate in the syncytiotrophoblast (Reichrtová et al., 1998). Accumulation of Pb leads to decreased cytochrome oxidase activity in placental cells (Reichrtová et al., 1998). Pb in maternal blood was significantly associated with a decreased uptake of calcium into isolated trophoblastic cells in the study by Lafond et al. (2004). A balance between heavy metal and selenium concentrations in maternal blood and placental tissue has suggested a protective role for selenium in restoring placental metabolic activities (Kantola et al., 2004). Therefore, in terms of adverse effects at the fetoplacental unit or within the whole body, the question may not only be of the actual heavy metal concentration but also the ratio between heavy metal burden and the amount of protecting "scavengers" present in the tissue. These markers display great interindividual and regional variation (Kantola et al., 2000, 2004).

In BeWo choriocarcinoma cells, Cd inhibits the function of the efflux transporter ABCG2 (Kummu et al., 2012), which probably has a role in transplacental transfer of many xenobiotics (Vähäkangas and Myllynen, 2009). In rats, Cd accumulates in placenta resulting in trophoblastic damage, which decreases the nutrient and oxygen transfer to fetus (Levin et al., 1981). Also in humans, Cd has been shown to accumulate in placenta (Korpela et al., 1986; Osman et al., 2000; Esteban-Vasallo et al., 2012). Kippler et al. (2010) found a link between placental Cd levels and impaired Zn transfer into the fetus, as well as increased amounts of metallothionein (MT) in placental tissue. Cd also inhibits placental transport of calcium. Furthermore, many other placental functions are disturbed by Cd as well, e.g., decidualization and hormonal balance (Gundacker and Hengstschläger, 2012). Toxicity of Hg to placenta includes at least disturbed amino acid transfer, placental oxygen consumption, enzyme activities, hormonal secretion, and membrane fluidity.

Organochlorine pesticides (OCPs) and other persistent organochlorine compounds have also been found in human placenta (Lopez-Espinosa et al., 2007; Bergonzi et al., 2011; Needham et al., 2011). The existence of OCPs in maternal blood, cord blood, and breast milk indicates that placenta, fetus, and infant are exposed to these chemicals (Al-Saleh et al., 2012; Dewan et al., 2013; Kezios et al., 2013). DDT (active p,p-DDT) is perhaps the best-studied OCP; p,p-DDT as well as its metabolites p,p-DDE and p,p-DDD have been detected in placenta, cord blood, and maternal blood (Al-Saleh et al., 2012) and can be used to evaluate exposure. Environmental pollutants, such as polybrominated diphenyl ethers (PBDEs) and polychlorinated biphenyls (PCBs), have been detected in placental tissue (Ma et al., 2012; Virtanen et al., 2012). Additionally, PBDEs, PCBs, hydroxylated metabolites of PCBs (polychlorobiphenylols; OH-PCBs), and pentachlorophenol have been detected in cord blood, maternal plasma, and breast milk (Guvenius et al., 2003; Park et al., 2009) indicating exposure of the placenta as well.

Environmental endocrine disruptors have been associated with various reproductive abnormalities (Robins et al., 2011). The plasticizer bisphenol A (BPA) has been detected in human cord blood, indicating that it crosses human placenta (Schönfelder et al., 2002; Lee et al., 2008; Wan et al., 2010). It has also been detected in human term placenta (Schönfelder et al., 2002) and various human biological fluids, such as maternal blood (Lee et al., 2008; Padmanabhan et al., 2008; Wan et al., 2010) and amniotic fluid (Yamada et al., 2002). In amniotic fluid five times higher concentrations of BPA have been measured than of other body fluids (Ikezuki et al., 2002). Also BPA metabolites have been detected in both maternal and cord sera, levels being higher in cord serum (Liu et al., 2017).

PLACENTAL FUNCTIONS AND MOLECULAR PATHWAYS INVOLVED IN TOXICITY

Xenobiotic Metabolism and Its Regulation in Human Placenta

While the liver plays a major role in xenobiotic metabolism, all tissues, including placenta, express at least some of the xenobiotic-metabolizing enzymes, although at a much lower level (for reviews, see Pasanen, 1999; Myllynen et al., 2007, 2009; Prouillac and Lecoeur, 2010). In the beginning of pregnancy placenta expresses mRNA of a wider variety of CYPs and conjugating enzymes (e.g., the UGT1A family) than at term (for a review see Myllynen et al., 2007). Functionally the most important of the xenobiotic-metabolizing CYP enzymes in placenta is CYP1A1. It is inducible, e.g., by cigarette smoking (Pasanen et al., 1988; Hakkola et al., 1996).

Kinetics of drugs and drug-metabolizing capacity as well as metabolizing enzyme profiles vary depending on the stage of pregnancy (Dickmann and Isoherranen, 2013), also in the placenta (Pasanen, 1999; Myllynen et al., 2005, 2007). Although the CYP-mediated metabolism in placenta is much more restricted than in the liver, a number of drugs and foreign chemicals are metabolized and activated. CYP1A1 activity is very low, but inducible, e.g., by cigarette smoking. In addition to CYP1A1, CYP19A1 (aromatase) has toxicological relevance and is also induced by xenobiotics. Both CYP1A1 and CYP19A1 are expressed and functional throughout the pregnancy. Other CYP enzymes expressed at protein level in human placenta include CYP11A1, CYP2J2, CYP2R1, CYP27B1, and CYP24A1. Even the CYP2 members in the placenta are evolutionally old with only physiological substrates, and most placental CYPs are linked to steroid metabolism (Storvik et al., 2014). Of the CYP enzymes, CYP19A1 is the most highly expressed. In the placenta CYP enzymes including CYP1A1 have unique regulation mechanisms and expression patterns different from those in the liver. Aromatase is nearly undetectable in normal hepatocytes (Granata et al., 2009), but increased expression and activity have been reported in diseased liver cells (Hata et al., 2013).

In addition to CYP1A1 and some other functionalization reactions, all the main conjugation activities are also expressed in human placenta but with different isoenzyme expression profiles than in the liver or kidney. All the most important phase 2 enzyme activities (glutathione transferase, epoxide hydrolase, N-acetyltransferase, sulfotransferases and UDP-glucuronyl transferase) are also found in human placental tissue (Pasanen and Pelkonen, 1990; Paakki et al., 2000a; b; Collier et al., 2000). Protein expression of the conjugating enzymes may also in some cases respond to chemical stress (Paakki et al., 2000a; b). However, both functionalization and conjugation activities are more specifically associated with endocrine functions of placenta (Pasanen, 1999).

In human placenta, CYP1A1 catalyzing the AHH activity (aryl hydrocarbon hydroxylase, named according to the first activity found for this enzyme) is the only functionally active "traditional" xenobiotic targeting CYP form. Toxicologically CYP1A1 is interesting because it activates several procarcinogens and proteratogens (Stejskalova and Pavek, 2011). Half a century ago, Welch et al. (1968) demonstrated that placental monooxygenase enzymes can metabolize BP in vitro. Later it was shown that BP is metabolized by the placental CYP-dependent system to its ultimate carcinogenic metabolite, BP-7,8-diol-9,10-epoxide capable of DNA binding, and that the DNA binding also correlates with AHH activity (Namkung and Juchau, 1980; Vähäkangas et al., 1989; Lagueux et al., 1999; Carlberg et al., 2000; Obolenskaya et al., 2010; Stejskalova and Pavek, 2011).

CYP19A1 in placenta is needed for estradiol production, and maintenance of pregnancy, as well as sexual differentiation of gonads and brain of the fetus. It catalyzes reactions producing estradiol from testosterone, estrone from androstenedione, and estriol from 16α-hydroxylated dehydroepiandrosterone. Interestingly, in addition to the production of estrogens from androgen derivates, CYP19A1 also metabolizes xenobiotics because it possesses a wider substrate-binding pocket than its vital function suggests. In addition, CYP19A1 can metabolize certain xenobiotics, such as buprenorphine (Deshmukh et al., 2003), and is itself inhibited by xenobiotics (Table 18.2).

CYP19A1 is very disproportionally highly expressed in placenta, compared to the liver (GEO DataSet Record GDS3113). Furthermore, among the mammalian CYP enzymes, CYP19A1 is the only one with placenta as the tissue of major expression. This high basal CYP19A1 transcription in placenta is determined by a promoter that is specific to humans (Bulun et al., 2003).

Human placental microsomal fraction, choriocarcinoma cells (e.g., JEG-3, BeWo), trophoblastic primary cells, or cells expressing recombinant CYP19A1 protein have been used to test the specificity and metabolic inhibitory potential on placental aromatase of several chemical structures (Table 18.2). In addition to classical steroids, and indole and azole-based aromatase inhibitors (Leze et al., 2004; Yahiaoui et al., 2004), several other chemical structures are capable of docking to the active center or prosthetic structures of CYP19A1 and influence the catalytic activity of the protein. For instance, BPA can directly inhibit the catalytic activity of CYP19A1 and downregulate the expression via the promoter I.1 (Watanabe et al., 2012). Aflatoxin B1 (AFB1) is a substrate for CYP19A1, and additionally affects CYP19A1 gene expression by secondary messengers (Storvik et al., 2011; Huuskonen et al., 2013).

The regulatory machineries driving the expression of CYPs in placenta mainly respond to maternal health status (Hakkola et al., 1996). The effect of maternal health may be complex and affect the enzyme activities in yet unknown ways. For example, oxidized LDLs and their PPARgamma and LXR ligands may decrease placental CYP expression due to an effect on placental development during the first two semesters (Pavan et al., 2004a; b). The best characterized regulators of CYP expression in placenta are the hormonal signals, hypoxia, proinflammatory cytokines, and certain xenobiotic compounds mimicking endogenous compounds.

Human placenta does not express most hepatic transcription factors (Storvik et al., 2014). For example,

TABLE 18.2 The Effect of Chemicals and Stress Factors on Human Placental CYP19A1

Compound	Response	Placental Test System	References
Aflatoxin B1	Substrate (aflatoxicol metabolite), induction	JEG-3 cells, microsomes	Sawada et al., 1993; Storvik et al., 2011; Huuskonen et al., 2013
Acetofenate	Inhibitor	JEG-3 cells	Chen et al., 2012
Atrazine	Inhibitor	Microsomes	Benachour et al., 2007
Benzofuran derivatives	Inhibitor	Microsomes	Saberi et al., 2006
Bisphenol A	Inhibition, downregulation via estrogen receptor	JEG-3 cells, microsomes	Benachour et al., 2007; Huang and Leung, 2009; Watanabe et al., 2012; Nativelle-Serpentini et al., 2003; Kwintkiewicz et al., 2010
	Inhibition through induction of PPARγ	Extrapolation on data from ovarian granulosa cells	
Buprenorphine	Inhibitor, metabolism Norbuprenorphine production	Microsomes	Fokina et al., 2011; Zharikova et al., 2006; Deshmukh et al., 2003
Calcitriol	Induction through atypical vitamin-D receptor pathway or through PKA-related mechanism	JEG-3 cells	Sun et al., 1998
		Trophoblasts	Barrera et al., 2007
Chrysin	Inhibitor	Microsomes, recombinant CYP19A1	Edmunds et al., 2005
Chlordecone	Inhibitor	Microsomes	Benachour et al., 2007
Cortisol	Inductions through PKA-related mechanism, or through glucocorticoid receptor	Syncytiotrophoblast	Wang et al., 2012
DDE	Inhibitor	Microsomes	Benachour et al., 2007
Diethylstilbestrol	Inhibitor	Microsomes	Benachour et al., 2007
Endoxifen	Inhibitor	Microsomes, recombinant CYP19A1	Lu et al., 2012a,b
Norendoxifen		Microsomes, recombinant CYP19A1	Lu et al., 2012a
N-desmethyl-tamoxifen		Microsomes, recombinant CYP19A1	
Forscolin	Induction via cAMP	JEG-3	Harada et al., 2003
Glucocorticoid (antepartum)	Downregulation	Placental mRNA at term	Paakki et al., 2000a
Glyphosate	Inhibitor	JEG-3 cells, recombinant CYP19A1	Richard et al., 2005
Levo-alpha-acetylmethanol	Metabolism	Microsomes, recombinant CYP19A1	Deshmukh et al., 2004
Lindane	Inhibitor	JEG-3 cells	Nativelle-Serpentini et al., 2003
Methadone	Substrate, mechanism-based inhibitor	Microsomes, recombinant CYP19A1	Nanovskaya et al., 2004; Hieronymus et al., 2006; Lu et al., 2010
Naringenin	Inhibitor	Microsomes, recombinant CYP19A1	Edmunds et al., 2005
Non−dioxin-like PCBs	Inhibitor	Microsomes	Antunes-Fernandes et al., 2011
Nonylphenol	Inhibitor	Microsomes	Benachour et al., 2007

TABLE 18.2 The Effect of Chemicals and Stress Factors on Human Placental CYP19A1—cont'd

Compound	Response	Placental Test System	References
Parabens	Inhibitor	Microsomes	van Meeuwen et al., 2008
Polybrominated diphenyl ethers—OH-derivatives	Inhibitor	Microsomes	Cantón et al., 2008
Retinol, all-trans retinoic acid	Inhibitor	JEG-3 cells	Ciolino et al., 2011
Serotonin	Induction	BeWo/JEG-3 cells	Klempan et al., 2011
Triptolide	Downregulation; multiple possible mechanisms	Placental microsomes, JEG-3 cells	Zhang et al., 2011a
Tributyltin oxide	Inhibitor	Microsomes	Benachour et al., 2007
Stress Factor	**Response**	**Placental Test System**	**Reference**
Hypoxia	Downregulation via HIF-1α controlled ERRγ expression	Primary cytotrophoblasts	Kumar and Mendelson, 2011
of C/EBP-beta related transcription	Induction	Based on promoter structure and active binding site	Bamberger et al., 2004
Estrogen receptor antagonist	Prevention of trophoblast differentiation related induction of CYP19	Primary cytotrophoblasts	Kumar et al., 2009
5-HT(2A) receptor activation	Induction through PKC-related mechanism	JEG-3 cells	Klempan et al., 2011
NR3B2 (ERRγ) activation	High basal CYP19A1 expression in primary trophoblasts (also when compared to JEG-3 or BeWo cells)	Primary cytotrophoblasts;	Kumar and Mendelson, 2011; Storvik et al., 2014
TGF-β	Downregulation through SMAD complex—related mechanism	JEG-3 cells	Zhou et al., 2009
Maternal smoking	Downregulation	In vivo exposed placental tissue. mRNA and enzymatic activity at term	Huuskonen et al., 2008

CAR (NR1I3), which is one of the major inducers of liver enzymes (Wallace and Redinbo, 2013), is barely detectable in human placenta, as are also PPARα (NR1C1), PXR (NR1I2), and HNF4α (NR2A1). On the other hand, placenta does express several other nuclear factors, such as cortisol receptor (NR3C1), estrogen receptors, ERRγ, and PPARγ, LXRα (NR1H3), SHR (NR0B2), and VDR (NR1I1) (Pavek and Dvorak, 2008). In addition, the aryl hydrocarbon receptor (AHR; NR3C1) is also present in very high levels, being able to induce CYP1A1 in a classical way, but also to interact with other nuclear factors (Huuskonen et al., 2008).

The expression of the CYP19A1 gene is regulated by tissue-specific promoters upstream of tissue-specific first exons (Mendelson and Kamat, 2007). Of the existing 10 or more promoters of the gene, 3 are reported to be active in human placenta. The activation of the placenta-specific constitutively active promoter I.1 leads to induction of CYP19A1 during pregnancy, which then elevates the estrogen levels in pregnant women up to 1000 times the baseline (Bulun et al., 2003). This increase of CYP19A1 expression occurs as the cytotrophoblasts differentiate into syncytiotrophoblast in developing placenta (Mendelson and Kamat, 2007). In addition to promoter I.1, promoters I.2 and I.8 have been described to be practically placenta specific (Demura et al., 2008).

CYP11A1, the key enzyme in steroid production, producing pregnenolone from cholesterol, is present in placenta at high level. The regulation of CYP11A1 in placenta is different from most other tissues, not regulated by the binding site shared by steroidogenic factor 1 (NR5A1) and liver receptor homolog-1 (NR5A2). However, the transcription factors AP1 and SP1, together with certain tissue-specific factors, may regulate the expression of CYP11A1 in placenta (Shih et al., 2011).

CYP2J2 is expressed in placenta at a moderate level (Pavek and Dvorak, 2008). It is regulated at least by proinflammatory signals, and it produces epoxyeicosatrienoic acid derivatives from arachidonic acid. These compounds are reported to be increased in preeclampsia (PE) and have cardiovascular effects (Herse et al., 2012). The regulation of CYP2J2 expression is linked to intracellular signaling pathways leading to AP1 activation in stress conditions (Cui et al., 2010).

The regulatory pathways and CYP enzymes required in vitamin D metabolism are also expressed in placenta (Storvik et al., 2014). These include CYP2R1, which converts calciferol to calcidiol, CYP27B1 catalyzing metabolism further to calcitriol, and CYP24A1 catabolizing calcitriol. All the abovementioned CYPs are affected in preeclampsia, probably through hypoxia or oxidative stress (Ma et al., 2012).

Placental Transporters

For transport of nutrients and other compounds, there is a rather wide variety of transporter proteins expressed in placental cells, both trophoblastic and endothelial cells (for reviews see, e.g., Vähäkangas and Myllynen, 2009; Staud et al., 2012; Iqbal et al., 2012). Transporters expressed in the placenta may facilitate transfer in both the maternal and fetal directions depending on the localization and function of the transporter protein. Drug transporters in human placenta include ABC and SLC (solute carrier protein) transporter families. Of the ABC transporters ABCB1 (P-glycoprotein or p-gp, MDR1), ABCG2 (BCRP or breast cancer resistance protein), and several ABCC (multidrug resistance-associated proteins or MRP) transporters are found in human placenta. Other transporters with xenobiotics as substrates include organic anion/cation transporters. Accumulating evidence suggests that these transporters modify also transplacental transfer of pharmaceutical drugs.

Toxicologically the most studied are the ABC transporters, which in human are efflux transporters. Of the ABC transporters, ABCG2/BCRP, ABCB1/p-gp, and several of the ABCC/MRP-proteins reside in the brush border apical membrane of syncytiotrophoblast-facing maternal blood. These transporters have a potential protective role against xenobiotics because they transport drugs and compounds back to maternal blood from syncytiotrophoblast. The ABC transporters are expressed in human placenta through the pregnancy and the expression level depends on the gestational stage. The expression of ABCB1/p-gp protein has been reported to decrease toward the end of pregnancy (Matthias et al., 2005). Results on the expression of ABCG2/BCRP in human placenta have been discrepant (Matthias et al., 2005; Meyer zu Schwabedissen et al., 2006; Yeboah et al., 2006). ABCC/MRP transporters have been reported to decrease toward the end of pregnancy (Meyer zu Schwabedissen et al., 2006).

Xenobiotics may be substrates for placental transporters but they may also interfere with transporter function or affect the expression level of transporters (for reviews see Vähäkangas and Myllynen, 2009; Staud et al., 2012; Iqbal et al., 2012). Although a number of studies suggest that transporter proteins have a role in transplacental toxicokinetics, much less is known about interactions of these transporters with other xenobiotics such as environmental contaminants, food-borne carcinogens, or abused substances (Table 18.3). However, studies in animal models and cells originating from other tissues suggest that a number of environmental contaminants in fact interact also with drug transporters (Carew and Leslie, 2010; Gundacker et al., 2010; Yang et al., 2010; Pacyniak et al., 2011). For instance, PBDE congeners seem to be substrates for OATPs (Pacyniak et al., 2011) and perfluorinated compounds for OAT4 and URAT1 (Yang et al., 2010). Interestingly, there is also experimental evidence suggesting that metals such as Pb, Cd, and arsenic (As) interact with ABC transporters (see e.g., Carew and Leslie, 2010; Gundacker et al., 2010; Thévenod, 2010; Kummu et al., 2012). ABCC1, ABCC2, and ABCB1 may participate in metal efflux from the cells (Gundacker et al., 2010; Thévenod, 2010; Carew and Leslie, 2010), whereas Cd seems to inhibit ABCG2 transporter function (Kummu et al., 2012).

In addition to "drug transporters," placenta also displays high activities of a variety of transporters for nutrient transfer (for a review see, e.g., Carter, 2012). Five types of amino acid transporters have been reported in human placenta: sodium-coupled neutral amino acid transporters, high affinity glutamate and neutral amino acid transporters, sodium- and chlorine-dependent transporters, cationic amino acid transporters, and glycoprotein-associated amino acid transporters. Human placenta expresses also several glucose transporters in microvillous and basal membranes of the syncytiotrophoblast as well as in capillary endothelium. Activity of the placental glucose and amino acid transport systems is influenced by gestational age and a range of environmental factors including heat stress, hypoxia, and under- and overnutrition, as well as exposure to hormones. Implications of interference with these transporters by chemicals have also been published. Nicotine, cocaine, or their combination inhibit several amino acid transporters (Pastrakuljic et al., 2000), which is a probable mechanism for the known effects on fetal growth. Abused substances including amphetamine and delta-9-tetrahydrocannabinol have been shown to decrease

TABLE 18.3 Examples of Interactions Between Environmental Contaminants, Food-Borne Carcinogens, and Abused Substances With Transporter Proteins in Human Placental Tissue

Xenobiotic	Transporter	Interaction	Placental Model	References
PhIP	ABCG2	Substrate, decreased placental transfer	Ex vivo perfused human placenta	Myllynen et al., 2008
IQ	ABCG2	No effect on placental transfer	Ex vivo perfused human placenta	Immonen et al., 2010
Cadmium	ABCG2	Inhibits ABCG2 function	BeWo cells	Kummu et al., 2012
Bisphenol A	ABCB1	ABCB1 inhibition	BeWo cells	Jin and Audus, 2005
Zearalenone	ABCB1, ABCC1, ABCC2, ABCG2	Induction of ABCB1, ABCC1, ABCC2, and ABCG2 mRNA expression	BeWo cells	Prouillac et al., 2009
Nicotine, cocaine	Amino acid transporters	Inhibition	Ex vivo perfused human placenta	Pastrakuljic et al., 2000
Smoking	ABCB1 and ABCG2	No effect on activity or expression	human placenta	Kolwankar et al., 2005
Aflatoxin B1	ABCG2, ABCC2, hOAT4	Induction in mRNA and/or protein expression	JEG-3 cells	Huuskonen et al., 2013
1-methyl-4-phenylpyridinium (MPP(+))	SERT	Uptake of MPP(+)	JAr cells	Martel and Keating, 2003
As(III)	ABCC2	Downregulation	Chorion cells	Yoshino et al., 2011
Methadone, cocaine, heroin	ABCB1	Increased protein expression	Ex vivo perfused human placenta	Malek et al., 2009

the uptake of glucose in BeWo choriocarcinoma cells (Araújo et al., 2008). In a murine model Xu et al. (2016) showed Cd-induced downregulation of the glucose transporter GLUT3 in placenta by hypermethylation of GLUT3 promoter through upregulation of DNA methyltransferases.

Placental Hormone Production

One of the most important functions of placenta is the production of steroid, protein, and peptide hormones (Table 18.4). Hormones, such as progesterone, estrogen, human chorionic gonadotropin (hCG), placental growth hormone (PGH), and human placental lactogen (hPL) are important for maintaining pregnancy and to support fetal development (Druckmann and Druckmann, 2005; Murphy et al., 2006; Cole, 2010; Newbern and Freemark, 2011). PGH and hPL have also an important role in regulating maternal insulin balance (Newbern and Freemark, 2011). More recently, leptin has appeared as an important regulator of placental growth and development (Maymo et al., 2011; Schanton et al., 2018). It induces proliferation and survival and inhibits apoptosis of trophoblastic cells. It is also important for implantation of embryo and fetal growth. Both leptin itself and all isoforms of its receptors are expressed in human placenta. Leptin secretion is induced transcriptionally by hCG. Other hormones are inactivated in the placenta, for instance prostaglandins, catecholamines, glucocorticoids, and thyroxine (Fowden and Forhead, 2004). Placental 11β-hydoxysteroid dehydrogenase type 2 (11βHSD2) inactivates glucocorticoids and limits feto-placental exposure to the higher maternal glucocorticoid concentrations (Seckl and Meaney, 2004).

The presence of hCG in urine or blood is used as an indicator of pregnancy because it is produced throughout pregnancy. It promotes progesterone production and fusion of cytotrophoblastic cells and their differentiation into syncytiotrophoblast at the beginning of pregnancy (Cole, 2010). Progesterone, on the other hand, maintains pregnancy by preventing the maternal immune system from attacking the fetus and stimulates maternal food intake (Druckmann and Druckmann, 2005; Newbern and Freemark, 2011). Placenta replaces the ovary as the source of progesterone during early pregnancy after about 8 weeks. The progesterone production continues to increase throughout the pregnancy. Placenta uses mainly maternal cholesterol as a precursor in progesterone biosynthesis. First cholesterol is converted in mitochondria to pregnenolone, which is further metabolized to progesterone by 3beta-hydroxysteroid dehydrogenase (Tuckey, 2005).

Placental syncytiotrophoblast synthesizes large amounts of estrogens using steroidal precursors from

TABLE 18.4 Examples of Human Placental Protein and Peptide Hormones

Hormone	Main Functions During Pregnancy
Human placental lactogen (hPL)	Induces maternal insulin resistance and facilitates the mobilization of maternal nutrients for fetal growth promotes growth of maternal tissues
Human chorionic gonadotropin (hCG)	Affects fetal and uterine growth
	Induces leptin expression in trophoblastic cells
	Promotes the maintenance of the corpus luteum during the early pregnancy
	Promotes villous trophoblast tissue formation and the development and growth of uterine spiral arteries
Adrenocorticotropin (ACTH)	Stimulates the release of cortisol and other steroids
Growth hormone variant (hGH-V)	Participates in regulation of maternal and fetal metabolism and the growth and development of the fetus.
Parathyroid hormone related protein (PTH-rP)	Regulates fetal mineral homeostasis, stimulates placental calcium (and possibly magnesium) transfer and affects mineralization of the skeleton
Relaxin	Regulates biochemical processes involved in remodeling the extracellular matrix of the cervix and vagina during pregnancy and rupture of the fetal membranes at term. Promotes uterine and placental growth and influences vascular development and proliferation in the endometrium
Inhibins	Inhibit FSH secretion, inhibit steroidogenesis and production of hCG by the syncytiotrophoblast, involved in the control of the fetomaternal communication required to maintain pregnancy
Activins	Stimulate production and/or secretion of hCG, human placental lactogen, progesterone, and estradiol, potentially involved in cytotrophoblast fusion and syncytialization. Immunomodulatory functions
Gonadotropin releasing hormone (GnRH)	Regulates hCG production, Inhibits the formation of progesterone and estrogen
Placental corticotropin-releasing hormone (CRH)	Stimulates adrenocorticotropin (ACTH) release. Modulates glucose transporter (GLUT) proteins in placental tissue. Role in the timing of birth
Somatostatin	Inhibits hormone secretion and nutrient transport
Ghrelin	Stimulates growth hormone release
Leptin	Regulates implantation as well as fetal and placental growth and development. In placenta regulates survival, proliferation, angiogenesis, and immunomodulation of trophoblast

maternal and fetal blood. C19-steroids dehydroepiandrosterone (DHEA) and its sulfate metabolite dehydroepiandrosterone sulfate (DHEA-S), originating from both maternal and fetal circulations, serve as precursors for estrogen biosynthesis. In human placenta, neither cholesterol nor progesterone can serve as a precursor for estrogen synthesis due to the lack of CYP17 enzyme. Conversion of C19-steroids to estrogens is catalyzed by placental aromatase (CYP19) (Mendelson and Kamat, 2007), which is known to be affected by a number of xenobiotics (Table 18.2). Estrogens produced in the syncytium preferentially enter maternal circulation. Estrogenic hormones increase uteroplacental blood flow, enhance steroid production, stimulate the development of mammary glands, and increase prostaglandin synthesis (Sadler, 2004; Newbern and Freemark, 2011).

Environmental contaminants and lifestyle-associated factors such as smoking and alcohol consumption affect placental hormone production (Table 18.5). Most of the studies have been carried out using in vitro cultured primary trophoblasts or the three most commonly used choriocarcinoma cell lines (BeWo, JEG-3, JAr) and the most commonly studied end point has been hCG production. In addition, hormone concentrations have been measured from placental tissue as well as from maternal and cord blood samples. For instance, smoking has been associated with decreased hCG and PGH secretion. On the other hand, the results regarding leptin and hPL secretion are controversial (Boyce et al., 1975; Spellacy et al., 1977; Mochizuki et al., 1984; Helland et al., 2001; Stasenko et al., 2010). Concentration measurements of hormones from maternal blood can easily be conducted throughout the pregnancy if a suitable biomarker for exposure or placental toxicity can be established.

Molecular Stress Protein Pathways Involved in Toxicity of Human Placenta

Any proteins, such as structural proteins, receptors, enzymes, transporters, and hormones can be targets of toxicity in the placenta. Relatively few studies are available on toxic mechanisms of environmental contaminants in human placenta or placental cells, but some changes in proteins associated with exposure to toxic agents and/or placental pathology have been clearly identified. From a biomarker point of view, these are naturally of the most immediate interest. Also, various environmental factors can affect the same signal transduction pathways, which are involved in some pathological conditions during pregnancy, such as intrauterine growth restriction (IUGR), PE, and HELLP

TABLE 18.5 Examples of Effects of Environmental Contaminants, Phytoestrogens, and Smoking on Human Placental Hormone Production

Compound	Effect on Hormone Secretion	Placental Model	References
p-Nonylphenol	β-hCG secretion ↑	Placental explants	Bechi et al., 2006
p-Nonylphenol	β-hCG secretion ↓	BeWo cells	Bechi et al., 2013
Bromodichloromethane (BDCM)	Immunoreactive CG, bioactive CG secretion ↓	Term primary trophoblasts	Chen et al., 2003; Chen et al., 2004
1,1,1,-Trichloro-2,2-bis(p-chlorophenyl) ethane (DDT)	hCG secretion ↓, Estradiol secretion↓	JEG-3 cells	Wójtowicz et al., 2007a,b
1,1,1,-Trichloro-2,2-bis(p-chlorophenyl) ethane (DDT)	hCG secretion ↓, Progesterone short term ↑, long term↓	Placental explants	Wójtowicz et al., 2008
1,1,-Dichloro-2,2-bis(p-chlorophenyl) ethylene (DDE)	hCG secretion↓, Progesterone secretion ↑, Estradiol secretion↓	JEG-3 cells	Wójtowicz et al., 2007a,b
1,1,-Dichloro-2,2-bis(p-chlorophenyl) ethylene (DDE)	hCG secretion ↓ Progesterone short term ↑, long term↓	Placental explants	Wójtowicz et al., 2008
2,3,7,8-Tetrachlorodibenzo-p-dioxin (TCDD)	Immunoreactive CG ↑, Bioactive CG ↔, Estradiol ↑/↓Progesterone ↓	Primary trophoblasts	Chen et al., 2003; Augustowska et al., 2003a,b
2,3,7,8-Tetrachlorodibenzo-p-dioxin (TCDD)	hCG secretion↓	JEG-3 cells	Augustowska et al., 2007
PCDDs/PCDFs	Progesterone ↔Estradiol ↑	Primary trophoblasts	Augustowska et al., 2003a,b
PCDDs/PCDFs	hCG secretion ↓	JEG-3 cells	Augustowska et al., 2007
Tributyltin (TBT)	hCG secretion ↑	JAr, JEG-3, and BeWo cells	Nakanishi et al. (2002)
Triphenyltin (TPT)	hCG secretion ↑	JAr, JEG-3, and BeWo cells	Nakanishi et al. (2002)
Chlorpyrifos (CPF)	hCG secretion ↑, Estradiol ↔, Progesterone ↔	JEG-3 cells	Ridano et al., 2012
Acetofenate (AF)	Progesterone ↑	JEG-3 cells	Chen et al., 2012
Bisphenol A (1 nM)	hCG secretion ↑	Placental explants	Mørck et al., 2010
Bisphenol A (30–125 μM)	hCG secretion ↓	BeWo cells	Mørck et al., 2010
H_2O_2	hCG secretion ↓	JEG-3 cells	McAleer and Tuan, 2001b
Diethylstilbestrol (DES)	Estradiol ↔	Term explants	Ling et al., 1984
Delor 103 (low chlorinated biphenyls)	Estradiol ↑	Placental explants	Grabic et al., 2006
Delor 106 (high chlorinated biphenyls)	Estradiol ↑	Placental explants	Grabic et al., 2006
Smoking	hPL concentration in serum ↓	In vivo	Boyce et al., 1975; Mochizuki et al., 1984
Smoking	hPL concentration in maternal serum ↑	Serum concentration during last month of pregnancy	Spellacy et al., 1977
Smoking	hCG concentration in serum↓	In vivo	Bernstein et al., 1989
Smoking	Cathechol estrogen formation ↑	Human placental tissue	Chao et al., 1981
Smoking/Cadmium	Leptin mRNA expression ↓	Human placental tissue	Stasenko et al., 2010
Smoking/Cadmium	Placental progesterone content ↓	Human placental tissue	Piasek et al., 2001, 2002
Smoking	No effect on leptin concentration	Maternal plasma at 18 and 35 wk, cord blood	Helland et al., 2001
Smoking	CRH concentration ↑	Maternal plasma at 24–26 wk	Kramer et al., 2010

Continued

TABLE 18.5 Examples of Effects of Environmental Contaminants, Phytoestrogens, and Smoking on Human Placental Hormone Production—cont'd

Compound	Effect on Hormone Secretion	Placental Model	References
Smoking	Cord blood ghrelin ↑ and PGH ↓	Blood concentration measurement	Bouhours-Nouet et al., 2006
Smoking	Maternal PGH ↓	Blood concentration measurement	Coutant et al., 2001
Cadmium	Progesterone release ↓	Human trophoblast cells	Jolibois et al., 1999
Cadmium	Progesterone synthesis ↑, hCG production initially ↓, followed by a recovery during which hCG ↑	JAr cells	Powlin et al., 1997
Cadmium	hCG production ↓	JEG-3 cells	McAleer and Tuan, 2001a
Cadmium	hCG production ↑	Placental explants	Boadi et al., 1992
Nicotine, cotinine, anabasine	Androstenedione and testosterone conversion to estrogen ↓	Choriocarcinoma cells	Barbieri et al., 1986
Nicotine	No effect on leptin mRNA levels	BeWo cells	Reseland et al., 2005
Genistein, daidzein	hCG production ↓, progesterone production ↓, Estrogen production↑	Primary trophoblasts	Jeschke et al., 2005; Richter et al., 2009
Genistein, daidzein	hCG production↑ (low doses) progesterone production ↓ (high doses)	BeWo and JEG-3 cells	Plessow et al., 2003
Genistein	EGF-induced hCG secretion ↓	JAr	Baker et al., 1998
Phytoestrogen extract	Estradiol production ↑, Progesterone production ↓	JEG-3 cells	Matscheski et al., 2006
Zearalenone	hCG production ↑, ↔	BeWo cells	Prouillac et al., 2009, 2012
Ethanol	hCG production ↑	Human placental trophoblasts	Karl et al., 1998

CRH, corticotropin releasing hormone; *EGF*, epidermal growth factor; *hCG*, human chorionic gonadotropin; *hPL*, human placental lactogen; *PGH*, placental growth hormone.

(hemolysis, elevated liver enzymes, and low platelet) syndrome.

Apoptosis is an important mechanism of toxic cell death. For instance BPA, an endocrine disruptor, has been shown to increase procaspase-3 cleavage to caspase-3, a key event in apoptosis induction, in human first trimester placental explants (Mørck et al., 2010). One of the key proteins in apoptosis induction is the tumor suppressor protein p53. It has an important role in cell toxicity because, in addition to apoptosis, it regulates various other cellular processes including the cell cycle, and it is commonly activated by chemical stress (Vähäkangas, 2003).

There are only a few in vitro studies in human placental cells available that have tried to pursue the role of p53 and related signal transduction pathways in toxicity caused by common environmental contaminants. For instance, data on 2,3,7,8-tetrachlorodibenzo-p-dioxin (TCDD) toxicity in human placental cells are still discrepant. Chen et al. (2010) showed that TCDD damages mitochondria in human trophoblastic JAr cells and induces p53-mediated apoptosis associated with Bax induction, cytochrome *c* release, and caspase-3 activation. On the contrary, TCDD or BP did not cause apoptosis in JEG-3 cells, another human trophoblastic cell line, in the study by Drukteinis et al. (2005). In these cells, TCDD has also no effect on cell proliferation while BP inhibits it (Zhang and Shiverick, 1997; Drukteinis et al., 2005). BP-induced inhibition may be associated with downregulation of epidermal growth factor receptors (EGFRs) (Zhang and Shiverick, 1997) and cell cycle arrest by activation of p53 through Ser-15 phosphorylation. Consequently, there is induction of the p53 target p21 with reduction in the expression of cyclin-dependent kinase 1 (CDK1) (Drukteinis et al., 2005). Preimplantation factor (PIF), secreted by viable embryos inhibits trophoblastic apoptosis through the p53 pathway and is lower in placentas from PE and IUGR pregnancies (Moindjie et al., 2016). Thus, the p53 protein seems to be involved in pathological conditions during pregnancy related to placental function. As with many other proteins, the level of p53 protein changes during pregnancy (Rolfo et al., 2012) and according to our own unpublished studies there is a very high interindividual variation in the level of p53 protein in human placenta. In cultured primary cytotrophoblasts isolated from normal human placenta, the expression of p53 has been shown to be higher under hypoxia (2% O_2) than under normal oxygen concentration (20%) (Hung et al., 2012). Recently, it has been shown that in human placentas associated with IUGR, the levels of p53 protein are higher than in placentas after normal pregnancy (Hung et al., 2012; Rolfo et al., 2012). In contrast to IUGR, Rolfo et al. (2012) observed that the amount of p53 protein is lower in PE placentas than in normal placentas. It is a well-known fact that smoking is associated with IUGR (e.g., Campbell et al., 2012) and increases the risk of IUGR by ethanol (Aliyu et al., 2009). The interesting inverse association between smoking and PE, related to changed balance of vasoactive factors, e.g., the soluble vascular endothelial growth factor receptor, fms-like tyrosine kinase-1 (sFlt-1), and placental growth factor (PlGF) by cigarette smoke condensate, nicotine and cotinine (Mehendale et al., 2007; Romani et al., 2011), seems to be dependent on the stage of pregnancy (Engel et al., 2013). Although the data are still discrepant, in many studies IUGR has been associated also with other chemical and environmental exposures, such as persistent organochlorine compounds (Bergonzi et al., 2011 and references within), air pollution particulates and nitrogen oxides (van den Hooven et al., 2012 and references within), endocrine disruptors (Meeker, 2012), and laboratory work (Halliday-Bell et al., 2010). Air pollution has also been shown to change the expression of sFlt-1 and PIGF in human placenta (van den Hooven et al., 2012).

Menai et al. (2012) suggested that one of the mechanisms by which tobacco smoking causes fetal growth restriction is accumulation and transplacental transfer of Cd. Heavy metals such as Hg, Pb, and Cd cross the human placenta, and Cd in particular, also accumulate in placenta (Esteban-Vasallo et al., 2012 and references within). Many studies indicate that smoking increases Cd levels in human placenta and this is associated with increased expression of metallothioneins (MTs) (Ronco et al., 2005, 2006; Sorkun et al., 2007). Ronco et al. (2006) specified that smoking induces especially the MT-2 isoform in human term placenta. MTs are probably the most important metal-binding proteins but also can act as antioxidants (Gundacker and Hengstschläger, 2012). Induction of MTs by Cd and Zn has been shown earlier also in primary human trophoblasts (Lehman and Poisner, 1984) and trophoblastic JAr cells (Wade et al., 1986; Powlin et al., 1997).

Cd is toxic in human placental trophoblastic cell lines (Valbonesi et al., 2008; Kummu et al., 2012). Cell death by Cd was associated with phosphorylation of ERK1/2, JNK1/2, and p38 mitogen activated protein kinases (MAPKs) in human immortalized HTR-8/SVneo trophoblasts (Valbonesi et al., 2008). These proteins are important regulators of various cellular processes such as proliferation, survival, apoptosis, and differentiation (for an extensive review see Cargnello and Roux, 2011). Also other agents may affect MAPK signaling pathways in placental cells. Saulsbury et al. (2008) suggest that p38 MAPK has a protective role against chlorpyrifos in human trophoblastic JAr cells because inhibition of the p38 MAPK pathway increases cytotoxicity caused by this organophosphate pesticide. The importance of p38 as a regulator of normal functioning of placenta is emphasized by the finding that the expression of p38 is reduced in placentas of women having HELLP syndrome during pregnancy (Corradetti et al., 2010). HELLP results from impaired trophoblast invasion and is present in about 10%–20% of women with severe PE (Jebbink et al., 2012). Migration and invasion of extravillous trophoblasts are key steps in the normal development of placenta (Knöfler, 2010). It is possible that this process is disturbed by toxic environmental compounds, such as Cd. Alvarez and Chakraborty (2011) showed that Cd, at concentrations that have no effect on cell viability, prevents migration of immortalized human HTR-8/SVneo trophoblastic cells. This may be due to actin cytoskeletal disorganization via caspase activation.

The signaling pathway of growth arrest and DNA damage-inducible 45 (Gadd45a), a stress sensor member of the Gadd45 family of genes, seems to be an important mediator in pathogenesis of human PE (Xiong et al., 2009, 2013; Luo et al., 2011). Gadd45a can be induced by various environmental stresses such as radiation, hypoxia, and oxygen radicals (Siafakas and Richardson, 2009). Luo et al. (2011) showed that Gadd45a expression and activation of p38 MAPK is higher in preeclamptic placenta than in placentas from pregnancy without any complications. This is supported by Xiong et al. (2013) who showed that in explants of human term placenta, various stresses associated with PE, including angiotensin II, hypoxia, and cytokines, increase the expression of Gadd45a and the release of sFlt-I. Similar results have been observed in rats with induced PE (Uddin et al., 2011). Increased levels of sFlt-1 and soluble endoglin (sEng) have been measured in serum of women with PE (Luo et al., 2011). It has been suggested that sFlt-1 and PlGF can be used as predictors of PE (Leaños-Miranda et al., 2012). The induction of Gadd45 is associated with activation of two MAPKs: p38 and JNK (Xiong et al., 2013). Also other factors, such as activin A and 8-isoprostane levels, are increased in the circulation of women with PE (Mandang et al., 2007). They are also produced more in human placental explants exposed to oxidative stress.

Possibilities for Studying Pathways and Proteins Involved in Toxicity of Human Placenta

To reveal human placental toxicity and the mechanisms behind helpful biomarker development, relevant models have to be utilized because of the uniqueness of human placenta when compared to other species (Benirschke et al., 2006). All existing models have restrictions and some are very difficult in practice to set up and manage. Trophoblastic cancer cell lines originating from human choriocarcinoma, such as BeWo, JEG-3, and JAr, have appeared very useful for mechanistic studies (Vähäkangas and Myllynen, 2006; Myllynen and Vähäkangas, 2013). In addition to trophoblastic cell lines, changes in cellular macromolecules and pathways as a result or mechanism of toxicity can be studied in cultured placental tissue explants or primary trophoblasts isolated from newly born placentas. Acute responses to toxic agents can also be studied in human placental perfusion where placental tissue can be retained viable for hours. If placental tissue is used for experimental purposes, production of hCG can be used as a marker of viability of the tissue (Partanen et al., 2010; Woo et al., 2012).

Ex vivo perfusion of a human placental cotyledon has appeared very useful to study transplacental transfer of chemical compounds, such as drugs and environmental contaminants (Schneider et al., 1972; Ala-Kokko et al., 2000; Vähäkangas and Myllynen, 2006: Myren et al., 2007; Hutson et al., 2011; Myllynen and Vähäkangas, 2013). In addition, the role of various transporters in transplacental transfer can be studied in placental perfusion. Specific inhibitors of transporters combined with a radioactively labeled compound can be utilized to find out whether a certain transporter is involved in the transfer (e.g., Myllynen et al., 2008; Karttunen et al., 2010; Woo et al., 2012). Radioactively labeled compounds are most widely used because radioactivity can be easily and sensitively measured using scintillation counting. Mechanisms of transplacental transfer can be modeled also by using a Transwell system with two chambers separated by a layer of trophoblastic cells forming tight junctions (Woo et al., 2012). Only a few studies exist to date and the cell line mostly used in Transwell studies is the human BeWo b30 cell line. Single nucleotide polymorphisms (SNPs) and other mechanisms of individual genetic variation can affect the activity of metabolizing enzymes and transporters, increasing or decreasing toxicity (Vähäkangas and Myllynen, 2009). SNPs in placental transporters can be functionally significant as shown in placental perfusion studies. As an example, Rahi et al. (2007) have shown in a perfusion model that the ABCB1/p-gp polymorphism 3435C/T in exon 26 affects transplacental transfer of the antipsychotic drug quetiapine.

In placental tissue or cells of placental origin, expression of various types of proteins, such as stress proteins (e.g., p53), cell cycle regulators (e.g., p21 and CDK1), and other signal transduction proteins (e.g., Wnt-signaling proteins β-catenin and E-cadherin), metabolizing enzymes (e.g., CYP1A1), and transporters (e.g., ABCC2/MRP2 and ABCG2/BCRP), has been shown using various immunological methods (Drukteinis et al., 2005; Hnat et al., 2005; Nakanishi et al., 2005; Myllynen et al., 2008; Storvik et al., 2011; Tsang et al., 2012; Woo et al., 2012; Huuskonen et al., 2013). Xenobiotics can change the expression level of proteins or their post-translational modifications. It is important to remember that expression of proteins in choriocarcinoma cell lines does not reflect in all cases the in vivo situation in the placenta (Vähäkangas and Myllynen, 2006).

In addition to proteins, corresponding gene expression at the mRNA level can be studied by reverse transcription polymerase chain reaction (RT-PCR) or quantitative real-time reverse transcription PCR (qRT-PCR) (Myllynen et al., 2008; Avissar-Whiting et al., 2010; Suter et al., 2010; Tsang et al., 2012; Huuskonen et al., 2013). Modern high-throughput microarray methodologies provide possibilities for gene expression profiling and "fishing" for mechanistically interesting candidate biomarkers (Huuskonen et al., 2008; Avissar-Whiting et al., 2010; Maccani et al., 2010). The omics techniques using various commercially available platforms enable analysis of the whole genome, transcriptome, proteome, or part of the epigenome (e.g., all known microRNAs) at the same time (for a review see Fowler, 2012). These methods are expensive and require special expertise, and studies pursuing their use in the development of biomarkers are in general at a very early stage of development. As to placental toxicity, very few studies exist to date.

POTENTIAL BIOMARKERS OF EXPOSURE AND TOXICITY IN HUMAN PLACENTA

Toxic Compounds and Their Metabolites as Biomarkers

Chemicals as well as their metabolites can be measured from placental tissue, cord blood, and fetal specimens (Table 18.1) and can be used as biomarkers of placental and fetal exposures. Theoretically they can also be pursued as biomarkers of placental and fetal toxicity. The ideal specimen probably varies based on the timing and duration of the exposure (Colby, 2017). Meconium, the stool produced during gestation, is usually used to assess longer term exposures because it reflects exposures occurring from the end of the second trimester through the third trimester. Cord tissue reflects more recent exposures than meconium, including exposures that happen just prior to or during delivery.

OCPs and other persistent organic pollutants, and their metabolites, have been detected in human placenta and can be used to demonstrate exposure. Cholinesterase activity in both maternal blood and placental tissue is lower in mothers living on Spanish farms, indicating potential placental toxicity of organophosphate pesticides (Vera et al., 2012). Decreased placental and/or maternal blood acetylcholinesterase activity may thus serve as an exposure biomarker of placental exposure to organophosphates.

Metabolites of ethanol have been suggested as biomarkers of ethanol exposure. Ethyl glucuronide (EtG), a primary metabolite of ethanol, has been detected in first trimester placental tissue (Morini et al., 2011) and is formed in perfused placenta (Matlow et al., 2013). In addition, another metabolite of ethanol, ethyl sulfate (EtS), has also been detected in first trimester placental tissue (Morini et al., 2011). Both are detectable in meconium of infants born to mothers who have consumed excessive amounts of alcohol during pregnancy (Pichini et al., 2009). Fatty acid ethyl esters (FAEE) are formed after ethanol consumption from free fatty acids and lipids in blood and Bakdash et al. (2010) have detected FAEE in meconium. They suggest that the combination of FAEE and metabolites of ethanol (e.g., EtG) in meconium increases the accuracy of fetal alcohol exposure assessment.

Cotinine is a major metabolite of nicotine and is used as a biomarker of nicotine exposure. It can be measured in several tissues or biological matrices including placenta, meconium, and cord blood (Ostrea et al., 1998; Vyhlidal et al., 2012). Cotinine levels in placenta correlate with the gene expression of xenobiotic-metabolizing enzymes (mRNA of CYP1A1 and CYP1B1) in fetal lung and liver tissue. Thus, placental cotinine is considered to be a reliable biomarker of fetal nicotine exposure (Vyhlidal et al., 2012). Also, another nicotine metabolite, trans-3′-hydroxycotinine, can be measured in meconium as a biomarker of fetal exposure (Gray et al., 2010). Additionally, it has been suggested that maternal or neonatal cadmium blood levels could be used as a potential biomarker for tobacco toxicity on fetal development. Cd, a toxic heavy metal present in large amounts in cigarettes, is accumulated in placental tissue (Al-Saleh et al., 2011) and transferred through the placenta (Butler Walker et al., 2006; Al-Saleh et al., 2011). Its levels in placenta, maternal blood, and urine are negatively correlated with birth weight (Kippler et al., 2010; Menai et al., 2012).

Levels of heavy metals in placental tissue prove the exposure of the placenta and have been suggested as markers of fetal exposure as well (Gundacker and Hengstschläger, 2012). However, at least in some cases, other markers are more useful for fetal exposure. Pb can be detected in placenta (Yoshida, 2002) and cord blood (Ding et al., 2013; Wu et al., 2013), but maternal blood levels have been suggested to be a more reliable indicator of fetal exposure (Iyengar and Rapp, 2001). The chemical form of Hg affects its cellular uptake and transfer through the placenta: Hg vapor and the methylated form cross the placenta easily (Ask et al., 2002), but inorganic Hg is more likely to be trapped in placental tissue (Yoshida, 2002). Human hair (from both mother and fetus) seems to be a more reliable matrix for the evaluation of mercury exposure than placenta (Iyengar and Rapp, 2001). As to arsenic, its levels in placenta correlate with maternal urinary levels and thus use of both urine and placenta could increase reliability when evaluating fetal exposure to arsenic.

Illicit drugs and opioids as well as their metabolites can be measured in placenta, meconium, and other biological specimens and be used as possible biomarkers of fetal exposure (Gray et al., 2010). Methadone and its metabolite 2-ethylidene-1,5-dimethyl-3,3-diphenylpyrrolidine have been detected in placenta, umbilical cord, cord blood (de Castro et al., 2011, 2013), and meconium (Ostrea et al., 1998; Gray et al., 2010). However, there are contradictory results as to whether their presence correlates with fetal birth outcome (e.g., birth weight, length, head circumference) or the severity of neonatal abstinence syndrome (Gray et al., 2010; de Castro et al., 2011). Of the other opioids, morphine, codeine, and hydromorphone have been detected in meconium, which is positively correlated with their presence in maternal urine (Ostrea et al., 1998; Gray et al., 2010). The presence of a heroin metabolite, 6-monoacetylmorphine, was detected in meconium in one study (Pichini et al., 2009), but not in another (Gray et al., 2010). Metabolites of cocaine have been detected in placenta, umbilical cord, and meconium: ecgonine methyl ester and hydroxybenzoylecgonine (BE-OH) in placenta (de Castro et al., 2013); benzoylecgonine in umbilical cord (de Castro et al., 2013); and benzoylecgonine and m-hydroxybenzoylecgonine in meconium (Ostrea et al., 1998; Gray et al., 2010). The presence of cocaine and its metabolites benzoylecgonine and m-hydroxybenzoylecgonine in meconium is associated with maternal urine positive for cocaine (Ostrea et al.,

1998; Gray et al., 2010). Cannabinoids can be also measured in meconium to assess fetal exposure (Bar-Oz et al., 2003).

Aflatoxins have been detected in maternal and cord blood as well as in breast milk of mothers living in contaminated areas, proving that fetus and infants are exposed to this toxic and carcinogenic mycotoxin (De Vries et al., 1989; Denning et al., 1990; Hsieh and Hsieh, 1993; Abdulrazzaq et al., 2002). Another mycotoxin, ochratoxin A, has been detected in maternal and fetal serum, cord blood, and also in breast milk, indicating early exposure of the fetus and placenta also to this mycotoxin (Breitholtz-Emanuelsson et al., 1993; Zimmerli and Dick, 1995; Rosner et al., 2000; Postupolski et al., 2006; Galvano et al., 2008; Biasucci et al., 2011).

PAHs have been detected in placenta and cord blood, indicating exposure at early stages (Yu et al., 2011; Al-Saleh et al., 2013). However, because they require metabolic activation to DNA-binding–activated metabolites, their DNA-adducts as biologically relevant biomarkers have been studied much more than the compounds themselves.

Placental DNA Adducts as Biomarkers

Metabolic activation is a key step for the toxicity of most genotoxic carcinogens (Miller and Miller, 1975; Pelkonen and Vähäkangas, 1980). These toxic metabolites, e.g., epoxides or methylcarbonium ions, may covalently bind to DNA and induce mutations (Fig. 18.3) (Dipple, 1995; Klaassen, 2001). Therefore, DNA adducts are biomarkers of an early, detectable, and critical step of carcinogenesis and can be used as biomarkers of exposure to reactive metabolites of genotoxic carcinogens. The predominant binding site in DNA is the guanine base, but compounds may also bind to other DNA bases (Beach and Gupta, 1992; Dipple, 1995; Poirier et al., 2000). Several organs, including placenta, contain capacities to repair covalent DNA adducts by, for example, nucleotide excision repair (Cheng et al., 1999). However, DNA adduct repair is substance- and organ-specific. Placental DNA adducts can be analyzed from in vivo exposed placental tissue, as well as from experimentally exposed placental tissue, trophoblastic cell lines, and primary trophoblasts(Table 18.6).

Since 1980s, many studies have demonstrated PAH-DNA adducts in human placental tissue (e.g., Manchester et al., 1988; Hansen et al., 1993; Arnould et al., 1997; Sram et al., 1999; Perera et al., 1999; Sanyal et al., 2007; Pratt et al., 2011a,b; Dodd-Butera et al., 2017), or in DNA incubated with placental microsomes (Vaught et al., 1979; Pelkonen and Saarni, 1980; Vähäkangas et al., 1989). PAH exposure from smoking (Everson et al., 1988; Hansen et al., 1992, 1993; Gallagher et al., 1994; Daube et al., 1997) and air pollution (Reddy et al., 1990; Manchester et al., 1992; Topinka et al., 1997;

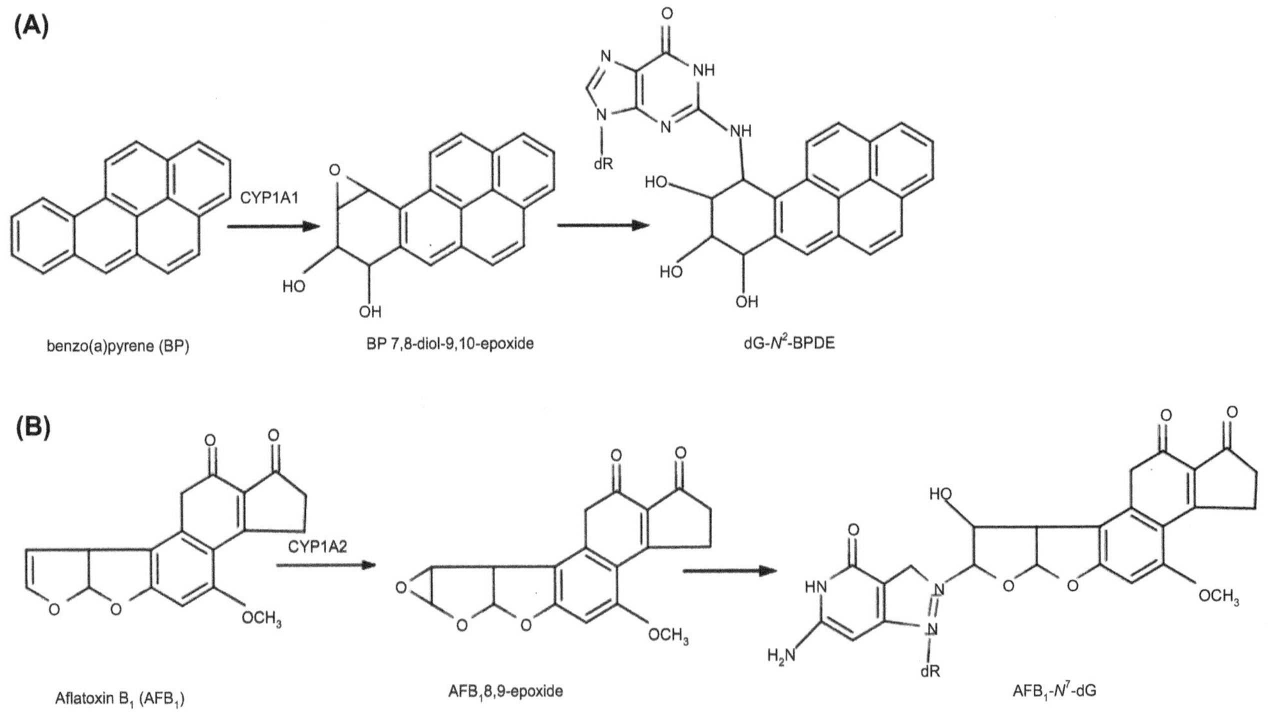

FIGURE 18.3 Benzo(a)pyrene and aflatoxin B1 DNA-adduct formation.

TABLE 18.6 Examples of DNA Adducts Found in Human Placenta In Vivo, Formed in Human Placental Perfusion Ex Vivo, and Catalyzed In Vitro by Human Placental Microsomes

Research Setting	DNA-Adducts	References
In vivo adducts	PAH adducts related to smoking	Everson et al., 1988; Hansen et al., 1992, 1993; Gallagher et al., 1994; Daube et al., 1997; Lagueux et al., 1999
	PAH adducts related to, e.g., air pollution	Hatch et al., 1990; Reddy et al., 1990; Manchester et al., 1992; Topinka et al., 1997; Whyatt et al., 1998; Marafie et al., 2000; Obolenskaya et al., 2010
	Benzo(a)pyrene-derived adducts	Everson et al., 1986; Manchester et al., 1988; Vo-Dinh et al., 1991; Sanyal et al., 1994; Arnould et al., 1997; Sanyal et al., 2007; Pratt et al., 2011a,b
	Aflatoxin B1–derived adducts	Hsieh and Hsieh, 1993
	Etheno-derived adducts	Chen et al., 1999; Doerge et al., 2000
	Organochlorine-derived adduct	Lagueux et al., 1999
Ex vivo adducts	Benzo(a)pyrene-derived adducts	Karttunen et al., 2010
	Aflatoxin B1–derived adducts	Partanen et al., 2010
In vitro adducts	Benzo(a)pyrene-derived adducts	Vaught et al., 1979; Pelkonen and Saarni, 1980; Vähäkangas et al., 1989; Prahalad et al., 1999
	Acrolein and crotonaldehyde adducts	Chen and Lin, 2009
	Formaldehyde-derived adducts	Zhong and Hee, 2009
	Carbamazepine and oxcarbazepine DNA binding	Myllynen et al., 1998

Whyatt et al., 1998; Sram et al., 1999) is associated with increased incidence of DNA adducts in human placenta, suggesting the transfer of pollutants and carcinogens in cigarette smoke from maternal to fetal tissues. PAH-DNA adducts seem to be localized especially in the cytotrophoblasts and syncytiotrophoblast knots lining the chorionic villi (Pratt et al., 2011a,b).

The presence of BPDE-DNA adducts in human placenta perfused with BP (Karttunen et al., 2010) indicates that PAHs can be metabolized in vivo in placental tissue. Trophoblastic cells contain metabolizing CYP enzymes and have been shown to contain the highest concentrations of PAH-DNA adducts among placental cells. PAH-DNA adducts have been proposed to be indicative for, or act as biomarkers of, harmful fetal outcomes, miscarriages, and placental pathologies (Perera et al., 1998, 1999; 2002; Obolenskaya et al., 2010). That nutritional status of mothers can affect the DNA-adduct levels, is implicated by the study of Dodd-Butera et al. (2017), who showed an inverse correlation between maternal serum folate levels and total PAH-DNA adduct levels. It has been suggested that the combination of placental PAH-DNA-adducts, placental CYP1A1 induction, and placental glutathione status could serve as a general biomarker for environmental chemical stress (Carlberg et al., 2000; Obolenskaya et al., 2010).

Aflatoxin-albumin adducts have been detected in both maternal and cord blood (Hsieh and Hsieh, 1993; Turner et al., 2007) and AFB1-DNA adducts in placental tissue and cord blood (Hsieh and Hsieh, 1993). AFB1 can be metabolized to its carcinogenic metabolite, aflatoxicol, by placental tissue fractions (Partanen et al., 2010). Some studies have tried to detect placental DNA adducts after exposure to compounds requiring CYP2E1 activation. However, after placental exposure to such genotoxic compounds, e.g., nitrosodimethylamine or acrylamide, no DNA adducts in placental tissue were found (Annola et al., 2008, 2009).

DNA adducts can be analyzed by several techniques (Table 18.7). The most common technique, due to its sensitivity and versatility, used in various laboratories for detecting DNA adducts in human placenta or other tissues, is the ^{32}P-postlabeling method originally developed by Randerath et al. in 1981 (Randerath et al., 1981) and further developed and used by many other groups (e.g., Spencer-Beach et al., 1996; for reviews see Jones, 2012; Phillips, 2013). The first demonstration of smoking-related bulky DNA adducts in human placenta, which was later confirmed by the same and other groups, was presented by Randerath's group in 1986 using this method (Everson et al., 1986, 1988; Reddy et al., 1990; Hansen et al., 1992; Gallagher et al., 1993; Daube et al., 1997; Pratt et al., 2011a,b). Other

TABLE 18.7 Advantages and Limitations of Analytical Methods for Measuring DNA and Protein Adducts

Method	Advantages	Limitation	References
Immunoassays	Reliable and inexpensive	Poor specificity, high amount of DNA needed	Farmer et al., 2005
^{32}P-postlabeling	High sensitivity	The efficiency of phosphorylation may be unknown or uncontrolled	Randerath and Randerath, 1994
LC-MS/MS or GC-MS	High specificity and sensitivity, possibility for chemical characterization of the adducts	Several pretreatments such as derivatization, vaporization, and ionization	Poirier and Weston, 1996; Farmer and Singh, 2008
Fluorescence spectroscopy, e.g., SFS	High specificity and sensitivity, rapid, and inexpensive	Requirement of a fluorescent adduct	Vähäkangas et al., 1985
Atomic absorbance spectrometry	High sensitivity to detect metal ions in DNA	Detection of metal ions with electrochemical conductance, there may be problems finding proper chromatographic separation for selective detection	Phillips, 2005
Analysis of radioactivity in DNA	Simple and inexpensive	Experimental only; does not tell about the types of adducts; requires radioactively labeled chemical	Myllynen et al., 1998

SFS, synchronous fluorescence spectrophotometry.

methods used to detect DNA adducts are immunoassays (Poirier et al., 1993), synchronous fluorescence spectrophotometry (Vähäkangas et al., 1985), LC- or GC/MS (Shuker et al., 1993; Farmer et al., 2005), atomic absorbance spectrometry (Reed et al., 1988), and electrochemical conductance (Floyd et al., 1986). Because of labor intensity, price, and other restrictions, none of the above methods is used in routine exposure analysis in any tissues. One of the latest techniques described in the literature is immunohistochemistry using the Automated Cellular Imaging System (Pratt et al., 2011a,b).

All in all, there are several robust and reliable analytical possibilities for studying placental adducts, which, however, at this stage are not practical in routine exposure or effect assessment. Studies of placental DNA adducts have mainly focused on PAH-compounds (e.g., BP and its derivates) or AFB1, as they are metabolized by CYP1A1 found in human placenta.

Xenobiotic-Metabolizing Enzymes, Placental Hormones, and Stress-Related Proteins as Biomarkers of Exposure and/or Toxicity in Placenta

As in other organs, many proteins in the placenta react with toxic compounds. One of the best studied groups so far is the xenobiotic metabolizing enzymes. Their activities may be modified in human placenta by maternal drug therapies and environmental and occupational factors (Myllynen et al., 2007). Most of the existing information concerns the CYP1A1 protein, the induction of which can be regarded as a general marker of exposure to xenobiotics. Also aromatase activity responds to environmental compounds, especially to those with hormonal activity.

Maternal cigarette smoking and environmental PAH exposure (Pasanen, 1999; Huuskonen et al., 2008; Suter et al., 2010), drug abuse (Paakki et al., 2000a,b), organochlorine exposure (Lagueux et al., 1999), and radiation (Obolenskaya et al., 2010) have been associated with CYP1A1 induction (for reviews see Pasanen et al., 1990; Collier et al., 2002, 2003). BP increases the expression of CYP1A1 protein in human placental JEG-3 cells (Drukteinis et al., 2005). CYP1A1 is one of the key enzymes in metabolic activation of PAH compounds (Xue and Warshawsky, 2005) and its activity is regulated by AhR (Stejskalova and Pavek, 2011). Also other environmental contaminants have been shown to increase the expression of CYP1A1 in human placental cells. TCDD increased CYP1A1 at mRNA (Tsang et al., 2012) and protein (Augustowska et al., 2007) levels in JEG-3 cells and at mRNA level in BeWo cells (Tsang et al., 2012). A mixture of PCDD and PCDF increases CYP1A1 protein levels in JEG-3 cells (Augustowska et al., 2007). In these cells, AFB1 induces CYP1A1 at mRNA level (Storvik et al., 2011; Huuskonen et al., 2013). Smoking modifies the levels of many placental transcription factors and proteins. With simultaneous CYP1A1 and 4B1 induction, smoking significantly upregulated the proteins SERPINA1, EFHD1, and KRT8 and downregulated SERPINB2, FGA, and HBB. Transcript expression of CYP1A1 and CYP4B1 were induced, whereas mRNA of HSD17B2, NFKB, and TGFB1 were repressed by smoking (Huuskonen et al., 2016). Clinical significance of these wide-spread changes by smoking in placenta need to be studied further.

Maternal glucocorticoid therapy suppresses placental xenobiotic metabolizing enzymes and aromatase activity (Paakki, 2000a). Interestingly, results on placental conjugating enzyme responses to exogenous chemical stress (cigarette smoking, environmental PAH compounds) are contradictory (Pasanen et al., 1990; Paakki, 2000b; Collier et al., 2002; Obolenskaya et al., 2010). This implicates that placental conjugation capacity does not detoxify all harmful reactive chemicals/intermediates that may target the fetoplacental unit.

The pregnancy-associated plasma protein-A (PAPP-A) is a product of the placenta and decidua and is secreted into the maternal circulation during human pregnancy. PAPP-A has recently been identified as an IGF binding protein-4 protease. Although PAPP-A is commonly analyzed clinically, there is very limited information available on the association of PAPP-A with environmental exposures. PAPP-A values have been reported to be somewhat lower in the blood of smoking than in nonsmoking mothers (Chelchowska et al., 2008), but the finding was not reproduced in another study (Ball et al., 2013). PAPP-A and hCG measured from maternal blood are used in prenatal screening for Down's syndrome. According to a relatively recent metaanalysis of these proteins in the first trimester, an unexplained low PAPP-A and/or a low hCG are associated with an increased frequency of adverse obstetrical outcomes (Gagnon and Wilson, 2008). Furthermore, during the second trimester, an unexplained elevation of maternal serum alfa-fetoprotein (AFP), hCG, and/or inhibin-A, or a decreased level of maternal serum AFP and/or unconjugated estriol, seem to be associated with an increased frequency of adverse obstetrical outcomes (Gagnon and Wilson, 2008). Maternal circulating inhibin A is mainly produced by the placental trophoblasts during pregnancy.

Production of hormones, such as progesterone and hCG, is a potential target of toxicity in placenta (Kawai et al., 2002; Nakanishi et al., 2005; Augustowska et al., 2007; Wójtowicz et al., 2007a,b; Stasenko et al., 2010). Many environmental endocrinologically active contaminants can affect the production or metabolism of hormones in placental cells (Augustowska et al., 2007; Storvik et al., 2011; Huuskonen et al., 2013) as shown in the studies on hCG in various placental models (Table 18.5). Different models may provide contradictory results as with Cd (Boadi et al., 1992; McAleer and Tuan, 2001b). Cd decreases progesterone secretion in cultured human trophoblasts isolated from term placentas (Kawai et al., 2002). This is associated with a decline in the mRNA expression of P450 cholesterol side-chain cleavage ($P450_{scc}$) and 3β-hydroxysteroid dehydrogenase (3β-HSD) enzymes involved in placental steroidogenesis. However, as to the activity of these enzymes, Cd inhibited only the activity of $P450_{scc}$. AFB1 increased the expression mRNA of steroid-metabolizing HSD enzymes 3B1 and 17B1 (Huuskonen et al., 2013). In addition, AFB1 induces CYP19A1 mRNA (Storvik et al., 2011; Huuskonen et al., 2013) and aromatase activity in JEG-3 cells (Storvik et al., 2011). Another class of stable environmental contaminants with endocrine disruption potential is PCBs. PCB28, a non−dioxin-like PCB (NDL-PCB) compound, and also hydroxyl-metabolites of NDL-PCB180, inhibit aromatase activity in human placental microsomes (Antunes-Fernandes et al., 2011).

Oxidative stress is currently believed to have a significant role in the pathophysiology of many complications of human pregnancy such as spontaneous miscarriage, early onset PE, and IUGR (Burton and Jauniaux, 2011). Also, gestational diabetes has been associated with enhanced placental oxidative stress (Gauster et al., 2012). Several environmental exposures have been shown to induce placental oxidative stress. Increased lipid peroxidation (LPO) in human placenta has been measured as a marker of oxidative stress after metal exposure. Levels of many metals (Cd, Zn, Ni, Pb, Cr) are positively correlated with LPO (Serafim et al., 2012). Also, TCDD, endrin, and lindane increase LPO in the placenta (Hassoun and Stohs, 1996). TCDD, endrin, and lindane also increase superoxide anion production and single strand breaks in DNA (Hassoun and Stohs, 1996). Smoking increases oxidative stress in the placenta, as shown by the increased levels of hydroxyl-2-nonenal (4-HNE) and 8-hydroxyguanosine (8-OHdG) (Sbrana et al., 2011). Furthermore, ethanol increases oxidative stress in in vitro cultured villous tissues as measured by nitrotyrosine, hydroxyl-2-nonenal, and 8-hydroxyguanosine staining (Kay et al., 2006).

Circulating cell-free nucleic acids analyzed from maternal plasma and serum are new promising potential biomarkers. Several studies have suggested that changes in the amount of cell-free fetal (cff) DNA may serve as an indicator of developing PE (for a review see Hahn et al., 2011). The increased levels of cffDNA before the onset of symptoms in PE may be due to hypoxia/reoxygenation within the intervillous space leading to oxidative stress in the tissue and increased placental apoptosis and necrosis. In addition, cell-free placental RNA is another potentially useful biomarker for PE. However, cffDNA or cell-free placental RNA as biomarkers of environmental compounds causing oxidative stress in the placenta awaits further studies.

To assess adverse effects of metal exposure in placenta, biological responses in placental tissue to metals have been tested as potential biomarkers of exposure and early effects. Some enzymes, such as ALAD, a key enzyme in hemoglobin synthesis, are inhibited by environmental contaminants and can thus be used as potential biomarkers (Serafim et al., 2012). For example, levels of the metals Cd, Cu, Zn, Ni, Pb, and Cr are negatively correlated with ALAD. Monoamine oxidase (MAO) activity in placenta is induced by metals in experimental animals and thus it may be a potential biomarker also in humans. In human placenta, MAO activity is positively correlated with manganese (Mn) concentrations in maternal and cord blood and has been suggested as a potential marker of Mn toxicity in human placenta (Abdelouahab et al., 2010).

Metallothioneins (MTs) are cysteine-rich proteins that are involved in metal metabolism. MT expression in placenta may serve as a biomarker for metal exposure (Sorkun et al., 2007; Zhang et al., 2011b; Serafim et al., 2012). In a recent Portuguese study, levels of many metals (Cd, Zn, Ni, Pb) in placenta were positively correlated with elevated levels of MTs when metal levels were within the range found in most European countries (Serafim et al., 2012). MT expression was also associated with cadmium but not lead concentration in placenta in a highly polluted area in Guiyu, China (Zhang et al., 2011b). These results suggest that induction of MT in placenta is an indicator of heavy metal exposure during pregnancy. In the study by Zhang et al., Cd exposure was also associated with lower expression of S100P. The S100 proteins, including S100A, -B, and -P, are Ca^{2+}-binding proteins, which can also bind other divalent ions. S100P is worth further studies as a potential biological indicator of Cd exposure (Zhang et al., 2011b).

Epigenetic Changes in Human Placenta as Potential Biomarkers of Toxicity

Epigenetics deals with molecular changes that affect gene and mRNA activity but are not due to changes in the genetic code (genetic changes) (Novakovic and Saffery, 2012). The most studied of the epigenetic phenomena are DNA methylation and modifications in histone proteins that regulate transcription. More recently, regulation of translation at the level of mRNA by a variety of noncoding RNA molecules has been revealed as a new important level of epigenetic regulation of protein expression. Noncoding RNAs have their own genes but are not translated into proteins.

It is by now well known that environmental chemicals can induce a variety of epigenetic changes (for reviews see Hou et al., 2012; Kim et al., 2012), and such changes are also linked to various adverse reproductive conditions, such as IUGR (Robins et al., 2011). Arai et al. (2011) studied in murine embryonal stem cells the effect of 25 environmental chemicals in relevant concentrations found in human cord blood serum, and found many of them capable of inducing changes in heterochromatin structure and DNA methylation. However, further research is required to establish the persistence and potential accumulation of epigenetic changes (Hou et al., 2012). It is also important to note that epigenetic changes are tissue- and cell-type specific and due to the novelty of the field a lot of discrepant data exist. Thus, the studies on how epigenetic changes can serve as biomarkers of toxicity in placenta are at an early stage, and studies on potential biomarkers have so far mainly concentrated on disease diagnosis and prognosis (Maccani and Marsit, 2009; Novakovic and Saffery, 2012; Fu et al., 2013).

In human placental epigenetics, considerable variation between placentas has been found, putatively due to environmental effects in addition to internal reasons (Yuen and Robinson, 2011). Such adaptation of the placenta according to changes in the maternal physiology and other environmental influences is naturally important for the optimal support of fetal health (Mouillet et al., 2011). Epigenetic alterations in the placenta are probably better tolerated than in the fetus, and placenta has been suggested to protect the fetus from environmental epigenetic influences (Yuen and Robinson, 2011).

What makes the situation difficult from the biomarker point of view is that, in addition to developmental, sex, and interindividual variation in placental epigenetics (Novakovic and Saffery, 2012), there is also intraplacental variation suggested to reflect localized effects in the placenta and its environment in the uterus (Yuen and Robinson, 2011). Grigoriu et al. (2011) compared methylation patterns in human placental trophoblastic cells and fibroblasts and showed differences between these cell types. They also concluded that the methylation profile of placenta is mainly due to the methylation pattern of cytotrophoblasts. Especially, in line with the extensive proliferation of cytotrophoblasts, promoters of many tumor suppressor genes were hypermethylated in the placenta.

In connection with environmental influences, the best studied are methylation changes, both global methylation and methylation of the promoter of some specific genes. Typically, methylation of CpG islands in the promoter region of a gene inhibits its transcription. Methylation patterns vary between different types of tissues and are susceptible to environmental influences

(Christensen et al., 2009). In human placenta, global methylation is lower than in other tissues (2.5%–3% vs. 4%–5%) (Novakovic and Saffery, 2012), for instance lower than methylation of maternal blood (Chu et al., 2011). Chu et al. (2011) found in the placenta tissue-specific differentially methylated regions, which were enriched within genes having tissue-specific functions. Methylation is dependent on placental development, so that global methylation increases by gestation (Novakovic et al., 2011). Sitras et al. (2012) have found specific gene groups, e.g., genes related to cell proliferation, upregulated in first trimester and other groups (e.g., genes related to G-protein–mediated signaling, ion transport, and neuronal activities) upregulated in third trimester placentas. Variation between the placentas increases also toward the end of pregnancy, implicating increasing impact by environmental factors. Both maternal nutrition and smoking have been shown to affect placental gene methylation (Table 18.8). For instance, CYP1A1 activity and protein induced by cigarette smoke, BP, or other PAH-like compounds seem to be associated with hypomethylation of regulatory elements of the CYP1A1 gene (Suter et al., 2010).

Another important level of epigenetic regulation is the microRNAs (miRNAs), small noncoding RNA molecules, which, by binding to mRNA, regulate translation, RNA processing, and RNA stability (Rosario et al., 2012). Micro RNAs regulate all the different cellular functions related to placental development, including trophoblast differentiation, proliferation, apoptosis, invasion, and angiogenesis, and are thus essential for proper development and function of the placenta (Fu et al., 2013). Aberrations in the miRNA expression are associated with some placental pathologies such as preterm labor (Fallen et al., 2018; for reviews see Chen and Wang, 2013; Fu et al., 2013).

TABLE 18.8 Examples of Epigenetic Changes in Human Placenta or Human Primary Trophoblasts Associated With Environmental Influences

Chemical/Exposure	Model	Epigenetic Changes	References
Tobacco smoking	In vivo exposure; human placental tissue	Downregulation of microRNA (miRs-16, 21 and 146a)	Maccani et al., 2010
Tobacco smoking	In vivo exposure; human placental tissue	Hypomethylation of CYP1A1 promoter region	Suter et al., 2010
Tobacco smoking	In vivo exposure; human placental tissue	Methylation changes, especially in genes GTF2H2C and GTF2H2D	Chhabra et al., 2014
Hypoxia	Primary human trophoblasts	Downregulation of miR-424	Mouillet et al., 2015
Hypoxia	Primary human cytotrophoblasts	Hypermethylation of functionally relevant genes including CP, ITGA5, SOD2, XDH, and ZNF2	Yuen et al., 2013
Air pollution	In vivo exposure, placental tissue analyzed for chemicals	Positive correlations: PBDE 209 with miR-188-5p PCBs with miR-1537 Cd with miR-1537 Negative correlation: PBDE 99 with let-7c	Li et al., 2015
Air pollution, PM2.5	Estimated in vivo exposure, placental tissue	Decrease of methylation of leptin gene promoter	Saenen et al., 2017
Phthalates	In vivo exposure, human placental tissue	Upregulation of lncRNAs	Machtinger et al., 2018
Lead	Human cord blood	Global hypomethylation	Pilsner et al., 2009
Mercury	In vivo exposure, human placenta	Hypomethylation of EMID2 gene	Maccani et al., 2015
Assisted reproductive technologies	Chorionic villous samples	Hypermethylation of imprinted genes; hypomethylation of nonimprinted genes	Zechner et al., 2010
In vitro fertilization	Placenta, cord blood	Higher mean methylation in placenta, lower in cord blood	Katari et al., 2009

Considering the fact that placenta expresses a wide variety of miRNAs, more than 600 identified so far, some of them unique to placenta (Chen and Wang, 2013), chemicals that affect placental miRNAs may be associated with placental toxicity (Mouillet et al., 2011). An important primate-specific genomic cluster in chromosome 19 (chromosomal region 19q13.4 or C19MC) seems to give rise to the most abundant placenta-specific miRNAs. It is transcribed as one large transcript that is processed into mature miRNAs. Among placental cells, C19MC is exclusively expressed in trophoblast-derived cells. Interestingly, C19MC is also expressed in human embryonic stem cells, with reduction of the expression by differentiation. Furthermore, aberrant expression of miRNAs encoded by this locus is found in some aggressive cancers. All these data implicate the important role of the miRNAs from C19MC in proliferation and differentiation. Some miRNAs are also abundantly expressed in placental mesenchymal stromal cells and some of the placenta-specific miRNAs are also found in maternal blood during pregnancy (Chen and Wang, 2013) implicating possibilities for biomarker development, and also for placental toxicity.

Hypoxia is related to placental pathology, such as preeclampsia, and many placental miRNAs are differentially expressed in preeclamptic placentas (Mouillet et al., 2011). MiR-210 is upregulated by both physiological and malignant hypoxia, and it is also significantly increased in preeclamptic placenta. MiR-424 is highly expressed in the placenta and is downregulated by hypoxia (Mouillet et al., 2013). MiR-205, which is specific to epithelium and linked to epithelial–mesenchymal transition and cancer, is upregulated in primary human trophoblasts by hypoxia (Mouillet et al., 2010b). Many drugs and chemicals (e.g., tobacco smoking; Zdravkovic et al., 2005) are known to induce hypoxia in the placenta. Thus such chemicals most probably also affect hypoxia-related miRNAs. Whether such miRNAs could serve as biomarkers for the exposure or effects of hypoxia-inducing chemicals remains to be studied.

Toxicologically, an interesting miRNA is miR-146a, which is downregulated in SV-40 transformed human trophoblasts by several environmental chemicals, for instance nicotine and BP (Maccani et al., 2010). It is also downregulated in placentas from smoking mothers, making it a candidate biomarker for the effects of smoking-related chemicals (Maccani et al., 2010). The fact that the same miR-146a is strongly induced by BPA (Avissar-Whiting et al., 2010) implicates further studies on the specificity of its downregulation as a biomarker of smoking-related toxicity in the placenta. MiR-146a is also responsive to folate deficiency and As. Its expression is related to cancer, with tumor suppressive or oncogenic effects, depending on the type of cancer. Its overexpression in the first trimester villous cells 3A increased proliferation of the cells as well as their sensitivity to bleomycin, a DNA-damaging cancer drug (Avissar-Whiting et al., 2010). All in all, miR-146a is an attractive candidate biomarker for environmental influences in the placenta. Other miRNAs downregulated by smoking, miR-16 and miR-21 (Maccani et al., 2010), are also clearly and markedly reduced in babies with a low birth weight (Maccani et al., 2011).

Dependence of miRNA changes on the developmental stage of the placenta was implicated by Avissar-Whiting et al. (2010) who used human placental cell lines, representing first and third trimester placentas. Although miR-146a was upregulated in the cells from first trimester, its expression was not detectable in the cell line from third trimester placenta.

MicroRNAs are found in blood within various microparticles or bound to nucleophosmin 1, which putatively protects them from ribonucleases (see Mouillet et al., 2011). During pregnancy, C19CM-derived miRNAs are found consistently in maternal blood and increased in fetal growth restriction (Mouillet et al., 2010a). Whether such changes are reflected in fetal blood is not yet known. In any case, it is an attractive idea to study the placenta-specific circulating miRNAs as possible biomarkers for placental and fetal toxicity. At this point, the whole field of epigenetics and, especially, miRNAs, is under lively development including the methodologies. Mouillet et al. (2011) justly remind us of the difficulties in normalization of the miRNA array data and poor concordance of the results by high-throughput platforms with RT-qPCR measurements.

USE OF PLACENTA IN REGULATORY TOXICOLOGY—CONSIDERATIONS BY ECVAM AND OTHER ORGANIZATIONS

Obviously, because developmental toxicity is such an important part of toxicity testing, and exposure of the developing individual is an important factor in the outcome, methods to measure placental transfer of foreign chemicals have been considered in the regulatory circles for quite some time.

OECD Guidelines

Reproductive and developmental toxicity studies have been regulated for a long time by the OECD guidelines. These studies are performed using experimental animals, rats, and rabbits, and, consequently due to interspecies differences, extrapolation of the results presents particular problems. In in vivo animal studies it is difficult to evaluate the contribution of the placental transfer or toxicity in the final developmental outcome. Obviously placenta itself may be the target of xenobiotics, resulting in disturbances in placental function or penetration. Low placental weight or other toxicity to placenta may have direct effects on the condition of pups. However, placental toxicity cannot explain all the effects, e.g., serious malformations observed in pups.

However, it is of interest that even if placental transfer controls the access of a chemical to the fetus and thus may influence the fetal adverse outcomes, only rather short sections concerning placenta can be found in appropriate OECD guidelines. Placental weights and sometimes placental condition are examined in the standard TG 414 test (developmental toxicity study in mice) as an additional end point. In TG 416 (two-generation reproductive toxicity study in mice), related to dosing, it is stated: "Information regarding placental distribution should be considered when adjusting the gavage dose based on weight."

The OECD TG 443 (Extended One-Generation Reproductive Toxicity Study in mice) contains a couple of statements about placenta. In relation to dosing, the following statement concerns placental transfer: "If TK data indicate a low placental transfer of the test substance, the gavage dose during the last week of pregnancy may have to be adjusted to prevent administration of an excessively toxic dose to the dam." Related to evaluation of the test results: "The physicochemical properties of the test substance, and when available, TK (toxicokinetic) data, including placental transfer and milk excretion, should be taken into consideration when evaluating the test results." However, it is not clear from where the TK data are supposed to come.

Recently, the OECD has developed so called Adverse Outcome Pathways (AOP), to increase knowledge of how chemicals induce adverse effects in humans and wildlife (https://aopwiki.org/aops). Also placenta is included in several OECD AOP proposals, as one of the events involved in the pathways, especially those related to vascular function/angiogenesis and induction of developmental toxicity; see, e.g., AOP 151: AhR signaling leading to placental vascular disruption and AOP 43: Disruption of VEGFR Signaling Leading to Developmental Defects. There is still room for identifying or reporting more AOPs where placenta plays a role in induction of an adverse effect.

Role of European Center for the Validation of Alternative Methods

ECVAM (European Center for the Validation of Alternative Methods)[1], which is responsible for the development and validation of alternative methods to replace in vivo animal tests, has considered placental models in its surveys and reviews on developmental toxicology. Moreover, there is a currently ongoing validation process for several methods to study adverse effects of xenobiotics on the placenta.

In 2005, ECVAM published an extensive survey on alternative methods available or under investigation for all major toxicities, including a chapter on reproductive and developmental toxicity (Bremer et al., 2005). The survey contains a short section on placental toxicity, including the placental barrier, and mentions both trophoblast cell lines and the human placental perfusion system. The chapter on toxicokinetics and metabolism (Coecke et al., 2005) includes some sections on bioaccumulation and distribution of chemicals across membranes and barriers, but it does not directly address the placental barrier. However, from the toxicokinetics point of view, it was envisaged that placental transfer studies would estimate the exposure of the fetus to a chemical under study. From the placental toxicity perspective, placental studies would reveal direct toxic effects of a compound under study on placental biochemistry, morphology, and functional status.

Adler et al. (2011) published an extensive survey to assess the status and prospects of alternative methods to animal testing for five toxicological areas (toxicokinetics, repeated dose toxicity, carcinogenicity, skin sensitization, and reproductive toxicity). This review contained two assays on placenta or placenta-derived cells, which were regarded as promising tools to study some aspects of fetal exposure to chemicals and mechanistic background for kinetics and toxicities. They were also seen potentially contributing as 3R tools[2] to OECD reproductive toxicity guidelines.

One of the aims of the EU Integrated Project ReProTect was to develop and optimize in vitro tests in reproductive toxicology, including tests using human placental models, to detect key events and mechanisms and to use these tests as building blocks in the integrated testing strategies (Bouvier d'Yvoire et al., 2012). Studies pursuing prevalidation of human placental perfusion (Myllynen et al., 2010; Mose et al., 2012; Karttunen et al., 2015) have indicated good intra- and interlaboratory comparison and transferability of the system.

[1] Currently EURL ECVAM (European Reference Laboratory, etc.).

[2] 3R is an acronym for refinement, reduction, and replacement of use of animals in toxicity studies.

Experimental Methods Concerning Placenta

Regarding the human placental perfusion assay, an ex vivo human placental perfusion provides an opportunity to carry out research without ethical difficulties. Other advantages of the placental perfusion method include the retention of in vivo placental organization and assessment of binding to placental tissue (Vähäkangas and Myllynen, 2006; Mose et al., 2008; Myllynen and Vähäkangas, 2013). However, the application of this assay is limited due to placenta to placenta variations and the limited relevance of the term placenta for the period of embryonic development. Because of the complexity of the assay, it is not applicable to routine testing of high numbers of test compounds. The trophoblast cell assay serves as an in vitro model of the rate-limiting barrier to maternal–fetal exchange and also for basic cellular and molecular investigations on toxicity mechanisms (Mørck et al., 2010).

It is clear that studies on isolated perfused human placenta as well as on properly characterized placental cells and cell lines would provide crucial information about the exposure of the fetus and placental toxicity and, in addition, basic characterization of enzymes, transporters, and targets of toxicity in the placenta. The use of human placenta obviates one of the most important problems in extrapolation and prediction of toxicity to humans, i.e., interspecies differences. The use of placental cells (and placentas if technically feasible and necessary) from animal species used for developmental toxicity studies would provide background for comparisons and extrapolation between animals and humans.

On the basis of the above argumentation, it would be of considerable advantage to include the option of placental studies in guidelines for reproductive and developmental toxicity studies.

According to the current views of regulatory toxicology, placental perfusion or other in vitro/ex vivo placental test methods can only be used in a tiered strategy and/or test battery. However, regulatory toxicology in general is shifting more toward the tiered or integrated testing strategies to minimize the use of animals and to get a more comprehensive picture of toxicity of the compounds. Therefore, as an example, the human placental perfusion study may have a key role in the integrated testing strategy, especially in a case when the compound is not transferred through the human placenta. It may thus prevent the need for conducting standard animal testing for the purpose of predicting developmental toxicity in humans.

CONCLUDING REMARKS AND FUTURE DIRECTIONS

A biomarker can be defined as a chemical, biochemical, or molecular entity that is, preferably quantitatively, associated with exposure, tissue change, or a condition. For the development of a biomarker one needs a measurable molecule (or a group of molecules), a method to analyze it with, and a matrix to measure it from. The method of analysis of the putative biomarker and its usefulness as a biomarker in humans have to be validated separately (Vähäkangas, 2008). This means that we need to know that what we measure is real, and thereafter, that it works in the selected matrix and, finally, in in vivo situations. Before these validation steps it would be best to talk of a potential biomarker. In the literature, there is unfortunately a lot of discrepancy concerning the validation.

Placenta has appeared potentially useful not only as a matrix for placental exposure and toxicity, but also for fetal, and even general, exposure and toxicity. New exciting possibilities for biomarkers indicating placental toxicity are being searched for among newly identified small noncoding RNAs and by high-throughput methodologies, such as genomics platforms for mRNA profiles. Their usefulness, especially for regulatory toxicology, remains to be studied.

1. Biomarkers of placental exposure and toxicity: The presence of a xenobiotic, e.g., heavy metal or PCB, or their metabolites or DNA-adducts in the placenta is a very clear proof of placental exposure. To develop biomarkers for placental toxicity is a more complex issue. Exposure or even accumulation of a xenobiotic in the placenta does not necessarily mean that it is toxic to the placenta. A change in placental biochemistry, e.g., in the level of certain protein, should preferably be mechanistically linked to the toxic effect in the placenta. A well-known indication of placental exposure to cigarette smoke or PAH compounds is the classical triad: induction of CYP1A1, resulting in increased amount of PAH-DNA adducts in placental tissue, followed by glutathione depletion and increased oxidative stress (Obolenskaya et al., 2010). This glutathione depletion may not only be restricted to chemical compounds but can reflect also radioactive-induced stress (Obolenskaya et al., 2010). Additionally, increased PAH load has resulted in human placenta in decreased aromatase activity (Kitawaki et al., 1993; Huuskonen et al., 2008). However, these connections to maternal overall hormonal balance have not been evaluated prospectively. Endocrine disruption by various environmental contaminants or even medicines is an important toxic end point in placenta.

Better knowledge of the pathways involved, including metabolizing enzymes, in endocrine disruption may, therefore, help to develop biomarkers of placental toxicity. Changes in CYP19 mRNA and protein can already be regarded as a qualitative, general biomarker of exposure to endocrine disrupters.

2. Placenta as a matrix of biomarkers for fetal toxicity: Because placenta is mainly of fetal origin, it is feasible to expect an association between placental and fetal exposure and toxicity. Indeed, in some cases a correlation between placental and cord blood and/or meconium levels of xenobiotics has been shown. The level of PAH-DNA adducts in both placental tissue and cord blood is higher than in maternal blood, and the level of these adducts correlates with fetotoxicity (Perera et al., 1999).

3. Placenta as a matrix of biomarkers for general toxicity: Binding to macromolecules of genotoxic compounds in the placenta is a clear sign of genotoxic potential in human tissue, and implicates adducts and genotoxic effects also in other tissues. Heavy metals in the placenta serve as biomarkers of metal exposure. They induce expression of MTs. MT in placenta is the best candidate as a placental biomarker of exposure to cadmium, and putatively other heavy metals during pregnancy, apart from showing the heavy metals themselves. Some old biomarkers seem to be relevant also when analyzed from placental tissue. Organophosphate pesticides decrease blood cholinesterase activity. Such a decrease can be seen after the exposure also in placental tissue, which may serve as a general biomarker. Recent examples of changes in placental proteome due to smoking, and miRNAs and methylation due to environmental chemicals, strengthen the hypothesis of the role of placenta as a biomonitoring target to explore placental changes as indicators of the well being of fetus or child. From the regulatory point of view in the future there are some critical issues to be resolved. Placenta has the same role throughout various species: nutrition, energy supply to fetus, barrier for inflammatory diseases, hormonal production, and other "housekeeping" functions. However, it does exhibit a huge species-dependent variation in its organization, anatomy, and histology that has never been taken into account when interpreting data obtained in animal studies. Obviously, such differences between the species are probably reflected in placental mechanistically relevant cellular responses. A good example is transport of immunoglobulin fractions from maternal to fetal circulation, which varies significantly between the species (Pentsuk and van der Laan, 2009).

References

Abdelouahab, N., Huel, G., Suvorov, A., 2010. Monoamine oxidase activity in placenta in relation to manganese, cadmium, lead, and mercury at delivery. Neurotoxicol. Teratol. 32, 256–261.

Abdulrazzaq, Y.M., Osman, N., Ibrahim, A., 2002. Fetal exposure to aflatoxins in the United Arab Emirates. Ann. Trop. Paediatr. 22, 3.

Acevedo, C.G., Huambachano, A., Perez, E., 1997. Effect of ethanol on human placental transport and metabolism of adenosine. Placenta 18, 387–392.

Adler, S., Basketter, D., Creton, S., 2011. Alternative (non-animal) methods for cosmetics testing: current status and future prospects. Arch. Toxicol. 85, 367–485.

Ala-Kokko, T.I., Myllynen, P., Vähäkangas, K., 2000. *Ex vivo* perfusion of the human placental cotyledon: implications for anesthetic pharmacology. Int. J. Obstet. Anesth. 9, 26–38.

Aliyu, M.H., Wilson, R.E., Zoorob, R., 2009. Prenatal alcohol consumption and fetal growth restriction: potentiation effect by concomitant smoking. Nicotine Tob. Res. 11, 36–43.

Al-Saleh, I., Al-Doush, A., Alsabbaheen, D., et al., 2012. Levels of DDT and its metabolites in placenta, maternal and cord blood and their potential influence on neonatal anthropometric measures. Sci. Total Environ. 416, 62–74.

Al-Saleh, I., Alsabbahen, A., Shinwari, N., 2013. Polycyclic aromatic hydrocarbons (PAHs) as determinants of various anthropometric measures of birth outcome. Sci. Total Environ. 444C, 565–578.

Al-Saleh, I., Shinwari, N., Mashhour, A., et al., 2011. Heavy metals (lead, cadmium and mercury) in maternal, cord blood and placenta of healthy women. Int. J. Hyg Environ. Health 214, 79–101.

Alvarez, M.M., Chakraborty, C., 2011. Cadmium inhibits motility factor-dependent migration of human trophoblast cells. Toxicol. Vitro 25, 1926–1933.

Annola, K., Heikkinen, A.T., Partanen, H., Woodhouse, H., Segerbäck, D., Vähäkangas, K., 2009. Transplacental transfer of nitrosodimethylamine in perfused human placenta. Placenta 30, 277–283.

Annola, K., Karttunen, V., Keski-Rahkonen, P., Myllynen, P., Segerbäck, D., Heinonen, S., Vähäkangas, K., 2008. Transplacental transfer of acrylamide and glycidamide are comparabale to that of antipyrine in perfused human placenta. Toxicol. Lett. 182, 50–56.

Antunes-Fernandes, E.C., Bovee, T.-F., Daamen, F.E., 2011. Some OH-PCBs are more potent inhibitors of aromatase activity and (anti-) glucocorticoids than non-dioxin like (NDL)-PCBs and $MeSO_2$-PCBs. Toxicol. Lett. 206, 158–165.

Arai, Y., Ohgane, J., Yagi, S., 2011. Epigenetic assessment of environmental chemicals detected in maternal peripheral and cord blood samples. J. Reprod. Dev. 57, 507–517.

Araújo, J.R., Gonçalves, P., Martel, F., 2008. Modulation of glucose uptake in a human choriocarcinoma cell line (BeWo) by dietary bioactive compounds and drugs of abuse. J. Biochem. 144, 177–186.

Arnould, J.P., Verhoest, P., Bach, V., et al., 1997. Detection of benzo[a]pyrene-DNA adducts in human placenta and umbilical cord blood. Hum. Exp. Toxicol. 16, 716–721.

Ask, K., Akesson, A., Berglund, M., Vahter, M., 2002. Inorganic mercury and methylmercury in placentas of Swedish women. Environ. Health Perspect. 110, 523–526.

Augustowska, K., Gregoraszczuk, E.L., Milewicz, T., 2003a. Effects of dioxin (2,3,7,8-TCDD) and PCDDs/PCDFs congeners mixture on steroidogenesis in human placenta tissue culture. Endocr. Regul. 37, 11–19.

Augustowska, K., Gregoraszczuk, E.L., Grochowalski, A., 2003b. Comparison of accumulation and altered steroid secretion by placental tissue treated with TCDD and natural mixture of PCDDs-PCDFs. Reproduction 126, 681–687.

Augustowska, K., Magnowska, Z., Kapiszewska, M., et al., 2007. Is the natural PCDD/PCDF mixture toxic for human placental JEG-3 cell line? The action of the toxicants on hormonal profile, CYP1A1 activity, DNA damage and cell apoptosis. Hum. Exp. Toxicol. 26, 407–417.

Avissar-Whiting, M., Veiga, K.R., Uhl, K.M., 2010. Bisphenol A exposure leads to specific microRNA alterations in placental cells. Reprod. Toxicol. 29, 401–406.

Bakdash, A., Burger, P., Goecke, T.W., 2010. Quantification of fatty acid ethyl esters (FAEE) and ethyl glucuronide (EtG) in meconium from newborns for detection of alcohol abuse in a maternal health evaluation study. Anal. Bioanal. Chem. 396, 2469–2477.

Baker, V.L., Murai, J.T., Taylor, R.N., 1998. Downregulation of protein kinase C by phorbol ester increases expression of epidermal growth factor receptors in transformed trophoblasts and amplifies human chorionic gonadotropin production. Placenta 19, 475–482.

Ball, S., Ekelund, C., Wright, D., 2013. Danish Fetal Medicine Study Group. Temporal effects of maternal and pregnancy characteristics on serum pregnancy-associated plasma protein-A and free β-human chorionic gonadotropin at 7–14 weeks' gestation. Ultrasound Obstet. Gynecol. 41, 33–39.

Bamberger, A.M., Makrigiannakis, A., Schroder, M., 2004. Expression pattern of the CCAAT/enhancer-binding proteins C/EBP-alpha, C/EBP-beta and C/EBP-delta in the human placenta. Virchows Arch. 444, 149–152.

Barbieri, R.L., Gochberg, J., Ryan, K.J., 1986. Nicotine, cotinine, and anabasine inhibit aromatase in human trophoblast in vitro. J. Clin. Invest. 77, 1727–1733.

Bar-Oz, B., Klein, J., Karaskov, T., Koren, G., 2003. Comparison of meconium and neonatal hair analysis for detection of gestational exposure to drugs of abuse. Arch. Dis. Child. Fetal Neonatal Ed. 88, F98–F100.

Barrera, D., Avila, E., Hernandez, G., 2007. Estradiol and progesterone synthesis in human placenta is stimulated by calcitriol. T. J. Steroid Biochem. Mol. Biol. 103, 529–532.

Beach, A.C., Gupta, R.C., 1992. Human biomonitoring and the ^{32}P-postlabelling assay. Carcinogenesis 13, 1053–1074.

Bechi, N., Ietta, F., Romagnoli, R., et al., 2006. Estrogen-like response to p-nonylphenol in human first trimester placenta and BeWo choriocarcinoma cells. Toxicol. Sci. 93, 75–81.

Bechi, N., Sorda, G., Spagnoletti, A., et al., 2013. Toxicity assessment on trophoblast cells for some environment polluting chemicals and 17β-estradiol. Toxicol. Vitro 27, 995–1000.

Benachour, N., Moslemi, S., Sipahutar, H., et al., 2007. Cytotoxic effects and aromatase inhibition by xenobiotic endocrine disrupters alone and in combination. Toxicol. Appl. Pharmacol. 222, 129–140.

Benirschke, K., Kaufmann, P., Baergen, R., 2006. Pathology of the Human Placenta. Springer, 2006.

Bergonzi, R., De Palma, G., Specchia, C., 2011. Persistent organochlorine compounds in fetal and maternal tissues: evaluation of their potential influence on several indicators of fetal growth and health. Sci. Total Environ. 409, 2888–2893.

Berlin, I., Heilbronner, C., Georgieu, S., et al., 2010. Newborns' cord blood plasma cotinine concentrations are similar to that of their delivering smoking mothers. Drug Alcohol Depend. 107, 250–252.

Bernstein, L., Pike, M.C., Lobo, R.A., 1989. Cigarette smoking in pregnancy results in marked decrease in maternal hCG and oestradiol levels. Br. J. Obstet. Gynaecol. 96, 92–96.

Biasucci, G., Calabrese, G., Di Giuseppe, R., 2011. The presence of ochratoxin A in cord serum and in human milk and its correspondence with maternal dietary habits. Eur. J. Nutr. 50, 211–218.

Boadi, W.Y., Shurtz-Swirski, R., Barnea, E.R., 1992. Secretion of human chorionic gonadotropin in superfused young placental tissue exposed to cadmium. Arch. Toxicol. 66, 95–99.

Bouhours-Nouet, N., Boux de Casson, F., Rouleau, S., 2006. Maternal and cord blood ghrelin in the pregnancies of smoking mothers: possible markers of nutrient availability for the fetus. Horm. Res. 66, 6–12.

Bouvier d'Yvoire, M., Bremer, S., Casati, S., 2012. ECVAM and new technologies for toxicity testing. Adv. Exp. Med. Biol. 745, 154–180.

Boyce, A., Schwartz, D., Hubert, C., et al., 1975. Smoking, human placental lactogen and birth weight. Br. J. Obstet. Gynaecol. 82, 964–967.

Breitholtz-Emanuelsson, A., Olsen, M., Oskarsson, A., et al., 1993. Ochratoxin A in cow's milk and in human milk with corresponding human blood samples. J. AOAC Int. 76, 842–846.

Bremer, S., Cortvrindt, R., Daston, G., 2005. Reproductive and developmental toxicity. Altern. Lab. Anim. 33, 183–209.

Bulun, E., Sebastian, S., Takayama, K., et al., 2003. The human CYP19 (aromatase P450) gene: update on physiologic roles and genomic organization of promoters. J. Steroid Biochem. Mol. Biol. 86, 219–224.

Burd, L., Roberts, D., Olson, M., et al., 2007. Ethanol and the placenta: a review. J. Matern. Fetal Neonatal Med. 20, 361–375.

Burton, G.J., Jauniaux, E., 2011. Oxidative stress. Best Pract. Res. Clin. Obstet. Gynaecol. 25, 287–299.

Burton, G.J., Fowden, A.L., Thornburg, K.L., 2016. Placental origins of chronic disease. Physiol. Rev. 96, 1509–1565.

Butler Walker, J., Houseman, J., Seddon, L., 2006. Maternal and umbilical cord blood levels of mercury, lead, cadmium, and essential trace elements in Arctic Canada. Environ. Res. 100, 295–318.

Campbell, M.K., Cartier, S., Xie, B., 2012. Determinants of small for gestational age birth at term. Paediatr. Perinat. Epidemiol. 26, 525–533.

Cantón, R.F., Scholten, D.E., Marsh, G., et al., 2008. Inhibition of human placental aromatase activity by hydroxylated polybrominated diphenyl ethers (OH-PBDEs). Toxicol. Appl. Pharmacol. 227, 68–75.

Cao, X.L., Zhang, J., Goodyer, C.G., 2012. Bisphenol A in human placental and fetal liver tissues collected from Greater Montreal area (Quebec) during 1998–2008. Chemosphere 89, 505–511.

Carew, M.W., Leslie, E.M., 2010. Selenium-dependent and -independent transport of arsenic by the human multidrug resistance protein 2 (MRP2/ABCC2): implications for the mutual detoxification of arsenic and selenium. Carcinogenesis 31, 1450–1455.

Cargnello, M., Roux, P.P., 2011. Activation and function of the MAPKs and their substrates, the MAPK-activated protein kinases. Microbiol. Mol. Biol. Rev. 75, 50–83.

Carlberg, C.E., Möller, L., Paakki, P., 2000. DNA adducts in human placenta as biomarkers for environmental pollution, analysed by the 32P-HPLC method. Biomarkers 5, 182–191.

Carter, A.M., 2012. Evolution of placental function in mammals: the molecular basis of gas and nutrient transfer, hormone secretion, and immune responses. Physiol. Rev. 92, 1543–1576.

Chhabra, D., Sharma, S., Kho, A.T., et al., 2014. Fetal lung and placental methylation is associated with in utero nicotine exposure. Epigenetics 9, 1473–1484.

Chao, S.T., Omiecinski, C.J., Namkung, M.J., 1981. Catechol estrogen formation in placental and fetal tissues of humans, macaques, rats and rabbits. Dev. Pharmacol. Ther. 2, 1–16.

Chełchowska, M., Gajewska, M.J., Ambroszkiewicz, J., 2008. The influence of lead on concentration of the pregnancy-associated plasma protein A (PAPP-A) in pregnant women smoking tobacco-preliminary study. Przegl. Lek. 65, 470–473.

Chen, H.J., Chiang, L.C., Tseng, M.C., 1999. Detection and quantification of 1,N(6)-ethenoadenine in human placental DNA by mass spectrometry. Chem. Res. Toxicol. 12, 1119–1126.

Chen, J., Douglas, G.C., Thirkill, T.L., 2003. Effect of bromodichloromethane on chorionic gonadotrophin secretion by human placental trophoblast cultures. Toxicol. Sci. 76, 75–82.

Chen, S., Liao, T., Wei, Y., et al., 2010. Endocrine disruptor, dioxin (TCDD)-induced mitochondrial dysfunction and apoptosis in human trophoblast-like JAR cells. Mol. Hum. Reprod. 16, 361–372.

Chen, H.J., Lin, W.P., 2009. Simultaneous quantification of 1,N2-propano-2′-deoxyguanosine adducts derived from acrolein and crotonaldehyde in human placenta and leukocytes by isotope dilution nanoflow LC nanospray ionization tandem mass spectrometry. Anal. Chem. 81, 9812–9818.

Chen, J., Thirkill, T.L., Lohstroh, P.N., 2004. Bromodichloromethane inhibits human placental trophoblast differentiation. Toxicol. Sci. 78, 166–174.

Chen, D.B., Wang, W., 2013. Human placental microRNAs and preeclampsia. Biol. Reprod. 88 (5), 130.

Chen, F., Zhang, Q., Wang, C., et al., 2012. Enantioselectivity in estrogenicity of the organochlorine insecticide acetofenate in human trophoblast and MCF-7 cells. Reprod. Toxicol. 33, 53–59.

Cheng, L., Guan, Y., Li, L., 1999. Expression in normal human tissues of five nucleotide excision repair genes measured simultaneously by multiplex reverse transcription-polymerase chain reaction. Cancer Epidemiol. Biomark. Prev. 9, 801–807.

Christensen, B.C., Houseman, E.A., Marsit, C.J., 2009. Aging and environmental exposures alter tissue-specific DNA methylation dependent upon CpG island context. PLoS Genet. 5, e1000602.

Chu, T., Handley, D., Bunce, K., 2011. Structural and regulatory characterization of the placental epigenome at its maternal interface. PLoS One 6, e14723.

Ciolino, H.P., Dai, Z., Nair, V., 2011. Retinol inhibits aromatase activity and expression in vitro. J. Nutr. Biochem. 22, 522–526.

Coecke, S., Blaauboer, B.J., Elaut, G., 2005. Toxicokinetics and metabolism. Altern. Lab. Anim. 33, 147–175.

Colby, J.M., 2017. Comparison of umbilical cord tissue and meconium for the confirmation of in utero drug exposure. Clin. Biochem. 50 (13–14), 784–790.

Cole, L.A., 2010. Biological functions of hCG and hCG-related molecules. Reprod. Biol. Endocrinol. 8, 102.

Collier, A.C., Ganley, N.A., Tingle, M.D., 2002. UDP-glucuronosyltransferase activity, expression and cellular localization in human placenta at term. Biochem. Pharmacol. 63, 409–419.

Collier, A.C., Helliwell, R.J., Keelan, J.A., 2003. 3′-Azido-3′-deoxythymidine (AZT) induces apoptosis and alters metabolic enzyme activity in human placenta. Toxicol. Appl. Pharmacol. 192, 164–173.

Collier, A.C., Tingle, M.D., Keelan, J.A., et al., 2000. A highly sensitive fluorescent microplate method for the determination of UDP-glucuronosyl transferase activity in tissues and placental cell lines. Drug Metab. Dispos. 28, 1184–1186.

Corradetti, A., Saccucci, F., Emanuelli, M., 2010. The role of 38α mitogen-activated protein kinase gene in the HELLP syndrome. Cell Stress Chaperones 15, 95–100.

Coutant, R., Boux de Casson, F., Douay, O., 2001. Relationships between placental GH concentration and maternal smoking, newborn gender, and maternal leptin: possible implications for birth weight. J. Clin. Endocrinol. Metab. 86, 4854–4859.

Crinnion, W.J., 2009. Maternal levels of xenobiotics that affect fetal development and childhood health. Altern. Med. Rev. 14, 212–222.

Cui, P.H., Lee, A.C., Zhou, F., et al., 2010. Impaired transactivation of the human CYP2J2 arachidonic acid epoxygenase gene in HepG2 cells subjected to nitrative stress. Br. J. Pharmacol. 159, 1440–1449.

Daube, H., Scherer, G., Riedel, K., 1997. DNA adducts in human placenta in relation to tobacco smoke exposure and plasma antioxidant status. J. Canc. Res. Clin. Oncol. 123, 141–151.

de Castro, A., Díaz, A., Piñeiro, B., 2013. Simultaneous determination of opiates, methadone, amphetamines, cocaine, and metabolites in human placenta and umbilical cord by LC-MS/MS. Anal. Bioanal. Chem. 405, 4295–4305.

de Castro, A., Jones, H.E., Johnson, R.E., 2011. Methadone, cocaine, opiates, and metabolite disposition in umbilical cord and correlations to maternal methadone dose and neonatal outcomes. Ther. Drug Monit. 33, 443–452.

Demura, M., Reierstad, S., Innes, J.E., et al., 2008. Novel promoter I.8 and promoter usage in the CYP19 (aromatase) gene. Reprod. Sci. 15, 1044–1053.

Denning, D.W., Allen, R., Wilkinson, A.P., et al., 1990. Transplacental transfer of aflatoxin in humans. Carcinogenesis 11, 1033–1035.

Deshmukh, S.V., Nanovskaya, T.N., Ahmed, M.S., 2003. Aromatase is the major enzyme metabolizing buprenorphine in human placenta. J. Pharmacol. Exp. Therapeut. 306, 1099–1105.

Deshmukh, S.V., Nanovskaya, T.N., Hankins, G.D., et al., 2004. N-demethylation of levo-alpha-acetylmethadol by human placental aromatase. Biochem. Pharmacol. 67, 885–892.

De Vries, H.R., Maxwell, S.M., Hendrickse, R.G., 1989. Foetal and neonatal exposure to aflatoxins. Acta Paediatr. Scand. 78, 373–378.

Dewan, P., Jain, V., Gupta, P., et al., 2013. Organochlorine pesticide residues in maternal blood, cord blood, placenta, and breastmilk and their relation to birth size. Chemosphere 90, 1704–1710.

Dickmann, L.J., Isoherranen, N., 2013. Quantitative prediction of CYP2B6 induction by estradiol during pregnancy: potential explanation for increased methadone clearance during pregnancy. Drug Metab. Dispos. 41, 270–274.

Ding, G., Cui, C., Chen, L., 2013. Prenatal low-level mercury exposure and neonatal anthropometry in rural northern China. Chemosphere 92, 1085–1089.

Dipple, A., 1995. DNA adducts of chemical carcinogens. Carcinogenesis 16, 437–441.

Dodd-Butera, T., Quintana, P.J., Ramirez-Zetina, M., et al., 2017. Placental biomarkers of PAH exposure and glutathione-S-transferase biotransformation enzymes in an obstetric population from Tijuana, Baja California, Mexico. Environ. Res. 152, 360–368.

Doerge, D.R., Churchwell, M.I., Fang, J.L., et al., 2000. Quantification of etheno-DNA adducts using liquid chromatography, on-line sample processing, and electrospray tandem mass spectrometry. Chem. Res. Toxicol. 12, 1259–1264.

Druckmann, R., Druckmann, M.A., 2005. Progesterone and the immunology of pregnancy. J. Steroid Biochem. Mol. Biol. 97, 389–396.

Drukteinis, J.S., Medrano, T., Ablordeppey, E.A., et al., 2005. Benzo[a] pyrene, but not 2,3,7,8-TCDD, induces G2/M cell cycle arrest, $p21^{CIP1}$ and p53 phosphorylation in human choriocarcinoma JEG-3 cells: a distinct signaling pathway. Placenta 26, S87–S95.

Edmunds, K.M., Holloway, A.C., Crankshaw, D.J., et al., 2005. The effects of dietary phytoestrogens on aromatase activity in human endometrial stromal cells. Reprod. Nutr. Dev. 45, 709–720.

El Marroun, H., Tiemeier, H., Steegers, E.A., et al., 2009. Intrauterine cannabis exposure affects fetal growth trajectories: the Generation R Study. J. Am. Acad. Child Adolesc. Psychiatr. 48, 1173–1181.

Engel, S.M., Scher, E., Wallenstein, S., et al., 2013. Maternal active and passive smoking and hypertensive disorders of pregnancy: risk with trimester-specific exposures. Epidemiology 24, 379–386.

England, L.J., Aagaard, K., Bloch, M., et al., 2017. Developmental toxicity of nicotine: a transdisciplinary synthesis and implications for emerging tobacco products. Neurosci. Biobehav. Rev. 72, 176–189.

Esteban-Vasallo, M.D., Aragonés, N., Pollan, M., et al., 2012. Mercury, cadmium, and lead levels in human placenta: a systematic review. Environ. Health Perspect. 12, 1369–1377.

Everson, R.B., Randerath, E., Santella, R.M., et al., 1988. Quantitative associations between DNA damage in human placenta and maternal smoking and birth weight. J. Natl. Cancer Inst. 80, 567–576.

Everson, R.B., Randerath, E., Santella, R.M., et al., 1986. Detection of smoking-related covalent DNA adducts in human placenta. Science 4733, 54–57.

Fallen, S., Baxter, D., Wu, X., et al., May 2018. Extracellular vesicle RNAs reflect placenta dysfunction and are a biomarker source for preterm labour. J. Cell Mol. Med. 22, 2760–2773.

Farmer, P.B., Singh, R., 2008. Use of DNA adducts to identify human health risk from exposure to hazardous environmental pollutants: the increasing role of mass spectrometry in assessing biologically effective doses of genotoxic carcinogens. Mutat. Res. 659, 68–76.

Farmer, P.B., Brown, K., Tompkins, E., et al., 2005. DNA adducts: mass spectrometry methods and future prospects. Toxicol. Appl. Pharmacol. 207, 293–301.

Fisher, S.E., Karl, P.I., 1988. Maternal ethanol use and selective fetal malnutrition. Recent Dev. Alcohol 6, 277–289.

Floyd, R.A., Watson, J.J., Wong, P.K., et al., 1986. Hydroxyl free radical adduct of deoxyguanosine: sensitive detection and mechanisms of formation. Free Radic. Res. Commun. 1, 163–172.

Fokina, V.M., Patrikeeva, S.L., Zharikova, O.L., et al., 2011. Transplacental transfer and metabolism of buprenorphine in preterm human placenta. Am. J. Perinatol. 28, 25–32.

Fowden, A.L., Forhead, A.J., 2004. Endocrine mechanisms of intrauterine programming. Reproduction 127, 515–526.

Fowler, B.A., 2012. Biomarkers in toxicology and risk assessment. Exper. Suppl. (Basel) 101, 459–470.

Fu, G., Brkić, J., Hayder, H., et al., 2013. MicroRNAs in human placental development and pregnancy complications. Int. J. Mol. Sci. 14, 5519–5544.

Gagnon, A., Wilson, R.D., Society of Obstetricians and Gynecologists of Canada Genetics Committee, 2008. Obstetrical complications associated with abnormal maternal serum markers analytes. J. Obstet. Gynaecol. Can. 30, 918–949.

Gallagher, J., Mumford, J., Li, X., et al., 1993. DNA adduct profiles and levels in placenta, blood and lung in relation to cigarette smoking and smoky coal emissions. IARC Sci. Publ. 124, 283–292.

Gallagher, J.E., Everson, R.B., Lewtas, J., et al., 1994. Comparison of DNA adduct levels in human placenta from polychlorinated biphenyl exposed women and smokers in which CYP 1A1 levels are similarly elevated. Teratog. Carcinog. Mutagen. 14, 183–192.

Galvano, F., Pietri, A., Bertuzzi, T., et al., 2008. Maternal dietary habits and mycotoxin occurrence in human mature milk. Mol. Nutr. Food Res. 52, 496–501.

Gauster, M., Desoye, G., Tötsch, M., et al., 2012. The placenta and gestational diabetes mellitus. Curr. Diabetes Rep. 12, 16–23.

Grabic, R., Hansen, L.G., Ptak, A., et al., 2006. Differential accumulation of low-chlorinated (Delor 103) and high-chlorinated (Delor 106) biphenyls in human placental tissue and opposite effects on conversion of DHEA to E2. Chemosphere 62, 573–580.

Granata, O.M., Cocciadifero, L., Campisi, I., et al., 2009. Androgen metabolism and biotransformation in nontumoral and malignant human liver tissues and cells. J. Steroid Biochem. Mol. Biol. 113, 290–295.

Gray, T.R., Choo, R.E., Concheiro, M., et al., 2010. Prenatal methadone exposure, meconium biomarker concentrations and neonatal abstinence syndrome. Addiction 105, 2151–2159.

Grigoriu, A., Ferreira, J.C., Choufani, S., et al., 2011. Cell specific patterns of methylation in the human placenta. Epigenetics 6, 368–379.

Gundacker, C., Gencik, M., Hengstschläger, M., 2010. The relevance of the individual genetic background for the toxicokinetics of two significant neurodevelopmental toxicants: mercury and lead. Mutat. Res. Rev. Mutat. Res. 705, 130–140.

Gundacker, C., Hengstschläger, M., 2012. The role of the placenta in fetal exposure to heavy metals. WMW (Wien. Med. Wochenschr.) 162, 201–206.

Guvenius, D.M., Aronsson, A., Ekman-Ordeberg, G., et al., 2003. Human prenatal and postnatal exposure to polybrominated diphenyl ethers, polychlorinated biphenyls, polychlorobiphenylols, and pentachlorophenol. Environ. Health Perspect. 111, 1235–1241.

Hahn, S., Rusterholz, C., Hösli, I., et al., 2011. Cell-free nucleic acids as potential markers for preeclampsia. Placenta 32, S17–S20.

Hakkola, J., Pasanen, M., Hukkanen, J., Pelkonen, O., et al., 1996. Expression of xenobiotic-metabolizing cytochrome P450 forms in human full-term placenta. Biochem. Pharmacol. 51, 403–411.

Halkoaho, A., Pietilä, A.M., Dumez, B., et al., 2010. Ethical aspects of human placental perfusion: interview of the mothers donating placenta. Placenta 31, 686–690.

Halliday-Bell, J.A., Quansah, R., Gissler, M., et al., 2010. Laboratory work and adverse pregnancy outcomes. Occup. Med. 60, 310–313.

Hansen, C., Asmussen, I., Autrup, H., 1993. Detection of carcinogen-DNA adducts in human fetal tissues by the ^{32}P-postlabeling procedure. Environ. Health Perspect. 99, 229–231.

Hansen, C., Sørensen, L.D., Asmussen, I., et al., 1992. Transplacental exposure to tobacco smoke in human-adduct formation in placenta and umbilical cord blood vessels. Teratog. Carcinog. Mutagen. 2, 51–60.

Harada, N., Yoshimura, N., Honda, S., 2003. Unique regulation of expression of human aromatase in the placenta. J. Steroid Biochem. Mol. Biol. 86, 327–334.

Hassoun, E.A., Stohs, S.J., 1996. TCDD, endrin and lindane induced oxidative stress in fetal and placental tissues of C57BL/6J and DBA/2J mice. Comp. Biochem. Physiol. C Pharmacol. Toxicol. Endocrinol. 115, 11–18.

Hata, S., Miki, Y., Saito, R., et al., 2013. Aromatase in human liver and its diseases. Cancer Med. 2, 305–315.

Hatch, M.C., Warburton, D., Santella, R.M., 1990. Polycyclic aromatic hydrocarbon-DNA adducts in spontaneously aborted fetal tissue. Carcinogenesis 9, 1673–1675.

Helland, I.B., Reseland, J.E., Saugstad, O.D., et al., 2001. Smoking related to plasma leptin concentration in pregnant women and their newborn infants. Acta Paediatr. 90, 282–287.

Herse, F., Lamarca, B., Hubel, C.A., et al., 2012. Cytochrome P450 subfamily 2J polypeptide 2 expression and circulating epoxyeicosatrienoic metabolites in preeclampsia. Circulation 126, 2990–2999.

Hieronymus, T.L., Nanovskaya, T.N., Deshmukh, S.V., et al., 2006. Methadone metabolism by early gestational age placentas. Am. J. Perinatol. 23, 287–294.

Hnat, M.D., Meadows, J.W., Brockman, D.E., et al., 2005. Heat shock protein-70 and 4-hydroxy-2-nonenal adduct in human placental villous tissue of normotensive preeclamptic and intrauterine growth restricted pregnancies. Am. J. Obstet. Gynecol. 193, 836–840.

Hou, L., Zhang, X., Wang, D., et al., 2012. Environmental chemical exposures and human epigenetics. Int. J. Epidemiol. 41, 79–105.

Hsieh, L.L., Hsieh, T.T., 1993. Detection of aflatoxin B1-DNA adducts in human placenta and cord blood. Cancer Res. 53, 1278–1280.

Huang, H., Leung, L.K., 2009. Bisphenol A downregulates CYP19 transcription in JEG-3 cells. Toxicol. Lett. 189, 248–252.

Hung, T.H., Chen, S.F., Lo, L.M., et al., 2012. Increased autophagy in placentas of intrauterine growth-restricted pregnancies. PLoS One 7 (7), e40957.

Hutson, J.R., Garcia-Bournissen, F., Davis, A., et al., 2011. The human placental perfusion model: a systematic review and development of a model to predict *in vivo* transfer of therapeutic drugs. Clin. Pharmacol. Ther. 90 (1), 67–76.

Huuskonen, P., Myllynen, P., Storvik, M., et al., 2013. The effects of aflatoxin B1 on transporters and steroid metabolizing enzymes in JEG-3 cells. Toxicol. Lett. 218 (3), 200–206.

Huuskonen, P., Storvik, M., Reinisalo, M., et al., 2008. Microarray analysis of the global alterations in the gene expression in the placentas from cigarette-smoking mothers. Clin. Pharmacol. Ther. 83, 542–550.

Huuskonen, P., Storvik, M., Amezaga, M.R., et al., 2016. Proteomics in placenta from cigarette smoking mothers. Reprod. Toxicol. 63, 22–31.

IARC, 2010. Monographs on the Evaluation of the Carcinogenic Risk of Chemicals to Humans: Some Non-heterocyclic Polycyclic Aromatic Hydrocarbons and Some Related Exposures, vol. 92. International Agency for Research on Cancer, Lyon, France.

Ikezuki, Y., Tsutsumi, O., Takai, Y., et al., 2002. Determination of bisphenol A concentrations in human biological fluids reveals significant early prenatal exposure. Hum. Reprod. 17, 2839–2841.

Immonen, E., Kummu, M., Petsalo, A., et al., 2010. Toxicokinetics of the food-toxin IQ in human placental perfusion is not affected by ABCG2 or xenobiotic metabolism. Placenta 31, 641–648.

Iqbal, M., Audette, M.C., Petropoulos, S., et al., 2012. Placental drug transporters and their role in fetal protection. Placenta 33 (3), 137–142.

Iyengar, G.V., Rapp, A., 2001. Human placenta as a 'dual' biomarker for monitoring fetal and maternal environment with special reference to potentially toxic trace elements. Part 3: toxic trace elements in placenta and placenta as a biomarker for these elements. Sci. Total Environ. 280 (1–3), 221–238.

Jaraczewska, K., Lulek, J., Covaci, A., et al., 2006. Distribution of polychlorinated biphenyls, organochlorine pesticides and polybrominated diphenyl ethers in human umbilical cord serum, maternal serum and milk from Wielkopolska region, Poland. Sci. Total Environ. 372, 20–31.

Jebbink, J., Wolters, A., Fernando, F., et al., 2012. Molecular genetics of preeclampsia and HELLP syndrome – a review. Biochim. Biophys. Acta 1822 (12), 1960–1969.

Jeschke, U., Briese, V., Richter, D.U., et al., 2005. Effects of phytoestrogens genistein and daidzein on production of human chorionic gonadotropin in term trophoblast cells in vitro. Gynecol. Endocrinol. 21, 180–184.

Jiménez-Díaz, I., Zafra-Gómez, A., Ballesteros, O., et al., 2010. Determination of Bisphenol A and its chlorinated derivatives in placental tissue samples by liquid chromatography-tandem mass spectrometry. J. Chromatogr. B 878 (32), 3363–3369.

Jin, H., Audus, K.L., 2005. Effect of bisphenol A on drug efflux in BeWo, a human trophoblast-like cell line. Placenta 26, S96–S103.

Jolibois Jr., L.S., Shi, W., George, W.J., et al., 1999. Cadmium accumulation and effects on progesterone release by cultured human trophoblast cells. Reprod. Toxicol. 13, 215–221.

Jones, N.J., 2012. ^{32}P-postlabelling for the sensitive detection of DNA adducts. Meth. Mol. Biol. 817, 183–206.

Jonsyn, F.E., Maxwell, S.M., Hendrickse, R.G., 1995. Human fetal exposure to ochratoxin A and aflatoxins. Ann. Trop. Paediatr. 15, 3–9.

Kantola, M., Purkunen, R., Kroger, P., et al., 2000. Accumulation of cadmium, zinc, and copper in maternal blood and developmental placental tissue: differences between Finland, Estonia, and St. Petersburg. Environ. Res. 83, 54–66.

Kantola, M., Purkunen, R., Kroger, P., et al., 2004. Selenium in pregnancy: is selenium an active defective ion against environmental chemical stress? Environ. Res. 96, 51–61.

Karl, P.I., Divald, A., Diehl, A.M., et al., 1998. Altered cyclic AMP-dependent human chorionic gonadotropin production in cultured human placental trophoblasts exposed to ethanol. Biochem. Pharmacol. 55, 45–51.

Karttunen, V., Myllynen, P., Prochazka, G., et al., 2010. Placental transfer and DNA binding of benzo(a)pyrene in human placental perfusion. Toxicol. Lett. 197, 75–81.

Karttunen, V., Sahlman, H., Repo, J.K., et al., 2015. Criteria and challenges of the human placental perfusion - data from a large series of perfusions. Toxicol. Vitro 29, 1482–1491.

Katari, S., Turan, N., Bibikova, M., et al., 2009. DNA methylation and gene expression differences in children conceived in vitro or in vivo. Hum. Mol. Genet. 18, 3769–3778.

Kawai, M., Swan, K.F., Green, A.E., et al., 2002. Placental endocrine disruption induced by cadmium: effects on P450 cholesterol side-chain cleavage and 3β-hydroxysteroid dehydrogenase enzymes in cultured human trophoblasts. Biol. Reprod. 67, 178–183.

Kay, H.H., Grindle, K.M., Magness, R.R., 2000. Ethanol exposure induces oxidative stress and impairs nitric oxide availability in the human placental villi: a possible mechanism of toxicity. Am. J. Obstet. Gynecol. 182, 682–688.

Kay, H.H., Tsoi, S., Grindle, K., et al., 2006. Markers of oxidative stress in placental villi exposed to ethanol. J. Soc. Gynecol. Invest. 13, 118–121.

Kezios, K.L., Liu, X., Cirillo, P.M., et al., 2013. Dichlorodiphenyltrichloroethane (DDT), DDT metabolites and pregnancy outcomes. Reprod. Toxicol. 35, 156–164.

Kim, M., Bae, M., Na, H., et al., 2012. Environmental toxicants–induced epigenetic alterations and their reversers. J. Environ. Sci. 30, 323–367.

Kippler, M., Hoque, A.M., Raqib, R., et al., 2010. Accumulation of cadmium in human placenta interacts with the transport of micronutrients to the fetus. Toxicol. Lett. 192, 162–168.

Kitawaki, J., Inoue, S., Tamura, T., et al., 1993. Cigarette smoking during pregnancy lowers aromatase cytochrome P-450 in the human placenta. J. Steroid Biochem. Mol. Biol. 45, 485–491.

Klaassen, C.D., 2001. Casarett & Doull's Toxicology: The Basic Science of Poisons, sixth ed. McGraw-Hill, New York, USA.

Klempan, T., Hudon-Thibeault, A.A., Oufkir, T., et al., 2011. Stimulation of serotonergic 5-HT2A receptor signaling increases placental aromatase (CYP19) activity and expression in BeWo and JEG-3 human choriocarcinoma cells. Placenta 32, 651–656.

Knöfler, M., 2010. Critical growth factors and signaling pathway controlling human trophoblast invasion. Int. J. Dev. Biol. 54, 269–280.

Kolwankar, D., Glover, D.D., Ware, J.A., et al., 2005. Expression and function of ABCB1 and ABCG2 in human placental tissue. Drug Metab. Dispos. 33, 524–529.

Korpela, H., Loueniva, R., Yrjanheikki, E., et al., 1986. Lead and cadmium concentrations in maternal and umbilical cord blood, amniotic fluid, placenta, and amniotic membranes. Am. J. Obstet. Gynecol. 155, 1086–1089.

Kramer, M.S., Lydon, J., Séguin, L., et al., 2010. Non-stress-related factors associated with maternal corticotrophin-releasing hormone (CRH) concentration. Paediatr. Perinat. Epidemiol. 24, 390–397.

Kumar, P., Mendelson, C.R., 2011. Estrogen-related receptor gamma (ERRgamma) mediates oxygen-dependent induction of aromatase (CYP19) gene expression during human trophoblast differentiation. Mol. Endocrinol. 25, 1513–1526.

Kumar, P., Kamat, A., Mendelson, C.R., 2009. Estrogen receptor alpha (ERalpha) mediates stimulatory effects of estrogen on aromatase (CYP19) gene expression in human placenta. Mol. Endocrinol. 23, 784–793.

Kummu, M., Sieppi, E., Wallin, K., et al., 2012. Cadmium inhibits ABCG2 transporter function in BeWo choriocarcinoma cell and MCF-7 cells overexpressing ABCG2. Placenta 33, 859–865.

Kwintkiewicz, J., Nishi, Y., Yanase, T., et al., 2010. Peroxisome proliferator-activated receptor-gamma mediates bisphenol A inhibition of FSH-stimulated IGF-1, aromatase, and estradiol in human granulosa cells. Environ. Health Perspect. 118, 400–406.

Lafond, J., Hamel, A., Takser, L., et al., 2004. Low environmental contamination by lead in pregnant women: effect on calcium transfer in human placental syncytiotrophoblasts. J. Toxicol. Environ. Health A 67, 1069–1079.

Lagueux, J., Pereg, D., Ayotte, P., et al., 1999. Cytochrome P450 CYP1A1 enzyme activity and DNA adducts in placenta of women environmentally exposed to organochlorines. Environ. Res. 80, 369–382.

Leaños-Miranda, A., Campos-Galicia, I., Isordia-Salas, I., et al., 2012. Changes in circulating concentrations of soluble fms-like tyrosine kinase-1 and placental growth factor measured by automated electrochemiluminescence immunoassays methods are predictors of preeclampsia. J. Hypert. 30, 2173–2181.

Lee, Y.J., Ryu, H.Y., Kim, H.K., et al., 2008. Maternal and fetal exposure to bisphenol A in Korea. Reprod. Toxicol. 25, 413–419.

Lehman, L.D., Poisner, A.M., 1984. Induction of metallothione in synthesis in cultured human trophoblasts by cadmium and zinc. J. Toxicol. Environ. Health 14, 419–432.

Levin, A.A., Plautz, J.R., di Sant'Agnese, P.A., et al., 1981. Cadmium: placental mechanisms of fetal toxicity. Placenta Suppl. 3, 303–318.

Lézé, M.P., Le Borgne, M., Marchand, P., et al., 2004. 2- and 3-[(aryl)(azolyl)methyl]indoles as potential non-steroidal aromatase inhibitors. J. Enzym. Inhib. Med. Chem. 19, 549–557.

Li, Q., Kappil, M.A., Li, A., Dassanayake, P.S., et al., 2015. Exploring the associations between microRNA expression profiles and environmental pollutants in human placenta from the National Children's Study (NCS). Epigenetics 10, 793–802.

Ling, W.Y., Wrixon, W.W., Acorn, T.P., 1984. Stimulation of estradiol production from estrone-3-sulfate by 5 alpha-dihydrotestosterone in cultured human placental explants. J. Steroid Biochem. 21, 653–657.

Liu, J., Li, J., Wu, Y., et al., 2017. Bisphenol a metabolites and bisphenol S in paired maternal and cord serum. Environ. Sci. Technol. 51, 2456–2463.

Lopez-Espinosa, M.J., Granada, A., Carreno, J., et al., 2007. Organochlorine pesticides in placentas from Southern Spain and some related factors. Placenta 28, 631–638.

Lu, W.J., Bies, R., Kamden, L.K., et al., 2010. Methadone: a substrate and mechanism-based inhibitor of CYP19 (aromatase). Drug Metab. Dispos. 38, 1308–1313.

Lu, W.J., Desta, Z., Flockhart, D.A., 2012a. Tamoxifen metabolites as active inhibitors of aromatase in the treatment of breast cancer. Breast Cancer Res. Treat. 131, 473–481.

Lu, W.J., Xu, C., Pei, Z., et al., 2012b. The tamoxifen metabolite norendoxifen is a potent and selective inhibitor of aromatase (CYP19) and a potential lead compound for novel therapeutic agents. Breast Cancer Res. Treat. 133, 99–109.

Luo, X., Yao, Z., Qi, H., et al., 2011. Gadd45α as an upstream signaling molecule of p38 MAPK triggers oxidative stress-induced sFlt-1 and sEng uprgulation in preeclampsia. Cell Tissue Res. 344, 551–565.

Ma, J., Qiu, X., Ren, A., et al., 2012. Using placenta to evaluate the polychlorinated biphenyls (PCBs) and polybrominated diphenyl ethers (PBDEs) exposure of fetus in a region with high prevalence of neural tube defects. Ecotoxicol. Environ. Saf. 86, 141–146.

Maccani, J.Z., Koestler, D.C., Lester, B., et al., 2015. Placental DNA methylation related to both infant toenail mercury and adverse neurobehavioral outcomes. Environ. Health Perspect. 123, 723–729.

Maccani, M.A., Avissar-Whiting, M., Banister, C.E., et al., 2010. Maternal cigarette smoking during pregnancy is associated with downregulation of miR-16, miR-21, and miR-146a in the placenta. Epigenetics 5, 583–589.

Maccani, M.A., Padbury, J.F., Marsit, C.J., 2011. miR-16 and miR-21 expression in the placenta is associated with fetal growth. PLoS One 6, e21210.

Maccani, M.A., Marsit, C.J., 2009. Epigenetics in the placenta. Am. J. Reprod. Immunol. 62, 78–89.

Machaalani, R., Ghazavi, E., Hinton, T., et al., 2014. Cigarette smoking during pregnancy regulates the expression of specific nicotinic acetylcholine receptor (nAChR) subunits in the human placenta. Toxicol. Appl. Pharmacol. 276, 204–212.

Machtinger, R., Zhong, J., Mansur, A., et al., 2018. Placental lncRNA expression is associated with prenatal phthalate exposure. Toxicol. Sci. 163 (1), 116–122.

Malek, A., Obrist, C., Wenzinger, S., et al., 2009. The impact of cocaine and heroin on the placental transfer of methadone. Reprod. Biol. Endocrinol. 7, 61.

Manchester, D.K., Bowman, E.D., Parker, N.B., et al., 1992. Determinants of polycyclic aromatic hydrocarbon-DNA adducts in human placenta. Cancer Res. 52, 1499–1503.

Manchester, D.K., Weston, A., Choi, J.S., et al., 1988. Detection of benzo[a]pyrene diol epoxide-DNA adducts in human placenta. Proc. Natl. Acad. Sci. U. S. A. 85, 9243–9247.

Mandang, S., Manuelpillai, U., Wallace, E.M., 2007. Oxidative stress increases placental and endothelial cell activin A secretion. J. Endocrinol. 192, 485–493.

Marafie, E.M., Marafie, I., Emery, S.J., et al., 2000. Biomonitoring the human population exposed to pollution from the oil fires in Kuwait: analysis of placental tissue using ^{32}P-postlabeling. Environ. Mol. Mutagen. 4, 274–282.

Martel, F., Keating, E., 2003. Uptake of 1-methyl-4-phenylpyridinium (MPP+) by the JAR human placental choriocarcinoma cell line: comparison with 5-hydroxytryptamine. Placenta 24, 361–369.

Matlow, J.N., Lubetsky, A., Aleksa, K., et al., 2013. The transfer of ethyl glucuronide across the dually perfused human placenta. Placenta 34, 369–373.

Matscheski, A., Richter, D.U., Hartmann, A.M., et al., 2006. Effects of phytoestrogen extracts isolated from rye, green and yellow pea seeds on hormone production and proliferation of trophoblast tumor cells Jeg3. Horm. Res. 65, 276–288.

Matthias, A.A., Hitti, J., Unadkat, J.D., 2005. P-glycoprotein and breast cancer resistance protein expression in human placentae of various gestational ages. Am. J. Physiol. Regul. Integr. Comp. Physiol. 289, R963–R969.

Maymó, J.L., Pérez, A., Gambino, Y., et al., 2011. Review: leptin gene expression in the placenta–regulation of a key hormone in trophoblast proliferation and survival. Placenta *32* (Suppl. 2), S146–S153.

McAleer, M.F., Tuan, R.S., 2001a. Metallothionein overexpression in human trophoblastic cells protects against cadmium-induced apoptosis. Vitro Mol. Toxicol. 14, 25–42.

McAleer, M.F., Tuan, R.S., 2001b. Metallothionein protects against severe oxidative stress-induced apoptosis of human trophoblastic cells. Vitro Mol. Toxicol. 14, 219–231.

Medina, A.E., 2011. Fetal alcohol spectrum disorders and abnormal neuronal plasticity. Neuroscientist 17, 274–287.

Meeker, J.D., 2012. Exposure to environmental endocrine disruptors and child development. Arch. Pediatr. Adolesc. Med. 166, 952–958.

Mehendale, R., Hibbard, J., Fazleabas, A., et al., 2007. Placental angiogenesis markers sFlt-1 and PlGF: response to cigarette smoke. Am. J. Obstet. Gynecol. 197, 363.e1-5.

Menai, M., Heude, B., Slama, R., et al., 2012. Association between maternal blood cadmium during pregnancy and birth weight and the risk of fetal growth restriction: the EDEN mother-child cohort study. Reprod. Toxicol. 34, 622–627.

Mendelson, C.R., Kamat, A., 2007. Mechanisms in the regulation of aromatase in developing ovary and placenta. J. Steroid Biochem. Mol. Biol. 106, 62–70.

Meyer zu Schwabedissen, H.E., Grube, M., Dreisbach, A., et al., 2006. Epidermal growth factor-mediated activation of the map kinase cascade results in altered expression and function of ABCG2 (BCRP). Drug Metab. Dispos. 34, 524–533.

Miller, J.A., Miller, E.C., 1975. Metabolic activation and reactivity of chemical carcinogens. Mutat. Res. 33, 25–26.

Miller, K.P., Ramos, K.S., 2001. Impact of cellular metabolism on the biological effects of benzo[a]pyrene and related hydrocarbons. Drug Metab. Rev. 33, 1–35.

Mochizuki, M., Maruo, T., Masuko, K., et al., 1984. Effects of smoking on fetoplacental-maternal system during pregnancy. Am. J. Obstet. Gynecol. 149, 413–420.

Moindjie, H., Santos, E.D., Gouesse, R.J., et al., 2016. Preimplantation factor is an anti-apoptotic effector in human trophoblasts involving p53 signaling pathway. Cell Death Dis. 7, e2504.
Mørck, T.J., Sorda, G., Bechi, N., et al., 2010. Placental transport and in vitro effects of Bisphenol A. Reprod. Toxicol. 30, 131–137.
Morini, L., Falcón, M., Pichini, S., et al., 2011. Ethyl-glucuronide and ethyl-sulfate in placental and fetal tissues by liquid chromatography coupled with tandem mass spectrometry. Anal. Biochem. 418, 30–36.
Mose, T., Kjaerstad, M.B., Mathiesen, L., et al., 2008. Placental passage of benzoic acid, caffeine, and glyphosate in an ex vivo human perfusion system. J. Toxicol. Environ. Health A 71, 984–991.
Mose, T., Mathiesen, L., Karttunen, V., et al., 2012. Meta-analysis of data from human ex vivo placental perfusion studies on genotoxic and immunotoxic agents within the integrated European project NewGeneris. Placenta 33, 433–439.
Mouillet, J.F., Chu, T., Hubel, C.A., et al., 2010a. The levels of hypoxia-regulated microRNAs in plasma of pregnant women with fetal growth restriction. Placenta 31, 781–784.
Mouillet, J.F., Chu, T., Nelson, D.M., et al., 2010b. MiR-205 silences MED1 in hypoxic primary human trophoblasts. FASEB J. 24, 2030–2039.
Mouillet, J.F., Chu, T., Sadovsky, Y., 2011. Expression patterns of placental microRNAs. Birth Defects Res. A Clin. Mol. Teratol. 91, 737–743.
Mouillet, J.F., Donker, R.B., Mishima, T., et al., 2013. The unique expression and function of miR-424 in human placental trophoblasts. Biol. Reprod. 89, 25.
Mouillet, J.F., Ouyang, Y., Coyne, C.B., et al., 2015. MicroRNAs in placental health and disease. Am. J. Obstet. Gynecol. 213 (4 Suppl.), S163–S172.
Murphy, V.E., Smith, R., Giles, W.B., et al., 2006. Endocrine regulation of human fetal growth: the role of the mother, placenta, and fetus. Endocr. Rev. 27, 141–169.
Myllynen, P., Vähäkangas, K., 2013. Placental transfer and metabolism: an overview of the experimental models utilizing human placental tissue. Toxicol. Vitro 27, 507–512.
Myllynen, P., Immonen, E., Kummu, M., et al., 2009. Developmental expression of drug metabolizing enzymes and transporter proteins in human placenta and fetal tissues. Expert Opin. Drug Metabol. Toxicol. 5, 1483–1499.
Myllynen, P., Kummu, M., Kangas, T., et al., 2008. ABCG2/BCRP decreases the transfer of a food-born chemical carcinogen, 2-amino-1-methyl-6-phenylimidazol[4,5-b]pyridine (PhIP) in perfused term human placenta. Toxicol. Appl. Pharmacol. 232, 210–217.
Myllynen, P., Mathiesen, L., Weimer, M., et al., 2010. Preliminary inter-laboratory comparison of the ex vivo dual human placental perfusion system. Reprod. Toxicol. 30, 94–102.
Myllynen, P., Pasanen, M., Pelkonen, O., 2005. Human placenta: a human organ for developmental toxicology research and biomonitoring. Placenta 26, 361–371.
Myllynen, P., Pasanen, M., Vähäkangas, K., 2007. The fate and effects of xenobiotics in human placenta. Expert Opin. Drug Metabol. Toxicol. 3, 331–346.
Myllynen, P., Pienimäki, P., Raunio, H., et al., 1998. Microsomal metabolism of carbamazepine and oxcarbazepine in liver and placenta. Hum. Exp. Toxicol. 12, 668–676.
Myöhänen, K., Vähäkangas, K., 2012. Foetal exposure to food and environmental carcinogens in human beings. Basic Clin. Pharmacol. Toxicol. 110, 101–112.
Myren, M., Mose, T., Mathiesen, L., et al., 2007. The human placenta – an alternative for studying foetal exposure. Toxicol. Vitro 21, 1332–1340.
Nakanishi, T., Kohroki, J., Suzuki, S., et al., 2002. Trialkyltin compounds enhance human CG secretion and aromatase activity in human placental choriocarcinoma cells. J. Clin. Endocrinol. Metab. 87, 2830–2837.
Nakanishi, T., Nishikawa, J., Hiromori, Y., et al., 2005. Trialkyltin compounds bind retinoid X receptor to alter human placental endocrine functions. Mol. Endocrinol. 19, 2502–2516.
Namkung, M.J., Juchau, M.R., 1980. On the capacity of human placental enzymes to catalyze the formation of diols from benzo [a]pyrene. Toxicol. Appl. Pharmacol. 55, 253–259.
Nanovskaya, T.N., Deshmukh, S.V., Nekhayeva, I.A., et al., 2004. Methadone metabolism by human placenta. Biochem. Pharmacol. 68 (2004), 583–591.
Nativelle-Serpentini, C., Richard, S., Séralini, G.E., et al., 2003. Aromatase activity modulation by lindane and bisphenol-A in human placental JEG-3 and transfected kidney E293 cells. Toxicol. Vitro 17, 413–422.
Needham, L.L., Grandjean, P., Heinzow, B., et al., 2011. Partition of environmental chemicals between maternal and fetal blood and tissues. Environ. Sci. Technol. 45, 1121–1126.
Newbern, D., Freemark, M., 2011. Placental hormones and the control of maternal metabolism and fetal growth. Curr. Opin. Endocrinol. Diabetes Obes. 18, 409–416.
Novakovic, B., Saffery, R., 2012. The ever growing complexity of placental epigenetics – role in adverse pregnancy outcomes and fetal programming. Placenta 33, 959–970.
Novakovic, B., Yuen, R.K., Gordon, L., et al., 2011. Evidence for widespread changes in promoter methylation profile in human placenta in response to increasing gestational age and environmental/stochastic factors. BMC Genom. 12, 529.
Obolenskaya, M.Y., Teplyuk, N.M., Divi, R.L., et al., 2010. Human placental glutathione S-transferase activity and polycyclic aromatic hydrocarbon DNA adducts as biomarkers for environmental oxidative stress in placentas from pregnant women living in radioactivity- and chemically-polluted regions. Toxicol. Lett. 196, 80–86.
Osman, K., Akesson, A., Berglund, M., et al., 2000. Toxic and essential elements in placentas of Swedish women. Clin. Biochem. 33, 131–138.
Ostrea Jr., E.M., Matias, O., Keane, C., et al., 1998. Spectrum of gestational exposure to illicit drugs and other xenobiotic agents in newborn infants by meconium analysis. J. Pediatr. 133, 513–515.
Paakki, P., Kirkinen, P., Helin, H., et al., 2000a. Antepartum glucocorticoid therapy suppresses human placental xenobiotic and steroid metabolizing enzymes. Placenta 21, 241–246.
Paakki, P., Stockmann, H., Kantola, M., et al., 2000b. Maternal drug abuse and human term placental xenobiotic and steroid metabolizing enzymes in vitro. Environ. Health Perspect. 108 (2000), 141–145.
Pacyniak, E., Hagenbuch, B., Klaassen, C.D., et al., 2011. Organic anion transporting polypeptides in the hepatic uptake of PBDE congeners in mice. Toxicol. Appl. Pharmacol. 257, 23–31.
Padmanabhan, V., Siefert, K., Ransom, S., et al., 2008. Maternal bisphenol-A levels at delivery: a looming problem? J. Perinatol. 28, 258–263.
Park, H.Y., Park, J.S., Sovcikova, E., et al., 2009. Exposure to hydroxylated polychlorinated biphenyls (OH-PCBs) in the prenatal period and subsequent neurodevelopment in eastern Slovakia. Environ. Health Perspect. 117, 1600–1606.
Partanen, H.A., El-Nezami, H.S., Leppänen, J.M., et al., 2010. Aflatoxin B1 transfer and metabolism in human placenta. Toxicol. Sci. 113, 216–225.
Pasanen, M., 1999. The expression and regulation of drug metabolism in human placenta. Adv. Drug Deliv. Rev. 38, 81–97.

Pasanen, M., Pelkonen, O., 1990. Xenobiotic and steroid-metabolizing monooxygenases catalyzed by cytochrome P450 and glutathione S-transferase conjugations in the human placenta and their relationships to maternal cigarette smoking. Placenta 11, 75–85.

Pasanen, M., Haaparanta, T., Sundin, M., et al., 1990. Immunochemical and molecular biological studies on human placental cigarette smoke-inducible cytochrome P-450-dependent monooxygenase activities. Toxicology 62, 175–187.

Pasanen, M., Stenback, F., Park, S.S., et al., 1988. Immunohistochemical detection of human placental cytochrome P-450-associated monooxygenase system inducible by maternal cigarette smoking. Placenta 9, 267–275.

Pastrakuljic, A., Derewlany, L.O., Knie, B., et al., 2000. The effects of cocaine and nicotine on amino acid transport across the human placental cotyledon perfused in vitro. J. Pharmacol. Exp. Therapeut. 294, 141–146.

Pavan, L., Hermouet, A., Tsatsaris, V., 2004a. Lipids from oxidized low-density lipoprotein modulate human trophoblast invasion: involvement of nuclear liver X receptors. Endocrinology 145, 4583–4591.

Pavan, L., Tsatsaris, V., Hermouet, A., 2004b. Oxidized low-density lipoproteins inhibit trophoblastic cell invasion. J. Clin. Endocrinol. Metab. 89, 1969–1972.

Pavek, P., Dvorak, Z., 2008. Xenobiotic-induced transcriptional regulation of xenobiotic metabolizing enzymes of the cytochrome P450 superfamily in human extrahepatic tissues. Curr. Drug Metabol. 9, 129–143.

Pedersen, M., von Stedingk, H., Botsivali, M., 2012. Birth weight, head circumference, and prenatal exposure to acrylamide from maternal diet: the European prospective mother-child study (NewGeneris). Environ. Health Perspect. 120, 739–1745.

Pelkonen, O., Saarni, H., 1980. Unusual patterns of benzo[a]pyrene metabolites and DNA-benzo[a]pyrene adducts produced by human placental microsomes in vitro. Chem. Biol. Interact. 3, 287–296.

Pelkonen, O., Vähäkangas, K., 1980. Metabolic activation and inactivation of chemical carcinogens. J. Toxicol. Environ. Health 6, 989–999.

Pentsuk, N., van der Laan, J.W., 2009. An interspecies comparison of placental antibody transfer: new insights into developmental toxicity testing of monoclonal antibodies. Birth Defects Res. B Dev. Reprod. Toxicol. 86, 328–344.

Perera, F., Hemminki, K., Jedrychowski, W., 2002. *In utero* DNA damage from environmental pollution is associated with somatic gene mutation in newborns. Cancer Epidem. Biomark. Res. 11, 1134–1137.

Perera, F.P., Jedrychowski, W., Rauh, V., et al., 1999. Molecular epidemiologic research on the effects of environmental pollutants on the fetus. Environ. Health Perspect. 107, 451–460.

Perera, F.P., Whyatt, R.M., Jedrychowski, W., 1998. Recent developments in molecular epidemiology: a study of the effects of environmental polycyclic aromatic hydrocarbons on birth outcomes in Poland. Am. J. Epidemiol. 147, 309–314.

Perera, F.P., Rauh, V., Whyatt, R.M., 2005a. A summary of recent findings on birth outcomes and developmental effects of prenatal ETS, PAH, and pesticide exposures. Neurotoxicology 26, 573–587.

Perera, F.P., Tang, D., Rauh, V., 2005b. Relationships among polycyclic aromatic hydrocarbon-DNA adducts, proximity to the World Trade Center, and effects on fetal growth. Environ. Health Perspect. 113, 1062–1067.

Phillips, D.H., 2002. Smoking-related DNA and protein adducts in human tissues. Carcinogenesis 23, 1979–2004.

Phillips, D.H., 2005. DNA adducts as markers of exposure and risk. Mutat. Res. 1 (2), 284–292.

Phillips, D.H., 2013. On the origins and development of the (32)P-postlabelling assay for carcinogen-DNA adducts. Cancer Lett. 334, 5–9.

Piasek, M., Blanusa, M., Kostial, K., et al., 2001. Placental cadmium and progesterone concentrations in cigarette smokers. Reprod. Toxicol. 15, 673–681.

Piasek, M., Laskey, J.W., Kostial, K., et al., 2002. Assessment of steroid disruption using cultures of whole ovary and/or placenta in rat and in human placental tissue. Int. Arch. Occup. Environ. Health 75 (Suppl.), S36–S44.

Pichini, S., Puig, C., Zuccaro, P., 2005. Assessment of exposure to opiates and cocaine during pregnancy in a Mediterranean city: preliminary results of the "Meconium Project". Forensic Sci. Int. 153, 59–65.

Pichini, S., Morini, L., Marchei, E., 2009. Ethylglucuronide and ethylsulfate in meconium to assess gestational ethanol exposure: preliminary results in two Mediterranean cohorts. Can. J. Clin. Pharmacol. 16, e370–e375.

Pilsner, J.R., Hu, H., Ettinger, A., 2009. Influence of prenatal lead exposure on genomic methylation of cord blood DNA. Environ. Health Perspect. 117, 1466–1471.

Plessow, D., Waldschläger, J., Richter, D.U., 2003. Effects of phytoestrogens on the trophoblast tumour cell lines BeWo and Jeg3. Anticancer Res. 23, 1081–1086.

Poirier, M.C., Weston, A., 1996. Human DNA adduct measurements: state of the art. Environ. Health Perspect. 104, 883–893.

Poirier, M.C., Reed, E., Shamkhani, H., et al., 1993. Platinum drug-DNA interactions in human tissues measured by cisplatin-DNA enzyme-linked immunosorbent assay and atomic absorbance spectroscopy. Environ. Health Perspect. 99, 149–154.

Poirier, M.C., Santella, R.M., Weston, A., 2000. Carcinogen macromolecular adducts and their measurement. Carcinogenesis 21, 353–359.

Postupolski, J., Karlowski, K., Kubik, P., 2006. Ochratoxin a in maternal and foetal blood and in maternal milk. Rocz. Panstw. Zakl. Hig. 57, 23–30.

Powlin, S.S., Keng, P.C., Miller, R.K., 1997. Toxicity of cadmium in human trophoblast cells (JAr choriocarcinoma): role of calmodulin and the calmodulin inhibitor, zaldaride maleate. Toxicol. Appl. Pharmacol. 144, 225–234.

Prahalad, A.K., Manchester, D.K., Hsu, I.C., et al., 1999. Human placental microsomal activation and DNA adduction by air pollutants. Bull. Environ. Contam. Toxicol. 62, 93–100.

Pratt, M.M., John, K., MacLean, A.B., et al., 2011a. Polycyclic aromatic hydrocarbon (PAH) exposure and DNA adduct semi-quantitation in archived human tissues. Int. J. Environ. Res. Publ. Health 8, 2675–2691.

Pratt, M.M., King, L.C., Adams, L.D., et al., 2011b. Assessment of multiple types of DNA damage in human placentas from smoking and nonsmoking women in the Czech Republic. Environ. Mol. Mutagen. 52, 58–68.

Prouillac, C., Lecoeur, S., 2010. The role of the placenta in fetal exposure to xenobiotics: importance of membrane transporters and human models for transfer studies. Drug Metab. Dispos. 38, 1623–1635.

Prouillac, C., Videmann, B., Mazallon, M., et al., 2009. Induction of cells differentiation and ABC transporters expression by a myco-estrogen, zearalenone, in human choriocarcinoma cell line (BeWo). Toxicology 263, 100–107.

Prouillac, C., Koraichi, F., Videmann, B., et al., 2012. In vitro toxicological effects of estrogenic mycotoxins on human placental cells: structure activity relationships. Toxicol. Appl. Pharmacol. 259, 366–375.

Rahi, M., Heikkinen, T., Härtter, S., et al., 2007. Placental transfer of quetiapine in relation to P-glycoprotein activity. J. Psychopharmacol. 21, 751–756.

Rama Sastry, B.V., Hemontolor, M.E., Olenick, M., 1999. Prostaglandin E2 in human placenta: its vascular effects and activation of prostaglandin E2 formation by nicotine and cotinine. Pharmacology 58, 70–86.

Ramirez, G.B., Cruz, M.C., Pagulayan, O., et al., 2000. The Tagum study I: analysis and clinical correlates of mercury in maternal and cord blood, breast milk, meconium, and infants' hair. Pediatrics 106, 774–781.

Randerath, K., Randerath, E., 1994. ^{32}P-postlabelling methods for DNA adduct detection: overview and critical evaluation. Drug Metab. Rev. 26, 67–85.

Randerath, K., Reddy, M.V., Gupta, R.C., 1981. 32P-labeling test for DNA damage. Proc. Natl. Acad. Sci. U.S.A. 78, 6126–6129.

Ranjit, N., Siefert, K., Padmanabhan, V., 2010. Bisphenol-A and disparities in birth outcomes: a review and directions for future research. J. Perinatol. 30, 2–9.

Reddy, M.V., Kenny, P.C., Randerath, K., 1990. ^{32}P-assay of DNA adducts in white blood cells and placentas of pregnant women: lack of residential wood combustion-related adducts but presence of tissue-specific endogenous adducts. Teratog. Carcinog. Mutagen. 10, 373–384.

Reed, E., Ozols, R.F., Tarone, R., et al., 1988. The measurement of cisplatin-DNA adduct levels in testicular cancer patients. Carcinogenesis 9, 1909–1911.

Reichrtova, E., Dorociak, F., Palkovicova, L., 1998. Sites of lead and nickel accumulation in the placental tissue. Hum. Exp. Toxicol. 17, 176–181.

Reseland, J.E., Mundal, H.H., Hollung, K., et al., 2005. Cigarette smoking may reduce plasma leptin concentration via catecholamines. Prostagl. Leukot. Essent. Fat. Acids 73, 43–49.

Richard, S., Moslemi, S., Sipahutar, H., et al., 2005. Differential effects of glyphosate and roundup on human placental cells and aromatase. Environ. Health Perspect. 113, 716–720.

Richter, D.U., Mylonas, I., Toth, B., et al., 2009. Effects of phytoestrogens genistein and daidzein on progesterone and estrogen (estradiol) production of human term trophoblast cells in vitro. Gynecol. Endocrinol. 25, 32–38.

Ridano, M.E., Racca, A.C., Flores-Martín, J., 2012. Chlorpyrifos modifies the expression of genes involved in human placental function. Reprod. Toxicol. 33, 331–338.

Robins, J.C., Marsit, C.J., Padbury, J.F., et al., 2011. Endocrine disruptors, environmental oxygen, epigenetics and pregnancy. Front. Biosci. 3, 690–700.

Rolfo, A., Garcia, J., Todros, T., et al., 2012. The double life of MULE in preeclamptic and IUGR placentae. Cell Death Dis. 3, e305.

Romani, F., Lanzone, A., Tropea, A., et al., 2011. Nicotine and cotinine affect the release of vasoactive factors by trophoblast cells and human umbilical vein endothelial cells. Placenta 32, 153–160.

Ronco, A.M., Arguello, G., Suazo, M., et al., 2005. Increased levels of metallothionein in placenta of smokers. Toxicology 208, 133–139.

Ronco, A.M., Garrido, F., Llanos, M.N., 2006. Smoking specifically induces metallothionein-2 isoform in human placenta at term. Toxicology 223, 46–53.

Rosario, F.J., Sadovsky, Y., Jansson, T., 2012. Gene targeting in primary human trophoblasts. Placenta 33, 754–762.

Rosner, H., Rohrmann, B., Peiker, G., 2000. Ochratoxin A in human serum. Arch. Leb. 51, 104–107.

Saberi, M.R., Vinh, T.K., Yee, S.W., et al., 2006. Potent CYP19 (aromatase) 1-[(benzofuran-2-yl)(phenylmethyl)pyridine, -imidazole, and -triazole inhibitors: synthesis and biological evaluation. J. Med. Chem. 49, 1016–1022.

Sadler, T.W., 2004. Langman's Medical Embryology, ninth ed. Lippincott Williams & Wilkins, Baltimore, USA.

Saenen, N.D., Vrijens, K., Janssen, B.G., et al., 2017. Lower placental leptin promoter methylation in association with fine particulate matter air pollution during pregnancy and placental nitrosative stress at birth in the ENVIRONAGE cohort. Environ. Health Perspect. 125, 262–268.

Sanyal, M.K., Li, Y.L., Belanger, K., 1994. Metabolism of polynuclear aromatic hydrocarbon in human term placenta influenced by cigarette smoke exposure. Reprod. Toxicol. 5, 411–418.

Sanyal, M.K., Mercan, D., Belanger, K., et al., 2007. DNA adducts in human placenta exposed to ambient environment and passive cigarette smoke during pregnancy. Birth Defects Res. A Clin. Mol. Teratol. 79, 289–294.

Saulsbury, M.D., Heyliger, S.O., Wang, K., et al., 2008. Characterization of chlorpyrifos-induced apoptosis in placental cells. Toxicology 244, 98–110.

Sawada, M., Kitamura, R., Norose, T., et al., 1993. Metabolic activation of aflatoxin B1 by human placental microsomes. J. Toxicol. Sci. 18, 129–132.

Sbrana, E., Suter, M.A., Abramovici, A.R., et al., 2011. Maternal tobacco use is associated with increased markers of oxidative stress in the placenta. Am. J. Obstet. Gynecol. 205, 246.e1-7.

Schanton, M., Maymó, J.L., Pérez-Pérez, A., et al., 2018. Involvement of leptin in the molecular physiology of the placenta. Reproduction 155, R1–R12.

Schempf, A.H., 2007. Illicit drug use and neonatal outcomes: a critical review. Obstet. Gynecol. Surv. 62, 749–757.

Schneider, H., Panigel, M., Dancis, J., 1972. Transfer across the perfused human placenta of antipyrine, sodium and leucine. Am. J. Obstet. Gynecol. 114, 822–828.

Schönfelder, G., Wittfoht, W., Hopp, H., et al., 2002. Parent bisphenol A accumulation in the human maternal-fetal-placental unit. Environ. Health Perspect. 110, A703–A707.

Seckl, J.R., Meaney, M.J., 2004. Glucocorticoid programming. Ann. N. Y. Acad. Sci. 1032, 63–84.

Sengpiel, V., Elind, E., Bacelis, J., et al., 2013. Maternal caffeine intake during pregnancy is associated with birth weight but not with gestational length: results from a large prospective observational cohort study. BMC Med. 11, 42.

Serafim, A., Company, R., Lopes, B., et al., 2012. Assessment of essential and nonessential metals and different metal exposure biomarkers in the human placenta in a population from the south of Portugal. J. Toxicol. Environ. Health A 75, 867–877.

Sexton, K., Salinas, J.J., McDonald, T.J., et al., 2011. Polycyclic aromatic hydrocarbons in maternal and umbilical cord blood from pregnant Hispanic women living in Brownsville, Texas. Int. J. Environ. Res. Publ. Health 8, 3365–3379.

Shih, M.C., Chiu, Y.N., Hu, M.C., et al., 2011. Regulation of steroid production: analysis of Cyp11a1 promoter. Mol. Cell. Endocrinol. 336, 80–84.

Shuker, D.E., Prevost, V., Friesen, M.D., et al., 1993. Urinary markers for measuring exposure to endogenous and exogenous alkylating agents and precursors. Environ. Health Perspect. 99, 33–37.

Siafakas, A.R., Richardson, D.R., 2009. Growth arrest and DNA damage-45 alpha (GADD45α). Int. J. Biochem. Cell Biol. 41, 986–989.

Sitras, V., Fenton, C., Paulssen, R., et al., 2012. Differences in gene expression between first and third trimester human placenta: a microarray study. PLoS One 7, e33294.

Sorkun, H.C., Bir, F., Akbulut, M., et al., 2007. The effects of air pollution and smoking on placental cadmium, zinc concentration and metallothionein expression. Toxicology 238, 15–22.

Spellacy, W.N., Buhi, W.C., Birk, S.A., 1977. The effect of smoking on serum human placental lactogen levels. Am. J. Obstet. Gynecol. 127, 232–234.

Spencer-Beach, G.G., Beach, A.C., Gupta, R.C., 1996. High-resolution anion-exchange and partition thin-layer chromatography for complex mixtures of 32P-postlabeled DNA adducts. J. Chromatogr. B Biomed. Appl. 677, 265–273.

Srám, R.J., Binková, B., Rössner, P., et al., 1999. Adverse reproductive outcomes from exposure to environmental mutagens. Mutat. Res. 1–2, 203–215.

Stasenko, S., Bradford, E.M., Piasek, M., et al., 2010. Metals in human placenta: focus on the effects of cadmium on steroid hormones and leptin. J. Appl. Toxicol. 30, 242–253.
Staud, F., Cerveny, L., Ceckova, M., 2012. Pharmacotherapy in pregnancy; effect of ABC and SLC transporters on drug transport across the placenta and fetal drug exposure. J. Drug Target. 20, 736–763.
Stejskalova, L., Pavek, P., 2011. The function of cytochrome P450 1A1 enzyme (CYP1A1) and aryl hydrocarbon receptor (AhR) in the placenta. Curr. Pharmaceut. Biotechnol. 12, 715–730.
Storvik, M., Huuskonen, P., Kyllönen, T., et al., 2011. Aflatoxin B1 – a potential endocrine disruptor – up-regulates CYP19A1 in JEG-3 cells. Toxicol. Lett. 202, 161–167.
Storvik, M., Huuskonen, P., Pehkonen, P., et al., 2014. The unique characteristics of the transcription of hormonal metabolism enzymes in placenta. Reprod. Toxicol. 47, 9–14.
Sun, T., Zhao, Y., Mangelsdorf, D.J., et al., 1998. Characterization of a region upstream of exon I.1 of the human CYP19 (aromatase) gene that mediates regulation by retinoids in human choriocarcinoma cells. Endocrinology 139, 1684–1691.
Suter, M., Abramovici, A., Showalter, L., et al., 2010. In utero tobacco exposure epigenetically modifies placental CYP1A1 expression. Metabolism 59, 1481–1490.
Suzuki, G., Nakano, M., Nakano, S., 2005. Distribution of PCDDs/PCDFs and Co-PCBs in human maternal blood, cord blood, placenta, milk, and adipose tissue: dioxins showing high toxic equivalence factor accumulate in the placenta. Biosci. Biotechnol. Biochem. 69, 1836–1847.
Taylor, S.M., Heron, A.E., Cannell, G.R., et al., 1994. Pressor effect of ethanol in the isolated perfused human placental lobule. Eur. J. Pharmacol. 270, 371–374.
Thévenod, F., 2010. Catch me if you can! Novel aspects of cadmium transport in mammalian cells. Biometals 23, 857–875.
Topinka, J., Binkova, B., Mrackova, G., et al., 1997. DNA adducts in human placenta as related to air pollution and to GSTM1 genotype. Mutat. Res. 390, 59–68.
Tsang, H., Cheung, T.Y., Kodithuwakku, S.P., et al., 2012. 2,3,7,8-Tetrachlorodibenzo-p-dioxin (TCDD) suppresses spheroids attachment on endometrial epithelial cells through the down-regulation of the Wnt-signaling pathway. Reprod. Toxicol. 33, 60–66.
Tsang, H.L., Wu, S., Leung, C.K., et al., 2011. Body burden of POPs of Hong Kong residents, based on human milk, maternal and cord serum. Environ. Int. 37, 142–151.
Tuckey, R.C., 2005. Progesterone synthesis by the human placenta. Placenta 26, 273–281.
Turner, P.C., Collinson, A.C., Cheung, Y.B., et al., 2007. Aflatoxin exposure in utero causes growth faltering in Gambian infants. Int. J. Epidemiol. 36, 1119–1125.
Uddin, M.N., Horvat, D., DeMorrow, S., et al., 2011. Marinobufagenin is an upstream modulator of Gadd45a stress signaling in preeclampsia. Biochim. Biophys. Acta 1812, 49–58.
Unuvar, E., Ahmadov, H., Kiziler, A.R., et al., 2007. Mercury levels in cord blood and meconium of healthy newborns and venous blood of their mothers: clinical, prospective cohort study. Sci. Total Environ. 374, 60–70.
Vähäkangas, K., 2003. Molecular epidemiology of human cancer risk: gene-environment interactions and p53 mutation spectrum in human lung cancer. Meth. Mol. Med. 74, 43–45.
Vähäkangas, K., 2008. Molecular epidemiology and ethics. Biomarkers of disease susceptibility. In: Wild, C., Vineis, P., Garte, S. (Eds.), Molecular Epidemiology of Chronic Diseases. John Wiley & Sons, Ltd, pp. 279–295.
Vähäkangas, K., Myllynen, P., 2006. Experimental methods to study human transplacental exposure to genotoxic agents. Mutat. Res. 608, 129–135.
Vähäkangas, K., Myllynen, P., 2009. Drug transporters in the human blood-placental barrier. Br. J. Pharmacol. 158, 665–678.
Vähäkangas, K., Haugen, A., Harris, C.C., 1985. An applied synchronous fluorescence spectrophotometric assay to study benzo[a]pyrene-diolepoxide-DNA adducts. Carcinogenesis 6, 1109–1115.
Vähäkangas, K., Raunio, H., Pasanen, M., et al., 1989. Comparison of the formation of benzo[a]pyrene diolepoxide-DNA adducts in vitro by rat and human microsomes: evidence for the involvement of P-450IA1 and P-450IA2. J. Biochem. Toxicol. 4, 79–86.
Vähäkangas, K.H., Veid, J., Karttunen, V., et al., 2011. The significance of ABC transporters in human placenta for the exposure of the fetus to xenobiotics. In: Gupta, R. (Ed.), Reproductive and Developmental Toxicology. Academic Press/Elsevier, Amsterdam, pp. 1051–1065.
Valbonesi, P., Ricci, L., Franzellitti, S., et al., 2008. Effects of cadmium on MAPK signaling pathways and HSP70 expression in a human trophoblast cell line. Placenta 29, 725–733.
van den Hooven, E.H., Pierik, F.H., de Kluizenaar, Y., et al., 2012. Air pollution exposure and markers of placental growth and function: the generation R study. Environ. Health Perspect. 120, 1753–1759.
van Meeuwen, J.A., van Son, O., Piersma, A.H., et al., 2008. Aromatase inhibiting and combined estrogenic effects of parabens and estrogenic effects of other additives in cosmetics. Toxicol. Appl. Pharmacol. 230, 372–382.
Vaught, J.B., Gurtoo, H.L., Parker, N.B., et al., 1979. Effect of smoking on benzo(a)pyrene metabolism by human placental microsomes. Cancer Res. 839, 3177–3183.
Vera, B., Santa Cruz, S., Magnarelli, G., 2012. Plasma cholinesterase and carboxylesterase activities and nuclear and mitochondrial lipid composition of human placenta associated with maternal exposure to pesticides. Reprod. Toxicol. 34, 402–407.
Virtanen, H.E., Koskenniemi, J.J., Sundqvist, E., et al., 2012. Associations between congenital cryptorchidism in newborn boys and levels of dioxins and PCBs in placenta. Int. J. Androl. 35, 283–293.
Vo-Dinh, T., Alarie, J.P., Johnson, R.W., et al., 1991. Evaluation of the fiber-optic antibody-based fluoroimmunosensor for DNA adducts in human placenta samples. Clin. Chem. 37, 532–535.
von Stedingk, H., Vikström, A.C., Rydberg, P., et al., 2011. Analysis of hemoglobin adducts from acrylamide, glycidamide, and ethylene oxide in paired mother/cord blood samples from Denmark. Chem. Res. Toxicol. 24, 1957–1965.
Vyhlidal, C.A., Riffel, A.K., Haley, K.J., et al., 2012. Cotinine in human placenta predicts induction of gene expression in fetal tissues. Drug Metab. Dispos. 41, 305–311.
Wade, J.V., Agrawal, P.R., Poisner, A.M., 1986. Induction of metallothionein in a human trophoblast cell line by cadmium and zinc. Life Sci. 39, 1361–1366.
Wagner, C.L., Katikaneni, L.D., Cox, T.H., et al., 1998. The impact of prenatal drug exposure on the neonate. Obstet. Gynecol. Clin. N. Am. 25, 169–194.
Wallace, B.D., Redinbo, M.R., 2013. Xenobiotic-sensing nuclear receptors involved in drug metabolism: a structural perspective. Drug Metab. Rev. 45, 79–100.
Wan, Y., Choi, K., Kim, S., et al., 2010. Wiseman S, Hydroxylated polybrominated diphenyl ethers and bisphenol A in pregnant women and their matching fetuses: placental transfer and potential risks. Environ. Sci. Technol. 44, 5233–5239.
Wang, W., Li, J., Ge, Y., et al., 2012. Cortisol induces aromatase expression in human placental syncytiotrophoblasts through the cAMP/Sp1 pathway. Endocrinology 153, 2012–2022.
Watanabe, M., Ohno, S., Nakajin, S., 2012. Effects of bisphenol A on the expression of cytochrome P450 aromatase (CYP19) in human fetal osteoblastic and granulosa cell-like cell lines. Toxicol. Lett. 210, 95–99.
Welch, R.M., Harrison, Y.E., Conney, A.H., et al., 1968. Cigarette smoking: stimulatory effect on metabolism of 3,4-benzpyrene by enzymes in human placenta. Science 160, 541–542.

Whyatt, R.M., Bell, D.A., Jedrychowski, W., et al., 1998. Polycyclic aromatic hydrocarbon-DNA adducts in human placenta and modulation by CYP1A1 induction and genotype. Carcinogenesis 19, 1389–1392.

Whyatt, R.M., Barr, D.B., Camann, D.E., et al., 2003. Contemporary-use pesticides in personal air samples during pregnancy and blood samples at delivery among urban minority mothers and newborns. Environ. Health Perspect. 111, 749–756.

Whyatt, R.M., Camann, D., Perera, F.P., et al., 2005. Biomarkers in assessing residential insecticide exposures during pregnancy and effects on fetal growth. Toxicol. Appl. Pharmacol. 206, 246–254.

Wojtowicz, A.K., Augustowska, K., Gregoraszczuk, E.L., 2007. The short- and long-term effects of two isomers of DDT and their metabolite DDE on hormone secretion and survival of human choriocarcinoma JEG-3 cells. Pharmacol. Rep. 59, 224–232.

Wójtowicz, A.K., Milewicz, T., Gregoraszczuk, E.Ł., 2007. DDT and its metabolite DDE alter steroid hormone secretion in human term placental explants by regulation of aromatase activity. Toxicol. Lett. 173, 24–30.

Wójtowicz, A.K., Milewicz, T., Gregoraszczuk, E.Ł., 2008. Time-dependent action of DDT (1,1,1-trichloro-2,2-bis(p-chlorophenyl) ethane) and its metabolite DDE (1,1-dichloro-2,2-bis(p-chlorophenyl)ethylene) on human chorionic gonadotropin and progesterone secretion. Gynecol. Endocrinol. 24, 54–58.

Woo, C.S., Partanen, H., Myllynen, P., et al., 2012. Fate of the teratogenic and carcinogenic ochratoxin A in human perfused placenta. Toxicol. Lett. 208, 92–99.

Wood, R.J., 2009. Manganese and birth outcome. Nutr. Rev. 67, 416–420.

Wu, M., Yan, C., Xu, J., et al., 2013. Umbilical cord blood mercury levels in China. J. Environ. Sci. 25, 386–392.

Xiong, Y., Liebermann, D.A., Holtzman, E.J., et al., 2013. Preeclampsia-associated stresses activate Gadd45a signaling and sFlt-1 in placental explants. J. Cell. Physiol. 228, 362–370.

Xiong, Y., Liebermann, D.A., Tront, J.S., et al., 2009. Gadd45a stress signaling regulates sFlt-I expression in preeclampsia. J. Cell. Physiol. 220, 632–639.

Xu, P., Wu, Z., Xi, Y., et al., Nov 30 2016. Epigenetic regulation of placental glucose transporters mediates maternal cadmium-induced fetal growth restriction. Toxicology 372, 34–41 pii: S0300–483X(16)30255.

Xue, W., Warshawsky, D., 2005. Metabolic activation of polycyclic and heterocyclic aromatic hydrocarbons and DNA damage: a review. Toxicol. Appl. Pharmacol. 206, 73–93.

Yahiaoui, S., Pouget, C., Fagnere, C., et al., 2004. Synthesis and evaluation of 4-triazolylflavans as new aromatase inhibitors. Bioorg. Med. Chem. Lett 14, 5215–5218.

Yamada, H., Furuta, I., Kato, E.H., et al., 2002. Maternal serum and amniotic fluid bisphenol A concentrations in the early second trimester. Reprod. Toxicol. 16, 735–739.

Yang, C.H., Glover, K.P., Han, X., 2010. Characterization of cellular uptake of perfluorooctanoate via organic anion-transporting polypeptide 1A2, organic anion transporter 4, and urate transporter 1 for their potential roles in mediating human renal reabsorption of perfluorocarboxylates. Toxicol. Sci. 117, 294–302.

Yeboah, D., Sun, M., Kingdom, D., et al., 2006. Expression of breast cancer resistance protein (BCRP/ABCG2) in human placenta throughout gestation and at term before and after labor. Can. J. Physiol. Pharmacol. 84, 1251–1258.

Yoshida, M., 2002. Placental to fetal transfer of mercury and fetotoxicity. Tohoku J. Exp. Med. 196, 79–88.

Yoshino, Y., Yuan, B., Kaise, T., et al., 2011. Contribution of aquaporin 9 and multidrug resistance-associated protein 2 to differential sensitivity to arsenite between primary cultured chorion and amnion cells prepared from human fetal membranes. Toxicol. Appl. Pharmacol. 257, 198–208.

Yu, Y., Wang, X., Wang, B., et al., 2011. Polycyclic aromatic hydrocarbon residues in human milk, placenta, and umbilical cord blood in Beijing, China. Environ. Sci. Technol. 45, 10235–10242.

Yuen, R.K.C., Robinson, W.P., 2011. Review: a high capacity of the human placenta for genetic and epigenetic variation: implications for assessing pregnancy outcome. Placenta Suppl. 2 (25), S136–S141.

Yuen, R.K., Chen, B., Blair, J.D., et al., 2013. Hypoxia alters the epigenetic profile in cultured human placental trophoblasts. Epigenetics 8, 192–202.

Zdravkovic, T., Genbacev, O., McMaster, M.T., et al., 2005. The adverse effects of maternal smoking on the human placenta: a review. Placenta 26 (Suppl. A), S81–S86.

Zechner, U., Pliushch, G., Schneider, E., et al., 2010. Quantitative methylation analysis of developmentally important genes in human pregnancy losses after ART and spontaneous conception. Mol. Hum. Reprod. 16, 704–713.

Zhang, L., Shiverick, K.T., 1997. Benzo(a)pyrene, but not 2,3,7,8-tetrachlorodibenzo-p-dioxin, alters cell proliferation and c-Myc and growth factor expression in human placental choriocarcinoma JEG-3 cells. Biochem. Biophys. Res. Commun. 231, 117–120.

Zhang, J., Liu, L., Mu, X., et al., 2011a. Effect of triptolide on aromatase activity in human placental microsomes and human placental JEG-3 cells. Arzneimittelforschung 61, 727–733.

Zhang, Q., Zhou, T., Xu, X., et al., 2011b. Downregulation of placental S100P is associated with cadmium exposure in Guiyu, an e-waste recycling town in China. Sci. Total Environ. 410–411, 53–58.

Zharikova, O.L., Deshmukh, S.V., Nanovskaya, T.N., et al., 2006. The effect of methadone and buprenorphine on human placental aromatase. Biochem. Pharmacol. 71, 1255–1264.

Zhong, W., Hee, S.S., 2009. Comparison of UV, fluorescence, and electrochemical detectors for the analysis of formaldehydeinduced DNA adducts. J. Anal. Toxicol. 29, 182–187.

Zhou, H., Fu, G., Yu, H., et al., 2009. Transforming growth factor-beta inhibits aromatase gene transcription in human trophoblast cells via the Smad2 signaling pathway. Reprod. Biol. Endocrinol. 7, 146.

Zhou, J., Liu, F., Yu, L., et al., 2018. nAChRs-ERK1/2-Egr-1 signaling participates in the developmental toxicity of nicotine by epigenetically down-regulating placental 11β-HSD2. Toxicol. Appl. Pharmacol. 344, 1–12.

Zimmerli, B., Dick, R., 1995. Determination of ochratoxin A at the ppt level in human blood, serum, milk and some foodstuffs by high-performance liquid chromatography with enhanced fluorescence detection and immunoaffinity column cleanup: methodology and Swiss data. J. Chromatogr. B Biomed. Appl. 666, 85–99.

CHAPTER

19

Early Biomarkers of Acute and Chronic Pancreatitis

Bhupendra S. Kaphalia

Department of Pathology, University of Texas Medical Branch, Galveston, TX, United States

INTRODUCTION

Early detection and prevention of pancreatic injury progressing to inflammation and fibrosis, maybe pancreatic cancer (PC), should be critical for the success of both clinical and surgical interventional therapies. Although a significant literature available on biomarkers of pancreatic diseases is focused on various forms of PC, identification of biomarkers for early pancreatic injury and precancerous stages should be the key to early prevention and therapy. Therefore, summarizing reliable and probable biomarkers identified for acute and chronic pancreatic toxicities/pancreatitis rather than PC is the focus of this chapter.

Both biliary duct disease and chronic alcohol abuse constitute 70%–75% of the etiologies of acute and chronic pancreatitis; remaining one-third may be idiopathic including those caused by chemical exposure, pesticide and metal poisoning, infections (viral, bacterial, and fungal), autoimmune and anatomical conditions, and the use of certain drugs. Congenital abnormalities, generally rare and asymptomatic, are related to two important events during embryological development of the gland, rotation and fusion. Anatomical disorders, deposition of cholesterol and related substances, obstruction of pancreatic duct due to tumor, and cyst formation could be rare (Lankisch and Banks, 1998; Kaphalia, 2011).

Pancreatic toxicity generally characterized by dysregulation of lipid metabolism and edema in early reversible stages followed by massive necrosis resulting into inflammation with or without fibrosis (scaring of the tissue) at the advanced stages. Irrespective of etiological agents, sequel of pancreatic injury follows a similar pattern for most of the etiologies. Some patients with pancreatitis can also develop PC. Overall, chronic pancreatitis and PC are serious diseases with great social, economic, and psychological impact and significant morbidity. The prognosis of pancreatitis and PC is very poor, and many patients die before reaching the clinical stage of the disease. Therefore, identification of biomarkers for prepancreatitis and postpancreatitis stages could be important for clinical interventions.

ANATOMICAL, PHYSIOLOGICAL, AND METABOLIC CONSIDERATIONS FOR PANCREATIC INJURY

Pancreas is an important digestive and glandular organ in vertebrates consisting of endocrine and exocrine components. About 85% of the gland consists of exocrine pancreas populated by acinar cells and a few duct cells. While endocrine pancreas secretes insulin hormone needed to control the body's glucose, exocrine pancreas synthesizes and stores zymogens (inactive form of digestive enzymes such as trypsinogen) required for the digestion of food. The digestive enzymes are secreted in the upper part of the intestine and activated by a gut hormone (enterokinase) for the digestion of food as detailed in several text books and reviews. An activation of zymogens within exocrine acinar cells in the gland triggers massive necrosis followed by inflammation with or without fibrosis, commonly termed as pancreatitis, and classified as acute pancreatitis and chronic pancreatitis.

ACUTE PANCREATITIS

Clinically, acute pancreatitis is characterized by an acute abdominal pain accompanied by elevated

Biomarkers in Toxicology, Second Edition
https://doi.org/10.1016/B978-0-12-814655-2.00019-0

pancreatic enzymes in the blood or urine, or both. Acute pancreatitis could be a single episode, or it may reoccur and is induced by the release of activated proteases from the pancreatic acinar cells and is often hemorrhagic. It occurs suddenly and lasts for a short period of time and usually resolves by itself in most of the cases. Severity of acute pancreatitis depends on histologic evaluation and may vary from mild (fat necrosis with enlarged gland [edema]) to severe (large confluent foci fat necrosis around the pancreas, focal hemorrhage, and acinar cell necrosis).

CHRONIC PANCREATITIS

Chronic or calcifying pancreatitis is a continuing inflammatory response characterized by severe morphological changes (such as irregular sclerosis and permanent loss of exocrine parenchyma), which may be focal, segmental, or diffused. Clinically, chronic pancreatitis is characterized by recurrent or persisting abdominal pain, although chronic pancreatitis may also present without pain. Chronic pancreatitis does not resolve by itself and could progress to a slow destruction of the pancreatic gland. Irrespective of etiology, the clinical pattern of chronic pancreatitis is characterized by the recurrent episodes of acute pancreatitis in early stages followed by pancreatic insufficiency, steatorrhea, pancreatic calcification and may be diabetes mellitus at the chronic stage. However, an intra-acinar activation of zymogens in the gland itself is the primary cause of pancreatic injury and pancreatitis (Lankisch and Banks, 1998).

BIOMOLECULAR BASIS OF PANCREATITIS

A large number of digestive enzymes are synthesized and stored by exocrine pancreas to be released into the upper part of intestine for digestion. Metabolic composition of the pancreas, particularly exocrine pancreas, has been previously detailed (Kaphalia, 2011). Activation of zymogens within the acinar cells is potentially damaging to the gland and causes autodigestion of exocrine pancreas and surrounding tissue. This process involves several extra-acinar cellular events, both in the pancreas and elsewhere in the body, due to the generation of inflammatory mediators such as cytokines, chemokines, and growth factors. It is also generally known that lipid degradation byproducts produced due to oxidative stress are involved in various target organ pathogenesis. Clinically, disease can be evident by noninvasive tests and by conventional physical and clinical symptoms only after inflammation (pancreatitis) has occurred due to a massive necrotic cell death. Having both exocrine and endocrine components in the pancreatic organ system, many patients with chronic pancreatitis also present overt diabetes mellitus as manifestation of either exocrine or endocrine pancreatic insufficiencies, or both. It is possible to have combined biomarkers of endocrine and exocrine pancreatic injury to evaluate chronic and advanced stages of pancreatitis and fibrosis, which involve both components of the gland. Therefore, identification and development of biomarkers of early pancreatic injury even before inflammation should have great translational and clinical benefits for interventional therapies.

Like other key gastrointestinal organs, exocrine pancreas is also metabolically an active organ, and expresses various phase I and phase II enzymes [oxidative, reductive and conjugation enzymes including fatty acid ethyl ester (FAEE) synthase] (Kaphalia, 2011). Such metabolic organization of the gland might also bioactivate and/or biotransform drugs and chemicals damaging to the gland itself. Therefore, certain proteins and peptides specific to the exocrine pancreas secreted in to the blood and/or excreted via urine during injury can be reliable and specific biomarkers of pancreatic injury/toxicity/inflammation.

BIOMARKERS OF ACUTE AND CHRONIC PANCREATITIS

Clinically, two types of biomarkers described are (1) those specific and related to target organ function detected in the blood and excreted through urine are termed as endogenous and (2) various chemical(s)/agent(s) and their metabolite(s) and infectious agents involved in initiation and progression of the disease are called exogenous biomarkers. Except for known cases of chemical poisoning and history of abuse (alcohol or drug) or occupational exposure, it is generally very difficult to pin point the agent responsible for the injury. In such cases patient information could be critical. Global approaches such as proteomics, metabolomics, and genomics could be important for biomarker discovery, but such approaches are generally not cost-effective and may require confirmation using Reverse transcription-polymerase chain reaction (RT-PCR), Liquid chromatography-mass spectrometry (LC-MS), and/or Gas chromatography-mass spectrometry (GC-MS) analyses.

A biomarker candidate should be measurable via noninvasive methods. Several such matrices as plasma, serum, urine, saliva, hair, feces, or sweat can be used for identification and levels of the etiologic agent(s) and its metabolite(s). Several biomarker candidates identified in experimental acute pancreatitis models can be developed using translational research approaches. Physical examination, imaging (endoscopic ultrasound, chest radiography, and barium sulfate X-rays), direct pancreatic function tests, and analysis of serum/plasma/urine for markers of pancreatic injury/pancreatitis should be gold standard for routine and less-expensive diagnosis (Table 19.1).

ABDOMINAL IMAGING TECHNIQUES

Abdominal ultrasound can be done with routine evaluation of acute pancreatitis; however, severity of pancreatitis can rarely be ascertained by the abdominal ultrasound. A less-invasive endoscopic ultrasonography and computed tomography (CT) and more invasive endoscopic retrograde cholangiopancreatography (ERCP) are better diagnostics for the gallstones in the common bile duct. ERCP is used primarily to diagnose and treat conditions of the bile ducts and main pancreatic duct, including gallstones, although the development of safer and relatively noninvasive investigations such as magnetic resonance cholangiopancreatography (MRCP), and endoscopic ultrasound has meant that ERCP is now rarely performed without therapeutic intent. Multimodal imaging systems such as single-photon emission computed tomography (SPECT)–CT, positron emission tomography (PET)–CT, and PET–magnetic resonance imaging (MRI) combine the best of both conventional diagnostic and molecular modalities, with the high resolution and high sensitivity of molecular imaging paired with anatomical or functional information of conventional diagnostic imaging. Currently, endoscopic ultrasound, CT, MRI, and ultrasonography form the conventional core imaging methodologies for pancreatic diseases (Dimastromatteo et al., 2017). Therefore, patient history and imaging results should be meaningful along with markers of pancreatic function tests in the plasma, urine, and/or saliva keeping in view the acute and chronic nature of the disease.

FUNCTIONAL BIOMARKERS OF ENDOCRINE AND EXOCRINE PANCREATITIS

Most available biomarkers of pancreatic injury are limited to acute pancreatitis. Markers of chronic pancreatitis necessarily do not correspond to those identified for acute pancreatitis. Elevated titers of autoantibodies directed against amylase α-2A are suggested a novel specific serologic biomarker to help identify patients at risk for autoimmune pancreatitis (AIP) and

TABLE 19.1 Biomarkers of Acute and Chronic Pancreatitis

Matrix	Biomarkers and/or Biomarker Candidates
Serum/plasma	**Indirect pancreatic functional biomarkers** Amylase, isoamylase, and total amylase Lipase Carbohydrate-deficient transferrin Trypsinogen activation products • Trypsinogen activation peptide (TAP) • Trypsin-1 and 2 and their conjugates with α-antitrypsin (trypsin-1-AAT and trypsin-2-AAT)
	C-reactive protein (CRP) Elastase-1 and phospholipase A_2 Procalcitonin (precursor of calcitonin) Proteins (proteomics, 2-dimensional gel electrophoresis (DGE), and MALDI-TOF/TOF) Lipids (lipidomics, NMR, GC-MS, and LC-MS) Metabolites (small-molecule metabolomics using LC-MS and GC-MS) Inflammatory cytokines and chemokines microRNAs (miR-216a and miR-217) Peptides (RA1609 and RT2864) IgG4 d-Dimer and angiopoietin-2
	Direct analysis of suspected etiologic agent(s) Alcohol and alcohol metabolites (FAEEs, EtS, EtG, and PEt), pesticides and metal poisoning and suicide cases, drugs and occupational exposure to chemicals), and infectious agents
Urinary	Amylase, isoamylase, TAP, CAPAP, phospholipase A_2
Fecal	Elastase-1
Pancreatic juice	Cathepsins B, L, and S (lysosomal hydrolases) (CAPAP)
Pancreatic tissue specimens	Differential expression of genes, proteins and microRNA, and altered metabolomics, inflammatory cytokines and chemokines, and growth factors
Breath analysis	Breath analysis for H_2S, N_2O, and 66U substance
Direct visualization	**Imaging techniques** Endoscopic ultrasound, CT, ERCP, MRI, SPECT-CT, PET-CT, PET-MRI

Note: The biomarkers of acute pancreatitis may not be applicable for chronic pancreatitis, which can be confirmed by imaging techniques (ultrasound, CT, MRCP, and/or ERCP) in conjunction with clinical symptoms consistent with chronic pancreatitis.
CAPAP, carboxypeptidase activation peptide; *CT*, computed tomography; *EtG*, ethyl glucuronide; *EtS*, ethyl sulfate; *FAEE*, fatty acid ethyl ester; *ERCP*, endoscopic retrograde cholangiopancreatography; *MRCP*, magnetic resonance cholangiopancreatography; *MRI*, magnetic resonance imaging; *PEt*, phosphatidylethanol; *SPECT*, single-photon emission computed tomography; *PET*, positron emission tomography; *TAP*, trypsinogen activation peptide.

fulminant type 1 diabetes for early prevention and therapy (Wiley and Pietropaolo, 2009). However, low serum levels of CD44, CD44v6, and neopterin as indicators of immune dysfunction in chronic pancreatitis could not be specific or selective (Schlosser et al., 2001). Similarly, oxidative stress–related lipid peroxidation products proposed as biomarkers of acute pancreatitis may have limited success due to lack of specificity (Col et al., 2010). Correlation between lipid peroxidation products such as malondialdehyde in plasma/serum/urine with severity of acute pancreatitis need to be established. A number of biomarkers of pancreatitis (acute or chronic) identified in the plasma/serum are described as following:

AMYLASE

All amylases (glycoside hydrolases) catalyze the breakdown of complex sugars (such as starch) and act at α-1,4-glycosidic bonds. Pancreatic and salivary amylase can be of several orders of magnitude than those in other tissues/organs such as lungs, tears, sweat, and human milk. However, amylases can also be found as minor genetic variants. α-Amylase, faster acting than other isozymes, is the major form present in both human pancreas and salivary glands with a half-life time in the serum of ~10 h. The levels can return to the normal within 24 h after pancreatic injury is resolved and/or if enzyme is not being added to the serum (Nord et al., 1973).

Amylase levels can show a variable response depending on the etiologic agent(s) involved in the pancreatitis, and its lower levels have been reported on admission in patients with alcohol-induced acute pancreatitis. About 50% of the patients with abdominal pain and hyperlipidemia considered to have acute pancreatitis as confirmed by CT show no significant elevation of either serum or urinary amylase levels (Toskes, 1990). Therefore, sensitivity of serum amylase estimation compromises the gold standard for the diagnosis of acute pancreatitis and pancreatic trauma (Lankisch and Banks, 1998; Moridani and Bromberg, 2003; Herman et al., 2011). Total serum amylase levels can be altered by changes in either pancreatic or salivary amylases (Tietz, 1988). Although measuring total amylase after inhibition of salivary amylase is suggested being more accurate diagnostic marker, the assay adds up more steps and cost of the test. Even, reducing the amylase ordering has been suggested for the emergency departments (Volz et al., 2010). Macroamylasemia due to binding of amylase with serum globulin can also mislead the amylase levels (Wilding et al., 1964). The precision of total amylase versus pancreatic amylase appears to be identical (Moridani and Bromberg, 2003). Therefore, a straightforward approach to measure total amylase is cost-effective and saves time.

Elevated urinary amylase is also a sensitive indicator of acute pancreatitis and increased in almost 95% of patients with pancreatitis and remains elevated longer than the serum amylase activity. An elevated urinary amylase (>3 folds) than the upper limit of normal is a clear diagnosis of acute pancreatitis. However, elevated urine amylase can also be seen in salivary gland disease, bowel perforation, and ketoacidosis (Lankisch and Banks, 1998). An abnormal urinary amylase–creatinine clearance ratio can be a proof of acute pancreatitis. Although urinary amylase has not been widely used, elevated urinary amylase and serum amylase both can be considered as diagnostic biomarkers of acute pancreatitis.

ISOAMYLASE

Any of several isoenzymes of α-amylase catalyzes the hydrolysis of 1,6-α-glycosidic branch linkages in glycogen and amylopectin and their β-limit dextrins. The ratio of the amounts of pancreatic and salivary isoamylases in the urine is proposed as an index of insufficient exocrine pancreatic functions (Hobbs et al., 1967). However, urinary isoamylase levels have not been used by the clinics so far.

LIPASE

Unlike amylase, lipase mainly is of pancreatic origin; a small amount can also be produced in the liver, stomach, and tongue. Lipase catalyzes the breakdown and hydrolysis of fats and acts at specific position on the glycerol backbone of lipid substrate (A1, A2, or A3). Pancreatic lipase is the main enzyme that converts triglycerides to monoglycerides and fatty acids. As compared with serum amylase and other markers so far been used, serum lipase appears to be a sensitive marker for acute pancreatitis. Serum lipase is highly sensitive, can be detected for several days, and is sensitive even with normal amylase levels. Most hospitals use serum lipase for the diagnosis of acute pancreatitis, and its specificity is considered to be excellent and can significantly reduce the cost to the hospitals (Ismail and Bhayana, 2017). Evidence-based guidelines also recommend the use of serum lipase over amylase, although both lack ability to distinguish severity and etiology of acute pancreatitis. However, the diagnosis of acute pancreatitis may require at least two or three diagnostic criteria such as characteristic abdominal pain, elevated serum lipase and/or amylase, and radiological evidence of pancreatitis.

CARBOHYDRATE-DEFICIENT TRANSFERRIN

Transferrin, a serum protein, is a polypeptide with two *N*-linked polysaccharide chains, which are branched with sialic acid (monosaccharide carbohydrate) residues. Transferrin carries iron through the bloodstream to the bone marrow, as well as to the liver and spleen. Correlation between carbohydrate-deficient transferrin (CDT) and mean corpuscular volume has been reported statistically significant with excessive alcohol consumption (Basterra et al., 2001; Aparicio et al., 2001; Sharpe, 2001). Elevated CDT found in the blood of chronic alcoholics can also be seen in various medical conditions (ARIP Laboratories, 2010). The CDT levels are shown to be correlated with alcohol consumption and possibly to acute pancreatitis in alcoholic patients (Jaakkola et al., 1994). Overall, CDT appears to have some clinical utility but could not be used by clinics for the diagnosis of acute or chronic pancreatitis, probably due to specificity and cost of analysis reasons.

ACTIVATION BY-PRODUCTS OF PANCREATIC TRYPSINOGEN AND CARBOXYPEPTIDASE B

Proenzyme trypsinogen occurs in two major isoenzymes in humans (trypsinogen 1, i.e., cationic trypsinogen form, and trypsinogen 2, i.e., anionic trypsinogen form), which are activated to trypsin by enterokinase (EK) in the gut for protein digestion. However, activation of trypsinogen within acinar cell plays a key role in pathogenesis of acute pancreatitis because conversion of trypsinogen triggers a cascade of reactions that activates the remaining zymogens resulting in autodigestion of the surrounding tissue (Rinderknecht, 1986). Trypsin rapidly gets inactivated by 1-α-antitrypsin (AAT), a neutralization process. High serum concentrations of trypsin-1-AAT and trypsin-2-AAT have been reported in acute pancreatitis (Borgström and Ohlsson, 1978; Hedström et al., 1994).

Trypsinogen activation peptide (TAP) is a by-product of trypsinogen activation process. Trypsin is a serine protease found in the digestive system of many vertebrates, where it hydrolyzes proteins at carboxyl side of the amino acids, lysine or arginine. Trypsin produced due to premature trypsinogen activation in the pancreas ideally fulfills criterion of organ specificity of a biomarker. The activation peptide at the amino terminus of vertebrate trypsinogen contains the sequence Asp-Asp-Asp-Asp-Lys (D_4K, highly conserved during vertebrate evolution) as the carboxy-terminal moiety. These peptides are generally liberated in the proximal small intestine during digestion, after recognition and cleavage at the lysine carbonyl by a small gut hormone, EK.

Intra-acinar activation of trypsinogen releases TAP into circulation, which is rapidly cleared by kidneys for excretion in urine because of its small size. TAP can be quantitated by an enzyme-linked immunosorbent assay (ELISA) in the plasma and urine using specific antibody to TAP (Wu et al., 2008). Only early acute or severe pancreatitis as compared with mild acute pancreatitis correlated with serum and urinary levels of TAP, which are rapidly depleted to undetectable levels. A quick collection and analysis of samples needs attention and results may differ depending on etiology of acute pancreatitis (for e.g., alcohol- versus gallstone-induced [Lempinen et al., 2003]). Serum TAP appears to have close correlation to severity of pancreatic injury and can be used as an early marker of acute pancreatitis.

As compared with serum TAP, urinary TAP has more consistent relationship with early acute pancreatitis. It has been reported that plasma TAP increased immediately after the induction of pancreatitis as compared with delayed excretion of TAP in the urine by several hours in experimental pancreatitis model (Wang et al., 2001). About 30% of the patients with acute pancreatitis had normal TAP values on admission (Gudgeon et al., 1990). There are contradictory reports regarding the utility of serum and urinary TAP, and its levels can remain undetected in mild cases of acute pancreatitis. Therefore, TAP levels only reflect severe acute pancreatitis (Kylänpää-Bäck et al., 2002; Johnson et al., 2004). However, the methods of analysis of trypsin and TAP are time-consuming and difficult and also not cost-effective for the routine clinical use.

Likewise TAP, procarboxypeptidase B has an activation peptide, carboxypeptidase activation peptide (CAPAP), which is a larger peptide and more stable in the serum and urine than other peptides (Burgos et al., 1991; Abu Hilal et al., 2007; Deng et al., 2015). Serum and urinary levels of CAPAP were found to correlate well with severe acute pancreatitis based on a study conducted in a small number of patients (Appelros et al., 2001; Pezzilli et al., 2000). Both TAP and CAPAP could serve the markers of early and severe form of acute pancreatitis at the time of admission to hospitals. Both serum and urinary CAPAP have the potential to act as a stratification marker on admission in predicting severity of acute pancreatitis. Therefore, utility of both markers needs to be tested in larger cohorts.

ELASTASE AND PHOSPHOLIPASE A2

Pancreatic elastase is a form of elastase and a subfamily of serine proteases (referring to chymotrypsin-like elastase family, member 3B [CELA3B]) that hydrolyze many proteins in addition to elastin. Human pancreatic elastase-1 (E1) is quite stable and remains undegraded during intestinal transit. Therefore its concentration in feces reflects exocrine pancreatic function. During an inflammation of the pancreas, E1 is released into the bloodstream. Thus the quantification of pancreatic elastase-1 in serum and feces allows diagnosis or exclusion of acute pancreatitis (Domínguez-Muñoz et al., 2006; Naruse et al., 2006). Urinary excretion of elastase-1 has been shown to increase in all patients with chronic pancreatic disease regardless of neoplastic or inflammatory nature of the disease (Fabris et al., 1989). Unfortunately, determination of elastase-1 does not provide additional advantage over combination of lipase and amylase for the diagnosis of acute pancreatitis and may not be a specific biomarker candidate.

Phospholipase A2 specifically recognizes the sn-2 acyl bond of phospholipids and catalytically hydrolyzes the bond releasing arachidonic acid and lysophospholipids. However, phospholipase A2 can be found in several other organs and induced in chronic liver cirrhosis thus lacking target organ specificity (Vishwanath et al., 1996). Urinary excretion of phospholipase A2 in chronic pancreatic diseases particularly during relapse and in other physiological conditions has been reported (Fabris et al., 1992). Like trypsin and TAP, the elastase and phospholipase A2 increase in serum and feces in acute pancreatitis but could not be used as conventional diagnostic biomarkers for acute pancreatitis probably due to cost considerations and intricate analysis protocols.

PROCALCITONIN

Procalcitonin is a precursor of calcitonin and serum procalcitonin as measured using chemoluminescent immunoassay was found to be a simple promising biomarker with accuracy similar to acute physiology and chronic health examination (APACHE)-II score, in predicting severity of acute pancreatitis (Woo et al., 2011). However, precalcitonin can be useful only in predicting the severity of acute pancreatitis and comparable to other biomarkers such as APACHE-II, C-reactive protein (CRP), lactate dehydrogenase (LDH), and urea.

C-REACTIVE PROTEIN AND PANCREATITIS-ASSOCIATED PROTEIN

CRP, an acute-phase proteins primarily synthesized in the liver, is a class of proteins whose plasma concentrations increase (positive acute-phase proteins) or decrease (negative acute-phase proteins) in response to inflammation. Determination of the severity of acute pancreatitis is difficult in the early phase after onset and often encounters difficulties in making decisions to initiate intensive care during the early phase. CRP values increase significantly in early stages of necrotic pancreatitis (Imamura et al., 2002). It is an important prognostic marker of pancreatic necrosis with the highest sensitivity and negative prognostic value given the cutoff is at 110 mg/L. The patients with CRP values below 110 mg/L are at low risk to develop pancreatic necrosis (Barauskas et al., 2004). Measurement of acute-phase proteins, especially CRP, is a useful marker of inflammation in both medical and veterinary clinical pathology. It correlates with the erythrocyte sedimentation rate. However, CRP is a marker of inflammation (proinflammatory cytokines), and higher CRP may also indicate liver failure (Ananian et al., 2005).

Pancreatitis-associated protein (PAP), an acute-phase protein such as CRP, is overexpressed in acute pancreatitis and detected in pancreatic juice from 49 patients with chronic pancreatitis. However, higher level of PAP also detected in 15 control subjects discredited the significance of the PAP as a reliable biomarker of chronic pancreatitis (Motoo et al., 2001).

Serum CRP, procalcitonin, and LDH have been used for the diagnosis of pancreatic necrosis, but any conclusions regarding the diagnostic test accuracy could not be arrived due to methodological deficiencies and uncertainty of the results (Komolafe et al., 2017). Summary of available data on diagnostic usefulness of markers of endothelial dysfunction and activated coagulation in early prediction of severe acute pancreatitis has been reviewed using the results of experimental studies and clinical trials targeting coagulation–inflammation interactions in severe acute pancreatitis. Among laboratory tests, d-dimer and angiopoietin-2 measurements have been found useful predictors of severe acute pancreatitis (Dumnicka et al., 2017).

CATHEPSINS

Pancreatitis is shown to be associated with increase in active form of cathepsins B, L, and S, which regulate premature activation of trypsinogen within pancreatic acinar cells (Lyo et al., 2012). Cathepsins are lysosomal proteases and postulated to play a role in the initiation of pancreatitis (van Acker et al., 2006). The levels of aforementioned cathepsins have been shown to increase in the pancreatic juice from patients with chronic pancreatitis. In case of ethanol-induced pancreatic injury, cathepsins respond to FAEEs (nonoxidative metabolites of ethanol-induced pancreatic lysosomal fragility) (Haber et al., 1993). However, lack of specificity and analysis of cathepsins mostly using fluorimetric are simply not cost-effective (Barrett, 1980).

PEPTIDES

Walgren et al. (2007) identified two peptide markers (RA1609 and RT2864) of pancreatic toxicity that are target organ specific using a model pancreatic toxin cyanohydroxybutene. These markers were verified in the serum of patients diagnosed with pancreatitis, and changes in serum RA1609 and RT2864 were indicative of and specific to pancreatitis. The human sera analyzed in this study were from patients with pancreatitis, which was not defined as acute or chronic. Moreover, clinical use of such markers has yet to be established for a routine analysis.

INFLAMMATORY CYTOKINES AND CHEMOKINES

Necrosis of exocrine pancreas, the primary cause of acute and chronic pancreatitis, can be identified by contrast-enhanced computerized tomography and magnetic resonance imaging. Massive necrotic cell death follows strong inflammatory responses and infection. Proinflammatory cytokines and chemokines such TNF, IL1α, IL1β, IL6, IL8, IL12, and MCP-1 have been identified to serve as biomarkers of acute and chronic pancreatitis (Papachristou, 2008; Achur et al., 2010). However, ability to predict which patients will develop severe disease is limited. Large prospective studies are still needed to address these questions by identifying risk factors for acute pancreatitis and serum biomarkers of severe disease conditions.

Correlation of specific serum inflammatory cytokines and chemokines increased or decreased with severity of pancreatitis need to be established before embarking on cytokines and chemokines as potential biomarkers of acute or chronic pancreatitis. Chronic pancreatitis is chronic inflammatory disease. However, a caution should be exercised to use certain or combination of inflammatory cytokines and chemokines in view of broad inflammatory response in chronic diseases of diverse nature. Secondary target organ toxicity is also one of the potential interference in such determinations.

Various laboratory markers for predicting acute pancreatitis including cytokines, activation peptides of pancreatic proteases, antiproteases, adhesion molecules, and several leucocyte-derived enzymes have been reviewed time and again, but due to their low accuracy, use of cumbersome laboratory analysis techniques, and high cost of analysis, a quest for finding new markers of acute pancreatitis continues.

BIOMARKERS OF ENDOCRINE AND AUTOIMMUNE PANCREATITIS

Endocrine functional impairment with insulin resistance can be used as an index of endocrine pancreatitis (Wu et al., 2011). Type 1 diabetes mellitus (loss of insulin-secreting capacity) is now classified as autoimmune (Type 1A) or idiopathic (Type 1B). Presence or absence of glutamic acid decarboxylase antibodies can differentiate type 1 diabetes. Some patients with idiopathic type 1 diabetes may also have nonautoimmune and remarkably abrupt onset and high serum pancreatic enzyme levels (Imagawa et al., 2000). Presence of elevated titers of autoantibodies directed to amylase α-2A may represent a novel specific biomarker of AIP and fulminant type 1 diabetes. The characteristic features of AIP are increased IgG4, lymphoplasmacytic sclerosing pancreatitis (abundant infiltration of IgG4+ plasmacytes and lymphocytes, storiform fibrosis, and obliterative phlebitis), and the involvement of other organs indicating a complex pathologic mechanism and biomarker identification (Okazaki and Uchida, 2015). Generally, endocrine pancreatic gland is well preserved even after an acute pancreatic insult (Bhopale et al., 2017b). However, glucagon, insulin, somatostatin, and polypeptides produced by endocrine pancreatic gland have not been target as reliable markers of acute endocrine pancreatic injury other than pancreatic endocrine insufficiency. A correlation between urinary p-aminobenzoic acid (PABA, a split product of n-benzoyl-1-tyrosyl-p-aminobenzoic acid) excretion and plasma glucagon concentration suggests that in chronic pancreatitis there is collateral impairment of exocrine and endocrine functions (Keller et al., 1984).

ETHANOL METABOLITES AND CONJUGATES

Ethanol is one of the major etiologic agents for acute and chronic alcoholic pancreatitis after biliary duct disease (Lankisch and Banks, 1998; Kaphalia, 2011). Acetaldehyde, an oxidative metabolite of ethanol, could not go beyond oxidative stress as a global toxicity endpoint. On the other hand, nonoxidative metabolism of ethanol to FAEEs is most prevalent ethanol metabolism in the pancreas frequently damaged during chronic alcohol abuse (Laposata and Lange, 1986). These esters are lipophilic and shown to cause selective pancreatic acinar cell toxicity (Werner et al., 1997; Kaphalia and Ansari, 2001; Gukovskaya et al., 2002; Vonlaufen et al., 2007; Wu et al., 2008). The plasma/serum levels of FAEEs are generally high and correlate well with blood alcohol concentration particularly during chronic alcohol abuse (Laposata and

Lange, 1986; Kaphalia et al., 2004). Our studies in the deer mouse model of chronic ethanol feeding have shown that increased levels of plasma FAEE correlate well with those in the pancreas (Kaphalia et al., 2010). However, an interrelationship between plasma/serum levels of FAEEs and severity of pancreatitis in patients with acute or chronic pancreatitis needs to be established.

Other nonoxidative metabolite of ethanol, which correlates with blood alcohol levels in the blood, is phosphatidylethanol (PEt) formed by phospholipase D–catalyzed reaction (Aradottir et al., 2006). A good clinical efficiency of PEt and its pancreatic toxicity has not been demonstrated so far for detecting chronic heavy drinking (Viel et al., 2012). Ethyl glucuronide (EtG) and ethyl sulfate (EtS) levels are typically used to determine whether an individual has been exposed to alcohol (ethanol) (Nanau and Neuman, 2015). Accuracy of the test is greater when both EtG and EtS are included. EtG in contrast to EtS may also be produced in vitro when ethanol-producing bacteria are present in the urine specimen; so detection of EtS indicates in vivo alcohol presence, thereby improving the specificity of the test. An EtG concentration >500 ng/mL is considered positive for alcohol presence only if the EtS concentration is ≥100 ng/mL. EtG may be present in the urine for up to 80 h after ethanol ingestion. EtS may be detectable for 24 h or more after the ethanol ingestion. The validity of the markers depends on the cutoff used for the test and the patient's metabolic rate and the time and duration of alcohol exposure. Although chronic alcohol abuse is one of the major causes of acute and chronic pancreatitis, the ethanol metabolites as markers can be limited to ethanol exposure, which can also be supported by the patient history of alcohol abuse and used in combination with some functional markers such as lipase and patient symptoms.

DIFFERENTIALLY ALTERED PROTEINS, METABOLITES, SMALL MOLECULES, AND MICRORNA

Direct functional tests can be conducted by identifying the proteins, metabolites (small molecule), and microRNAs differentially altered in the pancreatic juice and duodenal contents from proteins with acute or chronic pancreatitis. Omic and array research is a systemic global approach and provides a unique "fingerprint," which varies with time or stresses, that a cell or organism undergoes. High-resolution ^{1}H-NMR analysis of plasma and urine samples and data analysis as determined by principal component analysis using metabolomic modeling of L-arginine–induced exocrine pancreatitis suggest metabolomics a valuable approach, which can distinguish patients with pancreatitis from healthy controls (Bohus et al., 2008; Lusczek et al., 2013). Lipid profiling of serum and pancreatic fluid conducted in chronic pancreatitis using the LC-ESI-MS/MS system in five controls and six patients of mild and five of severe chronic pancreatitis was shown to increase the levels of oxidized fatty acid products suggesting their utility as biomarker candidates for the diagnosis of chronic pancreatitis (Stevens et al., 2012). However, the specificity and analysis cost-effectiveness of such biomarkers need to be established.

Identification of proteins has been done only in the pancreas in acute pancreatitis models (Zhang et al., 2012; Fétaud et al., 2008: García-Hernández et al., 2012). Therefore, significant efforts should be directed for profiling both plasma and pancreatic proteins differentially expressed in acute and chronic pancreatitis in animal models followed by their validation in the plasma of patients with acute or chronic pancreatitis. Proteomic profiling using SELDI-TOF-MS has been done in the serum as a predictor of severity of acute pancreatitis from 21 patients with mild and 7 patients with severe acute pancreatitis (Papachristou et al., 2007). Various other proteomic studies (separation of proteins by 2-dimensional polyacrylamide gel electrophoresis and identification by MALDI-TOF/TOF) done in the pancreatic fluid, plasma, serum, and pancreatic tissue from patients with chronic pancreatitis provide significant lead to further pursue the studies in larger cohorts (Chen et al., 2007; Hartmann et al., 2007). Recent papers from our laboratory have identified a number of proteins differentially expressed in the pancreas and plasma of a chronic ethanol feeding model of hepatic alcohol dehydrogenase (ADH)–deficient (ADH^-) deer mice (Bhopale et al., 2017a,b). Proteins differentially expressed in the pancreatic tissue of ethanol-fed mice are mainly amylase 2b precursor, 60-kD heat shock proteins, and those involved in ATP synthesis and blood pressure. Plasma proteins differentially expressed were mainly serine protease inhibitor A3A precursor, creatine kinase M-type, and transport protein, apolipoprotein E. These findings also suggest that chronic ethanol feeding does not provide a plasma profile that can be found for acute pancreatitis. While gene profiling studies can also identify biomarker candidates, studies are limited to target organ and in experimental models. Pathway analysis and system's biology approaches generally done to infer relationships between genes, protein, and small molecules (metabolomics) are important for identification of biomarkers and may be a future direction in the biomarker field. Therefore, novel concepts such as genetic, transcriptomic, and proteomic profiling as systems approach to pancreatitis or PC along with functional imaging for identification of specific disease pattern need to be introduced at least to find out the target molecules of interest.

MicroRNAs (miRs) are functional, 22 nt, noncoding RNAs that negatively regulate gene expression and play a role in the initiation and progression of certain diseases. However, identity of differentially altered target microRNA should be confirmed. Hierarchical clustering and principal component analysis of the data sets can distinguish pancreatic tissue of patients with pancreatitis from normal control tissue. Only a handful of studies are available to identify microRNAs as biomarkers in the experimental rat model of acute or chronic pancreatitis and in PC (Bloomston et al., 2007; Lee et al., 2007; Kong et al., 2010). Kong et al. (2010) identified plasma miR-216a as a potential marker of pancreatic injury in the rat model of acute pancreatitis. Like amylase and lipase, increased expression of miR-216a in the plasma was restricted to acute pancreatitis at 24 h after the insult. Serum miR-216a and miR-217 in rats fed with exocrine pancreatic toxicants (cerulean or 1-cyano-2-hydroxy-3-butene) are reported sensitive and specific biomarkers of acute exocrine pancreatic toxicity as compared with amylase and lipase and may add value to the measurement of classical pancreatic biomarkers (Wang et al., 2017b). Both miR-216a and miR-217 are claimed to be of pancreatic origin, correlated better with microscopic findings within the exocrine pancreas. Conversely, neither microRNA was increased in rats administered a proprietary molecule known to cause a lesion at the pancreatic endocrine–exocrine interface or in rats administered an established renal toxicant. Circulating miRNAs such as miR-488-5p and miR-938-5p have been suggested to have an important role in discriminating pancreatic ductal adenocarcinoma (PDAC) from healthy subjects and those with chronic pancreatitis (Ebrahimi et al., 2016). The miRNA–mRNA regulating network in chronic pancreatitis has been identified based on significant functional expression (Wang et al., 2017a). However, such miRNA-based markers need to be confirmed in patients with acute and chronic pancreatitis. In one study, two dozens of microRNAs have been shown to be differentially altered in the pancreatic tissue from patients with chronic pancreatitis ($n = 42$) as compared with that in adjacent normal tissue as control (Bloomston et al., 2007). These studies are very preliminary and need further investigations to establish their utility as biomarker candidates in clinical diagnostics. Even if microRNA microarray technologies identify some promising biomarkers of disease/injury, cost and precision are other factors to be considered.

COMPARISON OF BIOMARKERS BETWEEN ACUTE AND CHRONIC PANCREATITIS

Basic laboratory tests for diagnosing acute and chronic pancreatitis are vastly different because former one is short term largely indicated by significant increases for amylase, isoamylase, lipase, TAP, trypsin, elastase-1, phospholipase A2, CDT, CRP, PAP, CAPAP, procalcitonin in serum, plasma and/or urine, and other markers such as cathepsins in pancreatic juice. The indirect pancreatic function tests in serum or urine can be evident in early stages of chronic pancreatitis, but a number of biomarkers as mentioned previously may be in the normal range or fluctuate depending on the chronicity of the pancreatitis because slow destruction of the exocrine gland results in exocrine pancreatic insufficiency. Therefore, testing exocrine pancreatic insufficiency is still recommended when chronic pancreatitis is suspected. When the pancreatic injury progress to fibrosis following the activation of stellate cells as often noticed in chronic pancreatitis, a shift in cellular phenotype may also bring new markers of injury in picture. Although several tests including secretin test, Cerulean test, and Lundh test have been described in detail by Lankisch and Banks (1998), fecal tests for trypsin, chymotrypsin, and elastase-1 can also provide some clues for presence of chronic pancreatitis. Most of methods used are either titrimetric estimation or ELISA. Moreover, fecal chymotrypsin estimation is not novel. Most of these tests are not even sensitive to diagnose mild-to-moderate exocrine pancreatic insufficiency. Fecal fat estimation, particularly when stimulated lipase and protease output fall below 10% of the normal, is a safe biomarker for steatorrhea (DiMagno et al., 1973). However, fecal fat contents were of no value for differential diagnosis of pancreatic and nonpancreatic steatorrhea (Roberts et al., 1986).

Pancreas-specific chymotrypsin splits N-benzyl-L-tyrosyl-p-aminobenzoic acid to p-aminobenzoic acid (NBT–PABA) in the duodenum. Estimation of p-aminobenzoic acid (PABA) cannot be true because some gut bacteria also splits NBT–PABA causing a false normal test results. Chronic pancreatitis can be diagnosed by serum, urinary, and fecal biomarkers of pancreatic functions in conjunction with patient history and imaging (CT, MRCP, ERCP) results. In view of available biomarkers and nature of pathology, a single biomarker should not be sufficient to diagnose chronic pancreatitis. Identification and development of metabolomic-, proteomic-, and microRNA-based biomarkers should be undertaken in experimental models followed by in patients at various stages of acute and chronic pancreatitis.

BIOMARKERS FOR EARLY DETECTION OF PANCREATIC CANCER

Although focus of this chapter is biomarkers of pancreatitis (acute and chronic), early detection of PC is also critical for the success of interventional and surgical therapies. PC is most devastating and fatal disease without any overt symptoms at initial stages. Most common exocrine PCs are PDAC followed by adenosquamous carcinomas and broad-based pancreatic cystic neoplasma. However, pancreatic neuroendocrine tumors constitute a small number of cases ~1% of the all PCs. Endocrine pancreatic tumors have been variously called islet cell tumors, pancreas endocrine tumors (PETs), and pancreatic neuroendocrine tumors (PNETs). On the other hand, serum IgG4 was found to have high specificity and relatively low sensitivity in differential diagnosis of AIP from PC using meta-analysis of 11 studies comprising 523 AIP and 771 PC patients (Dai et al., 2018).

The annual clinically recognized incidence is low, about five per one million person/years. However, autopsy studies incidentally identify PETs in up to 1.5%, most of which would remain inert and asymptomatic. The majority of PNETs are usually categorized as benign, but the definition of malignancy in pancreas endocrine tumors has been ambiguous. The more aggressive endocrine PCs are known as pancreatic neuroendocrine carcinomas (PNECs). Similarly, there as likely been a degree of admixture of PNEC and extrapulmonary small cell carcinoma. Most of biomarkers of PCs, identified for PDAC, are carcinoembryonic, carbohydrate antigen (CA19-9), and mucin family (MUC), the later being most commonly used along with circulating tumor cells, markers in pancreatic juice, and signaling pathway have been extensively reviewed (Tanase et al., 2009; Poruk et al., 2013; Chakraborty et al., 2011). Elevated CA19-9 by itself is sufficient for differentiating pancreatic carcinoma and chronic pancreatitis. Early detection of PC offers promise of improved mortality rates through surgical resection. Therefore, biomarkers should be determined for the early stages of PC to improve the prognosis and mortality rates by using intervention therapies. For further details on biomarkers of PC, readers are referred to Chapter 49. In fact, a large number of miRNAs have been identified as putative biomarkers and therapeutic targets for an early detection of PC (Gayral et al., 2014; Calvano et al., 2016; Young et al., 2018). Innovative initiative and new approaches need to be undertaken to identify and validate the biomarkers for an early detection of PC (Kenner et al., 2017). It is expected that an early detection of PC can be very cost-effective with greater prognostic and clinical outcomes.

BREATHE ANALYSIS

Human-exhaled breath contains many molecules either present as gases or occurring in a soluble form in the vapor of the breath and can rapidly and easily assess the presence of volatile compounds (hydrogen sulfide, nitric oxide, and molecular mass 66 u substance) in the exhaled breath of chronic pancreatitis patients (Morselli-Labate et al., 2007). Although study was conducted in 11 patients with chronic pancreatitis and 31 healthy subjects and appears novel, no significant differences were found between patients and healthy controls.

CONCLUDING REMARKS AND FUTURE DIRECTIONS

Reliable and sensitive biomarkers of acute and chronic pancreatitis and PCs are needed, as current available biomarkers have limitations and lack specificity. Serum total amylase and lipase in combination or individually remains to be a gold standard for functional clinical diagnostics of acute pancreatitis if the test is conducted on admission of the patients to the hospital. Definitely, lipase assay has advantage over amylase in clinical diagnostics and has more specificity for diagnosing acute pancreatitis. However, imaging technologies are strong support for diagnosis of both acute and chronic pancreatitis. Identification and development of biomarkers of chronic pancreatitis remains to be investigated and is challenging because several indirect pancreatic functional biomarkers are inconclusive primarily because of impaired secretory components of exocrine pancreas due to slow destruction of the gland in chronic pancreatitis, particularly at later stages. The focus of future research lies in identification of molecular biomarkers by utilizing proteomic, metabolomic, and circulating microRNA array technologies by using very well-defined patient populations at various stages of the disease.

References

Abu Hilal, M., Ung, C.T., Westlake, S., et al., 2007. Carboxypeptidase-B activation peptide, a marker of pancreatic acinar injury, but not L-selectin, a marker of neutrophil activation, predicts severity of acute pancreatitis. J. Gastroenterol. Hepatol. 22 (3), 349–354.

Achur, R.N., Freeman, W.M., Vrana, K.E., 2010. Circulating cytokines as biomarkers of alcohol abuse and alcoholism. J. Neuroimmune Pharmacol. 5 (1), 83–91.

Ananian, P., Hardwigsen, J., Bernard, D., et al., 2005. Serum acute-phase protein level as indicator for liver failure after liver resection. Hepatogastroenterology 52 (63), 857–861.

Aparicio, J.R., Viedma, J.A., Aparisi, L., et al., 2001. Usefulness of carbohydrate-deficient transferrin and trypsin activity in the diagnosis of acute alcoholic pancreatitis. Am. J. Gastroenterol. 96 (6), 1777–1781.

Appelros, S., Petersson, U., Toh, S., et al., 2001. Activation peptide of carboxypeptidase B and anionic trypsinogen as early predictors of the severity of acute pancreatitis. Br. J. Surg. 88 (2), 216–221.

Aradottir, S., Asanovska, G., Gjerss, S., et al., 2006. Phosphatidylethanol (PEth) concentrations in blood are correlated to reported alcohol intake in alcohol-dependent patients. Alcohol Alcohol 41 (4), 431–437.

ARIP Laboratories, July 2010. Carbohydrate-Deficient Transferrin (CDT) for Alcohol Use. For Detection of Sustained Alcohol Intake, p. 2.

Barauskas, G., Svagzdys, S., Maleckas, A., 2004. C-reactive protein in early prediction of pancreatic necrosis. Medicina 40 (2), 135–140.

Barrett, A.J., 1980. Fluorimetric assays for cathepsin B and cathepsin H with methylcoumarylamide substrates. Biochem. J. 187 (3), 909–912.

Basterra, G., Casi, M.A., Alcorta, P., et al., 2001. Is carbohydrates-deficient transferrin the best test of the alcoholic etiology in acute pancreatitis? Rev. Esp. Enferm. Dig. 93 (8), 529–534.

Bhopale, K.K., Amer, S.M., Kaphalia, L., et al., 2017a. Proteomic profiling of liver and plasma in chronic ethanol feeding model of hepatic alcohol dehydrogenase-deficient deer mice. Alcohol Clin. Exp. Res. 41 (10), 1675–1685.

Bhopale, K.K., Amer, S.M., Kaphalia, L., et al., 2017b. Proteins differentially expressed in the pancreas of hepatic alcohol dehydrogenase-deficient Deer mice fed ethanol for 3 months. Pancreas 46 (6), 806–812.

Bloomston, M., Frankel, W.L., Petrocca, F., et al., 2007. MicroRNA expression patterns to differentiate pancreatic adenocarcinoma from normal pancreas and chronic pancreatitis. J. Am. Med. Assoc. 297 (17), 1901–1908.

Bohus, E., Coen, M., Keun, H.C., et al., 2008. Temporal metabonomic modeling of l-arginine-induced exocrine pancreatitis. J. Proteome Res. 7 (10), 4435–4445.

Borgström, A., Ohlsson, K., 1978. Immunoreactive trypsin in serum and peritoneal fluid in acute pancreatitis. Hoppe Seylers Z. Physiol. Chem. 359 (6), 677–681.

Burgos, F.J., Salvà, M., Villegas, V., et al., 1991. Analysis of the activation process of porcine procarboxypeptidase B and determination of the sequence of its activation segment. Biochemistry 30 (16), 4082–4089.

Calvano, J., Edwards, G., Hixson, C., et al., 2016. Serum microRNAs-217 and -375 as biomarkers of acute pancreatic injury in rats. Toxicology 368–369, 1–9.

Chakraborty, S., Baine, M.J., Sasson, A.R., et al., 2011. Current status of molecular markers for early detection of sporadic pancreatic cancer. Biochim. Biophys. Acta 1815 (1), 44–64.

Chen, R., Brentnall, T.A., Pan, S., et al., 2007. Quantitative proteomics analysis reveals that proteins differentially expressed in chronic pancreatitis are also frequently involved in pancreatic cancer. Mol. Cell. Proteomics 1331–1342.

Col, C., Dinler, K., Hasdemir, O., et al., 2010. Oxidative stress and lipid peroxidation products: effect of pinealectomy or exogenous melatonin injections on biomarkers of tissue damage during acute pancreatitis. Hepatobiliary Pancreat. Dis. Int. 9 (1), 78–82.

Dai, C., Cao, Q., Jiang, M., et al., 2018. Serum immunoglobulin G4 in discriminating autoimmune pancreatitis from pancreatic cancer: a diagnostic meta-analysis. Pancreas 47 (3), 280–284.

Deng, L., Wang, L., Yong, F., et al., 2015. Prediction of the severity of acute pancreatitis on admission by carboxypeptidase-B activation peptide: a systematic review and meta-analysis. Clin. Biochem. 48 (10–11), 740–746.

DiMagno, E.P., Go, V.L., Summerskill, W.H., 1973. Relations between pancreatic enzyme outputs and malabsorption in severe pancreatic insufficiency. N. Engl. J. Med. 288 (16), 813–815.

Dimastromatteo, J., Brentnall, T., Kelly, K.A., 2017. Imaging in pancreatic disease. Nat. Rev. Gastroenterol. Hepatol. 14, 97–109.

Domínguez-Muñoz, J.E., Villanueva, A., Lariño, J., et al., 2006. Accuracy of plasma levels of polymorphonuclear elastase as early prognostic marker of acute pancreatitis in routine clinical conditions. Eur. J. Gastroenterol. Hepatol. 18 (1), 79–83.

Dumnicka, P., Maduzia, D., Ceranowicz, P., et al., 2017. The interplay between inflammation, coagulation and endothelial injury in the early phase of acute pancreatitis: clinical implications. Int. J. Mol. Sci. 18 (2) pii: E354.

Ebrahimi, S., Hosseini, M., Ghasemi, F., et al., 2016. Circulating microRNAs as potential diagnostic, prognostic and therapeutic targets in pancreatic cancer. Curr. Pharmaceut. Des. 22 (42), 6444–6450.

Fabris, C., Basso, D., Benini, L., et al., 1989. Urinary elastase 1 in chronic pancreatic disease. Enzyme 42 (2), 80–86.

Fabris, C., Basso, D., Panozzo, M.P., et al., 1992. Urinary phospholipase A2 excretion in chronic pancreatic diseases. Int. J. Pancreatol. 11 (3), 179–184.

Fétaud, V., Frossard, J.L., Farina, A., et al., 2008. Proteomic profiling in an animal model of acute pancreatitis. Proteomics 8 (17), 3621–3631.

García-Hernández, V., Sánchez-Bernal, C., Sarmiento, N., et al., 2012. Proteomic analysis of the soluble and the lysosomal+mitochondrial fractions from rat pancreas: implications for cerulein-induced acute pancreatitis. Biochim. Biophys. Acta 1824 (9), 1058–1067.

Gayral, M., Jo, S., Hanoun, N., et al., 2014. MicroRNAs as emerging biomarkers and therapeutic targets for pancreatic cancer. World J. Gastroenterol. 20 (32), 11199–11209.

Gudgeon, A.M., Heath, D.I., Hurley, P., et al., 1990. Trypsinogen activation peptides assay in the early prediction of severity of acute pancreatitis. Lancet 335 (8680), 4–8.

Gukovskaya, A.S., Mouria, M., Gukovsky, I., et al., 2002. Ethanol metabolism and transcription factor activation in pancreatic acinar cells in rats. Gastroenterology 122, 106–118.

Haber, P.S., Wilson, J.S., Apte, M.V., et al., 1993. Fatty acid ethyl esters increase rat pancreatic lysosomal fragility. J. Lab. Clin. Med. 121 (6), 759–764.

Hartmann, D., Felix, K., Ehmann, M., et al., 2007. Protein expression profiling reveals distinctive changes in serum proteins associated with chronic pancreatitis. Pancreas 35 (4), 334–342.

Hedström, J., Leinonen, J., Sainio, V., et al., 1994. Time-resolved immunofluorometric assay of trypsin-2 complexed with alpha 1-antitrypsin in serum. Clin. Chem. 40 (9), 1761–1765.

Herman, R., Guire, K.E., Burd, R.S., et al., 2011. Utility of amylase and lipase as predictors of grade of injury or outcomes in pediatric patients with pancreatic trauma. J. Pediatr. Surg. 46 (5), 923–926.

Hobbs, J.R., Aw, S.E., Wootton, I.D., 1967. Urinary isoamylases in the diagnosis of chronic pancreatitis. Gut 8 (4), 402–407.

Imagawa, A., Hanafusa, T., Miyagawa, J., et al., 2000. A novel subtype of type 1 diabetes mellitus characterized by a rapid onset and an absence of diabetes-related antibodies. Osaka IDDM Study Group. N. Engl. J. Med. 342 (5), 301–307.

Imamura, T., Tanaka, S., Yoshida, H., et al., 2002. Significance of measurement of high-sensitivity C-reactive protein in acute pancreatitis. J. Gastroenterol. 37 (11), 935–938.

Ismail, O.Z., Bhayana, V., 2017. Lipase or amylase for the diagnosis of acute pancreatitis? Clin. Biochem. 50 (18), 1275–1280.

Jaakkola, M., Sillanaukee, P., Löf, K., et al., 1994. Blood tests for detection of alcoholic cause of acute pancreatitis. Lancet 343 (8909), 1328–1329.

Johnson, C.D., Lempinen, M., Imrie, C.W., et al., 2004. Urinary trypsinogen activation peptide as a marker of severe acute pancreatitis. Br. J. Surg. 91 (8), 1027–1033.

Kaphalia, B.S., 2011. Pancreatic toxicology. In: Nriagu, J.O. (Ed.), Encyclopedia of Environmental Health, vol. 4. Elsevier, , Burlington, pp. 315–324.

Kaphalia, B.S., Ansari, G.A.S., 2001. Fatty acid ethyl esters and ethanol-induced pancreatic injury. Cell. Mol. Biol. 47, OL173–OL179.

Kaphalia, B.S., Bhopale, K.K., Kondraganti, S., et al., 2010. Pancreatic injury in hepatic alcohol dehydrogenase-deficient deer mice after subchronic exposure to ethanol. Toxicol. Appl. Pharmacol. 246, 154–162.

Kaphalia, B.S., Cai, P., Khan, M.F., et al., 2004. Fatty acid ethyl esters: markers of alcohol abuse and alcoholism. Alcohol 34 (2–3), 151–158.

Keller, U., Szöllösy, E., Varga, L., et al., 1984. Pancreatic glucagon secretion and exocrine function (BT-PABA test) in chronic pancreatitis. Dig. Dis. Sci. 29 (9), 853–857.

Kenner, B.J., Go, V.L.W., Chari, S.T., et al., 2017. Early detection of pancreatic cancer: the role of industry in the development of biomarkers. Pancreas 46 (10), 1238–1241.

Komolafe, O., Pereira, S.P., Davidson, B.R., et al., 2017. Serum C-reactive protein, procalcitonin, and lactate dehydrogenase for the diagnosis of pancreatic necrosis. Cochrane Database Syst. Rev. 4, CD012645.

Kong, X.Y., Du, Y.Q., Li, L., et al., 2010. Plasma miR-216a as a potential marker of pancreatic injury in a rat model of acute pancreatitis. World J. Gastroenterol. 16 (36), 4599–4604.

Kylänpää-Bäck, M.L., Kemppainen, E., Puolakkainen, P., 2002. Trypsin-based laboratory methods and carboxypeptidase activation peptide in acute pancreatitis. JOP 3 (2), 34–48.

Lankisch, P.G., Banks, P.A., 1998. Pancreatitis. Springer-Verlag, Berlin, Heidelberg, New York, p. 377.

Laposata, E.A., Lange, L.G., 1986. Presence of nonoxidative ethanol metabolism in human organs commonly damaged by ethanol abuse. Science 231 (4737), 497–499.

Lee, E.J., Gusev, Y., Jiang, J., et al., 2007. Expression profiling identifies microRNA signature in pancreatic cancer. Int. J. Cancer 120 (5), 1046–1054.

Lempinen, M., Stenman, U.H., Puolakkainen, P., et al., 2003. Sequential changes in pancreatic markers in acute pancreatitis. Scand. J. Gastroenterol. 38 (6), 666–675.

Lusczek, E.R., Paulo, J.A., Saltzman, J.R., et al., 2013. Urinary 1H-NMR metabolomics can distinguish pancreatitis patients from healthy controls. JOP 14 (2), 161–170.

Lyo, V., Cattaruzza, F., Kim, T.N., et al., 2012. Active cathepsins B, L, and S in murine and human pancreatitis. Am. J. Physiol. Gastrointest. Liver Physiol. 303 (8), G894–G903.

Moridani, M.Y., Bromberg, I.L., 2003. Lipase and pancreatic amylase versus total amylase as biomarkers of pancreatitis: an analytical investigation. Clin. Biochem. 36 (1), 31–33.

Morselli-Labate, A.M., Fantini, L., Pezzilli, R., 2007. Hydrogen sulfide, nitric oxide and a molecular mass 66 u substance in the exhaled breath of chronic pancreatitis patients. Pancreatology 7 (5–6), 497–504.

Motoo, Y., Watanabe, H., Yamaguchi, Y., et al., 2001. Pancreatitis-associated protein levels in pancreatic juice from patients with pancreatic diseases. Pancreatology 1 (1), 43–47.

Nanau, R.M., Neuman, M.G., 2015. Biomolecules and biomarkers used in diagnosis of alcohol drinking and in monitoring therapeutic interventions. Biomolecules 5, 1339–1385.

Naruse, S., Ishiguro, H., Ko, S.B., et al., 2006. Fecal pancreatic elastase: a reproducible marker for severe exocrine pancreatic insufficiency. J. Gastroenterol. 41 (9), 901–908.

Nord, H.J., Weis, H.J., Cölle, H., 1973. [Metabolism of serum amylase]. Verh. Dtsch. Ges. Inn. Med. 79, 868–870.

Okazaki, K., Uchida, K., 2015. Autoimmune pancreatitis: the past, present, and future. Pancreas 44 (7), 1006–1016.

Papachristou, G.I., November 2008. Prediction of severe acute pancreatitis: current knowledge and novel insights. World J. Gastroenterol. 4 (41), 6273–6275.

Papachristou, G.I., Malehorn, D.E., Lamb, J., et al., 2007. Serum proteomic patterns as a predictor of severity in acute pancreatitis. Pancreatology 7 (4), 317–324.

Pezzilli, R., Morselli-Labate, A.M., Barbieri, A.R., et al., 2000. Clinical usefulness of the serum carboxypeptidase B activation peptide in acute pancreatitis. JOP 1 (3), 58–68.

Poruk, K.E., Gay, D.Z., Brown, K., et al., 2013. The clinical utility of CA 19-9 in pancreatic adenocarcinoma: diagnostic and prognostic updates. Curr. Mol. Med. 13 (3), 340–351.

Rinderknecht, H., 1986. Activation of pancreatic zymogens. Normal activation, premature intrapancreatic activation, protective mechanisms against inappropriate activation. Dig. Dis. Sci. 31 (3), 314–321.

Roberts, I.M., Poturich, C., Wald, A., 1986. Utility of fecal fat concentrations as screening test in pancreatic insufficiency. Dig. Dis. Sci. 31 (10), 1021–1024.

Schlosser, W., Gansauge, F., Schlosser, S., et al., 2001. Low serum levels of CD44, CD44v6, and neopterin indicate immune dysfunction in chronic pancreatitis. Pancreas 23 (4), 335–340.

Sharpe, P.C., 2001. Biochemical detection and monitoring of alcohol abuse and abstinence. Ann. Clin. Biochem. 38 (Pt 6), 652–664.

Stevens, T., Berk, M.P., Lopez, R., et al., 2012. Lipidomic profiling of serum and pancreatic fluid in chronic pancreatitis. Pancreas 41 (4), 518–522.

Tanase, C.P., Neagu, M., Albulescu, R., et al., 2009. Biomarkers in the diagnosis and early detection of pancreatic cancer. Expert Opin. Med. Diagn. 3 (5), 533–546.

Tietz, N.W., 1988. Amylase measurements in serum–old myths die hard. J. Clin. Chem. Clin. Biochem. 26 (5), 251–253.

Toskes, P.P., 1990. Hyperlipidemic pancreatitis. Gastroenterol. Clin. N. Am. 19 (4), 783–791.

van Acker, G.J., Perides, G., Steer, M.L., 2006. Co-localization hypothesis: a mechanism for the intrapancreatic activation of digestive enzymes during the early phases of acute pancreatitis. World J. Gastroenterol. 12, 1985–1990.

Viel, G., Boscolo-Berto, R., Cecchetto, G., et al., 2012. Phosphatidylethanol in blood as a marker of chronic alcohol use: a systematic review and meta-analysis. Int. J. Mol. Sci. 13 (11), 14788–14812.

Vishwanath, B.S., Frey, F.J., Escher, G., et al., 1996. Liver cirrhosis induces renal and liver phospholipase A2 activity in rats. J. Clin. Invest. 98 (2), 365–371.

Volz, K.A., McGillicuddy, D.C., Horowitz, G.L., et al., 2010. Eliminating amylase testing from the evaluation of pancreatitis in the emergency department. West. J. Emerg. Med. 11 (4), 344–347.

Vonlaufen, A., Wilson, J.S., Pirola, R.C., et al., 2007. Role of alcohol metabolism in chronic pancreatitis. Alcohol Res. Health 30, 48–54.

Walgren, J.L., Mitchell, M.D., Whiteley, L.O., et al., 2007. Identification of novel peptide safety markers for exocrine pancreatic toxicity induced by cyanohydroxybutene. Toxicol. Sci. 96, 174–183.

Wang, D., Xin, L., Lin, J.H., et al., 2017a. Identifying miRNA-mRNA regulation network of chronic pancreatitis based on the significant functional expression. Medicine 96 (21), e6668.

Wang, J., Huang, W., Thibault, S., et al., 2017b. Evaluation of miR-216a and miR-217 as potential biomarkers of acute exocrine pancreatic toxicity in rats. Toxicol. Pathol. 45 (2), 321–334.

Wang, Y., Naruse, S., Kitagawa, M., et al., 2001. Urinary excretion of trypsinogen activation peptide (TAP) in taurocholate-induced pancreatitis in rats. Pancreas 22 (1), 24–27.

Werner, J., Laposata, M., Fernández-del Castillo, C., et al., 1997. Pancreatic injury in rats induced by fatty acid ethyl ester, a nonoxidative metabolite of alcohol. Gastroenterology 113 (1), 286–294.

Wiley, J.W., Pietropaolo, M., 2009. Autoimmune pancreatitis: the emerging role of serologic biomarkers. Diabetes 58 (3), 520–522.

Wilding, P., Cooke, W.T., Nicholson, G.I., 1964. Globulin-bound amylase, A cause of persistently elevated levels in serum. Ann. Intern. Med. 60, 1053–1059.
Woo, S.M., Noh, M.H., Kim, B.G., et al., 2011. Comparison of serum procalcitonin with Ranson, APACHE-II, Glasgow and Balthazar CT severity index scores in predicting severity of acute pancreatitis. Korean J. Gastroenterol. 8 (1), 31–37.
Wu, H., Bhopale, K.K., Ansari, G.A.S., et al., 2008. Ethanol-induced cytotoxicity in rat pancreatic acinar AR42J cells: role of fatty acid ethyl esters. Alcohol Alcohol 43, 1–8.
Wu, D., Xu, Y., Zeng, Y., et al., 2011. Endocrine pancreatic function changes after acute pancreatitis. Pancreas 40 (7), 1006–1011.
Young, M.R., Wagner, P.D., Ghosh, S., et al., 2018. Validation of biomarkers for early detection of pancreatic cancer: summary of the alliance of pancreatic cancer consortia for biomarkers for early detection workshop. Pancreas 47 (2), 135–141.
Zhang, W., Zhao, Y., Zeng, Y., et al., 2012. Hyperlipidemic versus normal-lipid acute necrotic pancreatitis: proteomic analysis using an animal model. Pancreas 41 (2), 317–322.

CHAPTER

20

Skeletal Muscle Toxicity Biomarkers

Ramesh C. Gupta[1], *Robin B. Doss*[1], *Wolf-D. Dettbarn*[2], *Dejan Milatovic*[3]

[1]Toxicology Department, Breathitt Veterinary Center, Murray State University, Hopkinsville, Kentucky, United States; [2]Vanderbilt University, Nashville, TN, United States; [3]Charlottesville, VA, United States

INTRODUCTION

The skeletal muscles account for nearly half of mammalian body weight. The muscles are targets of toxicity for many drugs and chemicals for three major reasons. First, the muscles receive proportionately large amounts of toxicants; second, muscle metabolism is highly active; and third, the muscles are enriched with many reactive and binding sites, such as receptors, neurotransmitters, neurotropic and growth factors, enzymes, and other proteins. In addition, the muscles have drug transporter systems. Chemicals from various classes interact with these critical molecules and exert their toxic effects. Under strenuous conditions, such as chemical intoxication, skeletal muscles generate excess reactive oxygen species (ROS)/reactive nitrogen species (RNS), thereby causing oxidative/nitrosative stress leading to muscle toxicity. Chemicals of different classes can induce apoptosis, necrosis, or autophagy, resulting in muscle weakness, loss of muscle mass, atrophy, myopathy, or paralysis. Currently, drug-induced myopathy is among the most common causes of muscle disease. The target molecules are often used as biomarkers of exposure and/or effects of chemical/drug toxicities. In addition to drugs and chemicals, muscle toxicity occurs because of envenomations by poisonous snakes, scorpions, and other spiders. The distinct features of slow and fast muscles are the most fascinating aspects of skeletal muscle research because they can respond differently to each chemical. This chapter describes biomarkers of exposure and effects of drugs, chemicals, and biotoxins in skeletal muscles of experimental animals and humans and their relationship to clinical perspectives.

MUSCLE TYPES

Originally, muscles were characterized as having three different myofibers: type I myofibers (red, slow and aerobic muscles), and types IIA and IIB (white, fast and glycolytic muscles). Fiber type analysis by actomyosin ATPase reaction shows that the soleus is composed predominately of type I fibers with a few type IIA and IIB fibers. In contrast, the extensor digitorum longus (EDL) is composed predominately of type IIA and IIB fibers, with a few type I fibers. In the rat, total fiber numbers are approximately 1800 for soleus and 2500 for EDL (Gupta et al., 1985, 1986). Recently, types IC, IIC, IIAC, and IIAB, which have intermediate myosin ATPase–staining characteristics, have been identified in human muscles. Recently seven human muscle fiber types have been identified based on myosin ATPase–histochemical staining (from slowest to fastest): types I, IC, IIC, IIAC, IIA, IIAB, and IIB. Effects of toxic chemicals have rarely been studied on these fiber types, but chemicals such as organophosphate (OP) pesticides and nerve agents are known to have different effects on specific fiber types. The fast and slow muscles differ qualitatively and quantitatively in terms of biochemical constituents, and accordingly, these muscles are affected differently by various chemicals.

PESTICIDES

Cholinesterase Inhibitors

Next to the brain, skeletal muscles are major targets for the toxicity of OP and carbamate (CM) compounds. Understanding the skeletal muscle system in the context of OP/CM poisoning is interesting, yet very complex because muscles containing different fiber types often respond differently, even to the same OP or CM compound (Gupta et al., 2015). Skeletal muscles contain cholinergic as well as noncholinergic elements, which are directly or indirectly modulated by OP/CM compounds, and this influences overall toxicity outcome accordingly.

Biomarkers in Toxicology, Second Edition
https://doi.org/10.1016/B978-0-12-814655-2.00020-7

Behavioral Effects

In general, experimental animals acutely intoxicated with an OP (pesticide or nerve agent) or CM pesticide exhibit the onset of toxicity, such as salivation, lacrimation, muscle tremors, and fasciculations within 5–15 min. The signs of maximal severity are observed within 30–60 min and usually last for 2–6 h with increasing propensity. Thereafter, the intensity is reduced to a mild form and toxicity may still be observed after 24 h of exposure. With some OP insecticides, such as chlorpyrifos or parathion, occurrence of toxic signs is delayed. Toxic signs are less severe and last for a shorter period with CMs than with OPs. It needs to be pointed out that some OPs and CMs produce pronounced CNS effects, whereas others produce peripheral effects (Gupta et al., 1986, 1987a,b, 2015; Misulis et al., 1987; Gupta and Kadel, 1989, 1991).

Cholinergic Effects

All skeletal muscles contain major elements (a neurotransmitter ACh, an enzyme acetylcholinesterase (AChE) that hydrolyzes ACh, and an enzyme ChAT that synthesizes ACh) of the cholinergic system, but quantities of these elements vary from muscle to muscle. The fast fiber–containing muscle (EDL) has greater values of these cholinergic elements than the slow fiber muscle soleus or the mixed fiber muscle diaphragm (Gupta et al., 2015). OP and CM compounds exert muscle toxicity primarily by virtue of AChE inhibition at the neuromuscular junction (NMJ). Normally at the cholinergic synapse, AChE is essentially required for the removal of ACh from the synaptic cleft. But a marked inhibition of AChE by OPs or CMs profoundly modifies neuromuscular transmission by accumulated ACh, causing muscle hyperactivity, fasciculations, weakness, and muscle cell death by necrosis or apoptosis. Time-course studies of AChE inhibition in skeletal muscles have been reported for many OPs and CMs (Gupta et al., 1985, 1986, 1987a,b, 1991a,b, 2015; Gupta and Kadel, 1991; Thiermann et al., 2005). Evidently, AChE inactivation occurs to a varying degree in slow and fast muscles by each OP/CM compound. For example, Gupta et al. (1987a) demonstrated that selective inhibition of AChE activity in skeletal muscles was apparent within 1 h of soman administration (100 μg/kg, sc) in rats when soleus showed the maximum inhibition (87%), whereas EDL showed the least inhibition (47%). EDL-AChE was also less inhibited by other OP nerve agents, such as sarin, tabun, and VX (Gupta et al., 1987b, 1991a,b). However, with DFP (1.5 mg/kg, sc) AChE inhibition was maximal in EDL (80%), followed by diaphragm, and least in soleus (63%) (Gupta et al., 1986). It needs to be mentioned that AChE inhibition of 70% or greater is required to cause muscle hyperactivity. The observable toxic effects usually do not persist for more than 4–6 h, whereas recovery of AChE activity occurs at a much slower rate (i.e., days or weeks) depending on the muscle and the AChE inhibitor.

Skeletal muscles have also been characterized for molecular forms of AChE. Qualitative and quantitative variations exist among different species, as well as young versus adult (Massoulie and Bon, 1982; Barnard et al., 1984; Patterson et al., 1987, 1988). In the rat EDL, the G_1 (4S), G_4 (10S), and A_{12} (16S) molecular forms are predominant, whereas in soleus and diaphragm a fourth major molecular form is also present, the A_8 (12S). In rat soleus, the majority of the total AChE activity is contributed by the 12S and 16S forms, whereas in the diaphragm it is from 4S and 10S and in the EDL it is from 4S (Patterson et al., 1987; Gupta et al., 2015). The 16S form is found in high concentration at the endplates, and it is thought to be involved in neuromuscular transmission. The different molecular forms of AChE in soleus and EDL have apparent K_m values similar to that previously found in the diaphragm muscle. There appears to be no difference between catalytic sites of the molecular forms of AChE in fast EDL and slow soleus muscles, despite the different molecular form patterns and activity in these muscles.

After 1 h of soman administration (100 μg/kg, sc), all molecular forms of AChE were reduced to less than 10% of control (100%) in soleus and diaphragm. In EDL the 16S form, mainly localized at the NMJ, was not affected at this time, whereas the 10S form was completely inhibited and the 4S form was reduced to 50%. Further inhibition was seen after 24 h but even then the 16S form was least inhibited in EDL. In similar experiments, Gupta et al. (1986, 1987b, 1991) also reported differential inhibition of AChE and its molecular forms in various skeletal muscles of rats intoxicated with DFP (1.5 mg/kg, sc), tabun (200 μg/kg, sc), sarin (110 μg/kg, sc), and VX (12 μg/kg, sc).

Petrov et al. (2011) demonstrated different sensitivities of rat skeletal muscles to an anti-ChE agent, alkylammonium derivatives of 6-methyluracil (ADEMS) because of different sensitivities of molecular forms. The rat respiratory muscle diaphragm had markedly lower sensitivity than the locomotor muscle EDL to ADEMS. The intermuscular difference in sensitivity to ADEMS is partly explained by a higher level of mRNA and activity of 1,3-bis[5(diethyl-o-nitrobenzylammonium)pentyl]-6-methyluracildibromide (C-547)–resistant BuChE in the diaphragm. Moreover, diaphragm AChE was found to be more than 20 times less sensitive to C-547 than that from the EDL. The A_{12} form present in muscles appeared more sensitive to C-547. Some of the intermuscular differences are explained based on AChE and BuChE mRNA levels. For example, the ratio of AChE and BuChE mRNA level in the EDL was

4.6 and in the diaphragm it was 1. It appears that BuChE might be more significant for ACh hydrolysis in the diaphragm than in the EDL.

Morbidity and mortality associated with OP or CM intoxication is due to the effects on skeletal muscles in general and muscles of respiration in particular. Deaths from an overdose of OPs and CMs are due in part to respiratory paralysis by depolarizing neuromuscular blockade (Misulis et al., 1987; Besser et al., 1989; Karalliedde and Henry, 1993).

Inhibition of AChE and BuChE in skeletal muscles can be used as biomarkers of exposure and effect of OPs and CMs. But interpretation should be made with great caution, and factors such as species, sensitivity of AChE and BuChE to its inhibitor in a specific muscle, and different sensitivities of molecular forms of AChE and BuChE in that muscle should be taken into account. Following acute exposure, OPs and CMs have not been shown to alter the levels of ChAT or ACh in skeletal muscles.

Noncholinergic Effects

Excitotoxicity and Oxidative/Nitrosative Stress

Excitotoxicity caused by OPs and CMs is mediated by overstimulation of nicotinic acetylcholine (nACh) and NMDA receptors, leading to increased Ca^{2+} uptake (Gupta et al., 1986; Inns et al., 1990; Karalliedde and Henry, 1993). Lipid peroxidation, mitochondrial dysfunction/damage, loss of energy metabolites, reduction of cytochrome-c oxidase (COx) activity, and increased xanthine oxidase (XO) activity support the contention that AChE inhibitor–induced hyperactivity causes muscle injury by excessive formation of ROS and RNS (Yang and Dettbarn, 1998; Gupta and Goad, 2000; Jeyarasasingam et al., 2000; Gupta et al., 2002; Milatovic et al., 2000, 2005, 2006). Yang and Dettbarn (1998) demonstrated that in rats within 30–60 min of DFP injection (1.5 mg/kg, sc) COx activity was reduced by 56%, XO activity was increased by 56%, and a significant muscle fiber breakdown had occurred in the diaphragm. These findings suggested that ROS is produced during DFP-induced muscle hyperactivity, initiating lipid peroxidation and subsequent myopathy.

Acute toxic effects of AChE inhibitors on skeletal muscles are thought to involve oxidative and nitrosative stresses with increased generation of free radicals, such as ROS and RNS. Muscle hyperactivity with its increased oxygen and energy consumptions appears to be the primary cause of oxidative stress. Control values were not significantly different in soleus and EDL muscles of rats for biomarkers of oxidative stress, F_2-isoprostanes (F_2-IsoPs, 1.142 ± 0.027 and 1.177 ± 0.092 ng/g), nitrosative stress, and citrulline (469.7 ± 31.8 and 417.8 ± 18.5 nmol/g). However, the values were different for biomarkers of mitochondrial dysfunction (ATP, 3.66 ± 0.11 and 5.85 ± 0.14 μmol/g; and phosphocreatine, 7.91 ± 0.26 and 13.14 ± 0.31 μmol/g, in soleus and EDL, respectively).

Rats acutely intoxicated with a CM pesticide carbofuran (1.5 mg/kg, sc) showed signs of maximal toxicity including muscle hyperactivity within 60 min of exposure. At this time, F_2-IsoPs (177% and 153%) and citrulline (267% and 304%) levels were significantly increased (Figs. 20.1 and 20.2, respectively), whereas ATP (46% and 43%) levels were decreased in soleus and EDL (Milatovic et al., 2005). With this level of ATP depletion, muscle fiber damage is likely to occur as a result of metabolic catastrophe.

In a similar study, Gupta et al. (2002) determined citrulline, ATP, and phosphocreatine (PCr) levels in skeletal muscles of rats acutely intoxicated with DFP (1.5 mg/kg, sc). Within 15 min of DFP exposure, with an onset of fasciculations, citrulline levels were significantly elevated in the muscles (soleus, EDL, and

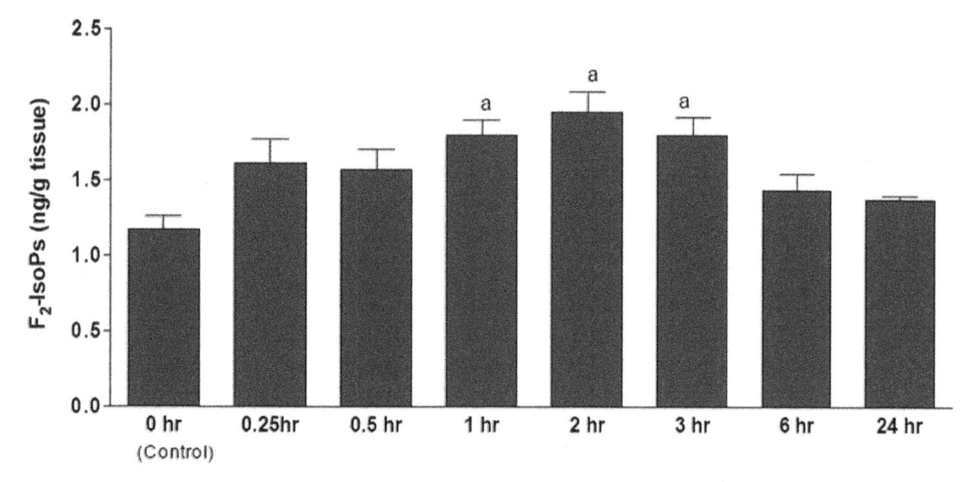

FIGURE 20.1 Effect of carbofuran (1.5 mg/kg, sc) on F_2-IsoPs levels in extensor digitorum longus (EDL) muscle of rats. Values are Means ± SEM (n = 4–6). [a]Significant difference between control and carbofuran-treated rats ($P < .05$).

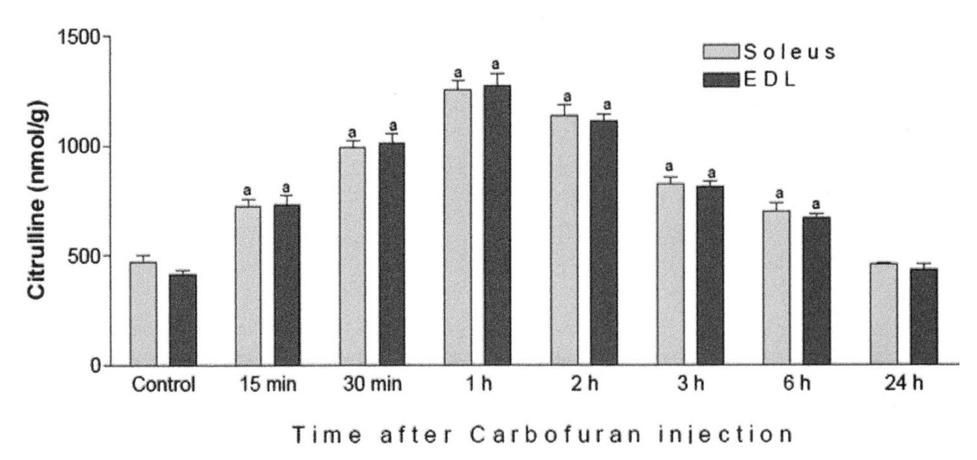

FIGURE 20.2 Effect of carbofuran (1.5 mg/kg, sc) on citrulline levels in soleus and extensor digitorum longus (EDL) muscles of rats. Values are Means ± SEM (n = 4–6). [a]Significant difference between control and carbofuran-treated rats ($P < .05$).

diaphragm). Maximum increases in citrulline (272%–288%) were noted 60 min after DFP injection. At this time point, AChE activity was reduced by 90%–96% (soleus < diaphragm < EDL). The levels of ATP and PCr were maximally reduced (30%–43%) and total adenine nucleotides (TAN) and total creatine compounds (TCC) showed declines. The findings revealed that the increase in NOS activity and nitric oxide (NO) was greater than the decrease of ATP and PCr.

NO has been linked to many pathological conditions, including deleterious effects of ChE inhibitors in skeletal muscles (Jeyarasasingam et al., 2000; Gupta et al., 2002; Milatovic et al., 2005, 2006). During DFP-induced muscle hyperactivity, the increased Ca^{2+} influx is associated with nicotinic ACh receptor (nAChR) activation at the NMJ. This finding was supported by previous studies showing that increased muscle contractility generates significantly greater quantities of ROS and RNS (Yang and Dettbarn, 1996; Clanton et al., 1999). Increased muscle contractility may also involve the activation of glutamate machinery (glutamate, glutamate receptors, and glutamate transporters) present at the NMJ mediated through increased calcium and NO. Increased NO exerts cellular toxicity primarily by depleting energy stores through a couple of mechanisms: (1) prolonging poly-(ADP-ribose) polymerase (PARP) activation and (2) inhibiting enzymes involved in mitochondrial respiration (cytochrome oxidase, aconitase, and mitochondrial CK) and glycolysis (phosphofructokinase). NO directly and specifically inhibits mitochondrial respiration by competing with molecular O_2 for binding to COx, thereby causing inhibition of ATP synthesis. Even at a nanomolar concentration, NO can directly inhibit COx activity, and cells producing large quantities of NO can inhibit their own respiration as well as the respiration of neighboring cells. Yang and Dettbarn (1998) found a 50% reduction of COx activity in diaphragm muscle caused by DFP-induced muscle hyperactivity, which could be due to the inhibitory effect of NO, the magnitude of which was increased by a factor of threefold. In essence, inhibition of COx appears to be the primary mechanism for ATP depletion because COx is the terminal and rate-limiting enzyme of the mitochondrial respiratory chain, which generates ATP by oxidative phosphorylation.

NO appears to regulate the coupling between energy supply and demand; it increases skeletal muscle blood flow and glucose transport and inhibits glycolysis and mitochondrial respiration (Kaminski and Andrade, 2001). The findings of Fukushima et al. (1997) and Gultekin et al. (2000) further supported the contention that the AChE-inhibiting OP insecticides cause lipid peroxidation and inhibition of oxidative phosphorylation. Gupta et al. (2002) demonstrated that the degree of NO elevation (threefold) was much greater than the declines of high-energy phosphates (30%–40%), suggesting the NO/NOS system to be a much more sensitive marker of nitrosative injury in skeletal muscles. Other factors, in addition to NO elevation, that contribute to the energy loss may include damage to mitochondria by AChE inhibitors (Laskowski et al., 1975; Gupta et al., 1999) and a higher rate of ATP utilization needed to generate NAD in the ADP-ribosylation of nuclear proteins, resulting in a decline of TAN and TCC.

OP-/CM-induced changes, such as increases in XO, F_2-IsoPs, and citrulline and decreases in COx, ATP, and PCr in skeletal muscles, can be used as biomarkers of oxidative stress and mitochondrial dysfunction.

Creatine Kinase and Creatine Kinase Isoenzymes

Creatine kinase (CK) catalyzes the synthesis of ATP and phosphocreatine (PCr) in a reversible Lohmann reaction (Creatine + ATP ↔ PCr + ADP). Data presented in Table 20.1 show the normal distribution of CK and its isoenzymes in the skeletal muscles of untreated control rats. Maximal CK activity was found in the EDL,

TABLE 20.1 Normal Distribution of CK and CK Isoenzymes in Skeletal Muscles and Serum of Rats

		CK Isoenzymes		
	Total CK	CK-BB (CK-1)	CK-MB (CK-2)	CK-MM (CK-3)
Soleus	2,062,800 ± 71,065 (100)	ND	ND	2,062,800 ± 71,065 (100)
EDL	4,659,125 ± 185,583 (100)[a,b]	ND	ND	4,659,125 ± 185,583 (100)[a,b]
Diaphragm	3,018,240 ± 110,777 (100)	ND	ND	3,018,240 ± 110,777 (100)
Serum	3769 ± 240 (100)	576 ± 45 (15.3)	148 ± 17 (3.9)	3086 ± 209 (80.8)

Values expressed in terms of IU/L are presented as Means ± SEM (n = 4–6). Numbers in parentheses are percentages of isoenzymes to total CK activity (100%). *CK*, creatine kinase; *EDL*, extensor digitorum longus; *ND*, none detected.
[a]*Significant difference between EDL and soleus (P <.001).*
[b]*Significant difference between EDL and diaphragm (P <.001).*

followed by diaphragm, with lowest activity in the soleus. The levels of energy metabolites (ATP and PCr) are reflected in these muscles according to the levels of CK (e.g., EDL having higher CK level had greater energy metabolites). Furthermore, during normal activity, fast-twitch fibers fire over a short period at higher rates than slow-twitch fibers, and for that reason, the EDL needs a greater amount of ATP and PCr (immediate precursor for ATP synthesis) than the soleus.

Electrophoretic separation of CK isoenzymes in all three muscles revealed the existence of only the CK-MM isoenzyme (Gupta et al., 1994). Further separation of the CK-MM isoenzyme for subforms showed only CK-MM3 in all three muscles. Serum had very little CK activity compared to muscles; however, the CK consisted of three distinct isoenzymes: CK-BB (15.3%), CK-MB (3.9%), and CK-MM (81.8%). Further separation of CK-MM isoenzyme by electrophoresis revealed the presence of three subforms: CK-MM1 (6.3%), CK-MM2 (24%), and CK-MM3 (69.7%). It is well known that the CK-MM3 subform is secreted from muscles into the plasma, where it converts into the MM2 and MM1 subforms by carboxypeptidase-N2 (CPN2).

Rats acutely intoxicated with a CM insecticide carbofuran (1.5 mg/kg, sc) showed significant increase in CK activity in diaphragm, a decrease in soleus, and no change in EDL, at the time of maximal severity (1 h post exposure). At the same time, activities of CK and all three CK isoenzymes were significantly elevated in serum. Carbofuran toxicity caused a shift in the serum CK-MM subform (i.e., higher sequential conversions of CK-MM3 subform to CK-MM2 and CK-MM2 to CK-MM1) possibly due to enhanced CPN2 activity.

Lactate Dehydrogenase and Lactate Dehydrogenase Isoenzymes

Lactate dehydrogenase (LDH) catalyzes the synthesis of lactate and pyruvate in a reversible reaction and is commonly used as a biomarker of cell damage or death. Normal distribution of LDH and its isoenzymes in skeletal muscles of rats revealed that in controls, LDH activity was found to be highest in the EDL, followed by diaphragm, and lowest in the soleus (Table 20.2). The enzyme activity was meager in serum compared with that of muscles (Gupta et al., 1994). It was interesting to note that all three muscles and serum

TABLE 20.2 Normal Distribution of LDH and LDH Isoenzymes in Skeletal Muscles and Serum of Rats

		LDH Isoenzymes				
	Total LDH	LDH-1 HHHH	LDH-2 HHHM	LDH-3 HHMM	LDH-4 HMMM	LDH-5 MMMM
Soleus	72,720 ± 2484 (100)	14,962 ± 476 (20.7)	19,054 ± 617 (26.3)	16,118 ± 692 (22.2)	14,254 ± 1332 (19.5)	8316 ± 890 (11.3)
EDL	207,300 ± 22,945 (100)[a,b]	7290 ± 1890 (3.5)[a,b]	12,300 ± 2033 (5.9)[a,b]	18,105 ± 1498 (8.7)	40,420 ± 4100 (19.5)[a,b]	129,870 ± 15,298 (62.6)[a,b]
Diaphragm	129,120 ± 4828 (100)	17,020 ± 504 (13.2)	17,498 ± 571 (13.5)	16,190 ± 780 (12.6)	26,738 ± 1080 (27.7)	51,680 ± 3081 (40.0)
Serum	740 ± 28 (100)	9.8 ± 1.4 (1.3)	17.4 ± 2.0 (2.3)	10.2 ± 1.2 (1.4)	38.0 ± 3.8 (5.2)	643 ± 65 (87.0)

Values expressed in terms of IU/L are presented as Means ± SEM (n = 4–6). Numbers in parentheses are percentages of isoenzymes to total LDH activity (100%). *EDL*, extensor digitorum longus; *LDH*, lactate dehydrogenase.
[a]*Significant difference between EDL and soleus (P <.01).*
[b]*Significant difference between EDL and diaphragm (P <.01).*

contained all five LDH isoenzymes, although with varying quantities. Each muscle presented a characteristic LDH isoenzyme pattern. For example, EDL contained a large proportion of LDH-5 (62.6%) and very little LDH-1 (3.5%). Soleus, on the other hand, had a very small amount of LDH-5 (11.3%) and about 20% each of the other four isoenzymes (LDH-1 through LDH-4). In general, values of LDH isoenzymes in diaphragm were intermediate to the EDL and soleus. Diaphragm had predominantly LDH-5 and LDH-4 (40% and 27.7%, respectively). In serum, isoenzyme LDH-5 was 87% of the total LDH activity (100%).

Carbofuran acute toxicity caused significant enhancement of LDH activity in EDL, diaphragm, and serum. Marked elevation of all five isoenzymes in serum occurred with maximum increases in LDH-1 and LDH-4 (threefold). Unlike serum, muscle LDH isoenzymes depicted variable patterns with carbofuran toxicity. A significant decline in ATP appears to be the mechanism involved in leakage of cytoplasmic/mitochondrial enzymes into circulation (Gupta et al., 1994). The fact that ATP is essential for retaining intracellular enzymes by maintaining cell membrane permeability and integrity was strongly reinforced by earlier investigations, demonstrating that depletion of ATP results in cell membrane damage, and ATP can be used as a biomarker of cell damage/death (Andreoli, 1993).

Histopathological Alterations

In general, the greater the AChE inhibition (during the initial 24–48 h period), the higher the number of lesions found in skeletal muscles (Gupta et al., 2015). The time courses of AChE inhibition and number of necrotic lesions in skeletal muscles of rats intoxicated acutely with DFP (1.5 mg/kg, sc) are presented in Table 20.3. With acute AChE inhibitors, the earliest lesions are focal areas of abnormality in the subjunctional area of the muscle fiber (Laskowski et al., 1977; Patterson et al., 1987, 1988). These focal changes progress to a breakdown of subjunctional fiber architecture followed by phagocytosis. Longitudinal sections reveal that the necrosis affects only a small segment of fiber lengths. During later stages, progressively greater lengths of muscle fibers are affected. Serial cross-sections of 10-μm thickness indicate that the number of lesions correlates with the greatest density of endplates. A significant increase in blood CK activity coincides with appearance of myonecrosis, indicating destruction of the muscle membrane (Gupta et al., 1991a,b, 1994, 2015). CMs, such as aldicarb and carbofuran, caused cell swelling, loss of myofibrillar structure, cytoplasmic clearing, and necrosis. Electron microscopic changes included mild focal loss of myofibrillar structure, intermyofibrillar clearing (fluid accumulation), and swelling of mitochondria (Fig. 20.3) (Gupta et al., 1999). Recently, Holovská et al. (2017) reported that a CM insecticide bendiocarb induced ultrastructural alterations in gastrocnemius muscles of rabbits receiving daily treatment (5 mg/kg body wt, po) for 20 days. Electron microscopy revealed dilatation of sarcoplasmic reticulum and myofilament disorganization. The most important alteration was the disruption of the sarcomeres due to the lysis of both thick and thin myofilaments. Muscle fibers also showed prominent mitochondrial swelling, and many mitochondria lacked cristae thus appearing as large membrane-bound cytoplasmic vesicles.

Despite the diversity in structures of the AChE inhibitors, the induced myopathic changes are the same, suggesting a common mechanism, i.e., muscle hyperactivity (Adler et al., 1992; Gupta and Dettbarn, 1992; Gupta et al., 2015; Jeyarasasingam et al., 2000). This mechanism is an excess of ACh and its prolonged interactions with nAChRs are associated with an excess of Ca^{+2}

TABLE 20.3 AChE Inhibition and Number of Necrotic Fibers Following a Single Sublethal Injection of DFP (1.5 mg/kg, sc)

	Soleus		EDL		Diaphragm	
Time After DFP Administration	**AChE**	**Lesions**	**AChE**	**Lesions**	**AChE**	**Lesions**
Control	100%	0	100%	0	100%	0
1 day	24%	27 ± 6[a]	17%	64 ± 11[a]	13%	308 ± 58[a]
2 day	24%	39 ± 9[a]	14%	71 ± 12[a]	13%	333 ± 63[a]
3 day	39%	5 ± 1[a]	31%	49 ± 5[a]	25%	174 ± 35[a]
7 day	65%	0	53%	0	69%	0

Values of AChE activity are expressed as % remaining activity of control (100%).
Numbers of lesions are presented per cross section from the mid-belly region of muscle.
Data are presented as the means ± SEM (n = 5–10).
EDL, extensor digitorum longus.
[a]*Significant difference between control and DFP-treated rats (P < .05).*

FIGURE 20.3 Electron micrograph of diaphragm muscle from aldicarb-treated (0.4 mg/kg, ip) rat showing mild focal loss of myofibrillar structure (*solid arrow*), intermyofibrillar clearing (fluid accumulation) (*arrowhead*), and swelling of mitochondria (*open arrow*).

and not a direct action of these inhibitors on the muscle (Gupta et al., 1986; Inns et al., 1990; Bright et al., 1991; Karalliedde and Henry, 1993).

Strychnine

Strychnine, obtained from the plants *Strychnos nux-vomica* or *Strychnos ignatii*, has been used as a rodenticide for more than a century. It has been involved in accidental poisonings in humans and malicious poisonings in animals. In small amounts, strychnine is known to be added to "street drugs," such as LSD, heroin, cocaine and others.

Strychnine is a potent convulsant. The convulsant action of strychnine is because of interference with the postsynaptic inhibition that is mediated by the amino acid glycine. Glycine is an inhibitory transmitter to motor neurons and interneurons in the spinal cord. Strychnine acts as a selective competitive antagonist to block the inhibitory effects of glycine at the glycine receptors. Studies suggest that strychnine and glycine interact with the same receptor (a ligand-gated chloride channel) but at different sites. There is also evidence of an increase in brain levels of glutamic acid, an amino acid that acts as a transmitter for excitatory nerve impulses that excites muscle contraction. The result of these effects is that skeletal muscles become hyperexcitable.

Roy et al. (2012) showed embryotoxicity of strychnine in zebrafish with long-term behavioral impairment in adults, possibly because of synaptic changes. Polymerase chain reaction analysis revealed indications that strychnine exposure altered expression of some genes related to glycinergic, GABAergic, and glutamatergic neuronal synapses during embryonic development.

The LD_{50} of strychnine in most species lies in the range of 0.4–3 mg/kg body wt (Gupta, 2018). Following the ingestion of strychnine, symptoms of poisoning usually appear within 15–60 min. Strychnine has profound toxic effects on the skeletal muscles. The initial clinical signs of poisoning include nervousness, restlessness, twitching of the muscles, and stiffness of the neck. As poisoning progresses, the muscular twitching becomes more pronounced and tetanic convulsions suddenly appear in all the striated muscles. The limbs are extended and the neck is curved to opisthotonus. As death approaches, the convulsions follow one another with increased rapidity, severity, and duration. Death results from asphyxia because of prolonged paralysis of the respiratory muscles, especially the diaphragm (Patocka, 2015). The main morphologic change following prolonged muscle contraction is rhabdomyolysis. Release of intracellular proteins may lead to myoglobinuric nephrosis, which in turn may result in renal failure for those who survive the acute tetanic phase. The presence of typical strychnine crystals can be identified using microscopy. Of course, chemical identification and confirmation of strychnine in the gastrointestinal content or body tissues/fluids by GC-MS is considered the best biomarker of strychnine toxicity (Gupta, 2018). The other characteristic laboratory findings, which can be used as secondary biomarkers of strychnine poisoning, include elevation of serum enzymes (GOT, CK, and LDH), lactic acidosis, hyperkalemia, and leukocytosis.

Paraquat

Paraquat, a commonly used herbicide, is known to cause myopathy in humans examined at both biopsy and necropsy. A few cases of paraquat poisoning and myopathy are described here in brief. Dinis-Oliveira et al. (2008) reviewed the toxicity of this herbicide in detail. Koppel et al. (1994) reported extensive myonecrosis in a postmortem specimen of intercostal muscle from a 52-year-old woman who had ingested an unknown dose of paraquat. A patient who died 5 days after ingestion of paraquat showed paraquat levels higher in the skeletal muscle, and an increase of CK levels appeared on the fourth day after admission. Tabata et al. (1999) observed degeneration of skeletal muscles (mainly rectus abdominis, psoas major, and diaphragm). Plasma CK values were highest on day 5 (1700 mU/mL) and 900 mU/mL on day 8. In an in vitro study, paraquat was shown to damage the mitochondrial membrane, disrupt oxidative phosphorylation, and halt the electron transport chain (Yamamoto et al., 1987). The presence of paraquat in muscle or body fluids can be used as biomarker of paraquat exposure and its toxic effects. Toxicity and biomarkers of paraquat and related herbicides are described in detail in Chapter 27 of this book.

METALS

Some metals are highly toxic to the biological system, specifically targeting skeletal muscle. Skeletal muscle regeneration is a known adaptive response to muscle injury that occurs through a process of myofiber degeneration, inflammation, and new fiber formation, resulting from satellite cell proliferation and differentiation (Kawai et al., 1990; Hurme and Kalimo, 1992). However, fibrosis that also develops during the healing process may hinder muscle regeneration. Yen et al. (2010) investigated the effects of arsenic trioxide (As_2O_3) on the myogenic differentiation of myoblasts in vitro (C2C12 myoblasts and primary mouse and human myoblasts) and skeletal muscle regeneration in vivo using a mouse model. Myogenesis involves three major steps: (1) withdrawal of myoblasts from the cell cycle, (2) subsequent expression of myotubes-specific genes, and (3) formation of multinucleated myotubes. Also, Akt is an important signaling pathway involved in myogenesis. Yen et al. (2010) demonstrated that low-dose As_2O_3 (0.1−0.5 μM) dose-dependently inhibited in vitro skeletal muscle differentiation as assessed by myogenin and myosin heavy chain expression, CK activity, and formation of multinucleated myotubes without apparent effects on cell viability. As_2O_3 significantly and dose-dependently inhibited skeletal myogenic differentiation (myogenin expression) by inhibiting phosphorylation of Akt protein and its downstream targets, mTOR and p70s6k. As_2O_3 also suppressed skeletal muscle differentiation and regeneration after glycerol-induced myopathy in the soleus in an in vivo mouse skeletal muscle regeneration model. Diaz-Villasenor et al. (2007) reported that arsenic exposure alters the expression or activity of different genes and proteins expressed in skeletal muscle and fat tissue. However, the precise mechanism of action of arsenic on skeletal myogenic differentiation remains unknown.

Muscle toxicity can also be induced by other metals, such as mercury, cadmium, lead, etc. (Gasparik et al., 2010; Tymoshenko et al., 2016). Recently, Tymoshenko et al. (2016) reported that high concentrations of Cu, Pb, Fe, Zn, Cr, and Mn caused swelling and activation of sclerotic and atrophic processes in the striated muscles of rats. These authors also observed enlarged muscle fibers and layers of connective tissue, deformation of their shapes, homogenous sarcoplasma, and hypochromic nuclei. The ultramicroscopic findings revealed multiple-dot hemorrhages with irregular cross section, deformed myofibril and hydropic and swollen mitochondria with destroyed inner membrane and deformed cristae.

Metals residue and microscopic and ultramicroscopic changes in the skeletal muscles usually serve as biomarkers of exposure and effects, respectively.

THERAPEUTIC DRUGS

Drugs of many classes are known to induce myotoxicity directly or indirectly, and their mechanisms of action are different and poorly understood. Drug-induced myopathy ranges from myalgias to chronic myopathy with severe weakness or massive rhabdomyolysis.

Lipid-Lowering Drugs

Currently, more than 25 million Americans use statins. Statins are 3-hydroxy-3-methylglutaryl-CoA reductase inhibitors, which attenuate cholesterol biosynthesis and are therefore widely used to manage hypercholesterolemia (Di Stasi et al., 2010). These medications have a number of additional important pleiotropic properties, such as improving endothelial function, enhancing the stability of atherosclerotic plaques, decreasing oxidative stress and inflammation, and inhibiting the thrombogenic response (Koch, 2012). The antiinflammatory effects of statins are thought to be neuroprotective. Furthermore, statins have beneficial extrahepatic effects on the immune system, CNS, and bone (Liao and Laufs, 2005). Statins are classified into two groups based on their solubility: lipophilic (e.g., atorvastatin, cerivastatin, fluvastatin, lovastatin, pitavastatin, and simvastatin) and hydrophilic (e.g., pravastatin and rosuvastatin). Each statin can be further classified into its lactone or acid forms, and both forms for all statins interconvert to achieve equilibrium in vivo (Skottheim et al., 2008). Lipophilic statins can pass through the cell membrane by passive diffusion and therefore can be widely distributed in a variety of cells and tissues, whereas hydrophilic statins are membrane-impermeable and therefore require drug transporters to enter cells. Statin transporters (Oatp1a4 and Oatp2b1) can carry both hydrophilic and lipophilic statins across the membranes.

Statins are generally well tolerated but can produce a variety of skeletal muscle-associated dose-dependent adverse effects (Schachter, 2005; Westwood et al., 2005; Sakamoto and Kimura, 2013; Norata et al., 2014). The major adverse effects of statins can range from muscle pain, cramping, soreness, fatigue, weakness, and myotoxicity, including rhabdomyolysis, myalgia, muscle fiber breakdown, and myositis (Mukhtar and Reckless, 2005; Di Stasi et al., 2010; Kurnik et al., 2012; Sakamoto and Kimura, 2013). In rare cases, statins can cause rapid muscle breakdown and rhabdomyolysis that can lead to death (Thompson et al., 2003; Antons et al., 2006).

Statin-induced myopathy appears to be multifactorial, including impaired signal transduction, cell trafficking, gene transcription, structural protein formation and regulation, and oxidative phosphorylation (Oh et al., 2007; Klopstock, 2008; Norata et al., 2014). Abnormal fat oxidation or mitochondrial dysfunction may be the primary mechanism underlying statin-induced myopathy (Kaufmann et al., 2006; Phillips and Haas, 2008). Other possible mechanisms are reduced sarcolemmal cholesterol and isoprenoids involved in muscle fiber apoptosis (Dirks and Jones, 2006; Klopstock, 2008; Sakamoto and Kimura, 2013; Norata et al., 2014). Lipophilic statins exert a concentration-dependent adverse effect on muscle cell viability and promote cell disruption via proteolysis and apoptosis. Also, lactone forms of statins exert more extensive myopathic effects compared with their respective acid forms. The higher myopathic potential of the lactone forms of statins may be due to their lipophilic nature, which increases their passive transport across the muscle membrane, yielding a greater propensity for muscle damage (Thompson et al., 2003).

In vitro studies confirm that lipophilic statins have greater myopathic effects than hydrophilic statins (Kobayashi et al., 2008; Vaklavas et al., 2008). Kaufmann et al. (2006) investigated mitochondrial toxicity of four lipophilic statins (cerivastatin, fluvastatin, atorvastatin, and simvastatin) and one hydrophilic statin (pravastatin). In L6 cells (rat skeletal muscle cell line), the four lipophilic statins (100 μmol/L) induced death in 27% −49% of the cells. Pravastatin was not toxic up to 1 mmol/L. Cerivastatin, fluvastatin, and atorvastatin (100 μmol/L) decreased the mitochondrial membrane potential by 49%−65%, whereas simvastatin and pravastatin were less toxic. In isolated rat skeletal muscle mitochondria, all statins, except pravastatin, decreased glutamate-driven state 3 respiration and respiratory control ratio. Beta-oxidation was decreased by 88% −96% in the presence of 100 μmol/L of the lipophilic statins, but only at higher concentrations by pravastatin. Lipophilic statins caused mitochondrial swelling, cytochrome c release, and DNA fragmentation in L6 cells.

Kaufmann et al. (2006) suggested that statin-induced myopathy may be attributed to its effect on coenzyme Q10 (CoQ10), which is a component of the inner mitochondrial membrane required for oxidative phosphorylation. By inhibiting CoQ10 biosynthesis, statins impair oxidative phosphorylation and energy production (Marcoff and Thompson, 2007; Vaklavas et al., 2008). It needs to be pointed out that statin intolerance in humans may be associated with a genomic variation in the CoQ10 pathway (Oh et al., 2007).

Hanai et al. (2007) reported that lovastatin induced the expression of atrogin-1 (a key gene involved in skeletal muscle atrophy) in humans with statin myopathy, in zebrafish embryos, and in murine skeletal muscle cells in vitro. Based on in vitro animal and human experiments, there is evidence that statins induce muscle damage by induction of the muscle-specific E3 ubiquitin ligase atrogin-1. Findings revealed that atrogin-1 may be a critical mediator of the muscle damage induced by statins.

Interestingly, Sakamoto et al. (2008) noted a discrepancy in the adverse effects of statins between clinical reports and in vitro studies using skeletal muscle cell models. In an in vitro study, Sakamoto and Kimura (2013) highlighted that statins cause depletion of isoprenoids and inactivation of small GTPases (especially Rab GTPases), which are critical for statin-induced myotoxicity, and can be used as biomarkers. In clinical reports, both lipophilic and hydrophilic statins caused myotoxicity, whereas in in vitro experiments using cell lines of myoblasts, lipophilic but not hydrophilic statins exerted myotoxicity. In the skeletal myofibers, both pravastatin and fluvastatin induced vacuolation and cell death, whereas in the mononuclear cells only fluvastatin, but not pravastatin, was toxic. mRNA of the organic anion—transporting polypeptides Oatp1a4 and Oatp2b1 were expressed in the skeletal myofibers, but not in mononucleate cells. The statin transporters Oatp1a4 and Oatp2b1 are expressed in rat skeletal myofibers but not in satellite cells, fibroblasts, or L6 myoblasts. This explains why hydrophilic pravastatin affects skeletal muscle but not skeletal myoblasts.

Creatine kinase (CK) has commonly been used as a biomarker of statin-related myopathy (Thompson et al., 2006). But in some cases of statin-myopathy, serum CK is found to be normal (Phillips et al., 2002). Therefore, CK level alone as a biomarker can be misleading, as elevated CK levels can occur without myopathic effects and are often seen as a result of exercise (Di Stasi et al., 2010). Therefore, in statin therapy, CK results should be interpreted with great caution. Elevated myoglobin is another biomarker used to identify damage to myocytes and often accompanies elevated circulatory CK levels (Thompson et al., 2003). Statins are also known to cause elevation of transaminases in the absence of muscle symptoms. Slade et al. (2006) reported elevation of phosphodiesters in myopathy with statin use because of accelerated myocytes membrane turnover or reduction in cholesterol synthesis. Thus, testing of phosphodiesters may assist physicians in identifying those patients who may have adverse effects due to statin use.

Dobkin (2005) suggested evaluation of skeletal muscle composition and function to assess the presence of myopathy in statin users. Muscle biopsies are an invasive procedure that may be used in research to assess histochemical and morphological alterations, but they are not clinical tests for myopathy (Di Stasi et al., 2010). Biopsies from statin users showed myopathic effects, including diffuse lipid droplet accumulation vacuoles, cytochrome oxidase–negative myofibers, and an increased number of ragged-red fibers (Phillips et al., 2002). Dobkin (2005) suggested measurement of muscle performance, alternative to muscle biopsies, to identify the myopathy. Findings revealed functional lower-extremity weakness of the hip flexors and abductors related to myopathy, possibly independent of CK level. This noninvasive approach for muscle performance testing can be an effective measure to identify and track muscle function and recovery after statin-induced myopathy.

Drugs Used Against AIDS

Patients infected with human immunodeficiency virus (HIV) are often more susceptible to myopathy because of a variety of disease- and treatment-related reasons (Kenaston et al., 2009). Zidovudine (AZT) has been used in patients with AIDS for more than three decades. The mechanism of action of AZT is inhibition of nucleoside reverse transcriptase. AZT acts as a false substrate for the mitochondrial DNA polymerase, thereby reducing mitochondrial DNA synthesis and causing mitochondrial myopathy. Although the exact mechanism of muscle toxicity is not known, patients may develop myopathy as myalgia, fatigue, muscle wasting, and diffuse weakness with a proximal predominance, along with an elevated CK. Pathological findings include muscle fiber necrosis and vacuoles with ragged-red fibers. Electron microscopy findings include changes in mitochondria, myofilaments, and tubules (Misulis et al., 2000). Unfortunately, a clear differentiation of AIDS myopathy from AZT myopathy is not always possible. Lee et al. (2012) reported skeletal muscle toxicity in HIV-infected adults treated with raltegravir. Findings revealed isolated CK elevation, myalgia without motor weakness, proximal myopathy on physical examination, and rhabdomyolysis.

Antiinflammatory Drugs

Increased levels of proinflammatory cytokines, both in the circulation and the skeletal muscle, have been attributed a key role in the pathophysiology of cachexia and muscle wasting (Fanzani et al., 2012). In this context, a more general antiinflammatory approach, as well as the use of so-called "immunomodulators," could be of interest. Thalidomide, pentoxyfylline, and statins exert antiinflammatory effects.

Antitumor and Anticancer Drugs

Several studies have attempted to identify the molecular mechanisms responsible for toxicity induced by antitumor and anticancer drugs. Doxorubicin (Dox), a potent antitumor agent used in cancer treatment, is myotoxic and results in significant reductions in skeletal muscle mass and function. It is established that Dox-induced toxicity is associated with an increased generation of ROS and oxidative damage within muscle fibers. Smuder et al. (2011) showed increased oxidative stress and activation of cellular proteases (calpain and caspase-3) in skeletal muscles of rats treated with Dox. Calpain is a Ca-dependent cysteine protease that promotes muscle atrophy by cleaving structural proteins. Increased ROS can lead to elevated levels of cytosolic Ca, which leads to calpain activation (Goll et al., 2003). In an in vitro study, Green and Leeuwenburgh (2002) observed that Dox administered to isolated mitochondria caused an increase in superoxide production, followed by increases in caspase-3 activity and cytochrome *c* efflux in H9C2 myocyte cells. Dox-induced cell death occurred because of mitochondrial dysfunction and apoptosis. Vincristine can cause neuropathy and myopathy, but its mechanism for myopathy is not clearly understood.

Various novel biomarkers, such as skeletal troponin I (sTnI), myosin light chain 3 (Myl3), CK M isoform (CKm), and fatty acid binding protein 3 (Fabp3) have been suggested in drug-induced skeletal muscle injury as they are more sensitive and tissue specific than conventional biomarkers, such as CK, AST, and LDH (Tonomura et al., 2012; EMA, 2015; FDA, 2015; Burch et al., 2016; Vlasakova et al., 2016; Goldstein, 2017).

DRUGS OF ABUSE

Alcohol, cocaine, opiates (heroin or pentazocine), and many other drugs of abuse are known to cause muscle damage by different mechanisms. In alcoholic binges, multiple mechanisms, including hypokalemia, hypophosphatemia, coma, or agitation, may combine to produce muscle damage. Alcohol also induces increases in ROS and RNS, thereby causing mitochondrial dysfunction. Drugs, such as ethanol, heroin, amphetamines, cocaine, and other sedatives, antidepressants, and stimulants are known to cause muscle damage by rhabdomyolysis (Richards, 2000; Chabria, 2006; Töth and Varga, 2009). The most serious sequela of rhabdomyolysis in alcoholics is renal failure. Muscle necrosis results in the release of large amounts of enzymes into the blood, such as CK, LDH, and others. Myoglobin is greatly increased in the blood and is excreted in the urine (Misulis et al., 2000). Rhabdomyolysis seen in association with heroin and barbiturate abuse is probably related to local pressure necrosis from the patient's own body while the patient is obtunded or unconscious. Rhabdomyolysis has also been associated with phencyclidine, possibly because of muscle damage caused by the use of restraints in hyperactive–combative patients. The acute dystonic reaction induced by phencyclidine may also cause muscle damage. Amphetamine abuse has been associated with rhabdomyolysis, and hyperpyrexia and increased muscular activity seem to cause muscle damage. Cocaine-induced rhabdomyolysis could be due to the sympathetic overactivity leading to muscular hyperactivity causing microinfarctions with skeletal muscle necrosis (Lombard et al., 1988). Elevated CK activity and myoglobinuria support the finding of cocaine-induced rhabdomyolysis and can be used as biomarkers of muscle toxicity.

Intramuscular injections, particularly repeated injections of opiates, may cause muscle damage with a fibrotic reaction that can result in muscle contractures (von Kemp et al., 1989). Subcutaneous administration of barbiturates may also cause focal necrosis of the adjacent muscles.

VENOMS AND ZOOTOXINS

Envenoming by snakebite, scorpion sting, and other zootoxins constitutes a significant health problem in many parts of the world, especially in developing countries. Most venoms and poisons are mixtures of a variety of chemical compounds that often act synergistically to produce their toxic effects (Gwaltney-Brant et al., 2018). Recent studies suggest that venoms and poisons from zootoxins cause tissue damage at the site of a bite or sting, but they can affect various vital organs, such as those in the nervous, cardiovascular, and reproductive and developmental systems (Del Brutto and Del Brutto, 2012; Gwaltney-Brant, 2017; Spyres et al., 2018).

Snake Toxins

More than 5 million people are bitten by snakes every year, resulting in more than 2 million cases of envenoming and 20,000–125,000 deaths (Kasturiratne et al., 2008). Venomous snakes are found in the families of *Colubridae*, *Crotalidae*, *Elapidae*, *Hydrophidae*, *Laticaudidae*, and *Viperidae* (Gwaltney-Brant, 2017; Gwaltney-Brant et al., 2018). Snake venoms are known to contain a complex mixture of neurotoxins, myotoxins, and coagulant toxins affecting the coagulation cascade and neuromuscular transmission, as well as causing neurological complications.

Metalloproteinases, serine proteases, and C-type lentins (common in viper and colubrid venoms) have anticoagulant or procoagulant activity and may be either agonists or antagonists of platelet aggregation. As a result, ischemic or hemorrhagic strokes may occur. In contrast, the venom of elapids is rich in phospholipase A2 and three-finger proteins, which are potent neurotoxins affecting neuromuscular transmission at either presynaptic or postsynaptic levels. Presynaptic-acting neurotoxins (β-neurotoxins) inhibit the release of ACh from the presynaptic terminals. Examples of β-neurotoxins are taipoxin, paradoxin, trimucrotoxin, viperotoxin, textilotoxin, and crotoxin. On the other hand, postsynaptic-acting neurotoxins (α-neurotoxins) are three-finger protein complexes, and these cause a reversible blockage of ACh receptors by exerting a curare-like mechanism of action (Del Brutto and Del Brutto, 2012). The best-characterized α-neurotoxins are irditoxins. These neurotoxins do not cross the blood–brain barrier (BBB), but they cause muscle paralysis by affecting neuromuscular transmission at either presynaptic or postsynaptic levels (Lewis and Gutmann, 2004). Some snake venoms contain both α- and β-neurotoxins and produce a complex blockage of neuromuscular transmission.

Mamba snake venom contains a mixture of neurotoxic compounds, including postsynaptic cholinoreceptor α neurotoxins, dendrotoxins, fasciculins, and muscarinic toxins (Anadon et al., 2015). Effects at the NMJ include AChE inhibition by fasciculins (named as they produced long-lasting fasciculations in lab animals) and increased presynaptic release of ACh by dendrotoxins (polypeptides that facilitate ACh release in response to nerve stimulation). In addition, mamba venom is very rich in ACh (6–24 mg/g), which facilitates neuromuscular transmission.

Fasciculins belong to the structural family of three-fingered toxins closely related by ~6,750 Da peptides. Fasciculins (Fasciculin 1–4) are proteinic AChE inhibitors as they bind to a peripheral regulatory anionic site of AChE in a noncompetitive and irreversible manner in mammals, especially at NMJ. Administration of fasciculin 1 and fasciculin 2 to mice at doses of 1–3 mg/kg and 0.05–2 mg/kg, ip, respectively, caused severe, generalized, and long-lasting fasciculations (5–7 h), followed by a gradual recovery to normal behavior. In vitro preincubation with fasciculins at concentrations of 0.01 μg/mL inhibited brain and muscle AChE up to 80%. Histochemical assay of AChE showed an almost complete disappearance of the black-brown precipitate at the neuromuscular endplate after in vitro incubation with fasciculins. The dendrotoxins exert toxicity by inhibiting voltage-dependent K^+ channels in motor nerve terminals, and they facilitate ACh release at the NMJ. Postsynaptic toxins present in mamba venom bind to and block nAChRs. The muscarinic toxins present in mamba venom are small (7 kDa) proteins that selectively bind to mAChRs. Recently, about 12 muscarinic toxins have been isolated and they differ in affinity for muscarinic receptor subtypes.

Snakebite envenoming has been characterized by complex pathological and pathophysiological profiles that include prominent local tissue damage, i.e., necrosis, hemorrhage, blistering and edema, and systemic alterations, such as bleeding, coagulopathy, cardiovascular shock, and renal failure (Gutiérrez et al., 2009a; b; Hernandez et al., 2011; Ranawaka et al., 2013; Da Silva et al., 2018; Teixera et al., 2018). Local tissue damage leading to necrosis of the skin and subcutaneous tissue is particularly relevant because it is frequently followed by poor tissue regeneration, with the occurrence of permanent sequelae associated with tissue loss and dysfunction. Local tissue damage in snakebite envenoming is due to the action of myotoxic phospholipase A_2 (PLA_{2s}), hemorrhagic snake venom metalloproteinases (SVMPs), and possible involvement of other components, such as hyaluronidases and nonhemorrhagic SVMPs (Gutiérrez et al., 2009a; Gutiérrez and Rucavado, 2000; Gutiérrez and Ownby, 2003; Teixera et al., 2018). The SVMPs are large multidomain proteins, which are classified as P-I, P-II, and P-III based on the presence or absence of additional nonproteinase domains (Fox and Serrano, 2005). The SVMPs are also classified as adamalysins that, alongside the matrixins, astacins and serralysins, make up the large metzincin gene superfamily (Casewell, 2012).

Cytotoxic enzymes (PLA_{2s} and metalloproteinases) activate proinflammatory mechanisms that cause edema, blister formation, and local tissue necrosis, which are a common occurrence at the site of a venomous snakebite. These enzymes also favor the release of bradykinin, prostaglandin, cytokines, and sympathomimetic amines, which are responsible for the pain experienced by the victims (Teixeira et al., 2009, 2018). Intense local pain, paresthesia, and local necrotic symptoms are more severe after viper, coral, and colubrid bites (Teixeira et al., 2009, 2018; Da Silva et al., 2018).

Chu et al. (2010) described venom ophthalmia as a syndrome characterized by ocular pain, hyperemia, blepharitis, and corneal erosions, which may occur when the venom of spitting cobras enters the eye of the victim. The myotoxins and neurotoxins can cause muscle weakness and paralysis by blocking the synaptic transmission at either presynaptic or postsynaptic levels (Lewis and Gutmann, 2004; Ranawaka et al., 2013). The first detectable signs of paralysis are palpebral ptosis and external ophthalmoplegia because ocular muscles are more susceptible to this neuromuscular blockage (Sanmuganathan, 1998). In the next few hours, facial and neck muscles become affected, followed by respiratory and limb muscles (Del Brutto and Del Brutto, 2012). The most common electromyographic (EMG) findings in these patients include prolongation of motor distal latencies and reduction in the amplitude of compound muscle action potentials (Singh et al., 1999). Lewis and Gutmann (2004) reported that generalized myokymia may occur after the bite of some rattlesnake species. The syndrome of continuous and spontaneous muscular activity resembling fasciculations and EMG finding of repetitive firing of motor union potentials was possibly due to a particular neurotoxin acting on voltage-gated K^+ channels, thus increasing nerve excitability.

Snake venom also negatively impact skeletal muscle regeneration. Skeletal muscle regeneration is a complex process, which involves the interaction of myogenic cells, resident cells, inflammatory cells, blood vessels, nerves, and extracellular matrix (Cicillot and Schiaffino, 2010). Injury to muscle fibers by snake venom leads to the activation of a population of quiescent myogenic cells, satellite cells located at the periphery of the fibers between muscle sarcolemma and the basement membrane (Cicillot and Schiaffino, 2010; Hernandez et al., 2011).

These authors and others have described two different patterns of muscle regeneration after snakebite-induced myonecrosis. When tissue is affected by venoms of toxins that cause muscle necrosis but do not affect the integrity of blood vessels, such as after injection of isolated myotoxic phospholipase A_2, regeneration proceeds successfully (Harris, 2003). When venoms affect both muscle fibers and the microvasculature, including hemorrhage, the regenerative process is impaired, with substitution of muscle tissue by fibrosis in some areas, and with the presence of regenerating fibers of reduced diameter (Gutiérrez et al., 1984; Salvini et al., 2001).

In an experimental study conducted in mice (CD1), Hernandez et al. (2011) showed that skeletal muscle regeneration is partially deficient after acute tissue damage induced by the injection of *Bothrops asper* venom by a hemorrhagic SVMP. A successful regenerative response occurs after myonecrosis induced by a myotoxin, which affects muscle fibers, but not the microvasculature nor the basement membranes of muscle fibers and nerves. Injection of the venom, at a dose that induces prominent hemorrhage, resulted in a drastic and rapid drop in the number of capillary vessels in gastrocnemius muscle, as evaluated by both the capillary/muscle fiber ratio and the number of capillaries per tissue area. These results supported the hypothesis that SVMP-induced basement membrane damage, in microvessels, muscle fibers, and nerves, is the main culprit for the poor regenerative outcome in the mouse muscle model. Fibrosis substitutes muscle tissue in some areas, and the diameter of regenerating fibers is abnormally small. Myotoxin, SVMP BaP1, and venom affected the integrity of axons in intramuscular nerves. In addition to poor muscle and microvascular regeneration, axonal regeneration is impaired. These experimental observations provided mechanistic insights and corroborated with clinical findings of poor muscle, nerve, and vasculature-regenerative outcome.

Identification and quantification of snake venom toxins, EMG findings, and biochemical and histopathological findings can be used as biomarkers of muscle, cardiac, and renal injuries inflicted by snakebites (Teixeira et al., 2009, 2018; Hernandez et al., 2011; Casewell, 2012; Del Brutto and Del Brutto, 2012; Pycroft et al., 2012; Ranawaka et al., 2013).

Spider Venom

More than 42,700 extant species of spiders have been described, and an even greater number remain to be characterized (Klint et al., 2012). Spider venoms are complex chemical cocktails, but the major components are small, disulfide-rich peptides (Kumar et al., 2011; Peng et al., 2014; Gwaltney-Brant et al., 2018). Because single venom can contain as many as 1000 peptides, over 10 million bioactive peptides are likely to be present in the venoms of spiders (Escoubas et al., 2006). The spider-venom peptides that have been described to date are detailed in ArachnoServer, a manually created database that provides information about proteinaceous toxins from spiders (Herzig et al., 2011).

Arachnids (spiders and scorpions), such as the brown recluse, the redbacked spider, and the mouse spider, often exert more neurotoxic effects on their victims; aches and acute rhabdomyolysis along with some necrotic lesions have been described in some cases (Gala and Katelaris, 1992). Rader et al. (2012) described positive and negative examination features for brown recluse envenomation and noted that spider bites predominantly occurred during April–October.

Spider-venom peptides produce toxicity by targeting Na_V channels. The term NaScTx is employed for scorpion toxins and NaSpTx is for spider venoms, and these toxins target Na_V channels (Rodriguez de la Vega and Passani, 2005; Klint et al., 2012; Peng et al., 2014). Kumar et al. (2011) described the structure of BTK-2 toxin, which is a novel $hK_V1.1$ inhibiting scorpion toxin. There are 12 different families of spider-venom peptides that target Na_V channels. Scorpion α and β toxins interact with voltage-gated Na channels (Na_Vs) at two pharmacologically distinct sites (Gurevitz, 2012). α Toxins bind at receptor site-3 and inhibit channel inactivation, whereas β toxins bind at receptor site-4 and shift the voltage-dependent activation toward more hyperpolarizing potentials. The two toxin classes are subdivided into distinct pharmacological groups according to their binding preferences and ability to compete for receptor sites at Na_V subtypes.

Black widow spiders are known to contain α latrotoxin. The toxin induces a pore in the myocytes membrane that is permeable to Na^+ and K^+, leading to membrane depolarization. Initial effects are muscle spasms, followed by depletion of ACh, and death occurs because of respiratory failure. In an electrophysiological study, Peng et al. (2014) found that black widow spiderling extract (10 μg/mL) could completely block the neuromuscular transmission in isolated mouse nerve-hemidiaphragm preparations within 21 ± 1.5 min, and 100 μg/mL extract could inhibit a certain percentage of voltage-activated Na^+, K^+, and Ca^{2+} channel currents in rat dorsal ganglion neurons. Recently, O'Connor et al. (2018) described clinical manifestations, such as pain, paresthesia, rhabdomyolysis, and cardiac abnormalities in humans after envenomation with bark scorpion (*Centruroides sculpturatus*). In an experimental study, Heidarpour et al. (2012) demonstrated that scorpion venom when injected in mice (5 mg/kg, sc) caused histopathological alterations in the heart and kidney.

BOTULINUM TOXIN

Botulinum neurotoxins (BoNTs) are produced primarily by *Clostridium botulinum* under anaerobic conditions; however, other *Clostridial* species, such as *Clostridium barati*, *Clostridium butyricum*, and *Clostridium argentinense* are also capable of producing neurotoxins. There are seven immunologically distinct serotypes of BoNT, and they are designated alphabetically, A to G (Simpson, 1981). Serotypes A, B, C1, and D have been associated with outbreaks of botulism in domestic animals, livestock, poultry, and wildlife, whereas serotypes A, B, E, and rarely F are known to cause diseases in humans (Cope, 2018). The active neurotoxin has a molecular mass of 150 kDa and exists as a polypeptide di-chain molecule. The di-chain consists of a heavy chain (HC) linked by a single disulfide bond to a light chain (LC). The 100-kDa HC is responsible for membrane targeting and cellular uptake, whereas the 50-kDa light chain mediates its intracellular action. The neurotoxin molecule can be further divided both structurally and functionally into three domains. The HC contains both a binding domain and a translocation domain, whereas the smaller LC contains a catalytically active domain. The LC of each toxin functions as a zinc-dependent endoprotease, cleaving at least one of three soluble *N*-ethylmaleimide–sensitive fusion (NSF) protein attachment receptor (SNARE) proteins involved in neurotransmitter release.

BoNTs selectively target peripheral cholinergic terminals because these toxins are too large to cross the BBB. Cope (2018) described that the mechanism of action of BoNT is a multistep process, which interrupts normal vesicular release of ACh from the presynaptic motor nerve terminals. The primary target site is the NMJ, where BoNTs act as potent inhibitors of synaptic transmission in skeletal muscles. In the first stage, the toxins must bind, via their HCs, to protein receptors (and gangliosides) located on the plasma membrane of the motor nerve terminal. The second stage involves receptor-mediated uptake of the toxin into endosomes (e.g., endocytosis). The third stage occurs within the endosome, where the disulfide linkage between the toxin heavy and light chains is reduced, and a subsequent drop in endosomal pH promotes a conformational change in the toxin molecule. Following the translocation to the cytosol, the toxin light chain is free to act on the intracellular target. During the final stage, the neurotoxins enzymatically cleave one of the three specific proteins found within the presynaptic terminal. These three proteins, SNAP-25, synaptobrevin, and syntaxin, known collectively as SNARE (soluble *N*-ethylmaleimide sensitive factor attachment protein receptor) proteins, are necessary for neurotransmitter release. Proteolysis of any of the SNARE proteins by BoNT either destabilizes or prevents the formation of functional SNARE complexes, inhibiting vesicular fusion and neurotransmitter release.

The catalytic active sites of the different BoNT serotypes vary slightly, giving each serotype both substrate and cleavage site specificity. Toxin serotypes A, C, and F cleave SNAP-25; serotypes B, D, F, and G cleave synaptobrevin; and serotype C cleaves syntaxin. BoNT prevents docking of synaptic vesicles with presynaptic plasma membrane by selective proteolysis of synaptic proteins. Neurotoxin-mediated cleavage of any of these substrates disrupts the processes involved in the exocytotic release of ACh and leads to descending flaccid paralysis of the affected skeletal muscles, starting usually in the bulbar musculature to involve deficits in sight, speech, and swallowing. Paralysis eventually progresses beyond cranial nerve palsies to include generalized muscle weakness and loss of critical accessory muscles of respiration. If untreated, death is inevitable from airway obstruction secondary to paralysis of pharyngeal, diaphragm, and accessory respiratory muscles, as well as loss of the protective gag reflex.

Exposure to BoNTs can produce lethal diseases in humans and animals (Anderson and Hilmas, 2015; Pirazzini et al., 2017; Cope, 2018). BoNT is the most potent biological toxin ever encountered, with lethal doses as low as 0.03 ng/kg body wt depending on both toxin type and animal species. In humans, six different forms of botulism have been described. These include: (1) foodborne botulism, (2) infant botulism, (3) wound botulism, (4) an adult form of infant botulism, (5) inadvertent systemic botulism, and (6) inhalation botulism. Clinically, botulism is recognized as a lower motor neuron disease resulting in progressive flaccid paralysis. Although deficits in somatic neuromuscular transmission are the most prominent effects, motor deficits in cranial nerve function, as well as the autonomic nervous system, have also been described. In animals, paresis begins in the hind limbs and progresses cranially, often resulting in quadriplegia and recumbency. In both humans and animals, death may result from respiratory muscle paralysis.

The onset of symptoms in botulism depends on the amount of toxin ingested or inhaled and the related kinetics of absorption (Anderson and Hilmas, 2015; Cope, 2018). Time to onset of clinical signs can range from as soon as 2 h to as long as 8 days, although symptoms typically appear between 12 and 72 h after consumption of toxin-contaminated food. It is important to mention that regardless of the exposure route, once BoNTs reach the circulation or lymphatic system, they lead to inhibition of the release of ACh from peripheral cholinergic nerve terminals, resulting in flaccid muscle paralysis. Of course, the specific paralytic profiles associated with each of the BoNTs are typically attributed to their unique proteolytic activities within the nerve terminal. These activities are known to be mediated by the LC components of the various neurotoxins. It is noteworthy that the muscle paralytic profile may vary for different serotypes and between animal species.

The mouse bioassay model is considered the gold standard for diagnosing botulism. The mouse LD_{50} still quantitates the purity of BoNT batches and is the basis of the international standard used in serum neutralization assay of BoNT antitoxin. The mouse phrenic nerve-hemidiaphragm assay has been used to measure the effect of BoNT on skeletal muscle contraction, and the doses necessary for inhibition are well characterized. Characteristic muscle paralysis is of some value in the diagnosis of botulism. Enzyme-linked immunosorbent assay (ELISA) is widely used to identify BoNTs. Recently, LC-MS has been used for identification, confirmation, and quantification of BoNTs. Identification and confirmation of BoNTs are the best biomarkers of botulism. For details on the pharmacology, toxicology, and biomarkers of BoNT, readers are referred to Anderson and Hilmas (2015), Pirazzini et al. (2017), and Cope (2018).

MYOTOXIC PLANTS

Muscle toxicity by poisonous plants is more common in domestic animals because of their food habits. Some plant toxins directly affect the muscular system, whereas others primarily affect organs such as the liver and nervous system, with secondary effects noted in the muscles. Plant toxins are known to exert a variety of toxic effects, including muscle weakness, lameness, myodegeneration, myopathy, etc. A list of plants and their toxins that are known to exert myotoxicity is provided in Table 20.4. Some of these plant toxins can cause minor pain or muscle weakness, whereas others can cause paralysis. In general, identification of a plant-specific toxin, elevated serum CK activity, and myoglobinuria are commonly used as biomarkers of muscle toxicity. In addition, microscopic changes in skeletal muscles are characteristic findings.

Plant-induced muscle toxicity and its biomarkers are discussed in detail in Chapter-37 of this book.

TABLE 20.4 Muscle Toxicity Caused by Plant Toxins

Plant	Toxin	Toxic Effects
Tansy Ragwort (*Senecio jacobea*)	Not identified	Muscle weakness and lameness
Day-Blooming Jessamine (*Cestrum diurnum*)	Similar to 1,25-dihydroxy-cholecalciferol	Calcinosis and lameness in horses
Golden Chain Tree (*Laburnum anagyroides*)	Not identified	Myodegeneration
Scotch Broom (*Cytisus scoparius*)	Quinolizidine alkaloids (sparteine, isosparteine, cytisine)	Myodegeneration
Coyotillo (*Karwinskia humboldtiana*)	Not identified	Myodegeneration
Mountain Thermopsis (*Thermopsis montana*)	Quinolizidine alkaloids (thermopsine, cytisine, *N*-methylcytisine, anagyrine)	Myodegeneration and muscle tremors
Coffee Weed/Sickle Pod Senna (*Cassia* spp.)	Not identified	Myodegeneration
Black Walnut (*Juglans nigra*)	Juglone	Laminitis
Hoary Alyssum (*Berteroa incana*)	Not identified	Laminitis and limb edema
Flatweed/Cat's Ears (*Hypochaeris radicata*)	Not identified	Lameness
Lupines (*Lupinus* spp.)	Anagyrine and ammodendrine	Muscular weakness and fasciculations
Poison Hemlock (*Conium maculatum*)	γ-Coniceine	Muscle fasciculations
Water Hemlock (*Cicuta* spp.)	Cicutoxin	Muscular tremors and myodegeneration
White Snakeroot (*Eupatorium rogosum*)	Tremetol	Muscle tremors
Rayless Goldenrod (*Haplopappus heterophyllus*)	Tremetone	Muscle tremors

CONCLUDING REMARKS AND FUTURE DIRECTIONS

The skeletal muscles account for nearly half of mammalian body weight, and they are the targets of toxicity for many drugs of use and abuse, pesticides and biotoxins. Muscles have different fiber type compositions and are enriched with binding and reactive sites, such as receptors, neurotropic factors, neurotransmitters, enzymes, drug transporter systems, etc. Currently, drug-induced myopathy is among the most common causes of muscle disease. Muscle toxicity can range from minor pain to something as severe as paralysis. For some chemicals, such as OP pesticides and nerve agents, the mechanism of muscle toxicity is well understood, whereas for other chemicals, drugs, and biotoxins, it is poorly understood. Chemicals can cause muscle cell death due to apoptosis, necrosis or autophagy. The presence of chemicals or their metabolites can be used as biomarkers of exposure, whereas histochemical and biochemical endpoints can be used as biomarkers of muscle toxicity. Recently, many new biomarkers of muscle toxicity have been added to the list of conventional biomarkers. Future studies will unravel novel mechanisms of action and biomarkers of toxicity, which will be more sensitive and organ specific.

References

Adler, M., Hinman, D., Hudson, C.S., 1992. Role of muscle fasciculations in the generation of myopathies in mammalian skeletal muscle. Brain Res. Bull. 29, 179–187.

Anadon, A., Martínez-Larrañaga, M.R., Valerio, L.G., 2015. Onchidal and fasciculins. In: Gupta, R.C. (Ed.), Handbook of Toxicology of Chemical Warfare Agents, second ed. Academic Press/Elsevier, Amsterdam, pp. 411–420.

Anderson, J., Hilmas, C.J., 2015. Botulinum toxin. In: Gupta, R.C. (Ed.), Handbook of Toxicology of Chemical Warfare Agents, second ed. Academic Press/Elsevier, Amsterdam, pp. 361–385.

Andreoli, S.P., 1993. ATP depletion and cell injury: what is the relationship? J. Lab. Clin. Med. 122, 232–233.

Antons, K.A., Williams, C.D., Baker, S.K., Phillips, P.S., 2006. Clinical perspectives of statin-induced rhabdomyolysis. Am. J. Med. 119, 400–409.

Barnard, E.A., Lai, J., Pizzey, J., 1984. Synaptic and extra-synaptic forms of acetylcholinesterase in skeletal muscles: variation with fiber type and functional considerations. In: Serratrice, G. (Ed.), Neuromuscular Diseases. Raven Press, Baltimore, pp. 455–463.

Besser, R., Gutman, L., Weilemann, L.S., 1989. Inactivation of end-plate acetylcholinesterase during the course of organophosphate intoxications. Arch. Toxicol. 63, 412–415.

Bright, J.E., Inns, R.H., Tuckwell, N.J., Griffiths, G.D., Marrs, T.C., 1991. A histochemical study of changes observed in the mouse diaphragm after organophosphate poisoning. Hum. Exp. Toxicol. 10, 9–14.

Burch, P.M., Hall, D.G., Walker, E.G., et al., 2016. Evaluation of the relative performance of drug-induced skeletal muscle injury biomarkers in rats. Toxicol. Sci. 150 (1), 247–256.

Casewell, N.R., 2012. On the ancestral recruitment of metalloproteinases into the venom of snakes. Toxicon 60, 449–454.

Chabria, S.B., 2006. Rhabdomyolysis: a manifestation of cyclobenzaprine toxicity. Case report. J. Occup. Med. Toxicol. 1, 16. https://doi.org/10.186/1745-6673-1-16.

Chu, E.R., Winstein, S.A., White, J., Warrell, D.A., 2010. Venom ophthalmia caused by venom of the spitting elapids and other snakes: report of ten cases with review of epidemiology, clinical features, pathophysiology and management. Toxicon 56, 259–272.

Cicillot, S., Schiaffino, S., 2010. Regeneration of mammalian skeletal muscle. Basic mechanisms and clinical implications. Curr. Pharmaceut. Des. 16, 906–914.

Clanton, T.L., Zuo, L., Klawitter, P., 1999. Oxidants and skeletal muscle function: physiologic and pathophysiologic implications. Proc. Soc. Exp. Biol. Med. 222, 253–262.

Cope, R.B., 2018. Botulinum neurotoxins. In: Gupta, R.C. (Ed.), Veterinary Toxicology: Basic and Clinical Principles, third ed. Academic Press/Elsevier, Amsterdam, pp. 743–757.

Da Silva, I.M., Bernal, J.C., Bisneto, P.F.G., et al., 2018. Snakebite by *Micrurus averyi* (Schmidt, 1939) in the Brazilian Amazon basin: case report. Toxicon 141, 51–54.

Del Brutto, O.H., Del Brutto, V.J., 2012. Neurological complications of venomous snake bites: a review. Acta Neurol. Scand. 125, 363–372.

Di Stasi, S.L., MacLeod, T.D., Winters, J.D., Binder-MacLeod, S.A., 2010. Effects of statins on skeletal muscle: a perspective for physical therapies. Phys. Ther. J 90, 1530–1542.

Diaz-Villasenor, A., Burns, A.L., Hiriart, M., et al., 2007. Arsenic-induced alteration in the expression of genes related to type 2 diabetes mellitus. Toxicol. Appl. Pharmacol. 225, 123–133.

Dinis-Oliveira, R.J., Duarte, J.A., Sanchez-Navarro, A., et al., 2008. Paraquat poisoning: mechanisms of lung toxicity, clinical features, and treatment. Crit. Rev. Toxicol. 38, 13–71.

Dirks, A.J., Jones, K.M., 2006. Statin-induced apoptosis and skeletal myopathy. Am. J. Physiol. 291, C1208–C1212.

Dobkin, B.H., 2005. Underappreciated statin-induced myopathic weakness causes disability. Neurorehabil. Neural Repair 19, 259–263.

European Medicines Agency (EMA), 2015. Letter of Support for Skeletal Muscle Injury Biomarkers. Available at: http://www.ema.europa.eu/docs/en_GB/document_library/Other/2015/03/WC500184458.pdf.

Escoubas, P., Sollod, B., King, G.F., 2006. Venom landscapes: mining the complexity of spider venoms via a combined cDNA and mass spectrometric approach. Toxicol 47, 650–663.

Fanzani, A., Conraads, V.M., Penna, F., Martinet, W., 2012. Molecular and cellular mechanisms of skeletal muscle atrophy: an update. J. Cachexia Sarcop Muscle 3, 163–179.

Food and Drug Administration (FDA), 2015. Biomarker Letter of Support for Plasma/serum Myl3, Skeletal Muscle Troponin I (STnI), Fabp3, and Creatine Kinase, Muscle Type (CK-M, the Homodimer CK-MM). Available at: http://www.fda.gov/downloads/Drugs/DevelopmentApprovalProcess/DrugDevelopmentToolsQualificationProgram/UCM432653.pdf.

Fox, J.W., Serrano, S.M.T., 2005. Structural considerations of the snake venom metalloproteinases, key members of the M12 reprolysin family of metalloproteinases. Toxicon 45, 969–985.

Fukushima, T., Hojo, N., Isobe, A., et al., 1997. Effects of organophosphorus compounds on fatty acid compositions and oxidative phosphorylation system in the brain of rats. Exp. Toxicol. Pathol. 49, 381–386.

Gala, S., Katelaris, C.H., 1992. Rhabdomyolysis due to redback spider envenomation. Med. J. Aust. 157, 66.

Gasparik, J., Vladarova, D., Capcarova, M., et al., 2010. Concentration of lead, cadmium, mercury and arsenic in leg skeletal muscles of three species of wild birds. J. Environ. Sci. Health, Part A 45 (7), 818–823.

Goldstein, R.A., 2017. Skeletal muscle injury biomarkers: assay qualification efforts and translation to the clinic. Toxicol. Pathol. 45 (7), 943–951.

Goll, D.E., Thompson, V.F., Li, H., et al., 2003. The calpain system. Physiol. Rev. 83, 731–801.

Green, P.S., Leeuwenburgh, C., 2002. Mitochondrial dysfunction is an early indicator of doxorubicin-induced apoptosis. Biochim. Biophys. Acta 1588, 94–101.

Gultekin, F., Ozturk, M., Akdogan, M., 2000. The effect of organophosphate insecticide chlorpyrifos-ethyl on lipid peroxidation and antioxidant enzymes (*in vitro*). Arch. Toxicol. 74, 533–538.

Gupta, R.C., 2018. Non-anticoagulant rodenticides. In: Gupta, R.C. (Ed.), Veterinary Toxicology: Basic and Clinical Principles. Academic Press/Elsevier, Amsterdam, pp. 613–626.

Gupta, R.C., Dettbarn, W.-D., 1992. Potential of memantine, *d*-tubocurarine, and atropine in preventing acute toxic myopathy induced by organophosphate nerve agents: soman, sarin, tabun and VX. Neurotoxicology 13, 649–662.

Gupta, R.C., Goad, J.T., 2000. Role of high-energy phosphates and their metabolites in protection of carbofuran-induced biochemical changes in diaphragm muscle by memantine. Arch. Toxicol. 74, 13–20.

Gupta, R.C., Kadel, W.L., 1989. Concerted role of carboxylesterases in the potentiation of carbofuran toxicity by iso-OPMA pretreatment. J. Toxicol. Environ. Health 26, 447–457.

Gupta, R.C., Kadel, W.L., 1991. Novel effects of memantine in antagonizing acute aldicarb toxicity: mechanistic and applied considerations. Drug Dev. Res. 24, 329–341.

Gupta, R.C., Patterson, G.T., Dettbarn, W.-D., 1985. Mechanisms involved in the development of tolerance to DFP toxicity. Fundam. Appl. Toxicol. 5, S17–S28.

Gupta, R.C., Patterson, G.T., Dettbarn, W.-D., 1986. Mechanisms of toxicity and tolerance to diisopropylphosphorofluoridate at the neuromuscular junction of the rat. Toxicol. Appl. Pharmacol. 84, 541–550.

Gupta, R.C., Patterson, G.T., Dettbarn, W.-D., 1987a. Biochemical and histochemical alterations following acute soman intoxication in the rat. Toxicol. Appl. Pharmacol. 87, 393–402.

Gupta, R.C., Patterson, G.T., Dettbarn, W.-D., 1987b. Acute tabun toxicity: biochemical and histochemical consequences in brain and skeletal muscles of rat. Toxicology 46, 329–341.

Gupta, R.C., Patterson, G.T., Dettbarn, W.-D., 1991a. Comparison of cholinergic and neuromuscular toxicity following acute exposure to sarin and VX in rat. Fundam. Appl. Toxicol. 16, 449–458.

Gupta, R.C., Goad, J.T., Kadel, W.L., 1991b. Carbofuran-induced alterations (*in vivo*) in high-energy phosphates, creatine kinase (CK) and CK isoenzymes. Arch. Toxicol. 65, 304–310.

Gupta, R.C., Goad, J.T., Kadel, W.L., 1994. Cholinergic and noncholinergic changes in skeletal muscles by carbofuran and methyl parathion. J. Toxicol. Environ. Health 43, 291–304.

Gupta, R.C., Dettbarn, W.-D., Sanecki, R.K., Goad, J.T., 1999. Biochemical and microscopic changes in diaphragm muscle by acute carbamate toxicity [abstract]. Toxicol. Sci. 48, 188.

Gupta, R.C., Milatovic, D., Dettbarn, W.-D., 2002. Involvement of nitric oxide in myotoxicity produced by diisopropylphosphorofluoridate (DFP)-induced muscle hyperactivity. Arch. Toxicol. 76, 715–726.

Gupta, R.C., Zaja-Milatovic, S., Dettbarn, W.-D., Malik, J.K., 2015. Skeletal muscle. In: Gupta, R.C. (Ed.), Handbook of Toxicology of Chemical Warfare Agents, second ed. Academic Press/Elsevier, Amsterdam, pp. 577–597.

Gurevitz, M., 2012. Mapping of scorpion toxin receptor sites at voltage-gated sodium channels. Toxicon 60, 502–511.

Gutiérrez, J.M., Ownby, C.L., 2003. Skeletal muscle degeneration induced by venom phospholipases A_2: insights into the mechanisms of local and systemic myotoxicity. Toxicon 42, 915–931.

Gutiérrez, J.M., Rucavado, A., 2000. Snake venom metalloproteinases: their role in the pathogenesis of local tissue damage. Biochimie 82, 841–850.

Gutiérrez, J.M., Ownby, C.L., Odell, G.V., 1984. Skeletal muscle regeneration after myonecrosis induced by crude venom and a myotoxin from the snake *Bothrops asper* (Fer-de-Lance). Toxicon 22, 719–731.

Gutiérrez, J.M., Rucavado, A., Chaves, F., et al., 2009a. Experimental pathology of local tissue damage induced by *Bothrops asper* snake venom. Toxicon 54, 958–975.

Gutiérrez, J.M., Escalante, T., Rucavado, A., 2009b. Experimental pathophysiology of systemic alterations induced by *Bothrops asper* snake venom. Toxicon 54, 976–987.

Gwaltney-Brant, S.M., 2017. Zootoxins. In: Gupta, R.C. (Ed.), Reproductive and Developmental Toxicology, second ed. Academic Press/Elsevier, Amsterdam, pp. 963–972.

Gwaltney-Brant, S.M., Dunayer, E., Youssef, H., 2018. Terrestrial zootoxins. In: Gupta, R.C. (Ed.), Veterinary Toxicology: Basic and Clinical Principles, third ed. Academic Press/Elsevier, Amsterdam, pp. 781–801.

Hanai, J.-I., Cao, P., Tanksale, P., Imamura, S., et al., 2007. The muscle –specific ubiquitin ligase atrogin-1/MAFbx mediates statin-induced muscle toxicity. J. Clin. Invest. 117, 3940–3951.

Harris, J.B., 2003. Myotoxic phospholipase A2 and the regeneration of skeletal muscles. Toxicon 42, 933–945.

Heidarpour, M., Ennaifer, E., Ahari, H., et al., 2012. Histopathological changes by *Hemiscorpius lepturus* scorpion in mice. Toxicon 59, 373–378.

Hernandez, R., Cabalceta, C., Saravia-Otten, P., et al., 2011. Poor regenerative outcome after skeletal muscle necrosis induced by *Bothrops asper* venom: alterations in microvasculature and nerves. PLoS One 6 (5), e19834, 1-11.

Herzig, V., Wood, D.L.A., Newell, F., et al., 2011. ArachnoServer 2.0, an updated online resource for spider toxin sequences and structures. Nucleic Acids Res. 39, D653–D657.

Holovská, K., Almášiová, V., Tarabová, L., et al., 2017. Effect of the bendiocarb on the ultrastructure of rabbit skeletal muscle. Acta Vet. BRNO 86, 219–222.

Hurme, T., Kalimo, H., 1992. Activation of myogenic precursor cells after muscle injury. Med. Sci. Sports Exerc. 24, 197–205.

Inns, R.H., Tuckwell, N.J., Bright, J.E., Marrs, T.C., 1990. Histochemical demonstration of calcium accumulation in muscle fibers after experimental organophosphate poisoning. Hum. Exp. Toxicol. 9, 245–250.

Jeyarasasingam, G., Yeluashvili, M., Quik, M., 2000. Nitric oxide is involved in acetylcholinesterase inhibitor-induced myopathy in rats. J. Pharmacol. Exp. Ther. 295, 314–320.

Kaminski, H.J., Andrade, F.H., 2001. Nitric oxide: biological effects on muscle and role in muscle diseases. Neuromuscular Disord. 11, 517–524.

Karalliedde, L., Henry, J.A., 1993. Effects of organophosphates on skeletal muscle. Hum. Exp. Toxicol. 12, 289–296.

Kasturiratne, A., Wickremasinghe, A.R., De Silva, N., et al., 2008. The global burden of snakebite: a literature analysis and modeling based on regional estimates of envenoming and deaths. PLoS Med. 5, e218.

Kaufmann, P., Török, M., Zahno, A., et al., 2006. Toxicity of statins on rat skeletal muscle mitochondria. Cell. Mol. Life Sci. 63, 2415–2425.

Kawai, H., Nishino, H., Kusaka, K., et al., 1990. Experimental glycerol myopathy: a histological study. Acta Neuropathol. 80, 192–197.

Kenaston, M.A., Abramson, E.M., Pfeiffer, M.E., Mills, E.M., 2009. Toxicology of skeletal muscle. In: Ballantyne, B., Marrs, T.C., Syversen, T. (Eds.), General and Applied Toxicology, third ed. John Wiley & Sons, Chichester, pp. 1457–1489.

Klint, J.K., Sebastian, S., Rupasinghe, D.B., et al., 2012. Spider-venom peptides that target voltage-gated sodium channels: pharmacological tools and potential therapeutic leads. Toxicon 60, 478–491.

Klopstock, T., 2008. Drug-induced myopathies. Curr. Opin. Neurol. 21, 590–595.

Kobayashi, M., Chisaki, I., Narumi, K., et al., 2008. Association between risk of myopathy and cholesterol-lowering effect: a comparison of all statins. Life Sci. 82, 969–975.

Koch, C.G., 2012. Statin therapy. Curr. Pharmaceut. Design 18, 6284–6290.

Koppel, C., van Wissmann, C., Barckow, D., Kenig, T., et al., 1994. Inhaled nitric oxide in advanced paraquat intoxication. J. Toxicol. Clin. Toxicol. 32, 205–214.

Kumar, G.S., Upadhyay, S., Mathew, M.K., Sharma, S.P., 2011. Solution structure of BTK-2, a novel $hK_v1.1$ inhibiting scorpion toxin, from the eastern Indian scorpion *Mesobuthus tumulus*. Biochim. Biophys. Acta 1814, 459–469.

Kurnik, D., Hochman, I., Vesterman-Landes, J., et al., 2012. Muscle pain and serum creatine kinase are not associated with low serum 25(OH) vitamin D levels in patients receiving statins. Clin. Endocrinol. 77, 36–41.

Laskowski, M.B., Olson, W.H., Dettbarn, W.-D., 1975. Ultrastructural changes at the motor endplate produced by an irreversible cholinesterase inhibition. Exp. Neurol. 47, 290–306.

Laskowski, M.B., Olson, W.H., Dettbarn, W.-D., 1977. An ultrastructural study of the initial myopathic changes produced by paraoxon. Exp. Neurol. 57, 13–33.

Lee, F.J., Aim, J., Bloch, M., et al., 2012. Skeletal muscle toxicity associated with raltegravir-based combination antiretroviral therapy in HIV-infected adults. In: 14th Intl Workshop Co-Morb Adverse Drug React HIV, Washington DC, July 19-21, 2012.

Lewis, R.L., Gutmann, L., 2004. Snake venoms and the neuromuscular junction. Semin. Neurol. 24, 175–179.

Liao, J.K., Laufs, U., 2005. Pleiotropic effects of statins. Pharmacol. Toxicol. 45, 89–118.

Lombard, J., Wong, B., Young, J.H., 1988. Acute renal failure due to rhabdomyolysis associated with cocaine toxicity. West. J. Med. 148, 466–468.

Marcoff, L., Thompson, P.D., 2007. The role of co-enzyme Q10 in statin-associated myopathy: a systemic review. J. Am. Coll. Cardiol. 49, 2231–2237.

Massoulie, J., Bon, S., 1982. The molecular forms of cholinesterase and acetylcholinesterase in vertebrates. Annu. Rev. Neurosci. 5, 57–106.

Milatovic, D., Zivin, M., Hustedt, E., Dettbarn, W.-D., 2000. Spin trapping agent phenyl-*N-tert*-butyl-nitrone prevents diisopropylphosphorofluoridate-induced excitotoxicity in skeletal muscle of the rat. Neurosci. Lett. 278, 25–28.

Milatovic, D., Gupta, R.C., Dekundy, A., et al., 2005. Carbofuran-induced oxidative stress in slow and fast skeletal muscles: prevention by memantine and atropine. Toxicology 208, 13–24.

Milatovic, D., Gupta, R.C., Aschner, M., 2006. Anticholinesterase toxicity and oxidative stress. Sci. World J 6, 295–310.

Misulis, K.E., Clinton, M.E., Dettbarn, W.-D., Gupta, R.C., 1987. Differences in central and peripheral neural actions between soman and diisopropyl fluorophosphate, organophosphorus inhibitors of cholinesterase. Toxicol. Appl. Pharmacol. 89, 391–398.

Misulis, K.E., Clinton, M.E., Dettbarn, W.-D., 2000. Toxicology of skeletal muscle. In: Ballantyne, B., Marrs, T.C., Syversen, T. (Eds.), General and Applied Toxicology, second ed. Macmillan Reference Ltd, London, pp. 937–964.

Mukhtar, R.Y., Reckless, J.P., 2005. Statin-induced myositis: a commonly encountered or rare side effect? Curr. Opin. Lipidol. 16, 640–647.
Norata, G.D., Tibolla, G., Catapano, A.L., 2014. Statins and skeletal muscles toxicity: from clinical trials to everyday practice. Pharmacol. Res. 88, 107–113.
Oh, J., Ban, M.R., Miski, B.A., et al., 2007. Genetic determinants of statin intolerance. Lipids Health Dis. 6, 7.
O'Connor, A.D., Padilla-Jones, A., Ruha, A.-M., 2018. Severe bark scorpion envenomation in adults. Clin. Toxicol. 56 (3), 170–174.
Patocka, J., 2015. Strychnine. In: Gupta, R.C. (Ed.), Handbook of Toxicology of Chemical Warfare Agents. Academic Press/Elsevier, Amsterdam, pp. 215–222.
Patterson, G.T., Gupta, R.C., Dettbarn, W.-D., 1987. Diversity of molecular form patterns of acetylcholinesterase in skeletal muscle of rat. Asia Pac. J. Pharmacol. 2, 265–273.
Patterson, G.T., Gupta, R.C., Misulis, K.E., Dettbarn, W.-D., 1988. Prevention of diisopropylphosphorofluoridate (DFP)-induced skeletal muscle fiber lesions in rat. Toxicology 48, 237–244.
Peng, X., Zhang, Y., Liu, J., et al., 2014. Physiological and biochemical analysis to reveal the molecular basis for Black Widow spiderling toxicity. J. Biochem. Mol. Toxicol. 28 (5), 198–205.
Petrov, K.A., Yagodina, L.O., Valeeva, G.R., et al., 2011. Different sensitivities of rat skeletal muscles and brain to novel anticholinesterase agents, alkylammonium derivatives of 6-methyluracil (ADEMS). Br. J. Pharmacol. 163, 732–744.
Phillips, P.S., Haas, R.H., 2008. Statin myopathy as a metabolic muscle disease. Expert Rev. Cardiovasc Ther. 6, 971–978.
Phillips, P.S., Haas, R.H., Bannykh, S., et al., 2002. Statin-associated myopathy with normal creatine kinase levels. Ann. Intern. Med. 137, 581–585.
Pirazzini, M., Rossetto, O., Eleopra, R., Montecucco, C., 2017. Botulinum neurotoxins: biology, pharmacology, and toxicology. Pharmacol. Rev. 69, 200–235.
Pycroft, K., Fry, B.G., Isbister, G.K., et al., 2012. Toxinology of venoms from five Australian lesser known elapid snakes. Basic Clin. Pharmacol. Toxicol. 111, 268–274.
Rader, R.K., Stoecker, W.V., Malters, J.M., et al., 2012. Seasonality of brown recluse populations is reflected by numbers of brown recluse envenomation. Toxicon 60, 1–3.
Ranawaka, U.K., Lalloo, D.G., de Silva, H.J., 2013. Neurotoxicity in snakebite- the limits of our knowledge. PLoS Negl. Trop. Dis. 7 (10), e2302.
Richards, J.R., 2000. Rhabdomyolysis and drugs of abuse. J. Emerg. Med. 19, 51–56.
Rodriguez de la Vega, R.C., Passani, L.D., 2005. Overview of scorpion toxins specific for Na^{+} channels and related peptides: biodiversity, structure-function relationships and evolution. Toxicon 46, 831–844.
Roy, N.M., Arpie, B., Lugo, J., Linney, E., et al., 2012. Brief embryonic strychnine exposure in zebrafish causes long-term adult behavioral impairment with indications of embryonic synaptic changes. Neurotoxicol. Teratol. 34, 587–591.
Sakamoto, K., Kimura, J., 2013. Mechanism of statin-induced rhabdomyolysis. J. Pharmacol. Sci. 123, 289–294.
Sakamoto, K., Mikami, H., Kimura, J., 2008. Involvement of organic anion transporting polypeptides in the toxicity of hydrophilic pravastatin and lipophilic fluvastatin in rat skeletal myofibers. Br. J. Pharmacol. 154, 1482–1490.
Salvini, T.F., Belluzzo, S.S., Selistre de Araujo, H.S., Souza, D.H., 2001. Regeneration and change of muscle fiber types after injury induced by a hemorrhagic fraction isolated from *Agkistrodon contortrix laticinctus*. Toxicon 39, 641–649.
Sanmuganathan, P.S., 1998. Myasthenic syndrome of snake bite envenoming: a clinical and neurophysiological study. Postgrad. Med. J. 74, 596–597.
Schachter, M., 2005. Chemical, pharmacokinetic and pharmacodynamic properties of statins: an update. Fundam. Clin. Pharmacol. 19, 117–125.
Simpson, L.L., 1981. The origin, structure, and pharmacological activity of botulinum toxin. Pharmacol. Rev. 33, 155–188.
Singh, G., Pannu, H.S., Chawla, P.S., Malhotra, S., 1999. Neuromuscular transmission failure due to common krait (*Bungarus caeruleaus*) envenomation. Muscle Nerve 22, 1637–1643.
Skottheim, I.B., Gedde-Dahl, A., Hejazifar, S., et al., 2008. Statin induced myotoxicity: the lactone forms are more potent than the acid forms in human skeletal muscle cells in vitro. Eur. J. Pharm. Sci. 33, 317–325.
Slade, J.M., Delano, M.C., Meyer, R.A., 2006. Elevated skeletal muscle phosphodiesters in adults using statin medications. Muscle Nerve 34, 782–784.
Smuder, A.J., Kavazis, A.N., Min, K., Powers, S.K., 2011. Exercise protects against doxorubicin-induced oxidative stress and proteolysis in skeletal muscle. J. Appl. Physiol. 110, 935–942.
Spyres, M.B., Ruha, A.-M., Kleinschmidt, K., et al., 2018. Epidemiology and clinical outcomes of snakebite in the elderly: a ToxIC database study. Clin. Toxicol. 56 (2), 108–112.
Tabata, N., Morita, M., Mimasaka, S., et al., 1999. Paraquat myopathy: report on two suicide cases. Forensic Sci. Int. 100, 117–126.
Teixeira, C., Cury, V., Moreira, V., et al., 2009. Inflammation induced by *Bothrops asper* venom. Toxicon 54, 988–997.
Teixera, L.F., de Carvalho, L.H., de Castro, O.B., et al., 2018. Local and systemic effects of BdipTX-1, a Lys-49 phospholipase A(2) isolated from *Bothrops diporus* snake venom. Toxicon 141, 55–64.
Thiermann, H., Szinicz, L., Eyer, P., et al., 2005. Correlation between red blood cell acetylcholinesterase activity and neuromuscular transmission in organophosphate poisoning. Chem. Biol. Interact. 157–158, 345–347.
Thompson, P.D., Clarkson, P., Karas, R.H., 2003. Statin-associated myopathy. J. Am. Med. Assoc. 289, 1681–1690.
Thompson, P.D., Clarkson, P.M., Rosenson, R.S., 2006. National lipid association statin safety task force muscle safety expert panel. An assessment of statin safety by muscle experts. Am. J. Cardiol. 97, 69C–76C.
Tonomura, Y., Matsushima, S., Kashiwagi, E., et al., 2012. Biomarker panel of cardiac and skeletal muscle troponins, fatty acid binding protein 3 and myosin light chain 3 for the accurate diagnosis of cardiotoxicity and musculoskeletal toxicity in rats. Toxicology 302, 179–189.
Tóth, A.R., Varga, T., 2009. Myocardium and striated muscle damage caused by licit or illicit drugs. Legal Med. 11, S484–S487.
Tymoshenko, A., Tkach, G., Sikora, V., et al., 2016. The microscopic and ultramicroscopic changes in the skeletal muscles, caused by heavy metal salts. Interv. Med. Appl. Sci. 8 (2), 82–88.
Vaklavas, C., Chatzizisis, Y.S., Ziakas, A., et al., 2008. Molecular basis of statin-associated myopathy. Atheroscerosis 202, 18–28.
Vlasakova, K., Lane, P., Michna, L., et al., 2016. Response to novel skeletal muscle biomarkers in dogs to drug-induced skeletal muscle injury or sustained endurance exercise. Toxicol. Sci. 156 (2), 422–427.
von Kemp, K., Herregodts, P., Duynslaeger, L., et al., 1989. Muscular fibrosis due to chronic intramuscular administration of narcotic analgesics. Acta Clin. Belg. 44, 383.
Westwood, F.R., Bigley, A., Randall, K., et al., 2005. Statin-induced muscle necrosis in the rat: distribution, development, and fiber selectivity. Toxicol. Pathol. 33, 246–257.

Yamamoto, T., Anno, M., Sato, T., 1987. Effects of paraquat on mitochondria of rat skeletal muscle. Comp. Biochem. Physiol. C 86 (2), 375–378.

Yang, Z.P., Dettbarn, W.-D., 1996. Diisopropylphosphorofluoridate induced cholinergic hyperactivity and lipid peroxidation. Toxicol. Appl. Pharmacol. 138, 48–53.

Yang, Z.P., Dettbarn, W.-D., 1998. Lipid peroxidation and changes in cytochrome c oxidase and xanthine oxidase activity in organophosphorus anticholinesterase induced myopathy. J. Physiol. 92, 157–161.

Yen, Y.-P., Tsai, K.-S., Chen, Y.-W., et al., 2010. Arsenic inhibits differentiation and muscle regeneration. Environ. Health Perspect. 118, 949–956.

CHAPTER

21

Ocular Biomarkers in Diseases and Toxicities

George A. Kontadakis[1], Domniki Fragou[2], Argyro Plaka[1], Antonio F. Hernández[5], George D. Kymionis[4], Aristidis M. Tsatsakis[3]

[1]Laboratory of Optics and Vision and Ophthalmology Department, University of Crete, Heraklion, Greece; [2]Laboratory of Forensic Medicine & Toxicology, School of Medicine, Aristotle University of Thessaloniki, Greece; [3]Center of Toxicology Science & Research, Medical School, University of Crete, Heraklion, Greece; [4]Jules Gonin Eye Hospital, Faculty of Biology and Medicine, University of Lausanne, Lausanne, Switzerland; [5]Department of Legal Medicine and Toxicology, University of Granada School of Medicine, Granada, Spain

INTRODUCTION

The human eye serves as the receptor of light, giving rise to the sense of seeing after the cortical processing of neuronal signals originating from ocular reception. To achieve reception of light and transform it into the signal that is sent to the cortex, a combination of fine-tuned and organized tissues is utilized, tissues that combine to form the structure of the human eye. Thus, ocular histology includes all three types of tissues—muscle, nerve, and epithelial—at a very high level of organization. Light enters the eye through the cornea, which is a transparent tissue responsible for 70% of the refractive power of the eye. The cornea is composed of epithelium externally, the corneal stroma (the basic part of its structure, consisting of organized collagen fibrils and proteoglycans), and the endothelium, which maintains the corneal humidity and transparency. The cornea is the front part of the bulbar wall and the rest of it comprises the sclera, a nontransparent collagen tissue. The cornea is covered by the tear film, which is composed of three layers: the lipid layer externally, the aqueous layer, and the mucus layer internally. Each is produced by different types of glands. Behind the cornea the anterior chamber is filled with aqueous humor, which preserves the intraocular pressure. It is produced by the ciliary body and removed mainly through the trabecular meshwork in the angle of the anterior chamber. The anterior chamber is separated from the posterior by the iris and the ciliary body. The pupil is the iris diaphragm that serves to regulate the quantity of light that passes through to the crystalline lens and then to the posterior segment of the eye, where light meets the retina. The crystalline lens is responsible for 30% of the refractive power of the eye in relaxed status and adds the needed power for accommodation of near vision in pre-presbyopic individuals. Behind the crystalline lens, the eye is filled with the vitreous body. In contact with the vitreous body is the retina (Fig. 21.1). The retina contains the photoreceptors that transform light to neural signal, which, after a complex path of intraretinal transformations, travels to the visual cortex through the optic nerve and the rest of the optic pathway. This chapter describes the biomarkers of ocular diseases and toxicities.

OCULAR BIOMARKERS

A large number of diseases with an acute or chronic course and variable involvement of different ocular tissues may be present in ocular pathology. Symptoms of ocular disease vary from subjects' discomfort to deterioration of ocular refractive function or permanent visual loss due to damage in the retinal-neuronal pathway.

Recently, several new methods for the identification and quantification of molecules in biological tissues have brought about the potential to identify molecular entities that may serve as potentially useful biomarkers in ophthalmology. Biomarkers may be helpful in (1) early detection of a disease, (2) may aid in predicting severity of disease, (3) may predict the rate of disease

Biomarkers in Toxicology, Second Edition
https://doi.org/10.1016/B978-0-12-814655-2.00021-9

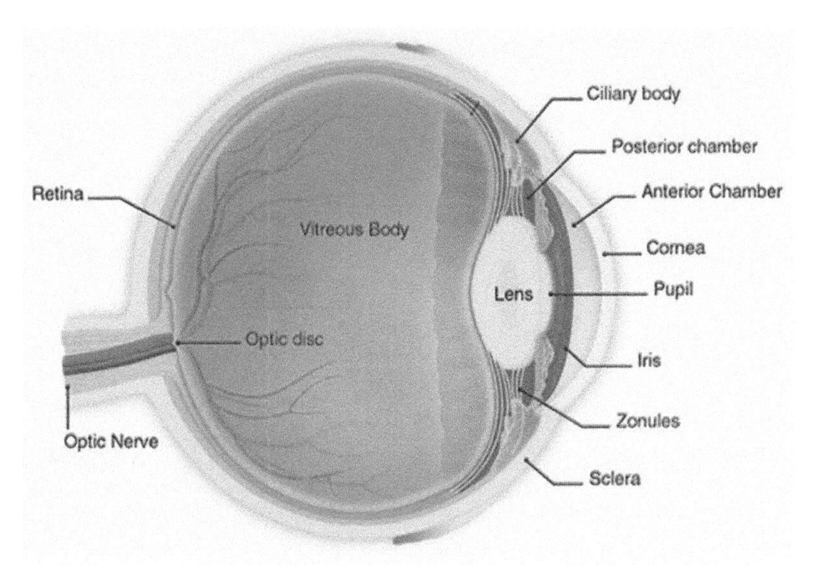

FIGURE 21.1 Overview of ocular anatomy.

progression, and (4) may serve as predictors of response to treatment.

Use of these molecular markers in the clinical setting appears to be a promising option. However, there are still many limitations, such as lack of common procedures for proper banking of biological tissues, as well as standardized methods and criteria in different studies. New studies are continually identifying biomarkers and seeking to standardize their values in health and disease.

In this chapter, we focus on molecular biomarkers that are involved in the diagnosis and management of several ocular pathologies and in ocular toxicity derived from agents prescribed for other systemic conditions.

Molecular Biomarkers in Ocular Surface Disease

Dry eye disease (DED) is an inflammatory disorder of the lacrimal functional unit of multifactorial origin leading to chronic ocular surface discomfort. Dry eye is a disease of the ocular surface characterized by a loss of homeostasis of the tear film that is accompanied by ocular symptoms, in which tear film instability and hyperosmolarity, ocular surface inflammation and damage, and neurosensory abnormalities play etiological roles (Craig et al., 2017). It is a significant burden for ophthalmic health care because of both its high prevalence and its capacity to affect a patients' quality of life. Because of its different origins and the variability of a patients' subjective response to clinically observed disease parameters, it is not easy to determine objective parameters for the evaluation of DED.

The pathophysiology of DED includes two distinct pathways, each with several possible initial causes that result in the same disease entity. One of the two is the aqueous deficient dry eye, where the production of tear is less than needed. The other is evaporative dry eye, where there is an increased evaporation of tears. In both conditions, there is increased osmolarity of the tear film and increased inflammation of the ocular surface, as well as patient discomfort. Proteomic analysis of the tear film of such patients has helped to identify molecular parameters that may serve as biomarkers for disease severity or predisposition to the disease.

Immune processes have been demonstrated to play a key role in dry eye. Cytokines, growth factors, and their receptors have been extensively studied in tears, corneal tissue, and conjunctival tissue. TNF-a, IL-1, and IL-6 are found to be increased in dry eye samples and are also correlated with clinical parameters. Boehm et al. (2011) have published a cytokine profile of dry eye patients' tears. They found increased tear levels of IL-1b, IL-6, IL-8, TNF-a, and IFN-g and increased expression of lipocalin, cystatin SN, and a-1 antitrypsin in dry eye patients. They also found correlations with the clinical parameters of DED. In this study, the findings were not significant in patients with evaporative dry eye, in contrast to other reports that found increased levels in such patients (Enriquez-de-Salamanca et al., 2010). These differences may be attributed to differences in methods of material collection and variations in the classification of patients and controls between studies. In a study by VanDerMeid et al. (2012), the tear levels of IL-1a, IL-1b, IL-6, IL-8, and TNF-a correlated inversely with Schirmer values; IL-1a, IL-6, and TNF-a also correlated directly with tear osmolarity. VanDerMeid et al. (2012) studied the correlations of dry eye clinical parameters (Schirmer's test, tear break up time (TBUT), tear osmolarity, and ocular surface disease index) with tear

MMP-9, MMP-1, MMP-2, MMP-7, and MMP-10 levels in a healthy volunteer group and found positive correlations of MMP levels with increasing disease severity according to each parameter. Chotikavanich et al. (2009) showed that tear MMP-9 activity was significantly higher in dry eye patients and proposed MMP-9 as a useful biomarker for dry eye. In addition, tear MMP-9 activity correlated strongly with clinical parameters, such as symptom severity score, topographic surface regularity index scores, and corneal and conjunctival fluorescein staining scores. Diurnal variation of these molecules and age should always be taken into account if these parameters are to be considered as biomarkers. Other associations with specific biomarkers have been established in recent studies of tear proteome in aqueous deficient versus evaporative dry eye patients. Such proteins include proline-rich protein 4 (which seems to be downregulated), mammaglobin B, lipophilin A, and calgranulin S100A8 that were found to be increased in these patients (Boehm et al., 2013; Perumal et al., 2016). Additionally, an association of dry eye with plasma levels of androgens has been established (Vehof et al., 2017).

Another agent that has been studied in dry eye patients is ocular mucin (MUC) levels. Recently, Corrales et al. (2011) presented differences in MUC-1, MUC-2, MUC-4, MUC-5ACn, and MUC-16 conjunctival levels in dry eye patients compared with controls. The lipid profile of dry eye patients is of high importance because of the role of the lipid layer in tear film stabilization and protection from evaporation. Changes in structure of lipids have been identified in Meibomian glands and tear film of patients with evaporative dry eye (Lam et al., 2011, 2014).

Another category of potential biomarkers for ocular surface disease is neuromediators associated with the functionality of ocular surface innervation. Corneal and ocular surface innervation is a part of the lacrimal functional unit, and this function is impaired in dry eye, as shown by studies of corneal sensitivity. Tear levels of some neuromediators have been proposed as potential useful markers of dry eye severity. Substance P, calcitonin gene-related peptide, neuropeptide Y, vasoactive intestinal peptide, and nerve growth factor tear levels have been correlated with clinical parameters of dry eye (Lambiase et al., 2011).

Molecular Biomarkers in Keratoconus

Keratoconus is a noninflammatory disorder that leads to corneal thinning, protrusion, and irregular astigmatism (corneal ectasia) (Fig. 21.2). In its end stages, corneal scarring occurs and vision is significantly impaired. It affects young individuals, mainly during puberty, and progression occurs until 30–40 years of age. Currently, diagnosis of the disease is done clinically with specific imaging examinations. Timely diagnosis is very important because of the possibility for stabilization treatment before the disease becomes visually impairing. In addition, diagnosis of subclinical cases or those predisposed to the disease is also very important because these patients are not suitable for refractive surgery and may develop corneal ectasia after being treated with laser surgery for their refractive error.

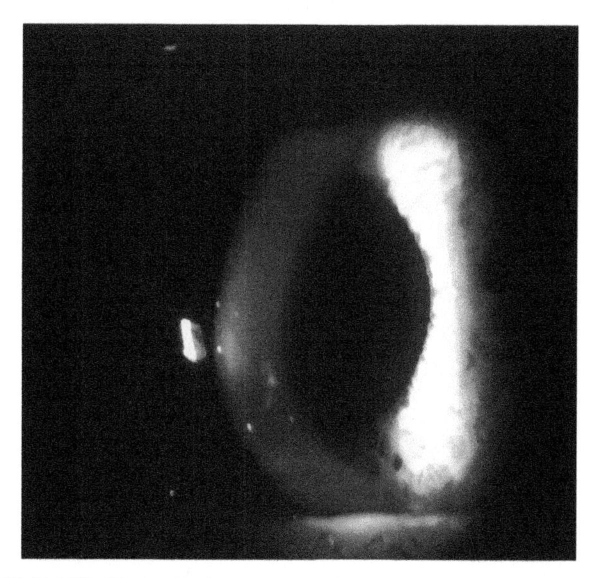

FIGURE 21.2 Slit lamp image of a patient with keratoconus.

Although it is considered a noninflammatory disorder, a chronic low-grade inflammation may play a role in the pathophysiology of keratoconus. Recent studies indicate that certain inflammatory molecules are elevated in tears from patients with keratoconus. Lema and Duran (2005) found that levels of IL-6, TNF-a, and MMP-9 are higher in keratoconus in comparison to controls. In addition, in patients with unilateral keratoconus (clinically evident keratoconus in one eye and subclinical in the fellow eye), IL-6 and TNF-a were increased in the tears of both keratoconus eyes and fellow eyes (Lema et al., 2009). Pannebaker et al. (2010) showed that the level of MMP-1 was increased in keratoconus. Other molecules found to be expressed differently in keratoconus corneas are keratins and mammaglobin B (Pannebaker et al., 2010), zinc-a2-glycoprotein, lactoferrin and immunoglobulin kappa chain (Lema et al., 2009), and α-enolase and β-actin (Srivastava et al., 2006). A recent study of tear film of patients with keratoconus suggested gross cystic disease fluid protein-15 or prolactin-inducible protein as a biomarker for keratoconus (Priyadarsini et al., 2014).

Molecular Biomarkers in Glaucoma

Glaucoma is a vision-threatening disease affecting a large percent of the population worldwide. Glaucoma progression leads to loss of peripheral vision (visual field) and eventually blindness. One in 40 adults over the age of 40 has glaucoma with loss of visual function, which equates to 60 million people worldwide being affected and 8.4 million being bilaterally blind. Increased intraocular pressure is the main cause of glaucoma, and this is caused mainly by obstruction in the outflow of aqueous humor. Timely diagnosis and initiation of intraocular pressure (IOP)-lowering medication is of great importance for prevention of visual loss. Follow-up of glaucoma patients is done mainly by IOP monitoring, visual fields sensitivity testing, and imaging of the optic nerve. The use of molecular biomarkers for the early diagnosis and the follow-up of patients with glaucoma is a field of great scientific interest and research. Biomarkers are identified in blood or ocular fluids such as aqueous humor and tears and show up- or downregulation in patients with glaucoma. The development of clinically useful biomarkers in glaucoma is currently an area of active investigation. Identification of such biomarkers offers insight into disease progression mechanisms. Weinstein et al. (1996) showed an association of decreased peripheral blood lymphocyte 3α-hydroxysteroid dehydrogenase (3α-HSD) activity with glaucoma. The authors concluded that the reduced levels of 3α-HSD activity in the peripheral blood lymphocytes may serve as a marker for primary open angle glaucoma (POAG) or those at risk of developing the disease.

Maruyama et al. (2002) suggested that a serum autoantibody against neuron-specific enolase (NSE) may be clinically useful for diagnosing early stages of glaucoma and monitoring glaucoma progression of normal tension glaucoma (NTG). According to the authors, the anti-NSE antibody titers were relatively higher in patients with visual field deterioration than in those without it.

Another molecule that may serve as a biomarker for early detection of glaucoma is the brain-derived neurotrophic factor (BDNF), which contributes to the building up and preserving of neurons. BDNF has been found significantly decreased in the tears of normal-tension glaucoma patients and open-angle glaucoma patients in comparison to controls (Ghaffariyeh et al., 2011).

Other studies suggest that CD44 content in ocular tissue may represent a biomarker for glaucoma (Mokbel et al., 2010). CD44 is a transmembrane protein and the principal receptor of the glycosaminoglycan, hyaluronan. In the same study, erythropoietin was also recognized as a biomarker for glaucoma.

Recent studies have revealed that the levels of homocysteine and hydroxyproline are significantly higher in the aqueous humor of patients with glaucoma than in controls (Ghanem et al., 2012).

Proteomic studies of aqueous humor and serum of glaucoma patients have found association of proteins related to inflammation and oxidative stress (Izzotti et al., 2010; Kaeslin et al., 2016; Gonzalez-Iglesias et al., 2014).

Currently, research is demonstrating several molecular biomarkers that are identified as possible indicators for the severity of glaucoma, the monitoring of the response to treatment, and also the identification of patients at high risk to develop glaucoma. Future research may provide valuable tools for clinical use in the treatment of glaucoma.

Molecular Biomarkers in Retinal Disease

Retinal pathology is one of the most intriguing parts of ophthalmology. Fundus disease may lead to irreversible visual impairment in several cases, and early diagnosis may help in the prevention or the protection of patients from disease progression. Diagnosis of retinal disorders is mainly from a clinical fundoscopic examination and specialized imaging examinations such as optical coherence tomography, fluorescent angiography, and indocyanine angiography. The most common disorders are age-related macular degeneration (ARMD) and diabetic retinopathy, which are included in the leading causes of blindness in the developed world.

Molecular Biomarkers in Age-Related Macular Degeneration

Currently research is targeted on identifying molecular agents that may contribute to the pathogenesis or the progression of ARMD. A recent study by Kim et al. (2012) discovered that in the aqueous humor of patients with ARMD, the protein composition was different from that in controls.

Recently, elevated levels of CXCL10 were reported in the sera and choroid of individuals with ARMD, and elevated intraocular CCL2 levels were observed in neovascular ARMD (Mo et al., 2010; Jonas et al., 2010). Newman et al. (2012) showed that all ARMD phenotypes in the retinal pigment epithelium (RPE)—choroid are associated with elevated expression of all, or a subset of, the following chemokines: CXCL1, CXCL2, CXCL9, CXCL10, CXCL11, CCL2, and CCL8. In addition, the upregulation of immunoglobulin genes supports an adaptive, autoimmune response in ARMD that is consistent with previous reports of immunoglobulins in drusen and drusen-associated RPE, as well as anticarboxyethylpyrrole adduct antibodies and antiretinal antigen autoantibodies in ARMD sera. Additionally, upregulation of immune response, complement system, and protease activity has been found in vitreous humor of ARMD patients, and upregulation of inflammation and apoptosis in aqueous humor (Koss et al., 2014; Yao et al., 2013).

Molecular Biomarkers in Ocular Oncology

There are several studies that attempt to connect specific biomarkers with ocular cancers. Retinoblastoma (RB) is a malignant tumor of the retina that affects children. In the study of Beta et al. (2013), miRNAs in the serum of children with RB were compared with those in normal age-matched serum. Expression of the oncogenic miRNAs, miR-17, miR-18a, and miR-20a by qRT-PCR was significant in the serum of the RB samples exploring the potential of serum miRNAs identification as noninvasive diagnosis. Researchers concluded that the identified miRNAs and their corresponding target genes could give insights into potential biomarkers and key events involved in the RB pathway.

Uveal and conjunctival melanomas (Fig. 21.3) have also been related to potential biomarkers. The Collaborative Ocular Oncology group (Onken et al., 2012) evaluated 459 patients for the prognostic performance of a 15-gene expression profiling (GEP) assay that assigns primary posterior uveal melanomas to prognostic subgroups: class 1 (low metastatic risk) and class 2 (high metastatic risk). They assumed that the GEP assay had a high technical success rate providing a highly significant improvement in prognostic accuracy over clinical TNM classification. Zoroquiain et al. (2012) analyzed the expression of p16ink4a (p16) in conjunctival melanocytic lesions in an immunohistochemical study and concluded that it appears to be a good marker to differentiate nevi and primary acquired melanoses (PAMs) from melanomas. Errington et al. (2012) found that uveal and conjunctival melanomas consistently expressed high levels of gp100, Melan-A/MART1, and tyrosinase, which are differentiation antigens. The relation of a stem cell marker, CD133, with uveal melanoma was explored by Thill et al. (2011). Differential expression of CD133 splice variants was found in the iris, ciliary body, retina, and retinal pigment epithelium/choroid as well as in the uveal melanoma cell lines.

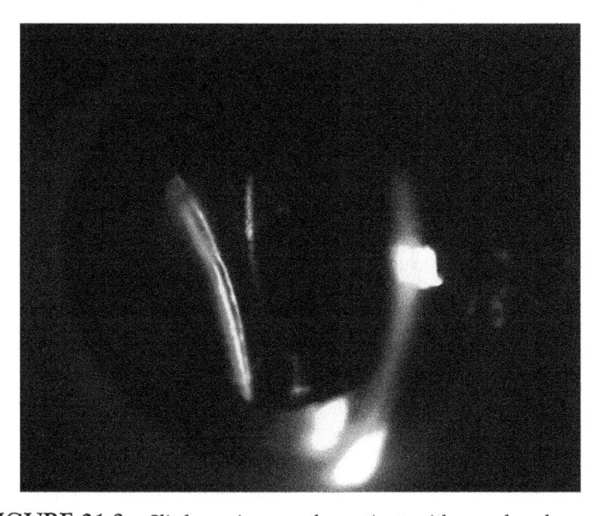

FIGURE 21.3 Slit lamp image of a patient with uveal melanoma.

Noninvasive biomarkers may potentially be used for the early diagnosis of ocular tumors and follow-up treatment. More experimental studies should be carried out to isolate more biomarkers, to evaluate their functional properties, and to explore possible therapeutic approaches.

SYSTEMIC AGENTS AND OCULAR TOXICITY

Many antidepressants, antipsychotics, anti-Parkinson drugs, antihistamines, anticonvulsants, decongestants, and beta-blockers, along with hormone replacement therapy, have been shown to cause dry eye symptoms as an adverse toxic effect. Herbal products with the same effect include niacin, kava, echinacea, and anticholinergic alkaloids (Askeroglu et al., 2013). Moreover, corneal toxicity can be caused in a number of individuals by fluoroquinolones, nonsteroidal antiinflammatory eye drops, glaucoma eye drops, preservatives in eye drops, aminoglycosides, chemotherapeutic medications, topical anesthetics, cyclooxygenase-2 inhibitors, bisphosphonates, retinoids, topical steroids, topical iodine, and some herbal medications such as black mustard, chamomile, cypress spurge, goa powder, and psyllium and also by the milky latex of the plant *Asclepias tuberosa* (Fraunfelder, 2006; Mikkelsen et al., 2017). Aminoglycosides, in particular, have been shown to cause vision loss, optic atrophy, glaucoma, and pigmentary degeneration (Hancock et al., 2005). Dexamethasone, fluocinolone, and triamcinolone are well-known corticosteroids that are very toxic to the retina at high doses and can even cause cataract and glaucoma (Penha et al., 2010). Retinal toxicity may also be caused by intravitreal use of the antibiotic amikacin (Widmer and Helbig, 2006) and intracameral injection of cefuroxime (Çiftçi et al., 2014).

Hydroxychloroquine and chloroquine have been used as antimalarial agents and for the treatment of dermatological and rheumatological diseases. The administration of these drugs can lead to a wide range of toxic ocular effects such as retinal damage, pigmentary retinopathy, keratopathy, corneal deposits, cataract, photophobia, ocular muscle imbalance, and loss of peripheral and night visions (Tzekov, 2005; Tehrani et al., 2008; Tailor et al., 2012; Telek et al., 2017). Ocular quinine toxicity should also be taken into serious consideration when treating vulnerable patient groups such as sleepwalkers, the elderly, and alcohol, and other drug abusers because of accidental overdose leading to loss of vision (Sinha and Al Husainy, 2013).

Since the 1960s, ethambutol hydrochloride has been known to cause dose- and duration-dependent ocular toxicity and, in particular, optic neuritis. The drug is used for the treatment of tuberculosis and its most common ocular adverse effect is retrobulbar neuritis (Chan and Kwok, 2006).

Low-dose fludarabine has been reported to cause hallucinations, visual changes, blurred vision, and even blindness. On the other hand, the ocular effects of high-dose fludarabine include hallucinations, blurred vision, amaurosis, bilateral papillitis, and, in some cases, cortical blindness (Ding et al., 2008).

Oral and inhaled steroids can cause cataract or glaucoma after chronic use or in individuals with high susceptibility. Oral antihistamine drugs used for the treatment of allergies can cause tear film dysfunction (dry eye syndrome) or conjunctival hyperreactivity (Bielory, 2006).

Antineoplastic agents used in chemotherapies have been proven to cause tear film changes and mucositis involving the conjunctival mucosa (Chaves et al., 2007). Cisplatin, tamoxifen, and interferons, all being chemotherapeutic agents, can cause vision loss; and cisplatin can also cause retinal neovascularization as well as blurred vision, papilledema, and optic neuritis (Kwan et al., 2006; Omoti and Omoti, 2006; Gul Baykalir et al., 2017). Moreover, zoledronic acid used in prostate cancer treatment can cause bilateral retrobulbar optic neuropathy in rare situations (Lavado et al., 2017), whereas cancer patients who undergo whole-brain radiotherapy might develop dry eye syndrome (Nanda et al., 2017). Melanoma patients treated with trametinib and dabrafenib developed reduced vision due to serous neuroretinal detachment of the fovea and should therefore be closely monitored throughout their treatment (Sarny et al., 2017).

In high concentrations, ornithine can cause retinal toxicity when administered as a supplement in patients with gyrate atrophy of the retina and choroid (Hayasaka et al., 2011). Atropine, which is used as a mydriatic in eye clinics, may cause corneal cytotoxicity and blurred vision by promoting apoptosis and cell cycle arrest (Tian et al., 2015). Topiramate toxicity was observed in two women who received the drug for the treatment of recurrent headaches and migraines. Both women are presented with macular folds associated with angle-closure glaucoma. The symptoms disappeared after discontinuation of the drug (Kumar et al., 2006).

Both macular and retinal lesions may appear after intravitreal administration of mepivacaine and adrenaline, such as in the case of local anesthesia for palpebral repair surgery (López-Herrero et al., 2017).

Deferoxamine, a chelating agent of iron and aluminum ions, can cause ocular retinal toxicity by damaging the retinal pigment epithelium. Ophthalmologic monitoring may be required when treating hematological or kidney conditions with deferoxamine (Szwarcberg et al., 1997). More recently a case of pseudovitelliform maculopathy was reported in a patient on deferoxamine treatment for 5 years. The maculopathy worsened when deferasirox, an alternative chelating agent, was administered (Bui et al., 2017).

Eye drops have been known to cause ocular toxicity after chronic use, sometimes caused by the preservatives included in the formulation. Therefore, symptoms such as cell loss; structural changes in the conjunctival epithelium and the corneal endothelium; and epithelium, fibrosis, and chronic inflammation of the subconjunctiva; and dry eye syndrome may occur (Huber-van der Velden et al., 2012). For instance, in experiments carried out in rabbits it was shown that when benzalkonium chloride, the most common preservative used in ocular drugs, was applied topically, the whole cornea was impaired and at high doses the barrier integrity of the corneal endothelium was disrupted (Chen et al., 2011).

Indocyanine green, a dye used in medical diagnostics, can cause retinal toxicity in high doses via degeneration of retinal layers and Müller cell dysfunction (Sato et al., 2002).

DED has also been reported after ciguatera fish poisoning in a 47-year-old woman. The patient had consumed Spanish mackerel that contained the ciguatera toxin (Sheck and Wilson, 2010).

Lead poisoning can also cause retinal toxicity as shown in a 35-year-old woman who suffered from loss of vision in the right eye (Gilhotra et al., 2007). Another heavy metal, iron, may play a role in retinal and macular degenerations, glaucoma, and cataract (He et al., 2007). In vitro experiments carried out in retinal pigment epithelium cells have shown that cadmium can cause disruption of the membrane integrity, alter cell morphology, and decrease the survival of the cells and could therefore play a role in age-related retinal disease in smokers (Wills et al., 2008). Cobalt toxicity following hip implant has also been reported involving degenerative alterations of the photoreceptor–retinal pigment epithelium complex with coroidal infarction and paracentral scotomas (Ng et al., 2013).

The toxic effects of crack cocaine abuse include corneal disturbances. For instance, a crack cocaine abuse case has been reported in which the patient had stromal ulceration and corneal epithelial disruption (Pilon and Scheiffle, 2006).

Mustard gas can affect the eyes, especially in cases of chronic involvement. Symptoms include dry eye, limbal stem cell deficiency, limbal ischemia, chronic blepharitis, aberrant conjunctival vessels, meibomian gland dysfunction, corneal neovascularization, corneal irregularity, thinning and scarring, and lipid and amyloid depositions (Baradaran-Rafii et al., 2011).

Ocular toxicity of hydrogen peroxide after habitual use as an eye wash was reported in a patient who was hospitalized for inflammation and scarring of the cornea and conjunctiva (Memarzadeh et al., 1993).

Organophosphate and organochloride compounds used widely as pesticides can cause ocular toxicity. Chronic exposure to those pesticides may lead to keratectasia and corneal neovascularization, which can result in blindness (Sanyal et al., 2017). Paraquat, a common herbicide, can cause ocular surface destruction via topical exposure (Vlahos et al., 1993).

Light or electromagnetic radiation can cause retinal damage although the eye has adaptive mechanisms to avoid possible damage (Youssef et al., 2011).

Methanol poisoning by ingestion has led to ocular damage in three patients. Retinal and cystoid macular edema was observed with paracentral scotomas, decreased retinal sensitivity, and enlargement of one blind spot as well as bilateral multifocal retinal pigment epithelial detachment (McKellar et al., 1992; Ranjan et al., 2014).

CONCLUDING REMARKS AND FUTURE DIRECTIONS

Several molecules that may be helpful in the diagnosis of ocular disease and also several systemic agents that may cause significant damage in the ocular tissue have been identified. Research in the field of biomarkers for ocular pathology is expanding, and it is possible that soon those molecules may serve as biomarkers for early diagnosis, treatment monitoring, and patient follow-up.

References

Askeroglu, U., Alleyne, B., Guyuron, B., 2013. Pharmaceutical and herbal products that may contribute to dry eyes. Plast. Reconstr. Surg. 131, 159–167.

Baradaran-Rafii, A., Eslani, M., Tseng, S.C., 2011. Sulfur mustard-induced ocular surface disorders. Ocul. Surf. 9, 163–178.

Beta, M., Venkatesan, N., Vasudevan, M., 2013. Identification and in silico analysis of retinoblastoma serum microRNA profile and gene targets towards prediction of novel serum biomarkers. Bioinform. Biol. Insights 7, 21–34.

Bielory, L., 2006. Ocular toxicity of systemic asthma and allergy treatments. Curr. Allergy Asthma Rep. 6, 299–305.

Boehm, N., Riechardt, A.I., Wiegand, M., 2011. Proinflammatory cytokine profiling of tears from dry eye patients by means of antibody microarrays. Invest. Ophthalmol. Vis. Sci. 52, 7725–7730.

Boehm, N., Funke, S., Wiegand, M., Wehrwein, N., Pfeiffer, N., Grus, F.H., 2013. Alterations in the tear proteome of dry eye patients—a matter of the clinical phenotype. Invest. Ophthalmol. Vis. Sci. 2385–2392.

Bui, K.M., Sadda, S.R., Salehi-Had, H., 2017. Pseudovitelliform maculopathy associated with deferoxamine toxicity: multimodal imaging and electrophysiology of a rare entity. Digit. J. Ophthalmol. 23 (1), 11–15.

Chan, R.Y., Kwok, A.K., 2006. Ocular toxicity of ethambutol. Hong Kong Med. J. 12, 56–60.

Chaves, A.P., Gomes, J.A., Höfling-Lima, A.L., 2007. Ocular changes induced by chemotherapy. Arq. Bras. Oftalmol. 70, 718–725.

Chen, W., Li, Z., Hu, J., 2011. Corneal alternations induced by topical application of benzalkonium chloride in rabbit. PLoS One 6, 10.

Chotikavanich, S., de Paiva, C.S., Li de, Q., 2009. Production and activity of matrix metalloproteinase-9 on the ocular surface increase in dysfunctional tear syndrome. Invest. Ophthalmol. Vis. Sci. 50, 3203–3209.

Çiftçi, S., Çiftçi, L., Dağ, U., 2014. Hemorrhagic retinal infarction due to inadvertent overdose of cefuroxime in cases of complicated cataract surgery: retrospective case series. Am. J. Ophthalmol. 157 (2), 421–425.

Corrales, R.M., Narayanan, S., Fernandez, I., 2011. Ocular mucin gene expression levels as biomarkers for the diagnosis of dry eye syndrome. Invest. Ophthalmol. Vis. Sci. 52, 8363–8369.

Craig, J.P., Nichols, K.K., Akpek, E.K., et al., 2017. TFOS DEWS II definition and classification report. Ocul. Surf. 276–283.

Ding, X., Herzlich, A.A., Bishop, R., et al., 2008. Ocular toxicity of fludarabine: a purine analog. Expert Rev. Ophthalmol. 3, 97–109.

Enriquez-de-Salamanca, A., Castellanos, E., Stern, M.E., 2010. Tear cytokine and chemokine analysis and clinical correlations in evaporative-type dry eye disease. Mol. Vis. 16, 862–873.

Errington, J.A., Conway, R.M., Walsh-Conway, N., 2012. Expression of cancer-testis antigens (MAGE-A1, MAGE-A3/6, MAGE-A4, MAGE-C1 and NY-ESO-1) in primary human uveal and conjunctival melanoma. Br. J. Ophthalmol. 96, 451–458.

Fraunfelder, F.W., 2006. Corneal toxicity from topical ocular and systemic medications. Cornea 25, 1133–1138.

Ghaffariyeh, A., Honarpisheh, N., Heidari, M.H., et al., 2011. Brain-derived neurotrophic factor as a biomarker in primary open-angle glaucoma. Optom. Vis. Sci. 88, 80–85.

Ghanem, A.A., Mady, S.M., El Awady, H.E., Arafa, L.F., 2012. Homocysteine and hydroxyproline levels in patients with primary open-angle glaucoma. Curr. Eye Res. 37, 712–718.

Gilhotra, J.S., Von Lany, H., Sharp, D.M., 2007. Retinal lead toxicity. Indian J. Ophthalmol. 55, 152–154.

Gonzalez-Iglesias, H., Alvarez, L., Garcia, M., et al., 2014. Comparative proteomic study in serum of patients with primary open-angle glaucoma and pseudoexfoliation glaucoma. J. Proteomics 65–78.

Gul Baykalir, B., Ciftci, O., Cetin, A., et al., 2017. The protective effect of fish oil against cisplatin induced eye damage in rats. Cutan. Ocul. Toxicol. 25, 1–6.

Hancock, H.A., Guidry, C., Read, R.W., et al., 2005. Acute aminoglycoside retinal toxicity in vivo and in vitro. Invest. Ophthalmol. Vis. Sci. 46, 4804–4808.

Hayasaka, S., Kodama, T., Ohira, A., 2011. Retinal risks of high-dose ornithine supplements: a review. Br. J. Nutr. 106, 801–811.

He, X., Hahn, P., Iacovelli, J., 2007. Iron homeostasis and toxicity in retinal degeneration. Prog. Retin. Eye Res. 26, 649–673.

Huber-van der Velden, K.K., Thieme, H., Eichhorn, M., 2012. Morphological alterations induced by preservatives in eye drops. Ophthalmologe 109, 1077–1081.

Izzotti, A., Longobardi, M., Cartiglia, C., Sacca, S.C., 2010. Proteome alterations in primary open angle glaucoma aqueous humor. J. Proteome Res. 4831–4838.

Jonas, J.B., Tao, Y., Neumaier, M., Findeisen, P., 2010. Monocyte chemoattractant protein 1, intercellular adhesion molecule 1, and vascular cell adhesion molecule 1 in exudative age-related macular degeneration. Arch. Ophthalmol. 128, 1281–1286.

Kaeslin, M.A., Killer, H.E., Fuhrer, C.A., et al., 2016. Changes to the Aqueous Humor Proteome during Glaucoma. PLoS One 11 (10) e0165314.

Kim, T.W., Kang, J.W., Ahn, J., 2012. Proteomic analysis of the aqueous humor in age-related macular degeneration (AMD) patients. J. Proteome Res. 11, 4034–4043.

Koss, M.J., Hoffmann, J., Nguyen, N., et al., 2014. Proteomics of vitreous humor of patients with exudative age-related macular degeneration. PLoS One e96895.

Kumar, M., Kesarwani, S., Rao, A., Garnaik, A., 2006. Macular folds: an unusual association in topiramate toxicity. Klin. Monbl. Augenheilkd. 223, 456–458.

Kwan, A.S., Sahu, A., Palexes, G., 2006. Retinal ischemia with neovascularization in cisplatin related retinal toxicity. Am. J. Ophthalmol. 141, 196–197.

Lam, S.M., Tong, L., Reux, B., et al., 2014. Lipidomic analysis of human tear fluid reveals structure-specific lipid alterations in dry eye syndrome. J. Lipid Res. 299–306.

Lam, S.M., Tong, L., Yong, S.S., et al., 2011. Meibum lipid composition in Asians with dry eye disease. PLoS One e24339.

Lambiase, A., Micera, A., Sacchetti, M., 2011. Alterations of tear neuromediators in dry eye disease. Arch. Ophthalmol. 129, 981–986.

Lavado, F.M., Prieto, M.P., Osorio, M.R.R., et al., 2017. Bilateral retrobulbar optic neuropathy as the only sign of zoledronic acid toxicity. J. Clin. Neurosci. 44, 243–245.

Lema, I., Duran, J.A., 2005. Inflammatory molecules in the tears of patients with keratoconus. Ophthalmology 112, 654–659.

Lema, I., Sobrino, T., Duran, J.A., et al., 2009. Subclinical keratoconus and inflammatory molecules from tears. Br. J. Ophthalmol. 93, 820–824.

López-Herrero, F., Sánchez-Vicente, J.L., Monge-Esquivel, J., et al., 2017. Retinal toxicity after accidental intravitreal injection of mepivacaine and adrenaline. Arch. Soc. Esp. Oftalmol. 6691 (17), 30209–30215.

Maruyama, I., Ikeda, Y., Nakazawa, M., Ohguro, H., 2002. Clinical roles of serum autoantibody against neuron-specific enolase in glaucoma patients. Tohoku J. Exp. Med. 197, 125–132.

McKellar, M.J., Hidajat, R.R., Elder, M.J., 1992. Acute ocular methanol toxicity: clinical and electrophysiological features. Ophthalmic Surg. 23, 622–624.

Memarzadeh, F., Shamie, N., Gaster, R.N., et al., 1993. Corneal and conjunctival toxicity from hydrogen peroxide: a patient with chronic self-induced injury. Aust. N. Z. J. Ophthalmol. 21, 187–190.

Mikkelsen, L.H., Hamoudi, H., Gül, C.A., Heegaard, S., 2017. Corneal toxicity following exposure to *Asclepias tuberosa*. Open Ophthalmol. J. 11, 1–4.

Mo, F.M., Proia, A.D., Hohnson, W.H., et al., 2010. Interferon γ-inducible protein-10 and eotaxin as biomarkers in age-related macular degeneration. Invest. Ophthalmol. Vis. Sci. 51, 4226–4236.

Mokbel, T.H., Ghanem, A.A., Kishk, H., et al., 2010. Erythropoietin and soluble CD44 levels in patients with primary open-angle glaucoma. Clin. Exp. Ophthalmol. 38, 560–565.

Nanda, T., Wu, C.C., Campbell, A.A., et al., 2017. Risk of dry eye syndrome in patients treated with whole-brain radiotherapy. Med. Dosim. 42 (4), 357–362.

Newman, A.M., Gallo, N.B., Hancox, L.S., 2012. Systems-level analysis of age-related macular degeneration reveals global biomarkers and phenotype-specific functional networks. Genome Med. 4, 16.

Ng, S.K., Ebneter, A., Gilhotra, J.S., 2013. Hip-implant related chorioretinal cobalt toxicity. Indian J. Ophthalmol. 61, 35–37.

Omoti, A.E., Omoti, C.E., 2006. Ocular toxicity of systemic anticancer chemotherapy. Pharm. Pract. 4, 55–59.

Onken, M.D., Worley, L.A., Char, D.H., Augsburger, J.J., Correa, Z.M., et al., 2012. Collaborative Ocular Oncology Group report number 1: prospective validation of a multi-gene prognostic assay in uveal melanoma. Ophthalmology 119, 1596–1603.

Pannebaker, C., Chandler, H.L., Nichols, J.J., 2010. Tear proteomics in keratoconus. Mol. Vis. 16, 1949–1957.

Penha, F.M., Rodrigues, E.B., Maia, M., 2010. Retinal and ocular toxicity in ocular application of drugs and chemicals – part II: retinal toxicity of current and new drugs. Ophthalmic Res. 44, 205–224.

Perumal, N., Funke, S., Pfeiffer, N., Grus, F.H., 2016. Proteomics analysis of human tears from aqueous-deficient and evaporative dry eye patients. Sci. Rep. 29629.

Pilon, A.F., Scheiffle, J., 2006. Ulcerative keratitis associated with crack-cocaine abuse. Cont. Lens Anterior Eye 29, 263–267.

Priyadarsini, S., Hjortdal, J., Sarker-Nag, A., et al., 2014. Gross cystic disease fluid protein-15/prolactin-inducible protein as a biomarker for keratoconus disease. PLoS One e113310.

Ranjan, R., Kushwaha, R., Gupta, R.C., Khan, P., 2014. An unusual case of bilateral multifocal retinal pigment epithelial detachment with methanol-induced optic neuritis. J. Med. Toxicol. 10 (1), 57–60.

Sanyal, S., Law, A., Law, S., 2017. Chronic pesticide exposure and consequential keratectasia and corneal neovascularisation. Exp. Eye Res. 164, 1–7.

Sarny, S., Neumayer, M., Kofler, J., El-Shabrawi, Y., 2017. Ocular toxicity due to trametinib and dabrafenib. BMC Ophthalmol. 17 (1), 146.

Sato, Y., Tomita, H., Sugano, E., 2002. Evaluation of indocyanine green toxicity to rat retinas. J. Fr. Ophtalmol. 25, 609–614.

Sheck, L., Wilson, G.A., 2010. Dry eye following ciguatera fish poisoning. Clin. Exp. Ophthalmol. 38, 315–317.

Sinha, A., Al Husainy, S., 2013. Ocular quinine toxicity in a sleepwalker. BMJ Case Rep. 1–3.

Srivastava, O.P., Chandrasekaran, D., Pfister, R.R., 2006. Molecular changes in selected epithelial proteins in human keratoconus corneas compared to normal corneas. Mol. Vis. 12, 1615–1625.

Szwarcberg, J., Mack, G., Flament, J., 1997. Ocular toxicity of deferoxamine: description and analysis of three observations. Aust. N. Z. J. Ophthalmol. 25, 225–230.

Tailor, R., Elaraoud, I., Good, P., et al., 2012. A case of severe hydroxychloroquine-induced retinal toxicity in a patient with recent onset of renal impairment: a review of the literature on the use of hydroxychloroquine in renal impairment. Case Rep. Ophthalmol. Med. 2012, 182747.

Tehrani, R., Ostrowski, R.A., Hariman, R., Jay, W.M., 2008. Ocular toxicity of hydroxychloroquine. Semin. Ophthalmol. 23, 201–209.

Telek, H.H., Yesilirmak, N., Sungur, G., et al., 2017. Retinal toxicity related to hydroxychloroquine in patients with systemic lupus erythematosus and rheumatoid arthritis. Doc. Ophthalmol. 135 (3), 187–194.

Thill, M., Berna, M.J., Grierson, R., 2011. Expression of CD133 and other putative stem cell markers in uveal melanoma. Melanoma Res. 21, 405–416.

Tian, C.L., Wen, Q., Fan, T.J., 2015. Cytotoxicity of atropine to human corneal epithelial cells by inducing cell cycle arrest and mitochondrion-dependent apoptosis. Exp. Toxicol. Pathol. 67 (10), 517–524.

Tzekov, R., 2005. Ocular toxicity due to chloroquine and hydroxychloroquine: electrophysiological and visual function correlates. Doc. Ophthalmol. 110, 111–120.

VanDerMeid, K.R., Su, S.P., Ward, K.W., Zhang, J.Z., 2012. Correlation of tear inflammatory cytokines and matrix metalloproteinases with four dry eye diagnostic tests. Invest. Ophthalmol. Vis. Sci. 53, 1512–1518.

Vehof, J., Hysi, P.G., Hammond, C.J., 2017. A metabolome-wide study of dry eye disease reveals serum androgens as biomarkers. Ophthalmology 505–511.

Vlahos, K., Goggin, M., Coster, D., 1993. Paraquat causes chronic ocular surface toxicity. Aust. N. Z. J. Ophthalmol. 21, 187–190.

Weinstein, B.I., Iyer, R.B., Binstock, J.M., 1996. Decreased 3 alpha-hydroxysteroid dehydrogenase activity in peripheral blood lymphocytes from patients with primary open angle glaucoma. Exp. Eye Res. 62, 39–45.

Widmer, S., Helbig, H., 2006. Presumed macular toxicity of intravitreal antibiotics. Ophthalmologica 220, 153–158.

Wills, N.K., Ramanujam, V.M., Chang, J., 2008. Cadmium accumulation in the human retina: effects of age, gender, and cellular toxicity. Exp. Eye Res. 86, 41–51.

Yao, J., Liu, X., Yang, Q., Zhuang, M., Wang, F., Chen, X., et al., 2013. Proteomic analysis of the aqueous humor in patients with wet age-related macular degeneration. Proteomics Clin. Appl. 550–560.

Youssef, P.N., Sheibani, N., Albert, D.M., 2011. Retinal light toxicity. Eye (Lond) 25, 1–14.

Zoroquiain, P., Fernandes, B.F., Gonzalez, S., 2012. p16ink4a expression in benign and malignant melanocytic conjunctival lesions. Int. J. Surg. Pathol. 20, 240–245.

CHAPTER

22

Biomarkers of Ototoxicity

Antonio F. Hernández[1], *Aristidis M. Tsatsakis*[2], *George A. Kontadakis*[3]

[1]Department of Legal Medicine and Toxicology, University of Granada School of Medicine, Granada, Spain;
[2]Center of Toxicology Science & Research, Medical School, University of Crete, Heraklion, Greece;
[3]Laboratory of Optics and Vision and Ophthalmology Department, University of Crete, Heraklion, Greece

INTRODUCTION

Any chemical or drug with the potential to cause toxic damage to the structures of the inner ear, including the cochlea and vestibular system (responsible for sound detection and balance, respectively), is considered ototoxic. The main chemicals associated with ototoxicity include specific classes of drugs (e.g., antimicrobials and antineoplastic agents) and industrial chemicals (organic solvents and heavy metals). Although ototoxic chemicals can cause cochleotoxicity or vestibulotoxicity, some of them can cause both, and the resultant toxicity can be transient or irreversible, depending on the compound and the length of exposure. Cochleotoxicity is manifested by hearing loss, difficulty with speech understanding in the presence of background noise, and tinnitus (ringing in the ears). In turn, vestibulotoxicity may exhibit dizziness, dysequilibrium (loss of balance) or ataxic gait, and nystagmus (involuntary abnormal eye movements). The functional and molecular biomarkers used to assess ototoxicity along with the underlying toxic mechanisms are addressed in this chapter.

STRUCTURE AND FUNCTION OF THE NORMAL AUDITORY SYSTEM

Anatomic Structure of the Human Auditory System

The auditory system consists of three components: the outer, middle, and inner ear, all of which work together to transfer sounds from the environment to the brain (Table 22.1, Fig. 22.1). The **outer ear** includes the pinna and the ear (or auditory) canal. The pinna, also called auricle, is a concave cartilaginous structure, which collects and directs sound waves into the ear canal toward the tympanic membrane (also called eardrum), thus increasing the sound pressure. The **middle ear** is composed of the eardrum and a cavity that contains the ossicular chain. This cavity is actually an extension of the nasopharynx via the eustachian tube, which acts as an air pressure equalizer for an optimum transmission of sound and also ventilates the middle ear (Fig. 22.1). The eardrum vibrates back and forth in response to air pressure changes caused by sound waves. The eardrum motion travels across the middle ear cavity via the three middle ear bones (or ossicles) called malleus (hammer), incus (anvil), and stapes (stirrup). The ossicles chain act thus as a lever, converting the lower-pressure sound vibrations of the tympanic membrane into higher-pressure sound vibrations at the oval window, a small membrane in the wall of the cochlea. This higher pressure is necessary because the structure beyond the oval window (inner ear) contains liquid rather than air as occurred at the tympanic membrane (Ervin, 2017).

The **inner ear** is composed of the sensory organ for hearing (the cochlea), as well as for balance (the vestibular system). The hearing part of the inner ear is the cochlea, a snail-shaped structure, which in cross section has three fluid-filled compartments divided by two tight junction-coupled cellular barriers, the basilar membrane and Reissner's membrane (Fig. 22.2). The two outer compartments are the scala tympani and the scala vestibule, whereas the central compartment is called scala media or cochlear duct (Ervin, 2017; Jiang et al., 2017). The vestibular apparatus is composed of three semicircular canals and the utricle and saccule, whose function is to maintain balance, regardless of head position or gravity, in conjunction with eye movement and somatosensory input (Ervin, 2017). The vestibular sensory

Biomarkers in Toxicology, Second Edition
https://doi.org/10.1016/B978-0-12-814655-2.00022-0

TABLE 22.1 Anatomic Structure of the Auditory System

Outer ear	Pinna (auricle)
	Ear canal (auditory canal)
Middle ear	Tympanic membrane (eardrum)
	Middle ear cavity (tympanic cavity)
	Ossicular chain (malleus, incus, stapes)
	Eustachian tube
Inner ear	Cochlea
	• Scala tympani
	• Scala media
	• Scala vestibuli
	• Organ of Corti
	• Stria vascularis
	Vestibulum (Vestibular system)
	• Semicircular canals
	• Utricule
	• Saccule

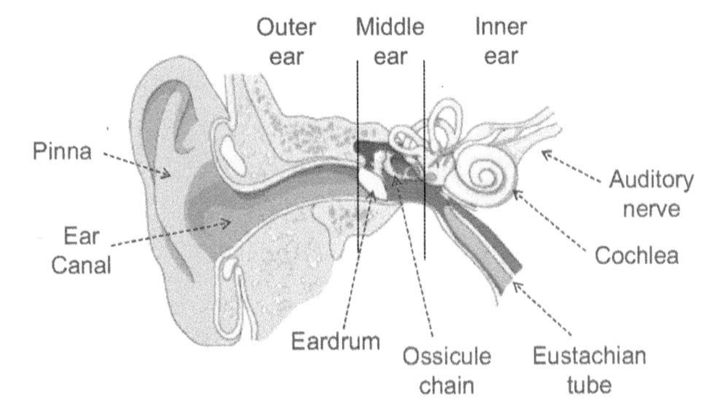

FIGURE 22.1 Cross section of the outer, middle, and inner ear. *Adapted from Ervin, S.E., 2017. Assessment Tools: Introduction to the Anatomy and Physiology of the Auditory System. Workplace INTEGRA, Inc., Greensboro, North Carolina.*

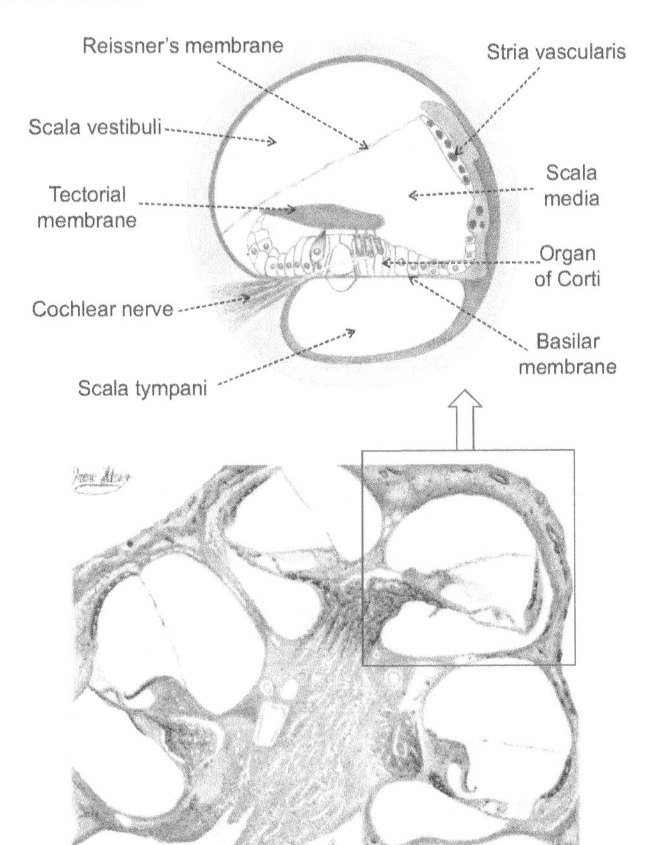

FIGURE 22.2 Structure of the cochlea (bottom) and schematic cross section of a cochlear canal (top) showing the organ of Corti.

epithelia, which control balance and spatial orientation, possess hair cells and supporting cells arrayed in the cristae of the semicircular canals and the maculae of the utricle and saccule (Brigande, 2017).

The organ of Corti, situated on the basilar membrane of the middle compartment, consists of sensory hair cells, which are the receptors for hearing, and adjacent supporting cells coupled together by apical tight junctions. There are two types of hair cells, the sensory inner hair cells (arranged in a single row) and outer hair cells (arranged in three rows) that amplify the sound-induced mechanical vibrations of the basilar membrane. Hair cells have cilia whose ends make contact with the tectorial membrane. Each hair cell is connected to a nerve fiber (afferent for inner hair cells and efferent for outer hair cells) that relays various impulses to the cochlear branch of the auditory nerve (Zdebik et al., 2009). Supporting cells, particularly pillar and Deiters' cells, provide a structural scaffold to enable mechanical stimulation of sensory hair cells and also play a central role in the primary repair response of hair cell loss (Wan et al., 2013). The organ of Corti transforms mechanical waves to electric signals in neurons, which are further transmitted to the brain.

The scala media is filled with endolymph whose composition is maintained by the epithelial cells of the stria vascularis in the lateral wall of the scala media, which consists of numerous capillary loops and small blood vessels (Zdebik et al., 2009). Endolymph provides metabolic support for the organ of Corti and contributes to the generation of the electrochemical (endocochlear) potential responsible for transduction of sound by sensory hair cells (Cunningham and Tucci, 2017). The endocochlear potential is generated by high K^+ concentrations in the endolymph because of $Na^+-K^+-Cl^-$ cotransporters, Na^+-K^+-ATPases, and rectifying potassium channels (Jiang et al., 2017).

At the base of the cochlea, the perilymph of the scala vestibuli contacts the oval window, and the perilymph of the scala tympani contacts the round window. The chemical difference between the endolymph (high concentration of K^+ and a low concentration of Na^+, similar to intracellular fluid) and perilymph (low in K^+ and high in Na^+, similar to interstitial fluid) is important for the function of the inner ear (Sprung et al., 2003).

Movement of the oval window creates motion in the cochlear fluid and along the basilar membrane. This motion excites frequency-specific areas of the organ of Corti. Although the apical portion of the basilar membrane transfers lower-frequency impulses, the basal end relays higher-frequency impulses (Ervin, 2017).

Although the ears are fully formed at birth, maturation of neuronal pathways and auditory structures continues during infancy and early childhood, making young children particularly vulnerable to the ototoxic effects of chemicals (Landier, 2016).

Hearing Mechanisms

In the human ear, a sound wave is transmitted through four separate mediums along the auditory system before a sound is perceived: the outer ear (air), middle ear (mechanical), inner ear (liquid), and the brain (neural). Air-transmitted sound waves are directed toward the delicate hearing mechanisms with the help of the outer ear, first by the pinna, which gently funnels sound waves into the ear canal that amplifies and direct sound toward the middle ear. When air movement strikes the tympanic membrane, this moves and sets the three little bones of the ossicular chain into motion. At this point, the energy generated through the sound waves is transformed into mechanical energy. This is then transmitted to the inner ear (the cochlea) via hydraulic waves that stimulate the mechanosensory hair cells of the organ of Corti. On stimulation, these cells convert sound energy into electrical activity in nerve fibers by releasing neurotransmitters that cause the neurons of the spiral ganglion to transmit the neural impulses through the VIII cranial nerve and various nuclei in the brain stem to the auditory cortex in the temporal lobe of the brain (Ervin, 2017).

Sound pressure waves entering the cochlea tonotopically vibrate the basilar membrane, deflecting the stereocilia projecting from the apices of outer hair cells into endolymph. As the stereocilia are deflected, mechanoelectrical transduction (MET) channels located at the tips of the stereocilia are opened, allowing K^+ to enter the cells. This K^+ influx opens voltage-dependent Ca^{++} channels and depolarizes outer hair cells, which change in length. This change originates motion of the basilar membrane, which in turn causes the associated inner hair cell to depolarize. This depolarization opens voltage-dependent Ca^{++} channels and initiates an electric potential that ultimately causes neurotransmitter release at the base of internal hair cell (known as ribbon synapses), eliciting an action potential in the dendrites of the spiral ganglion neurons. Since glutamate is the neurotransmitter released at the synapses between inner hair cells and spiral ganglion neurons, the postsynaptic afferent neurons represent a candidate target for excitotoxic damage (Brigande, 2017; Watson et al., 2010).

OVERVIEW ON OTOTOXICITY

Ototoxicity is a neurosensorial damage consisting of transient or definitive disturbances of auditory and/or vestibular function induced by chemicals targeting the cochlea and vestibular cells, respectively. Thus, both cochleotoxicant and vestibulotoxicant chemicals can be considered as ototoxicants (Campo et al., 2009).

Although sensory hearing loss is the result of damage to the organ of Corti or the stria vascularis, neural hearing loss is the result of loss or dysfunction of spiral ganglion neurons or of more proximal auditory structures (Cunningham and Tucci, 2017). Cochlear dysfunction spans from a slight increase of the hearing threshold, only detectable through audiometry, to complete deafness. Hearing loss can occur along with tinnitus (ringing in the ears). Clinically cochlear damage appears sooner than vestibular damage, which can even be severe before the onset of vertigo. Vestibular damages can go undetected especially if the damage development is slow and progressive and bilateral so that the actual extent of vestibular damage is hard to assess (Cianfrone et al., 2011). The main clinical features of cochlear and vestibular toxicity are shown in Table 22.2.

The first clinical manifestation of cochlear toxicity is the presence of tinnitus, usually bilateral, of sudden onset and great intensity, which may or may not evolve to hearing loss. In its initial stage the damage is limited to the acute frequencies (4000–8000 Hz), without affecting the conversational hearing, although subjects usually have a sensation of noise and auditory dullness.

TABLE 22.2 Characteristic Clinical Symptoms and Signs of Cochlear and Vestibular Toxicity

Cochlear Toxicity	Vestibular Toxicity
Tinnitus	Headache, nausea, vomiting
High-frequency hearing loss	Vertigo, loss of balance when closing the eyes
Low-frequency hearing loss	Loss of balance in ambulation
Deafness (may appear several weeks after stopping treatment)	Nystagmus

At this phase the damage is usually reversible. In its advanced stage the cochlear toxicity affects the inner hair cells of the cochlear apex affecting the most severe frequencies and conversational hearing. At this point the auditory deficit can be partially reversible or permanent. In parallel to the cochlear damage, vestibular toxicity usually develops, manifested by symptoms such as vertigo, nausea, dizziness, and nystagmus; however, these symptoms are often masked by visual compensatory and proprioceptive mechanisms, which is why the extent of this type of ototoxicity is difficult to assess, especially if it develops slowly and progressively (Casselbrant and Mandel, 2005; Cianfrone et al., 2011).

Hearing loss can be subdivided into three types: conductive, sensorineural or central hearing loss. However, sensorineural hearing loss is the most common one, usually caused by the functional impairment or loss of hair cells. These cells can be damaged by a variety of factors including genetic disorders, infectious diseases, overexposure to intense sound, and certain chemicals. However, cochlear hair cells cannot regenerate following injury because they lost their ability to proliferate during embryogenesis (Yamahara et al., 2015).

CLASSIFICATION OF CHEMICALS CAUSING OTOTOXICITY

Most ototoxic substances can be categorized as pharmaceuticals or workplace chemicals (Table 22.3). Apart from these agents, the most deleterious factor to hearing loss is occupational exposure to noise. Furthermore, the combination of chemicals and noise has been shown to be synergistic in their damaging effects to the hearing/balance mechanisms (Morata, 2007). After exposure to ototoxic chemicals there is a prolonged susceptibility period (up to 1 year) during which the ear may be more vulnerable to an increased damage.

Pharmaceuticals

The reported prevalence of ototoxicity in patients who have received potentially ototoxic therapy ranges from 4% to 90%, depending on factors such as age of patients, drug used, cumulative dose, and administration techniques. Considerable interindividual variability in the prevalence and severity of ototoxicity has been observed among patients receiving similar treatment, suggesting genetic susceptibility as a risk factor (Landier, 2016). Ototoxic chemicals may cause permanent disability because the damaged hair cells cannot regenerate or can do so to a very limited extent (Sedó-Cabezón et al., 2014).

TABLE 22.3 Chemicals Capable of Producing Ototoxicity

Group	Pharmaceuticals
Antimicrobials:	
• Aminoglucosides	Gentamicin, neomicine, streptomycin
• Glyopeptide antibiotics	Vancomicine
• Macrolides	Erythromicin
Antineoplastic	Cisplantin
	Vinca alkaloids (vincristine)
Loop diuretics	Furosemide, etacrinic acid
Analgesic—antipyretic	Salicylates (aspirin) and other nonsteroidal antiinflammatory (NSAIs)
Antipaludics	Quinines (chloroquine)
Antiepileptics	Carbamazepine, phenytoin, valproate, lamotrigine, gabapentin, vigabatrin, and oxcarbazepine
	Industrial Chemicals
Solvents	Butanol
	Carbon disulphide
	Ethanol
	Ethylbenzene
	n-heptane
	n-hexane
	Perchloroethylene
	Styrene
	Toluene
	Trichloroethylene
	Xylenes
Metals	Arsenic
	Lead
	Manganese
	Mercury
	Organotins (trimethyltin, triethyltin)
Others	Acrylonitrile
	Carbon monoxide
	Hydrogen cyanide
	Organophosphates
	Paraquat

Antibiotics

Aminoglycosides

Aminoglycosides are a group of antibiotics used to treat infections caused by Gram-negative bacteria and mycobacteria. Clinically used aminoglycosides can be primarily cochleotoxic (amikacin and kanamycin) or primarily vestibulotoxic (gentamicin, streptomycin, and tobramycin) (Selimoglu, 2007).

Aminoglycosides enter the scala media via the Reissner's membrane and the stria vascularis and progressively accumulate in the endolymph of the inner ear from where they can enter into the cochlear hair cells via endocytosis or the MET channel pore (Karasawa and Steyger, 2015). Aminoglycosides have greater access to cochlear hair cells than to vestibular hair cells, which are tightly embedded and completely surrounded by supporting cells (Ding et al., 2016). Aminoglycosides may also decrease the number of neurons of the vestibular system, either as a result of direct toxicity or secondarily to hair cell loss (Ishiyama et al., 2005). One of the proposed mechanisms of aminoglycoside-induced ototoxicity is the excessive formation of reactive oxygen species (ROS) followed by apoptosis of sensory hair cells (Rybak and Ramkumar, 2007). Likewise, aminoglycosides can interact with transition metals such as iron and copper, which are redox active and potentiate the formation of free radicals, thus producing oxidative cell damage (Sedó-Cabezón et al., 2014).

Macrolides

Macrolide antibiotics are widely used to treat a variety of conditions because of their broad spectrum of activity, particularly in penicillin-allergic patients. However, macrolides may elicit ototoxicity as an adverse effect, which may occur at standard oral doses and through multiple mechanisms. These include inhibition of ion transport in the stria vascularis and affectation of the auditory nerve, cochlear nucleus, and superior olivary complex. Some patients may have lower thresholds for penetration of erythromycin into the inner ear or increased sensitivity to erythromycin within the inner ear fluid. Previous exposures or the presence of an underlying hearing impairment can increase the risk of developing hearing loss from macrolides (Ikeda et al., 2018).

Salicylates and Nonsteroidal Antiinflammatory Drugs

Sodium salicylate, the active component of aspirin, is widely used for its antiinflammatory, antipyretic, and analgesic effects and also for its inhibitory effect on platelet aggregation. Salicylic acid is one of the earliest known ototoxic chemicals; at high doses it produces temporary tinnitus, vertigo, and sensorineural hearing loss. Salicylate quickly enters the cochlea where the perilymph concentration is similar to that in serum. Hearing loss is typically mild to moderate and bilaterally symmetric. Recovery usually occurs 24–72 h after cessation of the drug (de Almeida-Silva et al., 2011; Sheppard et al., 2015).

The underlying ototoxic mechanisms remain unclear but appear to involve changes in cochlear blood flow and reduction of electromotility of the outer hair cells. Inhibition of cyclooxygenase results in a reduction of cochlear blood flow. Salicylate does not alter the number or distribution of hair cells, but it competitively binds and inhibits prestin, a cochlear protein expressed by the outer hair cells, which reduces the electromotility of these cells and produces hearing loss and tinnitus (Zhang et al., 2014). Salicylate also blocks the potassium channel KCNQ4 in the cochlea, leading to outer hair cell degeneration (Hoshino et al., 2010).

Inner hair cells in the cochlea are also involved in the pathological mechanism of salicylate-induced tinnitus. This mechanism involves activation of cochlear N-methyl-D-aspartate receptors, tumor necrosis factor (TNF)−α, and interleukin (IL)-1β in the cochlea hair cells. The alteration of RIBEYE expression (a specific presynaptic protein) could be one possible mechanism underlying the change in the morphology of ribbon synapses and salicylate-induced tinnitus (Zhang et al., 2014).

Long-term treatment with high doses of salicylate can exert neurotoxic degeneration and apoptosis of spiral ganglion neurons and impair auditory neural activity of the cochlea. Changes in the expression of apoptotic genes, particularly among members of the TNF family, appear to play an important role in the degeneration (Wei et al., 2010).

In addition to salicylate, nonsteroidal antiinflammatory drugs (NSAIDs) at high doses may transiently cause tinnitus and mild to moderate hearing loss. NSAIDs impair the active process of the cochlea, the mechanosensory function of the outer hair cells, and affect peripheral and central auditory neurons. However, the endocochlear potential, an indicator of the function of the stria vascularis, is not affected after NSAID treatment (Hoshino et al., 2010).

Antineoplasic Drugs

Within the group of antineoplastic drugs, cisplatin shows the highest risk of ototoxic effects, although vinca alkaloids (e.g., vincristine) have also been involved.

Cisplatin

Platinum-based drugs (cisplatin, carboplatin, and oxaliplatin) are used for the treatment of solid tumors, such as ovarian, lung, head and neck, and testicular cancer. However, dose-limiting side effects, such as ototoxicity, neurotoxicity, and nephrotoxicity, are generally shown by the majority of patients (Sheth et al., 2017). Cisplatin can enter the cochlea through various mechanisms. Copper transporter 1 (Ctr1) is a major copper influx transporter and has been shown to mediate the uptake of cisplatin, carboplatin, and oxaliplatin. The

Ctr1 transporter is expressed by hair cells, the spiral ganglion neurons, and the stria vascularis. The uptake of cisplatin into cochlear cells is also mediated by organic cation transporters (OCTs), particularly the isoform OCT2, which is expressed in the organ of Corti and stria vascularis. Cytosolic uptake of cisplatin can also be facilitated by functional MET channels and transient receptor potential vanilloid 1 (TRPV1) channels (Sheth et al., 2017; Waissbluth and Daniel, 2013).

The degree of cisplatin-induced ototoxicity has been associated with cumulative doses (usually greater than 400 mg/m^2) and dose intensity, although considerable interindividual variability exists, with hearing threshold alterations being observed in 28%–77% of patients. The ototoxicity of cisplatin has a more pronounced effect in children, where the prevalence may be greater than 60% and with more severe symptoms (Karasawa and Steyger, 2015).

Platinum-based drugs, once activated, cross-link two purine bases of DNA by reacting with nitrogens at position seven, thus interfering with cell division and transcription. The resulting DNA irreversible changes inhibit tumor cell division and activate the apoptotic process, which may be also responsible for hearing loss if cochlear hair cells are affected (Starobova and Vetter, 2017). Cisplatin can also break double-stranded DNA in strial marginal and cochlear hair cells that are transcriptionally active, disrupting cellular physiology and inducing cytotoxicity (Karasawa and Steyger, 2015).

As occurred with aminoglycosides, platinum-induced ototoxicity is characterized by the production of toxic levels of ROS within the cochlea, with destruction of hair cells and damage to the stria vascularis and spiral ganglion cells. Given the tonotopic arrangement of cochlear hair cells, the initial insult begins at the base of the cochlea, in which high-frequency sounds are processed, and further drug administration results in damage that progresses toward the cochlear apex, in which lower (speech frequency) sounds are processed. Because the destroyed cochlear sensory hair cells cannot regenerate, the sensorineural hearing loss is almost always bilateral and irreversible and also can be accompanied by tinnitus and vertigo (Landier, 2016).

Vincristine

Vincristine is commonly used in the treatment of solid tumors, hematological tumors, and nonmalignant hematologic disorders. Vincristine is thought to be uptaken by cells through a carrier-mediated transport mechanism. Similar to other vinca alkaloids, vincristine binds to the β-subunit of tubulin and inhibits microtubule formation, leading to depolymerization of the microtubule and mitotic arrest at metaphase or cell death. Peripheral sensorimotor and autonomic neuropathy represents an important dose-limiting side effect that may occur from binding with tubulin and further disruption of microtubule assembly, axonal transport, and secretory functions, resulting in primary axonal degeneration (Kuruvilla et al., 2009). Experimental studies indicate that vinca alkaloids cause dose-related degeneration of hair cells. Although low and moderate doses of vincristine ($\leq$1.5 mg/m^2) do not have any significant hearing impairment, bilateral sensorineural hearing loss has been observed in adult patients receiving high doses of vincristine (2–2.5 mg/m^2) (Riga et al., 2006).

Loop Diuretics

Loop diuretics are a group of drugs that inhibit renal reabsorption of Na^+, Cl^-, and K^+ in the ascending loop of Henle. They include furosemide, bumetanide, ethacrynic acid, and torasemide, which are often used to treat kidney insufficiency, hypertension, and congestive heart failure. The hearing loss induced by loop diuretics is bilateral and usually reversible that lasts only during the treatment, unless these drugs are administered to patients with severe acute or chronic renal failure or in combination with other ototoxic drugs (e.g., cisplatin or aminoglycosides). In these cases the degree of hearing loss worsens and becomes permanent (Campo et al., 2013).

Although the tight junctions in the blood–cochlea barrier prevent toxic molecules from entering cochlea, when diuretics induce a transient ischemia, the barrier is temporarily disrupted allowing the entry of toxic chemicals. Loop diuretics interfere with strial adenylate cyclase and Na^+,K^+–ATPase and inhibit the Na^+–K^+–$2Cl^-$ cotransporter in the stria vascularis. Because renin is present in cochlear pericytes surrounding stria arterioles, diuretics may induce local vasoconstriction by renin secretion and angiotensin formation, resulting in a reduced blood flow of the vessels supplying the lateral wall. This mechanism parallels the effect of diuretics on kidney and ear. Ultimately, the ionic gradients between the endolymph and the perilymph are disturbed, resulting in a dose-dependent reduction of endocochlear potential, which affects sensorial transduction and produces a sudden high-frequency hearing loss (Campo et al., 2013; Ding et al., 2016).

Antimalarial Drugs

Although antimalarial medications such as quinine and chloroquine may cause sensorineural hearing loss and tinnitus after prolonged therapy in high doses, these effects are rare and usually reversible. A large dose of quinine can also lead to reversible hearing loss and tinnitus. Vertigo may appear as associated vestibular toxicity. Quinine's manifestation of ototoxicity is similar to that of salicylate, but the mechanisms of toxicity differ. These mechanisms are likely the result of an ischemic process that affects cochlear sensory hair cells, causes atrophy of stria vascularis, and decreases neuronal population (Bortoli and Santiago, 2007).

Phosphodiesterase Type 5 Inhibitors

Phosphodiesterase type 5 (PDE-5) inhibitors are prescribed for erectile dysfunction and pulmonary arterial hypertension. Sildenafil, and to a lesser extent vardenafil and tadalafil, have been associated with sudden sensorineural hearing loss in a large population-based study (McGwin, 2010). However, results from animal studies are controversial. Sildenafil inhibits the PDE-5, the enzyme that degrades cyclic guanosine monophosphate (cGMP), leading to an increase in cGMP and enhanced vasodilation and decreased vascular resistance. However, the question of why these drugs cause hearing loss is unclear. Alterations in the nitric oxide–cyclic guanosine monophosphate (NO/cGMP) pathway, including downstream second messenger molecules, have been previously involved in the development of hearing loss, including cisplatin and aminoglycoside ototoxicity (Maddox et al., 2009).

Antiepileptic Drugs

Data from experimental, cross-sectional, and prospective studies have shown evidence for the deleterious effect of some antiepileptic drugs on the auditory and vestibular systems. Long-term treatment with some of these drugs (e.g., carbamazepine, phenytoin, valproate, lamotrigine, gabapentin, vigabatrin, and oxcarbazepine), even at therapeutic doses, may result in sensorineural hearing loss, tinnitus, dizziness, and abnormalities in brain stem auditory evoked potentials, indicating auditory and central and/or peripheral vestibular dysfunctions (Hamed, 2017).

Industrial Chemicals

Workplace chemicals with ototoxic potential include industrial solvents (e.g., toluene, styrene, and trichloroethylene) and synthetic intermediates. These compounds are toxic to both the auditory and vestibular sensory systems, with their main targets being the hair cells (Sedó-Cabezón et al., 2014). The effects of these chemicals on ear function can be aggravated by noise, which remains a well-recognized cause of hearing impairment and poses an additive risk to hearing (Campo et al., 2013).

Among industrial solvents, styrene is one of the most ototoxic agents (Gagnaire and Langlais, 2005). Styrene is used in the production of plastics, synthetic rubbers, resins, insulating materials, and protective surface coatings. Although occupational studies on styrene-associated hearing impairment have led to contradictory results, experimental studies in animals showed that styrene can induce hearing loss. However, a threshold for ototoxicity has been identified (600–700 ppm for 4 weeks of exposure) (Chen et al., 2007). In contrast to most ototoxic drugs, which initially affect the base of the cochlea, styrene and other ototoxic solvents disrupt cochlear outer hair cells starting from the middle turn, thereby leading to hearing loss in the mid-frequency range (Crofton et al., 1994). However, supporting cells (e.g., Deiters' cells) appear to be the most vulnerable target of styrene in the cochlea, with inner hair cells being relatively insensitive to styrene exposure (Chen et al., 2008). Histochemical studies on styrene-induced apoptosis indicated that both mitochondrial-dependent pathway and death receptor–dependent pathway were involved in the styrene-induced hair cell death (Chen et al., 2007).

A systematic review of epidemiological studies confirmed the association between occupational exposure to diverse aromatic solvents and an increased risk of developing hearing loss, even at low concentrations (Odds ratio (OR) 2.05, 95% confidence interval (CI) 1.4–2.9). In addition, noise exacerbated the extent of hearing loss of those exposures as subjects concurrently exposed to solvents mixture and noise had an higher risk of hearing loss (OR 2.95, 95% CI 2.1–4.2) (Hormozi et al., 2017). Another study confirmed this finding and found that the combined exposure to a solvent mixture and noise increased the risk of hearing loss from 1.70 after 3 years of exposure to 8.25 after 12 years (Kaufman et al., 2005). The latency period depends on the characteristic of the exposure rather than on the type of ototoxic agent, although in the case of exposure to styrene the impairment may arise after 3 years of exposure (Johnson et al., 2006).

The cochleotoxic effects of aromatic solvents (styrene, toluene, and others) have been demonstrated in animal experiments, with longer exposures causing irreversible hearing impairments (Campo et al., 2013). Aromatic solvents can induce hearing loss by affecting hair cells of the organ of Corti, either directly after acute exposures or by the formation of chemically reactive intermediates after long-term exposures. These intermediates include ROS, which may disturb the transmembrane flux of K^+, with the accumulation of this cation into the outer channel of the organ of Corti causing outer hair cell cytotoxicity (Campo et al., 2013; Chen et al., 2007).

Exposure to high concentrations of trichloroethylene can disrupt cochlear sensory hair and spiral ganglion cells as well as the auditory nerve pathways within the cochlea. Conversely, the solvents carbon disulfide and n-hexane have been associated with retrocochlear dysfunction, as experimental studies have shown disturbances in the auditory nervous pathway beyond the cochlea (Campo et al., 2013).

Nitriles are industrial chemicals used as solvents (acetonitrile), monomers for polyacrylonitrile (acrylonitrile) and for the production of melamine resins (benzonitrile). These chemicals have shown cochleotoxic effects only in experimental animals, and hearing loss may occur with simultaneous exposure to acrylonitrile and low noise intensity (Campo et al., 2013).

Carbon monoxide and cyanides are asphyxiants that induce ototoxicity as a result of hypoxia within the cochlea. Experimental studies indicate that exposures to low concentrations of these asphyxiants impair cochlear function and induce reversible auditory effects, particularly affecting high-frequency tones. Although cyanides affect the stria vascularis, carbon monoxide induces glutamatergic excitotoxicity. Furthermore, both agents can potentiate permanent noise-induced hearing loss in both animals and humans (Campo et al., 2013).

Epidemiological and experimental studies have shown that long-term exposure to lead, mercury, and manganese, among other metal compounds, may damage different structural elements of the cochlea resulting in sensorineural hearing loss (Roth and Salvi, 2016). Workers chronically exposed to lead had impaired conduction in the auditory nerve and the auditory pathway in the lower brain system (Campo et al., 2013). Trimethyltin and triethyltin are also capable of inducing a dose-dependent hearing impairment in experimental animals as a result of increased calcium concentration in outer hair cells and spiral ganglion cells, with further disruption of the functioning of these cells (Liu and Fechter, 1996). Likewise, in vitro and in vivo studies indicate that manganese accumulates in the cochlea and that it alone may be ototoxic, although the combined exposure to manganese plus noise increases the risk of hearing loss over that caused by noise alone (Muthaiah et al., 2016).

Nonetheless, the putative hearing loss of prolonged (over)exposures to industrial chemicals may be confounded by exposure to high noise levels at work as noise by itself is known to cause hearing loss.

FACTORS MODIFYING CHEMICAL OTOTOXICITY

A number of factors contribute to an increased risk of ototoxicity of chemicals, including exposure to high concentrations of pharmaceutical or industrial agents, concurrent or consecutive exposure to several ototoxic chemicals, exposure to noise, preexisting hearing loss, genetic susceptibility, and age (Bauman, 2003). The administration of ototoxic drugs, particularly aminoglucosides, during pregnancy has been associated with impaired hearing, with a particularly sensitive phase up to approximately the fourth month of gestation (Mylonas, 2011). Young children and adults beyond 60 years are also more vulnerable to permanent hearing loss and/or balance problems. Certain predisposing diseases, such as liver or kidney failure, which cause the accumulation of certain active ingredients, may facilitate the appearance of ototoxicity. These circumstances may worsen the degree of chemical-induced ototoxicity (Bauman, 2003).

Considerable interindividual variability in the prevalence and severity of ototoxicity has been observed among patients receiving similar treatment with known ototoxic agents. Although some individuals remain unaffected at high cumulative doses, others experience severe damage at low doses. This interindividual variation can be partially accounted for by genetic susceptibility to the ototoxic effects of chemicals (Landier, 2016). Certain mitochondrial chromosome mutations can represent one of the genetic factors underlying vulnerability to aminoglycosides. The A1555G mutation (substitution of a guanine with an adenine) located on the mitochondrial RNA12S has been related to aminoglycoside ototoxicity. The fact that the mutated human form A1555G is very similar to the bacterial target for the aminoglycosides (ribosomal 16S RNA) may explain the occurrence of hearing loss even at low dosages of the drug. It has been estimated that 17% of the subjects eliciting aminoglycoside ototoxic effects harbor such mutation (Cianfrone et al., 2011).

Toxic interactions between ototoxic drugs are well known, particularly the potentiation of the ototoxic effect of aminoglycosides by loop diuretics. The mechanisms underlying this interaction can be related to the damage produced by loop diuretics on the tight cell junctions in the blood vessels of the stria vascularis, which disrupt the blood–cochlear barrier temporarily. This damage changes the positive endocochlear potential into a negative one, leading to a massive entry of the positively charged aminoglycosides into the endolymph, followed by hair cell loss and further permanent hearing loss (Ding et al., 2016; Liu et al., 2011).

On the other hand, interactions can be observed on simultaneous exposure to ototoxic chemicals and noise, such that moderate levels of noise can potentiate auditory dysfunction. Hearing loss can thus be greater than the sum of the individual effects. For instance, the lowest level of solvent exposure needed to elicit an auditory effect is reduced when individuals are co-exposed to other stressors, such as noise or carbon monoxide. This means that health-based reference values set for chemical-induced ototoxicity may not protect against hearing loss in the presence of high-level noise (Morata and Johnson, 2012).

UNDERLYING MECHANISMS OF OTOTOXICITY

Ototoxic compounds typically damage both the auditory sensory and the vestibular systems, as their main targets are the hair cells. As mentioned above, these are mechanosensory cells responsible for the transduction of either sound waves and linear and rotational movements of the head. Although sensorineural hearing loss occurs mainly as a consequence of degeneration and apoptosis of auditory hair cells, afferent dendrite terminals and the corresponding spiral ganglion neurons may also be damaged and undergo apoptosis. This damage can occur either primarily, before hair cell death, or secondarily, because of loss of trophic support after hair cell degeneration (Sedó-Cabezón et al., 2014).

After ototoxic chemicals reach the endolymph, they enter into hair cells and induce necrotic or apoptotic cell death through different mechanisms (Fig. 22.3). Nevertheless, some common features are shared between damaging mechanisms induced by these chemicals, noise and aging. For instance, the final common pathways usually involve oxidative stress and free radical–mediated damage (Niihori et al., 2015). The main toxic mechanisms include disruption of mitochondrial protein synthesis, formation of free radicals, activation of mitogen-activated protein kinases (MAPK, such as c-Jun N-terminal kinase—JNK), disruption of phosphoinositide homeostasis, mitochondrial dysfunction, and activation of caspases and nucleases (Leitner et al., 2011; Wang et al., 2007). Ultimately, activation of proapoptotic signaling pathways leads to apoptotic hair cells death (Fig. 22.3).

Chemical- or noise-induced hearing loss can be driven by the formation of free radicals, including ROS and reactive nitrogen species (RNS), within hair cells. Excessive production of these highly reactive molecules overwhelm the antioxidant defense mechanisms of the cochlea, causing a cascade of events eventually leading to apoptosis of hair cells and further sensorineural hearing loss. ROS identified in cochlear tissue are derived from hair cells mitochondria and include hydroxyl radicals, superoxide anions, and hydrogen peroxide (Falasca et al., 2017). ROS can form hydrogen peroxide (H_2O_2) or react with NO resulting in peroxynitrite formation ($ONOO^-$). Peroxynitrite can interact with proteins and form nitrotyrosine, whereas hydrogen peroxide reacts with iron to form the highly reactive hydroxyl radical (•OH). This radical then reacts with polyunsaturated fatty acids in cell membranes, produces lipid peroxidation, and generates the highly toxic aldehyde 4-hydroxynonenal (4-HNE) (Waissbluth and Daniel, 2013). Although ROS generation induces signal transduction cascades leading to apoptosis or necrosis, how ROS induces these signaling pathways remains unclear. One possibility is that increased levels of 4-HNE can induce intracellular Ca^{2+} influx, which triggers apoptosis of cochlear hair cells (Karasawa and Steyger, 2015). Ultimately, the imbalance between ROS and RNS together with reduced intrinsic antioxidant defenses represents the main elements underlying hair cell death through either apoptosis or necrosis.

The aquated forms of cisplatin are also highly reactive, particularly with cytoplasmic thiol-containing molecules such as the antioxidant glutathione and the antioxidant enzymes glutathione S-transferase,

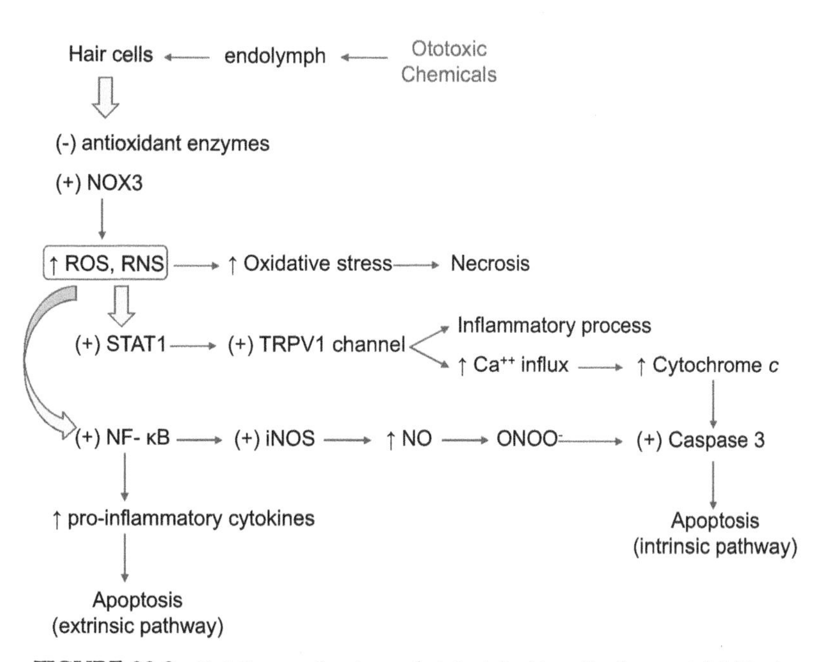

FIGURE 22.3 Putative mechanisms of ototoxicity (+, activation; −, inhibition).

glutathione peroxidase, and superoxide dismutase (Karasawa and Steyger, 2015; Waissbluth and Daniel, 2013). Once inhibited, these antioxidant defenses can no longer scavenge and neutralize reactive molecules, and hence the cellular redox status is shifted, leading to toxic levels of ROS and RNS and activation of the intrinsic apoptotic pathway (Fig. 22.3). Upregulation of antioxidant pathway activity, such as glutathione-*S*-transferases that are expressed in the mammalian cochlea, can protect against ototoxicity (Travis et al., 2014).

A major source of ROS in the cochlea is the NADPH oxidase 3 (NOX3), an isoform highly expressed in the organ of Corti and spiral ganglion, which is activated by cisplatin. The generation of ROS promotes activation of signal transducer and activator of transcription 1 (STAT1) leading to both inflammatory and proapoptotic actions in the cochlea. STAT1 stimulates the inflammatory process by activation of TRPV1 channels. The induction of TRPV1 also increases calcium inflow into cells, with massive cytochrome *c* release and further activation of caspase-3, thus triggering the intrinsic apoptotic pathway (Rybak et al., 2012; Sheth et al., 2017).

Another pathway leading to cochlea inflammation is activation of nuclear transcription factor-kappa B (NF-κB) through the production of ROS, as occurred with cisplatin (Chung et al., 2008). NF-κB upregulates the expression of proinflammatory cytokines such as IL-1β and TNF-α. TNF-α can, in turn, activate NF-κB resulting in a positive feedback loop that increases the inflammatory cascade (Waissbluth and Daniel, 2013). This local inflammation can activate the extrinsic signaling pathway leading to apoptosis (Fig. 22.3). NF-κB also increases the expression of inducible nitric oxide synthase (iNOS), leading to NO production that further reacts with superoxide to form peroxynitrite radical. This highly toxic molecule can activate caspase-3, which triggers apoptosis in the stria vascularis that eventually results in auditory dysfunction (Rybak et al., 2007).

The potassium channel KCNQ4, expressed in the mammalian cochlea, is another target for ototoxic chemicals. In fact, KCNQ4 mutations cause human hereditary hearing loss (O'Sullivan et al., 2017). Activation of this channel is associated with an outer hair cell potassium current (IK,n) and subsequent depolarization of outer hair cells (Hoshino et al., 2010). Salicylates cause a concentration-dependent (and reversible) reduction in IK,n (Wu et al., 2010). As the activity of KCNQ4 is linked to phosphatidylinositol 4,5-bisphosphate (PIP2) metabolism, aminoglycosides may indirectly cause KCNQ4 dysfunction by disrupting PIP2 homeostasis and in turn induce outer hair cell death (O'Sullivan et al., 2017).

The stria vascularis hosts the major blood–labyrinth barrier, a highly specialized capillary network that controls exchanges between blood and the interstitial space in the cochlea. Damage to strial cells in the cochlear lateral wall may breakdown the blood–labyrinth barrier and cause loss of the endolymphatic potential (Karasawa and Steyger, 2015; O'Sullivan et al., 2017). Some ototoxic chemicals (e.g., cisplatin) have shown to reduce the endocochlear potential in animal experiments, suggesting dysfunction of the stria vascularis. Aminoglycosides do not induce changes in endocochlear potentials during early phases but may do so after extensive treatment, leading to significant atrophy of the stria vascularis (Schacht et al., 2012).

Because hair cell synapses with the afferent spiral ganglion neurons are glutamatergic, these neurons can undergo excitotoxic damage. Excitotoxicity can be explained because under toxic stress hair cells have a limited capacity for regulating glutamate release and reuptake, and the excess of glutamate released can damage the afferent cochlear neurons as occurs with exposure to carbon monoxide (Sedó-Cabezón et al., 2014).

From the mechanistic data mentioned above it can be inferred that some molecules formed in the cochlea might be used as biomarkers of ototoxicty. These include molecular markers of oxidative stress (e.g., nitrotyrosine, a marker of NO production; or 4-HNE, a marker of cell membrane lipoperoxidation), antioxidant enzymes, cytoplasmic cytochrome *c*, iNOS and caspase-3, and inflammatory biomarkers (cytokines such as TNF-α and IL-1β). However, all these molecules are indirect biomarkers, as they lack specificity for cochlear damage.

FUNCTIONAL AND MOLECULAR BIOMARKERS FOR THE EVALUATION OF OTOTOXICITY

Early detection of ototoxic damage can improve the treatment outcome through minimizing hearing loss progression. The best way to detect ototoxicity is by assessing auditory and vestibular function directly. However, this section will address only the function of the auditory system, as this is more often assessed in clinical practice and occupational health services after exposure to ototoxic chemicals.

Functional Assessment of the Hearing Damage

Clinical evaluation of the hearing mechanisms can be performed with techniques that assess the functional integrity of the entire auditory system, from the tympanic membrane to cerebral cortex. The function of the auditory system can be assessed by using tests of mechanical sound transmission (middle ear function), neural sound transmission (cochlear function), and speech discrimination ability (central integration). In practice, the peripheral auditory pathway can be assessed by using pure-tone audiometry (PTA), high-frequency audiometry, or speech audiometry. On the other hand, the assessment of damage to the Central Auditory Nervous System can be made through electroacoustic tests (immitanciometry, immittance decay, otoacoustic emissions), electrophysiological tests (brain stem auditory evoked potential, auditory cortical response), and central auditory processing (Hoth and Baljić, 2017).

Pure-Tone Audiometric Testing

PTA is the standard test used for assessing hearing loss and usually involves the presentation of sounds across a wide range of frequencies in a sound-attenuating test booth. This method can detect significant increases in hearing threshold, consistent with ototoxicity criteria. Besides, PTA makes distinction between sensorineural (cochlear and central auditory pathways) and conductive (outer and middle ear) type of hearing loss. PTA obtains a representation of the lowest sound intensities that can be heard across the frequency spectrum (Lustig, 2010). Threshold shifts in patients with sensorineural hearing loss typically begin in the high-frequency range (>4 kHz) and generally progress to the lower frequencies in a staggered pattern over time with continued exposure to the ototoxic agent (Landier, 2016). Most audiometric evaluations used for monitoring chemical-induced hearing loss are limited to the range of speech spectrum frequencies (0.25–8 kHz). However, such testing would fail to detect hearing loss in the ultrahigh frequencies from 10 to 20 kHz. Expanding audiometry into this range (high-frequency audiometry) would allow to detect a substantial number of cases of hearing loss that otherwise would be missed (Singh-Chauhan et al., 2011). Monitoring at the high-frequency limit of hearing provides the earliest indication of ototoxicity, consistent with the systematic progression of cochlear damage from base to apex found in animal models (Fausti et al., 1999).

Although PTA can help determine the nature and degree of the hearing loss, pure-tone thresholds do not provide qualitative information on sound perception. This is important because individuals with hearing loss can hear but cannot understand. Speech audiometry provides this additional qualitative hearing assessment. The speech reception threshold is the lowest level at which an individual can repeat a word, a measure that correlates closely with pure-tone thresholds (Lustig, 2010).

An ototoxic hearing loss can effectively be determined by comparison of audiograms from before (baseline testing) and after the administration of a drug or exposure to an ototoxic chemical (Cianfrone et al., 2011). In most scales, hearing loss is assigned a grade ranging from 0 (normal hearing, or clinically insignificant loss) to 4 (severe or profound hearing loss) (Landier, 2016).

Distortion Product Otoacoustic Emission

PTA remains the most commonly used clinical test to measure the extent of hearing loss; however, it is time-consuming and cannot discriminate between cochlear and retrocochlear hearing loss. In case of central hearing loss, a PTA can indicate normal hearing, but a person can still have difficulty understanding speech, particularly in background noise, making it difficult to hold a conversation. Thus, there is a need to use tests that evaluate the auditory system more comprehensively, from the cochlea to the higher auditory pathways. These tests may help differentiate between the individual (or the combined) effects of ototoxicants and noise on hearing. One of these complementary tests is the use of distortion product otoacoustic emission (DPOAE), a sensitive tool that can detect early ototoxic changes in the cochlea. Otoacoustic emissions are sound signals resulting from the mechanical action of outer hair cells in the organ of Corti. Because DPOAE is less time-consuming, it would allow clinical monitoring of more individuals. The inclusion of DPOAEs in a battery of tests to assess hearing would facilitate the distinction between sensory and neural hearing disorders (Campo et al., 2013; Md Daud et al., 2014).

The diagnostic value of otoacoustic emissions lies in their association with normal outer hair cell function, which is usually indicative of normal hearing. They are stable over time within individual ears and their high test–retest reliability allows them to be used as an effective monitoring examination. Moreover, otoacoustic emissions are noninvasive, non–time-consuming, easy to perform, and do not require the individual's active participation (e.g., can be used in little children) (Riga et al., 2006). DPOAE are objective, noninvasive measures that may be more sensitive to initial changes in auditory function than conventional auditory testing (Landier, 2016). Overall, the objective otoacoustic emission data largely parallel the subjective audiometric data (Gürkov et al., 2008).

Acoustic Immitance Test

This testing objectively measures the ease with which sound is transmitted through the tympanic membrane and middle ear. It consists of tympanometry and the acoustic reflex. Tympanometry measures the changes that occur in the tympanic membrane and ossicular chain as a result of a change in air pressure in the ear canal. The acoustic reflex (or stapedius muscle reflex) is measured by sensing the altered compliance of the tympanic membrane in response to a loud auditory signal (Lustig, 2010).

Acoustic Brain Stem Response Testing

Acoustic brain stem response testing (ABR), also referred to as brain stem auditory evoked responses or brain stem auditory evoked potential, is an electrical test used to measure cochlear and retrocochlear function. This test allows the retrocochlear auditory pathways to be measured, that is, those pathways between the cochlea and auditory cortex. The two main parameters analyzed are wave amplitude and latency changes. If an abnormal ABR is found, further evaluation with magnetic resonance imaging is warranted (Lustig, 2010). Because ABR is a nonbehavioral test sensitive to ototoxicity, it can be applied to little children or individuals unable to cooperate with standard behavioral testing (PTA) (Landier, 2016).

Although the audiological monitoring of individuals exposed to ototoxic chemicals is now standardized, a battery of diagnostic methods are currently available to confirm balance disorders (e.g., dynamic posturography) and to assess the labyrinthine damage, e.g., Halmagyi test, videonystagmographic record of spontaneous nystagmus, and caloric tests (Sánchez-Sellero and Soto-Varela, 2016). In addition, the video head impulse test and vestibular evoked myogenic potentials seem to be suitable procedures for objective assessment of suspected vestibulotoxic effects. The early detection of such effects has important implications for the prevention of balance disorders (Walther et al., 2015).

Molecular Biomarkers

Serum biomarkers are useful molecular indicators that can be used to assess normal or abnormal biological processes in response to pharmacological or industrial chemicals. They can also provide information on progression of disease. To be useful, biomarkers must be present in easily accessible body fluids, easily quantifiable and sensitive enough to detect the biological state, or outcome of interest, and specific enough to discern a normal from a diseased state (Naples et al., 2018).

Despite the wide array of molecular biomarkers available in clinical practice to confirm many common diseases, blood tests for detecting hearing loss and vertigo are not yet clinically available. Currently, hearing loss can only be diagnosed through hearing tests such as those mentioned in the previous section. However, there is no way to detect hearing loss at its earliest stages, with prevention and intervention options being limited at the time of diagnosis when hearing loss has developed. Recently a few blood tests have been considered to be helpful, as occurred with connexin 26/30 (proteins for contacting adjacent cells) and pendrin (an anion exchanger that elevates the endolymphatic pH), which are evaluated for genetic hearing loss. Another example is measurement of nonspecific inflammatory markers, such as heat shock proteins, for the diagnosis of autoimmune inner ear disease. These tests generally have limited clinical applications, are not specific to the inner ear, and consequently have variable sensitivity and specificity (Liba et al., 2017).

However, there are several unique proteins specific to the inner ear with specialized functions that can be detected in minute quantities in the blood, which could be useful as otologic biomarkers. Because blood levels of these proteins correlate with inner ear disorders, they can potentially serve as biomarkers that help improve the early detection and diagnosis of hearing loss or vertigo. This is the case of otolin-1, a secreted glycoprotein expressed only in the inner ear, which increases in blood from patients with benign paroxysmal positional vertigo (Tabtabai et al., 2017). Besides, blood levels of prestin, an outer hair cell–specific protein, can be a useful biomarker for quantifying the extent of sensorineural hearing loss. In a rat model of noise-induced hearing loss, reduced hair cell counts and DPOAE levels were observed in the setting of decreased prestin levels (Parham and Dyhrfjeld-Johnsen, 2016). Prestin is likely not produced in major solid organs because its concentration in these organs does not exceed those in blood. Because prestin is released by injured outer hair cells, changes in blood prestin levels are linked to hearing loss; even before this loss it can be measured by functional tests. Nonetheless, if ototoxicity biomarkers are to be used into clinical practice, experimental validation is necessary before being translated to the clinical setting. In addition, biomarker levels should be sensitive to interventions aimed at ameliorating ototoxic damage (Liba et al., 2017; Naples et al., 2018).

Proteomic analysis of inner ear sensory epithelia has led to the identification of three major processes for normal hair cell physiology: energy metabolism, signal transduction, and cell cytoskeleton. Proteins involved in these processes can be exploited to develop new biomarkers and gain insight into pathophysiological mechanisms underlying ototoxicity (Alawieh et al., 2015).

Alternative Models

Damage to sensory hair cells in the inner ear can lead to permanent hearing or balance deficits, as regeneration of the inner ear sensory epithelia does not occur in mammals. Because zebrafish and other nonmammalian vertebrates have the ability to regenerate sensory hair cells, these models will allow for a better understanding of the molecular and cellular bases for this regenerative ability (Lush and Piotrowski, 2014).

Fish have neurosensory hair cells on their body surfaces, which can detect changes in movement and vibration from the surrounding water. These hair cells are structurally similar to those found within the human inner ear, making zebrafish a suitable model to study inner ear dysfunction. Because of their superficial location, zebrafish hair cells can be experimentally damaged by adding exogenous toxic agents. For normal zebrafish, the induction of water flow results in a predictable "head-to-current" swimming behavior called rheotaxis (Niihori et al., 2015). The addition of aminoglycoside toxins to water has caused damage to zebrafish lateral line hair cells and negatively impacts rheotaxis behavior in a dose-dependent manner (Suli et al., 2012).

CONCLUDING REMARKS AND FUTURE DIRECTIONS

The assessment of inner ear function remains the gold standard for functional clinical diagnostics of hearing loss. At present, there are no molecular biomarkers of ototoxicity that can accurately track, precede, or reliably correlate with predicted and clinical/functional evidence of inner ear damage. This is due to a number of critical issues, such as the minute size of the inner ear, the small number of hair cells (which lost their ability to proliferate in the early postnatal period), the poor accessibility of the inner ear, the small volume of endolymph fluid that needs much more sensitive techniques to detect extremely low amounts of diverse molecules, and the complex features of cochleovestibular diseases. Identification and development of novel molecular markers of early cochlear and vestibular impairment remain a research challenge because standard functional tests lack predictivity and show abnormal results once inner ear is damaged, particularly at later stages and many times irreversibly.

Although significant efforts have been made to develop new biomarkers for other target organs, these efforts have been limited in the area of ototoxicity. Moreover, the few biomarkers available are not suitable to be appropriately used for clinical and nonclinical research applications. Attempts are currently ongoing to qualify reliable and sensitive biomarkers of cochlear and vestibular damage to be applicable for diagnostic and research use in humans and laboratory animals. One of the key new directions of novel development in ototoxicity biomarker research is the understanding of omic-based biomarkers, which would also allow pathophysiological mechanisms to be identified. Future research efforts should utilize proteomic, metabolomic, and miRNA array technologies in well-defined patient populations to identify accurate and sensitive molecular biomarkers. This approach will enable the development of reliable profiles to identify at-risk individuals and to implement personalized strategies such as the use of otoprotective agents or avoidance of ototoxic drugs in susceptible individuals to decrease the prevalence of ototoxicity and improve health-related quality of life. The maintenance of hair cell and supporting cell numbers after cochlear injury is therefore important for the treatment of sensorineural hearing loss. To achieve such treatment, protection and/or regeneration of hair cells is necessary.

References

Alawieh, A., Mondello, S., Kobeissy, F., et al., 2015. Proteomics studies in inner ear disorders: pathophysiology and biomarkers. Expert Rev. Proteomics 12, 185–196.

Bauman, N.G., 2003. Ototoxic Drugs Exposed, second ed. GuidePost Publications, Stewartstown, PA.

Bortoli, R., Santiago, M., 2007. Chloroquine ototoxicity. Clin. Rheumatol. 26, 1809–1810.

Brigande, J.V., 2017. Hearing in the mouse of Usher. Nat. Biotechnol. 35, 216–218.

Campo, P., Maguin, K., Gabriel, S., et al., 2009. Combined Exposure to Noise and Ototoxic Substances. European Agency for Safety and Health at Work (EU–OSHA), Luxembourg.

Campo, P., Morata, T.C., Hong, O., 2013. Chemical exposure and hearing loss. Dis. Mon. 59, 119–138.

Casselbrant, M.L., Mandel, E.M., 2005. Balance disorders in children. Neurol. Clin. 23, 807–829.

Chen, G.D., Chi, L.H., Kostyniak, P.J., et al., 2007. Styrene induced alterations in biomarkers of exposure and effects in the cochlea: mechanisms of hearing loss. Toxicol. Sci. 98, 167–177.

Chen, G.D., Tanaka, C., Henderson, D., 2008. Relation between outer hair cell loss and hearing loss in rats exposed to styrene. Hear Res 243, 28–34.

Chung, W.H., Boo, S.H., Chung, M.K., 2008. Proapoptotic effects of NF-kappaB on cisplatin-induced cell death in auditory cell line. Acta Otolaryngol 128, 1063–1070.

Cianfrone, G., Pentangelo, D., Cianfrone, F., et al., 2011. Pharmacological drugs inducing ototoxicity, vestibular symptoms and tinnitus: a reasoned and updated guide. Eur. Rev. Med. Pharmacol. Sci. 15, 601–636.

Crofton, K.M., Lassiter, T.L., Rebert, C.S., 1994. Solvent-induced ototoxicity in rats: an atypical selective mid-frequency hearing deficit. Hear Res 80, 25–30.

Cunningham, L.L., Tucci, D.L., 2017. Hearing loss in adults. N. Engl. J. Med. 377, 2465–2473.

de Almeida-Silva, I., de Oliveira, J.A., Rossato, M., et al., 2011. Spontaneous reversibility of damage to outer hair cells after sodium salicylate induced ototoxicity. J. Laryngol. Otol. 125, 786–794.

Ding, D., Liu, H., Qi, W., et al., 2016. Ototoxic effects and mechanisms of loop diuretics. J. Otol. 11, 145–156.

Ervin, S.E., 2017. Assessment Tools: Introduction to the Anatomy and Physiology of the Auditory System. Workplace INTEGRA, Inc., Greensboro, North Carolina. http://www.workplaceintegra.com/hearing-articles/Ear-anatomy.html.

Falasca, V., Greco, A., Ralli, M., 2017. Noise induced hearing loss: the role of oxidative stress. Otolaryngol. Open J. SE (5), S1–S5.

Fausti, S.A., Henry, J.A., Helt, W.J., et al., 1999. An individualized, sensitive frequency range for early detection of ototoxicity. Ear Hear. 20, 497–505.

Gagnaire, F., Langlais, C., 2005. Relative ototoxicity of 21 aromatic solvents. Arch Toxicol 79, 346–354.

Gürkov, R., Eshetu, T., Miranda, I.B., et al., 2008. Ototoxicity of artemether/lumefantrine in the treatment of falciparum malaria: a randomized trial. Malar. J. 7, 179.

Hamed, S.A., 2017. .The auditory and vestibular toxicities induced by antiepileptic drugs. Expert Opin. Drug Saf. 16, 1281–1294.

Hormozi, M., Ansari-Moghaddam, A., Mirzaei, R., et al., 2017. The risk of hearing loss associated with occupational exposure to organic solvents mixture with and without concurrent noise exposure: a systematic review and meta-analysis. Int. J. Occup. Med. Environ. Health 30, 521–535.

Hoshino, T., Tabuchi, K., Hara, A., 2010. Effects of NSAIDs on the inner ear: possible involvement in cochlear protection. Pharmaceuticals (Basel) 3, 1286–1295.

Hoth, S., Baljić, I., 2017. Current audiological diagnostics. GMS Curr. Top. Otorhinolaryngol. Head Neck Surg. 16. Doc09.

Ikeda, A.K., Prince, A.A., Chen, J.X., et al., 2018. Macrolide-associated sensorineural hearing loss: a systematic review. Laryngoscope 128, 228–236.

Ishiyama, G., Finn, M., Lopez, I., et al., 2005. Unbiased quantification of Scarpa's ganglion neurons in aminoglycoside ototoxicity. J Vestib Res 15, 197–202.

Jiang, M., Karasawa, T., Steyger, P.S., 2017. Aminoglycoside-induced cochleotoxicity: a review. Front. Cell. Neurosci. 11, 308.

Johnson, A.C., Morata, T.C., Lindblad, A.C., et al., 2006. Audiological findings in workers exposed to styrene alone or in concert with noise. Noise Health 8, 45–57.

Karasawa, T., Steyger, P.S., 2015. An integrated view of cisplatin-induced nephrotoxicity and ototoxicity. Toxicol. Lett. 237, 219–227.

Kaufman, L.R., Lemasters, G.K., Olsen, D.M., et al., 2005. Effects of concurrent noise and jet fuel exposure on hearing loss. J. Occup. Environ. Med. 47, 212–218.

Kuruvilla, G., Perry, S., Wilson, B., et al., 2009. The natural history of vincristine-induced laryngeal paralysis in children. Arch. Otolaryngol. Head Neck Surg. 135, 101–105.

Landier, W., 2016. Ototoxicity and cancer therapy. Cancer 122, 1647–1658.

Leitner, M.G., Halaszovich, C.R., Oliver, D., 2011. Aminoglycosides inhibit KCNQ4 channels in cochlear outer hair cells via depletion of phosphatidylinositol(4,5)bisphosphate. Mol. Pharmacol. 79, 51–60.

Liba, B., Naples, J., Bezyk, E., et al., 2017. Changes in serum prestin concentration after exposure to cisplatin. Otol. Neurotol. 38, e501–e505.

Liu, Y., Fechter, L.D., 1996. Comparison of the effects of trimethyltin on the intracellular calcium levels in spiral ganglion cells and outer hair cells. Acta Otolaryngol. 116, 417–421.

Liu, H., Ding, D.L., Jiang, H.Y., et al., 2011. Ototoxic destruction by coadministration of kanamycin and ethacrynic acid in rats. J. Zhejiang Univ. Sci. B 12, 853–861.

Lush, M.E., Piotrowski, T., 2014. Sensory hair cell regeneration in the zebrafish lateral line. Dev. Dyn. 243, 1187–1202.

Lustig, L.R., 2010. The clinical cochlea. In: Fuchs, P. (Ed.), Oxford Handbook of Auditory Science: The Ear. Oxford University Press, pp. 15–48.

Maddox, P.T., Saunders, J., Chandrasekhar, S.S., 2009. Sudden hearing loss from PDE-5 inhibitors: a possible cellular stress etiology. Laryngoscope 119, 1586–1589.

McGwin, G., 2010. Phosphodiesterase type 5 inhibitor use and hearing impairment. Arch. Otolaryngol. Head Neck Surg. 136, 488–492.

Md Daud, M.K., Mohamadl, H., Haron, A., et al., 2014. Ototoxicity screening of patients treated with streptomycin using distortion product otoacoustic emissions. B-ENT 10, 53–58.

Morata, T.C., 2007. Promoting hearing health and the combined risk of noise-induced hearing loss and ototoxicity. Audiol. Med. 5, 33–40.

Morata, T.C., Johnson, A.C., 2012. Effect of exposure to chemicals on noise-induced hearing loss. In: Le Prell, C.G., Henderson, D., Fay, R.R., Popper, A.N. (Eds.), Noise-Induced Hearing Loss: Scientific Advances. Springer, New York, pp. 223–254.

Muthaiah, V.P.K., Chen, G.D., Ding, D., et al., 2016. Effect of manganese and manganese plus noise on auditory function and cochlear structures. Neurotoxicology 55, 65–73.

Mylonas, I., 2011. Antibiotic chemotherapy during pregnancy and lactation period: aspects for consideration. Arch. Gynecol. Obstet. 283, 7–18.

Naples, J., Cox, R., Bonaiuto, G., Parham, K., 2018. Prestin as an otologic biomarker of cisplatin ototoxicity in a Guinea pig model. Otolaryngol. Head Neck Surg. 158, 541–546.

Niihori, M., Platto, T., Igarashi, S., et al., 2015. Zebrafish swimming behavior as a biomarker for ototoxicity-induced hair cell damage: a high-throughput drug development platform targeting hearing loss. Transl. Res. 166, 440–450.

O'Sullivan, M.E., Perez, A., Lin, R., et al., 2017. Towards the prevention of aminoglycoside-related hearing loss. Front. Cell. Neurosci. 11, 325.

Parham, K., Dyhrfjeld-Johnsen, J., 2016. Outer hair cell molecular protein, prestin, as a serum biomarker for hearing loss: proof of concept. Otol. Neurotol. 37, 1217–1222.

Riga, M., Psarommatis, I., Korres, S., et al., 2006. The effect of treatment with vincristine on transient evoked and distortion product otoacoustic emissions. Int. J. Pediatr. Otorhinolaryngol. 70, 1003–1008.

Roth, J.A., Salvi, R., 2016. Ototoxicity of divalent metals. Neurotox. Res. 30, 268–282.

Rybak, L.P., Ramkumar, V., 2007. Ototoxicity. Kidney Int 72, 931–935.

Rybak, L.P., Whitworth, C.A., Mukherjea, D., Ramkumar, V., 2007. Mechanisms of cisplatin-induced ototoxicity and prevention. Hear Res 226, 157–167.

Rybak, L.P., Mukherjea, D., Jajoo, S., et al., 2012. siRNA-mediated knock-down of NOX3: therapy for hearing loss? Cell. Mol. Life Sci. 69, 2429–2434.

Sánchez-Sellero, I., Soto-Varela, A., 2016. Instability due to drug-induced vestibulotoxicity. J. Int. Adv. Otol. 12, 202–207.

Schacht, J., Talaska, A.E., Rybak, L.P., 2012. Cisplatin and aminoglycoside antibiotics: hearing loss and its prevention. Anat. Rec. (Hoboken) 295, 1837–1850.

Sedó-Cabezón, L., Boadas-Vaello, P., Soler-Martín, C., et al., 2014. Vestibular damage in chronic ototoxicity: a mini-review. Neurotoxicology 43, 21–27.

Selimoglu, E., 2007. Aminoglycoside-induced ototoxicity. Curr. Pharm. Des. 13, 119–126.

Sheppard, A.M., Chen, G.D., Salvi, R., 2015. Potassium ion channel openers, Maxipost and Retigabine, protect against peripheral salicylate ototoxicity in rats. Hear. Res. 327, 1–8.

Sheth, S., Mukherjea, D., Rybak, L.P., et al., 2017. Mechanisms of cisplatin-induced ototoxicity and otoprotection. Front. Cell. Neurosci. 11, 338.

Singh-Chauhan, R., Saxena, R.K., Varshey, S., 2011. The role of ultrahigh-frequency audiometry in the early detection of systemic drug-induced hearing loss. Ear Nose Throat J. 90, 218–222.

Sprung, J., Bourke, D.L., Contreras, M.G., et al., 2003. Perioperative hearing impairment. Anesthesiology 98, 241–257.

Starobova, H., Vetter, I., 2017. Pathophysiology of chemotherapy-induced peripheral neuropathy. Front. Mol. Neurosci. 10, 174.

Suli, A., Watson, G.M., Rubel, E.W., et al., 2012. Rheotaxis in larval zebrafish is mediated by lateral line mechanosensory hair cells. PLoS One 7, e29727.

Tabtabai, R., Haynes, L., Kuchel, G.A., et al., 2017. Age-related increase in blood levels of otolin-1 in humans. Otol. Neurotol. 38, 865–869.

Travis, L.B., Fossa, S.D., Sesso, H.D., et al., 2014. Chemotherapy-induced peripheral neurotoxicity and ototoxicity: new paradigms for translational genomics. J. Natl. Cancer Inst. 106 (5) pii: dju044.

Waissbluth, S., Daniel, S.J., 2013. Cisplatin-induced ototoxicity: transporters playing a role in cisplatin toxicity. Hear. Res. 299, 37–45.

Walther, L.E., Hülse, R., Lauer, K., et al., 2015. Current aspects of ototoxicity: local ototoxic effects, diagnosis, prevention, and treatment. HNO 63, 383–392.

Wan, G., Corfas, G., Stone, J.S., 2013. Inner ear supporting cells: rethinking the silent majority. Semin. Cell Dev. Biol. 24, 448–459.

Wang, J., Ruel, J., Ladrech, S., et al., 2007. Inhibition of the c-Jun N-terminal kinase-mediated mitochondrial cell death pathway restores auditory function in sound-exposed animals. Mol. Pharmacol. 71, 654–666.

Watson, C., Kirkcaldie, M., Paxinos, G., 2010. The Brain: An Introduction to Functional Neuroanatomy. Elsevier, Amsterdam, pp. 87–89.

Wei, L., Ding, D., Salvi, R., 2010. Salicylate-induced degeneration of cochlea spiral ganglion neurons-apoptosis signaling. Neuroscience 168, 288–299.

Wu, T., Lv, P., Kim, H.J., et al., 2010. Effect of salicylate on KCNQ4 of the Guinea pig outer hair cell. J. Neurophysiol. 103, 1969–1977.

Yamahara, K., Yamamoto, N., Nakagawa, T., et al., 2015. Insulin-like growth factor 1: a novel treatment for the protection or regeneration of cochlear hair cells. Hear. Res. 330 (Pt A), 2–9.

Zdebik, A.A., Wangemann, P., Jentsch, T.J., 2009. Potassium ion movement in the inner ear: insights from genetic disease and mouse models. Physiology (Bethesda) 24, 307–316.

Zhang, F.Y., Xue, Y.X., Liu, W.J., et al., 2014. Changes in the numbers of ribbon synapses and expression of RIBEYE in salicylate-induced tinnitus. Cell. Physiol. Biochem. 34, 753–767.

CHAPTER

23

Blood and Bone Marrow Toxicity Biomarkers

Sharon Gwaltney-Brant
Veterinary Information Network, Mahomet, IL, United States

INTRODUCTION

The blood and bone marrow are vital to survival, providing oxygen and nutrients to tissues, protecting the host from exogenous and endogenous invaders, providing the conduit for multisystemic communications, and preventing undue loss of bodily fluids. Because of its widespread influence within the body, toxic insults to the bone marrow or its daughter cells can result in severe consequences such as hypoxia, overwhelming infections, malignant neoplasia, or hemorrhage. The hematopoietic system contains the most mitotically active cells in the body, making it a prime target for toxicants that attack rapidly dividing cells.

The ability to quickly and effectively detect and measure toxicant-induced injury to hematopoietic or mature blood cells may allow for the institution of measures to mitigate the damage caused by the toxicant. Utilizing biomarkers as tools to predict the potential risks to the hematopoietic system posed by xenobiotics can enable us to preferentially select those compounds showing the least adverse effects on the body.

HEMATOPOIETIC SYSTEM

The hematopoietic system is composed of the bone marrow and the various cells that it produces. The production of blood cells is highly regimented and complex, involving innumerable cytokines, chemokines, neuropeptides, enzymes, and other chemical mediators. Although much has been learned about the intricate interactions among these players, there is still much that is not known about the mechanisms of hematopoietic cell proliferation, differentiation, maturation, and function.

Bone Marrow

In adults, hematopoietic marrow is concentrated in the spine, pelvis, sternum, ribs, calvarium, and proximal ends of the limb bones (Valli, 2007). Tucked away in the protective casing of cortical and cancellous bone, marrow is composed of hematopoietic cells, adipose tissue, and adjacent supportive cells and tissues. The microenvironment produced by the unique endosteal blood flow patterns, including bone—bone marrow portal capillary systems, provides the appropriate milieu for the proliferation, differentiation, and maturation of cellular components of the blood. Stromal stem cells give rise to adipocytes, osteoblasts, chondroblasts, and reticular cells that produce the structural scaffolding of the marrow and that secrete soluble mediators essential for the maintenance, differentiation, and growth of hematopoietic stem cells. In immature and young animals bone marrow is red, reflecting the high hematopoietic activity (Stockham and Scott, 2008). With age, red marrow is replaced by yellow marrow, which is composed of adipose tissue that differs from fat of other body sites in that marrow fat is more resistant to lipolysis in response to starvation (Valli, 2007). Conversion of yellow marrow back to red marrow may occur associated with pathologic states that stimulate hematopoiesis (e.g., anemia) and is most common in areas with higher blood flow such as endosteal surfaces.

Hematopoiesis

Hematopoiesis in the fetus occurs in multiple organs besides the bone marrow, including the thymic anlage, primordial lymph nodes, liver, spleen, kidney, and adrenals (Valli, 2007). At birth hematopoiesis is restricted primarily to the marrow, although occasional areas of extramedullary hematopoiesis may be found in the spleen or liver. In adults, extramedullary hematopoiesis (EMH) may occur in association with hematological disorders when bone marrow hematopoiesis is insufficient or ineffective. Typical sites of EMH include liver, spleen, lymph nodes, and paravertebral areas with the

Biomarkers in Toxicology, Second Edition
https://doi.org/10.1016/B978-0-12-814655-2.00023-2

intraspinal canal, presacral region, nasopharynx, and paranasal sinuses being less common locations for EMH (Sohawon et al., 2012). Although hematopoietic stem and progenitor cells capable of producing hematopoietic cells in vitro have been found in the stromal vascular fraction of adult adipose tissue, spontaneous EMH in adipose tissue is rare (Han et al., 2010).

Hematopoietic Cells

Hematopoiesis begins with the pluripotential stem cell, a primitive cell with almost unlimited capacity for self-renewal and the ability to differentiate into any of the blood cell lines. Growth factors and interleukins within the marrow microenvironment orchestrate the growth and differentiation of stem cells. With each level of differentiation, the commitment of the cell to a single cell type becomes more and more entrenched (See Fig. 23.1). In the event of an unusual demand for one particular cell type (e.g., erythrocytes in an anemic patient), production of other cell lines (e.g., neutrophils) will be reduced (Valli, 2007). Ontogenic relationships can also influence the relative production of cell lines; for example, erythroid and megakaryocytic cells derive from the same precursor cell lineage, and stimuli that increase production of erythrocytes frequently result in concurrent increases in platelet numbers. As an understanding of differentiation, growth, and kinetics of the cells and soluble mediators is essential to interpretation of toxicant-induced bone marrow and blood cell injury, a brief description of the myeloid cell lines follows.

Erythrocytes

Erythrocytes comprise up to 45% of the circulating blood volume and are vital to the transport of oxygen from the lungs to the tissues, as well as the transfer of carbon dioxide from the tissues to the lungs for

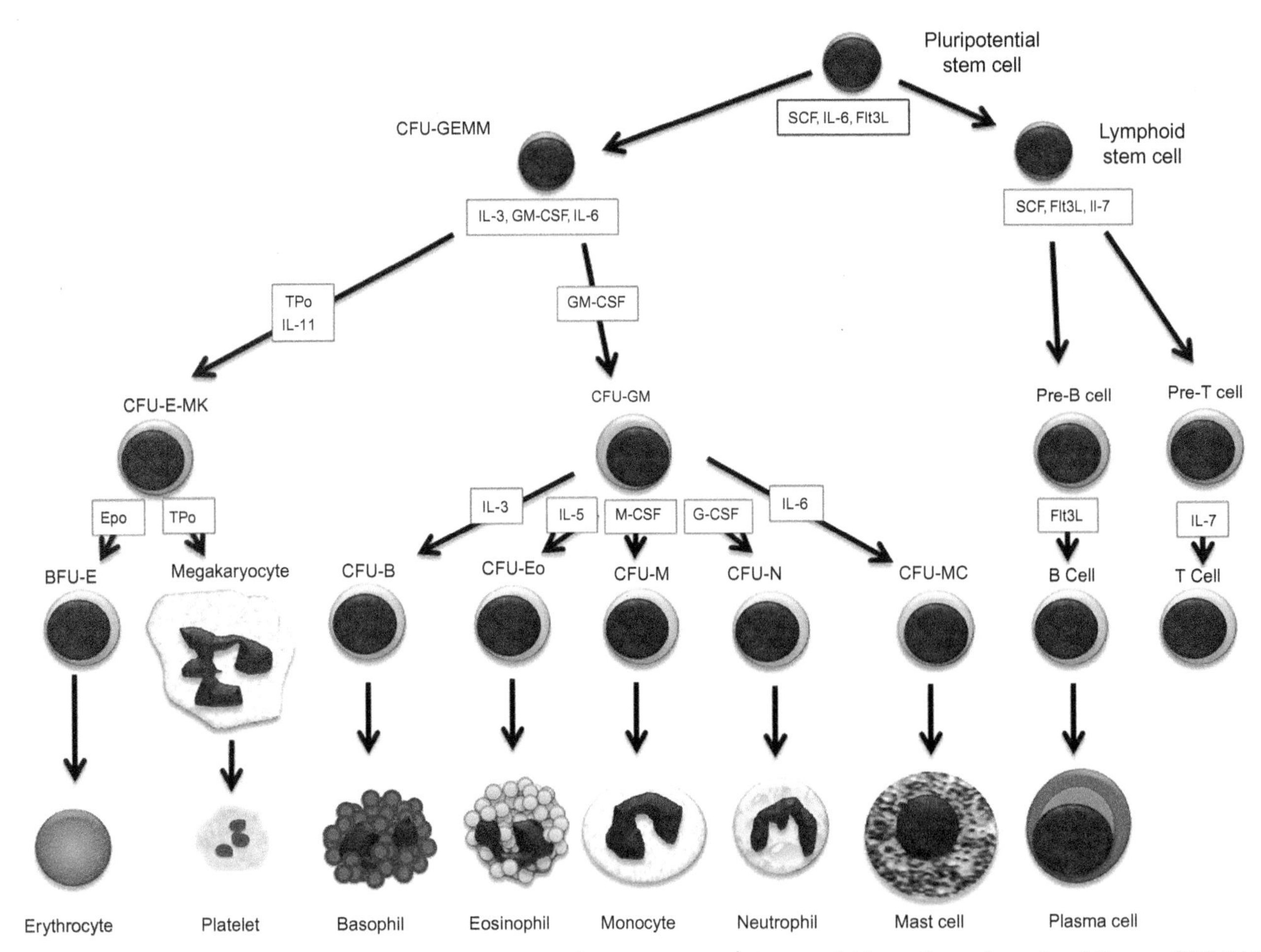

FIGURE 23.1 Differentiation of mammalian hematopoietic cells. Boxes contain primary soluble mediators for each cell lineage. *BFU-E*, blast forming unit, erythrocyte; *CFU-B*, colony forming unit, basophil; *CFU-E-MK*, colony forming unit erythrocyte, megakaryocyte; *CFU-Eo*, colony forming unit, eosinophil; *CFU-GEMM*, colony forming unit granulocyte, erythrocyte, macrophage, megakaryocyte; *CFU-GM*, colony forming unit granulocyte; *CFU-M*, colony forming unit, monocyte; *CFU-MC*, colony forming unit, mast cell; *SCF*, stem cell factor; *CFU-N*, colony forming unit, neutrophil; *Epo*, erythropoietin; *Flt3L*, Flt3 ligand; *G-CSF*, granulocyte colony-stimulating factor; *GM-CSF*, granulocyte-macrophage colony-stimulating factor; *IL*, interleukin; *M-CSF*, monocyte/macrophage colony-stimulating factor; *TPo*, thrombopoietin.

TABLE 23.1 Species Variation in Life Spans of Erythrocytes

Species	Average RBC Life Span (days)
Cat	70[a]
Cattle	150[a]
Dog	100[a]
Horse	150[a]
Human	120[a]
Rabbit	55[b]
Rat	58[b]

[a]*Stockham and Scott (2008).*
[b]*Rodnan et al. (1957).*

exhalation (Bloom and Brandt, 2008). Erythrocytes are also important in maintaining homeostatic blood pH, clearing of immune complexes and complement fragments, and regulation of blood flow to tissues. Stimulation of the common myeloid stem cell by interleukin-3 (IL-3), IL-6, IL-11, granulocyte colony-stimulating factor, and thrombopoietin results in the differentiation of the myeloid stem cell into the erythroid/megakaryocytic precursor cell (Aster, 2005). In the presence of erythropoietin, which is produced in the fetal liver and adult kidney, further differentiation into an erythrocyte colony forming unit (CFU) occurs, resulting in the development of erythroblast, which synthesizes and accumulates hemoglobin and eventually extrudes the nucleus, forming a reticulocyte. Reticulocytes retain stainable RNA and ribosomes which give the cells their name; as the RNA and ribosomes are lost, the reticulocytes become mature erythrocytes and are released into the circulation. Erythropoiesis requires approximately 4 days to complete. Toxicants that interfere with cell differentiation, proliferation, or growth can alter erythrocyte production, resulting in anemia or polycythemia. Toxicants that alter hemoglobin synthesis (e.g., lead) result in erythrocytes with reduced oxygen-carrying capacity and increased membrane fragility (Thompson, 2018).

Erythrocyte life spans vary with species (Table 23.1) and are related to the level of oxygen-derived radicals formed and the efficiency of the intrinsic erythrocyte antioxidant systems for each species (Kurata et al., 1993). Toxicants that deplete or damage erythrocyte antioxidant systems or increase free radical formation can shorten the erythrocyte life span.

Platelets

Platelets function in the formation of the hemostatic plug to mitigate hemorrhage from vascular damage. Platelets also play roles in wound healing and inflammation, primarily through communication with inflammatory cells via soluble mediators. Platelets are not cells, but rather cytoplasmic fragments derived from megakaryocytes. Megakaryocytes are formed from the erythroid/megakaryocytic precursor cell under the continued influence of thrombopoietin. It takes about 4 days for platelets to be produced and, once released into the blood, platelets have a life span of 5–10 days (Valli, 2007).

Monocytes

Monocytes account for ~5% of the total circulating leukocyte population and they are the precursors to tissue macrophages (Tizard, 2013). As part of the innate immune system macrophages function as phagocytes, removing pathogens and necrotic debris from sites of inflammation. Acting as antigen-presenting cells, macrophages also function in the acquired (adaptive) immune response. Under stimulation by IL-3, IL-6, granulocyte-macrophage colony-stimulating factor, and finally macrophage colony-stimulating factor, the myeloid stem cell gives rise to the monocyte precursor and ultimately differentiates into the monocyte. Once released into the bloodstream, monocytes circulate for approximately 3 days before entering tissues where they can replicate or differentiate into macrophages (Tizard, 2013). Tissue macrophages have extremely variable life spans, but generally are considered long-lived cells.

Neutrophils

The neutrophil is the most abundant leukocyte in circulation, comprising up to 75% of circulating white blood cells; up to two-thirds of the hematopoietic output of the bone marrow is composed of neutrophils (Tizard, 2013). Neutrophils are the first line of defense in innate immunity as they act as phagocytes and also have direct killing capabilities. Neutrophils share a common precursor cell with monocytes and differentiate under the influence of granulocyte stimulating factor. Neutrophils take approximately 6 days to be produced and have a relatively short life span of less than 12 h after entering the blood.

Eosinophils

Eosinophils have phagocytic and bactericidal capabilities, play a role in immune defense against parasites, and inactivate mediators released from mast cells (Stockham and Scott, 2008). Eosinophil precursors differentiate from myeloid precursor cells stimulated by IL-3, IL-5, IL-6, and granulocyte-macrophage colony-stimulating factor; further IL-5 exposure stimulates their differentiation into eosinophiloblasts which then mature into eosinophils. After minutes to hours in the blood, eosinophils migrate into tissues where they may persist for up to a few weeks.

Basophils

Basophils are the least abundant granulocyte population as they account for less than 1% of circulating leukocytes. Basophils play a role in immediate hypersensitivity disorders, as well as atopy, allergic contact dermatitis, and possibly autoimmune diseases such as systemic lupus erythematosus (Siracusa et al., 2011). Basophils can act as antigen-presenting cells, and they have antiparasitic functions similar to eosinophils. Basophils differentiate from myeloid stem cells under the influence primarily of IL-3 and other, as yet unidentified cytokines. Once mature, basophils have an estimated life span of 60–70 h (Siracusa et al., 2011).

Lymphocytes

Circulating lymphocytes account for 20%–35% of the leukocyte population. Lymphocytes differentiate from the common lymphoid precursor cell under the influence of IL-7, Flt3 ligand, and stem cell factor; further exposure to IL-7 stimulates maturation of T cells within the thymus, whereas Flt3 ligand stimulates B cell maturation in the bone marrow. Refer to Chapter 24, "Immunotoxicity Biomarkers," in this book for more detailed descriptions of lymphoid cells.

Soluble Mediators

A large number of soluble mediators have been identified that exert tremendous influence in the proliferation, differentiation, growth, maturation, and function of cells of the bone marrow and blood (Fig. 23.1). These include growth factors, colony-stimulating factors, interleukins, chemokines, and neuropeptides. Stem cell factor, an activator of the receptor tyrosin kinase c-Kit, is crucial for the initiation of normal hematopoiesis and, along with IL-6 and Flt3 ligand, mediates the commitment of pluripotential cells to lymphoid or myeloid lineages (Lennartsson and Ronnstrand, 2012). Flt3 ligand is another hematopoietic cytokine that interacts with tyrosine kinase III receptors essential for the initiation, expansion, and maintenance of hematopoiesis (Wodnar-Filipowicz, 2003). Expression of Flt3 receptors is limited to hematopoietic cells lacking lineage-specific markers, so they are found on the most primitive (i.e., least differentiated) hematopoietic cells. Stem cell factor and Flt3 ligand act synergistically with a wide range of specific colony-stimulating factors, growth factors, and interleukins (especially IL-3, IL-5, IL-6, IL-7, IL-11, and IL-15) to stimulate proliferation and mediate the commitment of cells to their respective lineages (Lyman, 1995).

MECHANISMS OF HEMATOTOXICITY

With so many cells, mediators, and enzymes involved in the production and maintenance of the bone marrow and blood, there are countless potential mechanisms by which toxicants can exert their adverse effects. Within the hematopoietic stem cell population only a fraction of cells is undergoing replication, with the majority of cells in a resting phase that protects them from toxic insult from external forces that target rapidly dividing cells (e.g., ionizing radiation, chemotherapeutic agents). Some toxicants such as lindane are directly cytotoxic to hematopoietic progenitor cells regardless of the phase of cell cycle, causing necrosis of cells with subsequent myelosuppression (Parent-Massin et al., 1994).

Direct damage to mature cells by toxicants can lead to hemolytic anemia, thrombocytopathy, or depressed immune function due to depletion of erythrocytes, platelets, or leukocytes, respectively. (Rebar, 1993) Indirect injury to blood or bone marrow cells can occur through a variety of means, including stimulation of immune responses against cell membranes or other structures, interference with cell surface receptors preventing normal cell function, and interference with cytokines or other soluble mediators necessary for hematopoietic cell proliferation, differentiation, maturation, or maintenance. For instance, the antineoplastic paclitaxel has been shown to alter the bone marrow microenvironment, which decreases the sensitivity of late erythroid progenitors to erythropoietin, resulting in depletion of erythroid precursors during late erythropoiesis (Juanisti et al., 2001). Toxicants may interfere with enzyme systems essential for normal cell function by inhibition of enzymes or interference with cofactors, such as occurs with interference with folate or cyanocobalamin by ethanol (Bloom and Brandt, 2008), and with decreased erythrocyte δ-aminolevulinic acid dehydrogenase activity in lead toxicosis (Jangid et al., 2012; Feska et al., 2012). Toxicant-induced alteration of enzymes involved in hemoglobin synthesis can result in mature erythrocytes with reduced oxygen-carrying capacity due to deficient hemoglobin levels. Platelet function can be permanently disabled by inhibitors of cyclooxygenase such as aspirin (Hall and Mazer, 2011). Inhibition of vitamin K epoxide reductase by coumadin-based anticoagulants (e.g., warfarin) results in depletion of vitamin K-dependent clotting factors II, VI, IX, and X, resulting in coagulopathy. Mature erythrocytes with normal hemoglobin levels can have their oxygen-carrying capacity altered by toxicants such as carbon monoxide or methemoglobin-inducing agents (e.g., nitrites).

BIOMARKERS OF HEMATOTOXICITY

A variety of assays are used clinically and nonclinically to evaluate the status of the bone marrow and blood (Table 23.2). Although newer technologies such as flow cytometry and automated cell counters have improved the speed and efficiency of some of the benchmark assays, microscopic examination of these tissues is still necessary for full evaluation of bone marrow and blood cell status. Biomarkers of leukocyte toxicity are discussed in Chapter 22, "Immunotoxicity Biomarkers."

MARKERS OF HEMATOPOIETIC/ HEMATOLOGIC TOXICITY

Complete Blood Count

The complete blood count (CBC) is the easiest, quickest, and most cost-effective method to get a rapid snapshot of the status of the blood and bone marrow, making it the most efficient means of screening the blood for evidence of toxicant-related injury (Adewoyin and Nwogoh, 2014). At a minimum, the CBC should include the following parameters: hematocrit (also called packed cell volume; Hct or PCV), mean cell volume (MCV), mean cell hemoglobin concentration (MCHC), total white blood cell count (WBC), differential white blood cell count, platelet count, and evaluation of stained blood smears.

Although modern blood analyzers can perform the cell counts, manual evaluation of the stained blood smear is necessary to verify accuracy of the counts when pathologic conditions are present. For example, nucleated red blood cells present in the circulation of lead poisoning cases may be read out by the machine as white blood cells. Similarly, morphologic aberrations due to toxicants, such as the presence of Heinz bodies in oxidant-induced hemolytic anemias or basophilic stippling in lead toxicosis, will need to be evaluated visually (Table 23.3). Platelet clumping may result in falsely lowered manual and automated platelet counts, again emphasizing the need for light microscopic confirmation. Abnormalities in hematocrit can include polycythemia, such as is seen with cobalt (Simonsen et al., 2012). Anemia is the more common hematocrit abnormality and may be the result of increased erythrocyte loss, such as in hemolytic anemia induced by oxidants, or decreased production due to bone marrow suppression. Mean cell volume is a measure of the size of erythrocytes and may be decreased in chronic lead toxicosis, as may the MCHC, resulting in a microcytic, hypochromic anemia.

Markers of Erythrocyte Toxicity

Biomarkers of erythrocyte toxicity include quantitative and qualitative evaluation of erythrocytes for abnormalities in number and/or morphology and evaluation for presence of abnormal compounds. Packed cell volume, or hematocrit (PCV, Hct), is the value representing the percentage of erythrocytes in the blood. Increases in PCV can occur following exposure to bone marrow stimulants such as erythropoietin or with exposure to toxicants that increase erythropoietin levels (e.g., cobalt) (Simonsen et al., 2012). Decreased PCV can occur following exposure that causes damage to mature erythrocytes or that damages or inhibits replication of erythrocyte progenitors within the bone marrow. Erythrocytic toxicants may cause identifiable alteration in erythrocyte morphology, which can give clues as to the type of toxicant that caused the damage (Table 23.3). For instance, lead toxicosis is sometimes associated with the presence of relatively large numbers of nucleated red blood cells or with basophilic stippling of erythrocytes, and echinocytosis is a prominent feature in many envenomations (Flachsenberger et al., 1995). Similarly, the presence of methemoglobin and/or Heinz bodies in erythrocytes is suggestive of injury due to oxidative compounds such as nitrites, chlorates, and phenols. In addition to reduced erythrocyte numbers and morphologic alterations, oxidative erythrocyte injury resulting in hemolysis will cause elevations in total serum bilirubin secondary to hemolysis.

Some toxicants cause no visible morphological change to red blood cells, but instead alter oxygen-carrying capacity to such a degree as to cause life-threatening hypoxia to the patient. Carbon monoxide binds to hemoglobin with high affinity, shifting the oxygen dissociation curve to the left, preventing oxygen delivery to tissues, and leaving no apparent morphological change in the erythrocyte, although grossly visible cherry red mucous membranes and blood are clues to the presence of hyperoxygenated blood (Guzman, 2012). Carboxyhemoglobin levels can be quickly measured in most hospital settings; levels over 20% are generally associated with signs of toxicosis, including shortness of breath, headache, and dizziness, whereas levels over 50% can be lethal.

Lead poisoning is a significant concern in many parts of the world because of past and/or current use of lead-based paints and gasoline (Liu et al., 2008). Lead alters erythrocyte function through interference with heme synthesis. Inhibition of δ-aminolevulinic acid dehydratase and ferrochetolase interferes with the insertion of lead into the protoporphyrin ring during heme

TABLE 23.2 Biomarkers of Hematotoxicity (See Also Table 22.1 for Additional Leukocytes Biomarkers)

Assay	Matrix	Endpoint	Example Toxicant
BIOMARKERS OF TOXIC EFFECTS ON BONE MARROW			
Bone marrow evaluation	Bone marrow	Altered M:E[a]	Alkylating agents (↑), azathioprine (↑), phenols (↓)
		Cellularity ↓	Busulfan
		Cellularity ↑	Cobalt
Morphologic alterations	Methotrexate (Megaloblastosis)		
Progenitor cell colony formation	Bone marrow, blood	Colony formation ↓	Clopidogrel, lindane
Metabonomics profiles	Urine	Altered metabolite profiles	Benzene (biomarker of exposure)
Proteomic profiles	Serum, blood	Altered protein profiles	Benzene
Toxicogenomic assays	Blood cells, bone marrow	Altered gene expression	Benzene, cisplatin, carboplatin
Glycophorin A gene loss mutation assay	Erythrocytes	Increased	Benzene
BIOMARKERS OF TOXIC EFFECTS ON HEMOGLOBIN			
δ-Aminolevulinic acid levels	Urine	Increased levels	Lead
δ-Aminolevulinic acid dehydratase activity	Blood	Decreased activity	Lead
Mean cell hemoglobin concentration (MCHC)	Blood	Decreased levels	Lead
Coproporphyrin	Urine	Increased levels	Lead
Zinc protoporphyrin	Blood	Increased levels	Lead
BIOMARKERS OF TOXIC EFFECTS ON ERYTHROCYTES			
Hematocrit/Packed cell volume	Blood	Increased Hct	Cobalt
	Decreased Hct	Oxidants, lindane	
Mean cell volume (MCV)	Blood	Decreased MCV	Lead
Reticulocyte count	Blood	Increased	Oxidants causing hemolysis (e.g., arsine)
Carboxyhemoglobin level	Blood	Increased	Carbon monoxide
Methemoglobin level	Blood	Increased	Oxidants (e.g., methylene blue)
Heinz body preparation	Blood	Increased	Oxidants (e.g., nitrites)
BIOMARKERS OF TOXIC EFFECTS ON THROMBOCYTES			
Bleeding time (BT)	In vivo assay	Prolonged	Aspirin
Platelet count	Blood	Decreased	Heparin
Direct platelet assays	Blood	Decreased activity	Aspirin
BIOMARKERS OF TOXIC EFFECTS ON LEUKOCYTES			
Leukocyte count	Blood	Increased neutrophils	Corticosteroids
		Decreased neutrophils	Phenothiazine
		Increased eosinophils	L-tryptophan
		Decreased eosinophils	Corticosteroids
		Increased monocytes	Corticosteroids
		Increased basophils	Allergens

TABLE 23.2 Biomarkers of Hematotoxicity (See Also Table 22.1 for Additional Leukocytes Biomarkers)—cont'd

Assay	Matrix	Endpoint	Example Toxicant
BIOMARKERS OF TOXIC EFFECTS ON HEMOSTASIS			
Activated partial thromboplastin time (APTT)	Blood	Prolonged APTT	Warfarin and related anticoagulants
Coagulation factor assays	Blood	↓ Factor V	Streptomycin, Penicillins
		↓ Factor VIII	Nitrofurazone
		↓ von Willebrand factor	Ciprofloxacin
Prothrombin time (PT, OSPT)	Blood	Prolonged PT	Warfarin and related anticoagulants
Proteins induced by vitamin K antagonism (PIVKA)	Blood	Increased PIVKA	Warfarin and related anticoagulants

[a]*Myeloid:Erythroid ratio.*

TABLE 23.3 Toxicant-Induced Erythrocyte Abnormalities

Erythrocyte Abnormality	Description	Significance	Example Toxicant
Basophilic stippling	Dark blue to purple dots or specks	Represents aggregated ribosomes	Lead-induced inhibition of pyrimidine 5′-nucleotidase results in decreased RNA degradation
Eccentrocyte	Eccentric dense staining hemoglobin with adjacent clear crescent or edge	Represents fusion of membranes damaged by oxidants	Nitrites
Echinocyte	Sharp, spiny membrane projections	Represents alterations to lipid membrane	Crotalid venom
Heinz body	Pale, rounded protruding defect in membrane; dark blue with new methylene blue stain	Represents precipitated hemoglobin due to oxidative injury	Nitrites, methylene blue
Nucleated erythrocyte	Presence of dark nucleus in red blood cell; basophilic tint to cytoplasm	Represents accelerated erythropoiesis and early release from bone marrow	Lead
Reticulocyte	Aggregated or punctated basophilic staining of cytoplasm	Represents residual RNA; accelerated erythropoiesis	Oxidant-induced hemolytic anemia
Siderotic granules	Fine granular basophilic inclusions	Represents iron accumulation in damaged mitochondria	Lead
Spherocyte	Decreased central pallor, decreased diameter	Represents membrane loss	Snake envenomations

formation (Sakai, 1995). Biomarkers of lead toxicosis include elevations in urinary δ-aminolevulinic acid and coproporphyrin levels, increased zinc protoporphyrin levels in the blood, decreased δ-aminolevulinic acid dehydratase activity, and decreased levels of erythrocyte hemoglobin (Jangid et al., 2012).

Markers of Platelet Toxicity

Toxic effects on platelets can be reflected by decreased platelet numbers and/or decreased platelet function, both of which can lead to increased incidence of bleeding due to loss of a critical hemostatic "plug" in the face of vascular injury. Decreased platelet numbers are readily identified during routine CBC analysis, although spurious thrombocytopenia can occur due to platelet clumping within the blood sample (Stockham and Scott, 2008). Bone marrow analysis may be helpful in determining if decreased platelet numbers are due to increased consumption or decreased megakaryocyte production. Decreased platelet function, such as occurs with cyclooxygenase inhibitors such as aspirin, can be measured using direct assays of platelet function. Bleeding time is a quick and rudimentary test of platelet function, which is often done as a prescreening test to determine if further platelet and/or coagulation assays are indicated. A small cut is made in the skin with a lancet or needle and the amount of time that it takes for the bleeding to cease is measured.

MARKERS OF BONE MARROW TOXICITY

Bone Marrow Evaluation

Evaluation of bone marrow can be done using histopathology, cytology, or flow cytometry. Histopathological evaluation of hematoxylin and eosin-stained (H&E) sections of bone marrow tissue may be used as an initial screening tool but has some limitations, so it is prudent to prepare bone marrow smears at the time of autopsy or biopsy for cytological examination in case further evaluation is needed (Elmore, 2006). Histopathology can provide estimates of cellular density, myeloid/erythroid (M:E) ratios, amount of hemosiderin present, and abnormalities in the numbers of megakaryocytes, adipocytes, and bone marrow stromal cells. Additionally, abnormalities such as necrosis, hemorrhage, fibrosis, granulomas, neoplasia, or alterations in endosteum, bone, and vasculature can be detected (Reagan et al., 2011). Because many of the mononuclear cells in bone marrow "all look alike" under H&E, differentiating the lymphoid lineage cells from early myeloid precursors can be difficult.

Cytological examination of Romanowsky-stained bone marrow smears may be more rewarding in cases where there is a need to differentiate early hematopoietic precursors, to correlate changes in peripheral cell numbers with bone marrow hyper- or hypocellularity, to investigate suspected abnormal erythropoiesis, to differentiate lymphoid from erythroid precursors, or to determine maturation indices of one or more cell lines (Elmore, 2006). Flow cytometry may be used in addition to or instead of cytology to further characterize alterations in bone marrow cells. However, although flow cytometry can readily classify cells into the major cell lineages and produce rapid counts of large numbers of cells, it does not evaluate cellular morphology; for this reason, flow cytometric methods are generally used in addition to either cytology or histopathology (Reagan et al., 2011).

Bone marrow evaluation generally starts with estimation of the relative cellularity and the M:E ratio, which compares the relative proportions of myeloid cells to erythrocytic cells (Table 23.4). Generalized hypocellularity of bone marrow can occur due to infection, irradiation, or exposure to various drugs or toxicants such as benzene, cephalosporins, chloramphenicol, or phenylbutazone (Bloom and Brandt, 2008; Bloom et al., 1987). Because immature hematopoietic cells are difficult to differentiate, the M:E ratio obtained by visual inspection is a relatively subjective estimation (Stockham and Scott, 2008). In most healthy mammals, the M:E ratio trends toward slightly >1, i.e., a somewhat greater proportion of myeloid cells (Elmore, 2006). Increases in M:E ratio can be due to increases in myeloid cells or decreases in erythrocytic cells, which can be determined by estimating the overall cellularity of the marrow. Hypercellular marrow with increased M:E ratio suggests myeloid hyperplasia, whereas hypocellularity of hematopoietic cells and increased M:E ratio suggests erythroid hypoplasia. The most common cause of myeloid hyperplasia is increased demand during times of acute or chronic inflammation, which oftentimes will also be reflected in the results of a CBC. Erythroid hypoplasia can occur in response to drugs such as isoniazid or azathioprine (Thompson and Gales, 1996).

Decreases in M:E ratio can occur during times of erythroid hyperplasia, such as occurs in response to acute hemorrhagic anemias or cobalt exposure, or to myeloid hypoplasia, such as may occur with carbamazepine or clozapine (Simonsen et al., 2012; Bloom and Brandt, 2008).

TABLE 23.4 Bone Marrow Evaluation, Assessments, and Potential Causative Agents

M:E Ratio	Cellularity	Assessment	Example Causes
Decreased	Hypercellular	Erythroid hyperplasia	Erythropoietin therapy or excess Hemorrhagic anemia Hemolytic anemia: e.g., acetanilide, nitrofurantoin, phenol, sulfonamides Cobalt
Decreased	Hypocellular	Granulocytic hypoplasia	Alkylating agents, benzene, carbamazepine, cisplatin, clozapine, isoniazid, lindane, methimazole, nitrosourea
Normal	Hypocellular	Bone marrow hypoplasia	Alkylating agents, benzene, chloramphenicol, chlorinated hydrocarbons, diclofenac, mycotoxins, sulfonamide
Increased	Hypercellular	Granulocytic hyperplasia	Inflammation leukomogenic response: e.g., alkylating agents, azathioprine, benzene (chronic), bleomycin, high-dose ionizing radiation, procarbazine
Increased	Hypocellular	Erythroid hypoplasia	Chronic renal disease (decreased erythropoietin) azathioprine, estrogen, isoniazid, phenytoin

Bone marrow examination should include evaluation of all of the major cellular components for anomalies in morphology. Toxicants such as ethanol, lead, or isoniazid can cause defects in synthesis of the porphyrin ring, resulting in failure of iron to incorporate into heme and leading to precipitation of iron within erythrocyte mitochondria, resulting in sideroblastic anemia (Bloom and Brandt, 2008). Megaloblastic anemia develops when folate or vitamin B_{12} deficiencies occur either due to nutritional deficiency or due to effects of toxicants such as phenytoin, phenobarbital, and sulfasalazine. In megaloblastic anemia, asynchronous development of nucleus and cytoplasm in erythroid precursors results in cells with abundant cytoplasm and immature, enlarged nuclei displaying exaggerated chromatin patterns (Stockham and Scott, 2008).

Progenitor Cell Colony Formation

Evaluation of hematopoietic cells in early differentiation stages is difficult by means of light microscopy and flow cytometry due to lack of distinctive cell markers. Bone marrow and blood cells cultured in a semisolid methylcellulose–based medium containing appropriate growth factors will proliferate and differentiate into colonies of maturing hematopoietic cells (Wognum et al., 2013; Masenini et al., 2012). Classification and enumeration of the colonies is performed by light microscopy or flow cytometry and allows for the quantification of erythroid, myeloid, lymphoid, and megakaryocytic cell lineages to detect toxicant-induced increases or decreases in specific hematopoietic cell lines.

MARKERS OF HEMOSTATIC TOXICITY

Prevention of excessive loss of blood is essential to maintain the integrity of the hematologic system; an elaborate and overlapping coagulation system, involving platelets and clotting factors, has evolved to ensure prompt repair of vascular leakage. Hemostasis involves sequential activation of serine proteases, culminating in the formation of fibrin clots to seal defects in the vasculature (Fig. 23.2).

Although hypercoagulable states can occur secondary to exposure to some chemotherapeutic agents (e.g., tamoxifen), most alterations in hemostatic ability are due to deficiencies in one or more components of the coagulation cascade resulting in ineffective coagulation and predisposing to uncontrolled hemorrhage. Decreases in coagulation factors can occur due to reduced synthesis or increased clearance. Perhaps the most common cause of decreased coagulation factor synthesis is due to vitamin K deficiency induced by anticoagulants such as warfarin and related compounds such as diphacinone and brodifacoum. These anticoagulants inhibit vitamin K epoxide reductase, a vital enzyme in the regeneration of vitamin K, resulting in an inability to generate vitamin K-dependent coagulation factors II, VII, IX, and X (Stafford, 2005). As these coagulation factors are depleted, coagulation becomes impaired and incomplete coagulation proteins are released into the circulation. Defects in coagulation can be measured using several assays, including activated partial thromboplastin time (APTT), prothrombin time (PT), proteins induced by vitamin K antagonism or absence (PIVKA), and measurement of individual coagulation factors. The APTT, PT, and PIVKA assays are the most frequently used for detecting early anticoagulant-induced coagulopathy (Stockham and Scott, 2008). The APTT and PT assays are performed by incubating test plasma with partial thromboplastin (APTT) or thromboplastin (PT) along with other intermediates and measuring the time until clot formation occurs. The PIVKA assay used in human diagnostic laboratories directly measures the incompletely carboxylated coagulation factors, whereas the PIVKA assay used in veterinary medicine (Owren's thrombotest) is essentially a modified PT assay (Stockham and Scott, 2008). Individual coagulation factors can be measured, and although these assays are more commonly used for suspected inherited coagulation factor deficiencies (e.g., hemophilia), certain xenobiotics can result in deficiencies of individual coagulation factors; for instance, ciprofloxacin

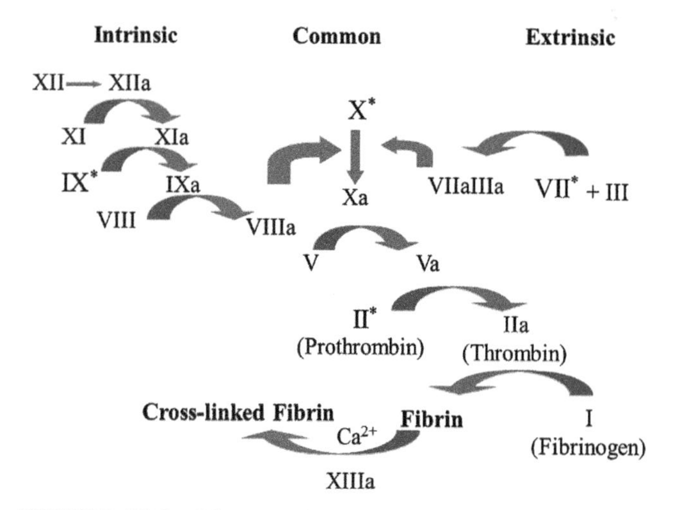

FIGURE 23.2 Schematic diagram of the coagulation cascade. Coagulation initiates with activation of intrinsic factor XII to XIIa and/or activation of the combined factors VII and III to VIIaIIIa. XIIa initiates a series of factor activations culminating in production of activated factor VIII. Factors VIIaIIIa and/or VIIa then activate factor X, triggering activation of factor V, which triggers the conversion of prothrombin to thrombin. Thrombin acts on fibrinogen to form fibrin. Enzymatic action by activated factor XIII in presence of ionized calcium results in cross-linkage of fibrin strands. *indicates vitamin K-dependent factors.

has been associated with transient acquired von Willebrand syndrome due to increased proteolysis of von Willebrand factor (Michiels et al., 2011).

CONCLUDING REMARKS AND FUTURE DIRECTIONS

Our understanding of the intricate connections responsible for the normal functioning of the blood and bone marrow has expanded tremendously over the last few decades. With this expanding knowledge has come the realization that this delicately balanced system is highly susceptible to injury induced by xenobiotics such as pharmaceuticals and environmental toxicants. Toxicant-induced bone marrow injury can have serious repercussions throughout the body and can put the host at risk of infection, hypoxic injury, or uncontrolled hemorrhage. This ripple effect underscores the need to determine that new and existing pharmaceuticals and environmental compounds pose minimal risk to blood or bone marrow components. Future research to further our knowledge of the mechanisms of toxicity of hematopoietic tissues will enable the development of pharmaceuticals, pesticides, and other compounds that pose less of a risk to humans, other animals, and the environment. Newer technologies in research such as toxicogenomics, proteomics, and metabonomics will allow further progress in mechanistic investigation and risk assessment.

Toxicogenomic profiles have been proposed as potentially useful biomarkers for exposure to various toxicants. For instance, circulating reticulocyte expression of the hemoglobin beta chain complex, aminolevulinic acid synthase 2, and cell division cycle 25 homolog B genes was altered following exposure of rats to myelosuppressive agents such as linezolid, cisplatin, and carboplatin, suggesting that this pattern of gene expression may pose a useful biomarker for myelosuppressive anemia in rats (Uehara et al., 2011). Microarray evaluation of the hepatic expression of six genes (Alas2, beta-glo, Eraf, Hmox1, Lgals3, and Rhced) showed high negative correlation with red blood cell counts and high positive correlation with total serum bilirubin levels in rats with drug-induced hemolysis, suggesting that these genes may be useful biomarkers for hemolytic anemia (Rokushima et al., 2007). Further validation of these and other toxicogenomic profiles will be required to determine if similar changes in gene expression are shared among other mammalian species. Benzene is a potent myelotoxic compound associated with aplastic anemia, acute myeloid leukemia, and other blood disorders in humans (Zhang et al., 2010). In humans, increased levels of gene duplication mutations in glycophorin A have been proposed as a potential biomarker for cumulative exposure to benzene, although sensitivity issues have been raised (Rothman et al., 1995; Smith and Rothman, 2000).

Proteomics is the branch of toxicogenomics that studies alterations in protein levels and posttranslational protein modifications that result from altered gene expression caused by exposure to toxicants (Joo et al., 2003). Two proteins, platelet factor 4 and connective tissue activating peptide III, have been found to be consistently downregulated in patients exposed to benzene when compared with control individuals and are being investigated as potential biomarkers for the early biologic effects of benzene (Zhang et al., 2010). Similarly, downregulation of platelet basic protein and apolipoprotein B100 has been detected in humans with hematotoxicity due to exposure to benzene (Huang et al., 2012). Plasma haptoglobin levels in the plasma of pancreatic cancer patients have been shown to be correlated to risk of hematologic adverse events from the chemotherapeutic agent gemcitabine (Matsubara et al., 2009) Further work is needed to determine if these potential biomarkers will prove to be reliable in determining risks of benzene exposure.

Metabonomics is a means of metabolic profiling to determine biological markers for mechanistic and diagnostic study in toxicology, pharmacology, and biomedicine. Benzene and its metabolites have been found in breath, blood, and urine, and levels in these matrices have been used as biomarkers for exposure and risk assessment in humans (Weisel, 2010). In mice, metabonomic profiles of benzene metabolites in urine have been further utilized as a sensitive tool to detect benzene-induced toxicity (Sun et al., 2012).

References

Adewoyin, A.S., Nwogoh, B., 2014. Peripheral blood film—a review. Ann. Ib. Postgrad. Med. 12 (2), 71–79.

Aster, J.C., 2005. Red blood cell and bleeding disorders. In: Kumar, V., Abbas, A.K., Fausto, N. (Eds.), Robbins and Cotran Pathologic Basis of Disease, seventh ed. Saunders/Elsevier, Philadelphia, PA.

Bloom, J.C., Brandt, J.T., 2008. Toxic responses of the blood. In: Klaassen, C.D. (Ed.), Casarett & Doull's Toxicology: The Basic Science of Poisons, seventh ed. McGraw-Hill Companies, New York, NY.

Bloom, J.C., Lewis, H.B., Sellers, T.S., et al., 1987. The hematologic effects of cefonicid and cefazedone in the dog: a potential model of cephalosporin hematotoxicity in man. Toxicol. Appl. Pharmacol. 90 (1), 135–142.

Elmore, S.A., 2006. Enhanced histopathology of the bone marrow. Toxicol. Pathol. 34 (5), 666–686.

Feska, L.R., Oliveira, E., Trombini, T., et al., 2012. Pyruvate kinase activity and δ-aminolevulinic acid dehydratease activity as biomarkers of toxicity in workers exposed to lead. Arch. Environ. Contam. Toxicol. 63 (3), 453–460.

Flachsenberger, W., Leight, C.M., Mirtschin, P.J., 1995. Spheroechinocytosis of human red blood cells caused by snake, red-back spider, bee and blue ringed octops and its inhibition by snake sera. Toxicon 33 (6), 791–797.

Guzman, J.A., 2012. Carbon monoxide poisoning. Crit. Care Clin. 28 (4), 537–548.
Hall, R., Mazer, C.D., 2011. Antiplatelet drugs: a review of their pharmacology and management in the perioperative period. Anesth. Analg. 112 (2), 292–318.
Han, J., Koh, Y.J., Moon, H.R., et al., 2010. Adipose tissue is an extramedullary reservoir for functional hematopoietic stem and progenitor cells. Blood 115 (5), 957–964.
Huang, Z., Wang, H., Huang, H., et al., 2012. iTRAQ-based proteomic profiling of human serum reveals down-regulation of platelet basic protein and apolipoprotein B100 in patients with hematotoxicity induced by chronic occupational benzene exposure. Toxicology 291 (1–3), 56–64.
Jangid, A.B., John, P.J., Yadav, D., et al., 2012. Impact of chronic lead exposure on selected biological markers. Int. J. Clin. Biochem. 27 (1), 83–89.
Joo, W.A., Kang, M.J., Son, W., et al., 2003. Monitoring protein expression by proteomics: human plasma exposed to benzene. Proteomics 3, 2402–2411.
Juanisti, J.A., Aguirre, M.V., Carmuega, R.J., et al., 2001. Hematotoxicity induced by paclitaxel: in vitro and in vivo assays during normal murine hematopoietic recovery. Methods Find. Exp. Clin. Pharmacol. 23 (4), 161–167.
Kurata, M., Suzuki, M., Agar, N.S., 1993. Antioxidant systems and erythrocyte life-span in mammals. Comp. Biochem. Physiol. B 106 (3), 477–487.
Lennartsson, J., Ronnstrand, L., 2012. Stem cell factor receptor/c-Kit: from basic science to clinical applications. Physiol. Rev. 92 (4), 1619–1649.
Liu, J., Goyer, R.A., Waalkes, M.P., 2008. Toxic effects of metals. In: Klaassen, C.D. (Ed.), Casarett & Doull's Toxicology: The Basic Science of Poisons, seventh ed. McGraw-Hill Companies, New York, NY.
Lyman, S.D., 1995. Biology of flt3 ligand and receptor. Int. J. Hematol. 62 (2), 63–73.
Masenini, S., Donzelli, M., Taegtmeyer, A.B., et al., 2012. Toxicity of clopidogrel and ticlopidine on human myeloid progenitor cells: importance of metabolites. Toxicology 229 (2–3), 139–145.
Matsubara, J., Orno, M., Negishi, A., et al., 2009. Identification of a predictive biomarker for hematologic toxicities of gemcitabine. J. Clin. Oncol. 27, 2261–2268.
Michiels, J.J., Budde, U., van der Planken, M., et al., 2011. Acquired von Willebrand syndromes: clinical features, aetiology, pathophysiology, classification and management. Best Pract Res Clin Haematol 14 (2), 401–436.
Parent-Massin, D., Thouvenot, D., Rio, B., et al., 1994. Lindane hematotoxicity confirmed by *in vitro* tests on human and rat progenitors. Hum. Exp. Toxicol. 13 (2), 103–106.
Reagan, W.J., Irizarry-Rovira, A., Poitout-Belissent, F., et al., 2011. Best practices for evaluation of bone marrow in nonclinical toxicity studies. Toxicol. Pathol. 39, 435–448.
Rebar, A.H., 1993. General responses of the bone marrow to injury. Toxicol. Pathol. 21 (2), 118–129.
Rodnan, G.P., Ebaugh Jr., F.G., Fox, M.R., 1957. The life span of the red blood cell and the red blood cell volume in the chicken, pigeon and duck as estimated by the use of Na2Cr51O4, with observations on red cell turnover rate in the mammal, bird and reptile. Blood. 12 (4), 355–366.
Rokushima, M., Omi, K., Araki, A., et al., 2007. A toxicogenomic approach revealed hepatic gene expression changes mechanistically linked to drug-induced hemolytic anemia. Toxicol. Sci. 95 (2), 474–484.
Rothman, N., Haas, R., Hayes, R.B., et al., 1995. Benzene induces gene-duplicating but not gene-inactivating mutation at the glycophorin A locus in exposed humans. Proc. Natl. Acad. Sci. U.S.A. 92, 2069–4071.
Sakai, T., 1995. Reviews on biochemical markers of lead exposure with special emphasis on heme and nucleotide metabolisms. Sangyo Eiseigaku Zasshi 37 (2), 99–112.
Simonsen, L.O., Harbak, H., Bennekou, P., 2012. Cobalt metabolism and toxicology—a brief update. Sci. Total Environ. 432, 210–215.
Siracusa, M.C., Comeau, M.R., Artis, D., 2011. New insights into basophil biology: initiators, regulators and effectors of type 2 inflammation. Ann. N. Y. Acad. Sci. 1217, 166–177.
Smith, M.T., Rothman, N., 2000. Biomarkers in the molecular epidemiology of benzene-exposed workers. J. Toxicol. Environ. Health Part A 61, 439–445.
Sohawon, D., Lau, K.K., Bowden, D.K., 2012. Extra-medullary haematopoiesis: a pictorial review of its typical and atypical locations. J. Med. Imag. Rad. Oncol. 56 (5), 538–544.
Stafford, D.W., 2005. The vitamin K cycle. J. Thromb. Hemost. 3, 1873–1878.
Stockham, S.L., Scott, M.A., 2008. Fundamentals of Veterinary Clinical Pathology, second ed. Blackwell Publishing, Ames, IA.
Sun, R., Zhan, J., Xiong, M., et al., 2012. Metabonomics biomarkers for subacute toxicity screening for benzene exposure in mice. J. Toxicol. Environ. Health 75 (18), 1163–1173.
Thompson, L.J., 2018. Lead. In: Gupta, R.C. (Ed.), Veterinary Toxicology: Basic and Clinical Principles, third ed. Academic PressElsevier, pp. 439–443.
Thompson, D.F., Gales, M.A., 1996. Drug-induced pure red cell aplasia. Pharmacotherapy 16 (6), 1002–1008.
Tizard, I.R., 2013. Veterinary Immunology, ninth ed. Elsevier, St. Louis, MO.
Uehara, T., Kondo, C., Yamate, J., et al., 2011. A toxicogenomic approach for identifying biomarkers for myelosuppressive anemia in rats. Toxicology 282 (3), 139–145.
Valli, V.E.O., 2007. Hematopoietic system. In: Maxie, M.G. (Ed.), Jubb, Kennedy and Palmer's Pathology of Domestic Animals, fifth ed. Saunders/Elsevier, St. Louis, MO.
Weisel, C.P., 2010. Benzene exposure: an overview of monitoring methods and their findings. Chem. Biol. Interact. 184 (1–2), 58–66.
Wodnar-Filipowicz, A., 2003. Flt3 ligand: role in control of hematopoietic and immune functions of the bone marrow. Physiology 18 (6), 247–251.
Wognum, B., Yaun, N., Lai, B., et al., 2013. Colony forming cell assays for human hematopoietic progenitor cells. Methods Mol. Biol. 946, 267–283.
Zhang, L., McHale, C.M., Rothman, N., et al., 2010. Systems biology of human benzene exposure. Chem. Biol. Interact. 84 (1–2), 86–93.

CHAPTER

24

Immunotoxicity Biomarkers

Sharon Gwaltney-Brant

Veterinary Information Network, Mahomet, IL, United States

INTRODUCTION

Employing a variety of cell types (e.g., granulocytes, macrophages, lymphocytes) and soluble mediators (e.g., interleukins, interferons, tumor necrosis factors), the immune system identifies, processes, and destroys foreign invaders (e.g., bacteria, viruses, fungi) and endogenous neoplasms. The immune system can distinguish between nonself (foreign) and selfantigens, effectively defending the body against foreign and domestic invaders that would threaten the homeostatic stability of the host. Immune responses result in the destruction and/or sequestration of nonself invaders to limit their ability to injure the host.

The complex interplay between the various components of the immune system achieve, under normal conditions, a delicate balance of physiological reactions that maintain the integrity and health of the host. Disruption of this delicate balance can result in a system-wide malfunction that can have deleterious effects on the host. A wide variety of toxicants can act to alter the immune system in such a way that the normal immune response is attenuated, intensified, or directed against the body's own tissues. Immunosuppression may result in more frequent and/or more severe infections, infections by organisms with normally low pathogenicity (opportunistic infections), or increased incidence of neoplasia (Descotes et al., 1995). Immunostimulation may result in increased incidence of allergic responses or autoimmunity.

THE IMMUNE SYSTEM

Innate immunity is the first line of defense against nonself invaders, and it includes physical and biochemical barriers such as skin and mucosal barriers, mucosal cilia and secretions, physiological alterations (e.g., fever), and cellular interactions. Innate immune responses are tasked with preventing pathogens (e.g., microbes, neoplastic cells) from gaining a foothold in the body, or, failing that, to hold off the pathogenic assault until reinforcements, in the form of cells of the adaptive immune response, can be recruited to assist in eliminating the threat. Adaptive (acquired) immunity encompasses the production of an immune response against nonself and utilizes specialized populations of cells to identify, isolate, and process antigens and initiate the appropriate agonistic response. Acquired immunity requires memory of previous exposure to trigger a response against a specific invader. At the heart of the immune system is a complex mixture of lymphoid organs, specialized cell types, and soluble mediators that interact in a complex and highly regulated fashion to generate appropriate responses to exogenous and endogenous threats to the integrity of the body.

LYMPHOID ORGANS AND TISSUES

Lymphoid organs form the stationary framework in which the migratory immune cells are generated, mature, and interact with antigens (Kaminski et al., 2008). Immune cells arise from the differentiation of self-renewing pluripotent stem cells within the bone marrow (Fig. 24.1). During differentiation, cells become committed to three distinct cell lineages—null cells, lymphoid precursors or myeloid precursors—from which the various immune cells arise. Lymphoid precursors further commit to either B or T cell differentiation, whereupon T cell precursors migrate to the thymus where they are programmed to distinguish between self and nonself.

Because of their roles in the production of B and T cells, the thymus and bone marrow are considered primary lymphoid organs. Secondary lymphoid organs include lymph nodes and spleen, which filter lymph and blood, respectively, and where naïve B and T cells are introduced to antigens. Tertiary lymphoid tissues/organs include Peyer's patches and surface-associated

Biomarkers in Toxicology, Second Edition
https://doi.org/10.1016/B978-0-12-814655-2.00024-4

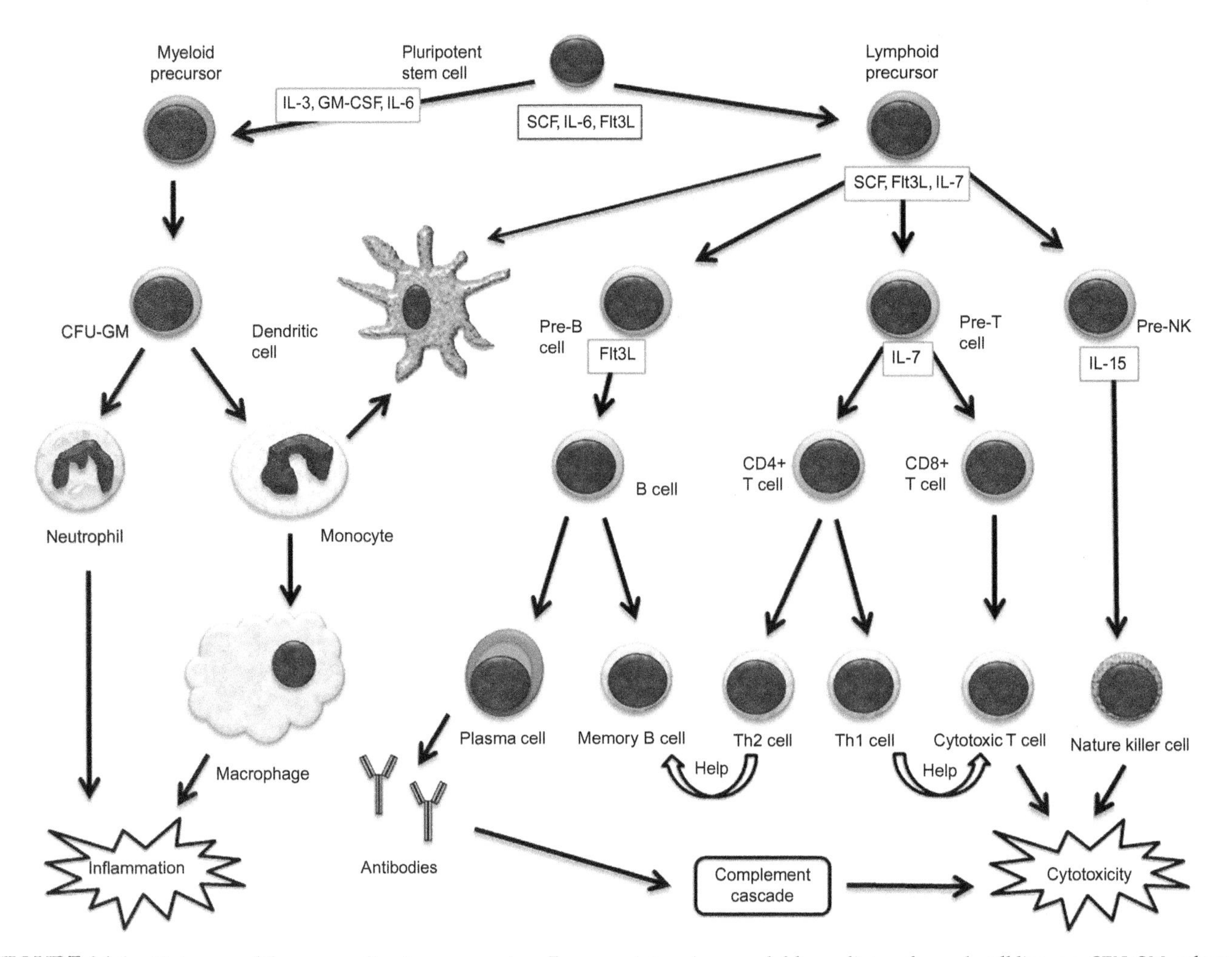

FIGURE 24.1 Ontogeny of the mammalian immune system. Boxes contain primary soluble mediators for each cell lineage. *CFU-GM*, colony-forming unit granulocyte; *Flt3L*, Flt3 ligand; *GM-CSF*, granulocyte-monocyte colony-stimulating factor; *IL*, interleukin; *Pre-B cell*, B cell precursor; *Pre-NK cell*, NK precursor; *Pre-T cell*, T cell precursor; *SCF*, stem cell factor. *From Stockham, S.L., Scott, M.A., 2008.* Fundamentals of Veterinary Clinical Pathology, *second ed. Blackwell Publishing, Ames, IA; Tizard (2013).*

lymphoid aggregates, such as bronchial-associated lymphoid tissue and gut-associated lymphoid tissue. Tertiary lymphoid tissues are the sites where memory and effector cells perform immunoregulatory, as well as immunologic, functions.

CELLS OF THE IMMUNE SYSTEM

Leukocytes, or white blood cells, are the cells involved in mediating immune reactions within the body. Whether producing antibodies, phagocytosing foreign invaders such as bacteria, presenting antigens for stimulation of humoral or cell-mediated immunity, or directly attacking neoplastic or virally infected cells, leukocytes are the workhorses of the immune system. Although independent cells, leukocytes are intimately linked via a chemical network of cytokines and chemokines that regulate the movement and activity of immune and inflammatory cells within the body.

B Cell Lymphocytes

B cells are produced and mature in the bone marrow. The role of the B cell lineage in adaptive immunity is to produce and secrete antibodies in response to binding of soluble antigen to receptors on the B cell membrane surface (Dean et al., 2008). Following antigenic stimulation, interaction between B cells and T helper cells triggers the differentiation of B cells into antibody-secreting plasma cells or into memory B cells, which provide for a rapid antibody response on re-exposure to those antigens at some future time. Five classes of antibody (immunoglobulin, Ig) are produced by plasma cells: IgA, IgD, Ig E, IgG, and IgM. Each antibody class has distinct biological functions in immune responses. Immunoglobulin M secretion predominates in the early stages of an initial immune response, while IgG is primarily secreted during secondary immune responses. Immunoglobulin D functions as a lymphocyte receptor, while IgA is the primary antibody associated with seromucous secretions and IgE mediates allergic reactions. Antibodies bind specific antigens on invading organisms, resulting in direct neutralization or in opsonization that facilitates phagocytosis. Antibodies may also coat target cells, triggering cytotoxic responses from Fc receptor–bearing natural killer cells. Complement fixation by IgG and IgM triggers cytotoxicity and phagocytosis of foreign invaders.

T Cell Lymphocytes

T cells are produced in the bone marrow and migrate to the thymus, where they mature. In the thymus, T cells are programmed to distinguish between antigens belonging to the host ("self") and those foreign to the host ("nonself") (Kaminski et al., 2008). T cells differentiate into either cytotoxic T cells or helper T cells, and, like B cells, T cells develop into memory cells that recognize specific antigens. However, unlike B cells, which can be activated by binding directly to soluble antigen, T cells require the intervention of antigen-presenting cells (APCs) such as macrophages or dendritic cells to become activated.

T helper (Th) cells function in a primarily regulatory role, facilitating the activities of B cells and other T cells through interactions between surface receptors and secretion of immunoregulatory cytokines (Dean et al., 2008). Several subtypes of Th cells have been identified based on differences in cytokine production and cell types impacted. Th1 cells are produced in response to interleukin (IL) 12, secrete interferon-γ, and are instrumental in the generation and maturation of cytotoxic T cells. Th2 cells develop in response to IL-4 and are major producers of IL-4, IL-5, and IL-13. These cells are essential in humoral responses and defense against parasites, and are also involved in allergic responses and asthma. Alterations in Th ratios have been identified with exposure to several toxicants; for example, lead (Pb) alters the balance of Th1 and Th2 cells, resulting in decreased Th1 (cell-mediated) response (DeWitt et al., 2011). Th3 cells are members of a group of regulatory T cells (Tregs) that suppress responses to self-antigens, as well as attenuating responses mediated by Th1 and Th2 cells (Corsini et al., 2011). Th17 cells develop in response to IL-6 and IL-13 and produce IL-17 and IL-21 (Mishra and Sopori, 2012). Th17 cells are important in recruitment of neutrophils and macrophages to areas of infection, and they have been implicated as principal cells involved in autoimmune and inflammatory diseases.

Cytotoxic T cells are effector cells that release specialized proteins that result in death and lysis of the target cell (Janeway et al., 2001). Recognition of specific antigens on target cells results in the release of granules from cytotoxic T cells. Within the granules are polymers of the protein perforin and proteases collectively referred to as granzymes. Polymerization of perforin forms a cylindrical structure that is inserted into the target cell membrane, allowing water, solutes, and granzymes to enter the cell. Within the cell, granzymes initiate an enzyme cascade leading to activation of nucleases and subsequent programmed cell death. Cytotoxic T cells also have a regulatory role by mediating suppression of B and T cell activity (Dean et al., 2008).

Neutrophils

Neutrophils (polymorphonuclear neutrophils, PMNs) derive from the granulocyte lineage of the bone marrow and are the most abundant leukocyte type in mammals. Abundant in the blood, neutrophils are the first cell type to arrive at an area of acute inflammation caused by infection or injury. Cytokines released by neutrophils at the inflammatory site recruit further immune cells, amplifying the inflammatory response. Toll-like receptors (TLRs), type I transmembrane proteins, within the neutrophil membrane are responsible for recognition of microbial-associated molecular patterns (MAMPs) such as bacterial lipoproteins, lipopolysaccharide, and peptidoglycan (Mishra and Sopori, 2012). Binding of TLRs to these MAMP ligands triggers the production of immunostimulatory and proinflammatory mediators. Neutrophils have phagocytic activity and can directly kill or induce apoptosis in target cells through the release of reactive oxygen species (Kaminski et al., 2008). Extracellular killing of foreign pathogens by neutrophils is accomplished by the ejection of cytoplasmic contents, nuclear histones, and chromatin in the form of a mesh that ensnares microbes which are subsequently killed by proteins and histones within the matrix of the neutrophil extracellular trap, as this mechanism has been termed (Gray et al., 2013). Recruitment of other immune cells such as Th1 cells and B cells to sites of inflammation is mediated by release of cytokines from activated neutrophils.

Macrophages

Derived from circulating blood monocytes, macrophages are phagocytic cells that reside in a variety of tissues, including liver (Kupffer cells), lung (pulmonary alveolar macrophages or PAMs, pulmonary interstitial macrophages or PIMs), skin (Langerhans cells), and brain (microglia) (Dean et al., 2008). Like neutrophils, macrophages possess TLRs that aid in recognition of microbes, and they function in both innate and adaptive immunity. Besides phagocytosis of pathogens, macrophages can act as APCs to T cells, secrete proinflammatory cytokines, and remove necrotic debris from inflammatory sites. Macrophages facilitate muscle regeneration through the secretion of substances that stimulate muscle proliferation, repair, and regeneration. In addition to their beneficial actions, macrophages can at times contribute to the progression and/or severity of certain disease conditions. Phagocytosis of pathogens or particles resistant to degradation can lead to chronic granulomatous diseases such as tuberculosis and asbestosis (Dean et al., 2008; Mishra and Sopori, 2012). Macrophages have also been implicated in enhancing the growth of certain tumors by establishing tumor promoting microenvironments, promoting angiogenesis, promoting metastasis, and suppressing antitumor immune response.

Natural Killer Cells

Natural killer cells (NK) are non-B, non-T lymphocytes that do not require specific antigenic activation to detect and kill pathogens, making NK a component of innate immunity. NK are important factors in destroying virus-infected cells and limiting the progression and spread of tumors. NK cells are located predominantly in the blood, lymphoid organs, liver, and lung. NK exert their cytolytic effects using perforin and granzymes in a manner similar to cytotoxic T cells (Kaminski et al., 2008). Although NK are considered important components of innate immunity, they are also involved in the initiation and regulation of acquired immunity through their response to and secretion of immunoregulatory cytokines (Dean et al., 2008).

Mast Cells

Mast cells originate in the bone marrow and are released into the blood as promastocytes, which then enter surface tissues with frequent exposure to potential pathogens or allergens such as the skin, respiratory tract, and alimentary tract (Galli et al., 2008). Once in their resident tissues, the promastocytes further differentiate and mature into recognizable mast cells. Mast cells play important roles in both innate and acquired immune responses. Mast cells can be activated by a variety of stimuli, including IgE, specific antigens, physical injury, chemical injury, or activated complement proteins (Prussin and Metcalfe, 2003). Mast cells function to destroy pathogens, increase vascular permeability, stimulate bronchial smooth muscle contraction, and degrade toxic endogenous peptides or exogenous venom components (Galli et al., 2008). Immunoregulatory functions of mast cells include recruitment of other immune cells, enhancement of antigen presentation to T cells, promotion of B cell IgE production, promotion of Th2 cell responses, and promotion of production of chemokines and cytokines of airway smooth muscle. Mast cells also serve to suppress sensitization for contact hypersensitivity, suppress cytokine production by T cells and monocytes, suppress proinflammatory cytokine and chemokine production by keratinocytes, and enhance the ability of dendritic cells to reduce T cell proliferation and cytokine production.

SOLUBLE MEDIATORS OF IMMUNITY

Cytokines

Cytokines are glycoproteins that play crucial roles in the regulation of both innate and acquired immune responses. Cytokines mediate cell-to-cell interactions and are important regulators of immune cell proliferation, differentiation, maturation, activation, migration, inhibition, and effector functions (Kaminski et al., 2008). Unlike hormones that act on cells of distant organs, cytokines are produced and act locally and they are rapidly cleared by the circulation; these factors play important roles in restricting their activity to the immediate environment. Th cells produce a large proportion of cytokines, but many other cells, both immune and nonimmune, are also capable of producing cytokines. A plethora of cytokines have been identified, and they can be roughly placed (with overlap in many cases) into five major classes based on function: interleukins, colony-stimulating factors, interferons, tumor necrosis factors, and hematopoietins (Dean et al., 2008). Interleukins are predominantly immunoregulatory compounds that act in stimulatory or inhibitory fashion on immune cells. Colony-stimulating factors are involved in the proliferation and differentiation of leukocyte progenitors. Interferons are primarily involved in regulating antiviral activity of immune cells. Tumor necrosis factors have antitumor activity and are involved in triggering apoptosis in target cells. Hematopoietins are involved in regulating bone marrow and production of hematopoietic cells. In addition to these major classes, there are many other cytokines with varied functions in immunoregulation that do not fit into these classes.

Chemokines

Chemokines are small peptide molecules that are produced by most cells of the body and that have a variety of vital biological functions both within and outside the immune system. Immunoregulatory functions of chemokines include mediation of allergic inflammation, leukocyte function, and Th1:Th2 ratios. Chemokines play important roles in autoimmunity and hypersensitivity disorders.

Other Soluble Mediators

Acute phase proteins include C-reactive protein, serum amyloid A, and serum amyloid P, which contribute to acute response to infection by binding pathogens such as bacteria and facilitating complement activation (Kaminski et al., 2008). The complement cascade is a system of globulins that ultimately forms a membrane attack complex capable of cytolytic action on membranes bound by antibodies. Reactive oxygen species released from neutrophils and macrophages as part of the killing response to microbes contribute to the inflammatory process. The cytolytic effects of granzyme and perforin have been previously discussed. Lymphoid cells possess receptors for various hormones, neuropeptides, and neurotransmitters, indicating that the immune system is able to respond to a variety of endocrine and paracrine signals from the endocrine and nervous systems that can attenuate or enhance the immune response. This neuroendocrine communication works both ways, as some immune cells are capable of secreting hormones and neurotransmitters that directly influence endocrine and nervous system functions.

MECHANISMS OF IMMUNOTOXICITY

It is apparent from the previous section that the immune response is not, as once thought, analogous to scattered bands of soldiers (immune cells) that patrol the body, burning and pillaging with abandon all in sight when they happen on foreign invaders (pathogens). Instead, in the healthy host, immune responses act like a highly disciplined army that uses sophisticated communication and tracking systems (soluble mediators, neuroendocrine input) to focus its efforts on removing foreign invaders in as specific and coordinated a manner as possible so as to minimize collateral damage to the host. Given the large number of players and tight regulatory requirements in the immune system arena, it should not be surprising that there are many opportunities for derangements in the system to develop because of interference of any one player by external forces such as toxicants. Additionally, because of the intricate connections between the immune system and other organ systems, toxicant-induced damage to nonimmune organs may result in altered immune responses both locally and throughout the body. Alterations in immune function may result in immunosuppression or excessive immunostimulation, either of which is detrimental to the overall health of the host.

The means by which toxicants may alter immune integrity are many and varied and can include direct and indirect mechanisms (Mishra and Sopori, 2012). Toxicant-induced damage to any single immune cell lineage can reduce the number and viability of that type of immune cell; because of the intricate intercellular communication, loss of one cell type will adversely affect other cells as well (e.g., loss of APCs can decrease T cell responses). Toxicants may inhibit protein synthesis by altering the structure or function of DNA, RNA, or synthetic enzymes, resulting in decreased immune cell proliferation and maturation, decreased antibody production, and decreased production of soluble mediators; the net result will be immunosuppression. Some toxicants inhibit enzymes or enzyme systems, interfering with normal cellular function; for example, organophosphate insecticides are thought to induce lymphocyte membrane dysfunction through inhibition of membrane esterases (Galloway and Handy, 2003). Compounds that alter immune cell membrane and receptor structures can interfere with signal transduction, cell-to-cell communication, degranulation, response to cytokines, and antibody secretion. Toxicants may bind to immune cell membrane receptors, resulting in aberrant cell signaling that can over- or understimulate cell responses. Toxicants may indirectly alter immune function through the triggering of stress responses that stimulate cortisol release, which can result in glucocorticoid-induced apoptosis of lymphocytes (Lill-Elghanian et al., 2002). Interaction of toxicants with cellular macromolecules can result in hapten formation, triggering hypersensitivity reactions. Similarly, toxicant-induced tissue damage can alter body proteins, leading to the development of autoantibodies and autoimmune disease.

BIOMARKERS

The field of immunotoxicology has undergone tremendous growth over the last 25 years as knowledge of the immune system, and its susceptibility to exogenous forces, has expanded (House and Selgrade, 2010). Identification of immune-mediated disorders within the human population has been continuously rising over the last several decades, and immune-mediated disease is a significant public health concern (Hochstenbach et al., 2010). With the realization that many pharmaceuticals and environmental toxicants have the potential to trigger immune-mediated diseases and with the rapid growth in development of immunomodulating drugs to treat conditions such as rheumatoid arthritis, there is increasing need for development of biomarkers that can identify exposure to, and injury by, xenobiotics that can have deleterious effects on the immune system. The ideal biomarker should show clinical relevance, reflect the molecular or biochemical basis of the disease process, show sensitivity and specificity, be reliable and minimally invasive, and have simplicity in use and application (Duramad and Holland, 2011). Unfortunately, an ideal biomarker panel that would reliably predict immunotoxicity of various compounds is yet to be developed. The paucity of adequate immunotoxicity tests has been recognized as a serious deficiency in the development of pharmaceuticals, evaluation of environmental toxicants, formulation of risk assessments for individuals and populations, and diagnosis of immune-mediated disorders (Duramad et al., 2007; Piccotti et al., 2009; Hochstenbach et al., 2010; Kawabata and Evans, 2012; Luebke, 2012).

The general strategy in evaluation of immunotoxicity is to apply a tiered approach, generally beginning with screening tests (tier 1) and following through with more comprehensive testing (tier 2) to confirm the nature of the immune defect (Dean et al., 2008). This approach is appropriate for both clinical and nonclinical evaluation for immunotoxicity, although the lack of validation of some tests can limit their use in clinical practice. Table 24.1 lists some biomarkers used in the assessment of immunotoxicology.

BIOMARKERS OF IMMUNE STATUS

Cell Counts

The standard hemogram, composed of a complete blood cell count with differential leukocyte count, is the most basic, rapid, and least expensive means of screening for adverse effects on the immune system. Elevations in total white blood cell counts can be indicative of inflammatory or neoplastic disorders; physiologic leukocytosis can occur as a result of acute or chronic stress. Elevations in neutrophils can occur following exposure to exogenous glucocorticoids and lithium and in early estrogen toxicosis (dogs) (Stockham and Scott, 2008; Petrini and Azzara, 2012). Toxicant-induced decreases in total leukocytes or neutrophils may be seen with bone marrow suppressants such as chemotherapeutic drugs, chloramphenicol, and sulfonamides, as well as compounds that cause destruction of mature granulocytes such as phenothiazines (Aster, 2005). Lymphocytosis may occur because of chronic infection or allergic reactions. Lymphopenia, as defined as absolute lymphocyte count $<1500/mm^3$, may occur because of severe stress, malnutrition, autoimmune disease, and hematopoietic malignancy (Dean et al., 2008). Monocytosis may occur secondary to stress or exogenous corticosteroid administration. Peripheral blood eosinophilia occurs with parasitism and hypothyroidism, and it is a significant feature of the eosinophilia–myalgia syndrome that has been associated with ingestion of the nutritional supplement L-tryptophan (Fenstrom, 2012).

Organ Weight

Determination of lymphoid organ weight has been a standard component of immunotoxicity testing during preclinical safety testing of pharmaceuticals and safety assessments of pesticides, industrial chemicals, and environmental contaminants. For obvious reasons, this type of immunologic screening is only suited to research studies.

Histopathology

Histomorphological examination of lymphoid organs is useful in assessing immunotoxicity of a xenobiotic. Microscopic determination of differences in cell density among the various lymphoid organs and immunohistochemical staining to differentiate various lymphoid cells can help identify which immune cell lineages are most severely affected. Accuracy of histomorphological scoring of immunotoxicity depends on a consistent and stringent scoring system to maximize specificity. Enhanced histopathological methods should be utilized to maximize accuracy, which includes identifying the separate compartments that support specific immune functions within each lymphoid organ, evaluating each compartment individually, and using semiquantitative descriptive (rather than interpretative) terminology to characterize any changes (Elmore, 2012). Within the spleen, five endpoints are evaluated: number of germinal centers, cellularity of periarteriolar lymphatic sheaths, lymphoid follicles, marginal zone, and red pulp. Lymph nodes have four endpoints that must be evaluated: cellularity of follicles, paracortical areas, medullary cords, and sinuses, while the thymus has three endpoints: corticomedullary ratio and cellularity of cortex and medulla. Although most commonly utilized for preclinical/premarketing safety assessments, histopathological evaluation of bone marrow and lymphoid tissue is also used clinically as part of the evaluation of the immune status of individual patients.

Antibody Assays

Serum levels of immunoglobulins have been used frequently in epidemiological studies and a variety of assay methods are readily available, including radioimmunoassay, enzyme immunoelectrodiffusion, enzyme-linked immunosorbent assay, single radial diffusion, double diffusion in agar gel, laser nephelometry, and immunoturbidimetric assay (Dean et al., 2008; Mali et al., 2009). The use of appropriate reference ("normal") ranges is vital because serum immunoglobulin levels can vary due to a variety of factors, including sex, age, ethnicity, nutritional status, geographical location, gender, and environment.

TABLE 24.1 Biomarkers Used in the Assessment of Immunotoxicology

Assay	Matrix	Endpoint	Example Toxicants
Biomarkers of Immune Status			
Leukocyte count	Blood	Total Leukocytes ↑	Cortisol, Lithium
		Total Leukocytes ↓	Chloramphenicol, Chemotherapeutics
		Neutrophils ↑	Cortisol, Estrogen (dogs, early), Lithium
		Neutrophils ↓	Phenothiazine
		Lymphocytes ↑	Allergens
		Lymphocytes ↓	Lithium
		Monocyte ↑	Cortisol
		Eosinophil ↑	L-Tryptophan
Organ Weight	Spleen, Thymus, Lymph Node	Organ Weight ↑	Immunostimulants
		Organ Weight ↓	Immunolytics
Enhanced histopathology	Tissues	Compartment cellularity ↑	Immunostimulants
		Compartment cellularity ↓, Apoptosis ↑	Immunosuppressants, Immunolytics
Antibody Assays	Serum, Secretions, Exudates	IgA, IgD, IgG, IgE, IgM ↓	B cell depressants
T-Dependent Antibody Response	Sensitized Spleen Cells	Plaque formation ↓	Dibenzo[def,p]chrysene, PAH[a], Indomethacin (rats)
Delayed-Type Hypersensitivity Response	*In vivo* (skin)	Induration	PPD tuberculin, *Toxicodendron*
Lymphocyte Phenotyping	Blood	Altered CD ratios	Lead, Beryllium
Natural Killer Cell Activity	Isolated lymphocytes	Cytolytic activity ↓	
Phagocytosis, chemotaxis, killing	Peritoneal Macrophages, Blood Neutrophils	Decreased adherence, chemotaxis, and/or phagocytosis	Aluminum trichloride, Deoxynivalenol
Proliferation Tests	Isolated lymphocytes	↓ T cell proliferation	Arsenic trioxide, Tributyltin
		↑ T cell proliferation	Beryllium
		↓ B cell proliferation	Prostaglandin
		↑ B cell proliferation	Lipopolysaccharide
Cytokine Profiling	Serum, Cell Supernatant, Saliva, Sputum	Measured levels	Cigarette smoke, VOC[b]
Urinary Leukotriene E(4)	Urine	Measured levels	Tobacco Smoke, Air Pollution
Biomarkers of Hypersensitivity			
Local Lymph Node Assay	Lymph Node	Radioactivity counts	Contact allergens
Human Repeat-Insult Patch Test	*In vivo* (Skin)	Erythema, irritation	Contact allergens
IgE Levels	Serum	Binding to antigen	Various allergens

Continued

TABLE 24.1 Biomarkers Used in the Assessment of Immunotoxicology—cont'd

Assay	Matrix	Endpoint	Example Toxicants
Biomarkers of Autoimmunity			
Autoantibodies	Serum, Inflammatory exudate	Measured levels	Mercury[c]
Proteomics	Plasma, Serum, Saliva, Sputum, Urine, Local Fluids, Tissues	Protein expression profile	N/A

[a]*PAH = polycyclic aromatic hydrocarbons*
[b]*VOC = volatile organic compounds*
[c]*Gardner RM, Nyland JF, Silva IA, Ventura AM, de Souza JM, Silbergeld EK (2009) Mercury exposure, serum antinuclear/antinucleolar antibodies and serum cytokine levels in mining populations in Amazonian Brazil: a cross-sectional study.* Envron Res ***110(4)****:345-354.*

In experimental animal studies, humoral responses to sheep erythrocytes have been used to evaluate compounds for immunotoxicity—following exposure to the compound in question, test animals are inoculated with sheep erythrocytes, which generally trigger robust innate and adaptive immune responses (Van Loveren et al., 2001). Rhesus monkeys exposed to polychlorinated biphenyls (PCBs) had reduced antibody responses to sheep erythrocytes, suggesting a depressant effect of the PCBs on the monkeys' immune systems.

Although antibody levels have been used for population studies in humans, generally to estimate exposure to infectious agents, results of studies attempting to document altered responses to vaccination (and thus altered immune function) in humans exposed to putative toxicants have been equivocal. For instance, cigarette smokers had suboptimal immune responses to vaccination with hepatitis B vaccine (Roome et al., 1993), yet had increased levels of antibodies to influenza following vaccination (Mancini et al., 1998). Furthermore, immunoglobulin levels do not necessarily correlate with protective antibody responses following vaccination. Exposure to immunotoxicants may result in decreased levels of one or more class of immunoglobulin. However, it is unlikely that measurements of serum immunoglobulin levels provide the level of sensitivity needed to detect mild to moderate toxicant-induced alterations in immune function within a population (Luster et al., 2005).

T-dependent Antibody Response

The T-dependent antibody response (TDAR) assay evaluates the ability to produce IgM antibodies following exposure to potentially immunotoxic compounds (White et al., 2010). Following exposure to the test compound, spleen cells are isolated from the test animal and incubated in a matrix of agar, sheep erythrocytes, and Guinea pig serum. Lysis of erythrocytes around IgM secreting antigen-specific plasma cells results in formation of clear plaques in the agar, which are then counted. The TDAR is a sensitive assay for measuring the effects of xenobiotics on the humoral immune system. Feeding pregnant rats indomethacin in late gestation can result in offspring that have suppressed TDAR because of alteration in release of Th2 cytokines (Kushima et al., 2009). Dibenzo[def,p]chrysene is a polycyclic aromatic hydrocarbon that is a potent carcinogen and that has been associated with persistent suppression of TDARs (Lauer et al., 2013).

Delayed-type Hypersensitivity Response

The delayed-type hypersensitivity (DTH) assay measures cell-mediated immunity, including ability of memory T cells to recognize foreign antigen, proliferate, migrate to the site of antigen, and secrete cytokines and chemokines, as well as ability of monocytes to migrate to the area (Kaminski et al., 2008). A classic example of the DTH assay is the PPD tuberculin test to detect individuals sensitized by *Mycobacterium* spp (Rosenstreich, 1993). The sensitized host has antigen inoculated intradermally, and the inoculated site is evaluated at 48 and 72 h for erythema, swelling, and edema.

Lymphocyte Phenotyping

Study of cell-surface markers (CDs) on immune cells using flow cytometry has greatly enhanced the ability to study lymphocyte development, activation, and function (Lappin and Black, 2003; Shipkova and Wiedland, 2012). In humans, B cells are identified by CD19 and T cells possess CD3. Using multiple fluorochromes to stain immune cells has resulted in the identification of a variety of cell subtypes and allows the study of the responses of the various subtypes to xenobiotics. Flow cytometry also allows for the study of numerous cytokines simultaneously, giving a better picture of the interactions between immune cells and soluble mediators. Using flow cytometry, changes in numbers of subpopulations of immune cells in response to a xenobiotic can be detected as can alterations in the amounts and types of cytokines elaborated in response to an immunotoxicant. For example, individuals occupationally exposed to lead had decreases in $CD3^+$ cells (i.e., B cells) and a decreased $CD4^+$ to $CD8^+$ cell ratio (Garcia-Leston et al., 2012). Evaluation of the number of beryllium-responsive $CD4^+$ in the blood has been proposed as a biomarker to distinguish patients with chronic beryllium disease from those who have been sensitized to beryllium but have not developed the pulmonary form of beryllium-induced disease (Martin et al., 2011).

Natural Killer Cell Activity

NK activity is measured by culturing lymphoid cells with radiolabeled tumor cells known to be susceptible to NK-mediated cytolysis. Quantitation of radioactivity in the supernatant indicates the degree of NK activity as the lysed tumor cells release their radiolabel. A technique using flow cytometry has been developed to quantitate NK activity and has the advantage of avoiding the use of radioactive isotopes (Kim et al., 2007).

Phagocytosis and Chemotaxis

Measurement of chemotaxis by immune cells generally involves incubating the cells in question in a diffusion chamber or on a porous substrate with a chemoattractant gradient used to direct chemotaxis. Soluble mediators elaborated by the cells may be measured and adhesion and chemotaxis evaluated. Failure of cells to adhere, migrate appropriately, or elaborate appropriate soluble mediators suggests impairment in immune function. Phagocytic activity of macrophages and neutrophils is measured in vitro by isolating peritoneal macrophages or peripheral blood neutrophils and incubating them with foreign antigen (usually radiolabeled) and then examining them for the presence of the marker antigen. A similar in vivo test involves injecting the antigen into the animal, which is subsequently killed and tissues examined for the presence of the marker antigen within phagocytes. Toxicant-induced deficiencies in phagocytosis ± adhesion can be evaluated. Defects in adherence, chemotaxis, and phagocytosis have been documented in macrophages exposed to aluminum and neutrophils exposed to deoxynivalenol (Hu et al., 2011; Gauthier et al., 2013).

Proliferation Tests

Proliferation tests measure the ability of B or T cells to proliferate in response to stimulation by antigen. B cells are stimulated by anti-CD19 antibody in the presence of IL-4 or with lipopolysaccharide, a B cell mitogen. T cells are incubated with anti-CD3 antibody in the presence of IL-2 or with either concanavalin A or phytohemagglutinin, T cell mitogens. Response is measured by the uptake of ^{3}H-thymidine into DNA of the cultured cells. Little to no response of the lymphocytes following antigenic stimulation suggests that the immune response has been suppressed. Arsenic and tributyltin have been shown to cause decreased T cell proliferation, while prostaglandin E has been associated with depressed B cell proliferation (Kurland et al., 1977; Frouin et al., 2008; Morzadec et al., 2012). Increased T cell proliferation occurs in beryllium disease, and microbial lipopolysaccharide is associated with increased B cell proliferation (Mond and Brunswick, 2003; Martin et al., 2011).

Cytokine Profiling

As previously mentioned, cytokines are crucial molecules in facilitating the proliferation, maturation, and function of immune cells, especially T cells. Cytokine levels can be measured by ELISA or flow cytometry in serum, plasma, bronchioalveolar fluid, or in supernatant from cultured cells. Cytokine profiling entails measuring the levels of a battery of cytokines simultaneously to provide insight into how immunotoxicants exert their effects on immune function. Through study of cytokine profiles, it has been determined that Th1 cytokines (IL-2, IFN-γ, TNF-β) are associated with contact sensitization, whereas Th2 cytokines (IL-4, IL-5, IL-6, IL-10) are associated with respiratory sensitization (Kaminski et al., 2008). Neonatal humans have predominantly Th2 cytokine profiles that gradually progress toward a balanced Th1/Th2 ratio (Luster et al., 2005). Exposure to toxicants that delay or prevent this balance from being achieved can predispose to immune dysfunction; for instance, sustained Th2 cytokine profiles increase the risk for allergic disease. Examples of environmental toxicants that can increase T2 cytokine production include maternal cigarette smoking and indoor volatile organic compounds. Cytokine profiling has been proposed for detecting immunotoxicity induced by nanoparticles used for therapeutic or diagnostic purposes (Elsabahy and Wooley, 2013).

Cytokine profiling does have some limitations as an immunotoxicology biomarker. Baseline data may not exist for a particular population (e.g., healthy children), and there are multiple genetic, physiological, and environmental factors that can contribute to significant individual variability in response (Luster et al., 2005). Lack of standardization of methods can make it difficult to make direct comparisons between studies if, for instance, one study used ELISA supernatant from in vitro stimulated peripheral blood cells, while the other used flow cytometric evaluation of intracellular cytokines while a third study used polymerase chain reaction to measure cytokine RNA rather than the proteins themselves. Finally, due to the transient nature of cytokines, a single profile is no more than a snapshot in time that may or may not reflect the response that is occurring in the target organ.

Other Soluble Mediators

Numerous endogenous compounds are being investigated as potential markers for immune disorders. Eosinophil-derived neurotoxin, an eosinophil degranulation product, has been proposed as a novel marker to diagnose, treat, and monitor allergic disease (Kim, 2013). Urinary leukotriene E(4) is being investigated as a biomarker of environmental exposure to tobacco smoke and air pollution, as well as a predictor of risk for asthma exacerbations related to tobacco smoke (Rabinovitch, 2012). Neuropteron, produced by monocytes and macrophages on activation, has been used as a measure of activation of the cell-mediated immune response (Garcia-Leston et al., 2012).

BIOMARKERS OF HYPERSENSITIVITY

Local Lymph Node Assay

The murine local lymph node assay (LLNA) has become the standard nonclinical test for testing the potential for chemical sensitization by xenobiotics (Anderson et al., 2011; Peiser et al., 2012). The LLNA has largely replaced the Guinea pig–based tests (i.e., Guinea pig maximization assay and Buehler assay) that had been used historically to test for chemical sensitization. The candidate substance is applied topically to mice once daily for three consecutive days, and on day 6 the mice are injected intravenously with ^{3}H-thymidine. Hours later, the mice are sacrificed and lymph nodes draining the site of chemical application are evaluated for radioactivity, which will correlate with degree of lymphocyte proliferation. Because lymphocyte proliferation has been shown to be related to chemical sensitization and the measured degree of response to the LLNA can be correlated to an objective measure of relative allergenic potency of a compound, the LLNA has been investigated as a potential tool to use in human quantitative risk assessment.

Limiting factors that need to be considered when using the LLNA include the fact that the assay measures reactions following acute (3-day duration) exposures without providing information on any differences in response with long-term or intermittent exposure. Additionally, interspecies differences need to be considered, as rodents have increased skin penetration of chemicals compared with humans. False positives in the LLNA may occur when nonspecific skin irritation, rather than chemical sensitization, occurs; tools used to attempt to filter out this complicating factor have been developed for use in research but they have not been validated for the purpose of risk assessment.

Human Repeat-Insult Patch Test

The human repeat-insult patch test is a test of skin sensitization that is primarily intended to confirm safety of topically applied compounds rather than identify skin sensitization (as with the Guinea pig and murine tests) (Kimber et al., 2001). Ethical issues with the use of human subjects prohibit or restrict the use of this method in many countries. An occlusive patch containing test material is applied to the upper arm or back of a human volunteer and maintained in place for 24 h after which the patch is removed and the skin evaluated for erythema and edema (Hayes et al., 2008). After a 24-h rest, a second patch is applied to the same site (Voss-Griffith test) or a new site (Draize test); the process is completed until a total of 9 (Voss-Griffith) or 10 patches (Draize) have been applied. Ten to fourteen - days after removal of the last patch, a challenge patch is applied for 24 h and the sites examined at 0, 48, and 96 h for evidence of reaction.

Immunoglobulin E Levels

Measurement of total serum IgE antibody levels can be used as a core screen for allergic disease, although tests for IgE response against a battery of allergens in a multiallergen screen are considered the core biomarker that allows characterization of allergic status of the individual (Szefler et al., 2012). A variety of assays have been developed in which serum from the patient is placed on the test substrate containing allergens and a colorimetric change occurs when IgE in the serum binds the allergen, indicating sensitization of the patient to the allergen. In vivo measurement of allergen-specific IgE is also available as "skin testing." Allergens are introduced to the skin via skin prick or intradermal injection and the resulting wheal and flare reaction identifies those allergens to which IgE is present.

BIOMARKERS OF AUTOIMMUNITY

Loss of "self" recognition by the immune system can lead to the development of antibodies against normal body components such as basement membranes or individual cell types (e.g., erythrocytes, keratinocytes). Toxicants can initiate this loss of self-recognition through tissue damage that alters cellular proteins, thereby triggering an autoimmune response (Mishra and Sopori, 2012). Autoimmunity can develop against a huge range of cellular and extracellular components, so it is not surprising that a single test for autoimmunity does not exist. Instead, diagnosis of autoimmunity generally relies on the detection of specific autoantibodies, and oftentimes an array of tests is required before the correct autoantigen is recognized (Lemmark, 2001). An ideal autoimmune biomarker would be measurable from serum or plasma and would reflect the real-time severity of the disease as it fluctuates during disease progression and therapy.

Currently, biomarker discovery efforts are focused on proteins that reflect the autoimmune process, namely degradation products arising from destruction of affected tissues, enzymes that are involved in tissue degradation, and cytokines or other proteins associated with immune activation (Prince, 2005). Many of these proteins are restricted to the anatomical site of production (e.g., synovial cavities in cases of autoimmune arthritis), which can make obtaining samples problematic. Proteomic platforms utilizing mass spectrometry or gel electrophoresis coupled with liquid chromatography have been investigated to identify and quantify proteins thought to contribute to autoimmune disease (Gibson et al., 2010). Although unique, single-protein biomarkers are unlikely to be found for most autoimmune diseases, characterization of the patterns of multiple proteins expressed during different autoimmune processes may provide unique profiles that aid in the diagnosis, prognostication, and monitoring of various autoimmune disorders (Ademowo et al., 2013). An advantage with using the proteomic methods is the ability to develop profiles for a variety of matrices, including plasma, serum, saliva, sputum, urine, local fluids (e.g., synovial fluid, cerebrospinal fluid), and tissues.

CONCLUDING REMARKS AND FUTURE DIRECTIONS

The field of immunotoxicology has expanded greatly in the past three decades with the growing understanding of the many functions and vulnerabilities of the immune system. No longer merely a "bug killer," the immune system is now known to be a complex, highly regulated, and delicate entity tasked with preserving the integrity of the host against outside forces. Increased understanding of how xenobiotics can impact the function of the immune system has led to development of assays to identify the effects of immunotoxicants and to study the mechanisms of immunotoxicity and develop rational risk assessments for the wide range of xenobiotics to which we are exposed on a daily basis. Immunotoxicology can be seen to be in its infancy, and future discoveries will undoubtedly further advance our knowledge and ability to predict, prevent, and treat immune-mediated disorders.

The development of in vitro or in silico (computer model) alternatives to the use of live animal research is an area of study that is receiving a great deal of attention within the scientific community (Corsini and Roggen, 2009). Although currently available methods fall short of being able to totally replace live animals for some vital parameters (e.g., carcinogenicity, reproductive toxicity), some progress is being made in the development of in vitro assays and in silico models as first-line screening tools for contact sensitizers. The myeloid U973 skin sensitization test and human cell line activation test (h-CLAT) have been accepted for prevalidation as alternatives to the LLNA for measuring skin sensitization (Anderson et al., 2011).

Immunotoxicogenomics is an emerging field of study that explores the potential to use analysis of gene expression in both the study of immunotoxicants and the assessment of relative risks of toxicants to the immune system (Luebke et al., 2006). Microarray analysis is already being used to investigate pathway-level changes that lead to altered immune responses, and the use of microarray analysis of genomes to screen for potential toxic effects, determine mechanisms of toxicant action, and provide data germane to the risk assessment process is already under investigation in other areas of toxicology; application to immunotoxicology is in progress. Using whole genome gene expression in human peripheral blood mononuclear cells, Hochstenbach et al.(2010) were able to identify a set of 48 genes that could distinguish a group of immunotoxic compounds from nonimmunotoxic compounds. Many barriers (e.g., cross-species extrapolation, correlating genotypic pattern with phenotypic expression, etc.) must be overcome before the marriage of immunotoxicology and genomics can be expected to produce reliable and useful means of evaluating compounds for immunotoxic potential, determining hazard assessments, and contributing to our understanding of immune function, but the field of immunotoxicogenomics holds much promise.

References

Ademowo, O.S., Staunton, L., FitzGerald, O., et al., 2013. Biomarkers of inflammatory arthritis and proteomics. In: Stanilova, S. (Ed.), Genes and Autoimmunity - Intracellular Signaling and Microbiome Contribution. InTech, ISBN 978-953-51-1028-6. https://doi.org/10.5772/54218.

Anderson, S.E., Siegel, P.D., Meade, B.J., 2011. The LLNA: a brief review of recent advances and limitations. J. Allergy 2011, 424203.

Aster, J.C., 2005. Diseases of white blood cells, lymph nodes, spleen and thymus. In: Kumar, V., Abbas, A.K., Fausto, N. (Eds.), Robbins and Cotran Qathologic Basis of Disease, seventh ed. Elsevier/Saunders, Philadelphia, PA.

Corsini, E., Oukka, M., Pieters, R., et al., 2011. Alterations in regulatory T cells: rediscovered pathways in immunotoxicology. J. Immunotoxicol. 8 (4), 251–257.

Corsini, E., Roggen, E.L., 2009. Immunotoxicology: opportunities for non-animal test development. Alt. Lab. Anim. 37 (4), 387–397.

Dean, J.H., House, R.V., Luster, M.I., 2008. Immunotoxicology: effects of and response to drugs and chemicals. In: Hayes, A.H. (Ed.), Principles and Methods of Toxicology, fifth ed. Informa Healthcare USA, New York, NY.

Descotes, J., Nicolas, B., Vial, T., 1995. Assessment of immunotoxic effects in humans. Clin. Chem. 41 (12), 1870–1873.

DeWitt, J.C., Peden-Adams, M.M., Keil, D.E., et al., 2011. Current status of developmental immunotoxicity: early-life patterns and testing. Toxciol. Pathol. 40, 230–236.

Duramad, P., Holland, N.T., 2011. Biomarkers of immunotoxicity for environmental and public health research. Int. J. Environ. Res. Publ. Health 8, 1388–1401.

Duramad, P., Tager, I.B., Holland, N.T., 2007. Cytokines and other immunological biomarkers in children's environmental health studies. Toxicol. Lett. 172 (1–2), 48–59.

Elmore, S.A., 2012. Enhanced histopathology of the immune system: a review and update. Toxicol. Pathol. 40, 148–156.

Elsabahy, M., Wooley, K.L., 2013. Cytokines as biomarkers of nanoparticle immunotoxicity. Chem. Soc. Rev. 42 (12), 5552–5576.

Fenstrom, J.D., 2012. Effects and side effects associated with the non-nutritional use of tryptophan by humans. J. Nutr. 142 (12), 2236S–2244S.

Frouin, H., Lebeuf, M., Saint-Louis, R., et al., 2008. Toxic effects of tributyltin and its metabolites on harbor seal (*Phoca vitulina*). immune cells in vitro. Aquat. Toxicol. 90 (3), 243–251.

Galli, S.J., Grimbaldeston, M., Tsai, M., 2008. Immunomodulatory mast cells: negative, as well as positive, regulators of innate and acquired immunity. Nat. Rev. Immunol. 8 (6), 478–486.

Galloway, T., Handy, R., 2003. Immunotoxicity of organophosphorus pesticides. Ecotoxicology 12, 345–363.

Garcia-Leston, J., Roma Torres, J., Mayan, O., et al., 2012. Assessment of immunotoxicity parameters in individuals occupationally exposed to lead. J. Toxicol. Environ. Health A 75, 807–818.

Gardner, R.M., Nyland, J.F., Silva, I.A., et al., 2010. Mercury exposure, serum antinuclear/antinucleolar antibodies, and serum cytokine levels in mining populations in Amazonian Brazil: a cross-sectional study. Environ. Res. 110 (4), 345–354.

Gauthier, T., Wache, Y., Laffitte, J., et al., 2013. Deoxynivalenol impairs the immune functions of neutrophils. Mol. Nutr. Food Res. https://doi.org/10.1002/mnfr.201200755 [PubMed 23427020].

Gibson, D.S., Banha, J., Penque, D., et al., 2010. Diagnostic and prognostic biomarker discovery strategies for autoimmune disorders. J. Proteomics 73, 1045–1060.

Gray, R.D., Lucas, C.D., Mackellar, A., et al., 2013. Activated of conventional protein kinase C (PKC). is critical in the generation of human neutrophil extracellular traps. J. Inflamm. 10 (1), 12.

Hayes, B.B., Patrick, E., Maibach, H.J., 2008. Dermatotoxicology. In: Hayes, A.H. (Ed.), Principles and Methods of Toxicology, fifth ed. Informa Healthcare USA, New York, NY.

Hochstenbach, K., van Leeuwen, D.M., Gmeunder, H., et al., 2010. Transcriptomic profile indicative of immunotoxic exposure: in vitro studies in peripheral blood mononuclear cells. Toxicol. Sci. 118 (1), 19–30.

House, R.V., Selgrade, M.J., 2010. A quarter-century of immunotoxicology: looking back, looking forward. Toxicol. Sci. 118 (1), 1–3.

Hu, C., Li, J., Zhu, Y., et al., 2011. Effects of aluminum exposure on the adherence, chemotaxis and phagocytosis capacity of peritoneal macrophages in rats. Biol. Trace Elem. Res. 144 (1–3), 1032–1038.

Janeway Jr., C.A., Travers, P., Walport, M., et al., 2001. Immunobiology: The Immune System in Health and Disease, fifth ed. Garland Science, New York.

Kaminski, N.E., Faubert Kaplan, B.L., Holsapple, M.P., 2008. Toxic responses of the immune system. In: Klaassen, C.D. (Ed.), Casarett & Doull's Toxicology: The Basic Science of Poisons, seventh ed. McGraw-Hill Companies, New York, NY.

Kawabata, T.T., Evans, E.W., 2012. Development of immunotoxicity testing strategies for immunomodulatory drugs. Toxicol. Pathol. 40, 288–293.

Kim, C.K., 2013. Eosinophil-derived neurotoxin: a novel biomarker for diagnosis and monitoring of asthma. Korean J. Pediatr. 56 (1), 8–12.

Kim, G.G., Donnenberg, V.S., Donnenberg, A.D., et al., 2007. A novel multiparametric flow cytometry-based cytotoxicity assay simultaneously immunophenotypes effector cells: comparisons to a 4 h ^{51}Cr-release assay. J. Immunol. Meth. 325 (1–2), 51–66.

Kimber, I., Basketter, D.A., Berthold, K., et al., 2001. Skin sensitization testing in potency and risk assessment. Toxicol. Sci. 59, 198–208.

Kurland, J.I., Kincade, P.W., Moore, M.A.S., 1977. Regulation of B-lymphocyte clonal proliferation by stimulatory and inhibitory macrophage-derived factors. J. Exp. Med. 146, 1420–1435.

Kushima, K., Sakuma, S., Furusawa, S., et al., 2009. Prenatal administration of indomethacin modulates Th2 cytokines in juvenile rats. Toxicol. Lett. 185 (1), 31–37.

Lappin, P.B., Black, L.E., 2003. Immune modulator studies in primates: the utility of flow cytometry and immunohistochemistry in the identification and characterization of immunotoxicity. Toxicol. Pathol. 31 (Suppl.), 111–118.

Lauer, F.T., Walker, M.K., Burchiel, S.W., 2013. Dibenzo[def,p]chrysene (DBC) suppresses antibody formation in spleen cells following oral exposures of mice. J. Toxicol. Environ. Health A 76 (1), 16–24.

Lemmark, A., 2001. Autoimmune diseases: are markers ready for prediction? J. Clin. Invest. 108 (8), 1091–1096.

Lill-Elghanian, D., Schwartz, K., King, L., et al., 2002. Glucocorticoid-induced apoptosis in early B cells from human bone marrow. Exp. Biol. Med. 227 (9), 763–770.

Luebke, R., 2012. Immunotoxicant screening and prioritization in the twenty-first century. Toxicol. Pathol. 40, 294–299.

Luebke, R.W., Holsapple, M.P., Ladicss, G.S., et al., 2006. Immunotoxicogenomics: the potential of genomics technology in the immunotoxicity risk assessment process. Toxicol. Sci. 94 (1), 22–27.

Luster, M.I., Johnson, V.J., Yucesoy, B., 2005. Biomarkers to assess potential developmental immunotoxicity in children. Toxicol. Appl. Pharmacol. 206, 229–236.

Mali, B., Arbruster, D., Serediak, E., et al., 2009. Comparison of immunoturbidimetric and immunonephelometric assays for specific proteins. Clin. Biochem. 42 (15), 1568–1571.

Mancini, D.A., Mendonca, R.M., Mendonca, R.Z., et al., 1998. Immune response to vaccine against influenza in smokers, non-smokers, and in individuals holding respiratory complications. Boll. Chim. Farm. 137, 21–25.

Martin, A.K., Mack, D.G., Falta, M.T., et al., 2011. Beryllium-specific $CD4^+$ T cells in blood as a biomarker of disease progression. J. Allergy Clin. Immunol. 128 (5), 1100–1106.
Mishra, N.C., Sopori, M.L., 2012. Immunotoxicity. In: Gupta, R.C. (Ed.), Veterinary Toxicology: Basic and Clinical Principles, second ed. Academic Press, San Diego, CA.
Mond, J.J., Brunswick, M., 2003. Proliferative assays for B cell function. Curr. Protoc. Immunol. (Chapter 3:Unit 3.10).
Morzadec, C., Bouezzedine, F., Macoch, M., et al., 2012. Inorganic arsenic impairs proliferation and cytokine expression in human primary T lymphocytes. Toxicology 300 (1–2), 45–56.
Peiser, M., Traulau, T., Heidler, J., et al., 2012. Allergic contact dermatitis: epidemiology, molecular mechanisms, in vitro methods and regulatory aspects. Cell. Mol. Life Sci. 69, 763–781.
Petrini, M., Azzara, A., 2012. Lithium in the treatment of neutropenia. Curr. Opin. Hematol. 19 (1), 52–57.
Piccotti, J.R., Lebrec, H.N., Evans, E., et al., 2009. Summary of a workshop on nonclinical and clinical immunotoxicity assessment of immunomodulatory drugs. J. Immunotoxicol. 6 (1), 1–10.
Prince, H.E., 2005. Biomarkers for diagnosing and monitoring autoimmune diseases. Biomarkers 10 (Suppl. 1), S44–S49.
Prussin, C., Metcalfe, D.D., 2003. IgE, mast cells, basophils, and eosinophils. J. Allergy Clin. Immunol. 111 (2 Suppl.), S486–S494.
Rabinovitch, N., 2012. Urinary leukotriene E4 as a biomarker of exposure, susceptibility and risk in asthma. Immunol. Allergy Clin. North Am. 32 (3), 433–445.
Roome, J., Walsh, S.J., Carter, M.L., et al., 1993. Hepatitis B vaccine responsiveness in Connecticut public safety personnel. J. Am. Med. Assoc. 270, 2931–2934.
Rosenstreich, D.L., 1993. Evaluation of delayed hypersensitivity: from PPD to poison ivy. Allergy Proc. 14 (6), 395–400.
Shipkova, M., Wieland, E., 2012. Surface markers of lymphocyte activation and markers of cell proliferation. Clin. Chim. Acta. 413 (17–18), 1338–1349.
Stockham, S.L., Scott, M.A., 2008. Fundamentals of Veterinary Clinical Pathology, second ed. Blackwell Publishing, Ames, IA.
Szefler, S.J., Wenzel, S., Brown, R., et al., 2012. Asthma outcomes: biomarkers. J. Allergy Clin. Immunol. 129 (3 Suppl.), S9–S23.
Tizard, I.R, 2013. Veterinary Immunology, ninth ed. Elsevier. ISBN: 978-1-4557-0362-3.
Van Loveren, H., van Amsterdam, J.G.C., Vandebriel, R.J., et al., 2001. Vaccine-induced antibody responses as parameters of the influence of endogenous and environmental factors. Environ. Health Perspect. 109 (8), 757–765.
White, K.L., Musgrove, D.L., Brown, R.D., 2010. The sheep erythrocyte T-dependent antibody response (TDAR). Methods Mol. Biol. 598, 173–184.

PART III

CHEMICAL AGENTS, SOLVENTS AND GASES TOXICITY BIOMARKERS

CHAPTER

25

Bisphenol A Biomarkers and Biomonitoring

P. Cohn[1], A.M. Fan[2]

[1]New Jersey Department of Health (retired), Trenton, NJ, United States; [2]California Environmental Protection Agency (retired), Oakland/Sacramento, CA, United States

INTRODUCTION

Bisphenol A (BPA), 2,2-bis (4-hydroxy-phenyl) propane, has been receiving increasing attention because of evidence on its reproductive toxicity and endocrine disruption effects in laboratory studies, the growing literature correlating environmental BPA exposure to adverse effects in humans, its high production volume, its widespread human exposure, and the potential diverse types of human health effects in addition to reproductive toxicity and endocrine disruption (Fan et al., 2017).

The widespread use of BPA in the production of polycarbonate plastics and epoxy resins used in many consumer products, including packaged foods, has led to its extensive distribution and contamination in the environment and nearly universal exposure to humans in developed countries. Due to the instability of the polymers, BPA leaches out because of heating or the acidity/alkalinity of the food. Importantly, maternal exposure to BPA results in embryos and newborns receiving BPA via placental transfer and breast milk.

The evidence available from laboratory animal studies is that unconjugated BPA (uBPA) has estrogenic activities of concern at levels found in human environmental exposure (Richter et al., 2007; Vandenberg et al., 2013), whereas BPA conjugated with glucuronide or sulfate is inactive. Most of the conjugation activity occurs in the liver, and blood flow from the intestinal tract goes first to the liver. Dermal absorption and direct absorption in the mouth do not pass first through the liver. The concern about individual variation in the level of uBPA informs one of the key biomarker issues in the next section.

Although the European Food Safety Authority (EFSA) based its temporary tolerable daily intake of 4 μg/kg body wt/day on increased kidney weight in male mice (EFSA, 2015), mammary gland, liver, various endocrine functions, and metabolism have also been shown to be affected (Vandenberg et al., 2013). As part of an update on BPA toxicology, the US National Toxicology Program (NTP) recently announced the peer review of the Draft NTP Research Report on the Consortium Linking Academic and Regulatory Insights on BPA Toxicity (CLARITY-BPA) Core Study (NTP, 2018). The draft report is on perinatal and chronic extended-dose-range study of BPA in rats. This study is part of the CLARITY-BPA, a research program between the NIEHS, NTP, and the National Center for Toxicological Research (NCTR) of the US Food and Drug Administration (FDA). CLARITY-BPA was developed to bridge guideline-compliant research conducted at the FDA with hypothesis-based research investigations conducted by academia on the toxicity of BPA. The final conclusions are expected in 2019, which will be based on the integration of the Core Study and Grantee Study findings.

In considering the toxicity and adverse health effects of BPA exposure, a particular concern is the possibility of low-dose effect and that nonmonotonicity of BPA would occur at doses lower than those used in regulatory toxicity studies, as demonstrated in nonmonotonic dose–response curves (NMDRC). The NMDRC is characterized by a change in the sign of the slope, and the function can take on the shape of a U, the shape of an inverted U, or another shape that has more than one inflection point. There are uncertainties requiring additional research on effects on mammary gland, and reproductive, neurobehavioral, immune, and metabolic systems at low levels. The causes of the U-shaped curve can include the inhibition/activation of different subsets of cellular receptors, one or more enzymes, and nuclear transcription factors and proteasomes as the dosage is changed from low to high.

Biomarkers in Toxicology, Second Edition
https://doi.org/10.1016/B978-0-12-814655-2.00025-6

The environmental and human health concerns have led to bans or restrictions in use in different countries. The preponderance of international regulatory bodies evaluated and determined that there is no health risk in the general population from average BPA exposure, and that existing exposures are below the health guidance value of 4 μg/kg body wt/day established by the EFSA (2015). However, other opinions have expressed concerns about its potential human health hazard and uncertainties relating to the toxicity and exposure of BPA. These aspects in toxicology and risk assessment of BPA have been described in an earlier discussion (Fan et al., 2017), which should be read as a supplementary compendium to this chapter to serve as a background to the present discussion on biomarkers with a focus on urinary BPA biomarkers. Readers are also referred to the discussion in an earlier chapter on biomarkers in toxicology, risk assessment, and chemical regulations for a discussion of the different types of biomarkers (i.e., biomarkers of exposure, biomarkers of effect, biomarkers of susceptibility) and some of the chemical samples presented (Fan, 2014).

The present discussion will focus on review of recent reports (last 10 years or relevant as appropriate) on the following: (1) studies that investigate the prevalence of BPA exposure in humans, using urinary BPA as biomarker, including several major population studies (e.g., biomonitoring studies such as the United States [US] National Health and Nutrition Examination Surveys [NHANES], Canadian Health Measures Survey [CHMS], Generation R study in the Netherlands, Norwegian Mother and Child Cohort Study [MoBa] for the general and certain specific populations); and (2) selected studies on findings on urinary BPA in pregnant women, children, and mother–child pairs, along with observations of neurobehavioral effects in some of the studies. Children are more susceptible to the effects of environmental toxicants under certain conditions especially in the early months of life (Marty, 2015). They have immature metabolic pathways and differences in ability to metabolize, detoxify, and eliminate toxins compared to adults. These led to a particular interest in BPA, which has been shown to have endocrine activities, reproductive toxicity, and adverse effects on the young.

BISPHENOL A BIOMARKER ISSUES IN BIOMONITORING

Environmental chemicals that have entered the human body may leave markers reflecting the exposure. The marker may be the chemical itself or a derivative of the chemical, measured in the biological matrix, but it may also be some change in the body resulting from the agent or its breakdown product(s). Biomonitoring programs provide an important source of human exposure data to environmental chemicals, including BPA.

BPA has been measured as total BPA, uBPA, BPA glucuronide, and BPA sulfate in human urines. In humans and other primates, orally administered BPA is rapidly absorbed and transformed to BPA glucuronide during first pass metabolism in the gut wall and the liver and a small amount of BPA is converted to a sulfate conjugate (WHO, 2009). More than 80% of orally administered BPA is cleared from the body in 5 h. The conjugated forms of BPA are devoid of estrogenic activity. BPA metabolism is influenced by age, gender, route of exposure, and physiological state (e.g., pregnant vs. nonpregnant). There are variations in metabolism due to genetic polymorphisms. There are also mechanisms to deconjugate BPA (Ginsberg and Rice, 2009), and these could be affected by polymorphisms in individuals. Some of these could affect only certain tissues, but not be obvious by measuring urinary BPA.

Urinary BPA has been the preferred biomarker in BPA studies because of its practicality. Calafat et al. (2015) discussed considerations that are essential when evaluating exposure to nonpersistent, semivolatile environmental chemicals such as phthalates and phenols (e.g., BPA) and concluded that the biological and technical factors strongly support urine as the optimal matrix for measuring nonpersistent, semivolatile, hydrophilic environmental agents. The factors include representativeness of usual personal exposures and not recent extraneous exposures; minimization of contamination arising from collection, sampling, or analysis procedures; pharmacokinetics (e.g., fast metabolism and urine is the compartment with the highest concentrations of metabolites); and knowledge of intraindividual reliability over the biologic window of interest. Potential external contamination sources include sampling containers, preservatives, air, or dust (the latter reported by Longnecker et al. (2013)).

There are various approaches to correct for urine dilution to avoid overestimation with highly concentrated urine samples and underestimation with dilute samples. Creatinine adjustment is widely used for urinary chemical testing and has been shown to vary by factors such as age, gender, and race/ethnicity. The easiest calculation is by reporting a chemical by μg/g of creatinine, but inclusion of creatinine in multivariate statistical regression is also used. The other method is to directly adjust for urine dilution/concentration by the specific gravity (SG) of the urine sample.

Vandenberg et al. (2010a,b) examined more than 80 published human biomonitoring studies that measured BPA concentrations in human tissues, urine, blood, and other fluids, along with two toxicokinetic studies of human BPA metabolism. The biomonitoring studies examined included measurements in thousands of individuals from many developed and some developing countries, and these studies overwhelmingly detected BPA in individual adults, adolescents, and children. uBPA or free BPA was routinely detected in blood in the ng/mL range, and conjugated BPA was routinely detected in the vast majority of urine samples (also in the ng/mL range). These data indicate that the general population is exposed to BPA and is at risk from internal exposure to uBPA. The studies provide data that account for all sources and routes of exposures. Although oral exposure has been thought to be the major source, not all sources for BPA have been identified. The authors found that (1) the large number of biomonitoring studies are highly consistent and therefore reliable; (2) the detection rates and concentrations of BPA in urine and blood of environmentally exposed individuals are remarkably similar in studies performed in many laboratories using a variety of techniques, including highly accurate and sensitive methods as used in the US Centers for Disease Control and Prevention's (CDC) national biomonitoring programs (Calafat et al., 2005; Calafat and Needham, 2008); and (3) there is no evidence to suggest that these studies have poor quality control (e.g., contamination from collection materials, breakdown of conjugates during storage, inadequate blanks). Overall, the reproducibility of these results showed that internal doses in humans for uBPA have a central measure of the distribution in the 0.5–3 ng/mL (equivalent to μg/L) range, but as much as 10 ng/mL.

Vandenberg et al. (2010a,b) also evaluated two toxicokinetic studies conducted in 15 adults administered BPA (Volkel et al., 2002, 2005) that suggested human uBPA exposure is negligible, and determined that the studies had significant deficiencies and are directly contradicted by environmental exposure studies. Other studies described in this chapter have provided additional extensive evidence of exposure.

The level of systemic BPA exposure, as represented by serum or plasma, may differ from exposure as represented by urinary BPA (Ginsberg and Rice, 2009; Vandenberg et al., 2014b; vom Saal and Welshons, 2014; Stahlhut et al., 2016), particularly in infants because of low activity of metabolic enzymes (Ginsberg and Rice, 2009). A remaining issue is the practicality of blood sampling in epidemiology. It is possible to reliably measure the various forms of BPA in blood (Vandenberg et al., 2014a), but there remains a debate about practicality (Calafat et al., 2015, 2016). Veiga-Lopez et al. (2015) conducted a fetal growth study based on maternal and cord plasma (see below in the Mother-Infant/Child Pairs section) following the round robin laboratory methods discussed by Vandenberg et al. (2014a). The observed median uBPA in first and third trimester BPA was 0.66 and 3.0 μg/L plasma, respectively, whereas the umbilical cord plasma median was 0.19 μg/L. The respective 75th percentile levels were 1.5, 16.6, and 2.8 μg/L, and the respective 95th percentile levels were 32, 59, and 14.8 μg/L. The total BPA levels were only moderately higher, by factors of 5%–20%, because of the low level of glucuronidation. The limit of detection for both uBPA and glucuronidated BPA was 0.02 μg/L.

Gerona et al. (2016) noted that most biomonitoring studies to date have estimated prenatal exposure to total BPA indirectly, using enzyme hydrolysis to cleave glucuronide and sulfate bonds with BPA, then quantifying uBPA. They used the direct analytical method, LC-MS/MS, to avoid the possibility of incomplete enzyme hydrolysis at higher BPA levels in a study of predominately low-income pregnant women (n = 112) in their second trimester. They found universal and high exposure to uBPA and its metabolites: median concentrations were 0.25, 2.6, and 0.31 μg/g for uBPA, BPA glucuronide, and BPA sulfate, respectively. Median total BPA, 4.0 μg/g, was roughly twice that measured in US pregnant women in other studies. The 75th percentile level in this study was 13.9 μg/g. On average, total BPA consisted of 71% BPA in glucuronide form, 15% BPA in sulfate form, and 14% in uBPA.

As BPA has a short average half-life of less than 6 h, a single measurement would be useful only if the environmental exposure is relatively constant over weeks or months, and issues of representativeness of longer-term exposures, temporal reliability, and stability after storage need to be considered. Nepomnaschy et al. (2009) evaluated the stability of BPA in urinary specimens after 22–24 years of storage and measured within-person temporal variability in urinary BPA. They concluded that the similar levels found in later sampling (see Section NHANES, below) provide indirect evidence that BPA is relatively stable during long-term freezer storage. These findings suggest that developmental effects of BPA exposure could be investigated with measurements from stored urine. Jusko et al. (2014) noted that exposure to BPA is episodic, varying in dose and duration. Because of its short half-life, an estimate of exposure based on a spot urine specimen is likely to misclassify individuals with regard to their average long-term exposure, for example, over 9 months of pregnancy. They evaluated the reproducibility of BPA concentration in serial urine specimens taken during pregnancy and noted that results indicate a high degree of within-person variability, which presents challenges for designing well-powered epidemiologic studies.

EXPOSURE AND SELECTED OUTCOME STUDIES

Analysis of health effects is an important use of surveys such as NHANES, although in general the epidemiological design is cross-sectional (exposure and outcome are measured at the same time) and, therefore, cannot be easily used for determining causality. Although questionnaire data on past use of tobacco and alcohol are valid retrospective data, chemicals that are quickly excreted, such as BPA, represent only a moment in time. Prospective studies on pregnant women and their children are presented in this chapter (in the subsections on Mother-Infant/Child Pairs and on Children, below). Urinary BPA in gestational samples taken during regular clinical visits have been compared to later outcomes, such as fetal growth, thyroid, neurodevelopment, and respiratory outcomes, which represent the largest number of studies, summarized in Tables 25.2–25.5. Other studies have also been published on other endocrine effects, metabolism, liver function, immune function, oxidative/nitrosative effects, and epigenetics, but are not discussed in this chapter.

LARGE GENERAL POPULATION BIOMONITORING PROGRAMS

The following presents a brief description of some of the large biomonitoring programs that measured BPA in human fluids. The review is not meant to be comprehensive. Studies usually included other potential endocrine disruptors, such as phthalates, but the focus was placed on BPA. Summary of BPA levels are shown in Table 25.1.

National Health and Nutrition Examination Survey, United States

In the United States, the Division of Laboratory Sciences of the CDC operates the National Biomonitoring Program (NBP), which also assesses the nutrition status of the US population (CDC, 2018a). Biomonitoring data are used to determine the presence and concentrations of chemicals in human blood, urine, breast milk, and saliva and to track exposure trends and impacts of public health programs.

The NBP measures more than 300 environmental chemicals and nutritional indicators. The data and biomonitoring research are used to develop various studies and publications, including two key reports: (1) the National Report on Human Exposure to Environmental Chemicals, which is a series of ongoing assessments of the US population's exposure to environmental chemicals; and (2) the National Report on Biochemical Indicators of Diet and Nutrition in the US population, which is a series of publications that provide ongoing assessment of the U.S. population's nutrition status. The NHANES is a program conducted by the National Center for Health Statistics of studies designed to assess the health and nutritional status of the noninstitutionalized civilian resident population of the United States (CDC, 2018b). The survey combines interviews and physical examinations. This information is used to estimate the prevalence of various diseases and conditions and to provide information for use in planning health policy. Urine BPA has been measured by NHANES in each 2-year cycle since 2003.

In 2005, Calafat et al. (2005) provided the first reference range of human internal dose levels of BPA in a demographically diverse human population. BPA was measured in archived urine samples from a reference population of 394 adults in the United States selected from the Third NHANES (NHANES III, 1988–94) callback cohort, a nonrepresentative subset of NHANES III composed of approximately 1000 adults. The urine samples were all spot samples, collected at different times throughout the day.

The concentration ranges of BPA were similar to those observed in other human populations. BPA was detected in 95% of the samples examined at concentrations $\geq$0.1 μg/L urine; the geometric mean (GM) and median concentrations were 1.33 μg/L (1.36 μg/g creatinine in urine, abbreviated below as μg/g) and 1.28 μg/L (1.32 μg/g), respectively; the 95th percentile concentration was 5.18 μg/L (7.95 μg/g). Gender and age group (<50 or $\geq$50 years of age) were not statistically significant predictors, but rural versus urban residence was. The adjusted GM of BPA for an urban resident (1.21 μg/L) was significantly lower than that for a rural resident (1.56 μg/L).

Since 1999, NHANES has been conducted continuously with approximately 10,000 respondents from 30 locations in each 2-year data cycle. BPA is measured in 1/3 of the samples (Parker et al., 2018). This study is limited to adults aged 20 and over with complete data on BPA, creatinine, race/ethnicity, gender, age, poverty status, urbanization, education, and cigarette smoking (6608). Summary results for 2003–12 NHANES showed that the estimated median of BPA (μg/L) by gender and race/Hispanic origin, for adults, are as follows: male—non-Hispanic white, 2.00; Mexican American, 2.00; non-Hispanic black, 2.70; females—non-Hispanic white, 1.6; Mexican American, 1.9; non-Hispanic black, 2.8.

In a second study, Calafat and Needham (2008) assessed exposure to BPA in the US general population. Total (free plus conjugated) urinary concentrations of BPA were measured in 2517 participants $\geq$6 years of age in the 2003–04 NHANES study. Urine

TABLE 25.1 Urinary Bisphenol A (BPA) Occurrence in the General Population and in Pregnant Women

Study	Cohort, Location	Years	Age (years)/ Gestation (weeks)	N	Concentration, Creatinine-Adjusted	Statistical Analysis/Comment
Calafat et al. (2008)	NHANES[a] subset, USA	2003–04			LSGM[p]	
			≥6 years	2514	2.6 μg/g	
			6–11 years	314	4.3 μg/g	
			12–19 years	713	2.8 μg/g	
			20–59 years	950	2.4 μg/g	
			≥60 years	537	2.3 μg/g	
				1296 ♀	2.8 μg/g	
				1228 ♂	2.4 μg/g	
CDC (2018a)		2009–10	>6 years	1469	8.0 μg/g	
Ye et al. (2008)	MoBa[b] subset Norway	1999–2008	Pregnant 17–18 weeks	110 of 65000	Mean 5.9 μg/g	Possible external contamination
Braun et al. (2011a)	HOME,[c] Cincinnati, OH, USA	2003–06	Pregnant	389	GM[q]	ICC = 0.10 (for unadj = 0.25) Correlation of log10 BPA
			16 weeks		1.7 μg/g	Week 16 versus 26: R = 0.12 (unadjusted BPA R = 0.28)
			26 weeks		2.0 μg/g	Week 16 versus birth: R = 0.06 (unadjusted BPA R = 0.21)
			Birth		2.0 μg/g	Week 26 versus birth: R = 0.12 (unadjusted BPA R = 0.28)
Casas et al. (2015, 2016), Gascon et al. (2015)	INMA,[d] Sabadell, Spain	2004–06	Pregnant 12 and 32 weeks	462–470	Median 2.4 μg/g	ICC = 0.14 (week 12 vs. week 32)
Braun et al. (2012)	EARTH,[e] Boston, MA, USA	2004–09	Pregnant	137 ≥ 2 samples	GM 1.5 μg/L	ICC = 0.12
Perera et al. (2012)	CCCEH,[f] New York City	1998–2003	Pregnant	361	GM	African American and Dominican
					1.96 μg/L	
			3–4 year old		3.94 μg/L (SG-adjusted)	
Harley et al. (2013)	CHAMACOS,[g] Salinas, CA, USA	1999–2000	Pregnant	292	GM	Week 12 versus 32: R = 0.25
			12 and 32 weeks		1.1 μg/g	
					(SG-adjusted)	
			5 year old		2.5 μg/g	
Jusko et al. (2014)	Gen-R[h] subset, Rotterdam, the Netherlands	2002–05	Pregnant	80 randomly selected from 1024 w/ complete sampling	Median	ICC = 0.32
			All weeks		2.9 μg/g	
			13 weeks		3.0 μg/g	
			20 weeks		3.0 μg/g	
			30 weeks		2.6 μg/g	
Philippat et al. (2014)	EDEN,[i] France	2003–06	Pregnant 23–29 weeks	520	Median 2.4 μg/L	
Lee et al. (2014)	MOCEH[j]	2006	Pregnant trimester 3	757	GM 1.9 μg/g	

Continued

TABLE 25.1 Urinary Bisphenol A (BPA) Occurrence in the General Population and in Pregnant Women—cont'd

Study	Cohort, Location	Years	Age (years)/ Gestation (weeks)	N	Concentration, Creatinine-Adjusted	Statistical Analysis/Comment
Kim et al. (2014)	School pupils, Seoul, South Korea		7–8 years	127	GM 1.02 μg/L	
Evans et al. (2014)	SFF-II,[k] multicenter in NY, USA	2002–05	Pregnant 10–39 years (mean 27)	157	Median 1.6 μg/g Mean 3.7 μg/g	
Arbuckle et al. (2014, 2015)	MIREC,[l] Canada	2008–11	Pregnant Trimester 1	1890	GM	
					0.91 μg/L (SG-adjusted)	
					1% uBPA at median BPA	
					13% uBPA at 95th percentile	
Gerona et al. (2016)	San Francisco, CA, USA	2009–11	Pregnant	112	Median 5.23 μg/g 14% uBPA	
Cantonwine et al. (2015)	Boston, MA and Philadelphia PA, USA	2006–08	Pregnant 10, 18, 26, 35 weeks	1695	GM 1.3–1.4 μg/g	Intersample time correlation Spearman R = 0.17–0.26, $P < .01$ ICC = 0.21
Park et al. (2016)	KoNEHS,[m] South Korea	2009–11	≥19 years	2044	GM 2.01 μg/g	
Health Canada (2017)	CHMS,[n] multicenter in Canada	2014–15		2560	GM	
			3–79 years		0.93 μg/L	
			3–5 years	511	2.0 μg/L	
			6–11 years	511	1.2 μg/L	
			12–19 years	505	0.83 μg/L	
Buckley et al. (2018)	MSCEHS[o]	1998–2002	Pregnant 25–40 weeks	159	Median 1.3 μg/L (regression-adjusted)	

[a]*National Health and Nutrition Examination Survey.*
[b]*Norwegian Mother and Child Cohort Study.*
[c]*Health Outcomes and Measures of the Environment.*
[d]*Infancia y Medio Ambiente (Childhood and Environment).*
[e]*Environment and Reproductive Health Study.*
[f]*Columbia Center for Children's Environmental Health.*
[g]*Center for the Health Assessment of Mothers and Children of Salinas.*
[h]*Generation - R.*
[i]*Etude des Déterminants pré et post natals du développement et de la santé de l'Enfant.*
[j]*Mothers and Children's Environmental Health.*
[k]*Study for Future Families II.*
[l]*Maternal-Infant Research on Environmental Chemicals.*
[m]*Korean National Environmental Health Survey.*
[n]*Canadian Health Measures Survey.*
[o]*Mount Sinai Children's Environmental Health Study.*
[p]*Least squares geometric mean.*
[q]*Geometric mean.*

concentrations of total BPA differed by race/ethnicity, age, gender, and household income. BPA was detected in 92.6% and 57.4% of the persons, respectively. Least square geometric mean (LSGM) concentrations of BPA were significantly lower in Mexican Americans than in non-Hispanic blacks and non-Hispanic whites. LSGM concentrations for non-Hispanic blacks and non-Hispanic whites were not statistically different. Females had statistically significantly higher BPA LSGM (2.8 μg/L) concentrations than males (2.4 μg/L). Children (6–11 years old) had higher concentrations (4.3 μg/L) than adolescents (2.8 μg/L), who in turn had higher concentrations than adults (2.4 μg/L). LSGM concentrations were lowest for participants in the high household income category (>$45,000/year).

Lakind and Naiman (2011) also estimated BPA daily intakes in the United States based on the 2005–06 NHANES urinary data on BPA and its metabolites. Excretion of ingested BPA into urine, mainly as the glucuronide conjugate, is essentially complete in 24 hr. Therefore, total urinary BPA in a 24-hr urine sample approximates the BPA intake from the previous 24 hr. Associations between urinary BPA levels and items from the NHANES questionnaire and examination data that specifically related to packaged food/drink consumption, smoking, and specific medical procedures were evaluated. The median daily BPA intake was approximately 34-ng/kg body wt/day for the overall population, and 54-ng/kg body wt/day for 6–11-year-olds, the youngest group. Intakes for men were statistically significantly higher than for women, and there was a significant decrease in daily BPA intake with increasing age. Estimates of daily BPA intake have decreased compared with those from the 2003–04 NHANES. Consumption of soda, school lunches, and meals prepared outside the home—but not bottled water or canned tuna—were associated with higher urinary BPA.

More recently, Lakind and Naiman (2015) compared the 2003–04 to 2011–12 NHANES to examine temporal trends and assess the resiliency of the previous reported associations. The median daily intake for the overall population was approximately 25 ng/kg day, which represents a gradual decrease in exposure.

Three approaches were assessed for their effect on trends and associations: (1) use of generic literature-based 24-h urine excretion volumes, (2) use of creatinine adjustments, and (3) use of individual urine flow rate data from NHANES. Estimates of associations between lifestyle/demographic/dietary factors and BPA exposure revealed inconsistencies related to both NHANES survey year and the three approaches listed above. The authors noted that these results demonstrate the difficulties in interpreting urinary BPA data, despite efforts to account for urine dilution. Stahlhut et al. (2009) investigated the relationship between urine BPA concentration and fasting time in 1469 adult participants in the 2003–04 NHANES. Because food is the predominant BPA exposure source, and BPA appears to be rapidly excreted, BPA levels in fasting individuals should decrease with increased fasting time. They estimated the BPA half-life for fasting times of <4.5 h, 4.5–8.5 h, and >8.5 h. Results showed that the overall half-life for the 0- to 24-hr interval was 43 h (95% confidence interval [CI], 26–119 h). Among those reporting fasting times of 4.5–8.5 h (n = 441), BPA declined significantly with fasting time, with a half-life of 4.1 h (95% CI, 2.6–10.6 h). However, within the fasting time intervals of 0–4.5 h (n = 129) and 8.5–24 h (n = 899), no appreciable decline was seen. Fasting time did not significantly predict highest (>12 ng/mL) or lowest (below limit of detection) BPA levels. Overall, BPA levels did not decline rapidly with fasting time in this sample. The authors noted that this suggests substantial nonfood exposure, accumulation in body tissues such as fat, or both. Of course, delayed uptake from the gut (due to fat content and differences in gut motility) and diurnal patterns of kidney function are also likely factors, even after exclusion of subjects using laxatives and antidiarrheal medications or drinking coffee, tea, or carbonated beverages or using mints or lozenges during the fasting period.

Nelson et al. (2012) analyzed data from the 2003–06 NHANES using multivariable linear regression to examine the association between urinary concentrations of BPA and multiple socioeconomic measures and race/ethnicity. BPA was inversely associated with family income, higher in people who reported very low food security and received emergency food assistance. This association was particularly strong in children: 6–11-year-olds whose families received emergency food had BPA levels 54% higher (95% CI, 13%–112%) than children of families who did not. Smaller and less consistent associations with education and occupation were observed. Mexican Americans had the lowest BPA concentrations of any racial/ethnic group. Income, education, occupation, and food security appear to capture different aspects of SEP that may be related to exposure to BPA and are not necessarily interchangeable as measures of SEP in environmental epidemiology studies. Differences by race/ethnicity were independent of SEP.

Silver et al. (2011) noted that previous studies using data from the NHANES showed an inconsistent association between prevalence of self-reported type-2 diabetes mellitus (T2DM) and urinary BPA. The authors also used hemoglobinA1c (HbA1c) as a surrogate for T2DM with a larger subset of NHANES. They analyzed data from 4389 adult from three NHANES cycles (2003–08). T2DM was defined as having a HbA1c ≥ 6.5% or use of diabetes medication. Multivariate analysis of the total sample revealed that a twofold increase in urinary BPA was associated with T2DM (OR = 1.08, 95% CI 1.02–1.16). However, when each NHANES cycle was examined individually, a statistically significant association was only found in the 2003/04 cycle, OR = 1.23.

Shankar and Teppala (2011) also examined the association between urinary BPA levels and diabetes mellitus (type 1) using the NHANES 2003–08. Results showed that positive association was observed, independent of confounding factors such as age, gender, race/ethnicity, body mass index (BMI), and serum cholesterol levels. Compared to quartile 1 (referent), the adjusted (age, gender, race/ethnicity, BMI, and serum cholesterol level) OR of diabetes associated with quartile 4 was 1.68 (1.22–2.30) (*P*-trend = .002). Prediabetes (5.7% –6.4% HbA1C or 100–125 mg/dL fasting blood glucose or 140–199 mg/dL 2-h blood glucose) was also associated (Sabanayagam et al., 2013). The adjusted odds ratio (AOR) was 1.34 (95% CI 1.03, 1.7) and was stronger among women and obese subjects.

Canadian Health Measures Survey

The Canadian Health Measures Survey (CHMS) is an ongoing national direct health measures survey (Health Canada, 2017). Statistics Canada, in partnership with Health Canada and the Public Health Agency of Canada, launched the CHMS in 2007 to collect health and wellness data and biological specimens from a nationally representative sample of Canadians. Biological specimens were analyzed for indicators of health status, chronic and infectious diseases, nutritional status, and environmental chemicals.

Four reports for four biennial cycles have been published on Human Biomonitoring of Environmental Chemicals in Canada in 2010, 2013, 2015, and 2017. Collection for cycle 5 began in January 2016 through the end of 2017. Data for Cycle 4 on 54 chemicals were collected from approximately 5700 Canadians aged 3–79 years at 16 sites in 10 provinces across Canada.

Selected findings on BPA concentrations expressed as the GMs of urine concentrations (μg/L) are shown as follows (adapted from Tables 9.1.1 and 9.1.2, Health Canada, 2017):

Total 3–79 years, N-2560/2559, 1.0 μg/L; 0.93 μg/L creatinine-adjusted
M 3–79 years, N-1273/1272, 1.2 μg/L; 0.92 μg/L creatinine-adjusted
F 3–37 years, N-1287, 0.92; 0.94 μg/L creatinine-adjusted
3–5 years, N-511, 1.2; 2.0 μg/L creatinine-adjusted
6–11 years, N-511/510, 1.1; 1.2 uμ/L creatinine-adjusted
12–19 years, N-505, 1.1; 0.83 μg/L creatinine-adjusted
20–39 years, N-362, 1.1; 0.91 μg/L creatinine-adjusted
46–59 years, N-311, 0.86; 0.78 μg/L creatinine-adjusted
60–79 years, N-360, 1.1; 1.0 μg/L creatinine-adjusted.

Generation R Study, The Netherlands

Various studies have reported on the Generation R study in Rotterdam, the Netherlands. The Generation R study biobank is a resource for epidemiological studies in children and their parents (Jaddoe et al., 2007). It is a population-based prospective cohort study in the Netherlands from fetal life until young adulthood. It focuses on four primary areas of research: (1) growth and physical development; (2) behavioral and cognitive development; (3) diseases in childhood; and (4) health and healthcare for pregnant women and children.

There are 9778 mothers enrolled (with a delivery date from April 2002 until January 2006) in the study. Prenatal and postnatal data collection has been conducted by physical examinations, questionnaires, interviews, ultrasound examinations, and biological samples. Major efforts have been conducted for collecting biological specimens including DNA, blood for phenotypes, and urine samples. Together with detailed phenotype measurements, these biological specimens form a unique resource for epidemiological studies focused on environmental exposures, genetic determinants, and their interactions in relation to growth, health, and development from fetal life onward.

Compared to other birth-cohort studies, the size of the Generation R study cohort is not larger, but the measurements are more detailed and the study group has a large ethnic variety.

An update on the Generation R study was provided by Kooijman et al. (2016). Response at baseline was 61%, and general follow-up rates until the age of 10 years were around 80%. Data collection in children and their parents include questionnaires, interviews, detailed physical and ultrasound examinations, behavioral observations, lung function, magnetic resonance imaging, and biological sampling. Genome and epigenome wide association screens are available.

Ye et al. (2008) reported on biological monitoring data from 100 pregnant women, a subset of participants in the Generation R study. The unadjusted and creatinine-adjusted concentrations were reported and compared to NHANES and other studies. The data showed similar exposure to BPA. Among the 100 pregnant women, BPA was detectable among 82% of samples. The creatinine-adjusted median concentration was 1.6 μg/g Cr, and the unadjusted median concentration was 1.2 μg/L. Urinary free BPA accounted for only a small percentage of total BPA.

Philips et al. (2018) examined first trimester bisphenol (BPX) and phthalate urine concentrations, including BPX and phthalate replacements, and found nutritional-, sociodemographic-, and lifestyle-related determinants. The authors measured first trimester BPX, phthalate, and creatinine urine concentrations in samples collected in 2004–05, with a median gestational age of 12.9 weeks in a population-based prospective cohort of 1396 mothers in the Netherlands. They examined associations of potential determinants with log-transformed BPX and phthalate concentrations. Outcomes were back-transformed. Nutritional analyses were performed in a subgroup of 642 Dutch participants only, as the Food Frequency Questionnaire was aimed at Dutch food patterns. Results showed that bisphenol A, bisphenol S (BPS), and bisphenol F (BPF) were detected in 79.2%, 67.8%, and 40.2% of the population, respectively. Bisphenol S and F are produced as alternatives to BPA, see Food Packaging and Bisphenol A Alternatives, below. The authors noted that bisphenol S and F exposure was highly prevalent in pregnant women in the Netherlands as early as 2004–05. Although associations of dietary and other key factors with BPX and phthalate concentrations were limited, adverse lifestyle factors including obesity and the lack of folic acid supplement use seem to be associated, likely as an indirect effect, with higher phthalate concentrations in pregnant women. The major limitation was the availability of only one urine sample per participant. The authors noted that previous studies among pregnant women generally reported higher levels of BPXs and phthalates to be associated with lower socioeconomic status, younger maternal age, and smoking (Arbuckle et al., 2014; 2015; Casas et al., 2013), likely due to food choices, but results are inconsistent (Arbuckle et al., 2014; Berman et al., 2014). Being overweight has also been suggested as a determinant of BPX and phthalate levels (Valvi et al., 2013). Likewise, this association is probably due to an indirect effect on food choices.

Jusko et al. (2014) determined the reproducibility of BPA using three urine specimens collected from 80 pregnant women selected as a random sample of the 1024 women with three urine specimens and complete data. Total maternal BPA concentrations were detectible in all specimens. On average (±SD), maternal urine specimens were collected at 13 (±1.7), 20 (±0.7), and 30 (±0.7) weeks of gestation, and median BPA concentrations were 1.1, 1.5, and 1.6 μg/L (3.0, 3.0, and 2.6 μg/g creatinine) across these three sampling periods. BPA concentrations in the present study were higher than those reported in a previous analysis of a subset of pregnant women in the Generation R study who had only one urine specimen collected after 20 weeks of pregnancy, likely owing to the greater proportion of older, Dutch, and better educated women participating in the present study, factors that have been also been associated with higher BPA concentrations in the Generation R study. In contrast, in the NHANES surveys BPA is higher in women of lower socioeconomic class (Calafat and Needham, 2008). To address the issue of potential contamination, uBPA concentrations in 26 samples having high total BPA were quantified, and all but one sample were found to have uBPA that comprised 1%–2% of the total BPA concentration.

Pair-wise measures of BPA concentrations were moderately correlated ($0.13 \leq$ rspearman$\leq$ 0.49), regardless of whether they were expressed on a volumetric or creatinine-adjusted basis. BPA measurements closer in time were more strongly correlated, although this was not true for the creatinine basis measures. The intraclass correlation coefficient (ICC, the percent of total variance explained by between-person variance) for the three sampling events were 0.32 (95% CI: 0.18, 0.46), and on a creatinine basis, 0.31 (95% CI: 0.16, 0.47), indicating low reproducibility.

Norwegian Mother and Child Cohort Study, Norway

The MoBa study was planned in the 1990s with the objective of testing specific etiological hypotheses by estimating the association between exposures (including genetic factors) and diseases, aiming at prevention and to collect data on as many relevant exposures and health outcomes as feasible (Magnus et al., 2006). It is a prospective birth cohort enrolling pregnant women throughout Norway from 1999 to 2008 (Rønningen et al., 2006). Biological material is to be collected from women in 100,000 pregnancies. In addition, cord blood from their newborn infants and samples from 80,000 fathers may result in more than 380,000 unique sample sets for long-term storage. In addition to biological materials, detailed questionnaire data are collected during pregnancy and as the child ages. Additional health outcome information is available through linkage to Norwegian national health registries. The overall aim is to study the effects of genetic and environmental factors on pregnancy outcomes and later health of the children as well as the parents. The study includes collection of questionnaires and registry information and aims at having DNA for all participants as well as biological samples from pregnant women to assess environmental exposures. The MoBa Biobank is charged with long-term storage of more than 380,000 biological samples from pregnant women, their partners, and their children for up to 100 years. Biological specimens include whole blood, plasma, DNA, and urine; samples are collected at 50 hospitals in Norway. The unique combination of biological specimens, DNA, and questionnaire data makes the MoBa Study a resource for many future investigations of the effects of genetic and environmental factors on pregnancy outcome and on human morbidity.

Ye et al. (2008) measured BPA in 10 pooled urine samples representing 110 pregnant women who participated in the MoBa study in 2004. Daily intakes were estimated from urinary data and compared with RfDs and TDIs. The authors found that MoBa women had a higher mean BPA concentration (4.50 μg/L) than the pregnant women in the Generation R study in the Netherlands and the NHANES in the United States. The higher levels of may have been due to consumption of canned food, especially fish/seafood.

In the MoBa study whereby over 65,000 urine specimens were collected from pregnant women, Longnecker et al. (2013) encountered unexpectedly high concentrations of total and free BPA, with 24% containing more than 10 ng/mL of total BPA, suggesting external contamination. Subsequent investigation of the source of free BPA in the specimens showed that under typical conditions used to collect urine specimens in an epidemiologic study of outpatients, contamination with BPA occurred by two separate mechanisms, one due to the preservative, and the other possibly via dust in clinics.

Other General Population Exposure Studies in Various Countries

Calafat et al. (2005) also presented briefly the urinary BPA levels in various populations in Southeast Asia reported by various researchers: (1) In Japan for 1992, 1999, 2002, and 2003, BPA was measured in urine samples from two different groups of university students in 1992 ($n = 50$, 92% male) and 1999 ($n = 56$, 87.5% male) with most of the BPA found in the urine as the glucuronide conjugate, and the median urinary BPA levels were approximately 2.2-fold higher than in 1999. (2) BPA glucuronide was detected in all of the urine samples collected from 48 female Japanese college students at concentrations ranging from 0.2 to 19.1 ng/mL, with a median level of 1.2 μg/L (0.77 μg/g). (3) In 2004, a median daily urinary excretion of BPA of 1.2 μg/day and an estimated maximum daily intake of BPA to be 0.23 μg/kg/day were reported based on the measurement of BPA in 24-hr urine samples collected from 36 men. This estimated intake was lower than the temporary TDI (10 μg/kg) set by the European Commission's Scientific Committee in Food in 2002. (4) In Korea for nonoccupational exposure assessment, urinary levels of BPA in a group of 73 adults showed a GM concentration of BPA of 9.54 μg/L (8.91 μg/g).

Park et al. (2016) reported that BPA urinary concentrations in Korea may be higher than those in other Asian countries and North America, but lower than or similar to those in European countries. The current study included a total of 2044 eligible adults of all ages and gender. The GM of preadjusted (adjusted) urinary BPA concentrations was 1.83 μg/L (2.01 μg/g) for subjects of all ages, and there was no statistically significant difference in BPA concentrations between males (1.90 μg/L, 1.87 μg/g) and females (1.76 μg/L, 2.16 μg/g).

In Sweden, phthalates and phenolic substances were investigated in urine samples from first-time mothers in Uppsala County, collected between 2009 and 2014 in the POPUP study (Persistent Organic Pollutants in Uppsala Primiparas) conducted by the Swedish National Food Agency (Gyllenhammar et al., 2017). The aim was to investigate if measures to decrease production and use of some of these nonpersistent chemicals have resulted in decreased human exposure, and to determine if exposure to replacement chemicals has increased. A total of 178 women were included in the data set; 30 women were sampled every year between 2009 and 2014 and the participating rate was 52%.

BPA showed a downward trend, with a half-life of 6.6 years, whereas BPF, identified as one of the substitutes for BPA, showed an increasing trend, with a doubling in mean concentrations every 3.8 years. The authors noted that declining temporal trends for BPA has also been seen in the US NHANES study between 2003 and 2012 (Lakind and Naiman, 2015) and also in young Swedish men between 2010 and 2013 (Jönsson et al., 2014). When compared with other studies, the authors indicated that median concentrations of BPA were in the same range as in young Swedish women and about 30% lower compared with Swedish mothers (Jönsson et al., 2014; Larsson et al., 2014). Concentrations of BPA found in the study were also lower compared to recent studies of 684 women in other European countries sampled in 2008 and 2011–12 (Covaci et al., 2015; Guidry et al., 2015).

PREGNANT WOMEN, MOTHER–INFANT/CHILD PAIRS, INFANTS, CHILDREN

Pregnant Women

Summary BPA levels for each study can also be found in Table 25.1.

Braun et al. (2011a) estimated the variability and predictors of serial urinary BPA concentrations taken during pregnancy (around 16 and 26 weeks of gestation) and at birth. Data were collected from pregnant women participating in the Health Outcomes and Measures of the Environment (HOME) Study, an ongoing prospective birth cohort in the Cincinnati, Ohio, metropolitan area designed to examine low-level environmental toxicant exposure. Beginning in March 2003, through 2006, women were identified from seven prenatal clinics associated with three hospitals. Eligibility criteria for enrollment in the study included ≤19 weeks of gestation, age ≥18 years, living in a house built before 1978, negative HIV status, and not taking medications for seizure or thyroid disorders. Letters were mailed to 5184 women who were ≥18 years of age and living in a house built before 1978. Of the 1263 eligible responding women, 468 enrolled in the study (37%), but 67 dropped out before delivery, and 3 had stillbirths. This analysis was further restricted to 389 mothers who delivered singleton children between March 2003 and January 2006.

Women provided three spot urine samples collected at approximately 16 and 26 weeks of gestation at prenatal care appointments and at birth. Results showed that the GM creatinine-standardized concentrations (micrograms per gram, μg/g) were 1.7 (16 weeks), 2.0 (26 weeks), and 2.0 (birth).

Log10-transformed urinary BPA concentrations were weakly correlated at 16 and 26 weeks ($r = 0.28$), 26 weeks and birth ($r = 0.28$), and 16 weeks and birth ($r = 0.21$). Correlations between creatinine-standardized urinary BPA concentrations were even lower at 16 and 26 weeks ($r = 0.12$), 26 weeks and birth ($r = 0.12$), and 16 weeks and birth ($r = 0.06$).

The ICC for serial urinary BPA measurements indicated poor reproducibility in analyses using unstandardized (ICC = 0.25) and creatinine-standardized concentrations (ICC = 0.10). Creatinine-standardized BPA concentrations exhibited low reproducibility (ICC = 0.11). The ICC was used to assess BPA variability and estimated associations between log10-transformed urinary BPA concentrations and demographic, occupational, dietary, and environmental factors, using mixed models. Only maternal education and occupation was associated. By occupation, cashiers had the highest BPA concentrations (GM: 2.8 μg/g) and serum cotinine, a tobacco smoke metabolite, was associated with higher BPA concentrations. In children almost 99% of BPA exposure was estimated to come from dietary sources. Dietary findings are further discussed below, in the Food packaging and BPA alternatives section. The authors noted that future studies should standardize or adjust for the timing of urine collection or measure BPA at multiple times to minimize biases due to within-day and within-woman variability of urinary BPA and creatinine concentrations.

Braun et al. (2012) also characterized the variability of urinary phthalate and BPA concentrations before and during pregnancy. No prior study had examined the same woman before and during pregnancy. Women 18–45 years of age were recruited from clinical partners seeking evaluation and treatment for infertility at the Massachusetts General Hospital Fertility Center in Boston between November 2004 and December 2009 for a prospective open-cohort study, the Environment and Reproductive Health (EARTH) Study. For BPA analyses, 137 women provided two or more urine samples before and during pregnancy. They were predominately white, highly educated, and had a mean (±SD) of 35 ± 4.1 years of age at enrollment. Before pregnancy, women provided between 2 and 13 urine samples (median, 3 samples) during a period of <1–110 weeks after enrollment (median, 12 weeks). During pregnancy, a median of three samples was collected. Results showed that urinary BPA concentrations were variable before and during pregnancy, ICC = 0.12, which is similar to observations in studies with multiple samples during pregnancy.

In the EDEN mother–child cohort (Etude des Déterminants pré et post natals du développement et de la santé de l'Enfant) recruited in Nancy and Poitiers, France, a subgroup of 587 pregnant women were tested once for urinary BPA and phthalates between the 23rd and 29th weeks of gestation during 2003–06 (Philippat et al., 2014). Median BPA adjusted by statistical regression for creatinine, as well as day and time of sampling, gestational age, duration of storage at room temperature, and year of analysis, was 2.4 μg/L.

Occupational and nonoccupational contact with paper receipts was positively associated with BPA in conjugated (glucuronidated + sulfated) form after adjustment for demographic characteristics. The author noted that recent consumption of foods and beverages likely to be contaminated with BPA was infrequent among participants.

Arbuckle et al. (2014, 2015) reported on free and conjugated forms of BPA during pregnancy in about 1890 first-trimester urine samples from the Maternal-Infant Research on Environmental Chemicals (MIREC) Study, a large multicenter cohort study of Canadian pregnant women. Study participants tended to be well educated, > 30 years of age, born in Canada, had never smoked, had an underweight or normal BMI prior to pregnancy, and had a parity of 0 or 1. Results showed that the glucuronides of BPA were the predominant forms measured (detected in 95% of samples), whereas the free forms were detected in 43% of samples. The GM of SG-adjusted urinary BPA glucuronide was 0.91 μg/L. On average, only about 1% of the total BPA was present in the unconjugated form. However, at the 95th percentile level of total BPA the percentage of uBPA versus total BPA was 12%. Significant predictors of BPA included maternal age <25, current smoking, low household income, and low education level. The results from this study represent the largest nation-level data on urinary concentrations of free and conjugated forms of BPA in pregnant women.

Teeguarden et al. (2016) evaluated 30 healthy pregnant women for total BPA exposure over a 30-h period comprising one-half day in the field and 1 day in a clinical setting. BPA and its metabolites were measured in serum and total BPA was measured in matching urine samples, including sampling outside the clinical setting. It was also intended to identify potential explanations for previously reported and anomalously high serum BPA concentrations.

The 30 pregnant volunteers were recruited for the study from the Salt Lake City, UT metropolitan area in 2014. Eligibility included no tobacco–alcohol–drug use, 18–40 (mean 26.5) years of age, and being at 15–35 (mean 23.7) weeks of a normal, low-risk pregnancy, with normal GI tract, kidney, and liver function, as determined by clinical laboratory tests, medical history, and physical examination. Volunteers were not taking medications that alter hepatic glucuronidation/sulfonation or renal elimination, were HIV- and hepatitis-free, and had not undergone dental procedures in the 2 days preceding the study. Two cohorts were recruited from a pool of 116 candidates. A 20-member BPA cohort with the highest potential for exposure to BPA was selected based on the screening survey to assess exposure to BPA by ingestion of canned foods and handling of cash register receipts.

Serum BPA levels were assessed in samples with a high corresponding urinary BPA result. The mean total exposure was similar to that in NHANES. Women working as cashiers did not have higher total BPA exposure. BPA was detected in some serum samples (0.25–0.51 ng/mL) but showed no relationship to total BPA in corresponding urine samples, no relationship to total BPA exposure, and had uBPA fractions of 60%–80%, consistent with established criteria for sample contamination. The authors concluded that typical exposures of North American pregnant women produce internal exposures to BPA in the picomolar range. They also suggested that previously published human biomonitoring studies of BPA in pregnant women that reported serum uBPA/total BPA ratios much greater than 15% are probably the result of contamination during sampling, and the reported concentrations appear unrelated to typical exposures in the general population of pregnant women. However, the findings, below, of Veiga-Lopez et al. (2015), following the round robin methods discussed by Vandenberg et al. (2014a), partly contradict this conclusion.

Mother–Infant/Child Pairs

The study of mother–infant/child pairs enables the prospective type of epidemiological study design. Prospective studies are a superior design because the exposure metrics are established before the outcome is measured. This helps to avoid misclassification of exposure and some aspects of confounding. The studies mostly examined fetal growth, neurodevelopment, and thyroid and respiratory outcomes, summarized in Tables 25.2–25.5.

A large number of studies has been published from the prospective HOME birth cohort (see above, in the section on studies of Pregnant Women), which measured a large number of chemicals. Spanier et al. (2012) compared BPA concentrations in serial maternal urine samples of 398 mother–infant pairs and assessed parent-reported child wheeze every 6 months for 3 years. Results showed that for 365 children for which both BPA and respiratory data were available, BPA was detected in 99% of maternal urine samples during pregnancy (16 and 26 weeks of gestation), with a mean prenatal creatinine-adjusted GM = 2.4 μg/g.

In multivariate analysis, a 1-unit increase in log-transformed creatinine-standardized mean prenatal BPA concentration was not significantly associated with child wheeze from birth to 3 years of age. Potential confounder data included socioeconomic factors, race/ethnicity, season, familial and child history of allergy, pet ownership, insect exposure, urban/suburban/rural residence, and maternal cotinine, a surrogate for tobacco use. Mean prenatal BPA above versus below the median was positively associated with wheeze at age 6 months and AOR of 2.3 (95% CI: 1.3, 4.1), but not at later (12, 18, 24, 30, and 36 months) time points. In secondary analyses evaluating associations of early and late prenatal BPA concentration separately, urinary BPA measured at 16 weeks gestation were associated with wheeze (AOR 1.2, 95% CI 0.99, 1.5), but not BPA from 26 weeks of gestation or at birth. In this regard it is of interest that BPA levels in the two gestational samples were only weakly correlated ($r = 0.18$, $P = .03$). Discussion presented below (Donohue et al., 2013) suggests that urinary levels may not be as representative at later gestational time points of BPA because of increasing retention in the fetoplacental unit as gestation nears term.

A follow-up study (Spanier et al., 2014) tracked the children through age 5 years, adding spirometry data for forced-expiratory-volume-in-1-second (FEV1) from testing at ages 4 (N = 155) and 5 (N = 193). There was a combined total of 208. Mean maternal BPA was associated with FEV1 at age 4 ($\beta = -14$, 95% CI, −24, −4, $P = .007$), but not at age 5. Child BPA at 4 and 5 was not associated with FEV1. The degree of association was similar with the 16-week and the 26-week gestational samples.

Yolton et al. (2011) examined the association of prenatal exposure to BPA in the HOME study with data from a Neonatal Intensive Care Unit Network Neurobehavioral

TABLE 25.2 Association of Gestational Urinary Bisphenol A (BPA) With Human Fetal Growth, as Measured at Birth

Study/Cohort	Design	Sampling Timeline	N	Test/Effect	Effect Difference
Snijder et al. (2013) Generation R, Netherlands	Maternal-child pairs (prospective-longitudinal)	Each trimester 2002–06	80 with all 3 samples	Fetal growth at birth >4.2 versus < 1.54 μg/g	
				Head circumference	−3.9 cm (−2.6 SD)
				Birth wt	−683 g (1.7 SD)
Philippat et al. (2012) EDEN,[a] France	Maternal-child pairs (prospective-longitudinal)	22–29 weeks gestation 2003–06	560	Fetal growth	NS
Veiga-Lopez et al. (2015) U Michigan Hospital, Ann Arbor, MI, USA	Maternal-child pairs (prospective-longitudinal)	First and third trimester **blood sampling**	80	Birth weight First trimester plasma BPA	
				Female pregnancy	−183 g/twofold uBPA[c] ($P < .005$) −607 g/10-fold uBPA ($P = .0093$)
				Male pregnancy	NS
				Third trimester Plasma BPA	
				Female pregnancy	96 g/twofold uBPA ($P < .05$) 318 g/10-fold uBPA ($P = .034$)
				Male pregnancy	NS
				Gestation time Third trimester plasma BPA	
				Female pregnancy	1.1 d/2-fold BPA
				Male pregnancy	NS
Casas et al. (2016) INMA-Sabadell,[b] Spain	Maternal-child pairs (prospective-longitudinal)	12, 32 weeks gestation 2004–06	470	Fetal growth, by ultrasound	NS
				Birth weight	
				Boys	−21.6 g/μg/g BPA (NS)
				Girls	−11.0 g/μg/g BPA (NS)

NS = not statistically significant; SD = standard deviation
[a]*Etude des Déterminants pré et post natals du développement et de la santé de l'Enfant.*
[b]*Infancia y Medio Ambiente.*
[c]*unconjugated BPA.*

Scale, assessed during a postnatal visit at 5 weeks old. They did not find evidence of an association among the 350 mother/infant pairs. Braun et al. (2009) examined the association between prenatal BPA exposure and behavior in 2-year-old children and investigated whether gender modified the association. Study data were obtained from the HOME Study cohort. Child behavior was assessed using the second edition of the Behavioral Assessment System for Children-2 (BASC-2). The association between prenatal BPA concentrations and BASC-2 scores was analyzed using linear regression. Median BPA concentrations among the 249 mothers included in this first set of analyses were 1.8 (16 weeks), 1.7 (26 weeks), and 1.3 (birth) ng/mL. Mean (±SD) BASC-2 externalizing and internalizing scores (see next paragraph) were 47.6 ± 7.8 and 44.8 ± 7.0, respectively. After adjustment for confounders (maternal age, race, education, mental health, household income, marital status, and a care-giving score), log10-transformed mean prenatal BPA concentrations were associated with externalizing scores, but only among females ($\beta = 6.0$; 95% CI, 0.1–12.0). In contrast, the trend for boys was inverse. Notably, the statistically nonsignificant trends for internalizing behavior

TABLE 25.3 Association of Gestational Urinary Bisphenol A (BPA) With Human Neonatal Thyroid Function

Study/Cohort	Design	Sampling Timeline	N/Age	Test/Effect	Effect Difference (95% CI)
Chevrier et al. (2013) CHAMACOS[a] Salinas, CA, USA	Maternal-child pairs (prospective-longitudinal)	12, 26 week gestation 1999–2000	476	TSH[c]	
				Boys, 12 + 26 week BPA	−9.9%/log2 (−16%, −3.5%)
				Boys, 12 week BPA	NS
				Boys, 26 week BPA	−9.3%/log2 (−16%, −1.7%)
				Girls	NS
				Other thyroid hormones	NS
Romano et al. (2015) HOME,[b] Cincinnati OH, USA	Maternal-child pairs (prospective-longitudinal)	16, 26 week gestation 2003–06	181	TSH	
				Boys	NS
				Girls, 26 week BPA	−43%/10-fold ↑ (−60%, −18%)
				Girls, 16 week BPA	NS/10-fold ↑
				Triiodothyronine	
				Girls, 16 + 26 week BPA	13%/10-fold ↑
				Other thyroid hormones	NS
Aung et al. (2017) Boston, MA and Philadelphia, PA	Maternal-child pairs (prospective-longitudinal)	10, 18, 26, 35 week gestation	116 preterm neonates	TSH	−8.2%/interquartile range
		2006–08	323 term neonates		

NS = not statistically significant; CI = confidence interval
[a]*Center for the Health Assessment of Mothers and Children of Salinas.*
[b]*Health Outcomes and Measures of the Environment.*
[c]*TSH.*

exhibited a similar pattern—with higher maternal BPA girls had a higher score and boys had a lower score. Compared with 26-week and birth concentrations, BPA concentrations collected around 16 weeks were more strongly associated with externalizing scores among all children ($\beta = 2.9$; 95% CI, 0.2–5.7), and this association was stronger in females ($\beta = 4.8$; 95% CI, 1.3–8.3) than in males. Among all children, measurements collected at $\leq$ 16 weeks showed a stronger association ($\beta = 5.1$; 95% CI, 1.5–8.6) with externalizing scores than did measurements taken at 17–21 weeks ($\beta = 0.6$; 95% CI, −2.9–4.1).

Externalizing and internalizing behaviors are problem behaviors, summarized as composite scores. The former includes outwardly directed behaviors, such as fighting, disobedience, stealing, and vandalism. The latter behaviors are focused inward, such as fear and withdrawal. A positive score means more problems. Braun et al. (2011b) continued their follow-up of the HOME study by examining behavior and executive function of 3-year-old children of 244 enrolled mothers. In addition to the three maternal spot urine samples, children's spot urine samples were collected at 1, 2, and 3 years of age during clinic or home visits, between 2004 and 2009. For statistical analysis, the authors characterized gestational and childhood BPA exposures by using the mean BPA concentrations of two urine samples from each mother and child. Behavior and executive function were measured by using the BASC-2 and the Behavior Rating Inventory of Executive Function-Preschool (BRIEF-P). BPA was detected in >97% of the gestational (median 2.0 μg/L) and childhood (median 4.1 μg/L) urine samples. Overall correlation of each individual's paired samples (i.e., early vs. later gestation, maternal vs. child, and the child's sequential samples) was weak: a Pearson R statistic < 0.25 or with creatinine-adjusted concentrations, a Pearson $R < 0.18$.

With adjustment for confounders, each 10-fold increase in gestational BPA concentrations was associated with more anxious and depressed behavior on the BASC-2 and poorer emotional control and inhibition on the BRIEF-P. The magnitude of the gestational BPA associations differed according to child gender; BASC-2 and BRIEF-P scores increased from 9 to 12 points among girls, but changes were null or negative among boys. Hyperactivity among girls was increased, $\beta = 9.1$

TABLE 25.4 Association of Urinary Bisphenol A (BPA) With Human Neurobehavioral Development

Study/Cohort	Design	Sampling Timeline	N/Age	Test/Effect	Slope or Rate Ratio (95% CI)
Braun et al. (2009) HOME[a] Study Cincinnati, OH, USA	Maternal-child pairs (prospective-longitudinal)	16, 26 week gestation 2004–06	249 2 year	*BASC-2*[h]	per log10 ↑ maternal BPA
				Externalizing behavior	β = 2.9 (0.2, 5.7)
				(≤16 week BPA only)	β = 5.1 (1.5, 8.6)
				Internalizing behavior	NS
Braun et al. (2011b) HOME Study Cincinnati, OH, USA	Maternal-child pairs (prospective-longitudinal)	16, 26 week gestation 2004–06 1,2,3 year 2004–09	244 3 year	*BASC-2*	per 10-fold ↑ maternal BPA
				Internalizing behavior	Primarily girls
				Hyperactivity, girls	β = 9.1 (3.1, 15); child BPA was NS
				BRIEF-P[i]	per 10-fold ↑ maternal BPA
				Emotional Control, girls	β = 9.1 (2.8, 15); child BPA was NS
Stacy et al. (2017) HOME Study Cincinnati, OH, USA	Maternal-child pairs (prospective-longitudinal)	16, 26 week gestation 2004–06 1–8 year 2004–15	228 8 year	*BASC-2*	per 10-fold ↑ maternal BPA
				Externalizing behavior	
				Boys	NS
				Girls	β = 6.2 (0.8, 11.6)/10-fold ↑ maternal BPA
				Boys	β = 3.9 (0.6, 7.2)/10-fold ↑ 8 year BPA
				Girls	NS
Harley et al. (2013) CHAMACOS[b] Salinas, CA, USA	Maternal-child pairs (prospective-longitudinal)	14, 26 week gestation 1999–2000	292 7–9 year	*BASC-2*	Continuous variable analysis
				Externalizing behavior	NS
		5 year 2004–09			
				Internalizing behavior	NS
Casas et al. (2015) INMA-Sabadell[c] Spain	Maternal-child pairs (prospective-longitudinal)	12, 32 weeks gestation 2004–06	382 1 year 365 4 year 361 7 year	Psychomotor	Third tertile versus First
				1 year	β = −4.3 (−8.1, −0.41)
				4 year	NS
				Hyperactivity	IRR[n] = 1.7 (1.1, 2.7) for 10-fold ↑ maternal BPA
				4 year	Primarily boys
				7 year	NS
Perera et al. (2012) CCCEH[d] New York City	Maternal-child pairs (prospective-longitudinal)	34 week gestation; 3–4 year 1998–2003	198 3–5 year	*CBCL*[j]	4th quartile versus lower quartiles
				Emotionally Reactive	
				Boys	RR[m] = 1.6 (1.1, 1.5)
				Anxious/Depressed	
				Girls	RR = 0.75 (0.57, 0.99)
				Aggressive Behavior	
				Boys	RR = 1.3 (1.1, 1.5)
				Girls	RR = 0.82 (0.70, 0.97)

Continued

TABLE 25.4 Association of Urinary Bisphenol A (BPA) With Human Neurobehavioral Development—cont'd

Study/Cohort	Design	Sampling Timeline	N/Age	Test/Effect	Slope or Rate Ratio (95% CI)
Evans et al. (2014) SFFII[e] New York multicenter	Maternal-child pairs (prospective-longitudinal)	10–39 week (mean = 27) gestation 2002–05	125 pairs with detected maternal BPA 6–10 year	*CBCL*	per natural log BPA
				Externalizing behavior	
				Boys	β = 0.35, *P* = .05
				Girls	NS
				Internalizing behavior	
				Boys	β = 0.45, *P* = .02
				Girls	NS
Philippat et al. (2017) EDEN,[f] France	Maternal-child pairs (prospective-longitudinal)	23–29 week gestation 2003–06	464 5–6 year	Hyperactivity - inattention	IRR = 1.08 (1.01, 1.14)
Nakiwala et al. (2018) EDEN,[f] France	Maternal-child pairs (prospective-longitudinal)	23–29 week gestation 2003–06	452 5–6 year	IQ (WPPSI-3[k])	NS (continuous variable analysis)
Tewar et al. (2016) 2003–04 NHANES[g]	Cross-sectional	8–15 year 2003–04	460 8–15 year	ADHD[l]	> versus < median BPA
				All children	AOR[o] = 5.7 (1.4, 86)
				Boys	AOR = 11 (1.4,20)
				Girls	AOR = 2.8 (0.4, 21)
Li et al. (2018) Guangzhou, China	case-control	6–12 year old not stated	465 6–12 year	ADHD boys and girls	
				quartile 2 versus 1	AOR = 1.8 (0.95, 3.4)
				quartile 3 versus 1	AOR = 7.4 (3.9, 14)
				quartile 4 versus 1	AOR = 9.4 (4.9, 18)
				quartiles 2, 3, 4 versus 1	
				Boys	AORs = 1.8, 9.1, 10
				Girls	AORs = 0.96, 3.4, 4.6

NS = not statistically significant.
[a]*Health Outcomes and Measures of the Environment.*
[b]*Center for the Health Assessment of Mothers and Children of Salinas.*
[c]*Infancia y Medio Ambiente.*
[d]*Columbia Center for Children's Environmental Health.*
[e]*Study for Future Families II.*
[f]*Etude des Déterminants pré et post natals du développement et de la santé de l'Enfant.*
[g]*National Health and Nutrition Examination Survey.*
[h]*Behavior Assessment System for Children, second Edition.*
[i]*Behavior Rating Inventory of Executive Function-Preschool.*
[j]*Child Behavior Checklist.*
[k]*Wechsler Preschool and Primary Scale of Intelligence.*
[l]*Attention-Deficit/Hyperactivity Disorder.*
[m]*Rate Ratio.*
[n]*Interquartile Rate Ratio.*
[o]*Adjusted Odds Ratio.*

(95% CI: 3.1, 15), but was decreased among boys, β = −6.3 (95% CI: −12, −0.6). Anxiety and depression were also associated with BPA among girls. Associations between childhood BPA exposure and neurobehavior were largely null and not modified by child gender. Overall, gestational BPA exposure affected behavioral and emotional regulation domains (anxiety, hyperactivity, emotional control, and behavioral inhibition) at 3 years of age, especially among girls, in a manner consistent with their previous findings.

A follow-up study of 228 mother–child pairs from the HOME Study (Stacy et al., 2016) found that each

TABLE 25.5 Association of Urinary Bisphenol A (BPA) With Human Respiratory Outcomes

Study/Cohort	Design	Sampling Timeline	N/Age	Test/Effect	Slope or Rate Ratio (95% CI)
Spanier et al. (2012) HOME[a] Study Cincinnati, OH, USA	Maternal-child pairs (prospective-longitudinal)	16, 26 week gestation 2004–06	398 every 6 mo 6 mo- 3 year	Mean prenatal BPA versus 6 mo wheeze, > versus < median BPA 16 week gestation BPA versus wheeze at 6 mo - 3 year, per log10 BPA 26 week gestation BPA versus wheeze at 6 mo - 3 year, per log10 BPA	AOR[g] = 2.3 (1.3, 4.1) AOR = 1.2 (1.0, 1.5) NS
Spanier et al. (2014) HOME Study Cincinnati, OH, USA	Maternal-child pairs (prospective-longitudinal)	16, 26 week gestation 2004–06; 1,2,3 year 2004–09	155 At 4 year 193 At 5 year	Mean prenatal BPA	
				FEV1 at 4 year, per 10-fold BPA	β = −14 (−24, −4)
				FEV1 at 5 year, per 10-fold BPA	NS
				Child BPA at 4 and 5 year	
				FEV1 at 4 and 5 year, per 10-fold BPA	NS
Donohue et al. (2013) CCCEH[b] New York City	Maternal-child pairs (prospective-longitudinal)	34 week gestation 1998–2006 3–7 year 2001–10	368–429 at 5 –7 year 474 at 5–7 year	Mean 3–7 year BPA	
				Wheeze, per log10 BPA	
				at 5 year	AOR = 1.5 (1.1, 2.0)
				at 6 year	AOR = 1.4 (1.0, 1.9)
				at 7 year	AOR = 1.4 (1.0, 2.0)
				Asthma, per log10 BPA	AOR = 1.6 (1.2, 2.1)
				prenatal BPA	
				Wheeze, by continuous BPA variable	
				at 5 year	AOR = 0.7 (0.5, 0.9)
				at 6,7 year	NS
				Asthma	NS
Kim et al. (2014) Korean elementary school	Longitudinal	7–8 year	127 7–12 year	Wheeze, per log10 increase of BPA	AOR = 2.5 (1.15,5.3)
				Asthma, per log10 increase of BPA	AOR = 2.1 (1.5, 3.0)
				Girls	AOR = 2.4 (2.2, 2.8)
Gascon et al. (2015) INMA-Sabadell[c] Spain	Maternal-child pairs (prospective-longitudinal)	12, 32 weeks gestation 2004–06	437-445 at 6 mo 385 at 4 year 361 at 7 year	Mean prenatal BPA	
				Wheeze, per 2-fold BPA	RR[h] = 1.2 (1.03, 1.4)
				Chest infection, per 2-fold BPA	RR = 1.15 (1.00, 1.3)
				Bronchitis, per 2-fold BPA	RR = 1.18 (1.01, 1.4)
				Asthma (tested only at 7 year), per 2-fold BPA	RR = 1.2 (0.95, 1.6)

Continued

TABLE 25.5 Association of Urinary Bisphenol A (BPA) With Human Respiratory Outcomes—cont'd

Study/Cohort	Design	Sampling Timeline	N/Age	Test/Effect	Slope or Rate Ratio (95% CI)
Vernet et al. (2017) EDEN,[d] France	Maternal-child pairs (prospective-longitudinal)	23–29 week gestation 2003–06	587	Wheeze, Bronchiitis, FEV1[f]	NS
				Asthma, per log10 BPA	HR[i] = 1.2 (0.97, 1.6)
Buckley et al. (2018) MSCEHS[e]	Maternal-child pairs (prospective-longitudinal)	25–40 week gestation 1998–2002	159 6–7 year	Asthma, per standard deviation of natural log BPA	
				Boys	AOR = 3.0 (1.4, 6.6)
				Girls	NS

NS, not statistically significant.
[a]*Health Outcomes and Measures of the Environment.*
[b]*Columbia Center for Children's Environmental Health.*
[c]*Infancia y Medio Ambiente.*
[d]*Etude des Déterminants pré et post natals du développement et de la santé de l'Enfant.*
[e]*Mount Sinai Children's Environmental Health Study.*
[f]*Spirometry data for forced expiratory volume in 1 s.*
[g]*Adjusted Odds Ratio.*
[h]*Rate Ratio.*
[i]*Hazard Ratio.*

10-fold increase in prenatal BPA was associated with more externalizing behaviors (see below) in 8-year-old girls ($\beta = 6.2$, 95% CI: 0.8, 11.6) on testing with the BASC-2, but not among boys. In contrast, externalizing behaviors in boys was associated with urinary BPA level at age 8 ($\beta = 3.9$, 95% CI: 0.6, 7.2), but not in girls.

Stacy et al. (2017) examined the patterns, variability, and predictors of urinary BPA concentrations in 337 mothers and children from the HOME Study. From 2003 to 2014 six urine samples were collected from children between 1 and 5 and at 8 years of age. Urinary BPA concentrations had a low degree of reproducibility (ICC < 0.2). Estimated daily intakes (EDIs) decreased with age. BPA concentrations were linked to consuming food stored or heated in plastic, consuming canned food and beverages, and handling cash register receipts. EDIs decreased as child age increased, with median intakes of 190 and 39 ng/kg/day for 1- and 8-year-olds, respectively.

Romano et al. (2015) observed that BPA in the 26th gestational week was associated with a 43% decrease of thyroid stimulating hormone (TSH) in the cord blood of female, but not male, neonates. BPA in the 16th gestational week was not associated with neonatal TSH. There was no effect on thyroxine (T4), but there was a marginal statistically significant association with triiodothyronine (T3) in girls. Chemical covariates unique to this study's analysis included iodine intake, polychlorinated biphenyls, and hexachlorobenzene.

The INMA study (Infancia y Medio Ambiente; Childhood and Environment) is a population-based birth cohort study that recruited 657 pregnant women in the Spanish region of Sabadell between 2004 and 2006 (Guxens et al., 2012). Pregnant women were recruited at their first routine prenatal care visit. Maternal urine samples were collected at 12 ± 1.7 and 32 ± 1.4 weeks of gestation. Total BPA was determined in 470 mother–child pairs. Data available for multivariate analysis included maternal age, socioeconomics, weight and height, parity, marital status, dietary information, use of household cleaning products, tobacco use, and season of birth, as well as first trimester blood mercury, total polychlorobiphenyls, and the DDT metabolite, DDE.

For the 462 children included in Gascon et al. (2015), reported below, the median maternal BPA was 2.4–2.5 μg/g and the interquartile level was 1.7–3.7 μg/g. The ICC for the two maternal samples was 0.15 for unadjusted BPA and 0.14 for adjusted BPA (Casas et al., 2016), indicating low reproducibility. The GM for the 37 pregnant mothers <25 years of age was 3.3 μg/g, whereas for the 83 pregnant mothers ≥35 years of age the GM was 2.2 μg/g.

Respiratory parameters in the children of the INMA-Sabadell cohort were studied by Gascon et al. (2015). Children were tested at 6 months (N = 437–445), 14 months (N = 424), 4 years (N = 385), and 7 years (N = 361). For each doubling in concentration of mean maternal urinary BPA the adjusted relative risks (RRs) over all ages combined increased for wheeze (RR, 1.20; 95% CI, 1.03–1.40; P = .02), chest infections (RR, 1.15; 95% CI, 1.00–1.32; P = .05), and bronchitis (RR, 1.18; 95% CI, 1.01–1.37; P = .04). Asthma, tested only at age 7, showed an association but not with statistical significance (RR, 1.21 (0.94–1.57; P = 0.14). These were slightly lower in a model that adjusted for other pollutants. The associations with early and late gestation BPA have shown different degrees of effect.

Casas et al. (2016) estimated growth curves for femur length, head circumference, abdominal circumference, and biparietal diameter and estimated fetal weight during pregnancy (weeks 12–20 and 20–34), and for birth weight, birth length, head circumference at birth, and placental weight in 470 births. The median maternal BPA level was 2.3 μg/g. This study, one of the first to combine repeat exposure biomarker measurements and multiple growth measures during pregnancy, found little statistically significant evidence of an effect on fetal growth. However, the trend for BPA was for lower birth weight (−21.6 g per μg/g BPA for girls and −11.0 g per μg/g for boys). Notably, the high intraindividual variability of the measurements meant likely exposure misclassification, probably random, thereby reducing the degree of observable linkage.

At age 4 (Valvi et al., 2013), maternal BPA was associated with increased waist circumference (β per log 10 μg/g = 0.28, 95% CI, 0.01, 0.57) and, with marginal statistical significance, BMI (β = 0.28, 95% CI, −0.06, 0.63).

Neurobehavioral testing (Casas et al., 2015) revealed that (1) psychomotor scores at age 1 were reduced in the highest tertile of creatinine-adjusted prenatal urinary BPA, but not at age 4, and (2) for each log10 μg/g of prenatal exposure there was increased incidence of hyperactivity, as reported by teachers using the Diagnostic and Statistical Manual of Mental Disorders-IV tool for Attention-Deficit/Hyperactivity Disorder (ADHD), when the children were 4 years old. The adjusted incidence rate ratio (IRR) for hyperactivity was 1.72 (95% CI: 1.08, 2.73) and was stronger in boys, 1.9 (95% CI; 1.01, 3.7), than in girls, 1.6 (95% CI: 0.79, 3.2). The IRR for inattention was elevated in boys (IRR = 1.4) but was not statistically significant. However, the IRR for second tertile versus the first tertile showed a statistically significant elevation, 1.6 (95% CI: 1.01, 2.4). When the children were tested again at the age of 7, there was no association of prenatal BPA with hyperactivity, as measured by parents with the Conner's Parent Rating Scales, nor were there links with externalizing and internalizing problems (Strengths and Difficulties Questionnaire). There were no significant associations with cognitive scores or their subscales. However, it must be noted that, ADHD is a complex disease and still not well understood, so there may be other important environmental chemicals (and other confounders) that were not included in the analysis. The relationships between early and late pregnancy samples and neurobehavioral testing were not analyzed separately.

Perera et al. (2012) examined the association between prenatal BPA exposure and child behavior, adjusting for postnatal BPA exposure and hypothesizing gender-specific effects. Participants in this study were mothers and their children in the Columbia Center for Children's Environmental Health (CCCEH) low income cohort, predominantly African-American and Dominican women, in New York City. Enrollment in 1998 was limited to women who were in the age range of 18–35 years, nonusers of tobacco products or illicit drugs, free of diabetes, hypertension, or known HIV, and who initiated prenatal care by the 20th week of pregnancy. Urine samples were collected at the 34th week of gestation on average and from children between 3 and 4 years of age.

Out of 337 mother–child pairs with prenatal BPA measurements and maternal prenatal questionnaire data, BPA was detectable in the urine of >90% of both mothers and children. The ranges of total SG-adjusted BPA in maternal and child urine samples were 0.24–38.53 μg/L and 0.42–73.50 μg/L, respectively, indicating a wide variation in exposure. The GMs for maternal and child urinary concentrations of BPA were 1.96 and 3.94 μg/L, respectively. No significant differences were observed between boys and girls in median and mean prenatal or postnatal BPA concentrations, whether SG-adjusted or not.

The authors assessed behavior of children between 3 and 5 years of age using the Child Behavior Checklist (CBCL) in 198 of these mother–child pairs (87 boys, 111 girls). Among boys, high prenatal BPA exposure (highest quartile vs. the three lower quartiles) was associated with significantly higher CBCL scores (more problems) on Emotionally Reactive (1.62 times greater; 95% CI: 1.13, 2.32) and Aggressive Behavior scales (1.29 times greater; 95% CI: 1.09, 1.53). Among girls, higher exposure was associated with lower scores on all syndromes, reaching statistical significance for Anxious/Depressed (0.75 times as high; 95% CI: 0.57, 0.99) and Aggressive Behavior (0.82 times as high; 95% CI: 0.70, 0.97) scales. These results suggest that prenatal exposure to BPA may affect child behavior, and differently among boys and girls. Among boys they found significant positive associations between prenatal BPA and CBCL scores of Emotionally Reactive and Aggressive Behavior. In contrast, among girls prenatal BPA was associated with significantly lower scores for the Aggressive Behavior and Anxious/Depressed syndromes. The authors noted that these findings are inconsistent with reports by Braun et al. (2009, 2011b), discussed above, who reported evidence of adverse effects predominantly in girls. The different ethnicity and socioeconomic status levels of the two sets of subjects provide a somewhat different neurobehavioral developmental setting that might contribute to the inconsistency.

Among the first studies of the fetal effect of BPA on asthma published on maternal–infant/child pairs, Donohue et al. (2013) looked at asthma in the CCCEH cohort. Sampling for urinary BPA was conducted during the third gestational trimester and in their children at ages 3, 5, and 7 years old. There was no correlation of BPA levels between gestational and child samples ($r \leq 0.02$) nor between samples taken at ages 3 and 5, but between ages 3 and 5 or 7 the r-statistic, 0.14–0.16, was significant. Wheeze was assessed with questionnaires at 5, 6, and 7 years of age, whereas an asthma assessment was conducted by a physician once between ages 5 and 12. The number of subjects with complete data for analysis ranged from 270 to 390. BPA levels at 3, 5, and 7 years were each associated with asthma, and the adjusted OR for mean child BPA was 1.6 (95% CI: 1.2, 2.1). The adjusted ORs were similar or lower for wheeze. Notably, third trimester maternal BPA was inversely associated with wheeze (AOR = 0.7, 95% CI: 0.5, 0.9).

The authors discussed a theory based on the pharmacokinetics of BPA in the pregnant mouse and measurements in human placenta in other studies. They noted that BPA was retained in the fetoplacental unit, which would limit the observable amount in the urine. If some pregnant women and fetuses displayed similar pharmacokinetics, it raises the possibility of significant misclassification of exposure, especially in the third trimester. This would be consistent with the finding above that 16th gestational week samples were more associated with childhood respiratory outcomes than 26th week samples (Spanier et al., 2012).

In the multicenter Study for Future Families II, initiated during 2002–05, maternal urinary samples were collected during a visit sometime in the 10–39 weeks (mean 27 weeks) of gestation timeframe (Evans et al., 2014). Maternal median BPA was 1.6 μg/g and mean BPA was 3.7 μg/g. Externalizing and internalizing behaviors measured with the CBCL in boys at aged 6–10 years was associated with maternal BPA among mother–child pairs in which the BPA was detectable (28 out of 153 pairs were excluded). The increases were small, but statistically significant, $\beta = 0.35$ ($P = .05$) and $\beta = 0.35$ ($P = .02$), respectively. Girls exhibited no associations with these measures.

Nested in the Generation R cohort, Snijder et al. (2013) analyzed a subset of 80 women for whom there was urinary BPA data from each trimester during pregnancy visits. The fetal growth rate was reduced in women with >4.2 μg/g BPA compared with women whose BPA was <1.54 μg/g. Weight at birth was reduced by 683 g and head circumference was reduced by 3.9 cm (See above for the analysis of BPA levels in this subset by Jusko et al. (2014)).

Maternal–child pairs in an agricultural area of California were the focus of the Center for the Health Assessment of Mothers and Children of Salinas study. Gestational urinary samples were taken at approximately 12 and 26 weeks during 1999–2000. The GM of creatinine-adjusted BPA was 1.1 μg/L and the two samples were only weakly correlated ($r = 0.25$). Maternal BPA, particularly the 26-week sample, was associated with lower TSH in neonatal boys (−9.9% per log2 unit increase of BPA), but not in girls (Chevrier et al., 2013). BPA at 26 weeks but not 12 weeks was also associated with lower total maternal thyroxine ($\beta = -0.13$ μg/dL per log2 unit) measured at 26 weeks. Iodine level was not a confounder.

BASC-2 internalizing behaviors tended to be increased with higher maternal BPA in 7-year-old boys but not girls in both mothers' and teachers' reports, though the score differences were not statistically significant (Harley et al., 2013). Early and late pregnancy samples were not examined separately. BPA was also measured in 5-year-olds (GM = 2.5 μg/L) and was associated with outcomes, primarily among girls (Harley et al., 2013). Analysis of reports on girls completed by mothers linked child BPA to several externalizing behaviors (score difference of 1.2, 95% CI: 0.3, 2.1) and two of its scales, including conduct problems (score difference of 1.8, 95% CI: 0.8, 2.9) and inattention (score difference of 0.9, 95% CI: 0.2, 1.6). Using a different tool, the Conners' ADHD/Diagnostic and Statistical Manual for Mental Disorders-IV, data generated by the mothers uncovered associations with inattention (score difference of 1.3, 95% CI: 0.4, 2.2), hyperactivity (score difference of 1.1, 95% CI: 0.1, 2.0), and ADHD (score difference of 1.3, 95% CI: 0.3, 2.3). Analysis of data from teacher reports also provided evidence of statistically significant associations among both boys and girls of BASC-2 internalizing behaviors (boys score difference of 1.8, 95% CI: 0.14, 3.1; girls score difference of 1.8, 95% CI: 0.1, 3.6) and its scales, especially anxiety and depression, but not externalizing behaviors. CADS components scored by teachers tended to be associated with girls. Inattention, hyperactivity, and ADHD score differences were 1.7, 0.4, and 1.1, respectively for boys and 1.0, 1.7, and 1.7 for girls.

The EDEN (Etude des Déterminants pré et post natals du développement et de la santé de l'Enfant) cohort examined mother–infant/child pairs from Poitiers and Nancy, France. Maternal urinary BPA sampled between gestational weeks 22 and 29 (N = 520) had 2.4 μg/L creatinine-adjusted BPA (Philippat et al., 2014). There was no association of urinary BPA with birth weight, but there was increased head circumference (Philippat et al., 2012). Children tended to have heavier weight at 12, 24 and 36 months but not with statistical significance (Philippat et al., 2014). This was similar to findings by Valvi et al. (2013) in the INMA-Sabadell cohort.

Follow-up (Philippat et al., 2017) found only a small effect due to urinary BPA from gestational weeks 23–29 on the internalizing behavior score at 3 years of age (IRR = 1.06, 95% CI: 1.00, 1.12, per doubling in BPA concentration, N = 457), especially for peer relationship problems, and on the externalizing behavior score, especially in the hyperactivity–inattention subscale at 5–6 years of age (IRR = 1.08, 95% CI: 1.01, 1.14, per doubling in BPA concentration). Dichotomizing the score data resulted in an OR for internalizing behavior of 1.2 with marginal statistical significance (95% CI: 0.96, 1.6) and an OR of 1.4 (95% CI: 1.08, 1.8) for hyperactivity–inattention. The data were provided by mothers (N = 464) who completed the Strength and Difficulties Questionnaire. Covariates included child age, maternal adiposity, parental education, breastfeeding duration, smoking, and maternal mental status. An IRR of 1.08 (95% CI: 1.01, 1.14) was reported. In addition, IQ was assessed in 5–6-year-old boys using the Wechsler Preschool and Primary Scale of Intelligence (Nakiwala et al., 2018). No association was observed.

Boys born to a subgroup of 587 women from the EDEN cohort in France were also examined at age 5 for respiratory health (Vernet et al., 2017). Comparison of FEV1 with log-transformed urinary BPA revealed a nonstatistically significant inverse relationship with FEV1 and a positive association with asthma (Hazard ratio = 1.2, 95% CI: 0.97, 1.55; third tertile HR = 1.4, 95% CI: 0.87, 2.2). There was no association with wheeze.

Preterm birth (<37 weeks) was studied by Cantonwine et al. (2015) in a study nested in a cohort enrolled during 2006–08 in Boston and Philadelphia. Urine samples were obtained over four visits (at 10th, 18th, 26th, and 35th weeks of gestation) from 1695 pregnant women. SG-corrected BPA indicated low stability across trimesters (ICC = 0.21; 95% CI: 0.16, 0.27), and ICCs were slightly higher in cases of preterm birth (0.25) compared with controls (0.19). Spearman correlations between study visits for BPA were low, 0.17–0.26 (all P-values $< .01$). The trend to preterm births with increasing gestational BPA was mostly not statistically significant, but notable, nonetheless, because the BPA from the fourth visit (33–38 weeks of gestation) was associated ($P < .01$) in all three statistical models with spontaneous preterm birth (N = 25; 298 controls). The AOR in the model with the most adjusting covariates, including total urinary phthalates, was 2.2 (95% CI: 1.3, 4.0).

Neonatal TSH was lower and thyroxine was increased with increased BPA, −8.2% and 4.8%, respectively, per natural log of the interquartile concentration (Aung et al., 2017). The association was primarily due to the gestational weeks of 26 and 35 samples, which are −13.5% and −17.3%. However, the conclusions are limited because this cohort was enriched with preterm births (116 preterm and 323 term neonates). The authors noted a general consistency between BPA and TSH in studies of men, nonpregnant women, and children.

Recently, Buckley et al. (2018) reported on the association of log-transformed urinary BPA in the third gestational trimester with previously diagnosed asthma in the respective 6–7-year-old boys (AOR = 3.0, 95% CI: 1.4, 6.6), but not girls, in the Mount Sinai Children's Environmental Health Study of 164 maternal–child pairs. The median unadjusted BPA concentration in the third trimester was 1.3 μg/L and the interquartile range was 0.6–2.3 μg/L.

A small blood sampling study of 80 maternal–infant pairs at the University of Michigan Hospital (Veiga-Lopez et al., 2015) found that birth weight of a female pregnancy was 183 g ($P > .005$) lower per twofold increase in plasma uBPA and 607 g ($P = .0093$) lower per 10-fold increase in uBPA, based on first trimester samples. The term uBPA had an opposite effect, a gain of 96 g ($P < .05$) per twofold increase of uBPA and 318 g ($P = .034$) per 10-fold increase, as well as an increased gestational age for female births, 1.1 days ($P = .06$) and 3.7 days ($P = .06$), respectively. There was no effect on male births. Notably, term uBPA was approximately twice as high as in the first trimester and was higher in 68% of term mothers. This may prove to be an example of a U-shaped dose–response curve. In addition, it is not clear from the literature whether the other studies analyzing birth weight and urinary BPA examined the different gestational weeks separately. If there was not a statistically significant difference, then, it may be that uBPA in blood is a more enlightening approach.

Children

The association between concurrently measured BPA and ADHD symptoms was examined in 460 8–15-year-old children, using data from the 2003–04 NHANES (Tewar et al., 2016). The prevalence of ADHD in the group was 7.1%. Among children with urinary BPA above the median, the prevalence was 11.2%, in contrast with 2.9% below the median. The AOR was 5.68 (95% CI: 1.6, 20), adjusted for child's age, family income, gender, race/ethnicity, urine creatinine, prenatal tobacco exposure, blood lead level (log-transformed), urine organophosphate metabolite level (log-transformed), and insurance status. For boys the AOR was 11 (95% CI: 1.4, 86), whereas in girls the AOR was 2.8 (95% CI: 0.4, 21).

Recently, the results of a case-control study of a group of 465 6–12-year-old children in Guangzhou, China showed that concurrently measured creatinine-adjusted urinary BPA was associated with ADHD in a concentration-dependent manner (Li et al., 2018). Study enrollment was composed of 215 children with ADHD and 250 controls. Comparing the second, third, and fourth quartiles to the first (referent) quartile of BPA, the AORs were, respectively, 1.8 (95% CI: 0.95, 3.4), 7.4 (95% CI: 3.9, 14), and 9.4 (95% CI: 4.9, 18). The effect was stronger in boys than girls, AOR of 1.8, 9.1, and 10.1, versus 0.96, 3.35, and 4.6, respectively, but only 27 of the 215 cases were girls. For cases the GM of creatinine-adjusted BPA was 4.63 μg/g versus 1.71 μg/g for controls. Like other studies (see above) this difference was larger in 6–9-year-old children, among whom the creatinine-adjusted BPA GM was 5.15 μg/g in cases versus 1.74 μg/g in controls. Consistent with these findings was the greater use of plastic food and beverage containers by ADHD children, but one cannot easily say whether higher exposure to BPA preceded the ADHD diagnosis because of the study design.

In a respiratory disease study, Kim et al. (2014) tested 127 children every 2 years at an elementary school in Seoul, South Korea, starting with urinary BPA measurement at age 7–8 and wheeze and asthma testing every 2 years, ending at age 11–12. Children with diagnosed asthma were excluded. The GM of BPA was 1.0 μg/L. A one-unit increase in log-transformed creatinine-adjusted BPA was associated with wheeze (AOR = 2.5, 95% CI: 1.15, 5.3) and asthma (HR = 2.1, 95% CI: 1.5, 3.0), but the association with asthma was primarily in girls (HR = 2.4, 95% CI: 2.2, 2.8). Covariates for analysis included parental asthma history, parental education level, tobacco use, pet ownership, and cockroach sensitization.

FOOD PACKAGING AND BISPHENOL A ALTERNATIVES

Volkel et al. (2011) determined both free and total BPA in urine samples of 1–2-month-old infants collected via urine bags. The infants were not exposed to chemicals via medical devices. Data were collected on infants after use of BPA-containing BPA-bottles in contrast to infants without use of PC-bottles. Free BPA was observed above the limit of quantitation (LOQ) in only 3 of 91 (3%) samples from 47 infants. Total BPA was observed in only 38 (42%) urine samples, at concentrations up to 18 μg/L. The highest concentration was just under the current EFSA guidance. Infants who were fed using baby bottles showed approximately twofold higher median levels of total BPA. The data also indicate substantial conjugation of BPA in infants.

It is interesting to note that BPA (and phthalates) exposures were substantially reduced when individuals' diets were restricted to food with limited packaging. Rudel et al. (2011) evaluated the contribution of food packaging to exposure, and measured urinary BPA and phthalate metabolites before, during, and after a "fresh foods" dietary intervention. Twenty participants in five families were selected based on self-reported use of canned and packaged foods. Participants ate their usual diet, followed by 3 days of "fresh foods" that were not canned or packaged in plastic, and then returned to their usual diet. Evening urine samples were collected over 8 days in January 2010 and were composited into preintervention, during intervention, and postintervention samples. Results showed that urine levels of BPA metabolites decreased significantly during the fresh foods intervention (e.g., BPA GM, 3.7 ng/mL preintervention vs. 1.2 ng/mL during intervention). The intervention reduced GM concentrations of BPA by 66% with the maxima reduced by 76%.

Morgan et al. (2011) quantified urinary total BPA in 81 children 23–64 months of age in Ohio over 48-h periods to quantify the relationship with overall exposure. Estimated median intake doses of BPA for these 81 children were 109 ng/kg/day (dietary ingestion), 0.06 ng/kg/day (nondietary ingestion), and 0.27 ng/kg/day (inhalation). The estimated median excreted amount of urinary BPA was 114 ng/kg/day, which closely matched exposure.

Consuming canned vegetables at least once a day was associated in the HOME study (Braun et al., 2012) with higher BPA concentrations (GM = 2.3 μg/g) compared with those consuming no canned vegetables (GM = 1.6 μg/g). BPA concentrations did not vary by consumption of fresh fruits and vegetables, canned fruit, or fresh and frozen fish. In the Spanish INMA-Sabadell cohort investigators (Casas et al., 2013) estimated that 21%–25% of the urinary BPA in the pregnant women was associated with consumption of canned fish (Casas et al., 2013).

Findings of BPA alternatives that have gradually replaced BPA have also been reported in urine, some data of which were included in studies described above. The substitution of BPA includes alternative BPXs such as bisphenol S (BPS; 4,4-sulfonyldiphenol), bisphenol B (2,2-bis[4-hydroxyphenyl]butane), BPF (4,4-dihydroxydiphenylmethane), and bisphenol AF (4,4-[hexa-fluoroisopropylidene] diphenol) (Usman and Ahmad, 2016). BPA-related compounds or BPXs consist of two phenol groups attached through a bridging carbon or other chemical structure. The para-hydroxy groups are critical in binding to estrogen receptor. Because BPXs are structurally similar to BPA, it is expected that they may also have the same toxicological effects on the biological system.

BPF and BPS are selected here as examples to examine in brief some of the findings related to these alternatives. BPF is absorbed by oral route and distributed to the whole organism, including the reproductive tract and the fetus by crossing the placental barrier, is mainly excreted through the urine, and has been shown to be as hormonally active as BPA. BPF has a broad range of industrial applications including manufacture of epoxy resins and polycarbonates (Usman and Ahmad, 2016). BPF residues have been found in foodstuffs packaged in containers with epoxy coatings, in the drinking water pumped out from the water pipes incidentally using bisphenol F diglycidyl ether–based epoxy resins for renovation; in vegetables, fish and seafood, meat and meat products, and beverages; and in indoor dust, and personal care products (PCPs). BPF was detected in 98% of the urine samples (LOQ of 0.03 μg/L) collected in the Swedish POPUP study during 2009–2014 with a median SG-adjusted concentration of 0.32 μg/L and with levels increasing between 2009 and 2014 (Gyllenhammar et al., 2017). As an environmental contaminant, BPF has been detected in environmental media such as surface water, sewage water, sediment samples, sewage sludge, liquid manure samples, compost water samples, and municipal landfill leachates.

BPS has hormonal potencies of the same magnitude as BPA. BPS use was increased in the production of epoxy resins, thermal papers, and infant feeding bottles after restrictions were put on BPA (Usman and Ahmad, 2016). It is more heat resistant and photoresistant than BPA. Studies showed that BPS has been detected in PCPs, foodstuffs (beverages, dairy products, fish and seafood, cereals, meat and meat products, and canned fruits and vegetables), indoor dust, river water, sewage, sediment, and sludge. The authors noted that moderate to high concentrations of BPS in the range of a few nanogram per gram to several milligrams per gram were detected in thermal receipts collected from various cities of United States, Korea, Vietnam, and Japan. The EDI through dermal absorption was 0.00418 mg/kg bw/day for the general population and 0.312 mg/kg bw/day for individuals who are occupationally exposed. BPS has a greater half-life and better dermal penetration than BPA and can lead to higher body burden in comparison to BPA. It is nondegradable in the aerobic conditions and likely to accumulate in the aquatic environment. Studies cited reported that (1) the detection rate of BPS in human matrices has been as high as 81% in urine and 3% in breast milk collected from United States and seven Asian countries; and (2) a mean EDI of 1.7 μg/day in Japan followed by China and the United States, 0.34 μg/day and 0.32 μg/day, respectively. Gyllenhammar et al. (2017) found BPS 68% of urine samples (Swedish POPUP study, see above) at an SG-adjusted median 0.048 μg/L.

CONCLUDING REMARKS AND FUTURE DIRECTIONS

BPA has been measured in human urine samples in various biomonitoring programs and research studies worldwide. An integrated approach that uses all data types along the environmental disease continuum is required for a complete understanding of the findings. Biomarkers may be particularly useful when they provide linkage to important exposure but must be measured in the correct matrix for exposure route/source of interest, and the analytical technology must be available, reliable, and reproducible, among other considerations. Nondetection may mean that the technology is limited in measuring low concentrations of specific chemicals, or that the exposure occurred at an earlier point in time, allowing for the chemical to be eliminated from the person's body before measurement took place. Although biomonitoring provides reference ranges, for most environmental chemicals there are no "standard ranges" or "safe ranges" established for biomarkers.

In using data on human exposure to environmental chemicals for performing risk assessment of such exposure, the information needed include an understanding of the pattern of exposure, toxicology and epidemiology databases, estimates of intake, measurements of exposure, use of biomarker data, and interpretation of the data and parameters to evaluate the associated public health implications and environmental management and regulatory decision-making. Therefore, for biomarkers to be useful for environmental (and occupational) health risk assessments, they have to be relevant and valid (WHO, 2001).

A good biomarker of exposure should be useful to predict adverse effects, rather than exposure levels. In addition, biomarkers must be validated before application in the risk assessment process, i.e., the relationship between the biomarker, the exposure, and the health outcome must be established. The general criteria for validating biomarkers include understanding the natural history, biological and temporal relevance, pharmacokinetics, background variability, dose response, and confounding factors (WHO, 2001). The validation process includes the critical assessment of different relationships: exposure–dose, dose response (effect), and effect–disease. Validation can focus on the demonstration of a correspondence between a biomarker of exposure and external exposure, or the assessment of their relationship with the concentration in the critical organ (e.g., concentration in the kidney vs. in blood or urine) or with critical effects (e.g., lead neurotoxicity vs. blood lead concentration).

For BPA, health reference values have been established based on animal data, but in the meantime, the results of a number of laboratory animal studies has not been sufficiently consistent. There are still uncertainties particularly on two aspects—low dose effects and NMDRC.

BPA has been measured as uBPA, conjugated BPA, and total BPA. As the conjugated forms are currently viewed to be the inactivated forms without endocrine activities, knowing the forms of BPA and the associated respective amounts present systemically in humans would be important for understanding the health implications of BPA in the body. Blood samples are a typical and valid approach to estimating systemic contaminant levels and uBPA is detectable in serum samples (Vandenberg et al., 2014a; vom Saal and Welshons, 2014), but it is not as practical as urine sampling.

A high degree of within-person variability has also been reported (Braun et al., 2011a; Jusko et al., 2014; Stacy et al., 2017). Lakind and Naiman (2015) noted the difficulties in interpreting urinary BPA data, despite efforts to account for urine dilution and translate spot sample data to daily exposure. In addition, the BPA levels measured in urine can vary by factors such as age, gender, and race/ethnicity.

A small but provocative study of blood plasma in mother–infant pairs (Veiga-Lopez et al., 2015) observed that gestational term uBPA was approximately twice as high as in the first trimester and was higher in 68% of term mothers. The results, if found to be accurate when repeated, may explain inconsistent findings within and between studies and suggest that the way forward in epidemiology of BPA should include blood sampling to better address different development time frames.

The implications of measurements of BPA concentrations in human tissues and their correlations to a dose–response relationship have yet to be clearly established. Likely, there are sensitive subsets of individuals and critical windows of effect during development, and the goal must be delineation of the mechanisms behind those subsets and time frames. Possibilities of mechanism include enzymatic and cellular receptor polymorphisms and epigenetic causes. Notably, behavior, cognitive ability, IQ, and ADHD are complex phenomena and still not well understood, so there may be other important environmental chemicals and other confounders that were not included. Other known critical pollutants also need to be included in the analysis, such as blood lead in the neurodevelopmental studies. However, the epidemiological studies point to a concern about neurodevelopmental and respiratory effects from fetal and childhood exposures.

References

Arbuckle, T.E., Davis, K., Marro, L., et al., 2014. Phthalate and bisphenol A exposure among pregnant women in Canada-results from the MIREC study. Environ. Int. 68, 55–65.

Arbuckle, T.E., Marro, L., Davis, K., et al., 2015. Exposure to free and conjugated forms of bisphenol A and triclosan among pregnant women in the MIREC cohort. Environ. Health Persp. 123, 277–284.

Aung, M.T., Johns, L.E., Ferguson, K.K., et al., 2017. Thyroid hormone parameters during pregnancy in relation to urinary bisphenol A concentrations: a repeated measures study. Environ. Int. 104, 33–40.

Berman, T., Goldsmith, R., Göen, T., et al., 2014. Demographic and dietary predictors of urinary bisphenol A concentrations in adults in Israel. Int. J. Hyg Environ. Health 217 (6), 638–644.

Braun, J.M., Kalkbrenner, A.E., Calafat, A.M., et al., 2011a. Variability and predictors of urinary bisphenol A concentrations during pregnancy. Environ. Health Persp. 119, 131–137.

Braun, J.M., Kalkbrenner, A.E., Calafat, A.M., et al., 2011b. Impact of early-life bisphenol A exposure on behavior and executive function in children. Pediatrics 128 (5), 873–882.

Braun, J.M., Smith, K.W., Williams, P.L., et al., 2012. Variability of urinary phthalate metabolite and bisphenol A concentrations before and during pregnancy. Environ. Health Persp. 120, 739–745.

Braun, J.M., Yolton, K., Dietrich, K.N., et al., 2009. Prenatal bisphenol A exposure and early childhood behavior. Environ. Health Persp. 117, 1945–1952.

Buckley, J.P., Quirós-Alcalá, L., Teitelbaum, S.L., et al., 2018. Associations of prenatal environmental phenol and phthalate biomarkers with respiratory and allergic diseases among children aged 6 and 7 years. Environ. Int. 115, 79–88.

Calafat, A.M., Kuklenyik, Z., Reidy, J.A., et al., 2005. Urinary concentrations of bisphenol A and 4-nonylphenol in a human reference population. Environ. Health Persp. 113, 391–395.

Calafat, A.M., Longnecker, M.P., Koch, H.M., et al., 2015. Optimal exposure biomarkers for nonpersistent chemicals in environmental epidemiology. Environ. Health Persp. 123 (7), A166–A168.

Calafat, A.M., Longnecker, M.P., Koch, H.M., et al., 2016. Response to "Comment on optimal exposure biomarkers for nonpersistent chemicals in environmental epidemiology". Environ. Health Persp. 124, A66–A67.

Calafat, A.M., Ye, X., Wong, L.-Y., et al., 2008. Exposure of the U.S. population to bisphenol A and 4-tertiary-octylphenol: 2003–2004. Environ. Health Persp. 116, 39–44.

Cantonwine, D.E., Ferguson, K.K., Mukherjee, B., et al., 2015. Urinary bisphenol A levels during pregnancy and risk of preterm birth. Environ. Health Persp. 123, 895–901.

Casas, M., Forns, J., Martínez, D., et al., 2015. Exposure to bisphenol A during pregnancy and child neuropsychological development in the INMA-Sabadell cohort. Environ. Res. 142, 671–679.

Casas, M., Valvi, D., Ballesteros-Gomez, A., et al., 2016. Exposure to bisphenol A and phthalates during pregnancy and ultrasound measures of fetal growth in the INMA-Sabadell cohort. Environ. Health Persp. 124, 521–528.

Casas, M, Valvi, D., Luque, N., et al., 2013. Dietary and sociodemographic determinants of bisphenol A urine concentrations in pregnant women and children. Environ. Int. 56, 10–18.

CDC, 2018a. Healthy People 2020. Chapter 12. Environmental Health. Centers for Disease Control and Prevention (CDC). https://www.cdc.gov/nchs/data/hpdata2020/HP2020MCR-C12-EH.pdf.

CDC, 2018b. National Health and Nutrition Examination Survey: Estimation Procedures, 2011–2014. In: Data Evaluation and Methods Research. Vital and Health Statistics Series 2, Number 177. Centers for Disease Control and Prevention (CDC), National Center for Health Statistics, Hyattsville, Maryland. DHHS Publication No. 2018–137.7.

Chevrier, J., Gunier, R.B., Bradman, A., et al., 2013. Maternal urinary bisphenol A during pregnancy and maternal and neonatal thyroid function in the CHAMACOS study. Environ. Health Persp. 121, 138–144.

Covaci, A., Den Hond, E., Geens, T., et al., 2015. Urinary BPA measurements in children and mothers from six European member states: overall results and determinants of exposure. Environ. Res. 141, 77–85.

Donohue, K.M., Miller, R.L., Perzanowski, M.S., et al., 2013. Prenatal and postnatal bisphenol A exposure and asthma development among inner-city children. J. Allergy Clin. Immunol. 131, 736–742.

EFSA, 2015. Scientific opinion on the risks to public health related to the presence of bisphenol A (BPA) in foodstuffs: executive summary. EFSA panel on food contact materials, enzymes, flavourings and processing aids (CEF). European Food Safety Authority, Parma, Italy. EFSA J. 13 (1), 3978 (1040 pages).

Evans, S.F., Kobrosli, R.W., Barrett, E.S., et al., 2014. Prenatal bisphenol A exposure and maternally reported behavior in boys and girls. Neurotoxicology 45, 91–99.

Fan, A.M., Chou, W.C., Lin, P.P., 2017. Toxicology and risk assessment of bisphenol A. In: Gupta, R.C. (Ed.), Developmental and Reproductive Toxicology, second ed. Academic Press/Elsevier, Amsterdam, pp. 765–795.

Fan, A.M., 2014. Biomarkers in toxicology, risk assessment and chemical regulations. In: Gupta, R.C. (Ed.), Biomarkers in Toxicology, first ed. Academic Press/Elsevier, Amsterdam, pp. 1057–1079.

Gascon, M., Casas, M., Morale,s, E., et al., 2015. Prenatal exposure to bisphenol A and phthalates and childhood respiratory tract infections and allergy. J. Allergy Clin. Immunol. 135, 370–378.

Gerona, R.R., Pan, J., Zota, A.R., et al., 2016. Direct measurement of bisphenol A (BPA), BPA glucuronide and BPA sulfate in a diverse and low-income population of pregnant women reveals high exposure, with potential implications for previous exposure estimates: a cross-sectional study. Environ. Health 15, 50.

Ginsberg, G., Rice, D.C., 2009. Does rapid metabolism ensure negligible risk from bisphenol A? Environ. Health Persp. 117, 1639–1643.

Guidry, V.T., Longnecker, M.P., Aase, H., et al., 2015. Measurement of total and free urinary phenol and paraben concentrations over the course of pregnancy: assessing reliability and contamination of specimens in the Norwegian mother and child cohort study. Environ. Health Persp. 123, 705–711.

Guxens, M., Ballester, F., Espada, M., et al., 2012. Cohort profile: the INMA—INfancia y Medio Ambiente — (environment and childhood) project. Int. J. Epidemiol. 41, 930–940.

Gyllenhammar, I., Glynn, A., Jönsson, B.A.G., et al., 2017. Diverging temporal trends of human exposure to bisphenols and plastizisers, such as phthalates, caused by substitution of legacy EDCs? Environ. Res. 153, 48–54.

Harley, K.G., Gunier, R.B., Kogut, K., et al., 2013. Prenatal and early childhood bisphenol A concentrations and behavior in school-aged children. Environ. Res. 126, 43–50.

Health Canada, 2017. Fourth Report on Human Biomonitoring of Environmental Chemicals in Canada. Results of the Canadian Health Measures Survey Cycle 4 (2014–2015).

Jaddoe, V.W., Bakker, R., van Duijn, C.M., et al., 2007. The Generation R Study Biobank: a resource for epidemiological studies in children and their parents. Eur. J. Epidemiol. 22, 917–923.

Jönsson, B.A.G., Amon, A., Lindh, C.H., 2014. Tidstrender för och halter av perfluorerade alkylsyror (PFAAs) i serum samt ftalatmetaboliter och alkylfenoler I urin hos unga svenska män och kvinnor - resultat från den fjärde uppföljningensundersökningen år 2013. *Avdelningen för Arbets-* och miljömedicin. Lunds universitet, Lund. Cited by Calafat et al., 2005.

Jusko, T.A., Shaw, P.A., Snijder, C.A., et al., 2014. Reproducibility of urinary bisphenol A concentrations measured during pregnancy in the Generation R Study. J. Expo. Sci. Environ. Epidemiol. 24 (5), 532–536.

Kim, K.N., Kim, J.H., Kwon, H.J., et al., 2014. Bisphenol A exposure and asthma development in school-age children: a longitudinal study. PLoS One 9, e111383.

Kooijman, M.N., Kruithof, C.J., van Duijn, C.M., et al., 2016. The Generation R Study: design and cohort update 2017. Eur. J. Epidemiol. 31, 1243–1246.

Lakind, J.S., Naiman, D.Q., 2015. Temporal trends in bisphenol A exposure in the United States from 2003 to 2012 and factors associated with BPA exposure: spot samples and urine dilution complicate data interpretation. Environ. Res. 142, 84–95.

Lakind, J.S., Naiman, D.Q., 2011. Daily intake of bisphenol A and potential sources of exposure: 2005–2006 National Health and Nutrition examination survey. J. Expo. Sci. Environ. Epidemiol. 21 (3), 272–279.

Larsson, K., Ljung Björklund, K., Palm, B., et al., 2014. Exposure determinants of phthalates, parabens, bisphenol A and triclosan in Swedish mothers and their children. Environ. Int. 73, 323–333.

Lee, B.E., Park, H., Hong, Y.C., et al., 2014. Prenatal bisphenol A and birth outcomes: MOCEH (mothers and Children's environmental Health) study. Int. J. Hyg Environ. Health 217 (2–3), 328–334.

Li, Y., Zhang, H., Kuang, H., et al., 2018. Relationship between bisphenol A exposure and attention-deficit/hyperactivity disorder: a case-control study for primary school children in Guangzhou, China. Environ. Pollut. 235, 141–149.

Longnecker, M.P., Harbak, K., Kissling, G.E., 2013. Reports from the field. The concentration of bisphenol A in urine is affected by specimen collection, a preservative, and handling. Environ. Res. 126, 211–214.

Magnus, P., Irgens, L.M., Haug, K., 2006. Cohort profile: the Norwegian mother and child cohort study (MoBa). Int. J. Epidemiol. 35, 1146–1150.

Marty, M., 2015. Consideration of infants and children in risk assessment. In: Fan, A.M., Khan, E.M., Alexeeff, G.V. (Eds.), Toxicology and Risk Assessment. Pan Stanford Publishing Pte. Ltd, pp. 93–133.

Morgan, M.K., Jones, P.A., Calafat, A.M., et al., 2011. Assessing the quantitative relationships between preschool children's exposures to bisphenol A by route and urinary biomonitoring. Environ. Sci. Technol. 45, 5309–5316.

Nakiwala, D., Pevre, H., Heude, B., et al., 2018. In-utero exposure to phenols and phthalates and the intelligence quotient of boys at 5 years. Environ. Health 17, 17.

NTP, 2018. Draft NTP Research Report on the CLARITY-BPA Core Study: A Perinatal and Chronic Extended-dose-range Study of Bisphenol A in Rats. NTP RR 9. Research Triangle Park, N.C. National Toxicology Program. (9), pp. 1–249.

Nelson, J.W., Scammell, M.K., Hatch, E.E., et al., 2012. Social disparities in exposures to bisphenol A and polyfluoroalkyl chemicals: a cross-sectional study within NHANES 2003–2006. Environ. Health 11 (10). Art. No 10.

Nepomnaschy, P.A., Baird, D.D., Weinberg, C.R., et al., 2009. Within-person variability in urinary bisphenol A concentrations: measurements from specimens after long-term frozen storage. Environ. Res. 109, 734–737.

Park, J.H., Hwang, M.S., Ko, A., et al., 2016. Risk assessment based on urinary bisphenol A levels in the general Korean population. Environ. Res. 150, 606–615.

Parker, J.D., Aoki, Y., Ingram, D., 2018. Does Creatinine Adjustment Method Affect Estimated BPA Levels? (Powerpoint Presentation). https://www.cdc.gov/nchs/ppt/nchs2015/parker_wednesday_brooksideab_dd5.pdf.

Perera, F., Vishnevetsky, J., Herbstman, J.B., et al., 2012. Prenatal bisphenol A exposure and child behavior in an inner-city cohort. Environ. Health Persp. 120, 1190–1194.

Philips, E.M., Jaddoe, V.W.V., Asimakopoulos, A., et al., 2018. Bisphenol and phthalate concentrations and its determinants among pregnant women in a population-based cohort in the Netherlands, 2004–5. Environ. Res. 161, 562–572.

Philippat, C., Botton, J., Calafat, A.M., 2014. Prenatal exposure to phenols and growth in boys. Epidemiology 25, 625–635.

Philippat, C., Mortamais, M., Chevrier, C., Petit, C., Calafat, A.M., Ye, X., et al., 2012. Exposure to phthalates and phenols during pregnancy and offspring size at birth. Environ. Health Persp. 120, 464–470.

Philippat, C., Nakiwala, D., Calafat, A.M., et al., 2017. Prenatal exposure to nonpersistent endocrine disruptors and behavior in boys at 3 and 5 years. Environ. Health Persp. 125, 097014.

Richter, C.A., Birnbaum, L.S., Farabollin,i, F., et al., 2007. *In vivo* effects of bisphenol A in laboratory rodent studies. Reprod. Toxicol. 24, 199–224.

Romano, M.E., Webster, G.M., Vuong, A.M., et al., 2015. Gestational urinary bisphenol A and maternal and newborn thyroid hormone concentrations: the HOME study. Environ. Res. 138, 453–460.

Rønningen, K.S., Paltiel, L., Meltzer, H.M., et al., 2006. The biobank of the Norwegian mother and child cohort study: a resource for the next 100 years. Eur. J. Epidemiol. 21, 619–625.

Rudel, R.A., Gray, J.M., Engel, C.L., et al., 2011. Food packaging and bisphenol A and bis(2-ethyhexyl) phthalate exposure: findings from a dietary intervention. Environ. Health Persp. 119, 914–920.

Sabanayagam, C., Teppala, S., Shankar, A., 2013. Relationship between urinary bisphenol A levels and prediabetes among subjects free of diabetes. Acta Diabetol. 50 (4), 625–631.

Shankar, A., Teppala, S., 2011. Relationship between urinary bisphenol A levels and diabetes mellitus. J. Clin. Endocrinol. Metab. 96, 3822–3826.

Silver, M.K., O'Neill, M.S., Sowers, M.R., et al., 2011. Urinary bisphenol A and type-2 diabetes in U.S. adults: data from NHANES 2003–2008. PLoS One 6 (10), e26868.

Snijder, C.A., Heederik, D., Pierik, F.H., et al., 2013. Fetal growth and prenatal exposure to bisphenol A: the Generation R Study. Environ. Health Persp. 121, 393–398.

Spanier, A.J., Kahn, R.S., Kunselman, A.R., et al., 2012. Prenatal exposure to bisphenol A and child wheeze from birth to 3 years of age. Environ. Health Persp. 119, 916–920.

Spanier, A.J., Kahn, R.S., Kunselman, A.R., Schaefer, E.W., et al., 2014. Bisphenol A exposure and the development of wheeze and lung function in children through age 5 years. JAMA Pediatr. 168, 1131–1137.

Stacy, S.L., Eliot, M., Calafat, A.M., et al., 2016. Patterns, variability, and predictors of urinary bisphenol A concentrations during childhood. Environ. Sci. Technol. 50 (11), 5981–5990.

Stacy, S.L., Papadonatos, G.D., Calafat, A.M., et al., 2017. Early life bisphenol A exposure and neurobehavior at 8 years of age: identifying windows of heightened vulnerability. Environ. Int. 107, 258–265.

Stahlhut, R.W., Welshons, W.V., Swan, S.H., 2009. Bisphenol A data in NHANES suggest longer than expected half-life, substantial nonfood exposure, or both. Environ. Health Persp. 117, 784–789.

Stahlhut, R.W., van Breeman, R.B., Gerona, R.R., et al., 2016. Comment on "Optimal exposure biomarkers for nonpersistent chemicals in environmental epidemiology". Environ. Health Persp. 124, A66.

Teeguarden, J.G., Twaddle, N.C., Churchwell, M.I., et al., 2016. Urine and serum biomonitoring of exposure to environmental estrogens. I: bisphenol A in pregnant women. Food Chem. Toxicol. 92, 129–142.

Tewar, S., Auinger, P., Braun, J.M., et al., 2016. Association of bisphenol A exposure and attention-deficit/hyperactivity disorder in a national sample of U.S. children. Environ. Res. 150, 112–118.

Usman, A., Ahmad, M., 2016. From BPA to its analogues: is it a safe journey? Chemosphere 158, 131–142.

Valvi, D., Casas, M., Mendez, M.A., et al., 2013. Prenatal exposure to bisphenol A and phthalates and childhood respiratory tract infections and allergy. Epidemiology 24 (6), 791–799.

Vandenberg, L.N., Chahoud, I., Heindel, J.J., et al., 2010a. Urinary, circulating, and tissue biomonitoring studies indicate widespread exposure to bisphenol A. Environ. Health Persp. 118 (8), 1055–1070.

Vandenberg, L.N., Chahoud, I., Padmanabhan, V., et al., 2010b. Biomonitoring studies should be used by regulatory agencies to assess human exposure levels and safety of bisphenol A. Environ. Health Persp. 118 (8), 1051–1054.

Vandenberg, L.N., Ehrlich, S., Belcher, S.M., et al., 2013. Low dose effects of bisphenol A: an integrated review of *in vitro*, laboratory animal and epidemiology studies. Endocr. Disruptors 1, e25078.

Vandenberg, L.N., Gerona, R.R., Kannan, K., et al., 2014a. A round robin approach to the analysis of bisphenol a (BPA) in human blood samples. Environ. Health 13, 25–45.

Vandenberg, L.N., Welshons, W.V., Vom Saal, F.S., et al., 2014b. Should oral gavage be abandoned in toxicity testing of endocrine disruptors? Environ. Health 13, 46–53.

Veiga-Lopez, A., Kannan, K., Liao, C., et al., 2015. Gender-specific effects on gestational length and birth weight by early pregnancy BPA exposure. J. Clin. Endocrinol. Metab. 100 (11), E1394–E1403.

Vernet, C., Pin, I., Giorgis-Allemand, L., et al., 2017. *In utero* exposure to select phenols and phthalates and respiratory health in five-year-old boys: a prospective study. Environ. Health Persp. 125, 097006.

Volkel, W., Kiranoglu, M., Fromme, H., 2011. Determination of free and total bisphenol A in urine of infants. Environ. Res. 111, 143–148.

Volkel, W., Colnot, T., Csanady, G.A., et al., 2002. Metabolism and kinetics of bisphenol A in humans at low doses following oral administration. Chem. Res. Toxicol. 15, 1281–1287.

Volkel, W., Bittner, N., Dekant, W., 2005. Quantitation of bisphenol A and bisphenol A glucuronide in biological samples by high performance liquid chromatography-tandem mass spectrometry. Drug Metab. Dispos. 33, 1748–1757.

vom Saal, F.S., Welshons, W.V., 2014. Evidence that bisphenol A (BPA) can be accurately measured without contamination in human serum and urine, and that BPA causes numerous hazards from multiple routes of exposure. Mol. Cell. Endocrinol. 398, 101–113.

WHO, 2001. Environmental Health Criteria 222, Biomarkers in Risk Assessment: Validity and Validation. United Nations Environment Programme (UNEP)/International Labor Organization (ILO)/ World Health Organization (WHO), Geneva, Switzerland.

WHO, 2009. Bisphenol A (BPA). Current State of Knowledge and Future Actions by WHO and FAO. International Food Safety Authorities Network (INFOSAN Information Note No. 5/2009-Bisphenol A.

Ye, X., Pierik, F.H., Hauser, R., et al., 2008. Urinary metabolite concentrations of organophosphorous pesticides, bisphenol A, and phthalates among pregnant women in Rotterdam, the Netherlands: the Generation R Study. Environ. Res. 108, 260–267.

Yolton, K., Xu, Y., Strauss, D., et al., 2011. Prenatal exposure to bisphenol A and phthalates and infant neurobehavior. Neurotoxicol. Teratol. 33, 558–566.

CHAPTER

26

Insecticides

Ramesh C. Gupta[1], *Ida R. Miller Mukherjee*[2], *Jitendra K. Malik*[3], *Robin B. Doss*[1], *Wolf-D. Dettbarn*[4], *Dejan Milatovic*[5]

[1]Toxicology Department, Breathitt Veterinary Center, Murray State University, Hopkinsville, Kentucky, United States; [2]Institute of Psychiatry and Human Behavior, Bambolim, Goa, India; [3]Division of Pharmacology and Toxicology, Indian Veterinary Research Institute, Izatnagar, Bareilly, Uttar Pradesh, India; [4]Vanderbilt University, Nashville, TN, United States; [5]Charlottesville, VA, United States

INTRODUCTION

Insecticides have been used around the world for centuries to control insects. People are exposed to insecticides in their homes, offices, gardens, workplaces, aircrafts, on military uniforms, or through trace contaminants in food. Insecticides are used most extensively in agriculture, horticulture, and forestry. As a result, farmers, farmhands, and their families are maximally exposed to these chemicals. Insecticides are also used to control vectors, such as mosquitoes and ticks that are involved in spreading public health diseases, such as malaria, West Nile disease, Lyme disease, and others. In developing countries, insecticides are often involved in suicide attempts in humans and malicious or accidental poisonings in pets, birds, and wildlife (Gupta, 2006; Satoh and Gupta, 2010; Lamb et al., 2016).

Insecticides constitute a large number of chemicals of different classes, and they not only exert toxicity in insects but also in vertebrate mammals through different mechanisms of action. Because of distinct differences in chemical structures, insecticides interact with different target and nontarget sites, including receptors, enzymes, and many other known and unknown molecules. Most insecticides are neurotoxicants as they target the nervous system, but they can also target other organs and body systems. The binding sites and adducts can be used as biomarkers of exposure and/or effects of insecticides. The insecticides are metabolized through different metabolic pathways and either parent compounds or their metabolites are often used as biomarkers of exposure. Insecticides can cause harmful health effects ranging from minor pain to death in nontarget mammalian (including humans), avian, and wildlife species.

Biomarkers of exposure from samples of human and animal tissues, fluids, and excreta offer qualitative and quantitative evidence of pesticide exposure (Barr and Buckley, 2011). These measurements are particularly useful in pesticides' toxicity because they can highlight population-based exposure trends and improve estimates of pesticide exposure and dose. Biomarkers of effects include measurements of biochemical, physiological, or behavioral alterations that are a consequence of pesticide exposure. Lastly, biomarkers of susceptibility are measurements of an individual's inherent ability to respond to insecticide exposures. These measurements include observations of molecular properties and functions, such as genetic polymorphisms and enzyme activities, which can affect the pharmacokinetics of insecticides, along with an individual's biochemical disposition toward disease progression and repair. Insecticide toxicity can be acute, subacute, or chronic, depending on the duration of exposure and the dose involved. Thus, selection of a sensitive, accurate, and validated biomarker of exposure, effects, and susceptibility appears to be a challenging task. In a very recent investigation, Appenzeller et al. (2017) highlighted that a specimen of hair can be as informative as plasma or urine for pesticide (organophosphates, carbamates [CMs], organochlorines, and pyrethroids) residue analysis for biomonitoring in epidemiological studies and

Biomarkers in Toxicology, Second Edition
https://doi.org/10.1016/B978-0-12-814655-2.00026-8

assessing adverse health effects. This chapter describes biomarkers of exposure, effects, and susceptibility of common insecticides in humans and animals.

ORGANOPHOSPHATES AND CARBAMATES

Organophosphate (OP) and CM insecticides are often discussed together as anticholinesterase (anti-ChE) agents because insecticides of both these classes inactivate the acetylcholinesterase (AChE) enzyme. AChE-inhibiting OPs are esters of phosphoric or phosphonic acid, while AChE-inhibiting CMs are esters of carbamic acid. Structurally, OPs are much more complex than CMs and are categorized into 13 different types, such as phosphates, phosphonates, phosphorothioates, and phosphoramidates. During the last half-century, anti-ChE insecticides have gained wide popularity around the world because of two major factors: (1) lack of residue persistence in the environment and in mammalian systems and (2) development of lesser resistance in insects compared with the organochlorine class of insecticides. Currently, hundreds of OPs and dozens of CMs are available on the market for their use as insecticides. Presently, OPs alone represent 50% of worldwide insecticide use. Due to their extreme toxicity and a lack of species selectivity, anti-ChE insecticides pose serious threats to the health of nontarget mammalian (including humans), wildlife, avian, and aquatic species. Depending on the dose and duration of exposure, these insecticides may adversely affect various body organs and systems (nervous system, skeletal muscles, digestive, cardiovascular, respiratory, ophthalmic, reproductive, endocrine, dermal, immune, and others) at cellular and molecular levels.

Mechanism of Toxicity

Mechanism of action differs substantially in acute, intermediate, and chronic toxicity of anti-ChE insecticides. Acute clinical signs of OPs and CMs are primarily attributed to AChE inactivation at synapses in the brain and at neuromuscular junctions in skeletal muscles (Pope, 1999; Gupta, 2006; Satoh and Gupta, 2010; Gupta and Milatovic, 2012; Heutelbeck et al., 2016). Physiologically, AChE is responsible for the hydrolysis of the neurotransmitter acetylcholine (ACh) and the termination of its biological activity within a microsecond. Both OPs and CMs react covalently with a serine residue in AChE in a similar manner to ACh. OPs and CMs inactivate AChE activity by phosphorylation and carbamylation, respectively, and differ quantitatively in rates of dephosphorylation and decarbamylation of inhibited AChE. It needs to be mentioned that the AChE enzyme can "age" with OPs and not with CMs. Following AChE inhibition, free ACh accumulates at the nerve endings of all cholinergic nerves and causes overstimulation of electrical activity. Inactivation of AChE activity >70% leads to a toxic level of ACh accumulation at central and peripheral sites. The molecular interactions between OPs and AChE have been studied in much more detail than those between CMs and AChE and have been described in previous publications (Timchalk, 2006; Gupta and Milatovic, 2012). Pavlovsky et al. (2003) demonstrated that pyridostigmine, by increasing free ACh, enhances glutamatergic transmission in hippocampal CA1 neurons. This mechanism offers an explanation as to how pyridostigmine and other AChE inhibitors, including OP nerve agents and pesticides, cause epileptic discharge and excitotoxic damage.

Evidence suggests that some of the AChE inhibitors directly interact with muscarinic ACh receptors (mAChRs) and nicotinic ACh receptors (nAChRs). The agonistic, antagonistic, potentiating, and inhibitory effects of AChE inhibitors on nAChRs were described by Smulders et al. (2003). These authors established that CM insecticides, which have a more potent interaction with nAChRs, are the less potent inhibitors of AChE; nAChRs are considered additional target sites, as they are involved in the toxicity of these insecticides.

In addition to AChE and ACh receptors, OPs and CMs bind to other serine-containing esterases (such as butyrylcholinesterase and carboxylesterases) and proteases in serum and tissues. In a series of experiments, Gupta et al. (1991a,b, 1993, 1994a,b) demonstrated significant changes in creatine kinase and lactic dehydrogenase, as well as changes in the isozymes of rats following an acute exposure to carbofuran and methyl parathion. Leakage of these enzymes into the serum from target tissues was due to carbofuran- or methyl parathion-induced depletion of ATP in tissues (Gupta et al., 2000, 2001a; b).

OP and CM insecticides also exert a myriad of toxic effects through multiple noncholinergic mechanisms soon after AChE inactivation (Gupta, 2004; Gupta et al., 2007, 2015; Zaja-Milatovic et al., 2009; Gupta and Milatovic, 2010, 2012; Terry, 2012). Among many noncholinergic mechanisms, activation of glutamatergic (NMDA receptors), adenosinergic, GABAergic, and monoaminergic systems appeared to be involved in the seizures and lethality associated with OP- or CM-induced poisonings (Solberg and Belkin, 1997; Choudhary et al., 2002; Dekundy and Kaminski, 2010; Slotkin and Seidler, 2012). In a very recent investigation, Umeda et al. (2018) suggested that carbofuran decreased glutamate receptor 2 (GluA2) protein expression and increased neuronal vulnerability to glutamate toxicity at concentrations that do not affect AChE activity.

OP- or CM-induced excitotoxicity for more than an hour can cause neurodegeneration and neuroinflammation in the cortex, amygdala, hippocampus, and other brain regions involved in initiation and propagation of convulsions and seizures. The early morphological lesions include dendritic swelling of pyramidal neurons in the CA1 sector of the hippocampus. The AChE inhibitor–induced neuronal cell death is a consequence of a series of extra- and intracellular events leading to the intracellular accumulation of Ca^{+2} ions and the generation of free radicals (Gupta et al., 2001a; b; Milatovic et al., 2010; Kazi and Oommen, 2012). OP- or CM-induced excessive production of free radicals causes oxidative and nitrosative stress to which the brain is especially vulnerable (Gupta et al., 2007, Zaja-Milatovic et al., 2009; Gupta and Milatovic, 2010, 2012). Lipids are readily attacked by free radicals, resulting in the formation of a number of peroxidation products, such as F_2-isoprostanes and F_4-neuroprostanes, which are formed nonenzymatically. F_4-neuroprostanes are specific biomarkers of oxidative damage to the neurons. These events, in addition to many more described elsewhere, cause irreversible destruction of cellular components, including proteins, DNA, and, particularly, mitochondria (Milatovic et al., 2006). Oxidative/nitrosative stress has been demonstrated to cause depletion of high-energy phosphates (ATP and phosphocreatine) by multiple mechanisms in the brain of rats acutely intoxicated with OP and CM insecticides (Gupta et al., 2001a; b; Gupta et al., 2007; Zaja-Milatovic et al., 2009; Zepeda-Arce et al., 2017).

As mentioned earlier, in addition to the nervous system and skeletal muscles, OPs and CMs adversely affect many other body organs and systems, and therefore additional mechanisms of action are involved in toxicity (Gupta, 2006; Satoh and Gupta, 2010; Gupta et al., 2017).

Biomarkers

Acute Toxicity

For the monitoring of insecticides, biomarkers can typically be divided into three categories: biomarkers of exposure, effect, and susceptibility. Biomarkers of exposure to OPs and CMs may include determination of residue of parent compounds and/or their metabolites and modified cells or their molecules (e.g., DNA and protein adducts) in biological tissue/fluids. Highly sophisticated and sensitive chromatographic, spectrometric, and other assays allow the confirmation and quantitation of OP/CM insecticides and their metabolites in body fluids and tissues at the ppm or ppb level. Jain (2006) described detailed aspects of sample selection, extraction, concentration, and analytical methods for the detection of various OPs and CMs and their metabolites. Because these insecticides are unstable and metabolize in body systems very rapidly, it is highly likely that most often metabolites will be detected and not the parent compounds. Some commonly used OPs and their major metabolites that are used as biomarkers of OP exposure are listed in Table 26.1.

TABLE 26.1 List of Some Common Organophosphate Insecticides and Their Metabolites

Parent Compound	Specific Metabolite(s)
Azinphos-methyl/-ethyl	1,2,3-Benzotriazin-4-one
Chlorfenvinphos	2-Chloro-1-(2,4-dichlorophenyl) vinyl ethyl hydrogen phosphate, 1-(2,4-dichlorophenyl)ethyl-β-D-glucuronic acid, 2,4-dichloromandelic acid, and 2,4-dichlorophenylethanediol glucuronide
Chlorpyrifos-methyl/-ethyl	Chlorpyrifos oxon and 3,5,6-trichloro-2-pyridinol
Coumaphos	3-Chloro-4-methyl-7-hydroxypyrimidine
Diazinon	Diazoxon and 2-isopropyl-4-methyl-6-hydroxypyrimidine
Fenitrothion	3-Methyl-4-nitrophenol
Isazofos, methyl/ethyl	5-Chloro-1,2-dihydro-1-isopropyl-[3H]-1,2,4-triazol-3-one
Malathion	Malaoxon, 2-[(dimethoxyphosphorothioyl) sulfanyl] succinic acid, malathion monocarboxylic acid, and malathion dicarboxylic acid
Methamidophos/acephate	*O,S*-Dimethyl hydrogen phosphorothioate
Parathion-methyl/-ethyl	Paraoxon, 4-nitrophenol, also known as *p*-nitrophenol
Pirimiphos-methyl	2-Diethylamino-6-methylpyrimidin-4-ol

Currently, a total of six dialkyl phosphate (DAP) metabolites are commonly identified as biomarkers of OP pesticide exposure and quantified in urine using gas chromatography–mass spectrometry (GC-MS) as biomarkers of OP pesticides exposure. Among these, there are three dimethyl phosphate metabolites (dimethylphosphate, dimethylthiophosphate, and dimethyldithiophosphate) and three diethyl phosphate metabolites (diethylphosphate, diethylthiophosphate, and diethyldithiophosphate). Maravgakis et al. (2012) demonstrated accumulation of diethylphosphates in rabbit hair as a biomarker of chronic exposure to diazinon and chlorpyrifos (CPF). Dimethyl phosphates are derived from pesticides such as dichlorvos, monocrotophos, and dicrotophos, and diethylphosphates are derived from pesticides such as chlorfenvinphos and paraoxon (Huen et al., 2012). These are the most commonly assayed metabolites for the exposure, biomonitoring, and risk assessment of OP pesticides (Duggan et al., 2003; Oulhote and Bouchard, 2013; Harley et al., 2016). Dulaurent et al. (2006) reported a liquid chromatography–mass spectrometry (LC-MS) method for simultaneous determination of six DAPs in urine. DAP concentrations provide nonspecific information about exposure to a class of OPs, rather than a specific OP compound. Currently, among OPs, the most common metabolite measured is 3,5,6-trichloropyridinol, a major metabolite of CPF (Smith et al., 2012; Morgan and Jones, 2013). Specific metabolites of malathion, such as malaoxon, malathion dicarboxylic acid, and α and β isomers of malathion monocarboxylic acid, are also measured. Martinez and Ballesteros (2012) determined chlorfenvinphos and its metabolites (2-chloro-1-(2,4-dichloro-phenyl)vinyl ethyl hydrogen phosphate; 1-(2,4-dichlorophenyl)ethyl-β-D-glucuronic acid; 2,4-dichloromandelic acid; and 2,4-dichlorophenylethanediol glucuronide) in various tissues and fluids using Gas Chromatograph-Flame Ionization Detector (GC-FID) and GC-MS. Other commonly determined metabolites of OPs include 2-isopropyl-4-methyl-6-hydroxypyrimidine, a metabolite of diazinon; 4-nitrophenol, a metabolite of parathion, methyl parathion, or o-ethyl o-4-nitrophenyl phenylphosphonothioate (EPN) (Barr and Buckley, 2011); and 3-methyl-4-nitrophenol, a metabolite of fenitrothion (Okamura et al., 2012). In a recent report, Appenzeller et al. (2017) suggested the use of hair analysis for residue determination of two commonly used OPs (diazinon and CPF) and metabolites (diethylphosphate, DEP; diethylthiophosphate, DETP; and trichloropyridinol, TCPy).

Residues of several CMs are measured in serum, plasma, or whole blood as biomarkers of exposure to CMs. Recently, Appenzeller et al. (2017) also used hair for the analysis of CMs. In general, CMs are unstable in blood, so their metabolites are often assayed. Metabolite(s) of carbaryl (1-naphthol), propoxur (2-isopropoxyphenol), and carbofuran (3-ketocarbofuran, 3-hydroxycarbofuran, and carbofuranphenol), methiocarb (methylthio-3,5-xylenol and methiocarb sulfoxide), and carbosulfan (carbofuran and 3-hydroxycarbofuran) are determined in serum or plasma for biomonitoring of CM insecticides (Gupta et al., 1994a,b; Tange et al., 2016; Zhang et al., 2018). Some commonly used CMs and their metabolites, which are used as biomarkers, are listed in Table 26.2.

Some CM insecticides commonly measured in urine include aldicarb, carbofuran, carbosulfan, and pirimicarb. Carbaryl exposure is usually estimated based on urinary measurements of 1-naphthol. However, 1-naphthol and 2-naphthol are also metabolites of naphthalene. Because the measurement of 1-naphthol does not distinguish these two pesticides, measurement of 4-hydroxycarbaryl glucuronide helps to circumvent this problem.

TABLE 26.2 List of Common Carbamate Insecticides and Their Metabolites

Parent Compound	Specific Metabolite(s)
Aldicarb	Aldicarb oxime, aldicarb sulfone, aldicarb sulfoxide, and aldicarb nitrile
Carbaryl	1-Naphthol and 4-hydroxycarbaryl glucuronide
Carbofuran	3-Hydroxycarbofuran, 3-ketocarbofuran, carbofuranphenol, 3-hydroxycarbofuranphenol, and carbofuran protein adducts
Carbosulfan	Carbofuran and 3-hydroxycarbofuran
Methiocarb	Methylthio-3,5-xylenol and methiocarb sulfoxide
Methomyl	Methomyl oxime
Propoxur	2-Isopropoxyphenol

In some studies, OP and CM insecticides and their metabolites have also been measured in saliva. The data can be used as biomarkers of exposure and for pharmacokinetics and pharmacodynamics (Timchalk, 2006, 2010; Timchalk et al., 2007; Smith et al., 2012). In two acute carbofuran poisoning cases, residues of carbofuran and its major metabolite 3-hydroxycarbofuran were detected in human hair (Dulaurent et al., 2011). Recently, in a rat model, Appenzeller et al. (2017) used hair analysis for residue determination of 2-isoprpoxyphenol (2-IPP) and carbofuranphenol. Novel biomarkers of exposure to OP insecticides, nerve agents, and CM insecticides form their adducts with serine of butyrylcholinesterase and tyrosine of albumin, transferrin, tubulin, keratin, and other proteins (Grigoryan et al., 2009; Li et al., 2009, 2010; Stefanidou et al., 2009; Read et al., 2010; Schopfer et al., 2010; Tacal and Lockridge, 2010; Rehman et al., 2016). These measurements can be directly related to the dose of an insecticide and are a function of insecticide exposure.

Blood AChE inhibition is still considered one of the most sensitive and reliable biomarkers of exposure to OP and CM insecticides. Scientific and regulatory communities have identified and recognized red blood cell (RBC) AChE inhibition as a sensitive biomarker of exposure to OPs and CMs because it serves as a sensitive surrogate endpoint for the inhibition of brain AChE. In California, regulators specify monitoring blood ChE when using pesticides with toxicities <50 mg/kg (Class I pesticides) and ≥ 50 and ≤ 500 mg/kg (Class II pesticides). Examples are aldicarb, azinphos-methyl (Class I pesticides), and malathion (Class II pesticide). In a recent experimental study, Reiss et al. (2012) stated that brain AChE inhibition is the most appropriate matrix for risk assessment of OP pesticides, such as CPF. Muttray et al. (2005) found that EEG is possibly more sensitive than ChE inhibition in farmers exposed to low doses of methyl parathion.

In recent years, egasyn-β-glucuronidase has also gained attention for being a sensitive biomarker of an acute OP exposure (Ueyama et al., 2010). In another study, liver prenylated methylated protein methyl esterase has been found as a sensitive enzyme in liver and brain to the exposure and effect of OPs. Perturbations in prenylated protein metabolism are involved in OP-induced noncholinergic toxicity, as prenylated proteins play a role in cell signaling, proliferation, differentiation, and apoptosis (Lamango, 2005). Yardan et al. (2013) suggested S100B protein as a potential biomarker in the assessment of clinical severity and prediction of mortality in acute OP poisoning. Alterations in these enzymes and proteins can be used as biomarkers of OP exposure or effects, but none of these can reveal the identity of a particular OP insecticide. Additionally, multiple noncholinergic mechanisms are involved in OP- and CM-induced developmental neurotoxicity that occurs at levels of OPs far below those that cause AChE inhibition (Slotkin, 2006; Bouchard et al., 2011; Slotkin and Seidler, 2012). Using an NMR-based metabonomic approach, Liang et al. (2012) reported alterations of urinary profiles of endogenous metabolites in rats subacutely exposed to propoxur. Interestingly, in many of the investigations cited here, noncholinergic alterations occurred at doses below those that induce AChE inhibition. These noncholinergic endpoints can also be used as biomarkers of OP- or CM-induced effects.

Intermediate Syndrome

Humans exposed to large doses of an OP (methamidophos, fenthion, dimethoate, monocrotophos, etc.) or a CM (carbofuran) insecticide can suffer from intermediate syndrome (IMS) after the acute cholinergic crisis. IMS usually occurs 24–96 h after exposure to the insecticide due to insufficient or lack of oxime therapy and severe and prolonged AChE inhibition. IMS is clearly a separate entity from acute toxicity and delayed neuropathy, and it is characterized by acute respiratory paresis and muscular weakness, primarily in the facial, neck, and proximal limb muscles. IMS is also accompanied by generalized weakness, cranial nerve palsies, depressed deep tendon reflexes, ptosis, and diplopia. These symptoms may last for several days or weeks, depending on the insecticide involved. It needs to be pointed out that despite severe AChE inhibition, muscle fasciculations and mAChRs-associated hypersecretory activities are absent. Multiple mechanisms appear to be involved in IMS, such as AChE inhibition, oxidative stress, nAChR mRNA expression, and changes in repetitive nerve stimulation (RNS). In addition, electrophysiological abnormalities occur at the neuromuscular junctions in patients with IMS. All these parameters can be used as biomarkers of IMS induced by OPs or CMs (De Bleecker, 2006).

Delayed Toxicity

Exposure to some OPs can cause another type of toxicity known as OP-induced delayed polyneuropathy (OPIDP). Symptoms of OPIDP appear after symptoms of acute cholinergic crisis and IMS subside, i.e., 2–3 weeks after a single-dose exposure. Symptoms of this syndrome include tingling of the hands and feet, followed by sensory loss, progressive muscle weakness, and flaccidity of the distal skeletal muscle of the lower and upper extremities, and ataxia. Pathogenesis of OPIDP involves phosphorylation (inhibition) of neuropathy target esterase (NTE) enzyme and its "aging" in peripheral nerves (Gupta and Milatovic, 2012). NTE is a large polypeptide of 1327 amino acids, a membrane-bound esterase with a molecular weight of 155 kDa. Its physiological role has not yet been established. Large epidemics of OPIDP occurred in the United States, Morocco, Fiji, and India from ingestion of triorthocresyl phosphate affecting tens of thousands of people. OPIDP has also been caused by certain OP pesticides, such as leptophos, dichlorvos, fenthion, trichloronat, trichlorfon, methamidophos, and CPF (Lotti and Moretto, 2005; Jokanović et al., 2011). Although the severity of OPIDP does not appear to correlate with the degree of aged NTE, NTE inhibition and its aging are still considered as the best understood mechanisms and biomarkers (Mangas et al., 2012; Heutelbeck et al., 2016).

Chronic Toxicity

Following chronic exposure to low levels of OPs, pesticide application workers, greenhouse workers, agricultural workers, and farm residents reveal a relatively consistent pattern of neurobehavioral deficits (Rohlman et al., 2011; Lein et al., 2012). Patients can suffer from chronic OP–induced neuropsychiatric disorder with symptoms of anxiety, depression, difficulty in concentrating, memory impairment, and others (Jamal et al., 2002; Pancetti et al., 2007; Ross et al., 2010; Chen, 2012; Payan-Renteria et al., 2012). Pancetti et al. (2007) described the role of an enzyme, acylpeptide hydrolase, in cognitive processes, which are usually compromised following chronic OP exposure. Mixers, loaders, applicators, and flaggers, but not field workers themselves, are tested for blood ChE activity, if they work with pesticides for 7 days or more within a 30-day period (Wilson et al., 2005). Some studies demonstrated a link between neurobehavioral performance and current biomarkers of OP exposure including blood ChE activity and urinary metabolites of OPs. Arguably, other studies suggested that these biomarkers are neither predictive nor diagnostic of the neurobehavioral effects of chronic pesticide exposure, and biomarkers of neuroinflammation and oxidative stress need to be included, in addition to blood ChE and urinary metabolites (Banks and Lein, 2012). Pancetti et al. (2007) described acylpeptide hydrolase as a more sensitive enzyme than AChE and therefore proposed it as a biomarker of OP exposure and associated cognitive deficits.

In addition, dystonic reactions, schizophrenia, cogwheel rigidity, choreoathetosis, and EEG changes have been reported with high-dose exposure. These extrapyramidal symptoms are thought to be due to the inhibition of the AChE in the human extrapyramidal area. Psychosis, delirium, aggression, hallucination, and depression may also be seen during recovery from the cholinergic syndrome. Schizophrenic and depressive reactions with severe memory impairment and difficulties in concentration are observed in workers exposed to these pesticides, and RBC AChE is used as a biomarker of these effects.

In an experimental study, Middlemore-Risher et al. (2010) observed that repeated exposures to low-level CPF caused impairments in sustained attention and increased impulsivity in rats. Speed et al. (2012) examined hippocampal synaptic transmission and pyramidal neuron synaptic spine density in mice treated with a non–signs-producing dose of CPF (5 mg/kg/day for 5 consecutive days) early (2–7 days) and late (3 months) after the last injection. Increased synaptic transmission was found in the CA3–CA1 region of the hippocampus of CPF-treated mice 2–7 days after the last injection. In contrast, 3 months after CPF treatment, a 50% reduction in synaptic transmission occurred due to a 50% decrease in CA1 pyramidal neuron synaptic spine density. Findings suggested that progressive synaptic abnormalities occurred leading to persistent brain damage, despite the absence of cholinergic toxicity. Rats acutely intoxicated with a signs-producing dose of carbofuran (1.5 mg/kg, sc) or DFP (1.5 mg/kg, sc) showed marked decreases in dendritic length and spine density in the CA1 region of hippocampal pyramidal neurons because of oxidative/nitrosative stress and depletion of high-energy phosphates (Gupta et al., 2007; Zaja-Milatovic et al., 2009). Both AChE inhibitors also caused significant increases in PGE2, a marker of neuroinflammation. Of course, cytokines and C-reactive protein can also be used as biomarkers of inflammation (Lein et al., 2012).

OPs and CMs produce a variety of reproductive and developmental toxicity in humans and animals. Following in utero exposure to OP/CM insecticides, brain damage can occur in children in an absence of cholinergic toxicity. Biomarkers for reproductive and developmental effects of these insecticides are described in detail by Gupta et al. (2017).

In an electrophysiological study, Nio and Breton (1994) investigated the effects of paraoxon and physostigmine on rabbit pyramidal cells firing pattern and hippocampal theta rhythm. The results revealed that paraoxon and physostigmine have a rather similar influence on the septohippocampal pathway and also suggested that paraoxon could act within local hippocampal circuitry through other systems than the cholinergic system exclusively. In similar studies, Desi and Nagymajtenyi (1999), Papp et al. (2004), and Narahashi (2006) investigated electrophysiological effects of various OPs and CMs, suggesting the involvement of noncholinergic mechanisms. In a preliminary study, De Luca et al. (2006) showed the electromyographic signal as a presymptomatic indicator of OPs in the body, using diisopropyl fluorophosphate (DFP) as a test substance. These parameters can be used as biomarkers of morphological and electrophysiological changes in central nervous system (CNS) and peripheral nervous system (PNS) and early indicators of behavioral alterations.

A human genetic polymorphism of paraoxonase-1 (PON-1) in detoxification of several OP insecticides, including the active metabolite of CPF, CPF oxon, is well established, resulting in the expression of PON-1 activity within a segment of the population. PON-1 and cytochrome P450 can be used as biomarkers of genetic polymorphism in regard to OP toxicity (Costa et al., 2006; Furlong et al., 2006; Huen et al., 2012; Lein et al., 2012). Huen et al. (2012) found that, compared with their mothers, newborns have much lower quantities of the detoxifying PON-1 enzyme, suggesting that infants may be especially vulnerable to OP pesticide exposure.

CHLORINATED HYDROCARBONS

Chlorinated hydrocarbon or organochlorine insecticides are classified into three groups: (1) dichlorodiphenylethanes (dichlorodiphenyltrichloroethane [DDT], dicofol, methoxychlor, and perthane); (2) hexachlorocyclohexanes (benzene hexachloride, chlordane, lindane, mirex, and toxaphene); and (3) chlorinated cyclodienes (aldrin, dieldrin, endrin, endosulfan, and heptachlor). The first use of the chlorinated hydrocarbons or organochlorines was for dielectrics and as fire retardants. The first use of these compounds as insecticides occurred when benzene was added to liquid chlorine in the field and it was noted that the product killed insects. Cyclodienes, such as aldrin and dieldrin, used as insecticides, became available for use in the 1940s.

DDT became available during World War II and was used extensively as an insecticide worldwide. Because of their persistence in the environment and in biological systems, most insecticides of this group have been eliminated from use today. Endosulfan and lindane are the most biodegradable organochlorines and are still used today.

Mechanism of Toxicity

There are at least two different mechanisms of action for organochlorine insecticides (Narahashi, 1987; Ensley, 2018a). DDT type insecticides affect the peripheral nerves and brain by slowing down Na^+ influx and inhibiting K^+ efflux, resulting in excess intracellular K^+ in the neuron, which partially depolarizes the cell. The threshold for another action potential is decreased leading to premature depolarization of the neuron. The aryl hydrocarbons and cyclodienes, in addition to decreasing action potentials, may inhibit the postsynaptic binding of GABA (Bloomquist and Soderlund, 1985). The cyclodiene-induced hyperactivity of the CNS and convulsions can be explained based on their structural resemblance to the GABA receptor antagonist picrotoxin. The cyclodienes act by competitive inhibition of the binding of GABA, an inhibitory neurotransmitter at its receptor, causing stimulation of the neurons. Both $GABA_A$ and $GABA_B$ receptors play a vital role in mammalian toxicity. $GABA_A$ receptors present in the mammalian synapse are ligand-gated chloride ion channels. In the human brain, the $GABA_A$ receptor consists of four or five hydrophobic domains, namely M_1, M_2, M_3, M_4, and M_5. The five M_2 domains are arranged to form a 5.6 Å diameter ion channel. $GABA_B$ receptors present in mammals are coupled to calcium and potassium channels and the action of this neurotransmitter is mediated by G proteins. Following its release in the synapse, GABA diffuses to the presynaptic terminal of another nerve, where GABA binds to its $GABA_A$ receptors. The binding of GABA to its receptor causes chloride ions to enter the synapse, leading to hyperpolarization of the terminal, which inhibits the release of other neurotransmitters. Due to such inhibition, postsynaptic stimulation of other nerves by other neurotransmitters, such as ACh, is reduced. As a consequence of inhibition of GABA, there is no synaptic downregulation and there can be excessive release of other neurotransmitters. CNS symptoms in animals poisoned by chlorinated cyclodienes include tremors, convulsions, ataxia, and changes in EEG patterns. The CNS symptoms could be due either to (1) inhibition of the Na^+/K^+-ATPase or the Ca^{2+}/Mg^{2+}-ATPase activity, which can then interfere with nerve action or release of neurotransmitters, or (2) inhibition of the GABA receptor function. Mladenovic et al. (2010) reported that lipid peroxidation may contribute to the neurotoxic effects of lindane in early acute intoxication and that behavioral manifestations correlate with lipid peroxidation in the rat brain hippocampus, which is one of the sites for initiation and propagation of seizures.

The nervous system of the developing organism appears to be more vulnerable to the toxicity of organochlorine insecticides, and multiple mechanisms are involved. DDT and related compounds produce a direct effect on the motor fibers and on the motor area of the cerebral cortex or they act as endocrine disruptors in the hypothalamic–hypophysis–thyroid axis. Neonatal exposure to DDT causes a significant reduction in the density of muscarinic receptors in the cerebral cortex of mice. These cholinergic receptors have a direct involvement in the process of neuronal excitement and inhibition. Following chronic exposure, DDT causes disruption of the thyroid system. DDT or its metabolites alter the production of thyroidal hormone and its availability to target tissues, given that the insecticide blocks glucuronidation. Thyroid hormone is known to play a pivotal role in the development of the cerebral cortex. This hormone serves as a signal for neuronal differentiation and maturation and participates in neuronal migration and proliferation, synaptogenesis, and myelinization.

Developmental exposure to dieldrin has been shown to alter the dopamine system and increase neurotoxicity in an animal model of Parkinson's disease. The dopamine transporter (DAT) and the vesicular monoamine transporter 2 (VMAT2) play an integral role in maintaining dopamine homeostasis, and the alteration of their levels during development could result in increased vulnerability of dopamine neurons later in life. Prenatal exposure of mice during gestation and lactation to low levels of dieldrin (0.3, 1, or 3 mg/kg every 3 days) caused a dose-dependent increase in protein and mRNA levels of DAT and VMAT2 in their offspring at 12 weeks of age (Richardson et al., 2006). In a recent investigation, Kamel et al. (2012) suggested that amyotrophic lateral sclerosis risk may be associated with use of organochlorine pesticides.

Biomarkers

Residues of organochlorine insecticides and their metabolites can be detected at ppb levels in human and animal tissues and fluids (blood, urine, and milk). Recently, Appenzeller et al. (2017) also suggested hair analysis for determination of organochlorines (γ-hexachlorocyclohexane (HCH), β-HCH, β-endosulfan, p,p′-DDT, p,p′-DDE, p,p′-DDD, dieldrin, and pentachlorophenol). GC and GC-MS are most commonly used to confirm and quantitate the residues of organochlorines and their metabolites. Biomonitoring studies have revealed that these insecticides are widely distributed throughout the body and deposit in fat and fatty tissue. Residues have been found in various human reproductive tissues, including amniotic fluid, blood stream, maternal blood, umbilical cord blood, breast milk, colostrum, placenta, semen, and urine (Malik et al., 2017). Studies suggest that the organochlorine endosulfan causes AChE inhibition, neurotoxicity, and brain function impairment in rats, rabbits, and zebrafish (Mor and Ozmen, 2010; Silva de Assis et al., 2011; Pereira et al., 2012).

In general, residues of organochlorine insecticides and/or their metabolites in blood/serum/plasma, fat, and urine are used as biomarkers of exposure. In the case of endosulfan, AChE inhibition can also be used as a biomarker of toxicity.

PYRETHRINS AND PYRETHROIDS

Pyrethrins and pyrethroids (synthetic pyrethrins) have a wide variety of applications in agriculture, public and animal health, and residential settings throughout the world. Pyrethroid insecticides are also used to disinfect commercial aircraft and military uniforms (Wei et al., 2012; Proctor et al., 2014). Pyrethrins were first developed as insecticides from extracts of the flower heads of *Chrysanthemum cinerariaefolium*. There are six naturally occurring pyrethrins (pyrethrins I and II, cinerins I and II, and jasmolins I and II). Because of their rapid decomposition by light and air, synthetic derivatives were developed and are commonly referred to as pyrethroids (Anadon et al., 2009). The first-generation pyrethroids were developed in the 1960s, including bioallethrin, tetramethrin, resmethrin, and bioresmethrin. By 1974, a second generation of more persistent compounds was developed, including permethrin, cypermethrin, and deltamethrin. These insecticides have significantly greater mammalian toxicity compared with first-generation pyrethroids. Later, some other pyrethroids, such as fenvalerate, lambda-cyhalothrin, and β-cyfluthrin, were discovered. Because of their high insecticidal potency, broad-spectrum activity, relatively low mammalian toxicity, lack of environmental persistence, and less insect resistance, pyrethroids have gained much success in the recent past and now account for more than 25% of the global insecticide market.

Using a recent nomenclature, pyrethroids are of two types and produce two syndromes through multiple mechanisms of action. Type I pyrethroids are those that lack α-cyano-3-phenoxybenzyl moiety and give rise to the tremor syndrome (T syndrome). A few examples of type I are pyrethrin I, allethrin, bioallethrin, tetramethrin, resmethrin, cismethrin, phenothrin, and permethrin. These insecticides cause severe fine tremors, marked reflex hyperexcitability, sympathetic activation, and paresthesia (with dermal exposure). Type II pyrethroids are those that contain α-cyano-3-phenoxybenzyl moiety and cause the choreoathetosis/salivation (CS) syndrome. A few examples of type II pyrethroids include cyphenothrin, cypermethrin, deltamethrin, fenvalerate, cyfluthrin, cyhalothrin, flucynthrate, and esfenvalerate. These insecticides produce profuse watery salivation, coarse tremors, increased extensor tone, moderate reflex hyperexcitability, sympathetic activation, choreoathetosis, seizures, and paresthesia (dermal exposure).

The clinical signs of type I and type II pyrethroids appear to be largely independent of the route of administration. *Trans-* and *cis*-isomers of fluorocyphenothrin confusingly produce type I and type II effects, respectively. Some pyrethroids, such as fenpropathrin, have been classified as hybrids and appear to show a mixture of type I and II effects (T and CS). All pyrethroids cause a reduction in motor activity. Reduced locomotor activity in the rat after oral gavage dosing has been used to quantify pyrethroid neurotoxicity. Unfortunately, the effect is nonspecific and does not readily allow conclusions to be made about mechanism of action (Gammon et al., 2012). Overall, type I pyrethroids with a *trans*-substituted acid moiety have a much lower mammalian toxicity than the corresponding *cis* isomers. This is partly due to the more rapid ester hydrolysis of *trans* rather than *cis* isomers. For type II pyrethroids, there is much less difference in toxicity between *trans* and *cis* isomers, suggesting a similar target site for type II pyrethroids.

Both pyrethroid classes have a similar range of mammalian toxicity, but for commercial pesticides, type II pyrethroids such as deltamethrin and cypermethrin are generally more toxic than type I pyrethroids such as permethrin (Ray and Forshaw, 2000). A survey of the general US population has revealed that exposure to pyrethroid insecticides is widespread and that children may have higher exposures than adolescents and adults (Barr et al., 2010). In general, young animals and children are more susceptible than adults to pyrethroid toxicity, and the focus of ongoing research is to explore to which degree children are more susceptible than adults. Flight attendants of commercial aircrafts disinfected with pyrethroids have complained of irritation of the skin and mucosa, sore throat, vomiting, abdominal pain, headache, dizziness, and nausea (Murawski, 2005; Sutton et al., 2007; Wei et al., 2012). It is important to mention that pyrethrins and pyrethroids also exert reproductive and developmental toxicity in animals and humans (Malik et al., 2017; Martin-Reina et al., 2017; Saito et al., 2017).

Mechanism of Toxicity

The primary effects of pyrethroids in mammalian species include neurotoxicity and neurobehavioral toxicity, which result primarily from hyperexcitation of the nervous system. Type I syndrome involves action in the CNS and PNS, whereas type II syndrome (CS) involves primarily the CNS. Hyperexcitation of the nervous system is caused by repetitive firing and depolarization of the nervous system. The primary site of action of pyrethroids is the sodium channels of cells, but they also affect chloride and calcium channels. These insecticides slow down the opening and closing of the sodium channels, ultimately leading to the excitation of the cell. An increase of sodium in the sodium channels results in a cell which is in a stable and hyperexcitable state. The duration of the sodium action potential is much shorter for type I pyrethroids than for type II. While type I pyrethroids result in primarily repetitive charges, cell membrane depolarization is the main mechanism of toxic action exerted by type II pyrethroids. The effect of type I pyrethroids on sodium channels in nerve membranes is similar to those produced by DDT-type insecticides. The direct action of pyrethroids on sensory nerve endings leads to paresthesia, as they cause repetitive firing (Ensley, 2018b).

If not all, most of the neurotoxic effects of pyrethroids are due to their interaction at voltage-gated sodium channels. The degree of hyperexcitability is dose-related, but the nature of this excitability is pyrethroid structure–dependent. Insect sodium channels are hundred times more sensitive than rat brain sodium channels. The lower sensitivity of mammalian sodium channel isoforms compared with insect channels to pyrethroids offers a mechanistic explanation for its species selective toxicity. Of course, it has become clear that besides insensitive mammalian sodium channels (e.g., Na_V 1.2 and Na_V 1.7), more sensitive isoforms also exist (e.g., Na_V 1.3, Na_V 1.6, or Na_V 1.8). McCavera and Soderlund (2012) provided direct evidence for the preferential binding of deltamethrin and tefluthrin (but not S-bioallethrin) to Na_V 1.6Q3 channels in the open state and implied that the pyrethroid receptor of resting and open channels occupies different conformations that exhibit distinct structure–activity relationships. Interestingly, the Na_V 1.6 isoform is not only a target for developmental neurotoxic effects but also a likely target for the central neurotoxic effects of pyrethroids in adult animals.

Voltage-gated chloride and calcium channels have been implicated as additional sites of action (Soderlund, 2012). Voltage-sensitive chloride channels are found in nerves, muscles, and salivary glands and are modulated by protein kinase C. The function of chloride channels is to control cell excitability. The decrease in the chloride-open channel state produced by type II pyrethroids serves to increase excitability and therefore to synergize pyrethroid actions on the sodium channel. While calcium channels may contribute in some way to the action of at least some pyrethroids, there is no basis at present to identify effects on calcium channels as an essential mechanism of pyrethroid intoxication (Shafer and Meyer, 2004). Gammon et al. (2012) suggested that different clinical signs associated with type I and type II syndromes result from effects on different channel isoforms or a combination of them. Interestingly, Ali (2012) demonstrated involvement of oxidative stress in pathogenesis of neurotoxicity by type II pyrethroid lambda-cyhalothrin in rats. For further details on molecular mechanisms in neurotoxicity and neurobehavioral toxicity of pyrethrins and pyrethroids, readers are referred to Wolansky and Harrill (2007), Soderlund (2012), and Van Thriel et al. (2012).

Recently, Lei et al. (2017) investigated the inhibitory effects of six commonly used pyrethroids (fenpropathrin, cyhalothrin, deltamethrin, fenvalerate, *tau*-fluvalinate, and permethrin) on two major human carboxylesterases (CES1 and CES2). Deltamethrin was found to be a strong inhibitor of CES1. Although its toxicological significance is yet to be determined, it could be used as a surrogate marker of deltamethrin exposure.

Biomarkers

Pyrethrins and pyrethroids are structurally diverse, and understanding their chemistry and toxicology plays a vital role in the development of biomarkers. Significant advancements in analytical chemistry and toxicology have led to the development of biomarkers to assess biomonitoring in the environment and exposure in the general population (Sudakin, 2006; Barr et al., 2010; Gammon et al., 2012; Furlong et al., 2017; Yoshida, 2017). Future challenges in the application of these biomarkers in epidemiological studies are being explored, as there is a need for improved understanding of the toxicokinetics and pharmacodynamics of pyrethroids in mammalian species, including humans.

Pyrethroids are of low to moderate toxicity because of their moderate absorption (40%–60%) and rapid metabolism following oral administration (Anand et al., 2006; Barr et al., 2010; Gammon et al., 2012; Furlong et al., 2017; Saito et al., 2017). Oxidases and esterases, primarily in the liver, metabolize pyrethroids at varying rates. The most rapidly metabolized pyrethroids have the lowest toxicity. Based on experimental data, Gammon et al. (2012) described that brain or blood plasma concentrations of parent pyrethroids correlate with acute toxicity and that metabolites (especially hydrolytic products) generally have little or no effect on neurotoxicity. In the context of biomarkers, most studies have investigated urine as the analytical matrix (Barr et al., 2010; Weilgomas, 2013; Weilgomas et al., 2013; Furlong et al., 2017; Oulhote and Bouchard, 2013; Yoshida, 2017). Leng et al. (1996) and Leng and Gries (2005) detected metabolites of some pyrethrins and pyrethroids (permethrin, cypermethrin, cyfluthrin, and deltamethrin), using GC-MS, in the urine of pesticide applicators. Using the same methodology, Wei et al. (2012) assayed metabolites of these pyrethroids in the urine of commercial flight attendants. Elflein et al. (2003) also used GC-MS to detect metabolites of some other commonly used pyrethroids, including allethrin, resmethrin, phenothrin, and tetramethrin in human urine. Recently, Yoshida (2017) developed an analytical method to detect 11 pyrethroid metabolites in urine using GC-MS. High-performance liquid chromatography (HPLC)– and LC-MS–based methods have also been employed to determine the residues of pyrethroids and their metabolites in body tissues and fluids (Baker et al., 2004; Anand et al., 2006; Kim et al., 2006; Barr et al., 2010). Ishibashi et al. (2012) developed a high-throughput assay for cypermethrin and tralomethrin using supercritical fluid chromatography–tandem mass spectrometry.

Metabolites of some of the commonly used pyrethrins/pyrethroids determined in urine as biomarkers are summarized in Table 26.3.

Barr et al. (2010) reported that 3-phenoxybenzoic acid (3-PBA), a metabolite common with many pyrethroid insecticides, was detected in >70% of urine samples tested in the United States. Non-Hispanic blacks had significantly higher 3-PBA concentrations than non-Hispanic whites and Mexican Americans, and children had significantly higher concentrations of 3-PBA than adolescents and adults. *Cis-* and *trans*-(2,2-dichlorovinyl)-2,2-dimethylcyclopropane-1-carboxylic acid (*cis-* and *trans*-Cl2CA) were highly correlated with each other and with 3-PBA, suggesting that 3-PBA was primarily derived from exposure to permethrin, cypermethrin, or their degradates. Wei et al. (2012) found that the flight attendants working on pyrethroid-disinsected commercial aircraft had significantly higher concentrations of 3-PBA and *cis-* and *trans*-Cl2CA in the postflight urine samples than those working on nondisinsected aircrafts and the general US population. Increase of pyrethroid metabolites in the preflight urine samples suggested an elevated body burden from a long-term exposure for those flight attendants routinely working on pyrethroid-treated aircraft. Interestingly, flight attendants working on international flights connected to Australia had higher urinary levels of 3-PBA and *cis-* and *trans*-Cl2CA than those on either domestic or other international flights flying between Asia, Europe, and North America. At veterinary diagnostic labs, residues of pyrethrins/pyrethroids have also been detected in the brains of cats that died from overexposure to these insecticides.

TABLE 26.3 Metabolites of Pyrethroids in Urine Used as Biomarkers for an Exposure to Pyrethroid Insecticides

Parent Compound	Specific Metabolite(s)
Allethrin	*Trans*-chrysanthemumdicarboxylic acid
Cyfluthrin	4-Fluoro-3-phenoxybenzoic acid, 3-phenoxybenzoic acid (3-PBA), *cis-* and *trans*-3-(2,2-dichlorovinyl)-2,2-dimethylcyclopropane carboxylic acid
Lambda-cyhalothrin	*Cis*-3-(2-chloro-3,3,3-trifluoroprop-1-en-1-yl)-2,2-dimethylcyclopropane carboxylic acid and 3-PBA
Cypermethrin	*Trans*-chrysanthemumdicarboxylic acid, *cis-* and *trans*-3-(2,2-dichlorovinyl)-2,2-dimethylcyclopropane carboxylic acid; 3-PBA
Deltamethrin	*Cis-* and *trans*-3-(2,2-dichlorovinyl)-2,2-dimethylcyclopropane carboxylic acid; *cis*-3-(2,2-dibromovinyl)-2,2-dimethylcyclopropane carboxylic acid; 3-PBA
Permethrin	*Cis-* and *trans*-3-(2,2-dichlorovinyl)-2,2-dimethylcyclopropane carboxylic acid; 3-PBA
Phenothrin	*Trans*-chrysanthemumdicarboxylic acid
Pyrethrum	*Trans*-chrysanthemumdicarboxylic acid
Resmethrin	*Trans*-chrysanthemumdicarboxylic acid
Tetramethrin	*Trans*-chrysanthemumdicarboxylic acid

Adapted from Malik, J.K., Aggarwal, M., Kalpana, S., and Gupta, R.C., 2017. Chlorinated hydrocarbons and pyrethrins/pyrethroids. In: Gupta, R.C. (Ed.), Reproductive and Developmental Toxicology. second ed. Academic Press/Elsevier, Amsterdam, pp. 633–655; Khemiri, R., Côte J., Fetoui, H., et al, 2017. Documenting the kinetic time course of lambda-cyhalothrin metabolites in orally exposed volunteers for the interpretation of biomonitoring data. Toxicol Lett. 276, 115–121, and others.

Hughes et al. (2016) realized that humans are often exposed to multiple pyrethroids. Accordingly, they determined residue of several pyrethroids in blood and brain of rats and suggested that summed pyrethroid rat blood concentration could be used as a surrogate for brain concentration as an aid to study the neurotoxic effect of pyrethroids in a mixture. Tornero-Velez et al. (2012) proposed a pharmacokinetic model for *cis-* and *trans*-permethrin disposition in rats and humans with aggregate exposure application. In this investigation, the description of pharmacokinetics in humans was based on the properties of permethrin, Physiologically-Based Pharmacokinetic Modeling (PBPK) models of deltamethrin in rats, and permethrin in vitro clearance data. The model was adapted with a biomarker submodel to evaluate exposure estimation in probabilistic risk assessment applications. Starr et al. (2012) investigated cumulative risk assessment of pyrethroid pesticides (permethrin, cypermethrin, β-cyfluthrin, deltamethrin, and esfenvalerate) taking into account pharmacokinetics, pharmacodynamics, and neurobehavioral assays. Findings supported the additive model of pyrethroid effect on motor activity and suggested that variation in the neurotoxicity of individual pyrethroids is related to toxicodynamic rather than toxicokinetic differences.

In addition to residue detection of pyrethrins and pyrethroids and their metabolites, hematological, biochemical, and histopathological alterations have been reported in target and nontarget tissues, and these changes can be used as biomarkers of pyrethrin/pyrethroid toxicity (Sayim et al., 2005; Yavasoglu et al., 2006).

AMITRAZ

Amitraz is a triazapentadiene compound of the formamidine pesticide family, which has a chemical formula of $C_{19}H_{23}N_3$ and a molecular weight of 293.41. Amitraz is a broad-spectrum insecticide and acaricide used in agriculture, horticulture, and veterinary medicine throughout the world since 1974. Amitraz is used to control ticks, mites, lice, and many other pests on dogs, sheep, cattle, and pigs. It persists on hair and wool long enough to control all stages of the parasite. Amitraz poisoning is frequently encountered in dogs and cats. With dogs, poisoning is most often associated with accidental ingestion of the flea and tick collar, resulting in severe toxicity and sometimes fatal poisoning. In humans, poisoning occurs because of oral ingestion of amitraz (Shitole et al., 2010). The major signs of acute poisoning are nausea, vomiting, coma, somnolence, miosis or mydriasis, bradycardia, hypo- or hyperthermia, polyuria, and respiratory failure. Amitraz should also not be used in diabetic animals, as it adversely affects the levels of glucose and insulin (Hsu and Schaffer, 1988).

Mechanism of Toxicity

Amitraz kills parasites by paralyzing their nervous system and sharp barbed mouth parts, thereby causing them to detach from animals. Unlike in insects, amitraz has been found to cause toxicity in animals by stimulating α_2-adrenergic receptors (α_2-AR), resulting in impairment of consciousness, respiratory depression, convulsions, bradycardia, hypotension, and hyperglycemia. Hypothermia occurs due to the inhibitory effect of amitraz on prostaglandin E2 synthesis. Amitraz acts centrally to influence blood pressure and heart rate by α_2-AR agonism, which causes a reduction in peripheral sympathetic tone (Cullen and Reynoldson, 1990). Marafon et al. (2010) observed significant declines in the heart rate and respiration rate of cats intoxicated with amitraz. Electrocardiography on an amitraz-poisoned English bulldog revealed prolonged QT intervals (Malmasi and Ghaffari, 2010). In the peripheral vasculature, both α_1- and α_2-AR are involved in the vasopressor action of amitraz, which results in hypotension. It is suggested that the central α_2-AR agonist activity of amitraz is responsible for CNS depression (Cullen and Reynoldson, 1990). Amitraz is also a potent inhibitor of the enzyme monoamine oxidase (MAO), which is responsible for degrading neurotransmitters (norepinephrine and serotonin), resulting in neurotoxicity and behavioral toxicity (Moser, 1991).

Poisoning is most often encountered in dogs and cats and is generally acute in nature (Filazi and Yurdakok-Dikmen, 2018). Onset of clinical signs is noted within 30 min to 2 h after ingestion. Common clinical signs include gastrointestinal (GI) disturbance (i.e., prolonged gastric transit time), CNS and respiratory depression, bradycardia, hypotension, and hypothermia. Death usually occurs due to respiratory failure.

Biochemical changes include hyperglycemia, glucosuria, suppressed insulin release, and elevation of liver transaminases activities. Histopathological changes may include hepatic and renal cortical necrosis, hemorrhage, and renal failure. Clinical symptoms reported in humans were giddiness, vomiting, drowsiness, and gastric dilatation (Shitole et al., 2010). Unconscious patients' CT brain scans revealed brain edema. For a detailed mechanism of amitraz toxicity, refer to del Pino et al. (2015).

Biomarkers

Amitraz is a highly lipid-soluble compound that is rapidly absorbed following oral ingestion or dermal application, thus making exposure potentially dangerous for animals and humans. In the stomach, amitraz can be metabolized to as many as six different metabolites and some of them are potentially toxic. The two major metabolites of amitraz are 2,4-dimethylformamide and *N*-(2,4-dimethylphenyl)-*N*′-methylformamide. The former metabolite is a relatively weak methemoglobin former in dogs and humans. These metabolites are further catabolized to 2,4-dimethylaniline and ultimately to 4-amino-3-methylbenzoic acid, which is the principal metabolite found in the liver and urine. A pharmacokinetic study in dogs revealed that amitraz is significantly absorbed and has a long elimination half-life, which is responsible for most of the observed clinical signs (Hugnet et al., 1996). Marafon et al. (2010) reported similar findings in cats. In ponies and sheep, amitraz has a brief half-life after intravenous (IV) administration because it is hydrolyzed in the blood by formaminidases (Pass and Mogg, 1995).

The proposed biomarkers of amitraz exposure and its toxicosis include detection of the residue of amitraz and its major metabolites in the plasma of poisoned animals or humans using GC with a nitrogen–phosphorus detector or thermionic specific detector (Marafon et al., 2010) or HPLC with a UV detector (Hugnet et al., 1996). Clinical signs of amitraz toxicity associated with α_2-AR agonism and biochemical and histopathological alterations can be used as biomarkers of amitraz toxicity. In a chronic study in rats, amitraz was reported to be embryotoxic and a teratogen at a maternally toxic dose of >10 mg/kg/day (Kim et al., 2007), and the no-observed-adverse-effect level (NOAEL) of amitraz for both dams and embryo fetal development was estimated to be 3 mg/kg/day. For further details on amitraz toxicity, see Filazi and Yurdakok-Dikmen (2018).

NEONICOTINOIDS

Neonicotinoids are a new class of insecticides, which includes imidacloprid, acetamiprid, thiacloprid, dinotefuran, nitenpyram, thiamethoxam, and clothianidin. These products are commonly used in agriculture and veterinary medicine. Neonicotinoids have a high target specificity to insects, a relatively low risk for nontarget mammalian species and the environment, and versatility in application methods (Ensley, 2018c). Among all neonicotinoids, imidacloprid is the most studied compound used as an insecticide for dermal application on animals, for grub control, and as an insecticide for crop protection. The neonicotinoids class accounts for about 20% of the current global insecticide market.

Mechanism of Toxicity

The neonicotinoids act on postsynaptic nicotinic acetylcholine receptors (nAChRs). In insects, these receptors are located entirely in the CNS. Mammalian tissue also contains multiple nAChRs subtypes, which are formed from different combinations of nine α, four β, and γ, δ, and ε subunits. In insects, there are binding sites in addition to nAChRs for neonicotinoids that are suggested to be used as mechanism-based biomarkers of exposure and effects (Badiou-Bénéteau et al., 2012). The selective toxicity of neonicotinoids in insects and mammals is attributed largely to the differential sensitivity of the insect and vertebrate nAChR subtypes. In insects, neonicotinoids act on at least three different subtypes of nAChRs and cause a biphasic response, i.e., an initial increase in the frequency of spontaneous discharge followed by a complete block to nerve propagation. In mammals, nAChRs are located in the brain, autonomic ganglia, skeletal muscle, and spinal cord. Neonicotinoids have a much lower activity in vertebrates as compared with insects because of the different binding properties of the various receptor subtypes. Tomizawa and Casida (2011) suggested that the high affinity for neonicotinoids at the insect nAChR is related to a single dominant binding orientation, whereas relative insensitivity at vertebrate nAChRs is caused by multiple binding confirmations in the agonist-binding pocket. Acute toxicity of the neonicotinoids in mammals is related to the potency at the α_7 nicotinic receptor subtype with the activity at the α_4, β_2, α_3, and α_1 receptors having a decreased effect on toxicity. Toxicity in mammals involves complex interactions at multiple receptor sites with some of the receptor types even having a combination of agonist and antagonist effects on the synapse (Ensley, 2018b). Furthermore, neonicotinoids have relatively poor penetration of the blood–brain barrier (BBB) in mammals (Rose, 2012).

Following an acute exposure to imidacloprid, clinical signs and symptoms can be observed as early as 15 min, with recovery within 8–24 h. At low doses, neonicotinoids cause decreased activity and tremors, mydriasis or miosis, and incoordination; and at higher doses, they cause hypothermia, staggering gait, salivation, trembling, and spasms. At lethal doses, death is observed within 4 h. Humans, after deliberately ingesting imidacloprid, show symptoms of aspiration pneumonia, CNS effects (agitation, confusion, and coma), and respiratory failure. Cardiovascular effects include tachycardia, palpitation, and ventricular fibrillation. It needs to be emphasized that nicotinoid metabolites derived from some neonicotinoids (e.g., imidacloprid and thiacloprid) are more potent vertebrate nAChR agonists than their parent molecules and are likely to be more toxic (Rose, 2012). Additional mechanisms have been described in neonicotinoids-induced toxic effects, such as hepatotoxicity (Swenson and Casida, 2013; Arfat et al., 2014), nephrotoxicity (Arfat et al., 2014), respiratory toxicity (Pandit et al., 2016), hyperglycemia (Khalil et al., 2017), genotoxicity (Stivaktakis et al., 2016), endocrine disruption (Pandey and Mohanty, 2015), and obesity (Park et al., 2013).

Biomarkers

Pharmacokinetic studies of imidacloprid revealed its rapid and complete absorption and distribution following oral administration in rats. Peak plasma levels were achieved within 2 h and the initial and terminal half-lives in plasma were about 3 h and 26–118 h, respectively. The highest tissue residues after 48 h were found in the liver, kidney, lung, and skin. Low concentrations of imidacloprid were found in the brain due to poor penetration of the BBB, and a low concentration was found in fat due to low lipophilicity. Elimination of 90% of the dose occurred within 24 h (75%–80% in the urine and the remainder in the feces originating from biliary excretion). Ensley (2018c) has described two pathways for imidacloprid metabolism in mammals. The first pathway involves oxidative cleavage of imidacloprid to imidazolidine and 6-chloronicotinic acid. The imidazolidine moiety is excreted in the urine. The 6-chloronicotinic acid is further degraded by glutathione conjugation to a derivative of mercapturic acid and then to methyl mercaptonicotinic acid. The mercaptonicotinic acid is then conjugated with glycine to form a hippuric acid conjugate, which is excreted. The second pathway involves hydroxylation of the imidazolidine ring, with two metabolites (5-hydroxyimidacloprid and 4-hydroxyimidacloprid) detected in the urine.

Residue detection of neonicotinoids and their metabolites in the body tissue/fluids serves as a biomarker of exposure. HPLC coupled with a UV or MS detector is commonly used to detect the residue of neonicotinoids and their metabolites. Signs and symptoms of toxicity and histopathological changes can also be used as biomarkers of neonicotinoid toxicity.

FIPRONIL

Fipronil is an insecticide of the phenylpyrazoles class and an active ingredient of one of the popular ectoparasiticide veterinary products, Frontline. Frontline is commonly used on dogs and cats to kill fleas and all stages of ticks, which may carry Lyme disease, and mites. Fipronil is also formulated as insect bait for roaches, ants, and termites; as a spray for pets; and as a granular form for turf and golf courses. The United States Environmental Protection Agency (USEPA) has determined fipronil to be safe for use on dogs and cats, with no harm to humans who handle these pets. Most of the time poisoning cases of fipronil occur in dogs and cats due to accidental ingestion of the product Frontline. In humans, poisoning is mainly due to accidental ingestions or suicide attempts. In agriculture, fipronil is widely used for soil treatment, seed coating, and crop protection.

Mechanism of Toxicity

The mechanism of action of fipronil is better understood in insects than it is in mammals. In insects, fipronil or its major metabolite (fipronil sulfone) noncompetitively binds to $GABA_A$-gated chloride channels, thereby blocking the inhibitory action of $GABA_A$ in the CNS. This leads to hyperexcitation at low doses and paralysis and death at higher doses. Fipronil exhibits a >500-fold selective toxicity to insects over mammals, primarily because of affinity differences in receptor binding between insect and mammalian receptors. In other words, fipronil binds more tightly to $GABA_A$ receptors in insects than in mammals. Fipronil is more selective at this receptor through the β3 subunit in insects than in mammals. This selectivity is less pronounced with fipronil metabolites (sulfone and desulfinyl). It needs to be emphasized that fipronil sulfone is rapidly formed in humans and experimental animals and can persist much longer than fipronil; therefore the toxicological effects are also likely due to the sulfone metabolite. The toxicity of another metabolite, fipronil desulfinyl, is qualitatively similar to that of fipronil, but the dose–effect curve for neurotoxic effects appears to be steeper for fipronil desulfinyl. In addition, fipronil desulfinyl appears to have a much greater affinity to bind to sites in the chloride ion channel of the rat brain GABA receptor. This finding appears to be consistent with the greater toxicity of fipronil desulfinyl in the CNS of mammals. Narahashi et al. (2010) has explained the mechanism of action of fipronil in insects and mammals in detail.

In laboratory animals, fipronil administration by the oral route can produce the signs of neurotoxicity, including convulsions, tremors, abnormal gait, and hunched posture. Similar signs can be produced following inhalation exposure. Poisoned dogs and cats usually show signs of tremors, convulsions, seizures, and death (Gupta and Anadon, 2018). Following dermal exposure, fipronil toxicity is more pronounced in rabbits than in rats, mice, and dogs. Humans exposed to fipronil by ingestion may show symptoms of headache, tonic–clonic convulsions, seizures, paresthesia, pneumonia, and death. Neurotoxic symptoms of fipronil poisoning in humans are typically associated with the antagonism of central GABA receptors.

It has been suggested that fipronil is a developmental neurotoxicant. In an in vitro study using differentiating N2a neuroblastoma cells, Sidiropoulou et al. (2011) demonstrated that fipronil caused severe disruption of the developmentally important ERK (½) MAP kinase signal transduction pathway, as evidenced by significant reductions in the activation state of MAP kinase (MEK [½]), and particularly ERK (½). These findings supported the contention that fipronil is a developmental neurotoxicant and unrelated to GABA receptor inhibition. Recently, Roques et al. (2012) demonstrated that fipronil and fipronil sulfone induced thyroid disruption in rats. However, fipronil sulfone has greater potential than fipronil for thyroid disruption because it persists much longer in the organism than fipronil itself.

Recent reports indicate that fipronil and its metabolite(s) can cause toxicity in liver, kidney, and other vital organs by dysregulating mitochondrial bioenergetics (by inhibiting mitochondrial respiratory chain), calcium homeostasis, oxidative and nitrosative stress, as well as causing damage to DNA and proteins (reviewed in Wang et al., 2016, Gupta and Anadon, 2018). Cell death can occur due to apoptosis or autophagy.

Biomarkers

After topical application of Frontline, fipronil spreads and sequesters in the lipids of the skin and hair follicles and continues to be released onto the skin and coat, resulting in long-lasting residual activity against fleas and ticks. Jennings et al. (2002) reported the maximum concentration of fipronil on the canine hair coat 24 h after a single application of Frontline Top Spot. With a descending concentration trend, fipronil residue was detected on dog's hair coat for a period of 30 days. Fipronil is metabolized to fipronil sulfone in the liver by cytochrome P450 (Roques et al., 2012). Fipronil and its metabolites (primarily sulfone) can persist in the tissues, particularly in fat and fatty tissues, for 1 week after treatment. The long half-life (150–245 h) of fipronil in blood may reflect a slow release of metabolites from fat. Pharmacokinetic studies in rats suggest that fipronil excretes mainly in the feces (45%–75%) and very little in the urine (5%–25%). Detection and confirmation of residue of fipronil and its metabolites is usually performed using GC-MS.

Detection of residue of fipronil or its metabolites in the body tissue, urine, feces, or on skin or hair can be used as biomarkers of fipronil exposure. Clinical signs and symptoms and pathological changes in liver are not specific and are of little value in terms of toxicological biomarkers. For further details on toxicity of fipronil, readers are referred to Gupta and Anadon (2018).

IVERMECTIN AND SELAMECTIN

Ivermectin is a semisynthetic macrocyclic lactone (ML), which was first isolated from *Streptomyces avermitilis*. Ivermectin is a mixture of two homologs (80% B_{1a} and 20% B_{1b}), and it was introduced to the market as abamectin in 1981, a potent antiparasitic animal health drug. The drug is approved at a very low dosage for the control of parasites in many animal species (cattle, sheep, swine, horses, dogs, and cats) but not approved in lactating cows, sheep, and goats. Although ivermectin poisoning can occur in any animal species, dogs of certain breeds (Collies, Old English sheepdog, German Shepherd, and others), Murray Grey cattle, and young animals (with less developed BBB) are particularly more sensitive. In Japan, Crump and Ōmura (2011) described ivermectin as a "wonder drug" in humans against onchocerciasis and lymphatic filariasis. In agriculture, ivermectin is used for its insecticidal, miticidal, and acaricidal activities.

Selamectin is a novel semisynthetic avermectin, which is marketed as Revolution for topical application on dogs and cats 6 weeks of age and older. Selamectin is used to kill fleas, ticks, and ear mites in dogs and cats. With ivermectin, selamectin, and other MLs (doramectin, eprinomectin, milbemycin, and moxidectin), poisoning in animals may occur because of overdosage or accidental exposure.

Mechanism of Toxicity

Ivermectin

Ivermectin is effective against nematodes and arthropods but not against cestodes and trematodes. This is because ivermectin acts as a GABA receptor agonist, and cestodes and trematodes lack a GABA system. In mammals, a GABA system is present only in the CNS. Ivermectin exerts toxicity by blocking the postsynaptic transmission of nerve impulses by potentiating the release and binding of GABA, thus blocking GABA-mediated transmission. In general, ivermectin does not cross the BBB in most animal species. The defective p-glycoprotein transporter (ABCB1) in the BBB has been found in at least 11 breeds of dogs (including Collies) and in many other animals. Therefore, when the p-glycoprotein transporter is defective or overwhelmed by ivermectin overdosage, the integrity of the BBB is compromised, leading to toxicosis in the animal. Some adverse effects of ivermectin in dogs, horses, and cattle also appear to be due to GABA-mediated cholinergic effects.

Both homologs of ivermectin are neurotoxicants and equally potent. The clinical signs and symptoms of ivermectin toxicosis, primarily involving the CNS, have been described in ivermectin-sensitive dogs. Further support for CNS involvement was the finding that the ivermectin concentrations in the brain were higher in dogs with p-glycoprotein defect displaying symptoms of ivermectin toxicosis than in nonsensitive dogs (Pullium et al., 1985). Clinical signs include dehydration, bradycardia, respiratory depression, cyanosis, mydriasis, and a diminished gag reflex. Toxic signs may also include vomiting, ataxia, tremors, hypersalivation, coma, and death. Ivermectin should not be given to kittens, as it can cross the BBB. At higher doses, other animals show similar signs of ivermectin toxicosis. In an experimental study in rats, abamectin has been found to cause testicular damage, thereby impairing male fertility (Celic-Ozenci et al., 2011).

Selamectin

In insects and parasites, selamectin binds to glutamate-gated chloride channels in the nervous system, causing them to remain open. This causes chloride ions to continuously flow into the nerve cell, changing the charge of the cell membrane. The continuous flow of chloride ions blocks neurotransmission, and transmission of stimuli to the muscle is prevented. Selamectin binding is irreversible, thereby causing prolonged channel opening and permanent hyperpolarization. Selamectin has no such effect in the mammalian nervous system and, therefore, it is considered to be a safe insecticide. However, in acute cases, signs of toxicoses may include hair loss at the site of application, vomiting, diarrhea with or without blood, lethargy, salivation, tachypnea, pruritus, urticaria, erythema, ataxia, and fever. In rare instances, seizures followed by death occur with overt acute overdose. In our study, dogs treated with a single topical application of Revolution (6 mg/kg body wt) showed no signs of any poisoning, though detectable residue of selamectin persisted on the skin and hair coat for 1 month (Gupta et al., 2005). For more details on the toxicity of ivermectin, selamectin, and other MLs, readers are referred to Gwaltney-Brant et al. (2018).

Biomarkers

In general, these insecticides are well absorbed, distributed widely throughout the body (higher concentrations in fat and fatty tissues), and tend to have long tissue residence times. They undergo some metabolism and are excreted unchanged in the feces via the bile. Because of the enterohepatic recycling, the half-lives of these compounds are in the range of days to weeks. Interestingly, residues of these compounds are excreted in the milk. Residue analysis of ivermectin, selamectin, or any other MLs, along with clinical signs, can serve as biomarkers of exposure and effects. These compounds are analyzed using HPLC with UV, fluorescence, or photodiode array detector (Anastaseo et al., 2002; Gupta et al., 2005; Gwaltney-Brant et al., 2018).

ROTENONE

Rotenone is one of the naturally occurring insecticides present in a number of plants of the *Derris*, *Lonchocarpus*, *Tephrosia*, and *Mundulea* species. It has a molecular formula of $C_{23}H_{22}O_6$ and a molecular weight of 394.42. Rotenone is used worldwide because it has broad-spectrum insecticidal, acaricidal, and other pesticidal properties. Its formulations include crystalline preparations (about 95%), emulsifiable solutions (about 50%), and dust (0.75%). In veterinary medicine, rotenone is used in powder form to control parasitic mites on chicken and other fowl, and lice and ticks on dogs, cats, and horses. Rotenone dust is also used to control beetles and aphids on vegetables, fruits, berries, and flowers. Rotenone emulsions are used for eliminating unwanted fish in the management of bodies of water. It is also formulated, along with other pesticides, such as carbaryl, pyrethrins, piperonyl butoxide, and lindane, in products to control insects, mites, ticks, lice, spiders, and undesirable fish.

According to a survey conducted by the USEPA in 1990, rotenone was found to be one of the pesticides most commonly used in and around the home. Rotenone is a very safe compound when properly used, but in higher doses it is toxic to humans, animals, and fish. WHO classifies rotenone as a moderately hazardous Class II pesticide. It has been involved in suicide attempts, in which acute congestive heart failure was the characteristic feature at autopsy. During the past decade, rotenone has received enormous attention because of its link to Parkinson's disease (Pan-Montojo et al., 2010; Xiong et al., 2012; Terron et al., 2018). Currently, rotenone is used extensively by researchers as an experimental drug to cause mitochondrial complex I inhibition and reproduce Parkinsonian motor deficits in animal models (Drolet et al., 2009; Xiong et al., 2012; Terron et al., 2018).

Mechanism of Toxicity

In insects, rotenone is both a contact and a systemic insecticide. Rotenone inhibits the transfer of electrons from Fe—S centers in complex I to ubiquinone in the electron transport chain. This prevents NADH from being converted into usable cellular energy, i.e., ATP. In mammals and fish, rotenone inhibits the oxidation of NADH to NAD, thereby blocking the oxidation of NAD and the substrates such as glutamate, α-ketoglutarate, and pyruvate. Rotenone causes inhibition of mitochondrial respiratory chain complex I, which can cause oxidative stress and lead to selective degeneration of striatal-nigral dopamine neurons. Besides complex I inhibition, mitochondrial dysfunction, nitrosative stress, increased nitric oxide and malondialdehyde levels, impaired proteostasis, aggregation of α-synuclein and polyubiquitin, activation of astrocytes and microglial cells, neuroinflammatory reaction, glutamate excitotoxicity, and degeneration of dopaminergic neurons in rotenone evoked Parkinsonism (reviewed in Xiong et al., 2012; Terron et al., 2018). In an in vitro study, Mundhall et al. (2012) demonstrated that 30 min superfusion of horizontal slices of rat midbrain with 100 nM rotenone caused significant injury to tyrosine hydroxylase—positive proximal dendrites in dorsal and ventral regions of the substantia nigra zona impacta and ventral tegmental area. Rotenone toxicity has been studied using various in vitro and in vivo models (reviewed in Xiong et al., 2012; Terron et al., 2018).

Rotenone exerts selective toxicity, as it is highly toxic to fish because of its rapid absorption from the GI tract in comparison with mammalian species in which it is poorly absorbed. The selective toxicity of rotenone in insects and fish versus mammals can also be explained by its metabolism. Rotenone converts to highly toxic metabolites in large quantities in insects and fish, while it converts to nontoxic metabolites in mammals.

Rotenone has been proven to be a neurotoxicant in all species tested. In mammals, acute rotenone exposure can produce vomiting, incoordination, muscle tremors, and clonic convulsions. Cardiovascular effects include tachycardia, hypotension, and impaired myocardial contractility. Death occurs due to cardiorespiratory failure. Chronic exposure to rotenone exerts Parkinson's disease–like neuropathology in experimental lab animals (Drolet et al., 2009; Pan-Montojo et al., 2010). A NOAEL of 0.4 mg/kg/day has been determined in rats and dogs. For further details on rotenone toxicity, readers are referred to Gupta (2012).

Biomarkers

In animals, rotenone has been found to be hundred times more toxic via the IV route than by the oral route because of its poor absorption from the GI tract. Rotenone is metabolized in the liver by NADP-linked hepatic microsomal enzymes. Several metabolites of rotenone have been identified as rotenoids, such as rotenolone I and II, hydroxyl and dihydroxyrotenone, etc. Approximately 20% of a rotenone dose is excreted in the urine within 24 h of oral administration in rats and mice. Unabsorbed rotenone from the GI tract is excreted in the feces. Residue detection of rotenone and/or its metabolites in blood, urine, feces, or liver can serve as a biomarker of rotenone exposure. Rotenone residue can be determined using HPLC with a fluorescence detector or Liquid Chromatograph-Mass Spectrometer-Mass Spectrometer (LC-MS-MS) (Caboni et al., 2008). Characteristic toxicological symptoms and histopathological changes can be used as biomarkers of rotenone toxicity.

CONCLUDING REMARKS AND FUTURE DIRECTIONS

The majority of insecticides are of chemical origin, and the rest are of biological origin. They are used worldwide in agriculture, horticulture, forestry, residential areas, gardens, homes, offices, and aircrafts. Insecticides are of diverse chemical structures, and therefore their mechanism of action, pharmacokinetics, and toxicity vary significantly. Most insecticides are neurotoxicants in insects and nontarget mammalian (including humans) species, wildlife, and aquatic species. In nontarget species, the insecticides appear to be of lesser toxicity than that in insects, but their toxic effects involve other body organs and systems, in addition to the nervous system. In general, detection of the residue of insecticides and/or their metabolites in body fluids (urine, blood serum, plasma, and milk), tissue, or hair is often used as a biomarker of exposure. Alterations in behavioral, biochemical, molecular, and histopathological endpoints are used as biomarkers of effects. In insecticide toxicity, very little is understood regarding biomarkers of genetic susceptibility and this needs to be explored in future studies. Future research will explore novel biomarkers with greater sensitivity, reliability, and reproducibility, preferably nondestructible and noninvasive for chemical toxicity and chemical-related long-term illnesses, such as neurodegenerative, metabolic, and carcinogenesis.

References

Ali, Z.Y., 2012. Neurotoxic effects of lambda-cyhalothrin, a synthetic pyrethroid pesticide: involvement of oxidative stress and protective role of antioxidant mixture. New York Sci. J. 5, 93–103.

Anadon, A., Martinez-Larranaga, M.R., Martinez, M.A., 2009. Use and abuse of pyrethrins and synthetic pyrethroids in veterinary medicine. Vet. J. 182, 7–20.

Anand, S.S., Bruckner, J.V., Haines, W.T., 2006. Characterization of deltamethrin metabolism by rat plasma and liver microsomes. Toxicol. Appl. Pharmacol. 212, 156–166.

Anastaseo, A., Asposito, M., Amorena, 2002. Residue study of ivermectin in plasma, milk, and mozzarella cheese following subcutaneous administration to buffalo (*Bubalus bubalis*). J. Agric. Food Chem. 50, 5244–5245.

Appenzeller, B.M.R., Hardy, E.M., Grova, N., et al., 2017. Hair analysis for the biomonitoring of pesticide exposure: comparison with blood and urine in a rat model. Arch. Toxicol. 91, 2813–2825.

Arfat, Y., Mahmood, N., Tahir, M.U., et al., 2014. Effect of imidacloprid on hepatotoxicity and nephrotoxicity in male albino mice. Toxicol. Rep. 1, 554–561.

Badiou-Bénéteau, A., Carvalho, S.M., Brunet, J.-L., 2012. Development of biomarkers of exposure to xenobiotics in the honey bee *Apis mellifera*: application to the systemic insecticide thiamethoxam. Ecotoxicol. Environ. Saf. 82, 22–31.

Baker, S.E., Olsson, A.O., Barr, D.B., 2004. Isotope dilution high-performance liquid chromatography-tendem mass spectrometry method for quantifying urinary metabolites of synthetic pyrethroid insecticides. Arch. Environ. Contam. Toxicol. 46, 281–288.

Banks, C.N., Lein, P.J., 2012. A review of experimental evidence linking neurotoxic organophosphorus compounds and inflammation. Neurotoxicology 33, 575–584.

Barr, D.B., Buckley, B., 2011. In vivo biomarkers and biomonitoring in reproductive and developmental toxicity. In: Gupta, R.C. (Ed.), Reproductive and Developmental Toxicology. Academic Press/Elsevier, Amsterdam, pp. 253–265.

Barr, D.B., Olsson, A.O., Wong, L.Y., et al., 2010. Urinary concentrations of metabolites of pyrethroid insecticides in the general US population: national health and nutrition examination survey 1999–2002. Environ. Health Perspect. 118, 742–748.

Bloomquist, J.R., Soderlund, D.M., 1985. Neurotoxic insecticides inhibit GABA-dependent chloride uptake by mouse brain vesicles. Biochem. Biophys. Res. Commun. 133, 37–43.

Bouchard, M.F., Chevrier, J., Harley, K.G., et al., 2011. Prenatal exposure to organophosphate pesticides and IQ in 7-year old children. Environ. Health Perspect. 119, 1189–1195.

Caboni, P., Sarais, G., Vargiu, S., et al., 2008. LC-MS-MS determination of rotenone, deguelin, and rotenolone in human serum. Chromatographia 68, 739–745.

Celic-Ozenci, C., Tasatargil, A., Tekcan, M., 2011. Effects of abamectin exposure on male fertility in rats: potential role of oxidative stress-mediated poly(ADP-ribose) polymerase (PARP) activation. Regul. Toxicol. Pharmacol. 61, 310–317.

Chen, A., 2012. Organophosphate-induced brain damage: mechanisms, neuropsychiatric and neuronal consequences, and potential therapeutic strategies. Neurotoxicology 33, 391–400.

Choudhary, S., Raheja, G., Gupta, V., Gill, K.D., 2002. Possible involvement of dopaminergic neurotransmitter system in dichlorvos induced delayed neurotoxicity. J. Biochem. Mol. Biol. Biophys. 6, 29–36.

Costa, L.G., Cole, T.B., Vitalone, A., et al., 2006. Paraoxon polymorphisms and toxicity of cholinesterase inhibitors. In: Gupta, R.C. (Ed.), Toxicology of Organophosphate and Carbamate Compounds. Academic Press/Elsevier, Amsterdam, pp. 247–256.

Crump, A., Ōmura, S., 2011. Ivermectin, 'wonder drug' from Japan: the human use perspective. Proc. Jpn. Acad. Ser. B 87, 13–28.

Cullen, L.K., Reynoldson, J.A., 1990. Central and peripheral alpha-adrenoceptor actions of amitraz in the dog. J. Vet. Pharmacol. Ther. 13, 86–92.

De Bleecker, J.L., 2006. Intermediate syndrome in organophosphate poisoning. In: Gupta, R.C. (Ed.), Toxicology of Organophosphate and Carbamate Compounds. Academic Press/Elsevier, Amsterdam, pp. 371–380.

De Luca, C.J., Buccafusco, J.J., Roy, S.H., 2006. The electromyographic signal as a presymptomatic indicator of organophosphates in the body. Muscle Nerve 33, 369–376.

del Pino, J., Moyano-Cires, P.V., Anadon, M.J., et al., 2015. Molecular mechanisms of amitraz mammalian toxicity: a comprehensive review of existing data. Chem. Res. Toxicol. 28, 1073–1094.

Dekundy, A., Kaminski, R.M., 2010. Central mechanisms of seizures and lethality following anticholinesterase pesticide exposure. In: Satoh, T., Gupta, R.C. (Eds.), Anticholinesterase Pesticides: Metabolism, Neurotoxicity, and Epidemiology. John Wiley & Sons, Hoboken, pp. 149–164.

Desi, I., Nagymajtenyi, L., 1999. Electrophysiological biomarkers of an organophosphorus pesticide, dichlorvos. Toxicol. Lett. 107, 55–64.

Drolet, R.E., Cannon, J.R., Montero, L., et al., 2009. Chronic rotenone exposure reproduces Parkinson's disease gastrointestinal neuropathology. Neurobiol. Dis. 36, 96–102.

Duggan, A., Charnley, G., Chen, W., 2003. Di-alkyl phosphate biomonitoring data: assessing cumulative exposure to organophosphate pesticides. Regul. Toxicol. Pharmacol. 37, 382–395.

Dulaurent, S., Saint-Marcoux, F., Marquet, P., et al., 2006. Simultaneous determination of six dialkylphosphates in urine by liquid chromatography tendem mass spectrometry. J. Chromatogr. B 831, 223–229.

Dulaurent, S., Gaulier, J.M., Zouaoui, K., 2011. Surprising hair analysis results following acute carbofuran intoxication. Forensic Sci. Int. 212, e10–e14.

Elflein, L., Berger-Preiss, E., Preiss, A., 2003. Human biomonitoring of pyrethrum and pyrethroid insecticides used indoors: determination of the metabolites E-cis/trans-chrysanthemumdicarboxylic acid in human urine by gas chromatography–mass spectrometry with negative chemical ionization. J. Chromatogr. B Technol. Biomed. Life Sci. 795, 195–207.

Ensley, S.M., 2018a. Organochlorines. In: Gupta, R.C. (Ed.), Veterinary Toxicology: Basic and Clinical Principles, third ed. Academic Press/Elsevier, Amsterdam, pp. 509–513.

Ensley, S.M., 2018b. Pyrethrins and pyrethroids. In: Gupta, R.C. (Ed.), Veterinary Toxicology: Basic and Clinical Principles, third ed. Academic Press/Elsevier, Amsterdam, pp. 515–519.

Ensley, S.M., 2018c. Neonicotinoids. In: Gupta, R.C. (Ed.), Veterinary Toxicology: Basic and Clinical Principles, third ed. Academic Press/Elsevier, Amsterdam, pp. 521–523.

Filazi, A., Yurdakok-Dikmen, B., 2018. Amitraz. In: Gupta, R.C. (Ed.), Vetrinary Toxicology: Basic and Clinical Principles, third ed. Academic Press/Elsevier, Amsterdam, pp. 525–531.

Furlong, C.C., Holland, N., Richter, N., 2006. PON1 status of farmworker mothers and children as a predictor of organophosphate sensitivity. Pharmacogen Genomics 16 (2006), 183–190.

Furlong, M.A., Barr, D.B., Wolff, M.S., et al., 2017. Prenatal exposure to pyrethroid pesticides and childhood behavior and executive functioning. Neurotoxicology 62, 231–238.

Gammon, D.W., Chandrasekaran, A., ElNaggar, S.F., 2012. Comparative metabolism and toxicology of pyrethroids in mammals. In: Marrs, T.C. (Ed.), Mammalian Toxicology of Insecticides. RSC Publishing, Cambridge, pp. 137–183.

Grigoryan, H., Li, B., Anderson, E.K., et al., 2009. Covalent binding of the organophosphorus agent FP-biotin to tyrosine in eight proteins that have no active site serine. Chem. Biol. Interact. 180, 492–498.

Gupta, R.C., 2004. Brain regional heterogeneity and toxicological mechanisms of organophosphates and carbamates. Toxicol. Mech. Meth. 14, 1–41.

Gupta, R.C. (Ed.), 2006. Toxicology of Organophosphate and Carbamate Compounds. Academic Press/Elsevier, Amsterdam, pp. 1–763.

Gupta, R.C., 2012. Rotenone. In: Gupta, R.C. (Ed.), Veterinary Toxicology: Basic and Clinical Principles. Academic Press/Elsevier, Amsterdam, pp. 620–623.

Gupta, R.C., Milatovic, S., Dettbarn, W.-D., et al., 2007. Neuronal oxidative injury and dendritic damage induced by carbofuran: protection by memantine. Toxicol. Appl. Pharmacol. 219, 97–105.

Gupta, R.C., Milatovic, D., 2010. Oxidative injury and neurodegeneration by OPs: protection by NMDA receptor antagonists and antioxidants. In: Weissman, B.A., Raveh, L. (Eds.), The Neurochemical Consequences of Organophosphate Poisoning in the CNS. Transworld Network, Trivendrum, pp. 19–39.

Gupta, R.C., Goad, J.T., Kadel, W.L., 1991a. In vivo alterations by carbofuran in high energy phosphates, creatine kinase (CK) and CK isoenzymes. Arch. Toxicol. 65 (4), 304–310.

Gupta, R.C., Goad, J.T., Kadel, W.L., 1991b. In vivo alterations in lactate dehydrogenase (LDH) and LDH isoenzymes patterns by acute carbofuran (Furadan) intoxication. Arch. Environ. Contam. Toxicol. 21, 263–269.

Gupta, R.C., Goad, J.T., Kadel, W.L., 1993. Protection and reversal by memantine and atropine of carbofuran-induced changes in biomarkers. Drug Dev. Res. 28, 153–160.

Gupta, R.C., Goad, J.T., Kadel, W.L., 1994a. Cholinergic and noncholinergic changes in skeletal muscles by carbofuran and methyl parathion. J. Toxicol. Environ. Health 43 (1994), 291–304.

Gupta, R.C., Goad, J.T., Kadel, W.L., 1994b. In vivo acute effects of carbofuran on protein, lipid and lipoproteins in rat liver and serum. J. Toxicol. Environ. Health 42, 451–462.

Gupta, R.C., Goad, J.T., Milatovic, D., Dettbarn, W.-D., 2000. Cholinergic and noncholinergic brain biomarkers of insecticides exposure and effects. Hum. Exp. Toxicol. 19, 1–12.

Gupta, R.C., Milatovic, D., Dettbarn, W.-D., 2001a. Nitric oxide (NO) modulates high-energy phosphates in rat brain regions with DFP or carbofuran: prevention by PBN or vitamin E. Arch. Toxicol. 75, 346–356.

Gupta, R.C., Milatovic, D., Dettbarn, W.-D., 2001b. Protection by antioxidants of DFP- or carbofuran-induced depletion of high-energy phosphates and their metabolites in rat brain regions. Neurotoxicology 22, 217–282.

Gupta, R.C., Masthay, M.B., Canerdy, T.D., et al., 2005. Human exposure to selamectin from dogs treated with Revolution™. Methodological consideration for selamectin isolation. Toxicol. Mech. Meth. 15, 317–321.

Gupta, R.C., Miller Mukherjee, I.R., Doss, R.B., et al., 2017. Organophosphates and carbamates. In: Gupta, R.C. (Ed.), Reproductive and Developmental Toxicology, second ed. Academic Press/Elsevier, Amsterdam, pp. 609–631.

Gupta, R.C., Milatovic, D., 2012. Toxicology of organophosphates and carbamates. In: Marrs, T.C. (Ed.), Mammalian Toxicology of Insecticides. RSC Publishing, Cambridge, pp. 104–136.

Gupta, R.C., Zaja-Milatovic, S., Dettbarn, W.-D., Malik, J.K., 2015. Skeletal muscle. In: Gupta, R.C. (Ed.), Handbook of Toxicology of Chemical Warfare Agents. Academic Press/Elsevier, Amsterdam, pp. 577–597.

Gupta, R.C., Anadon, A., 2018. Fipronil. In: Gupta, R.C. (Ed.), Vetrinary Toxicology: Basic and Clinical Principles, third ed. Academic Press/Elsevier, Amsterdam, pp. 533–538.

Gwaltney-Brant, S.M., DeClementi, C., Gupta, R.C., 2018. Macrocyclic lactone endectocides. In: Gupta, R.C. (Ed.), Veterinary Toxicology: Basic and Clinical Principles. Academic Press/Elsevier, Amsterdam, pp. 539–550.

Harley, K.G., Engel, S.M., Vedar, M.G., et al., 2016. Prenatal exposure to organophosphorus pesticides and fetal growth: pooled results from four longitudinal birth cohort studies. Environ. Health Perspect. 124, 1084–1092.

Heutelbeck, A.R.R., Bornemann, C., Lange, M., et al., 2016. Acetylcholinesterase and neuropathy target esterase activities in 11 cases of symptomatic flight crew members after fume events. J. Toxicol. Environ. Health Part A 79, 1050–1056.

Hsu, W.H., Schaffer, D.O., 1988. Effects of topical application of amitraz on plasma glucose and insulin concentrations in dogs. Am. J. Vet. Res. 49, 130–131.

Huen, K., Bradman, A., Harley, K., 2012. Organophosphate pesticide levels in blood and urine of women and newborns living in an agricultural community. Environ. Res. 117, 8–16.

Hughes, M.F., Ross, D.G., Starr, J.M., et al., 2016. Environmentally relevant pyrethroid mixtures: a study on the correlation of blood and brain concentrations of a mixture of pyrethroid insecticides to motor activity in the rat. Toxicology 359, 19–28.

Hugnet, C., Buronfusse, F., Pineau, X., 1996. Toxicity and kinetics of amitraz in dogs. Am. J. Vet. Res. 57, 1506–1510.

Ishibashi, M., Ando, T., Sakai, M., 2012. High-throughput simultaneous analysis of pesticides by supercritical fluid chromatography/tandem mass spectrometry. J. Chromatogr. A 1266, 143–148.

Jain, A.V., 2006. Analysis of organophosphate and carbamate pesticides and anticholinesterase therapeutic agents. In: Gupta, R.C. (Ed.), Toxicology of Organophosphate and Carbamate Compounds. Academic Press/Elsevier, Amsterdam, pp. 681–701.

Jamal, G.A., Hansen, S., Julu, P.O., 2002. Low level exposures to organophosphorus esters may cause neurotoxicity. Toxicology 181, 23–33.

Jennings, K.A., Canerdy, T.D., Keller, R.J., et al., 2002. Human exposure to fipronil from dogs treated with frontline. Vet. Hum. Toxicol. 44, 301–303.

Jokanović, M., Kosanović, M., Brkić, D., et al., 2011. Organophosphate induced delayed polyneuropathy in man: an overview. Clin. Neurol. Neurosurg. 113, 7–10.

Kamel, F., Umbach, D.M., Bedlack, R.S., Richards, M., 2012. Pesticide exposure and amyotrophic lateral sclerosis. Neurotoxicology 33, 457–462.

Kazi, A.I., Oommen, A., 2012. Monocrotophos induced oxidative damage associates with severe acetylcholinesterase inhibition in rat brain. Neurotoxicology 33, 156–161.

Khalil, S.R., Awad, A., Mohammed, H.H., et al., 2017. Imidacloprid insecticide exposure induces stress and disrupts glucose homeostasis in male rats. Env. Toxicol. Pharmacol. 55, 165–174.

Khemiri, R., Côte, J., Fetoui, H., et al., 2017. Documenting the kinetic time course of lambda-cyhalothrin metabolites in orally exposed volunteers for the interpretation of biomonitoring data. Toxicol. Lett. 276, 115–121.

Kim, K.B., Bartlett, M.G., Anand, S.S., et al., 2006. Rapid determination of the synthetic pyrethroid insecticide, deltamethrin, in rat plasma and tissues by HPLC. J. Chromatogr. B 834, 141–148.

Kim, J.C., Shin, J.Y., Yang, Y.S., Shin, D.H., 2007. Evaluation of developmental toxicity of amitraz in Sprague-Dawley rats. Arch. Environ. Contam. Toxicol. 52, 137–144.

Lamango, N.S., 2005. Liver prenylated methylated protein methyl esterase is an organophosphate-sensitive enzyme. J. Biochem. Mol. Toxicol. 19, 347–357.

Lamb, T., Selvarajah, L.R., Mohamed, F., et al., 2016. High lethality and minimal variation after acute self-poisoning with carbamate insecticides in Sri Lanka-implications for global suicide prevention. Clin. Toxicol. 54 (8), 624–631.

Lei, W., Wang, D.-D., Dou, T.-Y., et al., 2017. Assessment of the inhibitory effects of pyrethroids against human carboxylesterases. Toxicol. Appl. Pharmacol. 321, 48–56.

Lein, P.J., Bonner, M.R., Farahat, F.M., 2012. Experimental strategy for translational studies of organophosphorus pesticide neurotoxicity based on real-world occupational exposures to chlorpyrifos. Neurotoxicology 33, 660–668.

Leng, G., Gries, W., 2005. Simultaneous determination of pyrethroid and pyrethrin metabolites in human urine by gas chromatography-high resolution mass spectrometry. Chromatogr. B Technol. Biomed. Life Sci. 814, 285–294.

Leng, G., Kuehn, K.H., Idel, H., 1996. Biological monitoring of pyrethroid metabolites in urine of pest control operators. Toxicol. Lett. 88, 215–220.

Li, B., Ricordel, I., Schopfer, L.M., Baud, F., 2009. Dichlorvos, chlorpyrifos oxon and aldicarb adducts of butyrylcholinesterase, detected by mass spectrometry in human plasma following deliberate overdose. J. Appl. Toxicol. 30, 559–565.

Li, B., Ricordel, I., Schopfer, L.M., Baud, F., 2010. Detection of adduct on tyrosine 411 of albumin in humans poisoned by dichlorvos. Toxicol Sci. 116, 23–31.

Liang, Y.-J., Wang, H.-P., Yang, L., et al., 2012. Metabonomic responses in rat urine following subacute exposure to propoxur. Int. J. Toxicol. 31, 287–293.

Lotti, M., Moretto, A., 2005. Organophosphate-induced delayed polyneuropathy. Toxicol Rev. 24 (2005), 37–49.

Malik, J.K., Aggarwal, M., Kalpana, S., Gupta, R.C., 2017. Chlorinated hydrocarbons and pyrethrins/pyrethroids. In: Gupta, R.C. (Ed.), Reproductive and Developmental Toxicology, second ed. Academic Press/Elsevier, Amsterdam, pp. 633–655.

Malmasi, A., Ghaffari, S., 2010. Electrocardiographic abnormalities in an English bulldog with amitraz toxicity. Comp. Clin. Pathol. 19, 103–105.

Mangas, I., Vilanova, E., Estévez, J., 2012. NTE and non-NTE esterases in brain membrane: kinetic characterization with organophosphates. Toxicology 297, 17–25.

Marafon, C.M., Delfim, C.I.G., Valadao, C.A.A., et al., 2010. Analysis of amitraz in cats by gas chromatography. J. Vet. Pharmacol. Ther. 33, 411–414.

Martin-Reina, J., Duarte, J.A., Cerrillos, L., et al., 2017. Insecticide reproductive toxicity profile: organophosphate, carbamate and pyrethroids. J Toxins 4 (1), 1–7.

Martinez, M.A., Ballesteros, S., 2012. Two suicidal fatalities due to the ingestion of chlorfenvinphos formulations: simultaneous

determination of the pesticide and the petroleum distillates in tissues by gas chromatography-flame-ionization detection and gas chromatography-mass spectrometry. J. Anal. Toxicol. 36, 44–51.
Maravgakis, G., Tzatzarakis, M.N., Alegakis, A.K., et al., 2012. Diethyl phosphates accumulation in rabbit's hair as an indicator of long term exposure to diazinon and chlorpyrifos. Forensic Sci. Int. 218 (1–3), 106–110.
McCavera, S.J., Soderlund, D.M., 2012. Differential state-dependent modification of inactivation-deficient Na_V 1.6 sodium channels by the pyrethroid insecticides S-bioallethrin, tefluthrin and deltamethrin. Neurotoxicology 33, 384–390.
Middlemore-Risher, M.L., Buccafusco, J.J., Terry, A.V., 2010. Repeated exposures to low-level chlorpyrifos results in impairments in sustained attention and increased impulsivity in rats. Neurotoxicol. Teratol. 32, 415–424.
Milatovic, D., Gupta, R.C., Aschner, M., 2006. Anticholinesterase toxicity and oxidative stress. Sci. World J. 6, 295–310.
Milatovic, D., Aschner, M., Gupta, R.C., Zaja-Milatovic, S., et al., 2010. Involvement of oxidative stress in anticholinesterase pesticide toxicity. In: Satoh, T., Gupta, R.C. (Eds.), Anticholinesterase Pesticides: Metabolism, Neurotoxicity, and Epidemiology. John Wiley and Sons, Hoboken, pp. 135–147.
Mladenovic, D., Djuric, D., Petronijevic, N., 2010. The correlation between lipid peroxidation in different brain regions and the severity of lindane-induced seizures in rats. Mol. Cell. Biochem. 333, 243–250.
Mor, F., Ozmen, O., 2010. Endosulfan-induced neurotoxicity and serum acetylcholinesterase inhibition in rabbits: the protective effect of vit C. Pest. Biochem. Physiol. 96, 108–112.
Morgan, M.K., Jones, P.A., 2013. Dietary predictors of young children's exposure to current-use pesticides using urinary biomonitoring. Food Chem. Toxicol. 62, 131–141.
Moser, V.C., 1991. Investigations of amitraz neurotoxicity in rats. IV. Assessment of toxicity syndrome using a functional observational battery. Toxicol Sci. 17, 7–16.
Mundhall, A.C., Wu, Y.-N., Belknap, J.K., 2012. NMDA alters rotenone toxicity in rat substantia nigra zona impacta and ventral tegmental area dopamine neurons. Neurotoxicology 33, 429–435.
Murawski, J., 2005. Insecticide Use in Occupied Area of Aircraft. Springer-Verlag, Berlin, Germany.
Muttray, A., Spelmeyer, U., Degirmenci, M., 2005. Acute effects of low doses of methyl parathion on human EEG. Environ. Toxicol. Pharmacol. 19, 477–483.
Narahashi, T., 1987. Nerve membrane ion channels as the target site of environmental toxicants. Environ. Health Perspect. 71, 25–29.
Narahashi, T., 2006. Electrophysiological mechanisms in neurotoxicity of organophosphates and carbamates. In: Gupta, R.C. (Ed.), Toxicology of Organophosphate and Carbamate Compounds. Academic Press/Elsevier, Amsterdam, pp. 339–346.
Narahashi, T., Zhao, X., Ikeda, T., et al., 2010. Glutamate-activated chloride channels: unique fipronil targets present in insects but not in mammals. Pest. Biochem. Physiol. 97, 149–152.
Nio, J., Breton, P., 1994. Effects of organophosphates on rabbit pyramidal cells firing pattern and hippocampal theta rhythm. Brain Res. Bull. 33, 241–248.
Okamura, A., Saito, I., Ueyama, J., 2012. New analytical method for sensitive quantification of urinary 3-methyl-4-nitrophenol to assess fenitrothion exposure in general population and occupational sprayers. Toxicol. Lett. 210, 220–224.
Oulhote, Y., Bouchard, M.F., 2013. Urinary metabolites of organophosphate and pyrethroid pesticides and behavioral problems in Canadian children. Environ. Health Perspect. 121, 1378–1384.
Pan-Montojo, F., Anichtchik, O., Dening, Y., 2010. Progression of Parkinson's disease pathology is reproduced by intragastric administration of rotenone in mice. PLoS One 5, e8762.
Pancetti, F., Olmos, C., Dagnino-Subiabre, A., et al., 2007. Noncholinesterase effects induced by organophosphate pesticides and their relationship to cognitive processes: implications for the action of acylpeptide hydrolase. J. Toxicol. Environ. Health Part B 10, 630.
Pandey, S.P., Mohanty, B., 2015. The neonicotinoid pesticide imidacloprid and the dithiocarbamate fungicide mancozeb disrupt the pituitary-thyroid axis of a wildlife bird. Chemosphere 122, 227–234.
Papp, A., Pecze, L., Vezer, T., 2004. Comparison of the effect of subacute organophosphate exposure on the cortical and peripheral evoked activity in rats. Pest. Biochem. Physiol. 79, 94–100.
Park, Y., Kim, Y., Kim, J., et al., 2013. Imidacloprid, a neonicotinoid insecticide, potentiates adipogeness in 3T3-L1 adipocytes. J. Agric. Food Chem. 61 (1), 255–259.
Pass, M.A., Mogg, T.D., 1995. Pharmacokinetics and metabolism of amitraz in ponies and sheep. J. Vet. Pharmacol. Ther. 18 (1995), 210–215.
Pandit, A.A., Choudhary, S., Singh, R.B., et al., 2016. Imidacloprid induced histomorphological changes and expression of TLR-4 and TNFα in lung. Pest. Biochem. Physiol. 131, 9–17.
Pavlovsky, L., Browne, R.O., Friedman, A., 2003. Pyridostigmine enhances glutamatergic transmission in hippocampal CA1 neurons. Exp. Neurol. 179, 181–187.
Payan-Renteria, R., Garibay-Chavez, G., Rangel-Ascencio, R., 2012. Effect of chronic pesticide exposure in farm workers of a Mexico community. Arch. Environ. Occup. Health 67, 22–30.
Pereira, V.M., Bortolotto, J.W., Kist, L.W., 2012. Endosulfan exposure inhibits brain AChE activity and impairs swimming performance in adult Zebrafish (*Danio rerio*). Neurotoxicology 33, 469–475.
Pope, C.N., 1999. Organophosphorus pesticides: do they all have the same mechanism of toxicity? J. Toxicol. Environ. Health Part B 2 (1999), 161–181.
Proctor, S.P., Maule, A.L., Heaton, K.J., et al., 2014. Permethrin exposure from fabric-treated military uniforms under different wear-time scenarios. J. Expo. Sci. Environ. Epidemiol. 24 (6), 572–578.
Pullium, J.D., Seward, R.L., Henry, R.T., et al., 1985. Investigating ivermectin toxicity in collies. Vet. Med. 80, 33–40.
Ray, D.E., Forshaw, P.J., 2000. Pyrethroid insecticides: poisoning syndromes, synergies, and therapy. Clin. Toxicol. 38, 95–101.
Read, R.W., Riches, J.R., Stevens, J.A., et al., 2010. Biomarkers of organophosphorus nerve agent exposure: comparison of phosphylated butyrylcholinesterase and phosphylated albumin after oxime therapy. Arch. Toxicol. 84 (2010), 25–36.
Rehman, T., Khan, M.M., Shad, M.A., et al., 2016. Detection of carbofuran-protein adducts in serum of occupationally exposed pesticide factory workers in Pakistan. Chem. Res. Toxicol. 29, 1720–1728.
Reiss, R., Neal, B., Lamb, J.C., Juberg, D.R., 2012. Acetylcholinesterase inhibition dose-response modeling for chlorpyrifos and chlorpyrifos-oxon. Regul. Toxicol. Pharmacol. 63, 124–131.
Richardson, J.R., Caudle, W.M., Wang, M., 2006. Developmental exposure to the pesticide dieldrin alters the dopamine system and increases neurotoxicity in an animal model of Parkinson's disease. FASEB J. 20, 1695–1697.
Rohlman, D.S., Anger, W.K., Lein, P.J., 2011. Correlating neurobehavioral performance with biomarkers of organophosphorus pesticide exposure. Neurotoxicology 32, 268–276.
Roques, B.B., Lacroix, M.Z., Puel, S., et al., 2012. Cyp450-dependent biotransformation of the insecticide fipronil into fipronil sulfone can mediate fipronil-induced thyroid disruption in rats. Toxicol Sci. 127, 29–41.
Rose, P.H., 2012. Nicotine and neonicotinoids. In: Marrs, T.C. (Ed.), Mammalian Toxicology of Insecticides. RSC Publishing, Cambridge, pp. 184–220.

Ross, S.J.M., Brewin, C.R., Curran, H.V., 2010. Neuropsychological and psychiatric functioning in sheep farmers exposed to low levels of organophosphate pesticides. Neurotoxicol. Teratol. 32, 452–459.

Saito, H., Hara, K., Tanemura, K., 2017. Prenatal and postnatal exposure to low levels of permethrin exerts reproductive effects in male mice. Reprod. Toxicol. 74, 108–115.

Satoh, T., Gupta, R.C. (Eds.), 2010. Anticholinesterase Pesticides: Metabolism, Neurotoxicity, and Epidemiology. John Wiley and Sons, Hoboken, pp. 1–625.

Sayim, F., Yavasoglu, N.Y.K., Uyanikgil, Y., 2005. Neurotoxic effects of cypermethrin in wistar rats: a hematological, biochemical and histopathological study. J Health Sci. 51, 300–307.

Schopfer, L.M., Grigoryan, H., Li, B., et al., 2010. Mass spectral characterization of organophosphate-labeled, tyrosine-containing peptides: characteristic mass fragments and a new binding motif for organophosphates. J. Chromatogr. 8, 1297–1311.

Shafer, T.J., Meyer, D.A., 2004. Effects of pyrethroids on voltage-sensitive calcium channels: a critical evaluation of strengths, weaknesses, data needs, and relationship to assessment of cumulative neurotoxicity. Toxicol. Appl. Pharmacol. 196, 303–318.

Shitole, D.G., Kulkarni, R.S., Sathe, S.S., et al., 2010. Amitraz poisoning – an unusual pesticide poisoning. J. Assoc. Phys. India 58, 317–319.

Sidiropoulou, E., Sachana, M., Flaskos, J., 2011. Fipronil interferes with the differentiation of mouse N2a neuroblastoma cells. Toxicol. Lett. 201, 86–91.

Silva de Assis, H.C., Nicaretta, L., Marques, M.C.A., 2011. Anticholinesterase activity of endosulfan in Wistar rats. Bull. Environ. Contam. Toxicol. 86, 368–372.

Slotkin, T.A., 2006. Developmental neurotoxicity of organophosphates: a case study of chlorpyrifos. In: Gupta, R.C. (Ed.), Toxicology of Organophosphate and Carbamate Compounds. Academic Press/Elsevier, Amsterdam, pp. 293–314.

Slotkin, T.A., Seidler, F.J., 2012. Developmental neurotoxicity of organophosphates targets cell cycle and apoptosis, revealed by transcriptional profiles in vivo and in vitro. Neurotoxicol. Teratol. 34, 232–241.

Smith, J.N., Wang, J., Lin, Y., et al., 2012. Pharmacokinetics and pharmacodynamics of chlorpyrifos and 3,5,6-trichloro-2-pyridinol in rat saliva after chlorpyrifos administration. Toxicol Sci. 130, 245–256.

Smulders, C.J.G.M., Bueters, T.J.H., Van Kleef, R.G.D.M., et al., 2003. Selective effects of carbamate pesticides on rat neuronal nicotinic acetylcholine receptors and rat brain acetylcholinesterase. Toxicol. Appl. Pharmacol. 193, 139–146.

Soderlund, D.M., 2012. Molecular mechanisms of pyrethroid insecticide neurotoxicity: recent advances. Arch. Toxicol. 86, 165–181.

Solberg, Y., Belkin, M., 1997. The role of excitotoxicity in organophosphorus nerve agents central poisoning. Trends Pharmacol. 18, 183–185.

Speed, H.E., Blaiss, C.A., Kim, A., 2012. Delayed reduction of hippocampal synaptic transmission and spines following exposure to repeated subclinical doses of organophosphorus pesticide in adult mice. Toxicol Sci. 125, 196–208.

Starr, J.M., Scollon, E.J., Hughes, M.F., 2012. Environmentally relevant mixtures in cumulative assessment: an acute study of toxicokinetics and effects on motor activity in rats exposed to a mixture of pyrethroids. Toxicol Sci. 130, 309–318.

Stefanidou, M., Athanaselis, S., Spiliopoulou, H., 2009. Butyrylcholinesterase: biomarker for exposure to organophosphorus insecticides. Int. Med. J. 39, 57–60.

Stivaktakis, P.D., Kavvalakis, M.P., Tzatzarakis, M.N., et al., 2016. Long-term exposure of rabbits to imidacloprid as quantified in blood induces geneotoxic effect. Chemosphere 149, 108–113.

Sudakin, D.L., 2006. Pyrethroid insecticides: advances and challenges in biomonitoring. Clin. Toxicol. 44, 31–37.

Sutton, P.M., Vergara, X., Beckman, J., et al., 2007. Pesticide illness among flight attendants due to aircraft disinsection. Am. J. Ind. Med. 50, 345–356.

Swenson, T.L., Casida, J.E., 2013. Neonicotinoid formaldehyde generators: possible mechanism of mouse-specific hepatotoxicity/hepatocarcinogenicity of thiamethoxam. Toxicol. Lett. 216, 139–145.

Tacal, O., Lockridge, O., 2010. Methamidophos, dichlorvos, *o*-methoate and diazinon pesticides used in Turkey make a covalent bond with butyrylcholinesterase detected by mass spectrometry. J. Appl. Toxicol. 30, 469–475.

Tange, S., Fujimoto, N., Uramaru, N., et al., 2016. In vitro metabolism of methiocarb and carbaryl in rats, and its effect on their estrogenic and antiandrogenic activities. Environ. Toxicol. Pharmacol. 41, 289–297.

Terron, A., Bal-Price, A., Paini, A., et al., 2018. An adverse outcome pathway for parkinsonian motor deficits associated with mitochondrial complex I inhibition. Arch. Toxicol. 92, 41–82.

Terry, A.V., 2012. Functional consequences of repeated organophosphate exposure: potential non-cholinergic mechanisms. Pharmacol. Toxicol. 134, 355–365.

Timchalk, C., 2006. Physiologically based pharmacokinetic modeling of organophosphorus and carbamate pesticides. In: Gupta, R.C. (Ed.), Toxicology of Organophosphate and Carbamate Compounds. Academic Press/Elsevier, Amsterdam, pp. 103–125.

Timchalk, C., Campbell, J.A., Liu, G., et al., 2007. Development of a non-invasive biomonitoring approach to determine exposure to the organophosphorus insecticide chlorpyrifos in rat saliva. Toxicol. Appl. Pharmacol. 219, 217–225.

Timchalk, C., 2010. Biomonitoring of pesticides: pharmacokinetics of organophosphorus and carbamate insecticides. In: Satoh, T., Gupta, R.C. (Eds.), Anticholinesterase Pesticides: Metabolism, Neurotoxicity, and Epidemiology. John Wiley & Sons, Hoboken, pp. 267–288.

Tomizawa, M., Casida, J.E., 2011. Unique neonicotinoid binding confirmations conferring selective receptor interactions. J. Agric. Food Chem. 59, 2825–2828.

Tornero-Velez, R., Davis, J., Scollon, E.J., 2012. A pharmacokinetic model of *cis*- and *trans*-permethrin disposition in rats and humans with aggregate exposure application. Toxicol Sci. 130, 33–47.

Ueyama, J., Satoh, T., Kondo, T., et al., 2010. β-Glucuronidase activity is a sensitive biomarker to assess low-level organophosphorus insecticide exposure. Toxicol. Lett. 193, 115–119.

Umeda, K., Miyara, M., Ishida, K., et al., 2018. Carbofuran causes neuronal vulnerability to glutamate by decreasing GluA2 protein levels in rat primary cortical neurons. Arch. Toxicol. 92, 401–409.

Van Thriel, C., Hengstler, J.G., Marchan, R., 2012. Pyrethroid insecticide neurotoxicity. Arch. Toxicol. 86, 141–142.

Wang, X., Martínez, M.A., Wu, Q., et al., 2016. Fipronil insecticide toxicity: oxidative stress and metabolism. Crit. Rev. Toxicol. 46, 1–24.

Wei, B., Mohan, K.R., Weisel, C.P., 2012. Exposure of flight attendants to pyrethroid insecticides on commercial flights: urinary metabolite levels and implications. Int. J. Hyg Environ. Health 215, 465–473.

Weilgomas, B., 2013. Variability of urinary excretion of pyrethroid metabolites in seven persons over seven consecutive days-Implications for observational studies. Toxicol. Lett. 221, 15–22.

Weilgomas, B., Nahorski, W., Czarnowski, W., 2013. Urinary concentrations of pyrethroid metabolites in the convenience sample of an urban population of Northern Poland. Int. J. Hyg Environ. Health 216, 295–300.

Wilson, B.W., Arrieta, D.E., Henderson, J.D., 2005. Monitoring cholinesterases to detect pesticide exposure. Chem. Biol. Interact. 157–158, 253–256.

Wolansky, M.J., Harrill, J.A., 2007. Neurobehavioral toxicology of pyrethroid insecticides in adult animals: a critical review. Neurotoxicol. Teratol. 30, 55–78.

Xiong, N., Long, X., Xiong, J., 2012. Mitochondrial complex I inhibitor rotenone-induced toxicity and its potential mechanisms in Parkinson's disease model. Crit. Rev. Toxicol. 42, 613–632.

Yardan, T., Baydin, A., Acar, E., et al., 2013. The role of serum cholinesterase activity and S100B protein in the evaluation of organophosphate poisoning. Hum. Exp. Toxicol. 32 (10), 1081–1088.

Yavasoglu, A., Sayim, F., Uyanikgil, Y., 2006. The pyrethroid cypermethrin-induced biochemical and histological alterations in rat liver. J Health Sci. 52 (2006), 774–780.

Yoshida, T., 2017. Analytical method for pyrethroid metabolites in urine of the non-occupationally exposed population by gas chromatography/mass spectrometry. J. Chromatogr. Sci. 55 (9), 873–881.

Zaja-Milatovic, S., Gupta, R.C., Aschner, M., Milatovic, D., 2009. Protection of DFP-induced oxidative damage and neurodegeneration by antioxidants and NMDA receptor antagonist. Toxicol. Appl. Pharmacol. 240 (2009), 219–225.

Zepeda-Arce, R., Benitez-Trinidad, A., Medina-Díaz, I.M., et al., 2017. Oxidative stress and genetic damage among workers exposed primarily to organophosphate and pyrethroid pesticides. Environ. Toxicol. 32, 1754–1764.

Zhang, J., Guo, J., Lu, D., et al., 2018. Maternal urinary carbofuranphenol levels before delivery and birth outcomes in Sheyang birth cohort. Sci. Total Environ. 625, 1667–1672.

CHAPTER

27

Herbicides and Fungicides

P.K. Gupta

Director Toxicology Consulting Group, Former Principal Scientist and Chief Division of Pharmacology and Toxicology, IVRI, Advisor, World Health Organization (Geneva), Bareilly, India

INTRODUCTION

Ever since the dawn of civilization, it has been the major task of man to engage in a continuous endeavor to improve his living conditions. One of the main tasks in which human beings have been engaged is securing relief from hunger, one of the basic needs. Food production capacity is faced with an ever-growing number of challenges, including a world population expected to grow to nearly 10 billion by 2050 and a falling ratio of arable land to population. Food plays a vital and strategic role in a growing global population, but food production is encountering limits. For example, there is a limit to new areas to cultivate; therefore, we must increase agricultural production from the areas available. In fact, herbicides have significantly increased crop yields by reducing the strong competition of weeds with important and essential food crops. Along with improved crop varieties, both herbicides and fungicides have increased crop yields, decreased food costs, and enhanced the appearance of food. These benefits include increased crop and livestock yields, improved food safety, human health, quality of life and longevity, and reduced drudgery, energy use, and environmental degradation. A complex matrix of benefit interactions is explored for a range of beneficiaries at three main levels—local, national, and global—and in three main domains: social, economic, and environmental (Cooper and Dobson, 2007). Although the use of biopesticides is increasing (Gupta, 2006a), the use of chemical pesticides is still the mainstay in large-scale control of most insects, weeds, fungi, and other pests of economic or public health importance. Among pesticides, herbicides and fungicides have found extensive use in the eradication of unwanted plants and control of plant diseases (Gupta, 1989, 2004, 2010a, 2016).

On the other side, acute and long-term exposures to these chemicals and residues that remain on foods have given rise to many health effect issues (Gupta, 1986, 1988; 2006b). The toxic effects of some of these pesticide chemicals on animals (Gupta, 2018a,b,c,d) and their effects on developmental and reproductive abnormalities have been reviewed previously (Gupta, 2017). However, many point to health or environmental problems from accidental or deliberate exposure to pesticides. Because of the wide environmental dispersive uses of pesticides, regulatory agencies around the world have implemented intensive and robust product registration requirements that require state-of-the-art animal and environmental species testing that assures the safe use of modern pesticides. When pesticides are used rationally and carefully, in conjunction with other technologies in integrated pest management systems, it is more likely that their use will be justifiable (Cooper and Dobson, 2007).

The application of biological markers has been considered a major opportunity for assessing exposure and risk in determining health outcomes that are associated with exposures and for evaluating the success of risk mitigation strategies (Gupta, 2010b). Biomonitoring is an important tool that can be used to evaluate human exposure by measuring the levels of herbicides and fungicides and/or their metabolites or altered biological structures or functions in biological specimens or tissue. These measurements in biological media reflect human exposure (not the risk) to these chemicals through all relevant routes and can therefore be used to monitor aggregate and cumulative exposures (Barr et al., 1999; Barr and Angerer, 2006). Thus, with the availability of validated biomarker data, there are opportunities to reconstruct the exposure of an individual(s) at risk (Sachana et al., 2017). However, careful interpretation

Biomarkers in Toxicology, Second Edition
https://doi.org/10.1016/B978-0-12-814655-2.00027-X

of biomarkers and biomonitoring data related to toxicant exposures is necessary to accurately assess human exposure to chemicals and the associated health risks. Biomonitoring studies should be carefully designed to minimize inappropriate interpretation.

BACKGROUND

It is well known that human biomonitoring provides an efficient way to identify and quantify exposure to chemical substances. Once the toxicological effects have been identified in animal studies (and the mechanism of toxic effects has been elucidated), an adequate margin of safety is established by comparing with human exposure. Herbicides and fungicides are the most widely and commonly used pesticides worldwide. As such there is a likelihood of higher risk of exposure to these hazardous chemicals that may lead to adverse effects on health (Kapka-Skrzypczak et al., 2011). In view of these findings, the detection of populations at risk constitutes a very important topic. Therefore, the main focus of this chapter is to summarize the latest information on biomarkers and other widely used biomonitoring methods for studies on herbicides and fungicides or their metabolites levels, so as to assess the health risks associated with these chemicals.

TOXICOKINETICS

Toxicokinetics is associated with the absorption, distribution, metabolism, and excretion (ADME) of drugs and xenobiotics. Toxicokinetic studies provide important data on the amount of toxicant delivered to a target, as well as species-, age- and gender-specific influences on ADME properties. Human beings are exposed to herbicides or fungicides of different chemical classes. These chemicals may be ingested through food pesticide residues or absorbed through the skin or the respiratory system. Different factors regulate their ADME (Gupta, 2014). In general, liver is the primary site for biotransformation and may include activation and detoxification reactions through the cytochrome P450-dependent monooxygenase system, the flavin-containing monooxygenase, esterases, and a variety of transferases, most notably the glutathione (GSH) S-transferases (Hodgson and Meyer, 1997). Liver can serve as an important "first-pass metabolism" barrier to systemic distribution of pesticides, and particularly so with low-dose real-world oral (dietary residue) exposures encountered for most pesticides. The measurements of specific biomarkers after the absorption step, or during each subsequent step of the ADME process, are used to assess exposure by estimating the internal dose, which is defined as the amount of chemical absorbed into the body after an exposure has occurred. Biomonitoring of exposure to pesticides involves the measurement of an exposure biomarker, which can be pesticide(s), its metabolite(s), or reaction product(s) in biological media such as urine, blood or blood components, exhaled air, hair or nails, and tissues (Barr et al., 2006; Ngo et al., 2010). The knowledge of toxicokinetic values is, therefore, important to determine their biomarkers and for biomonitoring the chemicals in various body tissues or metrics used for this purpose. This also helps in the design of urine or other biological sampling time, which is of great relevance in a biomonitoring study, and sufficient collection time (i.e., spot samples vs. longer term continuous collection) will differ depending on the pesticide and other factors such as the route of exposure and chemical form. For example, in the case of herbicides such as 2,4-D, where they are essentially rapidly excreted unchanged in human urine, collection of a 24 h urine sample provides a very accurate estimate of total exposed dose of 2,4-D. Therefore, in planning biomonitoring, appropriate timings of the sampling are very important. This procedure can be difficult to perform among farm workers or pesticide sprayers because they are constantly undergoing occupational exposure, and this poses logistical and organizational problems with collecting samples (Kimata et al., 2009), and the procedure becomes more complex when the human and environmental biomarkers are the same for a particular chemical. Some of the herbicides or fungicides are very popular because they do not accumulate appreciably in humans and are rapidly metabolized and excreted in the urine, which means that the urinary metabolites of those pesticides can be measured up to a limited period only after an exposure has occurred. These measurements represent only a snapshot in time; thus, only exposures that occurred during the previous few hours or days can be captured. In that case, a single urinary measurement may not reflect the average exposure (Barr and Angerer, 2006). However, in the case of chronic exposure to those pesticides having prolonged systemic half-lives, urinary elimination may reach a steady state, which means that the chemical or metabolite present in the urine stays at a relatively constant level and reflects the average exposure. The specificity of urinary measurements is equally as high as blood measurements, but only in the case when the parent compound is excreted in urine (e.g., 2,4-dichlorophenoxyacetic acid (2,4-D), glyphosate, sulfonyl ureas). In another study, it was shown that variability in detection for each urinary metabolite within and across individuals indicates that any single measure of urinary metabolites cannot be considered a credible indicator of exposure for an individual. Furthermore, exposure estimation based on a urinary metabolite collected at a single time in an agricultural season is not a good indicator of population pesticide exposure, especially if the systemic half-life is very short. Results from these studies suggest that to provide a reliable characterization of pesticide exposure it is necessary to measure numerous pesticide urinary metabolites and collect multiple samples from each participant across a single agricultural season (Arcury et al., 2009). Therefore, ADME information for a chemical helps in careful design of biomonitoring studies.

MECHANISM OF ACTION

There are a number of biochemical changes that may act as indicators of biological response, such as enzyme changes due to exposure to herbicides and fungicides, which may be used as an internal marker of exposure; however, these types of biomarkers must be considered in the context of dose response, i.e., high-dose animal biomarkers are unlikely to be useful for low-dose human exposures. Likewise, specific biological effects of a chemical can be measured very early at the cellular level and may serve as indicators of exposure to that particular chemical. In most situations, it is difficult to pinpoint the exact mechanism of action. The mechanism of action of phenoxy derivatives, triazines, triazolopyrimidines, imidazolinones, dinitroaniline, and many other classes of herbicides is not precisely known in humans, especially at environmental exposure dose levels. However, at high-dose levels, phenoxy compounds are known to depress ribonuclease synthesis, uncouple oxidative phosphorylation, and increase the number of hepatic peroxisomes in animal studies. Oxadiazon causes hepatic porphyria in both mice and rats. The phenylurea herbicides linuron and monuron are rodent liver carcinogens. Chloroacetanilide and metolachlor have shown weak hepatocarcinogenicity in female rats and are nongenotoxic, suggesting a tumor-promoting action. The dinitro compounds markedly stimulate respiration while simultaneously impairing adenosine triphosphate synthesis. The main toxic action is uncoupling of oxidative phosphorylation, converting all cellular energy into the form of heat and causing extreme hyperthermia. In addition, the gut flora in ruminants is able to further reduce the dinitro compounds to diamine metabolites, which are capable of inducing methemoglobinemia (Gupta, 2018b). The available information on substituted anilines indicates that there is a nongenotoxic mechanism of action and lack of relevance to humans for the nasal turbinate, stomach, and/or thyroid oncogenic effects produced in rats. Glyphosate, a member of the phosphonomethyl amino acid group, selectively inhibits the enzyme 5-enolpyruvylshikimate 3-phosphate synthetase. The enzyme plays a key role in the biosynthesis of the intermediate, chorismate, which is necessary for the synthesis of the essential amino acids phenylalanine, tyrosine, and tryptophan. This aromatic amino acid biosynthesis pathway is found in plants, as well as in fungi and bacteria, but not in insects, birds, fish, mammals, and humans, thus providing a specific selective toxicity to plant species (Franz et al., 1997); this highlights that if an herbicide mode of action operates by a pathway not present in mammals, then this type of mode of action provides no basis for establishing a human biomarker. In mammalian systems, glyphosate can cause disruptions of mitochondrial and membrane integrity, and endocrine, reproductive, and developmental systems (Gasnier et al., 2009; Williams et al., 2012). Glyphosate can also induce apoptosis and necrosis in human umbilical, embryonic, placental, and testicular cells, and a testosterone decrease (Benachour and Seralini, 2009; Clair et al., 2012). Another compound of this group, glufosinate, acts by inhibiting the enzyme glutamine synthetase in animals. Glutamine synthetase in mammals is involved in ammonia homeostasis in many organs and the glutamine–glutamate shunt between γ-aminobutyrate and glutamate in the central nervous system (CNS). However, the enzyme normally works at a small fraction of its capacity, and considerable inhibition is required in mammals before ammonia levels increase. Hence, this does not lead to a problem in ammonia metabolism; the mammals obviously compensate by using other metabolic pathways (Hack et al., 1994; JMPR, 2000). Therefore, knowledge of mechanism of action may provide insight into selection of potential human biomarkers; however, particular attention must be directed at understanding dose response for such biomarkers to be effectively translated to actual real-world human applications. High-dose animal biomarkers are unlikely to be useful for real-world applications.

BIOMARKERS AND BIOMONITORING OF EXPOSURE

The expansion of knowledge at the molecular and cellular level has provided an opportunity to consider the addition of a myriad of new endpoints to toxicological evaluations utilizing the in vitro and in vivo animal models for better characterization of hazard potential of a chemical. This includes an array of new molecular biomarkers that have received substantial attention. Although biomarkers are frequently discussed as new approaches, it is well known to the clinician and toxicologist that biomarkers have been used in human medicine for centuries. In most cases, urine is the matrix for pesticides used for biomonitoring. A few biomarkers of herbicides and fungicides or their metabolites, along with selectivity of the measurement, are summarized in Table 27.1.

In some cases, the biomarkers to be measured are present in body fluids other than urine, such as exhaled breath or tears, which can serve as an indicator of exposure or, even, dose of a toxicant. In other cases, for example, the biomarker is an indicator of a disease process, such as individuals being evaluated for prostate cancer based on an elevated level of prostate-specific antigen in serum samples. New biomarkers of exposure will continue to be proposed. For each potential biomarker of exposure, it will be necessary to conduct experiments to validate the utility of the biomarker. The potential list of biomarkers for toxic responses is seemingly endless. Therefore, the application of biological markers has been considered as a major opportunity for exposure assessment to identify populations exposed to herbicides and fungicides. The biological

TABLE 27.1 Biomarkers of Selective Herbicides and Fungicides or Their Metabolites and Specificity of the Measurement

		Potential Biomarker[a]	Specificity[b]
Herbicide Class	**Herbicide**		
Phenoxy acid derivative	2,4-D	2,4-D	1
	2,4,5-T	2,4,5-T	1
	MCPA	MCPA	1
Triazines and triazoles	Atrazine	Atrazine and atrazine mercapturate	2,3
Phenyhirea herbicide	Diuron	Diuron and *N*-(3,4-dichlorophenyl)-urea	1,2
Protoporphyrinogen oxidase inhibiting herbicide		Porphyrin accumulation	3
Substituted anilines	Alachlor	Alachlor and alachlor mercapturate	2
	Metolachlor	Metolachlor and metolachlor mercapturate	2
Bipyridyl derivatives	Paraquat	Paraquat	1
	Diquat	Diquat	1
	Imazapyr	Imazapyr	1
Imidazolinones	Imazapic	Imazapic	1
	Imazethapyr	Imazethapyr	1
	Imazamox	Imazamox	1
	Imazaquin	Imazaquin	1
	Imazamethabenz-methyl	Imazamethabenz-methyl and Ester to imazamethabenz	1,2
Fungicide Class	**Fungicide**		
Dialkyldithiocarbamates		Common metabolite ETU	3,4
Anilinopyrimidines	Cyprodinil	4-cydopropyl-5-hydroxy-6-methyl-*N*-(4-hydroxy)-phenyl-2-pyrimidinamine	2
Chloroalkylthiodicarboximides (phthaiimides)	Captan	1,2,3,6-tetra-hydrophthalimide (THPI) and Thiazolidine-2-thione-4-carboxylic acid (TTCA)	2,3
	Folpet	THPA, phthalimide, and phthalic acid	2,3
Halogenated substituted monocyclic aromatics	Chlorothalonil	Thiol-derived metabolities	2

[a]*Urine is the matrix for analysis unless otherwise stated*
[b]*1, Parent pesticide measured; 2, Metabolite or contaminant measured but is reflective of the parent pesticide; 3, Reflective of class exposure to pesticides; 4, Reflective of exposure to pesticide and nonpesticide chemicals (e.g., other chemicals or degradates).*

markers can include a broad range of techniques that are designed to provide information on "an exogenous substance or its metabolite or a product of an interaction between a xenobiotic agent and some target molecule or cell that is measured in a compartment within an organism." Thus, biomarker measurements can provide quantitative information that can identify prior exposure or link exposures to an adverse effect/disease outcome. In most cases of exposure, the type of techniques needed do not exist, especially those needed to provide a direct linkage to disease. Such a situation is also true for various herbicides and other chemicals because we still only touch the tip of the iceberg of available active ingredients that can be measured (Barr and Angerer, 2006). A special challenge relates to recognizing the dynamics of the toxicokinetics and their mechanisms of action and establishing quantitative relationships between exposure and dose at any particular time over the course of the intoxication. Therefore, it is pertinent to discuss these topics, which are very important for biomarkers and biomonitoring of chemicals.

Herbicides

Herbicides are phytotoxic chemicals used for controlling various weeds. They have variable degrees of specificity. For example, paraquat kills all green plants, whereas phenoxy compounds are specific for certain groups of broadleaf plants. The worldwide consumption of herbicides is almost 48% of the total pesticide usage. Herbicides represent a plethora of chemically diverse structures and are inclusive of modes of action that are unique to plants or overlap with human biology. Most of the human or animal health problems that result from exposure to herbicides are due to their improper use or careless disposal of containers. Very few cases occur when these chemicals are used properly. However, there is increased concern about the effects of herbicides on human health because of runoff from agricultural applications and entrance into the drinking water supply (Gupta, 1986).

A large number of chemicals are used as herbicides (Gupta, 2018b) but, due to limited space, only those herbicides which are in active use and for which sufficient information on biomonitoring and biomarkers is available are discussed in greater detail in this chapter.

Phenoxy Acid Derivatives

This class of herbicides has been in continuous, extensive, and uninterrupted use since 1947. A combination of 2,4-D and 2,4,5-T, popularly known as "Agent Orange," was teratogenic in mouse. In the rat, this combination did not result in malformations, but behavioral effects were seen postnatally in some offspring. Because it was contaminated with the highly toxic and persistent 2,3,7,8-tetrachlorodibenzo-p-dioxin, along with other chlorinated dioxins and furans, 2,4,5-T has been banned for most applications. Although 2,4-D contains very small amounts of persistent chlorinated dioxins and furans, it is not expected to produce any adverse effects. In its ester or salt forms, it is commonly found in home and garden stores in combination with other herbicides such as dicamba or mecoprop for application on lawns. Absorption of 2,4-D occurs rapidly from the gastrointestinal tract (GIT), and plasma half-lives range from 3.5 to 18 h approximately. Dermal absorption was reported to occur rapidly but usually is less than 6%. The compound is extensively protein bound in vivo and distributed to the liver, kidneys, and lungs. 2,4-D has also been reported to cross the placental barrier in laboratory animals and pigs. 2,4-D is not metabolized to reactive intermediates, does not accumulate in tissues, and is excreted predominantly as the parent compound in urine. However, the rate of excretion via urine is inversely proportional to dose. 2,4-D has been detected in milk of lactating rats dosed with 2,4-D. The salts and esters of 2,4-D undergo acid and/or enzymatic hydrolysis to form 2,4-D acid, and possibly small amounts are conjugated with glycine or taurine. Excretion can be markedly enhanced by ion trapping using alkaline agents, as most of these herbicides are organic acids (Gupta, 2018b).

In addition to 2,4-D, mecoprop, MCPA (2-methyl-4-chloro-phenoxyacetic acid), and 2,4,5-T have been measured in urine (Kennepohl et al., 2010). Their mechanism of action indicates that they act as peroxisome proliferators. In addition, biomonitoring data for 2,4-D in urine samples are now available from a number of studies of both the general population (including preschool-age children) and farm applicators and their family members. Such data provide an integrated measure of absorbed dose from all pathways and routes of exposure (Alexander et al., 2007).

Using various toxicological information and monitoring data available in the literature, the hazards of 2,4-D were recently assessed by the US Environmental Protection Agency and the Canadian Pest Management Regulatory Agency (EPA, 2004; PMRA, 2007). The US EPA-derived reference doses (RfDs) for acute and chronic exposure to 2,4-D are based on external exposure metrics (administered dose), which are not directly useful for evaluating biomonitoring data. However, Biomonitoring Equivalent (BE) values (BEs are defined as an estimate of pesticide or its key metabolite in urine and/or blood of humans exposed to a regulatory standard health-protective dose, e.g., RfD) corresponding to RfDs for acute and chronic exposure scenarios are now available (Aylward and Hays, 2008). The use of BEs is a more scientific tool for performing exposure risk assessment while comparing with the biomonitoring data, as has been done for 2,4-D in Figs. 27.1 and 27.2, respectively, for the general population and field applicators (Aylward et al., 2010).

The available biomonitoring data for 2,4-D from the United States and Canada were compared with BE values. The biomonitoring data indicated margins of safety (ratio of BE value to biomarker concentration) of ~200 at the central tendency and 50 at the extremes in the general population. Median exposures for applicators and their family members during periods of use appear to be well within acute exposure guidance values. Biomonitoring data from these studies indicate that current exposures to 2,4-D are below applicable exposure guidance values (Hays et al., 2012). This example demonstrates the value of biomonitoring data in assessing population exposures in the context of existing risk assessments using the BE approach.

In conclusion, the extensive database of metabolic, toxicological, and epidemiological studies on 2,4-D has provided no evidence that 2,4-D poses any adverse health risk to humans. Furthermore, there is no credible or plausible evidence for any increased risk of cancer or noncancer outcomes in humans associated with 2,4-D exposure (Burns and Swaen, 2012).

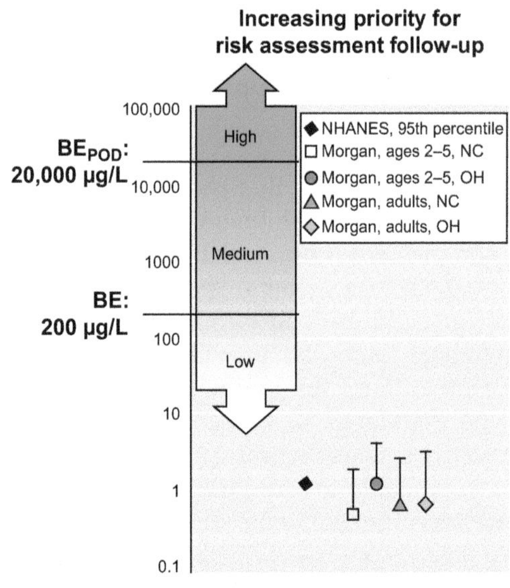

FIGURE 27.1 Urinary 2,4-D concentrations (μg/L) in general population studies presented in the context of the BE value corresponding to the US EPA RfD for general population chronic exposures. The symbol for data from NHANES (CDC, 2005) represents the 95th percentile for all tested participants (median values were below the LOD). The symbols for data from Morgan et al. (2008) (in key, Morgan) represent the median values for the children and adults from two states; bars extend to the 95th percentile for each group. The shaded regions represent concentration ranges associated with low, medium, and high priority for risk assessment follow-up based on the criteria described in the BE communications guidelines (LaKind et al., 2008) (Aylward et al., 2010).

Triazines and Triazoles

Another class of pesticides extensively used as herbicides are the triazines and triazoles. These herbicides have been extensively used in agriculture in the United States and other parts of the world for more than 50 years. These herbicides include both the asymmetrical and symmetrical triazines. Examples of symmetrical triazines are chloro-S-triazines (simazine, atrazine, propazine and cyanazine); the thiomethyl-S-triazines (ametryn, prometryn, terbutryn); and the methoxy-S-triazine (prometon). The commonly used asymmetrical triazine is metribuzin. These herbicides have low oral toxicity and are unlikely to pose acute hazards in normal use and do not produce developmental toxicity, except for ametryn, metribuzin, atrazine, and cyanazine, which may be slightly to moderately hazardous. Atrazine induced no structural malformations in either rats or rabbits. Another herbicide, simazine, provided contradictory results of developmental toxicity in different studies carried out in different species using different dose levels (Dilley et al., 1977).

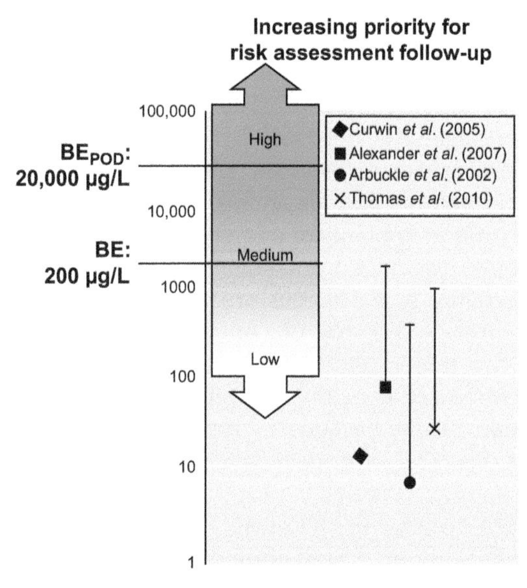

FIGURE 27.2 Urinary 2,4-D concentrations (μg/L) in applicators on the day after application of 2,4-D presented in the context of the human-equivalent BEPOD and target BE values associated with the occupational risk assessment in the United States (EPA, 2004). Symbols represent the median (or, in the case of Curwin et al., 2005 and Thomas et al., 2010, the geometric mean), and the bars extend to the maximum measured value in each study (not reported for Curwin et al., 2005). The description of shaded regions is shown by abbreviations: *BEPOD*, Chronic exposure at the human equivalent (Aylward et al., 2010); *POD*, Point of departure.

Extensive use of atrazine has resulted in widespread contamination of surface, ground-, and drinking water with this herbicide and its dealkylated and hydroxylated degradation products (Konstantinou et al., 2006). For this reason, atrazine has progressively been restricted or banned in the European countries over the last 20 years. The structurally related but more hydrophobic terbuthylazine is often used as a substitute. Contrary to the European countries, the United States continues to heavily use atrazine (LeBaron et al., 2008). Some of the symmetrical triazines are reported to produce mammary tumors specifically in the Sprague–Dawley strain of rats; however, a recent review on mode of action indicates that these tumors are of nonrelevance to humans (Breckenridge et al., 2010). Recent reviews on toxicology data from animal studies and human epidemiological data concluded that there is no scientific basis for inferring the existence of a causal relationship between triazine exposure and the occurrence of cancer in humans (Jowa and Howd, 2011; Sathiakumar et al., 2011). The metabolic studies of atrazine revealed that chemical structures of the compound are permutations of alkyl substituted 2,4-diamines of chlorotriazine. On entering the body, they are metabolized via the glutathione detoxification pathway or by simple dealkylation. For glutathione detoxification, the chlorine atom on the triazine herbicide is subject to an enzymatic catalyzed substitution by the free −SH on the internal cysteine residue of the glutathione tripeptide. The terminal peptides are enzymatically cleaved and the cysteine is N-acetylated. The mercapturate and dealkylation metabolites are then excreted into the urine. These metabolites are not specific for a single triazine but provide class exposure information. Triazines, as parent molecules, can also be measured as the intact pesticide in blood products (Sachana et al., 2017). The proposed metabolic pathways for atrazine are presented in Fig. 27.3 (Barr et al., 2007a,b). Animal versus human urine and/or blood metabolites from different studies are summarized in Table 27.2. Atrazine mercapturate appeared to be the major metabolite in humans after dermal or oral exposure to atrazine, which seems not to be the case in animal studies.

This table highlights the importance of understanding metabolic pathways of a compound in humans for selecting appropriate biomarkers for exposure. In a recent biomonitoring study for atrazine, atrazine mercapturate (Fig. 27.4) accounted for a major proportion of metabolites found in urine of atrazine applicators and measuring atrazine mercapturate plus atrazine could be used as a specific biomarker for the atrazine exposure (Mendas et al., 2012).

Another compound of this group is cyanazine. Cyanazine is more acutely toxic and resulted in developmental

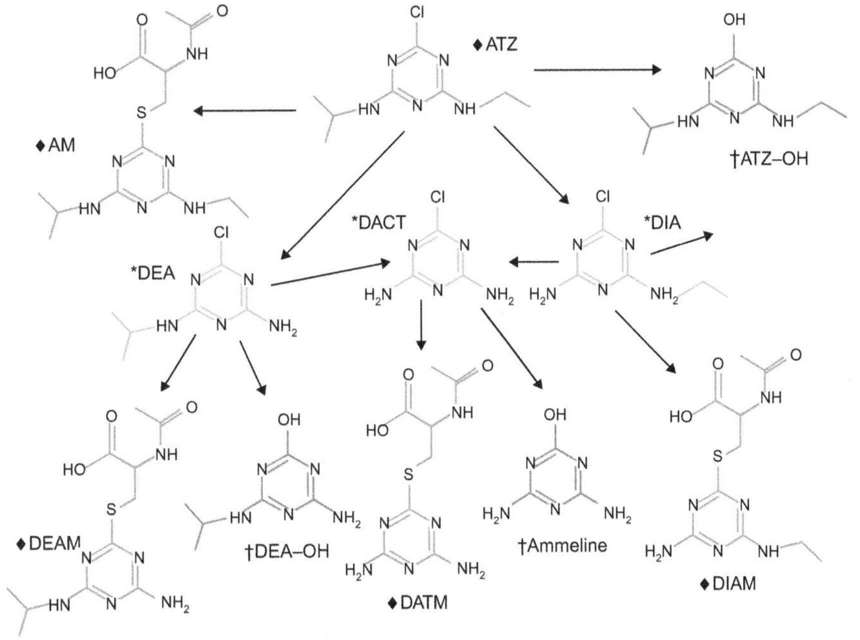

FIGURE 27.3 Proposed metabolism of ATZ. ATZ is shown in black; dealkylated metabolites are shown by *; hydroxylated metabolites are shown by †; and glutathione-derived mercapturic acid metabolites are shown by abbreviations: *AM*, atrazine mercapturate; *ATZ*, atrazine; *ATZ-OH*, hydroxyatrazine; *DACT*, diaminochlorotriazine; *DATM*, diaminotriazine mercapturate; *DEA*, desethylatrazine; *DEAM*, desethylatrazine mercapturate; *DEA-OH*, hydroxydesethylatrazine; *DIA*, desisopropyl atrazine; *DIAM*, desisopropyl atrazine mercapturate (Barr et al., 2007a,b).

TABLE 27.2 Summary of Animal and Human Metabolite Studies[a]

Study	Model	ATZ	AM	DEA	DACT	DIA	DAtM	ATZ'OH	Ammeline
Novartis[b]	Rat	ND	ND	Minor	1	Minor	1	ND	ND
Barr (Barr et al., 2007)	Rat	ND	ND	1	ND	1	ND	1	1
Bradwav[b]	Rat	NA	NA	1	1	1	NA	1	1
Erickson[b]	Swine	NA	NA	1	NA	NA	NA	NA	NA
Novartis[b]	Monkey	ND	3	2	1	4	After iv	ND	ND
Catenacci (Catenacci et al., 1993)	Human	Minor	NA	Minor	1	Minor	NA	NA	NA
Lucas (Lucas et al., 1993)	Human (D)	NA	1	Minor	Minor	Minor	NA	NA	NA
Buchholz (Buchholz et al., 1999)	Human (D)	NA	1	NA	2?	Minor	2?	NA	NA
Perry (Perry et al., 2000a,b)	Human	NA	1	2	NA	NA	NA	NA	NA
Catenacci (Catenacci et al., 2002)	Human	Minor	NA	2	1	Minor	NA	NA	NA
Mendas (Mendas et al., 2012)	Human	ND	1	NA	NA	NA	NA	NA	NA

ATZ, atrazine; *AM*, atrazine mercapturate; *DEA*, desethylatrazine; *DACT*, diaminochlorotriazine; *DIA*, desisopropyl atrazine; *DATM*, diaminotriazine mercapturate; *ATZ-OH*, hydroxyatrazine; *ND*, not detected; NA, not measured or applicable to the study; *?*, metabolite with tentative identification; *(D)*, dermal exposure. "After iv" indicates that metabolite was seen only after iv administration of atrazine.

[a]*1, a major metabolite; 2, a less abundant metabolite; minor, minor metabolite identified.*

[b]*Information obtained from documentation of internal studies conducted at Novartis. (Modified from Barr et al., 1997.)*

toxicity, presumably because of the presence of the cyano moiety. Effects noted at doses that were toxic to the mothers were cyclopia and diaphragmatic hernia in rabbits and an apparent increase in the incidences of skeletal variations in rats (Hodgson and Meyer, 1997; Gupta, 2018b).

Phenylurea Herbicides

Ureas and thioureas are a group of herbicides used for general weed control in agricultural and nonagricultural practices. The first urea herbicide, N, N-dimethyl-N0-(4-chlorophenyl)-urea was introduced in 1952 by DuPont under the common name of monuron. In subsequent years, many more derivatives of this class of compounds have been marketed. There are more than 20 polyureas available under different names such as diuron, fluometuron, isoproturon, linuron, buturon, chlorbromuron, chlortoluron, chloroxuron, difenoxuron, fenuron, methiuron, metobromuron, metoxuron, monuron, neburon, parafluron, siduron, tebuthiuron, tetrafluron, and thidiazuron. Diuron, the most common phenylurea, has been among the top 10 pesticides in use in the United States. The acute toxicity potential for all phenylureas appears to be low, but at high acute or repeated exposures, neurobehavioral alterations, body weight reductions, hematotoxicity, and hepatotoxicity have been reported. In general, ureas and thioureas do not cause developmental and reproductive toxicity, but monolinuron, linuron, and buturon are known to cause some teratogenic abnormalities in experimental animals. Linuron produced a high incidence of malformations in rat fetuses when given by gavage, but the chemical had no teratogenic potential in the rabbit under dietary regimen. A related chemical, monolinuron, caused cleft palate in mice (Matthiaschk and Roll, 1977; Liu, 2010).

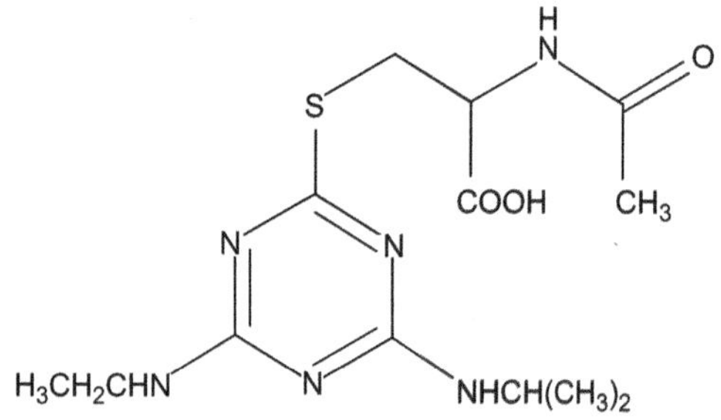

FIGURE 27.4 Chemical structure of atrazine mercapturate (N-acetyl-S-{4-(ethylamino)-6-[(1-methylethyl)amino]-1,3,5-triazine- 2-yl}-L-cysteine).

Absorption and excretion studies have revealed that polyureas are readily absorbed through the GI tract in rats and dogs and are mainly metabolized by dealkalization of the urea methyl groups (Boehme and Ernst, 1965). Hydrolysis of diuron to 3,4-dichloroaniline and oxidation to 3,4-dichlorophenol, as well as dihydroxylation at carbon 2 and/or 6 of the benzene ring, have also been reported. The predominant metabolite of diuron in urine is N-(3,4-dichlorophenyl)-urea. Diuron is partially excreted unchanged in feces and urine. The storage of diuron does not occur in tissues (Hodge et al., 1968; Hodgson and Meyer, 1997).

Protoporphyrinogen Oxidase–Inhibiting Herbicides

Protoporphyrinogen oxidase (Protox)–inhibiting herbicides have been used since the 1960s and now represent a relatively large and growing segment of the herbicide market. Nitrofen was the first Protox-inhibiting herbicide to be introduced for commercial use, in 1964. This diphenyl ether (DPE) herbicide was eventually recognized as a relatively weak inhibitor of Protox but was a lead compound of an entire class of structurally related herbicides that were much more active. Subsequently several DPE herbicides have been successfully commercialized (Nandihalli et al., 1992; Anderson et al., 1994). Nitrofen is a developmental toxicant and produced varying results in different species. It produced a high incidence of diaphragmatic hernias and harderian gland alterations in mice following oral treatment, and hydronephrosis and respiratory difficulties in rats. In hamsters, the compound when administered orally during gestation resulted in abnormal development of the para- and mesonephric ducts and ureteric bud, which was occasionally accompanied by renal agenesis in female offspring and predominantly left-sided agenesis of vas or epididymis and seminal vesicles in males. It was concluded that the teratogenic activity of nitrofen is mediated by alterations in maternal and fetal thyroid hormone status (Manson et al., 1984; Dayan and Duke, 2010).

After the first generation of Protox inhibitors (with the exception of oxadiazon), which was based on the DPE, numerous other non–oxygen-bridged compounds (non-DPE Protox inhibitors) with the same site of action (carfentrazone, JV 485, and oxadiargyl) have been commercialized. These compounds have little acute toxicity and are unlikely to pose any acute hazard in normal use. The developmental toxicity studies conducted on rats and rabbits indicate that the majority of the compounds did not show any reproductive, developmental, or teratogenic abnormalities, except at very high doses that elicit maternal toxicity. The developmental toxicity correlates with Protox accumulation (JMPR, 2004).

The Protox class of oxidase inhibitor herbicides is either not readily absorbed and/or are rapidly degraded by metabolism and/or excreted. The mammalian metabolites are similar to photochemical degradation products. In mammals, there are remarkable species differences in the levels of porphyrin accumulation resulting from exposure to Protox inhibitors. There is no bioaccumulation risk in animals. Metabolism of Protox inhibitors has been studied in a number of species, including rats, rabbits, goats, sheep, cattle, and chicken. In general, the metabolic degradation of these compounds by animals includes nitroreduction, deesterification, and conjugation to GSH, cysteine, and carbohydrates. Most of the metabolites are excreted in urine, with small amounts excreted in feces and milk. In chickens, ~95% of the metabolites are eliminated in excreta, with small amounts (0.09%) eliminated in the eggs (Hunt et al., 1977; Leung et al., 1991).

Substituted Anilines

Substituted aniline, an alachlor herbicide, was registered and introduced in 1967 for the preplant or preemergent control of a broad spectrum of grass, sedge, and broadleaf weeds (Heydens et al., 2010). The commonly used herbicides are alachlor, acetochlor, butachlor, metolachlor, and propachlor. This class of herbicides is slightly hazardous, except butachlor, which is not likely to pose any hazard. Based on various toxicity studies, the data support grouping alachlor, acetochlor, and butachlor with respect to a common mechanism of toxicity for nasal turbinate and thyroid tumors, and alachlor and butachlor for stomach tumors. The results from the reproductive and developmental toxicity studies indicate that these herbicides are not reproductive toxicants, developmental toxicants, or teratogens. However, in the rabbit study with butachlor, a slight increase in postimplantation loss and decreased fetal weights were observed at maternally toxic doses, but at dietary concentrations over two successive generations, no adverse effects on reproductive performance or pup survival was observed (Wilson and Takei, 1999).

Substituted anilines are well absorbed in rats orally. The dermal penetration in monkeys is relatively slow. The metabolism of alachlor in rats is complex because of extensive biliary excretion, intestinal microbial metabolism, and enterohepatic circulation of metabolites. The main route of excretion is urine and feces, and nearly 90% of the dose is eliminated in 10 days. The metabolism in rats and mice is similar; however, there are significant quantitative differences between the two species. These herbicides undergo *o*-demethylation with the release of formaldehyde; however, the relationship between metabolism-mediated formaldehyde release and nasal tumors induced was not established. In contrast, alachlor is metabolized in monkeys to a limited number of GSH and glucuronide conjugates, which are excreted primarily via kidney. Excretion in monkeys is more rapid than rodents with ~90% being excreted in the urine within 48 h. Alachlor metabolites undergo biliary excretion and hepatic circulation in rodents, whereas biliary excretion is limited in monkeys. In rats, acetochlor is rapidly metabolized to several polar metabolites and more than 95% are quickly excreted in urine and feces. The metabolites are the result of the mercapturic acid pathway formed by initial GSH conjugation. As in case of alachlor, acetochlor, butachlor, and propachlor also undergo glucuronide/glutathione conjugation and enter the hepatic circulation, leading to the formation of tertiary amide methylsulfide metabolite, which further undergo metabolism in liver and nasal tissue to form the putative carcinogen, diethyl quinoneimine (DEIQ) in rats (Feng et al., 1990; Heydens et al., 2010). In humans, the major urinary metabolites of alachlor, metolachlor, and acetachlor have been identified as their mercapturates. Alachlor mercapturate equivalent and metolachlor mercapturate equivalent were detected in more than half (n = 15) of the applicator preseason urine samples and in about 2% in the reference population (n = 46), whereas during the spraying season, the geometric mean amount of alachlor mercapturate equivalent and metolachlor mercapturate equivalent excreted in 24 h was 17 and 22 nmol, respectively. In addition, some of the intact chloroacetanilide herbicides have been measured in serum and plasma (Driskell et al., 1996; Driskell and Hill, 1997; Barr et al., 2007a,b). From these studies, it may be concluded that these metabolites are not inconsistent with the suggested metabolic pathways; however, the parent, plus the respective mercapturate metabolite in urine, can be used as a biomarker to monitor human exposure to substituted anilines (Sachana et al., 2017).

Bipyridyl Derivatives

This chemical class of herbicides includes paraquat and diquat. Paraquat was first described as a chemical redox indicator in 1882 by Weidel and Russo. In 1933, Michaelis and Hill discovered its redox properties and called the compound methyl viologen. It became commercially available as a herbicide in 1962 (Smith, 1997). It is a fast-acting, nonselective contact herbicide, absorbed by the foliage with some translocation in the xylem. Paraquat is the most toxic of the commonly used herbicides and the toxicity varies in different animals depending on the formulation and species used.

The compound is apparently nonteratogenic under standard testing regimens. High doses injected into pregnant rats and mice on various gestation days caused significant maternal toxicity but did not produce teratogenic effects (Bus and Gibson, 1975). At lower doses, gross soft tissue and slight increases in skeletal anomalies have been reported. In rats, costal cartilage malformations were reported. Diquat had no effect on fertility, was not teratogenic, and only produced fetotoxicity at doses that were maternally toxic. In the multigenerational study, cataract has been observed in rats at high doses (Lock and Wilks, 2010).

Paraquat is rapidly but incompletely absorbed from the GIT of laboratory animals and humans, with plasma concentration of 30–90 min, and poorly absorbed through contact skin. It has been reported that the dog absorbs more paraquat compared with rats, resulting in greater susceptibility of dogs toward paraquat toxicity. Distribution studies show higher radioactivity in Type I and Type II epithelial cells and the Clara cells of the rodent and human lungs, which are the major target cells for paraquat toxicity (Smith, 1997; Dinis-Oliveira et al., 2008). Paraquat is taken up into the brain via the neural amino acid transporter and thus may be a factor in the etiology of Parkinson's disease. The amount of paraquat excretion in feces corresponds to 60%–70% of the ingested dose. Paraquat is very poorly metabolized, and the bulk is excreted unchanged in the urine and feces (Van Dijck et al., 1975). The mechanism of action of paraquat and diquat is very similar at the molecular level and involves cyclic reduction–oxidation reactions, which produce reactive oxygen species and depletion of NADPH. However, the critical target organ differs for the two compounds, so the mammalian toxicity is quite different.

Although both herbicides affect kidneys, paraquat is selectively taken up in the lungs. Paraquat causes pulmonary lesions as a result of type I and type II pneumocytosis. The primary event in the mechanism of toxicity within cells is paraquat's ability to undergo a single electron reduction from the cation to form a free radical that is stable in the absence of oxygen. If oxygen is present, a concomitant reduction of oxygen takes place to form superoxide anion (O_2^-). Superoxide radical, in turn, is nonenzymatically converted to singlet oxygen, which attacks polyunsaturated lipids associated with cell membranes to form lipid hydroperoxides. Lipid hydroperoxides are normally converted to nontoxic lipid alcohols by the selenium-containing GSH-dependent enzyme, GSH peroxidase. Selenium deficiency, deficiency of GSH, or excess lipid hydroperoxides allow the lipid hydroperoxides to form lipid-free radicals. Lipid hydroperoxides are unstable in the presence of trace amounts of transition metal ions and decompose to free radicals, which in turn cause further peroxidation of polyunsaturated lipid in a process that is slowed by vitamin E. Peroxidation of the membranes could in turn cause cellular dysfunction and hence lead to cell damage or death, as summarized in Fig. 27.5 (Smith, 1997). For further details on mechanisms, toxicokinetics, clinical features, and treatment of paraquat poisoning, readers are referred to Dinis-Oliveira et al. (2008).

Unlike paraquat, diquat does not accumulate in the lungs; however, its presence is observed in the liver, kidney, plasma, and adrenal gland. Diquat does not enter the brain (Rose et al., 1976). Following oral administration, 90%–98% of the dose is eliminated via the urine. Metabolism studies indicate some unidentified metabolites of diquat in the urine of rabbits and guinea pigs. In rats, diquat monopyridone has been identified

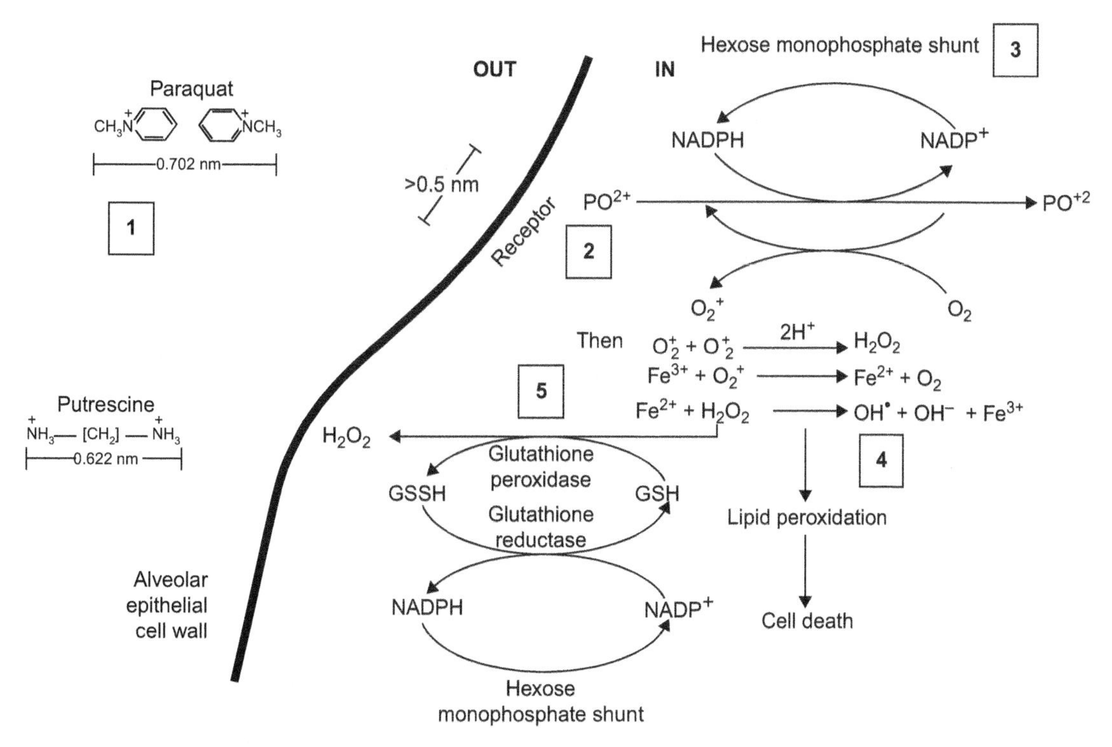

FIGURE 27.5 Schematic representation of mechanism of toxicity of paraquat. (1) Structure of paraquat and putrescine; (2) putative accumulation receptor; (3) redox cycling of paraquat, utilizing NADPH; (4) formation of hydroxyl radical (OH^o), leading to lipid peroxidation; and (5) detoxication of H_2O_2 via GSH reductase peroxidase couple, utilizing NADPH. *Reproduced with permission from Smith, L.L., 1997. Paraquat. In: Sipes, I.G., McQueen, C.A., and Gandolfi, A.J. (Eds.), Comprehensive Toxicology: Toxicology of the Respiratory System, vol. 8. Pergamon, New York, pp. 581–589.*

in the feces, at about 5% of an oral dose, whereas diquat-dipyridone has been detected in urine. These results indicate that diquat is probably metabolized by GI bacteria (Daniel and Gage, 1966; JMPR, 1993).

Amides and Acetamides

The commonly used herbicides include bensulide, dimethenamid-P, and propanil. They are slightly to moderately hazardous in normal use. Dimethenamid is a racemic mixture of the M and P stereoisomers, whereas the P isomer has useful herbicidal activity. Dimethenamid can reduce fetal body weights but is not teratogenic (Gupta, 2018b). Dimethenamid is slowly but well absorbed after oral administration (90% in rats) and is extensively metabolized in rats. The maximum concentration in blood is not achieved until about 72 h. Excretion is primarily via bile. By 168 h after treatment, an average of 90% of the administered dose is eliminated. In rats, the concentration of radioactivity in blood decreases more slowly than in other tissues and is associated with specific binding to globin (not in humans). Metabolism is primarily via the GSH conjugation pathway, but racemic dimethenamid is also metabolized by cytochrome P450 enzymes via reductive dechlorination, oxidation, hydroxylation, *o*-demethylation, and cyclization pathways, as well as conjugation with glucuronic acid. Unchanged dimethenamid in excreta accounts for only 1%–2% of the administered dose; more than 40 metabolites have been detected (JMPR, 2005).

Imidazolinones

Imidazolinone herbicide was discovered in the 1970s, with the first US patent awarded in 1980 for imazamethabenz-methyl. Imidazolinone herbicides include imazapyr, imazapic, imazethapyr, imazamox, and imazaquin. These herbicides are relatively nontoxic. The relatively high ADIs suggest that this class of compounds does not pose a concern for chronic dietary exposure to humans. Imidazolinone herbicides are not reproductive toxicants, developmental toxicants, or teratogens. Following single oral doses of imazapyr, imazapic, imazethapyr, imazamox, or imazaquin, these imidazolinone herbicides are rapidly absorbed and excreted. Although imazamethabenz-methyl is also rapidly absorbed in the rat, greater than 60% becomes metabolized via hydrolysis of the ester to imazamethabenz acid, which is rapidly excreted in the urine within 24 h. The presence of a significant amount of unchanged parent or metabolite in the urine within 24–48 h indicates a low potential for bioaccumulation of parent compound or acid metabolite in mammalian tissues (Hess et al., 2010).

Triazolopyrimidine

Triazolopyrimidine herbicides include cloransulam-methyl, diclosulam, florasulammethyl, flumetsulam, and metosulam. In general, they exhibit very low mammalian toxicity. In repeat-dose toxicity studies, the liver and kidneys have been identified as target organs with effects that were often adaptive in nature. This group of herbicides did not affect reproduction or fetal development or multigenerational reproduction. No evidence of maternal toxicity, embryo-fetotoxicity, or teratogenicity has been observed in experimental animals (EPA, 1997a,b; Billington et al., 2010). Triazolopyrimidines are rapidly absorbed, and the excretion is mainly through urine with small amounts being excreted in feces. They have a low potential for bioaccumulation. The only metabolite present is the 4-OH phenyl derivative and/or oxidation product. In metosulam toxicity, demethylation of the 3-methyl moiety of the phenyl ring and of the 3-methoxy moiety of the pyridine ring and other conjugation products of the parent material has been observed. Other minor metabolites include hydroxylated products of the pyridine ring, though the position of hydroxylation has not been identified. Because of rapid elimination, there is little potential for accumulation in the tissues (Billington et al., 2010).

Phosphonomethyl Amino Acids or Inhibitors of Aromatic Acid Biosynthesis

Inhibitors of aromatic acid biosynthesis (organic phosphorus), broad-spectrum, nonselective, postemergent, systemic herbicides with activity on essentially all annual and perennial plants have been developed. Monsanto discovered the herbicidal properties of glyphosate in 1970, and the first commercial formulation was introduced in 1974, under the brand "Roundup." Another compound, glufosinate (Basta), is marketed as the isopropyl amine. Trimethylsulfonium salts of glyphosate and the ammonium salt of glufosinate have low acute oral toxicity in mice and rats and are unlikely to pose acute hazard in normal use. No adverse effects on reproduction or development toxicity have been reported. These compounds are poorly absorbed both orally and via the dermal route. These herbicides are rapidly eliminated, and they are neither biotransformed nor accumulated in tissues. More than 70% of an orally administered dose of glyphosate is rapidly eliminated through feces and 20% through urine. The main metabolite of glyphosate is aminomethylphosphonic acid (AMPA); the AMPA is of no greater toxicological concern than its parent compound (JMPR, 2004).

Others

Carbamates, thiocarbamates, and dithiocarbamate compounds include derivatives of carbarnic acid (asulam, barban, chlorpropham, chlorbufam, karbutilate, and phenmedipham), derivatives of thiocarbarnic acid (butylate, cycloate, diallate, EPTC, molinate, and triallate), and derivatives of dithiocarbarnic acid (methamsodium). These herbicides have low-to-moderate toxicity in rats and do not pose acute hazards, but chlorpropham at high oral doses given during the organogenesis period caused malformations and various types of development toxicity (Gupta, 2018b).

There are several other herbicides that may cause various types of severe abnormalities or lead to developmental and reproductive abnormalities in animals. As such, most of them either have been banned or have restricted use. For example, aminotriazole or amitrole causes fetal thyroid lesions, as it is also an antithyroid agent (Shalette et al., 1963). Butiphos was reported to be teratogenic in the rabbit but was not so in the rat (Mirkhamidova et al., 1981). Chloridazon caused only resorptions in the rat but caused rib and tail anomalies in hamster fetuses of several litters. Prometryn produced head, limb, and tail defects in rat fetuses following daily administration during gestation (Schardein, 2000). Dinitrophenol compounds such as DNP (2,4-dinitrophenol), DNOC, dinoseb, and dinoterb are highly hazardous to animals and may cause developmental toxicity. Dinoseb is known to have teratogenic effects in several species. The compound when given by gavage to rats at maternally toxic doses reduced fetal body weight and increased the frequency of extra ribs. In rabbits, after dermal exposure, eye defects and neural malformations accompanied by maternal toxicity have been reported. With dinitro compounds, abortions have been reported in sows (Lorgue et al., 1996). Most of these chemicals have been banned from use in the United States. Dinoterb, another chemical of the same group, induced skeletal malformations by both oral and dermal administration in the rat, and skeletal, jaw, head, and visceral malformations in the rabbit (Schardein, 2000).

Compound such as astridiphane, a dinitroaniline compound, was a potent developmental toxicant in the mouse, and induced cleft palate and other toxicity at maternally toxic doses, but under the same conditions in the rat the compound increased the frequency of minor skeletal variations (Hanley et al., 1987).

Fungicides

Fungicides are agents that are used to prevent or eradicate fungal infections from plants or seeds. In agriculture, they are used to protect tubers, fruits, and vegetables during storage or are applied directly to ornamental plants, trees, field crops, cereals, and turf grasses. Numerous substances having widely varying chemical constituents are used as fungicides. Fungicides have been classified according to chemical structures or have been categorized agriculturally and horticulturally according to the mode of action (Ballantyne, 2004). According to the mode of application, fungicides are grouped as foliar, soil, and dressing fungicides. Foliar fungicides are applied as liquids or powders to the aerial green parts of plants, producing a protective barrier on the cuticular surface and systemic toxicity in the developing fungus. Soil fungicides are applied as liquids, dry powders, or granules, acting either through the vapor phase or by systemic properties. Dressing fungicides are applied to the postharvest crop as liquids or dry powders to prevent fungal infestation, particularly if stored under less than optimum conditions of temperature and humidity. With a few exceptions, most of the newly developed chemicals have a low order of toxicity to mammals. Public concern has focused on the positive mutagenicity tests obtained with some fungicides and the predictive possibility of both teratogenic and carcinogenic potential. The quantity of fungicides used on major crops is estimated to have increased 2.3-fold between 1964 and 1997. Use of inorganics such as sulfur, lime, copper, and mercury compounds has declined since the 1960s, but captan, chlorothalonils, and other organic materials account for 90% of fungicide use. Newer groups, such as benzimidazoles, conazoles, dicarboximides, and metal organic compounds, account for ~10% of fungicide use (Osteen and Padgitt, 2002).

Dialkyldithiocarbamates

Ethylenebisdithiocarbamate (EBDCs) fungicides are commonly used to control about 400 fungal pathogens on more than 100 crops. All members have an ethylenebisdithiocarbamate backbone, with different metals associated with the individual compounds. Out of five (mancozeb, maneb, metiram, zineb, and nabam), mancozeb, maneb, and metiram are the most widely used EBDCs. Zineb is used to a lesser degree, and nabam is no longer used in agriculture. The EBDCs have very low acute toxicity by the oral, dermal, and respiratory routes. Mancozeb, maneb, and metiram are unlikely to present an acute exposure hazard under conditions of normal use (Freudenthal et al., 1977).

Toxicokinetics and metabolism of mancozeb, maneb, and metiram in laboratory animals have indicated that the EBDCs are only partially absorbed, then rapidly metabolized, and excreted with no evidence of long-term bioaccumulation. Absorption of oral doses is rapid. Most of the administered dose is excreted within 24 h, with about half eliminated in the urine and half in the feces. Biliary excretion is minimal, indicating that only ~50% of oral doses are absorbed. Only low level residues are found in tissues, principally in the thyroid.

Ethylenethiourea (ETU) is the major metabolite. On average ~7.5% of an EBDC dose administered to rats is metabolized to ETU on a weight basis. The spectrum of metabolites produced in laboratory animals points to two common metabolic pathways (Fig. 27.6), both leading ultimately to the formation of glycine and incorporation into natural products. In the predominant pathway quantitatively, the dithiocarbamate linkages are hydrolyzed to produce ethylenediamine (EDA) directly, and EDA is oxidized to glycine, joining the intermediary metabolic pool at this point. The other pathway is responsible for the toxic effects of the EBDCs and involves oxidation to ethylenebisisothiocyanate sulfide and then to ETU, various derivatives of ETU, and ethyleneurea (EU) before rejoining the main pathway with conversion to EDA, glycine, and other natural products (Hurt et al., 2010).

ETU reversibly inhibits thyroid peroxidase-catalyzed iodination and coupling of tyrosine residues into the thyroid hormone. The correlation of thyroid hormonal changes with hypertrophy, hyperplasia, and neoplasia has been clearly demonstrated in specific studies of ETU in rats (Chhabra et al., 1992).

Based on various parameters, using individual toxicology databases for the respective EBDCs, mancozeb, maneb, and metiram, and their ETU metabolite, indicate an overall NOAEL of 5.0 mg/kg bw/day for the group (Fig. 27.7). Application of a standard 100-fold overall safety factor leads to a recommended acceptable daily intake (ADI) of 0.05 mg/kg/day. The factors discussed above for ETU, and the unusual reliability of the database, reflecting the combined results of four individually comprehensive databases, confirm that no additional uncertainty factors need be applied. This aggregate assessment produces a comparable, if slightly higher, estimate of the ADI than the FAO/WHO recommendation. The WHO panel reviewed the data for each of the actives individually and established ADIs of 0.05 mg/kg/day for mancozeb and maneb, and 0.03 mg/kg/day for metiram, resulting in the allocation of a group ADI of 0.03 mg/kg/day for the EBDCs collectively, including mancozeb, maneb, metiram, and zineb. The basis for the establishment of a group ADI was the similarity of the chemical structures of the EBDCs, the comparable toxicological profile of the EBDCs based on the toxic effects of ETU, and the fact that the parent EBDC residues cannot be differentiated using presently available regulatory analytical procedures (FAO/WHO, 1994). Because of their importance to worldwide agriculture, the EBDCs and ETU have been thoroughly tested over many years. Collectively, the data demonstrate that the use of the

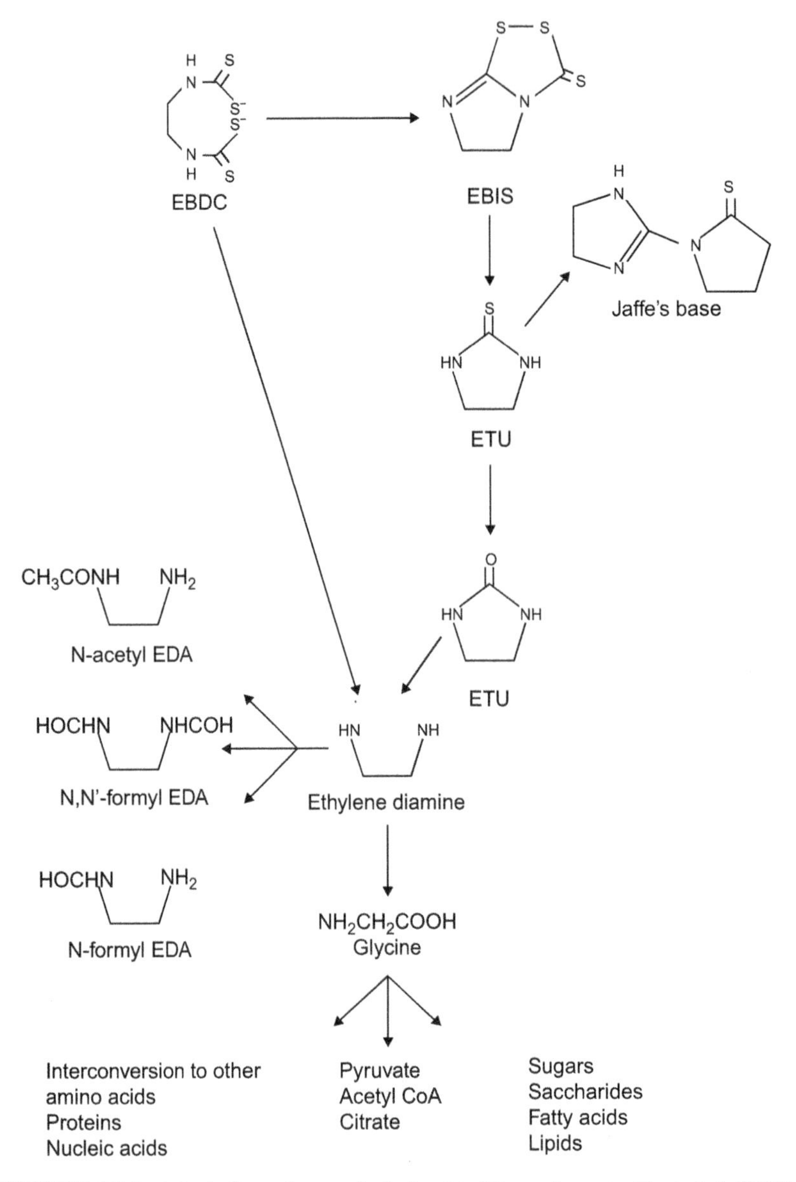

FIGURE 27.6 Metabolic pathway of ethylenebisdithiocarbamate. Hurt et al. (2010).

EBDCs results in essentially negligible exposure to consumers and low risk to farm workers, production workers, and people who are exposed through recreational activities (Hurt et al., 2010).

As ETU is the major and common metabolite of EBDCs, it can be used as a biomarker for EBDC exposures. Colosio et al. (2002) used ETU and mancozeb as an indicator of mancozeb exposure to vineyard workers. In this study, urine levels of ETU were significantly increased in workers after the work shift. However, the number of workers considered in this study was low. In another study with a comparatively larger group of workers, the authors concluded that, altogether, the differences between exposed and controls were not consistently correlated to any clinical impairment and suggested that the seasonal application of mancozeb does not pose a significant health risk to exposed subjects (Colosio et al., 2002).

Anilinopyrimidines

Anilinopyrimidines are a new chemical class of fungicides that are highly active against a broad range of fungi. The anilinopyrimidine class of fungicides includes cyprodinil, mepanipyrim, and pyrimethanil. The compounds have low toxicity and are unlikely to present acute hazards in normal use. Cyprodinil

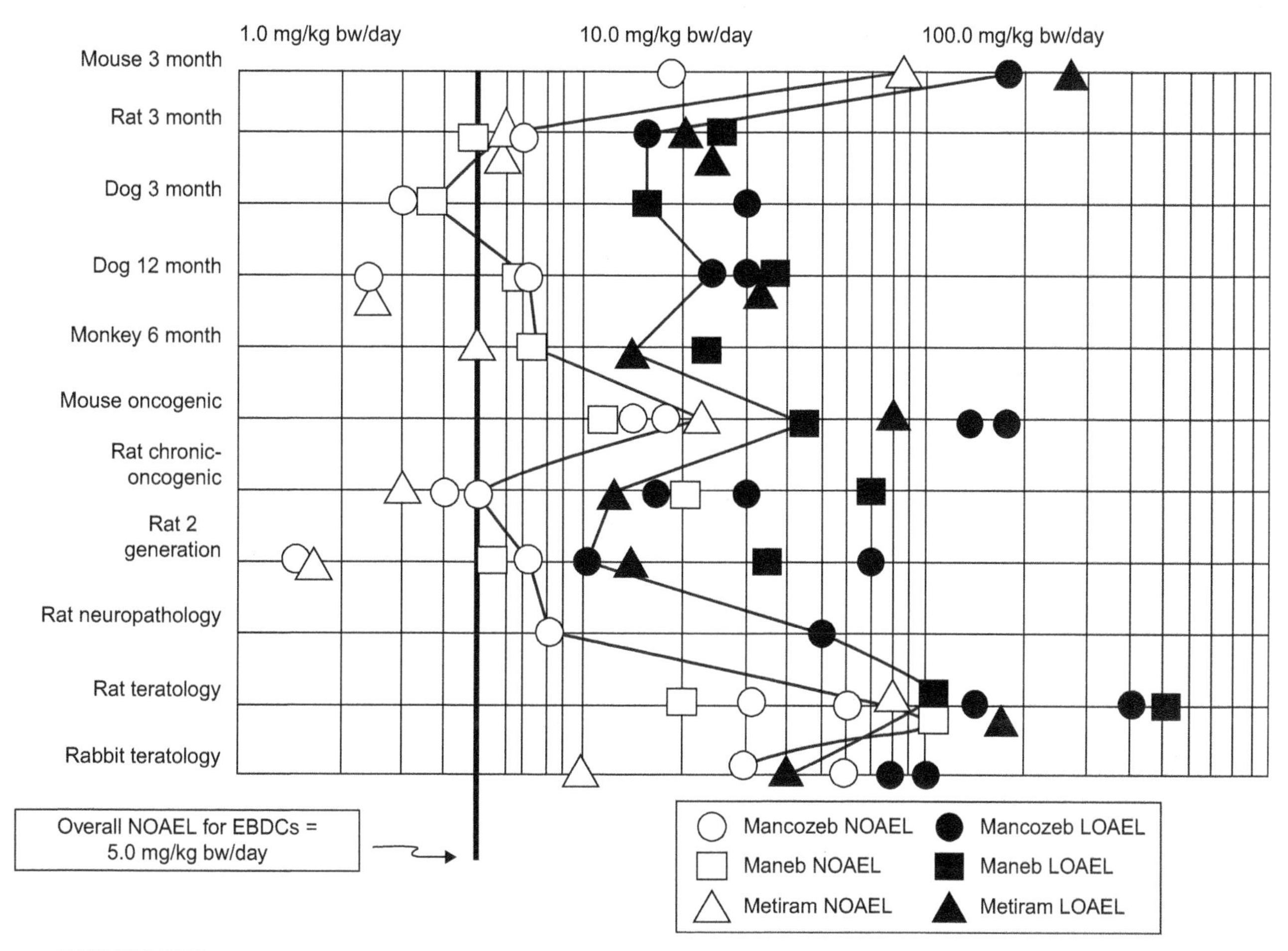

FIGURE 27.7 Summary of critical doses (NOAELs) for different EBDCs from various toxicology studies. Hurt et al. (2010).

produces hepatomegaly with hepatocellular hypertrophy and increased thyroid weights associated with follicular cell hypertrophy and hypochromasia in rats. Mepanipyrim causes hepatocellular fatty vacuolation and lipofuscin deposition in Kupffer cells and hepatocytes of dogs, whereas such changes are not observed in cyprodinil-treated rats (Terada et al., 1998). Pyrimethanil produces thyroid follicular cell tumors in rats and enhancement of hepatic thyroid hormone metabolism, which may be responsible for thyroid tumorigenesis (Hurlety, 1998). The findings in the thyroid were considered to be secondary to liver changes. Enhancement of hepatic thyroid hormone metabolism and excretion are considered to be the mode of action of thyroid tumorigenesis. Cyprodinil induced microsomal protein and cytochrome P450 contents along with ethoxyresorufin O-deethylase, pentoxyresorufin O-depentylase, and lauric acid 11- and 12-hydroxylase, and cytosolic glutathione S-transferase activities in rats. Cyprodinil and mepanipyrim induce the opposite effects on liver and blood lipid parameters in rats. In general, anilinopyrimidines do not have adverse effects on developmental toxicity. They are neither genotoxic nor have any carcinogenic potential (Waechter et al., 2010).

In rats, cyprodinil was almost completely metabolized. No unchanged parent molecule could be found in urine, whereas minor amounts of unchanged cyprodinil were found in feces. Most of the administered cyprodinil was metabolized by sequential oxidation of the phenyl and pyrimidine ring. The major phase 1 metabolite was identified as 4-cyclopropyl-5-hydroxy-6-methyl-N-(4-hydroxy)-phenyl-2-pyrimidinamine (metabolite 2). This metabolite was excreted in the urine as β-glucuronic acid conjugate, as well as mono- and disulfuric acid conjugates. Although female rats formed the monosulfate almost exclusively, the males excreted equal amounts of the mono- and disulfate. Further oxidation of the methyl group led to the formation of 4-cyclopropyl-5-hydroxy-6-hydroxymethyl-N-(4-hydroxy)-phenyl-2-pyrimidinamine (metabolite 3), which was excreted in the urine in unconjugated form.

Alternative pathways proceeded either by sequential oxidation of the phenyl ring to the 4-hydroxy and 3,4-dihydroxy derivatives (metabolites 4 and 5), followed by oxidation of the methyl group (metabolite 6), or started with hydroxylation of the methyl group as the first oxidation step (metabolite 9). Urinary and biliary metabolites were found to be conjugated with β-glucuronic acid and sulfuric acid. The major metabolites identified in feces were the 5-hydroxypyrimidine derivative of cyprodinil (metabolite 1) and metabolite 4. Metabolites 1 and 4 also were present in conjugated form in urine and bile.

Two additional metabolites were found in liver and/or kidney tissue but not in excreta. Metabolite 7 was identified as ring-hydroxylated N-phenyl-guanidine, a breakdown product of the pyrimidine ring moiety. Metabolite 8 (i.e., 4-cyclopropyl-6-methyl-pyrimidine-2-ylamine) demonstrated a cleavage of the parent molecule between the pyrimidine and the phenyl ring. Metabolite 7 was found exclusively in the liver, where it represented the major metabolite. Minor amounts of metabolite 8 were found in the liver and kidneys. It seems the metabolism of cyprodinil in humans has not been documented in the scientific literature.

The metabolism of cyprodinil in rats is summarized in Fig. 27.8.

Chloroalkylthiodicarboximides (Phthalimides)

This class of chemicals contains broad-spectrum fungicides (captan, folpet, captafol, etc.) used as surface protectants on many crops. Captan and folpet have been in use for over 55 years. They are usually nontoxic to mammals. Some compounds of this class cause developmental effects, whereas others do not, perhaps because of, and/or masked by, maternal toxicity and possible nutritional deficits (Costa, 1997). Captan induces hyperplasia of the crypt cells. Following treatment with folpet, immune function is reduced, villi length is reduced, and crypt compartments are expanded, thereby reducing the villi-to-crypt ratio in

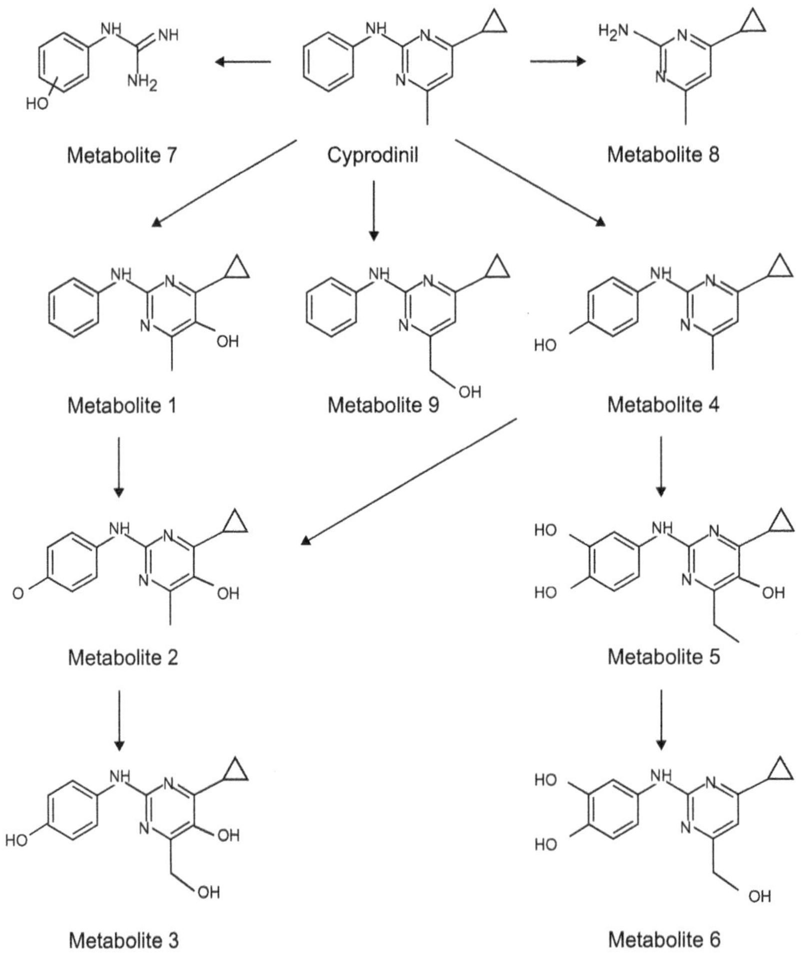

FIGURE 27.8 Phase I metabolism of cyprodinil in rats. Waechter et al. (2010).

mice (Tinston, 1995; Waterson, 1995). Captafol differs from captan and folpet in a number of ways, including structure and chemical activity. Both of them have low acute toxicity. They are not carcinogenic, mutagenic, or teratogenic. They are neither selective developmental toxicants nor reproductive toxicants. They are irritant to mucus membranes, especially of skin after repeated exposures (Gordon, 2010). Recently, the National Toxicology Program (NTP) judged that captafol is reasonably anticipated to be a human carcinogen based on sufficient evidence of carcinogenicity from studies in experimental animals and supporting data on mechanisms of carcinogenesis (NTP, 2011).

The captan is rapidly degraded to 1,2,3,6-tetrahydrophthalimide (THPI) and thiophosgene (via thiocarbonyl chloride) in the stomach before reaching the duodenum. THPI has a half-life of 1–4 s, and thiophosgene is detoxified by reaction with cysteine or glutathione and is rapidly excreted. The metabolic pathway of captan is presented in Fig. 27.9.

No captan is detected in blood or urine. It is therefore unlikely that these compounds or even thiophosgene would survive long enough to reach systemic targets such as the liver, uterus, or testes. Because of rapid elimination, meat, milk, or eggs from livestock/poultry would be devoid of the parent materials. Humans appear to metabolize captan in a similar manner to other mammals (Krieger and Thongsinthusak, 1993). THPI and thiazolidine-2-thione-4-carboxylic acid (TTCA) have been used as biomarkers in urine for captan exposure via either oral or dermal routes in animals and humans (van Welie et al., 1991; Heredia-Ortiz and

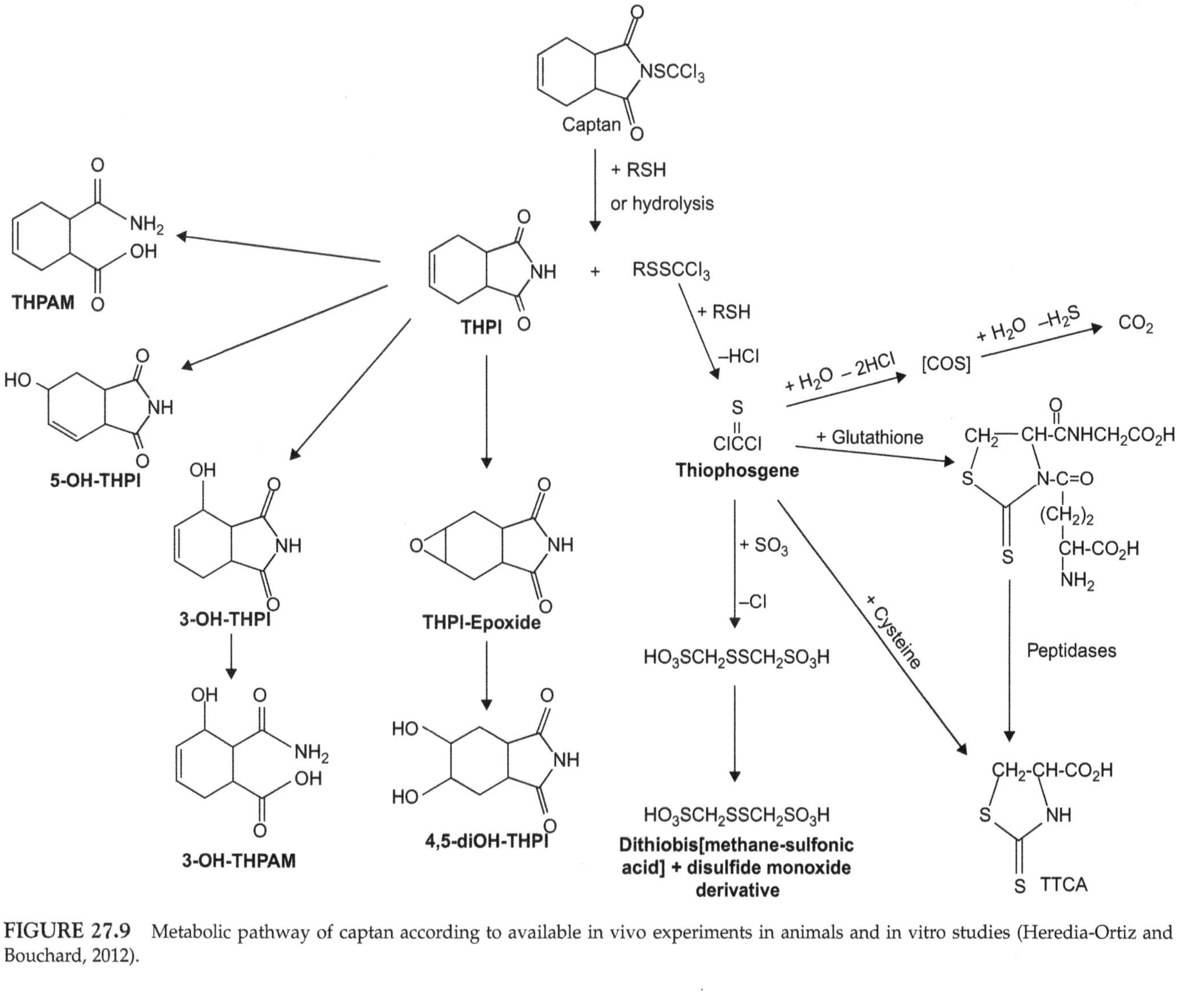

FIGURE 27.9 Metabolic pathway of captan according to available in vivo experiments in animals and in vitro studies (Heredia-Ortiz and Bouchard, 2012).

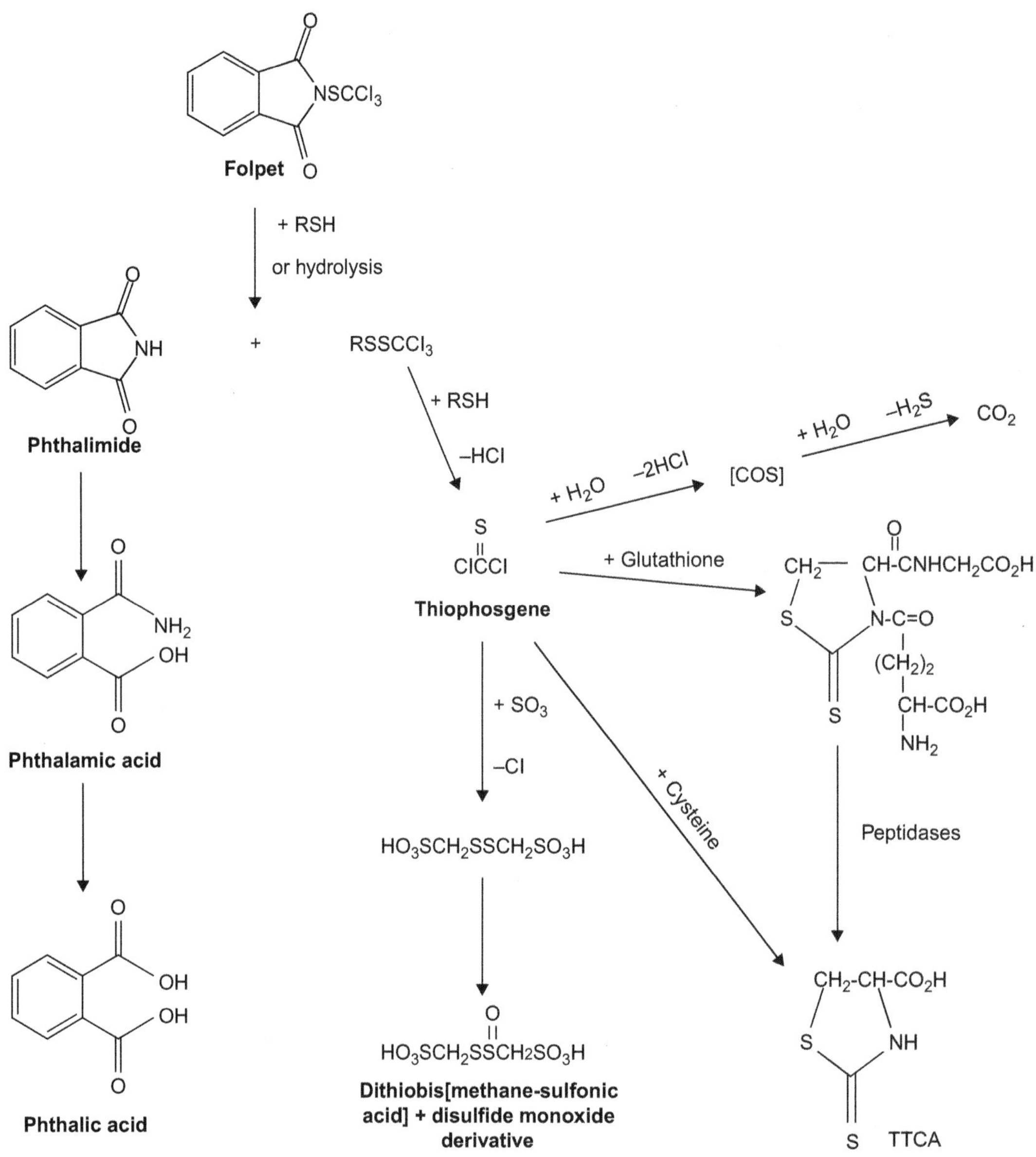

FIGURE 27.10 Metabolic pathway of folpet according to available in vivo experiments in animals and in vitro studies (Heredia-Ortiz et al., 2011).

Bouchard, 2012; Berthet et al., 2012a). Similarly, along with THPI, phthalimide (PI) and phthalic acid (PA) could be used as biomarkers for human exposure to folpet (Heredia-Ortiz et al., 2011; Berthet et al., 2012b). As stated earlier, these compounds degrade extremely rapidly when thiols are present (Fig. 27.10). In human blood, captan's t1/2 is less than 1 s and folpet's is less than 5 s. Thiophosgene, the reactive degradate that is formed from the trichloromethylthio side chain, reacts not only with thiols but also with other functional groups and its t1/2 is less than 0.6 s. Due to rapid degradation systemic exposure to captan, folpet, or their common degradate, thiophosgene, is absent. This, along with the low estimated dermal absorption rate of 0.5% per hour, assures that adverse systemic risk in agricultural workers is absent (Gordon, 2010).

Halogenated Substituted Monocyclic Aromatics

This class of chemicals includes chlorothalonil, dicloran, HCB, quintozene, PCP, dichlorophen, dinocap, tecnazene, and chloroneb. Chlorothalonil is a nontoxic halogenated benzonitrile fungicide with broad spectrum activity against vegetable, ornamental, orchard, and turf diseases. Repeated administration of chlorothalonil causes hyperplasia in the forestomach of rats and mice.

In dogs, there is no evidence of either neoplastic development or the occurrence of preneoplastic lesions in the kidney or stomach. The absence of stomach lesions in dogs is attributable to the anatomical differences between rodents and dogs (dogs do not possess a forestomach). Continued administration of chlorothalonil leads to the development of a regenerative hyperplasia within the renal proximal tubular epithelium. Continued regenerative hyperplasia ultimately results in progression of the kidney lesion to tubular adenoma and carcinoma. Initial cytotoxicity and regenerative hyperplasia within the proximal tubular epithelium are essential prerequisites for subsequent tumor development. The proposed mode of action for the induction of renal toxicity in rodents is outlined in Fig. 27.11. Human relevance to these stomach and kidney tumors is discussed in a review (Wilkinson and Killeen, 1996). Considering that chlorothalonil has been marketed for about 30 years, the reports of adverse effects are very rare.

Chlorothalonil is a reactive molecule toward thio(-SH) groups. It is a soft electrophile with a preference for sulfur nucleophiles rather than nitrogen/oxygen nucleophiles. Such chemicals tend to show reactivity toward protein containing critical S electrophiles rather than toward DNA (containing critical O and N nucleophiles). Following oral administration thiol-derived metabolites were identified in urine; only 3% of the dose appeared in urine with lower proportions excreted as thiol-derived metabolites, indicating that gut microflora may play a role in the disposition and metabolism of chlorothalonil in the rat. At least 80% of the administered dose has been shown to be excreted in feces within 96 h. Bile cannulation studies have confirmed that chlorothalonil undergoes enterohepatic circulation in the rat. A relatively high concentration of about 0.1% was observed in rat kidneys, one of the target organs for chlorothalonil toxicity (Parsons, 2010).

Mechanistic studies have been conducted to determine if a relationship exists between the ability to excrete urinary thiol-derived metabolites of chlorothalonil and to induce renal toxicity. Inhibition of γ-glutamyltranspeptidase using Acivicin (Syngenta, 1985) and renal organic anion transport using probenecid (Syngenta, 1986) decreased urinary thiol-derived metabolite excretion in rats. Administration of the monoglutathione conjugate to rats was shown to produce a qualitatively similar pattern of metabolite excretion to

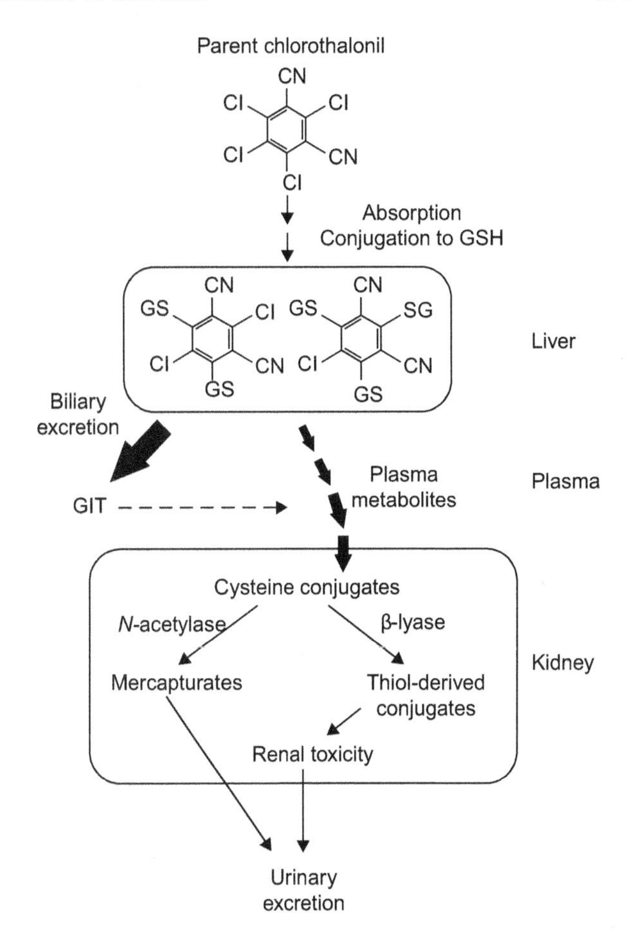

FIGURE 27.11 Schematic outlining potential pathways of chlorothalonil metabolism in the rat that leads to formation of toxic metabolites within the kidney. Following absorption from the GI tract, chlorothalonil is conjugated to glutathione in the liver. Further metabolic processing results in the formation of cysteine conjugates that may be detoxified via N-acetylase or activated to toxic thiol-derived species. *GIT*, gastrointestinal tract; *GSH*, glutathione. *Reproduced with permission from Parsons, P.P., 2010. Mammalian toxicokinetics and toxicity of chlorothalonil. In: Krieger, R. (Ed.), Hayes' Handbook of Pesticide Toxicology, third ed. vol. 2. Elsevier, New York, NY, pp. 1951–1966.*

that seen following administration of chlorothalonil itself (Syngenta, 1987a,b). These studies indicate that excretion of thiol-derived metabolites of chlorothalonil requires glutathione conjugation and then subsequent enzymatic processing of glutathione-derived conjugates that are selectively accumulated within the kidney. By analogy with other chemicals that undergo extensive glutathione conjugation, it is reasonable to presume that metabolism proceeds via cysteine conjugates and N-acetyl cysteine conjugates ("mercapturates") as outlined in Fig. 27.12 (Parsons, 2010).

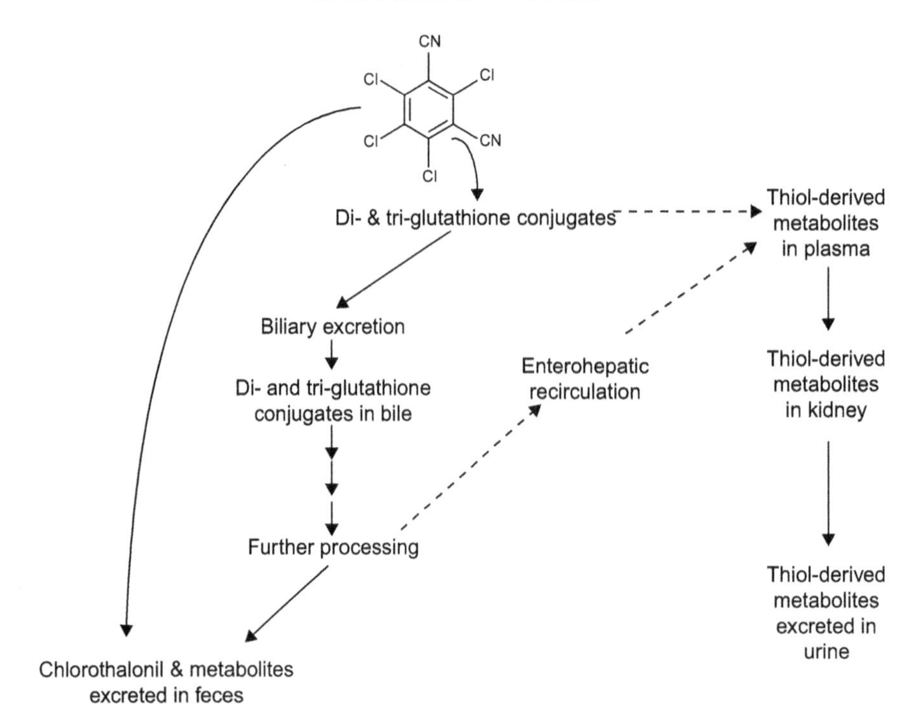

FIGURE 27.12 Diagram illustrating proposed metabolism of chlorothalonil following oral administration to rats. Broken lines indicate multistage events involving several enzymatic steps and transport processes. *Reproduced with permission from Parsons, P.P., 2010. Mammalian toxicokinetics and toxicity of chlorothalonil. In: Krieger, R. (Ed.), Hayes' Handbook of Pesticide Toxicology, third ed. vol. 2. Elsevier, New York, NY, pp. 1951–1966.*

CONCLUDING REMARKS AND FUTURE DIRECTIONS

To recognize whether a subject is exposed to herbicides and fungicides, or even accidental poisoning, standardized analytical procedures for diagnostic investigation of biological materials have become established. Accurately assessing exposure to these chemicals is critical in determining health outcomes that are associated with exposures and for evaluating the success of risk mitigation strategies. Biomonitoring is an important tool that can be used to evaluate human exposure to such xenobiotics by measuring the levels of these chemicals, their metabolites, or altered biological structures or functions in biological specimens or tissue. These measurements in biological media reflect human exposure to xenobiotics through all relevant routes and can therefore be used to monitor aggregate and cumulative exposures. Therefore, biomonitoring is a systematic continuous or repetitive activity of collection of biological samples for analysis of concentrations of xenobiotics, their metabolites, or specific nonadverse biological effect parameters for immediate application, with the objective to assess exposure and health risk to exposed subjects.

Although the biomonitoring data are not without their limitations, they still remain a viable tool and there is still much work left to be done. With existing data gaps, we may have more uncertain estimates of risks or health outcomes, but the biomonitoring data at least give us a starting point for our endeavors. We must continue and strive hard to fill in these gaps so that we can better interpret the data we generate.

References

Alexander, B.H., Mandel, J.S., Baker, B., et al., 2007. Biomonitoring of 2,4-dichlorophenoxyacetic acid exposure and dose in farm families. Environ. Health Perspect. 115, 370–376.

Anderson, R.J., Norris, A.E., Hess, F.D., 1994. Synthetic organic chemicals that act through the porphyrin pathway. Am. Chem. Soc. Symp. Ser. 559, 18–33.

Arcury, T.A., Grzywacz, J.G., Isom, S., et al., 2009. Seasonal variation in the measurement of urinary pesticide metabolites among Latino farm workers in eastern North Carolina. Int. J. Occup. Environ. Health 15, 339–350.

Aylward, L.L., Hays, S.M., 2008. Biomonitoring equivalents (BE) dossier for 2,4-dichlorophenoxyacetic acid (2,4-D) (CAS No. 94-75-7). Regul. Toxicol. Pharmacol. 51, S37–S48.

Aylward, L.L., Morgan, M.K., Arbuckle, T.E., et al., 2010. Biomonitoring data for 2,4-dichlorophenoxyacetic acid in the US and Canada: interpretation in a public health risk assessment context using Biomonitoring Equivalents. Environ. Health Perspect. 118, 177–181.

Ballantyne, B., 2004. Toxicology of fungicides. In: Marrs, T.C., Bryan, B. (Eds.), Pesticide Toxicology and International Regulation. Wiley, Hoboken, pp. 194–303.

Barr, D.B., Angerer, J., 2006. Potential uses of biomonitoring data: a case study using the organophosphorus pesticides chlorpyrifos and malathion. Environ. Health Perspect. 114, 1763–1769.

Barr, D.B., Panuwet, P., Nguyen, J.V., et al., 1997. Assessing exposure to atrazine and its metabolites using biomonitoring. Environ. Health Perspect. 115, 1474–1478.

Barr, D.B., Barr, J.R., Driskell, W.J., et al., 1999. Strategies for biological monitoring of exposure for contemporary-use pesticides. Toxicol. Ind. Health 15, 168–179.

Barr, D.B., Thomas, K., Curwin, B., et al., 2006. Biomonitoring of exposure in farm worker studies. Environ. Health Perspect. 114, 936–942.

Barr, D.B., Hines, C.J., Olsson, A.O., et al., 2007a. Identification of human urinary metabolites of acetochlor in exposed herbicide applicators by high-performance liquid chromatography-tandem mass spectrometry. J. Expo. Sci. Environ. Epidemiol. 17, 559–566.

Barr, D.B., Panuwet, P., Nguyen, J.V., et al., 2007b. Assessing exposure to atrazine and its metabolites using biomonitoring. Environ. Health Perspect. 115, 1474–1478.

Benachour, N., Seralini, G.-E., 2009. Glyphosate formulations induce apoptosis and necrosis in human umbilical, embryonic, and placental cells. Chem. Res. Toxicol. 22, 97–105.

Berthet, A., Bouchard, M., Danuser, B., 2012a. Toxicokineticsof captan and folpet biomarkers in orally exposed volunteers. J. Appl. Toxicol. 32, 194–201.

Berthet, A., Heredia-Ortiz, R., Vernez, D., et al., 2012b. A detailed urinary excretion time course study of captan and folpet biomarkers in workers for the estimation of dose, main route-of-entry and most appropriate sampling and analysis strategies. Ann. Occup. Hyg. 56, 815–828.

Billington, R., Gehen, S.C., Hanley Jr., T.R., 2010. Toxicology of triazolopyrimidine herbicides. In: Krieger, R. (Ed.), Hayes' Handbook of Pesticide Toxicology, third ed., vol. 2. Elsevier, San Diego, CA, pp. 1865–1885.

Boehme, C., Ernst, W., 1965. The mechanism of urea-herbicides in the rat: diuron and linuron. Food. Cosmet. Toxicol. 3, 797–802 (in German).

Breckenridge, C.B., Eldridge, J.C., Stevens, J.T., et al., 2010. Symmetric triazine herbicides: a review of regulatory toxicity endpoints. In: Krieger, R. (Ed.), Hayes' Handbook of Pesticide Toxicology, third ed., vol. 2. Elsevier, San Diego, CA, pp. 1711–1723.

Buchholz, B.A., Fultz, E., Haack, K.W., et al., 1999. HLPC accelerator MS measurements of atrazine metabolites in human urine after dermal exposure. Anal. Chem. 71, 3519–3525.

Burns, C.J., Swaen, G.M.H., 2012. Review of 2,4-dichlorophenoxyacetic acid (2,4-D) biomonitoring and epidemiology. Crit. Rev. Toxicol. 42, 768–786.

Bus, J.S., Gibson, J.E., 1975. Postnatal toxicity of chronically administered paraquat in mice and interactions with oxygen and bromobenzene. Toxicol. Appl. Pharmacol. 33, 461–470.

Catenacci, G., Barbieri, F., Bersani, M., et al., 1993. Biological monitoring of human exposure to atrazine. Toxicol. Lett. 69, 217–222.

Catenacci, G., Colli, G., Verni, P., et al., 2002. Environmental and biological monitoring of occupational exposure in atrazine formulating plant. G Ital. Med. Lav. Ergon. 24, 35–42.

CDC (Centers for Disease Control and Prevention), 2005. Third National Report on Human Exposure to Environmental Chemicals. NCEH Pub. No. 05-0570. National Center for Environmental Health, Division of Laboratory Sciences, Atlanta, GA, pp. 30341–33724.

Chhabra, R.S., Eustis, S., Haseman, J.K., et al., 1992. Comparative carcinogenicity of ethylene thiourea with or without perinatal exposure in rats and mice. Fund. Appl. Toxicol. 18, 405–417.

Clair, E., Mesnage, R., Travert, C., Seralini, G.-E., 2012. A glyphosate-based herbicide induced necrosis and apoptosis in mature rat testicular cells in vitro, and testosterone decrease at lower levels. Toxicol. In Vitro 26, 269–279.

Colosio, C., Fustinoni, S., Birindelli, S., et al., 2002. Ethylenethiourea in urine as an indicator of exposure to mancozeb in vineyard workers. Toxicol. Lett. 134, 133–140.

Cooper, J., Dobson, H., 2007. The benefits of pesticides to mankind and the environment. Crop Protect. 26, 1337–1348.

Costa, L.G., 1997. Basic toxicology of pesticides. Occup. Med. - State of the Art Rev. 12, 251–268.

Curwin, B.D., Hein, M.J., Sanderson, W.T., et al., 2005. Urinary and hand wipe pesticide levels among farmers and nonfarmers in Iowa. J. Expo. Anal. Environ. Epidemiol. 15, 500–508.

Daniel, J.W., Gage, J.C., 1966. Absorption and excretion of diquat and paraquat in rats. Br. J. Ind. Med. 23, 133–136.

Dayan, F.E., Duke, S.O., 2010. Protophyrinogen oxidase-inhibiting herbicides. In: Krieger, R. (Ed.), Hayes' Handbook of Pesticide Toxicology, third ed. Elsevier, San Diego, CA, pp. 1733–1751.

Dilley, J.V., Chernoff, N., Kay, D., et al., 1977. Inhalation teratology studies of five chemicals in rats. Toxicol. Appl. Pharmacol. 41, 196.

Dinis-Oliveira, R.J., Duarte, J.A., Sanchez-Navarro, A., et al., 2008. Paraquat poisoning: mechanisms of lung toxicity, clinical features, and treatment. Crit. Rev. Toxicol. 38, 13–71.

Driskell, W.J., Hill Jr., R.H., 1997. Identification of a major human urinary metabolite of metolachlor by LC-MS/MS. Bull. Environ. Contam. Toxicol. 58, 929–933.

Driskell, W.J., Hill Jr., R.H., Shealy, D.B., et al., 1996. Identification of a major human urinary metabolite of alachlor by LC-MS/MS. Bull. Environ. Contam. Toxicol. 56, 853–859.

EPA, 1997a. Cloransulam-methyl: Pesticide Fact Sheet. OPPTS 7501C.

EPA, 1997b. Cloransulam-methyl: pesticide tolerances. 40 CR 180. Fed. Regist. 62, 182.

EPA, 2004. May 1, 2004 memorandum (W J Hazel et al.) regarding "HED's Human Health Risk Assessment for the Reregistration Eligibility Decision (RED)-PC Code 030001-DP Barcode D287199" (EPA Office of Pesticide Programs).

FAO/WHO, 1994. Ethylenethiourea, ethylenebisdithiocarbamates, mancozeb, maneb, metiram, zineb. In: Report of the Joint Meeting of the FAO Panel of Experts on Pesticide Residues in Food and the Environment and the WHO Expert Group on Pesticide Residues, Geneva, 20–29 September 1993. Food and Agricultural Organization of the United Nations Plant Production and Protection, Paper 122.

Feng, P., Wilson, A., McClanahan, R., et al., 1990. Metabolism of alachlor by rat and mouse liver and nasal turbinate tissues. Drug Metab. Dispos. 18, 373–377.

Franz, J.E., Mato, M.K., Sikorski, J.A., 1997. Glyphosate: A Unique Global Herbicide, ACS Monograph No. 189. American Chemical Society, Washington, DC.

Freudenthal, R.I., Kerchner, G., Persing, R., Baron, R.L., 1977. Dietary subacute toxicity of ethylene thiourea in the laboratory rat. J. Environ. Pathol. Toxicol. 1, 147–161.

Gasnier, C., Dumont, C., Benachour, N., et al., 2009. Glyphosate-based herbicides are toxic and endocrine disruptors in human cell lines. Toxicology 262, 184–191.

Gordon, E.B., 2010. Captan and folpet. In: Krieger, R. (Ed.), Hayes' Handbook of Pesticide Toxicology, third ed., vol. 2. Elsevier, New York, NY, pp. 1915–1949.

Gupta, P.K., 1986. Pesticides in the Indian Environment. Interprint, New Delhi.

Gupta, P.K., 1988. Veterinary Toxicology. Cosmo, New Delhi.

Gupta, P.K., 1989. Pesticide production in India: an overview. In: Mishra, P.C. (Ed.), Soil Pollution and Soil Organisms. Ashish, New Delhi, pp. 1–16.

Gupta, P.K., 2004. Pesticide exposure -Indian scene. Toxicology 198, 83–90.

Gupta, P.K., 2006a. Status of biopesticides -Indian scene. Toxicol. Int. 13, 643–654.

Gupta, P.K., 2006b. WHO/FAO guidelines for cholinesterase inhibiting pesticide residues in food. In: Gupta, R.C. (Ed.), Toxicology of Organophosphate and Carbamate Compounds. Elsevier, Amsterdam, pp. 643–654.

Gupta, P.K., 2010a. Pesticides. In: Gupta, P.K. (Ed.), Modern Toxicology: Adverse Effects of Xenobiotics, vol. 2. Pharma Med Press/BSP, Hyderabad, India, pp. 1–60.

Gupta, P.K., 2010b. Epidemiological studies of anticholinesterase pesticides in India. In: Satoh, T., Gupta, R.C. (Eds.), Anticholinesterase Pesticides: Metabolism, Neurotoxicity, and Epidemiology. John Wiley & Sons, Hoboken, pp. 417–431.

Gupta, P.K., 2014. Essential Concepts in Toxicology. Pharma Med Press/BSP, Hyderabad, India.

Gupta, P.K., 2016. Fundamental of Toxicology: Essential Concepts Sand Applications. Elsevier-BSP, Elsevier/Academic Press, USA, p. 398.

Gupta, P.K., 2017. Herbicides and fungicides. In: Gupta, R.C. (Ed.), Reproductive and Developmental Toxicology, second ed. Academic Press/Elsevier, Amsterdam, pp. 657–679.

Gupta, P.K., 2018a. Epidemiology of animal poisonings in Asia. In: Gupta, R.C. (Ed.), Veterinary Toxicology: Basic and Clinical Principles, third ed. Academic Press/Elsevier, Amsterdam, pp. 57–69.

Gupta, P.K., 2018b. Toxicity of herbicides. In: Gupta, R.C. (Ed.), Veterinary Toxicology- Basic and Clinical Principals, 3rd ed. Elsevier, USA, pp. 553–568.

Gupta, P.K., 2018c. Toxicity of fungicides. In: Gupta, R.C. (Ed.), Veterinary Toxicology: Basic and Clinical Principles, third ed. Academic Press/Elsevier, Amsterdam, pp. 569–580.

Gupta, P.K., 2018d. Illustrative Toxicology. Academic Press/Elsevier, Amsterdam.

Hack, R., Ebert, E., Ehling, G., 1994. Glufosinate ammonium-Some aspects of its mode of action in mammals. Food Chem. Toxicol. 32, 461–470.

Hanley, T.R., John-Greene, J.A., Hayes, W.C., Rao, K.S., 1987. Embryotoxicity and fetotoxicity of orally administered tridiphane in mice and rats. Fund. Appl. Toxicol. 8, 179–187.

Hays, S.M., Aylward, L.L., Driver, J., et al., 2012. 2,4-D Exposure and risk assessment: comparison of external dose and biomonitoring based approaches. Regul. Toxicol. Pharmacol. 64, 481–489.

Heredia-Ortiz, R., Bouchard, M., 2012. Toxicokinetic modeling of captan fungicide and its tetrahydrophthalimide biomarker of exposure in humans. Toxicol. Lett. 213, 27–34.

Heredia-Ortiz, R., Berthet, A., Bouchard, M., 2011. Toxicokinetic modeling of folpet fungicide and its ring biomarkers of exposure in humans. J. Appl. Toxicol. https://doi.org/10.1002/jat.17822011. wileyonlinelibrary.com.

Hess, F.G., Harris, J.E., Pendino, K., Ponnock, K., 2010. Imidazolinones. In: Krieger, R. (Ed.), Hayes' Handbook of Pesticide Toxicology, third ed., vol. 2. Elsevier, San Diego, CA, pp. 1853–1863.

Heydens, W.F., Lamb, I.C., Wilson, A.G.E., 2010. Chloracetanilides. In: Krieger, R. (Ed.), Hayes' Handbook of Pesticide Toxicology, third ed., vol. 2. Elsevier, San Diego, CA, pp. 1753–1769.

Hodge, H.C., Downs, W.L., Panner, B.S., et al., 1968. Oral toxicity and metabolism of diuron (N-3,4-dichlorophenyl-N9, N9-dimethylurea) in rats and dogs. Food. Cosmet. Toxicol. 5, 513–531.

Hodgson, E., Meyer, S.A., 1997. Pesticides. In: Sipes, I.G., McQueen, C.A., Gandolfi, A.J. (Eds.), Comprehensive Toxicology: Hepatic and Gastrointestinal Toxicology, vol. 9. Pergamon, New York, pp. 369–387.

Hunt, L.M., Chamberlain, W.F., Gilbert, B.N., et al., 1977. Absorption, excretion, and metabolism of nitrofen by a sheep. J. Agric. Food Chem. 25, 1062–1065.

Hurlety, P.M., 1998. Mode of carcinogenic action of pesticides inducing thyroid follicular cell tumors in rodents. Environ. Health Perspect. 106, 437–445.

Hurt, S., Ollinger, J., Arce, G., et al., 2010. Dialkylthiocarbamates (EBDCs). In: Krieger, R. (Ed.), Hayes' Handbook of Pesticide Toxicology, third ed., vol. 2. Elsevier, San Diego, CA, pp. 1689–1710.

JMPR, Joint FAO/WHO Meeting on Pesticide Residues, 1993. Pesticide residues in food. Evaluation: Part II. Toxicological. In: Joint Meeting of the FAO Panel of Experts on Pesticide Residues in Food and the Environment and a WHO Expert Group on Pesticide Residues, WHO/PCS/94.4. World Health Organization, Geneva.

JMPR, Joint FAO/WHO Meeting on Pesticide Residues, 2000. Pesticide residues in food Evaluation: Part II. Toxicological. In: Joint Meeting of the FAO Panel of Experts on Pesticide Residues in Food and the Environment and a WHO Expert Group on Pesticide Residues, WHO/PCS/01.3. World Health Organization, Geneva.

JMPR, Joint FAO/WHO Meeting on Pesticide Residues, 2004. Pesticide residues in food. In: Report of the Joint Meeting of the FAO Panel of Experts on Pesticide Residues in Food and the Environment and a WHO Expert Group on Pesticide Residues, FAO Plant Production and Protection Paper, 178. Food and Agriculture Organization, Rome.

JMPR, Joint FAO/WHO Meeting on Pesticide Residues, 2005. Pesticide residues in food. In: Report of the Joint Meeting of the FAO Panel of Experts on Pesticide Residues in Food and the Environment and a WHO Expert Group on Pesticide Residues, FAO Plant Production and Protection Paper, 179. Food and Agriculture Organization, Rome.

Jowa, L., Howd, R., 2011. Should atrazine and related chlorotriazines be considered carcinogenic for human health risk assessment? J. Environ. Sci. Health C Environ. Carcinog. Ecotoxicol. Rev. 29, 91–144.

Kapka-Skrzypczak, L., Cyranka, M., Skrzypczak, M., Kruszewski, M., 2011. Biomonitoring and biomarkers of organophosphate pesticides exposure-state of the art. Ann. Agric. Environ. Med. 18, 294–303.

Kennepohl, E., Munro, I.C., Bus, J.S., 2010. Phenoxy herbicides (2,4-D). In: Krieger, R. (Ed.), Hayes Handbook of Pesticide Toxicology, third ed., vol. 2. Elsevier, San Diego, CA, pp. 1829–1847.

Kimata, A., Kondo, T., Ueyama, J., et al., 2009. Relationship between urinary pesticide metabolites and pest control operation among occupational pesticide sprayers. J. Occup. Health 51, 100–105.

Konstantinou, I.K., Hela, D.G., Albanis, T.A., 2006. The status of pesticides pollution in surface waters (rivers and Lakes) of Greece. Part 1. Review on occurrence and levels. Environ. Pollut. 141, 555–570.

Krieger, R.I., Thongsinthusak, T., 1993. Captan metabolism in humans yields two biomarkers, tetrahydrophthalimide (THPI) and thiazolidine-2-thione-4-carboxylic acid (TTCA), in urine. Drug Chem. Toxicol. 16, 207–225.

LaKind, J.S., Aylward, L.L., Brunk, C., et al., 2008. Guidelines for the communication of biomonitoring equivalents: report from the biomonitoring equivalents expert workshop. Regul. Toxicol. Pharmacol. 51, S16–S26.

LeBaron, H.M., McFarland, J.E., Burnside, O.C. (Eds.), 2008. The Triazine Herbicides. 50 Years Revolutionizing Agriculture. Elsevier, Amsterdam.

Leung, L.Y., Lyga, J.W., Robinson, R.A., 1991. Mechanism and distribution of the experimental triazolinone herbicide sulfentrazone in the rat, goat and hen. J. Agric. Food Chem. 39, 1509–1514.

Liu, J., 2010. Phenylurea herbicides. In: Krieger, R. (Ed.), Hayes' Handbook of Pesticide Toxicology, third ed., vol. 2. Elsevier, San Diego, CA, pp. 1725–1731.

Lock, E.A., Wilks, M.F., 2010. Paraquat. In: Krieger, R. (Ed.), Hayes' Handbook of Pesticide Toxicology, third ed., vol. 2. Elsevier, San Diego, CA, pp. 1771–1827.

Lorgue, G., Lechenet, J., Riviere, A., 1996. Clinical Veterinary Toxicology (English Version by M.J. Chapman). Blackwell, Oxford, UK.
Lucas, A.D., Jones, A.D., Goodrow, M.H., et al., 1993. Determination of atrazine metabolites in human urine: development of a biomarker of exposure. Chem. Res. Toxicol. 6, 107–116.
Manson, J.M., Brown, T.J., Baldwin, D.M., 1984. Teratogenicity of nitrofen (2,4-dinitro-40-nitrodiphenyl ether) and thyroid function in rat. Toxicol. Appl. Pharmacol. 73, 323–335.
Matthiaschk, G., Roll, R., 1977. Studies on the embryotoxicity of monolinuron and buturon in NMBI-mice. Arch. Toxicol. 38, 261–274.
Mendas, G., Vuletic, M., Galic, N., Drevenkar, V., 2012. Urinary metabolites as biomarkers of human exposure to atrazine: atrazine mercapturate in agricultural workers. Toxicol. Lett. 210, 174–181.
Millburn, P., 1975. Excretion of xenobiotics compounds in bile. In: Taylor, W. (Ed.), The Hepatobiology System: Fundamental and Pathological Mechanisms. Plenum, New York.
Mirkhamidova, P., Mirakhmedov, A.K., Sagatova, G.A., et al., 1981. Effect of butiphos on the structure and function of the liver in rabbit embryos. Uzb. Biol. Zh. 5, 45–47.
Morgan, M.K., Sheldon, L.S., Thomas, K.W., et al., 2008. Adult and children's exposure to 2,4-D from multiple sources and pathways. J. Expo. Sci. Environ. Epidemiol. 18, 486–494.
Nandihalli, U.B., Duke, M.V., Duke, S.O., 1992. Quantitative structure activity relationships of protoporphyrinogen oxidase inhibiting diphenyl ether herbicides. Pestic. Biochem. Physiol. 43, 193–211.
Ngo, M.A., O'Malley, M., Maibach, H.I., 2010. Percutaneous absorption and exposure assessment of pesticides. J. Appl. Toxicol. 30, 91–114.
NTP, 2011. Captafol. Rep. Carcinog. 12, 83–86.
Osteen, C.D., Padgitt, M., 2002. Economic issues of agricultural pesticide use and policy in the United States. In: Wheeler, W.B. (Ed.), Pesticides in Agriculture and the Environment. Dekker, New York, pp. 59–95.
Parsons, P.P., 2010. Mammalian toxicokinetics and toxicity of chlorothalonil. In: Krieger, R. (Ed.), Hayes' Handbook of Pesticide Toxicology, third ed., vol. 2. Elsevier, New York, NY, pp. 1951–1966.
Perry, M., Christiani, D., Dagenhart, D., et al., 2000a. Urinary biomarkers of atrazine exposure among farm pesticide applicators. Ann. Epidemiol. 10, 479.
Perry, M.J., Christiani, D.C., Mathew, J., et al., 2000b. Urinalysis of atrazine exposure in farm pesticide applicators. Toxicol. Ind. Health 16, 285–290.
PMRA (Pest Management Regulatory Agency), 2007. Proposed Acceptability for PACR2007 06. Continuing Registration Reevaluation of the Agricultural, Forestry, Aquatic and Industrial Site Uses of (2,4-Dichlorophenoxy)acetic Acid [2,4-D]. publications.gc.ca/collections/collection.../H113-18-2007-6E.pdf.
Rose, M.S., Lock, E.A., Smith, L.L., Wyatt, I., 1976. Paraquat accumulation. Tissue and species specificity. Biochem. Pharmacol. 25, 419–423.
Sachana, M., Flaskos, J., Hargreaves, A., 2017. In vitro biomarkers of developmental neurotoxicity. In: Gupta, R.C. (Ed.), Reproductive and Developmental Toxicology, second ed. Academic Press/Elsevier, Amsterdam, pp. 255–288.
Sathiakumar, N., MacLennan, P.A., Mandel, J., Delzell, E., 2011. A review of epidemiologic studies of triazine herbicides and cancer. Crit. Rev. Toxicol. 41, 1–34.
Schardein, J.L., 2000. Chemically Induced Birth Defects, third ed. Dekker, New York.
Shalette, M.I., Cotes, N., Goldsmith, E.D., 1963. Effects of 3-amino-1,2,4-triazole treatment during pregnancy on the development and structure of the thyroid of the fetal rat. Anat. Rec. 145, 284.
Smith, L.L., 1997. Paraquat. In: Sipes, I.G., McQueen, C.A., Gandolfi, A.J. (Eds.), Comprehensive Toxicology: Toxicology of the Respiratory System, vol. 8. Pergamon, New York, pp. 581–589.
Syngenta, 1985. Pilot study for the determination of the effects of probenecid pre-treatment on urinary metabolites and excretion of 14C-Chlorothalonil (14C-SDS-2787) following oral administration to male Sprague_Dawley rats Unpublished Syngenta study by Savides, MC, et al., Rep. 621–624 AM-85-0035-001.
Syngenta, 1986. Pilot study of the effect of the gamma-glutamyl transpeptidase inhibitor, AT-125 on the metabolism of 14CChlorothalonil Unpublished Syngenta study by Marciniszyn, JP, et al. Rep. 1376-86-0072-AM-002.
Syngenta, 1987a. Analysis of urine samples from a 90-day feeding yes no study in rats with the monoglutathione conjugate of Chlorothalonil (T-117–11) Unpublished Syngenta study by Mead, RL, et al., Rep.1108-85-0078-TX-006.
Syngenta, 1987b. Analysis of urine samples from a 90-day feeding study in rats with Chlorothalonil (T-117–11) Unpublished Syngenta study by Mead, RL, et al., Rep. 1115-85-0079-TX-005.
Terada, M., Mizuhashi, F., Tomita, T., et al., 1998. Mepanipyrim induced fatty liver in rats but not in mice and dogs. J. Toxicol. Sci. 23, 223–234.
Thomas, K.W., Dosemeci, M., Hoppin, J.A., et al., 2010. Urinary biomarker, dermal, and air measurement results for 2,4-D and chlorpyrifos farm applicators in the Agricultural Health Study. J. Expo. Sci. Environ. Epidemiol. 20, 119–134.
Tinston, D.J., 1995. Captan: Investigation of Duodenal Hyperplastia in Mice. Report CTL/4532. Central Toxicology Laboratory, Alderley Park, UK.
Van Dijck, A., Macs, R.A., Drost, R.H., et al., 1975. Paraquat poisoning in man. Arch. Toxicol. 35, 129–136.
van Welie, R.T.H., van Duyn, P., Lamme, E.K., et al., 1991. Determination of tetrahydrophthalimide and 2-thiothiazolidine- 4-carboxylic acid, urinary metabolites of the fungicide captan, in rats and humans. Int. Arch. Occup. Environ. Health 63, 181–186.
Waechter, F., Weber, E., Herner, T., May-Hertl, U., 2010. Cyprodinil: a fungicide of the anilinopyrimidine class. In: Krieger, R. (Ed.), Hayes' Handbook of Pesticide Toxicology, third ed., vol. 2. Elsevier, New York, NY, pp. 1903–1913.
Waterson, L., 1995. Folpet: Investigation of the Effects on the Duodenum of Male Mice after Dietary Administration for 28 Days with Recovery. Report MBS 45/943003. Huntingdon Research Centre, Huntingdon, UK.
Wilkinson, C.F., Killeen, J.C., 1996. A mechanistic interpretation of the oncogenicity of chlorothalonil in rodents and an assessment of human relevance. Regul. Toxicol. Pharmacol. 24, 69–84.
Williams, A.L., Watson, R.E., DeSesso, J.M., 2012. Developmental and reproductive outcomes in humans and animals after glyphosate exposure: a critical analysis. J. Toxicol. Environ. Health B Crit. Rev. 15, 39–96.
Wilson, A.G.E., Takei, A.S., 1999. Summary of toxicology studies with butachlor. J. Pestic. Sci. 25, 75–83.

CHAPTER

28

Polychlorinated Biphenyls, Polybrominated Biphenyls, and Brominated Flame Retardants

Prasada Rao S. Kodavanti[1], *Bommanna G. Loganathan*[2]

[1]Neurotoxicology Branch, Toxicity Assessment Division, National Health and Environmental Effects Research Laboratory, Office of Research and Development, U.S. Environmental Protection Agency, Research Triangle Park, NC, United States; [2]Department of Chemistry and Watershed Studies Institute, Murray State University, Murray, KY, United States

INTRODUCTION

Polychlorinated biphenyls (PCBs), polybrominated biphenyls (PBBs), and brominated flame retardants (BFRs) belong to a group of chemicals known as organohalogens. Organohalogens are organic compounds that contain chlorine, bromine, or fluorine atoms and the molecules are named as chlorinated, brominated, or fluorinated hydrocarbons, respectively. These compounds share common characteristics such as persistence in the environment, bioaccumulate in living organisms, long-range transport beyond the geographical regions of their use, and long-term health effects in wildlife and humans. Even though some of these compounds (e.g., PCBs and PBBs) have been banned or severely restricted in use in developed countries for more than three decades, they are still found in every component of the global ecosystem and pose a threat to human health (Kodavanti and Loganathan, 2017). Following the ban on production and usage, levels of regulated organohalogens have declined in the environment (Loganathan and Lam, 2012; Loganathan et al., 2016a,b). However, newly discovered organohalogens such as BFRs have taken their place and continue to be manufactured for commercial use. BFRs are widely used as flame retardants in a variety of consumer products and are considered to be emerging chemicals of concern with regard to human health and sustainability of the ecosystem (Guo et al., 2012; Kodavanti and Loganathan, 2016). These persistent organic chemicals have been shown to cause a variety of effects in humans including neurotoxic effects in children (Berghuis et al., 2015). This chapter focuses on physicochemical properties, environmental contamination and human exposure, health effects, and mode of action with particular emphasis on biomarkers of exposure and effects for two well-known and already regulated persistent chemical groups such as PCBs and PBBs, and the emerging persistent chemical group such as BFRs, which are partially regulated in some countries.

POLYCHLORINATED BIPHENYLS

The commercial production of PCBs began in 1929. PCBs were introduced under a number of trade names and sold in many different countries. Aroclor was the most common trade name for PCBs in the United States. PCB mixtures were named according to their chlorine content, Aroclor 1254 containing 54% chlorine by weight, and Aroclor 1260 containing 60%. PCB mixtures were produced for a variety of uses, including fluids in electrical transformers and capacitors, heat transfer fluids, hydraulic fluids, lubricating and cutting oils, and as additives in plastics, paints, copying paper, printing inks, adhesives, and sealants.

Physicochemical Properties and Environmental Levels

PCBs are colorless to light yellow, have no odor, and are a tasteless oily liquid or solids (Table 28.1). Some PCBs are volatile and may exist as a vapor in air. The physicochemical properties of PCBs depend on the number and positions of chlorine atoms in the biphenyl

Biomarkers in Toxicology, Second Edition
https://doi.org/10.1016/B978-0-12-814655-2.00028-1

TABLE 28.1 Physical and Chemical Properties of Polychlorinated biphenyls (PCBs), polybrominated biphenyls (PBBs), and Polybrominated Dipheny Ethers (PBDEs)

Property	PCBs	PBB (Hexa-, Octa-, and Deca-BB	PBDEs (Penta, Octa, and Deca)
CAS Numbers	Aroclor 1016–12674-11-2 Aroclor 1254–11097-69-1 Aroclor 1260–11096-82-5	HexaBB – 36355-01-8 OctaBB – 27858-07-7 DecaBB – 13654-09-6	Penta – 32534-81-9 Octa – 32536-52-0 Deca – 1163-19-5
Physical state at room temperature	Oil to viscous liquid	White solid	Viscous liquid to powder
Molecular weight (g/mol)	188 to 498	627 to 943	Up to 959.22
Water solubility (μg/L at 25°C)	0.0027 to 0.59	3 to 30	<0.1 to 13.3
Boiling point (°C)	275 to 450	Not applicable	300 to >400
Melting point (°C)	1	72 to 386	−7 to 306
Vapor pressure at 25°C (mm Hg)	10^{-12} to 10^{-4}	5.2×10^{-8}	2.2×10^{-7} to 3.47×10^{-8}
Octanol-water partition coefficient (log *Kow*)	4.7 to 6.8	5.53 to 9.10	6.265 to 6.97
Soil organic carbon-water coefficient (*Koc*)	No data	3.33 to 3.87 (HexaBB)	4.89 to 6.80
Henry's Law Constant (atm m^3/mol)	2.9×10^{-4} to 4.6×10^{-3}	1.38×10^{-6} to 5.7×10^{-3}	1.2×10^{-5} to 4.4×10^{-8}

Note: g/mol: gram per mole; μg/L: microgram per liter; °C: degrees Celcius; mm Hg: millimeters of mercury; atm m^3/mol: atmosphere cubic meters per mole. *ATSDR (2000). Toxicological Profile for Polychlorinated Biphenyls (PCBs). US Department of Health and Human Services. Agency for Toxic Substances and Disease Registry, ATSDR (2004). Public Health Statement: Polybrominated Biphenyls. US Department of Health and Human Services. Agency for Toxic Substances and Disease Registry; De Wit, C.A. (2002). An overview of brominated flame retardants in the environment. Chemosphere 46: 583–624.De Wit, 2002*

rings (Fig. 28.1). PCBs resist both acids and alkalis and are thermally stable. These properties made them useful in a wide variety of industrial applications but contribute to their persistence in the environment. Generally, PCBs are relatively insoluble in water, with solubility decreasing with increasing chlorination (Table 28.1). However, PCB congeners are readily soluble in nonpolar organic solvents and biological lipids. PCBs entered the earth's ecosphere primarily during their manufacture and use in a variety of applications. PCBs also entered the environment from accidental spills and leaks during the transport of PCB-containing materials.

Once in the environment, PCBs do not readily break down and therefore remain for very long periods. PCBs can easily cycle between air, water, and soil. For example, PCBs can enter the air by evaporation from both water and soil. In air, PCBs can be carried long distances from where these compounds are released and have been found in snow and seawater as far away as the Arctic and Antarctic (Macdonald et al., 2000; Corsolini, 2012). This has resulted in extended contamination and distribution throughout the global environment. Because of their persistent and lipophilic properties, PCBs bioaccumulate in various lower trophic organisms plankton, moving up the food chain through plankton, bivalve mollusks, fish, reptiles, marine mammals, birds, terrestrial mammals and including humans (Loganathan and Kannan, 1994; Loganathan and Lam, 2012). In concurrence with these findings, these compounds have been detected in fish and other food products. The primary route of human exposure to PCBs is through consumption of contaminated foods such as freshwater fish, dairy products, and meat. Unexpectedly, high levels of these compounds have been found in human adipose tissues, and blubber of marine mammals (Loganathan et al., 1993, Loganathan and Kannan, 1994).

Human Health Effects and Modes of Action

Information on health effects of PCBs first became available from studies of people accidentally exposed by consumption of contaminated rice oil in Japan in 1968 ("Yusho" incident) and Taiwan in 1979 ("Yu-Cheng" incident). Major symptoms of Yusho disease included acne form eruptions, pigmentation of the skin, nails, and conjunctivae, increased discharge from the eyes, and numbness of the limbs (Yusho Support Center, 2007). PCB exposures have been associated with low birth weight and learning and behavioral deficits in children of women who consumed PCB-contaminated fish from Lake Michigan (Jacobson and Jacobson, 1996). Health effects that have been associated with general environmental exposure to PCBs in humans and/or animals include histopathological and metabolic changes in liver, thyroid, dermal and ocular tissues/organs, immunological alterations, neurodevelopmental

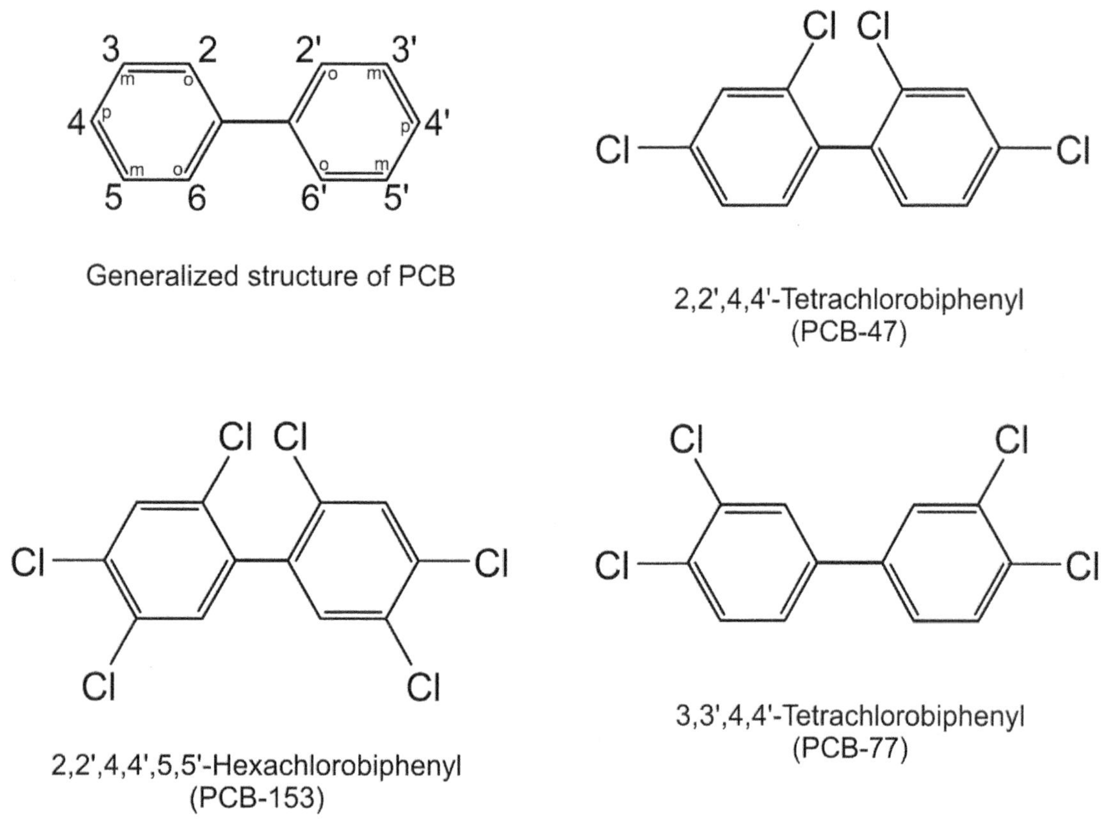

Generalized structure of PCB

2,2',4,4'-Tetrachlorobiphenyl (PCB-47)

2,2',4,4',5,5'-Hexachlorobiphenyl (PCB-153)

3,3',4,4'-Tetrachlorobiphenyl (PCB-77)

FIGURE 28.1 Generalized structure of polychlorinated biphenyls (PCBs) and structures of selected PCB congeners. The numbers in the generalized structure indicate the position for chlorines. The letters (*o*), (*m*), and (*p*) indicate *ortho-*, *meta-*, and *para-*substitutions for chlorine side groups. Predominant non–dioxin-like (PCB 47 and PCB 153) and dioxin-like (PCB 77) are shown here as examples.

changes, reduced birth weight, reproductive toxicity, and cancer. Our studies have shown neurobehavioral changes in adult animals and in animals that were exposed to PCBs during development (Kodavanti et al., 2000, 2010). Animals that were prenatally, postnatally, or perinatally exposed to PCBs showed many of the behavioral characteristics indicative of attention deficit hyperactivity disorder.

Apart from effects on motor activity, PCBs have been shown to decrease cognitive function in rats, nonhuman primates, and mice. Gender differences have been reported in a study where developmental exposure to PCBs significantly interfered with acquisition of the spatial alternation task in females, but not in males. Studies also indicated that perinatal exposure to Aroclor 1254 impaired radial arm maze performance in male rats and affected long-term potentiation in the dentate gyrus of the hippocampus. In nonhuman primates, developmental exposure to Aroclor 1016 or Aroclor 1248 resulted in long-term changes in cognitive function (Kodavanti et al., 2008). Other effects seen with developmental exposure to PCBs include sensory deficits. The most significant effect seen with developmental exposure to PCBs was deficits in hearing; the auditory threshold at low frequency (1 kHz) was significantly increased. PCB exposure during development has no significant effect on visual-, somatosensory-, or peripheral nerve-evoked potentials. These animal studies clearly indicated that developmental exposure to PCBs resulted in cognitive dysfunction, motor deficits, and hearing loss.

With regard to mode of action, coplanar PCBs (chlorine substitutions in *para* and *meta* positions in the biphenyl molecule) seem to elicit dioxin-like toxicity via AhR-mediated toxicity leading to the development of cancer. However, planar non–dioxin-like PCBs seem to exert neurotoxicity through their effects on thyroid hormones (THs) and intracellular signaling processes (Kodavanti and Tilson, 1997; see section "Perturbed Calcium Homeostasis and Kinase Signaling as Biomarkers of Effect" for details on the mode of action).

POLYBROMINATED BIPHENYLS

PBBs are a class of brominated hydrocarbons consisting of a central biphenyl structure to which 1–10 bromine atoms are attached (see Fig. 28.2). PBBs are a synthetic chemical mixture, not found in nature. Commercial scale production of PBBs was carried out using

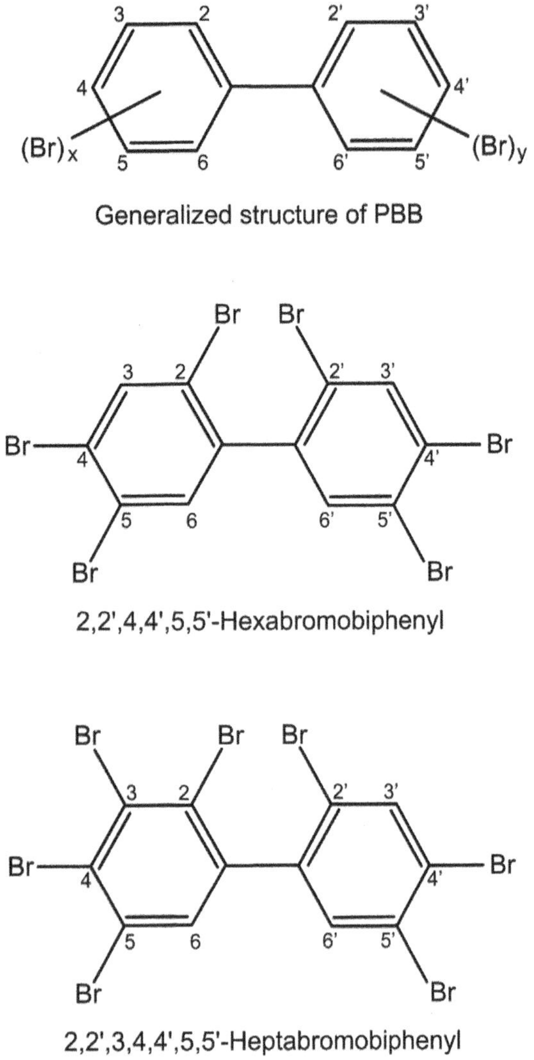

Generalized structure of PBB

2,2',4,4',5,5'-Hexabromobiphenyl

2,2',3,4,4',5,5'-Heptabromobiphenyl

FIGURE 28.2 Structural features of polybrominated biphenyls (PBBs). Generalized structure and selected PBB congeners of scientific interest are presented here. The numbers in the generalized structure indicate the position for bromines.

bromination of biphenyl, a process that results in a smaller number of product mixtures than the corresponding chlorination seen in PCBs (Sundström et al., 1976). Worldwide production of PBBs was estimated in 1994 to be 11,000 tons. However, the production statistics were not available in some countries known to have produced this compound (IPCS, 1994). PBBs were used as fire retardant additives in plastics that were used in a variety of consumer products, including furniture, textiles, electronic devices such as computer monitors and televisions, plastic foams, and other household products (ATSDR, 2004; US EPA, 2012). Because of their stability, lipophilic properties, and toxic biological effects, the production of PBBs has been banned in the United States since 1973 (US EPA, 2012). According to Hardy (2002), PBBs were manufactured in the United Kingdom until 1977, in Germany until the mid-1980s and in France until 2000. It has not been possible to find information on production volumes in Europe. In the United States, total production of PBBs during 1970–76 was estimated to be 6000 tons, of which 98% was FireMaster FF-1 and FireMaster BP-6.

Physicochemical Properties

PBBs are colorless to off-white solids. Based on the number and position of the bromine atoms attached to the two phenyl rings, there are 209 possible individual compounds, referred to as PBB congeners. PBBs have same number of congeners and the bromine substitution pattern and are identical to the PCB congeners (Ballschmiter et al., 1993; Kodavanti and Loganathan, 2016). Therefore, the PBBs share the same IUPAC numbering system as used for PCBs. PBB congeners have a molar mass range from 233 g/mole (for monoBBs) to 943 g/mole (decaBBs) (EFSA, 2010). PBBs are lipophilic and chemically stable compounds with low water solubility and low vapor pressures, which decreases with increasing degree of bromination. The octanol–water partition coefficients (log *Kow*) and volatility also vary between congeners (Table 28.1). These values are much higher than PCBs and this may contribute to the higher environmental mobility of PBBs as compared with PCBs. In general, PBBs are chemically recalcitrant and are resistant to heat, acids, bases, and oxidation (EFSA, 2010). These unique properties make PBBs highly desirable for industries that produce a variety of consumer products. The same properties make these compounds highly persistent in the environment and bioaccumulative in organisms and contribute to their toxic effects.

PBBs have been detected in air, dust, soil, fish and other seafood, meat and meat products, and milk and dairy products. PBBs were released into the air during manufacture and during incineration of PBB-containing materials. Because PBB-containing materials are no longer in circulation, it is unlikely that incineration is a significant source of PBBs into the atmosphere in recent years. The major former sources of PBBs in soil were production operations and disposal of PBB-containing products. At one point, PBB concentrations in soils near bagging and loading areas of the Michigan Chemical Corporations were 3500 and 2500 mg/kg, respectively (Di Carlo et al., 1978). As studies have shown that root uptake of PBBs in plants is extremely low, the intake by grazing animals might be due to soil ingestion during grazing (EFSA, 2010). The contamination levels of PBBs in biota are generally low. Gotsch et al. (2004) reported <10 ng/g wet weight of PBB-153 in egg sample of a white-tailed eagle in Norway. Von der Recke and Veter (2008) studied PBBs in the blubber of seals and harbor porpoises from the North Sea, the Baltic Sea, and the coastal waters of Iceland and North America. The authors observed hexaBBs dominated in most samples, followed by pentaBBs, heptaBBs, and octaBBs. NonaBBs and DecaBBs were not detected. Jaspers et al. (2005) investigated the occurrence of BB-153 in 39 egg samples from little owls (*Athene noctua)* from Belgium. They reported the concentrations of BB-153 ranged from 1 to 6 ng/g fat (mean 2 ng/g fat, $n = 39$ samples). For comparison, the corresponding levels of polybrominated dipheny ethers (PBDEs) were 10–94 ng/g fat (mean 32 ng/g fat). Recently, detailed studies conducted by the European Food Safety Authority (EFSA, 2010) reported levels of PBBs in fish and other seafood, meat and meat products, milk and dairy products. They found the highest frequency of detectable levels with the lowest proportion of nondetects in the food category of "fish and other seafood (including amphibians, reptiles, snails, and insects)". In a specific study on "fish meat," 9 out of 10 analyzed congeners (BB-15, 49-, 52-, 77-, 80-, 101-, 126-, 153-, and 169) showed increased fat content corresponding with increasing PBB contamination levels (except for BB-209). Shen et al. (2008) reported PBB contamination of human milk from Denmark and Finland indicating that BB-153 is the most frequently found and most abundant congener, with mean levels of 200 pg/g fat ($n = 65$) and 134 pg/g ($n = 65$) fat, respectively. These studies provide evidence of PBBs' widespread contamination and exposure to wildlife and humans.

Human Health Effects and Modes of Action

Humans are exposed to PBBs mainly via consumption of contaminated foods such as fish, dairy, and meat products. As PBBs are lipophilic, these compounds dissolve and accumulate in fat. PBBs have been detected in breast milk of exposed mothers, making mothers' milk a source of exposure to infants. PBBs are known to cross the placenta and to reach the fetuses before birth (ATSDR, 2004). Shen et al. (2008) analyzed PBBs in Danish and Finnish placenta samples and reported over 13 PBB congeners in the samples. BB-153 and BB-155 were the two most commonly detected congeners (detected in 100% and 77% of the samples, respectively). Mean concentrations of BB-153 were 304 and 83 pg/g fat for Danish ($n = 168$) and Finnish ($n = 112$) samples, respectively. The Michigan long-term study (Joseph et al., 2009) reported serum PBB concentrations in 145 mother/child pairs. In serum samples ($n = 112$) from mothers, collected between 1976 and 1979, PBB concentrations ranged from <LOD to 933 μg/L (median 2 μg/L), where LOD is the lowest detectable level of <1 μg/L. Analysis of serum samples of children ($n = 145$), born between 1973 and 1982, during infancy to 17 years old, showed that 73% of children had PBB levels <1 μg/L (LOD), whereas the remaining 27% of children had PBB concentrations at or >LOD (median 2.9 μg/L). In an another study, Wang et al. (2010) correlated serum PBB levels with TH levels in occupationally exposed people from electronic waste recycling and dismantling activity sites and in nonoccupationally exposed people. Median range of the sum of BB-77, BB-103, and BB-209 was 510.35 (<LOQ–4695) ng/g fat for the occupationally exposed group ($n = 239$). For people living around the electronic waste recycling site (nonoccupational exposure group, $n = 93$), the concentrations were 614 (<LOQ–4050) ng/g fat. For the control (people working in green plantations) group ($n = 116$), the serum contained 246 (<LOQ–5833) ng/g fat. The authors noted, in all three groups, the congener BB-77 was found to be the most abundant with median levels of 340, 440, 180 ng/g fat, respectively, for the three groups. The occurrence of PBBs was also investigated in tissues taken from cancer patients (Zhao et al., 2009). The authors measured 23 PBB congeners in the kidney ($n = 19$), liver ($n = 55$), and lungs ($n = 7$) tissue samples from cancer patients living near electronic waste-dismantling sites. Median (range) of concentrations for the total PBBs was 194 (39–1140), 193 (42–560), and 150 (102–677) ng/g fat for the kidney, liver, and lungs tissue samples, respectively. Fifty-nine percent of the total PBB was contributed by the lower brominated congeners such as PBB-2, BB-10, PBB-15, and PBB-30, whereas PBB-153 alone accounted for over 12% of the total.

PBB congeners have been reported to elicit liver toxicity, liver hyperplasia, and interference with thyroid function, cause endocrine-mediated reproductive effects, and have immunotoxicity, teratogenicity, and developmental and neurobehavioral effects (Roth and Wilks, 2014). Because of their structural similarity with PCBs, PBBs share many structure activity relationships and toxicological properties. Safe (1984) stated that binding to the AhR, induction of AhR-mediated gene expression, and subsequent dioxin-like behavior is the major toxic mode of action of non-*ortho* PBBs (BB-77, BB-126, and BB-169), and mono-*ortho*-brominated congeners in PBB mixtures. Molecular mechanisms are discussed in detail in the section "Induction of Cytochrome P450 Enzymes as a Biomarker of Exposure and Effect."

BROMINATED FLAME RETARDANTS

In the early 1970s, accidental brominated biphenyls (PBBs) poisoning occurred on Michigan farms. Inadvertent mixing of PBBs (trade name: FireMaster) in the production of feed (trade name: NutriMaster) for dairy cattle resulted in the destruction of 1.5 million chickens,

5 million eggs, 29,800 cattle, 34,000 pounds of milk products, 5920 hogs, and 1470 sheep, which cost approximately 75–100 million US dollars (Fries, 1985). After this poisoning episode, PBBs were removed from the market and, as previously mentioned, the United States has banned PBBs since 1973 (US EPA, 2012). Among the BFRs still in the market are brominated bisphenols, diphenyl ethers, and cyclododecanes (Fig. 28.3). These three classes are still produced in large volumes (Shaw and Kannan, 2009) and it is estimated that over 1 million metric tons of PBDEs have been produced. Although the manufacture of penta- and octa-bromo mixtures has ceased, the production of decabromodiphenyl ether (decaBDE) is still continuing in some countries (Guo et al., 2012). DecaBDE was banned in Sweden in 2007, followed by partial bans in four US states (Washington, Maine, Oregon, and Vermont), the European Union in 2008, and Canada in 2009. The REACH (Registration, Evaluation, Authorization, and Restriction of Chemical substances) program in the European Union announced on February 17, 2011 was the banning of hexabromocyclododecane (HBCD) for all polystyrene used in building insulation (http://ec.europa.eu/environment/chemicals/reach/reach_intro.htm).

PBDEs and other brominated and chlorinated flame retardants still in use are HBCD, decabromodiphenyl

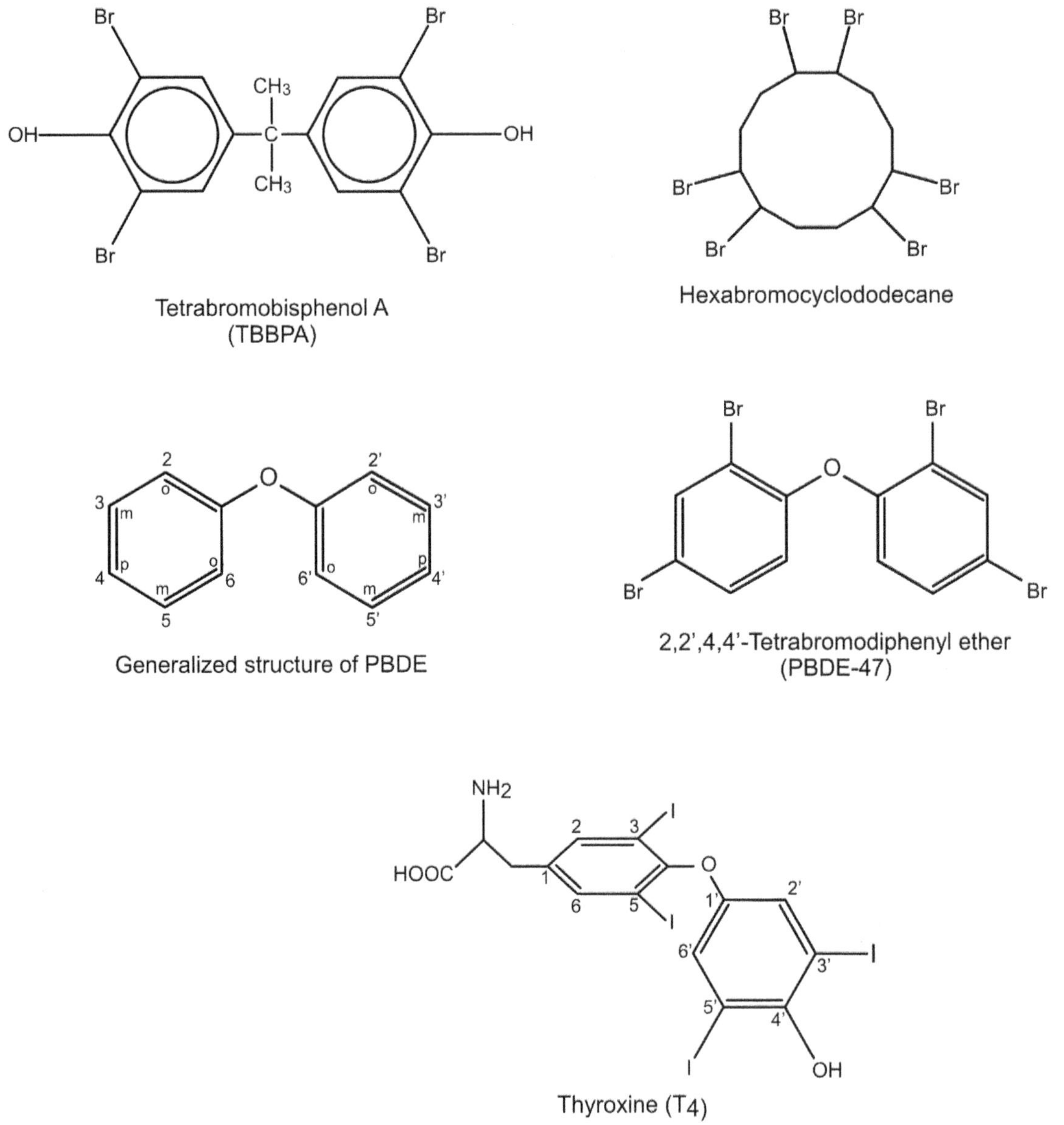

FIGURE 28.3 Chemical structures comparing selected brominated flame retardants (BFRs) with thyroxine (T_4). The numbers in the generalized structure of polybrominated dipheny ethers (PBDE) indicate the position of bromine. The letters (*o*), (*m*), and (*p*) indicate *ortho-*, *meta-*, and *para*-substitutions for bromine side groups.

ether, tetrabromobisphenol-A (TBBPA), tris(1-chloro-2-propyl) phosphate (TCPP), tris(2-chloroethyl)phosphate, and dechlorane plus. These compounds are used to meet fire safety standards for furniture, textiles, polyurethane foam, plastics used in electric and electronic equipment, printed circuit boards, curtains, carpets, etc. (Alaee et al., 2003; Stapleton et al., 2009; de Wit et al., 2010; Guo et al., 2012). The high production volume and the structural similarities of these brominated chemicals to other well-known toxic environmental contaminants such as DDTs and PCBs are the main concerns for environmental and human/animal health. Furthermore, polybrominated dioxins (PBDDs)/dibenzofurans (PBDFs), formed during heating or incineration of BFRs, have toxicological profiles similar to those of their chlorinated homologs (Birnbaum et al., 2003; DiGangi et al., 2010) but are more toxic than PBDEs. Like other organohalogens, BFRs are ubiquitous in the environment, bioaccumulate, and are toxic to animals and humans (Dye et al., 2007; Kodavanti et al., 2008; Kierkegaard et al., 2009; Guo et al., 2012; Shaw et al., 2012).

Low levels of PBDDs/PBDFs detected in the environmental samples suggest relatively lower exposure of biota (fish) and humans to these compounds.

PBDEs constitute an important group of flame retardants. PBDEs are added to consumer products so the products will not catch fire or will burn more slowly if exposed to flame or heat. PBDEs are added to plastics, upholstery, fabrics, and foams and are in common products such as computers, television sets, mobile phones, furniture, and carpet pads. In contrast to PCBs, PBDEs are currently being produced and used in household goods. PBDEs are primarily indoor pollutants. PBDEs leach into the environment when household wastes decompose in landfills or are incompletely incinerated. PBDE concentrations are rapidly increasing in the global environment and in human blood, breast milk, and liver. Although these chemicals are ubiquitous in the environment and bioaccumulate in wildlife and humans, potential toxic properties are still under investigation (Kodavanti, 2005; Shaw et al., 2010). New alternative flame retardants emerged to replace PBDEs, and Jin et al. (2016) reported species-specific accumulation of PBDEs and other emerging flame retardants in several species of birds from Korea. Human health concerns stem from the fact that PBDEs are persistent, bioaccumulative, and structurally similar to PCBs (see Figs. 28.2 and 28.3).

Physicochemical Properties

Tetrabromobisphenol-A (TBBPA; Fig. 28.3), the highest volume flame retardant worldwide, is primarily (90%) a reactive BFR covalently bound to the polymer structure and less likely to be released into the environment than are additive flame retardants (Birnbaum and Bergman, 2010). It is used mainly for the production of circuit board polymers. TBBPA is highly lipophilic (log $K_{ow} = 4.5$) and has low water solubility (0.72 mg/mL). TBBPA has been measured in the air (Zweidinger et al., 1979), soil, and sediment (Watanabe et al., 1983) but is generally not found in water samples. TBBPA is found in eggs of birds, human milk, and umbilical cord serum. TBBPA derivatives such as ethers are reported to be biologically active, which also may lead to adverse health effects (Birnbaum and Bergman, 2010).

Hexabromocyclododecane (HBCD; Fig. 28.3) is a nonaromatic brominated cyclic alkane with a molecular weight of 641.7 and is mainly used as an additive flame retardant in thermoplastic polymers with final applications in styrene resins (National Research Council, 2000). Like other BFRs, HBCD is highly lipophilic with a log K_{ow} of 5.6 and has low water solubility (0.0034 mg/L) (MacGregor and Nixon, 1997). The melting point is at 185–195°C and vapor pressure is 4.7×10^{-7} mm Hg. Studies have shown that HBCD is highly persistent, with a half-life of 3 days in air and 2025 days in water (Lyman et al., 1990), and is bioaccumulative with a bioconcentration factor of ~18,100 in fathead minnows (Veith and Defoe, 1979).

PBDEs comprise two phenyl rings linked by an oxygen molecule (thus the designation as "ether"; Fig. 28.3). The phenyl rings may have 1 to 10 bromine atoms, leading to formation of 209 possible congeners. PBDEs take the form of viscous liquid to powder and they are hydrophobic in nature (see additional details in Table 28.1).

The exact identity and pattern of various congeners in various commercial mixtures depends on the manufacturer and the specific product. Among these, the commercial "penta" mixture generally consists of PBDE congeners 99 (pentaBDE) and 47 (tetraBDE) as the major constituents (Fig. 28.3), which make up about 70% of the mixture (Huber and Ballschmiter, 2001). PBDE congener 100 (pentaBDE) is present at less than 10%, with PBDE congeners 153 and 154 (hexaBDEs) at less than 5% each. The commercial "octa" mixture is 10%–12% hexaBDE, 43%–44% heptaBDE, 31%–35% octa-BDE, 9%–11% nona-BDE, and 0%–1% decaBDE. The "deca" commercial mixture consists of 98% decaBDE, with a small percentage of nona-BDEs (World Health Organization, 1994; Hardy, 2000).

Human Health Effects and Modes of Action

TBBPA is of high ecotoxicologic concern because of its acute and chronic toxicity in several biota (US EPA, 2008; Lyche et al., 2015). Although rapidly metabolized by mammalian liver and eliminated in bile, urine, and feces

(Schauer et al., 2006), TBBPA has been detected in various environmental media and biota, including air, soils, water, sediment, and bird muscle, from electronic waste regions of China (Shi et al., 2009; Liu et al., 2016) and in bottlenose dolphins and bull sharks from the Florida Coast (Johnson-Restripo et al., 2008). TBBPA is cytotoxic, immunotoxic, a TH agonist and has the potential to disrupt estrogen signaling (Birnbaum and Staskal, 2004). TBBPA has also been shown to be toxic in rat brain cells in vitro where it causes oxidative stress and calcium influx and inhibits dopamine uptake (Reistad et al., 2007). Recent in vivo studies indicated that neonatal TBBPA exposure causes hearing deficits in rat offspring, similar to those observed following developmental exposure to PCBs (Lilienthal et al., 2008). Nakajima et al. (2009) correlated behavioral alterations following acute treatment with the presence of TBBPA in brain. Disruption of TH homeostasis is proposed to be the primary toxic effect of TBBPA and other BFRs. TBBPA has a closer structural relation to thyroxine (T_4) than PCBs and binds to transthyretin (TTR) with greater affinity than that of the naturally occurring ligand, T_4 (Meerts et al., 2000). The detailed mechanism by which BFRs can disrupt TH homeostasis is illustrated in the next section (Fig. 28.4).

Studies showed that HBCD, the second most used flame retardant, also has endocrine disrupting and reproductive effects (Birnbaum and Staskal, 2004). Many effects of HBCDs seem to occur during development. During developmental exposure, HBCDs have been shown to decrease bone density and retinoids and enhance immune response to sheep red blood cells (van der Ven et al., 2009). HBCD isomers are endocrine disruptors with antiandrogenic properties that inhibit aromatase and interact with steroid hormone receptors (Hamers et al., 2006). Like other BFRs, HBCDs may disrupt TH homeostasis resulting in decreased T_4 levels and increased thyroid-stimulating hormone (TSH) (Ema et al., 2007). Recent studies indicated that low-dose HBCD potentially can disrupt TH hormone receptor–mediated transactivation and impairs cerebellar Purkinje cell dendritogenesis (Ibhazehiebo et al., 2011).

Exposure to PBDEs in domestic/pet animals and humans may occur via multiple sources (air, water, dust, and food; Sjodin et al., 2008; Kodavanti and Loganathan, 2016). There are a few reports indicating high levels of serum PBDEs in household cats because of their high exposure to house dust (Dye et al., 2007). Studies show levels of PBDEs in animal, and human tissues have increased exponentially since the 1970s in the United States, Canada, and Sweden (Schecter et al., 2005; Guo et al., 2012). Elevated levels of PBDEs in North America were attributed to the greater use of the penta BDE mixture compared with the rest of the world. Like other lipophilic compounds, PBDEs readily cross the placenta into the fetus, providing an opportunity for PBDEs to interfere with human and animal developmental processes. Epidemiological studies indicate that childhood exposure to PBDEs results in neurodevelopmental changes in two cognitive assessments at 6 years of age (Chevrier et al., 2016; see review Linares et al., 2015). Fernie et al. (2017) reported spatiotemporal patterns and relationships among the diet, circulating THs, and exposure to flame retardants in an apex avian predator, the peregrine falcon. The results indicate that rural nestlings had significantly lower circulating TH levels, α-tocopherol, and oxidative status, but higher retinol levels and PBDE-153 when compared with the urban nestlings.

Laboratory studies with animals have indicated that commercial PBDE mixtures and the individual PBDE congeners that compose them affect the nervous, endocrine, reproductive, and immune systems. Regarding neurotoxic effects, several studies indicated that PBDEs, along with HBCDs, cause permanent aberrations in spontaneous behavior and habituation capability in mice after a single exposure at postnatal day (PND) 10 (a period of rapid brain growth and development). It is interesting to note that the effects seen on this behavioral paradigm with PBDEs are identical to those produced by PCBs (Eriksson and Fredriksson, 1996). Recent studies indicated that developmental exposure to a commercial pentabrominated mixture (DE-71) resulted in multiple effects. Although maternal or male body weights were not altered, female offspring were smaller compared with controls from PND 35–60. Exposure to DE-71 also resulted in accumulation of PBDE congeners in various tissues, including the brain, suggesting that PBDEs cross blood–placenta and blood–brain barriers (BBBs). Subtle changes in some parameters of neurobehavior, dramatic changes in circulating TH levels, and changes in both male and female reproductive endpoints were also observed (Kodavanti et al., 2010). In female Wistar rats exposed to DE71 (*penta*-PBDE mixture), deficits in reference memory have been described without alteration of working memory and hyperactivity (de-Miranda et al., 2016).

In addition to the effects on TH, there is evidence that PBDEs affect the cholinergic neurotransmitter system (Viberg et al., 2003a,b), with roles in memory and motor function. The effects of several PBDE congeners have been compared to PCBs for their ability to affect intracellular signaling in a cerebellar (brain) culture system (Kodavanti and Ward, 2005; Fan et al., 2010). The Ca/protein kinase C signaling pathways have been proposed as an additional mechanism of neurotoxicity for a number of chemicals, including PCBs and PBDEs. The order of potency for their effects on intracellular signaling was DE-71 (a commercial mixture of tetra-, penta-, and hexa-BDEs)>PBDE 47 > PBDE 100 > PBDE

Thyroid Hormone Disruption as a Biomarker of Exposure and Effect

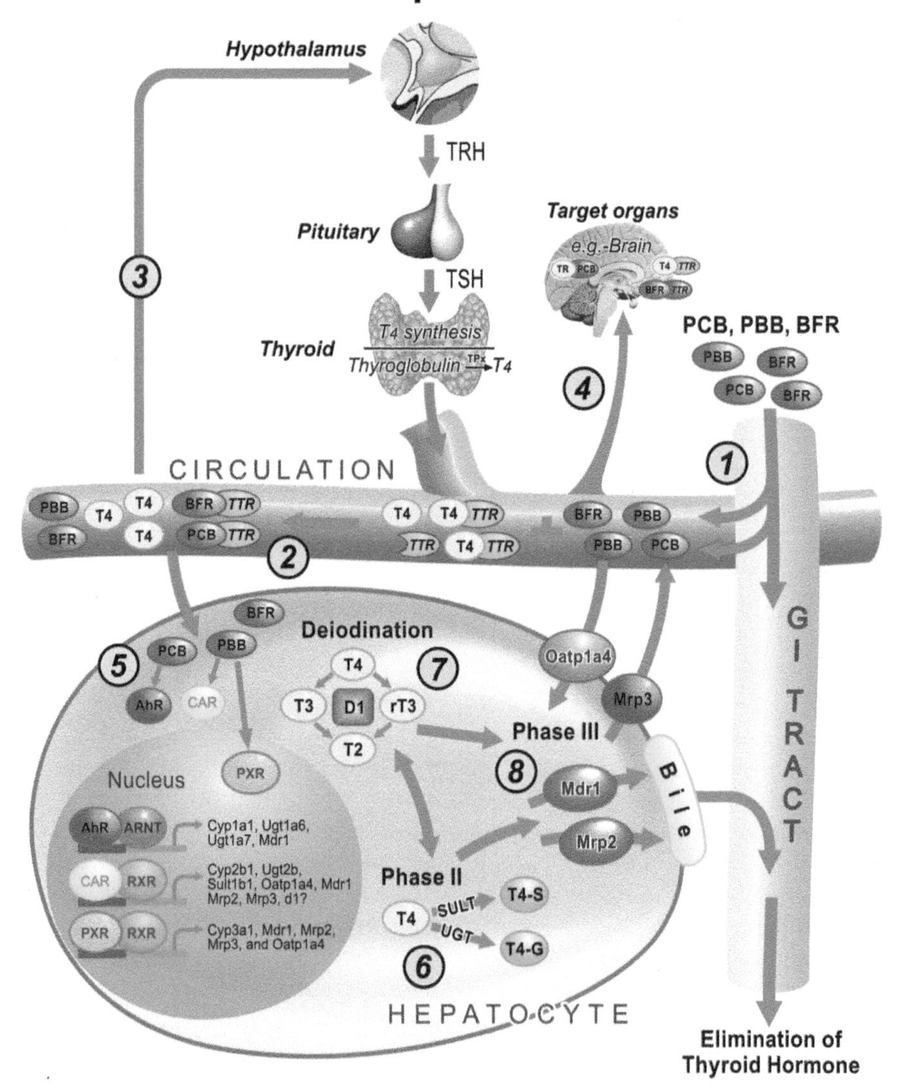

FIGURE 28.4 Schematic representation of thyroid hormone disruption as a biomarker of exposure and effect for polychlorinated biphenyls (PCBs), polybrominated biphenyls (PBBs), and brominated flame retardants (BFRs). Sequential processes by which these compounds may disrupt thyroid hormone homeostasis are illustrated by numbers. (1) PCBs, PBBs, and BFRs enter the circulation from gastrointestinal (GI) tract following exposure. (2) Parent compounds or hydroxylated metabolites can displace thyroxine (T_4) from serum binding proteins such as transthyretin (TTR). The resulting free T_4 will be eliminated following hepatic metabolism. (3) Reduced circulating T_4 levels stimulate the hypothalamic–pituitary axis to synthesize and secrete more T_4 by the thyroid gland. (4) Chemicals that were bound to TTR, along with T_4, will reach target organs, including the brain. (5) These chemicals activate nuclear receptors in liver cells (hepatocytes) initiating transcription of xenobiotic metabolizing enzymes (XMEs) for T_4 elimination. (6) XMEs consequently conjugate T_4 by phase II enzymes, uridine diphosphate glucuronyl transferase (UGT), and sulfotransferase (SULT). (7) Hepatic deiodinase (DI) deiodinates T_4 to its metabolites. (8) Influx transporters (Oatp1a4) further increase the T_4 uptake and metabolism. Efflux transporters eliminate T_4 or its conjugates from hepatocytes into either the serum (Mrp3) or the bile (Mrp2). These compounds do not have any effect on T_4 synthesis in the thyroid gland. *Adapted from Szabo D.T., Richardson V.M., Ross D.G., et al., 2009. Effects of perinatal PBDE exposure on hepatic phase I, phase II, phase III, and deiodinase 1 gene expression involved in thyroid hormone metabolism in male rat pups. Toxicol. Sci. 107, 27–39; Kodavanti P.R.S., Curras-Collazo M.C., 2010. Neuroendocrine actions of organohalogens: thyroid hormones, arginine vasopressin, and neuroplasticity. Front. Neuroendocrinol. 31, 479–496.*

99. On a molar basis, DE-71 was equipotent with Aroclor 1254; the most widely used commercial PCB mixture. A Swedish study found that PBDE 99 and PCB 52 produced effects on behavior when given together but not at the same dose given alone (Eriksson et al., 2006). The results suggest that there is little difference in neurotoxic potency between PBDEs and PCBs, and that effects of PCBs and PBDEs are additive. This implies that body burdens of PCBs and PBDEs in humans may be combined during the risk assessment of these compounds.

THYROID HORMONE DISRUPTION AS A BIOMARKER OF EXPOSURE AND EFFECT

PBDEs affect the TH system both in vivo and in vitro. In vivo studies show reductions in serum thyroxin (T_4) levels following exposure to PBDEs, both after acute and subacute exposure (Fowles et al., 1994) and after developmental exposure (Kodavanti et al., 2010). The negative correlation between these organohalogens and serum T4 levels has also been demonstrated in a North American human cohort (Makey et al., 2016). Hallgren et al. (2001) showed that PCBs are more potent compounds in reducing circulating T_4 levels than are PBDEs. Interaction studies with human TTR showed that PBDEs have to undergo metabolic activation before the compounds are able to competitively inhibit the binding of T_4 (Hamers et al., 2006). This interaction between PBDEs and TTR, however, appears to vary with species because several PBDE congeners (BDE-47, BDE-99) bind the piscine form of TTR with high affinity (Morgado et al., 2007). These species differences in PBDE interaction with TTR are likely due to differences in the structural properties of mammalian and piscine TTRs (Eneqvist et al., 2004). Zhou et al. (2001) concluded that short-term exposure to some commercial PBDE mixtures interferes with the TH system via upregulation of uridine diphosphate glucuronyl transferases (UGTs) and TH metabolizing enzymes. Richardson et al. (2008) indicated that although there was induction of hepatic UGTs, the decreases in circulating T_4 levels by PBDE 47 exposure might involve other mechanisms. In support of alternative mechanisms, several other reports indicated a good correlation between the degree of TH reduction by PCBs and PBDEs with a decrease in the ex vivo binding of ^{125}I-labeled T_4 to TTR and a lack of correlation with UGT induction (Hallgren et al., 2001; Hallgren and Darnerud, 2002). Further studies indicated that decabromodiphenyl ether (PBDE 209) decreased serum triiodothyronine (T_3) levels, but not T_4 levels, in the absence of any induction of UGTs (Tseng et al., 2008). These studies with PBDEs are similar to the reports on PCBs and support the conclusion that UGTs may not play a significant role in decreasing circulating TH levels by these groups of chemicals. Displacement of T_4 and binding of PCBs or PBDEs with the TTR transport protein might be a critical event in decreasing circulating TH levels.

Szabo et al. (2009) conducted extensive studies on different mechanisms involved in TH disruption after exposure to the PBDE mixture, DE-71. Developmental exposure resulted in significant increases both in hepatic cytochrome P450 enzyme activities and gene expression with significant decreases in hepatic deiodinase I (D1) activity and gene expression. The results from this study indicated that deiodination, active transport, and sulfation, in addition to glucuronidation, may be involved in the disruption of TH homeostasis (Szabo et al., 2009, Fig. 28.4). In addition to the effects on circulating TH levels, PBDEs also interfere with TH receptor (TR)–mediated TH signaling. Schriks et al. (2007) reported that hydroxyl-PBDEs increased T_3-induced thyroid receptor (TR)α-activation, but not T_3-induced TRβ-activation while 2,2′,3,3′,4,4′,5,5′,6-brominated diphenyl ether (PBDE 206) was antagonistic on both TRs. Lema et al. (2008) recently reported depressed plasma T_4 levels in flathead minnows exposed to PBDE 47 and this was accompanied by elevated mRNA levels for TSHβ in the pituitary. PBDE-47 exposure also elevated transcription of TRα in the brain of females and decreased mRNA levels for TRβ in the brain of both sexes of fathead minnows (Lema et al., 2008). The PBDE effects on TR would have physiological implications such as alterations in the development of neuronal cells as reported by Schreiber et al. (2010).

Based on the literature, the possible mechanism(s) of disruption of TH by exposure to PCBs, PBBs, and BFRs are shown in Fig. 28.4. These chemicals enter circulation through gastric absorption. T_4 is synthesized and released into circulation by the thyroid gland. These compounds did not interfere with the synthesis of T_4 by this gland. In circulation, PBDEs displace T_4 from serum binding proteins such as TTR. The resulting free T4 released from TTR will be subjected to hepatic metabolism and elimination.

The reductions in circulating T_4 levels increase TSH production via reduced negative feedback on the hypothalamic–pituitary axis (Lema et al., 2008), which induces increased synthesis and secretion of T_4 by the thyroid gland. PBDEs bound to TTR along with T_4 will reach target organs, including the brain, to elicit their effects. PBDEs activate nuclear receptors initiating transcription of xenobiotic metabolizing enzymes (XMEs) in the liver for T_4 elimination. XMEs consequently conjugate T_4 by phase II enzymes, UGT, and sulfotransferase (SULT). Deiodinase 1 (D1) can deiodinate T_4 to its metabolites. Influx transporters (Oatp1a4) further increase the T_4 uptake for metabolism. Efflux transporters eliminate T_4 or its conjugates from hepatocytes either into the serum (Mrp3) or the bile (Mrp2). Thus, TH disruption by PBDEs involves multiple mechanisms, including phase II glucuronidation and sulfation, TTR displacement, decreased hepatic deiodinase I activity, and increases in hepatic transporter phase III elimination (Fig. 28.4).

PERTURBED CALCIUM HOMEOSTASIS AND KINASE SIGNALING AS BIOMARKERS OF EFFECT

Studies with prototypic *ortho* (2,2′-dichlorobiphenyl, DCB) and non-*ortho* (3,3′,4,4′,5-pentachlorobiphenyl; PeCB or 4,4′-DCB) PCBs followed by extensive structure–activity relationships with more than 35 PCB congeners indicated alterations in intracellular calcium signaling. Increases in free Ca^{2+} ($[Ca^{2+}]_i$) with a fluorescent dye (Fluo-3AM) were reported for the first time; the *ortho*-substituted 2,2′-DCB was more effective than the non-*ortho*-substituted 3,3′,4,4′,5-PeCB (Kodavanti et al., 1993). The increase in $[Ca^{2+}]_i$ was slow, and a steady rise was observed with time (Kodavanti et al., 1993). The follow-up studies confirmed these observations in cerebellar granule neurons (Mundy et al., 1999; Bemis and Seegal, 2000), cortical neurons (Inglefield et al., 2002), and in human granulocytes (Voie and Fonnum, 1998). Studies characterizing the mechanisms of $[Ca^{2+}]_i$ increase indicated that 2,2′-DCB was an inhibitor of $^{45}Ca^{2+}$ uptake by mitochondria and microsomes with IC_{50} (concentration which inhibits control activity by 50%) values of 6–8 μM. 3,3′,4,4′,5-PeCB inhibited Ca^{2+} sequestration, but the effects were much less than those produced by equivalent concentrations of 2,2′-DCB. Synaptosomal Ca^{2+}-ATPase, involved in the Ca^{2+} extrusion process, was only inhibited by 2,2′-DCB, but not by 3,3′,4,4′,5-PeCB (Kodavanti et al., 1993). Further structure–activity relationship (SAR) studies indicated that congeners that are noncoplanar inhibited $^{45}Ca^{2+}$-uptake by microsomes and mitochondria, whereas coplanar congeners did not (Kodavanti et al., 1996a).

The disruption of Ca^{2+} homeostasis may have a significant effect on other signal transduction pathways (e.g., inositol phosphate [IP] and AA second messengers) regulated or modulated by Ca^{2+}. The congener 2,2′-DCB, but not 3,3′,4,4′,5-PeCB, affected basal and carbachol (CB)-stimulated IP accumulation in cerebellar granule cells (Kodavanti et al., 1994). Further studies indicated that any modulation of CB-stimulated IP accumulation is due to Ca^{2+} overload, but not due to activation of PKC activity (Kodavanti et al., 1994). AA is released intracellularly following activation of membrane phospholipases, and AA is an important second messenger in releasing Ca^{2+} from endoplasmic reticulum (Striggow and Ehrlich, 1997). Aroclor 1254 (a commercial mixture of PCBs) and 2,2′-DCB increased $[^3H]$-AA release in cerebellar granule cells, whereas 4,4-DCB did not (Kodavanti and Derr-Yellin, 1999); this is in agreement with previous structure–activity relationship studies on Ca^{2+} buffering and PKC translocation (Kodavanti and Tilson, 1997). A similar increase in $[^3H]$-AA was observed with structurally similar chemicals such as PBDE mixtures (Kodavanti and Derr-Yellin, 2002). Further studies indicated that the $[^3H]$-AA release caused by these chemicals could be due to activation of both Ca^{2+}-dependent and -independent phospholipase A_2 (PLA_2).

One of the downstream effects of perturbed Ca^{2+}-homeostasis is translocation of PKC from the cytosol to the membrane where it is activated (Trilivas and Brown, 1989). $[^3H]$-Phorbol ester ($[^3H]$-PDBu) binding has been used as an indicator of PKC translocation. The congener 2,2′-DCB increased $[^3H]$-PDBu binding in a concentration-dependent manner in cerebellar granule cells, whereas 3,3′,4,4′,5-PeCB had no effect on in concentrations up to 100 μM. The effect of 2,2′-DCB was time-dependent and also dependent on the presence of external Ca^{2+} in the medium. Sphingosine, a PKC translocation blocker, prevented 2,2′-DCB-induced increases in $[^3H]$-PDBu binding (Kodavanti and Tilson, 2000). Experiments with several pharmacological agents revealed that the effects are additive with glutamate, and none of the channel (glutamate, calcium, and sodium) antagonists blocked the response of 2,2′-DCB (Kodavanti et al., 1994).

Immunoblots of PKC-alpha and epsilon indicated that noncoplanar ortho-PCB decreased the cytosolic form and increased the membrane form significantly at 25 μM (Yang and Kodavanti, 2001). Subsequent SAR studies indicated that congeners that are noncoplanar increased PKC translocation, whereas coplanar congeners did not (for review, see Kodavanti and Tilson, 1997, 2000). This was further strengthened by observations with structurally similar chemicals such as polychlorinated diphenyl ethers (Kodavanti et al., 1996b). Nitric oxide (NO), which is produced by NOS, is a gaseous neurotransmitter. NO has an important role as a retrograde messenger in LTP, learning and memory processes, and endocrine function (Schuman and Madison, 1994). The congener 2,2′-DCB, but not 4,4′-DCB, inhibited both cytosolic (nNOS) and membrane (eNOS) forms of NOS (Sharma and Kodavanti, 2002).

These in vitro studies clearly demonstrated that second messenger systems, involved in the development of the nervous system, LTP, and learning and memory, are sensitive targets for the *ortho*-substituted PCBs and related chemicals. Fig. 28.5 illustrates the intracellular signaling events affected by these chemicals (ortho-PCBs and commercial PCB mixtures) at low micromolar concentrations and shorter exposure periods, where cytotoxicity is not evident. These signaling pathways include calcium homeostasis and PKC translocation. The rise of intracellular free Ca^{2+} is slow but steady following exposure. This free Ca^{2+} rise could be due to

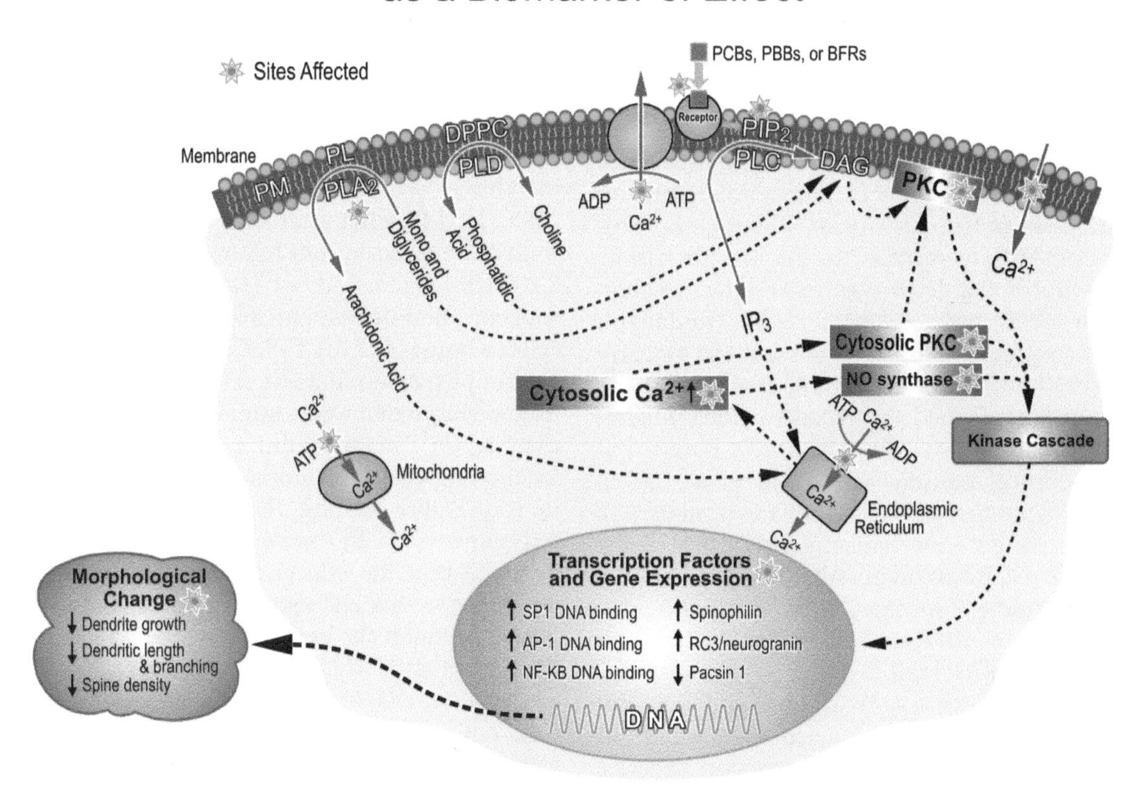

FIGURE 28.5 Schematic representation of calcium and kinase signaling as a biomarker of effect for polychlorinated biphenyls (PCBs), polybrominated biphenyls (PBBs), and brominated flame retardants (BFRs). The processes by which these compounds disrupt calcium homeostasis and kinase signaling are as follows. First, chemicals bind to the cell surface receptors and activate membrane phospholipases such as phospholipase C (PLC), phospholipase A2 (PLA_2), and phospholipase D (PLD). This will result in several second messengers such as arachidonic acid and inositol trisphosphate (IP_3), which will release calcium from intracellular stores such as endoplasmic reticulum. Following blockage of calcium sequestration mechanisms in mitochondria and endoplasmic reticulum, cytosolic free calcium levels rise. Increased cytosolic calcium translocates protein kinases from cytosol to the membrane where they are activated. This will result in the activation of kinase cascade triggering transcription of genes, which will result in a morphological change. *Adapted from Kodavanti, 2004.*

increased calcium influx, inhibited Ca^{2+} buffering mechanisms, and/or calcium release from intracellular stores by the products of membrane phospholipases. This increase in free Ca^{2+} could cause translocation of PKC. The coplanar non–ortho-PCBs have marginal effects on calcium homeostasis and no effects on PKC translocation. Literature reports indicate that at slightly higher concentrations, commercial PCB mixtures (Aroclors 1221 and 1254) have been seen to alter neurite outgrowth in PC12 cells (Angus and Contreras, 1995) and in hypothalamic cells (Gore et al., 2002). The possible mode of action for this structural change could be due to changes in intracellular signaling by these chemicals.

In vivo effects of PCBs have been studied with a commercial PCB mixture, Aroclor 1254, given orally from gestational day 6 through PND 21. Aroclor 1254 treatment did not alter maternal body weight or percent mortality, but caused a small and transient decrease in body weight gain of offspring. Both calcium homeostasis and PKC activities were significantly affected following developmental exposure to Aroclor 1254 (Kodavanti et al., 2000). Developmental exposure to PCBs also caused significant hypothyroxinemia and age-dependent alterations in the translocation of PKC isozymes; the effects were significant at PND 14 (Yang et al., 2003). Immunoblot analysis of PKC-alpha (α), -gamma (γ), and -epsilon (ε) from both cerebellum and hippocampus revealed that developmental exposure to Aroclor1254 caused a significant decrease in cytosolic fraction and an increase in particulate fraction. For some isozymes, the ratio between the two fractions was increased in a dose-dependent manner. Thus, the patterns of subcellular distributions of PKC isoforms following a developmental PCB exposure were PKC isozyme- and developmental stage-specific. The changes in PKC and other second messengers were associated with changes in transcription factors such as Sp1 and NF-κB, indicating changes in gene expression following developmental exposure to PCBs (Riyaz Basha et al., 2006). Considering the significant role of PKC signaling in motor behavior, learning, and memory, it is suggested that altered subcellular distribution of PKC isoforms at critical periods of brain development may be associated with activation of transcription factors and subsequent gene expression and may be a possible mechanism of PCB-induced neurotoxic effects. PKCα, γ, and ε may be among the target molecules implicated in PCB-induced neurological impairments during developmental exposure.

Further studies focused on the structural outcome for changes in the intracellular signaling pathway following developmental PCB exposure. Detailed brain morphometric evaluation was performed by measuring neuronal branching and spine density. Developmental exposure to PCBs affected normal dendritic development of Purkinje cells and CA1 pyramids (Lein et al., 2007). The branching area was significantly smaller in the PCB-exposed rats. When the rats became adults, there was continued neurostructural disruption of the CA1 dendritic arbor following PCB exposure, however, the branching area of the Purkinje cells returned to normal level. Developmental exposure to PCBs also resulted in a significantly smaller spine density in hippocampus, but not in cerebellum. This dysmorphic cytoarchitecture could be the structural basis for long-lasting neurocognitive deficits in PCB-exposed rats (Lein et al., 2007).

Previously, Pruitt et al. (1999) reported a reduced growth of intra- and infrapyramidal mossy fibers following developmental exposure to PCBs. These studies indicate that developmental exposure to a PCB mixture resulted in altered cellular distribution of PKC isoforms, which can subsequently disrupt the normal maintenance of signal transduction in developing neurons. The perturbations in intracellular signaling events could lead to structural changes in the brain. These findings suggest that altered subcellular distribution of PKC isoforms may be a possible mode of action for PCB-induced neurotoxicity. Fig. 28.5 illustrates the intracellular signaling events, transcription factors, and brain morphometry affected by these chemicals.

INDUCTION OF CYTOCHROME P450 ENZYMES AS A BIOMARKER OF EXPOSURE AND EFFECT

The adaptive response found in most organisms for a chemical insult is the induction and expression of genes for xenobiotic metabolizing enzymes (Denison and Nagym, 2003). One of the most potent and comprehensively studied properties of PCBs is their effect on mixed function oxidase (MFO) activity. The MFO system is located in the endoplasmic reticulum of many mammalian cells (Fig. 28.6). It has its terminal oxidase cytochrome P450 (CYP), which is responsible for the metabolism of many xenobiotics and endogenous compounds (Conney, 1967). Many drugs, pesticides, and herbicides cause changes in the activity of the MFO system by induction of cytochrome P450 and the enzyme activity associated with it (Parke, 1975). There are two major classes of cytochrome P450 inducing agents: (1) phenobarbital (PB) type and (2) 3-methylcholanthrene (3MC) type, which induces cytochrome P-448. Aroclor 1254 has been reported to have the inducing properties of both PB and 3MC (Alvares et al., 1973).

The suggestion that PBBs have similar toxic properties to PCBs was supported by the study by Dent et al. (1976) that dietary exposure to a mixture of PBBs containing mostly hexaBB induced the activities of the monoxygenase enzymes in rat liver microsomes. It has been reported that BB-52 is a weak "phenobarbital-like" inducer of CYP enzymes (Robertson et al., 1983). The authors further affirmed that BB-52 is an agonist of constitutive androstane receptor (CAR)/pregnane X receptor (PXR) transcription factors, responsible for gene expression of CYP2B, CYP3A, and other drug metabolizing enzymes and adverse effects connected with incorrect activation of CAR and PXR. Arneric et al. (1980) observed that metabolism of progesterone was accelerated in liver microsomes containing CYP enzymes from rats exposed to various concentrations of PBBs; simulation of hydroxylation of progesterone in positions 16α- and 6β-resembled the effects of phenobarbital, a prototype inducer of CAR- and PXR-dependent gene expression. The authors' observation confirmed induction effects of non–dioxin-like PBBs; certain effects of PBBs on the endocrine system are probably a consequence of enhanced steroid sex hormone catabolism, which might have led to a decrease in progesterone, 17β-estradiol, and testosterone levels, as well as THs.

Coplanar congeners such as non-*ortho* and mono-*ortho*-substituted PBBs elicited AhR-mediated gene expression and benzo[a]pyrene hydroxylase activity and depletion of thymus in immature male Wistar rats (Robertson et al., 1982, 1983). The authors further confirmed the AhR inducing potency of non-*ortho* PBBs using a rat hepatoma H4IIE Luc cell line with luciferase reporter system (DR CALLUX). BB-77, BB-126, and BB-169 were shown to be potent inducers of AhR-dependent luciferase and 7-ethoxyresorufin-O-deethylase activities (Fig. 28.6) and were calculated relative to 2,3,7,8-TCDD as a reference toxicant. The toxic equivalent (TEF) values were 0.0800, 0.16, and 0.0047, respectively (Behmisch et al., 2003). The TEF values were comparable with TEFs for coplanar PCBs.

Chen et al. (1992) studied PBB congeners for their ability to reduce serum retinol levels in male

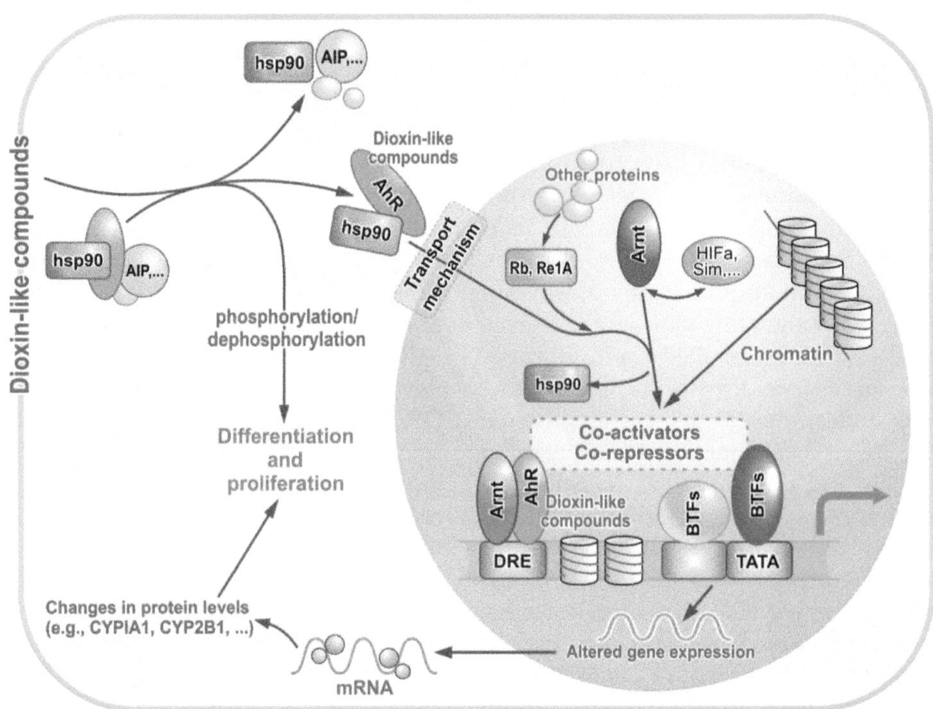

FIGURE 28.6 Schematic representation of functioning of the AhR (aryl hydrocarbon receptor) pathway. After entering the cell, the dioxin-like compounds bind to a protein complex in the cytoplasm consisting of AhR, Hsp90, AIP. On ligand binding AIP is released, exposing nuclear localization signal on AhR and leading to translocation of AhR from the cytoplasm to the nucleus. Within the nucleus, Hsp90 is released and AhR heterodimerizes with the Aryl Receptor Nuclear Translocator (ARNT). The AhR–ARNT complex then binds to multiple enhancer elements in the promotor region of the responsive genes in the AhR battery such as CYP1A. *Adapted from Denison, M.S., Nagym S.R., 2003. Activation of the aryl hydrocarebon receptor by structurally diverse exogeneous and endogenous chemicals. Annu. Rev. Pharmacol. Toxicol. 43, 309334.*

Sprague–Dawley rats. The authors noted an interesting observation that BB-169 had no effects on retinol levels, suggesting that a mechanism other than AhR activation is involved in this adverse effect. BB-77 and BB-126 significantly induced the retinyl ester hydrolase activity in the liver. Furthermore, high dietary levels of retinyl acetate produced some inhibitory effect on the promotion of hepatic altered foci induced by BB-169 in initiated rats (Rezabek et al., 1989). These studies showed that PBBs induce impairment of vitamin A and possibly also retinoid signaling and suggested that these effects result from different modes of action.

CONCLUDING REMARKS AND FUTURE DIRECTIONS

Human activities are responsible for the environmental input of many classes of chemicals through industrial, agricultural, and domestic sources. Biomonitoring studies have clearly shown a wide distribution of PCBs, PBBs, and PBDEs in the biosphere, resulting in exposures to almost all trophic level organisms with consequent adverse biological effects. This chapter presents biomarkers of response to environmental stressors, specifically in relation to the man-made synthetic organic chemicals PCBs, PBBs, and BFRs. Although other biochemical effects such as oxidative stress (Lee and Opanashuk, 2004; Mutlu et al., 2016) and neurotransmitter changes (Bell, 2014; Costa et al., 2014) have been reported for these chemicals, this chapter concentrates on the current three best prospects of early biomarkers of toxicity. Thyroid hormone disruption, perturbed calcium homeostasis and kinase signaling, and induction of cytochrome P450 enzymes are described as potential biomarkers of exposure and effect in exposed organisms. Data show that toxicity to these compounds appears at the subcellular level before being observed at individual or population levels. Further research, particularly for the BFRs, on initiating events, common molecular mechanisms, and the consequences of exposure, will improve our ability to determine earlier and better biomarkers of exposure. The relevant use of early biomarkers will improve our ability to assess both ecosystem and human risk.

Acknowledgments

The authors thank Dr. Aramandla Ramesh of Meharry Medical College, Nashville, TN, USA and Dr. Baki Sadi of Health Canada, Ottawa, ON, Canada for their excellent comments on an earlier version of this chapter. The contents of this article have been reviewed by the National Health and Environmental Effects Research Laboratory of the US Environmental Protection Agency and approved for publication. Approval does neither signify that the contents necessarily reflect the views and policies of the Agency nor does mention of trade names or commercial products constitute endorsement or recommendation for use.

References

Alaee, M., Arias, P., Sjodin, A., Bergman, A., 2003. An overview of commercially used brominated flame retardants, their applications, their use patterns in different countries/regions and possible modes of release. Environ. Int. 29, 683–689.

Alvares, A.P., Bickers, D.R., Kappas, A., 1973. Polychlorinated biphenyls: a new type of inducer of cytochrome P450 in the liver. Proc. Natl. Acad. Sci. Unit. States Am. 70, 1321.

Angus, W.G., Contreras, M.L., 1995. Aroclor 1254 alters the binding of ^{125}I-labeled nerve growth factor in PC12 cells. Neurosci. Lett. 191, 23–26.

Arneric, S.P., McCormac, K.M., Braselton Jr., W.E., Hook, J.B., 1980. Altered metabolism of progesterone by hepatic microsomes from rats following dietary exposure to polybrominated biphenyls. Toxicol. Appl. Pharmacol. 54, 187–196.

ATSDR, 2000. Toxicological Profile for Polychlorinated Biphenyls (PCBs). US Department of Health and Human Services. Agency for Toxic Substances and Disease Registry.

ATSDR, 2004. Public Health Statement: Polybrominated Biphenyls. US Department of Health and Human Services. Agency for Toxic Substances and Disease Registry.

Ballschmiter, K., Mennel, A., Buyten, J., 1993. Long-chain alkyl-polysiloxanes as nonpolar stationary phases in capillary gas-chromatography. Fresen. J. Anal. Chem. 346, 396–402.

Behmisch, P.A., Hosoe, K., Safe, S., 2003. Brominated dioxin-like compounds in vitro assessment in comparison to classical dioxin-like compounds and other poly aromatic compounds. Environ. Int. 29, 861–877.

Bell, M.R., 2014. Endocrine-disrupting actions of PCBs on brain development and social and reproductive behaviors. Curr. Opin. Pharmacol. 19, 134–144.

Bemis, J.C., Seegal, R.F., 2000. Polychlorinated biphenyls and methylmercury alter intracellular calcium concentrations in rat cerebellar granule cells. Neurotoxicology 21, 1123–1134.

Berghuis, S.A., Bos, A.F., Sauer, P.J., Roze, E., 2015. Developmental neurotoxicity of persistent organic pollutants: an update on childhood outcome. Archives Toxicol. 89 (5), 687–709.

Birnbaum, L.S., Staskal, D.F., 2004. Brominated flame retardants: cause for concern? Environ. Health Perspect. 112, 9–17.

Birnbaum, L.S., Bergman, A., 2010. Brominated and chlorinated flame retardants: The San Antonio Statement. Environ. Health Perspect. 118, A514–A515.

Birnbaum, L.S., Staskal, D.F., Diliberto, J.J., 2003. Health effects of polybrominated dibenzo-p-dioxins (PBDDs) and dibenzofurans (PBDFs). Environ. Int. 29, 855–860.

Chen, L.C., Berberian, I., Koch, B., et al., 1992. Polychlorinated and polybrominated congeners and retinoid levels in rat tissues: structure-activity relationships. Toxicol. Appl. Pharmacol. 11, 47–55.

Chevrier, C., Warembourg, C., Le Maner-Idrissi, G., et al., 2016. Childhood exposure to polybrominated diphenyl ethers and neurodevelopment at six years of age. NeuroToxicology 54, 81–88.

Conney, A.H., 1967. Pharmacological implications of microsomal enzyme induction. Pharmacol. Rev. 19, 317.

Corsolini, S., 2012. Contamination profile and temporal trend of POPs in Antarctic biota. In: Loganathan, B.G., Lam, P.K.S. (Eds.), Global Contamination Trends of Persistent Organic Chemicals. CRC Press, Taylor and Francis Group, pp. 571–628.

Costa, L.G., de Laat, R., Tagliaferri, S., Pellacani, C., 2014. A mechanistic view of polybrominated diphenyl ether (PBDE) developmental neurotoxicity. Toxicol. Lett. 230 (2), 282–294.

de-Miranda, A.S., Kuriyama, S.N., da-Silva, C.S., et al., 2016. Thyroid hormone disruption and cognitive impairment in rats exposed to PBDE during postnatal development. Reprod. Toxicol. 63, 114–124.

De Wit, C.A., 2002. An overview of brominated flame retardants in the environment. Chemosphere 46, 583–624.

De Wit, C.A., Herzke, D., Vorkamp, K., 2010. Brominated flame retardants in the Arctic environment- trends and new candidates. Sci. Total Environ. 408, 2885–2918.

Denison, M.S., Nagym, S.R., 2003. Activation of the aryl hydrocarebon receptor by structurally diverse exogeneous and endogenous chemicals. Annu. Rev. Pharmacol. Toxicol. 43, 309334.

Dent, J.G., Netter, K.J., Gibson, J.E., 1976. Effects of chronic administration of polybrominated biphenyls in parameters associated with hepatic drug metabolism. Res. Commun. Chem. Pathol. Pharmacol. 13, 75–82.

Di Carlo, F.J., Seifter, J., DeCarlo, V.J., 1978. Assessment of the hazards of polybrominated biphenyls. Environ. Health Perspect. 23, 350–365.

DiGangi, J., Blum, A., Bergman, A., et al., 2010. San Antonio statement on brominated and chlorinated flame retardants. Environ. Health Perspect. 118, A516–A518.

Dye, J.A., Venier, M., Zhu, L., et al., 2007. Elevated PBDE levels in pet cats: sentinels for humans? Environ. Sci. Technol. 41, 6350–6356.

EFSA, 2010. Scientific opinion on polybrominated biphenyls (PBBs) in food. EFSA J. 8 (10), 1789, 1–151.

Ema, M., Fujii, S., Hirata-Koizumi, M., Matsumoto, M., 2007. Two-generation reproductive toxicity study of the flame retardant hexabromocyclododecane in rats. Reprod. Toxicol. 25, 335–351.

Eneqvist, T., Lundberg, E., Karlsson, A., et al., 2004. High resolution crystal structures of piscine transthyretin reveal different binding modes for triiodothyronine and thyroxine. J. Biol. Chem. 279, 26411–26416.

Eriksson, P., Fredriksson, A., 1996. Developmental neurotoxicity of four ortho-substituted polychlorinated biphenyls in the neonatal mouse. Environ. Toxicol. Pharmacol. 1, 155–165.

Eriksson, P., Fischer, C., Fredriksson, A., 2006. Polybrominated diphenyl ethers, a group of brominated flame retardants, can interact with polychlorinated biphenyls in enhancing developmental neurobehavioral defects. Toxicol. Sci. 94, 302–309.

Fan, C.-Y., Besas, J., Kodavanti, P.R.S., 2010. Changes in mitogen-activated protein kinase in cerebellar granule neurons by polybrominated diphenyl ethers and polychlorinated biphenyls. Toxicol. Appl. Pharmacol. 245, 1–8.

Fernie, K., Chabot, D., Champoux, L., et al., 2017. Spatiotemporal patterns and relationships among the diet, biochemistry, and exposure to flame retardants in an apex avian predator, the peregrine falcon. Environ. Res. 158, 43–53.

Fowles, J.R., Fairbrother, A., Baecher-Steppan, L., Kerkvliet, N.I., 1994. Immunologic and endocrine effects of the flame-retardant pentabromodiphenyl ether (DE-71) in C57BL/6J mice. Toxicology 86, 49–61.

Fries, G.F., 1985. The PBB episode in Michigan: an overall appraisal. Crit. Rev. Toxicol. 16, 105–156.

Gore, A.C., Wu, T.J., Oung, T., et al., 2002. A novel mechanism for endocrine-disrupting effects of polychlorinated biphenyls: direct effects on gonadotropin-releasing hormone neurons. J. Neuroendocrinol. 14, 814–823.

Gotsch, A., Mariussen, E., van der Recke, R., et al., 2004. Enantioselective separation of atropisomeric PBB 132 and PBB 149 in extracts from white-tailed sea eagle. BFR Proceedings 351–354.

Guo, Y., Shaw, S., Kannan, K., 2012. Spatial and temporal trends of polybrominated diphenyl ethers. In: Loganathan, B., Lam, P.K.S. (Eds.), Global Contamination Trends of Persistent Organic Chemicals. CRC Press, Boca Raton, Florida, USA.

Hallgren, S., Darnerud, P.O., 2002. Polybrominated diphenyl ethers (PBDEs), polychlorinated biphneyls (PCBs) and chlorinated paraffins (CPs) in rats – testing interactions and mechanisms for thyroid hormone effects. Toxicology 177, 227–243.

Hallgren, S., Sinjari, T., Hakansson, H., Darnerud, H.O., 2001. Effects of polybrominated diphenyl ethers (PBDEs) and polychlorinated biphenyls (PCBs) on thyroid hormone and vitamin A levels in rats and mice. Arch. Toxicol. 75, 200–208.

Hamers, T., Kamstra, J.H., Sonneveld, E., et al., 2006. In vitro profiling of the endocrine-disrupting potency of brominated flame retardants. Toxicol. Sci. 92, 157–173.

Hardy, M.L., 2002. A comparison of the properties of the major commercial PBDPO/PBDE product to those of major PBB and PCB products. Chemosphere 46, 717–728.

Hardy, M., 2000. Distribution of decabromobiphenyl oxide in the environment. Organohalogen Compd. 47, 237–240.

Huber, S., Ballschmiter, K., 2001. Characterisation of five technical mixtures of brominated flame retardants. Fersen. J. Anal. Chem. 371, 882–890.

Ibhazehiebo, K., Iwasaki, T., Shimokawa, N., Koibuchi, N., 2011. 1,2,5,6,9,10-αHexabromocyclododecane (HBCD) impairs thyroid hormone-induced dendrite arborization of Purkinje cells and suppresses thyroid hormone receptor-mediated transcription. Cerebellum 10, 22–31.

Inglefield, J.R., Mundy, W.R., Meacham, C.A., Shafer, T.J., 2002. Identification of calcium-dependent and -independent signalling pathways involved in polychlorinated biphenyl-induced cyclic AMP-responsive element-binding protein phosphorylation in developing cortical neurons. Neuroscience 115, 559–573.

IPCS (International Program on Chemical Safety), 1994. Brominated Diphenyl Ethers. Environmental Health Criteria. 162. World Health Organization, Geneva.

Jacobson, J.L., Jacobson, S.W., 1996. Dose-response in perinatal exposure to polychlorinated biphenyls (PCBs): the Michigan and North Carolina cohort studies. Toxicol. Ind. Health 12, 435–445.

Jaspers, V., Covaci, A., Maervoet, J., et al., 2005. Brominated flame retardants and organochlorine pollutants in eggs of little owls (*Athene noctua*) from Belgium. Environ. Pollut. 13, 81–88.

Jin, X., Lee, S., Jeong, Y., et al., 2016. Species-specific accumulation of polybrominated diphenyl ethers (PBDEs) and other emerging flame retardants in several species of birds from Korea. Environ. Pol. 219, 191–200.

Johnson-Restripo, B., Adams, D.H., Kannan, K., 2008. Tetrabromobisphenol A (TBBPA) and hexabromocyclododecanes (HBCDs) in tissues of humans, dolphins, and sharks from the United States. Chemosphere 70, 1935–1944.

Joseph, A.D., Terrell, M.L., Small, C.M., et al., 2009. Assessing intergenerational transfer of a brominated flame retardant. J. Environ. Monit. 11, 802–807.

Kierkegaard, A., De Wit, C., Asplund, L., et al., 2009. A mass balance of tri-hexabromodiphenyl ethers in lactating cows. Environ. Sci. Technol. 43, 2602–2607.

Kodavanti, P.R.S., 2004. Intracellular Signaling and Developmental Neurotoxicity. In: Zawia, N.H. (Ed.), Molecular Neurotoxicology: Environmental Agents and Transcription-Transduction Coupling. CRC Press, Boca Raton, FL, pp. 151–182.

Kodavanti, P., 2005. Neurotoxicity of persistent organic pollutants: possible mode(s) of action and further considerations. Dose Response 3, 273–305.

Kodavanti, P.R., Tilson, H.A., 1997. Structure-activity relationships of potentially neurotoxic PCB congeners in the rat. NeuroToxicology 18, 425–442.

Kodavanti, P.R.S., Tilson, H.A., 2000. Neurochemical effects of environmental chemicals: in vitro and in vivo correlations on second messenger pathways. Ann. N. Y. Acad. Sci. 919, 97–105.

Kodavanti, P.R.S., Derr-Yellin, E.C., 1999. Activation of calcium-dependent and -independent phospholipase A_2 by non-coplanar polychlorinated biphenyls in rat cerebellar granule neurons. Organohalogen Compd. 42, 449–453.

Kodavanti, P.R.S., Derr-Yellin, E.C., 2002. Differential effects of polybrominated diphenyl ethers and polychlorinated biphenyls on $[^3H]$arachidonic acid release in rat cerebellar granule neurons. Toxicol. Sci. 68, 451–457.

Kodavanti, P.R.S., Loganathan, B.G., 2016. Brominated flame retardants: spatial and temporal trends in the environment and biota from the Pacific Basin Countries. In: Loganathan, B.G., Khim, J.S., Kodavanti, P.R., Masunaga, S. (Eds.), Persistent Organic Chemicals in the Environment: Status and Trends in the Pacific Basin Countries, ACS Symposium Series, vol. 1244. American Chemical Society and Oxford University Press, pp. 21–48.

Kodavanti, P.R.S., Loganathan, B.G., 2017. Organohalogen pollutants and human health. In: Quah, S.R., Cockerham, W.C. (Eds.), The International Encyclopedia of Public Health, second ed., vol. 5. Academic Press, Oxford, pp. 359–366.

Kodavanti, P.R.S., Ward, T.R., 2005. Differential effects of commercial polybrominated diphenyl ether and polychlorinated biphenyl mixture on intracellular signaling in rat brain in vitro. Toxicol. Sci. 85, 952–962.

Kodavanti, P.R.S., Curras-Collazo, M.C., 2010. Neuroendocrine actions of organohalogens: thyroid hormones, arginine vasopressin, and neuroplasticity. Front. Neuroendocrinol. 31, 479–496.

Kodavanti, P.R.S., Shin, D., Tilson, H.A., Harry, G.J., 1993. Comparative effects of two polychlorinated biphenyl congeners on calcium homeostasis in rat cerebellar granule cells. Toxicol. Appl. Pharmacol. 123, 97–106.

Kodavanti, P.R.S., Shafer, T.J., Ward, T.R., et al., 1994. Differential effects of polychlorinated biphenyl congeners on phosphoinositide hydrolysis and protein kinase C translocation in rat cerebellar granule cells. Brain Res. 662, 75–82.

Kodavanti, P.R.S., Ward, T.R., McKinney, J.D., Tilson, H.A., 1996a. Inhibition of microsomal and mitochondrial Ca^{2+} sequestration in rat cerebellum by polychlorinated biphenyl mixtures and congeners: structure-activity relationships. Arch. Toxicol. 70, 150–157.

Kodavanti, P.R.S., Ward, T.R., McKinney, J.D., et al., 1996b. Increased $[^3H]$phorbol ester binding in rat cerebellar granule cells and inhibition of $^{45}Ca^{2+}$ sequestration by polychlorinated diphenyl ether congeners and analogs: structure-activity relationships. Toxicol. Appl. Pharmacol. 138, 251–261.

Kodavanti, P.R.S., Mundy, W.R., Derr-Yellin, E.C., Tilson, H.A., 2000. Developmental exposure to Aroclor 1254 alters calcium buffering and protein kinase C activity in the brain. Toxicol. Sci. 54 (Suppl.), 76–77.

Kodavanti, P.R.S., Senthilkumar, K., Loganathan, B.G., 2008. Organohalogen pollutants and human health. In: Heggenhougen, K., Quah, S. (Eds.), International Encyclopedia and Public Health, vol. 4. Academic Press, San Diego, pp. 686–693.

Kodavanti, P.R.S., Coburn, C.G., Moser, V.C., et al., 2010. Developmental exposure to a commercial PBDE mixture, DE-71: neurobehavioral, hormonal, and reproductive effects. Toxicol. Sci. 116, 297–312.

Lee, D.W., Opanashuk, L.A., 2004. Polychlorinated biphenyl mixture Aroclor 1254-induced oxidative stress plays a role in dopaminergic cell injury. Neurotoxicology 25, 925–939.

Lein, P.J., Yang, D., Bachstetter, A.D., et al., 2007. Ontogenetic alterations in molecular and structural correlates of dendritic growth after developmental exposure to polychlorinated biphenyls. Environ. Health Perspect. 115, 556–563.

Lema, S.C., Dickey, J.T., Schultz, I.R., Swanson, P., 2008. Dietary exposure to 2,2′,4,4′-tetrabromodiphenyl ether (PBDE 47) alters thyroid

status and thyroid hormone-regulated gene transcription in the pituitary and brain. Environ. Health Perspect. 116, 1694–1699.
Lilienthal, H., Verwer, C.M., van der Ven, L.T., et al., 2008. Exposure to tetra bromobisphenol A (TBBPA) in Wistar rats: neurobehavioral effects in offspring from a one-generation reproduction study. Toxicology 246, 45–54.
Linares, V., Bellés, M., Domingo, J.L., 2015. Human exposure to PBDE and critical evaluation of health hazards. Arch. Toxicol. 89 (3), 335–356.
Liu, K., Li, J., Yan, S., et al., 2016. A review of status of tetrabromobisphenol A (TBBPA) in China. Chemosphere 148, 8–20.
Loganathan, B.G., Lam, P.K.S. (Eds.), 2012. Global Contamination Trends of Persistent Organic Chemicals. CRC Press, Taylor and Francis Group, p. 638.
Loganathan, B.G., Tanabe, S., Hidaka, Y., et al., 1993. Temporal trends of persistent organochlorine residues in human adipose tissue from Japan, 1928–1985. Environ. Pollut. 81, 31–39.
Loganathan, B.G., Kannan, K., 1994. Global organochlorine contamination: an Overview. AMBIO 23, 187–191.
Loganathan, B.G., Khim, J.S., Kodavanti, P.R., Masunaga, S. (Eds.), 2016a. Persistent Organic Chemicals in the Environment: Status and Trends in the Pacific Basin Countries I. ACS Symposium Series, vol. 1243. American Chem. Soc, Washington DC, p. 245.
Loganathan, B.G., Khim, J.S., Kodavanti, P.R., Masunaga, S. (Eds.), 2016b. Persistent Organic Chemicals in the Environment: Status and Trends in the Pacific Basin Countries II. ACS Symposium Series, vol. 1244. American Chemical Society, Washington DC, p. 264.
Lyche, J.L., Rosseland, C., Berge, G., Polder, A., 2015. Human health risk associated with brominated flame-retardants (BFRs). Environ. Int. 74, 170–180.
Lyman, W.J., Reehl, W.F., Rosenblatt, D.H., 1990. Handbook of chemical property estimation methods: Environmental behavior of organic compounds. United States Department of Energy, Office of Scientific and Technical Information. Biblio. 6902382 Oak Ridge, TN., USA.
Macdonald, R.W., Eisenreich, S.J., Bidleman, T.F., et al., 2000. Case studies on persistence and long-range transport of persistent organic pollutants. In: Klecka, G., Boethling, B., Franklin, J., et al. (Eds.), Evaluation of Persistence and Long-range Transport of Organic Chemicals in the Environment. Special Publication of Society of Environmental Toxcology and Chemistry. SETAC Press, pp. 245–314.
MacGregor, J.A., Nixon, W.B., 1997. Hexabromocyclododecane (HBCD): Determination of n-Octanol/Water Partition Coefficient. Wilflife International LTD 439C-104. Brominated Flame Retardant Industry Panel, Chemical Manufacturers Association, Arlington, VA.
Makey, C.M., McClean, M.D., Braverman, L.E., et al., 2016. Polybrominated diphenyl ether exposure and thyroid function tests in North American Adults. Environ. Health Perspect. 124 (4), 420.
Meerts, I.A., van Zanden, J.J., Luijks, E.A.C., et al., 2000. Potent competitive interactions of some brominated flame retardants and related compounds with human transthyretin in vitro. Toxicol. Sci. 56, 95–104.
Morgado, I., Hamers, T., Van der Ven, L., Powers, D.M., 2007. Disruption of thyroid hormone binding to sea bream recombinant transthyretin by ioxinyl and polybrominated diphenyl ethers. Chemosphere 69, 155–163.
Mundy, W.R., Shafer, T.J., Tilson, H.A., Kodavanti, P.R.S., 1999. Extracellular calcium is required for the polychlorinated biphenyl-induced increase of intracellular free calcium levels in cerebellar granule cell culture. Toxicology 136, 27–39.
Mutlu, E., Gao, L., Collins, L.B., et al., 2016. Polychlorinated biphenyls induce oxidative DNA adducts in female Sprague–Dawley rats. Chem. Res. Toxicol. 29 (8), 1335–1344.
Nakajima, A., Saigusa, D., Tetsu, N., et al., 2009. Neurobehavioral effects of tetrabromobisphenol A, a brominated flame retardant, in mice. Toxicol. Lett. 189, 78–83.
National Research Council, 2000. Toxicological Risks of Selected Flame-retardant Chemicals. The National Academies Press, Washington, DC.
Parke, D.V., 1975. Induction of drug-metabolizing enzymes. In: Parke, D.V. (Ed.), Enzyme Induction. Plenum Press, London, New York.
Pruitt, D.L., Meserve, L.A., Bingman, V.P., 1999. Reduced growth of intra- and infra-pyramidal mossy fibers is produced by continuous exposure to polychlorinated biphenyl. Toxicology 138, 11–17.
Reistad, T., Mariussen, E., Fonnum, F., 2007. In vitro toxicity of tetrabromobisphenol A on cerebellar granule cells: cell death, free radical formation, and calcium influx and extracellular glutamate. Toxicol. Sci. 96, 268–278.
Rezabek, M.S., Sleight, S.D., Jensen, P.K., Aust, S.D., 1989. Effects of dierary retinyl acetate on the promotion of hepatic enzyme-altered food by olybrominated biphenyls in initiated rats. Food Clin. Toxicol. 27, 539–544.
Richardson, V.M., Staskal, D.F., Ross, D.G., et al., 2008. Possible mechanisms of thyroid hormone disruption in mice by BDE 47, a major polybrominated diphenyl ether congener. Toxicol. Appl. Pharmacol. 226, 244–250.
Riyaz Basha, Md, Braddy, N.S., Zawia, N.H., Kodavanti, P.R.S., 2006. Ontogenetic alterations in prototypical transcription factors in the rat cerebellum and hippocampus following perinatal exposure to a commercial PCB mixture. Neurotoxicology 27, 118–124.
Robertson, L.W., Parkinson, A., Campbell, M.A., Safe, S., 1982. Polybrominated biphenyls as aryl hydrocarbon hydroxylase inducers: structure activity correlations. Chem. Biol. Interact. 42, 53–66.
Robertson, L.W., Andres, I.L., Safe, S.H., Lovering, S.L., 1983. Toxicity of 3, 3′, 4, 4′- and 2,2′, 5,5′-tetrabromobiphenyl: correlation of activity with aryl hydrocarbon hydroxylase induction and lack of protection by antioxidants. J. Toxicol. Environ. Health 11, 81–91.
Roth, N., Wilks, M.F., 2014. Neurodevelopmental and neurobehavioural effects of polybrominated and perfluorinated chemicals: a systematic review of the epidemiological literature using a quality assessment scheme. Toxicol. Lett. 230 (2), 271–281.
Safe, S., 1984. Polychlorinated biphenyls (PCBs) and polybrominated biphenyls (PBBs): biochemistry, toxicology, and mode of action. Crit. Rev. Toxicol. 13, 319–395.
Schauer, U.M., Volkel, W., Dekant, W., 2006. Toxicokinetics of tetrabromobisphenol a in humans and rats after oral administration. Toxicol. Sci. 91, 49–58.
Schecter, A., Papke, O., Tung, K., et al., 2005. Polybrominated diphenyl ether flame retardants in the U.S. population: current levels, temporal trends, and comparison with dioxins, dibenzofurans, and polychlorinated biphenyls. J. Occup. Environ. Med. 47, 199–211.
Schriks, M., Roessig, J.M., Murk, A.J., Furlow, J.D., 2007. Thyroid hormone receptor isoform selectivity of thyroid hormone disrupting compounds quantified with an in vitro reporter gene assay. Environ. Toxicol. Pharmacol. 23, 302–307.
Schreiber, T., Gassmann, K., Gotz, C., et al., 2010. Polybrominated diphenyl ethers induce developmental neurotoxicity in a human in vitro model: evidence for endocrine disruption. Environ. Health Perspect. 118, 572–578.
Sharma, R., Kodavanti, P.R.S., 2002. In vitro effects of polychlorinated biphenyls and hydroxy metabolites on nitric oxide synthases in rat brain. Toxicol. Appl. Pharmacol. 178, 127–136.
Shaw, S.D., Kannan, K., 2009. Polybrominated diphenyl ethers in marine ecosystems of the American continents: foresight from current knowledge. Rev. Environ. Health 24, 157–229.
Shaw, S.D., Blum, A., Weber, R., et al., 2010. Halogenated flame retardants: do the fire safety benefits justify the health and environmental risks? Rev. Environ. Health 25 (4), 261–305.
Shaw, S., Berger, M.L., Kannan, K., 2012. Status and trends of POPs in harbor seals form the Northwest Atlantic. In: Loganathan, B.G.,

Lam, P.K.S. (Eds.), Global Contamination Trends of Persistent Organic Chemicals. CRC Press, pp. 505–546.

Shen, H., Main, K.M., Anderson, A.M., et al., 2008. Concentrations of persistent organochlorine compounds in human milk and placenta are higher in Denmark than in Finland. Human Reporduction 23, 201–210.

Shi, T., Chen, S.-J., Luo, X.-J., et al., 2009. Occurrence of brominated flame retardants other than polybrominated diphenyl ethers in environmental and biota samples from southern China. Chemosphere 74, 910–916.

Schuman, E.M., Madison, D.V., 1994. Nitric oxide and synaptic function. Annu. Rev. Neurosci. 17, 153–183.

Sjodin, A., Papke, O., McGahee, E., et al., 2008. Concentrations of polybrominated diphenyl ethers (PBDEs) in household dust from various countries. Chemosphere 73 (1 Suppl.), S131–S136.

Stapleton, H.M., Klosterhaus, S., Eagle, S., et al., 2009. Detection of organophosphate flame retardants in furniture foam and U.S. House dust. Environ. Sci. Technol. 43, 7490–7495.

Striggow, F., Ehrlich, B.E., 1997. Regulation of intracellular calcium release channel function by arachidonic acid and leukotriene B_4. Biochem. Biophys. Res. Comm 237, 413–418.

Sundström, G., Hutzinger, O., Safe, S., Zitko, V., 1976. The synthesis and gas chromatographic properties of bromobiphenyls. Sci. Total Environ. 6, 15–29.

Szabo, D.T., Richardson, V.M., Ross, D.G., et al., 2009. Effects of perinatal PBDE exposure on hepatic phase I, phase II, phase III, and deiodinase 1 gene expression involved in thyroid hormone metabolism in male rat pups. Toxicol. Sci. 107, 27–39.

Trilivas, I., Brown, J.H., 1989. Increases in intracellular Ca^{2+} regulate the binding of $[^3H]$phorbol 12,13-dibutyrate to intact 1321N1 astrocytoma cells. J. Biol. Chem. 264, 3102–3107.

Tseng, L.H., Li, M.H., Tsai, S.S., et al., 2008. Developmental exposure to decabromodiphenyl ether (PBDE 209): effects on thyroid hormone and hepatic enzyme activity in male mouse offspring. Chemosphere 70, 640–647.

US EPA, 2008. Flame Retardants in Printed Circuit Boards. http://www.epa.gov/dfeprojects/pcb/full_report_pcb_flame_retardants_report_draft_11_10_08 to_e.pdf.

US EPA, 2012. Technical Fact Sheet- Polybrominated Diphenyl Ethers (PBDEs) and Polybrominated Biphenyls (PBBs). EPA-505-F-11–1007. May 2012.

van der Ven, L.T., van de Kuil, T., Leonards, P.E., et al., 2009. Endocrine effects of hexabromocyclododecane (HBCD) in a one-generation reproduction study in Wistar rats. Toxicol. Lett. 185, 51–62.

Veith, G.D., Defoe, D.L., 1979. Measuring and estimating the bioconcentration factor of chemicals in fish. J. Fish. Res. Board Can. 36, 1040–1048.

Viberg, H., Fredriksson, A., Eriksson, P., 2003a. Neonatal exposure to polybrominated diphenyl ether (PBDE 153) disrupts spontaneous behaviour, impairs learning and memory, and decreases hippocampal cholinergic receptors in adult mice. Toxicol. Appl. Pharmacol. 192, 95–106.

Viberg, H., Fredriksson, A., Jakobsson, E., et al., 2003b. Neurobehavioral derangements in adult mice receiving decabrominated diphenyl ether (PBDE 209) during a defined period of neonatal brain development. Toxicol. Sci. 76, 112–120.

Voie, O.A., Fonnum, F., 1998. *Ortho*-substituted polychlorinated biphenyls elevate intracellular $[Ca^{2+}]$ in human granulocytes. Environ. Toxicol. Pharmacol. 5, 105–112.

Von der Recke, R., Veter, W., 2008. Congener pattern of hexabromobiphenyls in marine biota from different provinces. Sci. Total Environ. 393, 358–366.

Wang, H., Zhang, Y., Liu, Q., et al., 2010. Examining the relationship between brominated flame retardants (BFR) exposure and changes of thyroid hormone levels around e-waste dismantling sites. Interntl. J. Hygiene Environ. Health 213, 369–380.

Watanabe, I., Kashimoto, T., Tatsukawa, R., 1983. The flame retardant tetrabromobisphenol A and its metabolite found in river and marine sediments in Japan. Chemosphere 12, 1533–1539.

World Health Organization, 1994. Environmental Health Criteria: Brominated Diphenyl Ethers, vol. 162. World Health Organization, Geneva, Switzerland.

Yang, J.-H., Kodavanti, P.R.S., 2001. Possible molecular targets of halogenated aromatic hydrocarbons in neuronal cells. Biochem. Biophys. Res. Comm 280, 1372–1377.

Yang, J.-H., Derr-Yellin, E.C., Kodavanti, P.R.S., 2003. Alterations in brain protein kinase C isoforms following developmental exposure to a polychlorinated biphenyl mixture. Mol. Brain Res. 111, 123–135.

Yusho Support Center, 2007. Left Behind the Yusho. A Report by Yusho Support Center, p. 77.

Zhao, G., Wang, Z., Zhuo, H., Zhao, A., 2009. Burdens of PBBs, PBDEs and PCBs in tissues of the cancer patients in the e-waste disassembly sites in Zheijang, China. Sci. Total Environ. 407, 4831–4837.

Zhou, T., Ross, D.G., DeVito, M.J., Crofton, K.M., 2001. Effects of short-term in vivo exposure to polybrominated diphenyl ethers on thyroid hormones and hepatic enzyme activities in weanling rats. Toxicol. Sci. 61, 76–82.

Zweidinger, R.A., Cooper, S.D., Erickson, M.D., et al., 1979. Sampling and analysis for semi volatile brominated organics in ambient air. In: Schuetzle, D. (Ed.), Monitoring Toxic Substances, ACS Symposium Series, vol. 94. American Chemical Society, Washington, DC, pp. 217–231.

CHAPTER

29

Polycyclic Aromatic Hydrocarbons

Leah D. Banks[1], *Kelly L. Harris*[1], *Kenneth J. Harris*[1], *Jane A. Mantey*[1], *Darryl B. Hood*[2], *Anthony E. Archibong*[3], *Aramandla Ramesh*[1]

[1]Department of Biochemistry, Cancer Biology, Neuroscience & Pharmacology, Meharry Medical College, Nashville, TN, United States; [2]College of Public Health, Ohio State University, Columbus, OH, United States; [3]Department of Microbiology, Immunology & Physiology, Meharry Medical College, Nashville, TN, United States

INTRODUCTION

Cancer risk assessment based on the 2-year rodent tumor bioassay is becoming obsolete because of the complexities associated with polycyclic aromatic hydrocarbon (PAH) exposures from occupational, dietary, and environmental settings. In addition to the costs and time factors involved, the relationship at best infers the relationship between the measured event (PAH exposure) and the tumor response. This approach, however, misses several endpoints that represent the various causal pathways involved in carcinogenesis. Additionally, the bioassay results could predict risk in the case of individuals but not at a population level. Therefore, to predict "at risk populations" from PAH exposure, appropriate biomarkers need to be employed for rapid screening of vulnerable populations and deploy appropriate preventive measures.

Biomarkers are cellular, biochemical, or molecular alterations that are measureable in biological media such as human tissues, cells, or fluids (Hulka, 1990). This definition has been extended to embrace biological parameters that could be measured and evaluated as an indicator of normal biological and pathogenic processes (Naylor, 2003). Biomarkers could be indicators of exposure, effect, or susceptibility (Links et al., 1995). These three categories of biomarkers are explained below (Links and Groopman, 2010). In the context of toxicology, biomarker of exposure indicates previous exposure to an environmental toxicant. This biomarker could be an exogenous chemical and interactive product (formed between a toxicant and endogenous compound) that could change the identity or status of target molecule. On the other hand, the biomarker of effect reports the nature and magnitude of biological (functional) response on exposure to an environmental pollutant. The third category, biomarker of susceptibility, outlines the heightened sensitivity of a subpopulation to the effects of a xenobiotic.

Biomarkers have power to be indispensable tools for preventing environmentally induced disease. The rising incidence of human exposure to persistent environmental pollutants necessitated use of biomarkers for disease detection and prevention at an early stage (Suk and Wilson, 2002). One such family of compounds that garnered a great deal of interest in the last century is PAHs. Being products of incomplete combustion they are prevalent in several environmental media (IARC, 2010; Ramesh et al., 2004, 2011, 2012) and through long-range atmospheric transport are carried over to places far away from point source areas (Shen et al., 2012). In addition, these toxicants are released from automobile exhaust, cigarette smoke, and industrial emissions. Several diseases such as cancer (lung, breast, colon, prostate), neurotoxicity, atherosclerosis, and infertility have been attributed to PAH exposure (IARC, 2010; Ramesh et al., 2011). Over the years, several biomarkers have been employed to detect susceptible populations from PAH-induced diseases. Some of the commonly employed biomarkers of PAH exposure, effect, and susceptibility are presented schematically in Fig. 29.1.

BIOMARKERS OF EXPOSURE

Hydroxylated Metabolites

As products of incomplete combustion, PAHs are prevalent in the environment, and exposure of people to PAHs in domestic, outdoor, and occupational settings

Biomarkers in Toxicology, Second Edition
https://doi.org/10.1016/B978-0-12-814655-2.00029-3

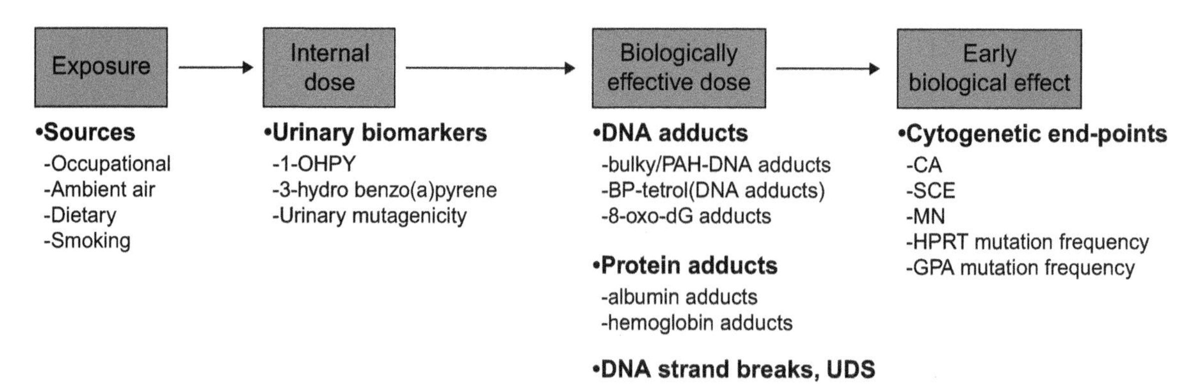

FIGURE 29.1 Overview of PAH exposures and biomarkers. *1-OH-PY*, 1-hydroxypyrene; *8-oxo-dG*, 8-oxo-2′-deoxyguanosine; *CA*, chromosome aberrations; *GPA*, glycophorin A; *HPRT*, hypoxanthine-guanine phosphoribosyltransferase; *MN*, micronuclei; *PAH*, polycyclic aromatic hydrocarbon; *SCE*, sister chromatid exchange. *Adopted from* Gyorffy E., Anna, L., Kovács, K., et al., 2008. Correlation between biomarkers of human exposure to genotoxins with focus on carcinogen-DNA adducts. Mutagenesis 23, 1–18; *with permission from Oxford University Press, UK.*

is inevitable. Hydroxylated metabolites of PAHs (OH-PAHs) generated through biotransformation have widely been used as biomarkers of exposure. Urinary concentrations of these metabolites have been employed in several studies. It is reported that urinary 1-hydroxypyrene (1-OHP) concentrations are higher for outdoor workers exposed to polluted air in urban setting than their unexposed counterparts (Ciarrocca et al., 2014). Similar observations were made in volunteers experimentally exposed to wood smoke (Li et al., 2016). Interestingly, the urinary concentrations of 1-nitropyrene are elevated in the United States–Mexico border residents, who tend to cross the border frequently, which is indicative of the higher urban pollution south of the border (Galaviz et al., 2017). Increased concentrations of 1-OHP glucuronide were reported in inner city Baltimore children from the home of cigarette smokers and children who more time spent outdoors (Peters et al., 2017). Children and elderly residents living near emission sources such as oil refineries and coal-fired power plants were also found to have elevated urinary concentrations of 1-OHP (Chen et al., 2017). Furthermore, restaurant workers exposed to fumes from repeated frying oil and restaurant waste oil were found to have higher urinary concentrations of 1-OHP (Ke et al., 2016). Another instance of occupation exposure to PAHs was reported in metallurgy workers in aluminum electrode production plant, who had high levels of urinary PAHs, the half-lives of which were estimated to range from 12 to 18 h (Lutier et al., 2016).

DNA Adducts

PAHs undergo biotransformation. As a result of this process, highly reactive metabolites are generated, which interact with cellular DNA and form PAH-DNA adducts. Measurement of these adducts serve an important purpose of human biomonitoring for exposure to PAH carcinogenesis and are also used as tools in molecular epidemiology studies (Farmer et al., 2003; Perera, 2000, 2011).

PAH-DNA adducts have been used as biomarkers to assess human exposure to mixtures of toxicants in Poland (Perera, 2000). Peripheral blood samples collected from residents of high-exposure regions revealed an association between environmental pollution and significant increases in levels of PAH-DNA adduct, sister chromatid exchanges (SCEs), chromosomal aberrations, and c-ras oncogene expression.

Another molecular epidemiology study that deserves mention was conducted by Taioli et al. (2007). The PAH-DNA adducts and oxidative DNA damage was measured in PAH-exposed populations from Prague (Czech Republic), Kosice (Slovakia), and Sofia (Bulgaria). The occupationally exposed individuals (policeman and bus drivers) were subjected to personal exposure monitoring, and the exposure of general population was monitored by measuring the levels of carcinogenic PAHs (c-PAHs) in ambient air. In the occupational category, policemen were exposed to PAHs to a greater extent than bus drivers. The average personal exposure to c-PAHs was found to be highest in Bulgaria, followed by Slovakia and the Czech Republic.

Further studies in this direction were undertaken by the SRAM research group in the Czech Republic (Rossner et al., 2013a). This study compared exposure of general public to contaminated air in Ostrava, a heavily polluted region, and Prague, a relatively clean region. The bulky PAH-DNA adducts in peripheral blood lymphocytes were used as biomarkers. A significant correlation between personal exposure to benzo(a)pyrene and BaP-DNA adduct levels were found in subjects from Ostrava, but not Prague.

Several analytical techniques have been employed for detection and quantitation of PAH-DNA adducts. The sources, types of samples used for analysis, and the methods used are summarized in Table 29.1.

TABLE 29.1 Sample Matrices and Methods Employed for Biomonitoring of Polycyclic Aromatic Hydrocarbon (PAH) DNA Adducts in Susceptible Populations

Exposure Route	Biological Matrix	Population	Analytical Method	References
Inhalation (cigarette smoke)	White blood cells	Healthy smokers and nonsmokers	Enzyme linked immunosorbent assay (ELISA)	Perera et al. (1987)
Inhalation (cigarette smoke)	White blood cells	Lung cancer patients	^{32}P-postlabeling	van Schooten et al. (1992)
Inhalation (cigarette smoke)	Peripheral leukocytes	Lung cancer patients	ELISA	Tang et al. (1995)
Inhalation (cigarette smoke)	Oral cavity (mouth floor and buccal mucosa)	Healthy subjects	Immunoperoxidase assay	Besaratinia et al. (2000)
Inhalation (cigarette smoke)	White blood cells	Smokers	Immunoassay	Funck-Brentano et al. (2006)
Inhalation (cigarette smoke)	Cervical mucosa	HPV-infected patients	Chemiluminescence assay	Pratt et al. (2007)
Inhalation (environmental tobacco smoke)	Maternal and umbilical cord blood	Pregnant mothers	?	Perera et al. (2007)
Inhalation (polluted air)	Placenta	Pregnant women from chemical and radioactive polluted areas	Chemiluminescence immune assay	Obolenskaya et al. (2010)
Inhalation (occupational)	White blood cells	Coke oven workers	ELISA	van Schooten et al. (1990)
Inhalation (occupational)	White blood cells	Iron foundry workers	ELISA	Santella et al. (1993)
Inhalation (occupational)	Peripheral blood lymphocytes	Aluminum plant workers	ELISA	Schoket et al. (1999)
Inhalation (occupational)	Peripheral blood lymphocytes	Coke oven workers	^{32}P-postlabeling High performance liquid chromatography (HPLC)	Brescia et al. (1999)
Inhalation (occupational)	Lymphocytes	Coke oven, aluminum plant workers, chimney sweeps	HPLC	Pavanello et al. (1999)
Dermal (coal tar)	White blood cells	Psoriasis patients	ELISA	Santella et al. (1995)
Dietary (charbroiled beef eaters)	White blood cells	Fire fighters who consumed barbecued food	ELISA	Rothman et al. (1993)
Dietary (charbroiled beef eaters)	White blood cells	Healthy subjects	ELISA	Kang et al. (1995)
Dietary (red meat eaters)	Leukocytes	Nonsmokers	Chemiluminescence assay	Gunter et al. (2007)
None specified	Colon tissue	Colon cancer patients	^{32}P-postlabeling	Al-Saleh et al. (2008)
None specified	Breast tissue	Breast cancer	ELISA	Sagiv et al. (2009)
None specified	Follicular cells	Infertile women	^{32}P-postlabeling	Al-Saleh et al. (2010)
None specified	Spermatozoa	Infertile men	Immunofluorescence assay	Ji et al. (2010)
None specified	Prostate	General population	Immunohistochemistry	Tang et al. (2013)

Protein Adducts

The formation of protein adducts is fundamentally similar to that of DNA adducts, i.e., reaction of a diol epoxide metabolite with a protein molecule (Autrup et al., 1999). Because of this property, protein adducts have been used as surrogates for DNA adducts, as metabolites of PAH carcinogens bind to both DNA and protein with similar dose-response kinetics (Poirier et al., 2000; Links and Groopman, 2010). Of all the proteins, hemoglobin and albumin have become the proteins of choice for molecular dosimetry studies. The characteristics that made protein adducts preferable over DNA adducts are that protein adducts are relatively more stable and are not removed by repair processes unlike DNA. Although protein adducts, when compared with DNA adducts, are considered as precise dosimeters (Links and Groopman, 2010), interpretation of results from protein adduct studies is limited by the techniques employed, which vary highly among laboratories (Castaño-Vinyals et al., 2004). Additionally, compared with DNA adducts, not many studies have measured protein adducts, and the data on PAH-protein adducts in humans are inconclusive owing to lack of sensibility in the assays used (Käfferlein et al., 2010).

Hemoglobin adduct concentrations of benzo(a)pyrene diol epoxide (BPDE) were reported to be high in newspaper vendors exposed to traffic exhaust (Pastorelli et al., 1996), compared with nonsmokers. On the other hand, BaP-albumin adducts did not show any significant difference between iron foundry workers and nonsmokers when serum samples were evaluated as an exposure measure (Omland et al., 1994). Similar results were reported from coke oven plant workers and nonoccupationally exposed individuals in an industrial area in Poland (Kure et al., 1997). The techniques employed for detection and quantitation of PAH-protein (albumin and/or globin) adducts for biomonitoring purposes are summarized in Table 29.2.

OTHER BIOMARKERS OF EXPOSURE

Aside from hydroxylated metabolites and PAH-DNA adducts, serum levels of liver enzymes such as alanine aminotransferase, aspartate aminotransferase, and superoxide dismutase, which are associated with liver function, have been used as indicators of PAH exposure in workers from occupational settings such as coke oven plants (Wu et al., 1997), brick kilns (Kamal et al., 2014), and petrochemical industry (Min et al., 2015). Additionally, in some instances markers of inflammation such as C-reactive protein and activated leukocyte cell adhesion molecule showed a positive correlation with hydroxylated metabolites of PAHs in coke oven workers (Yang et al., 2016). On the other hand, some studies do not provide conclusive evidence to establish a relationship between PAH exposure and some of these biomarkers (Clark et al., 2012; Alhamdow et al., 2017)). There are other caveats that limit the utilitarian value of the above-mentioned markers, which include the exposure dose, duration (time spent on assigned task), and other confounding factors such as smoking, diet, and BMI status. Therefore, the lack of robustness associated with these markers discourages their use for routine monitoring purposes.

BIOMARKERS OF EFFECT

Genetic Alterations

Some of the cytogenetic biomarkers that are currently being used as biomarkers of effect include chromosome

TABLE 29.2 Sample Matrices and Methods Employed for Biomonitoring of PAH-Protein Adducts in Susceptible Populations

Exposure Route	Biological Matrix	Population	Analytical Method	References
Dermal (coal tar)	White blood cells	Psoriasis patients	ELISA	Santella et al. (1995)
Inhalation (occupational)	Globin	Auto mechanics	Gas chromatography-mass spectrometry (GC-MS)	Nielsen et al. (1996)
Inhalation (occupational)	Serum albumin	Coke oven plant workers	ELISA	Kure et al. (1997)
Inhalation (occupational)	Serum albumin	Bus drivers, mail delivery workers	ELISA	Autrup et al. (1999)
Inhalation (occupational)	Globin	Auto mechanics	GC-MS	Melikian et al. (1999)
Inhalation (occupational)	Serum albumin	Traffic police officers	ELISA	Ruchirawat et al. (2002)
Inhalation (occupational)	Peripheral blood lymphocytes	Coke oven plant workers	HPLC	Wang et al. (2007)
Inhalation (occupational)	Peripheral blood lymphocytes	Coke oven plant workers	HPLC	Huang et al. (2012)
Diet and inhalation (smoking)	Globin and albumin	Healthy subjects	GC-MS	Scherer et al. (2000)

aberrations (CAs), SCEs, and micronuclei (MN). The CAs are characterized by structural alterations and rearrangements in chromosomes. Gu et al. (2008) observed BPDE (a carcinogenic metabolite of BaP)—induced chromosome 9p21 aberrations in bladder cancer cases. These studies suggest 9p21 as a marker for PAH-induced bladder cancers. The above-mentioned BaP metabolite has also been implicated in chromosome 3p deletions, which serves as a marker of an individual's susceptibility to renal cell carcinoma (Zhu et al., 2008).

MN are formed after mitosis from lagging chromatids. Cells bearing MN if enter mitosis produce daughter cells without MN. The formation of MN is linked to genetic damage. Increased frequencies of MN were found in coal-tar workers, who may have been exposed to PAHs in tar (Giri et al., 2012). In addition to occupational exposures, MN levels were monitored in mothers and newborns from a region in Czech Republic impacted by severe air pollution, whose BaP levels were found to be elevated in air samples (Rossnerova et al., 2011).

SCEs are exchanges of DNA segments between sister chromatids of a duplicated metaphase chromosome. Asphalt workers exposed to bitumen fumes registered a significant increase in SCEs (Karaman and Pirim, 2009). In addition, results of a metaanalysis study revealed statistically significant SCE frequencies in peripheral blood lymphocytes and peoples occupational exposure to PAHs (Wang et al., 2012a,b).

Glycophorin A (GPA) assay, which detects and quantifies erythrocytes with variant phenotypes at the autosomal locus, is a marker for population exposure to genotoxicants. A slight increase in GPA mutations was noticed in iron foundry workers exposed to PAHs (Perera et al., 1993). Additionally, a strong association between urinary PAH metabolites and GPA mutation frequencies was seen in incineration workers (Lee et al., 2002a,b).

The hypoxanthine-guanine phosphoribosyltransferase (HPRT) is used to denote transgene expression in T lymphocytes. Smoking (rich source of PAHs)—associated mutations at the HPRT locus was deleted in T lymphocytes of lung cancer patients (Hackman et al., 2000). Results of a metaanalysis study also revealed that children who are exposed to mothers' smoke registered a higher frequency of HPRT mutations as well (Neri et al., 2006).

In addition to chromosomal alterations, oncogenes and tumor suppressor genes have been used as markers of PAH exposure. Alguacil et al. (2003) used k-ras mutations to find out whether occupational exposure to PAHs had any effect on exocrine pancreatic cancer in factory workers. A weak association was noticed which cannot provide a definitive link between exposure and disease etiology.

It could be possible that some k-ras mutations are present at relatively high frequencies in some human tissues, but the methods applied may not have been able to detect those. Using an allele-specific competitive blocker polymerase chain reaction assay, Parsons et al. (2010) found k-ras mutations in normal human colon mucosa samples at high frequencies. A dose response for k-ras mutations caused by BaP has already been reported (Meng et al., 2010). Given the fact that dietary intake of PAHs contributes to colon cancers, oncomutations could be used as potential biomarkers for measuring PAH exposures in humans from occupational, environmental, and dietary settings.

The tumor suppressor gene p53 has also been used as a biomarker of PAH exposure. This gene regulates cell proliferation, differentiation, apoptosis, and DNA repair (Levine, 1997). Mordukhovich et al. (2010) reported an association between PAH exposure and p53 mutations subgroups in participants from the Long Island Breast Cancer Study Project. Other studies also have found positive association between PAH sources and p53 mutations in colon and lung cancers (Diergaarde et al., 2003; Harty et al., 1996).

Rossner et al. (2013b) studied the biomarkers of effect in Czech population exposed to PAHs through inhalation of polluted air in a highly contaminated region (Ostrava) and clean region (Prague). Samples collected from exposed humans were analyzed for oxidative stress markers (8-oxo-7,8-dihydro-2′-deoxyguanosine [8-oxodG], 15-F(2t)-isoprostane [15-F2t-ISOP]) and cytogenetic parameters (stable and unstable chromosomal aberrations). Lipid peroxidation measured by isoprostanes was elevated in subjects from Ostrava region compared with inhabitants of Prague. Other markers such as urinary excretion of 8-oxodG and unstable chromosomal aberrations were mostly comparable in both locations.

BIOMARKERS OF SUSCEPTIBILITY

Polymorphisms in PAH-metabolizing enzymes serve as markers of susceptibility. The PAH bioactivation enzyme, cytochrome P4501A1 (CYP1A1) gene polymorphism, was recorded in coal-tar workers (Giri et al., 2012). Gene polymorphisms in another CYP isozyme CYP1B1 and the antioxidant enzyme, catalase, were linked to BaP-DNA adduct levels in human lymphocytes (Schults et al., 2013). A strong association between PAH exposure during pregnancy and CYP1B1 polymorphisms leads to deficits in cognitive development of African American, Dominican, and Caucasian children (Wang et al., 2010). Also of interest was the finding that children living in the surrounding of a petrochemical factory in Mexico showed CYP1A1 and glutathione-S-transferase M1 (GSTM1) polymorphisms, increased DNA damage, and 1-OHP concentrations making them vulnerable to PAH exposure (Sanchez-Guerra et al., 2012).

Polymorphisms in CYP genes from 500 patients with colon cancer showed CYP1A2, CYP2E1, and microsomal epoxide hydrolase (mEH) (Kiss et al., 2007), indicating the susceptibility of individuals with CYP and mEH polymorphisms to colon cancer as these genes control PAH metabolism. Another study reported polymorphisms in CYP1A2, CYP2E1, CYP1B1, and CYP2C9 genes, which showed a strong association with red or processed meat consumption, a rich source of PAHs (Küry et al., 2007).

PAHs adsorbed onto particles in polluted air have been implicated in cardiovascular diseases and lung cancers in exposed populations. Binkova et al. (2007) found a strong association between PAH-DNA adducts levels and polymorphisms of DNA repair gene xeroderma pigmentosum group D (XPD) and the detoxification enzyme GSTM1 in population exposed to polluted air in Prague. This enzyme has also been suggested as a marker to monitor metabolic activation of PAHs in smokers (Wang et al., 2012b). Polymorphisms of another isozyme of GST family, the GSTP1, have been implicated in reduced fetal growth. These studies suggest the potential use of GSTP1 as a marker to detect risks arising from exposure to PAHs by mothers from susceptible populations (Duarte-Salles et al., 2012). Additionally, polymorphisms in genes coding for other detoxification enzymes such as GSTM1 and GSTT1 detected in smokers also showed a strong association with colorectal adenomas (Moore et al., 2005).

Another PAH detoxification enzyme, the UDP-glucuronosyltransferase (UGT) polymorphisms, has been associated with a reduced elimination of PAHs from the body. An increased risk of colon cancer was predicted in a population-based study whose subjects had UGT1A1 polymorphisms (Girard et al., 2008). Because intake of PAHs through diet has been involved in colon cancer causation (reviewed in Diggs et al., 2011), functional polymorphisms in other PAH detoxification genes hold lot of promise as markers not only for colon but also for other gastrointestinal tract cancers.

In addition to drug metabolizing enzymes, polymorphisms in aryl hydrocarbon receptor (AhR) gene is another biomarker, worthy of consideration. Gu et al. (2011) reported AhR polymorphisms in idiopathic infertile male subjects. These polymorphisms were found to be associated with reduced sperm concentration, increase in sperm DNA fragmentation, and BPDE-DNA adduct levels in spermatozoa.

Not only are polymorphisms in PAH activation and detoxification enzymes but also polymorphisms in DNA repair genes serve as useful biomarkers. In addition to DNA repair gene XPD (Binkova et al., 2007), polymorphisms exist in other DNA repair enzymes as well. An association among polymorphisms in DNA repair gene, the DNA methyltransferase 1 and 3B and DNA strand breaks were reported by Leng et al. (2008). These findings suggest that variants in haplotypes of these genes could be used as markers to predict cancer susceptibility. Polymorphisms in ataxia telangiectasia (ATM) lead to impairment in repair of chromosomal damage and cell cycle control in people exposed to toxicants. Wang et al. (2011) found that in workers occupationally exposed to PAHs, the ATM polymorphisms were associated with susceptibility to DNA repair capacity.

CONCLUDING REMARKS AND FUTURE DIRECTIONS

One of the caveats in using biomarkers is categorizing the exposure scenarios. Sometimes exposure misclassification cannot be avoided. This situation may lead to employing an incorrect biomarker that may not be able to capture the functional relationship between exposure and effect. In such situations, ex vivo studies are useful to control exposure settings so that gene–gene interactions could be studied well and help validating the biomarkers.

Thus far, several biomarkers that denote exposure, effect, and susceptibility have been proposed and studies involving these for most part yielded definitive information. At the same time, some studies were inconclusive because of the lack of correlation between biomarkers. In some studies, the PAH-DNA adducts concentrations failed to show either a correlation with urinary metabolites of PAHs or with DNA repair enzymes. To fully exploit the potential of biomarkers in PAH-induced disease diagnosis and intervention strategies, correlation among biomarkers of exposure and effect and effect and susceptibility needs to be investigated. This approach could be integrated into the exposome concept to address environmental and human health issues arising from PAH exposures.

Additionally, bioinformatics and systems biology resources could be exploited to advance the field further. Proteomic and genomic approaches could be used to identify additional biomarkers of PAH exposure, effect, or susceptibility. Data from these studies could facilitate the discovery and validation of new biomarkers for PAH-induced cardiovascular, neurological, reproductive, and developmental disorders and cancer.

Acknowledgments

The authors gratefully acknowledge the research funding from the National Institutes of Health through grant numbers 1RO1CA142845-04 (Ramesh), U54RR026140 (Archibong), 5G12RR03032 (Archibong and Ramesh), R56ES017448, U54NS041071 (Hood), S11ES014156 (Hood,

Ramesh and Archibong), 5T32HL007735-20 (Banks), 1F31ESO2407901 (Kelly Harris), Southern Regional Education Board, Atlanta (Kelly Harris), 5R25GM059994-13 (Kenneth Harris), and Title III grant from the US Department of Education (Mantey).

References

Alguacil, J., Porta, M., Kaapinen, T., et al., 2003. PANKRAS II Study Group. Occupational exposure to dyes, metals, polycyclic aromatic hydrocarbons and other agents and k-ras activation in humans exocrine pancreatic cancer. Int. J. Cancer 107, 635–641.

Alhamdow, A., Lindh, C., Albin, M., et al., 2017. Early markers of cardiovascular disease are associated with occupational exposure to polycyclic aromatic hydrocarbons. Sci. Rep. 7, 9426.

Al-Saleh, I., Arif, J., El-Doush, I., et al., 2008. Carcinogen DNA adducts and the risk of colon cancer: case-control study. Biomarkers 13, 201–216.

Al-Saleh, I., El-Doush, I., Arif, J., et al., 2010. Levels of DNA adducts in the blood and follicular fluid of women undergoing *in vitro* fertilization treatment and its correlation with the pregnancy outcome. Bull. Environ. Contam. Toxicol. 84, 23–28.

Autrup, H., Daneshavar, B., Dragsted, L.O., et al., 1999. Biomarkers of exposure to ambient air pollution-comparison of carcinogen-DNA adduct levels with other exposure markers and markers for oxidative stress. Environ. Health Perspect. 107, 233–238.

Besaratinia, A., Van Straaten, H.W., Godschalk, R.W., et al., 2000. Immunoperoxidase detection of polycyclic aromatic hydrocarbon-DNA adducts in mouth floor and buccal mucosa cells of smokers and nonsmokers. Environ. Mol. Mutagen. 36, 127–133.

Binkova, B., Chvatalova, I., Lnenickova, Z., et al., 2007. PAH-DNA adducts in environmentally exposed population in relation to metabolic and DNA repair gene polymorphisms. Mutat. Res. 620, 49–61.

Brescia, G., Celotti, L., Clonfero, E., et al., 1999. The influence of cytochrome P450 1A1 and glutathione S-transferase M1 genotypes on biomarker levels in coke-oven workers. Arch. Toxicol. 73, 431–439.

Castaño-Vinyals, G., D'Errico, A., Malats, N., et al., 2004. Biomarkers of exposure to polycyclic aromatic hydrocarbons from environmental air pollution. Occup. Environ. Med. 61, e12.

Chen, C.S., Yuan, T.H., Shie, R.H., et al., 2017. Linking sources to early effects by profiling urine metabolome of residents living near oil refineries and coal-fired power plants. Environ. Int. 102, 87–96.

Ciarrocca, M., Rosati, M.V., Tomei, F., et al., 2014. Is urinary 1-hydroxypyrene a valid biomarker for exposure to air pollution in outdoor workers? A meta-analysis. J. Expo. Sci. Environ. Epidemiol. 24, 17–26.

Clark 3rd, J.D., Serdar, B., Lee, D.J., et al., 2012. Exposure to polycyclic aromatic hydrocarbons and serum inflammatory markers of cardiovascular disease. Environ. Res. 117, 132–137.

Diergaarde, B., Vricling, A., Van Kraats, A.A., et al., 2003. Cigarette smoking and genetic alterations in sporadic colon carcinomas. Carcinogenesis 24, 565–571.

Diggs, D.L., Huderson, A.C., Harris, K.L., et al., 2011. Polycyclic aromatic hydrocarbons and digestive tract cancers: a perspective. J. Environ. Sci. Health C Environ. Carcinog. Ecotoxicol. Rev. 29, 324–357.

Duarte-Salles, T., Mendez, M.A., Morales, E., et al., 2012. Dietary benzo(a)pyrene and fetal growth: effect modification by vitamin C intake and glutathione S-transferase P1 polymorphism. Environ. Int. 45, 1–8.

Farmer, P.B., Singh, R., Kaur, B., et al., 2003. Molecular epidemiology studies of carcinogenic environmental pollutants. Effects of polycyclic aromatic hydrocarbons (PAHs) in environmental pollution on exogenous and oxidative DNA damage. Mutat. Res. 544, 397–402.

Funck-Brentano, C., Raphaël, M., Lafontaine, M., et al., 2006. Effects of type of smoking (pipe, cigars or cigarettes) on biological indices of tobacco exposure and toxicity. Lung Cancer 54, 11–18.

Galaviz, V.E., Quintana, P.J., Yost, M.G., et al., 2017. Urinary metabolites of 1-nitropyrene in US-Mexico border residents who frequently cross the San Ysidro Port of Entry. J. Expo. Sci. Environ. Epidemiol. 27, 84–89.

Girard, H., Butler, L.M., Villeneuve, L., et al., 2008. UGT1A1 and UGT1A9 functional variants, meat intake, and colon cancer, among Caucasians and African-Americans. Mutat. Res. 644, 56–63.

Giri, S.K., Yadav, A., Kumar, A., et al., 2012. CYP1A1 gene polymorphisms: modulator of genetic damage in coal-tar workers. Asian Pac. J. Cancer Prev. APJCP 13, 3409–3416.

Gu, J., Horikawa, Y., Chen, M., et al., 2008. Benzo(a)pyrene diol epoxide-induced chromosome 9p21 aberrations are associated with increased risk of bladder cancer. Cancer Epidemiol. Biomarkers Prev. 17, 2445–2450.

Gu, A., Ji, G., Long, Y., et al., 2011. Assessment of an association between an aryl hydrocarbon receptor gene (AHR) polymorphism and risk of male infertility. Toxicol. Sci. 122, 415–421.

Gunter, M.J., Divi, R.L., Kulldorff, M., et al., 2007. Leukocyte polycyclic aromatic hydrocarbon-DNA adducts formation and colorectal adenoma. Carcinogenesis 28, 1426–1429.

Hackman, P., Hou, S.M., Nyberg, F., et al., 2000. Mutational spectra at the hypoxanthine-guanine phosphoribosyltransferase (HPRT) locus in T-lymphocytes of nonsmoking and smoking lung cancer patients. Mutat. Res. 468, 45–61.

Harty, L.C., Guinee Jr., D.G., Travis, W.D., et al., 1996. p53 mutations and occupational exposures in a surgical surgical series of lung cancers. Cancer Epidemiol. Biomarkers Prev. 5, 997–1003.

Huang, G., Guo, H., Wu, T., 2012. Genetic variations of CYP2B6 gene were associated with plasma BPDE-Alb adducts and DNA damage levels in coke oven workers. Toxicol. Lett. 211, 232–238.

Hulka, B.S., 1990. Overview of biological markers. In: Hulka, B.S., Griffith, J.D., Wilcosky, T.C. (Eds.), Biological Markers in Epidemiology. Oxford University Press, New York, pp. 3–15.

IARC, 2010. Some non-heterocyclic polycyclic aromatic hydrocarbons and some related exposures. IARC Monographs on the Evaluation of Carcinogenic Risks to Humans 92, 861. Lyon: France.

Ji, G., Gu, A., Zhou, Y., et al., 2010. Interactions between exposure to environmental polycyclic aromatic hydrocarbons and DNA repair gene polymorphisms on bulky DNA adducts in human sperm. PLoS One 5.

Käfferlein, H.U., Marczynski, B., Mensing, T., et al., 2010. Albumin and hemoglobin adducts of benzo[a]pyrene in humans- analytical methods, exposure assessment, and recommendations for future directions. Crit. Rev. Toxicol. 40, 126–150.

Kamal, A., Malik, R.N., Martellini, T., et al., 2014. PAH exposure biomarkers are associated with clinico-chemical changes in the brick kiln workers in Pakistan. Sci. Total Environ. 490, 521–527.

Kang, D.H., Rothman, N., Poirier, M.C., et al., 1995. Interindividual differences in the concentration of 1-hydroxypyrene-glucuronide in urine and polycyclic aromatic hydrocarbon-DNA adducts in peripheral white blood cells after charbroiled beef consumption. Carcinogenesis 16, 1079–1085.

Karaman, A., Pirim, I., 2009. Exposure to bitumen fumes and genotoxic effects on Turkish asphalt workers. Clin. Toxicol. 47, 321–326.

Ke, Y., Huang, L., Xia, J., et al., 2016. Comparative study of oxidative stress biomarkers in urine of cooks exposed to three types of cooking-related particles. Toxicol. Lett. 255, 36–42.

Kiss, I., Orsós, Z., Gombos, K., et al., 2007. Association between allelic polymorphisms of metabolizing enzymes (CYP1A1, CYP1A2, CYP2E1, mEH) and occurrence of colorectal cancer in Hungary. Anticancer Res. 27, 2931–2937.

Kure, E.H., Andreassen, A., Ovrebø, S., et al., 1997. Benzo(a)pyrene-albumin adducts in humans exposed to polycyclic aromatic hydrocarbons in an industrial area of Poland. Occup. Environ. Med. 54, 662–666.

Küry, S., Buecher, B., Robiou-du-Pont, S., et al., 2007. Combinations of cytochrome P450 gene polymorphisms enhancing the risk for sporadic colorectal cancer related to red meat consumption. Cancer Epidemiol. Biomarkers Prev. 16, 1460–1467.

Lee, K.H., Lee, J., Ha, M., et al., 2002a. Influence of polymorphism of GSTM1 gene on association between glycophorin a mutant frequency and urinary PAH metabolites in incineration workers. J. Toxicol. Environ. Health 65, 355–363.

Lee, J., Kang, D., Lee, K.H., et al., 2002b. Influence of GSTM1 genotype on association between aromatic DNA adducts and urinary PAH metabolites in incineration workers. Mutat. Res. 514, 213–221.

Leng, S., Stidley, C.A., Bernauer, A.M., et al., 2008. Haplotypes of DNMT1 and DNMT3B are associated with mutagen sensitivity induced by benzo[a]pyrene diol epoxide among smokers. Carcinogenesis 29, 1380–1385.

Levine, A.J., 1997. p53, the cellular gate keeper for growth and cell division. Cell 88, 323–331.

Li, Z., Trinidad, D., Pittman, E.N., et al., 2016. Urinary polycyclic aromatic hydrocarbon metabolites as biomarkers to woodsmoke exposure - results from a controlled exposure study. J. Expo. Sci. Environ. Epidemiol. 26, 241–248.

Links, J.M., Kensler, T.W., Groopman, J.D., 1995. Biomarkers and mechanistic approaches in environmental epidemiology. Annu. Rev. Publ. Health 16, 83–103.

Links, J.M., Groopman, J.D., 2010. Biomarkers of exposure, effect, and susceptibility. In: McQueen, C.A., Bond, J., Ramos, K., Lamb, J., Guengerich, F.P., Lawrence, D., Walker, M., Campen, M., Schnellmann, R., Yost, G.S., Roth, R.A., Ganey, P., Hooser, S., Richburg, J., Hoyer, P., Knudsen, T., Daston, G., Philbert, M., Roberts, R. (Eds.), Comprehensive Toxicology. Elsevier, Amsterdam.

Lutier, S., Maître, A., Bonneterre, V., et al., 2016. Urinary elimination kinetics of 3-hydroxybenzo(a)pyrene and 1-hydroxypyrene of workers in a prebake aluminum electrode production plant: evaluation of diuresis correction methods for routine biological monitoring. Environ. Res. 147, 469–479.

Melikian, A.A., Malpure, S., John, A., et al., 1999. Determination of hemoglobin and serum albumin adducts of benzo[a]pyrene by gas chromatography-mass spectrometry in humans and their relations to exposure and to other biological markers. Polycycl. Aromat. Comp. 17, 125–134.

Meng, F., Knapp, G.W., Green, T., et al., 2010. K-ras mutant fraction in A/J mouse lung increases as a function of benzo(a)pyrene dose. Environ. Mol. Mutagen. 51, 146–155.

Min, Y.S., Lim, H.S., Kim, H., 2015. Biomarkers for polycyclic aromatic hydrocarbons and serum liver enzymes. Am. J. Ind. Med. 58, 764–772.

Moore, L.E., Huang, W.Y., Chatterjee, N., et al., 2005. GSTM1, GSTT1, and GSTP1 polymorphisms and risk of advanced colorectal adenoma. Cancer Epidemiol. Biomarkers Prev. 14, 1823–1827.

Mordukhovich, I., Rossner Jr., P., Terry, M.B., et al., 2010. Associations between polycyclic aromatic hydrocarbons-related exposures and p53 mutations in breast tumors. Environ. Health Perspect. 118, 511–518.

Naylor, S., 2003. Biomarkers: current perspectives and future prospects. Expert Rev. Mol. Diagn 3, 525–529.

Neri, M., Ugolini, D., Bonassi, S., et al., 2006. Children's exposure to environmental pollutants and biomarkers of genetic damage. II. Results of a comprehensive literature search and meta-analysis. Mutat. Res. 612, 14–39.

Nielsen, P.S., Andreassen, A., Farmer, P.B., et al., 1996. Biomonitoring of diesel exhaust-exposed workers. DNA and hemoglobin adducts and urinary 1-hydroxypyrene as markers of exposure. Toxicol. Lett. 86, 27–37.

Obolenskaya, M.Y., Teplyuk, N.M., Divi, R.L., et al., 2010. Human placental glutathione S-transferase activity and polycyclic aromatic hydrocarbon DNA adducts as biomarkers for environmental oxidative stress in placentas from pregnant women living in radioactivity- and chemically-polluted regions. Toxicol. Lett. 196, 80–86.

Omland, O., Sherson, D., Hansen, A.M., et al., 1994. Exposure of iron foundry workers to polycyclic aromatic hydrocarbons: benzo(a) pyrene- albumin adducts and 1-hydroxypyrene as biomarkers for exposure. Occup. Environ. Med. 51, 513–518.

Parsons, B.L., Myers, M.B., Meng, F., et al., 2010. Oncomutations as biomarkers of cancer risk. Environ. Mol. Mutagen. 51, 836–850.

Pastorelli, R., Resano, J., Guanci, M., et al., 1996. Hemoglobin adducts of benzo(a)pyrene diol epoxide in newspaper vendors: association with traffic exhaust. Carcinogenesis 17, 2389–2394.

Pavanello, S., Favretto, D., Brugnone, F., et al., 1999. HPLC/fluorescence determination of anti-BPDE-DNA adducts in mononuclear white blood cells from PAH-exposed humans. Carcinogenesis 20, 431–435.

Perera, F.P., 2000. Molecular epidemiology: on the path to prevention? J. Natl. Cancer Inst. 92, 602–612.

Perera, F.P., 2011. Molecular epidemiology, prenatal exposure and prevention of cancer. Environ. Health 10 (Suppl. 1), S5.

Perera, F., Santella, R.M., Brenner, D., et al., 1987. DNA adducts, protein adducts, and sister chromatid exchange in cigarette smokers and nonsmokers. J. Natl. Cancer Inst. 79, 449–456.

Perera, F.P., Tang, D.L., O'Neill, J.P., et al., 1993. HPRT and glycophorin A mutations in foundry workers: relationship to PAH exposure and to PAH-DNA adducts. Carcinogenesis 14, 969–973.

Perera, F.P., Tang, D., Rauh, V., 2007. Relationship between polycyclic aromatic hydrocarbon-DNA adducts, environmental tobacco smoke, and child development in the World Trade Center cohort. Environ. Health Perspect. 115, 1497–1502.

Peters, K.O., Williams, A.L., Abubaker, S., et al., 2017. Predictors of polycyclic aromatic hydrocarbon exposure and internal dose in inner city Baltimore children. J. Expo. Sci. Environ. Epidemiol. 27, 290–298.

Poirier, M.C., Santella, R.M., Weston, A., 2000. Carcinogen macromolecular adducts and their measurement. Carcinogenesis 21, 353–359.

Pratt, M.M., Sirajuddin, P., Poirier, M.C., et al., 2007. Polycyclic aromatic hydrocarbon-DNA adducts in cervix of women infected with carcinogenic human papillomavirus types: an immunohistochemistry study. Mutat. Res. 624, 114–123.

Ramesh, A., Walker, S.A., Hood, D.B., et al., 2004. Bioavailability and risk assessment of orally ingested polycyclic aromatic hydrocarbons. Int. J. Toxicol. 23, 301–333.

Ramesh, A., Archibong, A., Hood, D.B., et al., 2011. Global environmental distribution and human health effects of polycyclic aromatic hydrocarbons. In: Loganathan, B.G., Lam, P.K.-S. (Eds.), Global Contamination Trends of Persistent Organic Chemicals. Taylor & Francis Publishers, Boca Raton, Florida, pp. 95–124.

Ramesh, A., Archibong, A.E., Huderson, A.C., et al., 2012. Polycyclic aromatic hydrocarbons. In: Gupta, R.C. (Ed.), Veterinary Toxicology: Basic and Clinical Principles. Academic Press/Elsevier, Amsterdam, pp. 797–809.

Rossner Jr., P., Svecova, V., Schmuczerova, J., et al., 2013a. Analysis of biomarkers in a Czech population exposed to heavy air pollution. Part I: bulky DNA adducts. Mutagenesis 28, 89–95.

Rossner Jr., P., Rossnerova, A., Spatova, M., et al., 2013b. Analysis of biomarkers in a Czech population exposed to heavy air pollution. Part II: chromosomal aberrations and oxidative stress. Mutagenesis 28, 97–106.

Rossnerova, A., Spatova, M., Pastorkova, A., et al., 2011. Micronuclei levels in mothers and their newborns from regions with different types of air pollution. Mutat. Res. 715, 72–78.

Rothman, N., Correa-Villaseñor, A., Ford, D.P., et al., 1993. Contribution of occupation and diet to white blood cell polycyclic aromatic hydrocarbon-DNA adducts in wildland firefighters. Cancer Epidemiol. Biomarkers Prev. 2, 341–347.

Ruchirawat, M., Mahidol, C., Tangjarukij, C., et al., 2002. Exposure to genotoxins present in ambient air in Bangkok, Thailand—particle associated polycyclic aromatic hydrocarbons and biomarkers. Sci. Total Environ. 287, 121–132.

Sagiv, S.K., Gaudet, M.M., Eng, S.M., 2009. Polycyclic aromatic hydrocarbon-DNA adducts and survival among women with breast cancer. Environ. Res. 109, 287–291.

Sánchez-Guerra, M., Pelallo-Martínez, N., Díaz-Barriga, F., et al., 2012. Environmental polycyclic aromatic hydrocarbon (PAH) exposure and DNA damage in Mexican children. Mutat. Res. 742, 66–71.

Santella, R.M., Hemminki, K., Tang, D.L., et al., 1993. Polycyclic aromatic hydrocarbon-DNA adducts in white blood cells and urinary 1- hydroxypyrene in foundry workers. Cancer Epidemiol. Biomarkers Prev. 2, 59–62.

Santella, R.M., Perera, F.P., Young, T.L., et al., 1995. Polycyclic aromatic hydrocarbon-DNA and protein adducts in coal tar treated patients and controls and their relationship to glutathione S-transferase genotype. Mutat. Res. 334, 117–124.

Scherer, G., Frank, S., Riedel, K., et al., 2000. Biomonitoring of exposure to polycyclic aromatic hydrocarbons of nonoccupationally exposed persons. Cancer Epidemiol. Biomarkers Prev. 9, 373–380.

Schoket, B., Poirier, M.C., Mayer, G., et al., 1999. Biomonitoring of human genotoxicity induced by complex occupational exposures. Mutat. Res. 445, 193–203.

Schults, M.A., Chiu, R.K., Nagle, P.W., et al., 2013. Genetic polymorphisms in catalase and CYP1B1 determine DNA adduct formation by benzo(a)pyrene ex vivo. Mutagenesis 28, 181–185.

Shen, G., Tao, S., Wei, S., et al., 2012. Emissions of parent, nitro, and oxygenated polycyclic aromatic hydrocarbons from residential wood combustion in rural China. Environ. Sci. Technol. 46, 8123–8130.

Suk, W.A., Wilson, S.H., 2002. Overview and future of molecular biomarkers of exposure and early disease in environmental health. In: Wilson, S.H., Suk, W.A. (Eds.), Biomarkers of Environmentally Associated Disease. Lewis Publishers, Boca Raton, pp. 3–15.

Taioli, E., Sram, R.J., Garte, S., et al., 2007. Effects of polycyclic aromatic hydrocarbons (PAHs) in environmental pollution on exogenous and oxidative DNA damage (EXPAH project): description of the population under study. Mutat. Res. 620, 1–6.

Tang, D., Santella, R.M., Blackwood, A.M., et al., 1995. A molecular epidemiological case-control study of lung cancer. Cancer Epidemiol. Biomarkers Prev. 4, 341–346.

Tang, D., Kryvenko, O.N., Wang, Y., et al., 2013. Elevated polycyclic aromatic hydrocarbon-DNA adducts in benign prostate and risk of prostate cancer in African Americans. Carcinogenesis 34, 113–120.

van Schooten, F.J., Hillebrand, M.J., van Leeuwen, F.E., et al., 1990. Polycyclic aromatic hydrocarbon-DNA adducts in lung tissue from lung cancer patients. Carcinogenesis 11, 1677–1681.

van Schooten, F.J., Hillebrand, M.J., van Leeuwen, F.E., et al., 1992. Polycyclic aromatic hydrocarbon-DNA adducts in white blood cells from lung cancer patients: no correlation with adduct levels in lung. Carcinogenesis 13, 987–993.

Wang, H., Chen, W., Zheng, H., et al., 2007. Association between plasma BPDE-Alb adduct concentrations and DNA damage of peripheral blood lymphocytes among coke oven workers. Occup. Environ. Med. 64, 753–758.

Wang, S., Chanock, S., Tang, D., et al., 2010. Effect of gene-environment interactions on mental development in African American, Dominican, and Caucasian mothers and newborns. Ann. Hum. Genet. 74, 46–56.

Wang, Y., Cheng, J., Li, D., et al., 2011. Modulation of DNA repair capacity by ataxia telangiectasia mutated gene polymorphisms among polycyclic aromatic hydrocarbons-exposed workers. Toxicol. Sci. 124, 99–108.

Wang, Y., Yang, H., Li, L., et al., 2012a. Biomarkers of chromosomal damage in peripheral blood lymphocytes induced by polycyclic aromatic hydrocarbons: a meta-analysis. Int. Arch. Occup. Environ. Health 85, 13–25.

Wang, J., Zhong, Y., Carmella, S.G., et al., 2012b. Phenanthrene metabolism in smokers: use of a two-step diagnostic plot approach to identify subjects with extensive metabolic activation. J. Pharmacol. Exp. Therapeut. 342, 750–760.

Wu, M.T., Kelsey, K.T., Mao, I.F., et al., 1997. Elevated serum liver enzymes in coke oven and by-product workers. J. Occup. Environ. Med. 39, 527–533.

Yang, B., Deng, Q., Zhang, W., et al., 2016. Exposure to polycyclic aromatic hydrocarbons, plasma cytokines, and heart rate variability. Sci. Rep. 6, 19272.

Zhu, Y., Horikawa, Y., Yang, H., et al., 2008. BPDE induced lymphocytic chromosome 3p deletions may predict renal cell carcinoma risk. J. Urol. 179, 2416–2421.

CHAPTER

30

Metals

Swaran J.S. Flora, Abha Sharma

National Institute of Pharmaceutical Education and Research, Raebareli, U.P., India

INTRODUCTION

The term biomarker is used to measure the interaction between a biological system and an environmental agent, which may be chemical, physical, or biological. They form the essential end points of cascade of events involved in metal exposure and progression of related disease, analysis of which may contribute significantly to the successful risk management against heavy metal toxicities. This chapter provides some recent updates of biomarkers and their role in determining exposure to some of the specific metals. It will also provide a comprehensive account of molecular- and cellular-based biomarkers of toxicity.

CLASSIFICATION OF BIOMARKERS

Broadly biomarkers could be categorized into three categories:

1. ***Biomarkers of exposure***. These are directly related to the measurement of dose and function of a metal exposure. They can be further subdivided into the following:
   - *Molecular lesions* include formation of DNA and protein adducts, glycosylated hemoglobin, and amino acid conjugates through induction of electrophilic chemicals, chromosomal aberrations, and micronuclei following exposure to metals.
   - *Endogenously produced biomolecules* include production of various chemicals and enzymes in response to metal intoxication.
   - *Cellular/tissue changes* include alterations in sperm motility, sperm count, macrophage activity, red blood cell counts, and lymphocyte ratios.
   - *Unchanged exogenous agents* include heavy metals.
2. ***Biomarkers of effects***. These biomarkers are also known as "response biomarkers" include measurements of biochemical, physiological, or behavioral alterations as end results of metal exposure.
3. ***Biomarkers of susceptibility***. These biomarkers are strong indicators of the natural characteristics of an organism, which possibly contribute toward making them more susceptible to the effects of an exposure to a particular toxicant. These biomarkers can be genetic such as chromosomal aberrations, polymorphisms, etc.

SELECTION OF AN IDEAL BIOMARKER: BENEFITS AND DRAWBACKS

Choice or selection of biomarkers needs to be appropriate to attain accurate results. An ideal biomarker should possess certain specified characteristics such as:

1. Sample collection and analysis should be reliable and easy.
2. Use of biomarker is ethically proven.
3. Biomarker should be disease and organ specific and relevant intervention can be considered.
4. It should reflect a subclinical and reversible change.

BIOMONITORING OF EXPOSURE TO HEAVY METALS

Heavy metals are known to cause oxidative deterioration of biomolecules by initiating free radical mediated chain reaction resulting in lipid peroxidation, protein oxidation, and oxidation of nucleic acids such as DNA and RNA. Exposures to metals must be recognized promptly, and affected individuals should be evaluated and managed without delay.

Biomarkers in Toxicology, Second Edition
https://doi.org/10.1016/B978-0-12-814655-2.00030-X

LEAD

Inorganic Pb is considered to be one of the oldest occupational toxicants and is widely used in industries and life because of its favorable properties such as malleability, resistance to corrosion, and low melting point. Paints are the primary source of Pb exposure and the major source of Pb toxicity. Formation of reactive oxygen species (ROS) and changes in glutathione (GSH) as well as other antioxidant enzyme activities contribute equally to oxidative stress in lead intoxication.

Biomarkers of Exposure

1. ***Blood lead levels:*** Blood Pb (BPb) level (mainly erythrocyte lead) is the one of the primary and most verifiable biomarker used for the assessment of Pb exposure and toxicity. Up to 50% of inhaled Pb is transferred to bloodstream in adult humans, measurement of which reflects both recent and past exposures. Mobilization of Pb from bone into the blood contributes significantly (45%–55%) in enhancing the blood Pb levels, even in persons without excessive exposure to Pb. Evaluation of blood zinc protoporphyrin (ZPP) level can be used for determining Pb exposure (Yang et al., 2015).
2. ***Plasma lead levels:*** Plasma Pb levels (PPb) have also been associated with toxic effects of Pb as plasma fraction is rapidly exchangeable in the blood. Easy exchange of lead from bone to plasma occurs via exchangeable pool, whereas lead can move from nonexchangeable pool to the surface only when bone is actively being resorbed. Pb exposure affects plasma viscosity and also disturbs the agreeability and deformability of erythrocyte (Kasperczyk et al., 2015).
3. ***Urine lead levels:*** Spot urine specimen is subjected to large biological variations rendering it unreliable along with precipitation of urate salts in urine, which serves as another complicating factor in the analysis. Inorganic Pb remains unmetabolized and is excreted unchanged, primarily in the urine. Urine analysis is a noninvasive method and is favored for long-term biomonitoring and is acceptable in Pb workers as biomarker (Sommar et al., 2014).
4. ***Lead levels in hair and nails:*** Hair as biomarker of exposure has several advantages over other substrates, such as easy and noninvasive collection, minimal cost, inert, and easy storage and transportation for analysis. However, the ability to distinguish between endogenous and exogenous Pb is a major problem, and various geographical and ecological factors affect Pb distribution in hair. Techniques such as atomic absorption spectrometry can be used for the quantitative determination of Pb in fingernail, which indicates its environmental exposure. Another study showed that nail is not a reliable biomarker of exposure for women using surma (Ikegami et al., 2016). Toenails are preferred because they are comparatively less affected by various environmental contaminants, reliable, and noninvasive (Bakri et al., 2017).
5. ***Bone and teeth lead levels:*** Studies suggest the cumulative measurement of bone Pb levels as the most important determinant of some forms of toxicity. There is a strong association between bone Pb levels and blood Pb levels of adults exposed to lead (Behinaein et al., 2017). However, an appropriate selection of the bone type for analysis of Pb levels is necessary as different types of bones have different characteristics that potentially contribute toward their efficacy to absorb Pb. Though analysis of bone Pb levels has been recommended as one of the effective biomarker, still further research is warranted. X-ray fluoroscopy is considered among the most consistent and sensitive tools for assessing this. Compared to bones, teeth have been shown to be a superior indicator of Pb exposure as they also provide information about prenatal exposure and hence could be valuable in the understanding of various embryonic anomalies. The estimation of Pb in dental surface enamel by graphite furnace atomic absorption spectrometry may be used as a suitable biomarker of environmental exposure to Pb (De Oliveira et al., 2017).

Biomarkers of Effects

General signs and symptoms of Pb poisoning may include abdominal pain, anorexia, nausea and constipation, headache, joint and muscle pain, difficulties with concentration and memory, sleep disturbances, anemia with basophilic stippling, peripheral neuropathy and nephropathy. Some of the major biomarkers of effect or response biomarkers associated with Pb intoxication are discussed below:

1. ***Systemic manifestations:*** Lead poisoning has been associated with various hematological, gastrointestinal, and neurological dysfunctions, resulting in progression of diverse adverse effects and diseases.
   a. ***Hematological manifestations:*** Lead-induced hematological effects are mainly due to reduced life span of erythrocytes and inhibition of hemoglobin synthesis. These effects result in anemia, which may be mild, hyper, chronic, or microcytic. The adverse effects of lead appear even with blood concentration as low as 10 μg/dL. Basophilic stippling and premature erythrocyte hemolysis are the two common and promising biomarkers exhibiting hematological

manifestations on Pb exposure. Impairment of the activity of pyrimidine 5′-nucleotidase takes place, which increases the concentration of pyrimidine nucleotides in red blood cells, preventing maturation of erythroid elements and eventually leading to anemia presented in two case reports (Warang et al., 2017).

b. ***Gastrointestinal manifestations:*** Case reports indicate severe gastrointestinal dysfunction such as abdominal cramps, constipation, dilated stomach descending into the pelvis, small bowel distension, nausea, abdominal pain, belching, heart burn, loss of appetite hunger pain, liver damage as shown by liver biopsy (Verheij et al., 2009).

c. ***Neurological manifestations:*** Lead mimics calcium ions, which grants it the potential to cross blood–brain barrier and interferes with the normal development of a child's brain and nervous system. High levels of lead are associated with impaired cognitive function and various neuropsychiatric symptoms such as short-term memory and attention deficit disorders. Lead level in blood and urine in pediatric patients with attention deficit hyperactivity disorder was determined with the outcome that it is correlated inversely to neurological development in the early 7 years of life (Sánchez-Villegas et al., 2014).

d. ***Renal manifestations:*** Chronic exposure results in renal ganglioside alterations with urinary microalbumin excretion; however, these symptoms are variable and lack specificity. Chronic exposure is associated with tubule interstitial nephritis and progressive deterioration of renal function. The effect of low environmental exposure of Pb on kidney can be identified by monitoring the activity of lactate dehydrogenase. Proteinuria was also found in higher levels in Pb-exposed people (Cabral et al., 2012).

e. ***Cardiovascular manifestations:*** Association between Pb exposure and coronary heart disease, heart rate variability, and death from stroke has also been indicated; however, there are limited reports and scientific evidence to support this finding. High levels of serum asymmetric dimethylarginine, adipocyte fatty acid–binding protein, adiponectin, and chemerin have been reported recently in women living in Pb-contaminated area of Mexico (Ochoa-Martinez et al., 2017).

f. ***Reproductive manifestations:*** Lead affects both male and female reproductive systems and has been associated with decreased sperm count and motility in men and miscarriage, prematurity, low birth weight, and problems with development during childhood in women. *In utero* exposure of fetus to Pb might be due to the mobilization of Pb from mother's bones. Moreover, increased calcium intake may further mitigate this phenomenon (Pant et al., 2014).

g. ***Toxic manifestations on bones and teeth:*** Tooth decay, missed and filled teeth, dental abrasions, and significant increase in the prevalence of periodontal diseases such as gingivitis and periodontitis have been linked to Pb exposure. Use of combined laser ablation lead isotope and histological analysis method for the determination of Pb level in children's dentine and enamel of deciduous teeth furnish information about timing, duration, and source of Pb exposure during fetal development and early childhood (Shepherd et al., 2016). Presence of "Lead lines" along metaphyses of long bones and the margin of flat bones is also an important biomarker. The effect of Pb on bones of female rats suggests that it can damage bone biomechanics. Therefore, this finding would relate with the increases in the risk of bone fracture in elderly women exposed to Pb (Lee et al., 2016).

h. ***Respiratory manifestations:*** Inflammatory biomarkers in blood (WBC counts, PLA2 activity, and total protein levels) of Pb-exposed guinea pigs were examined and signified that inhalation exposure to lead acetate may lead to asthma-like disease (Farkhondeh et al., 2013).

i. ***DNA manifestations:*** Techniques such as comet, micronucleus, and chromosomal aberration tests are applied to identify the genetic damage occurred in heavy metal–exposed population. Elevated mean percent tail DNA, frequency of chromosomal aberration, and micronucleus in peripheral blood lymphocytes as well as in buccal epithelial cells were found in occupationally Pb-exposed worker (Chinde et al., 2014).

2. ***Biomarkers of oxidative damage***

a. ***ALAD-δ-aminolevulinic acid dehydratase:*** Several enzymes/enzymatic processes responsible for heme synthesis can be used as potential biomarkers for the lead toxicity, primarily being δ-aminolevulinic acid dehydratase (δ-ALAD), involved in catalyzing the condensation of two molecules of ALA to porphobilinogen. Blood Pb levels >20 μg/dL strongly inhibit the activity of ALAD, which makes it the most sensitive measurable biological index of lead toxicity. However, there are two limitations in the determination of ALAD activity: (1) wide range of normal activity, and (2) unstable nature of the activity during storage (La-Llave-León et al., 2017). 8-Hydroxy-2′-deoxyguanosine as a sensitive and noninvasive biomarker of Pb induced oxidative DNA damage (Pawlas et al., 2017).

b. ***ALA-U, ALA-B, ALA-P:*** ALA is synthesized in mitochondria from glycine and succinyl-CoA by ALA synthetase (ALAS). Inhibition of ALAD results in activation of ALAS, which further results in ALA accumulation in blood, plasma, and urine, thereby making ALA a critical biomarker of early biological effect of Pb also (Saxena and Flora, 2004). Effect of lead on bone marrow is more precisely depicted by increased ALA in plasma or blood than urine, although ALA-U has been a recommended biomarker of Pb exposure. Lead interrupts erythrocytes energy homeostasis, resulted with decreased glucose metabolism, which lead to cell dysfunction in Pb-exposed people (Feksa et al., 2012).

c. ***Zinc Protoporphyrin:*** ZPP concentrations are widely used as biomarkers for lead toxicity. Ferrochelatase is another crucial enzyme that catalyzes the insertion of iron into protoporphyrin IX. However, in the cases of Pb toxicity, this enzyme is inhibited and the pathway is interrupted. Also, if sufficient iron is not available, Zn is substituted for Fe, resulting in increased ZPP levels. This alteration is an important diagnostic feature coexisting with a limitation that these elevations do not appear in the blood until Pb levels reach 35 μg/dL. The three enzymes of heme biosynthesis, i.e., ALAD, coproporphyrin oxidase, and ferrochelatase, are affected by Pb exposure resulting in decreased level of erythropoietin, consequently reducing hemoglobin synthesis (Mazumdar et al., 2017).

d. ***Pyrimidine nucleotidase and nicotinamide adenine dinucleotide synthetase:*** Decreased activities of pyrimidine nucleotidase and nicotinamide adenine dinucleotide synthetase in blood also serve as useful biomarkers in humans for diagnosing lead intoxications (Kim et al., 2002).

Biomarkers of Susceptibility

1. ***Cytogenetic markers:*** Sister chromatic exchanges (SCEs), high-SCE frequency cells (HFCs), and DNA–protein cross-links (DPCs) have also been shown to be reliable markers for biomonitoring lead exposure. Genotoxicity of Pb was measured using SCE, erythrocyte ALAD, urinary delta-aminolevulinic acid (U-ALA), and BPb in workers occupationally exposed to Pb (Sanders et al., 2009).
2. ***Biomarkers of gene expression***
   a. ***ALAD, VDR, and HFE polymorphism:*** Several biomarkers at the molecular level also exist, which help in estimation of Pb in exposed individuals. ALAD G177C polymorphism yields two alleles, ALAD1 and ALAD2. Higher blood Pb levels are observed in individuals carrying two copies of ALAD2 allele; however, hemopoiesis and severe effects of intoxication in brain and bone are observed in ALAD 1–1 homozygotic individuals. The role of ALAD genotype polymorphism in Pb toxicity has been studied in Pb-exposed and -unexposed workers. They found higher level of B-Pb and U-Pb in ALAD2 carriers than in ALAD1 homozygotes. P-Pb values were found to be comparable in both ALAD genotypes, but ALAD1 homozygotes are more prone to kidney toxicity (Tian et al., 2013).
   b. ***Apoptotic gene expression:*** Mitochondrion plays an extremely crucial role in Pb-induced apoptosis and is dependent on mitochondrial permeability transition pore opening as studied in rat proximal tubular cells (Liu et al., 2016). Apoptotic signals triggered by Pb are stimulated by caspase cascade and extracellular signal–regulated kinase dephosphorylation pathway in hepatic stem cells of adult rat (Agarwal et al., 2009).

ARSENIC

Arsenic (As) is a highly toxic metalloid, which poses severe effects, and its use as a deadly poison has been known and reported for many years. Arsenic in its free form generates free radicals resulting in lipid peroxidation, depletion of antioxidant enzymes, and DNA damage, thereby establishing oxidative stress as the major mechanism of As-induced toxicity and carcinogenicity (Flora, 2011).

Biomarkers of Exposure

Assessment and quantification of As exposure usually involves the measurement of As levels in environmental media (water and soil) and/or the use of biomarkers of exposure (blood, urine, and nails). Analysis of total As in blood, hair, or nails, and total or speciated metabolites of As in urine is the most common biomarker of As exposure.

1. ***Blood arsenic levels:*** Blood As levels better indicate recent and relatively high-dose exposures. Arsenic is cleared from blood within a few hours after it is absorbed. Blood As levels, however, may not be very reliable and promising biomarker of exposure because of its rapid clearance, particularly in cases of intoxication with very low levels of inorganic As. Examination of porphyrin profile provides an indication of As exposure. Thus it can be a useful biomarker of As exposure (Hall et al., 2006).

2. *Urine arsenic levels:* Clouded urine is frequently seen in most severely As-intoxicated patients as absorbed As if primarily excreted through urine. Generally, urinary As is analyzed as total or speciated; however, during inorganic As exposure determination of both speciation and total urinary As contents plays a critical role. Although estimation of urinary As is a consistent biomarker of exposure, it poses few drawbacks, such as accurate sample collection time, volume of urine to be voided, etc. Arsenic can be determined by measuring it and its metabolite (monomethylated and dimethylated metabolites) concentrations in urine of exposed people as a biomarker of exposure (Laine et al., 2015).
3. *Arsenic levels in hairs and nails:* Hair and nails are also preferable samples to detect and quantify exposure to As, as absorbed arsenic accumulates in both hair and nails and its elevated levels are noted within few weeks after acute poisoning. This elevation may account to binding of As to sulfhydryl groups in keratin, and because hair and nails grow slowly, their analysis may give an indication of past As exposure. Arsenic content in both toenail and fingernail has been significantly elevated among individuals consuming contaminated well water and was reported to be significantly correlated with hair As content.

Biomarkers of Effects

General symptoms of As poisoning that develop gradually include headaches, confusion, severe diarrhea, drowsiness, blood in the urine, cramping muscles, hair loss, and stomach pain. Lungs, skin, kidney, and liver are the major target organs of arsenic toxicity, and coma leading to death is the ultimate consequence. Moreover, As poisoning has also been associated with increased risk of heart disease, cancer, stroke, chronic lower respiratory diseases, and diabetes. Such clinical manifestations form some of the important biomarkers of effect.

1. *Systemic manifestations*:
   a. *Skin manifestations:* Keratosis, hyperkeratosis, leukomelanosis, depigmentation, and melanosis are the characteristic skin lesions associated with As toxicity. They are typically exhibited by diffuse thickening of palms and soles, alone or in combination with nodules, and presence of numerous rounded hyperpigmented macules in the form of raindrops in the body. In severe cases, cracks and fissures may be seen in the soles. Leucomelanosis is another common skin lesion, consisting of hypopigmented macules with a spotty white appearance (Zhang et al., 2014).
   b. *Respiratory manifestations:* Scientific evidence from various studies/reports suggests the possible role of As poisoning in the occurrence of lung disease. Chronic respiratory disease in the form of chronic cough or chronic bronchitis because of prolonged drinking of As-contaminated water has been observed. In addition, reduced pulmonary function and features of restrictive lung disease and combined obstructive and restrictive lung disease were noted in individuals exposed to As (Steinmaus et al., 2016).
   c. *Gastrointestinal manifestations:* Chronic As toxicity has been reported to cause symptoms such as dyspepsia, abdominal pain, gastroenteritis, nausea, diarrhea, and anorexia indicating dysfunctioning of gastrointestinal system (Ahmad et al., 1997).
   d. *Hepatic manifestations:* Liver damage such as portal hypertension, liver fibrosis, hepatomegaly, and intrahepatic portal vein obstruction is associated with arsenic exposure. Liver cirrhosis was observed as one of the major outcome following medication with inorganic As compounds. Various studies show that As toxicity is predominantly associated with hepatomegaly, in which hepatic fibrosis is the predominant lesion. Dysregulated lipid metabolism, oxidative stress, altered cytokine profiles, and enhanced inflammation may be involved in arsenic-induced liver injury (Arteel, 2015).
   e. *Cardiovascular manifestations:* Exposure to As has been related to an increased incidence of cardiovascular diseases, especially Black Foot Disease, ischemic heart disease, and hypertension. Soluble thrombomodulin could be a specific and stable biomarker associated with chronic As exposure (Hasibuzzaman et al., 2017). Arsenic exposure was also found to be associated with increased blood pressure in pregnant women, which could lead to risk of cardiovascular disease (Farzan et al., 2015).
   f. *Neurological manifestations:* Peripheral neuropathy, paresthesia, increased incidence of cerebrovascular disease, sleep disturbances, weakness, cognitive and memory impairment, mood disorders, headache, and vertigo are some of the major outcomes of chronic arsenicosis as observed in adults and children exposed to As from air and water as evident by epidemiological studies (Tsuji et al., 2015).
   g. *Hematological manifestations and induction of diabetes mellitus:* Various hematological abnormalities have been reported in cases of acute and chronic As poisoning. A characteristic pattern of anemia and its accumulation in erythrocytes,

leucopoenia, neutrophil depletion, and thrombocytopenia were observed (Heck et al., 2008). Also, significantly increased prevalence of diabetes mellitus was reported due to drinking As-contaminated water.

h. ***Carcinogenesis and other effects:*** Skin cancers are one among the most prevalent cancers due to chronic As exposure and might arise in the hyperkeratotic areas or nonkeratotic areas of the trunk, extremities, or hand. Apart from skin carcinomas, urinary bladder, liver, and lung cancers are also reported due to chronic exposure to As. Increased level of serum vascular endothelial growth factor may cause angiogenesis in As-induced cancers (Rahman et al., 2015a,b).

2. ***Biomarkers of DNA damage:*** Excessive free radical generation has been linked to enhanced carcinogenesis/mutagenesis (Flora et al., 2008). Arsenic-generated free radicals have been shown to induce degradation of DNA resulting in malfunctioning of DNA repair enzymes.
   a. ***8-Hydroxyguanine, 8-hydroxyguanosine, and 8-hydroxy-2-deoxyguanosine formation:*** For free radical attack, guanine is the most sensitive base present in DNA, which gets converted to various markers such as 8-hydroxyguanine, 8-hydroxyguanosine, and 8-hydroxy-2-deoxyguanosine (8-OHdG) after oxidation. Among all other markers, 8-OHdG adduct is one of the most abundant base modifications excreted in urine, and it is regarded as a suitable oxidative DNA damage biomarker because of its easy collection. 8-Oxoadenine, the oxidized form of adenine, has also been detected in urine of As-exposed animals. All these markers are difficult to detect through existing analytical methods as they transform into stable end products. Still techniques such as single cell gel electrophoresis or comet assay has been extensively used for this purpose (Azqueta et al., 2015). Another recent technique employing immunospin trapping as in the ODD test detects As-induced DNA damage more efficiently. Maternal exposure to As leads to DNA damage that could be identified by measuring 8-oxo-7, 8-dihydro-2′-deoxyguanosine and N7-methylguanosine, which serve as biomarkers. The latter biomarker could be used for monitoring fetal health linked to As exposure (Chou et al., 2014).
3. ***Biomarkers of oxidative damage:*** At the biochemical level, increase in ROS levels and unsaturated reactive aldehydes, such as malondialdehyde (MDA), 4-Hydroxynonenal (HNE), and 2-propenal (acrolein) and isoprostanes are one among the first response to As-induced oxidative stress, levels of which can be measured as indirect indicators of oxidative injury to blood and soft tissues (Chouhan and Flora, 2010). Arsenic also alters antioxidative enzymes such as malondialdehyde, catalase, GSH, and glutathione peroxidase and elevation in hydroxyl radical (Guo et al., 2015).
   a. ***δ-Aminolevulinic acid dehydratase (ALAD):*** Arsenic has strong affinity for nucleophilic ligands, which results in direct inhibition of heme biosynthesis pathway through metal–mercaptide bond formation with the functional sulfhydryl groups. δ-Aminolevulinic acid dehydratase is found to be highly sensitive to the presence of As by virtue of its sulfhydryl moiety. Dose-dependent increase in urinary excretion of uroporphyrin and coproporphyrin was observed following As exposure. Heme oxygenase too has been proposed as a biomarker of response to environmental As exposure (Flora et al., 2007). The role of ALAD genotype in determining Pb toxicity in children has been studied, which suggests that ALAD1-2/2-2 genotype may have lower level of U-ALA in environmentally Pb-exposed children (Tasmin et al., 2015).
   b. ***Antioxidant enzymes:*** Antioxidant activity of enzymes such as SOD, CAT, glutathione-6-phosphate dehydrogenase, glutathione S-transferase (GST), GSH reductase, and GPx is directly or inversely proportional to As toxicity depending on the dose and time of exposure (Dobrakowski et al., 2016). After evaluating all pulmonary biomarker-based alterations, GST has also been proposed as a potential biomarker for As toxicity. However, the techniques introduced to measure the oxidative stress have not yet been fully exploited to monitor the status of exposed populations.

Biomarkers of Susceptibility

1. ***Cytogenetic markers:*** Sister Chromatic Exchange (SCE) is considered to be one of the early biomarker of As-induced genotoxicity. Along with SCE, presence of micronuclei (MN) in isolated bladder and buccal cells has been considered as possible target tissues for direct exposure to As from drinking water. Reports indicate an increase in MN in lymphocytes in individuals chronically exposed to As via drinking water or occupational exposure (Guzman et al., 2017).
2. ***Biomarkers of gene expression:*** Growing evidence indicates the association of As with alterations in the repair of oxidative damage, measurement of which holds great promise as a susceptibility biomarker

of oxidative DNA damage in populations exposed to As.

a. ***OGG1 polymorphism:*** At the gene level, chronic arsenicosis is linked with reduced expression of 8-oxoguanine DNA glycosylase (OGG1) gene encoding 8-oxoguanine DNA glycosylase, primary enzyme responsible for the removal of 8-oxoguanine from DNA (Fig. 30.1). Microbiomes present in human gut, skin, and respiratory tract have genes that are induced by As ppb level exposure and precede detoxification process. Therefore, microbiomes gene expression could be taken as biomarker of exposure (Rosen, 2015).
b. ***Polymorphism in genes involved in DNA repair pathways:*** A strong correlation was observed between the urinary levels of 8-oxo-dGuo and monomethylarsonic acid. The ability of As to interfere with various DNA-repairing enzymes can be attributed to the fact that As reacts with thiol groups of zinc binding structures present in Nucleotide excision repair (NER) enzymes, such as the Zn finger proteins XPA and XPD. It has been observed that chronic As exposure leads to *Ogg1* deficiency, which aggravates the normal cellular repair function process (Bach et al., 2015).
c. ***Genes involved in apoptosis:*** Generation of ROS, accumulation of calcium, deficiency of p53 and intracellular GSH, Akt/mTOR signaling pathway inhibition, and ROS/JNK activation are the events that collectively lead to apoptosis during As toxicity; and all these genotoxic and cytotoxic effects, both in vivo and in vitro, can be determined through single cell electrophoresis (comet assay) (Wang et al., 2017). Thus, various biomarkers exist for the rapid and efficient diagnosis of As toxicity, employing of which will definitely yield some positive results. However, in spite of enough scientific awareness, As toxicity continues at an alarming level, thereby proving to be a menace, which could be possibly reduced by increasing awareness, decreasing exposure to the toxicant, and better utilization of these biomarkers.

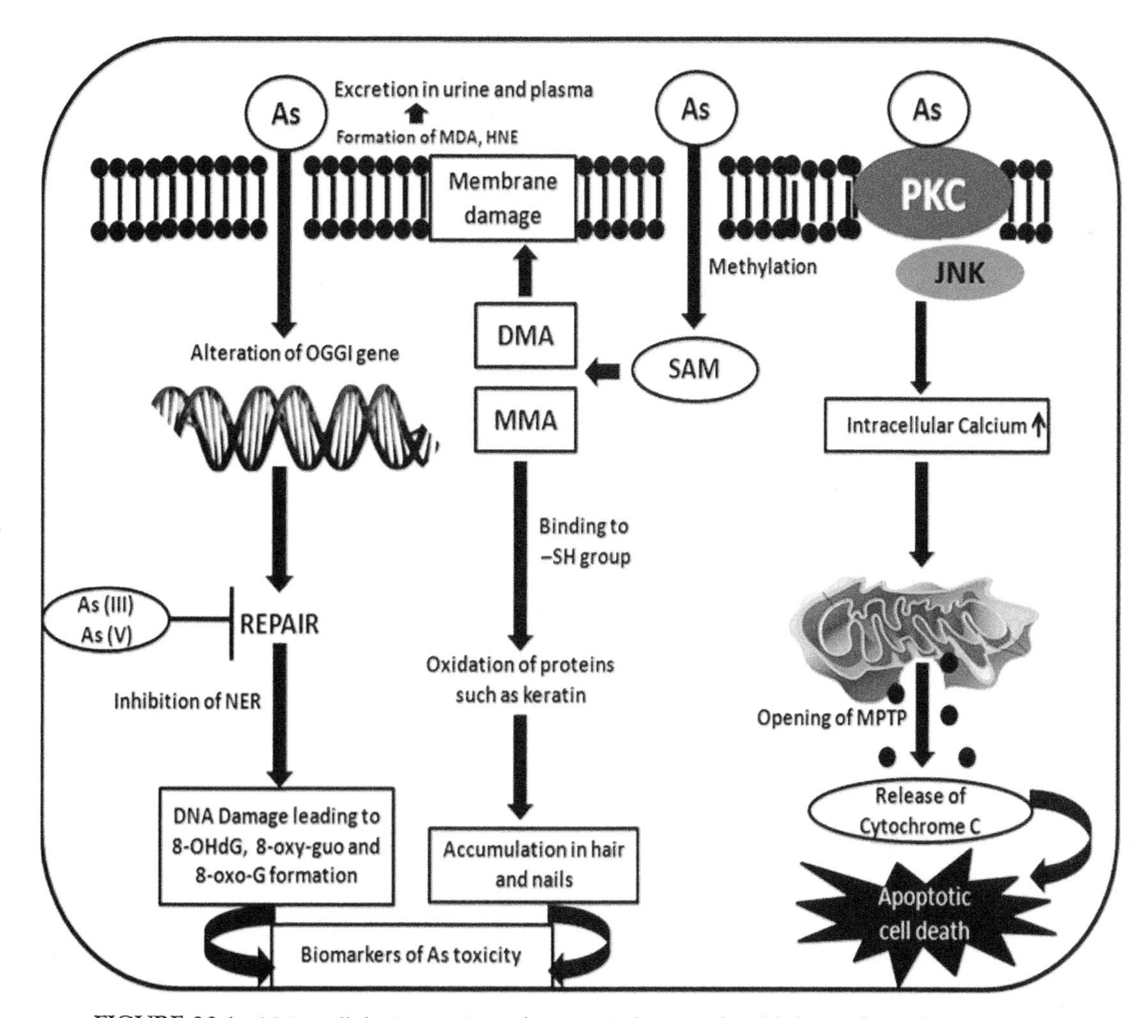

FIGURE 30.1 Major cellular interactions of arsenic, indicating plausible biomarkers of arsenic toxicity.

MERCURY

Mercury (Hg) is involved in causing serious public health disaster in Minamata Bay, Japan (Flora et al., 2008). Its toxic manifestations are also elicited in response to its binding with primary and secondary amine, amide, carboxyl, and phosphoryl groups.

Biomarkers of Exposure

Hg exposure could be complicated by the enhanced latency period of several weeks to months between the exposure and the development of clinical symptoms. Moreover, early symptoms are also hard to recognize. There exist several biomarkers of exposure for Hg poisoning, determining of which helps in early diagnosis and treatment. Analysis of Hg levels in blood, hair, and urine reflects recent exposure to the toxicant but do not correlate with total body burden. The presence of micronuclei has been detected in lymphocytes of exposed individuals and serves as an extremely important biomarker of exposure.

1. ***Blood mercury levels:*** Blood mercury level is determined using whole blood analysis and provides accurate information only about recent exposure, which renders it unsuitable for indicating prenatal exposure during early periods of pregnancy. Specifically alimental organic Hg and short-term exposure to mercury vapor are exhibited by the whole blood analysis, with organic Hg localized in erythrocytes. Thus, separate analysis of whole blood, erythrocytes, and plasma is recommended to analyze the presence of various species of Hg. Normally, the quotient of Hg content in erythrocytes and plasma is 2:1. A new method is developed and validated to measure MeHg exposure in newborns using dried blood spots (Basu et al., 2017).
2. ***Urine mercury levels:*** Analysis of urine samples for both mercury and creatinine determination also provides a much more valid index of recent exposure than the blood Hg level. A new method is developed for the determination of urine Hg. Samples were prepared using techniques such as vortex-assisted ionic liquid dispersive liquid–liquid microextraction and microvolume back-extraction, and for detection voltammetric analysis was performed using screen-printed electrodes modified with gold nanoparticles. The Hg threshold level in urine is 10–20 μg/L set by World Health Organization. The detection limit of this method was from 1.1 to 1.3 μg/L statistically (Fernández et al., 2016).
3. ***Mercury levels in hair, nails, breast milk, and umbilical cord blood:*** Mercury levels can be determined in hair, and total amount of this toxicant in both hair and blood correlates well with the total methyl mercuric content in the exposed individuals. Hair measurement provides an indirect noninvasive method of measuring Hg levels in the brain. Prenatal Hg exposure can also be assessed by analysis of umbilical cord tissue or umbilical cord blood, and maternal blood and maternal hair segments (Sakamoto et al., 2016). However, in case of exposure to mercuric vapors, hair is not a reliable biomarker. Along with all these, toenail and breast milk are also considered to be important biomarkers of MeHg toxicity. Hair samples of dentists were analyzed to determine the index of mercury exposure as they use dental amalgam. The hair sample showed 10-ppb mercury level, which is considered normal, so it is not taken into account of an occupational hazard to dentists (Wijesekara et al., 2017).

Limitations of Biomarkers of Exposure

All of the abovementioned biomarkers may surely help in rapid and accurate detection of Hg intoxications, but their selection poses certain difficulties influenced by number of factors that include (1) difficulty in obtaining blood samples post exposure; (2) variation in toxicokinetic, retention, and deposition; (3) extrapolation of low-level exposure data from animals to humans; (4) human exposure to other metal intoxicants, such as As, Pb, etc., in conjunction with Hg; and (5) difficulty in speciation, and it is essential to use certified reference material for quality control and assessment.

Biomarkers of Effects

In humans, Hg toxicity has been implicated in the development of various diseases and pathological conditions, resulting in dysfunctioning of vital organ systems of the body such as nervous system. There exist various biomarkers of effect or response biomarkers to detect systemic manifestations associated with mercury intoxication.

1. ***Systemic manifestations:***
   a. ***Neurological manifestations:*** Mercury poisoning has been reported to cause various neurodegenerative diseases such as amyotrophic lateral sclerosis, Alzheimer's diseases, and Parkinson's disease, exhibiting its strong neurotoxic potential. Its neurotoxic properties received labels such as "hatter's shakes" in which the individual experiences excessive tremors and psychological disorders such as being hyperirritable, blushing easily, having a labile temperament, being depressed or despondent, suffering from insomnia, and suffering from

fatigue. Moreover, mercuric chloride was found to increase free radical production in synaptosomal fractions suggesting oxidative damage and contributing significantly to neurotoxicity and neurodegeneration. Brain stem auditory evoked responses were measured as biomarkers of subtle Hg-induced neurological impairments. The people consuming methyl mercury via fish are at risk to suffer with amyotrophic lateral sclerosis (Andrew et al., 2018).

b. ***Reproductive manifestations:*** Mercury intoxication has been involved in causing various reproductive dysfunctions such as reduced number of sperms, testicular atrophy, reduced size of infants in one birth, reduced survival rate of fetuses, and fetal deformity. Inorganic Hg exposure impaired gonads of zebrafish, and the mechanism involved is an alternation of sex hormone levels that leads to disrupt transcription of hypothalamic-pituitary-gonadal (HPG) axis–related genes (Zhang et al., 2016).

c. ***Cardiovascular manifestations:*** Mercury exposure increases blood pressure as it accumulates in heart, following which are abnormal heartbeats and myocarditis. It has been reported to promote platelet aggregation and blood coagulation, myocardial infarction, altering heart rhythm and function, reduced sympathetic and parasympathetic nerve functions of the heart, and sclerosis of the arteries (Gribble et al., 2015).

d. ***Immunological manifestations:*** Occupational mercury exposures among miners are more at risk of autoimmune disorder. A report suggests that low level of organic Hg exposure is associated to subclinical autoimmunity among reproductive age females (Somers et al., 2015). Animal studies indicate adverse effect of Hg on immune system. However, fewer studies represented the immunotoxicity on human health. The difference may be due to dose variations, exposure route, and dissimilarity between the immune system of animal and human (Gardner and Nyland, 2016).

e. ***Nephrotoxic manifestations:*** Glomerular deposits of IgG1 and IgG4 are considered an important marker of mercury-induced nephropathy; however, their distribution and deposition varies according to Th1 and Th2 immune responses. An experimental study was conducted to determine the effect of Hg on rat kidney. Aldo-keto reductase and glutathione S-transferase pi (GSTP1) are identified as new biomarkers that could be utilized for evaluating nephrotoxicity (Shin et al., 2017).

2. ***Biomarkers of oxidative damage***

a. ***Urinary markers:*** Exposure to mercury elicits a specific change in the pattern of porphyrin excretion, characterized by increased urinary concentrations of coproporphyrin and pentacarboxyl porphyrin. Elevation in urinary NAG concentration is also associated with the direct poisoning of Hg compounds. Urinary porphyrins (coproporphyrin and precoproporhyrin) may also be developed as possible biomarkers of Hg toxicity in children with autism spectrum disorder (Khaled et al., 2016).

b. ***Metallothionein:*** Metallothionein, a low molecular weight (6–7 KDa) protein, is one of the major molecules playing critical role in scavenging and reducing the toxic effects of Hg. Metallothionein and genetic polymorphisms may affect urine and hair mercury concentrations. Glutathione-S-transferase (*GSTT1* and the *GSTM1*) genes deletion may be the risk factor for enhanced susceptibility to mercury exposure (Wang et al., 2012).

Biomarkers of Susceptibility

1. ***Cytogenetic markers:*** Determining alterations at molecular level due to metal intoxication forms another important category of biomarkers, helping in precise identification and diagnosis. Metal-specific lymphocytes act as biomarkers of sensitivity among patients suffering from health problems due to dental amalgams. Respiratory burst and chemotaxis in polymorphonuclear leukocytes have also been used to assess the extent of Hg exposure.
2. ***Biomarkers of apoptotic gene expression:*** Mercury significantly alters the expression of various genes involved in cell survival and apoptosis, proving it to be a potent genotoxin. It inhibits the activity of NF-Kβ, thereby enhancing the sensitivity of renal cells to apoptotic stimuli. Mercury-induced ROS generation leads to p38, caspase, and TNF-α activation; alteration of calcium homeostasis; and inflammatory cytokine gene expression that has a possible role in regulating both apoptosis and necrosis. Thioredoxin, an antioxidant system, oxidized by Hg compounds leads to increased ASK-1 phosphorylation and number of apoptotic cells along with activation of caspase-3 (Branco et al., 2017). Disorders of apoptosis and cell accumulation may play critical role in enhancing Hg-induced afflictions; however, their rapid diagnosis and measurement through biomonitoring may help in overcoming lethal consequences. Development of potent and novel biomarkers and proper utilization of

existing techniques will surely lead to reduction in cases of Hg intoxication.

CADMIUM

Cadmium (Cd) is recognized as a potential occupational health hazard because of its widespread and extensive dissemination in the environment. The famous itai-itai ("ouch-ouch") disease of Japan characterized by multiple fracture, distortion of the long bones in skeleton, and severe pain in the joints and spine was a result of consuming Cd-polluted rice.

Biomarkers of Exposure

1. ***Blood cadmium levels:*** Estimation of blood Cd levels is one of the major and promising biomarker to detect Cd toxicity and has also been shown to be an excellent indicator of Cd body burden. However, levels of Cd in blood mainly reflect recent exposure (Friedman et al., 2006).
2. ***Urine cadmium levels:*** Urinary excretion of Cd itself also serves for a good purpose and is regarded as a marker of both Cd exposure and proximal tubule injury. During exposure the concentration of Cd in the epithelial cells rise until the all the cells die and slough off into the urine. At this point urinary excretion of Cd increases markedly. Cd level in urine is an indicator of a long-term exposure because of its good to excellent temporal stability in urine. The urinary Cd is being affected by factors such as changes in smoking habits and increased excretion of proteins in some diseases (Vacchi-Suzzi et al., 2016) (Fig. 30.2).

Biomarkers of Effects

1. ***Systemic manifestations:*** "The Cd blues," often collectively referred to in this way, show flu-like symptoms such as chills, fever, and muscle ache and are induced on acute exposure to Cd. Further symptoms may lead to inflammation, cough, dryness, irritation of the nose and throat, headache, dizziness, weakness, fever, chills, and chest pain.

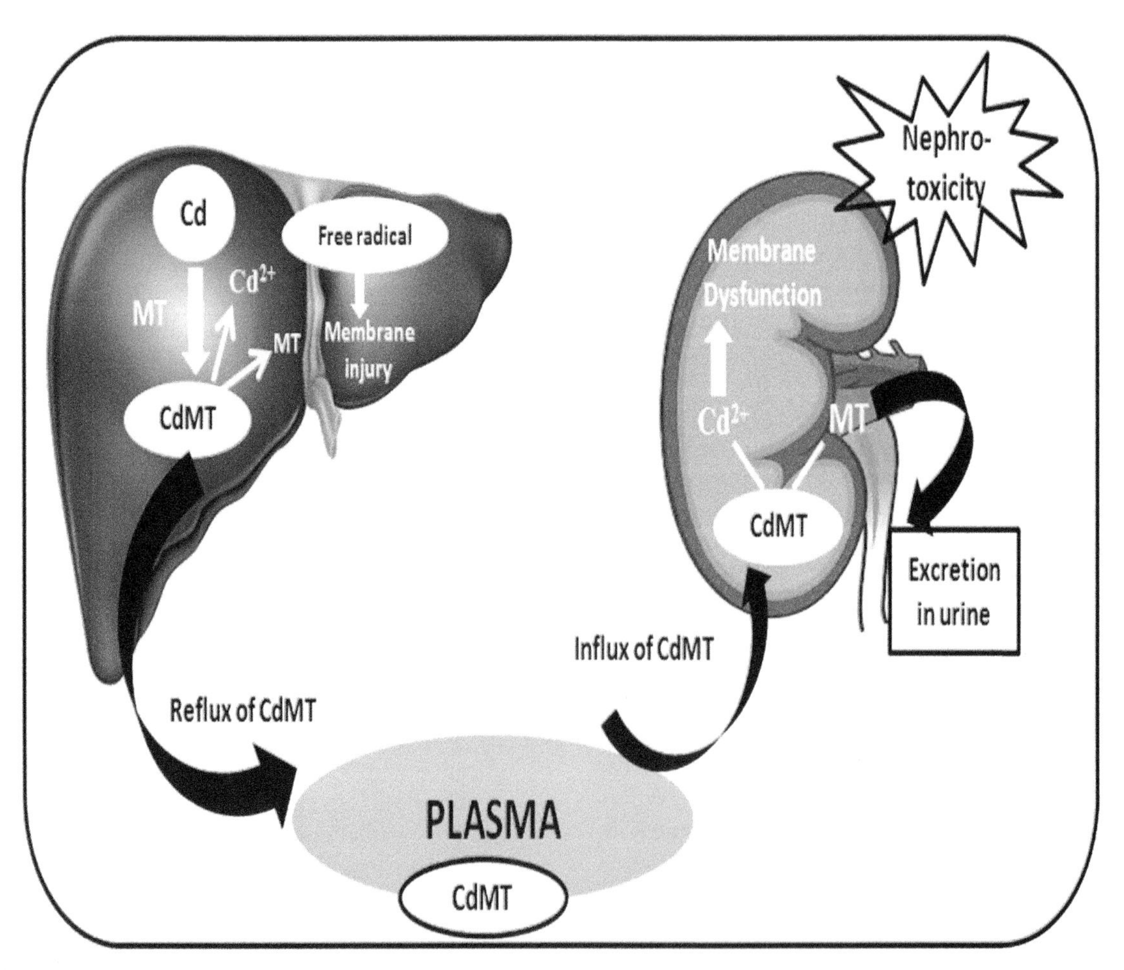

FIGURE 30.2 Induction of metallothionein by cadmium in liver and kidney, followed by excretion in urine, a major biomarker of Cd toxicity.

2. ***Renal manifestations:*** Kidney is reported to be the critical target organ as it is the main storage organ for Cd. However, Cd exerts its toxicity only when critical threshold is reached, which is estimated to be 150–200 ppm. Measuring urinary cystatin C concentration is being evaluated as biomarker of various types of ischemic and nephrotoxic renal injury (Prozialeck et al., 2016).
3. ***Bone manifestations:*** Increased risk of bone fractures, osteomalacia, pain in the back and in the extremities, difficulties in walking, and pain on bone pressure are some of the common response biomarkers associated with Cd toxicity. Effect of Cd on subchondral and interradicular bone of experimental rats was studied, and it resulted in decrease in bone volume and increase in tibial yellow bone marrow (Rodríguez and Mandalunis, 2016).
4. ***Carcinogenesis:*** Increased risk of prostrate, renal, breast, and lung cancers has been associated with occupational exposure to Cd. Cd exposure increases the risk of breast cancer (Larsson et al., 2015).
5. ***Other manifestations:*** Cadmium has been reported to cause impairment of pulmonary function suggestive of obstructive syndrome, and no adverse effects on hepatic, nervous, and reproductive systems has been documented so far. However, at the same time, enhanced Cd levels in blood were associated with a modest elevation in blood pressure levels. The effect of Cd exposure on pregnant women was found to be associated with decreased gestational age and increased preterm birth (Yang et al., 2016).

1. ***Urinary markers:***
   a. ***Metallothionein:*** Urinary levels of metallothionein have been used both as a marker of Cd exposure and renal injury. For example, identifying critical level of urinary metallothionein indicates the onset of toxic manifestations. It plays a key role in transporting Cd to the epithelial cells of the proximal tubule and is a specific metal-binding protein that makes it a superior biomarker. It was also found that genetic variation in the metallothionein gene region and metal-regulatory transcription factor 1 affects urinary Cd level (Adams et al., 2015). Urinary Cd showed good to excellent temporal stability in both sampling methods, i.e., spot urine or first morning void sampling. However, some factors such as increase of protein excretion and change in smoking habits may bring short-term changes in urinary Cd levels (Vacchi-Suzzi et al., 2016).
   b. ***β-2 microglobulin and Clara cell protein-16 (CC-16):*** Urinary biomarkers are considered to be the best and most promising in the diagnosis of Cd-induced nephrotoxicity. Some of the urinary biomarkers that have been used for this purpose include Cd-binding protein metallothionein and low molecular weight proteins such as β-2 microglobulin, Clara cell protein-16 (CC-16), and Cd itself. These are some of the effective biomarkers for renal dysfunction. β_2-Microglobulin and CC-16 are low–molecular weight proteins, filtered at the glomerulus and reabsorbed by the proximal tubule. In addition, Cd-induced renal injury can be identified by these existing markers only its late stages and by the time they are detected injury to the kidney becomes irreversible and untreatable. This evokes the need for better and early biomarkers of Cd-induced kidney injury. A study suggests that Cd-exposed rats showed increase level of urinary cystatin C because of disruption of megalin-mediated uptake of cystatin C by epithelial cells of the proximal tubule that could be an early biomarker for renal damage (Prozialeck et al., 2016).
   c. ***Kidney injury molecule-1:*** Another sensitive marker of cadmium-induced proximal tubular injury includes kidney injury molecule-1 (Kim-1). Elevated levels of Kim-1 appears in urine only after 6 weeks of Cd intoxication, whereas classic signs of Cd-induced proximal injury, which include overt polyuria and proteinuria, become evident only after 9–10 weeks. The ecto domain of Kim-1 is shed into the urine as a result of renal injury induced by a variety of agents including cisplatin, Hg, and chromium. It offers several advantages over the existing urinary biomarkers: (1) it is a more specific marker of Cd-induced proximal tubular injury because of specific production by injured renal cells and then shed into urine, (2) it is highly conserved and upregulated after renal injury in variety of species including mice, rats, nonhuman primates, and humans, (3) almost nil expression is observed in noninjured kidney, making it relatively easy to detect in cases of Cd-induced renal injury, and (4) it is highly stable in urine, without any requirement of special preservatives, even at the temperature of −80°C for months. Recent study suggests that Kim-1 might serve as a sensitive biomarker to assess kidney injury in children due to Cd exposure (Cárdenas-González et al., 2016).
   d. ***NAG, alanine aminopeptidase, and α-GST:*** Proximal tubule–derived enzymes NAG, alanine aminopeptidase (AAP), and α-GST are the other biomarkers, which have been studied. Reports

indicate that NAG and α-GST outperformed traditional markers of Cd toxicity such as β_2-microglobulin.
2. ***Biomarkers of DNA damage:*** Apart from all these urinary metallothionein mRNA, 8-OH-dG, albumin, and creatinine were also found to be sensitive markers of acute Cd-induced testicular damage and dysfunction. Measurement of 8-OHdG adduct in urine samples of exposed individuals has been used as an early marker of oxidative DNA damage, evaluated by the comet assay, immunoperoxidase staining coupled with a monoclonal antibody, and cytogenetic tests (Filipič and Hei, 2004).

Biomarkers of Susceptibility

1. ***Biomarkers of gene expression:*** In Cd-related carcinogenicity, various regulatory genes are activated, such as immediate early response genes (IEGs). Overexpressions of IEGs constitutively stimulate cell proliferation and induction of carcinogenesis. Cadmium-induced carcinogenicity is also found to be induced by expression of several stress response genes such as metallothionein (MT), genes for encoding heat-shock proteins (HSPs), oxidative stress response, and synthesis of GSH. Identification of these biomarkers is involved in detection of both early and late stages of Cd poisoning.
2. ***Biomarkers of apoptotic genes:*** Cadmium-induced cellular responses include variety of molecular markers that can be identified for the better and rapid identification of Cd-induced toxic manifestations. Some of them include, identifying release of cytochrome C from mitochondria, caspase-3 activation, intracellular GSH oxidation, inhibition in the expression of Bcl_2 and p53, and involvement of Fas-FADD caspase 8 pathways and MT-3 in inducing apoptosis in renal tubular cells. cDNA microarray and quantitative real-time PCR results revealed decreased expression of proapoptotic and DNA repair genes, suggesting enhanced carcinogenic potential of Cd. lncRNA-MALAT1 was reported as a novel biomarker of Cd exposure. It was observed that Cd toxicity increases MALAT1 expression, which regulates apoptosis, proliferation, and cell cycle progression (Huang et al., 2017a,b).

Thus, as evident from the reports available there exists various biomarkers for the effective diagnosis of Cd intoxications; however, few limitations exists, which has to be overcome to develop better and much more efficient biomarkers. Development of molecular biomarkers is the need of the hour, with a continuous increase in the number of metal-poisoning cases.

CHROMIUM

Chromium (Cr III and Cr VI) has widespread industrial applications such as chrome plating, welding inox steel and other special steels, painting, leather tanning, and wood preserving. Cr (VI) is classified as Group I human carcinogen and transported inside the cells through anion channels as chromate.

Biomarkers of Exposure

Analysis of Cr levels in various body fluids, such as blood, serum, and urine, also significantly determines exposure to Cr along with a measurement of the internalized dose of Cr and serves as important biomarkers of exposure.

1. ***Blood, serum, and urine chromium levels:*** Blood and urine are prominently used for the biological monitoring of workers exposed to Cr. The levels of indicators for nasal injury, genetic damage, and micronucleus were determined in low and high Cr-exposed workers; this indicates that blood Cr could be used as a biomarker for occupationally exposed people (Li et al., 2016). Recent study suggests that measuring serum MDA level in occupationally exposed workers could be a reliable biomarker of its exposure (Mozafari et al., 2016).
2. ***Exhaled breath condensate chromium levels:*** Analysis of exhaled breath condensate (EBC) is a novel approach being utilized currently to quantify both internal dose of toxicant inhaled and the level of free radical production (Caglieri et al., 2006). EBC seems to be suitable biomarkers for assessing inflammation and enhanced oxidative stress and can be efficiently used to investigate lung tissue levels of Cr (VI). Also, magnetic resonance imaging (MRI) is considered a good choice for monitoring of internalized doses of Cr in lung tissues.

Biomarker of Effects

1. ***Systemic manifestations:*** There are various health hazards and systemic effects associated with Cr exposure, which serves as important biomarkers of effect or response biomarkers for effective diagnosis of Cr poisoning.
   a. ***Respiratory manifestations:*** The adverse effect of Cr depends on exact dose, duration, and the specific compound involved. Higher incidences of nasal mucosa injury; dyspnea; urticaria; angioedema; pharyngitis; rhinitis; polyps of the upper respiratory tract; cough; phlegm; wheezing and shortness of breath; nasal allergy; nasal septal perforation; throat irritation; bronchial asthma;

alteration in lung function; and nose, throat, and lung tumors; tracheobronchitis; and bronchospasm have also been observed in workers exposed occupationally to chromium (Jamal et al., 2017).

b. ***Skin manifestations:*** Dermal contact with chromium compounds is known to cause various skin manifestations such as irritant and allergic contact dermatitis, localized erythematous or vesicular lesions at points of contact or generalized eczematous dermatitis, dryness, erythema, fissuring, papules, scaling, and swelling. Penetration of the skin by chromium leads to painless erosive ulceration known as "chrome holes" (Buters and Biedermann, 2017).

c. ***Carcinogenesis:*** Increased risk of respiratory system, lung, nasal, gastrointestinal, and sinus cancers has been shown to exist in workers exposed to Cr (VI). In addition, the carcinogenicity appears to be associated more with the inhalation of Cr (VI) compounds (Welling et al., 2015).

d. ***Developmental manifestation:*** Chromium exposure to pregnant women is a threat because of its potential health effects on susceptible embryos such as premature rupture of membranes, and there may be increased risk of delivering low birth-weight infants. The adverse effect of Cr on developing embryo may differ by infant gender (Huang et al., 2017a,b).

2. ***Biomarkers for oxidative damage:*** Although the mechanism of Cr (VI)-induced toxicity is not clear, it is believed that oxidative stress plays an important role, suggesting MDA as a prominent biomarker for occupational Cr exposure.

a. ***Biomarkers for DNA damage: 8-hydroxydeoxyguanosine and 8-Oxo-G:*** Measurement of oxidative DNA base modifications such as 8-hydroxydeoxyguanosine (8-OH-dG) and 8-Oxo-G plays vital role in the identification of Cr-induced DNA damage, thus serving as a key biomarker relevant to carcinogenesis. Estimation of 8-Oxo-G may not be wholly reliable and assessment of further oxidized products of 8-Oxo-G should be considered while evaluating chromate-induced DNA damage and mutagenesis. Increase in DNA strand breaks as estimated by comet assay is also suggested to be a valid biomarker for assessing genotoxic response to Cr compounds. Male workers occupationally exposed to Cr (VI) were investigated to find out the biomarker of oxidative DNA damage and lipid peroxidation. Exposure to Cr(VI) was linked with increase in the level of urinary 8-OHdG and MDA (Pan et al., 2018).

Biomarkers of Susceptibility

1. Cytogenetic markers

a. ***DNA–protein cross-links:*** Quantifying DNA–protein cross-links (DPC) in peripheral blood mononuclear cells of human subjects has also been used as a biomonitoring tool. This technique provides a sufficiently sensitive approach to determine DNA damage in Cr-exposed populations. Unique kinetic principles involved in the formation of this adduct points toward the ability to detect low levels of Cr exposure. Few advantages including the cost effectiveness of cross-link assay and multiple analyses of samples led to establishment of DPC measurements as one of the popular method for rapid screening of DNA damage among potentially exposed populations (Macfie et al., 2009).

b. ***Sister chromatic exchange (SCE:*** Blood may be analyzed for the presence of SCEs (intrachromosomal rearrangements, SCEs) in lymphocytes. Lymphocytes play an important role in biomonitoring as they come into contact with Cr(VI) while migrating through the blood, where uptake of Cr(VI) that has leached from the lungs takes place. But at the same time they seem to be inappropriate for the monitoring of early biological effects and effective dose of exposure. However, lymphocytes may be useful in monitoring exposure to Cr at high levels (Mamyrbaev et al., 2015).

2. ***Biomarkers of gene expression and apoptosis:*** Hypoxia-induced factor HIF-1 plays an important role in Cr-induced toxic manifestations. It is known and reported to modulate the expression of several cancer-related genes such as heme oxygenase1 and vascular endothelial growth factor by regulating oxygen homeostasis (Rana, 2008). Both p53-dependent and p53-independent pathways have been shown to play major role in Cr-induced apoptosis (Fig. 30.3). Generation of ROS by Cr (VI) activates upstream kinases resulting in phosphorylation of p53 and further DNA damage (Gavin et al., 2007).

THALLIUM

Thallium (Tl) is widely distributed in the environment in the earth's crust and is present at very low concentrations in soil and sulfide-based minerals. It is used in small amounts in pharmaceutical and electronics manufacturing, the latter being the current major industrial consumer of Tl (Karbowska, 2016). Interaction of Tl

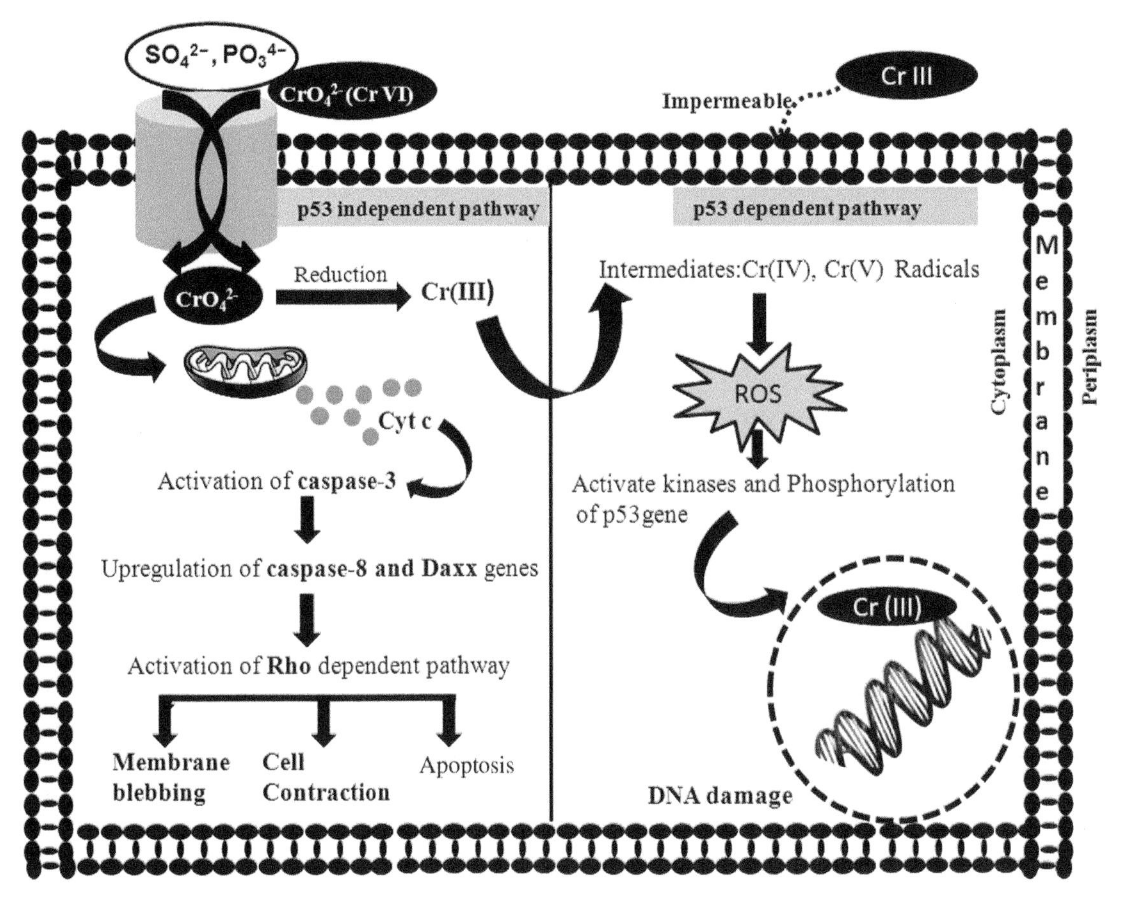

FIGURE 30.3 Mechanism of Cr induced genotoxicity and associated biomarkers of susceptibility.

with cells at different level is considered as a major mechanism involved in eliciting toxic manifestations. Person consuming thallium-contaminated alternative medicine suffered from diffuse alopecia with gastrointestinal and neurologic symptoms (Senthilkumaran et al., 2017).

Biomarkers of Exposure

Analysis of various body fluids, such as blood, serum, urine, and breast milk is a common biomarker employed for detection of exposure to Tl.

1. ***Blood thallium levels:*** Disappearance of Tl from the blood is observed within several days, but it is not considered to be a reliable biomarker as it reflects only recent exposures. It can readily cross the placental barrier and also excreted into the breast milk; however, elimination from the body tissues occur slowly through urine and feces (Blanchardon et al., 2005).
2. ***Urinary thallium levels:*** Urinary Tl levels reflect recent exposure. Appropriate methodology for analysis of urine or blood TI includes atomic absorption spectroscopy and inductively coupled plasma emission spectroscopy. Measurement of Tl from 24-h urine collection is the method of choice for diagnosis. A study demonstrated that qualitative Tl urinary assay is as important as quantitative assay in determining Tl poisoning that could be used in determination of Tl exposure in persons using drugs such as heroin and cocaine (Ghaderi et al., 2017).
3. ***Presence of thallium in hair samples:*** A good correlation was found between urinary and hair Tl concentrations, when both the samples were collected simultaneously. Elevated Tl levels in hair can be observed as early as 2–3 weeks after intoxication (Misra et al., 2003). Besides blood and urine, thallium also concentrates in hair, making it an efficient biomarker of Tl toxicity.

Biomarkers of Effects

Clinical manifestations of Tl poisoning can be characterized as acute, subchronic, and chronic, depending on the dose of Tl, route, and duration of exposure.

1. ***Systemic manifestations:*** Tl exhibits adverse effects on various organ systems, which serve as useful response biomarkers for the diagnosis of Tl intoxication.

a. ***Dermal and ocular effects:*** Complete hair loss along with hair discoloration usually occurs after a month of exposure, and characteristic pattern of black pigmentation at the roots of scalp hair may also be observed. Thallium interferes with normal process of cutaneous keratinization, parakeratosis, dilated hair follicles filled with keratin and necrotic sebaceous materials, mild epidermal atrophy, and vacuolar degeneration of the basal layer. Exposure to Tl for about 1 month leads to formation of transverse white lines in the nail called as Aldrich-Mees' lines or leukonychia striata (Zhao et al., 2008). In early stages of poisoning, the optic disk reveals the typical picture of neuritis with ill-defined and red papillae, followed by the development of pale or white papillae as a result of atrophy of the optic nerve.

b. ***Respiratory manifestations:*** Acute ingestion of Tl was reported to cause severe lung damage resulting in diffuse alveolar damage with hyaline membrane formation and pulmonary edema. Bronchopneumonia was also observed.

c. ***Gastrointestinal manifestations:*** Gastrointestinal symptoms were reported to occur immediately within one to few hours after acute exposure and comprised of inflammatory reaction in the first exposed tissues or organs causing glossitis, pharyngoesophagitis, gastritis, enteritis, and colitis. After 3–4 days of intoxication nausea and vomiting occurs, followed by abdominal pain, severe constipation, and stomach and duodenal ulcers.

d. ***Cardiovascular manifestations:*** Thallium toxicity induces various cardiac complications, such as myocardial damage, electrocardiographic changes, decrease in mean arterial pressure and heart rate, sinus tachycardia, irregular pulse, hypertension, and angina-like pain. It is quite common for paralyzed patients, often short of breath already, to develop severe dyspnea and cyanosis, followed by death (Cvjetko et al., 2010).

e. ***Hepatic manifestations:*** Tl compounds could induce reactive oxygen species formation, membrane lipid peroxidation, and reduced GSH oxidation and collapse mitochondrial membrane potential. As liver mitochondria are a target of Tl toxicity, its mechanism of toxicity was investigated. It was found that Tl has disruptive effect on the mitochondrial respiratory complexes, which lead to the ATP depletion and reactive oxygen species formation (Eskandari et al., 2015).

f. ***Renal manifestations:*** Renal excretion of Tl sulfate is slow and may be detected as late as 2 months after ingestion. During the initial stage of exposure there is albuminuria, with erythrocytes, leucocytes, and cylindrical casts in the urinary sediment; however, in the later phase of serious illness, there is a fall in the concentrating ability of the kidneys, often with recovery afterward.

g. ***Neurological manifestations:*** In very serious or even fatal cases, true "pseudobulbar paralysis," with paralysis of the ocular muscles, facial paralysis, amblyopia, and paralysis of the recurrent nerve was observed. However, paralysis of the vagal nerve may possibly be the direct cause of death. Acute oral exposure to Tl causes disturbances of the peripheral and central nervous systems leading to symptoms such as ataxia, tremor, multiple cranial palsies, numbness of toes and fingers, "burning feet" phenomenon, muscle cramps, convulsions, and death. In chronic poisoning, ataxia and paresthesia may be the outstanding symptoms, which may eventually progress to peripheral neuropathy with weakness and atrophy of the associated musculature (Osorio-Rico et al., 2017).

h. ***Reproductive manifestations:*** Reproductive system is highly susceptible to Tl and leads to various adverse effects, such as decreased sperm motility, inhibition of β-glucuronidase activity, histopathological alterations of the testes, increased embryolethality, decreased libido and impotence in humans, and lower sexual activity in laboratory animals. Thallium has the ability to cross placental barrier as evident by skin and nail dystrophy, alopecia, and low body weights in newborns of intoxicated mothers (Xia et al., 2016).

i. ***Developmental manifestations:*** Tl exposure may result in achondroplasia, leg bone curvature, parrot beak deformity, microcephaly, and decreased fetal size. High-level exposure to Tl at the end of pregnancy leads to the death of fetuses (Mulkey and Oehme, 1993).

Biomarkers of Susceptibility

1. *Cytogenetic markers:* Chromosomal aberrations are considered to be sensitive biomarkers of genotoxic damage, such as DNA breakage. Thallium has been shown to induce chromosomal aberrations and SCEs following exposure. Thallium showed weak mutagenic effect both in vitro and in vivo. However, no effect of the induction of primary DNA and chromosomal damage was observed. Cellular and genotoxic effects of Tl were investigated in human peripheral lymphocytes by chromosome aberration assays, and it was discovered that Tl induced clastogenic and aneuploidogenic effects (Rodríguez-Mercado et al., 2017).

2. ***Biomarkers of gene expression and apoptosis:*** Exposure to Tl leads to opening of mitochondrial membrane pores and release of cytochrome C, caspase-3, and caspase-9, resulting in cell death. Estimation of these proteins proves to be of immense value and serves as potential biomarkers of toxicity. p53, p21, and Bcl_2 estimation also play an important role in detecting Tl-induced apoptosis as Tl inhibits cell cycle progression at G2/M phase by suppressing CDK activity through the p53-mediated induction of the CDK inhibitor p21(Cip1), which in turn triggers the activation of a mitochondrial pathway and Bcl-2 family, promoting the formation of the apoptosome and, consequently, apoptosis (Cvjetko et al., 2010).

MANGANESE

Chronic exposure to manganese (Mn) leads to neurodegenerative changes resembling Parkinson's disease such as movement disorder, referred to as "manganism." A meta-analysis study revealed that elevated Mn level in plasma, whole blood and low level in serum in people are at higher risk for neurological disorder (Du et al., 2018).

Biomarkers of Exposure

The measurement of Mn in biological media (blood, serum, plasma, urine; and less frequently in hair, nails, cerebrospinal fluid, feces, and saliva) have been suggested as biomarkers of exposure. Blood and urinary Mn are the most widely investigated biomarkers of exposure. Half-life of Mn in blood has been reported to be 10–42 days, whereas it is less than 30 h in urine.

1. ***Blood Mn levels:*** Elevated blood Mn concentration accompanied by a progressive, symmetric Parkinsonism (characterized by prolonged L-dopa responsiveness) and a reduced [^{18}F]FDOPA uptake by Positron emission tomography has been reported in few cases (Racette et al., 2005). Report suggests a link between Mn exposures with serum prolactin level that could be a diagnostic marker of its exposure (Tutkun et al., 2014). Investigation of Mn in blood cell and plasma revealed that the former could serve as a potential internal biomarker. However, further validation studies need to be performed for its establishment as biomarker (Ge et al., 2018).
2. ***Urinary Mn levels:*** Studies suggest urine Mn as a potential noninvasive marker of Mn intoxication. As Mn gets mainly excreted in feces and only a very small proportion in urine, measuring urinary Mn is of little interest as biomarker of current exposure (Hoet et al., 2012).
3. ***Manganese levels in hair, nails, and saliva:*** Manganese concentration in saliva is a noninvasive biomarker of exposure. However, reports suggest that Mn in saliva is not a more sensitive measure or better indication of Mn exposure. Hairs and nails Mn level on the other hand may reflect low-level exposure of Mn (Ntihabose et al., 2017). Estimation of Mn in teeth and hair indicate long-term low-level Mn exposure, which could be a reliable biomarker of exposure. Several studies suggested that hair Mn is a more reliable and applicable biomarker of Mn exposure in school-aged children (Liang et al., 2016). A recent study suggests that the human toenails and tooth dentin Mn levels could be sensitive, specific, and easy-to-acquire biomarker of Mn exposure (Ward et al., 2017).

Biomarkers of Effects

Manganism due to an increased Mn concentration in brain regions is a neurological syndrome similar to Parkinson's disease (Rivera-Mancía et al., 2011). General symptoms associated with Mn exposure include behavioral changes such as increased anxiety or nervousness, apathy, loss of memory, decreases in concentration, bradykinesia, less-frequent resting tremor, rigidity, frequent dystonia of the trunk and extremities, extrapyramidal movement disorder with characteristic "cock-walk," and difficulty in walking backward are observed (Olanow, 2004).

1. ***Nervous system:*** Manganese can readily cross the blood–brain barrier by facilitating diffusion, active transport, divalent metal transport 1 (DMT-1)–mediated transport, and transferrin-dependent transport mechanisms leading to accumulation of Mn in various brain regions (Au et al., 2008). Use of excessive amounts of Mn-containing herbals and supplement products by the people could lead to neurotoxicity (Schuh, 2016). MnB could be diagnosed by a sensitive, feasible, effective, and semiquantitative index known as Pallidal index (Li et al., 2014). A study was performed to evaluate the low-level exposure of Mn on central nervous system function. It was found that T1 signal intensity in various parts of the basal ganglia significantly increased with Mn exposure, suggesting T1-weighted MRI a possible biomarker of exposure to Mn (Baker et al., 2015). An effect on the levels of serum prolactin, luteinizing hormone, follicle-stimulating hormone, testosterone, and thyroid-stimulating hormone that consequently influence the metabolism of neuroendocrine hormones has been reported (Kim et al., 2007).

2. ***Cardiovascular effects:*** An epidemiological study conducted among ferroalloy workers reported decreased systolic blood pressure. Similar studies have also been reported where potential cardiotoxicity in 656 workers (547 males and 109 females) engaged in Mn milling, smelting, and sintering to Mn dioxide exposure has been reported (Jiang and Zheng, 2005).
3. ***Hematological effects:*** Chronic Mn poisoning caused alterations in erythropoiesis and granulocyte formation. An increased level of erythrocyte superdioxide dismutase and plasma malondialdehyde in men working in Mn smelters has been reported, whereas an elevated level of serum prolactin and the increased activity of lymphocytes Mn-dependent superoxide dismutase have been suggested as useful peripheral biomarkers of Mn exposure. Children with iron deficiency were linked with increase whole blood Mn concentrations (Smith et al., 2013).
4. ***Reproductive effects:*** Impotence and loss of libido are the two common problems among male workers exposed to Mn. Another study suggested that Mn affects pituitary gland function, which is reversible when exposure stops. Welders with lower inhibin B level indicate impairment of the testicular sertoli cells (Ellingsen et al., 2007). Consumption of Mn-contaminated drinking water by the pregnant women may impair fetal growth (Rahman et al., 2015a,b).
5. ***Immunological effects:*** Suppression of T and B lymphocytes, primarily CD4+ and CD8+ T-lymphocytes leading to the reductions in serum immunoglobulin G (IgG) and alterations in immune cell number have been reported following pulmonary exposure to welding fumes containing Mn particles (Antonini et al., 2012).

Biomarkers of Susceptibility

Altered RNA expression and cytotoxic markers: An increased expression of GS mRNA has been observed because of Mn-potentiated cellular overload of iron. Overexpression of GS mRNA reflects a change in intracellular levels of Mn(II). An increase in proinflammatory genes (tumor necrosis factor-α, iNOS) and activated inflammatory proteins (P-p38, P-ERK, and P-JNK) in primary rat glial cells has been reported after Mn exposure (Chen et al., 2006). Polymorphism has also been suggested as a genetic marker with no direct physiological relevance to Mn neurotoxicity.

Concurrent Exposure of Heavy Metals

Human exposure occurring simultaneously to more than one heavy metals such as Pb, Cd, As, Mn and MeHg, may produce additive or synergistic toxic effects due to their binding affinity towards same biological pathway such as NMDA receptor, $Na^{+}-K^{+}$ ATPase pump, biological Ca^{+2}, Glu neurotransmitter (Flora, 2016). They have common mechanism of toxicity such as oxidative stress, variation in hematopoietic system, and disturbance in neurotransmitter function. The toxicity of heavy metals and their correlation with disease in humans have been studied by gene expression profiling. Heavy metals exposure may express genes that are responsible for detoxifying toxic metals, enzymes metabolizing xenobiotics, and repair DNA. Heavy metal–exposed people are more prone to renal and urological diseases. These genes could be validated to establish them as early biomarkers for renal damage. The mode of action of concurrent metal exposure and its mechanisms to induce cognitive dysfunction have been reported (Karri et al., 2016). Study showed that urine excretion of ALA can be established as biomarker of exposure and may also assist in evaluating onset of neurological diseases linked with exposure to heavy metals (Andrade et al., 2015).

CURRENT CONCERNS AND BIOLOGICAL RELEVANCE

Biomarkers are one of the keys to quantify the interaction of metals and its potential impact on living organisms, including human beings. Previous reports suggest long history of use of urine and blood as established matrices for biomarkers, but the advent of molecular biology has opened up new vistas. Introduction of novel analytical techniques and instrumentation led to recognition of specific compounds of metals in biological samples such as fluids and tissues. Detection of speciation of metallic ions and complexes marks an important step because only specific species of a metal has the ability to be transported into cells and across the blood–brain and placental barriers. Data obtained from biomonitoring can be utilized for a variety of applications eventually leading to enhanced assessment of exposure mitigation strategies. As a common surveillance tool, biomonitoring helps in identifying baseline exposure levels, trends in exposure levels over time, and unique subpopulations with higher exposure levels. Various exposure and kinetic models have been developed to estimate the dose of exposure to the toxicant; however, results get greatly influenced by many sources of variability and uncertainty. Biomonitoring helps in eliminating those errors and has been recognized as a valuable quantitative tool to improve dose estimates, thus helping in improvement of the human health through early risk assessment of metals. In conjunction with the existing biomarkers and technology, the advent of "omic" molecular technologies (genomic, proteomic, and metabolomic) has led to enormous augmentation in the speed and precision of biomarker (endpoint) measurements. The better and in-depth understanding of these early molecular markers in relation to other toxicological endpoint responses hold great promise in the future for the formation of more effective health surveillance strategies.

CONCLUDING REMARKS AND FUTURE DIRECTIONS

Biological survey forms a close association with environmental survey in environmental epidemiologies. Biomonitoring is useful for the assessment of exposure, but few limitations exist, which interfere with their ability to completely exhibit their potential. There is an immense spur to implement more biomonitoring tools into the field of exposure science, for which traditional methods should be improved and supplemented with newer, noninvasive technologies resulting in more informed risk assessment and decision-making. Moreover, advanced techniques are required to elucidate the mechanisms underlying interactive effects of metal mixtures and developing a gene–environment link. At the same time, expanding the repertoire of available biomarkers of metal exposure, investigating and employing the same in well-designed models will prove to be of immense benefit in the evaluation of metal intoxication along with designing of suitable measures to minimize hazardous exposures.

References

Adams, S.V., Barrick, B., Christopher, E.P., et al., 2015. Genetic variation in metallothionein and metal-regulatory transcription factor 1 in relation to urinary cadmium, copper, and zinc. Toxicol. Appl. Pharmacol. 289 (3), 381–388.

Agarwal, S., Roy, S., Ray, A., et al., 2009. Arsenic trioxide and lead acetate induce apoptosis in adult rat hepatic stem cells. Cell Biol. Toxicol. 25 (4), 403.

Ahmad, S.A., Bandaranayake, D., Khan, A.W., et al., 1997. Arsenic contamination in ground water and arsenicosis in Bangladesh. Int. J. Environ. Health Res. 7 (4), 271–276.

Andrade, V., Mateus, M., Batoréu, M., et al., 2015. Lead, arsenic, and manganese metal mixture exposures: focus on biomarkers of effect. Biol. Trace Elem. Res. 166 (1), 13–23.

Andrew, A.S., Chen, C.Y., Caller, T.A., et al., 2018. Toenail mercury levels are associated with amyotrophic lateral sclerosis (ALS) risk. Muscle Nerve. https://doi.org/10.1002/mus.26055.

Antonini, J.M., Zeidler-Erdely, P.C., Young, S.-H., et al., 2012. Systemic immune cell response in rats after pulmonary exposure to manganese-containing particles collected from welding aerosols. J. Immunotoxicol. 9 (2), 184–192.

Arteel, G.E., 2015. Hepatotoxicity. In: Arsenic: Exposure Sources, Health Risks, and Mechanisms of Toxicity, pp. 249–265.

Au, C., Benedetto, A., Aschner, M., 2008. Manganese transport in eukaryotes: the role of DMT1. Neurotoxicology 29 (4), 569–576.

Azqueta, A., Langie, S., Collins, A., 2015. 30 years of the Comet Assay: An Overview with Some New Insights. Frontiers Media SA, ISBN 978-2-88919-649-4.

Bach, J., Peremartí, J., Annangi, B., et al., 2015. Reduced cellular DNA repair capacity after environmentally relevant arsenic exposure. Influence of Ogg1 deficiency. Mutat. Res. Fund Mol. Mech. Mutagen 779, 144–151.

Baker, M.G., Criswell, S.R., Racette, B.A., et al., 2015. Neurological outcomes associated with low-level manganese exposure in an inception cohort of asymptomatic welding trainees. Scand. J. Work. Environ. Health 41 (1), 94.

Bakri, S., Hariri, A., Ma'arop, N., Hussin, N., 2017. Toenail as non-invasive biomarker in metal toxicity measurement of welding fumes exposure-a review. In: Paper Presented at the IOP Conference Series: Materials Science and Engineering.

Basu, N., Eng, J.W., Perkins, M., et al., 2017. Development and application of a novel method to characterize methylmercury exposure in newborns using dried blood spots. Environ. Res. 159, 276–282.

Behinaein, S., Chettle, D., Fisher, M., et al., 2017. Age and sex influence on bone and blood lead concentrations in a cohort of the general population living in Toronto. Physiol. Meas. 38 (3), 431.

Blanchardon, E., Challeton-de Vathaire, C., Boisson, P., et al., 2005. Long term retention and excretion of 201Tl in a patient after myocardial perfusion imaging. Radiat. Protect. Dosim. 113 (1), 47–53.

Branco, V., Coppo, L., Solá, S., et al., 2017. Impaired cross-talk between the thioredoxin and glutathione systems is related to ASK-1 mediated apoptosis in neuronal cells exposed to mercury. Redox Biol. 13, 278–287.

Buters, J., Biedermann, T., 2017. Chromium (VI) contact dermatitis: getting closer to understanding the underlying mechanisms of toxicity and sensitization! J. Invest. Dermatol. 137 (2), 274–277.

Cabral, M., Dieme, D., Verdin, A., et al., 2012. Low-level environmental exposure to lead and renal adverse effects: a cross-sectional study in the population of children bordering the Mbeubeuss landfill near Dakar, Senegal. Hum. Exp. Toxicol. 31 (12), 1280–1291.

Caglieri, A., Goldoni, M., Acampa, O., et al., 2006. The effect of inhaled chromium on different exhaled breath condensate biomarkers among chrome-plating workers. Environ. Health Perspect. 114 (4), 542.

Cárdenas-González, M., Osorio-Yáñez, C., Gaspar-Ramírez, O., et al., 2016. Environmental exposure to arsenic and chromium in children is associated with kidney injury molecule-1. Environ. Res. 150, 653–662.

Chen, C.-J., Ou, Y.-C., Lin, S.-Y., et al., 2006. Manganese modulates pro-inflammatory gene expression in activated glia. Neurochem. Int. 49 (1), 62–71.

Chinde, S., Kumari, M., Devi, K.R., et al., 2014. Assessment of genotoxic effects of lead in occupationally exposed workers. Environ. Sci. Pollut. Res. 21 (19), 11469–11480.

Chou, W.-C., Chen, H.-Y., Wang, C.-J., et al., 2014. Maternal arsenic exposure and DNA damage biomarkers, and the associations with birth outcomes in a general population from Taiwan. PLoS One 9 (2), e86398.

Chouhan, S., Flora, S., 2010. Arsenic and fluoride: two major ground water pollutants. Indian J. Exp. Biol. 48 (7), 666–678.

Cvjetko, P., Cvjetko, I., Pavlica, M., 2010. Thallium toxicity in humans. Arch. Ind. Hyg. Toxicol. 61 (1), 111–119.

De Oliveira, V.L.F., Gerlach, R.F., Martins, L.C., et al., 2017. Dental enamel as biomarker for environmental contaminants in relevant industrialized estuary areas in São Paulo, Brazil. Environ. Sci. Pollut. Res. 24 (16), 14080–14090.

Dobrakowski, M., Pawlas, N., Hudziec, E., et al., 2016. Glutathione, glutathione-related enzymes, and oxidative stress in individuals with subacute occupational exposure to lead. Environ. Toxicol. Pharmacol. 45, 235–240.

Du, K., Liu, M.-Y., Pan, Y.-Z., et al., 2018. Association of circulating manganese levels with Parkinson's disease: a meta-analysis. Neurosci. Lett. 665, 92–98.
Ellingsen, D.G., Chashchin, V., Haug, E., et al., 2007. An epidemiological study of reproductive function biomarkers in male welders. Biomarkers 12 (5), 497–509.
Eskandari, M.R., Mashayekhi, V., Aslani, M., et al., 2015. Toxicity of thallium on isolated rat liver mitochondria: the role of oxidative stress and MPT pore opening. Environ. Toxicol. 30 (2), 232–241.
Farkhondeh, T., Boskabady, M.H., Koohi, M.K., et al., 2013. The effect of lead exposure on selected blood inflammatory biomarkers in guinea pigs. Cardiovasc. Haematol. Disord. - Drug Targets 13 (1), 45–49.
Farzan, S.F., Chen, Y., Wu, F., et al., 2015. Blood pressure changes in relation to arsenic exposure in a US pregnancy cohort. Environ. Health Perspect. 123 (10), 999.
Feksa, L.R., Oliveira, E., Trombini, T., et al., 2012. Pyruvate kinase activity and δ-aminolevulinic acid dehydratase activity as biomarkers of toxicity in workers exposed to lead. Arch. Environ. Contam. Toxicol. 63 (3), 453–460.
Fernández, E., Vidal, L., Costa-García, A., et al., 2016. Mercury determination in urine samples by gold nanostructured screen-printed carbon electrodes after vortex-assisted ionic liquid dispersive liquid–liquid microextraction. Anal. Chim. Acta 915, 49–55.
Filipič, M., Hei, T.K., 2004. Mutagenicity of cadmium in mammalian cells: implication of oxidative DNA damage. Mutat. Res. Fund Mol. Mech. Mutagen 546 (1), 81–91.
Flora, S., Flora, G., Saxena, G., et al., 2007. Arsenic and lead induced free radical generation and their reversibility following chelation. Cell. Mol. Biol. 53 (1), 26–47.
Flora, S., Mittal, M., Mehta, A., 2008. Heavy metal induced oxidative stress & its possible reversal by chelation therapy. Indian J. Med. Res. 128 (4), 501.
Flora, S.J., 2011. Arsenic-induced oxidative stress and its reversibility. Free Radic. Biol. Med. 51 (2), 257–281.
Flora, S.J., 2016. Arsenic and dichlorvos: Possible interaction between two environmental contaminants. J. Trace Elem. Med. Biol. 35, 43–60.
Friedman, L.S., Lukyanova, E.M., Kundiev, Y.I., et al., 2006. Anthropometric, environmental, and dietary predictors of elevated blood cadmium levels in Ukrainian children: Ukraine ELSPAC group. Environ. Res. 102 (1), 83–89.
Gardner, R.M., Nyland, J.F., 2016. Immunotoxic effects of mercury. In: Environmental Influences on the Immune System. Springer, pp. 273–302.
Gavin, I.M., Gillis, B., Arbieva, Z., et al., 2007. Identification of human cell responses to hexavalent chromium. Environ. Mol. Mutagen. 48 (8), 650–657.
Ge, X., Wang, F., Zhong, Y., et al., 2018. Manganese in blood cells as an exposure biomarker in manganese-exposed workers healthy cohort. J. Trace Elem. Med. Biol. 45, 41–47.
Ghaderi, A., Banafshe, H.R., Khodabandehlo, S., et al., 2017. Qualitative thallium urinary assays are almost as valuable as quantitative tests: implication for outpatient settings in low and middle income countries. Electron. Physician 9 (4), 4190.
Gribble, M.O., Cheng, A., Berger, R.D., et al., 2015. Mercury exposure and heart rate variability: a systematic review. Curr. Environ. Health Rep. 2 (3), 304–314.
Guo, Y., Zhao, P., Guo, G., et al., 2015. The role of oxidative stress in gastrointestinal tract tissues induced by arsenic toxicity in cocks. Biol. Trace Elem. Res. 168 (2), 490–499.
Guzmán, O.D.L., Salazar, R.C., Martínez, N.P., et al., 2017. Micronucleus in exfoliated buccal cells of children from durango, Mexico, exposed to arsenic through drinking water. Revis Int de Contam Amb 33 (2), 281–287.
Hall, M., Chen, Y., Ahsan, H., et al., 2006. Blood arsenic as a biomarker of arsenic exposure: results from a prospective study. Toxicology 225 (2–3), 225–233.
Hasibuzzaman, M.M., Hossain, S., Islam, M.S., et al., 2017. Association between arsenic exposure and soluble thrombomodulin: a cross sectional study in Bangladesh. PLoS One 12 (4), e0175154.
Heck, J.E., Chen, Y., Grann, V.R., et al., 2008. Arsenic exposure and anemia in Bangladesh: a population-based study. J. Occup. Environ. Med. 50 (1), 80–87.
Hoet, P., Vanmarcke, E., Geens, T., et al., 2012. Manganese in plasma: a promising biomarker of exposure to Mn in welders. A pilot study. Toxicol. Lett. 213 (1), 69–74.
Huang, Q., Lu, Q., Chen, B., et al., 2017a. LncRNA-MALAT1 as a novel biomarker of cadmium toxicity regulates cell proliferation and apoptosis. Toxicol. Res. 6 (3), 361–371.
Huang, S., Xia, W., Li, Y., et al., 2017b. Association between maternal urinary chromium and premature rupture of membranes in the healthy baby cohort study in China. Environ. Pollut. 230, 53–60.
Ikegami, A., Takagi, M., Fatmi, Z., et al., 2016. External lead contamination of women's nails by surma in Pakistan: is the biomarker reliable? Environ. Pollut. 218, 723–727.
Jamal, A., Mehmood, A., Putus, T., et al., 2017. Prevalence of respiratory symptoms, bronchial asthma and obstructive lung disease among tannery workers. Peertechz J. Environ. Sci. Toxicol. 2 (1), 033–042.
Jiang, Y., Zheng, W., 2005. Cardiovascular toxicities upon managanese exposure. Cardiovasc. Toxicol. 5 (4), 345–354.
Karbowska, B., 2016. Presence of thallium in the environment: sources of contaminations, distribution and monitoring methods. Environ. Monit. Assess. 188 (11), 640.
Karri, V., Schuhmacher, M., Kumar, V., 2016. Heavy metals (Pb, Cd, as and MeHg) as risk factors for cognitive dysfunction: a general review of metal mixture mechanism in brain. Environ. Toxicol. Pharmacol. 48, 203–213.
Kasperczyk, S., Słowińska-Łożyńska, L., Kasperczyk, A., et al., 2015. The effect of occupational lead exposure on lipid peroxidation, protein carbonylation, and plasma viscosity. Toxicol. Ind. Health 31 (12), 1165–1171.
Khaled, E.M., Meguid, N.A., Bjørklund, G., et al., 2016. Altered urinary porphyrins & mercury exposure as biomarkers for autism severity in Egyptian children with autism spectrum disorder. Metab. Brain Dis. 31 (6), 1419–1426.
Kim, E.A., Cheong, H.K., Joo, K.D., et al., 2007. Effect of manganese exposure on the neuroendocrine system in welders. Neurotoxicology 28 (2), 263–269.
Kim, Y., Yoo, C.I., Lee, C.R., et al., 2002. Evaluation of activity of erythrocyte pyrimidine 5′-nucleotidase (P5N) in lead exposed workers: with focus on the effect on hemoglobin. Ind. Health 40 (1), 23–27.
La-Llave-León, O., Méndez-Hernández, E.M., Castellanos-Juárez, F.X., et al., 2017. Association between blood lead levels and delta-aminolevulinic acid dehydratase in pregnant women. Int. J. Environ. Res. Publ. Health 14 (4), 432.
Laine, J.E., Bailey, K.A., Rubio-Andrade, M., et al., 2015. Maternal arsenic exposure, arsenic methylation efficiency, and birth outcomes in the biomarkers of exposure to arsenic (BEAR) pregnancy cohort in Mexico. Environ. Health Perspect. 123 (2), 186.
Larsson, S.C., Orsini, N., Wolk, A., 2015. Urinary cadmium concentration and risk of breast cancer: a systematic review and dose-response meta-analysis. Am. J. Epidemiol. 182 (5), 375–380.
Lee, C.M., Terrizzi, A.R., Bozzini, C., et al., 2016. Chronic lead poisoning magnifies bone detrimental effects in an ovariectomized rat model of postmenopausal osteoporosis. Exp. Toxicol. Pathol. 68 (1), 47–53.
Li, P., Li, Y., Zhang, J., et al., 2016. Establishment of a reference value for chromium in the blood for biological monitoring among occupational chromium workers. Toxicol. Ind. Health 32 (10), 1737–1744.
Li, S.-J., Jiang, L., Fu, X., et al., 2014. Pallidal index as biomarker of manganese brain accumulation and associated with manganese levels in blood: a meta-analysis. PLoS One 9 (4), e93900.

Liang, G., Zhang, L.E., Ma, S., et al., 2016. Manganese accumulation in hair and teeth as a biomarker of manganese exposure and neurotoxicity in rats. Environ. Sci. Pollut. Res. 23 (12), 12265–12271.

Liu, G., Wang, Z.K., Wang, Z.Y., et al., 2016. Mitochondrial permeability transition and its regulatory components are implicated in apoptosis of primary cultures of rat proximal tubular cells exposed to lead. Arch. Toxicol. 90 (5), 1193–1209.

Macfie, A., Hagan, E., Zhitkovich, A., 2009. Mechanism of DNA– Protein cross-linking by chromium. Chem. Res. Toxicol. 23 (2), 341–347.

Mamyrbaev, A.A., Dzharkenov, T.A., Imangazina, Z.A., et al., 2015. Mutagenic and carcinogenic actions of chromium and its compounds. Environ. Health Prev. Med. 20 (3), 159.

Mazumdar, I., Goswami, K., Ali, M.S., 2017. Status of serum calcium, vitamin D and parathyroid hormone and hematological indices among lead exposed jewelry workers in Dhaka, Bangladesh. Indian J. Clin. Biochem. 32 (1), 110–116.

Misra, U., Kalita, J., Yadav, R., et al., 2003. Thallium poisoning: emphasis on early diagnosis and response to haemodialysis. Postgrad. Med. J. 79 (928), 103–105.

Mozafari, P., Azari, M.R., Shokoohi, Y., et al., October 2016. Feasibility of biological effective monitoring of chrome electroplaters to chromium through analysis of serum malondialdehyde. Int. J. Occup. Environ. Med. 7 (4), 199–206.

Mulkey, J.P., Oehme, F.W., 1993. A review of thallium toxicity. Vet. Hum. Toxicol. 35 (5), 445–453.

Ntihabose, R., Surette, C., Foucher, D., et al., 2017. Assessment of saliva, hair and toenails as biomarkers of low level exposure to manganese from drinking water in children. Neurotoxicology 64, 126–133.

Ochoa-Martínez, Á.C., Cardona-Lozano, E.D., Carrizales-Yáñez, L., et al., 2017. Serum concentrations of new predictive cardiovascular disease biomarkers in Mexican women exposed to lead. Arch. Environ. Contam. Toxicol. 1–11.

Olanow, C.W., 2004. Manganese-induced parkinsonism and Parkinson's disease. Ann. N. Y. Acad. Sci. 1012 (1), 209–223.

Osorio-Rico, L., Santamaria, A., Galván-Arzate, S., 2017. Thallium toxicity: general issues, neurological symptoms, and neurotoxic mechanisms. In: Neurotoxicity of Metals. Springer, pp. 345–353.

Pan, C.-H., Jeng, H.A., Lai, C.-H., 2018. Biomarkers of oxidative stress in electroplating workers exposed to hexavalent chromium. J. Expo. Sci. Environ. Epidemiol. 28 (1), 76.

Pant, N., Kumar, G., Upadhyay, A., et al., 2014. Reproductive toxicity of lead, cadmium, and phthalate exposure in men. Environ. Sci. Pollut. Res. 21 (18), 11066–11074.

Pawlas, N., Olewińska, E., Markiewicz-Górka, I., et al., 2017. Oxidative damage of DNA in subjects occupationally exposed to lead. Mutagenesis 5, 6.

Prozialeck, W.C., VanDreel, A., Ackerman, C.D., et al., 2016. Evaluation of cystatin C as an early biomarker of cadmium nephrotoxicity in the rat. Biometals 29 (1), 131–146.

Racette, B.A., Antenor, J.A., McGee-Minnich, L., et al., 2005. [18F] FDOPA PET and clinical features in parkinsonism due to manganism. Mov. Disord. 20 (4), 492–496.

Rahman, M., Al Mamun, A., Karim, M.R., et al., 2015a. Associations of total arsenic in drinking water, hair and nails with serum vascular endothelial growth factor in arsenic-endemic individuals in Bangladesh. Chemosphere 120, 336–342.

Rahman, S.M., Kippler, M., Ahmed, S., et al., 2015b. Manganese exposure through drinking water during pregnancy and size at birth: a prospective cohort study. Reprod. Toxicol. 53, 68–74.

Rana, S.V.S., 2008. Metals and apoptosis: recent developments. J. Trace Elem. Med. Biol. 22 (4), 262–284.

Rivera-Mancía, S., Ríos, C., Montes, S., 2011. Manganese accumulation in the CNS and associated pathologies. Biometals 24 (5), 811–825.

Rodríguez-Mercado, J.J., Mosqueda-Tapia, G., Altamirano-Lozano, M.A., 2017. Genotoxicity assessment of human peripheral Lymphocytes induced by thallium (I) and thallium (III). Toxicol. Environ. Chem. 99 (5–6), 987–998.

Rodríguez, J., Mandalunis, P.M., 2016. Effect of cadmium on bone tissue in growing animals. Exp. Toxicol. Pathol. 68 (7), 391–397.

Rosen, B.P., 2015. Microbiome Gene Expression as a Biomarker of Arsenic Exposure. U.S. Patent Application 14/704,323.

Sakamoto, M., Murata, K., Domingo, J.L., et al., 2016. Implications of mercury concentrations in umbilical cord tissue in relation to maternal hair segments as biomarkers for prenatal exposure to methylmercury. Environ. Res. 149, 282–287.

Sánchez-Villegas, C.M., Cortés-Vargas, A., Hidalgo-Luna, R., et al., 2014. Blood and urine lead levels in children with attention deficit hyperactivity disorder. Rev. Med. Inst. Mex. Seguro Soc. 52 (1), 20–27.

Sanders, T., Liu, Y., Buchner, V., et al., 2009. Neurotoxic effects and biomarkers of lead exposure: a review. Rev. Environ. Health 24 (1), 15–46.

Saxena, G., Flora, S., 2004. Lead-induced oxidative stress and hematological alterations and their response to combined administration of calcium disodium EDTA with a thiol chelator in rats. J. Biochem. Mol. Toxicol. 18 (4), 221–233.

Schuh, M.J., 2016. Possible Parkinson's disease induced by chronic manganese supplement ingestion. Consult. Pharm. 31 (12), 698–703.

Senthilkumaran, S., Balamurugan, N., Jena, N.N., et al., 2017. Acute alopecia: evidence to thallium poisoning. Int. J. Trichol. 9 (1), 30.

Shepherd, T.J., Dirks, W., Roberts, N.M., et al., 2016. Tracing fetal and childhood exposure to lead using isotope analysis of deciduous teeth. Environ. Res. 146, 145–153.

Shin, Y., Kim, K., Kim, E., et al., 2017. Identification of aldo-keto reductase (AKR7A1) and glutathione S-transferase pi (GSTP1) as novel renal damage biomarkers following exposure to mercury. Hum. Exp. Toxicol. https://doi.org/10.1177/0960327117751234.

Smith, E.A., Newland, P., Bestwick, K.G., et al., 2013. Increased whole blood manganese concentrations observed in children with iron deficiency anaemia. J. Trace Elem. Med. Biol. 27 (1), 65–69.

Somers, E.C., Ganser, M.A., Warren, J.S., et al., 2015. Mercury exposure and antinuclear antibodies among females of reproductive age in the United States: NHANES. Environ. Health Perspect. 123 (8), 792.

Sommar, J.N., Hedmer, M., Lundh, T., et al., 2014. Investigation of lead concentrations in whole blood, plasma and urine as biomarkers for biological monitoring of lead exposure. J. Expo. Sci. Environ. Epidemiol. 24 (1), 51.

Steinmaus, C., Ferreccio, C., Acevedo, J., et al., 2016. High risks of lung disease associated with early-life and moderate lifetime arsenic exposure in northern Chile. Toxicol. Appl. Pharmacol. 313, 10–15.

Tasmin, S., Furusawa, H., Ahmad, S.A., et al., 2015. Delta-aminolevulinic acid dehydratase (ALAD) polymorphism in lead exposed Bangladeshi children and its effect on urinary aminolevulinic acid (ALA). Environ. Res. 36, 318–323.

Tian, L., Zheng, G., Sommar, J.N., et al., 2013. Lead concentration in plasma as a biomarker of exposure and risk, and modification of toxicity by δ-aminolevulinic acid dehydratase gene polymorphism. Toxicol. Lett. 221 (2), 102–109.

Tsuji, J.S., Garry, M.R., Perez, V., et al., 2015. Low-level arsenic exposure and developmental neurotoxicity in children: a systematic review and risk assessment. Toxicology 337, 91–107.

Tutkun, E., Abusoğlu, S., Yılmaz, H., et al., 2014. Prolactin levels in manganese-exposed male welders. Pituitary 17 (6), 564–568.

Vacchi-Suzzi, C., Kruse, D., Harrington, J., et al., 2016. Is urinary cadmium a biomarker of long-term exposure in humans? A review. Curr. Environ. Health Rep. 3 (4), 450–458.

Verheij, J., Voortman, J., van Nieuwkerk, C., et al., 2009. Hepatic morphopathologic findings of lead poisoning in a drug addict: a case report. J. Gastrointestin Liver Dis. 18 (2), 225–227.

Wang, G., Zhang, T., Sun, W., et al., 2017. Arsenic sulfide induces apoptosis and autophagy through the activation of ROS/JNK and suppression of Akt/mTOR signaling pathways in osteosarcoma. Free Radic. Biol. Med. 106, 24–37.

Wang, Y., Goodrich, J.M., Gillespie, B., et al., 2012. An investigation of modifying effects of metallothionein single-nucleotide polymorphisms on the association between mercury exposure and biomarker levels. Environ. Health Perspect. 120 (4), 530.

Warang, P., Roshan, C., Kedar, P., 2017. Lead poisoning induced severe hemolytic anemia, basophilic stippling, mimicking erythrocyte pyrimidine 5′-nucleotidase deficiency in beta thalassemia minor. J. Clin. Toxicol. 7, 346.

Ward, E.J., Edmondson, D.A., Nour, M.M., et al., 2017. Toenail manganese: a sensitive and specific biomarker of exposure to manganese in career welders. Ann. Work Exp. Health 62 (1), 101–111.

Welling, R., Beaumont, J.J., Petersen, S.J., et al., 2015. Chromium VI and stomach cancer: a meta-analysis of the current epidemiological evidence. Occup. Environ. Med. 72 (2), 151–159.

Wijesekara, L.A., Usoof, R., Gamage, S.S., et al., 2017. Mercury levels in hair samples of dentists: a comparative study in Sri Lanka. J. Invest. Clin. Dent. e12302. https://doi.org/10.1111/jicd.12302.

Xia, W., Du, X., Zheng, T., et al., 2016. A case–control study of prenatal thallium exposure and low birth weight in China. Environ. Health Perspect. 124 (1), 164.

Yang, H., Zhang, H., Zhou, Q., et al., 2015. Study on relationships between biomarkers in workers with low-level occupational lead exposure. Zhonghua Lao Dong Wei Sheng Zhi Ye Bing Za Zhi 33 (6), 403–408.

Yang, J., Huo, W., Zhang, B., et al., 2016. Maternal urinary cadmium concentrations in relation to preterm birth in the healthy baby cohort study in China. Environ. Int. 94, 300–306.

Zhang, Q.-F., Li, Y.-W., Liu, Z.-H., et al., 2016. Reproductive toxicity of inorganic mercury exposure in adult zebrafish: histological damage, oxidative stress, and alterations of sex hormone and gene expression in the hypothalamic-pituitary-gonadal axis. Aquat. Toxicol. 177, 417–424.

Zhang, Q., Wang, D., Zheng, Q., et al., 2014. Joint effects of urinary arsenic methylation capacity with potential modifiers on arsenicosis: a cross-sectional study from an endemic arsenism area in Huhhot Basin, northern China. Environ. Res. 132, 281–289.

Zhao, G., Ding, M., Zhang, B., et al., 2008. Clinical manifestations and management of acute thallium poisoning. Eur. Neurol. 60 (6), 292–297.

CHAPTER

31

Melamine

Karyn Bischoff

Cornell University, Diagnostic Toxicologist, New York State Animal Health Diagnostic Center, Ithaca NY, United States

INTRODUCTION AND HISTORICAL BACKGROUND

Melamine (1,3,5-triazine-2,4,6-triamine) has numerous uses in manufacturing. It has been used in yellow pigments, dyes, and inks and can be polymerized with formaldehyde to produce a variety of durable resins, adhesives, cleansers, and flame retardants. Cyanuric acid is used to stabilize chlorine in swimming pools. Protein in foods is estimated based on the nitrogen content; because melamine is 67% nitrogen by molecular weight, it has been added to foodstuffs (both human and animal) to increase the apparent protein content. Structural formulas of melamine and cyanuric acid are shown in Fig. 31.1.

Early in 2007, there were reports of inappetence, vomiting, polyuria, polydipsia, and lethargy in cats and dogs, including laboratory cats on a feeding trial for a commercial pet food (Cianciolo et al., 2008). The pet food was recalled on March 15 and analyzed for possible contaminants. Melamine was found, as were low concentrations of cyanuric acid, ammelide, and ammeline, which are intermediates in the production and degradation of melamine. The United States Food and Drug Administration (FDA) investigation determined that two common pet food ingredients, sold to manufacturers as wheat gluten and rice protein concentrate, were mislabeled by Chinese exporters to avoid inspection (Osborne et al., 2008). Samples of the imported wheat gluten contained 8.4% melamine, 5.3% cyanuric acid, and 2.3% and 1.7% ammelide and ammeline, respectively (Rumbeiha et al., 2010). Eventually, more than 150 pet food products were identified as containing contaminated ingredients and were recalled. Analysis revealed that these products contained up to ~3200 ppm melamine and 600 ppm cyanuric acid (Cianciolo et al., 2008; Skinner et al., 2010).

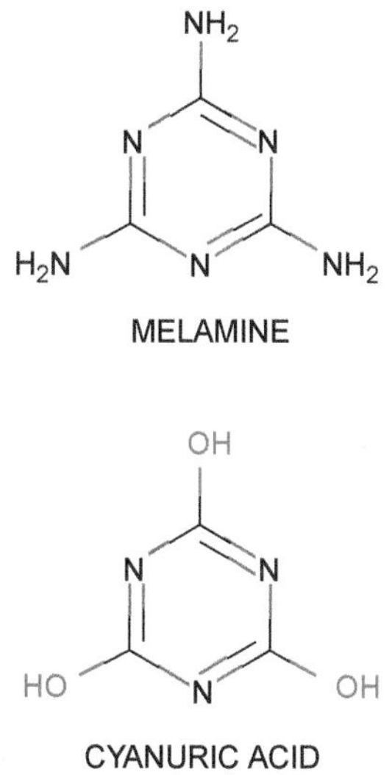

NH_2
N N
H_2N N NH_2
MELAMINE

OH
N N
HO N OH
CYANURIC ACID

FIGURE 31.1 Structural formulas of melamine and cyanuric acid.

Cats and dogs ingesting the contaminated food had evidence of renal failure. Cats had urine-specific gravities <1.035 and elevated serum urea nitrogen and creatinine concentrations. Urinalysis revealed the presence of circular green–brown crystals in urine sediment. Postmortem examination of animals that died typically noted bilateral renomegaly and evidence of uremia. Microscopic lesions were found primarily in the kidneys. Renal tubular necrosis was present, with evidence of rupture and regeneration. Distal convoluted tubules contained large golden-brown birefringent crystals (15–80 μm in diameter) with centrally radiating striations and smaller amorphous crystals (Cianciolo et al., 2008; Thompson et al., 2008). Crystals from kidneys and urine contained 70% cyanuric acid and 30% melamine based on infrared spectra (Osborne et al., 2008; Thompson et al., 2008).

Previous outbreaks of renal failure associated with pet foods and other animal feeds, including feeds for

Biomarkers in Toxicology, Second Edition
https://doi.org/10.1016/B978-0-12-814655-2.00031-1

poultry, swine, and raccoon dogs in the Republic of Korea, Japan, Thailand, Malaysia, Singapore, Taiwan, the Philippines, South Africa, Spain, China, and the United States were investigated and determined to be associated with melamine contamination (Acheson, 2007; Osborne et al., 2008; Reimschuessel et al., 2008, 2009; Bhalla et al., 2009; Gonzalez et al., 2009; Yhee et al., 2009; US FDA, 2010).

The 2007 pet food recall is considered a sentinel event by some (Osborne et al., 2008; Lewin-Smith et al., 2009). Indeed, 1 year later, melamine contamination of milk-based baby formula was found in China. Chinese authorities detected melamine concentrations between 2.5 and 2563 ppm in 13 commercial brands of milk powder and trace contamination in 9 others (Bhalla et al., 2009). Approximately 300,000 children may have been affected, more than 52,000 were hospitalized, and 6 died. Melamine-contaminated products were found in almost 70 countries, including the United States, and some children in Taiwan, Hong Kong, and Macau could have been clinically affected (Hau et al., 2009; Reimschuessel et al., 2009; Skinner et al., 2010).

Renal damage occurred in 0.61%–8.5% of exposed children, according to various estimates (Liu et al., 2010). Clinical signs of renal failure in children who consumed contaminated milk products included unexplained crying, anuria or polyuria, stranguria, hematuria, and the presence of stones in the urine, but many children were asymptomatic (Hau et al., 2009; Hu et al., 2010; Wen et al., 2010). The outbreak of melamine-induced nephropathy in children was different from the outbreak in companion animals and livestock in that cyanuric acid was not an important contaminant and was not required for crystal formation. Crystals in the infants contained melamine and uric acid at a molar ratio of 1:1 to 1:2 (Skinner et al., 2010; Wen et al., 2010).

Most affected children were between 6 and 18 months of age (Skinner et al., 2010; Wen et al., 2010). Clinical pathology findings in affected infants included elevated serum potassium, urea nitrogen, and creatinine concentrations (Sun et al., 2010). Hematuria and the presence of fan-shaped crystals were reported in urine samples and urine pH ranged from 5.0 to 7.5 (Hau et al., 2009). Renal calculi ranged in size from 2 to 18 mm in diameter, and approximately half were <5 mm in diameter. Diameter was dependent on the concentration of melamine in the diet but not on the duration of exposure (Hu et al., 2010). Renal calculi were usually bilateral; uroliths were also found in the ureter, unilaterally or bilaterally, and in the urinary bladder (Hau et al., 2009; Gao et al., 2010; Wen et al., 2010). A renal biopsy from an affected child showed glomerular sclerosis, swelling and necrosis of renal tubular epithelium, tubular dilation with the presence of material consistent with a stone, and an interstitial lymphoplasmacytic infiltrate. Tubular changes resolved after treatment (Sun et al., 2010).

TOXICOLOGY OF MELAMINE

Toxicity

As mentioned previously, until 2007 melamine was believed to have a relatively low toxicity based on animal studies. The oral LD_{50} of melamine is 3200 mg/kg in male rats, 3800 mg/kg in female rats, 3300 mg/kg in male mice, and 7000 mg/kg in female mice. Long-term administration of melamine to laboratory rodents at concentrations ranging from 0.225% to 0.9% of the diet produces urolithiasis (Melnick et al., 1984). A study involving dogs fed 125 mg/kg melamine reported crystalluria but no other adverse effects (Lipschitz and Stokey, 1945). Cyanuric acid has similarly low toxicity when given alone but is known to produce degenerative changes in the kidneys of guinea pigs when given at doses of 30 mg/kg body weight for 6 months, and in dogs and rats fed a diet containing 8% monosodium cyanurate (Canelli, 1974).

The combination of melamine and cyanuric acid is markedly more toxic to most animals than either compound when given alone. Cats fed diets containing 0.2% each of melamine, and cyanuric acid had evidence of acute renal failure within 48 h (Puschner et al., 2007). A pig fed 400 mg/kg each of melamine and cyanuric acid daily had transient bloody diarrhea within 24 h. Necropsy revealed perirenal edema and round golden-brown crystals with radiating striations in the kidneys. Similar lesions were present in tilapia, rainbow trout, and catfish dosed with 400 mg/kg each of melamine and cyanuric acid daily for 3 days (Reimschuessel et al., 2008).

Pharmacokinetics/Toxicokinetics

Melamine undergoes rapid renal elimination and is almost completely excreted within 1 day in monogastric mammals, though excretion is prolonged in ruminants (Dorne et al., 2012). Detectable melamine concentrations have been reported in edible tissues from animals, particularly the kidneys, for up to 4 days in lambs and up to 20 days in poultry (Lv et al., 2010; Bai et al., 2010). Melamine can be detected in cow's milk for up to 4 days post dosing (Shen et al., 2010).

Mechanism of Action

Cats and dogs ingesting contaminated food had evidence of renal failure. Although previous animal studies found that both melamine and cyanuric acid were relatively nontoxic when given individually, they caused crystal formation in renal tubules when given together. Melamine and cyanuric acid crystallize, forming a molecular lattice structure at pH = 5.8 (Osborne et al., 2008; Bhalla et al., 2009). Crystals form in the distal convoluted tubules of the kidneys and contain 70% cyanuric acid and 30% melamine based on infrared spectra results (Osborne et al., 2008; Thompson et al., 2008). Intratubular obstruction causing increased intrarenal pressure most likely contributes to the pathology.

After consuming contaminated formula for 3–6 months, children developed urolithiasis, which can result in obstruction and secondary renal changes. Tubular changes resolved after treatment (Sun et al., 2010). Although cyanuric acid did not contribute to the formation of urinary calculi in children during this incident, uroliths in children were produced by a similar interaction between melamine and uric acid (Gao et al., 2010). This interaction has also been reported in poultry (Bai et al., 2010). Humans and most other primates lack the enzyme uricase, which converts uric acid to allantoin (Reimschuessel et al., 2008). Compared to adults, human infants excrete between 5 and 8 times as much uric acid, possibly increasing their susceptibility to melamine toxicosis (Skinner et al., 2010). Urinary pH < 5.5 is associated with the formation of urate crystals and low pH and could contribute to melamine/urate stone formation (Gao et al., 2010).

Treatment

The basic treatment regimens for crystalluria and urolithiasis related to melamine ingestion include fluid therapy and supportive care in both veterinary and pediatric patients (Anon., 2007; Wen et al., 2010). Because low urinary pH is associated with crystal formation in infants, alkalization of the urine was used to maintain urine pH between 6.0 and 7.8 in affected children (Gao et al., 2010; Wen et al., 2010). Antispasmodic drugs such as anisodamine or atropine were given to facilitate excretion of uroliths and pain management was instituted (Bhalla et al., 2009; Wen et al., 2010). Most children recovered fully, but 12% were found to have renal abnormalities 6 months after treatment (Liu et al., 2010).

BIOMARKERS

There are relatively few published papers on biomarkers associated with exposure to melamine and related compounds. There have been studies in affected children, but most of those studies used biomarkers to explore the pathophysiology of kidney injury. Nonetheless, findings from these and other studies could be useful in diagnostics as specific, sensitive, early, noninvasive indicators of kidney damage associated with melamine exposure (Camacho et al., 2011; Gao et al., 2011; Zhang et al., 2012; Bandele et al., 2013).

Melamine + Cyanuric Acid Biomarkers

Serum chemistry findings in companion animals that were exposed to melamine and cyanuric acid in contaminated feed included indicators of renal failure such as increased blood urea nitrogen (BUN), creatinine, and phosphorus. Serum sodium and chloride concentrations were decreased. Other serum chemistry changes included increased cholesterol, triglycerides, and the liver enzyme γ-glutamyl transpeptidase (GGT) (Dobson et al., 2008; Chen et al., 2010). The presence of characteristic crystals was often noted on urinalysis.

Several rodent studies have looked for other markers associated with exposure to melamine + cyanuric acid. Zhang et al. (2012) fed diets containing 0–240 ppm melamine and equal concentrations of cyanuric acid to Fischer 344 (F344) rats. A low dose (120 ppm) resulted in increased urine concentrations of replication protein A-1 (RPA-1), though there was minimal evidence of renal injury on histopathology. RPA-1 indicates injury to renal distal convoluted tubules and collecting ducts. Bandele et al. (2013) did a similar study with F344 rats gavage fed 5 mg/kg/day melamine and an equal amount of cyanuric acid and found that the urine concentration of kidney injury molecule-1 (KIM-1) increased approximately 9- to 49-fold and clusterin increased about 8- to 24-fold. However, in rats given approximately equivalent daily doses, using dietary concentrations of 60 ppm of each contaminant, there was no elevation of RPA-1 or KIM-1. Increases in KIM-1 and clusterin were observed in gavage-fed rats on day 4, and peak concentrations were detected on day 14 of melamine + cyanuric acid exposure. Urine concentrations of both KIM-1 and clusterin declined somewhat by day 21. KIM-1 and clusterin are both indicators of renal response to injury. There was no change in serum BUN or creatinine concentrations in any of the rats. Silva et al. (2016) also looked at F344 rats and did find that BUN and creatinine were elevated in rats fed diets containing at least 240 ppm each of melamine and cyanuric acid for 28 days, but not in the lower dietary concentration groups. However, serum concentrations of some microRNAs, specifically microRNA-128-3p and microRNA-210-3p, were decreased at dietary concentrations of 180 ppm each of melamine and cyanuric acid, which were associated with histologic renal lesions.

Melamine appears to have effects on amino acid metabolism (Zhao and Lin, 2014). Schnackenberg et al. (2012)

looked specifically at hydroxyproline concentrations in the urine of F344 rats fed diets containing 0–240 ppm each of melamine and cyanuric acid. Hydroxyproline is a breakdown product of collagen; it is released during tissue injury and was used as an indicator for renal tubular damage. The increase in hydroxyproline was dose-dependent and correlated with histologic lesions of renal fibrosis. Urine hydroxyproline concentrations increased by day 4 in rats fed diets containing 240 ppm melamine + cyanuric acid, by day 14 in rats fed 180 ppm of each contaminant, and by day 28 in the group fed diets containing 120 ppm of each.

Xie et al. (2010) dosed Wistar rats for 15 days with gavage doses of 100–600 mg/kg/day melamine alone, or with 50 ppm melamine + 50 ppm cyanuric acid, and analyzed the urine using a variety of methods. An incomplete list of their findings is present in Table 31.1. These biomarkers were used to determine what metabolic pathways were affected by melamine and melamine + cyanuric acid. Again, amino acid metabolism was affected. Some metabolites were involved in tryptophan metabolism. These include 5-hydroxy-3-acetaldehyde, indolacetaldehyde, *N*-acetyl-5-hydroxytryptamine, and nicotinamide. Metabolites involved in tyrosine metabolism include tyramine, tyrosine, and dopa. Metabolites involved in arginine metabolism include L-arginine, urea, and guanidinoacetate. Cis-aconitate is involved in the citric acid cycle.

Kim et al. (2012) dosed male Sprague–Dawley rats with 63 mg/kg melamine and 6.3 mg/kg cyanuric acid for 30 days, then analyzed the kidneys for the biomarkers listed in Table 31.2. Decreased concentrations of asparagine, aspartate, glutathione, and alanine were associated with decreased resorptive ability of the renal proximal convoluted tubules. Medullary tissues had increased asparagine, choline, creatinine, cysteine, ethanolamine, isoleucine, glutamine, and myoinositol, but decreased phenylalanine and tyrosine. Renal cortex had increased isoleucine and decreased threonine and myoinositol. Increased concentrations of ethanol and lactic acid were attributed to inhibition of the citric acid cycle. High-mobility group protein B1 (HMGB1) is released as a proinflammatory mediator by necrotic cells. Increases in HMGB1 correlated with increases in BUN and serum creatinine. Increased netrin-1 is associated with toxic renal injury. Increases in netrin-1 were seen before changes in BUN and creatinine.

Camacho et al. (2011) and Bandele et al. (2013) looked at expression of genes involved in the regenerative response to injury in the kidney (Table 31.3). They fed diets containing 7–694 ppm each of melamine + cyanuric acid for 7 days or 60 ppm of each for 28 days to F344 rats. Results of these studies are listed in Table 31.3. The KIM-1 gene had the most dramatic upregulation, which correlated with dose and was not evident at subclinical doses. CLU, Spp1/osteopontin, and Lcn2 upregulation did not increase at subclinical doses either but increases were associated with a threshold dose of melamine + cyanuric acid and were not dose-dependent.

Pacini et al. (2014) looked at the effects of dietary melamine and cyanuric acid in trout. Crystals in the urinary tract were persistent at high dietary doses of melamine and cyanuric acid. Activity of catalase and glutathione S-transferase was elevated in kidneys of fish exposed to melamine and cyanuric acid or melamine alone, but not in fish fed a diet containing cyanuric acid alone.

TABLE 31.1 Biomarkers in the Urine of Wistar Rats Dosed With Melamine + Cyanuric Acid at Doses of 50 ppm Each or Melamine at Doses of 100, 300, and 600 mg/kg by Gavage for 15 days (Xie et al., 2010)

Biomarker	Concentration in Melamine + Cyanuric Acid Group	Concentration in Melamine Only Groups
5-Hydroxy-3-acetaldehyde	Decreased	Decreased
Indoleacetaldehyde	Increased	Increased
N-Acetyl-5-hydroxytryptamine	Decreased	Decreased
Nicotinamide	Decreased	
Tyramine	Increased	Increased
Tyrosine	Decreased	Decreased
Dopa	Decreased	
L-arginine	Decreased	
Urea	Increased	Increased
Guanidinoacetate	Decreased	Decreased
Cis-aconitate	Increased	Increased
Histidine	Increased	Increased
Leukotriene E4	Increased	Increased
Pipecolic acid	Increased	Increased
1-Methylguanine	Decreased	Decreased
(s)-3-Hydroxyisobutyrate	Increased	Increased
Riboflavin	Decreased	Decreased
dADP	Increased	Increased
Thymidine	Decreased	Decreased
Adenosine	Decreased	Decreased
3′,5′-cyclic adenosine	Decreased	Decreased
Pyridoxine	Decreased	Decreased

TABLE 31.2 Biomarkers in the Kidneys of F344 Rats Given Subclinical Doses of Melamine + Cyanuric Acid (5 mg/kg/day) in the Diet or via Gavage for 28 days (Kim et al., 2012)

Biomarker	Change in Concentration	Location
HMGB1	Increased	Kidney
Netrin-1	Increased	Kidney
Asparagine	Decreased	Renal cortex
Aspartate	Decreased	Renal cortex
Glutathione	Decreased	Renal cortex
Alanine	Decreased	Renal cortex
Hypoxanthine	Increased	Renal cortex
Isoleucine	Increased	Renal cortex
O-Phosphocholine	Increased	Renal cortex
Ethanol	Increased	Renal cortex
Lactate	Increased	Renal Cortex
N-Acetylglutarate	Decreased	Renal medulla
Phenylalanine	Decreased	Renal medulla
Taurine	Decreased	Renal medulla

TABLE 31.3 Gene Expression in Kidneys of Rats Dosed With Melamine and Cyanuric Acid at 7–694 ppm in the Diet for 7 days (Camacho et al., 2011) or 60 ppm in the Diet for 28 days (Bandele et al., 2013)

Biomarker	Regulation (Camacho et al., 2011)	Regulation (Bandele et al., 2013)	Correlated With Dose
KIM-1/Havcr1	Up 75–2500 X	Up 400–1500 X	Yes
CLU	Up	Up 60–90 X	No
Spp1/ osteopontin	Up	Up 60–90 X	No
Lcn2	Up	Up 60–90X	No
Timp-1	Up	Up 12 X	
Gapdh	Up		No

Melamine Only Biomarkers

Stones from the upper urinary tracts of children exposed to melamine were composed of melamine and uric acid at molar ratios between 1:1 and 1:2 (Chang et al., 2012; Liu et al., 2012). Uric acid was in the form of uric acid dihydrate and ammonium urate. No cyanuric acid or other melamine analog were detected. Uric acid is produced as the final step in the oxidation of purines in humans and other primates because in these species uric acid is produced as a final step in the oxidation of purines, whereas other mammals further oxidized uric acid to allantoin via uricase (Liu et al., 2012). Thus, serum uric acid concentrations in humans and related primates tend to be higher than those in other mammals. Serum uric acid was decreased in children with melamine-associated nephrolithiasis (Xie et al., 2010).

Persistent serum creatinine concentrations >7.1 μg/dL were associated with increased risk of stones. Some affected children had elevated total, direct, and indirect bilirubin and elevated alanine aminotransferase and aspartate aminotransaminase, both hepatic enzymes, in the serum (Hu et al., 2012). Urine was turbid and contained sediment, blood, and protein (Wang et al., 2009; Gao et al., 2011), but melamine was not always detectable in the urine of children with nephroliths (Wang et al., 2009). However, when Lin et al. (2013) analyzed urine in school-aged children for melamine with a spot test, it was present at small concentrations (mean 2.48 μg/mmol creatinine). They concluded that a spot test could be used to predict total excretion over the previous 24 h. The study also looked for β2-microglobulin, a biomarker for proximal tubular dysfunction, but it was not detected. Microalbumin and *N*-acetyl-β-glucosaminidase, other biomarkers for renal tubular injury, were present but did not correlate with melamine exposure.

Protein has a role in stone formation and was also present in small amounts in the melamine/uric acid stones from children. Protein can form the nucleus of the stone and help bind the crystals together. It has been proposed that cell membrane proteins can anchor the stone in place within the urinary tract. Examples of cell membrane proteins associated with uric acid/melamine stones include phosphatidylserine, CD44, osteopontin, and hyaluronan. Some metabolic and structural proteins, associated with leakage of cells during apoptosis or necrosis, were also found in the stones, as were inflammatory proteins such as α2-macroglobin, complement component 3, attractin, and fibronectin 1. Table 31.4 compares the proteins found in stones from children exposed to melamine to those found in stones from rats fed diets containing 2% melamine for 13 weeks.

Metabolites present in the urine of children exposed to melamine indicated an effect on the citric acid cycle (Zhao and Lin, 2014). Duan et al. (2011) analyzed urine from children with documented nephrolithiasis associated with either melamine exposure or other causes. The biomarkers they detected are listed in Table 31.5. Increased L-β-aspartyl-L-glycine, L-homocysteine sulfonic acid, hypoxanthine, and vanylglycol and decreased 5C-aglycone, proline, and 3-hydroxysebacic

TABLE 31.4 Proteins Isolated From Bladder Stones From Melamine-Exposed Humans and From Rats Fed 2% Melamine in the Diet for 13 Weeks (Liu et al., 2012)

Biomarker	Humans	Rats
Albumin	✓	✓
Transferrin	✓	✓
Macroglobulin	✓	✓
Immunoglobulin	✓	✓
Complement C3	✓	✓
Clusterin	✓	✓
Cathepson	✓	✓
Fibronectin	✓	✓
Plasminogen	✓	✓
Serum amyloid P	✓	✓
Apolipoprotein A	✓	✓
Apolipoprotein E	✓	✓
Hemoglobin	✓	✓
Kininogen-S-100	✓	✓
S100 calcium binding protein	✓	✓
Fibrinogen	✓	✓
A-actinin	✓	✓
Ceruloplasmin	✓	✓
Elongation factor 1	✓	✓
Tubulin	✓	✓
Histone H2B	✓	✓
Cytokeratin 7		✓
Cytokeratin 18		✓
Bcl-xl		✓
A-actin		✓

TABLE 31.5 Biomarkers Found in Urine From Two Groups of Children With Nephrolithiasis: Those Exposed to Melamine and Those Not Exposed to Melamine (Duan et al., 2011)

Biomarker	Change in Concentration	Melamine or Other Causes
L-beta-aspartyl-L-glycine	Increased	Melamine group only
5C-aglycone	Decreased	Melamine group only
Proline	Decreased	Melamine group only
3-hydroxysebacic acid	Decreased	Melamine group only
L-homocysteine sulfonic acid	Increased	Melamine group only
Hypoxanthine	Increased	Melamine group only
Vinylglycol	Increased	Melamine group only
Creatinine	Increased	Both groups
Uric acid	Increased	Both groups
7-methylguanine	Increased	Both groups

acid were associated with melamine exposure but not with other causes of nephrolithiasis. L-homocysteine sulfonic acid is a metabolite of homocysteine and a general indicator of renal disease. Hypoxanthine is a progenitor of uric acid and indicates uric acid accumulation in the tubules. Proline is an amino acid involved in glutamate metabolism and is suppressed by elevated uric acid. Uric acid accumulation in the renal tubules is also associated with decreased glycogen storage, for which 3-hydroxysebacic acid is a marker. 5C-aglycone is an indicator of decreased renal resorption of vitamin K. Wang et al. (2009) and Gao et al. (2011) also analyzed urine from children exposed to melamine, both before and after medical treatment. They determined that urine α1-microglobulin and β2-microglobulin, markers for resorptive function of the proximal convoluted tubules, were elevated at presentation. This elevation persisted for at least 3 months. Urine microalbumin, which indicates increased glomerular permeability, was mildly elevated and remained high for at least 3 months (Wang et al., 2009). Urine IgG concentrations are also indicative of glomerular damage. IgG elevations were associated with the presence of stones and remained elevated for more than 6 months. An indicator of damage to tubular epithelium, *N*-acetyl-β-D-glycosidase elevation correlated with younger patients and increasing stone diameter (Gao et al., 2011).

Hsieh et al. (2012) experimented with the effects of melamine on cultured human renal proximal tubular epithelium (HK-2 cells) at concentrations ranging from 100 to 1000 μg/mL to determine the effect on proteins, mRNA, and protein phosphorylation (Tables 31.6 and 31.8). Hsieh's group discovered downregulation of the antiapoptotic proteins Bcl-2 at moderate (500 μg/mL) and Bcl-x1 at high melamine concentrations (1000 μg/mL), as well as upregulation of proapoptotic proteins Bad and Bax at low (250 μg/mL) concentrations, resulting in increased apoptosis in the cells. Increased mRNA expression for vascular cell adhesion molecule-1 (VCAM-1) and transforming growth factor β1 (TGF-β1) was evident within 3 h of exposure to low (250 μg/mL) melamine concentrations. VCAM-1 is a chemoattractant for macrophages/monocytes, and TGF-β1 is a proapoptotic cytokine. Increased expression monocyte chemoattractant protein 1 (MCP-1) and the proinflammatory cytokine interleukin

TABLE 31.6 Protein Expression in Cultured Human Renal Proximal Tubular Cells (HK-2 Cells) Exposed to Concentrations of Melamine Ranging From 100 to 1000 μg/mL (Hsieh et al., 2012)

Biomarker	Regulation	Concentration of Melamine (μg/mL)
Bcl-2	Down	≥500
Bcl-x1	Down	1000
Bad	Up	≥250
Bax	Up	≥250
Cleaved caspase-3	Up	≥250
Cleaved caspase-9	Up	≥250

TABLE 31.7 Expression of mRNA in Cultured Human Renal Proximal Tubular Cells (HK-2 Cells) Exposed to Concentrations of Melamine Ranging From 100 to 1000 μg/mL (Hsieh et al., 2012)

mRNA	Regulation	Time Postexposure (h)	Concentration of Melamine (μg/mL)
IL-6	Up 200%	6–26	≥250
MCP-1	Up 150–300+%	6–24	≥125
VCAM-1	Up 130+%	3–24	≥250
TGF-β1	Up 200%	3–24	≥250
Fibronectin	>200%		≥250

TABLE 31.8 Phosphorylation of Inflammatory Mediators in Cultured Human Renal Proximal Tubular Cells (HK-2 Cells) Exposed to Concentrations of Melamine Ranging From 100 to 1000 μg/mL (Hsieh et al., 2012)

Inflammatory Mediator	Time Postexposure (min)	Dose-Dependent
44/42 MAPK	15–30	Yes
P38 MAPK	15–30	Yes
NF-κB	30	No

6 (IL-6) were evident a few hours later. Phosphorylation is another cell regulatory mechanism, and early phosphorylation of 44/42 mitogen-activated protein kinase (44/42 MAPK), P38 MAPK, and nuclear factor-κB (NF-κB), all of which regulate expression of proinflammatory genes, was noted.

Melamine was detected in the urine of 76% of the US population (Panuwet et al. 2012). Wu et al. (2015) looked at blood and urine from adults working in melamine manufacture. Melamine concentrations in urine were higher post- than preshift and decreased over the weekend. Creatinine, BUN, and uric acid were neither significantly elevated in melamine workers versus control workers from other fields, nor were the estimated glomerular filtration rate or estimated creatinine clearance rates significantly different. Microalbumin, *N*-acetyl *β*-D-glucosaminidase, and *β*2-microglobulin were measured in urine and compared to controls. *N*-Acetyl *β*-D-glucosaminidase concentrations increased with urinary melamine concentrations, and *β*2-microglobulin was more likely to be detectable in melamine workers. *N*-Acetyl *β*-D-glucosaminidase and *β*2-microglobulin are both biomarkers for early renal tubular damage. Microalbumin, a marker of glomerular damage, was not increased in melamine workers compared to controls.

Sun et al. (2012) did extensive work on biomarkers in urine, serum, and tissues of Wistar rats exposed to melamine at doses ranging from 250 to 1000 mg/kg/day (Tables 31.9 and 31.10). Succinate and citrate, involved in energy production via the citric acid cycle, were increased in the urine. Increased urinary taurine and creatinine were interpreted as general indicators of liver damage. BUN and creatinine are products of amino acid metabolism, and elevations in their concentrations are considered general indicators of renal insufficiency due to decreased glomerular filtration. Increases in dimethylglycine, *N*-acetylglycoprotein, taurine, very low–density lipoprotein (VLDL), low-density lipoprotein (LDL), and glucose were noted at the lowest melamine dose (250 mg/kg/day). Changes in VLDL, LDL, and glucose are possible indicators of alterations in carbohydrate metabolism. Increased serum choline, 3-hydroxybutyrate, and lactate were evident in the highest melamine dose (1000 mg/kg/day) group. All were considered indicative of a shift in metabolic pathways from glycolysis to lipolysis. Liver and kidney concentrations of choline were also increased. Concentrations of lactate, glutamate, and glucose were decreased in the liver of rats in the low-dose group, but increased in the kidneys, most likely due to a defect in tubular resorption.

Early et al. (2013) gave low (350 mg/kg/day) and high (1050 mg/kg/day) doses of melamine to Sprague–Dawley rats for 5 days and looked for changes in gene expression in renal tissue. They found upregulation of KIM-1, Clu, Spp1, A2m, Lcn2, Tnfrl2a, Gpnmb, and CD44 from 5- to 3000-fold, and downregulation of Tff3 two to 10-fold. Dysregulation correlated with histologic lesions in the kidneys.

Bandele et al. (2014) noted that the increased BUN and creatinine seen in rats given high doses

TABLE 31.9 Biomarkers Found in the Urine or Serum of Wistar Rats Exposed to Melamine at Doses of 250–1000 mg/kg/day (Sun et al., 2012)

Biomarker in Urine	Change	Biomarker in Serum	Change
Succinate	Increased	BUN	Increased
Citrate	Increased	Creatinine	Increased
Taurine	Increased (1000 ppm group)	Dimethylglycine	Increased
Creatine	Decreased (1000 ppm group)	VLDL	Increased
		LDL	Increased
		Taurine	Decreased
		Glucose	Decreased
		Choline	Increased (1000 ppm group)
		3-Hydroxybutyrate	Increased (1000 ppm group)
		Lactate	Decreased (1000 ppm group)

BUN, blood urea nitrogen; *LDL*, low-density lipoprotein; *VLDL*, very low–density lipoprotein.

TABLE 31.10 Biomarkers of Melamine Found in Liver or Kidney of Wistar Rats Exposed to Melamine at Doses of 250–1000 mg/kg/day (Sun et al., 2012)

Biomarker	Change in Liver	Change in Kidney
Choline	Increased (250–500 ppm groups)	Increased
Lactate	Decreased	Increased
Glutamate	Decreased (250–500 ppm groups)	Increased
Glucose	Decreased (250–500 ppm groups)	Increased
N-Acetylglycoprotein	Increased (1000 ppm group)	
Trimethylamine-N-oxide	Decreased (1000 ppm group)	
Creatinine	Increased (250–500 ppm groups)	
Amino acids		Increased
3-hydroxybutyrate		Decreased
Pyruvate		Decreased

(1000 mg/kg) of melamine were higher in pregnant versus nonpregnant females. (Increases were milder for cyanuric acid alone.) Serum uric acids were not altered. KIM-1, clusterin, and osteopontin increased in both pregnant and nonpregnant rats given melamine, but more rapidly in the pregnant rats.

CONCLUDING REMARKS AND FUTURE DIRECTIONS

Poor quality melamine containing cyanuric acid was fraudulently added to pet food ingredients by Chinese distributors and produced morbidity and death in dogs and cats in different areas of the world. This type of fraud had likely been ongoing for several years, but was first documented in 2007. The following year, it was discovered that a purer grade of melamine had been added to milk products, including baby formula, causing uroliths in infants and young children.

Melamine and cyanuric acid interact in the renal tubules forming crystals and producing renal failure, as was evident in dogs and cats during the 2007 pet food recalls in the United States. Although many cats and dogs died, many also responded to supportive care, which mostly consisted of fluid therapy. Melamine can also interact with uric acid, and this caused urolithiasis in young children in 2008. Fluid therapy and urine alkalization were the mainstay of therapy for these children and most survived.

There have been a few studies looking for biomarkers of melamine + cyanuric acid or melamine alone in vitro or in laboratory animals, and fewer studies of biomarkers for melamine in affected children. Most of the laboratory research thus far has been directed at clarifying the pathophysiology of melamine toxicosis. These studies suggest that the physiological disruptions caused by melamine and melamine + cyanuric acid affect amino acid and energy metabolism through multiple pathways. Furthermore, melamine is able to induce apoptosis in renal tubular epithelium in vitro. Some of the findings have potential uses in diagnostic testing for exposure to melamine or in differentiating melamine exposure from other causes of urolithiasis in children (Table 31.5). *N*-Acetyl-β-D glycosidase is of particular interest as an indicator of persistent urolithiasis. Upregulation of mRNA for TGF-β1, VCAM-1, MCP-1, and IL-6 was found to be an early indicator of melamine exposure in vitro, but it remains to be determined if the cytokines will be useful indicators in vivo.

References

Acheson, D., 2007. Importation of Contaminated Animal Feed Ingredients, Statement before the House Committee on Agriculture. Washington DC, United States, May 9, 2007.

Anon., 2007. Specialists confer about the pet food recall. J. Am. Vet. Med. Assoc. 233, 1603.

Bai, X., Bai, F., Zhang, K., 2010. Tissue deposition and residue depletion in laying hens exposed to melamine-contaminated diets. J. Agric. Food Chem. 58, 5414–5420.

Bandele, O., Camacho, L., Ferguson, M., 2013. Performance of urinary and gene expression biomarkers in detecting nephrotoxic effects of melamine and cyanuric acid following diverse scenarios of co-exposure. Food Chem. Toxicol. 51, 106–113.

Bandele, O.J., Stine, C.B., Ferguson, M., Black, T., Olejnik, N., Keltner, Z., Evans, E.R., Crosby, T.C., Reimschuessel, R., Sprando, R.L., 2014. Use of urinary renal biomarkers to evaluate the nephrotoxic effects of melamine or cyanuric acid in non-pregnant and pregnant rats. Food Chem. Toxicol. 74, 301–308.

Bhalla, V., Grimm, K.P., Chertow, G.M., Pao, A.C., 2009. Melamine nephrotoxicity: an emerging epidemic in an era of globalilzation. Kidney Int. 7, 774–779.

Camacho, L., Kelly, P., Beland, F.A., da Costa, G.G., 2011. Gene expression of biomarkers of nephrotoxicity in F344 rats co-exposed to melamine and cyanuric acid for seven days. Toxicol. Lett. 206, 166–171.

Canelli, E., 1974. Chemical, bacteriological, and toxicological properties of cyanuric acid and chlorinated isocyanurates as applied to swimming pool disinfection, a review. Am. J. Publ. Health 64, 155–162.

Chang, H., Shi, X., Shen, S., Wang, W., Yue, Z., 2012. Characterizatoin of melamine-associated urinary stones in children with consumption of melamine-contaminated infant formula. Clin Chem Acta 413, 985–991.

Chen, Y., Wenjun, Y., Wang, Z., 2010. Deposition of melamine in eggs from laying hens exposed to melamine contaminated feed. J. Agric. Food Chem. 58, 3512–3516.

Cianciolo, R.E., Bischoff, K., Ebel, J., 2008. Clinicopathologic, histologic, and toxicologic findings in 70 cats inadvertently exposed to pet food contaminated with melamine and cyanuric acid. J. Am. Vet. Med. Assoc. 333, 729–737.

Dobson, R.L., Motlagh, S., Quijano, M., 2008. Identification and characterization of toxicity of contaminants in pet food leading to an outbreak of renal toxicity in cats and dogs. Toxicol. Sci. 106, 251–262.

Dorne, J.L., Doerge, D.R., Vandenbroeck, M., 2012. Recent advances in the risk assessment of melamine and cyanuric acid in animal feed. Toxicol. Appl. Pharmacol. 42, 229–235.

Duan, H., Guan, J.N., Wu, Y., 2011. Identification of biomarkers for melamine-induced nephrolithiasis in young children based on ultra high performance liquid chromatography coupled to time-of-flight mass spectrometry (U-HPLC-Q-TOF/MS). J. Chromatogr. B 879, 3544–3550.

Early, R.J., Yu, H., Mu, X.P., Xu, H., Guo, L., Kong, Q., Zhou, J., He, B., Yang, X., Huang, H., Hu, E., Jiang, Y., 2013. Repeat oral dose toxicity stidies of melamine in rats and monkeys. Arch. Toxicol. 87, 517–527.

FDA, 2010. Melamine Pet Food Recall – Frequently Asked Questions February 2, 2010. Retrieved from United States Food and Drug Administration Department of Health and Human Services: http://www.fda.gov/animalveterinary/safetyhealth/RecallsWithdrawals/ucm129932.htm#AnimalFeed.

Gao, J., Shen, Y., Sun, N., 2010. Therapeutic effects of potassium sodium, hydrogen citrate on melamine-induced urinary calculi in China. Chin. Med. J. 123, 1112–1116.

Gao, J., Xu, H., Kuang, X.Y., 2011. Follow-up results of children with melamine induced urolithiasis: a prospective observational cohort study. World J. Pediatr. 7, 232–239.

Gonzalez, J., Puschner, B., Pérez, V., 2009. Nephrotoxicosis in Iberian piglets subsequent to exposure to melamine and derivatives in Spain between 2003 and 2006. J. Vet. Diagn. Invest. 21, 536–558.

Hau, A.K., Kwan, T.H., Kam-Tao, P., 2009. Melamine toxicity in the kidney. J. Am. Soc. Nephrol. 20, 245–250.

Hsieh, T.J., Tsai, Y.H., Hsieh, P.C., 2012. Melamine induces human renal proximal tubular cell injury via transforming growth factor-beta and oxidative stress. Toxicol. Sci. 130, 17–32.

Hu, P., Ling, L., Hu, B., Zhang, C.R., 2010. The size of melamine-induced stones is dependent on the melamine content of the formula fed, but not the duration of exposure. Pediatr. Nephrol. 25, 565–566.

Hu, P., Wang, J., Zhang, M., 2012. Liver involvement in melamine-associated nephrolithiasis. Arch. Iran. Med. 15, 247–248.

Kim, T.H., Ahn, M.R., Lim, H.Y., 2012. Evaluation of metabolomic profiling against renal toxicity in Sprague-Dawley rats treated with melamine and cyanuric acid. Arch. Toxicol. 86, 1885–1897.

Lewin-Smith, M.R., Kalasinsky, V., Mullick, F.G., Thompson, M.E., 2009. Melamine-containing crystals in the urinary tracts of domestic animals: sentinel event? Arch. Pathol. Lab Med. 133, 341–342.

Lin, Y.T., Tsai, M.T., Chen, Y.L., Cheng, C.M., Hung, C.C., Wu, C.F., Liu, C.C., Hsieh, T.J., Shiea, J., Chen, B.H., Wu, M.T., 2013. Can melamine levels in 1-spot overnight urine specimens predict the total previous 24-hour melamine excretion level in school children? Clin. Chem. Acta 420, 128–133.

Lipschitz, W.L., Stokey, E., 1945. The mode of action of three new diuretics: melamine, adenine, and formoguanamine. J. Phamacol. Exp. Ther. 82, 235–348.

Liu, J.M., Ren, A., Yang, L., 2010. Urinary tract abnormalities in Chinese rural children who consumed melamine-contaminated dairy products: a population-based scrreening and follow-up study. Can. Med. Assoc. J. 182, 439–443.

Liu, J.D., Liu, J.J., Yuang, J.H., 2012. Proteome of melamine urinary bladder stones and implications for stone formation. Toxicol. Lett. 212, 307–314.

Lv, X., Wang, J., Wu, L., 2010. Tissue deposition and residue depletion in lambs exposed to melamine and cyanuric acid-contaminated diets. J. Agric. Food Chem. 58, 943–948.

Melnick, R.L., Boorman, G.A., Haseman, J.K., Montali, R.J., Huff, J., 1984. Urolithiasis and bladder carcinogenicity of melamine in rodents. Toxicol. Appl. Pharmacol. 72, 292–303.

Osborne, C.A., Lulich, J.P., Ulrich, L.K., 2008. Melamine and cyanuric acid-induced crystaluria, uroliths, and nephrotoxicity in dogs and cats. Vet. Clin North Am. Small Animal 39, 1–14.

Pacini, N., Dörr, A.J.M., Elia, A.C., Scoparo, M., Abete, M.C., Prearo, M., 2014. Melamine-cyanurate complexes and oxidative stress markers in trout kidney following melamine and cyanuric acid long-term co-exposure and withdrawal. Fish Physiol. Biochem. 40 (214), 1609–1619.

Panuwet, P., Nguyen, J.V., Wade, E.L., D'souza, P.E., Ryan, P.B., Barr, D.B., 2012. Quantification of melamine in human urine using cation-exchange based high performance liquid chromatography tandem mass spectrometry. J. Chromatogr. B 887–888, 48–54.

Puschner, B., Poppenga, R., Lowenstine, L., Filigenzi, M.S., Pesavento, P.A., 2007. Assessment of melamine and cyanuric acid toxicity in cats. J. Vet. Diagn. Invest. 19, 616–624.

Reimschuessel, R., Evans, E., Andersen, W.C., 2009. Residue depletion of melamine and cysnuric acid in catfish and rainbow trout following oral administration. Vet. Pharmacol. Ther. 33, 172–182.

Reimschuessel, R., Gieseker, C.M., Miller, R.A., 2008. Evaluation of the renal effects of experimental feeding of melamine and cyanuric acid to fish and pigs. Am. J. Vet. Res. 69, 1217–1228.

Rumbeiha, W.K., Agnew, D., Maxie, G., 2010. Analysis of a survey database of pet food-induced poisoning in North America. J. Med. Toxicol. 6, 172–184.

Schnackenberg, L.K., Sun, J., Pence, L.M., 2012. Metabolomic evaluation of hydroxyproline as a potential marker of melamine and cyanuric acid nephrotoxicity in male and female Fischer F344 rats. Food Chem. Toxicol. 50, 3978–3983.

Shen, J.S., Wang, J.Q., Wei, H.Y., 2010. Transfer efficiency of melamine from feed to milk in lactating dairy cows fed with different doses of melamine. J. Dairy Sci. 93, 2060–2066.

Silva, C.S., Chang, C.W., Williams, D., Porter-Gill, P., Gamboa da Costa, G., Camacho, L., 2016. Effects of a 28-day co-exposure to melamine and cyanuric acid on levels of serum microRNAs in male and female Fisher 344 rats. Food Chem. Toxicol. 98, 11–16.

Skinner, C.G., Thomas, J.D., Osterloh, J.D., 2010. Melamine toxicity. J. Med. Toxicol. 6, 50–55.

Sun, N., Shen, Y., He, L.J., 2010. Histopathological features of the kidney after acute renal failure from melamine. N. Engl. J. Med. 362, 662.

Sun, Y.J., Wang, H.P., Liang, Y.J., 2012. An NMR-based metabolomic investigation of the subacute effects of melamine in rats. J. Proteome Res. 11, 2544–2550.

Thompson, M.E., Lewin-Smith, M.R., Kalasinsky, K., 2008. Characterization of melamine-containing and calciium oxalate crystals in three dogs with suspected pet food-induced nephrotoxicosis. Vet. Pathol. 55, 417–426.

Wang, I.J., Wu, Y.N., Wu, W.C., 2009. The association of clinical findings and exposure profiles with melamine associated urolithiasis. Arch. Dis. Childhood 94, 883–887.

Wen, J.G., Li, Z.Z., Zhang, H., 2010. Melamine related bilateral renal calculi in 50 children: single center experience in clinical diagnosis and treatment. J. Urol. 183, 1533–1538.

Wu, C.F., Peng, C.Y., Liu, C.C., Lin, W.Y., Pan, C.H., Cheng, C.M., Hsieh, H.M., Hsieh, T.J., Chen, B.H., Wu, M.T., 2015. Ambient melamine exposure and urinary biomarkers for early renal injury. J. Am. Soc. Nephrol. 26, 2821–2829.

Xie, G., Zheng, X., Qi, X., 2010. Metabolomic evaluation of melamine-induced acute renal toxicity in rats. J. Proteome Res. 9, 125–133.

Yhee, J.Y., Brown, C.A., Yu, C.H., 2009. Retrospective study of melamine/cyanuric acid-induced renal failure in dogs in Korea between 2003 and 2004. Vet. Pathol. 46, 348–354.

Zhang, Q., da Costa, G.G., Von Tungeln, L.S., 2012. Urinary biomarker detection of melamine- and cyanuric acid-induced kidney injury in rats. Toxicol. Sci. 129, 1–8.

Zhao, Y.Y., Lin, R.C., 2014. Metabolomics in nephrotoxicity. Adv. Clin. Chem. 65, 69–89.

CHAPTER

32

Biomarkers of Petroleum Products Toxicity

Robert W. Coppock[1], *Margitta M. Dziwenka*[2]

[1]Toxicologist and Associates, Ltd, Vegreville, AB, Canada; [2]Faculty of Medicine & Dentistry, Division of Health Sciences Laboratory Animal Services, Edmonton, AB, Canada

INTRODUCTION

Petroleum releases into the environment can occur as the crude product comes from the wellhead or as the refined products that are processed to meet chemical, engineering, automotive and other specifications. Crude petroleum exists as sweet crude petroleum with minimal sulfur content or as sour crude petroleum with low to high sulfur content (Coppock and Christian, 2018). It can exist as light to heavy crude petroleum and as bitumen. Because of a large number of causes, petroleum in its various crude and refined forms and production by-products such as fracturing flowback and produced waters can be released into the environment (He et al., 2017a). Although much focus is placed on maritime incidents, spills from pipelines and other equipment used to transport petroleum have multiple environmental and social impacts (Anon, 2012; Genereux et al., 2014). Crude petroleum is a complex mixture of compounds and even more complex when the possible congeners of the individual constituents are considered. Crude oil is estimated to contain 17,000 to 20,000 different chemicals with a large component of chemically unidentified mass. Abiotic chemical markers can be used to profile and trace petroleum in the environment and assess the exposure of biota to various forms of petroleum. For petroleum, biomarkers can be used to determine internal dose with biochemical and physiological and morphological responses and assess movement of petroleum hydrocarbons in the food web (Kroon et al., 2017; Sanni et al., 2017a). Biomarkers that are robust across different species are desirable for biomonitoring of undesirable effects (Sanni et al., 2017a,b). Interpreting biomarkers to assess the impacts of petroleum pollution on the life history of organisms is important for understanding and minimizing long-term impacts on biota.

CHEMICAL MARKERS

Recent advances in analytical chemistry are reducing the mass of unidentified compounds found in crude petroleum. Matching the profile of compounds, and identification of marker chemicals and isotopes are the backbone of forensic petroleum chemistry (Adhikari et al., 2017; Wang et al., 2018). Petroleum, once it is released into the environment, undergoes weathering processes. During weathering, petroleum dynamically changes in chemistry and toxicity (Coppock and Christian, 2018). These processes include evaporation, dissolution, photochemical reactions, biological degradation, emulsification, adsorption on suspended particles, and other physical–chemical activities that change its chemical composition and density, and generally concentrates polyaromatic molecules. There is evidence that the lower molecular weight and less substituted hydrocarbons are lost by first-order kinetics. The addition of dispersant can alter the weathering processes and enhance formation of microdroplets in the aquatic system and thereby alter toxicity. The fate and transport of spilled hydrocarbons can be traced using intrinsic marker chemicals that are resistant to weathering or unique profiles that are relatively unaffected by chemical changes that occur during weathering. Commonly used marker chemicals are the branched isoprenoid alkanes pristane and phytane, and the hopanes and steranes. Fingerprinting hydrocarbon groups and other indigenous compounds can identify the source of the crude oil. Biochemical versus petroleum sources of η-alkanes can be estimated using the carbon preference index (odd number of carbons/even number of hydrocarbons) and isotopic profiles. Similar methods can be applied for identification of biofuels. At the wellhead, petroleum is generally separated from water (production water), whereas flowback water is water returned

Biomarkers in Toxicology, Second Edition
https://doi.org/10.1016/B978-0-12-814655-2.00032-3

to surface during drilling and hydraulic fracturing. Flowback and production waters consist of a complex mixture of water, water-soluble hydrocarbons, oil minuscules, mineral salts, and chemicals used in fracking or well-maintenance chemicals including biocides. Produced and flowback waters from some formations may contain naturally occurring radioactive materials (NORMs), for example radium. With depletion, a petroleum formation generally increases in water production (water/oil ratio). The ion and chemical profile in flowback and produced waters can be unique, and they can be used as forensic markers in terrestrial spills and to provide evidence of groundwater contamination (DiGiulio and Jackson, 2016). Production water from offshore wells can be released directly into the marine environment. Production water contains hydrocarbons, alkylphenols, polycyclic aromatic hydrocarbons (PAHs, also known as polyaromatic hydrocarbons), and many other chemicals and these can be used as biomarkers.

Polycyclic Aromatic Hydrocarbons

Chemically, the PAHs consist of two or more fused aromatic rings. The PAHs can be used in fingerprinting for estimating the most likely source of petroleum. Environmental PAHs are from three primary sources: combustion of organic matter, waste materials, and fossil fuels, by biogenesis, and from crude and refined petroleum. Crude oil can contain 0.2 to 7% polyaromatic hydrocarbons. The PAHs in petroleum generally contain two or three aromatic rings, and refined petroleum can, depending on the distillation temperature, contain similar polyaromatic hydrocarbons plus other polyaromatics formed by the catalytic-cracking processes. Refined petroleum generally contains more alkyl PAHs. The PAHs formed during incomplete combustion of organic matter generally contain three or more aromatic rings. Generally there is an inverse relationship between combustion temperature and the formation of alkyl polycyclic aromatics in incomplete pyrolysis assemblages. The PAHs usually become concentrated as petroleum weathers.

Tissue and Body Fluid Levels of Petroleum Hydrocarbons

Bioavailability is the transfer of the petroleum hydrocarbons from the environment into the organism. Fauna can take up environmental petroleum hydrocarbons through a variety of mechanisms, and differences can occur across genera, species, and stages in life history. Differing sensitivities of aquatic organisms can be due to unique absorptive mechanisms. Water solubility and micronization of oil in water by turbulence and chemical dispersants are important factors in the internal dose acquired by aquatic organisms. Oil fouling by microdroplets of oil adhering to the chorion (acellular coat) of Atlantic haddock (*Melanogrammus aeglefinus*) eggs increases absorption of hydrocarbons that have low solubility in water (Sorensen et al., 2017). Oil droplets also reduce the absorption of lipids including cholesterol from the yolk sac, and there is also a positive relationship between oil fouling and cardiac defects. Copepods transfer PAHs from body fat to eggs (Hansen et al., 2017). For aquatic exposures, the time for equilibrium between the levels of petroleum hydrocarbons in body fluids and tissues and the levels in the environment is generally determined by route of exposure, time, and exposure concentration. Fish uptake of the C_4–C_7 alkylphenols with K_{ow} values less than 5 is generally directly from water and low uptake from feed (Sundt et al., 2009). The hydrocarbons that accumulate in tissues of a particular species generally are more resistant to metabolism. Organisms that occupy different ecological niches can be used to monitor the contamination of fauna by petroleum hydrocarbons. Dispersants may alter the uptake of petroleum hydrocarbons by altering bioavailability in a particular habitat. For liver metabolism, the first pass effect for clearance can occur following oral ingestion. In fish species studied, the alkylphenols are rapidly cleared from the body by the liver and excreted in the bile as conjugated and unconjugated metabolites. Bile is the body fluid of choice for estimating bioavailability of these hydrocarbons, and identification of alkylphenol metabolites in bile is considered a biomarker of exposure. Bivalve mollusks are a species that filter feeds and often concentrate both organic and inorganic chemicals in their tissues, which is a food safety issue in edible species. The location and feeding habits of fauna are a factor in their use as biomonitors. Caged blue mussels (*Mytilus edulis*) and semipermeable membrane devices can be used to concentrate polycyclic aromatics from seawater, and mussels generally are more efficient bioconcentrators than passive monitoring devices (Utvik et al., 1999). Mussels have been used as biomonitors of ecosystem health (Sundt et al., 2011; Marigomez et al., 2013). The PAHs found in tissues of aquatic organisms are generally proportioned to their abundance in the petroleum causing the incident. The polycyclic aromatic hydrocarbon level in tissues can be used as an indicator of pollution from petroleum or deposition from other sources. PAHs from spilled petroleum can bioaccumulate in fish, mussels, and copepods and follow the food chain to birds (Carls et al., 2015; Fernie et al., 2017; Sanni et al., 2017a; Willie et al., 2017; Agersted et al., 2018). Carps (*Cyprinus carpio*) accumulate two to three aromatic ring PAHs when fed a diet containing bunker A heavy oil. Nestling tree swallows (*Tachycineta bicolor*), from nests in the oil sands mining area of Alberta, have been shown to accumulate methylated dibenzothiophenes, parent PAHs, and alkyl PAHs (Fernie et al., 2017).

BIOCHEMICAL BIOMARKERS

CYP Biomarkers

The CYP (P450) family of enzymes has a broad spectrum of substrates and are important in Phase I oxidation, hydrolysis, and reduction biotransformation reactions (Burkina et al., 2017). The transcriptional upregulation of the CYP enzymes, especially CYP1 isoforms, occurs following exposure to polyaromatic hydrocarbons (PAHs) and other structurally related planar compounds, for example, polyhalogenated aromatic hydrocarbons. The CYP1 family is upregulated after the ligand attaches to the cytoplasmic aryl hydrocarbon receptor (AhR), and the ligand–AhR complex is translocated into the nucleus. The transcriptional changes by the activated AhR are ligand specific and generally are highly cell-type specific. In addition to upregulation of CYPs, the AhR activation can also disturb cell functions inclusive of embryologic molecular chemistry. In species exposed to petroleum, upregulation of the CYPs is generally considered to be a robust biomarker of exposure to PAHs (Martinez-Gomez et al., 2009; Alexander et al., 2017; Esler et al., 2017; Sanni et al., 2017a; b). There are genetic differences between animals for synthesis of CYP enzymes families. The CYPs can be measured by gene expression methods, measurement of the CYP proteins with monoclonal antibodies, other specific protein assay methods, and measuring enzyme activity using specific substrates. The microsomal CYP1 isoforms activity, using two different substrate methods, to measure the turnover for 7-ethoxyresorufin-O-deethylase (EROD) or benzo(a)pyrene by arylhydrocarbon hydroxylase (CYP1A) in liver and other tissues is generally accepted as a biomarker of response to xenobiotic exposure (Burkina et al., 2017). In copepods exposed to water-accommodated crude oil, the *Tk-CYP3024A3* and *Tj-CYP3024A2* genes were observed to be upregulated (Han et al., 2017). Swallow nestlings in the bitumen mining area of Alberta had increased activity (EROD substrate) of hepatic CYP1 (Li et al., 2017). In Japanese medaka, exposure to Cold Lake bitumen caused a monotonic increase in CYP1A mRNA (Madison et al., 2017). The CYP1 gene expression can be upregulated following exposure to petroleum without a corresponding increase in EROD biotransformation. There is limited evidence that upregulation of the AhR as measured by CYP1A induction increases the toxicity of crude oil to fish. Juvenile sockeye salmon were exposed to water-soluble fractions of natural gas condensate–diluted bitumen, and hepatic EROD biotransformation was increased (Alderman et al., 2017). In heart muscle, the transcription of aryl hydrocarbon hydroxylase had transient upregulation and transcript abundance of CYP1A mRNA is increased. The hepatic CYP1 biomarker has also been shown to be upregulated in fish exposed to produced water, fish and cormorants exposed to oil from the Deepwater Horizon blowout (MC252), swallows (EROD activity) nesting on reclaimed oil sands, and in cattle (EROD and AHH activities) experimentally administered crude oil (Khan et al., 1996; Smits et al., 2000; Alexander et al., 2017; Dubansky et al., 2017; He et al., 2017a; Li et al., 2017). Gill EROD biotransformation in Atlantic cod (*Gadus morhua*) was upregulated after 24 h of exposure to crude oil (0.025 w/w). The evaluation of CYP1 family as a biomonitor is not a specific response to petroleum-source PAHs, can be influenced by season and other factors, and requires careful data interpretation. Increased EROD biotransformation is an indicator of the environmental exposure to an AhR receptor agonist, and other parameters of intoxication may not significantly change. Other environmental contaminants with planar configuration, such as polyhalogenated aromatic hydrocarbons, can also upregulate CYP1A. Background levels of pollutants and species response variations can occur and must be accounted for in the design of the experimental or monitoring protocols that use upregulated CYPs as biomarkers. The EROD biotransformation can be used to monitor the exposure to PAHs, and the parent PAH may or may not be bioconcentrated in tissues because of biotransformation and elimination (Martinez-Gomez et al., 2009). The products of enzymatic activity of CYP1A can be both detoxication and toxication. In detoxication, the net effect includes formation of compounds that decrease toxicity, and generally these products are rapidly excreted. For toxication reactions, the end effect generally is the formation of highly reactive compounds reacting with DNA, proteins and other macromolecules, lipid peroxidation, and upregulated metabolic rates of endogenous compounds that can disrupt homeostasis including endocrine functions. The xenobiotic exposure–linked upregulation of CYP expression may change with species, sex, reproductive status, diet, starvation, and exposure conditions. A study in Atlantic cod larvae has shown that starvation increases CYP1A gene expressions by 6.4 times, and it is considered a magnitude less than gene expression after exposure to crude oil (Hansen et al., 2016). Exposure to multiple chemicals can result in complex interactions that may alter the expected values for CYP1 (He et al., 2017a,b). Spatial biomarker patterns can be used across species for biomonitoring oil spill impacts (Martinez-Gomez et al., 2009).

Biomarkers of Responses to Oxidate Stress

Oxidative stress can occur after exposure to petroleum. Parameters generally are enzymes and substrates in Phase II biotransformation, antioxidant enzymes, and oxidized indigenous substances (e.g., lipid peroxides). The activities of superoxide dismutase and glutathione-S-transferase activities (GST) in the digestive gland were elevated in the *Anomalocardia flexuosa* clam 48 h after pulse exposed in situ to low levels of diesel oil–contaminated sea water (Sardi et al., 2017). Responses can be conflicting between tissues. For example, in clams, the GST activity decreases in gills and increases in the digestive glands. Laughing gulls (*Leucophaeus atricilla*), in a laboratory study, were orally exposed to weathered MC252 in fish (Horak et al., 2017). Hepatic total glutathione, oxidized glutathione (GSSG), and reduced glutathione (rGSH) significantly increased as mean dose of oil increased, and a nonsignificant negative trend occurred in the rGSH:GSSG ratio. Sandpipers (*Calidris mauri*) orally exposed to MC252 had increased total antioxidant capacity (Bursian et al., 2017a). In gull and sandpiper studies, the birds did not have flight exercise and the associated oxidative stress. Rainbow trout (*Oncorhynchus mykiss*) embryos, exposed to hydraulic fracking and produced water, showed increased lipid peroxidation. They also showed increased mRNA abundance for UDP-glucoronosyl transferase, glutathione transferase, glutathione peroxidase, and superoxide dismutase (He et al., 2017b). Similar effects were observed in zebrafish (He et al., 2018). Lipid peroxidation was increased in earthworms (*Eudrilus eugeniae*) exposed to benzene, toluene, and xylene used as surrogate chemicals for spilled petroleum (Eseigbe et al., 2013). GST did not show a clear dose response in juvenile halibut (*Hippoglossus hippoglossus*), Atlantic salmon (*Salmo salar*), turbot (*Scophthalmus maximus*), and sprat (*Sprattus sprattus*) following exposure to dispersed crude oil (Sanni et al., 2017a). Across species, increased activity of GST may correlate with PAH levels or GST may decrease with exposures to PAHs. Exposure of zebrafish to flowback and produced waters altered the expression of genes related to oxidative stress (He et al., 2018). Cellular pathology can be used to evaluate the effects of oxidative stress. Studies using weathered crude oil in double-breasted cormorants (*Phalacrocorax auritus*) provided pathological evidence of oxidative damage in red blood cells (Harr et al., 2017a).

Genetic Biomarkers of Petroleum Exposure

The toxic effects of petroleum hydrocarbons and their metabolites on adverse outcomes from aberrant gene expression are being explored. Studies in embryos and larval forms are identifying petroleum hydrocarbon–linked changes in gene expression that lead to developmental defects. Analyses of gene molecular chemistry are showing that exposure to crude oil modifies gene expression in a manner that is predictive of adverse outcomes resulting in abnormal phenotypes and compromised chances for survival (Cherr et al., 2017; Incardona, 2017; Sorhus et al., 2017; Xu et al., 2017). Recent studies have shown that altered regulation of differentially expressed genes during embryogenesis and larval stages are linked to physiological and morphological defects in red drum (*Sciaenops ocellatus*), haddock, and zebrafish. Fertilized zebrafish eggs/larvae were exposed to organic extracts of hydraulic fracturing flowback and produced waters (He et al., 2018). There was a significant increase in the abundance of transcripts of genes related to biotransformation including nuclei receptors (AhR, and pregnane-x-receptor), Phase I enzymes (*cyp1a*, CYP1B, CYP1C1, CYP1C2, CYP2AA12, and CYP3A65), and Phase II enzyme (uridine 5′-diphospho-glucuronosyltransferase). A significant increased abundance was also observed in the genes that respond to oxidative stress and in endocrine-mediated genes. Other genes whose exposure to petroleum may uniquely change in developing fish embryos are the *ccbe1* of edema formation, cholesterol synthesis genes, and the hepatic expression of the *fgf7* fibroblast growth factor gene. DNA itself is a target for certain petroleum hydrocarbons and is damaged by multiple mechanisms. Briefly, DNA can be damaged by toxication reactions (generally Phase I) that form reactive chemicals that subsequently cause strand breaks, base modifications, intercalation of substances between double-stranded DNA strands, and the formation of DNA adducts. DNA strand breaks can be quantified by single-cell electrophoresis, adducts identified by ^{32}P post labeling, and chromosomal damage indexed on micronucleus formation. The comet assay can detect DNA damage in a few cells from most germinal and somatic tissues (Frenzilli et al., 2017). In aquatic organisms, the most sensitive tissues for DNA damage vary between species with gill and liver cells generally being considered the most sensitive. Rapid repair of DNA damage can make application of the comet time sensitive. The comet assay has been used in a large number of species including mollusks, toads, and a large number of fish species. Biomarkers have been developed to measure the potential for DNA damage to occur in vivo. Cultures of rainbow trout hepatocytes exposed to water from oil sands tailing pond water showed single and double DNA strand breaks, and this test is a sensitive in vivo biomarker of genetic toxicity. The PAHs in the tailing pond water or their metabolites in the hepatocytes likely caused the DNA damage. Laboratory studies in Atlantic cod using crude oil from the North Sea geologic formations have shown that 0.3 ppb total polycyclic aromatic hydrocarbons (measured value corresponding to 0.06 ppm crude oil) in sea water cause the formation of DNA adducts. A dose response increase in DNA adducts was observed in juvenile turbot, halibut, salmon (smoltified), and sprat exposed to sublethal levels of dispersed Arctic crude oil (Sanni et al., 2017a). The formation of DNA adducts can be correlated with the metabolites of polycyclic aromatic hydrocarbons identified in bile. Haddock is a fish species considered sensitive to genotoxic contaminants. DNA strand breaks, determined by comet assay of erythrocytes, have been observed in eelpout (*Zoarces viviparus*) following a harbor spill of bunker oil. Micronucleus formation in erythrocytes and liver of fish and in gill cells in oysters and mussels can be used as a parameter estimating chromosomal damage (Bolognesi et al., 2006). The micronucleus test is a biomarker of accumulated genetic damage that occurred in a cell. It is a common test used in a number of species including aquatic organisms and requires cell staining and light microscopy. Increased occurrences of micronuclei were observed in the gills of mussels (*Mytilus edulis*) collected before and after a spill of petroleum products that spread along the Lithuanian coast (Barsiene et al., 2012). The lymphocyte cytokinesis-block micronucleus assay, although not toxicant specific, is a sensitive biomarker for human exposure to petroleum, especially petroleum high in benzene (Angelini et al., 2016).

MORPHOLOGIC BIOMARKERS

Pathological changes in tissues and organs are used as biomarkers of petroleum toxicity. The tissues commonly examined by histopathology in most species are gills, heart, liver, intestines, kidney, and spleen. When coupled with enzymatic and other biomarkers, histopathology documents anatomical changes that can lead to pathophysiological dysfunctions and impaired health. The histopathology of the gill shows that the respiration, osmoregulation, and electrolyte balance may be compromised by petroleum hydrocarbons in water. Changes in the gill include edema and dilation of the secondary lamella. Proliferation of gill epithelium can occur with fusion of the secondary lamella. Increased severity of gill lesions can be observed with increasing concentrations of PAHs in sediment. Epithelial hyperplasia and fusing of lamella were reported in fish exposed to an environmental release of fracking flowback water. Necrosis and other lesions in the kidney can be linked with biliary levels of PAHs and their metabolites. An increase in the number of rodlet cells in the kidneys of bream (*Abramis brama*) can be observed after the fish were environmentally exposed to fuel oils. Hyperplasia of the intestinal mucous cells can also occur. Liver histopathology was used to assess the toxicity of Arabian crude oil and dispersed crude oil in juvenile rabbit fish (*Siganus canaliculatus*) (Agamy, 2012). Hepatic lesions evaluated were cytoplasmic vacuolization, megalocytosis, dispersed coagulative necrosis, lymphocytic infiltration, melanomacrophage aggregates, spongiosis hepatis, pericholangitis, and bile stenosis. In wild bream (*Abramis brama*) exposed to refined petroleum, hyperplasia of mucous cells in the skin was observed. The gills had lamellar capillary aneurysms, epithelial proliferation, and proliferation associated with fusion of the secondary lamella and mucous cell, which was hypertrophic. In a laboratory study, prespawning Atlantic cod were exposed in the laboratory to production water from North Sea oil wells (Sundt and Bjorkblom, 2011). In fish exposed to 0.2% production water (diluted with sea water), there was decreased ovarian ova development accompanied by atresia and inflammation in the ovary. Male fish exposed to production water were observed to have decreased mature sperm and increased spermatogonia and spermatocytes. In a study using mesocosms, fiddler crabs (*Uca longisignalis* and *U. panacea*) were exposed to Louisiana crude oil mixed in sediment (Franco et al., 2018). Microscopic examination of the hepatopancreas showed a threefold increase in the numbers of blister cells. This cellular change has also been reported in other species of crustaceans. Blister cells secrete digestive enzymes. Immunohistochemistry of killifish exposed in the wild to weathered crude oil (MC252) showed increased staining for CYP1A in gill pillar cells, head kidney distal and proximal tubular cells, and the vascular endothelial cells (Dubansky et al., 2017). Ingestion of fresh crude oil targets the lungs in cattle. This risk is exacerbated by the effects of volatile hydrocarbons on the nervous system and occurrence of emesis. Vacuolated pulmonary macrophages and hepatocytes generally occur. Double-breasted cormorants exposed to weathered crude oil had increased relative liver and kidney weights (Harr et al., 2017b). Orally exposed birds had renal histopathology consisting of squamous metaplasia and increased epithelial hypertrophy of the collecting ducts and renal tubules, and increased occurrences of mineralization. Orally dosed birds also had cytoplasmic vacuolization of exocrine pancreatic cells. Thyroid follicular hyperplasia was observed in birds dermally exposed to oil. The cardiac walls were thin with flabby musculature in all birds exposed to oil. Echocardiography showed likely oil-treatment linked dilative cardiomyopathy supporting pathophysiologic cardiac dysfunction (Harr et al., 2017c). Elevated thyroid hormones also have been observed in tree swallows nesting in active oil sands areas (Li et al., 2017).

Teratology

Abnormal organogenesis and early development can be a biomarker of eggs, embryos, and larvae being exposed to crude oil (Incardona et al., 2014; Cherr et al., 2017; Incardona, 2017; Alsaadi et al., 2018). The general syndrome observed in fish is fluid accumulation in the yolk sac and around the heart, abnormal myocardial development, craniofacial and other defects, and increased tissue staining for CYP1A (Hodson, 2017; Incardona, 2017; Sorhus et al., 2017). The currently identified teratogenic components in the crude oil hydrocarbon mixture are the planar PAHs and the dibenzothiophenes. The heart is a target at low PAHs concentrations in the developing fish embryo. Embryonic cardiotoxicity can occur with and without induction of the AhR, but induction of the AhR may contribute to cardiotoxicity. Crude petroleum and tricyclic PAHs cause abnormal cardiac rhythm. This effect is likely due to changes in ion gating, changes in membrane potentials, and subsequent alterations of the cardiac action potential. Abnormal heart contractility likely leads to abnormal cardiac development (proliferation and failed looping). There are interactions between physiological functions and the cascade of genetic events for normal development in proteins that regulate gating of calcium ions. The genes that control bone morphogenic proteins (BMPs) and genes regulating calcium ion homeostasis may be involved. In zebrafish, the mutant silent heart (*sih*) has decreased expression of troponin T and is a teratogenic syndrome very similar to that observed with PAH exposure. Atlantic haddocks have a greater sensitivity to crude petroleum than zebrafish and also have craniofacial terata following PAH exposure. Teratogenic effects were observed in red drum larvae exposed to ocean-weathered crude oil (Xu et al., 2017). Developmental abnormalities observed are decreased brain area and lens diameter, pericardial edema, and decreased body length. Diluted or pipeline bitumen is embryotoxic to fish and causes a condition in fish embryos that resembles blue sac disease (craniofacial malformations, pericardial and yolk sac edema, spinal curvatures, fin erosion, hemorrhaging, and induction of CYP1) (Alsaadi et al., 2018). A study in Japanese medaka showed that malformations increase with increasing total bitumen-linked PAHs levels in water (Madison et al., 2017). The malformations included craniofacial and yolk sac edema. The fish also had impairment of swim bladder inflation. Developmental toxicity of the organic chemical fraction of hydraulic fracturing flowback and produced waters was tested in zebrafish (He et al., 2018). Compounds in the organic extract included PAHs, alkyl PAHs, and other major substances including polyethylene glycols, alkyl ethoxylates, octylphenol ethoxylates, and high molecular weight (C_{49-79}) ethylene oxide polymeric material. A significant increase in spinal malformations and pericardial edema were observed. Exposure of fish to low transient levels of PAHs, which do not produce morphological changes, decreased swimming performance. Decreased physical performance likely compromises abilities to migrate, forage, and evade predators. The cardiac biomarker is linked to the collapse of the herring population in Prince William Sound after Exon Valdez oil spill (Cherr et al., 2017).

BIOMARKERS IN BIRDS

The toxicity of weathered MC252 sweet crude oil from the Deepwater Horizon blowout was studied in aquatic birds (Alexander et al., 2017; Bursian et al., 2017a,b; Cunningham et al., 2017; Harr et al., 2017a,b). Double-breasted cormorants were exposed, under laboratory conditions, to weathered MC252 crude oil by the oral and oiled feathers routes of exposure. Fish that had previously received a measured amount of oil were live fed to foraging birds to give a daily dose of 5 mL or 10 mL of oil/day. For feather–dermal exposure, 13 g of MC252 was applied to the breast and back feathers of the birds. Biochemical, hematological, histopathological, and clinical parameters were used as biomarkers. For hematology, anemia was observed in all of the cormorants exposed to MC252. The packed cell volume decreased in a dose-related manner, and Heinz bodies were observed in the red blood cells and reticulocytes were observed. Blood clotting times were prolonged. In cormorants orally exposed to oil in fish, there were significant decreases in plasma activities of aspartate amino transferase and gamma glutamyl transferase; decreases in the plasma levels of calcium, chloride, cholesterol, glucose, and total protein; and increases in plasma urea, uric acid, and phosphorus concentrations. Cormorants exposed by the oral and feather routes had an increase in the hepatic CYPs identified by an increase in biotransformation of EROD, benzyloxyresorufin O-debenzylase, methoxyresorufin O-demethylase, and pentoxyresorufin O-depentylase and an increase in CYP1A protein expression. Laughing gulls (*Leucophaeus atricilla*) were orally exposed to MC252 using the same methods as cormorants except the gulls were fed thawed frozen fish. The total hepatic glutathione/oxidized glutathione and reduced glutathione significantly increased with increasing mean oil dose. Plasma levels of 3-methyl histidine, an indicator of muscular wastage, were positively correlated with oil exposure. Exposure to crude oil can cause decreased red blood cell numbers and decreased hematocrit and is often accompanied by Heinz bodies being observed in the red blood cells. Plasma haptoglobin can decrease with increasing PAH exposure suggesting intravascular hemolysis. Plasma ferritin is increased with increasing PAH levels. Behavior of birds can be used as biomarkers. For example, birds that have oiled feathers can exhibit decreased wing performance and increased takeoff distances (Maggini et al., 2017a; b).

CONCLUDING REMARKS AND FUTURE DIRECTIONS

Diagnosis of petroleum intoxication is dependent on the use of chemical markers and biomarkers. Few biomarkers are specific for petroleum intoxication. Complicating factors in the diagnosis of petroleum intoxication from petroleum spills are natural systems that generally contain multiple anthropogenic pollutants. The measured response of in situ fauna to the parameters studied generally is not limited to the specific substances of interest but is the net response of the organism to all pollutants in its ecological niche. The logical and knowable use of chemical markers and biomarkers of petroleum intoxication can provide a diagnosis of exposure to petroleum hydrocarbons and a diagnosis of systemic harm. The biomarkers that are linked to understanding cause and effect with pathophysiology at the phenotypic and molecular levels can be directly linked to adverse outcomes that have negative effects on life history and population dynamics. A battery of selected biomarkers can be used to assess adverse effects of petroleum hydrocarbons across multiple genera and species of fauna and provide input for determining multiple species effects and relative risks. In the future, it is likely that real-time monitoring tools for identification of anomalies and changes in gene expression will enhance the current biomarker tools for understanding mechanisms and adverse effects from long-term exposures to petroleum hydrocarbons.

References

Adhikari, P.L., Wong, R.L., Overton, E.B., 2017. Application of enhanced gas chromatography/triple quadrupole mass spectrometry for monitoring petroleum weathering and forensic source fingerprinting in samples impacted by the Deepwater Horizon oil spill. Chemosphere 184, 939–950.

Agamy, E., 2012. Histopathological liver alterations in juvenile rabbit fish (*Siganus canaliculatus*) exposed to light Arabian crude oil, dispersed oil and dispersant. Ecotoxicol. Environ. Saf. 75, 171–179.

Agersted, M.D., Moller, E.F., Gustavson, K., 2018. Bioaccumulation of oil compounds in the high-arctic copepod *Calanus hyperboreus*. Aquat. Toxicol. 195, 8–14.

Alderman, S.L., Lin, F., Farrell, A.P., et al., 2017. Effects of diluted bitumen exposure on juvenile sockeye salmon: from cells to performance. Environ. Toxicol. Chem. 36, 354–360.

Alexander, C.R., Hooper, M.J., Cacela, D., et al., 2017. CYP1A protein expression and catalytic activity in double-crested cormorants experimentally exposed to Deepwater Horizon Mississippi Canyon 252 oil. Ecotoxicol. Environ. Saf. 142, 79–86.

Alsaadi, F., Hodson, P.V., Langlois, V.S., 2018. An embryonic field of study: the aquatic fate and toxicity of diluted bitumen. Bull Env Contam Toxicol 100, 8–13.

Angelini, S., Bermejo, J.L., Ravegnini, G., et al., 2016. Application of the lymphocyte cytokinesis-block micronucleus assay to populations exposed to petroleum and its derivatives: results from a systematic review and meta-analysis. Mutat. Res. Rev. 770, 58–72.

Anon, 2012. Pipeline Accident. Report Enbridge Incorporated Hazardous Liquid Pipeline Rupture and Release Marshall, Michigan. July 25, 2010. National Transportation Safety Board. Report No. NTSB/PAR-12/01 PB2012–916501 (Notation 8423, Adopted July 10, 2012). National Transportation Safety Board, Washington, DC.

Barsiene, J., Rybakovas, A., Garnaga, G., et al., 2012. Environmental genotoxicity and cytotoxicity studies in mussels before and after an oil spill at the marine oil terminal in the Baltic Sea. Environ. Mon. Assess. 184, 2067–2078.

Bolognesi, C., Perrone, E., Roggieri, P., et al., 2006. Bioindicators in monitoring long term genotoxic impact of oil spill: Haven case study. Mar. Environ. Res. 62 (Suppl.), S287–S291.

Burkina, V., Rasmussen, M.K., Pilipenko, N., et al., 2017. Comparison of xenobiotic-metabolising human, porcine, rodent, and piscine cytochrome P450. Toxicology 375, 10–27.

Bursian, S.J., Dean, K.M., Harr, K.E., et al., 2017a. Effect of oral exposure to artificially weathered Deepwater Horizon crude oil on blood chemistries, hepatic antioxidant enzyme activities, organ weights and histopathology in western sandpipers (*Calidris mauri*). Ecotoxicol. Environ. Saf. 146, 91–97.

Bursian, S.J., Alexander, C.R., Cacela, D., et al., 2017b. Reprint of: overview of avian toxicity studies for the Deepwater Horizon natural resource damage assessment. Ecotoxicol. Environ. Saf. 146, 4–10.

Carls, M.G., Larsen, M.L., Holland, L.G., 2015. Spilled oils: static mixtures or dynamic weathering and bioavailability? PLoS One 10, e0134448.

Cherr, G.N., Fairbairn, E., Whitehead, A., 2017. Impacts of petroleum-derived pollutants on fish development. Ann. Rev. Anim. Biosci. 5, 185–203.

Coppock, R.W., Christian, R.G., 2018. Petroleum. In: Gupta, R.C. (Ed.), Veterinary Toxicology Basic and Clinical Principles, third ed. Academic Press/Elsevier, Amsterdam, pp. 745–778.

Cunningham, F., Dean, K., Hanson-Dorr, K., et al., 2017. Reprint of: development of methods for avian oil toxicity studies using the double crested cormorant (*Phalacrocorax auritus*). Ecotoxicol. Environ. Saf. 146, 19–28.

DiGiulio, D.C., Jackson, R.B., 2016. Impact to underground sources of drinking water and domestic wells from production well stimulation and completion practices in the Pavillion, Wyoming, field. Environ. Sci. Technol. 50, 4524–4536.

Dubansky, B., Rice, C.D., Barrois, L.F., et al., 2017. Biomarkers of aryl-hydrocarbon receptor activity in gulf killifish (*Fundulus grandis*) from northern Gulf of Mexico marshes following the Deepwater Horizon oil spill. Arch. Environ. Contam. Toxicol. 73, 63–75.

Eseigbe, F.J., Doherty, V.F., Sogbanmu, T.O., et al., 2013. Histopathology alterations and lipid peroxidation as biomarkers of hydrocarbon-induced stress in earthworm, *Eudrilus eugeniae*. Environ. Monit. Assess. 185, 2189–2196.

Esler, D., Ballachey, B.E., Bowen, L., et al., 2017. Cessation of oil exposure in harlequin ducks after the Exxon Valdez oil spill: cytochrome P4501A biomarker evidence. Environ. Toxicol. Chem. 36, 1294–1300.

Fernie, K.J., Marteinson, S.C., Chen, D., et al., 2017. Elevated exposure, uptake and accumulation of polycyclic aromatic hydrocarbons by nestling tree swallows (*Tachycineta bicolor*) through multiple exposure routes in active mining-related areas of the Athabasca oil sands region. Sci. Total Environ. 624, 250–261.

Franco, M.E., Felgenhauer, B.E., Klerks, P.L., 2018. Crude oil toxicity to fiddler crabs (*Uca longisignalis* and *Uca panacea*) from the northern Gulf of Mexico: impacts on bioturbation, oxidative stress, and histology of the hepatopancreas. Environ. Toxicol. Chem. 37, 491–500.

Frenzilli, G., Bean, T.P., Lyons, B.P., 2017. The application of the Comet assay in aquatic environments. In: Dhawan, M., Anderson, D. (Eds.), Comet Assay in Toxicology, Issues in Toxicology, second ed., vol. 30. Royal Society of Chemistry, Cambridge, UK, pp. 354–368.

Genereux, M., Petit, G., Maltais, D., et al., 2014. The public health response during and after the Lac-Megantic train derailment tragedy: a case study. Disaster Health 2, 113–120.

Han, J., Kim, H.S., Kim, I.C., et al., 2017. Effects of water accommodated fractions (WAFs) of crude oil in two congeneric copepods *Tigriopus* sp. Ecotoxicol. Environ. Saf. 145, 511–517.

Hansen, B.H., Lie, K.K., Storseth, T.R., et al., 2016. Exposure of first-feeding cod larvae to dispersed crude oil results in similar transcriptional and metabolic responses as food deprivation. J. Toxicol. Environ. Health Part A 79, 558–571.

Hansen, B.H., Tarrant, A.M., Salaberria, I., et al., 2017. Maternal polycyclic aromatic hydrocarbon (PAH) transfer and effects on offspring of copepods exposed to dispersed oil with and without oil droplets. J. Toxicol. Environ. Health Part A 80, 881–894.

Harr, K.E., Cunningham, F.L., Pritsos, C.A., et al., 2017a. Weathered MC252 crude oil-induced anemia and abnormal erythroid morphology in double-crested cormorants (*Phalacrocorax auritus*) with light microscopic and ultrastructural description of Heinz bodies. Ecotoxicol. Environ. Saf. 146, 29–39.

Harr, K.E., Reavill, D.R., Bursian, S.J., et al., 2017b. Organ weights and histopathology of double-crested cormorants (*Phalacrocorax auritus*) dosed orally or dermally with artificially weathered Mississippi Canyon 252 crude oil. Ecotoxicol. Environ. Saf. 146, 52–61.

Harr, K.E., Rishniw, M., Rupp, T.L., et al., 2017c. Dermal exposure to weathered MC252 crude oil results in echocardiographically identifiable systolic myocardial dysfunction in double-crested cormorants (*Phalacrocorax auritus*). Ecotoxicol. Environ. Saf. 146, 76–82.

He, Y., Flynn, S.L., Folkerts, E.J., et al., 2017a. Chemical and toxicological characterizations of hydraulic fracturing flowback and produced water. Water Res. 114, 78–87.

He, Y., Folkerts, E.J., Zhang, Y., et al., 2017b. Effects on biotransformation, oxidative stress, and endocrine disruption in rainbow trout (*Oncorhynchus mykiss*) exposed to hydraulic fracturing flowback and produced water. Environ. Sci. Technol. 51, 940–947.

He, Y., Sun, C., Zhang, Y., et al., 2018. Developmental toxicity of the organic fraction from hydraulic fracturing flowback and produced waters to early life stages of zebrafish (*Danio rerio*). Environ. Sci. Technol. https://doi.org/10.1021/acs.est.7b06557 (in press).

Hodson, P.V., 2017. The toxicity to fish embryos of PAH in crude and refined oils. Arch. Environ. Contam. Toxicol. 73, 12–18.

Horak, K.E., Bursian, S.J., Ellis, C.K., et al., 2017. Toxic effects of orally ingested oil from the Deepwater Horizon spill on laughing gulls. Ecotoxicol. Environ. Saf. 146, 83–90.

Incardona, J.P., 2017. Molecular mechanisms of crude oil developmental toxicity in fish. Arch. Environ. Contam. Toxicol. 73, 19–32.

Incardona, J.P., Gardner, L.D., Linbo, T.L., et al., 2014. Deepwater Horizon crude oil impacts the developing hearts of large predatory pelagic fish. Proc. Natl. Acad. Sci. USA 111, E1510–E1518.

Khan, A.A., Coppock, R.W., Schuler, M.M., et al., 1996. Biochemical effects of Pembina Cardium crude oil exposure in cattle. Arch. Environ. Contam. Toxicol. 30, 349–355.

Kroon, F., Streten, C., Harries, S., 2017. A protocol for identifying suitable biomarkers to assess fish health: a systematic review. PLoS One 12, e0174762.

Li, C., Fu, L., Stafford, J., et al., 2017. The toxicity of oil sands process-affected water (OSPW): a critical review. Sci. Total Environ. 601–602, 1785–1802.

Madison, B.N., Hodson, P.V., Langlois, V.S., 2017. Cold Lake blend diluted bitumen toxicity to the early development of Japanese medaka. Environ. Pollut. 225, 579–586.

Maggini, I., Kennedy, L.V., Elliott, K.H., et al., 2017a. Trouble on takeoff: crude oil on feathers reduces escape performance of shorebirds. Ecotoxicol. Environ. Saf. 141, 171–177.

Maggini, I., Kennedy, L.V., Macmillan, A., et al., 2017b. Light oiling of feathers increases flight energy expenditure in a migratory shorebird. J. Exp. Biol. 220 (Part 13), 2372–2379.

Marigomez, I., Zorita, I., Izagirre, U., et al., 2013. Combined use of native and caged mussels to assess biological effects of pollution through the integrative biomarker approach. Aquat. Toxicol. 136–137C, 32–48.

Martinez-Gomez, C., Fernandez, B., Valdes, J., et al., 2009. Evaluation of three-year monitoring with biomarkers in fish following the Prestige oil spill (N Spain). Chemosphere 74, 613–620.

Sanni, S., Bjorkblom, C., Jonsson, H., et al., 2017a. I: Biomarker quantification in fish exposed to crude oil as input to species sensitivity distributions and threshold values for environmental monitoring. Mar. Environ. Res. 125, 10–24.

Sanni, S., Lyng, E., Pampanin, D.M., et al., 2017b. II. Species sensitivity distributions based on biomarkers and whole organism responses for integrated impact and risk assessment criteria. Mar. Environ. Res. 127, 11–23.

Sardi, A.E., Renaud, P.E., Morais, G.C., et al., 2017. Effects of an in situ diesel oil spill on oxidative stress in the clam *Anomalocardia flexuosa*. Environ. Pollut. 230, 891–901.

Smits, J.E., Wayland, M.E., Miller, M.J., et al., 2000. Reproductive, immune, and physiological end points in tree swallows on reclaimed oil sands mine sites. Environ. Toxicol. Chem. 19, 2951–2960.

Sorensen, L., Sorhus, E., Nordtug, T., et al., 2017. Oil droplet fouling and differential toxicokinetics of polycyclic aromatic hydrocarbons in embryos of Atlantic haddock and cod. PLoS One 12, 0180048.

Sorhus, E., Incardona, J.P., Furmanek, T., et al., 2017. Novel adverse outcome pathways revealed by chemical genetics in a developing marine fish. Elife 6.

Sundt, R.C., Bjorkblom, C., 2011. Effects of produced water on reproductive parameters in prespawning Atlantic cod (*Gadus morhua*). J. Toxicol. Environ. Health Part A 74, 543–554.

Sundt, R.C., Baussant, T., Beyer, J., 2009. Uptake and tissue distribution of C_4-C_7 alkylphenols in Atlantic cod (*Gadus morhua*): relevance for biomonitoring of produced water discharges from oil production. Mar. Pollut. Bull. 58, 72–79.

Sundt, R.C., Pampanin, D.M., Grung, M., et al., 2011. PAH body burden and biomarker responses in mussels (*Mytilus edulis*) exposed to produced water from a North Sea oil field: laboratory and field assessments. Mar. Pollut. Bull. 62, 1498–1505.

Utvik, T.I., Durell, G.S., Johnsen, S., 1999. Determining produced water originating polycyclic aromatic hydrocarbons in North Sea-waters: comparison of sampling techniques. Mar. Pollut. Bull. 38, 977–989.

Wang, Y., Liang, J., Wang, J., et al., 2018. Combining stable carbon isotope analysis and petroleum-fingerprinting to evaluate petroleum contamination in the Yanchang oilfield located on Loess Plateau in China. Environ. Sci. Pollut. Res. Int. 25, 2830–2841.

Willie, M., Esler, D., Boyd, W.S., et al., 2017. Spatial variation in polycyclic aromatic hydrocarbon exposure in Barrow's goldeneye (*Bucephala islandica*) in coastal British Columbia. Mar. Pollut. Bull. 118, 167–179.

Xu, E.G., Khursigara, A.J., Magnuson, J., et al., 2017. Larval red drum (*Sciaenops ocellatus*) sublethal exposure to weathered Deepwater Horizon crude oil: developmental and transcriptomic consequences. Environ. Sci. Technol. 51, 10162–10172.

CHAPTER

33

Biomarkers of Chemical Mixture Toxicity

Antonio F. Hernández[1], *Fernando Gil*[1], *Aristidis M. Tsatsakis*[2]

[1]Department of Legal Medicine and Toxicology, University of Granada School of Medicine, Granada, Spain; [2]Center of Toxicology Science & Research, Medical School, University of Crete, Heraklion, Greece

INTRODUCTION

Humans are continuously and increasingly exposed to low doses of a large number of chemicals through different sources and routes, such as food, drinking water, drugs, breathing air, and uptake through the skin. Many of these chemicals are persistent and may accumulate in animal tissues, may biomagnify in food webs, and can contribute to many chronic diseases, thus increasing the general population's concerns about the potential toxic effects of exposure to combined chemicals. Chemical mixtures may contain both known and unknown components, depending on whether they are intentionally or unintentionally formed; however, real-world mixtures tend to involve low doses of individual mixture components. The exposure scenario is characterized by exposure to all components of the mixture at the same time, either simultaneously or sequentially, and by a similar route as they occur together. Prediction of health risk from chemical mixture exposure is a complex issue that cannot be based only on the knowledge of individual chemicals because in such a case it would likely lead to wrong conclusions. The degree of exposure and the hazards associated with such combined exposures can be determined by risk assessment (Alexander et al., 2008; Crofton, 2009).

POTENTIAL CHEMICAL MIXTURES

Pesticides

Pesticides comprise of many different categories of chemicals and almost always occur in mixtures with each other in the environment or in foodstuffs. Multiple exposures to pesticides occur in occupational settings and from environmental and nutritional sources, with intensive agriculture workers being more heavily exposed to mixtures of pesticides. The type and severity of their adverse health effects are determined by the individual chemical category, the dose and the duration of exposure, and the exposure route. Because pesticides are often applied in mixtures to crops, their residues can be found in foods and drinking water. However, based on national and Europe-wide monitoring programs of pesticide residues in plant products, levels are infrequently above maximum residue limits and thus considerably below thresholds of concern (Alexander et al., 2008). Accordingly, the probability of experiencing adverse effects from combined exposures is considered to be rare, unless pesticides interact and produce synergism or potentiation (Hernández et al., 2013a, 2017), circumstances that could occur at doses above thresholds of concern. Moreover, finding a measurable amount of a pesticide (or metabolites) in a biological fluid or tissue does not mean that the levels of parent compounds or metabolites will have an adverse health effect (CDC, 2009).

Organophosphate (OP) and *N*-methylcarbamate insecticides are two of the most widely used pesticides that act by a similar mechanism of action, inhibition of acetylcholinesterase (AChE). Thus, it is important to evaluate the cumulative risk in the case of simultaneous exposure to mixtures of these compounds (Lowitt, 2006; Padilla, 2006). The nonspecific metabolites of OP pesticides, dialkylphosphates (DAPs), are the most commonly used indicators for the assessment of cumulative exposure to OPs in humans (Barr and Buckley, 2011; Kavvalakis and Tsatsakis, 2012) and may be useful for studies looking at epidemiological associations with adverse health outcomes. However, important limitations include a lack of specificity with respect to the OP from which they were derived, the occurrence of preformed DAPs in the environment, and the fact that DAP levels commonly encountered in

Biomarkers in Toxicology, Second Edition
https://doi.org/10.1016/B978-0-12-814655-2.00033-5

food and the environment are toxicologically irrelevant. For *N*-methylcarbamates, only acute exposure may need to be considered because of their reversible mode of action on AChE.

Organochlorinated Compounds

Various formerly used chlorinated pesticides (e.g., dichlorodiphenyltrichloroethane (DDT) and its major metabolite dichlorodiphenyldichloroethylene (DDE)), as well as the industrial chemicals polychlorinated biphenyls (PCBs) and the industrial waste polychlorinated dibenzodioxins (PCDDs), are ubiquitous environmental contaminants. As a result of their high chemical stability and environmental persistence, they may reach the trophic chain and bioaccumulate in fatty foods (González-Alzaga et al., 2018). However, the chlorinated pesticides and PCBs are no longer produced and the emissions of PCDDs into the environment have been significantly reduced over time. Thus, the levels of these persistent organochlorinated compounds in the environment and foods are generally decreasing (Alexander et al., 2008).

The general population is primarily exposed to dioxins and PCBs from diet (>90%), with foods from animal origin being the most important sources, particularly fatty fish (González-Alzaga et al., 2018). The toxic and biological effects of these compounds are mediated through the aryl hydrocarbon receptor (AhR). There is some evidence that subtle developmental effects caused by these organochlorinated compounds, alone or in combination, may occur at maternal body burdens that are only slightly higher than those expected from average daily intakes in Europe (EFSA, 2005).

Halogenated Flame Retardants

Brominated flame retardants, such as polybrominated diphenyl ethers (PBDEs), are known to leach from treated materials during their lifetime. These chemicals are also ubiquitous, persistent in the environment, bioaccumulative, and toxic to both humans and the environment. In contrast to the chlorinated organic compounds, the levels of PBDEs increased in the environment and in humans until 2004, when some technical mixtures were strictly banned in the United States and Europe, resulting in decreased levels in animals and humans. Despite the chemical structure of PBDEs resembling that of PCBs, they do not activate the cytosolic AhR, indicating a different mechanism of action from that of dioxin-like PCBs (Alexander et al., 2008). PBDEs may impair circulating hormone levels, decrease fertility, and interfere with neurodevelopment, presumably by inducing oxidative stress–related damage and interference with signal transduction (Costa et al., 2014).

Organophosphate flame retardants (OPFRs) were marketed as a less toxic and less bioaccumulative alternative than PBDEs. However, there are indications that OPFRs affect endocrine systems, producing sex-dependent effects on the hypothalamic–pituitary–gonad axis, and dysregulation of the thyroid hormone system. These chemicals can also affect several neurodevelopmental processes such as apoptosis, synaptogenesis, and neurite outgrowth (Dishaw et al., 2014).

Polycyclic Aromatic Hydrocarbons

Polycyclic aromatic hydrocarbons (PAHs) consist of a large group of several hundred chemically related organic compounds of varied toxicities that may contaminate food. PAH sources include domestic heating source systems, gasoline fuel exhaust, coal tar, asphalt, and cigarette smoke. These chemicals are usually found as a mixture containing two or more of them, some of which are persistent in the environment and resistant to biodegradation. They have been shown to cause carcinogenic and mutagenic effects and are potent immunosuppressants. Metabolites of pyrene and DNA adducts have been used as biomarkers of exposure to PAHs. Potential adverse birth outcomes have been described after PAH exposure during pregnancy (Ramesh and Archibong, 2011; Hood et al., 2011). In addition, the combined exposure of pregnant women to PAH and other common environmental pollutants, such as environmental tobacco smoke, might adversely affect fetal development (Polanska et al., 2014). As different PAHs probably have different mechanisms of action, a toxic equivalent factor principle is not suitable for cumulative risk assessment. The combined effect of PAHs is thus not always additive and, in some instances, synergistic and antagonistic effects have been observed. Benzo[*a*]pyrene (BaP) is the most potent carcinogenic substance among the PAHs, and as long as its relative contribution to total PAH is relatively constant in foods, BaP concentration is a good indicator of the total carcinogenic potency of a PAH mixture (Alexander et al., 2008).

Metals

Metals are potentially hazardous chemicals that occur in the environment as a result of natural geological conditions, industrial emissions, and mining activity. They show different chemical forms and bound states that play a major role in their toxicity and partially account for potential interactions with other chemicals. Besides, specific interactions between metals may account for by the differences between the observed effects of metal mixtures and additive predictions; however, significant deviations from additivity are not necessarily biologically relevant (Liu et al., 2017). Interactions can occur because of competition between divalent toxic and nontoxic metal cations for intestinal divalent metal transporters (DMT1) that mediate their absorption, leading to a reduced toxicity of heavy metals ingested. After absorption, metallothionein plays a crucial role in the homeostasis of metal elements because of the binding of its free thiol group to divalent ions such as Cd^{++}, Hg^{++}, Cu^{++}, Zn^{++}, and Pb^{++}. Metallothionein is also highly inducible, so previous exposure to toxic metals can reduce the adverse effect of a subsequent exposure (Alexander et al., 2008). The interaction of metal mixtures may differ across organs because of the varied ability to produce metallothionein, reduced glutathione, and heat shock proteins. More than an additive response may be observed under combined exposure to Pb, Cd, and As in experimental systems and human studies (Wang and Fowler, 2008).

BIOMONITORING FOR ASSESSING HUMAN EXPOSURE TO CHEMICAL MIXTURES

Human biomonitoring data and biomonitoring equivalents can be used to estimate absorbed doses of chemical compounds from all routes (Fig. 33.1). Biomonitoring is the direct measurement of a chemical or its metabolites in body fluids (e.g., urine, blood, or saliva), tissues (e.g., hair), or exhaled air that enables the assessment of people's exposure to toxic substances. These analytes are known as biological markers or biomarkers of exposure. Biomonitoring represents an actual measure of integrated exposure regardless of the route of exposure and allows for multiple chemical exposure assessments (Wang et al., 2010). Another important function of biomonitoring is the development of "reference ranges" that describe general population exposures to contaminants; however, health-based guidelines for their interpretation have not yet been established for many chemicals. Biomonitoring, as a measure of internal dose, is influenced by alterations of toxicokinetic processes resulting from interactions of the components of mixtures (CDC, 2009; SCHER, 2011). In contrast, biomonitoring equivalents are quantitative benchmarks of safe, or acceptable, concentrations of a chemical or its metabolite in biological specimens that are consistent with selected reference values, such as human threshold limits (Reference dose (RfD), tolerable daily intake (TDI) etc.), by taking into account the available toxicokinetic data of the chemical (Boogaard et al., 2011).

The chemical composition of a mixture is often unknown and the concentration of the individual components may vary with time and environmental conditions. Assessments of the exposure to a mixture generally use relevant available data, such as emissions data, measurement of the components (or a lead component) in the sample of interest, and biomarker information (SCHER, 2011). Biomarker identification and characterization is therefore critical for risk assessment. A framework for the assessment of combined exposures to multiple chemicals has been recommended by the WHO/IPCS using a tiered approach that involves stepwise considerations of both exposure and hazard with increasingly data-informed analyses (Gomes and Meek, 2009).

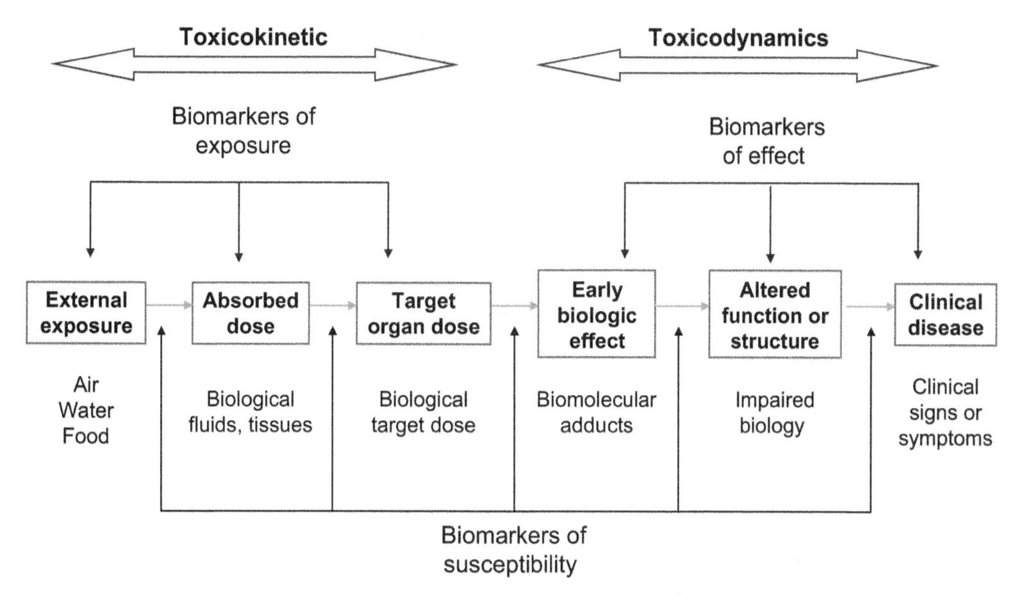

FIGURE 33.1 Paradigm of toxicity: continuum of events leading from external exposure to clinical disease. Role of biomonitoring to predict health consequences from the exposure.

Biological Specimens

Because of the relatively short biological half-life of nonpersistent chemicals in the body, the sampling period for the biological specimen to be analyzed can be important for the accuracy of representing the exposure to these chemicals. Physiological variations can affect the concentrations of certain chemicals and thus modify the sampling period. Blood and urine samples are the most widely used and accepted matrices for biomonitoring xenobiotic levels in the human body after occupational and environmental exposure. However, these biological fluids only reflect recent exposure (Gil and Hernández, 2009).

Blood

Blood is a complex matrix containing several components, such as proteins, lipids, and different types of cells. Biomonitoring of exposures to OP and *N*-methylcarbamate pesticides can easily be performed by using plasma and red blood cell cholinesterase activities (butyrylcholinesterase [BuChE] and AChE, respectively), which have become practical and useful tools for testing pesticide applicators to prevent overexposure to these compounds (Lessenger, 2005). Measurement of cholinesterase activity is a practical example of an integrated measure of a likely additive effect when there have been combined exposures to anticholinesterase compounds and would reflect any joint effects of these exposures (Alexander et al., 2008). Plasma BuChE activity is more sensitive to inhibition by some OPs than erythrocyte AChE activity and also recovers more rapidly following exposure. However, inter- and intraindividual variations in cholinesterase activities over time make this enzyme assay inaccurate at exposures causing up to 20% enzyme inhibition. In addition, it is not possible to elucidate which class of OP caused the inhibition, and certain OPs can induce adverse health effects without significant cholinesterase inhibition (Marsillach et al., 2013).

Persistent organic chemicals, such as the PCBs, PCDDs, polychlorinated dibenzofurans (PCDFs), organochlorine insecticides (DDT and its metabolite DDE), and brominated flame retardants, are chemicals readily measured in blood because of their partition into this matrix. The serum concentrations of these lipophilic chemicals can be affected by increased levels of serum triglyceride concentrations, as occurs transiently after a meal, because of the redistribution of these chemicals from adipose tissue. This variation can be corrected for by using lipid-adjusted units for these chemicals. The particular case of cord blood is of interest because it is indicative of fetal exposure, but a more sensitive analytical method is needed to measure these chemicals because cord blood has lower lipid concentration than maternal blood (Wang et al., 2010).

Heavy metals such as Pb, Cd, and Hg are typically measured in erythrocytes because of their affinity for thiol groups. They are found at increasing concentrations with advanced age as a result of their slow elimination from the body (Wang et al., 2010).

Urine

Chemicals metabolized to polar products are eliminated into the urine because of their increased hydrophilicity. Urine has the advantages of an easy collection and of higher concentrations of nonpersistent chemicals than has blood. Unlike blood, chemicals in the urine may not be at steady state because urine is an elimination pathway and its water content can vary throughout the day (Wang et al., 2010).

OP exposure can be assessed by analysis of the DAPs and leaving group metabolites in urine. DAP metabolites are produced during the hepatic and serum metabolism of many OPs and are eliminated in the urine. A leaving group, specific for each OP, is also generated during this metabolic process and excreted by the kidney. Urine DAP measurements do not provide specific information regarding the particular OP of exposure, as the metabolism of a given OP can release different DAPs, and the same DAPs can be released by different OPs. Moreover, detection of urinary DAPs or leaving groups cannot discriminate between OP exposure and environmental exposure to nontoxic OP degradation products ("preformed" DAPs in the environment), resulting in possible overestimation of the exposure. Furthermore, due to their short half-lives, they can be detected between 24 and 48 h following exposure and their levels can vary throughout the day. Thus, exposure assessment from a single urine sample may lead to conflicting results because of the variability of DAP excretion in humans. Regardless of these limitations, factors associated with greater DAP levels might be helpful for risk assessment, and measurement of DAPs in hair provides valuable information about chronic OP exposures (Barr and Angerer, 2006; Margariti and Tsatsakis (2009); Sudakin and Stone, 2011).

The presence of pyrethroid insecticide metabolites in the urine of the US population indicates widespread exposure to these widely used insecticides; 3-phenoxybenzoic acid (3-PBA), a common metabolite for many pyrethroids, has been detected in more than 70% of the samples of the 1999–2002 National Health and Nutrition Examination Survey, with children showing higher levels than adolescents and adults. Moreover, *cis*- and *trans*-(2,2-dichlorovinyl)-2,2-dimethylcyclopropane-1-carboxylic acids were highly correlated with 3-PBA, indicating that urinary 3-PBA was derived primarily from exposure to permethrin, cypermethrin, or their degradation products (Barr et al., 2010).

PAHs are converted by cytochrome P450 enzymes to hydroxylated metabolites, such as phenanthrols and 1-hydroxypyrene, which can be measured in urine. Although pyrene is commonly found in PAH mixtures, the extent of its contribution to the mixture can vary by time and setting. Thus, the measurement of 1-hydroxypyrene alone, despite the fact that it has been used widely as an indicator of recent exposure to PAH, may not accurately reflect the composition of the other PAHs in the mixture, particularly those with carcinogenic effects (Wang et al., 2010).

New Matrices for Assessing Cumulative Exposure

Alternative or complementary biological matrices are available for the detection of chemicals, such as human milk, saliva, adipose tissue, hair, meconium, and nail. The use of these matrices requires not only standardized collection, storage, processing, and analytical protocols but also understanding the absorption and metabolism of the chemicals of interest. The validation of analytical procedures is absolutely necessary for a proper implementation of nonconventional samples in biomonitoring programs (Gil and Hernández, 2015). Some matrices, such as hair and nails, are easily contaminated and difficult to collect in a standardized way. Others are difficult to obtain and available only in small amounts. Moreover, environmental chemicals are normally present in the biological matrix at trace levels, and thus, highly sensitive, specific, and selective multianalyte methods are required for the extraction, separation, and quantification of these chemicals (Calafat and Needham, 2009). The presence of a chemical compound in these matrices reflects an exposure, but given that many of them lack safe reference ranges, correlations with blood levels must be established to ensure that they are related to the total body burden (Esteban and Castaño, 2009).

Milk

Human milk is contaminated by lipophilic compounds as a result of long-term pollution by chemicals with long half-lives that tend to degrade slowly in the environment and to bioaccumulate in the food chain. Diet, particularly fatty fish, is a major factor that influences breast milk levels of persistent organohalogenated compounds. While organochlorine pesticide, PCB, and dioxin levels in breast milk have declined in countries where these chemicals have been banned or otherwise regulated, PBDE levels raised (Solomon and Weiss, 2002) until they experienced a modest decline after their ban (Bramwell et al., 2014). Human milk can be used to assess an infant's exposure to these chemicals from breastfeeding; however, the sampling of this matrix needs to be considered because the lipid content varies by the maturity of the milk. Although organic pollutants are the major chemicals measured in breast milk, heavy metals can also be determined in this matrix although they tend to accumulate more in blood (Esteban and Castaño, 2009).

Saliva

Saliva is a readily collectable material with potential diagnostic usefulness as a result of the presence of toxic compounds. However, saliva has several limitations as a suitable sample for analytical purposes, such as variation in flow rates, potential blood contamination during saliva collection, and the absence of reliable reference values for human populations (Gil and Hernández, 2015). In spite of these limitations, saliva has been accepted as a useful tool for biological monitoring of chemicals and may represent a complementary matrix to blood and urine for heavy metal analysis (Olmedo et al., 2010; Gil et al., 2011). Saliva sampling has also been shown to be suitable for measuring some pesticides rather than their metabolites in occupationally exposed adults and in children (Rodríguez et al., 2006).

Hair

Hair is a vehicle of excretion of substances from the human body and as such can indicate cumulative exposure to xenobiotics. Hair has been used to monitor occupational exposure to heavy metals, pesticides, PCDDs, PCBs, and, to a lesser extent, PAHs (Esteban and Castaño, 2009). Hair has also been proposed as an attractive choice for environmental health surveys of these and other environment chemicals. Human hair is a stable matrix, easily accessible for noninvasive sampling, and does not show storage changes for the period between sampling and analysis. Furthermore, it allows for repeated determinations over time. Although blood and urine concentrations clearly reflect recent exposure, hairs reflect past exposure, providing an average of their growth period. Limitations for hair include the potential for external contamination and the lack of sufficient information to define a normal range of reference typically found in the general population (Gil et al., 2011; Gil and Hernández, 2015).

Heavy metals are incorporated into the hair matrix by binding the thiol groups present in keratins. Their concentration varies significantly according to age, sex, hair color, hair care, smoking habits, and racial/ethnic factors. In the case of mercury, most of the levels measured in hair specimens represent exposure to organic mercury, particularly from fish consumption. Hence, hair is less useful for determining inorganic mercury. Nonetheless, the diverse sources of hair mercury can be individually identified by using high energy-resolution X-ray absorption spectroscopy as the different chemical forms of mercury bind in distinctive intermolecular configurations to hair proteins, as supported by molecular modeling (Manceau et al., 2016).

The detection of DAPs in human hair has been proposed as a useful tool for assessing either chronic exposure to OPs or past, acute poisoning if the appropriate hair segment is studied (Margariti and Tsatsakis, 2009; Tsatsakis et al., 2012). Recently, suitable analytical procedures have been reported for multiresidue pesticide analysis in human hair (Lehmann et al., 2018), which will allow for a better monitoring of long-term exposure to nonpersistent pesticides.

Meconium

The fetus can be exposed to different chemicals, including drugs of abuse, most of which are deposited and accumulated in the meconium, particularly during the second and third trimester of pregnancy. Meconium and placenta and cord blood are pertinent biological matrices for assessing the actual body burden of environmental contaminants in neonates. Meconium indicates cumulative exposure, whereas amniotic fluid is an indicator of fetal exposure to chemicals. Thus, meconium appears to be the best matrix for the determination of prenatal exposure to xenobiotics (Esteban and Castaño, 2009; Gil and Hernández, 2015).

For instance, measurement of DAP levels in meconium has been proposed as a potential biomarker for the assessment of OP exposure during pregnancy (Tsatsakis et al., 2009).

Biomolecular Adducts

The narrow sampling window for nonpersistent chemicals in blood can be extended by measuring biomolecular adducts whenever they are formed. The approximate duration of albumin and erythrocyte adducts is 21 days and 3 months, respectively. Because adducts last longer in blood than the parent chemical or metabolite, they can be used to assess for exposure when the sampling period is different from the time of exposure (Wang et al., 2010).

OP-modified enzymes remain longer in the blood than intact OPs, which are rapidly eliminated. Thus, they can be used as better biomarkers of exposure than the parent OPs. This is the case of OP-adducted BuChE and acylpeptide hydrolase (APH), which can be detected in plasma and red blood cells, respectively, several weeks following exposure because of their longer half-lives as compared with parent compounds or metabolites (11 days for BuChE and 33 days for APH). Moreover, both OP-adducted enzymes provide a longer window for detection than the determination of urinary metabolites or free parent OP compounds in blood (Marsillach et al., 2011).

The stable adducts formed on the active site of BuChE and APH following the aging process can be identified and characterized by liquid chromatography coupled with mass spectrometry (MS) and matrix-assisted laser desorption/ionization time-of-flight (John et al., 2008; Marsillach et al., 2013). In contrast to AChE, a baseline sample is not required with MS analyses for evaluation of OP exposures. OP agents may also bind to albumin, particularly on tyrosine-411, the most reactive residue of albumin. The OP—tyrosine bond is more stable and does not age like OPs adducted to active-site serine, allowing for a longer window of exposure detection and a more reliable identification of the OP involved in the exposure (Lockridge and Schopfer, 2010; Marsillach et al., 2013). OP-adducted albumin has been detected in serum from exposed humans even in the absence of AChE inhibition, and thus albumin has been proposed as a useful biomarker for OP exposure (Li et al., 2010).

Some PAH metabolites may also form adducts with biomolecules. Apart from the parent PAH or metabolite, the measurement of biomolecular adducts for some of these compounds in blood (e.g., binding of BaP to DNA, hemoglobin, or albumin) can extend the sampling window to assess for exposure and account for interindividual metabolic variations. Thus, determination of PAH-adduct levels is more informative than monitoring PAH exposures (Wang et al., 2010).

A comprehensive analysis of adducts (the so-called adductome) allows for the simultaneous detection of a variety of known and unknown adducts in DNA and protein samples by using MS-based methods. However, hemoglobin and serum albumin adducts are preferable over DNA adducts because of their greater abundance and half-life in blood.

DNA—protein cross-links are DNA adducts formed when proteins covalently bind to chromosomal DNA. These adducts can be induced by a variety of chemical and physical agents and have been associated with a number of pathophysiological conditions such as aging, cancer, heart disease, and neurodegenerative disorders (Groehler et al., 2017).

RISK ASSESSMENT OF COMBINED ACTIONS OF CHEMICAL MIXTURES

Because humans are often exposed to different chemical components at the same time, the possibility for combined effects of mixtures needs to be addressed in risk assessments for characterizing health risks for the particular exposure conditions. However, the complexity and variability of chemical mixtures and the paucity of data from studies using standard toxicological methods on combined actions of chemical mixtures challenge the risk assessment approach (Alexander et al., 2008). Evaluation of chemical mixture effects requires a systematic and integrated approach of in vivo, in vitro, and in silico data, together with systematic reviews (and metaanalysis, if available) of high-quality epidemiological studies (Hernández and Tsatsakis, 2017).

The prediction of toxic effects of a mixture is only possible with full knowledge of the individual components of the mixture, their mode of action, and dose-response curves. Nevertheless, this seldom occurs in practice. When information on the mode of action of the mixture components is unknown, the independent action concept is often used as default in human toxicology mixture assessments, although this approach may underestimate the toxicity. In turn, if the dose/concentration approach is applied, it may result in overprediction of toxicity (SCHER, 2011).

There are general principles of mixture toxicology describing how individual chemicals in a mixture affect one another: the concept of additivity (or no interaction) and interaction (Fig. 33.2). Additivity expectations can be derived from the concepts of dose addition and independent action, which assume that chemicals act by the same or different modes of action, respectively (Silins and Högberg, 2011). In this situation, also termed "non-interaction," the toxicity of a mixture resembles the effects expected to occur when all mixture components act without enhancing or diminishing their effects. By contrast, the term "interaction" indicates that the combined effects of two or more chemicals are stronger or weaker than would be expected on the basis of the additive effect, leading to synergism or antagonism, respectively. Interactions are difficult to predict because they are influenced by the relative dose levels, the route(s), timing and duration of exposure (including the biological persistence of the mixture components), and the biological targets (Alexander et al., 2008; Kortenkamp et al., 2009).

The nature of the combined effects of chemical mixtures varies significantly with dose. At exposures above the thresholds of effect for the individual components of the mixture, both toxicokinetic and toxicodynamic interactions can occur, thus making it difficult to predict the occurrence of adverse effects because of antagonism, potentiation, or synergism. Interactive effects giving rise to possible health concerns may start from concentrations around the lowest observed adverse effect level. Conversely, when exposures are below the respective dose thresholds for effect for each chemical in the mixture, the risk of combined adverse effects is much less likely to occur or is toxicologically negligible. For simple dissimilar action, individual components in the mixture are not assumed to contribute to the overall effect of the mixture if they are present at subthreshold levels of toxicity (Alexander et al., 2008; Mumtaz et al., 2010; SCHER, 2011).

Toxicokinetic interactions are a common cause of deviations from additivity. Examples are chemicals modifying the absorption, distribution, excretion, or metabolism of others. In contrast, toxicodynamic interactions occur at the cellular receptor/functional target level (Hernández et al., 2017). Competition of two components at the same target will usually result in addition

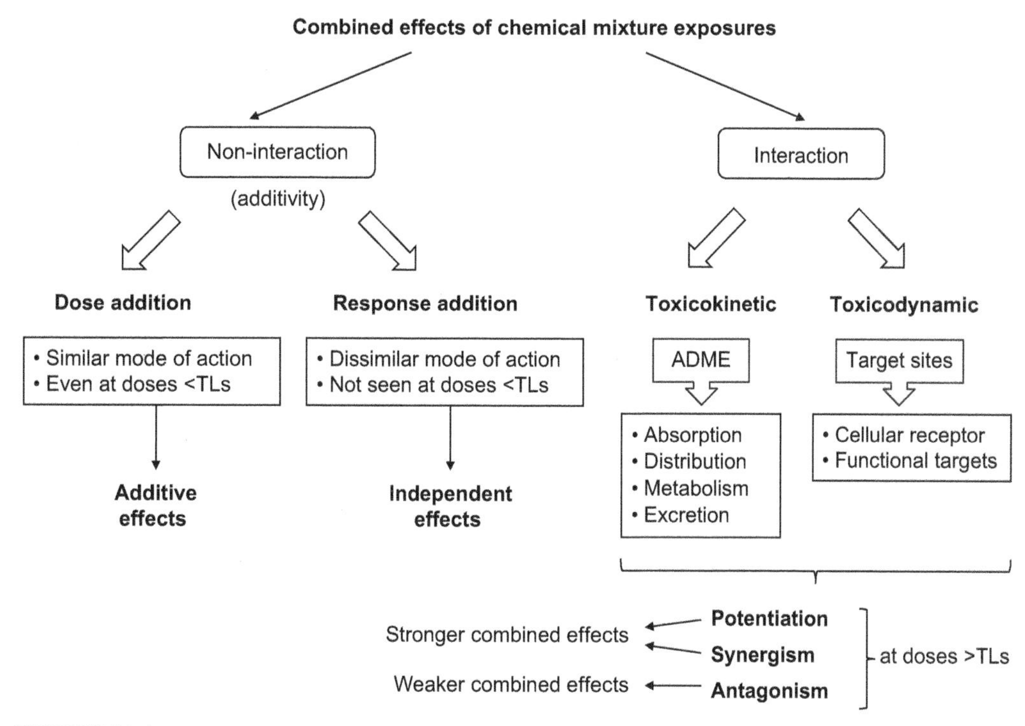

FIGURE 33.2 Type of combined effects of chemical mixtures exposure (TLs: threshold levels for toxicity).

of effects or antagonism (effect inhibition); it is very unlikely that this will lead to synergism. Weak agonists may function as antagonists by occupying the receptor and preventing the binding of a more potent ligand (Alexander et al., 2008; Wang et al., 2010).

Co-exposure to multiple metals may cause neurotoxic effects which are absent with exposure to a single metal at similar dose (Andrade et al., 2015). The additive toxicity approach has also been considered for assessing the combined effect of pesticides that have similar chemical structures and/or modes of toxic action. This is the case of concurrent multiple OP exposure, wherein a summation of the inhibitory effects of individual compounds on AChE activity is usually observed (Hernández et al., 2013a). A physiologically based pharmacokinetic model developed by Timchalk and Poet (2008) anticipated that at low environmentally relevant doses of the OP insecticides chlorpyrifos and diazinon, the cholinesterase inhibition would be dose additive. The assessment of the effect of a mixture of parathion, demeton-S-methyl, carbaryl, carbofuran, and aldicarb indicated an additive inhibitory effect on AChE activity (Mwila et al., 2013). Controversial findings have been reported for mixtures of several *N*-methylcarbamates (carbaryl, carbofuran, formetanate, methomyl, methiocarb, oxamyl, and propoxur) assayed in rats at equitoxic component doses and at environmental levels (Moser et al., 2012). In adult animals, the relative potency mixture showed dose additivity for erythrocyte AChE inhibition and motor activity, whereas brain AChE inhibition showed a modest greater-than-additive (synergistic) response but only at a middle dose. In rat pups, either dose-additive effects (brain AChE inhibition and motor activity) or slightly less-than-additive effects (erythrocyte AChE inhibition) were observed. At both ages, environmental levels of the chemical mixture showed greater-than-additive responses on the two cholinesterases and motor activity, with significant deviations from predicted at most doses tested (Moser et al., 2012).

Although additive effects imply that the chemicals in a mixture are acting by the same mechanism and at the same target, synergism can result from chemicals simultaneously acting in different toxicity pathways, thus magnifying their final toxic effect. Synergistic action in the carcinogenic process can be represented by the classical initiation–promotion model. Epidemiological evidence has indicated that cigarette smoke and inhaled arsenic exposure act synergistically to increase the incidence of lung cancer in smelter workers. By contrast, potentiation occurs when the toxicity of a chemical on a certain tissue or organ system is enhanced when given together with another chemical that alone does not have toxic effects on the same tissue or organ system (e.g., liver toxicity of carbon tetrachloride is enhanced with isopropanol) (Alexander et al., 2008).

BIOMARKERS OF TARGET ORGAN TOXICITY OF CHEMICAL MIXTURES

A biomarker of effect is usually a molecular indicator of a specific biological or disease state that can be measured in peripheral body fluids. Nevertheless, a broad concept would include any measurable change in a biological system or organism, including measured alterations in the structure or function of organs or tissues. Biomarkers of effect represent a valuable tool for assessing and controlling combined exposures from different routes and different compounds (Table 33.1). Recently, the approach of grouping chemicals focusing on common adverse outcomes could allow the use of shared effect biomarkers, even unspecific, as when used in combination they may reflect specific "signatures" of biochemical changes induced by a chemical mixture (Andrade et al., 2015). Early response biomarkers have the advantage of being interpreted within the context of integrated systems models, which connect these biomarkers to adverse outcomes of regulatory concern (DeBord et al., 2015).

A variety of strategies have been used for biomarker development, including transcriptional profiling, and proteomic and metabolomic approaches (Fig. 33.3). Examples of these markers are altered enzymatic activity, changes in protein expression or posttranslational modification, and altered gene expression, or protein or lipid metabolites, or a combination of these changes (Žurek and Fedora, 2012).

Biomarkers of Liver Damage

The liver plays a key role in the maintenance of the homeostasis of the organism and is the main organ responsible for detoxification of xenobiotics. As a result of Phase I microsomal metabolism, many chemicals undergo bioactivation instead of detoxification reactions, resulting in the generation of highly reactive metabolites. These molecular species may induce hepatotoxicity that can be assessed by measuring various peripheral proteins released in response to hepatocellular damage. However, the extremely large compensatory possibilities of the liver may account for the difficulty of proving a toxic effect in its early stage (Kaloyanova and Vergieva, 1987).

Conventional tests of hepatic damage provide information about the integrity of hepatocytes, such as serum levels of alanine and aspartate aminotransferases (ALT and AST, respectively), which are predominantly found in the periportal zone of the liver; however, ALT is considered as the gold standard biomarker of liver injury. The integrity of the biliary system is commonly assessed by measuring serum levels of bilirubin,

TABLE 33.1 Biomarkers for Assessing Target Organ Toxicity

Organ	Injury Location	Biomarker
Liver	Hepatocellular	ALT and AST (alanine and aspartate aminotransferases)
		GSTα (glutathione S-transferase alpha)
		Arginase I
		Ornithine carbamoyltransferase (OCT)
	Biliary system	GGT (γ-glutamyltransferase)
		ALP (alkaline phosphatase)
		5-Nucleotidase
		Bilirubin
	Biosynthesis	Albumin
		Prothrombin time
Kidney	Glomeruli	Urea
		Creatinine
		High molecular weight proteins (e.g., albumin[a])
		Cystatin C
		Glomerular filtration rate (GFR)
	Tubule	α_1-Microglobulin[a]
		β_2-Microglobulin[a]
		N-Acetylglucosaminidase (NAG)[a]
		Cystatin C[a]
		Kidney injury molecule-1 (KIM-1)[a]
		Neutrophil gelatinase–associated lipocalin (NGAL)
		GSTα[a] (for proximal tubule damage)
		GSTπ[a] (for distal tubule damage)
	Scarring	Collagen fragments[a] (e.g., fibronectin, laminin)
Brain	Neurons	Neuron-specific enolase (NSE)
		Hyperphosphorylated neurofilament (pNF-H)
		Ubiquitin C-terminal hydrolase L1 (UCH-L1)
		Microtubule-associated protein 2 (MAP-2)
		Microtubule-associated protein tau (MAPT)
		Translocator protein (TSPO)
	Glial cells	Glial fibrillary acidic protein (GFAP)
		Myelin basic protein (MBP)
		S100B protein
	Peripheral[b]	Erythrocyte AChE

Continued

TABLE 33.1 Biomarkers for Assessing Target Organ Toxicity—cont'd

Organ	Injury Location	Biomarker
Lung	Lung epithelia	CC16 (Clara cell protein 16)
		Krebs von den Lungen 6 (KL-6)
		Surfactant proteins (SP)
		Mucin 5B
		Cleaved cytokeratin 18 (cCK-18)
	Extracellular matrix remodeling	Matrix metalloproteases (MMP)
		Lysyl oxidase–like (LOXL)
	Immune dysfunction	Cytokines
		Chemokines
		T-cell activation

All biomarkers are determined in serum unless otherwise indicated.
[a]*Measured in urine.*
[b]*Surrogate biomarker for monitoring anticholinesterase pesticides exposure by measuring acetylcholinesterase activity in erythrocytes (red blood cells).*

γ-glutamyltransferase (GGT), and alkaline phosphatase (ALP) (Ozer et al., 2008; Gil and Hernández, 2009). Because of the lack of specificity of ALP, which can increase as a result of abnormalities of the biliary system and in certain skeletal disorders, either GGT or 5-nucleotidase can be assayed to clarify the origin of increased serum ALP. Increased serum bilirubin levels result from cholestatic and parenchymal liver diseases where the liver-blood cycling of conjugated bilirubin through membrane transporters is impaired. In turn,

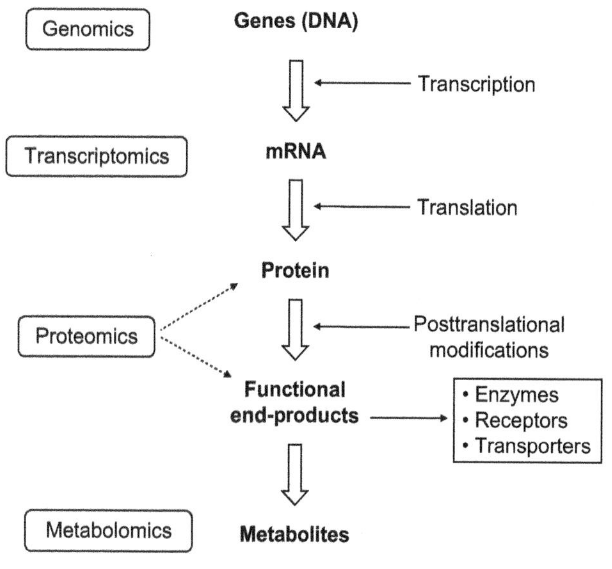

FIGURE 33.3 A functional -omics approach for biomarker development.

serum albumin and prothrombin time (which requires the presence of clotting factors produced in the liver) are the most commonly used tests to assess the biosynthetic capacity of the liver (Khalili et al., 2011).

Abnormalities in these liver function tests may be used to monitor liver damage from combined or mixed exposure. Additional information can be provided by more specific biochemical parameters not routinely used to assess drug/toxicant-induced liver injury because most of them involve ELISA measurement and a higher cost. Some examples are arginase I, ornithine carbamoyltransferase (OCT), glutathione S-transferase alpha (GSTα), and 4-hydroxyphenylpyruvate dioxygenase (Bailey et al., 2012). Arginase I catalyzes the catabolism of arginine to urea and ornithine and is highly liver specific, showing the earliest and greatest increase in serum after liver damage. OCT, like arginase I, is also involved in the urea cycle and its elevation in serum indicates damage to periportal hepatocytes (Ishikawa et al., 2003). From the eight GST classes (alpha, pi, mu, theta, kappa, sigma, omega, and zeta), the alpha GST class is more abundantly expressed in human liver and kidney. Unlike ALT and AST, GSTα is found in high concentration in centrilobular cells and is therefore more sensitive to injury in this metabolic zone of the liver. GSTα is released shortly after hepatocellular damage, leading to a greater increase in plasma levels than that found for ALT or AST (Singh et al., 2011).

Anand et al. (2005) examined the hepatotoxic outcome of secondary, tertiary, and quaternary mixtures of thioacetamide, allyl alcohol (a periportal hepatotoxicant), chloroform (a centrilobular hepatotoxicant), and trichloroethylene (that causes midzonal liver injury). These compounds are metabolized by the CYP2E1 enzyme and also exhibit nephrotoxic and carcinogenic potential in rodents. Higher ALT levels were observed without histopathological changes, similar to what has been described in chloroform alone studies, where slight plasma membrane injury was caused by its reactive metabolite phosgene. The observed subadditive toxicity was attributed to the fact that certain solvents (e.g., chloroform) enhance the elimination of others (e.g., trichloroethylene), leading to an antagonistic interaction (Anand et al., 2005). In a cross-sectional study a combined chemical exposure to lead and organic solvents under the permissive exposure level (indicating that it is not harmful to most workers) showed additive effects on the same target organ (liver), even if each single exposure occurs below the permissible exposure limit (Chang et al., 2013).

An increase in triglycerides and total cholesterol has been described as specific for toxic hepatitis (Kaloyanova and Vergieva, 1987; Provost et al., 2003) and, thus, they could also serve as markers of liver dysfunction. In fact, higher levels of triglycerides have been reported to reflect the total lifetime cumulative exposure to solvents (Silins and Högberg, 2011).

Kidney

The kidney is a target organ for many xenobiotics because of the high volume of blood flowing through it and the large capacity to concentrate toxins in the renal tubular system. Although the kidney is one of the main target organs for xenobiotic-induced toxicity, early detection of renal damage is difficult. Conventional biomarkers of kidney damage include serum creatinine and blood urea nitrogen (BUN); however, their levels increase when the kidney damage is greater than 50%, very likely indicating an irreversible progression. The increase in low molecular weight proteins in urine (such as α_1-microglobulin, β_2-microglobulin, or retinol-binding protein), assumed to be freely filtered, reflects tubular dysfunction and is an earlier and more sensitive biomarker of kidney tubular toxicity than is serum creatinine. Lysosome-derived enzymes, such as N-acetyl β-D-glucosaminidase (NAG) and β-galactosidase, are also used for monitoring proximal tubule insults because of their high activity in the kidney and its stability in urine (Gil and Hernández, 2009). Total urinary protein is a sensitive and early biomarker of changes in the glomerular filtration and the tubular reabsorption systems and has been used to assess progressive loss of kidney function (Goodsaid et al., 2009).

Several novel, sensitive, and tissue-specific biomarkers of nephrotoxicity have been proposed to improve detection of early acute kidney injury (AKI) following exposure to nephrotoxicants as compared with traditionally used biomarkers. They also enable the localization of the injury to a specific nephron site. Some examples are urinary kidney injury molecule-1 (KIM-1), neutrophil gelatinase–associated lipocalin (NGAL), interleukin-18 (IL-18), cystatin C, and GST, which are early diagnostic biomarkers of AKI; however, biomarkers such as NGAL, KIM-1, and cystatin C are also useful for chronic kidney disease (CKD). Biomarker combinations are required to increase diagnostic accuracy for either AKI or CKD detection (Wasung et al., 2015). Serum cystatin C has been considered a better endogenous marker of glomerular filtration rate (GFR) than serum creatinine in early stages of AKI and is less affected by age, gender, muscle mass, and ethnicity. Unlike creatinine, cystatin C is not secreted by the tubule, such that its appearance in urine indicates a tubular dysfunction with reabsorption failure (Charlton et al., 2014). Therefore, increased serum cystatin C is indicative of glomerular damage, while urinary levels indicate tubular damage. KIM-1 is a transmembrane renal tubular glycoprotein involved in immune regulation and in kidney tubular regeneration that is upregulated after ischemia- or nephrotoxicant-induced injury (Simsek et al., 2013). Although NGAL is an acute-phase reactant that is found to be elevated in inflammatory conditions, it is also released from lysosomes, brush border, and cytoplasm of proximal tubular epithelial cells (Wasung et al., 2015).

The assessment of nephrotoxicant-induced abnormal renal function might include the determination of the following biomarkers in serum or urine:

- Serum creatinine, BUN, and the GFR: These should be initially determined. As they lack enough sensitivity, much more sensitive biomarkers should be studied for assessing subclinical renal damage.
- Biomarkers of glomerular damage or dysfunction: Increased urine excretion of high molecular weight protein, such as albumin or transferrin, can be helpful for an early detection of glomerular barrier defect. Serum cystatin C correlates negatively with the GFR and is a much more sensitive marker for detecting early glomerular alterations.
- Biomarkers of tubular injury: Urinary cystatin C, KIM-1, α_1-microglobulin (which is more stable at acidic pH than β_2-microglobulin) and GSTα reflect a functional impairment of the proximal tubular reabsorption due to glomerular injury or direct tubular damage. Urine GSTπ is a biomarker useful for specifically assessing distal tubular damage.
- Biomarkers of recovery/repair: Urinary elevation of clusterin, a cytoprotective chaperone, indicates recovering of renal function from the renal injury.
- Biomarkers for scarring: Collagen fragments (e.g., type I collagen, fibronectin, laminin) are useful for monitoring the progression of glomerular damage and may also reflect the degree of tubulointerstitial fibrosis.

Differences in the excretion of these biomarkers reflect different stages of severity of renal function impairment. Under conditions of chronic solvent exposure, renal damage may remain clinically silent for many years owing to the large functional reserve capacity of the kidney. The degree of reversibility depends on the severity of the disease. At a later stage, the disease can be irreversible, leading to end-stage renal failure. It is therefore important to detect kidney defects as early as possible to prevent progression of the disease (Mangelsdorf, 2009).

Adverse effects induced by chronic exposure to Pb and Cd have been monitored by measuring urinary GSTα, a hallmark of early changes in the proximal tubular integrity, which showed a positive correlation with urinary levels of these heavy metals (Garçon et al., 2004). Combined exposure to As and Cd leading to increased urinary levels was followed by considerably enhanced urinary excretion of β_2-microglobulin or NAG than each exposure alone, indicating renal tubular damage (Nordberg, 2010). In addition, individuals showing lower Cd-induced metallothionein mRNA levels displayed higher β_2-microglobulin or NAG values at comparable levels of urinary Cd (Nordberg, 2010), indicating the protecting role of this inducible metal-binding protein.

Nervous System

Exposure to some chemicals may induce persistent changes in the nervous system by affecting different neuronal targets such as membranes, intracellular signaling, axonal transport, or energy supply. Chemicals in a mixture affecting any of these targets may interact to produce neuronal damage. Mature neurons have limited capability of regeneration and are therefore at greater risk for permanent damage after a toxic injury than many other cells in the body. This limitation is partially compensated for by the large reserve capacity of the central nervous system; only when the reserve decreases below a critical level does overt neurotoxicity ensue. On the other hand, the nervous system is particularly vulnerable to toxic compounds during the early period of development (Alexander et al., 2008).

Nonneurotoxic compounds can make the blood—brain barrier (BBB) more permeable to ionic compounds and thus may potentiate their toxicity. Likewise, biomarkers originating in the brain can gain access to the blood due to disruption of the BBB. Biomarkers that are measurable with minimally invasive techniques, such as biological fluid-based molecular markers and emerging neuroimaging methodologies, could be of utility both preclinically and clinically. Potential molecular markers include glial fibrillary acidic protein (GFAP, a biomarker of astrogliosis, which consists of a cellular reaction that indicates both neuronal and glial damage), ubiquitin C-terminal hydrolase L1 (UCH-L1, a biomarker of cell body injury), myelin basic protein (MBP, a biomarker of myelin disruption), microtubule-associated protein-2 (MAP-2, characteristic of dendritic injury), neurofilaments (NF, biomarkers of axonal injury), microtubule-associated protein tau (MAPT, biomarker of neurodegeneration/axonal injury), and translocator protein (TSPO, indicative of activated glia and also may signal neuroinflammation). In turn, the main neuroimaging methodologies are magnetic resonance imaging, magnetic resonance spectroscopy, and positron emission tomography (Roberts et al., 2015).

Neuronal tissue and glial-derived proteins are easily detectable in peripheral blood and thus may serve as surrogate markers of brain damage. NF, the most abundant protein of neurons and major components of the neuronal cytoskeleton, are released in large numbers when these cells are damaged. Because NFs are only found in neurons, their detection in blood points unambiguously to neuronal damage. NFs consist of three subunits, namely NF-light (L), NF-medium (M), and NF-heavy (H) based on their molecular weight. The function of NFs appears to be closely related to their phosphorylation status, and abnormally phosphorylated NFs in the cell bodies have been proposed as a common feature of neurodegenerative diseases. Hyperphosphorylated NFs may be a good serum biomarker for axonal injury and degeneration because of its stability within blood (Shaw, 2015).

When the central nervous system is injured in response to toxic substances or disease states, astroglial cells react to repair the damage. Astrocyte proliferation upregulates GFAP, an intermediate filament protein that provides support and strength to astroglial cells, which is partially released to blood where it can be measured (Žurek and Fedora, 2012).

Neuron-specific enolase (NSE, a glycolytic enzyme) and protein S100B (a calcium-binding protein abundant in astrocytes) have been shown to increase in serum after a number of cerebral insults, such as traumatic brain injury or stroke, resulting in structural damage. However, both NSE and S100B have limitations derived from false positive values (Žurek and Fedora, 2012).

Peripheral cells, including platelets, erythrocytes, lymphocytes, and fibroblasts, have been used to evaluate exposure to neurotoxic agents, mainly heavy metals, pesticides, and solvents (n-hexane and carbon disulfide). The best example is represented by the measurement of erythrocyte AChE following exposure to OP and *N*-methylcarbamate insecticides, as this enzyme better reflects brain AChE activity. Exposure to mixtures of these anticholinesterase pesticides has been assessed by using erythrocyte AChE inhibition as mentioned above.

Lungs

A number of molecules measured in serum, bronchoalveolar lavage fluid, or lung tissue can be used as biomarkers of lung toxicity and allow a better understanding of pathological processes. Possible biomarkers can be categorized in three groups: (1) biomarkers associated with alveolar epithelial cell dysfunction (e.g., Krebs von den Lungen 6—KL-6; surfactant proteins—SPs; mucin 5B; cleaved cytokeratin 18—cCK-18); (2) biomarkers associated with extracellular matrix remodeling and fibroproliferation (matrix metalloproteinases—MMPs; lysyl oxidase–like—LOXL; osteopontin); and (3) biomarkers related to immune dysfunction (e.g., cytokines, chemokines, and markers of T-cell activation) (Drakopanagiotakis et al., 2018).

Inhalation of toxic compounds may result in airway irritation featured by increased permeability, edema, and cell necrosis. Cytotoxic effects in the respiratory tract are often general and nonspecific and are related to water solubility of the toxic compound. Combined exposure to ozone and NO_2 has shown an interactive effect on lung function, leading at least to an additive irritation effect. Several cell types of the bronchial epithelium are extremely vulnerable to various types of injury. This is the case for Clara cells, which contain most of the lung CYP450 activity that confers a high xenobiotic metabolizing activity. Thus, these cells are particularly sensitive to toxic injury from inhaled xenobiotics that are metabolized by CYP450 enzymes (Gil and Hernández, 2009).

Clara cell protein (CC16, also referred to as CC10) is a low molecular weight protein that is secreted along the tracheobronchial tree, particularly in the terminal bronchioles where Clara cells are localized. It can be measured in serum, bronchoalveolar lavage fluid, and sputum as a peripheral lung marker for assessing the cellular integrity or the permeability of the lung epithelium. This protein is also involved in the protection of the respiratory tract against oxidative stress and inflammation. Acute exposures to pulmonary irritants can cause a transient increase in serum CC16 levels, which is not associated with impairment of pulmonary function (Broeckaert et al., 2000; Lakind et al., 2007).

Asphalt workers have shown increased mortality from respiratory diseases because of exposure to chemical mixtures of dust, oil mist, PAHs, and nitrogen dioxide. A significant increase in CC16 concentration has been reported in asphalt pavers exposed to chemical mixtures, particularly bitumen fume and vapor, and exhaust from engines. Thus, CC16 appears to be a useful biomarker for lung epithelial injury in workers exposed to lung toxic chemical mixtures (Ulvestad et al., 2007).

Particulate matter (PM) is a complex mixture of extremely small particles and liquid droplet present in the atmosphere and considered as air pollutants. The complex mixture consists of organic carbon, ammonium, nitrates, sulfates, mineral dust, trace elements, and water. PM can develop and exacerbate diverse respiratory diseases, such as asthma and chronic obstructive pulmonary disease. Besides, PM can induce oxidative stress and inflammation on respiratory airways and eventually can lead to cancer. A number of oxidative stress and inflammatory biomarkers have been considered as potentially helpful for assessing respiratory diseases induced by PM (Kim et al., 2017).

NONSPECIFIC BIOMARKERS OF TOXIC RESPONSE

Oxidative Stress

Many chemicals, either alone or in mixtures, may produce oxidative stress as a result of increasing the generation of free radicals, affecting the mitochondrial function or decreasing levels of antioxidants and antioxidant enzymes in blood and other tissues. The initiation of oxidative damage can be reversed by intracellular antioxidants or stimulation of antioxidant enzymes or repair systems. The increase in antioxidant levels leads to the scavenging of excess free radicals and contributes to a decrease in oxidative damage, whereas a decrease in antioxidants should lead to the opposite effect (Liu et al., 2000). A brief elevation in oxidative stress rapidly induces various antioxidant defenses, particularly antioxidant enzymes, such as superoxide dismutase, catalase, and glutathione peroxidase, which quickly reduce the cellular damage and limit the ability of testing methods to detect a change (Birben et al., 2012).

Assessment of oxidative stress involves the determination of biomarkers for oxidative damage, such as decreases in antioxidant concentrations or increases in oxidative metabolites. The latter can be monitored by measuring lipid peroxidation products (malondialdehyde), protein oxidation (protein carbonyl levels and glutamine synthetase activity), or endogenous antioxidants (ascorbic acid, α-tocopherol, reduced glutathione, total antioxidant capacity of serum, and ferric reducing ability of serum) (Liu et al., 2000). Biomarkers of oxidative damage of proteins (3-nitrotyrosine) and DNA (8-hydroxy-2′-deoxyguanosine) are also used for assessing oxidative stress; 3-nitrotyrosine is formed primarily from peroxynitrite, a damaging free radical generated from superoxide and nitric oxide. Other functional indicators of oxidative stress include relative levels of 3-chlorotyrosine and aconitase activity. Aconitase is highly sensitive to oxidative inactivation by the superoxide radicals produced in the mitochondria (Rose et al., 2012). However, changes in these biomarkers may not necessarily reflect a clinically significant or pathogenic event but merely indicate that the antioxidant defense system is functioning.

Oxidative stress is a sensitive endpoint for metal mixtures, very likely because mitochondria are common targets for certain metals that bind to thiol groups or displace essential metals in antioxidant enzymes. This is the case for Pb, Cd, Hg, and As, which produce alterations of target tissue concentrations of certain essential metals (e.g., Fe and Zn) or inhibit enzymes containing thiol groups (Wang and Fowler, 2008). Alterations in essential trace elements (Cu, Se, Zn) have been reported in workers exposed to metal mixtures (Pb and Cd) showing significant decreases in the Se-dependent enzyme glutathione peroxidase in both erythrocytes and serum (Wasowicz et al., 2001). Therefore, combined exposures to metal elements can modify the antioxidant capabilities of the blood.

Prooxidant/antioxidant alterations have been observed in subjects chronically exposed to airborne particles containing coal dust, usually by inducing enhanced lipid peroxidation together with a decline in antioxidants enzymes (Ávila Júnior et al., 2009). Nonenzymatic antioxidants, such as reduced glutathione and vitamin E, were depleted in subjects more exposed to air contamination, whereas enzymatic antioxidants were either induced (e.g., GST and catalase) or inhibited (superoxide dismutase and glutathione peroxidase). Besides workers, people living in the vicinity of the polluted mining areas also face oxidative stress and risk of coal mining–related diseases (Ávila Júnior et al., 2009).

In addition to their canonical toxicity mechanisms, OP, *N*-methylcarbamates, and organochlorine pesticides may induce cellular oxidative stress as a result of an enhanced generation of highly reactive molecules and/or reduced capacity of the antioxidant systems of the body, thus contributing to disruption of cell function (Dettbarn et al., 2006; Karami-Mohajeri and Abdollahi, 2011). Long-term pesticide exposure in intensive agriculture workers has been associated with reduced superoxide dismutase, glutathione reductase, and delta-aminolevulinic acid dehydratase activities (Hernández et al., 2013b). If these changes are sustained over time, there is an increased risk for developing a number of chronic and degenerative diseases wherein oxidative stress plays a pathogenic role.

DNA Damage and Genotoxicity

There is much knowledge about the genotoxic effects of individual chemicals but scarce information about possible interactions or combined actions among chemicals. Some xenobiotics, including PAHs, aromatic amines, heterocyclic amines, mycotoxins, alkylating agents, and nitrosamines, have reacting groups that can interact with nucleophilic targets, such as DNA, leading to genetic damage. However, genotoxic compounds may induce DNA damage by many other different mechanisms, including oxidative stress, thus making it difficult to predict the outcome of exposure to chemical mixtures. Besides, long-lived cells with high sensitivity to chemical agents may accumulate DNA damage over time, which, if not properly repaired, leads to mutations and ultimately to an increased risk of long-term effects such as inflammatory, carcinogenic, and reproductive disorders. Interactions that affect the bioavailability, metabolic activation or detoxification, or DNA binding or repair of DNA damage may influence the genotoxicity of a complex mixture. Initiators, promoters, and co-carcinogens all may act in concert to potentiate the final tumor outcome. However, combination effects at low and realistic human exposures, which are not possible to explore experimentally in animal studies, may be different from those obtained by experimentation at high doses (Alexander et al., 2008).

Different types of markers have been designed to assess the mutagenic (and possibly carcinogenic) response of certain chemicals. They are usually studied in peripheral cells, especially circulating lymphocytes or other cells readily accessible as the epithelial cells of the nasal or buccal mucosa. Although a positive response in these biomarkers does not always indicate increased risk of cancer, their early occurrence makes them useful for assessing toxic risks. The detection of double and single DNA strand breaks is relevant for effect monitoring of chemical mutagens and may serve as a surrogate for carcinogenesis. However, chromosomal changes are the most commonly studied biomarkers of genotoxicity. Stable chromosomal aberrations, especially long-lasting chromosomal rearrangements, have greater importance as indicators of carcinogenesis than reversible DNA strand breaks, which only reflect recent exposure. Indirect markers of chromosomal damage include sister chromatid exchange, induction of micronuclei (easily seen in enucleated cells such as circulating erythrocytes), and the comet assay, which is capable of detecting breakage of one or two strands of DNA (Greim, 2001). However, not every DNA-damaging chemical is potentially carcinogenic as it may be toxic to cells at high levels but does not pose a cancer risk in vivo. The lack of well-established dose—response relations between occupational or environmental exposures and the formation of strand breaks limits the applicability of these biomarkers (Angerer et al., 2007). Although DNA adducts have also been used to assess the genotoxic potential of chemicals (Gil and Hernández, 2009), they are considered biomarkers of exposure rather than biomarkers of effect.

PAHs may contribute to the genotoxicity and carcinogenicity of air pollution from urban environments. The exposed populations may also develop oxidative stress leading to oxidative DNA damage that results in increased levels of several markers of genotoxicity, including bulky DNA adducts, chromosomal aberrations, sister chromatid exchanges, and Ras oncogene overexpression (Farmer et al., 2003). A study conducted in children exposed to a mixture of pollutants from different sources found that the genotoxicity effects (evaluated by the comet assay) associated with exposure to low levels of benzene and Pb were enhanced by coexposure to high levels of PAHs (Pelallo-Martínez et al., 2014). An increase in genomic instability and cell death has been reported in the human buccal micronucleus cytome assay of workers occupationally exposed to coal residue mixtures as a consequence of oxidative damage (Rohr et al., 2013).

Occupational exposure to bitumen fumes in road paving workers has been associated with DNA damage assessed by the comet assay in blood mononuclear leukocytes (Lindberg et al., 2008). In this study, levels of DNA damage correlated with both urinary concentration of 1-hydroxypyrene and external air concentrations of bitumen fumes and vapors, thus supporting the genotoxic effects of this chemical mixture.

Combined exposures to metals have produced DNA strand breaks by a mechanism involving oxidative stress coupled with an inhibition of DNA damage reparation. Heavy metals such as Cd induce the generation of reactive oxygen species that can cause direct DNA damage. The urinary excretion of 8-hydroxyguanine is a reflection of an increased number of modified DNA bases. A synergistic effect of exposure to metals mixture on DNA oxidative stress may occur as a result of the combination of reactive oxygen species and impairment of DNA repair (Yoshioka et al., 2008).

CONCLUDING REMARKS AND FUTURE DIRECTIONS

Within mixture toxicity research, there is a need to identify the type and extent of interactions between the components of chemical mixtures and understand the underlying mechanisms. Experimental studies on whole mixtures should reflect real-life exposures, including their concentrations and component proportions, and should include several doses below the threshold level of exposure for individual compounds to assess for potential additive or supraadditive effects. The use of biomarkers of effect based on analogous endpoints of toxicity induced by individual chemicals in a mixture could be a useful tool for predicting and preventing the risk of toxicity. A better understanding mechanism underlying toxic interactions will provide a more robust basis for the development and selection of biomarkers of effect, which will be helpful for risk assessment and for the prevention of health risks in exposed populations. Advances in modern molecular biology techniques, such as genomics, proteomics, and metabolomics, as well as chip technologies, will enable the screening of effects of chemical mixtures at the molecular level and the development of more sensitive and specific methodologies for biological monitoring of combined exposures. Physiologically based toxicokinetic/toxicodynamic models will be of help for data-poor chemicals to estimate biomarker concentrations and mechanistically relevant effects in target organs. Further well-designed studies on the interactive toxicity of mixtures would broaden the scientific database on combined exposures of chemicals and would help to elucidate their impact on the health outcomes.

References

Alexander, J., Hetland, R.B., Vikse, R., et al., 2008. Opinion of the Scientific Steering Committee of the Norwegian Scientific Committee for Food Safety, 2008.

Anand, S.S., Mumtaz, M.M., Mehendale, H.M., 2005. Dose-dependent liver regeneration in chloroform, trichloroethylene and allyl alcohol ternary mixture hepatotoxicity in rats. Arch. Toxicol. 79, 671–682.

Andrade, V.M., Mateus, M.L., Batoréu, M.C., et al., 2015. Lead, arsenic, and manganese metal mixture exposures: focus on biomarkers of effect. Biol. Trace Elem. Res. 166, 13–23.

Angerer, J., Ewers, U., Wilhelm, M., 2007. Human biomonitoring: state of the art. Int. J. Hyg. Environ. Health 210, 201–228.

Ávila Júnior, S., Possamai, F.P., Budni, P., 2009. Occupational airborne contamination in south Brazil: 1. Oxidative stress detected in the blood of coal miners. Ecotoxicology 18, 1150–1157.

Bailey, W.J., Holder, D., Patel, H., Devlin, P., 2012. A performance evaluation of three drug-induced liver injury biomarkers in the rat: alpha-glutathione S-transferase, arginase 1, and 4-hydroxyphenylpyruvate dioxygenase. Toxicol. Sci. 130, 229–244.

Barr, D.B., Angerer, J., 2006. Potential uses of biomonitoring data: a case study using the organophosphorus pesticides chlorpyrifos and malathion. Environ. Health Perspect. 114, 1763–1769.

Barr, D.B., Buckley, B., 2011. In vivo biomarkers and biomonitoring in reproductive and developmental toxicity. In: Gupta, R.C. (Ed.), Reproductive and Developmental Toxicology. Academic Press/Elsevier, Amsterdam.

Barr, D.B., Olsson, A.O., Wong, L.Y., 2010. Urinary concentrations of metabolites of pyrethroid insecticides in the general U.S. population: National Health and Nutrition Examination Survey 1999–2002. Environ. Health Perspect. 118, 742–748.

Birben, E., Sahiner, U.M., Sackesen, C., et al., 2012. Oxidative stress and antioxidant defense. World Allergy Organ. J. 5, 9–19.

Boogaard, P.J., Hays, S.M., Aylward, L.L., 2011. Human biomonitoring as a pragmatic tool to support health risk management of chemicals – examples under the EU REACH programme. Regul. Toxicol. Pharmacol. 59, 125–132.

Bramwell, L., Fernandes, A., Rose, M., et al., 2014. PBDEs and PBBs in human serum and breast milk from cohabiting UK couples. Chemosphere 116, 67–74.

Broeckaert, F., Clippe, A., Knoops, B., et al., 2000. Clara cell secretory protein (CC16): features as a peripheral lung biomarker. Ann. N.Y. Acad. Sci. 923, 68–77.

Calafat, A.M., Needham, L.L., 2009. What additional factors beyond state-of-the-art analytical methods are needed for optimal generation and interpretation of biomonitoring data? Environ. Health Perspect. 117, 1481–1485.

CDC (Centers for Disease Control), 2009. Fourth National Report on Human Exposure to Environmental Chemicals. National Center for Environmental Health, Atlanta, Georgia. Available at: http://www.cdc.gov/exposurereport/pdf/FourthReport.pdf.

Chang, W.J., Joe, K.T., Park, H.Y., et al., 2013. The relationship of liver function tests to mixed exposure to lead and organic solvents. Ann. Occup. Environ. Med. 25, 5.

Charlton, J.R., Portilla, D., Okusa, M.D., 2014. A basic science view of acute kidney injury biomarkers. Nephrol. Dial. Transplant. 29, 1301–1311.

Costa, L.G., de Laat, R., Tagliaferri, S., et al., 2014. A mechanistic view of polybrominated diphenyl ether (PBDE) developmental neurotoxicity. Toxicol. Lett. 230, 282–294.

Crofton, K.M., 2009. WHO/IPCS international workshop on aggregate/cumulative risk assessment: overview. In: In IPCS-WHO Assessment of Combined Exposures to Multiple Chemicals-International Workshop, Harmonization Project Document 7, pp. 18–22.

DeBord, D.G., Burgoon, L., Edwards, S.W., et al., 2015. Systems biology and biomarkers of early effects for occupational exposure limit setting. J. Occup. Environ. Hyg. 12 (Suppl. 1), S41–S54.

Dettbarn, W.D., Milatovic, D., Gupta, R.C., 2006. Oxidative stress in anticholinesterase-induced excitotoxicity. In: Gupta, R.C. (Ed.), Toxicology of Organophosphate and Carbamate Compounds. Academic Press/Elsevier, Amsterdam, pp. 511–532.

Dishaw, L.V., Macaulay, L.J., Roberts, S.C., et al., 2014. Exposures, Mechanisms.

Drakopanagiotakis, F., Wujak, L., Wygrecka, M., Markart, P., 2018. Biomarkers in idiopathic pulmonary fibrosis. Matrix Biol. https://doi.org/10.1016/j.matbio.2018.01.023 (in press).

EFSA (European Food Safety Authority), 2005. Opinion of the Scientific Panel on Contaminants in the Food Chain on a request from the Commission related to the presence of non dioxin-like polychlorinated biphenyls (PCB) in feed and food. EFSA J. 284, 1–137.

Esteban, M., Castaño, A., 2009. Non-invasive matrices in human biomonitoring: a review. Environ. Int. 35, 438–449.

Farmer, P.B., Singh, R., Kaur, B., 2003. Molecular epidemiology studies of carcinogenic environmental pollutants. Effects of polycyclic aromatic hydrocarbons (PAHs) in environmental pollution on exogenous and oxidative DNA damage. Mutat. Res. 544, 397–402.

Garçon, G., Leleu, B., Zerimech, F., 2004. Biologic markers of oxidative stress and nephrotoxicity as studied in biomonitoring of adverse effects of occupational exposure to lead and cadmium. J. Occup. Environ. Med. 46, 1180–1186.

Gil, F., Hernández, A.F., 2009. Significance of biochemical markers in applied toxicology. In: Ballantyne, B., Marrs, T.C., Syversen, T. (Eds.), General and Applied Toxicology. John Wiley & Sons, Chichester (UK), pp. 847–857.

Gil, F., Hernández, A.F., Márquez, C., 2011. Biomonitorization of cadmium, chromium, manganese, nickel and lead in whole blood, urine, axillary hair and saliva in an occupationally exposed population. Sci. Total Environ. 409, 1172–1180.

Gil, F., Hernández, A.F., 2015. Toxicological importance of human biomonitoring of metallic and metalloid elements in different biological samples. Food Chem. Toxicol. 80, 287–297.

Gomes, J., Meek, B., 2009. Interactions between occupational and environmental factors in toxicology, hazard evaluation and risk assessment. In: Ballantyne, B., Marrs, T.C., Syversen, T. (Eds.), General and Applied Toxicology. John Wiley & Sons, Chichester (UK), pp. 2247–2264.

González-Alzaga, B., Lacasaña, M., Hernández, A.F., et al., 2018. Serum concentrations of organochlorine compounds and predictors of exposure in children living in agricultural communities from South-Eastern Spain. Environ. Pollut. 237, 685–694.

Goodsaid, F.M., Blank, M., Dieterle, F., 2009. Novel biomarkers of acute kidney toxicity. Clin. Pharmacol. Ther. 86, 490–496.

Greim, H., 2001. Endpoints and surrogates for use in population studies in toxicology. Toxicol. Lett. 120, 395–403.

Groehler 4th, A., Degner, A., Tretyakova, N.Y., 2017. Mass spectrometry-based tools to characterize DNA-protein cross-linking by bis-electrophiles. Basic Clin. Pharmacol. Toxicol. 121 (Suppl. 3), 63–77.

Hernández, A.F., Parrón, T., Tsatsakis, A.M., et al., 2013a. Toxic effects of pesticide mixtures at a molecular level: their relevance to human health. Toxicology 307, 136–145.

Hernández, A.F., Lacasaña, M., Gil, F., et al., 2013b. Evaluation of pesticide-induced oxidative stress from a gene-environment interaction perspective. Toxicology 307, 95–102.

Hernández, A.F., Tsatsakis, A.M., 2017. Human exposure to chemical mixtures: challenges for the integration of toxicology with epidemiology data in risk assessment. Food Chem. Toxicol. 103, 188–193.

Hernández, A.F., Gil, F., Lacasaña, M., 2017. Toxicological interactions of pesticide mixtures: an update. Arch. Toxicol. 9, 3211–3223.

Hood, D.B., Ramesh, A., Chirwa, S., 2011. Developmental toxicity of polycyclic aromatic hydrocarbons. In: Gupta, R.C. (Ed.), Reproductive and Developmental Toxicology. Academic Press/Elsevier, Amsterdam, pp. 593–606.

Ishikawa, H., Matsuzawa, T., Ohashi, K., et al., 2003. A novel method for measuring serum ornithine carbamoyltransferase. Ann. Clin. Biochem. 40, 264–268.

John, H., Worek, F., Thiermann, H., 2008. LC-MS-based procedures for monitoring of toxic organophosphorus compounds and verification of pesticide and nerve agent poisoning. Anal. Bioanal. Chem. 391, 97–116.

Kaloyanova, F., Vergieva, T., 1987. Estimation of human exposure to mixtures of chemicals: methods of clinical surveillance. In: Vouk, V.B., Batler, G.C., Upton, A.C., Parke, D.V., Asher, S.C. (Eds.), Methods for Assessing the Effects of Mixtures of Chemicals. John Wiley & Sons, Chichester, pp. 379–394.

Karami-Mohajeri, S., Abdollahi, M., 2011. Toxic influence of organophosphate, carbamate, and organochlorine pesticides on cellular metabolism of lipids, proteins, and carbohydrates: a systematic review. Hum. Exp. Toxicol. 30, 1119–1140.

Kavvalakis, M.P., Tsatsakis, A.M., 2012. The atlas of dialkylphosphates; assessment of cumulative human organophosphorus pesticides' exposure. Forensic Sci. Int. 218, 111–122.

Khalili, H., Dayyeh, B.A., Friedman, L.S., 2011. Assessment of liver function in clinical practice. In: Ginès, P., Kamath, P., Arroyo, V. (Eds.), Chronic Liver Failure. Clinical Gastroenterology. Humana Press, New York, pp. 47–76.

Kim, H.J., Choi, M.G., Park, M.K., Seo, Y.R., 2017. Predictive and prognostic biomarkers of respiratory diseases due to particulate matter exposure. J. Cancer Prev. 22, 6–15.

Kortenkamp, A., Backhaus, T., Faust, M., 2009. State of the Art on Mixture Toxicity. Report Available at: http://ec.europa.eu/environment/chemicals/effects/pdf/report_mixture_toxicity.pdf.

Lakind, J.S., Holgate, S.T., Ownby, D.R., 2007. A critical review of the use of Clara cell secretory protein (CC16) as a biomarker of acute or chronic pulmonary effects. Biomarkers 12, 445–467.

Lehmann, E., Oltramare, C., de Alencastro, L.F., 2018. Development of a modified QuEChERS method for multi-class pesticide analysis in human hair by GC-MS and UPLC-MS/MS. Anal. Chim. Acta 999, 87–98.

Lessenger, J.E., 2005. Fifteen years of experience in cholinesterase monitoring of insecticide applicators. J. Agromed. 10, 49–56.

Li, B., Ricordel, I., Schopfer, L.M., et al., 2010. Detection of adduct on tyrosine 411 of albumin in humans poisoned by dichlorvos. Toxicol. Sci. 116, 23–31.

Lindberg, H.K., Väänänen, V., Järventaus, H., et al., 2008. Genotoxic effects of fumes from asphalt modified with waste plastic and tall oil pitch. Mutat. Res. 653, 82–90.

Liu, J., Yeo, H.C., Overvik-Douki, E., et al., 2000. Chronically and acutely exercised rats: biomarkers of oxidative stress and endogenous antioxidants. J. Appl. Physiol. 89, 21–28.

Liu, Y., Vijver, M.G., Pan, B., et al., 2017. Toxicity models of metal mixtures established on the basis of "additivity" and "interactions". Front. Environ. Sci. Eng. 11, 10.

Lockridge, O., Schopfer, L.M., 2010. Review of tyrosine and lysine as new motifs for organophosphate binding to proteins that have no active site serine. Chem. Biol. Interact. 187, 344–348.

Lowitt, A.B., 2006. Federal regulations and risk assessment of organophosphate and carbamate pesticides. In: Gupta, R.C. (Ed.), Toxicology of Organophosphate and Carbamate Compounds. Academic Press/Elsevier, Amsterdam, pp. 617–632.

Manceau, A., Enescu, M., Simionovici, A., et al., 2016. Chemical forms of mercury in human hair reveal sources of exposure. Environ. Sci. Technol. 50, 10721–10729.

Mangelsdorf, I., 2009. Use of biomarkers of effect in the evaluation of cumulative exposure: nephrotoxicity of solvents at the workplace. In: IPCS-WHO Assessment of Combined Exposures to Multiple Chemicals. International Workshop, Harmonization Project Document 7, pp. 55–58.

Margariti, M.G., Tsatsakis, A.M., 2009. Analysis of dialkyl phosphate metabolites in hair using gas chromatography-mass spectrometry: a biomarker of chronic exposure to organophosphate pesticides. Biomarkers 14, 137–147.

Marsillach, J., Richter, R.J., Kim, J.H., 2011. Biomarkers of organophosphorus (OP) exposures in humans. Neurotoxicology 32, 656–660.

Marsillach, J., Costa, L.G., Furlong, C.E., 2013. Protein adducts as biomarkers of exposure to organophosphorus compounds. Toxicology 307, 46–54.

Moser, V.C., Padilla, S., Simmons, J.E., et al., 2012. Impact of chemical proportions on the acute neurotoxicity of a mixture of seven carbamates in preweanling and adult rats. Toxicol. Sci. 129, 126–134.

Mwila, K., Burton, M.H., Van Dyk, J.S., et al., 2013. The effect of mixtures of organophosphate and carbamate pesticides on acetylcholinesterase and application of chemometrics to identify pesticides in mixtures. Environ. Monit. Assess. 185, 2315–2327.

Mumtaz, M.M., Suk, W.A., Yang, R.S.H., 2010. Introduction to mixtures toxicology and risk assessment. In: Mumtaz, M.M. (Ed.), Principles and Practice of Mixtures Toxicology. Wiley-VCH Verlag GmbH & Co., Weinheim, Germany, pp. 1–25.

Nordberg, G.F., 2010. Biomarkers of exposure, effects and susceptibility in humans and their application in studies of interactions among metals in China. Toxicol. Lett. 192, 45–49.

Olmedo, P., Pla, A., Hernández, A.F., et al., 2010. Validation of a method to quantify chromium, cadmium, manganese, nickel and lead in human whole blood, urine, saliva and hair samples by electrothermal atomic absorption spectrometry. Anal. Chim. Acta 659, 60–67.

Ozer, J., Ratner, M., Shaw, M., et al., 2008. The current state of serum biomarkers of hepatotoxicity. Toxicology 245, 194–205.

Padilla, R.C., 2006. Cumulative effects of organophosphorus or carbamate pesticides. In: Gupta, R.C. (Ed.), Toxicology of Organophosphate and Carbamate Compounds. Academic Press/Elsevier, Amsterdam, pp. 607–615.

Pelallo-Martínez, N.A., Batres-Esquivel, L., Carrizales-Yáñez, L., et al., 2014. Genotoxic and hematological effects in children exposed to a chemical mixture in a petrochemical area in Mexico. Arch. Environ. Contam. Toxicol. 67, 1–8.

Polanska, K., Dettbarn, G., Jurewicz, J., et al., 2014. Effect of prenatal polycyclic aromatic hydrocarbons exposure on birth outcomes: the Polish mother and child cohort study. BioMed Res. Int. 2014, 408939.

Provost, J.P., Hanton, G., Le Net, J.L., 2003. Plasma triglycerides: an overlooked biomarker of hepatotoxicity in the rat. Comp. Clin. Pathol. 12, 95–101.

Ramesh, A., Archibong, A.E., 2011. Reproductive toxicity of polycyclic aromatic hydrocarbons: occupational relevance. In: Gupta, R.C. (Ed.), Reproductive and Developmental Toxicology. Academic Press/Elsevier, Amsterdam, pp. 577–591.

Roberts, R.A., Aschner, M., Calligaro, D., et al., 2015. Translational biomarkers of neurotoxicity: a Health and Environmental Sciences Institute perspective on the way forward. Toxicol. Sci. 148, 332–340.

Rodríguez, T., Younglove, L., Lu, C., Funez, A., et al., 2006. Biological monitoring of pesticide exposures among applicators and their children in Nicaragua. Int. J. Occup. Environ. Health 12, 312–320.

Rohr, P., da Silva, J., da Silva, F.R., et al., 2013. Evaluation of genetic damage in open-cast coal mine workers using the buccal micronucleus cytome assay. Environ. Mol. Mutagen. 54, 65–71.

Rose, S., Melnyk, S., Pavliv, O., et al., 2012. Evidence of oxidative damage and inflammation associated with low glutathione redox status in the autism brain. Transl. Psychiatry 2, e134.

SCHER (Scientific Committee on Health and Environmental Risks), 2011. Toxicity and the Assessment of Mixtures of Chemicals. Joint Opinion of DG SANCO SCs. European Commission.

Shaw, G., 2015. The use and potential of pNF-H as a general blood biomarker of axonal loss: an immediate application for CNS injury. In: Kobeissy, F.H. (Ed.), Brain Neurotrauma: Molecular, Neuropsychological, and Rehabilitation Aspects. CRC Press/Taylor & Francis, Boca Raton (FL). Available from: https://www.ncbi.nlm.nih.gov/books/NBK299212/ (Chapter 21).

Silins, I., Högberg, J., 2011. Combined toxic exposures and human health: biomarkers of exposure and effect. Int. J. Environ. Res. Publ. Health 8, 629–647.

Simsek, A., Tugcu, V., Tasci, A.I., 2013. New Biomarkers for the Quick Detection of Acute Kidney Injury. ISRN Nephrol. 394582.

Singh, S., Kumar, V., Thakur, S., et al., 2011. DNA damage and cholinesterase activity in occupational workers exposed to pesticides. Environ. Toxicol. Pharmacol. 31, 278–285.

Solomon, G.M., Weiss, P.M., 2002. Chemical contaminants in breast milk: time trends and regional variability. Environ. Health Perspect. 110, A339–A347.

Sudakin, D.L., Stone, D.L., 2011. Dialkyl phosphates as biomarkers of organophosphates: the current divide between epidemiology and clinical toxicology. Clin. Toxicol. 49, 771–781.

Timchalk, C., Poet, T.S., 2008. Development of a physiologically based pharmacokinetic and pharmacodynamic model to determine dosimetry and cholinesterase inhibition for a binary mixture of chlorpyrifos and diazinon in the rat. Neurotoxicology 29, 428–443.

Tsatsakis, A.M., Tzatzarakis, M.N., Koutroulakis, D., et al., 2009. Dialkyl phosphates in meconium as a biomarker of prenatal exposure to organophosphate pesticides: a study on pregnant women of rural areas in Crete, Greece. Xenobiotica 39, 364–373.

Tsatsakis, A.M., Tutudaki, M., Tzatzarakis, M.N., 2012. Is hair analysis for dialkyl phosphate metabolites a suitable biomarker for assessing past acute exposure to organophosphate pesticides? Hum. Exp. Toxicol. 31, 266–273.

Ulvestad, B., Randem, B.G., Andersson, L., et al., 2007. Clara cell protein as a biomarker for lung epithelial injury in asphalt workers. J. Occup. Environ. Med. 49, 1073–1078.

Wang, G., Fowler, B.A., 2008. Roles of biomarkers in evaluating interactions among mixtures of lead, cadmium and arsenic. Toxicol. Appl. Pharmacol. 233, 92–99.

Wang, R.Y., Ryan, P.B., Mumtaz, M., et al., 2010. Biomonitoring. In: Mumtaz, M.M. (Ed.), Principles and Practice of Mixtures Toxicology. Wiley-VCH Verlag GmbH & Co., Weinheim, Germany, pp. 569–593.

Wasowicz, W., Gromadzińska, J., Rydzyński, K., 2001. Blood concentration of essential trace elements and heavy metals in workers exposed to lead and cadmium. Int. J. Occup. Med. Environ. Health 14, 223–229.

Wasung, M.E., Chawla, L.S., Madero, M., 2015. Biomarkers of renal function, which and when? Clin. Chim. Acta 438, 35350–35357.

Yoshioka, N., Nakashima, H., Hosoda, K., 2008. Urinary excretion of an oxidative stress marker, 8-hydroxyguanine (8-OH-Gua), among nickel-cadmium battery workers. J. Occup. Health 50, 229–235.

Žurek, J., Fedora, M., 2012. The usefulness of S100B, NSE, GFAP, NF-H, secretagogin and Hsp70 as a predictive biomarker of outcome in children with traumatic brain injury. Acta Neurochir. 154, 93–103.

CHAPTER

34

Biomarkers of Toxic Solvents and Gases

Szabina A. Stice

Division of Biotechnology and GRAS Notice Review, Center for Food Safety and Applied Nutrition, US Food and Drug Administration, College Park, MD, United States

INTRODUCTION

Solvents are categorized as either organic or inorganic. Organic solvents can be further classified by their molecular structures and functional group(s) present (e.g., alcohols, ethers, ketones, aliphatic, and aromatic hydrocarbons). Solvents possess a wide range of toxicities depending on their number of carbon atoms, configuration, presence of double and triple bonds, and specific functional groups in the molecule, along with other factors. The toxicity of solvents can vary dramatically even within the same class. For example, while 2-hexanone, 3-heptanone, and 2-pentanone are all ketones with small differences in their structures, 2-hexanone is neurotoxic with a very low safe threshold value, 3-heptanone possesses only low neurotoxic potential, and 2-pentanone is not neurotoxic.

Solvents are widely used to dissolve, dilute, suspend, or extract other substances. They have various applications in the oil, gas, pharmaceutical, chemical, and other industries. These important chemicals are used as, or are used in, dry cleaning agents, nail polish removers, perfumes, detergents, spot removers, paints, varnishes, paint thinners, adhesives, drugs, and in numerous other products. Consequently, they may be commonplace in the work or home environments. The major routes of solvent exposure are the lungs (via inhalation), the gastrointestinal tract (via ingestion), and the skin. Exposure to gases may be accidental, intentional (abuse or chemical warfare agents), occupational, or environmental. During normal daily activities, nearly all of us are exposed to solvents and gases; exposure to airborne solvents is considered unavoidable (Bruckner and Warren, 2001).

As exposure may occur via multiple routes and various sources, air monitoring, a common way to estimate solvent and gas exposure, fails to provide information on the total exposure to humans and animals. On the other hand, biomarkers of toxic solvent and gas exposure enable us to assess total exposure in biological specimen(s). Furthermore, biological effects usually correlate better with biomarker levels than with air concentrations (Fiserova-Bergerova and Ogata, 1995). The specimens used most frequently for biomonitoring (measurement of a biomarker in a biological specimen) of toxic solvents and gases are blood, urine, and exhaled breath. As many solvents and gases have short residence times in the body, the sampling period for biological specimens must be carefully chosen to accurately represent exposure to these chemicals. The biomarker also must be chosen with care to ensure that it is a specific and sensitive marker of exposure to the toxic solvent or gas. Moreover, while some markers are indicators of recent exposures only, others can signal older exposures.

Early detection of biomarkers of exposure may enable us to either prevent the onset of a disease or halt its progression by avoiding further exposure to harmful chemicals and by monitoring the health of those exposed to toxic solvents and gases. Biomarkers of exposure include the measurement of the parent compound, its metabolite(s), protein adduct(s), or DNA adduct(s). Some markers may serve as biomarkers of both exposure and effects. Detection of biomarkers of effects aids us in early diagnosis and selection of effective treatment and may shed light on modes of toxic action. Biomarkers of effects include various changes at the cellular level (e.g., altered expression of metabolizing enzymes, enzyme inhibition, and altered gene expression patterns), preneoplastic changes, gene mutations, and chromosomal aberrations (Silins and Högberg, 2011).

While biomarkers of exposure are preferably, but not always, specific to the chemical of interest, biomarkers of effects are often not specific to a chemical (e.g., the

Biomarkers in Toxicology, Second Edition
https://doi.org/10.1016/B978-0-12-814655-2.00034-7

expression of the same metabolizing enzyme may be upregulated by various chemicals). Toxic effects and the mechanism of toxicity of a solvent or gas may be animal species- and/or strain-specific and depend on the route of exposure. When using residues of toxic chemicals or metabolic biomarkers to establish the extent of exposure, intra- and interspecies differences in absorption, distribution, metabolism, and/or elimination may pose significant challenges. Biomarkers that are common to test animals and humans are preferred for biomonitoring to ensure that biomarkers are predictive of effects in humans. Furthermore, as structurally related substances may have one or more common metabolites, metabolites used as biomarkers must be carefully selected; a multibiomarker approach should be employed instead of a single biomarker to detect exposure.

In reality, solvents, and chemicals in general, are often present as mixtures and not as single compounds. Some components of these mixtures may affect the toxicokinetics of other components, hence, biomarkers of exposure and effects (Hernández et al., 2014). Consequently, precautions should be taken when identifying biomarkers of mixtures or component(s) of mixtures. For a detailed discussion of biomarkers of mixtures, refer to Chapter 33. Although hundreds of solvents are in use, only a relatively small number of solvents and gases are discussed in this chapter because of space constraints and limited or unavailable biomarker data.

SOLVENTS

Solvents are used for a large number of applications. Historically, numerous chemicals or mixtures of chemicals have been used as dry cleaning solvents (e.g., benzene, chloroform, carbon tetrachloride, trichloroethylene, perchloroethylene, 1,1,1-trichloroethane, and liquid carbon dioxide) (SCRD, 2009). Acetone, ethyl acetate, and butyl acetate are utilized in nail polish removers. Ethanol is typically the solvent of choice in perfumes, but it is also employed as an excipient in medications. Many solvents, such as acetone, acetonitrile, benzene, and chloroform, are used in the pharmaceutical manufacturing process. In addition to the use of isopropyl alcohol (also known as rubbing alcohol) as a disinfectant, it is used as a solvent in cleaners along with 2-butoxyethanol, ethanol, and glycol ethers. Other common solvents include, but are not limited to, pentane, cyclopentane, hexane, cyclohexane, toluene, diethyl ether, dichloromethane, tetrahydrofuran, dimethyl sulfoxide, and methanol.

Many solvents, such as ethanol, diethyl ether, chloroform, gasoline, benzene, toluene, and acetone, are abused to achieve intoxication, often with dire health consequences. Benzene is known to be a human carcinogen based on sufficient evidence of carcinogenicity from human studies (NTP, 2016). Gasoline sniffing can cause nausea, vomiting, ataxia, dizziness, anesthesia, loss of consciousness, and even death (Poklis and Burkett, 1977). The primary target of toluene is the central nervous system (CNS). Although symptoms of CNS toxicity are reversible after acute inhalation, chronic exposure may result in irreversible damage to the CNS (Aydin et al., 2002). Acute and chronic exposures to toxic solvents are not unique to abusers; frequently they are the result of occupational exposure (Kamran and Bakshi, 1998; Filley et al., 2004). For instance, although toluene is a commonly inhaled product by solvent abusers, occupational exposure to this solvent also occurs in the dry cleaning, aviation, and chemical industries (Borne et al., 2005). Exposure to chloroform may occur due to both abuse and in occupational settings during the manufacture of this solvent. Workers at wastewater and other treatment plants can be exposed to high levels of chloroform (NTP, 2016).

If inhaled, ingested, or absorbed through the skin in large quantities, many solvents can cause CNS depression and produce symptoms of acute toxic exposure. Chronic exposures may lead to neurotoxicity, hepatotoxicity, nephrotoxicity, reproductive and developmental toxicity, cancer, or the development of other diseases. High levels of maternal solvent exposure have been reported to result in developmental disorders, fetal malformations, and even fetal death (Bowen, 2011). The physical and behavioral effects that may occur subsequent to elevated maternal solvent exposure are sometimes referred to as fetal solvent syndrome.

GASES

A number of gases, such as nitrous oxide (also known as laughing gas, used in medicine and dentistry and as an aerosol propellant), propane and butane (compressed to be used as fuels), and chlorofluorocarbons (in aerosols), are inhaled as recreational drugs. The major safety hazard of abused gases is risk of asphyxiation. Virtually every gas, excluding oxygen, can act as a simple asphyxiant by replacing oxygen from ambient air and interfering with oxygen delivery (Goldfranks, 2015). Unlike simple asphyxiants, chemical asphyxiants act by producing cellular hypoxia (interference with cellular respiration). Other gases exert toxicity by acting as irritants or as local or systemic toxins. Exposure to toxic gases (e.g., hydrogen sulfide, bromine, chlorine, and hydrogen fluoride) may also occur in occupational settings (Guidotti, 1994; Guillermo, 1994). Exposure to toxic gases may be the result of the intentional release of chemical warfare agents (e.g., chlorine and phosgene) or the consequence of accidental release (e.g., carbon monoxide [CO] [from the incomplete burning of fuel] and chlorine [from the inappropriate mixing of cleaning agents]). The accidental release of methyl isocyanate (liquid below but a gas above 39°C) in Bhopal, India, resulted in the immediate death of at least 3800 people in 1984 (Broughton, 2005). It also caused significant morbidity and early death for many more; the combined number of injured or dead is estimated to be over 250,000.

BIOMARKERS OF TOXIC SOLVENT AND GAS EXPOSURE

The Use of the Parent Compound as a Biomarker of Exposure

The measurements of solvents from biological samples have been used to monitor solvent exposures (Imbriani and Ghittori, 2005). The most commonly used biological sample is urine. The measurement of urinary excretion of unmodified solvents has numerous advantages such as, but not limited to, having a noninvasive sample collection procedure (compared with blood and tissue sampling), being specific and sensitive (able to detect low levels of exposure), and analytical techniques for their detection are widely available and/or easily developable (Gobba et al., 1997). Moreover, the correlation between environmental time-weighted average concentration (an index of the external dose of solvents) and the levels of unchanged solvents in the urine of individuals exposed via inhalation is generally good. The correlation between the ambient air and urine levels of selected toxic solvents and gases (as indicated by their r-values) is listed in Table 34.1.

The expired air (i.e., breath) is a mixture of dead space air and alveolar air (the air in the alveoli). Dead space air is exhaled before alveolar air. The alveoli are at the end

TABLE 34.1 Correlation Coefficients of the Ambient Air and Urine Levels of Toxic Solvents and Gases

Class of Solvent or Gas	Name of Solvent or Gas	r	References
Unsubstituted aromatic hydrocarbons	Benzene	0.50–0.97	Imbriani and Ghittori (2005) [a]
	Toluene	0.60–0.92	Imbriani and Ghittori (2005) [a] and Gobba et al. (1997) [a]
	Xylenes	0.64–0.97	Imbriani and Ghittori (2005) [a]
Unsubstituted nonaromatic hydrocarbons	n-Hexane	0.84	Imbriani et al. (1984)
	Cyclohexane	0.89	Ghittori et al. (1987)
	2-Methylpentane	0.86	Pezzagno et al. (1985)
Halogenated hydrocarbons	Tetrachloroethylene	0.67–0.88	Imbriani and Ghittori (2005) [a]
	Dichloromethane	0.86–0.92	Imbriani and Ghittori (2005) [a]
	Carbon tetrachloride	0.80	Ghittori et al. (1994)
	Methyl chloroform	0.93–0.95	Gobba et al. (1997) [a]
Used as inhaled anesthetics	Nitrous oxide	0.64–0.95	Imbriani and Ghittori (2005) [a]
	Halothane	0.81	Imbriani et al. (1995, 1998)
	Isoflurane	0.81	Imbriani et al. (1995, 1998)
	Enflurane	0.80	Imbriani et al. (1995, 1998)
	Desflurane	0.92	Alessio et al. (2003)
Ketones	Acetone	0.71–0.94	Imbriani and Ghittori (2005) [a]
	Methyl ethyl ketone	0.65–0.93	Gobba et al. (1997) [a]
	Methyl isobutyl ketone	0.90	Ogata et al. (1995)
Alcohols	Methanol	0.73–0.82	Imbriani and Ghittori (2005) [a]
	n-Butanol	0.73–0.82	Imbriani and Ghittori (2005) [a]
Ethers	Tetrahydrofuran	0.86–0.88	Ong et al. (1991)
Amides	Dimethylformamide	0.50	Lareo and Perbellini (1995)

[a]*Secondary references. See articles for multiple primary references for each substance.*

of the respiratory tree and are the site of gas exchange with blood. The concentration of volatile compounds in the end-tidal air (the last portion of expired air) is in constant equilibrium with that in the blood (Caro and Gallego, 2009). Consequently, the end-tidal air can be used to detect and measure exposure to volatile compounds. The concentrations of volatile organic solvents (e.g., toluene, ethyl acrylate, o-, m-, and p-xylene, ethylbenzene, benzene, and naphthalene) in the alveolar air and in the ambient air of workplaces showed good correlation, i.e., $r > 0.9$, for all 26 compounds studied by Caro and Gallego (2009). In another study, the correlation between the levels of 10 different solvents (e.g., dimethylformamide, acetone, methyl ethyl ketone, toluene, and n-hexane) in the alveolar air and the concentration in the environmental air of exposed workers was studied (Brugnone et al., 1980). The correlation coefficient between alveolar air levels and environmental air concentrations of these solvents ranged from 0.76 (for methyl ethyl ketone) to 0.98 (for cyclohexane). When the breath of dry cleaning workers was monitored for tetrachloroethylene exposure, a solvent also known as "dry cleaning fluid" because of its wide use in dry cleaning, it was found that exposure to tetrachloroethylene can be reliably monitored from breath (Droz and Guillemin, 1986). A correlation coefficient of 0.956, indicating strong correlation, was obtained for environmental air levels of "dry cleaning fluid" and its levels in breath. Numerous other articles discuss the use of breath for biomonitoring solvent and gas exposure (e.g., Apostoli et al., 1982; Brugnone et al., 1983, 1985; Fujino et al., 1992; Ghittori et al., 2004; Hamelin et al., 2004; Amorim and Cardeal, 2007). It is clear, based on these publications, that in addition to the use of urine for biomonitoring exposure to unmetabolized solvents and gases, breath can also be used to detect and measure exposure to these compounds. Like the collection of urine samples, the collection of breath samples is rapid and noninvasive. The correlation factors of ambient air and breath levels of selected toxic solvents are listed in Table 34.2.

In addition to urine and breath, blood may also be used for measuring biomarkers of toxic solvent exposure. Because it requires an invasive procedure to obtain and may expose those who collect it to blood-borne diseases, it is less preferred. Nonetheless, it can be as useful as urine or breath for biomonitoring. For example, in a study of toluene, the correlation coefficients for toluene levels in the air of occupational environment and concentrations of toluene in alveolar air, blood, and urine were comparable, i.e., 0.822, 0.850, and 0.846, respectively (Ghittori et al., 2004). Using blood for monitoring exposure in some cases may be superior and in other cases inferior compared with using breath or urine. In the case of acetone, monitoring exposure in blood was less useful than monitoring exposure using breath or urine. A search of the currently available literature yielded correlation coefficients of up to 0.94 for

TABLE 34.2 The Correlation Coefficients of the Ambient Air and Breath Levels of Toxic Solvents

Class of Solvent or Gas	Name of Solvent or Gas	r	References
Unsubstituted aromatic hydrocarbons	Benzene	0.82	Ghittori et al. (2004)
	Toluene	0.82–0.99	Brugnone et al. (1980), Ghittori et al. (2004), and Caro and Gallego (2009)
Unsubstituted nonaromatic hydrocarbons	n-Hexane	0.97	Brugnone et al. (1980)
	Cyclohexane	0.98	Brugnone et al. (1980)
	2-Methylpentane	0.97	Brugnone et al. (1980)
Halogenated hydrocarbons	Tetrachloroethylene	0.96	Droz and Guillemin (1986)
	Dichloromethane	0.98	McCammon et al. (1991)
	Chloroform	0.90–0.96	Caro and Gallego (2008)
Ketones	Acetone	0.71–0.82	Brugnone et al. (1980) and Fujino et al. (1992)
	Methyl ethyl ketone	0.76	Brugnone et al. (1980)
Alcohols	Isopropanol	0.92–0.95	Brugnone et al. (1983) and Ghimenti et al. (2015)
Ethers	Tetrahydrofuran	0.61	Ong et al. (1991)
Amides	Dimethylformamide	0.87	Brugnone et al. (1980)
Esters	Ethyl acetate	1	Caro and Gallego (2009)

monitoring acetone exposure in both breath and urine, while in blood the highest correlation coefficient found was 0.77 (Gobba et al., 1997; Lauwerys and Hoet, 2001).

The Use of Metabolites of Toxic Solvents or Gases as Biomarkers of Exposure

The metabolites of toxic solvents and gases can also be used to monitor exposures to these chemicals. Unfortunately, certain metabolites may form from more than one compound, rendering them nonspecific biomarkers of the parent compounds. For instance, n-hexane and 2-hexanone are both commonly used solvents that share the highly neurotoxic 2,5-hexanedione metabolite that is excreted in the urine and is used for monitoring exposure to these two solvents (Cardona et al., 1993; Lauwerys and Hoet, 2001; Prieto et al., 2003). Consequently, when the source of 2,5-hexanedione is unknown, determination of the parent compound from breath may be preferable (Periago et al., 1993; Hamelin et al., 2004). Other factors that may complicate the use of metabolites as biomarkers of exposure are changes in metabolism resulting from co-exposure to other solvents or gases or from metabolizing enzyme induction because of repeated exposure. In the case of n-hexane, after short-term, acute co-exposure with methyl ethyl ketone, a decrease in the concentration of 2,5-hexanedione was observed, whereas after repeated, chronic co-exposure with the methyl ethyl ketone, the urinary excretion of 2,5-hexanedione increased (Cardona et al., 1993; van Engelen et al., 1997). Repeated co-exposure of n-hexane with toluene tended to reduce and repeated co-exposure with acetone tended to increase the urinary excretion of 2,5-hexanedione (Cardona et al., 1993, 1996). While after co-exposure of n-hexane with other solvents the urinary concentration of 2,5-hexanedione does not necessarily reflect the exposure concentration of n-hexane, the 2,5-hexanedione level may be a useful marker of the assessment of these solvent mixtures' neurotoxic potential (Ichihara et al., 1998).

Isopropanol is another example where a metabolite could be used for biomonitoring. Not only was the correlation between alveolar and environmental concentrations of unmetabolized isopropanol found to be high (i.e., $r = 0.92$) but also the correlation between the concentrations of acetone (the chief metabolite of isopropanol) in the blood and the alveolar air and the alveolar isopropanol level (Brugnone et al., 1983). Additionally, the urinary acetone concentration was found to increase in proportion to the isopropanol exposure intensity ($r = 0.84$) (Kawai et al., 1990). Hence, in addition to isopropanol itself, acetone may be used as a biomarker of isopropanol exposure. Unfortunately, as acetone can also signal acetone exposure, caution must be exercised when it used as a biomarker of isopropanol exposure (Lauwerys and Hoet, 2001).

In addition to the parent compound, the metabolites of toluene, one of the most widely used organic solvents, are also used as biomarkers of toluene exposure (Cosnier et al., 2013, 2014). Numerous articles reported good correlation between alveolar air toluene concentrations and environmental air toluene concentrations (e.g., $r = 0.84$, 0.82, and 0.99 [Ghittori et al., 2004; Brugnone et al., 1980; Caro and Gallego, 2009, respectively]), showing that toluene is a good biomarker of toluene exposure. In addition, highly significant correlations between the urinary metabolites of toluene and the concentration of toluene in both blood and ambient air were found (Angerer and Krämer, 1996). The highest correlation coefficient was obtained for o-cresol, a metabolite of toluene, and toluene in ambient air. Therefore, in addition to employing toluene as a biomarker of toluene exposure, its o-cresol metabolite can also be used for this purpose. Unfortunately, when co-exposure to other solvents sharing similar metabolic pathways with toluene occurs (e.g., co-exposure with xylene), o-cresol levels, hence the correlation between o-cresol levels and exposure to toluene, can be significantly affected, rendering o-cresol to be unsuitable as a biomarker of toluene exposure (Tardif et al., 1998). Other metabolites of toluene (e.g., p-cresol and hippuric acid) are less suitable, or even outright unsuitable, as biomarkers of toluene exposure because of high and/or highly variable background levels from endogenous and dietary sources (Angerer and Krämer, 1996; Pierce et al., 1998; Tardif et al., 1998). For instance, at high toluene exposure levels, there was a good correlation ($r = 0.87$) between toluene exposure and urinary levels of hippuric acid; however, at low exposure concentrations (i.e., 10 ppm toluene), the excretion of hippuric acid was only slightly above the mean background level, rendering it unsuitable for biomonitoring low levels of toluene exposure (Tardif et al., 1998).

Benzene itself, just like toluene, can be its own biomarker of exposure. Like toluene, some of its metabolites are also suitable biomarkers of its exposure, while others are not. Because of their lack of specificity, phenol, hydroquinone, and catechol (metabolites of benzene) are not suitable biomarkers for the assessment of environmental or even occupational benzene exposure (Arnold et al., 2013). On the other hand, S-phenylmercapturic acid (SPMA), a minor metabolite of benzene, is generally regarded as a reliable and specific urinary biomarker of benzene exposure, as, so far, no background for SPMA has been found in nonexposed humans (Boogaard, 2009). However, the use of SPMA has its drawbacks. Like other mercapturates, it is unstable in alkaline urine, and as such, freezing or acidification of the sample is required. This sampling issue may potentially impact the interpretation of this biomarker. Furthermore, it is only a biomarker of recent occupational and nonoccupational benzene exposure but not mid- or long-term exposure, as SPMA has a relatively short mean half-life of 9–13 h and it does not bioaccumulate. Another metabolite of benzene, *trans, trans*-muconic acid (ttMA), is also used as an occupational biomarker of benzene exposure. A correlation of r = 0.816 was found for benzene levels in the breath zone of workers and levels of urinary ttMA (Inoue et al., 1989). As ttMA may have high and variable background levels present in the urine from smoking or dietary sources, it is not a sufficiently specific biomarker of low levels of benzene exposure (Pezzagno et al., 1999). Moreover, urinary ttMA levels were found to be suppressed in the case of co-exposure of toluene and benzene, further complicating the use of ttMA as a biomarker of benzene exposure (Inoue et al., 1989).

N,N-dimethylformamide (DMF) is a commonly employed solvent. As DMF is absorbed both via inhalation and through the skin, ambient air monitoring is not sufficient to assess individual body burden (Hennebrüder and Angerer, 2005). The urinary DMF metabolites N-methylformamide (NMF) and N-acetyl-S-(N-methylcarbamoyl)cysteine (AMCC) have been the most widely used biomarkers of DMF exposure (Wang et al., 2004; Imbriani et al., 2002a,b; Kim et al., 2004; Käfferlein et al., 2004). Unfortunately, co-exposure to other chemicals may affect the suitability of using DMF urinary metabolites as biomarkers of exposure; co-exposure of DMF with toluene was found to reduce the excretion of NMF in urine (Yang et al., 2000). Other examples of metabolites of some toxic solvents utilized as possible biomarkers of exposure are listed in Table 34.3.

The Use of Adducts as Biomarkers of Toxic Solvent and Gas Exposure

Parent compounds and/or their metabolites may form adducts with macromolecules such as protein or DNA. These adducts may also be used as biomarkers of exposure and/or effects. The human body is able to

TABLE 34.3 Metabolites of Toxic Solvents Used as Possible Biomarkers of Exposure

Exposure to	Biomarker Metabolite(s)	References
Benzene	S-Phenylmercapturic acid	Boogaard (2009), Inoue et al. (1989), and Pezzagno et al. (1999)
	tt-Muconic acid	
Toluene	o-Cresol	Angerer and Krämer (1996) and Tardif et al. (1998)
	Hippuric acid	
Xylenes	Methylhippuric acid	Jacobson and McLean (2003)
Ethylbenzene	Mandelic acid	Jang et al. (2000)
Chlorobenzene	4-Chlorocatechol	Ogata (1990)
	p-Chlorophenol	
Chloroethylene	Monochloroacetic acid	Monster (1986)
Trichloroethylene	Trichloroethanol	Lowry (1995)
	Trichloroacetic acid	
Tetrachloroethylene	Trichloroacetic acid	Monster (1986)
2-Ethoxyethanol	2-Ethoxyacetic acid	Lowry (1995)
Isopropanol	Acetone	Brugnone et al. (1983)
N,N-Dimethylformamide	N-Methylformamide	Wang et al. (2004), Imbriani et al. (2002a,b), Kim et al. (2004), and Käfferlein et al. (2004)
	N-Acetyl-S-(N-methylcarbomoyl)cysteine	
N,N-Dimethylacetamide	N-Methylacetamide	Perbellini et al. (2003)
Carbon disulfide	2-Thiothiazolidine-4-carboxylic acid	Simon et al. (1994)
n-Hexane	2,5-Hexanedione	Prieto et al. (2003)
2-Hexanone	2,5-Hexanedione	Lauwerys and Hoet (2001)
Methyl-t-butyl ether	t-Butyl alcohol	Buckley et al. (1997)

repair and eliminate DNA adducts. The rate and speed of repair of different DNA adducts are variable and often unknown. Hence, their use as biomarkers of exposure or effect may be limited (Needham et al., 2007; Pavanello and Lotti, 2014). On the other hand, hemoglobin (Hb) adducts, especially smaller ones, may not impact the life span of the erythrocyte. Furthermore, these adducts may accumulate in the body, and large quantities of Hb can be easily isolated from small volumes of blood for analysis. Consequently, Hb adducts can be utilized as biomarkers of exposure (Angerer et al., 2007). Protein (e.g., albumin) adducts are also used for biomonitoring. Albumin adducts are very useful for this purpose, as albumin is present in large quantities in the blood. Unlike DNA adducts, albumin adducts (like Hb adducts) are not prone to repair mechanisms and are chemically stable.

The above-mentioned adducts are also employed for biomonitoring exposure to toxic solvents and gases. An example of this is ethylene oxide, a reactive gas that induces nausea and vomiting, causes CNS depression, irritates the eyes, skin, and respiratory tract, and is mutagenic. Chronic exposure to it is associated with an increased risk of various cancers. It is commonly used for numerous industrial processes as an intermediate, as a sterilizing agent for medical supplies and foods, as a fumigant, and as an insecticide. In the case of this toxic gas, its 2-hydroxyethyl adduct with Hb can be used as a biomarker of ethylene oxide exposure (Csanády et al., 2000). Unfortunately, it is also a biomarker of ethylene exposure, as ethylene is biotransformed to ethylene oxide. The lack of specificity of biomarkers is also a problem in the case of CO. CO exposure is monitored with the help of carboxyhemoglobin (a complex of CO and Hb) levels in the blood. Unfortunately, this complex is not specific to CO; dichloromethane, an important but toxic solvent, and diiodomethane form the same complex with Hb as they are metabolized to CO (Boogaard, 2009).

Regarding DMF, in addition to its two urinary metabolites mentioned earlier (NMF and AMCC), its N-methylcarbamoyled-Hb-adduct can also be used for biomonitoring (Mráz et al., 2002). The DNA adduct of DMF (N^4-methylcarbamoylcytosine) was also reported as a possible biomarker to monitor occupational exposure to DMF using human urine (Hennebrüder and Angerer, 2005). According to Arnold et al. (2013), in addition to benzene and its metabolites, biological adducts of benzene may also be utilized as biomarkers of exposure. Multiple benzene metabolites (e.g., benzene oxide, benzoquinones, hydroquinone, and muconaldehyde) can bind to DNA. Unfortunately, while DNA adducts of benzene metabolites are potential biomarkers of benzene exposure, they cannot yet be employed as such in humans because of the lack of sensitive and specific analytical methods for their detection. Hb and plasma protein adducts of benzene oxide are also potential markers of benzene exposure, though Hb adducts might be diagnostically less sensitive than adducts of plasma proteins. Regrettably, currently available methods are not sensitive enough to monitor environmental benzene exposure using these Hb and plasma protein adducts, and they are also unsuitable for routine applications. Furthermore, the Hb and albumin adducts of 1,2- and 1,4-benzoquinone were found to be nonspecific biomarkers of benzene exposure. Nonetheless, unlike blood and urinary benzene and urinary SPMA levels, all of which are only suitable for detecting recent exposures, serum albumin adducts could provide a means of detecting older exposures because of their relatively long half-life ($\sim$21 days). Hence, further development of even more sensitive methods for the detection and quantitation of benzene oxide adducts is desirable. In addition, these methodologies must be automated for routine analysis.

The Use of Novel Biomarkers of Toxic Solvent and Gas Exposure

Alterations in gene expression can serve both as biomarkers of exposure and as early effects of exposure. An example is benzene, exposure to which is widespread in the population. Chronic industrial benzene exposure has been associated with an increased incidence of aplastic anemia (deficiency of platelets and red and white blood cells) because of bone marrow damage (Snyder, 2012). In addition, it is a leukemogen; the first reports of the association between benzene and leukemia are about a century old. The mechanisms of leukemogenesis and development of aplastic anemia have been the subjects of much research. Identification of biomarkers of effects often clarifies the mode of action. Furthermore, these biomarkers of effects could also serve as biomarkers of benzene exposure. Forrest et al. (2005) examined the effects of benzene exposure on peripheral blood mononuclear cell (PBMC) gene expression. Numerous genes were found to be affected by benzene, four of which were the most significantly and consistently affected and could be used as potential biomarkers of exposure. Two of these genes (CXCL16 and ZNF331) were consistently upregulated in exposed individuals, whereas two other genes (JUN and PF4) were consistently downregulated in exposed individuals.

Aberrant DNA methylation is common and has a role in the pathogenesis and prognosis of hematological cancers (Yang et al., 2014). Benzene is known to induce global DNA hypomethylation, and this may be a key mechanism underlying benzene hematotoxicity. Hence, examining DNA methylation profiles and alterations in mRNA expression in blood cells may serve as biomarkers for both benzene exposure and effect. DNA methylation profile from PBMCs of people chronically exposed to benzene was performed to identify new biomarkers of chronic benzene exposure. Aberrant hypomethylated STAT3 was found to be a potential biomarker of chronic benzene exposure. STAT3 was also shown to be upregulated. Activation of STAT3 is important in maintaining the characteristics of malignant tumors, and its inhibition may induce leukemic cell apoptosis, and as such, its potential in the treatment of certain types of leukemias is of interest. In summary, identification of novel biomarkers may serve as biomarkers of both exposure and effect and may help us understand the mechanisms of toxicity.

BIOMARKERS OF EFFECTS OF TOXIC SOLVENT AND GAS EXPOSURE

Biomarkers of effects reflect changes induced by toxic chemicals. Target organs of toxic solvents are the liver, kidney, and CNS, along with others. Multiple biomarkers of effects have been identified in these target organs. Unfortunately, as the selected examples below demonstrate, these biomarkers of effects are specific neither to a solvent nor to solvent or gas exposure. As biomarkers of effects are nonspecific in nature and humans are exposed to a variety of chemicals in their everyday lives that may or may not interact, relating these biomarkers of effects to a specific exposure is either challenging or outright impossible, and as such, they have limited use for health risk assessment (Boogaard, 2009).

The liver is an important target of many toxic solvents (e.g., DMF, carbon tetrachloride, dimethylacetamide, trichloroethylene, tetrachloroethylene, chloroform, and toluene). The liver transaminases alanine transaminase and aspartate transaminase are sensitive markers of liver injury and therefore can be used to monitor liver damage after solvent exposure (Silins and Högberg, 2011). Increased bilirubin, sorbitol dehydrogenase, gamma-glutamyl transpeptidase, and serum bile acid levels, along with other markers, may also signal liver injury after solvent exposure. Unfortunately, these biomarkers of effects are specific neither to a solvent nor to liver damage resulting from toxic solvent exposure and may be elevated because of various diseases and disorders (e.g., nonalcoholic fatty liver disease, alcoholic liver disease, medication-associated liver injury, viral hepatitis, and hemochromatosis).

The kidney is also a frequent target of toxic solvents. Numerous markers of kidney damage, such as increased urinary albumin and N-acetyl-beta-D-glucosaminidase levels, have been identified. In solvent (e.g., toluene or tetrachloroethene) exposed workers, albumin excretion exceeding the upper limit of the normal range may occur more frequently than in the corresponding control groups (Voss et al., 2005). Consequently, urinary albumin levels may be used to monitor effects of nephrotoxic solvents. However, this biomarker and other biomarkers of effects of nephrotoxic solvents are not specific to a solvent or to kidney injury due to solvent exposure. Nonetheless, periodic determination of urinary albumin levels is a useful tool for monitoring kidney damage in solvent-exposed workers. For a more detailed discussion on biomarkers of different target organs of chemicals, refer to the chapters in the *Systems Toxicity Biomarkers* section of this book.

INTERPRETATION AND USE OF BIOMONITORING DATA

The purpose of biomonitoring data is to assess exposure and, ultimately, health risk. Therefore, proper interpretation of the data is paramount. To interpret the data and derive quantitative reference or guidance values, knowledge of the quality of the analytical data and understanding of toxicokinetics of the toxic solvent or gas are necessary (Boogaard, 2009). Comprehension of the toxicokinetics of the substance of interest is essential, as some metabolites may form endogenously or from dietary ingredients or may originate from other chemicals that share the same metabolite with it. As the presence of a biomarker simply indicates exposure, one also needs to know the dose–response relationship to assess health risk; biomonitoring data must be correlated with toxicity data. Information on toxicity usually comes from animals, and sometimes humans, exposed to very high levels of toxic solvents or gases. These data are then extrapolated to lower exposures. Unfortunately, this method makes it hard to predict effects at low or very low levels of exposure. For a general discussion of how biomarkers fit into toxicological evaluation and risk assessment considerations and how risk assessment with the potential use of biomarkers is integrated into the development of chemical regulations, refer to Chapter 67.

CONCLUDING REMARKS AND FUTURE DIRECTIONS

Biomonitoring is a viable tool for assessing exposure, effects, and susceptibility. Nonetheless, it is not without limitations, and still much work is left to be done. Ongoing challenges include the development and validation of more specific and sensitive biomarkers of exposures and effects, as well as of sensitive, specific, and standardized analytical methods to detect and measure biomarkers and of finding ways to use these biomarkers for the prediction of disease risk or disease progression and outcome. Moreover, most of the currently used biomarkers of toxic solvent and gas exposure are only suitable for detecting fairly recent exposures. Finding biomarkers for detecting exposures from months, years, or decades ago is an area with much left to accomplish. In addition, exposure to mixtures, rather than to a single substance, is commonplace. Simultaneous exposure to multiple solvents and/or gases complicates biomonitoring of exposure and effects of a specific substance. More information is needed on co-exposure to multiple chemicals and their effects on specific biomarker(s) and the toxicity of a solvent or gas. Finally, ethical issues (such as confidentiality) in human biomonitoring also need to be addressed.

References

Alessio, A., Zadra, P., Negri, S., et al., 2003. Biological monitoring of occupational exposure to desflurane. G. Ital. Med. Lav. Ergon. 25 (2), 137–141.

Amorim, L.C.A., Cardeal, Z.D.L., 2007. Breath air analysis and its use as a biomarker in biological monitoring of occupational and environmental exposure to chemical agents. J. Chromatogr. B 853 (1), 1–9.

Angerer, J., Krämer, A., 1996. Occupational chronic exposure to organic solvents XVI. Ambient and biological monitoring of workers exposed to toluene. Int. Arch. Occup. Environ. Health 69 (2), 91–96.

Angerer, J., Ewers, U., Wilhelm, M., 2007. Human biomonitoring: state of the art. Int. J. Hyg. Environ. Health 210 (3), 201–228.

Apostoli, P., Brugnone, F., Perbellini, L., et al., 1982. Biomonitoring of occupational toluene exposure. Int. Arch. Occup. Environ. Health 50 (2), 153–168.

Arnold, S.M., Angerer, J., Boogaard, P.J., et al., 2013. The use of biomonitoring data in exposure and human health risk assessment: benzene case study. Crit. Rev. Toxicol. 43 (2), 119–153.

Aydin, K., Sencer, S., Demir, T., et al., 2002. Cranial MR findings in chronic toluene abuse by inhalation. Am. J. Neuroradiol. 23 (7), 1173–1179.

Boogaard, P.J., 2009. Biomonitoring of the workplace and environment. Gen. Appl. Syst. Toxicol. https://doi.org/10.1002/9780470744307.gat126.

Borne, J., Riascos, R., Cuellar, H., et al., 2005. Neuroimaging in drug and substance abuse part II: opioids and solvents. Top. Magn. Reson. Imaging 16 (3), 239–245.

Bowen, S.E., 2011. Two serious and challenging medical complications associated with volatile substance misuse: sudden sniffing death and fetal solvent syndrome. Subst. Use Misuse 46 (Suppl. 1), 68–72.

Broughton, E., 2005. The Bhopal disaster and its aftermath: a review. Environ. Health 4 (1), 6.

Bruckner, J.V., Warren, D.A., 2001. Toxic Effects of Solvents and Vapors. Casarett and Doull's Toxicology: The Basic Science of Poisons. McGraw-Hill, New York, pp. 871–872.

Brugnone, F., Perbellini, L., Gaffuri, E., Apostoli, P., 1980. Biomonitoring of industrial solvent exposures in workers' alveolar air. Int. Arch. Occup. Environ. Health 47 (3), 245–261.

Brugnone, F., Perbellini, L., Apostoli, P., et al., 1983. Isopropanol exposure: environmental and biological monitoring in a printing works. Occup. Environ. Med. 40 (2), 160–168.

Brugnone, F., Perbellini, L., Faccini, G., Pasini, F., 1985. Concentration of ethylene oxide in the alveolar air of occupationally exposed workers. Am. J. Ind. Med. 8 (1), 67–72.

Buckley, T.J., Prah, J.D., Ashley, D., et al., 1997. Body burden measurements and models to assess inhalation exposure to methyl tertiary butyl ether (MTBE). J. Air Waste Manag. Assoc. 47 (7), 739–752.

Cardona, A., Marhuenda, D., Martí, J., et al., 1993. Biological monitoring of occupational exposure ton-hexane by measurement of urinary 2, 5-hexanedione. Int. Arch. Occup. Environ. Health 65 (1), 71–74.

Cardona, A., Marhuenda, D., Prieto, M.J., et al., 1996. Behaviour of urinary 2, 5-hexanedione in occupational co-exposure to n-hexane and acetone. Int. Arch. Occup. Environ. Health 68 (2), 88–93.

Caro, J., Gallego, M., 2008. Alveolar air and urine analyses as biomarkers of exposure to trihalomethanes in an indoor swimming pool. Environ. Sci. Technol. 42 (13), 5002–5007.

Caro, J., Gallego, M., 2009. Environmental and biological monitoring of volatile organic compounds in the workplace. Chemosphere 77 (3), 426–433.

Cosnier, F., Cossec, B., Burgart, M., et al., 2013. Biomarkers of toluene exposure in rats: mercapturic acids versus traditional indicators (urinary hippuric acid and *o*-cresol and blood toluene). Xenobiotica 43 (8), 651–660.

Cosnier, F., Nunge, H., Brochard, C., et al., 2014. Impact of coexposure on toluene biomarkers in rats. Xenobiotica 44, 217–228.

Csanády, G.A., Denk, B., Pütz, C., et al., 2000. A physiological toxicokinetic model for exogenous and endogenous ethylene and ethylene oxide in rat, mouse, and human: formation of 2-hydroxyethyl adducts with hemoglobin and DNA. Toxicol. Appl. Pharmacol. 165 (1), 1–26.

Droz, P.O., Guillemin, M.P., 1986. Occupational exposure monitoring using breath analysis. J. Occup. Environ. Med. 28 (8), 593–602.

Filley, C.M., Halliday, W., Kleinschmidt-DeMasters, B.K., 2004. The effects of toluene on the central nervous system. J. Neuropathol. Exp. Neurol. 63 (1), 1–12.

Fiserova-Bergerova, V., Ogata, M., 1995. Biological monitoring of exposure to industrial chemicals. Biomark. Occup. Health Prog. Perspect. 89.

Forrest, M.S., Lan, Q., Hubbard, A.E., et al., 2005. Discovery of novel biomarkers by microarray analysis of peripheral blood mononuclear cell gene expression in benzene-exposed workers. Environ. Health Perspect. 113 (6), 801.

Fujino, A., Satoh, T., Takebayashi, T., et al., 1992. Biological monitoring of workers exposed to acetone in acetate fibre plants. Occup. Environ. Med. 49 (9), 654–657.

Ghimenti, S., Tabucchi, S., Bellagambi, F.G., et al., 2015. Determination of sevoflurane and isopropyl alcohol in exhaled breath by thermal desorption gas chromatography–mass spectrometry for exposure assessment of hospital staff. J. Pharmaceut. Biomed. Anal. 106, 218–223.

Ghittori, S., Imbriani, M., Pezzagno, G., Capodaglio, E., 1987. The urinary concentration of solvents as a biological indicator of exposure: proposal for the biological equivalent exposure limit for nine solvents. Am. Ind. Hyg. Assoc. J. 48 (9), 786–790.

Ghittori, S., Saretto, G., Saretto, G., 1994. Biological monitoring of workers exposed to carbon tetrachloride vapor. Appl. Occup. Environ. Hyg. 9 (5), 353–357.

Ghittori, S., Alessio, A., Negri, S., et al., 2004. A field method for sampling toluene in end-exhaled air, as a biomarker of occupational exposure: correlation with other exposure indices. Ind. Health 42 (2), 226–234.

Gobba, F., Ghittori, S., Imbriani, M., et al., 1997. The urinary excretion of solvents and gases for the biological monitoring of occupational exposure: a review. Sci. Total Environ. 199 (1–2), 3–12.

Goldfranks, L.R., 2015. Goldfrank's Toxicologic Emergencies. McGraw-Hill.

Guidotti, T.L., 1994. Occupational exposure to hydrogen sulfide in the sour gas industry: some unresolved issues. Int. Arch. Occup. Environ. Health 66 (3), 153–160.

Guillermo, A., 1994. Toxic gas inhalation. Clin. Pulm. Med. 1 (2), 84–92.

Hamelin, G., Truchon, G., Tardif, R., 2004. Comparison of unchanged n-hexane in alveolar air and 2, 5-hexanedione in urine for the biological monitoring of n-hexane exposure in human volunteers. Int. Arch. Occup. Environ. Health 77 (4), 264–270.

Hennebrüder, K., Angerer, J., 2005. Determination of DMF modified DNA base N 4-methylcarbamoylcytosine in human urine using off-line sample clean-up, two-dimensional LC and ESI-MS/MS detection. J. Chromatogr. B 822 (1), 124–132.

Hernández, A.F., Gil, F., Tsatsakis, A.M., 2014. Biomarkers of chemical mixture toxicity. In: Gupta, R.C. (Ed.), Biomarkers in Toxicology. Academic Press/Elsevier, Amsterdam, pp. 655–669.

Ichihara, G., Saito, I., Kamijima, M., et al., 1998. Urinary 2, 5-hexanedione increases with potentiation of neurotoxicity in chronic coexposure to n-hexane and methyl ethyl ketone. Int. Arch. Occup. Environ. Health 71 (2), 100–104.

Imbriani, M., Ghittori, S., Pezzagno, G., Capodaglio, E., 1984. n-Hexane urine elimination and weighted exposure concentration. Int. Arch. Occup. Environ. Health 55 (1), 33–41.

Imbriani, M., Ghittori, S., Pezzagno, G., Capodaglio, E., 1995. Anesthetic in urine as biological index of exposure in operating-room personnel. Journal of Toxicology and Environmental Health, Part A Current Issues 46 (2), 249–260.

Imbriani, M., Ghittori, S., Pezzagno, G., 1998. The biological monitoring of inhalation anaesthetics. G. Ital. Med. del Lav. Ergon. 20 (1), 44–49.

Imbriani, M., Negri, S., Ghittori, S., Maestri, L., 2002a. Measurement of urinary N-acetyl-S-(N-methylcarbamoyl) cysteine by high-performance liquid chromatography with direct ultraviolet detection. J. Chromatogr. B 778 (1), 231–236.

Imbriani, M., Maestri, L., Marraccini, P., et al., 2002b. Urinary determination of N-acetyl-S-(N-methylcarbamoyl) cysteine and N-methylformamide in workers exposed to N, N-dimethylformamide. Int. Arch. Occup. Environ. Health 75 (7), 445–452.

Imbriani, M., Ghittori, S., 2005. Gases and organic solvents in urine as biomarkers of occupational exposure: a review. Int. Arch. Occup. Environ. Health 78 (1), 1–19.

Inoue, O., Seiji, K., Nakatsuka, H., Watanabe, T., et al., 1989. Urinary t, t-muconic acid as an indicator of exposure to benzene. Occup. Environ. Med. 46 (2), 122–127.

Jacobson, G.A., McLean, S., 2003. Biological monitoring of low level occupational xylene exposure and the role of recent exposure. Ann. Occup. Hyg. 47 (4), 331–336.

Jang, J.Y., Droz, P.O., Kim, S., 2000. Biological monitoring of workers exposed to ethylbenzene and co-exposed to xylene. Int. Arch. Occup. Environ. Health 74 (1), 31–37.

Käfferlein, H.U., Mraz, J., Ferstl, C., Angerer, J., 2004. Analysis of metabolites of N, N-dimethylformamide in urine samples. International archives of occupational and environmental health 77 (6), 427–432.

Kamran, S., Bakshi, R., 1998. MRI in chronic toluene abuse: low signal in the cerebral cortex on T2-weighted images. Neuroradiology 40 (8), 519–521.

Kawai, T., Yasugi, T., Horiguchi, S.I., et al., 1990. Biological monitoring of occupational exposure to isopropyl alcohol vapor by urinalysis for acetone. Int. Arch. Occup. Environ. Health 62 (5), 409–413.

Kim, H.A., Kim, K., Heo, Y., et al., 2004. Biological monitoring of workers exposed to N, N-dimethylformamide in synthetic leather manufacturing factories in Korea. Int. Arch. Occup. Environ. Health 77 (2), 108–112.

Lareo, A.C., Perbellini, L., 1995. Biological monitoring of workers exposed to NN-dimethylformamide. Int. Arch. Occup. Environ. Health 67 (1), 47–52.

Lauwerys, R.R., Hoet, P., 2001. Industrial Chemical Exposure: Guidelines for Biological Monitoring. CRC Press.

Lowry, L.K., 1995. Role of biomarkers of exposure in the assessment of health risks. Toxicol. Lett. 77 (1–3), 31–38.

McCammon Jr., C.S., Glaser, R.A., Wells, V.E., et al., 1991. Exposure of workers engaged in furniture stripping to methylene chloride as determined by environmental and biological monitoring. Appl. Occup. Environ. Hyg. 6 (5), 371–379.

Monster, A.C., 1986. Biological monitoring of chlorinated hydrocarbon solvents. J. Occup. Environ. Med. 28 (8), 583–588.

Mráz, J., Dušková, Š., Gálová, E., et al., 2002. Improved gas chromatographic–mass spectrometric determination of the N-methylcarbamoyl adduct at the N-terminal valine of globin, a metabolic product of the solvent N, N-dimethylformamide. J. Chromatogr. B 778 (1), 357–365.

Needham, L.L., Calafat, A.M., Barr, D.B., 2007. Uses and issues of biomonitoring. Int. J. Hyg. Environ. Health 210 (3), 229–238.

NTP, 2016. National Toxicology Program, 14th Report on Carcinogens.

Ogata, M., 1990. Database for biological monitoring of aromatic solvents. In: Biological Monitoring of Exposure to Industrial Chemicals. Proceedings of the United States-Japan Cooperative Seminar on Biological Monitoring. ACGIH.

Ogata, M., Taguchi, T., Horike, T., 1995. Evaluation of exposure to solvents from their urinary excretions in workers coexposed to toluene, xylene, and methyl isobutyl ketone. Applied Occupational and Environmental Hygiene 10 (11), 913–920.

Ong, C.N., Chia, S.E., Phoon, W.H., Tan, K.T., 1991. Biological monitoring of occupational exposure to tetrahydrofuran. Occup. Environ. Med. 48 (9), 616–621.

Pavanello, S., Lotti, M., 2014. Biomonitoring exposures to carcinogens. In Biomarkers in Toxicology 785–798.

Perbellini, L., Princivalle, A., Caivano, M., Montagnani, R., 2003. Biological monitoring of occupational exposure to N, N-dimethylacetamide with identification of a new metabolite. Occup. Environ. Med. 60 (10), 746–751.

Periago, J.F., Cardona, A., Marhuenda, D., et al., 1993. Biological monitoring of occupational exposure to n-hexane by exhaled air analysis and urinalysis. Int. Arch. Occup. Environ. Health 65 (4), 275–278.

Pezzagno, G., Ghittori, S., Imbriani, M., Capodaglio, E., 1985. Eliminazione urinaria dei solventi durante esposizioni controllate: la loro concentrazione urinaria come indicatore biologico di esposizione. Atti 48, 511–522.

Pezzagno, G., Maestri, L., Fiorentino, M.L., 1999. Trans, trans-muconic acid, a biological indicator to low levels of environmental benzene: some aspects of its specificity. Am. J. Ind. Med. 35 (5), 511–518.

Pierce, C.H., Dills, R.L., Morgan, M.S., et al., 1998. Biological monitoring of controlled toluene exposure. Int. Arch. Occup. Environ. Health 71 (7), 433–444.

Poklis, A., Burkett, C.D., 1977. Gasoline sniffing: a review. Clin. Toxicol. 11 (1), 35–41.

Prieto, M.J., Marhuenda, D., Roel, J., Cardona, A., 2003. Free and total 2,5-hexanedione in biological monitoring of workers exposed to n-hexane in the shoe industry. Toxicol. Lett. 145 (3), 249–260.

SCRD, 2009. State Coalition for Remediation of Drycleaners. Chemicals Used in Drycleaning Operations.

Silins, I., Högberg, J., 2011. Combined toxic exposures and human health: biomarkers of exposure and effect. Int. J. Environ. Res. Public Health 8 (3), 629–647.

Simon, P., Nicot, T., Dieudonne, M., 1994. Dietary habits, a non-negligible source of 2-thiothiazolidine-4-carboxylic acid and possible overestimation of carbon disulfide exposure. Int. Arch. Occup. Environ. Health 66 (2), 85–90.

Snyder, R., 2012. Leukemia and benzene. Int. J. Environ. Res. Public Health 9 (8), 2875–2893.

Tardif, R., Truchon, G., Brodeur, J., 1998. Comparison of hippuric acid and o-cresol in urine and unchanged toluene in alveolar air for the biological monitoring of exposure to toluene in human volunteers. Appl. Occup. Environ. Hyg. 13 (2), 127–132.

van Engelen, J.G., Rebel-de Haan, W., Opdam, J.J., Mulder, G.J., 1997. Effect of coexposure to methyl ethyl ketone (MEK) on n-hexane toxicokinetics in human volunteers. Toxicol. Appl. Pharmacol. 144 (2), 385–395.

Voss, J.U., Roller, M., Brinkmann, E., Mangelsdorf, I., 2005. Nephrotoxicity of organic solvents: biomarkers for early detection. Int. Arch. Occup. Environ. Health 78 (6), 475–485.

Wang, V.S., Shih, T.S., Cheng, C.C., et al., 2004. Evaluation of current biological exposure index for occupational N, N-dimethylformamide exposure from synthetic leather workers. J. Occup. Environ. Med. 46 (7), 729–736.

Yang, J.S., Kim, E.A., Lee, M.Y., et al., 2000. Biological monitoring of occupational exposure to N, N-dimethylformamide—the effects of co-exposure to toluene or dermal exposure. Int. Arch. Occup. Environ. Health 73 (7), 463–470.

Yang, J., Bai, W., Niu, P., et al., 2014. Aberrant hypomethylated STAT3 was identified as a biomarker of chronic benzene poisoning through integrating DNA methylation and mRNA expression data. Exp. Mol. Pathol. 96 (3), 346–353.

PART IV

BIOTOXINS BIOMARKERS

CHAPTER

35

Freshwater Cyanotoxins

Gurjot Kaur

Human and Environmental Toxicology, Department of Biology, University of Konstanz, Konstanz, Germany

INTRODUCTION

Cyanobacteria, commonly known as blue-green algae, are photosynthetic prokaryotic bacteria that contribute to the normal microbial inhabitation on the surface of marine and freshwaters. Cyanobacteria are important primary producers in many freshwater ecosystems and are present in varying densities in most surface freshwaters worldwide. They may proliferate and form high-density blooms under certain conditions, particularly in abundance of specific nutrients. Cyanobacteria are of toxicological interest because many genera have the ability to produce toxins that affect aquatic organisms and terrestrial species such as wildlife, livestock, pets, and even humans on ingestion of contaminated water or seafood (Buratti et al., 2017).

The main energy source for photosynthesis in cyanobacteria is the green pigment chlorophyll-a. The name "blue-green algae" is derived from their ability to produce accessory photosynthetic blue pigments, phycobilins or phycocyanins, important for photon capture during photosynthesis. Red to brown accessory photosynthetic pigments, called phycoerythrins, are also produced, and depending on the levels of phycobilins and phycoerythrins, the colors of cyanobacteria vary between different shades of blue and green and, less commonly, brown to red. Being oxygenic phototrophic bacteria, cyanobacteria contain both type-1 and type-2 reaction centers or photosystems with a unique "phycobilisome" antenna network that feeds these reaction centers. Unlike plants and green algae, this special phycobilisome consists of 3-7 MDa large multimeric assemblies that associate with the surface of the chloroplast thylakoid membrane. The phycobilisome supplies excitation energy to both reaction centers by forming a megacomplex. Each phycobilisome is built of many units of chromophore-binding phycobiliproteins, phycocyanin, phycoerythrin, and allophycocyanin, and non-chromophore-binding linker proteins. The phycobiliproteins are responsible for the absorption of cyanobacteria at the 500–680 nm range and less commonly 710 nm due to specialized phycobilisomes (Liu et al., 2013).

The mere presence of a toxin-producing cyanobacterial species in surface water does not constitute a valid risk. Toxins are not always produced and may not be produced at toxicologically significant levels. Furthermore, toxin production is influenced by genetic and environmental factors. Although certain environmental features, including nutrient concentrations, water temperature and pH, play a role in triggering and maintaining toxin production, the critical parameters involved are not fully understood. Toxin production is also linked to cyanobacterial cell densities and growth rates, which are determined by the availability of nutrients, light intensity, water temperature, and interactions with other aquatic organisms. Important interactions include predation and competition for nutrients. Toxin production is generally more common during warmer summer weather but can occur at any time of the year. Occurrences of high densities of cyanobacteria, visible as discolorations of the water and/or the formation of algal scum, are often referred to as algal blooms. When algal blooms have the potential to cause harmful effects, they are referred to as harmful algal blooms (HABs). The incidence of HABs is recorded in many regions and is usually linked to human activities such as agriculture that results in excess nutrient inflow into surface water bodies. It can also be linked to the creation of lakes and ponds that provide suitable habitats for cyanobacteria. Following a toxic HAB formation, the risk of poisoning depends on the level of exposure. In the past, human exposures have occurred through drinking water, recreational water use, contaminated food or dietary supplements, and contaminated hemodialysis water. The toxins produced by freshwater cyanobacteria can be classified based on their toxic effects, their molecular structure, or their origins (Dai et al., 2016). Due to

Biomarkers in Toxicology, Second Edition
https://doi.org/10.1016/B978-0-12-814655-2.00035-9

this chapter's focus on biomarkers (Table 35.1), we will use a classification based on toxic effects.

HEPATOTOXINS

Microcystins

Introduction

Microcystins are cyclic heptapeptides and are arguably the most important cyanotoxins in terms of their worldwide impact on health and water quality. Microcystis spp., *Microcystis aeruginosa* in particular, are common and widespread producers of microcystins, but they can also be produced by several other genera of freshwater cyanobacteria, including *Anabaena, Planktothrix, Nostoc, Oscillatoria,* and *Anabaenopsis.* Microcystin concentrations are generally correlated with cyanobacterial cell density, with increasing correlation at high cell densities. People are often exposed when swimming, skiing, or boating in contaminated waters. Other routes of human exposure include drinking water, contaminated seafoods or nutritional supplements, and, in one tragic and lethal event, contaminated water in dialysis bags (Jochimsen et al., 1998). Terrestrial animals, including pets, livestock, and wildlife, are typically exposed when drinking from contaminated lakes and ponds. Aquatic species living in contaminated waters may also be affected. The ever-expanding number of new lakes and ponds created for water management, recreation, and drinking water for farm animals creates opportunities for microcystin production and exposure in susceptible populations. Potentially poisonous concentrations of microcystins are reached relatively frequently in lakes and ponds with high nutrient concentrations, and the frequency appears to be increasing in many regions due to the expansion of intensive agriculture, industrial development, and urbanization.

Chemistry

The cyclic heptapeptide structure of microcystins (Fig. 35.1) allows for considerable structural variability, and over 200 naturally occurring structural variants have been identified. Six amino acids, including four nonprotein and two protein, form a ring structure. One nonprotein amino acid, referred to as ADDA (3-amino-9-methoxy-2,6,8-trimethyl-10-phenyldeca-4,6-dienoic acid), forms a side chain. The ADDA side chain is consistent between microcystin variants and can be used to quantify microcystins independent of the variant type. The two protein amino acids, at positions 2 and 4 in the ring structure, contribute significantly to the structural variability. Microcystin variants are named based on the protein amino acids at these two variable positions. One of the most common variants

TABLE 35.1 Biomarkers of Exposure to Freshwater Cyanotoxins

Toxin	Specific Biomarkers	Nonspecific Biomarkers
Microcystin	Toxin isolation in tissues	Sphingolipid in liver
	Protein phosphatase (PP) inhibition	Bcl proteins in liver
	PP2A in liver	Lipid peroxidation
		Heat shock proteins
		8-Oxo-7,8-dihydro-2′-deoxyguanosine in liver
		Alkaline phosphatase in plasma
		Alanine transaminase in plasma
		Methemoglobinemia
		Hydroperoxidase
		Superoxide dismutase
		Muscle DNA/RNA ratio (chronic)
		Hormone levels (testosterone, estradiol)
		miRNA-122-5p
Cylindro-spermopsin	Toxin isolation in tissues	Urine protein
		Urine specific gravity
		Glutathione
		Lipid peroxidation
		Protein oxidation
		DNA oxidation
		Superoxide dismutase
		NADPH oxidase
		Catalase
		Gamma-glutamyl-cysteine synthetase
Anatoxin-a	Toxin isolation in tissues	Acetylcholinesterase (AChE)
		Lactate dehydrogenase
Anatoxin-a(s)	Toxin isolation in tissues	AChE
Saxitoxins	Toxin isolation in tissues	Glutathione peroxidase
		Lipid peroxidase
		Glutathione peroxidase
		Superoxide dismutase

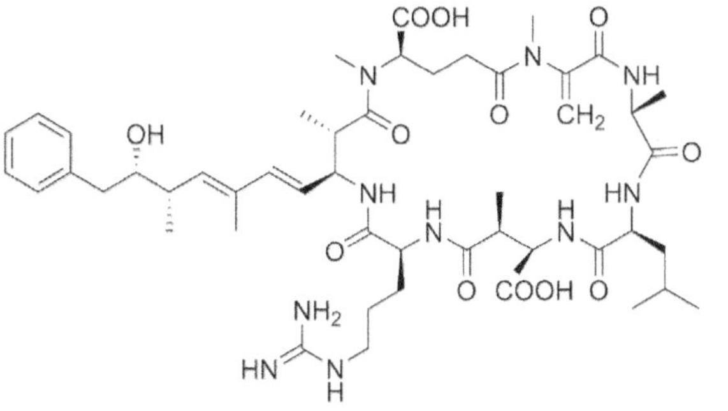

FIGURE 35.1 Microcystin chemical structure.

contains leucine and arginine at the protein amino acid positions and is referred to as microcystin-LR (Fig. 35.1). Microcystins are not excreted by cyanobacterial cells and are therefore strongly associated with cells as long as the cells remain intact. However, microcystins may be released into the surrounding water when cyanobacterial cells disintegrate. Microcystins are stable, with a typical environmental half-life of 10 weeks. The rate of breakdown is increased under direct sunlight, at high environmental temperatures (40°C), and at extremely low (<1) or high pH (>9) (Harada et al., 1996). Microcystin concentrations may be reduced by boiling or heating in a microwave oven. However, the concentration of free microcystins in muscle tissue from exposed fish may increase after boiling, due to the release of phosphatase-bound microcystins (Gutierrez-Praena et al., 2013).

Toxic Effects

Most of the available toxicological data on microcystins have been based on microcystin-LR. Other microcystin variants appear to be similar to microcystin-LR in their toxicological effects but different in potency. It should be noted that data used for determining toxicological relevance have been largely derived from experiments in rodents using intraperitoneal injection as the route of exposure using either extracts of complex crude mixtures derived from toxic algal blooms or purified components. Extrapolation between routes of exposure and between the toxicities of different complex mixtures makes the comparison of data from different studies challenging.

Many incidences of human exposure have come to light. One such incident is the exposure of 131 patients to microcystins in dialysis cartridges at a hemodialysis clinic in Caruaru, Brazil, in 1996. Eighty-nine percent of patients experienced early symptoms of toxicity such as visual disturbances, nausea, vomiting, and muscle weakness. Consequently, 100 patients developed acute liver failure and succumbed to death. It was conferred that this syndrome, now known as "Caruaru Syndrome," was a result of the dominant cyanobacteria population in water supply reservoir (Jochimsen et al., 1998).

The clinical effects of microcystin poisoning depend on the route and level of exposure and the mixture of components involved in the exposure. In typical exposures of mammals to toxic blooms, low to mild exposure levels are associated with irritant effects, resulting in inflammatory responses in the skin, respiratory system, and gastrointestinal system. Higher exposure levels, particularly oral exposures, result in liver damage and, if the liver damage is severe, liver failure (Briand et al., 2003). The earliest detectable signs of liver damage include liver swelling and increased concentrations of liver enzymes (alanine aminotransferase and aspartate aminotransferase) in the blood. Early symptoms typically appear within minutes to hours of exposure and include inappetence, depression, and vomiting, followed by diarrhea, which may become extreme and hemorrhagic. Inappetence and depression become progressively worse. The final stages of lethal poisoning may be associated with recumbency and coma (van der Merwe et al., 2012). An important difference between exposures in aquatic and terrestrial organisms is the chronic low-dose exposure in aquatic environments over days or weeks. In fish, exposures to microcystins cause cellular damage (particularly liver damage), similar to the effects seen in mammals (Tencalla and Dietrich, 1997). Mature fish are generally more resistant to microcystin toxicosis compared with juvenile fish. Although fish have the ability to avoid areas of accumulation of toxic algae, sublethal liver damage in fish is associated with accumulation of microcystins in fish food items, such as mussels, snails, and zooplankton (Malbrouck and Kestemont, 2006). Fishes are also susceptible to decreased water oxygen levels associated with the decay of algal scum, and this effect may also play a role in fish deaths (Ibelings and Havens, 2008).

Lower doses of microcystin-LR induce apoptosis, while higher concentrations induce necrosis in lymphocytes in a time- and concentration-dependent manner. Reorganization of the actin cytoskeleton, cell shrinkage, and filopodia disappearance was also observed in phagocytes, indicating impairment of the immune system (Rymuszka and Adaszek, 2013). Epidermal exposure to microcystin resulted in irritation of the skin. In cases of skin damage, faster penetration to deeper cell layers and absorption to systemic circulation were expected. Persistent toxic effects such as on keratinocytes migration and cytoskeleton were only observed after longer exposures to the skin (Kozdeba et al., 2014).

Cellular Transport

It is well known that microcystin transport into the cell is a prerequisite for the associated apical cytotoxicity. Following ingestion and release from cyanobacterial cells, microcystins are absorbed into the portal circulation from the small intestine via bile acid transporters in the intestinal wall. Microcystins are then accumulated in hepatocytes via organic anion transporting polypeptides (OATPs, OATP1B1 and OATP1B3) in hepatocyte membranes. In addition to the liver, human OATPs are also located in the blood-brain barrier (BBB) and kidney cells (Hagenbuch and Meier, 2004). At this stage, we must recall the aforementioned incidence of human exposure. Dialysis patients at a hemodialysis clinic in Caruaru, Brazil, who were exposed intravenously to varying concentrations of microcystin congeners developed immediate signs of neurotoxicity, e.g., dizziness, tinnitus, vertigo, with a later onset of hepatotoxicity and nephropathy leading finally to multiorgan failure (Jochimsen et al., 1998). Similar pathological changes have also been observed in mice and fish. Microcystins are transported across the BBB in an OATP-dependent manner and are able to induce neurotoxicity via inhibition of protein phosphatases (PPs) (Feurstein et al., 2009). It is no surprise that the presence of OATPs in proximal renal epithelial cells coupled with the kidney's role in toxin elimination predisposes renal cells in humans to nephrotoxicity. However, the exact OATP uptake kinetics is not established in the brain or kidney.

Secretion of biochemically active microcystin into the bile of rainbow trout corroborates an active efflux of microcystins from hepatocytes (Sahin et al., 1996). One report has linked efflux transporter in Zebrafish (zf-MDR3) to microcystin excretion (Lu et al., 2015). However, no information is yet available in humans as to which efflux transporters, if any, are important for microcystin efflux. The secretion of microcystin into bile also indicates the potential for enterohepatic recirculation of toxins, further endorsed by reports of microcystin conjugates (Sahin et al., 1996). Microcystin-glutathione (-GSH) and -cysteine (CYS) conjugates are reported to be the main routes of excretion of free microcystin after exposure in mice (Kondo et al., 1992) and aquatic organisms (Pflugmacher et al., 1998). Interestingly, it has to be noted that microcystin-GSH and microcystin-CYS conjugates show similar in vitro inhibition of PP1 and PP2A to that of microcystin-LR but lower toxicity when intratracheally administered in mice possibly because of lowered uptake into the liver or more rapid elimination into the bile. This indicates the presence of a detoxification mechanism to actively excrete microcystin conjugates out of the cell necessitating identification of involved efflux transporter(s).

Mechanism of Action

It is well established that cytotoxicity, after cellular uptake of microcystin congeners, microcystin-LR, microcystin-RR, microcystin-LW, and microcystin-LF, progresses through covalent binding and inhibition of PPs, mainly PP1 and PP2A. This, in turn, leads to an increase in cell phosphorylated-protein load and subsequent deregulation of fundamental cellular processes and apoptosis. The interaction occurs between the hydrophobic groove, the C-terminal groove, and the catalytic site of PP1 and the glutamate, methyl aspartate, and ADDA residues of the toxin (Goldberg et al., 1995) and covalently between the Sγ atom of cysteine-269 of PP2A and the terminal carbon atom of the Mdha side chain of the toxin (Xing et al., 2006). Microcystin-LR may also bind to ATP synthase, leading to hepatocyte apoptosis (Mikhailov et al., 2003). The microcystin-interacting PPs are ubiquitous and are found in all tissues and across species from mammals to plants to bacteria. In nature, PPs reverse the active state of kinases through the hydrolytic removal of the phosphoryl group from kinases. On inhibition, alterations in the cytoskeleton occur, such as hyperphosphorylation of cytokeratins, reorganization, disassembly of actin and microtubules, and ultimately disruption of cellular architecture (Huang et al., 2015; Zeng et al., 2015). PP2A inhibition also leads to MAPK pathway activation (ERK1/2, JNK, p38) and apoptosis. Microcystin-LR also impairs PP2A ability to bind tubulin and thus destabilizes the microtubules. Hepatotoxicity proceeds through changes in cytoskeletal architecture and cell viability by MAPK activation, hyperphosphorylation of p38, ezrin, and ERK1/2 (Komatsu et al., 2007).

The main pathway of microcystin toxicity occurs through oxidative stress. This oxidative stress is dose-dependent and proceeds with marked alterations in cytotoxicity markers such as Lactate dehydrogenase (LDH) leakage, lipid peroxidation, reactive oxygen species (ROS) production, and antioxidant enzymes. In addition, microcystin-LR induces JNK activation, which affects enzymes involved in energy metabolism and mitochondrial dysfunction, ultimately leading to hepatocyte apoptosis and oxidative liver injury. Microcystin-LR also induces intracellular ROS production due to a calcium-mediated loss of mitochondrial membrane potential. The differential toxicity of microcystin-LR and microcystin-RR has been linked to the differential expression of proinflammatory molecules, IL-6 and IL-8, wherein IL-8 production varies with the microcystin congener in intestinal cells. It has recently been shown in the liver of fish that microcystin-LR–mediated oxidative stress occurs with an increase in hydroxyl radicals and a subsequent induction of apoptosis-related genes (p38, JNKa, Bcl-2), wherein application of N-acetylcysteine provides protection against this stress. Research on mice has identified the role of Nrf2 and related antioxidant proteins, Keap1 and Cullin 3, as important protective proteins in oxidative stress mediated microcystin toxicity. In a normal cell, Keap1 regulates degradation and thus expression of Nrf2. During stress, Keap1 activity is disrupted, resulting in increased Nrf2 levels and activation of the oxidative defense system. In the absence of Nrf2 and Keap1 pathways, the microcystin-GSH conjugation pathway becomes an important pathway for cell protection (Valerio et al., 2016b).

Calcium has long been known to be important for microcystin-mediated toxicity. Its involvement may trigger mitochondrial dysfunction as discussed before (Ding et al., 2001). Ceramide has also been identified as another secondary messenger involved in microcystin toxicity (Li et al., 2012). Ceramide is known to be important for PP2A activity. It is involved in microcystin-induced toxicity through PP2A subunit regulation, subcellular localization and inhibition, and cytoskeletal destabilization. Desipramine, a ceramide synthase inhibitor, protects the cells against cytoskeletal destabilization, detachment, and apoptosis (Li et al., 2012). Nitric oxide production, another secondary messenger, has an emerging role in microcystin-associated toxicity (Ji et al., 2011).

As microcystin-LR crosses the BBB via brain-specific OATPs (Fischer et al., 2005), it induces hyperphosphorylation of *tau*, a neural microtubule-associated protein, in a biphasic and concentration-dependent manner, which in turn may lead to neuronal degeneration (Meng et al., 2015). ROS production is involved in this hyperphosphorylation and impairment of neurological functions. Microcystin also modulates the immune system. Microcystin-LR has been shown to induce Fas receptor, ligand, and NF-κB, important signals for apoptosis, and immune responses in cell culture models (Feng et al., 2011; Ji et al., 2011). Studies on the male reproductive system indicate a risk for toxicity (Chen et al., 2013a). Male mice have alterations in testes and epididymis after *M. aeruginosa* extract administration (Ding et al., 2006). Death of Leydig cells and subsequent decrease in testosterone production are also seen (Li et al., 2008). Similar cytotoxicity is observed in Sertoli cells (Chen et al., 2013b). Female mice display a reduction in ovary weight and loss of primordial follicle pool (Wu et al., 2014).

Microcystin-LR is now classified as a potential class 2B carcinogen to humans. Involvement of ERK1/2 pathway, increased expression of protooncogenes, c-fos, c-jun, and c-myc in mouse, and p53 after microcystin-LR-induced tumor is now shown. In addition, there are indications that microcystin-LR tumorigenesis is regulated by NF-κB, IFN-α, and TNF-α (Valerio et al., 2016b). Involvement of the glycolytic-oxidative-nitrosative stress pathway in microcystin-LR-induced toxicity in mice has been published, wherein coenzyme Q10 was capable of ameliorating the associated toxicity along with implications in cancer treatment (Lone et al., 2017). Increased thyroid hormones resulting in thyroid dysfunction was demonstrated in mice injected with microcystin-LR intraperitoneally. Disruption in glucose, triglyceride, and cholesterol levels concurrent with hyperphagia, polydipsia, and weight loss was also seen (Zhao et al., 2015). These are the first indications of endocrine toxicity in mammals and may have human relevance.

Biomarkers

Research on biomarkers of microcystin exposure has expanded in recent years, particularly in efforts to identify preclinical biomarkers of exposure, the detection and characterization of sublethal chronic exposures, and potential ecotoxicological impacts in aquatic species. Much of our understanding of the biomarkers of microcystin exposure has been derived from *in vivo* and *in vitro* laboratory studies using microcystis extracts, as opposed to pure toxins. Analysis of several studies demonstrated that crude extracts have toxicity characteristics that are different from pure microcystins (Falconer, 2007). The variability in the toxicological characteristics of complex mixtures associated with crude extracts, as is the case in most exposures, complicates the expression of biomarkers. However, PP inhibition and the detection of microcystins in tissues and body fluids serve as consistent and specific biomarkers of microcystin exposure. In the liver, markers of microcystin damage include increased sphingolipid concentrations, decreased PP2A levels, and decreased expression of Bcl-2 proteins (Billam et al., 2008).

Analysis of biomarkers in aquatic species revealed a high degree of variation between studies (Puschner and Humbert, 2007). Detoxification biomarkers such as reduced GSH and glutathione S-transferase (GST) were particularly inconsistent, presenting opposite expression trends in different studies. Biomarkers associated with oxidative stress, including superoxide dismutase, catalase, GSH peroxidase, GSH reductase, and lipid peroxidation, were more consistent in their trends. However, with the exception of lipid peroxidation, they were not associated with a sufficiently long half-life or systematic response. In a study using Japanese rice fish (*Oryzias latipes*) as a model, gender differences were observed in the toxic threshold to microcystin-LR, using PP inhibition and heat shock protein (Hsp60) as biomarkers (Deng et al., 2010). The DNA oxidative damage marker 8-oxo-7,8-dihydro-20-deoxyguanosine was found to be an indicator of microcystin-LR exposure in rat liver, both for *in vitro* and *in vivo* exposure, at levels that did not result in morphologically apparent cell damage (Maatouk et al., 2004). Chronic exposure to sublethal doses of microcystin-LR in mice indicated that nonspecific biomarkers of liver damage, such as increases in liver enzymes (alkaline phosphatase, aspartate transaminase, and alanine transaminase) in plasma, and indicators of oxidative stress, such as methemoglobin, hydrogen peroxides, and superoxide dismutase, are correlated with exposure (Sedan et al., 2013). Similar biomarker data showing involvement of oxidative stress were also shown in *Goodea gracilis*, fish found endemically in water bodies of Central Mexico, contaminated with microcystin-producing cyanobacteria (Olivares-Rubio et al., 2015). Sublethal exposures in Sacramento splittail (*Pogonichthys macrolepidotus*) were correlated with declining nutritional status, as indicated by a decreased RNA/DNA ratio in muscle (Acuna et al., 2012). A recent study on the water flea, *Daphnia magna*, showed that short-term microcystin exposure increased GST levels, whereas long-term exposure decreased levels. The same study also measured changes in antioxidant markers (catalase, Cu, Zn, and Mn superoxide dismutase) (Lyu et al., 2016). 17β-Estradiol, testosterone, 11-ketotestosterone, and follicle-stimulating hormone (FSH) levels were increased in the serum of females as compared with males in Zebrafish exposed to microcystin, although levels of testosterone, FSH, and luteinizing hormone were also changed in males (Liu et al., 2016). Proteomic profiling of *Saccharomyces cerevisiae* cells using 2D electrophoresis and Matrix-Assisted Laser Desorption/Ionization Time-of-Flight Mass Spectrometry (MALDI-TOF/TOF) after microcystin-LR exposure identified ~14 differentially expressed proteins that were involved in metabolism, genotoxicity, cytotoxicity, and stress response including PP1 and PP2A. Most genes were related to oxidative stress and apoptosis (Valerio et al., 2016a). Many recent studies have revealed the presence of huge amounts of microRNAs from damaged cells following a xenobiotic insult. An increase in miR-122-5p, liver-enriched microRNA, on microcystin-LR exposure in the blood plasma of whitefish *Coregonus lavaretus* was detected, wherein a rapid increase in miR-122-5p levels was observed after 8h, which continued till the end of the experiment. This has the potential to be a new generation diagnostic biomarker in fish that may help in noninvasive diagnosis of liver damage (Florczyk et al., 2016).

Cylindrospermopsin

Introduction

Cylindrospermopsin is an alkaloid cyanotoxin produced by several freshwater genera, including *Cylindrospermopsis, Aphanizomenon, Anabaena, Lyngbya, Umezakia*, and *Raphidiopsis* It is found worldwide in surface freshwaters and is a potential toxicant in drinking water supplies for a large human populace, as well as animals (Guzman-Guillen et al., 2013a). It was first reported after a hepatoenteritis outbreak due to a *Cylindrospermopsis raciborskii* bloom in a local drinking water supply in Australia in 1979 (Bourke et al., 1983).

Chemistry

Cylindrospermopsin is a polyketide alkaloid consisting of a tricyclic guanidine coupled with hydroxymethyluracil (Fig. 35.2). It is a highly water-soluble molecule because of its zwitterionic nature. There are two known structural variants: highly toxic 7-epicylindrospermopsin (Banker et al., 2001) and the relatively less toxic deoxycylindrospermopsin (Norris et al., 1999). Two more variants, 7-deoxy-desulfo-cylindrospermopsin and 7-deoxy-desulfo-12-acetyl-cylindrospermopsin, were later identified in the Thai strain of *C. raciborskii* (Wimmer et al., 2014). Cylindrospermopsin is resistant to high temperatures, sunlight, and pH extremes. Unlike microcystins, cylindrospermopsin is often secreted from cyanobacterial cells into the surrounding water (Rucker et al., 2007). It bioaccumulates, particularly in organisms, in lower level of the food chain, such as gastropods, bivalves, and crustaceans (Kinnear, 2010).

Toxic Effects

Impact on human health is asserted from many reported incidents. Mild skin irritation is usually reported. A major outbreak of poisoning in humans with 148 cases occurred in 1979 near a reservoir on Palm Island, Queensland, Australia. Unfortunately, a dense bloom of cylindrospermopsin-producing *C. raciborskii* was observed after a copper sulfate treatment of a reservoir. Eventually, algal cell lysis and discharge of harmful toxins occurred in the water. Prolific users living close to the reservoir were affected with a syndrome consisting of hepatic and kidney damage and severe gastroenteritis. Symptoms included hemorrhagic diarrhea, vomiting, fever, hepatomegaly, dehydration, electrolyte imbalances, acidosis, and hypovolemic shock. However, the potential role that the copper sulfate water treatment could have

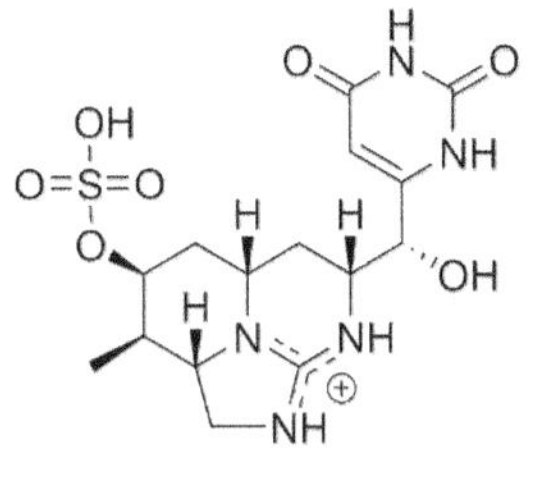

FIGURE 35.2 Cylindrospermopsin chemical structure.

played in the disease process remains uncertain. Intraperitoneal injection of extracts from *C. raciborskii* collected from the reservoir produced similar liver and kidney damage in mice (Griffiths and Saker, 2003).

Cylindrospermopsin was also isolated from water used in dialysis in Brazil, which caused liver failure in dialysis patients. However, the role of cylindrospermopsin in the disease process was not clear because the water was also contaminated with toxic concentrations of microcystin. *C. raciborskii* can cause mild skin irritation in some individuals, unfortunately, with no clear known mechanism, possibly because of unrelated components in animals and humans. While cylindrospermopsin outbreaks have been recently reported, there are many cases of co-occurrence with other cyanotoxins, making understanding of the relative contribution to symptoms difficult. Similar outbreak, popularly known as "Barcoo fever," was reported in Australian outback (Hayman, 1992).

Pure cylindrospermopsin injected into tilapia caused progressive tissue damage over a period of 5 days in the liver, kidney, heart, and gills (Gutierrez-Praena et al., 2013). In mice administered pure cylindrospermopsin, a dose-dependent increase in liver and kidney weight and alteration in hepatic and renal toxicity markers were observed after 11 weeks. Concomitant histopathological changes at the higher doses were also seen (Humpage and Falconer, 2003). Liver and kidney damages are consistently observed in laboratory rodents following exposure to acutely toxic doses of cylindrospermopsin-containing extracts. Typical liver pathology includes lipid infiltration and necrosis, mostly in the periacinar region of the prostate (Shaw et al., 2000). Dose-dependent DNA damage in mammalian cells is also observed, which can be prevented by CYP450 inhibitors, indicating an involvement of metabolic enzymes in progressing toxicity (Humpage et al., 2005). No reliable cylindrospermopsin-mediated human carcinogenicity, genotoxicity, and reproductive/developmental toxicity data are yet available.

Mechanism of Action

Previously, it was speculated that the main target tissue of purified cylindrospermopsin toxicosis is the liver, but lately the kidney has been shown to be the more sensitive target of toxicity. Four sequential phases of hepatocyte damage were identified by a time series analysis, including protein synthesis inhibition, membrane proliferation, lipid infiltration, and necrosis. Kidney pathology includes necrosis of the proximal tubules and protein accumulation in the distal tubules (Nair et al., 1999). Studies using crude extracts report higher potency and a wider range of effects compared with studies using purified cylindrospermopsin only, indicating that other components contribute to the toxic effects (Shaw et al., 2000).

Other tissues such as the thymus and heart are also affected (Terao et al., 1994). Cylindrospermopsin is a known potent inhibitor of protein synthesis in a concentration-dependent and irreversible manner, as confirmed both *in vitro* (Froscio et al., 2003) and *in vivo* (Terao et al., 1994), but the exact mechanism of action in different tissues is not fully elucidated. Interestingly, a decrease in cylindrospermopsin toxicity by the administration of cytochrome P450 inhibitors has been reported, suggesting an alternative toxicity mechanism possibly through metabolite formation. Cylindrospermopsin causes DNA fragmentation *in vitro* using metabolic activation by a cytochrome 450 enzyme as a necessary step (Bazin et al., 2010). New evidence suggests a cylindrospermopsin-induced upregulation of phase I enzymes (CYP1A1, CYP1B1, ALDH1A2, and CES2) and phase II enzymes (UGT1A6, UGT1A1, NAT1, and GSTM3) (Alja et al., 2013). Activation of p53 transcription factor and thus its target genes due to cylindrospermopsin stress has been detected in cell culture models (Bain et al., 2007). In addition, subtoxic concentrations of cylindrospermopsin are confirmed to modify protein levels involved in DNA repair and nucleosomal histones (Huguet et al., 2014). The role of oxidative stress is confirmed by increased ROS production, increased GSH content, and eventual apoptosis (Lopez-Alonso et al., 2013).

Biomarkers

Currently, satisfactory biomarkers for cylindrospermopsin toxicity are not known and only a tentative list of emerging biomarkers is provided. In a study of mice exposed to sublethal doses, no consistent changes were found in hematology or serum chemistry, indicators of liver and kidney damage. However, increased protein concentrations and changes in urine specific gravity were observed after 120 days of exposure (Humpage and Falconer, 2003).

On the contrary, change in several markers of oxidative stress indicates that oxidative stress plays a key role in the toxic effects seen in fish (Guzman-Guillen et al., 2013a). Chronic cylindrospermopsin exposure leads to a depletion of GSH, which precedes the appearance of clinical signs (Runnegar et al., 1994). Several other biomarkers of oxidative stress have been linked to the toxic effects of cylindrospermopsin, including the ratio of reduced GSH to oxidized GSH, lipid peroxidation, protein oxidation, DNA oxidation, and the activities of superoxide dismutase, NADPH oxidase, catalase, and gamma-glutamylcysteine synthetase (Guzman-Guillen et al., 2013b). Measurement of acetylcholinesterase (AChE) activity and lipid peroxidation levels are closely associated to cylindrospermopsin-associated neurotoxicity in tilapia fish. Histopathological changes in the brain, liver, kidney, intestine, and gills were also seen. The biochemical parameters inclusive oxidative stress biomarkers could be reversed by 3 days of depuration while the histopathological changes by 7 days of depuration (Guzman-Guillen et al., 2014). Cylindrospermopsin toxin markers have been recently detected using analytical pyrolysis (Py-GC/MS) and thermally assisted hydrolysis and methylation (TCh-GC/MS) coupled to GC/MS in environmental samples with an estimated 5 ppm detection threshold (Rios et al., 2014).

NEUROTOXINS

Anatoxin-a

Introduction

Anatoxin-a is a potent neurotoxin and is considered to be a major cyanotoxin of public health concern. It occurs worldwide in freshwaters and is produced by several genera of cyanobacteria, including *Anabaena, Aphanizomenon, Microcystis, Planktothrix, Raphidiopsis, Arthrospira, Cylindrospermum, Phormidium, Nostoc,* and *Oscillatoria*. Exposures occur mainly through consumption of contaminated drinking water but can also occur from recreational use of lakes and through contaminated dietary supplements (Osswald et al., 2007).

Chemistry

Anatoxin-a is a bicyclic amine alkaloid (Fig. 35.3). It contains a homotropane scaffold derived from glutamic acid. Stoichiometrically, it is chiral with two symmetric centers. Only the (+)-anatoxin-a enantiomeric form is naturally produced with the pKa of 9.4, indicating that it is mostly protonated under typical environmental pH conditions. Anatoxin-a is stable under sterile conditions but is susceptible to microbial biodegradation (Wonnacott and Gallagher, 2006). The half-life in a reservoir is reported to be 5 days under typical environmental pH conditions (Smith and Sutton, 1993). A structural analog, called homoanatoxin-a or methylene-anatoxin-a, has been isolated from *Oscillatoria formosa* (Skulberg et al., 1992). Small quantities of anatoxin-a are produced synthetically for use in acetylcholine receptor research.

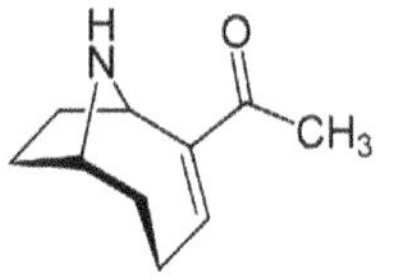

FIGURE 35.3 Anatoxin-a chemical structure.

Toxic Effects

Anatoxin-a was originally called Very Fast Death Factor because of its rapid lethal effects, within 2–7 min, in laboratory mice. Acute deaths following exposure to anatoxin-a have been recorded in multiple species, including dogs, cattle, and wildlife, and species differences in susceptibility to anatoxin-a have been observed even within animal types. Mallard ducks, for example, are more sensitive compared with ring-necked pheasants (Carmichael and Biggs, 1978). Anatoxin-a is rapidly absorbed from the gastrointestinal tract, as indicated by the rapidity of clinical effects following oral exposure. Clinical effects of poisoning may appear within minutes to hours after exposure and include loss of muscle coordination, muscle tremors, fasciculations, convulsions, and respiratory distress. The principal lethal effect is respiratory failure following loss of control over respiratory muscles (Osswald et al., 2007).

Mechanism of Action

Anatoxin-a is an agonist of peripheral and central acetylcholine receptors, with a 100-fold selectivity for nicotinic receptors over muscarinic receptors. It binds to the acetylcholine receptor at the same position as acetylcholine (pre- and postsynaptically), causing sodium/potassium ion channels to open and induce a depolarizing blockade. Anatoxin-a is more potent than acetylcholine or nicotine. Binding of anatoxin-a to the nicotinic acetylcholine receptors at neuromuscular junctions results in uncontrolled action potential propagation, which manifests clinically as uncoordinated muscle contraction, muscle fatigue, and paralysis. Cholinesterase does not break down anatoxin-a, leading to persistent muscle stimulation (Wonnacott and Gallagher, 2006). Stimulation of nicotinic receptors in the cardiovascular system causes increased heart rate and blood pressure (Sirén and Feurstein, 1990). Stimulation of presynaptic nicotinic receptors in the central nervous system by anatoxin-a may also cause the release of neurotransmitters such as dopamine, which could further increase the susceptibility of postsynaptic receptors to overstimulation (Wonnacott and Gallagher, 2006). Central nervous system receptors are, however, less sensitive to anatoxin-a when compared with peripheral receptors.

Biomarkers

The identification of biomarkers for anatoxin-a exposure, apart from the isolation of anatoxin-a from tissues, has remained elusive. Anatoxin-a appears to be a pure neurotoxin, with no characteristic blood chemistry changes and no gross or histological lesions associated with poisoning. The effect of sublethal doses of anatoxin-a, given intraperitoneally, on key enzyme activities (e.g., AChE, LDH, Isocitrate Dehydrogenase (IDH), Ethoxyresorufin-O-deethylase (EROD), GST) as biomarkers in muscle and liver of rainbow trout has been studied. An increase in AChE in muscle tissue and phase 1 and 2 enzymes, EROD and GST, in liver was observed. A similar increase in LDH activity, representative of anaerobic pathway induction, was noticed. All of these enzymes were upregulated to deal with the toxic response to anatoxin-a (Osswald et al., 2013). An analogous increase in AChE activity was observed in freshwater cladocerans following administration of *Anabaena spiroides* extract (Freitas et al., 2014).

Anatoxin-a(s)

Introduction

Anatoxin-a(s) is a natural organophosphate analog produced by cyanobacteria in the genus *Anabaena*. The (s) in the name of anatoxin-a(s) refers to salivation, which is a characteristic sign of poisoning observed in laboratory rodents following exposure. Anatoxin-a(s), although less frequently reported, has been detected in animal intoxications in Europe and the United States.

Chemistry

Structurally, anatoxin-a(s) is a cyclic N-hydroxyguanine with a phosphate ester moiety (Fig. 35.4) similar to typical synthetic organophosphate poisons. No structural variants have been described. Anatoxin-a(s) is rather unstable compared with other cyanotoxins (Matsunaga et al., 1989). Anatoxin-a(s) biosynthesis is not completely understood. Arginine is identified as a guanidine group precursor and erythro-4-hydroxyarginine as an intermediate product (Hemscheidt et al., 1995).

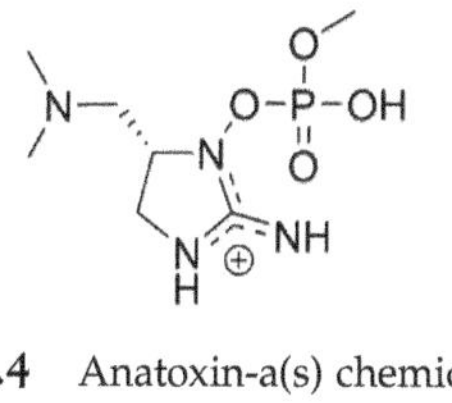

FIGURE 35.4 Anatoxin-a(s) chemical structure.

Toxic Effects

Anatoxin-a(s) poisoning is characterized by severe cholinergic stimulation of both nicotinic and muscarinic receptors. Clinical signs are indistinguishable from organophosphate poisoning. Muscarinic signs include salivation, lacrimation, urinary incontinence, and defecation, while nicotinic signs include muscle tremors, fasciculations, convulsions, and respiratory failure (Mahmood and Carmichael, 1986). The muscarinic effects of anatoxin-a(s) can be suppressed by atropine (Cook et al., 1990). Mice injected with *Anabaena flos-aquae* die because of respiratory arrest following convulsions within minutes of administration. Lowering of heart rate and blood pressure is also seen prior to decrease in respiratory function. As the AChE inhibition is irreversible, surviving mice show inhibition of erythrocyte cholinesterase for at least 8 days, indicating prolonged cholinesterase effects at sublethal doses (Cook et al., 1991).

Mechanism of Action

Similar to other organophosphate insecticides, anatoxin-a(s) is activated via oxidative metabolism and binds to AChE. It is a noncompetitive and irreversible inhibitor and, unlike the organophosphates, acts only on the peripheral nervous system in humans (Carmichael and Falconer, 1993). AChE is necessary for the inactivation of acetylcholine at the synaptic cleft hydrolysis. Inhibition of the enzyme therefore causes acetylcholine activity to build up at receptor sites, triggering excessive nicotinic and muscarinic receptor stimulation (Cook et al., 1990). In addition, as anatoxin-a(s) is a direct agonist at muscarinic receptors with indirect neuromuscular block, its effect can be temporarily blocked by atropine (Cook et al., 1991).

Biomarkers

As with other organophosphate poisonings, suppression of AChE in the brain and blood serves as a biomarker of exposure to anatoxin-a(s) (Freitas et al., 2016). No characteristic changes in blood chemistry and no gross or histological lesions have been described in mammals. Mouse bioassays may be used for anatoxin-a(s), though it has been difficult to establish adequate analytical methods such as LC-MS to determine environmental toxin levels due to a lack of stable and commercially available toxin standards.

Saxitoxins

Introduction

Saxitoxins are also known as paralytic shellfish poisons. Most human saxitoxin toxicoses have been associated with the ingestion of marine shellfish, which accumulate the saxitoxins produced by marine dinoflagellates. However, saxitoxins are also found in freshwaters, produced by cyanobacteria in the genera *Anabaena, Aphanizomenon, Planktothrix, Cylindrospermopsis, Lyngbya*, and *Scytonema* (Wiese et al., 2010).

Chemistry

Saxitoxins are highly polar, nonvolatile, tricyclic perhydropurine alkaloids (Fig. 35.5) derived from imidazoline guanidinium. Various structural substitutions produce at least 57 analog. Activity is mediated through positively charged guanidinium groups. Saxitoxins are heat stable and water soluble. They are tasteless and odorless and are not destroyed by cooking (Wiese et al., 2010). Saxitoxins can bioaccumulate in freshwater fish such as *tilapia* (Galvao et al., 2009).

Toxic Effects

The clinical presentation of saxitoxin poisoning varies depending on the level of exposure. The lag time between exposure and the appearance of clinical signs is highly variable and can range from minutes to as long as 72 h. At relatively low exposure levels, moderate paresthesias, often described as a tingling sensation, are experienced around the mouth and extremities. Larger exposures lead to a spreading numbness of the mouth, throat, and extremities. High exposure levels may also cause acute muscle paralysis and respiratory failure (Garcia et al., 2004).

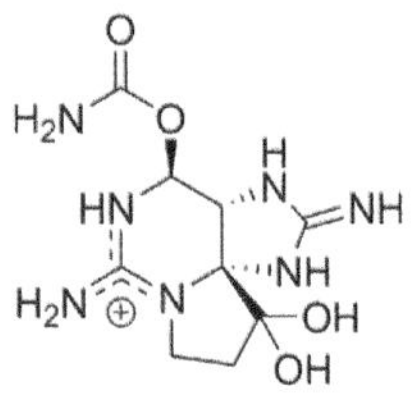

FIGURE 35.5 Saxitoxin chemical structure.

Mechanism of Action

Saxitoxins are selective but reversible blockers of voltage-gated sodium channel activity. They can cross the BBB and thus cause sodium channel blockage in the central nervous system, contributing to paralytic effects (Wiese et al., 2010). As these channels are important for propagation of action potentials in excitatory cells, binding of saxitoxin prevents ion flow (e.g., potassium) (Wang et al., 2003). Saxitoxin also blocks L-type calcium channels especially in the heart, resulting in early symptoms of a tingling sensation followed by paralysis and death by suffocation or cardiac arrest (Su et al., 2004).

Biomarkers

Exposure to saxitoxins is associated with ROS generation that can result in the activation of biomarkers of oxidative damage in fish, such as increased activities of GSH peroxidase, altered concentrations of lipid peroxidase, reduced GSH, and decreased activity of superoxide dismutase (Silva de Assis et al., 2013). Neuro-2A, a neuronal cell line-based bioassays, have attempted to provide reliable measurement of saxitoxin cytotoxicity using cell viability as a marker, although reliable indicators could not be observed with algal bioassays based on *Chlamydomonas reinhardtii*, which were tested in parallel (Perreault et al., 2011). However, as cytotoxicity is based on a myriad of factors, attempts have been made to determine functional endpoints. Microarray analysis did not produce gene expression as an important biomarker (Nicolas et al., 2015). A subsequent study to determine novel biomarkers in lower eukaryotic organisms identified multiple genes involved in copper and iron homeostasis and sulfur metabolism using quantitative real-time polymerase chain reaction. Generated expression profiles showed that genes tend to respond in a consistent manner. Important genes that undergo regulation after saxitoxin exposure include metallothioneins CUP1 and CRS5, ferric/cupric reductase FRE1, and the copper uptake transporter CTR1 (Cusick et al., 2009).

CONCLUDING REMARKS AND FUTURE DIRECTIONS

The hepatic and neurotoxic effects of common cyanotoxins are well understood down to the biochemical level with the exception of cylindrospermopsin. When new biomarkers for toxicity have emerged, the selection and interpretation of biomarkers presents a complex problem. This is partly because of the fact that some cyanotoxins, particularly anatoxin-a, have a mechanism of action that does not result in easily determinable biomarkers. In addition, natural exposures to cyanotoxins as algal extracts under laboratory conditions involve complex mixtures of biologically active compounds including, but not limited to, the cyanotoxins of primary interest. The toxicity of complex mixtures is variable and difficult to predict. Without well-documented models of complex mixture effects, the theoretical basis for predicting toxic effects and biomarker behavior is equally unclear. Increasing our understanding of the mixture effects created by co-exposure to multiple toxicants in association with cyanotoxins is therefore an important requirement. Finally, as most of the literature available is either from *in vitro* model systems or animal models, careful risk assessment and extrapolation to human disease form is mandatory.

References

Acuna, S., Deng, D.F., Lehman, P., et al., 2012. Sublethal dietary effects of *Microcystis* on *Sacramento splittail, Pogonichthys macrolepidotus*. Aquat. Toxicol. 110–111, 1–8.

Alja, S., Filipic, M., Novak, M., et al., 2013. Double strand breaks and cell-cycle arrest induced by the cyanobacterial toxin cylindrospermopsin in HepG2 cells. Mar. Drugs 11, 3077–3090.

Bain, P., Shaw, G., Patel, B., 2007. Induction of p53-regulated gene expression in human cell lines exposed to the cyanobacterial toxin cylindrospermopsin. J. Toxicol. Environ. Health 70, 1687–1693.

Banker, R., Carmeli, S., Werman, M., et al., 2001. Uracil moiety is required for toxicity of the cyanobacterial hepatotoxin cylindrospermopsin. J. Toxicol. Environ. Health 62, 281–288.

Bazin, E., Mourot, A., Humpage, A.R., et al., 2010. Genotoxicity of a freshwater cyanotoxin, cylindrospermopsin, in two human cell lines: Caco-2 and HepaRG. Environ. Mol. Mutagen. 51, 251–259.

Billam, M., Mukhi, S., Tang, L., et al., 2008. Toxic response indicators of microcystin-LR in F344 rats following a single-dose treatment. Toxicon 51, 1068–1080.

Bourke, A.T.C., Hawes, R.B., Neilson, A., et al., 1983. An outbreak of hepato-enteritis (the Palm Island mystery disease) possibly caused by algal intoxication. Toxicon 21, 45–48.

Briand, J.F., Jacquet, S., Bernard, C., et al., 2003. Health hazards for terrestrial vertebrates from toxic cyanobacteria in surface water ecosystems. Vet. Res. 34, 361–377.

Buratti, F.M., Manganelli, M., Vichi, S., et al., 2017. Cyanotoxins: producing organisms, occurrence, toxicity, mechanism of action and human health toxicological risk evaluation. Arch. Toxicol. 91, 1049–1130.

Carmichael, W.W., Biggs, D.F., 1978. Muscle sensitivity differences in two avian species to anatoxin-a produced by the freshwater cyanophyte Anabaena flos-aquae NRC-44-1. Can. J. Zool. 56, 510–512.

Carmichael, W.W., Falconer, I.R., 1993. Diseases related to freshwater blue-green algal toxins, and control measures. In: Algal Toxins in Seafood and Drinking Water. Academic Press, San Diego, pp. 187–209 (Chapter 12).

Chen, L., Zhang, X., Zhou, W., et al., 2013a. The interactive effects of cytoskeleton disruption and mitochondria dysfunction lead to reproductive toxicity induced by microcystin-LR. PLoS One 8, e53949.

Chen, Y., Zhou, Y., Wang, X., et al., 2013b. Microcystin-LR induces autophagy and apoptosis in rat Sertoli cells in vitro. Toxicon 76, 84–93.

Cook, W.O., Dahlem, A.M., Harlin, K.S., et al., 1991. Reversal of cholinesterase inhibition and clinical signs and the postmortem findings in mice after intraperitoneal administration of anatoxin-a(s), paraoxon or pyridostigmine. Vet. Hum. Toxicol. 33, 1–4.

Cook, W.O., Iwamoto, G.A., Schaeffer, D.J., et al., 1990. Pathophysiologic effects of anatoxin-a(s) in anaesthetized rats: the influence of atropine and artificial respiration. Pharmacol. Toxicol. 67, 151–155.

Cusick, K.D., Boyer, G.L., Wilhelm, S.W., et al., 2009. Transcriptional profiling of *Saccharomyces cerevisiae* upon exposure to saxitoxin. Environ. Sci. Technol. 43, 6039–6045.

Dai, R., Wang, P., Jia, P., et al., 2016. A review on factors affecting microcystins production by algae in aquatic environments. World J. Microbiol. Biotechnol. 32, 51.

Deng, D.F., Zheng, K., Teh, F.C., et al., 2010. Toxic threshold of dietary microcystin (-LR) for quart medaka. Toxicon 55, 787–794.

Ding, W.X., Shen, H.M., Ong, C.N., 2001. Pivotal role of mitochondrial Ca(2+) in microcystin-induced mitochondrial permeability transition in rat hepatocytes. Biochem. Biophys. Res. Commun. 285, 1155–1161.

Ding, X.S., Li, X.Y., Duan, H.Y., et al., 2006. Toxic effects of Microcystis cell extracts on the reproductive system of male mice. Toxicon 48, 973–979.

Falconer, I.R., 2007. Cyanobacterial toxins present in *Microcystis aeruginosa* extracts—more than microcystins! Toxicon 50, 585–588.

Feng, G., Abdalla, M., Li, Y., et al., 2011. NF-kappaB mediates the induction of Fas receptor and Fas ligand by microcystin-LR in HepG2 cells. Mol. Cell. Biochem. 352, 209–219.

Feurstein, D., Holst, K., Fischer, A., et al., 2009. Oatp-associated uptake and toxicity of microcystins in primary murine whole brain cells. Toxicol. Appl. Pharmacol. 234, 247–255.

Fischer, W.J., Altheimer, S., Cattori, V., et al., 2005. Organic anion transporting polypeptides expressed in liver and brain mediate uptake of microcystin. Toxicol. Appl. Pharmacol. 203, 257–263.

Florczyk, M., Brzuzan, P., Krom, J., et al., 2016. miR-122-5p as a plasma biomarker of liver injury in fish exposed to microcystin-LR. J. Fish Dis. 39, 741–751.

Freitas, E.C., Printes, L.B., Rocha, O., 2014. Acute effects of *Anabaena spiroides* extract and paraoxon-methyl on freshwater cladocerans from tropical and temperate regions: links between the ChE activity and survival and its implications for tropical ecotoxicological studies. Aquat. Toxicol. 146, 105–114.

Freitas, E.C., Printes, L.B., Rocha, O., 2016. Use of cholinesterase activity as an ecotoxicological marker to assess anatoxin-a(s) exposure: responses of two cladoceran species belonging to contrasting geographical regions. Harmful Algae 55, 150–162.

Froscio, S.M., Humpage, A.R., Burcham, P.C., et al., 2003. Cylindrospermopsin-induced protein synthesis inhibition and its dissociation from acute toxicity in mouse hepatocytes. Environ. Toxicol. 18, 243–251.

Galvao, J.A., Oetterer, M., Bittencourt-Oliveira Mdo, C., et al., 2009. Saxitoxins accumulation by freshwater tilapia (*Oreochromis niloticus*) for human consumption. Toxicon 54, 891–894.

Garcia, C., del Carmen Bravo, M., Lagos, M., et al., 2004. Paralytic shellfish poisoning: post-mortem analysis of tissue and body fluid samples from human victims in the Patagonia fjords. Toxicon 43, 149–158.

Goldberg, J., Huang, H.B., Kwon, Y.G., et al., 1995. Three-dimensional structure of the catalytic subunit of protein serine/threonine phosphatase-1. Nature 376, 745–753.

Griffiths, D.J., Saker, M.L., 2003. The Palm Island mystery disease 20 years on: a review of research on the cyanotoxin cylindrospermopsin. Environ. Toxicol. 18, 78–93.

Gutierrez-Praena, D., Jos, A., Pichardo, S., et al., 2013. Presence and bioaccumulation of microcystins and cylindrospermopsin in food and the effectiveness of some cooking techniques at decreasing their concentrations: a review. Food Chem. Toxicol. 53, 139–152.

Guzman-Guillen, R., Prieto, A.I., Moreno, I., et al., 2014. Effects of depuration on oxidative biomarkers in tilapia (*Oreochromis niloticus*) after subchronic exposure to cyanobacterium producing cylindrospermopsin. Aquat. Toxicol. 149, 40–49.

Guzman-Guillen, R., Prieto, A.I., Vasconcelos, V.M., et al., 2013a. Cyanobacterium producing cylindrospermopsin cause oxidative stress at environmentally relevant concentrations in sub-chronically exposed tilapia (*Oreochromis niloticus*). Chemosphere 90, 1184–1194.

Guzman-Guillen, R., Prieto, A.I., Vazquez, C.M., et al., 2013b. The protective role of l-carnitine against cylindrospermopsin-induced oxidative stress in tilapia (*Oreochromis niloticus*). Aquat. Toxicol. 132–133, 141–150.

Hagenbuch, B., Meier, P.J., 2004. Organic anion transporting polypeptides of the OATP/SLC21 family: phylogenetic classification as OATP/SLCO superfamily, new nomenclature and molecular/functional properties. Pflugers Arch. 447, 653–665.

Harada, K.I., Tsuji, K., Watanabe, M.F., Kondo, F., 1996. Stability of microcystins from cyanobacteria-III. Effect of pH and temperature. Phycologia 35, 83–88.

Hayman, J., 1992. Beyond the Barcoo-probable human tropical cyanobacterial poisoning in outback Australia. Med. J. Aust. 157, 794–796.

Hemscheidt, T., Burgoyne, D.L., Moore, R.E., 1995. Biosynthesis of anatoxin-a(s). (2S,4S)-4-hydroxyarginine as an intermediate. J. Chem. Soc. Chem. Commun. 2, 205–206.

Huang, X., Chen, L., Liu, W., et al., 2015. Involvement of oxidative stress and cytoskeletal disruption in microcystin-induced apoptosis in CIK cells. Aquat. Toxicol. 165, 41–50.

Huguet, A., Hatton, A., Villot, R., et al., 2014. Modulation of chromatin remodelling induced by the freshwater cyanotoxin cylindrospermopsin in human intestinal caco-2 cells. PLoS One 9, e99121.

Humpage, A.R., Falconer, I.R., 2003. Oral toxicity of the cyanobacterial toxin cylindrospermopsin in male Swiss albino mice: determination of no observed adverse effect level for deriving a drinking water guideline value. Environ. Toxicol. 18, 94–103.

Humpage, A.R., Fontaine, F., Froscio, S., et al., 2005. Cylindrospermopsin genotoxicity and cytotoxicity: role of cytochrome P-450 and oxidative stress. J. Toxicol. Environ. Health 68, 739–753.

Ibelings, B.W., Havens, K.E., 2008. Cyanobacterial toxins: a qualitative meta-analysis of concentrations, dosage and effects in freshwater, estuarine and marine biota. Adv. Exp. Med. Biol. 619, 675–732.

Ji, Y., Lu, G., Chen, G., et al., 2011. Microcystin-LR induces apoptosis via NF-kappaB/iNOS pathway in INS-1 cells. Int. J. Mol. Sci. 12, 4722–4734.

Jochimsen, E.M., Carmichael, W.W., An, J.S., et al., 1998. Liver failure and death after exposure to microcystins at a hemodialysis center in Brazil. N. Engl. J. Med. 338, 873–878.

Kinnear, S., 2010. Cylindrospermopsin: a decade of progress on bioaccumulation research. Mar. Drugs 8, 542–564.

Komatsu, M., Furukawa, T., Ikeda, R., et al., 2007. Involvement of mitogen-activated protein kinase signaling pathways in microcystin-LR-induced apoptosis after its selective uptake mediated by OATP1B1 and OATP1B3. Toxicol. Sci. 97, 407–416.

Kondo, F., Ikai, Y., Oka, H., et al., 1992. Formation, characterization, and toxicity of the glutathione and cysteine conjugates of toxic heptapeptide microcystins. Chem. Res. Toxicol. 5, 591–596.

Kozdeba, M., Borowczyk, J., Zimolag, E., et al., 2014. Microcystin-LR affects properties of human epidermal skin cells crucial for regenerative processes. Toxicon 80, 38–46.

Li, T., Ying, L., Wang, H., et al., 2012. Microcystin-LR induces ceramide to regulate PP2A and destabilize cytoskeleton in HEK293 cells. Toxicol. Sci. 128, 147–157.

Li, Y., Sheng, J., Sha, J., et al., 2008. The toxic effects of microcystin-LR on the reproductive system of male rats in vivo and in vitro. Reprod. Toxicol. 26, 239–245.

Liu, H., Zhang, H., Niedzwiedzki, D.M., et al., 2013. Phycobilisomes supply excitations to both photosystems in a megacomplex in cyanobacteria. Science 342, 1104–1107.

Liu, W., Chen, C., Chen, L., et al., 2016. Sex-dependent effects of microcystin-LR on hypothalamic-pituitary-gonad axis and gametogenesis of adult zebrafish. Sci. Rep. 6, 22819.

Lone, Y., Bhide, M., Koiri, R.K., 2017. Amelioratory effect of coenzyme Q10 on potential human carcinogen microcystin-LR induced toxicity in mice. Food Chem. Toxicol. 102, 176–185.

Lopez-Alonso, H., Rubiolo, J.A., Vega, F., et al., 2013. Protein synthesis inhibition and oxidative stress induced by cylindrospermopsin elicit apoptosis in primary rat hepatocytes. Chem. Res. Toxicol. 26, 203–212.

Lu, X., Long, Y., Sun, R., et al., 2015. Zebrafish Abcb4 is a potential efflux transporter of microcystin-LR. Comp. Biochem. Physiol. C Toxicol. Pharmacol. 167, 35–42.

Lyu, K., Gu, L., Li, B., et al., 2016. Stress-responsive expression of a glutathione S-transferase (delta) gene in waterflea *Daphnia magna* challenged by microcystin-producing and microcystin-free *Microcystis aeruginosa*. Harmful Algae 56, 1–8.

Maatouk, I., Bouaicha, N., Plessis, M.J., et al., 2004. Detection by 32P-postlabelling of 8-oxo-7,8-dihydro-2′-deoxyguanosine in DNA as biomarker of microcystin-LR- and nodularin-induced DNA damage in vitro in primary cultured rat hepatocytes and in vivo in rat liver. Mutat. Res. 564, 9–20.

Mahmood, N.A., Carmichael, W.W., 1986. The pharmacology of anatoxin-a(s), a neurotoxin produced by the freshwater cyanobacterium Anabaena flos-aquae NRC 525-17. Toxicon 24, 425–434.

Malbrouck, C., Kestemont, P., 2006. Effects of microcystins on fish. Environ. Toxicol. Chem. 25, 72–86.

Matsunaga, S., Moore, R.E., Niemczura, W.P., et al., 1989. Anatoxin-a(s), a potent anticholinesterase from Anabaena flos-aquae. J. Am. Chem. Soc. 111, 8021–8023.

Meng, G., Liu, J., Lin, S., et al., 2015. Microcystin-LR-caused ROS generation involved in p38 activation and tau hyperphosphorylation in neuroendocrine (PC12) cells. Environ. Toxicol. 30, 366–374.

Mikhailov, A., Harmala-Brasken, A.S., Hellman, J., et al., 2003. Identification of ATP-synthase as a novel intracellular target for microcystin-LR. Chem. Biol. Interact. 142, 223–237.

Nair, S.S., Leitch, J., Falconer, J., et al., 1999. Cardiac (n-3) non-esterified fatty acids are selectively increased in fish oil-fed pigs following myocardial ischemia. J. Nutr. 129, 1518–1523.

Nicolas, J., Bovee, T.F., Kamelia, L., et al., 2015. Exploration of new functional endpoints in neuro-2a cells for the detection of the marine biotoxins saxitoxin, palytoxin and tetrodotoxin. Toxicol. In Vitro 30, 341–347.

Norris, R.L., Eaglesham, G.K., Pierens, G., et al., 1999. Deoxycylindrospermopsin, an analog of cylindrospermopsin from *Cylindrospermopsis raciborskii*. Environ. Toxicol. 14, 163–165.

Olivares-Rubio, H.F., Martinez-Torres, M.L., Najera-Martinez, M., et al., 2015. Biomarkers involved in energy metabolism and oxidative stress response in the liver of *Goodea gracilis* Hubbs and Turner, 1939 exposed to the microcystin-producing *Microcystis aeruginosa* LB85 strain. Environ. Toxicol. 30, 1113–1124.

Osswald, J., Carvalho, A.P., Guimaraes, L., et al., 2013. Toxic effects of pure anatoxin-a on biomarkers of rainbow trout, *Oncorhynchus mykiss*. Toxicon 70, 162–169.

Osswald, J., Rellan, S., Gago, A., et al., 2007. Toxicology and detection methods of the alkaloid neurotoxin produced by cyanobacteria, anatoxin-a. Environ. Int. 33, 1070–1089.

Perreault, F., Matias, M.S., Melegari, S.P., et al., 2011. Investigation of animal and algal bioassays for reliable saxitoxin ecotoxicity and cytotoxicity risk evaluation. Ecotoxicol. Environ. Saf. 74, 1021–1026.

Pflugmacher, S., Wiegand, C., Oberemm, A., et al., 1998. Identification of an enzymatically formed glutathione conjugate of the cyanobacterial hepatotoxin microcystin-LR: the first step of detoxication. Biochim. Biophys. Acta 1425, 527–533.

Puschner, B., Humbert, J.-F., 2007. Cyanobacterial (blue-green algae) toxins. In: Gupta, R.C. (Ed.), Veterinary Toxicology: Basic and Clinical Principles. Academic Press, Elsevier, Amsterdam, pp. 714–724 (Chapter 59).

Rios, V., Prieto, A.I., Camean, A.M., et al., 2014. Detection of cylindrospermopsin toxin markers in cyanobacterial algal blooms using analytical pyrolysis (Py-GC/MS) and thermally-assisted hydrolysis and methylation (TCh-GC/MS). Chemosphere 108, 175–182.

Rucker, J., Stuken, A., Nixdorf, B., et al., 2007. Concentrations of particulate and dissolved cylindrospermopsin in 21 Aphanizomenon-dominated temperate lakes. Toxicon 50, 800–809.

Runnegar, M.T., Kong, S.M., Zhong, Y.Z., et al., 1994. The role of glutathione in the toxicity of a novel cyanobacterial alkaloid cylindrospermopsin in cultured rat hepatocytes. Biochem. Biophys. Res. Commun. 201, 235–241.

Rymuszka, A., Adaszek, L., 2013. Cytotoxic effects and changes in cytokine gene expression induced by microcystin-containing extract in fish immune cells—an in vitro and in vivo study. Fish Shellfish Immunol. 34, 1524–1532.

Sahin, A., Tencalla, F.G., Dietrich, D.R., et al., 1996. Biliary excretion of biochemically active cyanobacteria (blue-green algae) hepatotoxins in fish. Toxicology 106, 123–130.

Sedan, D., Giannuzzi, L., Rosso, L., et al., 2013. Biomarkers of prolonged exposure to microcystin-LR in mice. Toxicon 68, 9–17.

Shaw, G.R., Seawright, A.A., Moore, M.R., et al., 2000. Cylindrospermopsin, a cyanobacterial alkaloid: evaluation of its toxicologic activity. Ther. Drug Monit. 22, 89–92.

Silva de Assis, H.C., da Silva, C.A., Oba, E.T., et al., 2013. Hematologic and hepatic responses of the freshwater fish *Hoplias malabaricus* after saxitoxin exposure. Toxicon 66, 25–30.

Sirén, A.-L., Feurstein, G., 1990. Cardiovascular effects of anatoxin-A in the conscious rat. Toxicol. Appl. Pharmacol. 102, 91–100.

Skulberg, O.M., Skulberg, R., Carmichael, W.W., et al., 1992. Investigations of a neurotoxic oscillatorialean strain (Cyanophyceae) and its toxin. Isolation and characterization of homoanatoxin-a. Environ. Toxicol. Chem. 11, 321–329.

Smith, C., Sutton, A., 1993. The Persistence of Anatoxin-a in Reservoir Water. UK Report No. FR0427. Foundation for Water Research, Buckinghamshire, UK.

Su, Z., Sheets, M., Ishida, H., et al., 2004. Saxitoxin blocks L-type I_{Ca}. J. Pharmacol. Exp. Ther. 308, 324–329.

Tencalla, F., Dietrich, D., 1997. Biochemical characterization of microcystin toxicity in rainbow trout (*Oncorhynchus mykiss*). Toxicon 35, 583–595.

Terao, K., Ohmori, S., Igarashi, K., et al., 1994. Electron microscopic studies on experimental poisoning in mice induced by cylindrospermopsin isolated from blue-green alga *Umezakia natans*. Toxicon 32, 833–843.

Valerio, E., Campos, A., Osorio, H., et al., 2016a. Proteomic and real-time PCR analyses of *Saccharomyces cerevisiae* VL3 exposed to microcystin-LR reveals a set of protein alterations transversal to several eukaryotic models. Toxicon 112, 22–28.

Valerio, E., Vasconcelos, V., Campos, A., 2016b. New insights on the mode of action of microcystins in animal cells – a review. Mini Rev. Med. Chem. 16, 1032–1041.

van der Merwe, D., Sebbag, L., Nietfeld, J.C., et al., 2012. Investigation of a *Microcystis aeruginosa* cyanobacterial freshwater harmful algal bloom associated with acute microcystin toxicosis in a dog. J. Vet. Diagn. Invest. 24, 679–687.

Wang, J., Salata, J.J., Bennett, P.B., 2003. Saxitoxin is a gating modifier of HERG K+ channels. J. Gen. Physiol. 121, 583–598.

Wiese, M., D'Agostino, P.M., Mihali, T.K., et al., 2010. Neurotoxic alkaloids: saxitoxin and its analogs. Mar. Drugs 8, 2185–2211.

Wimmer, K.M., Strangman, W.K., Wright, J.L.C., 2014. 7-Deoxy-desulfo-cylindrospermopsin and 7-deoxy-desulfo-12-acetylcylindrospermopsin: two new cylindrospermopsin analogs isolated from a Thai strain of *Cylindrospermopsis raciborskii*. Harmful Algae 37, 203–206.

Wonnacott, S., Gallagher, T., 2006. The chemistry and pharmacology of anatoxin-a and related homotropanes with respect to nicotinic acetylcholine receptors. Mar. Drugs 4, 228–254.

Wu, J., Shao, S., Zhou, F., et al., 2014. Reproductive toxicity on female mice induced by microcystin-LR. Environ. Toxicol. Pharmacol. 37, 1–6.

Xing, Y., Xu, Y., Chen, Y., et al., 2006. Structure of protein phosphatase 2A core enzyme bound to tumor-inducing toxins. Cell 127, 341–353.

Zeng, J., Tu, W.W., Lazar, L., et al., 2015. Hyperphosphorylation of microfilament-associated proteins is involved in microcystin-LR-induced toxicity in HL7702 cells. Environ. Toxicol. 30, 981–988.

Zhao, Y., Xue, Q., Su, X., et al., 2015. Microcystin-LR induced thyroid dysfunction and metabolic disorders in mice. Toxicology 328, 135–141.

CHAPTER

36

Mycotoxins

Robert W. Coppock[1], *Margitta M. Dziwenka*[2]

[1]Toxicologist and Associates, Ltd, Vegreville, AB, Canada; [2]Faculty of Medicine & Dentistry, Division of Health Sciences Laboratory Animal Services, Edmonton, AB, Canada

INTRODUCTION

The term mycotoxin is generally restricted to chemicals synthesized by filamentous fungi. These chemicals, in concentrations that occur in nature, are toxic to vertebrates and other animals. Mycotoxins have variability in resisting degradation during the various heat- and fermentation-linked manufacturing processes. Milling of cereal grains can concentrate mycotoxins in the various fractions produced during milling operations. Mycotoxins can be present in raw commodities, processed raw ingredients, finished foods and feeds, and milling by-products. Animal source human foods can also contain mycotoxins. Any feedstuffs or foodstuffs that have supported the growth of toxigenic fungi can be contaminated with mycotoxins and they can be heterogeneously distributed in foodstuffs and feedstuffs. The FAO (2008) has estimated that 25% of the world food supply is contaminated with mycotoxins, and contamination may be substantially increased by regional adverse weather conditions and agricultural practices. Climate change is predicted to likely increase contamination of human and animal foods with mycotoxins (Medina et al., 2017). It is estimated that 5 billion people a year are exposed to mycotoxins (Pizzolato Montanha et al., 2018). When one mycotoxin is present, the risk of other mycotoxins being present is also increased (Lee and Ryu, 2017). A private laboratory reported that 72% of 17,316 samples were contaminated with one mycotoxin and 38% of the contaminated samples contained more than one mycotoxin (Streit et al., 2013). Foods can be contaminated by a succession of fungi that can produce multiple mycotoxins (Coppock and Jacobsen, 2009). In fact, the majority of mycotoxicoses linked to foodstuffs, that have been investigated in detail, predominantly show that the individual or animal was exposed to a mixture of mycotoxins. The most important recognized toxigenic fungi are in the genera *Aspergillus, Penicillium,* and *Fusarium* (Coppock and Jacobsen, 2009). The population structure of microorganisms in foodstuffs and feedstuffs is dependent on abiotic conditions of available water, temperature, and gas composition. Growth of pests and microorganisms can dynamically change these biotic conditions (Coppock and Jacobsen, 2009). For commodities, pest management before and after harvest, kernel damage during threshing and handling operations, and the presence of foreign material and debris in storage all contribute to increased microbial growth. Humans and animals are exposed to mycotoxins by the oral, inhalation, and dermal routes. Of these routes of exposure, oral and inhalation are the most important. Mycotoxins have diverse chemical structures and mechanisms of intoxication. All organ systems in the body can be targets for mycotoxins.

Biomarkers can be strategic tools to show exposure and adverse effects of exposure to mycotoxins. Biomarkers enable qualitative and quantitative assessment of exposure, assessment of internal dose, assessment of toxicological effects, and assessment of susceptibility. Proof that a diet containing mycotoxins has been consumed is proof of oral exposure, and quantitation of the mycotoxin(s) in foodstuff(s) provides input data to estimate the dose of mycotoxin(s) entering the body. Showing that ambient air is contaminated with mycotoxins and the proof that the contaminated air has been inhaled demonstrate exposure to mycotoxins. Documenting exposure to a mycotoxin can also be accomplished by assaying bodily tissues and fluids for the parent mycotoxin and its metabolites. The levels of mycotoxins and their metabolites inside the body are referred to as the internal dose. Biomarkers of the toxic effect can be specific adducts or some other chemical product showing upregulation or downregulation of critical biochemical chemical products or alteration of an enzyme system. The biological response to the

Biomarkers in Toxicology, Second Edition
https://doi.org/10.1016/B978-0-12-814655-2.00036-0

mycotoxin can be determined by histopathological procedures, clinicopathologic parameters, biochemical and clinical observed impairment of bodily function, changes in metabolic and system regulatory profiles, and medical observations that define a syndrome for a particular mycotoxicosis. Biomarkers of susceptibility can also exist and generally these are differences in enzyme activities in biotransformation pathways that exist between individuals, races, and species. If the excretion of the parent mycotoxin, its metabolites in body fluids including bile, etc., or levels in serum have been shown to be temporally correlated with external dose, these parameters can be used to estimate dietary levels. Internal dose can be correlated with adverse biological effects, adverse changes in metabolic profiles, endocrine functions and cell regulatory profiles, histopathology, etc., and can be used to estimate the damage that is occurring from exposure to mycotoxins. Levels of the parent mycotoxin or its metabolites in tissues document exposure.

AFLATOXINS

Background

It is estimated that at least 4.5 billion people worldwide are at risk of exposure to foodstuffs containing aflatoxins (AFs). Exposure to AFs is by the oral, inhalation, and dermal routes. The most commonly recognized aflatoxigenic fungi are *Aspergillus flavus*, *Aspergillus parasiticus*, and *Aspergillus nomius*, but other genera and species have been identified. These fungal species are also capable of forming other mycotoxins. For example, *A. flavus* also produces the mycotoxins cyclopiazonic acid, aspergillic acid, asperfuran, paspalinine, paspaline, and sterigmatocystin (STC). The general growth conditions for aflatoxigenic species require moisture contents in equilibrium at 80%–85% or more and temperatures of 13–42°C with optimum growth at 25–37°C (Coppock et al., 2018). The most common commodities contaminated with AFs are cereal grains, peanuts, nuts, oil seeds, spices, and condiments. Workers in feed mills, food processing facilities, poultry houses, grain storage and transportation operations, and textile works can be exposed to airborne AFs. Consumption of feedstuffs containing AFs can result in AF residues in edible animal products, including milk (Coppock et al., 2018). Dietary intake of AFs results in breast milk being contaminated with AFs, and for AFs and ochratoxin A (OTA), the milk/plasma ratio changes with the stage of lactation (Warth et al., 2016). The milk/plasma ratio of mycotoxins may increase because of the transfer of serum albumen in the mammary gland during early lactation. Breast milk can be considered the safer alternative for infants and young children because it contains a substantially less percentage of AFs than exist in available foodstuff consumed by the family. Essentially all fermented foods/feedstuffs and beverages can be contaminated with AFs if conditions favor the growth of aflatoxigenic fungi during fermentation and storage. AFs are classified as a human and animal carcinogen (Anon, 2012). It has been estimated that AFs are linked to 4.6%–28.2% of all global hepatocellular carcinomas, and at least 750,000 deaths from hepatocellular carcinomas occur each year (Livingstone et al., 2017). The AFs classified as carcinogens are AFB_1, AFG_1, and AFM_1 (Anon, 2012). Of these AFB_1 and AFM_1 are of the greatest concern for human exposure.

Chemistry and Legal Regulation of Aflatoxins

AFs have a difuranocoumarin chemical structure, and the primary AFs in order of decreasing toxicity are B_1, G_1, B_2, and G_2 (Coppock et al., 2018). AFs are heat stable and are not destroyed by cooking and processing of foodstuffs and feedstuffs. There are two chemical groups of AFs, the difurocoumarocyclopentenone series (denoted AFB_1, AFB_2, AFB_{2A}, AFM_1, AFM_2, AFM_{2A}, and aflatoxicol) and the difurocoumarolactone series (denoted AFG_1 and AFG_2). The "B" Group fluoresce blue in long-wavelength ultraviolet light and the "G" Group fluoresce green. The primary AFs of concern in feedstuffs are AFB_1, AFB_2, AFG_1, and AFG_2. Analytical results for AFs generally are the sum of the concentrations of AFB_1, AFB_2, AFG_1, and AFG_2. Generally, the levels of these four AFs are regulated under safe foods and feeds laws. AFB_1 is the most potent AF and this chemical form is generally the most abundant in feedstuffs and foodstuffs. Hydroxylated AF metabolites are excreted in milk and the important metabolites are AFM_1 and AFM_2 (Warth et al., 2016; Becker-Algeri et al., 2016). AFM_1 is the toxic metabolite of AFB_1, and AFM_2 is the hydroxylated form of AFB_2. Although AFM_1 and AFM_2 are commonly associated with milk and other edible animal products, aflatoxigenic fungi, when favorable growing conditions exist, can also produce these compounds in the processed and finished foods and feeds.

Toxicokinetics of Aflatoxins

The gastrointestinal tract in the rat rapidly absorbs AFs and there are data to support that other monogastric mammals have similar absorption kinetics. AFs are primarily found in portal blood with low levels in lymph drainage from the gastrointestinal tract (Coppock et al., 2018). Rats have been shown to absorb AFB_1 most efficiently from the duodenum and jejunum. The rate of AFB_1 absorption from the duodenum of rats in diestrus was greater than the rate of absorption of AFB_1 from the duodenum of rats in midlactation. High starch diet increases the absorption of AFB_1 in cattle (Pantaya et al., 2016). Young animals absorb AFs more efficiently than do older animals. Fasted human volunteers were each given low doses of ^{14}C-labeled AFB_1 (30 ng) in gelatin capsules (Jubert et al., 2009). The toxicokinetics fit a two-compartment model with peak plasma concentration occurring at approximately 1 h after administration. The first phase (α) half-life is 2.86 h and the β phase half-life is 64.5 h. The subjects excreted 29%–33% of the ^{14}C-AFB_1 dose within 72 h. AFs cross the placental barrier (Partanen et al., 2010). AFB_1 is absorbed through the rumen mucosa and is metabolized to AFM_1 in the rumen (Cook et al., 1986). Cattle were given a single oral dose of AFs from rice culture (42% AFB_1 and 27% AFB_2) in gelatin capsules. Thirty minutes later, AFB_1 and AFM_1 were observed in venous blood (jugular vein) and reached maximal levels at 4–8 h. AFs are distributed in the body and eliminated by first-order kinetics. The primary excretion route is bile, followed by urine, and parent AFs can be identified in urine. Milk is also an excretion route for AFs, and dietary AFB_1 to milk AFM_1 ranges from 0.09% to 0.43% (Warth et al., 2016; Becker-Algeri et al., 2016). The liver and kidney have the highest tissue concentrations of AFs. In a model using surgically obtained abdominal human skin, AFB_1 had a low level of absorption with a permeability coefficient of $Kp = 2.11 \times 10^{-4}$ cm/h (Boonen et al., 2012).

Aflatoxin Metabolism and Chemical Biomarkers of Aflatoxin Exposure

Differences in the biotransformation of AFs explain the majority of sensitivity observed between animal species and races and breeds within species (Dohnal et al., 2014). Diet also has an endogenous effect on the enzymes that biotransform AFs (Gross-Steinmeyer and Eaton, 2012). For AFB_1, the major biotransformation pathways are epoxidation, hydroxylation, O-dealkylation, and ketoreduction. The epoxidation toxication reaction for AFB_1 is the formation of the AFB_1-8,9-epoxide by the CYP1A2, CYP3A4, CYP3A5, and other CYP enzymes. AFs are not inducers of CYP enzymes, but dietary constituents, drugs, and other substances that upregulate or downregulate CYP1A2, CYP3A4, CYP3A5, and other CYP enzymes also increase or decrease the toxicity of AFB_1, respectively. The important detoxification route for AFB_1-8,9-epoxide is conjugation with glutathione and excretion as the AFB-mercapturic acid. Hydroxylation products of AFB_1 are AFM_1 and AFQ_1. The major metabolite from O-dealkylation is AFP_1 and ketoreduction is aflatoxicol. Of these biotransformation products, the AFB_1-8,9-epoxide is the most toxic followed by AFM_1. The AFB_1-8,9-epoxide binds to DNA to form the AFB-N7-guanine adduct (8,9-dihydro-8-(N7-guanyl)-9-hydroxyaflatoxin B_1). The AFB_1-N7-G adduct is unstable and rapidly depurinated and the AFB_1-N7-G is excreted in urine. The AFB_1-exo-8,9-epoxide is unstable and is hydrolyzed to AFB_1-8,9-dihydrodiol, which reacts with albumen to give the lysine adducts at Lys455 and Lys548 in bovine albumen and likely the same in human albumen (Sabbioni and Turesky, 2017). The AFB_1 DNA and albumen adducts plateaued in rats following 7 and 14 days of treatment, respectively. The half-life of the serum AFB_1-lysine adduct in rats is approximately 2.31 days and has first-order elimination kinetics (Qian et al., 2013). AFB_1 also forms stable noncovalent interactions with human, bovine, and rat serum albumen (Poor et al., 2017). Upregulation of hepatic miRNAs has been demonstrated in rats exposed to AFB_1 and maybe of initiation indicators of hepatocellular carcinoma type neoplasia (Livingstone et al., 2017). Upregulations observed by Livingstone et al. (2017) are miR-122-5p (5× increase), 34a-5p (13× increase), and 181c-3p (170× increase) and have biomarker potential. AFB_1 exposure causes a mutation in codon 249 (R249S) of TP53, and the R249S mutation of TP53 is a surrogate marker of AFB_1 exposure in humans with hepatocellular carcinoma (Nault, 2014). For dairy herds, the ratio of AFM_1 in milk to AFB_1 in diet can be used to predict reductions in AFM_1 when the diet is adjusted to reduce dietary levels (Coppock et al., 2018) Chemical biomarkers of AFs in humans are summarized in Table 36.1.

Histopathological Biomarkers of Exposure to Aflatoxins

AFs cause characteristic histopathology of the liver, and these lesions collectively are a biomarker of aflatoxicosis (Coppock et al., 2018). The lesions induced by AFs

TABLE 36.1 Biomarkers of Aflatoxin Exposure for Human Body Fluids

Urine	Blood	Milk
[a]AFB_1	[a]AFB_1	[a]AFB_1
[a]AFB_2	[a]AFB_2	[a]AFM_1
[a]AFG_1	[a]AFG_1	
AFG_2	AFG_2	
AFM_1	[a]AFM_1	
[a]AFQ_1	AFQ_1	
AFP_1	AFP_1	
[a]AFB_1-N^7-guanine	[b]AFB_1-albumen	
[a]Aflatoxin-mercapturic acid	[a]AFB_1-N^7-guanine	

[a]*Correlated with recent dietary intake.*
[b]*Indicative of dietary exposure and persists after exposure stops.*

vary with level and duration and level of exposure. The histopathology observed includes bile duct proliferation (hyperplasia), hepatocyte necrosis, and early fibrosis of the liver. The cytology of hepatocytes can be altered by abnormal variation in cell size by the formation of megalocytic and multinucleated cells. Species differences occur in the specific histopathology of AFs. Hyperplasia of the bile duct cells occurs rapidly in ducklings and may be present in horses, dogs, and chickens, and mild bile duct cell hyperplasia may be seen in cattle and pigs. The pattern of hepatocellular necrosis may vary with the species. In the rat, monkey, and duckling, the necrosis is reported to be periportal, whereas in cattle, pigs, horses, goats, and sheep, the pattern is reported to be centrilobular. In dogs, the lesion pattern may be either or both periportal or/and centrilobular, while in the rabbit the pattern is reported to be midzonal. In dogs, regeneration of hepatocytes may occur, and nodular hyperplasia may be present in turkeys, trout, and ducklings. Multinucleated hepatocytes have been observed in dogs, cattle, and other species. Reactive fibroblasts have been observed in dogs. Humans infected with hepatitis B virus appear to be at greater risk for hepatic neoplasia when they are exposed to AFs (Erkekoglu et al., 2017). Chronic liver injury and regenerative hyperplasia are considered critical for the development of liver neoplasia. Dividing liver cells is more likely to form DNA adducts.

STERIGMATOCYSTIN

STC is considered an emerging mycotoxin. The most common source of STC is *Aspergillus versicolor* (EFSA Panel, 2013). It is also produced by *A. flavus*, *A. parasiticus*, and *Aspergillus nidulans*, and in total, 55 toxigenic species are known to produce STC. STC is a precursor in the AFB_1 synthetic pathway, is structurally similar to AFB_1, and can also be produced as the end product by toxigenic fungi. The chemical name for STC is (3aR,12cS)-8-hydroxy-6-methoxy-3a,12c-dihydro-7H-furo[3′,2′:4,5]furo[2,3-c]xanthen-7-one. STC can exist as 1% of the biomass in *A. versicolor*. Microsomes form exo-STC-1,2-oxide that reacts with DNA to form the 1,2-dihydro-2-(N7-guanyl)-1-hydroxy STC adduct. STC is a carcinogen in rats and has been associated with stomach and liver cancer in humans. The glucuronide conjugate of STC is excreted in the urine of cattle consuming STC-contaminated rice straw supporting the growth of *Aspergillus niger*, and conjugated STC has been identified in the urine of cattle and horses consuming rice straw bedding (Fushimi et al., 2014; Takagi et al., 2018) Pathology of STC is hepatic necrosis, nuclear pleomorphism, multinucleated cells, and moderate biliary hyperplasia.

OCHRATOXINS

Production and Occurrence

OTA, B, and C are primarily produced by species of *Aspergillus* and *Penicillium* (Coppock and Jacobsen, 2009). The ochratoxins identified, in decreasing order of toxicity, are OTA, ochratoxin B, and ochratoxin C. Of these OTA is considered to be of clinical importance and is the most studied. For OTA production, generally, the *Penicillium* spp. are the most important toxigenic fungi in temperate climates and *Aspergillus* spp. are important in tropical climates (Wang et al., 2016). Chemically, ochratoxins consist of a polyketide-derived dihydroisocoumarin moiety linked at the 7-carboxy group to L-β-phenylalanine (Tao et al., 2018). The IUPAC nomenclature for OTA is (2S)-2-[[(3R)-5-chloro-8-hydroxy-3-methyl-1-oxo-3,4-dihydroisochromene-7-carbonyl]amino]-3-phenylpropanoic acid (NCBI, 2018). OTA can be produced in a variety of raw agricultural commodities, foodstuffs, fruits and nuts, raw materials used for beverages and spices, and finished beverages (FAO, 2008; Coppock and Jacobsen, 2009; Wang et al., 2016). OTA can also be present in meats as a residue from the food-producing animals being fed ochratoxin-contaminated feedstuffs or from fungal growth in processed and smoked meats and fish (Ringot et al., 2006; Pizzolato Montanha et al., 2018).

Toxicokinetics and Biotransformation of Ochratoxin A

OTA, in most species, is rapidly absorbed from the gastrointestinal tract and binds to serum proteins (Ringot et al., 2006). The pK_a for OTA is important for gastrointestinal absorption. The pK_a is 4.2–4.4 for the carboxyl group of the phenylalanine and 7.0–7.3 for the phenolic hydroxyl group of the isocoumarin moiety (Ringot et al., 2006). Passive diffusion, when in the nonionic and monoionic forms, is considered the predominant mechanism for absorption. OTA is fully protonated (nonionic) in acidic pH conditions. Likely due to decreasing ruminal pH, high starch diet increases the absorption of OTA in cattle (Pantaya et al., 2016). Diffusion of OTA, influenced by pH, is considered to be the primary mechanism for absorption in the stomach. Both diffusion and multidrug resistance-associated protein are likely important for absorption of OTA from the small intestine (Ringot et al., 2006; Vettorazzi et al., 2014). The efficiency of OTA absorption varies among species, and fish appear to have a low bioavailability of OTA (Bernhoft et al., 2017). OTA in ethanol had a high bioavailability in a human volunteer after oral administration on an empty stomach (Studer-Rohr et al., 2000).

Greater than 90% of the absorbed OTA in humans is bound to serum proteins, especially albumen and smaller proteins, and protein binding is almost identical to that of warfarin (Koszegi and Poor, 2016). The potential for protein binding is up to 99% and the nonsaturable range is 1–100 nmol. Protein binding of OTA varies among animals studied with fish, Atlantic salmon (*Salmo salar*), and carp (*Cyprinus carpio*), having the lowest level at 22% OTA bound to serum proteins (Bernhoft et al., 2017). The binding of OTA to plasma proteins in pigs, cattle, and humans is generally considered to be high. Of the species studied, humans have the longest serum half-life for OTA, estimated at 23.5–35.5 days, and carp the shortest at 0.68 h; the half-life of OTA in the serum or plasma of pigs and calves is estimated at 72 and 77.3 h, respectively (Sreemannarayana et al., 1988; Studer-Rohr et al., 2000; Ringot et al., 2006).

Human plasmas or sera levels of OTA have been reviewed and the levels vary for different geographic areas (Soto et al., 2016). There is some suggestion that OTA may be higher in the plasma of males than in the plasma of females. In a model using surgically obtained abdominal human skin, OTA was rapidly absorbed as shown by a permeability coefficient of $Kp = 8.20 \times 102^4$ cm/h (Boonen et al., 2012). In the majority of species, OTA has slow elimination. Urinary and fecal excretions are the primary methods by which OTA is cleared from the blood. OTA is excreted in bile and is reabsorbed from the gut with enterohepatic circulation occurring (Ringot et al., 2006). Excretion of OTA in bile is a major excretory route in laying hens for orally administered OTA (Armorini et al., 2015). In some species, hydrolysis of the ochratoxin conjugates by intestinal microorganisms is important in the hepatoenteric recycling. In goats and calves, fecal excretion can account for more than 50% of the dose administered, whereas in sheep less than 10% of the dose was excreted as ochratoxin-α (OTα) (Nip and Chu, 1979; Sreemannarayana et al., 1988). Renal excretion and secretion of ochratoxin occurs (Ringot et al., 2006; George et al., 2017). In the nephron, the organic anion transporters transport OTA from the glomerular filtrate into the tubular cells, and the multidrug resistance-associated protein 2 and other transporters may secrete OTA into the nephron (George et al., 2017). The toxicokinetics of OTA has been studied in sheep (Blank et al., 2003). Sheep were fed diets containing 0, 9.5, 19.0, and 28.5 OTA/kg body weight for 29 days, and the levels of OTA and metabolites were measured in rumen contents, serum, urine, and feces. Serum levels of OTA increased linearly with increased dose of OTA, and OTA and OTα in serum increased with time. OTA and OTα were identified in urine. Rumen flora and fauna metabolize OTA to OTα and diet has shown to alter the metabolism of OTA by rumen microorganisms (Xiao et al., 1991). Sheep on a high-roughage diet (hay) metabolized OTA in the rumen at a higher rate than sheep fed a diet high in grain. Sheep on a diet containing 70% concentrate and 30% grass silage and administered 9.5, 19.0, and 28.5 μg OTA/kg body weight consistently excreted 7%–8% of the dose in the urine predominantly as the OTα hydroxylated metabolite. The mean serum levels of OTA, in ascending order of dose, on the last day (day 29) of the study were 3.0, 6.7, and 9.9 ng/mL. OTA is absorbed from the lungs and occupational exposures occur from OTA-contaminated dust.

Important OTA biotransformation reactions are hydrolysis, hydroxylation, conjugation, and lactone opening (Heussner and Bingle, 2015; Koszegi and Poor, 2016). Microbes in the lower gut and rumen hydrolyze OTA to OTα, which is absorbed and excreted in urine. Cell cultures have been shown to produce OTα. The CYP enzymes hydroxylate OTA to 4S-OH-OTA, 4R-OH-OTA, 7′-OH-OTA, 9′-OH-OTA, 5′-OH-OTA, and ochratoxin B. Ochratoxin conjugates are excreted in bile and to a lesser extent in urine (Munoz et al., 2017). Lactone opening is considered to increase the toxicity of OTA and the lactone-open OTA has been identified in rat bile and chyme.

Chemical Biomarkers of Ochratoxin A

OTA crosses the placental barrier and at birth is present in fetal blood (Malir et al., 2013). In the species studied, OTA levels in maternal serum are not a good predictor of OTA levels in fetal serum at the time of birth. Finding OTA in the neonate before nursing, e.g. cord blood or other tissues, and in breast milk, are biomarkers of maternal to fetal and maternal to neonatal exposures, respectively. In horses, OTA has also been shown to cross the placental barrier (Minervini et al., 2013). OTA is also transferred from plasma to milk (Soto et al., 2016). Because OTA is bound to serum proteins, it is likely that both simple diffusion and transport mechanisms are involved in the shift of OTA from blood to milk. Colostrum is the highest in OTA and this may be due to the glandular transfer of OTA bound to plasma proteins into glandular colostrum (Munoz et al., 2014). The transfer rate of OTA from blood to milk may change with the percentage of OTA that is bound to serum proteins. The milk/blood ratio in rats has been shown to be 0.4–0.7 after the animals were administered 50 μg OTA/kg body weight 5 days a week for 21 days, and similar results were observed in a single oral dose study (Breitholtz-Emanuelsson et al., 1993). Soto et al. (2016) in a recent review reported the range of OTA in breast milk to be 0.002–13.1 ng/mL. Dietary habits of the mother influence the levels of OTA in breast milk, with consumption of foods high in OTA resulting in the higher levels in milk. Plasma/breast milk ratio has been reported at 0.40 ± 0.26 for colostrum, and after lactation onset (15 days–4 months) the plasma/milk ratio decreases to 0.15 ± 0.26 (Warth et al., 2016). Field sampling of cow milk has revealed the presence of OTA in milk sampled from tanker trucks and bulk tanks in different parts of Europe and market basket sampling in China. Hens fed a diet containing 5 ppm OTA had OTA appearing in the eggs after 5 days of feeding the contaminated diet and the level of OTA peaked in the eggs at 7.4 ± 1.03 ng/g on day 21 (Zahoor et al., 2012). Laying hens fed a diet containing 2 ppm OTA for 3 weeks did not have detectable levels (0.5 μg/kg) of OTA in their eggs (Denli et al., 2008). Other studies have shown that poultry diets with less than 2 ppm OTA do not result in detectable levels being transferred to eggs (Battacone et al., 2010).

Histopathologic Biomarkers of Exposure to Ochratoxins A

The kidney, in natural exposure conditions, is considered to be the primary target of OTA (Gupta et al., 2018). The renal toxicopathy of OTA has been reported for various species. In pigs, the temporal sequence of renal tubular histopathology is swelling of proximal tubular epithelial cells, appearance of eosinophilic hyaline droplets, condensation of the nuclei progressing to cellular necrosis, appearance of eosinophilic protein casts, inflammatory exudate, diffuse glomerular atrophy progressing to sclerosis and interstitial fibrosis (Stoev et al., 2012). In the rat, the pathogenesis of renal tubular histopathology is swelling of proximal tubular epithelial cells, appearance of eosinophilic hyaline droplets, condensation of the nuclei, and appearance of eosinophilic protein casts, progressing to cellular necrosis. Cell loss in the straight segment of proximal tubular epithelium can be accompanied by cell proliferation and nuclear enlargement, and these lesions progress to focal hyperplasia, tubular cell adenoma, and tubular cell carcinoma (Rached et al., 2007). OTA is a potent carcinogen in laboratory animals and a suspected human carcinogen (Anon, 1993). In the glomerulus, swelling and blebbing of the endothelial cells occur along with endothelial cellular necrosis. Interstitial nephropathy is also observed. Similar changes have been observed in poultry, rabbits, pigs, and dogs (Stoev et al., 2012; Prabu et al., 2013) and in intoxication with citrinin.

FUMONISINS

The fumonisins are primarily produced by *Fusarium verticillioides* (*Fusarium moniliforme*), *Fusarium proliferatum*, *Fusarium oxysporum*, *Fusarium globosum*, other *Fusarium* spp., and *Alternaria alternata*, which commonly infect maize and other foodstuffs (Coppock and Jacobsen, 2009; Scott, 2012; WHO, 2017). The occurrence of fumonisin in maize is important because maize is the third most important cereal grain for human consumption. Cereals in all land continents are contaminated with fumonisins (Lee and Ryu, 2017). Other foodstuffs that have been shown to contain fumonisin B_1 (FB_1) are garlic powder, black tea, peanuts, figs, soybeans, and cows' milk (Gazzotti et al., 2009; Scott, 2012). FB_1 has been shown to cause leukoencephalomalacia in horses, pulmonary edema in pigs, renal disease in sheep and other domestic species, and neoplasia in rats and mice. In humans, FB_1 is associated with neoplasia, stunting of children, and interference with folic acid uptake and therein associated with folic acid deficiency-linked birth defects (Coppock and Jacobsen, 2009; Gelderblom and Marasas, 2012).

Chemistry of Fumonisins

Fifteen different fumonisins have been described and by chemical properties grouped into categories A, B, C, and P (WHO, 2017). Of these categories groups A and B are considered toxic and the FB_1 group is considered to be the most toxic. There are at least 28 FB_1 isomers. The IUPAC name for FB_1 is (2S)-2-[2-[(5S,6R,7R,9R,11S,16R,18S,19S)-19-amino-6-[(3S)-3,4-dicarboxy-butanoyl]oxy-11,16,18-trihydroxy-5,9-dimethylicosan-7-yl]oxy-2-oxoethyl]butanedioic acid. FB_1 is structurally similar to sphingosine and inhibits sphinganine N-acetyltransferase (ceramide synthase), which is a group of enzyme for sphingolipid synthesis, and FB_1 is classed by IRAC as a Group 2B possible human carcinogen (Anon, 1993).

Toxicokinetics, Metabolism, and Chemical Biomarkers

FB_1 is a charged molecule and has a low level of gut absorption and a high rate of elimination. In the monogastric species studied, FB_1 has oral bioavailability, from 1% to 6% of the dose absorbed (Gelderblom and Marasas, 2012). The majority of kinetic studies (intravenous administration) suggest a two-compartment open model for the elimination of FB_1. Parenterally administered FB_1 has a half-life of 17 min in bile cannulated pigs, 18 min in rat, 40 min in vervet monkey, 49 min in laying hens, 85 min in turkey, and an estimated

128 min in human (Shephard et al., 2007; Tardieu et al., 2008). The half-life after oral administration of FB_1 is 214 min in turkeys (Tardieu et al., 2008). Approximately 6% of the absorbed FB_1 is bound to plasma proteins and other constituents. The absorbed FB_1 is rapidly cleared by the liver and excreted in the bile. There is evidence that hepatic-enteric circulation of FB_1 occurs. The majority of ingested FB_1 is eliminated in the feces and small amounts are eliminated in the urine (Shephard et al., 2007). FB_1 residues are found in the liver and kidney.

In vertebrates, muscle and fat have low levels of FB_1, whereas higher levels are found in the liver and kidney. Less than 1% of the FB_1 ingested is excreted in urine when humans consume biscuits and tortillas made from market basket flour contaminated with FB_1 (Riley et al., 2012). A study in South Africa provided estimates that 0.054% of FB_1 dietary intake was excreted in the urine, and urinary output was correlated with dietary intake (van der Westhuizen et al., 2011). In a model using surgically obtained abdominal human skin, FB_1 had a nondetectable permeability (Boonen et al., 2012). A long-term chronic study in nonhuman primates using culture material of *F. verticillioides* strain MRC 602 showed that FB_1 and metabolites were deposited in hair (Gelderblom et al., 2001; Sewram et al., 2003). Discarded barbershop hair was collected from a high maize consumption region of South Africa and assayed for FB_1 as a biomarker of chronic exposure to FB_1 (Sewram et al., 2003). The mean levels from the different geographic regions ranged from 22.2 to 33.0 μg/kg of hair. Using hair analyses and exposure data from a study in nonhuman primates, Sewram et al. (2003) estimated the chronic human exposure to be 18.4–27 μg of fumonisins/kg body weight. Another study concluded that FB_1 in human hair is a function of dietary intake of FB_1 (Bordin et al., 2015). A study in piglets fed diets containing 0, 3.1, 6.1, or 9.0 μg FB_1/g of feed for 28 days found a significant correlation between dietary FB_1 and the FB_1 incorporated into hair (Souto et al., 2017). FB_1 has been identified in market basket samples of milk. Eight out of 10 samples were positive for FB_1 and the values ranged from 0.26 to 0.43 μg/kg (Gazzotti et al., 2009). FB_1 is also present in breast milk from mothers consuming FB_1-contaminated foods, and breast milk is considered the safest infant food (Magoha et al., 2014; Warth et al., 2016). In de novo sphingolipid synthesis, sphinganine N-acetyltransferase (ceramide synthase) catalyzes acetylation of the sphingoid base sphinganine to form dihydroceramide. Changes in the sphingoid bases can occur before or at the onset of fumonisin toxicity. In horses and pigs, changes in the sphinganine/sphingosine ratio have been reported to occur in blood and urine before other indicators of intoxication occur (Bucci et al., 1998). Studies in poultry have shown increase in serum sphinganine/sphingosine ratio following exposure to FB_1 (Weibking et al., 1994; Tardieu et al., 2007, 2009). Studies in humans living in areas where FB_1 contamination of diets is endemic have shown that sphinganine/sphingosine ratios are not good predictors of individual dietary exposure to FB_1, but blood spot sphinganine 1-phosphate levels are positively correlated with individual urinary excretion of FB_1 (Xu et al., 2010; van der Westhuizen et al., 2013; Riley et al., 2015). Increased serum and tissue levels of sphinganine 1-phosphate in mice, rats, pigs, and horses have been proposed as a biomarker of exposure to FB_1 (Kim et al., 2006; Shephard et al., 2007).

Histopathological Biomarkers

Histopathology is an indicator of intoxication with FB_1. Apoptosis and dyssynchrony of the cell cycle may be an important pathological mechanism (Wang et al., 2013). FB_1-linked kidney lesions have been observed in essentially all animal species studied. In pigs, the renal lesions are pericapillary edema and granular degeneration of the proximal tubular cells, intermittent necrotic tubular epithelial cells, and granular casts (Colvin et al., 1993; Stoev et al., 2012). In nonhuman primates, renal lesions observed were cellular atrophy of the proximal tubular cells (Gelderblom et al., 2001). Degeneration of proximal tubular cells with individual cellular necrosis and apoptosis are observed and the lesion can continue into the outer strip of the medulla (Bucci et al., 1998). Urinary enzymes are considered a sensitive indicator of damage of the renal epithelium in rodents. There are few reports of their being studied and used as biomarkers of FB_1-linked renal damage in other species. FB_1 has been shown to be hepatotoxic in several species (Javed et al., 2005). The histopathology of hepatotoxicity is characterized in chickens by multifocal single-cell hepatocellular cytoplasmic swelling and fatty degeneration to pleomorphic vacuolated hepatocytes with marked hydropic degeneration and hepatocellular swelling in the higher dosage groups. Degenerative changes can occur in the biliary epithelial. Numerous mitotic figures in regenerative areas can be observed and periportal fibrosis can occur. In turkeys and ducks, the pathology is generalized hepatocellular hyperplasia. Liver pathology has been reported in pigs on a diet containing 190 ppm FB_1 fed for 83 days (Casteel et al., 1993). The observed liver pathology was hepatocellular necrosis occurring in the centrilobular to midzonal regions and diffuses hepatocellular vacuolation, biliary hyperplasia, and the presence of hyperplastic nodules. In field cases of FB_1 intoxication in pigs (dietary levels of 20–330 ppm

FB_1), pulmonary pathology is characterized by accumulation of acidophilic fibrillar material in alveoli, eosinophilic hyalinized alveolar capillary thrombi, and interlobular edema. Neurologic lesions in leukoencephalomalacia were characterized by degeneration of the white matter with inflammatory exudates and perifibrosis of blood vessels. Leukoencephalomalacia in horses is considered a biomarker of FB_1 intoxication (Maxie and Youssef, 2007).

DEOXYNIVALENOL TRICHOTHECENE

Deoxynivalenol (DON, vomitoxin) is one of the most prevalent mycotoxins in human foods and animal feeds. It is generally regarded as being heat stable, but there is evidence that some degradation may occur at bread-baking temperatures. DON is a member of the Group B trichothecene group of mycotoxins. It is a small sesquiterpenoid with an embedded epoxide at the 12–13 position. Chemically DON is identified as (3β,7α)-3,7,15-trihydroxy-12,13-epoxytrichothec-9-en-8-one. Important producers of DON are *Fusarium graminearum* (sexual state *Gibberella zeae*), which causes red ear rot of corn and *Fusarium* head blight (scab) of wheat and barley, and *Fusarium culmorum*, and *Fusarium pseudograminearum* (Coppock and Jacobsen, 2009). These species of *Fusarium* can also produce other trichothecenes and zearalenones (ZENs). The occurrences of DON contamination are linked to cereal grains being infected with species of *Fusarium*. Acetylated derivatives of DON (3-acetyl deoxynivalenol and 15-acetyl-deoxynivalenol) also can exist in foods and feeds (Payros et al., 2016). Plants deactivate DON by forming the DON-3β-D-glucopyranoside and this form is not detected by many analytical procedures. DON-3β-D-glucopyranoside is substantially less toxic than DON. Gut bacteria can hydrolyze these conjugates and liberate DON. DON is ribotoxic and inhibits protein synthesis and causes a cascade of cellular dysfunctions and oxidative stress. The gastrointestinal tract and immune, reproductive, endocrine, and nervous systems are targets for DON (Coppock et al., 1985; Maresca, 2013).

Toxicokinetics, Biotransformation, and Biomarkers

The toxicokinetics of DON varies with species, age, and sex (Payros et al., 2016). Rodent data show that males receiving the same oral dose of DON have higher plasma levels that females, and males are more sensitive to the toxic effects of DON (Clark et al., 2015). Absorption of DON varies from 7% in mature ruminants to 89% in chronically exposed piglets (Maresca, 2013; Payros et al., 2016). Pigs absorb 54%–89% of the DON ingested and there is evidence that absorption efficiency increases with chronic exposure possibly because of the toxic effects of DON on enterocytes (Goyarts and Danicke, 2006). Approximately 9% of the DON in blood is bound to plasma proteins (Meky et al., 2003). The intestinal cells, liver, and kidney detoxify absorbed DON by the formation of DON-glucuronides, and these are rapidly excreted predominantly in the urine with a small amount in the bile, and DON itself is also excreted in the urine. Microorganisms in the forestomach and gastrointestinal tract form the de-epoxide diene derivatives of DON and destruction of the 12–13 epoxide and greatly reduces toxicity. The 13-depoxy DON (DOM-1) microbial metabolite of DON has been regionally identified in human urine suggesting regional differences in gut microbial metabolism of DON, and there are also regional differences in the formation of DON conjugates (Chen et al., 2017). The DON-3-sulfate is excreted by several animal species and is likely excreted by humans. Very little parent DON is excreted in milk, and DOM-1 is excreted in milk and urine. Urinary excretion of DON-glucuronide has been studied as a biomarker of human and animal exposure and is considered to be a biomarker of exposure at least in moderately exposed populations (Meky et al., 2003; Thanner et al., 2016; Chen et al., 2017). The interval between food consumption and urine collection can affect the urinary level of DON and its metabolites.

ZEARALENONE

Background

ZEN is a mycotoxin with xenoestrogen activity that mimics 17β-estradiol and is considered an endocrine disruptor (Bandera et al., 2011; Kowalska et al., 2016). The IUPAC name for ZEN is (2E,11S)-15,17-dihydroxy-11-methyl-12-oxabicyclo[12.4.0] octadeca-1(14), 2,15,17-tetraene-7,13-dione, and it is also chemically described as 6-(10-hydroxy-6-oxo-trans-1-undecenyl)-β-resorcylic acid L-lactone. ZEN is produced by *F. graminearum*, *F. culmorum*, *Fusarium equiseti*, and *Fusarium cerealis* (*Fusarium crockwellense*) (Coppock and Jacobsen, 2009; Oldenburg et al., 2017). ZEN is a contaminate of foods, feeds, and beer and occurs in grain dust (Tangni and Pussemier, 2007). ZEN is gaining attention because of it being an endocrine disruptor, and some evidence that it is a growth–maturation disruptor in humans and may contribute to increased occurrences of breast cancer (Belhassen et al., 2015; Bandera et al., 2011).

Toxicokinetics, Biotransformation, and Biomarkers

The toxicokinetics and biotransformation of ZEN in food-producing animals has recently been reviewed (Danicke and Winkler, 2015). ZEN can be ingested as the aglycone and sulfate conjugates (ZEN-14-O-β-glucoside, ZEN-16-β-O-glucoside, and ZEN-14-sulfate). The aglycone forms of ZEN can be liberated by chyme-associated hydrolysis in the digestive tract. The aglycone forms of ZEN are not detected by many routine analytical procedures. The absolute absorption of ZEN is difficult to determine because of its rapid clearance by the liver and excretion into bile and biotransformation by gastrointestinal mucosal cells. Plasma levels of ZEN and its metabolites are low in poultry and rats after oral administration of ZEN (Danicke and Winkler, 2015; Devreese et al., 2015). This may be due to biotransformation in the gastrointestinal cells, hepatic clearance, and excretion in bile. Bolus doses of ZEN administered to poultry are primarily excreted in bile. The major pathways of ZEN biotransformation are hydroxylation and conjugation with glucuronic acid and sulfate. Oxidation of ZEN can occur by the CYP enzymes and in subsequent reactions can form reactive oxygen intermediates that form DNA adducts. Reduction of ZEN by 3α- and 3β-hydroxysteroid dehydrogenases forms α-zearalenol (α-ZEL) and β-zearalenol (β-ZEL), respectively.

Biotransformation of ZEN to α-ZEL is generally considered a toxication reaction because the estrogen receptors in many species have a 500-fold higher affinity for α-ZEL than ZEN. Conversely, the formation of β-ZEL is considered a detoxication reaction because there is, depending on the animal species, a 15-fold reduction in estrogenic activity. The rate of biotransformation of ZEN to α-ZEL and β-ZEL varies between species. ZEN, α-ZEL, and β-ZEL can be mutually transformed in hepatocytes. α-ZEL and β-ZEL, in the animal species studied, are the major metabolites of ZEN. ZEN, α-ZEL, and β-ZEL and their conjugates are excreted in bile. Bile levels can be 2.5 and 17 times higher than liver and urine, respectively, and bile is considered the best analyte for ZEN and metabolites including zeranol (α-zearalanol) (Lega et al., 2017). In chickens, oral exposure to ZEN results in ZEN, α-ZEL, and β-ZEL being excreted in the feces. In most food-producing species, ZEN, α-ZEL, and β-ZEL are excreted predominantly as glucuronide (α-ZEL-Glu, β-ZEL-Glu) and sulfate conjugates (Danicke and Winkler, 2015; Yang et al., 2017). In ZEN-exposed humans, pigs, and poultry, ZEN and α-ZEL and their conjugates are the main metabolites excreted. Cattle fed diets containing ZEN excrete ZEN and β-ZEL in their urine, and lactating cattle excrete β-ZEL, ZEN, and α-ZEL in milk. The milk/diet carryover rate is 0.008. Survey studies in horses suggest that ZEN is metabolized to β-ZEL in horses and ZEN, β-ZEL, and α-ZEL were identified, including glucuronide conjugates, in serum and urine (Schumann et al., 2016; Takagi et al., 2018). Humans excrete ZEN in breast milk. Humans excrete free ZEN, α-ZEL, and β-ZEL and ZEN-Glu, α-ZEL-Glu, and β-ZEL-Glu in urine with the glucuronide conjugates predominating (Mally et al., 2016). There appears to be regional differences in the biotransformation of ZEN by humans.

CONCLUDING REMARKS AND FUTURE DIRECTIONS

Worldwide, mycotoxins are an underappreciated cause of disease. Advances in diagnosis of mycotoxicoses are linked to advances in chemical, molecular, and histologic methods of diagnosis. The increased use of microchips to identify gene expression, protein signatures, and other molecular changes is advancing diagnosis of mycotoxicoses. Metabolomics has a high likelihood of being a sensitive method to assess the adverse metabolic effects of mycotoxins. Current studies are showing that adverse metabolic effects of chronic exposure to low levels of mycotoxins are underappreciated (Toda et al., 2017). Simplified analytical methods for simultaneously measuring multiple mycotoxins are showing that exposure to multiple mycotoxins is a frequent occurrence. Preventing the formation of mycotoxins by reducing conditions for mold growth and mycotoxin production is the best way forward.

References

Anon, 1993. Ochratoxin A, vol. 56. International Agency for Research on Cancer (IARC), p. 599.

Anon, 2012. Aflatoxins, 100F. International Agency for Research on Cancer (IARC), pp. 225–248.

Armorini, S., Al-Qudah, K.M., Altafini, A., et al., 2015. Biliary ochratoxin A as a biomarker of ochratoxin exposure in laying hens: an experimental study after administration of contaminated diets. Res. Vet. Sci. 100, 265–270.

Bandera, E.V., Chandran, U., Buckley, B., et al., 2011. Urinary mycoestrogens, body size and breast development in New Jersey girls. Sci. Total Environ. 409, 5221–5227.

Battacone, G., Nudda, A., Pulina, G., et al., 2010. Effects of ochratoxin A on livestock production. Toxins 2, 1796–1824.

Becker-Algeri, T.A., Castagnaro, D., de Bortoli, K., et al., 2016. Mycotoxins in bovine milk and dairy products: a review. J. Food Sci. 81, R544–R552.

Belhassen, H., Jimenez-Diaz, I., Arrebola, J.P., et al., 2015. Zearalenone and its metabolites in urine and breast cancer risk: a case-control study in Tunisia. Chemosphere 128, 1–6.

Bernhoft, A., Hogasen, H.R., Berntssen, M.H.G., et al., 2017. Tissue distribution and elimination of deoxynivalenol and ochratoxin A in dietary-exposed Atlantic salmon (*Salmo salar*). Food Addit. Contam. 34, 1211–1224.

Blank, R., Rolfs, J.P., Marquardt, R.R., et al., 2003. Effects of chronic ingestion of ochratoxin A on blood levels and excretion of the mycotoxin in sheep. J. Agric. Food Chem. 51, 6899–6905.

Boonen, J., Malysheva, S.V., De Saeger, S., et al., 2012. Human skin penetration of selected model mycotoxins. Toxicology 301, 21–32.

Bordin, K., Rottinghaus, G.E., Landers, B.R., et al., 2015. Evaluation of fumonisin exposure by determination of fumonisin B_1 in human hair and in Brazilian corn products. Food Control 53, 67–71.

Breitholtz-Emanuelsson, A., Palminger-Hallen, I., Wohlin, P.O., et al., 1993. Transfer of ochratoxin A from lactating rats to their offspring: a short-term study. Nat. Toxins 1, 347–352.

Bucci, T.J., Howard, P.C., Tolleson, W.H., et al., 1998. Renal effects of fumonisin mycotoxins in animals. Toxicol. Pathol. 26, 160–164.

Casteel, S.W., Turk, J.R., Cowart, R.P., et al., 1993. Chronic toxicity of fumonisin in weanling pigs. J. Vet. Diagn. Invest. 5, 413–417.

Chen, L., Yu, M., Wang, D., et al., 2017. Gender and geographical variability in the exposure pattern and metabolism of deoxynivalenol in humans: a review. J. Appl. Toxicol. 37, 60–70.

Clark, E.S., Flannery, B.M., Pestka, J.J., 2015. Murine anorectic response to deoxynivalenol (vomitoxin) is sex-dependent. Toxins 7, 2845–2859.

Colvin, B.M., Cooley, A.J., Beaver, R.W., 1993. Fumonisin toxicosis in swine: clinical and pathologic findings. J. Vet. Diagn. Invest. 5, 232–241.

Cook, W.O., Richard, J.L., Osweiler, G.D., et al., 1986. Clinical and pathologic changes in acute bovine aflatoxicosis: rumen motility and tissue and fluid concentrations of aflatoxins B_1 and M_1. Am. J. Vet. Res. 47, 1817–1825.

Coppock, R.W., Jacobsen, B.J., 2009. Mycotoxins in animal and human patients. Toxicol. Ind. Health 25, 637–655.

Coppock, R.W., Swanson, S.P., Gelberg, H.B., et al., 1985. Preliminary study of the pharmacokinetics and toxicopathy of deoxynivalenol (vomitoxin) in swine. Am. J. Vet. Res. 46, 169–174.

Coppock, R.W., Christian, R.G., Jacobsen, B.J., 2018. Aflatoxins. In: Gupta, R.C. (Ed.), Veterinary Toxicology Basic and Clinical Principles, third ed. Elsevier, Toronto, pp. 983–994.

Danicke, S., Winkler, J., 2015. Invited review: diagnosis of zearalenone (ZEN) exposure of farm animals and transfer of its residues into edible tissues (carry over). Food Chem. Toxicol. 84, 225–249.

Denli, M., Blandon, C.J., Guynot, M.E., et al., 2008. Efficacy of a new ochratoxin-binding agent (OcraTox) to counteract the deleterious effects of ochratoxin A in laying hens. Poultry Sci. 87, 2266–2272.

Devreese, M., Antonissen, G., Broekaert, N., et al., 2015. Comparative toxicokinetics, absolute oral bioavailability, and biotransformation of zearalenone in different poultry species. J. Agric. Food Chem. 63, 5092–5098.

Dohnal, V., Wu, Q., Kuca, K., et al., 2014. Metabolism of aflatoxins: key enzymes and interindividual as well as interspecies differences. Arch. Toxicol. 88, 1635–1644.

EFSA Panel, 2013. Scientific opinion on the risk for public and animal health related to the presence of sterigmatocystin in food and feed. EFSA J. 11, 3254.

Erkekoglu, P., Oral, D., Chao, M.W., et al., 2017. Hepatocellular carcinoma and possible chemical and biological causes: a review. J. Environ. Pathol. Toxicol. Oncol. 36, 171–190.

FAO, Food, Agriculture Organization of United Nations, 2008. Safety Evaluation of Certain Food Additives. In: Food Addit Series, vol. 59. WHO Press, Geniva.

Fushimi, Y., Takagi, M., Uno, S., et al., 2014. Measurement of sterigmatocystin concentrations in urine for monitoring the contamination of cattle feed. Toxins 6, 3117–3128.

Gazzotti, T., Serraino, A., Pagliuca, G., et al., 2009. Determination of fumonisin B_1 in bovine milk by LC-MS/MS. Food Control 20, 1171–1174.

Gelderblom, W.C., Marasas, W.F., 2012. Controversies in fumonisin mycotoxicology and risk assessment. Hum. Exp. Toxicol. 31, 215–235.

Gelderblom, W.C., Seier, J.V., Snijman, P.W., et al., 2001. Toxicity of culture material of *Fusarium verticillioides* strain MRC 826 to nonhuman primates. Environ. Health Perspect. 109 (Suppl 2), 267–276.

George, B., You, D., Joy, M.S., et al., 2017. Xenobiotic transporters and kidney injury. Adv. Drug Deliv. Rev. 116, 73–91.

Goyarts, T., Danicke, S., 2006. Bioavailability of the Fusarium toxin deoxynivalenol (DON) from naturally contaminated wheat for the pig. Toxicol. Lett. 163, 171–182.

Gross-Steinmeyer, K., Eaton, D.L., 2012. Dietary modulation of the biotransformation and genotoxicity of aflatoxin B(1). Toxicology 299, 69–79.

Gupta, R.C., Srivastava, A., Lall, R., 2018. Ochratoxin and citrinin. In: Gupta, R.C. (Ed.), Veterinary Toxicology Basic and Clinical Principles, third ed. Elsevier, Toronto, pp. 1019–1027.

Heussner, A.H., Bingle, L.E., 2015. Comparative ochratoxin toxicity: a review of the available data. Toxins 7, 4253–4282.

Javed, T., Bunte, R.M., Dombrink-Kurtzman, M.A., et al., 2005. Comparative pathologic changes in broiler chicks on feed amended with *Fusarium proliferatum* culture material or purified fumonisin B_1 and moniliformin. Mycopathologia 159, 553–564.

Jubert, C., Mata, J., Bench, G., et al., 2009. Effects of chlorophyll and chlorophyllin on low-dose aflatoxin B(1) pharmacokinetics in human volunteers. Canc. Prev. Res. 2, 1015–1022.

Kim, D.H., Yoo, H.S., Lee, Y.M., et al., 2006. Elevation of sphinganine 1-phosphate as a predictive biomarker for fumonisin exposure and toxicity in mice. J. Toxicol. Environ. Health Part A 69, 2071–2082.

Koszegi, T., Poor, M., 2016. Ochratoxin A: molecular interactions, mechanisms of toxicity and prevention at the molecular level. Toxins 8, 111.

Kowalska, K., Habrowska-Gorczynska, D.E., Piastowska-Ciesielska, A.W., et al., 2016. Zearalenone as an endocrine disruptor in humans. Environ. Toxicol. Pharmacol. 48, 141–149.

Lee, H.J., Ryu, D., 2017. Worldwide occurrence of mycotoxins in cereals and cereal-derived food products: public health perspectives of their co-occurrence. J. Agric. Food Chem. 65, 7034–7051.

Lega, F., Angeletti, R., Stella, R., et al., 2017. Abuse of anabolic agents in beef cattle: could bile be a possible alternative matrix? Food Chem. 229, 188–197.

Livingstone, M.C., Johnson, N.M., Roebuck, B.D., et al., 2017. Profound changes in miRNA expression during cancer initiation by aflatoxin B_1 and their abrogation by the chemopreventive triterpenoid CDDO-Im. Mol. Carcinog. 56, 2382–2390.

Magoha, H., De Meulenaer, B., Kimanya, M., et al., 2014. Fumonisin B_1 contamination in breast milk and its exposure in infants under 6 months of age in Rombo, Northern Tanzania. Food Chem. Toxicol. 74, 112–116.

Malir, F., Ostry, V., Pfohl-Leszkowicz, A., et al., 2013. Ochratoxin A: developmental and reproductive toxicity: an overview. Birth Defects Res. Part B 98, 493–502.

Mally, A., Solfrizzo, M., Degen, G.H., 2016. Biomonitoring of the mycotoxin zearalenone: current state-of-the art and application to human exposure assessment. Arch. Toxicol. 90, 1281–1292.

Maresca, M., 2013. From the gut to the brain: journey and pathophysiological effects of the food-associated trichothecene mycotoxin deoxynivalenol. Toxins 5, 784–820.

Maxie, G.M., Youssef, S., 2007. Nervous system. In: Maxie, G.M. (Ed.), Jubb, Kennedy, and Palmer's Pathology of Domestic Animals, , fifth ed.vol. 1. Elsevier, Toronto, pp. 1181–1199.
Medina, A., Akbar, A., Baazeem, A., et al., 2017. Climate change, food security and mycotoxins: do we know enough? Fungal Biol. Rev. 31, 143–154.
Meky, F.A., Turner, P.C., Ashcroft, A.E., et al., 2003. Development of a urinary biomarker of human exposure to deoxynivalenol. Food Chem. Toxicol. 41, 265–273.
Minervini, F., Giannoccaro, A., Nicassio, M., et al., 2013. First evidence of placental transfer of ochratoxin A in horses. Toxins 5, 84–92.
Munoz, K., Blaszkewicz, M., Campos, V., et al., 2014. Exposure of infants to ochratoxin A with breast milk. Arch. Toxicol. 88, 837–846.
Munoz, K., Cramer, B., Dopstadt, J., et al., 2017. Evidence of ochratoxin A conjugates in urine samples from infants and adults. Mycotoxin Res. 33, 39–47.
Nault, J.C., 2014. Pathogenesis of hepatocellular carcinoma according to aetiology. Best Pract. Res. Clin. Gastroenterol. 28, 937–947.
National Center for Biotechnology Information (NCBI), 2008. PubChem Compound Database; CID=442530. https://pubchem.ncbi.nlm.nih.gov/compound/442530.
Nip, W.K., Chu, F.S., 1979. Fate of ochratoxin A in goats. J. Environ. Sci. Health Part B 14, 319–333.
Oldenburg, E., Hoppner, F., Ellner, F., et al., 2017. Fusarium diseases of maize associated with mycotoxin contamination of agricultural products intended to be used for food and feed. Mycotoxin Res. 33, 167–182.
Pantaya, D., Morgavi, D.P., Silberberg, M., et al., 2016. Bioavailability of aflatoxin B_1 and ochratoxin A, but not fumonisin B_1 or deoxynivalenol, is increased in starch-induced low ruminal pH in nonlactating dairy cows. J. Dairy Sci. 99, 9759–9767.
Partanen, H.A., El-Nezami, H.S., Leppanen, J.M., et al., 2010. Aflatoxin B_1 transfer and metabolism in human placenta. Toxicol. Sci. 113, 216–225.
Payros, D., Alassane-Kpembi, I., Pierron, A., et al., 2016. Toxicology of deoxynivalenol and its acetylated and modified forms. Arch. Toxicol. 90, 2931–2957.
Pizzolato Montanha, F., Anater, A., Burchard, J.F., et al., 2018. Mycotoxins in dry-cured meats: a review. Food Chem. Toxicol. 111, 494–502.
Poor, M., Balint, M., Hetenyi, C., et al., 2017. Investigation of non-covalent interactions of aflatoxins (B_1, B_2, G_1, G_2, and M_1) with serum albumin. Toxins 9, 339.
Prabu, P.C., Dwivedi, P., Sharma, A.K., et al., 2013. Toxicopathological studies on the effects of aflatoxin B(1), ochratoxin A and their interaction in New Zealand white rabbits. Exp. Toxicol. Pathol. 65, 277–286.
Qian, G., Tang, L., Wang, F., et al., 2013. Physiologically based toxicokinetics of serum aflatoxin B_1-lysine adduct in F344 rats. Toxicology 303, 147–151.
Rached, E., Hard, G.C., Blumbach, K., et al., 2007. Ochratoxin A: 13-week oral toxicity and cell proliferation in male F344/n rats. Toxicol. Sci. 97, 288–298.
Riley, R.T., Torres, O., Showker, J.L., et al., 2012. The kinetics of urinary fumonisin B_1 excretion in humans consuming maize-based diets. Mol. Nutr. Food Res. 56, 1445–1455.
Riley, R.T., Torres, O., Matute, J., et al., 2015. Evidence for fumonisin inhibition of ceramide synthase in humans consuming maize-based foods and living in high exposure communities in Guatemala. Mol. Nutr. Food Res. 59, 2209–2224.
Ringot, D., Chango, A., Schneider, Y.J., et al., 2006. Toxicokinetics and toxicodynamics of ochratoxin A, an update. Chem. Biol. Interact. 159, 18–46.
Sabbioni, G., Turesky, R.J., 2017. Biomonitoring human albumin adducts: the past, the present, and the future. Chem. Res. Toxicol. 30, 332–366.
Schumann, B., Winkler, J., Mickenautsch, N., et al., 2016. Effects of deoxynivalenol (DON), zearalenone (ZEN), and related metabolites on equine peripheral blood mononuclear cells (PBMC) *in vitro* and background occurrence of these toxins in horses. Mycotoxin Res. 32, 153–161.
Scott, P.M., 2012. Recent research on fumonisins: a review. Food Addit. Contam. 29, 242–248.
Sewram, V., Mshicileli, N., Shephard, G.S., et al., 2003. Fumonisin mycotoxins in human hair. Biomarkers 8, 110–118.
Shephard, G.S., Van Der Westhuizen, L., Sewram, V., et al., 2007. Biomarkers of exposure to fumonisin mycotoxins: a review. Food Addit. Contam. 24, 1196–1201.
Soto, J.B., Ruiz, M.J., Manyes, L., et al., 2016. Blood, breast milk and urine: potential biomarkers of exposure and estimated daily intake of ochratoxin A: a review. Food Addit. Contam. 33, 313–328.
Souto, P., Jager, A.V., Tonin, F.G., et al., 2017. Determination of fumonisin B_1 levels in body fluids and hair from piglets fed fumonisin B_1-contaminated diets. Food Chem. Toxicol. 108, 1–9.
Sreemannarayana, O., Frohlich, A.A., Vitti, T.G., et al., 1988. Studies of the tolerance and disposition of ochratoxin A in young calves. J. Anim. Sci. 66, 1703–1711.
Stoev, S.D., Gundasheva, D., Zarkov, I., et al., 2012. Experimental mycotoxic nephropathy in pigs provoked by a mouldy diet containing ochratoxin A and fumonisin B_1. Exp. Toxicol. Pathol. 64, 733–741.
Streit, E., Naehrer, K., Rodrigues, I., et al., 2013. Mycotoxin occurrence in feed and feed raw materials worldwide: long-term analysis with special focus on Europe and Asia. J. Sci. Food Agric. 93, 2892–2899.
Studer-Rohr, I., Schlatter, J., Dietrich, D.R., 2000. Kinetic parameters and intraindividual fluctuations of ochratoxin A plasma levels in humans. Arch. Toxicol. 74, 499–510.
Takagi, M., Uno, S., Kokushi, E., et al., 2018. Measurement of urinary concentrations of the mycotoxins zearalenone and sterigmatocystin as biomarkers of exposure in mares. Reprod. Domest. Anim. 53, 68–73.
Tangni, E.K., Pussemier, L., 2007. Ergosterol and mycotoxins in grain dusts from fourteen Belgian cereal storages: a preliminary screening survey. J. Sci. Food Agric. 87, 1263–1270.
Tao, Y., Xie, S., Xu, F., et al., 2018. Ochratoxin A: toxicity, oxidative stress and metabolism. Food Chem. Toxicol. 112, 320–331.
Tardieu, D., Bailly, J.D., Skiba, F., et al., 2007. Chronic toxicity of fumonisins in turkeys. Poultry Sci. 86, 1887–1893.
Tardieu, D., Bailly, J.D., Skiba, F., et al., 2008. Toxicokinetics of fumonisin B_1 in Turkey poults and tissue persistence after exposure to a diet containing the maximum European tolerance for fumonisins in avian feeds. Food Chem. Toxicol. 46, 3213–3218.
Tardieu, D., Bailly, J.D., Benlashehr, I., et al., 2009. Tissue persistence of fumonisin B_1 in ducks and after exposure to a diet containing the maximum European tolerance for fumonisins in avian feeds. Chem. Biol. Interact. 182, 239–244.
Thanner, S., Czegledi, L., Schwartz-Zimmermann, H.E., et al., 2016. Urinary deoxynivalenol (DON) and zearalenone (ZEA) as biomarkers of DON and ZEA exposure of pigs. Mycotoxin Res. 32, 69–75.
Toda, K., Kokushi, E., Uno, S., et al., 2017. Gas chromatography-mass spectrometry for metabolite profiling of Japanese Black cattle naturally contaminated with zearalenone and sterigmatocystin. Toxins 21, 9.
van der Westhuizen, L., Shephard, G.S., Burger, H.M., et al., 2011. Fumonisin B_1 as a urinary biomarker of exposure in a maize intervention study among South African subsistence farmers. Cancer Epidemiol. Biomark. Prev. 20, 483–489.
van der Westhuizen, L., Shephard, G.S., Gelderblom, W.C.A., et al., 2013. Fumonisin biomarkers in maize eaters and implications for human disease. World Mycotoxin J. 6, 223–232.

Vettorazzi, A., Gonzalez-Penas, E., de Cerain, A.L., 2014. Ochratoxin A kinetics: a review of analytical methods and studies in rat model. Food Chem. Toxicol. 72, 273–288.

Wang, S.K., Liu, S., Yang, L.G., et al., 2013. Effect of fumonisin B_1 on the cell cycle of normal human liver cells. Mol. Med. Rep. 7, 1970–1976.

Wang, Y., Wang, L., Liu, F., et al., 2016. Ochratoxin A producing fungi, biosynthetic pathway and regulatory mechanisms. Toxins 8, 83.

Warth, B., Braun, D., Ezekiel, C.N., et al., 2016. Biomonitoring of mycotoxins in human breast milk: current state and future perspectives. Chem. Res. Toxicol. 29, 1087–1097.

Weibking, T.S., Ledoux, D.R., Bermudez, A.J., et al., 1994. Individual and combined effects of feeding *Fusarium moniliforme* culture material, containing known levels of fumonisin B_1, and aflatoxin B_1 in the young Turkey poult. Poultry Sci. 73, 1517–1525.

WHO, 2017. Evaluation of Certain Contaminants in Food. World Health Organization (WHO), Geneva, Switzerland (Report No. 1002).

Xiao, H., Marquardt, R.R., Frohlich, A.A., et al., 1991. Effect of a hay and a grain diet on the rate of hydrolysis of ochratoxin A in the rumen of sheep. J. Anim. Sci. 69, 3706–3714.

Xu, L., Cai, Q., Tang, L., et al., 2010. Evaluation of fumonisin biomarkers in a cross-sectional study with two high-risk populations in China. Food Addit. Contam. 27, 1161–1169.

Yang, S., Zhang, H., Sun, F., et al., 2017. Metabolic profile of zearalenone in liver microsomes from different species and its *in vivo* metabolism in rats and chickens using ultra high-pressure liquid chromatography-quadrupole/time-of-flight mass spectrometry. J. Agric. Food Chem. 65, 11292–11303.

Zahoor Ul, H., Khan, M.Z., Khan, A., et al., 2012. Effects of individual and combined administration of ochratoxin A and aflatoxin B_1 in tissues and eggs of White Leghorn breeder hens. J. Sci. Food Agric. 92, 1540–1544.

CHAPTER

37

Poisonous Plants: Biomarkers for Diagnosis

Kip E. Panter, Kevin D. Welch, Dale R. Gardner

United States Department of Agriculture, Agricultural Research Service, Poisonous Plant Research Laboratory, Logan, UT, United States

INTRODUCTION

Poisonous plants and the toxins they produce result in major economic losses to the livestock industries throughout the world. In the 17 western United States, it was estimated that more than $340 million in annual losses were attributed to poisonous plants (Nielsen et al., 1988; Nielsen and James, 1992). This cost estimate used 1989 figures and only considered death losses and measureable reproductive losses in cattle and sheep. A revised estimate of over $500 million annually using more current animal prices was recently reported (Holechek, 2002). Less obvious costs such as lost grazing opportunities, additional feed costs, increased health-care costs, management adjustments, increased culling costs, lost weight gains, delayed or failed reproduction, and the emotional stress accompanying many poisonous plant cases were not included in the Nielsen and James or Holechek analyses. When one considers these other costs, inflation, and current animal values, and when all pastures and ranges in the United States are factored in, the economic cost of poisonous plants to the livestock industry is significant. In addition, an often ignored cost is the environmental impact on plant biodiversity from invasive species, many of which are poisonous. These invasive and poisonous species are often aggressive invaders and reduce optimum utilization of private, federal, and state-managed forest, range, and pasture lands. This aspect alone has far-reaching implications, not only for livestock producers but also for many other segments of society.

Animals graze poisonous plants for a number of reasons. In some cases, it is a matter of survival. For example, in the arid and semiarid livestock-producing regions of the world, such as the western United States, regions of South Africa, Australia, China, South America, and others, browsing or grazing animals may have limited access to high-quality forage at certain times of the year and are forced to survive by grazing some poisonous species. In other instances, hay or harvested forages from areas where poisonous plants are abundant may be contaminated with a high percentage of poisonous plants, and when animals are fed contaminated hay, they may be poisoned. On many ranges, poisonous plants are a normal but small part of the animal's diet as they are highly nutritious, i.e., larkspurs, locoweeds, and lupines, to name a few.

Poisonous plant problems are often exacerbated during periods of below normal rainfall when the abundance of grasses is reduced. Frequently, the animal's diet selection changes during the season when grasses and palatable forbs mature and senesce; for example, the consumption of some poisonous plants such as lupines, locoweeds, or larkspurs, which stay green longer into the season, may increase as the season progresses. Pine needles and junipers are often consumed in the winter. In other instances, poisoning occurs early in the season when poisonous plants such as lupine or death camas have emerged ahead of grasses. To restate the obvious, poisoning by plants only occurs when animals eat too much too fast or graze it over extended periods. Therefore, management strategies utilizing multiple factors are required to minimize losses from poisonous plants.

Over the last 20 years, research at the USDA-ARS Poisonous Plant Research Laboratory in Logan, Utah, has provided information utilized by livestock producers and land managers to significantly reduce losses from poisonous plants and improve economic sustainability in rural communities. Part of this research has emphasized identifying biomarkers of poisoning for diagnostics and research. Although much more research is required, new information presented here will assist clinicians and diagnostic laboratories in diagnosing cases

Biomarkers in Toxicology, Second Edition
https://doi.org/10.1016/B978-0-12-814655-2.00037-2

of poisonous plant toxicoses. This chapter is not intended to be all inclusive but focuses on some of the most economically important and geographically widespread poisonous plants to livestock producers in the United States.

ASTRAGALUS AND *OXYTROPIS* SPECIES (LOCOWEEDS, NITROTOXIN SPP., AND SELENIUM ACCUMULATORS)

Of all the poisonous plants found on rangelands and pastures in the United States, the *Astragalus* and *Oxytropis* genera cause the greatest economic losses to the livestock industry in the western states (Graham et al., 2009; Cook et al., 2009c). There are three toxic syndromes associated with these species: (1) locoweed poisoning caused by species containing the indolizidine alkaloid swainsonine (~24 species); (2) species containing nitrotoxins (~356 species and varieties); and (3) selenium accumulators (~22 species).

Locoweeds

Locoweed poisoning was one of the first poisonous plant problems recognized by stockmen and was reported as early as 1873. There are about 24 known species of *Astragalus* and *Oxytropis* that contain swainsonine and have been implicated in livestock poisonings. The term "loco" is Spanish, meaning crazy, and colloquially describes the aberrant behavior of locoweed-poisoned animals. All species of *Astragalus* and *Oxytropis* containing swainsonine are collectively referred to as locoweeds.

Toxicology

There are numerous effects of locoweed on animals, but the classic syndrome from which the term "locoism" derived is one of the central neurological dysfunctions, resulting in aberrant and often aggressive and unpredictable behavior. The disease is chronic, developing after weeks of ingesting locoweeds, and begins with depression, dull-appearing eyes, and incoordination, progressing to uncharacteristic behavior, including aggression, staggering, solitary behavior, wasting, and eventually death if continued consumption is allowed. Other problems associated with locoweed ingestion include reproductive failure, abortion, birth defects, weight loss, and enhanced susceptibility to brisket disease at high elevations (Panter et al., 1999b).

Locoweed poisoning affects all animals but because of the transient nature of the poisoning, animals removed from the locoweed early in the toxicosis will recover most of their function and may be productive animals. In the final stages of locoism, central nervous system tissue shows swelling of axonal hillocks (meganeurites) and growth of new dendrites and synapses. This altered synaptic formation in nervous tissue in severely affected animals is permanent and may be the cause of some irreversible neurological signs. Because of neurological dysfunction and apparent permanence of some lesions in the nervous system, horses are believed to be unpredictable and therefore unsafe to use for riding or draft, but they may remain reproductively sound once they have recovered from the poisoning.

Toxin

The indolizidine alkaloid swainsonine **(1)** was first isolated from the Australian plant *Swainsona canescens* (Colegate et al., 1979) and then soon thereafter isolated with its *N*-oxide from locoweeds (Molyneux and James, 1982).

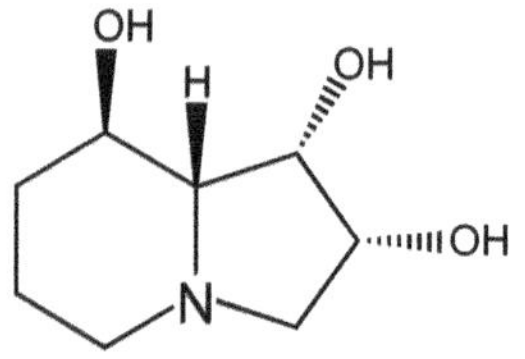

Swainsonine **(1)**

The locoweed poisoning syndrome is a lysosomal storage disease in which α-mannosidase is inhibited, resulting in prevention of hydrolysis of mannose-rich oligosaccharides in cells and accumulation of these oligosaccharides resulting in cellular dysfunction.

Pryor et al. (2009) discovered that swainsonine in *Astragalus* and *Oxytropis* species is produced by a fungal endophyte, *Undifilum oxytropis* (formerly called *Embellesia oxytropis*). A positive correlation was shown to exist between swainsonine concentrations found in the plant and concentrations of swainsonine produced by the endophytic fungus cultured from the same plant (Ralphs et al., 2008; Cook et al., 2009a). This same correlation was demonstrated for *Oxytropis glabra*, an important poisonous plant in the Inner Mongolia steppe (Ping et al., 2009).

Research results have shown that inhibition of α-mannosidase is relatively transient and quickly reversible once animals stop eating locoweed (Stegelmeier et al., 1994). Blood serum clearance of swainsonine is rapid (half-life of ~20 h); thus, the effects of locoweed should be reversible if tissue damage has not become extensive and permanent. This suggests that intermittent grazing of locoweed could be an effective means of reducing locoweed poisoning. There is also an apparent threshold dosage where severity of cell damage is more time-dependent than dosage-dependent. Once the threshold dosage is reached, which appears to be relatively low (0.35 mg/kg BW in the rat), eating more locoweed does not accelerate the toxicosis. Therefore, increasing animal numbers on loco pastures and reducing time of grazing is also a logical method to reduce economic impact yet utilize infested pastures.

Many locoweeds are biennials or perennials that flourish periodically under optimum environmental conditions. Historically, losses are regional and sporadic, with large regional economic impact. Individual cases of significant losses are frequent, and some historical cases are reported in James and Nielsen (1988).

Conditions of Grazing

In cows, preference to graze locoweed is relative to the amount and condition of other available forage. Many locoweeds are cool-season species that green up and start growth early in the spring, flower, set seed, and go dormant in summer, and then resume growth in fall. Livestock generally prefer the green, growing locoweeds to dormant grass. Sheep preferred the regrowth foliage of Green River milkvetch to dormant grasses during late fall and early winter on the desert range in eastern Utah (Ralphs et al., 2001). Horses selected green spotted locoweed instead of dormant grasses in the spring in Arizona (Pfister et al., 2003). Cattle readily grazed Wahweap milkvetch in proportion to its availability on desert winter range in southeastern Utah. In a series of grazing studies in northeast New Mexico, cattle readily grazed white locoweed in March–May but stopped grazing it in June as warm-season grasses became abundant and white locoweed matured and became coarse and rank. Stocker cattle grazed white-point locoweed on shortgrass prairies in May and early June, but weight losses continued throughout the summer, even though they had stopped eating locoweed. On mixed-grass prairies on the eastern foothills of the Rocky Mountains in northern Colorado, cattle ceased grazing white-point locoweed when it matured following flowering in mid-June and became rank and less palatable. However, they continued to graze it throughout the subsequent summer when abundant summer precipitation caused the locoweed to remain green and succulent (Ralphs et al., 2001).

Biomarkers of Poisoning (AST, GGTP, White Cell Count, Cellular Vacuolation, Serum Swainsonine, α-Mannosidase)

Clinically, locoweed poisoning in all classes of livestock and wildlife include depression, loss of appetite, weight loss followed by intention tremors, dull hair coat, proprioceptive deficits, aberrant behavior, abortion in pregnant animals, and eventually death with prolonged exposure. Evaluation of serum swainsonine versus serum alpha mannosidase will indicate recent exposure to locoweed; however, elimination kinetics of swainsonine is short ($t^{1/2}$ elimination of 20 h) and recovery rate of α-mannosidase is relative to disappearance of swainsonine. Therefore, serum evaluation of swainsonine or α-mannosidase is of limited value in diagnosis. Currently, serum biochemistry provides some support for a diagnosis as aspartate aminotransferase (AST), Lactic Acid Dehydrogenase (LDH), Gamma-Glutamyl Transferase (GGTP), and white blood cell counts are all elevated with locoweed poisoning. However, these are not pathognomonic to locoweed poisoning and without a history of locoweed ingestion would be of limited value as biomarkers. Currently, the best diagnosis for locoweed poisoning can only be obtained after necropsy and follow-up histopathology. Although there are few gross lesions seen in locoweed poisoned animals, there are many characteristic microscopic lesions. Most organ systems are affected; however, the nervous and endocrine systems are extremely sensitive and diseased cells of these organs are swollen and filled with dilated vacuoles described as cellular constipation. This cellular foamy vacuolation is readily observed in thyroid, pancreas, kidneys, testes, ovaries, and macrophages in nearly all tissues. Research at the Poisonous Plant Research Laboratory is focused on identifying selected abnormal proteins with a slow elimination kinetics as a biomarker to diagnose locoweed exposure and also as measure of prognosis for recovery.

Prevention of Poisoning and Management Recommendations

Prevention of poisoning remains a matter of management strategy adapted to individual grazing programs to minimize grazing of locoweed plants (Graham et al., 2009). Livestock should be denied access to locoweeds during critical periods when they are relatively more palatable than associated forages. On shortgrass prairies of northeastern New Mexico, stocker cattle should not be turned onto locoweed-infested rangelands until warm-season grasses start growth in late May or early June. Cattle on rangeland year-round should be removed from locoweed-infested areas in the spring when it is green and growing and warm-season grasses remain dormant. They can be returned to locoweed-infested pastures in summer when warm-season grasses are abundant.

Most locoweed species are endemic, growing only in certain habitats or on specific soils. Fences can be used to provide seasonal control of grazing. Reserving locoweed-free pastures for grazing during critical periods in spring and fall can prevent locoweed poisoning. Locoweed-free areas can be created by strategic herbicide use.

Currently, no broad management schemes or methods of treatment are known to generally prevent locoweed poisoning. Management strategies for individual operations have been developed once the grazing practices and options are identified, allowing utilization of the particular range and yet minimizing losses. It was determined that cattle generally rejected woolly loco even under extreme grazing conditions, but once they were forced to start eating it, they continued to graze it and became intoxicated. Ranchers should watch for these "loco eaters" and move them to locoweed-free pastures. Shortage of feed with high grazing pressure, social facilitation (loco eaters teaching nonloco eaters to accept loco), or supplementing with alfalfa hay or cubes may compel cattle to start grazing woolly locoweed. White-point locoweed is more palatable than woolly locoweed and is green before spring grasses begin to grow in northeastern New Mexico. Therefore, cattle readily graze white-point locoweed in early spring while grasses are dormant, but once green grass starts to grow, cattle switch off of locoweed. Recommendations include creating locoweed-free pastures through herbicide use, fencing, or selection of low locoweed-infested pastures for early spring grazing and also to provide a place to move the identified locoweed eaters. This practice appears to reduce the impact of locoweed on these ranges.

Grazing pressure can also force cattle to begin grazing locoweed when they run short of desirable forage (Ralphs et al., 1994). Ranchers should not overstock locoweed-infested ranges but, rather, should ensure adequate forage is always available. Improper use of some grazing systems may encourage livestock to graze locoweed.

In a grazing study comparing breeds, Brangus steers consumed more locoweed than did Hereford or Charolais steers (Ralphs et al., 1994). It was suggested that the gregarious nature of Brangus steers facilitated their acceptance of locoweed first among the other breeds. Observations of breed differences were acknowledged as early as 1909. C.D. Marsh suggested that black cattle and black-faced sheep were more likely to graze locoweed and become poisoned. Although this has not been experimentally substantiated, we do believe breed difference may play a role in an animal's propensity to graze locoweeds and become poisoned.

Nitro-Containing *Astragalus* (Milkvetches)

There are more than 260 species and varieties (356 taxa) of nitro-containing *Astragalus* in North America (Barneby, 1964; Welsh et al., 2007). They are frequently referred to as milkvetches, as are some of the other *Astragalus* species. Nitrotoxins are therefore the most common toxin in the *Astragalus*, followed by swainsonine (locoweeds), then selenium. Major livestock losses occur in many regions of the western United States. These plants are very diverse and concentrated on the deserts, foothills, and mountains of the west.

Toxicology

The nitro-containing *Astragalus* species cause acute and chronic poisoning in sheep, cattle, and horses. The acute form results in weakness, increased heart rate, respiratory distress, coma, and death. Although blood methemoglobin is high (induced from nitrotoxin metabolism to nitrites) and a contributing factor to the respiratory difficulties, administration of methylene blue in cattle does not prevent death. Therefore, the methemoglobinemia is apparently not the primary cause of death. The chronic form is the most frequent form of poisoning observed and follows a course of general weakness, incoordination, central nervous system involvement resulting in knuckling of the fetlocks, goose stepping, clicking of the hooves, and "cracker heels" progressing to paralysis and death. A respiratory syndrome is also present in the chronic and acute forms, with emphysema-like signs causing the animals to force respiration: "roaring disease." Sheep manifest the respiratory syndrome more than the central nervous syndrome and are more resistant to poisoning compared with cattle.

The toxic principles are β-d-glycosides of 3-nitro-1-propanol (NPOH) or 3-nitropropionic acid (NPA). The glycoside conversion occurs more readily in the ruminant because of the microflora and is apparently the reason for increased toxicity. The glycoside (miserotoxin **2**) is metabolized to the highly toxic NPOH in the gastrointestinal (GI) tract of ruminants (Williams et al., 1970). Thus, NPOH is absorbed in the gut and apparently converted to NPA by the liver. Further metabolism yields inorganic nitrite and an unidentified metabolite that may be involved in toxicity. It appears that NPOH is more rapidly absorbed from the gut than is NPA; therefore, forage containing the alcohol form is the most toxic.

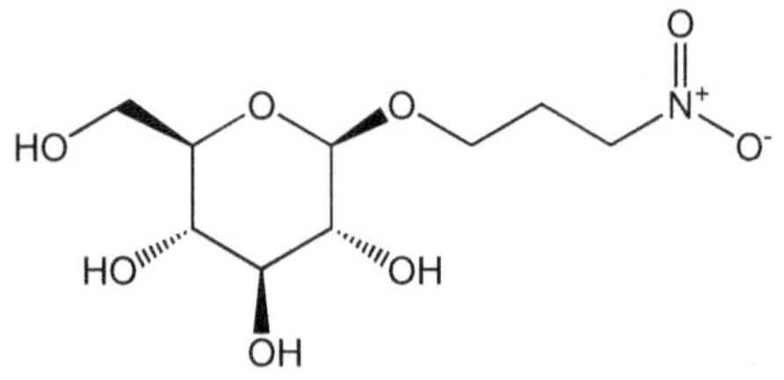

Miserotoxin (**2**)

Biomarkers of Poisoning (Clinical, Pathology, History of Ingestion, Methemoglobin)

The diagnosis of nitrotoxin-poisoned animals can be made by documenting exposure to the plant, identifying the characteristic clinical signs of poisoning, measuring serum 3-nitropropionic acid concentrations, and histological examination. Studies have shown that 3-nitropropionic acid is rapidly absorbed, distributed, and eliminated. Consequently serum 3-nitropropionic acid concentrations may be below detectable limits within days of removal from the plant. This limits the diagnostic usefulness of serum 3-nitropropionic acid detection, as none may be detected in the serum even though the animals are demonstrating severe clinical signs. There are no hematological or serum chemistry changes of diagnostic value. However, cattle and sheep have been shown to have elevated methemoglobin concentrations, in excess of 20%. The gross pathological changes that occur are not distinctive but include hepatic congestion, pulmonary emphysema and edema, excess pericardial fluid, and abomasal ulceration. Microscopic lesions are also nonspecific, with the exception of changes in the brains of cattle with chronic exposure. These changes include the presence of foci of necrosis in the thalamus, moderate necrosis of the Purkinje cells in cerebellar folia, spongiosis in the white matter of the globus pallidus, distension of the lateral ventricles, and wallerian degeneration of the ventral and lateral columns of the spinal cord and the sciatic nerve. Clinical signs of respiratory distress from the pulmonary emphysema "roaring disease" and rear leg paresis "cracker heels" would suggest 3-nitropropionic acid (milkvetch or miserotoxin) poisoning and should be considered in the differential diagnosis.

Prevention and Treatment

There is no preferred treatment for milkvetch poisoning, although treatment with methylene blue appears to reverse the methemoglobinemia but may not prevent death in cattle. Oxidation of NPOH to NPA is prevented if alcohol dehydrogenase is saturated with ethanol or inhibited with 4-methylpyrazole before NPOH is given. This suggests that NPOH is a good substrate for the enzyme, alcohol dehydrogenase. This information could be useful in acute cases; however, its value in treatment of poisoning in the field is limited.

Livestock losses can be reduced by decreasing the density of the *Astragalus* species with herbicides or avoiding grazing livestock on infested areas when the plant is most poisonous. Wasatch milkvetch contains the highest concentration of miserotoxin from bloom to immature seed pod stage of growth. Nitro compounds are found in all parts of the plant, but the leaves contain the highest concentration. Once the leaves begin to dry and lose their green color, the nitro levels drop very rapidly, and the plant is relatively nontoxic. However, the toxins in plants pressed green and preserved in herbaria appear to remain stable for years (Williams and Barneby, 1977). Herbicide treatment decreases the density of plants and also decreases the toxicity of the plants once they start to dry; therefore, spraying milkvetch appears to be the best method to reduce losses and still utilize infested ranges.

Seleniferous *Astragalus*

Approximately 22–24 species of *Astragalus* known to accumulate selenium (Se) have been identified (Rosenfeld and Beath, 1964; Welsh et al., 2007). These are less numerous and more geographically restricted than the nitro-containing species. Many of these species are referred to as Se-indicator plants because they only grow on soils high in bioavailable selenium; therefore, they are helpful in locating and identifying areas or soils high in selenium. The *Astragalus* species are deep-rooted plants and are thought to bring selenium (pump) from deeper soil profiles to the surface where shallow-rooted forbs and grasses will take up toxic levels. It is these facultative accumulators that create most of the subacute or chronic toxicity problems for livestock.

Toxicity

With selenium poisonings, one may observe acute, subacute, or chronic selenosis depending on the daily dose, Se type (organic vs. inorganic), and duration of exposure. Acute cases of selenium poisoning are rare and usually involve animals that have been exposed by one of four methods. (1) Forages that have superaccumulated selenium from seleniferous soils. (2) Environmental contamination from agricultural drain water, reclaimed soils from phosphate or ore mining, or from fly ash. (3) Accidental overdosing with pharmaceutical grade selenium such as Bo-Se in the treatment of white muscle disease. (4) Misformulated feed mixes. The signs of acute selenium poisoning include diarrhea, unusual postures, increased temperature and heart rate, dyspnea, tachypnea, respiratory distress, prostration, and death (Tiwary et al., 2006). Gross pathological findings are usually limited to pulmonary congestion, hemorrhage, and edema. Histologically, multifocal myocardial necrosis and pulmonary alveolar vasculitis are common (Tiwary et al., 2006).

Chronic selenium poisoning is most common and referred to as alkali disease because most areas with high concentrations of available selenium are alkaline in nature. Chronic selenosis occurs from prolonged ingestion of seleniferous forages containing 5–40 ppm

Se. Clinical signs include rough coat, hair or wool loss, poor growth, emaciation, abnormal hoof growth, lameness, dermatitis, and depressed reproduction (Rosenfeld and Beath, 1964; Raisbeck, 2000). In swine, a condition of paralysis (poliomyelomalacia or polioencephalomalacia) often occurs with cervical or lumbar involvement (Panter et al., 1996b). The description of a second chronic syndrome in cattle called "blind staggers" has been redefined and is now believed to be polioencephalomalacia induced by high sulfate water or high sulfate forage sources.

Selenium is found in soil and plants in both organic (**3**) and inorganic (**4**) forms. The organic forms are more bioavailable than the inorganic forms, resulting in higher animal tissue concentrations when administered at equivalent selenium doses (Tiwary et al., 2006; Davis et al. 2012). Although a dramatic difference in tissue selenium uptake between organic (selenomethionine) and plant (*Astragalus bisulcatus*) forms and inorganic (sodium selenate) forms occurs, the clinical and pathological syndromes are similar—i.e., poliomyelomalacia in pigs (Panter et al., 1996b) and pulmonary edema and hemorrhage in sheep and cattle (Tiwary et al., 2006; Davis et al. 2012).

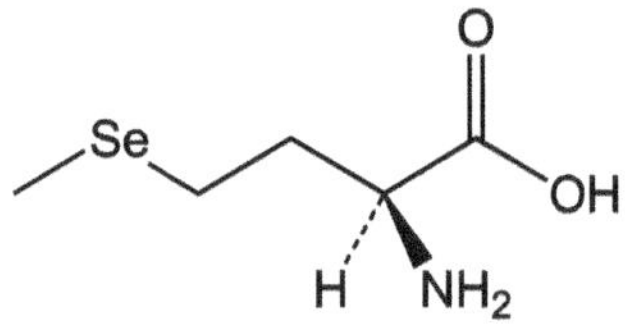

Selenomethionine (**3**)

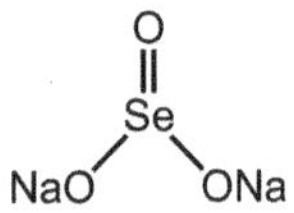

Sodium selenate (**4**)

Biomarkers of Poisoning (Liver Biopsy Se, Whole Blood Se, Enzyme, Hair, Mane, Tail, Hoof Samples)

A complete health history of the animals with excess selenium in the forages or feed supplementation providing the basis for suspecting selenium poisoning (acute or chronic) should be evaluated. Selenium is slowly excreted, requiring relatively long withdrawal times (weeks). Therefore, acute Se poisoning can be diagnosed by sampling numerous biological tissues, including whole blood, serum, liver biopsies, hair, tail, and hooves. Acute selenium poisoning can be easily diagnosed by sampling hair 3–4 weeks postexposure. Selenium appears to be excreted slightly slower from the liver; however, liver biopsies are a much more invasive procedure for diagnosis. Interestingly, selenium is excreted relatively slowly from the red blood cell, thus making whole blood a good tissue for analysis (Davis et al., 2012). The primary lesion in acute selenosis in cattle has been found to be myocardial necrosis with subsequent heart failure seen as passive congestion and centrilobular hepatic congestion with hepatocellular degeneration and necrosis. Consequently increases in serum creatine kinase and troponin can be used as biomarkers of acute selenosis. Chronic selenium poisoning results in bilateral hair loss over the withers of cattle and horses, tail and mane hair loss, overgrown coronary bands or separation and sloughing of the outer shell of the hooves, and an overall unkept appearance. In cattle excess selenium, or selenium deficiency, may reduce overall reproductive performance. In horses, sampling of the tail hair then cutting into 1 inch increments and performing selenium analysis on each of the 1 inch increments will provide a good history of selenium ingestion for up to a 3-year period depending on how long the tail hair is (Davis et al., 2014). In cattle, a hair sample should be clipped from the shoulder area close to the skin and divided into 2 or 3 samples based on hair length. This will provide a good evaluation of selenium status and an approximate time when selenium ingested. The 0.5 cm sample closest to the skin would indicate recent selenium exposure (0–7 days), whereas the center and outer samples would indicate more long-term exposure (Davis et al., 2012). In pigs, selenium poisoning (subacute and acute) manifests with cervical or lumbar paresis and lesions of polioencephalomalacia or poliomyelomalacia and chronic poisoning by hair loss, overgrown hooves, reduced growth rate, and overall poor or unkept appearance. In sheep, acute selenium poisoning is manifest by respiratory distress, depression, lethargy, and death within hours to a day or two after ingestion. There are few clinical signs associated with chronic selenium ingestion in sheep; however, reproductive performance may be impaired.

Prevention of Poisoning

There is no treatment for selenium poisoning except removal from the source, allowing spontaneous recovery in chronic cases. Treat overgrown hooves, i.e., trim horse's hooves frequently until normal hoof growth resumes (6–12 months). Monitoring soils in a particular area and understanding the plant communities can provide the management information to avoid poisoning. In areas where selenium is a problem, many ranchers have switched to grazing steers to avoid decreased reproductive efficiency in cows. Sheep appear to be more resistant to chronic selenosis compared with cattle and are better adapted for some of these ranges. However, sheep are sensitive to acute selenium poisoning, as was observed when a large number of sheep died within days after grazing on mine reclamation sites that contained very high soil and plant selenium concentrations (Panter, personal communications, 2004). Monitoring for selenium concentrations and forms in soil, vegetation, as well as animal tissues and hair, can help avoid poisoning incidents. Likewise, deficiency problems can be rapidly resolved with frequent monitoring and trace mineral supplementation.

LARKSPURS (*DELPHINIUM* SPP.)

There are more than 80 wild species of larkspurs in North America, classified into two general categories based primarily on mature plant height and distribution: low larkspurs and tall larkspurs. The larkspurs are a major cause of cattle losses on western ranges. As early as 1913, C.D. Marsh reported that more cattle deaths on western ranges are caused by larkspurs than by any other poisonous plant except locoweed. Large sporadic losses continue to occur on ranges with both low and tall larkspur species.

Toxicology

Larkspurs (*Delphinium* spp.) are a serious economic problem for cattle producers utilizing valley, foothill, and mountain rangelands in western North America. The toxicity of larkspur plants is due to norditerpenoid alkaloids, which occur as one of two chemical structural types—the *N*-(methylsuccinimido) anthranoyllycoctonine (MSAL) type (**5**) and the 7,8-methylenedioxylycoctonine (non-MSLA) type (**6**). Although the MSAL-type alkaloids are much more toxic (typically more than 20 times) (Panter et al., 2002), the non-MSAL-type alkaloids are generally more abundant in *Delphinium barbeyi* and *Delphinium occidentale* populations and will exacerbate the toxicoses (Gardner et al., 2002, Welch et al., 2010, 2012). The effects of 7, 8-methylenedioxylycoctonine-type diterpenoid alkaloids may enhance the overall toxicity of tall larkspur (Delphinium spp.) in cattle. Consequently, for a larkspur plant to be toxic to livestock, a sufficient quantity of MSAL-type alkaloids is required. However, MDL-type alkaloids appear to potentiate the overall toxicity of the MSAL-type alkaloids and should be considered when predicting potential toxicity of larkspur populations (Welch et al., 2010, 2012).

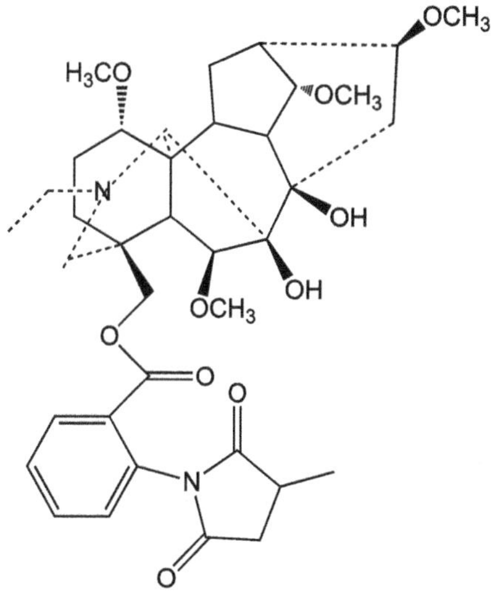

Methyllycaconitine (**5**)

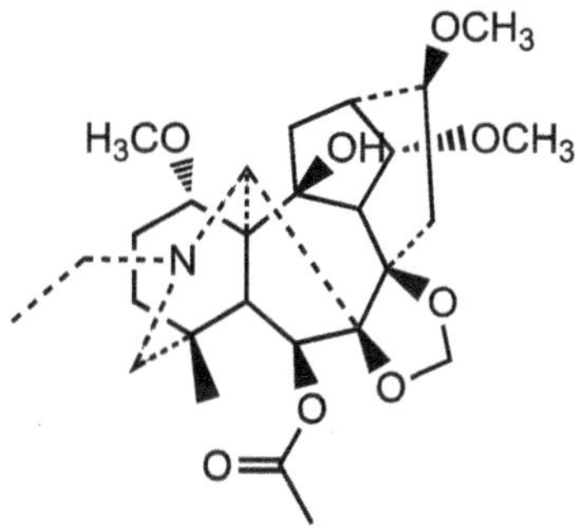

Deltaline (**6**)

Toxicity declines rapidly in tall larkspurs once pods begin to shatter. Measuring plant toxicity early in the growing season may allow prediction of grazing risk throughout the season (Ralphs et al., 2002). Because of the fact that the MSAL-type alkaloids are much more toxic than the MDL-type alkaloids, management recommendations for grazing cattle on larkspur-containing ranges are based primarily on the concentration of MSAL-type alkaloids in larkspur (Pfister et al., 2002; Ralphs et al., 2002).

Clinical signs of intoxication include muscular weakness and trembling, straddled stance, periodic collapse into sternal recumbency, respiratory difficulty, and death while in lateral recumbency. The primary result of larkspur toxicosis is neuromuscular paralysis from blockage at the postsynaptic neuromuscular junction (Benn and Jacyno, 1983). Cattle generally begin consuming tall larkspur after flowering racemes are elongated, and consumption increases as larkspur matures. Consumption usually peaks during the pod stage of growth in late summer, when cattle may eat large quantities (25%–30% of diet as herd average; >60% on some days by individual animals). Because larkspur toxicity generally declines throughout the growing season, and cattle tend to eat more larkspur after flowering, this risk period of greatest danger has been termed the "toxic window" (Pfister et al., 2002). This toxic window extends from the flower stage into the pod stage, or approximately 5 weeks, depending on temperature and elevation. Many ranchers typically defer grazing on tall larkspur-infested ranges until the flower stage to avoid death losses. This approach is dangerous as it places cattle in larkspur-infested pastures when risk of losses is high. This approach also wastes valuable, high-quality forage early in the season when risk of grazing is much lower. An additional 4–6 weeks of grazing may be obtained by grazing these ranges early, before larkspur elongates flowering racemes. The risk of losing cattle is low when grazing before flowering, even though larkspur is very toxic, because cattle prefer the highly palatable lush grasses. Once pods mature and begin to shatter, larkspur ranges can usually be grazed with impunity because pod toxicity declines rapidly, and leaf toxicity is low.

Managing cattle on low larkspur pastures is quite different from that on tall larkspur pastures. Not all low larkspurs are high risk, and alkaloid analyses should be done before grazing. On pastures where toxic species of low larkspur grow, delayed grazing until after pods shatter and plants begin to dry up is the only safe approach. Based on limited studies, cattle increase consumption of low larkspur after flowering, and increases in grazing pressure increase amounts of low larkspur eaten by cattle (Pfister et al., 2002).

Biomarkers of Poisoning (Blood Alkaloid, Liver Alkaloid, Plant Fragment, or Rumen Sample)

For diagnostic purposes, the toxic larkspur alkaloids can usually be detected in the whole blood, liver, kidney, sera, and rumen content of poisoned animals. Because of the acute nature of larkspur toxicosis, if the animals survive the initial intoxication they will recover and have no lasting effects. Consequently, in general, when a diagnosis of a larkspur-poisoned animal is needed, it is for a dead animal. The diagnosis of larkspur poisoning is currently made by documenting exposure to the plant, identifying the characteristic clinical signs, identifying larkspur in the rumen contents, and measuring serum alkaloid concentrations. Studies have shown that the larkspur alkaloids are rapidly absorbed, distributed, and eliminated. Serum alkaloid concentrations can be detected for approximately 7 days after intoxication. Because of the difficulty for most diagnostic laboratories to accurately identify plant parts in rumen contents and measure serum for larkspur alkaloids, experiments are being conducted to use PCR technology to identify larkspur plants in rumen contents. The use of a common and robust technology such as PCR will allow for most any veterinary diagnostic laboratory to determine if the plant was present in the rumen of the animal.

Prevention and Management of Poisoning

Grazing Management

A simple and low-risk grazing management scheme can often be used based simply on tall larkspur growth and phenology: (1) graze during early summer when sufficient grass is available until larkspur elongates flowering racemes (4–6 weeks, depending on elevation and weather); (2) remove livestock, or contend with potentially high risk from flowering to early pod stages of growth (4 or 5 weeks); and (3) graze with low risk during the late season when larkspur pods begin to shatter (4–6 weeks). This scheme can be refined substantially if livestock producers periodically obtain an estimate of the toxicity of tall larkspur, and if ranchers spend time periodically observing and documenting larkspur consumption by grazing cattle.

Management to reduce losses to low larkspur begins with recognition of the plant during spring. Vegetative low larkspur plants will typically begin growth before the major forage grasses. Low larkspur populations fluctuate with environmental conditions (Pfister, unpublished data). Risk of losing cattle is much higher during years with dense populations. During those years, recognizing the plant, and finding alternative pasture or waiting to graze infested pastures for 4–6 weeks until the low larkspur has dried up, will reduce losses. In addition, it is recommended that animals not be watered or provided mineral supplementation in areas that have high densities of larkspurs.

Drug Intervention

A variety of remedies have been applied in the field when ranchers find intoxicated animals, but most are without a solid scientific rationale. Treatment for overt poisoning is usually symptomatic, and recovery is often spontaneous if animals are not stressed further. Once the animal is observed showing muscular tremors, it should be allowed to drop back and proceed at its own pace. Poisoned animals should never be forced to continue moving because this will exacerbate the clinical effects and can result in death. Drugs that increase acetylcholine effectiveness at the neuromuscular junction have potential for reversing larkspur toxicosis or reducing susceptibility. The cholinergic drug physostigmine (0.08 mg/kg intravenous (i.v.)) has been successfully used under field and pen conditions to reverse clinical larkspur intoxication (Pfister et al., 1994). Similarly, i.v. administration of neostigmine (0.04 mg/kg) significantly reduced clinical signs in cattle (Green et al., 2009), and neostigmine administered intramuscularly at 0.02 mg/kg can be used as a rescue treatment for cattle that are recumbent. This reversal lasts approximately 2 h, and repeated injections of neostigmine are sometimes required. Under field conditions, neostigmine temporarily abates clinical signs and animals quickly (~15 min) become ambulatory. Depending on the larkspur dose, the intoxication may recur. The use of physostigmine-based treatments may aggravate losses in the absence of further treatment if intoxicated, yet ambulatory animals later develop increased muscular fatigue, dyspnea, and death.

LUPINES (*LUPINUS* SPP.)

The *Lupinus* genus contains more than 150 species of annual, perennial, or soft woody shrub lupines. More than 95 species occur in California alone. The lupines are rich in alkaloids, responsible for most of the toxic and teratogenic properties. There are domestic lupines that, through plant breeding, are low in alkaloid content and have been selected for ornamental purposes or for animal and human food. Only the range lupines known to cause poisoning or birth defects are discussed here.

Stockmen have long recognized the toxicity of lupines when livestock, particularly sheep, were poisoned in the late summer by the pods and seeds of lupine. Major losses in sheep were reported in the 1950s, and individual flock losses of hundreds and even thousands were reported. Lupines are also poisonous to other livestock, and field cases of poisoning in cattle, horses, and goats have been reported. However, the most recognized condition of lupine ingestion is the "crooked calf syndrome," a congenital condition in calves resulting in skeletal contracture-type malformations and cleft palate after their mothers have grazed lupines during sensitive periods of pregnancy (Panter et al., 1999a). The condition was first reported in 1959 and experimentally confirmed after large outbreaks in Oregon and Montana in 1967.

Toxicology

Most lupine species contain quinolizidine alkaloids, a few contain piperidine alkaloids, and some contain both. The specific alkaloids responsible for crooked calf syndrome are anagyrine **(7)**, ammodendrine **(8)**, *N*-acetylhystrine, and *N*-methyl ammodendrine. Hence, risk is based on chemical profile and the presence and concentration of these teratogenic alkaloids. It is known that chemical profile and concentration differ, resulting in changing levels of toxicity within and between species and populations. The chemical phenology has been studied in *Lupinus caudatus* and *Lupinus leucophyllus* (Lee et al., 2007). Total alkaloid concentration is high in the new early growth but diluted as the plant biomass increases. Pools of total alkaloids increase during the phenological growth stages and peak at the pod stage, concentrating in the pods. The teratogenic alkaloid anagyrine appears to be an end product in the biosynthetic pathway and accumulates in the floral parts and is stored in the seed. Following seed shatter, both concentration and pools of all alkaloids decline precipitously, leaving the senescent plant relatively nontoxic.

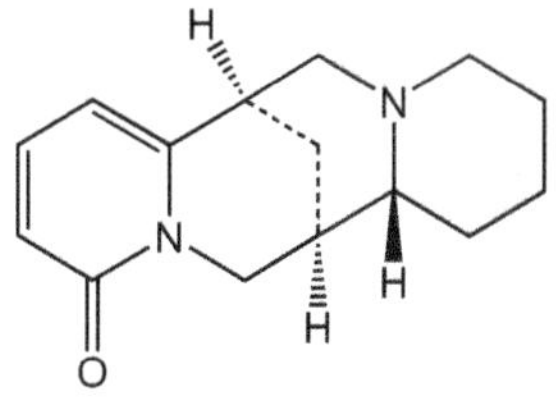

Anagyrine **(7)**

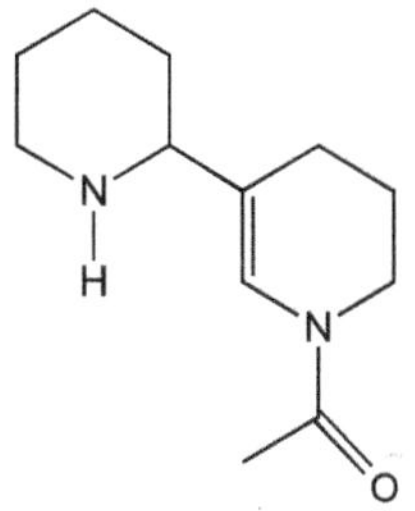

Ammodendrine **(8)**

The toxicity of lupines was first recognized in the early 1900s. Lupines can be toxic to sheep at 0.25%–1.5% of their body weight depending on alkaloid composition. A few cases of poisoning have occurred on young plants. Losses of 80–100 sheep in multiple bands have been reported during the past 5 years in Idaho and Wyoming (Panter, personal communication, 2005).

Poisoning by lupine plants should not be confused with lupinosis reported in Australia. Lupinosis is a condition that is entirely different, as it is a mycotoxicosis of livestock caused by toxins produced by the fungus *Phomopsis leptostromiformis*, which colonizes domestic lupine stubble. It affects livestock that graze lupine stubble and limits the use of this animal forage in Australia.

Historically, lupines were responsible for more sheep deaths than any other single plant in Montana, Idaho, and Utah. Most losses occurred from hungry sheep grazing seed pods. Poisoning occurred following trucking or trailing bands in late summer or fall, or after getting caught in early snowstorms that covered herbaceous vegetation. Hungry sheep nonselectively grazed lupine pods, which are high in alkaloids, and were poisoned. Large losses have also occurred when lupine hay harvested in the seed pod stage was fed in winter.

Lupine-induced crooked calf syndrome was first reported in 1959 and 1960 and experimentally confirmed in 1967 (Panter et al., 1999a). Crooked calf syndrome includes various skeletal contracture-type birth defects and occasionally cleft palate. The skeletal defects are similar to an inherited genetic condition reported in Charolais cattle. Based on epidemiologic evidence and chemical comparison of teratogenic and nonteratogenic lupines, the quinolizidine alkaloid anagyrine was determined to be the teratogen (Keeler, 1973). A second teratogen, a piperidine alkaloid, ammodendrine, found in *Lupinus formosus,* was also demonstrated to cause the condition (Keeler and Panter, 1989). Further research determined that the anagyrine-containing lupine only caused birth defects in cattle and did not affect sheep or goats; however, the piperidine alkaloid-containing lupine, *L. formosus,* induced similar birth defects in cattle and goats (Keeler and Panter, 1989).

Cattle Grazing

Different lupines produce varying toxic syndromes in livestock, apparently because the alkaloid profile varies remarkably among plant species and between populations. Season and environment influence alkaloid concentration in a given species of lupine. Generally, alkaloid content is highest in young plants and in mature seeds. Alkaloids are not lost on drying, so wild hay may be highly toxic if young lupine plants or seed pods are present. For many lupines, the time and degree of seeding vary from year to year. Most losses occur under conditions in which animals consume large amounts of plant or pods in a brief period, such as when they are being driven through an area of heavy lupine growth, unloaded into such an area, trailed through an area where the grass is covered by snow but the lupines are exposed, or when feeding lupine hay with pods. Most serious poisonings may occur in the late summer or early fall because lupine remains green after other forage has dried, and seed pods may be present. Once the poisonings were understood, the practice of harvesting lupine hay for winter sheep feed was discontinued.

Lupine is not very palatable to cattle, although it has been considered fair to good quality feed on some ranges that are heavily utilized. Its palatability or acceptability depends on availability and maturity of other forage. In a grazing study of velvet lupine (*L. leucophyllus*) on annual cheatgrass ranges in eastern Washington (Ralphs et al., 2006), cows selected lupine in July and August after cheatgrass dried and other forbs were depleted or matured and became less palatable. The deep-rooted lupine remained green and succulent longer into the summer than the other forage. Lupine was higher in crude protein and lower in fiber (NDF) than the other forages throughout the season (the crude protein level in foliage was 15%, and in seeds it was 36%).

The abundance of lupine is another factor influencing the amount of lupine consumed. Lupine population cycles are influenced by weather patterns. Catastrophic losses from lupine-induced crooked calves occurred in the Channel Scabland region of eastern Washington in 1997. Annual precipitation from 1995 to 1997 was 33% above average, precipitating a population outbreak of lupine throughout the region. The density of velvet lupine plants has declined since then (Ralphs, unpublished data), and the incidence of crooked calves has returned to what has become an acceptable tolerance of 1%–5% incidence.

Clinical signs of poisoning are of muscular weakness (neuromuscular blockade) beginning with nervousness, frequent urination and defecation, depression, frothing at the mouth, relaxation of the nictitating membrane, ataxia, muscular fasciculations, weakness, lethargy, collapse, sternal recumbency followed by lateral recumbency, respiratory failure, and death. Signs may appear within 15 min to 1 h after ingestion or as late as 24 h, depending on the amount and rate of ingestion. Death usually results from respiratory paralysis.

The incidence of crooked calves is variable geographically and from year to year within a given herd. Up to 100% of a given calf crop may be affected, and individuals may be more severely affected than others. Affected calves are generally born alive at full term. Dystocia may occur when calves are severely deformed and assistance is required, often resulting in cesarean section.

Arthrogryposis is the most common malformation observed and is often accompanied by one or more of the following: scoliosis, torticollis, kyphosis, or cleft palate. Elbow joints are often immobile because of malalignment of the ulna with the articular surfaces of the distal extremity of the humerus. The part of the limb distal to the elbow joint is often rotated laterally. In crooked calf syndrome, the osseous changes observed are permanent and generally become progressively worse as the calf grows and its limbs are subjected to greater load-bearing stress. Frequently, minor contractions such as "buck knees," often attributed to lupine, will resolve on their own, and the calf will appear relatively normal.

The sensitive gestational period in the pregnant cow for exposure is 40–70 days with periods extending to day 100 (Panter et al., 1997). The condition has been experimentally induced with dried ground lupine at 1 g plant/kg BW and with semipurified preparations of anagyrine (the apparent teratogen) at 30 mg anagyrine/kg BW fed daily from 30 to 70 days of gestation

(Keeler et al., 1976). The dose range of anagyrine to cause crooked calves is 6.5–11.9 mg/kg BW/day for 3 or 4 weeks during gestation days 40–70. Crooked calf disease has also been induced by feeding the piperidine alkaloid-containing lupine, *L. formosus* (Keeler and Panter, 1989). The teratogenic piperidines—ammodendrine, *N*-acetylhystrine, and *N*-methyl ammodendrine—are absorbed quickly after ingestion and can be detected in blood plasma by 0.5 h, with peak levels maintained for more than 24 h (Gardner and Panter, 1993). The mechanism of action has been determined to be an alkaloid-induced reduction in fetal movement by a neuromuscular blocking effect during the critical stages of gestation (Panter et al., 1990a). This inhibition of fetal movement is due to stimulation followed by desensitization of skeletal muscle-type nAChR (Lee et al., 2006). This mechanism is a common factor for multiple alkaloids found in many species of lupines, poison hemlock (*Conium*), and wild tree tobacco (*Nicotiana glauca*), and research using TE-671 cells that express human fetal muscle-type nAChR and SH-SY5Y cells that express human autonomic-type nAChR supports this mechanism (Green et al., 2010).

Biomarkers of Poisoning (Clinical Effects, History of Ingestion, Serum Analysis, Liver, Urine, Crooked Calf Syndrome (CCS))

There are few biomarkers for diagnosing lupine poisoning. However, there are clinical signs that should be considered in a differential diagnosis such as incoordination, nictitating membrane partially covering the eye, muscular weakness especially after mild exertion and death after a history of potential lupine exposure. Alkaloid screening of the serum, liver, rumen content, or urine of poisoned animals with positive analysis for quinolizidine or piperidine alkaloids is diagnostic, especially if a history of lupine exposure exists. When newborn calves are presented with skeletal contractures especially arthrogryposis, scoliosis, torticollis, kyphosis, and/or cleft palate, lupine ingestion during pregnancy should be suspected. Because of the acute nature of lupine toxicosis, if mature animals survive the initial intoxication they will recover and have no lasting effects. However, if a pregnant cow is poisoned between the GD 40 and 100, the calf will likely be born with skeletal contractures. The diagnosis of lupine poisoning is currently made by documenting exposure to the plant, identifying the characteristic clinical signs, identifying lupine in the rumen contents, and measuring serum alkaloid concentrations. Studies have shown that the lupine alkaloids are rapidly absorbed, distributed, and eliminated. Serum alkaloid concentrations can be detected for approximately 3–4 days after intoxication.

Prevention, Management, and Treatment

Keeler et al. (1977) proposed a simple management solution to prevent crooked calves: graze lupine-infested pastures so that the susceptible period of gestation (40–70 days) does not overlap the flower and pod stage of growth when anagyrine is highest. Ralphs et al. (2006) refined Keeler's recommendations to restrict access during the susceptible period of gestation, when anagyrine concentration is still high in the flower and pod stage, only when cattle are likely to eat lupine and in years when it is abundant. Panter et al. (2013) suggested that intermittent grazing between lupine-infested pastures and lupine-free pastures would allow the fetus to regain normal movement for a few days during the sensitive stage of gestation. It has been hypothesized that inhibited fetal movement over a prolonged period is required for severe malformations to occur (Panter et al., 1999a).

Lupines are easily controlled with 2,4-D-type broadleaf herbicides (Ralphs et al., 1991); however, herbicide treatment alone rarely provides long-term solutions to poisonous plant problems. Seed reserves in the soil will rapidly reestablish the stands if grazing management practices are not implemented.

Death losses in sheep can be reduced by recognizing the variability in lupine toxicity with stage of growth and the conditions under which animals graze the plant. Providing a choice of other quality forages usually prevents excess lupine grazing. The dangerous period of plant growth for sheep exists mainly with plants in the pod stage. The hazard increases if sheep are hungry, as is often the case with crowding, hauling, driving, or overgrazed conditions. The hazard is reduced or eliminated after seed pods shatter.

Treatment for overt poisoning is usually symptomatic, and recovery is often spontaneous if animals are not stressed further. Once the animal is observed showing muscular tremors, it should be allowed to drop back and proceed at its own pace. Poisoned animals should never be forced to continue moving because this will exacerbate the clinical effects and can result in death. The elimination of the toxic alkaloids in the urine is quite rapid ($t½ \sim 6.3–6.9$ h) and begins within hours of ingestion (Lopez-Ortiz et al., 2004). Therefore, allowing the animal to rest and move slowly will often result in full recovery within 24 h. There is no treatment for the malformations, and euthanasia is recommended for the serious skeletal defects and cleft palate. However, less severe contracture defects, particularly of the front legs (buck knees), will often resolve if the knee joint can be locked within 1 week after birth. If not, the defect generally becomes worse with growth and size, and although the animal will continue to grow, the front legs will break down until the animal is unable to survive.

POISON HEMLOCK (*CONIUM MACULATUM*)

Poison hemlock (*C. maculatum*) was introduced into the United States as an ornamental plant from Europe and has become naturalized throughout the United States (Kingsbury, 1964). The plant generally grows in waste places or habitats where there is adequate moisture to sustain populations. It is a biennial, however, where temperatures and adequate moisture allow seed germination in the fall the 2-year cycle for seed production may occur in one season. Poison hemlock has an interesting history as it is believed to be the plant from which a decoction (hemlock tea) was used to execute Socrates and others from that era. An interesting review describes the historical perspective of poison hemlock (Daughtery, 1995).

Toxicology

Eight piperidine alkaloids are known in poison hemlock, five of which are commonly discussed in the literature. Two alkaloids (coniine **(9)** and γ-coniceine **(10)**) are prevalent and likely responsible for toxicity and teratogenicity of the plant. γ-Coniceine is the predominant alkaloid in the early vegetative stage of plant growth and is a biochemical precursor to the other *Conium* alkaloids (Panter and Keeler, 1989). Coniine predominates in late growth and is found mainly in the seeds. γ-Coniceine is seven or eight times more toxic than coniine in mice (Lee et al., 2013). This makes the early growth plant most dangerous in the early spring, and the seedlings and regrowth again in the fall. This is also the time when green feed is limited to livestock and may impact their propensity to graze this plant. Frequently, poison hemlock encroaches into hay fields where it can be problematic in green chop or silage. Ensiling usually reduces toxicity; however, hot spots in silage pits increase risk of poisoning. Seeds, which are very toxic, can contaminate poultry and swine cereal grains (Panter and Keeler, 1989). Plants often lose their toxicity on drying, but seeds remain toxic as long as the seed coat is intact.

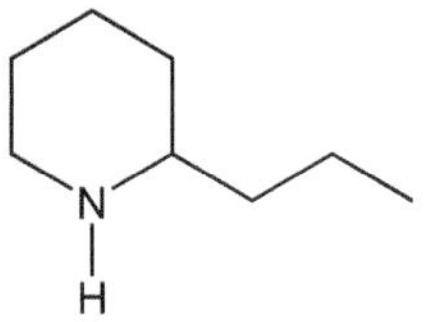

Coniine **(9)**

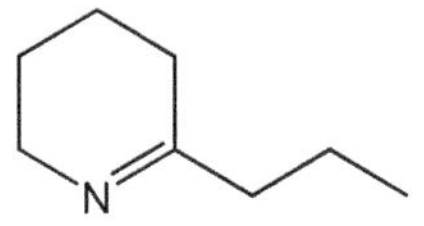

γ-Coniceine **(10)**

The clinical signs of toxicity are the same in all species and include initial stimulation (nervousness), resulting in frequent urination and defecation (no diarrhea), rapid pulse, temporarily impaired vision from the nictitating membrane covering the eyes, muscular weakness, muscle fasciculations, ataxia, incoordination followed by depression, recumbency, collapse, and death from respiratory failure (Panter et al., 1988). *Conium* plant and seed are teratogenic, causing contracture-type skeletal defects and cleft palate like those of lupine. Field cases of teratogenesis have been reported in cattle and swine and experimentally induced in cattle, swine, sheep, and goats (Panter et al., 1999a). Birth defects include arthrogryposis (twisting of front legs), scoliosis (deviation of spine), torticollis (twisted neck), and cleft palate. Field cases of skeletal defects and cleft palate in swine and cattle have been confirmed experimentally.

In cattle, the susceptible period for *Conium*-induced terata is the same as that described for lupine and is between days 40–70 of gestation. The defects, susceptible period of pregnancy, and probable mechanism of action are the same as those of crooked calf syndrome induced by lupines (Panter et al., 1999a). In brief, these alkaloids and their enantiomers in poison hemlock, lupines, and *N. glauca* were more effective in depolarizing the specialized cells TE-671, which express human fetal muscle-type nAChR, relative to SH-SY5Y, which predominantly express autonomic nAChRs, in a structure–activity relationship (Panter et al., 1990a; Lee et al., 2006, 2008; Green et al., 2010). In swine, sheep, and goats, the susceptible period of gestation is 30–60 days. Cleft palate has been induced in goats when plant, or toxins, was fed from 35 to 41 days of gestation (Panter and Keeler, 1992).

Field cases of poisoning have been reported in cattle, swine, horses, goats, elk, turkeys, quail, chickens, and Canada geese (Panter et al., 1999a). Poisoning in wild geese eating small seedlings in early spring was most recently reported (Panter, personal communication). Human cases of poisoning are occasionally reported in the literature, and a case of a child and his father mistakenly ingesting the plant has been reported. Field cases of teratogenesis have been reported in cattle and swine and experimentally induced in cattle, sheep, goats, and swine (Panter et al., 1990a). Pigs become habituated to poison hemlock, and if access to the plant is not limited, they will eat lethal amounts within a short time.

Biomarkers of Poisoning

There are no diagnostic lesions or serum biomarkers in poisoned animals, and diagnosis is based on clinical history of exposure and/or alkaloid (coniine, *N*-methyl coniine, or γ-coniceine) detection in the liver, urine, blood, or stomach contents. At necropsy, the presence of plant fragments in the rumen and a characteristic pungent odor in the contents with chemical confirmation of the alkaloids is diagnostic. Characteristic contracture-type skeletal birth defects or cleft palate in cattle, sheep, or goats similar to those produced by lupines should be included in the differential diagnosis.

Prevention and Treatment

Prevention of poisoning is based on recognizing the plant and its toxicity and avoidance of livestock exposure when hungry. If a lethal dose has not been ingested, the clinical signs will pass spontaneously, and a full recovery can be expected. Avoidance of stressing animals poisoned on *Conium* is recommended. However, if lethal doses have been ingested, supporting respiration, gastric lavage, and activated charcoal are recommended. Control of plants is easily accomplished using broadleaf herbicides; however, persistent control measures are recommended because seed reserves in the soil will quickly reestablish a population.

The mechanism of action of the *Conium* alkaloids is twofold. The most serious effect occurs at the neuromuscular junction, where they act as nondepolarizing blockers such as curare. Systemically, the toxins cause biphasic nicotinic effects, including salivation, mydriasis, and tachycardia, followed by bradycardia as a result of their action at the autonomic ganglia. The teratogenic effects are undoubtedly related to the neuromuscular effects on the fetus and have been shown to be related to reduction in fetal movement (Panter et al., 1990a). Likewise, cleft palate is caused by the tongue interfering in palate closure during reduced fetal movement and occurs during days 30–40 of gestation in swine, 32–41 days in goats, and 40–50 days in cattle (Panter and Keeler, 1992).

WATER HEMLOCK (CICUTA SPP.)

Water hemlock (*Cicuta* spp.) is among the most violently poisonous plants known to humans. It is often confused with poison hemlock because of its name, growth patterns, and appearance. However, there are distinct differences in morphology and habitat.

Toxicology

The primary toxic principle in water hemlock is a long-chain, highly unsaturated alcohol called cicutoxin (**11**). Water hemlock acts on the central nervous system inducing violent grand mal seizures and death from respiratory failure.

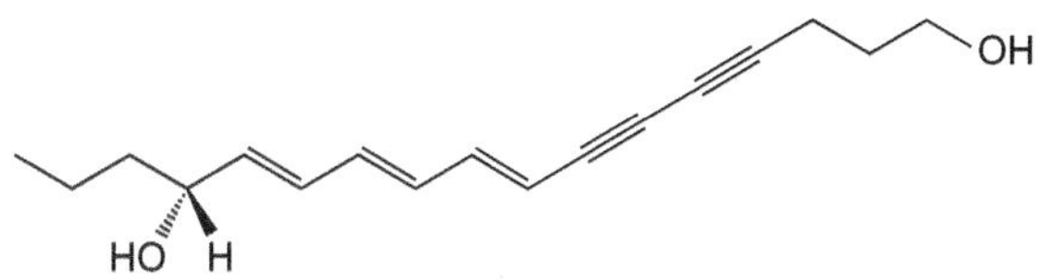

OH
HO H

Cicutoxin (**11**)

Tubers are the most toxic part of the plant, especially in early spring. The parsnip-like roots extending from the tuber are two to four times less toxic, and as the vegetative parts of the plant grow and mature, they become less toxic. Preliminary studies suggest that mature leaves and stems are much less toxic and are nontoxic after drying (Panter et al., 1988). Historically, water hemlock was believed to be most dangerous in early spring, and poisoning usually occurred when animals milled around in streambeds or sloughs and exposed tubers, which were then ingested. Although this is true, a recent case of poisoning and death in cattle after ingesting flower and green seed heads implicates this phenological stage as dangerous also (Panter et al., 2011). Chemical comparison of green seed and tubers combined with mouse bioassay studies showed that green seed is also toxic. Like tubers, the more mature vegetation, including leaves, flowers, and green seed heads, was palatable. The parsnip-like roots given free choice to hamsters showed that they were very palatable, and the hamsters actually preferred these roots to their normal laboratory chow. No signs of toxicoses were observed when hamsters had free choice to these roots. The roots are much less toxic compared with the tubers. Observations of cattle grazing early in spring suggest that the young shoots of water hemlock are very palatable because young plants growing in streambeds were frequently and extensively grazed without toxic sequelae (Panter, personal observation).

Clinical signs of poisoning appear within 10–15 min after ingestion and progress from nervousness, frothy salivation, ataxia, dyspnea, muscular tremors, and weakness to involuntary, spastic head and neck movements accompanied by rapid eye blinking and partial occlusion of the eyes from the nictitating membranes. This is quickly followed by collapse and intermittent grand mal seizures lasting 1 or 2 min each followed by relaxation periods of 8–10 min. Depending on the dosage, recovery may occur or seizures continue until death from exhaustion or respiratory failure. There appears to be a very narrow safety threshold in which a very small increase in dosage is all that is required to induce grand mal seizures (Panter et al., 1996a).

On necropsy, gross lesions are confined to pale areas in heart muscle and skeletal muscles, particularly the long digital extensor muscle groups (Panter et al., 1996a). Microscopic lesions include multifocal, subacute to chronic myocardial degeneration characterized by granular degeneration of myofiber cytoplasm necrosis and replacement fibrosis in the heart. These areas correspond to the pale areas observed grossly. There is bilateral symmetrical, subacute to chronic myofiber degeneration, and necrosis of the long digital extensor muscle groups. Clinical serum chemistry changes of elevated lactic dehydrogenase, aspartate aminotransferase, and creatine kinase occur in relation to severity of seizures. The extent of gross and microscopic lesions in muscles and elevated clinical chemistries are a result of the muscle damage from the severe seizures. In an experimental setting, administration of barbiturates at the onset of clinical signs prevented the seizures, death, and lesions in sheep. Interestingly, the lethal effects of a 3× lethal dose of water hemlock could be prevented with pentobarbital and no elevated serum chemistries or death occurred (Panter et al., 1996a).

Biomarkers of Poisoning (Blood Enzymes, CPK, AST, GGTP, Histological Lesions in the Heart, Long Muscles, Clinical Signs, History of Ingestion, Rumen Analysis)

Diagnosis can be made by documenting exposure to the plant, with positive identification of plant parts in the rumen, especially tubers or seeds. Additionally, cicutoxin may or may not be detected in stomach contents as the toxin is quite unstable and readily oxidized. Because of the acute nature of water hemlock poisoning, animals that survive the intoxication will not have any lasting effects, whereas severely poisoned animals are found dead, normally with indications of violent terminal grand mal seizures. Death normally occurs within 1–8 h after ingestion of the plant. In animals that do survive poisoning, elevations in serum LDH, AST, and CK from muscle damage induced by the seizures peaks 3 days after exposure then declines to normal levels within 8–10 days (Panter et al., 1996a). There are no consistent gross lesions, although bruises and lacerations may be evident on the skin, muscles, tongue, etc., resulting from the severe seizures. Small hemorrhages may be present in the long skeletal muscles, and white streaking may be evident on the surface of the heart. Microscopically, cardiac lesions, including multifocal pale areas in the left, right, and interventricular myocardium, may be evident with myocardial fibrosis. In the long digital extensor muscle groups there maybe acute to chronic, moderate, random myofiber necrosis with regeneration of skeletal myofibers. Because of the acute nature of water hemlock poisoning and narrow threshold between death and survival, successful treatment would require immediate response with appropriate drugs.

Prevention and Treatment

Prevention of poisoning is accomplished by recognizing the plant and avoiding exposing animals to it early in the spring or when in the flower/seed stage. Water hemlock is easily controlled with herbicides (2,4-D per manufacturer's specification); however, herbicide use is often restricted near natural water sources. If few plants are present, hand pulling (using appropriate gloves) may be accomplished using caution to discard tubers away from possible exposure to animals or humans.

Successful treatment with barbiturates or perhaps tranquilizers prevents death, lesion formation, and serum chemistry changes; however, treatment must be prompt (Panter et al., 1996a). This treatment has been successful in humans, but in animals it has never been demonstrated in the field and would require a veterinarian to respond very rapidly with appropriate drugs soon after the ingestion of the plant occurred.

PONDEROSA PINE NEEDLES (*PINUS* SPP.)

The needles of ponderosa pine have been known for years to induce abortion in pregnant cows when grazed during the last trimester of pregnancy (Gardner et al., 1999). Occasional toxicosis in pregnant cows occurs; however, cases of toxicosis in nonpregnant cows, steers, or bulls are not reported.

Toxicology

The toxin in ponderosa pine that induces abortion in cattle is the labdane resin acid isocupressic acid **(12)** (ICA; Gardner et al., 1994). Two related derivatives (succinyl ICA and acetyl ICA) also contribute to the induction of abortion after hydrolytic conversion to ICA in the rumen (Gardner et al., 1996). Other related labdane acids (agathic acid, imbricatoloic acid, and dihydroagathic acid) that are found in ponderosa pine needles at low levels may also contain abortifacient properties based on their similar chemical structure to ICA. Other genera and species have also been implicated in abortions, such as Monterey cypress (Parton et al., 1996), Korean pine (Kim et al., 2003), common juniper, lodgepole pine (Gardner et al., 1998), and other juniper species (Gardner et al., 2010; Welch et al., 2011a). Current research indicates that the concentration of labdane acids in ponderosa pine needles and western juniper bark is not uniform throughout the same tree, the concentrations can vary from location to location, and there is evidence for seasonal fluctuations as well (Cook et al., 2010, Welch et al., 2013a 2015).

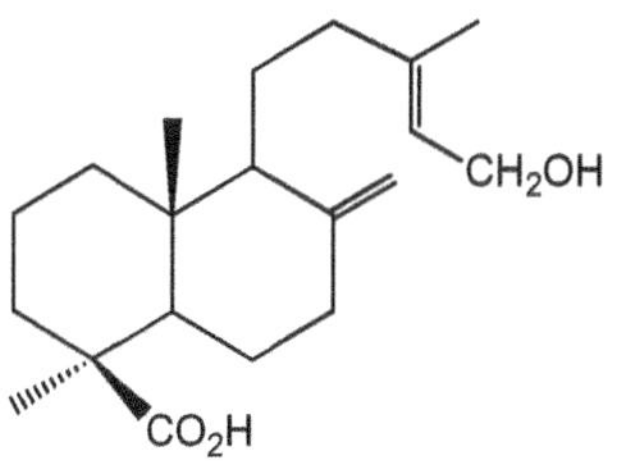

Isocupressic acid (**12**)

The primary toxicological effect of ponderosa pine needles in cattle is premature parturition and associated complications, such as retained fetal membranes, metritis, and occasional overt toxicosis and death (Gardner et al., 1999). Pine needle–induced abortion appears to mimic normal parturition except that it is premature. The abortions generally occur in the last trimester of pregnancy in the late fall, winter, or early spring. Abortions have been induced as early as 3 months of gestation and have been reported by ranchers to occur any time; however, the closer to the time of normal parturition that ingestion of pine needles occurs, the higher the risk of abortion. Abortions may occur following a single exposure to the needles, but results from controlled experiments indicate the highest incidence of abortion is in cows eating the needles over a period of 2–3 days. Abortions have been associated with grazing of green needles from trees, needles from slash piles following lumber activity, and dead, dry needles from the ground.

Abortions are generally characterized by weak uterine contractions, incomplete cervical dilation, dystocia, birth of weak but viable calves, agalactia, and retained fetal membranes (Gardner et al., 1999). Calves born after 255 days of gestation will often survive with extra care but need to be supplemented with colostrum and milk from other sources until the dam begins to lactate. Cows with retained fetal membranes may need antibiotic therapy to avoid uterine infections.

Pine needles will induce abortion in buffalo, but sheep, elk, and goats do not abort. Pine needles, pine bark, and new growth tips of branches are all abortifacient, and new growth tips are also toxic (Panter et al., 1990b). A separate toxic syndrome has been described in addition to abortion in which the abietane-type diterpene resin acids cause depression, feed refusal, weakness, neurological problems, and death. Specific compounds include abietic acid, dehydroabietic acid, and other related compounds (Stegelmeier et al., 1996). At 15%–30% of the diet, pine needles have been shown to alter rumen microflora and affect rumen fermentation (Pfister et al., 1992). Rumen stasis is part of the toxic syndrome (Stegelmeier et al., 1996).

Biomarkers of Poisoning (Premature Parturition; Live-Premature Calf; Retained Fetal Membrane; Agathic or Tetrahydroagathic Acid in Serum, Rumen Content, or Calf Tissues; History of Ingestion)

Diagnosis of pine needle–induced abortions is currently made by documenting exposure to needles, birth of the calf at least 2 weeks early, retained fetal membranes (RFM) in the cow, and agalactia. Recent studies have demonstrated that metabolites of ICA, including tetrahydroagathic acid, can be detected in the serum of the cow and calf for several days after parturition. Additional experiments are being conducted to determine what tissues/fluids in the calf, including dead calves, are optimum for a positive diagnosis. Isocupressic acid and related labdane diterpene acids found in pine needles are metabolized in both the rumen and the liver. The three major metabolites include agathic acid, dihydroagathic acid, and tetrahydoagathic acid (Gardner et al., 1999). Of these metabolites, tetrahydroagathic acid remains in the cow serum for the longest period and has been found to be the most relevant diagnostic component from abortion case samples submitted (Gardner, personal communications). There is research evidence that tetrahydoagathic acid is also found in fetal fluids (stomach and thoracic fluids; Snider et al., 2015). For diagnostic purposes, sera samples should be taken from the cow as soon as possible to optimize detection of the ICA metabolites. Additional samples from the aborted fetus should include blood (sera), stomach, and thoracic fluids.

Prevention and Treatment

The only recommendation to prevent pine needle abortion is to limit access of late term pregnant cows to pine trees. There is no known treatment for cattle once ingestion of pine needles has occurred. Open cows, steers, or bulls are apparently unaffected by pine needles; likewise, sheep, goats (pregnant or not), and horses can graze pine needles with impunity and experience no adverse effects. Supportive therapy (antibiotic treatment or uterine infusion for retained fetal membranes) is recommended for cows that have aborted, and intensive care of the calf may save its life. Grazing of pine needles intensifies during cold, inclement weather and if other forage is in short supply. In spring, before green grass is available, cows will leave bedding grounds in search of green forage and frequently graze green needles from low-hanging branches or old, dry needles from surrounding trees where the snow has melted. These cows are at risk and should be kept away from the pines. Research has also determined that cattle with low body condition are more likely to eat pine needles than cattle in adequate body condition (Pfister et al., 2008b). Consequently, it is recommended that pregnant cattle grazing in ponderosa pine areas be maintained in good body condition (Pfister et al., 2008). Anecdotal information suggests that pregnant llamas may be at risk from ingesting pine needles, but no research has been done to substantiate this (Panter, personal communications).

RAYLESS GOLDENROD (*ISOCOMA PLURIFLORA*)

White Snakeroot (*Eupatorium rugosum*)

Rayless goldenrod (*I. pluriflora*) and white snakeroot (*E. rugosum*) are toxic range plants of the southwestern and midwestern United States, respectively. The disease associated with toxicity has been referred to as "alkali disease" because originally it was associated with drinking alkali water. Currently, it is referred to as "milk sickness" or "trembles" (the same as white snakeroot in the midwest) because the toxin tremetone **(13)** is excreted in the milk and subsequently results in poisoning of humans and nursing offspring.

Toxicology

The toxic constituents of rayless goldenrod and white snakeroot are similar and the first reported toxin, tremetol, is actually a mixture of ketones and alcohols. Tremetone (**13**) (5-acetyl-2,3-dihydro-2-isopropenylbenzofuran) was thought to be the principle toxic factor; however, 11 different compounds have now been isolated and identified (Lee et al., 2010). The elucidation of different chemotypes of white snakeroot partially explains the sporadic and unpredictable toxicoses reported in livestock throughout the midwestern United States.

Tremetone **(13)**

Clinical signs of poisoning may occur after ingestion of 1%–15% BW during a 1- to 3-week period. Signs begin with depression or inactivity, followed by noticeable trembling of the fine muscles of the nose and legs. Most cases of poisoning reported constipation, nausea, vomition, rapid labored respiration, progressive muscular weakness, stiff gait, standing in a humped-up position, dribbling urine, inability to stand, coma, and death. Signs are similar in cattle, sheep, and goats. The disease is often more acute and severe in horses than in cattle, and horses may die of heart failure after subacute ingestion of white snakeroot and presumably rayless goldenrod. Cattle have also been poisoned on a related plant (*Isocoma acradenia*) in southern California (Galey et al., 1991). In this case, 21 of 60 cattle died and 15 of 60 were affected but recovered.

Biomarkers of Poisoning (Blood Enzymes, CPK, Troponin, Histology, Lesions in Heart, Long Muscles, Clinical Signs, History of Ingestion, Myonecrosis)

Diagnosis of rayless goldenrod poisoning is made by documenting exposure to the plant and identifying the characteristic clinical and pathological signs. The primary lesion and cause of the clinical trembles in rayless goldenrod intoxication in goats is skeletal muscle degeneration and necrosis. As these subtle lesions are most consistently seen in the quadriceps femoris and diaphragm at low dose, these may be the best tissues to examine. However, if there are clinical signs of poisoning and lesions are severe, the large appendicular muscles, such as the triceps brachii, biceps femoris, quadriceps femoris, and adductor, are likely to show significant histologic change. Higher doses produced myocardial lesions. Serum enzyme activities, including CK, LDH, AST, and alanine aminotransferase (ALT), have been shown to be significantly changed in poisoned animals. Although the mean cardiac troponin-I concentrations are not consistently elevated in different poisoned animals, individual animals do have elevated amounts. Creatine phosphokinase and ketones were elevated, and severe myonecrosis was described in the dead animals.

Prevention and Treatment

Rayless goldenrod is not readily palatable, and toxicity results from animals being forced to graze the plant because of lack of good quality forage. Avoiding overgrazing will usually minimize poisoning in livestock. White snakeroot is relatively palatable and may be ingested as part of the diet in cattle and horses.

Treatment is generally symptomatic and supportive, providing dry bedding, good shelter, and fresh feed and water. Activated charcoal and saline cathartic may be beneficial. Treatment may include fluids, B vitamins, ketosis therapy, and tube feeding. Hay and water should be placed within reach if the animal is recumbent. In lactating cows, frequent milking may facilitate a more rapid and complete recovery.

HALOGETON (*HALOGETON GLOMERATUS*)

Halogeton is an invasive, noxious, and poisonous weed introduced into the western United States from central Asia in the early 20th century. It was first collected along a railroad spur near Wells, Nevada, in 1934 and rapidly invaded 11.2 million acres of the cold deserts of the western United States (Young, 1999). There has been no appreciable spread since the 1980s because halogeton has filled all the suitable niches within its tolerance limits. It currently infests disturbed areas within the salt-desert shrub and sagebrush plant communities in the Great Basin, Colorado Plateau, and Wyoming's Red Desert physiographic provinces, which have 3–15 in of annual precipitation.

Halogeton's infamy began in the 1940s and 50s by causing large, catastrophic sheep losses. There were many instances of large dramatic losses; sometimes entire bands of sheep died overnight from halogeton poisoning. *Life* magazine ran a cover story titled "Stock Killing Weed" that focused national attention on halogeton (Young, 1999). Congress passed the Halogeton Act in 1952 with the intent to (1) detect the presence of halogeton; (2) determine its effects on livestock; and (3) control, suppress, and eradicate this stock-killing weed.

Federal research was reallocated from the Forest Service Experiment Stations to the Bureau of Plant Industries, creating a Range Research unit devoted specifically to "solving" the halogeton problem. It was realized that halogeton was not the problem but a symptom of a larger problem—that being degradation of desert rangelands (Young, 1999). It invaded disturbed sites where sheep congregated around railroad loading sites, trail heads, stock trails, road sides, and water holes. When hungry sheep were turned loose to graze, halogeton was the only feed available, and they consumed too much, too rapidly, and were poisoned.

Toxicology

The toxins are sodium and potassium oxalates (**14**), and *Halogeton* plants are frequently high (17%–29% total oxalates) in these oxalates in the fall and early winter resulting in death losses in sheep and cattle (Rood et al., 2014). Poisoning occurs when sheep consume more oxalates than the body can detoxify (James, 1999). Rumen microbes can adapt if animals are introduced slowly and prevented from eating too much halogeton.

Sodium oxalate (**14**)

Treatment of Poisoned Animals

Animals can be given excess water through a stomach tube to flush oxalates out in the urine, or supplemented with dicalcium phosphate in a drench to provide Ca that will combine with oxalates in the rumen and facilitate oxalate excretion. Intravenous injection of calcium gluconate can maintain blood Ca levels, but Ca oxalate crystals will continue to damage kidneys (James, 1999). Prevention is the key to avoid poisoning. Only hungry sheep are poisoned. Research has demonstrated that as little as 1 oz of soluble oxalates can be lethal to hungry sheep. Well-fed sheep grazing nutritious forage throughout the day can tolerate more than 4 oz of soluble oxalate. Sheep grazed in a desert plant community infested with halogeton consumed it in 5%–25% of their diets without ill effect. If other forage is available, they will likely not get a lethal dose. Historically, sheep were most often affected; however, multiple cases of cattle poisoning have recently been reported (J. Hall, personal communication, 2013).

Biomarkers of Poisoning (Blood or Ocular Calcium, Oxalate Analysis of Blood, History of Ingestion

Biomarkers of halogeton poisoning include serum hypocalcemia and oxalate crystals in the rumen and kidney. Diagnosis may be made by documenting exposure to the plant, identifying characteristic clinical and pathological signs of poisoning and histological examination. Blood calcium declines in poisoned animals, with concurrent increases in magnesium and phosphorus. There may also be an increase in serum LDH and Blood Urea Nitrogen (BUN), although these are not pathognomonic. Gross pathological lesions include the presence of fluid in the abdominal and chest cavities, as well as the pericardial sac. The lungs maybe filled with blood-tinged foam, and splotchy hemorrhages may be found on the surfaces of the heart, rumen, and other organs. The kidneys will be pale and swollen and contain oxalate crystals resembling glass shards and easily detected when the kidney tissue is cut with a knife. Microscopically the renal tubules will be filled with proteinaceous casts and calcium oxalate crystals, and the epithelium flattened and necrotic. There may also be edema and hemorrhage of the rumen wall and oxalate crystals present in the submucosa stomach compartment and in the walls of vessels.

Management to Prevent Poisoning

Never turn hungry sheep or cattle onto dense halogeton-infested sites. Provide supplemental feed and plenty of fresh water following trucking or trailing. Introduce sheep and cattle gradually to halogeton to allow rumen microbes to adjust. Some sheep producers graze their sheep on shadscale ranges (which contain low oxalate levels) before going into halogeton areas. Do not overgraze; maintain desert range in good condition. This prevents halogeton invasion and provides an alternative food source.

Herbicide control of halogeton is not recommended because the waxy surface of the leaves hinders absorption of most herbicides. More importantly, however, desirable desert shrubs are killed, leaving the site open for further invasion and degradation by halogeton and other invasive weeds.

PYRROLIZIDINE ALKALOID-CONTAINING PLANTS

Pyrrolizidine alkaloid (PA)—containing plants are numerous and worldwide in distribution and in toxic significance (Cheeke and Shull, 1985). Three plant families predominate in PA-producing genera and species: Compositae (*Senecio* spp.), Leguminosae (*Crotalaria* spp.), and Boraginaceae (*Heliotropium*, *Cynoglossum*, *Amsinckia*, *Echium*, and *Symphytum* spp.). Not all of these occur in the western United States.

Toxicology

More than 150 DHPAs (Dihydro Pyrrolizidine Alkaloids) have been identified, and their structural characteristics elucidated. The PAs contain the pyrrolizidine nucleus and can be represented by the basic structures of senecionine (**15**) and heliotrine (**16**). The toxic effects of all PAs are somewhat similar, although their potency varies because of their bioactivation in the liver to toxic metabolites called pyrroles (Fig. 37.1). These pyrroles are powerful alkylating agents that react with cellular proteins and cross-link DNA, resulting in cellular dysfunction, abnormal mitosis, and tissue necrosis. The primary effect is hepatic damage; however, many alkaloid and species-specific extrahepatic lesions have been described. Small amounts of pyrrole may enter the blood and be transported to other tissues, but there is debate on this issue because most pyrroles are highly reactive and not likely to make it into the circulation (Stegelmeier et al., 1999). When PA metabolites circulate, they probably do so as protein adducts that may be recycled. Some alkaloids (monocrotaline) may dissociate from their carrier proteins and damage other tissues such as the lungs. Pigs seem more prone to develop extrahepatic lesions.

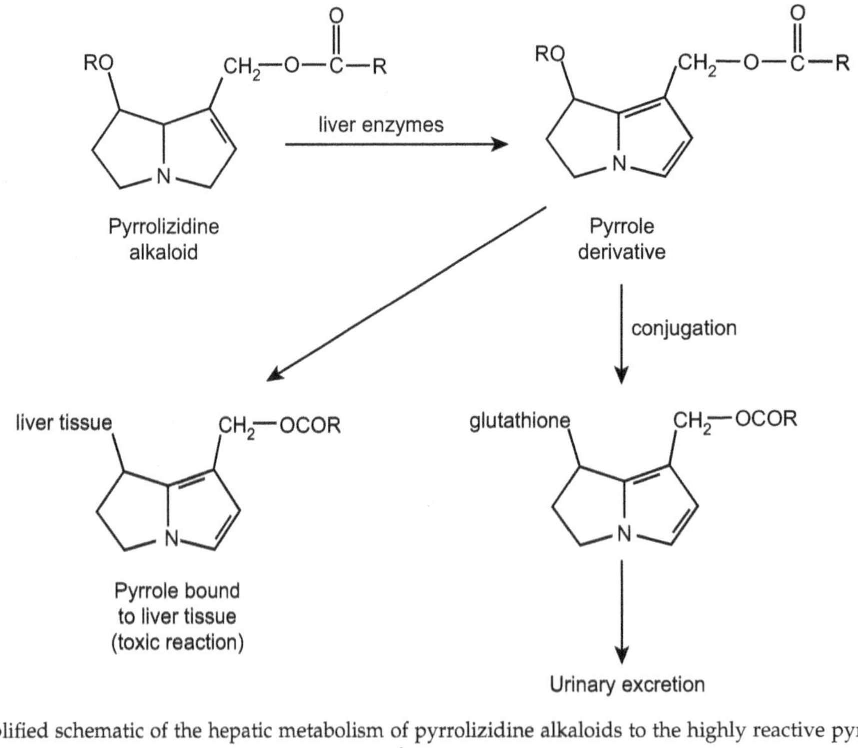

FIGURE 37.1 Simplified schematic of the hepatic metabolism of pyrrolizidine alkaloids to the highly reactive pyrroles that result in liver damage.

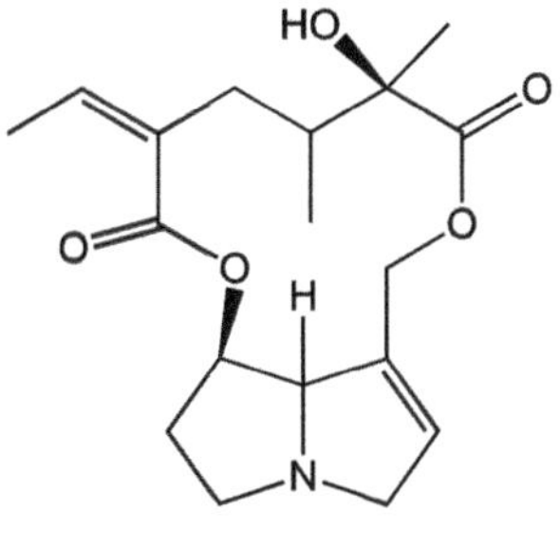

Senecionine (15)

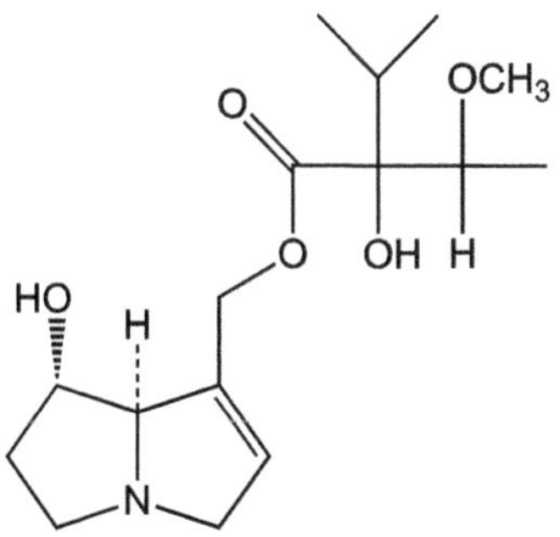

Heliotrine (16)

There are marked differences in susceptibility of livestock and laboratory animals to PA toxicosis. Cows are most sensitive, followed by horses, goats, and sheep, respectively. In small laboratory animals, rats are most sensitive, followed by rabbits, hamsters, guinea pigs, gerbils, and mice, respectively. Among avian species, chickens and turkeys are highly susceptible, whereas Japanese quail are resistant (Cheeke and Shull, 1985). Humans, especially fetuses, neonates, and children are particularly susceptible to DHPAs in herbal products, and many countries limit what a pregnant women should consume (Edgar et al., 2015).

Detoxification mechanisms of PAs generally involve the liver and GI tract. Evidence of ruminal detoxification in sheep suggests that this contributes to the reduced toxicity in that species. There are also substantial species-specific differences in the rate of PA metabolism. Both probably contribute to species susceptibility. For example, *Echium* and *Heliotropium* PAs are easily degraded by certain rumen microflora, but there is little evidence of ruminal degradation of *Senecio* PAs. The PAs in *Senecio* are macrocyclic esters of retronecine as opposed to the open esters found in heliotridine. Therefore, the reason for the difference in *Senecio* toxicity between sheep and cattle is unlikely to be the rumen detoxification but more likely differences in species-specific enzymatic activation of *Senecio* PAs. For example, in in vitro studies, retrorsine metabolism has been shown to be high in those species that are most susceptible and lowest in animals of least susceptibility (Brown et al., 2015). In addition, in vivo studies demonstrated that a higher pyrrole production rate occurred in cattle compared with sheep (Cheeke and Shull, 1985). Simple induction of liver microsomal enzymes by phenobarbitone increased pyrrole production and increased PA toxicity (LD_{50} in guinea pigs from $\approx$800 to 216 mg/kg). PA toxicity may disrupt other hepatic functions. Abnormal copper metabolism, clotting factors, NH_3 metabolism, protein metabolism, etc., may be affected in PA poisoning.

Biomarkers of Poisoning (Liver Enzymes, AST, LDH, GGTP, Cirrhosis, Histology, Pyrrole Analysis of Liver)

Toxicity of *Senecio*, *Heliotropium*, and *Echium* is largely confined to the liver, whereas *Crotalaria* will also cause significant lung damage. Typical histologic lesions are swelling of hepatocytes, hepatocyte necrosis, periportal necrosis, megalocytosis (enlarged parenchymal cells), karyomegaly (enlarged nuclei) fibrosis, bile duct proliferation, and vascular fibrosis and occlusion. Hepatic cells may be 10–30 times normal size, and DNA content may be 200 times normal (Brown et al., 2015).

In most species affected by PA poisoning, the liver becomes hard, fibrotic, and smaller. Because of decreased bile secretion, bilirubin levels in the blood rise, causing jaundice. Common clinical signs include ill thrift, depression, diarrhea, prolapsed rectum, ascites, edema in the GI tract, photosensitization, and aberrant behavior. In horses, "head pressing" or walking in straight lines regardless of obstacles in the path may occur. These neurological signs in horses are due to elevated blood ammonia from reduced liver function. PA poisoning may cause elevated blood ammonia, resulting in spongy degeneration of the central nervous system.

Elevated levels of serum enzymes such as ALT, AST, γ-glutamyl transferase (GGT), and alkaline phosphatase (ALP) are reported (Stegelmeier et al., 1999). Use of these tests for diagnosis is supportive but should not be relied on exclusively because they vary with animal species and other conditions. They may also be in the normal range even though liver damage has occurred, and they tend to be transient. Liver function tests such as bilirubin, bile acids, or sulfobromophthalein (BSP) clearance may be useful estimates of the extent of liver damage.

The sulfur-bound pyrrolic metabolites of PAs have been detected in tissues of poisoned animals after hydrolysis with an alcoholic silver nitrate solution and detection either by colorimetric methods or by GC-MS (Mattocks and Jukes, 1990; Seawright et al., 1991; Winter et al., 1993; Schoch et al., 2000). Mostly recently a modification of these methods has been used for detection of the sulfur bond metabolites using LC-MS (Lin et al., 2011; Brown et al., 2015). The pyrrolic metabolites also bind to DNA, which can be isolated, enzymatically hydrolyzed and then specific DNA base–pyrollic adducts detected by LC-MS (Fu et al., 2010).

Prevention and Treatment

There are no effective methods of prevention or treatment except avoidance of the plant and controlling plant populations with herbicides or biological control. Resistance to PA toxicosis in some species suggests that the possibility may exist to increase resistance to PAs. Dietary factors such as increased protein, particularly those high in sulfur amino acids, had minor protective effects in some species. Antioxidants such as Butylated Hydroxytoluene (BHT) and ethoxyquin induced increased detoxifying enzymes such as glutathione S-transferase and epoxide hydrolase (Mattocks and Jukes, 1990; Seawright et al., 1991; Winter et al., 1993). Zinc salts have been shown to provide some protection against hepatotoxicosis from sporidesmin or lupinosis in New Zealand and Australia, and zinc supplementation reduced toxicity in rats from *Senecio* alkaloids (Burrows and Tyrl, 2013; Knight and Walter, 2001).

Many of these plants were introduced either inadvertently or intentionally. Without natural predators to keep populations in check, they experienced explosive growth and distribution followed by epidemic proportions of toxicity. Introduction of biological controls and natural population controls has reduced many of the plant populations and thus toxicoses have declined. Sheep, a resistant species, have been used to graze plants, particularly *Senecio jacobaea*.

PHOTOSENSITIZING PLANTS

Numerous plants cause photosensitization resulting in lost production to livestock producers. Photosensitization is the development of abnormally high reactivity to ultraviolet radiation or natural sunlight in the skin or mucous membranes. Primarily induced in livestock by various poisonous plants, the syndrome in livestock has been defined as primary or secondary photosensitization.

Toxicology

Primary

In primary photosensitization, the photoreactive agent is absorbed directly from the plant and reaches the peripheral circulation and skin, where it reacts with the ultraviolet rays of the sun and results in sunburn, particularly of unprotected areas of the body. Hypericin (**17**) and fagopyrin are polyphenolic derivatives from St. John's wort and buckwheat, respectively, and are primary photodynamic agents (Cheeke and Shull, 1985). By definition, primary photosensitization does not induce hepatic damage. Most agents are ingested, but some may induce lesions through skin contact. Several of these plants are weedy in nature and can contaminate pastures and feed. Exposure to some plants is increasing as they are becoming widely used as herbal remedies and naturopathic treatments. In most cases, the photodynamic agent is absorbed from the digestive tract unchanged and reaches the skin in its "native" form (Stegelmeier, 2002).

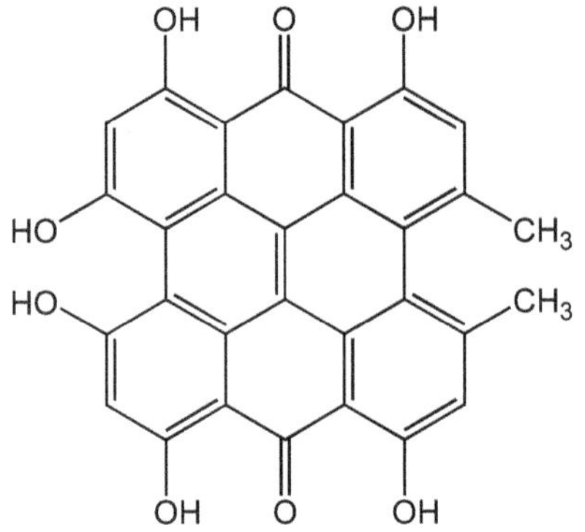

OH O OH HO CH_3 HO CH_3 OH O OH

Hypericin (**17**)

Secondary

In secondary or hepatogenous photosensitization, the photoreactive agent is phylloerythrin, a degradation product of chlorophyll. Phylloerythrin is produced in the stomach of animals, especially ruminants, and absorbed into the bloodstream. In normal animals, the hepatocytes conjugate phylloerythrin and excrete it in the bile. However, if the liver is damaged or bile secretion is impaired, phylloerythrin accumulates in the liver, the blood, and subsequently the skin, causing photosensitivity. This is the most common cause of photosensitization in livestock (Knight and Walter, 2001). Because chlorophyll is almost always present in the diet of livestock, the etiologic agent of secondary photosensitization is the hepatotoxic agent.

The dermatologic signs of photosensitization in livestock are similar regardless of the plant or toxicant involved. Degree or severity varies, depending on the amount of toxin or reactive phylloerythrin in the skin, degree of exposure to sunlight, and amount of normal physical photoprotection (hair, pigmentation, or shelter). First signs in most animals are restlessness or discomfort from irritated skin, followed by photophobia, squinting, tearing, erythema, itching, and sloughing of skin in exposed areas (i.e., lips, ears, eyelids, udder, external genitalia, or white pigmented areas) (Burrows and Tyrl, 2013). Swelling in the head and ears (edema) of sheep after ingestion of *Tetradymia* has been referred to as big head. It was determined that sheep grazing black sagebrush (*Artemesia nova*) before *Tetradymia* were three times more likely to develop this photosensitization (Johnson, 1974). Tissue sloughing and serum leakage may occur where tissue damage is extensive. Primary photosensitization rarely results in death. However, in secondary or hepatogenic photosensitization, the severity of liver damage and secondary metabolic and neurologic changes of hepatic failure may ultimately result in death. Recovery may leave sunburned animals debilitated from scar tissue formation and wool or hair loss.

Biomarkers of Poisoning (Bilirubin, Liver Enzymes, Skin Biopsies, History of Ingestion)

Identifying chronic hepatic disease is complicated because many of the serum markers for hepatic disease have returned to normal. As normal hepatocytes become replaced with fibrous connective tissue, there are fewer damaged cells to elevate serum enzymes. Percutaneous liver biopsies are invaluable in identifying and diagnosing these cases (Stegelmeier et al., 1999).

Plant-induced hepatopathy generally results in characteristic histologic lesions. For example, PAs generally cause bridging portal fibrosis with hepatocellular necrosis, biliary proliferation, and megalocytosis. *Panicum* and *Tribulus* species generally produce a crystalline cholangiohepatitis. Liver biopsy also provides prognostic information. The degree of damage is correlated directly with the animal's ability to compensate, recover, and provide useful production. Note that the liver reacts to insult in a limited number of ways, and most histologic changes are not pathognomonic. Hepatic cirrhosis (necrosis, fibrosis, and biliary proliferation) involves nonspecific changes that can be initiated by a variety of toxic and infectious agents (Stegelmeier et al., 1999).

Prevention and Treatment

Prevention of poisoning lies in controlling plants with photosensitizing potential and providing adequate quality forage to animals. Treatment after poisoning involves removing animals from sun exposure, treating areas of necrosis and sunburn, antibiotic therapy, and supplementing young animals when access to sunburned udders is prevented because of nursing discomfort to dams.

DEATH CAMAS

All death camas (*Zigadenus*) species are assumed to be toxic; however, variation in toxicity exists between species and even within species depending on season, climate, soils, and geographical locations. Poisoning in sheep, cattle, horses, pigs, fowl, and humans has been reported. Because of their grazing habits, the largest losses generally occur in sheep. Death camas is generally not palatable to livestock but is one of the earliest species to emerge in the spring. Poisoning most frequently occurs in spring when other more palatable forage is not available, or on overgrazed ranges where there is a lack of more desirable forage. Poisonings have resulted due to management errors in which hungry animals were placed in death camas–infested areas (Panter et al., 1987).

Toxicity of Death Camas to Livestock

The toxins in death camas are of the cevanine steroidal alkaloid type—i.e., zygacine **(18)**. Zygacine is a very potent compound with an i.v. LD_{50} of 2 mg/kg and an oral LD_{50} of 130 mg/kg in mice (Welch et al., 2011b). Clinical signs of toxicosis are similar in all livestock poisoned by *Zigadenus*, irrespective of the species of plant involved. Excessive salivation is noted first, with foamy froth around the nose and muzzle that persists, followed by nausea and occasionally vomition in ruminants (Panter et al., 1987). Intestinal peristalsis is dramatically increased, accompanied by frequent defecation and urination. Muscular weakness with accompanying ataxia, muscular fasciculations, prostration, and eventual death may follow. The pulse becomes rapid and weak, and the respiration rate increases but the amplitude is reduced. Some animals become cyanotic, and the spasmodic struggling for breath may be confused with convulsions.

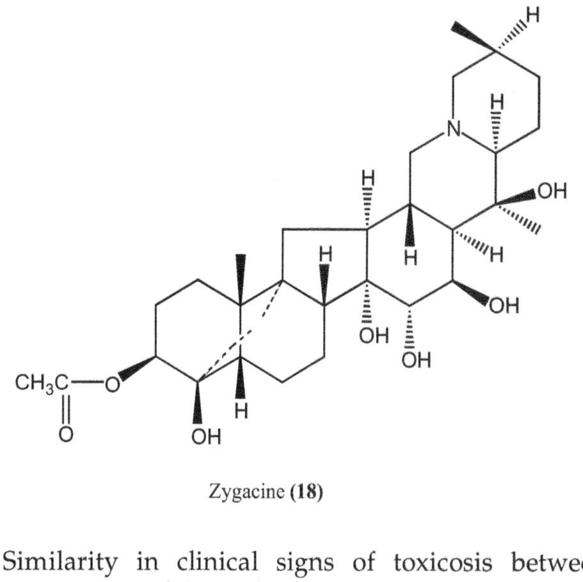

Zygacine **(18)**

Similarity in clinical signs of toxicosis between certain species of these plants suggests that the same alkaloids are present; however, differences in concentrations can explain the differences in relative toxicity of different species.

Biomarkers of Poisoning (Alkaloid Analysis of Serum, Liver, Pathology, History of Ingestion)

The heart fails before respiration, and at necropsy the heart is usually found in diastole. A comatose period may range from a few hours to a few days before death. Pathological lesions are those of pulmonary congestion. Gross lesions of sheep include severe pulmonary congestion, hemorrhage, edema, and subcutaneous hemorrhage in the thoracic regions. Microscopic lesions include severe pulmonary congestion with infiltration of red blood cells in the alveolar spaces and edema. Diagnosis of poisoning may be established by clinical signs of toxicosis, evidence of death camas being grazed, histopathological analysis of tissues from necropsied animals, and identification of death camas in the rumen or stomach contents (Panter et al., 1987). Zygacine has been detected in rumen content from field cases and in serum from sheep and mice experimentally dosed (Welch et al., 2011b, 2013b, 2016).

Management and Prevention

Conditions conducive to poisoning by death camas include driving animals through death camas–infested ranges; not allowing animals to graze selectively; unloading hungry animals in infested areas; lambing, bedding, watering, or salting livestock in death camas–infested areas; or placing animals on range where little forage is available. Poisoning generally occurs in the early spring when death camas is the first green forage available, and the young immature foliage is the most toxic. Single flock losses of 300–500 sheep have been reported (Panter et al., 1987). Three key factors contribute to sheep losses: (1) driving sheep through heavily infested areas of death camas when the sheep are hungry; (2) bedding sheep for the night near death camas–infested areas, providing immediate access to death camas the following morning; and (3) forcing sick sheep to travel will contribute to the stress, exacerbating the toxic effects and increasing the losses.

KNAPWEEDS: *CENTAUREA* SPP.

The knapweeds are a large group of plants with primarily noxious, invasive characteristics. Although this genus is not a great risk for livestock producers, a serious disease of horses called nigropallidal encephalomalacia warrants its inclusion in this chapter. There are 450–500 species of *Centaurea*, and 29 species have been described in North America (Burrows and Tyrl, 2013). Most of these have been introduced and have had a huge negative impact on rangelands in the western United States. Although most species are opportunists and will aggressively invade rangelands, especially those that have been overgrazed, burned, or disturbed, only two species are of any toxicologic significance—*Centaurea repens* (Russian knapweed) and *Centaurea solstitialis* (yellow star thistle).

Toxicology

The compounds isolated from knapweeds include a large class called sesquiterpene lactones. Although the putative toxin causing the neurological disease in horses has not been specifically identified, six of these compounds have been screened for cytotoxicity in an in vitro neuronal cell bioassay. The rank order of activity is repin (**19**) > subluteolide > janerin > cynaropicrin > acroptilin > solstitialin (Riopelle and Stevens, 1993). Toxicity of solstitialin A-13 acetate and cynaropicrin to primary cultures of fetal rat substantia nigra cells has been demonstrated. These sesquiterpene lactones are quite unstable, and it has been hypothesized that they are precursors to the ultimate neurotoxin. Also, there are aspartic and glutamic acids present in these plants, and they possess neuroexcitatory properties.

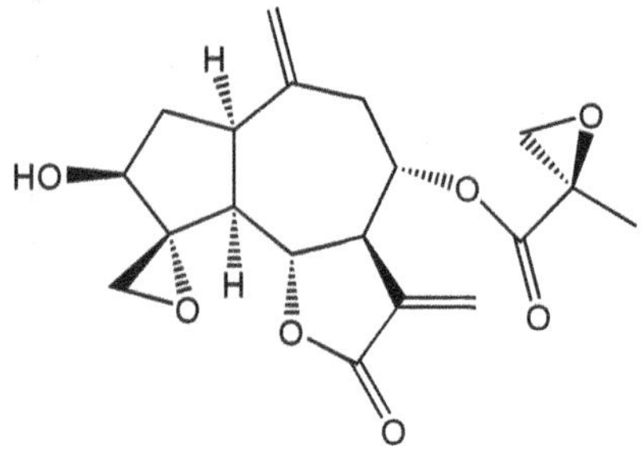

Repin (**19**)

Clinical Signs

Thus far, only yellow star thistle and Russian knapweed have been implicated in toxicoses in the United States, and only in horses (Panter, 1991). Apparently, ruminants are not affected, and the *Centaurea* spp. may be useful forage for sheep and goats. However, in other countries, toxicoses in ruminants have been reported. For example, in South Africa, *C. repens* fed to sheep at 600 g doses for 2 days caused an acute digestive upset and pulmonary edema and ascites. In Azerbaijan, *C. repens* is reported to cause a neurological disease in buffalo similar to that which has been described in horses. However, no neuropathology similar to that seen in horses was observed in the buffalo. Toxicity generally occurs in summer and fall when forage is depleted and horses are forced to graze less palatable species. Ingestion often occurs for several months or more before an abrupt onset of neurological dysfunction is observed.

Often, the disease progresses to dehydration, starvation, and bizarre behavior, including submergence of the head in water to allow water to flow into the esophagus or lapping water like a dog. *C. repens* appears to be more toxic than *C. solstitialis*, but prolonged ingestion is required by both before disease appears. The amount of plant ingested to induce the clinical effects is reported to be 60% or more of body weight for *C. repens* and 100% or more of body weight for *C. solstitialis*. Intermittent grazing can prevent disease, indicating that there is not a cumulative effect but, rather, a threshold must be exceeded before neurological signs are observed (Cordy, 1978). Once neurological signs are observed in horses, prognosis for recovery is poor and euthanasia should be considered.

Biomarkers of Poisoning (History of Ingestion, Pathology Brain Lesions, Horses Only)

Impaired eating and drinking are often the first observable signs. Depression and hypertonicity of the lips and tongue follow, and a constant chewing may be observed, hence the name "chewing disease." Abnormal tongue and lip postures may be observed, and other neurological signs include locomotor difficulties such as aimless walking, drowsy appearance, and inactivity with the head held low. The neurological disease is considered permanent, and although some improvement may be seen, difficulty eating and drinking may preclude long-term recovery.

The lesions are very specific and limited to the globus pallidus and the substantia nigra (nigropallidal encephalomalacia), where distinct pale yellowish to buff-colored foci or softening and cavitation are seen (Cordy, 1978). The lesions are typically bilateral and symmetrical. This specificity of the lesions for the basal ganglia has prompted more investigations into unraveling the mysteries of human diseases associated with dopaminergic pathways, such as Parkinson's or Huntington's disease, and tardive dyskinesia. This disease in horses is often called equine Parkinsonism. This unusual disease is manifest by an almost immediate onset after prolonged ingestion, suggesting an all-or-none type of acute neurological crisis. The lesions develop quickly and completely, and progressive stages of degeneration rarely occur except for some changes in the adjoining neurons adjacent to the necrotic foci in the globus pallidus and the pars reticularis of the substantia nigra (Cordy, 1978). Microscopically, there is extensive necrosis of neurons, glia, and capillaries within sharply defined margins of the involved brain centers. Occasionally, lesions may be observed in the gray and white matter of the brain.

Prevention and Treatment

Good veterinary care and supportive therapy, including good feed, easy access to water, supplemental vitamins, and good nursing care, is essential for survival. Treatment of the disease once it is manifest is not generally successful. However, in Argentina, affected horses have been treated with glutamine synthetase and a bovine brain ganglioside extract given daily intramuscularly for 1 month with some success (Selfero et al., 1989). When animals are first observed grazing *Centaurea* spp., they should be immediately removed to better pastures. Prevention of the disease is easily accomplished by knowing the plants that exist in one's pastures, by providing good quality and adequate amounts of forages and feed, and by frequent observation of one's animal's grazing patterns and behavior.

Control of plant invasion by good range/pasture management to prevent overgrazing and loss of other competitive grasses and forbs is important. Herbicide control is quite easily accomplished with broadleaf products, including 2,4-D, dicamba, and picloram. These plants are prolific seed producers, and follow-up treatment is required to eliminate the populations. Seeds are often distributed through contaminated hay or other feed sources, and initial populations often start near feed bunks and spread from there. Because of their morphology, size, and parachute-like structures, seeds are easily spread by wind and water. Understanding one's weeds and close monitoring of populations will help in the control of these highly invasive species.

CONCLUDING REMARKS AND FUTURE DIRECTIONS

Even with our ever-increasing knowledge about poisonous plants and their toxins, poisonings continue to occur, some catastrophic, on livestock operations. Poisoning in humans and companion animals from toxic plants also continues to be a significant risk, especially to pets and children. Identification of specific biomarkers of poisoning from plants is relatively new and is often limited to pathology after the fact or chemical analysis of biological fluids or tissues. Research at the Poisonous Plant Research Laboratory continues to emphasize methods and biomarkers that will improve early diagnosis and ultimately reduce losses from poisonous plants.

References

Barneby, R.C., 1964. Atlas of North American Astragalus: Parts I and II, vol. 13. Memoirs New York Botanical Garden, New York.

Benn, M.H., Jacyno, J.M., 1983. The toxicology and pharmacology of the diterpenoid alkaloids. In: Pelletier, S.W. (Ed.), Alkaloids: Chemical and Biological Perspectives. Wiley, New York, pp. 153–210.

Brown, A.W., Stegelmeier, B.L., Colegate, S.M., et al., 2015. The comparative toxicity of a reduced, crude comfrey (*Symphytum officianle*) alkaloid extract and the pure, comfrey-derived pyrrolizidine alkaloids, lycopsamine and intermedine in chicks (*Gallus gallus domesticus*). J. Appl. Toxicol. 36, 716–725.

Burrows, G.E., Tyrl, R.J., 2013. Toxic Plants of North America, second ed. John Wiley and Sons, Inc., Ames, IA.

Cheeke, P.R., Shull, L.R., 1985. Natural Toxicants in Feeds and Poisonous Plants. AVI, Westport, CT.

Colegate, S.M., Dorling, P.R., Huxtable, C.R., 1979. A spectroscopic investigation of swainsonine: an α-mannosidase inhibitor isolated from *Swainsona canescens*. Aust. J. Chem. 32, 2257–2264.

Cook, D., Gardner, D.R., Pfister, J.A., Panter, K.E., et al., 2010. Differences in ponderosa pine isocupressic acid concentrations across space and time. Rangelands 32, 14–17.

Cook, D., Gardner, D.R., Welch, K.D., et al., 2009a. Quantitative PCR method to measure the fungal endophyte in locoweeds. J. Agric. Food Chem. 57, 6050–6054.

Cook, D., Ralphs, M.H., Welch, K.D., Stegelmeier, B.L., 2009c. Locoweed poisoning in livestock. Rangelands 31 (1), 16–21.

Cordy, D.R., 1978. *Centaurea* species and equine nigropallidal encephalomalacia. In: Keeler, R.F., Van Kampen, K.R., James, L.F. (Eds.), Effects of Poisonous Plants on Livestock. Academic Press, New York, pp. 327–336.

Daugherty, C.G., 1995. The death of Socrates and the toxicology of hemlock. J. Med. Biogr. 3, 178–182.

Davis, T.Z., Stegelmeier, B.L., Hall, J.O., 2014. Analysis in horse hair as a means of evaluating selenium toxicoses and long-term exposures. J. Agric. Food Chem. 62, 7393–7397.

Davis, T.Z., Stegelmeier, B.L., Panter, K.E., et al., 2012. Toxicokinetics and pathology of plant-associated acute selenium toxicosis in steers. J. Vet. Diagn. Invest. 24 (2), 319–327.

Edgar, J.A., Molyneux, R.J., Colegate, S.M., 2015. Pyrrolizidine alkaloids: potential role in the etiology of cancers, pulmonary hypertension, congenital anomalies, and liver disease. Chem. Res. Toxicol. 28, 4–20.

Fu, P.P., Chou, M.W., Churchwell, M., et al., 2010. High-performance liquid chromatography electrospray ionization tandem mass spectrometry for the detection and quantitation of pyrrolizidine alkaloid-derived DNA adducts *in vitro* and *in vivo*. Chem. Res. Toxicol. 23, 637–652.

Galey, F.D., Hoffman, R., Maas, J., Barr, B., et al., 1991. Suspected *Haplopappus acradenius* toxicosis in beef heifers. In: Paper Presented at the 34th Annual Meeting of the American Association of Veterinary Laboratory Diagnosticians, October, San Diego.

Gardner, D.R., James, L.F., Molyneux, R.J., Panter, K.E., et al., 1994. Ponderosa pine needle-induced abortion in beef cattle: identification of isocupressic acid as the principal active compound. J. Agric. Food Chem. 42, 756–761.

Gardner, D.R., James, L.F., Panter, K.E., et al., 1999. Ponderosa pine and broom snakeweed: poisonous plants that affect livestock. J. Nat. Toxins 8, 27–34.

Gardner, D.R., Panter, K.E., 1993. Comparison of blood plasma alkaloid levels in cattle, sheep and goats fed *Lupinus caudatus*. J. Nat. Toxins 2, 1–11.

Gardner, D.R., Panter, K.E., James, L.F., Stegelmeier, B.L., 1998. Abortifacient effects of lodgepole pine (*Pinus contorta*) and common juniper (*Juniperus communis*) on cattle. Vet. Hum. Toxicol. 40 (5), 260–263.

Gardner, D.R., Panter, K.E., Molyneux, R.J., et al., 1996. Abortifacient activity in beef cattle of acetyl and succinylisocupressic acid from ponderosa pine. J. Agric. Food Chem. 44, 3257–3261.

Gardner, D.R., Panter, K.E., Stegelmeier, B.L., 2010. Implication of agathic acid from Utah juniper bark as an abortifacient compound in cattle. J. Appl. Toxicol. 30, 115–119.

Gardner, D.R., Ralphs, M.H., Turner, D.L., Welsh, S.L., 2002. Taxonomic implications of diterpene alkaloids in three toxic tall larkspur species (*Delphinium* spp.). Biochem. Syst. Ecol. 30, 77–90.

Graham, D., Creamer, R., Cook, D., et al., 2009. Solutions to locoweed poisoning in New Mexico and the western United States. Rangelands 31 (6), 3–8.

Green, B.T., Lee, S.T., Panter, K.E., et al., 2010. Actions of piperidine alkaloid teratogens at fetal nicotinic acetylcholine receptors. Neurotoxicol. Teratol. 32, 383–390.

Green, B.T., Pfister, J.A., Cook, D., et al., 2009. Effects of larkspur (*Delphinium barbeyi*) on heart rate and electrically evoked electromyographic response of the external anal sphincter in cattle. Am. J. Vet. Res. 70, 539–546.

Holechek, J.L., 2002. Do most livestock losses to poisonous plants result from "poor" range management? J. Range Manag. 55, 270–276.

James, L.F., 1999. Halogeton poisoning in livestock. J. Nat. Toxins 8, 395–403.

James, L.F., Nielsen, D.B., 1988. Locoweeds: assessment of the problem on western U.S. rangelands. In: James, L.F., Ralphs, M.H., Nielsen, D.B. (Eds.), The Ecology and Economic Impact of Poisonous Plants on Livestock Production. Westview, Boulder, CO, pp. 119–129.

Johnson, A.E., 1974. Experimental photosensitization and toxicity in sheep produced by *Tetradymia glabrata*. Can. J. Comp. Med. 33, 404–410.

Keeler, R.F., 1973. Lupin alkaloids from teratogenic and nonteratogenic lupines: 2. Identification of the major alkaloids by tandem gas chromatography–mass spectrometry in plants producing crooked calf disease. Teratology 7, 31–36.

Keeler, R.F., Cronin, E.H., Shupe, J.L., 1976. Lupin alkaloids from teratogenic and nonteratogenic lupine IV. Concentration of total alkaloids, individual major alkaloids and the teratogen anagyrine as a function of plant part and stage of growth and their relationship to crooked calf disease. J. Toxicol. Environ. Health 1, 899–908.

Keeler, R.F., James, L.F., Shupe, J.L., Van Kampen, K.R., 1977. Lupine-induced crooked calf disease and a management method to reduce incidence. J. Range Manag. 30, 97–102.

Keeler, R.F., Panter, K.E., 1989. Piperidine alkaloid composition and relation to crooked calf disease-inducing potential of *Lupinus formosus*. Teratology 40, 423–432.

Kim, I.-H., Choi, K.-C., An, B.-S., et al., 2003. Effect on abortion of feeding Korean pine needles to pregnant Korean native cows. Can. J. Vet. Res. 67, 194–197.

Kingsbury, J.M., 1964. Poisonous Plants of the United States and Canada. Prentice-Hall, Inc, Englewood Cliffs, NJ.

Knight, A.P., Walter, R.G., 2001. A Guide to Plant Poisoning of Animals in North America. Teton New Media, Jackson, WY.

Lee, S.T., Davis, T.Z., Gardner, D.R., et al., 2010. Tremetone and structurally related compounds in white snakeroot (*Ageratina altissima*): a plant associated with trembles and milk sickness. J. Agric. Food Chem. 58, 8560–8565.

Lee, S.T., Green, B.T., Welch, K.D., et al., 2013. Stereoselective potencies and relative toxicities of gamma-coniceine and *N*-methylconiine enantiomers. Chem. Res. Toxicol. 26, 616–621.

Lee, S.T., Green, B.T., Welch, K.D., et al., 2008. Stereoselective potencies and relative toxicities of coniine enantiomers. Chem. Res. Toxicol. 21, 2061–2064.

Lee, S.T., Panter, K.E., Gardner, D.R., et al., 2006. Relative toxicities and neuromuscular nicotinic receptor agonistic potencies of anabasine enantiomers and anabaseine. Neurotoxicol. Teratol. 28, 220–228.

Lee, S.T., Ralphs, M.H., Panter, K.E., et al., 2007. Alkaloid profiles, concentration and pools in velvet lupine (*Lupinus leucophyllus*) over the growing season. J. Chem. Ecol. 33, 75–84.

Lin, G., Wang, J.Y., Li, N., et al., 2011. Hepatic sinusoidal obstruction syndrome associated with consumption of *Gynura segetum*. J. Hepatol. 54, 666–673.

Lopez-Ortiz, S.L., Panter, K.E., Pfister, J.A., Launchbaugh, K.L., 2004. The effect of body condition on disposition of alkaloids from silvery lupine (*Lupinus argenteus* Pursh) in sheep. J. Anim. Sci. 82, 2798–2805.

Mattocks, A.R., Jukes, R., 1990. Recovery of the pyrrolic nucleus of pyrrolizidine alkaloid metabolites from sulphur conjugates in tissues and body fluids. Chem. Biol. Interact. 75, 225–239.

Molyneux, R.J., James, L.F., 1982. Loco intoxication: indolizidine alkaloids of spotted locoweed (*Astragalus lentiginosus*). Science 216, 190–191.

Nielsen, D.B., James, L.F., 1992. The economic impacts of livestock poisonings by plants. In: James, L.F., Keeler, R.F., Bailey, E.M., Cheeke, P.R., Hegarty, M.P. (Eds.), Poisonous Plants: Proceedings of the Third International Symposium. Iowa State University Press, Ames, IA, pp. 3–10.

Nielsen, D.B., Rimbey, N.R., James, L.F., 1988. Economic considerations of poisonous plants on livestock. In: James, L.F., Ralphs, M.H., Nielsen, D.B. (Eds.), The Ecology and Economic Impact of Poisonous Plants on Livestock Production. Westview, Boulder, CO, pp. 5–15.

Panter, K.E., 1991. Neurotoxicity of the knapweeds (*Centaurea* spp.) in horses. In: James, L.F., Evans, J.O., Ralphs, M.H., Child, R.D. (Eds.), Noxious Range Weeds, pp. 495–499.

Panter, K.E., Baker, D.C., Kechele, P.O., 1996a. Water hemlock (*Cicuta douglasii*) toxicoses in sheep: pathologic description and prevention of lesions and death. J. Vet. Diagn. Invest. 8, 474–480.

Panter, K.E., Bunch, T.D., Keeler, R.F., et al., 1990a. Multiple congenital contractures (MCC) and cleft palate induced in goats by ingestion of piperidine alkaloid-containing plants: reduction in fetal movement as the probable cause. Clin. Toxicol. 28, 69–83.

Panter, K.E., Gardner, D.R., Gay, C.C., et al., 1997. Observations of *Lupinus sulphureus*-induced crooked calf disease. J. Range Manag. 50, 587–592.

Panter, K.E., Gardner, D.R., Stegelmeier, B.L., et al., 2011. Water hemlock poisoning in cattle: ingestion of immature *Cicuta maculata* seed as the probable cause. Toxicon 57, 157–161.

Panter, K.E., Gay, C.C., Clinesmith, R., Platt, T.E., 2013. Management practices to reduce lupine-induced crooked calf syndrome in the northwest. Rangelands 35 (2), 12–16.

Panter, K.E., Hartley, W.J., James, L.F., et al., 1996b. Comparative toxicity of selenium from seleno-DL-methionine, sodium selenate, and *Astragalus bisulcatus* in pigs. Fund. Appl. Toxicol. 32, 217–223.

Panter, K.E., James, L.F., Gardner, D.R., 1999a. Lupines, poison-hemlock and *Nicotiana* spp.: toxicity and teratogenicity in livestock. J. Nat. Toxins 8, 117–134.

Panter, K.E., James, L.F., Short, R.E., et al., 1990b. Premature bovine parturition induced by ponderosa pine: effects of pine needles, bark, and branch tips. Cornell Vet. 80, 329–333.

Panter, K.E., James, L.F., Stegelmeier, B.L., et al., 1999b. Locoweeds: effects on reproduction in livestock. J. Nat. Toxins 8, 53–62.

Panter, K.E., Keeler, R.F., 1989. Piperidine alkaloids of poison hemlock (*Conium maculatum*). In: Cheeke, P.R. (Ed.), Toxicants of Plant Origin, Vol. I: Alkaloids. CRC Press, Boca Raton, FL, pp. 109–132.

Panter, K.E., Keeler, R.F., 1992. Induction of cleft palate in goats by *Nicotiana glauca* during a narrow gestational period and the relation to reduction in fetal movement. J. Nat. Toxins 1, 25–32.

Panter, K.E., Keeler, R.F., Baker, D.C., 1988. Toxicoses in livestock from the hemlocks (*Conium* and *Cicuta* spp.). J. Anim. Sci. 66, 2407–2413.

Panter, K.E., Manners, G.D., Stegelmeier, B.L., et al., 2002. Larkspur poisoning: toxicology and alkaloid structure-activity relationships. Biochem. Syst. Ecol. 30, 113–128.

Panter, K.E., Ralphs, M.H., Smart, R.A., Duelke, B., 1987. Death camas poisoning in sheep: a case report. Vet. Hum. Toxicol. 29, 45–48.

Parton, K., Gardner, D., William, N.B., 1996. Isocupressic acid, an abortifacient component of *Cupressus macrocarpa*. N. Z. Vet. J. 44, 109–111.

Pfister, J.A., Adams, D.C., Wiedmeier, R.D., Cates, R.G., 1992. Adverse effects of pine needles on aspects of digestive performance in cattle. J. Range Manag. 45, 528–533.

Pfister, J.A., Panter, K.E., Gardner, D.R., et al., 2008. Effect of body condition on consumption of pine needles (*Pinus ponderosa*) by beef cows. J. Anim. Sci. 86, 3608–3616.

Pfister, J.A., Panter, K.E., Manners, G.D., 1994. Effective dose in cattle of toxic alkaloids from tall larkspur (*Delphinium barbeyi*). Vet. Hum. Toxicol. 36, 10–11.

Pfister, J.A., Ralphs, M.H., Gardner, D.R., et al., 2002. Management of three toxic *Delphinium* species based on alkaloid concentration. Biochem. Syst. Ecol. 30, 129–138.

Pfister, J.A., Stegelmeier, B.L., Gardner, D.R., James, L.F., 2003. Grazing of spotted locoweed (*Astragalus lentiginosus*) by cattle and horses in Arizona. J. Anim. Sci. 81, 2285–2293.

Ping, L., Child, D., Meng-Li, Z., et al., 2009. Culture and identification of endophytic fungi from *Oxytropis glabra* DC. Acta Ecol. Sinica 20 (1), 53–58.

Pryor, B.M., Creamer, R., Shoemaker, R.A., et al., 2009. *Undifilum*, a new genus for endophytic *Embellesia oxytropis* and parasitic *Helminthosporium bornmuelleri* on legumes. Botany 87, 178–194.

Raisbeck, M.F., 2000. Selenosis. Vet. Clin. N. Am. Food Anim. Pract. 16, 465–480.

Ralphs, M.H., Creamer, R., Baucom, D., et al., 2008. Relationship between the endophyte *Embellisia* spp. and the toxic alkaloid swainsonine in major locoweed species (*Astragalus* and *Oxytropis*). J. Chem. Ecol. 34, 32–38.

Ralphs, M.H., Gardner, D.R., Turner, D.L., et al., 2002. Predicting toxicity of tall larkspur (*Delphinium barbeyi*): measurement of the variation in alkaloid concentration among plants and among years. J. Chem. Ecol. 28, 2327–2341.

Ralphs, M.H., Graham, D., James, L.F., 1994. Social facilitation influences cattle to graze locoweed. J. Range Manag. 47, 123–126.

Ralphs, M.H., Greathouse, G., Knight, A.P., et al., 2001. Cattle preference for Lambert locoweed over white locoweed throughout their phenological stages. J. Range Manag. 54, 265–268.

Ralphs, M.H., Panter, K.E., Gay, C.C., et al., 2006. Cattle consumption of velvet lupine (*Lupinus leucophyllus*) in the channel scablands of eastern Washington. J. Range Ecol. Manage. 59, 204–207.

Ralphs, M.H., Whitson, T.D., Ueckert, D.N., 1991. Herbicide control of poisonous plants. Rangelands 13 (2), 73–77.

Riopelle, R.J., Stevens, K.L., 1993. *In vitro* neurotoxicity bioassay: neurotoxicity of sesquiterpene lactones. In: Colegate, S.M., Molyneux, R.J. (Eds.), Bioactive Natural Products: Detection, Isolation, and Structural Determination. CRC Press, Boca Raton, FL, pp. 457–463.

Rood, K.A., Panter, K.E., Gardner, D.R., et al., 2014. Halogeton (*H. glomeratus*) poisoning in cattle: case report. Int. J. Poisonous Plant Res. 3 (1), 23–25.

Rosenfeld, I., Beath, O.A., 1964. Selenium: Geobotany, Biochemistry, Toxicity and Nutrition. Academic Press, New York.

Schoch, T.K., Gardner, D.R., Stegelmeier, B.L., 2000. GC/MS/MS detection of pyrrolic metabolites in animals poisoned with the pyrrolizidine alkaloid riddelline. J. Nat. Toxins 9, 197–206.

Seawright, A.A., Hrdicka, J., Wright, J.D., et al., 1991. The identification of hepatotoxic pyrrolizidine alkaloid exposure I horses by the demonstration of sulphur-bound pyrrolic metabolites on their hemoglobin. Vet. Hum. Toxicol. 33, 286–287.

Selfero, N.A., Merlassino, J.L., Audisio, S., 1989. Intoxication by *Centaurea solstitialis*: therapeutic evaluation. Therios 13, 42–44.

Snider, D.B., Gardner, D.R., Janke, B.H., Ensley, S.M., 2015. Pine needle abortion biomarker detected in bovine fetal fluids. J. Vet. Diagn. Invest. 27 (1), 74–79.

Stegelmeier, B.L., 2002. Equine photosensitization. Clin. Tech. Equine Pract. 1 (2), 81–88.

Stegelmeier, B.L., Edgar, J.A., Colegate, S.M., et al., 1999. Pyrrolizidine alkaloid plants, metabolism and toxicity. J. Nat. Toxins 8, 95–116.

Stegelmeier, B.L., Gardner, D.R., James, L.F., et al., 1996. The toxic and abortifacient effects of ponderosa pine. Vet. Pathol. 33, 22–28.

Stegelmeier, B.L., Ralphs, M.H., Gardner, D.R., et al., 1994. Serum α-mannosidase and the clinicopathologic alterations of locoweed (*Astragalus mollissimus*) intoxication in range cattle. J. Vet. Diagn. Invest. 6, 473–479.

Tiwary, A.K., Stegelmeier, B.L., Panter, K.E., et al., 2006. Comparative toxicosis of sodium selenite and selenomethionine in lambs. J. Vet. Diagn. Invest. 18, 61–70.

Welch, K.D., Cook, D., Gardner, D.R., et al., 2013a. A comparison of the abortifacient risk of Western Juniper trees in Oregon. Rangelands 35 (1), 40–44.

Welch, K.D., Gardner, D.R., Panter, K.E., et al., 2011a. Western juniper induced abortions in beef cattle. Int. J. Poisonous Plant Res. 1 (1), 72–79.

Welch, K.D., Green, B.T., Gardner, D.R., et al., 2010. Influence of 7,8-methylenedioxylycoctonine-type alkaloids on the toxic effects associated with ingestion of tall larkspur (*Delphinium* spp.) in cattle. Am. J. Vet. Res. 71, 487–492.

Welch, K.D., Green, B.T., Gardner, D.R., et al., 2012. The effect of 7,8-methylenedioxylycoctyonine-type diterpenoid alkaloids on the toxicity of tall larkspur (*Delphinium* spp.) in cattle. J. Anim. Sci. 90 (7), 2394–2401.

Welch, K.D., Green, B.T., Gardner, D.R., et al., 2013b. The effect of low larkspur (*Delphinium* spp.) co-administration on the acute toxicity of death camas (*Zigadenus* spp.) in sheep. Toxicon 76, 50–58.

Welch, K.D., Green, B.T., Gardner, D.R., et al., 2016. The effect of co-administration of death camas (*Zigadenus* spp.) and low larkspur (*Delphinium* spp.) in cattle. Toxins 8, 21–34.

Welch, K.D., Panter, K.E., Gardner, D.R., et al., 2011b. The acute toxicity of the death camas (*Zigadenus* spp.) alkaloid zygacine in mice, including the effect of methyllycaconitine co-administration on zygacine toxicity. J. Anim. Sci. 89, 1650–1657.

Welch, K.D., Parsons, C., Gardner, D.R., et al., 2015. Evaluation of the seasonal and annual abortifacient risk of western juniper trees on Oregon rangelands. Rangelands 37 (4), 139–143.

Welsh, S.L., Ralphs, M.H., Panter, K.E., et al., 2007. Locoweeds of North America. In: Panter, K.E., Wierenga, T.L., Pfister, J.A. (Eds.), Poisonous Plants: Global Research and Solutions. CAB International, Wallingford, UK, pp. 20–29.

Williams, M.C., Barneby, R.C., 1977. The occurrence of nitro-toxins in North American *Astragalus* (Fabaceae). Brittonia 29, 310–326.

Williams, M.C., Norris, F.A., Van Kampen, K.R., 1970. Metabolism of miserotoxin to 3-nitro-1-propanol in bovine and ovine ruminal fluids. Am. J. Vet. Res. 31, 259–262.

Winter, H., Seawright, A.A., Hrdlicka, J., et al., 1993. Pyrrolizidine alkaloid poisoning of yaks-diagnosis of pyrrolizidine alkaloid exposure by the demonstration of sulphur-conjugated pyrrolic metabolites of the alkaloid in circulating haemoglobin. Aust. Vet. J. 70, 312–313.

Young, J.A., 1999. Halogeton: A History of Mid-20th Century Range Conservation in the Intermountain Area. US Department of Agriculture, Agricultural Research Service, Washington, DC.

PART V

PHARMACEUTICALS AND NUTRACEUTICALS BIOMARKERS

CHAPTER

38

Biomarkers of Drug Toxicity and Safety Evaluation

Arturo Anadón, María Rosa Martínez-Larrañaga, Irma Ares, María Aránzazu Martínez

Department of Pharmacology and Toxicology, Faculty of Veterinary Medicine, Universidad Complutense de Madrid, Madrid, Spain

INTRODUCTION

The costs of drug research and development continue to increase, and the total number of approved drugs continues to decrease, down 50% since 1999. On the average, it takes 10–15 years to get one new drug from molecule to medicine because of the specific requirements for approval of new drug candidates. The low productivity crisis in pharmaceutical research and development suggests that the current paradigm of drug discovery and development used in the past 30 years is unsustainable. Some new approaches have improved the success rate; for example, approximately 50% of the drug failures were attributed to unacceptable pharmacokinetics (i.e., absorption, distribution, metabolism, and excretion) and toxicity. Hence, filtering and optimizing kinetics and toxicity properties in early stage of drug discovery are widely applied to reduce the high attrition rate. However, drug safety remains a major concern. The costs of bringing a single new drug to market now exceed $800 million (Adams and Brantner, 2010) as a consequence of substantial increases in the cost of clinical and preclinical studies, longer development times, and an increase in the number of failed drugs in late phases of drug development, in particular, phase III clinical trials, because of lack of efficacy and toxicity; furthermore, promising drug candidates are frequently pulled from development based on preclinical toxicities that are not readily monitored in humans.

Drug safety assessment is a complex process that involves many preclinical tests and clinical evaluations to ensure the highest level of drug safety profiles. Therefore, interpreting the interaction between drugs and genes, including drug–protein interactions, drug–microRNA (miRNA) associations, drug–single-nucleotide polymorphisms associations, among others events, is a critical stage to improve drug toxicity evaluation and enable precision medicine (Hong et al., 2010). These interactions or associations can appear at molecular, cellular, organ, and or organism levels.

Biomarkers are used in the drug development programs in both nonclinical species and in human beings to assess safety and efficacy of novel pharmacologically active substances. Therefore, development of new, predictive safety and efficacy biomarkers is expected to reduce the time and cost of drug development. Recent late-phase drug withdrawals for safety concerns, as experienced by a number of pharmaceutical companies, increase the pressure to change preclinical and early clinical safety testing. To improve drug safety and deliver better treatments to patients, it is critical to advance our understanding of mechanisms of toxicity and to apply new technologies, for example, molecular biomarkers, to the challenge of predicting drug safety and improving mechanism-based risk assessment to ensure the safe use of new pharmaceutical drugs. Industry and regulatory agencies expect that new and better toxicity biomarkers (also known as *safety biomarkers*) will play an important role in improving key aspects of the drug development process. Successful translation of many biomarkers into clinical practice is now increasing. However, as indicated previously, to be more efficient, considerable improvement in the use of novel technical tools (omics-based technology) for effective safety assessment is needed. Omics technology is now accepted as essential methodological part of systems or integrative biology, and it is possible to observe global changes of transcripts

Biomarkers in Toxicology, Second Edition
https://doi.org/10.1016/B978-0-12-814655-2.00038-4

(genomics), proteins (proteomics), or metabolites (metabolomics) in cellular systems or tissues.

To date in the pharmaceutical industry, safety biomarkers are applied both preclinically (for early detection of toxicity, the selection of the safest drug candidate, sensitive safety monitoring in regulatory toxicity studies, and selection of dosing regimens) and clinically (for translation into human studies, safety monitoring in late clinical phases, development of personalized health-care strategies, and postmarket safety surveillance) (Dieterle et al., 2010). In preclinical, the goal of using novel qualified predictive safety biomarkers is to assist in selecting drug candidates that are more likely to be tolerated in humans. Limitations of the current standards and a lack of adequate sensitive, specific, and easily accessible biomarkers in current clinical practice underscore the need for medical research aimed at developing novel safety biomarkers that are widely available. During lead discovery, a biomarker can assist in the screening of chemicals, based on the ability of the candidate molecule to modulate the activity of a specific process, such as a biochemical pathway.

Although biomarkers have been used in drug development and treatment of disease for a long time, the identification of new predictive safety and efficacy biomarkers is expected to reduce the time and cost of drug development. Biomarkers offer a means to affect rational drug design early in the development process and accelerate translation of drug development from animal to man. The analytical and clinical validity of biomarkers must be demonstrated using relevant clinical samples in drug development and clinical practice.

CLINICAL DRUG DEVELOPMENT

The objective of the clinical development is to bring about translation into human safety monitoring in late clinical phases, allowing the development of personalized medicine strategies and postmarket supporting safety. In clinical trials, biomarkers from blood and urine, together with a variety of imaging approaches, are often the only available sources to address concerns raised by histopathological analysis in preclinical safety studies. In clinical development, however, access to organ biopsies is in most cases not possible, and drug development decisions rely on conventional clinical diagnosis and biomarker measurements. It is important, therefore, that new safety biomarkers be able to reflect tissue changes that have historically been monitored using histopathology. Ideally, peripheral biomarkers could be correlated to the severity and time course of an organ-specific toxicity as detected by histopathology, show reversibility, and then be translated to early human studies.

Clinical drug development includes different segments: Phase I, II, and III clinical trials, pharmacokinetics, pharmacodynamics, and studies to investigate special toxicities. Phase IV clinical trials for postmarket drug safety surveillance are conducted to identify and evaluate the long-term effects of new drugs and treatments over a lengthy period for a greater number of patients.

An initial set of studies, typically studies of appropriate length by the route intended for humans, is performed in both a rodent (typically rat) and a nonrodent (usually a dog or a primate). This is required to support Phase I clinical testing. During Phase I trials, a drug is tested in healthy volunteers (almost always males) to assess indication of toxicity, safety in terms of pharmacokinetics, the mechanism characterization of toxicity, general biological effects, and possible side effects and to define the tolerance (maximum tolerated dose). Phase II trials are conducted in volunteers who have the disease to determine the safety of the drug, the correct dosage (therapeutic efficacy), more details on pharmacokinetics and metabolism, and the short-term side effects. Phase I and Phase II are considered to be early clinical development. Efficacy and safety biomarkers need to be identified as early as possible in drug development (e.g., in Phase II or earlier) to incorporate their proper use in the late-phase trials for clinical validation and qualification. Phase III clinical trials (or late clinical development) confirm efficacy seen in Phase II trials. In the Phase III trials (which are large, long, and expensive), the drug is typically tested in double-blind, placebo-controlled trials to demonstrate effectiveness and establish the side effects and adverse events associated with use. Patient groups selected for Phase III trials are large and homogeneous to produce proof of safety and efficacy of the human drug. By the time Phase III testing is completed, some additional preclinical safety tests must also generally be in hand; these include the three separate reproductive and developmental toxicity studies and carcinogenicity studies in both rats and mice (unless the period of therapeutic usage is intended to be very short). Some assessment of genetic toxicity will also be expected (Gad, 2009). Based on all the data submitted to regulatory authorities for review, a human medicine is either accepted or rejected for approval.

During early clinical development, when fewer numbers of subjects are studied, the value of pharmacokinetics/pharmacodynamics (PK/PD) modeling is in its ability to quantitatively relate the dose with exposure and changes in biomarkers that are causally linked to clinical outcomes. In the last years, PK/PD modeling and simulation has been routinely applied during medicine research and development to help in the design, analysis, and interpretation of clinical trial data. Thus, PK/PD reasoning is likely to be a core discipline at all stages of medicine research and development in the near future. Clinical development of potential medicines has undergone a drastic change in recent years; early-stage testing at present is geared toward "learning" about potential safety and efficacy dose-response of drugs and the later stage toward "confirming," in large trials, the acceptable safety and efficacy profile at one dose. Biomarkers, combined with PK/PD analysis, have played a key role in enabling this learn–confirm paradigm, which has made clinical development more efficient than before. In the next few years, drug discovery and preclinical development are also ready to undergo a similar change—experiments in highly unreliable animal models of diseases are likely to be replaced by ones more focused on learning about the target and disease pathways in animal models and human tissues. Biomeasures, combined with systems pharmacology analysis, will aid integration of the generated data into translatable models. Success of both these approaches will depend on the availability of reliable biomarker and biomeasure data such as -omics, receptor-level, and intracellular bioanalytical assays (Roskos et al., 2011). Systems pharmacology (i.e., pharmacological application of systems biology) integrates the classical pharmacology data with "omics" data, empowering construction of mechanism-based predictive models for application of biomarker-based drug safety assessment in drug discovery and development, as well as regulatory evaluation. The goal of biomarker-based drug safety assessment is to explore and characterize all biological mechanisms-mediating drug toxicity and side effects. Additionally, bioinformatics resources are of much interest for the development of predictive models for biomarker-based drug safety assessment. There are four main types of bioinformatics resources: (1) drug–target interactions, (2) drug side effects, (3) toxicogenomics, and (4) protein–protein interactions and biological pathways.

Application of Biomarkers in Postmarketing

If the drug is approved, additional postmarket or long after the drug approval (or Phase IV) studies may be required to ensure the product's safety in the general population. All approved human medicines are monitored by passive surveillance to determine adverse events and deviations in product quality. Serious and rare adverse effects of drugs are often observed only after marketing of the drug, as premarketing clinical trials are limited in the number of patients being studied. Information obtained postmarketing has resulted in recent labeling changes (e.g., carbamazepine [CBZ], warfarin).

The usefulness of a biomarker may also be discovered in studies performed in the Phase IV or during usual therapy. Adverse drug reactions (ADRs) are a major issue for clinical practice and include a wide variety of clinical toxic drug reactions. The term adverse reaction usually excludes nontherapeutic overdosage (i.e., accidental drug exposure) or lack of efficacy. According to the axiom of Paracelsus, all compounds are toxic at high doses and all are safe at very low doses (Borzelleca, 2000). Toxic and adverse effects of drugs may be classified as pharmacological, pathological, or genotoxic and their incidence and seriousness are related, at least over some range, to the concentration of the toxic chemical in the body. Many chemicals are not toxic themselves but are activated by transformation into toxic metabolites. The toxic response is then dependent on the balance of the rate at which the toxic metabolites are produced and destroyed.

Severe ADRs are often classified into two groups. The *first group* is dose-dependent and predictable on the basis of the drug's known pharmacological actions of the therapeutic drug (examples include bleeding induced by oral anticoagulants [i.e., warfarin], hypoglycemia induced by diabetic drugs, leukopenia induced by cytotoxic anticancer drugs, or convulsions produced by local anesthetics). These may occur within therapeutic dose ranges and those that are undesirable in a given therapeutic situation and often represent an exaggeration of the known primary and/or secondary pharmacology of the drug (i.e., side effects) or may occur with doses above the therapeutic range for a particular patient (i.e., overdosage toxicity). The phenotypes of the *second group* are not explained by the mode of action. These reactions are idiosyncratic, are unpredictable from the pharmacological action of the drug, and are not necessarily dose-dependent. This second group is responsible of nearly 10%–15% of the all ADRs, particularly idiosyncratic reactions, which include, for instance, severe cutaneous disorders, such as Stevens–Johnson syndrome (SJS) or toxic epidermal necrolysis (TEN), and injury to the liver caused by different drugs.

Genetic variants that cause susceptibility to a drug reaction loom large. The identification of such variants is expected to improve the management of patient care by determining which patients should avoid a specific drug and which patients should take a modified dose of the drug. Genes that are currently known to be associated with ADRs can be classified mainly into three categories: (1) drug-metabolizing enzymes, (2) drug transporters, and (3) human lymphocyte antigens (HLAs). Genes in the first two categories influence the PK/PD of drugs, and poor clearance of drugs from the body can increase the concentration of the drug to a toxic level and thus result in ADRs. For example, genetic variants in the genes encoding thiopurine *S*-methyltransferase and uridine diphosphate glucuronosyltransferase 1A1 are known to increase the risk of myelotoxicity associated with treatment with azathioprine and irinotecan, respectively (Chung et al., 2004). For the third category, the HLA-B*1502 allele is associated with serious dermatologic reactions (e.g., TEN and SJS to the tricyclic antidepressant drug CBZ, an anticonvulsant). CBZ developed SJS/TEN; studies illustrated a strong association between CBZ-induced SJS and the presence of the HLA-B*1502 allele, rendering this allele in a small patient from Taiwan a risk factor (i.e., a predictive biomarker) (Hung et al., 2006). For these drugs are recommended that the relevant genetic tests be performed before the drug is prescribed. Another example of pharmacogenomics and drug toxicity is with statins. A strong association has been described between myopathy and two tightly linked variants in the *SLCO1B1* gene in patients consuming simvastatin (Pasanen et al., 2006), as well as polymorphisms in some solute transporter genes associated with altered hepatic uptake of pravastatin, indicating that genetic variations in *SLCO1B1* might affect the variability of outcomes in patients treated with statins. A variant of *SLCO1B1* was found in a patient with pravastatin-induced myopathy (Morimoto et al., 2004). Warfarin is another example, in which a significant proportion of variation in the PK/PD between patients influences the anticoagulant response from a fixed dose. Variability has been linked to the genetic variants of CYP2C9 and VKORC1 (Schwarz et al., 2008).

The various circumstances of drug toxicity can be summarized as follows: (1) *toxicity on target* (or mechanism-based) toxicity (i.e., toxicity due to interaction of the drug with the same target that produces the desired pharmacological response); (2) *hypersensitivity and immune responses* (e.g., allergic responses); (3) *off-target toxicity* (i.e., the drug is not specific in its interactions; binding to an alternate target is the cause of toxicity). One example is the terfenadine, which binds to H_1 receptor, eliciting the desirable antihistaminic response and to the hERG channel, causing arrhythmias; (4) *bioactivation* (i.e., some drugs are converted to reactive metabolites); and (5) *idiosyncratic reactions* (Guencherich, 2011).

The context of idiosyncratic reactions is individual and they are usually rare ($1/10^3$–$1/10^4$). At least five theories have been proposed to explain these reactions: The *first theory* involves polymorphisms or rare alleles of metabolism enzymes. The concept is that sensitivity is due to lack of (or too much) metabolism of a drug, including a lack of detoxication (e.g., having ultrarapid metabolizers in the P450 2D6 group) and also being deficient in a glutathione (GSH) transferase or other enzymes to detoxify the medicinal product, occurring in approximately 1% of Caucasian population. The *second theory* is the hapten formation. Some individuals show more activation of a drug to yield a hapten and the variation in the immune systems of individuals will dictate that only a few will show this response (Uetrecht, 2007). The *third theory* is sometimes referred to as the inflammagen model (Ganey et al., 2004). It is theorized that the bioactivation and other events occur in many people and that inflammation (or other predisposing episodes) render only some individual more sensitive. The *fourth theory* is the danger hypothesis. Injured tissue produces danger signals (e.g., lipid oxidation products, cytokines) that evoke a toxic response, not the drug or its metabolites. The *fifth theory* is the pharmacological intervention model (Gerber and Pichler, 2004). In this model, drugs elicit immunological responses by reversible binding to protein (i.e., without covalent binding). One basis for this model is the formation of reactive metabolites.

Among ADRs, hepatotoxicity and nephrotoxicity are two major reasons that drugs are withdrawn postmarket in part because of the failure of animal safety studies to translate to human safety of a new chemical entity. Thus, there is need for new efficient biomarkers that can translate effectively. Additionally, personalized medicine biomarkers can be used to predict whether that individual will respond favorably or adversely to a drug or medical therapy. Metabolomics can play a role in providing translational biomarkers for organ injury and for predicting a susceptible subset of the population that should avoid specific drugs. Metabolomics refers to "the measurement of the metabolite pool that exists within a cell under a particular set of conditions" (Fiehn, 2002), and metabonomics describes "the quantitative measurement of the dynamic multiparametric metabolic response of living systems to pathophysiological stimuli or genetic modification" (Nicholson et al., 1999). Metabolomics and metabonomics are metabolic profiling methods that offer the opportunity to identify biomarkers or patterns of biomarker changes related to drug toxicity in biofluid samples, such as urine and blood, which can be collected with relative ease. Moreover, it is possible by measuring the gene expression patterns before and after compound treatment to study interactions between structure and activity of the whole genome and adverse biological effects with exogenous stressors (Aardema and MacGregor, 2002). Genes that consistently exhibit increased or decreased expression during these toxic responses in model systems serve as biomarkers to predict potential adverse preclinical outcomes. The overall aim of toxicogenomics analysis is the identification of the mode of action by which a compound induces a toxic/adverse effect (Edwards and Preston, 2008). In addition to the identification of characteristic molecular pathways, it has been reported that specific transcripts can be used as mechanistic or predictive biomarker for organ-specific toxic alterations (Zidek et al., 2007).

INTEGRATION AND USE OF SAFETY BIOMARKERS IN DRUG DEVELOPMENT

Safety biomarkers can be used to predict, detect, and monitor drug-induced toxicity during both preclinical studies and human trials. Unlike techniques for histopathology, blood and urine biomarkers are noninvasive but remain quantifiable and of translational value. The recognition by both the pharmaceutical industry and regulatory agencies that toxicological biomarkers may have much to offer in helping to bring safe drugs to the market being a reality. Biomarker measurements reflect the time course of an injury and provide information on the molecular mechanisms of toxicity (Boekelheide and Schuppe-Koistinen, 2012). In addition, biomarker measurements support target validation and proof of target, mechanism, and efficacy, and they are being developed first in preclinical animal models of disease; the majority of biomarker research is done in clinical trials that test cancer drugs, which represent the single largest therapeutic class of drug in development. There are actually a number of guidelines and recommendations from regulatory agencies on validation of bioanalytical methods, in particular for measuring drugs or metabolites, as well as published guidance from the US Food and Drug Administration (FDA) and European Medicines Agency (EMA) dedicated to genomic biomarker assays.

Biomarkers have been used in clinical diagnosis and drug development for decades to measure physiological parameters such as blood pressure, heart rate, body temperature, and urine color, as well as clinical chemistry parameters such as enzymes in blood and urine, serum creatinine levels, and urinary glucose. Examples are the appearance of glucose and albumin in urine with hyperglycemia as biomarkers of diabetes, and blood pressure and cholesterol reduction as biomarkers that have attained a status as surrogate endpoints for drug approval because they are clearly linked to mortality resulting from heart attack and stroke (Temple, 1999). Currently, biomarkers are being developed to identify patients at disease risk and to predict potential treatment responders, adverse event occurrences, and favorable clinical outcomes for many disease states, especially cancer.

Advances in molecular biology and analytical technologies have enabled the discovery of novel biomarkers that include individual genes, transcripts, proteins, and endogenous metabolites (including lipids) and their associated patterns of expression. The use of novel, but less well-established, pharmacodynamics biomarkers can further facilitation decision-making from discovery through preclinical development and into clinical trials, whereas rapid advances in genomics and proteomics have increased the discovery of new biomarkers and their value in drug development and treatment of disease. Other examples include the use of imaging patterns, electrical signals, and cell levels in body fluids to identify potential concerns (Boekelheide and Schuppe-Koistinen, 2012).

The effective integration of biomarkers into clinical development programs, with the objectives of enriching a responder population or identifying patients at risk for an adverse event, is essential; it may facilitate new medicinal product development and promote personalized medicine.

Application of Safety Biomarkers in Different Phases of Drug Development

During the discovery phase, there are several opportunities for in vivo toxicity assessment. Toxicological biomarkers are important in all phases of drug development, including discovery, preclinical assessment, clinical trials, and postmarket surveillance. During the toxicity studies, traditional biomarkers utilized for toxicity assessment, for instance, are physiological measures, clinical hematology and biochemical parameters, macroscopic and microscopic tissue assessment, and specific biomarkers of potential issues identified during target validation and through in vitro indicator assays, which can be estimated in the context of in vivo exposures. Additionally, there is an opportunity to assess the potential for unexpected compound-related toxicity in animal disease models, for which the systemic effects of a disease process may lead to aggravation of toxicity (Boelsterli, 2003).

Pathology, including all the aspects of anatomic (histopathology), clinical chemistry, and clinical pathology (hematology), is generally considered the single most significant portion of data to come out of systemic toxicity studies (particularly the repeat dose, with versions from 14 days to 2 years in duration). Anatomic pathology evaluations actually consist of three related sets of data (gross pathology observations, organ weights, and microscopic pathology) that are collected during termination of the study animals. At the end of the study, a number of tissues are collected during termination of all surviving animals (test and control). In toxicology, histopathology remains the "gold standard" for safety assessment during the preclinical phase of drug development, permitting researchers to identify the type, localization (i.e., target organ), and severity of organ lesions being a sensitive technique that conveys large amounts of information on the spatial and temporal relationship of toxicological processes. Organ weights and terminal body weights are recorded at study termination so that absolute and relative (to body weight) values can be statistically evaluated. The problem existing with histopathology as the only approach to evaluating tissue injury is that it is very costly in terms of animals and time and therefore interferes with the ability to extend drug screening. Bindhu et al. (2007) have provided a review of practices of how such organ weight information is evaluated and utilized in the overall evaluation of pathology and adverse effects. In general, with the exception of brain weights, relative (to body weight) changes are considered more relevant to identifying target organ toxicities. The evaluation of the pathological changes induced in laboratory animals by new drugs represents the cornerstone of their safety assessment before they can be first tried in patients. This preliminary assessment, which is based largely on conventional histopathological techniques, represents a major contribution to the development of new treatments for both human and animal diseases. The basic model of the dosing laboratory with various dose of new drug for increasing periods of time, accompanied by careful clinical observations, biochemical and hematological monitoring, followed by histopathological examination of the tissue, remains essentially unaltered.

Histopathology has particular limitations as a gold standard in the translation of preclinical animal-based safety studies to the clinical testing of a new drug. Although pharmaceutical companies rely on a variety of biochemical, mechanistic, molecular, and physiological datasets in the development of a new drug, histopathological analysis of animal studies remains a key contributor to the decision about safety, despite the known limitations of these data (Olson et al., 2000). Histopathology analysis lacks the quantitative objectivity inherent in biomarkers; while biomarkers are measured with the evolving new technologies, histopathology remains at the discretion of individual pathologists. Other considerations are the impact of variability in histopathology methods on biomarker qualification. Burkhardt et al. (2011) pointed out in a study documented by four different pathologists who evaluated slides (blinded and nonblinded) the interpathologist variation as a source of variability to alter the biomarker interpretation. The pathologist is required not only to evaluate alterations to organs and tissues and any relationship that they might have to drug treatment but also to assess the likely relevance any treatment-related finding they might have for patients (Gad, 2009).

The role of biomarkers in the preclinical safety gives the lie to the assessment of indication of toxicity, the mechanism characterization of toxicity, and defining the maximum tolerated dose. It is pivotal to determine the relevance of biomarkers in each preclinical species: the correlation of safety markers with PK/PD data allows the refinement of drug formulation, administration, and dosing/schedule schemes.

TRADITIONAL INDICATOR FOR DRUG TOXICITY ASSESSMENT

Traditional indicators of target organ toxicity used in drug preclinical safety studies consist of a battery of clinical pathology parameters in blood and urine coupled with histopathological examination of altered tissues. Table 38.1 lists the clinical pathology parameters often measured in multidose toxicity studies.

TABLE 38.1 Standard Clinical Pathology Measures in Fluids

Clinical Chemistry	Albumin, alkaline phosphatase, bilirubin, blood urea nitrogen, calcium phosphorus, chloride, creatinine, creatine kinase, gamma-glutamyltransferase, globulin, glucose, alanine aminotransferase, aspartate aminotransferase, lactate dehydrogenase, total bilirubin, total cholesterol, total protein, triglycerides, electrolytes (Na, K, Cl), glucose, bile acids, total CO_2.
Hematology	Red blood cell count, hemoglobin, hematocrit or packed cell volume, erythrocytic indices (mean corpuscular hemoglobin, mean corpuscular volume, mean corpuscular hemoglobin concentration, red cell distribution width), platelet count, mean platelet volume, prothrombin time, reticulocyte count, white cell count, white cell differential count, activated partial thromboplastin time.
Urinalysis	Color, clarity, volume, specific gravity or osmolality, pH, bilirubin, urobilinogen, glucose, occult blood, phosphorus, total protein, electrolytes (Na, K, Cl), nitrites, ketones, creatinine, microscopic sediment.

Several analytes and enzymes are used to assess the integrity and the function of organs (Lanning, 2006). However, it should be highlighted that changes in clinical pathology parameters must be interpreted in the context of understanding of pathophysiology and the known effects of the drugs tested. Some of these enzymes and analytes are briefly described in the following sections.

Alkaline Phosphatase

Alkaline phosphatase (AP) is composed of several isoenzymes that can hydrolyze organic phosphatase ester bonds in an alkaline medium, giving rise to an organic radical and inorganic phosphate. These isoenzymes are present in practically all tissues of the body, especially at or in the cell membranes; there are specific forms of AP in liver, bone, intestine, placenta, and kidney. However, the predominant forms present in normal serum are the liver and bone forms. In addition, normal young growing animals have serum AP activities up to three times higher than those in normal mature animals (Stockham and Scott, 2002). In addition, during pregnancy, serum AP levels rise two- to fourfold by the ninth month and return to normal by 21 days postpartum. It appears that the AP is associated with lipid transport in the intestine and liver and calcification in the bone. AP increases markedly in diseases that impair bile formation (cholestasis) and to a lesser extent in hepatocellular disease. The three major causes of high serum AP activity are induction of hepatic AP (i.e., intrahepatic or posthepatic cholestasis), induction of hepatic AP release (i.e., iatrogenic corticosteroids, phenobarbital, hyperadrenocorticism), and increase of osteoblastic activity (i.e., fractures).

Creatine Kinase

Creatine kinase (CK) is a leakage enzyme, and increased serum values occur with reversible and irreversible damage; high serum CK values are indicative of active or recent muscle injury or damage.

Aminotransferases

These enzymes, including alanine aminotransferase (ALT) and aspartate aminotransferase (AST), are blood indicators of hepatocyte damage, which have been used for decades in preclinical studies and for clinical monitoring of adverse effects. ALT is reliable for routine screening for liver injury or disease, whereas AST is present in skeletal muscle and heart and most commonly associated with their damage. AST is relatively nonspecific, but high levels indicate liver cell injury. These enzymes leak out of the hepatocyte cytosol when the plasma membrane permeability is altered. The magnitude of ALT increases in liver damage or disease is greater than that of AST. Serum aminotransferases and AP are common indicators of hepatocellular and hepatobiliary injury, respectively.

AST is not only found in the liver but also in the muscle. Therefore, in case of a compound suspected to create lesions of the muscle, other enzymes such as sorbitol dehydrogenase (SDH) (liver-specific) and CK (muscle and brain) could be used to differentiate the source(s) of AST. SDH is quite sensitive but is fairly unstable, so samples should be processed as quickly as possible.

Lactate Dehydrogenase

Lactate dehydrogenase (LDH) is a hydrogen transfer enzyme that is found in the cytoplasm of most of the cells of the body. LDH, commonly included in routine analysis, is insensitive as an indicator of hepatocellular injury. The peak of the increase in serum LDH values is usually observed within 48–72 h and tissue levels of LDH are about 500 times greater than normal serum levels. Causes of increased serum LDHs include muscle damage or necrosis, hemolysis, liver disease, renal tubular necrosis, pyelonephritis, and malignant neoplasia.

Bilirubin

Bilirubin is the yellow breakdown product of normal heme catabolism and is measured to assess liver damage. Hyperbilirubinemia results from increased bilirubin production, decreased liver uptake, or conjugation or decreases biliary excretion; it is often seen prior to increases in serum total bilirubin. Typically, increased total bilirubinemia is accompanied by enhanced Alanine aminotransferase (ALP) (Stockham and Scott, 2002). It is known that (1) increases in serum total bilirubin may be due to increased hemoglobin destruction (i.e., hemolytic hyperbilirubinemia) or liver damage (i.e., obstructive hyperbilirubinemia) and (2) increased serum total bilirubin is an early indicator of cholestasis or hepatopathy. Serum bilirubin may not be a particularly sensitive index of liver dysfunction or disease prognosis.

Creatinine and Blood Urea Nitrogen

Serum creatinine can provide similar information to blood urea nitrogen (BUN) in renal disease or postrenal obstruction or leakage. It is freely filtered through the glomerulus; however, small amounts are reabsorbed by renal tubules and secreted by the proximal tubules. Creatinine also estimates blood filtering capacity of kidney, as BUN does, but is more specific than BUN. Creatinine levels are also increased by reduced renal perfusion. Creatinine clearance may also be measured and is considered to be an accurate index of glomerular filtration rate (GFR) but BUN estimates blood filtering capacity of the kidneys (i.e., it is significantly elevated until kidney function is reduced 60%–75%). Serum creatinine and BUN are commonly used markers of the renal clearance of nitrogenous waste and thus of the kidney function, whereas creatinine serum levels are modulated by muscle mass, age, dehydration status, and extrarenal clearance (gut). The FDA (2008) accepted BUN and serum creatinine in rat kidney safety assessment studies that use histopathology as the gold standard.

A significant proportion of creatinine (up to 25% in healthy human) is secreted by the tubuli into urine and is not filtrated by the glomeruli. Furthermore, serum creatinine and urea levels are influenced by nonrenal sources (protein supplementation) and other adverse effects, including changes in protein synthesis and catabolic states resulting from lesions of the muscle. Urea is the primary end product of protein catabolism in most species and is freely filtered at the glomerulus, but it is equally freely diffusible out of the tubule, and a highly variable but significant amount (40%–70%) reenters the extracellular fluid (and ultimately the plasma). The rate of urea production does not constantly increase with a high-protein diet and with enhanced tissue breakdown due to hemorrhage, trauma, or glucocorticoid therapy; in addition, a chronic or an abruptly absent adequate food supply or intake increases BUN

(Urbschat et al., 2011). In comparison, a low-protein diet and/or advanced liver disease can lower the BUN without change in GFR, despite an already existing reduction in GFR (Proulx et al., 2005). The extent of urea reabsorption is in general in inverse proportion to tubular flow rates.

Plasma Proteins

The proteins as α- and β-proteins, albumin, and clotting factors (but not γ-globulin) are synthesized in the liver. Albumin is the most abundant protein in blood plasma (comprises about half of the blood serum protein) and is the major determinant of plasma oncotic pressure and transports numerous substances; its serum concentration is determined by (1) the relative rates of its synthesis and degradation or loss, (2) its distribution between the intra- and extravascular beds, and (3) the plasma volume. Albumin detects tubular injury and is most commonly used because it is the highest level protein in urine. Albumin is also the protein most likely to be found at increased levels early with a variety of renal disorders, in particular with glomerular injury, although it is not a specific marker for any one site. Normally, serum total protein concentration is in direct proportion to the serum albumin concentration. The size and negative electric charge of albumin exclude it from excretion in the glomerulus. These serum proteins and some others increase nonspecifically in response to tissue injury (e.g., inflammation) with the release of cytokines.

Total protein alterations are generally associated with decreased production (liver) or increased loss (kidney). Total protein (a classical biomarker for the glomerular filtration membrane) is the ensemble of all protein species measured together. Large and strongly charged proteins are not filtered by the glomerulus, but the small proteins go free through the glomerular barrier and become reabsorbed by the proximal tubular. Proteins appeared in the urine result from damaged nephrons. In urine, the abnormally high excretion of proteins is called proteinuria and has been highlighted in the literature as (1) a clinical prognostic marker, (2) a preclinical and clinical diagnostic marker to detect acute kidney injury (AKI), and (3) a factor predicting progressive loss of renal function in the context of a variety of diseases (Chesney et al., 1981; Diamond and Yoburn, 1982; Herget-Rosenthal et al., 2004a,b; Polkinghorne, 2006; Richmond et al., 1982). However, in the setting of early AKI, proteinuria does function as a suitable biomarker. This is strengthened by several aspects: classical AKI is not accompanied by proteinuria/albuminuria; therefore, proteinuria would not be a marker in the course of recovering from AKI (Urbschat et al., 2011). Alterations of the glomerular filtration barrier, such as damage of the glomerular podocytes, lead to leakage of plasma proteins into the ultrafiltrate (Schmid et al., 2003).

The normal glomerular filtrate contains 10 mg protein/L, but only approximately 1% is normally present in the urine because of the strong reabsorption capacity of the proximal tubule. If this reabsorption reaches a saturation point due to excessive glomerular damage/leakage with an often observed associated "poisoning" of the proximal tubules or if the tubular protein reabsorption complex is directly damaged by toxic agents, proteinuria can be observed despite normal GFRs (Guder and Hofmann, 1992).

A number of recent publications highlight the diagnostic power of total protein for AKI induced by nephrotoxicants, such as cisplatin, thalidomide, pamidronate, aminoglycosides, ifosfamide, doxorubicin, or nonsteroidal antiinflammatory drugs (Benoehr et al., 2005; Desikan et al., 2002; Guo and Nzerue, 2002; Koerbin et al., 2001; Zomas et al., 2000).

Total Protein

This is a classical biomarker for the glomerular filtration membrane measured in the urine. Alterations of the glomerular filtration barrier, such as the damage to the glomerular podocytes, result in leakage of plasma proteins into the ultrafiltrate. If the reabsorption of proteins reaches a saturation point due to excessive glomerular damage/leakage with an often observed associated "poisoning" of the proximal tubules or if the tubular protein reabsorption complex is directly damaged by toxic agents, proteinuria can be observed despite normal GFRs (Guder and Hofmann, 1992). Total protein is used in the AKI diagnostic for a known number of nephrotoxicant drugs such as cisplatin, thalidomide, pamidronate, aminoglycosides, ifosfamide, doxorubicin, or nonsteroidal antiinflammatory drugs.

Gamma-Glutamyltranspeptidase

Gamma-glutamyltransferase (GGT) is present in the liver, pancreas, and kidney. GGT is elevated in cholestasis, and the liver levels of this microsomal enzyme are increased in response to microsomal enzyme induction. Likewise, anticonvulsant drugs have been shown to elevate the GGT and AP activities in preclinical species and humans in the absence of histological evidence of hepatic injury (Mendis et al., 1993).

Hydroxybutyrate Dehydrogenase

Hydroxybutyrate dehydrogenase (HBDH) is a cardiac biochemical marker. The patient's high level of HBDH indicates myocardial infarction.

TABLE 38.2 Existing Biomarkers of Clinical Pathology

Biomarkers	Use	Limitations
Serum creatinine[a]	To detect kidney toxicity in preclinical and clinical studies; routine clinical care	Sensitivity and specificity: Poor sensitivity and specificity for detection of renal injury. Limitation for estimating glomerular filtration rate. Insensitive for detection of histological injury in preclinical toxicity studies and humans.
Blood urea nitrogen[a]	To measure renal function	Poor sensitivity and specificity for detection of renal injury. Limited ability to monitor drug toxicity in humans.
Urinalysis	To identify a renal toxicant	Relatively primitive and the reliability of the results correspond with the methods employed.

[a]*Not very sensitive or specific to detect acute kidney injury because of the fact that they are affected by many renal and extrarenal factors.*

Lactate Dehydrogenase

Lactate dehydrogenase (LDH) is a hydrogen transfer enzyme that is found in the cytoplasm of most of the cells of the body. The peak of the increase in serum LDH values is usually observed within 48–72 h. Causes of increased serum LDH comprise muscle and liver damage, renal tubular necrosis, pyelonephritis, and malignant neoplasia; this enzyme is not very specific unless isozymes are evaluated.

Sodium, Potassium, and Chloride

These ions are measured to assess the electrolyte status. Electrolyte gain or loss can occur in the gastrointestinal tract (e.g., dietary intake, loss of saliva, gut stasis, diarrhea, vomiting), kidney (e.g., lack of antidiuretic hormone, excess or lack of aldosterone, tubular disease), lung (e.g., hyperventilation, febrile episodes), and skin (e.g., sweat, febrile episodes). Sodium is the major cation of extracellular fluid and is essential in the maintenance of water distribution and osmotic pressure. In acute renal failure, the fractional excretion of sodium is the most accurate screening test to differentiate between prerenal and intrarenal disease origin (e.g., a value below 1% suggests prerenal disease) (Urbschat et al., 2011). Potassium is the major intracellular cation and is the critical ion in maintaining ionic gradient for neural impulse transmission. Chloride is the major extracellular anion and is regulated passively by gradients derived from active sodium transport across cell membranes; like sodium, chloride is involved in water distribution, osmotic pressure, and anion–cation balance in the extracellular fluid compartment. The serum chloride concentration is directly proportional to the sodium concentration.

Urinalysis

The analysis of urine should be performed in any toxicological study, particularly in rodents when the test material is suspected to be a renal toxicant. The urinalysis provides a unique opportunity to selectively and noninvasively sample a single target organ by examination of its end product. The results, which should be interpreted along with histopathology results, can be effective in determining whether the kidney is functioning properly or if its capacity has been overcome. As a minimum, the general appearance, specific gravity, acidity, protein, glucose, ketones, bilirubin, urobilinogen, and cellular content/morphology (microscopic) should be measured. The urine volume, combined with assessment of urine concentration (specific gravity or osmolality), can serve as an index to renal function. The uses and limitations of some existing biomarkers of clinical pathology are listed in Table 38.2.

BIOMARKERS OF DRUG LIVER TOXICITY

Drug-induced adverse liver effects as drug-induced liver injury (DILI) are frequent events that appeared in nonclinical and/or clinical studies during the development of new drug entities. The detection of hepatotoxicity alerts is a continuous process covering all drug development phases, and the identification of a hepatotoxic potential in nonclinical studies has frequently resulted in delayed or discontinued development of drug candidates. DILI is the most common adverse event causing withdrawal of approved drugs and postmarketing regulatory decisions and is associated with significant mortality. Metabolic activation of drugs is an important mechanism in DILI.

To date, standard nonclinical toxicity studies remain the cornerstone of the prediction of hepatotoxicity in humans. However, new approaches and the refinement of existing methods are necessary to improve prediction of DILI in humans. Various promising investigative approaches are currently being evaluated for potential screening purposes or use on a case-by-case basis, following hepatotoxicity alerts in standard nonclinical toxicity studies. Furthermore, a number of industry/academic/regulatory consortia are assessing the utility of new approaches. These consortia are among others operating under the Innovative Medicines Initiative (IMI) and the C-path and International Life Sciences Institute (ILSI)/Health and Environmental Sciences Institute (HESI) institutes. HESI qualified two biomarkers of drug-induced nephrotoxicity in rats, urinary clusterin and renal papillary antigen 1 (RPA-1), with FDA. RPA-1 was qualified by FDA for voluntary use in detecting acute drug-induced renal tubule alterations, particularly in the collecting duct, in male rats when used in conjunction with traditional clinical chemistry markers and histopathology in Good Laboratory Practice (GLP) toxicology studies for drugs for which there is previous preclinical evidence of drug-induced nephrotoxicity or where it is likely given the experience with other members of the pharmacologic class.

The liver is the primary responsible organ for drug metabolism using phase I and phase II and thus is exposed to drugs and metabolites rapidly after gastrointestinal absorption. Many liver cytochrome P450 enzymes can be either activated or suppressed by bioactive nutritional components or other drugs and these interactions can affect the drug metabolism and clearance from the liver. Hence, metabolic status may be a significant element in determining whether a toxic reaction occurs in the liver and if so, the capability of the patient to overcome the insult. The metabolism, which is intended to inactivate xenobiotics, can also lead to the formation of toxic, reactive metabolites, which can impair cellular functions. Reactive metabolites can bind to macromolecules and different structures within the cell, leading to loss of protein function, perturbation of lipid metabolism, DNA damage, and oxidative stress and for that several biomarkers have been proposed (Fig. 38.1). In addition, other consequences of xenobiotic hepatotoxicity are known, such as inhibition of hepatic transporters, inflammatory processes, bile duct hyperplasia, and hepatic tumors. Selective serum markers of apoptosis, necrosis, and inflammation would have benefit for differentiating the fundamental mechanisms of DILI.

The gold standards used to monitor hepatotoxicity are the serum liver transaminase enzymes ALT and AST. Changes in the ALT serum levels indicate hepatocyte necrosis with high sensitivity and fairly good specificity. The specificity and sensibility of ALT and AST are limited as there is a lack of correlation of liver enzyme changes and the observable histopathological damage. Therefore, the increases of serum ALT levels are not often correlated with liver histomorphologic alteration.

Some candidate drugs may cause isolated adverse liver effects in humans that are not predicted from nonclinical studies. This type of DILI is termed *idiosyncratic*. The mechanism of human idiosyncratic liver effects appears to include an interaction of genetic and nongenetic factors that are not reproduced in standard nonclinical toxicity studies (Ulrich, 2007; Walgren et al., 2005). Retrospective analysis of nonclinical data has provided no evidence that idiosyncratic drug-induced hepatotoxicity in humans could have been predicted from nonclinical toxicity data (Kaplowitz, 2005; Ong et al., 2007; Peters, 2005). Given the current understanding that idiosyncratic drug reactions are rare, human-specific, and most often dose-independent events, it should be emphasized that an improved prediction of idiosyncratic hepatotoxicity may not be achievable on the basis of nonclinical toxicity data. It is conceivable that improved detection and prediction of idiosyncratic DILI may be achieved in the future by new predictive biomarkers and/or in vitro and/or

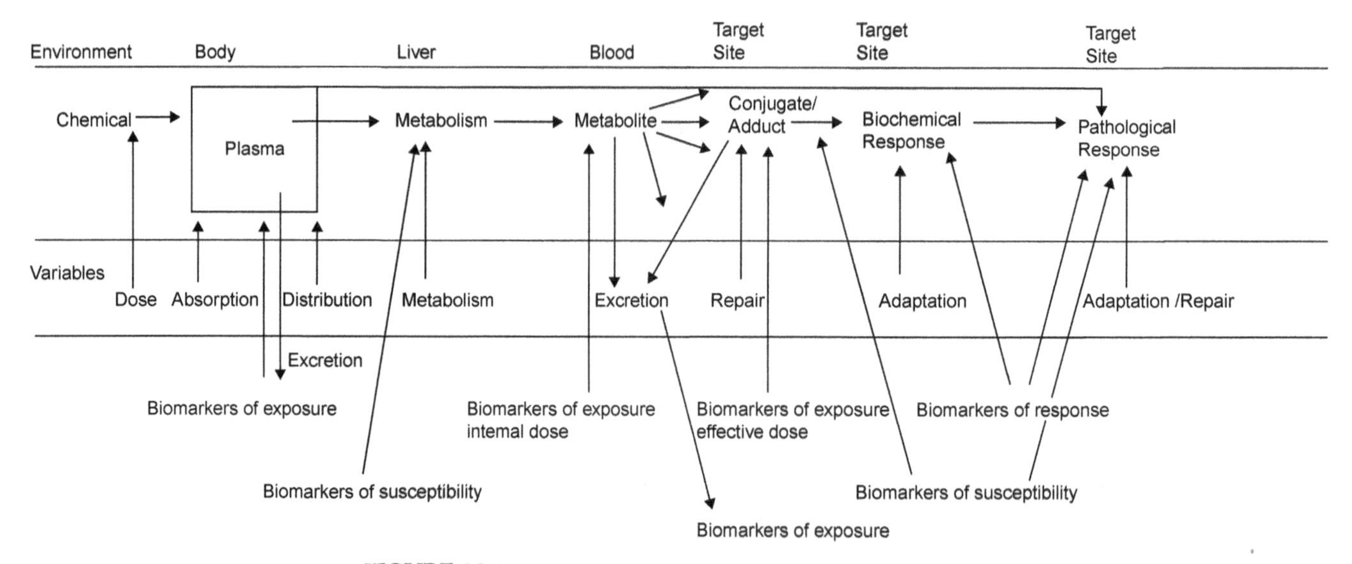

FIGURE 38.1 The biomarker paradigm for metabolism of drugs.

in vivo models, which are currently not available and/or validated. Approaches of individual companies and international collaborative research initiatives aiming at incorporating promising innovative tests into the regulatory requirements are continuing (Evans et al., 2004).

Biomarker for Monitoring Liver Injury

The best characteristics of hepatic damage biomarkers include organ specificity for liver, strong correlation with well-defined hepatic histomorphologic changes, outperformance of or added information to ALT and/or AST values, adaptation of screening assays to high-throughput modalities that are commercially available, and sample accessibility by noninvasive procedures such as blood collection (Ozer et al., 2008). Preclinical assays currently include some combination of serum ALT, AST, AP, GGT, SDH, total bilirubin, and bile acids among others.

ALT is a biomarker for hepatocyte injury, but its blood level can increase for a variety of reasons; thus, its value alone is not a reliable biomarker for liver injury. On that basis, the pharmaceutical industry and regulatory agencies have invested much effort toward discovering novel biomarkers that are organ-specific and benefit both preclinical and clinical studies.

It is known that blood-borne mRNA species were elevated at low doses of hepatotoxicants that did not cause changes in the classical biomarkers of liver injury, such as ALT and AST. Moreover, at higher doses, the mRNA species were elevated earlier than the classical biomarkers (Mendrick, 2011).

Alanine Aminotransferase and Its Two Isoforms

Alanine aminotransferase (ALT) is the clinical chemistry gold standard for detection of liver injury or hepatotoxic effects. Although the overall clinical utility of serum ALT measurements is exceptional, it does not always correlate well with preclinical histomorphologic data. Damaged hepatocytes release their contents including ALT and AST into the extracellular space. The ratio of serum AST to ALT can be used to differentiate liver damage from other organ damage (Ozer et al., 2008). ALT levels are greater than AST levels in certain types of chronic liver disease, for example, in hepatitis.

Two forms of ALT have been identified, ALT1 and ALT2, encoded by separate genes. In normal human tissue, high expression of ALT1 was found in liver, skeletal muscle, and kidney and low levels in heart muscle and not detectable in pancreas. High ALT2 activity was detected in heart and skeletal muscle, whereas no ALT2 expression was found in liver or kidney (Lindblom et al., 2007). As with the liver, the traditional markers of AST and CK lack both specificity and sensibility.

Humans express ALT1 and ALT2 and these two forms of the enzyme share approximately 70% identity (Yang et al., 2002). The cellular and tissue distribution of the different ALT proteins has not been characterized in humans, and their relative contribution to serum is unknown. As a result of immunoassay, it is possible to distinguish human ALT1 and ALT2.

Glutamate Dehydrogenase

Glutamate dehydrogenase (GLDH) activity is present in mitochondria and has a role in oxidative deamination of glutamate. Because it is an inducible enzyme, the utility of GLDH as a marker of liver cell necrosis has been questioned in humans during excessive alcohol intake. GLDH may appear in blood without appreciable evidence of hepatic necrosis either by cytoplasmic enzyme synthesis or mitochondrial material within released cytoplasmic membrane blebs (Solter, 2005). GLDH activity increases with hepatocellular damage as does the ALT. Moreover, unlike transaminases, GLDH activity is not inhibited by compounds that interfere with pyridoxal-5′-phosphate, such as isoniazid and lead (O'Brien et al., 2002). Serum GLDH activity is more liver-specific than transaminases and is not substantially affected by skeletal muscle damage.

Serum F Protein

Serum F protein is a 44-kDa protein that is produced in large amounts in liver and small quantities in kidney and is present at low serum concentrations in normal humans. The serum F protein was identified as 4-hydroxyphenylpyruvate dioxygenase, which is a key enzyme in tyrosine catabolism (Neve et al., 2003). The serum F protein concentration was a more sensitive and specific marker of liver damage than conventional liver function tests (AST, ALP, GGT activities) and showed a close correlation with the histological assessment of liver damage (Foster et al., 1989). Serum F protein can be measured by radioimmunoassay in a variety of human diseases and showed elevations in the serum of patients with hepatocellular damage (Foster et al., 1989). Serum F protein in patients taking CBZ and phenytoin as monotherapeutics, in patients receiving multiple drugs, and in patients taking sodium valproate were slightly elevated being an indicator of hepatocellular dysfunction associated with anticonvulsant therapy. Evidence indicates that serum F protein is not influenced by enzyme induction (Callaghan et al., 1994).

Arginase-I

Arginase is a hydrolase that catalyzes the catabolism of arginine to urea and ornithine. Cytosolic mitochondrial isozyme arginase-I is highly liver-specific, almost exclusively a periportal liver enzyme, making it a candidate biomarker for liver toxicity. The activity of arginase-I in serum is considered to be an exact test of liver function. However, the activity of arginase-II, localized in perivenous hepatocytes, is increased in human cirrhotic liver while arginase-I is decreased. It has been suggested that such changes in both the expression and activity of the two arginase isozymes in cirrhotic liver compared with control tissue may allow compensation of ammonia detoxification in various zones of hepatic acinus in cirrhotic livers (Chrzanowska et al., 2009).

In humans, liver graft function after transplantation is dependent on ischemia-reperfusion injury, toxicity of drugs (i.e., immunosuppressive agents and antimicrobials), and transplant rejection. Arginase I has been evaluated as a more specific test of liver function compared with traditional serum markers for this model situation. Serum arginase I reaches the maximum concentration 1 day after liver transplantation and showed powerful and significant correlation with serum AST and ALT activities (Ashamiss et al., 2004).

New Biomarkers for Monitoring Liver Injury

A number of promising biomarkers for liver injury with better sensitivity and specificity have been proposed, as described in the following sections:

Alpha-Glutathione S-transferase

Glutathione S-transferases (GSTs) are inducible phase II detoxification enzymes, and their action reduces drug toxicity and facilitates urinary excretion of metabolites. GST catalyzes the conjugation of GSH with reactive metabolites formed during phase I metabolism. There are four isozymes of GST (alpha, pi, mu, and theta), which are expressed in human and other mammals. The alpha-glutathione S-transferase (α-GST) class consists of two subunits (dimer) and is expressed in human hepatocytes where it accounts for approximately 5% of the soluble protein (Coles and Kadlubar, 2005). The diagnostic use of α-GST (detects necrosis in centrolobular region) to assess acute hepatotoxicity has been compared with serum aminotransferases. Unlike aminotransferases, which are principally found in the periportal hepatocytes, α-GST is also uniformly distributed throughout the centrilobular region of the liver. α-GST has a good correlation between its level and liver histopathological endpoints and rapidly decreases to baseline levels after cessation of the drug causing the injury (Higuchi et al., 2001). α-GST expression is confined to liver and kidney; however, renal injury does not cause increased serum but increased urine levels.

The human α-GST polymorphisms could modulate serum α-GST levels, so the lack of α-GST elevation with drug toxicant treatment might indicate an individual with polymorphism(s) in α-GST. The most frequently found polymorphism in human GST-α is in hGSTA1*B, which correlates with reduced liver expression levels as the SP1 promoter element is mutated (Coles et al., 2001). Its frequency is high (40% in Caucasians, 35% in Africans, and 15% in Asians), and therefore this polymorphism could have high impact on the values obtained in human clinical trials (Coles and Kadlubar, 2005). Changes in α-GST have to be determined relative to predose levels. α-GST was not qualified by FDA because it was found to either increase or decrease depending on the location of the renal injury, which might confound data interpretation.

Gamma-Glutamyl Transpeptidase

Gamma-glutamyl transpeptidase (GGT) activity is localized in liver, kidney, and pancreas tissues. The highest GGT concentration is in the kidney and pancreas followed by the liver. GGT has multiple functions including catalytic transfer of γ-glutamyl groups to amino acids and short peptides, and hydrolysis of GSH to a gamma-glutamyl moiety and cysteinylglycine in GSH and GSH conjugate catabolism (Csanaky and Gregus, 2005). GGT activity in the serum is a marker of hepatobiliary injury, especially cholestasis and biliary effects, although it is not only expressed in liver but also in kidney, pancreas, and bile ducts. This enzyme might be affected in severe ADRs.

Sorbitol Dehydrogenase

Sorbitol dehydrogenase (SDH) catalyzes the reversible oxidation-reduction of sorbitol, fructose, and NADH. It is widely distributed in tissues throughout the body and is primarily found in the cytoplasm and mitochondria of liver, kidney, and seminal vesicles. SDH is a specific indicator of acute hepatocellular injury; this can be useful in rodents and of value in humans (Khayrollah et al., 1982). SDH can be considered as a liver toxicity biomarker that has functions across preclinical species and it performs comparably to other liver enzymatic activities.

Purine Nucleotide Phosphorylase

Purine nucleotide phosphorylase (PNP) is a key enzyme involved in the purine pathway located in the cytoplasm of different cell types (e.g., endothelial cells, Kupffer cells, and hepatocytes) and is a leakage marker released into hepatic sinusoids during necrosis. It reversibly catalyzes the phosphorolysis of nucleosides to their respective bases and corresponding 1-(deoxy)-ribose-phosphate. PNP is found in a number of tissues, particularly in liver, and to a lesser extent in heart and muscle (Marrer and Dieterle, 2010) and could be modified in an ADR (Bantia et al., 1996). PNP detects hepatocyte necrosis.

Malate Dehydrogenase

Malate dehydrogenase (MDH) is involved in the Krebs cycle, catalyzing the reversible transformation of malate into oxaloacetate utilizing NAD^+. It is also a leakage marker released into the serum after tissue damage, being highly abundant in liver followed by heart, skeletal muscle, and brain. This enzyme is localized in the mitochondria (10%) and in the cytoplasm (90%) (Marrer and Dieterle, 2010). MDH activity has proven to be useful in staging liver diseases and is increased in cirrhotic patients (Misra et al., 1991).

Paraoxonase-1

Paraoxonase-1 (PON-1) is a high-density lipoprotein-associated esterase, which is involved in metabolism/detoxification of organophosphates (e.g., diazoxon and paraoxon) in the liver and protects low-density lipoproteins (LDL) from oxidative modification. PON-1 is reduced in serum in response to injury. PON-1 detects liver function marker. Reduced PON-1 levels in serum are a consequence of reduced PON-1 synthesis and secretion in response to an injured liver (Ferré et al., 2002). Studies indicate that plasma PON activity in humans exhibits a polymorphic distribution, with individuals showing a trimodal pattern with high, intermediate, or low PON activity. Gene frequencies for high or low metabolizers have also been shown to vary among groups of different ethnic or geographical origins. The molecular basis of the polymorphism has been associated with several mutations (Brophy et al., 2001). In a given population, mutations in PON-1 lead to variations in plasma PON-1 activity up to 40-fold and differences in PON-1 protein levels up to 13-fold are also present within a single PON1 genotype (Furlong et al., 2002). Because of polymorphisms, the utility of PON-1 activity appears to be greater for preclinical evaluation of hepatotoxicity compared to use in the clinic.

Biomarkers in Acetaminophen Hepatotoxicity

Acetaminophen is the most widely used and popular over-the-counter analgesic and antipyretic drugs in the world, which is safe when taken at therapeutic doses but produced hepatotoxicity and kidney damage following overdose. Many studies indicate that oxidative stress is involved in the various toxicities associated with acetaminophen (Wang et al., 2017). Acetaminophen contributes to around 50% of cases of all of drug-related acute liver failure in adults both in the United States and United Kingdom (UK), results in more than 200 deaths/year in the UK alone, and is a component in 40% of the 80,000 poisoning presentations to UK hospitals (Ostapowicz et al., 2002). An overall scheme of drug toxicity includes many aspects, some of which are related to metabolism. The hepatotoxicity of acetaminophen is initiated by *N*-acetyl-*p*-benzoquinone imine (NAPQI), a reactive metabolite formed by cytochrome P450-mediated (P450 2E1) process, which first depletes GSH and then covalently binds to cellular proteins (Fig. 38.2). The reactive iminoquinone product reacts with mitochondrial proteins and produces mitochondrial injury and reactive oxygen species (ROS), and the latter, in turn, activates the cytoplasmic signal

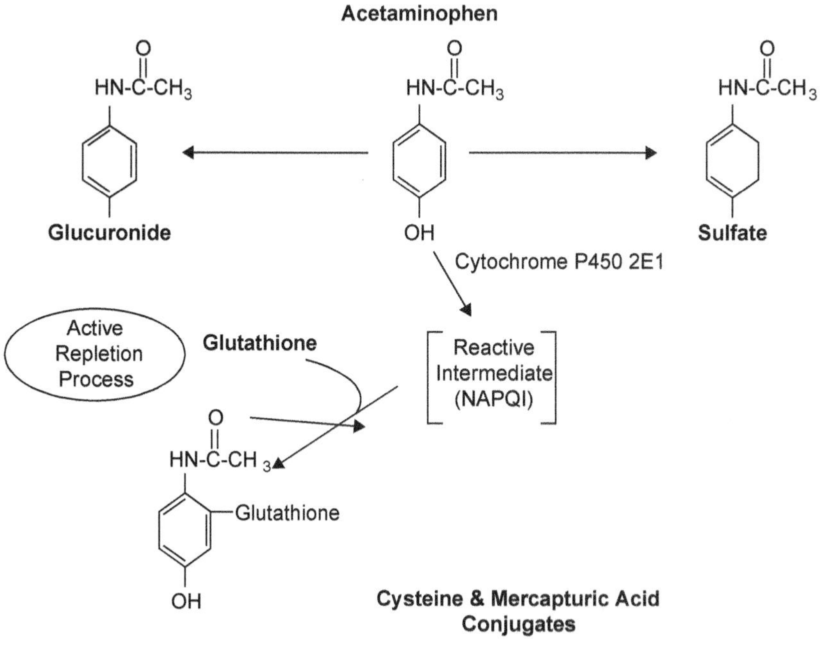

FIGURE 38.2 Metabolism of acetaminophen.

transduction pathway. Covalent binding has been studied in hepatotoxicity as a result of drug activated to reactive products. The degree of covalent binding is higher for hepatotoxic drugs than for nonhepatotoxic drugs; nevertheless, the variation in covalent binding is considerable for both these drug categories (Guengerich, 2011). GSH and taurine are sulfur-containing antioxidants that can reduce or prevent acetaminophen-induced drug toxicity by deactivating NAPQI, the toxic metabolite of acetaminophen. It was shown that urinary levels of creatine were increased on day 1 or day 2 in five high-dose liver toxicity studies and the taurine concentration in urine was increased in three of five liver toxicity studies, while GSH, *S*-adenosylmethionine was found to be reduced in four of five liver toxicity studies (Schnackenberg et al., 2009).

The comparison of toxicity of acetaminophen with its metaisomer, 3-hydroxyacetanilide, with respect to levels of total covalent binding both in vitro and in vivo, has also been studied (Roberts et al., 1990). However, different reactive intermediates are produced from acetaminophen and the *meta*isomer, an iminoquinone and *ortho*quinone, respectively. The early events lead to mitochondrial dysfunction with ROS and peroxynitrite formation (Jaeschke et al., 2003). Mitochondrial stress has since developed in terms of being a major aspect of drug toxicity, so some of the evidence suggests a combination of the drug (or drug metabolite) promoting oxidative stress (a "direct" effect) and alteration of signal transduction systems resulting in further loss of mitochondrial function (an "indirect" effect). It appeared that acetaminophen generated more mitochondrial binding and the *meta*congener more cytosolic binding (Jones et al., 2010).

The cellular events that link drug metabolism to clinical outcome are poorly understood. Selective serum markers of apoptosis, necrosis, and inflammation would be of benefit for differentiating the underlying mechanisms of DILI. Necrosis is the final and ultimate form of cell death. However, it has been suggested that intracellular events following acetaminophen metabolic activation can cause hepatocyte apoptosis. Hepatocyte apoptosis, necrosis, and innate immune activation have been defined as features of the toxicological response associated with the hepatotoxin acetaminophen. Circulating biomarkers that represent the different mechanistic aspects of acetaminophen hepatotoxicity may have utility in the clinical situation to provide information regarding the mechanism of acute liver failure.

The circulating miRNAs represent a class of liver-specific blood-based biomarkers during clinical acetaminophen hepatotoxicity (Starkey Lewis et al., 2011). Moreover, high mobility group box 1 protein (HMGB1) and keratin-18 (K18) have been reported as circulating mechanistic indicators of cell death balance in clinical acetaminophen hepatotoxicity (Craig et al., 2011; Antoine et al., 2012). Hypoacetylated HMGB1 (necrosis indicator), caspase-cleaved K18 (apoptosis indicator), and full-length K18 (necrosis indicator) present in serum showed strong correlation with the histological time course of cell death and was more sensitive than ALT. The hyperacetylated form of HMGB1 (inflammatory indicator) in serum indicated that hepatotoxicity was associated with an inflammatory response. The use of a combined molecular, proteomic, and histological approach to investigate the differing circulating forms of HMGB1 and K18 as potential serum-based proteins could define the multidimensional mechanistic aspects of the dynamics between apoptosis, necrosis, and inflammation during DILI (Antoine et al., 2009). Liquid chromatography-tandem mass_Spectrometry (LC/MS/MS), histological, and proteomic evaluation could be used for differentiating the circulating molecular forms of HMGB1 and K18.

High Mobility Group Protein B1

HMGB1 is a nuclear-binding protein that has proinflammatory activity and targets Toll-like receptors and the receptor for advanced glycation end products on target cells (Hori et al., 1995; Park et al., 2004). It is released in a hyperacetylated form with the involvement of distinct lysine residues from activated innate immune cells (Bonaldi et al., 2003) and passively in a hypoacetylated form by necrotic cells, while it is known not to be released by apoptotic or secondary necrotic cells (Scaffidi et al., 2002).

Caspase-Cleaved K18

Caspase-cleaved K18 is an early event in structural rearrangement during apoptosis. Full-length K18 released passively during necrotic cell death and fragmented K18 generated during apoptosis can be released into the blood and accumulate over time. Serum quantification of caspase-cleaved and full-length K18, which can represent markers of apoptosis and necrosis, has been used during pharmacodynamics of chemotherapeutic drug monitoring in patients and animal models (Cumming et al., 2008).

HMGB1 is passively released by cells undergoing necrosis and is actively secreted as an inflammatory mediator by monocytes and macrophages. Immunoassays directed toward the recognition of caspase-cleaved K18 (cK18; apoptosis) and full length K18 (FL-K18; necrosis) have been reported in clinical studies as biomarkers for the therapeutic drug monitoring of chemotherapeutic agents (Cummings et al., 2007) and for the quantification of hepatocyte cell death mode during liver disorders such as nonalcoholic steatohepatitis (Wieckowska et al., 2006).

Glutathione S-Transferase Alpha

The biomarker glutathione S-transferase alpha (GST-α) has been used in patients with self-administered acetaminophen overdose. In these situations, elevated serum GST-α levels were observed, as well as elevated serum F protein levels. GST-α, serum F protein, and ALT activity all showed elevations postadmission, which was consistent with liver biopsies indicating necrosis (Beckett et al., 1989).

Malate Dehydrogenase

Malate dehydrogenase (MDH) activity is one biochemical index of acetaminophen liver injury that coincided with histological evidence of necrosis (Zieve et al., 1985). The MDH activity increases in a manner correlated with morphological changes after administration of thioacetamide, diethylnitrosamine, and diethanolamine (Korsrud et al., 1972). MDH detects hepatocyte necrosis.

Cytochrome c

Cytochrome *c* is another potential mitochondrial marker of drug-induced injury which has been shown to reach peak elevations in serum along with ALT and AST by acetaminophen.

Biomarkers for Drug-Induced Liver Injury in Tuberculosis—Human Immunodeficiency Virus—Infected Patients

In patients with pulmonary tuberculosis, a significant improvement in oxidative stress and suppression of inflammatory response has been reported after the initial 2-month therapy. However, it has been shown that even after 6 months of successful chemotherapy, pulmonary tuberculosis is still associated with increased levels of circulating lipid peroxides and low plasma concentration of antioxidants. Before chemotherapy, mycobacteria induce ROS production by activating both mononuclear and polymorphonuclear phagocytes. Tuberculosis is, therefore, characterized by poor antioxidants defense that exposes the host to oxidative tissue damage.

The potentially hepatotoxic antituberculosis drugs from the first-line regimen are isoniazid, rifampicin, and pyrazinamide. The conditions that increase the risk of antituberculosis drug-induced hepatitis can be summarized as the following: malnutrition, excessive alcohol intake, aging, chronic hepatic diseases including viral infections (e.g., HIV/AIDS), female sex, ethnicity (Asians), concurrent administration of enzyme inducers, and inadequate compliance (McIlleron et al., 2007).

It has been described that tuberculosis is the most common opportunistic fatal infection in HIV-infected patients. Concomitant HIV and tuberculosis treatment is recommended in patients with low CD4 cell counts. The concurrent treatment with antiretroviral and antituberculosis drugs is complicated because of ADRs; in particular it exacerbates risk for DILI, and overlapping toxicity between drugs used to treat HIV and tuberculosis could also complicate the management. Although rifampicin and efavirenz are key drugs used for concomitant tuberculosis and HIV therapy, data on the concomitant use—related liver injury and biomarkers are limited, particularly from sub-Saharan Africa, a region highly affected by HIV/AIDS and tuberculosis, as is much of the remainder of the continent. In general, drugs used to treat HIV and tuberculosis infections are known to induce drug-metabolizing enzymes and transporter proteins. Induction might also lead to increased production of harmful reactive intermediates and ROS. There is high incidence of antitubercular and antiretroviral DILI in Ethiopian patients. Patient variability in systemic efavirenz exposure and pharmacogenetic variations in NAT2, CYP2B6, and ABCB1 genes determine susceptibility to DILI in tuberculosis—HIV co-infected patients. Close monitoring of plasma efavirenz level and liver enzymes during early therapy and/or genotyping practice in HIV clinics are recommended for early identification of patients at risk of DILI (Yimer et al., 2011).

BIOMARKER OF DRUG-INDUCED KIDNEY TOXICITY

Urine color, clarity, and volume, as well as specific gravity or osmolality, an index of urine concentration, and renal clearance measurements, are all useful marker guides to detect renal injury in experimental animals and humans (Sharratt and Frazer, 1963). Microscopic examination of urine sediment was introduced, in combination with the previous biomarkers, to look for cells, casts, crystals, and bacteria. The advent of dipstick tests for protein, glucose, ketones, hemoglobin, and bilirubin in urine also enabled rapid qualitative measurement. Two known functional serum biomarkers, serum creatinine and BUN, are commonly used to detect kidney toxicity in preclinical and clinical studies, as well as in routine clinical care. Both, however, have severe limitations relating to sensitivity and specificity.

The most efficient way to prevent or attenuate nephrotoxicity is to have sensitive and specific biomarkers. These biomarkers should be able to sensitively predict toxicity in preclinical models and clinical situations so that they can be used to efficiently guide drug developers to modify or discard the potential therapeutics and replace them with variants that affect the same target without the toxicity. However, it is important to recognize that safety concerns must always be incorporated into a general "risk—benefit" analysis and that toxicity of a drug does not necessarily mean that it should not be developed or approved. Some examples of nephrotoxic drugs that have provided a very high therapeutic benefit are the aminoglycoside antimicrobials, the cancer drug cisplatin, and the antiviral tenofovir. Drugs that compete with creatinine for renal transport pathways, in particular cimetidine, trimethoprim, and salicylates, will decrease the secretion component and elevate baseline plasma levels in the absence of a reduction in GFR. The enzyme used as a biomarker of nephron injury, β-galactosidase, is elevated following some glomerular toxicants before proteinuria is evident (Price, 1982).

The discovery of electrophoresis for the separation of proteins (i.e., small molecular weight urinary proteins) as markers of renal injury (e.g., β2-microglobulin and retinol-binding protein) has been very important. This was later on supplemented with analysis of enzymes and proteins in urine, such as LDH and AP. A number of known marketed drugs induce kidney injury, with the main mechanisms of nephrotoxicity being the following: vasoconstriction, altered intraglomerular hemodynamics, direct tubular cell toxicity, interstitial nephritis, crystal deposition, thrombotic microangiopathy, and osmotic nephrosis (Schetz et al., 2005).

Several alternatives to serum creatinine and BUN have been proposed in response to the urgent need for biomarkers that predict human nephrotoxicity in preclinical studies, allow more timely diagnosis of AKI in humans, and ideally localize the injury to a specific nephron site. Nowadays, there are new kidney safety biomarkers to detect AKI, such as kidney function biomarkers (serum cystatin C, detects GFR), de novo expression biomarkers (KIM-1 (urinary), detects proximal tubular injury and regeneration, neutrophil gelatinase–associated lipocalin (NGAL) (urine, serum), detects proximal tubular injury), leakage markers (NAG), or functional biomarkers for the glomerular filtration membrane (total protein) and tubular reabsorption processes (β2-microglobulin). Among the most important biomarkers are KIM-1, β2-macroglobulin and total protein (urinary) (detects glomerular injury and tubular reabsorption complex), urinary clusterin (detects tubular injury, with similar performance and characteristics to Kim-1), albumin (detects tubular injury), trefoil factor 3 (TFF3; detects tubular injury), and urinary cystatin C (detects glomerular injury with subsequent impairment of tubular reabsorption, with similar performance and characteristics to those of urinary β2-microglobulin) (Marrer and Dieterle, 2010).

Urinary kidney injury markers include molecule-1 (KIM-1), NGAL, interleukin-18 (IL-18), cystatin C, clusterin, fatty acid–binding protein liver type, and osteopontin. These biomarkers not only have the potential to both transform the way to detect and quantify nephrotoxicity and prevent the development and entry into the market of nephrotoxic drugs but also allow the continued development of potentially useful drugs that, without the help of biomarkers, would be erroneously believed to be toxic on the basis of a particular preclinical model. It is important to consider that biomarkers for one type of kidney toxicity may not be as useful in another, so a good biomarker for injury may not reliably indicate delayed repair; a biomarker that detects inflammation effectively may not be as sensitive as in detecting early proximal tubule toxicity in the absence of inflammation. A biomarker of injury might not detect a functional defect, such as is observed in Fanconi syndrome or nephrogenic diabetes insipidus, and a biomarker useful in an animal model may or may not be useful in the same way in humans. Another issue is whether panels of biomarkers will be more informative than a single biomarker. At first, this might seem logical because different biomarkers might be more sensitive or specific for different forms of injury. Nevertheless, if multiple biomarkers are used to detect a similar form of injury, an adjudication process will be necessary if the biomarkers suggest different outcomes.

Urinary β2-Microglobulin

Urinary β2-microglobulin is used to monitor tubular reabsorption. It has been shown that impairment of the tubular uptake causes increased excreted urinary β2-microglobulin levels. Two mechanisms have been identified that are involved in this impairment. Firstly, glomerular alterations, damages, and/or diseases allow higher molecular weight proteins to pass through the filtration membrane, causing a high protein load in the tubuli. As a consequence, proteins such as albumin compete for common transport mechanisms, decreasing the tubular uptake of β2-microglobulin and increasing the excretion into urine (Thielemans et al., 1994). Secondly, the tubular reabsorption complex is directly affected by treatment with drugs or different tubular diseases. Increases of urinary β2-microglobulin in the context of nephrotoxicity have been described in HIV patients treated with tenofovir disoproxil fumarate and other antiretroviral agents, patients with aminoglycoside treatments, cisplatin treatment regimens, and gold treatment (sodium aurothiomalate) for rheumatoid arthritis.

BIOMARKERS OF DRUG-INDUCED VASCULAR INJURY

Drug-induced vascular injury (DIVI) is a major concern not only preclinically but also clinically because of a total lack of diagnostic markers and gaps in fundamental science of pathogenesis, mechanism of injury, and development of vascular lesions. Clinical development of novel therapies can often be a danger when DIVI is observed in preclinical toxicity studies. DIVI is a vascular injury that develops acutely in response to drug administration and progresses to vascular inflammation. It is a common finding in preclinical toxicity studies involving vasoactive drugs that alter blood pressure and heart rate such as minoxidil, hydralazine, or milrinone. For these drugs, the decision to proceed from preclinical safety to clinical development was made based on measuring heart rate and blood pressure as surrogate markers in clinical trials. However, these two clinical endpoints are not significantly perturbed with other known DIVI drugs such as endothelin receptor antagonists. The observation that drugs can cause DIVI without hemodynamic changes has led the regulatory agencies to require new DIVI biomarkers for use in the clinical setting and stoppage of drug development by pharmaceutical companies when there is pathological evidence of DIVI in preclinical studies (Louden and Morgan, 2001).

DIVI as the previously defined preclinical type of lesions has not been clearly documented in the clinic. This introduces difficulties in the translational development of biomarkers, as it is unclear how biomarkers, which correlate with histopathology-confirmed preclinical vascular injury, can be clinically qualified. Nevertheless, investigational work has led to a few promising biomarkers, which still have to overcome a number of limitations.

Biomarkers for Monitoring Drug-Induced Vascular Injury

Traditionally, the regulatory authorities and pharmaceutical companies manage the potential risk of drug-induced vascular toxicity as the occurrence correlated with decreases in blood pressure, concomitant reflex tachycardia, and increases in coronary blood flow. By contrast, recent experiences with endothelin receptor antagonists suggest that vascular injury is not always associated with intense hypotension and marked reflex tachycardia. Localized vasodilatation, inability of the vasculature to maintain tone, and alterations in regional blood flow dynamics all play a key role in the pathogenesis of this vascular lesion (Brott et al., 2005). Hemodynamic parameters such as heart rate and mean arterial blood pressure are not satisfactory clinical surrogates for managing the potential risk of drug-induced vascular toxicity. Identification of biochemical diagnostic biomarkers with preclinical and clinical utility would be of value.

The first promising biomarker investigated preclinically and clinically is von Willebrand factor (vWF) and the vWF propeptide (vWFpp). vWF is a glycoprotein, synthesized by endothelial cells and megakaryocytes and released into the circulation, being a marker of endothelial cell injury. vWF is released on vascular damage and is transiently increased in plasma prior to morphological evidence of damage in dogs and rats treated with vascular toxicants, such as fenoldopam or potassium channel openers (Newsholme et al., 2000). The C-Path Institute's Predictive Safety Testing Consortium (PSTC) currently investigates the preclinical and clinical utility of this exploratory biomarker, as well as of vWFpp, to monitor DIVI. Thus far, a good correlation has been found between the amplitude of the transient change of vWF and the severity of progress of the lesions, but a time disconnect (early transient increase followed by return to baseline values despite lesions progression) puts in question the predictability, reliability, and general clinical applicability of circulation vWF as a diagnostic marker to monitor progressive vascular injury. The possible explanation for this "only" transient release is the trapping of a significant amount of the released vWF in the vessel wall and surrounding tissues. Therefore, plasma vWF probably represents an underestimation of the actual amount released from the damaged endothelial cells, especially for later time points. Circulating plasma vWF could still be a good marker of endothelial cell perturbation prior to morphological changes, but this excludes its use for monitoring the progression of lesions (Brott et al., 2005).

Another promising biomarker of vascular injury is more specific to endothelial and smooth muscle cell injury: caveolin-1 (Cav-1), the major structure protein of caveolae. Caveolae (little caves) were first described as flask-shaped plasma membrane invagination capable of transporting molecules across the endothelial barrier. They are located on the surface of endothelial cells and smooth muscle cells and contain both adenylyl cyclase and nitric oxide synthases (NOS). The caveolin family comprises at least three genes that are expressed heterogeneously in different cell types. Interestingly, adenylyl cyclase and NOS are co-localized with and functionally regulated by Cav-1, being a good biomarker of DIVI. Cav-1 has been reported to be a potential biomarker of vascular injury caused by certain vasodilator drugs (Brott et al., 2005).

Data suggest that Cav-1 is highly expressed in normal blood vessels but there is a decrease of Cav-1 expression specifically at the site of vascular damage, pointing to a possible release of Cav-1 from vessel walls into blood. Treatment of cancer cells with cytotoxic (antineoplastic) agents usually increases caveolin expression (Belanger et al., 2003). The C-Path Institute's PSTC currently investigates the preclinical and clinical utility of this exploratory biomarker, as well as of vWFpp to monitor DIVI (PSTC, 2008).

Vascular biomarkers and cytokines related with atherosclerosis in regularly treated and attack-free familial Mediterranean fever patients have been described. In a study where the examinations were performed during attack-free period, all patients received colchicine treatment, a first-line treatment agent, for disease control and for preventing and arresting renal amyloidosis; the antiinflammatory effect of colchicine is multifaceted and is mainly based on its antimitotic and antichemotactic activity on granulocytes and it reduces vascular injury parameters. IL-17 was significantly higher in familial Mediterranean fever patients, and the markers related to endothelial injury, including asymmetric dimethylarginine, osteoprotegerin, and thrombomodulin, were significantly downregulated (Pamuk et al., 2013). Colchicine treatment may have a negative effect (in favor of maintaining vascular homeostasis) on asymmetric dimethylarginine concentrations.

BIOMARKERS OF DRUG-INDUCED CARDIAC INJURY

Coronary artery disease and acute myocardial infarction (AMI) are the leading causes of death in developed and developing countries. Biomarkers of injury to the heart have only been developed in more recent times. The myocytes are the major cell in the heart, and the heart's purpose is to pump blood. Because myocytes essentially cannot be regenerated, if heart cells die, then cardiac function has a high probability of being impaired. When the cell dies, the proteins inside the cell will be released, with proteins in the cytoplasm leaving the cell more rapidly than the ones in membranes or fixed cell elements. The most sensitive markers should be those in the highest abundance in the cell, and as the major function of the heart is contraction, the proteins involved in contraction and producing energy to support it should be good candidates for biomarkers of cardiac injury, which could be detected in blood. One also has to consider the means by which markers can reach the blood stream. As occlusion of blood flow is the primary cause of myocardial infarction, most of the proteins reach the blood via the lymphatics, where they can be prone to degradation, leading to delayed appearance in the blood (Rosalki et al., 2004).

Monoclonal Antibodies

The discovery of monoclonal antibodies (Kohler and Milstein, 2005) revolutionized diagnostics for human diseases, including myocardial infarction with the development of a monoclonal antibody reactive only with CK-MB isoenzyme. This led to a quantitative test. The old gold standards myoglobin and CK-MB served as first-line cardiotoxicity markers, despite limited specificity and short serum half-life, until the recent addition of the cardiac troponins T (cTn-I and cTn-T), qualified and more specific biomarkers of cardiac injury (Rosalki et al., 2004).

Troponins

Troponin (TN) is a component of the heart muscle and its release is indicative of early events in heart tissue degeneration, necrosis, and myocyte damage and it is an obvious candidate biomarker. In humans, cardiac TNs, myocardial-specific structural proteins, are the preferred markers as the cornerstone for the diagnosis of myocardial infarction, and increased cardiac TN is defined as a measurement greater than the 99th percentile of an appropriate reference group. TN, however, offers far more than just improved diagnostic sensitivity and specificity (NTP, 2006). Several groups have demonstrated a powerful relationship between the increase in TN and the risk of mortality in patients presenting with a non–ST elevation acute coronary syndrome, i.e., without classical changes on the electrocardiogram consistent with an acute injury pattern. There are two specific types of TN, TnT and TnI. TN-T and QT prolongation have been used as biomarkers to evaluate potential for cardiac electrophysiology toxicities and prognosis after heart attack.

Cardiac TNs I and T have high sensitivity and specificity for damage to the heart muscle. Emergency patients with chest pain are systematically tested for elevated cardiac TNs levels to confirm a presumption of myocardial infarct. In AMI, TN levels rise as early as 3.5 h after the onset of chest pain. TNs are not only markers for myocardial infarction but also markers of heart muscle damage (i.e., detects necrosis of heart muscle) related to multiple causes, such as severe tachycardia, inflammatory conditions (myopericarditis), and different cardiomyopathies. Being leakage markers, elevations of TNs are only seen when necrosis of the heart muscle has occurred similarly to CK-MB and myoglobin. But in contrast to these two markers, TNs feature a longer half-life, increasing the window of detection after the primary insult. Cardiac TNs have low baseline levels supporting a sensitive detection. For example, increased cTn-I and cTn-T levels were found in rats just 2 h after administration of the cardiotoxicant isoprenaline (Bertinchant et al., 2000).

TNs are also used in nonclinical drug development studies in rats and dogs for three different context of uses (COUs): (1) when there is a "previous indication of cardiac structural damage with a particular drug" states that "cardiac troponin testing can help estimate a lowest toxic dose or a highest non-toxic dose to help choose doses for human testing," (2) when "there is known cardiac structural damage with a particular pharmacologic class of a drug and histopathologic analyses do not reveal structural damage" states that "circulating cardiac troponins may be used to support or refute the inference of low cardiotoxic potential," (3) when "unexpected cardiac structural toxicity is found in a nonclinical study" states that the "retroactive ('reflex') examination of serum or plasma from that study for cardiac troponins can be used to help determine an NOAEL or LOAEL." This retroactive study may then support utilization of this biomarker in future safety testing (FDA, 2017).

B-type Natriuretic Protein

B-type natriuretic protein (BNP) is a peptide hormone released from cardiomyocytes during ventricular stress. BNP is transcribed within the ventricular myocardium and has compensatory diuretic activity during heart failure. The synthesis of BNP occurs through a preprohormone—preproBNP—cleaved into proBNP. Once produced in myocardial cells, proBNP is in turn cleaved into an N-terminal fragment NT-proBNP and BNP fragment, which are logically secreted in a 1:1 ratio. Because BNP is primarily synthesized and secreted by myocytes, in ventricular myocardium, in proportion to the end diastolic pressure and ventricular volume/stretch, the measurement of plasma BNP concentration was introduced to the clinical laboratory as a test to identify cardiac failure (i.e., congestive heart failure) and related cardiac indications. The amino terminal cleavage equivalent (NT-proBNP) is now progressively replacing BNP, thanks to a better stability and the availability of more sensitive and robust assays (Rehman and Januzzi, 2008). Plasma BNP is found at considerable concentrations, and its clearance is mediated by endopeptidases leading to a half-life of about 5–10 min (Pemberton et al., 2000).

BNP is regarded as a cardiac-specific hormone, reflecting mechanical stress on the myocardium (Kramer and Milting, 2011). BNP is typically elevated when pathological findings are noted and thus was recommended as a biomarker to evaluate myocardial pressure and volume overload and is considered a strong "negative predictor," as a low value means volume parameters are normal. Increased BNP concentrations are an independent predictor of mortality in patients presenting with acute coronary syndromes. The group did not know whether an abnormal BNP level would constitute an adverse response. Although BNP protein assays based on serum samples are available for humans, in rodents the assay requires RNA extraction from the heart, which limits measurements at necropsy (NTP, 2006).

The prognostic utility of natriuretic peptides has been explored particularly in ischemic heart disease and conditions leading to right ventricular dysfunction. The limiting factor is the specificity of serum NT-proBNP related to its clearance only through the kidney, rendering kidney disease a cofactor.

Circulating Micro-RNAs

In AMI patients, earlier biomarker of the relative delay release time of troponin are demanded to reduce mortality by AMI. Recent studies have found that circulating micro-RNAs (miRNAs) are closely linked to myocardial injury and currently are emerging as sensitive biomarkers in plasma and serum. The increased levels of circulating miRNAs after, and parallel the extent of myocardial damage are estimated by cardiac troponin. More than 200 miRNAs exist in the heart; miR-1, miR-133, miR-208, miR-499, and other miRNAs are expressed in heart (Li et al., 2012). For details on miRNAs as biomarkers of cardiovascular toxicity and diseases, see Chapter 12 in this book.

The Heart Breakout Group recommended the routine inclusion of three biomarkers in the National Toxicology Program (NTP) subchronic studies: TN, α2-macroglobulin in the rat, and BNP in conjunction with ultrasound. All of these biomarkers are considered indicative of a disease process rather than predictive.

BIOMARKERS OF DRUG-INDUCED BRAIN INJURY

Neurological diseases create difficulties in both diagnosis and in the ability to objectively monitor responses to treatment, as few body fluid biomarkers exist and direct sampling of the brain is rarely advised. Thus, diagnosis of some diseases and brain injuries utilizes subjective clinical assessment. Signs of treatment effectiveness can take months and in many cases are based solely on subjective patient feedback. Thus, the possible utilization of genomics applied to peripheral blood cells in this area would fulfill a great clinical need. Much remains to be done in terms of reproducing the gene signatures in new biological samples, demonstrating, for example, disease exclusivity, before their usefulness can be ascertained.

One goal of personalized medicine is to improve the prediction of an individual patient's response to therapy. A pilot study examined valproic acid (VPA) or carbamazepine (CBZ) treatment in children with epilepsy using whole blood genomic expression, where epileptic patients could be distinguished as with and without CBZ treatment (Tang et al., 2004). It was found that gene expression patterns could discriminate between patients who were and were not responding to VPA treatment. There was little overlap between the

VPA and CBZ signatures, suggesting that there are drug-specific responses that can be detected. If such biomarker patterns can be replicated in new studies, they would be very useful for personalized medicine.

Translational biomarkers of mood disorders have also been identified (Le-Niculescu et al., 2008), studying the gene expression differences in whole blood samples between patients with bipolar disorder and in whole blood samples from mice models of bipolar disease. These authors correlated these results with previously published data from their laboratory on gene expression changes in the brains of mice in these models (Ogden et al., 2004) and data reported on gene expression levels in human postmortem brain from individuals with mood disorders; genetic links to this disorder were found. They selected the top 10 genes that could identify patients with high and low moods. Moreover, there have been biomarkers identified for the diagnosis of hallucinations (gene panel of seven) and for delusions (gene panel of 31) in psychotic patients; some overlap was found in the biomarkers for hallucinations, delusions, and mood state, but it was found that biomarkers for mood state tended to be regulated in the opposite direction than were seen in psychotic patients.

Another approach is to evaluate the individual patient before drug exposure to ascertain susceptibility to drug-induced adverse events. A study was performed in patients recruited in a clinical trial for an Alzheimer's drug, in which the transcriptome of peripheral blood mononuclear cells was examined prior to drug treatment. Two biomarker panels were recognized: one set that identified those individuals susceptible to an adverse event and a second panel that could predict efficacy of the drug (O'Toole et al., 2005).

Studies have discovered that the expression of miRNA may provide a new tool in detecting neurological disease, ADRs, and renal allograft rejection. A recent study examined the miRNA levels in the brain and in the blood of rats subjected to stroke (ischemic or hemorrhagic) and chemically induced seizures and reported unique miRNA patterns for each condition, with a correlation between the brain and blood patterns. The miRNAs most affected were miR-298, miR-155, and miR-362-3p (Liu et al., 2010). Another neurological condition, multiple sclerosis, was also taken into consideration for the miRNAs, and so not reduced blood levels of miR-17 and miR-20a were found in these patients. To perform this experiment, T cell knock-in and knockout models were used in vitro to determine the genes regulated by these miRNA species and they were found to be T-cell activation genes; these same genes were expressed at higher levels in the blood of patients with multiple sclerosis (Cox et al., 2010). In conclusion, the data suggests that these miRNA may be useful biomarkers.

BIOMARKERS IN DRUG SAFETY EVALUATION

Importance of Understanding Specific Mechanisms of Toxicity in Drug Development

Modern high-throughput technologies for transcripts, proteins, and endogenous metabolites offer a major opportunity to systematically identify sensitive and specific blood and urine safety biomarkers that could serve as an index of damage specific to each of the important tissues and organs. Today, it is technically feasible to identify tissue-specific markers through the use of these—omics technologies by integrating and correlating toxicology data. These changes in gene, protein, and metabolite expression are consistent, sensitive, and early and provide the molecular basis of drug-induced injuries. The genomic-derived candidate markers can then be localized to organs using in situ hybridization (ISH) and immunohistochemistry (IHC) for their proteins. The concerns at the transcript level are reflected on the protein level and thereafter secreted into body fluids where they can be quantified (Dix et al., 2007).

A greater understanding of the molecular basis of toxicity and its influence on disease and disease progression can play a major role in drug development outcomes. In the early phase of research, biomarkers support the mechanistic characterization of toxicity, show an early indication of toxicity, and help define the maximal tolerated dose. Biomarkers can be used in early or late drug development for enrichment of patient populations to increase the odds of detecting a phenotypic or clinical efficacy signal. For example, data from early clinical trials that enroll patients with poor metabolizer genotypes in early phases of clinical trials to evaluate dose—concentration—response relationships in patients with different genotypes can inform the study design of latter-phase clinical studies. In latter stages of development, stratification approaches might be employed for looking at response in subgroups of patients.

Safety biomarkers can also play an important role in deciding if candidate drugs are transferred from the preclinical to the clinical phase in the case where traditional clinical markers would not detect early-onset organ toxicity. If a pharmaceutical company can clearly show in preclinical studies that the novel biomarkers can be used to detect early toxicity, to monitor onset and reversibility, and to manage any potential adverse effect of a new drug with significant therapeutic potential, a clinical implementation strategy with these biomarkers can enable a clinical development program on a case-by-case basis (PSTC, 2008).

Biomarkers in Drug Safety Evaluation

The Biomarkers Definitions Working Group (BDWG) of the National Institute of Health (NIH) defined a biomarker as "a characteristic that is objectively measured and evaluated as an indicator of normal biological processes or pharmacological responses to a therapeutic intervention." To maintain definitional integrity, it has been agreed that the term *validation* will refer to the technical characterization and documentation of method performance, and the term *qualification* will refer to the evidentiary process of linking a biomarker with specific biological processes and clinical endpoints (BDWG, 2001). A molecular biomarker is defined as a specific subset of small molecule (metabolite) or large molecule (protein or DNA) biomarkers that may be discovered using genomic technology, as opposed to clinical biomarkers arising from imaging technology (Lewin and Weiner, 2004).

Molecular profiling (the application of genomic, transcriptomic, proteomic, and metabolomics technologies to the development of in vivo molecular biomarker) of a patient's phenotype utilizes molecular biomarkers of the genome, transcriptome, proteome, and metabolome to provide a molecular basis for the normal or healthy patient phenotype in a particular environment and thereby provides many opportunities for improving the choice of therapy and the prediction of drug response and disease pathology (Koch, 2004). Alteration of molecular biomarker in the diseased state compared with the normal state (characterized by the normal molecular phenotype) provides a molecular characterization of the disease phenotype, which is utilized in the diagnosis.

Biomarkers are considered to be parameters of injury or toxicity to help diagnose or monitor disease, but the biomarkers present in blood and urine give no information on the variability or range of response from individual cells or areas within a tissue. Biomarkers that represent highly sensitive and specific indicators of disease pathways have been used as substitutes for outcomes in clinical trials when evidence indicates that they predict clinical risk or benefit. In consequence, useful biomarkers can be quantitated and should provide information about normal or pathobiological states (Lock and Bonventre, 2008). The most reliable way to assess the clinical impact of a therapeutic intervention is through its effect on well-defined clinical endpoints (e.g., survival, myocardial infarction, stroke). A clinical endpoint is also a characteristic or variable that reflects how a patient feels or functions or how long a patient survives (Frank and Hargreaves, 2003).

Certain biomarkers have been used as substitutes for clinical endpoints (i.e., the most credible characteristic used in the assessment of the benefits and risks of a therapeutic intervention in randomized clinical trials) as a basis for drug approval. The term *surrogate endpoint* is defined as a biomarker that is intended to substitute for a clinical endpoint. A surrogate endpoint is expected to predict clinical benefit (or harm or lack of benefit or harm) based on epidemiologic data.

A conceptual paradigm has been developed to show the relationship of a biomarker to a clinical endpoint and the application of the biomarker as surrogate endpoint in the evaluation of therapeutic interventions (BDWG, 2001) (Fig. 38.1). The model also shows that biomarkers can be useful in the assessment of safety, as well as efficacy (MacGregor et al., 1995; Santella, 1997). Some biomarkers (e.g., blood pressure) may have dual functions in assessing efficacy (i.e., benefit) and safety (i.e., risk). Characterization of a biomarker as a surrogate endpoint requires it to be "reasonably likely, based on epidemiologic, therapeutic, pathophysiologic or other evidence, to predict clinical benefit" as indicated in the Food and Drug Modernization Act of 1997. The utility of a biomarker as a surrogate endpoint (i.e., a biomarker that is intended to substitute for a clinical endpoint) requires demonstration of its accuracy (the correlation of the measure with the clinical endpoint) and precision (the reproducibility of the measure) (Fig. 38.3).

Biomarkers have been used in drug development and treatment monitoring for a long time and quantitatively measure perturbations to the system anywhere in the causal path between drug administration and clinical effect.

Likewise, the concept of bridging biomarkers of toxicity would include not only those biomarkers that span use during in-life studies in preclinical species and also in humans but also in vitro or ex vivo study results used as surrogate biomarkers that accurately "predict" a toxic outcome in preclinical studies and/or humans (early-phase clinical trial to establish the "proof of concept") or to establish in vitro–in vivo correlations. On the other hand, there is wide consensus in the scientific community that biomarkers are and will be of utility as estimating tools in improving risk assessment of drugs. For the assessment of nephrotoxicity in vitro, different test systems are available, such as isolated perfused kidneys, cut slices, and isolated fragments and cells (glomeruli and tubules), which are relatively close to the in vivo situation, although the isolation and cultivation is complex and the time of cultivation and treatment is hardy restricting (Fuchs and Hewitt, 2011).

The essential need for a public process biomarker validation is addressed by the European Commission that started development of the IMI draft (IMI, 2005) and the process defined by the recommendations from the European Centre for the Validation of Alternative Methods (ECVAM), which include the development of in vitro models employing cells and tissues from rats

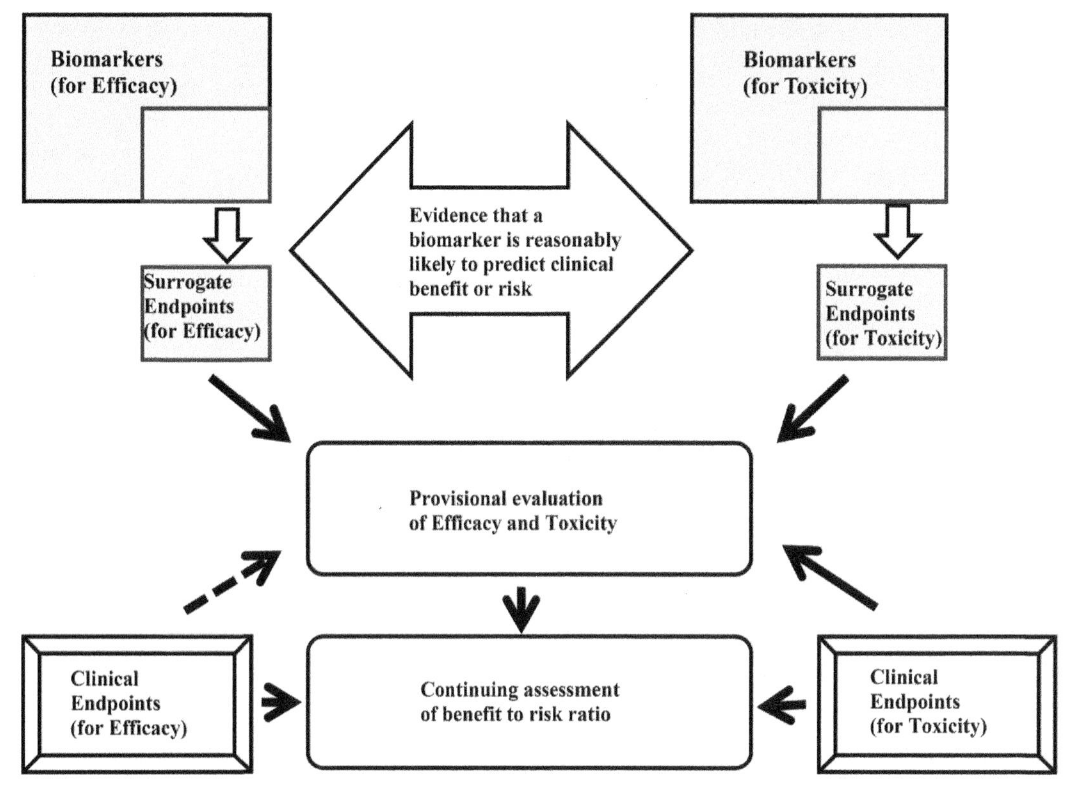

FIGURE 38.3 The biomarker paradigm for drug development (BDWG, 2001).

and mice, so that the in vitro and in vivo data can be compared and the possibility for extrapolating in vitro results to the in vivo situation in man can be determined. IMI (2005) defined desirable characteristics for a biomarker of drug safety assessment: (1) specific for certain types of injury; (2) indicates injury in a variety of experimental species and humans; (3) can be used to bridge across nonclinical/preclinical studies to clinical and surveillance types of studies; (4) more effective at indicating injury than any other biomarker currently used; (5) used instead of classic biomarkers, not in addition; (6) can be easily measured (in real time) even at a later stage (measurement is not strongly time dependent); (7) more reproducible and sensitive than the original toxicity endpoint it would replace; and (8) reduces the number of individuals tested (animals or humans). These IMI definitions are closely reflected in the work of the PSTC of the Critical Path (C-Path) Institute.

Following BDWG (2001), biomarkers have many other important applications in disease detection and monitoring of health status. These applications include the following: (1) as a diagnostic tool for the identification of those patients with a disease or abnormal condition (e.g., elevated blood glucose concentration for the diagnosis of diabetes mellitus); (2) as a tool for staging of disease (e.g., measurements of carcinoembryonic antigen-125 for various cancers) or classification of the extent of disease (e.g., prostate-specific antigen [PSA] concentration in blood used in cancer detection) to reflect extent of tumor growth and metastasis and α-fetoprotein (AFM) for hepatocellular carcinoma; PSA and AFM have never been formally validated or tested in prospective trials; (3) as an indicator of disease prognosis (e.g., anatomic measurement of tumor shrinkage of certain cancers); and (4) for prediction and monitoring of clinical response to an intervention (e.g., blood cholesterol concentrations for determination of the risk of heart disease).

Safety Biomarkers for Preclinical and/or Clinical Perspectives

Kidney

To date, AKI, also referred to as "acute renal failure," is associated to a high mortality rate of up to 80% (Han and Bonventre, 2004). It has been highlighted that nephrotoxicity contributes to 8%–60% of all cases of AKI, with the mechanisms of nephrotoxicity involving vasoconstriction, altered intraglomerular hemodynamics, direct tubular cell toxicity, interstitial nephritis, crystal deposition, thrombotic microangiopathy, and osmotic nephrosis (Schetz et al., 2005).

The traditional tests to determine kidney toxicity, such as BUN and serum creatinine, are functional markers (i.e., constant production by the body and a clearance via the kidney) and are not optimal for early detection or localization of kidney damage, being significantly upregulated when up to two-third of the nephron's function has already been lost. Determination of the creatinine clearance measurement is accomplished by measurement of creatinine in blood and in 24 h collection urine. Creatinine, a muscle-derived by-product of creatine metabolism, has several significant advantages as a plasma marker of GFR over urea: it is synthesized and delivered to the plasma at a fairly consistent rate. Serum creatinine, as reference parameter for kidney function, must be regarded with caution because of the following features: (1) serum creatinine concentrations can vary widely within age, gender, muscle mass, muscle metabolism, overall body weight, nutrition situation, and hydration status; (2) serum creatinine concentrations may not change until a significant amount of kidney function has already been lost, meaning that renal injury is already present or occurs before serum creatinine is elevated; (3) at lower rates of glomerular filtration, the amount of tubular secretion of creatinine results in overestimation of renal function (the latter capacity of the kidneys to excrete creatinine is hardly predictable for the individual; it also depends on some medication interfering with tubular creatinine transport); (4) during acute changes in glomerular filtration, serum creatinine does not accurately describe kidney function until a steady-state equilibrium has been reached, which may require several days (Nguyen and Devarajan, 2008). In the short run, BUN and serum creatinine showed poor sensitivity and specificity for detection of renal injury (Star, 1998). Serum creatinine and GFR estimation can only be used in patients with stable or chronically altered kidney dysfunction. For that reason, efforts have been done to find out new urinary biomarker.

Kidney Injury Molecule-1

Kidney injury molecule-1 (KIM-1) is a type 1 transmembrane protein containing a 6-cystein immunoglobulin-like domain and a mucin domain that is not detectable in normal kidney, tissue, or urine. KIM-1 is a biomarker because transmembrane protein is expressed by tubule epithelial cells in response to injury. KIM-1 is sensitive and specific, has no interferences with pathologies unrelated to kidney, and is highly stable in urine. It exhibits homology to a monkey gene, hepatitis A virus cell receptor 1 (HAVcr-1), and therefore is sometimes referred to as HAVcr-1. KIM-1 mRNA and protein are expressed at a very low level in normal kidney and increase dramatically in injured kidney, particularly in differentiated proximal tubule cells in human and rodent kidney after AKI, ischemia, drug-induced nephrotoxicity, renal cell carcinoma, tubule-interstitial fibrosis, and inflammation; acute and chronic kidney disease all induce KIM-1 gene and protein expressions.

β2-Microglobulin

Urinary β2-microglobulin is a single polypeptide chain of 12 kDa, which is readily filtered by the glomeruli and almost completely reabsorbed and metabolized by the tubuli; therefore, it is used to monitoring tubular reabsorption. Only 0.3% of the β2-microglobulin filtered by the glomeruli is normally excreted into urine. It has been shown that impairment of the tubular uptake causes increased excreted urinary β2-microglobulin concentration levels, up to several hundred-folds. Two identified mechanisms have been involved in this impairment. Firstly, glomerular alterations, damages, and/or diseases allow higher molecular weight proteins to pass through the filtration membrane, causing a high protein load in the tubuli. Consequently, proteins such as albumin compete for common transport mechanisms, decreasing the tubular uptake of β2-microglobulin and increasing the excretion into urine (Thielemans et al., 1994). Secondly, the tubular reabsorption complex is directly impacted by treatment with drugs or different tubular diseases. Increases of urinary β2-microglobulin in the context of nephrotoxicity have been described in more than 200 peer-reviewed publications, such as HIV patients treated with tenofovir disoproxil fumarate and other antiretroviral agents (Gatanaga et al., 2006), patients with aminoglycoside treatments (Schentag and Plaut, 1980), cisplatin treatment regimens (Tirelli et al., 1985), and gold treatment (sodium aurothiomalate) for rheumatoid arthritis (Latt et al., 1981), to name a few examples. In human urine, α-1 and β2-microglobulins are commonly used, due to their relative instability in acid or contaminated urine; in particular, α-1 microglobulin is present in fairly high abundance in normal urine, shows robust elevation with tubular disease, and is slightly more stable to degradation than β2-microglobulin, making it the preferred marker of tubular malfunction in human bioassays (Marchewka et al., 2001; Price, 2002).

Cystatin C

Serum cystatin C is a low molecular protein with a molecular weight of ~13 kDa. Cystatin C is a 120 amino acid nonglycosylated basic cysteine protease inhibitor widely expressed by all nucleated cells in the body. Urinary cystatin C detects glomerular injury with subsequent impairment of tubular reabsorption, aw well as with similar performance and characteristics as urinary β2-microglobulin. Cystatin C is directly and freely filtered by the glomerulus, is reabsorbed completely, is not secreted by the tubule, and is considered an ideal marker for GFR. Urinary excretion of the low molecular weight protein cystatin C, which is an endogenous marker of renal dysfunction, correlates with the severity of acute tubular damage. The blood levels of cystatin C are not significantly affected by age, gender, race, or overall muscle mass. It is a marker for the estimation of glomerular function (e.g., in cachectic patients or early AKI, where serum creatinine could underestimate the true renal function) (Urbschat et al., 2011). Cystatin C assay has been compared with serum creatinine measurements in human beings, which is identical in sensitivity. Cystatin C assays are all antibody-based and used against human cystatin C but have antigenic cross-reactivity across species.

Clusterin

Clusterin is involved in cell adhesion, membrane recycling, tissue remodeling, cell cycle regulation apoptosis, and DNA repair. Clusterin is overexpressed in human renal diseases, in particular in renal ischemia, unilateral urethral obstruction, or in response to various nephrotoxins. Urinary clusterin detects tubular injury, with similar performance and characteristics to KIM-1.

Trefoil Factor 3

Trefoil factor 3 (TFF3) a new promising kidney biomarker, detects tubular injury for acute drug-induced renal toxicity (EMEA, 2008).

Overall, the novel biomarkers were more sensitive and specific for kidney injury when compared with BUN and creatinine. Novel and existing biomarkers such as KIM-1, albumin, clusterin, and TFF3 were approved by FDA for use as biomarkers of drug-induced acute kidney tubule alterations in GLP rat studies used to support clinical trials. Total protein, β2 microglobulin, and cystatin C were approved by FDA for use as biomarkers of drug-induced acute glomerular alterations, damage, and/or impairment of kidney tubular reabsorption in GLP rat studies used to support clinical trials (Sauer and Porter, 2018.

Other Kidney Biomarkers

Several protein biomarkers have been assayed as noninvasive indicators of kidney injury.

Lipocalin-2

Lipocalin-2, also referred to as NGAL, is a small 25 kDa secreted protein (belongs to lipocalin family) bound to gelatinases from neutrophils and epithelial cells, including those of the proximal tubule. It is expressed and secreted in kidney epithelial cells only during inflammation as cell protection mechanism, especially in damaged proximal tubuli to induce reepithelialization. mRNA and protein levels are extremely rapidly increased within 3 h of renal ischemic injury in animals and humans. NGAL was found to be a useful early predictor in contrast to KIM-1. A possible issue with NGAL is its specificity; this marker is induced in serum by bacterial infections, chronic obstructive pulmonary disease, and liver injury (Xu and Venge, 2000). Urinary and plasma NGAL have also been proven to predict contrast-induced AKI with a high diagnostic power just 2 h after contrast administration (Bachorzewska-Gajewska et al., 2007; Hirsch et al., 2007) and, similarly, to IL-18, increased NGAL levels in urine collected the day after kidney transplantation could predict recipients with subsequent delayed graft function and who needed dialysis (Parikh et al., 2006). NGAL was identified as one of the fastest upregulated genes in the early phase of the postischemic mouse kidney (Supavekin et al., 2003) being detected in the very first urine sample within 2 h following ischemia and displaying increased levels correlating with the duration of ischemia. NGAL was found to be a useful early predictor of AKI, with urine or plasma/serum NGAL levels functioning as well. Additionally, NGAL level had prognostic value for clinical endpoints, such as initiation of dialysis and mortality (Haase et al., 2009).

N-Acetyl-β-D-Glucosaminidase

N-Acetyl-β-D-glucosaminidase (NAG) is a lysosomal enzyme of 140 kDa with two isoforms (A and B) predominantly found in proximal tubules cells so that increased activity of this enzyme in the urine suggests tubular injury to cells and thus can serve as a specific urinary marker for the tubular cells. Because of its high molecular weight, NAG is not filtered by glomeruli and its excretion into urine correlates with increased tubular lysosomal activity, tubular cell injury (leakage), and indirectly with increased proteinuria being a leakage marker. NAG has been used for long time as a marker because of its localization in the proximal tubules. In the course of kidney diseases, AKI, and treatment with nephrotoxic compounds, urinary NAG levels remain persistently elevated before rises of serum creatinine and BUN (Price, 1992). The increase in urinary NAG activity indicates damage to tubular cells, although it can also reflect increased lysosomal activity without cellular damage (Liangos et al., 2007). However, the use of NAG remains limited because its urinary excretion is also elevated in glomerular diseases (e.g., diabetic nephropathy) (Marchewka et al., 2001).

Interleukin-18

Interleukin-18 (IL-18) is a proinflammatory cytokine that belongs to the IL-1 superfamily and is produced by macrophages and other cells. Apart from its physiological role, IL-18 is also able to induce severe inflammatory reactions, which suggests its role in certain inflammatory disorders. IL-18 is constitutively expressed in the intercalated cell of the late distal convoluted tubule, the connecting tubule, and the collecting duct of the healthy human kidney. Moreover, these cells contain three major components required for the release of the active proinflammatory cytokine IL-18, namely pro-IL-18, P2X7, and the intracellular cysteine protease caspase-1, which converts the preform of IL-18 to its active form, which then exits the cell and may enter the urine, for example, in AKI. Furthermore, urinary IL-18 was significantly upregulated prior to the increase in serum creatinine in patients with acute respiratory distress syndrome who developed AKI, predicting mortality at the time of mechanical ventilation (Parikh et al., 2006). Plasma IL-18 levels are known to be increased in various pathophysiological states, such as inflammatory arthritis, inflammatory bowel diseases, systemic lupus erythematosus, psoriasis, hepatitis, and multiple sclerosis. Thus, although this cytokine seems to be a candidate biomarker in the setting of AKI, its proinflammatory properties and its upregulation in inflammatory disease may limit its application in terms of sensitivity and specificity.

Matrix Metalloproteinase-9

Matrix metalloproteinase-9 (MMP-9) is elevated in postischemic kidney tissue in an animal model. Although it has not been evaluated as a urinary biomarker for AKI, it seems to be an early predictive urinary biomarker of ischemia AKI in children and adult patients after cardiac surgery (Han et al., 2008). These authors examine (1) the utility of urinary biomarkers MMM-9, NAG, and KIM-1 (singly or in combination) for detection of AKI in a cross-sectional study in adults and (2) the temporal expression pattern of urinary biomarkers before the development of AKI in a prospective case–control study in children.

Netrin-1

Netrin-1, a laminin-related neuronal guidance molecule, is highly expressed and excreted in the urine after AKI in animals. Urinary netrin-1 excretion was increased dramatically in 13 patients with AKI, whereas no changes were detected in 6 healthy volunteer urine samples. Therefore, urinary netrin-1 might be a promising early upregulated biomarker for detection of renal injury (Reeves et al., 2008).

Monocyte Chemotactic Peptide-1

Monocyte chemotactic peptide-1 (MCP-1) is a potent chemokine originated by renal cells and acting as mediator in acute ischemic and toxic kidney injury. In the mouse model, MCP-1 protein and its mRNA increased in intrarenal injury more than those of NGAL. In prerenal and postrenal injury, NGAL and MCP-1 gene expressions increased in a comparable way. In patients with AKI, MPC-1 and mRNA levels are increased in urine without overlap in the absolute urine. On that basis, MPC-1 might be a biomarker of AKI, possibly giving complementary information to the NGAL analysis (Munshi et al., 2011).

Liver

The liver is a complex organ with interdependent metabolic, excretory, and defense functions. The liver is the primary organ for drug metabolism and thus is exposed to drugs and metabolites rapidly after gastrointestinal absorption. The gold standards used preclinically and clinically to monitor hepatotoxicity are the liver enzymes ALT and AST measured in blood. Changes in the ALT serum levels point to hepatocyte necrosis with high sensitivity and fairly good specificity. Unfortunately, too high sensitivity is not an advantage here. Often serum ALT levels increase in the absence of histomorphologic alteration in the liver. This emphasizes the potential for other sources of serum ALT activity. Nevertheless, ALT is still more liver-specific, as it is found in highest concentrations in the cytosol of hepatocytes, whereas AST serum activity is also found in blood cells, skeletal muscle, and heart besides of hepatocytes (Hunt et al., 2007). As a result, increased levels of transaminases may be caused by liver-independent processes such as muscle injury, cardiac events, changes of body mass, blood-related pathologies, or pancreatitis.

Heart

The discovery of monoclonal antibodies (Kohler and Milstein, 1975) revolutionized diagnostics for human diseases, including myocardial infarction with the development of a monoclonal antibody reactive only with CK-MB isoenzyme. This led to a quantitative test. A search for more specific markers of cardiac injury focused on the cTn1 and cTnT, which are currently used (Rosalki et al., 2004; FDA, 2012).

The Heart Breakout Group recommended the routine inclusion of three biomarkers in NTP subchronic studies: TN, α_2-macroglobulin in the rat, and BNP in conjunction with ultrasound. All of these biomarkers are considered indicative of a disease process rather than predictive.

α_2-Macroglobulin

α_2-Macroglobulin in the rat, analogous to human C-reactive protein (CRP), was recommended even though this marker is not cardiac-specific because the investigators felt it was important to have an indication of systemic inflammation. α_2-Macroglobulin was suggested as a way to address potential effects on the vasculature.

Like BNP, α_2-macroglobulin is considered to be a "negative" predictor such that a normal value indicates the absence of systemic inflammation and absence of vascular injury. An elevated α_2-macroglobulin would probably be associated with vascular injury, including inflammation (NTP, 2006).

Ultrasound

Ultrasound can be used to identify an adverse effect. Ultrasound is noninvasive technique that requires a live animal and has been recommended for suspected cardiotoxicants (NTP, 2006).

Others Novel Biomarkers

Other novel biochemical plasma biomarkers described for terminal heart failure patients, which might predict an advanced mortality risk or even recovery, are described in the following sections.

C-Reactive Protein

C-reactive protein (CRP) is an acute-phase reactant and has been recommended as a predictive biochemical biomarker for risk of coronary disease. CRP is not a specific biomarker of inflammation and correlates very poorly with extent of atherosclerosis throughout angiography (biomarker of risk). CRP contributes to risk assessment independently of serum cholesterol measures and in particular is consistent with histopathology findings of active inflammatory cells in plaque (Albert et al., 2001). Other inflammatory biomarkers, such as lipoprotein-associated phospholipase A_2, secretory type II phospholipase A_2, oxidized LDL, myeloperoxidase, and growth differentiation factor, are under investigation.

Galectin-3

Galectin-3 (also known as Mac-2, CBP-35, or LBP) is a β-galactoside–binding protein with a preference for lactose and N-acetyllactosamine (Ochieng et al., 2004). There are a number of cell types that express galectin-3 such as neutrophils, macrophages, and mast cells, and lung, stomach, colon, uterus, and ovary cells (Kim et al., 2007). Myocardial galectin-3 was found to be increased in those rats, which progressed to heart failure (Sharma et al., 2004), and therefore is used as a biomarker for prediction of mortality in severe heart failure even under mechanical circulatory support (Kramer and Milting, 2011).

Matrix Metalloproteinases

Matrix metalloproteinases (MMPs) constitute a group of zinc- and calcium-dependent endopeptidases, which are also involved in pathomechanisms of cardiovascular, inflammatory, and oncological diseases (Manicone and McGuire, 2008). More than 20 MMPs have been described, with MMP-2 (gelatinase A) and MMP-9 (gelatinase B) being the most relevant MMPs in cardiovascular disease. The proteolytic activity of MMP-2 and MMP-9 is regulated by cleavage of the propeptide and through inhibition of the active enzyme by endogenous inhibitors of MMPs (tissue inhibitors of metalloproteinases [TIMP] or α_2-macroglobulins) (Clark et al., 2008).

Tissue Inhibitors of Metalloproteinases

The tissue inhibitors of metalloproteinases (TIMP) family is comprised of four members. TIMPs bind MMPs in a stoichiometric 1:1 ratio and thereby block access of substrates to the catalytic domain of the endopeptidases. TIMP-1, TIMP-2, and TIMP-4 are secreted proteins, whereas TIMP-3, as a membrane bound TIMP, is restricted to the extracellular matrix. In the context of secreted biomarkers in heart failure, TIMP-1 and TIMP-4 are the most frequently discussed TIMP family members. The highest TIMP-1 levels have been found in ischemic cardiomyopathy.

Lung

The NTP sponsored a workshop to identified additional biomarkers of lung disease and lipid metabolism (i.e., metabolic syndrome), which could be assessed routinely in NTP subchronic toxicology studies (NTP, 2006).

In pulmonology, one of the biomarkers used is the partial pressure of oxygen (acute lung injury and acute respiratory distress syndrome are defined by the difference between the alveolar and arterial concentration of oxygen). This biomarker reflects the critical life-sustaining function of the organ (delivery of oxygen to the circulation to support aerobic metabolism) (Waikar et al., 2009). Additionally, the Lung Breakout Group considered and discussed a variety of potential approaches for collecting biomarkers ranging from lavage fluid analysis, respiratory function, enhanced tissue pathology, and imaging to gene analysis and proteomics. Of these, the Lung Breakout Group felt three assays would be most useful for the NTP: (1) bronchoalveolar lavage analysis, especially appropriate for obtaining cell counts and differentials. Molecular biomarkers that can be obtained through lavage analysis are, for example, chemokines, cytokines, antioxidants, and albumin, among others, but the Breakout Group did not make specific recommendations. Instead, they recommended the NTP select a panel of markers to evaluate processes such as immunity (innate and acquired) and inflammation; (2) histopathology markers, reflecting lung injury, inflammation, apoptosis, repair, and other events either early or late in the disease process. The Breakout Group suggested the trichrome and Periodic acid–Schiff stains and Ki67 protein for cell proliferation, conducting immune histochemistry for proteins assessed in bronchoalveolar lavage analysis; and (3) gene expression analysis. The Breakout Group recommended that the NTP explore the use of gene expression analysis not as a routine measure, but on a more limited basis.

Lipid/Carbohydrate Metabolism

The three highest priority biomarkers identified by the lipid/carbohydrate metabolism Breakout Group were serum cholesterol/triglycerides, insulin, and GSH. Liver histological analysis was recommended to separate microvesicular and macrovesicular fatty changes to distinguish different pathological processes. Moreover, insulin was recommended because it is considered a better marker of insulin resistance than glucose, but it was suggested to consider the measure of glucose bound to hemoglobin in red blood cells (hemoglobin A1C or HbA1C) or fructosamine. Regarding metabolic disorders, the GSH measurement was recommended as a marker of whole body oxidative stress. Other biomarkers proposed were sterol regulatory element binding proteins in the liver (if histochemical techniques can be developed) indicating early events in cholesterol (SREBP 2) and fatty acid (SREBP 1) synthesis and liver triglycerides. The measures of inflammation (i.e., tumor necrosis factor alpha and IL-6) could be useful as a routine measurement or in special studies, even though changes are not specific to perturbations in lipid and carbohydrate metabolism.

Central Nervous System

It is now stated that a triplet of cerebrospinal fluid (CSF) biomarker (total tau, phosphorylated-tau, and the 42 amino acid fragment of amyloid beta [Aβ]) reflect core neuropathological features of Alzheimer's disease (AD) and contribute diagnostically relevant information if measured in a proper manner. The cause of AD, a serious neurodegenerative disease, is currently unknown but pathological, genetic, and nonclinical evidence suggests that Aβ peptides and specifically the highly amyloidogenic isoform Aβ42 (with 42 residues) are involved in the pathogenesis of AD. CSF biomarker signature based on a low Aβ1-42 and a high T-tau and/or positron emission tomography (PET)-amyloid imaging positive/negative can be useful to identify patients with clinical diagnosis of mild to moderate AD who are at increased risk to have an underlying AD neuropathology for the purposes of enriching a clinical trial population.

The EMA (2011a) published a positive qualification opinion capacitating the use of low CSF Aβ42 and high CSF tau as enrichment tools for clinical studies of amyloid-targeted therapies in predementia AD. There were two purposes of that qualification. The first purpose was to seek a broadened use of CSF biomarkers as tools to enrich clinical trials in AD dementia patients (mild to moderately severe) with neuropathology most likely to benefit from treatment with amyloid-modulating therapies. The second purpose was to support qualification of PET-amyloid imaging as a second biomarker to be used as an enrichment tool in clinical studies of amyloid-targeted therapies in patients with predementia AD and in patients with mild to moderately severe AD. In patients with minimal cognitive impairment (MCI), a positive CSF biomarker signature based on a low Aβ1-42 and a high T-tau is predictive of evolution to AD dementia type. ELISA methods to measure this CSF biomarker's signature are commercially available but the process of measurement is also complex. The CSF biomarker signature based on a low Aβ1-42 and a high-tau qualifies for identifying MCI patients as close as possible to the prodromal stage of AD, as defined by the Dubois criteria (Dubois et al., 2005).

There are several types of biomarkers relevant for central nervous system diseases drug development. There are five primary types of biomarkers relevant to AD drug development: (1) Brain imaging: *Structure* (magnetic resonance imaging (MRI); diffusion tensor imaging: white matter tracts, cortical thickness mapping); *function* (fluorodeoxyglucose PET); functional MRI; MRI arterial spin labeling; single-photon emission computed tomography (SPECT) of cerebral blood flow; dopamine transporter SPECT; *molecular and chemical constituents* (amyloid PET, MR spectroscopy); (2) electrophysiologic measures: electroencephalography, evoked potentials; (3) plasma, serum, and CSF (amyloid-related measures [Aβ40, Aβ42, other Aβ species]; inflammatory markers [cytokines]; oxidation markers [isoprostanes]; other serum and CSF measures; amyloid synthesis/clearance with stable isotope-labeled kinetics); (4) "omics" platforms with microarray and spectroscopic determination of multiple gene, protein, lipid, metabolite, or other measures combined with advanced informatics required to interpret the study results; and (5) genetics (disease-related [e.g., apolipoprotein genotype]; pharmacogenetics [e.g., CYP enzyme genotypes]) (Cummings et al., 2013). The biomarker IL-18 has also been found to increase the AD-associated Aβ production in human neuron cells.

Genetic and Genomic Biomarkers

Genetic biomarkers (i.e., genetic variation) for drug efficacy and adverse effects have traditionally been derived from two sources: (1) genetic variation in the target on which the drug acts and (2) variation in enzymes that metabolize the drug. The genetic constitution can influence an individual's sensitivity toward exogenous substances, drugs in particular. Biomarkers that predict risk for disease can identify patients who will have a greater than average benefit from therapy. This creates a new opportunity to enrich clinical trials with patients who are likely to have more events and to achieve earlier drug approval. Markers that predict risk of cardiovascular, thrombotic, and liver diseases may also identify a subset of individuals at substantially elevated risk for adverse drug effects (White et al., 2006).

In principle, identifying genetic risk factors for severe ADRs, particularly idiosyncratic reactions (i.e., severe cutaneous disorders and DILI caused by several drugs), could significantly decrease the incidence rate of ADRs and improve the process of drug development. The genetic basis of ADRs can be categorized into two groups. The first group involves genes that drive pharmacological mechanism (drug targets, drug metabolizing enzymes, and drug transporters (Nakamura, 2008)). The second category involves the immune system; one important molecule for ADRs associated with immune reactions is the HLA, which plays a key role in initiation of immune responses and killing target cells by presenting antigens to the T-cell receptor (Rudolph et al., 2006).

The HLA gene region codes for three classical class I (HLA-A, HLA-B, and HLA-C) and three class II (HLA-DR, HLA-DP, and HLA-DQ) antigens. Class I antigens are recognized by cytotoxic $CD8^+$ T cells and class II by $CD4^+$ T cells. Of the highly polymorphic HLA genes, HLA-B is the most polymorphic with over 800 variants reported in the human genome (Horton et al., 2004). HLA genes within each class encode structurally similar but distinct HLA proteins that bind and present HLA-type–specific antigenic peptides to T-cell receptors (Horton et al., 2004; Rudolph et al., 2006). HLA disease associations that are related to genes with immunological and inflammatory functions have been identified in many autoimmune and inflammatory conditions. Among these idiosyncratic ADRs, the usefulness of abacavir HLA-genetic biomarker (HLA-B*5701) (i.e., induced hypersensitivity reaction in HIV-infected patients) has been confirmed in Caucasians from several prospective studies, such as the PREDICT-1 study (Mallal et al., 2008). This finding has been successfully applied in clinical practice, and a prescreening for HLA-B*5701 before treatment with abacavir has shown a reduction in the occurrence of abacavir-associated hypersensitivity reactions. The association of CBZ-induced SJS/TEN and an HLA-genetic biomarker (HLA-B*1502) in Han Chinese is extremely high compared with other drugs (Chung et al., 2004). Therefore, HLA-B*1502 screening is recommended for CBZ in clinical practice by the US FDA, and HLA-B*5701 screening is recommended for abacavir by the US FDA and EMA. Before treatment with CBZ or abacavir, HLA analysis should be performed to exclude HLA-B*1502 or HLA-B*5701 unless the patient is from a population that shows extremely low frequency of these HLA types. Such exclusion of patients from treatment with causative drugs would markedly reduce the possibility of severe ADRs and prevent overestimation of severe cutaneous ADRs that could otherwise result in excessive discontinuation of treatment (Hung et al., 2006).

Other biomarkers of drug toxicity of genetic variation in human are dihydropyrimidine dehydrogenase, thiopurine methyltransferase, CYP2C19, the *bcr-abl* translocation, Factor V Leiden, and APOE4 (White et al., 2006).

Classic examples of genomic biomarkers (i.e., genetic variation) for drug efficacy include genetic variation in the drug target (including its expression level) and drug-metabolizing enzymes. There are numerous metabolic biomarker tests that are on the market as FDA-approved or laboratory-developed tests. Recent US FDA approvals of tests for cytochrome P-450, CYP2D6 and CYP2C19, and uridine diphosphate glucuronosyltransferase (UGT)1A1 have given regulatory endorsement to biomarkers that can improve drug safety by identifying individuals at risk for drug toxicity. The test results are indicative of whether the patient is an ultrarapid, extensive, intermediate, or poor metabolizer of a drug that is substrate for CYP2D6 or CYP2C19. This type of genotyping knowledge may assist the treating clinician in selecting the right dose for a given patient to achieve target systemic drug exposure. UGT1A1 genotype information is of the utility in the treatment use of irinotecan, a drug used to treat colon cancer and small cell lung cancer.

Oncotype DX predicts the risk of women with node-negative, estrogen receptor–positive (ER^+) breast cancer of experiencing cancer recurrence 10 year following diagnosis and also predicts the extent of benefit with chemotherapy. A similar test is the MammaPrint approved by the FDA. Genetic tests recently approved by the FDA are tests for the polymorphisms in two enzymes, CYP2C9 and VKORC1, for determining an optimal starting dose for warfarin therapy.

Biomarker Validation and Qualification

Most safety biomarkers have been used in safety assessment for many years without being formally qualified, and many of them show deficiencies with respect to sensitivity, specificity, and predictive value (Boekelheide and Schuppe-Koistinen, 2012). With that in mind, during development of biomarkers, the following steps should be followed: (1) identification, (2) preclinical qualification, and (3) clinical qualification and diagnose use (Marrer and Dieterle, 2010). These are described in more detail in the following sections.

Identification of Safety Biomarkers

Preclinical screens of histopathology and informative biomarkers are essential in transitioning new drugs into human testing. For this to take place, the peripheral biomarkers must reflect histopathological changes or events. The safety biomarker study should be oriented toward identification biomarkers that have the greatest measured peripherally in body fluids (i.e., serum, plasma, and urine) when beginning the investigations of toxicity.

At present, identification of the biomarkers needs toxicogenomics, which implies the integration and correlation of data from toxicology studies. Toxicogenomics relates to substance-induced changes of the transcriptome to identify cellular and subcellular mechanisms. Perturbation of the expression profile of specific genes leads to pathological outcomes. Taking the examples of nephrotoxic drugs, a number of genes such as clusterin, NGAL, and KIM-1 are strongly and coherently deregulated in the kidney and correlate well with histopathologically confirmed lesions in the tubules (Marrer and Dieterle, 2010). The genomic-derived candidates can then be localized in the organ of interest by ISH and IHC for its protein.

The perturbation at the mRNA level of a perfect safety biomarker candidate is reflected in its protein level and subsequent secretion into periphery. A protein modulated by toxicity can be identified by proteomics (measured total cellular protein and determine the posttranslational modification and fate of proteins) in the tissue or peripheral fluids. Data from proteins can be more relevant that the information from genomic analysis, due to the fact that the changes in mRNA levels do not necessarily correlate with changes in protein expression.

Once this identification of protein biomarker candidates is done, it is followed by the development of immunoassays for routine determinations as to which methods need validation in terms of specificity, selectivity, accuracy, precision, sensitivity, robustness, linearity and parallelism, and dynamic range (using a definition of an upper and lower limit of detection and quantification). Traditional methods used to characterize and quantify biomarkers from preclinical and clinical studies are chromatography and mass spectrometry.

Preclinical Qualification

The regulatory agencies have initiated the creation of various consortia that are working toward the identification and characterization of exploratory biomarkers to qualify them for a specific use. Currently, collaborations exist between pharmaceutical companies, regulatory agencies (US FDA and EMA), and academia for qualifying biomarkers, providing a model that investigates and identifies reliable safety markers for preclinical applications. An example of such a collaboration was initiated by the Nephrotoxicity Working Group of the PSTC, which was created as part of the FDA's C-Path Initiative in 2007. PSTC was initially established with the primary aim of qualifying novel nonclinical safety biomarkers for use in nonclinical safety assessment studies to support drug development. As discussed elsewhere, most novel safety biomarkers being evaluated by PSTC have been correlated with histopathological changes in target organs and are designed to substitute for histopathological evaluation in its absence (Dieterle et al., 2010). The PSTC nonclinical qualification strategy is to use specific prototypical organ toxicants in rats, dogs, and nonhuman primates to define correlations between novel biomarker response and histopathological changes. Currently, the primary target organs of interest for PSTC are the liver, kidney, testis, pancreas, heart, skeletal muscle, and vascular system. Regulatory qualification is the formal regulatory endorsement or acceptance of a drug development tool for a specific COU.

Qualification of novel nonclinical biomarkers for use in animal models provides a regulatory endorsed tool to assess the safety and efficacy of new drugs prior to clinical development with the next step being translation of the preclinical biomarker to a qualified clinical biomarker (Sauer and Porter, 2018).

Renal biomarkers have been divided into clinical application of tubular markers and glomerular markers. Other groups are presently involved in qualifying biomarkers to detect hepatotoxicity, vascular injury, nongenotoxic carcinogens, and myopathy. Thus, each group is dealing with qualification of a biomarker within a certain context. A marker may be useful in one clinical setting—for example, patients in an intensive care unit or patients with sepsis or after cardiac surgery—but not in another. The very best biomarkers will be useful in many situations (Lock and Bonventre, 2008).

Many biomarker candidates have failed to show sufficient specificity and sensitivity for clinical use, although other promising candidates have emerged recently. In the Nephrotoxicity Working Group of the PSTC, performance of each new biomarker versus the accepted standards of BUN and serum creatinine is evaluated by comparison of the area under the curve (AUC) of the receiver operating characteristic (ROC) analysis for each new biomarker with the similar data obtained for BUN and creatinine. ROC curves provide a comprehensive and visually attractive way to summarize the accuracy of predictions. Each point on the curve represents the true-positive rate and false-positive rate associated with a particular test value. ROC curves have been generated both from data merged from all positive histopathology scores as a benchmark for renal injury for all studies by study site, as histopathology is widely regarded as the "gold standard" for animal studies, and for data from subset ranges of these scores. AUC provides a useful metric to compare different tests (indicator variables). While an AUC value close to 1 indicates an excellent diagnostic test, a curve that lies close to the diagonal (AUC = 0.5) has no information content and thus no diagnostic practicability. Finally, in confirmatory studies, the best cut-off value is determined for each biomarker based on the best ratio between sensitivity and specificity.

The Nephrotoxicity Working Group of PSTC selected 23 urinary biomarkers and evaluated the utility of most promising biomarkers in several rat models of kidney injury, which were performed under GLP regulations. A first set of new nephrotoxicity biomarkers is currently being evaluated in rats using known nephrotoxicants (e.g., gentamicin, cisplatin). These biomarkers reflect toxicity in different anatomical regions of the kidney and are intended to provide earlier warning signs of drug-induced toxicity. Seven renal safety biomarkers (KIM-1, albumin, total protein, β2-microglobulin, cystatin C, clusterin, and TFF3) were qualified for limited use in nonclinical and clinical drug development following submission of drug toxicity studies and analyses of biomarker performance for the US FDA and EMA by PSTC Nephrotoxicity Working Group. The ROC curves have been performed for the complete KIM-1 and albumin data and also for the complete KIM-1, clusterin, total protein, β-2 microglobulin, and cystatin data. The FDA and the EMA, respectively, have concluded that seven urinary biomarkers of kidney injury submitted by the C-Path Institute of PSTC are considered qualified for particular uses in regulatory decision-making. The first four can be included as biomarkers of drug-induced acute kidney tubular alterations in GLP rat studies to support clinical trials, whereas the last three can be included as biomarkers for acute glomerular alteration/damage and/or impairment of kidney tubular reabsorption (Dieterle et al., 2010). The current gold standard biomarkers, serum creatinine and BUN, are insensitive and appear only in advanced stages of kidney injury (Waikar et al., 2009).

A number of accessible protein biomarkers, with a high probability of success in diagnosing nephrotoxicity in rats and monkeys, are available as exploratory biomarkers. These include (in addition to the widely known biomarkers such as urinary albumin, urinary total protein, urinary β2-microglobulin) the Kim-1, clusterin, and urinary cystatin C, TFF3, as well as changes in the differential expression of other genes included in a toxicogenomic signature for nephrotoxicity. These seven kidney biomarkers correlate to either tubular histomorphologic alterations or to glomerulopathy with functional tubular involvement and "add information" to serum creatinine and BUN.

To support the use of biomarkers for translational contexts to allow the translation of a drug into human studies, which caused renal injury in preclinical GLP studies, the clinical data were accompanied by peer-reviewed experts supporting their clinical utility. As a result, the EMA and FDA concluded that (1) in the context of nonclinical drug development, the seven urinary kidney biomarkers submitted were considered acceptable for the detection of acute drug-induced nephrotoxicity, either tubular or glomerular with associated tubular involvement. They provide additional and complementary information to BUN and serum creatinine to correlate with histopathological alterations considered to be the gold standard. Additional data on the correlation between the biomarkers and the evolution and reversibility of AKI and information on species-specificity are needed. In addition, further knowledge on species-specificity is required. However, it was recognized that KIM-1, β-2 microglobulin, and total protein are considered acceptable for preclinical context. (2) In the context of clinical drug development, it was recognized that it is worthwhile to further explore, in early clinical trials, the potential of KIM-1, albumin, total protein, β2-microglobulin, urinary clusterin, urinary TFF3, and urinary cystatin C as clinical biomarkers for acute drug-induced kidney (DIKI) injury.

Until further data are available to correlate these biomarkers with the evolution of the nephrotoxic alterations and their reversibility, their general use for monitoring nephrotoxicity in the clinical setting cannot be qualified. The use of these renal biomarkers in clinical trials is to be considered on a case-by-case basis to gather further data to qualify their usefulness in monitoring drug-induced renal toxicity in man.

The regular qualification of biomarkers is actually supported by a formal biomarker qualification process, which was established by the EMA (formerly EMEA) (scientific advice procedures) in the course of this formal biomarker submission. The EMA issued a draft guidance document for pharmacogenetics in 2005 (EMEA, 2005). The EMA and FDA guidelines for voluntary genomic data submission, both the documents, contribute to minimize risks of creating inadvertent obstacles to the use of genomic technology. The EMA issued other documents related with pharmacogenomic biomarkers as well (EMA, 2007; EMA, 2011a; b).

Clinical Qualification and Diagnostic Use

The clinical qualification of safety biomarkers for wide contexts of use is faced with additional challenges compared with preclinical qualifications. In addition to the absence of histopathology evaluation as gold standard, the many challenges include clinical endpoints that are often nonperfect gold standards, less-controlled heterogeneous populations with sometimes underlying chronic diseases and co-medications, low incidences rates of drug-induced toxicities in clinical studies, ethical concerns, logistic challenges, and many more (Marrer and Dieterle, 2010). In addition to the PSTC consortium, to help define the operational approach for the qualification of safety biomarkers, the European IMI Strategic Research Agenda issued a draft by the European Commission (IMI, 2005) safer and faster evidence-based translation consortium (SAFE-T), which has established a generic qualification strategy for new translational safety biomarkers (TSBMs) that will allow early identification, assessment, and management of drug-induced injuries throughout the research and development process. The SAFE-T consortium is a public–private partnership comprising 20 participants from the pharmaceutical industry, small to medium enterprises, academic institutions, and clinical units of excellence with representatives from the EMA and FDA as external observers and advisor.

The clinical studies in SAFE-T are to be conducted by applying all relevant national and European Union (EU) legislation and regulations. The collection, use, and retention of blood samples will be compliant with Directive 95/46/EC of the European Parliament and of the Council of October 24, 1995, on the protection of individuals with regard to the processing of personal data and on the free movement of such data (OJ L 281, 23.11.95). The Charter of Fundamental Rights of the EU dedicates certain political, social, and economic rights for EU citizens and residents into EU law (drafted by the European Convention and solemnly proclaimed on December 7, 2000, by the European Parliament, the Council of Ministers, and the European Commission; its then legal status was uncertain and it did not have full legal effect until the entry into force of the Treaty of Lisbon on December 1, 2009), and Directive 2001/20/EC of the European Parliament and of the Council of April 4, 2001, on the approximation of the laws, regulations, and administrative provisions of the member states relating to the implementation of good clinical practice in the conduct of clinical trials on medicinal products for human use (OJ L 121, 1.5.2001).

The SAFE-T consortium proposes a generic qualification strategy for TSBMs, outlining proposals on how to generate sufficient preclinical and clinical evidence to qualify new TSBM for regulatory decision-making in defined contexts. The experience gained during the course of the SAFE-T project for three organ toxicities will be integrated into improvements for this initial generic approach. Each work program DIKI, DILI, and DIVI will follow the general qualification plan of a two-stage program of studies: initial exploratory followed by confirmatory studies. The biomarker profile for DILI is considered exploratory preclinical and clinical and for DIVI, exploratory (Matheis et al., 2011).

As stated previously, most safety biomarkers in use today have not been formally qualified. There is no systematic scientific qualification strategy in place allowing for the accumulation of sufficient clinical and biological evidence for the acceptance of biomarkers independent of a specific drug. Regulatory agencies have established submission procedures for the regulatory review and endorsement of biomarker qualification, but they have not yet defined the scientific standards and approaches needed. The SAFE-T consortium has proposed principles of a research plan to develop biomarkers for regulatory decision-making in specific circumstances (Matheis et al., 2011) (see Fig. 38.4). This type of endorsement is known as "regulatory qualification"; that is, the formal regulatory endorsement or acceptance of a drug development tool for a specific COU.

The first submission of kidney safety biomarkers by the PSTC (Dieterle et al., 2010) also opened the door to a new framework of fit-for-purpose qualification of biomarkers instead of having an absolute "all or nothing" qualification. With more data and evidence, the limited context can be extended. This principle has been referred to as an "incremental," "progressive," or "rolling" qualification. Thus, the SAFE-T consortium proposes broad principles of a research plan including assay method validation and biomarker qualification to develop biomarkers for regulatory decision-making in specific contexts. While clinical diseases have a long history, systematic efforts to discover and validate markers of adverse exposure to or effects of exogenous chemical agents (i.e., toxicological biomarkers) are generally more recent. Most safety biomarkers have been used in safety assessment for decades without being formally qualified, and many of them show significant deficiencies regarding sensitivity, specificity, and predictive value. They provide little molecular understanding of the mechanisms of toxicity and their deficiencies contribute to safety-related drug attrition and withdrawals (Kola and Landis, 2004).

Translational Safety Biomarker

Translational research is the research process that investigates and translates nonclinical research results into clinical applications and tests their safety and efficacy in a Phase 1 clinical trial. Traditionally, nonclinical drug toxicity can be investigated using cellular tests, cell culture, and tissue culture models to explore

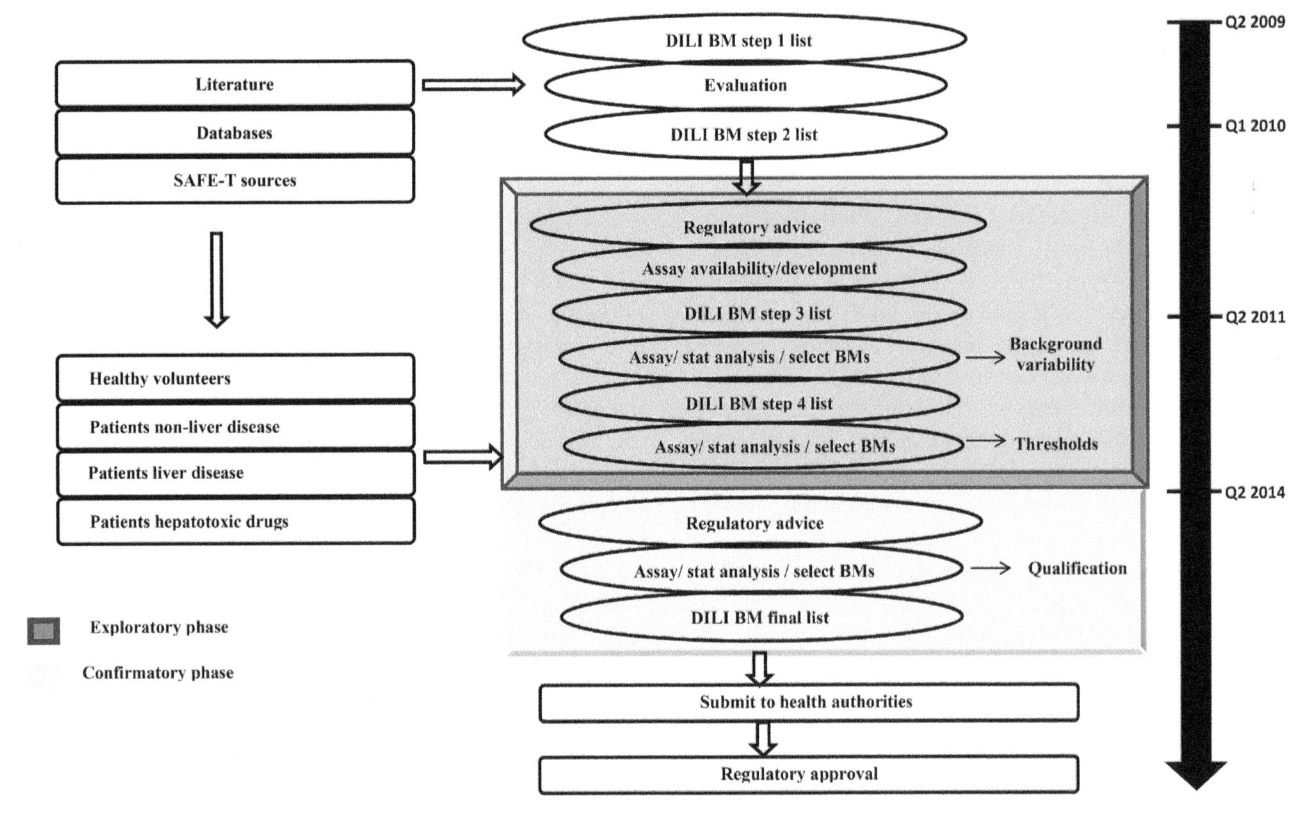

FIGURE 38.4 Biomarker qualification process proposed by safer and faster evidence-based translation (SAFE-T) consortium.

molecular, cellular, tissue, and even organ effects in vitro. Moreover, animal models provide needed information at the organism level. The future of preclinical biomarker qualification describes the translational role that preclinical biomarkers play in the qualification process for clinical biomarkers. One characteristic of the safety biomarkers is that it is noninvasive and translates between species. Biomarkers may have their greatest value in early efficacy and safety evaluations such as in vitro studies in tissue samples, in vivo studies in animal models, and early-phase clinical trials to establish "proof of concept." To improve these outcomes, a systemic and dynamic understanding of physiological and pathophysiological processes is needed, along with expanded application of translational biomarkers for drug efficacy and tissue and organ injury. TSBMs can predict, detect, and monitor drug-induced toxicity during human trials for testing drugs and are needed to assess whether toxicities observed in laboratory animal studies are relevant to humans at therapeutic doses.

In the case of studies in experimental animals, the biomarker must also be the one that is relevant to humans. The most valuable biomarkers are those that can be used in animals and humans. These "translational biomarkers" can be rigorously studied in animals, thereby establishing well-defined relationships between biomarker levels and tissue histopathology. One of the most notable challenges in assessing drug toxicity in humans is that we do not have tools capable of predicting toxicity across species borders (Bonventre et al., 2010). The dependence of preclinical screening on histopathology and weakly informative biomarkers causes considerable delays and inefficiency in transitioning new drugs into human testing. Therefore, delays confirmation of the safety and effectiveness of new therapies is needed (Warnock and Peck, 2010). In humans the transitional biomarkers are of particular interest, as they can be used to monitor safety and efficacy in clinical trials, in particular when the capacity to obtain tissue is severely restricted for performing histopathology. For example, liver biopsies are performed by percutaneous or transvenous routes and they have limitations on the procedure and relative contraindications.

One example of the importance of translation research is translational neuroscience, which provides a framework for advancing development of new therapies for AD patients. This translational neuroscience includes new preclinical models that may better predict human efficacy and safety, leading to improved clinical trial designs and outcomes that will accelerate drug development and the use of biomarkers for more rapidly providing information regarding the effects of drugs on the underlying disease biology. The NIH is responding to the crisis in drug development by funding translational research. The formation of the National Center for Advancing Translational Science (NCATS) is one milestone in reorganizing the NIH to orient more toward public–private partnerships and product development (Collins, 2011). NCATS has resources to support drug discovery and advance promising compounds through preclinical development, including assay development and high-throughput screening, synthesis, formulation, pharmacokinetics, toxicology, medicinal chemistry, molecular libraries probe production, genomics, interference RNA, tissue chips for drug screening, and technologies for identifying and validating drug targets.

CONCLUDING REMARKS AND FUTURE DIRECTIONS

Unexpected drug toxicity is still a major reason for pharmaceutical companies withdrawing drugs from the marketplace. Among ADRs, hepatotoxicity and nephrotoxicity are two major causes for drugs being withdrawn postmarket, in part because of the failure of animal safety studies to translate to human safety of a new chemical entity. Two examples of high risk for drug adverse reactions are the acetaminophen hepatotoxicity and the high incidence of antitubercular and antiretroviral DILI (McIlleron et al., 2007). In both clinical situations, biomarkers hold much interest. In the first condition, the circulating miRNAs represent a class of liver specific blood-based biomarkers during clinical acetaminophen hepatotoxicity. In the second condition, the pharmacogenetic variations in genes determine susceptibility to DILI in tuberculosis/HIV coinfected patients. Other important drug adverse reactions are related to vessels and heart, as well as brain injuries. Pharmaceutical industries and regulatory agencies expect that new and better safety biomarkers will play an important role in improving key aspects of the drug development process, as well as shortening the time-consuming process and reducing the cost. Thus, there is a need for new efficient biomarkers with enough sensitivity and specificity to translate effectively from the preclinical phase to clinical phase. One characteristic of the biomarkers is that they can be noninvasive and can translate between species. The TSBMs can predict, detect, and monitor drug-induced toxicity during human trials for testing drugs and are needed to assess whether toxicities observed in laboratory animal studies are relevant to humans at therapeutic doses. The most valuable biomarkers are those that can be used in animals and humans. The safety biomarkers to be used in safety assessment should be formally qualified and validated and are important in all phases of drug development, including discovery, preclinical assessment, clinical trials, and postmarket surveillance. Biomarkers from blood and urine could replace traditional histopathological evaluation to determine adverse responses in several organs, tissues, and systems such as kidney, liver, heart, vascular vessels, lung, central nervous system, and lipid/carbohydrate metabolism. New technologies and molecular approaches, including molecular profiling and molecular pathology with rapid detection of urine and blood as a complement to the classical toolbox, promise a future of further discoveries and improvements in the safety biomarker field and for the first time have been accepted by the regulatory agencies. There are high demands for ideal properties and roles of biomarkers in several pathological conditions. Biomarkers should be organ-specific in addition to having rapid execution and being inexpensive, noninvasive, reliable, and capable of an early detection. Personalized medicine biomarkers can be of utility to predict whether an individual will respond adversely to drug therapy. For some drugs, it is recommended to use genetic tests before the drug is prescribed. Studies of disease could discover biomarkers that can also be used to identify drug-induced injury (in particular, liver, kidney, vessels, heart, and brain injuries). To optimize clinical utility of biomarkers, body fluids are preferred over direct tissue sampling. The possible utilization of genomics applied to peripheral blood cells in the area of disease and brain injuries would fulfill a great medical clinical need. Much remains to be done in terms of reproducing the gene signatures in new biological samples, demonstrating, for example, disease exclusivity, before their usefulness can be discerned. Studies have discovered that the expression of miRNA may provide a new tool in detecting ADRs.

References

Aardema, M.J., MacGregor, J.T., 2002. Toxicology and genetic toxicology in the new era of "toxicogenomics": impact of "-omics" technologies. Mutat. Res. 499, 13–25.

Adams, C.P., Brantner, V.V., 2010. Spending on new drug development. Health Econ. 19, 130–141.

Albert, M.A., Danielson, E., Rifai, N., et al., 2001. Effects of statin therapy on C-reactive protein levels. The pravastatin inflammatory/CRP evaluation (PRINCE). A randomized trial and cohort study. J. Am. Med. Assoc. 286, 64–70.

Antoine, D.J., Williams, D.P., Kipar, A., et al., 2009. High-mobility group box-1 protein and keratin-18, circulating serum proteins informative of acetaminophen-induced necrosis and apoptosis in vivo. Toxicol. Sci. 112, 521–531.

Antoine, D.J., Jenkins, R.E., Dear, J.W., et al., 2012. Molecular forms of HMGB1 and keratin-18 as mechanistic biomarkers for mode of cell death and prognosis during clinical acetaminophen hepatotoxicity. J. Hepatol. 56, 1070–1079.

Ashamiss, F., Wierzbicki, Z., Chrzanowska, A., et al., 2004. Clinical significance of arginase after liver transplantation. Ann. Transplant. 9, 58–60.

BDWG, 2001. Biomarkers and surrogate endpoints: preferred definitions and conceptual framework. Clin. Pharmacol. Ther. 69, 89–95.

Bachorzewska-Gajewska, H., Malyszko, J., Sitniewska, E., et al., 2007. Could neutrophil-gelatinase associated lipocalin and cystatin C predict the development of contrast-induced nephropathy after percutaneous coronary interventions in patients with stable angina and normal serum creatinine values? Kidney Blood Press. Res. 30 (6), 408–415.

Bantia, S., Montgomery, J.A., Johnson, H.G., et al., 1996. In vivo and in vitro pharmacologic activity of the purine nucleoside phosphorylase inhibitor BCX-34: the role of GTP and dGTP. Immunopharmacology 35, 53–63.

Beckett, G.J., Foster, G.R., Hussey, A.J., et al., 1989. Plasma glutathione-S-transferase and F protein are more sensitive than alanine aminotransferase as markers of paracetamol (acetaminophen)-induced liver damage. Clin. Chem. 35, 2186–2189.

Belanger, M.M., Roussel, E., Couer, J., 2003. Up-regulation of caveolin expression by cytotoxic agents in drug-sensitive cancer cells. Anticancer Drug 14, 281–287.

Benoehr, P., Krueth, P., Bokemeyer, C., et al., 2005. Nephroprotection by theophylline in patients with cisplatin chemotherapy: a randomized, single-blinded, placebo-controlled trial. J. Am. Soc. Nephrol. 16, 452–458.

Bertinchant, J.P., Robert, E., Polge, A., et al., 2000. Comparison of the diagnostic value of cardiac troponin I and T determinations for detecting early myocardial damage and the relationship with histological findings after isoprenaline-induced cardiac injury in rats. Clin. Chim. Acta 298 (1–2), 13–28.

Bindhu, M., Yano, B., Sekkers, R.S., et al., 2007. Evaluation of organ weights for rodent and non-rodent toxicity studies: a review of regulatory guidelines and a survey of current practices. Toxicol. Pathol. 35, 742–750.

Boekelheide, K., Schuppe-Koistinen, I., 2012. SOT/EUROTOX debate: biomarkers from blood and urine will replace traditional histopathological evaluation to determine adverse responses. Toxicol. Sci. 129, 249–255.

Boelsterli, U.A., 2003. Animal model of human disease in drug safety assessment. Toxicol. Sci. 28, 109–121.

Bonaldi, T., Talamo, F., Scaffidi, P., et al., 2003. Monocytic cells hyperacetylate chromatin protein HMGB1 to redirect it towards secretion. EMBO J. 22, 5551–5560.

Bonventre, J.V., Vaidya, V.S., Schmouder, R., et al., 2010. Next- generation biomarkers for detecting kidney toxicity. Nat. Biotechnol. 28, 436–440.

Borzelleca, J.F., 2000. Paracelsus: herald of modern toxicology. Toxicol. Sci. 53, 2–4.

Brophy, V.H., Hastings, M.D., Clendenning, J.B., et al., 2001. Polymorphisms in the human paraoxonase (PON1) promoter. Pharmacogenetics 11, 77–84.

Brott, D., Gould, S., Jones, H., et al., 2005. Biomarkers of drug-induced vascular injury. Toxicol. Appl. Pharmacol. 207, S441–S445.

Burkhardt, J.E., Pandher, K., Solter, P.F., et al., 2011. Recommendations for the evaluation of pathology data in nonclinical safety biomarkers qualification studies. Toxicol. Pathol. 39, 1129–1137.

Callaghan, N., Majeed, T., O'Connell, A., et al., 1994. A comparative study of serum F protein and other liver function tests as an index of hepatocellular damage in epileptic patients. Acta Neurol. Scand. 89, 237–241.

Chrzanowska, A., Gajewska, B., Barańczyk-Kuźma, A., 2009. Arginase isoenzymes in human cirrhotic liver. Acta Biochem. Pol. 56, 465–469.

Chesney, P.J., Davis, J.P., Purdy, W.K., et al., 1981. Clinical manifestations of toxic shock syndrome. J. Am. Med. Assoc. 246, 741–748.

Chung, W.H., Hung, S.I., Hong, H.S., et al., 2004. Medical genetics: a marker for Stevens-Johnson syndrome. Nature 428, 486.

Clark, I.M., Swingler, T.E., Sampieri, C.L., et al., 2008. The regulation of matrix metalloproteinases and their inhibitors. Int. J. Biochem. Cell Biol. 40, 1362–1378.

Coles, B.F., Kadlubar, F.F., 2005. Human alpha class glutathione-S-transferases: genetic polymorphism, expression, and susceptibility to disease. Methods Enzymol. 401, 9–42.

Coles, B.F., Morel, F., Rauch, C., et al., 2001. Effect of polymorphism in the human glutathione-S-transferase A1 promoter on hepatic GSTA1 and GSTA2 expression. Pharmacogenetics 11, 663–669.

Collins, F.S., 2011. Reengineering translational science: the time is right. Sci. Transl. Med. 3 (90), 90cm17.

Cox, M.B., Cairns, M.J., Gandhi, K.S., et al., 2010. MicroRNAs miR-17 and miR-20a inhibit T cell activation genes and are under-expressed in MS whole blood. PLoS One 5, e12132. https://doi.org/10.1371/journal.pone.0012132.

Craig, D.G., Lee, P., Pryde, E.A., et al., 2011. Circulating apoptotic and necrotic cell death markers in patients with acute liver injury. Liver Int. 31 (8), 1127–1136.

Csanaky, I., Gregus, Z., 2005. Role of glutathione in reduction of arsenate and of gamma-glutamyltranspeptidase in disposition of arsenite in rats. Toxicology 207, 91–104.

Cummings, J., Ranson, M., Butt, F., et al., 2007. Qualification of M30 and M65 ELISAs as surrogate biomarkers of cell death: long term antigen stability in cancer patient plasma. Cancer Chemother. Pharmacol. 60, 921–924.

Cummings, J., Ward, T.H., Greystoke, A., et al., 2008. Biomarker method validation in anticancer drug development. Br. J. Pharmacol. 153, 646–656.

Cummings, J.L., Banks, S.J., Gary, R.K., et al., 2013. Alzheimer's disease drug development: translational neuroscience strategies. CNS Spectr. 18, 199–208.

Desikan, R., Veksler, Y., Raza, S., et al., 2002. Nephrotic proteinuria associated with high-dose pamidronate in multiple myeloma. Br. J. Haematol. 119 (2), 496–499.

Diamond, J.R., Yoburn, D.C., 1982. Nonoliguric acute renal failure. Arch. Intern. Med. 142 (10), 1882–1984.

Dieterle, F., Sistare, F., Goodsaid, F., et al., 2010a. Renal biomarker qualification submission: a dialog between the FDA_EMEA and predictive safety testing consortium. Nat. Biotechnol. 28, 455–462.

Dieterle, F., Perentes, E., Cordier, A., et al., 2010b. Urinary clusterin, cystatin C [beta]2-microglobulin and total protein as markers to detect druginduced kidney injury. Nat. Biotechnol. 28, 463–469.

Dix, D.J., Houck, K.A., Martin, M.T., et al., 2007. The ToxCast program for prioritizing toxicity testing of environmental chemicals. Toxicol. Sci. 95, 5–12.

Dubois, B., Feldman, H.H., Jacova, C., et al., 2005. Research criteria for the diagnosis of Alzheimer's disease: 327 revising the NINCDS-ADRDA criteria. Lancet Neurol. 6, 734–746.

Edwards, S.W., Preston, R.J., 2008. Systems biology and mode of action based risk assessment. Toxicol. Sci. 106, 312–318.

EMA, November 15, 2007. Committee for Medicinal Products for Human Use (CHMP). Reflection Paper on Pharmacogenomics Samples, Testing and Data Handling. EMEA/CHMP/PGxWP/201914/2006.

EMA, February 10, 2011a. Qualification Opinion of Alzheimer's Disease Novel Methodologies/biomarkers for BMS-708163. EMA/CHMP/SAWP/102001/2011.

EMA, June 9, 2011b. Reflection Paper on Methodological Issues Associated with Pharmacogenomics Biomarkers in Relation to Clinical Development and Patient Selection. EMA/446337/2011.

EMEA, March 17, 2005. Guidelines on Pharmacogenetics Briefing Meetings. EMEA/CHMP/2022704.

EMEA, 2008. Committee for Medicinal Products for Human Use (CHMP). Final Conclusion on the Pilot Joint EMEA/FDA VXDS Experience on Qualification of Nephrotoxicity Biomarkers. EMEA/679719/2008, Rev. 1.

Evans, D.C., Watt, A.P., Nicoll-Griffith, D.A., et al., 2004. Drug-protein adducts: an industry perspective on minimizing the potential for drug bioactivation in drug discovery and development. Chem. Res. Toxicol. 17, 3–16.

FDA, 2008. FDA Qualification of Seven Biomarkers of Drug-Induced Nephrotoxicity in Rats. In: www.fda.gov/downloads/Drugs/developmentAoorivalProcess/DrugDevelopmentToolsQualificationProgram/UCM285031.pdf.

FDA, 2012. Biomarker Qualification Decision: Preclinical CTnT and CTnI. In: www.fda.gov/downloads/Drugs/DevelopmentApprovalProcess/DrugDevelopmentToolsQualification.Program/UCM294644.pdf.

FDA, 2017. Biomarker Qualification Decision: Preclinical CTnT and CTnl. Development approval process/drug development tools qualification program/UCM294644.

Ferré, N., Camps, J., Prats, E., et al., 2002. Serum paraoxonase activity: a new additional test for the improved evaluation of chronic liver damage. Clin. Chem. 48, 261–268.

Fiehn, O., 2002. Metabolomics – the link between genotypes and phenotypes. Plant Mol. Biol. 48, 155–171.

Frank, R., Hargreaves, R., 2003. Clinical biomarkers in drug discovery and development. Nat. Rev. 2, 566–580.

Foster, G.R., Goldin, R.D., Oliveira, D.B., 1989. Serum F protein: a new sensitive and specific test of hepatocellular damage. Clin. Chim. Acta 184, 85–92.

Fuchs, T.C., Hewitt, P., 2011. Biomarkers for drug-induced renal damage and nephrotoxicity. An overview for applied toxicology. AAPS J. 13, 615–631.

Furlong, C.E., Cole, T.B., Jarvik, G.P., et al., 2002. Pharmacogenomic considerations of the paraoxonase polymorphisms. Pharmacogenomics 3, 341–348.

Gad, S.C., 2009. Drug Safety Evaluation, second ed. John Wiley and Sons, Inc, Hoboken, New Jersey.

Ganey, P.E., Luyendyk, J.P., Maddox, J.F., et al., 2004. Adverse hepatic drug reactions: inflammatory episodes as consequence and contributor. Chem. Biol. Interact. 150, 35–51.

Gatanaga, H., Tachikawa, N., Kikuchi, Y., et al., 2006. Urinary beta2-microglobulin as a possible sensitive marker for renal injury caused by tenofovir disoproxil fumarate. AIDS Res. Hum. Retroviruses 22, 744–748.

Gerber, B.O., Pichler, W.J., 2004. Cellular mechanisms of T cell mediated drug hypersensitivity. Curr. Opin. Immunol. 16, 732–737.

Guder, W.G., Hofmann, W., 1992. Markers for the diagnosis and monitoring of renal tubular lesions. Clin. Nephrol. 38, S3–S7.

Guengerich, F.P., 2011. Mechanisms of drug toxicity and relevance to pharmaceutical development. Drug Metabol. Pharmacokinet. 26 (1), 3–14.

Guo, X., Nzerue, C., 2002. How to prevent, recognize, and treat drug-induced nephrotoxicity. Cleve. Clin. J. Med. 69, 289–297.

Haase, M., Bellomo, R., Devarajan, P., et al., 2009. Accuracy of neutrophil gelatinase-associated lipocalin (NGAL) in diagnosis and prognosis in acute kidney injury: a systemic review and meta-analysis. Am. J. Kidney Dis. 54, 1012–1024.

Han, W.K., Bonventre, J.V., 2004. Biologic markers for early detection of acute kidney injury. Curr. Opin. Crit. Care 10, 476–482.

Han, W.K., Waikar, S.S., Johnson, A., et al., 2008. Urinary biomarkers in the early diagnosis of acute kidney injury. Kidney Int. 73 (7), 863–869.

Herget-Rosenthal, S., Marggraf, G., Hüsing, J., et al., 2004a. Early detection of acute renal failure by serum cystatin C. Kidney Int. 66, 1115–1122.

Herget-Rosenthal, S., Poppen, D., Hüsing, J., et al., 2004b. Prognostic value of tubular proteinuria and enzymuria in nonoliguric acute tubular necrosis. Clin. Chem. 50, 552–558.

Higuchi, H., Adachi, Y., Wada, H., et al., 2001. Comparison of plasma alpha glutathione S-transferase concentrations during and after low-flow sevoflurane or isoflurane anaesthesia. Acta Anaesthesiol. Scand. 45, 1226–1229.

Hirsch, R., Dent, C., Pfriem, H., et al., 2007. NGAL is an early predictive biomarker of contrast-induced nephropathy in children. Pediatr. Nephrol. 22, 2089–2095.

Hong, H., Goodsaid, F., Shi, L., et al., 2010. Molecular biomarkers: a US FDA effort. Biomarkers Med. 4 (2), 215–225.

Hori, O., Brett, J., Slattery, T., et al., 1995. The receptor for advanced glycation end products (RAGE) is a cellular binding site for amphoterin: mediation of neurite outgrowth and co-expression of rage and amphoterin in the developing nervous system. J. Biol. Chem. 270, 25752–25761.

Horton, R., Wilming, L., Rand, V., et al., 2004. Gene map of the extended human MHC. Nat. Rev. Genet. 5, 889–899.

Hung, S.I., Chung, W.H., Jee, S.H., et al., 2006. Genetic susceptibility to carbamazepine-induced cutaneous adverse drug reactions. Pharmacogenet. Genomics 16, 297–306.

Hunt, C.M., Papay, J.I., Edwards, R.I., et al., 2007. Monitoring liver safety in drug development: the GSK experience. Regul. Toxicol. Pharmacol. 49, 90–100.

IMI, July 26, 2005. The Innovative Medicines Initiative Strategic Research Agenda. Creating Biomedical R&D Leadership for Europe to Benefit Patients and Society. Draft document. European Commission.

Jaeschke, H., Knight, T.R., Bajt, M.L., 2003. The role of oxidant stress and reactive nitrogen species in acetaminophen hepatotoxicity. Toxicol. Lett. 144, 279–288.

Jones, D.P., Lemasters, J.J., Han, D., et al., 2010. Mechanism of pathogenesis in drug hepatotoxicity putting the stress on mitochondria. Mol. Interv. 10, 98–111.

Kaplowitz, N., 2005. Idiosyncratic drug hepatotoxicity. Nat. Rev. Drug Discov. 4, 489–499.

Khayrollah, A.A., Al-Tamer, Y.Y., Taka, M., et al., 1982. Serum alcohol dehydrogenase activity in liver diseases. Ann. Clin. Biochem. 19, 35–42.

Kim, H., Lee, J., Hyun, J.W., et al., 2007. Expression and immunohistochemical localization of galectin-3 in various mouse tissues. Cell Biol. Int. 31, 655–662.

Koch, W.H., 2004. Technology platforms for pharmacogenomic diagnostic assays. Nat. Rev. Drug Discov. 3, 749–761.

Koerbin, G., Taylor, L., Dutton, J., et al., 2001. Aminoglycoside interference with the Dade Behring pyrogallol red-molybdate method for the measurement of total urine protein. Clin. Chem. 47, 2183–2184.

Kohler, G., Milstein, C., 1975. Continuous cultures of fused cells secreting antibodies of predefined specificity. Nature 256, 495–497.

Köhler, G., Milstein, C., 2005. Continuous cultures of fused cells secreting antibody of predefined specificity. 1975. J. Immunol. 174, 2453–2455.

Kola, I., Landis, J., 2004. Can the pharmaceutical industry reduce attrition rates ? Nat. Rev. Drug Discov. 3 (8), 711.

Kramer, F., Milting, H., 2011. Novel biomarkers in human terminal heart failure and under mechanical circulatory support. Biomarkers 16, S31–S41.

Korsrud, G.O., Grice, H.C., McLaughlan, J.M., 1972. Sensitivity of several serum enzymes in detecting carbon tetrachloride-induced liver damage in rats. Toxicol. Appl. Pharmacol. 22, 474–483.

Lanning, L.L., 2006. Toxicologic pathology assessment. In: Jacobaon-Kram, D., Keller, K.A. (Eds.), Toxicological Testing Handbook: Principles, Applications, and Data Interpretation, second ed. Informa Healthcare, New York, pp. 439–464.

Latt, D., Weiss, J.B., Jayson, M.I., 1981. Beta 2-microglobulin levels in serum and urine of rheumatoid arthritis patients on gold therapy. Ann. Rheum. Dis. 40, 157–160.

Le Niculescu, H., Kurian, S.M., Yehyawi, N., et al., 2008. Identifying blood biomarkers for mood disorders using convergent functional genomics. Mol. Psychiatry 14, 156–174.

Lewin, D.A., Weiner, M.P., 2004. Molecular biomarkers in drug development. Drug Discov. Today 9, 976–983.

Li, C., Pei, F., Zhu, X., et al., 2012. Circulating microRNAs as novel and sensitive biomarkers of acute myocardial infarction. Clin. Biochem. 45, 727–732.

Liangos, O., Perianayagam, M.C., Vaidya, V.S., et al., 2007. Urinary *N*-acetyl-beta-(d)-glucosaminidase activity and kidney injury molecule-1 level are associated with adverse outcomes in acute renal failure. J. Am. Soc. Nephrol. 18, 904–912.

Lindblom, P., Rafter, I., Copley, C., et al., 2007. Isoforms of alanine aminotransferases in human tissues and serum—differential tissue expression using novel antibodies. Arch. Biochem. Biophys. 466 (1), 66–77.

Liu, D.Z., Tian, Y., Ander, B.P., et al., 2010. Brain and blood microRNA expression profiling of ischemic stroke, intracerebral hemorrhage, and kainate seizures. J. Cerebr. Blood Flow Metabol. 30, 92–101.

Lock, E.A., Bonventre, J.V., 2008. Biomarkers in translation; past, present and future. Toxicology 245, 163–166.

Louden, C., Morgan, D.G., 2001. Pathology and pathophysiology od drug-induced arterial injury in laboratory animals and its implications on the evaluation of novel chemical entities for human clinical trials. Pharmacol. Toxicol. 89, 158–170.

MacGregor, J.T., Farr, S., Tucker, J.D., et al., 1995. New molecular endpoints and methods for routine toxicity testing. Fundam. Appl. Toxicol. 26, 156–173.

Mallal, S., Phillips, E., Carosi, G., et al., 2008. HLA-B*5701 screening for hypersensitivity to abacavir. N. Engl. J. Med. 358, 568–579.

Manicone, A.M., McGuire, J.K., 2008. Matrix metalloproteinases as modulators of inflammation. Semin. Cell Dev. Biol. 19, 34–41.

McIlleron, H., Meintjes, G., Burman, W.J., et al., 2007. Complications of antiretroviral therapy in patients with tuberculosis: drug interactions, toxicity and immune reconstitution inflammatory syndrome. J. Infect. Dis. 196, S63–S75.

Marchewka, Z., Kuzniar, J., Dlugosz, A., 2001. Enzymuria and beta2-microglobulinuria in the assessment of the influence of proteinuria on the progression of glomerulopathies. Int. Urol. Nephrol. 33, 673–676.

Marrer, E., Dieterle, F., 2010. Impact of biomarker development on drug safety assessment. Toxicol. Appl. Pharmacol. 243, 167–179.

Matheis, K., Laurie, D., Andriamandroso, C., et al., 2011. A generic operational strategy to qualify translational safety biomarkers. Drug Discov. Today 16, 600–608.

Mendis, G.P., Gibberd, F.B., Hunt, H.A., 1993. Plasma activities of hepatic enzymes in patients on anticonvulsant therapy. Seizure 2, 319–323.

Mendrick, D.L., 2011. Transcriptional profiling to identify biomarkers of disease and drug response. Pharmacogenomics 12, 235–249.

Misra, M.K., Khanna, A.K., Sharma, R., et al., 1991. Serum malate dehydrogenase (MDH) in portal hypertension—its value as a diagnostic and prognostic indicator. Indian J. Med. Sci. 45, 31–34.

Morimoto, K., Oishi, T., Ueda, S., et al., 2004. A novel variant allele of OATP-C (SLCO1B1) found in a Japanese patient with pravastatin-induced myopathy. Drug Metab. Pharmacokinet. 19, 453–455.

Munshi, R., Johnson, A., Siew, E.D., et al., 2011. MCP-1 gene activation marks acute kidney injury. J. Am. Soc. Nephrol. 22, 165–175.

Nakamura, Y., 2008. Pharmacogenomics and drug toxicity. N. Engl. J. Med. 359, 856–858.

Neve, S., Aarenstrup, L., Tornehave, D., et al., 2003. Tissue distribution, intracellular localization and proteolytic processing of rat 4-hydroxyphenylpyruvate dioxygenase. Cell Biol. Int. 27, 611–624.

Newsholme, S.J., Thudium, D.T., Gossett, K.A., et al., 2000. Evaluation of plasma von Willebrand factor as a biomarker for acute arterial damage in rats. Toxicol. Pathol. 28, 688–693.

Nguyen, M.T., Devarajan, P., 2008. Biomarkers for the early detection of acute kidney injury. Pediatr. Nephrol. 23, 2151–2157.

Nicholson, J.K., Lindon, J.C., Holmes, E., 1999. Metabonomics: understanding the metabolic responses of living systems to pathophysiological stimuli via multivariate statistical analysis of biological NMR spectroscopic data. Xenobiotica 29, 1181–1189.

NTP, 2006. Biomarkers for Toxicology Studies. September 20–21, 2006 at the NIEHS.

O'Brien, P.J., Slaughter, M.R., Polley, S.R., et al., 2002. Advantages of glutamate dehydrogenase as a blood biomarker of acute hepatic injury in rats. Lab. Anim. 36, 313–321.

Ochieng, J., Furtak, V., Lukyanov, P., 2004. Extracellular functions of galectin-3. Glycoconj. J. 19, 527–535.

Ogden, C.A., Rich, M.E., Schork, N.J., et al., 2004. Candidate genes, pathways and mechanisms for bipolar (manic-depressive) and related disorders: an expanded convergent functional genomics approach. Mol. Psychiatry 9, 1007–1029.

Olson, H., Betton, G., Robinson, D., et al., 2000. Concordance of the toxicity of pharmaceuticals in humans and in animals. Regul. Toxicol. Pharmacol. 32, 56–67.

Ong, M.M.K., Latchoumycandane, C., Boelsterli, U.A., 2007. Troglitazone-induced hepatic necrosis in an animal model of silent genetic mitochondrial abnormalities. Toxicol. Sci. 97 (1), 205–213.

Ostapowicz, G., Fontana, R.J., Schiodt, F.V., et al., 2002. Results of a prospective study of acute liver failure at 17 tertiary care centers in the United States. Ann. Intern. Med. 137, 947–954.

O'Toole, M., Janszen, D.B., Slonim, D.K., et al., 2005. Risk factors associated with b-amyloid(1–42) immunotherapy in preimmunization gene expression patterns of blood cells. Arch. Neurol. 62, 1531–1536.

Ozer, J., Ratner, M., Shaw, M., et al., 2008. The current state of serum biomarkers of hepatotoxicity. Toxicology 245, 194–205.

Pamuk, B.O., Sari, I., Kozaci, D.L., 2013. Evaluation of circulating endothelial biomarkers in familial Mediterranean fever. Rheumatol. Int. 33, 1967–1972.

Park, J.S., Svetkauskaite, D., He, Q., et al., 2004. Involvement of toll-like receptors 2 and 4 in cellular activation by high mobility group box 1 protein. J. Biol. Chem. 279, 7370–7377.

Parikh, C.R., Jani, A., Mishra, J., et al., 2006. Urine NGAL and IL-18 are predictive biomarkers for delayed graft function following kidney transplantation. Am. J. Transplant. 6, 1639–1645.

Pasanen, M.K., Neuvonen, M., Neuvonen, P.J., et al., 2006. SLCO1B1polymorphism markedly affects the pharmacokinetics of simvastatin acid. Pharmacogenet. Genomics 16, 873–879.

Pemberton, C.J., Johnson, M.L., Yandle, T.G., et al., 2000. Deconvolution analysis of cardiac natriuretic peptides during acute volume overload. Hypertension 36, 355–359.

Peters, T.S., 2005. Do preclinical strategies help predict human hepatotoxic potentials. Toxicol. Pathol. 33, 146–154.

Polkinghorne, K.R., 2006. Detection and measurement of urinary protein. Curr. Opin. Nephrol. Hypertens. 15, 625–630.

Price, R.G., 1982. Urinary enzymes, nephrotoxicity and renal disease. Toxicology 23, 99–134.

Price, R.G., 1992. The role of NAG (*N*-acetyl-beta-D-glucosaminidase) in the diagnosis of kidney disease including the monitoring of nephrotoxicity. Clin. Nephrol. 38, S14–S19.

Price, R.G., 2002. Early markers of nephrotoxicity. Comp. Clin. Pathol. 11, 2–7.

Proulx, N.L., Akbari, A., Garg, A.X., et al., 2005. Measured creatinine clearance from timed urine collections substantially overestimates glomerular filtration rate in patients with liver cirrhosis: a systematic review and individual patient meta-analysis. Nephrol. Dial. Transplant. 20, 1617–1622.

PSTC, 2008. FDA and EMEA Conclude That New Renal Safety Biomarkers Are Qualified for Specific Regulatory Purposes [serial online]. Available at: http://www.cpath .org/pdf/PSTC_nephro_VXDS_summary_final.pdf.

Reeves, W.B., Kwon, O., Ramesh, G., 2008. Netrin-1 and kidney injury. II. Netrin-1 is an early biomarker of acute kidney injury. Am. J. Physiol. Ren. Physiol. 294, F731–F738.

Rehman, S.U., Januzzi, J.L., 2008. Natriuretic peptide testing in primary care. Curr. Cardiol. Rev. 4, 300–308.

Roberts, A.S., Price, V.F., Jollow, D.J., 1990. Acetaminophen structure-toxicity studies: in vivo covalent binding of a non-hepatotoxic analog, 3-hydroxyacetanilide. Toxicol. Appl. Pharmacol. 105, 195–208.

Rosalki, S.B., Roberts, R., Katus, H.A., et al., 2004. Cardiac biomarkers for detection of myocardial infarction: perspectives from past to present. Clin. Chem. 50, 2205–2213.

Richmond, J.M., Sibbald, W.J., Linton, A.M., et al., 1982. Patterns of urinary protein excretion in patients with sepsis. Nephron 31, 219–223.

Roskos, L.K., Schneider, A., Vainshtein, I., et al., 2011. PK-PD modeling of protein drugs: implications in assay development. Bioanalysis 3, 659–675.

Rudolph, M.G., Stanfield, R.L., Wilson, I.A., 2006. How TCRs bind MHCs, peptides, and coreceptors. Annu. Rev. Immunol. 24, 419–466.

Santella, R.M., 1997. DNA damage as an intermediate biomarker in intervention studies. Proc. Soc. Exp. Biol. Med. 216, 166–171.

Sauer, J.-M., Porter, A.C., 2018. Preclinical biomarker qualification. Exp. Biol. Med. 243 (3), 222–227.

Scaffidi, P., Misteli, T., Bianchi, M.E., 2002. Release of chromatin protein HMGB1 by necrotic cells triggers inflammation. Nature 418, 191–195.

Schetz, M., Dasta, J., Goldstein, S., et al., 2005. Drug-induced acute kidney injury. Curr. Opin. Crit. Care 11, 555–565.

Schnackenberg, L.K., Chen, M., Sun, J., et al., 2009. Evaluations of the trans-sulfuration pathway in multiple liver toxicity studies. Toxicol. Appl. Pharmacol. 235, 25–32.

Schentag, J.J., Plaut, M.E., 1980. Patterns of urinary beta 2-microglobulin excretion by patients treated with aminoglycosides. Kidney Int. 17, 654–661.

Schmid, H., Henger, A., Cohen, C.D., et al., 2003. Gene expression profiles of podocyte-associated molecules as diagnostic markers in acquired proteinuric diseases. J. Am. Soc. Nephrol. 14 (11), 2958–2966.

Schwarz, U.I., Ritchle, M.D., Bradford, Y., et al., 2008. Genetic determinants of response to warfarin during initial anticoagulation. N. Engl. J. Med. 358, 999–1008.

Sharma, UC., Pokharel, S., van Brakel, T.J., van Berlo, J.H., Cleutjens, J.P.M., Schroen, B., André, S., Crijns, H.J.G.M., Gabius, H-J., Maessen, J., Pinto, Y.M., 2004. Galectin-3 marks activated macrophages in failure-prone hypertrophied hearts and contributes to cardiac dysfunction. Circulation 110, 3121–3128.

Sharratt, M., Frazer, A.C., 1963. The sensitivity of function tests in detecting renal damage in the rat. Toxicol. Appl. Pharmacol. 5, 36–48.

Solter, P.F., 2005. Clinical pathology approaches to hepatic injury. Toxicol. Pathol. 33, 9–16.

Star, R.A., 1998. Treatment of acute renal failure. Kidney Int. 54, 1817–1831.

Starkey Lewis, P., Dear, J., Platt, V., et al., 2011. Circulating microRNAs as potential markers of human drug induced liver injury. Hepatology 54, 1767–1776.

Stockham, S.L., Scott, M.A., 2002. Fundamentals of Veterinary Pathology. Iowa State Press, Ames, Iowa.

Supavekin, S., Zhang, W., Kucherlapati, R., et al., 2003. Differential gene expression following early renal ischemia/reperfusion. Kidney Int. 63, 1714–1724.

Tang, Y., Glauser, T.A., Gilbert, D.L., et al., 2004. Valproic acid blood genomic expression patterns in children with epilepsy – a pilot study. Acta Neurol. Scand. 109, 159–168.

Thielemans, N., Lauwerys, R., Bernard, A., 1994. Competition between albumin and lowmolecular-weight proteins for renal tubular uptake in experimental nephropathies. Nephron 66, 453–458.

Temple, R., 1999. Are surrogate markers adequate to assess cardiovascular disease drugs? J. Am. Med. Assoc. 282, 790–795.

Tirelli, A.S., Colombo, N., Cavanna, G., et al., 1985. Follow-up study of enzymuria and beta 2 microglobulinuria during cis-platinum treatment. Eur. J. Clin. Pharmacol. 29, 313–318.

Uetrecht, J., 2007. Idiosyncratic drug reactions: current understanding. Annu. Rev. Pharmacol. Toxicol. 21, 84–92.

Ulrich, R.G., 2007. Idiosyncratic toxicity: a convergence of risk factors. Annu. Rev. Med. 58, 17–34.

Urbschat, A., Obermüller, N., Haferkamp, A., 2011. Biomarkers of kidney injury. Biomarkers 16, S22–S30.

Waikar, S.S., Betensky, R.A., Bonventre, J.V., 2009. Creatinine as the gold standard for kidney injury biomarker studies? Nephrol. Dial. Transplant. 24, 3263–3265.

Walgren, J.L., Mitchell, M.D., Thompson, D.C., 2005. Role of metabolism in drug-induced idiosyncratic hepatotoxicity. Crit. Rev. Toxicol. 35, 325–361.

Wang, X., Wu, Q., Liu, A., et al., 2017. Paracetamol: overdose-induced oxidative stress toxicity, metabolism, and protective effects of various compounds in vivo and in vitro. Drug Metab. Rev. 49 (4), 395–437.

Warnock, D.G., Peck, C.C., 2010. A roadmap for biomarker qualification. Nat. Biotechnol. 28, 444–445.

Wieckowska, A., Zein, N.N., Yerian, L.M., et al., 2006. In vivo assessment of liver cell apoptosis as a novel biomarker of disease severity in nonalcoholic fatty liver disease. Hepatology 44, 27–33.

White, T.J., Clark, A.G., Broder, S., 2006. Genome-based biomarkers for adverse drug effects, patient enrichment and prediction of drug response, and their incorporation into clinical trial design. Per. Med. 3, 177–185.

Xu, S., Venge, P., 2000. Lipocalins as biochemical markers of disease. Biochim. Biophys. Acta 1482, 298–307.

Yang, R.Z., Blaileanu, G., Hansen, B.C., et al., 2002. cDNA cloning, genomic structure, chromosomal mapping, and functional expression of a novel human alanine aminotransferase. Genomics 79, 445–450.

Yimer, G., Ueda, N., Habtewold, A., et al., 2011. Pharmacogenetic & pharmacokinetic biomarker for Efavirenz based ARV and rifampicin based anti-TB drug induced liver injury in TB-HIV infected patiens. PLoS One 6, e27810.

Zidek, N., Hellmann, J., Kramer, P.J., Hewitt, P.G., 2007. Acute hepatotoxicity: a predictive model based on focused illumina microarrays. Toxicol. Sci. 99, 289–302.

Zieve, L., Anderson, W., Dozeman, R., et al., 1985. Acetaminophen liver injury: sequential changes in two biochemical indices of regeneration and their relationship to histologic alterations. J. Lab. Clin. Med. 105, 619–624.

Zomas, A., Anagnostopoulos, N., Dimopoulos, M.A., 2000. Successful treatment of multiple myeloma relapsing after high-dose therapy and autologous transplantation with thalidomide as a single agent. Bone Marrow Transplant. 25 (12), 1319–1320.

C H A P T E R

39

Risk Assessment, Regulation, and the Role of Biomarkers for the Evaluation of Dietary Ingredients Present in Dietary Supplements

Sandra A. James-Yi[1], *Corey J. Hilmas*[2], *Daniel S. Fabricant*[3]

[1]Associate Principle Scientist, Product Safety/Nutritional Toxicology, Mary Kay Inc., Addison, Texas, United States; [2]Senior Vice President of Scientific and Regulatory Affairs, Natural Products Association, Washington D.C., United States; [3]President and CEO, Natural Products Association, Washington D.C., United States

INTRODUCTION

Dietary supplements in the United States are an FDA-regulated commodity used by an increasing number of individuals on a daily basis. This topic brings up a wide range of opinions from both supplement users and nonusers alike, in regard to regulation, safety, and efficacy. There are reports in the literature suggesting that the use of certain dietary supplement products and ingredients, specifically, those with a known or predictable pharmacological activity (nutraceuticals), have the potential for adverse events even if a preemptive risk assessment is performed prior to market distribution. Despite the long-held standing that a particular dietary ingredient is "safe" due to a substantial history of traditional use, information and research substantiating present-day usage and form may be lacking.

There are considerable drivers in the marketplace that contribute to or potentially compound risk. These include a public perception that these products are inherently safe because they are advertized as "all-natural"; have a history of traditional use but in a different form or manufactured by a different process; and are sold to the consumer as a capsule or tablet, similar to prescription drugs they would get from a medical professional, and thus deemed to have undergone the same process, scrutiny, and guidance as pharmaceuticals prior to release in the marketplace or there may be the belief that they have passed regulatory standards for product safety or efficacy, by a regulatory body such as FDA. In addition, because of the passage of time between the original use of a supplement and present day suggested use, and the evolution of modern-day medicine, there is a significant loss of historical knowledge and familiarity with many products, their nutritive value, and recommended usages within the traditional medical community.

Because of customer demand, many health-care practitioners are becoming more familiar with many of the dietary supplements on the market. The industry has grown at such a rapid pace that all of the products and combination products that are available to the average consumer may be difficult for medical practitioners and the general public to follow. Inherent in the science of nutrition is the difficulty in developing and performing robust studies. The scale with which products are coming onto the market makes this even more of a challenge for health-care professionals, industry, government regulators, and consumers alike. Communication is essential between health-care practitioners and their patients with respect to what is known about individual and combination dietary supplement products and the applicable cautions that need to be tailored to an individual's specific needs.

A contributing factor to the risk associated with the use of dietary supplements is consumer misinformation or lack of education for their intended use. Consumers of supplements often "self-prescribe" over-the-counter dietary supplements based on information from web-based public domains, bloggers, media outlets, word-of-mouth within their communities, manufacturers, and distributors of products.

Biomarkers in Toxicology, Second Edition
https://doi.org/10.1016/B978-0-12-814655-2.00039-6

The goal of this chapter is to review the process of assessing risk associated with the use of dietary supplements; the role that biomarkers play in the risk assessment and identification of hazards associated with the use of dietary supplements by different populations of users; and how science informs FDA regulatory decisions with respect to FDA's regulatory tools for application to dietary supplements. Although not exhaustive, listed below are factors to take into consideration when evaluating comprehensive safety profiles for dietary supplements as part of the process of risk assessment and hazard identification. Although the goal of this chapter is to address the biomarkers of physiological and biochemical changes corresponding with or directly caused by dietary supplements and ingredients with pharmacological activity (nutraceuticals) or toxicity, this subject is immense and incorporates all of the chapters in this text as they apply to biomarkers of multiple organ system functions. Dietary supplements are commonly mixtures and many times they are derived from the extrapolation of botanicals used in traditional or folk medicine for modern-day applications. As such they will have multiple target sites for activity that cannot be easily condensed into a simple algorithm. Factors that need to be considered by toxicologists in performing risk assessments for these products include but are not limited to the following:

- **Inappropriate application of the known pharmacology** of an ingredient or supplement for uses that differ from their traditional or historical applications. For example, ephedra has an extensive history of use for treatment of clinical symptoms associated with respiratory disease. Its traditional use was never intended to be for weight loss. Thus, the known pharmacology of the ingredient, especially safety and efficacy, was not applicable to products marketed or intended to be used for weight loss.
- **Changes in the duration of use** of a dietary supplement from the traditional posology of acute or intermittent usage in folk medicine to an intended current usage of prolonged, continuous, or chronic usage, typical of nutritional supplementation to an individual's diet.
- **Changes in dose** causing a larger, possibly excessive dose to be administered to achieve some particular result(s). Maximum levels of product, close to the tolerable upper limit for administration, may be delivered to consumers to make sure that a product on a store shelf has a labeled amount of constituent ingredients at the time of purchase and use. Some products that may have a short shelf-life of active constituent(s) have resulted in industry maximizing the amount per serving for dosing, based on tolerable, upper limits for a reasonable expectation of safe use rather than the minimal amount used to effect as is practiced within the pharmaceutical industry.
- **Changes in processing** that are significantly different from traditional or well-established methods. Herbs that historically have been consumed in one specific form, or in a particular cultural tradition for use, may now be processed or manufactured differently for the mass market. An example would be a botanical supplement that historically used a particular part of the plant for the basis of a tea preparation in folk medicine, may now be extracted with an organic solvent to extract and concentrate a specific, physiologically active component of the same plant. This significantly changes the chemical profile being delivered to consumers. The inherent synergist or antagonistic activity of constituents in the original tea may be gone and the resulting, concentrated extract may have an inherently different activity, posology, and risk assessment for a reasonable expectation for safe use within the general population for its intended use.
- **Concurrent use** of dietary supplements by consumers taken in conjunction with pharmaceuticals, especially drugs with a narrow therapeutic index. The medicinal drugs and supplements may have similar mechanistic actions or pathways of metabolism. Health-care professionals and the general population may not be educated adequately to think of dietary supplements contributing to potential problems with this type of polypharmacy. The concurrent use of multiple dietary supplements or herbal medicines with additive, synergistic, potentiating, or antagonistic pharmacological and toxicological activity can have serious consequences in individuals being treated for other medical conditions.

As science-based professionals, there needs to be a heightened level of communication between regulatory, academia, industry, and health-care professions, with the goal of identifying patterns of hazard or risk associated with the use of dietary supplements. The consumer rightfully assumes and expects that all stakeholders interested in dietary supplement safety have done due diligence prior to a product reaching the market. A reasonable expectation of product safety standards must align with the precautionary principle used by toxicologists involved in providing risk assessments to prevent potential harm in a very dynamic, demanding, lucrative, and growing marketplace.

OVERVIEW OF REGULATION OF DIETARY SUPPLEMENTS IN THE UNITED STATES

Federal Food, Drug, and Cosmetic Act

Dietary supplements are governed by four major sections of the Federal Food, Drug, and Cosmetic Act (Federal FD&C Act).

- Sec. 201 [21 U.S.C. 321] DEFINITIONS (ff)
- Sec. 402 [21 U.S.C. 342] ADULTERATED FOOD (f) and (g)
- Sec. 403 [21 U.S.C. 343] MISBRANDED FOOD
- Sec. 413 [21 U.S.C. 350b] NEW DIETARY INGREDIENTS

Dietary supplements are regulated under the Federal FD&C Act (the Act), as amended by the Dietary Supplement Health and Education Act (DSHEA) of 1994, and under the Fair Packaging and Labeling Act. DSHEA established a framework for regulating the safety of dietary supplements. This gave regulatory authority to the FDA to take action if there are product safety concerns (Hobson, 2009). Implementing regulations for certain provisions of DSHEA can be found in Title 21 of the Code of Federal Regulations (21 CFR). The term *dietary supplement* as defined in section 201(ff) of the Federal FD&C Act (the Act) is a product (other than tobacco) intended to supplement the diet that contains one or more of certain dietary ingredients, such as a vitamin, mineral, herb, other botanicals, amino acid, or dietary substance for use by man to supplement the diet by increasing the total dietary intake, of a concentrate, metabolite, constituent, extract, or combination of the preceding ingredients. Section 201(ff) of the Act further limits the term *dietary supplement* to mean products that are intended for ingestion in a form described in section 411(c)(1)(B)(i) of the Act (tablet, capsule, powder, softgel, gelcap, or liquid), which are not represented as conventional food, or as the sole item of a meal or of the diet, and that are specifically labeled as dietary supplements. According to 201(ff), "[e]xcept for purposes of section 201(g), a dietary supplement shall be deemed to be a food within the meaning of this Act." Based on case law, dietary supplements are not only regulated as food but must also be ingested, where "ingestion" means to take into the stomach and gastrointestinal tract by the oral route of administration having first passed through the mouth.

The term *dietary supplement* does not include products that are approved drugs, certified antibiotics, licensed biologics, or products that are authorized for investigation as a new drug, antibiotic, or biologic (and for which substantial clinical investigations have been instituted and for which the existence of such investigations have been made public), unless the product was marketed as a dietary supplement or as a food before it was approved as a drug, antibiotic, or biologic, or, in the case of investigational products, before the public disclosure of such investigations.

Dietary supplements may not contain dietary ingredients that present a significant or unreasonable risk of illness or injury as set forth in section 402(f) of the Act, or that may render the supplement injurious to health (see section 402(a)(1) of the Act).

Under section 413 of the Act, a manufacturer of a dietary supplement that is or that contains a new dietary ingredient (NDI) (an ingredient that was not marketed in the United States before October 15, 1994) must submit to FDA, at least 75 days before marketing, information substantiating the manufacturer's conclusions that use of the dietary supplement will be expected to be reasonably safe for its intended use. Alternatively, a manufacturer of a dietary supplement that is or contains an NDI may petition FDA to issue an order prescribing the conditions under which such a dietary supplement will be expected to be reasonably safe. Nevertheless, it is still the responsibility of the manufacturer to ensure that a dietary ingredient used in a dietary supplement is safe for its intended use.

It is important to remember that there is no authoritative "list" of substances that were marketed in the United States as ingredients in dietary supplements or foods before October 15, 1994. Therefore, the manufacturer must determine if an ingredient may be a "new dietary ingredient" under the Act for which a notification is required. If a manufacturer does not have a basis to conclude that a notification is not required, and they introduce a product containing the ingredient into interstate commerce, they do so at the risk that it may be considered adulterated as a matter of law. In the absence of evidence that a dietary ingredient does not require a notification to be submitted for it, manufacturers may wish to make the required notification to avoid uncertainty. In the September 23, 1997 Federal Register, FDA published a final rule promulgating a procedural regulation in 21 CFR 190.6 for making the required notification.

DSHEA was followed by the Dietary Supplement and Nonprescription Drug Consumer Protection Act (Public Law 109−462), which was signed into law on December 22, 2006 and became effective on December 22, 2007. This law amended the Federal FD&C Act with respect to reporting of serious adverse events related to dietary supplements and nonprescription drugs. The law has four major provisions: requiring (1) the collection of all adverse event reports by manufacturers, distributors, and retailers of dietary supplements; (2) the reporting of serious adverse event reports to the FDA; (3) firms to maintain records of reports of all adverse events from use of the dietary supplement(s) such that FDA is allowed to audit those records; and (4) that dietary supplement labels provide information to facilitate the reporting of serious adverse events associated with their use by consumers.

Lastly, Public Law P.L. 111–353, the FDA Food Safety Modernization Act, was signed into law on January 4, 2011. The major elements of the law include: prevention-based controls, inspection, and compliance to hold industry accountable; greater tools for increased food safety; mandatory recall authority for all food products; and enhanced partnerships among food safety agencies.

Regulatory Challenges

There is no requirement for the manufacturer of a dietary supplement to provide FDA with evidence of a product's efficacy. The manufacturer also need not provide FDA with evidence of ingredient safety prior to marketing, unless the product contains an NDI that has not been part of the food supply as an article used for food or is an article of food that has been modified in a form that is constitutionally, chemically altered.

FDA's Division of Dietary Supplement Program is tasked with the challenge of regulating a progressively expanding portfolio of dietary ingredients and finished products that are marketed as dietary supplements for a wide array of uses. When DSHEA amended the act in 1994, there were approximately 4000 products on the market; as of a 2009 Nielsen market survey, FDA estimated that there were 85,000 products on the market and this number is continuing to grow globally. With such growth the agency has encountered a number of unforeseen challenges. For example, FDA has found that dietary supplements marketed for sexual enhancement, weight loss, body building, and "calming" influences, can contain active pharmaceutical ingredients, analog of approved drugs, and other compounds, such as anabolic steroids, that do not qualify as dietary ingredients. Finding and testing these supplements for hidden, undeclared ingredients and removing those products from the market are a top priority for the Agency.

FDA's regulation of dietary supplements is particularly challenging given the seemingly endless volume of products that potentially transgress current regulations and the painstaking and costly process required by the government agency in proving that a violation exists. FDA continues to see dietary supplements marketed with ingredients that have been banned (i.e., ephedrine alkaloids and kratom) as well as dietary ingredients for which a notification to support the reasonable expectation of safety as an NDI was never submitted or containing an active ingredient that is undeclared on the labeling.

After identifying one of numerous potentially violative products, FDA is responsible for analyzing the ingredients of a product to determine whether a specific product is not in compliance. FDA conducts the laboratory analysis to prove the presence or absence of a product's ingredients that may compromise its safety. Even if the labeling of a particular dietary supplement identifies an ingredient suspected to render the product violative, the process requiring confirmation of a violation must be instituted, which is rigorous, extremely time-consuming, and expensive. When an ingredient that may render a product violative is adequately identified, FDA must then determine whether the ingredient is a dietary ingredient through an extensive search of the available scientific literature. Once it is determined whether the ingredient is a dietary ingredient, an examination of the product's other ingredients, labeling, and other promotional material is required to determine its proper regulatory status (e.g., unapproved new drug or an adulterated dietary supplement, i.e., nondeclared NDI).

Dietary Ingredient Safety

The Federal FD&C Act (the Act) prohibits the distribution of adulterated foods in interstate commerce. Under section 402(f)(1)(A) of the Act (21 U.S.C. 342(f)(1)(A)), a food is considered adulterated if it is a dietary supplement containing a dietary ingredient that presents a significant or unreasonable risk of illness or injury under the conditions of use recommended or suggested on the labeling, or if there are no stated conditions of use suggested or recommended in the labeling, under conditions of ordinary use. FDA bears the burden of establishing that the product presents a significant or unreasonable risk. Because dietary supplements are presumed to be safe, FDA's evaluation of whether a dietary supplement presents a significant or unreasonable risk generally takes place after the product is already on the market, with the exception of products that contain certain NDIs. Whether there is suspicion of a banned ingredient, concern over combinations of dietary ingredients, or safety concerns over an NDI, the use of biomarkers is crucial in determining whether dietary ingredients are adulterated. Although this list is not all-inclusive, Table 39.1 highlights several dietary ingredients of concern; the biomarkers that should be analyzed in the ingredient for quality control and may represent a component for possible risk and the known primary target organ systems associated with exposure causing clinical symptoms associated with the adverse event that is known to occur after ingestion. In the case of NDIs, manufacturers must demonstrate

TABLE 39.1 Dietary Ingredients of Concern, Their Biomarkers of Toxicity, and Target Organs Affected as a Result of Their Chronic Consumption in Dietary Supplements

Ingredient of Concern in a Dietary Supplement	Biomarker of Toxicity: Toxicant/Toxin, Drug, Metabolite	Target Organ System
Aristolochia spp. *and Asarum* spp.	Aristolochic acid	Renal/Urinary tract, liver
Blue-green algae	Microcystin	Hepatic
Bovine Source	Prohibited cattle material (21 CFR 189.5)	Central Nervous System
Butterbur (*Petasites hybridus*) root	Pyrrolizidine alkaloids	Hepatic, Cardiovascular
Comfrey (*Symphytum officinale* L)	Pyrrolizidine alkaloids	Hepatic, Cardiovascular
DMAA (1,3–dimethylamylamine)	1,3-Dimethylamylamine	Cardiovascular, Respiratory, Central and Peripheral Nervous system, Gastrointestinal
Ephedra	Ephedrine alkaloids	Cardiovascular, Central, and Peripheral Nervous System
Hydroxycut (previous formulation)	Unknown	Cardiovascular, Hepatic, Central Nervous System, Skeletal Muscle (Rhabdomyolysis), Renal
Kava (*Piper methysticum*)	Unknown	Hepatic
Lichen (*Usnea*)	Usnic acid	Hepatic
Kratom (*Mitragyna speciosa*)	Mitragynine, 7-hydroxymitragynine	Central Nervous and Respiratory System Depression
Rauwolfia	Reserpine	Central and Peripheral Nervous System, Cardiovascular
Red yeast rice	Lovastatin	Skeletal muscle toxicity, cardiovascular
Silver (listed as colloidal, ionic, native, alginate, protein, etc.)	Silver: Consumer advisory—risk of argyria at levels greater than RfD: 5 μg/kg BW/day	Dermal, Ophthalmologic (Note: All organ systems can be noticeably affected)
Desiccated thyroid glandulars	Thyroid hormone (9 CFR 310.15)	All organ systems

the process by which particular biomarkers of toxicity are absent within the process of notification to the FDA for an NDI.

As part of the process of determining tolerance and safe usage of a dietary supplement, a risk assessment is performed by a qualified toxicologist experienced in nutritional product safety. Because so little information and independent, peer-reviewed research is available for many of the products coming on to the market, the toxicologist responsible for the risk assessment needs to be armed with an extensive knowledge base and resources for the interpretation of a broad range of credible, scientific, primary, peer-reviewed data, including but not limited to in vitro, in vivo, and molecular studies as well as human clinical trials, case reports, and a myriad of other scientific and nonscientific, nutritional, historical, and environmental data to support the generation of safety profiles of dietary ingredient(s) for a reasonable expectation of tolerance and safety under the conditions of their intended use.

The Risk Assessment

In general, acceptable and recommended dietary daily intakes for the majority of vitamin and mineral exposures in the general population of healthy men, women, children, infants, and pregnant and/or lactating women have been well characterized. These guidelines are by no means static but are part of a dynamic process of characterization that continues to develop as methods of analyzing nutritional science, toxicology, and risk assessment develops among various subpopulations. For the purposes of this discussion on risk assessment, and the use of biomarkers of exposure and effect for the majority of newly marketed dietary supplements, the focus of this discussion will be on the role of biomarkers within the context of risk assessment of botanically derived dietary supplements.

Human populations are inherently outbred. The complexity of a "general" population and individual variation is currently being developed with customized tools and models provided by global genomics, epigenomics, proteomics, metabolomics, and big data such as Read-Across (Berggren et al., 2015). A case in point is the investigation and discovery of the various functions and perturbations of the microbiome and the use of prebiotics, probiotics, and synbiotics as dietary supplements. Biomarkers are objective measures of biochemical pathways, systemic organ function, internal and external perturbations of environmental markers, and chemical fingerprints of mixtures, plants and a multitude of other identifiers that may be present in natural or synthetic biological systems. Targeted research is just beginning to answer questions about the multisystemic, gender, environmental, cultural, and ancestral effects that nutrition and microbes contribute to a person's health.

A single, botanical, dietary ingredient (BDI) is a complex mixture of chemical compounds. Many of the products marketed as dietary supplements have multiple BDIs in formulations that have the potential for a multitude of additive, synergistic, potentiating, and/or antagonistic properties. In addition, constituents of these products are being grown, manufactured, and transported from vendors from all over the world. The ability to provide a reasonable expectation of safe use of these constituents by assessing quality, purity, stability, identity, and consistency of product, free from contamination or adulteration, under Good Manufacturing Practices, is vitally important for the protection of the consumer for the products intended use. Assessment of biomarkers and the chemical characterization or fingerprint of chemical constituents are vitally important to the process of public safety procedures and risk comparisons between BDIs. This is very important as new products come to market that may not be not in their natural state but in a chemical or physically modified form.

Although there is very little information or guidance specifically addressing the risk assessment of dietary supplements in the peer-reviewed literature, the nutritional toxicologist can follow the basic principles of risk assessment, weight of evidence, and the precautionary principle when assessing dietary supplements.

There are four basic components to risk assessment: *Hazard Identification, Hazard Characterization, Exposure potential, and Risk Characterization* (Boobis, 2007). The following is an overview of the steps that form this process for the evaluation of dietary supplements.

Hazard Identification

Hazard identification is performed to identify the range of possible intrinsic capabilities of a specific dietary supplement, its dietary ingredients (BDIs), and the characterized chemical constituents of the ingredients for the potential to cause harm. Because it is the responsibility of the manufacturer to ensure that a dietary ingredient used in a dietary supplement is safe for its intended use prior to distribution, industry must do due diligence to identify possible hazards that may be inherent to a BDI before it is exposed to the general population. It must also establish guidelines and biomarkers of botanicals for quality control. Hazard identification forms the basis for the development of the product label. This is a diagnostic challenge for botanically based products. Before a BDI product is introduced for development, there is usually some history of use. Important features associated with an investigation of a botanical are the region or country of origin; botanical nomenclature, plant genetic identification, and chemotype; original cultural purpose for use and dosing regimen; original methods of processing a BDI such as seasonal conditions for harvesting, part of plant used, processing, extraction methods, and storage. It is within this context for comparison that the risk characterization can be performed in relation to the suggested label usage under development for a specific BDI.

Because there is typically evidence of use of the BDI of interest or comparable products that have either recent or historical use, a comprehensive search of the literature for medical case reports of toxicity, clinical research, animal studies, in vitro analysis and any other pertinent mechanistic or molecular data of the BDI, or a specific chemical component of the BDI, is undertaken. Interpretation of the quality of data is highly dependent on the ability of the assessor(s) to have the experience to integrate the information to produce a toxicological narrative of risk. Experience with the principles of toxicological assessment, pathology, animal and human physiology and medicine, pharmacology, biochemistry, nutrition, and research study development and the principles of epidemiology and public health, all factor in to the creation of a robust risk assessment. In addition, it is very important to understand the origin of the data. Questions should be asked as to who owns the original study data? Who owns the laboratory or Contract Research Organization doing the study, the analysis, and the interpretation of the data? What was the initial incentive or purpose for the research project or clinical study? Was there a null hypothesis or a priori theoretical proposal or was there another motivation for conducting the study with expected outcomes and usage for the data?

Hazard Characterization

Hazard characterization is based on many components including but not limited to the results of cluster analysis and correlation of multiple biomarkers associated with human clinical studies, animal studies, molecular and mechanistic analysis, toxicokinetic and toxicodynamic information, clinical pathologic and histopathologic testing and documentation of adverse events occurring within the context of various studies. Results are compared for consistency in physiological effect across studies of similar BDIs and their chemical components. There is a cumulative need for multiple tests to assess toxicity and predictive signatures (Waters and Merrick, 2009). Conservation along with corroboration of more than one functional pathway that merges within and between various animal physiological systems is necessary to establish health-based guidance values such as the upper limit of tolerance for various populations of consumers.

Negative outcome data are critical to identifying a lack of physiological changes that have the possibility of being associated with an adverse effect. Consistent with quality research, randomized, blinded, controlled studies utilizing credible, robust, and objective biomarkers associated with comprehensive baseline serum chemistries, physical examinations, health status of the population(s) tested, exposure data, ADME, and evaluation procedures for long-term chronic effects and reversibility are warranted. Many clinical studies associated with the use of BDI's tend to focus on the beneficial effects of a given product utilizing a wide variety of disparate, subjective outcomes. The studies tend to lack a standardized platform of objective, repeatable criteria for comparative analysis between studies. A standardized set of biomarkers indicating the presence or absence of adverse events is an integral function of robust clinical research studies and needs to be better established for research specifically relating to the use of BDI's. Typical of most past and current studies, if adverse effects are considered by the researcher to have not occurred, there will be a short statement of "No adverse events were recorded." This is insufficient. Procedural guidelines, a priori, for how the adverse events were collected, recorded, and the criteria for evaluation are critical to the interpretation of a study and a possible adverse incident. Did these events occur in more than one individual? Were they recurring? Were there variables associated with subpopulations including race, gender, and life-stage that will help in identifying individual variation in intake and sensitivity (Boobis, 2007)? Were they self-reported or were they conducted by trained personnel capable of obtaining objective assessment? This process is very fluid. As more information surrounding the use (and misuse) of BDIs in the public sector is garnered, epidemiological data for long- and short-term exposure outcomes and postmarketing data generated by academia, government and industry, in addition to the regulatory guidelines established by the FDA for adverse event reporting, will become critical for future product development and quality control.

Exposure Assessment

Risk is a function of hazard and exposure. Outside of acceptable or estimated daily intake for consumption of components of BDIs that may occur in a specific population's normal diet on a local basis, global sourcing and production of BDIs in dietary supplements carries significant challenges to manufacturers and distributers. Supply-chain exposure to contaminants in the field, residual solvents after manufacture, and contamination or economically motivated adulteration of product at various stages of production can occur. Close associations with suppliers and supply chain-management in conjunction with standard markers for the presence or absence of toxicants are key to decreasing exposure to contamination and subsequently, risk. As an example, the raw, unprocessed butterbur plant (*Petasites hybridus* L.) contains pyrrolizidine alkaloids (PAs), which have been associated with hepatobiliary veno-occlusive disease and neoplasia (Danesch and Rittinghausen, 2003; Sadler et al., 2007; Burrows and Tyrl, 2013). Butterbur has been used for hundreds of years as an analgesic and antiinflammatory remedy. Currently it is being used as adjunct support for more chronic conditions such as migraine headaches, allergic rhinitis, and asthma. The pharmacologically active substances are believed to be the sesquiterpene esters, petasin, and furanopetasin (Anderson et al., 2009; Chizzola et al., 2006). During processing PAs are removed. Standard, preemptive, operating procedures and specifications need to be in place for quality analysis and control indicating the absence of PAs at critical points of production prior to distribution to the consumer to assure its absence for chronic use that may result in more severe, debilitating, and life-threatening conditions.

Risk Characterization

Because of limited available data available on many dietary supplements, toxicologists find themselves quite often in the position of proving a negative, that the dietary supplement or ingredient under investigation will not be associated with any serious adverse events to support a reasonable expectation of safety. A conservative approach is always warranted in an environment where there may be competing objectives for production and release of a dietary supplement product to the market. Toxicologists, charged with completing a risk assessment on a NDI or supplement, will need to rely on the weight of evidence from known data in conjunction with the precautionary principle when evaluating product. In some cases it is quite possible and appropriate for the toxicologist to conclude that there is insufficient evidence to produce a tolerable upper limit or observable safe level for consumption of a dietary ingredient that would have a reasonable expectation of safety under the conditions of its intended use.

BIOMARKERS OF TOXICITY: DIETARY INGREDIENTS

Biomarkers Associated with the Presence of Banned Ingredients

Banned Dietary Ingredients Causing Neurostimulation—Ephedrine Alkaloids

Dietary ingredients are banned if they present an unreasonable risk of illness or injury. In the *Federal Register* of February 11, 2004 (69 FR 6788), FDA interpreted and applied the "unreasonable risk" standard in a final rule declaring dietary supplements containing ephedrine alkaloids to be adulterated because they present an unreasonable risk of illness or injury. As the rule explained, "unreasonable risk" implies a risk–benefit calculation that weighs a product's risks against its benefits under the conditions of use recommended or suggested in the product's labeling, or if the labeling is silent, under ordinary conditions of use. The Secretary of Health and Human Services also has authority under the statute to act where a product poses an "imminent hazard" to public health. The risk–benefit calculation is also true for multiple or combinations of dietary ingredients with common mechanisms and/or modes of action.

Ephedrine alkaloids are chemical stimulants that were originally found to naturally occur in some botanicals. In the 10 years prior to the published Ephedrine Final Rule of 2004, dietary supplements containing ephedrine alkaloids were labeled and marketed with claims for weight loss, energy, or enhancement of athletic performance. Today, FDA still encounters dietary supplement products containing ephedrine alkaloids in the form of raw botanicals imported from foreign suppliers intended for use in dietary supplements or destined for production of methamphetamine.

Ephedrine alkaloids are members of a large family of pharmacologically active compounds known as phenylethylamines or sympathomimetics. They include pseudoephedrine (a decongestant), phenylephrine (cold tablets, nasal sprays, and hemorrhoid treatment), phenylpropanolamine (PPA) (for treatment of urinary incontinence) and amphetamine-based products (for treatment of narcolepsy and attention deficit disorder). Their mechanism of action is due to the release of catecholamines (dopamine and norepinephrine) and serotonin. Primary effects associated with toxicity are due to the activation of both α- and β-adrenergic receptors causing serious adverse effects in primarily the central nervous and cardiovascular systems. Clinical symptoms that have been associated with toxicity due to phenylethylamines are hallucination, hyperactivity, agitation, tremors, seizures, sleep disorders, vasoconstriction, arrhythmias, hypertension, platelet aggregation, disseminated intravascular coagulopathy, hyperthermia, tachypnea, hyperinsulinemia, cardiovascular collapse, cerebrovascular hemorrhage/infarction, multiorgan failure, and death.

Ephedrine alkaloids are primarily derived from raw material and extracts from the plants *Ephedra sinica* Stapf, *Ephedra equisetina* Bunge, *Ephedra intermedia* var. *tibetica* Staph, and *Ephedra distachya* L. (the *Ephedras*) (Betz and Tab, 1995; World Health Organization, 1999). Other botanical sources found to contain ephedrine alkaloids include *Sida cordifolia* L. and *Pinellia ternate* (Thunb.) Makino (Ghosal et al., 1975; Oshio et al., 1978). Ma huang, *Ephedra*, Chinese *Ephedra*, and epitonin are several names used for the botanical sources. The plants were mostly found in desert regions of China and Mongolia and were the original source of ephedrine and pseudoephedrine prior to the early 1900s.

Other common names that have been used for the various plants that contain ephedrine alkaloids include sea grape, yellow horse, joint fir, popotillo, and country mallow. Although ephedrine is the primary alkaloid isolated from these plants, other closely related chemicals associated with these botanicals can also be used as biomarkers indicating contamination or adulteration of a dietary supplement. They include norephedrine, methylephedrine, norpseudoephedrine, and methylpseudoephedrine (Chen and Schmidt, 1930; Mahuang, 1987; Karch, 1996; Bruneton, 1995; Cui et al., 1991).

Banned Dietary Ingredients Causing Neurosuppression—**Mitragyna speciosa** *(Kratom)*

Mitragyna speciosa and extracts of its leaf are collectively referred to as kratom or by other synonyms and pseudonyms, including *Nauclea speciosa*, biak-biak, cratom, gratom, kakuam, katawn, kedemba, ketum, krathom, mambog, madat, maeng da, mitragynine extract, *Mitragyna javanica*, red vein/white vein, thang, ithang, and thom. *M. speciosa* or kratom is a botanical ingredient that would generally qualify as a dietary ingredient under section 201(ff) of the Federal FD&C Act (the Act) [U.S.C. 321(ff)]. The leaf and extracts are usually imported with a product code 54 as they are almost always intended as dietary supplements for ingestion. Kratom is obtained overseas from the dried leaves of a tree that grows most commonly in Thailand, Malaysia, Indonesia, Sumatra, Bali, and Vietnam. The whole leaf, crushed leaf, powder forms, and extracts are shipped directly to consumers for oral consumption or indirectly through US distributors in the United States.

Dry leaves of kratom are typically ingested as a tea or the dried leaves are crushed into smaller particulates for ingestion in foods that mask kratom's bitter taste. The powders, dry leaves, and crushed leaf forms can also be ingested by either directly placing them in blank capsules or wrapping them in bathroom tissue squares.

After kratom is placed in bathroom tissue, the tissue is twisted at both ends to contain the material, and then placed into the mouth and swallowed. This process is termed "parachuting." The raw botanical material is also incorporated into finished dietary supplements.

M. speciosa gained use among American consumers in the belief that it will relieve pain, anxiety, and depression and treat the symptoms of opioid addiction and for its euphoric properties for recreational use. It contains two potent alkaloids, mitragynine and 7-hydroxymitragynine. Both are biomarkers for *M. speciosa* toxicity. In low concentrations, these alkaloids produce euphoria, energy, and other stimulating effects. In higher concentrations, they are sedating and possess antinociceptive properties. Based on scientific research and modeling by the FDA, they have concluded that the compounds in Kratom have structural similarities with other opioid analgesics and exert effects on mu, delta (Thongpradichote et al., 1998), and kappa opioid receptors (Boyer et al., 2008; Dale et al., 2012), and they are now classified as opioids (FDA, 2018a).

There are safety concerns regarding the dietary ingredient kratom that support detaining this ingredient and legal enforcement against any dietary supplements known to contain kratom. These products are considered adulterated under section 402(f)(1)(B) of the Act (21 U.S.C. 342(f)(1)(B)). FDA's review of publicly available information regarding kratom concluded that there is not a reasonable assurance that the ingredient does not present a significant or unreasonable risk of injury to consumers.

The scientific literature has disclosed serious concerns regarding the toxicity of kratom in multiple organ systems. Consumption of kratom can lead to respiratory depression, nervousness, agitation, aggression, sleeplessness, hallucinations, delusions, tremors, loss of libido, constipation, skin hyperpigmentation, nausea, vomiting, addiction, and severe withdrawal signs and symptoms after refraining from use.

Detention and refusals of kratom imports do not rely on the provision in section 413 of the Act that deems a dietary supplement containing an NDI to be adulterated unless the manufacturer or distributor has filed an NDI notification with FDA at least 75 days before marketing (21 U.S.C. 350b).[1] Instead, kratom detentions and refusals rely on the adulteration standard in section 402(f)(1)(B) of the Act, which provides that a food is adulterated if it contains an NDI for which there is inadequate information to provide a reasonable assurance that any such ingredient does not present a significant or unreasonable risk of illness or injury. In general, FDA would not be inclined to support the detention of bulk NDIs unless it had identified concrete safety concerns. Reports of addiction clinics dedicated to kratom abuse and emergency room visits for kratom intoxication have increased over the past 6 months. Kratom is an emerging issue of public health that should be taken seriously. From 2010 to 2015, calls to poison control centers have increased 10-fold (FDA, 2017) along with at least 36 deaths associated with its use (FDA, 2017). Based on this information the FDA has exercised its jurisdiction to seize imports and enforce voluntary destruction by companies that had imported kratom for the intent of distribution to consumers (FDA, 2017; 2018b).

Biomarkers Associated with the Presence of Endocrine Disruptors

Dessicated Thyroid Glandulars

Dietary supplements containing animal-derived, thyroid glandular tissue are currently marketed for weight loss, body building, nutritional support, and altering thyroid hormone levels. Marketing such a dietary supplement to explicitly or implicitly treat hypothyroidism, with respect to an individual's thyroid stimulating hormone (TSH) or thyroid hormone level, would be a disease claim as per Title 21 CFR 101.93(g). The public health issue for using such a product, even for weight loss, is the tendency for consumers to incorrectly self-medicate themselves to achieve weight loss by jump-starting a perceived underactive thyroid.

Dietary supplements are considered food under the Federal FD&C Act (the Act). Animal-derived thyroid tissue is declared on many dietary supplement product labels as a dietary ingredient, but it is not. In contrast to FDA-approved thyroid-containing products, which are synthetic drugs approved for defined medical conditions, at consistently defined and approved concentrations and dosages of administration, desiccated thyroid supplements are primarily derived from the thyroid glands of slaughterhouse animals. The thyroid tissue that is ingested, even if it is cooked, or frozen prior to cooking, contains variable levels of the thyroid hormones, levothyroxine (T_4), and levothyronine (T_3). Both forms of thyroid hormone are active after ingestion and absorption into the general circulation with the potential of resulting in multisystemic effects (Table 39.2) that can result in the clinical syndrome known as thyrotoxicosis factitia (Bouchard, 2015).

[1] There is an exception to the notification requirement for dietary supplements that contain only dietary ingredients that have been present in the food supply as articles used for food without chemical alteration. 21 U.S.C. 350b(a)(1).

TABLE 39.2 Clinical Signs and Symptoms of Hyperthyroidism due to Consumption of Thyroid Glandular Ingredients Contained in Dietary Supplements (Bouchard, 2015)

Organ Systems Affected by Exogenous Thyroid Hormone Toxicity	Clinical Sign or Symptom
Ocular	Alterations in vision, photophobia, eye irritation, exophthalmos
Musculoskeletal system	Fatigue, muscle weakness, exertional intolerance, rhabdomyolysis, accelerated osteoporosis
Respiratory system	Dyspnea
Nervous system	Irritability, anxiety, behavioral changes, mental disturbance/psychosis, insomnia, eating disorders, tremor, paralysis, coma
Cardiovascular system	Palpitations, tachydysrhythmias (atrial fibrillation/flutter, extra systoles, high-output congestive heart failure), widened pulse pressure (elevated systolic, decreased diastolic), left ventricular hypertrophy, focal myocarditis, cardiac failure, pulmonary hypertension, angina pectoris, thromboembolism, myocardial infarction, stroke, sudden death
Gastrointestinal system	Gastrointestinal disturbance (nausea, vomiting, diarrhea)
Reproductive system	Impaired fertility and fetal development
Metabolic	Hyperthermia, heat intolerance, increased diaphoresis, weight loss, hypothyroidism
Dermal	Delayed palmar desquamation

Historically, ingestion of animal-derived thyroid gland has been used by for a variety of cosmetic and medical purposes. Ancient Egyptians used them as a beauty aid. Enlarged, goiterous necks were considered fashionable in women (Bouchard, 2015). Up until the 1950s thyroid hormone from animal tissue was used to treat conditions such as hypothyroidism (Bouchard, 2015). Currently, thyroid hormone derived from animal tissue is being marketed as an aid for weight loss and as a stimulant. Its use has been fraught with adverse events. In some cases consumption has resulted in severe, clinical, multisystemic symptoms, and even sudden death (Bouchard, 2015) (Table 39.2).

Thyroid tissue as an ingredient invariably contains an unknown amount and ratio of thyroid hormones that have the potential to result in hyperthyroidism. Thus, in accordance with 9 CFR 310.15(a)[2], livestock thyroid glands and laryngeal muscle tissue shall not be used for human food. Accordingly, thyroid tissue does not appear to fit within any category of "dietary ingredient" in 201(ff)(1). It is not considered a "dietary substance" because it is not part of usual food or drink for humans and even if thyroid tissue could be identified as a "dietary ingredient," it would fall under the category of "adulterated" in accordance with Section 402(f)(1)(A) of the Federal FD&C Act (the Act) [U.S.C. 342(f)(1)(A)] because it presents a significant or unreasonable risk of illness or injury.

Contamination of dietary supplements by thyroid tissue is identified by analysis of product for the presence of the biomarkers thyroxine (T_4) and triiodothyronine (T_3). Clinical symptoms of thyrotoxicosis factitia are not readily identifiable through blood analysis of either thyroxine (T_4) and/or triiodothyronine (T_3) due to the fact that even in the presence of normal or low levels, severe adverse events associated with toxic exposure can occur (Bouchard, 2015).

Diagnosis of exposure of an individual from either acute or chronic consumption of excessive thyroid hormones can be identified primarily through blood analysis for the suppression of circulating TSH. Any argument to support allowing residual or trace amounts of thyroid hormone in a dietary ingredient because it is present in many common foods that we normally eat is not a valid argument. First, consumption of meat containing bovine thyroid tissue has been shown to lead to thyrotoxicosis (Hedberg et al., 1987; Kinney et al., 1988). Second, the mere fact that a substance exists in minuscule quantities in a food humans typically consume does not make that substance a "dietary" one. There are many substances found in typical human foods that people avoid eating, and which could never be termed a "dietary ingredient." Examples of substances people may unintentionally ingest include pesticides and heavy metals. Just because they may be there in trace amounts from environmental contamination does not make them a dietary ingredient.

Supplementing the diet with oral thyroid hormone from a glandular supplement is not needed or recommended in normal, euthyroid individuals. Medical literature contains numerous case reports of adverse events in individuals who ingested herbalal supplements adulterated with thyroid hormone.

[2] Under 21 CFR 316.3(b)(2), active moiety means "the molecule or ion, excluding those appended portions of the molecule that cause the drug to be an ester, salt (including a salt with hydrogen or coordination bonds), or other noncovalent derivative (such as a complex, chelate, or clathrate) of the molecule, responsible for the physiological or pharmacological action of the drug substance."

Many describe life-threatening events. In April 2006, French health authorities reported an incident causing one death and several hospital intensive care unit admissions after taking a "slimming aid" consisting of powdered thyroid extract (Thyroid Extract, 2006). All had symptoms of thyrotoxicosis, including heart palpitations and altered consciousness.

Another case series described five individuals who took Enzo-caps, which claimed to be a natural food product containing papaya, garlic, and kelp, as a weight loss adjunct. All five patients suffered from symptoms suggestive of hyperthyroidism. Each tablet was found to contain up to 112 μg of T_4 and 11.1 μg of T_3. It was later found to also contain a sympathomimetic agent and diuretic. For comparison, the mean dosage of levothyroxine for treatment of diagnosed, hypothyroid, adult patients is 1.7 μg per kg body weight per day, for a typical daily dose of 100–150 μg (Bouchard, 2015; Braunstein et al., 1986). Depending on age, subpopulations can have enhanced sensitivity to thyroid hormone. An average daily dosage for an elderly, adult, hypothyroid patient is 12.5–50 μg/day (Bouchard, 2015).

Thyroid glandular tissue is just one example of the importance of analysis for certain biomarkers as part of a standard process of manufacturing for identification of contamination of raw materials and end-products, by even minute amounts of endocrine disrupting agents. Standard operating procedures for identification of these types of adulterants cannot be overemphasized as companies perform the due diligence necessary to protect consumers in a global environment.

Biomarkers Associated With the Presence of Carcinogens

Aristolochic Acid

Aristolochic acid refers to a family of nitrophenanthrene compounds found concentrated in the roots/rhizomes of plants of the family Aristolochiaceae, and in particular in the genus *Aristolochia*. Lower concentrations are found in the fruit and leaves. There are many other genera in the family Aristolochiaceae that can be found distributed around the world. Many are used as medicinal plants. These include species of plants in the genera *Asarum, Bragantia, Stephania, Clematis, Akebia, Cocculus, Diploclisia, Saussurea, Menispermum*, and *Sinomenium*, as well as Mu tong, Fang ji, Guang fang ji, Fang chi, Kan-Mokutsu (Japanese), and Mokutsu (Japanese).

In the latter part of the 20th century reports started to appear for serious adverse events associated with renal failure and renal fibrosis occurring in individuals ingesting herbal weight loss aids. The herbal product was labeled to have contained *Stephania tetrandra*, a botanical not known to contain aristolochic acid, but was later found to be inadvertently substituted with the botanical *Aristolochia fangchi*. Subsequent reports confirmed that the renal lesions were related to exposure due to ingestion of products containing aristolochic acids (Vanhaelen et al., 1994; Schmeiser et al., 1996; Nortier et al., 2000; Muniz-Martinez et al., 2002).

Similar diagnoses were reported elsewhere in Europe, Asia, and the United States. In July 1999, two cases of nephropathy, associated with the use of Chinese botanical preparations, were reported from the United Kingdom. Both of these patients had ingested botanical preparations for the treatment of "eczema." Renal biopsies showed a pattern of extensive tubular loss, most prominent in the outer cortex, and severe interstitial fibrosis. These pathological features are typical of what has now come to be called "Chinese herb nephropathy" due to the toxicity associated with ingestion of aristolochic acid.

In 2001, after reports of rapidly developing renal failure and urothelial carcinoma associated with the use of herbal preparations containing aristolochic acid containing dietary ingredients, manufacturers and distributors of supplements were advised to remove these products from the market. Public health statements were released, advising consumers not to consume these supplements (Schwetz, 2001). An FDA import alert (IA #54-10) called for the detention of any botanical, drug, or other products found to contain the more than the 69 types of aristolochic acids; herbs known to contain aristolochic acids; and/or herbal preparations that had been found to have been replaced by herbs containing aristolochic acids. At that time it was clear that consumption of herbal products containing aristolochic acids, both intentionally and through apparent misbranding, was associated with a rapid onset of renal interstitial fibrosis, renal failure, and an increased incidence of urothelial carcinomas. The rapid onset, severity, and irreversibility of the damage suggested that detection of any amount of aristolochic acid should be viewed as a potential health risk.

The carcinogenic potential of aristolochic acid started to appear in the 1980s. In addition to renal damage, safety studies reported an increased incidence in pathologic lesions indicating aristolochic acid–induced carcinogenicity in rodents orally administered aristolochic acid (Mengs et al., 1982). The rodents developed multiple cancers that included lymphoma, as well as cancers in the kidney, bladder, stomach, and lung. Short-term studies resulted in tumors in the urinary tract in addition to the other renal pathological changes indicative of Chinese herbal nephropathy (Mengs et al., 1982; Mengs, 1987; Mengs and Stotzem, 1993).

The precipitating event for aristolochic acid was an outbreak in Belgium. In 1992, a Belgian cohort of 70 patients, known to have consumed a slimming regimen containing a powdered extract of several Chinese herbs, was found to have interstitial renal fibrosis, with 30 individuals progressing to end-stage renal disease (Vanherweghem et al., 1993; Cosyns et al., 1994; Depierreux et al., 1994; Vanherweghem, 1998). The diagnosis was made based on an unusual and distinctive pattern of renal nephropathy and the finding of biomarkers for aristolochic acid–DNA adducts in the urothelium indicating exposure and potential toxicity. The renal fibrosis occurred 12–24 months after the initial injury with an increased incidence of urothelial cancer being associated with this pattern of renal nephropathy (Hung, 2015; Debelle et al., 2008).

The International Agency for Research on Cancer (IARC), an arm of the World Health Organization, would later conclude that there was sufficient evidence to determine that aristolochic acid–containing herbs are carcinogenic in humans, thus classifying them as Group I carcinogens (IARC, 2002).

To this day, reports from around the world, and especially in east Asia, continue to link rapid onset of renal failure and/or renal cancer with consumption of teas brewed from the ground root of these botanicals (Gillerot et al., 2001; Tanaka et al., 2001; Krumme et al., 2001; Cronin et al., 2002; Yang et al., 2003; Lee et al., 2004).

Pyrrolizidine Alkaloids—*Symphytum* Spp. (Common Name, Comfrey)

In June 2001, FDA published an Advisory to dietary supplement manufacturers to remove comfrey as an ingredient in dietary supplements. The Advisory alerted the supplement industry to the available scientific information that firmly establishes that dietary supplements that contain comfrey or any other source of PAs are adulterated under the Act. Also, the Agency stressed that manufacturers need to identify and report adverse events associated with any product that contains an ingredient that has the potential to contain PAs. The Agency cited serious adverse health effects and opined on the presence of PAs as potent hepatotoxins.

Comfrey, like a number of other plants (e.g., *Senecio* species), contain PAs. The toxicity of PAs to humans is well documented (Huxtable and Cheeke, 1989; Winship, 1991; McDermott and Ridker, 1990; Burrows and Tyrl, 2013). FDA's position with regard to comfrey is that the PAs present in comfrey are toxic chemicals that can lead to serious adverse health effects if taken into the body. It is clear that oral exposure is potentially hazardous. It is equally clear that PAs are harmful when allowed to enter the body through broken skin, or other nonoral routes, such as suppositories. The Federal Trade Commission has limited them to external use only. Accordingly, FDA believes that external use of comfrey products, containing PAs, present minimal risks to consumers when such products are not used as suppositories and are not applied to broken skin.

Hepatic veno-occlusive disease and neoplasia as well as pulmonary hypertension are the major documented forms of injury to humans (Burrows and Tyrl, 2013) from chronic or excessive intake of PA-containing herbals. Laboratory animal studies, epidemiological studies, and veterinary case reports suggest that the toxic effects are much broader. Animal exposure has resulted in pulmonary, kidney, and gastrointestinal pathologies including cancer (Burrows and Tyrl, 2013).

Four countries, the United Kingdom, Australia, Canada, and Germany, have restricted the availability of products containing comfrey, and other countries permit use of comfrey only under a physician's prescription.

There is variability in the levels of PAs in various species of plants, thus, the concerns FDA has about a comfrey-containing dietary supplement or ingredient depends on the exact species identified. Identification of this biomarker is a crucial component for the development, manufacture, and distribution of a dietary supplement derived from plants that have the potential to containing PAs as part of the process for quality control. This is especially important in light of the potential for chronic exposure to result in hepatic lesions, including cancer, if due diligence is not adhered to by the manufacturer prior to marketing.

The FDA ruling banned internal use of *Symphytum officinale* L. (common comfrey), *Symphytum asperum* Lepech (rough or prickly comfrey), and *Symphytum xuplandicum* Nyman (Russian comfrey), as well as any other plant/substance containing PAs. While FDA did not examine the safety of other comfrey species such as *Symphytum tuberosum* L. (tuberous comfrey), which is suggested to contain negligible amounts of PAs, FDA and potential manufacturers of products would rely on the presence of the PA to determine whether this species is to be permitted to be used as a dietary ingredient. Only after such sampling and analysis of this biomarker of toxicity, would they be in a position to provide an opinion on whether its use in dietary supplements would not present a significant or unreasonable risk of illness or injury under the conditions of use recommended on the label. Because the burden and responsibility for assuring that such a product is not adulterated under the Act lies with the manufacturer, and with the associated costs involved during the product life cycle, alternative species that are known to not have an association with the presence of pyrrolizidine alkaloids should take precedence rather than risking the potential for future contamination of product.

Although FDA continues to voice its concern about the safety of dietary supplements containing comfrey, this does not mean that FDA prohibits the marketing of dietary supplements that contain comfrey. Although dietary supplements do not require premarket approval or approval from FDA before marketing and distribution, the manufacturer is responsible for determining that its products are safe under the Act. However, the Agency is unaware of conclusive scientific data that would establish that a dietary supplement containing comfrey would be safe. Therefore, for a manufacturer interested in marketing a comfrey-containing dietary supplement, it would be required to submit information in the form of a NDI notification that forms the basis by which it establishes that such a product is safe within the meaning of the Act. A manufacturer would then have to demonstrate that the comfrey-containing ingredient did not contain the toxic PA biomarker as part of the NDI notification, as well as chemistry and additional safety data to market such a product as a dietary supplement for its intended use.

BIOMARKERS OF TOXICITY: NEW DIETARY INGREDIENTS

Premarket Notification of a New Dietary Ingredient

The Act requires a premarket safety notification for NDIs introduced into commerce following the passage of the DSHEA of 1994. Under section 413(c) of the Act (21 U.S.C. 350b(c)), an "NDI" is a dietary ingredient that was not marketed in the United States before October 15, 1994. A dietary supplement that contains an NDI is deemed to be adulterated unless one of the following two conditions are met: (1) The dietary supplement must contain only dietary ingredients that have been present in the food supply as an article used for food in a form in which the food has not been chemically altered; or (2) there must be a history of use or other evidence of safety establishing that the dietary ingredient when used under the conditions recommended or suggested in the labeling of the dietary supplement will reasonably be expected to be safe, and the manufacturer or distributor of the dietary supplement containing the NDI must provide FDA with information, including any citation to published articles, which forms the basis on which the manufacturer or distributor has concluded that a dietary supplement containing such dietary ingredient will reasonably be expected to be safe.

FDA has established requirements for premarket notification of NDIs in 21 CFR § 190.6, based on authority granted in sections 201(ff), 301, 402, 413, and 701 of the Federal FD&C Act (21 U.S.C. 321(ff), 331, 342, 350b, 371). 21 CFR 190.6(a) which states that at least 75 days before introducing or delivering for introduction into interstate commerce a dietary supplement that contains an NDI that has not been present in the food supply as an article used for food in a form in which the food has not been chemically altered, the manufacturer or distributor of that supplement, or of the NDI, shall submit to the Office of Nutritional Products, Labeling and Dietary Supplements, Center for Food Safety and Applied Nutrition information including any citation to published articles that is the basis on which the manufacturer or distributor has concluded that a dietary supplement containing such dietary ingredient will reasonably be expected to be safe. That notification should include pertinent information that satisfies the regulatory requirements established under 21 CFR 190.6, documentation demonstrating clear identification of the test article, history of use, and weight of evidence for a reasonable expectation of safety.

If a dietary supplement containing an NDI is subject to the notification requirement and this requirement is not met, or if there is no history of use or other evidence of safety establishing a reasonable expectation of safety, the dietary supplement is deemed to be adulterated under section 402(f)(1)(B) of the act because there is inadequate information to provide reasonable assurance that the product does not present a significant or unreasonable risk of illness or injury.

Within this context, biomarkers are used at many critical points of the product life cycle to establish safety. This includes but is not limited to identification of possible contamination or adulteration of product at any point in the process for quality control; physiological effects of use of product through the evaluation of animal and human research; a comprehensive evaluation of the peer-reviewed literature concerning any evidence to substantiate use and safety of product; the establishment of labeling including cautions of use for specific subpopulations of individuals; epidemiological evaluation of use and historical and ancillary use of product.

Premarket Notification for New Dietary Supplement as Enhancers of Metabolism, Energy, and Weight Loss

Usnic Acid

On November 20, 2001, US FDA warned consumers to stop using LipoKinetix, a dietary supplement marketed for weight loss, because it was associated with a number of serious adverse event reports involving hepatotoxicity (Favreau et al., 2002) including acute hepatitis, irreversible liver injury, liver failure, and eventual transplant. US FDA reported on six persons who developed acute hepatitis and/or liver failure while using LipoKinetix. The injuries reported to FDA occurred in persons between 20 and 32 years of age, and no other cause for liver disease was identified in the cohort. In all cases, no preexisting medical condition that would predispose the consumer to liver injury was identified. Liver injury occurred fairly rapidly, between 2 weeks and 3 months, after starting LipoKinetix. Although the product contained multiple dietary ingredients such as norephedrine (PPA), caffeine, yohimbine, diiodothyronine and sodium usniate, usnic acid was the dietary ingredient suspected to have led to hepatotoxicity in the adverse event reports.

Usnic acid is typically derived from botanicals of *Usnea* spp. (Ingólfsdóttir, 2002), but it is also present in other genera of lichens, including *Alectoria*, *Cladonia*, *Evernia*, *Lecanora*, and *Ramalina*. Usnic acid is a complex dibenzofuran derivative, produced naturally by certain lichen species. In recent years, usnic acid and its salt form, sodium usniate, have been marketed in the United States as an ingredient in dietary supplement products, mostly with claims and marketing to promote weight reduction, enhancement of metabolism, and inhibition of bacteria.

Although lichens containing *Usnea* have been formulated into topical products and used as traditional medicines in other countries, FDA is unaware of any evidence that usnic acid was marketed before October 15, 1994. Similar to many NDIs on the market, FDA has not received any notification for usnic acid or usniate as an NDI for which a premarket notification pursuant to 21 U.S.C. 50b(a)(2) is a requirement. The NDI notification process, a premarket gate to ensure reasonable expectation of safety, is designed to pick up toxicological signals in the data before the dietary ingredient contained in the dietary supplement reaches the market. The notification process can only work in a premarket capacity if notifications for NDIs are submitted to FDA.

Lichens have evolved an innate capacity to survive extreme environmental conditions. They excrete bioweathering and bioactive secondary metabolites (e.g., usnic acid), which provide chemical protection from invading biologicals such as viruses, bacteria, protozoa, competing fungi, algae, plants, and animal predators. Although lichens produce a vast array of secondary metabolites, representing diverse classes of chemical compounds (e.g., lactones, aromatic compounds, quinines, terpenes, diphenyl esters, dibenzofuranes, and aliphatic acids) (Fiedler et al., 1986; Huneck and Yoshimura, 1996; Huneck, 1999), the most characterized and studied is the polycyclic usnic acid (Correche et al., 1998). *Usnea* content varies depending on the lichen species and region. *Alectoria* spp. are known to contain up to 6% usnic acid, whereas the thallus of *Usnea laevis* from the Venezuelan Andes contains approximately half that amount (Marcano et al., 1999). Although *Usnea* spp. synthesize and excrete usnic acid in response to a toxic environment, usnic acid is a useful biomarker for assessing pollution as well as toxicity in dietary supplements.

Although adverse events suggest that usnic acid is the culprit for liver toxicity, we do not have a definitive mechanism for how this occurs. Subchronic and chronic toxicology studies in rodents and clinical studies in humans are absent. What is known regarding usnic acid toxicity is related to its apparent mechanism of action to uncouple oxidative phosphorylation in mitochondria (Johnson et al., 1950; Abo-Khatwa et al., 1996; Cardarelli et al., 1997). This mechanism could account for its ability to kill microorganisms (Lauterwein et al., 1995), accelerate metabolism, cause weight loss, and induce liver toxicity (Krähenbühl, 2001; Sonko et al., 2011; Yellapu et al., 2011; Sahu et al., 2012; Liu et al., 2012; Moreira et al., 2013). In vitro analysis to screen for mitochondrial damage should be a consideration for research and development of products with similar claims. Biomarkers indicative of mitochondrial toxicity should be taken under consideration as a required step in the premarket analysis for justification of safety of any weight loss or energy products considering the possible seriousness of adverse events that might result due to "enhanced" metabolism.

BIOMARKERS OF TOXICITY: NEW DIETARY INGREDIENTS

Active Moiety of a New Dietary Ingredient

Red Yeast Rice—Monacolin K and Statin Activity

Another issue of concern is the physiological effect and regulatory status of the relevant article or active moiety[2] in any marketed dietary supplement or NDI. Some NDI notifications received by the FDA concern ingredients that do not meet the statutory definition of a dietary ingredient and are therefore excluded as dietary ingredients under the U.S.C. 321(ff)(3)(B)[3], section 201(ff)(3)(B) of the Federal FD&C Act (The Act) (21 U.S.C. § 321(ff)(3)(B)). Such dietary supplements may not include articles approved as a new drug or authorized for investigation as a new drug under section 505 of the Act, unless the article was marketed as a dietary supplement or food before its approval as a drug.

The case law clarifying DSHEA regarding this issue involved the dietary supplement product called

[3] The term "dietary supplement" is defined in 21 U.S.C. 321(ff). Under 21 U.S.C. § 321(ff)(3)(B), dietary supplements may not include articles approved as a new drug under 21 U.S.C. § 355 (section 505 of the Act), unless the article was marketed as a dietary supplement or food before its approval as a drug.

Cholestin (red yeast rice) marketed by Pharmanex (35F. Supp. 1341, 2001). Cholestin was a "traditional" Asian product known alternatively as "red yeast rice," "Hong Qu," or "Red Koji." Red yeast rice has been used both in traditional Asian cuisine and medicine. Products that are marketed as dietary supplements that contain red yeast rice, like all dietary supplements, must meet the requirements set forth in the Federal FD&C Act, as amended by the DSHEA of 1994. Red yeast rice (*Monascus purpureus*) contain monacolins that are fungal, secondary metabolites capable of inhibiting 3-hydroxy-3-methylglutaryl-CoA (HMG CoA) reductase, which is involved in the synthetic pathway for cholesterol production. They are more commonly known as Statins. By definition, the presence of biomarkers for monacolins in red yeast rice would not permit its use as a dietary ingredient in a dietary supplement. The term "dietary supplement" as defined in 21 U.S.C. 321(ff) contains a number of exclusions. For example, under section 321(ff)(3)(B)(i), an article that is approved as a new drug under 21 U.S.C. 355 is excluded from being a dietary supplement unless it was marketed as a dietary supplement or as a food before such approval. It is this provision of the FD&C Act that bears directly on the legal status of Pharmanex's product that was the subject of an FDA enforcement action on Cholestin and certain other products that contain red yeast rice, specifically sold as dietary supplements.

In 1998, FDA issued an administrative proceeding stating that Pharmanex's product, Cholestin,[4] which was promoted as a dietary supplement intended to reduce cholesterol levels, was not a dietary supplement, but instead is considered an unapproved drug under the FD&C Act (Pharmanex, Inc, 1998). FDA based its decision, in part, on the fact that the red yeast rice in Cholestin was not simply red yeast rice that traditionally had been used as food. Instead, the agency concluded that Pharmanex had taken several actions in the marketing and manufacturing of Cholestin such that the relevant ingredient in the product, Monacolin K, otherwise known as lovastatin, an approved drug in the product Mevacor, was the basis for their product claims. This prohibited Cholestin from being marketed as a dietary supplement under the exclusion clause in section 321(ff)(3)(B)(i). Traditional red yeast rice does not contain lovastatin in measurable concentrations. The fact that Pharmanex took steps to artificially induce and enhance lovastatin production in its red yeast rice product and promoted its product to reduce blood cholesterol levels caused the product to be a drug under the FD&C Act.

The FDA's final determination was that lovastatin was not marketed as a dietary supplement prior to its approval as a new drug and Pharmanex failed to show that lovastatin was previously marketed as a dietary supplement, food, or component of a food; therefore, Cholestin did not qualify as a dietary supplement.

All red yeast rice products manufactured with the purpose of increasing levels of monacolins are considered a threat to public health. Their chronic use comes with the potential for adverse events such as myopathies as well as the potential to cause damage to the kidneys with chronic usage. There is also added risk for individuals who are taking other statin medications or concurrent dietary supplements such as niacin that can potentiate the risk and incidence of adverse reactions.

On February 16, 1999, the United States District Court for the District of Utah set aside the FDA's May 20, 1998 administrative determination that Cholestin is a drug. The United States Court of Appeals for the 10th Circuit reversed the District Court's decision on July 21, 2000 and remanded the case back to the District Court for consideration of record-based issues not reached by the lower court in its original decision. On March 30, 2001, the United States District Court for the District of Utah issued a Memorandum Decision and Order on the remaining record-based issues. The District Court affirmed the FDA's administrative decision that Cholestin is a drug, not a dietary supplement. Taken together, the courts' decisions in the Pharmanex litigation means that red yeast rice products containing lovastatin are unapproved new drugs in violation of the FD&C Act and may not be marketed as dietary supplements (Pharmanex, Inc. v. Shalala, 2001).

The decision in the Pharmanex case does not, however, prohibit the marketing of all red yeast rice-containing products as dietary supplements. The decision only limits the marketing of products as dietary supplements if they contain substances that are excluded from the dietary supplement definition. The agency has not objected to the marketing of dietary supplements containing red yeast rice if they do not contain lovastatin or other substances that are approved drugs. The Congressional intent of the exclusion clause that is the basis of the agency's decision on Cholestin was to protect the research development incentives for new drugs to treat serious diseases.

There is no "action level" for lovastatin, or any other substance excluded under section 201(ff)(3)(B), which

[4] Cholestin consists of the yeast *Monascus purpureus* when fermented on premium rice powder. The fermentation of the rice with this yeast, under certain conditions, produces a product that contains lovastatin, the active ingredient in the prescription cholesterol-lowering drug Mevacor.

would trigger action by FDA. Rather, as was the case with Cholestin, FDA would look at the following:

- level of a substance in a proposed dietary ingredient
- consider the manufacturing, processing, and composition and determine how it compares to traditionally prepared dietary ingredients
- provision for evidence of intended use or claims for the product to determine whether a given product violates the Act

BIOMARKERS OF TOXICITY: NEW DIETARY INGREDIENTS

Safety Assessment of Multiple Dietary Ingredients in a Dietary Supplement

As with other dietary supplements to be marketed, those containing multiple dietary ingredients, must substantiate through a history of use or other evidence of safety, that under the conditions recommended or suggested in the labeling of the dietary supplement, the combination of dietary ingredients will have a reasonable expectation of safety. As with other new ingredients, at least 75 days before being introduced or delivered for introduction into interstate commerce, the manufacturer or distributor of the dietary ingredient or dietary supplement must provide the Secretary (and by delegation, FDA) with information, including any citation to published articles, which forms the basis on which the manufacturer or distributor has concluded that a dietary supplement containing such dietary ingredients will not pose an unreasonable risk for illness or injury to the general public. This is especially true when the ingredients achieve the same effect through either different or similar mechanisms of action.

New Dietary Ingredients (Supplementation with Multiple, Combinations of Dietary Ingredients) Rauwolfia and Yohimbe Bark Extract

Pharmaceutical and dietary ingredients are capable of chemical and physiological interactions that have the potential to increase the incidence of adverse events in individuals. Interactions can lead to antagonism or potentiation of drug activity and/or exacerbate underlying medical conditions that may be present with or without other diagnosed systemic disease.

Because of an assumption of safe use of dietary supplements in the general, healthy population, a lack of knowledge concerning the interactions of many dietary ingredients and the belief by many consumers that because many of the products are "natural" or of plant origin, that they can be assumed to be safe, self-medication of multiple products is one of the most common precipitating circumstances surrounding adverse events and product complaints.

Drug interactions can lead to changes in the activity of one or more dietary ingredients with concurrent use or a similar situation can occur when multiple dietary ingredients are placed in the same dietary supplement. This is significant if the dietary ingredients within the dietary supplements have similar pharmacological effects. For example, both *Rauwolfia*, containing the active ingredient, reserpine, and yohimbe bark extract, containing the α-2 adrenergic antagonist, yohimbine, can lower blood pressure. When combined in a dietary supplement, there exists the potential for an additive or synergistic response. This may go beyond permitted structure–function claims on dietary supplement products for maintaining blood pressure in the normal range. Combination products have the potential to precipitate serious adverse events such as hypotension, dizziness, reflex tachycardia, and heart palpitations, especially in individuals who may be taking other medications with a similar mechanism of action or who have other systemic, underlying health conditions.

Reserpine, an alkaloid isolated from plants in the genera *Rauwolfia*, was one of the first drugs developed for treatment of high blood pressure. Reserpine is an FDA-approved prescription medication for the treatment of hypertension. The prescribed, adult, label dosage for reserpine, for use as an antihypertensive agent, is 0.1–0.25 mg per individual per day.[5] According to a recent study, reserpine can be found in *Rauwolfia serpentina* at a concentration of 0.1442% (Kumar et al., 2010). A dietary supplement product, labeled to contain 50 mg of *R. serpentina* in each serving (1 capsule) with directions to take two capsules per day, could potentially contain upward of 0.0721 mg reserpine (0.1442% of 50 mg) per capsule and lead to a total daily exposure level of 0.144 mg reserpine. This is in the range for an amount that would be indicated in a prescription medication for an individual with hypertension.

As an FDA-approved prescription medication, reserpine contains potential risks and toxicological effects that restricts its use to uncomplicated hypertensive patients. Reserpine is classified as a pregnancy category C. When reserpine is administered parenterally it has been shown to be teratogenic in rats and to have an embryocidal effect in guinea pigs. It is unclear how reserpine affects the dietary supplement consumer who is normotensive. Furthermore, the US National

[5] Reserpine [package insert]. Princeton (NJ): Sandoz, 2011.

Toxicology Program has considered reserpine to be a "probable cancer-causing substance" and "reasonably anticipated to be a human carcinogen" based on sufficient evidence of carcinogenicity from studies in experimental animals (NTP, 2011). These risks are unreasonable for a dietary supplement that must have a reasonable expectation of safe use, in a healthy adult, in the general population.

The package insert states that concomitant use of reserpine with other antihypertensive agents necessitates careful titration of dosage with each agent. This is because hypotension, hypothermia, central respiratory depression, and bradycardia may develop in cases of overdose of reserpine or synergistic action of reserpine with other antihypertensives. Thus, the combination of reserpine with any other antihypertensive botanical such as yohimbe bark extract can precipitate an adverse event. Their combination would necessitate a separate NDI notification to address this impact in the normotensive consumer. In contrast, the NDI *Rauwolfia vomitoria* (NDI 013) contains the botanical with the reserpine component removed. To receive an acknowledgment from FDA at this time, all *Rauwolfia* spp.—derived botanicals, for use in dietary supplements, must demonstrate that reserpine has been removed from the end-product prior to distribution to consumers.

This example illustrates the risks that can result when dietary ingredients with similar pharmacological actions are combined in the same product without submitting a premarket NDI; without appropriate warnings to the consumer of toxicological effects that can potentially develop and without premarket analysis for biomarkers of physiologically active ingredients that should not be present in the final product for distribution.

New Dietary Ingredients (Supplementation with Multiple, Combinations of Dietary Ingredients)—Galantamine and Huperzine A (Cholinesterase Inhibitors)

In a race to market, galantamine and huperzine A, both reversible cholinesterase inhibitors, were acknowledged dietary ingredients through NDI notification before they were investigational new drugs. Therefore, they are permitted to be used in dietary supplements under the conditions of use and consumption level as described in the "acknowledged" NDI notifications. FDA has concerns with increases in serving levels for galantamine and huperzine A whether they are used together or alone. Going forward, a manufacturer or distributor would be obliged to submit an NDI with safety evidence if they were to market the same ingredient (manufactured using the same process); if they wanted to market new products with higher serving levels of ingredients; if they wanted to produce dietary supplement combinations with other dietary ingredients (i.e., huperzine A combined with galantamine); or if they developed new technologies to manufacture the dietary ingredient. In other words, any change in formula and/or chemical composition would elicit a need to submit an NDI prior to distribution of a new product.

The chemical structures for galantamine and huperzine A are provided in Fig. 39.1. Galantamine and huperzine A are reversible inhibitors of the enzyme acetylcholinesterase (AChE) (Taylor et al., 1996). Because their acute toxicity is due to the inhibition of AChE, measurement of AChE activity in red blood cells (whole blood analysis) is a critical biomarker for monitoring the toxicity of these dietary ingredients to adequately ensure a reasonable expectation of safety under the ordinary conditions of use for the product. AChE serves to limit the duration of the neurotransmitter acetylcholine (ACh) at muscarinic and nicotinic cholinergic receptors and therefore prevents its accumulation at synaptic junctions (Hilmas et al., 2009; Williams and Hilmas, 2010). Exposure to AChE inhibitors, both reversible and irreversible, results in symptoms indicative of widespread overstimulation. Fig. 39.2 illustrates the target organs, symptoms, and clinical signs that are associated with overstimulation. These symptoms include bronchoconstriction, increases in tracheobronchial secretion, lacrimation, increased urination, increased gastrointestinal motility, diarrhea, emesis, muscle weakness, diaphoresis, decreased heart rate, and central nervous system signs such as dizziness, blurred vision, mental confusion, respiratory depression, tremors, seizures, paralysis, and coma.

Galantamine and huperzine A have a very rapid on-rate and slow off-rate or release. Their kinetic profiles indicate predicted tight interactions with AChE in docking models. Best-fit docking models for both compounds involve different functional groups. Fig. 39.3 illustrates the three-dimensional docking of galantamine to both human and *Torpedo californica* forms of AChE. Fig. 39.4 illustrates three-dimensional docking of huperzine A to human AChE. Thus, they have been used in as prophylactic and postexposure therapeutic agents in preclinical studies to protect rodents against muscular paralysis and seizures induced by nerve agents.

Safety evidence in an NDI notification would have to demonstrate that AChE inhibition was not cumulative as a result of chronic (daily) consumption. Red blood cell, muscle, and brain cholinesterase activity have all been used, primarily in in vivo and in vitro studies, to measure and evaluate peripheral and central neurotoxicity as a result of galantamine and huperzine A administration. These activities should correlate to the whole

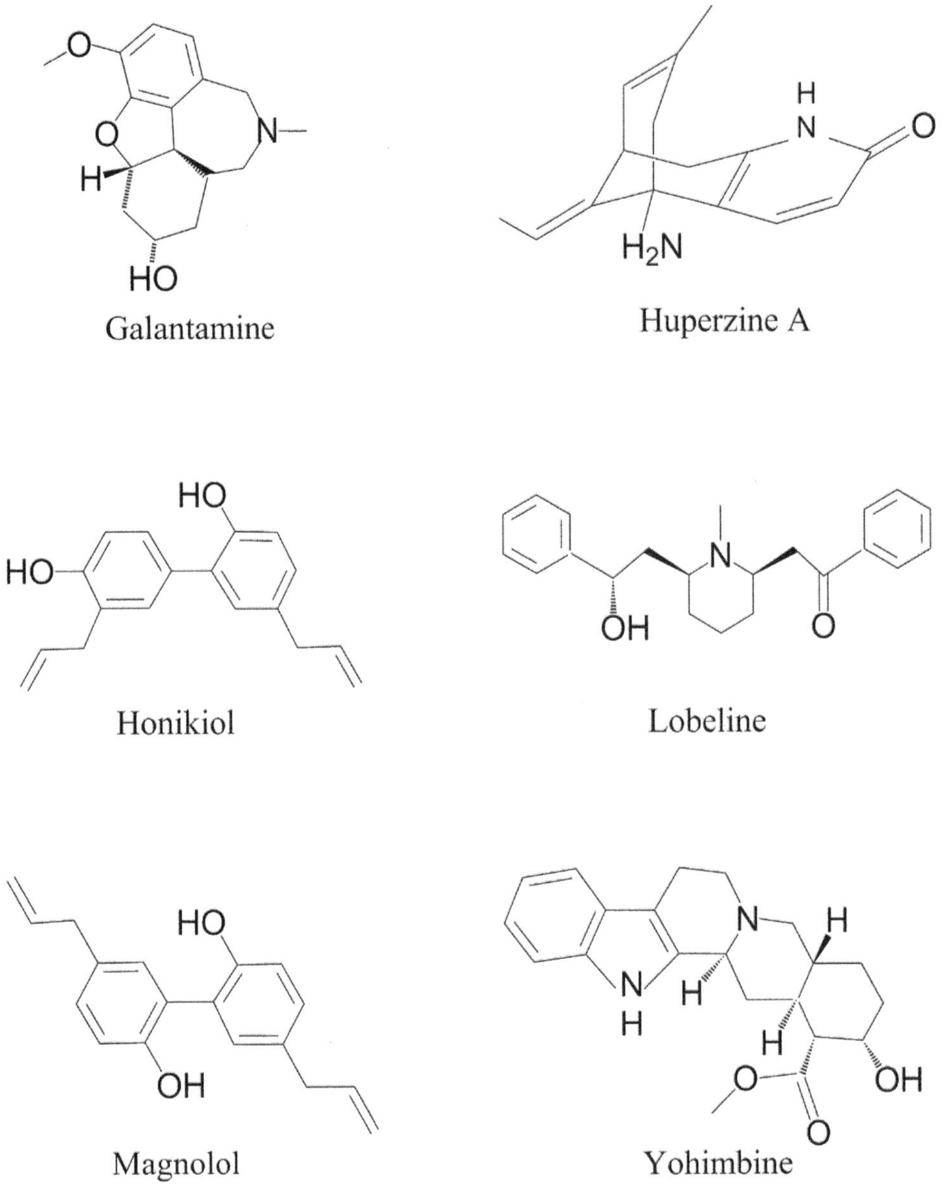

Galantamine

Huperzine A

Honikiol

Lobeline

Magnolol

Yohimbine

FIGURE 39.1 Chemical structures of galantamine, huperzine A, honokiol, lobeline, magnolol, and yohimbine. *The authors would like to acknowledge Dr. Mariton Dos Santos for providing this illustration.*

blood or plasma levels of galantamine or huperzine A in the animal model (Steiner et al., 2012).

There are other botanicals that contain reversible cholinesterase inhibitors. An example is honokiol and magnolol, present in *Magnolia* spp. (Fig. 39.1). The combination of an extract of *Magnolia* with galantamine or huperzine A may pose significant synergistic activity. Another is Lobeline (Fig. 39.1), a nicotinic agonist that may also act synergistically to produce a cholinergic crisis if combined with galantamine, huperzine A, or any other ingredient with AChE-inhibiting activity.

Appropriate safety studies should measure and evaluate AChE activity as a biomarker of toxicity when these ingredients or others are suspected of combined nicotinic and muscarinic AChE activity prior to their use in a dietary supplement product.

CONCLUDING REMARKS AND FUTURE DIRECTIONS

Dietary supplements are becoming more ubiquitous in the market and are becoming accepted in the general public as important components of many foods and beverages. Some may call them functional foods, others may call them nutraceuticals, dietary supplements, botanicals, or natural products to support health. With the

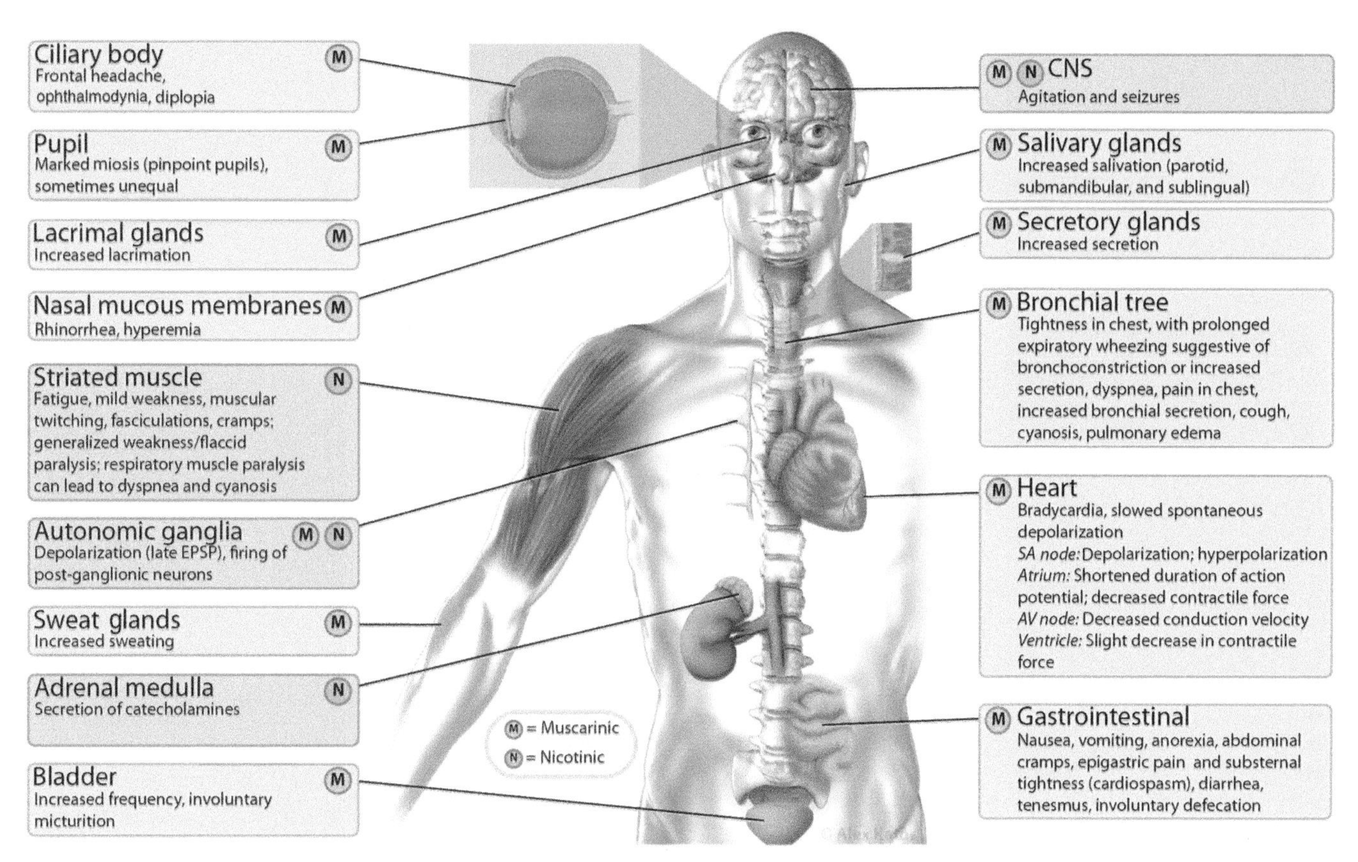

FIGURE 39.2 Overstimulation of muscarinic and nicotinic cholinergic systems due to inhibition of acetylcholinesterase (AChE) inhibition. Some dietary ingredients inhibit AChE, resulting in overstimulation by the neurotransmitter ACh at both muscarinic and nicotinic receptors located in various organ systems throughout the body. This figure illustrates the various physiological markers of toxicity which result from cholinergic activation in the presence of AChE inhibition. *The authors would like to acknowledge Alexandre Katos for providing this illustration.*

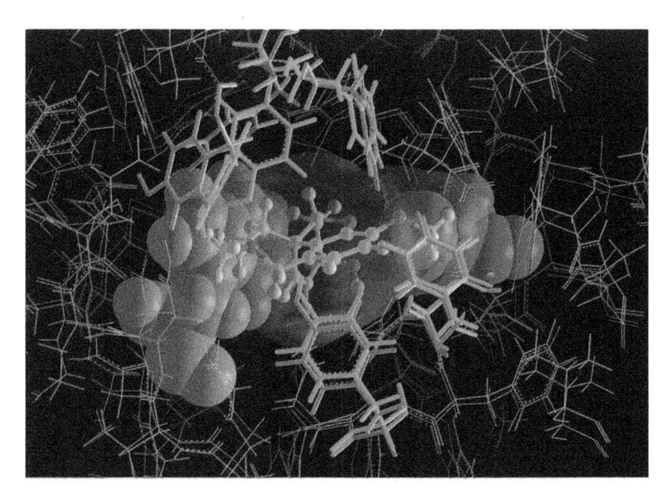

FIGURE 39.3 Three-dimensional docking of galantamine into AChE from human and *Torpedo californica*. The most appropriate portions of human AChE are displayed. The ball display on the left-hand side is the active site tryptophan, and the balls of AChE on the right-hand side represent the active site serine. Galantamine, represented in ball-and-stick configuration, is shown by atom type and has a transparent solvent accessible surface to demonstrate that it is filling a pocket. This docking was performed using Insight II software by aligning the backbones of the two AChEs.

increasing costs of health care, many people are looking for ways to lead healthier lifestyles, and whether good or bad, they are listening to broad avenues of media and marketing about the pros and cons of these products. Education of the public is important but at the same time the majority of individuals do not have the background in science and nutrition to adequately understand the complexity of what they are ingesting. Thus, it is vitally important for the medical community, industry, government regulators, academia, and media outlets to work together to make sure that only high-quality products are manufactured, marketed, and distributed properly in the general marketplace.

This chapter uses examples of dietary supplements and their ingredients that have been or are currently still available to consumers in a review for the use of biomarkers as they apply to various stages of the life cycle of these products.

An additional goal of this chapter is to remedy the misconception that dietary supplements are not regulated. They are. It should be stated that they are regulated not as a pharmaceutical where a lack of efficacy is considered an adverse event for its intended use but

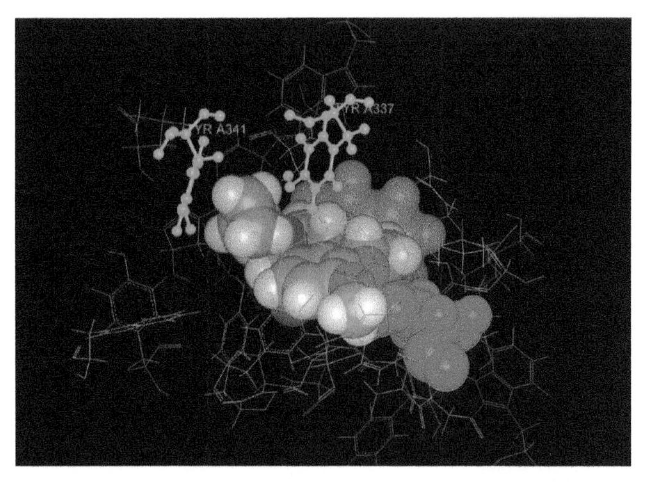

FIGURE 39.4 Three-dimensional docking of huperzine A, a known anticholinesterase alkaloid, into human AChE. The three-dimensional rendering shows a best-fit model for docking of huperzine A functional groups with the two tyrosine functional groups (Tyr 337 and Tyr 341) on human AChE. Although these tyrosines are normally not involved in GAL docking, they may play a role in huperzine A docking. For reference, the active site tryptophan of AChE is oriented in the background; the active site serine is on the right-hand side. Huperzine A is colored by atom using a space-filling ball model configuration to show it is filling a pocket, rather than the ball-and-stick configuration with transparent solvent accessible surface as shown in the previous figure. The two tyrosines are in a ball-and-stick model configuration. This docking was performed using Insight II software.

are more closely aligned to the guidance's found for historical and NDIs that may be found in food products for human consumption.

Information has also been provided for how biomarkers inform worldwide regulatory decisions; the development of risk assessments for the protection of public health; and for industry to be able to manufacture and distribute dietary supplements with an appropriate weight of evidence to support an expectation of tolerance and safety for the general population of consumers for their proper, intended use.

References

Abo-Khatwa, A.N., al-Robai, A.A., al-Jawhari, D.A., 1996. Lichen acids as uncouplers of oxidative phosphorylation of mouse-liver mitochondria. Nat. Toxins 4, 96–102.

Anderson, N., Meier, T., Borlak, C., 2009. Toxicogenomics applied to cultures of human hepatocytes enabled an identification of novel *Petasites hybridus* extracts for the treatment of migraine with improved hepatobiliary safety. Toxicol. Sci. 112 (2), 507–520.

Berggren, E., Amcoff, P., Benigni, R., et al., 2015. Chemical safety assessment using read-across: assessing the use of novel testing methods to strengthen the evidence base for decision making. Environ. Health Perspect. 123 (12), 1232–1240.

Betz, J.M., Tab, F., 1995. Review of plant chemistry: Alkaloids of Ma Huang (Ephedra spp.), Food and Drug Administration, Briefing materials for Food Advisory Committee Special Working Group on Foods Containing Ephedrine Alkaloids. Center for Food Safety and Applied Nutrition, pp. 1–14.

Boobis, A.R., 2007. Risk Assessment of dietary supplements. Novartis Found. Symp. 282, 3–28.

Bouchard, N.C., 2015. Thyroid and anti-thyroid medications. In: Hoffman, R.S., Howland, M.A., Lewin, N.A., Nelson, L.S., Goldfrank, L.R. (Eds.), Goldfrank's Toxicologic Emergencies, tenth ed. McGraw-Hill Education, New York, New York, pp. 763–771.

Boyer, E.W., Babu, K.M., Adkins, J.E., et al., 2008. Self-treatment of opioid withdrawal using kratom (*Mitragyna speciosa* korth). Addiction 103, 1048–1050.

Braunstein, G.D., Koblin, R., Sugawara, M., et al., 1986. Unintentional thyrotoxicosis factitia due to a diet pill. West. J. Med. 145, 388–391.

Bruneton, J., 1995. Phenethylamines. In: Bruneton, J. (Ed.), Pharmacognosy, Phytochemistry, Medicinal Plants. New York. Laviosier Publishing, New York.

Burrows, G.E., Tyrl, R.J., 2013. Toxic Plants of North America. Senecio spp. (pp. 202–214) and Symphytum spp. (pp. 275–277), second ed. Wiley-Blackwell, Hoboken, New Jersey.

Cardarelli, M., Serino, G., Campanella, L., 1997. Antimitotic effects of usnic acid on different biological systems. Cell. Mol. Life Sci. 53, 667–672.

Chen, K.K., Schmidt, C.F., 1930. Ephedrine and related substances. Medicine 9, 1–117.

Chizzola, R., Langer, T., Franz, C., 2006. An approach to the inheritance of the sesquiterpene chemotypes within *Petasites hybridus*. Planta Med. 72, 1254–1256.

Correche, E.R., Carrasco, M., Escudero, M.E., 1998. Study of the cytotoxic and antimicrobial activities of usnic acid and derivates. Fitoterapia 69, 493–501.

Cosyns, J.P., Jadoul, M., Squifflet, J.P., 1994. Chinese herbs nephropathy: a clue to Balkan endemic nephropathy? Kidney Int. 45, 1680–1688.

Cronin, A.J., Maidment, G., Cook, T., 2002. Aristolochic acid as a causative factor in a case of Chinese herbal nephropathy. Nephrol. Dial. Transplant. 17, 524–525.

Cui, J., Zho, T., Zhang, J., Lou, Z., 1991. Analysis of alkaloids in Chinese *Ephedra* species by gas chromatographic methods. Phytochem. Anal. 2, 116–119.

Dale, O., Ma, G., Gemelli, C., 2012. Effects of mitragynine and its derivatives on human opioid receptors (delta, kappa, and mu). Planta Med. Congress Abstract 78, 91.

Danesch, U., Rittinghausen, R., 2003. Safety of a patented special butterbur root extract for migraine prevention. Headache 43, 76–78.

Debelle, F.D., Vanherweghem, J.L., Nortier, J.L., 2008. Aristolochic acid nephropathy: a worldwide problem. Kidney Int. 74, 158–169.

Depierreux, M., Van Damme, B., Vanden Houte, K., et al., 1994. Pathologic aspects of a newly described nephropathy related to the prolonged use of Chinese herbs. Am. J. Kidney Dis. 24, 172–180.

Favreau, J.T., Ryu, M.L., Braunstein, G., 2002. Severe hepatotoxicity associated with the dietary supplement LipoKinetix. Ann. Intern. Med. 136, 590–595.

Fiedler, P., Gambaro, V., Garbarino, J.A., et al., 1986. Epiphorellic acids 1 and 2, two diaryl ethers from the lichen Cornicularia epiphorella. Phytochemistry 25, 461–465.

Food, Drug Administration (FDA), 2017. FDA statement: Statement from FDA Commissioner Scott Gottlieb, M.D. On FDA Advisory about Deadly Risks Associated with Kratom. https://www.fda.gov/NewsEvents/Newsroom/PressAnnouncements/ucm584970.htm.

Food, Drug Administration (FDA), 2018a. FDA News Release: FDA Oversees Destruction and Recall of Kratom Products; and Reiterates its Concerns on Risks Associated with This Opioid. https://www.fda.gov/newsevents/newsroom/pressannouncements/ucm597649.htm.

Food, Drug Administration (FDA), 2018b. Public Health Focus: FDA and Kratom. https://www.fda.gov/NewsEvents/PublicHealthFocus/ucm584952.htm.

Ghosal, S., Chauhan, R.B., Mehta, R., 1975. Alkaloids of Sida cordifolia. Phytochem. Rep. 14, 830–832.

Gillerot, G., Jadoul, M., Arlt, V.M., 2001. Aristolochic acid nephropathy in a Chinese patient: time to abandon the term "Chinese herbs nephropathy"? Am. J. Kidney Dis. 38, E26.

Hedberg, C.W., Fishbein, D.B., Janssen, R.S., 1987. An outbreak of thyrotoxicosis caused by the consumption of bovine thyroid gland in ground beef. New. Eng. J. Med. 316, 993–998.

Hilmas, C.J., Poole, M.J., Finneran, K., et al., 2009. Galantamine is a novel post-exposure therapeutic against lethal VX challenge. Toxicol. Appl. Pharmacol. 240, 166–173.

Hobson, D.W., 2009. Basic toxicological issues in product-safety evaluations. In: Ballantyne, B., Marrs, T.C., Syversen, T. (Eds.), General and Applied Toxicology, third ed., vol. 5. John Wiley and Sons, Ltd, West Sussex, United Kingdom, p. 2656.

Huneck, S., 1999. The significance of lichens and their metabolites. Naturwissenschaften 86, 559–570.

Huneck, S., Yoshimura, Y., 1996. Identification of Lichen Substances. Springer, New York.

Hung, O., 2015. Herbal preparations. In: Hoffman, R.S., Howland, M.A., Lewin, N.A., Nelson, L.S., Goldfrank, L.R. (Eds.), Goldfrank's Toxicologic Emergencies, tenth ed. New York: McGraw-Hill Education, New York, pp. 598–615.

Huxtable, R.J., Cheeke, P.J., 1989. Human health implications of pyrrolizidine alkaloids and herbs containing them. In: Cheeke, P.R. (Ed.), Toxicants of Plant Origin. CRC Press, Inc, Boca Raton, Florida, pp. 41–86.

Ingólfsdóttir, K., 2002. Usnic acid. Phytochemistry 61, 729–736.

International Agency for Research on Cancer (IARC), 2002. Some Traditional Herbal Medicines, Some Mycotoxins, Naphthalene and Styrene. Aristolochia Species and Aristolochic Acids. IARC Monographs on the Evaluation of Carcinogenic Risk to Humans, vol. 82. WHO, Lyon, pp. 69–128.

Johnson, R.B., Feldott, G., Lardy, H.A., 1950. The mode of action of the antibiotic, usnic acid. Arch. Biochem. 28, 317–323.

Karch, S.B., 1996. Other naturally occurring stimulants. In: Karch, S.B. (Ed.), The Pathology of Drug Abuse. CRC Press, Boca Raton, Florida, pp. 177–198.

Kinney, J.S., Hurwitz, E.S., Fishbein, D.B., 1988. Community outbreak of thyrotoxicosis: epidemiology, immunogenetic characteristics and long-term outcome. Am. J. Med. 84, 10–18.

Krähenbühl, S., 2001. Mitochondria: important target for drug toxicity? J. Hepatol. 34, 334–336.

Krumme, B., Endmeir, R., Vanhaelen, M., et al., 2001. Reversible Fanconi syndrome after ingestion of a Chinese herbal 'remedy' containing aristolochic acid. Nephrol. Dial. Transplant. 16, 400–402.

Kumar, V.H., Nirmala Shashidhara, S., et al., 2010. Reserpine content of *Rauwolfia serpentina* in response to geographical variation. Int. J. Pharma Bio Sci. 1, 429–434.

Lauterwein, M., Oethinger, M., Belsner, K., et al., 1995. In vitro activities of the lichen secondary metabolites vulpinic acid, (+)-usnic acid, and (−)-usnic acid against aerobic and anaerobic microorganisms. Antimicrob. Agents Chemother. 39, 2541–2543.

Lee, S., Lee, T., Lee, B., 2004. Fanconi's syndrome and subsequent progressive renal failure caused by a Chinese herb containing aristolochic acid. Nephrology 9, 126–129.

Liu, Q., Zhao, X., Lu, X., et al., 2012. Proteomic study on usnic-acid-induced hepatotoxicity in rats. J. Agric. Food Chem. 60, 7312–7317.

Mahuang (Appendix: Mahuanggen), Chang, H.M., But, P.P.-H., 1987. Pharmacology and applications of Chinese materia medica. In: Chang, H.-M., But, P.P.-H. (Eds.), Pharmacology and Applications of Chinese Materia Medica. World Scientific Publishing Co. Pte. Ltd, Singapore, China, pp. 1119–1124.

Marcano, V., Rodriguez-Alcocer, V., Moralez-Méndez, A., 1999. Occurrence of usnic acid in *Usnea laevis* Nylander (lichenized ascomycetes) from the Venezuelan Andes. J. Ethnopharmacol. 66, 343–346.

McDermott, W.V., Ridker, P.M., 1990. The Budd-Chiari syndrome and hepatic veno-occlusive disease: recognition and treatment. Arch. Surg. 125, 525–527.

Mengs, U., 1987. Acute toxicity of aristolochic acid in rodents. Arch. Toxicol. 59, 328–331.

Mengs, U., Stotzem, C.D., 1993. Renal toxicity of aristolochic acid in rats as an example of nephrotoxicity testing in routine toxicology. Arch. Toxicol. 67, 307–311.

Mengs, U., Lang, W., Poch, J.-H., 1982. The carcinogenic action of aristolochic acid in rats. Arch. Toxicol. 51, 107–119.

Moreira, C.T., Oliveira, A.L., Comar, J.F., et al., 2013. Harmful effects of usnic acid on hepatic metabolism. Chem. Biol. Interact. 203, 502–511.

Muniz-Martinez, M.C., Nortier, J.L., Vereerstraeten, P., et al., 2002. Progression rate of Chinese-herb nephropathy: impact of *Aristolochia fangchi* ingested dose. Nephrol. Dial. Transplant. 17, 1–5.

National Toxicology Program (NTP), 2011. Reserpine. Report on Carcinogens, twelfth ed. Department of Health and Human Services, Public Health Service, National Toxicology Program, Research Triangle Park, NC: U.S, pp. 370–371.

Nortier, J.L., Martinez, M.C., Schmeiser, H.H., 2000. Urothelial carcinoma associated with the use of a Chinese herb (*Aristolochia fangchi*). N. Engl. J. Med. 342, 1686–1692.

Oshio, H., Tsukui, M., Matsuoka, T., 1978. Isolation of *l*-Ephedrine from 'Pinelliae tuber'. Chem. Pharm. Bull. (Tokyo) 26, 2096–2097.

Pharmanex Inc, 1998. Administrative Proceeding. Docket No. 97P-0441; Final Decision. May 20, 1998 letter to Stuart M. Pape. Found in. http://www.fda.gov/ohrms/dockets/dockets/97p0441/ans0002.pdf.

Pharmanex Inc. v. Shalala, (2001). No. 2:97CV262K, 2001 WL 741419 (D. Utah Mar.30, 2001). Found in 35F. Supp. 1341, 2001. Case No. 2: 97CV72K. Pharmanex, Inc. v. Shalala, 35F. Supp. 1341 (2001 WL 741419 (2001 U.S. District Court Utah).

Sadler, C., Vanderjagt, L., Vohra, S., 2007. Complementary, holistic, and integrative medicine: butterbur. Pediatr. Rev. 28, 235–238.

Sahu, S.C., O'Donnell, M.W., Sprando, R.L., 2012. Interactive toxicity of usnic acid and lipopolysaccharides in human liver HepG2 cells. J. Appl. Toxicol. 32, 739–749.

Schmeiser, H.H., Bieler, C.A., Wiessler, M., 1996. Detection of DNA adducts formed by aristolochic acid in renal tissue from patients with Chinese herbs nephropathy. Cancer Res. 56, 2025–2028.

Schwetz, B.A., 2001. From the food and drug administration. J. Am. Med. Assoc. 285, 2705.

Sonko, B.J., Schmitt, T.C., Guo, L., 2011. Assessment of usnic acid toxicity in rat primary hepatocytes using ^{13}C isotopomer distribution analysis of lactate, glutamate and glucose. Food Chem. Toxicol. 49, 2968–2974.

Steiner, W.E., Pikalov, I.A., Williams, P.T., et al., 2012. An extraction assay analysis for galanthamine in Guinea pig plasma and its application to nerve agent countermeasures. J. Anal. Bioanal. Tech. 3, 1–5.

Tanaka, A., Nishida, R., Yoshida, T., 2001. Outbreak of Chinese herb nephropathy in Japan: are there any differences from Belgium? Intern. Med. 40, 296–300.

Taylor, P., Hardman, J.G., Limbird, L.E., et al., 1996. Anticholinesterase agents. In: Hardman, J.G., Limbird, L.E., Molinoff, P.B., Ruddon, R.W., Bilman, A.G. (Eds.), The Pharmacological Basis of Therapeutics. McGraw-Hill, New York, New York, pp. 131–149.

Thongpradichote, S., Matsumoto, K., Tohda, M., 1998. Identification of opioid receptor subtypes in antinociceptive actions of supraspinally-administered mitragynine in mice. Life Sci. 62, 1371–1378.

Vanhaelen, M., Vanhaelen-Fastre, R., But, P., et al., 1994. Identification of aristolochic acid in Chinese herbs (Letter to the editor). Lancet 343, 174.

Vanherweghem, J.L., 1998. Misuse of herbal remedies: the case of an outbreak of terminal renal failure in Belgium (Chinese herbs nephropathy). J. Alt. Complement Med. 4, 9–13.

Vanherweghem, J.L., Depierreux, M., Tielemans, C., 1993. Rapidly progressive interstitial renal fibrosis in young women: association with slimming regimen including Chinese herbs. Lancet 341, 387–391.
Waters, M.D., Merrick, B., 2009. Toxicogenomics and the evolution of systems toxicology. In: Ballantyne, B., Marrs, T.C., Syversen, T. (Eds.), General and Applied Toxicology, , third ed.vol. 1. John Wiley and Sons, Ltd, West Sussex, United Kingdom, pp. 511–537.
Williams, P.T., Hilmas, C.J., 2010. Cholinergic effects on ocular flutter in Guinea pigs following nerve agent exposure: a review. J. Med. CBR Def. 8, 1–13.
Winship, K.A., 1991. Toxicity of comfrey. Adverse Drug React. Toxicol. Rev. 10, 47–59.
World Health Organization (WHO), 1999. Monographs on selected medicinal plants. Herba Ephedrae 145–153.
Yang, L., Zhou, X., Pang, N., 2003. Clinical analysis of two children with aristolochic acid nephropathy. Zhonghua Er Ke Za Zhi 41, 552–553.
Yellapu, R.K., Mittal, V., Grewal, P., et al., 2011. Acute liver failure caused by 'fat burners' and dietary supplements: a case report and literature review. Can. J. Gastroenterol. 25, 157–160.

CHAPTER

40

Nutriphenomics in Rodent Models: Impact of Dietary Choices on Toxicological Biomarkers

Michael A. Pellizzon, Matthew R. Ricci

Research Diets, Inc., New Brunswick, NJ, United States

INTRODUCTION

When designing an experiment involving animal models, there are many factors that will affect a given phenotype (i.e., observable or biochemical response), and these can be broken into two basic categories: the genotype of the animal and the environment. Rodent models such as mice and rats are the most commonly used animals in biomedical research because of their small size, low cost, the vast array of strains commercially available, and the ease of modifying their genetic background. There are many rodent models for studying chronic disease states such as metabolic disorders (i.e., obesity, nonalcoholic fatty liver disease (NAFLD)) or cancer, some which are driven by dietary means, genetic modification, or both. Therefore, many considerations need to be taken into account when choosing the proper rodent model for a given study. In some cases, phenotype is very predictable, but in others, rodents can have a variable response, including when presented with a dietary challenge (Enriori et al., 2007; Zhang et al., 2012).

Like genotype, most environmental factors, such as light/dark cycle, temperature, humidity, and cage density, and type are typically well controlled, but diet (unless being directly studied) is usually a last consideration and is either not disclosed or is underreported in experimental studies. Most of the time, terms such as "a standard diet" or "a normal diet" are the commonly used "descriptors" when defining a diet in methods sections of publications, and in many of these cases, it is likely that the diet was not carefully considered. The importance of diet in toxicological studies was reviewed quite some time ago by toxicologists (Greenfield and Briggs, 1971; Wise, 1982), but some toxicologists still have not embraced this concept.

As the researcher chooses a diet for their study, they should ask a number of questions. Is the formula "open" to the public? Do I have access to the exact composition of the diet? Is the diet variable from batch to batch? Can I easily modify the diet formula? Does the diet contain contaminants? By having an open and consistent formula that is easily modified as needed, it is then possible to effectively study how and why various dietary manipulations influence any given biological parameter and form valid conclusions from the data. The influence of dietary factors on an animal's phenotype is called *nutriphenomics* and will occur under any experimental condition, whether diet is intentionally being altered to influence an animal's phenotype or not. Although *nutriphenomics* is a term typically used to describe intentional diet-induced disease phenotypes, there are dietary factors that may affect certain toxicological biomarkers, which may not be noticed unless they are being directly studied. These factors include certain nutrients and non-nutrients that are found in commercially available diets. This chapter will provide an overview of dietary choices available, common diet-induced disease phenotypes, and factors in commonly used control diets that can affect *nutriphenomics* in rodent studies.

Dietary Choices

Rodent diets can be broken down into two basic categories: grain-based (GB) or (cereal-based) diets or purified diets. A third category is chemically defined diets, which include ingredients that are entirely synthetic, but it is not something that is commonly used in research mainly because of their extreme expense and will not be discussed.

Biomarkers in Toxicology, Second Edition
https://doi.org/10.1016/B978-0-12-814655-2.00040-2

Grain-Based Diets

GB diets are made with unrefined cereal grains or animal ingredients that contain multiple nutrients and non-nutrients (Fig. 40.1). GB diets typically include agricultural grade ingredients such as ground corn, ground wheat, wheat gluten, wheat middlings, barley, ground oats, soybean meal, alfalfa meal, and animal byproducts such as fish meal and porcine animal meal in varying proportions. Most GB diets are "closed formulas" or proprietary, and therefore the actual concentrations of these ingredients are not disclosed. Furthermore, GB diet companies will alter levels or sources of ingredients without disclosing these changes to researchers. These changes can be necessary to maintain a given nutrient concentration because nutrient levels in GB ingredients will vary with soil conditions, climate, and timing of harvest or sequence. However, and perhaps unwittingly, this can lead to changes in other nutrients and non-nutrients. For example, protein is typically monitored closely in GB diets, and a commonly used source of protein in these diets is soybean meal (around 50% protein). Besides protein, soybean meal also contains fiber, fat, minerals, and non-nutrients, such as phytoestrogens. Phytoestrogens, as their name implies, have similarities to the chemical structure of estrogen. Phytoestrogens in soybean meal are a complex mixture of isoflavones that are in several different chemical forms (daidzin, 6″-OAc daidzin, 6″-OMal daidzin, daidzein, genistin, 6″-OAc genistin, 6″-OMal genistin, genistein, glycitin, 6″-OMal glycitin, and glycitein), and each of these compounds is present either as β-glycosides (bound to glucose) or as aglycones (not bound to glucose). Phytoestrogens in soybean meal are mainly in the chemical forms genistin and daidzin (β-glycosides) and the levels of these phytoestrogens can change significantly based on soy variety, location of harvest, and time of year (Eldridge and Kwolek, 1983; Brown and Setchell, 2001). Therefore, as soybean meal is altered intentionally in some GB diets to maintain protein, this intentional adjustment could cause unintentional changes to levels of phytoestrogens (Thigpen et al., 1999; Jensen and Ritskes-Hoitinga, 2007). Some GB diets are "open formulas," and well-known examples of these include those developed by the National Institute of Health in the early 1970s (i.e., NIH-07, NIH-31). These "open" formulas will always contain a fixed level of ingredients, but it is difficult to maintain consistency in nutrient and non-nutrient contents of a given open- and fixed-formula GB diet from batch to batch as alluded to above (Rao and Knapka, 1987).

GB diet producers also offer "phytoestrogen-free" diets by replacing soybean meal or both soybean meal and alfalfa meal (contains the phytoestrogen coumestrol) with casein (containing around 87% protein) or combinations of other plant protein sources. Although this reduces phytoestrogen levels (Jensen and Ritskes-Hoitinga, 2007), there are still grains in GB diets that contain smaller amounts of phytoestrogens including alfalfa, wheat, barley, corn, and oats (Farnsworth et al., 1975) and potentially other endocrine disruptors or confounders present in other ingredients that could mask the influence of a toxicological phenotype of interest (Heindel and vom Saal, 2008). In fact, several other potential contaminants exist, including

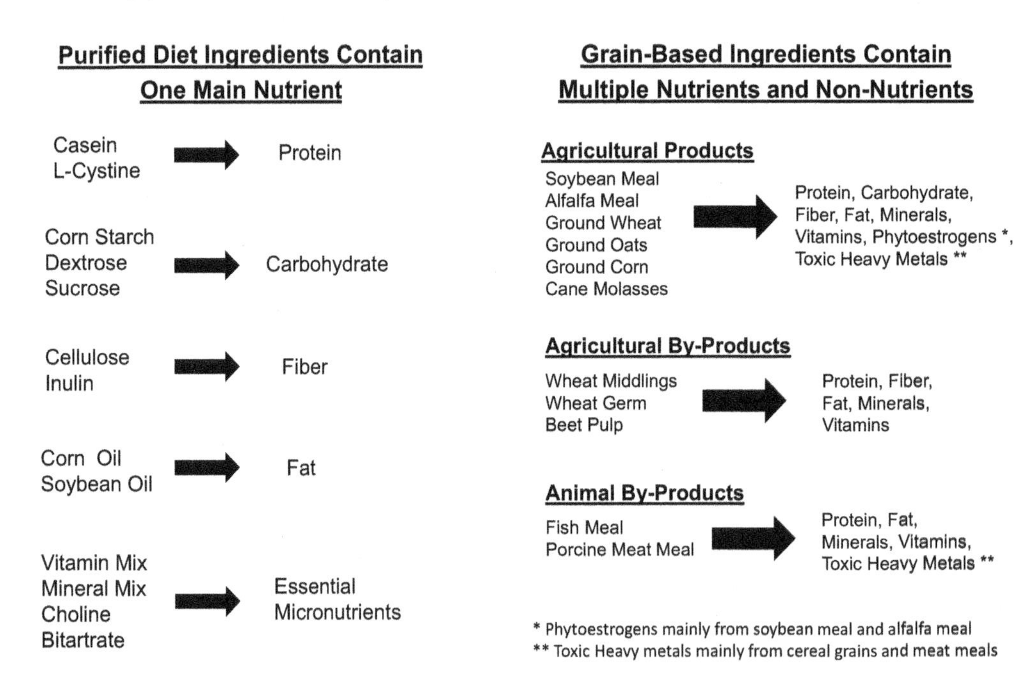

FIGURE 40.1 Common ingredients in purified diets and GB diets and their nutrient/nonnutrient contributions.

heavy metals such as arsenic (Kozul et al., 2008; Mesnage et al., 2015), endotoxins (Hrncir et al., 2008; Schwarzer et al., 2017), and pesticides and pollutants (Mesnage et al., 2015). Often these contaminants are present at biologically relevant levels (Kozul et al., 2008; Thigpen et al., 2013; Mesnage et al., 2015; Schwarzer et al., 2017). There is evidence that the levels of contaminants in GB diets vary across batches of the same diet (Greenman et al., 1980; Jensen and Ritskes-Hoitinga, 2007; Thigpen et al., 2007), which could lead to different findings depending on the batch used.

In addition to a diverse array of contaminants, GB diets contain both very high fiber contents and variable types of fiber because of the presence of multiple cereal grains. The diversity in the types of fibers present in GB diets has been described decades ago and showed that rat and mouse GB diets varied from 8% to 22% (total fiber) and proportions of different fiber types (i.e., pectin, hemicellulose, cellulose, and lignin) were highly variable among different diets (Wise and Gilburt, 1980). Therefore, GB diets contain fermentable fiber sources (i.e., soluble fibers such as pectin and to a partial degree from hemicellulose), which have the potential to modify the gut microbiome significantly. Thus, their use may limit our understanding of how microbiota affect overall health.

Purified Diets

Purified diets are open and fixed formulas made with refined ingredients, which contain one main nutrient and are commonly phytoestrogen-free, depending on the protein source. Given their open nature, the formulas are reportable and their nutrient contents are more easily defined than diets based on less refined grains and animal by-products. For example, casein is ~87% protein (~11% moisture), corn starch is ~88% carbohydrate (~11% moisture), soybean oil is 100% fat, cellulose is >99% insoluble fiber, and vitamin and mineral mixes are chemically defined ingredients. These ingredients contain minimal non-nutrients, and as such, it is possible to repeat a given nutrient composition of a particular diet from one lot to the next while limiting non-nutrients that can influence phenotype (Wise, 1982).

Purified diets have been around since the 1920s and were used to prove the essentiality of vitamins and minerals (Knapka, 1988). Until the 1970s, researchers made their own purified diets in the lab, and it was not uncommon for the formulas to vary from one researcher to the next, or for essential nutrients to be missing or perhaps simply unreported (Greenfield and Briggs, 1971). These dietary differences among studies in a given area of toxicology (or other fields) led to confounding factors and reduced the ability of collaborating researchers to draw conclusions. In an effort to reduce what was the variable influence of diet in toxicology studies, a committee of nutritional scientists known as the American Institute of Nutrition (AIN) set forth in 1973 to establish an open, fixed formula to meet requirements for growth, reproduction, and lactation of rats and mice (Bieri et al., 1977). Such a diet was considered important for toxicologists because it contained only a minimal level of non-nutrients that could influence the toxicological phenotype of a given rodent model. The initial formula, called the *AIN-76*, fulfilled the nutrient recommendations for rodents established in 1972 by the Committee on Animal Nutrition of the National Research Council (National Research Council, 1972). As researchers began using this diet, they noted that a high percentage of their rats were hemorrhaging either internally or externally; the problem was solved by adding 10-fold higher concentrations of menadione (vitamin K) (Roebuck et al., 1979). This revised formula was called the *AIN-76A* (Table 40.1), and additional suggestions to improve the formula were proposed in an editorial by Bieri in 1980 (Bieri, 1980). About a decade later, a new group of scientists formed another AIN committee over some of the nutritional concerns indicated by Bieri with the *AIN-76A* diet. This committee developed a new set of diets, which would become the *AIN-93* series diets (*AIN-93G* and *AIN-93M*, for growth and maintenance, respectively) (Reeves et al., 1993). Despite some improvements, studies with the *AIN-93* series diets along with certain formulation issues, such as a low phosphorus mineral mix that relies on casein to supply part of the phosphorus requirement, have suggested that more improvements are required (Lien et al., 2001). That being said, these purified diets provide toxicologists and others a phytoestrogen-free formula that has a consistent nutrient composition from batch to batch, is easy to report, and is easily modified to the researcher's advantage.

TABLE 40.1 The AIN-76A Rodent Diet

Ingredient	g	kcal
Casein	200	800
DL-Methionine	3	12
Corn Starch	150	600
Sucrose	500	2000
Cellulose	50	0
Corn Oil	50	450
Mineral Mix	35	0
Vitamin Mix	10	40
Choline Bitartrate	2	0
Total	**1000**	**3902**

Diet-Induced Metabolic Disorders

Much work has been done with respect to how diet influences disease phenotypes in rodent models, and purified diets have been crucial in this effort. In some cases, it is important to address the influence of an environmental toxicant or endocrine disruptor on a metabolic disorder.

High-Fat Diets for Diet-Induced Obesity Models

At the time of this writing, a Google Scholar search for "high-fat diet rat" and "high-fat diet mice" yielded over 359,000 articles, dating back to the 1940s. For much of this time, a high-fat diet (HFD) was made by adding various levels and types of fat to the available GB diet already in the animal facility. However, this is not ideal because (1) fat addition can lead to nutritional inadequacies (nutrient dilution) in the diet and (2) the use of GB diets has inherent drawbacks as discussed earlier (variable and unknown composition). As an example, phytoestrogens can reduce adiposity and improve insulin sensitivity in mice relative to diets with low phytoestrogens (Cederroth et al., 2007, 2008).

Because purified diets allow for the easy formulation of an HFD without nutrient dilution, it is no surprise that in the last ~20–25 years, purified ingredient HFDs have been widely used to study obesity and its associated comorbidities in rodents. There is a dose–response relationship between dietary fat and body weight and adiposity (Donovan et al., 2009; Jiang et al., 2009) and because animals fed a higher fat diet tend to gain more weight in a shorter period of time (which is often seen as cost savings to the researcher), the popularity of using diets containing 60% of calories from fat has grown as a standard method of promoting diet-induced obesity (DIO) in rodent models.

Although most rodents tend to become obese on an HFD, there are variable responses in body weight gain, glucose tolerance, triglycerides (TG), and other parameters depending on the strain and gender (Levin et al., 1997; Rossmeisl et al., 2003) and type of dietary fat (Ikemoto et al., 1996; Wang et al., 2002; Buettner et al., 2006). Outbred Sprague–Dawley and Wistar rats have a variable response to an HFD in that some animals become obese, whereas others remain as lean as low–fat diet fed rats (Farley et al., 2003). It is common to separate these rats (based on body weight gain) into DIO and diet-resistant (DR) groups (Chang et al., 1990; Farley et al., 2003; Levin and Dunn-Meynell, 2006). DIO and DR rats have been selectively bred over time such that the propensity to gain weight or resist obesity on an HFD is known in utero, allowing the researcher to look early in life (prior to the onset of obesity) for genetic traits that may later predispose them to their DIO or DR phenotypes (Levin et al., 1997; Ricci and Levin, 2003). Diet-induced obesity in Sprague–Dawley and Wistar rats can induce glucose intolerance (Drake et al., 2005; Laurent et al., 2007), fatty liver (Ross et al., 2012), inflammation (De Souza et al., 2005; de La Serre et al., 2010), cardiac dysfunction (Burgmaier et al., 2010), and increased muscle lipid deposition (Storlien et al., 1991) and affect development of offspring when fed to pregnant rats (White et al., 2009).

Mouse models of DIO are in more widespread use compared with those of rats, and the most commonly used mouse model of DIO is the C57BL/6 mouse, substrains of which are available from different breeders. Although it is rarely reported, there is some variability in weight gain in this strain, with one paper reporting that about 2/3 of the mice on an HFD became obese (Enriori et al., 2007). In C57BL/6 mice, HFDs induce obesity (Van Heek et al., 1997), cause insulin resistance (IR) (DeFuria et al., 2009), increase muscle triglyceride levels (Park et al., 2005), reduce glucose tolerance (Gallou-Kabani et al., 2007), cause inflammation (Weisberg et al., 2003), induce fatty liver (Ito et al., 2007a), change gut microflora populations (Ravussin et al., 2012), affect cognition (Pistell et al., 2010), and alter neurogenesis (Park et al., 2010), among other phenotypes. Other mouse strains such as the A/J or SWR/J are resistant to DIO (Surwit et al., 1995; Prpic et al., 2003). However, strains that may exhibit similar levels of obesity may have varied metabolic responses. For example, C57BL/6 mice are more glucose intolerant, compared with obese AKR mice that are more insulin resistant (Rossmeisl et al., 2003).

Diet-Induced Atherosclerosis/ Hypercholesterolemia Models

Western-type diets containing high levels of saturated fat and cholesterol are commonly used to "push" atherosclerosis risk factors (elevated total cholesterol [TC] and low-density lipoprotein cholesterol [LDL-C]) in certain rodent models such as mice, hamsters, and guinea pigs. Not surprisingly, different animal models require different diets.

Mice and Rats

Most wild-type mice and rats are not ideal models of cardiovascular disease research because they typically have very low levels of LDL-C and high levels of high-density lipoprotein cholesterol (HDL-C). This is in contrast to humans in whom the reverse is true. Although diets containing high levels of cholesterol and saturated fat (~0.5% cholesterol, ~40 kcal% fat) can increase TC, both LDL-C and HDL-C (Srivastava et al., 1991, 1992; Srivastava, 1994) increase, which limits atherosclerosis development (Getz and Reardon, 2006). The addition of the bile acid and cholic acid to diet (in combination with cholesterol) will increase LDL-C (Nishina et al.,

1990, 1993; Zulet et al., 1999; Jeong et al., 2005; Yokozawa et al., 2006) by both facilitating fat and cholesterol absorption and reducing conversion of cholesterol to bile acids (Horton et al., 1995; Ando et al., 2005). However, cholic acid can also independently influence genes that regulate lipoprotein metabolism and inflammation and reduce plasma TG and HDL-C (Nishina et al., 1990; Ando et al., 2005; Getz and Reardon, 2006) and reduces body weight gain (without changing food intake), glucose intolerance, and plasma insulin either with or without the presence of cholesterol (Ikemoto et al., 1997). Therefore data from cholic acid–containing diets should be interpreted carefully.

Genetically modified mice such as those with mutations that slow the removal of cholesterol from the blood have led to more "human-like" phenotypes and can show significant elevations in circulating LDL-C and atherosclerotic lesions, especially when dietary cholesterol is added. Some of these knockout mouse models (such as the LDL receptor knockout and the apolipoprotein E knockout [apoE KO]) can be very responsive after 12 weeks on a high cholesterol diet (0.15%–1.25% cholesterol) (Lichtman et al., 1999; Collins et al., 2001; Joseph et al., 2002). Lesion development is very dramatic in apoE KO mice fed a Western-type diet, and beginning stages of atherosclerosis (i.e., fatty streak lesions) can be found at 6 weeks (Nakashima et al., 1994). A combination of multiple genetic modifications (e.g., apoE + matrix metalloproteinase double KO) with a high-fat lard-based diet with cholesterol (0.15%) was found to increase plaque rupture after only 8 weeks (Johnson et al., 2005). With these mouse models, the main influence on atherosclerosis is dietary cholesterol rather than the level of fat (Davis et al., 2001; Wu et al., 2006), but certain threshold levels of dietary cholesterol may exist, at least within the context of a low-fat purified diet (Teupser et al., 2003). Very high–fat diets (i.e., 60 kcal% fat) are capable of inducing some atherosclerosis (Subramanian et al., 2008; King et al., 2009), and the fatty acid profile and carbohydrate form (i.e., fructose, sucrose) can be manipulated to modify the atherosclerosis phenotype to the researchers advantage (Merat et al., 1999; Collins et al., 2001; Merkel et al., 2001).

Hamsters

Although hamsters typically have a high percentage of HDL-C when fed a low-fat/low-cholesterol diet (similar to rats and mice), dietary cholesterol alone will increase LDL-C, and like humans, saturated fat can increase these levels further (Otto et al., 1995; Alexaki et al., 2004). Diets containing high levels of saturated fat and cholesterol can drive the initial stages of atherosclerosis (fatty streaks, foam cells) in as little as 6 weeks (Kahlon et al., 1996). Even a diet high in saturated fat without added cholesterol can increase aortic cholesterol accumulation compared with a lower saturated fat diet with added cholesterol (Alexaki et al., 2004). The different response of the hamster (compared with the rat or mouse) to dietary cholesterol is related to different mechanisms of cholesterol processing by the liver (Dietschy et al., 1993; Horton et al., 1995; Khosla and Sundram, 1996). With the hamster, it is also important to consider the source of dietary protein because hamsters fed casein- and lactalbumin-based diets had higher levels of LDL-C and atherosclerosis than those fed an equal amount of soy protein, and like humans, males may be more susceptible than females (Blair et al., 2002).

Guinea Pigs

Unlike other wild-type rodents, guinea pigs have a similar cholesterol profile and lipoprotein metabolism to humans (more LDL-C vs. HDL-C) when maintained on a low-fat/low-cholesterol diet (Fernandez and Volek, 2006). Like hamsters, diets high in saturated fat can elevate TC and LDL-C levels and a diet with added cholesterol (at least up to 0.3%, w/w) can cause further elevations in LDL-C and induce atherosclerotic lesions (i.e., fatty streaks) after 12 weeks (Lin et al., 1994; Cos et al., 2001; Zern et al., 2003). Carbohydrate/fat ratio is also important to atherosclerosis development, as high-cholesterol diets that are high in carbohydrate and moderate in fat are more capable of promoting atherosclerosis (likely via increased number of smaller LDL particles) than those low in carbohydrate but very high in fat (Torres-Gonzalez et al., 2006). Also, plasma TC levels are higher with casein versus soy protein (Fernandez et al., 1999) and with sucrose versus starch (Fernandez et al., 1996), in the context of a cholesterol-containing diet. The sensitivity of LDL-C to dietary manipulation with minimal change to HDL-C highlights the value of the guinea pig for studies examining the influence of drug therapies on lowering LDL-C (Conde et al., 1996; Aggarwal et al., 2005).

High-Fructose/Sucrose Diets for Hypertriglyceridemia and Insulin Resistance

In humans and rodents, the presence of higher levels of dietary carbohydrate as fructose or sucrose is capable of causing hypertriglyceridemia (hyperTG) and IR via increasing lipogenesis and glucose production in the liver (Daly et al., 1997; Basciano et al., 2005). Sprague–Dawley and Wistar rats are both established models of sucrose-induced IR and hyperTG (Pagliassotti et al., 1996, 2000), which can develop as quickly as 2 weeks in rats fed mostly sucrose (i.e., 68 kcal%) relative to corn starch (Pagliassotti et al., 1996). The fructose component of sucrose is largely responsible for the hyperTG and IR produced by high-sucrose diets (Sleder et al., 1980; Thorburn et al., 1989; Thresher et al., 2000).

Although very high levels of sucrose or fructose are commonly used for driving IR in rodents, even lower levels of dietary sucrose (17% of energy) can cause this after chronic feeding periods (i.e., 30 weeks) (Pagliassotti and Prach, 1995). High-fructose diets can also raise liver TG levels and induce liver inflammation (Kawasaki et al., 2009) as well as adipose tissue and kidney inflammation (Oudot et al., 2013). Furthermore, gender is important in the development of sucrose-induced IR and hyperTG in rats as females (unlike males) are typically not responsive to elevations in dietary sucrose (Horton et al., 1997). Unlike HFD, high sucrose or fructose diets promote marginal weight gain in rats, and this typically requires a prolonged period of time and a significantly greater energy intake (Chicco et al., 2003).

Similar to rats, hamsters are sensitive to metabolic alterations by high-fructose diets (~60% of energy) and quickly develop IR and hyperTG only after 2 weeks (Kasim-Karakas et al., 1996; Taghibiglou et al., 2000), but they may not be as sensitive to sucrose-induced hyperTG and IR (Kasim-Karakas et al., 1996), suggesting that the level of dietary fructose is important for this model. The addition of dietary cholesterol (0.25%) allows for the simultaneous development of hypercholesterolemia, greater IR, and hyperTG in this model compared to fructose alone (Basciano et al., 2009). The ability to induce these three metabolic disturbances together makes the hamster an attractive model for studying these conditions in one model without the need to alter background genetics.

In contrast to rats and hamsters, mice are used less frequently as a model for sucrose/fructose-induced phenotypes, as the commonly used C57BL/6 strain either does not develop IR or develops the phenotype more slowly (Nagata et al., 2004; Sumiyoshi et al., 2006). For example, glucose intolerance (attributed to reduced pancreatic insulin secretion) develops in C57BL/6 mice fed a high-sucrose diet over the course of 10–55 weeks (Sumiyoshi et al., 2006). However, the mouse genome is much easier to manipulate than that of the rat and in certain cases can cause knockout models to have hyperTG with high dietary fructose (Hecker et al., 2012).

Nonalcoholic Fatty Liver Disease

NAFLD encompasses a spectrum of disease states, from steatosis (fatty liver) to nonalcoholic steatohepatitis (also called NASH; steatosis with inflammatory changes), which may be followed by progression to fibrosis, cirrhosis and hepatocellular carcinoma (Zafrani, 2004). There are several different dietary approaches to induce NAFLD. These different dietary approaches produce different severities of disease along the NAFLD spectrum and likely work by different mechanisms.

Of the dietary approaches discussed here, methionine–choline deficient (MCD) diets produce the most severe phenotype in the shortest timeframe. Used for over 40 years, MCD diets will quickly induce measurable hepatic steatosis (mainly macrovesicular) in rodents by 2–4 weeks, and this progresses to inflammation and fibrosis shortly thereafter (Weltman et al., 1996; Sahai et al., 2004). Fat levels in MCD diets can vary, though they typically contain about 20% fat by energy (most "control" diets, purified ingredient or grain based, tend to be ~10% by energy). The mechanism for steatosis includes increased hepatic fatty acid uptake, impaired very low density lipoprotein secretion (due to lack of phosphatidyl choline synthesis), and increased fatty acid transport proteins (Kulinski et al., 2004; Rinella and Green, 2004). MCD diet–induced NASH is reversible in rats by switching to a diet with sufficient methionine and choline (Mu et al., 2010). The disadvantage of using MCD diets is that they induce rapid weight loss (due to a vastly lower caloric intake) and the rodents do not become insulin resistant (Kirsch et al., 2003; Rinella and Green, 2004), unlike the typical obese, insulin-resistant human NAFLD patient.

Similar to MCD diets, choline deficient (CD) diets also tend to contain higher levels of fat, though it is often difficult to know the specifics because, unfortunately, authors rarely publish the details of the diets. CD diets induce steatosis, inflammation, and fibrosis over 10 weeks without any difference in body weight compared with the control group (Fujita et al., 2010). This lack of weight loss makes CD diets more appealing to some researchers. If weight gain is desired, a CD HFD (i.e., 45 kcal% fat) can increase body weight of C57BL/6 mice further than a matched low-fat diet (10 kcal% fat) with or without choline after 8 weeks, but mice remain glucose tolerant (Raubenheimer et al., 2006). Even longer feeding periods of up to 1 year have found that a CD 45 kcal% fat diet (same as Raubenheimer et al.) can drive hepatocarcinoma in mice relative to those simply fed a choline-sufficient 45 kcal% fat diet (Wolf et al., 2014). The mechanisms involved with liver fat accumulation may be different from those at work during MCD diet feeding (Kulinski et al., 2004). When both CD and MCD diets were fed to rats for 7 weeks, the MCD diet group had higher scores of liver inflammation and steatosis than the CD group. However, the CD fed rats gained weight, were insulin resistant, and had higher plasma lipids than the MCD group (Veteläinen et al., 2007).

As discussed earlier, HFDs are well known to increase body weight and body fat and induce IR in rodent models. HFD (with sufficient methionine and choline) can also increase liver fat levels quite rapidly (within days) as well as hepatic IR before significant increases in peripheral fat deposition occur (Samuel et al., 2004). Chronically, HFDs-induced liver fat accumulation may not follow a linear progression and liver fat levels may actually decrease and then increase again during prolonged HFDs feeding (Gauthier et al., 2006). When fed for equal lengths of time, HFD feeding results in 10-fold lower liver fat levels compared to what accumulates on an MCD diet (Romestaing et al., 2007). In general, HFD feeding does not produce liver fibrosis but only mild steatosis as compared with MCD diets (Anstee and Goldin, 2006), thus highlighting an important difference between these dietary regimes. It is important to remember that the term "HFDs" encompasses a wide variety of diet formulas that can be expected to alter the liver phenotype in various ways.

In a dietary combination approach, Matsumoto et al. (2013) used a high fat (60% of energy), CD diet containing only 0.1% methionine by weight (most purified ingredient diets based on casein contain 5–8 times this amount). Fatty liver, markers of liver injury, and inflammation were increased after 1–3 weeks, and after an initial loss of body weight, the C57BL/6 mice gained weight at a trajectory similar to control animals (Matsumoto et al., 2013). In contrast, A/J mice exhibited reduced weight and inflammation with no evidence of fibrosis (Matsumoto et al., 2013). Increasing methionine from 0.1% to 0.2% in the context of a 45 kcal% fat diet increased body weight gain and still allowed for NASH to develop after 12 weeks in C57BL/6 mice (Chiba et al., 2016). Thus it seems that HFDs containing lower than normal levels of methionine and choline allow for the development of NASH without massive weight loss. This idea of modifying so-called "standard" HFDs is powerful because it allows the researcher to "fine-tune" the phenotype to meet their needs.

Combination HFDs with 40 kcal% fat using a vegetable-based shortening (combination of palm oil and partially hydrogenated soybean oil), 2% cholesterol, and 20% as fructose source have been used often by those interested in developing both NASH and metabolic disease. Both lard or the shortening were capable of increasing intrahepatic lipid levels in ob/ob knockout mice, which are genetically obese and already have a higher intrahepatic lipid level on a low-fat diet, and a similar effect was found in C57BL/6 mice, but the shortening was found to elevate the lipid level even more than lard in both these mouse models after 12 (for ob/ob) and 16 (for C57BL/6) weeks (Trevaskis et al., 2012). To develop more advanced NASH including fibrosis, it is necessary to feed C57BL/6 mice up to 30 weeks (Trevaskis et al., 2012). Therefore, prolonged feeding is necessary to drive a combination of fibrosis with metabolic disorders in wild-type mice.

Diets High in Sodium (and Fructose) for Hypertension

Diet-induced hypertension is possible in both rats and mice though there are more published papers using rats, perhaps because of their larger size, the amount of physiological data available, and robust blood pressure response that some strains present.

The main dietary contributor to diet-induced hypertension is the level of NaCl. Levels of NaCl well above what are normally found in purified (~0.1%) or GB diets (~0.3–0.4%) are required to raise blood pressure in most rodent models. The Dahl salt-sensitive (SS) rat shows a significant rise in blood pressure within 2–4 weeks after being fed a purified diet containing 8% NaCl (Ogihara et al., 2002; Karmakar et al., 2011), though lower levels of NaCl (4%) will still raise blood pressure (Konda et al., 2006) albeit more slowly (Owens, 2006). This rise in blood pressure can be attenuated by supplementing the diet with extra potassium (Ogihara et al., 2002).

Aside from the NaCl level itself, the background diet can also affect the hypertension phenotype. When either low (0.4%) or high (4%) NaCl was added to both a GB diet and a purified diet, Dahl SS rats that were fed the purified diet had higher blood pressure and more renal damage compared with GB diet–fed rats (Mattson et al., 2004). Given the many differences between GB and purified diets, it is difficult to know exactly which components were responsible. The possibilities include the source of protein (Nevala et al., 2000), carbohydrate (Buñag et al., 1983), fat (Zhang et al., 1999), fiber (Preuss et al., 1995), and/or the level of minerals such as potassium (Ogihara et al., 2002), all of which can greatly differ between GB and purified diets.

Aside from salt-sensitive rats, outbred rat strains such as the Sprague–Dawley (which are in widespread use for obesity research) can develop hypertension. When fed an 8% NaCl diet, hypertension develops, but this usually occurs over a longer time period and to a lesser magnitude compared with Dahl SS rats (Thierry-Palmer et al., 2010). Sprague–Dawley rats can also develop hypertension as they become obese on an HFD (Dobrian et al., 2000). Diets high in fructose (around 60% of calories) but with normal levels of NaCl (0.1%) can induce metabolic abnormalities, including increased blood pressure (Vasudevan et al., 2005; Sánchez-Lozada et al., 2007) and kidney damage in both Sprague–Dawley and Wistar rats (Hwang et al., 1987; Vasudevan et al., 2005; Sánchez-Lozada et al., 2007). The IR induced by such high-fructose diets (Hwang et al., 1987) is believed to play a causal role in the development of hypertension (DeFronzo, 1981).

Even in a spontaneous rat model of hypertension (such as the spontaneously hypertensive rat [SHR], which will develop hypertension on a variety of diets), diet can be used to modify the onset or degree of this disease. For example, dietary supplementation with antioxidants (such as vitamins E and C) can lower blood pressure in stroke-prone SHR (Noguchi et al., 2004), as can dietary calcium supplementation (Sallinen et al., 1996). As mentioned earlier, the mouse is not as widely used for the study of diet-induced hypertension. Inbred mice such as the C57BL/6 can develop elevated blood pressure on purified diets high in NaCl (8%), though this appears to be on the order of several months (Yu et al., 2004).

Although C57BL/6 mice are sensitive to diet-induced obesity when fed an HFD, mean arterial blood pressure was slightly, but significantly increased relative to a low-fat diet (around 3 mm Hg) after 7 days of feeding a very HFD (i.e., 60 kcal% fat, mainly lard). The addition of 5% NaCl to the HFD increased blood pressure by 4.5 mm Hg compared to baseline measures on an HFD with normal NaCl levels (i.e., 0.3%) and was also increased (by 2.1 mm Hg) compared to those fed a low-fat diet with similar 5% NaCl levels (Nizar et al., 2016). From a mechanism standpoint, they observed that after being exposed to an acute NaCl load, high fat feeding reduced sodium excretion compared with low fat feeding.

Metabolic Disease Development and the Importance of Fiber Type

Although increasing dietary fat in the diet is a major factor affecting weight gain and metabolic disease development in rodents, the underlying agents that may be key to driving these effects are the type of fiber and the gut microbiome. Over 2000 years ago, Hippocrates proposed a link between gut health and disease development ("All disease starts in the gut"), and recent studies in rodent models have shown that there is merit to this blanket statement when it comes to metabolic disease and possibly other diseases. The presence of the 100 trillion bacteria in our lower gastrointestinal tract (made up of 500–1000 different species) has important implications in this link between the gut and metabolic disease development. These bacteria are "fed" with the fiber present in our diets, in particular, soluble fiber (i.e., prebiotic fibers), and in turn this fiber can affect bacterial composition rapidly, leading to changes in gut health. Replacement of cellulose (i.e., insoluble fiber typically used in purified diets) with soluble fibers (such as inulin) in a HFD has been found to have a dramatic effect on gut health, including reduced adiposity, improved gut morphology, improved insulin sensitivity, better glucose tolerance, increased beneficial bacteria and SCFAs, and reduced gap junctions between enterocytes, the latter of which reduces permeability and slows transfer of bacterial substances such as lipopolysaccharides thus limiting low-grade inflammation. All of these effects are preceded by changes in gut microbiota (Chassaing et al., 2015, 2017; Zou et al., 2017). These effects by inulin on gut and metabolic health were found to be mediated by interleukin (IL)-22 rather than changes in short-chain fatty acids in one study (Zou et al., 2017). IL-22 is produced by immune cells to promote both epithelial cell proliferation and induce antimicrobials, which provides a direct link to how fiber-induced changes in gut microbiota influence low-grade inflammation and metabolic disease risk (Zou et al., 2017). However, the particular diet may play some role in this process as Brooks et al. reported that short-chain fatty acid production was important to the beneficial influence of soluble fiber on reducing adiposity and liver triglyceride levels when animals were fed a lower level of inulin (7.5 g%) in the context of a high-fat (21 g % as milk fat)/high- sucrose (34 g%) diet (Brooks et al., 2017). Therefore, it's important to consider the diet background when forming conclusions regarding how fiber is mediating its effect.

Know Your Control Diet

The choice of the control diet used in a diet-induced metabolic disease study can profoundly affect data interpretation. When a study is performed using a GB diet as a "control" diet for a purified HFD, it is not possible to know whether the differences between the HFD and the GB diet are driven by the higher level of fat or because of the other factors that differ between the diets. To know that the observed differences are due to the high fat content, it is important to have a low-fat diet that matches the nutrient sources in the HFD (except for the fat and carbohydrate levels). This is easy to do with purified ingredients because each ingredient contains one main nutrient. However, all too often do we find that a GB diet has been used as the "control" diet for a purified HFD (Warden and Fisler, 2008; Pellizzon and Ricci, 2018). Using a GB diet as a comparator diet can lead to a severe misinterpretation of how a purified HFD influences the gut morphology and microbiota. This was highlighted in recent studies that demonstrated that mice fed a high-fat purified diet (i.e., 60 kcal% fat, mainly lard) had a reduced cecum and colon size relative to those from mice fed a GB diet, whereas there was no difference in gut morphology in the matched low-fat purified diet (10 kcal% fat, increased carbohydrate in place of lard) (Chassaing et al., 2015; Dalby et al., 2017). Furthermore, the same high-fat purified diet altered ileal and cecal microbiota profiles significantly from those fed the GB diet, whereas those fed a matched low-fat purified diet were similar to the high-fat fed group (Dalby et al., 2017). Although fat level is an important variable for driving adiposity, fiber type can mediate effects on gut morphology and microbiota in rodents, which can in turn alter metabolic disease development by an HFD. Therefore, it is important to choose the control diet carefully and consider the fiber type when designing any metabolic disease studies (Pellizzon and Ricci, 2018).

Potential Effects of Grain-Based Diets and Low-Fat Purified Diets on the Rodent Phenotype

Similar to the importance of choosing the proper control diet, choosing the diet in any study should not be overlooked as this may lead to significant misinterpretation of data. Like mice with a certain genetic background, each diet background is unique and there is no such thing as a "standard chow", a common reference made for the diet used in methods sections of publications. In fact, this vague term and the lack of reporting about the particular diet being fed suggest a general lack of consideration for the diet background in most studies, which is perplexing, given how dietary factors have been shown to alter study outcomes in so many cases. Both GB and purified low-fat diets (such as the AIN diets) can alter an animal's health status (either unfavorably or favorably) and interact with, attenuate, or completely mask the influence of a toxicological compound of interest on a given phenotype. Therefore, the details of the diet used (i.e., diet number, formulation, if available) should be reported in any study for the scientific community to critically evaluate the validity of the findings.

Grain-Based Diets

Phytoestrogens and Development/Maturation

As discussed previously, GB diets can contain soybean and alfalfa meals, which are known to contain biologically relevant and variable levels of phytoestrogens including genistin, daidzin, glyceitin, genistein, daidzein, and coumestrol. In contrast, purified diets such as the *AIN-76A* and *AIN-93* series diets contain no phytoestrogens. When ingested, these phytoestrogens undergo biotransformation into more absorbable forms such that genistin and daidzin are converted into the isoflavones genistein and daidzein. Daidzein can undergo further metabolism by resident bacteria to S-(-) equol and enter the circulation. This latter form, equol, is found in highest concentrations in the serum and urine of adult rats and mice consuming soybean meal–containing GB diets, and it remains high in newborn pups after in utero exposure to these diets (Brown and Setchell, 2001). As mentioned earlier, different GB diet batches (i.e., same diet, but different mill dates) can vary in isoflavone concentrations three- to fourfold, and this can have a concentration-dependent influence on circulating phytoestrogens levels (Thigpen et al., 2004). However, these circulating levels can differ significantly in mice fed different GB diets, even when dietary isoflavones levels are similar (Brown and Setchell, 2001) suggesting that it is difficult to predict circulating isoflavone levels based on the concentration in the diet. Once in the circulation, these phytoestrogens are free to target estrogen receptors, in particular estrogen receptor β, but with a lower affinity compared to estradiol (Oseni et al., 2008). The effects can be either pro- or antiestrogenic, depending on the stage of life. Examples of a proestrogenic effect are found during development and maturation and include a dose-dependent reduction in timing of vaginal opening (VO) and an increase in uterine weights when rats and mice have prepubertal exposure to phytoestrogens (Thigpen et al., 2002, 2007; Heindel and vom Saal, 2008). Examples of antiestrogenic effects include the antagonistic effect of soy isoflavones on endogenous estrogens in certain cancers as discussed in the next section. The levels of phytoestrogens in GB diets are also high enough to negate effects of endogenously administered endocrine disruptors (such as bisphenol A) on DNA methylation and oocyte development, and their variability may contribute to inconsistencies in data where such compounds are studied during earlier stages of life (Thigpen et al., 2013).

Phytoestrogens and Cancer

Aside from altering pubertal onset, GB diets have the potential to inhibit carcinogenesis, and the presence of many plant-based compounds such as phytoestrogens is likely key to the reduced frequency and slower onset of tumors in certain cancer models relative to phytoestrogen-free diets such as the *AIN-76A* (Fullerton et al., 1991, 1992). There can be several mechanisms by which phytoestrogens play a role in carcinogenesis, but one is likely through their ability to bind estrogen receptors (Nikov et al., 2000). Their action may be either pro- or anticarcinogenic depending on age, mode of cancer induction, phytoestrogen dose, and rodent model being studied (Bouker and Hilakivi-Clarke, 2000). Genistein can dose-dependently increase mammary tumor area in estrogen-sensitive ovariectomized mice (Allred et al., 2001), but in contrast, early life (gestation and lactation) exposure to genistein can dose-dependently *reduce* tumor formation in a carcinogen-induced mammary cancer model with intact ovaries (Fritz et al., 1998). In a prostate cancer mouse model (transgenic mice with prostatic adenocarcinoma), genistein at levels found in GB diets can reduce prostate weight of mice with adenomas and improve survival (Mentor-Marcel et al., 2005). Phytoestrogens can, depending on the dose, either inhibit or promote the efficacy of drugs called selective estrogen receptor modulators (SERMs). For example, in a transgenic mouse model of mammary carcinogenesis, lower doses of isoflavones (211 mg/kg diet) abrogated tamoxifen's ability to reduce carcinogenesis relative to either higher concentrations of isoflavones (491 mg/kg) or phytoestrogen-free diets (GB diet or purified) based on casein (Liu et al., 2005). Therefore in studies evaluating the influence of SERMs on tumors, it is important to eliminate exposure to phytoestrogens or any other compounds in diets that can bind estrogen receptors.

GB diets that contain low levels of phytoestrogens are available but it is not possible to truly control other factors within these GB diets, given each ingredient contains multiple nutrients and non-nutrients. Aside from offering a clean background and consistent composition that is phytoestrogen free, the refined nature of the ingredients in purified diets allows for easy customizations, which provide researchers an additional advantage in the cancer field. Making select changes to various nutrients such as increasing the level of fat calories in place of carbohydrate, reducing fiber, and certain vitamins (i.e., lower folate) and minerals (i.e., lower calcium) can increase tumor incidence in wild-type mice without carcinogen exposure given a long enough timeframe (Newmark et al., 2001; Yang et al., 2008). Therefore, purified diets allow the researcher to conduct studies with animals that may be more relevant to humans in addition to minimizing the presence of estrogenic compounds that can influence carcinogenesis.

Arsenic and Heavy Metals

GB diets are commonly based on cereal grains such as ground wheat, ground corn, and ground oats and meat meals such as fish meal, all of which can contain varying levels of toxic heavy metals (Newberne, 1975; Greenman et al., 1980; Wise, 1982). It has been suggested that the levels of these heavy metals in rodent GB diets are not high enough to affect animals from a disease standpoint (Greenman et al., 1980). However, some toxic heavy metals such as arsenic can be found in these diets at biologically relevant levels, and it is critical to consider potential dietary contributions especially when determining the effect of lower doses of heavy metals on the phenotype of rodent models. One study found that arsenic levels were quite high (390 ppb) and other heavy metals (i.e., cadmium, lead, nickel, and vanadium) were also present in a GB diet—all of which were virtually absent in the purified *AIN-76A diet* (Kozul et al., 2008). Of the 390-ppb arsenic, 56 ppb was considered inorganic arsenic, whereas none was detectable in the *AIN-76A*. This level of inorganic arsenic in this particular GB diet was higher than what is designated by the Enviromental Protection Agency as safe in drinking water (National Research Council, 1999, 2001), and therefore can severely compromise data from studies using GB diets to evaluate the influence of lower doses (i.e., 10–100 ppb) of arsenic (or other heavy metals). In fact, similar gene expression levels in both type I and type II metabolism genes (cytochrome P450 enzymes, involved in xenobiotic metabolism) as well as inflammatory genes were found in liver and lung of GB diet–fed mice with or without arsenic in water (10 or 100 ppb). In contrast clear differences in gene expression were found between dosed or control mice fed the *AIN-76A* (Kozul et al., 2008). Although arsenic in GB diet was at 4 to 40 times the level being studied and was likely to blame for the lack of expressional changes, it could be also that other factors in the diet were masking the effects.

Follow-up studies suggest that arsenic levels much lower than those found in GB diets can significantly influence an animal's phenotype. For example, 100-ppb arsenic in the water consumed by mice fed a purified diet can impair inflammatory factor activation and the immune response to infection (Kozul et al., 2009a,b). When these same investigators wanted to see how arsenic exposure early in life (i.e., during gestation and lactation) influenced the immune response in these mice, profound reductions in postnatal growth and development were observed with only 10-ppb arsenic, which was likely attributed to reduced energy from milk (i.e., lower fat content) of lactating dams fed arsenic (Kozul-Horvath et al., 2012).

Other Compounds or Contaminants

A list of phytoestrogens and toxic heavy metals in GB diets and how they may influence phenotypes are summarized in Fig. 40.2. In addition to these contaminants, other compounds that may be present in GB diets at varying concentrations include mycotoxins (zearalenone), herbicide and pesticide residues, polychlorinated dibenzo-p-dioxins (PCDDs), and dibenzofurans, which can influence various toxicological phenotypes by targeting estrogen receptors and the aryl hydrocarbon receptor (AhR) (Rao and Knapka, 1987; Schecter et al., 1996; Thigpen et al., 2004). These factors may work alone or together (additively or synergistically) with other contaminants previously discussed (i.e., phytoestrogens, heavy metals) to influence phenotype. For example, GB diets were found to activate AhR of cells that modulate immunity defense and detoxification in the intestine and throughout the body in mice (Ito et al., 2007b; Li et al., 2011). Although it is unknown what ligand(s) in GB diets affected the AhR, a similar effect on the AhR was found when mice were fed a purified diet containing a high concentration of a known ligand of the AhR (indole-3-carbinol) found in cruciferous vegetables. Although GB diets do not contain cruciferous vegetable sources, it may be that other known ligands of this receptor such as isoflavones and PCDDs known to be present in these diets could very well be responsible for these effects (Amakura et al., 2003).

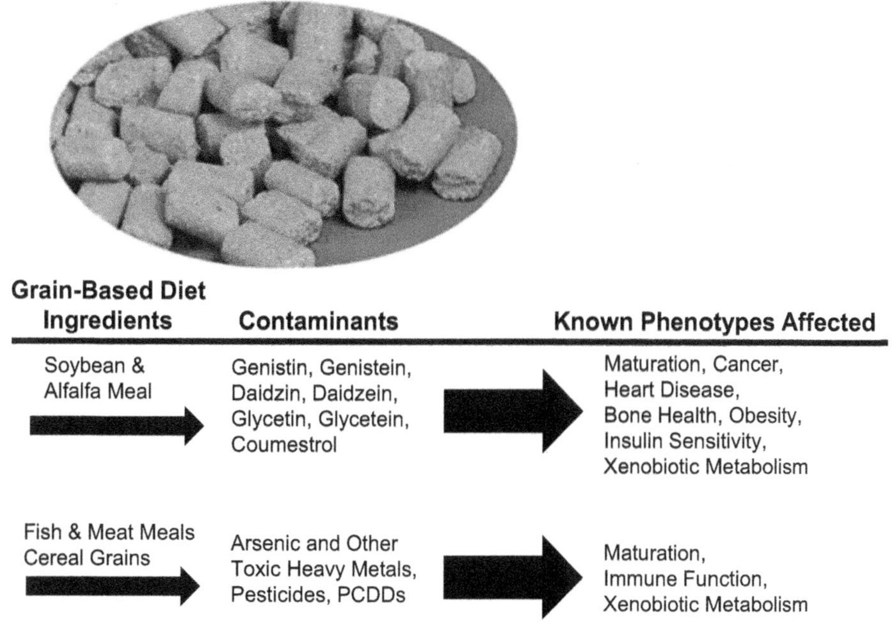

FIGURE 40.2 GB diet contaminants that can influence various phenotypes in rodents.

Purified Low-Fat Diets and Health Status of Rodents

As pointed out earlier, purified diets have been used in laboratory animal research since the 1920s and were essential for determining nutrient recommendations because researchers were able to remove specific vitamins and minerals to determine requirements. However, there are phenotypical responses to purified diets that may not be favorable, and in some cases, they have led to unfavorable health outcomes. Research has pointed out where the potential flaws may lie, which have led to improvements in the purified diet formulas in use today.

Kidney Calcinosis

The *AIN-76A* diet has a long history of use since its initial conception in 1976, but along the way, it has come under criticism with good reason. One initial observation found with feeding the *AIN-76A* was that it promoted nephrocalcinosis (or kidney calcinosis (KC)), particularly in rats. Nephrocalcinosis is an abnormal condition of the kidney in which deposits of calcium (Ca) form in the filtering units. This may reduce kidney function and eventually lead to significant damage and kidney stone formation. Although calcium and phosphorus levels were in the recommended range for rodents, studies done carefully were able to determine that a low Ca to phosphorus (P) molar ratio (0.75) in the *AIN-76A* was the culprit for KC in weanling female rats (Cockell et al., 2002; Cockell and Belonje, 2004). This phenotype develops very quickly (i.e., 2–4 weeks) and is irreversible in these rats (Ritskes-Hoitinga et al., 1989; Peterson et al., 1996; Matsuzaki et al., 2002; Cockell and Belonje, 2004). Other factors that may affect development of KC within this diet include the type of carbohydrate, fiber, and sulfur-containing amino acid (AA) supplement—specifically, fructose (Bergstra et al., 1993), insoluble fiber cellulose (Anderson et al., 1985), and DL-methionine (Reeves et al., 1993b). That said, it is important to point out also that the Ca to P ratio is likely most important to this phenotype given GB diets such as the NIH-07 or NIH-31M (Ca to P ratio = 0.9) were found to cause mild KC in young, female rats (Rao, 2002).

When the AIN committee met to formulate the *AIN-93* series diets, increasing the Ca to P ratio above 1 was at the top of their list (Reeves et al., 1993a). The *AIN-93* diets contain a Ca to P ratio of 1.3:1 and that, with a lower level of sucrose (10% vs. 50%) was able to "fix" the KC problem, but at the expense of having a mineral mix that is deficient in phosphorus and relying on casein (0.7% phosphorus) to fill the phosphorus void. Even still, a previous report suggested that some rats still have a mild degree of KC when fed the *AIN-93G* diet, suggesting this "fix" was not enough (Cockell and Belonje, 2004).

Pubertal Onset

Given what was discussed above regarding the influence of phytoestrogens on pubertal onset, one might speculate that a phytoestrogen-free diet would prevent precocious pubertal onset in rodents. However, a phytoestrogen-free diet such as the *AIN-76A* or *AIN-93G* diet can also reduce the timing of VO and increase uterine weight, even in some cases when compared with phytoestrogen-containing GB diets, depending on the animal model (Thigpen et al., 2002, 2007). This is thought to be due to the higher metabolizable energy (or caloric density) content of purified diets (3.9–4 kcal/g diet in *AIN-76A* and *AIN-93G*, respectively) compared with GB diets (3–3.5 kcal/g) (Thigpen et al., 2002, 2007). GB diets typically contain 15%–25% fiber (both insoluble and soluble) and around 3%–5% soluble fiber (Pellizzon and Ricci, unpublished observations), whereas the AIN diets contain 5% of it (only insoluble fiber). Therefore, although it is currently unknown what particular dietary factor(s) in purified diets is "estrogenic," it is possible that a reduced fiber content (and therefore higher metabolizable energy level) and perhaps even lack of soluble fiber are responsible and can easily be addressed by simply adding more fiber in purified diets.

Metabolic Effects

Aside from KC, the metabolic phenotype, including plasma and liver lipids, glucose tolerance, and blood pressure, can be significantly altered by purified diets and in some cases, not on purpose. For example, the *AIN-76A* and *AIN-93G* can have adverse effects on metabolic phenotypes in rodents, including increased body weight, blood and liver lipids, and blood pressure, compared to GB diets (Fullerton et al., 1992; Reeves et al., 1993b; Lien et al., 2001; Mattson et al., 2004). These diets contain sucrose in two concentrations (50% and 10% in the *AIN-76A* and *AIN-93G* diets, respectively), and high sucrose has been found to reduce glucose tolerance in C57BL/6 mice (Sumiyoshi et al., 2006) and induce IR, hyperTG, hepatic steatosis, and hypertension in rats (Hwang et al., 1987; Pagliassotti and Prach, 1995; Pagliassotti et al., 1996), even at lower concentrations, at least in rats (Thresher et al., 2000).

As mentioned above, fiber levels are also very low in purified diets compared with GB diets. Furthermore, purified diets typically have only insoluble fiber (cellulose), whereas GB diets have both insoluble and soluble fiber sources from cereal grains. Crude fiber levels of GB diets are typically listed at 5%–6%, but these diets typically contain four times more total dietary fiber than the crude fiber analyses indicate (Wise and Gilburt, 1980) (~18% insoluble and ~2–5% soluble). In comparison, the *AIN-76A* and *AIN-93G* contain 5% fiber by weight (50 g per 3902 kcals and 50 g per 4000 kcals for *AIN-76A* and *AIN-93G*, respectively) as cellulose (100% insoluble fiber). Modification of total fiber and fiber type in purified diets to be similar in composition to GB diets may lead to a healthier rodent phenotype. For example, addition of soluble fibers such as inulin can also reduce body weight, adiposity, and

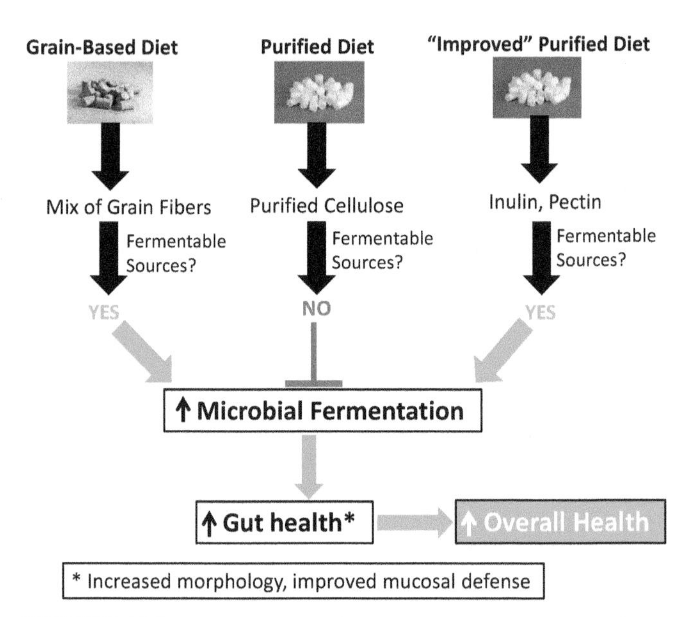

FIGURE 40.3 Current purified diets may be improved by adding soluble fiber sources.

blood and liver lipids and improve glucose tolerance of rodents, and these changes are likely due to increases in short-chain fatty acids produced by fermentation of inulin by gut microbiota (Delzenne et al., 2002, 2007). Other mechanisms likely exist, including more direct effects of microbiota to reduce inflammation, and increase epithelial cell formation and mucosal layer thickness for improved barrier function (Chassaing et al., 2017; Zou et al., 2017). Additional fiber (both nonfermentable and fermentable) can also influence risk for other chronic diseases including cancer (Jacobs, 1986; Taper and Roberfroid, 2002) and may improve life span in rodents fed purified diets, which is usually shorter than those fed GB diets (Fullerton et al., 1992). Therefore, an increase in total fiber level and addition of soluble fiber are warranted in purified diets for normal growth and health maintenance of rodents (Fig. 40.3).

CONCLUDING REMARKS AND FUTURE DIRECTIONS

The use of animals in research has provided us with an important means to understand more about our biology in general and the many toxicological factors in our environment that can affect our health. When designing any study using rodent models, diet needs to be considered, and one should ask three questions before making a decision: Can I report the diet? Can I repeat the diet? Can I revise the diet? Purified diets allow for each of these questions to be answered as "yes" and such diets are required in toxicological studies to minimize background contaminants.

References

Aggarwal, D., West, K.L., Zern, T.L., et al., 2005. JTT-130, a microsomal triglyceride transfer protein (MTP) inhibitor lowers plasma triglycerides and LDL cholesterol concentrations without increasing hepatic triglycerides in guinea pigs. BMC Cardiovasc. Disord. 5, 30.

Alexaki, A., Wilson, T.A., Atallah, M.T., et al., 2004. Hamsters fed diets high in saturated fat have increased cholesterol accumulation and cytokine production in the aortic arch compared with cholesterol-fed hamsters with moderately elevated plasma non-HDL cholesterol concentrations. J. Nutr. 134, 410–415.

Allred, C.D., Allred, K.F., Ju, Y.H., et al., 2001. Soy diets containing varying amounts of genistein stimulate growth of estrogen-dependent (MCF-7) tumors in a dose-dependent manner. Cancer Res. 61, 5045–5050.

Amakura, Y., Tsutsumi, T., Nakamura, M., et al., 2003. Activation of the aryl hydrocarbon receptor by some vegetable constituents determined using in vitro reporter gene assay. Biol. Pharm. Bull. 26, 532–539.

Anderson, R.L., Kanerva, R.L., Francis, W.R., 1985. Effect of dietary fiber composition on renal mineralization in female rats ingesting purified diet - AIN-76. Nutr. Rep. Int. 31, 1331–1339.

Ando, H., Tsuruoka, S., Yamamoto, H., et al., 2005. Regulation of cholesterol 7alpha-hydroxylase mRNA expression in C57BL/6 mice fed an atherogenic diet. Atherosclerosis 178, 265–269.

Anstee, Q.M., Goldin, R.D., 2006. Mouse models in non-alcoholic fatty liver disease and steatohepatitis research. Int. J. Exp. Pathol. 87, 1–16.

Basciano, H., Federico, L., Adeli, K., 2005. Fructose, insulin resistance, and metabolic dyslipidemia. Nutr. Metab. 2, 5.

Basciano, H., Miller, A.E., Naples, M., et al., 2009. Metabolic effects of dietary cholesterol in an animal model of insulin resistance and hepatic steatosis. Am. J. Physiol. Endocrinol. Metab. 297, E462–E473.

Bergstra, A.E., Lemmens, A.G., Beynen, A.C., 1993. Dietary fructose vs. glucose stimulates nephrocalcinogenesis in female rats. J. Nutr. 123, 1320–1327.

Bieri, J.G., 1980. Second report of the ad hoc committee on standards for nutritional studies. J. Nutr. 110, 1726.

Bieri, J.G., Stoewsand, G.S., Briggs, G.M., et al., 1977. Report of the American Institute of nutrition ad hoc committee on standards for nutritional studies. J. Nutr. 107, 1340–1348.

Blair, R.M., Appt, S.E., Bennetau-Pelissero, C., et al., 2002. Dietary soy and soy isoflavones have gender-specific effects on plasma lipids and isoflavones in golden Syrian f(1)b hybrid hamsters. J. Nutr. 132, 3585–3591.

Bouker, K.B., Hilakivi-Clarke, L., 2000. Genistein: does it prevent or promote breast cancer? Environ. Health Perspect. 108, 701–708.

Brooks, L., Viardot, A., Tsakmaki, A., et al., 2017. Fermentable carbohydrate stimulates FFAR2-dependent colonic PYY cell expansion to increase satiety. Mol. Metab. 6, 48–60.

Brown, N.M., Setchell, K.D., 2001. Animal models impacted by phytoestrogens in commercial chow: implications for pathways influenced by hormones. Lab. Invest. 81, 735–747.

Buettner, R., Parhofer, K., Woenckhaus, M., et al., 2006. Defining high-fat-diet rat models: metabolic and molecular effects of different fat types. J. Mol. Endocrinol. 36, 485. Soc Endocrinology.

Buñag, R.D., Tomita, T., Sasaki, S., 1983. Chronic sucrose ingestion induces mild hypertension and tachycardia in rats. Hypertension 5, 218–225.

Burgmaier, M., Sen, S., Philip, F., et al., 2010. Metabolic adaptation follows contractile dysfunction in the heart of obese Zucker rats fed a high-fat "Western" diet. Obesity 18, 1895–1901.

Cederroth, C.R., Vinciguerra, M., Gjinovci, A., et al., 2008. Dietary phytoestrogens activate AMP-activated protein kinase with improvement in lipid and glucose metabolism. Diabetes 57, 1176–1185.

Cederroth, C.R., Vinciguerra, M., Kühne, F., et al., 2007. A phytoestrogen-rich diet increases energy expenditure and decreases adiposity in mice. Environ. Health Perspect. 115, 1467–1473.

Chang, S.A.M., Lin, D., Peters, J.C., 1990. Metabolic differences between obesity-prone and obesity-resistant rats. Am. J. Physiol. 259, R1103–R1110.

Chassaing, B., Miles-Brown, J., Pellizzon, M., et al., 2015. Lack of soluble fiber drives diet-induced adiposity in mice. Am. J. Physiol. 309, G528–G541.

Chassaing, B., Vijay-Kumar, M., Gewirtz, A.T., 2017. How diet can impact gut microbiota to promote or endanger health. Curr. Opin. Gastroenterol. 33, 417–421.

Chiba, T., Suzuki, S., Sato, Y., et al., 2016. Evaluation of methionine content in a high-fat and choline-deficient diet on body weight gain and the development of non-alcoholic steatohepatitis in mice. PLoS One 11.

Chicco, A., D'Alessandro, M.E., Karabatas, L., et al., 2003. Muscle lipid metabolism and insulin secretion are altered in insulin-resistant rats fed a high sucrose diet. J. Nutr. 133, 127–133.

Cockell, K.A., Belonje, B., 2004. Nephrocalcinosis caused by dietary calcium:phosphorus imbalance in female rats develops rapidly and is irreversible. J. Nutr. 134, 637–640.

Cockell, K.A., L'Abbé, M.R., Belonje, B., 2002. The concentrations and ratio of dietary calcium and phosphorus influence development of nephrocalcinosis in female rats. J. Nutr. 132, 252–256.

Collins, A.R., Meehan, W.P., Kintscher, U., et al., 2001. Troglitazone inhibits formation of early atherosclerotic lesions in diabetic and nondiabetic low density lipoprotein receptor-deficient mice. Arterioscler. Thromb. Vasc. Biol. 21, 365–371.

Conde, K., Vergara-Jimenez, M., Krause, B.R., et al., 1996. Hypocholesterolemic actions of atorvastatin are associated with alterations on hepatic cholesterol metabolism and lipoprotein composition in the guinea pig. J. Lipid Res. 37, 2372–2382.

Cos, E., Ramjiganesh, T., Roy, S., et al., 2001. Soluble fiber and soybean protein reduce atherosclerotic lesions in guinea pigs. Sex and hormonal status determine lesion extension. Lipids 36, 1209–1216.

Dalby, M.J., Ross, A.W., Walker, A.W., et al., 2017. Dietary uncoupling of gut microbiota and energy harvesting from obesity and glucose tolerance in mice. Cell Rep. 21, 1521–1533 (Elsevier Company).

Daly, M.E., Vale, C., Walker, M., et al., 1997. Dietary carbohydrates and insulin sensitivity: a review of the evidence and clinical implications. Am. J. Clin. Nutr. 66, 1072–1085.

Davis, H.R., Compton, D.S., Hoos, L., et al., 2001. Ezetimibe, a potent cholesterol absorption inhibitor, inhibits the development of atherosclerosis in ApoE knockout mice. Arterioscler. Thromb. 2032–2038.

de La Serre, C.B., Ellis, C.L., Lee, J., et al., 2010. Propensity to high-fat diet-induced obesity in rats is associated with changes in the gut microbiota and gut inflammation. Am. J. Physiol. Gastrointest. Liver Physiol. 299, G440–G448.

De Souza, C.T., Araujo, E.P., Bordin, S., et al., 2005. Consumption of a fat-rich diet activates a proinflammatory response and induces insulin resistance in the hypothalamus. Endocrinology 146, 4192–4199.

DeFronzo, R.A., 1981. The effect of insulin on renal sodium metabolism. A review with clinical implications. Diabetologia 21, 165–171.

DeFuria, J., Bennett, G., Strissel, K.J., et al., 2009. Dietary blueberry attenuates whole-body insulin resistance in high fat-fed mice by reducing adipocyte death and its inflammatory sequelae. J. Nutr. 139, 1510–1516.

Delzenne, N.M., Cani, P.D., Neyrinck, A.M., 2007. Modulation of glucagon-like peptide 1 and energy metabolism by inulin and oligofructose: experimental data. J. Nutr. 137, 2547S–2551S.

Delzenne, N.M., Daubioul, C., Neyrinck, A., et al., 2002. Inulin and oligofructose modulate lipid metabolism in animals: review ofbiochemical events and future prospects. Br. J. Nutr. 87, S255–S259.

Dietschy, J.M., Turley, S.D., Spady, D.K., 1993. Role of liver in the maintenance of cholesterol and low density lipoprotein homeostasis in different animal species, including humans. J. Lipid Res. 34, 1637–1659.

Dobrian, A.D., Davies, M.J., Prewitt, R.L., et al., 2000. Development of hypertension in a rat model of diet-induced obesity. Hypertension 35, 1009–1015.

Donovan, M.J., Paulino, G., Raybould, H.E., 2009. Activation of hindbrain neurons in response to gastrointestinal lipid is attenuated by high fat, high energy diets in mice prone to diet-induced obesity. Brain Res. 1248, 136–140 (Elsevier B.V).

Drake, A.J., Livingstone, D.E.W., Andrew, R., et al., 2005. Reduced adipose glucocorticoid reactivation and increased hepatic glucocorticoid clearance as an early adaptation to high-fat feeding in Wistar rats. Endocrinology 146, 913–919.

Eldridge, A.C., Kwolek, W.F., 1983. Soybean isoflavones: effect of environment and variety on composition. J. Agric. Food Chem. 31, 394–396.

Enriori, P.J., Evans, A.E., Sinnayah, P., et al., 2007. Diet-induced obesity causes severe but reversible leptin resistance in arcuate melanocortin neurons. Cell Metabol. 5, 181–194 (Elsevier).

Farley, C., Cook, J.A., Spar, B.D., et al., 2003. Meal pattern analysis of diet-induced obesity in susceptible and resistant rats. Obesity 11, 845–851. Nature Publishing Group.

Farnsworth, N.R., Bingel, A.S., Cordell, G.A., et al., 1975. Potential value of plants as sources of new antifertility agents. J. Pharmacol. Sci. 64, 717–754.

Fernandez, M.L., Vergara-Jimenez, M., Conde, K., et al., 1996. Dietary carbohydrate type and fat amount alter VLDL and LDL metabolism in guinea pigs. J. Nutr. 126, 2494–2504.

Fernandez, M.L., Volek, J.S., 2006. Guinea pigs: a suitable animal model to study lipoprotein metabolism, atherosclerosis and inflammation. Nutr. Metab. 3, 17.

Fernandez, M.L., Wilson, T.A., Conde, K., et al., 1999. Hamsters and guinea pigs differ in their plasma lipoprotein cholesterol distribution when fed diets varying in animal protein, soluble fiber, or cholesterol content. J. Nutr. 129, 1323–1332.

Fritz, W.A., Coward, L., Wang, J., et al., 1998. Dietary genistein: perinatal mammary cancer prevention, bioavailability and toxicity testing in the rat. Carcinogenesis 19, 2151–2158.

Fujita, K., Nozaki, Y., Yoneda, M., et al., 2010. Nitric oxide plays a crucial role in the development/progression of nonalcoholic steatohepatitis in the choline-deficient, l-amino acid-defined diet-fed rat model. Alcohol Clin. Exp. Res. 34 (Suppl 1), S18–S24.

Fullerton, F.R., Greenman, D.L., Bucci, T., 1992. Effects of diet type on incidence of spontaneous and acetylaminofluorene-induced liver and bladder tumors in Balb/c mice fed AIN-76A versus NIH-07 diet. Fund. Appl. Toxicol. 18, 193–199.

Fullerton, F.R., Greenman, D.L., McCarty, C.C., et al., 1991. Increased incidence of spontaneous and 2-acetylaminofluorene-induced liver and bladder tumors in B6C3F1 mice fed AIN-76A diet versus NIH-07 diet. Fund. Appl. Toxicol. 16, 51–60.

Gallou-Kabani, C., Vigé, A., Gross, M.S., et al., 2007. C57BL/6J and A/J mice fed a high-fat diet delineate components of metabolic syndrome. Obesity 15, 1996–2005. Nature Publishing Group.

Gauthier, M.-S., Favier, R., Lavoie, J.-M., 2006. Time course of the development of non-alcoholic hepatic steatosis in response to high-fat diet-induced obesity in rats. Br. J. Nutr. 95, 273–281.

Getz, G.S., Reardon, C.A., 2006. Diet and murine atherosclerosis. Arterioscler. Thromb. Vasc. Biol. 26, 242–249.

Greenfield, H., Briggs, G.M., 1971. Nutritional methodology in metabolic research with rats. Annu. Rev. Biochem. 40, 549–572.

Greenman, D.L., Oller, W.L., Littlefield, N.A., et al., 1980. Commercial laboratory animal diets: toxicant and nutrient variability. J. Toxicol. Environ. Health 6, 235–246.

Hecker, P.A., Mapanga, R.F., Kimar, C.P., et al., 2012. Effects of glucose-6-phosphate dehydrogenase deficiency on the metabolic and cardiac responses to obesogenic or high-fructose diets. AJP Endocrinol Metab 303, E959–E972.

Heindel, J.J., vom Saal, F.S., 2008. Meeting report: batch-to-batch variability in estrogenic activity in commercial animal diets—importance and approaches for laboratory animal research. Environ. Health Perspect. 116, 389–393.

Horton, J.D., Cuthbert, J.A., Spady, D.K., 1995. Regulation of hepatic 7 alpha-hydroxylase expression and response to dietary cholesterol in the rat and hamster. J. Biol. Chem. 270, 5381–5387.

Horton, T.J., Gayles, E.C., Prach, P.A., et al., 1997. Female rats do not develop sucrose-induced insulin resistance. Am. J. Physiol. 272, R1571–R1576.

Hrncir, T., Stepankova, R., Kozakova, H., et al., 2008. Gut microbiota and lipopolysaccharide content of the diet influence development of regulatory T cells: studies in germ-free mice. BMC Immunol. 9, 65.

Hwang, I.S., Ho, H., Hoffman, B.B., et al., 1987. Fructose-induced insulin resistance and hypertension in rats. Hypertension 10, 512–516.

Ikemoto, S., Takahashi, M., Tsunoda, N., et al., 1996. High-fat diet-induced hyperglycemia and obesity in mice: differential effects of dietary oils. Metabolism 45, 1539–1546 (Elsevier).

Ikemoto, S., Takahashi, M., Tsunoda, N., et al., 1997. Cholate inhibits high-fat diet-induced hyperglycemia and obesity with acyl-CoA synthetase mRNA decrease. Am. J. Physiol. 37–45.

Ito, M., Suzuki, J., Tsujioka, S., et al., 2007a. Longitudinal analysis of murine steatohepatitis model induced by chronic exposure to high-fat diet. Hepatol. Res. 37, 50–57. Wiley Online Library.

Ito, S., Chen, C., Satoh, J., et al., 2007b. Dietary phytochemicals regulate whole-body CYP1A1 expression through an arylhydrocarbon receptor nuclear translocator-dependent system in gut. J. Clin. Invest. 117, 1940–1950.

Jacobs, L.R., 1986. Relationship between dietary fiber and cancer: metabolic, physiologic, and cellular mechanisms. Proc. Soc. Exp. Biol. Med. 183, 299–310. Royal Society of Medicine.

Jensen, M.N., Ritskes-Hoitinga, M., 2007. How isoflavone levels in common rodent diets can interfere with the value of animal models and with experimental results. Lab. Anim. 41, 1–18.

Jeong, W.-I., Jeong, D.-H., Do, S.-H., et al., 2005. Mild hepatic fibrosis in cholesterol and sodium cholate diet-fed rats. J. Vet. Med. Sci. 67, 235–242.

Jiang, L., Wang, Q., Yu, Y., et al., 2009. Leptin contributes to the adaptive responses of mice to high-fat diet intake through suppressing the lipogenic pathway. PLoS One 4, e6884.

Johnson, J.L., George, S.J., Newby, A.C., et al., 2005. Divergent effects of matrix metalloproteinases 3, 7, 9, and 12 on atherosclerotic plaque stability in mouse brachiocephalic arteries. Proc. Natl. Acad. Sci. U.S.A. 102, 15575–15580.

Joseph, S.B., McKilligin, E., Pei, L., et al., 2002. Synthetic LXR ligand inhibits the development of atherosclerosis in mice. Proc. Natl. Acad. Sci. U.S.A. 99, 7604–7609.

Kahlon, T., Chow, F.I., Irving, D.W., et al., 1996. Cholesterol response and foam cell formation in hamsters fed two levels of saturated fat and various levels of cholesterol. Nutr. Res. 16, 1353–1368.

Karmakar, S., Das, D., Maiti, A., et al., 2011. Black tea prevents high fat diet-induced non-alcoholic steatohepatitis. Phyther Res. 25, 1073–1081.

Kasim-Karakas, S.E., Vriend, H., Almario, R., et al., 1996. Effects of dietary carbohydrates on glucose and lipid metabolism in golden Syrian hamsters. J. Lab. Clin. Med. 128, 208–213.

Kawasaki, T., Igarashi, K., Koeda, T., et al., 2009. Rats fed fructose-enriched diets have characteristics of nonalcoholic hepatic steatosis. J. Nutr. 139, 2067–2071.

Khosla, P., Sundram, K., 1996. Effects of dietary fatty acid composition on plasma cholesterol. Prog. Lipid Res. 35, 93–132.

King, V.L., Hatch, N.W., Chan, H.W., et al., 2009. A murine model of obesity with accelerated atherosclerosis. Obesity 18, 35–41 (Nature Publishing Group).

Kirsch, R., Clarkson, V., Shephard, E.G., et al., 2003. Rodent nutritional model of non-alcoholic steatohepatitis: species, strain and sex difference studies. J. Gastroenterol. Hepatol. 18, 1272–1282.

Knapka, J.J., 1988. Animal diets for biomedical research. Lab. Anim. 17, 17–18.

Konda, T., Enomoto, A., Takahara, A., et al., 2006. Effects of L/N-type calcium channel antagonist, cilnidipine on progressive renal injuries in Dahl salt-sensitive rats. Biol. Pharm. Bull. 29, 933–937.

Kozul-Horvath, C.D., Zandbergen, F., Jackson, B.P., et al., 2012. Effects of low-dose drinking water arsenic on mouse fetal and postnatal growth and development. PLoS One 7, 1–9.

Kozul, C.D., Ely, K.H., Enelow, R.I., et al., 2009a. Low-dose arsenic compromises the immune response to Influenza A infection in vivo. Environ. Health Perspect. 117, 1441–1447.

Kozul, C.D., Hampton, T.H., Davey, J.C., et al., 2009b. Chronic exposure to arsenic in the drinking water alters the expression of immune response genes in mouse lung. Environ. Health Perspect. 117, 1108–1115.

Kozul, C.D., Nomikos Athena, P., Hampton, T.H., et al., May 28, 2008. Laboratory diet profoundly alters gene expression and confounds genomic analysis in mouse liver and lung. Chem. Biol. Interact. 173 (2), 129–140.

Kulinski, A., Vance, D.E., Vance, J.E., 2004. A choline-deficient diet in mice inhibits neither the CDP-choline pathway for phosphatidylcholine synthesis in hepatocytes nor apolipoprotein B secretion. J. Biol. Chem. 279, 23916–23924.

Laurent, D., Didier, L., Yerby, B., et al., 2007. Diet-induced modulation of mitochondrial activity in rat muscle. Am. J. Physiol. Endocrinol. Metab. 293, E1169–E1177.

Levin, B., Dunn-Meynell, A., Balkan, B., et al., 1997. Selective breeding diet-induced obesity obesity selective breeding for for diet-induced and resistance in Sprague-Dawley rats and resistance in rats. Am. J. Physiol. 273, R725–R730.

Levin, B.E., Dunn-Meynell, A.A., 2006. Differential effects of exercise on body weight gain and adiposity in obesity-prone and -resistant rats. Int. J. Obes. 30, 722–727.

Li, Y., Innocentin, S., Withers, D.R., et al., 2011. Exogenous stimuli maintain intraepithelial lymphocytes via aryl hydrocarbon receptor activation. Cell 147, 629–640.

Lichtman, A.H., Clinton, S.K., Iiyama, K., et al., 1999. Hyperlipidemia and atherosclerotic lesion development in LDL receptor-deficient mice fed defined semipurified diets with and without cholate. Arterioscler. Thromb. Vasc. Biol. 19, 1938–1944.

Lien, E.L., Boyle, F.G., Wrenn, J.M., et al., 2001. Comparison of AIN-76A and AIN-93G diets: a 13-week study in rats. Food Chem. Toxicol. 39, 385–392.

Lin, E.C., Fernandez, M.L., Tosca, M.A., et al., 1994. Regulation of hepatic LDL metabolism in the guinea pig by dietary fat and cholesterol. J. Lipid Res. 35, 446–457.

Liu, B., Edgerton, S., Yang, X., et al., 2005. Low-dose dietary phytoestrogen abrogates tamoxifen-associated mammary tumor prevention. Cancer Res. 65, 879–886.

Matsumoto, M., Hada, N., Sakamaki, Y., et al., 2013. An improved mouse model that rapidly develops fibrosis in non-alcoholic steatohepatitis. Int. J. Exp. Pathol. 94, 93–103.

Matsuzaki, H., Katsumata, S., Masuyama, R., et al., 2002. Sex differences in kidney mineral concentrations and urinary albumin excretion in rats given high-phosphorus feed. Biosci. Biotechnol. Biochem. 66, 1737–1739.

Mattson, D.L., Kunert, M.P., Kaldunski, M.L., et al., 2004. Influence of diet and genetics on hypertension and renal disease in Dahl salt-sensitive rats. Physiol. Genom. 16, 194–203.

Mentor-Marcel, R., Lamartiniere, C.A., Eltoum, I.A., et al., 2005. Dietary genistein improves survival and reduces expression of osteopontin in the prostate of transgenic mice with prostatic adenocarcinoma (TRAMP). J. Nutr. 135, 989–995.

Merat, S., Casanada, F., Sutphin, M., et al., 1999. Western-type diets induce insulin resistance and hyperinsulinemia in LDL receptor-deficient mice but do not increase aortic atherosclerosis compared with normoinsulinemic mice in which similar plasma cholesterol levels are achieved by a fructose-rich diet. Arterioscler. Thromb. Vasc. Biol. 19, 1223–1230.

Merkel, M., Velez-Carrasco, W., Hudgins, L.C., et al., 2001. Compared with saturated fatty acids, dietary monounsaturated fatty acids and carbohydrates increase atherosclerosis and VLDL cholesterol levels in LDL receptor-deficient, but not apolipoprotein E-deficient, mice. Proc. Natl. Acad. Sci. U.S.A. 98, 13294–13299.

Mesnage, R., Defarge, N., Rocque, L.M., et al., 2015. Laboratory rodent diets contain toxic levels of environmental contaminants: implications for regulatory tests. PLoS One 10, e0128429.

Mu, Y., Ogawa, T., Kawada, N., 2010. Reversibility of fibrosis, inflammation, and endoplasmic reticulum stress in the liver of rats fed a methionine-choline-deficient diet. Lab. Invest. 90, 245–256.

Nagata, R., Nishio, Y., Sekine, O., et al., 2004. Single nucleotide polymorphism (-468 Gly to A) at the promoter region of SREBP-1c associates with genetic defect of fructose-induced hepatic lipogenesis. J. Biol. Chem. 279, 29031–29042.

Nakashima, Y., Plump, A.S., Raines, E.W., et al., 1994. ApoE-deficient mice develop lesions of all phases of atherosclerosis throughout the arterial tree. Arterioscler. Thromb. 14, 133–140.

National Research Council, 1999. Arsenic in Drinking Water. Natl Acad Press, Washington, DC.

National Research Council, 2001. Arsenic in Drinking Water: 2001 Update. Natl Acad Press, Washington, DC.

National Research Council, 1972. Nutrient Requirements of Laboratory Animals, second ed. National Academy of Sciences, Washington, D.C.

Nevala, R., Vaskonen, T., Vehniäinen, J., et al., 2000. Soy based diet attenuates the development of hypertension when compared to casein based diet in spontaneously hypertensive rat. Life Sci. 66, 115–124.

Newberne, P.M., 1975. Influence on pharmacological experiments of chemicals and other factors in diets of laboratory animals. Food Sourc Incident Drug Expo 34, 209–218.

Newmark, H.L., Yang, K., Lipkin, M., et al., 2001. A Western-style diet induces benign and malignant neoplasms in the colon of normal C57Bl/6 mice. Carcinogenesis 22, 1871–1875.

Nikov, G.N., Hopkins, N.E., Boue, S., et al., 2000. Interactions of dietary estrogens with human estrogen receptors and the effect on estrogen receptor-estrogen response element complex formation. Environ. Health Perspect. 108, 867–872.

Nishina, P.M., Lowe, S., Verstuyft, J., et al., 1993. Effects of dietary fats from animal and plant sources on diet-induced fatty streak lesions in C57BL/6J mice. J. Lipid Res. 34, 1413–1422.

Nishina, P.M., Verstuyft, J., Paigen, B., 1990. Synthetic low and high fat diets for the study of atherosclerosis in the mouse. J. Lipid Res. 31, 859–869.

Nizar, J.M., Dong, W., McClellan, R.B., et al., 2016. Sodium-sensitive elevation in blood pressure is ENAC independent in diet-induced obesity and insulin resistance. Am. J. Physiol. https://doi.org/10.1152/ajprenal.00265.2015.
Noguchi, T., Ikeda, K., Sasaki, Y., et al., 2004. Effects of vitamin E and sesamin on hypertension and cerebral thrombogenesis in stroke-prone spontaneously hypertensive rats. Clin. Exp. Pharmacol. Physiol. 31 (Suppl 2), S24–S26.
Ogihara, T., Asano, T., Ando, K., et al., 2002. High-salt diet enhances insulin signaling and induces insulin resistance in Dahl salt-sensitive rats. Hypertension 40, 83–89.
Oseni, T., Patel, R., Pyle, J., et al., 2008. Selective estrogen receptor modulators and phytoestrogens. Planta Med. 74, 1656–1665.
Otto, J., Ordovas, J.M., Smith, D., et al., 1995. Lovastatin inhibits diet induced atherosclerosis in F1B golden Syrian hamsters. Atherosclerosis 114, 19–28.
Oudot, C., Lajoix, A.D., Jover, B., et al., 2013. Dietary sodium restriction prevents kidney damage in high fructose-fed rats. Kidney Int. 83, 674–683.
Owens, D., 2006. Surgically and chemically induced models of disease. In: Suckow, M.A., et al. (Eds.), The Laboratory Rat. Elsevier Academic Press, pp. 711–732.
Pagliassotti, M.J., Gayles, E.C., Podolin, D.A., et al., 2000. Developmental stage modifies diet-induced peripheral insulin resistance in rats. Am. J. Physiol. Regul. Integr. Comp. Physiol. 278, R66–R73.
Pagliassotti, M.J., Prach, P.A., 1995. Quantity of sucrose alters the tissue pattern and time course of insulin resistance in young rats. Am. J. Physiol. 269, R641–R646.
Pagliassotti, M.J., Prach, P.A., Koppenhafer, T.A., et al., 1996. Changes in insulin action, triglycerides, and lipid composition during sucrose feeding in rats. Am. J. Physiol. 271, R1319–R1326.
Park, H.R., Park, M., Choi, J., et al., 2010. A high-fat diet impairs neurogenesis: involvement of lipid peroxidation and brain-derived neurotrophic factor. Neurosci. Lett. 482, 235–239.
Park, S.-Y., Cho, Y.-R., Kim, H.-J., et al., 2005. Unraveling the temporal pattern of diet-induced insulin resistance in individual organs and cardiac dysfunction in C57BL/6 mice. Diabetes 54, 3530–3540.
Pellizzon, M.A., Ricci, M.R., 2018. The common use of improper control diets in diet-induced metabolic disease research confounds data interpretation: the fiber factor. Nutr. Metab. 15, 3.
Peterson, C.A., Baker, D.H., Erdman, J.W., 1996. Diet-induced nephrocalcinosis in female rats is irreversible and is induced primarily before the completion of adolescence. J. Nutr. 126, 259–265.
Pistell, P.J., Morrison, C.D., Gupta, S., et al., 2010. Cognitive impairment following high fat diet consumption is associated with brain inflammation. J. Neuroimmunol. 219, 25–32.
Preuss, H.G., Gondal, J.A., Bustos, E., et al., 1995. Effects of chromium and guar on sugar-induced hypertension in rats. Clin. Nephrol. 44, 170–177.
Prpic, V., Watson, P.M., Frampton, I.C., et al., 2003. Differential mechanisms and development of leptin resistance in A/J versus C57BL/6J mice during diet-induced obesity. Endocrinology 144, 1155–1163.
Rao, G.N., 2002. Diet and kidney diseases in rats. Toxicol. Pathol. 30, 651–656.
Rao, G.N., Knapka, J.J., 1987. Contaminant and nutrient concentrations of natural ingredient rat and mouse diet used in chemical toxicology studies. Fund. Appl. Toxicol. 9, 329–338.
Raubenheimer, P.J., Nyirenda, M.J., Walker, B.R., 2006. A choline-deficient diet exacerbates fatty liver but attenuates insulin resistance and glucose intolerance in mice fed a high-fat diet. Diabetes 55, 2015–2020.
Ravussin, Y., Koren, O., Spor, A., et al., 2012. Responses of gut microbiota to diet composition and weight loss in lean and obese mice. Obesity 20, 738–747.
Reeves, P.G., Nielsen, F.H., Fahey, G.C., 1993a. AIN-93 purified diets for laboratory rodents: final report of the American Institute of nutrition ad hoc writing committee on the reformulation of the AIN-76A rodent diet. J. Nutr. 123, 939–951.
Reeves, P.G., Rossow, K.L., Lindlauf, J., 1993b. Development and testing of the AIN-93 purified diets for rodents: results on growth, kidney calcification and bone mineralization in rats and mice. J. Nutr. 123, 1923–1931.
Ricci, M.R., Levin, B.E., 2003. Ontogeny of diet-induced obesity in selectively bred Sprague-Dawley rats. Am. J. Physiol. Integr. Comp. Physiol. 285, R610 (Am Physiological Soc).
Rinella, M.E., Green, R.M., 2004. The methionine-choline deficient dietary model of steatohepatitis does not exhibit insulin resistance. J. Hepatol. 40, 47–51.
Ritskes-Hoitinga, J., Lemmens, A.G., Beynen, A.C., 1989. Nutrition and kidney calcification in rats. Lab. Anim. 23, 313–318.
Roebuck, B.D., Wilpone, S.A., Fifield, D.S., et al., 1979. Letter to the editor, dear Dr. Hill. J. Nutr. 109, 924–925.
Romestaing, C., Piquet, M.-A., Bedu, E., Rouleau, V., Dautresme, M., Hourmand-Ollivier, I., Filippi, C., Duchamp, C., Sibille, B., 2007. Long term highly saturated fat diet does not induce NASH in Wistar rats. Nutr. Metab. 4, 4.
Ross, A.P., Bruggeman, E.C., Kasumu, A.W., et al., 2012. Non-alcoholic fatty liver disease impairs hippocampal-dependent memory in male rats. Physiol. Behav. 106, 133–141.
Rossmeisl, M., Rim, J.S., Koza, R.A., et al., 2003. Variation in type 2 diabetes-related traits in mouse strains susceptible to diet-induced obesity. Diabetes 52, 1958–1966.
Sahai, A., Malladi, P., Melin-Aldana, H., et al., 2004. Upregulation of osteopontin expression is involved in the development of nonalcoholic steatohepatitis in a dietary murine model. Am. J. Physiol. Gastrointest. Liver Physiol. 287, G264–G273.
Sallinen, K., Arvola, P., Wuorela, H., et al., 1996. High calcium diet reduces blood pressure in exercised and nonexercised hypertensive rats. Am. J. Hypertens. 9, 144–156.
Samuel, V.T., Liu, Z.-X., Qu, X., et al., 2004. Mechanism of hepatic insulin resistance in non-alcoholic fatty liver disease. J. Biol. Chem. 279, 32345–32353.
Sánchez-Lozada, L.G., Tapia, E., Jiménez, A., et al., 2007. Fructose-induced metabolic syndrome is associated with glomerular hypertension and renal microvascular damage in rats. Am. J. Physiol. Ren. Physiol. 292, F423–F429.
Schecter, A.J., Olson, J., Papke, O., 1996. Exposure of laboratory animals to polychlorinated dibenzodioxins and polychlorinated dibenzofurans from commerical rodent chow. Chemosphere 32, 501–508.
Schwarzer, M., Srutkova, D., Hermanova, P., et al., 2017. Diet matters: endotoxin in the diet impacts the level of allergic sensitization in germ-free mice. PLoS One 12, 1–15.
Sleder, J., Chen, Y.D., Cully, M.D., et al., 1980. Hyperinsulinemia in fructose-induced hypertriglyceridemia in the rat. Metabolism 29, 303–305.
Srivastava, R.A., 1994. Saturated fatty acid, but not cholesterol, regulates apolipoprotein AI gene expression by posttranscriptional mechanism. Biochem. Mol. Biol. Int. 34, 393–402.
Srivastava, R.A., Jiao, S., Tang, J.J., et al., 1991. In vivo regulation of low-density lipoprotein receptor and apolipoprotein B gene expressions by dietary fat and cholesterol in inbred strains of mice. Biochim. Biophys. Acta 1086, 29–43.
Srivastava, R.A., Tang, J., Krul, E.S., et al., 1992. Dietary fatty acids and dietary cholesterol differ in their effect on the in vivo regulation of apolipoprotein A-I and A-II gene expression in inbred strains of mice. Biochim. Biophys. Acta 1125, 251–261.

Storlien, L.H., Jenkins, A.B., Chisholm, D.J., et al., 1991. Influence of dietary fat composition on development of insulin resistance in rats. Relationship to muscle triglyceride and omega-3 fatty acids in muscle phospholipid. Diabetes 40, 280–289.

Subramanian, S., Han, C.Y., Chiba, T., et al., 2008. Dietary cholesterol worsens adipose tissue macrophage accumulation and atherosclerosis in obese LDL receptor-deficient mice. Arterioscler. Thromb. Vasc. Biol. 28, 685–691.

Sumiyoshi, M., Sakanaka, M., Kimura, Y., 2006. Chronic intake of high-fat and high-sucrose diets differentially affects glucose intolerance in mice. J. Nutr. 136, 582–587.

Surwit, R., Feinglos, M., Rodin, J., et al., 1995. Differential effects of fat and sucrose on the development of obesity and diabetes in C57BL/6J and A/J mice. Metabolism 44, 645–651 (Elsevier).

Taghibiglou, C., Carpentier, A., Van Iderstine, S.C., et al., 2000. Mechanisms of hepatic very low density lipoprotein overproduction in insulin resistance. Evidence for enhanced lipoprotein assembly, reduced intracellular ApoB degradation, and increased microsomal triglyceride transfer protein in a fructose-fed hamster model. J. Biol. Chem. 275, 8416–8425.

Taper, H.S., Roberfroid, M.B., 2002. Inulin/oligofructose and anticancer therapy. Br. J. Nutr. 87, S283–S286.

Teupser, D., Persky, A.D., Breslow, J.L., 2003. Induction of atherosclerosis by low-fat, semisynthetic diets in LDL receptor-deficient C57BL/6J and FVB/NJ mice: comparison of lesions of the aortic root, brachiocephalic artery, and whole aorta (en face measurement). Arterioscler. Thromb. Vasc. Biol. 23, 1907–1913.

Thierry-Palmer, M., Tewolde, T.K., Emmett, N.L., et al., 2010. High dietary salt does not significantly affect plasma 25-hydroxyvitamin D concentrations of Sprague Dawley rats. BMC Res. Notes 3, 332.

Thigpen, J., Setchell, K., Ahlmark, K., et al., 1999. Phytoestrogen content of purified, open- and closed-formula laboratory animal diets. Lab. Anim. Sci. 49, 530–536.

Thigpen, J.E., Haseman, J.K., Saunders, H., et al., 2002. Dietary factors affecting uterine weights of immature CD-1 mice used in uterotrophic bioassays. Canc. Detect. Prev. 26, 381–393.

Thigpen, J.E., Setchell, K.D., Kissling, G.E., et al., 2013. The estrogenic content of rodent diets, bedding, cages, and water bottles and its effect on bisphenol a studies. J. Am. Assoc. Lab. Anim. Sci. 52, 130–141.

Thigpen, J.E., Setchell, K.D.R., Padilla-Banks, E., et al., 2007. Variations in phytoestrogen content between different mill dates of the same diet produces significant differences in the time of vaginal opening in CD-1 mice and F344 rats but not in CD Sprague-Dawley rats. Environ. Health Perspect. 115, 1717–1726.

Thigpen, J.E., Setchell, K.D.R., Saunders, H., et al., 2004. Selecting the appropriate rodent diet for endocrine disruptor research and testing studies. ILAR J. 45, 401–416.

Thorburn, A.W., Storlien, L.H., Jenkins, A.B., et al., 1989. Fructose-induced in vivo insulin resistance and elevated plasma triglyceride levels in rats. Am. J. Clin. Nutr. 49, 1155. Am Soc Nutrition.

Thresher, J.S., Podolin, D.A., Wei, Y., et al., 2000. Comparison of the effects of sucrose and fructose on insulin action and glucose tolerance. Am. J. Physiol. Regul. Integr. Comp. Physiol. 279, R1334–R1340.

Torres-Gonzalez, M., Volek, J.S., Sharman, M., et al., 2006. Dietary carbohydrate and cholesterol influence the number of particles and distributions of lipoprotein subfractions in guinea pigs. J. Nutr. Biochem. 17, 773–779.

Trevaskis, J.L., Griffin, P.S., Wittmer, C., et al., 2012. Glucagon-like peptide-1 receptor agonism improves metabolic, biochemical, and histopathological indices of nonalcoholic steatohepatitis in mice. AJP Gastrointest. Liver Physiol. 302, G762–G772.

Van Heek, M., Compton, D.S., France, C.F., et al., 1997. Diet-induced obese mice develop peripheral, but not central, resistance to leptin. J. Clin. Invest. 99, 385–390.

Vasudevan, H., Xiang, H., McNeill, J.H., 2005. Differential regulation of insulin resistance and hypertension by sex hormones in fructose-fed male rats. Am. J. Physiol. Heart Circ. Physiol. 289, H1335–H1342.

Veteläinen, R., van Vliet, A., van Gulik, T.M., 2007. Essential pathogenic and metabolic differences in steatosis induced by choline or methione-choline deficient diets in a rat model. J. Gastroenterol. Hepatol. 22, 1526–1533.

Wang, H., Storlien, L.H., Huang, X.F., 2002. Effects of dietary fat types on body fatness, leptin, and ARC leptin receptor, NPY, and AgRP mRNA expression. Am. J. Physiol. Metab. 282, E1352. Am Physiological Soc.

Warden, C.H., Fisler, J.S., 2008. Comparisons of diets used in animal models of high-fat feeding. Cell Metabol. 7, 277.

Weisberg, S.P., McCann, D., Desai, M., et al., 2003. Obesity is associated with macrophage accumulation in adipose tissue. J. Clin. Invest. 112, 1796–1808.

Weltman, M.D., Farrell, G.C., Liddle, C., 1996. Increased hepatocyte CYP2E1 expression in a rat nutritional model of hepatic steatosis with inflammation. Gastroenterology 111, 1645–1653.

White, C.L., Purpera, M.N., Morrison, C.D., 2009. Maternal obesity is necessary for programming effect of high-fat diet on offspring. Am. J. Physiol. Regul. Integr. Comp. Physiol. 296, R1464–R1472.

Wise, A., 1982. Interaction of diet and toxicity—the future role of purified diet in toxicological research. Arch. Toxicol. 50, 287–299.

Wise, A., Gilburt, D.J., 1980. The variability of dietary fibre in laboratory animal diets and its relevance to the control of experimental conditions. Fd. Cosmet. Toxicol. 18, 643–648.

Wolf, M.J., Adili, A., Piotrowitz, K., et al., 2014. Metabolic activation of intrahepatic CD8+ T cells and NKT cells causes nonalcoholic steatohepatitis and liver cancer via cross-talk with hepatocytes. Canc. Cell 26, 549–564.

Wu, L., Vikramadithyan, R., Yu, S., et al., 2006. Addition of dietary fat to cholesterol in the diets of LDL receptor knockout mice: effects on plasma insulin, lipoproteins, and atherosclerosis. J. Lipid Res. 47, 2215–2222.

Yang, K., Kurihara, N., Fan, K., et al., 2008. Dietary induction of colonic tumors in a mouse model of sporadic colon cancer. Cancer Res. 68, 7803–7810.

Yokozawa, T., Cho, E.J., Sasaki, S., et al., 2006. The protective role of Chinese prescription Kangen-karyu extract on diet-induced hypercholesterolemia in rats. Biol. Pharm. Bull. 29, 760–765.

Yu, Q., Larson, D.F., Slayback, D., et al., 2004. Characterization of high-salt and high-fat diets on cardiac and vascular function in mice. Cardiovasc. Toxicol. 4, 37–46.

Zafrani, E.S., 2004. Non-alcoholic fatty liver disease: an emerging pathological spectrum. Eur. J. For. Pathol. 444, 3–12.

Zern, T.L., West, K.L., Fernandez, M.L., 2003. Grape polyphenols decrease plasma triglycerides and cholesterol accumulation in the aorta of ovariectomized guinea pigs. J. Nutr. 133, 2268–2272.

Zhang, H.Y., Reddy, S., Kotchen, T.A., 1999. A high sucrose, high linoleic acid diet potentiates hypertension in the Dahl salt sensitive rat. Am. J. Hypertens. 12, 183–187.

Zhang, L.-N., Morgan, D.G., Clapham, J.C., et al., 2012. Factors predicting nongenetic variability in body weight gain induced by a high-fat diet in inbred C57BL/6J mice. Obesity 20, 1179–1188.

Zou, J., Chassaing, B., Singh, V., Pellizzon, M., et al., 2017. Fiber-mediated nourishment of gut microbiota protects against diet-induced obesity by restoring IL-22-mediated colonic health. Cell Host Microbe. https://doi.org/10.1016/j.chom.2017.11.003.

Zulet, M.A., Barber, A., Garcin, H., et al., 1999. Alterations in carbohydrate and lipid metabolism induced by a diet rich in coconut oil and cholesterol in a rat model. J. Am. Coll. Nutr. 18, 36–42.

PART VI

NANOMATERIALS AND RADIATION

CHAPTER

41

Engineered Nanomaterials: Biomarkers of Exposure and Effect

Enrico Bergamaschi[1], *Mary Gulumian*[2], *Jun Kanno*[3], *Kai Savolainen*[4]

[1]Laboratory of Toxicology and Occupational Epidemiology, Department of Public Health Science and Pediatrics, University of Turin, Italy; [2]Toxicology Research Projects NIOH, School of Pathology, University of the Witwatersrand South Africa; [3]Japan Bioassay Research Center, Japan Organization of Occupational Health and Safety, Hadano, Japan; [4]Nanosafety Research Centre, Finnish Institute of Occupational Health, Helsinki, Finland

INTRODUCTION AND BACKGROUND

The use of engineered nanomaterials (ENM) and their applications has grown dramatically since the turn of the 21st century because of multiple technological benefits of the use of material at nanoscale. More than 1600 "nanoenabled" products in commerce, all required workers for that to happen (http://www.nanotechproject.org/cpi/), and, according to recent estimates, 6 million workers will be potentially exposed to ENM in 2020 (Roco, 2011).

Nanotechnologies is in fact an umbrella name for a great number of technologies that utilize material at nanoscale for different purposes. For these reasons, nanotechnologies have been recognized as highly cross-cutting technologies, whose products, based on the use of nanomaterials, utilize physical and chemical properties of ENM, different from their chemically identical bulk counterparts (Savolainen et al., 2010).

However, during the production and use of ENM, there is the chance of exposure of workers, consumers, and the environment (Savolainen et al., 2010; Kuhlbusch et al., 2011; Valsami-Jones and Lynch, 2015). The effects of such exposure cannot be predicted based on our current understanding of chemicals, given the fact that material at nanoscale has both a particulate identity and a molecular identity, which are responsible for the potential biological effects. A recent review has summarized work conducted in relation to exposure to ENMs in the workplace and processes that may lead to such exposure (Kuhlbusch et al., 2011). For example, it could be shown that exposure to TiO_2 nanomaterials may arise during cleaning and maintenance operations as well as in the case of a failure of normal operation (Plitzko, 2009), to TiO_2 and silver during the reactor and vacuum pump operations (Lee et al., 2011), to carbon nanotubes (CNTs) during spraying, preparation, ultrasonic dispersion, wafer heating, and opening the water bath cover (Lee et al., 2010), or to precious metal nanoparticles (NPs) in a furnace room and in an electrorefining area (Miller et al., 2010).

Although it has been argued that size as such does not cause harmful effects (European Commission, 2011), a number of studies have convincingly shown that ENMs cause toxic effects not induced by chemically identical but larger particles (Rossi et al., 2010; Palomäki et al., 2011; Catalán et al., 2012). These effects are likely due to exposure to the unique features of ENM, either to their intrinsic properties or to their small entity that allows them to reach targets not reachable by their larger, chemically identical counterparts (Kreyling et al., 2009). An important issue in this context may also be the biocorona formed to surround NPs, and larger particles, once they reach biological environments (Monopoli et al., 2012).

Another major challenge in the assessment of hazards of ENM to experimental animals, humans, and environmental species is, that being in a particulate form, the behavior of ENM differ dramatically from that of traditional soluble chemicals having an impact, not only on the kinetics of ENM in biological environments but also on their potentially harmful effects (Pietroiusti et al., 2018).

The properties of nanomaterials cause marked challenges to the assessment of hazards of ENM via the lungs, but also via other exposure routes. However, in the lungs, when the NPs most readily reach the body, they become

Biomarkers in Toxicology, Second Edition
https://doi.org/10.1016/B978-0-12-814655-2.00041-4

covered by biomolecules (Monopoli et al., 2012) rendering their kinetic behavior and effects more difficult to assess. The special features of the airways add to the complexity to the hazard assessment of ENM via the inhalational route. In general terms, assessing effects of ENM is demanding because the associations of harmful effects of ENM features (physicochemical and biological) are not well understood.

To this end, novel approaches for the prediction of material at nanoscale need to be developed (Kinaret et al., 2017a,b). There are currently a number of ongoing attempts to develop such hazard prediction tools and frameworks [www.nanosolutionsfp7.com/; www.guidenano.eu/; www.nanomile:eu-vri.eu/; www.sun-fp7.eu/]. Some of the known effects of ENM include those of titanium dioxide (National Institute of Occupational Safety and Health—NIOSH, 2011) and of metal oxides and metals (Saber et al., 2013), and carbon-containing materials induced pulmonary inflammation (Ryman-Rasmussen et al., 2009a; Palomäki et al., 2011; Mercer et al., 2013; Kinaret et al., 2017b). These effects include, among others, inflammation, granuloma formation, and fibrosis of the lungs (Rossi et al., 2010; Ryman-Rasmussen et al., 2009b; Saber et al., 2013). It has also been shown that both tangled and rigid rod-like CNTs (Mitsui-7) can reach the lung and subsequently the subpleural space and cause collagen deposition. Subpleural space is also the site of pulmonary mesothelioma initiation (Ryman-Rasmussen et al., 2009b; Mercer et al., 2013). However, only rigid, rod-like CNTs have been shown to induce mesothelioma in rodents (Takagi et al., 2012; Sargent et al., 2014). In addition, several fibrous and crystalline ENM have been shown to induce genotoxic effects in vivo and in vitro (Catalán et al., 2012; Kinaret et al., 2017a).

In the light of reports on the production of pleural effusion and pulmonary fibrosis following exposure to nanomaterials (Ryman-Rasmussen et al., 2009a,b), and also the carcinogenicity of CNTs (Chernova et al., 2017), search for biomarkers as tools to protect workers has been emphasized (Schulte and Hauser, 2012; Bergamaschi, 2012; Iavicoli et al., 2014; Bergamaschi et al., 2017). Being able to develop biomarkers of exposure to ENM would greatly increase the certainty of assessment of potential risks of exposure to ENM. In fact, it is quite obvious that the (ENM) biomarkers will become available when detailed molecular mechanisms of ENM-induced diseases can be better characterized. Until then, bridging data between in vitro and in vivo reactions are required. The understanding of possible diseases of particles is currently based on biokinetics (absorption, distribution, metabolism, excretion: ADME) viewed, with regard to the various ENM considering both their physicochemical characteristics and their interactions with biomolecules, and even considering particulate matter (PM) toxicity, such as asbestos. At the current state, these two areas are serving as starting points for knowledge and understanding within the new fields of molecular biology. Several topics have to be considered in the development of biomarkers of exposure for ENM, including foreign body recognition systems and/or immune systems, especially of innate nature. These studies are applicable to acute and chronic in vivo responses including virtually permanent deposition of PMs to the reticuloendothelial system (RES). In any case, for sound growth of the nanomaterial industry and protection of workers and users, promotion of new studies and usage of available data in a reasonable balance is practical and essential.

The development of novel biomarkers for ENM is also hampered by the lack of a systematic database on ENM toxicity, even though there is a plethora of detailed information on specific toxicity of several ENM. This renders ENM risk and safety assessment a challenge, especially because information on exposure to ENM in occupational setting or other environments is lacking for most of the materials (Kuhlbusch et al., 2011; Valsami-Jones and Lynch, 2015; Pietroiusti et al., 2018). This situation is reflected by the fact that there are no occupational exposure limits implemented for any of the ENM anywhere (Van Broekhuizen and Reijnders, 2011, Van Broekhuizen et al., 2012). It is not surprising that there are concerns regarding the safety of ENM in work places, consumer products, and the environment.

Epidemiological studies were crucial in identifying a correlation between exposure to different sizes of particles and fibers in causing a disease (Donaldson and Seaton, 2012). The search for biomarkers of exposure to, and effects of particles could only become possible once the elucidation of the molecular mechanisms involved in particle-induced diseases, including fibrosis, lung cancer, and mesothelioma (Mossmann, 2000; Grosse et al., 2014) had been established. Such correlations were identified later between ambient PM_{10} and ultrafine particles (UFP) and cardiovascular diseases (Dockery et al., 1993) or pulmonary function (Pope and Kanner, 1993). The relevance of oxidative and inflammatory markers due to exposure to particles and fibers and especially nanomaterials has been recently reviewed by Bergamaschi et al. (2017). In addition, in some cases, ions released from the NPs such as silver, gold, and iron can be measured in urine and in blood (Iavicoli et al., 2014).

From the outset, attempts have been directed to develop biomarkers of exposure to nanomaterials based on their ability to induce oxidative stress and inflammation (Gulumian et al., 2006; Johnston et al., 2013; Manke et al., 2013). Biomarkers of oxidative stress and inflammation have been shown to have an association with the biopersistence of particles and fibers (Searl et al., 1999) resulting in frustrated phagocytosis and oxidative cellular stress, especially in the lungs. The challenge that has remained in the development of biomarkers of exposure to nanomaterials and other particulates has been the lack of specificity toward ENM. Hence, so far these biomarkers seem to work at a group level in a given epidemiological study, but they seem not to be suitable for the assessment of exposure of a given worker to ENM because so many exposures to different kind of PM, including fungi, bacteria, wood dust, and man-made mineral fibers among others induce a similar response (e.g., Savolainen et al., 2010; Bergamaschi et al., 2017).

The search for appropriate biomarkers for exposure to NPs will be of great relevance to the need for such exposures to be detected early enough in the process of toxicity and also pathogenicity. Successful identification of such biomarkers will prevent the recurrence of the experience with larger particles where exposures continued unabated for very long periods until pathological changes were observed, producing diseases such as fibrosis and cancer. In addition, earlier lessons of the benefits of biomarkers for oxidative stress and inflammation might help also in the search for biomarkers for ENM. It is, though, of importance to remember the lack of specificity of such biomarkers, and for that reason also search for biomarkers with a better specificity would be of value (Palomäki et al., 2011; Kinaret et al., 2017a,b).

CLASSIFICATION AND CHARACTERISTICS OF NANOMATERIALS

According to the definition given in the technical report of the International Standards Organization (ISO) (ISO/TR 14786, 2014), nanomaterials may include nanosized objects with one or more external dimension in the nanoscale. Nanomaterials may therefore be distinguished by their shape as either NPs (all three dimensions in the nanoscale), nanofibers (two dimensions in the nanoscale, including nanowires, nanotubes, and nanorods), or nanoplates (one dimension in the nanoscale).

Because of the multiplicity in chemical composition, shape, and size of NPs, as well as multiplicity in the effects they produce in in vitro cell cultures and in vivo—in pulmonary, cardiac, reproductive, renal, and cutaneous systems (Kumar et al., 2012)—it will therefore be useful to classify NPs into general groups: carbon-based, inorganic metal, or metal oxides as well as organic NPs, with the hope that biomarkers that might be specific to a certain group but not to other groups of NPs may emerge .

The classification of nanomaterials may be made as carbon-based NPs that include fullerenes, carbon nanofibers, graphene, and carbon black (CB); inorganic metal-based NPs that include gold and silver NPs; and metal oxide NPs that include among others titanium dioxide, zinc oxide, and cerium oxide; quantum dots (QDs); and finally organic NPs that include organic polymers.

Carbon-Based Nanoparticles: Fullerenes

Fullerenes are carbon-based allotropes in a hexagonal network of carbon atoms, which may be in the form of a hollow sphere, ellipsoid, tube, or plane. When in spherical cages, they may contain between 28 to more than 100 carbon atoms, the most widely studied of which are those containing 60 carbon atoms (C60), first synthesized by Kroto et al. (1985). When in layers, they are called graphenes, and when in hollow cylinders they are called CNTs, which may be single layer (SWCNT), double layer (DWCNT), or multiple layers (MWCNT). Finally, CB, composed of partially amorphous graphitic material, is mostly spherical in shape (Aitken et al., 2004).

Inorganic Nanoparticles

Inorganic NPs are particles in nanometric dimensions composed of pure metals or metal oxide, or are of metallic composition. The most common examples include gold, silver, aluminum, titanium, silica, tungsten, manganese, copper, cerium, iron, molybdenum, and palladium NPs. They are also synthesized in various geometries, examples being spherical, nanoshells, nanorods, tripods, tetrapods, nanocages, and star-shaped nanorice-shaped gold NPs (Chen et al., 2003; Chen et al., 2005; Nehl et al., 2006; Wang et al., 2006; Murphy et al., 2008).

Generally, QDs are fabricated from groups II–VI or groups III–V elements of the periodic table (Aitken et al., 2004). Examples include indium phosphate (InP), indium arsenate (InAs), gallium arsenate (GaAs), gallium nitride (GaN), zinc sulfide (ZnS), zinc–selenium (ZnSe), cadmium–selenium (CdSe), and cadmium–tellurium (CdTe) metalloid cores (Hines and Guyot-Sionnest, 1996; Dabbousi et al., 1997). Newer, heavier structures (e.g., CdTe/CdSe, CdSe/ZnTe) and hybrids composed of lead–selenium (PbSe) have also been synthesized (Kim et al., 2003).

Organic Nanoparticles

The highly branched and symmetrical molecules known as dendrimers are the most recently recognized members of the polymer family. Their unique branched topologies give dendrimers properties that differ substantially from those of linear polymers. Dendrimers were first synthesized by Vögtle in 1978 (Buhleier et al., 1978). They are three-dimensional globular, monodisperse, highly branched polymers prepared in a series of repetitive reactions from simple branched monomer units emitted from a central core with an exterior corrugated surfacing whose size and shape can be precisely controlled. Dendrimers are fabricated from monomers using either convergent or divergent step-growth polymerization. Dendrimers' unique architecture enhances their ability to exhibit high functional structures (Zeng and Zimmerman, 1997; Pricl et al., 2003; Lee et al., 2005; Namazi and Adeli, 2005). Subsequently, a wide range of dendrimers of different structural classes are synthesized using divergent (built from the central core to the periphery) or convergent (built from the periphery toward the central core) strategies—using repeat units ranging from pure hydrocarbons to peptides, or coordination compounds.

Attempts to classify nanomaterials in their risk assessment for regulatory purposes have included those that are based on a set of performance metrics that measure both the toxicity and physicochemical characteristics of the original materials, as well as the expected environmental impacts through the product life cycle (Tervonen et al., 2008) or more recently based on features that control nanomaterial biological interactions (Castagnola et al., 2017) or on their structure–activity relationship (Gajewicz et al., 2018).

Toxicity Testing of Nanomaterials

The introduction of a radical overhaul for testing synthetic chemicals in animal models for their adverse effects on humans and the environment has recently been advocated. As such, design of integrated testing strategies has been proposed, using modern methods including cell culture techniques with the implementation of a combination of biochemical knowledge of cellular pathways with genomics, proteomics, and metabonomics (Hartung, 2009).

Although these in vitro techniques may be useful in the initial identification of the toxicity of NPs and in the elucidation of the mechanisms involved in their toxicity, long-term animal studies comparing the toxicity and carcinogenicity of certain types of NPs have been strongly advocated (Shi et al., 2013). The necessity for conducting such long-term studies has emanated from the fact that mineral particles and fibers in general and certain NPs in particular are shown to be biopersistent (Oberdörster, 2010).

In addition, these short-term biological tests are shown to produce false positives in the case of non-biopersistent fibers because although they may have effects in vitro, they do not persist long enough in the lungs for a sufficient dose to build up and produce effects in vivo (Donaldson and Tran, 2004). This possibility was confirmed with biopersistent and non-biopersistent nanofibers, where it was found that the biopersistence was influenced by both fiber dimensions and solubility (Searl et al., 1999). Moreover, concerns were raised as to the ability of short-term in vitro assays to accurately predict the in vivo effects of the functionalized products of inhaled CNTs (Zhang et al., 2013). However, once the functionalized side chains are removed and the biopersistent core structure is presented naked to the organism, both ex-A and ex-N NPs become identical (Sayes et al., 2007; Warheit et al., 2009).

DEFINITION AND MEANING OF BIOLOGICAL MONITORING AND ITS APPLICATION TO ENGINEERED NANOMATERIALS

Biological monitoring (BM) deals with the systematic or repetitive measurement of chemical or biochemical markers in fluids, tissues, or other accessible matrices from people exposed to or with past exposure to xenobiotics. BM can be used with the purpose of identifying potential hazards of new and emerging chemicals—potentially including ENMs—thus identifying groups at higher risk of health outcomes (Schulte and Hauser, 2012). The main objectives of such periodical measurements are (1) the assessment of individual or group exposure; (2) the identification of early, specific, nonadverse biological effect parameters that are indicative, if compared with adequate reference values, of an actual or potential condition leading to health damage; and, ultimately (3) the assessment of health risk to exposed subjects (Manno et al., 2010). Biomarkers are recommended for use in assessing the effects or exposure to harmful substances, specifically where there are low or intermittent levels of exposure, mixtures of toxicants that may act synergistically, or exposure resulting in disease with long latency period. Biomarkers are increasingly used as surrogate indicators of designated events in a biological system due to the inaccessibility of target organs; in spite of this limitation, it is thought that biomarkers are more directly related to the adverse effects that one attempts to prevent than any ambient measurement, and this supports the use of BM in risk assessment (Smolders et al., 2010). On the other hand, BM is more complex in terms of standardization and interpretative efforts as compared to ambient monitoring; the use of biomarkers requires a toxicological knowledge for their interpretation and ethical issues should be addressed as generally required in human studies.

Assessing even subtle health effects resulting from exposure to ENM is challenging for several reasons, including the heterogeneity of nano-objects in real life, and the lack of available tests with a known sensitivity and specificity to detect physiological and biological modifications clearly related to particle exposure.

Hazard studies have identified the most significant biological responses and target organ/systems affected by different ENM (Pappi et al., 2008; Aschberger et al.,

2010; Savolainen et al., 2010). Unfortunately, research over the past few years has not been conducted with the aim of identifying threshold exposures. The effects of ENM on human health are—to a large extent—unknown, and currently there is no report of any definitive human disease that is caused or worsened by ENM exposure. The current body of information on the human health effects potentially related to ENM comes from the studies on incidental NPs, air pollution epidemiology, and studies on occupational exposures with similarities to NPs such as welding fumes, ultrafine CB, or diesel exhausts (Madl and Pinkerton, 2009). Epidemiological studies have found hazardous respiratory effects from occupational exposure to some industrial processes involving generation of significant amounts of UFP, such as CB (Wellmann et al., 2006), fumed silica (Merget et al., 2002), metal oxides (Antonini, 2003; Luo et al., 2009), and fibers of concern.

The main effects attributed to ENM are (1) lung inflammation and fibrosis; (2) genotoxicity and DNA oxidative changes; (3) carcinogenic or procarcinogenic effects; and (4) vascular impairment resulting from endothelial activation, prothrombotic effects, and accelerated atherosclerosis. Although the respiratory and cardiovascular systems represent the main or the most studied targets of ENM, recent researches suggest that other organs can be indirectly affected by ENM exposure. Lung effects largely depend on the physicochemical characteristics and surface properties of instilled/inhaled particles. While the surface properties of insoluble particles are the main determinants of their interaction with biological systems, those ENM that rapidly dissolve into toxic ions lead readily to inflammation (Donaldson et al., 2013). Some ENM, such as titanium dioxide (TiO_2), copper oxide (CuO), ZnO and iron oxide NPs, cationic polystyrene, and C60 fullerene, have demonstrated prooxidative and proinflammatory properties both in vitro and in vivo, mainly related to their surface reactivity and chemical composition (Madl and Pinkerton, 2009), although not all NPs cause inflammation via a mechanism involving oxidative stress. The tissue response to CNTs following instillation or inhalation is characterized by transient inflammatory changes, oxidative stress, fibrosis (Shvedova et al., 2008), and cancer (Chernova et al., 2017).

Through a process of translocation across biological barriers, NPs can reach and deposit in secondary target organs where they may induce adverse biological reactions. Therefore, a correct assessment of NP-induced adverse effects should take into account the different aspects of toxicokinetics and tissues that may be targeted by NPs.

Cardiovascular effects have been assessed in workers handling TiO_2 nanomaterials, thus suggesting that exposure to particles with a diameter <300 nm might affect parasympathetic function leading to higher heart rate variability in workers (Ichihara et al., 2016). Acute exposure to TiO_2 NPs acutely alters cardiac excitability and increases the likelihood of arrhythmic events (Savi et al., 2014) and also induces myocarditis in mice (Hong et al., 2015).

Recent experimental findings, considering the role of new players in gut physiology (e.g., the microbiota), shed light on several outcomes of the interaction between ENM and gastrointestinal tract, fostering for long-term unpredictable consequences (Pietroiusti et al., 2017). Ruiz et al. (2017) have demonstrated that TiO_2 NPs exacerbate experimentally induced colitis and that TiO_2 in blood is significantly higher in human beings suffering from inflammatory bowel diseases during the acute phase of the disease. Few in vivo studies demonstrated that NPs may affect kidney function, leading to both tubular and glomerular changes (Iavicoli et al., 2016). The issue of potential effects of manufactured NPs on brain functions, given that NP's potential to induce oxidative stress, inflammation, cell death by apoptosis, or changes in the level of expression of certain neurotransmitters, has been recently reviewed (Bencsik et al., 2018).

Studies on the relationship between particulate pollutants (including elemental carbon) and health effects have generated a panel of circulating biomarkers reflecting inflammation end points, endothelial activation, platelet activation, oxidative damage to DNA and lipids, and antioxidant capacity (Loft et al., 2008; Møller and Loft, 2010).

Challenges of the Development of Biomarkers of Exposure to Engineered Nanomaterials Due to Their Biokinetics

Considering the behavior of ENM in a biological system, i.e., in a cell or an organ, as shown in Fig. 41.1, there are three issues of special interest: primary NPs, secondary aggregated NPs, and NPs firmly attached to the cellular matrix.

The first stage occurs before the exposure, the second at the site of absorption, and the third at the level of distribution, metabolism, and excretion and deposition at the potential final destiny of the particle, if any. At the third stage, the active dispersion mechanisms and attachment of the NP through bonding by the host with other insoluble foreign particles determine the final fate of the ENM. For the functionalized NPs, initial toxicity may depend largely on the functionalized side chain moieties that interact first with the organism's biomolecules. However, as shown in Fig. 41.2, metabolic activity of the host organism may eventually succeed in removing the side chains, leaving the biopersistent core structure, such as C_{60} for fullerene-based functionalized products.

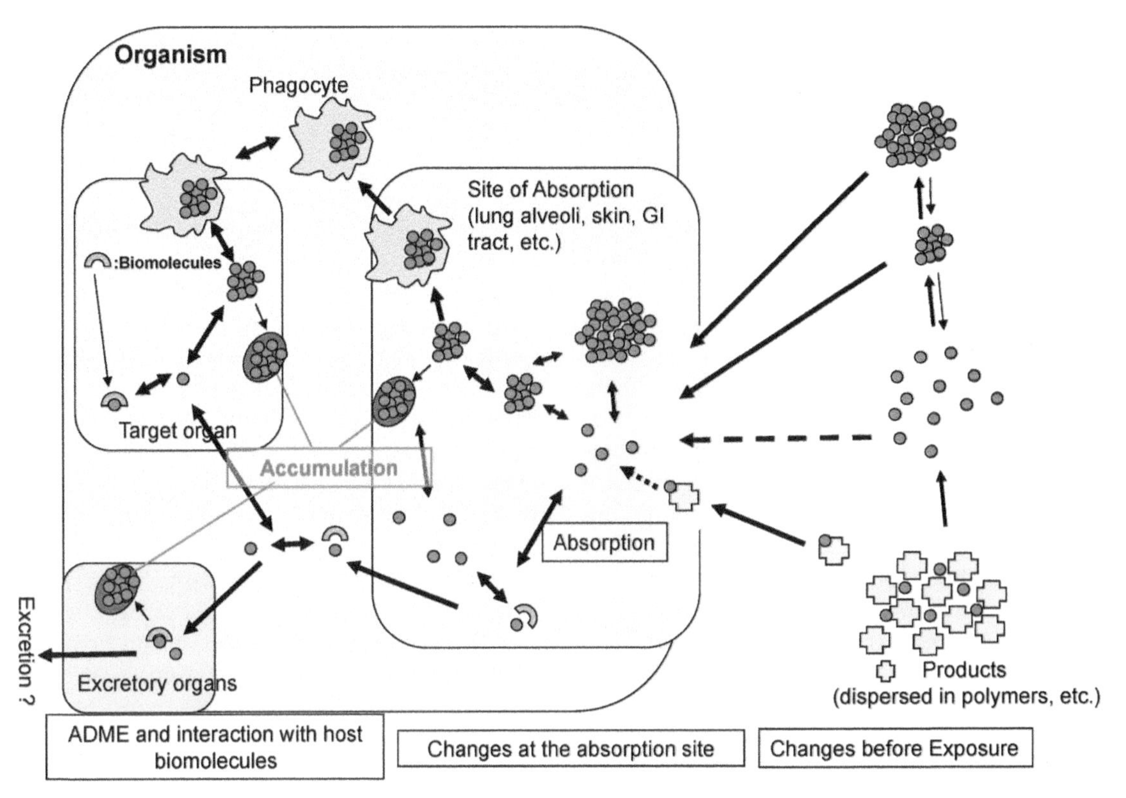

FIGURE 41.1 A pictorial representation of absorption, distribution, metabolism, excretion, and deposition of engineered nanomaterials in cells and tissues.

The diversity and complexity of an ever-increasing number of NM with varying physicochemical properties, and the complex and changeable nature of nano-biointeractions, make the biological behavior of NM not predictable on the basis of their inherent properties. Recent findings indicate that the physicochemical properties and the integrity of NPs can change dramatically following internalization by cells in vitro and in vivo, even for NPs with high colloidal stability such as polymer-coated gold NPs (Kreyling et al., 2015).

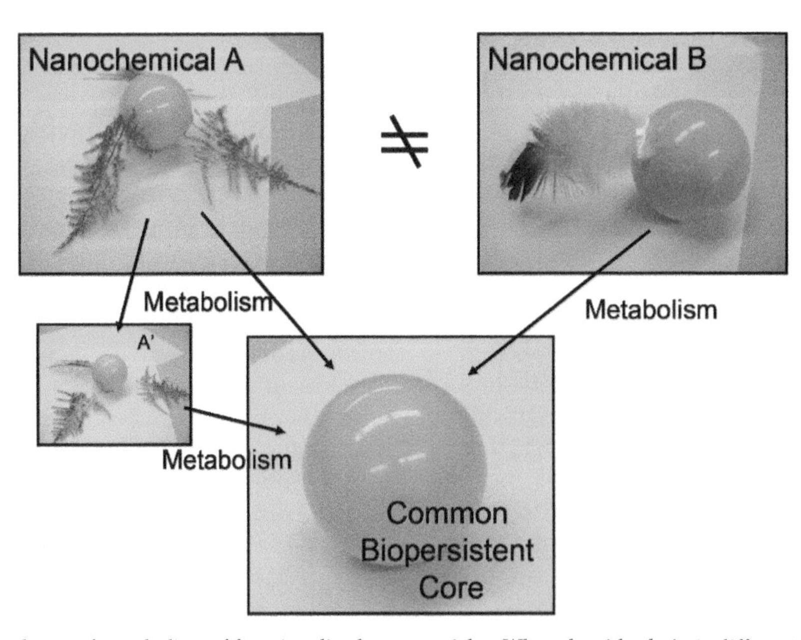

FIGURE 41.2 Symbolic scheme of metabolism of functionalized nanoparticles. When the side chain is different, then the molecules A and B are chemically and toxicologically different. However, once the functionalized side chains are removed and the biopersistent core structure is presented naked to the organism, both ex-A and ex-B nanoparticles become identical.

While the material-intrinsic properties determine the "synthetic identity," the context-dependent properties of the nanomaterial influence its "biological identity"—which is shaped, in part, by the adsorption of biomolecules that form a "corona" on the surface of NPs; the composition of this "bio-corona" depends on the particular biofluid and may exhibit dynamic changes as the NP crosses from one biological compartment to another (Monopoli et al., 2012; Fadeel et al., 2013). Moreover, particles of similar size and composition can raise qualitatively different effects in relation to their methods of synthesis, as for amorphous silica NPs (DiCristo et al., 2015).

Besides intrinsic characteristics of particles, the presence of contaminants can cause significant changes in their chemical identity, leading to increase in biological activity. This issue is clearly described by synergistic effect of NP and endotoxins on cytokine production and immune cells activation (Bianchi et al., 2015; Li and Boraschi, 2016). Such changes in surface characteristics and chemical identity have important implications for biomarker research because experimental setting may not be truly representative of particle–cell interactions in a real-life scenario. As a result, our current testing strategy is actualized by using simplified models, different dosing regimens, and exposure routes and so represents a compromise to understand what could happen in humans, which requires more caution in the interpretation of findings with consideration of the weight of evidence (Bergamaschi et al., 2015).

BIOLOGICAL INTERACTIONS RELEVANT TO BIOMARKERS OF EXPOSURE TO ENGINEERED NANOMATERIALS AT MOLECULAR, CELLULAR, AND ORGAN LEVEL

Biological interfaces of ENM are numerous and complex, existing both on the cell surface and inside the cell. When compared between primary (nonaggregated) and secondary (aggregated) NPs of the same origin, the target biomolecules may be different. In particular, the host defense mechanisms against the insoluble, i.e., biopersistent, NPs may be similar to those used by the cells against bacterial and viral infections. The most important differences between normal chemical toxicity and ENM toxicity are the activation of the innate immune system against foreign bodies, the subsequent influence on the acquired immune system, and the so-called indirect oxidative stress responses, which eventually lead to cell death (apoptosis and necrosis), cell proliferation and tissue modification (e.g., fibrosis), and indirect genotoxicity and carcinogenesis.

Studies on asbestos-induced noncancerous and cancerous diseases have revealed several key events at the molecular level in the cells. As for the morphological events, fiber carcinogenesis has mesothelial cells and phagocytic cells of the immune system as targets. The former target is subject to direct damage by the exposure, resulting in clastogenicity and indirect attack by phagocytes nearby. The indirect danger signal from the phagocytes protects damaged mesothelial cells from apoptosis, thus increasing the chances of the attacked cells to proceed to neoplastic forms. (Nagai and Toyokuni, 2010). The importance of the length and the aspect ratio (length/thickness) of the fiber in the frustrated phagocytosis has been thoroughly reported (Stanton et al., 1981; Pott et al., 1994: Roller et al., 1997). It is likely that the chronic active inflammatory lesions are more important than fibrous scar formation (Poland et al., 2008; Donaldson et al., 2010; Takagi et al., 2012). Condensation of secondary elements to the fibers, such as iron in the body for the Fenton reaction, is postulated as a direct mechanism or as an augmenting factor of fiber mesotheliomagenesis (Nagai et al., 2011).

Another classical example of particle-induced carcinogenesis is thorotrast (Mori et al., 1983). The size of thorium dioxide primary particles in emulsion, used as an X-ray contrast medium, is approximately 10 nm in diameter (Riedel et al., 1983). Primary particles are concentrated initially in the first line of macrophages to form larger clusters and when these macrophages die off by either alpha ray effect or by foreign body-based oxidative stress, the cluster is released to the intercellular space, and phagocytosed again by the second line of macrophage, resulting in the formation of larger clusters (Nishizawa et al., 1987). As a whole, the biological half-life time has been reported as over 400 years in some reports (Janower et al., 1968). It has been proposed that the carcinogenic effect is linearly proportional to radioactivity of the emulsion, indicating that the permanent retention of the particle in the RES is not responsible for the effect (Wesch et al., 1983). However, this event clearly shows that biopersistent ENM administered into the bloodstream or into a place where the particle eventually enters the bloodstream will be gradually concentrated into larger clusters and permanently trapped by the RES.

The recognition of foreign NM has been studied at the molecular level. The mechanisms of foreign body recognition share the same signal pathways and reactions to the recognition of bacteria and viruses, i.e., the innate immune system. NALP3/NLRP3 inflammasome was reported as a mediator of the sensing mechanism of foreign ENM via lysosomal damage and endogenous "danger" signals (Dostert et al., 2008; Palomäki et al., 2011). It seems important to identify both the direct oxidative stress by immediate responses to the ENM and the indirect IFN signaling-mediated oxidative stress for the understanding of ENM generated cell injury and cell growth that may eventually lead to tissue/organ damage and carcinogenesis (Mullan et al., 2005; Chakrabarti et al., 2011).

The precipitating types of adjuvant, such as aluminum salts, are reported to induce cell migration to the site of coadministered antigen and induce cell killing so that the mesh of DNA from the dead cell activates the immune activity via distinct signaling pathways (Marichal et al., 2011). Diesel exhaust and pollen allergy enhancement might be considered as a similar phenomenon to the ENM attribution to the immune system—more toward adverse immune reactions, such as allergy, persistent immune activation triggering autoimmune diseases, and immune suppression (anergy) (Siegel et al., 2004; Marichal et al., 2011). The finding that the double-stranded DNA from the host tissue may take part in enhancing immunization of an externally applied antigen may add some consideration to the lifelong deposition of biopersistent ENM and the induction of production of biomolecules that could serve as bioindicators of exposure to ENM.

Protein scaffolding and nanomaterials is another aspect of biological interaction. An example is the acceleration of bone-implant connection as a result of accelerated bone formation by CNTs used at the interface of the implant and grafted tissue (Saito et al., 2008; Usui et al., 2008). The authors claim that there is minimal inflammatory reaction at the interface. These findings address the issue of whether primary particles deposited in the mesenchymal connective tissue would stimulate noninflammatory reaction, such as direct stimulation of fibroblast or mesenchymal stem cells to produce progressive pathology such as fibrosis, angiogenesis, noninflammatory granulation tissue formation, or bone formation. The systemic distribution of single fibers we found after MWCNT intraperitoneal injection poses an unanswered question whether tissue reaction would eventually take place in those deposition sites.

Examples of Given Biomarkers

Mesothelin is reported to increase in the serum of human patients with mesothelioma (Imashimizu et al., 2011) and in a rat study where multiwall CNTs were given via intraperitoneal exposure (Sakamoto et al., 2010). MIP1alpha is also reported to be secreted by macrophages treated with titanium dioxide (Xu et al., 2010). There are many other putative markers reported in relation to toxicity of NPs, such as CD146 and IMP3 (insulin-like growth factor 2 mRNA-binding protein 3) (Okazaki et al., 2013). Many of them are waiting for further validation.

Histological biomarkers for MWCNT of asbestos size against mesotheliomagenesis include our finding that the chronic inflammatory lesion consisting of MWCNT-laden activated macrophages without acute inflammatory cell infiltration, granuloma formation, and fibrotic scar formation (Fig. 41.3), considered as a phenotype of frustrated phagocytosis, is important for mesotheliomagenesis (Takagi et al., 2012). This phenotype is considered to be supported by a study on SWCNT (Mangum et al., 2006). This finding may be in good correlation with the acute phase biomarkers posted by several research studies (Poland et al., 2008; Murray et al., 2012). For a morphological picture demonstrating the impact of MWCNT on the mesothelial lining in the peritoneal cavity in sensitive mice strain refer Fig. 41.4 (Takagi et al., 2012).

Biomarkers of Exposure

Once inhaled, NPs can deposit in lung cells including alveolar macrophages and translocate through or between epithelial and endothelial cells into the blood and lymph circulation, potentially reaching sensitive target sites including bone marrow, lymph nodes, spleen, heart, and central nervous system. Translocation to many organs by ENM has been demonstrated and quantified (Geiser and Kreyling, 2010; Holgate, 2010), but its clinical relevance is not known. The likelihood of being taken up can vary dramatically among exposure conditions and by the behavior of particle aerosols in which nano-objects are commonly present and taken up as agglomerates and/or aggregates of various aerodynamic diameter. The internal dose of common chemicals is usually assessed by measuring both the amount of the substance and/or its metabolites or as a product of interaction with biomolecules. An ideal biomarker of exposure should be specific for the exposure of interest and also detectable in small quantities, measurable by noninvasive techniques, and capable of providing a positive predictive value to a specific health status (Mutti, 2001). Toxicokinetic and toxicodynamic data, which are available for a few ENM (e.g., metal NPs, fullerenes, and single-walled CNT) suggest that translocation rates from the portal-of-entry to secondary organs are very low (Kreyling et al., 2002; Choi et al., 2010; Geiser and Kreyling, 2010). So far, few studies have provided quantitative demonstration of ENM translocation from the lung. Recent research undertook biokinetic analysis in rats, using near-infrared imaging, to follow the fate of intratracheally instilled NPs of various size, surface modification, and core composition. Choi et al. (2010) found that noncationic NPs smaller than 34 nm in diameter, which did not bind serum proteins, reached the regional lymph nodes quickly. However, larger NPs are consistently retained in the lungs. When the ENM

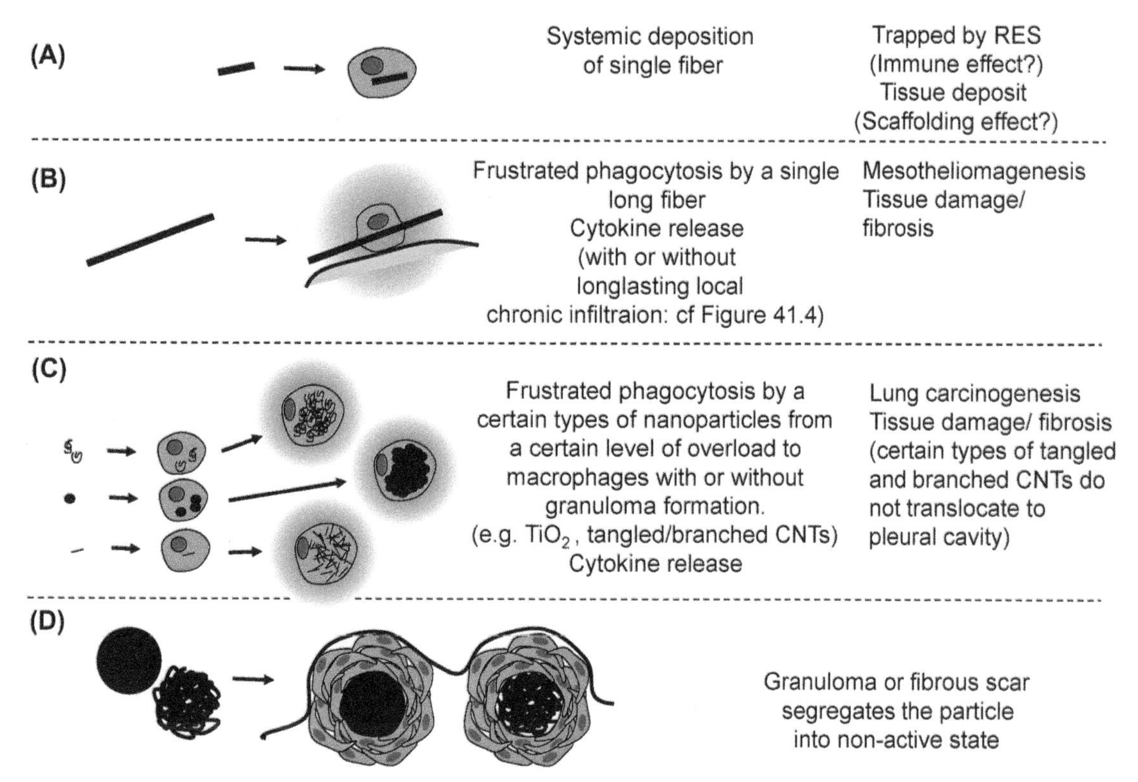

FIGURE 41.3 (A) A nanoparticle short or small enough to be phagocytosed by a single macrophage is easily removed from the initial exposure site, and either ends up trapped in the reticuloendothelial system or distributed systemically via bloodstream to renal glomeruli, choroid plexus, and other organs. (B) A relatively long fiber (cf. Stanton et al., 1981) such as asbestos, multiwall carbon nanotubes, and some single wall nanotubes are phagocytosed by more than one macrophage, resulting in the release of various cytokines and other inflammatory mediators without the formation of granulomas or scars. This state is reported to last for a considerably long period, resulting in mesotheliomagenesis in cases of the peritoneal/pleural cavity, and local fibrosis in pulmonary alveoli. (C) Certain types of nanomaterials, either primary or secondary particles, that are small enough to be phagocytized by a macrophage can induce frustrated phagocytosis from a certain level of overload to the macrophage. It is not well known what kind of condition leads to subsequent formation of foreign body granuloma. Lack of formation of such granuloma can be seen for a long period in some condition. In case of MWCNT, it is experienced that the fibers recovered from the pleural cavity are almost always the straight ones without branching. (D) Large aggregate and agglomerate are entrapped by a group of macrophages and result in formation of foreign body granuloma and eventually fibrous scarring. This reaction effectively segregates the foreign body from the inflammatory system and thus does not seem to contribute to mesotheliomagenesis and further expansion of diffuse fibrosis. These different types of engineered nanomaterials have a different potential for penetrating barriers and hence reaching different organs. They also have different ability to induce the synthesis of biomolecules that could serve as a biomarker of exposure to ENM.

diameter falls below 6 nm, and the particles are zwitterionic, about half of the NPs rapidly enter the bloodstream from alveolar airspace and are mostly cleared from the body by means of renal filtration. Thus, while the size seems of minor importance for the toxicity, it dramatically affects particle translocation and biokinetics.

Although some ENM have been demonstrated to easily cross biological barriers, the number of ENM entering the systemic circulation could be too low to lead to any significant internal dose, and hence not be detected in peripheral tissues in humans. There is a general consensus on the likelihood that particles whose aerodynamic diameter is between 0.1 and 1.0 μm reach the lower respiratory tract, but their deposition is low compared with particles with lower and higher aerodynamic diameter because of symmetric human lung morphology (Broday and Agnon, 2007). The daily dose of inhaled NP in humans is limited to a maximal NP aerosol concentration of $10^{12}/m^3$ (beyond that, coagulation rapidly occurs under ambient conditions), and the daily volume of air inhaled by an adult is $15\ m^3$. Assuming a particle deposition of 0.3, the daily deposition amounts to about 5×10^{12} NPs (Oberdörster, 2010). As a result, it may take years to reach an appreciable internal dose, whereas the biologically effective dose (i.e., the entity within any dose of particles in tissue that drives a critical pathophysiologically relevant form of toxicity, e.g., oxidative stress, inflammation, genotoxicity, or proliferation) may be achieved independently from the mass of inhaled particles (Donaldson et al., 2010).

Low-soluble ENM (e.g., CNT or TiO_2)—even though they can partially break down in contact with

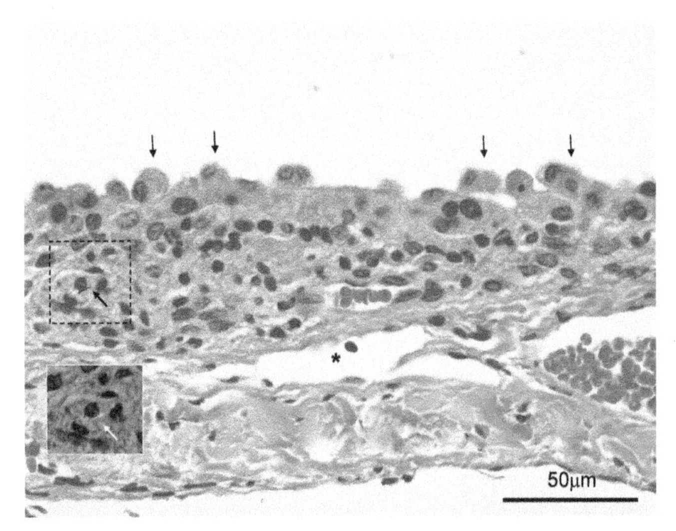

FIGURE 41.4 Atypical mesothelial hyperplasia of the tendinous portion of diaphragm of a mouse with 3 μg of MWCNT injected intraperitoneally (sampled at terminal sacrifice, i.e., 365 days after i.p. inoculation of the MWCNT (Takagi et al., 2012). (*Black arrows*: hobnail appearance of the atypically hyperplastic mesothelial cells. *Asterisk*: lymphatic drainage of the peritoneal cavity.) The polarized image is of the *dotted area*). (*White arrow*: an MWCNT fiber in a macrophage-like cell [birefringent]). *Cancer Science. Reproduced with permission.*

phagocytic cells—show a slow clearance leading to accumulation over time with exposure. For many particles, size and agglomeration state are influential in dictating toxicity; thus, particle dissolution and the release of ions—which would be expected to be greater for smaller particles—may affect organ distribution. Cho et al. (2013) studied the absorption, distribution, and excretion patterns of TiO_2 and ZnO NPs following oral administration. Zinc concentrations in blood, organs, and urine were higher than concentrations of titanium when NPs were administered orally for 13 weeks. The urine concentration of Ti in the TiO_2-treatment groups showed no significant differences compared with the control group; in contrast, the concentration of Zn in the urine of ZnO-treatment groups was significantly increased in the middle- and high-dose groups and showed positive trend dose-responses. In vivo, NPs are mostly retained in the liver, and fragments of the organic shell are excreted through the kidneys, probably due to proteolytic digestion (Kreyling et al., 2015).

Following liver accumulation, metal NPs show a protracted elimination and slow release of particles from the target organ into systemic circulation (Johnston et al., 2010), giving rise to detectable trace amounts in body fluids. Demonstration of translocation of ENM from lung into systemic circulation is theoretically possible for metallic NPs, which release metal ions or dissolve in biological media. Similar to metal species in welding fumes, metallic elements are measurable in blood and urine with appropriate analytical methods, giving an estimate of current or past exposure. For instance, Lee et al. (2012) measured silver concentration in blood and urine in Korean workers exposed to silver NPs, whereas in the study of Lee et al. (2015) molybdenum blood concentration was measured as a candidate biomarker for MWCNT exposure.

To estimate the dose retained at the portal of entry, it can be informative to assess the tissue dose reaching the target organ, e.g., the lung. Sampling of exhaled breath condensate (EBC) provides a matrix for the simultaneous monitoring of exposure and effects on target organ. As EBC mainly consists of water that is practically free of potentially interfering solutes, it is an ideal biological fluid for elemental determinations based on electrothermal atomic absorption spectroscopy or inductively coupled plasma-mass spectrometry. This novel approach has represented a significant advancement over the analysis of alternative media (blood, serum, urine, hair), which are not as reliable (owing to interfering substances in the complex matrix) and reflect systemic rather than lung (target tissue) levels, e.g., of pneumotoxic metallic elements (Goldoni et al., 2004). Particles of rutile and/or anatase were found in the EBC of exposed workers in 70% of the postshift samples, the mean concentration of titanium in production workers being 24.1 ± 1.8 μg/L, whereas in the research workers the values were below the limit of quantitation, e.g., 4.0 ± 0.2 μg/L. Thus, the concentration of titanium originating from TiO_2 in EBC might serve as a direct exposure marker in workers producing TiO_2 pigment; however, the stability of its concentrations in EBC not being influenced by current exposure support the use as a biomarkers of deposited dose following long-term exposure (Pelclova et al., 2015a).

Biomarkers of Effect

A biomarker of effect is defined as any measurable biochemical, physiological, or other alteration within an organism that, depending on magnitude, can be recognized as an established or potential health impairment or disease (Henderson et al., 1989). Biomarkers should not be considered as diagnostic tests, but rather as indicators reflecting early modifications preceding progressive structural or functional damage at the molecular, cellular, and tissue level, i.e., changes possibly leading to adverse effects but completely reversible on the removal from the exposure of concern. Biomarkers of effect can be used to evaluate whether a well-characterized exposure is associated with a shift in the distribution of relevant biochemical or functional end points, indicative of early changes in the target or critical organs/tissue. To assess such early events associated with exposure to ENM, the choice of potential biomarkers can give insights on local and systemic oxidative stress, systemic inflammation, and inflammatory response in target organs, as in respiratory and cardiovascular systems.

Biomarkers of Lung Inflammation and Systemic Effects

Because airway inflammation is the main outcome investigated following NM exposure, inflammatory biomarkers deserve greater importance for biomonitoring purposes. Breath analysis has been proposed as a noninvasive approach that allows the identification of the inflammatory and oxidative stress biomarkers involved in the pathogenesis of various clinical conditions (Montuschi, 2007) and for investigating occupational lung diseases (Corradi et al., 2010), exposure to welding fumes (Gube et al., 2010) or particles from aircraft engines (Desvergne et al., 2016). Besides macromolecules (e.g., DNA or RNA), an increasing panel of biomarkers reflecting oxidative stress and inflammatory pathways can be determined in EBC. Thiobarbituric acid reactive substances, such as MDA—a product of membrane lipoperoxidation—and 8-isoprostane—a peroxidation product of prostaglandin metabolism—can be quantified; proinflammatory cytokines, such as leukotrienes B4 (LTB4), can be determined as biomarkers of inflammation. Inflammatory response in the airways is characterized by an influx of neutrophils, whose activation is associated with a respiratory burst resulting in overproduction of hydrogen peroxide (H_2O_2), changes in pH, and depletion of the glutathione (GSH) pool.

Various classes of proteins and volatile compounds can be measured in EBC, including the saturated hydrocarbons and oxygen-containing substances formed during the fatty acid lipid peroxidation of cell membranes. A wide variety of carbonyl compounds are generated as secondary oxidation products during respiratory burst. In particular, saturated aldehydes, such as 4-hydroxy-trans-nonenal (HNE) and 4-hydroxy-trans-hexenal (HHE), are formed by the peroxidation of omega-3 and -6 fatty acids, the basic components of cell membrane phospholipids. EBC concentrations of MDA, HNE, and HHE were significantly higher in workers exposed to TiO_2 aerosols as compared with unexposed people (Pelclova et al., 2015b). Oxidative stress biomarkers (MDA, HNE, HHE, C6–C10, 8-isoprostane, 8-OHdG, 8-OHG, 5-OHMeU, 3-ClTyr, 3-NOTyr, o-Tyr, and C11) were elevated in the EBC of workers exposed to NPs during iron oxide pigment production compared with control subjects (Pelclova et al., 2016).

Recently, Lacombe et al. (2018) performed an in-depth proteomics characterization of EBC; a total of 229 unique proteins were identified in EBC among which 153 proteins were detected in both EBC pooled samples. A detailed bioinformatics analysis of these 153 proteins showed that most of the proteins identified corresponded to proteins secreted in the respiratory tract (lung, bronchi).

Breath can be analyzed also in the gaseous phase (as exhaled breath). For instance, nitric oxide (NO) is produced by all cellular components of pulmonary inflammation (macrophages, epithelial cells, mast cells, lymphocytes, and granulocytes). Studying NO exhaled from the lower airways offers a unique possibility to study features of pulmonary NO metabolism noninvasively in all states characterized by lung inflammation. In a follow-up study, Wu et al. (2014) found an increase in fractional exhaled nitric oxide (FENO) among workers handling nano-TiO_2 powders, but not in all of the NM exposed categories investigated; the values recorded were over the threshold of 35 ppb, suggesting a chronic airways inflammation. Differences in FENO between workers of an MWCNT facility and nonexposed, with no difference in lung function or the pneumoproteins have been found by Vlaanderen et al. (2017).

Pneumoproteins, such as Clara cell protein (CC16) and surfactant-associated protein B (SP-B) in the serum, have been validated as markers of alveolo–capillary barrier integrity/permeability in human studies on gaseous/particulate pollutants (Broekaert et al., 2000; Gulumian et al., 2006). There is evidence that acute exposures to certain pulmonary irritants can cause a transient increase in serum CC16 levels, and limited evidence also suggests that a transient increase in serum CC16 levels can be caused by a localized pulmonary inflammation without impairment of pulmonary function. The biological interpretation of chronic changes in serum CC16 is less clear, owing to chronobiological variability. In workers in indium tin oxide production plants, significant positive relationships were found between S-In and surfactant protein A (SP-A), and surfactant protein D (SP-D) levels, sensitive markers of interstitial lung disease. SP-A and SP-D levels were elevated significantly in the workers with moderately high indium exposure (Liu et al., 2012).

Early events at vascular level can be assessed by a panel of circulating biomarkers reflecting inflammation end points, platelet activation, and antioxidant capacity (as assessed by the activity of Cu/Zn-superoxide dismutase, and glutathione peroxidase-1). Inactivation of antioxidant enzymes within erythrocytes, plasma interleukin-6 (IL-6), and soluble tumor necrosis factor-receptor II (sTNF-RII), investigated during a longitudinal study, showed a positive association with vehicle emissions tracers (Delfino et al., 2010). High-sensitivity C-reactive protein in plasma, plasma fibrinogen, and IL-6 could represent good candidates for BM of ENM.

Acute phase proteins, such as C-reactive protein (CRP), are commonly used as biomarkers of (inflammatory) disease, and associations have also been shown between inflammatory biomarkers including CRP and exposure to PM air pollution. 70-nm silica nanoparticles (nSP70) induced a higher level of acute phase proteins such as haptoglobin, CRP, and serum amyloid A (SAA) than larger silica particles (diameter >100 nm) (Higashisaka et al., 2011). In addition, the level of these acute phase proteins was elevated in the plasma of mice after intranasal treatment with nSP30. The same authors identified hemopexin (another acute phase protein) as a potential biomarker for predicting the effects of silica NPs (Higashisaka et al., 2012).

Following intratracheal instillation of titanium dioxide (TiO_2), CB, diesel exhaust particles and CNTs, tissue mRNA expression of acute phase protein, and plasma levels of Serum Amyloid 3 were increased. Interestingly, these inflammatory biomarkers significantly correlated with the magnitude of neutrophilic influx in bronchoalveolar lavage fluid, thus suggesting that lung inflammation is associated with the expression of biomarkers predictive of cardiovascular outcomes (Saber et al., 2013).

Exposure to NPs could enhance the adhesion of endothelial cells and modify the membrane structure of vascular endothelium, which plays an important role in the regulation of fibrinolysis. Radomski et al. (2005) showed that both urban dusts and engineered carbon particles, such as CNT and CB—except C60CS—stimulated platelet aggregation and accelerated the rate of vascular thrombosis in rat carotid arteries with a similar rank order of efficacy. All particles resulted in upregulation of GPIIb/IIIa in platelets. In contrast, particles differentially affected the release of platelet granules, as well as the activity of thromboxane-, ADP-, matrix metalloproteinase-, and protein kinase C-dependent pathways of aggregation. Exposure to nano Ag (0.05–0.1 mg/kg i.v. or 5–10 mg/kg i.t. instillation) enhanced platelet aggregation and promoted venous thrombus formation in rats (Jun et al., 2011). Therefore, assessment of platelet aggregation and expression of GPIIb/IIIa in platelets may be useful to evaluate possible prothrombotic effects induced by NM. Early systemic prothrombotic effects induced by fine particle (<2.5 μm) exposure were detected by the quantification of levels of the plasminogen activator inhibitor-1 (PAI-1) (Kilinc et al., 2011). Elevated PAI-1 is closely associated with enhanced thrombosis by impairing fibrinolysis; metal NPs have proven to increase PAI-1 expression in endothelial cells in vitro (Yu et al., 2010).

Vesterdal et al. (2010) showed that exposure of young and aged apolipoprotein E knockout mice (apoE−/−) to CB (Printex 90, 14 nm average size) by intratracheal instillation, resulted in modest vasomotor impairment, with a lack of association with nitrosative stress (3-nitrotyrosine), and without increases in the expression of vascular adhesion molecule and intercellular adhesion molecule (ICAM-1) on endothelial cells or in plaque progression.

Interestingly, workers exposed to a mixture of NM showed statistically significant changes across exposure risk classes for high sensitive C reactive protein levels and inflammatory cell activation (increased ICAM-1 in macrophages), IL-6, and fibrinogen. Moreover, the depression of antioxidant enzymes, namely SOD and GPx, were associated with NM handling (Liou et al., 2012).

Erdely et al. (2011) demonstrated that the pulmonary exposure to CNT triggered the induction of primary cytokines such as IL-6 and IL-1β, which regulate multiple pathways of the inflammatory cascade as well as secondary inflammatory mediators, chemokines (CCL2, 4, 19, 22; CXCL1, 2), which directly regulate leukocyte recruitment to the inflammatory site. Iavicoli et al. (2018) investigated the adverse effects induced by subchronic intravenous administration of palladium NPs (PdNPs) on the immune system of female Wistar rats by evaluating cytokines (e.g., IL-1α, IL-2, IL-4, IL-6, IL-10, IL-12), the granulocyte–macrophage colony-stimulating factor , the INF-γ, and TNF-α serum levels at different dose levels (0, 0.012, 0.12, 1.2, and 12 μg PdNPs/kg b.w. till 60 days). Subchronic exposure to PdNPs induced a decreasing trend in serum levels in most of the cytokines investigated, with the highest concentration (12 μg/kg) determining significant inhibitory effects.

In a small cohort of workers occupationally exposed to MWCNTs, Shvedova et al. (2016) found significant changes in the ncRNA and mRNA expression profiles between exposed (inhalable concentration 14.42 ± 3.8 $\mu g/m^3$; respirable concentration 2.83 ± 0.6 $\mu g/m^3$ as elemental carbon concentrations in breathing zone samples from workers) and nonexposed workers. A dysregulation of profile of genes involved in cell cycle regulation/progression/control, apoptosis, cell proliferation, and carcinogenetic pathways was characterized in eight workers having direct contact with MWCNT-containing aerosols in the previous 6 months (Shvedova et al., 2016). In a concomitant study on the same group (Fatkhutdinova et al., 2016), it was found that exposure to MWCNTs in the order of 3 times above the recommended exposure level proposed by the NIOSH (1 $\mu g/m^3$) caused significant increase in IL-1β, IL-4, IL-5, IL6, TNF-α, inflammatory cytokines, and KL-6 in sputum samples. Moreover, the level of TGF-β1 was increased in serum obtained from young (<30 years old) exposed workers. In the serum samples, levels of IL-1β, IL-4, IL-10, and TNF-α were significantly elevated in the MWCNT exposed group.

Vlaanderen et al. (2017) assessed 51 immune markers and three pneumoproteins in serum among 22 workers of an MWCNT producing facility and 39 age- and gender-matched, unexposed controls, considering potentially confounding parameters (age, body mass index, smoking, and sex). These authors found significant upward trends for immune markers C–C motif ligand 20, basic fibroblast growth factor, and soluble IL-1 receptor II with increasing exposure to MWCNT.

Biomarkers of Oxidative DNA Damage and RNA Methylation

Panel studies and cross-sectional investigations on health effects of PM exposure have found consistent associations between exposure to combustion-derived particles and products of oxidative damage to DNA and lipids (Loft et al., 2008; Møller and Loft, 2010). Among the DNA oxidation products, 8-hydroxy-2-deoxyguanosine (8-OHdG) and 8-hydroxyguanosine (8-OHG), measured in DNA of peripheral blood cells and urine, have been the most studied. Base excision repair products of oxidative damage to DNA in urine seem to originate mostly from the oxidation of the deoxynucleotide pool and do not represent solely repairing/excretion of the oxidized-DNA guanine, but biomarkers of effective dose.

Urinary 8-OH-dG concentration has been investigated as a potential biomarker of oxidative stress in response to exposure to incidental NPs emitted from photocopiers, and it was found significantly increased as compared with background levels (Khatri et al., 2013). In EBC of workers occupationally exposed to TiO_2 aerosols, Pelclova et al. (2015b) found an increase in markers of oxidation of nucleic acids (including 8-hydroxy-2-deoxyguanosine [8-OHdG], 8-hydroxyguanosine [8-OHG], 5-hydroxymethyl uracil [5-OHMeU]). Conversely, in workers handling nanomaterials in 14 plants in Taiwan, 8-OH-dG urinary and plasma levels did not show significant differences compared with controls (Liou et al., 2012).

The DNA-damaging potential of many NPs of different composition (metals, metal oxides, silica, QDs, fullerenes, nanofibers, SW-, and MWCNT) has been demonstrated in vitro (Gonzalez et al., 2008; Singh et al., 2009; Magdolenova et al., 2013). NM can affect DNA also by direct mechanisms, including a mechanical interference with cellular and nuclear components, such as microtubules of the mitotic spindle (Gonzalez et al., 2008). Although interactions between the particles and the assay cannot be totally excluded, the use of Comet assay in human biomonitoring studies could provide valuable information for hazard identification of NM (Karlsson, 2010; Karlsson et al., 2015). Modified Single Cell Gel Electophoresis (SCGE) provides parallel information on oxidative DNA damage caused by NM, such as CNT, detecting oxidized/damaged pyrimidines and purines (Muller et al., 2008; Migliore at al., 2010), and revealing dose–effect and dose–response relationships between the mass concentration of NM and the frequency of micronucleated lymphocytes. Both tests have been applied in a cohort of nanomaterial workers in Taiwan, but there were no significant differences in changes between the exposed and control workers between baseline and the 6-month follow-up (Liao et al., 2014).

DNA methylation, a major genomic mechanism of gene expression control, can be affected by ROS, which are considered as one of the main cellular stressors generated by PM exposure as well as by some metals. Stoccoro et al. (2013) highlighted the ability of certain NPs to induce an impaired expression of genes involved in DNA methylation reactions leading to global DNA methylation changes, as well as changes of gene-specific methylation of tumor suppressor genes, inflammatory genes, and DNA repair genes, all potentially involved in cancer development. Moreover, some nanosized compounds are able to induce changes in the acetylation and methylation of histone tails, as well as microRNA (miRNAs) deregulated expression.

Brown et al. (2016) assessed the promoter methylation of inflammatory genes (IFN-γ and TNF-α) after MWCNT exposure and found a correlation between these changes and initial cytokine production. In addition, methylation of a gene involved in tissue fibrosis (Thy-1) was also altered in a way that matched collagen deposition. These authors also found that MWCNT exposure lead to DNA hypomethylation in the lung and blood, which coincided with disease development (i.e., fibrosis).

MiRNAs are noncoding small RNAs that regulate the expression of broad gene networks at the posttranscriptional level, interacting with several mRNA targets, and their use as possible biomarkers of the effects of acute and chronic environmental exposure has been suggested (Vrijens et al., 2015). Some nanosized compounds are able to induce selected miRNAs deregulated expression, and this may help in finding specific fingerprints. Ng et al. (2011) demonstrated that gold nanoparticles (AuNPs) altered the expression of 19 genes in human fetal lung fibroblasts, upregulating the miRNA-155 (miR-155) and downregulating the *PROS1* gene—a gene encoding a vitamin K-dependent plasma protein that functions as a cofactor for the anticoagulant protease, activated protein C to inhibit blood coagulation. Silencing of miR-155 established *PROS1* as its possible target gene. DNA methylation profiling analysis of the *PROS1* gene revealed no changes in the methylation status of this gene in AuNP-treated fibroblasts, whereas chromatin condensation and reorganization was observed in the nucleus of fibroblasts exposed to AuNPs.

Nagano et al. (2013) compared the effectiveness of serum levels of liver-specific or -enriched miRNAs (miR-122, miR-192, and miR-194) with that of conventional hepatic biomarkers (alanine aminotransferase and aspartate aminotransferase) as biomarkers for nSP70 induced liver damage in mice.

Toward Specific Biomarkers for Engineered Nanomaterials Exposure

Research of biomarkers reflecting exposure to certain particles and fibers of concern have already generated a large amount of data supporting the validity of intermediate end points to assess changes before clinically apparent disease occurs (Gulumian et al., 2006) and several biomarkers can reliably help in assessing exposure and effects of NM (Iavicoli et al., 2014; Bergamaschi et al., 2017). The challenge that remains for the use of biomarkers of exposure for NM is the lack of specificity of all the biomarkers developed so far that could hamper the applicability of biomonitoring to nano-object.

It should be recognized that the existence of nanospecific (i.e., size-dependent) effects is an arbitrary assumption because the threshold of 100 nm does not infer, per se, new properties, whereas a gradual change in surface reactivity could actually modulate biological interactions independently from size. Properties other than size, such as shape, surface area, surface charge, and reactivity, can more effectively describe the dose leading to biological effects, as in the classical particle and fiber toxicology science (Auffan et al., 2009; Fubini et al., 2010; Donaldson and Poland, 2013). This does not mean that there are no changes specifically induced by some particles at certain size thresholds, but these changes should be interpreted as mode of actions instead as "new" toxicological properties (Donaldson and Poland, 2013; Lynch et al., 2014). For instance, the more toxicity and DNA damaging potential of nanosized than microsized CuO particles in A549 cells has been attributed to the ability to deliver Cu^{2+} inside cells ("*Trojan horse*" effect). In contrast, the micrometer particles of TiO_2 caused more DNA damage compared to the NPs, which is likely explained by the crystal structures.

New achievements in biomarker discovery came from the study of the intercellular communication pathways (Valadi et al., 2007; Raposo and Stoorvogel, 2013; Lin et al., 2015). In vitro studies have revealed that cellular uptake of NPs provoked micro vesicular endosome (MVe) formation intracellularly; however, few studies have studied the biological consequences once the exosomes have been secreted extracellularly in vivo. On i.t. instillation (4 μg or 20 μg) with 43-nm diameter magnetic iron oxide nanoparticles (MIONs), Zhu et al. (2012) observed a dose-dependent generation of exosomes in the alveolar region of BALB/c mice. These exosomes were quickly eliminated from alveoli into systemic circulation and largely transferred their signals to the immune system, with activation of splenic T cells. Interestingly, the maximum dose used in this study is equal to a half-workday deposition mass in human lung tissue at the permissible exposure limit of iron oxide fume (at a concentration of 10 mg/m^3) suggested by the Occupational Safety and Health Standards (OSHA, USA). These findings suggest that respiratory exposure to MIONs can activate systemic T cells in susceptible individuals, through exosome-mediated signaling pathways.

Systems Toxicology, the integration of classical toxicology with quantitative analysis of large networks of molecular and functional changes, is also expected to provide information on the dynamic interaction between molecular components of biological systems and about the possible mechanisms, by assessing whether specific biological pathways are activated/perturbed by specific NM, thus identifying fingerprints and nano-specific end points useful for hazard identification and, ultimately, for risk assessment (RA) (Costa and Fadeel, 2016). Comparative proteomic studies have shown strong similarities in the pulmonary response to different ENM with known hazardous particles and fibers. One repeated aspiration study (4 μg per mouse, twice a week, for 3 weeks) allowed the identification of a pattern of 109 proteins representing cellular processes affected by both SWCNT and crocidolite asbestos; S100a9, a high-sensitivity marker of inflammation, can be proposed as a biomarker of human response to SWCNT exposure (Teeguarden et al., 2011). Mice exposed by pharyngeal aspiration to 40 μg CNT showed increased inflammatory blood gene expression and serum cytokines followed by an acute phase response (e.g., CRP, SAA-1, SAP). At 28 days, serum acute-phase proteins with immune function including complement C3, apolipoproteins A-I and A-II, and 1-macroglobulin were increased (Erdely et al., 2011). CNT exposure resulted in measurable systemic markers but lacked specificity to distinguish from other pulmonary exposures.

A toxicogenomic approach has been used in assessing specific mechanisms at the molecular level, identifying patterns of cellular perturbations in specific pathways, through identification and quantification of global shifts in gene expression in cell models challenged with ENM. Pulmonary exposure to CNT resulted in an elevated series of measurable potential biomarkers in blood, including genes expressed in the circulating blood cells and/or soluble proteins not unique to the type of particles (Erdely et al., 2009). In particular, MWCNT-induced gene upregulation of more than half of the tested genes in the lung related to inflammation, oxidative stress, coagulation, and tissue remodeling and to a significant increase in the circulating blood gene expression of several biomarkers of neutrophil response. Interestingly, several genes were activated in the circulating blood cells but not in the lung, at least at 4 h after exposure to MWCNT, e.g., osteopontin (a marker of early mesothelioma), colony stimulating factor-1 (CSF-1), and insulin growth factor receptor 1. Exposure to CNT also triggered the induction of primary cytokines such as IL-6 and IL-1b, which regulate multiple pathways of the inflammatory cascade as well as secondary inflammatory mediators, and chemokines, which directly regulate leukocyte recruitment to the inflammation site.

Guo et al. (2012) analyzed mRNA expression profiles in lungs of mice exposed to 0–80 μg of MWCNT by pharyngeal aspiration until 56 days postexposure and identified sets of genes associated with human lung cancer risk and progression with significant odds ratios. C57BL/6 mice exposed to 18, 54, and 162 μg Printex 90 carbon black nanoparticles (CBNP) showed perturbation of pathways, networks, and transcription factors of predicted phenotypes (e.g., pulmonary inflammation and genotoxicity) that correlated with dose and time. Comparison to inflammatory lung disease models (i.e., allergic airway inflammation, bacterial infection, and tissue injury and fibrosis) and human disease profiles revealed that induced gene expression changes in Printex 90 exposed mice were similar to those typical for pulmonary injury and fibrosis. Very similar fibrotic pathways were perturbed in CBNP-exposed mice and human fibrosis disease models, thus supporting the use of toxicogenomic profiles in human health risk assessment of NPs (Bourdon et al., 2013).

Palomäki et al. (2015) performed proteomics analyses of human macrophages exposed to tangled or rigid, long MWCNTs, or crocidolite asbestos, using hyphenated techniques and concluded that not all types of CNTs are as hazardous as asbestos fibers.

Kinaret et al. (2017b) have systematically investigated transcriptomic responses of the THP-1 macrophage cell line and lung tissues of mice, specifically induced by several carbon nanomaterials (CNMs). They observed only a minute overlap between the sets of intrinsic property-correlated genes at different exposure scenarios, suggesting specific transcriptional programs working in different exposure scenarios. However, when the effects of the CNM were investigated at the level of significantly altered molecular functions, a broader picture of substantial commonality emerged. As a result, in vitro exposures can efficiently recapitulate the complex molecular functions altered in vivo.

High-throughput (omics) methods, if applied in a rigorous manner, hold great promise for the development of (novel) biomarkers and biomarker signatures. In addition, achievements in this field may make more consistent the regulatory approach to hazard assessment based on dynamic adverse outcome pathway (AOP) models. (Vietti et al., 2016; Labib et al., 2016). With the aim to evaluate the application of global gene expression data in deriving pathway-based points of departure for MWCNT-induced lung fibrosis, as a noncancer end point of regulatory importance, Labib et al. (2016) showed an early perturbation of similar biological pathways regardless from MWCNT types across the doses and postexposure time points studied. The authors also showed that transcriptional benchmark dose (BMD) values for pathways associated with fibrosis (4.0–30.4 μg/mouse) were comparable to the BMDs derived by NIOSH for MWCNT-induced lung fibrotic lesions (namely, 21.0–27.1 μg/mouse), thus suggesting that transcriptomic data can be used to derive acceptable levels of exposure to NM in product development (Labib et al., 2016).

CONCLUDING REMARKS AND FUTURE DIRECTIONS

BM is an important component of the occupational and environmental health surveillance, especially when occupational and/or environmental exposure monitoring data are unavailable or difficult to obtain. The literature on short-term effects of air pollutants and the available literature on NM, suggest identifying multiple biomarkers—a biomarker profile—to assess both effects at the "portal of entry" (e.g., inflammatory changes, short-term respiratory changes, respiratory, eye or skin irritation) and systemic effects (e.g., heart-rate variability, platelet aggregation and prothrombotic changes, acute phase proteins) in susceptible subgroups of the general population exposed to incidental NPs or to a mixture of pollutants.

Table 41.1 summarizes a panel of biomarkers of exposure and effect potentially available for human biomonitoring studies aimed at assessing early effects and health outcomes.

In spite of the lack of validation of "nanospecific" biomarkers, it is proposed that at this stage that the sensitivity rather than the specificity of biomarkers should be privileged to identify potentially predictive biomarkers suggestive of the biological pathway and mechanistic changes underlying the causality of exposure conditions and association with hazards (e.g., at workplace) and also to identify predictive biomarkers (Bergamaschi et al., 2015). For the moment, all studies carried out so far (see the review by Liou et al., 2015) have used a cross-sectional design and consequently do not allow confirming the observed effects and understanding the dynamics of their occurrence and

TABLE 41.1 Appraisal of Biomarkers of Exposure or Effects Relating to Ultrafine Particle or Engineered Nanomaterial

Quality of Biomarker	
Biomarkers of exposure	• Exhaled particles and/or elements in EBC (estimate of the "deposited dose" or "target tissue dose") • Elements analysis in biological fluids (excretion, body burden) • Protein modification ("corona")
Biomarkers of effective dose/early effect	• Lipid peroxidation products in EBC or blood (MDA, TBARS, conjugated dienes, LTB4, F2- and 8-isoprostane) • DNA excision base products (8-OHdG, 8-oxo-Gua, 8-OHG) • Exhaled NO (FeNO) and nitrosative stress products (3-nitrotyrosine) • Carbonyl compounds (4-HNE, 4-HHE) in EBC • Serum pneumoproteins (CC16) • Platelet activation/aggregation and prothrombotic changes • Acute phase proteins: hsCRP, SAA, Haptoglobin, Hemopexin • IL-6 and sTNF-RII • Coagulation factors (fibrinogen, plasminogen activator inhibitor-1[PAI-1]) • Vascular adhesion molecules (VCAM-1) and intercellular adhesion molecule (ICAM-1)
Biomarkers reflecting alterations in cell structure/function	• Fibrogenic markers (KL-6 glycoprotein; MMP-1, MMP-7, MMP- 9) • Osteopontin (Early mesothelioma development) • Micronucleus • DNA strand breaks (Comet assay + FPG-ENDO III) • Epigenetic markers: DNA (hypo)methylation; MicroRNAs (miRNAs) • Extracellular micro- and nanovescicles, exosomes (EMVs' cargo characterization)

The table includes biomarkers validated in human studies on people exposed to different ultrafine or fine particulates or known fractions, and biomarkers specifically investigated in relation to ENM.
Note: *4-HNE*, 4-hydroxy-2-nonenal; *8-isoprostane*, 8-isoprostaglandin F2α; *8-OHdG*, 8-hydroxy-2′-deoxyguanosine; *8-OHG*, 8-hydroxyguanosine; *8-oxo-Gua*, 8-oxo-7,8-dihydroguanine; *CC16*, Clara cell protein; *EBC*, exhaled breath condensate; *FPG-ENDOIII*, lesions detected as sites in DNA sensitive to formamidopyrimidine DNA glycosylase and endonuclease III; *hsCRP*, high sensitivity C-reactive protein; *IL-6*, plasma Interleukin 6; *LTB4*, Leucotriene-B4; *MDA*, malondialdehyde; *NO*, nitric oxide; *SAA*, serum amyloid A; *sTNF-RII*, soluble tumor necrosis factor-receptor II; *TBARS*, thiobarbituric acid reactive substances.

duration. Consequently, this kind of study could only help to identify some "intermediate" biological changes of effects. Only the study of Lee et al. (2015) showed a reduction, though not clinically significant, in lung function parameters in a cohort of workers occupationally exposed to nanoscale CB. Longitudinal panel studies with repeated exposure and effect biomarker measurement are necessary to investigate whether this unspecific "intermediate" biological changes could be indicative or predictive of clinical effects and apply them in health surveillance programs. To provide a coherent approach and make future epidemiological research a reality, a well-defined framework is needed for the careful choice of materials, exposure characterization, identification of study populations, definition of health end points, and evaluation of the appropriateness of study designs, data collection and analysis, and interpretation of the results (Riediker et al., 2012). Moreover, the scientific, methodological, political, and regulatory issues that make epidemiological research in nanotechnology-exposed communities particularly complex. Standardization of data collection and harmonization of research protocols are needed to eliminate misclassification of exposures and health effects. Forming ENM worker cohorts from a combination of smaller cohorts and overcoming selection bias are also challenges (Guseva Canu et al., 2018).

In conclusion, BM of exposure and effect should represent a valuable component of an integrated strategy and a proactive approach to risk assessment and management. Hence, it is an opportunity for companies committed to the responsible development of nanotechnology and also an ethical obligation toward all populations of workers that everything is done to assure a safe working environment.

Acknowledgments

The research leading to this review has been developed under the EU FP7 NANOSOLUTIONS Project (Grant agreement 309329) and the EU Horizon 2020 BIORIMA Project (Grant agreement 760928).

References

Aitken, R.J., Creely, K.S., Tran, C.L., 2004. Nanoparticles: An Occupational Hygiene Review. Health and Safety Executive (HSE).
Antonini, J.M., 2003. Health effects of welding. Crit. Rev. Toxicol. 33, 61–103.
Aschberger, K., Johnston, H.J., Stone, V., 2010. Review of carbon nanotubes toxicity and exposure. Appraisal of human health risk assessment based on open literature. Crit. Rev. Toxicol. 40, 759–790.
Auffan, M., Rose, J., Bottero, J.Y., 2009. Towards a definition of inorganic nanoparticles from an environmental, health and safety perspective. Nat. Nanotechnol. 4, 634–641.
Bencsik, A., Lestaevel, P., Guseva Canu, I., 2018. Nano- and neurotoxicology: an emerging discipline. Prog. Neurobiol. 160, 45–63.

Bergamaschi, E., 2012. Human biomonitoring of engineered nanoparticles: an appraisal of critical issues and potential biomarkers. J. Nanomater. https://doi.org/10.1155/2012/564121.

Bergamaschi, E., Guseva-Canu, I., Prina-Mello, A., Magrini, A., 2017. Biomonitoring. In: Fadeel, B., Pietroiusti, A., Shvedova, A. (Eds.), Adverse Effects of Engineered Nanomaterials, second ed. Academic Press, London, England, pp. 225–260.

Bergamaschi, E., Poland, C., Guseva Canu, I., Prina-Mello, A., 2015. The role of biological monitoring in nano-safety. Nano Today 10, 274–277.

Bianchi, M.G., Allegri, M., Costa, A.L., et al., 2015. Titanium dioxide nanoparticles enhance macrophage activation by LPS through a TLR4-dependent intracellular pathway. Toxicol. Res. 4, 385–398.

Bourdon, J.A., Williams, A., Kuo, B., 2013. Gene expression profiling to identify potentially relevant disease outcomes and support human health risk assessment for carbon black nanoparticle exposure. Toxicology 303, 83–93.

Broeckaert, F., Arsalane, K., Hermans, C., et al., 2000. Serum Clara cell protein: a sensitive biomarker of increased lung epithelium permeability caused by ambient ozone. Environ. Health Perspect. 108 (6), 533–537.

Broday, D.M., Agnon, Y., 2007. Asymmetric human lung morphology induce particle deposition variation. J. Aero. Sci. 38, 701–718.

Brown, T.A., Lee, J.W., Holian, A., et al., 2016. Alterations in DNA methylation corresponding with lung inflammation and as a biomarker for disease development after MWCNT exposure. Nanotoxicology 10 (4), 453–461.

Buhleier, E., Wehner, W., Vogtle, F., 1978. Cascade and nonskid-chain-like synthesis of molecular cavity topologies. Synthesis 2, 155–158.

Castagnola, V., Cookman, J., de Araújo, J.M., et al., 2017. Towards a classification strategy for complex nanostructures. Nanoscale Horiz. 2, 187–198.

Catalán, J., Järventaus, H., Vippola, M., et al., 2012. Induction of chromosomal aberrations by carbon nanotubes and titanium dioxide nanoparticles in human lymphocytes in vitro. Nanotoxicology 6, 825–836.

Chakrabarti, A., Jha, B.K., Silverman, R.H., 2011. New insights into the role of RNase L in innate immunity. J. Interferon Cytokine Res. 31, 49–57.

Chen, J., Saeki, F., Wiley, B.J., 2005. Gold nanocages: engineering the structure for biomedical applications. Adv. Mater. 17, 2255–2261.

Chen, S., Wang, Z.L., Ballato, J., et al., 2003. Monopod, bipod, tripod, and tetrapod gold nanocrystals. J. Am. Chem. Soc. 125, 16186–16187.

Chernova, T., Murphy, F.A., Galavotti, S., et al., 2017. Long-fiber carbon nanotubes replicate asbestos-induced mesothelioma with disruption of the tumor suppressor gene Cdkn2a (Ink4a/Arf). Curr. Biol. 27, 3302–3314.

Cho, W.S., Kang, B.C., Lee, J.K., 2013. Comparative absorption, distribution, and excretion of titanium dioxide and zinc oxide nanoparticles after repeated oral administration. Part. Fibre Toxicol. 10, 9.

Choi, H.S., Ashitate, Y., Lee, J.H., 2010. Rapid translocation of nanoparticles from the lung airspaces to the body. Nat. Biotechnol. 28, 1300–1303.

Corradi, M., Gergelova, P., Mutti, A., 2010. Use of exhaled breath condensate to investigate occupational lung diseases. Curr. Opin. Allergy Clin. Immunol. 10, 93–98.

Costa, P.M., Fadeel, B., 2016. Emerging systems biology approaches in nanotoxicology: towards a mechanism-based understanding of nanomaterial hazard and risk. Toxicol. Appl. Pharmacol. 299, 101–111.

Dabbousi, B.O., Rodriguez-Viejo, J., Mikulec, F.V., 1997. (CdSe)ZnS core–shell quantum dots: synthesis and characterization of a size series of highly luminescent nanocrystallites. J. Phys. Chem. B 101, 9463–9475.

Delfino, R.J., Staimer, N., Tjoa, T., 2010. Association of biomarkers of systemic inflammation with organic components and source tracers in quasi-ultrafine particles. Environ. Health Perspect. 118, 756–762.

Desvergne, C.M., Dubosson, M., Touri, L., et al., 2016. Assessment of nanoparticles and metal exposure of airport workers using exhaled breath condensate. J. Breath Res. 10, 036006.

Di Cristo, L., Movia, D., Bianchi, M.G., et al., 2015. Proinflammatory effects of pyrogenic and precipitated amorphous silica nanoparticles in innate immunity cells. Toxicol. Sci. 150 (1), 40–53.

Dockery, D.W., Pope, C.A., Xu, X., 1993. An association between air pollution and mortality in six U.S. cities. N. Engl. J. Med. 329, 1753–1759.

Donaldson, K., Poland, C.A., 2013. Nanotoxicity: challenging the myth of nano-specific toxicity. Curr. Opin. Biotechnol. 24, 724–734.

Donaldson, K., Seaton, A., 2012. A short history of the toxicology of inhaled particles. Part. Fibre Toxicol. 9, 13.

Donaldson, K., Tran, C.L., 2004. An introduction to the short-term toxicology of respirable industrial fibres. Mutat. Res. 553, 5–9.

Donaldson, K., Murphy, F.A., Duffin, R., Poland, C.A., 2010. Asbestos, carbon nanotubes and the pleural mesothelium: a review of the hypothesis regarding the role of long fibre retention in the parietal pleura, inflammation and mesothelioma. Part. Fibre Toxicol. 7, 5.

Donaldson, K., Schinwald, A., Murphy, F., 2013. The biologically effective dose in inhalation nanotoxicology. Acc. Chem. Res. 46, 723–732.

Dostert, C., Pétrilli, V., Van Bruggen, R., 2008. Innate immune activation through Nalp3 inflammasome sensing of asbestos and silica. Science 320, 674–677.

Erdely, A., Hulderman, T., Salmen, R., 2009. Cross-talk between lung and systemic circulation during carbon nanotube respiratory exposure: potential biomarkers. Nano Lett. 9, 36–43.

Erdely, A., Liston, A., Salmen-Muniz, R., et al., 2011. Identification of systemic markers from a pulmonary carbon nanotube exposure. J. Occup. Environ. Med. 53 (6), S80–S86.

European Commission (EC), 2011. Recommendations. Commission recommendation of 18 October 2011 on the definition of nanomaterial. Brussels, Belgium Off. J. Eur. Union L 275/38.

Fadeel, B., Feliu, N., Vogt, C., et al., 2013. Bridge over troubled waters: understanding the synthetic and biological identities of engineered nanomaterials. WIREs Nanomed. Nanobiotechnol. 5, 111–129.

Fatkhutdinova, L.M., Khaliullin, T.O., Vasil'yeva, O.L., et al., 2016. Fibrosis biomarkers in workers exposed to MWCNTs. Toxicol. Appl. Pharmacol. 15, 299–305.

Fubini, B., Ghiazza, M., Fenoglio, I., 2010. Physico-chemical features of engineered nanoparticles relevant to their toxicity. Nanotoxicology 4, 347–363.

Gajewicz, A., Puzyn, T., Odziomek, K., et al., 2018. Decision tree models to classify nanomaterials according to the DF4nanoGrouping scheme. Nanotoxicology 12, 1–17.

Geiser, M., Kreyling, W., 2010. Deposition and biokinetics of inhaled nanoparticles. Part. Fibre Toxicol. 7, 2–17.

Goldoni, M., Catalani, S., De Palma, G., et al., 2004. Exhaled breath condensate as a suitable matrix to assess lung dose and effects in workers exposed to cobalt and tungsten. Environ. Health Perspect. 112, 1293–1298.

Gonzalez, L., Lison, D., Kirsch-Volders, M., 2008. Genotoxicity of engineered nanomaterials: a critical review. Nanotoxicology 2, 252–273.

Grosse, Y., Guyton, K.Z., Lauby-Secretan, B., et al., 2014. Carcinogenicity of fluoro-edenite, silicon carbide fibres and whiskers, and carbon nanotubes. Lancet Oncol. 15, 1427–1428.

Gube, M., Ebel, J., Brand, P., et al., 2010. Biological effect markers in exhaled breath condensate and biomonitoring in welders: impact of smoking and protection equipment. Int. Arch. Occup. Environ. Health 83, 803–811.

Gulumian, M., Borm, P.J., Vallyathan, V., 2006. Mechanistically identified suitable biomarkers of exposure, effect, and susceptibility for silicosis and coal-worker's pneumoconiosis: a comprehensive review. J. Toxicol. Environ. Health 9, 357–395.

Guo, N.L., Wan, Y.W., Denvir, J., et al., 2012. Multi-walled carbon nanotube-induced gene signatures in the mouse lung: potential predictive value for human lung cancer risk and prognosis. J. Toxicol. Environ. Health 75, 1129–1153.

Guseva Canu, I., Schulte, P.A., Riediker, M., et al., 2018. Methodological, political and legal issues in the assessment of the effects of nanotechnology on human health. J. Epidemiol. Community Health 72 (2), 148–153.

Hartung, T., 2009. Toxicology for the twenty-first century. Nature 460, 208–212.

Henderson, R.F., Bechtold, W.E., Bond, J.A., Sun, J.D., 1989. The use of biological markers in toxicology. Crit. Rev. Toxicol. 20, 65–82.

Higashisaka, K., Yoshioka, Y., Yamashita, Y., et al., 2011. Acute phase proteins as biomarkers for predicting the exposure and toxicity of nanomaterials. Biomaterials 32, 3–9.

Higashisaka, K., Yoshioka, Y., Yamashita, K., et al., 2012. Hemopexin as biomarkers for analyzing the biological responses associated with exposure to silica nanoparticles. Nanoscale Res. Lett. 7, 555.

Hines, M.A., Guyot-Sionnest, P., 1996. Synthesis and characterization of strongly luminescing ZnS-capped CdSe nanocrystals. J. Phys. Chem. B 100, 468–471.

Holgate, S., 2010. Exposure, uptake distribution and toxicity of nanomaterials in humans. J. Biomed. Nanotechnol. 6, 1–19.

Hong, F., Wang, L., Yu, X., et al., 2015. Toxicological effect of TiO_2 nanoparticle-induced myocarditis in mice. Nanoscale Res. Lett. 10, 326.

Iavicoli, I., Leso, V., Manno, M., Schulte, P.A., 2014. Biomarkers of nanomaterial exposure and effect: current status. J. Nanopart. Res. 16, 2302.

Iavicoli, I., Fontana, L., Leso, V., et al., 2018. Subchronic exposure to palladium nanoparticles affects serum levels of cytokines in female Wistar rats. Hum. Exp. Toxicol. 37 (3), 309–320.

Iavicoli, I., Fontana, L., Nordberg, G., 2016. The effects of nanoparticles on the renal system. Crit. Rev. Toxicol. 46 (6). https://doi.org/10.1080/10408444.2016.1181047.

Ichihara, S., Li, W., Omura, S., et al., 2016. Exposure assessment and heart rate variability monitoring in workers handling titanium dioxide particles: a pilot study. J. Nanopart. Res. 18, 52.

Imashimizu, K., Shiomi, K., Maeda, M., et al., 2011. Feasibility of large-scale screening using N-ERC/mesothelin levels in the blood for the early diagnosis of malignant mesothelioma. Exp. Ther. Med. 2, 409–411.

ISO/TR 14786, 2014. Nanotechnologies - Considerations for the Development of Chemical Nomenclature for Selected Nano-objects.

Janower, M.L., Sidel, V.W., Baker, W.H., et al., 1968. Late clinical and laboratory manifestations of thorotrast administration in cerebral arteriography: a follow-up study of thirty patients. N. Engl. J. Med. 279, 186–189.

Johnston, H.J., Hutchison, G., Christensen, F.M., et al., 2010. A review of the in vivo and in vitro toxicity of silver and gold particulates particle attributes and biological mechanisms responsible for the observed toxicity. Crit. Rev. Toxicol. 40, 328–346.

Johnston, H., Pojana, G., Zuin, S., et al., 2013. Engineered nanomaterial risk. Lessons learnt from completed nanotoxicology studies: potential solutions to current and future challenges. Crit. Rev. Toxicol. 43, 1–20.

Jun, E.A., Lim, K.M., Kim, K., et al., 2011. Silver nanoparticles enhance thrombus formation through increased platelet aggregation and procoagulant activity. Nanotoxicology 5 (2), 157–167.

Karlsson, H.L., 2010. The comet assay in nanotoxicology research. Anal. Bioanal. Chem. 398, 651–666.

Karlsson, H.L., Di Bucchianico, S., Collins, A.R., Dusinska, M., 2015. Can the comet assay be used reliably to detect nanoparticle-induced genotoxicity? Environ. Mol. Mutagen. 56 (2), 82–96.

Khatri, M., Bello, D., Pal, A.K., et al., 2013. Evaluation of cytotoxic, genotoxic and inflammatory responses of nanoparticles from photocopiers in three human cell lines. Part. Fibre Toxicol. 10, 42. https://doi.org/10.1186/1743-8977-10-42.

Kilinç, E., Schulz, H., Kuiper, J.A.J.M., et al., 2011. The procoagulant effects of fine particulate matter in vivo. Part. Fibre Toxicol. 8, 12.

Kim, S., Fisher, B., Eisler, H.J., et al., 2003. Type-II quantum dots: CdTe/CdSe(core/shell) and CdSe/ZnTe(core/shell) heterostructures. J. Am. Chem. Soc. 125, 11466–11467.

Kinaret, P., Ilves, M., Fortino, V., et al., 2017a. Inhalation and oropharyngeal aspiration exposure to rod-like carbon nanotubes induce similar airway inflammation and biological responses in mouse lungs. ACS Nano 11 (1), 291–303.

Kinaret, P., Marwah, V., Fortino, V., et al., 2017b. Network analysis reveals similar transcriptomic responses to intrinsic properties of carbon nanomaterials *in vitro* and *in vivo*. ACS Nano 11 (4), 3786–3796.

Kreyling, W.G., Semmler-Behnke, M., Erbe, F., et al., 2002. Translocation of ultrafine insoluble iridium particles from lung epithelium to extrapulmonary organs is size dependent but very low. J. Toxicol. Environ. Health 65 (20), 1513–1530.

Kreyling, W.G., Semmler-Behnke, M., Seitz, J., et al., 2009. Size dependence of the translocation of inhaled iridium and carbon nanoparticle aggregates from the lung of rats to the blood and secondary target organs. Inhal. Toxicol. 21 (Suppl. 1), 55–60.

Kreyling, W.G., Abdelmonem, A.M., Ali, Z., et al., 2015. In vivo integrity of polymer-coated gold nanoparticles. Nat. Nanotechnol. 10 (7), 619–623.

Kroto, H.W., Heath, J.H., O'Brien, S.C., et al., 1985. Smalley, C_{60}: Buckminsterfullerene. Nature 318, 162–163.

Kuhlbusch, T.A., Asbach, C., Fissan, H., et al., 2011. Nanoparticle exposure at nanotechnology workplaces: a review. Part. Fibre Toxicol. 8, 22.

Kumar, V., Kumari, A., Guleria, P., et al., 2012. Evaluating the toxicity of selected types of nanochemicals. Rev. Environ. Contam. Toxicol. 215, 39–121.

Labib, S., Williams, A., Yauk, C.L., et al., 2016. Nano-risk Science: application of toxicogenomics in an adverse outcome pathway framework for risk assessment of multi-walled carbon nanotubes. Part. Fibre Toxicol. 13, 15. https://doi.org/10.1186/s12989-016-0125-9.

Lacombe, M., Desvergne, C.M., Combes, F.L., et al., 2018. Proteomic characterization of human exhaled breath condensate. J. Breath Res. 12, 021001.

Lee, J.S., Choi, Y.C., Shin, J.H., et al., 2015. Health surveillance study of workers who manufacture multi-walled carbon nanotubes. Nanotoxicology 9, 802–811.

Lee, C.C., MacKay, J.A., Fréchet, J.M., Szoka, F.C., 2005. Designing dendrimers for biological applications. Nat. Biotechnol. 23, 1517–1526.

Lee, J.H., Kwon, M., Ji, J.H., et al., 2011. Exposure assessment of workplaces manufacturing nanosized TiO_2 and silver. Inhal. Toxicol. 23, 226–236.

Lee, J.H., Lee, S.B., Bae, G.N., et al., 2010. Exposure assessment of carbon nanotube manufacturing workplaces. Inhal. Toxicol. 22, 369–381.

Lee, J.H., Mun, J., Park, J.D., Yu, I.J., 2012. A health surveillance case study on workers who manufacture silver nanomaterials. Nanotoxicology 6, 667–669.

Li, Y., Boraschi, D., 2016. Endotoxin contamination: a key element in the interpretation of nanosafety studies. Nanomedicine 11 (3), 269–287.

Liao, H.Y., Chung, Y.T., Tsou, T.C., et al., 2014. Six-month follow-up study of health markers of nanomaterials among workers handling engineered nanomaterials. Nanotoxicology 8, 100–110.

Lin, J., Li, J., Huang, B., et al., 2015. Exosomes: novel biomarkers for clinical diagnosis. Sci. World J. 2015, 657086.

Liou, H.S., Tsai, C., Pelclova, D., et al., 2015. Assessing the first wave of epidemiological studies of nanomaterial workers. J. Nanopart. Res. 17, 413.

Liou, S.H., Tsou, T.C., Wang, S.L., et al., 2012. Epidemiological study of health hazards among workers handling engineered nanomaterials. J. Nanopart. Res. 14, 878–885.

Liu, H.H., Chen, C.Y., Chen, G.I., et al., 2012. Relationship between indium exposure and oxidative damage in workers in indium tin oxide production plants. Int. Arch. Occup. Environ. Health 85, 447–453.

Loft, S., Danielsen, P.O., Mikkelsen, L., et al., 2008. Biomarkers of oxidative damage to DNA and repair. Biochem. Soc. Trans. 36, 1071–1076.

Luo, J.C., Hsu, K.H., Shen, W.S., et al., 2009. Inflammatory responses and oxidative stress from metal fume exposure in automobile welders. J. Occup. Environ. Med. 51, 95–103.

Lynch, I., Weiss, C., Valsami-Jones, E., 2014. A strategy for grouping of nanomaterials based on key physico-chemical descriptors as a basis for safer-by-design NMs. Nano Today 9, 266–270.

Madl, A.K., Pinkerton, K.E., 2009. Health effects of inhaled engineered and incidental nanoparticles. Crit. Rev. Toxicol. 39, 629–658.

Magdolenova, Z., Collins, A., Kumar, A., et al., 2013. Mechanisms of genotoxicity. A review of in vitro and in vivo studies with engineered nanoparticles. Nanotoxicology 8 (3), 233–278.

Mangum, J.B., Turpin, E.A., Antao-Menezes, A., 2006. Single-walled carbon nanotube (SWCNT)-induced interstitial fibrosis in the lungs of rats is associated with increased levels of PDGF mRNA and the formation of unique intercellular carbon structures that bridge alveolar macrophages in situ. Part. Fibre Toxicol. 3, 15.

Manke, A., Wang, L., Rojanasakul, Y., 2013. Mechanisms of nanoparticle-induced oxidative stress and toxicity. BioMed Res. Int. 2013, 942916.

Manno, M., Viau, C., Cocker, J., et al., 2010. Biomonitoring for occupational health risk assessment (BOHRA). Toxicol. Lett. 192, 3–16.

Marichal, T., Ohata, K., Bedoret, D., et al., 2011. DNA released from dying host cells mediates aluminum adjuvant activity. Nat. Med. 17, 996–1002.

Mercer, R.R., Scabilloni, J.F., Hubbs, A.F., et al., 2013. Distribution and fibrotic response following inhalation exposure to multi-walled carbon nanotubes. Part. Fibre Toxicol. 10, 33.

Merget, R., Bauer, T., Küpper, H.U., et al., 2002. Health hazards due to the inhalation of amorphous silica. Arch. Toxicol. 75, 625–634.

Migliore, L., Saracino, D., Bonelli, A., et al., 2010. Carbon nanotubes induce oxidative DNA damage in RAW 264.7 cells. Environ. Mol. Mutagen. 51, 294–303.

Miller, A., Drake, P.L., Hintz, P., et al., 2010. Characterizing exposures to airborne metals and nanoparticle emissions in a refinery. Ann. Occup. Hyg. 54, 504–513.

Møller, P., Loft, S., 2010. Oxidative damage to DNA and lipids as biomarkers of exposure to air pollution. Environ. Health Perspect. 118, 1126–1136.

Monopoli, M.P., Aberg, C., Salvati, A., Dawson, K., 2012. A biomolecular coronas provide the biological identity of nanosized materials. Nat. Nanotechnol. 7 (12), 779–786.

Montuschi, P., 2007. Analysis of exhaled breath condensate in respiratory medicine. Methodological aspects and potential clinical applications. Ther. Adv. Respir. Dis. 1, 5–23.

Mori, T., Kato, Y., Kumatori, T., et al., 1983. Epidemiological follow-up study of Japanese Thorotrast cases - 1980. Health Phys. 44 (1), 261–272.

Mossman, B.T., 2000. Mechanisms of action of poorly soluble particulates in overload-related lung pathology. Inhal. Toxicol. 12, 141–148.

Mullan, P.B., Hosey, A.M., Buckley, N.E., et al., 2005. The 2,5 oligoadenylate synthetase/RNaseL pathway is a novel effector of BRCA1- and interferon-gamma-mediated apoptosis. Oncogene 24, 5492–5501.

Muller, J., Huaux, F., Fonseca, A., 2008. Structural defects play a major role in the acute lung toxicity of multiwall carbon nanotubes: toxicological aspects. Chem. Res. Toxicol. 21, 1698–1705.

Murphy, C.J., Gole, A.M., Stone, J.W., et al., 2008. Gold nanoparticles in biology: beyond toxicity to cellular imaging. Acc. Chem. Res. 41, 1721–1730.

Murray, A.R., Kisin, E.R., Tkach, A.V., et al., 2012. Factoring-in agglomeration of carbon nanotubes and nanofibers for better prediction of their toxicity versus asbestos. Part. Fibre Toxicol. 9, 10.

Mutti, A., 2001. Biomarkers of Exposure and Effect for Non Carcinogenic End-points. International Programme on Chemical Safety. Environmental Health Criteria 222, Biomarkers in Risk Assessment: Validity and Validation, 104. World Health Organization, Geneva.

Nagai, H., Toyokuni, S., 2010. Biopersistent fiber-induced inflammation and carcinogenesis: lessons learned from asbestos toward safety of fibrous nanomaterials. Arch. Biochem. Biophys. 502, 1–7.

Nagai, H., Ishihara, T., Lee, W.H., et al., 2011. Asbestos surface provides a niche for oxidative modification. Cancer Sci. 102, 2118–2125.

Nagano, T., Higashisaka, K., Kunieda, A., et al., 2013. Liver-specific microRNAs as biomarkers of nanomaterial-induced liver damage. Nanotechnology 24, 405102.

Namazi, H., Adeli, M., 2005. Dendrimers of citric acid and poly(ethylene glycol) as the new drug-delivery agents. Biomaterials 26, 1175–1183.

Nehl, C.L., Liao, H., Hafner, J.H., et al., 2006. Optical properties of star-shaped gold nanoparticles. Nano Lett. 6, 683–688.

Ng, C.T., Dheen, S.T., Yip, W.C., et al., 2011. The induction of epigenetic regulation of PROS1 gene in lung fibroblasts by gold nanoparticles and implications for potential lung injury. Biomaterials 32, 7609–7615.

NIOSH, 2011. Current Intelligence Bulletin 63: Occupational Exposure to Titanium Dioxide. U.S. Department of Health and Human Services, Centers for Disease Control and Prevention, National Institute for Occupational Safety and Health, DHHS (NIOSH), Cincinnati, OH. Publication No. 201–160.

Nishizawa, K., Kamiya, Y., Kaneko, M., 1987. Possible mechanism for the formation of tumors by thorotrast based on crystallographic characterization of thorotrast particles in tissues. J. Clin. Biochem. Nutr. 3, 241–250.

Oberdörster, G., 2010. Safety assessment for nanotechnology and nanomedicine: concepts of nanotoxicology. J. Intern. Med. 267, 89–105.

Okazaki, Y., Nagai, H., Chew, S.H., 2013. CD146 and IMP3 predict prognosis of asbestos-induced rat mesothelioma. Cancer Sci. 104, 989–995.

Palomäki, J., Välimäki, E., Sund, J., et al., 2011. Long, needle-like carbon nanotubes and asbestos activate the NLRP3 inflammasome through a similar mechanism. ACS Nano 5 (9), 6861–6870.

Palomäki, J., Sund, J., Vippola, M., et al., 2015. A secretomics analysis reveals major differences in the macrophage responses towards different types of carbon nanotubes. Nanotoxicology 9 (6), 719–728.

Pappi, T., Schiffmann, D., Weiss, D., et al., 2008. Human health implications of nanomaterial exposure. Nanotoxicology 2, 9–27.

Pelclova, D., Barosova, H., Kukutschova, J., et al., 2015a. Raman microspectroscopy of exhaled breath condensate and urine in workers exposed to fine and nano TiO_2 particles: a cross-sectional study. J. Breath Res. 9 (3), 036008.

Pelclova, D., Zdimal, V., Fenclova, Z., et al., 2015b. Markers of oxidative damage of nucleic acids and proteins among workers exposed to TiO2 (nano)particles. Occup. Environ. Med. 73 (2), 110–118.

Pelclova, D., Zdimal, V., Kacer, P., et al., 2016. Oxidative stress markers are elevated in exhaled breath condensate of workers exposed to nanoparticles during iron oxide pigment production. J. Breath Res. 10 (1), 016004.

Pietroiusti, A., Bergamaschi, E., Campagna, M., et al., 2017. The unrecognized occupational relevance of the interaction between engineered nanomaterials and the gastro-intestinal tract: a consensus paper from a multidisciplinary working group. Part. Fibre Toxicol. 14 (1), 47.

Pietroiusti, A., Stockmann-Juvala, H., Lucaroni, F., Savolainen, K., 2018. Nanomaterial exposure, toxicity, and impact on human health. WIREs Nanomed. Nanobiotechnol. e1513.

Plitzko, S., 2009. Workplace exposure to engineered nanoparticles. Inhal. Toxicol. 21 (S1), 25–29.

Poland, C.A., Duffin, R., Kinloch, I., et al., 2008. Carbon nanotubes introduced into the abdominal cavity of mice show asbestos-like pathogenicity in a pilot study. Nat. Nanotechnol. 3, 423–428.

Pope, C.A., Kanner, R.E., 1993. Acute effects of PM_{10} pollution on pulmonary function of smokers with mild to moderate chronic obstructive pulmonary disease. Am. Rev. Respir. Dis. 147 (6 Pt 1), 1336–1340.

Pott, F., Roller, M., Kamino, K., et al., 1994. Significance of durability of mineral fibers for their toxicity and carcinogenic potency in the abdominal cavity of rats in comparison with the low sensitivity of inhalation studies. Environ. Health Perspect. 102 (S5), 145–150.

Pricl, S., Fermeglia, M., Ferrone, M., et al., 2003. Scaling properties in the molecular structure of three dimensional, nanosized phenylene-based dendrimers as studied by atomistic molecular dynamics simulations. Carbon 41, 2269–2283.

Radomski, A., Jurasz, P., Alonso-Escolano, D., et al., 2005. Nanoparticle-induced platelet aggregation and vascular thrombosis. Br. J. Pharmacol. 146, 882–893.

Raposo, G., Stoorvogel, W., 2013. Extracellular vesicles: exosomes, microvesicles, and friends. J. Cell Biol. 200 (4), 373–383.

Riedel, W., Dalheimer, A., Said, M., et al., 1983. Recent results of the German Thorotrast study - dose relevant physical and biological properties of Thorotrast equivalent colloids. Health Phys. 44 (S1), 293–298.

Riediker, M., Schubauer-Berigan, M., Brouwer, D.H., et al., 2012. A roadmap towards a globally harmonized approach for occupational health surveillance and epidemiology in nanomaterial workers. J. Occup. Environ. Med. 54, 1214–1223.

Roco, M.C., 2011. The long view of nanotechnology development: the National Nanotechnology Initiative at 10 years. J. Nanopart. Res. 13, 427–445.

Roller, M., Pott, F., Kamino, K., et al., 1997. Dose–response relationship of fibrous dusts in intraperitoneal studies. Environ. Health Perspect. 105 (S5), 1253–1256.

Rossi, E.M., Pylkkänen, L., Koivisto, A.J., Nykäsenoja, H., Wolff, H., Savolainen, K., Alenius, H., 2010. Inhalation exposure to nanosized and fine TiO2 particles inhibits features of allergic asthma in a murine model. Part. Fibre Toxicol. 7, 35.

Ruiz, P.A., Morón, B., Becker, H.M., et al., 2017. Titanium dioxide nanoparticles exacerbate DSS-induced colitis: role of the NLRP3 inflammasome. Gut 66, 1216–1224.

Ryman-Rasmussen, J.P., Cesta, M.F., Brody, A.R., et al., 2009a. Inhaled carbon nanotubes reach the subpleural tissue in mice. Nat. Nanotechnol. 4 (11), 747–751.

Ryman-Rasmussen, J.P., Tewksbury, E.W., Moss, O.R., et al., 2009b. Inhaled multiwalled carbon nanotubes potentiate airway fibrosis in murine allergic asthma. Am. J. Respir. Cell Mol. Biol. 40 (3), 349–358.

Saber, A.T., Lamson, J.S., Jacobsen, N.R., et al., 2013. Particle-induced pulmonary acute phase response correlates with neutrophil influx linking inhaled particles and cardiovascular risk. PLoS One 8 (7), e69020.

Saito, N., Usui, Y., Aoki, K., et al., 2008. Carbon nanotubes for biomaterials in contact with bone. Curr. Med. Chem. 15, 523–527.

Sakamoto, Y., Dai, N., Hagiwara, Y., et al., 2010. Serum level of expressed in renal carcinoma (ERC)/mesothelin in rats with mesothelial proliferative lesions induced by multi-wall carbon nanotube (MWCNT). J. Toxicol. Sci. 35, 265–270.

Sargent, L.M., Porter, D.W., Staska, L.M., et al., 2014. Promotion of lung adenocarcinoma following inhalation exposure to multi-walled carbon nanotubes. Part. Fibre Toxicol. 11, 3.

Savi, M., Rossi, S., Bocchi, L., et al., 2014. Titanium dioxide nanoparticles promote arrhythmias via a direct interaction with rat cardiac tissue. Part. Fibre Toxicol. 11, 63.

Savolainen, K., Alenius, H., Norppa, H., 2010. Risk assessment of engineered nanomaterials and nanotechnologies: a review. Toxicology 269, 92–104.

Sayes, C.M., Reed, K.L., Warheit, D.B., 2007. Assessing toxicity of fine and nanoparticles: comparing *in vitro* measurements to *in vivo* pulmonary toxicity profiles. Toxicol. Sci. 97, 163–180.

Schulte, P.A., Hauser, J.E., 2012. The use of biomarkers in occupational health research, practice, and policy. Toxicol. Lett. 213, 91–99.

Searl, A., Buchanan, A., Cullen, R.T., 1999. Biopersistence and durability of nine mineral fibre types in rat lungs over 12 months. Ann. Occup. Hyg. 43, 143–153.

Shi, H., Magaye, R., Castranova, V., et al., 2013. Titanium dioxide nanoparticles: a review of current toxicological data. Part. Fibre Toxicol. 10, 15.

Shvedova, A.A., Kisin, E., Murray, A.R., et al., 2008. Inhalation vs. aspiration of singlewalled carbon nanotubes in C57BL/6 mice: inflammation, fibrosis, oxidative stress, and mutagenesis. Am. J. Physiol. Lung Cell Mol. Physiol. 295, L552–L565.

Shvedova, A.A., Yanamala, N., Kisin, E.R., et al., 2016. Integrated analysis of dysregulated ncRNA and mRNA expression profiles in humans exposed to carbon nanotubes. PLoS One. https://doi.org/10.1371/journal.pone.0150628.

Siegel, P.D., Saxena, R.K., Saxena, Q.B., 2004. Effect of diesel exhaust particulate (DEP) on immune responses: contributions of particulate versus organic soluble components. J. Toxicol. Environ. Health 67, 221–231.

Singh, N., Manshian, B., Jenkins, G.J.S., et al., 2009. NanoGenotoxicology: the DNA damaging potential of engineered nanomaterials. Biomaterials 30, 3891–3914.

Smolders, R., Bartonova, A., Boogaard, P.J., 2010. The use of biomarkers for risk assessment: reporting from the INTARESE/ENVIRISK Workshop in Prague. Int. J. Hyg. Environ. Health 213, 395–400.

Stanton, M.F., Layard, M., Tegeris, A., 1981. Relation of particle dimension to carcinogenicity in amphibole asbestoses and other fibrous minerals. J. Natl. Cancer Inst. 67, 965–975.

Stoccoro, A., Karlsson, H.-L., Coppedè, F., et al., 2013. Epigenetic effects of nano-sized materials. Toxicology 313, 3–14.

Takagi, A., Hirose, A., Futakuchi, M., et al., 2012. Dose-dependent mesothelioma induction by intraperitoneal administration of multiwall carbon nanotubes in p53 heterozygous mice. Cancer Sci. 103, 1440–1444.

Teeguarden, J.G., Webb-Robertson, B.J., Waters, K.M., et al., 2011. Comparative proteomics and pulmonary toxicity of instilled single-walled carbon nanotubes, crocidolite asbestos, and ultrafine carbon black in mice. Toxicol. Sci. 120, 123–135.

Tervonen, T., Linkov, I., Figueira, J.R., et al., 2008. Risk-based classification system of nanomaterials. J. Nanopart. Res. https://doi.org/10.1007/s11051-008-9546-1.

Usui, Y., Aoki, K., Narita, N., 2008. Carbon nanotubes with high bone-tissue compatibility and bone-formation acceleration effects. Small 4, 240–246.

Valadi, H., Ekström, K., Bossios, A., et al., 2007. Exosome-mediated transfer of mRNAs and microRNAs is a novel mechanism of genetic exchange between cells. Nat. Cell Biol. 9, 654–659.

Valsami-Jones, E., Lynch, I., 2015. How safe are nanomaterials? Science 350 (6259), 388–389.

Van Broekhuizen, P., Reijnders, L., 2011. Building blocks for a precautionary approach to the use of nanomaterials: positions taken by trade unions and environmental NGOs in the European nanotechnologies debate. Risk Anal. 31 (10), 1646–1657.

Van Broekhuizen, P., van Veelen, W., Streekstra, W.H., et al., 2012. Exposure limits for nanoparticles: report of an international workshop on nano reference values. Ann. Occup. Hyg. 56 (5), 515–524.

Vesterdal, L.K., Folkmann, J.K., Jacobsen, N.R., et al., 2010. Pulmonary exposure to carbon black nanoparticles and vascular effects. Part. Fibre Toxicol. 7, 33.

Vietti, G., Lison, D., van den Brule, S., 2016. Mechanisms of lung fibrosis induced by carbon nanotubes: towards an Adverse Outcome Pathway (AOP). Part. Fibre Toxicol. 13, 11.

Vlaanderen, J., Pronk, A., Rothman, N., et al., 2017. A cross-sectional study of changes in markers of immunological effects and lung health due to exposure to multi-walled carbon nanotubes. Nanotoxicology 11 (3), 395–404.

Vrijens, K., Bollati, V., Nawrot, T.S., 2015. MicroRNAs as potential signatures of environmental exposure or effect: a systematic review. Environ. Health Perspect. 123 (5), 399–411.

Wang, H., Brandl, D.W., Le, F., et al., 2006. Nanorice: a hybrid plasmonic nanostructure. Nano Lett. 6, 827–832.

Warheit, D.B., Sayes, C.M., Reed, K.L., et al., 2009. Nanoscale and fine zinc oxide particles: can in vitro assays accurately forecast lung hazards following inhalation exposures? Environ. Sci. Technol. 43, 7939–7945.

Wellmann, J., Weiland, S.K., Neiteler, G., et al., 2006. Cancer mortality in German carbon black workers 1976–98. Occup. Environ. Med. 63, 513–521.

Wesch, H., van Kaick, G., Riedel, W., et al., 1983. Recent results of the German Thorotrast study – statistical evaluation of animal experiments with regard to the nonradiation effects in human thorotrastosis. Health Phys. 44 (S1), 317–321.

Wu, W.T., Liao, H.Y., Chung, Y.T., et al., 2014. Effect of nanoparticles exposure on fractional exhaled nitric oxide (FENO) in workers exposed to nanomaterials. Int. J. Mol. Sci. 15, 878–894.

Xu, J., Futakuchi, M., Iigo, M., et al., 2010. Involvement of macrophage inflammatory protein 1alpha (MIP1alpha) in promotion of rat lung and mammary carcinogenic activity of nanoscale titanium dioxide particles administered by intra-pulmonary spraying. Carcinogenesis 31, 927–935.

Yu, M., Mo, Y., Wan, R., et al., 2010. Regulation of plasminogen activator inhibitor-1 expression in endothelial cells with exposure to metal nanoparticles. Toxicol. Lett. 195, 82–89.

Zeng, F., Zimmerman, S.C., 1997. Dendrimers in supramolecular chemistry: from molecular recognition to self-assembly. Chem. Rev. 97, 1681–1712.

Zhang, Y., Deng, J., Zhang, Y., et al., 2013. Functionalized single-walled carbon nanotubes cause reversible acute lung injury and induce fibrosis in mice. J. Mol. Med. 91, 117–128.

Zhu, M., Li, Y., Shi, J., et al., 2012. Cellular responses to nanomaterials: exosomes as extrapulmonary signaling conveyors for nanoparticle-induced systemic immune activation. Small 8 (3), 404–412.

CHAPTER

42

Biomarkers of Exposure and Responses to Ionizing Radiation

Roger O. McClellan

Independent Advisor, Toxicology and Risk Analysis, Albuquerque, New Mexico, United States

INTRODUCTION

The health effects of ionizing radiation from external sources and internally deposited radionuclides have been more extensively investigated than any other hazardous agent or class of agents (McClellan, 2014). Thus, ionizing radiation is an ideal toxicant to evaluate relative to the concept of biomarkers. The previous edition of this book included a chapter entitled "Biomarkers of Radiation Injury and Response" (Ray et al., 2014). This chapter complements the earlier chapter and draws heavily on the author's experience from over 60 years of conducting, managing, and interpreting research on the health effects of radionuclides deposited internally following ingestion or inhalation exposure.

As I will relate later, the earliest knowledge of radiation-induced health effects came from human experience, initially ad hoc observations followed much later by more formal epidemiological studies. Epidemiological investigations are by their very nature studies of "human misfortune" in that people should never purposefully exposed to provide human evidence of the effects of radiation exposure, the populations studied have been inadvertently exposed. The research my colleagues and I conducted with controlled exposures of laboratory animals was intended to complement and extend the findings from epidemiological studies. Special attention in the controlled exposure animal studies was directed to gaining information on a range of radionuclides with different decay characteristics (beta- vs. alpha-emitters and half-life) and radiation dose patterns for which human data did not exist and was unlikely to be obtained.

As I will relate later, most of the experimental studies my colleagues and I planned and conducted involved multiple levels of exposure and intake. For every radionuclide and chemical form studied, parallel investigations were conducted to provide kinetic descriptions of the disposition of the radionuclide and allow estimation of the radiation dose delivered to various tissues over time following intake. In each life span study (LSS), a continuum of intakes and cumulative doses were evaluated. At the highest intakes and radiation doses, acute deaths were observed. At lower intakes and radiation doses, the health effects observed were debilitating and caused intermediate term deaths. At yet lower intakes and radiation doses, the effects were indistinguishable from the health effects seen in aging controls except the occurrence of neoplasms was elevated compared to controls. At the lowest intakes and radiation doses, the life span and causes of death in the radionuclide exposed animals, including occurrence of cancer, were not distinguishable from that observed in control animals.

The contents of this chapter build on two recent book chapters prepared by the author. There is a review of "Concepts in Veterinary Toxicology" (McClellan, 2018) and a comprehensive review of "Radiation Toxicity" (McClellan, 2014). In this chapter, the findings from both epidemiological studies and controlled exposure investigations in laboratory animals will be discussed in a "biomarker" context building on my earlier review of radiation toxicity. This context recognizes that the use of biomarkers is really an extension of the practice of medicine, from the time of its origins, with skillful interpretation of signs and symptoms used to diagnose disease and guide treatment. The author strongly endorses the use of biomarkers of both exposure and health effects to aid in understanding the health effects of radiation exposures from external sources and internally deposited radionuclides. Some radiation-induced effects are characterized by a threshold in the dose–response relationship with an increase in the frequency and severity of the health effect

Biomarkers in Toxicology, Second Edition
https://doi.org/10.1016/B978-0-12-814655-2.00042-6

as radiation doses are increased above a threshold level. Other effects, such as cancer, have a stochastic (random occurrence) relationship relative to dose and the incidence of the disease increases above background rates in proportion to the increase in radiation dose. The author offers words of caution on the use of biomarkers for interpreting and assigning attribution or causality to radiation exposure for effects such as cancer, that occur randomly, i.e., in a stochastic dose–response manner over and above their random occurrence in nonexposed individuals.

Knowledge available today is generally not adequate to accurately predict the occurrence of most cancers before they are clinically manifested. Excess Relative Risks (ERRs) for cancer occurrence are statistically estimated in epidemiological studies and in a few experimental studies and attributed to an etiological agent such as radiation exposure. However, current scientific knowledge is not sufficient to identify which of the cancer cases observed in a cohort are specifically caused by a specific etiological factor versus unknown causality. Unfortunately, most experimental studies are designed to test outcomes of a deterministic nature rather than a stochastic nature.

In my opinion, the advancement of the development and application of biomarkers, especially biomarkers of radiation exposure and the resultant radiation-attributable health effects, is going to require a rigorous, critical assessment of the current status of this specialized area. This will provide a basis for developing a pragmatic and strategic approach to the planning and conduct of future research building on our past collective experience. The development and application of biomarkers for ionizing radiation and other hazardous agents is at a critical juncture. New methods and approaches are needed! The quantitative nature of radiation science suggests that lessons learned and concepts developed with ionizing radiation as a hazardous agent may be useful applied to chemical agents.

KEY DEFINITIONS AND UNITS

Some key definitions are provided in Table 42.1 for keywords and phrases regularly used in radiation science to augment the Glossary provided for this textbook. A concise summary of key units used in radiation science and radiation protection is provided in Table 42.2. The earliest units, termed conventional, have ultimately been superseded by the newer International System of Units which were first developed in Europe and more slowly accepted in the United States. In general, when referencing previous published research, I will use the units used by the previous author. The reader also needs

TABLE 42.1 Key Definitions in Radiation Science and the Effects of Radiation on Health

Alpha-Particles	A charged particle consisting of two protons and two neutrons bound together into a particle identical to a Helium-4 nucleus.
Beta-Particles	A charged particle of very small mass emitted spontaneously from the nuclei of certain radioactive elements. Physically, a beta-particle is identical to an electron moving at high velocity.
Elements	One of the distinct, basic varieties of matter occurring in nature.
Fission	Electromagnetic radiation of variable energy originating in atomic nuclei. Physically, X-rays and gamma-rays are identical.
Fusion	The process whereby the nucleus of a particular heavy element following absorption of a neutron splits into two nuclei of lighter elements with release of substantial energy. Controlled and sustained fission occurs in a nuclear reactor. Fission of Pu^{239} and U^{235} is a key component of the detonation of nuclear weapons. The process whereby the nuclei of lighter elements, especially those of hydrogen (deuterium and tritium), combine to form the nucleus of heavier elements with the release of substantial energy. A thermonuclear weapon involves fission followed by fusion.
Gamma-Rays	Electromagnetic radiation of variable energy originating in atomic nuclei. Physically, X-rays and gamma-rays are identical.
Isotopes	Forms of the same element having identical chemical properties but differing in atomic mass (due to different numbers of neutrons in their respective nuclei) and nuclear properties (radioactivity).
Nuclide	An atomic species of an element distinguished by the composition of its nucleus, i.e., the number of protons and neutrons.
Radioactive Half-Life	The time required for the activity of a given radionuclide to decrease to half its initial value due to radioactive decay.
Radionuclide or Radioisotope	An isotope or nuclide that is unstable and will spontaneously decay by emission of alpha or beta-particles.
Transmutation **X-rays**	Conversion of one chemical element into another electromagnetic radiation of variable energy produced by slowing down fast electrons. X-rays can penetrate the body and be absorbed or pass through the body.

TABLE 42.2 International Units for Radiation and Radioactivity Compared With Old Units

International Units	Describes	Old Units
Becquerel (Bq)	Radioactivity, the spontaneous decay of atomic nuclei 1 Bq = 1 disintegration/s	Curie = 3.7 $X10^{10}$ Bq
Gray (Gy)	Dose to tissue 1 Gy = energy uptake of 1 J/kg	Rad = 10 mGy
Sievert (Sv)	Effective dose,[a] dose normalized to effects of gamma radiation by applying a Radiation-Weighting factor (WR) based on the Relative Biological Effectiveness (RBE) of the radiation of interest, to the absorbed dose in an organ or tissue to derive the equivalent dose. The effective dose is usually expressed as millisievert (mSv) 1 mSv = 10^{-3} Sv	Rem[b] = 10 mSv

[a] *Allows conversion of dose from gamma, beta, or alpha radiation to a standard unit.*
[b] *Rem, originally based on Roentgen Equivalent Man.*

to be mindful that different prefixes are used, such as with Curies or Becquerels, to span many orders of magnitude from micro to milli levels of the base Unit.

RADIATION PROTECTION SYSTEM

Over the past century, a systematic approach to radiation protection has developed. Its origins were international followed by the development of national systems. International guidance is currently provided by the International Commission on Radiological Protection (ICRP). In the United States, a congressional chartered organization, the National Council on Radiation Protection and Measurements (NCRP), provides guidance with individual federal, state, and local agencies having legislated responsibility and authority for radiation protection activities to protect workers and the public. I served as an elected member of the NCRP Council (1971–2001) and now serve as a Distinguished Emeritus Member. From time to time, in the United States, the National Academies, National Academy of Sciences, National Academy of Engineering, National Academy of Medicine (previously, the Institute of Medicine) through the National Research Council (NRC), also offers independent advice on radiation issues. I have served on many of those committees, including chairing the NRC Committee on Toxicology for 7 years, and I have been an elected member of the National Academy of Medicine since 1990.

The radiation protection system is uniquely different than the system used throughout the world for chemicals. By and large, the hazards and risks of chemicals are assessed and regulations developed chemical by chemical, a very tedious and demanding process requiring substantial information on each chemical. Very early in the radiation field, it became apparent that it would be very difficult to separately consider each type of radiation (X-rays, gamma-rays, neutrons, beta-particles, and alpha-particles) and each radionuclide (I^{131}, Sr^{90}, Ra^{226}, Pu^{239}, etc.) separately. Alternatively, a system for radiation was developed that recognized that all radiation acts via ionization of tissue in causing health effects. Thus, a holistic approach has developed based on absorbed tissue dose as the common metric irrespective of whether the ionizing radiation arises from external sources on internally-deposited radionuclides.

Current recommended dose limits are shown in Table 42.3.

Radiation-weighting factors (W_g) and tissue-weighting factors (W_t) that undergird the dose limits are shown in Tables 42.4 and 42.5. In the author's opinion, the limits shown in Table 42.3 are grounded in scientific judgment guided by risk considerations. However, they were not developed based on achieving specific limits on risk. Table 42.6 shows the detriment-adjusted nominal risk coefficients for stochastic effects (both cancer and heritable effects). One point to be made is that from 1990 to 2007, the estimates for heritable effects have been reduced, whereas the estimates for cancer have been slightly increased. As an aside, there was considerable concern beginning in the 1930s and continuing through the 1950s and 1960s for heritable effects. As new epidemiological and experimental evidence became available, it was possible to place heritable effects in better perspective. It is apparent the radiation protection system is highly quantitative and strongly grounded in a source-exposure-dose-response paradigm (Fig. 42.1). This paradigm undergirds the risk analysis system for both radiation and chemicals. An alternative presentation of the paradigm is shown in Fig. 42.2. This version emphasizes various elements of the pathogenesis of radiation-induced responses, including molecular and cellular processes that have emerged in recent decades. It is apparent that this paradigm will accommodate consideration of exposure from multiple sources to individuals or populations. Radiation dose is the control component of the paradigm for initiating consideration of responses to ionizing radiation, including an increase in overt diseases over and

TABLE 42.3 Recommended Dose Limits for Ionizing Radiation (mSv in a year)

Type of Limit	ICRP	NCRP
A. Occupational exposure Stochastic effects Effective dose limit (cumulative)	20 mSv/year averaged over 5 years, not to exceed 50 mSv in any 1 year	10 mSv × age
Annual	50 mSv/year	50 mSv/year
Deterministic effects Dose equivalent limits for tissues and organs (annual) Lens of eye Skin, hands, and feet	150 mSv/year 500 mSv/year	150 mSv/year 500 mSv/year
B. Embryo/fetus exposure Effective dose limit after pregnancy is declared	0.5 mSv/month	Total of 1 mSv to abdomen surface
C. Public exposure (annual) Effective dose limit, continuous or frequent exposure Effective dose limits, infrequent exposure Dose equivalent limits Lens of eye Skin and extremities	No distinction between frequent and infrequent—1 mSv/year 15 mSv/year 50 mSv/year	1 mSv/year 5 mSv/year 15 mSv/year 50 mSv/year
D. Negligible individual dose (annual)	No statement	0.01 mSv/year

ICRP (International Commission on Radiological Protection), 2007. The 2007 Recommendations of the International Commission on Radiological Protection, ICRP Publication 103 and NCRP (National Council on Radiation Protection and Measurements), 1993. Recommendations for Limits for Exposure to Ionizing Radiation, NCRP Report No. 116.

TABLE 42.4 Recommended Radiation-Weighting Factors (W_R)[a]

Radiation Type	Radiation-Weighting Factor (W_R)
Photons	1
Electrons and muons	1
Protons and charged pions	2
Alpha-particles, fission fragments, heavy ions	20
Neutrons	A continuous function of energy

[a]W_R values are used to adjust the absorbed dose in an organ or tissue to derive the equivalent dose.
From ICRP (International Commission on Radiological Protection), 2007. The 2007 Recommendations of the International Commission on Radiological Protection, ICRP Publication 103.

TABLE 42.5 Recommended Tissue-Weighting Factors (W_T)[a]

Tissue	W_T	Sum of W_T Values
Bone marrow (red), colon, lung, stomach, breast, remainder tissues[b]	0.12	0.72
Gonads	0.08	0.08
Bladder, esophagus, liver, thyroid	0.04	0.16
Bone surface, brain, salivary glands, skin	0.01	0.04
Total		1.00

[a]Used to derive the effective dose.
[b]Remainder tissues: adrenals, extrathoracic (ET) region, gall bladder, heart, kidneys, lymphatic nodes, muscle, oral mucosa, pancreas, prostate (♂), small intestine, spleen, thymus, uterus/cervix (♀).
ICRP (International Commission on Radiological Protection), 2007. The 2007 Recommendations of the International Commission on Radiological Protection, ICRP Publication 103.

above those caused by others, and frequently, as yet unknown risk factors.

SOURCES OF RADIATION EXPOSURE

This section briefly reviews the source of radiation exposure for the US population to provide a context for the rest of the chapter (Table 42.7). One overarching point needs to be made.

We all live and work in a "sea of radiation." About half of the average radiation exposure (3.1 mSv/year) comes from background. The other half (about 3.1 mSv/year) comes from anthropogenic, i.e., man-made sources, including medical uses of radiation. The anthropogenic radiation dose varies substantially between individuals because a substantial portion comes from medical procedures. From a personal view point,

TABLE 42.6 Detriment-Adjusted Nominal Risk Coefficients for Stochastic Effects After Exposure to Radiation at Low Dose Rate (10^{-2} per Sv)

	Cancer		Heritable Effects		Cancer Total Detriment	
Exposed Population	1990[a]	2007[b]	1990	2007	1990	2007
Whole	6.0	5.5	1.3	0.2	7.3	5.7
Adult	4.8	4.1	0.8	0.1	5.6	4.2

[a]*1990 cancer values based on fatal cancer risk weighted for nonfatal cancer, relative life years lost for fatal cancers, and life impairment for nonfatal cancer.*
[b]*2007 cancer values based on data on cancer incidence weighted for lethality and life impairment.*
Boecker, B.B., 1969. The metabolism of ^{137}Cs inhaled as ^{137}CsCl by the beagle dog. Proc. Soc. Exp. Biol. Med. 130(3), 966–971.

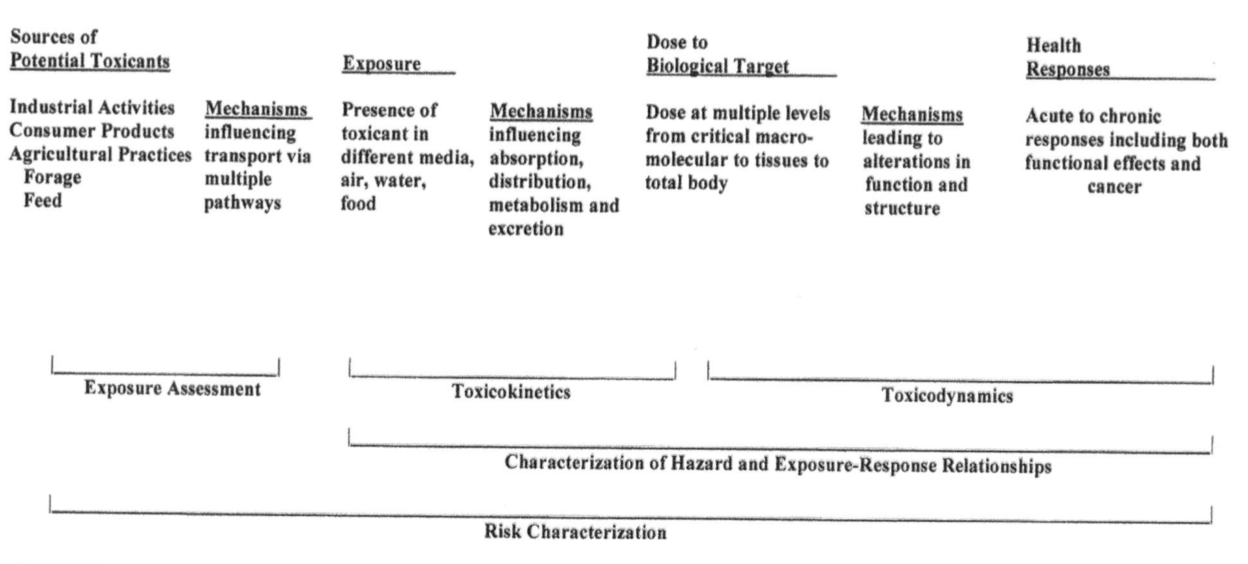

FIGURE 42.1 Critical linkages for integrating information from sources of toxicants to the development of adverse health effects.

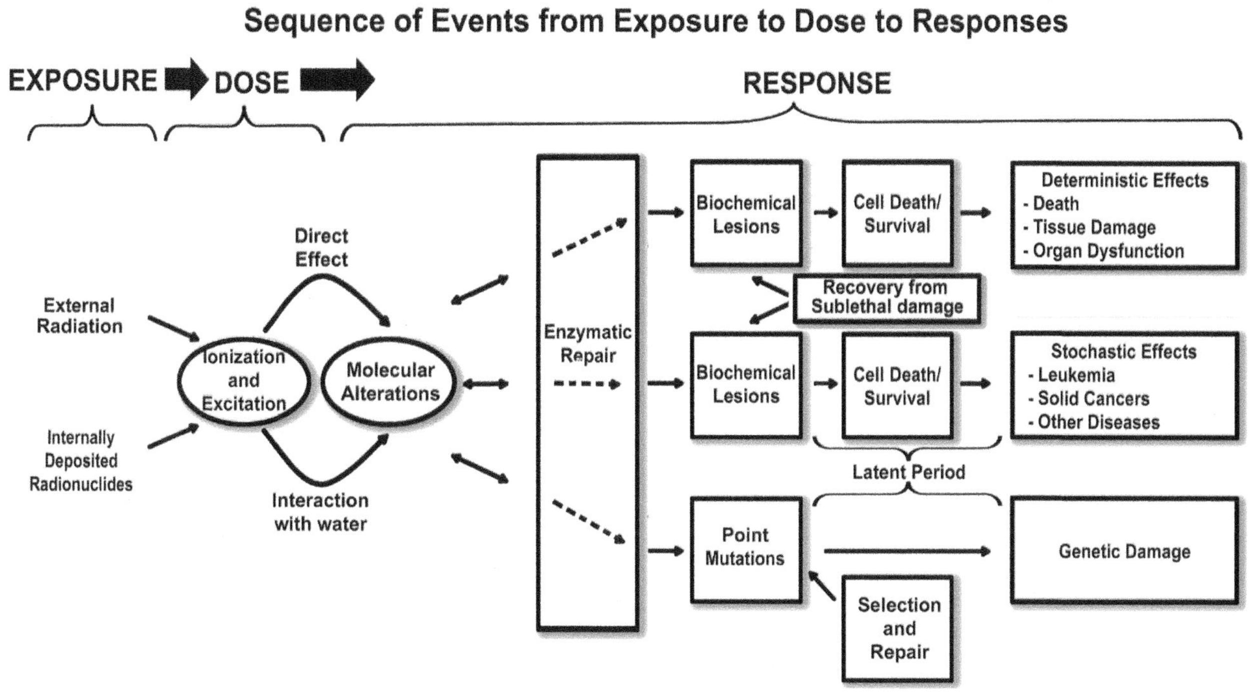

Sequence of Events from Exposure to Dose to Responses

EXPOSURE DOSE RESPONSE

External Radiation
Internally Deposited Radionuclides
Ionization and Excitation
Direct Effect
Interaction with water
Molecular Alterations
Enzymatic Repair
Biochemical Lesions
Cell Death/ Survival
Deterministic Effects
- Death
- Tissue Damage
- Organ Dysfunction
Recovery from Sublethal damage
Biochemical Lesions
Cell Death/ Survival
Stochastic Effects
- Leukemia
- Solid Cancers
- Other Diseases
Latent Period
Point Mutations
Genetic Damage
Selection and Repair

FIGURE 42.2 Conceptual framework linking events from exposure to external sources or internally deposited radioactivity to dose to biological responses (Bushong, 2013).

TABLE 42.7 Relative Contribution of Different Radiation Sources to the Average Radiation Dose Received by the US Public for 2006

Background	Millisieverts (mSv)	Percent
• Space radiation	0.31	5
• Internal radionuclides	0.31	5
• Terrestrial dose	0.19	3
• Radon and thoron	2.29	37
Background	3.10	
Anthropogenic		
• Computer tomography (CT)	1.49	24
• Nuclear medicine	0.74	12
• Interventional fluoroscopy	0.43	7
• Conventional radiography/fluoroscopy	0.31	5
• Consumer products	0.12	2
• Occupational	<0.1	<0.1
• Industrial	<0.1	<0.1
Anthropogenic	3.10	

Data from National Council on Radiation Protection and Measurements (NCRP), 2009. Report 160, Ionizing Radiation Exposure of the Population of the United States.

TABLE 42.8 Early History: Toxicity of Internally Deposited Radionuclides

1895	Roentgen Discovers X-Rays
1898:	Marie Curie discovers Ra^{226} and Po^{210}, coins "radioactivity"
1899:	Rutherford discovers alpha and beta-particles
1913:	Muller links radon exposure of miners with lung cancer
>1915:	Women painting luminescent dials develop bone disease
1932:	Lawrence builds cyclotron (UC Berkeley)
	(UC Berkeley), Joe Hamilton and others begin tracer studies
1938:	Strassmann and Hahn discover nuclear fission
1938:	Frisch and Meitner explain nuclear fission and daughter products
1940:	Seaborg discovers Pu^{239}
1941:	Manhattan Project begins under US Army
1942:	The Manhattan project becomes operational
	University of Chicago: Basic chemistry physics and biology
	New and unique facilities constructed
	Oak Ridge, TN—Test reactor and U^{235} Production
	Richland, WA—Reactors and separation facilities for Pu^{239} production
	Los Alamos—Design and fabrication of atomic bombs
1945:	Detonation of bombs
	July 16: Pu^{239} Bomb tested at Trinity site, Alamogordo, New Mexico
	August 6: U^{235} Bomb dropped on Hiroshima, Japan
	August 9: Pu^{239} Bomb dropped on Nagasaki, Japan
	(Blast overpressure, thermal, external radiation, and residual radionuclides)
1946:	Beginning of the U.S. Atomic Energy Commission which will later become the
	Energy Research and Development Administration which becomes the current
	Department of Energy

I am confident that every medical procedure involving radiation doses I have had has yielded a substantial benefit that far outweighed any hypothetical or estimated risk.

HISTORICAL PERSPECTIVE

Some key early events related to discovery of radiation phenomena and radiation-induced health effects are shown in Table 42.8. Investigations on the health effects of ionizing radiation began soon after William Conrad Roentgen's discovery of X-rays in 1895. Dermatological changes were soon observed in response to short-term high-dose external exposures to X-rays. The radiation dose–response-related nature of dermal changes was soon recognized with an apparent threshold that had to be exceeded before effects were observed. This phenomenon, the severity and frequency of adverse health effects being related to dose rate and total dose would later be observed for all the various health effects caused by radiation exposure. Within a short period of time, X-rays were being used routinely to produce radiographic images. The field of diagnostic radiology was underway.

As soon as the cell-killing effects of radiation were apparent, X-rays began to be used to treat superficial skin cancers and cancers located in deep tissues. In this same time period, naturally occurring radionuclides, such as Ra^{226} and Po^{210}, and their emissions were discovered. Marie Curie coined the term "radioactivity." It was quickly realized that radon from decay of Ra^{226} could be placed in capsules which could then be implanted in tumors to treat cancer. The field of radiation therapy was underway.

In the early 1900s, naturally occurring radioactive materials, such as alpha-particle emitting radium and thorium, were purified and added to phosphorescent

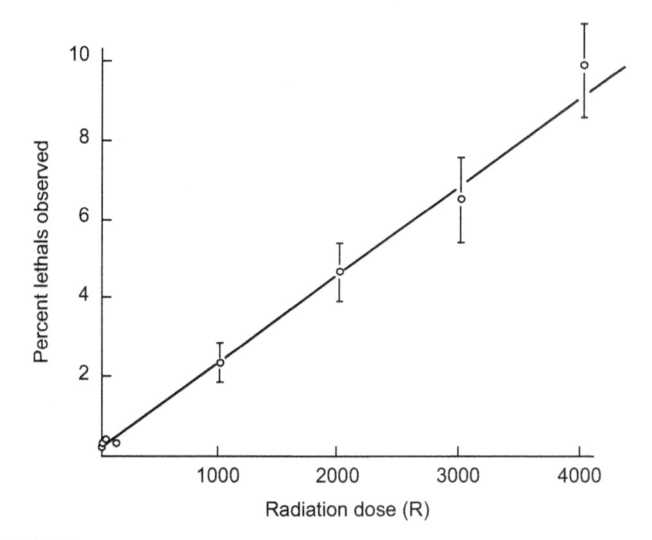

FIGURE 42.3 Dose–response relationship for induction of mutations in X-irradiated *Drosophila* (Note the substantial radiation doses studied).

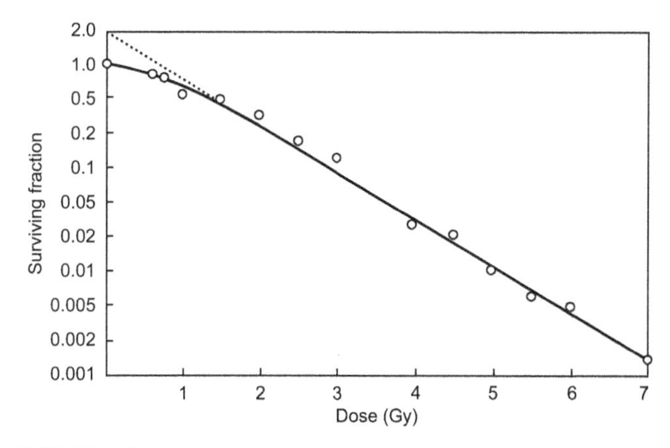

FIGURE 42.4 Survival curve for Hela cells in culture exposed to X-rays. *Adapted from Puck, T.T. and Markus, P.I. (1956). Action of x-rays on mammalian cells. J. Exp. Med. 103:653–666.*

paints to create luminous dials for clocks and instruments. The painting of dials was typically done by young women who would "tip" the brushes on their lips and inadvertently ingest the material. A recent book, "*The Radium Girls*," (Moore, 2016) provides vivid descriptions of working conditions and health outcomes in that time period. The Ra^{226} behaved like calcium and translocated to the skeleton, including the mandible. The women soon had a loss of teeth, damage to their mandibles, blood dyscrasias, and bone cancer. The carcinogenic potential of chronic radiation exposure was clearly evident. In retrospect, miners working in the mountains of central Europe several centuries earlier had likely experienced radiation-induced lung cancer related to chronic inhalation exposure to radon and its daughter products in the air arising from pitchblende containing ore.

In the 1920s and 1930s, x- and gamma-rays were used by numerous investigators to irradiate all manner of biological organisms from cells to mammals and plants and observe a range of biological effects. This included numerous studies to investigate hereditable effects and somatic effects in tissues resulting in a wide range of health effects. Again, radiation dose–response relationships were frequently documented over a range of doses although the emphasis was usually on observing some specific effects, thus, quite high dose rates and large total doses delivered over a relatively short period of time were frequently used. The results of an early study with *Drosophila* are shown in Fig. 42.3. To complement this demonstration of the mutagenic properties of radiation, shown in Fig. 42.4, is a survival curve for cells from research conducted in the 1950s by Puck and Markus (1956). Yet later, the concepts of mutations, cell proliferation, and cell killing were joined as carcinogenesis (Gray, 1965) (Fig. 42.5).

The 1930s were also a period of major developments in radiological science. The invention of the cyclotron in 1932 by the physicist, Ernest Lawrence, at the University of California–Berkeley, provided a method for creating numerous new radionuclides of naturally occurring elements and discovering new elements. The availability of new radionuclides, such as Fe^{59} and P^{32}, provided material for radiotracer studies in laboratory animals and human subjects. The application of these new tools in medicine was facilitated by the fact that Ernest Lawrence's brother, John Lawrence, was a physician at the University of California–Berkeley medical college. In retrospect, the early research of John Lawrence, his key collaborator, Joe Hamilton, and other colleagues was pioneering work with biomarkers. The discovery of fission by Strassman and Hahn (reported in a manuscript to *Naturwissenschaften*, January 6, 1939) and the interpretation of "nuclear fission" and "daughter products" by Frisch and Meitner (in a manuscript in Nature, February 11, 1939) would be of enormous significance. The use of the terms, fission and daughters, came from Frisch's experience in biology. The physical and biological sciences have been joined ever since in the study of radiological phenomena, including radiation-induced health effects.

One of the new elements discovered at Berkeley by a team led by Glenn Seaborg was Plutonium. Seaborg, with his junior colleagues, Joe Kennedy and Ed McMillan, and assistance from Emilo Segre, a research associate, and Art Wahl, a graduate student, first discovered Pu^{238} and then very soon thereafter Pu^{239} produced in the cyclotron. On March 28, 1941, they demonstrated that Pu^{239} undergoes fission with slow neutrons with a large probability. Thus, they revealed the potential for using Pu^{239} as an explosive ingredient in a nuclear or atomic bomb.

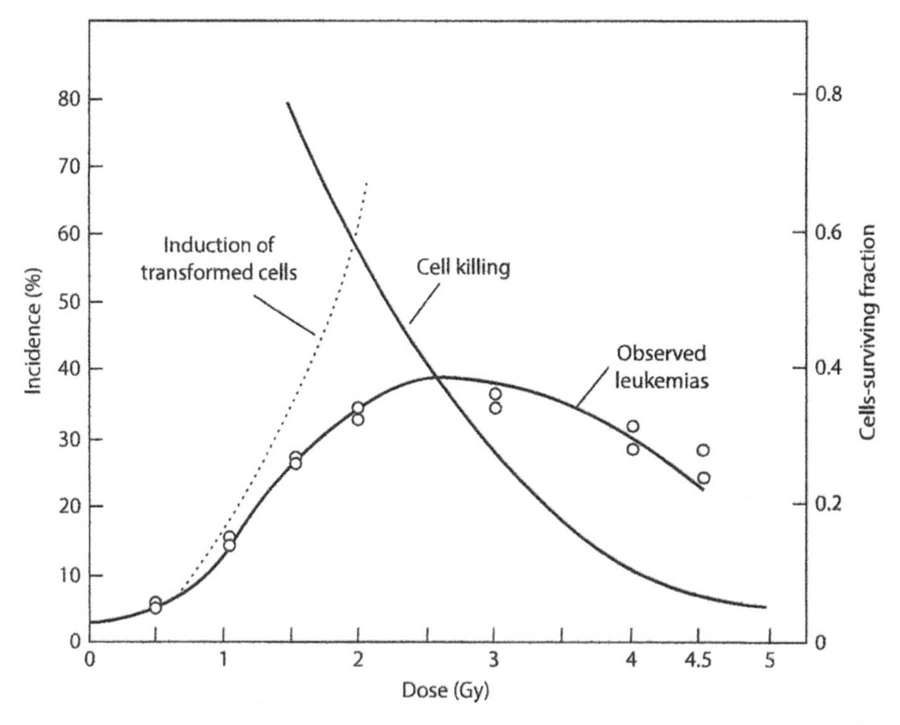

FIGURE 42.5 The postulated role of cell killing and induction of cell killing in producing leukemia in irradiated mice (Gray, 1965).

In 1941, as World War II was beginning in Europe and on the eve of the December 7, 1941, bombing of Pearl Harbor, Hawaii, a young radiation biologist, Paul S. Henshaw, published what I view as one of the earliest classic papers in radiation biology and radiation protection in Volume 1 of the Journal of the National Cancer Institute (Henshaw, 1941). The National Cancer Institute had only been started a short time earlier. This paper, which included the phrase "tolerance dose" in the title succinctly synthesized what was known at that time about radiation-induced effects. It included a very simple schematic (Fig. 42.6) that illustrated the now well-known two fundamentally different dose–response relationships for radiation exposure. One relationship has a "deterministic" form with a clear threshold. The second is a linear, no-threshold relationship with a "stochastic" form. In 1941, it was thought that acute and chronic health effects from intake of chemicals were generally of the "deterministic" form. This underlying assumption had already begun to be used in setting occupational exposure limits for chemicals, i.e., the setting threshold limit values.

The earliest radiation protection standards were also grounded in the "deterministic" or threshold dose–response relationship. An example is the exposition in the document entitled "The Tolerance Dose" authored by Cantril and Parker (1945) released on January 5, 1945. This document outlined the basis for the radiation protection program for workers at the Hanford, WA facilities. Three radiation tolerance levels were proposed

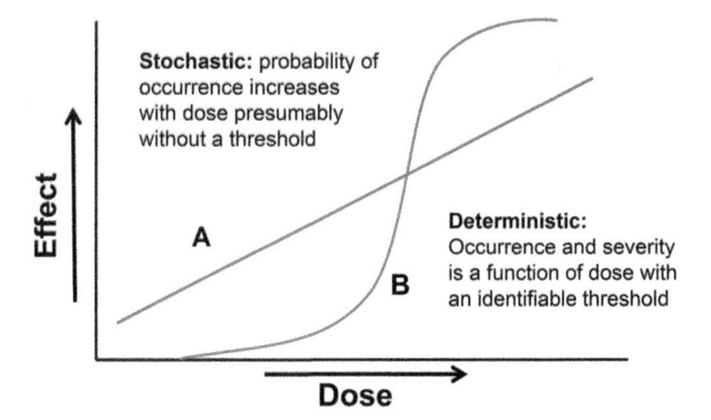

FIGURE 42.6 Schematic rendering of two fundamentally different dose–effect relationships for radiation. (Used in "The Tolerance Dose" by Cantril and Parker, January 5, 1945). *Adapted from Henshaw, P.S., 1941. Biologic significance of the tolerance dose in X-ray and radium protection. J. Natl. Cancer Inst. 1, 789–805.*

to protect workers: (1) 0.1 R/day for x- and gamma-rays; (2) 1×10^{-14} Ci radon/cc for radon in air; and (3) 0.1 μg Ra^{226} as a maximum amount deposited in the body. Cantril and Parker (1945) emphasized that all three tolerance doses were based on "human misfortune" and were uncertain, a point Parker would repeatedly make in later years. In particular, he noted major uncertainties in considering the effects of protracted exposure in contrast to brief exposures and related limitations in the human data base, especially in dealing with newly discovered elements and radionuclides, such as Pu^{239} and new applications of radiation. He also noted the possible role for the results of animal experiments. When doing so, he emphasized the importance of understanding the potential role of the results of the animal studies before initiating the studies. He wanted to be assured the animal studies met a clear need.

Personal Experience

The author's awareness of radiation issues and what is now known as nuclear energy dates to the summer of 1944. At that time, I moved from a small town on the prairies of southwest Minnesota, where I had lived on a farm in 1943–44 with my grandparents, to join my parents in Richland, WA. My parents had been working at Hanford in 1943–44 building unique facilities that were to produce a secret product. In August 1945, with the detonation on August 6, 1945, of a Uranium-235 fueled bomb over Hiroshima and on August 9, 1945, a Plutonium-239 fueled bomb over Nagasaki, Japan, it became known that Hanford's secret product was Pu-239. The feasibility of using Pu-239 as the fuel in a nuclear explosive device had been demonstrated in the "Trinity" tower test at Alamogordo, NM, on July 16, 1945. Robert Oppenheimer, who headed the Los Alamos, NM, bomb design and fabrication team was reported to have quoted a passage from Bhagavad Gita: "If the radiance of a thousand suns were to burst at once into the sky, that would be like the splendor of the Mighty One … I am become Death, the shatter of worlds." The world would never be the same, the Nuclear Age had begun. It soon became public knowledge that Hanford was a part of the Manhattan Project along with U-235 production facilities at Oak Ridge, TN, bomb design and fabrication activities at Los Alamos, NM, and basic science operations at Chicago, IL, and Berkeley, CA.

Useful references for the early years of the Nuclear Age are books by Hewlett and Anderson (1962), Hewlett and Duncan (1969), Szasa (1984), and Rhodes (1995). A very comprehensive text, "Radioactivity and Health: A History" (Stannard and Baalman, 1988) will be of special interest to biomedically oriented readers. J. Newell Stannard was a pioneering radiation toxicologist at the University of Rochester. The University of Rochester had a small but significant research program on the toxicity of inhaled uranium and Pu^{239} that was part of the Manhattan project. The associated graduate education program trained many radiation biologists and aerosol scientists, including many who ultimately would work at Hanford and Lovelace. Herbert M. Parker's Publications and Other Contributions to Radiological and Health Physics (Kathren et al., 1986) will be of interest to readers interested in early developments in radiation protection.

When I arrived in Richland in 1944, I was under age for radiation work so I missed out on helping Enrico Fermi start the Hanford 100B reactor on September 26, 1944, to irradiate uranium fuel to produce Pu-239 or to join workers in the Hanford facilities on December 26, 1944, to separate Pu-239 from the irradiated uranium fuel and fission products using a process developed by Glenn Seaborg. Instead, I started the third grade. The Hanford nuclear reactors were large blocks of carbon with horizontal tube channels. Uranium fuel elements clad with aluminum was placed in the tubes for irradiation under carefully controlled conditions. Purified cold water from the Columbia River flowed through the tube channels to cool the reactors. The sustained and controlled nuclear chain reaction produced neutrons and a plethora of fission products. This included Te^{131} that quickly decayed to I^{131} and large quantities of Sr^{90} and Cs^{137}. A small portion of the U^{238} atoms were transmutated to Np^{239} which then quickly decayed to Pu^{239}.

Quite by serendipity, Simon Cantril, a physician experienced in radiation oncology, and Herbert M. Parker, a radiological physicist, arrived at Hanford in the fall of 1944 and put pen to paper and documented the radiation protection program for Hanford as I noted earlier (Cantril and Parker, 1945). A key figure in that document was from the classic paper of Henshaw (1941) (Fig. 42.6). Mr. Parker was a leader in creating the field of radiation protection, originally code-named, "Health Physics," a name that stuck, much to Mr. Parker's chagrin. He was first and foremost a radiological or medical physicist, a skilled experimentalist and research team leader. Mr. Parker would later become one of my mentors in his role as head of the Hanford Research Laboratories, a position he held from 1956 until January 1965. I had the good fortune to work in those laboratories as a student intern in the summers of 1957, 1958, and 1959 and full-time from 1960 through 1965.

I did not recognize it at the time; however, in the late 1940s, as an elementary school student, I was introduced to the field of biomarkers. I noted that periodically a gray metal box would be left on the front door step of our home. I knew the box and contents must be related to my father's employment; the box had a HEW (for Hanford Engineering Works) label on it and my father's payroll number. When I inquired about the contents, my father said it contained his "pee bottles" and he needed to collect all his urine when at home to be added to the urine collected on the job. The urine was analyzed for Pu^{239} as a "biomarker" of his potential body burden of Pu-239. The initial "allowable skeletal burden" of Pu^{239} was derived from the human experience with Ra^{226}. My father was a technician in the Hanford facilities that separated Pu-239 from irradiated uranium fuel and large quantities of fission products and, thus, had the potential for exposure to Pu^{239}. Several decades later,

I learned that several dozen of the early Pu^{239} workers at the Los Alamos National Laboratory in World War II had joined together in what they called informally the IPPU Club (Voelz et al., 1997).

Let me leap ahead. In the 1980s, Glenn Seaborg visited the Lovelace Laboratories in New Mexico. During the course of his visit, we discussed studies we were conducting on inhaled Pu^{238} and Pu^{239}. He recalled that when he separated the first microgram quantities of Pu^{239} at the University of Chicago from fuel irradiated in a reactor at Clinton (Oak Ridge), TN, he was concerned about its potential toxicity. He was aware of the unfortunate experience of the Radium Dial painters and did not want to see it repeated with Pu workers. He immediately made arrangements to send a small sample of Pu^{239} to his colleague, Joe Hamilton at the University of California–Berkeley. Hamilton and his team injected Pu^{239} into rats to evaluate its disposition in the body. They found Pu^{239}, an actinide, behaved like the lanthanides and was deposited in both the skeleton and liver unlike the alkaline earth compounds, Ca and Ra, which primarily concentrated in the skeleton. These observations were key to developing and interpreting the urinary bioassay for Pu^{239}, one of the first applications of biomarkers in radiation protection starting in 1945.

In 1950, as a first-year high-school student, I enrolled in the traditional college preparation courses and, in addition, a course in Vocational Agriculture. The latter course had just been initiated in the Richland, WA high school. One of the attractions of the course was that every student also became a participant in a cooperative school farm. The farm had a variety of agricultural enterprises, including maintenance of a large flock of sheep as "off-site" controls for a large research study underway on the Hanford Site of the effects of chronic ingestion of I^{131}. There is not space here to provide supporting details but I think it is safe to assume the development of the Richland Vocational Agriculture program and School Farm was strongly encouraged by government authorities who, at the time, were conducting a large research program to understand the health effects of I^{131}.

Why study I^{131}? I have already noted it is a key constituent in freshly irradiated uranium fuel. On December 2–3, 1949, Hanford conducted a planned study, "Operation Green Run," releasing to the atmosphere 7000 to 12,000 Ci of I^{131}. The study did not go as planned! "Operation Green Run" was an attempt to understand what was happening at a facility the Soviet Union built that "copy-catted" Hanford. These facilities were located at Mayak, Russia, in the Ural Mountains. When the United States detected I^{131} in the atmosphere in 1949, it was assumed that Mayak had begun operations and was producing Pu^{239} for Russian bombs. The question was what does the detection of atmospheric I^{131} mean in terms of Mayak production of Pu-239? One approach was to assume the Soviets Union, like the United States in early 1945, was eager to produce Pu-239 and that they were processing "green fuel." "Green fuel" is fuel that had been recently irradiated and processed without much time postirradiation to "cool down." Thus, the "green fuel" contains substantial quantities of short-lived fission product radionuclides like I^{131} and its precursors. Very soon it was suggested that "Operation Green Run" be conducted to replicate Hanford operations in 1945 providing an "algorithm" for estimating Mayak production of Pu^{239}. This "experiment" would involve substantial downwind monitoring, including use of air craft with samplers as well-characterized "green" fuel was processed. The anticipated outcome was X quantity of I^{131} in the atmosphere from Y quantity of irradiated fuel yielding Z quantity of Pu-239. "Operation Green Run" provided the answer. However, because an atmospheric inversion set up soon after fuel reprocessing was initiated, eastern Washington was fumigated with I^{131}, physical half-life 8 days, for several days as the I^{131} contaminated cloud failed to dispense as expected. Jumping ahead, a detailed epidemiological study of the "Hanford Downwinders" did not identify any increase in thyroid cancer, over and above background rates, attributable to Hanford releases of I^{131} (Davis et al., 2004).

My experiences at the School Farm also introduced me to the field of biomarkers. In the early 1950s, nuclear weapons testing resulted in airborne radioactivity around the globe; three of the most prominent radionuclides were I^{131}, Cs^{137}, and Sr^{90}. Ingested Iodine-131 goes to the thyroid, Sr^{90} behaving like Ca goes to the skeleton, and Cs^{137} behaving like K is distributed in soft tissues throughout the body. The local grown forage fed the sheep was routinely monitored for I^{131} from fallout and the thyroids of the sheep routinely monitored for I^{131} using external ionization chamber monitors placed over the thyroids. I very quickly deduced that the Hanford scientists were interested in understanding the relationship between I^{131} intake, I^{131} in the thyroid, the estimated radiation dose to the thyroid and potential development of thyroid disease, including cancer. They wanted to make certain they knew the estimated radiation doses to the thyroid the off-site controls sheep were receiving from fallout I^{131}.

Working at the School Farm, I came in contact with a number of Hanford scientists, including Leo K. Bustad, a Veterinary scientist who would ultimately become my primary mentor. The experience as a high-school student introduced me to issues I would investigate later as a student and then as a staff scientist working in the Hanford Laboratories. The context for much of that research is shown in Fig. 42.7, which can be viewed as an extension of Fig. 42.1.

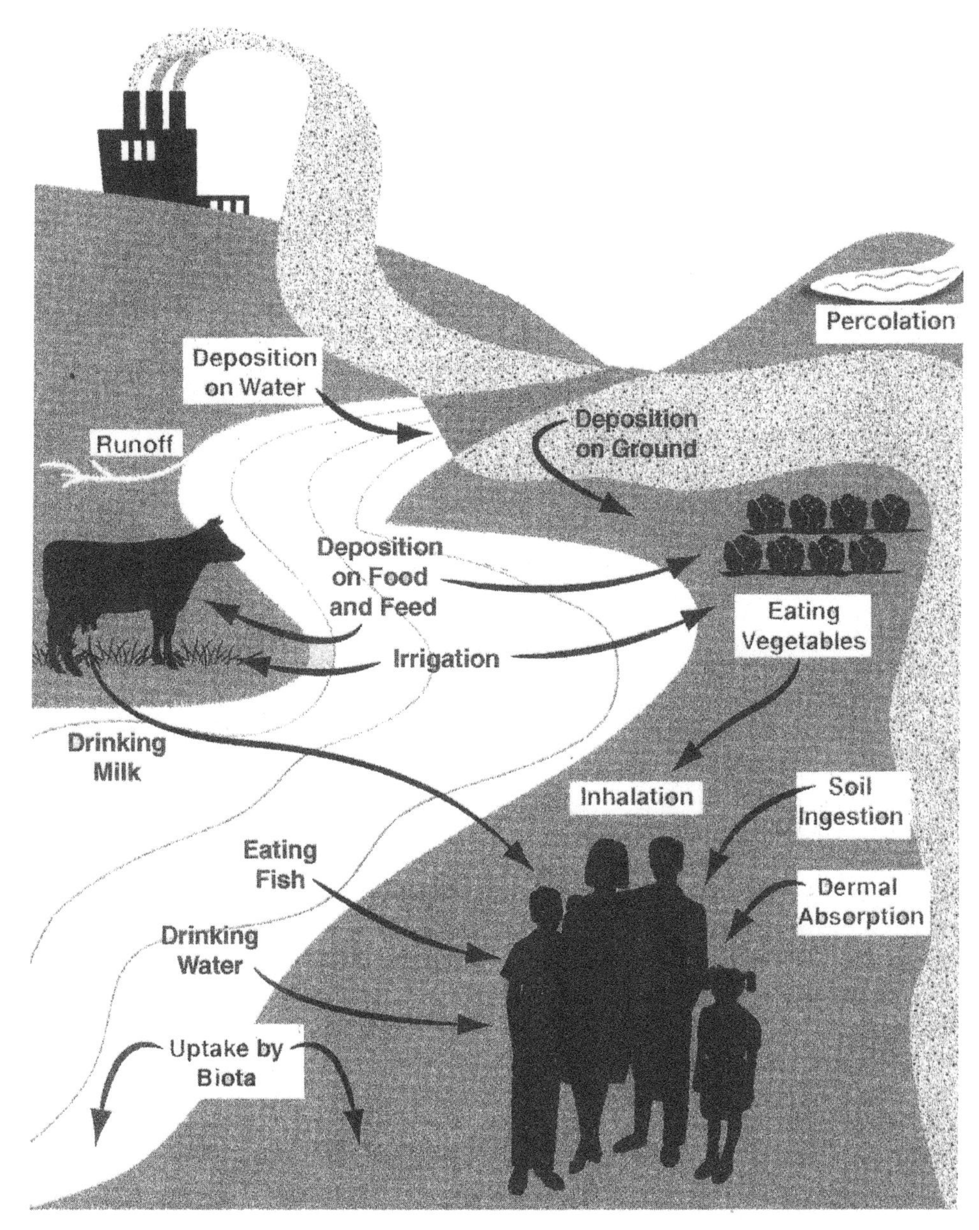

FIGURE 42.7 Schematic representation of pathways for radioactivity to reach humans (Paustenbach, 2001).

In 1954, I enrolled at Washington State College (now Washington State University) initially studying engineering and economics. In both of these fields, one quickly gains an appreciation for understanding both basic concepts and being quantitative. In 1956, I began my studies in Veterinary Medicine leading to receipt of my Doctor of Veterinary Medicine degree in 1960. In studying medicine, one quickly gains an appreciation for the importance of learning overarching concepts and synthesizing information. As one progresses into clinical medicine, the importance of understanding the role of symptoms and signs in diagnosing and treating diseases becomes apparent. One soon comes to appreciate the difference between symptoms, what the patient is experiencing, versus signs which the diagnostician observes or measures. It also becomes apparent that single signs are rarely pathognomonic of a specific disease. The diagnosis of common diseases usually involves an understanding of a combination of signs and symptoms. A single sign, some of which would clearly qualify for what we now call "biomarkers," is rarely uniquely related to a single disease. The study of infectious diseases leads students to think about exposure to single agents leading to specific diseases, essentially a deterministic relationship. Later, as students encounter more complex diseases, such as cancer (in reality a family of diseases), it becomes apparent that the underlying etiology is frequently multifactorial and the pathogenesis of the disease is quite complex. I will return to this topic later.

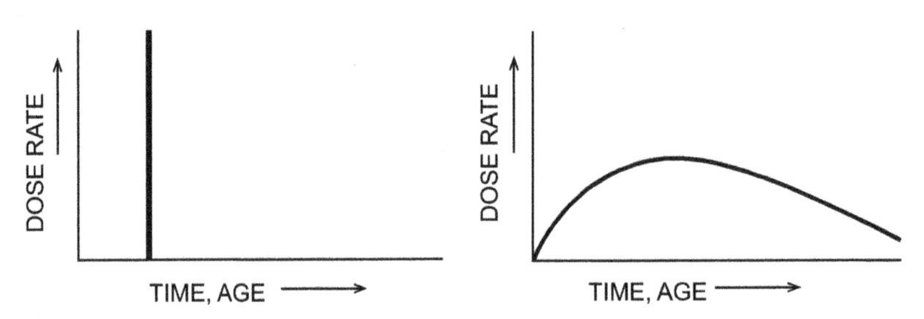

FIGURE 42.8 External brief radiation exposures and resulting tissue dose occur simultaneously, whereas the intake of radioactivity results in a tissue dose in a protracted tissue dose.

In the summers of 1957 and 1959, working with Bustad, I conducted research using sheep to compare the effects of brief X-irradiation of the thyroid using an external X-ray beam with the protracted irradiation of the thyroid from the beta-particle emitting I^{131} deposited in the thyroid. The brief exposures to high-dose rate X-rays were much more effective in producing pathological changes than the protracted beta-particle irradiation from the I^{131} decaying in the thyroid. This was a clear example that dose rate matters (Fig. 42.8).

In 1957, as I began my summer internship, Bustad suggested I prepare and present a few brief seminars to junior technicians and animal caretakers to acquaint them with the basic concepts of radiation biology. In retrospect, he was a very positive manipulator. What better way to have me learn basic radiation biology! He gave me a set of reprints that he thought would be helpful to me in preparing the lectures. Included in the stack were copies of the classic paper by Henshaw (1941) and "The Tolerance Dose" document by Cantril and Parker (1945) that I noted earlier which by then had been declassified. Both contained Fig. 42.6. Ironically, the use of the linear, no-threshold relationship is still being debated three quarters of a century later (NCRP, 2018).

In the summer of 1958, I investigated the uptake of various radionuclides, including Zn^{65}, from fresh nuclear reactor effluent water by bean plants, recall Fig. 42.7. I soon became aware of the importance of creating a "source-intake-body burden/tissue distribution-tissue dose-tissue response" paradigm to understand the role of any specific radionuclide toxicant, such as I^{131}, in causing disease. The potential for biomarkers at several steps was immediately obvious.

In 1960, I joined the Hanford Laboratories as a full-time staff scientist. I continued conducting research on I^{131}. This included participation in a landmark study to simulate the effects of an accidental release of I^{131}. Recall the "Green Run" release. In our new study, dairy cows were fed I^{131}, their thyroids and milk monitored for I^{131} and human volunteers drank aliquots of the milk and their thyroids monitored for I^{131} (See Fig. 42.9). The measurements of I^{131} in the cow's thyroid and milk and human thyroid certainly qualified as biomarkers. As an aside, the radiation dose to the thyroids of the human volunteers was less than received with typical radiographic medical procedures. The interrelationships which were observed proved valuable in predicting potential outcomes following reactor accidents such as those that happened at Windscale (October 10, 1957) in Britain, Chernobyl (April 26, 1986) in the Ukraine, and Fukushima (March 11, 2011) in Japan. Of special note, regulatory officials in Japan, mindful of this data, took action to limit distribution and intake of I^{131}-contaminated milk, thereby minimizing radiation exposure of children and preventing the occurrence of thyroid disease. Intake of I^{131}-contaminated milk was not effectively controlled after the Chernobyl accident, and an increase in thyroid cancer was observed in individuals irradiated as children, in some regions downwind (Lubin et al., 2017). This outcome was consistent with the results of the extensive studies conducted at Hanford with I^{131} (McClellan, 1995).

My primary responsibility at Hanford in the early 1960s was conduct of a large-scale study in miniature pigs ingesting Sr^{90}. The health end points of concern, based on knowledge that Sr^{90} concentrated in bone resulting in irradiation of bone and bone marrow, were induction of bone cancer and hematopoietic tissue dyscrasia. The study was started with three levels of Sr^{90} intake; 1, 5, and 25 μCi of Sr^{90}/day and a control group. When health effects were not observed in the first year, levels of 125, 625, and 3100 μCi Sr^{90}/day were added. In retrospect, the latter two levels were probably overzealous additions—I was a young toxicologist eager to identify effects! It would take more experience before I would understand the need for long-term observations, especially when studying induction of cancer, and the importance of understanding exposure—dose—response relationships. In the early 1960s, scientists were just beginning to think about integrating mutational events with cell killing (recall Fig. 42.5).

The Sr^{90} pig study provided numerous opportunities to characterize potential biomarkers. All the pigs were routinely monitored in a whole-body counter to follow build up of Sr^{90} in the skeleton, routine radiographs were taken of the skeleton to detect any potential tumors, and routine hematological assessments were made anticipating the occurrence of leukemia

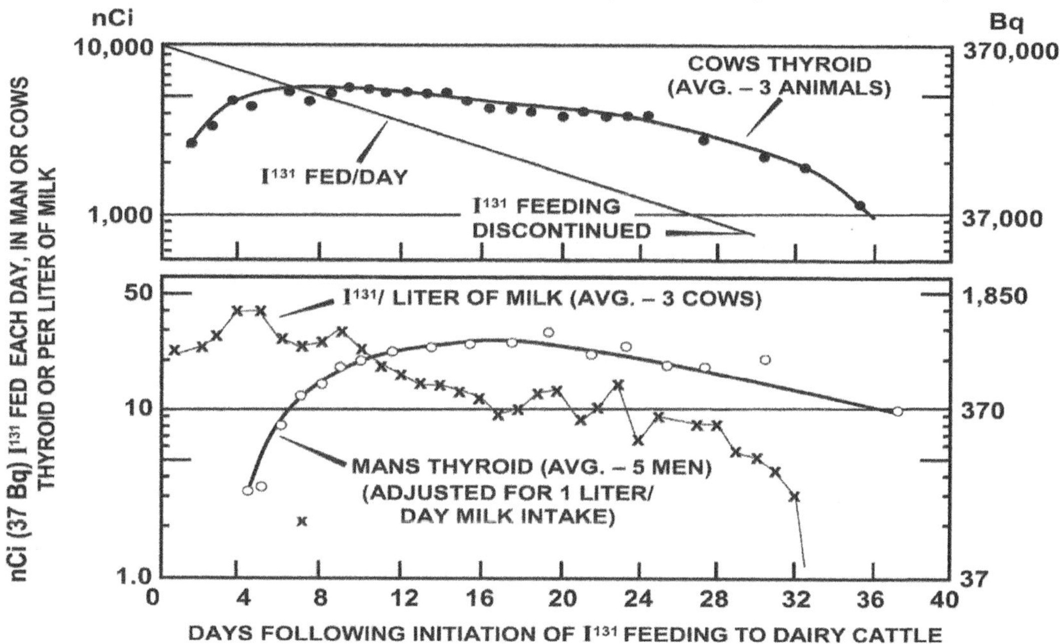

FIGURE 42.9 Interrelationships for radioiodine (I^{131}): Intake by cows, concentration in cow's thyroid and milk, and human thyroid. *Adapted from Bustad, L.K., McClellan, R.O., Garner, R.J., 1965. The significance of radionuclide contamination in ruminants. In: Physiology of Digestion in the Ruminant. Butterworth, Washington, DC, pp. 131–143.*

(McClellan et al., 1962, 1963a,b; McClellan and Bustad, 1963, 1964; McClellan, 1966). In addition, a number of special studies were conducted to assess bone marrow function, especially the production and survival of erythrocytes and leukocytes (McClellan, 1963; McClellan et al., 1963b, 1964).

One special study is worthy of review. In the fall of 1960, I visited the University of Chicago. I left my visit to the University of Chicago with two new friends, John Doull and Eugene Goldwasser. John would become my mentor in toxicology (McClellan, 2017). Goldwasser, a biochemist, explained the basic concepts of erythropoiesis to me and gave me a small sample of partially purified erythropoietin, the hormone regulating red blood cell production, which he had just discovered. It would take a number of years before human erythropoietin was purified (Miyaka et al., 1977; Goldwasser, 1996). As an aside, the University of Chicago failed to file a patent for this discovery which would later be exploited by others for commercial development. As Goldwasser later related to me, the proceeds from a patent on erythropoietin, if it had been filed and awarded, would have supported his laboratory and that of many colleagues in an elegant manner for their total career.

On returning to Hanford, I immediately set about developing an assay system for erythropoietin using polycythemic mice, created using simulated high altitude exposure to produce hypoxia. The hypoxic mice begin producing red blood cells ultimately achieving packed cell volumes of 70% or greater. Removed from the high altitude chamber, red cell production ceased. They were now candidates for assaying for erythropoietin. This assay was used to conduct studies to see if I could identify erythropoietin in pigs as an early marker of bone marrow damage. As expected, I did detect circulating erythropoietin in the serum of pigs following removal of large quantities of blood validating the assay system. I also detected elevated levels of erythropoietin in pigs that received large doses of radiation to the bone marrow from ingested Sr^{90}. As customary, I promptly published the results in the Hanford Biology Division Annual Report (McClellan, 1963). I was running fast! I did not take time to prepare a paper for publication in a peer-reviewed journal.

Fifteen years later, in the mid-1970s, I attended my first international conference that included scientists from mainland China. After I presented a lecture on the health effects of internally deposited radionuclides, a Chinese scientist asked what my thoughts were on use of erythropoietin as a biomarker for injury from Sr^{90}. I was puzzled by what prompted this very specific question and responded that I had done some preliminary research on this subject in the early 1960s. The individual who asked the question and I had a number of conversations at the conference; I was impressed by his substantial knowledge of my past research. When I returned to the United States, I learned my new friend was the senior-most Chinese official responsible for radiation health effects research and radiation protection across China. He was aware of every publication I and my colleagues ever wrote on radiation and its effects. I did arrange for him to visit our laboratory in Albuquerque, NM. He later arranged for me to visit his laboratory in Peking at the Institute of Radiation Medicine, which he headed, within the Chinese Academy of Military Medicine, as well as other Chinese laboratories. On one of my visits to China, my friend thanked me for the extensive animal studies we conducted on ingested and inhaled radionuclides. He noted the Chinese waited eagerly to see our reports and papers; it minimized their need to conduct similar research.

In 1965–66, I was temporarily assigned to the Division of Biology and Medicine of the US Atomic Energy Commission (AEC) (now the US Department of Energy) in Germantown, MD, a suburb of Washington, DC. At that time, the AEC had a very strong scientific research portfolio, including major research programs in the biomedical and environmental sciences. The Chair of the AEC at that time was Glenn Seaborg. My primary responsibilities were for oversight of the Agency's "internally deposited radionuclide research program" and a very limited program on chemical toxicity. Quite fortuitously for me, I shared an office with Paul Henshaw, the pioneering radiobiologist I mentioned earlier, recall the Henshaw (1941) paper. Paul and I had many conversations fueled by his recalling his participation in research at Oak Ridge as part of the Manhattan Project. He also recalled he was one of the first civilian scientists to visit Hiroshima and Nagasaki following the atomic bombings and had a role in setting up the Atomic Bomb Casualty Commission (ABCC). The AEC assignment was valuable in giving me a contextual understanding of important societal issues concerning radiation that had the potential for resolution through scientific knowledge acquired from well-planned research. My field of vision was broadened to include concerns related to the use of Pu^{238} as a thermal source in space nuclear power systems, the use of nuclear propulsion in naval vessels, and development of conventional uranium-fueled reactors as well as Plutonium Recycle Reactors and Breeder Reactors. The importance of using a "source-intake-dose-mechanisms of action-attributable disease" paradigm as illustrated in Figs. 42.1 and 42.2 was reinforced and the need for considering both exposed individuals and populations became even more apparent.

In the fall of 1966, I joined the Lovelace Foundation for Medical Education and Research in Albuquerque, NM, with responsibility for a research program directed at acquiring information needed to improve the estimates of the consequences of a catastrophic nuclear reactor accident (U.S.AEC, 1957). In a sense, the Institute's mission was to understand the potential consequences of the kind of reactor accident with release of large quantities of radioactive material to the atmosphere as would occur at Chernobyl in the Ukraine on April 26, 1986.

The scope of the Lovelace research program has been described in detail in previous publications (McClellan et al., 1986; McClellan, 2014). Suffice it to note here that numerous life span studies were conducted with fission product radionuclides (Y^{90}, Y^{91}, Cs^{137}, Ce^{144}, and Sr^{90}) that emit beta-particles and gamma-rays as they decay with different half-lives and alpha-particle emitting Pu^{238} and Pu^{239}. Every major study consisted of two phases. Both phases used aerosols that were very well characterized as to their chemical composition and particle size, both real and aerodynamic size. In the first phase, animals were exposed briefly to a well-characterized radioactive aerosol to simulate an accidental intake and then animals followed for variable periods of time before they were euthanized and tissues collected for detailed radiochemical analysis. The goal of these studies was to create a kinetic model to dynamically simulate the retention and tissue distribution of the radionuclide and ultimately serve as the basis for calculating the radiation dose to important organs (see Figs. 42.1 and 42.2).

The second phase involved an evaluation of exposure-related health effects in life span studies. The most significant findings were from the extensive studies using Beagle dogs. The animals were observed clinically for up to the normal life span of control Beagle dogs, about 15 years. Initially, very frequent clinical examinations were conducted with associated hematological and serum biochemical profiles assessed. It soon became apparent that the aggressive schedule of observations was useful in documenting the early and intermediate term deaths. However, they yielded limited scientific information to predict the occurrence of late-occurring deaths. Indeed, it soon became apparent that the best prognostic information on illness in an experimental animal, including the occurrence of cancer, came from the observations of the animal caretakers who had regular contact with the animals. My physician colleagues note this situation is similar to that which they frequently encounter with their human patients.

The major finding from the studies with experiments with Beagle dogs exposed to radionuclides and conducted at Lovelace, the University of California–Davis, University of Utah and Hanford (Pacific Northwest Laboratory) was the marked influence of radiation quality (low-LET radiation vs. alpha radiation) and the radiation dose pattern on life span and cancer induction (McClellan, 2014). Alpha irradiation from Pu^{238} or Pu^{239} was 7.5–12.4 times more potent than beta radiation from Y^{91}, Ce^{144}, and Sr^{90} inhaled in fused aluminosilicate particles for causing lung cancer. A combined

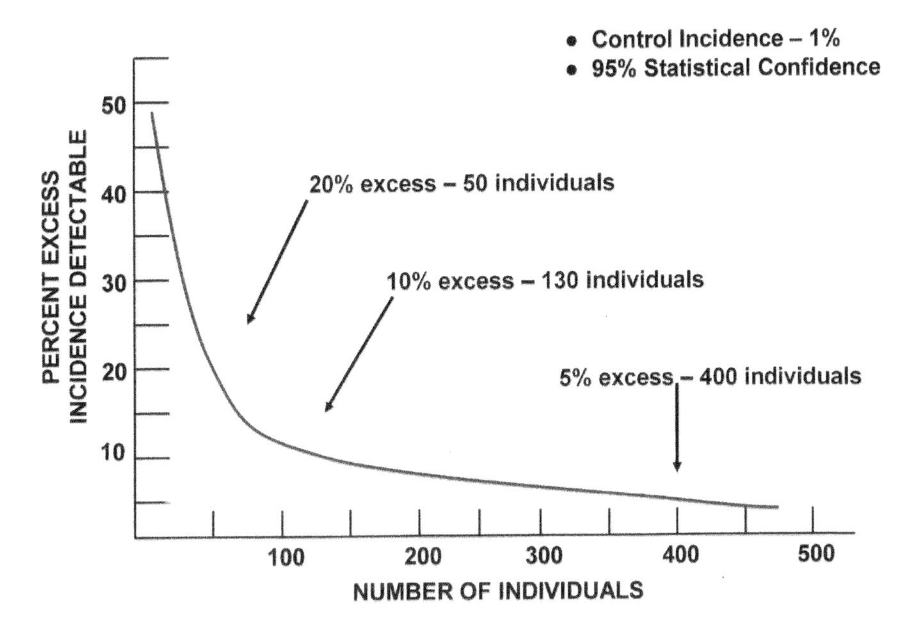

FIGURE 42.10 Relationships between number of subjects required to detect excess risk and the level of detectable excess risk.

Dose, Dose Rate Effectiveness Factor (DDREF) of 2.0 is currently recommended by the ICRP. The information from the dog studies needs to be carefully considered in revision and development of a separate Dose Rate Effectiveness and Dose Effectiveness Factors for use in radiation protection.

As an aside, it is worthy of note that the dog studies with inhaled and ingested radionuclides were not designed to provide support for or against use of the LNT model for radiation-induced cancer. This is the case because of the statistical limitations of the studies based on the size of the populations studied (Fig. 42.10).

Despite the substantial investments made in the last half century in cancer research, few specific markers of prognostic value for detecting cancer have yet been developed. Perhaps the best known is prostate-specific antigen (PSA) for prostate cancer. In my opinion, I am alive today because my prostate cancer was diagnosed 20 years ago based on an elevated PSA measurement which prompted a needle biopsy followed by a radical surgical prostatectomy by a highly skilled surgeon. Fortunately, it was surgically removed before it could metastasize to other sites. Ironically, today routine PSA screening of men over 50 years of age remains controversial with ongoing debate of the benefits and risks of screening of individuals and populations at different ages. It is not enough to have biomarkers available; it is crucial they be used in a thoughtful manner. Let me emphasize that PSA is a useful biomarker of prostate cancer even though we do not yet know the specific etiology of prostate cancer. The challenges of finding biomarkers of pending cancer or even clinically evident cancer are very substantial. The prospect of identifying the etiology (for example, radiation) of a specific cancer even after diagnosis is even more daunting. I will return to that topic later.

BIOMONITORING OF EXPOSURE TO RADIATION

Many individuals and government organizations have noted the need to develop biomarkers of radiation exposure calling attention to the potential for radiation disasters from a nuclear detonation or accident (Ray et al., 2014; Pellmar and Rockwell, 2005). I share these concerns; however, I quickly note that before additional investments are made in this area, the potential funders and researchers are strongly encouraged to review all the previous research and related findings conducted in this area going back to 1946. Many approaches have been tried with very modest success. Indeed, post September 11, 2001, many of the same old approaches dating to the late 1940s and early 1950s were frequently tried again with limited awareness of the past experiences. I am particularly concerned that many government agencies and individual scientists find it convenient to build support for research on any aspect of radiation using widespread "radiation phobia" rather than by conducting a critical assessment of what has been done in the past and is required to develop a strategic assessment of needs and plan of action.

Many of these individuals do not appear to understand why nuclear weapons are of concern as weapons of mass destruction. It is not that they are great sources of radiation! The primary reason is blast overpressure with its tremendous destructive impact, including its lethal and debilitating effects on people, followed by thermal injury, and then external radiation exposure. Radiation exposure, via inhalation or ingestion of radionuclides from a nuclear detonation, is by comparison with blast, thermal or direct radiation effects, a minor concern. For those not familiar with blast effects, I encourage them to read the classic document, "The Effects of Nuclear Weapons" (Glasstone and Dolan, 1977).

Concern for the health effects of radioactive materials purposefully dispersed is another matter. In my opinion, concerns for such potential releases resulting in overt radiation-attributable health effects have been overstated. The real concern for such accidents is dealing with the fear created by widespread radioactive contamination and potential radiation exposures at even low levels. Yet radiation phobia continues to sell research programs.

Radiation-Attributable Disease

Let us now turn attention to radiation-attributable disease, first addressing the acute radiation syndrome (ARS) followed by consideration of late-occurring health effects focusing on cancer. Mettler and Upton (2008) in their book, "*Medical Effects of Ionizing Radiation*," provide an excellent review of the medical consequences of radiation exposure from the effects of nuclear detonations as occurred at Hiroshima and Nagasaki and nuclear accidents such as Chernobyl. Tables 42.9–42.11 provide useful summaries of symptoms and signs of ARS.

Many of the signs are indeed biomarkers of acute disease. Their use represents good old-fashioned medical practices to diagnose and treat diseases.

The chapter by Ray et al. (2014) in the earlier edition of this textbook reviews a number of more sophisticated biomarker measurements that may complement traditional, time-tested biomarkers and be useful in the triage of patients and the diagnosis and treatment of exposed individuals. However, I would argue that none of these sophisticated biomarkers offers as much information as one gains with a traditional hematologic evaluation. I am equally pessimistic as to the likelihood of developing a magic pharmacologic pill for treating the ARS, a very complex syndrome.

Late-Occurring Health Effects Attributable to Radiation Exposure

A very large body of information from human populations exists on late-occurring health effects attributable to radiation exposure from both external sources and internally deposited radionuclides, i.e., studies of "human misfortune." The human data have been reviewed by Mettler and Upton (2008), Shore (2013), and, very recently, by the National NCRP (NCRP, 2018). McClellan (2014) reviewed the human data and also considered in some detail the information on inhaled and ingested radionuclides from life span studies in Beagle dogs.

Some of the major human populations studied following external exposure to X-radiation or gamma radiation are shown in Tables 42.12 and 42.13. Key studies in Beagle dogs are shown in Table 42.14. These studies provide clear evidence that the most significant late-occurring radiation-induced effects are cancers. These cancers occur in many organs of the body with whole-body irradiation and in the most heavily irradiated organs with partial body exposure or internally deposited radionuclides.

The most substantial data base available on late-occurring effects of radiation has been developed on the Japanese A-Bomb survivors. A cooperative program by the governments of Japan and the United States was initiated soon after World War II. The goals were (1) to provide medical care for the Japanese populations exposed at Hiroshima and Nagasaki and (2) conduct research to gain insights into the long-term effects of brief external exposure to radiation. The initial efforts were carried out by the ABCC which later became the Radiation Effects Research Foundation (RERF). Funding from the United States originally came from the US Atomic Energy Commission (now the US Department of Energy) budget and passed through the US National Academics of Science/NRC to RERF. It is important to note that substantial effort has been expanded to estimate the radiation exposure and organ doses for each individual in the ABCC/RERF cohorts. These estimates were developed taking account of precisely where each individual was when the detonations occurred and any shielding between them and the detonation (Cullins et al., 2006; Young and Bennett, 2006).

There are three ABCC/RERF cohorts. The LSS cohort includes 120,321 individuals: 82,214 in Hiroshima and 38,107 in Nagasaki. A second group is a cohort of about 3300 individuals who received their exposure in utero as a result of their mother's whole-body exposure. Information on this cohort has provided valuable insights into the incidence of malformation, growth retardation, microcephaly, and mental retardation related to radiation exposure. A third group is a cohort of 77,000 children born to survivors of the atomic bombing. Observations on this group provide valuable data for assessing heritable effects (Neel et al., 1990; Satoh et al., 1996). Grant et al. (2015) has reported on the risk of death among these children of atomic bomb survivors.

The first report on mortality in the LSS cohort was published by Beebe et al. (1962). The 14th in the series was published by Ozasa et al. (2012, 2013). A more recent publication by Grant et al. (2015) provides

TABLE 42.9 Prodromal Phase of Acute Radiation Sickness (ARS)

	Degree of ARS and Approximate Dose of Acute WBE (Gy)				
Symptoms and Medical Response	**Mild (1–2 Gy)**	**Moderate (2–4 Gy)**	**Severe (4–6 Gy)**	**Very Severe (6–8 Gy)**	**Lethal[a] (>8 Gy)**
Vomiting onset	2 h after exposure or later	1–2 h after exposure	Earlier than 1 h after exposure	Earlier than 30 min after exposure	Earlier than 10 min after exposure
% of incidence	10–50	70–90	100	100	100
Diarrhea	None	None	Mild	Heavy	Heavy
Onset	—	—	3–8 h	1–3 h	Within min or 1 h
% of incidence	—	—	<10	>10	Almost 100
Headache	Slight	Mild	Moderate	Severe	Severe
Onset	—	—	4–24 h	80	1–2 h
% of incidence	—	—	50	May be altered	80–90
Consciousness	Unaffected	Unaffected	Unaffected		Unconsciousness (may last sec/min)
Onset	—	—	—	—	Sec/min
% of incidence	—	—	—	—	100 (at >50 Gy)
Body temperature	Normal	Increased	Fever	High fever	High fever
Onset	—	1–3 h	1–2 h	<1 h	<1 h
% of incidence	—	30–80	80–100	100	100
Medical Response	Occupation observation	Observation in general hospital, treatment in specialized hospital if needed	Treatment in specialized hospital	Treatment in specialized hospital	Palliative treatment (symptomatic only)

WHE, whole-body exposure.
[a]*With appropriate supportive therapy individuals may survive for 6–12 months with whole-body doses as high as 12 Gy.*
Adapted from International Atomic Energy Agency (IAEA), 1998. Diagnosis and Treatment of Radiation Injuries. Safety Report Series #2, IAEA, Vienna.

TABLE 42.10 Latent Phase of Acute Radiation Sickness (ARS)

	Degree of ARS and Approximate Dose of Acute WBE (Gy)[a]				
	Mild (1–2 Gy)	**Moderate (2–4 Gy)**	**Severe (4–6 Gy)**	**Very Severe (6–8 Gy)**	**Lethal (>8 Gy)**
Lymphocytes (G/L) (days 3–6)	0.8–1.5	0.5–0.8	0.3–0.5	1.0–0.3	0.0–0.1
Granulocytes (G/L)	>2.0	1.5–2.0	1.0–1.5	≤0.5	≤0.1
Diarrhea	None	None	Rare	Appears on days 6–9	Appears on days 4–5
Epilation	None	Moderate beginning on day 15 or later	Moderate, beginning on day 11–21	Complete earlier than day 11	Complete earlier than day 10
Latency period (days)	21–35	18–28	8–18	7 or less	None
Medical response	Hospitalization not necessary	Hospitalization recommended	Hospitalization necessary	Hospitalization urgently necessary	Symptomatic treatment only

[a]*G/L gigaliter.* WBE, *whole-body exposure.*
Adapted from International Atomic Energy Agency (IAEA), 1998. Diagnosis and Treatment of Radiation Injuries. Safety Report Series #2, IAEA, Vienna.

TABLE 42.11 Findings of Critical Phase of Acute Radiation Sickness (ARS) Following Whole-Body Exposure (WBE)

	Degree of ARS and Approximate Dose of Acute WBE (Gy)[a]				
	Mild (1–2 Gy)	Moderate (2–4 Gy)	Severe (4–6 Gy)	Very Severe (6–8 Gy)	Lethal (>8 Gy)
Lymphocytes (G/L)	0.8–1.5	0.5–0.8	0.3–0.5	0.1–0.3	0.0–0.1
Platelets (G/L)	60–100 10%–25%	10–60 25%–40%	25–35 40%–80%	15–25 60%–80%	<20 80-100%[b]
Clinical manifestations	Fatigue, weakness	Fever, infections, bleeding, weakness, epilation	High fever, infections, bleeding, epilation	High fever, diarrhea, vomiting, dizziness and disorientation, hypotension	High fever, diarrhea, unconsciousness
Lethality (%)	0	0–50 Onset 6–8 wk	20–70 Onset 4–8 wk	50–100 Onset 1–2 wk	100 1–2 wk
Medical response	Prophylactic	Special prophylactic treatment from days 14–20; isolation from days 10–20	Special prophylactic treatment from days 7–10; isolation from the beginning	Special treatment from the first day; isolation from the beginning	Symptomatic only

[a] *1 Gy = 100 rad.*
[b] *In very severe cases, with a dose >50 Gy, deaths precedes cytopenia.*
Adapted from International Atomic Energy Agency (IAEA), 1998. Diagnosis and Treatment of Radiation Injuries. Safety Report Series #2, IAEA, Vienna.

TABLE 42.12 Human Populations Studied Following External Exposure to X- or Gamma Radiation

Population	Effect	Key Reference
Atomic bomb survivors	Cancer and other diseases	Ozasa et al. (2012)
Prenatal irradiation	Leukemia Cancer	Doll and Wakeford (1997)
Ankylosing spondylitis patients	Cancer	Court Brown and Doll (1965)
Radiologists	Leukemia	Matanowski et al. (1975) Smith and Doll (1981)
Thymic enlargement	Thyroid cancer	Shore et al. (1993)
Tinea capitis	Thyroid cancer	Shore et al. (2003) Ron et al. (1989)

TABLE 42.13 Key Human Populations Studied Following Intake of Radioactivity

Population	Radiation Source	Target Organ	Key References
Uranium miners	Radon and daughters	Lung	BEIR VI (1999)
Radium dial painters	^{226}Ra and ^{228}Ra	Skeleton	Evans et al. (1972) Rowland (1994) Taylor et al. (1989)
Thorotrast patients	ThO_2	Liver andand spleen	Olsen et al. (1990) Taylor et al. (1989)
Thyrotoxicosis patients	^{131}I	Thyroid	Dobyns et al. (1974) Ron et al. (1998) Holm et al. (1980)
Thyroid diagnosis patients	^{131}I	Thyroid	Holm et al. (1989)
Marshall Islanders	$^{131,132,133,135}I$	Thyroid	Conrad et al. (1980)
Mayak workers	^{239}Pu	Skeleton	Solkolnikov et al., 2008

TABLE 42.14 Major Life span Studies With Internally Deposited Radionuclides in Beagle Dogs With Multiple Exposure Levels and Life span Observation

Laboratory	Radionuclides	Route of Administration
University of Utah	Ra^{226}, Ra^{228}, Th^{228}, Pu^{239}, Am^{241}, Sr^{90}	Single intravenous injection
UC-Davis	Ra226 Sr^{90}	Multiple intravenous injections Chronic ingestion
Hanford	$Pu^{238}PuO_2$, $Ra^{239}O_2$, Pu^{239} $(NO_2)4$	Inhalation of polydisperse aerosols
Lovelace	Y^{90}, Y^{91}, Ce^{144} and Sr^{90} Chloride $Cs^{137}Cl_2$ Y^{90}, Y^{91}, Ce^{144} and Sr^{90} in fused aluminosilicate particles $Pu^{238}O_2$ and $Pu^{239}O_2$	Inhalation of polydisperse aerosol Intravenous injection Inhalation of polydisperse aerosol Inhalation of monodisperse particles

information on a longer-term follow-up from 1958 through 2009. Some detailed results from 14th mortality update will be reviewed here. The observed and excess deaths for the LSS cohort through 2003 are shown in Table 42.15. The estimates of ERR for the various causes of death are shown in Fig. 42.11. It is of interest to note that of the 86,601 individuals alive and enrolled in the cohort in 1950, a total of 46,614 individuals or 53.8% of the individuals had died by 2003 with 46.2% were surviving. The 10,929 cancer deaths represented 23.4% of the total number of deaths through 2003. In aggregate, it was estimated, using statistical methods, that 527

TABLE 42.15 Observed and Estimated Excess Deaths in Cancer and Noncancer Diseases in the ABCC/RERF Life Span Study (LSS) cohort

			Solid Cancer			Noncancer diseases[b]		
Colon Dose (Gy)	**Number of Subjects**	**Person Years**	**Number of Deaths**	**Number of Excess cases[a]**	**Attributable Fraction (%)**	**Number of Deaths**	**Number of Excess cases[b]**	**Attributable Fraction (%)**
<0.005	38,509	1,465,240	4621	2	0	15,906	1	0
0.005-	29,961	1,143,900	3653	49	1.3	12,304	36	0.3
0.1-	5974	226,914	789	46	5.8	2504	36	1.4
0.2-	6356	239,273	870	109	12.5	2736	82	3.0
0.5-	3424	129,333	519	128	24.7	1357	86	6.3
1-	1763	66,602	353	123	34.8	657	76	11.6
2+	624	22,947	124	70	56.5	221	36	16.3
Total	86,611	3,294,210	10,929	527	4.8	35,685	353	1.0

[a]*Based on the ERR model was defined as the linear model with effect modification:* $\gamma_o(c,s,b.a)[1 + \beta_1 d \bullet exp(\tau e + \upsilon\ ln(a)) \bullet (1\ \sigma\ s)]$.
[b]*Non-neoplastic blood diseases were excluded from noncancer diseases.*
Ozasa et al., 2012. Studies of the mortality of atomic bomb survivors, report 14, 1950–2003: an overview of cancer and noncancer diseases. Radiat. Res. 177, 229–243.

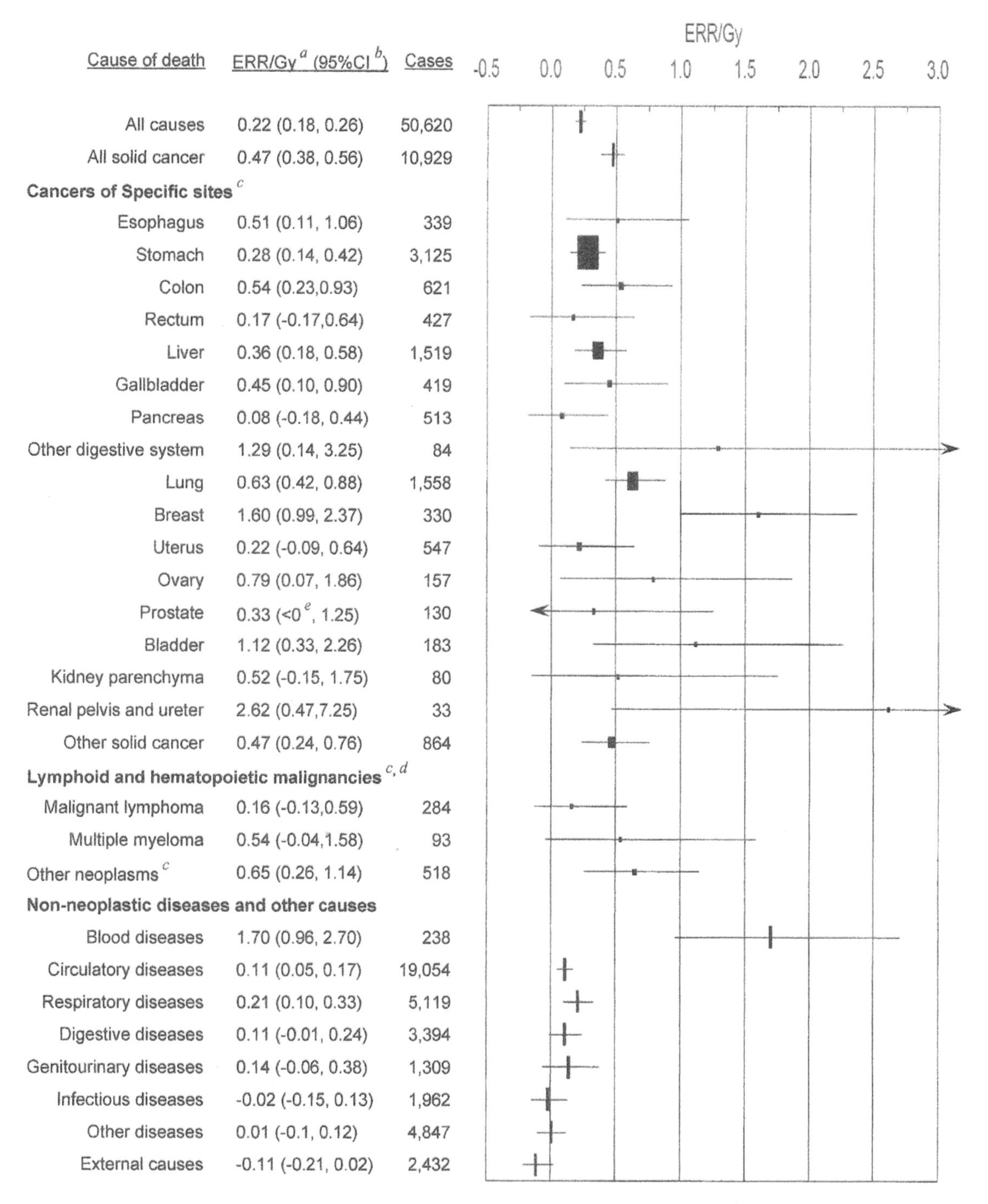

FIGURE 42.11 Estimates of excess relative risk (ERR) per Gy and 95% CI for major causes of death. [a]ERR was estimated using the linear dose model, in which city, sex, age at exposure, and attained age were included in the background rates, but not allowing radiation effect modification by those factors. [b]Confidence interval. *Horizontal bars* show 95% confidence intervals. [c]The size of plots for site-specific cancers was proportional to the number of cases. [d]ERR (95% CI) of leukemia was 3.1 (1.8, 4.3) at 1 Gy and 0.15 (−0.01, 031) at 0.1 Gy based on a linear-quadratic model with 318 cases (not displayed in the figure). [e]The lower limit of 95% CI was lower than zero, but not specified by calculation. (1950–2003). *Ozasa, K., Shimizu, Y., Suyama, A., et al. (2012). Studies of the mortality of atomic bomb survivors, report 14, 1950–2003: An overview of cancer and noncancer diseases. Radiat. Res. 177:229–243. © 2019 Radiation Research Society.*

excess cancer deaths among 10,929 deaths from solid cancers were attributable to radiation exposure. Most of these were in groups with colon doses of over 0.1 Gy and over half were in groups with colon doses of over 0.5 Gy. Contrary to common beliefs among the public and many scientists, radiation is not a very potent carcinogen. I strongly suspect most readers of this chapter are surprised by the small number of excess cancers observed to date in the Japanese A-Bomb survivors.

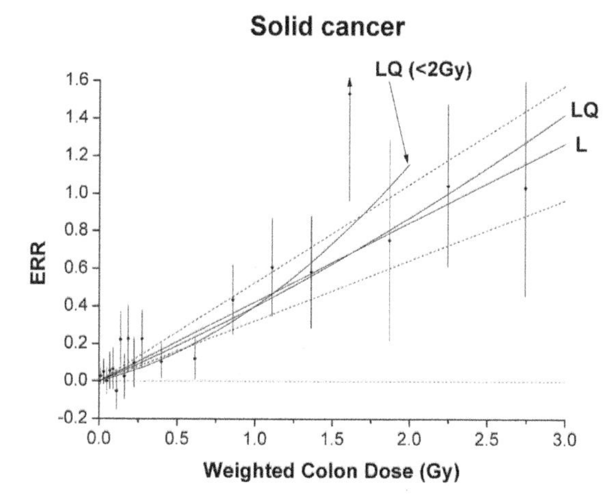

FIGURE 42.12 Excess relative risk (ERR) for all solid cancers in relation to radiation exposure. The *black circles* represent ERR, and the *bars* are the 95% Confidence Interval for the dose categories together with trend estimates based on linear (L) with 95% Confidence Intervals (*dotted lines*) and Linear–Quadratic (L–Q) models using the full dose range and LQ model for data restricted to dose < 2 Gy. *Ozasa, K., Shimizu, Y., Suyama, A., et al. (2012). Studies of the mortality of atomic bomb survivors, report 14, 1950–2003: An overview of cancer and noncancer diseases. Radiat. Res. 177:229–243. © 2019 Radiation Research Society.*

It is important to note the 527 excess deaths attributable to radiation exposure were estimated using statistical methods. At this time there are no scientific methods for distinguishing which of the cancer cases was caused by radiation exposure versus other causes. To be blunt, none of the cancers carries a sign—"This cancer was caused by radiation!" If biomarkers had existed for identification of developing cancer, they would have been very useful for diagnostic purposes and early treatment. However, it is very unlikely that any of these prospective biomarkers would have identified which of the about 1 in 20 cancers attributable to radiation exposure was actually caused by radiation exposure.

Alternative presentations of the relationship between ERR and dose in the LSS are shown in Figs. 42.12 and 42.13. It is always important to carefully examine the scale on figures. This is a good example of why that is important. Fig. 42.12 at a glance appears to show a linear, no-threshold dose–response relationship, and Fig. 42.13 appears to illustrate a clear threshold. Now note the difference in the scales. Fig. 42.12 has an arithmetic horizontal scale and Fig. 42.13 is logarithmic. When the difference in scales is considered, it is apparent that in both figures, using the same data, with alternative dose–response models, the results are not remarkably different. How many angels can dance on the head of a pin? It depends on the size of the pin! These two figures are germane to the argument of whether the cancer response per unit dose follows a linear no-threshold relationship or not. Recall that Henshaw (1941) raised this issue (Fig. 42.4). Three quarters of a century later, the issue is still being debated. Some individuals, such as Clarke (2008), a former chair of the International Commission on Radiation Protection, have argued that the specific shape of the curve relating excess cancer risk to exposure levels approaching background is irrelevant to the setting of radiation protection standards (Fig. 42.14).

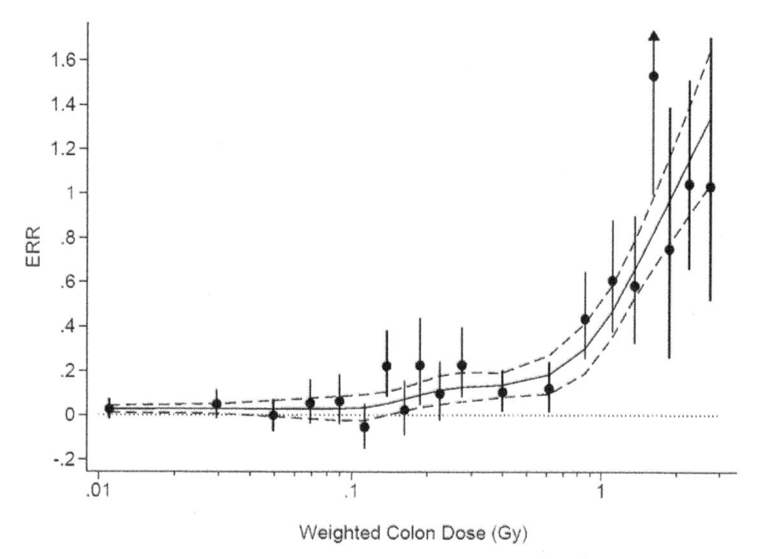

FIGURE 42.13 Excess relative risk (ERR) for all solid cancer in atomic bomb survivors in relation to radiation exposure. The *black circles* and *error bars* represent ERR and 95% CIs for the dose categories. *Solid line*—fit to the ERR data using a multiple linear regression in which weighted log colon dose was entered into the model using a restricted cubic spline transformation with five knots. Regression weights were equal to the inverse of the variance of the point estimates. *Dashed lines* are 95% CI of the fit. Figure from performing analysis equivalent to (Doss et al., 2012) with the corrected data in Ozasa et al. (2013). *Doss, M., Egleston, B.L. & Litwin, S. Comments on "Studies of the mortality of atomic bomb survivors, report 14, 1950–2003: an overview of cancer and noncancer diseases" (Radiat Res 2012; 177:229–243). Radiat Res 178, 244–245, (2012). © 2019 Radiation Research Society.*

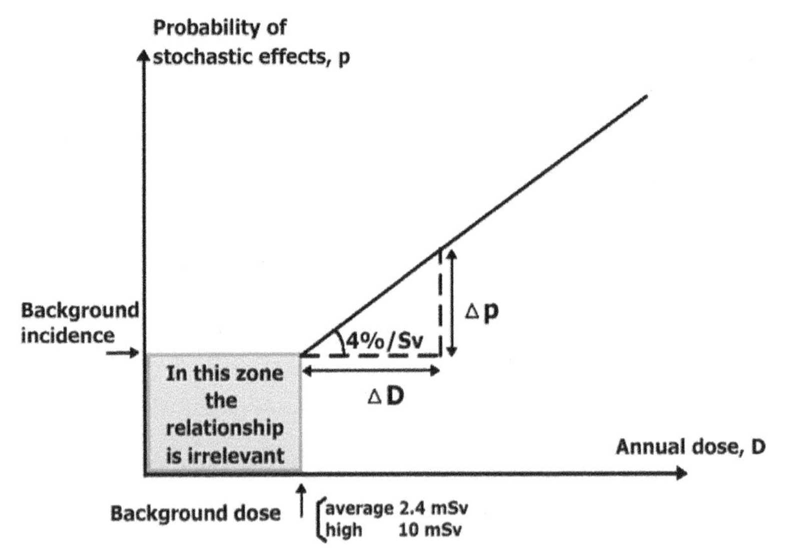

FIGURE 42.14 Region of the dose–response curve when risk factors apply (Clarke, 2008).

In the United States, the NCRP has recently released a commentary addressing implications of recent epidemiologic studies for the use of the linear, non-threshold model for radiation protection (NCRP, 2018). This very comprehensive report evaluated 29 major populations with published reports on cancer and radiation exposure based on (1) dosimetry, (2) epidemiological, and (3) statistical considerations to assess their (4) overall support for using or not using a linear, non-threshold model for radiation protection purposes. The results are summarized in Table 42.16.

The NCRP Commentary concludes "that no alternative dose-response relationship appears more pragmatic or prudent for radiation protection purposes than the LNT model." I fully concur.

In my opinion, the continuation of the LNT debate over the past 75 years relates in part to the debate rarely being properly framed with several groups of scientists talking past each other. In some cases, individuals have entered the debate on the use of LNT models for setting radiation protection standards with a lack of understanding of how radiation protection standards are set. Few of the radiation protection standards or regulations are actually set based on achieving some specific risk target, i.e., less than one extra cancer over a lifetime in a population of 10,000 individuals. Rather, the numerical radiation protection standards were initially set based on scientific judgment and have evolved over time. Post hoc analysis can provide estimates of excess risk for a given numerical standard. However, those results should not be confused with setting a standard based on achieving a risk limit.

PATH FORWARD

Information from a NRC Biological Effects of Ionizing Radiation VII Report (NRC, 2006) can be used to illustrate some of the challenges faced in development of biomarkers of cancer, including those attributable to exposure to ionizing radiation. That report indicated "a comprehensive review of biological and biophysical data supports a 'linear-no-threshold' (LNT) risk model – that the risk of cancer proceeds in a linear fashion at lower doses without a threshold and that the lowest dose has the potential to cause a small increase in risk to humans." The report defined low doses as those in the range of near zero up to about 100 mSv (0.1 Sv) of low-LET radiation. The BEIR VII Committee was so bold as to suggest approximately one extra cancer would occur in 100 people from a single 100 mSv exposure to low-LET radiation in addition to the 42 cancers that would be diagnosed with cancers from causes unrelated to radiation (Fig. 42.15). In this figure, each of 100 hypothetical individuals is represented by a symbol. The 57 open circles are individuals that died without cancer. The 42 darkened circles are individuals that died with cancer attributed to causes other than radiation. The single star is an individual that died with cancer statistically attributed to radiation exposure. It is immediately clear from the figure that cancer is a very common disease. Most of these cancers occur late in life and, thus, as life spans are extended more cancers are likely to be observed. The competition between cancer, cardiovascular disease, and other causes of death continues! Most of the observed cancers caused the

TABLE 42.16 Ratings of the Quality of Cancer Studies Reviewed and Their Degree of Support for the LNT Model

Study (or Groups of studies)[a]	Dosimetry[b]	Epidemiology[b]	Statistics[b]	Support for LNT Model[c]
LSS, Japanese atomic bomb survivors (Grant et al., 2017)[d]	3	3	3	4
INWORKS (French, UK, US combined cohorts) (Richardson et al., 2015)	3	3	3	4
TB fluoroscopic examinations and breast cancer (Little and Boice, 2003)	3	3	2	4
Childhood Japan atomic bomb exposure (Preston et al., 2008*	3	3	3	4
Childhood thyroid cancer studies (Lubin et al., 2017; Ron et al., 1989)	3	3	3	4
Mayak nuclear workers (Sokolnikov et al., 2015)	2	2	3	3
Chernobyl fallout, Ukraine and Belarus thyroid cancer (Brenner et al., 2011; Zablotake et al., 2011)	3	2	2	3
Breast cancer studies, after childhood exposure (Eidemuller et al., 2015)	2	3	3	3
In utero exposures, Japan atomic bomb (Preston et al., 2008)	2	3	3	3
Techa River, nearby residents (Schonfeld et al., 2013)	2	2	2	2.5
In utero exposures, medical (Wakeford, 2008)	1	2	2	2.5
Japanese nuclear workers (Akiba and Mizuno, 2012)	2.5	2	3	2
Chernobyl cleanup workers, Russia (Kashcheev et al., 2015)	1	1.5	2	2
US radiologic technologists (Liu et al., 2014; Preston et al., 2016)[e]	1	2	2	2
Mound nuclear workers (Boice et al., 2014)	2	1.5	1.5	2
Rocketdyne nuclear workers (Boice et al., 2011)	2	2	2	2
French uranium processing workers (Zhivin et al., 2016)	2.5	3	1.5	2
Medical X-ray workers, China (Sun et al., 2016)	1.5	1.5	2	2[f]
Taiwan radio-contaminated buildings, residents (Hsieh et al., 2017)	2	1.5	1.5	2[f]
Background radiation levels and childhood leukemia (Kendall et al., 2013)	1.5	2	2	2
In utero exposures, Mayak and echa (Akleyev et al., 2016)	1	1.5	2	1
Hanford ^{131}I fallout study (Davis et al., 2004)	2	3	1.5	1
Kerala, India, HBRA (Nair et al., 2009)	2	2	1.5	1
Canadian nuclear workers (Zablotska et al., 2014)	2.5	3	3	1
US atomic veterans (Caldwell et al., 2016)	3	3	3	1
Yangjiang, China, HBRA (Tao et al., 2012)	1.5	1	1	1[f]
CT examinations of young persons (Pearce et al., 2012)	1	1.5	1.5	1[f]
Childhood medical X-rays and leukemia studies (aggregate of >10 studies) (Little, 1999; Wakeford, 2008)	1	2	1.5	1[f]
Nuclear weapons test fallout studies (aggregate of eight studies (Lyon et al., 2006)[g]	1.5	1	1.5	1[f]

[a]*A representative recent publication is listed for each study. Others are found in the text.*
[b]*Judged quality of the dosimetry, epidemiology, and statistics scores: 1 = weak-to-moderate, 2 = moderate, 2.5 = moderate-strong, 3 = strong.*
[c]*Ratings of the support for LNT: 1 = essentially no support (null or negative; or unreliable and inconclusive), 2 = weak-to-moderate support, 3 = moderate support, 4 = strong support. Study ratings were based on reported solid cancer (or close surrogates) risk unless noted otherwise.*
[d]*Studies excluded: ecological studies of risks around nuclear sites (no dosimetry and extremely low exposures); the 15-Country worker study and other studies that overlap with the more recent INWORKS; the MPS is not yet completed, but published components of it are included separately; studies of genetic effects (since no human heritable risks have been shown); studies of tissue reactions (because these are generally not believed to be LNT).*
[e]*The dosimetry used in the Preston et al. (2016) study of breast cancer, based on the Simon et al. (2014) dosimetry, is significantly improved and would be rated "2" compared with the dosimetry that was used in other published epidemiologic analyses of this cohort. However, since little other epidemiology has been published using the new dosimetry, for the purposes of this Commentary, these studies are limited in their support for LNT.*
[f]*Considered "weak" support or "inconclusive" primarily because of weaknesses in the epidemiology, dosimetry or statistical risk modeling. The other studies scored as 1 (no support) had reasonable methodologies but provided little or no support for the LNT model because their risk coefficients were essentially zero or negative.*
[g]*Fallout studies were included as a group (they mostly had little or poor dosimetry and many were studies of aggregates rather than individuals, however, the Hanford ^{131}I fallout study was of better quality, so was identified separately).*

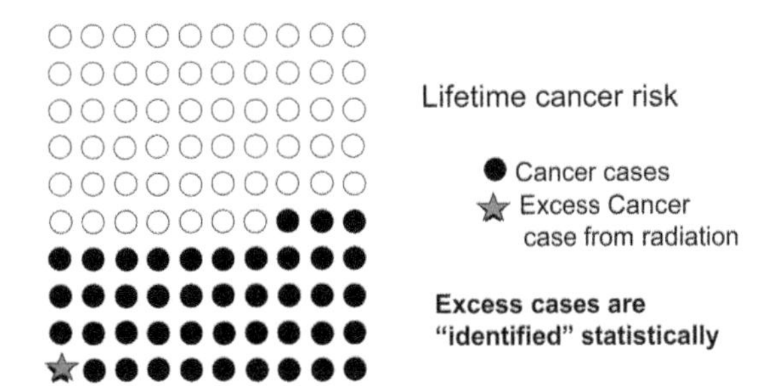

FIGURE 42.15 Linear, no-threshold exposure—response model; Science or Policy?

death of the individuals because cancer treatment in general is not highly effective, although it can extend the life span of some of the treated individual.

In my opinion, the NRC committee grossly understated the uncertainties in their estimates. Yes, the number of background cancer cases can be estimated, indeed, extrapolated from past experience with reasonable certainty. However, in my opinion, the extra case estimated from the baseline cancer rate and a highly uncertain estimate of the potency of radiation for projecting an ERR for single 100 mSv exposures of the population is a highly uncertain! Nonetheless, let us use the BEIR VII Committee results as a basis for considering where future investments in cancer and biomarkers research might yield the greatest benefits to Society-at-large.

It is obvious that any increase in knowledge that will aid in earlier diagnosis and treatment of the baseline cancer cases, irrespective of their etiology, will yield a positive net benefit to Society. By comparison, the development and use of any potential biomarkers to diagnose and treat cancers, the relatively rare cancers uniquely caused by radiation exposures below 0.1 mSv will have a trivial benefit to Society. The radiation-attributable cancer is only one of 43 cancers in the example. I expressed the view earlier that much of the radiation-related biomarker research had been spurred by "radiation phobia." The impact of "radiation phobia" including that generated by the BEIR VII report may have negative impacts if it discourages the use of medical procedures that involve exposure to ionizing radiation resulting in less effective medical care. I conclude that it is important for Society-at-large to focus scientific and financial resources on developing and using "biomarkers" for all cancers whose early diagnosis and treatment will yield the greatest benefit to Society.

A paper by Cohen et al. (2018) offers promise for development of useful "biomarkers" for early diagnose and treatment of cancer. Their large team reported a new single blood test, Cancer SEEK, that screens for cancers in the ovary, liver, stomach, pancreas, esophagus, colorectum, lung, or breast. Consideration of the occurrence of cancer in different organs in the A-Bomb survivors indicates that if this test can be validated by large population studies, it could be very useful. A large portion of the cancers in the A-Bomb survivors were in the eight organs studied by the Cohen team, see Fig. 42.11. Recalling the size and duration of the LSS cohort study helps one understand the challenge of validating Cancer SEEK, it will require the study of a very large population over a number of years.

At this juncture, let me note that starting in 1998, the US Department of Energy initiated a low dose radiation research program that continued for just over a decade. The origin of the program, its etiology, research content, and major findings have recently been reviewed by Brooks (2018), who served as Chief Scientist for the program. At its initiation, a decision was made to focus on doses of less than 0.1 Gy of low-LET ionizing radiation. Early in the program, a decision was made to focus on the use of recent advances in modern molecular biology. To a large extent, the research program consisted of individual investigator–originated research efforts on adaptive responses, genomic instability, and bystander effects in cells. Very few of the research projects include direct linkages to observations of cancer in humans or laboratory animals. The program generated many findings reported in peer-reviewed publications. Some of the observations made clearly qualify as biomarkers of radiation exposure and response. It is not clear if the collective findings have had significant impact on the central issue of the relationship between low doses of radiation, typically given over short periods of time, and the occurrence of excess cancer, over and above background incidence.

A survey of many of the papers cited by Brooks (2018) and the earlier radiation biomarker chapter by Ray et al. (2014) reveals that most of the past research, to better understand early markers of radiation-induced disease, has used experimental designs and related statistical approaches useful for "deterministic" or threshold dose–response relationships. The approaches, appropriate for stochastic responses, routinely used in epidemiological investigations, have rarely been used in the experimental studies. There appear to be a substantial "disconnect" between the experimental studies and the "stochastic" end point of primary concern—cancer in humans.

In designing future biomarker studies, it is important to recall that health end points, like cancer, occur in a time and dose-dependent manner following brief or protracted radiation exposure. There are many reasons to anticipate that biomarkers of intermediate events leading to cancer will also have time and dose-dependent relationships. Future studies need to be designed to identify these relationships. Studies of biomarkers of effects that utilize single doses and only a few observational times are unlikely to yield very much useful information. Unfortunately, in the past many studies have been conducted without taking account of these considerations.

One approach to bridging this gap may be increased use of biologically based models of carcinogenesis as pioneered by S. H. Moolgavkar (Moolgavkar, 2016, 1977, 1978, 1986, 1990, 1994, 2004; Moolgavkar and Knudson, 1981; Moolgavkar et al., 1999; Little et al., 2008). Such models allow the integration of multiple kinds of data in a quantitative manner with linkages to the end point of concern—cancer as it occurs in age-related fashion over the lifetime of individuals in populations. The world of experimental observations needs to be merged with the real world of epidemiological findings on human populations!

Although this chapter specifically addresses ionizing radiation, the lessons learned from experience with radiation should also be applied to chemical agents. Some individuals might argue that both radiation and chemicals may provide model systems for development of biomarkers of cancer. I find that argument unconvincing if it focuses on cancers produced by high dose exposures to either radiation or chemicals. These are the cancers that are most likely to have unique characteristics related to radiation and chemical exposure compared with cancers of unknown etiology.

We need to collectively develop a new strategic approach to planning and conduct of our biomarker research. With a very few exceptions, it is clear that today's science is not adequate to identify the etiology of the vast majority of cancers observed in populations in countries with well-developed economies such as the United States with populations that have long life spans which allow a large portion of the population to develop cancer. In my opinion, focusing research on radiation and chemical caused cancer is unlikely to help understand the real major factors that cause the cancers diagnosed in about 40% of the population over their lifetime.

Dedication and Acknowledgments

This chapter is dedicated to the memory of three of my mentors who had major influence on my career. Leo K. Bustad guided my early research on internally deposited radionuclides and encouraged development of my skills in comparative medicine and use of biomarkers. Herbert M. Parker encouraged me to recognize the role of issue-resolving research addressing major Societal needs. Furthermore, he taught me that science informs policy decision, and that scientists need to engage with the public in communicating science and the setting of policy and development of regulations. I was honored to be asked to give the 2018 Herbert M. Parker Memorial lectures at Washington State University-Tri-Cities Campus, Richland, WA, on April 24 and 25, 2018. This chapter, in part, builds on those lectures. John Doull, from 1960 when we first met until his death in 2017, mentored me on the art and science of toxicology. We had countless discussions over whether toxicology was a discipline (John's position) versus toxicology as an area of activity that uses the concepts and tools of many disciplines (my position). His impact on toxicology over 70 years was immense (McClellan, 2017).

I also gratefully acknowledge the pleasure I had working at Hanford and Lovelace with many skilled scientists from different disciplines, a cadre of dedicated technical support personnel and animal care staff, all of whom understood the value and essential role of team work in conducting research on radiation toxicity with results that have had major impact on Society (McClellan, 2014). I hope the historical orientation I used in this chapter will encourage others to delve into and understand our scientific past and build on it by working with others to create an improved science base to inform public policy decisions.

References

Akiba, S., Mizuno, S., 2012. The third analysis of cancer mortality among Japanese nuclear workers, 1991–2001: estimation of excess relative risk per radiation dose. J. Radiol. Prot. 32 (1), 73–83.

Akleyev, A., Deltour, I., Krestinina, L., Sokolnikov, M., Tsarev, y., Tolstykh, E., Schuz, J., 2016. Incidence and mortality of solid cancers in people exposed in utero to ionizing radiation: pooled analysis of two cohorts from the Southern Urals, Russia. PLoS One 11 (8), e0160372. https://doi.org/10.1371/journal.pone.0160372.

Beebe, G.W., Ishida, M., Jablon, S., 1962. Studies of the mortality of a-bomb survivors. I. Plan of study and mortality in the medical subsample (Selection 1) 1950-1958. Radiat. Res. 16, 253–280.

BEIR VI, 1999. Health Effects of Exposure to Radon: Committee on Health Risks of Exposure to Radon. National Research Council. National Academy Press, Washington, DC.

Boecker, B.B., 1969. The metabolism of ^{137}Cs inhaled as ^{137}CsCl by the beagle dog. Proc. Soc. Exp. Biol. Med. 130 (3), 966–971.

Boice Jr., J.D., Cohen, S.S., Mumma, M.T., Ellis, E.D., Eckerman, K.F., et al., 2011. Updated mortality analysis of radiation workers at Rocketdyne (Atomic International), 1948-2008. Radiat. Res. 176 (2), 244–258.

Boice Jr., J.D., 2014. The Miller Worker study comes to Oak Ridge. Health Phys. News XLII (7), 16–17.

Brenner, A.V., Tronko, M.D., Hatch, M., Bogdanova, T.L., Oliynik, V.A., et al., 2011. I-131 dose response for incident thyroid cancers in Ukraine related to the Chernobyl accident. Environ. Health Perspect. 119 (7), 933–939.

Brooks, A., 2018. Low Dose Radiation: The History of the U.S. Department of Energy Program. Washington State University Press, Pullman, WA, 316 pp.

Bushong, S.C., 2013. Radiological Science for Technologists: Physics, Biology and Protection, tenth ed. Elsevier/Mosby, St. Louis, MO.

Bustad, L.K., McClellan, R.O., Garner, R.J., 1965. The significance of radionuclide contamination in ruminants. In: Physiology of Digestion in the Ruminant. Butterworth, Washington, DC, pp. 131–143.

Caldwell, G.G., Zack, M.M., Mumma, M.T., Falk, H., Heath, C.W., Till, J.E., Chen, H., Boice, J.D., 2016. Mortality among military participants at the 1957 PLUMBBOB nuclear weapons test series and from leukemia among participants at the SMOKY test. J. Radiol. Prot. 36 (3), 474–489.

Cantril, S.T., Parker, H.M., 1945. The Tolerance Dose. Manhattan District Report. NDDC-1100, Oak Ridge, TN.

Clarke, R.H., 2008. The risks from exposure to ionizing radiation. In: Till, J.E., Grogan, H.A. (Eds.), Radiological Risk Assessment and Environmental Analysis. Oxford University Press, New York, NY, pp. 531–550.

Cohen, J.D., Lu, L., Wang, Y., Thoburn, C., et al., 2018. Detection and location of surgically resectable cancers with a multi-analytic blood test. Science 359 (6378), 926–930.

Conrad, R.A., et al., 1980. Review of Medical Findings in a Marshallese Population 26 years After Accidental Exposure to Radioactive Fallout. Brookhaven National Laboratory, Upton, NY.

Court Brown, W.M., Doll, R., 1965. Mortality from cancer and other causes after radiotherapy for ankylosing spondylitis. Br. Med. J. 2, 1327–1332.

Cullins, H.M., Fujita, S., Funamoto, S., Grant, C.J., Kerr, G.D., Preston, D.L., 2006. Dose estimation for atomic bomb survivor studies: its evolution and present status. Radiat. Res. 166 (1 Pt. 2), 219–254.

Davis, S., Kopecky, K.J., Hamilton, T.E., Onstad, L., 2004. Thyroid neoplasia, autoimmune thyroiditis, and hypothyroidism in persons exposed to iodine 131 from the Hanford Nuclear Site. J. Am. Med. Assoc. 292 (21), 2600–2613.

Dobyns, B.M., Sheline, G.E., Workman, J.B., Tompkins, E.A., McConahey, W.M., Becker, D.V., 1974. Malignant and benign neoplasms of the thyroid in patients treated for hyperthyroidism: a report of the cooperative thyrotoxicosis therapy follow-up study. J. Clin. Endocrinol. Metab. 38, 976–998.

Doll, R., Wakeford, R., 1997. Risk of childhood cancer from fetal irradiation. Br. J. Radiol. 70, 130–139.

Doss, M., Egleston, B.L., Litwin, S., 2012. Comments on "studies of the mortality of atomic bomb survivors, Report 14, 1950-2003: an overview of cancer and noncancer diseases. Radiat. Res. 177, 229–243.

Doss, M., 2013. Linear no-theshold model vs. radiation hormesis. Dose Response J. 11, 480–497.

Eidemuller, M., Holmberg, E., Jacob, P., Lundell, M., Karlsson, P., 2015. Breast cancer risk and possible mechanisms of radiation-induced genomic instability in the Swedish hemangioma cohort after reanalyzed dosimetry. Mutat. Res. 775, 1–9. https://doi.org/10.1016/j.mrfmmm.2015.03.002.

Evans, R.D., Keane, A.T., Shanahan, M.M., 1972. Radiogenic effects in man of long-term skeletal alpha-irradiation. In: Stover, B.J., Jee, W.S.S. (Eds.), Radiobiology of Plutonium. JW Press, Salt Lake City, UT, pp. 431–468.

Glasstone, S., Dolan, P.J., 1977. The Effects of Nuclear Weapons. U.S. Department of Defense and U.S. Department of Energy.

Goldwasser, E., 1996. Erythropoietin: a somewhat personal history. Perspect. Biol. Med. 40, 18–32.

Grant, E.J., Furukawa, K., Sakata, R., Sugiyama, H., Sadakane, A., et al., 2015. Risk of death among children of atomic bomb survivors after 62 years of follow-up: a cohort study. Lancet Oncol. 16 (13), 1316–1323.

Grant, E.J., Brenner, A., Sugiyama, H., Sakata, R., Sadakane, A., et al., 2017. Solid cancer incidence among the life span study of atomic bomb survivors: 1958-2009. Radiat. Res. 187)5), 513–527.

Gray, L.H., 1965. Radiation biology and cancer. Published for the University of Texas, M.D. Anderson Hospital and Tumor Institute. In: Cellular Radiation Biology: A Symposium Considering Radiation Effects in the Cell and Possible Implications for Cancer Therapy: A Collection of Papers. Lippincott, Williams and Wilkins, Philadelphia, pp. 8–25.

Henshaw, P.S., 1941. Biologic significance of the tolerance dose in x-ray and radium protection. J. Natl. Cancer Inst. 1, 789–805.

Hewlett, R.G., Anderson, O.E., 1962. The New World, 1939-1946, Volume 1. A History of the United States Atomic Energy Commission. The Pennsylvania State University Press, University Park, PA, USA and London.

Hewlett, R.G., Duncan, F., 1969. Atomic Shield, 1947/1952. Volume II, A History of the United States Atomic Energy Commission. The Pennsylvania State University Press, University Park, PA, USA and London.

Holm, L.M., Dahlquist, I., Israelsson, A., Lundell, G., 1980. Malignant thyroid tumors after 131I therapy: a retrospective cohort study. N. Engl. J. Med. 303 (4), 188–191.

Holm, L.M., Wiklund, K.E., Lundell, G.E., et al., 1989. Cancer risk in populations examined with diagnostic doses of ^{131}I. J. Natl. Cancer Inst. 81 (4), 302–306.

Hsieh, W.H., Lin, L.F., Ho, J.C., Chang, P.W., 2017. 30 years follow-up and increased risks of breast cancer and leukaemia after long-term low-dose-rate radiation exposure. Br. J. Cancer 117 (12), 1883–1887.

IAEC, 1998. Diagnosis and Treatment of Radiation Injuries. Safety Report Series #2. IAEA, Vienna.

ICRP (International Commission on Radiation Protection), 2007. The 2007 Recommendations of the International Commission on Radiological Protection. ICRP Publication 103. [Not in text].

Kashcheev, V.V., Chekin, S.y., Maksioutov, M.A., Tumanov, K.A., Kochergina, E.V., et al., 2015. Incidence and mortality of solid cancer among emergency workers of the Chernobyl accident: assessment of radiation risks for the follow-up period of 199–2009. Radiat. Environ. Biophys. 54 (1), 13–23.

Kathren, R.L., Baalman, R.W., Bair, W.J., 1986. Herbert M. Parker: Publications and Other Contributions to Radiological and Health Physics. Battelle Press, Columbus, OH and Richland, WA.

Kendall, G.M., Little, M.P., Wakeford, R., Bunch, K.J., Miles, J.C.H., et al., 2013. A record-based case-control study of natural background radiation and the incidence of childhood leukaemia and other cancers in Great Britain during 1980-2006. Leukemia 27 (1), 3–9.

Little, J., 1999. Ionizing radiation. In: Epidemiology of Childhood Cancer. IARC Scientific Publication 149 (International Agency for Research on Cancer, Lyon, France, pp. 90–147.

Little, M.P., Boice Jr., J.D., 2003. Analysis of breast cancer in the Massachusetts TB fluoroscopy cohort and in the Japanese A-bomb survivors, taking account of dosimetry error and curvature in the A-bomb dose response: absence of evidence of reduction in risk following fractionated irradiation. Int. J. Low Radiat. 11 (1), 88–101.

Little, M., Heidenreich, W., Moolgavkar, S.H., Schoelinbergr, H., Thomas, D.C., 2008. Sysems biological and mechanistic modelling of radiation-induced cancer. Radiat. Environ. Biophys. 47, 39–47.

Liu, J.J., Freedman, D.M., Little, M.P., Doody, M.M., Alexander, B.H., et al., 2014. Work history and mortality risks in 90,268 US radiological technologists. Occup. Environ. Med. 71 (12), 819–835.

Lubin, J.H., Adams, M.J., Shore, R., Holmberg, E., Schneider, A.B., Hawkins, M.M., Robison, L.L., Inskip, P.D., Lundell, M., Johannson, R., Kleinerman, R.A., Vathaire, D.E., Damaber, L., Sadetzki, S., Tucker, M., Sakata, R., Veiga, L.H.S., 2017. Thyroid cancer following childhood low dose radiation exposure: a pooled analysis of nine cohorts. J. Clin. Endocrinol. Metab. 102 (7), 2575–2583.

Lyon, J.L., Aldler, S.C., Stone, M.B., Scholl, A., Reading, J.C., et al., 2006. Thyroid disease associated with exposure to the Nevada Nuclear Weapons Test Site radiation: a reevaluation based on corrected dosimetry and examination data. Epidemiology 17 (6), 604–614.

Matanowski, G.M., Seltser, G., Sartwell, P.E., Diamond, E.L., Elliot, E.A., 1975. The current mortality rates of radiologists and other physician specialists: specific causes of death. Am. J. Epidemiol. 101, 199–210.

McClellan, R.O., Clarke, W.J., McKenney, J.R., Bustad, L.K., 1962. Preliminary observations on the biological effects of Sr-90 in miniature pigs. Am. J. Vet. Res. 23, 910–912.

McClellan, R.O., Bustad, L.K., 1963. Strontium-990 and calcium in milk of miniature igs. Int. J. Radiat. Biol. 6, 173–180.

McClellan, R.O., Casey, H.W., Hahn, F.F., 1963a. Erythrokinetics of miniature pigs ingesting Sr-90. In: Hanford Atomic Products Operation, Biology Division Annual Report, HW-76000, pp. 89–92.

McClellan, R.O., 1963. Erythropoietin in miniature swine ingesting Sr-90. In: Hanford Atomic Products Operation Biology Laboratory Annual Report for 1962, HW-76000, vol. 86, pp. 93–94.

McClellan, R.O., Kerr, M.E., Bustad, L.K., 1963b. Reproductive performance of miniature pigs ingesting Sr-90 daily. Nature 197, 670–671.

McClellan, R.O., Bustad, L.K., 1964. Toxicity of significant radionuclides in large animals. Ann. N.Y. Acad. Sci. 111 (2), 793–811.

McClellan, R.O., Vogt, G.S., Kane, R.E., Hahn, F.F., 1964. Endotoxin-induced neutrophil response in miniature pigs ingesting Sr-90 daily. Nature 201, 721–724.

McClellan, R.O., 1966. Hematopoietic tissue neoplasms in animals administered Sr-90. Health Phys. 12, 1362–1365.

McClellan, R.O., 1986. Twenty-five years of Lovelace research in inhalation toxicology. N. M. J. Sci. 26, 330–345.

McClellan, R.O., Boecker, B.B., Hahn, F.F., Muggenburg, B.A., 1986. Lovelace ITRI studies on the toxicity of inhaled radionuclides in Beagle dogs. In: Thompson, R.C., Mahaffe, J.A. (Eds.), Lifespan Radiation Effects Studies in Animals: What Can They Tell Us? Office of Scientific and Technical Information, U.S. Department of Energy, Washington, DC, pp. 74–96.

McClellan, R.O., 1995. Hanford animal studies of radioiodine. Radiat. Protect. Dosim. 60, 295–305.

McClellan, R.O., 2014. Radiation toxicity (Chapter 18). In: Wall ace Hayes, A., Kruger, C.L. (Eds.), Principles and Methods of Toxicology, sixth ed. Taylor and Francis, pp. 883–955.

McClellan, R.O., 2017. A tribute to John Doull, BS, PhD, MD scientist, physician, educator, communicator, advisor and above all, a gentleman and friend to many. Int. J. Toxicol. 36(4)) (277–286), 2017.

McClellan, R.O., 2018. Concepts in veterinary toxicology. In: Gupta, R.C. (Ed.), Veterinary Toxicology: Basic and Clinical Principles. Elsevier/Academic Press, London, UK.

Mettler Jr., F.A., Upton, A.C., 2008. Medical Effects of Ionizing Radiation, third ed. Elsevier, , Philadelphia.

Miyake, T., Kung, C.K., Goldwasser, E., 1977. Purification of human erythropoietin. J. Biol. Chem. 2523, 5558–5574.

Moolgavkar, S., 1977. The multi-stage theory of carcinogenesis. Int. J. Cancer 18, 730.

Moolgavkar, S., 1978. The multistage theory of carcinogenesis and the age distribution of cancer in man. JNCI 61, 49–52.

Moolgavkar, S.H., Knudson, A.G., 1981. Mutation and cancer: a model for human carcinogenesis. JNCI 66, 1037–1052.

Moolgavkar, S.H., 1986. Carcinogenesis modelling: from molecular biology to epidemiology. Annu. Rev. Publ. Health 7, 151–170.

Moolgavkar, S.H., Lubeck, G., 1990. Two-event model for carcinogenesis: biological, mathematical and statistical considerations. Risk Anal. 10, 323–341.

Moolgavkar, S.H., 1994. Biological models of carcinogenesis and quantitative cancer risk assessment: Guest Editorial. Risk Anal. 14, 879–882.

Moolgavkar, S.H., Krewski, D., Schwarz, M., 1999. Mechanisms of carcinogenesis and biologically-based models for quantitative estimation and prediction of cancer risk. In: Moolgavkar, S.H., Krewski, D., Zeise, L., Cardis, E., Moller, H. (Eds.), Quantitative Estimation and Prediciton of Cancer Risk. IARC Scientific Publications, pp. 179–238.

Moolgavkar, S.H., 2004. Fifty years of the multi-stage model: remarks on a landmark paper. Int. J. Epidemiol. 33, 1182–1183.

Moolgavkar, S.H., 2016. Commentary: Multistage carcinogenesis and epidemiological studies of cancer. Int. J. Epidemiol. 45, 645–649.

Moore, K., 2016. The Radium Girls: The Dark Story of American's Shining Women. Source Books, Naperville, IL.

Nair, R.R.K., Rajan, B., Akiba, S., Jayalekshmi, P., Nair, M.K., et al., 2009. Background radiation and cancer incidence in Kerala, India – Karunagappally cohort study. Health Phys. 98 (1), 55–66.

NCRP (National Council for Radiation Protection, 1993. Recommendations for Limits for Exposure to Ionizing Radiation. National Council on Radiation Protection Report No. 116.

NCRP (National Council for Radiation Protection, 2018. Applications of Recent Epidemiologic Studies for the Linear-Nonthreshold Model and Radiation Protection. NCRP Commentary No. 27. National Council on Radiation Protection and Measurements, Bethesda, MD, pp. 1–191.

Neel, J.V., Schull, J., Awa, A.A., et al., 1990. The children of parents exposed to atomic bombs: estimates of genetic doubling dose of radiation for humans. Am. J. Hum. Genet. 46, 1053–1072.

NRC (National Research Council). BEIR VII, 2006. Health Risks from Exposure to Low Levels of Ionizing Radiation, Phase 2. National Academies Press, Washington, DC.

Olsen, J.H., Andersson, M., Boice Jr., J.D., 1990. Thorotrast exposure and cancer risk. Health Phys. 58 (2), 222–233.

Ozasa, K., Shimizu, Y., Suyama, A., et al., 2012. Studies of the mortality of atomic bomb survivors, report 14, 1950-2003: an overview of cancer and noncancer diseases. Radiat. Res. 177, 229–243.

Ozasa, K., Shimizu, Y., Suyama, A., Kasagi, F., Soda, M., Grant, E.J., Sakata, R., Sugiyama, H., Kodama, K., 2013. Errata: studies of the mortality of atomic bomb survivors, report 14, 1950-2003: an overview of cancer and noncancer diseases. Radiat. Res. 179 (4), e40–e41.

Paustenbach, D.J., 2001. The practice of exposure assessment. In: Hayes, A.W. (Ed.), Principles and Methods of Toxicology. Taylor and Francis, Philadelphia, PA, pp. 387–448.

Pearce, M.S., Salotti, J.A., Little, M.P., McHugh, K., Lee, C., et al., 2012. Radiation exposure from CT scans in childhood and subsequent risk of leukaemia and brain tumours: a retrospective cohort study. Lancet 380 (9840), 499–505.

Pellmar, T.C., Rockwell, S., 2005. Priority list of research areas for radiological and nuclear threat countermeasures. Radiat. Res. 163, 115–123.

Preston, D.L., Cullings, H., Suyama, A., Funamoto, S., Nishi, N., et al., 2008. Solid cancer incidence in atomic bomb survivors exposed *in utero* or as young children. J. Natl. Cancer Inst. 100 (6), 428–436.

Preston, D.L., Kitahara, C.M., Freedman, D.M., Sigurdson, A.J., Simon, S.L., et al., 2016. Breast cancer risk and protracted low-to-moderate dose occupational radiation exposure in the US radiologic technologists cohort, 1983-2008. Br. J. Cancer 115 (9), 1105–1112.

Puck, T.T., Markus, P.I., 1956. Action of x-rays on mammalian cells. J. Exp. Med. 103, 653–666.

Ray, K., Hudak, K., Citrin, D., Stik, M., 2014. Biomarkers of radiation injury and response. In: Gupta, R. (Ed.), Biomarkers in Toxicology.

Rhodes, R., 1995. Dark Sun: The Making of the Hydrogen Bomb. Touchstone, New York, NY.

Richardson, D.B., Cardis, E., Daniels, R.D., Gillies, M., O'Hagan, J.A., et al., 2015. Risk of cancer from occupational exposure to ionizing radiation: retrospective cohort study of workers in France, the United Kingdom and the United States (INWORKS). Br. Med. J. 351, h5359. https://doi.org/10.1136/bmj.h5359.

Ron, E., Modan, B., Preston, D., Alfandary, E., Stovall, M., Boice Jr., J.D., 1989. Thyroid neoplasia following low-dose radiation in childhood. Radiat. Res. 120 (3), 516–531.

Ron, E., Doody, M.M., Becker, D.V., et al., 1998. Cancer mortality following treatment for adult hyperthyroidism: cooperative thyrotoxicosis therapy follow-up study group. J. Am. Med. Assoc. 280 (4), 347–355.

Rowland, R.E., 1994. Radium in Humans: A Review of U. Studies. Argonne National Laboratory, Argonne, IL. ANL/ER-3.

Satoh, C., Takahashi, N., Asakawa, J., Kodaire, M., Kuick, R., Hanash, S.M., Neel, J.V., 1996. Genetic analysis of children of atomic bomb survivors. Environ. Health Perspect. 104 (Suppl. 3), 511–519.

Schonfeld, S.J., Krestinina, L.Y., Epifanova, S., Degteva, M.O., Akleyev, A.V., Preston, D.L., 2013. Solid cancer mortality in the Techa River cohort (1950-2007). Radiat. Res. 179 (2), 183–189.

Schull, W.J., 1990. Song Among the Ruins. Harvard University Press, Cambridge, MA, USA and London.

Shore, R.E., Hildreth, N., Dvoretsky, P., Andresen, E., Moseson, M., Pasternack, B., 1993. Thyroid cancer among persons given x-ray treatment in infancy for an enlarged thymus gland. Am. J. Epidemiol. 137 (10), 1068–1080.

Shore, R.E., Moseson, M., Harley, N., Pasternack, B.S., 2003. Tumors and other diseases following childhood x-ray treatment for ringworm of the scalp (Tinea Capitis). Health Phys. 85 (4), 404–408.

Shore, R.E., 2013. Radiation impacts on human health: certain fuzzy and unknown. In: Presentation to the Annual Meeting of the NCRP, Bethesda, MD, March 11, 2013. Health Phys.

Smith, P.G., Doll, R., 1981. Mortality from cancer and all causes among British radiologists. Br. J. Radiol. 54 (639), 187–194.

Simon, S.L., Preston, D.L., Linet, M.S., Miller, J.S., Sigurdson, A.J., et al., 2014. Radiation organ doses received in a nationwide cohort of U.S. radiologic technologists: Methods and findings. Radiat. Res. 188 (5), 507–528.

Sokolnikov, M.E., Gilbert, E.S., Preston, D.L., et al., 2008. Lung, liver and cancer mortality in Mayak workers. Int. J. Cancer 123, 905–911.

Sokolnikov, M.E., Preston, D., Gilbert, E., Schonfeld, S., Koshurnikova, N., 2015. Radiation effects on mortality from solid cancers other than lung, liver, and bone cancer in the Mayak worker cohort: 1948-2008. PLoS One 10 (2), e0116684. https://doi.org/10.11371/journal.pone.0117784.

Stannard, J.N., Baalman, R.W., 1988. Radioactivity and Health: A History. Office of Scientific and Technical Information, Springfield, VA.

Sun, Z., Inskip, P.D., Wang, J., Kwon, D., Zhao, Y., Zhang, L., Wang, Q., Fan, S., 2016. Solid cancer incidence among Chinese medical diagnostic x-ray works, 1950-1995: estimation of radiation-related risks. Int. J. Cancer 138 (12), 2875–2883.

Szasa, F.M., 1984. The Day the Sun Rose Twice. University of New Mexico Press, Albuquerque, NM.

Tao, Z., Akiba, S., Zha, Y., Sun, Q., Zou, J., et al., 2012. Cancer and non-cancer mortality among inhabitants in the high background radiation area of Yangjiang, China (1979-1998). Health Phys. 102 (2), 173–181.

Taylor, D.M. (Ed.), 1989. Risks from Radium and Thorotrast. British Inst. of Radiology, London, UK.

Till, J.E., Grogan, H.A. (Eds.), 2008. Radiological Risk Assessment and Environmental Analysis. Oxford University Press, New York.

USAEC (U.S. Atomic Energy Commission), 1957. Theoretical Possibilities and Consequences of Major Accidents in Large Nuclear Power Plants. WASH-740. Prepared by Brookhaven National Laboratory.

Voelz, G.L., Lawrence, J.N.P., Johnson, E.R., 1997. Fifty years of plutonium exposure to the Manhattan Project workers: an update. Health Phys. 73, 611–619.

Wakeford, H., 2008. Childhood leukaemia following medical diagnostic exposure to ionizing radiation *in utero* or after birth. Radiat. Protect. Dosim. 132 (2), 166–174.

Young, R., Bennett, B. (Eds.), 2006. A Revised System for Atomic Bomb Survivor Dose Estimates. Radiation Effects Research Foundation, Hiroshima.

Zablotska, L.B., Ron, E., Rozhko, A.V., Hatch, M., Polyanskaya, O.N., et al., 2011. Thyroid cancer risk in Belarus among children and adolescents exposed to radioiodine after the Chernobyl accident. Br. J. Cancer 104 (1), 181–187.

Zablotska, L.B., Lane, R.S.D., Thompson, P.A., 2014. A reanalysis of cancer mortality in Canadian nuclear workers (1956-1994) based on revised exposure and cohort data. Br. J. Cancer 110 (1), 214–223.

Zhivin, S., Guseva Canu, I., Samson, E., Laurent, O., Grellier, J., Collomb, P., Zablotska, L.B., Laurier, D., 2016. Mortality (1968-2008) in a French cohort of uranium enrichment workers potentially exposed to rapidly soluble uranium compounds. Occup. Environ. Med. 73 (3), 167–174.

PART VII

CARCINOGENS BIOMONITORING AND CANCER BIOMARKERS

CHAPTER

43

Biomonitoring Exposures to Carcinogens

Sofia Pavanello, Marcello Lotti

Dipartimento di Scienze Cardio-Toraco-Vascolari e Sanità Pubblica, Università degli Studi Padova, Padova, Italy

INTRODUCTION

A wide array of analytical tools exists to assess biological parameters, referred to as biomarkers, which are used in clinical medicine as indicators of normal or pathological processes and pharmacological responses (Biomarkers Definitions Working Group, 2001). However, in environmental and occupational health the term biomarker is used somehow differently, indicating signaling events in biological systems or samples (NRC, 1987) that are aimed at preventing adverse effects of exposures (Angerer et al., 2007). A procedure known as human biomonitoring measures the size of human exposures to chemicals (Angerer et al., 2011), their biochemical effects (Mateuca et al., 2012), and individual susceptibility (Christiani et al., 2008). Data obtained from this procedure are used for a variety of purposes including risk assessment, the setting and the compliance to occupational and environmental limits of exposure, and in molecular epidemiology.

Biomonitoring of carcinogens, however, requires a unique conceptual framework because cancer is thought to be the long-term result of multiple causal interactions between environmental and genetic/epigenetic factors (Thomas, 2010). Factors involved in cancer causality include both external (pollutants, diet, drugs) and internal (hormones, human, and microbial metabolites) environments, collectively known as the exposome that vary both quantitatively and qualitatively throughout the lifespan (Rappaport and Smith, 2010). Exposome research was thought as a complement to the genome sequencing, but because of its complexity it is at an early stage of development (Dennis et al., 2017). Several genetic/epigenetic players are also involved in cancer initiation and development, and different pathways and sequence of events within variable time frames may be activated even in the same type of cancer (de Vogel et al., 2009). Mutations may occur on different components of a given signaling pathway, each mutation associated with a different cancer (Jakowlew, 2006), and players may even change their role as it occurs when a tumor suppressor gene switches into an oncogene (Abbas and Dutta, 2009) and when autophagy favors or hinders the carcinogenic process (Choi et al., 2013). Consequently, every individual tumor even of the same histopathological subtype as another tumor is distinct with respect to its genetic alterations and development, though the pathways affected in different tumors may be similar (Vogelstein et al., 2013), and the mutation pattern of individual cells may vary even within the same tumor (Gerlinger et al., 2012).

The variety and variability of environmental and genetic/epigenetic factors in the development of cancer may explain the large interindividual differences in susceptibility to the disease, but they also hamper the understanding of the significance of biomarkers used to monitor carcinogen exposures (Pavanello and Lotti, 2012). In fact none of these biomarkers, used either alone or in combination, are able to capture the complex dynamics of both exposures and carcinogenesis (Lahiri, 2011). This chapter summarizes biomarkers currently used for carcinogen biomonitoring and highlights their values and limitations, restricting the examples of exposures to those chemicals classified in groups 1 and 2A by the International Agency for Research on Cancer (IARC, 2011).

ASSESSING EXPOSURES

Chemicals and Metabolites

Current exposures to carcinogens, mainly at work places, are estimated by measuring either the chemical itself or its metabolite(s) in body fluids (Tan et al., 2012). Examples are many folds. Metals such as arsenic, cadmium, lead, and nickel are measured in urine and blood (Becker

Biomarkers in Toxicology, Second Edition
https://doi.org/10.1016/B978-0-12-814655-2.00043-8

et al., 2013; Wilhelm et al., 2004), the hydroxylated metabolites of pyrene, phenanthrene, and naphthalene (Campo et al., 2010; Pesch et al., 2011), and their hydroxyl-nitrated analog such as hydroxy-1-nitropyrene and hydroxy-N-acetyl-1-aminopyrene in urine (Miller-Schulze et al., 2013), dioxin and nondioxin-like polychlorinated biphenyl in plasma (Schettgen et al., 2012) and mother milk (Mannetje et al., 2012) lipids, and benzene in blood, urine, and expired air with its metabolites S-phenylmercapturic and t,t-muconic acids in urine (Egeghy et al., 2003; Weisel, 2010). A broad spectrum of carcinogenic aromatic amines has also been detected in urine samples such as *o*-toluidine (Kütting et al., 2009; Richter, 2015) and 4-aminobiphenyl (Ward et al., 1996; Riedel et al., 2006). However, these tests have little or no value in assessing the risk of developing cancer. When they have been used for that purpose, as for instance to assess lung cancer risk in smokers and health risk in benzene exposure, results appear either somehow obvious and redundant or difficult to be interpreted. For instance a prospective nested case-control study shows that urinary levels of 4-methylnitrosamino-1-(3-pyridyl)-1-butanol, a metabolite of the tobacco-specific compound 4-methylnitrosamino-1-(3-pyridyl)-1-butanone, and cotinine a metabolite of nicotine were both associated with risk of developing lung cancer in a dose-dependent manner (Yuan et al., 2009). However, individuals of the cohort that developed lung cancer had higher numbers of pack-years of smoking than those remaining cancer free, suggesting that the self-reported smoking history may be enough to assess cancer risk. Even when specific biomarkers of exposure and large amounts of data are available, such as in the case of benzene, the interpretation of biomonitoring data of exposure for human risk assessment presents several problems, as recently outlined (Arnold et al., 2013). These include issues associated with the short half-life of this chemical and gaps in knowledge regarding the relationship between the biomarker and subsequent toxic effects.

Undoubtedly, the ability to detect hazardous substances and their metabolites in body fluids now seems to far exceed our understanding of their biological significance. In this respect the relative role of these biomarkers, as compared with other information, has been defined within a framework for biomonitoring guidance values aimed at assessing risk (Bevan et al., 2012). At present, the body fluid concentrations of chemicals and metabolites may be useful to assess the hygienic conditions and practices at work place and also to comply with reference values and limits of exposure, when available. However, even for these purposes, caution should be exercised because gender differences may blur the picture as shown by urinary t,t-transmuconic acid in female workers, which was higher than that of male workers exposed to benzene at comparable levels of exposure (Moro et al., 2017).

DNA and Protein Adducts

Adduct formation is the result of the covalent binding between reactive electrophilic substances and the nucleophilic sites of DNA and proteins. The preferred sites for adduct formation are guanine with five nucleophilic sites on DNA and the sulfhydryl group of cystein, the nitrogen of histidin, and the N-terminal valine on proteins. DNA adducts, such as those detected with ^{32}P-postlabeling and immunoassay methods, have been measured in lymphocytes of workers with different exposures. However, these adducts are nonspecific because the responsible electrophilic substance cannot be identified. These studies include workers exposed to polycyclic aromatic hydrocarbons (PAHs) in iron foundries (Santella et al., 1993), coke oven plants (Brescia et al., 1999), aluminum plants (van Schooten et al., 1995), garages, car repair shops (Nielsen et al., 1996a), rubber industries (Peters et al., 2008), shipyards (Lee et al., 2003), and incineration facilities (Lee et al., 2002). In this way other exposures to PAHs have been assessed in bus drivers (Nielsen et al., 1996b; Hemminki et al., 1994), asphalt workers (McClean et al., 2007), policemen (Binkova et al., 2007), US army soldiers (Poirier et al., 1998), and dietary exposures as well (Kang et al., 1995).

Further examples of these nonspecific DNA adducts include those measured in subjects occupationally exposed to diesel exhaust (Scheepers et al., 2002), epichlorohydrin (Plna et al., 2000), and benzidine and benzidine-based dyes (Zhou et al., 1997). Another adduct, which is nonspecific because it is a biomarker of endogenous DNA oxidation, is the 8-hydroxy guanidine (Pilger and Rüdiger, 2006) that was measured in individuals exposed to PAHs (Marczynski et al., 2009; Chen et al., 2011; Huang et al., 2012), chromium (Zhang et al., 2011), polychlorinated dibenzodioxins, polychlorinated dibenzofuranes, polychlorinated biphenyls (Wen et al., 2008), and benzene (Yoshioka et al., 2008) and in smokers (Manini et al., 2010).

Other studies adopted chemical-specific methods such as mass spectrometry detection and include measurements of alkyl-DNA adducts such as N7, N2-alkylguanines (Boysen et al., 2009), O-6-methylguanine (Reh et al., 2000), N7-(2-hydroxyethyl)guanine (Huang et al., 2011), and etheno-adducts such as 1,N6-ethenoadenine, 3,N4-ethenocytosine, 1,N2–ethenoguanine, and N2,3-ethenoguanine (Bartsch and Nair, 2000). Examples of chemical-specific DNA adducts that have been measured in lymphocyte DNA and urine after exposures to benzo(a)pyrene and aflatoxin are reported in Table 43.1.

The major advantage of measuring chemical-specific DNA adducts is to offer an assessment of the dose of carcinogens close to the molecular target. However, it is unclear how accurate this assessment would be because rate and speed of repair of the various DNA adducts are unknown. In addition, because DNA adducts are measured in peripheral blood leukocytes they may not represent those formed at the DNA of target organs.

Adducts to hemoglobin and serum albumin have been measured after exposure to a variety of electrophilic carcinogens and have been recently reviewed (Sabbioni, 2017; Sabbioni and Turesky, 2017). Table 43.2 shows examples of these adducts that have been detected using chemical-specific methods. However, these also include nonspecific adducts because they are formed endogenously such as alkyl-adducts cleaved at the N-terminal valine of hemoglobin and chemical-specific adducts, both quantified by mass spectrometry. Advantages of measuring hemoglobin and serum albumin adducts include the amount of protein present in blood, which is larger than that of

TABLE 43.1 Examples of Specific DNA Adducts

Chemical	Adduct	Matrix	References
Benzo[*a*] pyrene	Anti-BPDE-DNA[a]	White blood cells	Rojas et al. (1995), Pavanello et al. (2010)
Benzo[*a*] pyrene	BP-6-*N*7Gua[b] and BP-6-*N*7Ade[c]	Urine	Casale et al. (2001), Chen et al. (2005)
Aflatoxin	AFB-*N*7-Gua[d]	Urine	Egner et al. (2006)

[a]*B[a]P-tetrol-I-1 (r-7,c-10,t-8,t-9-tetrahydroxy-7,8,9,10-tetrahydro-benzo[a] pyrene) released after acid hydrolysis of DNA.*
[b]*7-(Benzo[a]pyren-6-yl)guanine.*
[c]*7-(Benzo[a]pyren-6-yl)adenine.*
[d]*Aflatoxin-N7-guanine.*

TABLE 43.2 Examples of Chemical Specific Adducts on Hemoglobin and Albumin

Chemical	Hemoglobin-Adduct[a]	References
Ethylene, -oxide	Hydroxyethylvaline	Bailey et al. (1988), Angerer et al. (1998)
Butadiene, -oxide	N-(2-hydroxy-3-butenyl) valine	Begemann et al. (2001)
Formaldehyde	N-methylenvaline	Bono et al. (2012)
Acrylamide	N-2-carbamoylethylvaline	Bergmark (1997); Schettgen et al. (2002)
Glycidol	N-(2,3-dihydroxy-propyl)valine	Honda et al. (2011)
	Hemoglobin-Adduct[b]	
Benzene	Cysteinyl-benzene-oxide; cysteinyl-1,4-benzoquinone	Rappaport et al. (2005)
2-Nitrotoluene	Toluenamide-cysteine	Jones et al. (2005)
o-, p-, m-toluidine	o-, p-, m-toluidine-cysteine	Bryant et al. (1988), Riffelmann et al. (1995), Ward et al. (1996), Richter (2015)
2-Naphthylamine	2-Naphthylamide-cysteine	Bryant et al. (1988), Sabbioni (2017)
4-Aminobiphenyl	4-Aminobiphenylamide-cysteine	Bryant et al. (1988), Richter et al. (2001), Sabbioni (2017)
1-Nitro-pyrene	1-Amino-pyrene-cysteine	Zwirner-Baier and Neumann (1999), Sabbioni (2017)
4-(Methylnitrosamino)-1-(3-pyridyl)-1-butanone	4-(Methylnitrosamino)-1-(3-pyridyl)-1-butanol-cysteine and -hystidine	Hoffmann and Hecht (1985), Myers and Ali (2008)
Benzo[a]pyrene	7,8,9-Trihydroxy-r-7,f-8,-9,c-10-tetrahydrobenzo [a]pyren-10-yl-aspartic and -glutamic ester	Skipper et al. (1989)
	Albumin-Adduct[b]	
Aflatoxin-1	Aflatoxin-1-dialdehyde-lysine	Guengerich et al. (2002), McCoy et al. (2008)
Benzo[a]pyrene	7,8,9-Trihydroxy-r-7,t-8,t-9,c-10-tetrahydrobenzo [a]pyren-10-yl-histidine	Day et al. (1991)
Benzene	Cysteinyl-benzene-oxide and cysteinyl-1,4-benzoquinone	Rappaport et al. (2002), Lin et al. (2007)
4-Aminodiphenyl	Aminodiphenylamide-cysteine	Thier et al. (2001)

[a]*Adduct released after N-alkyl Edman reaction.*
[b]*Chemical is released after mild alkaline or acidic hydrolysis.*

nuclear DNA, and their stability as opposed to DNA repair capabilities. In the case of hemoglobin adducts, they represent a measure of short-term cumulative exposure, given the long life span of red blood cells (Tornqvist et al., 2002). Nevertheless, it should be made clear that these proteins do not represent the target of carcinogenic substances.

Finally, as for the concentrations of chemicals and their metabolites in body fluids, DNA and protein adducts represent snapshot measurements that are not representative of doses of carcinogens that may cause cancer after prolonged exposures. However, attempts have been recently made to extrapolate lifetime risks from spot biomarkers data sets using a statistical approach (Pleil and Sobus, 2013).

ASSESSING EFFECTS

Chromosomal Aberrations and Sister Chromatid Exchanges

Cytogenetic alterations in cultured peripheral blood lymphocytes, such as chromosomal aberrations (CAs) and sister chromatid exchanges (SCEs), have been used as biomarkers of exposure to and early effects of carcinogens. CAs are structural aberrations including chromosome-type and chromatid-type breaks and rearrangements, usually analyzed by light microscopy from Giemsa-stained metaphase specimens or by fluorescence microscopy using in situ hybridization (Albertini et al., 2000; Norppa, 2004). SCEs are interchanges of DNA replication products between sister chromatids thought to represent homologous recombination repair of DNA double strand breaks (Johnson and Jasin, 2000). SCEs are analyzed by fluorescence microscopy from second division metaphases after bromodeoxyuridine staining (Albertini et al., 2000; Norppa, 2004). Several studies reported increased frequencies of SCE (Miner et al., 1983; Bender et al., 1988; Kalina et al., 1998) and CAs (Bender et al., 1988; Reuterwall et al., 1991; Forni et al., 1996; Kalina et al., 1998) in workers exposed to PAHs. Few studies, where internal exposure to PAHs was quantified, reported rather small increases and high variability of SCE frequencies (Buchet et al., 1995; Popp et al., 1997; Siwińska et al., 2004; van Delft et al., 2001; van Hummelen et al., 1993). In addition, the study of Siwińska et al. (2004) measured CA frequencies but no increases were observed.

Because an increased number of CAs is a marker of genomic instability (Fenech, 2006), an event occurring at early stages of carcinogenesis, a large international prospective study was undertaken. Results show that an increased frequency of CAs (Bonassi et al., 2008), but not of SCEs (Norppa et al., 2006), in peripheral lymphocytes predicts cancer risk and, in particular, the risk of stomach cancer was associated with increased frequency of ring chromosomes. However, the association was weak and independent from known exposures to carcinogens, such as smoking and occupation. Several reasons might account for this lack of association considering that several confounding factors such as age (Bolognesi et al., 1997), sex (Bonassi et al., 1995), and smoking (Wang et al., 2012a,b) might influence the frequencies of SCE and CA in peripheral blood lymphocytes.

Micronuclei

Micronuclei (MN) are small, extranuclear bodies identifiable by light microscopy arising in cells dividing from fragments (without centromere) or whole chromosomes/chromatids lagging behind in anaphase. MN can result from a large spectrum of mechanisms including misattachment of tubulin, defects in kinetochore proteins or assembly, late replication, epigenetic modifications of histones, nucleoplasmic bridge formation, and gene amplification (Fenech et al., 2011). They may be formed spontaneously as a mechanism to remove extra DNA (Shimizu et al., 2007) or during normal development from progenitor cells (Peterson et al., 2008). They may also result from exposure to clastogens or aneugens. MN frequencies have been analyzed in various cell types (peripheral lymphocytes, erythrocytes, and exfoliated epithelial cells) and used to monitor genetic damage in humans. The cytokinesis-block micronucleus test in cultured human lymphocytes is the preferred method for measuring MN (Fenech et al., 2011). This method can also be used to measure nucleoplasmic bridges, a biomarker of dicentric chromosomes, and nuclear buds, biomarkers of gene amplification. The simplicity of MN assay led to its exponential use worldwide and guidelines are available for its uses in human biomonitoring (Albertini et al., 2000; Fenech, 2007).

MN frequencies are higher in women than men and increase with age (Kirsch-Volders et al., 2011). Examples of the uses of this biomarker assessing PAH and vinyl chloride exposures are given in Table 43.3. Although almost all studies on vinyl chloride workers show increased MN formation (Bolognesi et al., 2017), even after different cumulative exposures, the results on PAHs show that at certain low levels of exposures they are not associated with increased MN frequencies.

Increased MN frequencies in lymphocytes were found to be predictive of cancer risk in one large prospective study (Bonassi et al., 2007). Although the association was weak and independent from exposures

TABLE 43.3 Examples of Increased Micronuclei in White Blood Cells of Workers Exposed to Polycyclic Aromatic Hydrocarbons and Vinyl Chloride

Exposure to Polycyclic Aromatic Hydrocarbons (Urinary 1-Hydroxypyrene µmol/mol Creatinine)	Fold Increase Versus Controls	References
0.39 ± 0.9 (mean ± SD)	No increase	van Hummelen et al. (1993)
1.5 ± 4.7 (mean ± SD)	No increase	Buchet et al. (1995)
1.44 (0.040–3.75) (median [range])	1.1	Brescia et al. (1999)
1.04 ± 0.67 (mean ± SD)	No increase[a]	van Delft et al. (2001)
9.0 (5.4–16.0) (median [interquartile range])	2.0	Siwińska et al. (2004)
12.0 (10.4–13.9) (geometric mean [95% confidence interval])	2.3	Leng et al. (2004)
5.7 (1.4–12.0) (median [interquartile range])	3	Liu et al. (2006)
3.08 (0.41–7.48) (median [range])	4[a]	Pavanello et al. (2010)
Exposure to Vinyl Chloride (Estimated Cumulative Environmental Exposure)		
10.3–301 992.0 mg (range)	4.0	Wang et al. (2010a)
8866.56 (16.78–301.992) mg (median [range])	3.7	Ji et al. (2010)
21.809 (576–301.992) mg (median [range])	4.0	Wang et al. (2010b)
32.2 (1.33–845.5) ppm-year (median [range])	3.0	Wen-Bin et al. (2009)

[a]*Nonsmokers only.*

to carcinogens, such as smoking and occupation, this study is worth to be mentioned because it is one of the few that provides a prospective assessment of cancer biomarkers.

Comet Assay

The comet assay is a single-cell gel electrophoresis method for measuring DNA strand breaks in cells. Cells embedded in agarose on a microscope slide are lysed with detergent and high salt to form nucleoids containing supercoiled loops of DNA linked to the nuclear matrix. Subsequent electrophoresis at high pH results in structures resembling comets, observed by fluorescence microscopy. The intensity of the comet tail relative to the head reflects the number of DNA breaks. The likely explanation for comet tail formation is that loops containing a break lose their supercoiling and become free to extend toward the anode (Tice et al., 2000; Collins, 2004). Besides single-strand DNA breaks, alkali labile sites (apurinic/apyrimidic sites), cross-links, and incomplete DNA repair sites in individual cells can be detected. This technique has been further modified to detect specific classes of DNA adducts, e.g., thymidine dimers, oxidative damage by using lesion-specific antibodies and specific DNA repair enzymes (Lorenzo et al., 2013). The comet assay has been used in molecular epidemiology as a biomarker of effects of occupational exposures, and Table 43.4 shows examples of studies on white blood cells of workers with different exposures to carcinogens. Results appear inconsistent. In addition, several studies have been published where the effects of occupational exposures in the production and application of pesticides have been evaluated (Bhalli et al., 2006; Bian et al., 2004; Garaj-Vrhovac and Zeljezic, 2002; Paz-y-Miño et al., 2004), even if the qualitative and quantitative assessments of exposure are often missing. The comet assay can be applied to a broad spectrum of cells such as white blood, buccal, nasal, sperm, and exfoliated cells from urinary tract and lachrymal ducts, although technical variables should be carefully controlled both within and across laboratories. Besides exogenous exposures such as smoking, diet, and drinking habits another confounding factor is the basal level of DNA damage that is influenced by endogenous exposures such as oxidative stress.

Specific Gene Mutations

Mutant forms of the proteins p53 and p21 encoded by p53 tumor suppressor gene and Ki-ras oncogene are detectable in blood serum by specific monoclonal antibodies and have been used to monitor exposures to carcinogens (De Vivo et al., 1994; Trivers et al., 1995). Table 43.5 shows examples of measurements of mutated p53 and p21-ki-ras proteins in serum of workers exposed to vinyl chloride as compared with controls and across different exposures.

Although mutated p53 was found in liver angiosarcomas that rose in workers exposed to vinyl chloride (Hollstein et al., 1994), and 5 out of 15 workers affected by angiosarcoma were positive for serum-mutated p53, only 2 out of 77 exposed individuals were positive before the development of neoplasia. These results raise the question as to whether these are biomarkers of early effect or disease.

In general, the use of these biomarkers to monitor exposures to carcinogens underestimates the complexity of cancer development. Cancer cells acquire their

TABLE 43.4 Comet Assay: Examples of Tail Moment[a] Increases in White Blood Cells of Workers Exposed to Carcinogens

Chemical	Exposure	Fold Increase Versus Controls	References
Benzene	Urinary s-phenyl mercapturic acid: 5.10 (1.55–7.20) μg/g creatinine; median (range)	1.39[b]	Fracasso et al. (2010)
Butadiene	1.73 (0.024–23.0) mg/m^3; median (range)	No increase[b]	Srám et al. (1998)
Polycyclic aromatic hydrocarbons	Urinary 1-hydroxypyrene 1.04 ± 0.67 μmol/mol creatinine; mean ± SD	No increase[b]	van Delft et al. (2001)
	9.0 (0.1–35.0) μmol/mol creatinine; median (range)	3.12	Marczynski et al. (2002)
	9.0 (5.4–16.0) μmol/mol creatinine; median (interquartile range)	No increase[b]	Siwińska et al. (2004)
	12.0 (10.4–13.9) μmol/mol creatinine; geometric mean (95% confidence interval)	2.61	Leng et al. (2004)
	0.41 μmol/mol creatinine; geometric mean	1.03[b]	Cavallo et al. (2006)
Radiation	196 (0–1401) μsv; 1 month cumulative mean (range)	2.78	Wojewódzka et al. (1998)
	10 msv; 5 month cumulative mean	No increase[b]	Aka et al. (2004)
	82.3 (19.5–242.3) msv; 1 year cumulative mean (range)	1.23	Touil et al. (2002)
Piombo	Blood 39.63 ± 7.56 μg/100 mL; mean ± SD	1.51	Fracasso et al. (2002)
Chromium	Urinary 7.31 μg/g creatinine; mean	1.57	Gambelunghe et al. (2003)

[a]*Tail moment (TM) is defined as some measure of tail length multiplied by the fraction of DNA in the tail.*
[b]*Non smokers.*

characteristics using different strategies and players (Hanahan and Weinberg, 2011), and the time for a given gene alteration to occur varies during the development of the same cancer type (Guimaraes and Hainaut, 2002; Letouzé et al., 2017).

TABLE 43.5 Mutated p53 and p21-ki-ras in Serum of Workers Exposed to Vinyl Chloride

Marker	Fold Increase Versus Controls	Fold Increase in High Versus Low Exposed Workers (ppm years)	References
p53	2.4		Schindler et al. (2007)
p21-ki-ras	2.3		
p53		2.7 (>2000 vs. 100–1000)	Mocci and Nettuno (2006)
p53	3.6		John Luo et al. (2003)
p21-ki-ras	10		
p53		2.0 (>40 vs. <40)	Wong et al. (2002)
p53	6.4	1.7 (>5000 vs. <500)	Smith et al. (1998)
p21-ki-ras	49		De Vivo et al. (1994)

Telomere Shortening

Telomeres are DNA repeat sequences (TTAGGG) that, together with associated proteins, form a sheltering complex that caps chromosomal ends and protects their integrity (Chan and Blackburn, 2004). Because of limitation of DNA polymerase (telomerase), chromosomal stability is gradually lost as telomeres shorten with each round of cell division (Harley et al., 1990). Consequently, telomere length shortening is considered a cellular marker of biological aging (Mather et al., 2011).

Epidemiological retrospective (Broberg et al., 2005; Hou et al., 2009; Jang et al., 2008) and prospective (McGrath et al., 2007; Willeit et al., 2010) studies show that individuals with shorter telomeres in peripheral blood leukocytes have a higher risk of cancer.

Telomeres, as triple guanine–containing sequences, are highly sensitive to damage by oxidative stress (von Zglinicki, 2002), alkylation (Petersen et al., 1998), or ultraviolet irradiation (Oikawa et al., 2001). Telomere shortening have been associated with exposures to ionizing radiation (Ilyenko et al., 2011), alcohol (Pavanello et al., 2011), N-nitrosamines (Li et al., 2011a,b), PAHs (Pavanello et al., 2010), benzene (Hoxha et al., 2009), and polychlorinated biphenyls (Ziegler et al., 2017). Although the shortening can be further accelerated by internal stressors such as oxidative stress (von Zglinicki, 2002; Epel et al., 2004), its use is hampered for the assessment of occupational exposure effects.

Epigenetic Changes

The term "epigenetics" literally means "above the genetics," and it is generally used to refer to modification of gene expression without a change in the DNA sequence. Epigenetic changes include DNA methylation at the carbon-5 position of cytosine in CpG dinucleotides, changes of chromatin by posttranslational histone modifications, and miRNA expression (Pogribny and Rusyn, 2013). Epigenetic signatures can be transmitted through generations, may reflect the prenatal environmental exposure, and may persist when exposure ends.

A variety of environmental pollutant exposures is linked with epigenetic variations in blood leukocytes. These variations induced by human genotoxic carcinogens have been recently reviewed (Marrone et al., 2014; Chappell et al., 2016). Most studies assessed exposures to carcinogenic metals where, for instance, blood arsenic was associated with global hypermethylation (Pilsner et al., 2007), or hypomethylation at higher concentrations (Pilsner et al., 2011; Lambrou et al., 2012). Prenatal lead exposure was associated with increased methylation of long interspersed nuclear element-1 (LINE-1) and short interspersed repeated DNA elements (SINEs). The latter named Alu repeats represent a surrogate measure of global methylation (Pilsner et al., 2012).

Exposure to particulate matter rich in lead, cadmium, and chromium was associated with increased miR-222 and miR-21 expression (Bollati et al., 2010), whereas that rich in nickel and arsenic was positively correlated with both histone 3 lysine 4 dimethylation (H3K4me2) and histone 3 lysine 9 acetylation (H3K9ac) (Cantone et al., 2011). Occupational exposure to nickel was associated with increased histone 3 lysine 4 trimethylation (H3K4me3) and decreased histone 3 lysine 9 dimethylation (H3K9me2) (Arita et al., 2012).

Benzene exposures of gas-station attendants and police officers were associated with reduction of LINE-1 and Alu methylation. Benzene exposure was also associated with hypermethylation of *p15* gene and hypomethylation of the melanoma antigen family 1 (*MAGE-1*), a tumor-associated antigen in hematological malignancies (Bollati et al., 2007). PAH exposures were associated with hypermethylation of LINE1 and Alu (Pavanello et al., 2009) and *p53* hypomethylation (Pavanello et al., 2009; Alegría-Torres et al., 2013). An inverse linear relationship between plasma persistent organic pollutants and polychlorinated biphenyls and Alu methylation was observed in Inuit individuals that have the highest reported levels of persistent organic pollutants worldwide (Rusiecki et al., 2008). However, it is not clear how these associations between epigenetic modifications and exposures reflect cancer risk.

ASSESSING SUSCEPTIBILITY

The phenotypic variability of xenobiotic metabolism and DNA repair may confer susceptibility to cancer development. The genetic polymorphism of enzymes involved in these processes accounts for these variabilities. Thus, the polymorphisms of both phase I enzymes involved in bioactivation of carcinogens and phase II enzymes involved in detoxification modulate the size of dose at the target. The polymorphism of enzymes involved in DNA repair mechanisms influences the persistance of DNA damage, but the majority of associations between DNA repair gene variants and cancer risk have not been replicated (Vineis et al., 2009). Table 43.6 shows examples of phases I and II and of DNA repair polymorphic genes that have been used as biomarkers of susceptibility in biomonitoring studies.

These biomarkers of susceptibility have several limitations (Ketelslegers et al., 2008) when considering the multiplicity of metabolic pathways that take place concurrently at different rates. Therefore, by assessing one or fewer metabolic pathways the amount of ultimate carcinogen(s) can hardly be predicted. In addition, the risk conferred by single variants is quite small, as shown in genome-wide association studies (Manolio et al., 2009). Finally, genetic screening in occupational settings is ethically questionable considering that primary prevention should represent the option of choice (Christiani et al., 2008).

NEW SCENARIOS TO ASSESS EXPOSURES AND EFFECTS: THE OMICS APPROACH

Current research on biomarkers aimed at the identification of new ones is focused on no hypothesis-driven studies. Novel high-throughput techniques in combination with advanced biostatistics and bioinformatics tools (Baker, 2013) quantitate global sets of molecules (i.e., polymorphisms, chemicals, RNAs, proteins, etc.) in biological fluids allowing the identification of new biomarkers to be used in biomonitoring studies.

For instance, genome wide association studies (GWAS) explore genetic associations with diseases and successfully identified common single nucleotide polymorphisms (SNPs) for a wide variety of cancers (http://www.genome.gov/26525384). Thus, the results of GWAS may represent the basis for a comeback to a hypothesis driven biomarkers discovery. Similarly, an environment wide association study (EWAS) evaluates multiple environmental factors and their association

TABLE 43.6 Examples of Genetic Susceptibility as Assessed in Biomonitoring Studies

Function	Polymorphic Genes	Exposure	Endpoint	References
Phase I reactions	*ADH1*	Alcohol	Telomere Mean corpuscular volume of erythrocytes	Pavanello et al. (2011) Pavanello et al. (2012)
	ALDH2	Alcohol	Sister chromatid exchanges	Morimoto and Takeshita (1996)
		Vinyl chloride	Micronuclei; sister chromatid exchanges	Wong et al. (2003)
	CYP1A1	Polycyclic aromatic hydrocarbons	Urinary 1-pyrenol DNA adducts	Zhang et al. (2001) Rojas et al. (2000)
	CYP1A2	Aflatoxin	Urinary metabolites	Gross-Steinmeyer and Eaton (2012)
		Aromatic amines	4-Aminobiphenyl–hemoglobin adducts	Landi et al. (1996)
		Heterocyclic aromatic amines	Urinary 2-amino-3,8-dimethylimidazo [4,5-f] quinoxaline metabolites	Stillwell et al. (1997)
	CYP2A6	Nicotine	Urinary and plasma 3′-hydroxycotinine	Nagano et al. (2010)
	CYP2D6	Benzene	S-phenyl mercapturic acid	Rossi et al. (1999)
	CYP2E1	Vinyl chloride Polycyclic aromatic hydrocarbons	Sister chromatid exchanges Micronuclei Urinary naphthols	Wong et al. (2003) Ji et al. (2010) Nan et al. (2001)
Phase II reactions	*NAT2*	Aromatic amines	4-Aminobiphenyl–hemoglobin adducts Aromatic-DNA adducts	Landi et al. (1996) Godschalk et al. (2001)
	NAT2	Benzidine	Urinary benzidine and N-acetylated benzidine N-Acetylbenzidine-hemoglobin adduct	Rothman et al. (1996) Beyerbach et al. (2006)
	GSTM1	Polycyclic aromatic hydrocarbons	DNA adducts	Pavanello et al. (2005, 2008)
		Aromatic amines	Aromatic-DNA adducts	Liu et al. (2013)
		Aflatoxin	Aflatoxin-albumin adduct	Ahsan et al. (2001)
		Epichlorohydrin	Sister chromatid exchanges	Cheng et al. (1999)
	GSTT1	Ethylene oxide Acrylonitrile	Hydroxyethyl-valine	Yong et al. (2001) Fennell et al. (2000)
DNA repair	*XRCC1*	Aflatoxins Vinyl chloride	Aflatoxin-DNA adducts Micronuclei; sister chromatid exchanges	Lunn et al. (1999) Ji et al. (2010), Wang et al. (2010a)
	XPD, XPC	Polycyclic aromatic hydrocarbons	DNA adducts	Pavanello et al. (2005, 2008)

ADH1, alcohol dehydrogenase 1 (class I); *ALDH2*, aldehyde dehydrogenase 2 family; *CYP1A1*, cytochrome P450, family 1, subfamily A, polypeptide 1; *CYP1A2*, cytochrome P450, family 1, subfamily A, polypeptide 2; *CYP2A6*, cytochrome P450, family 2, subfamily A, polypeptide 6; *CYP2D6*-cytochrome P450, family 2, subfamily D, polypeptide 6; *CYP2E1*, cytochrome P450, family 2, subfamily E, polypeptide 1; *NAT2*, N-acetyltransferase 2 (arylamine N-acetyltransferase); *GSTM1*, glutathione S-transferase mu 1; *GSTT1*, glutathione S-transferase theta 1; *XRCC1*, X-ray repair complementing defective repair in Chinese hamster cells 1; *XPD*, xeroderma pigmentosum complementation group D; *XPC*, xeroderma pigmentosum, complementation group C.

with disease (Patel et al., 2010, 2012). The combining of EWAS with GWAS might allow a better understanding of gene–environment interactions in the development of cancer and the identification of biomarkers to monitor such interactions (Patel et al., 2013).

New opportunities may be derived from measuring the results of exposures from all sources both external and internal that occur throughout the life span. This approach named exposome by Wild (2005, 2012) is represented by the set of chemicals derived from sources outside genetic control that include diet, pathogens, microbiome, smoking, psychological stress, drugs, and pollution (Rappaport, 2012).

New technologies are not only opening new scenarios for biomarker discovery but also new challenges, which include the need of repeated measurements of global sets of biomarkers to be collected at different critical life stages. Only in this way can the dynamics of exposures and early and subsequent effects be captured.

Adductomic

The concept of "adductome," which represents all covalent adducts linked to nucleophilic sites of proteins and DNA, has recently paved the way for the characterization of all reactive electrophiles in biological samples (Rappaport et al., 2012; Carlsson and Törnqvist, 2017). Insufficient knowledge of DNA adduct fragmentation patterns and limited availability of DNA adduct standards currently act as the bottleneck for a full characterization of unknown DNA adducts. The research on adductome has mainly focused on hemoglobin and serum albumin adduct detection, because of the greater abundance and lack of damage repair on these proteins. An adductomic approach for the screening of adducts to N-terminal valine in hemoglobin of human blood samples has been developed for detecting exposure to food and endogenous agents in general population (Carlsson et al., 2014, 2015; Carlsson and Törnqvist, 2016). This approach was also applied to school-age children to measure adduct levels, originating from unspecified endogenous processes or exogenous exposure (Carlsson and Törnqvist, 2017). Two pilot studies instead report the analysis of albumin adductome in relation to tobacco smoke (Preston et al., 2017; Li et al., 2011a,b). Serum albumin Cys34 adductome approach was applied to investigate exposure to high levels of indoor combustion products (Lu et al., 2017) and to tobacco smoke (Grigoryan et al., 2016). These pilot studies are still based on few dozen subjects and further research is required to enlarge the spectrum of the adducts to be identified.

Transcriptomics

Transcriptomics has become a powerful tool for studying genome-wide responses to environmental exposures in human populations. Genome-wide approaches, based on microarrays and more recently on next-generation sequencing (NGS), have been used to gain a comprehensive assessment of the gene expression signature associated with exposures. Transcriptome modifications associated with specific exposures have been detected in target (e.g., airway epithelium) and "surrogate" tissues (e.g., cells from peripheral blood leucocytes and umbilical cord blood). Exposures include benzene (McHale et al., 2009, 2010, 2011; Forrest et al., 2005), diesel exhaust (Peretz et al., 2007), welding fumes (Wang et al., 2005), dioxin (McHale et al., 2007), arsenic (Argos et al., 2006), and acrylamide (Hochstenbach et al., 2012). In particular, Wang et al. (2005), in a pre- and post-exposure study changes were detected in the transcriptome of blood leucocytes of welders as soon as 6 h after occupational exposure to heavy metal fumes. The differentially expressed genes mainly involved in processes related to inflammation, oxidative stress, cell signaling, cell cycle, and apoptosis. Particular attention has been paid to the effect of occupational exposure to benzene on the transcriptome of blood leukocytes in groups of shoe factories and refineries (Forrest et al., 2005; McHale et al., 2009, 2010; 2011). The analysis conducted on two microarray platforms (Affimetrix and Illumina) reveals a large number of differently expressed genes linked to immune and inflammatory responses. The transcriptome analysis is also sensitive to low benzene exposures (<10 ppm) (McHale et al., 2011) where a leukemia-associated signature was identified (Li et al., 2014). Several studies have also investigated the impact of tobacco smoke on the transcriptome of peripheral blood leukocytes (Wright et al., 2012; Lampe et al., 2004; van Leeuwen et al., 2007) and airway epithelial cells (Hackett et al., 2012; Beane et al., 2007, 2011; Steiling et al., 2009; Spira et al., 2004; Tilley et al., 2011). Transcriptomic patterns associated with tobacco smoke are frequently related to inflammatory processes and oxidative stress. Similar transcriptomic profiles have also been revealed after occupational exposure to welding and diesel exhaust fumes (Wang et al., 2005; Peretz et al., 2007). Moreover, many overexpressed genes in smokers return to normal expression levels a few weeks after smoking cessation (Beane et al., 2007; Zhang et al., 2008). This suggests that transcriptome changes may be reversible signatures. This, together with experimental problems and difficulties in interpretation of subtle changes of low-dose effects, limits the value of transcriptomics in individual studies and for cross-study comparisons.

Epigenomic

The introduction of Infinium Human Methylation Bead Chip technology (27 and 450K) has allowed the analysis of the DNA methylation status throughout all the genome, which is defined as epigenome-wide analysis. DNA regions, most susceptible to environmental exposures, are the promoter regions in transposable elements of some housekeeping genes that are adjacent to genes with metastable epialleles and the regulatory elements of imprinted genes. In fact, these target regions, rich in CpG dinucleotide sequences, are preferential site of DNA methylation.

Most attention has been paid on exploring the relationship between epigenome-wide changes and smoking exposure in several cross-sectional studies (Zeilinger et al., 2013, Breitling et al., 2011, Hillemacher et al., 2008, Shenker et al., 2013, Joehanes et al., 2016, Lee and Pausova, 2013). In particular, changes in aryl hydrocarbon signaling pathway (AHRR, receptor signaling involved in detoxification of the components of tobacco smoke) have been coherently detected

(Monick et al., 2012; Joubert et al., 2012), but other end points showed inconsistent results.

Cross-sectional studies on birth cohorts examined umbilical cord blood epigenome in relation to lead (Wu et al., 2017), arsenic (Green et al., 2016; Cardenas et al., 2015), and mercury exposures (Cardenas et al., 2017). Epigenome analysis seems to represent more stable signatures of environmental exposure than changes in the transcriptome. However, it is still unclear how these epigenetic modifications are related to the complexity of human exposures and how they may predict cancer risk.

Metabolomic

Metabolomic is an analytical approach that aims to detect and quantify the metabolome, which is the sum of all low molecular weight metabolites (molecules <1 kDa) present in biological samples. About 20,000 metabolites are known in humans (Wishart et al., 2018). Metabolome can be measured in blood, urine, or tissues (Holmes et al., 2008).

Over the last years, an increased application of metabolomics, for the identification of disease biomarkers, reveals large interindividual differences in relation to environmental factors, linked to particular time of an individual's life (Assfalg et al., 2008), different living regions (Holmes et al., 2008; Saadatian-Elahi et al., 2009), and genetic factors.

The untargeted metabolomic approach is mainly used to identify metabolites over- or underexpressed in individuals with different exposures. In this approach, hundreds to thousands of metabolites are detected in biological samples. Metabolites are then identified using multivariate statistical methods and their chemical structure is determined by comparison with those stored in large metabolite databases (Dunn et al., 2011).

Metabolomics have been used to characterize the effect of occupational and smoking exposures. In particular, alterations of amino acid and lipid metabolism, subsequent to occupational exposure to welding fumes (Kuo et al., 2012; Wei et al., 2013; Wang et al., 2012a), polyvinyl chloride (Guardiola et al., 2016), smelter fumes rich in arsenic, lead, and chromium (Dudka et al., 2014), and environmental exposures to dioxin (Saberi Hosnijeh et al., 2013; Jeanneret et al., 2014) and tobacco smoke (Wang-Sattler et al., 2008) have been reported.

Metabolomic techniques appear to be the most omics-versatile approach in which metabolic profiles from samples collected under diverse conditions (e.g., different exposures, treatments, or physiologic states) may be systematically compared. However, there are still many uncertainties in terms of experimental design for larger scale analysis and computational methods for characterization of metabolites that hinder the identification to what extent such metabolic changes associate with a given individual exposure.

CONCLUDING REMARKS AND FUTURE DIRECTIONS

The practice of biomonitoring carcinogen exposures appears to have limited value. Certainly, the clinical significance of exposure and effect of biomonitoring is unknown because these tests are not predictive of cancer development. As far as their preventive uses, these are largely postulated given the paucity of prospective validation studies. However, tests to assess the size of exposures may be used to monitor hygienic conditions at work place. Difficult and also questionable in some respect is the use of biomonitoring data for risk assessment. Finally, interventions of primary prevention as opposed to biomonitoring individual susceptibility should be preferred, in particular in occupational settings. Indeed, a new era for biomarker discovery may result from the application of new technologies.

References

Abbas, T., Dutta, A., 2009. p21 in cancer: intricate networks and multiple activities. Nat. Rev. Cancer 9 (6), 400–414.

Ahsan, H., Wang, L.Y., Chen, C.J., et al., 2001. Variability in aflatoxin-albumin adduct levels and effects of hepatitis B and C virus infection and glutathione S-transferase M1 and T1 genotype. Environ. Health Perspect. 109 (8), 833–837.

Aka, P., Mateuca, R., Buchet, J.P., et al., 2004. Are genetic polymorphisms in OGG1, XRCC1, and XRCC3 genes predictive for the DNA strand break repair phenotype and genotoxicity in workers exposed to low dose ionizing radiations? Mutat. Res. 556 (1–2), 169–181.

Albertini, R.J., Anderson, D., Douglas, G.R., et al., 2000. IPCS guidelines for the monitoring of genotoxic effects of carcinogens in humans. International Programme on Chemical Safety. Mutat. Res. 463 (2), 111–172.

Alegría-Torres, J.A., Barretta, F., Batres-Esquivel, L.E., et al., 2013. Epigenetic markers of exposure to polycyclic aromatic hydrocarbons in Mexican brickmakers: a pilot study. Chemosphere 91 (4), 475–480.

Angerer, J., Aylward, L.L., Hays, S.M., et al., September 2011. Human biomonitoring assessment values: approaches and data requirements. Int. J. Hyg. Environ. Health 214 (5), 348–360.

Angerer, J., Bader, M., Krämer, A., 1998. Ambient and biochemical effect monitoring of workers exposed to ethylene oxide. Int. Arch. Occup. Environ. Health 71 (1), 14–18.

Angerer, J., Ewers, U., Wilhelm, M., 2007. Human biomonitoring: state of the art. Int. J. Hyg. Environ. Health 210 (3–4), 201–228.

Argos, M., Kibriya, M.G., Parvez, F., et al., 2006. Gene expression profiles in peripheral lymphocytes by arsenic exposure and skin lesion status in a Bangladeshi population. Cancer Epidemiol. Biomarkers 15 (7), 1367–1375.

Arita, A., Niu, J., Qu, Q., et al., 2012. Global levels of histone modifications in peripheral blood mononuclear cells of subjects with exposure to nickel. Environ. Health Perspect. 120 (2), 198–203.

Arnold, S.M., Angerer, J., Boogaard, P.J., et al., 2013. The use of biomonitoring data in exposure and human health risk assessment: benzene case study. Crit. Rev. Toxicol. 43 (2), 119–153.

Assfalg, M., Bertini, I., Colangiuli, D., et al., 2008. Evidence of different metabolic phenotypes in humans. Proc. Natl. Acad. Sci. 105 (5), 1420–1424.

Bailey, E., Brooks, A.G., Dollery, C.T., et al., 1988. Hydroxyethylvaline adduct formation in haemoglobin as a biological monitor of cigarette smoke intake. Arch. Toxicol. 62 (4), 247–253.

Baker, M., 2013. Big biology: the 'omes puzzle. Nature 494 (7438), 416–419.

Bartsch, H., Nair, J., 2000. Ultrasensitive and specific detection methods for exocyclic DNA adducts: markers for lipid peroxidation and oxidative stress. Toxicology 153 (1–3), 105–114.

Beane, J., Sebastiani, P., Liu, G., et al., 2007. Reversible and permanent effects of tobacco smoke exposure on airway epithelial gene expression. Genome Biol. 8 (9), R201.

Beane, J., Vick, J., Schembri, F., et al., 2011. Characterizing the impact of smoking and lung cancer on the airway transcriptome using RNA-Seq. Cancer Prev. Res. 4 (6), 803–817.

Becker, K., Schroeter-Kermani, C., Seiwert, M., et al., 2013. German health-related environmental monitoring: assessing time trends of the general population's exposure to heavy metals. Int. J. Hyg. Environ. Health 216 (3), 250–254.

Begemann, P., Srám, R.J., Neumann, H.G., 2001. Hemoglobin adducts of epoxybutene in workers occupationally exposed to 1,3-butadiene. Arch. Toxicol. 74 (11), 680–687.

Bender, M.A., Leonard, R.C., White Jr., O., et al., 1988. Chromosomal aberrations and sister-chromatid exchanges in lymphocytes from coke oven workers. Mutat. Res. 206 (1), 11–16.

Bergmark, E., 1997. Hemoglobin adducts of acrylamide and acrylonitrile in laboratory workers, smokers and nonsmokers. Chem. Res. Toxicol. 10 (1), 78–84.

Bevan, R., Angerer, J., Cocker, J., et al., 2012. Framework for the development and application of environmental biological monitoring guidance values. Regul. Toxicol. Pharmacol. 63 (3), 453–460.

Beyerbach, A., Rothman, N., Bhatnagar, V.K., et al., 2006. Hemoglobin adducts in workers exposed to benzidine and azo dyes. Carcinogenesis 27 (8), 1600–1606.

Bhalli, J.A., Khan, Q.M., Nasim, A., 2006. DNA damage in Pakistani pesticide-manufacturing workers assayed using the Comet assay. Environ. Mol. Mutagen. 47 (8), 587–593.

Bian, Q., Xu, L.C., Wang, S.L., et al., 2004. Study on the relation between occupational fenvalerate exposure and spermatozoa DNA damage of pesticide factory workers. Occup. Environ. Med. 61 (12), 999–1005.

Binkova, B., Chvatalova, I., Lnenickova, Z., et al., 2007. PAH-DNA adducts in environmentally exposed population in relation to metabolic and DNA repair gene polymorphisms. Mutat. Res. 620 (1–2), 49–61.

Biomarkers Definitions Working Group, 2001. Biomarkers and surrogate endpoints: preferred definitions and conceptual framework. Clin. Pharmacol. Ther. 69 (3), 89–95.

Bollati, V., Baccarelli, A., Hou, L., et al., 2007. Changes in DNA methylation patterns in subjects exposed to low-dose benzene. Cancer Res. 67 (3), 876–880.

Bollati, V., Marinelli, B., Apostoli, P., et al., 2010. Exposure to metal-rich particulate matter modifies the expression of candidate microRNAs in peripheral blood leukocytes. Environ. Health Perspect. 118 (6), 763–768.

Bolognesi, C., Abbondandolo, A., Barale, R., et al., 1997. Age-related increase of baseline frequencies of sister chromatid exchanges, chromosome aberrations, and micronuclei in human lymphocytes. Cancer Epidemiol. Biomarkers Prev. 6 (4), 249–256.

Bolognesi, C., Bruzzone, M., Ceppi, M., et al., 2017. The lymphocyte cytokinesis block micronucleus test in human populations occupationally exposed to vinyl chloride: a systematic review and meta-analysis. Mutat. Res. 774, 1–11.

Bonassi, S., Bolognesi, C., Abbondandolo, A., et al., 1995. Influence of sex on cytogenetic end points: evidence from a large human sample and review of the literature. Cancer Epidemiol. Biomarkers Prev. 4 (6), 671–679.

Bonassi, S., Norppa, H., Ceppi, M., et al., 2008. Chromosomal aberration frequency in lymphocytes predicts the risk of cancer: results from a pooled cohort study of 22,358 subjects in 11 countries. Carcinogenesis 29 (6), 1178–1183.

Bonassi, S., Znaor, A., Ceppi, M., et al., 2007. An increased micronucleus frequency in peripheral blood lymphocytes predicts the risk of cancer in humans. Carcinogenesis 28 (3), 625–631.

Bono, R., Romanazzi, V., Pirro, V., et al., 2012. Formaldehyde and tobacco smoke as alkylating agents: the formation of N-methylenvaline in pathologists and in plastic laminate workers. Sci. Total Environ. 414, 701–707.

Boysen, G., Pachkowski, B.F., Nakamura, J., et al., 2009. The formation and biological significance of N7-guanine adducts. Mutat. Res. 678 (2), 76–94.

Breitling, L.P., Yang, R., Korn, B., et al., 2011. Tobacco-smoking-related differential DNA methylation: 27K discovery and replication. Am. J. Hum. Genet. 88 (4), 450–457.

Brescia, G., Celotti, L., Clonfero, E., et al., 1999. The influence of cytochrome P450 1A1 and glutathione S-transferase M1 genotypes on biomarker levels in coke-oven workers. Arch. Toxicol. 73 (8–9), 431–439.

Broberg, K., Bjork, J., Paulsson, K., et al., 2005. Constitutional short telomeres are strong genetic susceptibility markers for bladder cancer. Carcinogenesis 26 (7), 1263–1271.

Bryant, M.S., Vineis, P., Skipper, P.L., et al., 1988. Hemoglobin adducts of aromatic amines: associations with smoking status and type of tobacco. Proc. Natl. Acad. Sci. U. S. A. 85 (24), 9788–9791.

Buchet, J.P., Ferreira Jr., M., Burrion, J.B., et al., 1995. Tumor markers in serum, polyamines and modified nucleosides in urine, and cytogenetic aberrations in lymphocytes of workers exposed to polycyclic aromatic hydrocarbons. Am. J. Ind. Med. 27 (4), 523–543.

Campo, L., Rossella, F., Pavanello, S., et al., 2010. Urinary profiles to assess polycyclic aromatic hydrocarbons exposure in coke-oven workers. Toxicol. Lett. 192 (1), 72–78.

Cantone, L., Nordio, F., Hou, L., et al., 2011. Inhalable metal-rich air particles and histone H3K4 dimethylation and H3K9 acetylation in a cross-sectional study of steel workers. Environ. Health Perspect. 119 (7), 964–969.

Cardenas, A., Houseman, E.A., Baccarelli, A.A., et al., 2015. In utero arsenic exposure and epigenome-wide associations in placenta, umbilical artery, and human umbilical vein endothelial cells. Epigenetics 10 (11), 1054–1063.

Cardenas, A., Rifas-Shiman, S.L., Agha, G., et al., 2017. Persistent DNA methylation changes associated with prenatal mercury exposure and cognitive performance during childhood. Sci. Rep. 7 (1), 288.

Carlsson, H., Motwani, H.V., Osterman Golkar, S., et al., 2015. Characterization of a hemoglobin adduct from ethyl vinyl ketone detected in human blood samples. Chem. Res. Toxicol. 28 (11), 2120–2129.

Carlsson, H., Törnqvist, M., 2016. Strategy for identifying unknown hemoglobin adducts using adductome LC-MS/MS data: identification of adducts corresponding to acrylic acid, glyoxal, methylglyoxal, and 1-octen-3-one. Food Chem. Toxicol. 92, 94–103.

Carlsson, H., Törnqvist, M., 2017. An adductomic approach to identify electrophiles in vivo. Basic Clin. Pharmacol. Toxicol. 3, 44–54.

Carlsson, H., von Stedingk, H., Nilsson, U., et al., 2014. LC–MS/MS screening strategy for unknown adducts to N-terminal valine in hemoglobin applied to smokers and nonsmokers. Chem. Res. Toxicol. 27 (12), 2062–2070.

Casale, G.P., Singhal, M., Bhattacharya, S., et al., 2001. Detection and quantification of depurinated benzo[a]pyrene-adducted DNA bases in the urine of cigarette smokers and women exposed to household coal smoke. Chem. Res. Toxicol. 14 (2), 192–201.

Cavallo, D., Ursini, C.L., Bavazzano, P., et al., 2006. Sister chromatid exchange and oxidative DNA damage in paving workers exposed to PAHs. Ann. Occup. Hyg. 50 (3), 211–218.

Chan, S.R., Blackburn, E.H., 2004. Telomeres and telomerase. Philos. Trans. R. Soc. Lond. B Biol. Sci. 359 (1441), 109–121.

Chappell, G., Pogribny, I.P., Guyton, K.Z., et al., 2016. Epigenetic alterations induced by genotoxic occupational and environmental human chemical carcinogens: a systematic literature review. Mutat. Res. Rev. 768, 27–45.

Chen, M., Ho, C.W., Huang, Y.C., et al., 2011. Glycine N-methyltransferase affects urinary 1-hydroxypyrene and 8-hydroxy-2′-deoxyguanosine levels after PAH exposure. J. Occup. Environ. Med. 53 (7), 812–819.

Chen, Y.L., Wang, C.J., Wu, K.Y., 2005. Analysis of N7-(benzo[a]pyrene-6-yl)guanine in urine using two-step solid-phase extraction and isotope dilution with liquid chromatography/tandem mass spectrometry. Rapid Commun. Mass Spectrom. 19 (7), 893–898.

Cheng, T.J., Hwang, S.J., Kuo, H.W., et al., 1999. Exposure to epichlorohydrin and dimethylformamide, glutathione S-transferases and sister chromatid exchange frequencies in peripheral lymphocytes. Arch. Toxicol. 73 (4–5), 282–287.

Choi, A.M., Ryter, S.W., Levine, B., 2013. Autophagy in human health and disease. N. Engl. J. Med. 368 (19), 1845–1846.

Christiani, D.C., Mehta, A.J., Yu, C.L., 2008. Genetic susceptibility to occupational exposures. Occup. Environ. Med. 65 (6), 430–436.

Collins, A.R., 2004. The comet assay for DNA damage and repair: principles, applications, and limitations. Mol. Biotechnol. 26 (3), 249–261.

Day, B.W., Skipper, P.L., Zaia, J., et al., 1991. Benzo[a]pyrene anti-diol epoxide covalently modifies human serum albumin carboxylate side chains and imidazole side chain of histidine146. J. Am. Chem. Soc. 113, 8505–8509.

De Vivo, I., Marion, M.J., Smith, S.J., et al., 1994. Mutant c-Ki-ras p21 protein in chemical carcinogenesis in humans exposed to vinyl chloride. Cancer Causes Control 5 (3), 273–278.

de Vogel, S., Weijenberg, M.P., Herman, J.G., et al., 2009. MGMT and MLH1 promoter methylation versus APC, KRAS and BRAF gene mutations in colorectal cancer: indications for distinct pathways and sequence of events. Ann. Oncol. 20 (7), 1216–1222.

Dennis, K.K., Marder, E., Balshaw, D.M., et al., 2017. Biomonitoring in the era of the exposome. Environ. Health Perspect. 125, 502–510.

Dudka, I., Kossowska, B., Senhadri, H., et al., 2014. Metabonomic analysis of serum of workers occupationally exposed to arsenic, cadmium and lead for biomarker research: a preliminary study. Environ. Int. 68, 71–81.

Dunn, W.B., Broadhurst, D., Begley, P., et al., 2011. Procedures for large-scale metabolic profiling of serum and plasma using gas chromatography and liquid chromatography coupled to mass spectrometry. Nat. Protoc. 6 (7), 1060–1083.

Egeghy, P.P., Hauf-Cabalo, L., Gibson, R., et al., 2003. Benzene and naphthalene in air and breath as indicators of exposure to jet fuel. Occup. Environ. Med. 60 (12), 969–976.

Egner, P.A., Groopman, J.D., Wang, J.S., et al., 2006. Quantification of aflatoxin-B1-N7-Guanine in human urine by high-performance liquid chromatography and isotope dilution tandem mass spectrometry. Chem. Res. Toxicol. 19 (9), 1191–1195.

Epel, E.S., Blackburn, E.H., Lin, J., et al., 2004. Accelerated telomere shortening in response to life stress. Proc. Natl. Acad. Sci. U. S. A. 101 (49), 17312–17315.

Fenech, M., Holland, N., Zeiger, E., et al., 2011. The HUMN and HUMNxL international collaboration projects on human micronucleus assays in lymphocytes and buccal cells–past, present and future. Mutagenesis 26 (1), 239–245.

Fenech, M., 2006. Cytokinesis-block micronucleus assay evolves into a "cytome" assay of chromosomal instability, mitotic dysfunction and cell death. Mutat. Res. 600 (1–2), 58–66.

Fenech, M., 2007. Cytokinesis-block micronucleus cytome assay. Nat. Protoc. 2 (5), 1084–1104.

Fennell, T.R., MacNeela, J.P., Morris, R.W., et al., 2000. Hemoglobin adducts from acrylonitrile and ethylene oxide in cigarette smokers: effects of glutathione S-transferase T1-null and M1-null genotypes. Cancer Epidemiol. Biomarkers Prev. 9 (7), 705–712.

Forni, A., Guanti, G., Bukvic, N., et al., 1996. Cytogenetic studies in coke oven workers. Toxicol. Lett. 88 (1–3), 185–189.

Forrest, M.S., Lan, Q., Hubbard, A.E., et al., 2005. Discovery of novel biomarkers by microarray analysis of peripheral blood mononuclear cell gene expression in benzene-exposed workers. Environ. Health Perspect. 113 (6), 801–807.

Fracasso, M.E., Doria, D., Bartolucci, G.B., et al., 2010. Low air levels of benzene: correlation between biomarkers of exposure and genotoxic effects. Toxicol. Lett. 192 (1), 22–28.

Fracasso, M.E., Perbellini, L., Soldà, S., et al., 2002. Lead induced DNA strand breaks in lymphocytes of exposed workers: role of reactive oxygen species and protein kinase C. Mutat. Res. 515 (1–2), 159–169.

Gambelunghe, A., Piccinini, R., Ambrogi, M., et al., 2003. Primary DNA damage in chrome-plating workers. Toxicology 188 (2–3), 187–195.

Garaj-Vrhovac, V., Zeljezic, D., 2002. Assessment of genome damage in a population of Croatian workers employed in pesticide production by chromosomal aberration analysis, micronucleus assay and Comet assay. J. Appl. Toxicol. 22 (4), 249–255.

Gerlinger, M., Rowan, A.J., Horswell, S., et al., 2012. Intratumor heterogeneity and branched evolution revealed by multiregion sequencing. N. Engl. J. Med. 366 (10), 883–892.

Godschalk, R.W., Dallinga, J.W., Wikman, H., et al., 2001. Modulation of DNA and protein adducts in smokers by genetic polymorphisms in GSTM1,GSTT1, NAT1 and NAT2. Pharmacogenetics 11 (5), 389–398.

Green, B.B., Karagas, M.R., Punshon, T., et al., 2016. Epigenome-wide assessment of DNA methylation in the placenta and arsenic exposure in the New Hampshire birth cohort study (USA). Environ. Health Perspect. 124 (8), 1253–1260.

Grigoryan, H., Edmands, W., Lu, S.S., et al., 2016. Adductomics pipeline for untargeted analysis of modifications to Cys34 of human serum albumin. Anal. Chem. 88 (21), 10504–10512.

Gross-Steinmeyer, K., Eaton, D.L., 2012. Dietary modulation of the biotransformation and genotoxicity of aflatoxin B(1). Toxicology 299 (2–3), 69–79.

Guardiola, J.J., Beier, J.I., Falkner, K.C., et al., 2016. Occupational exposures at a polyvinyl chloride production facility are associated with significant changes to the plasma metabolome. Toxicol. Appl. Pharmacol. 313, 47–56.

Guengerich, F.P., Arneson, K.O., Williams, K.M., et al., 2002. Reaction of aflatoxin B1 oxidation products with lysine. Chem. Res. Toxicol. 15 (6), 780–792.

Guimaraes, D.P., Hainaut, P., 2002. TP53: a key gene in human cancer. Biochimie 84 (1), 83–93.

Hackett, N.R., Butler, M.W., Shaykhiev, R., et al., 2012. RNA-Seq quantification of the human small airway epithelium transcriptome. BMC Genom. 13, 82.

Hanahan, D., Weinberg, R.A., 2011. Hallmarks of cancer: the next generation. Cell 144 (5), 646–674.

Harley, C.B., Futcher, A.B., Greider, C.W., 1990. Telomeres shorten during ageing of human fibroblasts. Nature 345 (6274), 458–460.

Hemminki, K., Zhang, L.F., Krüger, J., et al., 1994. Exposure of bus and taxi drivers to urban air pollutants as measured by DNA and protein adducts. Toxicol. Lett. 72 (1–3), 171–174.

Hillemacher, T., Frieling, H., Moskau, S., et al., 2008. Global DNA methylation is influenced by smoking behaviour. Eur. Neuropsychopharmacol 18 (4), 295–298.

Hochstenbach, K., van Leeuwen, D.M., Gmuender, H., et al., 2012. Global gene expression analysis in cord blood reveals gender-specific differences in response to carcinogenic exposure in utero. Cancer Epidemiol. Biomarkers 21 (10), 1756–1767.

Hoffmann, D.S., Hecht, S.S., 1985. Nicotine-derived N-nitrosamines and tobacco-related cancer: current status and future directions. Cancer Res. 45 (3), 935–944.

Hollstein, M., Marion, M.J., Lehman, T., et al., 1994. p53 mutations at A: T base pairs in angiosarcomas of vinyl chloride-exposed factory workers. Carcinogenesis 15 (1), 1–3.

Holmes, E., Wilson, I.D., Nicholson, J.K., 2008. Metabolic phenotyping in health and disease. Cell 134 (5), 714–717.

Honda, H., Onishi, M., Fujii, K., et al., 2011. Measurement of glycidol hemoglobin adducts in humans who ingest edible oil containing small amounts of glycidol fatty acid esters. Food Chem. Toxicol. 49 (10), 2536–2540.

Hou, L., Savage, S.A., Blaser, M.J., et al., 2009. Telomere length in peripheral leukocyte DNA and gastric cancer risk. Cancer Epidemiol. Biomarkers Prev. 18 (11), 3103–3109.

Hoxha, M., Dioni, L., Bonzini, M., et al., 2009. Association between leukocyte telomere shortening and exposure to traffic pollution: a cross-sectional study on traffic officers and indoor office workers. Environ. Health 8, 41.

Huang, C.C., Wu, C.F., Shih, W.C., et al., 2011. Comparative analysis of urinary N7-(2-hydroxyethyl)guanine for ethylene oxide- and non-exposed workers. Toxicol. Lett. 202 (3), 237–243.

Huang, H.B., Lai, C.H., Chen, G.W., et al., 2012. Traffic-related air pollution and DNA damage: a longitudinal study in Taiwanese traffic conductors. PLoS One 7 (5), e37412.

IARC, 2011. IARC Monographs on the Evaluation of Carcinogenic Risks to Humans. In: A Review of Human Carcinogens, vol. 100. International Agency for Research on Cancer, Lyon, France. http://monographs.iarc.fr/ENG/Monographs/PDFs/index.php.

Ilyenko, I., Lyaskivska, O., Bazyka, D., 2011. Analysis of relative telomere length and apoptosis in humans exposed to ionising radiation. Exp. Oncol. 33 (4), 235–238.

Jakowlew, S.B., 2006. Transforming growth factor-β in cancer and metastasis. Cancer Metastasis Rev. 25 (3), 435–457.

Jang, J.S., Choi, Y.Y., Lee, W.K., et al., 2008. Telomere length and the risk of lung cancer. Cancer Sci. 99 (7), 1385–1389.

Jeanneret, F., Boccard, J., Badoud, F., et al., 2014. Human urinary biomarkers of dioxin exposure: analysis by metabolomics and biologically driven data dimensionality reduction. Toxicol. Lett. 230 (2), 234–243.

Ji, F., Wang, W., Xia, Z.L., et al., 2010. Prevalence and persistence of chromosomal damage and susceptible genotypes of metabolic and DNA repair genes in Chinese vinyl chloride-exposed workers. Carcinogenesis 31 (4), 648–653.

Joehanes, R., Just, A.C., Marioni, R.E., et al., 2016. Epigenetic signatures of cigarette smoking. Circ. Cardiovasc. Genet. 9 (5), 436–447.

John Luo, J.C., Cheng, T.J., Du, C.L., et al., 2003. Molecular epidemiology of plasma oncoproteins in vinyl chloride monomer workers in Taiwan. Cancer Detect Prev. 27 (2), 94–101.

Johnson, R.D., Jasin, M., 2000. Sister chromatid gene conversion is a prominent double-strand break repair pathway in mammalian cells. EMBO J. 19 (13), 3398–3407.

Jones, C.R., Liu, Y.Y., Sepai, O., et al., 2005. Hemoglobin adducts in workers exposed to nitrotoluenes. Carcinogenesis 26 (1), 133–143.

Joubert, B.R., Håberg, S.E., Nilsen, R.M., et al., 2012. 450K epigenome-wide scan identifies differential DNA methylation in newborns related to maternal smoking during pregnancy. Environ. Health Perspect. 120 (10), 1425–1431.

Kalina, I., Brezáni, P., Gajdosová, D., et al., 1998. Cytogenetic monitoring in coke oven workers. Mutat. Res. 417 (1), 9–17.

Kang, D.H., Rothman, N., Poirier, M.C., et al., 1995. Interindividual differences in the concentration of 1-hydroxypyrene-glucuronide in urine and polycyclic aromatic hydrocarbon-DNA adducts in peripheral white blood cells after charbroiled beef consumption. Carcinogenesis 16 (5), 1079–1085.

Ketelslegers, H.B., Gottschalk, R.W.H., Koppen, G., et al., 2008. Multiplex genotyping as a biomarker for susceptibility to carcinogenic exposure in the FLEHS biomonitoring study. Cancer Epidemiol. Biomarkers Prev. 17 (8), 1902–1912.

Kirsch-Volders, M., Plas, G., Elhajouji, A., et al., 2011. The in vitro MN assay in 2011: origin and fate, biological significance, protocols, high throughput methodologies and toxicological relevance. Arch. Toxicol. 85 (8), 873–899.

Kuo, C.H., Wang, K.C., Tian, T.F., et al., 2012. Metabolomic characterization of laborers exposed to welding fumes. Chem. Res. Toxicol. 25 (3), 676–686.

Kütting, B., Göen, T., Schwegler, U., et al., 2009. Monoarylamines in the general population—a cross-sectional population-based study including 1004 Bavarian subjects. Int. J. Hyg. Environ. Health 212 (3), 298–309.

Lahiri, D.K., 2011. An integrated approach to genome studies. Science 331 (6014), 147.

Lambrou, A., Baccarelli, A., Wright, R.O., et al., 2012. Arsenic exposure and DNA methylation among elderly men. Epidemiology 23 (5), 668–676.

Lampe, J.W., Stepaniants, S.B., Mao, M., et al., 2004. Signatures of environmental exposures using peripheral leukocyte gene expression: tobacco smoke. Cancer Epidemiol. Biomarkers 13 (3), 445–453.

Landi, M.T., Zocchetti, C., Bernucci, I., et al., 1996. Cytochrome P4501A2: enzyme induction and genetic control in determining 4-aminobiphenyl-hemoglobin adduct levels. Cancer Epidemiol. Biomarkers Prev. 5 (9), 693–698.

Lee, J., Kang, D., Lee, K.H., et al., 2002. Influence of GSTM1 genotype on association between aromatic DNA adducts and urinary PAH metabolites in incineration workers. Mutat. Res. 514 (1–2), 213–221.

Lee, K.H., Ichiba, M., Zhang, J., et al., 2003. Multiple biomarkers study in painters in a shipyard in Korea. Mutat. Res. 540 (1), 89–98.

Lee, K.W., Pausova, Z., 2013. Cigarette smoking and DNA methylation. Front. Genet. 4, 1323.

Leng, S., Cheng, J., Pan, Z., et al., 2004. Associations between XRCC1 and ERCC2 polymorphisms and DNA damage in peripheral blood lymphocyte among coke oven workers. Biomarkers 9 (4–5), 395–406.

Letouzé, E., Shinde, J., Renault, V., et al., 2017. Mutational signatures reveal the dynamic interplay of risk factors and cellular processes during liver tumorigenesis. Nat. Commun. 8, 1315.

Li, H., Grigoryan, H., Funk, W.E., et al., 2011a. Profiling Cys34 adducts of human serum albumin by fixed-step selected reaction monitoring. Mol. Cell. Proteomics 10 (3). https://doi.org/10.1074/mcp.M110.004606.

Li, H., Jönsson, B.A., Lindh, C.H., et al., 2011b. N-nitrosamines are associated with shorter telomere length. Scand. J. Work. Environ. Health 37 (4), 316–324.

Li, K., Jing, Y., Yang, C., et al., 2014. Increased leukemia-associated gene expression in benzene-exposed workers. Sci. Rep. 4, 5369.

Lin, Y.S., Vermeulen, R., Tsai, C.H., et al., 2007. Albumin adducts of electrophilic benzene metabolites in benzene-exposed and control workers. Environ. Health Perspect. 115 (1), 28–34.

Liu, A.L., Lu, W.Q., Wang, Z.Z., et al., 2006. Elevated levels of urinary 8-hydroxy-2-deoxyguanosine, lymphocytic micronuclei, and serum glutathione S-transferase in workers exposed to coke oven emissions. Environ. Health Perspect. 114 (5), 673–677.

Liu, M., Chen, L., Zhou, R., et al., 2013. Association between GSTM1 polymorphism and DNA adduct concentration in the occupational workers exposed to PAHs: a meta-analysis. Gene 519 (1), 71–76.

Lorenzo, Y., Costa, S., Collins, A.R., et al., 2013. The comet assay, DNA damage, DNA repair and cytotoxicity: hedgehogs are not always dead. Mutagenesis 28 (4), 427–432.

Lu, S.S., Grigoryan, H., Edmands, W.M., et al., 2017. Profiling the serum albumin Cys34 adductome of solid fuel users in Xuanwei and Fuyuan, China. Environ. Sci. Technol. 51 (1), 46–57.

Lunn, R.M., Langlois, R.G., Hsieh, L.L., et al., 1999. XRCC1 polymorphisms: effects on aflatoxin B1-DNA adducts and glycophorin A variant frequency. Cancer Res. 59 (11), 2557–2561.

Manini, P., De Palma, G., Andreoli, R., et al., 2010. Occupational exposure to low levels of benzene: biomarkers of exposure and nucleic acid oxidation and their modulation by polymorphic xenobiotic metabolizing enzymes. Toxicol. Lett. 193 (3), 229–235.

Mannetje, A., Coakley, J., Mueller, J.F., et al., 2012. Partitioning of persistent organic pollutants (POPs) between human serum and breast milk: a literature review. Chemosphere 89 (8), 911–918.

Manolio, T.A., Collins, F.S., Cox, N.J., et al., 2009. Finding the missing heritability of complex diseases. Nature 461 (7265), 747–753.

Marczynski, B., Pesch, B., Wilhelm, M., et al., 2009. Occupational exposure to polycyclic aromatic hydrocarbons and DNA damage by industry: a nationwide study in Germany. Arch. Toxicol. 83 (10), 947–957.

Marczynski, B., Rihs, H.P., Rossbach, B., et al., 2002. Analysis of 8-oxo-7,8-dihydro-2′-deoxyguanosine and DNA strand breaks in white blood cells of occupationally exposed workers: comparison with ambient monitoring, urinary metabolites and enzyme polymorphisms. Carcinogenesis 23 (2), 273–281.

Marrone, A.K., Beland, F.A., Pogribny, I.P., 2014. Noncoding RNA response to xenobiotic exposure: an indicator of toxicity and carcinogenicity. Expert Opin. Drug Metabol. Toxicol. 10, 1409–1422.

Mateuca, R.A., Decordier, I., Kirsch-Volders, M., 2012. Cytogenetic methods in human biomonitoring: principles and uses. Methods Mol. Biol. 817, 305–334.

Mather, K.A., Jorm, A.F., Parslow, R.A., et al., 2011. Is telomere length a biomarker of aging? A review. J. Gerontol. A Biol. Sci. Med. Sci. 66 (2), 202–213.

McClean, M.D., Wiencke, J.K., Kelsey, K.T., et al., 2007. DNA adducts among asphalt paving workers. Ann. Occup. Hyg. 51 (1), 27–34.

McCoy, L.F., Scholl, P.F., Sutcliffe, A.E., et al., 2008. Human aflatoxin albumin adducts quantitatively compared by ELISA, HPLC with fluorescence detection, and HPLC with isotope dilution mass spectrometry. Cancer Epidemiol. Biomark. Prev. 17 (7), 1653–1657.

McGrath, M., Wong, J.Y., Michaud, D., et al., 2007. Telomere length, cigarette smoking, and bladder cancer risk in men and women. Cancer Epidemiol. Biomarkers Prev. 16 (4), 815–819.

McHale, C.M., Zhang, L., Lan, Q., et al., 2011. Global gene expression profiling of a population exposed to a range of benzene levels. Environ. Health Perspect. 119, 628–634.

McHale, C.M., Zhang, L., Hubbard, A.E., et al., 2010. Toxicogenomic profiling of chemically exposed humans in risk assessment. Mutat. Res. 705, 172–183.

McHale, C.M., Zhang, L., Lan, Q., et al., 2009. Changes in the peripheral blood transcriptome associated with occupational benzene exposure identified by cross-comparison on two microarray platforms. Genomics 93, 343–349.

McHale, C.M., Zhang, L., Hubbard, A.E., et al., 2007. Microarray analysis of gene expression in peripheral blood mononuclear cells from dioxin-exposed human subjects. Toxicology 229, 101–113.

Miller-Schulze, J.P., Paulsen, M., Kameda, T., et al., 2013. Evaluation of urinary metabolites of 1-nitropyrene as biomarkers for exposure to diesel exhaust in taxi drivers of Shenyang, China. J. Expo. Sci. Environ. Epidemiol. 23 (2), 170–175.

Miner, J.K., Rom, W.N., Livingston, G.K., et al., 1983. Lymphocyte sister chromatid exchange (SCE) frequencies in coke oven workers. J. Occup. Med. 25 (1), 30–33.

Mocci, F., Nettuno, M., 2006. Plasma mutant-p53 protein and anti-p53 antibody as a marker: an experience in vinyl chloride workers in Italy. J. Occup. Environ. Med. 48 (2), 158–164.

Monick, M.M., Beach, S.R., Plume, J., et al., 2012. Coordinated changes in AHRR methylation in lymphoblasts and pulmonary macrophages from smokers. Am. J. Med. Genet. B Neuropsychiatr. Genet. 159B (2), 141–151.

Morimoto, K., Takeshita, T., 1996. Low Km aldehyde dehydrogenase (ALDH2) polymorphism, alcohol-drinking behavior, and chromosome alterations in peripheral lymphocytes. Environ. Health Perspect. 104 (Suppl. 3), 563–567.

Moro, A.M., Brucker, N., Charão, M.F., et al., 2017. Biomonitoring of gasoline station attendants exposed to benzene: effect of gender. Mutat. Res. 813, 1–9.

Myers, S.R., Ali, M.Y., 2008. Haemoglobin adducts as biomarkers of exposure to tobacco-related nitrosamines. Biomarkers 13 (2), 145–159.

Nagano, T., Shimizu, M., Kiyotani, K., et al., 2010. Biomonitoring of urinary cotinine concentrations associated with plasma levels of nicotine metabolites after daily cigarette smoking in a male Japanese population. Int. J. Environ. Res. Publ. Health 7 (7), 2953–2964.

Nan, H.M., Kim, H., Lim, H.S., et al., 2001. Effects of occupation, lifestyle and genetic polymorphisms of CYP1A1, CYP2E1, GSTM1 and GSTT1 on urinary 1-hydroxypyrene and 2-naphthol concentrations. Carcinogenesis 22 (5), 787–793.

Nielsen, P.S., Andreassen, A., Farmer, P.B., et al., 1996a. Biomonitoring of diesel exhaust-exposed workers. DNA and hemoglobin adducts and urinary 1-hydroxypyrene as markers of exposure. Toxicol. Lett. 86 (1), 27–37.

Nielsen, P.S., de Pater, N., Okkels, H., et al., 1996b. Environmental air pollution and DNA adducts in Copenhagen bus drivers—effect of GSTM1 and NAT2 genotypes on adduct levels. Carcinogenesis 17 (5), 1021–1027.

Norppa, H., Bonassi, S., Hansteen, I.L., et al., 2006. Chromosomal aberrations and SCEs as biomarkers of cancer risk. Mutat. Res. 600 (1–2), 37–45.

Norppa, H., 2004. Cytogenetic biomarkers and genetic polymorphisms. Toxicol. Lett. 149 (1–3), 309–334.

NRC, 1987. Committee on biological markers of the National Research Council. Biological markers in environmental health research. Eviron. Health Perspect. 74, 3–9.

Oikawa, S., Tada-Oikawa, S., Kawanishi, S., 2001. Site-specific DNA damage at the GGG sequence by UVA involves acceleration of telomere shortening. Biochemistry 40 (15), 4763–4768.

Patel, C.J., Bhattacharya, J., Butte, A.J., 2010. An Environment-Wide Association Study (EWAS) on type 2 diabetes mellitus. PLoS One 5 (5), e10746.

Patel, C.J., Chen, R., Kodama, K., et al., 2013. Systematic identification of interaction effects between genome- and environment-wide associations in type 2 diabetes mellitus. Hum. Genet. 132 (5), 495–508.

Patel, C.J., Cullen, M.R., Ioannidis, J.P., et al., 2012. Systematic evaluation of environmental factors: persistent pollutants and nutrients correlated with serum lipid levels. Int. J. Epidemiol. 41 (3), 828–843.

Pavanello, S., Bollati, V., Pesatori, A.C., et al., 2009. Global and gene-specific promoter methylation changes are related to anti-B[a]PDE-DNA adduct levels and influence micronuclei levels in polycyclic aromatic hydrocarbon-exposed individuals. Int. J. Cancer 125 (7), 1692–1697.

Pavanello, S., Hoxha, M., Dioni, L., et al., 2011. Shortened telomeres in individuals with abuse in alcohol consumption. Int. J. Cancer 129 (4), 983–992.

Pavanello, S., Lotti, M., 2012. Biological monitoring of carcinogens: current status and perspectives. Arch. Toxicol. 86 (4), 535–541.

Pavanello, S., Pesatori, A.C., Dioni, L., et al., 2010. Shorter telomere length in peripheral blood lymphocytes of workers exposed to polycyclic aromatic hydrocarbons. Carcinogenesis 31 (2), 216–221.

Pavanello, S., Pulliero, A., Clonfero, E., 2008. Influence of GSTM1 null and low repair XPC PAT+ on anti-B[a]PDE-DNA adduct in mononuclear white blood cells of subjects low exposed to PAHs through smoking and diet. Mutat. Res. 638 (1–2), 195–204.

Pavanello, S., Pulliero, A., Siwinska, E., et al., 2005. Reduced nucleotide excision repair and GSTM1-null genotypes influence anti-B[a]PDE-DNA adduct levels in mononuclear white blood cells of highly PAH-exposed coke oven workers. Carcinogenesis 26 (1), 169–175.

Pavanello, S., Snenghi, R., Nalesso, A., et al., 2012. Alcohol drinking, mean corpuscular volume of erythrocytes, and alcohol metabolic genotypes in drunk drivers. Alcohol 46 (1), 61–68.

Paz-y-Miño, C., Arévalo, M., Sanchez, M.E., et al., 2004. Chromosome and DNA damage analysis in individuals occupationally exposed to pesticides with relation to genetic polymorphism for CYP 1A1 gene in Ecuador. Mutat. Res. 562 (1–2), 77–89.

Peretz, A., Peck, E.C., Bammler, T.K., et al., 2007. Diesel exhaust inhalation and assessment of peripheral blood mononuclear cell gene transcription effects: an exploratory study of healthy human volunteers. Inhal. Toxicol. 19 (14), 1107–1119.

Pesch, B., Spickenheuer, A., Kendzia, B., et al., 2011. Urinary metabolites of polycyclic aromatic hydrocarbons in workers exposed to vapours and aerosols of bitumen. Arch. Toxicol. 85 (Suppl. 1), S29–S39.

Peters, S., Talaska, G., Jönsson, B.A., et al., 2008. Polycyclic aromatic hydrocarbon exposure, urinary mutagenicity, and DNA adducts in rubber manufacturing workers. Cancer Epidemiol. Biomarkers Prev. 17 (6), 1452–1459.

Petersen, S., Saretzki, G., von Zglinicki, T., 1998. Preferential accumulation of single-stranded regions in telomeres of human fibroblasts. Exp. Cell Res. 239 (1), 152–160.

Peterson, S.E., Westra, J.W., Paczkowski, C.M., et al., 2008. Chromosomal mosaicism in neural stem cells. Methods Mol. Biol. 438, 197–204.

Pilger, A., Rüdiger, H.W., 2006. 8-Hydroxy-2′-deoxyguanosine as a marker of oxidative DNA damage related to occupational and environmental exposures. Int. Arch. Occup. Environ. Health 80 (1), 1–15.

Pilsner, J.R., Hall, M.N., Liu, X., et al., 2011. Associations of plasma selenium with arsenic and genomic methylation of leukocyte DNA in Bangladesh. Environ. Health Perspect. 119 (1), 113–118.

Pilsner, J.R., Hall, M.N., Liu, X., et al., 2012. Influence of prenatal arsenic exposure and newborn sex on global methylation of cord blood DNA. PLoS One 7 (5), e37147.

Pilsner, J.R., Liu, X., Ahsan, H., et al., 2007. Genomic methylation of peripheral blood leukocyte DNA: influences of arsenic and folate in Bangladeshi adults. Am. J. Clin. Nutr. 86 (4), 1179–1186.

Pleil, J.D., Sobus, J.R., 2013. Estimating lifetime risk from spot biomarker data and interclass correlation coefficients (ICC). J. Toxicol. Environ. Health 76 (12), 747–766.

Plna, K., Osterman-Golkar, S., Nogradi, E., et al., 2000. 32P-post-labelling of 7-(3-chloro-2-hydroxypropyl)guanine in white blood cells of workers occupationally exposed to epichlorohydrin. Carcinogenesis 21 (2), 275–280.

Pogribny, I.P., Rusyn, I., 2013. Environmental toxicants, epigenetics, and cancer. Adv. Exp. Med. Biol. 754, 215–232.

Poirier, M.C., Weston, A., Schoket, B., et al., 1998. Biomonitoring of United States Army soldiers serving in Kuwait in 1991. Cancer Epidemiol. Biomarkers Prev. 7 (6), 545–551.

Popp, W., Vahrenholz, C., Schell, C., et al., 1997. DNA single strand breakage, DNA adducts, and sister chromatid exchange in lymphocytes and phenanthrene and pyrene metabolites in urine of coke oven workers. Occup. Environ. Med. 54 (3), 176–183.

Preston, G.W., Plusquin, M., Sozeri, O., et al., 2017. Refinement of a methodology for untargeted detection of serum albumin adducts in human populations. Chem. Res. Toxicol. 30 (12), 2120–2129.

Rappaport, S.M., Li, H., Grigoryan, H., et al., 2012. Adductomics: characterizing exposures to reactive electrophiles. Toxicol. Lett. 213 (1), 83–90.

Rappaport, S.M., Smith, M.T., 2010. Epidemiology. Environment and disease risks. Science 330 (6003), 460–461.

Rappaport, S.M., Waidyanatha, S., Qu, Q., et al., 2002. Albumin adducts of benzene oxide and 1,4-benzoquinone as measures of human benzene metabolism. Cancer Res. 62 (5), 1330–1337.

Rappaport, S.M., Waidyanatha, S., Yeowell-O'Connell, K., et al., 2005. Protein adducts as biomarkers of human benzene metabolism. Chem. Biol. Interact. 153–154, 103–109.

Rappaport, S.M., 2012. Biomarkers intersect with the exposome. Biomarkers 17 (6), 483–489.

Reh, B.D., DeBord, D.G., Butler, M.A., et al., 2000. O(6)-methylguanine DNA adducts associated with occupational nitrosamine exposure. Carcinogenesis 21 (1), 29–33.

Reuterwall, C., Aringer, L., Elinder, C.G., et al., 1991. Assessment of genotoxic exposure in Swedish coke-oven work by different methods of biological monitoring. Scand. J. Work. Environ. Health 17 (2), 123–132.

Richter, E., Rösler, S., Scherer, G., et al., 2001. Haemoglobin adducts from aromatic amines in children in relation to area of residence and exposure to environmental tobacco smoke. Int. Arch. Occup. Environ. Health 74 (6), 421–428.

Richter, E., 2015. Biomonitoring of human exposure to arylamines. Front. Biosci. 7, 222–238.

Riedel, K., Scherer, G., Engl, J., et al., 2006. Determination of three carcinogenic aromatic amines in urine of smokers and nonsmokers. J. Anal. Toxicol. 30 (3), 187–195.

Riffelmann, M., Müller, G., Schmieding, W., et al., 1995. Biomonitoring of urinary aromatic amines and arylamine hemoglobin adducts in exposed workers and nonexposed control persons. Int. Arch. Occup. Environ. Health 68 (1), 36–43.

Rojas, M., Alexandrov, K., Auburtin, G., et al., 1995. Anti-benzo[a]pyrene diolepoxide–DNA adduct levels in peripheral mononuclear cells from coke oven workers and the enhancing effect of smoking. Carcinogenesis 16 (6), 1373–1376.

Rojas, M., Cascorbi, I., Alexandrov, K., et al., 2000. Modulation of benzo[a]pyrene diolepoxide-DNA adduct levels in human white blood cells by CYP1A1, GSTM1 and GSTT1 polymorphism. Carcinogenesis 21 (1), 35–41.

Rossi, A.M., Guarnieri, C., Rovesti, S., et al., 1999. Genetic polymorphisms influence variability in benzene metabolism in humans. Pharmacogenetics 9 (4), 445–451.

Rothman, N., Bhatnagar, V.K., Hayes, R.B., et al., 1996. The impact of interindividual variation in NAT2 activity on benzidine urinary metabolites and urothelial DNA adducts in exposed workers. Proc. Natl. Acad. Sci. U. S. A. 93 (10), 5084–5089.

Rusiecki, J.A., Baccarelli, A., Bollati, V., et al., 2008. Greenlandic Inuit Global DNA hypomethylation is associated with high serum-persistent organic pollutants in. Environ. Health Perspect. 116 (11), 1547–1552.

Saadatian-Elahi, M., Slimani, N., Chajès, V., et al., 2009. Plasma phospholipid fatty acid profiles and their association with food intakes: results from a cross-sectional study within the European prospective investigation into cancer and nutrition. Am. J. Clin. Nutr. 89 (1), 331–346.

Sabbioni, G., Turesky, R.J., 2017. Biomonitoring human albumin adducts: the past, the present and the future. Chem. Res. Toxicol. 30, 332–366.

Sabbioni, G., 2017. Hemoglobin adducts and urinary metabolites of arylamines and nitroarenes. Chem. Res. Toxicol. 30, 1733–1766.

Saberi Hosnijeh, F., Pechlivanis, A., Keun, H.C., et al., 2013. Serum metabolomic pertubations among workers exposed to 2,3,7,8-tetrachlorodibenzo-p-dioxin (TCDD). Environ. Mol. Mutagen. 54 (7), 558–565.

Santella, R.M., Hemminki, K., Tang, D.L., et al., 1993. Polycyclic aromatic hydrocarbon-DNA adducts in white blood cells and urinary 1-hydroxypyrene in foundry workers. Cancer Epidemiol. Biomarkers Prev. 2 (1), 59–62.

Scheepers, P.T., Coggon, D., Knudsen, L.E., et al., 2002. BIOMarkers for occupational diesel exhaust exposure monitoring (BIOMODEM)–a study in underground mining. Toxicol. Lett. 134 (1–3), 305–317.

Schettgen, T., Broding, H.C., Angerer, J., et al., 2002. Hemoglobin adducts of ethylene oxide, propylene oxide, acrylonitrile and acrylamide-biomarkers in occupational and environmental medicine. Toxicol. Lett. 134 (1–3), 65–70.

Schettgen, T., Gube, M., Esser, A., et al., 2012. Plasma polychlorinated biphenyls (PCB) levels of workers in a transformer recycling company, their family members, and employees of surrounding companies. J. Toxicol. Environ. Health 75 (8–10), 414–422.

Schindler, J., Li, Y., Marion, M.J., et al., 2007. The effect of genetic polymorphisms in the vinyl chloride metabolic pathway on mutagenic risk. J. Hum. Genet. 52 (5), 448–455.

Shenker, N.S., Ueland, P.M., Polidoro, S., et al., 2013. DNA methylation as a long-term biomarker of exposure to tobacco smoke. Epidemiology 24 (5), 712–716.

Shimizu, N., Misaka, N., Utani, K., 2007. Nonselective DNA damage induced by a replication inhibitor results in the selective elimination of extrachromosomal double minutes from human cancer cells. Genes Chromosomes Cancer 46 (10), 865–874.

Siwińska, E., Mielzyńska, D., Kapka, L., 2004. Association between urinary 1-hydroxypyrene and genotoxic effects in coke oven workers. Occup. Environ. Med. 61 (3), e10.

Skipper, P.L., Naylor, S., Gan, L.S., et al., 1989. Origin of tetrahydrotetrols derived from human hemoglobin adducts of benzo[a] pyrene. Chem. Res. Toxicol. 2 (5), 280–281.

Smith, S.J., Li, Y., Whitley, R., et al., 1998. Molecular epidemiology of p53 protein mutations in workers exposed to vinyl chloride. Am. J. Epidemiol. 147 (3), 302–308.

Spira, A., Beane, J., Shah, V., et al., 2004. Effects of cigarette smoke on the human airway epithelial cell transcriptome. Proc. Natl. Acad. Sci. 101 (27), 10143–10148.

Srám, R.J., Rössner, P., Peltonen, K., et al., 1998. Chromosomal aberrations, sister-chromatid exchanges, cells with high frequency of SCE, micronuclei and comet assay parameters in 1,3-butadiene-exposed workers. Mutat. Res. 419 (1–3), 145–154.

Steiling, K., Kadar, A.Y., Bergerat, A., et al., 2009. Comparison of proteomic and transcriptomic profiles in the bronchial airway epithelium of current and never smokers. PLoS One 4 (4), e5043.

Stillwell, W.G., Kidd, L.C., Wishnok, J.S., et al., 1997. Urinary excretion of unmetabolized and phase II conjugates of 2-amino-1-methyl-6-phenylimidazo[4,5-b]pyridine and 2-amino-3,8-dimethylimidazo [4,5-f]quinoxaline in humans: relationship to cytochrome P4501A2 and N-acetyltransferase activity. Cancer Res. 57 (16), 3457–3464.

Tan, Y.M., Sobus, J., Chang, D., et al., 2012. Reconstructing human exposures using biomarkers and other "clues". J. Toxicol. Environ. Health B Crit. Rev. 15 (1), 22–38.

Thier, R., Lewalter, J., Selinski, S., et al., 2001. Biological monitoring in workers in a nitrobenzene reduction plant: haemoglobin versus serum albumin adducts. Int. Arch. Occup. Environ. Health 74 (7), 483–488.

Thomas, D., 2010. Gene–environment-wide association studies: emerging approaches. Nat. Rev. Genet. 11 (4), 259–272.

Tice, R.R., Agurell, E., Anderson, D., et al., 2000. Single cell gel/comet assay: guidelines for in vitro and in vivo genetic toxicology testing. Environ. Mol. Mutagen. 35 (3), 206–221.

Tilley, A.E., O'Connor, T.P., Hackett, N.R., et al., 2011. Biologic phenotyping of the human small airway epithelial response to cigarette smoking. PLoS One 6 (7), e22798.

Tornqvist, M., Fred, C., Haglund, J., et al., 2002. Protein adducts: quantitative and qualitative aspects of their formation, analysis and applications. J. Chromatogr. B Anal. Technol. Biomed. Life Sci. 778, 279–308.

Touil, N., Aka, P.V., Buchet, J.P., et al., 2002. Assessment of genotoxic effects related to chronic low level exposure to ionizing radiation using biomarkers for DNA damage and repair. Mutagenesis 17 (3), 223–232.

Trivers, G.E., Cawley, H.L., DeBenedetti, V.M., et al., 1995. Anti-p53 antibodies in sera of workers occupationally exposed to vinyl chloride. J. Natl. Cancer Inst. 87 (18), 1400–1407.

van Delft, J.H., Steenwinkel, M.S., van Asten, J.G., et al., 2001. Biological monitoring the exposure to polycyclic aromatic hydrocarbons of coke oven workers in relation to smoking and genetic polymorphisms for GSTM1 and GSTT1. Ann. Occup. Hyg. 45 (5), 395–408.

van Hummelen, P., Gennart, J.P., Buchet, J.P., et al., 1993. Biological markers in PAH exposed workers and controls. Mutat. Res. 300 (3–4), 231–239.

van Leeuwen, D.M., van Agen, E., Gottschalk, R.W., et al., 2007. Cigarette smoke-induced differential gene expression in blood cells from monozygotic twin pairs. Carcinogenesis 28 (3), 691–697.

van Schooten, F.J., Jongeneelen, F.J., Hillebrand, M.J., et al., 1995. Polycyclic aromatic hydrocarbon-DNA adducts in white blood cell DNA and 1-hydroxypyrene in the urine from aluminum workers: relation with job category and synergistic effect of smoking. Cancer Epidemiol. Biomarkers Prev. 4 (1), 69–77.

Vineis, P., Manuguerra, M., Kavvoura, F.K., et al., 2009. A field synopsis on low-penetrance variants in DNA repair genes and cancer susceptibility. J. Natl. Cancer Inst. 101 (1), 24–36.

Vogelstein, B., Papadopoulos, N., Velculescu, V.E., et al., 2013. Cancer genome landscapes. Science 339 (6127), 1546–1558.

von Zglinicki, T., 2002. Oxidative stress shortens telomeres. Trends Biochem. Sci. 27, 339–344.

Wang, K.C., Kuo, C.H., Tian, T.F., et al., 2012a. Metabolomic characterization of laborers exposed to welding fumes. Chem. Res. Toxicol. 25 (3), 676–686.

Wang, Q., Ji, F., Sun, Y., et al., 2010a. Genetic polymorphisms of XRCC1, HOGG1 and MGMT and micronucleus occurrence in Chinese vinyl chloride-exposed workers. Carcinogenesis 31 (6), 1068–1073.

Wang, W., Qiu, Y.L., Ji, F., et al., 2010b. Genetic polymorphisms in metabolizing enzymes and susceptibility of chromosomal damage induced by vinyl chloride monomer in a Chinese worker population. J. Occup. Environ. Med. 52 (2), 163–168.

Wang, Y., Yang, H., Li, L., et al., 2012b. Biomarkers of chromosomal damage in peripheral blood lymphocytes induced by polycyclic aromatic hydrocarbons: a meta-analysis. Int. Arch. Occup. Environ. Health 85 (1), 13–25.

Wang, Z., Neuburg, D., Li, C., et al., 2005. Global gene expression profiling in whole-blood samples from individuals exposed to metal fumes. Environ. Health Perspect. 113 (2), 233–241.

Wang-Sattler, R., Yu, Y., Mittelstrass, K., et al., 2008. Metabolic profiling reveals distinct variations linked to nicotine consumption in humans—first results from the KORA study. PLoS One 3 (12), e3863.

Ward, E.M., Sabbioni, G., DeBord, D.G., et al., 1996. Monitoring of aromatic amine exposures in workers at a chemical plant with a known bladder cancer excess. J. Natl. Cancer Inst. 88 (15), 1046–1052.

Wei, Y., Wang, Z., Chang, C.Y., et al., 2013. Global metabolomic profiling reveals an association of metal fume exposure and plasma unsaturated fatty acids. PLoS One 8 (10), e77413.

Weisel, C.P., 2010. Benzene exposure: an overview of monitoring methods and their findings. Chem. Biol. Interact. 184 (1–2), 58–66.

Wen, S., Yang, F.X., Gong, Y., et al., 2008. Elevated levels of urinary 8-hydroxy-2′-deoxyguanosine in male electrical and electronic equipment dismantling workers exposed to high concentrations of polychlorinated dibenzo-p-dioxins and dibenzofurans, polybrominated diphenyl ethers, and polychlorinated biphenyls. Environ. Sci. Technol. 42 (11), 4202–4208.

Wen-Bin, M., Wei, W., Yu-Lan, Q., et al., 2009. Micronucleus occurrence related to base excision repair gene polymorphisms in Chinese workers occupationally exposed to vinyl chloride monomer. J. Occup. Environ. Med. 51 (5), 578–585.

Wild, C.P., 2005. Complementing the genome with an "exposome": the outstanding challenge of environmental exposure measurement in molecular epidemiology. Cancer Epidemiol. Biomarkers Prev. 14 (8), 1847–1850.

Wild, C.P., 2012. The exposome: from concept to utility. Int. J. Epidemiol. 41 (1), 24–32.

Wilhelm, M., Ewers, U., Schulz, C., 2004. Revised and new reference values for some trace elements in blood and urine for human biomonitoring in environmental medicine. Int. J. Hyg. Environ. Health 207 (1), 69–73.

Willeit, P., Willeit, J., Mayr, A., et al., 2010. Telomere length and risk of incident cancer and cancer mortality. J. Am. Med. Assoc. 304, 69–75.

Wishart, D.S., Feunang, Y.D., Marcu, A., et al., 2018. HMDB 4.0: the human metabolome database for 2018. Nucleic Acids Res. 46 (D1), D608–D617.

Wojewódzka, M., Kruszewski, M., Iwaneñko, T., et al., 1998. Application of the comet assay for monitoring DNA damage in workers exposed to chronic low-dose irradiation. I. Strand breakage. Mutat. Res. 416 (1–2), 21–35.

Wong, R.H., Du, C.L., Wang, J.D., et al., 2002. XRCC1 and CYP2E1 polymorphisms as susceptibility factors of plasma mutant p53 protein and anti-p53 antibody expression in vinyl chloride monomer-exposed polyvinyl chloride workers. Cancer Epidemiol. Biomarkers Prev. 11 (5), 475–482.

Wong, R.H., Wang, J.D., Hsieh, L.L., et al., 2003. XRCC1, CYP2E1 and ALDH2 genetic polymorphisms and sister chromatid exchange frequency alterations amongst vinyl chloride monomer-exposed polyvinyl chloride workers. Arch. Toxicol. 77 (8), 433–440.

Wright, W.R., Parzych, K., Crawford, D., et al., 2012. Inflammatory transcriptome profiling of human monocytes exposed acutely to cigarette smoke. PLoS One 7, e30120.

Wu, S., Hivert, M.F., Cardenas, A., et al., 2017. Exposure to low levels of lead in utero and umbilical cord blood DNA methylation in project viva: an Epigenome-Wide Association Study. Environ. Health Perspect. 125 (8), 087019.

Yong, L.C., Schulte, P.A., Wiencke, J.K., et al., 2001. Hemoglobin adducts and sister chromatid exchanges in hospital workers exposed to ethylene oxide: effects of glutathione S-transferase T1 and M1 genotypes. Cancer Epidemiol. Biomarkers Prev. 10 (5), 539–550.

Yoshioka, N., Nakashima, H., Hosoda, K., et al., 2008. Urinary excretion of an oxidative stress marker, 8-hydroxyguanine (8-OH-Gua), among nickel-cadmium battery workers. J. Occup. Health 50 (3), 229–235.

Yuan, J.M., Koh, W.P., Murphy, S.E., et al., 2009. Urinary levels of tobacco-specific nitrosamine metabolites in relation to lung cancer development in two prospective cohorts of cigarette smokers. Cancer Res. 69 (7), 2990–2995.

Zeilinger, S., Kuhnel, B., Klopp, N., et al., 2013. Tobacco smoking leads to extensive genome-wide changes in DNA methylation. PLoS One 8, e63812.

Zhang, J., Ichiba, M., Hara, K., et al., 2001. Urinary 1-hydroxypyrene in coke oven workers relative to exposure, alcohol consumption, and metabolic enzymes. Occup. Environ. Med. 58 (11), 716–721.

Zhang, L., Lee, J.J., Tang, H., et al., 2008. Impact of smoking cessation on global gene expression in the bronchial epithelium of chronic smokers. Cancer Prev. Res. 1, 112–118.

Zhang, X.H., Zhang, X., Wang, X.C., et al., 2011. Chronic occupational exposure to hexavalent chromium causes DNA damage in electroplating workers. BMC Publ. Health 11, 224.

Zhou, Q., Talaska, G., Jaeger, M., et al., 1997. Benzidine-DNA adduct levels in human peripheral white blood cells significantly correlate with levels in exfoliated urothelial cells. Mutat. Res. 393 (3), 199–205.

Ziegler, S., Schettgen, T., Beier, F., et al., 2017. Accelerated telomere shortening in peripheral blood lymphocytes after occupational polychlorinated biphenyls exposure. Arch. Toxicol. 91, 289–300.

Zwirner-Baier, I., Neumann, H.G., 1999. Polycyclic nitroarenes (nitro-PAHs) as biomarkers of exposure to diesel exhaust. Mutat. Res. 441 (1), 135–144.

CHAPTER

44

Genotoxicity Biomarkers: Molecular Basis of Genetic Variability and Susceptibility

Szabina A. Stice[1,a], *Sudheer R. Beedanagari*[2,b], *Suryanarayana V. Vulimiri*[3,c], *Sneha P. Bhatia*[4,b], *Brinda Mahadevan*[5,b]

[1]Division of Biotechnology and GRAS Notice Review, Center for Food Safety and Applied Nutrition, US Food and Drug Administration, College Park, MD, United States; [2]Bristol Myers Squibb, New Brunswick, NJ, United States; [3]National Center for Environmental Assessment, Environmental Protection Agency (EPA), Washington DC, United States; [4]Research Institute for Fragrance Materials, NJ, United States; [5]Medical Safety & Surveillance, Abbott Laboratories, Columbus, OH, United States

INTRODUCTION

Genetic toxicology elucidates the effects of chemical and physical agents on genetic material or deoxyribonucleic acid (DNA) causing alteration(s) in the structural integrity of the genome. Evaluation of genetic damage by using a combination of in silico, in vitro, and in vivo assays would enable us to utilize a weight of evidence approach in reaching a conclusion regarding the mutagenic or carcinogenic potential of a chemical. Genetic damage is a natural phenomenon occurring endogenously in the cells as a result of oxidative respiration and metabolism, but this damage is continually being repaired by a variety of DNA repair processes present in all cells. However, when the genetic damage becomes overwhelming for the cellular machinery to handle, genetic lesions accumulate that could lead to a wide variety of adverse outcomes, including apoptosis, altered cell division or proliferation, and DNA replication and mutagenesis that may result in carcinogenesis. Carcinogenesis is a multistep process, but broadly it is a three-step process, consisting of initiation, promotion, and progression. DNA damage plays an important role in the initiation step of carcinogenesis, where a carcinogen or its activated metabolite covalently binds to DNA forming a DNA adduct, which may be repaired or, following cell division, may be fixed as a mutation, leading to a heritable alteration in the DNA sequence.

Genetic toxicology encompasses the study of a wide variety of DNA damage end points, including, but not limited to mutagenicity, clastogenicity, and aneugenicity. Mutagenicity refers to the ability of an agent to alter the nucleotide sequence of DNA. Mutations are categorized as base-pair substitutions, frameshifts, additions, and deletions of nucleotide base(s) or chromosomes. Mutations are heritable and hence may have a permanent or long-lasting adverse effect on the health of an organism. Although mutations naturally arise during evolution or DNA replication, most of them occur in areas of DNA that do not exert any harmful effects to the organism. However, if mutations occur in the coding regions (e.g., genes) or in the noncoding (e.g., regulatory regions of genes), then they may lead to adverse effects due to altered gene expression patterns or formation of nonfunctional

[a] The opinions expressed in this article are the author's personal opinions and they do not necessarily reflect that of FDA, DHHS, or the Federal Government

[b] The views expressed in this chapter are those of the authors and do not necessarily reflect the views or policies of the authors' representative institutes

[c] The views expressed in this chapter are those of the author and do not necessarily reflect the views or policies of the U.S. Environmental Protection Agency

Biomarkers in Toxicology, Second Edition
https://doi.org/10.1016/B978-0-12-814655-2.00044-X

or altered proteins and enzymes, thus impacting the functional activities of the cells. Not all mutations in the coding region lead to adverse effects. In the case of silent mutations, the base substitution in a codon will either not result in a change of the amino acid the codon codes for or even if it results in an amino acid change, it will not affect the functionality of the protein product.

Clastogenicity refers to the ability of an agent to cause disruptions or breakages of chromosomes that lead to sections of the chromosome being rearranged, deleted, or added (i.e., structural alteration of chromosome). Aneugenicity reflects the ability of an agent to give rise to a daughter cell that has an abnormal number of chromosomes (i.e., numerical alteration of chromosome). It occurs due to errors in chromosomal segregation, altered tubulin polymerization, and spindle apparatus instability. Aneugenicity may result in polysomy or monosomy. In polysomy, the organism has at least one more chromosome than normal. Polysomy is found in many diseases such as Down syndrome (trisomy of chromosome 21) and Cat eye syndrome (the short arm and a small section of the long arm are present either three [trisomy] or four [tetrasomy] times). In monosomy, either only one chromosome from a pair is present or only a portion of the chromosome has one pair and the rest has two (partial monosomy). Sufferers of Turner syndrome completely or partially miss one of the X chromosomes.

A wide variety of in silico, in vitro, and in vivo assays are used to predict the three common types of genotoxic damage: mutagenicity, clastogenicity, and aneugenicity. In vitro assays in general are designed to achieve high sensitivity in detecting DNA damage caused by chemicals or agents. High sensitivity of these in vitro assays often leads to decreased specificity in predicting the carcinogenic potential of chemicals. In vivo genotoxic assays on the other hand have high specificity, but tend to exhibit low sensitivity in predicting the carcinogenic potential of chemicals. Therefore, to increase the specificity and sensitivity of detecting DNA damage, a combination of in vitro and in vivo genotoxicity assays is used, commonly referred to as the "battery approach," for predicting the carcinogenicity potential of a test article in humans. The use of a genotoxic battery approach has been proven to have high sensitivity and specificity (~85%–90%) (Kirkland et al., 2005).

The main emphasis of this chapter has been placed on providing a general overview of currently available genotoxic biomarkers. It is beyond the scope of this chapter to discuss carcinogenesis and relevant predictors/assays, but these are covered elsewhere in other chapters of this textbook. Here we discuss the genotoxicity assays that can be used to monitor and measure biomarkers of DNA damage. In addition, the mechanisms of different types of DNA damage, and the molecular basis of genetic variability and susceptibility are also elaborated with examples. Comprehensive details on the conduct of commonly used genetic toxicology assays can be reviewed by accessing the Organization for Economic Cooperation and Development (OECD) guidelines (http://www.oecd.org/chemicalsafety/testing/).

GENOTOXIC BIOMARKER DETECTION METHODS

In Silico Approaches

As indicated earlier and discussed next in this chapter, there are several in vitro and in vivo genotoxicity testing assays that are available to aid in identifying genotoxic damage. However, the global call for reduced animal usage with emphasis on the three Rs (Reduce, Refine, and Replace) and the phasing out of animal testing has led to rapid expansion and evolution of a myriad of in vitro and in silico genetic toxicology screening assays. The objective of in silico screening techniques is to apply mathematical, statistical, chemical, and computational tools to the existing genetic toxicology data from the literature and to develop efficient and effective genotoxicity predictive models. These predictive models can be used to screen large numbers of untested new chemicals to predict their genotoxic potential. In silico genotoxicity screening methods offer a fast, reliable, and economical solution for preliminary screening of large numbers of chemicals. The in silico modeling has its roots in Structure–Activity Relationship (SAR). The earliest examples of SAR were the Ashby–Tennant alerts, which associated certain chemical structures with mutation, based on the expert knowledge and experience of these scientists. Others have continued to expand the list of structural alerts associated with DNA reactivity based on known chemical reactivity and expert knowledge. Later, an alternative approach was used based on quantifying structure–activity relationships or QSARs using statistical correlation and quantum chemical properties. QSAR models use linear free energy relationships to describe reaction rates and equilibrium. There are a wide variety of in silico tools that have been packaged into software programs capable of predicting a variety of genotoxic end points, but the best validated are models predicting bacterial mutagenicity end points. These in silico tools can be divided into three main categories: databases, virtual screening programs, and prediction systems.

The primary step in in silico assessment of a chemical is to perform a search in public databases to determine if any genotoxicity data exist for the chemical under investigation. The databases collect and store genotoxicity data obtained from multiple sources, including published literature, public data sources, and proprietary in-house historical data from within a testing laboratory obtained over the years of conducting genotoxicity testing. The databases are either open to public use free of cost or available commercially, where the users pay service charges to

TABLE 44.1 Freely Available In Silico Tools for Genotoxicity and Carcinogenicity Prediction

Software Name	End point(s) Predicted
Toxicity Estimation Software Tool (TEST) https://www.epa.gov/chemical-research/toxicity-estimation-software-tool-test	Mutagenicity
Lazar Toxicity Predictions http://lazar.in-silico.de	Mutagenicity, carcinogenicity
VEGA https://www.vegahub.eu/	Mutagenicity, carcinogenicity
ToxTree https://eurl-ecvam.jrc.ec.europa.eu/laboratories-research/predictive_toxicology/qsar_tools/toxtree	Mutagenicity, carcinogenicity
OncoLogic https://www.epa.gov/tsca-screening-tools/oncologictm-computer-system-evaluate-carcinogenic-potential-chemicals	Carcinogenicity
QSAR Toolbox www.qsartoolbox.org	Mutagenicity, carcinogenicity

access them. Some examples of free databases are the Toxicology Data Network of the US National Library of Medicine, the Aggregated Computational Toxicology Online Resource of the Environmental Protection Agency, the Carcinogenic Potency Database, and the National Toxicology Program of the US Department of Health and Human Services. Examples of commercial databases include PharmaPendium, SciFinder, and STN (See Table 44.1 and Box 44.1 for additional database information.). There are some tools that serve the dual purpose of a database and prediction system; examples include Derek Nexus, Vitic Nexus, Leadscope, and so forth.

Prediction systems are software programs that enable us to make predictions of species and assay-specific end points, such as in vivo mouse micronucleus test and in vitro chromosome aberration test, mouse lymphoma assay, and micronucleus (MN) test in several species. These software programs use different types of algorithms to compute the propensity of a specific chemical to induce genotoxicity. The algorithms for these systems can be divided into three main categories: knowledge-based, statistical methods, and read-across systems. Knowledge-based predictive systems focus on organizing relevant experimental genotoxicity data as knowledge guidelines or rules in a computer program. The in silico software compares the structure of the chemical of interest/investigation with the chemicals in the knowledge library of the database. For example, one of the widely used commercial knowledge-based predictive systems is Deductive Estimation of Risk from Existing Knowledge (Derek); version 14 comprises 82 rules or alerts for mutagenicity. The Derek rules are written and maintained by experts from the nonprofit organization Lhasa Limited (www.lhasalimited.org). There are also several free open-source knowledge-based in silico predicting software packages, such as ToxTree (http://toxtree.sourceforge.net/), Bioclipse-DS (www.bioclipse.net), and OncoLogic (https://www.epa.gov/tsca-screening-tools/oncologictm-computer-system-evaluate-carcinogenic-potential-chemicals) (Table 44.1).

A structural fragment statistically related to the biological activity is called a *biophore*. Biophores of known genotoxic chemicals are compared in databases through biological activity or QSAR modeling. For example, presence of the N—N=O (nitrosamine) fragment is strongly statistically correlated with mutagenicity of both aromatic and aliphatic chemicals. Biophores or structure moieties that are related to toxic or therapeutic end points are referred to as "toxicophore" and "pharmacophore", respectively. If a biophore is present, then the software predicts the chemical as active, and if no biophore is found, the software predicts the chemical as inactive (Klopman et al., 1985; Rosenkranz and Klopman, 1990). Based on the distribution of a biophore and its correlation to toxicity, the statistical systems give a statistical value or predictive value of the biophore. For example, if the database contains 50 nitrosamines, and all of these are active, then presence of the N—N=O biophore is a very strong predictor of toxicity. On the other hand, if only half of the nitrosamines in the database are active and half are inactive, then the confidence that chemicals containing N—N=O are active will be only 50%. The term "structural alert" has also been applied in computational toxicology, meaning a chemical moiety that serves as an indicator of toxic outcome (Klopmann et al., 1985; Rosenkranz and Klopman, 1990). The two widely used commercial statistically fragment-based in silico systems are Leadscope (www.leadscope.com), whose QSAR models were developed by Informatics and Computational Safety Analysis Staff (ICSAS) of the Food and Drug Administration, and MultiCASE (www.multicase.com), whose QSAR models were based on proprietary statistical algorithms developed by Klopman and Rosenkranz (1984). There are also several free open source statistically fragment-based in silico predicting software packages, such as ToxTree (http://toxtree.sourceforge.net/) and Bioclipse-DS (www.bioclipse.net). In read-across systems, the end point result of one chemical is used to predict the end point of a structurally similar chemical. The QSAR toolbox (http://www.oecd.

BOX 44.1

FREELY AVAILABLE DATABASES WITH GENOTOXICITY AND CARCINOGENICITY INFORMATION

- ACToR (Environmental Protection Agency's [EPA] Aggregated Computational Toxicology Online Resource) (https://actor.epa.gov/actor/home.xhtml): Aggregates toxicity data from thousands of public sources on over 500,000 chemicals. It is also the warehouse for EPA web applications that can be used to explore and visualize complex computational toxicology information.
- NTP (National Toxicology Program) (https://ntp.niehs.nih.gov/): Contains data on subchronic, chronic, and in vitro and in vivo genotoxicity assays and carcinogenicity studies.
- CCRIS (Chemical Carcinogenesis Research Information System) (https://toxnet.nlm.nih.gov/newtoxnet/ccris.htm): Contains chemical records with carcinogenicity, mutagenicity, tumor promotion, and tumor inhibition test results.
- CPDB (Carcinogenicity Potency Database) (https://toxnet.nlm.nih.gov/cpdb/cpdb.html): Contains information on carcinogenicity studies on 1547 chemicals.
- GENE-TOX (Genetic Toxicology Databank) (https://toxnet.nlm.nih.gov/newtoxnet/genetox.htm): Provides genetic toxicology test data for more than 3000 chemicals from EPA.
- HSDB (Hazardous Substances databank) (https://toxnet.nlm.nih.gov/newtoxnet/hsdb.htm): It is a toxicology database containing over 5800 individual chemical records. Available data include genotoxicity and carcinogenicity test results.
- IRIS (Integrated Risk Information System, EPA) (https://cfpub.epa.gov/ncea/iris/search/basic/index.cfm): Contains descriptive and quantitative information related to human cancer and noncancer health effects that may result from exposure to 511 substances in the environment.
- TOXNET (Toxicology Data Network) (http://toxnet.nlm.nih.gov): It is a group of databases such as CCRIS, CPDB, GENE-TOX, HSDB, IRIS, etc.
- HPVIS (High Production Volume Information System, EPA) (https://ofmext.epa.gov/hpvis/HPVISlogon): Provides access to genetic toxicity, carcinogenicity, and other information on chemicals that are manufactured in exceptionally large amounts.
- EURL ECVAM (European Union Reference Laboratory for Alternatives to Animal Testing (https://eurl-ecvam.jrc.ec.europa.eu/databases/genotoxicity-carcinogenicity-db): A structured master database that compiles available genotoxicity and carcinogenicity data for Ames positive chemicals originating from different sources.
- ATSDR (Agency for Toxic Substances and Registry (https://www.atsdr.cdc.gov/toxprofiles/index.asp): Contains toxicological profiles, including mutagenicity, and carcinogenicity data, for hazardous substances.
- ECHA (European Chemicals Agency) (https://echa.europa.eu/information-on-chemicals): Contains information, including genotoxicity and carcinogenicity, on chemicals manufactured and imported in Europe.
- Tox Benchmark (http://doc.ml.tu-berlin.de/toxbenchmark/): Ames mutagenicity data set for 6512 chemicals.
- PharmaPendium (www.pharmapendium.com): Offers searchable Food and Drug Administration and European Medical Agency approval packages with excerpted preclinical, clinical, and postmarketing safety data into a single longitudinal database.
- INCHEM (International Programme on Chemical Safety) (http://www.inchem.org/): Consolidates information from a number of intergovernmental organizations on chemicals commonly used throughout the world, which may also occur as contaminants in the environment and food.

org/chemicalsafety/risk-assessment/oecd-qsar-toolbox.htm) and Analog Identification Method (https://www.epa.gov/tsca-screening-tools/analog-identification-methodology-aim-tool) are examples of in silico tools that can be used to find structurally similar chemicals.

Although the in silico systems provide an economical, faster, and easier screening of a large set of chemicals, they are far from being replacements of in vitro and in vivo genetic toxicology screening assays, due to lack of strong correlation between the in silico predictions and experimental outcomes. However, currently they serve as powerful tools for early genetic toxicology screening of chemicals, impurities, and degradants.

IN VITRO AND IN VIVO BIOMARKERS OF GENOTOXICITY

Gene Mutations in Prokaryotes

Assays for gene mutations can detect either reversion or forward mutation. Reversions are back mutations that restore gene function in a mutant (i.e., bring about a return to the wild type phenotype). In contrast, forward mutations are mutations in a wild-type gene. Bacterial forward mutation assays are less commonly used than reversion assays. The Ames test is an assay that measures bacterial revertant colony formation. The test was named to honor Bruce Ames, who first identified and reported the utility of this assay in detecting mutations in 1974 (Ames, 1974). Ames created a series of genetically modified strains of *Salmonella typhimurium* with mutations in the histidine gene such that these strains require histidine for growth. In the Ames assay, auxotrophic bacteria that require the amino acid histidine for growth are treated with a test chemical or vehicle or positive control on an agar plate consisting of growth media that lacks histidine. Therefore, only bacteria that have reverted back to wild type histidine locus are now capable of growth and formation of colonies. Increased numbers of revertant colonies in test chemical treated plates compared to concurrent untreated or vehicle control plates is an indication of the mutagenic potential of the test chemical (Fig. 44.1). As bacteria lack or have limited metabolic activation potential, the Ames assay is almost always carried out with and without exogenous metabolic activation to help distinguish chemicals that require metabolic activation for genotoxicity from those that have intrinsic genotoxic potential.

The Ames assay is the most commonly used assay to predict mutagenic potential of a test compound across several scientific disciplines and industries, because historically those chemicals shown to be positive in the Ames assay have also been shown to be positive in the in vivo rodent bioassay (~65% concordance), used to measure carcinogenic potential of a test chemical. The Ames assay is considered a gold standard assay to predict human carcinogenic potential of a chemical, by health regulatory authorities globally (Kirkland et al., 2005). The bacterial reversion assay is commonly carried out using several bacterial strains. Both *S. typhimurium* and *Escherichia coli* strains can detect base-pair substitution and frameshift mutations. These strains are genetically manipulated to increase their sensitivity to detect mutation frequency. A list of commonly used bacterial strains employed to detect reverse gene mutations and detailed assay guidelines can be obtained in OECD test 471 guideline (OECD 471, 1997; Mortelmans and Zeiger, 2000).

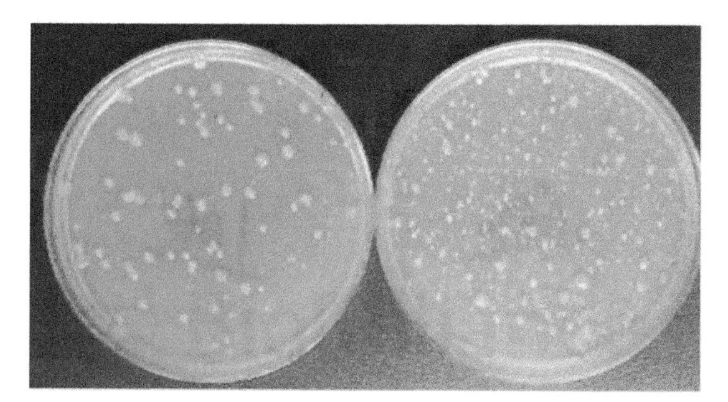

FIGURE 44.1 Ames assay plates, showing the increase in revertant colony formation in mutagenic chemical treated plates, relative to vehicle control plates. *Images courtesy of James Wojciechowski and Andrew Henwood, Bristol Myers Squibb, NJ, USA.*

Micronucleus Formation

Increased incidence of micronuclei formation serves as a good biomarker for genotoxic damage. A MN, also known as Howell–Jolly body, is often a broken chromatid/chromosome fragment or rarely a whole chromatid/chromosome that remains outside the nucleus after cell division. Micronuclei formed can be of various sizes, but typically they vary from 1/10th to 1/100th the size of the original nucleus (see Fig. 44.2). Acentric chromatid/chromosome fragments are usually the result of unrepaired or misrepaired DNA breaks. An MN consisting of an entire chromatid(s)/chromosome(s) arises due to deficiency in chromosome segregation. For a thorough review of the mechanisms behind MN formation, see the review by Luzhna et al. (2013).

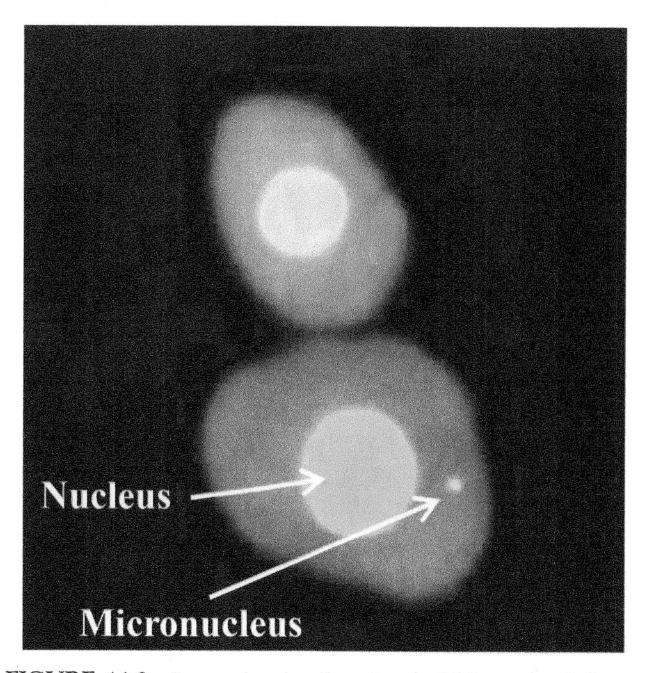

FIGURE 44.2 Image showing the micronuclei formation in hepatocytes. *Images courtesy of Rie Takashima, Mitsubishi Chemical Medience Corporation, Japan.*

The formation of MN can be readily measured both in vitro in different cell types (in vitro MN assay) and in vivo samples (in vivo MN assay) including peripheral blood, bone marrow, and tissues of rodents and other species using MN assay. Although in principle the in vitro MN assay can be performed in any cell type (primary or immortalized cells), the most commonly used cell types are Chinese hamster ovary (CHO) strain -K1 and -WBL cells, Chinese hamster V79 lung epithelial cells, L5178Y mouse lymphoma cells, TK6 human lymphoblastoid cells, and primary human lymphocytes. The in vitro MN assay is typically carried out with and without exogenous metabolic activation. For the in vivo MN assay, bone marrow or peripheral blood samples from rodents treated with multiple doses of test compound, vehicle, and appropriate positive control are analyzed for the formation of MN. Cell division (one mitosis) is required for the formation of micronuclei, and use of cytochalasin B (CytoB) blocks the cells in the binucleated stage with micronuclei; hence this is called the cytochalasin-blocked micronucleus assay. Cells treated with CytoB undergo karyokinesis but not cytokinesis, resulting in binucleated cells, so by evaluating binucleated cells only those cells that have undergone one cell replication are evaluated. MN formation can be due to either a clastogenic and/or an aneugenic DNA damaging event. However, MN formed by clastogens lack centromeres, whereas those micronuclei resulting from loss of an entire chromatid or chromosome (aneugenic event) contain a centromere. Hence, by centromere staining using various methods, such as fluorescence in situ hybridization (FISH) or other immunofluorescence methods, one can determine if micronuclei are a result of clastogenic or aneugenic events (Degrassi and Tanzarella, 1988; Miller and Adler, 1990; De Stoppelaar et al., 1997). Although yet debated, the size of MN can also serve as an indicator of an aneugenic or a clastogenic event, as the size of MN formed due to aneugenic events is generally larger than that resulting from a clastogenic event (Hashimoto et al., 2010). Advanced and high throughput MN detection technologies, such as flow cytometry and imaging-based technologies, make this a popular biomarker to measure genotoxic damage. More details on the conduct, analysis, and interpretation of in vitro and in vivo MN assays can be obtained in the OECD test guidelines 487 and 474, respectively (Krishna and Hayashi, 2000; Hayashi et al., 2000; Kirsch-Volders et al., 2003; Lorge et al., 2006; OECD Test 487, 2016g; OECD Test 474, 2016b).

Chromosomal Aberrations

One of the traditional and direct cytogenetic methods to detect genotoxic damage is based on a direct observation of the changes in the chromosomes after exposure to a chemical insult in cell populations under a microscope. Chromosomal aberrations (CAs) can be measured by the CA assay, which is commonly used in both clinical and nonclinical settings. Different types of CA are accumulated due to genotoxic damage such as insertions, deletions, translocations, endoreduplication, polyploidy, chromatid aberrations, and chromosomal gain and chromosomal loss. Depending on the relative abundance of a particular type of CA in comparison with other types of aberrations occurring due to treatment with a chemical, it is possible to understand the mechanism of action of the test chemical; for example, an abnormal number of chromosomes suggests aneuploidy, rather than clastogenicity. Although numerical changes can be observed in the CA assay, the MN assay is far more sensitive to chemicals inducing this type of DNA damage. CA studies can be carried out in both in vitro and in vivo test models. In vitro CA studies may be conducted with primary or immortalized cells, but regardless of cells used all require addition of an exogenous S9 liver metabolic fraction to enable detection of chemicals requiring metabolism to the ultimate genotoxicant. For in vivo CA studies, bone marrow or peripheral blood samples from rodents are commonly used. The main advantage of treating intact animals is that they include mammalian metabolism and DNA repair.

CA are usually studied in cell populations at the metaphase stage of the cell division, as chromosomes are condensed and damage is easily visualized at this stage. The cell populations can be arrested in metaphase by treatment with chemicals such as colcemid. Similar to MN detection, in principle the CA assay can be performed in any cell type. However, CHO cells are most commonly used in the in vitro studies because of the large size of chromosomes, fewer chromosomal numbers, a stable and well-defined karyotype, and a relatively shorter cell division interval (Fig. 44.3). Detection of the CA assay is a direct subjective method of measuring genotoxicity levels, not amenable to automation, and requires skilled cytogeneticists to analyze and score the CA. Invention of new technologies, such as fluorescence in situ hybridization (FISH) and chromosome painting, allows scientists to identify CA with relative ease and also to determine insertions and translocations more precisely, which is not possible with traditional CA assay. More details on the conduct, analysis, and interpretation of the CA assay data can be inferred from Bignold (2009) and OECD tests 473 (2016a, in vitro mammalian CA) and 475 (2016c, in vivo mammalian CA).

Comet Formation

The Comet assay, also known as the single cell gel electrophoresis assay, detects single or double strand breaks measured at the individual cell level. This assay

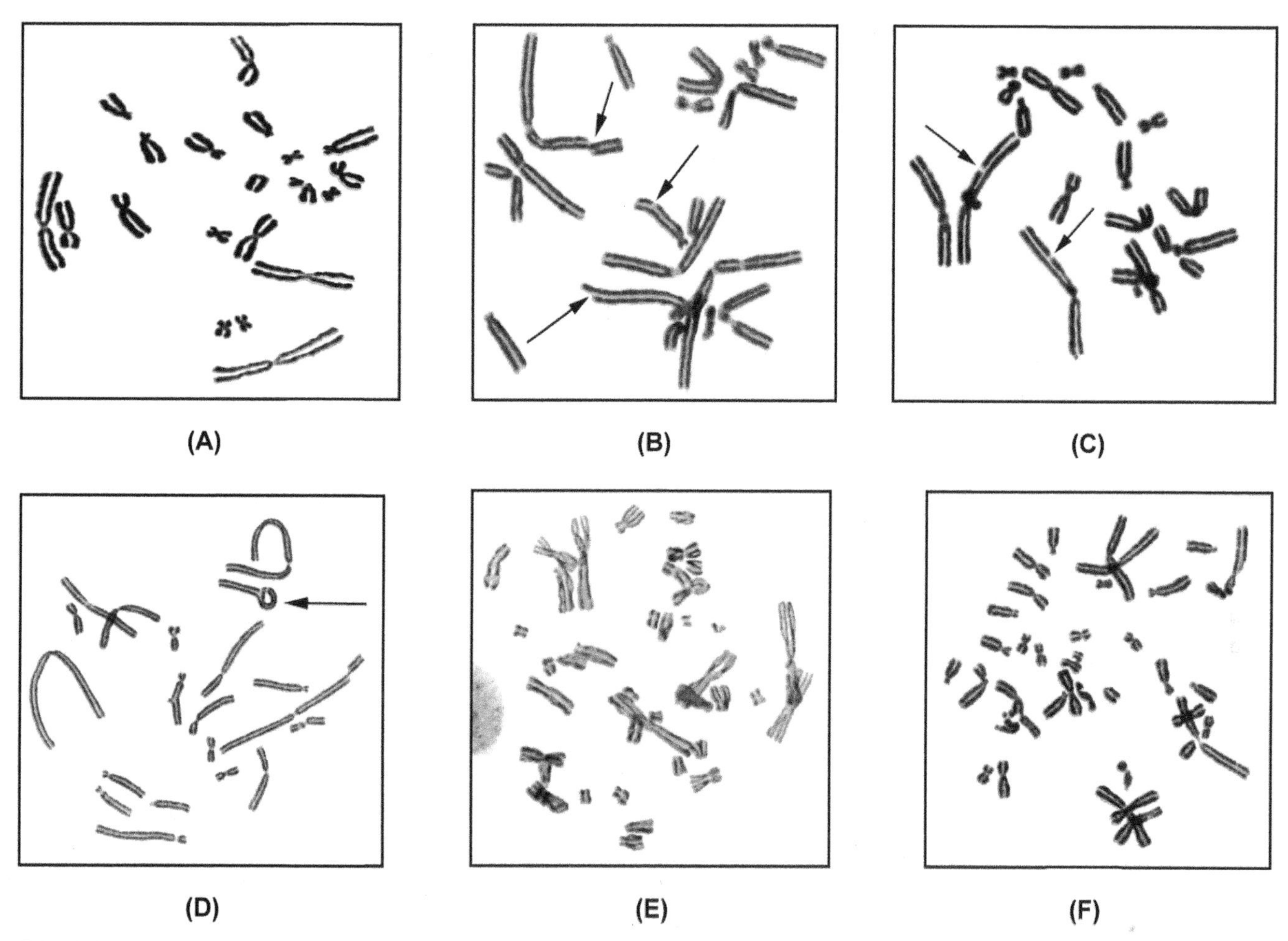

FIGURE 44.3 Images showing different chromosomal aberrations in CHO cells: normal (A), chromatid breaks (B), chromatid gaps (C), chromosome intrachange (D), endoreduplication (E), and polyploidy (F). *Images courtesy of James Wojciechowski and Louis Nimzroff, Bristol Myers Squibb, NJ, USA.*

was first developed by Ostling and Johansson in 1984 and was later revised by Singh et al. in 1988. With the advent of recent technological innovations such as fluorescent DNA stains and automated comet scoring for analysis, comet assay has emerged as a popular assay to detect DNA damage. Unlike the other assays described above, the comet assay measures transient genetic damage that is not a fixed change to the DNA. Strand breaks may be repaired, hence the requirement for very short time intervals between the final treatment and damage evaluation. The comet assay may be conducted in vitro using single cells from immortalized cell lines or in vivo for any tissue that can be dispersed to a single cell suspension.

At the molecular level, the formation of "comets" in the DNA of cells on genotoxic insult can be visualized through the method of gel electrophoresis and indicates DNA strand breaks, as the damaged DNA migrates at a different rate than nondamaged DNA during electrophoresis. In the comet assay, when a damaged DNA–containing single cell suspension embedded in low melting agarose is electrophoresed, the damaged DNA migrates away from the undamaged DNA–containing nucleoid body, resembling the structure of a comet, hence the name comet assay. In the comet structure, the undamaged DNA nucleoid part is referred to as the "head" and the trailing damaged DNA streak is referred to as the "tail" (Fig. 44.4). The percentage of DNA in the tail is directly proportional to the percentage of DNA damage that has occurred in a particular cell. Thus, by counting a representative sample of ~100–300 cells per tissue, it is possible to arrive at the average percentage of DNA damage accumulated in a particular tissue due to genotoxic stress.

The in vitro comet assay can be performed in any rodent or human cancer cell lines and human lymphocytes. It is typically carried out with and without exogenous metabolic activation, usually by supplementing with induced rat liver S9 fraction as indicated in other assays detailed earlier in this chapter. Although in principle any tissue can be used for in vivo comet assay analysis, high blood circulating organs such as

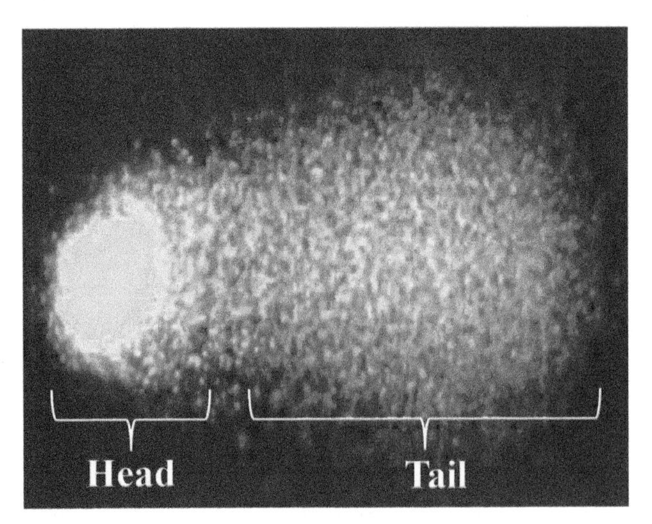

FIGURE 44.4 Image showing the comet formation comprising the head (undamaged DNA) and tail (damaged DNA). *Images courtesy of James Wojciechowski and Dan Roberts, Bristol Myers Squibb, NJ, USA.*

liver, kidney, and spleen are the target organ tissues that are analyzed. Also, if the route of administration is oral, then the site of contact or exposed tissue such as stomach mucosal cells are commonly used for comet analyses. More details on the conduct, analysis, and interpretation of the comet assay data can be inferred from Tice et al. (2000), Azqueta et al. (2011), Burlinson (2012), and OECD test 489 (2016h).

Toxicogenomic Signatures

Toxicogenomic studies in relation to genetic toxicology utilize the comprehensive gene expression data to identify gene expression signatures that strongly correlate with genetic toxicity. The use of these approaches has a long history and due to rapid development in gene expression technologies such as microarrays, real-time PCR, RNA sequencing, and other bioinformatics tools, the field of toxicogenomics has emerged as an area that may have immense potential for genotoxicity testing. In toxicogenomic studies, changes in gene expression patterns in response to genotoxic stress are most commonly studied, but changes in protein expression or metabolite formation can also be investigated. In addition, toxicogenomic approaches have the potential to overcome some of the limitations posed by traditional genotoxicity testing approaches, such as use of a limited number of animals, especially when changes in a limited set of genes are being investigated. Also, toxicogenomic approaches are more amenable for high throughput and high content screening than are traditional methods, thus aiding in reducing the costs associated with genetic toxicity screening and risk assessment.

Several investigators to date have reported different toxicogenomic signatures both in vitro using rodent and human cancer cell lines and in vivo in rodent models, to predict carcinogenicity in rodents and/or in humans and to distinguish genotoxic and nongenotoxic agents (Thybaud et al., 2007; Ellinger-Ziegelbauer et al., 2009; Fielden et al., 2011; Thomas and Waters, 2016; Li et al., 2015, 2017). Although toxicogenomic signatures have been proposed to differentiate genotoxins versus nongenotoxins, aneugens versus clastogens, carcinogens versus noncarcinogens, genotoxic carcinogens versus nongenotoxic carcinogens, etc., no general consensus exists on their usage in genetic toxicology testing. This is primarily because of diverse test models and gene expression analysis and bioinformatics technologies used to arrive at these toxicogenomic signatures, independently by different groups. Public repositories of gene expression studies include the Gene Expression Omnibus of the National Center for Biotechnology Information (https://www.ncbi.nlm.nih.gov/geo/) and the ArrayExpress of the European Bioinformatics Institute (https://www.ebi.ac.uk/arrayexpress/). Additionally, TOXsIgN (TOXicogenomic sIgNature) serves as a free cross-species repository for toxicogenomic signatures (http://toxsign.genouest.org/app/#/). The Comparative Toxicogenomics Database (CTD) is a public website that is maintained by the Department of Biological Sciences at North Carolina State University (http://ctdbase.org/). One of the aims of CTD is to advance the understanding of toxicogenomics and contains data on chemical–gene interactions and chemical–disease, gene–disease, and chemical–phenotype associations.

Other Genotoxic Biomarker Detection Methods

Other genotoxic biomarker detection methods include in vitro mammalian cell gene mutation assays (OECD test 476, 2016d), in vivo transgenic rodent somatic and germ cell gene mutation assays (OECD test 488, 2013), in vivo mammalian spermatogonial chromosome aberration test (OECD test 483, 2016f), in vivo mouse heritable translocation assay (OECD test 485, 1986), and the in vivo rodent dominant lethal test (OECD test 478, 2016e). Guidelines for performing these assays can be found in the above OECD test guidelines along with other genetic toxicology test guidelines (https://www.oecd-ilibrary.org/).

BIOMARKERS AND MECHANISM OF ACTION

Biomarkers of exposure are considered measures of internal dose. The genetic toxicology assays elaborated on in the previous sections of this chapter provide a general standard battery of genotoxic biomarkers available for the evaluation of functional cellular and molecular alterations on chemical exposure. Because changes in cellular molecules are thought to precede toxic outcomes, appropriate changes serve as early, sensitive indicators of potential genotoxicity (inclusive of mutagenicity and clastogenicity). Therefore, application of genetic techniques and biomarkers to provide an understanding of the effects of chemicals on cellular alterations is crucial. In this section, we will elaborate on how genotoxic biomarkers would facilitate better understanding of gene mutations, structural, and numerical chromosome changes that induce changes to genetic material (DNA).

Gene Mutations and Their Mechanism of Action

Mutation by definition refers to a permanent change in the nucleotide sequence of the genome of an organism. Mutation can be beneficial or detrimental to the cell. In this section, detrimental mutations are reviewed. Mutations in the DNA sequence may trigger a cascade of reactions that can adversely impact cellular events such as replication, transcription, or translation, possibly eventually resulting in cell death. Mutations can be classified into two main types: acquired (or somatic) and hereditary (or germline mutation). Acquired mutations occur in certain, but not all, cells of an organism as a result of DNA damage due to environmental agents such as nuclear radiation, UV exposure, chemical exposure (exogenous or endogenous), or errors during the DNA replication or transcription process. Elegant repair mechanisms exist to revert DNA damage caused by various mutagens and to repair errors that occur during replication or transcription; however, these repair mechanisms have their limitations and are not able to sufficiently restore the damage that is caused by potent mutagens. Acquired mutations in somatic cells cannot be passed down to the next generation. When a mutation is present in the germ cells (eggs or sperm), it can be passed down to the next generation. This type of mutation will be present in every cell of the offspring and is called a hereditary mutation.

Gene mutation refers to a permanent alteration in the DNA sequence of a gene. This change can range in size from a base to a large segment of the DNA that spans across multiple genes. The DNA sequence of a gene can be altered by missense mutation, nonsense mutation, insertion, deletion, frameshift mutation, and repeat expansion.

Chemical mutagens induce mutagenicity through various mechanisms. For example, 2-amino purine has a structure that is similar to that of the nitrogenous base adenine and, during the DNA replication, it forms a base analog. Then there are other mutagens that function as intercalating agents. An intercalating agent inserts itself between the strands of dsDNA and causes the DNA to unwind, thereby inhibiting mechanisms involved in replication and transcription.

The first step in identifying the genotoxic and carcinogenic potential of various agents is the conduction of genotoxicity assays. Among the various assays, as alluded to earlier in this chapter, the Ames test is the very first assay used as the point of detection to eliminate potential mutagenic/carcinogenic chemicals. All chemicals identified as mutagens in an Ames test may not necessarily test positive for carcinogenicity (Eastmond et al., 2009; Claxton et al., 2010). For example, chemicals that contain the nitrate functional group may test positive in the Ames test because nitrate compounds may generate the signaling molecule nitric oxide. It is important to understand, as described earlier in this chapter, that in an Ames test, the tester strains that are used are mutated and are unable to grow without histidine. Mutagenic chemicals would revert these mutated strains back to the wild type and enable them to grow in absence of histidine. Sodium azide is used as a positive control for *S. typhimurium* strains TA100 and TA1535 in absence of S9. Strains TA1535 and TA100 carry a *hisG46* marker that results from the substitution of amino acid leucine (GAG/CTC) by a proline (GGG/CCC). Sodium azide reverts the mutation to the wild-type by inducing point mutations at the GC base pair sites (Mortelmans and Zeiger, 2000). In bacteria, sodium azide is converted to the mutagenic compound azidoalanine by the bacterial enzyme *O*-acetylserine(thio)lyase. However, in a rodent carcinogenicity assay, sodium azide is determined to be negative. Hence, the mutagenic effects observed in *S. typhimurium* or *E. coli* are irrelevant for evaluation of the genotoxicity in humans. This is an interesting example of a chemical that induces strong mutagenic response in the Ames test, but is negative in the in vivo rodent carcinogenicity assay.

2-Nitrofluorene is used as positive control for *S. typhimurium* strains TA98 and TA1538 in absence of S9. 2-Nitrofluorene is converted to N-hydroxy-2-aminofluorene by nitroreductase activity. During this conversion, hydroxylamine intermediates form, which results in esterification to electrophiles and formation of DNA adducts. The mutagenic potency of nitrated fluorenes increases on addition of nitro groups (McCoy et al., 1981). Strains TA1538 and TA98 carry a frameshift mutation that affects the reading frame near the –C–G–C–G–C–G–C–G– sequence. 2-Nitrofluorene induces frameshift mutations and reverts these strains to the wild-type state (Mortelmans and Zeiger, 2000). Various aromatic nitroso derivatives such as 4-nitro-*o*-phenylenediamine also induce frameshift mutation in a similar manner. Hence, 4-nitro-*o*-phenylenediamine is often used as a positive control for strains TA98 and TA1538.

Most often, chemicals would cause a mutagenic effect only in the presence of metabolic factors (S9) because the presence of metabolic factors could result in the formation of mutagenic metabolites. For this reason, 2-aminoanthracene, an aromatic amine, is commonly used as a positive control in the Ames assay for all the bacterial strains only in presence of metabolic factors.

Chromosomal Aberrations and Their Mechanisms of Action

There is accumulating evidence that MN and CAs can independently aid in the prediction of risk of cancer development (Hagmar et al., 1998a; b; 2004; Bonassi et al., 2000; Boffetta et al., 2006; Bonassi et al., 2007). Often, in addition to gene mutations, these genotoxicity end points aid in the interpretation of chemical exposures and cancer risk. Hence, genotoxicity biomonitoring studies using MN or CA formation are widely undertaken. Recently, MN and nuclear abnormalities tests were used as biomarkers in the fish *Oreochromis niloticus* from a polluted pond in winter, spring, and summer, wherein fish collected at the polluted pond exhibited higher rates of DNA damage, as indicated by increased rates of MN and numerical aberrations in erythrocytes when compared to the negative control (Seriani et al., 2012). To provide comprehensive coverage of the mutagenic potential of a chemical, it is important to generate information on three types of genetic damage, namely gene mutation, changes to chromosome structure (i.e., clastogenicity), and number (i.e., aneuploidy) to provide comprehensive coverage of the mutagenic potential of a chemical. A clastogen is a chemical that can cause breaks in chromosomes, leading to sections of the chromosome being deleted, added, or rearranged. Known clastogens include acridine yellow, benzene, ethylene oxide, arsenic, phosphine mimosine, vincristine, etc. An aneugenic agent, on the other hand, is a chemical that promotes aneuploidy in cells by affecting normal cell division and migration of chromosomes during mitosis (or meiosis). Thus, aneuploidy is defined as an abnormal number of chromosomes in any given homologous pair of chromosomes. A few known examples of chemicals that are aneugenic are vinblastine, griseofulvin, colcemid, and mitomycin C. In general, several in vitro and in vivo mammalian cell assays described earlier can be performed to assess the potential of a chemical to cause chromosomal damage. Elaborated below are genotoxicity biomarkers and their mechanism of action, with a few examples.

Most of the mechanisms of CAs involve the concepts of clastogens directly acting on DNA to produce strand breaks, and subsequently, the survival of these directly caused DNA strand breaks—or misrepairs of them—to metaphase when they appear as chromosomal "breaks" or translocations. Chemicals such as mitomycin C, cadmium chloride, and 5-fluorouracil that are direct-acting clastogens have been shown to be positive in the MN assay (Shi et al., 2009) in addition to other known indirect clastogens, such as benzo[a]pyrene, cyclophosphamide, and 2-aminoanthracene. Many cytotoxic cancer therapies, such as cyclophosphamide, are positive in the in vivo MN assay by their mechanism of action. The main effect of cyclophosphamide is due to its metabolite phosphoramide mustard. This metabolite is only formed in cells that have low levels of aldehyde dehydrogenase. Phosphoramide mustard forms DNA crosslinks both between and within DNA strands at guanine N-7 positions (known as interstrand and intrastrand cross linkages, respectively). This, being irreversible, leads to cell death.

Induction of MN in vivo could occur by non−DNA-damaging mechanisms and the identification of a structurally diverse class of compounds that appear to bind noncovalently to DNA and that could enhance the DNA nicking activity, resulting in micronuclei induction, has been reviewed in Snyder (2010). A classic example of the mechanism of noncovalent DNA binding would be topoisomerase inhibitors such as etoposide. Etoposide is a chemotherapeutic drug that forms a ternary complex with DNA and the topoisomerase II enzyme (which aids in DNA unwinding), prevents religation of the DNA strands, and by doing so induces DNA strand breaks that cause errors in DNA synthesis and promote apoptosis of the cancer cell (Hande, 1998). Other examples are drugs that induce hypothermia (morphine, chlorpromazine), hyperthermia, or increased erythropoiesis following bone marrow toxicity or direct stimulation of precursor cells to divide.

Aneugens such as vinblastine, vincristine, colchicine, and nocodazole have been shown to induce micronuclei in the MN assay. Studies reviewed indicate that these chemicals interact with a number of cell components and processes such as synthesis and functioning of the spindle fibers, the activity of the centrosome, and modification of centromeres, resulting in the production of aneuploid progeny cells (Parry et al., 2002). A well-studied example of a chemical that induces aneuploidy is vincristine. It is a dimeric alkaloid, a mitotic inhibitor, and is used in cancer chemotherapy. Vincristine binds to tubulin dimers, inhibiting assembly of microtubule structures. Disruption of the microtubules arrests mitosis in metaphase, therefore affecting all rapidly dividing cell types including cancer cells; it has been used widely to treat pediatric leukemias. The genotoxicity and cytogenetic effects of vincristine at different genetic end points have been demonstrated using biomarkers such as the comet assay and MN assay, and a gene mutation assay in occupational exposure settings (Hongping et al., 2006).

Taken together, both the in vitro and in vivo genetic toxicology assays have proven invaluable in understanding the mechanism of action in a biological system and serve as essential biomarkers of genotoxicity.

MOLECULAR BASIS OF GENETIC VARIABILITY AND SUSCEPTIBILITY

Genotype is the sum total of all the genetic information of an individual. In a population, individuals differ from one another with regard to their genotypes. These differences in the genetic traits are often referred to as genetic variability. Although the variation between individuals constitutes genetic variability, the variability within the population is genetic diversity. Genetic variability or variation in genetic traits in response to the genetic and environmental influences determines the susceptibility of the individual or the population to toxicant injury or disease. Genetic susceptibility represents an increased likelihood of developing a disease, otherwise known as genetic predisposition.

Genetic Variability

Genetic variability represents the genetic differences within or between populations. Several possible factors, including gene flow due to population migration, homologous recombination or crossing over during meiosis, polyploidy, and mutations, might contribute to the genetic variability in the population. Of these possibilities, mutations make the primary source for genetic variation. Mutations affecting a single nucleotide (point mutation), whereby new alleles (alternative forms of the same gene or locus) are formed, constitute a common source of genetic variation. On the other hand, chromosomal mutations move the alleles around, either on to different loci on the same chromosome or a different chromosome by homologous recombination or translocation resulting in genetic variation, but this form is less common.

Gene mutations and epimutations (those that affect gene expression, but do not alter gene sequence) are also markers of genetic variability, which may have an impact on the population in different ways. A positive impact may enhance the well-being of the individual, and it is likely to be propagated in the population through the natural selection process, whereas the negative impact may be concealed if the mutation is lethal; the neutral mutations will not cause an impact. Both endogenous and exogenous factors contribute to genetic variations. Endogenous sources of genetic variations include heritable traits, life stage, and aging, whereas exogenous factors include preexisting health conditions, coexposure to chemicals, and dietary and psychosocial stress factors.

Examples of common genetic alterations or variants include Short Tandem Repeats (STRs), Single Nucleotide Polymorphisms (SNPs), and Copy Number Variants (CNVs). The STRs are also referred to as microsatellites or simple sequence repeats, and there are 1–6 or more base pairs of repetitive DNA sequences that are considered molecular markers of genetic variation. STR or microsatellite scans could help researchers in identifying the candidate genes involved in genetic susceptibility to a particular disease. Minisatellites are 10–60 or more base pairs of repetitive DNA sequences. Microsatellites and minisatellites together are classified as variable number of tandem repeats. CNV is a structural variation wherein an abnormal number of copies of a gene or a fragment of the gene occurs in the genome of an individual as a result of either duplication or deletion of a part of genome. Inherently, the human genome shows interindividual variability in the copy number of DNA sequences. Two types of CNVs exist in the human population; they are either inherited or arise de novo in the genome of an individual. CNVs play an important role not only in population diversity but also in human disease. CNVs have been associated with neuropsychiatric diseases such as autism, schizophrenia, and Prader–Willi in addition to a wide variety of other diseases such as Crohn's disease, psoriasis, and systemic lupus erythematosus (Girirajan et al., 2011; Thapar and Cooper, 2013; Ionita-Laza et al., 2009).

Genetic alterations or variations commonly occur in organisms due to two reasons: allelic variations and CAs. Allelic variations represent mutations in a given gene locus, whereas CAs represent structural damage to chromosomes or chromatids affecting multiple genes or loci. Exposure to environmental agents and drugs or pharmaceuticals can cause genetic alterations, which can affect both somatic and germ cells in the form of somatic and germline mutations, respectively. In somatic cells, genetic alterations can induce immediate effects, such as cell death or a slow transformation of cells into cancer cells. Usually these hazardous cellular events are preceded by accumulation of genotoxic or cytogenetic damage that can be monitored or detected using biomarkers such as mutations, MN formation, sister chromatid exchanges, CAs, and DNA repair. In contrast, heritable germline mutations occur in germ cells that are passed on to the offspring, such as chromosome aberrations, leading to birth defects and genetic diseases.

Genotoxic effects such as mutations in somatic cells precede the genetic changes following exposure to environmental agents. Several biomarkers of exposure and genotoxic effect have been reported from somatic cells in a number of studies due to their relative abundance in number compared to germ cells (Bickham et al., 2000). However, when the mutations occur in germ cells, those mutations are more detrimental than somatic mutations, as they can be inherited, thus making the investigation or studying of heritable mutations more critical for hazard identification and risk assessment. Biomarkers such as chromosomal rearrangements or abnormal number of chromosomes (e.g., trisomy, monosomy) are often associated with certain birth defects and genetic diseases that could be used for the prediction of certain disease outcomes by screening pregnant women. In this context, the affected population is likely to have low viability, life expectancy, and fertility because the induced heritable mutations are likely to cause several medical complications in the organisms harboring these genetic mutations. Also, the influence of genetic and environmental factors affecting the spontaneous formation of cytogenetic biomarkers of genotoxicity in humans can be measured. For example, in twins of different age groups, spontaneous micronuclei formation has been shown to be increased with age, particularly in females.

Epigenetic changes or modifications involve modifying DNA base pairs or proteins rather than changing the genome sequence itself, as is the case in mutations. Epigenetic changes include both heritable and nonheritable changes. Most common epigenetic mechanisms include DNA methylation, histone protein modifications (methylation, phosphorylation, acetylation, ubiquitylation, and sumoylation), and noncoding RNAs (miRNAs, picoRNAs, etc.). Epigenetic mechanisms regulate or modulate gene expression both in transient and permanent manners in response to endogenous and exogenous environmental stress factors (Hernandez and Singleton, 2012). Epigenetic mechanisms are highly dynamic in nature, thus facilitating modulation of gene and protein expression and are greatly influenced by environmental, dietary, race, gender, and age factors of the organism under study. DNA damage or genotoxic stress is shown to induce the miRNA family miR34 through the p53 pathway, which includes three miRNAs: miRNA-34a, miRNA-34b, and miRNA-34c (He et al., 2007). Also, miR-125b and miR-504 are shown to negatively regulate p53 expression, a key regulator of the genotoxic damage response pathway (Le et al., 2009; Hu et al., 2010). A few more examples of direct regulation of cytochrome P450 (CYP) enzymes by miRNAs that affect the toxicokinetics of drugs include miR-27b for CYP1B1 and CYP3A4 and miR-378 for CYP2E1. SNPs in miRNAs are another regulatory possibility affecting the toxicokinetics of drugs. Other sources of genetic variability include homologous recombination, translocation, and polyploidy.

With the completion of the Human Genome Project in 2003, common genetic variants that contribute to the susceptibility and progression of disease and the variability in the response to treatment have been mapped using state-of-the-art methods in genotyping and sequencing (Abramowicz, 2003). High-throughput screening techniques, such as next-generation DNA sequencing (NGS) and other wide-range genetic screening technologies, will facilitate identification of sources for genetic variability among populations. Understanding the relationship between genetic variation and biological function is essential for unraveling the critical aspects of the biology, evolution, and pathophysiology of humans. In terms of disease susceptibility, two kinds of genetic variants are known. The first one is the rare variants that drastically affect simple traits (e.g., cystic fibrosis), whereas the second variant is a common one having mild impact on complex traits (e.g., diabetes). However, there are a great many unknowns between these two categories that are yet to be identified by the scientific community. To investigate the data gaps and explore the association between phenotype and genotype, the 1000 genomes project was initiated in 2008. This is an international research effort aimed at measuring the human genetic variations in 1000 unrelated individuals, by performing genome-wide sequencing of 1000 individual genomes using NGS platforms (1000 Genomes Project Consortium, 2010). In a pilot study of the 1000 genomes project, they identified 250–300 loss-of-function variants in annotated genes. Also, the genes related to human disease susceptibility were compiled in the HapMap (Haplotype Map) project, which was coordinated by a group of countries including the United States, United Kingdom, China, Canada, Japan, and Nigeria. From a limited set of DNA samples, the HapMap project together with the Human Genome Project and the SNP Consortium identified 10 million common SNPs. Information on the linkage disequilibrium patterns of these SNPs was employed in conducting genome-wide association studies (GWAS); thereby several hundreds of genomic loci affecting human diseases were identified (Buchanan et al., 2012).

Susceptibility

Susceptibility involves an increased likelihood of developing a given disease due to heritable changes such as gene mutations; it is also referred to as genetic predisposition. Genetic polymorphisms in drug-metabolizing enzymes play an important role in the susceptibility or sensitivity of individuals to environmental chemicals. Genetic markers such as SNPs are correlated with genetic variation in a population. The SNPs are inherited as a cluster, which is referred to as a haplotype. Influence of genetic variation on response to pharmaceutical agents or drugs, determined by SNPs or gene expression, is referred to as pharmacogenomics. Based on their location in the genome, the SNPs are classified into three types: (1) SNPs within the coding region of the genes that are referred to as coding SNPs (cSNPs), which may or may not cause changes in amino acids; (2) SNPs in noncoding sequences that are upstream of regulatory regions and introns of the genes, referred to as perigenic SNPs (pSNPs) and which may not be functional; and (3) SNPs in intergenic (between genes) sequences that are referred to as coding intergenic SNPs (iSNPs). In general, cSNPs are more detrimental to organisms than pSNPs or iSNPs. However, the cSNPs may not always have a detrimental impact because single nucleotide change may not always result in a change in amino acid due to the degenerative nature of the genetic code. Variations in the SNPs in the population can be related to the differences in susceptibility of individuals to diseases, pathogens, or toxicant exposure. SNPs have been used as high-resolution markers in genetic mapping using GWAS, a concept in which several genetic variants are examined to see whether they have an association with a particular trait. SNPs differ at a single nucleotide and often have two alleles in the population. The linkage disequilibrium pattern, i.e., a nonrandom association of two alleles at two or more loci, determines the number of SNPs required (Christiani et al., 2008).

Susceptibility of individuals to genotoxic environmental pollutant or toxicant exposures could also be affected by polymorphisms in phase I and phase II xenobiotic metabolizing enzymes. As reviewed recently by Zanger and Schwab (2013), the impact of genomic markers on cytochrome P450 enzymes has been in the order of CYP2D6>CYP2C19~CYP2A6>CYP2B6 >CYP2C9>CYP3A4/5 and this study focused on common variants, but not rare variants. In addition, DNA repair enzymes belonging to different repair pathways, such as base excision repair, nucleotide excision repair, mismatch repair, and recombination repair, may also contribute to the genetic variation and susceptibility of individuals towards DNA damage and carcinogenesis. Many of these sets of enzymes act by forming protein–protein interactions leading to large enzyme complexes, and the fidelity and functional ability of these enzymes can be affected by polymorphisms that might result in altered ability of the repair enzymes to form a functional enzyme complex (Miller et al., 2001).

Similarly to the case of genetic variability, genetic effects and epigenetic mechanisms also play a significant role in regulating the genetic susceptibility of an individual or organism. For example, CYP1A1 gene promoter methylation in the human lung tissue of nonsmokers is higher compared to that in heavy smokers (Anttila et al., 2003). Environmental factors can also influence CYP polymorphism in smokers and nonsmokers.

CONCLUDING REMARKS AND FUTURE DIRECTIONS

Screening for genotoxic damage using several in silico, in vitro, and in vivo screening assays is an important part of pharmaceutical and chemical safety testing across a wide range of industries. Existing genotoxic battery testing that investigates and monitors biomarkers of genotoxic damage, such as increased bacterial revertant colonies, MN formation, comet formation, and CAs, are of tremendous help in genotoxic hazard identification and risk assessment. However, currently there are no sensitive and specific short term in vivo assays to evaluate carcinogenicity of chemicals caused via nongenotoxic or epigenetic mechanisms and only a limited number of sensitive in vitro assays. In vitro assays to assess both genotoxic and nongenotoxic carcinogens include Syrian Hamster Embryo cell transformation assay (SHE CTA) and Bhas 42 CTA (OECD Guidance Document, 2015; OECD Draft Guidance Document, 2015). The 2-year in vivo rodent bioassay that is used currently to evaluate carcinogenicity of chemicals is extremely time-consuming and cost prohibitive for screening a plethora of potential genotoxic chemicals. In this regard, cutting-edge technologies such as toxicogenomics, high content imaging, FISH, and high throughput approaches, such as NGS technologies, would be valuable to gain insights into genotoxic, nongenotoxic, and epigenetic mechanisms that play an important role in carcinogenesis, genetic variability, and susceptibility.

References

1000 Genomes Project Consortium, 2010. A map of human genome variation from population-scale sequencing. Nature 467 (7319), 1061.

Abramowicz, M., 2003. The human genome project in retrospect. Adv. Genet. 50, 231.

Ames, B.N., 1974. A combined bacterial and liver test system for detection and classification of carcinogens as mutagens. Genetics 78 (1), 91.

Anttila, S., Hakkola, J., Tuominen, P., et al., 2003. Methylation of cytochrome P4501A1 promoter in the lung is associated with tobacco smoking. Cancer Res. 63 (24), 8623–8628.

Azqueta, A., Gutzkow, K.B., Brunborg, G., et al., 2011. Towards a more reliable comet assay: optimising agarose concentration, unwinding time and electrophoresis conditions. Mutat. Res. Genet. Toxicol. Environ. Mutagen. 724 (1), 41–45.

Bickham, J.W., Sandhu, S., Hebert, P.D., et al., 2000. Effects of chemical contaminants on genetic diversity in natural populations: implications for biomonitoring and ecotoxicology. Mutat. Res. Rev. Mutat. Res. 463 (1), 33–51.

Bignold, L.P., 2009. Mechanisms of clastogen-induced chromosomal aberrations: a critical review and description of a model based on failures of tethering of DNA strand ends to strand-breaking enzymes. Mutat. Res. Rev. Mutat. Res. 681 (2), 271–298.

Boffetta, P., van der Hel, O., Norppa, H., et al., 2006. Chromosomal aberrations and cancer risk: results of a cohort study from Central Europe. Am. J. Epidemiol. 165 (1), 36–43.

Bonassi, S., Hagmar, L., Strömberg, U., et al., 2000. Chromosomal aberrations in lymphocytes predict human cancer independently of exposure to carcinogens. Cancer Res. 60 (6), 1619–1625.

Bonassi, S., Znaor, A., Ceppi, M., et al., 2007. An increased micronucleus frequency in peripheral blood lymphocytes predicts the risk of cancer in humans. Carcinogenesis 28 (3), 625–631.

Buchanan, C.C., Torstenson, E.S., Bush, W.S., et al., 2012. A comparison of cataloged variation between International HapMap Consortium and 1000 genomes project data. J. Am. Med. Inf. Assoc. 19 (2), 289–294.

Burlinson, B., 2012. The in vitro and in vivo comet assays. In: Genetic Toxicology. Springer, New York, NY, pp. 143–163.

Christiani, D.C., Mehta, A.J., Yu, C.L., 2008. Genetic susceptibility to occupational exposures. Occup. Environ. Med. 65 (6), 430–436.

Claxton, L.D., Umbuzeiro, G.D.A., DeMarini, D.M., 2010. The Salmonella mutagenicity assay: the stethoscope of genetic toxicology for the 21st century. Environ. Health Perspect. 118 (11), 1515.

de Stoppelaar, J.M., de Roos, B., Mohn, G.R., et al., 1997. Analysis of DES-induced micronuclei in binucleated rat fibroblasts: comparison between FISH with a rat satellite I probe and immunocytochemical staining with CREST serum. Mutat. Res. Genet. Toxicol. Environ. Mutagen. 392 (1), 139–149.

Degrassi, F., Tanzarella, C., 1988. Immunofluorescent staining of kinetochores in micronuclei: a new assay for the detection of aneuploidy. Mut. Res./Environ. Mutagen. Relat. Subj. 203 (5), 339–345.

Eastmond, D.A., Hartwig, A., Anderson, D., et al., 2009. Mutagenicity testing for chemical risk assessment: update of the WHO/IPCS Harmonized Scheme. Mutagenesis 24 (4), 341–349.

Ellinger-Ziegelbauer, H., Fostel, J.M., Aruga, C., et al., 2009. Characterization and interlaboratory comparison of a gene expression signature for differentiating genotoxic mechanisms. Toxicol. Sci. 110 (2), 341–352.

Fielden, M.R., Adai, A., Dunn, R.T., et al., 2011. Development and evaluation of a genomic signature for the prediction and mechanistic assessment of nongenotoxic hepatocarcinogens in the rat. Toxicol. Sci. 124 (1), 54–74.

Girirajan, S., Campbell, C.D., Eichler, E.E., 2011. Human copy number variation and complex genetic disease. Annu. Rev. Genet. 45, 203–226.

Hagmar, L., Bonassi, S., Strömberg, U., et al., 1998a. Chromosomal aberrations in lymphocytes predict human cancer: a report from the European Study group on Cytogenetic biomarkers and health (ESCH). Cancer Res. 58 (18), 4117–4121.

Hagmar, L., Bonassi, S., Strömberg, U., et al., 1998b. Cancer predictive value of cytogenetic markers used in occupational health surveillance programs. In: Genes and Environment in Cancer. Springer, Berlin, Heidelberg, pp. 177–184.

Hagmar, L., Strömberg, U., Bonassi, S., et al., 2004. Impact of types of lymphocyte chromosomal aberrations on human cancer risk: results from Nordic and Italian cohorts. Cancer Res. 64 (6), 2258–2263.

Hande, K.R., 1998. Etoposide: four decades of development of a topoisomerase II inhibitor. Eur. J. Cancer 34 (10), 1514–1521.

Hashimoto, K., Nakajima, Y., Matsumura, S., Chatani, F., 2010. An in vitro micronucleus assay with size-classified micronucleus counting to discriminate aneugens from clastogens. Toxicol In Vitro 24 (1), 208–216.

Hayashi, M., MacGregor, J.T., Gatehouse, D.G., et al., 2000. In vivo rodent erythrocyte micronucleus assay. II. Some aspects of protocol design including repeated treatments, integration with toxicity testing, and automated scoring. Environ. Mol. Mutagen. 35 (3), 234–252.

He, L., He, X., Lim, L.P., et al., 2007. A microRNA component of the p53 tumour suppressor network. Nature 447 (7148), 1130.

Hernandez, D.G., Singleton, A.B., 2012. Using DNA methylation to understand biological consequences of genetic variability. Neurodegener. Dis. 9 (2), 53–59.

Hongping, D., Jianlin, L., Meibian, Z., et al., 2006. Detecting the cytogenetic effects in workers occupationally exposed to vincristine with four genetic tests. Mutat. Res. Fundam. Mol. Mech. Mutagen. 599 (1), 152–159.

Hu, W., Chan, C.S., Wu, R., et al., 2010. Negative regulation of tumor suppressor p53 by microRNA miR-504. Mol. Cell 38 (5), 689–699.

Ionita-Laza, I., Rogers, A.J., Lange, C., et al., 2009. Genetic association analysis of copy-number variation (CNV) in human disease pathogenesis. Genomics 93 (1), 22–26.

Kirkland, D., Aardema, M., Henderson, L., et al., 2005. Evaluation of the ability of a battery of three in vitro genotoxicity tests to discriminate rodent carcinogens and non-carcinogens: I. Sensitivity, specificity and relative predictivity. Mutat. Res. Genet. Toxicol. Environ. Mutagen. 584 (1), 1–256.

Kirsch-Volders, M., Sofuni, T., Aardema, M., et al., 2003. Report from the in vitro micronucleus assay working group. Mutat. Res. Genet. Toxicol. Environ. Mutagen. 540 (2), 153–163.

Klopman, G., Rosenkranz, H.S., 1984. Structural requirements for the mutagenicity of environmental nitroarenes. Mutat. Res. Fundam. Mol. Mech. Mutagen. 126 (3), 227–238.

Klopman, G., Frierson, M.R., Rosenkranz, H.S., 1985. Computer analysis of toxicological data bases: mutagenicity of aromatic amines in Salmonella tester strains. Environ. Mutagen. 7 (5), 625–644.

Krishna, G., Hayashi, M., 2000. In vivo rodent micronucleus assay: protocol, conduct and data interpretation. Mutat. Res. Genet. Toxicol. Environ. Mutagen. 455 (1), 155–166.

Le, M.T., Teh, C., Shyh-Chang, N., et al., 2009. MicroRNA-125b is a novel negative regulator of p53. Gene Dev. 23 (7), 862–876.

Li, H.H., Hyduke, D.R., Chen, R., et al., 2015. Development of a toxicogenomics signature for genotoxicity using a dose-optimization and informatics strategy in human cells. Environ. Mol. Mutagen. 56 (6), 505–519.

Li, H.H., Chen, R., Hyduke, D.R., et al., 2017. Development and validation of a high-throughput transcriptomic biomarker to address 21st century genetic toxicology needs. Proc. Natl. Acad. Sci. Unit. States Am. 114 (51), E10881–E10889.

Lorge, E., Thybaud, V., Aardema, M.J., et al., 2006. SFTG international collaborative study on in vitro micronucleus test: I. General conditions and overall conclusions of the study. Mutat. Res. Genet. Toxicol. Environ. Mutagen. 607 (1), 13–36.

Luzhna, L., Kathiria, P., Kovalchuk, O., 2013. Micronuclei in genotoxicity assessment: from genetics to epigenetics and beyond. Front. Genet. 4, 131.

McCoy, E.C., Rosenkranz, E.J., Rosenkranz, H.S., et al., 1981. Nitrated fluorene derivatives are potent frameshift mutagens. Mutat. Res. Genet. Toxicol. 90 (1), 11–20.

Miller, B.M., Adler, I.D., 1990. Application of antikinetochore antibody staining (CREST staining) to micronuclei in erythrocytes induced in vivo. Mutagenesis 5 (4), 411–415.

Miller, I.I.I.M.C., Mohrenweiser, H.W., Bell, D.A., 2001. Genetic variability in susceptibility and response to toxicants. Toxicol. Lett. 120 (1–3), 269–280.

Mortelmans, K., Zeiger, E., 2000. The Ames Salmonella/microsome mutagenicity assay. Mutat. Res. Fundam. Mol. Mech. Mutagen. 455 (1), 29–60.

Organisation for Economic Co-operation and Development (OECD), 1986. Test No. 485: Genetic Toxicology, Mouse Heritable Translocation Assay, OECD Guidelines for the Testing of Chemicals,

Section 4. OECD Publishing, Paris. https://doi.org/10.1787/9789264071506-en.
Organisation for Economic Co-operation and Development (OECD), 1997. Test No. 471: Bacterial Reverse Mutation Test, OECD Guidelines for the Testing of Chemicals, Section 4. OECD Publishing, Paris. https://doi.org/10.1787/9789264071247-en.
Organisation for Economic Co-operation and Development (OECD), 2013. Test No. 488: Transgenic Rodent Somatic and Germ Cell Gene Mutation Assays, OECD Guidelines for the Testing of Chemicals, Section 4. OECD Publishing, Paris. https://doi.org/10.1787/9789264203907-en.
Organisation for Economic Co-operation and Development (OECD), 2015a. Guidance Document on the In Vitro Syrian Hamster Embryo (SHE) Cell Transformation Assay. Series on Testing & Assessment No. 214. ENV/JM/MONO, 18.
Organisation for Economic Co-operation and Development (OECD), 2015b. Hatano Research Institute (HRI) Draft guidance Document for in vitro Bhas 42 cell transformation assay. Retrieved July 12, 2018, from: http://www.oecd.org/env/ehs/testing/Bhas 42 CTA GD after 3rd comments-F_CLEAN.pdf.
Organisation for Economic Co-operation and Development (OECD), 2016a. Test No. 473: In Vitro Mammalian Chromosomal Aberration Test, OECD Guidelines for the Testing of Chemicals, Section 4. OECD Publishing, Paris. https://doi.org/10.1787/9789264264649-en.
Organisation for Economic Co-operation and Development (OECD), 2016b. Test No. 474: Mammalian Erythrocyte Micronucleus Test, OECD Guidelines for the Testing of Chemicals, Section 4. OECD Publishing, Paris. https://doi.org/10.1787/9789264264762-en.
Organisation for Economic Co-operation and Development (OECD), 2016c. Test No. 475: Mammalian Bone Marrow Chromosomal Aberration Test, OECD Guidelines for the Testing of Chemicals, Section 4. OECD Publishing, Paris. https://doi.org/10.1787/9789264264786-en.
Organisation for Economic Co-operation and Development (OECD), 2016d. Test No. 476: In Vitro Mammalian Cell Gene Mutation Tests Using the Hprt and Xprt Genes, OECD Guidelines for the Testing of Chemicals, Section 4. OECD Publishing, Paris. https://doi.org/10.1787/9789264264809-en.
Organisation for Economic Co-operation and Development (OECD), 2016e. Test No. 478: Rodent Dominant Lethal Test, OECD Guidelines for the Testing of Chemicals, Section 4. OECD Publishing, Paris. https://doi.org/10.1787/9789264264823-en.
Organisation for Economic Co-operation and Development (OECD), 2016f. Test No. 483: Mammalian Spermatogonial Chromosomal Aberration Test, OECD Guidelines for the Testing of Chemicals, Section 4. OECD Publishing, Paris. https://doi.org/10.1787/9789264264847-en.
Organisation for Economic Co-operation and Development (OECD), 2016g. Test No. 487: In Vitro Mammalian Cell Micronucleus Test, OECD Guidelines for the Testing of Chemicals, Section 4. OECD Publishing, Paris. https://doi.org/10.1787/9789264264861-en.
Organisation for Economic Co-operation and Development (OECD), 2016h. Test No. 489: In Vivo Mammalian Alkaline Comet Assay, OECD Guidelines for the Testing of Chemicals, Section 4. OECD Publishing, Paris. https://doi.org/10.1787/9789264264885-en.
Parry, E.M., Parry, J.M., Corso, C., et al., 2002. Detection and characterization of mechanisms of action of aneugenic chemicals. Mutagenesis 17 (6), 509–521.
Rosenkranz, H.S., Klopman, G., 1990. Structural basis of carcinogenicity in rodents of genotoxicants and non-genotoxicants. Mutat. Res. Fundam. Mol. Mech. Mutagen. 228 (2), 105–124.
Seriani, R., Abessa, D., Kirschbaum, A.A., et al., 2012. Water toxicity and cyto-genotoxicity biomarkers in the fish *Oreochromis niloticus* (Cichlidae). Ecotoxicol. Environ. Contam. 7 (2).
Shi, J., Bezabhie, R., Szkudlinska, A., 2009. Further evaluation of a flow cytometric in vitro micronucleus assay in CHO-K1 cells: a reliable platform that detects micronuclei and discriminates apoptotic bodies. Mutagenesis 25 (1), 33–40.
Snyder, R.D., 2010. Possible structural and functional determinants contributing to the clastogenicity of pharmaceuticals. Environ. Mol. Mutagen. 51 (8–9), 800–814.
Thapar, A., Cooper, M., 2013. Copy number variation: what is it and what has it told us about child psychiatric disorders? J. Am. Acad. Child Adolesc. Psychiatr. 52 (8), 772.
Thomas, R.S., Waters, M.D. (Eds.), 2016. Toxicogenomics in Predictive Carcinogenicity. Royal Society of Chemistry. https://doi.org/10.1039/9781782624059.
Thybaud, V., Le Fevre, A.C., Boitier, E., 2007. Application of toxicogenomics to genetic toxicology risk assessment. Environ. Mol. Mutagen. 48 (5), 369–379.
Tice, R.R., Agurell, E., Anderson, D., et al., 2000. Single cell gel/comet assay: guidelines for in vitro and in vivo genetic toxicology testing. Environ. Mol. Mutagen. 35 (3), 206–221.
Zanger, U.M., Schwab, M., 2013. Cytochrome P450 enzymes in drug metabolism: regulation of gene expression, enzyme activities, and impact of genetic variation. Pharmacol. Ther. 138 (1), 103–141.

CHAPTER

45

Epigenetic Biomarkers in Toxicology

Anirudh J. Chintalapati, Frank A. Barile

St. John's University, College of Pharmacy and Health Sciences, Department of Pharmaceutical Sciences, Toxicology Division, New York, United States

INTRODUCTION

Epigenetics was first coined in 1942 by Conrad Hal Waddington, who used it to describe how genes and their products interact to produce diversity of phenotypes (Jablonka and Lamb, 2002). Since then, the definition has evolved to signify heritable changes in gene expression, which do not involve alterations in DNA sequence (Roloff and Nuber, 2005). Major epigenetic (EG) mechanisms include DNA methylation, covalent posttranslational modifications (PTMs) of histone proteins, and noncoding RNAs (ncRNAs). These mechanisms mediate transcriptional alterations to orchestrate processes necessary to maintain chromatin structure and differential patterns of gene expression. DNA methylation is covalent addition of methyl groups to cytosine of CpG dinucleotides (cytosine-5′−phosphate bond−guanine), which is implicated in gene silencing. PTMs of histone proteins, including acetylation, methylation, phosphorylation, ubiquitination, and sumoylation, create dynamic states of chromatin that influence gene expression (Gadhia and Barile, 2014). NcRNAs are associated with nonprotein coding segments of the mammalian genome, which regulates and influences several EG modifications in addition to regulation of developmental processes, such as embryogenesis and tissue morphogenesis (Sauvageau *et al.*, 2013; Grotece *et al.*, 2013). These and other accompanying EG modifications play important roles in processes of cellular development, imprinting, cellular differentiation, X-chromosome inactivation, and transposon silencing (Barile, 2013). Furthermore, tracing subcellular EG mechanistic changes provide novel biomarkers for drug classification, disease diagnosis, and/or prognosis and endow viable therapeutic targets. For example, endocrine disruptors vinclozolin and methoxychlor induce alterations in DNA methylation pattern in germline and are classified as EG reproductive/developmental toxicants (Anway *et al.*, 2005; Autrup *et al.*, 2015).

Although it is well established that the science of genetics modulates etiology of disease, especially cancer, awareness of EG etiology of disease states has more recently been the subject of increasing scrutiny. The Basis of diseases that are implicated to involve EG causes includes mechanisms that do not require genetic mutations. These EG mechanisms incorporate aberrations in DNA methylation and/or improper histone modifications, ostensibly resulting in abnormal gene expression and malfunction of normal EG progression (Feinberg, 2007). Such alterations are mainly conveyed through either loss of function or gain of function of EG modifiers, resulting in an abnormal phenotype. The following discussion describes critical functions of EG regulators and consequences of their malfunction.

DNA METHYLATION

DNA methylation refers to the covalent addition of a methyl group donated by cofactor *S*-adenosyl methionine (SAM) to carbon-5 of cytosine of CpG dinucleotides, catalyzed via action of DNA methyltransferases. Product of this reaction, 5-methylcytosine (5-mc), is capable of pairing with guanine without compromising DNA integrity (Goodman and Watson, 2002). Termed as a "chemically improbable reaction" (Chen *et al.*, 1991), DNA methylation is only possible by base flipping, or *eversion*, of target cytosine away from the inner helix (Klimasauskas *et al.*, 1994; Cheng and Roberts, 2001; Goll and Bestor, 2005). As a stable, highly conserved, but reversible modification, DNA methylation is most notably involved in repression of genomic activity, although some examples of genetic activation

Biomarkers in Toxicology, Second Edition
https://doi.org/10.1016/B978-0-12-814655-2.00045-1

are noted. For instance, CpG island methylation at gene promoter region of FoxA2, a transcriptional factor responsible for endodermal development and pancreatic beta-cell gene expression, resulted in transcriptional activation of FoxA2 gene expression (Halpern et al., 2014). In addition, DNA methylation also has fundamental roles in processes of X-chromosome inactivation, imprinting, embryogenesis, and silencing of transposons. Loss of conserved patterns, resulting in either hypo- or hypermethylation, results in abnormal gene expression whose consequences are implicated in carcinogenesis.

Pathways of DNA Methylation

Two modes of DNA methylation are now widely accepted: *de novo* and maintenance methylation. *De novo* methylation occurs when new patterns of DNA methylation are established, whereas maintenance methylation involves conservation of existing patterns during DNA replication. How *de novo* methylation is prompted is incompletely understood. However, Klose and Bird (2006) postulate that *de novo* methyltransferases target their substrates via transcriptional repressors or other factors that aim at specific sequences of DNA. Alternatively, maintenance methylation relies on compliment of each methylated parent strand formed during replication; cytosine, however, as opposed to 5-mc, is incorporated into regions that should be methylated, resulting in hemimethylated DNA (Goodman and Watson, 2002). DNA methyltransferases then transfer methyl groups to cytosine of newly formed compliment strands, thus conserving the methylation pattern and completing replication processes (Goodman and Watson, 2002).

Demethylation is a biological method by which methyl groups are removed from cytosine residues, generally resulting in gene reactivity. Since there are no known active demethylases, demethylation is proposed to occur by both **passive** and **active** reactions (Ooi and Bestor, 2008). **Passive** DNA methylation is described as failure to methylate DNA following its replication as a function of deficit or malfunction of DNA methyltransferase 1 (DNMT1). Alternatively, **active** demethylation occurs by two different mechanisms. Hydrolytic deamination of 5-mc in DNA to thymine is suggested as one mechanism. This base substitution, when repaired properly, is replaced by an unmethylated cytosine. Conversion of 5-mc to 5-hydroxymethylcytosine (5-hmC), 5-formylcytosine, and 5-carboxylcytosine by ten-eleven translocation enzymes (TET) is a second pathway for demethylation (Tammen *et al.*, 2012). Both passive and active reactions result in loss of DNA methylation patterns.

TABLE 45.1 DNA Methyltransferases (DNMTs)

Name	Description
DNMT1	Preference for hemimethylated DNA; maintenance methyltransferase; some de novo methylation activity
DNMT3a	De novo methyltransferase
DNMT3b	De novo methyltransferase
DNMT3L	Catalytically inactive; required cofactor for DNMT3a and DNMT3b activity
DNMT1o	Oocyte-specific form of DNMT1
DNMT2	Not required for either de novo or maintenance methylation

DNA Methyltransferases

Several forms of DNA methyltransferases have been identified and are outlined in Table 45.1. The First purified and cloned eukaryotic DNA methyltransferase was DNA (cytosine-5)-methyltransferase (Bestor *et al.*, 1988), later termed DNMT1. DNMT1 is usually referred to as maintenance methyltransferase as it has a higher preference for hemimethylated DNA (Pradhan *et al.*, 1999; Hermann *et al.*, 2004). DNMT1 is targeted to replication fork by specific factors, such as UHRF1 (ubiquitin-like, containing PHD and RING finger domains protein 1), which recognize hemimethylated DNA (Bostick *et al.*, 2007; Sharif *et al.*, 2007). DNMT1o is an oocyte-specific truncated form of DNMT1 found in very high levels in oocytes (Goll and Bestor, 2005). Other DNA methyltransferases include DNMT3a and DNMT3b, predominant *de novo* methyltransferases (Okano *et al.*, 1999). Their activity is linked with DNMT3-like protein (DNMT3L), an inactive regulatory cofactor that enhances *de novo* methylation (Suetake *et al.*, 2004). DNMT3b is also responsible for methylation of satellite DNA and pericentromeric regions of chromosomes (Okano *et al.*, 1999). For most of the last three decades, biochemical analyses of DNMT2 enzymes that contain motifs similar to DNA methyltransferases demonstrated very weak, albeit no specific, DNA methyltransferase activity (Okano *et al.*, 1998a,b). More recent data confirm RNA methylation activity of DNMT2 at cytosine38 (C38) of $tRNAAsp^{GTC}$ in mice, *Drosophila melanogaster* and *Arabidopsis thaliana*. Current advances in sequence alignment analysis of methyltransferase catalytic domains suggest that eukaryotic DNMT1, DNMT2, and DNMT3 probably evolved from common ancestral DNMT2-like RNA methyltransferases. Progressive evolution steered DNMT1 and -3 to shift DNA to RNA, while action of DNMT2 remained conserved (Goll *et al.*, 2006; Jeltsch *et al.*, 2017).

Transcriptional Repression and DNA Methyl-Binding Proteins

Transcriptional silencing elicited by DNA methylation occurs by two different mechanisms—through "removal of a permissive environment" or by "acquisition of a repressive environment" (Prokhortchouk and Defossez, 2008).

In the "removal of a permissive environment," a protruding methyl group is oriented away from the major groove of DNA, thus interfering with transcription factor (TF) binding (Klose and Bird, 2006; Perera and Herbstman, 2011). In a second mechanism, "acquisition of a repressive environment" involves transcriptional repressors that recognize methylated DNA. The repressors, or DNA methyl-binding proteins (MBPs), bind and prevent accessibility to DNA, thus obscuring these areas from TFs. In addition, DNA MBPs recruit transcriptional corepressors that create repressive chromatin, thus solidifying gene suppression (Robertson and Jones, 2000; Klose and Bird, 2006). In the latter environment, DNA methylation is viewed as initiator of process, while activity of repressors and corepressors facilitates transcriptional silencing (Jones and Baylin, 2002). Active genes, however, can also be methylated, but the degree of methylation is an important consideration (Zemach *et al.*, 2010).

Table 45.2 lists three families of DNA MBPs: (1) methyl-CpG-binding domain (MBD)–containing family; (2) zinc finger proteins; and (3) SET and RING finger associated (SRA) domain-containing proteins (Prokhortchouk and Defossez, 2008; Sasai and Defossez, 2009). MBDs prevent binding of TFs and associate with corepressor complexes to enhance gene repression; for example, *MeCP2* (MBD-containing protein 2) associates with histone deacetylase (HDAC)–containing corepressor *SIN3* (Jones *et al.*, 1998; Nan *et al.*, 1998). Similar to MBDs, zinc finger protein *Kaiso* mediates gene silencing by linking with another deacetylase-containing corepressor complex, *N-CoR* (Yoon *et al.*, 2003). Other zinc finger proteins (*ZBTB4* and *ZBTB38*) are capable of binding to single CpG dinucleotides to mediate transcriptional repression (Filion *et al.*, 2006). SRA domain-containing proteins *UHRF1* (ICBP90) and *UHRF2* (NIRF) recognize and join with hemimethylated DNA (Mori *et al.*, 2012). Interestingly, *UHRF1* is essential for maintenance methylation (Bostick *et al.*, 2007; Sharif *et al.*, 2007).

HISTONE MODIFICATIONS

Within the beukaryotic nucleus, DNA is packaged by histones; i.e., proteins, which together with DNA, form the nucleosome (Barile, 2013). Nucleosome is a fundamental repeat unit of chromatin and consists of a 147-base pair DNA wrapped around two of each of core histones—H2A, H2B, H3, and H4—at a rate of 1.8 turns. Together with DNA, core histones thus comprise the histone octamer structure. Linker histone, H1, along with a string of DNA, connects neighboring nucleosomes (Luger *et al.*, 1997). Core histones contain protruding N-terminal tails that participate in PTMs, including acetylation, methylation, phosphorylation, ubiquitination, and sumoylation, and are summarized in Table 45.3.

PTMs affect histone/DNA, histone/histone, and histone/protein interactions, all of which influence chromatin structure and activity. According to the "histone code" hypothesis, cross talk between varying PTMs orchestrates DNA functions, such as gene transcription, replication, and repair (Strahl and Allis, 2000). Different PTMs activate or repress these functions depending on location and type of modification and presence of adjacent modifications (Kouzarides, 2007). PTMs are also site-specific, interdependent, and contribute to organization of chromatin (Lachner and Jenuwein, 2002).

Chromatin is found in two distinct forms in eukaryotic cells: heterochromatin and euchromatin. Heterochromatin, or condensed, closed form of chromatin, is transcriptionally silent, whereas euchromatin is opened, less condensed form, which allows for both active and silent

TABLE 45.2 DNA Methyl-Binding Proteins

Family	Members	Description
MBD	MBD1, MBD2, MBD4, MeCP2	Methyl-CpG-binding domain-containing; associate with corepressor complexes; involved in transcriptional repression[a,b]
Zinc finger	Kaiso, ZBTB4, ZBTB38	Zinc finger domain-containing; associate with corepressor complexes; involved in transcriptional repression[c]
SRA	UHRF1 (ICBP90), UHRF2 (NIRF)	SRA domain-containing; preference for hemimethylated DNA[d]

MBD, methyl-binding domain; *MeCP2*, methyl-CpG-binding protein 2; *SRA*, SET and RING finger associated domain; *UHRF*, ubiquitin-like, containing PHD and RING finger domains; *ZBTB*, zinc finger and BTB domain. Alternate acronyms are provided in parenthesis.

[a] *Jones et al. (1998).*

[b] *Nan et al. (1998).*

[c] *Yoon et al. (2003).*

[d] *Mori et al. (2012).*

TABLE 45.3 Posttranslational Histone Modifications

Type of Modification	Key Mediators	Required Cofactor/Substrate	Antagonizers (Erasers)
Acetylation	Histone acetyltransferases	Acetyl coenzyme A	Histone deacetylases
Methylation (arginine)	Protein arginine methyltransferases	*S*-Adenosyl methionine (SAM)	JMJD6; demethylimination
Methylation (lysine)	Histone lysine methyltransferases	SAM	Lysine demethylases
Phosphorylation	Serine/threonine and tyrosine kinases	ATP	Phosphatases
Ubiquitination	E1-activating, E2-conjugating, E3-ligating ubiquitin enzymes	Ubiquitin	Deubiquitinases
Sumoylation	E1-activating, E2-conjugating, E3-ligating SUMO enzymes	SUMO1, SUMO2, SUMO3	SUMO-cleaving isopeptidases

ATP, adenosine triphosphate; *SUMO*, small ubiquitin-related modifier.

gene transcription. This accounts for an important role of euchromatin in DNA repair and replication (Roloff and Nuber, 2005; Kouzarides, 2007). In general, heterochromatin is associated with low levels of histone acetylation but high levels of histone methylation (Kouzarides, 2007). Active transcription within euchromatin is supported by high levels of histone acetylation and trimethylation, whereas transcriptionally inactive euchromatin associates with low levels of histone acetylation, methylation, and phosphorylation (Kouzarides, 2007).

Histone Acetylation

Acetylation occurs in part through counterbalancing activities of histone acetyltransferases (HATs) and HDACs. HATs are generally accepted as transcriptional activators as their activity weakens interactions between histones and DNA (Robertson and Jones, 2000). HAT activity promotes hyperacetylation, thus stimulating an open chromatin accompanied by increased transcriptional movement (Roth *et al.*, 2001). Conversely, HDACs are transcriptional repressors that reverse HAT activity, consequently restoring histone/DNA interactions that support repressive chromatin (Robertson and Jones, 2000). Most importantly, acetylation not only disrupts histone/DNA interactions but also initiates histone/histone and histone/regulatory protein exchanges, all of which have weighty effects on DNA transcription (Roth *et al.*, 2001). Both HATs and HDACs often exist within multiprotein complexes (Bannister and Kouzarides, 2011).

Using acetyl coenzyme A as a cofactor, HAT catalyzes transfer of an acetyl group to lysine (K) side chains, neutralizing lysine's positive charge (Bannister and Kouzarides, 2011). HAT exerts its modifications to weaken interaction between histones and DNA, essentially by disrupting electrostatic interactions, namely between positive charge of lysine residues and negative charge of DNA (Sterner and Berger, 2000; Bannister and Kouzarides, 2011). The outcome is the production of a relaxed, open state of chromatin that is permissive to TF binding and transcriptional activation. Two major classes of HATs are categorized: type-A and type-B. Originally, HATs were defined by location—nuclear HATs were designated type-A, while cytoplasmic HATs were assigned to type-B (Roth *et al.*, 2001). However, increasing evidence suggests that HATs are not confined to one or other compartment (Parthun, 2007). Only known type-B HAT is HAT1, which performs acetylation of newly synthesized histones required for nuclear transport and deposition (Parthun, 2007; Shahbazian and Grunstein, 2007). Type-A HATs include *CBP/p300*, *GNAT*, and *MYST* families, which are classified and listed in Table 45.4, and are

TABLE 45.4 Histone Acetyltransferases

Family	Members	Histone Targets
CBP/p300	CBP/p300	H2A, H2B, H3, H4
GNAT	GCN5	H3, H4
	PCAF	H3, H4
	HAT1	H4
MYST	TIP60	H2A, H3, H4
	HBO1 (MYST2)	H4
	MOZ (MYST3)	H3
	MORF (MYST4)	H2A, H3, H4
	MOF (MYST1)	H4

CBP, CREB-binding protein; *GCN*, general control nonrepressed; *GNAT*, Gcn5-related N-acetyltransferase; *HAT*, histone acetyltransferase; *HBO*, HAT bound to ORC; *MOF*, males absent on first; *MORF*, MOZ-related factor; *MOZ*, monocytic leukemia zinc finger protein; *PCAF*, p300/CREB-binding protein-associated factor; *TIP*, HIV Tat-interacting protein. Alternate acronyms are provided in parentheses.
Adapted from Sterner, D.E., Berger, S.L., 2000. Acetylation of histones and transcription-related factors. Microbiol. Mol. Biol. Rev. 64, 435–459; Avvakumov and Côté, 2007; Bhaumik, S.R., Smith, E., Shilatifard, A., 2007. Covalent modifications of histones during development and disease pathogenesis. Nat. Struct. Mol. Biol. 14, 1008–1016.

TABLE 45.5 Histone Deacetylases (HDACs)

Class	Members	Description
I	HDAC1, HDAC2, HDAC3, HDAC8	Components of multiprotein nuclear complexes required for transcriptional repression (except HDAC8)[a]
IIa	HDAC4, HDAC5, HDAC7, HDAC9	Regulate nuclear-cytoplasmic shuttling; contain 14-3-3 binding domain[a]
IIb	HDAC6, HDAC10	HDAC6 is a major cytoplasmic deacetylase; HDAC10 associates with other HDACs[b]
III	SIRT1, SIRT2, SIRT3, SIRT5	NAD+-dependent; regulate cellular responses to stress[c]
IV	HDAC11	Structurally more related to class I than II HDACs[b]

NAD, nicotinamide adenine dinucleotide; *SIRT*, sirtuin.
[a]*Witt et al. (2009).*
[b]*de Ruijter et al. (2003).*
[c]*Saunders and Verdin (2007).*

based on sequence homology and structure (Bannister and Kouzarides, 2011).

HDACs, in contrast, remove acetyl groups from lysine residues, restoring lysine's positive charge. Deacetylation is characterized by transcriptional repression. Four major classes of HDACs account for most of enzymatic action: class I, class II, class III (more commonly referred to as "sirtuins"), and class IV HDACs (Table 45.5). Class I, II, and IV HDACs require Zn^{2+} to catalyze removal of acetyl group, while sirtuins necessitate NAD^{+} as a cofactor (Saunders and Verdin, 2007; Witt *et al.*, 2009). Enzymes are involved in many chromatin repressive complexes (De Ruijter *et al.*, 2003; Kouzarides, 2007). For instance, HDAC1 (class I) is associated with *Sin3* corepressor complex, which is recruited to methylated DNA by DNA MBP *MeCP2*, the product of which is formation of hypoacetylated chromatin and gene repression (Jones *et al.*, 1998; Nan *et al.*, 1998).

Histone Methylation

Histone methylation occurs on lysine (mono-, di-, or trimethylation) or arginine (mono- or symmetric/asymmetric dimethylation) residues. Depending on site and degree of methylation, lysine methylation either activates or represses transcription, whereas arginine methylation is generally an activating mark (Kouzarides, 2002; Völkel and Angrand, 2007). Three major protein families are responsible for histone methylation: protein arginine methyltransferases (*PRMTs*), *SET* domain-containing subfamilies, and *Dot1/DOT1L* (Völkel and Angrand, 2007).

TABLE 45.6 Histone Arginine Methyltransferases

Type	Function	Members	Target Residues	Activity
I	Monomethylation and asymmetrical dimethylation	PRMT1	H4R3	Transcriptional activation
		PRMT4 (CARM1)	H3R17, H3R26	Transcriptional activation
		PRMT6	H3R2	Transcriptional repression
II	Monomethylation and symmetrical dimethylation	PRMT5 (JBP1)	H4R3, H3R8	Transcriptional repression

PRMT, protein arginine methyltransferase. Alternate acronyms are provided in parenthesis.
Adapted from Di Lorenzo, A., Bedford, M.T., 2011. Histone arginine methylation. FEBS Lett. 585, 2024–2031.

Table 45.6 assembles histone PRMTs as type I or type II enzymes, based on their ability to perform monomethylation and asymmetric (type I) or symmetric (type II) dimethylation (Kouzarides, 2002; Bedford and Clarke, 2009; Bannister and Kouzarides, 2011). Addition of methyl group(s) from donor (SAM) by PRMTs to ω-guanidino moiety of arginine (R) residues disrupts hydrogen bond formation between arginine and nearby acceptors, while increasing residue's affinity for aromatic rings (Chen *et al.*, 1999). Common arginine methylation sites include R2, R17, and R26 of H3 and R3 of H4, resulting in transcriptional activation or repression, depending on site of methylation (Di Lorenzo and Bedford, 2011).

Jumonji C (JmjC) domain-containing iron- and 2-oxoglutarate-dependent dioxygenase (*JMJD6*) is the only known arginine demethylase (Chang *et al.*, 2007). It reverses dimethyl H3R2 (H3R2me2) and mono- and dimethyl H4R3 marks (H4R3me1 and H4R3me2), respectively (Chang *et al.*, 2007). Other mechanisms of demethylation involve demethylimination; i.e., conversion of arginine and monomethyl-arginine to citrulline by enzymatic activity of *PADI4/PAD4* (peptidylarginine deiminase 4), which prevents or reverses arginine methylation and, hence, transcriptional activation (Kouzarides, 2007; Völkel and Angrand, 2007).

Using SAM as a methyl donor, histone lysine methyltransferases (HKMTs) introduce methyl groups to ε-amino group of lysine (K), resulting in enhanced or repressed transcription (Bannister and Kouzarides, 2011). Unlike acetylation, methylation does not result in any change to oxidation states; instead it increases hydrophobic properties (Rice and Allis, 2001). Most HKMTs are generally site specific, modifying corresponding lysine residues on specific histones (Kouzarides, 2007). All HKMTs contain catalytically active SET domain, except for *Dot1/DOT1L* (Völkel and

TABLE 45.7 Histone Lysine Methyltransferases

Family	Structural Features	Representative Members	Target Residues
SET1	SET domain followed by a post-SET region	MLL, MLL2, MLL3, MLL4, SET1A, SET1B	H3K4
SUV39	SET domain flanked by pre- and post-SET regions	SUV39H1, SUV39H2, SETDB1 (ESET), EHMT1 (GLP1)	H3K9
		EHMT2 (G9a; Bat8)	H3K9; H3K27
SET2	SET domain flanked by AWS and post-SET regions	NSD1	H3K36; H4K20
		NSD2	H3K36
SUV4-20	SET domain followed by a post-SET region	SUV4-20h1, SUV4-20h2	H4K20
PRDM (RIZ)	SET domain and C2H2 zinc finger domain	PRDM2 (RIZ1)	H3K9
Dot1/DOT1L	Lacks SET domain	DOT1L	H3K79
EZ	SET domain preceded by pre-SET region	EZH1, EZH2	H3K27
SYMD	SET domain and MYND zinc finger domain	SYMD3	H3K4
Others	SET domain and MORN domain	SET7/9	H3K4

C2H2, two cysteine, two histidine; *DOT1l*, DOT1-like; *EHMT*, euchromatic histone lysine-N-methyltransferase; *EZH*, enhancer of zeste homolog; *MLL*, mixed lineage leukemia; *MORN*, membrane occupation and recognition nexus repeat; *MYND*, myeloid, Nervy, DEAF-1; *NSD*, nuclear receptor binding SET domain; *PRDM*, PR domain-containing; *SET*, Su(var)3-9, Enhancer of zeste (E(z)), and trithorax (trx); *SUV*, suppressor of variegation; *SYMD*, SET and MYND domain-containing. Alternate acronyms are provided in parenthesis.

Adapted from Kouzarides, T., 2002. Histone methylation in transcriptional control. Curr. Opin. Genet. Dev. 12, 198–209; Kouzarides, T., 2007. Chromatin modifications and their function. Cell 128, 639–705; Völkel, P., Angrand, P., 2007. Control of histone lysine methylation in EG regulation. Biochimie 89, 1–20.

Angrand, 2007; Bannister and Kouzarides, 2011). SET domain, or *Su(var)3-9, enhancer of zeste (E(z))*, and *trithorax (trx)* domain, is required for functional histone lysine methylation.

Table 45.7 summarizes properties of HKMT families, including SUV39 family, whose members, SUV39H1 and SUV39H2, were the first mammalian HKMTs identified (Völkel and Angrand, 2007). Additional HKMTs consist of SET1, SET2, EZH, SMYD, and PRDM families, as well as other SET-containing proteins, such as SET7/9 and SET8 (PR-SET7), and non–SET-containing protein, Dot1/DOT1L (Völkel and Angrand, 2007). Lysine methylation occurs on residues K4, K9, K27, K36, and K79 of H3 and K20 of H4; H3K9, H3K27, and H4K20 methylation results in transcriptional repression, while H3K4, H3K36, and H3K79 methylation causes transcriptional activation (Table 45.8) (Gadhia *et al.*, 2012; Lachner and Jenuwein, 2002; Völkel and Angrand, 2007).

Table 45.9 illustrates families of histone lysine methylases and demethylases whose activities are antagonistic.

TABLE 45.8 Association of Histone Lysine Methylation Marks

Transcriptional Activation[a]	Transcriptional Repression[a]
H3K4me2/3; H3K9me1; H3K27me1; H3K36me3; H3K79me2/3; H2BK5me1	H3K9me3; H3K27me3

[a]*Varier and Timmers (2011).*

Of these, lysine-specific demethylase 1 (LSD1) was the first to be discovered (amine oxidase). When combined with repressor, Co-REST, the complex activates demethylation of both H3K4me1 and -me2, which results in transcriptional repression (Kouzarides, 2007; Völkel and Angrand, 2007). Mechanism of LSD1 demethylation involves oxidation of an amine to form an unstable imine intermediate that ultimately regenerates unmethylated lysine residues (Völkel and Angrand, 2007). LSD1 also demethylates H3K27me1 and -me2, during interactions with androgen receptor, which is presumably the basis for transcriptional activation of androgen-responsive genes (Kouzarides, 2007; Völkel and Angrand, 2007). Other lysine demethylases include *JmjC* domain-containing families, hydroxylases *JHDM1*, *JHDM2*, and *JHDM3*, which require *JmjC* domains for their catalytic activity (Klose *et al.*, 2006).

Histone Phosphorylation

Important for transcription and cell cycle progression, phosphorylation is dependent on upstream signal transduction cascades (Baek, 2011). When activated, these signaling pathways regulate gene transcription and alter chromatin states. This is accomplished by stimulating phosphorylation of histones, TFs, and transcriptional coregulators (Baek, 2011). Phosphorylation involves activity of kinases that transfer phosphate groups from ATP to hydroxyl groups of target residue,

TABLE 45.9 Histone Demethylases

Family	Member(s)	Target Residues
HISTONE ARGININE DEMETHYLASE		
	Member	**Target Residues**[a]
	JMJD6	H3R2me2 H4R3me2/1
HISTONE LYSINE DEMETHYLASES		
Family	**Members**	**Target Residues**[b]
LSD1	LSD1 (BHC110; AOF2)	H3K4me2/1 H3K9me2
JHDM1	JHDM1A, JHDM1B	H3K36me2/1
JHDM2	JHDM2A (JMJD1A), JHDM2B (JMJD1B), JHDM2C (JMJD1C)	H3K9me3/2
JHDM3	JHDM3A (JMJD2A), JHDM3B (JMJD2B), JHDM3B (JMJD2C; GASC1), JHDM3D (JMJD2D)	H3K9me3/2 H3K36me3/2

JHDM, JmjC domain-containing histone demethylase; *JMJD*, JmjC domain-containing; *LSD*, lysine-specific demethylase. Alternate acronyms are provided in parenthesis.
[a]*Chang et al. (2007).*
[b]*Bhaumik et al. (2007).*

TABLE 45.10 Histone Phosphorylation

Target	Mediator(s)	Significance
H2BS14	MST1	Apoptotic chromatin condensation
H3S10	IKKα, MSK1/2, PIM1, PKB (AKT), RSK2	Transcriptional activation
	Aurora B	Mitotic chromatin condensation
H3S28	MSK1/2	Transcriptional activation
	Aurora B	Mitotic chromatin condensation
H3T11	CHK1	GCN5 recruitment, leading to acetylation of H3K9 and H3K14
H3Y41	JAK2	Transcriptional activation

CHK, checkpoint kinase; *IKK*, IκB kinase; *JAK*, Janus kinase; *MSK*, mitogen- and stress-activated protein kinase; *MST1*, mammalian sterile-20-like kinase; *PKB*, protein kinase B; *PIM*, protooncogene serine/threonine-protein kinase; *RSK*, p90 ribosomal S6 kinase.
Adapted from Baek, S.H., 2011. When signaling kinases meet histones and histone modifiers in nucleus. Mol. Cell 42, 274–284.

resulting in net formation of negative charges (Bannister and Kouzarides, 2011). As presented in Table 45.10, kinases modify histone serine and threonine residues (serine/threonine kinases), as well as tyrosine residues (tyrosine kinases), generally leading to transcriptional activation or chromatin condensation (Baek, 2011). For example, phosphorylation of H3S10 by a variety of serine/threonine kinases (RSK2, MSK1/2, protein kinase B [PKB] [ATK]) is a frontrunner for transcriptional activation (Baek, 2011). Alternatively, during mitosis, chromatin condensation is facilitated by phosphorylation of H3S10 and H3S28 by *Aurora B*, a process necessary for cell cycle progression (Sugiyama *et al.*, 2002; Bhaumik *et al.*, 2007; Baek, 2011).

In addition to specific TFs, kinases directly or indirectly target other histone modifiers, such as HATs (CREB, CBP, and GCN5), HDACs (SIRT1 and HDAC5), and HKMTs (EZH2), thus modulating their activity (Baek, 2011). For instance, direct phosphorylation of EZH2 by PKB (AKT) inhibits lysine methyltransferase activity, producing a decrease in H3K27 trimethylation (Baek, 2011). In addition, GCN5 is indirectly recruited following phosphorylation of H3T11, establishing acetylation of neighboring H3K9 and H3K14 residues (Shimada *et al.*, 2008); phosphorylation of H3S10, however, has a contrasting effect, causing hypoacetylation of same residues (Edmondson *et al.*, 2002). Activity of kinases is reversed by phosphatases, such as effects of serine/threonine phosphatases, PP1 and PP2A, the net reaction of which is dephosphorylation necessary for firm control of chromatin condensation (Sugiyama *et al.*, 2002).

Histone Ubiquitination and Sumoylation

In addition to mediating protein degradation, protein trafficking, stress responses, and cell cycle progression, ubiquitination also regulates gene transcription via histones (Pickart, 2001; Zhang, 2003). Ubiquitination of histones, unlike other proteins, is not linked to degradation; instead it dictates transcription in a site-specific manner by impacting higher-order chromatin structure (it is a bulky modification) (Zhang, 2003; Bhaumik *et al.*, 2007). Consequently the pathway is capable of recruiting other regulatory molecules that control transcription, thus enhancing or interfering with other neighboring histone modifications (Zhang, 2003).

Ubiquitination is the covalent addition of ubiquitin (Ub) via an isopeptide bond between C-terminal glycine of Ub and ε-amino group of lysine of either target proteins (i.e., histones) or other Ub molecules (Pickart, 2001; Zhang, 2003). Furthermore, ubiquitination is likely to occur by presence of multiple lysine residues found within Ub, allowing for formation of poly-Ub proteins (Zhang, 2003). Target residues include H2AK119 (mono- or poly-Ub), a biomarker associated with transcriptional repression, and H2BK120 (mono-Ub), a biomarker required for initiation and elongation of transcription (Zhang, 2003; Kouzarides, 2007).

Ubiquitination is performed by sequential activity of one E1-activating enzyme, varying E2-conjugating enzymes, and target-specific, RING finger domain-containing E3 ligases (Pickart, 2001; Zhang, 2003; Bhaumik *et al.*, 2007). For example, RING1, a polycomb group (PcG) protein, is an E3 ligase that performs monoubiquitination of H2AK119 during gene silencing (Bhaumik *et al.*, 2007; Kouzarides, 2007). Deubiquitination, or removal of ubiquitin groups, occurs via activity of deubiquitinases (isopeptidases), such as 2A-DUB, which mediates deubiquitination of H2AK119 (Zhang, 2003; Bhaumik *et al.*, 2007).

Sumoylation, like ubiquitination, is covalent addition of a small protein (SUMO) to lysine of a target protein via an isopeptide bond. SUMO, or "small ubiquitin-related modifier," is an ubiquitin-like protein distantly related to ubiquitin (Melchior, 2000). Members of SUMO family include SUMO1 (sentrin), SUMO2 (SMT3a; sentrin-3), and SUMO3 (SMT3b; sentrin-2), which are conjugated to target proteins by sequential activity of SUMO-activating enzyme (E1), -conjugating enzyme (E2), and -ligase (E3) (Johnson, 2004). Sumoylation is a reversible process, where removal of SUMO groups is performed by a group of isopeptidases—i.e., Ulp domain-containing SUMO-cleaving enzymes, such as *SENP1* (Johnson, 2004).

SUMO modifications regulate transcriptional processes and nuclear transport, as well as promote assembly of multiprotein complexes (Johnson, 2004). In addition, SUMO, which conjugates same lysine targets as ubiquitin, competes with and blocks ubiquitin binding at these sites (Johnson, 2004). Many targets of SUMO are involved in regulation of gene transcription, including transcriptional activators, coactivators, repressors, and corepressors, where sumoylation generally induces transcriptional repression (Johnson, 2004). For instance, SUMO1 is required for repressive activity of HDAC1, HDAC4, and p300 (David *et al.*, 2002; Kirsh *et al.*, 2002; Girdwood *et al.*, 2003).

TABLE 45.11 Chromatin Remodeling: Polycomb and Trithorax Group Complexes

Complex	Main Components	Function
POLYCOMB GROUP		
PRC2	EZH2, EED, RBBP7 (RBAP46), RBBP4/RBAP48, SUZ12[a,b,c]	Trimethylation of H3K27
PRC1	BMI1, CBX2, CBX4, CBX6, CBX7, CBX8, MEL18 (PCGF2), NSPC1 (PCGF1), PHC1-3, RING1, RING1B (RNF2)[b,c]	Recognizes H3K27me3; monoubiquitination of H2AK119
TRITHORAX GROUP		
MLL	MLL, MLL2, MLL3, MLL5, ASH1L, ASH2L, HCF1, MEN1, WDR5[c]	Trimethylation of H3K4

ASH, absent small or homeotic; *BMI1*, B lymphoma Mo-MLV insertion region 1; *CBX*, chromobox protein homologue; *EED*, embryonic ectoderm development; *EZH*, enhancer of zeste homologue; *HCF1*, host cell factor 1; *MEL18*, melanoma nuclear protein 18; *MEN1*, multiple endocrine neoplasia; *NSPC1*, nervous system Polycomb 1; *MLL*, mixed linage leukemia; *PHC*, polyhomeotic homologue; *PRC*, polycomb repressive complex; *RBBP*, retinoblastoma binding protein; *RING*, ring finger protein; *SUZ12*, suppressor of zeste homologue; *WDR5*, WD repeat domain 5. Some alternate acronyms are provided in parenthesis.

[a]*Schuettengruber et al. (2007).*

[b]*Simon, J.A., Kingston, R.E., 2009. Mechanisms of polycomb gene silencing: knowns and unknowns. Nat. Rev. Mol. Cell Biol. 10, 697–708.*

[c]*Mills (2010).*

POLYCOMB AND TRITHORAX PROTEINS

PcG and trithorax group (TrxG) proteins are evolutionarily conserved multiprotein complexes that regulate chromatin states in an opposing manner (Mills, 2010). More specifically, Table 45.11 outlines activities of PcG protein complexes that result in transcriptional repression, whereas those of TrxG promote transcriptional activation (Schuettengruber *et al.*, 2007; Mills, 2010). Both PcG and TrxG protein complexes are recruited to chromatin by DNA regulatory elements, where they form covalent modifications of histones (Mills, 2010; Tammen *et al.*, 2012).

Two main PcG protein complexes are reported: polycomb repressor complexes 2 (PRC2) and 1 (PRC1). Catalytic activity of PRC2 resides in its SET domain-containing subunit, EZH2 (Schuettengruber et al., 2007; Mills, 2010). Like other SET enzymes, EZH2 is responsible for histone lysine methylation; more specifically, it causes trimethylation of H3K27, a repressive mark (Schuettengruber *et al.*, 2007; Mills, 2010). EZH2 activity is dependent on the presence of other PRC2 subunits, such as EED and SUZ12 (Mills, 2010). Conversely, PRC1 recognizes H3K27me3 mark, leading to further PcG-dependent chromatin silencing through monoubiquitination of H2AK119. This activity is supported through action of its RING1 subunit (Kouzarides, 2007; Tammen *et al.*, 2012), a mark of which prevents RNA polymerase activity, thus halting gene transcription (Tammen *et al.*, 2012).

TrxG protein complexes are involved in several histone-modifying reactions. Mixed lineage leukemia (MLL) complex or mixed lineage leukemia (Tables 45.7 and 45.11) is one possible pathway, while ATP-dependent chromatin remodeling, such as execution by switch/sucrose nonfermentable (SWI/SNF), ISWI (protein **ISWI** or **imitation SWI** of *drosophila melanogaster*), and CHD (Chromodomain Helicase DNA-binding family of proteins)complexes (Mills, 2010), is another. Histone-modifying activity of MLL complexes is dependent on catalytic activity of SET domain-containing MLL subunit (MLL, MLL2, MLL3, or MLL5; Table 45.11), which results in trimethylation of H3K4 and subsequent transcriptional activation (Mills, 2010). Interestingly, as acetylation is an activating mark and prevents H3K27 trimethylation, MLL complex also recruits HATs, such as MYST1 (Dou et al., 2005), further promoting its transcription-enhancing activity (Mills, 2010).

NONCODING RNA

The significance of mRNA, tRNA, and rRNA in transcription of DNA, followed by translation and protein synthesis, and their associated molecular machinery, has been well established for several decades. Remarkably, genome-wide analysis in eukaryotes reveals that only 1%–2% of a transcribed genome is required to be functionally active and to encode protein; the transcribed, nonprotein coding segment of DNA is referred to as "junk DNA," and its nontranslatable product is labeled ncRNA (Frith *et al.*, 2005; Kaikkonen *et al.*, 2011). Exploration of the biological significance of both junk DNA and ncRNAs started in early 1990s leading to the discovery of several types of ncRNAs, which are primarily categorized according to the variable lengths formed by their constituent nucleotides. Although microRNAs (miRNAs) are the most studied variants among ncRNAs, other representatives such as small nucleolar RNAs, transcribed ultraconserved regions, PIWI-interacting RNAs (piRNAs), and a diverse group of long noncoding RNAs (lncRNAs) have been discovered and are known for their differential modulatory effects on genomic transcription and translation (Kaikkonen *et al.*, 2011). Table 45.12 outlines the variety of ncRNAs, as well as their length and frequency of occurrence in the human genome.

The biological reputation of miRNAs is attributed to their ability to specifically target mRNA, the basis of which leads to posttranscriptional gene silencing and suppression of translation. Biogenesis of miRNAs is mediated by transcription of miRNA genes by pol II promoter to produce a primary transcript of miRNA (Khandelwal *et al.*, 2015). The primary transcript produced is the precursor of miRNA constituting several thousands of nucleotides in length, which, when cleaved by nuclear Drosha/DGCR8 complex, forms long precursor miRNA (pre-miRNA) spanning hundreds of nucleotides. Pre-miRNA is then exported to the cytoplasm via Ran-GTP/exportin-5 complex and is further cleaved by Dicer/TRBP (TAR RNA-binding protein) complex to form mature miRNA, with a final length of 19–24 nucleotides. Mature miRNA combines with Argonaute proteins, which materializes into RNA-induced silencing complex. The completely formed complex recognizes and targets mRNA eventually causing transcriptional repression (He and Hannon, 2004). In contrast, biogenesis of piRNAs is Dicer-independent and is functionally involved in maintenance of genomic stability in germline cells. PiRNAs form complexes with PIWI proteins and suppress transposable element expression by either cleavage or progressive DNA methylation, rendering gene silencing (Aravin *et al.*, 2007).

TABLE 45.12 Classification of Noncoding RNAs

Type	Length (nt)	Function	Number in Humans	Example
tiRNAs	17–18	Transcription regulation	>5000	CAP1 gene and cAMP pathway
miRNAs	19–24	Target and inactivate mRNA causing transcriptional repression	>1500	miR-34b/c, miR-200
piRNAs	26–31	DNA methylation and transposon repression	~23,000	Targeting LINE1 and IAP elements
TSSa RNAs	20–90	Transcription regulation	>10,000	RNF12 and CCDC52 gene regulation
PASRs	22–200	Gene promoter region alterations	>10,000	Associated with protein coding genes
PROMPTs	<200	Transcription regulation	Unknown	EXT1 and RBM 39 gene processing
snoRNAs	60–300	rRNA alterations	>300	SNORD
T-UCRs	>200	mRNA and transcriptional regulation	>350	uc.338, uc.160+
NATs	>200	Gene expression regulation; RNA editing/stability	Unknown	p15-NAT
eRNAs	>200	Enhance gene transcription	Unknown	P53 derived eRNAs
lincRNAs	>200	Chromatin complexation	>1000	HOTAIR, HOTTIP

CAP1, adenylyl cyclase-associated protein 1; *CCDC52*, coiled-coil domain containing 52; *eRNAs*, enhancer-derived RNAs; *EXT*, exostosin 1; *HOTAIR*, homebox transcript antisense RNA; *HOTTIP*, HOXA distal transcript antisense RNA; *lincRNAs*, large intergenic noncoding RNAs; *LINE1*, long interspersed element 1; *miRNAs*, microRNAs; *NATs*, natural antisense transcripts; *PASRs*, promoter-associated small RNAs; *piRNAs*, PIWI-interacting RNAs; *PROMPTS*, promoter upstream transcripts; *RBM 39*, RNA-binding motif protein 39; *RNF12*, ring finger protein 12; *snoRNAa*, small nucleolar RNAs; SNORD, C/D box snoRNAs; *tiRNAs*, transcription initiation RNAs; *TSSa RNAs*, transcription start site–associated RNAs; *T-UCRs*, transcribed ultraconserved regions.
Adapted and modified from Esteller, M., 2011. Non-coding RNAs in human disease. Nat. Rev. Genet. 12(12), 861.

LncRNAs are diverse within the class with more than 200 bp length, accompanied by an ability to inhibit transcriptional regulation during gene expression using multiple mechanisms. From an EG perspective, lncRNAs are postulated to recruit chromatin remodeling complexes to specific gene loci resulting in progressive transcriptional manipulations (Gupta *et al.*, 2010). Furthermore, analysis of lncRNAs as a function of its gene and associated EG modifications indicated that the effects of these alterations are similar to those of protein encoding genes. Case in point—the presence of the transcriptionally active biomarker modifications, H3K4me3 and H3K36me3, in genes encoding lncRNAs resulted in their own overexpression, a situation analogous to genes that encode proteins (Sati *et al.*, 2012). However, limited evidence suggests an inverse relationship of EG modifications in lncRNAs coding genes contrary to mRNA coding genes. This phenomenon is embodied in a study displaying hypermethylation (transcriptional repressor) of the transcription start site of lncRNA coding genes having no significant effect on their expression levels (Sati *et al.*, 2012).

EPIGENETICS AND DISEASE

The field of EGs is now associated with an abundance of biomarkers, and the mechanisms of numerous diseases have been correlated with it. Consequently, biomarkers offer significant opportunities to understand the progression of pathologies and disease states. The role of biomarkers thus encourages appreciation of the mechanisms associated with the disease state, aids in the diagnoses of pathologies, and is involved in identifying interactions of therapeutic modalities with pathological consequences. In addition, biomarkers are involved in aberrations associated with DNA methylation, disruption of chromatin structure, or defects in "gain- or loss of function" of histone-modifying enzymes, all of which adversely affect gene expression (Feinberg, 2007). Thus, EG dysfunction results in loss of imprinting (LOI), DNA hypo- and hypermethylation, abnormal histone modifications, and generation of ncRNAs, frequently resulting in inactivation, or overexpression, of specific genomic loci (Herceg and Hainaut, 2007). Accordingly, EG dysregulation is implicated in generation or development of cancer, autoimmune diseases, and metabolic disorders (Calabro *et al.*, 2008).

Abnormalities of DNA Methylation

Abnormal regulation of DNA methylation is implicated in the origin of many genetic diseases, presumably a result of mutations in genes whose protein products have essential functions in DNA methylation and recognition. As such they are noted in immunodeficiency and centromere instability, including facial anomalies syndrome (ICF) and Rett syndrome, respectively. ICF is a rare autosomal recessive disorder characterized by hypomethylation of pericentromeric repeats. Its origin is a point mutation in the gene encoding DNMT3b, resulting in abnormal methylation (Xu *et al.*, 1999). The mutation adversely affects function of PWWP domain, responsible for normal chromatin binding (Klose and Bird, 2006). Alternatively, DNA methylation occurs normally in Rett syndrome, but the characteristic EG predicament is displayed by a defect in the ability to recognize methyl marks (Bienvenu and Chelly, 2006; Feinberg, 2007). This neurodegenerative disease is triggered by mutations in DNA MBD and transcription repression domain of the gene that encodes MeCP2, thus interfering with its proper functioning (Amir *et al.*, 1999).

Aberrations in DNA methylation patterns, such as in those required for imprinting, are also implicated in a variety of diseases. Regulated by EG processes during normal development, imprinting or silencing of either maternal or paternal alleles of particular genes is critical for proper cell functioning. When imprinting fails, the result is biallelic expression of certain monoallelic genes, initiating abnormal gene overexpression (Herceg and Hainaut, 2007). Diseases attributable to imprinting failure include Beckwith–Wiedemann syndrome (characterized by Wilms' tumors), Prader–Willi syndrome, and Angelman syndrome (Feinberg, 2007). In addition, LOI, i.e., either activation of a normally silent imprinted allele or silencing of a typically expressed imprinted allele, is implicated in cancer development, such as neuroblastoma and invasive breast cancer (Pedersen *et al.*, 1999; Astuti *et al.*, 2005; Feinberg, 2007).

As observed in LOI, DNA hypo- or hypermethylation results in aberrant gene expression and its pathological consequences. Hypomethylation occurs when established methylation patterns are lost through demethylation. Conversely, hypermethylation results from introduction of new methylation patterns where previously there may have not been an impact. Global hypomethylation is a general hallmark of cancer and is associated with aberrant activation of protooncogenes and chromosome instability (Jones and Baylin, 2002; Gopalakrishnan *et al.*, 2008; Herceg and Hainaut, 2007). Gene-specific hypermethylation is also a characteristic of many cancers, frequently occurring within promoter regions of tumor suppressor genes (Jones and Baylin, 2002; Feinberg, 2007).

Abnormalities in Histone Modifications

Abnormalities in histone modifications result from either loss-of-function or gain-of-function mutations in regulators. For instance, Rubinstein–Taybi syndrome is instigated through microdeletions, translocations, and point mutations in the gene that encodes for acetyltransferase coactivator CBP (Petrij *et al.*, 1995). Other examples of disorders resulting from histone methyltransferase malfunction include Sotos syndrome and Wolf–Hirschhorn syndrome, represented by mutations in genes encoding NSD1 and NSD2, respectively (Bergemann *et al.*, 2005; Feinberg, 2007). Sotos syndrome is characterized by a triad of Wilms' tumor, leukemia, and tissue overgrowth, while Wolf–Hirschhorn syndrome presents with mental retardation, epilepsy, and cranio-facial dysgenesis (Bergemann *et al.*, 2005; Feinberg, 2007).

As increasing evidence of EG mechanistic changes in multiple disease conditions is discovered, selected examples of cancer and autoimmune diseases, such as systemic lupus erythematosus (SLE), type 1 diabetes mellitus (T1-DM), and metabolic syndrome (MS), are further discussed in this context.

Epigenetics and Cancer

Cancer cells exploit innate metabolic pathways for their survival and hyperproliferation. This metabolic dysregulation associated with tumor microenvironment leads to an anomalous EG regulation pattern that subsequently disrupts cellular homeostasis (Shen and Laird, 2013). A pathological hallmark of cancer, consequentially, is disruption of genetic mechanisms by an assortment of mutations, copy number alterations, and impairment of DNA repair mechanisms, which ultimately result in extensive microsatellite instability throughout the genome (Hanahan and Weinberg, 2011). Mutations in EG regulators include writers—DNA methyltransferases, histone methyltransferases, HATs; readers—bromodomain, chromodomain, tudor protein; erasers—HDACs, histone demethylases; and chromatin remodelers—SWI/SNF complex; all of which result in abnormal processing patterns of the genetic code (Simó-Riudalbas and Esteller, 2014; Soshnev *et al.*, 2016).

Approximately 40% of CpG islands occur at promoter regions of human genes spanning at the 5′-end of their regulatory regions (Goll and Bestor, 2005; Gopalakrishnan *et al.*, 2008). Genome-wide analysis of several cancers has demonstrated a 10% increase in progressive methylation of normally unmethylated promoter CpG of tumor suppressor genes rendering them silenced (Simó-Riudalbas and Esteller, 2014). This phenomenon of methylation associated with underexpression of demethylation enzymes, such as TET methylcytosine dioxygenase family, activation-induced cytidine deaminase, and thymine DNA glycosylase, results in perpetual inactivation of tumor suppressor genes. Hypermethylation of promoter region CpG islands of tumor suppressor genes and subsequent gene silencing leads to the inability to regulate DNA repair and cell cycle progression (Mulero-Navarro and Esteller, 2008). Examples of gene-specific hypermethylation of tumor suppressor genes include *RB* in retinoblastoma, $p15^{INK4B}$, in particular forms of leukemia, and *BRAC1* in breast and ovarian cancer (Sakai *et al.*, 1991; Herman *et al.*, 1996; Esteller *et al.*, 2000). In addition, it is important to note that 5-mc, a methyl mark, is an endogenous mutagen. During spontaneous deamination, 5-mc is converted to thymine, a substitution that is more difficult to detect by DNA repair mechanisms as compared with deamination of cytosine, to form uracil (Robertson and Jones, 2000). This process has been implicated in mutagenesis of tumor suppressor genes, such as *TP53*, the gene that encodes p53. Point mutations in locations of known methylated cytosines in *TP53* result in amino acid surrogates in its protein product, affecting abnormal function and disease (Rideout III *et al.*, 1990).

Substantial variability in the genome and epigenome exists between and within specific cancer types. For instance, one of the challenges in breast cancer therapeutics is the high degree of heterogeneity exhibited by different breast tumors. Notably, a higher frequency of DNA methylation is observed in luminal (ER+) breast tumors compared with nonluminal (ER−) subtypes. (Interestingly, breast cancers are ER+ in the initial stages of the disease and later progress to more aggressive ER− breast cancers.) Hypermethylation of genes that encode ER has been proposed as a mechanism of silencing initially overexpressed ER receptors, which then progress to ER− breast cancer (Benevolenskaya *et al.*, 2016). Table 45.13 summarizes a selected list of drugs which exploit EG modifications as their therapeutic targets in cancer treatment.

HATs are also linked to fusion proteins caused by chromosomal translocations (Herceg and Hainaut, 2007). As an illustration, in acute myeloid leukemia, CBP was found to form fusion protein with MOZ, which causes global loss of acetylation (Borrow *et al.*, 1996; Fraga *et al.*, 2005). Another leukemia-related fusion protein, MORF-CBP, also results in a loss of acetylation (Fraga *et al.*, 2005). Mutations in genes encoding CBP coactivator, p300, result in truncated forms of protein and are also coupled to colorectal and breast cancers (Gayther *et al.*, 2000). MLL is implicated in human lymphoid and myeloid leukemias and in myelodysplastic syndrome (Varier and Timmers, 2011). Chromosomal aberrations of gene encoding MLL result in over 50 different types of translocations (Varier and Timmers, 2011). In addition, overexpression of EZH2, whose consequences are responsible for increased gene repression, is a hallmark of metastatic prostate cancer (Feinberg,

TABLE 45.13 Epigenetic Classification of Drugs in Cancer Therapeutics

Epigenetic Class of Drug	Compound	Target; Subclass	Clinical Stage of Development; Sponsor	Indication
DNA methylation inhibitors	5-Azacytidine	DNMT1	Approved; Celgene	MDS
	Decitabine	DNMT1	Approved; Otsuka, Janssen	MDS
Histone methyltransferase inhibitors	EPZ-5676	DOT1L	Clinical trials; Epizyme Inc.	Hematological malignancy
	EPZ-6438	EZH2	Clinical trials; Epizyme Inc.	Lymphoma and solid tumors
Histone demethylase inhibitors	Tranylcypromine analog	HDM	Clinical trials; Oryzon genomics	Leukemia
Histone acetyltransferase inhibitors	C646	p300	Preclinical trails; John Hopkins University	Multiple tumor types
Histone deacetylase inhibitors	Vorinostat	HDACs	Approved; Merck	CTCL
	Pracinostat	HDACs	Approved; MEI pharma	AML
	Panobinostat	HDACs; Hydroxamic acid derivatives	Approved; Novartis	Multiple myeloma
	Romidepsin	HDACs; Cyclic peptides	Approved; Gloucester pharmaceuticals	CTCL, PTCL

AML, acute myeloid leukemia; *CTCL*, cutaneous T cell lymphoma; *DOT1L*, disruptor of telomeric silencing 1-like histone H3K79 methyltransferase; *EZH2*, enhancer of zeste homolog 2; *MDS*, myelodysplastic syndrome; *p300*, E1A-binding protein; *PTCL*, peripheral T cell lymphoma.

Adapted from Biswas, S., Rao, C.M., 2017. Epigenetics in cancer: fundamentals and beyond. Pharmacol. Ther. 173, 118–134; Bennett, R.L., Licht, J.D., 2018. Targeting epigenetics in cancer. Ann. Rev. Pharmacol. Toxicol. 58, 187–207.

2007). Another overexpressed polycomb-related factor, BMI1, is observed in colorectal and liver carcinomas, non-small cell lung cancer, and mantle cell lymphoma (Valk-Lingbeek *et al.*, 2004).

Epigenetically altered genes and their expression can be modified by partial to complete reversal of undesired aberrant EG modifications, which makes them ideal targets for cancer therapeutics.

Epigenetics and Autoimmune Disease

Systemic Lupus Erythemosus

Historically, evidence has accumulated demonstrating the involvement of EG alterations in pathogenesis of autoimmune diseases. Richardson (1986) first demonstrated activation of autologous macrophages resulting in development of an autoimmune lupus-like condition when the patient was treated with the demethylation agent, 5-azacytidine (Richardson, *et al.*, 2004). Table 45.14 summarizes characteristic EG marks associated with several genes regulating immunity.

SLE is a chronic, systemic, autoimmune disease with diverse clinical manifestations affecting multiple organs. Pathological manifestations of SLE are mediated by production of autoantibodies that activate the complement cascade and trigger cytotoxic hypersensitivity in systemic circulation. Clinical evidence suggests that CD4+ T cells are globally hypomethylated in SLE and inhibitors of DNA methylation, such as procainamide and hydralazine, and induce lupus-like disease in mice (Cornacchia *et al.*, 1988). Furthermore, patients treated with high-dose hydralazine for longer durations exhibited lupus-like symptoms that waned on termination of the drug. This phenomenon of hydralazine-induced T cell hypomethylation in SLE is caused by inhibition of extracellular signal–regulated kinase pathway (Gorelik and Richardson, 2010). Promoter region hypomethylation and hence overexpression of select genes in T cells, including perforin, CD70, and CD11a, are linked with SLE. The results lead to CD4+ T cell–induced monocyte death, hyperstimulation of B cells, and disease-induced flares (Lu *et al.*, 2002; Richardson *et al.*, 2004; Kaplan *et al.*, 2004). In addition, elevation in global 5-hmC levels and increased expression of TET2 and TET3, which collectively convert methylcytosine to hydroxymethylcytosine in CD4+ T cells, is associated with active transcription and overexpression of select immune-related genes, notably, SOCS1, NR2F6, and IL15RA. The results lead to hyperstimulation of the immune system (Zhao *et al.*, 2016). Inactivation of HAT affected severe lupus-like disease in mice, progressing to glomerulonephritis and premature death (Forster *et al.*, 2007). In vivo studies demonstrate global histone H3 and H4 hypoacetylation and H3K9 hypomethylation

TABLE 45.14 Epigenetic Marks Associated With Genes Regulating Immune Cells

Type of Cell	Gene	Function	Characteristic Epigenetic Modification
T cell	CD3d[a]	T cell development	Active, H3 acetylation
	Bcl11b[a]	T cell development	
	Hhex[a]	Hematopoietic progenitor differentiation	Repressive, H3K27me3
	Lmo2[a]	Hematopoietic progenitor differentiation	
	Tbet[b]	Transcriptional regulator of Th1 subset of CD4+ T cells	**Naive state** - Hypermethylation with associated repressive histone modifications **Immune-activated state** - Hypomethylation and loss of histone repressive marks
	Gata3[b]	Transcriptional regulator of Th2 subset of CD4+ T cells	
	Rorgt[b]	Transcriptional regulator of Th17 subset of CD4+ T cells	
	FOXP3[c]	Transcription factor (TF) of immunosuppressive Treg	DNA hypomethylation
Macrophages	Pu. 1[d]	TF for macrophage lineage determination	Recruitment of TFs favoring open chromatin structure
Dendritic cells	CD209[e]	Essential factor for dendritic cell trafficking	Active, H3K9ac in CD209 gene; loss of histone repressive marks; Promoter region hypomethylation
B cells	PAX5[a]	Early B cell development	Hypomethylation before development of B cells; Hypermethylation upon development and maturation of B cells
	Pu. 1[d]		
	MB1[e]		
	Blimp-1[f]	Regulation of plasma cell differentiation	Variable histone active and repressive marks

Bcl11b, B cell lymphoma/leukemia 11B; *Blimp-1*, B-lymphocyte–induced maturation protein 1; *CD209*, dendritic cell–specific intracellular adhesion molecules-3 grabbing nonintegrin; *CD3d*, T cell surface glycoprotein CD3 delta chain; *FOXP3*, forkhead box protein P3; *Gata3*, GATA binding protein 3; *Hhex*, hematopoietically expressed homeobox; *Lmo2*, LIM domain only protein 2; *MB1*, membrane bound immunoglobulin associated protein; *PAX5*, paired box homeotic gene 5; *Pu. 1*, Spi-1 protooncogene; *Rorgt*, RAR related orphan receptor gamma; *H3*, Histone 3; *H3K27me3*, histone 3 lysine 27 trimethyl modification; *H3K9ac*, histone 3 lysine 9 acetyl modification.

[a] *Rodriguez et al. (2015).*
[b] *Wang et al. (2015).*
[c] *Huehn and Beyer (2015).*
[d] *Kapellos and Iqbal (2016).*
[e] *Suárez-Álvarez et al. (2013).*
[f] *Lee et al. (2003).*

in lupus-prone mice (Hu *et al.*, 2008). Additionally, significant improvement in glomerulonephritis and splenomegaly was observed in lupus-prone mice when treated with HDAC inhibitors, including trichostatin A or suberoylanilide hydroxamic acid (Mishra *et al.*, 2003; Reilly *et al.*, 2011).

Type 1 Diabetes Mellitus

T1-DM, traditionally identified with insulin-dependent diabetes, is characterized by autoimmune destruction of insulin producing β cells in islets of Langerhans in the pancreas. The outcome results in deficiency of insulin in systemic circulation, leading to hyperglycemia, ketoacidosis, and other micro/macrovascular complications (Eisenbarth, 2005). The autoimmune reaction involved is elicited by hyperproliferation of B cells, CD4+ and CD8+ T cells, and autoantibodies against β cells causing their progressive destruction (Eisenbarth, 2005). T1-DM–mediated microvascular complications, such as retinopathy and neuropathy, are associated with concomitant elevation of H3K9ac levels at the promoter region of proinflammatory signaling molecule NF-kB (Miao *et al.*, 2014). Furthermore, genome-wide analysis of DNA methylation in T1-DM patients identified differential global methylation levels that can serve as a prospective biomarker for diabetic complications (Gautier *et al.*, 2015).

Epigenetics and Metabolic Syndrome

Studies concerning metabolic diseases and associated changes in variable regions of the mammalian epigenome, compared with healthy controls, demonstrated several alterations of EG fingerprints. MS is a classic example of metabolic disease characterized by a varied combination of multiple etiologies such as hypertension, dyslipidemia, obesity, and type 2 diabetes mellitus (Kaur, 2014). Increased expression of suppressor of cytokine signaling 3 (SOCS3) gene is linked to elevated risk of MS; DNA methylation analyses of obese subjects demonstrate hypomethylation and subsequent overexpression of SOCS3 gene (Ali *et al.*, 2016). MiRNAs miR-33, -34a, and -22 are considered biologically significant in regulating lipid metabolism and are proposed to serve as valuable EG biomarkers to monitor MS (Yang *et al.*, 2015).

CONCLUDING REMARKS AND FUTURE DIRECTIONS

EG mechanisms, such as DNA methylation, covalent PTMs of histones, and ncRNAs, play important roles in normal cellular processes. Dysfunction of such mechanisms, in the absence of induced mutations, results in abnormal gene expression and progression of pathologies. As EG changes typically occur early in a disease process, aberrant EG regulators are ideal targets for disease identification, diagnosis, monitoring, and follow-up treatment (Herceg and Hainaut, 2007). These regulators also act as valuable biomarkers in global public health risk assessment in the detection and prevention of diseases, notwithstanding the benefit of calculating treatment response (Mulero-Navarro and Esteller, 2008).

References

Ali, O., Cerjak, D., Kent Jr., J.W., et al., 2016. Methylation of SOCS3 is inversely associated with metabolic syndrome in an epigenome-wide association study of obesity. Epigenetics 11 (9), 699–707.

Amir, R.E., Van den Veyver, I.B., Wan, M., et al., 1999. Rett syndrome is caused by mutations in X-linked *MECP2*, encoding methyl-CpG-binding. Nat. Genet. 23, 185–188.

Anway, M.D., Cupp, A.S., Uzumcu, M., et al., 2005. Epigenetic transgenerational actions of endocrine disruptors and male fertility. Science 308 (5727), 1466–1469.

Aravin, A.A., Sachidanandam, R., Girard, A., et al., 2007. Developmentally regulated piRNA clusters implicate MILI in transposon control. Science 316 (5825), 744–747.

Astuti, D., Latif, F., Wagner, K., et al., 2005. Epigenetic alteration at *DLKI-GTL2* imprinted domain in human neoplasia: analysis of neuroblastoma, phaeochromocytoma, and Wilms' tumour. Br. J. Cancer 92, 1574–1580.

Autrup, H., Barile, F.A., Blaauboer, B.J., et al., 2015. Principles of pharmacology and toxicology also govern effects of chemicals on the endocrine system. Toxicol. Sci. 146 (1), 11–15.

Avvakumov, N., Côté, J., 2007. Functions of myst family histone acetyltransferases and their link to disease. Subcell. Biochem 41, 295–317.

Baek, S.H., 2011. When signaling kinases meet histones and histone modifiers in nucleus. Mol. Cell 42, 274–284.

Bannister, A.J., Kouzarides, T., 2011. Regulation of chromatin by histone modifications. Cell Res. 21, 381–395.

Barile, F.A., 2013. Chapter 20. Toxicogenomics and EG testing in vitro. In: Principles of Toxicology Testing, second ed. CRC Press/Taylor & Francis, Boca Raton, FL, USA, p. 276.

Bedford, M.T., Clarke, S.G., 2009. Protein arginine methylation in mammals: who, what, and why. Mol. Cell 33, 1–13.

Benevolenskaya, E.V., Islam, A.B., Ahsan, H., et al., 2016. DNA methylation and hormone receptor status in breast cancer. Clin. Epigenet. 8 (1), 17.

Bennett, R.L., Licht, J.D., 2018. Targeting epigenetics in cancer. Annu. Rev. Pharmacol. Toxicol. 58, 187–207.

Bergemann, A.D., Cole, F., Hisrchhorn, K., 2005. The etiology of Wolf-Hirschhorn syndrome. Trends Genet. 21, 188–195.

Bestor, T., Laudano, A., Mattaliano, R., et al., 1988. Cloning and sequencing of a cDNA encoding DNA methyltransferase of mouse cells. The carboxyl-terminal domain of mammalian enzymes is related to bacterial restriction methyltransferases. J. Mol. Biol. 203, 971–983.

Bhaumik, S.R., Smith, E., Shilatifard, A., 2007. Covalent modifications of histones during development and disease pathogenesis. Nat. Struct. Mol. Biol. 14, 1008–1016.

Bienvenu, T., Chelly, J., 2006. Molecular genetics of Rett syndrome: when DNA methylation goes unrecognized. Nat. Rev. Genet. 7, 415–426.

Biswas, S., Rao, C.M., 2017. Epigenetics in cancer: fundamentals and beyond. Pharmacol. Ther. 173, 118–134.

Borrow, J., Stanton Jr., V.P., Andresen, J.M., et al., 1996. The translocation t(8;16)(p11;p13) of acute myeloid leukaemia fuses to the CREB-binding protein. Nat. Genet. 14, 33–41.

Bostick, M., Kim, J., Estève, P., et al., 2007. UHRF1 plays a role in maintaining DNA methylation in mammalian cells. Science 317, 1760–1764.

Calabro, A.R., Konsoula, R., Barile, F.A., 2008. Evaluation of *in vitro* cytotoxicity and paracellular permeability of intact monolayers with mouse embryonic stem cells. Toxicol. In Vitro 22 (5), 1273–1284.

Chang, B., Chen, Y., Zhao, Y., et al., 2007. JMJD6 is a histone arginine demethylase. Science 318, 444–447.

Chen, D., Ma, H., Hong, H., et al., 1999. Regulation of transcription by protein methyltransferases. Science 284, 2174–2177.

Chen, L., MacMillan, A.M., Chang, W., et al., 1991. Direct identification of the active-site nucleophile in a DNA (cytosine-5)-methyltransferase. Biochemistry 30, 11018–11025.

Cheng, X., Roberts, R.J., 2001. AdoMet-dependent methylation, DNA methyltransferases and base flipping. Nucleic Acids Res. 29, 3784–3795.

Cornacchia, E., Golbus, J., Maybaum, J., et al., 1988. Hydralazine and procainamide inhibit T cell DNA methylation and induce autoreactivity. J. Immunol. 140 (7), 2197–2200.

David, G., Neptune, M.A., DePinho, R.A., 2002. SUMO-1 modification of histone deacetylase 1 (HDAC1) modulates its biological activities. J. Biol. Chem. 277, 23658–23663.

de Ruijter, A.J.M., van Gennip, A.H., Caron, H.N., Kemp, S., et al., 2003. Histone deacetylase (HDACs): characterization of classical HDAC family. Biochem. J. 370, 737–749.

Di Lorenzo, A., Bedford, M.T., 2011. Histone arginine methylation. FEBS Lett. 585, 2024–2031.

Dou, Y., Milne, T.A., Tackett, A.J., et al., 2005. Physical association and coordinate function of H3K4 methyltransferase MLL1 and H4 K16 acetyltransferase MOF. Cell 121, 873–885.

Edmondson, D.G., Davie, J.K., Zhou, J., et al., 2002. Site-specific loss of acetylation upon phosphorylation of H3. J. Biol. Chem. 277, 29496–29502.

Eisenbarth, G.S., 2005. Type 1 diabetes mellitus. In: Joslin's Diabetes Mellitus, vol. 14, pp. 399–424.

Esteller, M., Maria Silva, J., Dominguez, G., et al., 2000. Promoter hypermethylation and BRCA1 inactivation in sporadic breast and ovarian tumors. J. Natl. Cancer Inst. 92, 564–569.

Esteller, M., 2011. Non-coding RNAs in human disease. Nat. Rev. Genet. 12 (12), 861.

Feinberg, A.P., 2007. Phenotypic plasticity and epigenetics of human disease. Nature 447, 433–440.

Filion, G.J., Zhenilo, S., Salozhin, S., Yamada, D., Prokhortchouk, E., Defossez, P.A., 2006. A family of human zinc finger proteins that bind methylated DNA and repress transcription. Molecular and Cell Biol. 26 (1), 169–181.

Forster, N., Gallinat, S., Jablonska, J., et al., 2007. p300 protein acetyltransferase activity suppresses systemic lupus erythematosus-like autoimmune disease in mice. J. Immunol. 178 (11), 6941–6948.

Fraga, M.F., Ballestar, E., Villar-Garea, A., et al., 2005. Loss of acetylation at Lys16 and trimethylation at Lys20 of histone H4 is a common hallmark of human cancer. Nat. Genet. 37, 391–400.

Frith, M.C., Pheasant, M., Mattick, J.S., 2005. Genomics: Amazing Complexity of Human Transcriptome.

Gadhia, S.R., Barile, F.A., 2014. Epigenetic modeling and stem cells in toxicology testing. In: Handbook of Nanotoxicology, Nanomedicine and Stem Cell Use in Toxicology, pp. 359–378.

Gadhia, S.R., Calabro, A.R., Barile, F.A., 2012. Trace metals alter DNA repair and histone modification pathways concurrently in mouse embryonic stem cells. Toxicol. Lett. 212 (2), 169–179.

Gautier, J.F., Porcher, R., Khalil, C.A., et al., 2015. Kidney dysfunction in adult offspring exposed in utero to type 1 diabetes is associated with alterations in genome-wide DNA methylation. PLoS One 10 (8), e0134654.

Gayther, S.A., Batley, S.J., Linger, L., et al., 2000. Mutations truncating EP300 acetylase in human cancers. Nat. Genet. 24, 300–303.

Girdwood, D., Bumpass, D., Vaughan, O.A., et al., 2003. p300 transcriptional repression is mediated by SUMO modification. Mol. Cell 11, 1043–1054.

Goll, M.G., Bestor, T.H., 2005. Eukaryotic cytosine methyltransferases. Ann. Rev. Biochem. 74, 481–514.

Goll, M.G., Kirpekar, F., Maggert, K.A., et al., 2006. Methylation of tRNAAsp by DNA methyltransferase homolog Dnmt2. Science 311 (5759), 395–398.

Goodman, J.I., Watson, R.E., 2002. Altered DNA methylation: a secondary mechanism involved in carcinogenesis. Ann. Rev. Pharmocol. 42, 501–525.

Gopalakrishnan, S., Van Emburgh, B.O., Robertson, K.D., 2008. DNA methylation in development and human disease. Mutat. Res. Fund Mol. M 647, 30–38.

Gorelik, G., Richardson, B., 2010. Key role of ERK pathway signaling in lupus. Autoimmunity 43 (1), 17–22.

Grote, P., Wittler, L., Hendrix, D., et al., 2013. Tissue-specific lncRNA Fendrr is an essential regulator of heart and body wall development in mouse. Dev. Cell 24 (2), 206–214.

Gupta, R.A., Shah, N., Wang, K.C., et al., 2010. Long non-coding RNA HOTAIR reprograms chromatin state to promote cancer metastasis. Nature 464 (7291), 1071.

Halpern, K.B., Vana, T., Walker, M.D., 2014. Paradoxical role of DNA methylation in activation of FoxA2 gene expression during endoderm development. J. Biol. Chem. 289 (34), 23882–23892.

Hanahan, D., Weinberg, R.A., 2011. Hallmarks of cancer: next generation. Cell 144 (5), 646–674.

He, L., Hannon, G.J., 2004. MicroRNAs: small RNAs with a big role in gene regulation. Nat. Rev. Genet. 5 (7), 522.

Herceg, Z., Hainaut, P., 2007. Genetic and EG alterations as biomarkers for cancer detection, diagnosis and prognosis. Mol. Oncol. 1, 26–41.

Herman, J.G., Jen, J., Merlo, A., et al., 1996. Hypermethylation-associated inactivation indicates a tumor suppress role for p15INK4B1. Cancer Res. 56, 722–727.

Hermann, A., Goyal, R., Jeltsch, A., 2004. Dnmt1 DNA-(cytosine-C5)-methyltransferase methylates DNA processively with high preference for hemimethylated target sites. J. Biol. Chem. 279, 48350–48359.

Hu, N., Qiu, X., Luo, Y., et al., 2008. Abnormal histone modification patterns in lupus CD4+ T cells. J. Rheumatol. 35 (5), 804–810.

Huehn, J., Beyer, M., February 2015. Epigenetic and transcriptional control of Foxp3+ regulatory T cells. Semin. Immunol. 27 (1), 10–18. Academic Press.

Jablonka, E., Lamb, M.L., 2002. The changing concept of epigenetics. Ann. N. Y. Acad. Sci. 981, 82–96.

Jeltsch, A., Ehrenhofer-Murray, A., Jurkowski, T.P., et al., 2017. Mechanism and biological role of Dnmt2 in nucleic acid methylation. RNA Biol. 14 (9), 1108–1123.

Johnson, E.S., 2004. Protein modification by SUMO. Annu. Rev. Biochem. 73, 355–382.

Jones, P.A., Baylin, S.B., 2002. The fundamental role of EG events in cancer. Nat. Rev. Genet. 3, 415–428.

Jones, P.L., Veenstra, G.J.C., Wade, P.A., et al., 1998. Methylated DNA and MeCP2 recruit histone deacetylase to repress transcription. Nat. Genet. 19, 187–191.

Kaikkonen, M.U., Lam, M.T., Glass, C.K., 2011. Non-coding RNAs as regulators of gene expression and epigenetics. Cardiovasc. Res. 90 (3), 430–440.
Kapellos, T.S., Iqbal, A.J., 2016. Epigenetic control of macrophage polarisation and soluble mediator gene expression during inflammation. Mediat. Inflamm. 2016.
Kaplan, M.J., Lu, Q., Wu, A., et al., 2004. Demethylation of promoter regulatory elements contributes to perforin overexpression in CD4+ lupus T cells. J. Immunol. 172 (6), 3652–3661.
Kaur, J., 2014. A comprehensive review on metabolic syndrome. Cardiol. Res. Pract. 2014.
Khandelwal, A., Bacolla, A., Vasquez, K.M., et al., 2015. Long non-coding RNA: a new paradigm for lung cancer. Mol. Carcinog. 54 (11), 1235–1251.
Kirsh, O., Seeler, J.S., Pichler, A., et al., 2002. SUMO E3 ligase RanBP2 promotes modification of HDAC4 deacetylase. EMBO J. 21, 2682–2691.
Klimasauskas, S., Kumar, S., Roberts, R.J., et al., 1994. Hhal methyltransferase flips its target base out of DNA helix. Cell 76, 357–369.
Klose, R.J., Bird, A.P., 2006. Genomic DNA methylation: mark and its mediators. Trends Biochem. Sci. 31, 89–97.
Klose, R.J., Kallin, E.M., Zhang, Y., 2006. JmjC-domain-containing proteins and histone demethylation. Nat. Rev. Genet. 7, 715–727.
Kouzarides, T., 2002. Histone methylation in transcriptional control. Curr. Opin. Genet. Dev. 12, 198–209.
Kouzarides, T., 2007. Chromatin modifications and their function. Cell 128, 639–705.
Lachner, M., Jenuwein, T., 2002. Many faces of histone lysine methylation. Curr. Opin. Cell Biol. 14, 286–298.
Lee, S.C., Bottaro, A., Insel, R.A., 2003. Activation of terminal B cell differentiation by inhibition of histone deacetylation. Mol. Immunol. 39 (15), 923–932.
Lu, Q., Kaplan, M., Ray, D., et al., 2002. Demethylation of ITGAL (CD11a) regulatory sequences in systemic lupus erythematosus. Arthritis Rheum. 46 (5), 1282–1291.
Luger, K., Mäder, A.W., Richmond, R.K., et al., 1997. Crystal structure of nucleosome core particle at 2.8 Å resolution. Nature 389, 251–260.
Melchior, F., 2000. SUMO-Nonclassical ubiquitin. Annu. Rev. Cell Dev. Biol. 16, 591–626.
Miao, F., Chen, Z., Genuth, S., et al., 2014. Evaluating role of EG histone modifications in metabolic memory of type 1 diabetes. Diabetes. https://doi.org/10.2337/db13-1251.
Mills, A.A., 2010. Throwing cancer switch: reciprocal roles of polycomb and trithorax proteins. Nat. Rev. Cancer 10, 669–682.
Mishra, N., Reilly, C.M., Brown, D.R., et al., 2003. Histone deacetylase inhibitors modulate renal disease in MRL-lpr/lpr mouse. J. Clin. Invest. 111 (4), 539–552.
Mori, T., Ikeda, D.D., Yamaguchi, Y., et al., 2012. NIRF/UHRF2 occupies a central position in cell cycle network and allows coupling with EG landscape. FEBS Lett. 586, 1570–1583.
Mulero-Navarro, S., Esteller, M., 2008. Epigenetic biomarkers for human cancer: time is now. Crit. Rev. Oncol. Hematol. 68, 1–11.
Nan, X., Ng, H., Johnson, C.A., et al., 1998. Transcriptional repression by methyl-CpG-binding protein MeCP2 involves a histone deacetylase complex. Nature 393, 386–389.
Okano, M., Xie, S., Li, E., 1998a. Dnmt2 is not required for *de novo* and maintenance methylation of viral DNA in embryonic stem cells. Nucleic Acids Res. 26, 2536–2540.
Okano, M., Xie, S., Li, E., 1998b. Cloning and characterization of a family of novel mammalian DNA (cytosine-5) methyltransferases. Nat. Genet. 19, 219–220.
Okano, M., Bell, D.W., Haber, D.A., 1999. DNA methyltransferases Dnmt3a and Dnmt3b are essential for de novo methylation and mammalian development. Cell 99, 247–257.
Ooi, S.K.T., Bestor, T.H., 2008. Colorful history of active DNA demethylation. Cell 133, 1145–1148.
Parthun, M.R., 2007. Hat1: emerging cellular roles of a type B histone acetyltransferase. Oncogene 26, 5319–5328.
Pedersen, I.S., Dervan, P.A., Broderick, D., et al., 1999. Frequent loss of imprinting of *PEG1/MEST* in invasive break cancer. Cancer Res. 59, 5449–5451.
Petrij, F., Giles, R.H., Dauwerse, H.G., et al., 1995. Rubinstein-Taybi syndrome caused by mutations in transcriptional co-activator CBP. Nature 376, 348–351.
Perera, F., Herbstman, J., 2011. Prenatal environmental exposures, epigenetics, and disease. Reprod. Toxicol. 31, 363–373.
Pickart, C.M., 2001. Mechanisms underlying ubiquitination. Annu. Rev. Biochem. 70, 503–533.
Pradhan, S., Bacolla, A., Wells, R.D., et al., 1999. Recombinant human DNA (cytosine-5) methyltransferase. I. Expression, purification, and comparison of de novo and maintenance methylation. J. Biol. Chem. 274, 33002–33010.
Prokhortchouk, E., Defossez, P., 2008. The cell biology of DNA methylation in mammals. Biochim. Biophys. Acta 1783, 2167–2173.
Reilly, C.M., Regna, N., Mishra, N., 2011. HDAC inhibition in lupus models. Mol. Med. 17 (5–6), 417.
Rice, J.C., Allis, C.D., 2001. Histone methylation versus histone acetylation: new insights into EG regulation. Curr. Opin. Cell Biol. 13, 263–273.
Richardson, B., 1986. Effect of an inhibitor of DNA methylation on T cells. II. 5-Azacytidine induces self-reactivity in antigen-specific T4+ cells. Hum. Immunol. 17 (4), 456–470.
Richardson, B., Ray, D., Yung, R., 2004. Murine models of lupus induced by hypomethylated T cells. In: Autoimmunity. Humana Press, pp. 285–294.
Rideout III, W.M., Coetzee, G.A., Olumi, A.F., et al., 1990. 5-Methylcytosine as an endogenous mutagen in human LDL receptor and p53 genes. Science 249, 1288–1290.
Robertson, K.D., Jones, P.A., 2000. DNA methylation: past, present and future directions. Carcinogenesis 21, 461–467.
Rodriguez, R.M., Lopez-Larrea, C., Suarez-Alvarez, B., 2015. Epigenetic dynamics during CD4+ T cells lineage commitment. Int. J. Biochem. Cell Biol. 67, 75–85.
Roloff, T.C., Nuber, U.A., 2005. Chromatin, epigenetics and stem cells. Eur. J. Cell Biol. 84, 123–135.
Roth, S.Y., Denu, J.M., Allis, C.D., 2001. Histone acetyltransferases. Annu. Rev. Biochem. 70, 81–120.
Sakai, T., Ohtani, N., McGee, T.L., Robbins, P.D., Dryja, T.P., 1991. Oncogenic germ-line mutations in Sp1 and ATF sites in the human retinoblastoma gene. Nature 353 (6339), 83–86.
Sasai, N., Defossez, P., 2009. Many paths to one goal? proteins that recognize methylated DNA in eukaryotes. Int. J. Dev. Biol. 53, 323–334.
Sati, S., Ghosh, S., Jain, V., et al., 2012. Genome-wide analysis reveals distinct patterns of EG features in long non-coding RNA loci. Nucleic Acids Res. 40 (20), 10018–10031.
Saunders, L.R., Verdin, E., 2007. Sirtuins: critical regulators at crossroads between cancer and aging. Oncogene 26, 5489–5504.
Sauvageau, M., Goff, L.A., Lodato, S., et al., 2013. Multiple knockout mouse models reveal lincRNAs are required for life and brain development. Elife 2, e01749.
Schuettengruber, B., Chourrout, D., Vervoort, M., et al., 2007. Genome regulation by polycomb and trithorax proteins. Cell 128, 735–745.
Shahbazian, M.D., Grunstein, M., 2007. Functions of site-specific histone acetylation and deacetylation. Annu. Rev. Biochem. 76, 75–100.
Sharif, J., Muto, M., Takebayashi, S., et al., 2007. SRA protein Np95 mediates EG inheritance by recruiting Dnmt1 to methylate DNA. Nature 450, 908–912.

Shen, H., Laird, P.W., 2013. Interplay between cancer genome and epigenome. Cell 153 (1), 38–55.

Shimada, M., Niida, H., Zineldeen, D.H., et al., 2008. Chk1 is a histone H3 threonine 11 kinase that regulates DNA damage-induced transcriptional repression. Cell 132, 221–232.

Simó-Riudalbas, L., Esteller, M., 2014. Cancer genomics identifies disrupted EG genes. Hum. Genet. 133 (6), 713–725.

Soshnev, A.A., Josefowicz, S.Z., Allis, C.D., 2016. Greater than sum of parts: complexity of dynamic epigenome. Mol. Cell 62 (5), 681–694.

Sterner, D.E., Berger, S.L., 2000. Acetylation of histones and transcription-related factors. Microbiol. Mol. Biol. Rev. 64, 435–459.

Strahl, B.D., Allis, D., 2000. Language of covalent histone modifications. Nature 403, 41–45.

Suárez-Álvarez, B., Baragaño Raneros, A., Ortega, F., et al., 2013. Epigenetic modulation of immune function: a potential target for tolerance. Epigenetics 8 (7), 694–702.

Suetake, I., Shinozaki, F., Miyagawa, J., et al., 2004. DMT3L stimulated DNA methylation activity of Dnmt3a and Dnmt3b through a direct interaction. J. Biol. Chem. 279, 27816–27823.

Sugiyama, K., Sugiura, K., Hara, T., et al., 2002. Aurora-B associated protein phosphatases as negative regulators of kinase activation. Oncogene 21, 3103–3111.

Tammen, S.A., Frisco, S., Choi, S., 2012. Epigenetics: link between nature and nurture. Mol. Aspect. Med. https://doi.org/10.1016/j.mam.2012.07.018.

Valk-Lingbeek, M.E., Bruggeman, S.W.M., van Lohuizen, M., 2004. Stem cells and cancer: the polycomb connection. Cell 118, 409–418.

Varier, R.A., Timmers, H.T.M., 2011. Histone lysine methylation and demethylation pathways in cancer. Biochim. Biophys. Acta 1815, 75–89.

Völkel, P., Angrand, P., 2007. Control of histone lysine methylation in EG regulation. Biochimie 89, 1–20.

Wang, C., Collins, M., Kuchroo, V.K., 2015. Effector T cell differentiation: are master regulators of effector T cells still masters? Curr. Opin. Immunol. 37, 6–10.

Witt, O., Deubzer, H.E., Milde, T., et al., 2009. HDAC family: what are cancer relevant targets? Cancer Lett. 277, 8–21.

Xu, G., Bestor, T.H., Bourc'his, D., et al., 1999. Chromosome instability and immunodeficiency syndrome caused by mutations in a DNA methyltransferase gene. Nature 402, 187–191.

Yang, Z., Cappello, T., Wang, L., 2015. Emerging role of microRNAs in lipid metabolism. Acta Pharm. Sin. B 5 (2), 145–150.

Yoon, H., Chan, D.W., Reynold, A.B., et al., 2003. N-CoR mediates DNA methylation-dependent repression through a methyl CpG binding protein Kaiso. Mol. Cell 12, 723–734.

Zemach, A., McDaniel, I.E., Silva, P., et al., 2010. Genome-wide evolutionary analysis of eukaryotic DNA methylation. Science 328, 916–919.

Zhang, Y., 2003. Transcriptional regulation by histone ubiquitination and deubiquitination. Gene Dev. 17, 2733–2740.

Zhao, M., Wang, J., Liao, W., et al., 2016. Increased 5-hydroxymethylcytosine in CD4+ T cells in systemic lupus erythematosus. J. Autoimmun. 69, 64–73.

CHAPTER

46

Risk Factors as Biomarkers of Susceptibility in Breast Cancer

Carolina Negrei[1], *Bianca Galateanu*[2]

[1]Departament of Toxicology, Faculty of Pharmacy, "Carol Davila" University of Medicine and Pharmacy, Bucharest, Romania; [2]Department of Biochemistry and Molecular Biology, University of Bucharest, Bucharest, Romania

INTRODUCTION

Epidemiology

Over the entire span of their lives women, and postmenopausal women especially, have an 11% risk of developing breast cancer, making this the most widespread malignancy among women. The current worldwide trend in incidence rate is higher in developed countries, but changes in lifestyle may lead to increases in less developed countries as well (Tao et al., 2015).

Pathology

Adenocarcinomas are the most type of common cancer, almost excluding lymphoma and sarcoma pathology. Both noninvasive and invasive breast tumors usually originate in the lobules and ducts, and their specific morphology becomes evident via microscopic examination, which provides decisive histological data to distinguish lobular from ductal carcinomas. According to present evidence, the site of origin for both types is considered to be the terminal duct lobuloalveolar unit (Russo and Russo, 1999).

Usually determined by mammography (Ernster et al., 2002), atypical ductal hyperplasia leading to in situ breast carcinomas do not pervade the basement membrane but may be considered as early precursors of invasive types (Kuerer et al., 2009). This has been successfully shown in a number of progression models for normal breast tissue.

Between ductal and lobular invasive types, the histological category most commonly diagnosed (approximately 80% of all cases) is ductal carcinoma. Invasive lobular carcinoma occurs much less often (Li et al., 2003), mostly in elderly women and, despite a specific metastatic pattern, it has a similar prognosis for other ages (Arpino et al., 2004). Other well-defined morphologic but more infrequent subtypes providing better prognosis may include Paget and adenoid cystic, papillary, medullary, tubular, and mucinous carcinomas.

Risk Factors

Among the most significant risk factors for developing breast cancer, inherent predisposing factors have to be taken into account. Factors such as gender, age, and history of benign or malignant breast disease in both patients and their family (more significantly in a first-degree relative) may be broad predictors. Although gender would seem irrelevant as a breast cancer risk factor, because of the difference between males and females in exposure to hormones, male breast cancer cases used to be very rare, but they are becoming more frequent of late. The risk of developing breast cancer is mainly age dependent as a consequence of genetic and epigenetic changes (Fraga et al., 2007). Thus, there is low risk but more aggressive progression before the age of 25, and the risk increases significantly in the third decade. Diagnosis is largely established prior to menopause, possibly indicative of a hormonal status association (Tao et al., 2015).

Several more specific predisposing factors stand out, the most important of which are hormonal and reproductive particulars related to estrogen exposure (age at first period, first pregnancy and menopause, high levels of endogenous sex steroid hormone, lifetime use of hormonal drugs), breast density, and genetic and epigenetic alterations. Breast density relates to the amount of mammary gland stromal and epithelial cells, and higher density arises from more extensive fibrous and glandular

Biomarkers in Toxicology, Second Edition
https://doi.org/10.1016/B978-0-12-814655-2.00046-3

tissue as opposed to adipose tissue. This is a significant (four times higher than average) risk for both breast cancer occurrence and more aggressive tumors.

Further factors of individual risk are proliferative breast lesions with an excessive number of cells, some of which display morphological abnormalities. At the same time, behavioral and lifestyle factors (weight, diet, exercise, and use of alcohol) should not be overlooked (Veronesi et al., 2005). Risk factors have a cumulative effect on overall risk.

Life Stage–Related Risks

In addition to genetic and environmental risks, there is a certain increased vulnerability to cancer depending on the person's development of life phase, when the tissue is more susceptible to the occurrence of alterations under the influence of epigenetic factors. These factors are likely to trigger changes in the underlying body structure and functions as well as cellular modifications. For breast development, important stages include the fetal stage, puberty with its marked growth, and late pregnancy with its precipitated preparation for breastfeeding. In the fetal phase, the more pronounced vulnerability is the result of the immaturity of most protection fostering mechanisms such as DNA repair, functioning of the immune system, enzymes and liver metabolism, and incomplete formation of the blood–brain barrier (BBB) to ensure protection against external aggression (Fenton et al., 2012). Developmental stage–related susceptibility may be determined by biomarkers as well, and recent research has focused on rodent models to offer evidence of potential epigenetic or underlying genetic risk factors. Thus, the following have been deemed of significance and they have been presented in order of relevant life stage importance.

Gestation and Development of the Fetus

Summarized in the theory of "fetal basis of adult disease," the first 3 months of pregnancy in particular and the entire pre-birth period are characterized by quick cellular division and growth, susceptible to the influence of environmental toxicity or modified hormone (such as estrogen) action (Birnbaum and Fenton, 2003), which can alter underlying tissue structure, function, and programming (Fenton et al., 2012).

The risk of cancer later in life has been suggested to be influenced by endogenous factors, such as altered hormonal exposure as indicated by weight at birth, gestational age, birth order, twins and maternal age, and preeclampsia (Park et al., 2008). The influence of elevated estrogen levels during pregnancy on cancer risk later on is also apparent from exposure during pregnancy to exogenous agents capable of endocrinal disruption (such as diethylstilbestrol or phytoestrogens) (Hoover et al., 2011).

Puberty

As a period of rapid and substantial growth, puberty is the second life stage with an increased susceptibility to breast cancer. On an intrinsic level, puberty is normally governed by two endocrine processes: first, development of the hypothalamic–pituitary–gonadal axis (gonadarche) and secondly, maturation of the hypothalamic–pituitary–adrenal axis and the ability to produce and secrete androgens (adrenarche) (Fenton et al., 2012; Wan et al., 2012). This accounts for the capacity of disturbances in either or both axes to determine modified pubertal timing.

The onset of puberty is indicated by physical markers such as breast development, the so-called "growth spurt," growth of pubic hair, and the first menstrual cycle and its development towards regularity (Fenton et al., 2012). Breast, bone, and brain development are under the control of estrogens (reviewed in Fenton et al., 2012). However, early onset of puberty may be the result of disruptions with major effects on cancer risk.

Precocious puberty may be determined by hormonal profiling and the appearance of sudden breast development or pubic hair growth (Carel et al., 2004), together with an assessment of bone age and performance of pelvic ultrasound checks. This is usually identified in girls under 8 years of age. Elevated breast density is a very relevant risk factor for breast cancer (higher breast tissue density means five times more likely to develop cancer) (IBCERCC, 2013). Early breast development and increased numbers of terminal duct lobular units increase vulnerability to exposure to carcinogenic factors (Fenton, 2006; Biro et al., 2010).

Precocious onset of menarche is also a risk factor for breast cancer (D'Aloisio et al., 2013), so much so that a woman's risk is approximately 10% lower for each year of menarche delay (Biro et al., 2010). According to recent studies, an association has been established between early onset of menarche and vitamin D deficiency [defined as 25-hydroxyvitamin D [25(OH)D] blood level<20 ng/mL (IOM, 2011)], which makes vitamin D deficiency an important factor in breast cancer risk. Other intrinsic factors include height and obesity. In a pubertal pattern of growth, height is in direct relation to estrogen as well as a sex steroid–dependent increase in growth hormone and production of insulin-like growth factor-I (Carel et al., 2004), making it a definite risk factor for breast cancer. Obesity, on the other hand, as indicated by body mass index (BMI) is a risk factor for breast cancer in young girls, as associated with precocious breast development (Biro et al., 2003; Wan et al., 2012) triggered by increased body fat (Biro et al., 2003), and women after menopause. Exogenous agents, such as exposure to radiation and environmental compounds, directly determine DNA mutation, subsequently altered mammary gland environment, and deregulation of the mammary stem cell (IBCERCC, 2013).

Susceptible stages in adulthood are pregnancy and lactation because hormonal developments and changes in the morphology of terminal duct lobular units (Kobayashi et al., 2012; Faupel-Badger et al., 2013) may also increase cancer risk.

Pregnancy

In epidemiological studies, pregnancy has been shown to generally lower the breast cancer risk for the longer term, despite its transitory short-term effect of increasing post-birth risk due to the stimulation of malignant cell transformation. Thus, multiple births are a strong protective factor for breast cancer risk (Faupel-Badger et al., 2013).

Lactation and Breastfeeding

Milk secretion and breastfeeding involve all developments turning the mammary epithelium into a mature milk-producing gland (Faupel-Badger et al., 2013) after birth (Neville et al., 2001). According to research, breastfeeding and particularly sustained lactation can be a protective factor as far as breast cancer is concerned (Bernier et al., 2000; Kobayashi et al., 2012), possibly due to a decrease in estrogen exposure resulting from fewer menstrual cycles.

Family History

A family history of breast cancer is a predictor of likely disease, even though if present in first- or second-degree relatives. When diagnosis of the relative in question occurs at age 50 or in case of multiple relatives affected by the disease (Pharoah et al., 1997).

Genetic Modifications

Despite the fact that testing for different genetic biomarkers is comparatively easy to perform, the challenge lies in the ability to consider these biomarkers in their context, making efforts toward predictive genetic testing while not overlooking the medical, social, and psychological implications of results, whether positive, negative, or pending (Walsh et al., 2016).

History of Benign Conditions

A personal history of nonproliferative benign conditions such as fibroadenomas and cysts is not a marker of high breast cancer risk (Page et al., 2003; Dupont et al., 1993). However, the risk rises by 1.5–1.9 with proliferative benign conditions, such as papilloma, hyperplasia, or radial scar, and 4–6 times for atypical lobular or ductal hyperplasia.

Prognostic Factors

Although not thoroughly clarified and therefore not formally endorsed for overall clinical use, criteria such as the stage (extent) of the disease, tumor grade, expression of estrogen receptors, expression of progesterone receptors, and human epidermal receptor-2/neu are among the prognostic factors used to determine the likely progression of the disease (Harris et al., 2007; Bast et al., 2001). Given its importance among prognostic factors, staging is currently achieved based on the TNM system (Singletary et al., 2002), as it focuses on indicators of clinical importance such as the size of the tumor (T), involvement level of lymph nodes (N), and presence of distant metastases (M).

Although not predictive of response to therapy, the above characteristics are strong individual predictors of future progress of the disease, with tumor size indicating death independently from the other criteria. Generally, poorer prognosis is evidenced by larger tumor size, greater number of affected lymph nodes (Cianfrocca and Goldstein, 2004), and presence of distant metastasis (with 2 years median overall survival time) (Giordano et al., 2004). Another prognosis factor, tumor grade, is established as high, moderate, or low, with high-grade disease characterized by faster progression and extreme potential for spreading.

Currently, grading is mainly achieved using the Nottingham histological grade system (Elston and Ellis, 1991), based on criteria referring to glandular formation and cell-related aspects such as size, shape, proliferation rate, and pattern. As far as disease markers are concerned, the most important to examine as far as prognosis are the expressions of the estrogen and progesterone receptors, both belonging to the steroid hormone receptor group.

Generally, expression of the estrogen and/or the progesterone receptors is predictive of improved response to hormonal therapy and better survival. Expression of both receptor types indicates up to 80% effective response. Expression of the estrogen receptor only predicts 30% responsiveness, whereas absence of either receptor expressions indicates nonresponsiveness to estrogen receptor modulator therapy (Lapidus et al., 1998).

An additional marker to examine is the human epidermal growth factor receptor 2 (HER2/neu) of the HER receptor tyrosine kinases. This is overexpressed in 30% of breast cancers, which are more aggressive and show poor prognosis. The drug of choice for the treatment of HER2/neu overexpressing breast cancer cases is trastuzumab (Menard et al., 2003).

Other gene expression profiles have recently been researched to help predict responsiveness to therapy and the likelihood of recurrence and response to treatment (e.g., assays such as Oncotype DX, the MammaPrint test, the Rotterdam Signature, and the Breast Cancer Gene Expression Ratio) (Harris et al., 2007).

Detection and Screening

The current means used for breast cancer detection and screening are physical examination combined with imaging tools such as ultrasonography, mammography, magnetic resonance imaging (MRI), and positron emission tomography, as well as analysis of tumor markers.

Tumor Marker Analysis

The term "tumor marker" refers to features distinguishing the tumor from normal tissue, which are measurable and/or visible outcomes of tumourigenesis (Levenson, 2007).

Biomarkers' efficacy depends directly on the degree of their clinical sensitivity (the ratio of subjects with confirmed disease with positive test results) and specificity (the ratio of healthy control subjects with negative test results) (Pepe et al., 2001).

The main characteristics of effective detection biomarkers include: (1) none or minimal invasive, (2) use of only small specimen amounts, (3) site-specificity expressed in the ability to discern between nonmalignant and malignant events at the level of the same tissue/organ, (4) enhanced specificity allowing for limitation of false-positive results, (5) ease of performance, (6) cost-effectiveness, and (7) independence from the observer (Levenson, 2007).

In relation to breast cancer, researchers have examined a range of analytes of biological molecules with the potential to act as biomarkers for the disease. Therefore, various carbohydrates, lipids, polyamines, proteins, DNA, and RNA have been studied in breast tissue, plasma, serum, and ductal lavage fluid or nipple aspirate fluid (Levenson, 2007). The study of proteins as biomarkers (proteomics) applies biochemical analysis to proteins and focuses on the function of expressed genes. Thus, proteomics is able to render an accurate and dynamic account not only of the cell intrinsic genetic program but also of the impact of its proximal environment. This ability of proteomics to relate gene sequence to cellular physiology suggests the usefulness of biomarker discovery for complementing the genome with the proteome.

In addition, with the aid of recent techniques such as matrix-assisted laser desorption/ionization time-of-flight mass spectrometry (MALDI-TOF MS) and surface-enhanced laser desorption/ionization (SELDI-TOF MS), high-throughput analysis of the proteome has become feasible. However, the tumor marker value of proteins remains to be clarified through comprehensive prospective clinical studies, and the only cancer biomarker with proven effectiveness is the prostate-specific antigen screening for prostate cancer (Lane et al., 2010).

Among proteins with a substantial value for breast cancer diagnosis, prognosis, or prediction, one should mention CA27-29 and CA15-3 (Gast et al., 2009; Duffy, 2006) or the extracellular mucin 1 (MUC1) protein (antigens for MUC1). MUC1 overexpression and aberrant glycosylation are markers for several cancer types. Recent studies have focused on the capacity of autoantibodies to aberrant O-glycoforms of MUC1 to operate as diagnostic biomarkers. Further sensitive biomarkers for effective early detection of breast cancer have been found in cancer-associated immunoglobulin G autoantibodies in breast cancer patients' serum against various aberrant O-glycopeptide epitopes derived from MUC1 (Wandall et al., 2010).

The technique of measuring gene expression from available sequence information (genomics) allows for determination of an expression profile representing cell function and phenotype (transcriptome). Such molecular signatures of potential biomarker value for early detection have recently come within reach by means of such technologies as multiplex polymerase chain reaction (PCR) (Srinivas et al., 2001) and cDNA oligonucleotide arrays.

In addition, according to certain studies, it seems that plasma mRNA may be used as a highly sensitive tumor marker thus allowing earlier cancer detection (Silva et al., 2002). However, research needs to be continued to resolve the problem of RNA stability in the bloodstream, more so in the presence of high serum ribonuclease levels found in cancer patients (Levenson, 2007).

Testing for Genetic and Molecular Changes

Genetic testing for mutations in breast cancer susceptibility genes such as BRCA1, BRCA2 (conducted so far in over one million individuals), and other genes is one prototype for the incorporation of genomics into personalized medicine practice. It has proved effective for both improvement of specific strategies for screening and prevention and use as a marker in targeted therapy. However, close compliance with the principles of genetic counseling has become even more important with the increasingly rapid development of molecular sequencing for the purposes of successful targeting of therapies. At the same time, it is key to establish whether genomic analysis is conducted to determine inherited susceptibilities or whether its primary purpose is genomic analysis of the tumor.

Approximately 30% of high-risk breast cancer families and around 15% of breast cancer familial relative risk (i.e., the disease risk for an affected individual's relative/the disease risk for the general population) may be attributed to pathogenic mutations (King, 2014;

Bahcall, 2013; Antoniou et al., 2008; Rebbeck et al., 2011). However, the difficulty arises to contextualize the risk of disease of inherited mutations and sequence variants in BRCA1/BRCA2.

Syndromes Predisposing to Breast Cancer

Hereditary Breast and Ovarian Syndrome

Given the numerous cases of worldwide breast cancer obviously developed as the result of an inherited predisposition, the identification of mutated genes as genetic risk biomarkers markedly increases in relevance. Thus, genetic testing becomes all the more important and worthwhile, so much so as early detection results in a more than 90% cure rate. In that respect, as highlighted above, BRCA1 and BRCA2 predominate among breast cancer susceptibility genes. BRCA1 is a large gene, consisting of 24 exons, out of which 22 are coding; the remaining two are noncoding. The location of BRCA1 is chromosome 17.

BRCA2 consists of 27 exons and is located on chromosome 13 (genenames.org).

Both genes are involved in DNA response to damage and in homologous recombination (Venkitaraman, 220).

Families with a history of both male and female breast cancers show higher pretest probability for BRCA2 testing, whereas families with both ovarian and breast cancers present higher pretest probability for BRCA1 testing (Frank et al., 2002). According to epidemiologic studies, mutations in BRCA1 determine a 70% lifetime breast cancer risk by age 70 (Whittemore et al., 1997; Claus et al., 1991, 1993, 1996) and even 90% for certain families with frequent early onset of ovarian or breast cancers (Ford et al., 1994).

High Penetrance Genes Predisposing to Breast Cancer

Actual development of cancer depends on the contribution of factors such as the particular type of constitutional aberration in BRCA1 or BRCA2, occurrence of modifying genomic alterations, and influence of the environment. Susceptibility, however, exists mostly as the result of genomic aberrations consisting of premature truncations of the BRCA1 and BRCA2 proteins by frameshift or nonsense mutations.

Among the more than 2000 distinct rare variants of BRCA1 and BRCA2, there are missense mutations, intronic changes, and small in-frame deletions and insertions that have been reported (Breast Cancer Information Core; www.research.nhgri.nih.gov/bic). BRCA1 main domains are located in the RING finger, and BRCT domains and their involvement are key for DNA repair function.

The location of pathogenic, highly penetrant missense mutations in BRCA2 involves the DNA binding domain in particular (Guidugli et al., 2013, 2014). In BRCA1 and BRCA2 there occur extensive genomic structural variations or rearrangements, representing 14% of mutations and 2.6% of mutations, respectively. The difference in the volume of structural variations between BRCA1 and BRCA2 may be attributed to the numerous Alu repeats characterizing the genomic region of BRCA1 gene location (Judkins et al., 2012).

In relation to BRCA1 and BRCA2, "founder" (population specific) mutations have been identified, among which the mutations occurring in the eastern European Jewish population have been well studied and documented. Thus, BRCA1 is the site of two mutations (5382insC and 185delG) and BRCA2 undergoes one mutation (6174delT) and up to 3% of this population are carriers of such a founder mutation (Offit et al., 1996; Szabo and King, 1997; Thorlacius et al., 1997). Given that 5% of all mutations in BRCA1 and BRCA2 in breast cancer patients belonging to this population group display nonfounder mutations, reflex full gene sequencing may be necessary in the case of negative results (Szabo and King, 1997; Thorlacius et al., 1997).

Carriers identified from population studies show a lower degree of disease penetrance as compared with carriers evidenced by means of kindred-based studies.

Because of an almost 60% breast cancer risk and a 40% ovarian cancer likelihood by age 70, identified BRCA1 mutation carriers are encouraged to be regularly screened. On the other hand, the likelihood of developing the disease by the same age in carriers of the BRCA2 mutation is 49% for breast cancer and 18% for ovarian cancer (Chen and Parmigiani, 2007). However, a certain degree of risk variability has been observed between the two types of cancer, explained in part, in respect to the risk for ovarian cancer, of genotype–phenotype correlations suggested by statistical data, highlighting a connection between specific phenotypes and the location of BRCA1/BRCA2 mutations. This has prompted the assumption that frameshift- and nonsense-type mutations occurring in each distal and proximal genomic region associate with lower ovarian cancer risk compared to analogous mutations by the center of either coding sequence (also known as "ovarian cancer cluster regions") (Gayther et al., 1996, 1997). From the perspective of the mutation location, statistic cohort data gathered from BRCA1/BRCA2 mutation carriers in the records of the CIMBA group (Consortium of Investigators of Modifiers of BRCA1/BRCA2) highlight relative decreases in breast cancer risk and increases in ovarian cancer risk concerning mutations in each gene central region. At the same time, there seems to be a higher breast cancer risk concerning mutations in the 3′ and 5′ regions in each gene.

An additional clarification for this risk variability comes from genome-wide association studies, which have prompted for the action of common genetic modifiers of ovarian and breast cancer risks in BRCA1/BRCA2 mutation carriers (Antoniou et al., 2010; Gaudet et al., 2010; Rebbeck et al., 2011; Wang et al., 2010).

The risk derived from the genomic location of BRCA1/BRCA2 mutations combined with that triggered by modifier genes differentiates between BRCA1 mutation carriers and BRCA2 mutation carriers at highest risk; thus, by age 80, the likelihood in the former carrier group to develop breast cancer is >81% and for ovarian cancer it is >63% (Gaudet et al., 2013).

If considered in the context of further risk-modifying variables observed in carriers of the BRCA1/BRCA2 mutation, mutation location and modifier genes are currently emerging as significant genomic biomarkers, whose worth may reside in their prospective capacity to better estimate the risk itself and to provide improved prognosis of disease behavior. An additional value may lie in profiling the phenotype of hereditary disease (such as the estrogen receptor status). This has been suggested by the observed tendency of tumors in BRCA1 mutation carrier breast cancer patients toward more aggressive disease features (Bignon et al., 1995; Jacquemier et al., 1995; Johannsson et al., 1997; Robson et al., 1998). However, accurate interpretation of genetic biomarkers for inherited risk needs to also take note of variants of unknown significance (e.g., small in-frame deletion/insertion variants, as well as intronic and missense ones).

Classification of these variants in all genes as neutral/low effect (Lindor et al., 2012) or pathogenic is a difficult task in clinical genetic testing. In the absence of an established database, such classification used animal or in silico models in an attempt to as nearly as possible infer the behavior of the disease in humans from patterns of development profiled in animals is difficult. It is also challenging to predict the functional influence of encountered variants on the foundation of the structure and/or conservation of amino acids. More recently, efforts have converged toward initiation of a database, which is designed for organization and oversight of clinical information on such variants (Clinvar—www.ncbi.nlm.nih.gov/clinvar).

Consistent work by the International Evidence-Based Network for the Interpretation of Germline Mutant Alleles (ENIGMA) Consortium has focused on BRCA1 and BRCA2, furthered by the Global Alliance for Genomics and with the aim of establishing an international database of BRCA1/BRCA2 variants. In addition to the database of variants of unknown significance as a tool for more accurate interpretation of genetic biomarkers, research has also endeavored to design quantitative risk prediction methods as necessary algorithms for assessment of such variants' likelihood for pathogenicity. Thus, for each variant envisaged, work involves measurement of family disease history, conservation, the pathology of the tumor and RNA splicing effects (Guidugli et al., 2013; Spurdle et al., 2012; Iversen et al., 2011; Lindor et al., 2012), and combined evolutionary sequence conservation (Guidugli et al., 2013, 2014; Lindor et al., 2012, Tavtigian et al., 2008, Goldgar et al., 2004). However, the need to compensate for the lack of statistical power concerning individual or rare variants has prompted development and use of high-throughput quantitative cell-based in vitro assays, aimed at providing a reliable estimate of their outcome in relation to BRCA1 and BRCA2 protein functions. Specificity and sensitivity for the variant specific biomarker VUS (Guidugli et al., 2013) or variants of unknown significance are assessed with acknowledged controls of normal versus pathogenic mutations.

Hypomorphic mutations preserve protein activity to some extent, but they also present particular difficulties in gene variant interpretation. Identification and clinical validation of variants or biomarkers in genes predisposing to breast cancer, which feature more moderate risk, allow for the design of increasingly personalized strategies for surveillance and prevention of disease.

Therapeutic Implications of Genetic Biomarkers

Clinical management of individuals with BRCA1 and BRCA2 mutation has been adjusted from the perspective of genetic testing. Genetic testing is a valuable source of information for evidence-based medical decisions. Identification of BRCA mutations requires constant surveillance of the breast, achieved by means of the carrier's self-examination, accompanied by clinical examination, and sustained and validated by mammography, sonography, or MRI (Burke et al., 1997; Kriege et al., 2004; Morris et al., 2003), which should be performed annually as early as the age of 25 in women at hereditary risk (Mettler et al., 1996).

Mammography and MRI, however, show different levels of sensitivity in the case of BRCA mutation carriers, with the former missing as many as 29% of new tumors thus making MRI the standard one. Breast cancer risk for BRCA1 and BRCA 2 mutation carriers may also be significantly (>90%) decreased by mastectomy, but this is comparatively infrequent. Chemoprevention is a further means of breast cancer risk reduction in BRCA1 and BRCA2 mutation carriers as well, and studies are underway to develop more advanced options than tamoxifen.

The standard current means to reduce risk in such mutation carriers is salpingo-oophorectomy, with effectiveness rates between 80% and 96% for ovarian cancer (Kauff et al., 2002; Rebbeck et al., 2002; Finch et al.,

2006) and approximately 50% for breast cancer. This is most likely due to its capacity to induce menopause (Kauff et al., 2002; Rebbeck et al., 2002; Eisen et al., 2005), resulting in a 60% reduction of overall mortality in these patients (Domchek et al., 2010). Several drugs have proven effectiveness against BRCA1 and BRCA2 cancers, such as platinum chemotherapy and poly(-ADP-ribose) polymerase inhibitors.

Other Highly Penetrant Breast Cancer Predisposing Genes

One important outcome of the development of sequencing technologies is the discovery of further predisposing genes, such as TP53 and CDH1. Although rare, genomic alterations in the TP53 gene encoding the tumor suppressor protein p. TP53 is involved in Li-Fraumeni syndrome and a comparatively increased risk of breast cancer (Hisada et al., 1998). Negative/Indeterminate results in tests for BRCA1 and BRCA2 mutations testing and a marked history of cancer in the family require testing for TP53 mutation. In addition to the diffuse gastric cancer risk, CDH1 mutation carriers also face a 40%–50% risk of lobular breast cancer (http://www.nccn.org/professionals/physician_gls/pdf/genetics_screening.pdf). Further, highly penetrant breast cancer predisposing genes PTEN and STK11 determine overt phenotypes such as PTEN hamartoma tumor syndromes and the Peutz-Jeghers syndrome in respective patients of marked breast cancer risk (Amos et al., 2004; Boardman et al., 1998, 2000a; b).

Moderate Penetrance Breast Cancer Genes

Carriers of moderately penetrant breast cancer predisposing genes such as CHEK2, ATM, PALB2, BRIP1, RAD51C, RAD51D, BARD1 should be subjected to screening in line with the patient's family and personal histories.

Low Penetrance Breast Cancer Polygenes

Breast cancer risk may also be weakly increased by common genetic variants (e.g., TGFBR2, MYC, TET2) identified in genome-wide association studies in 76 loci (Couch et al., 2014; Michailidou et al., 2013; Maxwell and Nathanson, 2013). Although still unclear for the most part, one of their mechanisms of action is via gene transcription. Among relevant signatures, single nucleotide polymorphisms, microsatellite instability, and epigenetic changes (e.g., changes in DNA methylation) should be mentioned.

Epigenetic changes such as DNA methylation involve markers regulating gene expression, but these do not modify the original DNA sequence. Such markers may undergo modifications themselves and may be inherited. As a gene expression regulator, DNA methylation can silence genes for tumor suppression and thus bear significant influence on tumorigenesis.

The value of biomarkers for DNA methylation relies on the similarity of primary tumor and blood plasma methylation patterns (Hoque et al., 2006) in the very incipient stages of breast tumor progression (van Hoesel et al., 2013; Radpour et al., 2011; Fabian et al., 2005; Wong et al., 2011; Yan et al., 2006). The proven high sensitivity (>90%) of plasma measurement of DNA methylation of certain genes renders it an effective prospective means for screening.

DNA Methylation, Definitions, and Measurement Methods

Altered gene expression is mainly triggered by microRNA expression, DNA methylation levels, and histone modifications. Epigenetic phenomena for which DNA methylation is key are cell development and differentiation, genomic imprinting and silencing of transposable elements, and inactivation of the X-chromosome. As representative of DNA methylation, methylated cytosine 5-methylcytosine (5mC), found in approximately 4% of cytosines, is an outcome of adding a methyl group to the 5′ position of cytosine mainly in CpG sequences. As a result of hypomethylation, total 5mC levels are known to be lower in tumors compared with neighboring tissues (reviewed in Robertson, 2005). This DNA loss of methylation occurs mainly in repetitive DNA elements, resulting in their reactivation and enhanced aberrant recombination, rendering them genomically unstable.

Gene-specific hypomethylation is also possible and leads to repeated expression of affected genes. Hypomethylation is a modification occurring early in cancer. The opposite phenomenon, gene-specific increase in epigenetic methylation (hypermethylation), more commonly occurs in the CpG island promoters. Its capacity to lead to gene inactivation is relevant for cancer as it also affects tumor suppressor genes, as well as mutation. Numerous genes feature hypermethylated CpG island promoters in breast cancer, which are involved in regulation of the cell cycle, in DNA repair, in the remodeling of chromatin, in cell signaling, and in transcription and tumor cell invasion and apoptosis.

An additional important type of methylated cytosine is 5-hydroxymethylcytosine (5hmC), even if at significantly lower levels. The 5hmC cytosine results from oxidation of 5mC by Tet enzymes (Kohli and Zhang, 2013). This pathway of oxidation can continue, rendering 5-carboxylcytosine and 5-formylcytosine, as substrates for the DNA repair enzyme thymidine–DNA glycosylase, emerging as one mechanism for methyl group removal from cytosine.

IMPACT OF METHYLATION BIOMARKERS

DNA Methylation Markers and Primary Prevention

In the context of increasing research efforts for cancer prevention, studies have also been directed toward assessment of the potential of DNA methylation as a biomarker for cancer risk evaluation. From the various studies conducted, some have aimed to examine the correlation between breast cancer risk and DNA methylation, both gene specific and global (reviewed in Terry et al., 2011; Brennan and Flanagan, 2012).

For instance, risk assessment studies were performed on white blood cell DNA, concluding on significant association of both hyper- and hypo-global methylation with breast cancer, highlighting the potential as a biomarker for breast cancer risk global DNA methylation.

However, there are certain limitations at this time to the approach of DNA methylation as cancer risk predictor, such as a certain degree of uncertainty derived from the potential for resulting data to be in fact an outcome of differences in cell populations or a response to underlying disease. In addition, DNA methylation can be influenced by other factors as well, such as genetics and age, to which lifestyle and environmental factors may be added (e.g., diet, smoking, air pollution, heavy metals, stress) (Terry et al., 2011; Bakulski and Fallin, 2014). Therefore, evaluation of the value of DNA methylation as a predictor for cancer requires larger population studies, allowing for extended follow-up, as well as serial blood collections (Pepe et al., 2001).

DNA Methylation Markers for Secondary Prevention and Early Detection

Screening tools for diagnosis and to determine treatment and prognosis are proposed as secondary preventive means. Despite its proven practical usefulness in effectively reducing breast cancer mortality for women at medium risk, regular screening by mammography (Mandelblatt et al., 2009; Webb et al., 2014) is generally not entirely satisfactory in specificity and sensitivity and even more limited in the case of younger females or those with dense breasts (Elmore et al., 1998; Alagaratnam and Wong, 1985; Moss, 2004; Qaseem et al., 2007). Such limitations as well as the complexity of the disease make a single-marker approach insufficient for effective early detection. The basic principle underlying the current general strategy in the management of cancer is to improve screening sensitivity and specificity by actively promoting the use of multiple various markers combined with different risk factors, adjusted for by reliable and validated statistical models (Wald et al., 1999).

Therefore, more effective early detection requires both discovery of additional markers and more precise assessment and stratification of risk. In that respect, biomarkers as additional screening means for early detection are useful for high-risk groups, as they have the ability to identify the disease in the absence of symptoms as well as in cases with normal results by mammography and breast examination (Evron et al., 2001). One such tool consists of plasma/serum cancer screening biomarkers such as the CA-125 for ovarian cancer.

A screening tool not used very often for secondary prevention is the category of blood biomarkers such as circulating cell-free DNA in the plasma, released from apoptotic and necrotic cells in the tumor. A number of studies have concluded that circulating DNA levels are higher in the presence of cancer. The potential usefulness of this blood-based biomarker has been further supported by the fact that, though initially examined for mutations, methylation patterns were discovered in circulating DNA that were similar to those of primary tumor ones (Gormally et al., 2007; Wang et al., 2010; Van De Voorde et al., 2012; Suijkerbuijk et al., 2011).

Methylation of plasma DNA has been found to possess early detection marker potential (van Hoesel et al., 2013) resulting from studies showing DNA hypermethylation of certain biomarkers (e.g., RARβ2, RASSF1A) in early breast cancer stages. Several specific characteristics support the biomarker potential of DNA methylation. Thus, promoter hypermethylation is more frequent than mutations and occurs early in breast tumorigenesis (Hoque et al., 2009; Lewis et al., 2005; Pasquali et al., 2007). In addition, aberrant DNA methylation in malignant cells also occurs in the immediate tissue. There are also technical aspects contributing to the usefulness of DNA methylation as a biomarker. Given its stability and as such the possibility for PCR amplification, analysis of aberrations only requires small amounts of DNA in comparison to gene expression profiling (Sidransky, 1997). Moreover, hypermethylated sequences are easier to detect compared to genetic alterations due to the positive signal they form against an unmethylated background. Furthermore, current research is being conducted on multiple genes such as RASSF1A, CDH1, BRCA1, APC, GSTP1, RARβ (reviewed in Ma et al., 2013, Van De Voorde et al., 2012, Suijkerbuijk et al., 2011).

For adequate sensitivity and specificity in the detection of breast cancer, a range of epigenetic markers should be used (Van De Voorde et al., 2012).

Results from research showing the presence of aberrant promoter hypermethylation in serum or plasma DNA in high-risk cases for a substantial amount of time prior to diagnosis further suggest its usefulness as a screening tool of plasma DNA methylation markers. In fact, there is increasing evidence that plasma DNA methylation of certain genes allows for over 90% sensitivity for breast cancer detection. One important benefit of blood markers for more effective screening is its use as an alternative or additional option to MRI and mammography, especially in cases where these screening tools have shown less sensitivity, and to reduce cumulative radiation.

DNA Methylation Markers and Tertiary Prevention and Role in Prognosis

Despite abundant information on DNA methylation in tissue samples at diagnosis, research of the correlation between DNA methylation patterns on diagnosis and subsequent prognosis and overall survival is scarce. Results of all recent studies that initiated on correlations between DNA methylation and breast cancer have converged toward cumulatively demonstrating the importance of methylation biomarkers present in tissue, plasma, and serum samples for secondary and tertiary prevention, as well as for prognosis of disease behavior and therapy outcome for the patient.

Further studies, especially cohort studies, are needed to accurately clarify the prognostic contribution of DNA methylation biomarkers, particularly in samples harvested at baseline. Continued extensive research is a necessary means to demonstrate the usefulness for screening and prognosis in general practice of complementing standard clinical markers measured at diagnosis (stage, grade, tumor size, molecular subtype) with assessment of methylation biomarkers.

To conclude, growing information on all aspects of the correlation between incidence of breast cancer and DNA methylation markers has not been fully clarified. Prospective research is required to eliminate temporality-related uncertainties resulting from the retrospective character of most studies thus far, which has made it difficult to determine whether aberrant methylation is a cause of or a consequence of cancer or its treatment. Future research also needs to ascertain, by repeated measurement, the potential for environmental factors to alter levels of DNA methylation markers, as well as the capacity of such modifications to influence the risk itself.

Finally, the panel of studied gene targets should be diversified.

MALE BREAST CANCER SUSCEPTIBILITY FACTORS

Epidemiology

Male breast cancer accounts for less than 1% of all breast cancers and only 0.2% of all cancers in males (ACS, 2013).

Pathology

Although rare, the glandular and adipose tissues of the breast in adult males can grow and proliferate as a result of a process similar to that occurring in females subject to high estrogen and progesterone levels, ending in disruption of the estrogen-to-androgen ratio (Johnson and Murad, 2009; Dickson, 2012). Growth of adipose tissue and a decrease of testosterone production thus inclining the estrogen:androgen ratio toward estrogen is the primary cause of breast cancer in men of age >65 (Niewoehner and Schorer, 2008).

In approximately 85% of cases, male breast cancer appears as a subareolar, unilateral thickening of the breast and is usually painless and may involve nipple retraction, ulceration, or discharge (Johansen Taber et al., 2010; Zygogianni et al., 2012).

Risk Factors

Major risk factors for male breast cancer include excess estrogen:androgen ratio, family disease history, inclusion in a particular ethnic group, occurrence of specific gene mutations, and environmental factors (Niewoehner and Schorer, 2008; Johansen Taber et al., 2010; Zygogianni et al., 2012; Ruddy and Winer, 2013). One of the most important causes of higher estrogen: androgen ratio underlying male breast cancer is the Klinefelter syndrome (XXY sex chromosomes), which is also characterized by more estrogen and progesterone receptors in the breast tissue (Aksglaede et al., 2013). Both of these features pose a very significant breast cancer risk (20–30 times normal) (Brinton, 2011).

Among other causes of excess estrogen:androgen ratio potentially leading to male breast cancer are low levels of androgen determined by such testicular anomalies as orchitis or cryptorchidism, elevated estrogen and androgen caused by congenital adrenal hyperplasia or Leydig cell tumor, therapy with exogenous estrogen, increased estradiol levels triggered by abnormal aromatase function or Sertoli cell tumor, abnormal levels of sex hormone binding globulin, hyperthyroidism, liver or renal disease, obesity, environmental antiandrogenic and proestrogenic factors (e.g., excessive heat), and carcinogenic compounds and radiation (Niewoehner and Schorer, 2008; Johansen Taber et al., 2010;

Zygogianni et al., 2012; Ruddy and Winer, 2013). Given their shared risk factors, gynecomastia may be considered a marker of breast cancer risk.

Recently, a new cause of gynecomastia has been added to the established one (high effect of estrogens or the low effect of androgens, i.e., use of estrogen receptor binding drugs e.g., digitalis, diazepam, and ketoconazole) and exogenous compounds (marijuana) (Niewoehner and Schorer, 2008; Swerdloff and Ng, 2011).

Another important risk factor for male breast cancer is mutation in the BRCA2 gene (accounting for 5%–10% likelihood) and, to a lesser extent, the BRCA1 gene (Ruddy and Winer, 2013). Because of a current lack of awareness of this disease in men, screening and breast examination for males is infrequent, which sharply diminishes the possibility for early detection. In over 50% of cases, tumors of the male breast are diagnosed at stage II or more, whereas this only happens in about 35% of cases in women (Johansen Taber et al., 2010; Zygogianni et al., 2012).

Cancer Aggressiveness Risk Factors

Accumulating evidence from clinical and epidemiological studies increasingly demonstrating the diversity of risk factors for cancer occurrence and for life-threatening cancer, as well as the complexity of their interplay, suggests a potential difference between prediagnostic risk factors for cancer occurrence and those for cancer death (Autier, 2012; Barnett et al., 2008), even in the same organ. This renders the significance of finding the so-called "cancer aggressiveness risk factor," consisting of individual-specific genetic, lifestyle, and/or environmental features or of measuring nontumor biomarkers to determine individuals who are more likely to develop aggressive, life-threatening cancers. From this perspective, it has been determined that even if highly instrumental in breast cancer risk, reproductive factors are weak for breast cancer mortality (Barnett et al., 2008). Though modest as breast cancer risk triggers in post-menopausal women, adiposity increases the risk for breast cancer death in women before menopause (Loi et al., 2005). The example of fertility may be added to these, which is generally a weak risk for breast cancer but an aggravating one for cancer death in women aged 40 and above and in those who give birth (Daling et al., 2002).

Impact of Cancer Aggressiveness Risk Factors for Patient Management and Health Policies

The identification of cancer aggressiveness risk factors is a most important tool for risk stratification and therefore proves useful for medical practice and health policies. Primary prevention would acquire the necessary base to target efforts in high-risk individuals by providing personalized counseling. Furthermore, given the basic purpose of screening to detect cancer early and thus prevent lethal disease outcomes, application of cancer aggressiveness risk factors would decrease harm from unnecessary screening in less aggressive cancer risk cases. By differentiating between subjects at higher risk of cancer death and those at mere higher risk of cancer occurrence, screening would thus become more cost-effective and improve participation due to better patient awareness.

At the same time, use of such tools would improve cancer patient stratification and referral, distinguishing between absence of life-threatening risk, where active surveillance is the adequate approach, and presence of life-threatening risk, requiring immediate treatment. A further benefit would impact decisions for therapeutic options, warning against cancer aggressiveness risk factors leading to likelihood of relapse despite an ostensibly favorable prognosis and endorsing more resolute management of such patients.

Lastly, discovery of such risk factors would greatly assist in clarification of biological mechanisms leading to poor prognosis for progression, relapse, and lethal cancer outcome. In the context of cancer overdiagnosis and resulting overtreatment with unnecessarily aggressive therapy, relevance of cancer aggressiveness risk factors for medical practice lies in their potential to prioritize referral to diagnosis and specialized care. By eliminating pseudocancers and borderline or in situ cancers from therapy, their use can increase cost-effectiveness of both screening and therapy and reduce associated harm.

Further research in this respect would prove highly beneficial for the design of screening policies, chemoprevention strategies, and making better-informed decisions on treatment options based on more accurate evaluation of relapse risks and would avoid unnecessary treatment.

References

Aksglaede, L., Link, K., Giwercman, A., 2013. 47,XXY Klinefelter syndrome: clinical characteristics and age-specific recommendations for medical management. Am. J. Med. Genet. C. Semin. Med. Genet. 163, 55–63.

Alagaratnam, T.T., Wong, J., 1985. Limitations of mammography in Chinese females. Clin. Radiol. 36 (2), 175–177.

American Cancer Society (ACS), 2013. Breast Cancer. Available at: www.cancer.org/cancer/breastcancer/index.

Amos, C.I., Keitheri-Cheteri, M.B., Sabripour, M., et al., 2004. Genotype-phenotype correlations in Peutz-Jeghers syndrome. J. Med. Genet. 41 (5), 327–333.

Antoniou, A.C., Beesley, J., McGuffog, L., et al., 2010. Common breast cancer susceptibility alleles and the risk of breast cancer for BRCA1 and BRCA2 mutation carriers: implications for risk prediction. Cancer Res. 70 (23), 9742–9754.

Antoniou, A.C., Cunningham, A.P., Peto, J., et al., 2008. The BOADICEA model of genetic susceptibility to breast and ovarian cancers: updates and extensions. Br. J. Cancer 98 (8), 1457–1466.

Arpino, G., Bardou, V.J., Clark, G.M., Elledge, R.M., 2004. Infiltrating lobular carcinoma of the breast: tumor characteristics and clinical outcome. Breast Cancer Res. 6 (3), R149–R156.

Autier, P., 2012. Risk factors for breast cancer for women aged 40 to 49 years. Ann. Intern. Med. 157 (7), 529.

Bahcall, O.G., 2013. iCOGS collection provides a collaborative model. Foreword. Nat. Genet. 45, 343.

Bakulski, K.M., Fallin, M.D., 2014. Epigenetic epidemiology: promises for public health research. Environ. Mol. Mutagen. 55 (3), 171–183.

Barnett, G.C., Shah, M., Redman, K., et al., 2008. Risk factors for the incidence of breast cancer: do they affect survival from the disease? J. Clin. Oncol. 26 (20), 3310–3316.

Bast Jr., R.C., Ravdin, P., Hayes, D.F., et al., 2001. 2000 update of recommendations for the use of tumor markers in breast and colorectal cancer: clinical practice guidelines of the American Society of Clinical Oncology. J. Clin. Oncol. 19 (6), 1865–1878.

Bernier, M.O., Plu-Bureau, G., Bossard, N., et al., 2000. Breastfeeding and risk of breast cancer: a metaanalysis of published studies. Hum. Reprod. Update 6, 374–386.

Bignon, Y.J., Fonck, Y., Chassagne, M.C., 1995. Histoprognostic grade in tumours from families with hereditary predisposition to breast cancer. Lancet 346 (8969), 258.

Birnbaum, L.S., Fenton, S.E., 2003. Cancer and developmental exposure to endocrine disruptors. Environ. Health Perspect. 111, 389–394.

Biro, F.M., Galvez, M.P., Greenspan, L.C., 2010. Pubertal assessment method and baseline characteristics in a mixed longitudinal study of girls. Pediatrics 126, e583–e590.

Biro, F.M., Lucky, A.W., Simbartl, L.A., 2003. Pubertal maturation in girls and the relationship to anthropometric changes: pathways through puberty. J. Pediatr. 142, 643–646.

Boardman, L.A., Thibodeau, S.N., Schaid, D.J., et al., 1998. Increased risk for cancer in patients with the Peutz-Jeghers syndrome. Ann. Intern. Med. 128 (11), 896–899.

Boardman, L.A., Couch, F.J., Burgart, L.J., et al., 2000a. Genetic heterogeneity in Peutz-Jeghers syndrome. Hum. Mutat. 16 (1), 23–30.

Boardman, L.A., Pittelkow, M.R., Couch, F.J., et al., 2000b. Association of Peutz-Jeghers-like mucocutaneous pigmentation with breast and gynecologic carcinomas in women. Medicine 79 (5), 293–298.

Brennan, K., Flanagan, J.M., 2012. Is there a link between genome-wide hypomethylation in blood and cancer risk? Cancer Prev. Res. 5 (12), 1345–1357.

Brinton, L.A., 2011. Breast cancer risk among patients with Klinefelter syndrome. Acta Paediatr. 100, 814–818.

Burke, W., Petersen, G., Lynch, P., et al., 1997. Recommendations for follow-up care of individuals with an inherited predisposition to cancer. I. Hereditary nonpolyposis colon cancer. Cancer Genetics Studies Consortium. J. Am. Med. Assoc. 277 (11), 915–919.

Carel, J.C., Lahlou, N., Roger, M., Chaussain, J.L., 2004. Precocious puberty and statural growth. Hum. Reprod. Update 10, 135–147.

Chen, S., Parmigiani, G., 2007. Meta-analysis of BRCA1 and BRCA2 penetrance. J. Clin. Oncol. 25 (11), 1329–1333.

Cianfrocca, M., Goldstein, L.J., 2004. Prognostic and predictive factors in early-stage breast cancer. Oncol. 9 (6), 606–616.

Claus, E.B., Risch, N., Thompson, W.D., 1991. Genetic analysis of breast cancer in the cancer and steroid hormone study. Am. J. Hum. Genet. 48 (2), 232–242.

Claus, E.B., Risch, N., Thompson, W.D., 1993. The calculation of breast cancer risk for women with a first degree family history of ovarian cancer. Breast Cancer Res. Treat. 28 (2), 115–120.

Claus, E.B., Schildkraut, J.M., Thompson, W.D., Risch, N.J., 1996. The genetic attributable risk of breast and ovarian cancer. Cancer 77 (11), 2318–2324.

Couch, F.J., Nathanson, K.L., Offit, K., 2014. Two decades after BRCA: setting paradigms in personalized cancer care and prevention. Science 343 (6178), 1466–1470.

D'Aloisio, A.A., Deroo, L.A., Baird, D.D., Weinberg, C.R., Sandler, D.R., 2013. Prenatal and infant exposures and age at menarche. Epidemiology 24, 277–284.

Daling, J.R., Malone, K.E., Doody, D.R., et al., 2002. The relation of reproductive factors to mortality from breast cancer. Cancer Epidemiol. Biomarkers Prev. 11 (3), 235–241.

Dickson, G., 2012. Gynecomastia. Am. Fam. Physician 85, 716–722.

Domchek, S.M., Friebel, T.M., Singer, C.F., et al., 2010. Association of risk-reducing surgery in BRCA1 or BRCA2 mutation carriers with cancer risk and mortality. J. Am. Med. Assoc. 304 (9), 967–975.

Duffy, M.J., 2006. Serum tumor markers in breast cancer: are they of clinical value? Clin. Chem. 52 (3), 345–351.

Dupont, W.D., Parl, F.F., Hartmann, W.H., et al., 1993. Breast cancer risk associated with proliferative breast disease and atypical hyperplasia. Cancer 71 (4), 1258–1265.

Eisen, A., Lubinski, J., Klijn, J., et al., 2005. Breast cancer risk following bilateral oophorectomy in BRCA1 and BRCA2 mutation carriers: an international case-control study. J. Clin. Oncol. 23 (30), 7491–7496.

Elmore, J.G., Barton, M.B., Moceri, V.M., et al., 1998. Ten-year risk of false positive screening mammograms and clinical breast examinations. N. Engl. J. Med. 338 (16), 1089–1096.

Elston, C.W., Ellis, I.O., 1991. Pathological prognostic factors in breast cancer. I. The value of histological grade in breast cancer: experience from a large study with long-term follow-up. Histopathology 19 (5), 403–410.

Ernster, V.L., Ballard-Barbash, R., Barlow, W.E., et al., 2002. Detection of ductal carcinoma in situ in women undergoing screening mammography. J. Natl. Cancer Inst. 94 (20), 1546–1554.

Evron, E., Dooley, W.C., Umbricht, C.B., et al., 2001. Detection of breast cancer cells in ductal lavage fluid by methylation-specific PCR. Lancet 357 (9265), 1335–1336.

Fabian, C.J., Kimler, B.F., Mayo, M.S., Khan, S.A., 2005. Breast-tissue sampling for risk assessment and prevention. Endocr. Relat. Canc. 12 (2), 185–213.

Faupel-Badger, J.M., Arcaro, K.F., Balkam, J.J., 2013. Postpartum remodeling, lactation, and breast cancer risk: summary of a National Cancer Institute-sponsored workshop. J. Natl. Cancer Inst. 105, 166–174.

Fenton, S.E., Reed, C., Newbold, R.R., 2012. Perinatal environmental exposures affect mammary development, function, and cancer risk in adulthood. Annu. Rev. Pharmacol. Toxicol. 52, 455–479.

Fenton, S.E., 2006. Endocrine-disrupting compounds and mammary gland development: early exposure and later life consequences. Endocrinology 147, S18–S24.

Finch, A., Beiner, M., Lubinski, J., et al., 2006. Salpingo-oophorectomy and the risk of ovarian, fallopian tube, and peritoneal cancers in women with a BRCA1 or BRCA2 mutation. J. Am. Med. Assoc. 296 (2), 185–192.

Ford, D., Easton, D.F., Bishop, D.T., et al., 1994. Risks of cancer in BRCA1-mutation carriers. Breast Cancer Linkage Consortium. Lancet 343 (8899), 692–695.

Fraga, M.F., Agrelo, R., Esteller, M., 2007. Cross-talk between aging and cancer: the epigenetic language. Ann. N. Y. Acad. Sci. 1100, 60–74.

Frank, T.S., Deffenbaugh, A.M., Reid, J.E., et al., 2002. Clinical characteristics of individuals with germline mutations in BRCA1 and BRCA2: analysis of 10,000 individuals. J. Clin. Oncol. 20 (6), 1480–1490.

Gast, M.C., Schellens, J.H., Beijnen, J.H., 2009. Clinical proteomics in breast cancer: a review. Breast Cancer Res. Treat. 116 (1), 17–29.

Gaudet, M.M., Kirchhoff, T., Green, T., et al., 2010. Common genetic variants and modification of penetrance of BRCA2-associated breast cancer. PLoS Genet. 6 (10), e1001183.

Gaudet, M.M., Kuchenbaecker, K.B., Vijai, J., et al., 2013. Identification of a BRCA2-specific modifier locus at 6p24 related to breast cancer risk. PLoS Genet. 9 (3), e1003173.

Gayther, S.A., Harrington, P., Russell, P., et al., 1996. Rapid detection of regionally clustered germ-line BRCA1 mutations by multiplex heteroduplex analysis. UKCCCR Familial Ovarian Cancer Study Group. Am. J. Hum. Genet. 58 (3), 451–456.

Gayther, S.A., Harrington, P., Russell, P., et al., 1997. Frequently occurring germ-line mutations of the BRCA1 gene in ovarian cancer families from Russia. Am. J. Hum. Genet. 60 (5), 1239–1242.

Giordano, S.H., Buzdar, A.U., Smith, T.L., et al., 2004. Is breast cancer survival improving? Cancer 100 (1), 44–52.

Goldgar, D.E., Easton, D.F., Deffenbaugh, A.M., et al., 2004. Integrated evaluation of DNA sequence variants of unknown clinical significance: application to BRCA1 and BRCA2. Am. J. Hum. Genet. 75 (4), 535–544.

Gormally, E., Caboux, E., Vineis, P., Hainaut, P., 2007. Circulating free DNA in plasma or serum as biomarker of carcinogenesis: practical aspects and biological significance. Mutat. Res. 635, 105–117.

Guidugli, L., Carreira, A., Caputo, S.M., et al., 2014. Functional assays for analysis of variants of uncertain significance in BRCA2. Hum. Mutat. 35 (2), 151–164.

Guidugli, L., Pankratz, V.S., Singh, N., et al., 2013. A classification model for BRCA2 DNA binding domain missense variants based on homology-directed repair activity. Cancer Res. 73 (1), 265–275.

Harris, L., Fritsche, H., Mennel, R., et al., 2007. American Society of Clinical Oncology 2007 update of recommendations for the use of tumor markers in breast cancer. J. Clin. Oncol. 25 (33), 5287–5312.

Hisada, M., Garber, J.E., Fung, C.Y., et al., 1998. Multiple primary cancers in families with Li-Fraumeni syndrome. J. Natl. Cancer Inst. 90 (8), 606–611.

Hoover, R.N., Hyer, M., Pfeiffer, R.M., 2011. Adverse health outcomes in women exposed in utero to diethylstilbestrol. N. Engl. J. Med. 365, 1304–1314.

Hoque, M.O., Feng, Q., Toure, P., et al., 2006. Detection of aberrant methylation of four genes in plasma DNA for the Wdetection of breast cancer. J. Clin. Oncol. 24 (26), 4262–4269.

Hoque, M.O., Prencipe, M., Poeta, M.L., et al., 2009. Changes in CpG islands promoter methylation patterns during ductal breast carcinoma progression. Cancer Epidemiol. Biomark. Prev. 18 (10), 2694–2700.

Interagency Breast Cancer and Environmental Research Coordinating Committee (IBCERCC), 2013. Breast Cancer and the Environment: Prioritizing Prevention. Available at: www.niehs.nih.gov/about/assets/docs/ibcercc_full_508.pdf.

IOM (Institute of Medicine), 2011. Dietary Reference Intakes for Calcium and Vitamin D. Committee to Review Dietary Reference Intakes for Calcium and Vitamin D. The National Academies Press, Washington, DC.

Iversen Jr., E.S., Couch, F.J., Goldgar, D.E., et al., 2011. A computational method to classify variants of uncertain significance using functional assay data with application to BRCA1. Cancer Epidemiol. Biomark. Prev. 20 (6), 1078–1088.

Jacquemier, J., Eisinger, F., Birnbaum, D., Sobol, H., 1995. Histoprognostic grade in BRCA1-associated breast cancer. Lancet 345 (8963), 1503.

Johannsson, O.T., Idvall, I., Anderson, C., et al., 1997. Tumour biological features of BRCA1-induced breast and ovarian cancer. Eur. J. Cancer 33 (3), 362–371.

Johansen Taber, K.A., Morisy, L.R., Osbahr, A.J., Dickinson, B.D., 2010. Male breast cancer: risk factors, diagnosis, and management. Oncol. Rep. 24, 1115–1120.

Johnson, R.E., Murad, M.H., 2009. Gynecomastia: pathophysiology, evaluation, and management. Mayo Clin. Proc. 84, 1010–1015.

Judkins, T., Rosenthal, E., Arnell, C., et al., 2012. Clinical significance of large rearrangements in BRCA1 and BRCA2. Cancer 118 (21), 5210–5216.

Kauff, N.D., Satagopan, J.M., Robson, M.E., et al., 2002. Risk-reducing salpingooophorectomy in women with a BRCA1 or BRCA2 mutation. N. Engl. J. Med. 346 (21), 1609–1615.

King, M.-C., 2014. The race' to clone BRCA1. Science 343, 1462–1465.

Kobayashi, S., Sugiura, H., Ando, Y., 2012. Reproductive history and breast cancer risk. Breast Canc. 19, 302–308.

Kohli, R.M., Zhang, Y., 2013. TET enzymes, TDG and the dynamics of DNA demethylation. Nature 502 (7472), 472–479.

Kriege, M., Brekelmans, C.T., Boetes, C., et al., 2004. Efficacy of MRI and mammography for breast-cancer screening in women with a familial or genetic predisposition. N. Engl. J. Med. 351 (5), 427–437.

Kuerer, H.M., Albarracin, C.T., Yang, W.T., et al., 2009. Ductal carcinoma in situ: state of the science and roadmap to advance the field. J. Clin. Oncol. 27 (2), 279–288.

Lane, J.A., Hamdy, F.C., Martin, R.M., et al., 2010. Latest results from the UK trials evaluating prostate cancer screening and treatment: the CAP and ProtecT studies. Eur. J. Cancer 46 (17), 3095–3101.

Lapidus, R.G., Nass, S.J., Davidson, N.E., 1998. The loss of estrogen and progesterone receptor gene expression in human breast cancer. J. Mammary Gland Biol. Neoplasia 3 (1), 85–94.

Levenson, V.V., 2007. Biomarkers for early detection of breast cancer: what, when, and where? Biochim. Biophys. Acta 1770 (6), 847–856.

Lewis, C.M., Cler, L.R., Bu, D.-W., et al., 2005. Promoter hypermethylation in benign breast epithelium in relation to predicted breast cancer risk. Clin. Cancer Res. 11 (1), 166–172.

Li, C.I., Anderson, B.O., Daling, J.R., Moe, R.E., 2003. Trends in incidence rates of invasive lobular and ductal breast carcinoma. JAMA 289 (11), 1421–1424.

Lindor, N.M., Guidugli, L., Wang, X., et al., 2012. A review of a multifactorial probability-based model for classification of BRCA1 and BRCA2 variants of uncertain significance (VUS). Hum. Mutat. 33 (1), 8–21.

Loi, S., Milne, R.L., Friedlander, M.L., et al., 2005. Obesity and outcomes in premenopausal and postmenopausal breast cancer. Cancer Epidemiol. Biomarkers Prev. 14 (7), 1686–1691.

Ma, Y., Wang, X., Jin, H., 2013. Methylated DNA and microRNA in body fluids as biomarkers for cancer detection. Int. J. Mol. Sci. 14 (5), 10307–10331.

Mandelblatt, J.S., Cronin, K.A., Bailey, S., et al., 2009. Effects of mammography screening under different screening schedules: model estimates of potential benefits and harms. Ann. Intern. Med. 151 (10), 738–747.

Maxwell, K.N., Nathanson, K.L., 2013. Common breast cancer risk variants in the post-COGS era: a comprehensive review. Breast Cancer Res. 15 (6), 212.

Menard, S., Pupa, S.M., Campiglio, M., Tagliabue, E., 2003. Biologic and therapeutic role of HER2 in cancer. Oncogene 22 (42), 6570–6578.

Mettler, F.A., Upton, A.C., Kelsey, C.A., et al., 1996. Benefits versus risks from mammography: a critical reassessment. Cancer 77 (5), 903–909.

Michailidou, K., Hall, P., Gonzalez-Neira, A., et al., 2013. Large-scale genotyping identifies 41 new loci associated with breast cancer risk. Nat. Genet. 45 (4), 353–361, 361e351–352.

Morris, E.A., Liberman, L., Ballon, D.J., et al., 2003. MRI of occult breast carcinoma in a high-risk population. Am. J. Roentgenol. 181 (3), 619–626.

Moss, S., 2004. Should women under 50 be screened for breast cancer? Br. J. Cancer 91 (3), 413–417.

Neville, M.C., Morton, J., Umemura, S., 2001. Lactogenesis: the transition from pregnancy to lactation. Pediatr. Clin. North Am. 48, 35–52.

Niewoehner, C.B., Schorer, A.E., 2008. Gynaecomastia and breast cancer in men. BMJ 336, 709–713.

Offit, K., Gilewski, T., McGuire, P., et al., 1996. Germline BRCA1 185delAG mutations in Jewish women with breast cancer. Lancet 347 (9016), 1643–1645.

Page, D.L., Schuyler, P.A., Dupont, W.D., et al., 2003. Atypical lobular hyperplasia as a unilateral predictor of breast cancer risk: a retrospective cohort study. Lancet 361 (9352), 125–129.

Park, S.K., Kang, D., McGlynn, K.A., 2008. Intrauterine environments and breast cancer risk: meta-analysis and systematic review. Breast Cancer Res. 10, R8.

Pasquali, L., Bedeir, A., Ringquist, S., et al., 2007. Quantification of CpG island methylation in progressive breast lesions from normal to invasive carcinoma. Cancer Lett. 257 (1), 136–144.

Pepe, M.S., Etzioni, R., Feng, Z., et al., 2001. Phases of biomarker development for early detection of cancer. J. Natl. Cancer Inst. 93 (14), 1054–1061.

Pharoah, P.D., Day, N.E., Duffy, S., et al., 1997. Family history and the risk of breast cancer: a systematic review and meta-analysis. Int. J. Cancer 71 (5), 800–809.

Qaseem, A., Snow, V., Sherif, K., et al., 2007. Screening mammography for women 40 to 49 years of age: a clinical practice guideline from the American College of Physicians. Ann. Intern. Med. 146 (7), 511–515.

Radpour, R., Barekati, Z., Kohler, C., et al., 2011. Hypermethylation of tumor suppressor genes involved in critical regulatory pathways for developing a blood-based test in breast cancer. PLoS One 6 (1), e16080.

Rebbeck, T.R., Lynch, H.T., Neuhausen, S.L., et al., 2002. Prophylactic oophorectomy in carriers of BRCA1 or BRCA2 mutations. N. Engl. J. Med. 346 (21), 1616–1622.

Rebbeck, T.R., Mitra, N., Domchek, S.M., et al., 2011. Modification of BRCA1-associated breast and ovarian cancer risk by BRCA1-interacting genes. Cancer Res. 71 (17), 5792–5805.

Robertson, K.D., 2005. DNA methylation and human disease. Nat. Rev. Genet. 6 (8), 597–610.

Robson, M., Rajan, P., Rosen, P.P., et al., 1998. BRCA-associated breast cancer: absence of a characteristic immunophenotype. Cancer Res. 58 (9), 1839–1842.

Ruddy, K.J., Winer, E.P., 2013. Male breast cancer: risk factors, biology, diagnosis, treatment and survivorship. Ann. Oncol. 24, 1434–1443.

Russo, J., Russo, I.H., 1999. Cellular basis of breast cancer susceptibility. Oncol. Res. 11 (4), 169–178.

Sidransky, D., 1997. Nucleic acid-based methods for the detection of cancer. Science 278 (5340), 1054–1058.

Silva, J., Silva, J.M., Garcia, V., et al., 2002. RNA is more sensitive than DNA in identification of breast cancer patients bearing tumor nucleic acids in plasma. Genes Chromosomes Cancer 35 (4), 375–376.

Singletary, S.E., Allred, C., Ashley, P., et al., 2002. Revision of the American Joint Committee on Cancer staging system for breast cancer. J. Clin. Oncol. 20 (17), 3628–3636.

Spurdle, A.B., Whiley, P.J., Thompson, B., et al., 2012. BRCA1 R1699Q variant displaying ambiguous functional abrogation confers intermediate breast and ovarian cancer risk. J. Med. Genet. 49 (8), 525–532.

Srinivas, P.R., Kramer, B.S., Srivastava, S., 2001. Trends in biomarker research for cancer detection. Lancet Oncol. 2 (11), 698–704.

Suijkerbuijk, K.P.M., van Diest, P.J., van der Wall, E., 2011. Improving early breast cancer detection: focus on methylation. Ann. Oncol. 22 (1), 24–29.

Swerdloff, R.S., Ng, J.C.M., 2011. Gynecomastia: Etiology, Diagnosis, and Treatment. Available at: www.endotext.org/male/male14/male14.html.

Szabo, C.I., King, M.C., 1997. Population genetics of BRCA1 and BRCA2. Am. J. Hum. Genet. 60 (5), 1013–1020.

Tao, Z., Shi, A., Lu, C., et al., 2015. Breast Cancer: epidemiology and etiology. Cell Biochem. Biophys. 72 (2), 333–338.

Tavtigian, S.V., Byrnes, G.B., Goldgar, D.E., Thomas, A., 2008. Classification of rare missense substitutions, using risk surfaces, with genetic- and molecular-epidemiology applications. Hum. Mutat. 29 (11), 1342–1354.

Terry, M.B., Delgado-Cruzata, L., Vin-Raviv, N., et al., 2011. DNA methylation in white blood cells: association with risk factors in epidemiologic studies. Epigenetics 6 (7), 828–837.

Thorlacius, S., Sigurdsson, S., Bjarnadottir, H., et al., 1997. Study of a single BRCA2 mutation with high Carrier frequency in a small population. Am. J. Hum. Genet. 60 (5), 1079–1084.

Van De Voorde, L., Speeckaert, R., Van Gestel, D., et al., 2012. DNA methylation-based biomarkers in serum of patients with breast cancer. Mutat. Res. 751 (2), 304–325.

van Hoesel, A.Q., Sato, Y., Elashoff, D.A., et al., 2013. Assessment of DNA methylation status in early stages of breast cancer development. Br. J. Cancer 108 (10), 2033–2038.

Veronesi, U., Boyle, P., Goldhirsch, A., et al., 2005. Breast cancer. Lancet 365 (9472), 1727–1741.

Wald, N.J., Hackshaw, A.K., Frost, C.D., 1999. When can a risk factor be used as a worthwhile screening test? BMJ 319 (7224), 1562–1565.

Walsh, M.F., Nathanson, K.L., Couch, F.J., Offit, K., 2016. Genomic biomarkers for breast cancer risk. Adv. Exp. Med. Biol. 882, 1–32.

Wan, W., Deng, X., Archer, K.J., Sun, S.S., 2012. Pubertal pathways and the relationship to anthropometric changes in childhood: the Fels longitudinal study. Open J. Pediatr. 2, 9.

Wandall, H.H., Blixt, O., Tarp, M.A., et al., 2010. Cancer biomarkers defined by autoantibody signatures to aberrant Oglycopeptide epitopes. Cancer Res. 70 (4), 1306–1313.

Wang, X., Pankratz, V.S., Fredericksen, Z., et al., 2010. Common variants associated with breast cancer in genomewide association studies are modifiers of breast cancer risk in BRCA1 and BRCA2 mutation carriers. Hum. Mol. Genet. 19 (14), 2886–2897.

Webb, M.L., Cady, B., Michaelson, J.S., et al., 2014. A failure analysis of invasive breast cancer. Cancer 120 (18), 2839–2846.

Whittemore, A.S., Gong, G., Itnyre, J., 1997. Prevalence and contribution of BRCA1 mutations in breast cancer and ovarian cancer: results from three U.S. population-based case-control studies of ovarian cancer. Am. J. Hum. Genet. 60 (3), 496–504.

Wong, E.M., Southey, M.C., Fox, S.B., et al., 2011. Constitutional methylation of the BRCA1 promoter is specifically associated with BRCA1 mutation-associated pathology in early-onset breast cancer. Cancer Prev. Res. 4 (1), 23–33.

Yan, P.S., Venkataramu, C., Ibrahim, A., et al., 2006. Mapping geographic zones of cancer risk with epigenetic biomarkers in normal breast tissue. Clin. Cancer Res. 12 (22), 6626–6636.

Zygogianni, A.G., Kyrgias, G., Gennatas, C., 2012. Male breast carcinoma: epidemiology, risk factors and current therapeutic approaches. Asian Pac. J. Cancer Prev 13, 15–19.

CHAPTER

47

Pancreatic and Ovarian Cancer Biomarkers

George Georgiadis[1,a], Charalampos Belantis[1,a], Charalampos Mamoulakis[1], John Tsiaoussis[2], Wallace A. Hayes[3], Aristidis M. Tsatsakis[4]

[1]Department of Urology, University General Hospital of Heraklion, University of Crete, Medical School, Heraklion, Crete, Greece; [2]Laboratory of Anatomy-Histology-Embryology, Medical School, University of Crete, Heraklion, Crete, Greece; [3]University of South Florida College of Public Health, Tampa, FL, United States; Michigan State University, East Lansing, MI, United States; [4]Center of Toxicology Science & Research, Medical School, University of Crete, Heraklion, Greece

INTRODUCTION

The challenge of caring for patients diagnosed with malignancies of the pancreas or ovaries remains daunting despite steady advances in risk assessment and treatment for these conditions. The prospect of curative treatment for both of these cancers is restricted to those patients who present with the earliest stages of disease; however, effective and practical methods of identifying these patients remain elusive. Clinical management of the majority of patients with pancreatic cancer (PC) or ovarian cancer (OC), for whom surgical resection is either not indicated or only indicated as a palliative measure, largely relies on systemic chemotherapy as the primary therapeutic option. Outcomes in the systemic treatment of PC and OC could be significantly improved with the use of biomarkers, aiding in the design of personalized treatment options offering the greatest chance of favorable responses. This chapter will provide an overview of recent and ongoing trends related to the use of biomarkers for diagnosis, prognosis and prediction of response to therapy in PC and OC with an emphasis on protein biomarkers measured in body fluids (Table 47.1). A discussion of currently used clinical biomarkers within each disease setting will be provided, followed by a review of the most promising recent findings related to novel biomarker development.

CURRENTLY USED CLINICAL BIOMARKERS

Cancer Antigen 19-9 in Pancreatic Cancer

The mucin-associated sialylated Lewis (a) antigen CA 19-9 is a biomarker of pancreatic duct adenocarcinoma (PDAC) with limited clinical utility in the screening setting. CA 19-9 has demonstrated modest effectiveness in screening of symptomatic individuals with a median sensitivity of 79% (range 70%–90%) and median specificity of 82% (range 68%–91%); however, it appears to be ineffective in screening of asymptomatic subjects (Goonetilleke and Siriwardena, 2007). The principal limitations include its frequent elevation in pancreatitis and obstructive jaundice that frequently cooccur with PC and a variety of other malignant and benign conditions. CA 19-9 is also severely limited in detecting many early stage malignancies and is unsuitable for use in the estimated 5%–10% of patients, who are carriers of the Lewis-negative genotype and develop tumors that do not express the antigen (Locker et al., 2006). The clinical usefulness of CA 19-9 is improved in patient cohorts enriched for pancreatobiliary disease (Pleskow et al., 1989).

Preoperative CA 19-9 levels offer modest prognostic value in localized and advanced PDAC, but are not accurately or reproducibly correlated with disease

[a] Authors with equal contribution.

Biomarkers in Toxicology, Second Edition
https://doi.org/10.1016/B978-0-12-814655-2.00047-5

TABLE 47.1 Overview of Diagnostic, Prognostic, and Predictive Biomarkers in Pancreatic and Ovarian Cancer

	Pancreatic Cancer	Ovarian Cancer
Diagnostic biomarkers	**Protein biomarker combinations**	**Protein biomarker combinations**
	Individual biomarkers CA 19-9 (symptomatic patients), OPN, TIMP-1, CEA, ICAM-1, osteoprotegerin, miRNAs, cell-free DNA (KRAS)	**Individual biomarkers** CA-125/HE4/LPA/mesothelin/PRSS8/tissue kallikreins/FOLR1/miRNAs/gene mutations
		FDA approved triage tools ROMA/ROCA/OVA1
Prognostic biomarkers	**Gene signatures**	**Gene signatures**
	Gene mutations *DPC4*	**Genetic markers** *RAB25, FGF1, ING-5*
	Protein biomarkers CA 19-9 (resectability, post-chemo), MUC1, MUC2, mesothelin, HuR, TgII, HES-1, HEY-1	**Protein biomarkers** OPN, kallikreins, cyclin E, VEGF, bikunin, relaxin, high progesterone and estrogen receptors, galectins, $CD8^{+}/CD20^{+}$, vasohibin-1, cyclin E, Ki67, E-cadherin)
Predictive biomarkers	**Protein biomarkers** HuR, miRNAs, SPARC, Hent1	**Gene signatures**
	Genetic markers *BRCA1, BRCA2, PALB*	**Protein biomarkers** kallikrein, bikunin, ABCF2, YB-1, miRNAs, long noncoding RNAs, IL-37, galectins, ERCC1, TUBB3, PGP, PDGFR-beta VEGFR2

clinical manifestations (Brody et al., 2011). Several studies observed Cox proportional hazard ratios for CA 19-9 comparable to lymph node and margin status (Barton et al., 2009; Hartwig et al., 2011). However, preoperative CA 19-9 levels with a cutoff value of 130 U/mL offer some indication of whether PDAC is resectable (Maithel et al., 2008). Postoperative levels of CA 19-9 offer much more substantial information regarding prognosis, with low, normalized levels indicating improved overall survival and a failure to normalize associated with median survival of <1 year. Nevertheless, normalization of CA 19-9 levels may take 3–6 months, representing a significant window of uncertainty for patients and physicians.

CA 19-9 is weakly prognostic before treatment with chemotherapy or radiotherapy and considerably more informative as a marker of response. Similar to observations regarding postoperative CA 19-9 in resectable PDAC, normalization of biomarker levels or failure to normalize is significantly associated with improved or diminished survival, respectively. The predictive performance of CA 19-9 is greatly limited by the variability among and within individuals diagnosed with PDAC. CA 19-9 levels for an individual patient may fluctuate considerably over time, resulting in an inaccurate impression of treatment response or disease progression that leads to considerable patient anxiety.

Cancer Antigen-125 in Ovarian Cancer

CA-125 is the most widely used OC biomarker to date. It is a high-molecular-weight membrane glycoprotein, encoded by the MUC16 gene. The precise function of CA-125 in normal physiology/development of malignancy has yet to be identified. The cutoff normal value of 35 KU/L has been determined from the distribution of healthy individuals to include 99% of the population.

The initial finding of CA-125 levels >35 U/mL in ~83% of patients with advanced epithelial OC and in only 1%–2% of the normal population led to investigation of its use as a biomarker for early detection of OC (Bast et al., 1983; Zurawski et al., 1988). However, there are weaknesses that make its use as an early detection biomarker for OC equivocal. It is absent in 20% of OC because of nonepithelial carcinoma and elevated in numerous malignant and benign conditions such as endometrial cancer, endocervical cancer, PC, breast cancer, lymphoma, lung cancer, peritonitis, endometriosis, liver cirrhosis, pancreatitis, and acute pelvic inflammatory disease. Factors such as race/ethnicity, age, hysterectomy, smoking history, and obesity may influence CA-125 level. Despite these well-recognized limitations, CA-125 remains the most widely studied serum biomarker for OC. It is currently FDA approved for disease monitoring in all patients diagnosed with epithelial

OC and it is used concurrently with transvaginal ultrasonography (TVS) in women with hereditary breast OC syndrome every 6 months after the age of 30 or 5–10 years before the earlier presentation age of cancer in their family.

The best available protocol for early detection of OC to date, a combination of screening for elevated CA-125 and TVS in the presence of elevated CA-125, does not meet the stringent criteria for cost-effectiveness espoused by the US Preventive Services Task Force. As a result, no professional group currently recommends screening for OC in the general population (Grossman et al., 2018). Despite the stand-alone biomarker weaknesses, CA-125 has been used together with other biomarkers such as the circulating receptor of alpha-fetoprotein to improving the sensitivity of the former for stages I/II and III/IV from 58.1% to 76% and 79.6% to 88.2%, respectively (Tcherkassova et al., 2011).

In the PLCO cancer screening trial, 78,216 healthy women between the ages of 55 and 74 were randomly assigned either to undergo annual CA-125 testing plus TVS or to receive "usual care." The positive predictive value of screening was 1.0%–1.3% during a 4-year period. However, 72% of detected cases were stage III or IV, indicating that screening does not result in stage shift. The PLCO project team concluded that the CA-125/TVS screening approach does not reduce disease-specific mortality in comparison to usual care but increases invasive medical procedures and associated harms (Pinsky et al., 2016). In the UKCTOCS, 202,638 postmenopausal women between the ages of 50 and 74 at average risk for OC were randomly assigned to annual pelvic examination (control arm), TVS alone (TVS arm), or measurement of CA-125 plus TVS in cases of elevated CA-125 level (multimodality arm) (Jacobs et al., 2016). Compared to the TVS arm, the multimodality arm showed a significantly greater specificity (99.8% vs. 98.2%) and a higher positive predictive value (35.1% vs. 2.8%); sensitivity did not differ significantly between arms. Both TVS and multimodality arms demonstrated a higher proportion of stage I–II cancers. Despite the initial analysis, the 7–14 year follow-up showed a significant mortality reduction in the multimodality arm when prevalent cases were excluded (Jacobs et al., 2016).

Following technological advances of the modern era, the use of computerized algorithms has managed to enhance the sensitivity of CA-125. Risk of Ovarian Cancer Algorithm (ROCA) uses patient's age and CA-125 to stratify them into low, intermediate, and high risk for OC, enabling clinicians to proceed with further interventions. It has been shown that ROCA improves CA-125 sensitivity to 86% in preclinical detection (Skates, 2012).

CA-125 has also been evaluated as a prognostic biomarker in OC both postoperatively and during disease remission. Following cytoreduction, a steep decline is an indication of optimal surgical outcomes and a failure to demonstrate such a response is associated with suboptimal cytoreduction and increased risk of relapse (Yoo et al., 2008; Zivanovic et al., 2009). Clinical post-treatment remission is defined by imaging and normalization of CA-125 levels. Persistently abnormal levels following treatment indicate a high risk of recurrence within 6 months and a high likelihood of disease-related death. CA-125 retains prognostic value even in those patients who achieve a clinical remission with levels below 35 U/mL. Elevated levels within this normal range are associated with reduced overall and progression-free survival (Juretzka et al., 2007), whereas serially increasing values within the normal range are associated with increased risk of recurrence (Karam and Karlan, 2010). Current utility of CA-125 predominantly resides in monitoring patients in clinical remission for CA-125 levels, which become elevated out of the normal range preceding clinical and/or radiological indications of recurrence by 2–4 months (Bridgewater et al., 1999; Rustin et al., 1996, 2001; Tuxen et al., 2000, 2002; Vergote et al., 2000). However, clinical management of asymptomatic patients with elevated CA-125 levels remains controversial and a significant source of anxiety for patients.

CA-125 has demonstrated some promise as a predictive biomarker related to optimal cytoreduction. Before surgical resection of OC, levels of CA-125 have been demonstrated to indirectly reflect the degree of tumor burden. It has been shown that CA-125 levels >500 U/mL are associated with a diminished likelihood of optimal cytoreduction (Diaz-Padilla et al., 2012). Given the high response rate typically observed following the administration of chemotherapy in epithelial OC, CA-125 is not widely utilized as a predictive biomarker in this setting. The use of neoadjuvant chemotherapy is a current topic of interest in the management of OC as a means of improving the potential of optimal cytoreduction in advanced disease. It has been recently suggested in this treatment setting that initial CA-125 levels >2000 U/mL predict improved progression-free survival (Kang et al., 2011).

DIAGNOSTIC BIOMARKERS

Pancreatic Cancer

CA 19-9 is ineffective in detecting PDAC in asymptomatic patients, whereas sensitivity and specificity are elevated in symptomatic ones. Following CA 19-9, carcinoembryonic antigen (CEA) is the second most common antigen currently used but sensitivity is low and inferior to CA 19-9 (Swords et al., 2016; Viterbo et al., 2016). Combined use of CA 19-9, CA 242, and CEA has been proposed. A metaanalysis has shown the highest sensitivity with the combination of CA 19-9 and CA 242, while the combination of CA 19-9, CA 242, and CEA provides the highest specificity (Zhang et al., 2015b).

For the early detection of PDAC, many circulating, cellular, gene, or molecular biomarkers have been studied. Attempts have been made to identify biomarkers in urine/pancreatic juice accompanied with endoscopic techniques. Studies using combination of markers, including CA 19-9, have been particularly successful in detecting early disease. Detection of circulating tumor cells in blood and peritoneal fluid using reverse transcription polymerase chain reaction has been proposed as a viable and sensitive method (Kelly et al., 2009).

An innovative approach on pathogenetic mechanisms of PDAC is the involvement of inflammation (Ahmad et al., 2014). Systemic inflammation markers such as neutrophil–lymphocyte ratio, platelet–lymphocyte ratio, and modified Glasgow Prognostic Score have been investigated (Martin et al., 2014). Neutrophil to lymphocyte ratio is calculated by dividing the neutrophil count by the lymphocyte count, with sensitivity increased at levels >5. This ratio is significantly higher in patients with PDAC (Garcea et al., 2011). Moreover, survival in advanced PDAC is strongly related to the levels of various systemic inflammation markers as they are calculated from routine blood tests. DJ-1 is a novel biomarker and when its serum levels were compared to those of CA 19-9, its area under the curve was higher (Chen et al., 2012). It has been also suggested that in 68.5% of patients with PDAC, DJ-1 correlates with tumor invasion, predicting a poor outcome. Other circulating biomarkers such as PAM4.Ic3b, REG4, serum phosphoprotein signal-regulated kinase (p-ERK1/2), and CEACAM1 need further validation before clinical use.

Several gene biomarkers have been studied in PDAC as it is known to be a genetically predisposed disease. Most commonly mutated genes are KRAS (K-Ras 2), CDKN2A/p16, TP53, and SMAD4/DPC4. PDAC has the highest amount of KRAS mutations, and >50% of patients exhibit this abnormality. Activated KRAS activates the RAF family kinase RAF1, BRAF, ARAF leading to uncontrolled cancer development. The KRAS mutation presents a critical event for PDAC for diagnosis and prognosis (Nio et al., 2001). P53 phosphoprotein is encoded by gene TP53, which is mutated in up to 50% of PDAC cases. As it is a tumor suppressor gene and activates genes due to cellular stress or DNA damage, it induces growth arrest (Moore et al., 2001). P16 is encoded by CDKN2A gene, whereas lack of P16 protein is associated with advanced disease and poor prognosis. Modification of CDKN2A causes melanoma and increased risk of PC (Duff and Clarke, 1998). DPC4/SMAD4 is a protein encoded by the SMAD4 gene located in chromosome 18. It has been found mutated in 55% of patients with PDAC (Duff and Clarke, 1998). Metastin is a gene product of the tumor suppressor gene KISS-1. PDAC expresses lower KISS-1 levels in comparison to normal pancreatic tissue; hence the presence of metastin correlates with longer survival (Masui et al., 2004). Phosphate and Tensin (PTEN) is a tumor suppressor gene encoded at chromosome 10q23.3. PTEN deletion has resulted in metaplasia leading to PDAC. In about 70% of PDAC, expression of PTEN is low or absent (Osayi et al., 2014).

Regarding molecular biomarkers, there is a high expression in all areas susceptible to PC (like liver, lymph node, and lung) of CXCL12, CXCR4 protein receptors of the CXC chemokine ligand 12. High expression of CXCR4 in resected PDAC is associated with shorter overall survival/distal metastases (Osayi et al., 2014). Cathepsin B (CTSB) is a lysosomal protease shown to degrade components of extracellular matrix and contributes to poor prognosis due to local invasion and metastasis promotion (Osayi et al., 2014). Another important factor in tumor invasion/metastases is angiogenesis. Several growth factors are implicated in promoting angiogenesis, the most important of which is the vascular endothelial growth factor (VEGF) that has been associated with PDAC metastases, especially to the liver. Epidermal growth factor receptor, a transmembrane glycoprotein, is overexpressed in 40%–70% of patient with PC (Jung et al., 2015; Yamanaka et al., 1993).

MicroRNAs (miRNAs) are noncoding molecules that aim to preserve normal cellular signals. They are involved in posttranscriptional gene regulation. Their utilization helps differentiate PDAC from chronic pancreatitis. Besides that, they appear to predict disease-free and overall survival. miRNA can act as a tumor suppressor or as an oncogene (Jamieson et al., 2012). miRs-21, miR-210, miR-155, and miR-196a are upregulated in PC tissue, representing possible novel diagnostic and prognostic factor (Ajit, 2012). miR-21 overexpression is reported as the lesion initiator in the context of pancreatic intraepithelial neoplasia causing tumor progression. The upregulation of miR-221 in PC leads to distant metastasis and unresectable tumors (Kishikawa et al., 2015).

Ovarian Cancer

Although CA-125 remains the most useful OC biomarker, numerous efforts and strategies aiming at utilizing CA-125 for screening purposes have not proven fruitful. A popular strategy has emerged within OC screening research wherein additional biomarkers are sought, which are capable of complementing the performance of CA-125 to augment sensitivity and specificity. For example, a study utilizing >2000 women identified a four-biomarker panel consisting of CA-125, human epididymis protein (HE4), CEA (a glycoprotein biomarker associated with colorectal cancer and also loosely associated with PC), and vascular cell adhesion molecule-1 (VCAM-1; a leukocyte adhesion factor involved in extravasation of white blood cells during an inflammatory response), which could differentiate early stage OC from controls with a sensitivity/specificity of 86/98 (Yurkovetsky et al., 2010). The combination of CA-125, transthyretin (TTR), and apolipoprotein A1 (ApoA1) has also demonstrated some usefulness in the early detection of OC (Kim et al., 2012; Su et al., 2007). TTR and ApoA1 are acute phase reactants produced in the liver and released in acute inflammation state. Each of the biomarkers used individually is unlikely to offer a high specificity given that they are likely to be elevated in response to a number of inflammatory conditions. However, their use in combination with CA-125 appears to increase their efficacy. The performance of this panel has been further enhanced with the addition of other protein biomarkers (H418, transferrin). The overall performance of these panels is comparable to that of the four-biomarker panel described above. Similar levels of performance have also been reported for several larger combinations of CA-125, tumor markers, inflammatory mediators, and acute phase reactants consisting of 5–11 proteins (Amonkar et al., 2009; Edgell et al., 2010; Skates et al., 2004). In all, the use of CA-125-based biomarker combinations has produced some encouraging results; however, none of the identified panels have yet to perform reproducibly at a level that would warrant widespread implementation.

HE4 is encoded by the WFDC2 gene and is a small secretory protein that is involved in the maturation of sperm. It is overexpressed in OC and is a FDA approved peptide for monitoring epithelial OC. HE4 is expressed in epithelial OC that do not express CA-125. HE4 on its own showed the highest sensitivity for stage I disease with a specificity of 45.9% at 95% sensitivity. It has been shown that using HE4 in a multivariate panel with CA-125 yields a significantly better specificity than any of the stand-alone uses.

Lysophosphatidic acid (LPA) is a small bioactive phospholipid that acts as an agent promoting cell proliferation, invasion smooth muscle cell contraction, cell migration, cell survival, wound healing, morphology alteration, and tissue differentiation. A metaanalysis of 19 controlled studies, including 2520 subjects, showed that plasma LPA was significantly higher for patients with OC compared with those with benign disease or healthy individuals (Li et al., 2015).

Mesothelin is a protein encoded by the MSLN gene, is present on normal mesothelial cells and is overexpressed in tumors such as OC, PDACs, and mesotheliomas. A recent metaanalysis, indicated 94% specificity for mesothelin and 62% sensitivity for OC, suggesting that although it cannot be utilized as a stand-alone biomarker, it might be combined with CA-125 or HE4 for better outcomes (Madeira et al., 2016).

Human prostasin is a glycosylphosphatidylinositol-anchored extracellular serine protease encoded by PRSS8 and is located on chromosome 16p11.2. It is present in the seminal fluid and apical surface of epithelial tissues (the lungs, kidney, liver, bronchi, colon, and salivary glands). It acts as a proteolytic activator of epithelial sodium channels. Raised expression of prostasin is associated with various malignancies, including OC. Combined with CA-125, the sensitivity and specificity for OC is 92% and 94%, respectively. Furthermore, Tamir which induces at least an 100-fold overexpression of the PRSS8 gene in all types and all stages of OC is a promising minimally invasive biomarker of OC but should not be used to differentiate stages (Tamir et al., 2016). PRSS8's diagnostic performance is also vastly improved when combined with OVA1 (Yip et al., 2011).

Human kallikreins are low-molecular-weight proteases encoded by a family of 15 genes located on chromosome 13q13.4. Plasma kallikrein is responsible for blood pressure regulation, tissue inflammation, and plasmin formation. Tissue kallikreins play a role in smooth muscle contraction, hormonal regulation, blood pressure control, vascular cell repair, and neoangiogenesis. These proteins influence the course of various types of cancers. Their role in cancer pathogenesis and progression is due to dysregulation of growth factors, cell adhesion molecules and cell surface receptors, as well as the degradation of extracellular proteins, effect that can induce a disruption of the physiological barriers against cancer.

In OC, KLK6 and KLK10 are elevated in serum. Moreover, in advanced stages, they correlate significantly with CA-125. KLK11 is elevated notably in patients with OC compared to patients with benign disease or healthy individuals. KLK6 has a better specificity than CA-125 for early detection and is not elevated in healthy individuals. Combined use of KLK6, KLK13, and CA-125 has a greater sensitivity for early stage diagnosis (White et al., 2009). A metaanalysis of five studies, including a total of 1150 patients, reported a sensitivity of 50%, a specificity of 91% with a positive and negative likelihood ratio of 7.2 and 0.51, respectively, whereas the combined use of KLK6 and CA-125 achieved superior diagnostic efficacy for advanced OC (Yang et al., 2016).

miRNAs (noncoding RNAs secreted from cells into the circulation that regulate gene expression by altering proliferation, differentiation, and apoptosis perform their action via translational inhibition or mRNA degradation; see above) act both as tumor suppressors or promoters. Tp53 regulates miRNAs. As Tp53 mutations are found in 96% of OC, it becomes apparent that there is dysregulation of the miRNAs. On the other hand miRNAs downregulate BRCA1 and 2, hence they can affect the onset of the cancerous process (Leung et al., 2016). It has been shown that miRNA 21, 141, 200a, 200b, 203, and 214 could play an important role in diagnosing OC as they are overexpressed. On the other hand, miRNA 199a, 140, 145, and 125b1 were found to be downregulated (Iorio et al., 2007). miRNAs found in the exosomes of individuals with malignant disease are significantly different from the miRNA profile of healthy individuals or patients with benign disease (Taylor and Gercel-Taylor, 2008). It has been reported that a combination of CA-125 and six miRNAs (200a-3p, 766-3p, 26a-5p, 142-3p, let 7d-5p, and 328-3p) achieves a sensitivity of 98.4% and a specificity of 95.6% (Yokoi et al., 2017). It has been suggested that it is possible to overcome the difficulty of distinguishing a malignant from a benign mass without surgical intervention by applying a combination of seven miRNAs with a 86.1% sensitivity and 83.3% specificity (Yokoi et al., 2017). It also has been suggested that detection of circulating miRNAs might contribute significantly to early detection, and therefore miRNAs could be used as prognostic biomarkers because over- or under-expression of certain of these proteins is linked to poor prognosis, recurrent, or even metastatic disease (S. Zhang et al., 2015a).

At least 10% of epithelial OCs are hereditary and in ~90% of those individuals, BRCA1 and 2 are the accountable cause, as well as RAD51D. Furthermore, Tp53 has been identified in 96% of the cases, mainly in high-grade tumors. On the other hand, KRAS and BRAF mutations are more common in low-grade subtypes. Moreover, mutations to DNA mismatch repair genes have also been found to be responsible for malignant transformation, such as hMLH 1 and 2 genes. Therefore, genetic testing for mutations might identify potentially high-risk patients before clinical presentation of the disease. Epigenetic changes such as DNA hyper- or hypomethylation that can deactivate tumor suppressor genes, also are accountable for genetic alterations, hence promotion of carcinogenesis.

Folate receptor alpha (FOLR1) is a membrane receptor that facilitates the transportation of folate into cells. Rapid mitosis and proliferation require high levels of folate to secure adequate DNA synthesis. Several studies have confirmed overexpression of FOLR1 in OC. It has been showed that FOLR1 is significantly elevated in the serum of patients with OC compared to patients with benign disease or healthy individuals (Leung et al., 2013). Furthermore, a strong synergetic effect of HE4 + FOLR1 and CA-125 + FOLR1 combinations has been demonstrated (Kurosaki et al., 2016; Leung et al., 2016).

Biomarkers in the Evaluation of a Pelvic Mass

The overall prevalence of pelvic abnormalities is estimated at 7% and it is expected that 5%–10% of American women will receive prophylactic surgery for suspected OC at some point in their lives (DiSaia and Creasman, 1997). The burden of early identification of potential OC falls predominantly on the obstetrician/gynecologist whose training in the management of cancer patients is usually limited. A series of diverse studies have demonstrated a decrease in the relative risk of reoperation, and increases in disease-free interval and overall survival for women operated on by gynecological oncologists compared with gynecologists and general surgeons (Bristow et al., 2002; Eisenkop et al., 1992; Tingulstad et al., 2003). In addition to family history, pelvic examination, ascites, and evidence of local or distant metastases, the CA-125 blood test is included in the standard criteria espoused by The Society of Gynecologic Oncology and the American College of Obstetrics and Gynecology regarding referral of a patient with a pelvic mass to a gynecological oncologist. This set of criteria has produced disappointing results in prospective studies, particularly those evaluating premenopausal women with early stage disease, providing sensitivity/specificity levels as low as 47%/77% (Dearking et al., 2007). Considerable effort has been focused on the identification of additional biomarkers capable of complementing the performance of CA-125. Several combinations have recently been approved for clinical use based on trial performance and these are discussed below. Each of the FDA approved tests is intended to aid referring physicians in choosing the most appropriate specialist for surgical intervention for patients already planning to undergo surgery.

A scoring model termed the Risk of Ovarian Malignancy Algorithm (ROMA) has been developed, which incorporates measurements of CA-125 and HE4 along with menopausal status to assign high or low risk of malignancy to a woman presenting with a pelvic mass. Based on the results of a prospective, multicenter, blinded clinical trial, ROMA was approved by the FDA for use in determining the risk of OC in pre- and postmenopausal women with a pelvic mass. The clinical impact of the ROMA test remains to be assessed as physicians independently evaluate its performance and make decisions regarding its implementation. The CA-125/HE4 combination and ROMA have received considerable attention in preclinical studies with mixed results. The superior diagnostic performance achieved through the use of the combination of both biomarkers over either biomarker used individually is widely agreed on.

However, despite the diagnostic utility of ROMA in comparison to more traditional methods of assessment, including imaging, physical exam, and an older diagnostic model, the risk of malignancy algorithm has remained controversial (Montagnana et al., 2011, 2017).

OVA1 is another multivariate index assay for identifying high-risk OCs by using five proteins: CA-125, ApoA1, transthyretin, beta 2 microglobulin, and transferrin. It uses two different immunoassay platforms, the Roche Elecsys for CA-125 and the Siemens BNII for the four other analytes. The results are calculated with the use of an algorithm, OvaCalc. Taking into account the menopausal status of the individual, it calculates a score. Having a score of more than 5 or more than 4.4 for premenopausal and postmenopausal women, respectively, shows a higher probability for malignant disease. OVA1 has been FDA approved since 2009 as an adjunct to diagnosis of adnexal masses when radiological and clinical indetermination exists. The test is not approved for screening individuals without an adnexal mass. It should neither be utilized if conservative management is favored by clinical assessment nor if a referral to a gynecologic oncologist is necessary as it has no value for the patient's future management. The panel was evaluated in a clinical trial that utilized immunoassays targeting each of the five markers in a set of 524 women diagnosed with a pelvic mass and recommended for surgery (Ueland et al., 2011; Ware Miller et al., 2011). When the OVA1 panel was substituted for CA-125 within the American College of Obstetricians and Gynecologists OC referral guidelines, it provided not only an increase in sensitivity and negative predictive value in comparison to CA-125 but also a decrease in specificity and positive predictive value. When the OVA1 test was added to a normal physician assessment, similar trends were observed. Regarding, its performance, the presurgery assessment and OVA1 score had 96% sensitivity, 35% specificity, 40% positive predictive value, and 95% negative predictive value, whereas the physician presurgery assessment arm had 75% sensitivity, 79% specificity, 62% positive predictive value, and 88% negative predictive value. Limitations for this test include rheumatoid factor >250 IU/mL or triglyceride levels >4.5 g/L, as well as diagnosis of malignancy in the last 5 years (Nolen and Lokshin, 2013).

PROGNOSTIC BIOMARKERS

Pancreatic Cancer

Indicators of prognosis in PC could inform the development of treatment plans tailored to optimize the curative and/or palliative benefits provided to individual patients. Prognostic biomarkers would assist physicians in determining the appropriateness of surgical resection, local radiation therapy, and systemic chemotherapy. Currently, CA 19-9 offers only a limited ability to aid in these types of decisions and other biomarkers commonly used in the monitoring of PC, including CEA and CA-125, have been hampered by low levels of disease specificity. A number of additional biomarkers, including genetic markers and proteins, have received attention as potential prognostic biomarkers of PC; however, none of these have yet performed at a level warranting clinical implementation.

Gene signatures are routinely used in the management of breast cancer, and similar tools have been sought in the setting of PC. In two notable studies, a protein signature consisting of 171 genes and a much smaller 6-gene signature were both shown to stratify patients on the basis of survival (Donahue et al., 2012; Stratford et al., 2010). On an individual basis, the mutation status of the *DPC4* gene, which encodes the transcription factor SMAD4, has been proposed as a means of differentiating patients likely to develop local versus distant recurrence following surgery. Such a role was supported by the results of two independent studies (Blackford et al., 2009; Iacobuzio-Donahue et al., 2009) and would suggest a method of identifying patients more likely to benefit from aggressive local therapy following surgery. However, more recent findings regarding SMAD4 expression in tissue obtained from PC patients indicated no difference between individuals demonstrating local versus distant recurrence (Winter et al., 2013), casting doubt on the usefulness of this biomarker.

A number of proteins have emerged as potential prognostic biomarkers for PC. Three such proteins, MUC1, MUC2, and mesothelin, were identified from among 13 PC-related proteins in a single study as markers capable of differentiating PC on the basis of tumor aggressiveness (Winter et al., 2012). MUC1 and MUC2 are membrane-bound proteins characterized by altered glycosylation patterns associated with a variety of malignancies. Mesothelin is a glycosylphosphatidylinositol-anchored cell surface protein that is overexpressed in mesothelioma, OC, and squamous cell carcinoma. OPN (osteopontin), another protein biomarker, which has shown promise in the early detection of PC, has also been implicated as a prognostic biomarker based on the observation that serum levels of OPN in PC patients are correlated with diminished survival (Poruk et al., 2013). Human antigen R (HuR) is an RNA binding protein involved in the regulation of gene expression in response to cellular stress. Upregulation of HuR has been linked to the development of malignancy and has also been shown to be inversely correlated with prognosis in PC (Denkert et al., 2006; Richards et al., 2010). Tissue transglutaminase (TgII) is an enzyme responsible for catalyzing calcium-dependent transamination reactions with a diverse and expanding repertoire of intra- and extracellular functions. Demonstrated and proposed roles for TgII in cancer development and progression include cell survival, drug resistance, and cell mobility and migration. TgII is currently the focus of considerable investigation as a prognostic marker and therapeutic target in PC (Mehta, 2009).

Notch cytoplasmic signaling and HES-1 or HEY-1 expression are significantly increased in metaplastic ductal epithelium and invasive PC. Immunohistochemistry of resected specimens revealed increased Notch1 in the nerves, Notch2 and Notch 3 in vascular smooth muscle, and Notch4 in vascular endothelium. These findings implicate Notch in both invasion and angiogenesis. HEY-1 nuclear expression independently, is not a good predictive marker of overall or disease-free survival, but is associated with aggressive tumor phenotypes (Gungor et al., 2011).

Ovarian Cancer

Similarly to the approach for PC, gene signatures have been investigated in the setting of OC, which might have a prognostic value in both pre- and postoperative management. Several gene signatures have been reported to stratify patients on the basis of overall survival (Bonome et al., 2008; Spentzos et al., 2004). The identification of gene signatures offers not only potential for development of prognostic tools but may also shed light on the biological mechanisms underlying disease development through closer examinations of the molecular markers included in each signature. Several individual genetic markers have been identified that offer prognostic information in OC, including *RAB25* and *FGF1*. *RAB25* encodes a cell membrane protein involved in pseudopod extension implicated in the promotion of cell proliferation, survival, and migration, whereas *FGF1* encodes a member of the fibroblast growth factor family, a network of potent mitogens involved in numerous biological processes. Increased levels of *RAB25* mRNA are associated with decreased disease-free survival (Cheng et al., 2004), whereas *FGF1* gene amplification is significantly associated with poor survival (Birrer et al., 2007). The Cancer Genome Atlas (TCGA) project has shown that TP53 mutation is present in almost all high-grade OC and that 20% of patients with high-grade carcinoma harbor BRCA1 mutation. In patients with TP53 and BRCA1, the prognosis is poor. Women harboring only BRCA1 mutation seem to have a survival benefit. This is due to the fact that BRCA patients present at a young age and BRCA genes are involved in DNA damage recognition/homologous repair, which means that when there is a BRCA mutation there is impaired homologous repair and chemotherapy has greater potential of destroying those cells (Rzepecka et al., 2017). ING5 gene expresses a protein, characterized as a tumor suppressor protein as it inhibits cell glucose metabolism, invasion, migration, transition and induces apoptosis. Downregulated ING5 is linked to poor prognosis because it is associated with vascular invasion, lymphatic invasion, and late International Federation of Gynecology and Obstetrics (FIGO) staging (Zheng et al., 2017).

In an attempt to identify prognostic biomarkers to guide clinical practice, numerous circulating and tumor-expressed proteins have been investigated. Circulating and tumor-associated levels of osteopontin have been reported to be positively correlated with recurrent disease, bulk disease, metastasis, and poorer prognosis among patients with advanced disease (Coticchia et al., 2008; Hu et al., 2015). High serum levels of relaxin and KLK10 have also been shown to be unfavorable prognostic factors (Guo et al., 2017). On the contrary, elevated serum levels of KLK11 have been shown to be an independent favorable predictor in advanced high-grade serous OC (Geng et al., 2017). High progesterone and estrogen receptors are associated with favorable outcome in endometrioid OC. Galectins have also been reported to have various effects on disease progression. Elevated galectin 1 and 7 levels in the cytoplasm have been reported as negative prognostic factors. Multiple immune markers have a prognostic value for OC because the protective role of the immune system is well established (Labrie et al., 2017). The presence of $CD8^{+}$ and $CD20^{+}$ tumor-infiltrating lymphocytes and HLA II predicts a favorable prognosis (Santoiemma et al., 2016). Vasohibin-1, a negative feedback inhibitor of angiogenesis induced by VEGF (a prognostic biomarker for OC negatively correlating with survival) in vascular endothelial cells has been associated with poor prognosis (Sano et al., 2017). A metaanalysis, showed that cyclin E (the main regulator for transition from G1 to S phase) is overexpressed, and is a poor prognostic factor for overall survival (Farley et al., 2003). Ki67, a nuclear protein essential for cellular proliferation, is an independent prognostic factor for poor overall survival in OC patients (Liu et al., 2012). Diminished levels of E-cadherin, a glycoprotein that mediates calcium-dependent cell adhesion and interacts with actin strengthening cell–cell interaction, have been reported to manifest poor prognosis and possible late/disseminated OC due to loss of cellular contacts allowing cancer progression/dissemination (Rosso et al., 2017).

PREDICTIVE BIOMARKERS

Pancreatic Cancer

The search for biomarkers predictive of therapy response in PC has been largely centered on cellular factors related to the transport and metabolism of gemcitabine, the current standard of care. Although a number of such factors have been evaluated (Brody et al., 2011), none has been deemed to be of sufficient quality for clinical use. The role of HuR in cellular handling of gemcitabine remains uncertain; however, this factor has demonstrated considerable promise as a predictive biomarker of the gemcitabine response. Although HuR overexpression is negatively correlated with overall prognosis in PC, it is positively correlated with prolonged survival following gemcitabine administration independent of tumor stage but its value remains to be validated in independent large trials (Costantino et al., 2009; Richards et al., 2010).

Mutations in several genes in terms of the Fanconi anemia DNA repair pathway have been associated with PC and may serve as predictive biomarkers for response to agents targeting DNA synthesis and repair, including poly ADP-ribose polymerase (PARP) inhibitors, platinum-based agents, or mitomycin C. Patients harboring mutations in *BRCA1*, *BRCA2*, and *PALB* have demonstrated encouraging responses to these agents (Lowery et al., 2011; Villarroel et al., 2011).

Receptor tyrosine kinase inhibitors represent a promising class of anticancer therapeutics currently in use or in clinical trials for a number of malignancies. One such agent targeting epidermal growth factor receptor, erlotinib, is currently approved for the treatment of metastatic PC in combination with gemcitabine. The introduction of erlotinib and other targeted agents in PC will require the identification of additional predictive biomarkers. Potential biomarkers may include soluble or membrane-bound receptors, or molecules which function within the signaling pathways of targeted receptors (Akt, PI3k, etc.)

The goal of the predictive biomarkers is to determine the response to a specific therapy. SPARC is a promising therapeutic biomarker in patient treated with combination nab-paclitaxel−gemcitabine chemotherapy. Patients with high-expression SPARC display increased overall survival (Von Hoff et al., 2011). Hent1 appears to play a role in the internalization of gemcitabine by PC cells and is an important factor for gemcitabine efficacy in patients with PC. Patients positive for Hent1 have longer median survival with gemcitabine therapy. Interestingly, this positive finding was not observed with other chemotherapy agents (Spratlin et al., 2004).The increased expression of some miRNAs predicts favorable response to gemcitabine treatment. They can be identified by mutational analysis on DNA or RNA and are commercially available. Some of them are miR-21, miR-200b, miR-29b, miR-221, and miR-494 (Karakatsanis et al., 2013).

Ovarian Cancer

A number of gene signatures and several protein biomarkers such as kallikrein 10 and bikunin have emerged as potential predictive biomarkers. In separate studies, elevated levels of each biomarker were found to correlate with poor response to chemotherapy (Borley et al., 2012; Costantino et al., 2009; Farley et al., 2003; Hefler et al., 2006; Liu et al., 2012; Lowery et al., 2011; Luo et al., 2001; Matsuzaki et al., 2005; Na et al., 2009; Villarroel et al., 2011). Elevated protein levels of ABCF2, a member of the ATP-binding cassette family of transmembrane transport proteins, were recently observed to correlate with chemoresistance in OC patients (Tsuda et al., 2005). miRNA let-7i has been found to have a predictive value as its levels are significantly reduced in chemotherapy-resistant patients, presaging a shorter progression-free survival (Yang et al., 2008). Another group of four miRNAs (miR-200b, miR-1274A, miR-141, and miR-200c) appears to have a predictive value as low levels of them are associated with better overall survival, with the latter able to give information regarding response to immunotherapy (Halvorsen et al., 2017). Moving to a different direction, it has been suggested that a long noncoding RNA signature held a rather high predictive accuracy of resistance to chemotherapy and was intimately associated with patients' progression-free survival (Wang et al., 2017). Another potential marker that could be implicated as an OC predictive marker and provide information about the outcome is interleukin-37, a novel antiinflammatory cytokine that is found in late-stage OC, lymph node progression, and residual disease postsurgical management and is associated with unfavorable outcome (Huo et al., 2017).

Furthermore, elevated galectin 1, 8, 9, levels are also associated with poor chemotherapy and other treatment outcomes, as well as a decline in overall survival and progression-free disease (Labrie et al., 2017). A recent study showed that three proteins, ERCC1, TUBB3 and PGP, had a significant prognostic and predictive value as they identified patients who had chemoresistance to a platinum/taxane combination (Herzog et al., 2016). Two preangiogenic factors, PDGFR-beta and VEGFR2, have been shown to have unfavorable predictive news regarding chemoresistance if elevated (Avril et al., 2017).

CONCLUDING REMARKS AND FUTURE DIRECTIONS

The diverse and unmet challenges associated with the management of patients diagnosed with PC and OC have prompted an immense amount of research into the development of clinical biomarkers as tools aimed at diagnosing, monitoring, and triaging patients more effectively. Despite this effort, only a few clinically useful biomarkers are currently available, notably, CA 19-9 for PC and CA-125 for OC. While each of these remains an important component of patient management in its respective disease setting, well-appreciated limitations are associated with each demand in the discovery and development of additional, more effective diagnostic, prognostic, and predictive biomarkers.

Research efforts aimed at the development of diagnostic biomarkers for use in screening for PC and OC are intense, diverse, innovative, and productive, resulting in numerous promising reports. Such reports have described the use of proteins, genes, miRNAs, proteomic signatures, and a variety of additional platforms. However, as alluded to by the disappointing results of the PLCO trial, population-based screening for either of these diseases does not appear to be on the horizon. Further research will be necessary to identify biomarkers capable of achieving sensitivity/specificity levels approaching 99% to overcome the challenge of disease rarity and establish a screening test that achieves acceptably low levels of false-positive results while detecting diseased individuals with high fidelity. Only then large screening trials such as the PLCO and UKCTOCS will be warranted to establish efficacy and increase the advancement of biomarker testing into clinical implementation.

The successful clinical implementation of the ROMA and OVA1 tests represents significant achievements within the field of pelvic mass management and also within the larger field of biomarker research. The next step in the evaluation of these tests will be the so-called Phase IV trials of postmarketing efficacy. Large studies of the use of these tests by general practitioners and gynecologists will be necessary to assess the impact on OC morbidity and mortality. Comparative effectiveness research should also be undertaken to directly compare the use of the ROMA and OVA1 tests to more precisely define the indications for each and maximize the benefits.

Regarding prognostic and predictive biomarkers, many preliminary reports have been described for each; however, relatively less has been achieved in the search for predictive biomarkers. Efforts should now be made to work beyond the preliminary stage and begin to standardize analysis platforms and sampling criteria to focus on the most promising and reproducible biomarkers.

Efficacy achieved in this manner should lead to advancement into larger trials and eventual progression into clinical implementation. In conclusion, the use of biomarkers in multiple aspects of the clinical management of PC and OC has become a priority for clinicians and researchers and is likely to increase steadily in the coming years. The advancement of novel biomarker tests into the clinic should lead to significant benefits in patient care.

References

Ahmad, J., Grimes, N., Farid, S., et al., 2014. Inflammatory response related scoring systems in assessing the prognosis of patients with pancreatic ductal adenocarcinoma: a systematic review. Hepatobiliary Pancreat. Dis. Int. 13, 474–481.

Ajit, S.K., 2012. Circulating microRNAs as biomarkers, therapeutic targets, and signaling molecules. Sensors 12, 3359–3369.

Amonkar, S.D., Bertenshaw, G.P., Chen, T.H., et al., 2009. Development and preliminary evaluation of a multivariate index assay for ovarian cancer. PLoS One 4, e4599.

Avril, S., Dincer, Y., Malinowsky, K., et al., 2017. Increased PDGFR-beta and VEGFR-2 protein levels are associated with resistance to platinum-based chemotherapy and adverse outcome of ovarian cancer patients. Oncotarget 8, 97851–97861.

Barton, J.G., Bois, J.P., Sarr, M.G., et al., 2009. Predictive and prognostic value of CA 19-9 in resected pancreatic adenocarcinoma. J. Gastrointest. Surg. 13, 2050–2058.

Bast Jr., R.C., Klug, T.L., St John, E., et al., 1983. A radioimmunoassay using a monoclonal antibody to monitor the course of epithelial ovarian cancer. N. Engl. J. Med. 309, 883–887.

Birrer, M.J., Johnson, M.E., Hao, K., et al., 2007. Whole genome oligonucleotide-based array comparative genomic hybridization analysis identified fibroblast growth factor 1 as a prognostic marker for advanced-stage serous ovarian adenocarcinomas. J. Clin. Oncol. 25, 2281–2287.

Blackford, A., Serrano, O.K., Wolfgang, C.L., et al., 2009. SMAD4 gene mutations are associated with poor prognosis in pancreatic cancer. Clin. Cancer Res. 15, 4674–4679.

Bonome, T., Levine, D.A., Shih, J., et al., 2008. A gene signature predicting for survival in suboptimally debulked patients with ovarian cancer. Cancer Res. 68, 5478–5486.

Borley, J., Wilhelm-Benartzi, C., Brown, R., et al., 2012. Does tumour biology determine surgical success in the treatment of epithelial ovarian cancer? A systematic literature review. Br. J. Cancer 107, 1069–1074.

Bridgewater, J.A., Nelstrop, A.E., Rustin, G.J., et al., 1999. Comparison of standard and CA-125 response criteria in patients with epithelial ovarian cancer treated with platinum or paclitaxel. J. Clin. Oncol. 17, 501–508.

Bristow, R.E., Tomacruz, R.S., Armstrong, D.K., et al., 2002. Survival effect of maximal cytoreductive surgery for advanced ovarian carcinoma during the platinum era: a meta-analysis. J. Clin. Oncol. 20, 1248–1259.

Brody, J.R., Witkiewicz, A.K., Yeo, C.J., 2011. The past, present, and future of biomarkers: a need for molecular beacons for the clinical management of pancreatic cancer. Adv. Surg. 45, 301–321.

Chen, Y., Kang, M., Lu, W., et al., 2012. DJ-1, a novel biomarker and a selected target gene for overcoming chemoresistance in pancreatic cancer. J. Cancer Res. Clin. Oncol. 138, 1463–1474.

Cheng, K.W., Lahad, J.P., Kuo, W.L., et al., 2004. The RAB25 small GTPase determines aggressiveness of ovarian and breast cancers. Nat. Med. 10, 1251–1256.

Costantino, C.L., Witkiewicz, A.K., Kuwano, Y., et al., 2009. The role of HuR in gemcitabine efficacy in pancreatic cancer: HuR Upregulates the expression of the gemcitabine metabolizing enzyme deoxycytidine kinase. Cancer Res. 69, 4567–4572.

Coticchia, C.M., Yang, J., Moses, M.A., 2008. Ovarian cancer biomarkers: current options and future promise. J. Natl. Compr. Canc. Netw. 6, 795–802.

Dearking, A.C., Aletti, G.D., McGree, M.E., et al., 2007. How relevant are ACOG and SGO guidelines for referral of adnexal mass? Obstet. Gynecol. 110, 841–848.

Denkert, C., Koch, I., von Keyserlingk, N., et al., 2006. Expression of the ELAV-like protein HuR in human colon cancer: association with tumor stage and cyclooxygenase-2. Mod. Pathol. 19, 1261–1269.

Diaz-Padilla, I., Razak, A.R., Minig, L., et al., 2012. Prognostic and predictive value of CA-125 in the primary treatment of epithelial ovarian cancer: potentials and pitfalls. Clin. Transl. Oncol. 14, 15–20.

DiSaia, P.J., Creasman, W.T., 1997. The adnexal mass and early ovarian cancer. Clin. Gynecol. Oncol. 253–281.

Donahue, T.R., Tran, L.M., Hill, R., et al., 2012. Integrative survival-based molecular profiling of human pancreatic cancer. Clin. Cancer Res. 18, 1352–1363.

Duff, E.K., Clarke, A.R., 1998. Smad4 (DPC4)–a potent tumour suppressor? Br. J. Cancer 78, 1615–1619.

Edgell, T., Martin-Roussety, G., Barker, G., et al., 2010. Phase II biomarker trial of a multimarker diagnostic for ovarian cancer. J. Cancer Res. Clin. Oncol. 136, 1079–1088.

Eisenkop, S.M., Spirtos, N.M., Montag, T.W., et al., 1992. The impact of subspecialty training on the management of advanced ovarian cancer. Gynecol. Oncol. 47, 203–209.

Farley, J., Smith, L.M., Darcy, K.M., et al., 2003. Cyclin E expression is a significant predictor of survival in advanced, suboptimally debulked ovarian epithelial cancers: a Gynecologic Oncology Group study. Cancer Res. 63, 1235–1241.

Garcea, G., Ladwa, N., Neal, C.P., et al., 2011. Preoperative neutrophil-to-lymphocyte ratio (NLR) is associated with reduced disease-free survival following curative resection of pancreatic adenocarcinoma. World J. Surg. 35, 868–872.

Geng, X., Liu, Y., Diersch, S., et al., 2017. Clinical relevance of kallikrein-related peptidase 9, 10, 11, and 15 mRNA expression in advanced high-grade serous ovarian cancer. PLoS One 12, e0186847.

Goonetilleke, K.S., Siriwardena, A.K., 2007. Systematic review of carbohydrate antigen (CA 19-9) as a biochemical marker in the diagnosis of pancreatic cancer. Eur. J. Surg. Oncol. 33, 266–270.

Grossman, D.C., Curry, S.J., Owens, D.K., et al., 2018. Screening for ovarian cancer: US preventive services task force recommendation statement. J. Am. Med. Assoc. 319, 588–594.

Gungor, C., Zander, H., Effenberger, K.E., et al., 2011. Notch signaling activated by replication stress-induced expression of midkine drives epithelial-mesenchymal transition and chemoresistance in pancreatic cancer. Cancer Res. 71, 5009–5019.

Guo, X., Liu, Y., Huang, X., et al., 2017. Serum relaxin as a diagnostic and prognostic marker in patients with epithelial ovarian cancer. Cancer Biomark. 21, 81–87.

Halvorsen, A.R., Kristensen, G., Embleton, A., et al., 2017. Evaluation of prognostic and predictive significance of circulating microRNAs in ovarian cancer patients. Dis. Markers 2017, 3098542.

Hartwig, W., Hackert, T., Hinz, U., et al., 2011. Pancreatic cancer surgery in the new millennium: better prediction of outcome. Ann. Surg. 254, 311–319.

Hefler, L.A., Zeillinger, R., Grimm, C., et al., 2006. Preoperative serum vascular endothelial growth factor as a prognostic parameter in ovarian cancer. Gynecol. Oncol. 103, 512–517.

Herzog, T.J., Spetzler, D., Xiao, N., et al., 2016. Impact of molecular profiling on overall survival of patients with advanced ovarian cancer. Oncotarget 7, 19840–19849.

Hu, Z.D., Wei, T.T., Yang, M., et al., 2015. Diagnostic value of osteopontin in ovarian cancer: a meta-analysis and systematic review. PLoS One 10, e0126444.

Huo, J., Hu, J., Liu, G., et al., 2017. Elevated serum interleukin-37 level is a predictive biomarker of poor prognosis in epithelial ovarian cancer patients. Arch. Gynecol. Obstet. 295, 459–465.

Iacobuzio-Donahue, C.A., Fu, B., Yachida, S., et al., 2009. DPC4 gene status of the primary carcinoma correlates with patterns of failure in patients with pancreatic cancer. J. Clin. Oncol. 27, 1806–1813.

Iorio, M.V., Visone, R., Di Leva, G., et al., 2007. MicroRNA signatures in human ovarian cancer. Cancer Res. 67, 8699–8707.

Jacobs, I.J., Menon, U., Ryan, A., et al., 2016. Ovarian cancer screening and mortality in the UK Collaborative Trial of Ovarian Cancer Screening (UKCTOCS): a randomised controlled trial. Lancet 387, 945–956.

Jamieson, N.B., Morran, D.C., Morton, J.P., et al., 2012. MicroRNA molecular profiles associated with diagnosis, clinicopathologic criteria, and overall survival in patients with resectable pancreatic ductal adenocarcinoma. Clin. Cancer Res. 18, 534–545.

Jung, H.Y., Fattet, L., Yang, J., 2015. Molecular pathways: linking tumor microenvironment to epithelial-mesenchymal transition in metastasis. Clin. Cancer Res. 21, 962–968.

Juretzka, M.M., Barakat, R.R., Chi, D.S., et al., 2007. CA125 level as a predictor of progression-free survival and overall survival in ovarian cancer patients with surgically defined disease status prior to the initiation of intraperitoneal consolidation therapy. Gynecol. Oncol. 104, 176–180.

Kang, S., Kim, T.J., Seo, S.S., et al., 2011. Interaction between preoperative CA-125 level and survival benefit of neoadjuvant chemotherapy in advanced epithelial ovarian cancer. Gynecol. Oncol. 120, 18–22.

Karakatsanis, A., Papaconstantinou, I., Gazouli, M., et al., 2013. Expression of microRNAs, miR-21, miR-31, miR-122, miR-145, miR-146a, miR-200c, miR-221, miR-222, and miR-223 inpatients with hepatocellular carcinoma or intrahepatic cholangiocarcinoma and its prognostic significance. Mol. Carcinog. 52, 297–303.

Karam, A.K., Karlan, B.Y., 2010. Ovarian cancer: the duplicity of CA125 measurement. Nat. Rev. Clin. Oncol. 7, 335–339.

Kelly, K.J., Wong, J., Gladdy, R., et al., 2009. Prognostic impact of RT-PCR-based detection of peritoneal micrometastases in patients with pancreatic cancer undergoing curative resection. Ann. Surg. Oncol. 16, 3333–3339.

Kim, Y.W., Bae, S.M., Lim, H., et al., 2012. Development of multiplexed bead-based immunoassays for the detection of early stage ovarian cancer using a combination of serum biomarkers. PLoS One 7, e44960.

Kishikawa, T., Otsuka, M., Ohno, M., et al., 2015. Circulating RNAs as new biomarkers for detecting pancreatic cancer. World J. Gastroenterol. 21, 8527–8540.

Kurosaki, A., Hasegawa, K., Kato, T., et al., 2016. Serum folate receptor alpha as a biomarker for ovarian cancer: implications for diagnosis, prognosis and predicting its local tumor expression. Int. J. Cancer 138, 1994–2002.

Labrie, M., De Araujo, L.O.F., Communal, L., et al., 2017. Tissue and plasma levels of galectins in patients with high grade serous ovarian carcinoma as new predictive biomarkers. Sci. Rep. 7, 13244.

Leung, F., Bernardini, M.Q., Brown, M.D., et al., 2016. Validation of a novel biomarker panel for the detection of ovarian cancer. Cancer Epidemiol. Biomark. Prev. 25, 1333–1340.

Leung, F., Dimitromanolakis, A., Kobayashi, H., et al., 2013. Folate-receptor 1 (FOLR1) protein is elevated in the serum of ovarian cancer patients. Clin. Biochem. 46, 1462–1468.

Li, Y.Y., Zhang, W.C., Zhang, J.L., et al., 2015. Plasma levels of lysophosphatidic acid in ovarian cancer versus controls: a meta-analysis. Lipids Health Dis. 14, 72.

Liu, P., Sun, Y.L., Du, J., et al., 2012. CD105/Ki67 coexpression correlates with tumor progression and poor prognosis in epithelial ovarian cancer. Int. J. Gynecol. Cancer 22, 586–592.

Locker, G.Y., Hamilton, S., Harris, J., et al., 2006. ASCO 2006 update of recommendations for the use of tumor markers in gastrointestinal cancer. J. Clin. Oncol. 24, 5313–5327.

Lowery, M.A., Kelsen, D.P., Stadler, Z.K., et al., 2011. An emerging entity: pancreatic adenocarcinoma associated with a known BRCA mutation: clinical descriptors, treatment implications, and future directions. Oncologist 16, 1397–1402.

Luo, L.Y., Katsaros, D., Scorilas, A., et al., 2001. Prognostic value of human kallikrein 10 expression in epithelial ovarian carcinoma. Clin. Cancer Res. 7, 2372–2379.

Madeira, K., Dondossola, E.R., Farias, B.F., et al., 2016. Mesothelin as a biomarker for ovarian carcinoma: a meta-analysis. An. Acad. Bras. Cienc. 88, 923–932.

Maithel, S.K., Maloney, S., Winston, C., et al., 2008. Preoperative CA 19-9 and the yield of staging laparoscopy in patients with radiographically resectable pancreatic adenocarcinoma. Ann. Surg. Oncol. 15, 3512–3520.

Martin, H.L., Ohara, K., Kiberu, A., et al., 2014. Prognostic value of systemic inflammation-based markers in advanced pancreatic cancer. Intern. Med. J. 44, 676–682.

Masui, T., Doi, R., Mori, T., et al., 2004. Metastin and its variant forms suppress migration of pancreatic cancer cells. Biochem. Biophys. Res. Commun. 315, 85–92.

Matsuzaki, H., Kobayashi, H., Yagyu, T., et al., 2005. Plasma bikunin as a favorable prognostic factor in ovarian cancer. J. Clin. Oncol. 23, 1463–1472.

Mehta, K., 2009. Biological and therapeutic significance of tissue transglutaminase in pancreatic cancer. Amino Acids 36, 709–716.

Montagnana, M., Benati, M., Danese, E., 2017. Circulating biomarkers in epithelial ovarian cancer diagnosis: from present to future perspective. Ann. Transl. Med. 5, 276.

Montagnana, M., Danese, E., Ruzzenente, O., et al., 2011. The ROMA (Risk of Ovarian Malignancy Algorithm) for estimating the risk of epithelial ovarian cancer in women presenting with pelvic mass: is it really useful? Clin. Chem. Lab. Med. 49, 521–525.

Moore, P.S., Sipos, B., Orlandini, S., et al., 2001. Genetic profile of 22 pancreatic carcinoma cell lines. Analysis of K-ras, p53, p16 and DPC4/Smad4. Virchows Arch. 439, 798–802.

Na, Y.J., Farley, J., Zeh, A., et al., 2009. Ovarian cancer: markers of response. Int. J. Gynecol. Cancer 19 (Suppl. 2), S21–S29.

Nio, Y., Dong, M., Iguchi, C., et al., 2001. Expression of Bcl-2 and p53 protein in resectable invasive ductal carcinoma of the pancreas: effects on clinical outcome and efficacy of adjuvant chemotherapy. J. Surg. Oncol. 76, 188–196.

Nolen, B.M., Lokshin, A.E., 2013. Biomarker testing for ovarian cancer: clinical utility of multiplex assays. Mol. Diagn. Ther. 17, 139–146.

Osayi, S.N., Bloomston, M., Schmidt, C.M., et al., 2014. Biomarkers as predictors of recurrence following curative resection for pancreatic ductal adenocarcinoma: a review. BioMed Res. Int. 2014, 468959.

Pinsky, P.F., Yu, K., Kramer, B.S., et al., 2016. Extended mortality results for ovarian cancer screening in the PLCO trial with median 15years follow-up. Gynecol. Oncol. 143, 270–275.

Pleskow, D.K., Berger, H.J., Gyves, J., et al., 1989. Evaluation of a serologic marker, CA19-9, in the diagnosis of pancreatic cancer. Ann. Intern. Med. 110, 704–709.

Poruk, K.E., Firpo, M.A., Scaife, C.L., et al., 2013. Serum osteopontin and tissue inhibitor of metalloproteinase 1 as diagnostic and prognostic biomarkers for pancreatic adenocarcinoma. Pancreas 42, 193–197.

Richards, N.G., Rittenhouse, D.W., Freydin, B., et al., 2010. HuR status is a powerful marker for prognosis and response to gemcitabine-based chemotherapy for resected pancreatic ductal adenocarcinoma patients. Ann. Surg. 252, 499–505 discussion 505–496.

Rosso, M., Majem, B., Devis, L., et al., 2017. E-cadherin: a determinant molecule associated with ovarian cancer progression, dissemination and aggressiveness. PLoS One 12, e0184439.

Rustin, G.J., Marples, M., Nelstrop, A.E., et al., 2001. Use of CA-125 to define progression of ovarian cancer in patients with persistently elevated levels. J. Clin. Oncol. 19, 4054–4057.

Rustin, G.J., Nelstrop, A.E., Tuxen, M.K., et al., 1996. Defining progression of ovarian carcinoma during follow-up according to CA 125: a North Thames Ovary Group Study. Ann. Oncol. 7, 361–364.

Rzepecka, I.K., Szafron, L.M., Stys, A., et al., 2017. Prognosis of patients with BRCA1-associated ovarian carcinomas depends on TP53 accumulation status in tumor cells. Gynecol. Oncol. 144, 369–376.

Sano, R., Kanomata, N., Suzuki, S., et al., 2017. Vasohibin-1 is a poor prognostic factor of ovarian carcinoma. Tohoku J. Exp. Med. 243, 107–114.

Santoiemma, P.P., Reyes, C., Wang, L.P., et al., 2016. Systematic evaluation of multiple immune markers reveals prognostic factors in ovarian cancer. Gynecol. Oncol. 143, 120–127.

Skates, S.J., 2012. Ovarian cancer screening: development of the risk of ovarian cancer algorithm (ROCA) and ROCA screening trials. Int. J. Gynecol. Cancer 22 (Suppl. 1), S24–S26.

Skates, S.J., Horick, N., Yu, Y., et al., 2004. Preoperative sensitivity and specificity for early-stage ovarian cancer when combining cancer antigen CA-125II, CA 15-3, CA 72-4, and macrophage colony-stimulating factor using mixtures of multivariate normal distributions. J. Clin. Oncol. 22, 4059–4066.

Spentzos, D., Levine, D.A., Ramoni, M.F., et al., 2004. Gene expression signature with independent prognostic significance in epithelial ovarian cancer. J. Clin. Oncol. 22, 4700–4710.

Spratlin, J., Sangha, R., Glubrecht, D., et al., 2004. The absence of human equilibrative nucleoside transporter 1 is associated with reduced survival in patients with gemcitabine-treated pancreas adenocarcinoma. Clin. Cancer Res. 10, 6956–6961.

Stratford, J.K., Bentrem, D.J., Anderson, J.M., et al., 2010. A six-gene signature predicts survival of patients with localized pancreatic ductal adenocarcinoma. PLoS Med. 7, e1000307.

Su, F., Lang, J., Kumar, A., et al., 2007. Validation of candidate serum ovarian cancer biomarkers for early detection. Biomark. Insights 2, 369–375.

Swords, D.S., Firpo, M.A., Scaife, C.L., et al., 2016. Biomarkers in pancreatic adenocarcinoma: current perspectives. OncoTargets Ther. 9, 7459–7467.

Tamir, A., Gangadharan, A., Balwani, S., et al., 2016. The serine protease prostasin (PRSS8) is a potential biomarker for early detection of ovarian cancer. J. Ovarian Res. 9, 20.

Taylor, D.D., Gercel-Taylor, C., 2008. MicroRNA signatures of tumor-derived exosomes as diagnostic biomarkers of ovarian cancer. Gynecol. Oncol. 110, 13–21.

Tcherkassova, J., Abramovich, C., Moro, R., et al., 2011. Combination of CA125 and RECAF biomarkers for early detection of ovarian cancer. Tumour Biol. 32, 831–838.

Tingulstad, S., Skjeldestad, F.E., Hagen, B., 2003. The effect of centralization of primary surgery on survival in ovarian cancer patients. Obstet. Gynecol. 102, 499–505.

Tsuda, H., Ito, Y.M., Ohashi, Y., et al., 2005. Identification of overexpression and amplification of ABCF2 in clear cell ovarian adenocarcinomas by cDNA microarray analyses. Clin. Cancer Res. 11, 6880–6888.

Tuxen, M.K., Soletormos, G., Dombernowsky, P., 2002. Serum tumor marker CA 125 for monitoring ovarian cancer during follow-up. Scand. J. Clin. Lab. Invest. 62, 177–188.

Tuxen, M.K., Soletormos, G., Rustin, G.J., et al., 2000. Biological variation and analytical imprecision of CA 125 in patients with ovarian cancer. Scand. J. Clin. Lab. Invest. 60, 713–721.

Ueland, F.R., Desimone, C.P., Seamon, L.G., et al., 2011. Effectiveness of a multivariate index assay in the preoperative assessment of ovarian tumors. Obstet. Gynecol. 117, 1289–1297.

Vergote, I., Rustin, G.J., Eisenhauer, E.A., et al., 2000. Re: new guidelines to evaluate the response to treatment in solid tumors [ovarian cancer]. Gynecologic Cancer Intergroup. J. Natl. Cancer Inst. 92, 1534–1535.

Villarroel, M.C., Rajeshkumar, N.V., Garrido-Laguna, I., et al., 2011. Personalizing cancer treatment in the age of global genomic analyses: PALB2 gene mutations and the response to DNA damaging agents in pancreatic cancer. Mol. Cancer Ther. 10, 3–8.

Viterbo, D., Gausman, V., Gonda, T., 2016. Diagnostic and therapeutic biomarkers in pancreaticobiliary malignancy. World J. Gastrointest. Endosc. 8, 128–142.

Von Hoff, D.D., Ramanathan, R.K., Borad, M.J., et al., 2011. Gemcitabine plus nab-paclitaxel is an active regimen in patients with advanced pancreatic cancer: a phase I/II trial. J. Clin. Oncol. 29, 4548–4554.

Wang, L., Hu, Y., Xiang, X., et al., 2017. Identification of long noncoding RNA signature for paclitaxel-resistant patients with advanced ovarian cancer. Oncotarget 8, 64191–64202.

Ware Miller, R., Smith, A., DeSimone, C.P., et al., 2011. Performance of the American College of Obstetricians and Gynecologists' ovarian tumor referral guidelines with a multivariate index assay. Obstet. Gynecol. 117, 1298–1306.

White, N.M., Mathews, M., Yousef, G.M., et al., 2009. Human kallikrein related peptidases 6 and 13 in combination withCA125 is a more sensitive test for ovarian cancer than CA125 alone. Cancer Biomark. 5, 279–287.

Winter, J.M., Tang, L.H., Klimstra, D.S., et al., 2012. A novel survival-based tissue microarray of pancreatic cancer validates MUC1 and mesothelin as biomarkers. PLoS One 7, e40157.

Winter, J.M., Yeo, C.J., Brody, J.R., 2013. Diagnostic, prognostic, and predictive biomarkers in pancreatic cancer. J. Surg. Oncol. 107, 15–22.

Yamanaka, Y., Friess, H., Kobrin, M.S., et al., 1993. Coexpression of epidermal growth factor receptor and ligands in human pancreatic cancer is associated with enhanced tumor aggressiveness. Anticancer Res. 13, 565–569.

Yang, F., Hu, Z.D., Chen, Y., et al., 2016. Diagnostic value of KLK6 as an ovarian cancer biomarker: a meta-analysis. Biomed. Rep. 4, 681–686.

Yang, N., Kaur, S., Volinia, S., et al., 2008. MicroRNA microarray identifies Let-7i as a novel biomarker and therapeutic target in human epithelial ovarian cancer. Cancer Res. 68, 10307–10314.

Yip, P., Chen, T.H., Seshaiah, P., et al., 2011. Comprehensive serum profiling for the discovery of epithelial ovarian cancer biomarkers. PLoS One 6, e29533.

Yokoi, A., Yoshioka, Y., Hirakawa, A., et al., 2017. A combination of circulating miRNAs for the early detection of ovarian cancer. Oncotarget 8, 89811–89823.

Yoo, S.C., Yoon, J.H., Lyu, M.O., et al., 2008. Significance of postoperative CA-125 decline after cytoreductive surgery in stage IIIC/IV ovarian cancer. J. Gynecol. Oncol. 19, 169–172.

Yurkovetsky, Z., Skates, S., Lomakin, A., et al., 2010. Development of a multimarker assay for early detection of ovarian cancer. J. Clin. Oncol. 28, 2159–2166.

Zhang, S., Lu, Z., Unruh, A.K., et al., 2015a. Clinically relevant microRNAs in ovarian cancer. Mol. Cancer Res. 13, 393–401.

Zhang, Y., Yang, J., Li, H., et al., 2015b. Tumor markers CA19-9, CA242 and CEA in the diagnosis of pancreatic cancer: a meta-analysis. Int. J. Clin. Exp. Med. 8, 11683–11691.

Zheng, H.C., Zhao, S., Song, Y., et al., 2017. The roles of ING5 expression in ovarian carcinogenesis and subsequent progression: a target of gene therapy. Oncotarget 8, 103449–103464.

Zivanovic, O., Sima, C.S., Iasonos, A., et al., 2009. Exploratory analysis of serum CA-125 response to surgery and the risk of relapse in patients with FIGO stage IIIC ovarian cancer. Gynecol. Oncol. 115, 209–214.

Zurawski Jr., V.R., Orjaseter, H., Andersen, A., et al., 1988. Elevated serum CA 125 levels prior to diagnosis of ovarian neoplasia: relevance for early detection of ovarian cancer. Int. J. Cancer 42, 677–680.

CHAPTER

48

Prostate Cancer Biomarkers

Charalampos Mamoulakis[1], *Charalampos Mavridis*[1], *George Georgiadis*[1], *Charalampos Belantis*[1], *Ioannis E. Zisis*[1], *Iordanis Skamagkas*[1], *Ioannis Heretis*[1], *Wallace A. Hayes*[2], *Aristidis M. Tsatsakis*[3]

[1]Department of Urology, University General Hospital of Heraklion, University of Crete, Medical School, Heraklion, Crete, Greece; [2]University of South Florida College of Public Health, Tampa, FL, United States; Michigan State University, East Lansing, MI, United States; [3]Center of Toxicology Science & Research, Medical School, University of Crete, Heraklion, Greece

INTRODUCTION

Prostate cancer (PCa) represents a major health issue with an incidence mainly dependent on age. It is the second most commonly diagnosed cancer in men (1.1 million diagnoses worldwide in 2012; 15% of all cancers diagnosed) (Ferlay et al., 2015). Detection on autopsy is similar worldwide (Haas et al., 2008) with prevalence at age <30 years of 5%, increasing by an odds ratio of 1.7 per decade, to 59% by age >79 years (Bell et al., 2015). However, the incidence varies widely (highest in Australia/New Zealand, Northern America, Western and Northern Europe; largely due to prostate-specific antigen (PSA) testing use and the aging population; low in Eastern and South Central Asia, while rates in Eastern and Southern Europe, which were low, have showed a steady increase) (Ferlay et al., 2015; Haas et al., 2008). On the other hand, mortality rates vary less (generally high in populations of African descent, intermediate in the United States, and very low in Asia) (Ferlay et al., 2015). Genetic factors are associated with the risk of (aggressive) PCa, but ongoing trials are needed to define the clinical applicability of screening for genetic susceptibility. A variety of exogenous/environmental factors may have an impact on the risk of progression, but no definitive recommendation can be provided for specific preventive or dietary measures to reduce the risk of developing PCa (Mottet et al., 2017).

The therapeutic success rate for PCa can be significantly improved if the disease is diagnosed early. In most cases, this disease can be treated effectively and even eradicated when detected at a very early stage. Thus, a successful therapy depends heavily on clinical indicators (biomarkers) for early detection and progression of the disease, as well as the prediction after the clinical intervention. However, there are a lack of reliable biomarkers to date that can specifically distinguish between those patients who should be treated adequately to stop the aggressive form of the disease and those who should avoid overtreatment of the indolent form. Current advances in molecular techniques have provided new tools to facilitate the discovery of new biomarkers for PCa. These emerging biomarkers will be critical in developing new clinically reliable indicators with high specificity for the diagnosis and prognosis of PCa. This chapter examines the current and emerging PCa biomarkers on their diagnostic and prognostic potential for prediction of disease status and response to drug- or treatment-related toxicities.

SCREENING AND EARLY DETECTION FOR PROSTATE CANCER

Population (mass) screening refers to the "systematic examination of asymptomatic men (at risk)" usually driven by health authorities, whereas early detection (opportunistic) testing refers to individualized findings driven by the individual being tested and/or his physician. Both strategies aim at the reduction of disease-specific mortality and quality of life maintenance.

PCa screening remains one of the most controversial topics in urology, but there is currently strong evidence against systematic population-based screening in all

Biomarkers in Toxicology, Second Edition
https://doi.org/10.1016/B978-0-12-814655-2.00048-7

countries, including Europe (Mottet et al., 2017). It has been recently shown from data of the European Randomized Study of Screening for Prostate Cancer (ERSPC) with follow-up of over 10 years that reduction in mortality remains unchanged with extended follow-up (Schroder et al., 2014). Screening for PCa has never been shown to be detrimental at the population level but its impact on overall quality of life remains unclear (Booth et al., 2014; Heijnsdijk et al., 2012; Vasarainen et al., 2013). An updated Cochrane review (Hayes and Barry, 2014) has showed that:

1. Screening is associated with increased diagnosis (risk ratio (RR): 1.3; 95% CI: 1.02–1.65), with detection of more localized and less advanced disease (RR: 1.79; 95% CI: 1.19–2.70; and RR: 0.80; 95% CI: 0.73–0.87, respectively).
2. Based on pooled results of four available randomized controlled trials (RCTs), no overall survival benefit has been detected (RR: 1.00; 95% CI: 0.96–1.03).
3. Based on pooled and individual results of five RCTs, randomizing >341,000 men, no PCa-specific survival benefit has been detected (RR: 1.00; 95% CI: 0.86–1.17).
4. Screening is associated with minor and major harms (overdiagnosis and overtreatment), despite the fact that biopsy is not associated with any mortality.

According to the European Association of Urology (EAU)-European Society for Radiotherapy and Oncology (ESTRO)-International Society of Geriatric Oncology (SIOG) guidelines on PCa (Mottet et al., 2017), an individualized risk-adapted strategy for early detection might be offered to a well-informed man with a good performance status and a life expectancy of at least 10–15 years. However, it is crucial to carefully detect patients likely to benefit most from individual early diagnosis, considering potential balances/harms because this approach may be associated with a substantial risk of overdiagnosis and the long-term survival/quality of life benefits of this approach needs to be proven at a population level. Men at increased risk of having PCa include those >50 years, those >45 years with positive paternal or maternal family history (Albright et al., 2015) and African Americans (Kamangar et al., 2006). Men with a serum PSA level >1 ng/mL at 40 years and >2 ng/mL at 60 years (Carlsson et al., 2014; Vickers et al., 2013) are also at increased risk of metastasis/death. Informed men requesting early diagnosis should be given a PSA test and undergo a digital rectal examination (DRE) but optimal intervals are unknown, whereas a risk-adapted strategy might be considered based on the initial PSA level (every 2 years for those initially at risk, or postponed up to 8 to 10 years in those not at risk). The age at which early diagnosis should be stopped is controversial, but men with <15 year life expectancy are unlikely to benefit (Mottet et al., 2017).

Several new biomarkers such as BRCA2 (a genetic abnormality associated with an increased risk shown prospectively (Bancroft et al., 2014; Gulati et al., 2017)), TMPRSS2-Erg fusion, PCA3 (Leyten et al., 2014; Vedder et al., 2014), or kallikreins as incorporated in the Phi or 4K score tests (Boegemann et al., 2016; Bryant et al., 2015) have been shown to add sensitivity/specificity on top of PSA, potentially avoiding unnecessary biopsies and lowering overdiagnosis. However, there is currently too limited data to base a recommendation on and further data are needed before such biological markers can be used in standard clinical practice (Mottet et al., 2017).

CONTEMPORARY CLINICAL BIOMARKERS IN PROSTATE CANCER

Prostate-Specific Antigen

PSA is a serine protease kallikrein family member, clustered in a 300-kb region encoded on chromosome 19q13.4. It is expressed both in normal and cancerous prostatic tissue. Therefore it constitutes an organ but not a cancer-specific biomarker that can be elevated in benign prostatic hyperplasia (BPH), prostatitis, and other nonmalignant conditions. Nevertheless, the use of PSA as a serum marker has revolutionized PCa diagnosis (Stamey et al., 1987). As an independent variable, it is a better predictor of PCa than DRE or transrectal ultrasonography (TRUS) of the prostate (Catalona et al., 2017). Currently, serum PSA is widely used in clinical practice for diagnostic, posttreatment prognostic, recurrence monitoring, and drug efficacy determining biomarker, but it is considered a suboptimal tool for screening, diagnosis, and prognosis due to limitations such as low specificity and its inability to identify indolent from aggressive tumors.

Two large population-based studies in Europe (ERSPC) and in the United States (Prostate, Lung, Colorectal, and Ovarian [PLCO] Cancer Screening Trial) have confirmed its modest clinical value as a screening tool for PCa (Auvinen et al., 2016; Bibbins-Domingo et al., 2017; Pinsky et al., 2017; Schroder et al., 2014). Higher serum PSA levels indicate a greater likelihood of PCa. Many men with low serum PSA levels may harbor PCa and therefore, an optimal threshold for detecting nonpalpable but clinically significant PCa is precluded (Table 48.1) (Mottet et al., 2017). Nomograms may help in predicting indolent PCa (Dong et al., 2008).

Efforts to improve the diagnostic performance of serum PSA have been extensively investigated using several modifications of PSA testing (Mottet et al., 2017). PSA density (level of serum PSA divided by the TRUS-determined prostate volume) has shown that the

TABLE 48.1 Risk of PCa in Patients With Serum PSA Levels up to 4 ng/mL

PSA (ng/mL)	PCa Gleason Score >7 (%)	PCa (%)
0.0–0.5	0.8	6.6
0.6–1.0	1.0	10.1
1.1–2.0	2.0	17.0
2.1–3.0	4.6	23.9
3.1–4.0	6.7	26.9

PCa, prostate cancer; *PSA*, prostate-specific antigen.

higher the PSA density is the more likely that PCa is clinically significant. PSA kinetics (PSA velocity; i.e., the absolute annual increase in serum PSA (ng/mL/year) and PSA doubling time; i.e., the exponential increase in serum PSA over time) may have a prognostic role in treatment but limited diagnostic value because of background noise (total prostate volume and BPH), different intervals between PSA determinations, and acceleration/deceleration of PSA velocity and PSA doubling time over time. Consequently, PSA kinetics do not provide additional information compared with PSA alone. Free/total PSA ratio stratifies the risk of PCa in men with a total serum PSA level of 4–10 ng/mL and negative DRE. The ratio is clinically useless if the total serum PSA level is >10 ng/mL or for following up PCa because it may be adversely affected by several factors (e.g., instability of free PSA at 4°C and room temperature, variable assay characteristics, and concomitant BPH in large prostates).

Additional Serum Testing

A few commercially available assays measure a panel of kallikreins in serum or plasma (Ashley et al., 2017; Saini, 2016). The most well-known and widely used are the prostate health index (PHI) test and the four kallikrein (4K) score test (Mottet et al., 2017). The former was approved by FDA in 2012. It is calculated for each patient as ([−2] proPSA isoform (p2PSA, one of the forms of free PSA)/freePSA) × $PSA^{1/2}$). The test was developed by Beckman Coulter in partnership with the NCI Early Detection Research Network and intends to distinguish cancerous and benign prostatic conditions in men aged ≥50 years with a normal DRE and total serum PSA levels of 4–10 ng/mL, thus determining the need of biopsy in such cases and, reducing unnecessary biopsies (Ashley et al., 2017; Saini, 2016). Several studies suggest that the use of PHI along with [−2] proPSA significantly improves PCa detection in cases with a Gleason score ≥7 and improves the accuracy of established PCa predictors at biopsy compared to PSA and free PSA.

The four kallikrein (4K) score test, offered by OPKO Health Inc., is available as a Clinical Laboratory Improvement Amendments–based laboratory developed test, i.e., commercially developed for PCa but not yet approved by the FDA (Ashley et al., 2017; Saini, 2016). It measures the blood plasma levels of a panel of four prostate-derived kallikrein proteins, i.e., total PSA, free PSA, intact PSA, and human kallikrein 2 (hK2), another kallikrein with 80% PSA homology. These biomarkers are combined with other parameters such as age, DRE status, and prior biopsy status to calculate the risk of pathologically insignificant versus aggressive PCa. The test assists to identify patients eligible for biopsy based on the probability of having aggressive PCa, and helps to avoid unnecessary biopsies in low-risk patients.

A few prospective multicenter studies have shown that both tests described above outperform the free/total PSA ratio for PCa detection, with an improved prediction of clinically significant PCa in cases with a serum PSA level between 2 and 10 ng/mL, whereas in a head-to-head comparison both tests perform equally (Mottet et al., 2017). PHI remains to be tested as a population-based screening tool for showing clinically meaningful benefit for screening that will lead to reduction of PCa mortality. The predictive ability of the 4K test appears to increase the diagnostic accuracy of detecting PCa using a subcohort of the ERSPC study, but a head-to-head comparison in large cohorts with serum PSA alone has not yet been studied. Nevertheless, according to the EAU-ESTRO-SIOG guidelines on PCa, before performing a prostate biopsy, a further risk assessment for asymptomatic men with a PSA level between 2 and 10 ng/mL, an additional serum or urine-based test (e.g., PHI, 4K score or Prostate cancer gene 3 [PCA3]; see below) among other tools, such as risk calculators or multiparametric magnetic resonance imaging (mpMRI), is recommended to avoid unnecessary biopsies (Mottet et al., 2017).

mpMRI is mainly recommended if clinical suspicion persists in spite of previous negative biopsy for deciding on a repeat systematic prostate biopsy with extra targeting of any lesions seen. On the other hand, despite the fact that additional information may be gained by additional serum-based, urine-based, or even tissue-based epigenetic tests (e.g., PHI, 4K score; PCA3 or ConfirmMDx [see below], respectively) the role of such tests in deciding whether to take a repeat biopsy in men with previous negative biopsy is uncertain and probably not cost-effective (Mottet et al., 2017).

Prostate Cancer Antigen 3

PCA3 or DD3 is a prostate-specific, noncoding mRNA biomarker with unknown function that is detectable in urine sediments obtained after three strokes of prostatic massage during DRE. It is encoded on chromosome 9q21–22; it is not expressed in normal prostate tissue and has very low expression in hyperplastic prostate but is highly expressed in PCa (Ashley et al., 2017; Saini, 2016). The commercially available Progensa urine test (Hologic, Marlborough, MA) is superior to total and percent-free PSA for detection of PCa in men with elevated PSA as it shows significant increases in the area under the receiver-operator characteristic curve for positive biopsies (Mottet et al., 2017). An optimal cutoff for urinary PCA3 levels have yet to be established for maximizing the tradeoff between clinical benefit versus overdiagnosis (Saini, 2016). Currently, the main indication for the Progensa test is to determine whether repeat biopsy is needed after an initially negative biopsy (Mottet et al., 2017), but its clinical effectiveness and cost-effectiveness is uncertain (Nicholson et al., 2015).

CANDIDATE BIOMARKERS FOR PROSTATE CANCER

Current advancements in proteomics, tissue microarray, DNA microarray, immunohistochemical staining, microRNA, single nucleotide polymorphism analysis for genetic variations, and other biotechnologies have paved the way and significantly increased the pace at which novel biomarkers are being discovered and developed. Using these methodologies, researchers have reported a number of biomarkers with great potential: These biomarkers are currently undergoing further investigation for validation. A few of the recently identified candidates that have generated some excitement for their potential as PCa biomarkers are discussed below.

ConfirmMDx Test

The ConfirmMDx (MDx Health Inc.) test, is a non-FDA approved test that detects an epigenetic field effect or "halo" associated with a cancerization process at the DNA methylation level in cells adjacent to cancer foci (Hammarsten et al., 2016; Saini, 2016). The concept is that benign prostatic tissue near a PCa focus shows distinct epigenetic alterations. Core specimens collected during a 12-core biopsy are used (minimum requirement of eight cores; from apex, mid- and left/right base, and two additional locations of the prostate). If PCa is missed at biopsy, demonstration of epigenetic changes in the adjacent benign tissue indicates the presence of carcinoma. The test quantifies the methylation level of the promoter regions of three genes (RASSF1, GSTP1, and APC) in benign prostatic tissue. A multicenter study concluded that repeat biopsy could be avoided in men with a negative test, namely methylation absent in all three markers (negative predictive value of 88%) (Partin et al., 2014). Because of limited available data, no recommendation can be made regarding its routine application (Mottet et al., 2017).

Cell Cycle Progression Signature Test

The CCP signature test (Prolaris; Myriad Genetics Inc.) is a genomic test for predicting PCa aggressiveness in conjunction with clinical parameters such as Gleason score and PSA. The test quantifies a patient's risk of disease progression and PCa-specific mortality using a gene-expression-based cell cycle progression (CCP) score. In biopsy-derived PCa tissue, the expression of a set of 31 cell cycle progression–associated genes and 15 housekeeping genes is measured to predict disease progression (Mottet et al., 2017; Saini, 2016). Expression of the cell cycle–related genes is correlated with PCa proliferation, serving as a risk stratification tool that enables a better treatment/monitoring strategy for patients at diagnosis. Low expression of these genes is associated with low risk of progression and those men may be candidates for active surveillance, whereas high expression is associated with higher risk of disease progression in men who may be treated. Potential utility of the CCP test has been evaluated in a US-based clinical setting (Shore et al., 2014). It was concluded that the CCP score adds meaningful new information to the risk assessment for localized PCa patients and that real-world use of the test is likely to lead to a change in treatment in a significant portion of tested patients, particularly by shifting patients toward more conservative management, which could reduce overtreatment of patients with less aggressive disease. Results of prospective multicenter studies are awaited before recommendation in routine application (Mottet et al., 2017).

Oncotype DX Test

Similarly, the Oncotype Dx (Genomic Health, Redwood City, CA) test is a multigene expression assay developed for formalin-fixed paraffin-embedded diagnostic prostate needle biopsies containing as little as 1 mm of PCa (Ashley et al., 2017; Saini, 2016). This RNA-based test reveals the underlying biology of the tumor to determine aggressiveness by assessing the activity of a set of 12 cancer-related genes involved in four different biological pathways (androgen pathway (AZGP1, KLK2, SRD5A2, RAM13C), proliferation (TPX2), cellular organization (FLNC, GSN, TPM2, GSTM2), and stromal response (BGN, COL1A1 and SFRP4)). Five reference genes are included to normalize data and control for variability. These measurements are algorithmically combined to calculate a Genomic Prostate Score. The assay has been validated as a predictor of aggressive PCa (Knezevic et al., 2013). Results of prospective multicenter studies are awaited before a recommendation can be made regarding their routine application (Mottet et al., 2017).

Transmembrane Protease Serine 2-Erythroblast Transformation Specific–Related Gene Fusion Rearrangement

Transmembrane protease serine 2, also known as TMPRSS2, is an androgen-regulated, type II transmembrane-bound serine protease that is locally expressed in the prostate and overexpressed in neoplastic prostate epithelium. TMPRSS2 was thought to play a role in PCa metastasis through the activation of protease-activated receptor-2 (PAR-2). An extensive study focusing on gene fusion transcripts in PCa identified the fusion between TMPRSS2 (located at 21q22.3) with the transcription factor genes ERG (erythrocyte transformation specific–related gene) (21q22.2) and ETV1 (7p21.1), both ETS family members. One TMPRSS2 allele loses its promoter, and one of the ERG alleles gains it, resulting in an over expression of ETS family members in the cancer cells (Mwamukonda et al., 2010). The gene fusion rearrangements between TMPRSS2 and ERG or ETV1 have been reported in several cancers, particularly in hematological malignancies. TMPRSS2-ERG is the most frequent oncogenic gene fusion rearrangement in PCa. It has been observed in almost half of PCa patients and detected in about one-quarter of patients with prostatic intraepithelial neoplasia. This TMPRSS2-ERG fusion is usually found in PCa tissue from men undergoing prostatectomy, and especially among men in North America with PCa on biopsy: However, it is not present in benign prostate biopsy. Another multicenter study showed that the TMPRSS2-ERG urine test could be used to select men with clinically significant PCa (Leyten et al., 2014). The TMPRSS2-ERG fusion can be detected in the urine after DRE with 37% sensitivity and 93% specificity. The addition of TMPRSS2-ERG detection also increased the sensitivity of the urine PCA3 test from 62% to 73% (Mosquera et al., 2009) and resulted in a greater prediction of positive tumors with a higher Gleason score. The results suggest that the combination of TMPRSS2-ERG detection with PCA3 can be very useful in accurately predicting PCa development during screening. TMPRSS2–ERG could be an useful tool for managing and stratifying patients with newly diagnosed PCa as those with fusion-negative TMPRSS2–ERG have worst outcomes (Stone, 2017).

Glutathione S-transferase P1

Glutathione S-transferase P1 (GSTP1) is a ubiquitous family of multifunctional enzymes that conjugate reactive substrates with reduced glutathione and is involved in cellular detoxification. Their role in protecting the cells from oxidative attack, and consequently being upregulated in the presence of free radicals, makes them a prime candidate for consideration as a cancer biomarker. Hypermethylation of the GSTP1 gene was detected in plasma samples from patients with PCa. Genomic DNA was analyzed using methylation-specific polymerase chain reaction technique. Analyzing the GSTP1 gene could discriminate between PCa and BPH patients (Dumache et al., 2014). With the help of polymerase chain reaction (PCR), the methylated GSTP1 gene can be detected in the urine of men who have undergone prostate biopsy (Gonzalgo et al., 2003). This implies the possible additional use of this biomarker in risk stratification of men undergoing prostate biopsy. GSTP1 displays several good characteristics that make it a viable biomarker. For instance, it is highly prevalent in the disease condition, and clinicians are able to measure quantitatively the methylation status of the gene in biopsy/prostatectomy tissues and in cells isolated from serum, urine, and seminal plasma. Furthermore, GSTP1 inactivation by hypermethylation leads to a cell's reduced ability to detoxify electrophilic compounds (such as carcinogens and cytotoxic drugs) by glutathione conjugation, a mechanism that accounts for increased cell vulnerability to carcinogenesis. Therefore, hypermethylation of the GSTP1 gene in prostate cells could also serve as a potential toxicity biomarker for carcinogens and response to cytotoxic anticancer drug treatment. If it is successfully validated, GSTP1 methylation testing of cells derived from samples of serum and urine may possess a significant clinical potential for early detection of PCa and post-treatment monitoring of the disease.

α-Methylacyl Coenzyme A Racemase

α-Methylacyl coenzyme A racemase (AMACR) is an enzyme localized to peroxisomes, is involved in fat metabolism and has been identified to function as a growth promoter, independent of androgens, in PCa. The AMACR gene is overexpressed in PCa tissue and is currently used as a highly specific tissue biomarker to aid in the pathological diagnosis of PCa (Rubin et al., 2002). The potential use of analyzing the levels of AMACR from urine and needle biopsy for PCa detection has been tested (Rogers et al., 2004). Considerably more enhanced sensitivity and specificity in PCa patients with mid-range PSA levels have been observed with AMACR antibodies in comparison to PSA testing alone. This demonstrates that AMACR can be useful in distinguishing control subjects from those with PCa. Some of the limitations of AMACR as a biomarker include the possibility of humoral response and production of endogenous AMACR antibody as a result of certain cancers other than prostate in patients suffering from autoimmune diseases. In addition, AMACR levels are commonly increased in patients with other urological disorders such as BPH. In spite of these restraints, the diagnostic capability for characterizing organ-confined and metastatic PCa is increased with the addition of the AMACR test to serum PSA testing. A promising new noninvasive screening test for PCa is to use quantitative reverse transcriptase PCR (RT-PCR) to identify the ratio of AMACR-to-PSA transcript (Zha et al., 2003). Furthermore, AMACR could be a potential therapeutic target. Use of AMACR inhibitors may be complimentary to androgen therapy (Carnell et al., 2013). Further testing is in progress to assess and possibly validate the prospective use of this serum biomarker.

Chromogranin A

Chromogranin A (GRN-A, also known as secretory protein I), part of the granin family of proteins, is an acidic protein that has been identified in all neuroendocrine cell types and is produced in larger amounts than other secreted proteins by those cells. The growth of prostate cells has been found to be regulated by peptides derived from GRN-A (Lee et al., 1994). Because it is produced and secreted by prostate cells, GRN-A has been examined for its diagnostic and prognostic values as a biomarker for PCa. Based on prior studies, one can use GRN-A to monitor the success of cancer treatment and to predict the outcome of diseases that are androgen-independent. These predictions would precede any indication of PSA progression and reveal that increased levels of GRN-A correlated with disease progression and diminished overall survival (Berruti et al., 2005). In addition, GRN-A levels could be a potential prognostic biomarker in castration-resistant PCa (CRPC) treated with enzalutamide or abiraterone and other new hormonal therapies (Burgio et al., 2014; Conteduca et al., 2014). Therefore, GRN-A may be very useful as a prognostic factor in patients with advanced PCa. However, limitations do exist. One noted weakness of using GRN-A as a biomarker is the fact that neuroendocrine cells do not reside in all prostate tumors. Another disadvantage is its inability to detect the disease at a very early stage. In spite of these shortcomings, GRN-A may possess clinical potential as a biomarker for progressive and recurrent PCa. More research is necessary to clearly determine the clinical value of GRN-A as a serum and tumor marker for PCa.

Prostate-Specific Membrane Antigen

Prostate-specific membrane antigen (PSMA) is a cell surface membrane protein that exhibits some enzymatic activity, although its biological role remains unclear. PSMA has been detected in prostate tissues, PCa cells, and serum. Higher levels of PSMA appear to correspond with poorer clinical outcomes. PSMA levels are higher in primary PCa and metastatic disease, with over 90% of the protein prevalent in diseased tissues. The serum levels of PSMA increase with age and are considerably higher in men above 50 years of age. Although levels of PSMA found in PCa patients are considerably higher than levels found in healthy men or those with BPH, and PSMA appears to be upregulated in patients following hormone deprivation therapy, concrete evidence has yet to validate that a relationship exists between high levels of serum PSMA and the aggressive disposition of the disease. In fact, some studies have observed a decrease in serum PSMA in advanced cases of the disease (Xiao et al., 2001). One of the shortcomings of using PSMA as a serum biomarker is that high PSMA levels have been noticed in the serum of breast cancer patients. This could make it difficult, in some cases, to accurately diagnose men with PCa. Another limitation previously alluded to is the fact that levels of serum PSMA tend to naturally increase with age, potentially complicating diagnoses sought in elder years. PSMA could be a therapeutic target for monoclonal antibodies in patients with PCa (Muthumani et al., 2017). PSMA also has a role in biochemically recurrent PCa (BCR). In fact the superiority of 68Ga-PSMA-PET/CT compared to other imaging techniques seems to be most marked at low PSA levels (Udovicich et al., 2017). More data are required to validate whether this biomarker is clinically acceptable for use in PCa detection, monitoring of treatment, or as an actual means of treatment.

Prostate Stem Cell Antigen

Prostate stem cell antigen (PSCA) is a membrane glycoprotein predominantly expressed in the prostate. Although the expression of PSCA has been shown to be upregulated in the majority of PCa, its biological role in PCa is tentative. Studies have implied that PSCA may be involved in various processes, including androgen-independent progression, metastasis, or signal transduction. In most PCa, a positive correlation was detected between increased levels of PSCA expression and higher Gleason grades, indicative of more an advanced tumor stage and poor overall survival (Li et al., 2017; Reiter et al., 1998). PSCA expression is also associated with seminal vesicle invasion and capsular invasion in PCa; these findings suggest that PSCA has potential as a therapeutic target. In addition, elevated PSCA could be a prognostic indicator for BCR after radical prostatectomy (RP) in patients with clinically localized prostatic cancer (Kim et al., 2017). Using human xenografts grown in SCID mice, researchers found that anti-PSCA monoclonal antibodies inhibited prostate tumor growth and metastasis formation (Ross et al., 2002), providing evidence for PSCA immunotherapy targeting as a potential treatment for PCa. Although PSCA displayed inferior sensitivity and considerable inability to distinguish between malignant and benign disease when juxtaposed against PSA and PSMA, the circulating prostate biomarker possesses the highest disease specificity and independent predictive value (Hara et al., 2002). Despite the research revelations about PSCA and its potential, no definitive conclusions have been reached in regard to it being a serum biomarker. Factors that militate against PSCA as a candidate for further development include an inadequate number of published studies supporting PSCA as a valuable clinical biomarker and the lack of better measuring techniques. Hence, the value of PSCA as a therapeutic target and its effectiveness as a clinical PCa marker must await more supportive evidence.

Early Prostate Cancer Antigen

Early PCa antigen (EPCA) is a nuclear matrix protein linked with nuclear transformations that occur in early PCa development. EPCA was found in PCa precursor lesions, specifically in prostatic intraepithelial neoplasia and proliferative inflammatory atrophy, as well as in PCa tissue. Also, the protein has been identified in men with a preliminary negative biopsy, but who later developed cancer (Dhir et al., 2004). EPCA cannot be linked with early carcinogenesis as no relationship could be found between EPCA and the disease stage or Gleason score after RP (Uetsuki et al., 2005). However, EPCA-2 could differentiate BPH from cancer in PCa suspects (Pourmand et al., 2016). Furthermore, studies have verified the potential diagnostic value of serum EPCA by demonstrating the ability of EPCA antibodies to recognize PCa. Although EPCA appears not to be present in patients devoid of PCa, it has been detected in surroundings free of, but adjacent to, the cancer (Dhir et al., 2004). More studies are needed to further characterize the protein as a suitable biomarker to diagnose PCa.

B7-H3

B7-H3 is the first immune molecule that possibly participates in the development of PCa and in predicting the recurrence and progression of cancer. B7-H3 is a member of the B7 family, a group of proteins that are important ligands interacting with known and unknown receptors to regulate the activation and function of T lymphocytes. The B7-H family proteins, including B7-H3, can both arrest cancer growth and shield cancers from the immune system by paralyzing immune cells, thereby exhibiting both an immune stimulatory and inhibitory role in cancer growth (Luo et al., 2004). More research is warranted, however, to understand how the immune system is affected by B7-H3. This information is critical and will help to establish the effectiveness of B7-H3 as a clinical biomarker of disease and target for therapy. In contrast to PSA, B7-H3 remains bound to the surface of normal prostate cells and the surfaces of premalignant and cancerous prostate cells that show no apparent indication of migration (metastatic ability); this attribute, in particular, makes B7-H3 an attractive therapeutic target and biomarker (Roth et al., 2007). One study revealed a link between a rising level of B7-H3 in PCa and adverse pathologic features of the disease. In a large cohort, B7-H3 expression is presented in most of the localized PCa cases. Increased levels of B7-H3 were associated with tumor aggressiveness and extent (Benzon et al., 2017). If the connection holds, B7-H3 may have the potential to independently predict PCa progression, making it useful as a diagnostic and prognostic biomarker to evaluate disease status and immunotherapeutic responses.

Sarcosine

Sarcosine, a N-methyl derivative of glycine, is a natural amino acid found in muscles and other body tissues. It is classified under the group collectively known as metabolites. Sarcosine was found to stimulate malignant growth of benign PCa cells has been suggested as a potential indicator of malignancy of PCa cells when detected in the urine (Sreekumar et al., 2009). Following screening of urine, blood, and tissues and profiling more than 1126 metabolites related to PCa, researchers were able to differentiate between benign prostate, clinically localized PCa, and metastatic disease based on the levels of sarcosine. Higher level of sarcosine were observed in invasive PCa cell lines compared with benign prostate epithelial cells (Sreekumar et al., 2009). Furthermore, it was reported that PCa invasion weakened when glycine-N-methyltransferase, the enzyme that catalyzes the production of sarcosine from glycine, was knocked down. In contrast, knocking down the enzyme responsible for sarcosine degradation or adding exogenous sarcosine stimulated an invasive phenotype in BPH cells. Together, these results suggest that sarcosine may be a vital metabolic intermediary propelling PCa cells toward invasion and aggressiveness (Couzin, 2009). The ultimate goal of diagnosis is to detect aggressive-type PCas at their premature stage. At the center of the contradicting scientific viewpoint regarding whether sarcosine is a better diagnostic biomarker or not than PSA at detecting aggressive PCa is the concern raised by the fact that several scientists, with personal investment interests in sarcosine use, have criticized for their possible motives. Another study reported that sarcosine in urine samples of patients with PCa was not a significant biomarker for cancer diagnosis (Gkotsos et al., 2017). Another potential role for sarcosine would be to study its usefulness in the diagnosis of PCa. If proven successful with further human use, this radiotracer containing c11-sarcosine might exemplify a new concept for PET imaging of cancer (Piert et al., 2017). Subsequently, further investigations must be performed on the metabolites with patient outcomes monitored long term to truly substantiate any correlation of sarcosine with patients who developed different forms of PCa.

Caveolin-1

Caveolin-1 (Cav-1), an integral membrane protein expressed in two isoforms (Cav-1α and Cav-1β), is a major component of caveolae membranes in vivo. It has been implicated in regulating several signaling pathways and mediating intracellular processes, specifically as a negative regulator in several mitogenic pathways and in oncogenesis. Cav-1 seems to function as a tumor suppressor protein at early stages of cancer progression. Studies of prostate tissue from men with only localized PCa indicate a significant decrease in levels of Cav-1. It was also discovered that the protein was absent in tumor tissue from men with metastatic PCa, and the reduced levels of Cav-1 were associated with a high Gleason score (Thompson et al., 2010). However, Cav-1 is also found to be upregulated in several metastatic and multidrug-resistant cancer cell lines and in some human tumor specimens (Shatz and Liscovitch, 2008). Cav-1 is secreted by PCa cells. Other studies have shown that this secreted protein can promote cell survival and angiogenic activities. Cav-1 has been reported to overexpress in PCa cells and is associated with the progression of the disease. Research conducted on stromal Cav-1 expression in patients with BPH, primary PCas, and PCa metastases revealed that almost all BPH samples showed an abundant stromal Cav-1 immunostaining, while a subset of samples with primary PCa had significantly decreased levels of stromal Cav-1. All metastatic tumors (either from lymph node or bone) lacked stromal Cav-1 staining (Di Vizio et al., 2009). The concentration of preoperative serum Cav-1 showed prognostic potential in patients undergoing radical prostatectomy (Thompson et al., 2010). Recent clinical studies have shown the possible role of Caveolin-1 as a biomarker for cancer prognosis during RT (Mahmood et al., 2016). Also Cav-1 could be a useful tool in management for patients in active surveillance or watchful waiting (Basourakos et al., 2018; Hammarsten et al., 2016). These findings suggest Cav-1 expression may be a useful prognostic marker for PCa.

Serum Calcium

PCa cells express calcium-sensing, G-protein-coupled receptors, which can be activated by extracellular calcium. These cells also express calcium-dependent potassium channels that regulate their proliferation by controlling the entry of calcium into the cells. High levels of total calcium in serum have been linked with greater risk of fatal PCa through growth and metastasis (Lallet-Daher et al., 2009). Independent investigations by separate groups confirmed the association between increased serum calcium and terminal PCa, suggesting that serum calcium is a promising prospective biomarker for the screening of fatal PCa (Schwartz, 2009).

Hypermethylation of PDZ, LIM Domain Protein 4 Gene, and PDLIM5

Hypermethylation of the PDLIM4 gene has been shown to be a sensitive molecular tool in detecting prostate tumorigenesis (Vanaja et al., 2006). PDLIM4 mRNA and protein levels were found to decrease in various PCa cell lines. PDLIM4 may function as a tumor suppressor by associating with actin in PCa cells, thereby controlling cell proliferation (Vanaja et al., 2009). PDLIM5 (PDZ and LIM domain 5) is a cytoplasmatic protein, which may be involved in oncogenesis of many types of cancer such as the lungs, thyroid, breast, and gastric (Edlund et al., 2012; Eeckhoute et al., 2006; Heiliger et al., 2012; Li et al., 2015b). Recent study showed that PDLIM5 was upregulated in PCa tissues (compared with normal prostate tissue). PDLIM5 could be an oncogene in PCa cells associated with progression, metastasis, and recurrence of cancer (Liu et al., 2017). These findings support the potential use of hypermethylated PDLIM4 and PDLIM5 as a biomarker of PCa.

Exosomes

Exosomes are nanometer-sized vesicles secreted by a broad range of normal and neoplastic cell types. They contain both functional mRNA and microRNA, called exosomal shuttle RNA, which are often transported from cell to cell where they can continue to be functional (Valadi et al., 2007). Exosomes are constituents of urine, with a degree of variability in urine specimens. Because they often carry genetic components that come directly from tumors, such vesicles may be a useful noninvasive source of markers in screening for renal diseases and PCa. A recent study reported the presence of PCA3 and TMPRSS2-ERG fusion, two known PCa biomarkers, in exosomes from urine samples of PCa patients (Nilsson et al., 2009), shedding more light on its clinical use. The presence and quantification of exosomes as a potential noninvasive source of tumor markers could be used to diagnose, monitor the status of PCa concerning drug resistance and metastasis. It may also serve as a target in cancer therapy (Pan et al., 2017).

Ki-67

Ki-67, a cell proliferation-associated marker (and probably the only one with an expression pattern under a level of cell cycle regulation) (Scholzen and Gerdes, 2000), has been described as one of the most promising biomarkers of PCa. Ki-67 has been suggested as a prolific predictive biomarker, for men who have low-grade, low-stage PCa and PSA relapse after radical -prostatectomy (Khatami et al., 2009). In a 6-year study, involving 808 patients diagnosed with PCa, an immunohistochemical assessment of Ki-67 expression was evaluated for its relationship to the specificity of the cancer and overall survival. Compared to information from the Gleason score and PSA, Ki-67 provided additional prognostic information (Khatami et al., 2009). In another study of a group of men treated with radiotherapy and androgen deprivation for PCa, Ki-67 expression levels in conjunction with MDM2 were found to be correlated to distant metastasis and survivability (Khor et al., 2009). A recent study enrolling 535 patients showed that Ki-67 expression was an independent biomarker for high Gleason grade, larger tumor size, and biochemical failure (Richardsen et al., 2017). Also a recent metaanalysis showed that low Ki-67 is associated with improved overall and disease-specific survivor. Furthermore, the review suggests a decreased risk of developing metastasis (Berlin et al., 2017). The results of this study mount evidence on Ki-67 robustness and clinical validity, warranting prospective validation accounting for well-established clinical prognostic factors Nevertheless, further studies will be needed to validate these results and explore the possibility of combining Ki-67 with existing prognostic tools as a powerful biomarker for localized PCa.

Golgi Phosphoprotein 2 and 3

Golgi phosphoprotein 2, GOLPH2, is a gene that codes for type II Golgi membrane antigen GOLPH2/GP73. It is usually expressed in various epithelial cells and is reported to be frequently overexpressed in PCa, although its function is currently unknown. Golgi phosphoprotein 3 (GOLPH3/GPP34) is also a membrane protein and the presence of the GOLPH3 gene was found in several human cancers, including prostate (Hua et al., 2012). The overexpression of GOLPH3 has been associated with cancer's aggressiveness and proliferation of PCa in androgen-independent cell lines (Dai et al., 2015; Li et al., 2015a). A study has observed a higher level of GOLPH2 in the Golgi apparatus in PCa cells compared with that of normal cells, thereby indicating that GOLPH2 can be used as a biomarker in distinguishing between normal cells and cancer cells (Wei et al., 2008). Another study revealed that an increase in the levels of GOLPH2 (along with some of the other biomarkers assayed) was not only a critical but also a better indicator for PCa than PSA or PCA3 alone (Laxman et al., 2008). A comparative study of GOLPH2 and AMACR in benign and malignant prostate lesions revealed that GOLPH2 expression was considerably higher in PCa tissues compared with normal tissues, and in about 90% of the cases studied, GOLPH2 protein was upregulated. Furthermore, this upregulation was noticed in about 85% of PCa cases that were negative for AMACR (Kristiansen et al., 2008) Also a recent study examined GOLPH2 as a potential target for cancer therapy. Particularly GOLPH2-regulated oncolytic adenovirus GD55 was showed to have antitumor effects on PCa cells (Ying et al., 2017). Another recent study about GOLPH3 expression in PCa correlated with the Gleason score and poor overall survival (Abd El-Maqsoud et al., 2016). Another study showed that patients with GOLPH3 expression presented shorter overall survival and biochemical recurrence-free survival. In consequence, GOLPH3 may be a prognostic biomarker for patients who have undergone radical prostatectomy (Zhang et al., 2015). These findings show that both GOLPH3 and GOLPH2 could be promising targets for new therapeutic strategies for the treatment of androgen-dependent PCa and castration-resistant PCa and as potential diagnostic–prognostic tools.

DAB2 Interacting Protein

DAB2-interacting protein (DAB2IP) is a Ras GTPase-activating protein that functions as a tumor suppressor. The human DAB2IP gene is frequently observed to be downregulated in PCa cell lines (Chen et al., 2003). Studies have shown that loss of expression of DAB2IP may be a result of altered epigenetic regulations such as DNA methylation and histone modification (Chen et al., 2005). The abnormal methylation in the promoter region of the DAB2IP gene has been reported to be responsible for transcriptional silencing and consequently performing a significant function in the progression of PCa (Chen et al., 2003). Also a loss of DAB2IP expression in CRPC cells was associated with chemoresistance (Wu et al., 2013). A link between a genetic variation in DAB2IP and the risk of aggressive PCa was reported (Duggan et al., 2007). This research indicates that DAB2IP protein, after further studies, can potentially be used as a very effective novel biomarker for PCa diagnosis.

Other Emerging Prostate Cancer Biomarkers

A recent study demonstrated that CCL11 (Eotaxin-1), a new diagnostic serum biomarker, helps distinguish between prostatic enlargement and PCa among men demonstrating low, but detectable, serum PSA values (Agarwal et al., 2013). The use of multigene signature-based prognosis, such as the triproliferation markers of Ki-67, TOP2A (DNA topoisomerase II, alpha), and E2F1, has shown potential to improve the prediction and treatment stratification of PCa (Malhotra et al., 2011). Another study used a 7-gene microarray panel of blood samples and demonstrated their value as potential blood-based biomarkers for a minimally invasive test for PCa aggressiveness (Liong et al., 2012). Other identified novel predictors of PCa include the ratio of serum testosterone to PSA (Gurbuz et al., 2012), microRNA expression profile (Yin et al., 2010), epigenetic biomarkers APC and RASSF1 (Van Neste et al., 2012), an epigenetic assay detecting methylation of GTSP1, RASSF1, and APC (SelectMDx) (Partin et al., 2016), prognostic tissue biomarkers Bcl-2 and p53, and blood- and urine-based biomarkers, including human glandular kallikrein 2 (hK2), TGF-β1, endoglin, IL-6, urokinase plasminogen activator (uPA), and its receptor (uPAR) (Shariat et al., 2011). In addition, the urine RNA marker SPINK1 was shown to serve as a promising alternative or addition to serum-based biomarkers (Roobol et al., 2011). Other promising biomarkers under investigation include ANPEP, ABL1, EFNA1, HSPB1, INMT, and TRIP13 (Larkin et al., 2012).

CONCLUDING REMARKS AND FUTURE DIRECTIONS

PCa is currently the most frequently diagnosed cancer in males and constitutes a major health issue in developed countries. On the other hand, the majority of PCa cases are not considered clinically significant and certainly not lethal. Thus, it is particularly important to balance the risks of overtreatment for indolent disease with undertreatment for a potentially aggressive tumor. The discrepancies in the case of PCa emphasize the need to identify reliable biomarkers for the early detection of the disease that have aggressive features and for early and radical intervention. The current clinical use of PSA toward this end is obviously not adequate because PSA is prostate-specific, but not a PCa-specific marker, as it is known to increase in other prostate diseases such as BPH and inflammation.

A critical point that has been reiterated is the fact that an ideal biomarker has to show a high level of specificity and sensitivity to prevent false-positive screening tests, which will create anxiety in patients and lead to more expensive and invasive testing. Thus far, studies, although inconclusive, have found that the likelihood of identifying a biomarker with such sensitivity and specificity remains low. Therefore, combining markers is thought to be the next best thing to improve the accuracy of diagnosing, treating, and surveillance of recurrence.

The technology used to discover potential biomarkers has advanced significantly in recent years. Recent gene expression analyses have revealed several promising biomarkers in the areas of epigenetics, genomics, and the transcriptome, some of which are currently under investigation as clinical tests for early detection and better prognostic prediction of PCa. Despite several biomarkers having displayed some potential in early phase studies, few so far appear likely to possess the appropriate level of sensitivity and specificity required to reliably determine the choice and course of therapeutic treatment for PCa. This may explain why only a small number of biomarkers are routinely validated for use in drug development or qualified for clinical applications, despite the apparent progress in this research field. Several limitations still exist with the current technology that hinders the discovery and development of new biomarkers for all forms of cancer, including PCa. Some of these impediments may be overcome through the development of new technologies and improved strategies. For example, one strategy proposed would pair the diagnostic test with the therapeutic agent. Another strategy calls for more attention to studies that can generate quantified biomarkers related to cell signaling pathways, as these biomarkers can be applied across a wide range of tumor types and diseases, as well as in different tests and drugs.

In conclusion, substantial discoveries still lie before us in this field, and methodologies for the clinical evaluation of existing and novel biomarkers need to be further explored. Although much could be gained from the discovery of novel biomarkers for early detection of PCa, prediction of the malignant potential of the disease, and guidance of individualized therapy for patients, the near future of PCa prognosis may eventually come to count on a few "elite club" biomarkers, which hopefully will accurately predict the incidence, stage, and progression of the disease, as well as reliably evaluate drug response and related toxicity. From a safety standpoint, biomarkers that can be measured and monitored for the toxicity are particularly useful in terms of minimizing toxicity of cancer drugs and therapy. Although extensive clinical validation of these novel biomarkers remains one of the most significant and daunting challenges, overcoming this impediment will by no means eliminate all the problems hampering the identification and development of biomarkers for this disease. However, in the process of searching for novel biomarkers, particularly those that help analyze risk stratification and predict toxicity of intervention, we may discover valuable insights into the mechanisms of PCa and its response to therapy that could in the long run lead to a cure of this disease.

References

Abd El-Maqsoud, N.M., Osman, N.A., Abd El-Hamid, A.M., et al., 2016. Golgi phosphoprotein-3 and Y-box-binding protein-1 are novel markers correlating with poor prognosis in prostate cancer. Clin. Genitourin. Cancer 14, e143–152.

Agarwal, M., He, C., Siddiqui, J., et al., 2013. CCL11 (eotaxin-1): a new diagnostic serum marker for prostate cancer. Prostate 73, 573–581.

Albright, F., Stephenson, R.A., Agarwal, N., et al., 2015. Prostate cancer risk prediction based on complete prostate cancer family history. Prostate 75, 390–398.

Ashley, V.A., Joseph, M.B., Kamlesh, K.Y., et al., 2017. The use of biomarkers in prostate cancer screening and treatment. Rev. Urol. 19, 221–234.

Auvinen, A., Moss, S.M., Tammela, T.L., et al., 2016. Absolute effect of prostate cancer screening: balance of benefits and harms by center within the European randomized study of prostate cancer screening. Clin. Cancer Res. 22, 243–249.

Bancroft, E.K., Page, E.C., Castro, E., et al., 2014. Targeted prostate cancer screening in BRCA1 and BRCA2 mutation carriers: results from the initial screening round of the IMPACT study. Eur. Urol. 66, 489–499.

Basourakos, S.P., Davis, J.W., Chapin, B.F., et al., 2018. Baseline and longitudinal plasma caveolin-1 level as a biomarker in active surveillance for early-stage prostate cancer. BJU Int. 121, 69–76.

Bell, K.J., Del Mar, C., Wright, G., et al., 2015. Prevalence of incidental prostate cancer: a systematic review of autopsy studies. Int. J. Cancer 137, 1749–1757.

Benzon, B., Zhao, S.G., Haffner, M.C., et al., 2017. Correlation of B7-H3 with androgen receptor, immune pathways and poor outcome in prostate cancer: an expression-based analysis. Prostate Cancer Prostatic Dis. 20, 28–35.

Berlin, A., Castro-Mesta, J.F., Rodriguez-Romo, L., et al., 2017. Prognostic role of Ki-67 score in localized prostate cancer: a systematic review and meta-analysis. Urol. Oncol. 35, 499–506.

Berruti, A., Mosca, A., Tucci, M., et al., 2005. Independent prognostic role of circulating chromogranin A in prostate cancer patients with hormone-refractory disease. Endocr. Relat. Cancer 12, 109–117.

Bibbins-Domingo, K., Grossman, D.C., Curry, S.J., 2017. The US preventive services task force 2017 draft recommendation statement on screening for prostate cancer: an invitation to review and comment. J. Am. Med. Assoc. 317, 1949–1950.

Boegemann, M., Stephan, C., Cammann, H., et al., 2016. The percentage of prostate-specific antigen (PSA) isoform [-2]proPSA and the Prostate Health Index improve the diagnostic accuracy for clinically relevant prostate cancer at initial and repeat biopsy compared with total PSA and percentage free PSA in men aged </=65 years. BJU Int. 117, 72–79.

Booth, N., Rissanen, P., Tammela, T.L., et al., 2014. Health-related quality of life in the Finnish trial of screening for prostate cancer. Eur. Urol. 65, 39–47.

Bryant, R.J., Sjoberg, D.D., Vickers, A.J., et al., 2015. Predicting high-grade cancer at ten-core prostate biopsy using four kallikrein markers measured in blood in the ProtecT study. J. Natl. Cancer Inst. 107.

Burgio, S.L., Conteduca, V., Menna, C., et al., 2014. Chromogranin A predicts outcome in prostate cancer patients treated with abiraterone. Endocr. Relat. Cancer 21, 487–493.

Carlsson, S., Assel, M., Sjoberg, D., et al., 2014. Influence of blood prostate specific antigen levels at age 60 on benefits and harms of prostate cancer screening: population based cohort study. BMJ 348, g2296.

Carnell, A.J., Kirk, R., Smith, M., et al., 2013. Inhibition of human alpha-methylacyl CoA racemase (AMACR): a target for prostate cancer. ChemMedChem 8, 1643–1647.

Catalona, W.J., Richie, J.P., Ahmann, F.R., et al., 2017. Comparison of digital rectal examination and serum prostate specific antigen in the early detection of prostate cancer: results of a multicenter clinical trial of 6,630 men. J. Urol. 197, S200–S207.

Chen, H., Toyooka, S., Gazdar, A.F., et al., 2003. Epigenetic regulation of a novel tumor suppressor gene (hDAB2IP) in prostate cancer cell lines. J. Biol. Chem. 278, 3121–3130.

Chen, H., Tu, S.W., Hsieh, J.T., 2005. Down-regulation of human DAB2IP gene expression mediated by polycomb Ezh2 complex and histone deacetylase in prostate cancer. J. Biol. Chem. 280, 22437–22444.

Conteduca, V., Burgio, S.L., Menna, C., et al., 2014. Chromogranin A is a potential prognostic marker in prostate cancer patients treated with enzalutamide. Prostate 74, 1691–1696.

Couzin, J., 2009. Biomarkers. Metabolite in urine may point to high-risk prostate cancer. Science 323, 865.

Dai, T., Zhang, D., Cai, M., et al., 2015. Golgi phosphoprotein 3 (GOLPH3) promotes hepatocellular carcinoma cell aggressiveness by activating the NF-kappaB pathway. J. Pathol. 235, 490–501.

Dhir, R., Vietmeier, B., Arlotti, J., et al., 2004. Early identification of individuals with prostate cancer in negative biopsies. J. Urol. 171, 1419–1423.

Di Vizio, D., Morello, M., Sotgia, F., et al., 2009. An absence of stromal caveolin-1 is associated with advanced prostate cancer, metastatic disease and epithelial Akt activation. Cell Cycle 8, 2420–2424.

Dong, F., Kattan, M.W., Steyerberg, E.W., et al., 2008. Validation of pretreatment nomograms for predicting indolent prostate cancer: efficacy in contemporary urological practice. J. Urol. 180, 150–154 discussion 154.

Duggan, D., Zheng, S.L., Knowlton, M., et al., 2007. Two genome-wide association studies of aggressive prostate cancer implicate putative prostate tumor suppressor gene DAB2IP. J. Natl. Cancer Inst. 99, 1836–1844.

Dumache, R., Puiu, M., Motoc, M., et al., 2014. Prostate cancer molecular detection in plasma samples by glutathione S-transferase P1 (GSTP1) methylation analysis. Clin. Lab. 60, 847–852.

Edlund, K., Lindskog, C., Saito, A., et al., 2012. CD99 is a novel prognostic stromal marker in non-small cell lung cancer. Int. J. Cancer 131, 2264–2273.

Eeckhoute, J., Carroll, J.S., Geistlinger, T.R., et al., 2006. A cell-type-specific transcriptional network required for estrogen regulation of cyclin D1 and cell cycle progression in breast cancer. Genes Dev. 20, 2513–2526.

Ferlay, J., Soerjomataram, I., Dikshit, R., et al., 2015. Cancer incidence and mortality worldwide: sources, methods and major patterns in GLOBOCAN 2012. Int. J. Cancer 136, E359–E386.

Gkotsos, G., Virgiliou, C., Lagoudaki, I., et al., 2017. The role of sarcosine, uracil, and kynurenic acid metabolism in urine for diagnosis and progression monitoring of prostate cancer. Metabolites 7.

Gonzalgo, M.L., Pavlovich, C.P., Lee, S.M., et al., 2003. Prostate cancer detection by GSTP1 methylation analysis of postbiopsy urine specimens. Clin. Cancer Res. 9, 2673–2677.

Gulati, R., Cheng, H.H., Lange, P.H., et al., 2017. Screening men at increased risk for prostate cancer diagnosis: model estimates of benefits and harms. Cancer Epidemiol. Biomark. Prev. 26, 222–227.

Gurbuz, C., Canat, L., Atis, G., et al., 2012. The role of serum testosterone to prostate-specific antigen ratio as a predictor of prostate cancer risk. Kaohsiung J. Med. Sci. 28, 649–653.

Haas, G.P., Delongchamps, N., Brawley, O.W., et al., 2008. The worldwide epidemiology of prostate cancer: perspectives from autopsy studies. Can. J. Urol. 15, 3866–3871.

Hammarsten, P., Dahl Scherdin, T., Hagglof, C., et al., 2016. High caveolin-1 expression in tumor stroma is associated with a favourable outcome in prostate cancer patients managed by watchful waiting. PLoS One 11, e0164016.

Hara, N., Kasahara, T., Kawasaki, T., et al., 2002. Reverse transcription-polymerase chain reaction detection of prostate-specific antigen, prostate-specific membrane antigen, and prostate stem cell antigen in one milliliter of peripheral blood: value for the staging of prostate cancer. Clin. Cancer Res. 8, 1794–1799.

Hayes, J.H., Barry, M.J., 2014. Screening for prostate cancer with the prostate-specific antigen test: a review of current evidence. J. Am. Med. Assoc. 311, 1143–1149.

Heijnsdijk, E.A., Wever, E.M., Auvinen, A., et al., 2012. Quality-of-life effects of prostate-specific antigen screening. N. Engl. J. Med. 367, 595–605.

Heiliger, K.J., Hess, J., Vitagliano, D., et al., 2012. Novel candidate genes of thyroid tumourigenesis identified in Trk-T1 transgenic mice. Endocr. Relat. Cancer 19, 409–421.

Hua, X., Yu, L., Pan, W., et al., 2012. Increased expression of Golgi phosphoprotein-3 is associated with tumor aggressiveness and poor prognosis of prostate cancer. Diagn. Pathol. 7, 127.

Kamangar, F., Dores, G.M., Anderson, W.F., 2006. Patterns of cancer incidence, mortality, and prevalence across five continents: defining priorities to reduce cancer disparities in different geographic regions of the world. J. Clin. Oncol. 24, 2137–2150.

Khatami, A., Hugosson, J., Wang, W., et al., 2009. Ki-67 in screen-detected, low-grade, low-stage prostate cancer, relation to prostate-specific antigen doubling time, Gleason score and prostate-specific antigen relapse after radical prostatectomy. Scand. J. Urol. Nephrol. 43, 12–18.

Khor, L.Y., Bae, K., Paulus, R., et al., 2009. MDM2 and Ki-67 predict for distant metastasis and mortality in men treated with radiotherapy and androgen deprivation for prostate cancer: RTOG 92-02. J. Clin. Oncol. 27, 3177–3184.

Kim, S.H., Park, W.S., Park, B.R., et al., 2017. PSCA, Cox-2, and Ki-67 are independent, predictive markers of biochemical recurrence in clinically localized prostate cancer: a retrospective study. Asian J. Androl. 19, 458–462.

Knezevic, D., Goddard, A.D., Natraj, N., et al., 2013. Analytical validation of the oncotype DX prostate cancer assay - a clinical RT-PCR assay optimized for prostate needle biopsies. BMC Genom. 14, 690.

Kristiansen, G., Fritzsche, F.R., Wassermann, K., et al., 2008. GOLPH2 protein expression as a novel tissue biomarker for prostate cancer: implications for tissue-based diagnostics. Br. J. Cancer 99, 939–948.

Lallet-Daher, H., Roudbaraki, M., Bavencoffe, A., et al., 2009. Intermediate-conductance Ca^{2+}-activated K^+ channels (IKCa1) regulate human prostate cancer cell proliferation through a close control of calcium entry. Oncogene 28, 1792–1806.

Larkin, S.E., Holmes, S., Cree, I.A., et al., 2012. Identification of markers of prostate cancer progression using candidate gene expression. Br. J. Cancer 106, 157–165.

Laxman, B., Morris, D.S., Yu, J., et al., 2008. A first-generation multiplex biomarker analysis of urine for the early detection of prostate cancer. Cancer Res. 68, 645–649.

Lee, W.H., Morton, R.A., Epstein, J.I., et al., 1994. Cytidine methylation of regulatory sequences near the pi-class glutathione S-transferase gene accompanies human prostatic carcinogenesis. Proc. Natl. Acad. Sci. U. S. A. 91, 11733–11737.

Leyten, G.H., Hessels, D., Jannink, S.A., et al., 2014. Prospective multicentre evaluation of PCA3 and TMPRSS2-ERG gene fusions as diagnostic and prognostic urinary biomarkers for prostate cancer. Eur. Urol. 65, 534–542.

Li, E., Liu, L., Li, F., et al., 2017. PSCA promotes prostate cancer proliferation and cell-cycle progression by up-regulating c-Myc. Prostate 77, 1563–1572.

Li, W., Guo, F., Gu, M., et al., 2015a. Increased expression of GOLPH3 is associated with the proliferation of prostate cancer. J. Cancer 6, 420–429.

Li, Y., Gao, Y., Xu, Y., et al., 2015b. si-RNA-mediated knockdown of PDLIM5 suppresses gastric cancer cell proliferation in vitro. Chem. Biol. Drug Des. 85, 447–453.

Liong, M.L., Lim, C.R., Yang, H., et al., 2012. Blood-based biomarkers of aggressive prostate cancer. PLoS One 7, e45802.

Liu, X., Chen, L., Huang, H., et al., 2017. High expression of PDLIM5 facilitates cell tumorigenesis and migration by maintaining AMPK activation in prostate cancer. Oncotarget 8, 98117–98134.

Luo, L., Chapoval, A.I., Flies, D.B., et al., 2004. B7-H3 enhances tumor immunity in vivo by costimulating rapid clonal expansion of antigen-specific CD8+ cytolytic T cells. J. Immunol. 173, 5445–5450.

Mahmood, J., Zaveri, S.R., Murti, S.C., et al., 2016. Caveolin-1: a novel prognostic biomarker of radioresistance in cancer. Int. J. Radiat. Biol. 92, 747–753.

Malhotra, S., Lapointe, J., Salari, K., et al., 2011. A tri-marker proliferation index predicts biochemical recurrence after surgery for prostate cancer. PLoS One 6, e20293.

Mosquera, J.M., Mehra, R., Regan, M.M., et al., 2009. Prevalence of TMPRSS2-ERG fusion prostate cancer among men undergoing prostate biopsy in the United States. Clin. Cancer Res. 15, 4706–4711.

Mottet, N., Bellmunt, J., Bolla, M., et al., 2017. EAU-ESTRO-SIOG guidelines on prostate cancer. Part 1: screening, diagnosis, and local treatment with curative intent. Eur. Urol. 71, 618–629.

Muthumani, K., Marnin, L., Kudchodkar, S.B., et al., 2017. Novel prostate cancer immunotherapy with a DNA-encoded anti-prostate-specific membrane antigen monoclonal antibody. Cancer Immunol. Immunother. 66, 1577–1588.

Mwamukonda, K., Chen, Y., Ravindranath, L., et al., 2010. Quantitative expression of TMPRSS2 transcript in prostate tumor cells reflects TMPRSS2-ERG fusion status. Prostate Cancer Prostatic Dis. 13, 47–51.

Nicholson, A., Mahon, J., Boland, A., et al., 2015. The clinical effectiveness and cost-effectiveness of the PROGENSA(R) prostate cancer antigen 3 assay and the Prostate Health Index in the diagnosis of prostate cancer: a systematic review and economic evaluation. Health Technol. Assess. 19 i–xxxi, 1–191.

Nilsson, J., Skog, J., Nordstrand, A., et al., 2009. Prostate cancer-derived urine exosomes: a novel approach to biomarkers for prostate cancer. Br. J. Cancer 100, 1603–1607.

Pan, J., Ding, M., Xu, K., et al., 2017. Exosomes in diagnosis and therapy of prostate cancer. Oncotarget 8, 97693–97700.

Partin, A.W., Van Neste, L., Klein, E.A., et al., 2014. Clinical validation of an epigenetic assay to predict negative histopathological results in repeat prostate biopsies. J. Urol. 192, 1081–1087.

Partin, A.W., Criekinge, W.V., Trock, B.J., et al., 2016. Clinical evaluation of an epigenetic assay to predict missed cancer in prostate biopsy specimens. Trans. Am. Clin. Climatol. Assoc. 127, 313–327.

Piert, M., Shao, X., Raffel, D., et al., 2017. Preclinical evaluation of (11)C-Sarcosine as a substrate of proton-coupled amino acid transporters and first human application in prostate cancer. J. Nucl. Med. 58, 1216–1223.

Pinsky, P.F., Prorok, P.C., Yu, K., et al., 2017. Extended mortality results for prostate cancer screening in the PLCO trial with median follow-up of 15 years. Cancer 123, 592–599.

Pourmand, G., Safavi, M., Ahmadi, A., et al., 2016. EPCA2.22: a silver lining for early diagnosis of prostate cancer. Urol. J. 13, 2845–2848.

Reiter, R.E., Gu, Z., Watabe, T., et al., 1998. Prostate stem cell antigen: a cell surface marker overexpressed in prostate cancer. Proc. Natl. Acad. Sci. U. S. A. 95, 1735–1740.

Richardsen, E., Andersen, S., Al-Saad, S., et al., 2017. Evaluation of the proliferation marker Ki-67 in a large prostatectomy cohort. PLoS One 12, e0186852.

Rogers, C.G., Yan, G., Zha, S., et al., 2004. Prostate cancer detection on urinalysis for alpha methylacyl coenzyme a racemase protein. J. Urol. 172, 1501–1503.

Roobol, M.J., Haese, A., Bjartell, A., 2011. Tumour markers in prostate cancer III: biomarkers in urine. Acta Oncol. 50 (Suppl. 1), 85–89.

Ross, S., Spencer, S.D., Holcomb, I., et al., 2002. Prostate stem cell antigen as therapy target: tissue expression and in vivo efficacy of an immunoconjugate. Cancer Res. 62, 2546–2553.

Roth, T.J., Sheinin, Y., Lohse, C.M., et al., 2007. B7-H3 ligand expression by prostate cancer: a novel marker of prognosis and potential target for therapy. Cancer Res. 67, 7893–7900.

Rubin, M.A., Zhou, M., Dhanasekaran, S.M., et al., 2002. alpha-Methylacyl coenzyme A racemase as a tissue biomarker for prostate cancer. J. Am. Med. Assoc. 287, 1662–1670.

Saini, S., 2016. PSA and beyond: alternative prostate cancer biomarkers. Cell. Oncol. 39, 97–106.

Scholzen, T., Gerdes, J., 2000. The Ki-67 protein: from the known and the unknown. J. Cell. Physiol. 182, 311–322.

Schroder, F.H., Hugosson, J., Roobol, M.J., et al., 2014. Screening and prostate cancer mortality: results of the European randomised study of screening for prostate cancer (ERSPC) at 13 years of follow-up. Lancet 384, 2027–2035.

Schwartz, G.G., 2009. Is serum calcium a biomarker of fatal prostate cancer? Future Oncol. 5, 577–580.

Shariat, S.F., Scherr, D.S., Gupta, A., et al., 2011. Emerging biomarkers for prostate cancer diagnosis, staging, and prognosis. Arch. Esp. Urol. 64, 681–694.

Shatz, M., Liscovitch, M., 2008. Caveolin-1: a tumor-promoting role in human cancer. Int. J. Radiat. Biol. 84, 177–189.

Shore, N., Concepcion, R., Saltzstein, D., et al., 2014. Clinical utility of a biopsy-based cell cycle gene expression assay in localized prostate cancer. Curr. Med. Res. Opin. 30, 547–553.

Sreekumar, A., Poisson, L.M., Rajendiran, T.M., et al., 2009. Metabolomic profiles delineate potential role for sarcosine in prostate cancer progression. Nature 457, 910–914.

Stamey, T.A., Yang, N., Hay, A.R., et al., 1987. Prostate-specific antigen as a serum marker for adenocarcinoma of the prostate. N. Engl. J. Med. 317, 909–916.

Stone, L., 2017. Prostate cancer: stuck in the middle: interstitial genes in TMPRSS2-ERG fusion. Nat. Rev. Urol. 15 (1).

Thompson, T.C., Tahir, S.A., Li, L., et al., 2010. The role of caveolin-1 in prostate cancer: clinical implications. Prostate Cancer Prostatic Dis. 13, 6–11.

Udovicich, C., Perera, M., Hofman, M.S., et al., 2017. (68)Ga-prostate-specific membrane antigen-positron emission tomography/computed tomography in advanced prostate cancer: current state and future trends. Prostate Int. 5, 125–129.

Uetsuki, H., Tsunemori, H., Taoka, R., et al., 2005. Expression of a novel biomarker, EPCA, in adenocarcinomas and precancerous lesions in the prostate. J. Urol. 174, 514–518.

Valadi, H., Ekstrom, K., Bossios, A., et al., 2007. Exosome-mediated transfer of mRNAs and microRNAs is a novel mechanism of genetic exchange between cells. Nat. Cell Biol. 9, 654–659.

Van Neste, L., Bigley, J., Toll, A., et al., 2012. A tissue biopsy-based epigenetic multiplex PCR assay for prostate cancer detection. BMC Urol. 12, 16.

Vanaja, D.K., Ballman, K.V., Morlan, B.W., et al., 2006. PDLIM4 repression by hypermethylation as a potential biomarker for prostate cancer. Clin. Cancer Res. 12, 1128–1136.

Vanaja, D.K., Grossmann, M.E., Cheville, J.C., et al., 2009. PDLIM4, an actin binding protein, suppresses prostate cancer cell growth. Cancer Invest. 27, 264–272.

Vasarainen, H., Malmi, H., Maattanen, L., et al., 2013. Effects of prostate cancer screening on health-related quality of life: results of the Finnish arm of the European randomized screening trial (ERSPC). Acta Oncol. 52, 1615–1621.

Vedder, M.M., de Bekker-Grob, E.W., Lilja, H.G., et al., 2014. The added value of percentage of free to total prostate-specific antigen, PCA3, and a kallikrein panel to the ERSPC risk calculator for prostate cancer in prescreened men. Eur. Urol. 66, 1109–1115.

Vickers, A.J., Ulmert, D., Sjoberg, D.D., et al., 2013. Strategy for detection of prostate cancer based on relation between prostate specific antigen at age 40-55 and long term risk of metastasis: case-control study. BMJ 346, f2023.

Wei, S., Dunn, T.A., Isaacs, W.B., et al., 2008. GOLPH2 and MYO6: putative prostate cancer markers localized to the Golgi apparatus. Prostate 68, 1387–1395.

Wu, K., Xie, D., Zou, Y., et al., 2013. The mechanism of DAB2IP in chemoresistance of prostate cancer cells. Clin. Cancer Res. 19, 4740–4749.

Xiao, Z., Adam, B.L., Cazares, L.H., et al., 2001. Quantitation of serum prostate-specific membrane antigen by a novel protein biochip immunoassay discriminates benign from malignant prostate disease. Cancer Res. 61, 6029–6033.

Yin, Y., Li, M., Li, H., et al., 2010. Expressions of 6 microRNAs in prostate cancer. Zhonghua Nan Ke Xue 16, 599–605.

Ying, C., Xiao, B.D., Qin, Y., et al., 2017. GOLPH2-regulated oncolytic adenovirus, GD55, exerts strong killing effect on human prostate cancer stem-like cells in vitro and in vivo. Acta Pharmacol. Sin. 39 (3).

Zha, S., Ferdinandusse, S., Denis, S., et al., 2003. Alpha-methylacyl-CoA racemase as an androgen-independent growth modifier in prostate cancer. Cancer Res. 63, 7365–7376.

Zhang, L., Guo, F., Gao, X., et al., 2015. Golgi phosphoprotein 3 expression predicts poor prognosis in patients with prostate cancer undergoing radical prostatectomy. Mol. Med. Rep. 12, 1298–1304.

PART VIII

DISEASE BIOMARKERS

CHAPTER

49

Biomarkers of Alzheimer's Disease

Jason Pitt

University of Evansville, Evansville, IN

INTRODUCTION

Alzheimer's disease (AD) is the most common form of dementia, affecting an estimated 5.4 million Americans in 2012, with this estimate rising to 11–16 million by 2050 (Alzheimer's Association, 2012). AD patients exhibit episodic memory impairments, judgment deficiencies, confusion, altered mood, and motor impairments at later stages. Although AD is diagnosed clinically, confirmation requires postmortem examination demonstrating brain atrophy and the positive identification of two AD neuropathological hallmarks: (1) *neurofibrillary tangles* that contain aggregated forms of the microtubule-associated protein tau and (2) *amyloid plaques* that are predominantly composed of the peptide Aβ. Brain atrophy involves the loss of neurons and synapses. The loss of synapses correlates highly with the memory dysfunction observed in patients. This synaptic and neuronal deterioration most likely results from synaptotoxic Aβ oligomers (AβOs), which appear to cause subtle forms of synaptic dysfunction through altered glutamatergic signaling (see LTP Impairments and Excitotoxicity Through Glutamatergic Dysfunction) and cell death through mechanisms involving tau dysfunction (see Posttranslational Modifications of Tau Cause Aggregation and Synaptic Starving).

Although no cure exists for AD, there are two types of drugs currently being used for treatment of AD. The *NMDA receptor antagonist* memantine (Namenda) prevents glutamate excitotoxicity caused by overactivation of the NMDA receptor (Chen et al., 1992). Although Namenda can slow the cognitive decline in patients with moderate to severe AD, Namenda neither reverses nor prevents cognitive impairments and offers no apparent benefit to patients with mild to moderate AD. The second class of AD therapeutics are the *cholinesterase inhibitors* (ChEIs) Aricept, Exelon, Razadyne, and Cognex. ChEIs elevate acetylcholine levels, which partially offsets the loss of cholinergic input to the cerebral cortex that occurs in AD. ChEIs are effective for mild to moderate AD, unlike Namenda, with clinical trials reporting slower rates of cognitive decline and functional improvements. As with Namenda, the long-term effectiveness of ChEIs is questionable.

Another promising therapy currently in Phase III clinical trials is Aβ immunization. In a case study, active immunization with aggregated Aβ (AN-1792) reduced plaque burden and the associated dystrophic neurites and reactive astrocytes in cortical areas, although neurofibrillary tangles were still present (Nicoll et al., 2003), suggesting that immunization may be more successful if employed early on in disease progression. However, aseptic meningoencephalitis in a subset of patients forced early termination of active immunization trials (Orgogozo et al., 2003). *Passive immunization*, in which anti-Aβ antibodies are administered directly to boost patient titers, shows promise in mouse models (Bard and Cannon, 2000). However, human trials have shown mixed results. In one trial, immunized patients had reduced deterioration of cognitive functions measured with a battery of neuropsychological tests, although no differences were observed with standard measures of cognitive function (e.g., the Alzheimer's Disease Assessment Scale—Cognitive subscale [ADAS-cog] and the Mini-Mental State Exam [MMSE]), and the trial was stopped because of meningoencephalitis in 18 of the 300 (6%) immunized patients (Gilman et al., 2005). Currently, no Aβ immunotherapy has been successfully implemented in human subjects (reviewed in Pul et al., 2011).

There is some debate regarding the reason for the clinical trial failures of Aβ immunotherapies. One possibility is that therapies are focusing on the wrong target (i.e., Aβ does not play a causative role in AD).

Biomarkers in Toxicology, Second Edition
https://doi.org/10.1016/B978-0-12-814655-2.00049-9

Based on the genetic risk factors for familial AD, this does not seem likely (see Genetic Risk Factors for AD). A more plausible explanation is that clinical trials are not started early enough to be effective. It is widely believed that after the onset of dementia, neural function has been compromised to a degree that is not likely to be reversed (Petersen, 2003). As an example, brain atrophy involves a significant loss of synapses and, more importantly, neurons that will not be replaced. No amount of pharmacological intervention is likely to reverse cognitive impairments in severely atrophied brains.

Because the prospect of treating AD after dementia onset appears grim, there is substantial interest in identifying at-risk, predemented individuals for therapeutic intervention. Several large-scale preventative clinical trials are underway to test whether preventive immunization can prevent familial AD (Lambracht-Washington and Rosenberg, 2013) is in preparation to study prevention of familial AD. However, because the vast majority of AD cases are sporadic, AD biomarker studies have focused largely on subjects with *mild cognitive impairment* (MCI), which is viewed as an intermediate stage between normal aging and AD. It is acknowledged that, although MCI is clearly associated with an elevated risk for developing AD, a diagnosis of MCI does not guarantee disease progression. Hence, identification of biomarkers that (1) predict conversion to AD and (2) identify underlying etiologies in MCI patients has considerable potential to improve therapeutic approaches, potentially reducing the cost and sample sizes of clinical trials by 60% and 67%, respectively. This chapter discusses current AD biomarkers that may fulfill these functions.

TABLE 49.1 Alzheimer's Linked Mutations

Gene	Mutations	Outcome
Amyloid precursor protein (APP; Gene ID 351)	K670N/M671L (Swedish)	Increased Aβ production through potentiated BACE1 activity; elevated risk of AD
	A673T	Decreased Aβ production through disrupted BACE1 activity; reduced risk of AD
	A692G (Flemish)	Increased Aβ production through disrupted α-secretase activity; elevated risk of AD
	E693G (Arctic)	Increased Aβ polymerization; elevated risk of AD
	I716V (Florida)	Increased $A\beta_{42}$:$A\beta_{40}$ ratio; elevated risk of AD
	V717I (London)	Increased $A\beta_{42}$:$A\beta_{40}$ ratio; elevated risk of AD
Presenilin 1 (PSEN1; Gene ID 5663)	150+ mutations, mostly missense	Increased $A\beta_{42}$:$A\beta_{40}$ ratio; elevated risk of AD
Presenilin 2 (PSEN2; Gene ID 5664)	10+ mutations	Increased $A\beta_{42}$:$A\beta_{40}$ ratio; elevated risk of AD

AD, Alzheimer's disease.

GENETIC RISK FACTORS FOR ALZHEIMER'S DISEASE

There is strong evidence that Aβ plays a critical role in initiating AD. Although roughly 99% of AD cases are sporadic, heritable forms of AD provide insights into disease etiology. Each of the genes identified in cases of familial AD is involved in the formation of Aβ, and the only known mutation that decreases the incidence of AD impairs Aβ production (Table 49.1). This section (1) explains the apparent causative role of Aβ in AD progression, (2) highlights the importance of soluble, oligomeric forms of Aβ, and (3) discusses AD risk factors in the context of Aβ-induced neural dysfunction.

Aβ accumulation is central to AD progression and involves proteolytic liberation from a longer parent peptide and rapid self-association. Aβ, the main component of amyloid plaques, is derived from the amyloid precursor protein (APP) through proteolytic cleavage by two enzymes, β- and γ-secretase. β-secretase (BACE1) cleaves the extracellular portion of APP between amino acids 671M and 672D. γ-Secretase is a multiprotein complex consisting of at least presenilin proteins, nicastrin, aph-1, and pen-2 that cleaves APP within the plasma membrane at variable sites to produce Aβ monomers ranging from 32 to 43 amino acids in length. $A\beta_{40}$ is the most common form of Aβ, while $A\beta_{42}$ is more prone to aggregate into oligomeric and fibrillar forms because of the additional hydrophobic amino acids. For this reason, circulating levels of $A\beta_{42}$ are more closely associated with AD progression, and AD biomarkers that target $A\beta_{42}$ have better prognostic value than those targeting $A\beta_{40}$ (see Decreased CSF $A\beta_{42}$).

Evidence for a causal role of Aβ in AD pathogenesis comes from the only genetic risk factor for sporadic AD—the presence of at least one apolipoprotein E4

(ApoE4) allele. Aβ is cleared from the brain through a variety of mechanisms, one of which is receptor-mediated removal by ApoE and its receptors. Aβ is bound by ApoE, either directly or indirectly, and brought into neurons and glia for proteolytic degradation or transported across the blood–brain barrier through interactions with a variety of ApoE receptors, including low-density lipoprotein receptor (LDLR), low-density lipoprotein receptor-related protein 1 (LRP1), and very low-density lipoprotein receptor (VLDLR). The ApoE gene has three polymorphic alleles (ApoE2, ApoE3, and ApoE4) that differ by a single amino acid. Although it is not clear exactly how ApoE4 promotes AD onset, findings indicate several possible mechanisms, including increased Aβ production and aggregation, decreased Aβ clearance, and synaptic impairments (reviewed in Bu, 2009). Regardless of the mechanism(s) involved, the presence of two ApoE4 alleles decreases the mean age of AD onset from 84 to 68 years of age and increases an individual's risk of developing AD by roughly fourfold (Corder et al., 1993).

Much stronger evidence for a causative role of Aβ in AD progression is provided by mutations associated with familial AD. Mutations responsible for familial AD occur either in APP or the presenilins and result in an increased production in $A\beta_{42}$. Interestingly, a recently identified mutation in APP (A673T) near the site of BACE1 cleavage decreases Aβ production, thereby protecting carriers against AD onset. Although still somewhat debated, it is relatively well accepted that insoluble amyloid plaques represent pathological end points, rather than disease-initiating agents: amyloid plaques are found in the brains of cognitively normal individuals (Morris et al., 1996), and AD can occur in the absence of apparent plaque formation compared with age-matched controls (Tomiyama et al., 2008). It is now widely accepted that soluble AβOs, rather than insoluble amyloid plaques, are the toxic-initiating factors in AD (reviewed in Wilcox et al., 2011).

MECHANISMS OF SYNAPTIC DYSFUNCTION AND NEURONAL LOSS

One of the currently hypothesized cellular mechanisms of memory storage is termed long-term potentiation (LTP). In LTP, there is an activity-dependent boost of excitatory glutamatergic transmission that persists for hours to days. In the canonical mechanism of LTP induction, calcium influx through the NMDA receptor leads to increased AMPA receptor conductance and content at postsynaptic sites. Because of its memory-relevant functions, altered glutamatergic signaling in the hippocampus is likely to account for at least some of the AD-associated memory deficits.

Long-Term Potentiation Impairments and Excitotoxicity Through Glutamatergic Dysfunction

AβOs bind at or near synapses and induce a biphasic alteration in glutamatergic signaling. Initially, AβOs potentiate glutamatergic signaling through activation of both ionotropic and metabotropic glutamate receptors. This leads to calcium dysregulation and inevitably loss of synapses. This is evidenced by experiments in transgenic AD mice which showed impaired synaptic transmission in the Schaffer collateral that coincides with a decreased AMPA:NMDA ratio. This early synaptic impairment can be relieved by the ionotropic glutamate receptor antagonist kynurenate, demonstrating that, early on, AβOs activate glutamatergic signaling.

After initiating synaptic loss through potentiated glutamatergic signaling, AβOs alter synaptic plasticity and decrease glutamatergic transmission. In a mouse model of AD, synaptic loss precedes impairments in LTP by roughly 1 month. Interestingly, while LTP is inhibited, mechanisms of long-term depression remain intact. Although a vast oversimplification, this suggests that AβOs make synapses unable to store memory (i.e., disrupt LTP formation) while leaving them fully capable of losing memories (i.e., LTP is unaffected). In addition to inhibiting LTP, AβOs also impair baseline transmission by reducing CaMKII-dependent AMPA receptor phosphorylation at a site that increases its single-channel conductance.

Although it is possible that AβOs potentiate ionotropic glutamate receptor activity directly, it may also be the case that AβOs increase glutamate receptor activation by inhibiting glutamate uptake. Indeed, there is evidence that AβOs impair glutamate uptake at hippocampal synapses. One study reported a 30% reduction of glutamate uptake in hippocampal synaptosomes in the presence of AβOs (Li et al., 2009). It should be noted that synaptosomes contain pre- and postsynaptic components, as well as associated astrocytic processes, so it is unclear whether AβOs are targeting neurons, astrocytes, or both. Based on work from isolated cell cultures, it seems that AβOs are capable of reducing glutamate uptake in both cell types. These data demonstrate multiple ways in which AβOs alter synaptic transmission: (1) potentiating glutamatergic signaling, (2) inducing synapse loss, (3) reducing glutamatergic signaling and impairing synaptic potentiation, and (4) preventing glutamate uptake in neurons and astrocytes.

Posttranslational Modifications of Tau Cause Aggregation and Synaptic Starving

Tau is a microtubule-associated protein found abundantly within axons, where it likely plays a role in controlling the rate of axonal transport and microtubule assembly. In AD, tau becomes hyperphosphorylated and subsequently cleaved by caspases, causing it to aggregate into neurofibrillary tangles (reviewed in Binder et al., 2005). Tau hyperphosphorylation is a useful AD biomarker (see Elevated CSF Total-Tau and Phospho-Tau) and is likely responsible for the redistribution of tau from axons to the somatodendritic compartment. This disturbance in tau localization and function disrupts microtubule transport of vesicles and energy sources, leading to synaptic and neuronal degeneration.

Although both tau and Aβ are critical players in AD-associated neuronal dysfunction, tau-mediated toxicity appears to act downstream of Aβ, as cultured neurons from tau knockout mice are resistant to Aβ-induced toxicity. These roles are supported by findings from AD biomarker studies, which suggest (1) that Aβ accumulation occurs early on in AD progression and (2) that the severity of neural damage correlates better with tau biomarkers than those for Aβ.

BIOMARKERS

A significant step toward identifying reliable AD biomarkers was taken by the Alzheimer's Disease Neuroimaging Initiative (ADNI), which hosts a public database of potential cerebrospinal fluid (CSF) and imaging biomarkers (UCLA LONI, 2013). By collecting biological and clinical measures in a standardized manner across time, ADNI is expected to provide a wealth of information regarding the validity of candidate biomarkers.

The standardized measures taken by ADNI help alleviate some of the shortcomings of many AD biomarker studies. Although not always feasible, studies should (1) test their prognostic value, (2) correlate with continuous measurements of disease progression, such as the ADAS-cog (the standard primary neuropsychological measure for AD trials) or the MMSE, (3) compare biomarker covariance and combinatorial utility, such as the p-tau$_{T181}$:Aβ$_{42}$ ratio (see Elevated CSF p-Tau$_1$:Aβ$_{42}$ Ratio), and (4) validate AD progression postmortem.

AD biomarker research is in its early stages, with studies characterizing biomarkers in select, at-risk populations. Many studies include MCI subjects as a sort of prodromal, predemented stage of AD. Although individuals with MCI have a higher annual rate of conversion to AD (6%–15%) than the general population (1%–2%) because MCI is of low specificity for AD, the inclusion of clinically diagnosed MCI patients in clinical trials as an intermediate phase along the AD spectrum is likely flawed. The National Institute on Aging/Alzheimer's Association Diagnostic Guidelines for AD has suggested the use of AD biomarkers to estimate whether MCI is likely to be "MCI due to AD":

MCI due to AD, Low Likelihood—MCI patients with negative biomarkers for both Aβ and neuronal injury.
MCI due to AD, Intermediate Likelihood—MCI patients with either (1) a positive Aβ biomarker or (2) a positive biomarker for neuronal injury.
MCI due to AD, High Likelihood—MCI patients with positive biomarkers for both Aβ and neuronal injury.

The National Institute on Aging/Alzheimer's Association Diagnostic Guidelines for AD also suggested classifying AD biomarkers into three groups: (1) biomarkers of Aβ deposition, (2) biomarkers of neuronal injury, and (3) associated biochemical change. However, some of the biomarkers of neuronal injury are likely to reflect dysfunction in glial cells, such as ^{18}F-2-deoxy-2-fluoro-D-glucose (FDG) positron emission tomography (PET) and single-photon emission computed tomography (SPECT) perfusion imaging. Furthermore, the distinction between biomarkers of neuronal injury and associated biochemical changes is not necessarily clear. For example, markers of synaptic damage (i.e., neuronal injury) are classified as associated biochemical changes. To simplify the matter, this author has chosen to organize AD biomarkers into two groups: (1) prognostic biomarkers (may predict AD onset) and (2) diagnostic biomarkers (reflect neural damage). All biomarker data are summarized in Table 49.2.

Prognostic Biomarkers

Biomarkers for Aβ and tau abnormalities are useful for predicting conversion to AD, especially when used together. The formation of Aβ plaques and tau tangles are the defining hallmarks of AD; biomarkers based on these likely reflect AD-associated pathology. Biomarkers of Aβ deposition are potentially useful predictive tools, as Aβ accumulates in the brain 10–20 years before clinical symptoms are evident (Price and Morris, 1999), whereas tau biomarkers are considered to reflect neuronal damage and are clearly associated with AD (Fig. 49.1).

Decreased Cerebrospinal Fluid Aβ$_{42}$

Aβ can be measured from plasma and CSF samples with minimally invasive procedures using an enzyme-linked immunosorbent assay (ELISA) to distinguish between Aβ peptides of different lengths. Here, the focus is on CSF Aβ$_{42}$ measurements because (1) Aβ$_{42}$ is considered to be more disease-relevant than Aβ$_{40}$ and (2) a large number of studies find that plasma Aβ levels do

TABLE 49.2 Selected Alzheimer's Disease Biomarkers With Highly Evidenced Prognostic and Diagnostic Value

Category	AD Pathology	Biomarker	Change in AD
1. Prognostic biomarkers	Increased amyloid burden	CSF $A\beta_{42}$	Decreased
		PiB-PET	Increased
	Increased tau (total-tau and phospho-tau)	CSF Tau	Increased
		CSF p-Tau (T181, S199, T231)	Increased
		CSF Tau:$A\beta_{42}$	Increased
2. Diagnostic biomarkers	Regional brain atrophy	CT and MRI	Regional atrophy, particularly in the hippocampus
	Hypometabolism	FDG-PET (glucose uptake)	Regional hypometabolism
	Altered cerebral blood flow	SPECT perfusion imaging	Regional alterations in cerebral perfusion
	Oxidative damage	Isoprostanes	Increased

AD, Alzheimer's disease; *CT*, computed tomography; *MRI*, magnetic resonance imaging; *PiB*, Pittsburgh compound B; *SPECT*, single-photon emission computed tomography.

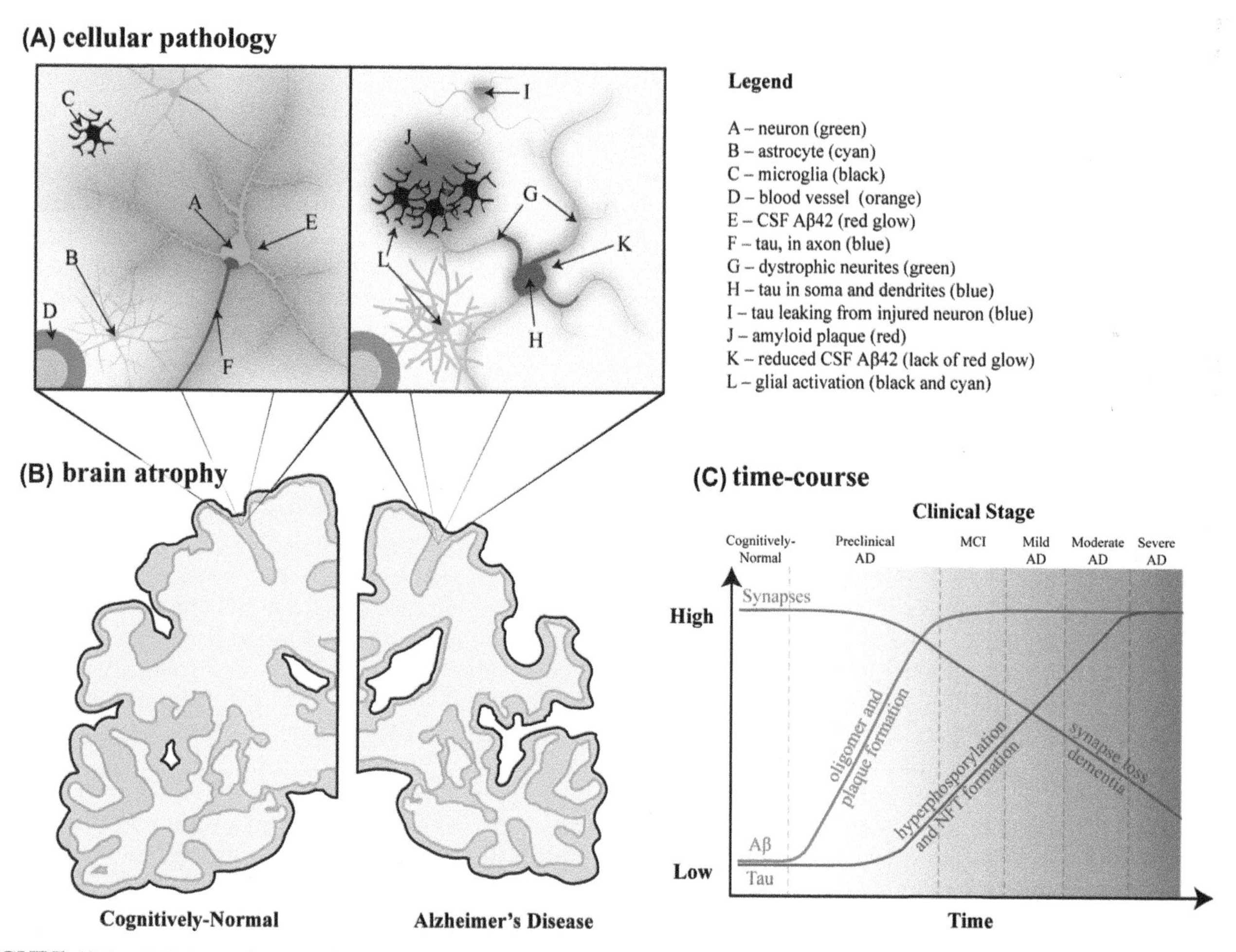

FIGURE 49.1 Cellular and temporal characteristics of Alzheimer's disease (AD) pathology. (A) Cellular pathology of AD. Active neurons (A) produce Aβ (E), which is released extracellularly. After release, Aβ is rapidly cleared by microglia ((C); the macrophages of the brain) and astrocytes (B), which contact blood vessels (D) to fulfill their protective and supportive functions. In healthy neurons (left), tau is localized in the axon (F). In AD (right), the overproduction of aggregation-prone forms of Aβ leads to several distinctive pathologies. First, tau relocates to the cell body and dendrites (H), leading to synaptic starving and neuritic dystrophy (G) and eventually cell death (I). Glial activation (L) is observed around Aβ plaques (J). The aggregation of Aβ into plaques acts as a sink, soaking up soluble Aβ from the CSF (K). (B) Significant brain atrophy occurs in AD. (C) Time course of AD. Aβ pathology precedes cognitive dysfunction by 10–20 years. Tau pathology follows, reflecting neural damage associated with compromised neural and synaptic integrity.

not associate with disease severity. While detection of other Aβ fragments through unbiased proteomic approaches has been used successfully to distinguish AD patients from nondemented controls, proteomics is not as well established as ELISA-based quantification of $A\beta_{42}$ (see "Concluding Remarks and Future Directions").

Although counterintuitive, reduced CSF levels of $A\beta_{42}$ are associated with an increased risk of developing AD. Indeed, CSF levels of $A\beta_{42}$ are decreased in AD patients compared with nondemented controls. Decreased CSF $A\beta_{42}$ also predicts conversion to AD in MCI patients. This drop in $A\beta_{42}$ is apparently caused by the formation of insoluble amyloid, throughout the course of AD, which acts as a sink for soluble Aβ, as CSF $A\beta_{42}$. This is supported by a negative correlation between CSF $A\beta_{42}$ and insoluble amyloid levels (Fagan et al., 2006). However, it is also possible that the self-association of Aβ into oligomeric toxins may preclude detection with antibodies raised against monomeric Aβ (see Concluding Remarks and Future Directions). Currently, detection of AβOs as an AD biomarker is not standard practice, although a negative correlation between the level of high molecular weight AβOs in CSF and performance on the MMSE were reported in patients with AD or MCI (Fukumoto et al., 2010).

Although CSF levels of $A\beta_{42}$ are reliable in many small-scale, single-site studies, multicenter studies revealed the unstable nature of Aβ. Because of the aggregation-prone nature of Aβ, it is a problematic peptide to keep in solution. Furthermore, the circadian fluctuations in CSF Aβ levels create an additional source of variability (Huang et al., 2012). Although CSF $A\beta_{42}$ levels are more prone to interpatient and intersite variability, they are still critical to the diagnosis of AD.

Increased Positron Emission Tomography Amyloid Imaging

One potential solution to the problem of measuring Aβ in CSF is to measure Aβ pathology in vivo. This is accomplished by the use of amyloid-sensitive PET ligands. The most successful amyloid imaging agent is ^{11}C-labeled Pittsburgh compound B (PiB), which binds to Aβ fibrils with nanomolar affinity ($K_d = 4.7$ nM) (Mathis et al., 2003). PiB binding is increased in AD patient brains, showing a similar pattern to that of amyloid detected postmortem. PiB binding has also been used to distinguish MCI patients that convert to AD—PiB levels correlate with reductions in CSF $A\beta_{42}$, elevated CSF tau and phospho-tau levels, episodic memory impairments, cerebral atrophy, and reduced cerebral glucose uptake (Forsberg et al., 2008).

It should be noted that amyloid imaging has more prognostic, rather than diagnostic, value. High PiB binding is observed in some cognitively normal individuals (Fagan et al., 2006), and while PiB levels remained stable across a 2-year longitudinal study of AD patients, cerebral glucose uptake continued to decline (Engler et al., 2006). This supports the idea that amyloid deposition plateaus before dementia becomes severe.

Elevated Cerebrospinal Fluid Total-Tau and Phospho-Tau

Tau aggregation is the second neuropathological hallmark of AD. It is thought that when neurons become sufficiently damaged, they leak tau into the extracellular milieu, although it is also possible that pathological forms of tau may spread between cells in a prion-like fashion (Clavaguera et al., 2009). Regardless of how tau enters the CSF, elevated CSF tau levels predict AD with specificity and sensitivity well above 80% in most studies. As with total-tau, levels of phosphorylated tau (e.g., T181, S199, and T231) are also elevated in AD and converting MCI patients.

Measuring CSF tau offers some advantages over Aβ measurements. The levels of total and phosphorylated tau in the CSF correlate better with brain atrophy (i.e., cell damage) than CSF Aβ or amyloid (Apostolova et al., 2010), CSF tau appears to be more stable than Aβ (Lewczuk et al., 2006), and CSF tau/p-tau may have better predictive value, at least in presymptomatic carriers of familial AD-associated mutations (Ringman et al., 2008). Despite these advantages, Aβ biomarkers are likely to have more prognostic value than tau biomarkers and even greater predictive power is achieved when CSF tau measurements are interpreted in conjunction with CSF Aβ.

Elevated Cerebrospinal Fluid p-tau:$A\beta_{42}$ Ratio

AD is characterized by significant synaptic and neuronal degeneration in the presence of amyloid plaques and tangles. Based on this, many investigators measure the ratio of CSF p-tau and $A\beta_{42}$ as a means to better distinguish AD from control subjects with specificity and sensitivity of 72%–94% and 83%–95%, respectively (De Meyer et al., 2010). When compared with measures of CSF tau, p-tau, or $A\beta_{42}$, CSF p-tau:$A\beta_{42}$ generally has greater predictive value (Herukka et al., 2005). The p-tau:$A\beta_{42}$ ratio can also be used to distinguish AD from other dementias, including semantic dementia, frontotemporal dementia, posterior cortical atrophy, and primary progressive nonfluent or logopenic aphasia (de Souza et al., 2011).

Elevated CSF tau:$A\beta_{42}$ also predicts conversion of MCI patients to AD. In one study, ADNI subjects with low CSF $A\beta_{42}$ (<192 pg/mL) and high p-tau$_{T181}$ (>23 pg/mL) were considered to have pathological CSF and had significantly reduced MMSE scores and elevated genetic associations for AD (Schott and ADNI Investigators, 2012). The predictive value of the CSF p-tau_{T181}:$A\beta_{42}$ ratio was tested on MCI converters and nonconverters from 198 MCI subjects from the ADNI trial (Samtani et al., 2013). In this study a cutoff value of −1.86 for log CSF p-tau_{181p}:$A\beta_{42}$ was used to distinguish pathological CSF (>−1.86; i.e., elevated p-tau_{181p} and reduced $A\beta_{42}$) and nonpathological CSF (≤−1.86) with a negative predictive value of 84% and a positive predictive value of 57%. This study suggests that CSF biomarkers, particularly p-tau:$A\beta_{42}$ ratios, can be used to estimate likelihood of disease progression in MCI patients.

MicroRNA Biomarkers

MicroRNAs (miRNAs) have received considerable attention as potential biomarkers for disease states for a few reasons. First, miRNAs regulate the expression of a variety of genes in eukaryotes—changes in the miRNA expression profile reflect changes in the overall gene expression profile. Second, miRNAs are secreted from cells, allowing them to be harvested from blood or CSF. Lastly, miRNA libraries are available to allow the screening of hundreds of miRNAs in a given sample.

A large number of miRNAs have been associated with AD. For example, Alzheimer's patients show altered expression of several miRNAs involved in the processing of tau and Aβ (reviewed in Femminella et al., 2015). If these miRNAs play a causative role in disease progression, as these data suggest, then they should be able to act as biomarkers of the disease. Although we can detect different levels of specific miRNAs in blood and brain samples from Alzheimer's patients, these changes in miRNA expression are not specific to AD. For example, miR-9 is one of the miRNAs found at altered levels in Alzheimer's patients—miR-9 levels are lowered in the serum and higher in the brain. MiR-9 expression is also deregulated in Huntington's patients, Parkinson's patients, and several different types of cancer. This lack of specificity is inherent in miRNAs, as they regulate a large number of genes. Given this, even sufficiently large miRNA profiles will not likely be used to make specific diagnoses without additional, specific readouts (e.g., CSF p-tau:$A\beta_{42}$ ratio).

Diagnostic Biomarkers

The underlying cause of AD dementia is multivariant, particularly in sporadic AD. Biomarkers of neural damage may provide insight into disease mechanisms, allowing clinicians to choose appropriate therapies when available. More importantly, biomarkers that estimate neural damage may distinguish MCI patients likely to remain stable or decline across time, which will aid in patient selection for clinical trials. However, biomarkers of neural damage are not likely to allow a definitive diagnosis of AD in the absence of positive Aβ and tau biomarkers.

Brain Atrophy

The progressive nature of AD dementia is a consequence of the severe synaptic and neuronal degeneration that occurs throughout the course of the disease. Given the characteristic pattern of tau neuropathology and concomitant neurodegeneration observed in AD patients, one would expect to observe associated decreases in brain volume throughout the course of the disease. Such measurements are attainable with computed tomography and magnetic resonance imaging (MRI), with MRI being the technique of choice in more recent work. A number of studies have shown that AD patients exhibit increased atrophy in medial temporal regions, in particular the hippocampal formation, as well as general brain atrophy. Measurements of hippocampal atrophy also support the idea that biomarkers of tau, rather than Aβ, are more closely associated with neural damage—the degree of hippocampal atrophy correlates better with tau than Aβ biomarkers (Apostolova et al., 2010). In addition to the regional differences noted above, there are also temporal differences. Longitudinal studies have shown that brain atrophy is higher in AD, occurring two- to tenfold more rapidly than in age-matched controls. Although structural MRI can clearly distinguish AD patients from age-matched controls, its ability to identify MCI patients who later convert to AD is not yet established (reviewed in Fayed et al., 2012).

^{18}F-2-Deoxy-2-Fluoro-D-Glucose Positron Emission Tomography

Cerebral metabolism is estimated using FDG, a radio-labeled glucose analog that is not metabolized because of the lack of a 2′ hydroxyl group. Using PET, the rate of FDG uptake can be measured (FDG-PET) as a surrogate measure of the cerebral metabolic rate of glucose consumption. In AD patients, decreased FDG-PET typically occurs in temporoparietal, posterior cingulate, and precuneus cortices (Minoshima et al., 1997; Hoffman et al., 2000; Sakamoto et al., 2002; Nestor et al., 2003). Severe cases of AD have additional hypometabolism in prefrontal cortices (Mosconi, 2005), which likely underlies disturbances in higher-level cognitive tasks. Decreased FDG-PET has also been reported to distinguish MCI patients who later convert to AD.

As noted previously, alterations in FDG-PET may reflect changes in neuronal and astrocytic function. It is estimated that astrocytic end-feet cover the vast majority (>99%) of the vascular surface in the brain, and astrocyte-derived signals can control local blood supply (Iadecola and Nedergaard, 2007). This leads to the speculation that changes in FDG-PET and SPECT (see below) observed in AD may result from altered astrocytic function, which in turn leads to neuronal impairments and atrophy.

Single-Photon Emission Computed Tomography Perfusion Imaging

SPECT has shown that regional reductions in cortical perfusion occur in AD. Areas with reduced perfusion show a similar pattern to that of FDG-PET, with AD patients having reduced perfusion in temporoparietal, posterior cingulate, and precuneus cortices. Hypoperfusion also correlates with the degree of parietotemporal atrophy and postmortem Braak staging (Jobst et al., 1992). SPECT also distinguishes MCI patients who convert to AD from those who remain stable (Hirao et al., 2005). Taken together, this work suggests that cortical hypoperfusion is a useful measure of AD progression that tracks neural damage along the course of the disease.

F_2-Isoprostanes

Isoprostanes are the result of lipid peroxidation and serve as markers of oxidative damage. F_2-isoprostanes are increased in CSF of AD patients, and this increase correlates with cognitive decline in APOE4 carriers (Duits et al., 2013), cortical atrophy (Montine et al., 1999), and dementia severity (Praticò et al., 2000). CSF F_2-isoprostanes may also have predictive value, as presymptomatic carriers with familial AD-associated mutations had significantly elevated CSF F_2-isoprostane levels (Ringman et al., 2008). Although far less work has focused on isoprostanes compared with more standard AD CSF biomarkers, such as tau or Aβ, this work suggests that CSF F_2-isoprostanes are potentially useful AD biomarkers.

CONCLUDING REMARKS AND FUTURE DIRECTIONS

The biomarkers discussed in this chapter are the most widely studied AD biomarkers. There are many other biomarkers that are newly emerging or of debatable prognostic or diagnostic value. One future approach that is particularly appealing is an unbiased proteomic analysis of CSF samples. Unlike antibody-based recognition, proteomic mass spectroscopy analysis allows the identification of many different proteins, as well as variants of the same protein. One study used proteomic methods to identify novel biomarkers to distinguish AD from cognitively normal controls and other neurodegenerative disorders (Hu et al., 2010). In a separate study, the pattern of Aβ fragments ($A\beta_{1-16}$, $A\beta_{1-33}$, $A\beta_{1-39}$, and $A\beta_{1-42}$) was found to be different in CSF from AD patients compared with nondemented controls (Portelius et al., 2006). However, due to technical limitations, proteomic analyses are not likely to be adopted as standard practice.

CSF is in direct contact with the brain and is rich in molecular components that are thought to reflect changes in brain function. In fact, CSF biomarkers have proven far more reliable than serum biomarkers for detecting changes that occur in AD. Aβ biomarkers are likely to be of great prognostic value, especially when coupled with tau biomarkers. Biomarkers of neural damage are more likely to aid in estimating disease progression, rather than distinguishing AD from other forms of dementia or predicting which MCI patients are likely to convert to AD (Small et al., 2006). Because of the unstable nature of Aβ and tau, the collection, handling, and storage of CSF samples is a major source of variability within and between study sites. This variability may be reduced by adopting the National Institute on Aging's *Biospecimen Best Practice Guidelines for the Alzheimer's Disease Centers* (National Institute on Aging, 2007). As of yet, there are no validated cutoff values for any AD biomarkers. A significant step in this direction was started by ADNI, which hosts a public database of AD biomarkers acquired across multiple sites (UCLA LONI, 2013). Earlier detection is expected to aid in clinical trial design and may allow individuals to better plan for their future. However, in the absence of effective therapies, a positive diagnosis may do more harm than good.

References

Alzheimer's Association, 2012. Alzheimer's disease facts and figures. Alzheimer's Dement. 8, 131–168.

Apostolova, L.G., Hwang, K.S., Andrawis, J.P., 2010. Alzheimer's disease neuroimaging initiative: 3D PIB and CSF biomarker associations with hippocampal atrophy in ADNI subjects. Neurobiol. Aging 31, 1284–1303.

Bard, C., Cannon, R., 2000. Barbour, Peripherally administered antibodies against amyloid beta-peptide enter the central nervous system and reduce pathology in a mouse model of Alzheimer disease. Nat. Med. 6, 916–919.

Binder, L.I., Guillozet-Bongaarts, A.L., Garcia-Sierra, F., Berry, R.W., 2005. Tau, tangles, and Alzheimer's disease. Biochim. Biophys. Acta 1739, 216–223.

Duits, F.H., Kester, M.I., Scheffer, P.G., Blankenstein, M.A., Scheltens, P., Teunissen, C.E., van der Flier, W.M., 2013. Increase in Cerebrospinal Fluid F 2-Isoprostanes is Related to Cognitive Decline in APOE ε4 Carriers. J. Alzheimers Dis. 36 (3), 563–570.

Bu, G., 2009. Apolipoprotein E and its receptors in Alzheimer's disease: pathways, pathogenesis and therapy. Nat. Rev. Neurosci. 10, 333–344.

Chen, H.S., Pellegrini, J.W., Aggarwal, S.K., 1992. Open-channel block of N-methyl-D-aspartate (NMDA) responses by memantine: therapeutic advantage against NMDA receptor-mediated neurotoxicity. J. Neurosci. 12, 4427–4436.

Clavaguera, F., Bolmont, T., Crowther, R.A., 2009. Transmission and spreading of tauopathy in transgenic mouse brain. Nat. Cell Biol. 11, 909–913.

Corder, E.H., Saunders, A.M., Strittmatter, W.J., 1993. Gene dose of apolipoprotein E type 4 allele and the risk of Alzheimer's disease in late onset families. Science 261, 921–923.

De Meyer, G., Shapiro, F., Vanderstichele, H., 2010. Alzheimer's disease neuroimaging initiative, 2010: diagnosis-independent Alzheimer disease biomarker signature in cognitively normal elderly people. Arch. Neurol. 67, 949–956.

de Souza, L.C., Lamari, F., Belliard, S., 2011. Cerebrospinal fluid biomarkers in the differential diagnosis of Alzheimer's disease from other cortical dementias. J. Neurol. Neurosurg. Psychiatr. 82, 240–246.

Engler, H., Forsberg, A., Almkvist, O., 2006. Two-year follow-up of amyloid deposition in patients with Alzheimer's disease. Brain 129, 2856–2866.

Fagan, A.M., Mintun, M.A., Mach, R.H., 2006. Inverse relation between in vivo amyloid imaging load and cerebrospinal fluid Abeta42 in humans. Ann. Neurol. 59, 512–519.

Fayed, N., Modrego, P.J., Salinas, G.R., Gazulla, J., 2012. Magnetic resonance imaging based clinical research in Alzheimer's disease. J. Alzheimers Dis. 31 (Suppl. 3), S5–S18.

Femminella, G.D., Ferrara, N., Rengo, G., 2015. The emerging role of microRNAs in Alzheimer's disease. Front. Physiol. 12, 40.

Forsberg, A., Engler, H., Almkvist, O., 2008. PET imaging of amyloid deposition in patients with mild cognitive impairment. Neurobiol. Aging 29, 1456–1465.

Fukumoto, H., Tokuda, T., Kasai, T., 2010. High-molecular-weight beta-amyloid oligomers are elevated in cerebrospinal fluid of Alzheimer patients. FASEB. J. 24, 2716–2726.

Gilman, S., Koller, M., Black, R.S., 2005. AN1792(QS-21)-201 Study Team, 2005: clinical effects of Abeta immunization (AN1792) in patients with AD in an interrupted trial. Neurology 64, 1553–1562.

Herukka, S.-K., Hallikainen, M., Soininen, H., Pirttilä, T., 2005. CSF Abeta42 and tau or phosphorylated tau and prediction of progressive mild cognitive impairment. Neurology 64, 1294–1297.

Hirao, K., Ohnishi, T., Hirata, Y., 2005. The prediction of rapid conversion to Alzheimer's disease in mild cognitive impairment using regional cerebral blood flow SPECT. Neuroimage 28, 1014–1021.

Hoffman, J.M., Welsh-Bohmer, K.A., Hanson, M., 2000. FDG PET imaging in patients with pathologically verified dementia. J. Nucl. Med. 41, 1920–1928.

Hu, W.T., Chen-Plotkin, A., Arnold, S.E., 2010. Novel CSF biomarkers for Alzheimer's disease and mild cognitive impairment. Acta Neuropathol. 119, 669–678.

Huang, Y., Potter, R., Sigurdson, W., 2012. Effects of age and amyloid deposition on Aβ dynamics in the human central nervous system. Arch. Neurol. 69, 51–58.

Iadecola, C., Nedergaard, M., 2007. Glial regulation of the cerebral microvasculature. Nat. Neurosci. 10, 1369–1376.

Jobst, K.A., Smith, A.D., Barker, C.S., 1992. Association of atrophy of the medial temporal lobe with reduced blood flow in the posterior parietotemporal cortex in patients with a clinical and pathological diagnosis of Alzheimer's disease. J. Neurol. Neurosurg. Psychiatr. 55, 190–194.

Lambracht-Washington, D., Rosenberg, R.N., 2013. Advances in the development of vaccines for Alzheimer's disease. Discov. Med. 15 (84), 319–326.

Lewczuk, P., Beck, G., Ganslandt, O., 2006. International quality control survey of neurochemical dementia diagnostics. Neurosci. Lett. 409, 1–4.

Li, S., Hong, S., Shepardson, N.E., 2009. Soluble oligomers of amyloid Beta protein facilitate hippocampal long-term depression by disrupting neuronal glutamate uptake. Neuron 62, 788–801.

Mathis, C.A., Wang, Y., Holt, D.P., 2003. Synthesis and evaluation of 11C-labeled 6-substituted 2-arylbenzothiazoles as amyloid imaging agents. J. Med. Chem. 46, 2740–2754.

Minoshima, S., Giordani, B., Berent, S., 1997. Metabolic reduction in the posterior cingulate cortex in very early Alzheimer's disease. Ann. Neurol. 42, 85–94.

Montine, T.J., Markesbery, W.R., Zackert, W., 1999. The magnitude of brain lipid peroxidation correlates with the extent of degeneration but not with density of neuritic plaques or neurofibrillary tangles or with APOE genotype in Alzheimer's disease patients. Am. J. Pathol. 155, 863–868.

Morris, J.C., Storandt, M., McKeel, D.W., 1996. Cerebral amyloid deposition and diffuse plaques in "normal" aging: evidence for presymptomatic and very mild Alzheimer's disease. Neurology 46, 707–719.

Mosconi, L., 2005. Brain glucose metabolism in the early and specific diagnosis of Alzheimer's disease. FDG-PET studies in MCI and AD. Eur. J. Nucl. Med. Mol. Imag. 32, 486–510.

National Institute on Aging, 2007. NIA Biospecimen Best Practice Guidelines for the Alzheimer's Disease Centers. Retrieved from https://www.alz.washington.edu/BiospecimenTaskForce.html.

Nestor, P.J., Fryer, T.D., Smielewski, P., Hodges, J.R., 2003. Limbic hypometabolism in Alzheimer's disease and mild cognitive impairment. Ann. Neurol. 54, 343–351.

Nicoll, J.A.R., Wilkinson, D., Holmes, C., 2003. Neuropathology of human Alzheimer disease after immunization with amyloid-beta peptide: a case report. Nat. Med. 9, 448–452.

Orgogozo, J.-M., Gilman, S., Dartigues, J.-F., 2003. Subacute meningoencephalitis in a subset of patients with AD after Abeta42 immunization. Neurology 61, 46–54.

Petersen, R.C., 2003. Mild cognitive impairment clinical trials. Nat. Rev. Drug Discov. 2, 646–653.

Portelius, E., Zetterberg, H., Andreasson, U., 2006. An Alzheimer's disease-specific beta-amyloid fragment signature in cerebrospinal fluid. Neurosci. Lett. 409, 215–219.

Praticò, D., Clark, C.M., Lee, V.M., 2000. Increased 8,12-iso-iPF2alpha-VI in Alzheimer's disease: correlation of a noninvasive index of lipid peroxidation with disease severity. Ann. Neurol. 48, 809–812.

Price, J.L., Morris, J.C., 1999. Tangles and plaques in nondemented aging and "preclinical" Alzheimer's disease. Ann. Neurol. 45, 358–368.

Pul, R., Dodel, R., Stangel, M., 2011. Antibody-based therapy in Alzheimer's disease. Exp. Opin. Biol. Ther. 11, 343–357.

Ringman, J.M., Younkin, S.G., Praticò, D., 2008. Biochemical markers in persons with preclinical familial Alzheimer disease. Neurology 71, 85–92.

Sakamoto, S., Ishii, K., Sasaki, M., 2002. Differences in cerebral metabolic impairment between early and late onset types of Alzheimer's disease. J. Neurol. Sci. 200, 27–32.

Samtani, M.N., Raghavan, N., Shi, Y., 2013. Alzheimer's Disease Neuroimaging Initiative, 2013: disease progression model in subjects with mild cognitive impairment from the Alzheimer's disease neuroimaging initiative: CSF biomarkers predict population subtypes. Br. J. Clin. Pharmacol. 75, 146–161.

Schott, J.M., ADNI, Investigators., 2012. Using CSF biomarkers to replicate genetic associations in Alzheimer's disease. Neurobiol. Aging 33 (7), 1486–e9.

Small, G.W., Kepe, V., Ercoli, L.M., 2006. PET of brain amyloid and tau in mild cognitive impairment. N. Engl. J. Med. 355, 2652–2663.

Tomiyama, T., Nagata, T., Shimada, H., Teraoka, R., Fukushima, A., Kanemitsu, H., Wada, Y., 2008. A new amyloid β variant favoring oligomerization in Alzheimer's-type dementia. Ann. Neurol. 63 (3), 377–387.

UCLA Laboratory of Neuro Imaging (LONI), 2013. Alzheimer's Disease Neuroimaging Initiative. Available from: http://adni.loni.usc.edu/.

Wilcox, K.C., Lacor, P.N., Pitt, J., Klein, W.L., 2011. Aβ oligomer-induced synapse degeneration in Alzheimer's disease. Cell. Mol. Neurobiol. 31, 939–948.

CHAPTER

50

Biomarkers of Parkinson's Disease

Huajun Jin, Arthi Kanthasamy, Vellareddy Anantharam, Anumantha G. Kanthasamy

Parkinson's Disorder Research Laboratory, Iowa Center for Advanced Neurotoxicology, Department of Biomedical Sciences, Iowa State University, Ames, IA, United States

Abbreviations

8-OHdG 8-hydroxy-2′-deoxyguanosine
8-OHG 8-hydroxydeoxyguanosine
AADC Aromatic amino acid decarboxylase
AD Alzheimer's disease
AR-JP Autosomal recessive juvenile-onset Parkinsonism
CBGD Corticobasal ganglionic degeneration
CNS Central nervous system
COX Cyclooxygenase
CSF Cerebrospinal fluid
DAT Dopamine transporter
DHTBZ ^{11}C-dihydrotetrabenazine
EPDA European Parkinson's Disease Association
FDOPA [^{18}F]-6-fluorodopa
fMRI functional magnetic resonance imaging
IFN-γ Interferon-γ
iNOS Inducible nitric oxide synthase
LB Lewy bodies
LBD Lewy body dementia
LN Lewy neurites
MSA Multiple system atrophy
PD Parkinson's disease
PET Positron emission tomography
PSP progressive supranuclear palsy
SNc Substantia nigra pars compacta
SNP Single-nucleotide polymorphism
SPECT Emission computed tomography
TGF-α Transforming growth factor-α
TNF-α Tumor necrosis factor-α
VMAT2 Vesicular monoamine transporter

INTRODUCTION

Parkinson's disease (PD) is a neurodegenerative disorder of the central nervous system (CNS) first described in the medical literature by British physician James Parkinson in 1817 (Parkinson, 1817; Goetz, 2011; Pan et al., 2018). According to data available from the European Parkinson's Disease Association, 6.3 million people have PD worldwide, making it the second most common neurodegenerative disease after Alzheimer's disease (AD) (Jankovic, 1988; Tanner, 1992). In the United States, this disease affects more than 1 million people at an estimated annual cost of $28 billion, and about 50,000–60,000 new cases are diagnosed annually. These figures are expected to increase substantially as the median age of the population continues to rise in the coming decades, creating an even greater public health issue (Olanow, 2004). Both PD prevalence and incidence are age-dependent, with approximately 1% of the population older than 65 years being affected, rising to 4%–5% of the population over 85 years (Van Den Eeden et al., 2003; Farrer, 2006). Typical clinical symptoms of PD include resting tremor, rigidity, bradykinesia, and postural abnormalities. As the disease progresses, patients often show reduced facial expressions, speak softly, and have difficulty chewing and swallowing. In addition to motor deficits, PD patients often exhibit a number of nonmotor symptoms as well, such as autonomic dysfunction, cognitive and neurobehavioral disorders, and sensory and sleep abnormalities (Jankovic, 2008). Many of these symptoms can arise prior to motor dysfunction (Abbott et al., 2005, 2007; Haehner et al., 2007; Odin et al., 2018).

The pathological hallmark of PD is progressive loss of dopaminergic neurons within the substantia nigra pars compacta (SNc), leading to a profound depletion of dopamine in the striatum, a central component of the basal ganglia responsible for the initiation and control of body movement. Such depletion is believed to underlie many of the clinical manifestations of PD (Crossman, 1989; Dunnett and Bjorklund, 1999; Zhang et al., 2000). The onset of disease symptoms is marked by a level of depigmentation in the SNc encompassing approximately 60% of the SNc dopaminergic neurons and by

Biomarkers in Toxicology, Second Edition
https://doi.org/10.1016/B978-0-12-814655-2.00050-5

about 80% of the putamen dopamine having been lost (Kirik et al., 1998). Another important pathological feature for PD is the formation in the neuron's cytoplasm of round eosinophilic inclusions (Lewy bodies [LB]) containing aggregates of many different proteins and lipids, and thread-like proteinaceous deposits within neurites (Lewy neurites) of surviving dopaminergic neurons (Werner et al., 2008).

Thus far, there is no cure for PD. All current treatments are symptomatic; none slow or prevent neuronal death progression in the dopaminergic system (Olanow, 2004; Obeso et al., 2010; Lang and Espay, 2018; Ullah and Khan, 2018). Current standard treatment therapy is based on levodopa, one of the intermediate molecules in the genesis of dopamine (Clarke, 2004). As a dopamine replacement therapy, levodopa is initially effective at improving PD symptoms for most patients, but with continued treatment, patients may develop dyskinesia, dystonia, and other motor complications, as well as nonmotor adverse effects. Thus, the beneficial effects of levodopa wear off over time, and its clinical efficacy gradually declines as the disease advances (Lewitt, 2008).

Early diagnosis or even preclinical identification of individuals prior to the onset of resting tremor is essential to optimally managing PD. However, clinical diagnosis, especially in the early stages of disease, can be challenging owing to its insidious, asymmetrical, and unilateral onset and lack of pathognomonic signs or symptoms (Brooks, 2002; Caslake et al., 2009; Scorza et al., 2018). A definite diagnosis requires the findings of neuronal loss and depigmentation in the SNc and the presence of LBs in the brain stem, all of which can only be documented postmortem (Farrer, 2006). Currently, making a diagnosis of PD is clinical in nature, mainly relying on the presence or absence of various clinical features (bradykinesia, resting tremor, muscle rigidity, and postural instability) as well as the experience of the treating physician, and frequently requires a long-term watchful waiting (Pahwa and Lyons, 2010; Jann, 2011; Crosiers et al., 2012; Marsili et al., 2018). In many cases, diagnostic accuracy, particularly in the early stages of disease, is quite low, even among neurological specialists. An early study pointed out a more than one-third rate of misdiagnosis in PD found at autopsy (Ward and Gibb, 1990). This appreciable misdiagnosis arises partly from the clinical and pathological overlaps between PD and other movement disorders. These atypical Parkinsonian disorders, including multiple system atrophy (MSA), progressive supranuclear palsy (PSP), corticobasal ganglionic degeneration, and Lewy body dementia, display specific features of PD in addition to other complex neurological symptoms resulting from neuronal damage to brain areas other than the nigrostriatal dopaminergic system. Differentiating PD from these disorders remains problematic, and as a result, an early PD diagnosis may be revised years later when atypical features indicating other forms of Parkinsonism are apparent.

Based on the aforementioned challenges in diagnosing PD, there is a critical need for robust biomarkers with the ability to confirm diagnosis in the early stages of disease. Such biomarkers are needed as well to optimize the development of drugs targeting these early stages. Furthermore, a validated biological marker that reflects PD severity/progression is also urgently required. Despite the continuing lack of reliable biomarkers for PD, extensive research has been concentrated on the development of neuropathological, biochemical, neuroimaging, molecular, and genetic biomarker candidates of the disease. This has created considerable excitement in recent years, leading to novel diagnostic approaches and contributing to the development of new drugs and therapeutic biological products. This chapter will focus on the most promising PD biomarkers.

GENETIC BIOMARKERS

Extensive genetic research during the past two decades has led to the identification of not only several monogenic forms of PD, but also numerous causative mutations in several genes that increase the risk of PD. Although monogenic forms of PD appear to be rare, accounting for about 30% of familial and 3%–5% of sporadic cases, understanding these genetic variations will aid in identifying the underlying disease mechanisms, thereby accelerating the search for disease-specific biomarkers. In the following section, we will discuss the most important genetic mutations.

α-Synuclein

The *SNCA* gene coding for the protein α-synuclein was the first gene implicated in the familial forms of PD when a missense mutation A53T within the gene was isolated from a large Italian–Greek family with autosomal dominant PD, characterized by early onset (50 years) and rapid disease progression (Polymeropoulos et al., 1996, 1997). Subsequent studies identified two further point mutations (A30P and E46K) in the *SNCA* gene in a German and Spanish family, respectively (Kruger et al., 1998; Zarranz et al., 2004). All these families had clinical and pathological features similar to those observed in sporadic PD and responded to levodopa medication, although some atypical phenotypes also have been observed. For example, cognitive decline and severe central hypoventilation have been noted in several

A53T-associated patients (Polymeropoulos et al., 1997; Spira et al., 2001), and interestingly, patients with the E46K mutation exhibit some clinical features typical of DLB in addition to Parkinsonism (Zarranz et al., 2004). Apart from these missense substitutions, genomic rearrangements, including duplication (Chartier-Harlin et al., 2004; Ibanez et al., 2004; Nishioka et al., 2006; Ahn et al., 2008) and triplication (Singleton et al., 2003) of the wild-type *SNCA* gene, were also reported to cause autosomal-dominantly inherited PD in several families. In contrast to families with gene triplications, who were affected in their thirties and often presented with severe phenotypes, such as rapid progression, early dementia, and reduced life span, the clinical phenotype in patients with *SNCA* duplications resembles more closely those of sporadic PD patients (Chartier-Harlin et al., 2004; Fuchs et al., 2007). Interestingly, a Rep1 microsatellite polymorphism located on the *SNCA* gene promoter (Maraganore et al., 2006) and several single-nucleotide polymorphisms at the 5′ and 3′ regions have been associated with higher risk for sporadic PD (Mueller et al., 2005; Mizuta et al., 2006; Winkler et al., 2007; Pankratz et al., 2009). Although the cases of familial PD associated with α-synuclein mutations are extremely rare (Lee and Trojanowski, 2006), a significant role for α-synuclein in the pathogenesis of PD is highlighted by the identification of α-synuclein as the major component of LBs in both sporadic and familial PD (Spillantini et al., 1997, 1998; Takeda et al., 1998; Bayer et al., 1999). Using α-synuclein as a biochemical marker of PD will be discussed at length in this chapter (see Section 3.1).

LRRK2

The PARK8 locus encompassing the *LRRK2* gene was initially mapped in a large Japanese family with late-onset autosomal dominant PD (Funayama et al., 2002). Subsequently, two groups concurrently identified mutations within the *LRRK2* gene as the causative gene for PARK8-linked familial PD (Paisan-Ruiz et al., 2004; Zimprich et al., 2004). Since then, six point mutations with definite pathogenicity (R1441C, R1441G, Y1699C, G2019S, I1122V, and I2020T) and numerous putative pathogenic mutations have been identified in the *LRRK2* gene in both familial and sporadic PD (Cookson, 2005; Funayama et al., 2005; Gilks et al., 2005; Nichols et al., 2005; Mata et al., 2006; Tomiyama et al., 2006; Lu and Tan, 2008; Hatano et al., 2009; Haugarvoll and Wszolek, 2009). *LRRK2* mutations are the most frequently known cause of autosomal dominant forms of familial PD (Klein and Schlossmacher, 2007; Mizuno et al., 2008). The known *LRRK2* variants are estimated to account for approximately 2% of sporadic and 10% of familial PD cases (Berg et al., 2005; Di Fonzo et al., 2005; Mata et al., 2006). In particular, the *LRRK2* G2019S mutation is the best studied and most frequent substitution in the Caucasian population, accounting for approximately 0.5%–2.0% of apparently sporadic and 5%–6% of familial PD cases (Di Fonzo et al., 2005; Farrer et al., 2005; Gilks et al., 2005; Kachergus et al., 2005; Nichols et al., 2005; Tomiyama et al., 2006). However, the G2019S mutation frequency appears to vary with ethnicity (Tan et al., 2005; Lesage et al., 2006), with an extremely high frequency (30%–40%) of familial and sporadic PD patients from North Africa (Lesage et al., 2006) and 10%–30% of Ashkenazi Jews (Ozelius et al., 2006), but very rarely found in Asia (Tan et al., 2005), South Africa (Okubadejo et al., 2008), and some European countries (Xiromerisiou et al., 2007). Additionally, two polymorphic mutations R1628P and G2385R have been found to confer susceptibility to PD in Asian populations (Funayama et al., 2007; Ross et al., 2008). The penetrance of G2019S-associated disease appears to be age-dependent (Kachergus et al., 2005), but variations have been subsequently reported (Lesage et al., 2005; Clark et al., 2006; Goldwurm et al., 2007). The clinical and neurochemical phenotypes of patients with *LRRK2* mutations usually resemble sporadic PD, with neuronal degeneration accompanied by LB and responsiveness to levodopa. However, the disease pathologies can be quite variable, even within the same family (Tan and Skipper, 2007). Recent evidence showed that LRRK2 could play a role in mediating neuroinflammation (Schapansky et al., 2015; Taymans et al., 2017; Chen et al., 2018). The expression of LRRK2 in microglia is higher compared with that in neurons (Schapansky et al., 2015), and LRRK2 inhibition can attenuate microglia-mediated inflammatory responses (Moehle et al., 2012). At the molecular level, LRRK2 was reported to regulate microglial activity through modulation of chemokine (C-X3-C) receptor 1- or NFκB-mediated signaling pathways (Russo et al., 2015; Ma et al., 2016).

Parkin

Mutations in the *parkin* gene were originally identified in Japanese families with autosomal recessive juvenile-onset Parkinsonism (Ishikawa and Tsuji, 1996; Kitada et al., 1998). Subsequent studies have identified a wide variety of *parkin* mutations in PD cases, including point mutations, exonic rearrangements, deletions, and duplications (Lucking et al., 2001; Tan and Skipper, 2007). To date, more than 100 different *parkin* mutations have since been identified (Tan and Skipper, 2007). *Parkin* mutations are the most commonly known cause of autosomal recessive early-onset PD (Mizuno et al., 2008; Hatano et al., 2009), accounting for about 50% of familial and 20% of sporadic early-onset PD cases (Lucking et al., 2000). In general, *parkin*-proven disease has typical signs of PD but with an earlier age of onset (typically before 40 years), dystonia at onset, a slower progression, and a dramatic response to levodopa manipulation (Lohmann et al., 2003). However, several mutations in *parkin* may lead to a clinical presentation indistinguishable from typical late-onset idiopathic PD (Abbas et al., 1999; Klein et al., 2000; Foroud et al., 2003; Oliveira et al., 2003; Hatano et al., 2009). The pathological features of *parkin*-associated Parkinsonism typically include a loss of nigral neurons and a moderate loss of neurons in the locus coeruleus region (Mori et al., 1998). However, LBs are usually absent (Mizuno et al., 2001; Mizuno et al., 2001; Mata et al., 2004). Nevertheless, for the sporadic forms of PD, *parkin* has been identified as a component of LB.

PINK1

PINK1 mutations were initially identified in a large Italian family with an autosomal recessive form of PD (Valente et al., 2001). Since then, more than 50 pathogenic *PINK1* mutations have been identified (Hatano et al., 2009). These mutations include point mutations, as well as insertions and deletions that result in frameshift and truncation of the protein (Atsumi et al., 2006; Exner et al., 2007). Mutations in the *PINK1* gene were estimated to account for 1%–8% of familial or early-onset PD (Klein and Schlossmacher, 2007), and as such, *PINK1* mutations are the second most commonly known cause of autosomal recessive PD, after *parkin* mutations (Hatano et al., 2004; Valente et al., 2004). The clinical phenotype of *PINK1*-associated PD resembles sporadic PD with rare atypical features such as dystonia at onset and dementia similar to those with *parkin* mutations (Hatano et al., 2004; Valente et al., 2004a, 2004b; Steinlechner et al., 2007). It is not clear whether LBs are present in this PINK1-linked disease, because no neuropathological examination of the homozygous pathogenic mutation has been reported (Hardy et al., 2009).

DJ-1

Mutations in *DJ-1* were first identified in one Dutch family with autosomal recessive early-onset PD (van Duijn et al., 2001). Additional mutations including missense, exonic deletions, and splice-site alterations were further identified (Bonifati et al., 2003, 2004; Hering et al., 2004). The *DJ-1* mutations are extremely rare, accounting for less than 1% of early-onset PD cases (Clark et al., 2004; Lockhart et al., 2004). In general, patients with *DJ-1* mutations exhibit a clinical presentation similar to that of *parkin* or *PINK1* mutation-associated Parkinsonism (Hatano et al., 2009). Like *PINK1* mutations, a neuropathological investigation has not yet been reported, and for this reason it is not clear whether the LB phenotype is present in this disorder (Hardy et al., 2009). Although DJ-1 is not an essential component of LBs, it appears to be consistently colocalized with neuronal tau-positive inclusions and glial cytoplasmic inclusions (Neumann et al., 2004), providing a link between DJ-1 and distinct neurodegenerative diseases. DJ-1 has been extensively evaluated as a biochemical marker of PD (see Section 3.2.1).

BIOCHEMICAL MARKERS

Although the specific etiology of PD remains largely unknown, extensive evidence suggests that several pathogenic pathways play a key role in the development of the disease, including oxidative and nitrosative stress, mitochondrial dysfunction, impairment of the ubiquitin-proteasome system, apoptosis, and inflammation. Therefore, characterizing body fluids like cerebrospinal fluid (CSF), urine, or blood or the tissue concentration profiles of selected key proteins or other molecules involved in PD pathogenesis is necessary if we are to uncover promising biomarkers for PD.

α-Synuclein and Abnormal Protein Accumulations

Impaired protein degradation, leading to the aggregation of misfolded, mutant, and toxic variants of proteins, plays a key role in the molecular pathogenesis of PD and other neurodegenerative diseases (Kopito, 2000; Goldberg, 2003; Ross and Poirier, 2004; Bentea et al., 2017; Sharma and Priya, 2017). As described previously, the presence of intracytoplasmic proteinaceous inclusions in the SNc, known as LBs, is one of the neuropathological hallmarks of PD. LBs are composed of a variety of free and ubiquitinated proteins, including α-synuclein, ubiquitin, parkin, proteasome subunits, UCHL1, torsin-A, synphilin-1, chaperons, and neurofilaments (Forno, 1996). During the past decade, significant efforts toward biomarker development in PD have been underway to validate certain proteins that are present in LBs and seem to be associated with abnormal protein aggregation, particularly α-synuclein.

α-Synuclein is a primary component of LB and neurites, and the formation of oligomeric aggregates of α-synuclein has long been associated with PD and other synucleinopathies, including DLB and MSA among others (Galvin et al., 2001; Marti et al., 2003). Levels of α-synuclein in the CNS are tightly regulated by a balance of α-synuclein synthesis, clearance, aggregation, and secretion (Lashuel et al., 2013). The mechanism of α-synuclein clearance remains elusive, but some reports suggest that α-synuclein is degraded mainly through the proteasomal system (Xilouri et al., 2013) and autophagy pathway via lysosomes (Lee et al., 2004; Xilouri et al., 2008). α-Synuclein at first was considered to be an exclusively intracellular protein due to its lack of endoplasmic reticular signal peptides. However, growing evidence suggests that both soluble monomeric and oligomeric forms of α-synuclein can be detected in the biological fluids, such as plasma and CSF, of healthy and diseased individuals (Borghi et al., 2000; El-Agnaf et al., 2003; Tokuda et al., 2006). Interestingly, more recent discoveries indicate that the toxic α-synuclein monomers, oligomers, and fibrils can propagate via the prion-like mechanism of cell-to-cell transfer via exosomes. These transferred proteins possess the capacity to seed aggregation in recipient cells, thereby spreading disease to other brain regions (Lee et al., 2008; Desplats et al., 2009; Hansen et al., 2011). As exosomes contain a lipid membrane with moieties similar to those seen on the plasma membrane of various cells, it is easy for a distant cell to either engulf these extracellular vesicles or simply allow entry via plasma membrane fusion and diffusion into the cell cytosol. This cell-to-cell transfer of aggregated α-synuclein proteins is thought to cause the progression of PD. A recent study suggests that the exosomal/total α-synuclein ratio in plasma is associated with disease severity in PD patients. CSF has long been known to be optimal for measuring brain metabolism in both healthy and unhealthy individuals due to the relatively free exchange of molecules between the brain and CSF (Anoop et al., 2010). Indeed, previous studies of CSF concentrations of biomarkers for AD, such as β-amyloid and tau, have displayed a good correlation with disease progression (Mattsson et al., 2009; Mattsson and Zetterberg, 2009). For these reasons, levels of α-synuclein in CSF have great promise as potential biomarkers for PD and other synucleinopathies. The possible presence of α-synuclein in CSF first came to light in 2000 when a method combining immunoprecipitation and immunoblotting was applied. In this study, however, no significant difference between PD and control subjects was observed (Borghi et al., 2000). More recently, several independent studies using either ELISA or bead-based Luminex assays have demonstrated significantly reduced levels of total CSF α-synuclein in PD patients compared to healthy controls (Tokuda et al., 2006; Mollenhauer et al., 2008, 2011; Hong et al., 2010; Kasuga et al., 2010; Parnetti et al., 2011). However, most of these studies did not highlight a correlation between disease severity/duration and total CSF α-synuclein levels, suggesting that total CSF α-synuclein may not be a sensitive biomarker for disease progression. Notably, several other investigations reported comparable levels of total CSF α-synuclein between control and PD patients (Ohrfelt et al., 2009). This discrepancy is possibly due to differences in sampling protocols and in methodological quantification of α-synuclein, as well as possible blood contamination of CSF samples during lumbar puncture. In addition to total α-synuclein, several recent studies also investigated the CSF concentration of oligomeric α-synuclein in patients with PD and in age-matched controls (Tokuda et al., 2010; Park et al., 2011; Sierks et al., 2011). Results demonstrated that levels of oligomeric α-synuclein in the CSF are significantly higher in patients with PD as compared to controls, suggesting that α-synuclein oligomers in CSF may be a promising biomarker for PD and other synucleinopathies, although further validation is required.

Besides CSF, detection of α-synuclein in blood as a potential PD biomarker also has been evaluated by several investigators, but the results have been variable and inconclusive, mainly due to the fact that more than 98% of blood α-synuclein is contained in red blood cells, and various in vitro or in vivo hemolytic changes are confounders in these studies. In some of these studies, increased levels of α-synuclein (El-Agnaf et al., 2006; Lee et al., 2006; Duran et al., 2010) or phosphorylated α-synuclein (Foulds et al., 2011) in plasma from PD patients have been detected relative to controls. In contrast, in other studies, comparable (Shi et al., 2010) or significantly decreased (Li et al., 2007; Laske et al., 2011) levels of plasma α-synuclein between groups was found. Interestingly, a recent study indicates that plasma α-synuclein can predict cognitive decline in PD (Lin et al., 2017). Additionally, several studies have detected α-synuclein in other tissue types, including skin (Hoepken et al., 2008) and saliva (Del Tredici et al., 2010; Devic et al., 2011). Skin and salivary biomarkers have received special attention because they are highly accessible and easily obtained. However, potential biomarkers based on skin and salivary α-synuclein remain preliminary and need further confirmation.

Mitochondrial Dysfunction and Oxidative Stress Markers

Over the last several decades, mitochondrial dysfunction and increased oxidative stress have gained wide acceptance as pathogenic pathways contributing to PD pathogenesis. There is considerable evidence for mitochondrial dysfunction and increased oxidative stress in the brains of PD patients. Impairment of complex I activity of the mitochondrial electron transport chain has been detected in the SNc, skeletal muscle, lymphocytes, and platelets of patients with PD (Mizuno al. 1989; Parker et al., 1989; Schapira et al., 1989; Yoshino et al., 1992; Barroso et al., 1993; Mann et al., 1994; Haas et al.,

1995; Penn et al., 1995; Blandini et al., 1998). Increased oxidation of complex I subunits and reduced rates of electron transfer through complex I, as well as misassembly of complex I, were also demonstrated in PD brains (Keeney et al., 2006). Moreover, increased mtDNA deletions were detected in nigral neurons in PD brains (Bender et al., 2006). Similarly, postmortem studies have consistently observed high levels of oxidation of lipids (Dexter et al., 1989), proteins (Yoritaka et al., 1996; Floor and Wetzel, 1998), and nucleic acids (Alam et al., 1997) in the SNc of sporadic PD brains. In addition, significant alterations of the antioxidant defense system, in particular reduced glutathione, are found in the SNc of PD patients (Sian et al., 1994). Thus, metabolites and proteins linked to mitochondrial dysfunction and oxidative stress are increasingly the focus of biomarker studies of PD.

DJ-1

DJ-1 is a ubiquitously expressed protein and its mutations are associated with an autosomal recessive form of PD and cancers. This protein is also believed to play an important role in sporadic PD, particularly owing to its antioxidative activity and associated neuroprotective function (Taira et al., 2004; Menzies et al., 2005; Yang et al., 2005). DJ-1's protective effect has also been linked to its ability to regulate gene transcription, stabilize protective proteins, or mediate cell signal pathways (Shendelman et al., 2004; Zhou et al., 2006; da Costa, 2007). In addition, several lines of evidence point to a mitochondrial regulation function for DJ-1 (Hao et al., 2010; Irrcher et al., 2010; Kamp et al., 2010; Thomas et al., 2011). DJ-1 can be secreted into biological fluids, making it a promising biological marker for PD. Several studies on DJ-1 concentrations in CSF and blood have been carried out; however, the results are mostly contradictory and inconclusive. Using a semiquantitative immunoblot approach, two initial studies by Waragai and colleagues showed that DJ-1 levels are significantly higher in the CSF (Waragai et al., 2006) and blood (Waragai et al., 2007) of PD patients compared to controls. By contrast, Maita et al. (2008) examined DJ-1 with an ELISA-based method, but failed to find a disease-related difference in blood levels of DJ-1. In a more recent study, described above for measuring CSF α-synuclein, CSF DJ-1 levels were found to be significantly reduced in PD cases compared to controls and AD cases (Hong et al., 2010). Notably, there was no correlation of DJ-1 with disease severity in this study. The same group also showed that DJ-1 levels in human blood did not differ between PD patients and controls (Shi et al., 2010). Similar to what was seen with α-synuclein, those divergent results might be due to variations in detection and sensitivity among different experimental systems as well as possible blood contamination of CSF.

More recently, using mass spectrometry analyses, Lin et al. (2012) detected oxidized forms of DJ-1 in blood as potential biomarkers of PD. Among the seven posttranslationally modified isoforms of DJ-1 identified in this study, two of 4-HNE oxidatively modified DJ-1 isoforms in blood were found to differ between PD patients and controls, suggesting that blood-oxidized DJ-1 may be a candidate biomarker of PD. This finding is of special interest because oxidative modifications of DJ-1 are indicative of increased oxidative stress and have been linked to sporadic PD and AD (Choi et al., 2006; Saito et al., 2009; Inden et al., 2011; Ren et al., 2011). Further studies measuring oxidized DJ-1 in CSF or blood are definitely warranted, particularly with the availability of specific antibodies recognizing oxidized DJ-1 (Ooe et al., 2006; Saito et al., 2009).

Urate

Urate (also known as uric acid) represents the main end product of purine metabolism and is present both intracellularly and in all biological fluids (Cipriani et al., 2010). It has been demonstrated to scavenge superoxide (Skinner et al., 1998), peroxynitrite (Whiteman et al., 2002), hydroxyl radical (Cohen et al., 1984; Aruoma and Halliwell, 1989; Kaur and Halliwell, 1990; Regoli and Winston, 1999), and singlet oxygen (Ames et al., 1981), as well as to form chelate complexes with metal ions (Davies et al., 1986), thereby serving as a major antioxidant in body fluids. With these antioxidant and metal chelation propensities, urate has been implicated to exert neuroprotective effects against oxidative damage in both cell culture (Guerreiro et al., 2009) and animal (Duan et al., 2002; Haberman et al., 2007) models of neurodegeneration. Many investigators have reported a significant correlation between urate and PD. The first evidence was reported by Church et al. (1994), who found that urate levels in the substantia nigra of PD were significantly lower compared to age-matched controls, and that adding urate to PD homogenates attenuated oxidative stress as indicated by reduced dopamine oxidation. Subsequently, four prospective cohort studies on blood urate and PD risk were carried out and similar results were obtained (Davis et al., 1996; de Lau et al., 2005; Weisskopf et al., 2007; Chen et al., 2009). They demonstrated that individuals with higher blood levels of urate have a significantly reduced risk of developing PD. This work strengthens the link between oxidative stress and PD pathogenesis and the rationale for considering urate as a potential biomarker for PD risk. Additionally, several studies tested the hypothesis that urate may also be a prognostic biomarker of clinical PD progression. The first prospective study of blood urate and PD progression demonstrated that patients with increasing baseline serum levels of urate progress markedly slower toward the disability end point that required dopamine treatment (Schwarzschild et al., 2008). Ascherio et al. (2009) confirmed the inverse association between baseline serum urate and clinical progression rate in an independent cohort with early PD. In this study, they also identified an inverse link between CSF urate and clinical PD progression. Although all these findings indicate that urate could be a promising marker of the pathophysiological changes underlying PD, the biomarker utility of urate for PD diagnosis and prognosis remains limited due to its largely nonspecific action as an antioxidant.

8-OHG and 8-OHdG

Oxidative damage to nucleic acids has often been used as a marker for various diseases involving abnormally high oxidative stress, such as certain cancers, cardiovascular and neurodegenerative diseases (Cooke et al., 2003; Nakabeppu et al., 2007; Moreira et al., 2008). Among multiple adducts of nucleoside oxidation, 8-hydroxydeoxyguanosine (8-OHG) and 8-hydroxy-2′-deoxyguanosine (8-OHdG) are two of the most common modifications of nucleic acids under conditions of oxidative stress (Fiala et al., 1989; Loft and Poulsen, 1996; Wamer and Wei, 1997). Using highly specific antibodies against 8-OHG and 8-OHdG, higher levels of 8-OHG/8-OHdG have been found in the caudate nuclei and SNc of PD patients compared to age-matched controls (Alam et al., 1997; Shimura-Miura et al., 1999; Zhang et al., 1999). Interestingly, several studies have demonstrated increased concentrations of 8-OHG/8-OHdG in the CSF of PD patients (Kikuchi et al., 2002; Abe et al., 2003; Michell et al., 2004). In another study investigating the link between urinary 8-OHG and PD progression, Sato et al. (2005) found that urinary 8-OHG levels in PD patients were associated with the stages of PD, suggesting that urinary 8-OHG may be a potential marker of PD progression, although it has yet to be replicated independently. With regard to serum 8-OHG/8-OHdG concentrations in PD patients, however, available data are highly variable and inconclusive (Kikuchi et al., 2002; Abe et al., 2003).

Neuroinflammation Markers

Growing evidence supports the involvement of neuroinflammation in the pathogenesis of PD and many different brain diseases (Mosley et al., 2006; Gao et al., 2008; Przedborski, 2010; Qian et al., 2010). Neuroinflammation in the brain is characterized by the activation of microglia and astrogliosis as well as the expression of major inflammatory mediators (Barnum and Tansey, 2012; Hirsch et al., 2012; Sekiyama et al., 2012; Taylor et al., 2013). The first evidence of neuroinflammation in PD came from postmortem studies, which showed microglial activation, astrogliosis, and lymphocytic infiltration in the nigra of PD patients. Epidemiological data indicated that nonsteroidal anti-inflammatory drugs might reduce the risk of PD (Chen et al., 2003). Interestingly, a plethora of proinflammatory cytokines such as interferon-γ (IFN-γ), interleukin1-β, 6, and 2 (IL1-β, IL-6, IL-2), tumor necrosis factor-α (TNF-α), transforming growth factor-α (TGF-α), and TGF-β1 (Boka et al., 1994; Mogi et al., 1994a, 1994b, 1994c; Mogi et al., 1995a, 1995b; Mogi et al., 1996a, 1996b; Hunot et al., 1999), as well as several inflammatory enzymes such as inducible nitric oxide synthase (iNOS), cyclooxygenase (COX-1 and COX-2), and gp91phox, the main component of NADPH-oxidase (Hunot et al., 1999; Knott et al., 1999; Teismann et al., 2003; Teismann et al., 2003a, 2003b; Wu et al., 2003; Przedborski, 2010), have also been found to be upregulated in PD brains. These findings raise the possibility that examining the CSF and serum concentrations of these inflammatory mediators may be a sensitive biomarker for PD. Indeed, several studies on CSF (Blum-Degen et al., 1995; Muller et al., 1998; Guo et al., 2009) and serum (Dufek et al., 2009; Scalzo et al., 2011) proinflammatory cytokines and chemokines have been published; however, these results remained largely inconclusive. In addition, the lack of disease specificity is a major drawback limiting the use of inflammation and glial activation markers in clinical trials and in practice.

Apoptosis Markers

Apoptosis has been widely implicated in PD for mediating dopaminergic neuron death, although it continues to be debated (Mattson, 2000; Vila and Przedborski, 2003). Initially, it was demonstrated that increased numbers of TUNEL-positive dopaminergic neurons exist in the postmortem brains of PD patients. Further studies showed the activation of different initiator and effector caspases, including caspase-8, -9, and -3 in the brains of PD patients (Hartmann et al., 2000, 2001; Viswanath et al., 2001), although other studies failed to find such activation (Banati et al., 1998; Jellinger, 2000). In MPTP-intoxicated animals, substantial evidence exists demonstrating all major events of an intrinsic apoptotic pathway, including cytochrome C release, caspase-3 activation, and further cell death after MPTP administration. Furthermore, there is abundant evidence that some of the genetic causes of PD, including α-synuclein (Manning-Bog et al., 2003; Sidhu et al., 2004; Chandra et al., 2005; Machida et al., 2005; Leng and Chuang, 2006), parkin (Darios et al., 2003; Jiang et al., 2004; Machida et al., 2005), PINK1 (Petit et al., 2005; Plun-Favreau et al., 2007), and DJ-1 (Canet-Aviles et al., 2004; Junn et al., 2005; Xu et al., 2005), are directly and primarily involved in the regulation of apoptotic pathways. So far, biomarker studies in this area remain rare. An early CSF study demonstrated reduced levels of CSF Annexin V in PD patients; however, their CSF Annexin V levels did not correlate with dementia, duration of symptoms, age, sex, or PD treatment (Vermes et al., 1999).

NEUROIMAGING MARKERS

The potential utility of structural and functional neuroimaging as biological markers for diagnosing PD and its progression has been extensively advanced in recent years (de la Fuente-Fernandez et al., 2011; Stoessl, 2011; Di Biasio et al., 2012). Various imaging techniques have been increasingly investigated, including single photon emission computed tomography (SPECT), positron emission tomography (PET), and functional magnetic resonance imaging (fMRI) (Godau et al., 2012; Seibyl et al., 2012). Based on a number of radiotracers whose uptake reflects the integrity of the nigrostriatal tract, both PET and SPECT imaging are able to detect several important PD pathological parameters, such as dopaminergic dysfunction, motor symptoms, and protein aggregation (Niethammer et al., 2012; Stoessl, 2012). Multiple radiotracers have been developed to evaluate the functionality of dopamine terminals in PD. *First*, [^{18}F]-6-fluorodopa (FDOPA) PET assesses the activity of aromatic amino acid decarboxylase (AADC). It has been demonstrated that striatal FDOPA binding is markedly reduced in patients suffering from PD and is inversely correlated with motor disability (Brooks et al., 1990; Eidelberg et al., 1990; Otsuka et al., 1996; Broussolle et al., 1999; Thobois et al., 2004). Furthermore, several lines of evidence indicate that this radiotracer can also be used to monitor the stages of dopaminergic degeneration in PD (Fearnley and Lees, 1991; Bruck et al., 2009; Pavese et al., 2011). *Second*, various tropane-based PET and SPECT tracers label dopamine transporters (DAT) on the presynaptic dopamine nerve terminals and thus directly determine the integrity of the nigrostriatal system in PD (Aquilonius, 1991; Marie et al., 1995). *Third*, the ^{11}C-dihydrotetrabenazine (DHTBZ) PET radiotracer examines vesicular monoamine transporter (VMAT2) density (Stoessl, 2007; de la Fuente-Fernandez et al., 2009). These DAT and VMAT2 radiotracers routinely showed reduced binding in PD patients and have been shown to correlate with disease progression (Vingerhoets et al., 1994; Booij et al., 1997; Lee et al., 2000; Hilker et al., 2005; Panzacchi et al., 2008). Dopaminergic imaging using PET and SPECT tracers has demonstrated enhanced sensitivities in differentiating clinically probable early PD from non-PD controls, thus showing promise in improving diagnostic accuracy (Marek et al., 1996; Seibyl et al., 1996; Meara et al., 1999; Pirker et al., 2002; Marshall et al., 2009). However, neither PET nor SPECT can distinguish between causes of dopamine deficiency and also differentiate PD from atypical Parkinsonism including MSA, CBD, and PSP. Additionally, fMRI imaging may be used to differentiate PD from atypical Parkinsonism (Gupta et al., 2010), but its use is not widely accepted. In general, all these brain imaging markers are able to monitor disease pathology and are intended as an adjunct to other diagnostic procedures. However, their use is primarily limited to research applications due to their cost and availability.

CONCLUDING REMARKS AND FUTURE DIRECTIONS

This chapter discusses promising biomarkers for PD. Overall, there is increasing evidence that a number of potential biochemical and genetic biomarkers show promise for earlier, accurate diagnosis of PD. Nonetheless, a major limitation for using these markers in clinical practice is their lack of specificity and sensitivity. On the other hand, functional imaging markers are more specific and sensitive, but their utility has not been widely accepted due to their high cost and limited access. Clearly, a panel of diagnostic assays combining multiple biomarkers from different categories and even within the same category, together with thorough neuroimaging testing and clinical work-up, may be more effective in diagnosing and tracking PD progression, as well as accelerating the pace of developing more meaningful disease-modifying therapies for PD.

Acknowledgments

The writing of this chapter was supported by the National Institutes of Health R01 grants, NS100090 and ES026892 to AGK, and NS088206 to AK. The W. Eugene and Linda Lloyd Endowed Chair and Eminent Scholar to AGK and the Salisbury Endowed Chair to AK are also acknowledged. We also acknowledge Scott and Nancy Ambrust Endowment support for our research. We thank Gary Zenitsky for assistance in the preparation of this chapter.

References

Abbas, N., Lucking, C.B., Ricard, S., et al., 1999. A wide variety of mutations in the parkin gene are responsible for autosomal recessive parkinsonism in Europe. French Parkinson's Disease Genetics Study Group and the European Consortium on Genetic Susceptibility in Parkinson's Disease. Hum. Mol. Genet. 8 (4), 567–574.

Abbott, R.D., Ross, G.W., Petrovitch, H., et al., 2007. Bowel movement frequency in late-life and incidental Lewy bodies. Mov. Disord. 22 (11), 1581–1586.

Abbott, R.D., Ross, G.W., White, L.R., et al., 2005. Excessive daytime sleepiness and subsequent development of Parkinson disease. Neurology 65 (9), 1442–1446.

Abe, T., Isobe, C., Murata, T., et al., 2003. Alteration of 8-hydroxyguanosine concentrations in the cerebrospinal fluid and serum from patients with Parkinson's disease. Neurosci. Lett. 336 (2), 105–108.
Ahn, T.B., Kim, S.Y., Kim, J.Y., et al., 2008. alpha-Synuclein gene duplication is present in sporadic Parkinson disease. Neurology 70 (1), 43–49.
Alam, Z.I., Jenner, A., Daniel, S.E., et al., 1997. Oxidative DNA damage in the parkinsonian brain: an apparent selective increase in 8-hydroxyguanine levels in substantia nigra. J. Neurochem. 69 (3), 1196–1203.
Ames, B.N., Cathcart, R., Schwiers, E., et al., 1981. Uric acid provides an antioxidant defense in humans against oxidant- and radical-caused aging and cancer: a hypothesis. Proc. Natl. Acad. Sci. U. S. A. 78 (11), 6858–6862.
Anoop, A., Singh, P.K., Jacob, R.S., et al., 2010. CSF biomarkers for Alzheimer's disease diagnosis. Int. J. Alzheimer's Dis. 2010.
Aquilonius, S.M., 1991. What has PET told us about Parkinson's disease? Acta Neurol. Scand. Suppl. 136, 37–39.
Aruoma, O.I., Halliwell, B., 1989. Inactivation of alpha 1-antiproteinase by hydroxyl radicals. The effect of uric acid. FEBS Lett. 244 (1), 76–80.
Ascherio, A., LeWitt, P.A., Xu, K., et al., 2009. Urate as a predictor of the rate of clinical decline in Parkinson disease. Arch. Neurol. 66 (12), 1460–1468.
Atsumi, M., Li, Y., Tomiyama, H., et al., 2006. A 62-year-old woman with early-onset Parkinson's disease associated with the PINKi gene deletion. Rinsho Shinkeigaku 46 (3), 199–202.
Banati, R.B., Daniel, S.E., Blunt, S.B., 1998. Glial pathology but absence of apoptotic nigral neurons in long-standing Parkinson's disease. Mov. Disord. 13 (2), 221–227.
Barnum, C.J., Tansey, M.G., 2012. Neuroinflammation and non-motor symptoms: the dark passenger of Parkinson's disease? Curr. Neurol. Neurosci. Rep. 12 (4), 350–358.
Barroso, N., Campos, Y., Huertas, R., et al., 1993. Respiratory chain enzyme activities in lymphocytes from untreated patients with Parkinson disease. Clin. Chem. 39 (4), 667–669.
Bayer, T.A., Jakala, P., Hartmann, T., et al., 1999. Alpha-synuclein accumulates in Lewy bodies in Parkinson's disease and dementia with Lewy bodies but not in Alzheimer's disease beta-amyloid plaque cores. Neurosci. Lett. 266 (3), 213–216.
Bender, A., Krishnan, K.J., Morris, C.M., et al., 2006. High levels of mitochondrial DNA deletions in substantia nigra neurons in aging and Parkinson disease. Nat. Genet. 38 (5), 515–517.
Bentea, E., Verbruggen, L., Massie, A., 2017. The proteasome inhibition model of Parkinson's disease. J. Parkinson's Dis. 7 (1), 31–63.
Berg, D., Schweitzer, K., Leitner, P., et al., 2005. Type and frequency of mutations in the LRRK2 gene in familial and sporadic Parkinson's disease. Brain 128 (Pt 12), 3000–3011.
Blandini, F., Nappi, G., Greenamyre, J.T., 1998. Quantitative study of mitochondrial complex I in platelets of parkinsonian patients. Mov. Disord. 13 (1), 11–15.
Blum-Degen, D., Muller, T., Kuhn, W., et al., 1995. Interleukin-1 beta and interleukin-6 are elevated in the cerebrospinal fluid of Alzheimer's and de novo Parkinson's disease patients. Neurosci. Lett. 202 (1–2), 17–20.
Boka, G., Anglade, P., Wallach, D., et al., 1994. Immunocytochemical analysis of tumor necrosis factor and its receptors in Parkinson's disease. Neurosci. Lett. 172 (1–2), 151–154.
Bonifati, V., Oostra, B.A., Heutink, P., 2004. Linking DJ-1 to neurodegeneration offers novel insights for understanding the pathogenesis of Parkinson's disease. J. Mol. Med. 82 (3), 163–174.
Bonifati, V., Rizzu, P., van Baren, M.J., et al., 2003. Mutations in the DJ-1 gene associated with autosomal recessive early-onset parkinsonism. Science 299 (5604), 256–259.
Booij, J., Korn, P., Linszen, D.H., et al., 1997. Assessment of endogenous dopamine release by methylphenidate challenge using iodine-123 iodobenzamide single-photon emission tomography. Eur. J. Nucl. Med. 24 (6), 674–677.
Borghi, R., Marchese, R., Negro, A., et al., 2000. Full length alpha-synuclein is present in cerebrospinal fluid from Parkinson's disease and normal subjects. Neurosci. Lett. 287 (1), 65–67.
Brooks, D.J., 2002. Diagnosis and management of atypical parkinsonian syndromes. J. Neurol. Neurosurg. Psychiatry 72 (Suppl. 1), I10–I16.
Brooks, D.J., Ibanez, V., Sawle, G.V., et al., 1990. Differing patterns of striatal 18F-dopa uptake in Parkinson's disease, multiple system atrophy, and progressive supranuclear palsy. Ann. Neurol. 28 (4), 547–555.
Broussolle, E., Dentresangle, C., Landais, P., et al., 1999. The relation of putamen and caudate nucleus 18F-Dopa uptake to motor and cognitive performances in Parkinson's disease. J. Neurol. Sci. 166 (2), 141–151.
Bruck, A., Aalto, S., Rauhala, E., et al., 2009. A follow-up study on 6-[18F]fluoro-L-dopa uptake in early Parkinson's disease shows nonlinear progression in the putamen. Mov. Disord. 24 (7), 1009–1015.
Canet-Aviles, R.M., Wilson, M.A., Miller, D.W., et al., 2004. The Parkinson's disease protein DJ-1 is neuroprotective due to cysteine-sulfinic acid-driven mitochondrial localization. Proc. Natl. Acad. Sci. U. S. A. 101 (24), 9103–9108.
Caslake, R., Taylor, K.S., Counsell, C.E., 2009. Parkinson's disease misdiagnosed as stroke. BMJ Case Rep. 2009.
Chandra, S., Gallardo, G., Fernandez-Chacon, R., et al., 2005. Alpha-synuclein cooperates with CSPalpha in preventing neurodegeneration. Cell 123 (3), 383–396.
Chartier-Harlin, M.C., Kachergus, J., Roumier, C., et al., 2004. Alpha-synuclein locus duplication as a cause of familial Parkinson's disease. Lancet 364 (9440), 1167–1169.
Chen, H., Mosley, T.H., Alonso, A., et al., 2009. Plasma urate and Parkinson's disease in the Atherosclerosis Risk in Communities (ARIC) study. Am. J. Epidemiol. 169 (9), 1064–1069.
Chen, H., Zhang, S.M., Hernan, M.A., et al., 2003. Nonsteroidal anti-inflammatory drugs and the risk of Parkinson disease. Arch. Neurol. 60 (8), 1059–1064.
Chen, J., Su, P., Luo, W., et al., 2018. Role of LRRK2 in manganese-induced neuroinflammation and microglial autophagy. Biochem. Biophys. Res. Commun. 498 (1), 171–177.
Choi, J., Sullards, M.C., Olzmann, J.A., et al., 2006. Oxidative damage of DJ-1 is linked to sporadic Parkinson and Alzheimer diseases. J. Biol. Chem. 281 (16), 10816–10824.
Church, W.H., Ward, V.L., 1994. Uric acid is reduced in the substantia nigra in Parkinson's disease: effect on dopamine oxidation. Brain Res. Bull. 33 (4), 419–425.
Cipriani, S., Chen, X., Schwarzschild, M.A., 2010. Urate: a novel biomarker of Parkinson's disease risk, diagnosis and prognosis. Biomark. Med. 4 (5), 701–712.
Clark, L.N., Afridi, S., Mejia-Santana, H., et al., 2004. Analysis of an early-onset Parkinson's disease cohort for DJ-1 mutations. Mov. Disord. 19 (7), 796–800.
Clark, L.N., Wang, Y., Karlins, E., et al., 2006. Frequency of LRRK2 mutations in early- and late-onset Parkinson disease. Neurology 67 (10), 1786–1791.
Clarke, C.E., 2004. Neuroprotection and pharmacotherapy for motor symptoms in Parkinson's disease. Lancet Neurol. 3 (8), 466–474.
Cohen, A.M., Aberdroth, R.E., Hochstein, P., 1984. Inhibition of free radical-induced DNA damage by uric acid. FEBS Lett. 174 (1), 147–150.
Cooke, M.S., Evans, M.D., Dizdaroglu, M., et al., 2003. Oxidative DNA damage: mechanisms, mutation, and disease. Faseb. J. 17 (10), 1195–1214.

Cookson, M.R., 2005. The biochemistry of Parkinson's disease. Annu. Rev. Biochem. 74, 29–52.
Crosiers, D., Pickut, B., Theuns, J., et al., 2012. Non-motor symptoms in a Flanders-Belgian population of 215 Parkinson's disease patients as assessed by the Non-Motor Symptoms Questionnaire. Am. J. Neurodegener. Dis. 1 (2), 160–167.
Crossman, A.R., 1989. Neural mechanisms in disorders of movement. Comp. Biochem. Physiol. A 93 (1), 141–149.
da Costa, C.A., 2007. DJ-1: a newcomer in Parkinson's disease pathology. Curr. Mol. Med. 7 (7), 650–657.
Darios, F., Corti, O., Lucking, C.B., et al., 2003. Parkin prevents mitochondrial swelling and cytochrome c release in mitochondria-dependent cell death. Hum. Mol. Genet. 12 (5), 517–526.
Davies, K.J., Sevanian, A., Muakkassah-Kelly, S.F., et al., 1986. Uric acid-iron ion complexes. A new aspect of the antioxidant functions of uric acid. Biochem. J. 235 (3), 747–754.
Davis, J.W., Grandinetti, A., Waslien, C.I., et al., 1996. Observations on serum uric acid levels and the risk of idiopathic Parkinson's disease. Am. J. Epidemiol. 144 (5), 480–484.
de la Fuente-Fernandez, R., Appel-Cresswell, S., Doudet, D.J., et al., 2011. Functional neuroimaging in Parkinson's disease. Expert Opin. Med. Diagn. 5 (2), 109–120.
de la Fuente-Fernandez, R., Sossi, V., McCormick, S., et al., 2009. Visualizing vesicular dopamine dynamics in Parkinson's disease. Synapse 63 (8), 713–716.
de Lau, L.M., Koudstaal, P.J., Hofman, A., et al., 2005. Serum uric acid levels and the risk of Parkinson disease. Ann. Neurol. 58 (5), 797–800.
Del Tredici, K., Hawkes, C.H., Ghebremedhin, E., et al., 2010. Lewy pathology in the submandibular gland of individuals with incidental Lewy body disease and sporadic Parkinson's disease. Acta Neuropathol. 119 (6), 703–713.
Desplats, P., Lee, H.J., Bae, E.J., et al., 2009. Inclusion formation and neuronal cell death through neuron-to-neuron transmission of alpha-synuclein. Proc. Natl. Acad. Sci. U. S. A. 106 (31), 13010–13015.
Devic, I., Hwang, H., Edgar, J.S., et al., 2011. Salivary alpha-synuclein and DJ-1: potential biomarkers for Parkinson's disease. Brain 134 (Pt 7), e178.
Dexter, D.T., Carter, C.J., Wells, F.R., et al., 1989. Basal lipid peroxidation in substantia nigra is increased in Parkinson's disease. J. Neurochem. 52 (2), 381–389.
Di Biasio, F., Vanacore, N., Fasano, A., et al., 2012. Neuropsychology, neuroimaging or motor phenotype in diagnosis of Parkinson's disease-dementia: which matters most? J. Neural. Transm. 119 (5), 597–604.
Di Fonzo, A., Rohe, C.F., Ferreira, J., et al., 2005. A frequent LRRK2 gene mutation associated with autosomal dominant Parkinson's disease. Lancet 365 (9457), 412–415.
Duan, W., Ladenheim, B., Cutler, R.G., et al., 2002. Dietary folate deficiency and elevated homocysteine levels endanger dopaminergic neurons in models of Parkinson's disease. J. Neurochem. 80 (1), 101–110.
Dufek, M., Hamanova, M., Lokaj, J., et al., 2009. Serum inflammatory biomarkers in Parkinson's disease. Parkinsonism. Relat. Disord. 15 (4), 318–320.
Dunnett, S.B., Bjorklund, A., 1999. Prospects for new restorative and neuroprotective treatments in Parkinson's disease. Nature 399 (6738 Suppl.), A32–A39.
Duran, R., Barrero, F.J., Morales, B., et al., 2010. Plasma alpha-synuclein in patients with Parkinson's disease with and without treatment. Mov. Disord. 25 (4), 489–493.
Eidelberg, D., Moeller, J.R., Dhawan, V., et al., 1990. The metabolic anatomy of Parkinson's disease: complementary [18F]fluorodeoxyglucose and [18F]fluorodopa positron emission tomographic studies. Mov. Disord. 5 (3), 203–213.
El-Agnaf, O.M., Salem, S.A., Paleologou, K.E., et al., 2003. Alpha-synuclein implicated in Parkinson's disease is present in extracellular biological fluids, including human plasma. Faseb. J. 17 (13), 1945–1947.
El-Agnaf, O.M., Salem, S.A., Paleologou, K.E., et al., 2006. Detection of oligomeric forms of alpha-synuclein protein in human plasma as a potential biomarker for Parkinson's disease. Faseb. J. 20 (3), 419–425.
Exner, N., Treske, B., Paquet, D., et al., 2007. Loss-of-function of human PINK1 results in mitochondrial pathology and can be rescued by parkin. J. Neurosci. 27 (45), 12413–12418.
Farrer, M., Stone, J., Mata, I.F., et al., 2005. LRRK2 mutations in Parkinson disease. Neurology 65 (5), 738–740.
Farrer, M.J., 2006. Genetics of Parkinson disease: paradigm shifts and future prospects. Nat. Rev. Genet. 7 (4), 306–318.
Fearnley, J.M., Lees, A.J., 1991. Ageing and Parkinson's disease: substantia nigra regional selectivity. Brain 114 (Pt 5), 2283–2301.
Fiala, E.S., Conaway, C.C., Mathis, J.E., 1989. Oxidative DNA and RNA damage in the livers of Sprague-Dawley rats treated with the hepatocarcinogen 2-nitropropane. Cancer. Res. 49 (20), 5518–5522.
Floor, E., Wetzel, M.G., 1998. Increased protein oxidation in human substantia nigra pars compacta in comparison with basal ganglia and prefrontal cortex measured with an improved dinitrophenylhydrazine assay. J. Neurochem. 70 (1), 268–275.
Forno, L.S., 1996. Neuropathology of Parkinson's disease. J. Neuropathol. Exp. Neurol. 55 (3), 259–272.
Foroud, T., Uniacke, S.K., Liu, L., et al., 2003. Heterozygosity for a mutation in the parkin gene leads to later onset Parkinson disease. Neurology 60 (5), 796–801.
Foulds, P.G., Mitchell, J.D., Parker, A., et al., 2011. Phosphorylated alpha-synuclein can be detected in blood plasma and is potentially a useful biomarker for Parkinson's disease. Faseb. J. 25 (12), 4127–4137.
Fuchs, J., Nilsson, C., Kachergus, J., et al., 2007. Phenotypic variation in a large Swedish pedigree due to SNCA duplication and triplication. Neurology 68 (12), 916–922.
Funayama, M., Hasegawa, K., Kowa, H., et al., 2002. A new locus for Parkinson's disease (PARK8) maps to chromosome 12p11.2-q13.1. Ann. Neurol. 51 (3), 296–301.
Funayama, M., Hasegawa, K., Ohta, E., et al., 2005. An LRRK2 mutation as a cause for the parkinsonism in the original PARK8 family. Ann. Neurol. 57 (6), 918–921.
Funayama, M., Li, Y., Tomiyama, H., et al., 2007. Leucine-rich repeat kinase 2 G2385R variant is a risk factor for Parkinson disease in Asian population. Neuroreport 18 (3), 273–275.
Galvin, J.E., Lee, V.M., Trojanowski, J.Q., 2001. Synucleinopathies: clinical and pathological implications. Arch. Neurol. 58 (2), 186–190.
Gao, H.M., Kotzbauer, P.T., Uryu, K., et al., 2008. Neuroinflammation and oxidation/nitration of alpha-synuclein linked to dopaminergic neurodegeneration. J. Neurosci. 28 (30), 7687–7698.
Gilks, W.P., Abou-Sleiman, P.M., Gandhi, S., et al., 2005. A common LRRK2 mutation in idiopathic Parkinson's disease. Lancet 365 (9457), 415–416.
Godau, J., Hussl, A., Lolekha, P., et al., 2012. Neuroimaging: current role in detecting pre-motor Parkinson's disease. Mov. Disord. 27 (5), 634–643.
Goetz, C.G., 2011. The history of Parkinson's disease: early clinical descriptions and neurological therapies. Cold. Spring. Harb. Perspect. Med. 1 (1), a008862.
Goldberg, A.L., 2003. Protein degradation and protection against misfolded or damaged proteins. Nature 426 (6968), 895–899.
Goldwurm, S., Zini, M., Mariani, L., et al., 2007. Evaluation of LRRK2 G2019S penetrance: relevance for genetic counseling in Parkinson disease. Neurology 68 (14), 1141–1143.
Guerreiro, S., Ponceau, A., Toulorge, D., et al., 2009. Protection of midbrain dopaminergic neurons by the end-product of purine

metabolism uric acid: potentiation by low-level depolarization. J. Neurochem. 109 (4), 1118–1128.
Guo, J., Sun, Z., Xiao, S., et al., 2009. Proteomic analysis of the cerebrospinal fluid of Parkinson's disease patients. Cell Res. 19 (12), 1401–1403.
Gupta, D., Saini, J., Kesavadas, C., et al., 2010. Utility of susceptibility-weighted MRI in differentiating Parkinson's disease and atypical parkinsonism. Neuroradiology 52 (12), 1087–1094.
Haas, R.H., Nasirian, F., Nakano, K., et al., 1995. Low platelet mitochondrial complex I and complex II/III activity in early untreated Parkinson's disease. Ann. Neurol. 37 (6), 714–722.
Haberman, F., Tang, S.C., Arumugam, T.V., et al., 2007. Soluble neuroprotective antioxidant uric acid analogs ameliorate ischemic brain injury in mice. NeuroMolecular Med. 9 (4), 315–323.
Haehner, A., Hummel, T., Hummel, C., et al., 2007. Olfactory loss may be a first sign of idiopathic Parkinson's disease. Mov. Disord. 22 (6), 839–842.
Hansen, C., Angot, E., Bergstrom, A.L., et al., 2011. alpha-Synuclein propagates from mouse brain to grafted dopaminergic neurons and seeds aggregation in cultured human cells. J. Clin. Invest. 121 (2), 715–725.
Hao, L.Y., Giasson, B.I., Bonini, N.M., 2010. DJ-1 is critical for mitochondrial function and rescues PINK1 loss of function. Proc. Natl. Acad. Sci. U. S. A. 107 (21), 9747–9752.
Hardy, J., Lewis, P., Revesz, T., et al., 2009. The genetics of Parkinson's syndromes: a critical review. Curr. Opin. Genet. Dev. 19 (3), 254–265.
Hartmann, A., Hunot, S., Michel, P.P., et al., 2000. Caspase-3: a vulnerability factor and final effector in apoptotic death of dopaminergic neurons in Parkinson's disease. Proc. Natl. Acad. Sci. U. S. A. 97 (6), 2875–2880.
Hartmann, A., Troadec, J.D., Hunot, S., et al., 2001. Caspase-8 is an effector in apoptotic death of dopaminergic neurons in Parkinson's disease, but pathway inhibition results in neuronal necrosis. J. Neurosci. 21 (7), 2247–2255.
Hatano, T., Kubo, S., Sato, S., et al., 2009. Pathogenesis of familial Parkinson's disease: new insights based on monogenic forms of Parkinson's disease. J. Neurochem. 111 (5), 1075–1093.
Hatano, Y., Li, Y., Sato, K., et al., 2004a. Novel PINK1 mutations in early-onset parkinsonism. Ann. Neurol. 56 (3), 424–427.
Hatano, Y., Sato, K., Elibol, B., et al., 2004b. PARK6-linked autosomal recessive early-onset parkinsonism in Asian populations. Neurology 63 (8), 1482–1485.
Haugarvoll, K., Wszolek, Z.K., 2009. "Clinical features of LRRK2 parkinsonism. Parkinsonism. Relat. Disord. 15 (Suppl. 3), S205–S208.
Hering, R., Strauss, K.M., Tao, X., et al., 2004. Novel homozygous p.E64D mutation in DJ1 in early onset Parkinson disease (PARK7). Hum. Mutat. 24 (4), 321–329.
Hilker, R., Schweitzer, K., Coburger, S., et al., 2005. Nonlinear progression of Parkinson disease as determined by serial positron emission tomographic imaging of striatal fluorodopa F 18 activity. Arch. Neurol. 62 (3), 378–382.
Hirsch, E.C., Vyas, S., Hunot, S., 2012. Neuroinflammation in Parkinson's disease. Parkinsonism. Relat. Disord. 18 (Suppl. 1), S210–S212.
Hoepken, H.H., Gispert, S., Azizov, M., et al., 2008. Parkinson patient fibroblasts show increased alpha-synuclein expression. Exp. Neurol. 212 (2), 307–313.
Hong, Z., Shi, M., Chung, K.A., et al., 2010. DJ-1 and alpha-synuclein in human cerebrospinal fluid as biomarkers of Parkinson's disease. Brain 133 (Pt 3), 713–726.
Hunot, S., Dugas, N., Faucheux, B., et al., 1999. FcepsilonRII/CD23 is expressed in Parkinson's disease and induces, in vitro, production of nitric oxide and tumor necrosis factor-alpha in glial cells. J. Neurosci. 19 (9), 3440–3447.
Ibanez, P., Bonnet, A.M., Debarges, B., et al., 2004. Causal relation between alpha-synuclein gene duplication and familial Parkinson's disease. Lancet 364 (9440), 1169–1171.
Inden, M., Kitamura, Y., Takahashi, K., et al., 2011. Protection against dopaminergic neurodegeneration in Parkinson's disease-model animals by a modulator of the oxidized form of DJ-1, a wild-type of familial Parkinson's disease-linked PARK7. J. Pharmacol. Sci. 117 (3), 189–203.
Irrcher, I., Aleyasin, H., Seifert, E.L., et al., 2010. Loss of the Parkinson's disease-linked gene DJ-1 perturbs mitochondrial dynamics. Hum. Mol. Genet. 19 (19), 3734–3746.
Ishikawa, A., Tsuji, S., 1996. Clinical analysis of 17 patients in 12 Japanese families with autosomal-recessive type juvenile parkinsonism. Neurology 47 (1), 160–166.
Jankovic, J., 1988. Parkinson's disease: recent advances in therapy. South. Med. J. 81 (8), 1021–1027.
Jankovic, J., 2008. Parkinson's disease: clinical features and diagnosis. J. Neurol. Neurosurg. Psychiatry 79 (4), 368–376.
Jann, M.W., 2011. Advanced strategies for treatment of Parkinson's disease: the role of early treatment. Am. J. Manag. Care 17 (Suppl. 12), S315–S321.
Jellinger, K.A., 2000. Cell death mechanisms in Parkinson's disease. J. Neural. Transm. 107 (1), 1–29.
Jiang, H., Ren, Y., Zhao, J., et al., 2004. Parkin protects human dopaminergic neuroblastoma cells against dopamine-induced apoptosis. Hum. Mol. Genet. 13 (16), 1745–1754.
Junn, E., Taniguchi, H., Jeong, B.S., et al., 2005. Interaction of DJ-1 with Daxx inhibits apoptosis signal-regulating kinase 1 activity and cell death. Proc. Natl. Acad. Sci. U. S. A. 102 (27), 9691–9696.
Kachergus, J., Mata, I.F., Hulihan, M., et al., 2005. Identification of a novel LRRK2 mutation linked to autosomal dominant parkinsonism: evidence of a common founder across European populations. Am. J. Hum. Genet. 76 (4), 672–680.
Kamp, F., Exner, N., Lutz, A.K., et al., 2010. Inhibition of mitochondrial fusion by alpha-synuclein is rescued by PINK1, Parkin and DJ-1. Embo. J. 29 (20), 3571–3589.
Kasuga, K., Tokutake, T., Ishikawa, A., et al., 2010. Differential levels of alpha-synuclein, beta-amyloid42 and tau in CSF between patients with dementia with Lewy bodies and Alzheimer's disease. J. Neurol. Neurosurg. Psychiatry 81 (6), 608–610.
Kaur, H., Halliwell, B., 1990. Action of biologically-relevant oxidizing species upon uric acid. Identification of uric acid oxidation products. Chem. Biol. Interact. 73 (2–3), 235–247.
Keeney, P.M., Xie, J., Capaldi, R.A., et al., 2006. Parkinson's disease brain mitochondrial complex I has oxidatively damaged subunits and is functionally impaired and misassembled. J. Neurosci. 26 (19), 5256–5264.
Kikuchi, A., Takeda, A., Onodera, H., et al., 2002. Systemic increase of oxidative nucleic acid damage in Parkinson's disease and multiple system atrophy. Neurobiol. Dis. 9 (2), 244–248.
Kirik, D., Rosenblad, C., Bjorklund, A., 1998. Characterization of behavioral and neurodegenerative changes following partial lesions of the nigrostriatal dopamine system induced by intrastriatal 6-hydroxydopamine in the rat. Exp. Neurol. 152 (2), 259–277.
Kitada, T., Asakawa, S., Hattori, N., et al., 1998. Mutations in the parkin gene cause autosomal recessive juvenile parkinsonism. Nature 392 (6676), 605–608.
Klein, C., Pramstaller, P.P., Kis, B., et al., 2000. Parkin deletions in a family with adult-onset, tremor-dominant parkinsonism: expanding the phenotype. Ann. Neurol. 48 (1), 65–71.
Klein, C., Schlossmacher, M.G., 2007. Parkinson disease, 10 years after its genetic revolution: multiple clues to a complex disorder. Neurology 69 (22), 2093–2104.
Knott, C., Wilkin, G.P., Stern, G., 1999. Astrocytes and microglia in the substantia nigra and caudate-putamen in Parkinson's disease. Parkinsonism. Relat. Disord. 5 (3), 115–122.

Kopito, R.R., 2000. Aggresomes, inclusion bodies and protein aggregation. Trends Cell Biol. 10 (12), 524–530.
Kruger, R., Kuhn, W., Muller, T., et al., 1998. Ala30Pro mutation in the gene encoding alpha-synuclein in Parkinson's disease. Nat. Genet. 18 (2), 106–108.
Lang, A.E., Espay, A.J., 2018. Disease modification in Parkinson's Disease: current approaches, challenges, and future considerations. Mov. Disord.
Lashuel, H.A., Overk, C.R., Oueslati, A., et al., 2013. The many faces of alpha-synuclein: from structure and toxicity to therapeutic target. Nat. Rev. Neurosci. 14 (1), 38–48.
Laske, C., Fallgatter, A.J., Stransky, E., et al., 2011. Decreased alpha-synuclein serum levels in patients with Lewy body dementia compared to Alzheimer's disease patients and control subjects. Dement. Geriatr. Cogn. Disord. 31 (6), 413–416.
Lee, C.S., Samii, A., Sossi, V., et al., 2000. In vivo positron emission tomographic evidence for compensatory changes in presynaptic dopaminergic nerve terminals in Parkinson's disease. Ann. Neurol. 47 (4), 493–503.
Lee, H.J., Khoshaghideh, F., Patel, S., et al., 2004. Clearance of alpha-synuclein oligomeric intermediates via the lysosomal degradation pathway. J. Neurosci. 24 (8), 1888–1896.
Lee, H.J., Suk, J.E., Bae, E.J., et al., 2008. Assembly-dependent endocytosis and clearance of extracellular alpha-synuclein. Int. J. Biochem. Cell Biol. 40 (9), 1835–1849.
Lee, P.H., Lee, G., Park, H.J., et al., 2006. The plasma alpha-synuclein levels in patients with Parkinson's disease and multiple system atrophy. J. Neural. Transm. 113 (10), 1435–1439.
Lee, V.M., Trojanowski, J.Q., 2006. Mechanisms of Parkinson's disease linked to pathological alpha-synuclein: new targets for drug discovery. Neuron 52 (1), 33–38.
Leng, Y., Chuang, D.M., 2006. Endogenous alpha-synuclein is induced by valproic acid through histone deacetylase inhibition and participates in neuroprotection against glutamate-induced excitotoxicity. J. Neurosci. 26 (28), 7502–7512.
Lesage, S., Durr, A., Tazir, M., et al., 2006. LRRK2 G2019S as a cause of Parkinson's disease in North African Arabs. N. Engl. J. Med. 354 (4), 422–423.
Lesage, S., Ibanez, P., Lohmann, E., et al., 2005. G2019S LRRK2 mutation in French and North African families with Parkinson's disease. Ann. Neurol. 58 (5), 784–787.
Lewitt, P.A., 2008. Levodopa for the treatment of Parkinson's disease. N. Engl. J. Med. 359 (23), 2468–2476.
Li, Q.X., Mok, S.S., Laughton, K.M., et al., 2007. Plasma alpha-synuclein is decreased in subjects with Parkinson's disease. Exp. Neurol. 204 (2), 583–588.
Lin, C.H., Yang, S.Y., Horng, H.E., et al., 2017. Plasma alpha-synuclein predicts cognitive decline in Parkinson's disease. J. Neurol. Neurosurg. Psychiatry 88 (10), 818–824.
Lin, X., Cook, T.J., Zabetian, C.P., et al., 2012. DJ-1 isoforms in whole blood as potential biomarkers of Parkinson disease. Sci. Rep. 2, 954.
Lockhart, P.J., Lincoln, S., Hulihan, M., et al., 2004. DJ-1 mutations are a rare cause of recessively inherited early onset parkinsonism mediated by loss of protein function. J. Med. Genet. 41 (3), e22.
Loft, S., Poulsen, H.E., 1996. Cancer risk and oxidative DNA damage in man. J. Mol. Med. (Berl.) 74 (6), 297–312.
Lohmann, E., Periquet, M., Bonifati, V., et al., 2003. How much phenotypic variation can be attributed to parkin genotype? Ann. Neurol. 54 (2), 176–185.
Lu, Y.W., Tan, E.K., 2008. Molecular biology changes associated with LRRK2 mutations in Parkinson's disease. J. Neurosci. Res. 86 (9), 1895–1901.
Lucking, C.B., Bonifati, V., Periquet, M., et al., 2001. Pseudo-dominant inheritance and exon 2 triplication in a family with parkin gene mutations. Neurology 57 (5), 924–927.
Lucking, C.B., Durr, A., Bonifati, V., et al., 2000. Association between early-onset Parkinson's disease and mutations in the parkin gene. N. Engl. J. Med. 342 (21), 1560–1567.
Ma, B., Xu, L., Pan, X., et al., 2016. LRRK2 modulates microglial activity through regulation of chemokine (C-X3-C) receptor 1 -mediated signalling pathways. Hum. Mol. Genet. 25 (16), 3515–3523.
Machida, Y., Chiba, T., Takayanagi, A., et al., 2005. Common anti-apoptotic roles of parkin and alpha-synuclein in human dopaminergic cells. Biochem. Biophys. Res. Commun. 332 (1), 233–240.
Maita, C., Tsuji, S., Yabe, I., et al., 2008. Secretion of DJ-1 into the serum of patients with Parkinson's disease. Neurosci. Lett. 431 (1), 86–89.
Mann, V.M., Cooper, J.M., Daniel, S.E., et al., 1994. Complex I, iron, and ferritin in Parkinson's disease substantia nigra. Ann. Neurol. 36 (6), 876–881.
Manning-Bog, A.B., McCormack, A.L., Purisai, M.G., et al., 2003. Alpha-synuclein overexpression protects against paraquat-induced neurodegeneration. J. Neurosci. 23 (8), 3095–3099.
Maraganore, D.M., de Andrade, M., Elbaz, A., et al., 2006. Collaborative analysis of alpha-synuclein gene promoter variability and Parkinson disease. Jama 296 (6), 661–670.
Marek, K.L., Seibyl, J.P., Zoghbi, S.S., et al., 1996. [123I] beta-CIT/SPECT imaging demonstrates bilateral loss of dopamine transporters in hemi-Parkinson's disease. Neurology 46 (1), 231–237.
Marie, R.M., Barre, L., Rioux, P., et al., 1995. PET imaging of neocortical monoaminergic terminals in Parkinson's disease. J. Neural. Transm. Park. Dis. Dement. Sect. 9 (1), 55–71.
Marshall, V.L., Reininger, C.B., Marquardt, M., et al., 2009. Parkinson's disease is overdiagnosed clinically at baseline in diagnostically uncertain cases: a 3-year European multicenter study with repeat [123I]FP-CIT SPECT. Mov. Disord. 24 (4), 500–508.
Marsili, L., Rizzo, G., Colosimo, C., 2018. Diagnostic criteria for Parkinson's disease: from James Parkinson to the concept of prodromal disease. Front. Neurol. 9, 156.
Marti, M.J., Tolosa, E., Campdelacreu, J., 2003. Clinical overview of the synucleinopathies. Mov. Disord. 18 (Suppl 6), S21–S27.
Mata, I.F., Lockhart, P.J., Farrer, M.J., 2004. Parkin genetics: one model for Parkinson's disease. Hum. Mol. Genet. 13 (Spec No 1), R127–R133.
Mata, I.F., Ross, O.A., Kachergus, J., et al., 2006a. LRRK2 mutations are a common cause of Parkinson's disease in Spain. Eur. J. Neurol. 13 (4), 391–394.
Mata, I.F., Wedemeyer, W.J., Farrer, M.J., et al., 2006b. LRRK2 in Parkinson's disease: protein domains and functional insights. Trends. Neurosci. 29 (5), 286–293.
Mattson, M.P., 2000. Apoptosis in neurodegenerative disorders. Nat. Rev. Mol. Cell Biol. 1 (2), 120–129.
Mattsson, N., Blennow, K., Zetterberg, H., 2009. CSF biomarkers: pinpointing Alzheimer pathogenesis. Ann. N. Y. Acad. Sci. 1180, 28–35.
Mattsson, N., Zetterberg, H., 2009. Alzheimer's disease and CSF biomarkers: key challenges for broad clinical applications. Biomark. Med. 3 (6), 735–737.
Meara, J., Bhowmick, B.K., Hobson, P., 1999. Accuracy of diagnosis in patients with presumed Parkinson's disease. Age Ageing 28 (2), 99–102.
Menzies, F.M., Yenisetti, S.C., Min, K.T., 2005. Roles of Drosophila DJ-1 in survival of dopaminergic neurons and oxidative stress. Curr. Biol. 15 (17), 1578–1582.
Michell, A.W., Lewis, S.J., Foltynie, T., et al., 2004. Biomarkers and Parkinson's disease. Brain 127 (Pt 8), 1693–1705.
Mizuno, Y., Hattori, N., Kitada, T., et al., 2001a. Familial Parkinson's disease. Alpha-synuclein and parkin. Adv. Neurol. 86, 13–21.
Mizuno, Y., Hattori, N., Kubo, S., et al., 2008. Progress in the pathogenesis and genetics of Parkinson's disease. Philos. Trans. R. Soc. Lond. B Biol. Sci. 363 (1500), 2215–2227.

Mizuno, Y., Hattori, N., Mori, H., et al., 2001b. Parkin and Parkinson's disease. Curr. Opin. Neurol. 14 (4), 477–482.

Mizuno, Y., Ohta, S., Tanaka, M., et al., 1989. Deficiencies in complex I subunits of the respiratory chain in Parkinson's disease. Biochem. Biophys. Res. Commun. 163 (3), 1450–1455.

Mizuta, I., Satake, W., Nakabayashi, Y., et al., 2006. Multiple candidate gene analysis identifies alpha-synuclein as a susceptibility gene for sporadic Parkinson's disease. Hum. Mol. Genet. 15 (7), 1151–1158.

Moehle, M.S., Webber, P.J., Tse, T., et al., 2012. LRRK2 inhibition attenuates microglial inflammatory responses. J. Neurosci. 32 (5), 1602–1611.

Mogi, M., Harada, M., Kondo, T., et al., 1995a. Transforming growth factor-beta 1 levels are elevated in the striatum and in ventricular cerebrospinal fluid in Parkinson's disease. Neurosci. Lett. 193 (2), 129–132.

Mogi, M., Harada, M., Kondo, T., et al., 1994a. Interleukin-1 beta, interleukin-6, epidermal growth factor and transforming growth factor-alpha are elevated in the brain from parkinsonian patients. Neurosci. Lett. 180 (2), 147–150.

Mogi, M., Harada, M., Kondo, T., et al., 1996a. Interleukin-2 but not basic fibroblast growth factor is elevated in parkinsonian brain. Short communication. J. Neural. Transm. 103 (8–9), 1077–1081.

Mogi, M., Harada, M., Narabayashi, H., et al., 1996b. Interleukin (IL)-1 beta, IL-2, IL-4, IL-6 and transforming growth factor-alpha levels are elevated in ventricular cerebrospinal fluid in juvenile parkinsonism and Parkinson's disease. Neurosci. Lett. 211 (1), 13–16.

Mogi, M., Harada, M., Riederer, P., et al., 1994b. Tumor necrosis factor-alpha (TNF-alpha) increases both in the brain and in the cerebrospinal fluid from parkinsonian patients. Neurosci. Lett. 165 (1–2), 208–210.

Mogi, M., Inagaki, H., Kojima, K., et al., 1995b. Transforming growth factor-alpha in human submandibular gland and saliva. J. Immunoassay 16 (4), 379–394.

Mogi, M., Kage, T., Chino, T., et al., 1994c. Increased beta 2-microglobulin in both parotid and submandibular/sublingual saliva from patients with Sjogren's syndrome. Arch. Oral Biol. 39 (10), 913–915.

Mollenhauer, B., Cullen, V., Kahn, I., et al., 2008. Direct quantification of CSF alpha-synuclein by ELISA and first cross-sectional study in patients with neurodegeneration. Exp. Neurol. 213 (2), 315–325.

Mollenhauer, B., Locascio, J.J., Schulz-Schaeffer, W., et al., 2011. alpha-Synuclein and tau concentrations in cerebrospinal fluid of patients presenting with parkinsonism: a cohort study. Lancet Neurol. 10 (3), 230–240.

Moreira, P.I., Nunomura, A., Nakamura, M., et al., 2008. Nucleic acid oxidation in Alzheimer disease. Free Radic. Biol. Med. 44 (8), 1493–1505.

Mori, H., Kondo, T., Yokochi, M., et al., 1998. Pathologic and biochemical studies of juvenile parkinsonism linked to chromosome 6q. Neurology 51 (3), 890–892.

Mosley, R.L., Benner, E.J., Kadiu, I., et al., 2006. Neuroinflammation, oxidative stress and the pathogenesis of Parkinson's disease. Clin. Neurosci. Res. 6 (5), 261–281.

Mueller, J.C., Fuchs, J., Hofer, A., et al., 2005. Multiple regions of alpha-synuclein are associated with Parkinson's disease. Ann. Neurol. 57 (4), 535–541.

Muller, T., Blum-Degen, D., Przuntek, H., et al., 1998. Interleukin-6 levels in cerebrospinal fluid inversely correlate to severity of Parkinson's disease. Acta Neurol. Scand. 98 (2), 142–144.

Nakabeppu, Y., Tsuchimoto, D., Yamaguchi, H., et al., 2007. Oxidative damage in nucleic acids and Parkinson's disease. J. Neurosci. Res. 85 (5), 919–934.

Neumann, M., Muller, V., Gorner, K., et al., 2004. Pathological properties of the Parkinson's disease-associated protein DJ-1 in alpha-synucleinopathies and tauopathies: relevance for multiple system atrophy and Pick's disease. Acta Neuropathol. 107 (6), 489–496.

Nichols, W.C., Pankratz, N., Hernandez, D., et al., 2005. Genetic screening for a single common LRRK2 mutation in familial Parkinson's disease. Lancet 365 (9457), 410–412.

Niethammer, M., Feigin, A., Eidelberg, D., 2012. Functional neuroimaging in Parkinson's disease. Cold. Spring. Harb. Perspect. Med. 2 (5), a009274.

Nishioka, K., Hayashi, S., Farrer, M.J., et al., 2006. Clinical heterogeneity of alpha-synuclein gene duplication in Parkinson's disease. Ann. Neurol. 59 (2), 298–309.

Obeso, J.A., Rodriguez-Oroz, M.C., Goetz, C.G., et al., 2010. Missing pieces in the Parkinson's disease puzzle. Nat. Med. 16 (6), 653–661.

Odin, P., Chaudhuri, K.R., Volkmann, J., et al., 2018. Viewpoint and practical recommendations from a movement disorder specialist panel on objective measurement in the clinical management of Parkinson's disease. NPJ. Parkinsons. Dis. 4, 14.

Ohrfelt, A., Grognet, P., Andreasen, N., et al., 2009. Cerebrospinal fluid alpha-synuclein in neurodegenerative disorders-a marker of synapse loss? Neurosci. Lett. 450 (3), 332–335.

Okubadejo, N., Britton, A., Crews, C., et al., 2008. Analysis of Nigerians with apparently sporadic Parkinson disease for mutations in LRRK2, PRKN and ATXN3. PLoS One 3 (10), e3421.

Olanow, C.W., 2004. The scientific basis for the current treatment of Parkinson's disease. Annu. Rev. Med. 55, 41–60.

Oliveira, S.A., Scott, W.K., Martin, E.R., et al., 2003. Parkin mutations and susceptibility alleles in late-onset Parkinson's disease. Ann. Neurol. 53 (5), 624–629.

Ooe, H., Iguchi-Ariga, S.M., Ariga, H., 2006. Establishment of specific antibodies that recognize C106-oxidized DJ-1. Neurosci. Lett. 404 (1–2), 166–169.

Otsuka, M., Ichiya, Y., Kuwabara, Y., et al., 1996. Differences in the reduced 18F-Dopa uptakes of the caudate and the putamen in Parkinson's disease: correlations with the three main symptoms. J. Neurol. Sci. 136 (1–2), 169–173.

Ozelius, L.J., Senthil, G., Saunders-Pullman, R., et al., 2006. LRRK2 G2019S as a cause of Parkinson's disease in Ashkenazi Jews. N. Engl. J. Med. 354 (4), 424–425.

Pahwa, R., Lyons, K.E., 2010. Early diagnosis of Parkinson's disease: recommendations from diagnostic clinical guidelines. Am. J. Manag. Care 16. Suppl Implications: S94–99.

Paisan-Ruiz, C., Jain, S., Evans, E.W., et al., 2004. Cloning of the gene containing mutations that cause PARK8-linked Parkinson's disease. Neuron 44 (4), 595–600.

Pan, P.Y., Zhu, Y., Shen, Y., et al., 2018. Crosstalk between presynaptic trafficking and autophagy in Parkinson's disease. Neurobiol. Dis.

Pankratz, N., Nichols, W.C., Elsaesser, V.E., et al., 2009. Alpha-synuclein and familial Parkinson's disease. Mov. Disord. 24 (8), 1125–1131.

Panzacchi, A., Moresco, R.M., Garibotto, V., et al., 2008. A voxel-based PET study of dopamine transporters in Parkinson's disease: relevance of age at onset. Neurobiol. Dis. 31 (1), 102–109.

Park, M.J., Cheon, S.M., Bae, H.R., et al., 2011. Elevated levels of alpha-synuclein oligomer in the cerebrospinal fluid of drug-naive patients with Parkinson's disease. J. Clin. Neurol. 7 (4), 215–222.

Parker Jr., W.D., Boyson, S.J., Parks, J.K., 1989. Abnormalities of the electron transport chain in idiopathic Parkinson's disease. Ann. Neurol. 26 (6), 719–723.

Parkinson, J., 1817. An Essay on the Shaking Palsy. Whittingham and Rowland for Sherwood. Neely and Jones, London.

Parnetti, L., Chiasserini, D., Bellomo, G., et al., 2011. Cerebrospinal fluid Tau/alpha-synuclein ratio in Parkinson's disease and degenerative dementias. Mov. Disord. 26 (8), 1428–1435.

Pavese, N., Rivero-Bosch, M., Lewis, S.J., et al., 2011. Progression of monoaminergic dysfunction in Parkinson's disease: a longitudinal 18F-dopa PET study. Neuroimage 56 (3), 1463–1468.

Penn, A.M., Roberts, T., Hodder, J., et al., 1995. Generalized mitochondrial dysfunction in Parkinson's disease detected by magnetic resonance spectroscopy of muscle. Neurology 45 (11), 2097–2099.

Petit, A., Kawarai, T., Paitel, E., et al., 2005. Wild-type PINK1 prevents basal and induced neuronal apoptosis, a protective effect abrogated by Parkinson disease-related mutations. J. Biol. Chem. 280 (40), 34025–34032.

Pirker, W., Djamshidian, S., Asenbaum, S., et al., 2002. Progression of dopaminergic degeneration in Parkinson's disease and atypical parkinsonism: a longitudinal beta-CIT SPECT study. Mov. Disord. 17 (1), 45–53.

Plun-Favreau, H., Klupsch, K., Moisoi, N., et al., 2007. The mitochondrial protease HtrA2 is regulated by Parkinson's disease-associated kinase PINK1. Nat. Cell Biol. 9 (11), 1243–1252.

Polymeropoulos, M.H., Higgins, J.J., Golbe, L.I., et al., 1996. Mapping of a gene for Parkinson's disease to chromosome 4q21-q23. Science 274 (5290), 1197–1199.

Polymeropoulos, M.H., Lavedan, C., Leroy, E., et al., 1997. Mutation in the alpha-synuclein gene identified in families with Parkinson's disease. Science 276 (5321), 2045–2047.

Przedborski, S., 2010. Inflammation and Parkinson's disease pathogenesis. Mov. Disord. 25 (Suppl 1), S55–S57.

Qian, L., Flood, P.M., Hong, J.S., 2010. Neuroinflammation is a key player in Parkinson's disease and a prime target for therapy. J. Neural. Transm. 117 (8), 971–979.

Regoli, F., Winston, G.W., 1999. Quantification of total oxidant scavenging capacity of antioxidants for peroxynitrite, peroxyl radicals, and hydroxyl radicals. Toxicol. Appl. Pharmacol. 156 (2), 96–105.

Ren, H., Fu, K., Wang, D., et al., 2011. Oxidized DJ-1 interacts with the mitochondrial protein BCL-XL. J. Biol. Chem. 286 (40), 35308–35317.

Ross, C.A., Poirier, M.A., 2004. Protein aggregation and neurodegenerative disease. Nat. Med. 10 Suppl, S10–S17.

Ross, O.A., Wu, Y.R., Lee, M.C., et al., 2008. Analysis of Lrrk2 R1628P as a risk factor for Parkinson's disease. Ann. Neurol. 64 (1), 88–92.

Russo, I., Berti, G., Plotegher, N., et al., 2015. Leucine-rich repeat kinase 2 positively regulates inflammation and down-regulates NF-kappaB p50 signaling in cultured microglia cells. J. Neuroinflammation 12, 230.

Saito, Y., Hamakubo, T., Yoshida, Y., et al., 2009. Preparation and application of monoclonal antibodies against oxidized DJ-1. Significant elevation of oxidized DJ-1 in erythrocytes of early-stage Parkinson disease patients. Neurosci. Lett. 465 (1), 1–5.

Sato, S., Mizuno, Y., Hattori, N., 2005. Urinary 8-hydroxydeoxyguanosine levels as a biomarker for progression of Parkinson disease. Neurology 64 (6), 1081–1083.

Scalzo, P., de Miranda, A.S., Guerra Amaral, D.C., et al., 2011. Serum levels of chemokines in Parkinson's disease. Neuroimmunomodulation 18 (4), 240–244.

Schapansky, J., Nardozzi, J.D., LaVoie, M.J., 2015. The complex relationships between microglia, alpha-synuclein, and LRRK2 in Parkinson's disease. Neuroscience 302, 74–88.

Schapira, A.H., Cooper, J.M., Dexter, D., et al., 1989. Mitochondrial complex I deficiency in Parkinson's disease. Lancet 1 (8649), 1269.

Schwarzschild, M.A., Schwid, S.R., Marek, K., et al., 2008. Serum urate as a predictor of clinical and radiographic progression in Parkinson disease. Arch. Neurol. 65 (6), 716–723.

Scorza, F.A., Fiorini, A.C., Scorza, C.A., et al., 2018. Cardiac abnormalities in Parkinson's disease and Parkinsonism. J. Clin. Neurosci.

Seibyl, J., Russell, D., Jennings, D., et al., 2012. Neuroimaging over the course of Parkinson's disease: from early detection of the at-risk patient to improving pharmacotherapy of later-stage disease. Semin. Nucl. Med. 42 (6), 406–414.

Seibyl, J.P., Marek, K., Innis, R.B., 1996. Images in neuroscience. Neuroimaging, XII. SPECT imaging of dopamine nerve terminals. Am. J. Psychiatry. 153 (9), 1131.

Sekiyama, K., Sugama, S., Fujita, M., et al., 2012. Neuroinflammation in Parkinson's disease and related disorders: a lesson from genetically manipulated mouse models of alpha-synucleinopathies. Parkinsons. Dis. 2012, 271732.

Sharma, S.K., Priya, S., 2017. Expanding role of molecular chaperones in regulating alpha-synuclein misfolding; implications in Parkinson's disease. Cell. Mol. Life Sci. 74 (4), 617–629.

Shendelman, S., Jonason, A., Martinat, C., et al., 2004. DJ-1 is a redox-dependent molecular chaperone that inhibits alpha-synuclein aggregate formation. PLoS Biol. 2 (11), e362.

Shi, M., Zabetian, C.P., Hancock, A.M., et al., 2010. Significance and confounders of peripheral DJ-1 and alpha-synuclein in Parkinson's disease. Neurosci. Lett. 480 (1), 78–82.

Shimura-Miura, H., Hattori, N., Kang, D., et al., 1999. Increased 8-oxo-dGTPase in the mitochondria of substantia nigral neurons in Parkinson's disease. Ann. Neurol. 46 (6), 920–924.

Sian, J., Dexter, D.T., Lees, A.J., et al., 1994. Alterations in glutathione levels in Parkinson's disease and other neurodegenerative disorders affecting basal ganglia. Ann. Neurol. 36 (3), 348–355.

Sidhu, A., Wersinger, C., Moussa, C.E., et al., 2004. The role of alpha-synuclein in both neuroprotection and neurodegeneration. Ann. N. Y. Acad. Sci. 1035, 250–270.

Sierks, M.R., Chatterjee, G., McGraw, C., et al., 2011. CSF levels of oligomeric alpha-synuclein and beta-amyloid as biomarkers for neurodegenerative disease. Integr. Biol. (Camb) 3 (12), 1188–1196.

Singleton, A.B., Farrer, M., Johnson, J., et al., 2003. alpha-Synuclein locus triplication causes Parkinson's disease. Science 302 (5646), 841.

Skinner, K.A., White, C.R., Patel, R., et al., 1998. Nitrosation of uric acid by peroxynitrite. Formation of a vasoactive nitric oxide donor. J. Biol. Chem. 273 (38), 24491–24497.

Spillantini, M.G., Crowther, R.A., Jakes, R., et al., 1998. alpha-Synuclein in filamentous inclusions of Lewy bodies from Parkinson's disease and dementia with lewy bodies. Proc. Natl. Acad. Sci. U. S. A. 95 (11), 6469–6473.

Spillantini, M.G., Schmidt, M.L., Lee, V.M., et al., 1997. Alpha-synuclein in lewy bodies. Nature 388 (6645), 839–840.

Spira, P.J., Sharpe, D.M., Halliday, G., et al., 2001. Clinical and pathological features of a Parkinsonian syndrome in a family with an Ala53Thr alpha-synuclein mutation. Ann. Neurol. 49 (3), 313–319.

Steinlechner, S., Stahlberg, J., Volkel, B., et al., 2007. Co-occurrence of affective and schizophrenia spectrum disorders with PINK1 mutations. J. Neurol. Neurosurg. Psychiatry 78 (5), 532–535.

Stoessl, A.J., 2007. Positron emission tomography in premotor Parkinson's disease. Parkinsonism. Relat. Disord. 13 (Suppl. 3), S421–S424.

Stoessl, A.J., 2011. Neuroimaging in Parkinson's disease. Neurotherapeutics 8 (1), 72–81.

Stoessl, A.J., 2012. Neuroimaging in Parkinson's disease: from pathology to diagnosis. Parkinsonism. Relat. Disord. 18 (Suppl. 1), S55–S59.

Taira, T., Saito, Y., Niki, T., et al., 2004. DJ-1 has a role in antioxidative stress to prevent cell death. EMBO Rep. 5 (2), 213–218.

Takeda, A., Mallory, M., Sundsmo, M., et al., 1998. Abnormal accumulation of NACP/alpha-synuclein in neurodegenerative disorders. Am. J. Pathol. 152 (2), 367–372.

Tan, E.K., Shen, H., Tan, L.C., et al., 2005. The G2019S LRRK2 mutation is uncommon in an Asian cohort of Parkinson's disease patients. Neurosci. Lett. 384 (3), 327–329.

Tan, E.K., Skipper, L.M., 2007. Pathogenic mutations in Parkinson disease. Hum. Mutat. 28 (7), 641–653.

Tanner, C.M., 1992. Epidemiology of Parkinson's disease. Neurol. Clin. 10 (2), 317–329.

Taylor, J.M., Main, B.S., Crack, P.J., 2013. Neuroinflammation and oxidative stress: Co-conspirators in the pathology of Parkinson's disease. Neurochem. Int. 62 (5), 803–819.

Taymans, J.M., Mutez, E., Drouyer, M., et al., 2017. LRRK2 detection in human biofluids: potential use as a Parkinson's disease biomarker? Biochem. Soc. Trans. 45 (1), 207–212.

Teismann, P., Tieu, K., Choi, D.K., et al., 2003a. Cyclooxygenase-2 is instrumental in Parkinson's disease neurodegeneration. Proc. Natl. Acad. Sci. U. S. A. 100 (9), 5473–5478.

Teismann, P., Vila, M., Choi, D.K., et al., 2003b. COX-2 and neurodegeneration in Parkinson's disease. Ann. N. Y. Acad. Sci. 991, 272–277.

Thobois, S., Jahanshahi, M., Pinto, S., et al., 2004. PET and SPECT functional imaging studies in Parkinsonian syndromes: from the lesion to its consequences. Neuroimage 23 (1), 1–16.

Thomas, K.J., McCoy, M.K., Blackinton, J., et al., 2011. DJ-1 acts in parallel to the PINK1/parkin pathway to control mitochondrial function and autophagy. Hum. Mol. Genet. 20 (1), 40–50.

Tokuda, T., Qureshi, M.M., Ardah, M.T., et al., 2010. Detection of elevated levels of alpha-synuclein oligomers in CSF from patients with Parkinson disease. Neurology 75 (20), 1766–1772.

Tokuda, T., Salem, S.A., Allsop, D., et al., 2006. Decreased alpha-synuclein in cerebrospinal fluid of aged individuals and subjects with Parkinson's disease. Biochem. Biophys. Res. Commun. 349 (1), 162–166.

Tomiyama, H., Li, Y., Funayama, M., et al., 2006. Clinicogenetic study of mutations in LRRK2 exon 41 in Parkinson's disease patients from 18 countries. Mov. Disord. 21 (8), 1102–1108.

Ullah, H., Khan, H., 2018. Anti-Parkinson potential of silymarin: mechanistic insight and therapeutic standing. Front. Pharmacol. 9, 422.

Valente, E.M., Abou-Sleiman, P.M., Caputo, V., et al., 2004a. Hereditary early-onset Parkinson's disease caused by mutations in PINK1. Science 304 (5674), 1158–1160.

Valente, E.M., Bentivoglio, A.R., Dixon, P.H., et al., 2001. Localization of a novel locus for autosomal recessive early-onset parkinsonism, PARK6, on human chromosome 1p35-p36. Am. J. Hum. Genet. 68 (4), 895–900.

Valente, E.M., Salvi, S., Ialongo, T., et al., 2004b. PINK1 mutations are associated with sporadic early-onset parkinsonism. Ann. Neurol. 56 (3), 336–341.

Van Den Eeden, S.K., Tanner, C.M., Bernstein, A.L., et al., 2003. Incidence of Parkinson's disease: variation by age, gender, and race/ethnicity. Am. J. Epidemiol. 157 (11), 1015–1022.

van Duijn, C.M., Dekker, M.C., Bonifati, V., et al., 2001. Park7, a novel locus for autosomal recessive early-onset parkinsonism, on chromosome 1p36. Am. J. Hum. Genet. 69 (3), 629–634.

Vermes, I., Steur, E.N., Reutelingsperger, C., et al., 1999. Decreased concentration of annexin V in parkinsonian cerebrospinal fluid: speculation on the underlying cause. Mov. Disord. 14 (6), 1008–1010.

Vila, M., Przedborski, S., 2003. Targeting programmed cell death in neurodegenerative diseases. Nat. Rev. Neurosci. 4 (5), 365–375.

Vingerhoets, F.J., Snow, B.J., Lee, C.S., et al., 1994. Longitudinal fluorodopa positron emission tomographic studies of the evolution of idiopathic parkinsonism. Ann. Neurol. 36 (5), 759–764.

Viswanath, V., Wu, Y., Boonplueang, R., et al., 2001. Caspase-9 activation results in downstream caspase-8 activation and bid cleavage in 1-methyl-4-phenyl-1,2,3,6-tetrahydropyridine-induced Parkinson's disease. J. Neurosci. 21 (24), 9519–9528.

Wamer, W.G., Wei, R.R., 1997. In vitro photooxidation of nucleic acids by ultraviolet A radiation. Photochem. Photobiol. 65 (3), 560–563.

Waragai, M., Nakai, M., Wei, J., et al., 2007. Plasma levels of DJ-1 as a possible marker for progression of sporadic Parkinson's disease. Neurosci. Lett. 425 (1), 18–22.

Waragai, M., Wei, J., Fujita, M., et al., 2006. Increased level of DJ-1 in the cerebrospinal fluids of sporadic Parkinson's disease. Biochem. Biophys. Res. Commun. 345 (3), 967–972.

Ward, C.D., Gibb, W.R., 1990. Research diagnostic criteria for Parkinson's disease. Adv. Neurol. 53, 245–249.

Weisskopf, M.G., O'Reilly, E., Chen, H., et al., 2007. Plasma urate and risk of Parkinson's disease. Am. J. Epidemiol. 166 (5), 561–567.

Werner, C.J., Heyny-von Haussen, R., Mall, G., et al., 2008. Proteome analysis of human substantia nigra in Parkinson's disease. Proteome Sci. 6, 8.

Whiteman, M., Ketsawatsakul, U., Halliwell, B., 2002. A reassessment of the peroxynitrite scavenging activity of uric acid. Ann. N. Y. Acad. Sci. 962, 242–259.

Winkler, S., Hagenah, J., Lincoln, S., et al., 2007. alpha-Synuclein and Parkinson disease susceptibility. Neurology 69 (18), 1745–1750.

Wu, D.C., Teismann, P., Tieu, K., et al., 2003. NADPH oxidase mediates oxidative stress in the 1-methyl-4-phenyl-1,2,3,6-tetrahydropyridine model of Parkinson's disease. Proc. Natl. Acad. Sci. U. S. A. 100 (10), 6145–6150.

Xilouri, M., Brekk, O.R., Stefanis, L., 2013. Alpha-synuclein and protein degradation systems: a reciprocal relationship. Mol. Neurobiol. 47 (2), 537–551.

Xilouri, M., Vogiatzi, T., Vekrellis, K., et al., 2008. alpha-synuclein degradation by autophagic pathways: a potential key to Parkinson's disease pathogenesis. Autophagy 4 (7), 917–919.

Xiromerisiou, G., Hadjigeorgiou, G.M., Gourbali, V., et al., 2007. Screening for SNCA and LRRK2 mutations in Greek sporadic and autosomal dominant Parkinson's disease: identification of two novel LRRK2 variants. Eur. J. Neurol. 14 (1), 7–11.

Xu, J., Zhong, N., Wang, H., et al., 2005. The Parkinson's disease-associated DJ-1 protein is a transcriptional co-activator that protects against neuronal apoptosis. Hum. Mol. Genet. 14 (9), 1231–1241.

Yang, Y., Gehrke, S., Haque, M.E., et al., 2005. Inactivation of Drosophila DJ-1 leads to impairments of oxidative stress response and phosphatidylinositol 3-kinase/Akt signaling. Proc. Natl. Acad. Sci. U. S. A. 102 (38), 13670–13675.

Yoritaka, A., Hattori, N., Uchida, K., et al., 1996. Immunohistochemical detection of 4-hydroxynonenal protein adducts in Parkinson disease. Proc. Natl. Acad. Sci. U. S. A. 93 (7), 2696–2701.

Yoshino, H., Nakagawa-Hattori, Y., Kondo, T., et al., 1992. Mitochondrial complex I and II activities of lymphocytes and platelets in Parkinson's disease. J. Neural. Transm. Park. Dis. Dement. Sect. 4 (1), 27–34.

Zarranz, J.J., Alegre, J., Gomez-Esteban, J.C., et al., 2004. The new mutation, E46K, of alpha-synuclein causes Parkinson and Lewy body dementia. Ann. Neurol. 55 (2), 164–173.

Zhang, J., Perry, G., Smith, M.A., et al., 1999. Parkinson's disease is associated with oxidative damage to cytoplasmic DNA and RNA in substantia nigra neurons. Am. J. Pathol. 154 (5), 1423–1429.

Zhang, Y., Dawson, V.L., Dawson, T.M., 2000. Oxidative stress and genetics in the pathogenesis of Parkinson's disease. Neurobiol. Dis. 7 (4), 240–250.

Zhou, W., Zhu, M., Wilson, M.A., et al., 2006. The oxidation state of DJ-1 regulates its chaperone activity toward alpha-synuclein. J. Mol. Biol. 356 (4), 1036–1048.

Zimprich, A., Biskup, S., Leitner, P., et al., 2004. Mutations in LRRK2 cause autosomal-dominant parkinsonism with pleomorphic pathology. Neuron 44 (4), 601–607.

CHAPTER

51

Biomarkers for Drugs of Abuse and Neuropsychiatric Disorders: Models and Mechanisms

Pushpinder Kaur Multani[1], *Nitin Saini*[1], *Ravneet Kaur*[2], *Vandana Saini*[3]
[1]Johnson & Johnson, Janssen Research & Development, Malvern, PA, United States;
[2]Aveley, Western Australia; [3]Washington, United States

INTRODUCTION

Drug abuse and mental health problems are common in the United States and worldwide. Most drugs of abuse exert their initial reinforcing effects by activating reward circuits in the brain and, although initial drug experimentation is largely a voluntary behavior, continued drug use impairs brain function by interfering with the capacity to exert self-control over drug-taking behaviors and rendering the brain more sensitive to stress and negative moods. Addiction changes the brain significantly, causing alterations in a person's personality and disturbing the normal functioning of the brain and leading to neuropsychiatric (NP) disorders. The shared neurobiological substrates (brain regions), overlapping genetic vulnerability, environmental stress, and developmental issues are some of the factors causing an individual to be susceptible to drug addiction, comorbid with NP illness such as depression, anxiety, and bipolar disorder. The mechanism of action of drugs of abuse is attributed toward their interaction with distinct neurotransmitter transporter, receptors signaling, epigenetic changes with upregulation or downregulation expression of genes implicated in neuroplasticity followed by interruption of the intracellular key proteins. This modify the information processing of neuronal circuits in the brain areas involved in reward, executive function, control, awareness, mood, and stress resulting in behavioral dysfunctions. This chapter is focused on the cellular and molecular mechanisms of drug of abuse, their involvement in NP disorders, and animal models and biomarkers of NP disorders.

DRUGS OF ABUSE

Addiction to drugs of abuse (psychostimulants) is a brain disease resulting in a loss of control over drug-taking or compulsive drug-seeking, despite noxious consequences (Nestler and Hyman, 2010). It is a serious social burden and has a destructive influence on the health and standard of living of an addicted person. Addiction to psychostimulants is characterized by at least three phases: (1) the continued use of a substance despite adverse consequences; (2) the diminishing effectiveness of the substance over time or the need for larger doses to achieve the same desired effect (tolerance); and (3) physical and psychological symptoms of distress, discomfort, or impairment on reduction or cessation of substance use (withdrawal) (DSM-IV, 1994). These three phases of addiction are associated with structural and functional changes in the brain that can be explained collectively as neuronal plasticity.

Psychostimulants such as cocaine, opioids, amphetamines, methamphetamine (Meth), methylphenidate, and other recreational drugs, such as 3,4-methylene dioxymethamphetamine (MDMA, "ecstasy"), cannabis, alcohol, morphine, and gamma-hydroxybutyric acid, influence the central nervous system (CNS) remarkably by acting on various molecular targets, resulting in neuroadaptive changes responsible for the dependence and addiction that they create. A psychostimulant is a substance that enhances arousal and euphoria acutely, and psychosis and addiction chronically.

At the cellular level, the drug of abuse upsets neuronal functions by imitating (opiate drugs—endogenous

Biomarkers in Toxicology, Second Edition
https://doi.org/10.1016/B978-0-12-814655-2.00051-7

opioids, nicotine—acetylcholine), stimulating (cocaine, amphetamines—dopamine [DA]; ecstasy—serotonin), or blocking (alcohol—glutamate [GLU]) the effects of the neurotransmitters.

Neuronal Basis of Drug Dependence

Essentially all drugs of abuse exert their acute reinforcing properties via the mesocorticolimbic DA pathway, encompassing DA neurons that originate in the ventral tegmental area (VTA) and project to the striatum and other limbic regions, including the prefrontal cortex (PFC), amygdala, and hippocampus. Specifically following brain circuits and structures mediate reward (nucleus accumbens [NAc] and ventral pallidum), memory and learning (amygdala and hippocampus), motivation/drive (orbitofrontal cortex [OFC] and subcallosal cortex), and control (PFC and anterior cingulate gyrus). These brain centers are innervated by dopaminergic and glutamatergic projections, mediating neuroadaptations in drug addiction. The reward pathway is acutely activated by all the drugs of abuse and mediates the reinforcing effects of these drugs.

Neurotoxicity of Psychostimulants

The known mechanism of psychostimulants is that these substances increase the extracellular level of several neurotransmitters including DA, serotonin (5-HT), and norepinephrine (NE) by competing with monoamine transporters (dopamine transporter [DAT]; serotonin transporter [SERT]; or norepinephrine transporter [NET] or on the vesicular monoamine transporter-1,2 [VMAT-1,VMAT-2], Multani et al., 2012) or their receptors, resulting in their rewarding effects and dependence.

AMPHETAMINE

Amphetamine, methamphetamine (Meth), methylenedioxy congeners, and MDMA, or ecstasy, are the best-known psychostimulants belonging to the class of "amphetamines" with disparate molecular sites of action in the CNS. Meth, due to its low cost and long duration of action, is the most abused drug (more than heroin and cocaine), used by more than 25 million people in the world; 5.2% of American adults have used Meth at least once, with populations in the Western, Southern, and Midwest states being the most affected in the United States (Rathinam et al., 2007). Meth abuse has reached epidemic proportions worldwide and is a major public concern as it is associated with deficits in attention, memory, and executive functions in humans. Norephedrine, hydroxyamphetamine, and hydroxynorephedrine are active metabolites of amphetamine, and *N*-methyl-α-methyldopamine α-methyldopamine and 6-hydroxy-α-methyldopamine are active metabolites of ecstasy (or MDMA) and act as neurotoxicants in the brain. The NP complications might, in part, be related to drug-induced neurotoxic effects, demonstrated by damage to dopaminergic and serotonergic terminals, induced oxidative stress, activation of transcription factors, DNA damage, excitotoxicity, blood–brain barrier (BBB) breakdown, and neuronal apoptosis, as well as activation of astroglial and microglial cells in the brain.

The potential clinical neuropathological consequences of Meth abuse include agitation, anxiety, aggressive behaviors, paranoia, hypertension, hyperthermia in acute users, and deficits in attention, working memory, and executive functions in chronic psychostimulant abusers. In addition to its effect on the brain, ingestions of large doses of Meth cause life-threatening hyperthermia (above 41°C), cardiac arrhythmias, heart attacks, cerebrovascular hemorrhages, strokes, seizures, renal failure, and liver failure. Here we attempt to discuss some of the mechanisms that underlie Meth-induced neurodegenerative effects.

Oxidative Stress Due to High Dopamine Content in Methamphetamine-Induced Neurotoxicity

Meth toxic effects are due to its similar chemical structure to DA, by virtue of which Meth has high affinity for the DAT (Iversen, 2006), allowing its entry into dopaminergic neurons, replacing endogenous DA, and depleting vesicular stores of DA by release from synaptic vesicles into the cytoplasm and by reverse transport into the synaptic cleft, thereby promoting high DA levels in the brain. The neurotoxic effects due to DA release by Meth are due to the formation as well as redox cycling of DA quinones and superoxide radicals within nerve terminals. Meth intoxication induces increased DA levels and its metabolism by monoamine oxidase increases hydrogen peroxide production, which further has the capacity to interact with metal ions to form very toxic hydroxyl radicals. Various studies have shown that Meth neurotoxicity is further increased by shifting the balance between reactive oxygen species (ROS) production and the reduced capacity of the antioxidant enzyme system to scavenge ROS. The excessive generation of ROS due to Meth intoxication has the ability to damage the building blocks of the cell, such as lipids, proteins, and mitochondrial and nuclear DNA (Potashkin and Meredith, 2006). Meth reduces the levels of glutathione and antioxidant enzymes and increases lipid peroxidation and the formation of protein carbonyls. The use of antioxidants selenium, melatonin, ascorbic acid, vitamin E, the free radical scavenger, edaravone, and 7-nitroindazole completely blocks the formation of 3-nitrotyrosine (3-NT) striatal DA, 5-HT depletion, and tyrosine hydroxylase (TH), providing evidence of NO- and peroxynitrite-based mechanisms underlying Meth-induced monoaminergic neurotoxicity.

Fos B Transcription Factor Is Responsible for the Synaptic Modification Following Meth Abuse

Gene expression changes in the mesocorticolimbic system of an abused brain modify synaptic transmissions that underlie some of the enduring behavioral characteristics of addiction. Activation and suppression of transcription factors and epigenetic mechanisms and induction of noncoding RNAs are some of the mechanisms through which drugs of abuse contribute their long-lasting changes in the brain. ΔFosB is one such transcription factor, a truncated splice variant of the FosB gene, and it shares homology with other Fos family members including c-Fos, FosB, Fra1, and Fra2, which all heterodimerize with Jun family proteins (c-Jun, Jun B, or Jun D) to form activator protein-1 (AP-1) transcription factors. Meth effects are in part mediated by the activation of AP-1 transcription factors including upregulation of c-Jun, c-Fos, Jun B, and Jun D expression. c-Fos heterozygote mice showed substantial neuropathology in terms of DA axonal degeneration, neuronal cell death, and decreased basal levels of cell adhesion receptors integrins following Meth administration, supporting a protective role of c-Fos as well as integrins against Meth damage (Gilcrease, 2007).

Meth Neurotoxicity Involving DNA Damage

Meth-induced oxidative stress leads to DNA damage by the formation of oxidative DNA products such as 7,8-dihydro-8-oxoguanine. The threat of this oxidative DNA damage involves mutations and transcriptional delay contributing to teratogenic effects because developmental events must occur within a specific time window and any delays may result in structural or functional abnormalities of the developing brain (Pastoriza-Gallego et al., 2007). Cell cycle arrest or apoptosis resulting from oxidative DNA damage may also play a role in developmental abnormalities, which account for developmental deficits observed in children born to Meth-abusing mothers. Global gene expressions hint that Meth administration causes changes in the expression of a number of genes that participate in DNA repair, including multifunctional DNA repair enzyme (APEX), polymerase β (PolB), and ligase1 (LIG1).

Meth Excitotoxicity Is Mediated by Glutamate Release and Activation of Glutamate Receptors

GLU is the most abundant excitatory amino acid in the CNS and is capable of producing neuronal damage. MDMA administration in animals induces neuroadaptive changes in gene transcript expressions of glutamatergic N-methyl-D-aspartate (NMDA) and AMPA receptor subunits, metabotropic receptors, and transporters in cortex, caudate putamen, hippocampus, and hypothalamus brain regions, regulating the reward-related associative learning, cognition, memory, and neuroendocrine functions. In addition, NMDA GLU receptor antagonist MK-801 can attenuate Meth-induced dopaminergic toxicity (Battaglia et al., 2002). Memantine, an antagonist of NMDA-receptor and alpha-7 nACh receptors, prevents MDMA-induced serotonergic neurotoxicity in rats (Chipana et al., 2008). Knockout mice deficient in either neuronal nitric oxide synthase or inducible nitric oxide synthase (iNOS) showed protection against psychostimulant-induced damage to monoaminergic axons, indicating the GLU/NO pathway as a strong player in Meth neurotoxicity.

Meth Neurotoxicity Is the Result of Blood–Brain Barrier Dysfunction and Hyperthermia

Meth crosses the BBB easily. BBB disruption is investigated by using protein tracers, such as Evans Blue, iodine, and albumin; immunohistochemistry IgGI in the cortex, hippocampus, thalamus, hypothalamus, cerebellum, and amygdala revealed the effect of Meth neurotoxicity. The appearance of albumin immunoreactivity in the neuropil and leakage of serum albumin into the brain tissue had been observed as a consequence of Meth-induced BBB breakdown, further leading to neuronal damage, myelin degeneration, reactive astrocytosis in the parietal and occipital cortices, extensive degeneration of pyramidal cells, and activation of microglia in the amygdala and hippocampus of rats (Sharma et al., 2007). The neurotoxic effects of Meth-induced BBB damage has been linked to hyperthermia, as Meth causes dose-dependent temperature increases, which was generally correlative in different brain sites (NAcc, hippocampus), temporal muscle, and body core. The acute hyperthermia produced by MDMA in laboratory experimental animals is one of the few effects that can be directly compared to humans. Antioxidant H-290/51 pretreatment prevented hyperthermia, neuronal damage, myelin degradation, glial response, and leakage of serum albumin into brain tissue, establishing the role of free radicals in BBB damage and oxidative stress in Meth neurotoxicity.

Mitochondrial and Endoplasmic Reticulum—Dependent Death Pathway in Meth-Induced Apoptosis

Mitochondria are common targets of ROS and reactive nitrogen species, and GLU-mediated excitotoxicity produced by Meth. Mitochondrial dysfunction due to depletion of ATP, increased energy use, and inhibition of succinate dehydrogenase (complex II of the electron transport chain, ETC) and cytochrome oxidase activity (Complex IV of the ETC) play an important role in mediating the prooxidant and neurotoxic effects of the substituted Meth neurotoxicity. The two major apoptosis-triggering signaling pathways, (1) the mitochondrial (intrinsic) and (2) the death receptor pathways (extrinsic), contribute to the neurotoxic effects of Meth.

In the intrinsic or mitochondrial pathway, apoptosis is triggered by intracellular stress signals, mainly associated with outer mitochondrial membrane permeabilization (MMP). MMP may occur through an increase in ROS production at the mitochondria, favoring permeabilization of the inner mitochondrial membrane, decrease of mitochondrial potential, and, hence, the complete loss of mitochondrial function. Most morphological changes of apoptotic cells are caused by cysteine proteases, termed cysteinyl aspartate–specific proteases or caspases, which are activated by intrinsic and extrinsic pathways and cleave a restricted group of target substrates after an aspartate residue. However, MMP may also be triggered by proapoptotic members of the Bcl-2 family, which facilitate the formation of specific release channels in the mitochondrial outer membrane.

Meth-induced upregulation of proapoptotic proteins BAX and BID, and decreases in antiapoptotic proteins Bcl-2 and Bcl-XL, are accompanied by the release of mitochondrial proteins, cytochrome *c*, and apoptosis inducing factor Smac/DIABLO into the cytosol, which participates in the execution of Meth-induced apoptosis, resulting in the downstream activation of caspases 9 and 3, and disruption of various cellular proteins (Jayanthi et al., 2004).

Not only do the mitochondria experience the havoc of Meth-induced oxidative stress, cellular organelles, such as the endoplasmic reticulum (ER) regulating synthesis, folding, and transport of proteins, and the main intracellular store for Ca^{2+} can contribute to neuronal cell death. Meth-induced activation of calpain (a calcium-responsive cytosolic protease) is preceded by the generation of ROS in the ER and imbalances in Ca^{2+} homeostasis appear to conduce ER-dependent apoptosis. Calpain-mediated proteolysis of the cytoskeletal protein spectrin and the microtubule protein tau in the rat cortex, hippocampus, and striatum has been observed following Meth exposure. Calpain inhibitors attenuate Meth-induced spectrin and tau proteolysis neuronal death, strongly implicating ER stress and calpain activation in the mechanisms of Meth neuronal degeneration.

Neuroinflammatory Mechanism in Meth-Induced Neurotoxicity

The role of inflammation in the neurotoxicity of Meth is contributed to by an increase in the activation of microglia. The resident immune cells—microglia—within the CNS protect the brain against any injury and toxic damage (Raivich, 2005). Under normal physiological conditions, microglia typically exist in a resting state characterized by ramified morphology, continuously scanning the neuronal environment. However, on toxic insult, microglia are readily activated, transforming from a resting ramified state into an amoeboid morphology, which is necessary for immune response and neuron survival, but overactivation of microglia results in the secretion of a variety of reactive species including proinflammatory cytokines, prostaglandins, nitric oxide, and superoxide, all of which can damage neuronal tissue. It has been noted that the activation of microglia occurs prior to the damage of dopaminergic neurons and axons by Meth. The NMDA-receptor antagonist MK-801, dextromethorphan, and antiinflammatory drug ketoprofen inhibit microglial activation. Microglial activation can be used as a selective biomarker for neuronal axonal damage following Meth exposure. Meth-induced increase in the levels of tumor necrosis factor-α (TNF-α), IL-1β, and microglial activation describe the neuroinflammatory mechanism of Meth neurotoxicity.

COCAINE

Cocaine is a crystalline tropane alkaloid obtained from the leaves of the coca plant. Cocaine comprises nearly 1% of the coca leaf when the plant is grown in high altitudes, and people living in the Andes Mountains have chewed these leaves for at least 1200 years. Cocaine metabolism gives rise to active compounds that cross the BBB, namely benzoylecgonine, norcocaine, and cocaethylene. Cocaine and norcocaine are cationic amphiphiles, which penetrate into acidic intracellular organelles such as mitochondria, thereby leading to the neurotoxic effects of cocaine. Repeated cocaine administration in animals reduces DA, which provides an explanation for depression, irritability, anxiety, and suicide in cocaine-addicted individuals.

Epigenetic Regulation (ΔFos-B and Cyclic Adenosine Monophosphate Response Element Binding Protein) in Cocaine-Induced Neuropsychiatric Disorders

ΔFos and cyclic adenosine monophosphate (cAMP) response element binding protein (CREB) are the two main transcription factors that induce gene expression following cocaine exposure. Transcription factors can recruit chromatin-modifying enzymes executing long-lasting changes in neuronal circuits responsible for NP disorders. Epigenetic regulation of transcription is a coordinated interplay of mechanisms involving chromatin-modifying enzymes, nucleosome remodeling complexes, histone variant incorporation (transcriptionally active state), and DNA methylation (transcriptional repression), which translates incoming signaling events by altering chromatin structure in a specific and precise manner that in turn regulates the genes carrying out neuroadaptive changes following cocaine abuse. Cocaine promotes alterations in histone acetylation, phosphorylation, and methylation levels, as well as DNA methylation levels in the NAc mediating drug–induced behaviors. Cocaine increases H4 acetylation at the proximal promoters of the immediate early genes c-Fos, Fos-B (in NAc), and of cdk5 and brain-derived neurotrophic factor (BDNF) genes (in PFC) required for the development of cocaine-induced behaviors (Renthal et al., 2009). Cocaine-induced ΔFosB accumulation acts as a molecular "switch" in the transition from drug abuse to addiction due to the longer persistence of activated ΔFos-B.

Acute or chronic cocaine administration leads to an increase in the phosphorylation of CREB within the PFC, NAc, dorsal striatum, and amygdala. Once phosphorylated, CREB dimers, bound to specific CRE (cAMP response element) sites on target genes such as BDNF, corticotrophin releasing factor (CRF), and dynorphin produce behavioral, learning, and memory effects.

Cocaine-Induced Changes in Synaptic Plasticity Is Mediated Through Brain-Derived Neurotrophic Factor

BDNF is the most prevalent neurotropin in the brain and is involved in the regulation of activity-dependent synaptic plasticity, survival, and differentiation of neurons. BDNF is synthetized as pro-BDNF (32 kDa) and secreted as a mixture of pro-BDNF and mature BDNF (14 kDa); it plays a vital role in learning and memory cognitive functions (Bekinschtein et al., 2008).

BDNF is an endogenous ligand for tropomyosin-receptor kinase (Trk) receptors. BDNF binding at the cell surface initiates autophosphorylation of the tropomyosin receptor kinase B (TrkB) receptor through one of at least three signaling cascades: (1) the mitogen-activated protein (MAP) kinase pathway, which has been implicated in neuronal differentiation and neurite outgrowth; (2) the PI3-kinase pathway, which enables cell survival; and (3) The PLCγ pathway, involved in synaptic plasticity and transmission. These cascades converge on nuclear transcription factors such as CREB and ELK to regulate gene expression and induce neuroadaptations. Cocaine-induced BDNF exonIV expression within brain regions is associated with reward circuitry (Boulle et al., 2012). The promoter region of exonIV contains specific binding sites for the transcriptional regulators cAMP response element-binding (CREB) and methyl CpG binding protein 2 (MeCP2), both of which have established associations with regulation of gene expression at the epigenetic level. Cocaine increases BDNF messenger RNA (mRNA) in the PFC, and NAc is associated with increases in cocaine craving, suggesting a causal relationship between BDNF and drug-mediated responses. Increased BDNF expression is associated with enhanced cocaine cravings, and impaired BDNF signaling is associated with a variety of neurological and psychiatric conditions, including depression, schizophrenia (SZ), obsessive–compulsive disorder, dementia, anorexia, and epilepsy.

OPIOIDS

Opium has been known for centuries for its medicinal and recreational use. The active ingredient of opium is morphine, which is used for the treatment of acute and chronic pain. In the market, opioids are known by various names such as codeine, morphine, and tebanine (natural opioids); methadone (synthetic opioid); and heroin, oxymorphone, and idrossimorfon (semisynthetic opioids). The opioid system carries out the nociceptive (pain) processing, and modulation of gastrointestinal, endocrine, and cognitive functions. The opioid system comprises various receptors (OP) and they are named based on the targets of opioids: mu, KOP (kappa), DOP (delta), and NOP for the nociceptin orphanin FQ peptide receptor. The beneficial and the addictive effects of opioid use are evident from their interaction with endogenous opioids: endorphins, enkephalins, dynorphins, and endomorphins. The interaction of exogenous opioid with endogenous opioid and their receptors (mostly MOP) is responsible for the aberrant synaptic plasticity underlying addiction. μ-Receptors mediate analgesia, euphoria, physical dependence; K-receptors mediate sedation analgesia and mitosis; and δ-receptors mediate analgesia and release of growth hormones.

Opioid Actions Are Orchestrated Through Inhibitory G Protein (Gi/Go)

All of the opioid receptors are G-protein-coupled receptors (GPCRs) characterized by seven transmembrane domains (Koneru et al., 2009). Opioids convey their message to their effector protein via inhibitory G-protein (Gi/Go). Opiates disturb G-protein concentration and expression in discrete brain regions, resulting in adaptive changes in multiple intracellular signaling, accounting for opiate tolerance. Activation of opioid receptors with morphine mainly leads to: (1) closing of voltage sensitive calcium channels; (2) stimulation of potassium efflux leading to hyperpolarization; and (3) reduced cAMP production via inhibition of adenylyl cyclase, followed by reduced neuronal cell excitability and nerve impulses along with inhibition of neurotransmitters release.

Intracellular Targets of Opioid Activation: cAMP, Ca^{2+}, MAPK, CREB, AP-1 Transcription Factors

The slowing down of neurological functions after opioids is due to inhibition of Ca^{2+} current, increase in potassium conductance, and inhibition of formation of cAMP. In addition to this, opioids also display excitatory activity through increased mobilization of Ca^{2+} from its intracellular stores as a result of stimulation of inositol lipid hydrolysis and production of IP3 and diacylglycerol (DAG) (Cunha-Oliveira et al., 2008). Opioids activate the MAPK/extracellular signal-regulated kinase (ERK) pathway integrating signals from second messenger systems, such as Ca^{2+} and DAG, and control activity-dependent regulation of neuronal functions, neuronal differentiation, and survival. MAPK/ERK phosphorylation is PKC/CAMK-II dependent (acute opioid exposure) and protein kinase A (PKA) dependent (chronic opioid exposure), and it has the ability to activate downstream transcription factors, such as Ca^{2+}/CREB, which regulate numerous genes such as c-Fos, CRF, TH, and BDNF, and enkephalins and dynorphin. The 32-kDa dopamine- and adenosine 3′,5′-monophosphate-regulated phosphoprotein plays an extensive role in the pathogenesis of opioid addiction by acting as a mediator of the activity of the ERK signaling cascades, and it represents a thrilling hub for drug-induced changes in long-term synaptic plasticity.

Neurotoxic Effects of Opioids Linked with Oxidative Stress, Mitochondrial Dysfunction, and Apoptosis

Opioids decrease enzymatic cellular activity of superoxide dismutase (SOD), catalase, and glutathione peroxidase (GPX) and reduce glutathione levels. Lipid peroxidation and oxidative DNA and protein damage are the consequences of increased oxidative stress in opioid abuse and constitute a fundamental component of neurotoxicity. Neuronal dysfunction due to apoptotic cell death is a strong feature of opioid neurotoxicity observed in mouse cortex and hippocampus, rat spinal cord in fetal human neurons, and rat cortical neurons. Opioids cause release of cytochrome *c*, activate caspase, decrease the Bcl-2/Bax ratio, and upregulate proapoptotic proteins Fas, FasL, and Bad (Tramullas et al., 2008), causing loss of mitochondrial membrane potential, and thus implicating mitochondria as a substantial component of opiate-related neurotoxic effects.

Cognitive Deficits During Opioid Addiction Are Due to Inhibition of Neurogenesis

Decreased hippocampal neurogenesis due to opioids affects the ability of the brain to build new neurons. In adult mammalian brain there are two main neurogenic regions, the subventricular zone (SVZ), and the subgranular zone (SGZ) of the dentate gyrus in the hippocampus. Morphine and heroin decrease SGZ and SVZ neurogenesis (Eisch and Harburg, 2006), resulting in long-lasting effects on learning, memory, and cognition.

CANNABINOIDS/MARIJUANA

The derivative (marijuana) of the *Cannabis sativa* plant has been known for recreational and therapeutic uses for centuries; 3.3% of young adults (19–28 years of age) use marijuana on a daily basis and 54% of people aged between 26 and 34 have used it at least once, making marijuana the most commonly used drug of abuse in the United States. The Δ9-tetrahydrocannabinol metabolite of marijuana is the major psychoactive compound, causing its euphoric effects in the brain by acting on several neurotransmitters, including DA, γ-aminobutyric acid (GABA), GLU, serotonin, and noradrenaline. Cannabinoids, by acting collectively as a retrograde messenger, target both excitatory as well as inhibitory synapses (Alger, 2012). The biological action of marijuana and its endogenous ligands (endocannabinoids [eCBs]: made by the human body) occurs through cannabinoid receptors (CBRs: CB1R and CB2R), so-called reward pathways implicated in the neuropathophysiology of addiction. In addition to the euphoric effects of marijuana, increased appetite, anxiety, paranoia, and reversible cognitive impairment with deficits in memory and learning are also NP features of cannabis toxicity.

Cannabinoid Signaling Is Mediated by Cannabinoid Receptors and Endocannabinoids

Three major components of cannabinoid (CB) signaling consist of (1) Gi/Go protein-coupled receptors (GPCRs), mainly cannabinoid type 1 and type 2 receptors (CB1R and CB2R); (2) the endogenous cannabinoid ligands (eCBs): anandamide (AEA) and 2-arachidonoylglycerol (2-AG); and (3) synthetic, degradative enzymes and transporters modulating eCB expression and their action at receptors. The retrograde mechanism by which eCBs regulate synaptic function involves the backward flow of eCB postsynaptically to the presynaptic CB1 receptors and inhibits neurotransmitter release.

Endogenous Cannabinoid Ligand Interaction With Cannabinoid Type 1 Receptor Regulates Expression of Genes Involved in Long-Term Depression

CB1R is highly expressed in the cortex, substantia nigra pars reticulata, globus pallidus, cerebellum, hippocampus, and brainstem, thereby modulating a wide range of neural functions, such as motor activity, learning, and memory. CB1R, located presynaptically on GABA and GLU neurons, decreases their release by inhibition of presynaptic Ca^{2+} influx through voltage-gated Ca^{2+} channels (VGCCs). CB1R activation is coupled with inhibition of adenylyl cyclase, downregulation of the cAMP/PKA pathway, and activation of the MAP kinase/ERK pathway, thereby regulating expression of genes involved in the reinforcement behavior of marijuana use. eCBs are the best characterized retrograde messenger for carrying out neurological actions of endocannabinoids. Formation of 2-AG occurs through the influx of Ca^{2+} via VGCCs and by activating the enzyme phospholipase Cb (PLCb), which triggers short- and long-term plasticity (Hashimotodani et al., 2007). eCBs interaction with CB1R is capable of inducing presynaptic forms of long-term depression (eCB-LTD) at both excitatory and inhibitory synapses. Marijuana alters eCB levels, resulting in the disruption of memory and learning associated with marijuana use.

ALCOHOL

Interaction between genetic, environmental, and neurobiological factors can produce great risks for developing chronic alcoholism. Initial recreational use of alcohol can later on become alcohol dependence, featuring craving a strong desire to drink, impaired control, and withdrawal symptoms such as nausea, sweating, and anxiety. Despite the serious health and social concerns, people consume alcohol because of their vulnerability to relapse. The principal psychoactive ingredient of alcoholic beverages is ethanol, which is responsible for a wide range of clinical symptoms associated with alcohol consumption.

Modulation of γ-Aminobutyric Acid Type A and N-Methyl-D-aspartate Receptors by Ethanol Contributes to Anxiety, Arousal, and Dysphoria-Like Symptoms

The neurotransmitter system including DA, serotonin, GABA, glutamic acid, adenosine, and neuropeptides are neurobiological substrates underlying short- and long-term adaptations to ethanol dependence. Specifically, ethanol is a positive allosteric modulator of GABA receptors (i.e., reinforces inhibitory neurotransmission) and negative allosteric modulator of NMDA glutamatergic receptors (i.e., inhibition of excitatory neurotransmission). The acute sedative effects of ethanol are therefore due to the potentiation of GABAergic inhibition with long-term adaptive changes in the NMDA glutamatergic receptors, promoting alcohol dependence. Upregulation of the glutamatergic system and a downregulation of the GABAergic system due to chronic exposure to ethanol lead to compensatory changes in the expression of ethanol-responsive ion channels and receptors. This explains the consequences such as arousal, anxiety, and sleeplessness, along with the persistence of negative craving during alcohol withdrawal. In alcoholics, there is considerable interaction of opioid systems, as ethanol releases endorphins, endogenous opioid peptides, in the brain, thereby indirectly activating the dopaminergic reinforcement/reward system. Furthermore, ethanol disrupts the cholinergic system by activation of nicotinic acetylcholine receptors on dopaminergic neurons, leading to excitotoxic damage of cholinergic functions and to psychiatric disorders.

Astrocytes Are the Main Site of Ethanol-Induced Oxidative Stress and Neurotoxicity

Astrocytes represent the major cellular site of ethanol metabolism within the brain, and protect neurons from ethanol-induced oxidative stress. In the nervous tissue, ethanol can be metabolized by catalase, cytochrome P450, and alcohol dehydrogenase, and it upregulates antioxidant enzymatic activities of SOD, catalase, GPX, and heat shock proteins such as HSP70, protecting astrocytes from oxidative damage (Henderson, 2006). Ethanol in astrocytes with other cell types impairs cellular redox status, cell growth, and differentiation, interfering with the stimulatory effect of trophic factors or altering the expression of cytoskeletal proteins consequently perturbing the neuron–glia interactions and generating developmental defects in the brain (Rathinam et al., 2007; González et al., 2007). ROS generation by ethanol inside mitochondria impairs oxidative phosphorylation, ion channel modification, lipid peroxidation, and causes DNA damage, thus activating cellular death pathways responsible for alcohol-related NP disorders. Ethanol's ability to form inflammatory molecules in astrocytes occurs through nuclear factor kappa B (NF-κB) activation involved in the upregulation of cyclooxygenase 2 (COX-2) and iNOS expression.

Generation of Proinflammatory Molecules Activates Mitogen-Activated Protein Kinase Signaling by Ethanol

Glial cell activation by ethanol generates inflammatory molecules such as iNOS, COX-2, and IL-1β, which leads to signaling of MAP kinases, including ERK1/2, p-38, and JNK, causing caspase-3 activation and apoptosis. Imbalance in calcium homeostasis as a result of ethanol exposure induces nuclear fragmentation, DNA laddering, cell contraction, membrane blebbing, and chromatin condensation of astrocytes.

Abused Substances and Neuropsychiatric Disorders

Long-term exposure to drugs of abuse and being dependent on them involve changes in cognitive and executive functioning (high order skills, selective attention, inhibition, and flexibility). As discussed previously, the rewarding and reinforcing effects of these drugs are due to their interaction with the VTA, striatum, PFC (motivation and cognitive control), amygdala, and hippocampus (memory). These neuronal circuits show changes in the gene and protein levels of synaptic architecture in various NP diseases such as SZ or mood disorders (bipolar disorder), anxiety, and depression, explaining the comorbidity of mental illness as well as substance use disorder (SUD) (Fig. 51.1). The lifetime prevalence rate of any bipolar disorder with SUD is 47.3%, bipolar I disorder and any SUD, 60.3%, and major depression and alcohol dependence comorbidity, 40.3%. NP disorders such as SZ, major depression disorder (MDD), bipolar disorder (BPD), and obsessive–compulsive disorder account for one of the top category of mental ailments worldwide. According to the WHO report, by the year 2020, 4 out of 10 diseases will be NP-related diseases, with depression ranking first for both sexes and BPD ranking seventh (WHO, 2009). The critical current challenge in NP research is the development of novel interventions due to the complex pathophysiology of the NP diseases, which arise from the high level of etiologic heterogeneity and involvement of several multifactorial environmental and genetic factors. The following section of the chapter gives a brief overview of the clinical symptoms and underlying cellular mechanisms of the three main NP diseases: SZ, MDD, and bipolar disorder (BPD).

SCHIZOPHRENIA

The unknown etiology of SZ makes it a very complex psychiatric disorder, affecting 0.5%–1% of the world's population. The clinical symptoms of SZ are presented in three ways (Lewis and Sweet, 2009). Positive or psychotic symptoms include delusions, hallucinations, thought disorder, and abnormal psychomotor activity. Negative symptoms include social withdrawal, impairments in initiative and motivation, a reduced capacity to recognize and express emotional states, and poverty in the amount or content of speech. Cognitive symptoms include disturbances in selective attention, working memory, executive control, episodic memory, language comprehension, and social emotional processing. Schizophrenic individuals also experience severe depression and suicidal thoughts, abuse drugs excessively, and carry a high risk of cardiovascular disease and diabetes.

The most prominent features of SZ are as follows:

- **SZ has a strong genetic component**: Adoption, family, and twin studies strongly document heritability of SZ. The risk of developing SZ in first-degree relatives of a proband is approximately 10%, as opposed to 3% in second-degree relatives. Genome-wide linkage studies have identified at least 15 susceptible genetic loci with weak to moderate linkages to SZ including DISC1, Neuregulin 1, Dysbindin, D-amino acid oxidase activator, and catechol-*o*-methyl transferase.
- **Environmental and nongenetic factors are linked to SZ**: Season of birth, geographical variation, substance abuse, uterine environment, prenatal stress, prenatal exposure to environmental toxicants, family size, and availability of adequate maternal diet are contributing factors in the development of SZ.

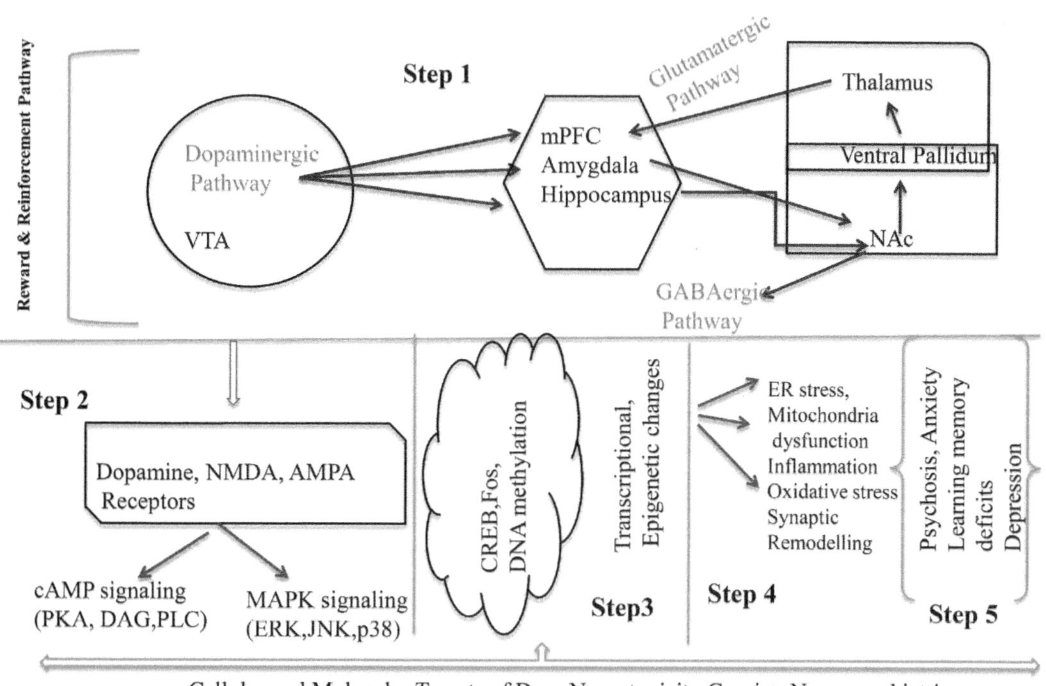

FIGURE 51.1 **Relationship between drug of abuse and neuropsychiatric disorders. Step 1**: The main brain circuits and structures implicated in the effects of the drugs are the ones that mediate reward (nucleus accumbens [NAc] and ventral pallidum), memory and learning (amygdala and hippocampus), and control (prefrontal cortex). These brain centers are innervated by dopaminergic and glutamatergic projections, mediating neuroadaptations in drug addiction. The reward pathway is acutely activated by all the drugs of abuse and mediates the reinforcing effects of these drugs. Arrows indicate innervations from one brain region to another.**Step 2:** Neurotoxicity by drugs of abuse upsets neuronal functions by imitating, stimulating, or blocking monoamine transporters (DAT; SERT; or NET) or their receptors, resulting in the activation of MAPK signaling and the cAMP pathway. **Step 3:** Activation of MAPK and cAMP pathway transduces signals to nuclear protein (CREB) and immediate early genes (fos) in the nucleus for the long-lasting changes through transcriptional and epigenetic mechanisms. **Step 4:** Protein and gene expression changes lead to structural and functional changes in the mitochondria and endoplasmic reticulum causing oxidative stress, DNA damage, apoptosis, and synaptic remodeling by altering neurotrophic levels. **Step 5:** The combined effect of step 1 through step 4 is the manifestations of neuropsychiatric symptoms such as anxiety, mood disorder, schizophrenia, and learning memory deficits.

- **Neuroanatomical alteration in SZ**: Schizophrenic PFC and hippocampus undergo substantial neuroanatomical changes. The brain and gray matter volume is reduced, whereas third and fourth ventricular volume is increased.
- **Neuronal circuits undergo functional changes in SZ**: Changes in neural substrates such as AMPA and NMDA receptors and their affiliated signaling, trafficking, scaffolding, and anchoring proteins impact learning, memory, and other cognitive functions that underlie neuroplasticity. Disruption of both NMDA and AMPA receptor subtypes, with change in binding sites as well as altered stoichiometry of subunit composition contribute to psychopathology in SZ. The expression of NMDA-associated protein postsynaptic density (PSD95, 95 kDa), SAP102 (synapse associated protein (102 kDa), neurofilament-L in the thalamus, occipital cortex, PFC, and hippocampus are found to be altered in schizophrenic postmortem brains. These NMDA-associated proteins facilitate the interface of NMDA receptors and intracellular signaling processes. AMPA receptor–related molecules such as SAP97 (synapse associated protein, 97 kDa), and NSF (*N*-ethylmaleimide-sensitive fusion protein) expression is changed in SZ. NSF and SAP97 are components of the intracellular machinery that mobilize AMPA receptors to the plasma membrane from constitutive and newly synthesized pools of receptors, respectively. Disordered signaling due to alteration in the NMDA and AMPA receptor associated molecules is one of the potential causes of impaired neuroplasticity in SZ. The reduced transcription of synaptic proteins suggests that SZ is a disorder of neuronal miswiring.
- **Disrupted synaptic transmission in SZ**: Magnetic resonance spectroscopy of schizophrenics showed glutamatergic abnormalities due to decreased levels of the amino acid N-acetyl aspartic acid in the PFC, temporal cortex, thalamus, basal ganglia, and cerebellum. Schizophrenic brain also has reduced expression of DA receptor subtypes or dopaminergic

synaptic function, reduced GABA synthesis, and GLU receptor hypofunction.

- **Cytokines in SZ**: Dysregulation in the cellular and humoral immune responses by shifting a type 2 immune response (Th2) to a type 1 (Th1) response is seen in SZ. The levels of interferon-γ gene, Th1 proinflammatory cytokine, IL-1β, TNF-α, IL-6, IL-2, IL-10, and IL-4, critical antiinflammatory cytokines, are elevated in the SZ disorder.
- **MAP kinase signaling proteins are altered in SZ:** ERK, JNK, and p38 are the three MAP kinase groups that modify protein and gene expression in neurons following exposure to abused drugs. Because of the cardinal role of MAP kinase in mediating synaptic plasticity, learning, and memory, alterations in the expression and/or function of various intermediates of MAP kinase cascades are involved in the neuropathophysiological events such as apoptosis in SZ. The altered protein expression of ERK, JNK, and p38 MAP kinase and their downstream gene target, such as Elk-1, CREB, ATF-2, c-Jun, and c-Fos, increases in the cerebellum and thalamus in SZ (Todorova et al., 2003).
- **Mitochondrial functional and structural changes account for multifactorial inheritability in SZ:** Morphological deformation and a reduction in the number and density of mitochondria, followed by changes in the protein as well as mRNA levels of the mitochondrial oxidative phosphorylation system (OX-PHOS), cytochrome *c* oxidase (complex IV), and NADH ubiquinone dehydrogenase (complex I), indicate that mitochondrial dysfunction constitutes an important component of impaired brain energy metabolism, developmental aberrations, abnormal neurotransmission, and neuronal connectivity, thus accounting for the multifactorial inheritability in SZ.

Bipolar Disorder/Mood Disorder

Bipolar disorder (BD) is a mood disorder characterized by recurrent episodes of mania and depression. The major depressive disorder (unipolar disorder), depression, and bipolar disorder are the most commonly diagnosed mood disorders. Elated mood and hyperactivity along with diminished sleep are the features of the manic episode, whereas the depressive episode is characterized by depressed mood, loss of interest, appetite loss, sleep disturbance, psychomotor retardation, feelings of worthlessness, and suicidal ideation. It is estimated that 12%–17% of individuals suffer from mood disorders at some point in their lifetime, with treatment costing the United States alone 100 billion dollars each year. The clinical symptoms of BD suggest that it is a neuroprogressive disorder with changes in tissue damage, and neuronal structure and functions that form the neural substrate for the regulation of mood.

The mechanisms of BD include the following:

- **Genetic heritability is the major risk factor for BD**: The most established finding is the familial nature of mood disorders, evidenced by the role of genetic factors in twin studies. Genetic factors are liable for 37%–75% of major depressive disorder and 60%–85% for bipolar disorder (Smoller and Finn, 2003). In twin studies, monozygotic twins have two- to threefold higher concordance for major depressive disorder than dizygotic twins and four- to eightfold higher concordance in BD. Genetic studies have implicated chromosomes 18, 11, and X.
- **Structural and functional CNS changes in BD**: Structural abnormalities in BD are reflective of ventricular enlargement and progressive loss of gray matter thickness. Voxel-based morphometry studies of gray matter in BD found gray matter reduction in left rostral anterior cingulate cortex (ACC) and right frontoinsular cortex, brain areas linked to executive control and emotional processing (Bora et al., 2010). Genetic studies indicate that individuals with BD and Val66Met BDNF genotype had significantly lower anterior cingulate volumes than those with a homozygous Val/Val genotype.
- **High levels of cytokines are associated with BD**: Inflammation in BD is characterized by increased plasma levels of proinflammatory cytokines, such as interleukin-IL-6 and tumor necrosis factor-α; increased acute phase protein levels, including haptoglobin, CRP, and complement factors, C3C or C4 concentrations; and increased IL-1β, NF-κB, and IL-1RA protein and mRNA levels in postmortem frontal cortex of bipolar patients. High levels of cytokines are associated with depression, mania, and cognitive decline in BD (Goldstein et al., 2009). Gene association studies showed -511C/T polymorphism of IL-1β with gray matter deficits in bipolar patients.
- **Oxidative stress and mitochondrial dysfunction in BD:** Oxidative stress and mitochondrial dysfunction is the fundamental abnormality in BD. High rates of comorbidity of BD with mitochondrial diseases, impaired brain energy metabolism, the effects of mood stabilizers on mitochondria, increased mitochondrial DNA deletion, mitochondrial DNA mutations/polymorphisms, and downregulation of mitochondrial gene expression in BD suggest that mitochondrial impairment is a critical component of the pathophysiology of BD. Reduced activity of mitochondrial ETC complex I, and altered expression

of subunits from mitochondrial ETC complex I, also contribute to the mitochondrial dysfunction in BD. Genome-wide linkage scans suggested linkage of chromosome 19p13 in BD, where the majority of complex I subunits genes are located. Pharmacological use of glutathione precursor and the free radical scavenger N-acetyl cysteine reduces depressive symptoms and improves functioning and quality of life in BD.

- **cAMP second messenger system is crucial in BD:** The cAMP actions are carried out through serotonergic (5-HT) and adrenergic G-protein-coupled receptors (GPCR) and are thought to play a central role in mood regulation. These receptors bind to stimulatory Gs and inhibitory Gi/Go proteins, activating adenylyl cyclase (AC), the enzyme that catalyzes the formation of cAMP from ATP. cAMP regulates many cellular functions of CNS, including metabolism and gene transcription, through the activation of PKA, which further processes the upstream information by phosphorylating the cellular substrates, such as cAMP regulatory element binding protein (CREB), leading to transcription of potentially relevant CRE-containing genes such as BDNF that regulate cell proliferation, development, plasticity, and survival of neurons. Various tricyclic antidepressants increase the expression of CRE-containing genes, such as CREB and BDNF, suggesting that the cAMP system is downregulated in depressed subjects and pharmacological upregulation of cAMP signaling produces protective effects. Mood stabilizers such as lithium elevate basal AC activity, and VPA decreases the cAMP response to stimuli.

Anxiety Disorders

The development of a mild form of anxiety and fear are crucial components of adaptive changes for the overall behavioral and autonomic "stress" response to dangerous situations that threaten to perturb homeostasis. Anxiety disorders have a lifetime prevalence of 15%–20%. The most common is generalized anxiety disorder, with lifetime prevalence close to 5%; social anxiety prevalence ranges from 2% to 14%, panic disorder ranges from 2% to 4%; and posttraumatic stress disorder has a prevalence of 8%. Any peripheral adverse stimuli activating sympathetic activity has the potential to activate brain areas, such as the locus coeruleus, amygdala, hypothalamus, and hippocampus, which are of prime importance in the formation of anxiety and fear. Various biochemical and cellular studies have identified that centrally acting neurotransmitters (DA, 5-HT, GABA, NE), corticotrophin releasing hormone, neuropeptide Y, cholecystokin, and substance P and their relevant neural circuits are implicated in the development of anxiety. A substantial number of transgenic studies investigating anxiety-like behavior have identified the key intracellular signaling enzymes, including calcium/calmodulin-dependent protein kinase-II (CAMK-II) α, adenylate cyclase type VIII, and PKCε and γ. The knockout of PKCε and γ provides resistance to anxiety and inhibition of PKC has anxiolytic effects in humans. Inositol, the building block of the PI signaling pathway, has been examined as a potential anxiolytic in the clinical trials of pain and depression.

ANIMAL MODELS IN NEUROPSYCHIATRY

Animal models have increased our understanding of psychiatric diseases tremendously. Animal models are designed based on behavior and psychopharmacology theories, mimicking a human condition, and the substantive contribution of animal models to fundamental aspects of human psychopathology is quite convincing. Two major goals of psychiatry for which animal models are designed are: (1) to know the underlying mechanisms of psychiatric diseases, and (2) to accurately determine the behavioral and pharmacological treatments of psychiatric diseases. To achieve these goals, the following approaches are applied:

1. Genetics: Selective breeding, random mutation and screening, transgenic animals (for example, knockouts, knockins, and overexpression), viral-mediated gene delivery to brain.
2. Pharmacology: Administration of neurotransmitter agonist or antagonist.
3. Environmental: Chronic social stress (adult or during development) and chronic physical stress.
4. Electrical stimulation and lesions: Brain stimulation, including optogenetic approaches and anatomical lesions.

The validity of psychiatric animal models in mimicking human symptoms is still being investigated. There are four measures that describe validity: construct validity, face validity, content validity, and predictive validity.

1. Construct validity is "the degree to which a score can be interpreted as representing the intended underlying construct" (Cook and Beckman, 2006). Construct validity in animal models incorporates a conceptual similarity to the cause of human disease.
2. Face validity is "the extent to which an animal model recapitulates important features of a human disease" (Nestler and Hyman, 2010).

3. Content validity "evaluates whether all dimensions of the target are measured by the given test or recapitulated by the given model" (Nestler and Hyman, 2010).
4. Predictive validity is "measure of agreement between results obtained by a new instrument or model and the results obtained from previously accredited tools or models (Gold Standards)." The predictive validity is often quantified by the use of statistical correlation coefficients between two sets of measurements for the same population. Predictive validity in animal models "incorporates specificity of responses to treatments that are effective in human disease" (Nestler and Hyman, 2010).

Besides validity, the reliability, efficacy, and practicality of animal models should also be evaluated and considered for a particular animal model to support clinically productive hypotheses.

Animal Models of Depression

"Learned helplessness" or "behavioral despair" have been used to demonstrate how animals stop struggling in an inescapable situation, and thus model depression. The most employed behavioral tests in labs to study depression are the "forced swim test" (FST) and "tail suspension test" (TST).

1. **FST:** Based on the fact that animals subjected to an inescapable situation of being located in water for an unlimited time, will eventually stop swimming (Porsolt et al., 1977).
2. **TST:** Based on the fact that mice under inescapable stress of being suspended become immobile (Steru et al., 1985). FST and TST are used for primary antidepressant screening and they mimic acute stress only.
3. **The foot shock escape:** This is another test that has been proposed as a model of learned hopelessness in mice in which an animal can escape an environmental stressor (Fukui et al., 2007). This test has a limitation, as it displays symptomatic overlap between depression and anxiety manifestations.
4. **Chronic social defeat test (CSDF):** CSDF is another advanced behavioral model of learned helplessness in which long experience of defeat in daily social confrontations can lead to patterns of passive submissive behavior (Venzala et al., 2012). In one form of this paradigm, a mouse is repeatedly subjected to a more dominant mouse, creating chronic stress. As opposed to FST and TST, which mimic acute stress only, CSDF overcomes this problem, mimicking chronic stress, as the majority of stress stimuli in humans leading to psychopathology are of a social nature, and the CSDF model has an advantage over other models in its resemblance to human stress-related psychopathologies.
5. **Olfactory bulbectomy:** Bilateral ablation of olfactory lobes is the most recommended fit model of depression. Olfactory ablation disrupts the limbic–hypothalamic axis with neuroendocrine changes compatible with depression in humans. Bilateral ablation of the olfactory lobes of rats results in behavioral changes, gender recognition, and social dominance among male rodents. Validity instruments have shown a significant overlap between these symptoms in rats and patients with major depression.

Animal Models of Anxiety

Anxiety and fear emotions are essential components of endurance in a living species. Adverse life events shape anxiety-related pathologies, such as generalized anxiety, panic disorders, or social phobia, which reflect extremes in trait anxiety. Animal models of anxiety are required to expose the neuroendocrine, neurochemical, and neurogenetic mechanisms of complex behavioral phenotypes for potential targets for psychotherapy. The following animal models are the best studied for anxiety traits:

1. **Fear-potentiated startle (FPS):** The FPS test includes two steps in which an intrinsically aversive unconditioned stimulus United States (foot shock) is paired with a conditioned stimulus (CS, e.g., light) initially. In the next step, the CS is subsequently used to elicit a startle reflex. Fear-potentiated startle is defined as greater startle amplitude in the presence of the CS (vs. absence). The acoustic startle response (ASR) is enhanced in certain anxiety disorders (Bijlsma et al., 2010).
2. **Contextual fear conditioning:** The animal is placed in a novel environment where an aversive stimulus is provided. When the animal is returned to the environment again, it may demonstrate a fear response if it remembers the environment. Freezing (a period of watchful immobility except for respiration) is measured as an indicator of fear.
3. **Cued fear conditioning:** Cued fear conditioning is theoretically similar to contextual conditioning, with an exception that a cue is added as a CS to exclude the effect of context itself. This is because the context (environment) is not as accurate as a shock to elicit the fear responses. Moreover, to separate contextual effects from conditioning induced by the cue, some investigators preexpose the animals to the context without the cue. On a second exposure to the context, the cue is administered to evaluate the amount of fear.

4. **Open-field test:** Unconditioned fear and anxiety is assessed by open field test. The standard open field is a brightly lit, circular arena with opaque walls and a floor scored in a grid. The animal is placed on a small stage in the open field. The test examines the internal dilemma between the innate fear of being in the central open area versus the tendency to explore new environments. When they are anxious, animals prefer to stay close to the walls (thigmotaxis).
5. **Elevated plus maze:** The maze is a cross-shaped device with two open and two closed arms elevated above the floor. The elevated plus maze examines the conflict between desire to explore a novel environment and the tendency to escape from the elevated open arms of the maze. When anxious, the natural tendency of rodents is to prefer enclosed dark spaces. An increase in proportion of open arm activity (duration and/or entries) reflects less anxiety (Carter et al., 2011).
6. **Light–dark transition:** The concept of the light–dark transition test is the same as for the elevated plus maze. The apparatus consists of a dark chamber and a brightly illuminated chamber. This forces the animal to choose between exploring the brightly lit area of the open field or staying in the dark area. When anxious, the mice tend to stay in the dark area. Anxiety-reducing drugs increase the number of entries into the bright chamber and the duration of time spent there (Savignac et al., 2011).
7. **Social interaction test:** This test is designed to see the natural reaction of an animal meeting a stranger animal. This test provides a choice between going into an empty chamber or a chamber containing a stranger mouse. Various behavior measures such as sniffing, following, allogrooming, biting, and mounting are considered as dependent variables. Other investigators may compare time spent by the animal examining an inanimate object to time spent interacting with the stranger (Flint and Shifman, 2008).
8. **Vogel conflict test:** The Vogel conflict test has been widely used for detecting anxiolytic-like effects of drugs. In this test water-deprived animals are given water. After a short period of drinking, all or some licks on the drinking are punished by mild electrical shocks leading to a significant reduction of water consumption in deprived animals. There are differences in water deprivation, water access periods, and punishment schedules, and drinking responses are reestablished by anxiolytic drugs (Gleason and Witkin, 2007).
9. **Defensive probe burying:** This test measures the effects of an aversive stimulus and anxiolytic drugs on behavior. A stationary electrified probe is used for shock, following which, rats exhibit varying periods of inactivity, immobility (a passive response), and bouts of burying behavior (an active response) consisting of vigorous burrowing, shoveling, and flicking movements with the forepaws and head to displace the cage-bedding material toward the probe.

Animal Models of Schizophrenia

SZ, with its positive, negative, and cognitive deficit symptoms, is one of the most difficult psychiatric disorders to model. Delusion-like symptoms of SZ have not yet been modeled, but others such as deficits in sensorimotor gating, working memory, and social recognition have been modeled in mice. A majority of the animal models of SZ are based on the induction of abnormal behaviors by psychostimulants and hallucinogens.

1. **Psychostimulant model:** Amphetamine exposure is the most widely used drug-induced model of SZ and exhibits similarities between the effects of amphetamine in humans and the symptoms of SZ (face validity). Amphetamine releases DA by acting directly on presynaptic dopaminergic terminals. Animals receiving amphetamine challenge exhibit excessive locomotor activation; this is an excellent measure of face validity given that individuals with SZ have excessive striatal DA release in response to an amphetamine challenge compared with healthy controls.
2. **Sensorimotor gating model:** The impairments in gating or filtering, leading to sensory overload and cognitive fragmentation, are prominent clinical features of SZ; this is termed as *sensorimotor gating*. The model of sensorimotor gating, such as prepulse inhibition (PPI), which measures sensorimotor gating, is the attenuation of the startle reflex following a sudden intense startling stimulus when a weaker sensory stimulus is previously applied. PPI provides an opportunity for normal individuals to filter or block most of the unnecessary sensory and cognitive stimuli on a daily basis, which reflects a protective function; lack of this ability is often seen in patients with SZ. A slight variation of PPI is the ASR, observed following the presentation of a sudden loud noise. ASR measures startle activity by coupling the startle platform to a 0020 strain gauge transducer. Habituation to the startle response reflecting abnormal stimulus processing is a clinical feature observed in SZ. Another model of sensory gating is latent inhibition (Li), a measure of reduced learning or attentional dysfunction. It is based on the hypothesis that nonreinforced, incidental stimulus exposure disregard the stimulus, which suggests inhibition of memory formation. Li has reasonable construct validity, as attention and information processing is disrupted in SZ.

3. **Social interaction model:** The core negative symptom in SZ is the lack of socialization in the affected individuals. A social interaction test model in animals is used to see the interaction between the mouse and an unfamiliar stranger mouse in an open field. The unfamiliar mouse moves freely (Miyakawa et al., 2003) or is contained in an enclosure (Sankoorikal et al., 2006).

BIOMARKERS OF BIPOLAR DISORDER AND SCHIZOPHRENIA

The quest for identification of biomarkers involved in psychiatric disorders has been going on for many years. According to National Institutes of Health "a biomarker is a characteristic that is objectively measured and evaluated as an indicator of normal biological and pathogenic processes, or pharmacological responses to a therapeutic intervention." The task of identifying a biomarker for NP disorder is quite daunting because of the overlap and complexity of symptoms between disorders. The diagnosis of most NP disorders is based on self-report and behavioral observations, so the strategies to improve therapeutic selectivity, which largely rely on the identification of biomarkers, will lead to improved diagnosis and potentially pave the way for more effective treatment of patients.

Biomarkers for bipolar disorder (BD) have been categorized into neurobiological factors, peripheral markers, and genetic variables.

1. Neurobiological factors: Earlier studies revealed that biological factors such as expansion of the third and lateral ventricles, white matter hyperintensities, condensed gray matter in the hippocampus and cerebellum, and enlarged size of the amygdala indicate structural abnormalities in bipolar disease. Individuals with hyperactivation of the frontal polar cortex and increased memory load are more prone to develop BD, making these two factors markers for BD (Thermenos et al., 2010).
2. Peripheral markers: Serial monitoring of peripheral markers could be useful as a simple, rapid, low cost, and dependable method for prediction and treatment outcome in bipolar disease. The markers that have been assessed in BD patients include BDNF, glial involvement, biochemical markers—inflammation (pro- and antiinflammatory cytokines)—and oxidative stress and mitochondrial dysfunction.
   - BDNF: BDNF, a 27 kDa polypeptide synthesized from proBDF, belongs to neutrophins, which are homodimeric proteins. It is a significant mediator for cell survival and functions by binding to TrkB, so it may result in advancement of bipolar disease (Hashimoto, 2010). Studies have shown that BDNF 196G/A gene polymorphism results in hippocampal activation, decreased hippocampal N-acetyl aspartate and periodic memory suggesting its involvement in bipolar disease. Its serum level was reported to be decreased in depressed, hyper, and unmedicated patients in the late stage of illness. Further study revealed decreased levels of mRNAs of lymphocyte-derived BDNF and of protein BDNF levels in platelets of hyper patients. Therefore, BDNF plays an important role in BD pathogenesis.
   - Glial involvement: Astrocytes secrete calcium-binding protein, S100B, which causes paracrine and autocrine effects on neurons. Glia S100B toxicity is directly proportional to its concentration in serum (Andreazza et al., 2007). It is also an important marker in various psychiatric disorders such as SZ and major depression. Studies have reported that during mood variations, the activity of astrocytes is altered, resulting in increased extracellular levels of S100B. Higher serum levels of this protein during depressive and hyper mood cause brain damage. Therefore, the alteration of glial and astrocyte cells in mood disorders may be involved in the pathogenesis of BD.
   - Inflammatory factors: Various studies have shown that proinflammatory cytokines IL-2, IL-6, IL-8 (interleukins) and TNF-α were found to be increased in depressive and late stages of BD patients as compared to controls. This is in conformity with the results that showed that IL-6 and TNF-α mRNA levels were increased in BD patients with hyper state of illness. It has been postulated by proteomics study that BD involves inflammatory pathways of IL-13, TNF-α, and apolipoprotein A1 (Herberth et al., 2011). One of the studies also revealed that high sensitivity C-reactive protein is associated with worsening of the hyper state in youngsters with BD (Berk et al., 2011). Antiinflammatory cytokine IL-10 was found to be decreased in the hyper state. Therefore, evaluation of proinflammatory factors may enlighten the role of these important markers in the progression of BD.
   - Oxidative stress: Oxidative stress is caused by an imbalance between ROS and antioxidants. Malondialdehyde (MDA), the end product of lipid peroxidation, can be produced from free radical attack on polyunsaturated fatty acids (PUFAs) (Andreazza et al., 2008). It is a marker of oxidative stress, which can be evaluated through the levels of thiobarbituric acid reactive substances (TBARS). Oxidative stress causes disruption in lipid messenger signaling systems and alters the conformation of proteins. An increment in TBARS levels has been postulated in BD patients in hyper, depressive, and reduction stages leading to

continuous oxidative damage to the lipid structure during BD. Therefore, TBARS can be used to categorize BD in different stages, with higher levels of TBARS during later stages of BD. The other free radical that has been extensively studied is nitric oxide. Nitric oxide (NO) is derived from L-arginine through the Ca^{2+}-dependent enzyme, nitric oxide synthase (NOS). It is an essential physiological signaling molecule that mediates various cellular functions, but also induces cytotoxic and mutagenic effects when present in excess. Studies have shown that persons with BD have higher levels of GLU, glutamate + glutamine, and lactate in the PFC (Frye et al., 2007). Increased GLU levels leads to an increase in Ca^{2+} levels, which activates NOS. NOS forms nitric oxide, which disturbs normal protein function through the following equations:

$$\bullet NO + \bullet O_2^- \rightarrow ONOO^- \text{ (neurotoxic)}$$

$$(\bullet NO, \text{Nitric oxide}) + (\bullet O_2, \text{Superoxide})$$

$$\rightarrow (ONOO^-, \text{Peroxynitrite})$$

$$ONOO^- + \text{Protein} - SH \rightarrow \text{Disturbed protein functions}$$

$$(\text{Protein} - SH, \text{thiol groups of amino acids and proteins})$$

Therefore, oxidative stress causes DNA fragmentation in the ACC and lymphocytes resulting in telomere shortening, a biomarker in BD patients.

3. Genetic variables: The Alpha 1C subunit of the L-type voltage-gated calcium channel (CACNA1C) and ankyrin 3 (ANK3) are the two genetic markers that are the risk loci in BD. CACNA1C facilitates different calcium-dependent processes such as regulation of dendritic calcium influx in neurons. Increase in gray matter volume and intensity in the right amygdala and hypothalamus has been found to be due to CACNA1C rs1006737 risk allele (G to A) (Perrier et al., 2011). ANK3, ankyrin protein, modifies the neuronal sodium channel activity and plays an important role in cell motility, stimulation, and proliferation, and maintenance of specialized membrane domain. Therefore, several genetic studies confirmed these genes in modulating pathways involved in the pathogenesis of BD.

BIOMARKERS OF SCHIZOPHRENIA

Diagnosis of SZ is still complicated because of its unknown pathogenesis and etiology. The current criteria for its diagnosis include positive symptoms (hallucinations, delusions), negative symptoms (depression, lack of motivation), and cognitive impairment reported by patients, mental status examination, and clinical presentation. Because of advances in genetics, genomics, and proteomics, the overall burden of SZ can be reduced in early stages by discovering new biomarkers for it (Pillai and Buckley, 2012). The biomarkers for SZ can be categorized as *trait markers*, *peripheral biomarkers*, and *proteomic candidate markers*. These three categories are described as follows:

1. **Trait markers**: Trait markers include endophenotypes that are examined during SZ. Endophenotypes are assumed not only to be due to genetic abnormalities, but also to nongenetic factors and interaction between environmental stressors and a genetic disposition; for example, interaction between fetal hypoxia and predisposing genes has been studied in the pathogenesis of SZ (Pillai and Buckley, 2012). The different endophenotypes are cognitive measures, neurophysiological measures, and structural and functional abnormalities. Cognitive measures include selective attentions, working memory, face memory, and emotion processing. Neurophysiological measures include PPI of startle reflex, eye tracking dysfunctions, and mismatch negativity, whereas neuroimaging includes frontal hypoactivation in response to intellectual tasks, magnetic resonance imaging, and whole brain nonlinear pattern classifications. These endophenotypes, when studied together in different combinations, resulted in improved prognostic outcomes as compared to single endophenotypes (Greenwood et al., 2011). Therefore, trait markers can be assessed during the progression of SZ.
2. **Peripheral biomarkers:** These play an important role in SZ. Some of the useful peripheral biomarkers are blood-based biomarkers, such as inflammation, microRNA (miRNA), oxidative stress, and neurotransmitters and neuronal signaling molecules, such as BDNF.
   a. Blood-based biomarkers, which are analyzed in serum/plasma using a multi-analyte profiling (MAP) platform, signify the role of immunological and inflammatory processes during the first stage of SZ. The innate immune system consists of humoral and cellular components. Humoral components are activated by the complement system and acute phase proteins (APPs), which participate in systemic nonspecific reactions in response to disturbances in homeostasis called acute phase response. During stress, proinflammatory cytokines IL-6 and TNF-α are secreted in the blood, causing variations in the adrenocorticotrophic hormone (ACTH) and

glucocorticoid levels by hyperactivating the hypothalamic–pituitary–adrenal axis (Bradley and Dinan, 2010), complement system stimulation, and alterations in the APPs levels, inducing behavioral neuroendocrine and metabolic changes during SZ. The cellular component is activated by monocytes, macrophages, granulocytes, and natural killer cells. Activated T and B lymphocytes of the cellular immune system produce active immunotransmitters IL-2, IL-12, IL-18, and TNF-α, which reveal that the innate immune response is activated against stressors in SZ.

b. **miRNA:** This plays a regulatory role in SZ by inhibiting the translation of mRNA, which makes it one of the possible biomarkers for this disease. Postmortem brain studies have shown that around 30 miRNA expressions were varied in patients with SZ (Beveridge et al., 2010). Further studies revealed that the gene expression between CNS tissue and peripheral blood is thought to be the same, as peripheral leukocyte gene expressions are affected by the CNS through cytokines, neurotransmitters, and hormones. Altered miRNA expressions of blood mononuclear leukocytes were found in patients with negative symptoms, neurocognitive performance scores, and event-related potentials. This is in concordance with the study that showed that seven miRNA gene expressions were altered in leukocytes, with hsa-miR-34a most differentially expressed in SZ patients (Lai et al., 2011). Therefore, miRNA can be assessed as a noninvasive biomarker for SZ.

c. **Oxidative stress:** Free radicals generated during oxidative stress mainly affect brain because of high levels of PUFAs, relatively low levels of antioxidants, and high metal content and oxygen utilization in the brain. Oxidative stress in the form of lipid peroxidation and a compromised antioxidant system plays a key role in the pathophysiology of SZ (Ciobica et al., 2011). The products of lipid peroxidation, which were assessed during SZ, are MDA in the form of TBARS and 4-hydroxy nonenal (HNE), a highly cytotoxic reactive α,β-aldehyde. Neuronal functions are disturbed by TBARS at high levels, which change the membrane fluidity or alter the membrane receptors. Studies have examined the increased levels of 4-HNE and MDA in leukocytes, erythrocytes, and plasma in SZ patients as compared to controls. The results on antioxidant levels in SZ are ambiguous. The variations in antioxidant levels may be due to illness duration, symptom severity, and therapeutic features. Some authors have shown no difference in SOD levels, whereas some have shown both decrease and increase in SOD levels. Similar results were reported for GPX with increased and decreased levels in SZ patients.

d. **BDNF:** BDNF is one of the significant and most studied markers in SZ, which is expressed in the hippocampus, cerebral cortex, basal forebrain, striatum, and hypothalamus and regulates neuronal structure and function along with their growth, differentiation, and survival, such as GABA neurons. Different studies have reported the decrease in BDNF levels of cerebrospinal fluid and serum in first episode SZ patients (Thompson Ray et al., 2011). Similar results were observed in the PFC of SZ patients with decreased BDNF levels. BDNF knockout mice were found to have deficits in spatial learning and memory as BDNF facilitates the strengthening of hippocampal synapses associated with learning and memory. Therefore, BDNF can be considered as a potential biomarker of SZ.

3. **Proteomic analysis:** Dysfunctions of oligodendrocytes and glycolysis were observed by proteomic studies in the mediodorsal thalamus of SZ patients (Martins-de-Souza et al., 2010). Oligodendrocytes form myelin around axons and also maintain myelin in the CNS. It has been found that sphingomyelin and galactocerebrosides 1 and 2 (myelin membrane components) were decreased along with upregulation of myelin basic protein and myelin oligodendrocyte glycoprotein in the thalamus of SZ patients. Glycolysis alterations were studied by differential expression of enzymes such as triose phosphate isomerase and glyceraldehyde phosphate dehydrogenase (GAPDH) in SZ patients. Consequent studies of glycolysis alterations further revealed lower levels of pyruvate (end product of glycolysis) in SZ. Because of variations in glucose metabolism, the levels of NADPH, essential for oxidative phosphorylation, were affected. NADPH levels are regulated by proteins transketolase, μ crystallin, glutathione s-transferase, and GAPDH, which were found to be increased in SZ patients.

CONCLUDING REMARKS AND FUTURE DIRECTIONS

Addiction to drugs of abuse adversely affects the CNS by altering neurotransmitter release and their actions on their respective receptors. Chronic abuse of amphetamine, cocaine, cannabis, and alcohol is associated with deficits in attention, memory, and executive functions, suggesting their neurotoxicity to brain. Studies on the neurotoxic effect of abused drugs have identified cellular and molecular pathways involving oxidative stress, activation of transcription factors, DNA damage, excitotoxicity, BBB breakdown, microglial activation, and various apoptotic pathways through mitochondrial dysfunction and inhibition of neurogenesis. The comorbidity among drug addiction and NP disorders is because both share common neural substrates and brain areas that underlie depression, anxiety, and SZ. Genetic and pharmacological manipulations have made animal models available to understand the causes of NP symptoms. Blood-based biomarkers, such as BDNF, 4-HNE, and MDA in leukocytes and erythrocytes, and those in genetic studies, have given us more hope to diagnose and treat NP diseases effectively.

Despite the significant strides made in understanding the neurotoxic consequences of drug addiction and NP diseases, due to overlapping symptoms and shared pathways, significant challenges still exist for the complete spectrum of treatment of pathophysiology associated with NP disorders. Future studies require successful translational research on animal models, which can illuminate the path to disease mechanisms and their therapeutic targets.

References

Alger, B.E., 2012. Endocannabinoids at the synapse a decade after the dies mirabilis: what we still do not know. J. Physiol. 590, 2203–2212.

Andreazza, A.C., Cassini, C., Rosa, A.R., Leite, M.C., 2007. Serum S100B and antioxidant enzymes in bipolar patients. J. Psychiatr. Res. 41, 523–529.

Andreazza, A.C., Kauer-Sant'Anna, M., Frey, B.N., 2008. Oxidative stress markers in bipolar disorder: a meta-analysis. J. Affect. Disord. 111, 135–144.

Battaglia, G., Fornai, F., Busceti, C.L., 2002. Selective blockade of mGlu5 metabotropic glutamate receptors is protective against methamphetamine neurotoxicity. J. Neurosci. 22, 2135–2141.

Bekinschtein, P., Cammarota, M., Izquierdo, I., Medina, J.H., 2008. BDNF and memory formation and storage. Neuroscientist 4, 147–156.

Berk, M., Kapczinski, F., Andreazza, A.C., 2011. Pathways underlying neuroprogression in bipolar disorder: focus on inflammation, oxidative stress and neurotrophic factor. Neurosci. Biobehav. Rev. 35, 804–817.

Beveridge, N.J., Gardiner, E., Carroll, A.P., et al., 2010. Schizophrenia is associated with an increase in cortical microRNA biogenesis. Mol. Psychiatry 15, 1176–1189.

Bijlsma, E.Y., de Jongh, R., Olivier, B., Groenink, L., 2010. Fear-potentiated startle, but not light-enhanced startle, is enhanced by anxiogenic drugs. Pharmacol. Biochem. Behav. 96, 24–31.

Bora, E., Fornito, A., Yucel, M., Pantelis, C., 2010. Voxelwise meta-analysis of gray matter abnormalities in bipolar disorder. Biol. Psychiatry 67, 1097–1105.

Boulle, F., van den Hove, D.L., Jakob, S.B., 2012. Epigenetic regulation of the BDNF gene: implications for psychiatric disorders. Mol. Psychiatry 17, 584–596.

Bradley, A.J., Dinan, T.G., 2010. A systematic review of hypothalamic-pituitary-adrenal axis function in schizophrenia: implications for mortality. J. Psychopharmacol. 24, 91–118.

Carter, M.D., Shah, C.R., Muller, C.L., 2011. Absence of preference for social novelty and increased grooming in integrin beta3 knockout mice: initial studies and future directions. Autism Res. 4, 57–67.

Chipana, C., Torres, I., Camarasa, J., Pubill, D., Escubedo, E., 2008. Memantine protects against amphetamine derivatives-induced neurotoxic damage in rodents. Neuropharmacology 54, 1254–1263.

Ciobica, A., Padurariu, M., Dobrin, I., et al., 2011. Oxidative stress in schizophrenia – focusing on the main markers. Psychiatr. Danub. 23, 237–245.

Cook, D.A., Beckman, T.J., 2006. Current concepts in validity and 1061 reliability for psychometric instruments: theory and application. Am. J. Med. 119.

Cunha-Oliveira, T., Rego, A.C., Oliveira, C.R., 2008. Cellular and molecular mechanisms involved in the neurotoxicity of opioid and psychostimulant drugs. Brain Res. Rev. 58, 192–208.

Diagnostic and Statistical Manual of Mental Disorders, fourth ed., 1994. American Psychiatric Association, Washington, DC.

Eisch, A.J., Harburg, G.C., 2006. Opiates, psychostimulants, and adult hippocampal neurogenesis: insights for addiction and stem cell biology. Hippocampus 16, 271–286.

Flint, J., Shifman, S., 2008. Animal models of psychiatric disease. Curr. Opin. Genet. Dev. 18, 235–240.

Frye, M.A., Watzl, J., Banakar, S., 2007. Increased anterior cingulate/medial prefrontal cortical glutamate and creatine in bipolar depression. Neuropsychopharmacology 32, 2490–2499.

Fukui, M., Rodriguiz, R.M., Zhou, J., 2007. Vmat2 heterozygous mutant mice display a depressive-like phenotype. J. Neurosci. 27, 10520–10529.

Gilcrease, M.Z., 2007. Integrin signaling in epithelial cells. Cancer Lett. 247, 1–25.

Gleason, S.D., Witkin, J.M., 2007. A parametric analysis of punishment frequency as a determinant of the response to chlordiazepoxide in the Vogel conflict test in rats. Pharmacol. Biochem. Behav. 87, 380–385.

Goldstein, B.I., Kemp, D.E., Soczynska, J.K., McIntyre, R.S., 2009. Inflammation and the phenomenology, pathophysiology, comorbidity, and treatment of bipolar disorder: a systematic review of the literature. J. Clin. Psychiatry 70, 1078–1090.

González, A., Pariente, J.A., Salido, G.M., 2007. Ethanol stimulates ROS generation by mitochondria through Ca^{2+} mobilization and increases GFAP content in rat hippocampal astrocytes. Brain Res. 1178, 28–37.

Greenwood, T.A., Lazzeroni, L.C., Murray, S.S., 2011. Analysis of 94 candidate genes and 12 endophenotypes for schizophrenia from the Consortium on the Genetics of Schizophrenia. Am. J. Psychiatry 168, 930–946.

Hashimoto, K., 2010. Brain-derived neurotrophic factor as a biomarker for mood disorders: an historical overview and future directions. Psychiatry Clin. Neurosci. 64, 341–357.

Hashimotodani, Y., Ohno-Shosaku, T., Kano, M., 2007. Ca^{2+}-assisted receptor-driven endocannabinoid release: mechanisms that associate presynaptic and postsynaptic activities. Curr. Opin. Neurobiol. 17, 360–365.

Henderson, G.I., 2006. Astrocyte control of fetal cortical neuron glutathione homeostasis: up-regulation by ethanol. J. Neurochem. 96, 1289–1300.

Herberth, M., Koethe, D., Levin, Y., 2011. Peripheral profiling analysis for bipolar disorder reveals markers associated with reduced cell survival. Proteomics 11, 94–105.

Iversen, L., 2006. Neurotransmitter transporters and their impact on the development of psychopharmacology. Br. J. Pharmacol. 147, S82–S88.

Jayanthi, S., Deng, X., Noailles, P.A., et al., 2004. Methamphetamine induces neuronal apoptosis via cross-talks between endoplasmic reticulum and mitochondria- dependent death cascades. FASEB J. 18, 238–251.

Koneru, A., Satyanarayana, S., Rizwan, S., 2009. Endogenous opioids: their physiological role and receptors. Global J. Pharmacol. 3, 149–153.

Lai, C.-Y., Yu, S.-L., Hsieh, M.H., 2011. MicroRNA expression aberration as potential peripheral blood biomarkers for Schizophrenia. PLoS One 6, e21635.

Lewis, D.A., Sweet, R.A., 2009. Schizophrenia from a neural circuitry perspective: advancing toward rational pharmacological therapies. J. Clin. Invest. 119, 706–716.

Martins-de-Souza, D., Maccarrone, G., Wobrock, T., Zerr, I., 2010. Proteome analysis of the thalamus and cerebrospinal fluid reveals glycolysis dysfunction and potential biomarkers candidates for schizophrenia. J. Psychiatr. Res. 44, 1176–1189.

Miyakawa, T., Leiter, L.M., Gerber, D.J., 2003. Conditional calcineurin knockout mice exhibit multiple abnormal behaviors related to schizophrenia. Proc. Natl. Acad. Sci. U.S.A. 100, 8987–8992.

Multani, P.K., Hodge, R., Estévez, M.A., 2012. VMAT1 deletion causes neuronal loss in the hippocampus and neurocognitive deficits in spatial discrimination. Neuroscience 29, 32–44.

Nestler, E.J., Hyman, S.E., 2010. Animal models of neuropsychiatric disorders. Nat. Neurosci. 13, 1161–1169.

Pastoriza-Gallego, M., Armier, J., Sarasin, A., 2007. Transcription through 8-oxoguanine in DNA repair-proficient and Csb(−)/Ogg1(−) DNA repair-deficient mouse embryonic fibroblasts is dependent upon promoter strength and sequence context. Mutagenesis 22, 343–351.

Perrier, E., Pompei, F., Ruberto, G., 2011. Initial evidence for the role of CACNA1C on subcortical brain morphology in patients with bipolar disorder. Eur. Psychiatry 26, 135–137.

Pillai, A., Buckley, P.F., 2012. Reliable biomarkers and predictors of schizophrenia and its treatment. Psychiatr. Clin. N. Am. 35, 645–659.

Porsolt, R.D., Bertin, A., Jalfre, M., 1977. Behavioral despair in mice: a primary screening test for antidepressants. Arch. Int. Pharmacodyn. Ther. 229, 327–336.

Potashkin, J.A., Meredith, G.E., 2006. The role of oxidative stress in the dysregulation of gene expression and protein metabolism in neurodegenerative disease. Antioxid. Redox Signal. 8, 144–151.

Raivich, G., 2005. Like cops on the beat: the active role of resting microglia. Trends Neurosci. 28, 571–573.

Rathinam, M.L., Watts, L.T., Stark, A.A., 2007. Why do we need an addiction supplement focused on methamphetamine? Addiction 102, 1–4.

Renthal, W., Kumar, A., Xiao, G., 2009. Genome-wide analysis of chromatin regulation by cocaine reveals a role for sirtuins. Neuron 62, 335–348.

Sankoorikal, G.M., Kaercher, K.A., Boon, C.J., et al., 2006. A mouse model system for genetic analysis of sociability: C57BL/6J versus BALB/cJ inbred mouse strains. Biol. Psychiatry 59, 415–423.

Savignac, H.M., Dinan, T.G., Cryan, J.F., 2011. Resistance to early-life stress in mice: effects of genetic background and stress duration. Front. Behav. Neurosci. 5, 13.

Sharma, H.S., Sjoquist, P.O., Ali, S.F., 2007. Drugs of abuse-induced hyperthermia, blood-brain barrier dysfunction and neurotoxicity: neuroprotective effects of a new antioxidant compound H-290/51. Curr. Pharm. Des. 13, 1903–1923.

Smoller, J.W., Finn, C.T., 2003. Family, twin, and adoption studies of bipolar disorder. Am. J. Med. Genet. Part C: Semin. Med. Genet. 123C, 48–58.

Steru, L., Chermat, R., Thierry, B., Simon, P., 1985. The tail suspension test: a new method for screening antidepressants in mice. Psychopharmacology 85, 367–370.

Thermenos, H.W., Goldstein, J.M., Milanovic, S.M., 2010. An fMRI study of working memory in persons with bipolar disorder or at genetic risk for bipolar disorder. Am. J. Med. Genet. Part B: Neuropsychiatr. Genet. 153B, 120–131.

Thompson Ray, M., Weickert, C.S., Wyatt, E., 2011. Decreased BDNF, trkB-TK1 and GAD67 mRNA expression in the hippocampus of individuals with schizophrenia and mood disorders. J. Psychiatry Neurosci. 36, 195–203.

Todorova, V.K., Elbein, A.D., Kyosseva, S.V., 2003. Increased expression of c-Jun transcription factor in cerebellar vermis of patients with schizophrenia. Neuropsychopharmacology 28, 1506–1514.

Tramullas, M., Martinez-Cue, C., Hurle, M.A., 2008. Chronic administration of heroin to mice produces up-regulation of brain apoptosis-related proteins and impairs spatial learning and memory. Neuropharmacology 54, 640–652.

Venzala, E., Garcia-Garcia, A.L., Elizalde, N., Delagrange, P., Tordera, R.M., 2012. Chronic social defeat stress model: behavioral features, antidepressant action, and interaction with biological risk factors. Psychopharmacology 224, 313–325.

World Health Organization (WHO), 2009. The Global Burden of Disease: 2004 Update. World Health Organization, Geneva.

CHAPTER

52

Osteoarthritis Biomarkers

Ramesh C. Gupta[1], *Ajay Srivastava*[2], *Rajiv Lall*[2], *Anita Sinha*[2]

[1]Toxicology Department, Breathitt Veterinary Center, Murray State University, Hopkinsville, Kentucky, United States; [2]Vets Plus Inc., Menomonie, WI, United States

Abbreviations

ADAMTS A disintegrin and metalloprotease with thrombospondin motifs
AMPK AMP-activated protein kinase
COMP Cartilage oligomeric matrix protein
K-L Grade Kellgren–Lawrence Grade
LACS Local-area cartilage segmentation
MMPs Matrix metalloproteinases
MMP-3/-9 Matrix metalloproteinases-3/-9
sRAGE Soluble receptor for advanced glycation end products
SIRT1 Sirtuin 1
TSG-6 Tumor necrosis factor-stimulated gene-6

INTRODUCTION

Osteoarthritis (OA) is an inflammatory joint disease characterized by chronic and progressive cartilage degeneration, osteophyte formation, subchondral sclerosis, bone marrow lesions, hypertrophy of bone at the margin, and changes in the synovial membrane. OA is a disease of the entire joint and surrounding muscles, and it commonly affects humans, dogs, and horses. OA can affect any joint in the body, but more commonly the knee, hip, arm, and foot. Currently, worldwide estimates are that 9.6% of men and 18.0% of women over the age of 60 have symptomatic OA. Almost 80% of patients with OA have some degree of limitation of movement, and 25% cannot perform their major daily activities of life (Lourido et al., 2014). Aging is the main causative factor for OA, although many other factors are also involved, such as injury/trauma, excess or lack of exercise, dietary habits, obesity, joint malalignment (e.g., varus or valgus deformity), genetic predisposition, infection, etc. (Abramson and Attur, 2009; van Meurs, 2017). As a result of aging, the pathophysiology of OA is very complex, and multiple mechanisms and signaling pathways are involved in inflammation, pain, and cartilage degeneration/loss.

Biomarkers play a pivotal role in the diagnosis of OA, which is often based on clinical (synovial fluid, serum, and urine biomarkers), radiographic, and MRI assessments. During the past decade, significant progress has been made in identification, confirmation, and validation of biomarkers, some of which are used in early disease diagnosis, prognosis, drug development, and monitoring recovery progress during treatment. The application of proteomic technologies has generated several new nonconventional biomarkers (Patra and Sandell, 2011; Ruiz-Romero and Blanco, 2009, 2010; Mobasheri, 2012; Lourido et al., 2014; Sanchez et al., 2017). Although values of biomarkers in various aspects of OA have been recognized for a long time, quests for novel biomarkers will continue forever (Lotz et al., 2013; deBakker et al., 2017; van Meurs, 2017). This chapter describes contributing factors and pathophysiology of OA; classification of OA biomarkers; and recent developments in identification, qualification, and validation of biomarkers of OA in humans and animals.

PATHOPHYSIOLOGY AND SIGNALING PATHWAYS IN OSTEOARTHRITIS

OA has been recognized as a heterogeneous disease of the entire joint, encompassing multiple tissues, including cartilage, bone, adipose, and skeletal muscle, thereby causing remodeling and/or failure of the joint (Abramson and Attur, 2009; Thote et al., 2013; Yuan et al., 2014; Tonge et al., 2014; Svala et al., 2017). Degradation of articular cartilage is a hallmark of OA, which occurs in two phases: anabolic phase, in which chondrocytes attempt to repair the damaged extracellular matrix (ECM), and a catabolic phase, in which enzymes produced by different cells (including chondrocytes) digest the extracellular matrix (Svala et al., 2017). This catabolic

Biomarkers in Toxicology, Second Edition
https://doi.org/10.1016/B978-0-12-814655-2.00052-9

phase also includes inhibition of matrix synthesis, thereby ensuing accelerated erosion of the cartilage.

In the early stages of OA, progressive depletion of the cartilage proteoglycan leads to a net loss of matrix from the cartilage (Thote et al., 2013). Normally, the joint cartilage consists of 5% chondrocytes and 95% ECM. In OA, chondrocytes serve as mechanosensors and osmosensors, altering their metabolism in response to physicochemical changes in the microenvironment. OA is characterized by the degradation of cartilage matrix components, including cartilage-specific type II collagen and the proteoglycan aggrecan, ultimately resulting in the loss of cartilage structure and function. Breakdown and deterioration of the cartilage have been correlated with increased activities of certain enzymes, including matrix metalloproteinases (MMPs). Fig. 52.1 shows an OA joint with thinning and destruction of the cartilage and cartilage remnants in joint space.

In OA cartilage, decreases in the number of chondrocytes and in their ability to regenerate the ECM in response to stress has been described (Portal-Núñez et al., 2016). OA chondrocytes show a senescence secretory phenotype (SSP) exhibiting overproduction of cytokines (IL-1 and IL-6), matrix metalloproteinases (e.g., MMP-3, MMP-8, and MMP-13), and growth factors (e.g., epidermal growth factor). Of the three major MMPs that degrade native collagen, MMP-13 appears to be the most important in OA because it preferentially degrades type II collagen, and it has been shown that expression of MMP-13 greatly increases in OA (Abramson and Attur, 2009). During cartilage degeneration in OA, the inflammatory processes cause excess production of ROS, RNS, oxygen, and PGE2, and as a result, their increased levels are found within the joint (Bakker et al., 2017; Wan and Zhao, 2017). Evidence suggests that in chondrocytes, excessive ROS production leads to hyperperoxidation, protein carbonylation, and DNA damage, which all play a major role in the induction of SSP (Portal-Núñez et al., 2016). Activity of proinflammatory cytokine IL-1β can be detected in synovial fluid from OA joints where it induces gene expression of matrix-degrading enzymes in chondrocytes. In the synovial fluid of OA patients, another proinflammatory molecule from natural killer cells has been shown to be expressed, i.e., protease granzyme A, which may contribute to chronic articular inflammation (Jaime et al., 2017).

Nuclear factor kappa B (NF-κB) appears to be a key transcription factor that drives the expression of OA-inducer genes in response to the activation of toll-like receptors (TLR) and interleukin receptors by ECM fragments and IL-1β. Proinflammatory cytokines, such as IL-1β, are produced in the OA-stimulated mitogen-activated protein kinase (MAPK) pathways through the extracellular signal-regulated kinases (ERK)½, p38 kinase, and c-Jun N terminal kinase (JNK). Elevated MAPK phosphorylation results in activation of transcription factors, which in turn upregulates the production of several molecules, such as MMPs and aggrecanases, which are responsible for matrix degradation. In OA, IL-1β induces activation of the NF-κB signaling pathway in human chondrocytes, thereby causing low-grade pain (Wan and Zhao, 2017). Sluzalska et al. (2017) demonstrated that NF-κB, p38, MAPK, and JNK signaling pathways are all involved in IL-1β–induced phospholipids biosynthesis (PLs). PLs are key molecules for cartilage maintenance, and their presence in synovial fluid plays a major role in joint lubrication to protect cartilage surfaces. Signaling pathways directly mediated by lipids and those that involve mammalian target of rapamycin (mTOR) pathways are reported both in normal and OA cartilage (Villalvilla et al., 2013).

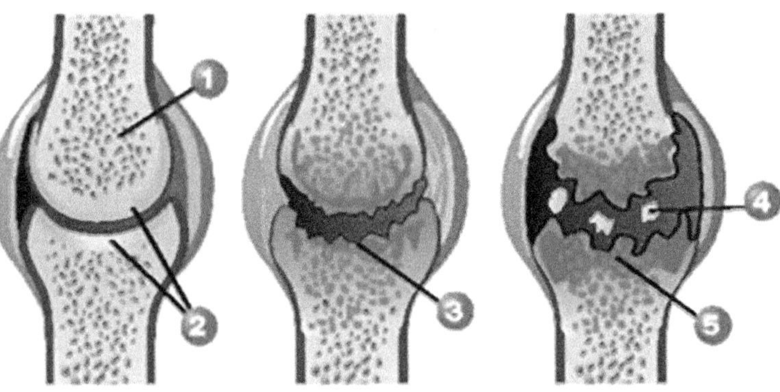

FIGURE 52.1 Osteoarthritis (OA) joint showing thinning of cartilage and cartilage remnants in joint space.

Sillat et al. (2013) demonstrated that mature chondrocytes express TLR1 and TLR2 and may react to cartilage matrix/chondrocyte-derived danger signals or degradation products, which leads to synthesis of proinflammatory cytokines; and that stimulates further TLR and cytokine expression, establishing a vicious circle. Evidence also suggests that hyaluronan and fibronectin may act as TLR pathway stimuli in development and progression of OA involving innate immune system (Scanzello et al., 2008). Furthermore, during the inflammatory response, dysregulation of AMPK and Sirtuin 1 in articular chondrocytes modulate intracellular energy metabolism (Liu-Bryan, 2015). These findings suggest that OA can act as an autoinflammatory disease and links the old mechanical wear-and-tear concept with modern biochemical views of OA. In essence, the chondrocyte itself is the earliest and most important inflammatory cell in OA.

In an interesting investigation, Koskinen et al. (2011) reported that adipocytokine leptin alone, and in combination with IL-1β, upregulated production of MMP-1, MMP-3, and MMP-13 in human OA cartilage. Effects of leptin on production of all three MMPs were mediated through TNF-κβ, protein kinase C, and MAPK pathways. Leptin concentrations in synovial fluid from OA patients correlated positively with MMP-1 and MMP-3 in synovial fluid and cartilage.

In a recent report, Ishimaru et al. (2014) delineated the role of chondroitin sulfate (CS) glycosyltransferase in cartilage growth and degeneration. By now, six CS glycosyltransferases [CS synthase 1 (CSS1/CHSY1); chondroitin polymerizing factor (CHPF); CS synthase 3 (CSS3); chondroitin polymerizing factor 2 (CHPF2); CS *N*-acetylgalactosaminyltransferase 1 (CSGALNACT1); and CS *N*-acetylgalactosaminyltransferase 2 (CSGALNACT2)] have been known to be involved in CS biosynthesis in mammals. The concentration and chain length of CS were found to be reduced closer to the more degraded cartilage. Inhibition of CS glycosyltransferase gene expression may reduce CS chain length, which may contribute to OA progression.

Yuan et al. (2014) described that a bone–cartilage modulatory pathway, transforming growth factor β (TGFβ), is required for the maintenance of metabolic homeostasis and structural integrity of healthy cartilage. The factor is highly expressed in normal cartilage, but almost absent in OA cartilage. Wang et al. (2017) further emphasized that TGFβ/ALK5 signaling maintains articular cartilage homeostasis, in part, by upregulating proteoglycan 4 (PRG4) expression through the PKA–CREB signaling pathway in articular chondrocytes. Loss/interruption of TGFβ signaling in cartilage results in loss of proteoglycans and cartilage degeneration. Thus, inhibition of endogenous TGFβ leads to increased damage to cartilage.

Wnt/β-catenin is another signaling pathway, which plays a critical modulatory role in maintaining the bone–cartilage biochemical unit (reviewed in Gupta, 2016). Increased expression of Wnt signaling agonists, such as Wnt-induced signaling protein-1, Wnt-16, and Wnt-28, have been reported in human OA cartilage and may contribute to cartilage degradation by upregulating MMPs and aggrecanases (Blom et al., 2009). Recently, Huang et al. (2017) reported that Wnt5a as a signaling pathway may contribute to OA by promoting MMP-1, MMP-3, and MMP-13 expression, as well as MMP-1 and MMP-13 protein productions in chondrocytes. Wnt5a activated beta-catenin–independent signaling, including calmodulin-dependent protein kinase II (CaMKII), JNK, p38, ERK½, p65, and Akt. Inhibition of JNK, p38, ERK, pl-3 kinase, and CaMKII by specific signaling inhibitors suppressed Wnt5a-mediated MMP-1 and MMP-13 production.

Involvement of the endocannabinoid system in OA and bone remodeling has been well reported (Idris and Ralston, 2012; La Porta et al., 2014; Valastro et al., 2017). To date, two cannabinoid receptors (CBR1 and CBR2) and two endogenous endocannabinoids (arachidonoylethanolamide and 2-arachidonoylglycerol) have been identified in synovial fluid (Valastro et al., 2017). Cannabinoid receptor activation can modulate inflammation and nociception in animal models of joint inflammation.

Recent evidence suggests that the initiation and progression of OA are contributed to various expression patterns and enhanced activation of histone deacetylases (HDACs) (Lu et al., 2014; Carpio and Westendorf, 2016). Overexpression of HDAC4 promotes matrix-degrading enzyme levels and enhances catabolic activity of chondrocytes in OA cartilage (Lu et al., 2014). Both HDAC1 and HDAC2 suppress the expression of genes encoding ECM, such as type II collagen (Cil2a1), aggrecan (ACAN), and cartilage oligomeric protein (COMP). It has also been demonstrated that HDAC inhibitors can reduce the expression of matrix-degrading enzymes in chondrocytes and fibroblasts (Young et al., 2005). Most recently, Mao et al. (2017) found that miR-92a-3p functions as a negative regulator by downregulating HDAC2 in both chondrogenesis and OA pathogenesis. To explore its potential application as a therapeutic agent, in vivo studies need to be carried out. In another study, it has been shown that miR-140 is expressed specifically in cartilage and regulates ECM-degrading enzymes (Si et al., 2017).

Recently, proteomic technologies have helped researchers in understanding novel insight into the mechanisms of human articular cartilage degradation and a number of new cartilage-characteristic proteins with possible biomarker value for early diagnosis and prognosis of OA (Ruiz-Romero and Blanco, 2009, 2010;

Lourido et al., 2014; Sanchez et al., 2017). Lourido et al. (2014) identified a panel of 76 proteins and found decreased osteoprotegerin and increased periostin release from OA cartilage. Increased levels of periostin were detected in the synovial fluid of OA patients.

Finally, several candidate genes encoding proteins of the extracellular matrix of the articular cartilage have been associated with early-onset OA (Holderbaum et al., 1999). Mutations in genes expressed in types II, IV, V, and VI collagens, as well as in COMP, have been noted. Candidate genes that are not structural proteins for OA have also been identified, such as frizzled-related protein 3, asporin, and von Willebrand factor gene (Reginato and Olsen, 2002; Ikegawa, 2007; Abramson and Attur, 2009).

By now, it is clear that it may take decades to fully understand the underlying pathophysiology of OA.

CLASSIFICATION OF OSTEOARTHRITIS BIOMARKERS

Biomarkers are urgently needed for the early detection of OA and monitoring its progression/reversal during treatment. Biomarkers of OA can be classified as markers of biochemical changes, inflammation, cartilage degeneration/loss, etc. (Altman et al., 1986; Bauer et al., 2006). In addition, the US Food and Drug Administration and European Medicines Agency have recently published guidelines recommending a higher level of integration of biomarkers in the development and testing of new drugs to advance decision-making on dosing, time and treatment effect, trial design, risk/benefit analysis, and personalized medicine (Bauer et al., 2006; Karsdal et al., 2009; Bay-Jensen et al., 2016). Currently, the board of directors (D-BOARD) consortium, an EU and PPP (public private partnership) initiative, is working on identification and validation of novel biomarkers and pushing those qualifications forward, by applying the BIPED (*B*urden of disease, *I*nvestigative, *P*rognostic, *E*fficacy of intervention, and *D*iagnostic) criteria for hypothesis testing (http://www.d-board.eu/dboard/index.aspx). Biomarkers of burden of disease are exemplified by the Kellgren and Lawrence radiographic grade (K-L grade), serum COMP, serum hyaluronan, and urinary carbonyl-terminal cross-linking telopeptide of type II collagen (CTXII). Prognostic biomarkers require longitudinal studies (prospective or retrospective) showing an increase in K-L grade by >1 and measurements of biochemical molecules/fragments of protein, DNA, and COMP (>1 unit in serum); CTX-II; serum hyaluronic acid; and pentosidine. In addition, these biomarkers include degree of joint space narrowing (JSN) or change in cartilage volume by MRI and correlated well with urinary CTX-II levels. Efficacy of intervention biomarkers may be measured prior to therapy to predict treatment efficacy or measure short-term changes that occur as a result of pharmacologic or other interventions (e.g., measuring concentrations of a biomarker of cartilage degradation, such as CTXII). Some examples of diagnostic biomarkers include COMP, CTXII, NTX-I. For detailed BIPED classification and examples of biomarkers, see Rousseau and Delmas (2007). Recently, Bay-Jensen et al. (2016) classified OA biomarkers into three broad categories: (1) inflammatory biomarkers, such as cytokines, chemokines, or cell type markers relevant to OA pathology; (2) biomarkers reflecting the turnover of the ECM of OA cartilage; and (3) biomarkers that target autoantibodies, signaling molecules, or growth factors (discussed above in pathophysiology). It is noteworthy that a biomarker can fall into more than one category.

BIOMARKERS OF COMFORT, MOBILITY, FUNCTION, AND INFLAMMATION AND PAIN

The diagnosis of OA often relies on clinical and radiological examinations of late and irreversible stages. Sensitive serum biomarkers specific for early stages (potentially reversible) of OA are lacking. It is suggested that combining biochemical markers with tissue and cell imaging techniques and bioinformatics (i.e., machine learning, clustering, and data visualization) may facilitate the development of biomarkers enabling earlier detection of OA (Mobasheri, 2012; Mobasheri et al., 2017).

Humans

In OA in humans, several scoring systems are used to measure joint mobility, flexibility, and pain, such as Knee Osteoarthritis Scoring System (KOSS), modified Knee Injury and Osteoarthritis Index (mKOOS) global score, Clinical American College of Rheumatism (ACR) criteria, Whole-Organ Magnetic Resonance Imaging Score (WORMS) of the knee, Boston Leeds Osteoarthritis Knee Score (BLOKS), MRI Osteoarthritis Knee Score (MOAKS), modified Western Ontario and McMaster Universities Arthritis Index (mWOMAC) subscale, and Visual Analog Scales (VAS) (Kornaat et al., 2005; Peterfy et al., 2004; Hunter et al., 2008, 2011; Belo et al., 2009; Runhaar et al., 2014). The global mKOOS test is composed of four categories (symptoms, stiffness, discomfort, and function/daily living activities). Additionally, mWOMAC subscale scores consist of 24 items divided into three subscales assessing discomfort (5 items), stiffness (2 items), and physical function (17 items). Additionally, self-reported feelings of joint health, comfort, mobility, and function are the primary outcome measures of OA (Lopez et al., 2017).

OA severity is often determined using weight-bearing anteroposterior radiographs of the affected joints, which are evaluated according to the Kellgren and Lawrence classification. The grade of OA is described as follows: Grade 0, no radiographic findings of OA; Grade 1, minute osteophytes of doubtful clinical significance; Grade 2, definite osteophytes with unimpaired joint space; Grade 3, definite osteophytes with moderate JSN; and Grade 4, definite osteophytes with severe JSN and subchondral sclerosis.

Animals

Dogs and horses are evaluated for overall pain, pain upon limb manipulation, and exercise-associated lameness using the Glasgow scoring system on a monthly basis for a study period of at least 4–5 months (Gupta et al., 2009, 2012). Overall pain is graded on a scale of 0–10: 0, no pain; 5, moderate pain; and 10, severe and constant pain. Pain upon limb manipulation is evaluated during the extension and flexion of all four limbs for a period of several minutes. Pain level is graded on a scale of 0–4: 0, no pain; 1, mild; 2, moderate; 3, severe; and 4, severe and constant. Pain and lameness are measured after physical exercise and graded on a scale of 0–4: 0, no pain; 1, mild; 2, moderate; 3, severe; and 4, severe and constant. In horses, pain is measured using similar criteria and scales. Additionally, flexibility and range of motion in the affected joints are measured using a Goniometer (May et al., 2015). Hielm-Björkman et al. (2009) used a five-point scale to measure pain in OA dogs [0, no sign of pain; 1, mild pain (dog turns head in recognition); 2, moderate pain (dog pulls limb away); 3, severe pain (dog vocalizes or becomes aggressive); and 4, extreme pain (dog does not allow palpation)].

In dogs, Ground Force Plate (GFP) (Kistler Instrument, Amherst, NY, USA) is utilized to quantitatively measure the lameness-associated pain in each leg of each dog (Gupta et al., 2012). The Kistler's GFP system consists of plates, lasers, and a computer. The GFP measures two major parameters: (1) peak vertical force or g force (Newton/Kg body weight), and (2) impulse area (Newton/Kg body weight). In a similar way, GFP can quantitatively measure the pain level in horses.

Observation of Ortolani and Cranial Tibial Drawer Examination

Along with the external evaluations of pain, such as gait, or lameness, other options may also be performed during the physical examination. The Ortolani Maneuver is a common test performed on canines that are predisposed to hip dysplasia such as German shepherds or larger breed canines (Siberian huskies, Rottweilers, Newfoundlands, and Labradors). Ortolani is performed on the hip joint by flexing the knee and hip to 90 degrees, placing the index finger on the greater trochanters, and abducting the hip (Ortolani's sign, 2007). As the hip is abducted, or moved away from the body, a positive Ortolani will be shown with a "clunk" sound or feeling as the femoral head relocates anteriorly to the acetabulum, or hip socket. The Ortolani is typically used on patients that have been sedated to receive a true positive or negative sign. In moderate arthritis, dogs may exhibit a negative Ortolani sign.

Cranial tibial drawer is another test that can be performed on physical examination to indicate arthritic changes and diagnose the rupture of the cranial cruciate ligament (CCL). In this procedure, the canine is in a laterally recumbency with the veterinarian located behind the patient. The thumb of one hand is placed on the caudal aspect of the femoral condular region, and the index finger of the same hand is placed over the patella. The thumb of the other hand is placed on the head of the fibula, and the index finger is placed on the tibial crest (Devine, 1993). A positive tibial drawer is elicited with the ability to move the tibia cranially or forward in respect to the fixed femur.

Currently, experimental/small animals are heavily used to unravel pathways of pain and develop models to measure OA-associated pain (electrophysiology, von Frey hair algesiometry, acoustic frequency and duration recording for vocalization measurement, and nerve injury marker activating transcription factor-3 (Malfait et al., 2013)).

OSTEOARTHRITIS BIOMARKERS IN SERUM, SYNOVIAL FLUID, AND URINE

There are many biomarkers of inflammation and cartilage degeneration associated with OA that can be determined in synovial fluid, serum, or urine, and they have potential as a diagnostic utility. Synovial changes in OA are regarded by many as a secondary response to the degradation of cartilage (Felson, 2013) though there are others who advocate them as a primary driver for OA, which may be partly responsible for pain and disease progression (Benito et al., 2005; Berenbaum, 2013). Many of the biomarkers of OA are listed in Table 52.1; and some of these are described in brief below.

Biomarkers of Inflammation

Biomarkers of joint inflammation can be detected in serum, synovial fluid, and urine much earlier than irreversible joint damage and radiographic changes (Das et al., 2015).

TABLE 52.1 Biomarkers of Osteoarthritis (OA)

	Biomarkers	References
OA-associated inflammation and pain	Cytokines (IL-1β, IL-6, IL-8, IL-10, IL-17, and IL-22), adipocytokines/adipokines (leptin, resistin, and visfatin)	Stannus et al. (2010), Koskinen et al. (2011), Berry et al. (2011), Goldring and Otero (2011), Francin et al. (2014), Daghestani and Kraus (2015), Deligne et al. (2015), Fowler-Brown et al. (2015), Bay-Jensen et al. (2016), Wan and Zhao (2017)
	Tumor necrosis factor-α (TNF-α)	Stannus et al. (2010), Jaime et al. (2017), Wan and Zhao (2017)
	Bradykinin B_1 receptor	Duclos et al. (2016)
	Chemokines (C-X-C and C−C)	Abramson and Attur (2009), Endres et al. (2010), Daghestani and Kraus (2015)
	Toll-like receptors (TLR1 and TLR2)	Scanzello et al. (2008), Sillat et al. (2013)
	Neuropeptides	Zhang et al. (2013), Heikkilä et al. (2017)
	Activating transcription factor-3 (AFT-3)	Malfait et al. (2013)
	Prostaglandin E2 (PGE2)	Lin et al. (2006), Heikkilä et al. (2017)
	Fractalkine (CX3CL1)	Huo et al. (2015), Bay-Jensen et al. (2016)
	Nuclear factor-kappa B (NF-κB)	Shakibaei et al. (2007), Goldring and Otero (2011), Rasheed et al. (2016)
	C-reactive protein (CRP)	Daghestani and Kraus (2015), Bay-Jensen et al. (2016), Hillstrom et al. (2016)
	Coll2-1, and Coll2-1 NO2	Henroitin et al. (2013, 2014)
	sRAGE	Bierhaus et al. (2006), Chayanupatkul and Honsawek (2010)
	Protease granzyme A	Jaime et al. (2017)
	TSG-6	Maier et al. (1996), Bardos et al. (2001), Bayliss et al. (2001), Milner and Day (2003), Wisniewski et al. (2014)
	AMPK and SIRT1	Liu-Bryan (2015)
	Erythrocyte sedimentation rate (ESR)	Murdock et al. (2016)
	Uric acid	Denoble et al. (2011)
OA-associated cartilage degeneration/loss	COMP, COMP neoepitope, and ykl-40	Petersson et al. (1998), Clark et al. (1999), Williams and Spector (2008), Benedetti et al. (2010), Nagala et al. (2012), Shahi et al. (2013), Sharif et al. (2004), Varma and Dalal (2013), Das et al. (2015), Skiöldebrand et al. (2017)
	Cartilage collagenases	Ehrlich et al. (1978), Goldring and Otero (2011)
	Bone sialoprotein (BSP)	Petersson et al. (1998)
	MMPs (MMP-1, MMP-3, MMP-10, MMP-13)	Pelletier et al. (2010), Koskinen et al. (2011), Hsueh et al. (2014)
	C-telopeptide of type I collagen (CTXI)	Wu et al. (2017)
	N-telopeptide of type I collagen (NTXI)	Wu et al. (2017)
	N-terminal procollagen III propeptides (PIIINP)	Wu et al. (2017)
	Type II C-telopeptide	Jung et al. (2004)
	Type II collagen	Bay-Jensen et al. (2016)
	Type II collagen neoepitope	Bay-Jensen et al. (2016)
	Coll2-1 (a type II collagen fragment)	Henroitin et al. (2013)
	Hyaluronic acid/Hyaluronan	Plickert et al. (2013), Singh et al. (2015), Das et al. (2015)
	Adiponectin	Francin et al. (2014)

TABLE 52.1 Biomarkers of Osteoarthritis (OA)—cont'd

	Biomarkers	References
	Aggrecan neoepitope, aggrecan, and aggrecanase 1 (ADAMTS-4) and aggrecanase 2 (ADAMTS-5)	Dufield et al. (2010), Swearingen et al. (2010a,b), Goldring and Otero (2011), Hsueh et al. (2014), Li et al. (2014); Peffers et al. (2014)
	Biglycan, decorin, and matrilin-1	Poole et al. (1996), Wiberg et al. (2003), Peffers et al. (2014)
	Ghrelin	Wu et al. (2017)
	Clusterin and Lubricin (proteoglycan 4)	Swan et al. (2011), Elsaid et al. (2012), Ritter et al. (2014), Svala et al. (2017)
	Fibulin-3 peptides (Fib3-1 and Fib3-2)	Henroitin et el (2012)
	Follistatin-like protein 1 (FSTL1)	Wang et al. (2011)
	Fibromodulin and fibronectin	Hsueh et al. (2014), Peffers et al. (2014)
	Uric acid	Denoble et al. (2011)
	Radiographic biomarkers (K-L Grade Criteria)	Kellgren and Lawrence (1957), Bauer et al. (2006), Mobasheri (2012), Hall et al. (2014), Andronescu et al. (2015), Palmer et al. (2017), Ramírez-Flores et al. (2017)
	MRI- and ultrasound-biomarkers	Peterfy et al. (2004), Gamero et al. (2005), Garvican et al. (2010a,b), Guermazi et al. (2013), Eckstein et al. (2014), Hall et al. (2014), Peffers et al. (2014), Nieminen et al. (2017), Roemer et al. (2014), Palmer et al. (2017), Ramírez-Flores et al. (2017)
	Cartilage thickness and cartilage damage score	Bagi et al. (2017), Wirth et al. (2017)
	Cartilage volume, using local-area cartilage segmentation (LACS) software method	Schaefert et al. (2017)
Proteomic and metabolomic analysis for OA biomarkers	Cartilage	Guo et al. (2008), Castro-Perez et al. (2010), Lourido et al. (2014), Sanchez et al. (2017)
	Synovial fluid	Ruiz-Romero and Blanco (2009, 2010), Han et al. (2012)
	Serum/plasma	Castro-Perez et al. (2010), Zhai et al. (2010)
	Urine	Hsueh et al. (2014)
MicroRNAs in OA	miR-9, miR-16, miR-22, miR-33a; miR-92a-3p; miR-98; miR-370; miR-140; miR-146a, miR-222, miR-373; miR-16-5p; miR-26a-5p; miR-634; and others	Li et al. (2011), Beyer et al. (2015), Kostopoulou et al. (2015), Li et al. (2015), Song et al. (2015), Cui et al. (2016), Rasheed et al. (2016), Wang et al. (2016), Cong et al. (2017), Mao et al. (2017), Si et al. (2017), van Meurs (2017)

Cytokines/Adipocytokines, Chemokines, and Neuropeptides

In a number of studies, profiles of cytokines/adipocytokines, leptins (Stannus et al., 2010; Koskinen et al., 2011; Berry et al., 2011; Goldring and Otero, 2011; Deligne et al., 2015; Fowler-Brown et al., 2015; Bay-Jensen et al., 2016; Hillstrom et al., 2016; Wan and Zhao., 2017), chemokines (Endres et al., 2010; Daghestani and Kraus, 2015; Bay-Jensen et al., 2016), and neuropeptides (Zhang et al., 2013) have been described in serum and/or synovial fluid in relation to OA inflammation.

Prostaglandin E2 and EP4 Receptor

Prostaglandin E2 (PGE2), a proinflammatory mediator, is produced from arachidonic acid by cyclooxygenase (COX) enzymes. PGE2 plays a pivotal role in the development of joint inflammation and pain in arthritis by binding of PGE2 to the EP4 receptor, one of the four G-proteins (Lin et al., 2006).

Fractalkine

Fractalkine (CX3CL1) plays an important role in inflammation and chronic pain, and its levels were found to be high in serum and synovial fluid and correlated with WOMAC pain and WOMAC total scores (Huo et al., 2015).

Erythrocyte Sedimentation Rate

In a clinical setting, an increase in erythrocyte sedimentation rate is commonly used as a biomarker of inflammation in OA patients (Murdock et al., 2016).

Biomarkers of Osteoarthritis Progression and Cartilage Degeneration

There is a great need to identify biomarkers of OA progression because the disease process is highly unpredictable. Biomarkers of cartilage degeneration (COMP, MMPs, clusterin, lubricin, bone sialoprotein, *N*-terminal propeptides of procollagen type I (P1NP), and β−C-terminal telopeptide (β-CTX), and many others) can be detected in synovial fluid, serum, and other bodily fluids (Petersson et al., 1998; Abramson and Attur, 2009). Noncollagenous and nonproteoglycan components of cartilage can also be detected in body fluids following their release due to cartilage turnover (Garvican et al., 2010a,b; Abramson and Attur, 2009).

Early detection of OA and OA progression can be evaluated macroscopically, histologically, immunohistochemically, and via imaging (MRI). In recent years, MRI has demonstrated its relevance for the evaluation of structural changes during the development and progression of knee OA (Guermazi et al., 2013). In a recent investigation, Nieminen et al. (2017) demonstrated that histopathological information relevant to OA can be reliably obtained from contrast-enhanced micro-computed tomography (CE mu CT images). This new grading system may be used as a reference for 3D imaging and analysis techniques intended for volumetric evaluation of OA pathology in research and clinical applications. Availability of new technologies and tools for improved morphological and pathophysiological understanding of OA-related changes in joints are an aid to early diagnosis and prognosis and in the monitoring of treatment (Boesen et al., 2016; Wyatt et al., 2017). Some important biomarkers of OA progression are discussed below.

Cartilage Oligomeric Matrix Protein (COMP)

COMP is a pentameric noncollagenous glycoprotein (435,000 Da) belonging to the heterogeneous family of thrombospondin, which can bind to collagen types I, II, and IX. It has been reported that COMP can bind up to five collagen molecules, thereby retaining them in close proximity. By this process, COMP facilitates the collagen−collagen interactions and microfibril formation. It is reported that COMP is mainly produced by articular chondrocytes, and COMP levels in serum and synovial fluid might be related to cartilage degeneration (Clark et al., 1999; Williams and Spector, 2008; Benedetti et al., 2010). Das et al. (2015) identified COMP as a potential biomarker of OA, which has shown significant clinical promise as a tool for early detection, therapeutic monitoring, prognostication, and drug development. Increase in serum COMP level by > 1 unit increased the probability of radiographic progression by 15% (Sharif et al., 2004). The findings of Varma and Dalal (2013) and Shahi et al. (2013) suggested that serum COMP level, along with clinical profile including family history, joint injury and other visible symptoms, and radiographs, could be used as valuable biomarkers for early diagnosis, prognosis, and management of OA.

In a most recent investigation, Skiöldebrand et al. (2017) identified and quantified a unique COMP neoepitope in the synovial fluid from joints of healthy horses and those with different stages of OA. The findings revealed that the increase in the COMP neoepitope in the synovial fluid from horses with acute lameness suggests that this neoepitope has the potential to be a novel candidate biomarker for the early molecular changes in articular cartilage associated with OA.

Matrix Metalloproteinases (MMPs)

MMPs, also known as "Matrixins," are capable of degrading ECM proteins. It is well established that a number of MMPs (MMP-1, MMP-3, MMP-10, and MMP-13) are significantly involved in degeneration/loss of cartilage in OA (Pelletier et al., 2010; Koskinen et al., 2011). Increased levels of some of the MMPs in synovial fluid and serum have been correlated with cartilage degeneration and therefore have potential to serve as biomarkers of OA progression.

Hyaluronan

Hyaluronic acid (HA; hyaluronate, hyaluronan) is an anionic, nonsulfated glycosaminoglycan and is an important component of articular cartilage present as a coat around each chondrocyte. Molecular weight (MW) of HA can range from 5000 to 20,000,000 DA. The average MW of HA in human synovial fluid is 3−4 million DA. HA is synthesized by a class of integral membrane proteins called hyaluronan synthases (HAS), of which vertebrates have three types (HAS1, HAS2, and HAS3). HA is degraded by a family of enzymes called hyaluronidases. In a number of studies, HA has been recognized as a biomarker of OA (Venable et al., 2008; Plickert et al., 2013; Das et al., 2015; Singh et al., 2015). For further details on HA in relation to OA, readers can referred Gupta (2016).

Lubricin

The glycoprotein lubricin (also known as proteoglycan 4) is secreted by chondrocytes, synoviocytes, and meniscus cells. It contributes to the boundary lubrication of the articular cartilage surface and is found in synovial fluid. As a boundary lubricant, lubricin reduces the coefficient of friction of the articular surface. Svala et al. (2017) demonstrated for the first time that the reduced sialation of lubricin in synovial fluid from OA joints may affect the boundary lubrication of the superficial layer of articular cartilage and could be one of the early events in the progression of OA. Reduced expression of lubricin in synovial synoviocytes and superficial zone chondrocytes with subsequent loss of joint lubrication are early events in the inflammation associated with OA. Using mass spectrometry, Ritter et al. (2014) identified several proteins (including clusterin and lubricin) in human plasma as predictors of OA progression.

Intra-articular injections of lubricin were reported to improve damaged cartilage and preserve viability of superficial zone chondrocytes (Swan et al., 2011; Elsaid et al., 2012).

Follistatin-Like Protein 1 (FSTL 1)

FSTL 1 is a glycoprotein implicated in OA pathogenesis, which secretes into synovial fluid and serum from OA articular cartilage. Wang et al. (2011) found that the FSTL 1 mRNA and protein levels were substantially elevated in the synovial tissues from OA patients. The FSTL 1 expression was strong in the cytoplasm of the synovial and capillary endothelial cells of the synovial tissues, but it was weak in the chondrocytes of the articular cartilage from OA patients. Interestingly, the serum and synovial fluid FSTL 1 levels were maximally higher in female OA patients than in male patients. The serum FSTL 1 levels of OA patients had significant correlations with K-L grade, JSN, WOMAC stiffness scale, and the WOMAC function subscale. The findings suggested that FSTL 1 could be a biomarker from synovium and might represent a biomarker of the burden of disease (severity of joint damage). However, FSTL 1 has two limitations: (1) it is not specific to OA, as it plays roles in kidney, heart, and other organs; and (2) it poorly correlates with chondrocytes of the articular cartilage. Furthermore, there are conflicting reports that FSTL 1 has potential preventive effects on joint destruction by inhibiting the production of MMPs and cytokines both in synovial cells in vitro and in mouse models in vivo (Tanaka et al., 2003; Kawabata et al., 2004).

Ghrelin

Ghrelin is a peptide hormone, which is involved in energy homeostasis, autoimmunity, inflammation, and many other physiological/pathophysiological processes. Wu et al. (2017) found that significant increases in serum levels of ghrelin were associated with increased knee symptoms, intrapatellar fat pad (IPFP) signal intensity alteration, and serum levels of MMP-3, MMP-13, *N*-telopeptide of type I collagen (NTXI), and *N*-terminal procollagen III propeptide (PIIINP), suggesting that ghrelin may have a role in knee OA.

TSG-6

Tumor necrosis factor-stimulated gene-6 (TSG-6) is a member of the hyaluronan-binding proteins. TSG-6 plays a role in joint inflammation because of its inducible expression by the proinflammatory cytokines IL-1β and TNF-α in human synoviocytes and chondrocytes, and the presence of increased levels of this protein in the synovial fluids and cartilage of a majority of arthritis patients has been reported (Bayliss et al., 2001; Milner and Day, 2003; Wisniewski and Vilcek, 2004). TSG-6 serves as a biomarker of OA progression and is of particular interest for aiding development of disease modifying OA drug (DMOAD) and in identifying high-risk patients who would benefit mostly from the use of DMOAD (Kraus et al., 2011; Wisniewski et al., 2014). Potent antiinflammatory and chondroprotective activities of this protein have also been demonstrated in experimental arthritis models (Bardos et al., 2001).

Cartilage Damage and Imaging Biomarkers

Although many joint structures are affected in OA, articular cartilage is one of the main tissues involved in the OA disease process (Roemer et al., 2014). Articular damage or loss in OA is detected by radiography and measuring decreases in joint space width (JSW) on the radiograph, the so-called "gold standard." However, radiographic evidence is seen only after significant cartilage degradation has already taken place (Mobasheri, 2012). Radiographic measures are less than adequate for multiple reasons. First, radiographs indicate changes in bone and only indirectly measure alterations in cartilage (Bauer et al., 2006). Second, the measurement of articular cartilage change, namely JSN, is itself confounded by meniscal cartilage lesions and meniscal extrusion. Third, bone marrow perturbations and synovial abnormalities may go undetected. Fourth, radiographic features characteristic of OA appear only after significant deterioration has occurred in both the hard and soft tissues within and around the joint and changes may occur relatively slowly. Fifth, radiographic features are usually poorly correlated with joint function.

Bauer et al. (2006) stated that imaging techniques might themselves be considered as biomarkers for the pathologic joint abnormalities that define OA. Imaging measures should be accurate, precise (reliable), specific, sensitive to a longitudinal change, and acceptable to regulatory agencies (Eckstein et al., 2014). Recently, Palmer et al. (2017) reported that delayed gadolinium magnetic resonance imaging of cartilage (dGEMRIC) could detect glycosaminoglycan loss in the cartilage of asymptomatic individuals with cam morphology. These investigators found a positive correlation between dGEMRIC and the magnitude of JSW narrowing. Baseline dGEMRIC is able to predict the development of radiographic OA. MRI offers the potential to identify patients who may benefit from early intervention to prevent the development of OA.

Recent studies demonstrated that articular cartilage thickness loss/cartilage degeneration score, and osteophyte formation, measured by MRI and μ-computed tomography were found to be associated with concurrent and subsequent radiographic progression and with concurrent symptomatic progression, and therefore they can be used as biomarkers of OA progression (Bagi et al., 2017; Wirth et al., 2017). Schaefert et al. (2017) validated a previously developed local-area cartilage segmentation software method to measure cartilage volume using MRI scans. Cartilage volume change was strongly correlated with radiographic and pain progressions. In canines, Ramírez-Flores et al. (2017) found a correlation between synovial fluid effusion and osteophytosis in the stifle joint, using the results of orthopedic, radiographic, ultrasonographic, and arthroscopic examinations. Inflammation, effusion, and capsular thickening can be clinically evident in some joints with OA and are more frequently observed using sensitive measures such as ultrasound and MRI (Hall et al., 2014).

Newly Identified Biomarkers

Currently, there is no universal biomarker for OA, and the available biomarkers lack specificity and sensitivity for early diagnosis and monitoring progression of OA. Therefore, new biomarkers are needed. New biomarkers are also needed to distinguish between catabolic and maintenance events because many existing biomarkers reflect normal cartilage turnover, tissue repair, or ECM remodeling. A new biomarker for diagnosing OA should be evaluated by comparison against an established gold standard in an appropriate spectrum of subjects. For OA, an accepted "gold standard" diagnostic test is the radiograph, and typically, K-L grade ≥ 2 is required for a diagnosis of OA (Altman et al., 1986).

De Senny et al. (2011) characterized four novel biomarkers in the serum of knee-OA patients (V65 vitronectin fragment, C3f peptide, CTAP-III, and m/z 3762 protein). Mobasheri (2012) and Mobasheri et al. (2017) recognized that enrichment of the deaminated epitope of D-COMP suggests that OA disease progression is associated with posttranslational modifications that may show specificity for particular joint sites. Fibulin-3 peptides (Fib3-1 and Fib3-2) have been proposed as potential biomarkers of OA along with FSTL 1 (described above), a new serum biomarker with the capacity to reflect the severity of joint damage. The "membrane attack complex (MAC)" component of complement has also been implicated in OA.

Recently, Mobasheri et al. (2017) identified new biomarkers based on novel technologies, such as "omics." These authors referred to several new biomarkers, such as adipocytokines including leptin and adiponectin. ADAM metallopeptidase with thrombospondin type 1 motif 4 (ADAMTS-4), aggrecan ARGS neoepitope fragment (ARGS) in synovial fluid, and plasma chemokine (CeC motif) ligand 3 (CCL3) were reported as potential new biomarkers of knee OA. New and refined proteomic technologies and novel assays including a fluoro-microbead guiding chip for measuring C-telopeptide of type II collagen (CTX-II) in serum and urine and a novel magnetic nanoparticle-based technology (termed magnetic capture) for collecting and concentrating CTX-II also have great potential to be biomarkers of OA.

DNA-Methylation and MicroRNAs (miR)

Recently, Rogers et al. (2015) examined the role of inflammation-related genes in OA from the perspective of genetics, epigenetics, and gene expression. There is no compelling evidence that DNA variation in inflammatory genes is an OA risk factor. However, there is compelling evidence that epigenetic effects involving inflammatory genes are a component of OA and that alteration in the expression of these genes is also highly relevant to the disease process. van Meurs (2017) stated that current technological advances in the area of "omics" have elevated multiple molecular levels, from detailed sequencing of the genome, to epigenetic markers, such as DNA-methylation and microRNAs (miRNAs), transcriptomics, and metabolomics. The objective is to identify biomarkers that predict OA's early onset and progression. miRNAs have great promise as biomarkers because they are relatively stable in circulation. The exact role of miRNAs in OA is still under investigation, but as such, they could potentially be good biomarkers (van Meurs, 2017). Published studies have focused on one or two miRNAs, based on the hypothesis that they target a gene or metabolism that plays an important role in OA pathogenesis. For example, miR-33a, which is known to be important in cholesterol metabolism (Kostopoulou et al., 2015), was found to also regulate cholesterol metabolism in chondrocytes through the TGF-β1/Akt/SREBP-2 pathway, as well as cholesterol efflux-related genes ABCA1 and ApoA1. miR-370 and miRNA-373 were found to regulate expression of SHMT-2 and MECP-2 in chondrocytes (Song et al., 2015). miR-16-5p has been shown to regulate SMAD5 expression in cartilage (Li et al., 2015) and miRNA-26a-5p was found to regulate iNOS expression in cartilage (Rasheed et al., 2016). miR-634 has been shown to target PIK3R1 in chondrocytes (Cui et al., 2016). Beyer et al. (2015) identified let-7e as a potential predictor for severe knee or hip OA, but an increase in circulating levels of let-7e could be due to other conditions, such as metabolic dysfunction or ischemic stroke.

miR-140 levels are significantly reduced in human OA cartilage–derived chondrocytes and synovial fluid. Overexpressing miR-140 in primary human chondrocytes promoted collagen II expression and inhibited MMP-13 and ADAMTS-5 expression (Si et al., 2017). These authors demonstrated that intra-articular injection of miR-140 can alleviate OA progression by modulating ECM homeostasis in rats and may have potential as a new therapy for OA. In an OA rat model, Wang et al. (2016) confirmed the effects of miR-98 on apoptosis in cartilage cells in vivo. MiR-98 expression is reduced in the cartilage cells of OA patients and the overexpression of miR-98 inhibits cartilage cell apoptosis, whereas inhibition of miR-98 leads to cartilage cell apoptosis.

Recently, Cong et al. (2017) identified 46 differentially expressed miRNAs involved in autophagy, inflammation, chondrocyte apoptosis, chondrocyte differentiation and homeostasis, chondrocyte metabolism, and degradation of the ECM. Additionally, these authors identified a wide range of miRNAs that have been shown to be differentially expressed in OA. The function of upregulated miRNAs primarily target transcription. These miRNAs may be useful as diagnostic biomarkers and/or may provide new therapeutic targets in OA.

CORRELATION OF CIRCULATORY BIOMARKERS WITH RADIOGRAPHIC/IMAGING BIOMARKERS AND SYMPTOMS OF OSTEOARTHRITIS

Although circulatory biomarkers, particularly of inflammation, are poorly correlated with radiographic or imaging evidence of OA, they are often used for early detection of this disease (Orita et al., 2011). Patra and Sandell (2011) reported that biochemical markers offer potential nonradiographic alternatives to detect early, nonsymptomatic OA. Accumulating evidence supports that COMP predicts MRI cartilage loss and appears as a useful biomarker of early OA (Williams and Spector, 2008). Combinations of cartilage-derived and bone-derived biomarkers have been used in the context of OA presence, severity, and bone turnover in cartilage integrity to assess the impact of treatment (Clark et al., 1999; Patra and Sandell, 2011). Circulatory biomarkers of inflammation have been used to profile OA progression attesting to the inflammation of OA joints. Stannus et al. (2010) found that the circulating levels of IL-6 and TNF-α are associated with knee radiographic OA and knee cartilage loss in older adults. Berry et al. (2011) investigated the relationship of serum markers of cartilage metabolism to imaging and clinical outcome measures of knee joint structures. Pelletier et al. (2010) reported a decrease in serum levels of MMPs in prediction of the disease-modifying effects of OA drugs assessed by quantitative MRI in patients with knee OA. Of course, there are many other studies that have described correlation between body fluid biomarkers and joint inflammation and cartilage degeneration/loss.

CONCLUDING REMARKS AND FUTURE DIRECTIONS

OA is a disease of the entire joint but primarily causes degeneration of articular cartilage. Currently, there are no reliable, quantifiable, and easily measured biomarkers that provide an earlier diagnosis, inform on the prognosis of disease, and monitor responses to therapeutic modalities. Available biomarkers also lack sensitivity and specificity. Biomarkers of both cartilage proteoglycan and noncollagenous, nonproteoglycan components of cartilage can be detected following their release as a result of turnover and disease. Therefore, continuing quests are underway to identify, test, validate, and qualify biomarkers of OA. Genetic and genomic biomarkers could be essential to stratify OA patients and assess their risk of developing OA later in life with different underlying driving etiologies (van Meurs, 2017). Many molecular "omic" advancements will be made in the near future to the genetic and epigenetic data. Imaging continues to play an important role in OA research, where several exciting new technologies and computer-aided analysis methods are emerging to complement the conventional imaging approaches. New biomarkers are needed to distinguish between catabolic and maintenance events because many existing biomarkers reflect normal cartilage turnover, tissue repair, or ECM remodeling. Circulatory microRNAs are not only offering early detection of disease but also providing basis for the development of novel targeted therapies for OA. There are still many challenges that we have yet to overcome in the near future.

Acknowledgments

The authors would like to thank Ms. Robin B. Doss and Ms. Denise M. Gupta for their technical assistance in preparation of this chapter.

References

Abramson, S.B., Attur, M., 2009. Developments in the scientific understanding of osteoarthritis. Arthritis Res. Ther. 11, 227.

Altman, R., Asch, E., Bloch, D., et al., 1986. Development of criteria for the classification and reporting of knee osteoarthritis. Arthritis Rheum. 29 (8), 1039–1049.

Andronescu, A., Kelly, L., Kearney, M.T., et al., 2015. Associations between early radiographic and computed tomographic measures and canine hip joint osteoarthritis at maturity. Am. J. Vet. Res. 76 (1), 19–27.

Bagi, C.M., Berryman, E.R., Teo, S., et al., 2017. Oral administration of undenatured native chicken type II collagen (UC-II) diminished deterioration of articular cartilage in a rat model of osteoarthritis (OA). Osteoarthr. Cartil. 25, 2080–2090.

Bakker, B., Eijkel, G.B., Heeren, R.M.A., et al., 2017. Oxygen-dependent lipid profiles of three-dimensional cultured human chondrocytes revealed by MALDI-MSI. Anal. Chem. 89, 9438–9444.

Bardos, T., Kamath, R.V., Mikecz, K., et al., 2001. Anti-inflammatory and chondroprotective effect of TSG-6 (tumor necrosis factor-alpha-stimulated gene-6) in murine models of experimental arthritis. Am. J. Pathol. 159, 1711–1721.

Bauer, D.C., Hunter, D.J., Abramson, S.B., et al., 2006. Classification of osteoarthritis biomarkers: a proposed approach. Osteoarthr. Cartil. 14, 723–727.

Bay-Jensen, A.C., Reker, D., Kjelgaard-Petersen, C.F., et al., 2016. Osteoarthritis year in review 2015: soluble biomarkers and the BIPED criteria. Osteoarthr. Cartil. 24, 9–20.

Bayliss, M.T., Howat, S.L., Dudhia, J., et al., 2001. Up-regulation and differential expression of the hyaluronan-binding protein TSG-6 in cartilage and synovium in rheumatoid arthritis and osteoarthritis. Osteoarthr. Cartil. 9, 42–48.

Belo, J.N., Berger, M.Y., Koes, B.W., et al., 2009. The prognostic value of the clinical ACR classification criteria of knee osteoarthritis for persisting knee complaints and increase of disability in general practice. Osteoarthr. Cartil. 17, 1288–1292.

Benedetti, S., Canino, C., Tonti, G., et al., 2010. Biomarkers of oxidation, inflammation and cartilage degradation in osteoarthritis patients undergoing sulfur-based spa therapies. Clin. Biochem. 43, 973–978.

Benito, M.J., Veale, D.J., Fitzgerald, O., et al., 2005. Synovial tissue inflammation in early and late osteoarthritis. Ann. Rheum. Dis. 64, 1263–1267.

Berenbaum, F., 2013. Osteoarthritis as an inflammatory disease (osteoarthritis is not osteoarthrosis!). Osteoarthr. Cartil. 21, 16–21.

Berry, P.A., Jones, S.W., Cicuttini, F.M., et al., 2011. Temporal relationship between serum adipokines, biomarkers of bone and cartilage turnover, and cartilage volume loss in a population with clinical knee osteoarthritis. Arthritis Rheum. 63, 700–707.

Beyer, C., Zampetaki, A., Lin, N.Y., et al., 2015. Signature of circulating microRNAs in osteoarthritis. Ann. Rheum. Dis. 74 (3), e18.

Bierhaus, A., Stern, D.M., Nawroth, P.P., 2006. Rage in inflammation: a new therapeutic target? Curr. Opin. Invest. Drugs 7, 985–991.

Blom, A.B., Brockbank, S.M., van Lent, P.L., et al., 2009. Involvement of the Wnt signaling pathway in experimental and human osteoarthritis: prominent role of Wnt-induced signaling protein 1. Arthritis Rheum. 60, 501–512.

Boesen, M., Ellegaard, K., Henriksen, M., et al., 2016. Osteoarthritis year in review 2016: imaging. Osteoarthritis Cartil. 25 (2), 216–226.

Carpio, L.R., Westendorf, J.J., 2016. Histone deacetylases in cartilage homeostasis and osteoarthritis. Curr. Rheumatol. 18, 52.

Castro-Perez, J.M., Kamphorst, J., DeGroot, J., et al., 2010. Comprehensive LC-MS lipidomic analysis using a shotgun approach and its application to biomarker detection and identification in osteoarthritis patients. J. Proteome Res. 9, 2377–2389.

Chayanupatkul, M., Honsawek, S., 2010. Soluble receptor for advanced glycation end products (sRAGE) in plasma and synovial fluid is inversely associated with disease severity of knee osteoarthritis. Clin. Biochem. 43, 1133–1137.

Clark, A.G., Jordan, J.M., Vilim, V., et al., 1999. Serum cartilage oligometric matrix protein reflects osteoarthritis presence and severity: the Johnston County Osteoarthritis Project. Arthritis Rheum. 42 (11), 2356–2364.

Cong, L., Zhu, Y., Tu, G., 2017. A bioinformatics analysis of microRNAs role in osteoarthritis. Osteoarthr. Cartil. 25 (8), 1362–1371.

Cui, X., Wang, S., Cai, H., et al., 2016. Overexpression of microRNA-634 suppresses survival and matrix synthesis of human osteoarthritis chondrocytes by targeting PIK3R1. Sci. Rep. 6, 23117.

Daghestani, H.N., Kraus, V.B., 2015. Inflammatory biomarkers in osteoarthritis. Osteoarthr. Cartil. 23, 1890–1896.

Das, B.R., Roy, A., Khan, F.R., 2015. Cartilage oligometric matrix protein in monitoring and prognostication of osteoarthritis and its utility in drug development. Perspect. Clin. Res. 6 (1), 4–9.

deBakker, E., Stroobants, V., VanDael, F., et al., 2017. Canine synovial fluid biomarkers for early detection and monitoring of osteoarthritis. Vet. Rec. 180, 328. https://doi.org/10.1136/vr.103982.

De Senny, D., Sharif, M., Fillet, M., et al., 2011. Discovery and biochemical characterization of four novel biomarkers for osteoarthritis. Ann. Rheum. Dis. 70, 1144–1152.

Deligne, C., Casulli, S., Pigenet, A., et al., 2015. Differential expression of interleukin-17 and interleukin-22 in inflamed and non-inflamed synovium from osteoarthritis patients. Osteoarthr. Cartil. 23, 1843–1852.

Denoble, A.E., Huffman, K.M., Stabler, T.V., et al., 2011. Uric acid is a danger signal of increasing risk for osteoarthritis through inflammasome activation. Proc. Natl. Acad. Sci. U.S.A. 108, 2088–2093.

Duclos, B.A., Rugg, C.A., White, J., et al., 2016. Pharmacological evaluation of a selective bradykinin B_1 antagonist in a canine model of arthritis. J. Vet. Pharmacol. Ther. 40, 70–76.

Devine, S.B., 1993. Cranial tibial thrust: a primary force in the canine stifle. J. Am. Vet. Med. Assoc. 183 (4), 456–459.

Dufield, D.R., Nemirovskiy, O.V., Jennings, M.G., et al., 2010. An immunoaffinity liquid chromatography-tandem mass spectrometry assay for detection of endogenous aggrecan fragments in biological fluids: use as a biomarker for aggrecanase activity and cartilage degradation. Anal. Biochem. 406, 113–123.

Eckstein, F., Guermazi, A., Gold, G., et al., 2014. Imaging of cartilage and bone: promises and pitfalls in clinical trials of osteoarthritis. Osteoarthr. Cartil. 22, 1516–1532.

Ehrlich, M.G., Houle, P.A., Vigliani, G., et al., 1978. Correlation between articular cartilage collagenase activity and osteoarthritis. Arthritis Rheum. 21 (7), 761–766.

Elsaid, K.A., Zhang, L., Waller, K., et al., 2012. The impact of forced joint exercise on lubricin biosynthesis from articular cartilage following ACL transection and intra-articular lubricin's effect in exercised joints following ACL transection. Osteoarthr. Cartil. 20, 940–948.

Endres, M., Andreas, K., Kalwitz, G., et al., 2010. Chemokine profile of synovial fluid from normal, osteoarthritis and rheumatoid arthritis patients: CCL25, CXCL10 and XCL1 recruit human subchondral mesenchymal progenitor cells. Osteoarthr. Cartil. 18 (11), 1458–1466.

Felson, D.T., 2013. Osteoarthritis as a disease of mechanics. Osteoarthr. Cartil. 21, 10–15.

Fowler-Brown, A., Kim, D.H., Shi, L., et al., 2015. The mediating effect of leptin on the relationship between body weight and knee osteoarthritis in older adults. Arthritis Rheum. 67, 169–175.

Francin, P.J., Abot, A., Guillaume, C., et al., 2014. Association between adiponectin and cartilage degradation in human osteoarthritis. Osteoarthr. Cartil. 22 (3), 519–526.

Garnero, P., Peterfy, C., Zaim, S., et al., 2005. Bone marrow abnormalities on magnetic resonance imaging are associated with type II collagen degradation in knee osteoarthritis: a three-month longitudinal study. Arthritis Rheum. 52 (9), 2822–2829.

Garvican, E.R., Vaughan-Thomas, A., Innes, J.F., Clegg, P.D., 2010a. Biomarkers of cartilage-turnover. Part 1: markers of collagen degradation and synthesis. Vet. J. 185, 36–42.

Garvican, E.R., Vaughan-Thomas, A., Clegg, P.D., Innes, J.F., 2010b. Biomarkers of cartilage-turnover. Part 2: noncollagenous markers. Vet. J. 185, 43–49.

Goldring, M.B., Otero, M., 2011. Inflammation in osteoarthritis. Curr. Opin. Rheumatol. 23 (5), 471–478.

Guermazi, A., Roemer, F.W., Haugen, I.K., et al., 2013. MRI-based semiquantitative scoring of joint pathology in osteoarthritis. Nat. Rev. Rheumatol. 9, 236–251.

Guo, D., Tan, W., Wang, F., et al., 2008. Proteomic analysis of human articular cartilage: identification of differentially expressed proteins in knee osteoarthritis. Joint Bone Spine 75, 439–444.

Gupta, R.C., Canerdy, T.D., Skaggs, P., et al., 2009. Therapeutic efficacy of undenatured type-II collagen (UC-II) in comparison to glucosamine and chondroitin in arthritic horses. J. Vet. Pharmacol. Ther. 32, 577–584.

Gupta, R.C., Canerdy, T.D., Lindley, J., et al., 2012. Comparative therapeutic efficacy and safety of type-II collagen (UC-II), glucosamine and chondroitin in arthritic dogs: pain evaluation by ground force plate. Anim. Physiol. Anim. Nutr. 96, 770–777.

Gupta, R.C., 2016. Nutraceuticals in arthritis. In: Gupta, R.C. (Ed.), Nutraceuticals: Efficacy, Safety and Toxicity. Academic Press/Elsevier, Amsterdam, pp. 161–176.

Hall, M., Doherty, S., Courtney, P., et al., 2014. Synovial pathology detected on ultrasound correlates with the severity of radiographic knee osteoarthritis more than with symptoms. Osteoarthr. Cartil. 22, 1627–1633.

Han, M.Y., Dai, J.J., Zhang, Y., et al., 2012. Identification of osteoarthritis biomarkers by proteomic analysis of synovial fluid. J. Int. Med. Res. 40, 2243–2250.

Heikkilä, H.M., Hielm-Björkman, A.K., Innes, J.F., et al., 2017. The effect of intra-articular botulinum toxin A on substance P, prostaglandin E2, and tumor necrosis factor alpha in the canine osteoarthritic joint. BMC Vet. Res. 13, 74.

Henrotin, Y., Gharibi, M., Mazzucchelli, G., et al., 2012. Fibulin 3 peptides Fib3-1 and Fib3-2 are potential biomarkers of osteoarthritis. Arthritis Rheum. 64, 2260–2267.

Henrotin, Y., Chevalier, X., Deberg, M., et al., 2013. Early decrease of serum biomarkers of type II collagen degradation (Coll2-1) and joint inflammation (Coll2-1 NO2) by hyaluronic acid intra-articular injections in patients with knee osteoarthritis: a research study part of the Biovisco study. J. Orthop. Res. 31, 901–907.

Henrotin, Y., Gharbi, M., Dierckxsens, Y., et al., 2014. Decrease of a specific biomarker of collagen degradation in osteoarthritis, Coll2-1, by treatment with highly bioavailable curcumin during an exploratory clinical trial. BMC Compl. Altern. Med. 14, 159.

Hielm-Björkman, A.K., Rita, H., Tulamo, R.M., 2009. Psychometric testing of the Helsinki chronic pain index by completion of a questionnaire in Finnish by owners of dogs with chronic signs of pain caused by osteoarthritis. Am. J. Vet. Res. 70, 727–734.

Hillstrom, A., Bylin, J., Hagman, R., et al., 2016. Measurement of serum C-reactive protein concentration for discriminating between suppurative arthritis and osteoarthritis in dogs. BMC Vet. Res. 12 https://doi.org/10.1186/s12917-016-0868-4.

Holderbaum, D., Haqqi, T.M., Moskowitz, R.W., 1999. Genetics and osteoarthritis: exposing the iceberg. Arthritis Rheum. 42, 397–405.

Hsueh, M.-F., Önnerfjord, P., Kraus, V.B., 2014. Biomarkers and proteomic analysis of osteoarthritis. Matrix Biol. 39, 56–66.

Huang, G., Chubinskaya, S., Liao, W., et al., 2017. Wnt5a induces catabolic signaling and matrix metalloproteinase production in human articular chondrocytes. Osteoarthr. Cartil. 25 (9), 1505–1515.

Hunter, D.J., Lo, G.H., Gale, D., et al., 2008. The reliability of a new scoring system for knee oateoarthritis MRI and the validity of bone marrow lesion assessment: BLOKS (Boston Leeds Osteoarthritis Knee Score). Ann. Rheum. Dis. 67, 206–211.

Hunter, D.J., Guermazi, A., Lo, G.H., et al., 2011. Evolution of semi-quantitative whole joint assessment of knee OA: MOAKS (MRI Osteoarthritis Knee Score). Osteoarthr. Cartil. 19, 990–1002.

Huo, L.W., Ye, Y.L., Wang, G.W., et al., 2015. Fractalkine (CXCL1): a biomarker reflecting symptomatic severity in patients with knee osteoarthritis. J. Invest. Med. 63, 626–631.

Idris, A.I., Ralston, S.H., 2012. Role of cannabinoids in regulation of bone remodeling. Front. Endocrinol. 3, 136.

Ikegawa, S., 2007. New gene associations in osteoarthritis: what do they provide, and where do we go? Curr. Opin. Rheumatol. 19, 429–434.

Ishimaru, D., Sugiara, N., Akiyama, H., et al., 2014. Alterations in the chondroitin sulfate chain in human osteoarthritic cartilage of the knee. Osteoarthr. Cartil. 22, 250–258.

Jaime, P., García-Guerrero, N., Estella, R., et al., 2017. $CD56^+/CD16^-$ natural killer cells expressing the inflammatory protease Granzyme A are enriched in synovial fluid from patients with osteoarthritis. Osteoarthr. Cartil. 25, 1708–1718.

Jung, M., Christgau, S., Lukoschek, M., et al., 2004. Increased urinary concentration of collagen type II C-telopeptide fragments in patients with osteoarthritis. Pathobiology 71 (2), 70–76.

Karsdal, M.A., Henriksen, K., Leaming, D.J., et al., 2009. Biochemical markers and the FDA critical path: how biomarkers may contribute to the understanding of pathophysiology and provide unique and necessary tools for drug development. Biomarkers 14, 181–202.

Kawabata, D., Tanaka, M., Fujii, T., et al., 2004. Ameliorative effects of follistatin-related protein/TSC-36/FSTL 1 on joint inflammation in a mouse model of arthritis. Arthritis Rheum. 50, 660–668.

Kellgren, J.H., Lawrence, J.S., 1957. Radiological assessment of osteoarthrosis. Ann. Rheum. Dis. 16, 494–502.

Kornaat, P.R., Ceulemans, R.Y., Kroon, H.M., et al., 2005. MRI assessment of knee osteoarthritis: Knee Osteoarthritis Scoring System (KOSS)- inter-observer and intra-observer reproducibility of a compartment-based scoring system. Skeletal Radiol. 34, 95–102.

Koskinen, A., Vuolteenaho, K., Nieminen, R., et al., 2011. Leptin enhances MMP-1, MMP-3 and MMP-13 production in human osteoarthritic cartilage and correlates with MMP-1 and MMP-3 in synovial fluid from OA patients. Clin. Exp. Rheumatol. 29 (1), 57–64.

Kostopoulou, F., Malizos, K.N., Papathanasiou, I., et al., 2015. MicroRNA-33a regulates cholesterol synthesis and cholesterol efflux-related genes in osteoarthritic chondrocytes. Arthritis Res. Ther. 17, 42.

Kraus, V.B., Burnett, B., Coindreau, J., et al., 2011. Application of biomarkers in the development of drugs intended for the treatment of osteoarthritis. Osteoarthr. Cartil. 19, 515–542.

La Porta, C., Bura, S.A., Negrete, R., et al., 2014. Involvement of the endocannabinoid system in osteoarthritis pain. Eur. J. Neurosci. 39, 485–500.

Li, X., Gibson, G., Kim, J.-S., et al., 2011. MicroRNA-146a is linked to pain-related pathophysiology of osteoarthritis. Gene 480, 34–41.

Li, W., Du, C., Wang, H., et al., 2014. Increased serum ADAMTS-4 in knee osteoarthritis: a potential indicator for the diagnosis of osteoarthritis in early stages. Genet. Mol. Res. 13, 9642–9649.

Li, L., Jia, J., Liu, X., et al., 2015. MicroRNA-16-5p controls development of osteoarthritis by targeting SMAD3 in chondrocytes. Curr. Pharm. Des. 21 (35), 5160–5167.

Lin, C.-R., Amaya, F., Barrett, L., et al., 2006. Prostaglandin E2 receptor EP4 contributes to inflammatory pain hypersensitivity. J. Pharmacol. Exp. Ther. 319, 1096–1103.

Liu-Bryan, R., 2015. Inflammation and intracellular metabolism: new targets in OA. Osteoarthr. Cartil. 23, 1835–1842.

Lopez, H.L., Habowski, S.M., Sandrock, J.E., 2017. Effects of dietary supplementation with a standardized aqueous extract of *Terminalia chebula* fruit (AyuFlex®) on joint mobility, comfort, and functional capacity in healthy overweight subjects: a randomized placebo-controlled clinical trial. BMC Compl. Altern. Med. 17, 475.

Lotz, M., Martel-Pelletier, J., Christiansen, C., et al., 2013. Value of biomarkers in osteoarthritis: current status and perspectives. Ann. Rheum. Dis. 72 (11), 1756–1763.

Lourido, L., Calamia, V., Mateos, J., et al., 2014. Quantitative proteomic profiling of human articular cartilage degradation in osteoarthritis. J. Proteome Res. 13, 6096–6106.

Lu, J., Sun, Y., Ge, Q., et al., 2014. Histone deacetylase 4 alters cartilage homeostasis in human osteoarthritis. BMC Muscoskel. Disord. 15, 438.

Maier, R., Wisniewski, H.-G., Vilcek, J., et al., 1996. TSG-6 expression in human articular chondrocytes. Arthritis Rheum. 39 (4), 552–559.

Malfait, A.M., Little, C.B., McDougall, J.J., 2013. A commentary on modelling osteoarthritis pain in small animals. Osteoarthr. Cartil. 21, 1316–1326.

Mao, G., Zhang, Z., Huang, Z., et al., 2017. MicroRNA-92a-3p regulates the expression of cartilage-specific genes by directly targeting histone deacetylase 2 in chondrogenesis and degradation. Osteoarthr. Cartil. 25, 521–532.

May, K., Gupta, R.C., Miller, J., et al., 2015. Therapeutic efficacy and safety evaluation of a novel chromium supplement (Crominex® +3-) in moderately arthritic horses. Jacobs J. Vet. Sci. Res. 2 (1), 014.

Milner, C.M., Day, A.J., 2003. TSG-6: a multifunctional protein associated with inflammation. J. Cell Sci. 116, 1863–1873.

Mobasheri, A., 2012. Osteoarthritis year 2012 in review: biomarkers. Osteoarthr. Cartil. 20, 1451–1464.

Mobasheri, A., Bay-Jensen, A.C., van Spil, W.E., et al., 2017. Osteoarthritis year in review 2016: biomarkers (biochemical markers). Osteoarthr. Cartil. 25 (2), 199–208.

Murdock, N., Gupta, R.C., Vega, N., et al., 2016. Evaluation of *Terminalia chebula* extract for anti-arthritic efficacy and safety in osteoarthritic dogs. Vet. Sci. Technol. 7 (1), 1–8.

Nagala, I.A., Taher, A.A., Ibrahim, E.M., et al., 2012. Evaluation of the role of cartilage oligomeric matrix protein and YKL-40 as biomarkers in knee osteoarthritis patients. Nat. Sci. 10, 43.

Nieminen, H.J., Gahunia, H.K., Pritzker, K.P.H., et al., 2017. 3D histopathological grading of osteochondral tissue using contrast-enhanced micro-computed tomography. Osteoarthr. Cartil. 25 (10), 1680–1689.

Orita, S., Koshi, T., Mitsuka, T., et al., 2011. Associations between proinflammatory cytokines in the synovial fluid and radiographic grading and pain-related scores in 47 consecutive patients with osteoarthritis of the knee. BMC Muscoskel. Disord. 12, 144.

Ortolani's sign, 2007. Saunders Comprehensive Veterinary Dictionary, third ed. Elsevier, Inc. Retrieved from: http://medical-dictionary.thefreedictionary.com/Ortolani_s_sign.

Palmer, A., Fernquest, S., Rombach, I., et al., 2017. Diagnostic and prognostic value of delayed gadolinium enhanced magnetic resonance imaging of cartilage (dGEMRIC) in early osteoarthritis of the hip. Osteoarthr. Cartil. 25 (9), 1468–1477.

Peffers, M.J., Cillero-Pastor, B., Eijkel, G.B., et al., 2014. Matrix assisted laser desorption ionization mass spectrometry imaging identifies markers of ageing and osteoarthritic cartilage. Arthritis Res. Ther. 16, R110.

Petra, D., Sandell, L., 2011. Recent advances in biomarkers in osteoarthritis. Curr. Opin. Rheumatol. 23 (5), 465–470.

Pelletier, J.P., Raynauld, J.P., Caron, J., et al., 2010. Decrease in serum level of matrix metalloproteinases is predictive of the disease-modifying effect of osteoarthritis drugs assessed by quantitative MRI in patients with knee osteoarthritis. Ann. Rheum. Dis. 69, 2095–2101.

Petersson, I.F., Boegard, T., Svensson, B., et al., 1998. Changes in cartilage and bone metabolism identified by serum markers in early osteoarthritis of the knee joint. Br. J. Rheumatol. 37 (1), 46–50.

Peterfy, C.G., Guermazi, A., Zaim, S., et al., 2004. Whole-organ magnetic resonance imaging score (WORMS) of the knee in osteoarthritis. Osteoarthr. Cartil. 12, 177–190.

Plickert, H., Bondzio, A., Einspanier, R., et al., 2013. Hyaluronic acid concentrations in synovial fluid of dogs with different stages of osteoarthritis. Res. Vet. Sci. 94, 728–734.

Poole, A.R., Rosenberg, L.C., Reiner, A., et al., 1996. Contents and distributions of the proteoglycans (decorin and biglycan) in normal and osteoarthritic human articular cartilage. J. Orthop. Res. 14, 681–689.

Portal-Núñez, S., Esbrit, P., Alcaraz, M.J., et al., 2016. Oxidative stress, autophagy, epigenetic changes and regulation by miRNA as potential therapeutic targets in osteoarthritis. Biochem. Pharmacol. 108, 1–10.

Ramírez-Flores, G.I., Angel-Caraza, J.D., Quijano-Hernández, I.A., et al., 2017. Correlation between osteoarthritic changes in the stifle joint in dogs and the results of orthopedic, radiographic, ultrasonographic and arthroscopic examinations. Vet. Res. Commun. 41, 129–137.

Rasheed, Z., Al-Shobaili, H.A., Rasheed, N., et al., 2016. MicroRNA-26a-5p regulates the expression of inducible nitric oxide synthase via activation of NF-kappaB pathway in human osteoarthritis chondrocytes. Arch. Biochem. Biophys. 594, 61–67.

Reginato, A.M., Olsen, B.R., 2002. The role of structural genes in the pathogenesis of osteoarthritic disorders. Arthritis Res. 4, 337–345.

Ritter, S.Y., Collins, J., Krastins, B., et al., 2014. Mass spectrometry assays of plasma biomarkers to predict radiographic progression of knee osteoarthritis. Arthritis Res. Ther. 16 (5), 456.

Roemer, F.W., Guermazi, A., Trattnig, S., et al., 2014. Whole joint MRI assessment of surgical cartilage repair of the knee: cartilage repair osteoarthritis knee score (CROAKS). Osteoarthr. Cartil. 22, 779–799.

Rogers, E.L., Reynard, L.N., Loughlin, J., 2015. The role of inflammation-related genes in osteoarthritis. Osteoarthr. Cartil. 23, 1933–1938.

Rousseau, J.C., Delmas, P.D., 2007. Biologic markers in osteoarthritis. Nat. Clin. Pract. Rheumatol. 3, 347–356.

Ruiz-Romero, C., Blanco, F.J., 2009. The role of proteomics in osteoarthritis pathogenesis research. Osteoarthr. Cartil. 10, 543–556.

Ruiz-Romero, C., Blanco, F.J., 2010. Proteomics role in the search for improved diagnosis, prognosis and treatment of osteoarthritis. Osteoarthr. Cartil. 18, 500–509.

Runhaar, J., Schiphof, D., van Meer, B., et al., 2014. How to define subregional osteoarthritis progression using semi-quantitative MRI osteoarthritis knee score (MOAKS). Osteoarthr. Cartil. 22, 1533–1536.

Sanchez, C., Bay-Jensen, A.C., Pap, T., et al., 2017. Chondrocyte secretome: a source of novel insights and exploratory biomarkers of osteoarthritis. Osteoarthr. Cartil. 25 (8), 1199–1209.

Scanzello, C.R., Plaas, A., Crow, M.K., 2008. Innate immune system activation in osteoarthritis: is osteoarthritis a chronic wound? Curr. Opin. Rheumatol. 20 (5), 565–572.

Schaefert, L.F., Sury, M., Yin, M., et al., 2017. Quantitative measurement of medial femoral knee cartilage volume-Analysis of the OA biomarkers consortium FNIH study cohort. Osteoarthr. Cartil. 25 (7), 1107–1113.

Shahi, U., Shahi, N.T., Khanna, V., et al., 2013. Cartilage oligomeric matrix protein: a potential diagnostic, prognostic and therapeutic biomarker of knee osteoarthritis. Orthoped. Res. Soc. Congr. Abstract.

Shakibaei, M., John, T., Schulze-Tanzil, G., et al., 2007. Suppression of NF-kappaB activation by curcumin leads to inhibition of expression of cyclooxygenase-2 and matrix metalloproteinase -9 in human articular chondrocytes: implications for the treatment of osteoarthritis. Biochem. Pharmacol. 73, 1434–1445.

Sharif, M., Kirwan, J.R., Elson, C.J., et al., 2004. Suggestion of nonlinear or phasic progression of knee osteoarthritis based on measurements of serum cartilage oligomeric matrix protein levels over five years. Arthritis Rheum. 50 (8), 2479–2488.

Si, H.B., Zeng, Y., Liu, S.Y., et al., 2017. Intra-articular injection of microRNA-140 (miRNA-140) alleviates osteoarthritis (OA) progression by modulating extracellular matrix (ECM) homeostasis in rats. Osteoarthr. Cartil. 25 (10), 1698–1707.

Sillat, T., Barreto, G., Clarijs, P., et al., 2013. Toll-like receptors in human chondrocytes and osteoarthritic cartilage. Acta Orthop. 84 (6), 585–592.

Singh, S., Kumar, D., Kumar, S., et al., 2015. Cartilage oligomeric matrix protein (COMP) and hyaluronic acid (HA): Diagnostic biomarkers of knee osteoarthritis. MOJ Orthoped. Rheumatol. 2 (2), 00044.

Skiöldebrand, E., Ekman, S., Hultén, L.M., et al., 2017. Cartilage oligomeric protein neoepitope in the synovial fluid of horses with acute lameness: a new biomarker for the early stages of osteoarthritis. Equine Vet. J. 49, 662–667.

Sluzalska, K.D., Liebisch, G., Lochnit, G., et al., 2017. Interleukin-1 beta affects the phospholipid biosynthesis of fibroblast-like synoviocytes from human osteoarthritic knee joints. Osteoarthr. Cartil. 25 (11), 1890–1899.

Song, J., Kim, D., Chung, C.H., Jin, E.J., 2015. miR-370 and miR-373 regulate the pathogenesis of osteoarthritis by modulating one-carbon metabolism via SHMT-2 and MECP-2, respectively. Aging Cell 14 (5), 826–837.

Stannus, O., Jones, G., Cicuttini, F., et al., 2010. Circulating levels of IL-6 and TNF-alpha are associated with knee radiographic osteoarthritis and knee cartilage loss in older adults. Osteoarthr. Cartil. 18, 1441–1447.

Svala, E., Jin, C., Rüetschi, U., et al., 2017. Characterization of lubricin in synovial fluid from horses with osteoarthritis. Equine Vet. J. 49, 116–123.

Swan, D.A., Hendren, R.B., Radin, E.L., et al., 2011. The lubricating activity of synovial fluid glycoproteins. Arthritis Rheum. 24, 22–30.

Swearingen, C.A., Chambers, M.G., Lin, C., et al., 2010a. A short-term pharmacodynamic model for monitoring aggrecanase activity: injection of monosodium iodoacetate (MIA) in rats and assessment of aggrecan neoepitope release in synovial fluid using novel ELISAs. Osteoarthr. Cartil. 18, 1159–1166.

Swearingen, C.A., Carpenter, J.W., Siegel, R., et al., 2010b. Developmental of a novel clinical biomarker assay to detect and quantify aggrecanase-generated aggrecan fragments in human synovial fluid, serum and urine. Osteoarthr. Cartil. 18, 1150–1158.

Tanaka, M., Ozaki, S., Kawabata, D., et al., 2003. Potential preventive effects of follistatin-related protein/TSC-36 on joint destruction and antagonistic modulation of its autoantibodies in rheumatoid arthritis. Int. Immunol. 15, 71–77.

Thote, T., Lin, A.S.P., Raji, Y., et al., 2013. Localized 3D analysis of cartilage composition and morphology in small animal models of joint degeneration. Osteoarthr. Cartil. 21, 1132–1141.

Tonge, D.P., Pearson, M.J., Jones, S.W., 2014. The hallmarks of osteoarthritis and the potential to develop personalized disease-modifying pharmacological therapeutics. Osteoarthr. Cartil. 22, 609–621.

Valastro, C., Campanile, D., Marinaro, M., et al., 2017. Characterization of endocannabinoids and related acylethanolamides in the synovial fluid of dogs with osteoarthritis: a pilot study. BMC Vet. Res. 13, 309.

van Meurs, J.B.J., 2017. Osteoarthritis year in review 2016: genetics, genomics and epigenetics. Osteoarthr. Cartil. 25, 181–189.

Varma, P., Dalal, K., 2013. Serum cartilage oligomeric matrix protein (COMP) in knee osteoarthritis: a novel diagnostic and prognostic biomarker. J. Orthop. Res. 31 (7), 999–1006.

Venable, R.O., Stoker, A.M., Cook, C.R., et al., 2008. Examination of synovial fluid hyaluronan quantity and quality in stifle joints of dogs with osteoarthritis. Am. J. Vet. Res. 69 (12), 1569–1573.

Villalvilla, A., Gómez, R., Largo, R., et al., 2013. Lipid transport and metabolism in healthy and osteoarthritic cartilage. Int. J. Mol. Sci. 14 (10), 20793–20808.

Wan, Z.-H., Zhao, Q., 2017. Gypenoside inhibits interleukin-1β-induced inflammatory response in human osteoarthritis chondrocytes. J. Biochem. Mol. Toxicol. e21926.

Wang, Y., Li, D., Xu, N., et al., 2011. Follistatin-like protein 1: a serum biochemical marker reflecting the severity of joint damage in patients with osteoarthritis. Arthritis Res. Ther. 13, R193.

Wang, G.L., Wu, Y.B., Liu, J.T., Li, C.Y., 2016. Upregulation of miR-98 inhibits apoptosis in cartilage cells in osteoarthritis. Genet. Test. Mol. Biomarkers 20 (11), 645–653.

Wang, Q., Tan, Q.Y., Xu, W., et al., 2017. Cartilage-specific deletion of Alk5 gene results in a progressive osteoarthritis-like phenotype in mice. Osteoarthr. Cartil. 25 (11), 1868–1879.

Wiberg, C., Klatt, A.R., Wagener, R., et al., 2003. Complexes of matrilin-1 and biglycan or decorin connect collagen VI microfibrils to both collagen II and aggrecan. J. Biol. Chem. 278, 37698–37704.

Williams, F.M.K., Spector, T.D., 2008. Biomarkers in osteoarthritis. Arthritis Res. Ther. 10, 101.

Wirth, W., Hunter, D.J., Nevitt, M.C., et al., 2017. Predictive and concurrent validity of cartilage thickness change as a marker of knee osteoarthritis progression: data from the osteoarthritis initiative. Osteoarthr. Cartil. 25, 2063–2071.

Wisniewski, H.G., Vilcek, J., 2004. Cytokine-induced gene expression at the crossroads of innate immunity, inflammation and fertility: TSG-6 and PTX3/TSG-14. Cytokine Growth Factor Rev. 15, 129–146.

Wisniewski, H.-G., Colón, E., Liublinska, V., et al., 2014. TSG-6 activity as a novel biomarker of progression in knee osteoarthritis. Osteoarthr. Cartil. 22, 235–241.

Wu, J., Wang, K., Xu, J., et al., 2017. Association between serum ghrelin and knee symptoms, joint structures and cartilage or bone biomarkers in patients with knee osteoarthritis. Osteoarthr. Cartil. 25 (9), 1428–1435.

Wyatt, L.A., Moreton, B.J., Mapp, P.I., et al., 2017. Histopathological subgroups in knee osteoarthritis. Osteoarthr. Cartil. 25 (1), 14–22.

Young, D.A., Lakey, R.L., Pennington, C.J., et al., 2005. Histone deacetylase inhibitors modulate metalloproteinase gene expression in chondrocytes and block cartilage resorption. Arthritis Res. Ther. 7, R503–R512.

Yuan, X.L., Meng, H.Y., Wang, Y.C., et al., 2014. Bone-cartilage interface crosstalk in osteoarthritis: potential pathways and future therapeutic strategies. Osteoarthr. Cartil. 22, 1077–1089.

Zhai, G., Wang-Sattler, R., Hart, D.J., et al., 2010. Serum branched-chain amino acid to histidine ratio: a novel metabolomic biomarker of knee osteoarthritis. Ann. Rheum. Dis. 69, 1227–1231.

Zhang, R.-X., Ren, K., Dubner, R., 2013. Osteoarthritis pain mechanisms: basic studies in animal models. Osteoarthr. Cartil. 21, 1308–1315.

CHAPTER

53

Pathological Biomarkers in Toxicology

Meliton N. Novilla[1,2], *Vincent P. Meador*[3], *Stewart B. Jacobson*[1], *Jessica S. Fortin*[4]

[1]Shin Nippon Biomedical Laboratories USA, Ltd, Everett, WA, United States; [2]School of Veterinary Medicine, Purdue University, West Lafayette, IN, United States; [3]Covance Laboratories, Inc., Global Pathology, Madison, WI, United States; [4]College of Veterinary Medicine, Michigan State University, East Lansing, MI, United States

INTRODUCTION

Simply defined, a biomarker is "a specific physical trait or measurable biologically produced change in the body connected with health or disease conditions" (Hunt, 2009). In toxicology practice, biomarkers are associated with some aspect of normal or abnormal biological function resulting from exposure to drugs, food additives, biopharmaceuticals, medical devices, environmental chemicals, and plants. From the three broad biomarker categories of exposure, effect, and susceptibility (Timbrell, 1998; Barr and Buckley, 2011), this chapter focuses on biomarkers of effect, limited to morphologic and clinicopathologic alterations in organs and tissues from which diagnoses are based. Brief descriptions of historical advancements are given, from the systematic practice of autopsies around 1346 and the invention of the light microscope in 1600 and its use in medicine beginning in the 1800s (Rosai, 1997). The 1800s to the recent past, considered to be the diagnostic era, was the period during which pathologists were principally involved in delivering diagnoses of disease conditions (Hunt, 2009). Pathologists arrived at their conclusions from the anamnesis, gross examination of specimens and later histopathologic examination after the light microscope was introduced into medical practice. From the 1960s to the 1990s, the inclusion of special pathology techniques, such as electron microscopy, confocal laser scanning microscopy, immunohistochemistry, and in situ hybridization, facilitated the transition of diagnostic pathology toward a prognostic and therapeutically oriented discipline (Rosai, 1997; Schnitt, 2003; Hunt, 2009). Now, sophisticated and novel technologies that relate to biomarkers of effect, exposure, and susceptibility to disease are becoming available, as evidenced by the many chapters in this book. These biomarkers aid in disease diagnosis, understanding pathogenesis, defining prognosis, and helping to guide selections of the best options for therapy.

This chapter provides examples of pathological biomarkers, with emphasis on morphologic (gross and microscopic) and clinicopathologic changes. Biomarker examples of the effects of investigational drugs, biologicals, medical devices, environmental contaminants, and poisonous plants are given when the biomarker has been validated and applied in preclinical and clinical studies conducted in humans and animals. However, because of space limitations, not all information can be included; hence, apologies are extended to colleagues whose work has been omitted or inadvertently missed.

DIAGNOSTIC PATHOLOGY

Among the most significant historical advances in the evolution of medicine was the systematic practice of the autopsy/necropsy beginning in the 14th century, followed by the use of the light microscope in the 20th century. Restrictions against opening the human body after death were eased during the Black Death pandemic (1347–1350). Prior to this development, dissection (vivisection) had been carried out since ancient times for reasons related to animistic and naturalistic philosophies rather than to determine the cause of the disease. Hippocrates (468–377 BC), founder of a school of medicine in ancient Greece, taught that disease resulted not from divine or supernatural origin but from natural causes (King and Meehan, 1973). Hippocrates was the first great naturalistic physician and his concepts influenced the course of scientific medicine ever after. However, it

Biomarkers in Toxicology, Second Edition
https://doi.org/10.1016/B978-0-12-814655-2.00053-0

would be almost 1800 years after his death before the autopsy found its place in medicine.

Prior to these milestone events, diagnoses were based on clinical signs and symptoms. It has been reported that before the first histologic examination of colon cancer in humans, diagnosis was based on clinical observations of bowel symptoms, cachexia, and abdominal mass, which were deemed sufficient to confirm the diagnosis, as late as 1990 (cited by Hunt, 2009). Many years elapsed before autopsies became accepted practice; hence the medical practitioner had to have observed the cardinal signs of inflammation: *rubor, tumor, calor, dolor et functio laesa*, learned to recognize gross abnormalities from apparently normal organs and tissues, and applied knowledge of prevailing maladies at the time.

Gross findings were considered in disease diagnosis after physicians performed autopsies on dying patients "to know more clearly the illnesses of their bodies" (cited by Rosai, 1997). The systematic performance of the autopsy ushered the birth of the discipline of pathology in human medicine and, ipso facto, in veterinary medicine. Rosai (1997) reported that those who practiced the discipline of pathology employed morphological techniques to explain symptoms and signs, determine the cause of death, guide therapy, and predict the evolution of disease. With the introduction of the light microscope and other diagnostic techniques in the 20th century, pathology practice evolved to its present state. For the toxicologic pathologist, adequate training and ever-increasing improvements in processes and procedures have resulted in high-quality tissue specimens, proper identification of significant histopathologic findings, and accurate, correctly interpreted diagnoses (Crissman et al., 2004; Van Tongeren et al., 2011).

According to Buck et al. (1973), the accurate diagnosis of toxicosis is based on history, clinical signs, postmortem findings, chemical analysis, and laboratory animal tests. Depending on data availability, diagnoses rendered may vary from suspected (presumptive or preliminary) to definitive or confirmatory as in the following examples.

Enzootic Hematuria

Some cattle at a ranch in Cotabato province, Mindanao Island in the Philippines were clinically observed to have been excreting bloody urine by the farm veterinarian. Leptospirosis was diagnosed earlier following the isolation of *Leptospira pomona* by the Animal Disease Diagnostic Laboratory, University of the Philippines College of Veterinary Medicine (UPCVM), Diliman, Quezon City. However, hematuria persisted even after treatment; hence, a team consisting of a microbiologist, parasitologist, and pathologist from the UPCVM were invited to conduct field necropsies and collect samples for additional tests. During the inspection and tour of the ranch, lush growths of bracken fern were found alongside pasture fences and hillsides on the ranch; hence, bracken fern intoxication was suspected. Fenced cattle are known to be exposed to toxins, such as ptaquiloside and other toxins, mutagens, and carcinogens, from consumption of tips of crosiers and young fronds of the bracken fern along with pasture grasses during grazing (Panter et al., 2011; Panter et al., 2012).

At necropsy, the urinary bladder in three of five animals contained residual reddish urine and multiple red nodules on the bladder mucosa. The nodules correlated histologically with hemorrhagic and proliferative lesions, consistent with those described in bracken fern intoxication (Pamukcu et al., 1976; Jones and Hunt, 1983). The red nodular growths on the urinary bladder mucosa were gross pathological biomarkers. Maxie (1985) reported that more than 90% of cattle with enzootic hematuria had urinary bladder tumors. Being associated with the clinical signs of bloody urination and probable prolonged ingestion of bracken fern, the bladder tumors confirmed the suspected diagnosis of bracken fern–induced toxicity in the cattle herd.

Clinicopathological findings that might be encountered are anemia, leucopenia, monocytosis, thrombocytopenia, hypergammaglobulinemia, microhematuria, and proteinuria (Perez-Alenza et al., 2006). Regarding *L. pomona* infection, Chirathaworn and Kongpan (2014) reported that several inflammatory mediators were higher in susceptible animals versus resistant hosts. Immune responses with cytokines/chemokines and serum proteins (TNFα, IL-6, IL-8, and PTX3) that were induced following *Leptospira* infection may be correlated positively with mortality and were suggested to be biomarkers for disease severity in human with leptospirosis.

There are several studies on the expression of biomarkers in bovine enzootic hematuria (BEH) bladder lesions to detect molecular changes in the different stages of the disease. Up to 13% of the bladder lesions were found to be nonneoplastic and only half of the neoplastic lesions were malignant. The urothelial carcinoma was the most frequent tumor type (Carvalho et al., 2006; Peixoto et al., 2003). Cytokeratin is a protein found in the intracytoplasmic cytoskeleton of epithelial cells commonly used as an immunohistochemistry marker to identify epithelial cells in normal or malignant state. Cytokeratin 7 (CK7) is expressed in the urothelium, and this protein is frequently used to confirm the urothelial origin of tumors (Moll, 1998). There is a conserved transmembrane protein across species namely uroplakins (UPS) that is expressed in the superficial urothelial cells. The marker uroplakin III (UPIII) is also routinely used to confirm the urothelial origin of tumors. However, in humans and domestic animals, a loss of expression can occur in high-grade urothelial tumors (Ambrosio et al., 2001; Ohtsuka et al., 2006; Ramos-Vara et al., 2003). An immunohistochemical study was performed on epithelial bladder tumor samples from 37 animals affected with BEH. In both high-grade and high-stage urothelial carcinomas, there was loss of UPIII and CK7 expression. Interestingly, there was immunoreactivity for p53 in high-grade and high-stage carcinomas (Cota et al., 2013).

Cyclopia

Consumption of another toxic plant, *Veratrum californicum*, has been associated with distinctive teratogenic abnormalities in ruminants (Burrows and Tyrl, 2001). Jervoline alkaloids in this plant such as cyclopamine, cycloposine, and jervine have been incriminated in producing teratogenic effects in the offspring of ewes and less commonly cows and goats (Panter et al., 2011). Pregnant ewes ingesting the plant from the 12th to the 14th day of gestation may have prolonged gestation due to the absence of the pituitary gland in the malformed offspring. The most striking malformation is partial or complete cyclopia in which one eye or two fused eyes are in a single orbit with a median skin protuberance above. Other abnormalities such as cleft palate and limb reductions have also been described. The gross findings and history of dietary exposure are presumptive diagnostic pathological biomarkers of the toxic plant's teratogenic effect.

Taxus (Yew) Poisoning

There are no characteristic gross or microscopic tissue alterations of Taxus (yew) toxicity. These evergreen plants are ornamental shrubs that contain toxic alkaloids (taxines) and irritant oils, which have been reported to cause sudden death in humans and a variety of animals. All parts of the plant except the aril are highly poisonous, with death attributed to the cardiotoxic effects of taxine alkaloids (Wilson and Hooser, 2018). In ruminants, nonspecific myocardial hemorrhages and focal nonsuppurative interstitial myocarditis have been reported, but diagnosis is usually based on a history of known or potential exposure to the yew plant. In the dairy cow ingesting trimmings of the plant, the presence of leaves, stems, twigs, and seeds or remnants thereof in the stomach contents is diagnostic and recognized as the gross pathological biomarker of yew toxicity. In horses where extensive mastication and digestion of the plant occur, stomach contents may be difficult to assess by the naked eye. Microscopic examination for plant parts and identification of taxine alkaloids in the stomach contents by the gas chromatography/mass spectrometry (GC/MS) method have been used to confirm diagnosis (Tiwary et al., 2005).

Cardiotoxic alkaloids from the yew leaves (*Taxus baccata*) have been identified and quantitated in perimortem samples of serum and gastric contents in human medicine (Musshoff et al., 1993; Stríbrný et al., 2010). The main substance identified by HPLC and GC-MS (gas chromatography/mass spectrometry) was 3,5-dimethoxyphenol, the aglycone of taxicatine. In cases of intoxication by yew, 3,5-dimethoxyphenol can be used as a marker.

Ergotism

Gross lesions occur on the extremities of cattle grazing on pastures infected with the fungi *Claviceps purpurea* and *Neotyphodium coenophialum* or fed grain contaminated with the ergot alkaloids, ergotamine, and ergovaline, respectively. One week after consumption of *C. purpurea*, swelling and redness of the extremities, particularly of the posterior limb, develop and the tips of the ear and tail have gangrene, which may slough off about 2 weeks following ingestion of the ergovaline-producing fungus; necrosis (dry gangrene) of the distal extremities is observed (Jones and Hunt, 1983). The gross observations are indicative of peripheral vasoconstriction induced by the ergot alkaloids produced by the fungal organisms.

Interestingly, an ELISA test to detect ergot alkaloids is available and offered by Randox Food Diagnostics. The Ergot Alkaloid ELISA kit is validated for flour and seed. The ELISA test offers excellent limits of detection for the toxin ergotamine at 1 ppb. GC/MS is routinely used for identification and confirmation of ergot alkaloids. For further details on toxicity and biomarkers of ergot alkaloids poisoning refer to Gupta et al. (2018).

Morphologic (Gross and Microscopic) and Clinical Pathology

Until the light microscope was incorporated into medical practice, gross pathology was the primary means to establish a diagnosis (Hunt, 2009). However, 200 years elapsed from the invention of the very first microscope in 1600 until it was incorporated into medical practice (Hagdu, 2002). Although Anton Van Leeuwenhoek (1632–1723) was the first to use the light microscope to study organisms and Robert Hooke (1635–1703) was the first to use the microscope in the practice of medicine, integration of microscopy into medical practice did not occur until Rudolf Virchow (1821–1902) introduced the light microscope in pathology, allowing the study of cellular and later subcellular events for a better understanding of disease. The use of the microscope by pathologists enhanced their skills and diagnostic acumen. With the microscope, identification of causative agents and pathognomonic lesions facilitated diagnosis. For instance, light microscopic identification of Negri bodies in neurons of a dog is pathognomonic for rabies. However, few pathognomonic lesions are encountered by the toxicologic pathologist in preclinical safety and toxicity studies.

Solar Injuries

Damage to the skin by ultraviolet (UV) light has resulted in sunburn, solar dermatitis, and neoplasia (Hargis and Ginn, 2006). Solar injuries begin clinically with the cardinal signs of inflammation: *rubor, tumor, calor, dolor et functio laesa*. Gross lesions of erythema, scaling, and crusting occur where there is little or no pigment or hair in the affected skin, whereas the haired or pigmented areas of the skin are unaffected. Microscopically, epidermal acanthosis and follicular hyperkeratosis (comedones) develop. Multiple layers of compacted stratum corneum may form cutaneous horns and the comedones can rupture (furunculosis), releasing follicular contents into the dermis. The endogenous foreign materials, including follicular stratum corneum, hair shafts, and sebum may get infected and cause an inflammatory response with secondary bacterial infection. The gross and microscopic findings and distribution pattern support the diagnosis of solar dermatitis.

Skin Photosensitization

In primary skin photosensitization, gross lesions occur in nonpigmented and lightly pigmented skin and poorly haired areas of the body that include erythema and edema, followed by blisters, exudation, and necrosis and sloughing of necrotic tissue (Hargis and Ginn, 2006). Histologically, there is coagulative epidermal necrosis, which may extend to the superficial dermis, follicular epithelium, and adnexal glands. Fibrinoid degeneration and thrombosis of dermal vessels result in infarction and sloughing of necrotic tissues and secondary bacterial infection. Skin damage occurs as a result of ingestion of preformed photodynamic substances (e.g., helianthrone, furocoumarin, psoralens, phytoalexins) contained in a variety of poisonous plants, or the administration of phenothiazine, which is converted to photoreactive metabolite in the intestinal tract. The photodynamic substances are absorbed into the skin where they react with UV light, releasing energy that produces reactive oxygen molecules, degranulation of mast cells, production of inflammatory mediators, and cell damage. The gross findings on the skin associated with exposure to photoreactive substances are considered to be the pathological biomarkers of primary photosensitization. The results of one research group suggest that changes of cell surface thiols and/or amines may be useful biomarkers to predict photosensitization potential of chemicals (Oeda et al., 2016). Further research is needed to find useful biomarkers for skin photosensitization.

Secondary (hepatogenous) photosensitization occurs primarily in herbivorous animals, but any animal with massive hepatic disorder given a chlorophyll-rich ration can be affected (Hargis and Ginn, 2006). The liver fails to excrete phylloerythrin, a chlorophyll breakdown product formed in the intestinal tract due to inherited hepatic defects, primary hepatocellular damage, or bile duct obstruction from toxic plants, pyrrolizidine alkaloids, or mycotoxins. For example, gross findings associated with the mycotoxin sporidesmin, produced by the fungus *Pithomyces chartarum*, included bile-stained liver with prominent bile ducts, dilated with bile and surrounded by periductal edema. In addition, cholestasis and facial eczema were also observed in affected sheep.

Skin Neoplasia

Chronic exposure to sunlight has been stipulated by Madewell (1981) as the best-known etiologic factor for many skin tumors in animals. After chronic exposure, the skin becomes wrinkled and thickened due to epidermal hyperplasia, fibrosis, and elastosis, which could progress to neoplasia, such as hemangiomas, hemangiosarcomas, and squamous cell carcinomas. The most frequently diagnosed malignant tumor in cattle is ocular squamous cell carcinoma. Although the cause is thought to be multifactorial, cattle with nonpigmented eyelids and conjunctiva are predominantly affected. Premalignant stages including plaques, keratomas, papillomas, and the squamous cell carcinomas are found in the junction of the cornea and sclera, the third eyelid, and on the margins of upper and lower eyelids.

Scrotal hemangioma in boars is rarely reported in North America but common in tropical countries. According to Szcech et al. (1973), the probable index case in the United States was a Chester White boar from a swine-breeding establishment in Indiana. Grossly, a hemorrhagic, 1.0×1.5 cm, dark blue and pedunculated wartlike mass was located at the junction of the scrotum and perineum. Following surgical excision and histopathology, the mass was diagnosed as a capillary hemangioma. Novilla (1989) reported a high number of scrotal capillary hemangiomas, 317 (31% of 1027 boars were affected) from 17 farms in the Philippines (Fig. 53.1). Light-skinned purebred boars had more and larger scrotal tumors than dark-skinned boars had. Two- or three-way crossbred boars tended to have growths in nonpigmented areas of the scrotal skin. Grossly, the

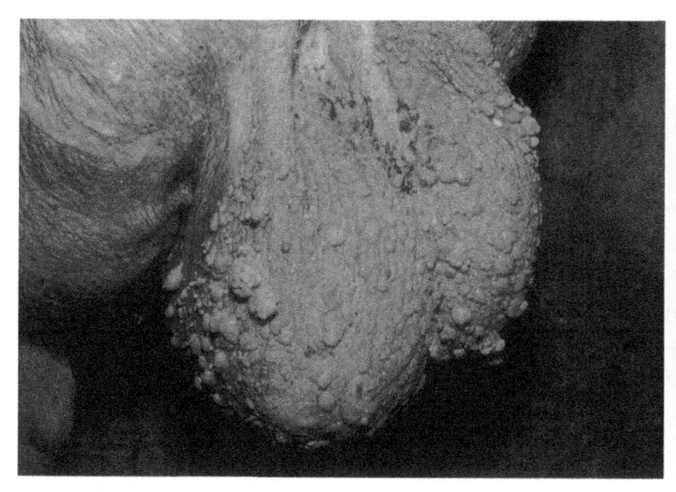

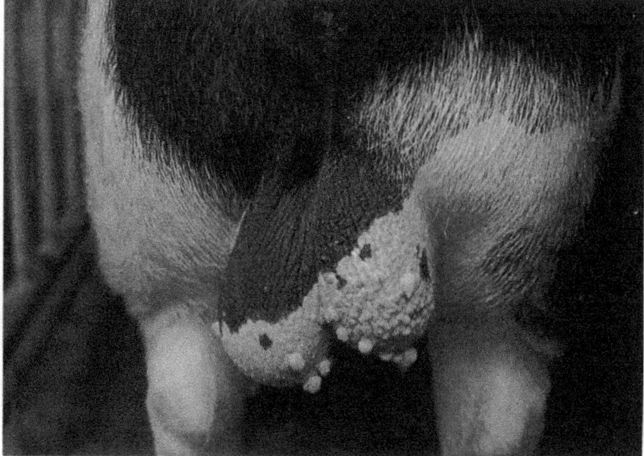

FIGURE 53.1 Scrotal capillary hemangiomas in Landrace Crossbred boars from the Philippines. *Courtesy of Dr. D. A. Novilla, USDA, FSIS Morristown, Tennessee.*

lesions varied from tiny reddish blotches or blebs to pedunculated warty and cauliflower-like masses.

Histologically, the dermis contained lobules of capillary-like vascular spaces lined by mostly endothelial-type cells supported by delicate connective tissue stroma. Gross and microscopic features were consistent with previous descriptions of scrotal hemangioma (Munro et al., 1982) and presumed to be due to exposure to sunlight.

According to Kycko and Reichert (2014), many candidate biomarkers were identified and proven to be not specific enough to be further investigated and developed as a diagnostic tool for skin cancer. Reports concerning proteomic biomarkers of spontaneously occurring tumors in animals provided preliminary results up to this point. Most studies resulted in protein changes involving a small number of cases, which limits the significance of the findings. The candidate biomarkers are usually nonspecific for a particular type of cancer and, in most cases, components of the acute phase response, which is not necessarily associated with cancer.

Iatrogenic Acromegaly in Dogs

Dogs administered progestogens indirectly develop a syndrome of growth hormone excess (La Perle and Capen, 2012). Injection of medroxyprogesterone acetate to prevent estrus in dogs has been shown to produce off-target adverse effects. There is expression of the growth hormone gene in the mammary gland, elevated circulating levels of growth hormone, and clinical manifestations of acromegaly. Affected animals have coarse facial features with markedly folded and thickened skin of the face, enlarged abdomen, and expansion of interdigital spaces.

According to Kooistra (2006), diagnosis of GH excess is established by measuring basal plasma GH levels. The basal plasma GH level in acromegalic animals in many instances exceeds the upper limit of the reference range. The diagnosis may be supported by the measurement of elevated plasma IGF-I levels. IGF-I is bound to proteins and for this reason its level is more stable versus GH. However, there is some overlap in plasma IGF-I levels between healthy animals and individuals with acromegaly.

Endocrine Alopecia

Endocrine alopecia may occur from iatrogenic hyperadrenocorticism or hyperestrogenism in dogs from administration of glucocorticoids or estrogens, respectively. Other reported causes are bilateral adrenocortical hyperplasia secondary to a pituitary tumor, functional adrenocortical hyperplastic nodule, or neoplasm inducing hyperadrenocorticism. Dogs with hyperadrenocorticism have truncal alopecia, sparing the head and limbs, distended abdomen, calcinosis cutis, and thin skin. Histologically, the epidermis, dermis, and hair follicles are atrophic with follicular hyperkeratosis and calcinosis cutis in hyperadrenocorticism.

Hyperestrogenism associated with ovarian cysts in females and functional Sertoli cell tumor of the testis in males also induces endocrine alopecia (Hargis and Ginn, 2006). In hypestrogenism, gross findings include symmetrical alopecia and hyperpigmentation over the posterior trunk and limbs (Fig. 53.2). In addition, male dogs may develop pendulous prepuce and enlargement of the nipples and prostate, whereas female dogs have enlarged vulva and abnormal estrus cycle. Epidermal and follicular hyperkeratosis and follicular atrophy are observed in hyperestrogenism. A history of iatrogenic exposure to glucocorticoids or estrogen together with clinical and morphologic findings supports a diagnosis of endocrine alopecia.

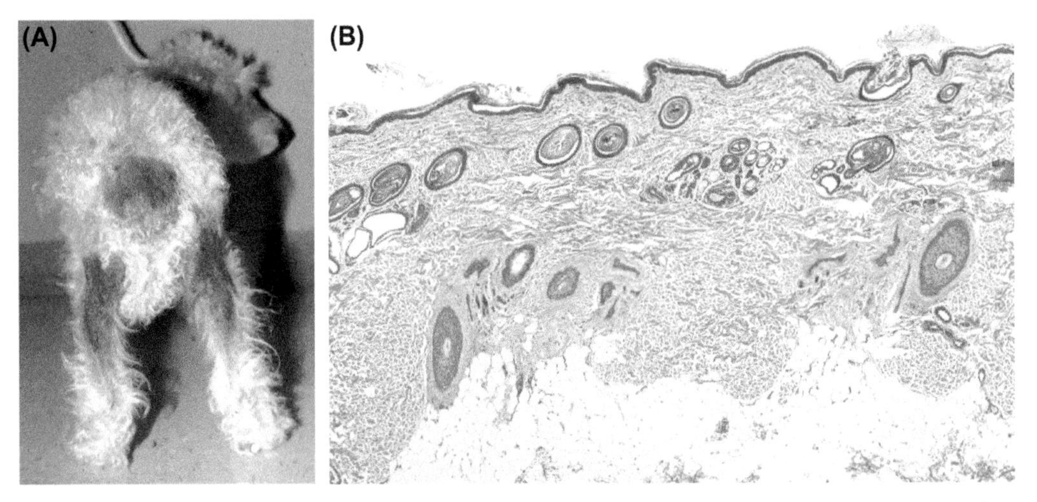

FIGURE 53.2 Hyperestrogenism, iatrogenic diethylstilbestrol, dog. *Reproduced from Pathologic Basis of Veterinary Disease (Fig. 17–69), fourth ed. p. 1223. Academic Press/Elsevier.*

Alopecia and Inhibition of Hair Growth (Mimosine Toxicity)

Fariñas (1951) reported the utilization of the legume *Leucaena glauca*, also known as *ipil-ipil*, Koa haole, or Santa Elena tree, as a fodder and pasture crop in the tropics. However, loss of hair has been reported after ingestion of seeds in native women and seeds and foliage in animals. The toxic nonprotein free amino acid, identified as mimosine or leucenol, was shown to act as a tyrosine analog capable of competitive inhibition of tyrosinase and inhibition of thyroxine decarboxylase (Crounce et al., 1962). Dietary administration of the purified toxic principle and ground seeds to mice produced hair growth inhibition and loss of hair, probably from inhibition of mitosis of hair follicles, especially in the matrix (Montagna and Yun, 1963). Growing (anagen) hairs were affected, and there was complete recovery following withdrawal from treatment. Grossly, the skin appeared thin and atrophic. Histologic findings of absent mitosis, destruction of matrix cells of the hair follicles, and extracellular melanin were observed. Morphologic findings together with a history of exposure to ipil-ipil seeds and foliage are suggestive of mimosine-induced alopecia.

Acute Bovine Pulmonary Edema and Emphysema

Acute bovine pulmonary edema and emphysema (*ABPE* or "fog fever") occurs in adult beef-type cattle grazing fall pastures on changing from sparse dry range to lush green pastures (Dungworth, 1985). Gross findings are observed primarily in the dorsocaudal lobes of the lungs, which are pale, soft, and rubbery in texture and moist on cut section. Microscopically, diffuse interstitial pneumonia with severe alveolar and interstitial edema and interlobular emphysema are observed. There is hyaline membrane formation within alveoli and hyperplasia of type II alveolar epithelial cells in animals that survive for several days.

ABPE is related to increased amounts of L-tryptophan in the ingested pasture grasses, which are metabolized to 3-methyl indole (3MI) in the rumen, absorbed into the bloodstream, and carried to the lung. In the lung, 3MI is converted by mixed function oxidases (MFOs) in Clara cells to a pneumotoxic compound that produces extensive necrosis of bronchiolar epithelial cells and type I pneumocytes and increases alveolar permeability leading to edema, interstitial pneumonia, and emphysema. Experimentally reducing the ruminal conversion of L-tryptophan to 3MI prevents the development of the condition. The history and morphologic findings support the presumptive diagnosis of ABPE.

Toxic Myopathy—Ionophore Toxic Syndrome

Ionophore toxic syndromes have resulted from overdosage, misuse, and drug interactions of feed additives, including formulations with monensin, lasalocid, salinomycin, narasin, maduramicin, laidlomycin, or semduramicin. Target organs damaged by toxic doses of ionophores include the heart and skeletal muscles in all species studied (reviewed by Novilla, 2012). The most important change is a toxic myopathy characterized by focal areas of degeneration, necrosis, and repair in cardiac and skeletal muscles with a variable inflammatory component (Novilla and Folkerts, 1986; Van Vleet et al., 1991). The development of muscle lesions varies among domestic species. The heart is primarily affected in horses, skeletal muscle in pigs and dogs, and there is about equal tissue predilection in rats, chickens, and cattle. In addition, neurotoxic effects have been reported for lasalocid (Shlosberg et al., 1985; Safran et al., 1993), narasin (Novilla et al., 1994), and salinomycin (Van der Linde-Sipman et al., 1999). Neuropathic changes occurred in peripheral nerves and the spinal cord. Focal swelling, fragmentation, loss of axons, and formation of digestion chambers filled with macrophages were observed in both sensory and motor nerves, and there was vacuolation with swelling, degeneration, and fragmentation of myelin sheaths and axons in the spinal cord.

It is not easy to diagnose ionophore toxicoses. Clinical signs and muscle lesions of monensin toxicoses are not pathognomonic. In the absence of proof of a gross feed mixing error with monensin (usually >5×), the diagnostic pathologist must go through a process of exclusion of potential causes of the lesion. Confirmatory diagnosis requires laboratory assays to determine the identity and amounts of the ionophore involved or concurrent use of an incompatible drug.

In humans, there are three publications of intentional exposure to monensin. According to Kouyoumdjian et al. (2001), a 17-year-old Brazilian male admitted ingesting monensin premix (Rumensin, exact amount unknown), probably to develop muscle. Instead he fell ill, was hospitalized, and died from acute rhabdomyolysis with renal failure. Although the amount of monensin ingested in this case was not estimated, in another case cited by the authors, two deaths among six people who consumed baked goods made with premix were attributed to monensin exposure of at least 10 times the optimum daily dose fed to cattle. In another report from Brazil, a 16-year-old farmworker who ingested approximately 500 mg of monensin (5 g of Rumensin 100 premix) "to become stronger" developed an early and severe rhabdomyolysis followed by acute renal failure, heart failure, and death (Caldeira et al., 2001). More recently, a 58-year-old man after ingesting 300 mg monensin (self-treatment) for suspected brain toxoplasmosis presented with 8 days of vomiting and abdominal pain had severe rhabdomyolysis without renal toxicity and survived (Blain et al., 2017). Echocardiography revealed decreased ejection fraction from 69% to 56% on his echocardiogram. Clinical pathology findings included decreased total CK (from peak above 100,000 to 5192 U/L) following 15 days of aggressive hydration and sodium bicarbonate therapy.

Consequent to ionophore-induced muscle damage, significant increases in enzymes of muscle origin occur (Amend et al., 1981); Van Vleet et al., 1983). Levels of aspartate transaminase, creatine phosphokinase, lactic dehydrogenase, alkaline phosphatase (ALP), blood urea nitrogen (BUN), and total bilirubin (TB) are elevated; calcium (Ca) and potassium (K) levels transiently decrease, whereas sodium (Na) levels are within normal limits. Cardiac troponins (both cTnI and cTnT) were reported to be highly sensitive and specific biomarker of myocardial injury in humans (O'Brien, 2008) and animals. The published range of cardiac troponin I (cTnI) in clinically normal horses was 0.0–0.06 ng/mL compared to that of 0.08–3.68 ng/mL found in six horses gavaged with a single dose of 1.0–1.5 mg monensin/kg body weight (Divers et al., 2009; Kraus et al., 2010). Because these monensin doses are close to the LD50 of 1.38 mg/kg body weight (Matsuoka et al., 1996), it was not surprising that the biomarker revealed the presence of the myocardial injury caused by toxic doses of monensin, as it would for any significant injury to heart muscle. However, the presence of heart injury is insufficient to confirm a diagnosis of monensin toxicity, so toxic exposure must be demonstrated.

Ionophore toxicoses may be suspected when there is a history of a feed-related problem in a group of animals; clinical signs of anorexia, diarrhea, labored breathing, depression, locomotor disorder, *colic*, recumbency, and death; lesions affecting heart and skeletal muscles; or congestive heart failure. The clinical signs and lesions induced by toxic levels of ionophores are not pathognomonic. However, a history of recent introduction of newly formulated feed or supplement to a herd or flock in which signs and lesions are present may cause one to suspect that acute intoxication has occurred. Dose and time factors influence the severity and distribution of lesions. Animals that die soon after exposure may not have muscle lesions discernible by light microscopy. Lesions are likely to be found in animals that survived longer than a week. Although a presumptive diagnosis of ionophore toxicosis can be made based on history, clinical signs, lesions, and considerations of the differential diagnosis, specific assays are needed for confirmatory diagnosis. Confirmatory diagnosis requires efficient laboratory assays to determine the identity and amounts of the ionophore involved and a thorough consideration of differential diagnoses. For suspected cases of monensin toxicity in horses, the exertional myopathies, such as equine rhabdomyolysis (Monday morning disease) and hyperkalemic periodic paralysis, plant poisoning from coffee senna and white snakeroot, blister beetle intoxication, colic, and laminitis (Whitlock et al., 1978; Amend et al. , 1981), should be ruled out as differential diagnoses. Complete herd history, clinical examination, successful supportive treatment, and necropsy may help differentiate these conditions from monensin and other ionophore toxicoses (Tables 53.1 and 53.2).

TABLE 53.1 Clinical Pathology Biomarkers of Ionophore Toxicoses

Elevated	Decreased	No Change
Aspartate transaminase	Calcium	Sodium
Creatine kinase	Potassium	
Lactic dehydrogenase		
Alkaline phosphatase		
Blood urea nitrogen		
Total bilirubin		
Cardiac troponin 1		

Reproduced from Veterinary Toxicology (Table 80.6 Some biomarkers of ionophore toxicity), third ed. Academic Press, Elsevier, p. 50.

Drug-Eluting Stents

Stents are commonly used to treat diseased or partially occluded coronary or peripheral arteries by acting as a scaffold to hold open the dilated segment of artery following balloon angioplasty. Early problems associated with use of bare metal stents included restenosis of the artery due to thrombosis and neointimal proliferation. Drug-eluting stents (DESs), first available in the United States in 2003, were developed to decrease restenosis and improve patient outcomes (Fig. 53.3).

DESs are polymer-coated stents impregnated with antiproliferative or antiinflammatory drugs. These stents release drug (eluate) following direct contact with the arterial wall. Common drugs used with DES include sirolimus and paclitaxel, both members of the taxine family used as immunosuppressive or oncologic chemotherapeutic agents. Specific biomarkers of toxicity include evaluation of vessel injury, inflammation, endothelialization, neointimal proliferation, coverage of stent struts, and stent apposition to the artery wall, which can be characterized microscopically with hemotoxylin and eosin, elastin, and trichrome stains. Standardized scoring systems are commonly used to evaluate these parameters. Neointimal proliferation can also be estimated with standard immunohistochemical methods for cell proliferation such as Brdu (bromodeoxyuridine), proliferating cell nuclear antigen (PCNA), or Ki-67.

TABLE 53.2 Summary of Cardiac and Skeletal Muscle Biomarkers

Biomarkers	Sources	Interpretation
LEAKAGE MARKERS		
Cardiac troponin 1 (cTn1)	Cardiac muscle	Cell injury/necrosis
Cardiac troponin 1 (cTnT)	Cardiac muscle	Cell injury/necrosis
Heart-type fatty acid binding protein (H-FABP or FABP3)	Cardiac muscle and skeletal muscle	Cell injury/necrosis
Myoglobin	Cardiac muscle and skeletal muscle	Cell injury/necrosis
Myosin light chains (Mlc)	Cardiac muscle and skeletal muscle	Cell injury/necrosis
Creatine kinase (CK)	Cardiac muscle and skeletal muscle, brain, GI tract	Cell injury/necrosis
CK-MM	Cardiac muscle and skeletal muscle	Cell injury/necrosis
CK-MB	Cardiac muscle and skeletal muscle	Cell injury/necrosis
Lactate dehydrogenase (LD)	All muscle types, liver, RBCs	Cell injury/necrosis
Aspartate aminotransferase (AST)	All muscle types, liver, RBCs	Cell injury/necrosis
Skeletal muscle troponin I (sTnI)	Skeletal muscle	Cell injury/necrosis
FUNCTIONAL MARKERS		
Atrial natriuretic peptide (ANP)	Primarily cardiac atria	Atrial wall stretch
Brain natriuretic peptide (BNP, proBNP, NT-proBNP)	Primarily cardiac ventricles	Ventricular wall stretch

Reproduced from Faqi, A.S. A Comprehensive Guide to Toxicology in Nonclinical Drug Development. Academic Press/Elsevier, p. 459.

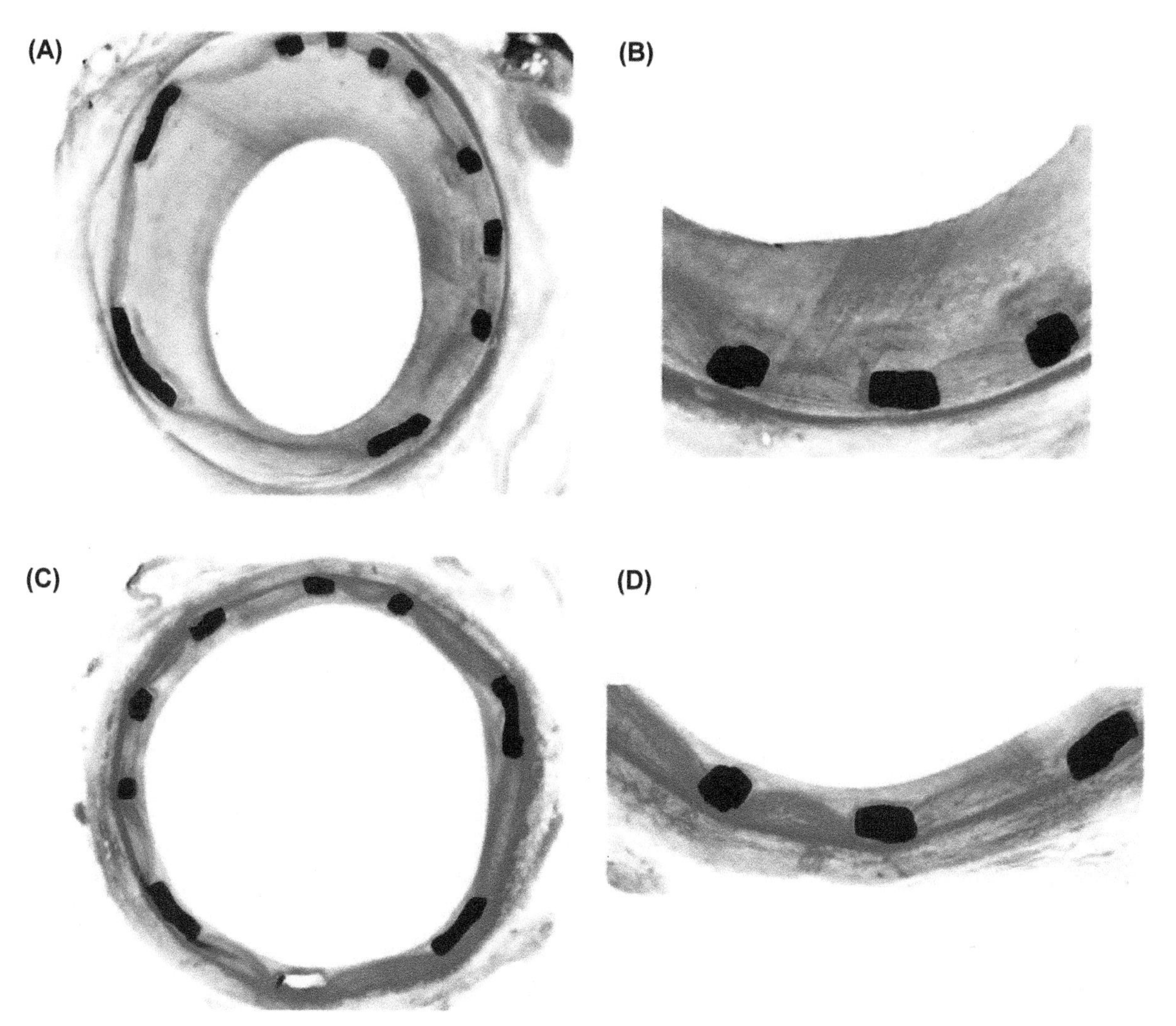

FIGURE 53.3 Actinomycin-D stent in rabbit iliac arteries. *From Schwartz et al., 2004. JACC 44, 1373–1385; photo(s) of drug-eluting stents from Dr. S. B. Jacobson.*

One study has indicated an association of restenosis occurrence after DES implantation with diabetes mellitus, circulating CD45-positive platelets, and neutrophil to lymphocyte ratio (Melnikov et al., 2017).

HEPATOPATHIES

Glycogen Hepatopathy

Glycogen accumulates in hepatocytes secondary to xenobiotics that increase glycogen synthesis or decrease glycogenolysis. Glucocorticoid analog, such as dexamethasone, are potent stimulators of glycogen synthesis, especially in dogs, causing centrilobular hepatocellular hypertrophy secondary to glycogen accumulation (Fig. 53.4). Glycogen accumulation extends into midzonal regions when the effect has greater magnitude. Liver size and weight are increased because of the increased hepatocellular glycogen accumulation. Ultrastructurally, hepatocytes have large accumulations of

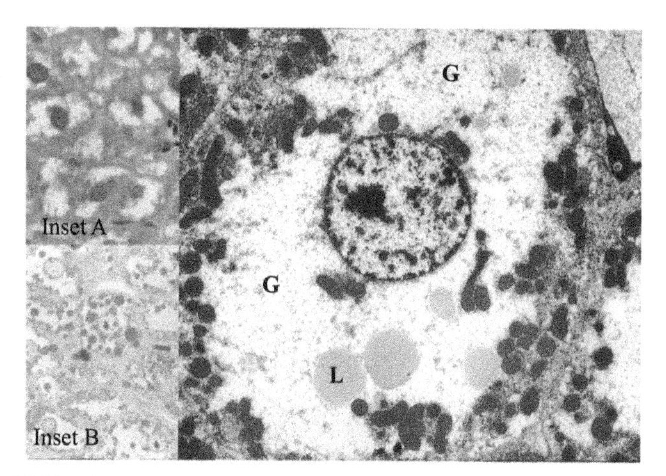

FIGURE 53.4 Glycogen hepatopathy caused by corticosteroid administration to a dog. *Courtesy of Dr. V. Meador, Covance Laboratories Global Pathology.*

glycogen free in the cytosol that displaces cytoplasmic components to the cell periphery. Interspersed in glycogen are lipid droplets.

Hepatotoxicity Secondary to Enzyme Induction

Proliferation of smooth endoplasmic reticulum (SER) is a well-documented change seen secondary to administration of xenobiotics that induce cytochrome P450 enzymes (Guengrich, 2007). Hepatocytes have a full complement of cytochrome P450s (CYP) because one of the liver's main functions is metabolism of foreign substances. Induction most commonly occurs in centrilobular hepatocytes. Not all xenobiotics that induce CYP cause proliferation of SER. Proliferation is dependent on the subset of CYP that are induced, as no single agent induces all CYPs. Cytochrome P450 enzymes are grouped into families (e.g., CYP1, CYP2, CYP3) with capital letters designating subfamilies (e.g., CYP1A, CYP1B, CYP1C) and numerals further identifying individual enzymes (e.g., CYP1A1). Not all CYP inducers cause SER proliferation and liver enlargement (Barka and Popper, 1967); however, SER proliferation is nearly pathognomonic for CYP induction.

Classic inducers of SER proliferation are phenobarbital and its analog. Phenobarbital is metabolized by, and induces, cytochrome P4502B, the major membrane protein of SER (2d). Proliferation can be rapid and dramatic, with morphologic changes occurring within 4 days. Liver weight and size are increased. By light microscopy in Hematoxylin and Eosin (HE) sections, centrilobular hepatocytes are hypertrophied because of increased amounts of homogeneous eosinophilic cytoplasm. SER proliferation begins in centrilobular hepatocytes (Fig. 53.5). With increasing magnitude, proliferation will extend into midzonal regions and rarely to periportal hepatocytes. Withdrawal of the inciting xenobiotic is accompanied by a rapid regression of changes (Cheville, 1994). Effete SER forms aggregates that are removed from the cell by autophagocytosis (Feldman et al., 1980).

Hepatotoxicity secondary to hepatocellular enzyme induction can occur through increased activation of xenobiotics to hepatotoxins (Zimmerman, 1999). Subsets of CYP inactivate xenobiotics while others can activate creating reactive electrophiles (Greaves, 2007). Aflatoxin B1, a mycotoxin, is activated to a number of metabolites, including exo-8,9-epoxide, a hepatocarcinogen. Acetaminophen, an analgesic, is activated to a reactive iminoquinone. Trichloroethylene (TCE), an industrial toxicant and previously used anesthetic, is activated to TCE oxide, which forms unstable protein adducts. Troglitazone, an antidiabetes drug, is metabolized to electrophilic reactive metabolites.

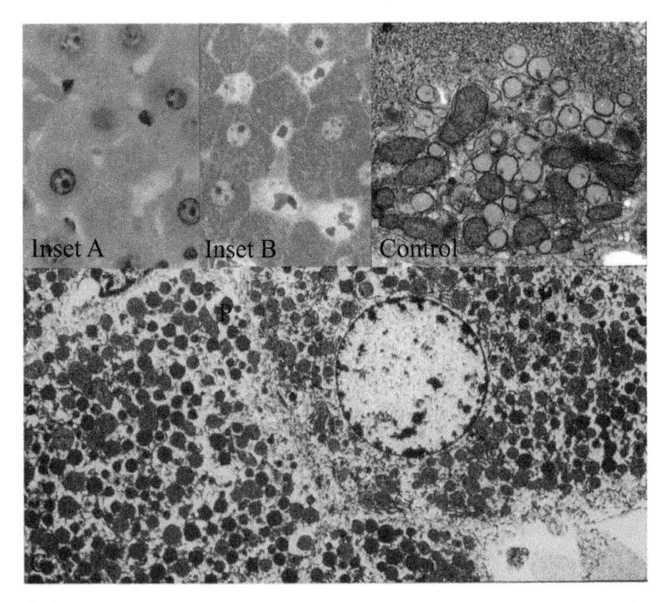

FIGURE 53.5 Smooth endoplasmic reticulum proliferation in hepatocytes. *Courtesy of Dr. V. Meador, Covance Laboratories Global Pathology.*

Peroxisome Proliferator-Activated Receptor Hepatopathy (Characteristic Lesion by Light and Electron Microscopy) in Rodents

Peroxisomes are identified primarily in liver and kidneys, and they are located in cytoplasm often adjacent to mitochondria. Drug-induced effects on peroxisomes have been reported to increase numbers and cause qualitative changes in appearance, but decrease in numbers has not been reported. Stimulation of the hepatocytes with peroxisome proliferator-activated receptor (PPAR) α causes proliferation of peroxisomes (Fig. 53.6). Drugs that induce CYP4A, e.g., PPARα, cause proliferation of hepatic peroxisomes. The PPARα receptor is the transcription factor that activates CYP4A. In general, the change is quantitative causing an increase in numbers, which is appreciable in the liver, but has also been reported in the kidney. By light microscopy, liver has centrilobular hepatocellular hypertrophy, with enlarged hepatocytes having a finely stippled eosinophilic granular appearance. Liver weight can be increased. Changes in appearance of peroxisomes have been reported, but apart from disappearance of the crystalline core, qualitative changes are seldom detected. There is species susceptibility to peroxisome proliferation: rodents are most susceptible, with dogs, nonhuman primates, humans, and guinea pigs being poorly susceptible.

For the purpose of this discussion, "disorder" is considered as any morphologic change of peroxisomes as evidenced by a quantitative and/or qualitative change compared to control. Compromise of

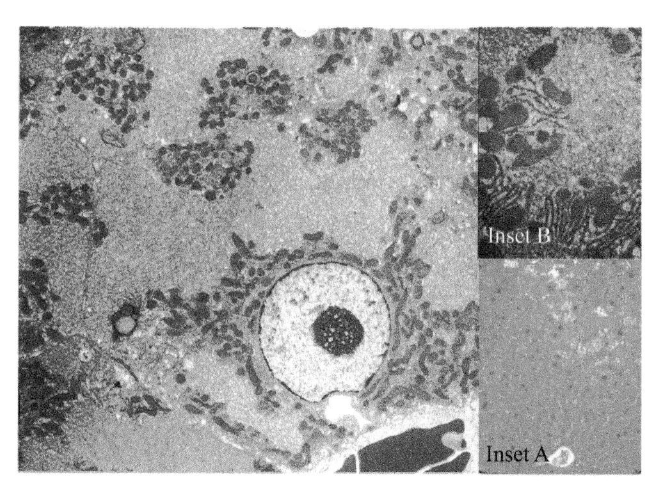

FIGURE 53.6 Peroxisome proliferation in a rat given peroxisome proliferator-activated receptor alpha. *Courtesy of Dr. V. Meador, Covance Laboratories Global Pathology.*

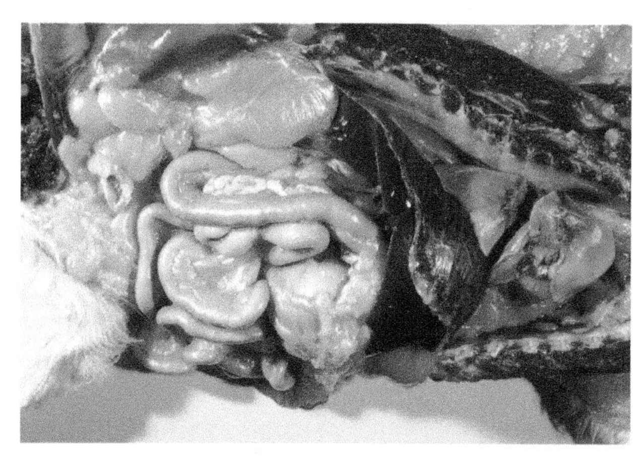

FIGURE 53.7 Icterus, hemolytic anemia, dog. *Reproduced from Pathologic Basis of Veterinary Disease (Fig. 1–72), fourth ed. p. 58. Academic Press/Elsevier.*

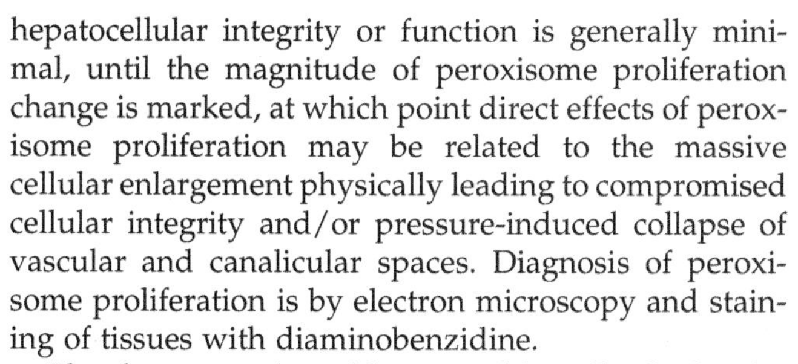

hepatocellular integrity or function is generally minimal, until the magnitude of peroxisome proliferation change is marked, at which point direct effects of peroxisome proliferation may be related to the massive cellular enlargement physically leading to compromised cellular integrity and/or pressure-induced collapse of vascular and canalicular spaces. Diagnosis of peroxisome proliferation is by electron microscopy and staining of tissues with diaminobenzidine.

The demonstration of lysosomal lamellar bodies by TEM might be considered as a pathologic biomarker of phospholipidosis. A readily accessible, reliable, and noninvasive clinical biomarker of phospholipidosis need to be found. Phospholipidosis results from a multiplicity of different molecules involving different tissues. One impediment is that each phospholipidosis-producing chemical must be evaluated independently (Chatman et al., 2009).

Iron Toxicosis

Iron-containing products can increase iron pigment in hepatocytes and/or Kupffer cells. Excessive iron supplementation in dogs and cats has resulted in hemochromatosis due to iron overload in the liver. Cases of iron poisoning have occurred in suckling pigs administered iron dextran intramuscularly and in newborn foals administered ferrous fumarate to prevent anemia (Cullen, 2006). In severe toxicosis, there is significant mortality due to massive hepatic necrosis and affected piglets die soon after injection. There is yellowish brown discoloration (icterus) of tissues, notably those near the injection site. Although the yellowish brown discoloration is considered a gross pathological biomarker in the case of iron dextran poisoning, icterus has been reported in other diseases and conditions with extensive liver damage; see Fig. 53.7.

Copper Poisoning in Sheep

In ruminants, especially sheep, accumulation of copper in the liver may occur over time because of poor regulation of copper storage. Copper toxicity in sheep results from: (1) dietary excess from contamination of pastures from sprays or fertilizer or access to copper-containing mineral blocks intended for cattle; (2) grazing on pasture grasses containing normal copper but inadequate levels of molybdenum, which antagonizes copper uptake by the liver; and (3) grazing on pastures with hepatotoxic plants such as *Heliotropium*, *Crotalaria*, and *Senecio*, which produce pyrrolizidine alkaloids (Cullen, 2006). Sudden release of the copper results in severe hemolytic anemia and extensive hepatic necrosis. The breakdown of erythrocytes results in release of hemoglobin, which stains the plasma pink and discolors the kidney dark blue and the urine dark red. Gross findings are usually characteristic and include generalized icterus, greatly enlarged "gunmetal blue" kidneys with hemorrhagic mottling, mildly enlarged, friable and yellowish liver, distended gall bladder with thick greenish bile, and enlarged brown to black spleen. Microscopically, there are multifocal areas of hepatocellular vacuolation and necrosis in the liver, and the kidney tubules have hyaline and coarsely granular hemoglobin casts resulting from intravascular hemolysis with glomerular filtration of hemoglobin into the uriniferous space.

NEPHROPATHIES

Polycystic Kidney Disease in Rats

Dietary administration of diphenylthiazole to rats for 4–8 weeks resulted in three- to fivefold bilateral and symmetrical enlargement of the kidneys due to marked cystic changes (Carone, 1986). The most prominent enlargement occurs in the outer medulla with radially arranged dilated collecting tubules and occasional dilation of cortical collecting and distal tubules. The histologic correlates include focal degeneration and necrosis and mild tubular dilation of collecting ducts at 2 weeks of feeding and marked cystic change at 4–8 weeks. Recovery from the distinctive morphologic changes occurred after diphenylthiazole treatment withdrawal of 4–8 weeks.

Urinary neutrophil gelatinase-associated lipocalin (NGAL) and IL-18 excretion have been shown to be slightly elevated in ADPKD. However, the increased level does not correlate with alterations in total kidney volume or kidney function. A potential explanation could be the discontinuity between individual cysts and the urinary collecting system in this pathology (Parikh et al., 2012).

Alpha 2-Microglobulin Nephropathy

Abnormal accumulation of alpha 2-microglobulin (alpha-2 mu-globulin) in kidney lysosomes of male rats has been described in nephropathy resulting from exposure to a variety of chemicals, including D-limonene, a constituent of orange juice. The nephropathy is characterized by distinctive cytoplasmic hyaline droplets in epithelial lining cells of proximal convoluted tubules and granular casts at the junction of the inner and outer stripe of the outer medulla and linear mineral deposits in the descending limb of the loop of Henle. Accumulation of hyaline droplets leads to tubular apoptosis followed later by regeneration, exacerbation of spontaneous renal disease, and tumor formation.

Tubulointerstitial Nephritis

A considerable number of xenobiotics can induce kidney injury. Most nephrotoxicants damage the renal tubules with many toxicants affecting specific segments. Microscopic findings include interstitial inflammation, edema, fibrosis, and atrophy. Acute kidney injury has been proposed to encompass the full spectrum of renal injury from minor elevations of serum chemistry values to anuric renal failure (Gwaltney-Brant, 2012). In drug-induced tubulointerstitial nephritis (DTIN) microscopic findings include interstitial inflammation, edema, fibrosis, and atrophy (Wu et al., 2010). Lack of correlation has limited the usefulness of established noninvasive diagnostic biomarkers in defining severity and progression of kidney injury. For instance, renal azotemia (increased BUN) occurs only after approximately 75% of the nephrons have lost function (Turk et al., 1997). According to Wu et al. (2010), a renal biopsy is the gold standard for diagnosis, prognosis, and selection of therapies for DTIN. Urinary monocyte chemotactic peptide-1 (MCP-1) levels correlated with the degree of severity of acute lesions in DTIN. The roles of **NGAL** and α1-microglobulin in chronic lesions need further evaluation (Wu et al., 2010), and to monitor the progression rate of drug-induced chronic DTIN, urinary *N*-acetyl-β-D-glucosaminidase (NAG) and matrix metalloproteinases (MMPs) 2 and 9 are two possible candidates to consider (Shi et al., 2013).

Ethylene Glycol (Oxalate) Poisoning

Toxic amounts of ethylene glycol, which is commonly used as antifreeze, get metabolized in the liver via alcohol dehydrogenase resulting in formation of oxalates and development of hypocalcemia. The latter occur from the formation of insoluble calcium oxalates that are widely deposited but most severely in the kidney, resulting in crystalluria, which is used as a biomarker for this toxicity. The precipitated calcium oxalates, seen better under polarized light as numerous pale yellow, birefringent (rosette-shaped) crystals in proximal convoluted tubules are consistent and characteristic in animals that die from ethylene glycol toxicity (Thrall et al., 1985; Stice et al., 2018). In addition to ethylene glycol, oxalate poisoning can result from ingestion of oxalate salts, plants with toxic levels of oxalates, and plants infected with fungi that produce oxalates. Grossly, the kidneys are pale, swollen, and feel gritty when cut. Microscopically, degeneration and necrosis of proximal convoluted tubules associated with pale yellow variable-shaped crystals forming sheaves, rosettes, or prisms within proximal tubular lumens, and cytoplasm of the lining of epithelial cells and interstitial spaces confirm the diagnosis of oxalate poisoning.

Melamine Poisoning

Melamine cyanurate crystals (MCA) precipitate in the kidney of dogs and cats ingesting commercial pet food contaminated with melamine and cyanuric acid (Bischoff, 2018). Grossly, there is hemorrhage and interstitial edema primarily of the medullary region and histologic correlates of tubular degeneration and necrosis of distal straight tubules associated with pale yellow to brown crystals that were fan to starburst to globular in appearance within the lumen of the distal straight and collecting tubules. The more downstream location and different morphology of the crystals is diagnostic of MCA poisoning.

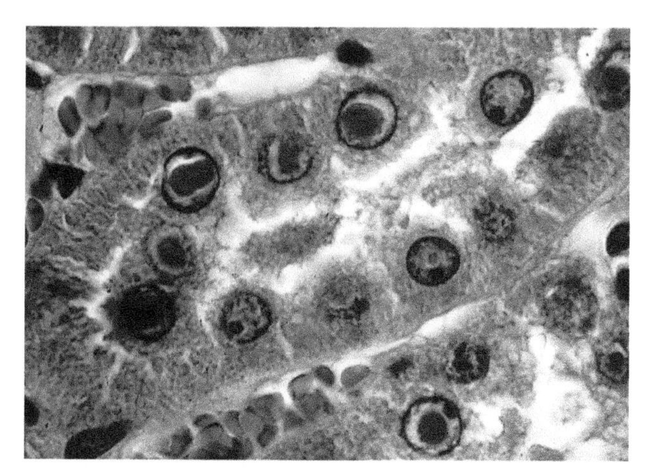

FIGURE 53.8 Lead-acid fast inclusion bodies in rat. *Reproduced from Pathologic Basis of Veterinary Disease (Fig. 11–41), fourth ed. p. 654. Academic Press/Elsevier.*

Lead Toxicity

In the kidney, toxic tubular necrosis is caused by several classes of naturally occurring and synthetic compounds, including heavy metals such as lead, cadmium, thallium, and inorganic mercury. The nephrotoxic heavy metal cannot be identified by renal lesions alone, except for lead. Exposure to old paints and batteries may result in lead toxicity in which renal tubular epithelial cells have irregularly shaped intranuclear inclusion bodies that are acid fast (Myers and McGavin, 2006; Fig. 53.8). The inclusion bodies, composed of lead protein complexes, are considered to be a diagnostic pathological biomarker for lead toxicosis.

Nephrotoxicity and Acute Renal Injury

One major factor that explains the distinct susceptibility of the kidney to injury is the exposure to a variety of exogenous compounds (e.g., aminoglycosides, cisplatin, radiocontrast agents) and endogenous compounds (e.g., free hemoglobin after hemolysis, free myoglobin after rhabdomyolysis) via the blood flow. Both kidneys receive roughly 25% of cardiac output. Heavy metals (e.g., chromium, lead, and mercury), analgesics (including acetaminophen and nonsteroidal antiinflammatory drugs), aminoglycoside antibiotics (e.g., neomycin, gentamicin), immunosuppressive agents (e.g., cyclosporine, tacrolimus), and chemotherapeutic agents (e.g., cisplatin, carboplatin) are known common nephrotoxic drugs that cause kidney damage (Schnellman, 2008; Pazhayattil and Shirali, 2014). For drug-induced renal dysfunction, the potential mechanisms underlying their toxicity include alterations in renal perfusion and glomerular filtration, tubular cell damage, and tubular obstructions.

To detect and monitor renal dysfunction, clinical pathology end points such as BUN, serum creatinine (sCr), phosphorus, and urine-specific gravity has been established and used over the past decades. However, these parameters are relatively insensitive indicators in chronic injury and exhibit alteration only when more than two-thirds nephrons have functional deficit (Stockham and Scott, 2008). In addition to BUN and sCr, there are seven renal injury biomarkers with improved sensitivity that was qualified by the FDA to monitor drug-induced renal injury in rats (Dieterle et al., 2010a,b). These seven renal injury biomarkers are kidney injury molecule-1, albumin, total protein, β2-microglobulin, cystatin C, clusterin, and trefoil factor-3 (Rosenberg and Paller, 1991; Witzgall et al., 1994; Ichimura et al., 1998; Han et al., 2002; Davis et al., 2004; Filler et al., 2005; Vaidya et al., 2006; Perez-Rojas et al., 2007; Bonventre et al., 2010; Dieterle et al., 2010a,b; Yu et al., 2010; Vlasakova et al., 2014). Clusterin and renal papillary antigen-1 have been used to detect acute drug-induced renal tubule alterations in rats. Several biomarkers such as NGAL (lipocalin-2) have been used in drug development settings (Ennulat and Adler, 2015; Mishra et al., 2003). For several decades, more traditional biomarkers employing enzymes that are released from damaged renal tubular cells such as ALP, lysosomal enzymes N-acetyl-β-glucosaminidase (NAG), and gamma glutamyltransferase (GGT) have been used as markers of renal injury in animal studies (Clemo (1998); Emeigh (2005); D'Amico and Bazzi (2003).

It is important to understand the relationship between biomarkers and microscopic pathologic changes on the kidney as assessed by light microscopy. Because of renal cell injury, enzymes such as BUN are elevated prerenal, renal, or post renal; but, BUN can also increase with diet, without renal injury. For the kidneys, histopathologic findings remain the gold standard by which biomarkers are validated for use (Table 53.3).

Polioencephalomalacia

Polioencephalomalacia is diagnosed at necropsy of cattle, sheep, and, less commonly, goats observed with clinical signs including depression, stupor, ataxia, head pressing, apparent cortical blindness, opisthotonos, convulsions, and recumbency with paddling of the limbs. Although primarily associated with thiamine deficiency, this may be a multifactorial metabolic disorder. Multiple dietary factors have been proposed, such as feeding high carbohydrate rations with little roughage (Zachary, 2006). In cattle, microcavitation of the deep cortical lamina adjacent to subcortical white matter is characteristic of chronic polioencephalomalacia.

TABLE 53.3 Summary of Renal Injury Biomarkers

Biomarkers	Sources	Interpretation
Kidney injury molecule 1 (KIM-1, TIM-1, HAVCR-1)[a]	Proximal tubules	Toxic, ischemic, septic renal tubular injury
Neutrophil gelatinase-associated lipocalin (NGAL/Lipocalin-2)[b]	Proximal and distal tubules, neutrophils, bone marrow	Renal tubular injury (urinary), inflammatory (blood)
Clusterin	Proximal and distal tubules	Toxic, ischemic, septic renal tubular injury
Cystatin C	All nucleated cells, renal tubules	Plasma levels inversely related to GFR; renal tubular injury (urine)
Alkaline phosphatase (ALP)	Renal tubules	Renal tubular injury
N-Acetyl-β-glucosaminidase (NAG)	Renal tubules	Renal tubular injury
Gamma glutamyltransferase (GGT)	Renal tubules	Renal tubular injury
Total protein	Plasma protein	Glomerular injury
Albumin (microalbumin)	Plasma protein	Glomerular injury
α1/β2-microglobulin	Plasma protein	Glomerular injury and/or proximal tubular injury

[a]*May not be useful in canines.*
[b]*False-positive increases can be induced in neutrophils by various stimuli (inflammation, infections, neoplasia).*
Reproduced from A Comprehensive Guide to Toxicology in Nonclinical Drug Development (Table 17.1), first ed. Academic Press/Elsevier, p. 457.

Results from a previous study suggest that gut fill, Cu status, and cytochrome *c* oxidase activity in brain tissue might be influenced and compromised by wet distillers grains with solubles when fed at 60% of diet dry matter in diets based on steam-flaked corn. It might imply a greater susceptibility to polioencephalomalacia (Ponce et al., 2014).

Sodium Chloride Toxicity

Sodium chloride toxicity or salt poisoning has been reported primarily in pigs, poultry, and occasionally in ruminants, horses, and dogs following increased intake of sodium chloride from feed rations or supplements with limited access to drinking water. Gross lesions include inconsistent meningeal congestion and edema, and transverse slices of fixed brain have shown subgross evidence of laminar necrosis. Histologically, laminar cerebrocortical necrosis accompanied by astrocytic swelling and infiltration of meningeal and cerebral blood vessels by eosinophils were observed in pigs. The latter was considered the microscopic pathology biomarker of salt poisoning.

Nigropallidal Encephalomalacia in Horses

Sharply demarcated foci of pale yellowish discoloration to buff-colored foci of softening or cavitation of the substantia nigra and the globus pallidus have been reported in horses grazing on yellow star thistle (*Centaurea solstitialis*) or Russian knapweed (*Acroptilon repens*) for long periods of time (Cordy, 1978). The gross lesions, which were usually bilateral, correlated histologically with necrosis and loss of neurons as well as necrosis of axons, glia, and blood vessels in the anterior regions of the globus pallidus and substantia nigra. In addition, localized encephalomalacia (necrosis) of the nucleus of the inferior colliculus (Fig. 53.9), the mesencephalic nucleus of the trigeminal nerve, and the dentate nucleus was also reported in horses fed dried Russian knapweed (Young et al., 1960). Morphologic findings

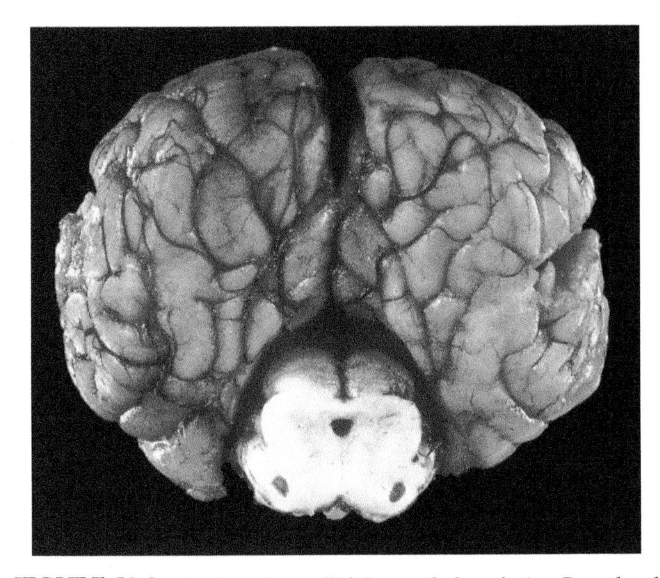

FIGURE 53.9 Equine nigropallidal encephalomalacia. *Reproduced from Pathologic Basis of Veterinary Disease (Fig. 11–79), fourth ed. p. 921. Academic Press/Elsevier.*

are distinctive for the neurotoxicity that develops in horses. A potential biomarker of nigropallidal encephalomalacia in horses would be in relation to oxidative stress. Results from a previous study suggest that glutathione (GSH) depletion by the sesquiterpene lactone repin may increase the susceptibility to oxidative damage in equine nigropallidal encephalomalacia (Tukov et al., 2004).

Phospholipidosis

Cationic amphophilic drugs (CADs) from different pharmacologic classes are known to induce accumulation of phospholipids and drugs within a variety of cells (Reasor, 1989). When taken up by the lysosomal system the CADs, such as chlorphentermine, amiodarone, and chloroquine, may raise intracellular pH and form complexes with polar lipids or mucopolysaccharides, which interfere with lysosomal enzyme activity, resulting in accumulation of polar lipids (Hein et al., 1990). By light microscopy, affected cells are foamy because of presence of lamellar inclusion bodies.

Wünschmann et al. (2006) reported that captive Humboldt penguins found dead after treatment for avian malaria with higher than usual doses of chloroquine had neuronal storage disease. Affected birds were found histologically to have moderate to marked vacuolation of Purkinje cells, neurons of the brainstem, and motor neurons of the spinal cord. By electron microscopy, the cytoplasmic vacuoles were multilayered lamellar structures by electron microscopy, characteristic of drug-induced lysosomal storage disease (Fig. 53.10). For chloroquine toxicity, gangliosides rather than phospholipids are primarily stored (Klinghardt et al., 1981). A tentative histopathologic diagnosis may be enhanced by Immunohistochemistry (IHC) with antibodies for autosomal membrane associated protein-2 (LAMP-2) and anti-ADRP for adipophilin. However, the widely accepted standard for a diagnosis of lysosomal storage disease is electron microscopic identification of concentric lamellar inclusion bodies in secondary lysosomes (Monteith et al., 2006).

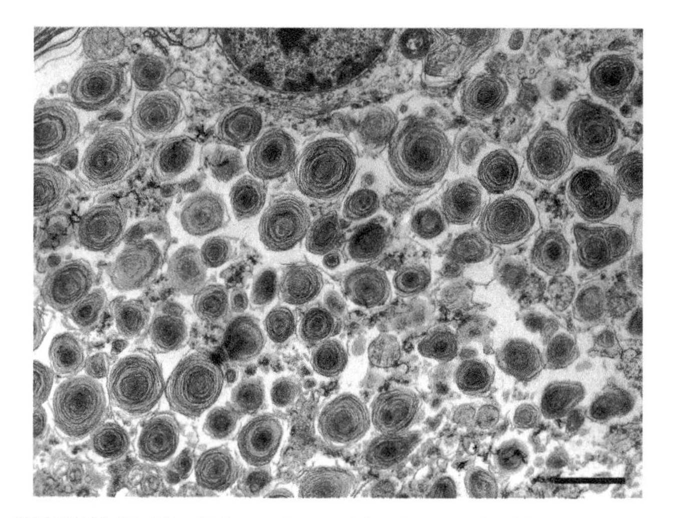

FIGURE 53.10 Chloroquine toxicity in penguin. Electron micrograph of a spinal cord motoneuron with distended perikaryon due to the presence of numerous multilayered concentric lamellar structures. Bar = 300 nm. *Courtesy of Dr. A. Wünschmann, College of Veterinary Medicine, University of Minnesota.*

The central nervous system that contains numerous cell bodies with specific functions is protected from xenobiotics by the blood–brain barrier (BBB); but, some molecules can cross BBB by nature of their physicochemical properties or design. As of this writing, "there are no routinely used molecular/chemical validated biomarkers to assess xenobiotic related injury to the nervous system."

CONCLUDING REMARKS AND FUTURE DIRECTIONS

This chapter focused on morphologic (gross and microscopic) and clinical pathology findings as pathological biomarkers of adverse effects of drugs, food additives, biopharmaceuticals, medical devices, environmental chemicals, and plants. Introduction of the autopsy/necropsy and the use of the light microscope to investigate diseases were important milestones that provided recognizable postmortem morphologic end points—the gross and microscopic pathological biomarkers. Later, integration of special morphologic pathology techniques resulted in increased accuracy of diagnoses reported accurately and interpreted correctly in a timely manner. Now there are validated biomarkers being used to fine tune diagnosis, prognosis, and therapy with more being discovered through "omics" technologies (Milburn et al., 2013). Because morphologic techniques remain typically the gold standard against which specificity and sensitivity of novel biomarkers are being validated, the toxicologic pathologist should be aware of and adapt to his/her expanding role. As per *Best Practices Guideline: Toxicologic Histopathology* (Crissman et al., 2004), the toxicologic pathologist must be qualified and possess credentials that document a high level of education, training, experience, and expertise and shall be provided well-prepared slides, laboratory test results, and other study information necessary to conduct an accurate pathologic evaluation. A draft guidance regarding the use of histopathology for biomarker qualification studies has been issued (Food and Drug Administration, 2011). Continued improvements in pathology processes and procedures will come from the alignment with *Best Practices Guidelines*. Taken together with the personal integrity of the toxicologic pathologist, this assures a high-quality pathology report, contributing to the final reports of R&D safety, toxicity, and efficacy studies for the benefit of man and animals.

References

Ambrosio, V., Borzacchiello, G., Bruno, F., et al., 2001. Uroplakin expression in the urothelial tumors of cows. Vet. Pathol. 38, 657–660.

Amend, J.F., Mallon, F.M., Wren, W.B., Ramos, A.S., 1981. Equine monensin toxicosis: some experimental clinicopathologic observations. Comp Cont Ed 11, S173–S183.

Barka, T., Popper, H., 1967. Liver enlargement and drug toxicity. Medicine 46, 103–117.

Barr, D.B., Buckley, B., 2011. *In vivo* biomarkers and biomonitoring in reproductive and developmental toxicity. In: Gupta, R.C. (Ed.), Reproductive and Developmental Toxicology, first ed. Academic Press/Elsevier, Amsterdam, pp. 253–265.

Bischoff, K., 2018. Melamine and cyanuric acid. In: Gupta, R.C. (Ed.), Veterinary Toxicology. Basic and Clinical Principles, third ed. Academic Press/Elsevier, Amsterdam, pp. 1067–1072.

Blain, M., Garrard, A., Poppenga, R., et al., 2017. Survival after severe rhabdomyolysis following monensin ingestion. J. Med. Toxicol. 13, 259–262.

Bonventre, J.V., Vaidya, V.S., Schmouder, R., et al., 2010. Next-generation biomarkers for detecting kidney toxicity. Nat. Biotechnol. 28, 436–440.

Buck, W.B., Osweiller, G.D., Van Gelder, G.D., 1973. Clinical and Diagnostic Veterinary Toxicology, second ed. Kendall/Hunt Publishing Co, Dubuque, Iowa.

Burrows, G.R., Tyrl, R.J., 2001. Toxic Plants of North America. Iowa State University Press, Ames, Iowa.

Caldeira, C., Neves, W.S., Cury, P.M., et al., 2001. Rhabdomyolysis, acute renal failure, and death after monensin ingestion. Am. J. Kidney Dis. 38, 1108–1112.

Carone, F.A., 1986. Diphenylthiazole-induced renal cystic disease, rat. In: Jones, T.C., Mohr, U., Hunt, R.D. (Eds.), Urinary System: Monographs on Pathology of Laboratory Animals. Springer-Verlag, New York, pp. 262–267.

Carvalho, T., Pinto, C., Peleteiro, M.C., 2006. Urinary bladder lesions in bovine enzootic haematuria. J. Comp. Pathol. 134, 336–346.

Chatman, L.A., Morton, D., Johnson, T.O., et al., 2009. A strategy for risk management of drug-induced phospholipidosis. Toxicol. Pathol. 37, 997–1005.

Clemo, F.A., 1998. Urinary enzyme evaluation of nephrotoxicity in the dog. Toxicol. Pathol. 26, 29–32.

Cheville, N.F., 1994. Blockade of metabolic pathways. In: Cheville, N.F. (Ed.), Ultrastructural Pathology: An Introduction to Interpretation. Iowa State University Press, Ames, Iowa, pp. 126–181.

Chirathaworn, C., Kongpan, S., 2014. Immune responses to Leptospira infection: roles as biomarkers for disease severity. Braz. J. Infect. Dis. 18, 77–81.

Cordy, D.R., 1978. *Centaurea solstitialis* and equine nigropallidal encephalomalacia. In: Keeler, R.F., VanKampen, K.R., James, L.F. (Eds.), Effects of Poisonous Plants on Livestock. Academic Press, NY, pp. 327–336.

Cota, J.B., Carvalho, T., Pinto, C., et al., 2013. Epithelial Urinary bladder tumors from cows with enzootic hematuria: structural and cell cycle-related protein expression. Vet. Pathol. 51, 749–754.

Crissman, J.W., Goodman, D.G., Hildebrandt, P.K., 2004. Best practices guideline: toxicologic histopathology. Toxicol. Pathol. 32, 126–131.

Crounce, R.G., Maxwell, J.D., Blank, H., 1962. Inhibition of hair growth by mimosine. Nature 194, 694–695.

Cullen, J.M., 2006. Liver, biliary system and exocrine pancreas. In: McGavin, M.D., Taylor, Z.M. (Eds.), Pathologic Basis of Veterinary Disease, fourth ed. Academic Press/Elsevier, Amsterdam, pp. 393–461.

D'Amico, G., Bazzi, C., 2003. Urinary protein and enzyme excretion as markers of tubular damage. Curr. Opin. Nephrol. Hypertens. 12, 639–643.

Davis 2nd, J.W., Goodsaid, F.M., Bral, C.M., et al., 2004. Quantitative gene expression analysis in a nonhuman primate model of antibiotic-induced nephrotoxicity. Toxicol. Appl. Pharmacol. 200, 16–26.

Dieterle, F., Perentes, E., Cordier, A., et al., 2010b. Urinary clusterin, cystatin C, beta2-microglobulin and total protein as markers to detect drug-induced kidney injury. Nat. Biotechnol. 28, 463–469.

Dieterle, F., Sistare, F., Goodsaid, F., et al., 2010b. Renal biomarker qualification submission: a dialog between the FDA-EMEA and predictive safety testing consortium. Nat. Biotechnol. 28, 455–462.

Divers, T.J., Kraus, M.S., Jesty, S.A., et al., 2009. Clinical findings and serum cardiac troponin I concentrations in horses after intragastric administration of sodium monensin. J. Vet. Diagn. Invest. 21, 338–343.

Dungworth, D.L., 1985. The respiratory system. In: Jubb, K.V.F., Kennedy, P.C., Palmer, N. (Eds.), Pathology of Domestic Animals, third ed. Academic Press, New York, pp. 413–556.

Emeigh Hart, S.G., 2005. Assessment of renal injury in vivo. J. Pharmacol. Toxicol. Meth. 52, 30–45.

Ennulat, D., Adler, S., 2015. Recent successes in the identification, development, and qualification of translational biomarkers: the next generation of kidney injury biomarkers. Toxicol. Pathol. 43, 62–69.

Fariñas, E.C., 1951. Ipil-ipil (L. glauca), the "alfalfa" of the tropics, its establishment, culture, and utilization as a fodder and pasture crop. Philipp. J. Anim. Ind. 19, 6584.

Feldman, D., Swarm, R.L., Becker, J., 1980. Elimination of excess smooth endoplasmic reticulum after phenobarbital administration. J. Histochem. Cytochem. 28, 997–1006.

Filler, G., Bokenkamp, A., Hofmann, W., et al., 2005. Cystatin C as a marker of GFR–history, indications, and future research. Clin. Biochem. 38, 1–8.

Food and Drug Administration, 2011. Use of histology in biomarker qualification studies. Draft Guidance for Industry. Available from: http://www.fda.gov/downloads/Drugs/GuidanceCompliance RegulatoryInformation/Guidances/UCM285297.pdf.

Gupta, R.C., Evans, T.E., Nicholson, S.S., 2018. Ergot and fescue toxicosis. In: Gupta, R.C. (Ed.), Veterinary Toxicology. Basic and Clinical Principles, third ed. Academic Press/Elsevier, Amsterdam, pp. 995–1001.

Greaves, P., 2007. Liver. In: Greaves, P. (Ed.), Histopathology of Preclinical Toxicity Studies, third ed. Elsevier, New York, pp. 457–504.

Guengerich, F.P., 2007. Cytochrome P450 activation of toxins and hepatotoxicity. In: Kaplowitz, N., Deleve, L.D. (Eds.), Drug-Induced Liver Disease, 2nd ed. Informa Healthcare, New York, pp. 13–32.

Gwaltney-Brant, S.M., 2012. Renal toxicity. In: Gupta, R.C. (Ed.), Veterinary Toxicology. Basic and Clinical Principles, third ed. Academic Press/Elsevier, Amsterdam, pp. 259–272.

Hagdu, S.I., 2002. The first use of the microscope in medicine. Ann. Clin. Lab. Sci. 32, 309–310.

Han, W.K., Bailly, V., Abichandani, R., et al., 2002. Kidney Injury Molecule-1 (KIM-1): a novel biomarker for human renal proximal tubule injury. Kidney Int. 62, 237–244.

Hargis, A.M., Ginn, P.E., 2006. The integument. In: McGavin, M.D., Taylor, Z.M. (Eds.), Pathologic Basis of Veterinary Disease, fourth ed. Academic Press/Elsevier, Amsterdam, pp. 1107–1261.

Hein, L., Lullmann-Rauch, R., Mohr, K., 1990. Human accumulation potential of xenobiotics: potential of catamphiphilic drugs to promote their accumulation via inducing lipidosis or mucopolysaccharidosis. Xenobiotica 20, 1259–1267.

Hunt, J.L., 2009. Biomarkers in anatomic pathology: adding value to diagnosis. Arch. Pathol. Lab. Anim. Med. 133, 532–536.

Ichimura, T., Bonventre, J.V., Bailly, V., et al., 1998. Kidney injury molecule-1 (KIM-1), a putative epithelial cell adhesion molecule containing a novel immunoglobulin domain, is up-regulated in renal cells after injury. J. Biol. Chem. 273, 4135–4142.

Jones, T.C., Hunt, R.D., 1983. Veterinary Pathology, fifth ed. Lea & Febiger, Philadelphia.

King, L.S., Meehan, M.C., 1973. A history of the autopsy: a review. Am. J. Pathol. 73, 514–544.

Klinghardt, G.W., Fredman, P., Svennerholm, L., 1981. Chloroquine intoxication induces ganglioside storage in nervous tissue: a chemical and histopathological study of brain, spinal cord, dorsal root ganglia, and retinal in the miniature pig. J. Neurochem. 37, 897–908.

Kooistra, H.S., 2006. Growth Hormone Disorders in Dogs. In: WSAVA World Congress Proceedings, Prague.

Kraus, M.S., Gelzer, J.S.A., Ducharme, N.G., et al., 2010. Measurement of plasma cardiac troponin I concentration by use of a point-of-care analyzer in clinically normal horses and horses with experimentally induced cardiac disease. Am. J. Vet. Res. 71, 55–59.

Kycko, A., Reichert, M., 2014. Proteomics in the search for biomarkers of animal cancer. Curr. Protein Pept. Sci. 15, 36–44.

La Perle, K., Capen, C.C., 2012. Endocrine system. In: McGavin, M.D., Taylor, Z.M. (Eds.), Pathologic Basis of Veterinary Disease, fourth ed. Academic Press/Elsevier, Amsterdam, pp. 1107–1261.

Madewell, B.R., 1981. Neoplasms in domestic animals: a review of experimental and spontaneous carcinogenesis. Yale J. Biol. Med. 54, 111–125.

Matsuoka, T., Novilla, M.N., Thomson, T.D., Donoho, A.L., 1996. Review of monensin toxicosis in horses. J. Equine Vet. Sci. 16, 8–15.

McGavin, M.D., Taylor, Z.M. (Eds.), 2006. Pathologic Basis of Veterinary Disease, fourth ed. Academic Press/Elsevier, Amsterdam, pp. 833–971.

Maxie, M.G., 1985. The urinary system. In: Jubb, K.V.F., Kennedy, P.C., Palmer, N. (Eds.), Pathology of Domestic Animals, third ed. Academic Press, New York, pp. 343–411.

Melnikov, I., Gabbasov, Z., Kozlov, S., et al., 2017. Assessment of novel biomarkers for restenosis occurrence after drug-eluting stent implantation in patients with stable coronary artery disease and type 2 diabetes mellitus. Atherosclerosis 263, 121–122.

Milburn, M.V., Ryals, J.A., Guo, L., 2013. Toxicometabolomiscs: Technology and Applications. Chapter 34. In: A Comprehensive Guide in Toxicology and Preclinical Drug Development. Academic Press/Elsevier, Amsterdam, pp. 807–825.

Mishra, J., Ma, Q., Prada, A., et al., 2003. Identification of neutrophil gelatinase-associated lipocalin as a novel early urinary biomarker for ischemic renal injury. J. Am. Soc. Nephrol. 14, 2534–2543.

Moll, R., 1998. Cytokeratins as markers of differentiation in the diagnosis of epithelial tumors. Subcell. Biochem. 31, 205–262.

Montagna, W., Yun, J.S., 1963. The effects of the seeds of *Leucaena glauca* on the hair follicles of the mouse. J. Invest. Dermatol. 40, 325–332.

Monteith, D.K., Morgan, R.E., Halstead, B., 2006. In vitro assays and biomarkers for drug-induced phospholipidosis. Expert Opin. Drug Metabol. Toxicol. 2, 687–696.

Munro, R., Head, K.W., Munro, H.M.C., 1982. Scrotal haemangiomas in boars. J. Comp. Pathol. 92, 109–115.

Musshoff, F., Jacob, B., Fowinkel, C., et al., 1993. Suicidal yew leaves ingestion — phloroglucindimethylether (3,5-dimethoxyphenol) as a marker for poisoning from *Taxus baccata*. Int. J. Leg. Med. 106, 45–50.

Myers, R.K., McGavin, M.D., 2006. Cellular and tissue responses to injury. In: McGavin, M.D., Taylor, Z.M. (Eds.), Pathologic Basis of Veterinary Disease, fourth ed. Academic Press/Elsevier, Amsterdam, pp. 3–62.

Novilla, D.A., 1989. Capillary Hemangiomas of the Scrotum in Philippine Boars. DVM Degree thesis. Araneta University, Malabon, Rizal, Philippines.

Novilla, M.N., 2012. Ionophores. In: Gupta, R.C. (Ed.), Veterinary Toxicology: Basic and Clinical Principles, second ed. Academic Press, New York, pp. 1281–1299.

Novilla, M.N., Folkerts, T.M., 1986. Ionophores: monensin, lasalocid, salinomycin, narasinCurrent. In: Howard, J.L. (Ed.), Current Veterinary Therapy – Food Animal Practice. Academic Press, New York, pp. 359–363.

Novilla, M.N., Owen, N.V., Todd, G.C., 1994. The comparative toxicology of narasin in laboratory animals. Vet. Hum. Toxicol. 36, 318–332.

O'Brien, P.J., 2008. Cardiac troponin is the most effective translational safety biomarker for myocardial injury in cardiotoxicity. Toxicology 245, 206–218.

Oeda, S., Hirota, M., Nishida, H., et al., 2016. Development of an in vitro photosensitization test based on changes of cell-surface thiols and amines as biomarkers: the photo-SH/NH_2 test. J. Toxicol. Sci. 41, 129–142.

Ohtsuka, Y., Kawakami, S., Fujii, Y., et al., 2006. Loss of uroplakin III expression is associated with a poor prognosis in patients with urothelial carcinoma of the upper urinary tract. BJU Int. 97, 1322–1326.

Pamukcu, A.M., Price, J.M., Bryan, J.T., 1976. Naturally occurring and bracken fern-induced urinary bladder tumors. Vet. Pathol. 13, 110–122.

Parikh, C.R., Dahl, N.K., Chapman, A.B., et al., 2012. Evaluation of urine biomarkers of kidney injury in polycystic kidney disease. Kidney Int. 81, 784–790.

Panter, K.E., Welch, K.D., Gardner, D.R., 2011. Toxic plants. In: Gupta, R.C. (Ed.), Reproductive and Developmental Toxicology. Academic Press/Elsevier, Amsterdam, pp. 689–705.

Panter, K.E., Welch, K.D., Gardner, D.R., 2012. Poisonous plants in the United States. In: Gupta, R.C. (Ed.), Veterinary Toxicology: Basic and Clinical Principles, 2nd edn. Academic Press, New York, pp. 1031–1079.

Pazhayattil, G.S., Shirali, A.C., 2014. Drug-induced impairment of renal function. Int. J. Nephrol. Renovasc. Dis. 7, 457–468.

Peixoto, P.V., França, T.N., Barros, C.S.L., et al., 2003. Histopathological aspects of bovine enzootic hematuria in Brazil. Pesqui. Vet. Bras. 23, 65–81.

Perez-Alenza, M.D., Blanco, J., Sardon, D., et al., 2006. A Clinicopathological findings in cattle exposed to chronic bracken fern toxicity. N. Z. Vet. J. 54, 185–192.

Perez-Rojas, J., Blanco, J.A., Cruz, C., et al., 2007. Mineralocorticoid receptor blockade confers renoprotection in preexisting chronic cyclosporine nephrotoxicity. Am. J. Physiol. Ren. Physiol. 292, F131–F139.

Ponce, C.H., Brown, M.S., Osterstock, J.B., et al., 2014. Effects of wet corn distillers grains with solubles on visceral organ mass, trace mineral status, and polioencephalomalacia biomarkers of individually-fed cattle. J. Anim. Sci. 92, 4034–4046.

Ramos-Vara, J.A., Miller, M.A., Boucher, M., et al., 2003. Immunohistochemical detection of uroplakin III, cytokeratin 7, and cytokeratin 20 in canine urothelial tumors. Vet. Pathol. 40, 55–62.

Reasor, M.J., 1989. A review of the biology and toxicological implications of the induction of lysosomal lamellar bodies of drugs. Toxicol. Appl. Pharmacol. 97, 47–56.

Rosai, J., 1997. Pathology: a historical opportunity. Am. J. Pathol. 151, 3–6.

Rosenberg, M.E., Paller, M.S., 1991. Differential gene expression in the recovery from ischemic renal injury. Kidney Int. 39, 1156–1161.

Safran, N., Aisenberg, I., Bark, I., 1993. Paralytic syndrome attributed to lasalocid residues in a commercial ration fed to dogs. J. Am. Vet. Med. Assoc. 202, 1273–1275.

Schnellman, R., 2008. Toxic responses of the kidney. In: Klaassen, C.D. (Ed.), Casarett and Doull's Toxicology: The Basic Science of Poisons, seventh ed. McGraw-Hill, New York, pp. 583–608.

Schnitt, S.J., 2003. Pharmacopathology: a new spin on an old idea. Am. J. Surg. Pathol. 7, 121–123.

Shi, Y., Su, T., Qu, L., et al., 2013. Evaluation of urinary biomarkers for the prognosis of drug-associated chronic tubulointerstitial nephritis. Am. J. Med. Sci. 346, 283–288.

Shlosberg, A., Weisman, Y., Klopper, U., et al., 1985. Neurotoxic action of lasalocid at high doses. Vet. Rec. 117, 394.

Stice, S., Thrall, M.A., Hamar, D.W., 2018. Alcohols and glycols. In: Gupta, R.C. (Ed.), Veterinary Toxicology. Basic and Clinical Principles, third ed. Academic Press/Elsevier, Amsterdam, pp. 647–657.

Stockham, S., Scott, M., 2008. Enzymes. In: Stockham, S.L., Scott, M.A. (Eds.), Fundamentals of Veterinary Clinical Pathology, second ed. Blackwell Publishing, Ames.

Stríbrný, J., Dogosi, M., Snupárek, Z., et al., 2010. 3,5-dimethoxyfenol–marker intoxication with *Taxus baccata*. Soud Lek. 55, 36–39.

Szcech, G., Carlton, W.W., Olander, H.J., 1973. Haemangioma on the scrotum in a Chester White boar. Can. Vet. J. 14, 16–18.

Thrall, M.A., Dial, S.M., Winder, D.M., 1985. Identification of calcium oxalate monohydrate crystals by x-ray diffraction in urine of ethylene glycol-intoxicated dogs. Vet. Pathol. 22, 625–628.

Timbrell, J.A., 1998. Biomarkers in toxicology. Toxicology 129, 1–12.

Tiwary, A.K., Puschner, B., Kinde, H., et al., 2005. Diagnosis of Taxus (Yew) poisoning in a horse. J. Vet. Diagn. Invest. 17, 252–255.

Tukov, F.F., Anand, S., Gadepalli, R.S., et al., 2004. Inactivation of the cytotoxic activity of repin, a sesquiterpene lactone from *Centaurea repens*. Chem. Res. Toxicol. 17, 1170–1176.

Turk, J.R., Casteel, R.W., Kaneko, J.J., et al., 1997. Clinical biochemistry in toxicology. In: Kaneko, J.J., Harvey, J.W., Bruss, M.L. (Eds.), Clinical Biochemistry Domestic Animals, fifth ed. Academic Press, NY, pp. 829–843.

Vaidya, V.S., Ramirez, V., Ichimura, T., et al., 2006. Urinary kidney injury molecule-1: a sensitive quantitative biomarker for early detection of kidney tubular injury. Am. J. Physiol. Ren. Physiol. 290, F517–F529.

Van der Linde-Sipman, J.S., Van den Ingh, T.S.G.A.M., Van Es, J.J., 1999. Salinomycin-induced polyneuropathy in cats: morphologic and epidemiologic data. Vet. Pathol. 36, 152–156.

Van Tongeren, S., Fagerland, J.A., Conner, M.W., et al., 2011. The role of the toxicologic pathologists in the biopharmaceutical industry. Int. J. Toxicol. 30, 568–582.

Van Vleet, J.F., Ferrans, V.J., Herman, E., 1991. Cardiovascular and skeletal muscle system. In: Hascheck, W.M., Rousseaux, C.G. (Eds.), Handbook of Toxicologic Pathology. Academic Press, San Diego, pp. 539–624.

Van Vleet, J.F., Amstutz, H.E., Weirich, W.E., Rebar, A.H., Ferrans, V.J., 1983. Clinical, clinicopathologic and pathologic alterations in acute monensin toxicosis in cattle. Am. J. Vet. Res. 44, 2133–2144.

Vlasakova, K., Erdos, Z., Troth, S.P., et al., 2014. Evaluation of the relative performance of 12 urinary biomarkers for renal safety across 22 rat sensitivity and specificity studies. Toxicol. Sci. 138, 3–20.

Whitlock, R.H., White, N.A., Rowland, G.N., Plue, R., 1978. Monensin toxicosis in horses; clinical manifestations. Proc. Am. Assoc. Equine Practnrs 24, 473–486.

Wilson, C.R., Hooser, S.B., 2018. Toxicity of yew (*Taxus* spp.) alkaloids. In: Gupta, R.C. (Ed.), Veterinary Toxicology: Basic and Clinical Principles, third ed. Academic Press/Elsevier, Amsterdam, pp. 947–953.

Witzgall, R., Brown, D., Schwarz, C., et al., 1994. Localization of proliferating cell nuclear antigen, vimentin, c-Fos, and clusterin in the postischemic kidney. Evidence for a heterogenous genetic response among nephron segments, and a large pool of mitotically active and dedifferentiated cells. J. Clin. Invest. 93, 2175–2188.

Wu, Y., Yang, L., Su, T., et al., 2010. Pathological significance of a panel of urinary biomarkers in patients with drug-induced tubulointerstitial nephritis. Clin. J. Am. Soc. Nephrol. 5, 1954–1959.

Wünschmann, A., Armien, A., Wallace, R., et al., 2006. Neuronal storage disease in a group of captive Humboldt penguins (*Sphenicus humboldti*). Vet. Pathol. 43, 1029–1033.

Young, S., Brown, W.W., Klinger, B., 1960. Nigropallidal encephalomalacia in horses fed Russian knapweed *Centaurea repens* L. Am. J. Vet. Res. 31, 1393–1404.

Yu, Y., Jin, H., Holder, D., et al., 2010. Urinary biomarkers trefoil factor 3 and albumin enable early detection of kidney tubular injury. Nat. Biotechnol. 28, 470–477.

Zachary, J.F., 2006. Nervous system. In: McGavin, M.D., Taylor, Z.M. (Eds.), Pathologic Basis of Veterinary Disease, fourth ed. Academic Press/Elsevier, Amsterdam, pp. 833–971.

Zimmerman, H.J., 1999. Hepatotoxicity: The Adverse Effects of Drugs and Other Chemicals on the Liver, second ed. Lippincott Williams & Wilkins, Philadelphia, pp. 111–145.

C H A P T E R

54

Oral Pathology Biomarkers

Anupama Mukherjee

Oral Pathology, Microbiology and Forensic Odontology, Goa Dental College and Hospital, Panaji, India

INTRODUCTION

The oral cavity is regarded as a window to the human body. The stomatognathic system plays a critical role in numerous physiological processes such as digestion, respiration, and speech. Subtle and early signs of various diseases often manifest in the oral tissues. This necessitates a detailed study of the oral cavity and its associated structures to identify and understand local and systemic disease processes.

Oral and maxillofacial pathology is a specialty of dentistry and pathology, which deals with the nature, identification, and management of diseases affecting the oral and maxillofacial regions. It is a science that investigates the causes, processes, and effects of these diseases. The practice of oral and maxillofacial pathology includes research along with diagnosis of diseases using clinical, radiographic, microscopic, biochemical, or other examinations and the subsequent management of patients.

Numerous developmental disturbances/disorders often involve the orofacial regions. These may be as a result of hereditary factors, administration of drugs, or even exposure to radiation during dental development. Additionally, the use of tobacco in a smoked or smokeless form or betel quid chewing in synergy with alcohol consumption, predispose an individual to oral potentially malignant disorders (OPMDs) and subsequently oral cancer (George et al., 2011; Sarode et al., 2012). A continuous and excessive exposure to inorganic substances such as fluoride, silica, asbestos, and heavy metals (lead, mercury, etc.) may manifest in the oral mucosa, as a sign of toxicity.

Various biomolecules (growth factors, enzymes, and interleukins) and the presence of toxicants can be assessed in oral samples. Saliva, a biofluid, exclusive to the oral cavity has emerged as a significant diagnostic tool (Langie et al., 2017). The altered concentrations or unusual presence of certain biomolecules in saliva are now being utilized as biomarkers for diagnosis and prognosis. Biomarkers of oral diseases are more routinely evaluated using tissue biopsies. The application of special techniques such as histochemistry and immunohistochemistry allow a more detailed evaluation of the proteomic and molecular profile of diseases. This chapter describes oral pathology biomarkers of select toxicants and diseases.

ANATOMICAL AND HISTOLOGICAL CONSIDERATIONS OF THE ORAL CAVITY

The stomatognathic system displays a harmonious coexistence of functional hard and soft tissues. The hard tissue components include the teeth, temporomandibular joint, mandible, and maxilla, whereas the soft tissue components include the oral mucosa (including tongue) and the salivary glands (Nanci, 2017). (Fig. 54.1).

Teeth

The teeth represent the exposed mineralized structures of the oral cavity. They play a vital role in mastication along with aesthetics and provide support to facial tissues such as the lips and cheeks. Teeth are composed of enamel, dentine, and cementum, which are mineralized, whereas the pulp accounts for the vital, soft, connective tissue enclosed within the tooth.

The oral mucosa refers to the soft tissue lining the oral cavity. Mucosal organization at the various anatomic sites of the oral cavity is categorized as (1) lining mucosa, (2) masticatory mucosa, and (3) specialized mucosa based on its regional variation in structure and function (Table 54.1).

The salivary glands produce a multifunctional fluid, saliva, which moistens the oral cavity. The predominant

Biomarkers in Toxicology, Second Edition
https://doi.org/10.1016/B978-0-12-814655-2.00054-2

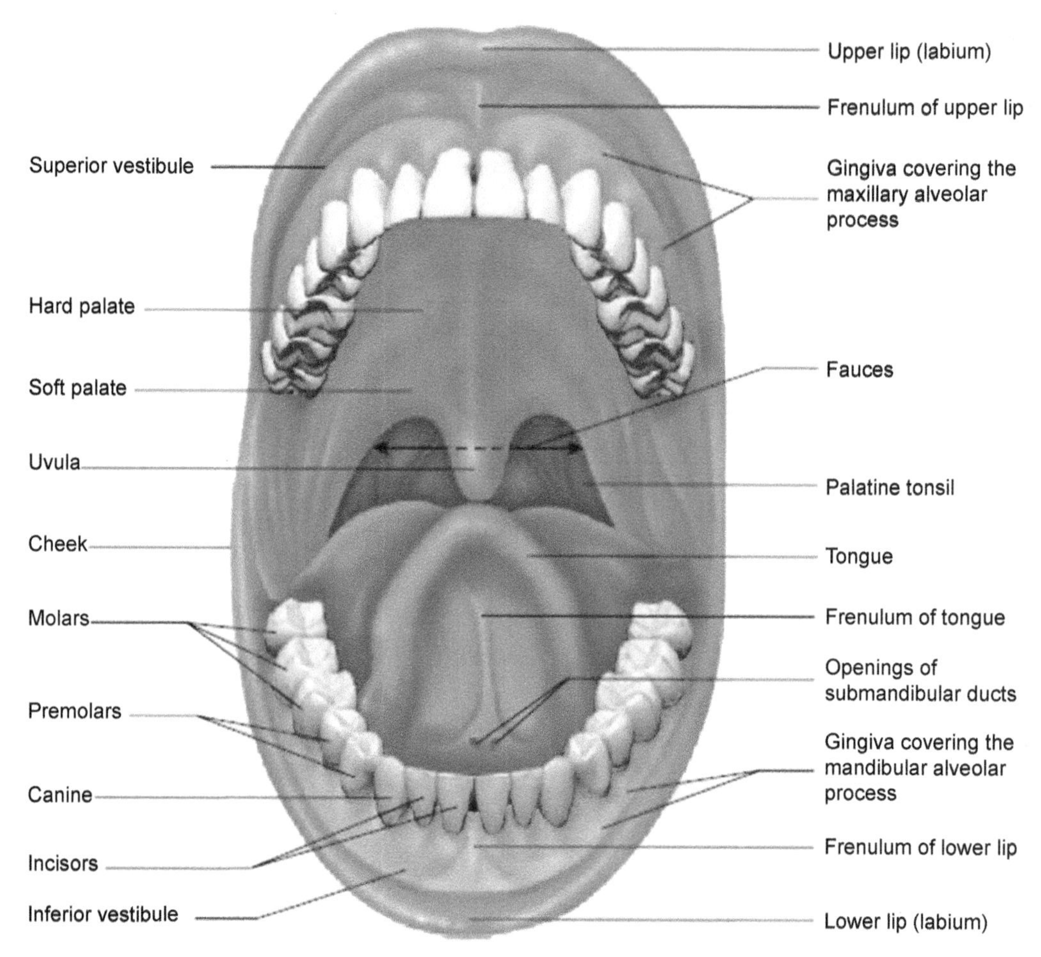

FIGURE 54.1 Shows the various anatomical regions of the oral cavity. *Image courtesy: Seeley–Stephens–Tate: Anatomy and Physiology, sixth ed., Part 4-Regulations and Maintenance, Chapter: 24 Digestive System, The McGraw–Hill Companies, 2004.*

functions of saliva include lubrication, mastication, gustation (taste) and digestion. It also plays an important role in speech, oral pH maintenance, and ion exchange between tooth surfaces, along with providing defense against early microbial invasions. Saliva also acts as a vehicle for numerous substances, such as vitamins, drugs, glucose and salts, facilitating their dispersion and rapid absorption by the oral mucosa.

HISTOLOGY OF ORAL MUCOSA

The oral mucosa comprises the oral epithelium and connective tissue. (Fig. 54.2). The epithelium is stratified squamous in nature and may be keratinized or nonkeratinized in different areas. It forms a physical barrier to the external environment. The epithelium is supported by the connective tissue, designated as the lamina propria and submucosa. The connective tissue comprises the vascular, muscular, and neural components. The submucosa in the oral cavity also shows the presence of specialized glandular structures. These are the serous or seromucous minor salivary gland acini. The organizational relations of the various tissue elements when viewed under a microscope are shown in Fig. 54.3. The basement membrane is a specialized zone that maintains the structural integrity of the epithelium, along with its intimate association with the underlying connective tissue. Histologically, it is observed as an interface between the basal layer of epithelial cells and the lamina propria (Nanci, 2017).

Thus the variable histomorphological features confer each oral site with a specialized function. The areas covered by the masticatory mucosa protect the underlying tissues from microtrauma and abrasions during mastication. The specialized mucosa on the dorsum of the tongue allows for taste perception. Based on the variable keratinization of the oral mucosa, some sites (sublingual, floor of the mouth) are more permeable as compared to others (palate). (Fig. 54.4). The presence of minor salivary gland acini, in conjunction with the major salivary glands, maintains a moistened mucosa. The water content of these secretions acts as a vehicle to facilitate intercellular transport and absorption (Nanci, 2017).

TABLE 54.1 A Comparative Overview of the Numerous Regional Variations of the Oral Mucosa

Region	Overlying Epithelium	Lamina Propria	Submucosa
LINING MUCOSA			
Soft palate	Thin (150 μm), nonkeratinized stratified squamous epithelium; few taste bud present.	Thick with numerous short papillae; elastic fibers forming an elastic lamina; highly vascular with well-developed capillary network	Diffuse tissue containing numerous minor salivary gland acini
Ventral surface of the tongue	Thin, nonkeratinized, stratified squamous epithelium.	Thin with numerous short papillae and some elastic fibers; a few minor salivary gland acini; capillary network in subpapillary layer.	Thin and irregular, where absent, mucosa is bound to connective tissue surrounding tongue musculature.
Floor of the mouth	Very thin (100 μm), nonkeratinized stratified squamous epithelium.	Short papillae; some elastic fibers; extensive vascular supply with short anastomosing capillary loops.	Loose fibrous connective tissue, containing fat and minor salivary glands.
Alveolar mucosa	Thin, nonkeratinized stratified squamous epithelium.	Short papillae, connective tissue containing many elastic fibers; capillary loops close to the surface.	Loose connective tissue containing thick elastic fibers attaching it to the periosteum of alveolar process; minor salivary gland acini.
Labial and Buccal mucosa	Very thick (500 μ), nonkeratinized stratified squamous epithelium.	Long, slender papillae; dense fibrous connective tissue containing collagen and some elastic fibers; rich vascular supply giving off anastomosing capillary loops into papillae.	Mucosa firmly attached to underlying muscle by collagen and elastin; dense collagenous connective tissue with fat, minor salivary glands, some sebaceous glands.
Lips: Vermilion zone	Thin, orthokeratinized, stratified squamous epithelium.	Numerous narrow papillae; capillary loops close to surface in papillary layer.	Mucosa firmly attached to underlying muscle; some sebaceous glands in vermilion border, minor salivary glands and fat in intermediate zone.
Lips: Intermediate zone	Thin, parakeratinized, stratified squamous epithelium.	Long irregular papillae; elastic and collagen fibers in connective tissue.	
MASTICATORY MUCOSA			
Gingiva	Thick (250 μm), orthokeratinized or parakeratinized, stratified squamous epithelium often showing stippled surface.	Long, narrow papillae, dense collagenous connective tissue; long capillary loops with numerous anastomoses.	No distinct layer, mucosa firmly attached by collagen fibers to cementum and periosteum of alveolar process.
Hard palate	Thick, orthokeratinized (parakeratinized in some parts), stratified squamous epithelium thrown into transverse palatine ridges (rugae).	Long papillae; thick, dense collagenous tissue, especially under rugae; moderate vascular supply with short capillary loops.	Dense collagenous connective tissue attaching mucosa to periosteum ("mucoperiosteum"), fat and minor salivary gland acini are packed into connective tissue in regions where mucosa overlies lateral palatine neurovascular bundles.
SPECIALIZED MUCOSA			
Dorsal surface of the tongue	Thick, keratinized, and nonkeratinized stratified squamous epithelium forming various types of papillae, some bearing taste buds.	Long papillae, minor salivary gland acini in posterior region; rich innervation, particularly near taste buds; capillary plexus in papillary layer, large vessels lying deeper.	No distinct layer; mucosa is bound to connective tissue surrounding musculature of tongue.

Courtesy: Antonio Nanci. Ten Cate's Oral Histology-Development, Structure and Function, ninth ed. Elsevier, Amsterdam.

Cellular permeability is attributed to cell structure and concentrations of substances in the cytoplasm. The lipid bilayer configuration of the cell membrane allows trafficking of lipid soluble substances. Special protein channels and transport proteins help carry water-soluble substances across the cell membrane. These structural characteristics allow for selective permeability across the membrane. The mechanisms by which substances travel across the cell membrane include (1) active transport, (2) passive diffusion, and

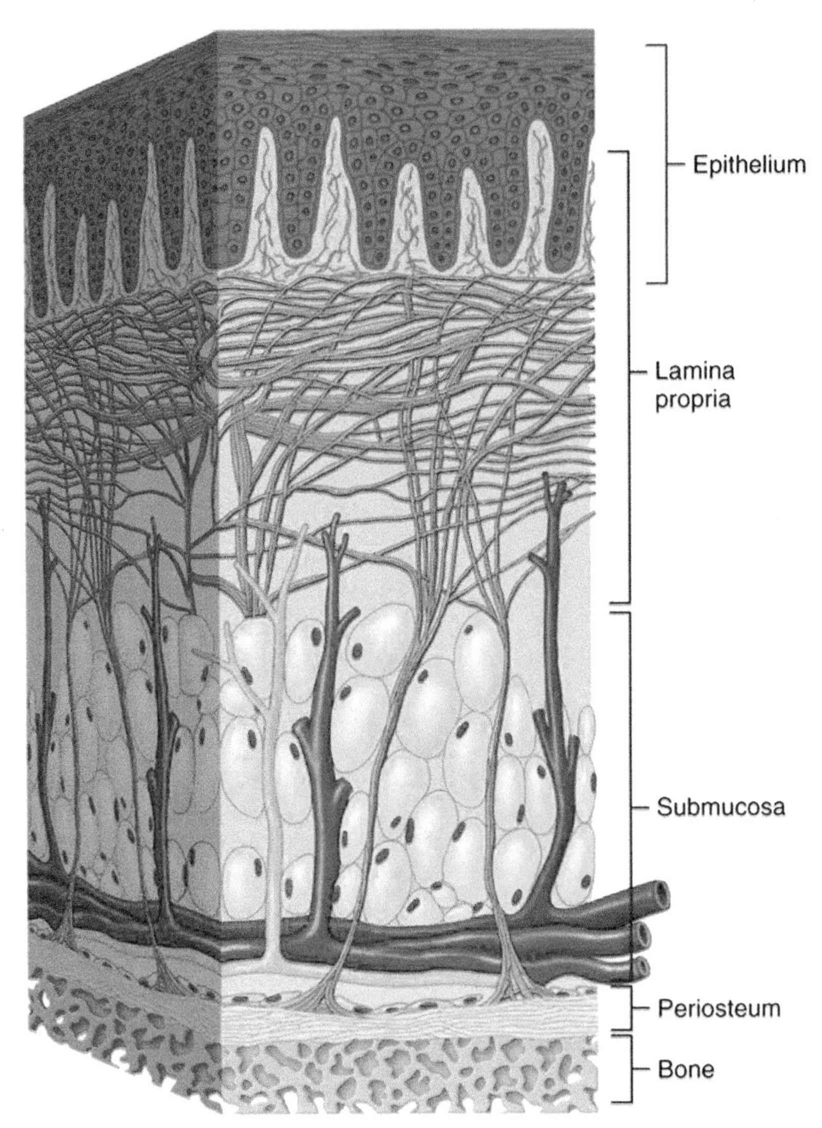

FIGURE 54.2 Schematic representation of the various components of the oral mucosa. *Image courtesy: Antonio Nanci, 2017. Ten Cate's Oral Histology-Development, Structure and Function, ninth ed.*

(3) facilitated diffusion. Occasionally, endocytosis may be demonstrated by a cell. However, this is not common in oral epithelial cells.

The stratified squamous epithelium lining the oral mucosa, like all epithelial cells, shows the presence of membrane surface specializations called cell junctions, which allow cell-to-cell communication along with the passage of substances through and between cells. Thus, epithelial cells can allow and promote the diffusion of gases, liquids, and nutrients. Epithelial cells demonstrate four major types of cell-cell junctions. The **tight junction** (zonula occludens) and the **adherent junction** (zonula adherens) are typically close together and each forms a continuous ribbon around the cell's apical end. Multiple ridges of the tight, occluding junctions prevent passive flow of material between the cells. The adhering junctions, usually located immediately below the tight junctions, serve to stabilize and strengthen circular bands around the cells and help hold the layer of cells together (Fig. 54.5).

Both desmosomes and gap junctions make spot-like plaques between two cells. Bound to intermediate filaments inside the cells, **desmosomes** form very strong attachment points and play a major role in maintaining the integrity of the epithelium. **Gap junctions**, which are patches of many **connexons** in the adjacent cell membranes, have little strength but serve as intercellular channels for the flow of molecules. The specific

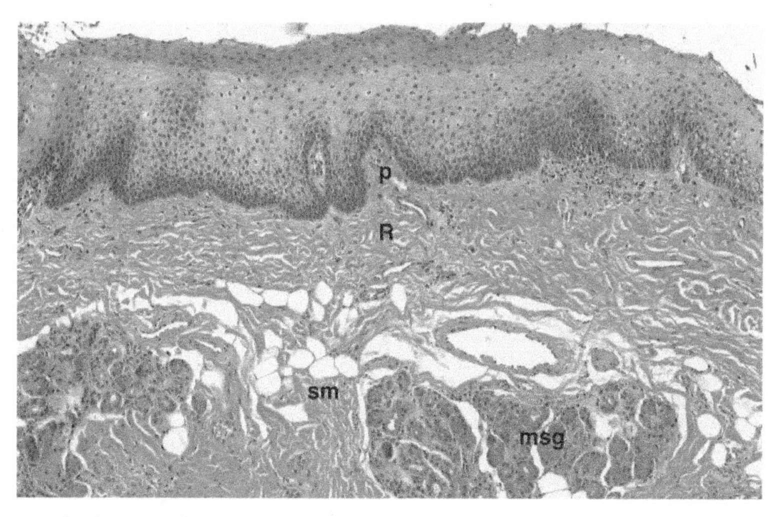

FIGURE 54.3 A pictomicrograph showing the tissue organization of the oral mucosa. *MSG*, minor salivary glands; *P*, connective tissue papillae; *R*, reticular lamina propria; *SM*, submucosa. *Image courtesy: Ellen Eisenberg, Easwar Natarajan, Bradley K. Formaker, 2018. Oral Mucosa and Mucosal Sensation, Department of Oral Health and Diagnostic Sciences, School of Dental Medicine, University of Connecticut, Storrs, CT, USA.*

arrangement of these junctions forms a "junctional complex" and defines the basolateral surface of a cell (Nanci, 2017).

ORAL PHYSIOLOGY AND PHARMACODYNAMICS

The oral cavity allows for two major routes of administration of drugs, oral (as when ingested) and mucosal. Saliva constantly bathes the oral mucosa and acts as a vehicle allowing dissolution of substances to pass across. Mucosal absorption bypasses first pass metabolism in the liver and directly delivers the drug/substance into the vascular system. Absorption by the oral mucosa occurs predominantly via passive diffusion into the lipid membrane.

Numerous studies have been initiated to understand the process of mucosal absorption, particularly for various drugs. A rapid absorption of drugs/substances via the mouth into the mucosal membranes does not imply that they are immediately transported to the systemic circulation. This is because the rate-limiting step could be the transport of molecules through the mucosal membranes into the systemic circulation. Mucosal absorption is usually rapid and short, depending largely on duration of contact and permeability of the mucosa. The sublingual area/floor of the mouth is most permeable followed by the buccal mucosa and palate. Mucosal permeability is attributed to relative thickness of epithelium, keratinization, and vascularity. The floor of the mouth is mostly vascular and thinly epithelialized. Thus, it demonstrates a rate of absorption about 3–10 times greater than that seen via other routes of administration (Narang and Sharma, 2011).

A key role, in mucosal absorption, is played by the secretory glands of the oral cavity—the major and minor salivary glands. Saliva comprises water, mucin, enzymes, and salts. The anatomical location of the sublingual salivary glands also contributes to maintaining a moist environment that can act as a vehicle for the dissolution and distribution of substances (Nanci, 2017). Certain factors that influence oral mucosal absorption include lipid solubility and permeability of solution (osmosis), molecular weight of substances, and the ionization state (pH).

The cells of the oral epithelium and epidermis are also capable of absorbing by endocytosis. These engulfed particles are usually too large to diffuse through its wall. It is unlikely that this mechanism is used across the entire stratified epithelium. It is also unlikely that active transport processes operate within the oral mucosa. Occasionally large inert substances (heavy metals/pigments) may accumulate in the fine capillaries and tissues present as clinical signs of toxicity or pigmentation.

BIOMARKERS IN THE ORAL CAVITY

Biomarkers have been defined as a biological characteristic that is objectively measured and evaluated as an indicator of normal biological/pathological processes or a response to a therapeutic intervention (Mayeux, 2004). The WHO has defined biomarkers as "almost any measurement reflecting an interaction between a biological system and a potential hazard, which may be chemical, physical, or biological. The measured response may be functional and physiological, biochemical at the cellular level, or a molecular interaction."

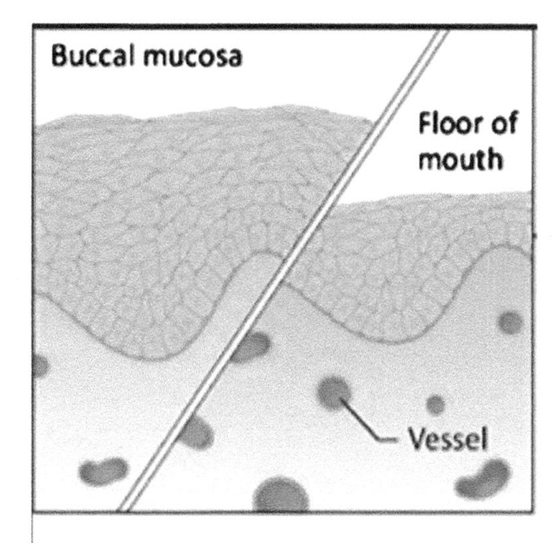

Masticatory mucosa

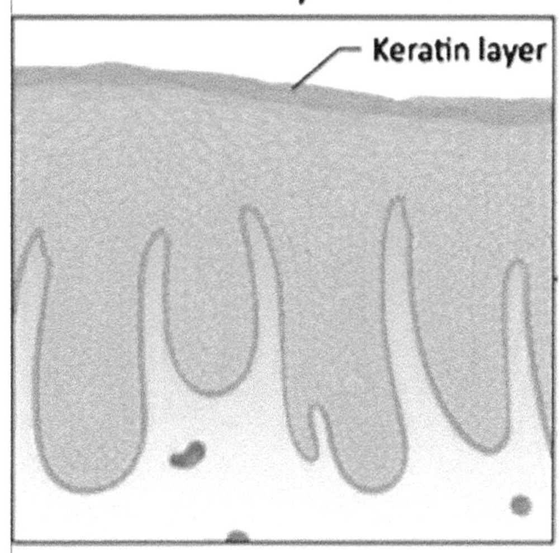

Tongue mucosa

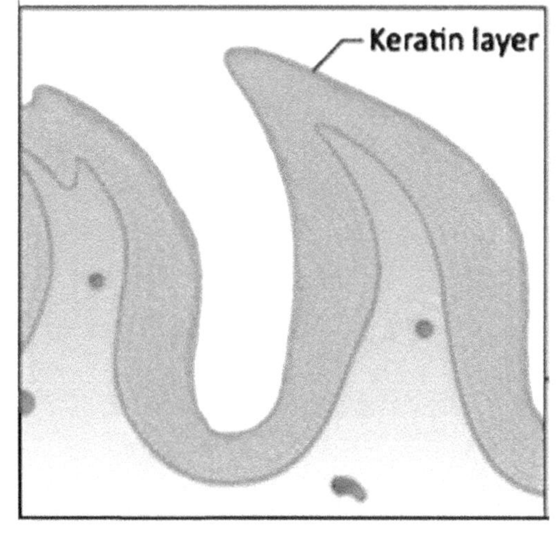

FIGURE 54.4 Schematic representation comparing the variable epithelial and connective tissue thickness at various oral sites. The buccal mucosa and floor of the mouth are richly vascular area. *Images courtesy: Niki M. Moutsopoulos, Joanne E. Konkel, 2018. Tissue-specific immunity at the oral mucosal barrier. Trends Immunol. 39 (4), 276–287.*

Biomarkers can be broadly categorized as biomarkers of exposure, which are used in risk prediction, and biomarkers of disease, which are used in screening, diagnosis, and monitoring of disease progression (Mayeux, 2004). Although drug and metal toxicity may manifest as clinically and histologically observable features, these are predominantly biomarkers of exposure/risk. For other oral diseases such as periodontitis, oral squamous cell carcinoma, pemphigus, pemphigoid, oral lichen planus (Sultan et al., 2014; Gopalakrishnan et al., 2016), etc., the evaluation of distinct histological features, proteomic profiles, and gene expression have been used as biomarkers for screening to assess disease progression and aggressiveness (Fernández-González et al., 2011; Mehdipour et al., 2018).

To simplify and categorize the various biomarkers observed in the oral cavity, a three-tier approach may be adopted. This would include a) clinical biomarkers, observable clinical features that reflect the risk/presence of a disease either systemically or locally in the oral cavity (brownish mottled teeth suggestive of fluorosis), and b) tissue biomarkers, the histopathological characteristics that are informative of the degree of variation from normal. They aid in providing a definitive diagnosis and help ascertain the severity or aggressiveness of the disease (keratin pearls as dyskeratosis as seen in squamous cell carcinoma (Gopalakrishnan et al., 2016)). c) Molecular biomarkers include molecular profiles or the expression of specific protein(s), the presence of which may be increased/decreased or be abnormal in a diseased state. These markers may also be used to monitor disease progression or for the evaluation of therapy (antibodies directed against DSG-3 and/or DSG-1 in pemphigus).

COMMON DRUGS AND TOXICANTS SHOWING ORAL MANIFESTATIONS

1. Drugs
   a. Tetracycline/Minocycline
   b. Bisphosphonates
   c. Hydantoin
2. Heavy Metals
   a. Mercury
   b. Silver
   c. Lead
   d. Arsenic
   e. Bismuth
3. Poisonings
   a. Cyanide
   b. Organophosphates
   c. Aluminum phosphide

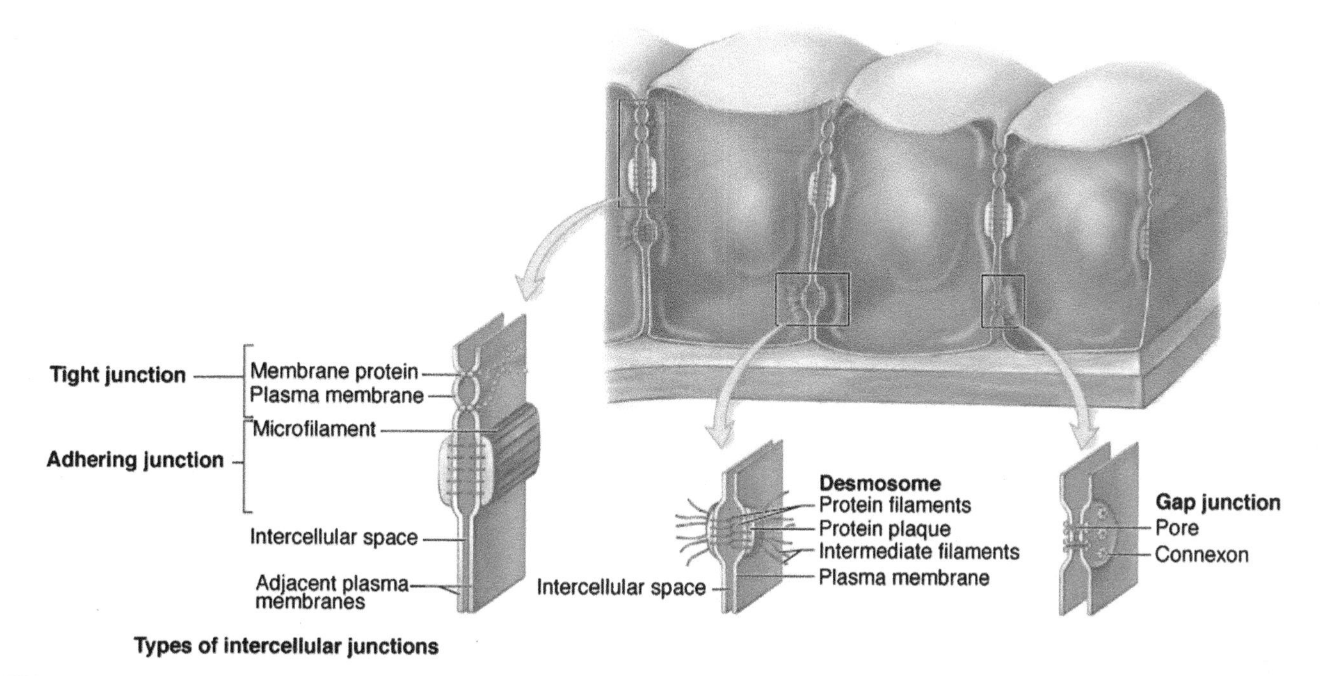

FIGURE 54.5 Schematic diagram showing the various cell junctions in epithelial cells. *Image courtesy: Mescher, A.L. Junqueira's Basic Histology: Text and Atlas, twelfth ed. http://www.accessmedicine.com.*

SELECT DISEASES OF THE ORAL CAVITY

1. Periodontal disease
2. Immunologically mediated oral diseases/ Autoimmune diseases
   a. Pemphigus
   b. Pemphigoid
   c. Oral Lichen Planus
3. Oral squamous cell carcinoma

ORAL BIOMARKERS OF EXPOSURE AND EFFECTS OF SELECT DRUGS/TOXICANTS

Tetracycline/Minocycline

Tetracycline is a broad-spectrum antibiotic introduced in 1948. It was routinely administered to children and adults for the treatment of common infections caused by Gram-positive bacteria. A few Gram-negative bacteria also showed sensitivity to the drug. Minocycline, a semisynthetic derivative of tetracycline, is commonly administered to treat acne. A prominent side effect of long-term administration of tetracycline is its incorporation into various mineralizing sites. This occurs because of the select affinity of tetracycline for calcium ions. The drug forms a chelate with the calcium hydroxyapatite crystals and is deposited into teeth and bones during mineralization. Minocycline causes staining via different mechanisms. The drug assumes a highly protein-bound state and preferentially binds to collagen-rich tissues such as teeth and bones. Alternatively, the chelation of a minocycline breakdown product similar to hemosiderin forms an insoluble complex with teeth (Raymond and Cook, 2015). The excretion of minocycline in the gingival cervical fluid and its subsequent oxidation when incorporated into enamel could also cause staining, which appears as a blackish discoloration. The administration of tetracycline and its derivatives to pregnant mothers (second and third trimester) or children (up to 7 years of age) results in staining of teeth and bones. The distribution and intensity of color vary based on the specific form of the drug along with the duration of exposure.

The occurrence of staining is noted in 3%–6% of minocycline-treated patients. Clinical biomarkers of tetracycline administration include the following:

- Yellowish or brownish-gray staining of deciduous and permanent teeth. Discoloration is more intense at the time of tooth eruption and gradually becomes browner. Occasionally a bluish discoloration has also been reported (Johnston, 2013).
- Chlortetracycline administration leads to brownish-gray color. Oxytetracycline leads to a yellowish color.
- Bright yellow fluorescence under ultraviolet light.
- Minocycline administration causes "black bones," "black or green roots," and a blue-gray to gray hue, darkening the crowns of permanent teeth.

Such teeth usually pose an esthetic challenge and are not functionally compromised. Various techniques such as composite restorations or dental veneers are often employed to improve patient aesthetics.

Mercury and Silver

Mercury is a glistening, odorless liquid that becomes a colorless and odorless gas when heated. Mercury is very toxic and exceedingly bioaccumulative. It is a major cause of acute heavy metal poisoning. Because mercury is ubiquitous in the environment, individuals are routinely exposed to it via air, water, and food. According to WHO, the maximum allowable environmental level of mercury is 50 μg/day in the workplace. Three major forms of mercury in descending order of toxicity are organomercury, mercury vapor, and inorganic mercury. Organomercury (methyl-, ethyl-forms) is well absorbed or may even be formed because of the interaction of gut flora with inorganic mercury.

The toxic effects of mercury are noted at a cellular level and eventually manifest clinically. Mercury affects cellular integrity and disrupts intracellular calcium homeostasis. Excessive mercury also interferes with transcription and translation, resulting in the disappearance of ribosomes and endoplasmic reticulum activity. These cell organelles are vital for cell functions (Jaishankar et al., 2014). Mercury is largely neurotoxic as it depletes glutathione and thiols. It also leads to the generation of reactive oxygen species (ROS) and oxidative stress. High levels of mercury in the blood can be nephrotoxic and neurotoxic. Severe mercury toxicity, particularly from methyl mercury is termed as the "Minamata disease". It is a severe neurological syndrome characterized by headache, loss of peripheral vision, numbness in hands and feet, general muscle weakness, and damage to hearing and speech. The disease was first noted in Japanese residents of the city of Minamata in 1956 (Yorifuji et al., 2017).

Silver is a heavy metal occurring naturally as a soft, glistening metal. It occurs as a powdery white (silver nitrate and silver chloride) or dark gray to black compound (silver sulfide and silver oxide). Exposure to silver occurs via jewelry, silverware, medication, printed photographic material, electronic equipment, and dental fillings with silver in its metallic form. Excessive exposure to silver results in argyria. A clinical biomarker would be a gray to blue-gray discoloration of the skin and body tissues. This phenomenon is also referred to as "ashen-Gray" discoloration.

Dental amalgam is an alloy of silver and mercury. It contains about 50% mercury, along with other toxic metals such as tin, copper, nickel, palladium, etc. Amalgam restorations continuously leak mercury due to its low vapor pressure and loss due to the galvanic action of mercury with dissimilar metals in the mouth. It has been reported that an amalgam filling releases about 0.8–0.10 μg of mercury vapor each day. Often, residual amalgam from a restoration may be lodged into the gingiva. This can result in a grayish-black discoloration of the localized area referred to as an "Amalgam tattoo." This poses no significant health hazard (Sivapathasundharam, 2016).

A key microscopic biomarker is the appearance of black, granular deposits of silver salt in the submucosa. This is a key feature to rule out localized pigmentation. The measurement of urinary mercury levels in adults and children has been used as a parameter in various studies aimed at understanding the toxic effects of dental amalgam. Results remain elusive because age of the patient, time of sampling, duration since filling, and even masticatory habits have shown to play a role.

Dental practitioners and dental assistants are at a high risk to develop mercury toxicity because of the use of amalgam in dental practice. Handling of mercury during restorative procedures, removal of old amalgam restorations, and mercury spills/open disposal jars pose a long-term health risk. Numerous studies have evaluated mercury levels in the blood, urine, hair, and nails of dental practitioners and reported higher values as compared with patients and nondental professionals.

Lead

Lead is a bluish-gray metallic element that is soft, ductile, durable, and heavy. It is used in a variety of products such as batteries, paint, solders, and ceramics. More recently the use of lead in pipes, gasoline, and cosmetics has been curbed owing to the propensity to cause toxicity. Exposure to lead can occur via ingestion, absorption (from skin), or inhalation. Lead poisoning (plumbism, saturnism, painter's colic) may manifest as acute or chronic poisoning. The severity of presentation depends on the amount of lead in the blood and tissues, as well as duration and amount of exposure. Organic forms of lead are more toxic than inorganic forms. The excess lead binds to the sulfhydryl groups of proteins. It also causes an inhibition in the synthesis of heme, an essential component of hemoglobin, myoglobin, and various cytochromes.

Plumbism can present in an acute or chronic form based on duration of exposure. Nausea, vomiting, diarrhea, muscle pain, abdominal pain, and weakness are noted in clinically acute lead poisoning. Headaches, fatigue, stupor, and anemia are noted due to chronic lead poisoning. Additionally, oral signs would include astringency, metallic taste in the mouth, and the appearance of "lead hue," a biomarker of chronic lead exposure. It appears as a characteristic bluish-gray line along the free gingiva around the tooth.

A tissue biomarker of plumbism is the perivascular deposit of lead sulfide in the submucosa and basement membrane zone. In tissue samples, lead can be demonstrated histochemically using the rhodizonate method. Blood levels of lead have often been used to aid in the diagnosis of lead toxicity/poisoning, which is a clinicopathological correlation. It is suggested that cognitive impairment occurs at blood lead levels of 5 μg/dL, particularly in children. At highly elevated blood lead levels of 70 μg/dL, encephalopathy, seizures, and even death may occur (Jaishankar et al., 2014; Ihmed, 2016).

In a study by Gardner et al. (2016), saliva has been reported as a reliable sample for lead evaluation when screening children for toxicity. Based on their study, saliva was reliable for lead concentrations of less than 5 μg/dL. Salivary samples with a concentration of lead greater than 1.40 μg/dL should be confirmed using blood evaluation.

The oral cavity provides subtle indications of toxicities and poisonings. Data regarding the correlation between the clinical biomarkers and serum and/or salivary levels of metals remain unexplored. Often in case of acute toxicity or poisoning, the systemic manifestations are overwhelming, and diagnosis based solely on oral biomarkers is not necessitated. In the case of inert metals, such as silver or exposure to amalgam, oral findings are often suggestive and diagnostic when correlated with history and examination.

BIOMARKERS OF SELECT DISEASES

Biomarkers have assumed a forefront in the assessment of numerous oral diseases. This approach allows them to be categorized as screening, diagnostic, and prognostic biomarkers. Biomarkers are evaluated using techniques ranging from routine histopathology to specialized techniques, such as immunohistochemistry, ELISA, immunofluorescence, and even fluorescence in situ hybridization (to detect gene fusions and characterize soft tissue tumors).

PERIODONTAL DISEASE

Periodontal disease refers to a microbial infection affecting the supporting structures of the tooth, that is, the gingiva, periodontal ligament, and alveolar bone. It predominantly affects the connective tissue structures and is clinically preceded by gingivitis. If not intercepted, periodontal disease progressively leads to loss of tooth support and eventually tooth loss. Periodontal disease affects almost 50% of the adult population in the United States. The clinically observable biomarkers of gingivitis include bleeding on probing the gingival crevice, along with redness and edema of gingival tissues (Taba et al., 2005).

The clinically evaluated diagnostic biomarkers for periodontitis include pocket formation due to loss of attachment at the dentogingival unit and gingival recession, which is the apical migration of the gingiva. These features are often accompanied by tooth mobility. A diagnostic radiographic finding of loss of alveolar bone is significant. Aggressive forms of periodontal disease can lead to early tooth loss resulting in masticatory difficulties and poor aesthetics. Over a long period, it may also render the jawbones unsupportive to rehabilitation techniques such as dental implants or prosthesis. Hence, methods for early detection and assessment of disease progression could decrease morbidity. Tools to measure periodontal disease at the clinical, cellular, and molecular level have been devised. Studies evaluating saliva and gingival crevicular fluid (GCF), a fluid found in the space between the tooth and gingiva, for potential biomarkers of periodontal disease have been carried out. Considering that periodontitis is characterized by connective tissue destruction, the levels of matrix metalloproteinases (MMP) in GCF have been evaluated. In particular, MMP-8 has demonstrated an 83% sensitivity and 96% specificity in differentiating periodontitis from gingivitis and other disease sites. A study by Holmlund et al. (2004) reported an increase in IL-1α, IL-1β, and IL-1 receptor antagonist, which are associated with bone resorption, particularly in samples from diseased sites.

It has been established that an increase in prostaglandins (PGE2), IL-1β, and TNF-α is associated with the severity of periodontal disease. A review by Loos and Tjoa (2000) identified GCF-related probable diagnostic biomarkers of periodontitis. This included alkaline phosphatase, β glucuronidase, cathepsin B, collagenase-2 matrix metalloproteinase (MMP-8), gelatinase (MMP-9), dipeptidyl peptidase (DPP) II and III, and elastase. Early detection could help address and arrest the disease in its initial stage, using suitable therapeutic regimens of antibiotics, oral prophylaxis, and root planning procedures.

IMMUNOLOGICALLY MEDIATED ORAL DISEASES

Autoimmune diseases encompass a group of diseases where an immunological dysfunction results in failure of the host to recognize and distinguish self- from non—self-tissues. Some conditions which present as mucocutaneous diseases, may manifest solely in the oral cavity, at a given time and pose a diagnostic challenge. In such a situation, the use of specific biomarkers can help eliminate other differential diagnoses and allow effective treatment planning. Some of the immunologically mediated oral diseases are listed below:

• Hypersensitive Reaction	• Reiter's Syndrome
• Pemphigus vulgaris	• Linear IgA
• Paraneoplastic pemphigus	• Epidermolysis bullosa
• Cicatricial or mucocutaneous pemphigoid	• Erythema multiforme
• Cutaneous, bullous pemphigoid	• Lupus erythematosus
• Lichen planus	• Systemic drug reaction
• Scleroderma	• Crest syndrome
• Bechet's syndrome	

PEMPHIGUS

Pemphigus is classified as a vesiculobullous lesion with an autoimmune pathogenesis. Pemphigus is a broad term that encompasses a number of mucocutaneous presentations. The term "Pemphigus" is derived from the Greek word "Pemphix" implying blister or bubble (Mignogna et al., 2009). Some cases that involve the skin and mucous membranes extensively may pose to be life threatening (Azizi and Lawaf, 2008).There are various subtypes of pemphigus, each of which are characterized by the loss of normal cell-to-cell adhesion secondary to the binding of autoantibodies. This results in the formation of an intraepithelial vesicle with acantholysis (loss of cellular cohesion) (Mignogna et al., 2009; Zunt, 1996).

The three major subsets of pemphigus are (1) pemphigus vulgaris, (2) pemphigus foliaceus, and (3) paraneoplastic pemphigus. Pemphigus vulgaris is a relatively rare disease with an onset around 50–60 years of age. It commonly affects the oral mucosa. Oral lesions precede cutaneous lesions and are often the first sign of disease. They are described as "first to show and last to go" because they are the most challenging to address with therapy.

Clinically, lesions are noted on the soft palate, buccal mucosa, lower lip, and ventral surface of the tongue. They appear as oral vesicles and bullae that readily rupture to give rise to multiple eroded ulcers that may bleed.

The condition is attributed to the abnormal production of autoantibodies directed toward the specific molecules desmoglein-3 (DSG-3) and desmoglein-1 (DSG-1). Both molecules are structural components of desmosomes which are a type of cell junction. (Azizi and Lawaf, 2008). The disruption of cell adhesion leads to blister formation. The expression of DSG-3 is seen in the parabasal cells of the oral epithelium and epidermis. DSG-1 is noted in the superficial layers of the epidermis and minimally expressed in the oral epithelium. Thus, patients with autoantibodies directed only toward DSG-1 shall exclusively demonstrate dermal lesions, whereas those with autoantibodies directed toward DSG-3 alone, or DSG-1 and DGS-3, shall demonstrate oral lesions (Mignogna et al., 2009).

PEMPHIGOID

Pemphigoid is also classified as a vesiculobullous lesion with an autoimmune pathogenesis. The term pemphigoid represents the similarity in its clinical presentation with pemphigus. Yet its histopathological features, immunological findings, and prognosis differ. Two distinct lesions, mucous membrane pemphigoid, and bullous pemphigoid are observed (Neville et al., 2015; Sivapathasundharam, 2016).

Mucous membrane pemphigoid or cicatricial pemphigoid refers to a heterogeneous group of blistering lesions caused by autoantibodies that are directed toward any of the numerous components of the basement membrane (Xu et al., 2013). It is usually seen in females 50–60 years of age. Intraoral lesions are seen more commonly as compared to bullous pemphigoid. Clinically, the appearance of bullae or vesicles is noted. They rupture to form ulcerated, denuded surfaces on the mucosa and may persist for weeks to months if left untreated. Other mucosal surfaces in addition to the oral cavity, such as the conjunctival, nasal, laryngeal, esophageal, and vaginal mucosa, may also be involved (Petruzzi, 2012; Xu et al., 2013). In some cases, dermal involvement is also seen. The clinical course in patients is usually progressive and protracted as compared to bullous pemphigoid (Zunt, 1996).

The appearance of denuded gingiva termed as "desquamated gingivitis" is often used as a clinical biomarker to differentiate this lesion from bullous pemphigoid, though it may also be seen in erosive lichen planus or pemphigus vulgaris. Bullous pemphigoid is described as the most commonly occurring autoimmune condition (Xu et al., 2013). It is characterized by the production of autoantibodies toward the components of the hemidesmosome cell surface specializations that attach the basal cells to the basement membrane.

Clinically, it is noted in older adults usually 60–80 years of age with no gender predilection. Involvement of the oral mucosa is not a very common finding. If present, oral mucosal lesions are usually noted as shallow erythematous ulcers with smooth and distinct margins. This is due to the rupture of the intraoral vesicle/bullae as a result of constant low-grade trauma

TABLE 54.2 Clinical and Tissue Biomarkers That Aid in Eliminating Some Confounding Vesiculobullous/Autoimmune Lesions

Condition	Mean Age	Sex Prediliction	Clinical Features	Histopathology
Pemphigus vulgaris	40–60 years	Equal	Vesicles, erosions, and ulcerations on any oral mucosal site or skin surface	Intraepithelial clefting
Paraneoplastic pemphigus	60–70 years	Equal	Vesicles, erosions, or ulceration on any mucosa or skin surface	Subepithelial and intraepithelial clefting
Mucous membrane pemphigoid	60–70 years	Female	Primarily mucosal lesions	Subepithelial clefting
Bullous Pemphigoid	70–80 years	Equal	Primarily skin lesions	Subepithelial clefting

Based on Xu, H.-H., Werth, V.P., Parisi, E., et al., 2013. Mucous membrane pemphigoid. Dent. Clin. North Am. 57 (4), 611–630; Neville, B., Damm, D.D., Allen, C. et al., 2015. Oral and Maxillofacial Pathology, fourth ed. Elsevier, Amsterdam; Sivapathasundharam, B., 2016. Shafer's Textbook of Oral Pathology. Elsevier, India.

within the oral cavity. Even though numerous clinical findings can be enlisted to differentiate pemphigus and pemphigoid, histopathological and immunological evaluations provide a more definitive diagnosis, particularly in cases that do not show a classical presentation. Clinical and tissue biomarkers that distinguish and aid in eliminating some confounding vesiculobullous/autoimmune lesions are summarized in Table 54.2.

ORAL LICHEN PLANUS

Lichen planus is a chronic inflammatory mucocutaneous condition. Oral lichen planus (OLP) refers to the oral presentation of the disease, and it may/may not be accompanied by skin lesions. Females in the fifth to sixth decade of life are more commonly affected (Fernández-González et al., 2011). A vast spectrum of clinical presentations are noted, such as unilateral or bilateral white striations, papules, or plaques on the buccal mucosa, labial mucosa, or tongue. The appearance of "striae of Wickham" is a pathognomic clinical feature of oral lichen planus. Lesions involving the gingiva cause inflammation and denudation of the surface epithelium (Kamath et al., 2015). This phenomenon is also referred to as desquamative gingivitis. Erythema, erosion, and blisters may or may not be present.

Numerous theories have been cited to understand the pathogenesis. OLP is regarded as a dysregulated T cell-mediated disorder to exogenous triggers or a dysregulated response to autologous keratinocyte antigens (autoimmune) (Fernández-González et al., 2011). A number of etiological agents have been proposed, such as:

1. local and systemic inducers of cell-mediated hypersensitivity,
2. stress,
3. autoimmune response to epithelial antigens or a dysregulated response to external antigens, and
4. viral infections.

Histopathologically, the pathognomonic features of lichen planus include liquefactive degeneration of basal cell layer, saw tooth rete pegs, a band of chronic inflammation in the sub-basilar region, along with supplementary findings such as hyaline/colloidal bodies and Max–Joseph cleft (Fernández-González et al., 2011; Kamath et al., 2015; Mehdipour et al., 2018).

Lichen planus is not a life-threatening disease, but it continues to be listed as a potentially malignant disorder (OPMD) (Sarode et al., 2012). This has been attributed to mixed reports of malignant transformation of OLP to oral squamous cell carcinoma (Fitzpatrick et al., 2014). The presence of chronic inflammation, as seen with OLP, has been recognized as a potential cause of malignant transformation.

Diagnosing OLP is a clinicopathological correlation and may pose a challenge because similar clinical features may also be noted in oral lichenoid reaction (an allergic response to drugs, restorative materials, betel quid chewing, etc.), pemphigus, pemphigoid, and leukoedema. In such cases, evaluation of biomarkers using immunofluorescence can help eliminate other possibilities.

Immunofluorescence is based on the principle of antigen–antibody reaction. The resulting complex is tagged with a fluorescent compound to allow visualization. Two major techniques are utilized, direct immunofluorescence (aims at detecting the presence of the antigen in the patient's biopsy sample) and indirect immunofluorescence (aims at detecting the presence of antigen-directed antibodies in the patient's serum). Because autoimmunity is a dysregulation of the immune system, antigen and antibody evaluations, along with the distribution of the immune complex, can be utilized as a specific diagnostic biomarker. The application of immunofluorescence in the diagnosis of immune-related

TABLE 54.3 Application of Immunofluorescence for the Diagnosis of Immune-Mediated Lesions

Lesion	Direct Immunofluorescence	Indirect Immunofluorescence
Pemphigus vulgaris	Intercellular deposition of IgG (IgG 1 and 4) throughout epidermis—chicken wire/fish net appearance	Intercellular circulating IgG autoantibodies that bind to epidermis in 80%–90% cases
Paraneoplastic pemphigus	IgG with or without C3 in an intercellular pattern. Granular or linear deposition of C3, IgG, and/or IgM along dermal–epidermal junction in minor cases	Both intercellular intraepidermal antibody deposition and along dermal–epidermal junction
Cicatricial pemphigoid	Linear deposits of C3, IgG, IgA at dermal epidermal junction—"shoreline pattern"	Circulating IgG autoantibodies directed against basement membrane one in 20% cases + circulating autoantibodies to laminin 5
Bullous pemphigoid	IgG (70%) and C3 (90%–100%) deposition at dermal–epidermal junction	Circulating autoantibodies against basement membrane zone
Erythema multiforme	Granular deposits of IgG, C3, IgM, and fibrinogen around dermal vessels or at dermal–epidermal junction	–
Lichen planus	Shaggy deposit at dermal–epidermal junction of IgM (within scattered cytoid bodies), C3 and IgG with fibrin deposition at basement membrane zone	–
Systemic lupus erythematosus	Deposition of IgG, IgM, or C3 in a shaggy band at basement membrane (positive Lupus band test)	–

conditions has gained universal acceptance (Anuradha et al., 2011; Rastogi et al., 2014) (Table 54.3).

ORAL SQUAMOUS CELL CARCINOMA

A tumor is defined as an abnormal mass of tissue, the growth of which exceeds and is uncoordinated with that of the normal tissues and persists in the same excessive manner after the cessation of the stimuli that evoked the change. Tumors can be benign or malignant based on their proliferation characteristics, local tissue invasion, and behavior. The head and neck region is afflicted by distinct tumor entities that may affect the face, oral cavity, nasal and paranasal structures, pharynx, and salivary glands. Oral squamous cell carcinoma (OSCC) accounts for almost 90% of all oral cancers, with over 300,000 new cases reported annually worldwide. OSCC is a malignant epithelial neoplasm affecting the mucosal lining of the oral cavity and oropharynx. Despite attempts aimed at early diagnosis and improving treatment and prognosis, OSCC continues to have a dismal 5-year survival rate of 55%. OSCC can affect the tongue, buccal mucosa, gingiva, palate, floor of the mouth, and lips in decreasing incidence. Most cases are detected at an advanced stage and therapeutic alternatives are expensive and disfiguring. OSCC was reported to be a "disease of the elderly" because it commonly affected males over the age of 50. More recently, a bimodal distribution of cases has been reported, some as early as in 27–30 year old individuals.

The use of tobacco in a smoked/smokeless form with or without the synergistic effects of alcohol consumption is a major risk factor to develop the disease. The malignant transformation of OPMDs (leukoplakia, oral submucous fibrosis, and oral lichen planus) (George et al., 2011; Sarode et al., 2012) continues to bewilder clinicians and researchers because of the unpredictable course of these diseases culminating into OSCC. Recently, the role of chronic inflammation due to microtrauma, chronic mucosal irritation, sharp tooth, or denture clasps has also been recognized as leading to nonhabit-associated OSCC.

The molecular aspects of OSCC reveal the derangement of basic cellular processes that govern DNA repair, cell proliferation, and cell survival. This is led by genetic alterations and mutations that dysregulate normal functions. Mutation in tumor suppressor genes-p53 and/or oncogenes such as epidermal growth factor receptor (EGFR), Bmi (B lymphoma Mo-MLV insertion region 1 homolog), c-myc, RAS result in uncontrolled proliferation and cell survival (Lavanya et al., 2016). The tumor also establishes a microenvironment conducive for its proliferative and metabolic needs. As the tumor thrives, the increase in vascular and lymphatic channels provides routes for metastasis to lymph nodes and distant organs. Thus, OSCC cannot be viewed solely as an epithelial pathology but more like a rouge organ.

Attempts to grade (degree of disorganization of epithelium) and stage (disease severity) the disease have been made so as to facilitate treatment planning and prognostication. Often, cases within the same stage/grade demonstrate a variable outcome. Hence, the accurate evaluation of tumor aggressiveness and behavior remains a challenge.

Biomarkers for tumors include histopathological characteristics (tissue markers) and tumor markers.

Routine histopathology remains the gold standard for diagnosis and can be regarded as a definitive diagnostic tissue biomarker. Other specialized techniques such as immunohistochemistry, ELISA, and immunofluorescence have been proposed to actively evaluate the various phases of tumor development from initiation, promotion, and progression (Lavanya et al., 2016). A tumor marker is a substance present in or produced by a tumor or the tumor's host in response to the presence of the tumor and can be used to differentiate a tumor from normal tissues or to determine the presence of a tumor based on its measurement in blood or secretions (Rivera et al., 2017).

Scully and Burkhardt (1993) proposed that tumor markers can be categorized based on (1) tissue and cell site specificity, (2) their interaction with tissues, and (3) their association with tumoral processes such as invasion and metastasis. The role of tumor markers as an aid for early diagnosis, disease prognostication, and for treatment evaluation has steadily evolved.

The application of tumor markers for OSCC attempts to address two major challenges:

- to predict the transformation of oral potentially malignant disorders (leukoplakia, erythroplakia, discoid lupus erythematosus, oral submucous fibrosis) to frank malignancy (OSCC), and
- to predict metastasis and understand tumor behavior and aggressiveness.

The evaluation of tumor tissue, patient blood, and salivary samples has been carried out in an attempt to unravel efficient biomarkers. Because a universally accepted panel of markers is yet to be proposed, we regard the listed markers as "potential tumor biomarkers." An enhanced expression of Bmi and survivin in OPMDs, which transformed to OSCC, has been demonstrated. Increased expression of transforming growth factor alpha (TGF-α), epidermal growth factor receptor (EGFR), and fibroblast growth factor receptor (FGFR) have been suggested as early markers for head and neck carcinogenesis (Feng et al., 2013).

The expression of cytokeratin (AE-1/3) can aid in characterizing the epithelial cell of origin in tumors, which are atypical or dedifferentiated. Cytokeratin-19 (CK-19) is associated with high-grade dysplasia and early OSCC, whereas cytokeratin-13 (CK-13) has been proposed as a prognostic marker to reveal metastasis and tumor progression (Frohwitter et al., 2016). The decrease in expression of cell adhesion molecule E-cadherin and an increase in N-cadherin is indicative of an invasive aggressive tumor.

To understand the stromal response, expression of vascular endothelial growth factor (VEGF), matrix metalloproteinases (MMP-2,-9,-13), alpha smooth muscle actin (α-SMA), interleukin-8 (IL-8), CD-163, and CD-44 have been evaluated in addition to over 35 other markers with promising results. Although studies have been carried out on tumor tissues, the salivary biomarkers that are significantly altered in OSCC patients are inhibitors of apoptosis (IAP), squamous cell carcinoma associated antigen (SCC−Ag), carcinoembryonic antigen (CEA), carcino-antigen (CA19-9), CA128, serum tumor marker (CA125), tissue polypeptide specific antigen (TPS), reactive nitrogen species (RNS) (Radhika et al., 2016) and 8-Oxo-2′-deoxyguanosine (8-OHdG), which are critical markers of DNA damage, lactate dehydrogenase (LDH), and salivary IgA(s-IgA), to name a few.

SALIVA: A HIDDEN PLETHORA OF BIOMARKERS

Saliva is the exclusive omnipresent biofluid of the oral cavity. It is the result of synthesis and secretion of products from the major and minor salivary glands. Saliva is attributed with distinct functions, as previously stated, and contributes toward maintaining oral health. In the event of oral disease infection, inflammatory or neoplastic, alterations in salivary constituents have been demonstrated. Over 100 potential salivary biomarkers have been reported. A variety of techniques such as liquid chromatography, gel and capillary electrophoresis, mass spectrometry, microbial cultures, immunoassays, and magnetic bead immunoprecipitation have been used to evaluate the proteomic, metabolomic, and microbiomic classes of salivary biomarkers (Wang et al., 2016).

Saliva is considered as the best diagnostic tool for periodontitis. A salivary sample can help ascertain the number and taxa of bacteria and help categorize the disease (Aimetti et al., 2012). A diagnosis of Leishmaniasis in asymptomatic patients can be established by detecting the *Leishmania siamensis* DNA in salivary samples via polymerase chain reaction (PCR) even before its appearance in blood (Siriyasatien et al., 2016). The detection of HIV antibodies in saliva has paved the way for rapid diagnostic kits allowing self-testing, prior to laboratory confirmation (Wang et al., 2016). OraQuick is an in-home oral swab-based testing kit, approved by the FDA for use in the United States (US Food and Drug Administration, 2017).

Dental caries is a microbial disease of the teeth, characterized by the irreversible loss of tooth structure due to demineralization of the inorganic component along with dissolution of the organic matrix constituting a tooth. Salivary samples from patients can be evaluated using specific tests referred to as caries activity tests. The *Lactobacillus* colony count test provides information regarding the microbial load of the cariogenic bacteria *Lactobacillus*. The Snyder test evaluates the ability of salivary microbes to form organic acids from a carbohydrate substrate, which eventually leads to demineralization of the tooth. Enamel solubility test and saliva flow rate are a few other saliva-based investigations that are employed as predictive biomarkers for dental caries (Wang et al., 2016). As a result of the evaluation, preventive measures may be adopted for highly susceptible individuals or the caries vaccine may be advised.

Antibodies (IgG) directed against dengue virus antigen, Ebola virus antigen, and *Plasmodium falciparum* antigens have also been detected in saliva. The application of these markers in a clinical setting warrants further investigation and validation. Numerous studies have investigated the correlation of serum and salivary glucose levels in diabetics (Gupta et al., 2015). An ongoing challenge of this correlation is to establish a noninvasive, rapid investigation to monitor the glycemic status of patients.

In an attempt to facilitate early diagnosis of OSCC, the evaluation of potential biomarkers such as cytokines, i.e., interleukin-6, interleukin-8, matrix metalloproteinases MMP1, MMP3, MMP9, TNF-α, vascular endothelial growth factor A (VEGF-A), fibroblast growth factors, and transferrin in salivary samples has revealed promising results (Radhika et al., 2016). The assessment of CD-59, profilin, and Mac-2–binding protein (M2BP) have demonstrated an 83% specificity and 90% sensitivity for OSCC detection (Wang et al., 2016). SaliMark OSCC is a recently developed noninvasive saliva-based evaluation to detect OSCC. An introduction of these evaluation techniques, as part of a routine work-up, needs further validation.

Investigating the role of saliva in the detection of distant malignancies is still under way. A study by Bigler et al. (2002) suggests that the response to chemotherapy in breast cancer could be measured by the expression of an oncogenic protein c-erbB-2 in saliva. It is evident that saliva holds numerous biomarkers that are yet to be assayed and correlated. (Fig. 54.6). The major challenges for investigations and research using saliva include (1) inherent variability of the sample due to dehydration, use of medication or salivary gland disorders, and (2) inconsistent sample collection techniques along with the confounding effects of diet, oral hygiene status, and contamination with oropharyngeal secretions with the samples collected. Yet, salivary diagnostics and investigations remain a promising avenue to be explored owing to the ease of sample collection, patient compliance, and substantial availability of sample, with minimal risks of collection site complications or infections. Thus, the assay of salivary biomarkers could be applied as predictive and diagnostic biomarkers, contributing toward disease prevention, early diagnosis, disease monitoring, and evaluation of treatment.

CONCLUDING REMARKS AND FUTURE DIRECTIONS

The exposure to any substance, organic or inorganic, of intrinsic or extrinsic origin, in an excessive amount, can pose to be a threat to normal vital functions. Often individuals may remain oblivious of the risk factors they are exposed to until the disease manifests. The advancements in laboratory techniques have facilitated the early diagnosis of hereditary and acquired pathologies, which would have otherwise presented only at an advanced stage of the disease. Thus health care services now focus not only on treatment but also on prevention and screening.

The use of biomarkers has been the key to detection and diagnosis since early medical practice, wherein clinical features of a disease were the only accessible and assessable biomarkers. Over the years, study of the biological response and involvement of organs and organ systems in various diseases has unfurled countless biomarkers ranging from tissue characteristics to the molecular signature of diseases.

The oral cavity provides a general impression of an individual's health status. This view is substantiated by the early and subtle hints of disease, which are often observable at this site. Although not many studies have focused on the oral signs of toxicity, it remains an intriguing area to be investigated because oral signs may precede the actual systemic manifestation of toxicity, thus allowing early intervention.

Oral diseases encompass conditions as basic as dental caries and as lethal as oral cancer. A spectrum of other pathologies ranging from hereditary conditions such as amelogenesis imperfecta and regional odontodysplasia to orofacial cysts such as dentigerous cysts, radicular cysts, odontogenic keratocyst, and even tumors such as

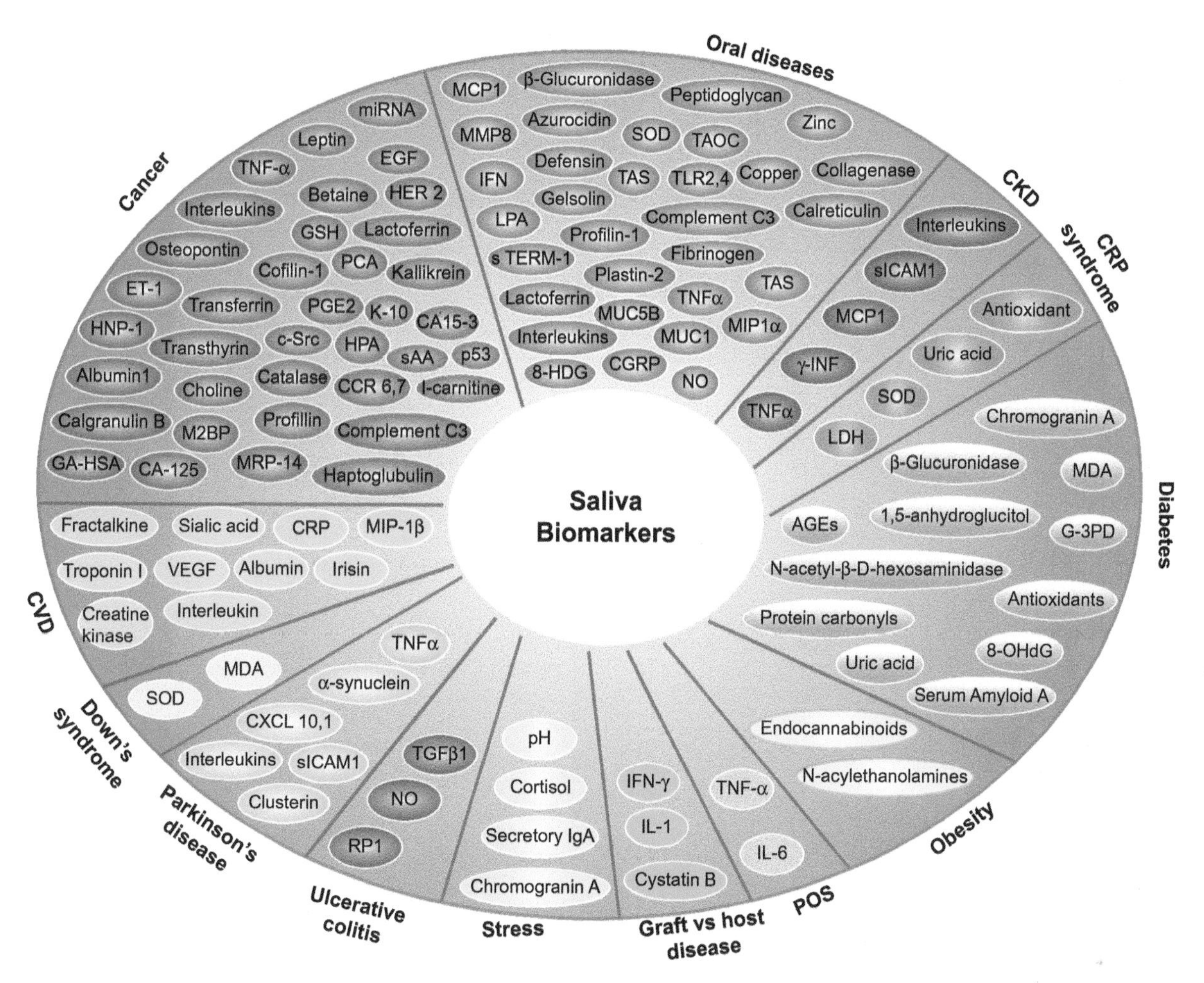

FIGURE 54.6 An overview of the various biomarkers that have been detected in saliva along with their association with specific disease processes. *Image courtesy: Prasad, S., Tyagi, A.K., Aggarwal, B.B., 2016. Detection of inflammatory biomarkers in saliva and urine: potential in diagnosis, prevention, and treatment for chronic diseases. Exp. Biol. Med. 241, 783–799.*

ameloblastoma, pleomorphic adenomas and calcifying epithelial odontogenic tumors are also encountered. The use of specific clinical (radiographic) and tissue biomarkers is often employed for diagnosis. Molecular biomarkers in oral diseases continue to be an evolving arena. The discussion of each pathology and its biomarkers is extensive and external to this chapter.

The sublingual route of drug administration has garnered much attention as an efficient means of systemic drug delivery. Various pharmaceuticals have developed drug preparations such as lozenges, oral sprays, gels, and dispersible tablets to best use this route.

What remains to be achieved is the proposal and acceptance of a standardized universal panel of biomarkers, which can be used to detect, diagnose, and evaluate efficacy of treatment for a particular disease. As newer facts regarding etiopathogenesis and pathophysiology are unfurled, novel biomarkers continue to emerge while many go on to become obsolete. Further studies to quantify the various organic and inorganic constituents of saliva, as affected by various diseases, could aid in devising a quick, noninvasive and simple screening test.

References

Anuradha, C.H., Malathi, N., Anandan, S., et al., 2011. Current concepts of immunofluorescence in oral mucocutaneous diseases. J. Oral Maxillofac. Pathol. 15 (3), 261–266.

Azizi, A., Lawaf, S., 2008. The management of oral pemphigus vulgaris with systemic corticosteroid and dapsone. J. Dent. Res. Dent. Clin. Dent. Prosp. 2 (1), 33–37.

Bigler, L.R., Streckfus, C.F., Copeland, L., et al., 2002. The potential use of saliva to detect recurrence of disease in women with breast carcinoma. J. Oral Pathol. Med. 31 (7), 421–431.

Eisenberg, E., Natarajan, E., Formaker, B.K., 2018. Oral Mucosa and Mucosal Sensation. Department of Oral Health and Diagnostic Sciences, School of Dental Medicine. University of Connecticut, Storrs, CT, USA.

Feng, J.Q., Xu, Z.Y., Shi, L.J., et al., 2013. Expression of cancer stem cell markers ALDH1 and Bmi1 in oral erythroplakia and the risk of oral cancer. J. Oral Pathol. Med. 42 (2), 148–153.

Fernández-González, F., Vásquez-Álvarez, R., Reboires-López, D., et al., 2011. Histopathological findings in oral lichen planus and their correlation with the clinical manifestations. Med. Oral Patol. Oral Cir. Bucal 16 (5), e641–e646.

Fitzpatrick, S.G., Hirsch, S.A., Gordon, S.C., 2014. The malignant transformation of oral lichen planus and oral lichenoid lesions: a systematic review. J. Am. Dental Ass. 145 (1), 45–56.

Frohwitter, G., Buerger, H., Van Diest, P.J., et al., 2016. Cytokeratin and protein expression patterns in squamous cell carcinoma of the oral cavity provide evidence for two distinct pathogenetic pathways. Oncol. Lett. 12 (1), 107–113.

Gardner, S.L., Geller, R.J., Hannigan, R., et al., 2016. Evaluating oral fluid as a screening tool for lead posioning. J. Anal. Toxicol. 40 (9), 744–748.

George, A., Sreenivasan, B.S., Sunil, s, et al., 2011. Potentially malignant disorders of oral cavity. Oral Maxillofac. Pathol. J. 2 (1), 95–100.

Gopalakrishnan, A., Balan, A., Kumar, N.R., et al., 2016. Malignant potential of oral lichen planus an analysis of literature over the past 20 years. Int. J. Appl. Dental Sci. 2 (2), 1–5.

Gupta, S., Sandhu, S.V., Bansal, H., et al., 2015. Comparison of salivary and serum glucose levels in diabetic patients. J. Diabetes Sci. Technol. 9 (1), 91–96.

Holmlund, A., Hänström, L., Lerner, U.H., 2004. Bone resorbing activity and cytokine levels in gingival crevicular fluid before and after treatment of periodontal disease. J. Clin. Periodontol. 31 (6), 475–482.

Ihmed, M.H.M., 2016. Heavy metal toxicity - metabolism, absorption, distribution, excretion and mechanism of toxicity for each of the metals. World News Nat. Sci. 4, 20–32.

Jaishankar, M., Testeen, T., Anbalagan, N., et al., 2014. Toxicity, mechanism and health effects of some heavy metals. Interdiscipl. Toxicol. 7 (2), 60–72.

Johnston, S., 2013. Feeling blue? Minocycline-induced staining of the teeth, oral mucosa, sclerae and ears – a case report. Br. Dental J. 215 (2), 71–73.

Kamath, V.V., Setlur, K., Yerlagudda, K., 2015. Oral lichenoid lesions - a review and update. Indian J. Dermatol. 60 (1), 102.

Langie, S.A.S., Moisse, M., Declerck, K., et al., 2017. Salivary DNA methylation profiling: aspects to consider for biomarker identification. Basic Clin. Pharmacol. Toxicol. 121, 93–101.

Lavanya, R., Mamatha, B., Waghray, S., et al., 2016. Role of tumor markers in oral cancer: an overview. Br. J. Med. Med. Res. 15 (7), 1–9.

Loos, B.G., Tjoa, S., 2000. Host-derived diagnostic markers for periodontitis: do they exist in gingival crevice fluid? Periodontology 39, 53–72.

Mayeux, R., 2004. Biomarkers: potential uses and limitations. J. Am. Soc. Exp. NeuroTher. 1, 182–188.

Mehdipour, M., Shahidi, M., Manifar, S., et al., 2018. Diagnostic and prognostic relevance of salivary microRNA-21, -125a,–31and -200a levels in patients with oral lichen planus - a short report. Cell. Oncol. https://doi.org/10.1007/s13402-018-0372-x.

Mignogna, M.D., Fortuna, G., Leuci, S., 2009. Oral pemphigus. Minerva Stomatol. 58 (10), 501–518.

Moutsopoulos, N.M., Konkel, J.E., 2018. Tissue-specific immunity at the oral mucosal barrier. Trends Immunol. 39 (4), 276–287.

Nanci, A., 2017. Ten Cate's Oral Histology- Development, Structure and Function, ninth ed. Elsevier, Amsterdam.

Narang, N., Sharma, J., 2011. Sublingual mucosa as a route for systemic drug delivery. Int. J. Pharm. Pharmaceut. Sci. 3 (2), 18–22.

Neville, B., Damm, D.D., Allen, C., et al., 2015. Oral and Maxillofacial Pathology, fourth ed. Elsevier, Amsterdam.

Petruzzi, M., 2012. Mucous membrane pemphigoid affecting the oral cavity: short review on etiopathogenesis, diagnosis and treatment. Immunopharmacol. Immunotoxicol. 34 (3), 363–367.

Prasad, S., Tyagi, A.K., Aggarwal, B.B., 2016. Detection of inflammatory biomarkers in saliva and urine: potential in diagnosis, prevention, and treatment for chronic diseases. Exp. Biol. Med. 241, 783–799.

Radhika, T., Jeddy, N., Nithya, S., et al., 2016. Salivary biomarkers in oral squamous cell carcinoma – an insight. J. Oral Biol. Craniofac. Res. 6 (1), S51–S54.

Rastogi, V., Sharma, R., Misra, S.R., et al., 2014. Diagnostic procedures for autoimmune vesiculobullous diseases: a review. J. Oral Maxillofac. Pathol. 18 (3), 390–397.

Raymond, J., Cook, D., 2015. Still leaving stains on teeth—the legacy of minocycline? Austr. Med. J. 8 (4), 139–142.

Rivera, C., Oliveira, A.K., Costa, R.A.P., et al., 2017. Prognostic biomarkers in oral squamous cell carcinoma: a systematic review. Oral Oncol. 72, 38–47.

Sarode, S.C., Sarode, G.S., Tupkari, J.V., 2012. Oral potentially malignant disorders: precising the definition. Oral Oncol. 48 (9), 759–760.

Scully, C., Burkhardt, A., 1993. Tissue markers of potentially malignant human oral epithelium lesions. J. Oral Pathol. Med. 22, 246–256.

Siriyasatien, P., Chusri, S., Kraivichian, K., et al., 2016. Early detection of novel *Leishmania* species DNA in the saliva of two HIV-infected patients. BMC Infect. Dis. 16, 89.

Sivapathasundharam, B., 2016. Shafer's Textbook of Oral Pathology. Elsevier, India.

Sultan, M.K., Sadatullah, S., Shaik, M.A., 2014. Have biomarkers made their mark? A brief review of dental biomarkers. J. Dent. Res. Rev. 1 (1), 37–41.

Taba Jr., M., Kinney, J., Kim, A.S., et al., 2005. Diagnostic biomarkers for oral and periodontal diseases. Dent. Clin. North Am. 49 (3), 551–571.

US Food and Drug Administration, 2017. Available from: https://www.fda.gov/MedicalDevices/ProductsandMedicalProcedures/DentalProducts/DentalAmalgam/ucm171094.htm.

Wang, A., Wang, C.P., Tu, M., et al., 2016. Oral biofluid biomarker research: current status and emerging frontiers. Diagnostics 6 (4), 1–15.

Xu, H.-H., Werth, V.P., Parisi, E., et al., 2013. Mucous membrane pemphigoid. Dent. Clin. North Am. 57 (4), 611–630.

Yorifuji, T., Kashima, S., Suryadhi, M.A.H., et al., 2017. Temporal trends of infant and birth outcomes in Minamata after severe methylmercury exposure. Env. Poll. 231 (Pt. 2), 1586–1592.

Zunt, S.L., 1996. Vesiculobullous disease of the oral cavity. Dermatol. Clin. 14 (2), 291–302.

PART IX

SPECIAL TOPICS

CHAPTER

55

Biomarkers of Mitochondrial Dysfunction and Toxicity

Carlos M. Palmeira, João S. Teodoro, Rui Silva, Anabela P. Rolo

Center for Neurosciences and Cell Biology of the University of Coimbra and Department of Life Sciences of the University of Coimbra Largo Marquês de Pombal, Coimbra, Portugal

INTRODUCTION

Mitochondria play a central role in the life and death of cells. They are not only the mere center for energy metabolism and ATP generation but are also the prime location for different catabolic and anabolic processes, calcium fluxes, and various signaling pathways, while also playing major role in cell life-defining processes such as apoptosis. Mitochondria maintain cellular homeostasis by interacting with reactive oxygen–nitrogen species and responding adequately to different stimuli. In this context, the interaction of pharmacological agents with mitochondria is an aspect of molecular biology that is too often disregarded, not only in terms of toxicology but also from a pharmaceutical point of view, especially when considering the potential therapeutic applications related to the modulation of mitochondrial activity.

Numerous works have shown that mitochondria are a major toxicological target, with their dysfunction being a major mechanism of drug-induced injury. The aim of this chapter is to highlight the role of mitochondria and the modulation of mitochondrial activities in pharmacology and toxicology and also to stress some of the potential therapeutic approaches.

In recent years, there has been extraordinary progress in mitochondrial science that has further outlined the critical role of these organelles in cell biology, pathophysiology, and the diagnosis and therapeutic treatment of different human diseases, such as ischemic diseases, diabetes, some forms of neurodegeneration, and cancer (Duchen, 2004b; Scatena et al., 2007; Giorgi et al., 2012). Mitochondrial physiology and pathophysiology is notably complex, and the role of mitochondria in bioenergetics is also linked, as mentioned earlier, to other essential functions, such as anabolic metabolism, the balance of redox potential, cell death and differentiation, and mitosis. In addition to these basic functions, mitochondria are associated with more specialized cell activities, including calcium homeostasis and thermogenesis, reactive oxygen species (ROS) and reactive nitrogen species signaling, maintenance of ion channels, and the transport of metabolites. Consequently, the basis of different congenital mitochondrial diseases on a molecular level is equally complex and heterogeneous, making mitochondrial pathophysiology difficult to investigate (Hamm-Alvarez and Cadenas, 2009; Cardoso et al., 2010).

This field is made even more challenging by recent evidence that suggests mitochondrial structure and function is dynamic. Specifically, mitochondria possess many interesting properties, such as the ability to fuse or divide, move along microtubules and microfilaments, or undergo turnover (Westermann, 2010; Michel et al., 2012), and these unique properties are often overlooked in research. Undoubtedly, much is still unknown about the mutual interactions between mitochondrial energetics, biogenesis, dynamics, and degradation (Detmer and Chan, 2007), and the contribution of these interactions to mitochondrial toxicology and pharmacology.

MITOCHONDRIAL FUNCTION: GENERAL OVERVIEW

The mitochondrion consists of four main structures or compartments: two membranes, the intermembrane space, and the matrix within the inner membrane. The mitochondrial outer membrane (MOM) separates the cytosol from the intermembrane space. The MOM is responsible for interfacing with the cytosol and its interactions with cytoskeletal elements, which are important for

Biomarkers in Toxicology, Second Edition
https://doi.org/10.1016/B978-0-12-814655-2.00055-4

the movement of mitochondria within a cell. This mobility is essential for the distribution of mitochondria during cell division and differentiation. The mitochondrial inner membrane (MIM) separates the intermembrane space from the matrix. The foldings of the MIM toward the matrix (cristae) serve to increase the surface area of this membrane. Mitochondria also move along intermediate actin filaments, using kinesin and dynein proteins. The MIM hosts the most important redox reactions converting the energy of nutrients into ATP. These reactions are catalyzed by the mitochondrial electron transport chain (ETC), which transports electrons from several substrates to oxygen, in the complex multistep process termed as *mitochondrial respiration*. According to the chemiosmotic theory, mitochondrial respiration generates a transmembrane potential ($\Delta\Psi_m$) across the inner membrane, which is used by ATP synthase to phosphorylate ADP. The MIM is normally virtually impermeable to protons and other ions, and this solute barrier function of the MIM is critical for energy transduction. Permeabilization of the MIM dissipates the $\Delta\Psi_m$ and thereby uncouples the process of respiration from the ATP synthase, halting mitochondrial ATP production (Kushnareva and Newmeyer, 2010). Hence, the free energy of respiration is used to pump protons from the matrix to the intermembrane space (IMS), establishing an electrochemical gradient. Because the MIM displays an extremely low passive permeability to protons, an electrochemical gradient ($\Delta\mu H^+$) is built across the membrane. The electrochemical gradient is the sum of two components: the proton concentration difference and the electrical potential difference across the membrane. The estimated magnitude of the proton electrochemical gradient is about −220 mV (negative inside), and under physiological conditions most of the gradient is in the form of electrical potential difference. The proton gradient is converted in ATP by the F_1F_0-ATP synthase. F_1F_0-ATP synthase couples the transport of the protons back to the matrix with the phosphorylation of ADP to ATP.

Inefficient electron transfer through complexes I–IV causes human diseases in part not only because of loss of energy generation capacity but also of insults to the various enzymes (particularly complexes I, II, and III) induce production of toxic ROS. Defects of complex V are also a cause of mitochondrial dysfunction (Schapira, 2006; Wu et al., 2010a; Abramov et al., 2011). It has also been reported that the deterioration of mitochondrial function underlies common metabolic-related diseases (Rolo and Palmeira, 2006; Palmeira et al., 2007; Turner and Heilbronn, 2008), and several studies have identified compromised oxidative metabolism, altered mitochondrial structure and dynamics, and impaired biogenesis and gene expression in insulin resistance or type 2 diabetes (T2DM) models (Cheng et al., 2010; Rolo et al., 2011; Gomes et al., 2012; Dela and Helge, 2013; Teodoro et al., 2013).

In addition to the process of ATP formation, mitochondria are highly dynamic organelles that have been implicated in the regulation of a great and increasing number of physiological processes. Cells need energy not only to support their vital functions but also to die gracefully, through programmed cell death, or apoptosis (Kushnareva and Newmeyer, 2010). Execution of an apoptotic program includes energy-dependent steps, including kinase signaling, formation of the apoptosome, and effector caspase activation. Furthermore, mitochondrial regulation is also present beyond cell death mechanisms. Indeed, besides oxidative ATP production, mitochondria assume other functions such as heme synthesis, β-oxidation of free fatty acids, metabolism of certain amino acids, production of free radical species, formation and export of Fe/S clusters, and iron metabolism and play a crucial role in calcium homeostasis (Duchen, 2004a; Michel et al., 2012). In addition, initially described as a key checkpoint of intrinsic programmed cell death, accumulating data point to mitochondria as a central platform involved in many cellular pathways, such as those recently highlighted participating in the innate immune response (West et al., 2011) or its lipidic contribution to autophagosomal membrane genesis during starvation-induced autophagy (Hailey et al., 2010). Still, the regulatory roles of mitochondria over normal physiology include the transduction pathway that underlies the secretion of insulin in response to glucose by pancreatic β-cells and in the evaluation of oxygen tension necessary for sensing oxygen in the carotid body and the pulmonary vasculature. Mitochondria also house key enzyme systems quite distinct from those required for intermediary metabolism—the rate-limiting enzymes in steroid biosynthesis and even the carbonic anhydrase required for acid secretion in the stomach (Duchen, 2004b). By accumulating calcium when cytosolic calcium levels are high, mitochondria play subtle roles in coordinating the complexities of intracellular calcium-signaling pathways, at least in some cell types, in which their contribution may be extremely important in the finer aspects of cell regulation. The physiological "uncoupling" of mitochondria plays a central role as a heat-generating mechanism in nonshivering thermogenesis in young and small mammals. It has also been suggested that the production of free radical species by mitochondria might play a key role as a signaling mechanism—for example, in the regulation of ion-channel activities and also in initiating cytoprotective mechanisms in stressed cells (Michel et al., 2012).

MITOCHONDRIAL TOXICITY

Mitochondrial dysfunction is a fundamental mechanism in the pathogenesis of several significant toxic effects in mammals, especially those associated with the liver, skeletal and cardiac muscle, and the central nervous system. These changes can also occur as part of the natural aging process and have been linked to cellular mechanisms in several human disease states including Parkinson's and Alzheimer's diseases, as well as ischemic perfusion injury and the effects of hyperglycemia in diabetes mellitus (Amacher, 2005).

Knowledge of the effects of xenobiotics on mitochondrial function has expanded to the point that chemical structure and properties can guide the pharmaceutical scientist in anticipating mitochondrial toxicity. Recognition that maintenance of the mitochondrial membrane potential is essential for normal mitochondrial function has resulted in the development of predictive cell-based or isolated mitochondrial assay systems for detecting these effects with new chemical entities. The homeostatic role of some uncoupling proteins, differences in mitochondrial sensitivity to toxicity, and the pivotal role of mitochondrial permeability transition (MPT) as the determinant of apoptotic cell death are factors that underlie the adverse effects of some drugs in mammalian systems. To preserve mitochondrial integrity in potential target organs during therapeutic regimens, a basic understanding of mitochondrial function and its monitoring in the drug development program are essential.

At the mitochondrial level, there are several potential drug targets that can lead to toxicity, but a real clinical counterpart has been demonstrated only for a few of them. Recently, antiviral nucleoside analog have shown mitochondrial toxicity through the inhibition of DNA polymerase gamma. Other drugs targeted to different components of the mitochondrial channels can disrupt ion homeostasis or affect the MPT pore. Many molecules are known as inhibitors of the mitochondrial ETC, interfering with one or more of the complexes in the respiratory chain. Some drugs, including nonsteroidal antiinflammatory drugs (NSAIDs), may lead to uncoupling of oxidative phosphorylation, whereas the mitochondrial toxicity of other drugs seems to depend on the production of free radicals, although this mechanism has yet to be clearly defined. Besides toxicity, other drugs have been targeted toward mitochondria to treat mitochondrial dysfunctions. A clear example is the recent development of drugs that target the mitochondria of cancer cells to trigger apoptosis or necrosis, thus promoting cell death and fighting cancer (Rohlena et al., 2011).

XENOBIOTICS AND MITOCHONDRIAL DYSFUNCTION

Mitochondria, because of their central role in metabolism and cell function, have been often used to assess chemical-induced toxicity. Organophosphorus (OPs) pesticides are a class of widely used pesticides in agriculture and in domestic uses. Mitochondria as a site of cellular oxygen consumption and energy production can be a target for OP poisoning as a noncholinergic mechanism of OPs toxicity (Gupta et al., 2001a,b; Karami-Mohajeri and Abdollahi, 2013). Some toxic effects of OPs arise from the dysfunction of mitochondrial oxidative phosphorylation through alteration of the activities of all respiratory complexes and disruption of the mitochondrial membrane. Reduction of ATP synthesis or induction of its hydrolysis can impair the cellular metabolism. The OPs perturb cellular and mitochondrial antioxidant defenses, ROS generation, and calcium uptake and promote oxidative and genotoxic damage triggering cell death via cytochrome *c* released from mitochondria and consequent activation of caspases. Mitochondrial dysfunction induced by OPs can be restored by use of antioxidants such as vitamin E and C, alpha-tocopherol, and electron donors and increasing the cytosolic ATP level. Moreover, other organophosphates have been reported to induce neuron apoptosis in hen spinal cords, which might be mediated by the activation of the mitochondrial apoptotic pathway, causing neuropathy (Zou et al., 2013).

Some compounds used as food additives for growth promotion, such as olaquindox, induce DNA damage and oxidative stress, causing apoptosis in liver cells through the mitochondrial pathway (Zou et al., 2011). More recently, there have been reports of an increase in the frequency of mitochondrial DNA (mtDNA) somatic mutations in lung tissues of fruit growers that had been exposed to pesticides multiple times via inhalation (Wang and Zhao, 2012). The mitochondrial genome is particularly prone to DNA damage, because of its limited DNA repair capabilities, lack of protective histone proteins, and the low tolerance of damaged DNA. Moreover, mitochondria are known to be the major source of reactive oxygen in most mammalian cell types, as well as a major target organelle for oxidative damage (Chomyn and Attardi, 2003). Mitochondrial superoxide and H_2O_2 can cause direct damage to mitochondrial proteins, resulting in nuclear and mitochondrial genotoxicity (Shen et al., 2005), and initiation of apoptosis.

It has been reported that dioxins cause sustained oxidative stress and damage in liver mitochondria from mice exposed to TCDD and in hepatocytes (Senft et al., 2002; Aly and Domènech, 2009); thus, mitochondria are also a direct target for dioxin-induced toxicity. In both hepatic mitochondria isolated from TCDD-treated mice and mitochondria incubated in vitro with TCDD, a number of functional alterations have been observed, including a defect in ATP synthesis and increased ROS production (Senft et al., 2002; Shen et al., 2005; Shertzer et al., 2006; Kopf and Walker, 2010). TCDD decreases hepatic ATP levels through changes in mitochondrial F_0F_1-ATP synthase and ubiquinone and generates mitochondrial oxidative DNA damage, which is exacerbated by decreasing mitochondrial reduced (active) glutathione and by inner membrane hyperpolarization. These mitochondrial effects of TCDD are also associated with altered expression of nuclearly encoded mitochondrial genes (Forgacs et al., 2010; Dere et al., 2011), as well as apoptosis induction involving calcium/calmodulin signaling (Kobayashi et al., 2009). In primary hepatocytes, TCDD has been shown to induce an oxidative stress response involving mitochondrial dysfunction (Aly and Domènech, 2009) and, in mice, exposure to TCDD causes a loss in mitochondrial membrane potential mediated by AhR-dependent production of ROS (Fisher et al., 2005).

Furthermore, previous studies have identified mitochondrial targets of environmental pollutants, namely ROS production and decreased ATP content (Shertzer et al., 2006). Consequently, maintenance of cellular function is strictly dependent on the existence of a healthy population of mitochondria, given that alterations of mitochondrial bioenergetic features by toxicants reduce energetic charge and may ultimately result in cell death. Some detrimental effects of dibenzofuran (DBF), a ubiquitous dioxin-like compound considered to be an environmental pollutant, have already been reported in lung mitochondria (Duarte et al., 2011) and lung cells (Duarte et al., 2012); in addition, some previous studies reported that environmental toxicants induce mitochondrial damage (Palmeira and Madeira, 1997), proving that some pollutants injure mitochondria directly. More recently, siRNA-mediated knockdown of the AhR in lung epithelial cells and fibroblasts was shown to increase sensitivity to smoke-induced apoptosis (Souza et al., 2013), and these effects involved mitochondrial dysfunction, decreased antioxidant enzymes, and oxidative stress.

MITOCHONDRIA AND DISEASE

In a clinical setting, research has shown a significant relationship between mitochondrial metabolic abnormalities and tumors found in renal carcinomas, glioblastomas, paragangliomas, or skin leiomyoma, which has led to the discovery of new genes, oncogenes, and oncometabolites involved in the regulation of cellular and mitochondrial energy production with a particular focus on reevaluating the Warburg effect (Fulda et al., 2010; Ralph et al., 2010; Solaini et al., 2011). Furthermore, the examination of rare neurological diseases, such as Charcot-Marie Tooth type 2a, autosomal dominant optic 53 atrophy, lethal mitochondrial and peroxisomal fission, and spastic paraplegia, has suggested the involvement of MFN2, OPA1, DRP1, or paraplegin in the auxiliary control of mitochondrial energy production (Benard et al., 2010; Du and Yan, 2010; Zhu, 2010). Advances in the understanding of mitochondrial apoptosis have suggested a supplementary role for Bcl-2 or Bax in the regulation of mitochondrial respiration and dynamics, which has led to the investigation of alternative mechanisms of energy regulation (Benard et al., 2010). In addition, different metabolic diseases, such as diabetes, obesity, and nonalcoholic fatty liver disease (NAFLD), and the more general metabolic syndrome underline the role of dysfunctional mitochondria in pathogenesis (Dalgaard, 2011; Rolo et al., 2011).

MITOCHONDRIAL DYSFUNCTION IN DIABETES

In a situation of excess nutrients, mitochondrial membrane potential ($\Delta\Psi_m$) can rise to abnormally high levels, with concomitant excessive reduction of the mitochondrial respiratory chain complexes. This occurs because of elevated levels of ATP and low levels of ADP, meaning that the membrane potential generated by the oxidation of substrates is not totally utilized and begins to build up. Although there is a normal buildup of membrane potential, when it reaches high enough values it can lead to extremely dangerous situations. Overreduction means that the electrons obtained from substrate oxidation can no longer reach molecular O_2 at complex IV or cytochrome *c* oxidase to generate H_2O. As the vectorial ejection of protons against their gradient is a requirement for electronic transport across the respiratory chain, given a high enough membrane potential, the electronic leap between each complex no longer carries enough energy to transport protons against their enlarged gradient, and for that reason electrons get "stuck" inside the respiratory chain. This is most dangerous, for these proteic complexes are in an altered, unstable conformational state, which they must abandon by getting rid of the electrons to anything that will take them. That turns out to be molecular O_2, resulting in the heightened generation of ROS. Given enough time, the abnormal ROS generation overwhelms natural antioxidant defenses and creates mitochondrial damage, further increasing their generation and leading ultimately to cellular and tissue dysfunctions (Fig. 55.1).

The increase in nutrients offered without elevated demand for ATP leads to the abnormal augment of membrane potential ($\Delta\Psi_m$). This, in turn, leads to increased ROS generation that, if prolonged enough, causes cellular and mitochondrial damage. By activating PGC-1α (by phosphorylation—AMPK and deacetylation—sirtuins), one can induce the activation of the mitochondrial biogenic program, leading to the generation of more mitochondria. More mitochondria allows for better

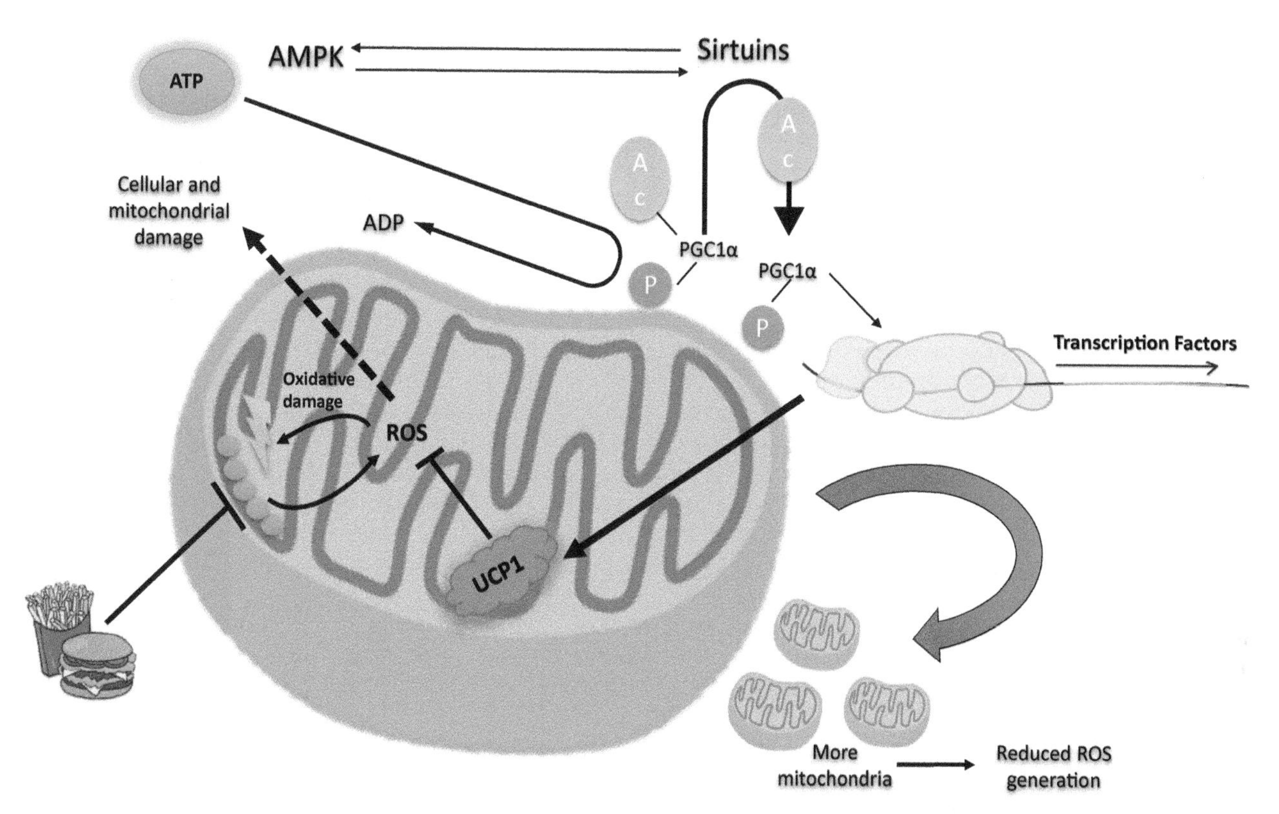

FIGURE 55.1 Role of mitochondria in diabetes and obesity. *AMPK*, AMP-activated protein kinase; *PGC1α*, peroxisome proliferator-activated receptor γ coactivator 1α; *ROS*,./ reactive oxygen species; *UCP1*, uncoupling protein 1.

handling of the excess nutrients, thus reducing ROS generation. Another way to reduce $\Delta\Psi_m$ is by mildly uncoupling the oxidative phosphorylation. UCP1 (and other members of the uncoupling protein family) accomplishes this by reducing $\Delta\Psi_m$ and generating heat, thus preventing ROS generation.

It was found that mitochondria from high-fat fed (HFD) rats were morphologically and structurally altered (Lieber et al., 2004; Kim et al., 2008). These studies demonstrate that there appears to exist a direct correlation between altered mitochondrial functionality and insulin resistance, obesity, and diabetes (Vial et al., 2010). As such, correct mitochondrial structure and function correction could lead to the unveiling of therapeutic strategies to treat obesity and diabetes.

The master regulator of mitochondrial biogenesis, the peroxisome proliferator–activated receptor γ (PPARγ) coactivator 1α (PGC1α), is of great necessity for the correct number, structure, and function of mitochondria. The regulation of PGC1α can occur by several means: its expression, phosphorylation, and acetylation status (Fernandez-Marcos and Auwerx, 2011), for example. The ones highlighted are particularly important, for they appear to be dependent on the cell's energetic status.

In fact, it has been shown that sirtuin 1 (SirT1) regulates PGC1α. Sirtuins are a class of NAD^+-dependent deacetylases and, as such, their activity on gene transcription can be classified as a nutrient-sensitive action. Sirtuins' activity as gene transcription modulators has been explored in various fields of investigation, from aging to obesity, diabetes, and Alzheimer's, to name a few (Yamamoto et al., 2007). SirT1 effects on metabolic regulation were found on SirT1-*null* mice, which have decreased insulin release, whereas overexpression of SirT1 has increased insulin response to glucose (Moynihan et al., 2005). Also, SirT1 deacetylates and thus activates PGC1α, correlating directly to improved metabolic status (Nemoto et al., 2005). Curiously, SirT1 decreases uncoupling protein 2 (UCP2) expression, resulting in increased mitochondrial coupling and thus reducing substrate utilization (Moynihan et al., 2005), which makes some sense because elevated NAD^+ levels activate SirT1 and, as such, the cell has energetic needs and should not waste $\Delta\Psi_m$. Conversely, PPARγ's expression is downregulated by SirT1, resulting in decreased adipogenesis and increased lipolysis (Picard and Auwerx, 2002). SirT1 activation of PGC1α in brown adipocytes leads to increased mitochondrial biogenesis and thus increased thermogenic dissipation of excess nutrients (Lagouge et al., 2006). For further reading on SirT1, readers are encouraged to read the excellent work by Yamamoto et al. (2007). These works are sometimes conflicting and make it difficult to understand the effects of SirT1 on metabolism. In fact, until very recently, the role of resveratrol (the most famous natural SirT1 activator) was still questioned and not fully understood (Hubbard et al., 2013). SirT1 is the most famous and most studied sirtuin, but it is not the only one. Another extremely important sirtuin is the normally mitochondrial native SirT3. Expression of SirT3, in white and brown adipose tissue (WAT and BAT, respectively), is induced by calorie restriction and cold exposure. One of the most interesting facts about SirT3 is that its constitutive expression causes increased levels of PGC1α and UCP1 expression, and as such it is a very attractive target for obesity reduction. Because it is a mitochondrial sirtuin, SirT3's activity leads to decreased acetylation of mitochondrial proteins, which invariably results in increased activity (Shi et al., 2005). We have recently shown that the isoquinoline alkaloid berberine is a potent inducer of SirT3's activity in high fat–fed rats, which at least partially explains this compound's potent anti-obesity effects (Teodoro et al., 2013). Because SirT1 is a NAD^+-dependent deacetylase, its activity is dependent on the cell's reductive status, and a high nutrient ambient leads to decreased SirT1 activity, where PGC1α remains acetylated and its activity diminished (Cantó and Auwerx, 2009). Another metabolic sensor, the AMP-activated protein kinase (AMPK), is activated in the presence of low energy levels, i.e., when the ATP levels are low (or, more appropriately, when AMP levels are high). It phosphorylates PGC1α, activating it to produce more mitochondria to try to elevate ATP levels (Cantó and Auwerx, 2009). As such, these sensors' activities on PGC1α were designed to counter situations of low energy stress and are completely shut down in obesity and diabetes. Consequently, the artificial induction of their activation can be considered a potential and extremely attractive therapeutic strategy, especially considering that oxidative phosphorylation inhibition is a hallmark of HFD and diabetic animals.

AMPK is an important metabolic sensor and regulator, being involved in, among other effects, glucose uptake, lipidic β-oxidation, and mitochondrial biogenesis. Its effects are present on several organs, from the liver to the brain, from WAT and BAT to skeletal muscle, i.e., all metabolic-relevant tissues (Winder et al., 2000). Because AMP activates AMPK, a rise in this adenosine nucleotide (with concomitant decrease in ATP) signals the cell to begin substrate oxidation processes to generate ATP. As paralleled by NAD^+ and SirT1, this (and subsequent downstream effects) can be explored (and has been extensively studied) for obesity management. Although SirT1 deacetylates proteins and histones, AMPK phosphorylates and alters proteins' activity (either increasing or decreasing) (Hardie et al., 2012). AMPK induces GluT1 activation and GluT4 migration to the cellular membrane and thus increased glucose uptake and oxidation (Barnes et al., 2002; Pehmøller et al., 2009). AMPK also mediates fatty acid uptake in cardiac cells (Habets et al., 2009), while improving their uptake and oxidation mainly by the inhibition of acetyl-CoA carboxylase and thus increasing mitochondrial import of fatty acids (Merrill et al., 1997) and by increasing the glycolytic rate (Marsin et al., 2002). Another key effect of AMPK is on mitochondrial biogenesis for, unsurprisingly, AMPK phosphorylates and activates PGC1α, thus increasing mitochondrial content, particularly in skeletal muscle (Winder et al., 2000). Finally, AMPK can also activate (and be activated by) SirT1, by increasing cellular NAD^+ levels (Cantó et al., 2010). As with sirtuins, these are just some effects of AMPK on metabolism, for it is also involved in many other biological functions. For further reading, we refer the reader to Hardie et al. (2012).

Oxidative stress also plays a major role in mitochondrial dysfunction in high-energy situations. In fact, as noted before, increased ROS generation is common in high-energy situations. The increased ROS generation, along with mitochondrial damage, also causes the activation of inflammatory pathways (as the *c*-Jun N-terminal kinase [JNK] and mitogen-activated protein kinase [MAPK]). These cause the inactivation of the insulin signaling pathway and loss of GluT4 (the insulin-sensitive glucose transporter) translocation to the cellular membrane (Qatanani and Lazar, 2007), increasing insulin resistance and thus exacerbating the problem. Also, because of the high energy levels, it comes as no surprise that the expression of proteins involved in lipid handling and mitochondrial lipid β-oxidation is diminished (Schreurs et al., 2010). Furthermore, ROS contribute to diminish glycolytic rates, as it is known that ROS inhibit the key glycolytic enzyme GAPDH (Du et al., 2000), which appears to be a self-defense mechanism against glucose damage (Rolo and Palmeira, 2006). Also, the persistent excess of nutrients leads to the maintenance of said inhibition and worsening of the situation. All of this contributes to increased lipid deposition inside cells, which affects not just mitochondrial function, but also the entire cell. Increasing the number of mitochondria is an attractive strategy for it leads to more units to carry the load of more nutrients, because not only would ROS generation be attenuated but also more antioxidant defenses would also be present. On the other hand, mildly uncoupling mitochondria leads to decreased ROS generation by the caloric dissipation of the electrochemical protonic gradient (Korshunov et al., 1997; Skulachev, 1998). There is already much work being conducted on both perspectives, both yielding very promising results (for further reading, please refer to Ren et al., 2010; Wu et al., 2013).

In terms of increasing mitochondrial numbers, there has been an enormous push toward research on brown adipose tissue since its recent discovery in human adults. Brown adipose tissue (BAT) evolved into the main source of heat generation in small mammals by having a high content of UCP1-expressing mitochondria (UCP1 can be and is considered a hallmark of BAT) (Cannon and Nedergaard, 2004). When activated, BAT generates a high metabolic rate, sustained by a rather large rate of substrate oxidation, which is obtained both from its lipid droplets and from circulation (Bartelt et al., 2011). As such, overactivation of BAT is, theoretically, a highly effective therapeutic strategy for obesity. In fact, we have recently shown just that, with the use of the bile acid chenodeoxycholic acid (CDCA) an obese phenotype can be normalized by elevating BAT UCP1 activity (Teodoro et al., 2014). It was thought that BAT was not present in adult humans, voiding such therapeutic approaches, but, as noticed before, it has been shown otherwise (Whittle, 2012). However, Vosselman et al. (2012) demonstrated that overstimulation of BAT in adult humans was hardly a valid strategy. As such, thermogenic therapy for obesity in adult humans could only be a failed idea, if not for the fact that adipocytes are highly plastic cells. This means that, given the right stimuli, white adipocytes can be, to some extent, converted into brown-like cells, the so-called "brite" or beige adipocytes (the opposite, i.e., the conversion of brown into white is also possible). As such, the next "big thing" in metabolic research is the conversion of white into brown adipocytes thus creating elevated basal metabolic rates, burning more fuel, and decreasing obesity. Most therapeutic strategies already studied and reported, which reduce adiposity in white adipocytes, produce metabolic alterations that are common to what is described to happen in "brite" induction—i.e., the activation of PPARα and induction of lipolysis, increased leptin release, and induction of mitochondrial biogenesis (Flachs et al., 2013). These alterations are usually associated with UCP1 induction and nonshivering thermogenesis. Curiously, UCP1-*null* mice are obesity-resistant when exposed to cold, but not at thermoneutrality (Anunciado-Koza et al., 2008). To make matters worse, the work by Nedergaard and Cannon (2013) skillfully argues that, despite the huge increase in UCP1 expression in WAT (arising from virtually zero), the overall contribution of these newly formed "brite" cells to the body's basal metabolic rate is negligible at best. As such, the authors propose that some other mechanism is responsible for the effects demonstrated in other works. Flachs et al. (2011) suggest that n-3 PUFA anti-obesogenic effects are not UCP1-dependent, which is also the case when combined with calorie restriction, but they are caused by increased cycling of triglyceride/free fatty acid cycle (TG/FFA cycle), a so-called metabolic futile cycle, for it consumes energy while not causing the generation of products. This cycle could be behind many anti-obesogenic effects of countless compounds, whose effects are clear, but whose mechanisms are not. Research into these metabolic pathways will probably become a hot topic for obesity research in the near future. We have shed some light into this matter, by demonstrating once more using CDCA that there is more than just thermogenic dissipation behind bile acids' anti-obesity effects, in particular, an acceleration of metabolic functions (Teodoro et al., 2016).

MITOCHONDRIAL DYSFUNCTION IN ISCHEMIA/REPERFUSION

Ischemia/reperfusion (I/R) injury is a phenomenon whereby damage to a hypoxic organ is accentuated following the return of the oxygen supply, and it has been recognized as a clinically important pathological disorder. I/R may occur in many clinical situations such as transplantation, resection, trauma, shock, hemorrhage, and thermal injury.

The mitochondrial function is impaired in I/R settings, leading to an alteration of energy metabolism. Ischemia leads to the cessation of oxidative phosphorylation, which causes tissue ATP and creatine phosphate concentrations to decrease with a simultaneous increase in ADP, AMP, and inorganic phosphate (Pi) concentrations. During ischemia, anaerobic glycolysis and ATP degradation produce H^+-maintaining mitochondrial membrane potential. As maintenance of ion gradients across the plasma membrane and between cellular compartments depends on ATP-driven reactions, metabolic disruption by injurious stresses may rapidly perturb cellular ion homeostasis.

During oxygen deprivation the intracellular H^+, Na^+, and Ca2$^+$ levels are elevated, inducing osmotic stress and causing mitochondrial damage. Intracellular H^+ accumulation activates the Na^+/H^+ exchanger, leading to Na^+ influx. Na^+ efflux is attenuated because the Na^+/K^+-ATPase is inhibited during ischemia. Therefore, Na^+/H^+ exchange activity leads to increasing intracellular Na^+ (Inserte et al., 2002, 2006; Murphy and Steenbergen, 2008). This augmentation of intracellular Na^+ during ischemia is accompanied by an increase in intracellular Ca^{2+} through reverse mode of the Na^+/Ca^{2+} exchanger. Although Na^+ overload stimulates Ca^{2+} influx by the Na^+/Ca^{2+} exchanger and depletion of ATP reduces Ca^{2+} uptake by the endoplasmic reticulum, the Ca^{2+} level is maintained as modest during ischemia because acidosis inhibits the Na^+/Ca^{2+} exchanger, and cytosolic Ca^{2+} is taken up by the mitochondria as long as its membrane potential is maintained. Influx of extracellular Ca^{2+} is responsible for irreversible cell injury as shown by studies in which removal of extracellular Ca^{2+} protects against various hepatotoxicants (Schanne et al., 1979; Farber, 1982).

ROS generation plays a major role in damaging the organ during ischemia and sensitizing it to reperfusion. The source of the ROS is uncertain, perhaps involving complexes I and III of the ETC of mitochondria, or perhaps xanthine/xanthine oxidase (X/XOD) acting on xanthine formed from the degradation of adenosine (AMP is slowly converted into adenosine and then inosine and xanthine through a purine degradation pathway). There is a gradual decline in cellular integrity as a consequence of combinative action of ATP depletion, elevated intracellular Ca^{2+}, and ROS. Thus, the ATP-dependent repair processes are incapable of operating. Maintenance of mitochondrial integrity is a critical determinant of cell outcome. As such, if mitochondria remain sufficiently intact to generate ATP after short periods of ischemia, tissue damage is reversible and can be repaired. But if the ischemia is more aggressive then recovery is not possible. The ATP restored during reperfusion will exacerbate the damage to the organ due to metabolic disorders that accumulate during ischemia leading to cell death. The increased Ca^{2+} and bursts of ROS generation are characteristics of reperfusion.

Probably the majority of ROS is formed by uncoupled mitochondria, mainly from mitochondrial complexes I and III of the ETC (Jaeschke and Mitchell, 1989; Turrens, 2003). When the respiratory chain is inhibited by absence of oxygen and then reexposed to oxygen, ubiquinone can become partially reduced to ubisemiquinone. It can then react with oxygen to generate superoxide that is reduced to hydrogen peroxide by superoxide dismutase. Hydrogen peroxide is removed by glutathione peroxidase or catalase, but if ferrous ions (or other transition metals such as copper) are present it will form the highly reactive hydroxyl radical through a Fenton reaction (Becker, 2004). In fact, mitochondrial lipids and proteins that are damaged during ischemia favor ROS generation during reperfusion (Inserte et al., 2002). ROS cause peroxidation of cardiolipin of the inner mitochondrial membrane, impairing electron flow through the ETC (Petrosillo et al., 2003; Paradies et al., 2004). Moreover, lipid peroxidation causes the release of reactive aldehydes such as 4-hydroxynonenal that alters membrane proteins (Echtay et al., 2003). ROS also have direct effects on several respiratory chain components and can cause inhibition of the ATP synthase and ANT (adenine nucleotide translocase). There is depletion in superoxide dismutase, glutathione peroxidase, and glutathione during reperfusion that enhances oxidative stress.

Mitochondria are the major target of ROS and Ca^{2+} overload. These agents are potent inducers of the MPT, resulting in mitochondrial-initiated cell death. A major consequence of MPT induction is inhibition of oxidative phosphorylation, which when unrestrained will lead to necrotic cell death. The permeability transition has also been pointed to as being involved in apoptosis, through the release of proapoptotic factors, such as cytochrome *c*, and other apoptosis-inducing factors into the cytosol (Forbes et al., 2001; Murata et al., 2001). In response to proapoptotic signals, Bax, a proapoptotic member of the Bcl-2 family, is translocated to the mitochondria and can form channels that allow the release of cytochrome *c* from the mitochondrial intermembrane space (Borutaite and Brown, 2003). In conditions of ATP depletion, apoptosis can deviate to necrosis (necroapoptosis). Changes in mitochondrial morphology achieved by fission and fusion may play an important role as a determinant of cell viability. It is important to understand the molecular mechanisms of mitochondrial dynamics and their relationship with ischemia-reperfusion injury.

Given the role of mitochondria in ischemia/reperfusion injury, strategies have been developed that focus on maintaining mitochondrial function and consequently reducing the damage. Perfusion with GSK-3β inhibitors reduces cell death induced by I/R (Tong et al., 2002; Gross et al., 2004; Pagel et al., 2006; Gomez et al., 2008). It is thought that the mechanism of protection elicited by GSK-3β inhibition is related to modulation of MPT, by interaction between GSK-3β and components of the MPT process (Juhaszova et al., 2008). Phospho-GSK-3β can bind to the ANT, voltage-dependent anion channel (VDAC), or Cyclophilin D (CypD) (Pastorino et al., 2005; Nishihara et al., 2007). Pretreatment with indirubin-3′-oxime (an inhibitor of GSK-3β) in conditions of hepatic I/R protects the liver by maintaining mitochondrial calcium homeostasis, thus preserving mitochondrial function and hepatic energetic balance (Varela et al., 2010). GSK-3β inactivation by indirubin-3′-oxime acts as pharmacological preconditioning, modulating the susceptibility to MPT induction and preserving mitochondrial function after I/R.

The suppression of the ANT-CypD interaction may contribute to the elevation of the threshold for MPT induction. CypD null mice mitochondria have been demonstrated to have higher Ca^{2+} buffering capacity, demonstrating a desensitization of these mitochondria to Ca^{2+}-induced MPT (Baines et al., 2005). Recently, a relationship was established between SirT3 and CypD: SirT3 deacetylates and inactivates CypD causing its dissociation from the ANT (Shulga et al., 2010). The decrease in SirT3 activity leads to increased activation of the MPT in response to Ca^{2+} increases, cardiac stress, and aging, resulting in a decline in cardiac function (Hafner et al., 2010). This ability to suppress MPT formation indicates SirT3 as a potential target for new drugs that protect against I/R.

Mitochondrial K-ATP channels are normally closed in vivo because of the inhibitory concentrations of ATP and ADP. Mitochondrial K-ATP–dependent matrix alkalization, preservation of mitochondrial volume, and the structure of the intermembrane space, as well as MPT inhibition, are involved in a preconditioning protective action (Andrukhiv et al., 2006; Costa et al., 2006; Costa and Garlid, 2008). Diazoxide is a selective mitochondrial K-ATP channel agonist that was previously shown to decrease I/R injury induced by orthotopic liver transplantation (Huet et al., 2004). The protective effects of diazoxide against hepatic I/R were dependent on Bcl-2 expression and also with the inhibition of mitochondrial cytochrome *c* release, being abolished by siRNA knockdown of Bcl-2 (Wu et al., 2010b). We have recently contributed to the demonstration of the essential role of mitochondrial function preservation in an I/R setting (Alexandrino et al., 2016) (Fig. 55.2).

The events of ischemia/reperfusion lead to a number of cellular alterations, with particular relevance for mitochondrial function. During ischemia, restriction of blood flow limits access to nutrients, ions and, most relevantly, oxygen. Because ATP requirements are maintained (and, in some cases, elevated), ATP generation drains the mitochondrial membrane potential. Eventually, the cell has to resort to anaerobic generation of ATP through glycolysis, which results in the accumulation of lactate and concomitant decrease in pH. Furthermore, ion exchanges are altered leading to intracellular Na^+ accumulation and Ca^{2+} overload (see full text for further details). Accompanying the decrease in mitochondrial function, there is a mild increase in ROS generation. If the ischemic event is prolonged and if the cell was not prepared for it (for example, by preconditioning it with pharmacological agents or with short, repetitive ischemia/reperfusion events), during reperfusion, the restoration of blood flow and particularly of oxygen restores mitochondrial activity but in a totally different setting. Mitochondrial environment and function are compromised because of alterations suffered during ischemia, and ROS generation is highly exacerbated, with resulting damage to mitochondrial components such as the members of the respiratory chain and cardiolipin, heightening the problem. All this leads to the induction of the mitochondrial permeability pore, with concomitant release of cytochrome *c* and other proapoptotic factors, which might lead to cell death. Various therapeutic agents have already been shown to be modulators of mitochondrial function and normalizers during ischemia/reperfusion. Of note, berberine leads to an overactivation of SirT3 and GSK-3β inhibitors such as indirubin-3′-oxime (see text for further details).

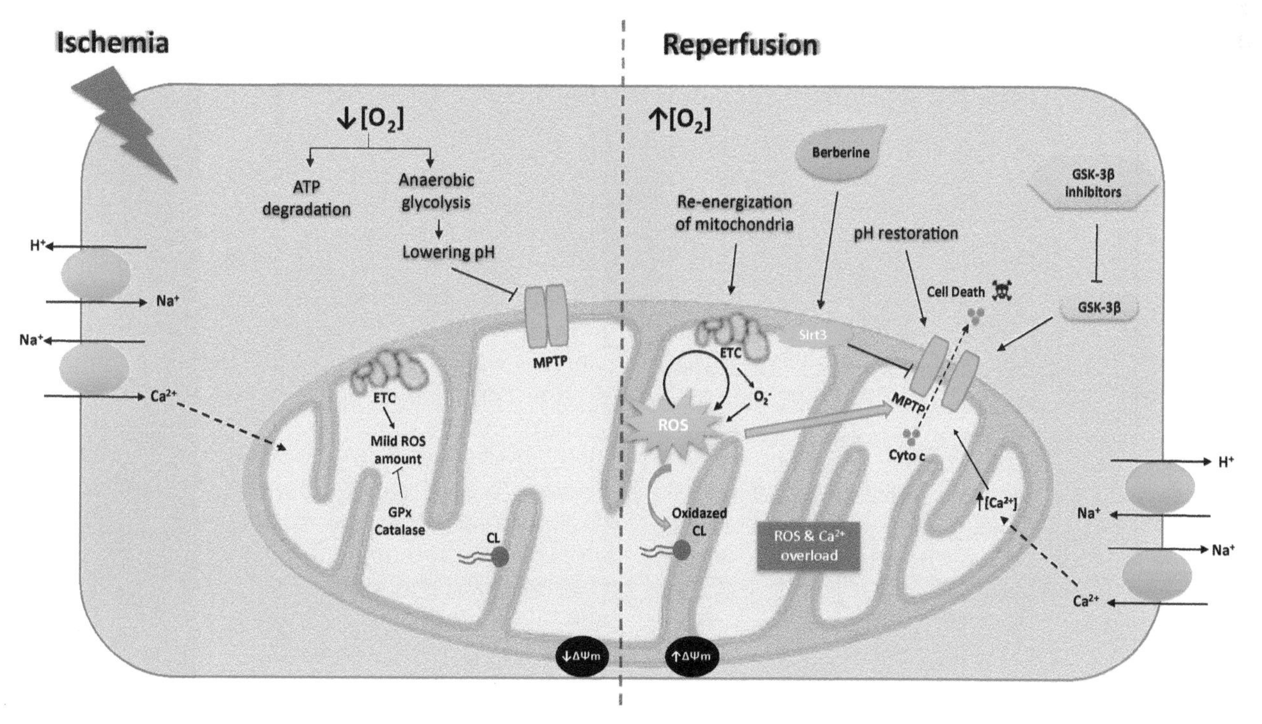

FIGURE 55.2 Role of mitochondria in Ischemia/Reperfusion injury. $\Delta\Psi_m$, mitochondrial membrane potential; *CL*, cardiolipin; *ETC*, electronic transport chain; *GPx*, glutathione peroxidase; *MPTP*, mitochondrial permeability transition pore; *ROS*, reactive oxygen species.

MITOCHONDRIAL DYSFUNCTION IN CANCER

As noted before, mitochondria are responsible for a wide array of reactions and phenomena in the cell. Energy (ATP) production, redox status maintenance, ROS generation, Ca^{2+} storage, and apoptosis control, to name a few, are all dependent on and/or take place in mitochondria. As such, it comes as no surprise to find that mitochondrial alterations and/or deregulation are involved in cancer development. In fact, alterations in biosynthetic pathways, cell signaling, or DNA replication can shift a cell from a quiescent to a proliferative status (Wallace, 2012). The idea of mitochondrial dysfunction in cancer is rather ancient, dating back to the description of the Warburg effect, meaning the realization that cancerous cells have increased lactic acid production in the presence of oxygen, the so-called aerobic glycolysis. This led to the idea that cancer cells have damaged mitochondria because pyruvate is converted into lactate, and glycolysis is the main source of ATP for the cancerous cell. This is further supported by the notion that the interior of tumors is a highly anoxic area, elevating the need for glycolytic-derived ATP and downplaying the need for a functional mitochondrial population. All these factors would come into play in creating a highly proliferative, malignant cell. As discussed below, a healthy mitochondrial population is as vital to a normal cell as it is to a cancerous one. It is noteworthy, however, that it is true that mtDNA is highly mutated in cancer cells and these cells have a higher glycolytic rate than do normal cells. It is the fact that mitochondria are not essential to the cancerous cell that is in dispute.

One way by which we can demonstrate that mitochondria are essential to cancerous cells comes from studies where the mtDNA was removed by ethidium bromide exposure, generating the so-called ρ^0 cells. These cells present a lower metabolic rate, resulting in decreased growth and tumor formation (Cavalli et al., 1997; Weinberg et al., 2010). In his seminal review, Wallace (2012) points to Tasmanian devils' and dogs' transmissible tumors, which cause mtDNA decay and which would have disappeared long ago, if not for periodic uptake of normal mtDNA from host cells.

Both somatic and germline mtDNA mutations have been reported in almost all cases of tumors and cancers. Despite the fact that some of these alterations could be considered normal heterogeny of the population, there are no doubts about the direct correlation of some of these mutations with a cancerous phenotype. Consequently, almost all cancerous cells demonstrate impaired OXPHOS activity, when compared with normal cells. As such, these alterations can be beneficial to the cancer cell, by promoting neoplastic alterations and allowing the cancer cell to adapt to a wide array of metabolic environments (Brandon et al., 2006). In fact, several mtDNA mutations have been demonstrated to be positively selected in cancer cells, in contrast to normal cells (Gasparre et al., 2008). But the presence of a mutation that will affect mtDNA and thus lead to a cancerous phenotype appears to not even be necessary, for it has been shown that the nuclear-encoded SUV3 RNA helicase is required for correct mitochondrial function and biogenesis. In fact, $SUV3^{+/+}$ can present the same altered mitochondrial phenotype of a heterozygous or double recessive individual, because of maternal inheritance of altered mitochondria (Chen et al., 2012). But if cancerous cells do indeed present mitochondrial alterations and impaired OXPHOS activity, then how is it possible that functional mitochondria are needed for the viability of said cells?

For one, mitochondrial activity can be altered by more than just mtDNA mutations. In fact, various nDNA mutations for mitochondrial proteins have been reported in various types of cancers. For example, mutations in the complex II proteins have been shown to lead to increased ROS generation, decreased succinate consumption, and decreased respiratory rates and thus a shift toward glycolysis, in a mechanism that appears to involve the stabilization of the hypoxia-inducible factor 1α, HIF1α (Wallace and Fan, 2010; Kurelac et al., 2011). In the same fashion, fumarate hydratase, an enzyme of the Krebs cycle, has also been shown to be mutated in various cancers, leading to increased succinate levels and thus the shift toward increased glycolytic metabolism but, apparently, without the involvement of HIF4α (Adam et al., 2011; Frezza et al., 2011). There are many other reported mutations in other mitochondrial proteins, for example isocitrate dehydrogenase (Ward et al., 2012), and complexes I, III, IV, and V of the respiratory chain (Wallace, 2012).

Mitochondria produce ROS as a by-product of their activity. It is estimated that roughly 5% of all oxygen consumed in a cell is not converted into water but rather into a reactive species. Although a by-product, it has been recently shown that ROS have a signaling activity (Lander, 1997; Devasagayam et al., 2004), resulting in their production being not only required but also needed (within certain limits). As such, a complex (and, for most cases, effective) antioxidant system is in place to reduce them and thus neutralize their activity. However, mitochondrial defects, metabolic imbalances, mutations, and lack of mtDNA histones all lead to increased mtDNA mutations and thus increased cancerous potential. ROS can be one of the most important causal factors for said mutations. In fact, it has been demonstrated that ROS generation is augmented by inhibition of the OXPHOS, in particular of the ATP synthase (Sánchez-Cenizo et al., 2010). In addition to contributing to increased ROS generation, this blockade of OXPHOS also leads to the already discussed metabolic shift toward glycolysis.

We have previously mentioned that the shift toward a greater dependence on glycolytic-derived ATP is a standard condition in the progression from a normal to a cancerous cell. This shift typically involves the overactivation of the PI3K–Akt signaling pathway (Jones and Thompson, 2009), which leads to increased expression of glucose transporters, glycolytic and lipogenic enzymes, and activation of the mechanistic target of rapamycin, mTOR (DeBerardinis et al., 2008), resulting in elevated rates of glycolysis and lactate generation. This pathway also inhibits fatty acid oxidation and prevents carbon flow toward the Krebs cycle (Gatenby and Gillies, 2004; DeBerardinis et al., 2008). Furthermore, by inhibition of the master regulator of mitochondrial biogenesis, PGC1α, this pathway leads to decreased mitochondrial respiratory activity and content, of both respiratory chain units and antioxidant defenses (Daitoku et al., 2003). As mentioned before, HIF1α stabilization is a recurring phenomenon in cancer cells, so it comes as no surprise to realize that HIF1α can impair mitochondrial biogenesis (Zhang et al., 2007). But why does the cancerous cell go to such an extent as to possibly limit its own ATP production? The answer is rather simple: the downplay of mitochondrial bioenergetics is a side effect of the need for the activation of glycolysis-parallel anabolic pathways. By increasing glycolysis, more carbon is shunted toward, for example, the pentose phosphate pathway for nucleotide synthesis and NADPH generation to combat oxidative stress (Grüning et al., 2011). Also, the increased generation of glycerol-3-phosphate in glycolysis leads to heightened lipogenesis (Esechie and Du, 2009), which requires mitochondrial-derived acetyl-CoA, supplied by a complex process that involves the aggressive oncogene Myc, in an anaplerotic refill of Krebs cycle intermediaries (DeBerardinis et al., 2007), making cancer cells dependent on glutamine for their survival (Wise et al., 2008). The antitumor p53 protein provides another proof of mitochondrial dependence of cancer cells. p53 inhibits glycolysis and diverts carbon toward the pentose phosphate pathway, leading to increased NADPH generation; in addition, its activation by telomere shortening prevents PGC1α-mediated mitochondrial biogenesis, thus leading to increased ROS generation and cellular senescence (Sahin and Depinho, 2010). As such, it is clear that, for a cancerous cell, a viable, efficient mitochondrial population is a vital requirement (Vander Heiden et al., 2009).

There are many other ways by which mitochondrial function is linked with cancer (masterfully reviewed in Wallace, 2012). If mitochondria are so vital for the cancer cell, then it is only reasonable to assume that targeting mitochondria to try to treat cancer is a viable therapeutic strategy. In fact, various works (Fulda et al., 2010; Hockenbery, 2010; Wenner, 2012) have focused on the subject. Therefore, we will briefly approach the thematic in these works.

Cancer trademarks, such as immortal potential, unresponsiveness to growth arrest signaling, increased anabolic metabolism, and decreased apoptosis and autophagy, have already been linked to mitochondria (Galluzzi et al., 2010). Despite the already discussed dependence on mitochondrial activity, cancer cells have structurally and functionally different mitochondria when compared with normal cells (Modica-Napolitano and Singh, 2004). One way to target mitochondria is by targeting its membrane potential ($\Delta\Psi_m$). Specific ANT ligands can lead to the induction of the MPT (Lehenkari et al., 2002; Don et al., 2003; Oudard et al., 2003) and could be a valid strategy because cancer cells should be more susceptible owing to having higher metabolic rates and higher Ca^{2+} loads. Also, the peripheral benzodiazepine receptor (PBR) is thought to be a component of the MPT (as well as ANT) and binds to the voltage-dependent anion channel (VDAC) and prevents MPT induction. Thus, it comes as no surprise that PBR is typically overexpressed in various cancers, as it blocks the antiapoptotic effect of the Bcl-2 protein family. As such, PBR ligands have shown antitumor activity (Decaudin et al., 2002). These are but a few examples of how reducing $\Delta\Psi_m$ and inducing apoptosis can be considered a valid and promising anticancer strategy. In fact, various compounds are already in test that target these described mechanisms.

It is evident that a compound or therapeutic strategy that would increase ROS generation or inhibit antioxidant defenses in cancer cells could be immensely helpful. Along these lines, there are already some promising results with ROS inducers (Sarin et al., 2006; Bey et al., 2007; Mehta et al., 2009) and antioxidant defense compounds (Alexandre et al., 2006; Trachootham et al., 2006; Dragovich et al., 2007), to name a few.

In the mitochondrial intermembrane space reside various proapoptotic factors (e.g., cytochrome *c*, cyt c), and their release is a common phenomenon in mitochondrial-dependent apoptosis, a phenomenon that usually involves the proapoptotic proteins BAX and BAK (Chipuk et al., 2006). As such, the modulation of the activities of the pro- and antiapoptotic family of proteins (for example, the Bcl-2 family) can provide interesting therapeutic approaches. The compounds ABT-737 (Oltersdorf et al., 2005) and ABT-263 (Tse et al., 2008) are just two examples.

Of course, targeting mitochondrial metabolic activity is also a strong possibility. Inhibition of the pyruvate dehydrogenase kinase (which blocks PD activity of converting pyruvate into acetyl-CoA) by dichloroacetate leads to increased ROS generation in cancerous, but not in normal cells (Bonnet et al., 2007). Also, inhibition of the expression of lactate dehydrogenase A has shown promising results, by impeding the conversion of pyruvate into lactate (Fantin et al., 2006). Pyruvate is also actively sent toward lipid synthesis, so the block of the key enzyme ATP citrate lyase has also shown promising results (Hatzivassiliou et al., 2005). These are only a few examples of how targeting mitochondrial metabolism is a valid therapeutic strategy to combat cancer.

There are many other compounds and therapeutic strategies and targets being currently tested against cancer, many of them also involving mitochondria. Therefore, it would be no surprise if the cure for cancer came from a strategy that involved mitochondria.

CONCLUDING REMARKS AND FUTURE DIRECTIONS

Considering that mitochondria play an essential role in cellular homeostasis and signaling processes, identifying biomarkers of mitochondrial dysfunction and toxicity is of fundamental relevance. Research into these mitochondrial targets is at present a hot topic in several diseases (diabetes, obesity, cancer, etc., to name just a few examples) and drug-induced toxicity. An improved understanding of the changes that occur at the mitochondrial level is essential to discover new therapeutic targets for mitochondria-related diseases and toxicity exposure.

References

Abramov, A.Y., Gegg, M., Grünewald, A., et al., 2011. Bioenergetic consequences of PINK1 mutations in Parkinson disease. PLoS One 6, e25622.

Adam, J., Hatipoglu, E., O'Flaherty, L., et al., 2011. Renal cyst formation in Fh1-deficient mice is independent of the Hif/Phd pathway: roles for fumarate in KEAP1 succination and Nrf2 signaling. Cancer Cell 20, 524–537.

Alexandre, J., Nicco, C., Chéreau, C., et al., 2006. Improvement of the therapeutic index of anticancer drugs by the superoxide dismutase mimic mangafodipir. J. Natl. Cancer Inst. 98, 236–244.

Alexandrino, H., Varela, A.T., Teodoro, J.S., et al., 2016. Mitochondrial bioenergetics and posthepatectomy liver dysfunction. Eur. J. Clin. Invest. 46, 627–635.

Aly, H.A.A., Domènech, Ò., 2009. Cytotoxicity and mitochondrial dysfunction of 2,3,7,8-tetrachlorodibenzo-p-dioxin (TCDD) in isolated rat hepatocytes. Toxicol. Lett. 191, 79–87.

Amacher, D.E., 2005. Drug-associated mitochondrial toxicity and its detection. Curr. Med. Chem. 12, 1829–1839.

Andrukhiv, A., Costa, A.D., West, I.C., et al., 2006. Opening mitoKATP increases superoxide generation from Complex I of the electron transport chain. Am. J. Physiol. 291, H2067–H2074.

Anunciado-Koza, R., Ukropec, J., Koza, R.A., et al., 2008. Inactivation of UCP1 and the glycerol phosphate cycle synergistically increases energy expenditure to resist diet-induced obesity. J. Biol. Chem. 283, 27688–27697.

Baines, C.P., Kaiser, R.A., Purcell, N.H., et al., 2005. Loss of cyclophilin D reveals a critical role for mitochondrial permeability transition in cell death. Nature 434, 658–662.

Barnes, K., Ingram, J.C., Porras, O.H., et al., 2002. Activation of GLUT1 by metabolic and osmotic stress: potential involvement of AMP-activated protein kinase (AMPK). J. Cell Sci. 115, 2433–2442.

Bartelt, A., Bruns, O.T., Reimer, R., et al., 2011. Brown adipose tissue activity controls triglyceride clearance. Nat. Med. 17, 200–205.

Becker, L.B., 2004. New concepts in reactive oxygen species and cardiovascular reperfusion physiology. Cardiovasc. Res. 61, 461–470.

Benard, G., Bellance, N., Jose, C., et al., 2010. Multi-site control and regulation of mitochondrial energy production. Biochim. Biophys. Acta 1797, 698–709.

Bey, E.A., Bentle, M.S., Reinicke, K.E., et al., 2007. An NQO1- and PARP-1-mediated cell death pathway induced in non-small-cell lung cancer cells by β-lapachone. Proc. Natl. Acad. Sci. U.S.A. 104, 11832–11837.

Bonnet, S., Archer, S.L., Allalunis-Turner, J., et al., 2007. A mitochondria-K^+ channel axis is suppressed in cancer and its normalization promotes apoptosis and inhibits cancer growth. Cancer Cell 11, 37–51.

Borutaite, V., Brown, G.C., 2003. Mitochondria in apoptosis of ischemic heart. FEBS Lett. 541, 1–5.

Brandon, M., Baldi, P., Wallace, D.C., 2006. Mitochondrial mutations in cancer. Oncogene 25, 4647–4662.

Cannon, B., Nedergaard, J., 2004. Brown adipose tissue: function and physiological significance. Physiol. Rev. 84, 277–359.

Cantó, C., Auwerx, J., 2009. PGC-1alpha, SIRT1 and AMPK, an energy sensing network that controls energy expenditure. Curr. Opin. Lipidol. 20, 98–105.

Cantó, C., Jiang, L.Q., Deshmukh, A.S., et al., 2010. Interdependence of AMPK and SIRT1 for metabolic adaptation to fasting and exercise in skeletal muscle. Cell Metabol. 11, 213–219.

Cardoso, A.R., Queliconi, B.B., Kowaltowski, A.J., 2010. Mitochondrial ion transport pathways: role in metabolic diseases. Biochim. Biophys. Acta 1797, 832–838.

Cavalli, L.R., Varella-Garcia, M., Liang, B.C., 1997. Diminished tumorigenic phenotype after depletion of mitochondrial DNA. Cell Growth Differ. 8, 1189–1198.

Chen, P.L., Chen, C.F., Chen, Y., et al., 2012. Mitochondrial genome instability resulting from SUV3 haploinsufficiency leads to tumorigenesis and shortened lifespan. Oncogene 32, 1193–1201.

Cheng, Z., Tseng, Y., White, M.F., 2010. Insulin signaling meets mitochondria in metabolism. Trends Endocrinol. Metabol. 21, 589–598.

Chipuk, J.E., Bouchier-Hayes, L., Green, D.R., 2006. Mitochondrial outer membrane permeabilization during apoptosis: the innocent bystander scenario. Cell Death Differ. 13, 1396–1402.

Chomyn, A., Attardi, G., 2003. MtDNA mutations in aging and apoptosis. Biochem. Biophys. Res. Commun. 304, 519–529.

Costa, A.D., Jakob, R., Costa, C.L., et al., 2006. The mechanism by which mitoKATP opening and H_2O_2 inhibit the mitochondrial permeability transition. J. Biol. Chem. 281, 20801–20808.

Costa, A.D.T., Garlid, K.D., 2008. Intramitochondrial signaling: interactions among mito KATP, PKCε, ROS, and MPT. Am. J. Physiol. 295, H874–H882.

Daitoku, H., Yamagata, K., Matsuzaki, H., et al., 2003. Regulation of PGC-1 promoter activity by protein kinase B and the forkhead transcription factor FKHR. Diabetes 52, 642–649.

Dalgaard, L.T., 2011. Genetic variance in uncoupling protein 2 in relation to obesity, Type 2 diabetes, and related metabolic traits: focus on the functional -866G>A Promoter Variant (rs659366). J. Obes. 2011, 340241.

DeBerardinis, R.J., Lum, J.J., Hatzivassiliou, G., et al., 2008. The biology of cancer: metabolic reprogramming fuels cell growth and proliferation. Cell Metabol. 7, 11–20.

DeBerardinis, R.J., Mancuso, A., Daikhin, E., et al., 2007. Beyond aerobic glycolysis: transformed cells can engage in glutamine metabolism that exceeds the requirement for protein and nucleotide synthesis. Proc. Natl. Acad. Sci. U.S.A. 104, 19345–19350.

Decaudin, D., Castedo, M., Nemati, F., Beurdeley-Thomas, A., et al., 2002. Peripheral benzodiazepine receptor ligands reverse apoptosis resistance of cancer cells in vitro and in vivo. Cancer Res. 62, 1388–1393.

Dela, F., Helge, J.W., 2013. Insulin resistance and mitochondrial function in skeletal muscle. Int. J. Biochem. Cell Biol. 45, 11–15.

Dere, E., Lee, A.W., Burgoon, L.D., Zacharewski, T.R., 2011. Differences in TCDD-elicited gene expression profiles in human HepG2, mouse Hepa1c1c7 and rat H4IIE hepatoma cells. BMC Genom. 12, 193.

Detmer, S.A., Chan, D.C., 2007. Functions and dysfunctions of mitochondrial dynamics. Nature 8, 870–879.

Devasagayam, T.P.A., Tilak, J.C., Boloor, K.K., et al., 2004. Free radicals and antioxidants in human health: current status and future prospects. J. Assoc. Phys. India 52, 794–804.

Don, A.S., Kisker, O., Dilda, P., et al., 2003. A peptide trivalent arsenical inhibits tumor angiogenesis by perturbing mitochondrial function in angiogenic endothelial cells. Cancer Cell 3, 497–509.

Dragovich, T., Gordon, M., Mendelson, D., et al., 2007. Phase I trial of imexon in patients with advanced malignancy. J. Clin. Oncol. 25, 1779–1784.

Du, H., Yan, S.S., 2010. Mitochondrial medicine for neurodegenerative diseases. Int. J. Biochem. Cell Biol. 42, 560–572.

Du, X., Edelstein, D., Rossetti, L., et al., 2000. Hyperglycemia-induced mitochondrial superoxide overproduction activates the hexosamine pathway and induces plasminogen activator inhibitor-1 expression by increasing Sp1 glycosylation. Proc. Natl. Acad. Sci. U.S.A. 97, 12222–12226.

Duarte, F.V., Simões, A.M., Teodoro, J.S., Rolo, A.P., Palmeira, C.M., 2011. Exposure to dibenzofuran affects lung mitochondrial function in vitro. Toxicol. Mech. Meth. 21, 571–576.

Duarte, F.V., Teodoro, J.S., Rolo, A.P., Palmeira, C.M., 2012. Exposure to dibenzofuran triggers autophagy in lung cells. Toxicol. Lett. 209, 35–42.

Duchen, M.R., 2004a. Mitochondria in health and disease: perspectives on a new mitochondrial biology. Mol. Aspect. Med. 25, 365–451.

Duchen, M.R., 2004b. Roles of mitochondria in health and disease. Diabetes 53, S96–S102.

Echtay, K.S., Esteves, T.C., Pakay, J.L., et al., 2003. A signalling role for 4-hydroxy-2-nonenal in regulation of mitochondrial uncoupling. EMBO J. 22, 4103–4110.

Esechie, A., Du, G., 2009. Increased lipogenesis in cancer. Commun. Integr. Biol. 2, 545–548.

Fantin, V.R., St-Pierre, J., Leder, P., 2006. Attenuation of LDH-A expression uncovers a link between glycolysis, mitochondrial physiology, and tumor maintenance. Cancer Cell 9, 425–434.

Farber, J.L., 1982. Membrane injury and calcium homeostasis in the pathogenesis of coagulative necrosis. Lab. Invest. 47, 114–123.

Fernandez-Marcos, P.J., Auwerx, J., 2011. Regulation of PGC-1α, a nodal regulator of mitochondrial biogenesis. Am. J. Clin. Nutr. 93, 884S–890S.

Fisher, M.T., Nagarkatti, M., Nagarkatti, P.S., 2005. Aryl hydrocarbon receptor-dependent induction of loss of mitochondrial membrane potential in epididydimal spermatozoa by 2,3,7,8-tetrachlorodibenzo-p-dioxin (TCDD). Toxicol. Lett. 157, 99–107.

Flachs, P., Rossmeisl, M., Kuda, O., Kopecky, J., 2013. Stimulation of mitochondrial oxidative capacity in white fat independent of UCP1: a key to lean phenotype. Biochim. Biophys. Acta 1831, 986–1003.

Flachs, P., Rühl, R., Hensler, M., et al., 2011. Synergistic induction of lipid catabolism and anti-inflammatory lipids in white fat of dietary obese mice in response to calorie restriction and n-3 fatty acids. Diabetologia 54, 2626–2638.

Forbes, R.A., Steenbergen, C., Murphy, E., 2001. Diazoxide-induced cardioprotection requires signaling through a redox-sensitive mechanism. Circ. Res. 88, 802–809.

Forgacs, A.L., Burgoon, L.D., Lynn, S.G., LaPres, J.J., Zacharewski, T., 2010. Effects of TCDD on the expression of nuclear encoded mitochondrial genes. Toxicol. Appl. Pharmacol. 246, 58–65.

Frezza, C., Zheng, L., Folger, O., Rajagopalan, K.N., et al., 2011. Haem oxygenase is synthetically lethal with the tumour suppressor fumarate hydratase. Nature 477, 225–228.

Fulda, S., Galluzzi, L., Kroemer, G., 2010. Targeting mitochondria for cancer therapy. Nat. Rev. Drug Discov. 9, 447–464.

Galluzzi, L., Morselli, E., Kepp, O., et al., 2010. Mitochondrial gateways to cancer. Mol. Aspect. Med. 31, 1–20.

Gasparre, G., Hervouet, E., de Laplanche, E., et al., 2008. Clonal expansion of mutated mitochondrial DNA is associated with tumor formation and complex I deficiency in the benign renal oncocytoma. Hum. Mol. Genet. 17, 986–995.

Gatenby, R.A., Gillies, R.J., 2004. Why do cancers have high aerobic glycolysis? Nat. Rev. Canc. 4, 891–899.

Giorgi, C., Agnoletto, C., Bononi, A., et al., 2012. Mitochondrial calcium homeostasis as potential target for mitochondrial medicine. Mitochondrion 12, 77–85.

Gomes, A.P., Duarte, F.V., Nunes, 2012. Berberine protects against high fat diet-induced dysfunction in muscle mitochondria by inducing SIRT1-dependent mitochondrial biogenesis. Biochim. Biophys. Acta 1822, 185–195.

Gomez, L., Paillard, M., Thibault, H., Derumeaux, G., Ovize, M., 2008. Inhibition of GSK3-beta by postconditioning is required to prevent opening of the mitochondrial permeability transition pore during reperfusion. Circulation 117, 2761–2768.

Gross, E.R., Hsu, A.K., Fross, G.J., 2004. Opioid-induced cardioprotection occurs via glycogen synthase kinase beta during reperfusion in intact rat hearts. Circ. Res. 94, 960–966.

Grüning, N.-M., Rinnerthaler, M., Bluemlein, K., et al., 2011. Pyruvate kinase triggers a metabolic feedback loop that controls redox metabolism in respiring cells. Cell Metabol. 14, 415–427.

Gupta, R.C., Milatovic, D., Dettbarn, W.-D., 2001a. Nitric oxide (NO) modulates high-energy phosphates in rat brain regions with DFP or carbofuran: prevention by PBN or vitamin E. Arch. Toxicol. 75, 346–356.

Gupta, R.C., Milatovic, D., Dettbarn, W.-D., 2001b. Depletion of energy metabolites following acetylcholinesterase inhibitor-induced status epilepticus: protection by antioxidants. Neurotoxicology 22, 271–282.

Habets, D.D.J., Coumans, W.A., Hasnaoui El, M., et al., 2009. Crucial role for LKB1 to AMPKalpha2 axis in the regulation of CD36-mediated long-chain fatty acid uptake into cardiomyocytes. Biochim. Biophys. Acta 1791, 212–219.

Hafner, A.V., Dai, J., Gomes, A.P., et al., 2010. Regulation of the mPTP by SIRT3-mediated deacetylation of CypD at lysine 166 suppresses age-related cardiac hypertrophy. Aging 2, 914–923.

Hailey, D.W., Rambold, A.S., Satpute-Krishnan, P., et al., 2010. Mitochondria supply membranes for autophagosome biogenesis during starvation. Cell 141, 656–667.

Hamm-Alvarez, S., Cadenas, E., 2009. Mitochondrial medicine and therapeutics, Part II. Preface. Adv. Drug Deliv. Rev. 61, 1233.

Hardie, D.G., Ross, F.A., Hawley, S.A., 2012. AMPK: a nutrient and energy sensor that maintains energy homeostasis. Nat. Rev. Mol. Cell Biol. 13, 251–262.

Hatzivassiliou, G., Zhao, F., Bauer, D.E., et al., 2005. ATP citrate lyase inhibition can suppress tumor cell growth. Cancer Cell 8, 311–321.

Hockenbery, D.M., 2010. Targeting mitochondria for cancer therapy. Environ. Mol. Mutagen 51, 476–489.

Hubbard, B.P., Gomes, A.P., Dai, H., et al., 2013. Evidence for a common mechanism of SIRT1 regulation by allosteric activators. Science 339, 1216–1219.

Huet, P.M., Nagaoka, M.R., Desbiens, G., et al., 2004. Sinusoidal endothelial cell and hepatocyte death following cold ischemia-arm reperfusion of the rat liver. Hepatology 39, 1110–1119.

Inserte, J., Gacrcia-Dorado, D., Hernando, V., Barba, I., Soler-Soler, J., 2006. Ischemic preconditioning prevents calpain-mediated impairment of Na+/K+-ATPase activity during early reperfusion. Cardiovasc. Res. 70, 364–373.

Inserte, J., Garcia-Dorado, D., Ruiz-Meana, M., et al., 2002. Effect of inhibition of Na^+/Ca^{2+} exchanger at the time of myocardial reperfusion on hypercontracture and cell death. Cardiovasc. Res. 55, 739–748.

Jaeschke, H., Mitchell, J.R., 1989. Mitochondria and xanthine oxidase both generate reactive oxygen species in isolated perfused rat liver after hypoxic injury. Biochem. Biophys. Res. Commun. 160, 140–147.

Jones, R.G., Thompson, C.B., 2009. Tumor suppressors and cell metabolism: a recipe for cancer growth. Genes Dev. 23, 537–548.

Juhaszova, M., Wang, S., Zorov, D.B., et al., 2008. The identity and regulation of the mitochondrial permeability transition pore: where the known meets the unknown. Ann. N.Y. Acad. Sci. 1123, 197–212.

Karami-Mohajeri, S., Abdollahi, M., 2013. Mitochondrial dysfunction and organophosphorus compounds. Toxicol. Appl. Pharmacol. 270, 39–44.

Kim, J., Wei, Y., Sowers, J., 2008. Role of mitochondrial dysfunction in insulin resistance. Circ. Res. 102, 401–414.

Kobayashi, D., Ahmed, S., Ishida, M., Kasai, S., Kikuchi, H., 2009. Calcium/calmodulin signaling elicits release of cytochrome c during 2,3,7,8-tetrachlorodibenzo-p-dioxin-induced apoptosis in the human lymphoblastic T-cell line. L-MAT Toxicol. 258, 25–32.

Kopf, P.G., Walker, M.K., 2010. 2,3,7,8-tetrachlorodibenzo-p-dioxin increases reactive oxygen species production in human endothelial cells via induction of cytochrome P4501A1. Toxicol. Appl. Pharmacol. 245, 91–99.

Korshunov, S.S., Skulachev, V.P., Starkov, A.A., 1997. High protonic potential actuates a mechanism of production of reactive oxygen species in mitochondria. FEBS Lett. 416, 15–18.

Kurelac, I., Romeo, G., Gasparre, G., 2011. Mitochondrial metabolism and cancer. Mitochondrion 11, 635–637.

Kushnareva, Y., Newmeyer, D.D., 2010. Bioenergetics and cell death. Ann. N.Y. Acad. Sci. 1201, 50–57.

Lagouge, M., Argmann, C., Gerhart-Hines, Z., et al., 2006. Resveratrol improves mitochondrial function and protects against metabolic disease by activating SIRT1 and PGC-1alpha. Cell 127, 1109–1122.

Lander, H.M., 1997. An essential role for free radicals and derived species in signal transduction. FASEB J. 11, 118–124.

Lehenkari, P.P., Kellinsalmi, M., Näpänkangas, J.P., et al., 2002. Further insight into mechanism of action of clodronate: inhibition of mitochondrial ADP/ATP translocase by a nonhydrolyzable, adenine-containing metabolite. Mol. Pharmacol. 61, 1255–1262.

Lieber, C.S., Leo, M.A., Mak, K.M., et al., 2004. Model of nonalcoholic steatohepatitis. Am. J. Clin. Nutr. 79, 502–509.

Marsin, A.-S., Bouzin, C., Bertrand, L., Hue, L., 2002. The stimulation of glycolysis by hypoxia in activated monocytes is mediated by AMP-activated protein kinase and inducible 6-phosphofructo-2-kinase. J. Biol. Chem. 277, 30778–30783.

Mehta, M.P., Shapiro, W.R., Phan, S.C., et al., 2009. Motexafin gadolinium combined with prompt whole brain radiotherapy prolongs time to neurologic progression in non-small-cell lung cancer patients with brain metastases: results of a phase III trial. Int. J. Radiat. Oncol. Biol. Phys. 73, 1069–1076.

Merrill, G.F., Kurth, E.J., Hardie, D.G., Winder, W.W., 1997. AICA riboside increases AMP-activated protein kinase, fatty acid oxidation, and glucose uptake in rat muscle. Am. J. Physiol. 273, E1107–E1112.

Michel, S., Wanet, A., De Pauw, A., et al., 2012. Crosstalk between mitochondrial (dys)function and mitochondrial abundance. J. Cell. Physiol. 227, 2297–2310.

Modica-Napolitano, J.S., Singh, K.K., 2004. Mitochondrial dysfunction in cancer. Mitochondrion 4, 755–762.

Moynihan, K.A., Grimm, A.A., Plueger, M.M., et al., 2005. Increased dosage of mammalian Sir2 in pancreatic beta cells enhances glucose-stimulated insulin secretion in mice. Cell Metabol. 2, 105–117.

Murata, M., Akao, M., O'Rourke, B., et al., 2001. Mitochondrial ATP-sensitive potassium channels attenuate matrix Ca2+ overload during simulated ischemia and reperfusion: possible mechanism of cardioprotection. Circ. Res. 89, 891–898.

Murphy, E., Steenbergen, C., 2008. Ion transport and energetics during cell death and protection. Physiology 2, 115–123.

Nedergaard, J., Cannon, B., 2013. UCP1 mRNA does not produce heat. Biochim. Biophys. Acta 1831, 943–949.

Nemoto, S., Fergusson, M.M., Finkel, T., 2005. SIRT1 functionally interacts with the metabolic regulator and transcriptional coactivator PGC-1{alpha}. J. Biol. Chem. 280, 16456–16460.

Nishihara, M., Miura, T., Miki, T., et al., 2007. Modulation of the mitochondrial permeability transition pore complex in GSK-3 β-mediated myocardial protection. J. Mol. Cell. Cardiol. 43, 564–570.

Oltersdorf, T., Elmore, S.W., Shoemaker, A.R., et al., 2005. An inhibitor of Bcl-2 family proteins induces regression of solid tumours. Nature 435, 677–681.

Oudard, S., Carpentier, A., Banu, E., et al., 2003. Phase II study of lonidamine and diazepam in the treatment of recurrent glioblastoma multiforme. J. Neurooncol. 63, 81–86.

Pagel, P.S., Krolikowski, J.G., Neff, D.A., et al., 2006. Inhibition of glycogen synthase kinases enhances isoflurane-induced protetion against myocardial infarction during early reperfusion in vivo. Anesth. Analg. 102, 1348–1354.

Palmeira, C.M., Madeira, V.M., 1997. Mercuric chloride toxicity in rat liver mitochondria and isolated hepatocytes. Environ. Toxicol. Pharmacol. 3, 229–235.

Palmeira, C.M., Rolo, A.P., Berthiaume, J., Bjork, J.A., Wallace, K.B., 2007. Hyperglycemia decreases mitochondrial function: the regulatory role of mitochondrial biogenesis. Toxicol. Appl. Pharmacol. 225, 214–220.

Paradies, G., Petrosillo, G., Pistolese, M., et al., 2004. Decrease in mitochondrial complex I activity in ischemic/reperfused rat heart – involvement of reactive oxygen species and cardiolipin. Circ. Res. 94, 53–59.

Pastorino, J.G., Hoek, J.B., Shulga, N., 2005. Activation of glycogen synthase kinase 3beta disrupts the binding of hexokinase II to mitochondria by phosphorylating voltage-dependent anion channel and potentiates chemotherapy-induced cytotoxicity. Cancer Res. 65, 10545–10554.

Pehmøller, C., Treebak, J.T., Birk, J.B., et al., 2009. Genetic disruption of AMPK signaling abolishes both contraction-and insulin-stimulated TBC1D1 phosphorylation and 14-3-3 binding in mouse skeletal muscle. Am. J. Physiol. Endocrinol. Metab. 297, E665–E675.

Petrosillo, G., Ruggiero, F.M., Di Venosa, N., et al., 2003. Decreased complex III activity in mitochondria isolated from rat heart subjected to ischemia and reperfusion: role of reactive oxygen species and cardiolipin. FASEB J. 17, U395–U413.

Picard, F., Auwerx, J., 2002. PPARγ and glucose homeostasis. Annu. Rev. Nutr. 22, 167–197.
Qatanani, M., Lazar, M.A., 2007. Mechanisms of obesity-associated insulin resistance: many choices on the menu. Genes Dev. 21, 1443–1455.
Ralph, S.J., Rodríguez-Enríquez, S., Neuzil, J., et al., 2010. Bioenergetic pathways in tumor mitochondria as targets for cancer therapy and the importance of the ROS-induced apoptotic trigger. Mol. Aspect. Med. 31, 29–59.
Ren, J., Pulakat, L., Whaley-Connell, A., Sowers, J.R., 2010. Mitochondrial biogenesis in the metabolic syndrome and cardiovascular disease. J. Mol. Med. 88, 993–1001.
Rohlena, J., Dong, L.-F., Ralph, S.J., Neuzil, J., 2011. Anticancer drugs targeting the mitochondrial electron transport chain. Antioxidants Redox Signal. 15, 2951–2974.
Rolo, A.P., Palmeira, C.M., 2006. Diabetes and mitochondrial function: role of hyperglycemia and oxidative stress. Toxicol. Appl. Pharmacol. 212, 167–178.
Rolo, A.P., Gomes, A.P., Palmeira, C.M., 2011. Regulation of mitochondrial biogenesis in metabolic syndrome. Curr. Drug Targets 12, 872–878.
Sahin, E., Depinho, R.A., 2010. Linking functional decline of telomeres, mitochondria and stem cells during ageing. Nature 464, 520–528.
Sánchez-Cenizo, L., Formentini, L., Aldea, M., et al., 2010. Up-regulation of the ATPase inhibitory factor 1 (IF1) of the mitochondrial H+-ATP synthase in human tumors mediates the metabolic shift of cancer cells to a Warburg phenotype. J. Biol. Chem. 285, 25308–25313.
Sarin, S.K., Kumar, M., Garg, S., et al., 2006. High dose vitamin K3 infusion in advanced hepatocellular carcinoma. J. Gastroenterol. Hepatol. 21, 1478–1482.
Scatena, R., Bottoni, P., Botta, G., et al., 2007. The role of mitochondria in pharmacotoxicology: a reevaluation of an old, newly emerging topic. Am. J. Cell Phys. 293, C12–C21.
Schanne, F.A.X., Kane, A.B., Young, E.A., Farber, J.L., 1979. Calcium dependence of toxic cell death: a final common pathway. Science 206, 700–702.
Schapira, A.H.V., 2006. Mitochondrial disease. Lancet 368, 70–82.
Schreurs, M., Kuipers, F., van der Leij, F.R., 2010. Regulatory enzymes of mitochondrial beta-oxidation as targets for treatment of the metabolic syndrome. Obes. Rev. 11, 380–388.
Senft, A.P., Dalton, T.P., Nebert, D.W., et al., 2002. Dioxin increases reactive oxygen production in mouse liver mitochondria. Toxicol. Appl. Pharmacol. 178, 15–21.
Shen, D., Dalton, T.P., Nebert, D.W., Shertzer, H.G., 2005. Glutathione redox state regulates mitochondrial reactive oxygen production. J. Biol. Chem. 280, 25305–25312.
Shertzer, H.G., Genter, M.B., Shen, D., et al., 2006. TCDD decreases ATP levels and increases reactive oxygen production through changes in mitochondrial F(0)F(1)-ATP synthase and ubiquinone. Toxicol. Appl. Pharmacol. 217, 363–374.
Shi, T., Wang, F., Stieren, E., Tong, Q., 2005. SIRT3, a mitochondrial sirtuin deacetylase, regulates mitochondrial function and thermogenesis in brown adipocytes. J. Biol. Chem. 280, 13560–13567.
Shulga, N., Wilson-Smith, R., Pastorino, J.G., 2010. Sirtuin-3 deacetylation of cyclophilin D induces dissociation of hexokinase II from the mitochondria. J. Cell Sci. 123, 894–902.
Skulachev, V.P., 1998. Uncoupling: new approaches to an old problem of bioenergetics. Biochim. Biophys. Acta 1363, 100–124.
Solaini, G., Sgarbi, G., Baracca, A., 2011. Oxidative phosphorylation in cancer cells. Biochim. Biophys. Acta 1807, 534–542.
Souza, A.O., Pereira, L.C.M., Oliveira, D.P., Dorta, D.J., 2013. BDE-99 congener induces cell death by apoptosis of human hepatoblastoma cell line – HepG2. Toxicol In Vitro 27, 580–587.
Teodoro, J.S., Duarte, F.V., Gomes, A.P., et al., 2013. Berberine reverts hepatic mitochondrial dysfunction in high-fat fed rats: a possible role for SirT3 activation. Mitochondrion 13, 637–646.
Teodoro, J.S., Zouhar, P., Flachs, P., et al., 2014. Enhancement of brown fat thermogenesis using chenodeoxycholic acid in mice. Int. J. Obes. 38, 1027–1034.
Teodoro, J.S., Rolo, A.P., Jarak, I., et al., 2016. The bile acid chenodeoxycholic acid directly modulates metabolic pathways in white adipose tissue in vitro: insight into how bile acids decrease obesity. NMR Biomed. 29, 1391–1402.
Tong, H., Imahashi, K., Steenbergen, C., Murphy, E., 2002. Phosphorylation of glycogen synthase kinase-3b beta during preconditioning through a phosphatidylinositol-3-kinase dependent pathway is cardioprotective. Circ. Res. 90, 377–379.
Trachootham, D., Zhou, Y., Zhang, H., et al., 2006. Selective killing of oncogenically transformed cells through a ROS-mediated mechanism by β-phenylethyl isothiocyanate. Cancer Cell 10, 241–252.
Tse, C., Shoemaker, A.R., Adickes, J., et al., 2008. ABT-263: a potent and orally bioavailable Bcl-2 family inhibitor. Cancer Res. 68, 3421–3428.
Turner, N., Heilbronn, L.K., 2008. Is mitochondrial dysfunction a cause of insulin resistance? Trends Endocrinol. Metabol. 19, 324–330.
Turrens, J.F., 2003. Mitochondrial formation of reactive oxygen species. J. Physiol. 552, 335–344.
Vander Heiden, M.G., Cantley, L.C., Thompson, C.B., 2009. Understanding the Warburg effect: the metabolic requirements of cell proliferation. Science 324, 1029–1033.
Varela, A.T., Simões, A.M., Teodoro, J.S., et al., 2010. Indirubin-3′-oxime prevents hepatic I/R damage by inhibiting GSK-3beta and mitochondrial permeability transition. Mitochondrion 10, 456–463.
Vial, G., Dubouchaud, H., Leverve, X.M., 2010. Liver mitochondria and insulin resistance. Acta Biochim. Pol. 57, 389–392.
Vosselman, M.J., der Lans van, A.A.J.J., Brans, B., et al., 2012. Systemic β-adrenergic stimulation of thermogenesis is not accompanied by brown adipose tissue activity in humans. Diabetes 61, 3106–3113.
Wallace, D.C., 2012. Mitochondria and cancer. Nat. Rev. Canc. 12, 685–698.
Wallace, D.C., Fan, W., 2010. Energetics, epigenetics, mitochondrial genetics. Mitochondrion 10, 12–31.
Wang, C.-Y., Zhao, Z.-B., 2012. Somatic mtDNA mutations in lung tissues of pesticide-exposed fruit growers. Toxicology 291, 51–55.
Ward, P.S., Cross, J.R., Lu, C., et al., 2012. Identification of additional IDH mutations associated with oncometabolite R(−)-2-hydroxyglutarate production. Oncogene 31, 2491–2498.
Weinberg, F., Hamanaka, R., Wheaton, W.W., et al., 2010. Mitochondrial metabolism and ROS generation are essential for Kras-mediated tumorigenicity. Proc. Natl. Acad. Sci. U.S.A. 107, 8788–8793.
Wenner, C.E., 2012. Targeting mitochondria as a therapeutic target in cancer. J. Cell. Physiol. 227, 450–456.
West, A.P., Shadel, G.S., Ghosh, S., 2011. Mitochondria in innate immune responses. Nat. Rev. Immunol. 11, 389–402.
Westermann, B., 2010. Mitochondrial fusion and fission in cell life and death. Nature 11, 872–884.
Whittle, A., 2012. Searching for ways to switch on brown fat: are we getting warmer? J. Mol. Endocrinol. 49, R79–R87.
Winder, W.W., Holmes, B.F., Rubink, D.S., et al., 2000. Activation of AMP-activated protein kinase increases mitochondrial enzymes in skeletal muscle. J. Appl. Physiol. 88, 2219–2226.
Wise, D.R., DeBerardinis, R.J., Mancuso, A., et al., 2008. Myc regulates a transcriptional program that stimulates mitochondrial glutaminolysis and leads to glutamine addiction. Proc. Natl. Acad. Sci. U.S.A. 105, 18782–18787. PMCID: PMC2596212.
Wu, J., Cohen, P., Spiegelman, B.M., 2013. Adaptive thermogenesis in adipocytes: is beige the new brown? Genes Dev. 27, 234–250.

Wu, Y.-T., Wu, S.-B., Lee, W.-Y., Wei, Y.-H., 2010a. Mitochondrial respiratory dysfunction-elicited oxidative stress and posttranslational protein modification in mitochondrial diseases. Ann. N.Y. Acad. Sci. 1201, 147–156.

Wu, Q., Tang, C., Zhang, Y.J., et al., 2010b. Diazoxide suppresses hepatic ischemia/reperfusion injury after mouse liver transplantation by a BCL-2-dependent mechanism. J. Surg. Res. 169, 155–166.

Yamamoto, H., Schoonjans, K., Auwerx, J., 2007. Sirtuin functions in health and disease. Mol. Endocrinol. 21, 1745–1755.

Zhang, H., Gao, P., Fukuda, R., et al., 2007. HIF-1 inhibits mitochondrial biogenesis and cellular respiration in VHL-deficient renal cell carcinoma by repression of C-MYC activity. Canc. Cell 11, 407–420.

Zhu, X., 2010. Mitochondrial dysfunction: mitochondrial diseases and pathways with a focus on neurodegeneration. Preface. Biochim Biophys Acta 1802, 1.

Zou, C., Kou, R., Gao, Y., et al., 2013. Activation of mitochondria-mediated apoptotic pathway in tri-ortho-cresyl phosphate-induced delayed neuropathy. Neurochem. Int. 67, 965–972.

Zou, J., Chen, Q., Jin, X., et al., 2011. Olaquindox induces apoptosis through the mitochondrial pathway in HepG2 cells. Toxicology 285, 104–113.

CHAPTER

56

Biomarkers of Blood–Brain Barrier Dysfunction

Rekha K. Gupta[1], *Ramesh C. Gupta*[2]

[1]School of Medicine, University of Louisville, Louisville, KY, United States; [2]Toxicology Department, Breathitt Veterinary Center, Murray State University, Hopkinsville, Kentucky, United States

INTRODUCTION

Out of all vital organs in the body, the central nervous system (CNS) is the most complex organ in terms of its structure and function. It is composed of certain interfaces, such as the blood–brain barrier (BBB), the blood–cerebrospinal fluid barrier (BCSFB), and the blood–spinal cord barrier. These can collectively be called blood–CNS barriers or brain barriers. Evidence for the existence of the BBB came from the observations (based on staining technique) of Nobel Laureate Paul Ehrlich in the early 20th century. In 1921, the findings of a Russian neurophysiologist, Lina Stern, further supported the existence of BBB and designated it as the "hematoencephalic" barrier. Finally, electron microscopic observations confirmed the presence of the BBB (tight junctions between the endothelial cells that form brain capillaries) (Abbot et al., 2006, 2010; Palmer, 2010). Deli et al. (2005) defined the BBB as a dynamic interaction between cerebral and endothelial cells constituting the anatomical basis of the BBB and other neighboring cells, such as astroglia, pericytes, perivascular microglia, and neurons. The cross talk between these cells endows endothelial cells with a unique BBB phenotype comprising not only the morphological barrier of endothelial tight junctions but also the biochemical (enzymatic and metabolic) barriers, as well as the uptake and efflux transport systems (Abbott et al., 2010).

The BBB is formed of extremely tight junctions offering a wide range of functions, such as transendothelial transport systems, enzymes, and regulations of leukocyte permeation, which thereby generates the transport, enzymatic, and immune regulatory functions of the BBB (Abbott and Friedman, 2012). Studies have also revealed important stages, cell types, and signaling pathways involved in BBB development.

Under physiological conditions, the BBB regulates the exchange of nutrients, waste, and immune cells between the blood and the nervous tissue of the CNS and is the most important component preserving CNS homeostasis and neuronal function (Abbott et al., 2010). The BBB also protects the brain from xenobiotics, pharmaceuticals, nutraceuticals, pathogens, and various cells, proteins, and neurotransmitters present in the blood. Dysfunctional blood–CNS barrier mechanisms contribute to the pathology of neurological conditions, ranging from trauma to neurodegenerative diseases (Erickson and Banks, 2013).

The brain barriers, particularly BBB, are very sensitive to the toxic insult of chemicals and biotoxins, and these barriers play pivotal roles in the initiation and progression of neurodegenerative diseases. It is noteworthy that unlike toxicants that compromise the integrity of BBB, compounds such as melatonin can promote BBB integrity via multiple molecular pathways (Jumnongprakhon et al., 2016). This chapter describes in brief the structure and function of brain barriers and the biomarkers of toxic effects of metals, pesticides, mycotoxins, drugs of abuse, and some diseases involving the CNS.

STRUCTURE AND FUNCTION OF BRAIN BARRIERS

The structure and function of brain barriers are very complex and maintain the neuronal microenvironment, playing pivotal roles in CNS homeostasis, fibrinolysis and coagulation, vasotonus regulation, and blood cell activation, and migration during physiological and pathological processes (Zheng et al., 2003). The BBB is also vital for protecting the CNS from systemic

Biomarkers in Toxicology, Second Edition
https://doi.org/10.1016/B978-0-12-814655-2.00056-6

perturbations and from elements of the peripheral immune system, so that any impairment of its integrity could have far-reaching consequences on the health of the CNS (Drouin-Ouellet et al., 2015). The BBB is formed by a complex cellular system of endothelial cells, astroglia, pericytes, perivascular macrophages, and a basal lamina (Bradbury, 1985; Abbott et al., 2006; Gupta et al., 2015). Endothelial cells regulate the selective transport and metabolism of substances from blood to brain as well as in the opposite direction from the parenchyma back to the systemic circulation. Astrocytes project their end feet tightly to the cerebral endothelial cells (CEC), influencing and conserving the barrier function of these cells (Abbott et al., 2006). The prominent role of astrocytes in supporting a healthy BBB may be the result of trophic factors secreted by astrocytes that nourish and regulate brain microvascular endothelial cells (BMVEC) (Ivey et al., 2009). CEC are embedded in the basal lamina together with pericytes and perivascular macrophages (De Vries et al., 1997; Abbott et al., 2006). Pericytes are characterized as contractile cells that surround the brain capillaries with long processes, and they play a role in controlling the growth of endothelial cells. Pericytes may influence the integrity of the capillaries and conserve the barrier function. The lumen of the cerebral capillaries is covered by CEC in which the functional and morphological basis of the BBB resides.

The BBB acts as a physical and metabolic barrier because a complex tight junction system between adjacent endothelial cells restricts most paracellular movement of ions and solutes across the brain endothelium (Partridge, 2002). Tight junctions of the BBB are composed of an intricate combination of at least three integral transmembrane proteins (claudins, junctional adhesion molecules (JAMs), and occludin), and cytoplasmic accessary proteins, such as zonula occludens (ZO-1, ZO-2, and AF6). Daneman et al. (2010) reported that a large number of transcripts are expressed and enriched at the BBB. These transcripts encode tight junction molecules, transporters, metabolic enzymes, signaling cascades, and proteins of unknown function. The best-characterized molecules at the BBB and tight junctions include claudin-5 and claudin-12, occludin, ZO-1 and ZO-2, marveld2, cingulin-like 1, jam4, and pard3. Also at the BBB, several signaling pathways have been identified and characterized, including serotonin receptor signaling, clathrin-mediated endocytosis signaling, dopamine receptor signaling, wnt/beta-catenin signaling, and several others.

In more than 99% of the brain capillaries, a BBB is present, but in some areas of the brain, a BCSFB can be found. This barrier is present in the circumventricular areas, such as the median eminence, pituitary, choroid plexus, subfornical organ, organum vasculosum of the lamina terminalis, and the area postrema. The BCSFB is not as strict as the BBB, but it does prevent the entrance of blood-borne compounds into the brain. Because the surface of the BBB is 5000-fold greater than that of the BCSFB, the main route of entry for compounds from plasma into the brain is via the brain capillaries (Partridge, 1986). Major tight junction proteins include occludin and claudin. Claudin-5 is a major functional constituent and a critical determinant of BBB paracellular permeability and charge-selective hydrophilic paracellular pores, whereas occludin enhances tight junction tightness. Recently, it has been reported that endophilin-1 regulates the expression of ZO-1 and occludin via the EGFR-JNK signaling pathway (Chen et al., 2015).

Several enzymes (monoamine oxidase A and B, catechol *O*-methyltransferase, epoxy hydrolase, endopeptidases, acetylcholinesterase (AChE), pseudocholinesterase, dopa decarboxylase, γ-glutamyl transpeptidase, etc.) present in endothelial cells are important elements of the BBB, constituting the so-called metabolic barrier. The role of the BBB as a metabolic barrier was further substantiated by the presence of mitochondria in CECs (Fenstermacher, 1989). Thus, the BBB is considered a physical as well as a metabolic barrier.

The brain barriers have been assigned many vital functions, and one of them is to provide required nutrients to the CNS. These essential nutrients are transported into the brain by means of selective/carrier mechanisms. Several transport systems have been characterized varying from passive transport (such as diffusion) to active and energy requiring processes (De Vries et al., 1997).

The diffusion of compounds across the plasma membranes of the endothelial cells of the BBB is dependent on the physicochemical properties, such as lipophilicity, molecular weight, electrical charge, and extent of ionization. Lipid-soluble compounds penetrate the brain barriers readily and equilibrate easily between blood and brain tissue. Many transporters are expressed at the BBB (Garrick et al., 2003; Zlokovic, 2008) and BCSFB (Herbert et al., 1986; Ghersi-Egea and Strazielle, 2002). Compounds that are a substrate for P-glycoprotein (Pgp) are less efficiently transported across the BBB. Specific carrier systems have been characterized: glucose transporter system (GLUT-1) for sugars and large neutral amino acids system for amino acids, in addition to carrier system for purine, nucleoside, thiamine, monocarboxylic acid, and thyroid hormone. The other two carrier transporters in the human BBB are ATP-binding cassette (ABC) transporters and solute carrier transporters (Geier et al., 2013).

Some transporters that are involved in drug distribution in the brain are also involved in drug efflux. The ABC efflux transporter Pgp has been demonstrated as a key element of the BBB that can actively transport a huge variety of lipophilic drugs out of the brain capillary endothelial cells (Löscher and Potschka, 2005). Drees et al. (2005) and Löscher and Potschka (2005) reported that the multidrug resistance proteins BMDP/ABCG2 and BCRP are also involved in efflux pump at the BBB.

In Vivo and *In Vitro* Models to Study the Blood–Brain Barrier

The movement of compounds from the circulating blood into the brain is strictly regulated by the brain capillary endothelial cells, which constitute the BBB. The importance of the BBB is not only in the passage of nutrients and toxicants but therapeutic drugs and antidotes as well. So, to understand the crossing of compounds of various classes, drugs, and antidotes, various in vivo and in vitro models have been developed (reviewed in Gupta et al., 2015).

In Vivo Model

Birngruber et al. (2013) reported a cerebral open flow microperfusion as a new membrane-free technique for measuring substance transport across the intact BBB in Sprague–Dawley rats. This in vivo technique is based on a probe that is inserted into the brain, thereby rupturing the BBB. The BBB is usually reestablished within 15 days, which then allows sampling of interstitial brain fluid under physiological conditions. This technique also allows monitoring of BBB permeability, which can be useful for measuring pharmacokinetics across the BBB and pharmacodynamics in the brain. Using tracers, such as Evans blue (EB), horseradish peroxidase, and [^{131}I]albumin, breakdown of the BBB in humans and experimental animals has been studied under many conditions, such as hypoglycemia, hypertension, seizures/convulsions, inflammation, etc. (Öztaş, 1996).

In Vitro Models

In vitro reconstituted models of the BBB from different mammalian species have been employed since the late 1970s. Bowman et al. (1983) introduced the first in vitro BBB filter model. An insert was made of nylon mesh and polycarbonate tubing, and bovine brain endothelial cells were seated on it for studying the effect of calcium-free medium and osmotic shock on sucrose flux. Since then, a variety of chambers and inserts from different materials and with diverse pore size have become commercially available. Garberg et al. (2005) used an in vitro model for BBB permeability based on the use of a continuous cell line and to investigate the specificity of this model. These authors developed a coculture procedure that mimics the in vivo situation by culturing brain capillary endothelial cells on one side of a filter and astrocytes on the other. Under these conditions, endothelial cells retain all the endothelial cell markers and the characteristics of the BBB, including tight junctions and enzymes (such as γ-glutamyl transpeptidase and monoamine oxidase (MAO)) activities.

Deli et al. (2005) presented permeability data from various in vitro BBB models by measuring transendothelial electrical resistance (TEER) and by calculation of permeability coefficients for paracellular or transendothelial tracers. These authors summarized the results of primary cultures of cerebral microvascular endothelial cells or immortalized cell lines from bovine, human, porcine, and rodent origin. They also described the effect of coculture with astroglia, neurons, mesenchymal cells, blood cells, and conditioned media, as well as the physiological influence of serum components, hormones, growth factors, lipids, and lipoproteins on the BBB function.

The strong correlation between the in vivo (Oldenhorf method) and in vitro (coculture) drug transport, the relative ease with which such cocultures can be produced in large quantities, and the reproducibility of the system, provide evidence for an efficient system for the screening of drugs that are active in the CNS (Dehouck et al., 1997). These authors suggested that the coculture method is a useful system for investigating passive diffusion, carrier-mediated transport, and P-glycoprotein (Pgp)–dependent drug transport. The in vitro permeabilities of propranolol and cyclosporine A were parallel with indications from in vivo extraction, showing that transporters and Pgp are expressed in the coculture system. For further details, readers are referred to Deli et al. (2005) who reviewed various in vitro models covering bovine, human, porcine, and rodent (murine and rat) brain endothelial cell–based systems. Bovine systems provide a high yield of brain endothelial cells sufficient for pharmacological screening, and they are widely used in basic as well as in applied research. Mouse brain yields the least endothelial cells compared to other species.

Some examples of the modulators of BBB permeability in *in vitro* models are (1) both cAMP elevator peptide hormone adrenomedulin and calcitonin gene-related peptide decrease paracellular permeability, (2) a glucocorticoid hormone, hydrocortisone, improves the barrier properties, (3) insulin exerts a tightening effect on tight junctions, and (4) catecholamines (adrenaline and noradrenaline) increase the sodium fluorescein flux.

In essence, in vitro models have been widely used in pharmacological research for screening drugs and drug candidate molecules for either modifying BBB permeability or investigating brain penetration (Deli et al., 2005). This area of research is very important for permeability screening during drug development in the pharmaceutical industry (Gerlach et al., 2018).

Toxicants Affecting the Central Nervous System Barriers

Metals

A number of metals (aluminum, copper, iron, lead, manganese, and mercury) are known to modulate directly or indirectly the structure, function, or permeability of the BBB and/or BCSFB. Effects of these metals on brain barriers and the brain are described here in brief (Table 56.1).

Aluminum

Aluminum (Al)-induced neurotoxicity has been known for more than a century. Al has also been involved in neurodegenerative diseases, such as Alzheimer's (AD), encephalopathy, and amyotrophic lateral sclerosis (ALS) (Delonche and Pages, 1997; Savory et al., 2006; Bondy, 2010; Garcia et al., 2010). In a pharmacokinetic study, Yokel (2012) mentioned that Al can enter the brain from blood by two routes: (1) through the BBB and (2) through the choroid plexus into the CSF of the ventricles within the brain and then into the brain. Al appears to bind with transferrin and citrate. Al concentration in serum can be 25-fold higher than in CSF. The glutamate transporter system Xc^- has been suggested to mediate brain Al citrate uptake. Al mainly deposits in the hippocampus, cortex, and amygdala, which are the areas of brain that are also rich in glutamatergic neurons and transferrin receptors.

Al is known to associate with many epithelia and endothelia in the BBB and may be responsible for compromising the properties and integrity of these membranes, thereby altering the barrier function (Vorbrodt et al., 1994). Al-related toxicological effects are observed on both sides of the BBB (Delonche and Pages, 1997). Al is known to cause development of neurofibrillary tangles and degeneration of cerebral neurons in laboratory animals (Zheng, 2001). Al directly affects neuronal pathways and appears to act as a direct BBB toxicant. Accumulation of Al in lysosomes (protease-rich vacuoles) and the hyperphosphorylation of neurofilaments are involved in the molecular mechanism of Al-induced neurotoxicity and neurodegenerative disease, such as AD.

Copper and Iron

Copper (Cu) is an essential element and involved in many vital biochemical reactions and physiological functions, such as in Cu-containing enzymes, angiogenesis, nerve myelination, and endorphin action (Choi and Zheng, 2009; Pal, 2014). Cu access in the body, due to mutation in the Cu-transporter ATP7B gene located on the long arm (q) of chromosome 13 (13Q 14.3), may lead to the neurologic disorder known as Wilson's disease. In contrast, Cu deficiency in liver and brain, due to a genetic disorder in the expression of the Cu-transporter ATP7A, may lead to Menkes disease. An imbalance in Cu homeostasis due to defective Cu transporters or any other reasons at the BBB may play a role in the pathogenesis of neurodegenerative diseases, such as AD, Parkinson's (PD), spongiform encephalopathies, and ALS.

In serum, about 65%–90% of Cu tightly binds with ceruloplasmin, and the rest loosely binds with albumin, transcuprein, and amino acids. It appears that ceruloplasmin sequesters Cu and thus tightly regulates the movement of Cu into the CSF (Choi and Zheng, 2009). Free Cu ions are the main species for Cu transport into the brain, and uptake and distribution of Cu varies between different brain regions. The BBB appears to serve as the main entrance for Cu to get access to brain parenchyma, whereas the BCSFB is more likely involved in the regulation of Cu homeostasis in the CSF. Uptake of Cu into the cells is mediated by two transporter proteins (Cu transporter 1 (Ctr1) and divalent metal transporter 1 (DMT1)). Inside the cells, ATP7A and ATP7B are Cu transport proteins that participate in Cu efflux (Choi and Zheng, 2009; Pal, 2014). Monnot et al. (2011) reported that Cu appears to enter the brain primarily via the BBB and is subsequently removed from the CSF by the BCSFB.

Iron (Fe) is a trace mineral that acts as both an electron acceptor and electron donor, and thereby it is essential to life (McCarthy and Kosman, 2015). Fe enters the brain by crossing the BBB and BCSFB. Fe is essential for normal brain function, as it is involved as a cofactor in myelination, mitochondrial energy generation, neurotransmission, oxygen transport, and cellular division. Fe is also involved in several neurological diseases, such as AD, PD, and aceruloplasminemia. There are two possible mechanisms for Fe uptake into BMVEC: one, which is referred to as transferrin-bound iron uptake involving transferrin (Tf) endocytosis, and the other is uptake of Fe from nontransferrin-bound iron. The latter process involves an Fe transporter, which is a divalent cation transporter such as DMT1.

High intracellular Fe concentration is toxic to cells. The endothelial cells of the BBB sequester Fe from the blood at their apical surface and release Fe into the brain at their basolateral surface. Ferroportin is the only known mammalian cellular Fe exporter.

Monnot et al. (2011) investigated how Fe levels regulated brain Cu levels transport through BBB and BCSFB

TABLE 56.1 Toxicants That Can Cross the Brain Barriers May Cause Dysfunction/Damage to Barriers

Toxicants		Biomarkers	References
Metals	Aluminum (Al)	Decreased serum inorganic phosphorus, neurofibrillary tangles development, cerebral neurons degeneration, neuropeptide delta sleep-inducing peptide (DSIP), transferrin, disruption of BBB, and increased BBB permeability	Vorbrodt et al. (1994), Zheng (2001), and Yokel RA (2012)
	Copper (Cu)	High Cu in Serum/plasma and urine, ceruloplasmin, transporters in BBB and BCSFB, and Cu chaperone for SOD protein	Zheng et al. (2003), Choi and Zheng (2009), and Pal (2014)
	Iron (Fe)	Ferritin, transferrin, amyloid-β precursor protein, increased amyloid-β aggregates, ferroportin, ceruloplasmin, zyklopen, and hephaestin	Zheng et al. (2003) and McCarthy and Kosman (2015)
	Lead (Pb)	Increased BBB permeability, increased $A\beta_{1-40}$ in choroid plexus, protein kinase C (PKC-zeta), reduction of tight junction protein (occludin), and tyrosine kinase Src, and reduction of transthyretin in CSF	Laterra et al. (1992), Bradbury and Deane (1993), Zhao et al. (1998), Zheng et al. (2003), Behl et al. (2009), Song et al. (2014), and Gupta et al. (2015)
	Manganese (Mn)	Blood and saliva Mn levels, and transferrin receptor in BBB and BCSFB	Aschner and Gannon (1994), Zheng (2001), Crossgove et al. (2003), Zheng et al. (2003), Milatovic et al. (2011), and Ge et al. (2018)
	Mercury (Hg)	Hg concentration, and BBB disruption	Chang and Hartman (1972), Aschner and Clarkson (1988), and Zheng et al. (2003)
	Zinc (Zn)	Increased BBB permeability, BBB breakdown, increased Zn and Na concentrations, Cu, Fe, and Mg deficiency in brain, inhibition of Na/K-ATPase	Yorulmaz et al. (2013) and Giacconi et al. (2017)
Pesticides	Organophosphate pesticides and nerve agents	OP residue, increased BBB permeability, AChE inhibition, BuChE inhibition, increased ACh level, seizures, BBB dysfunction, edema, and neuronal damage/loss	Ashani and Catravas (1981), Drewes and Singh (1985), Petrali et al. (1985), Carpentier et al. (1990), Sinha and Shukla (2003), Bhavari and Reddy (2005), Zaja-Milatovic et al. (2009), Mercey et al. (2012), and Martin-Reina et al. (2017)
	Carbamates(including pyridostigmine bromide)	Carbamate residue, AChE inhibition, BuChE inhibition, increased ACh level, seizures, and increased BBB permeability, and neuronal damage/loss	Ropp et al. (2008), Gupta et al. (2007), Amourette et al. (2009), and Martin-Reina et al. (2017)
	Chlorinated hydrocarbons	Increased BBB permeability	Sinha et al. (2003) and Malik et al. (2017)
	Pyrethrins/pyrethroids	Increased BBB permeability, BBB dysfunction, and pyrethrin/pyrethroid residue	Sinha et al. (2004), Amaraneni et al. (2016), and Martin-Reina et al. (2017)
	Neonicotinoids (Imidacloprid)	Imidacloprid residue	Rose (2012)
	Nicotine	Nicotine residue	Rose (2012)
	Rotenone	Mitochondrial complex I inhibition, autoradiographic findings, mitochondrial dysfunction, decreased ATP level, degeneration of dopaminergic neurons, and Parkinsonian motor deficits	Higgins and Greenamyre (1996), Gupta (2012), Heinz et al. (2017), and Terron et al. (2018)
	Herbicides (paraquat)	Paraquat level, reduced tyrosine hydroxylase, and dopamine level	Shimizu et al. (2001) and Prasad et al. (2009)
	PCBs/Dioxin/TCDD		Eum et al. (2008)

Continued

TABLE 56.1 Toxicants That Can Cross the Brain Barriers May Cause Dysfunction/Damage to Barriers—cont'd

Toxicants		Biomarkers	References
Mycotoxins/ Biotoxins/Endotoxin	Aflatoxins	Reduced tight junction proteins (ZO-1, ZO-2, and AF6) AFB1 concentration in brain, AFB1-DNA (AFB1-N7-guanine) adducts, and HBMEC death	Qureshi et al. (2015)
	Deoxynivalenol (vomitoxin)	BBB and BCSFB impairment, reduced BBB integrity, increased BBB permeability, reduced TEER, increased MAPK, decreased claudin-3 and claudin-4, and increased lactate dehydrogenase release	Prelusky et al. (1990), Maresca (2013), and Behrens et al. (2015)
	T-2 and HT-2 toxins	Increased BBB permeability, reduced protein synthesis, decreased MAO activity, and enhanced MMP-9	Wang et al. (1998), Ravindran et al. (2011), and Weidner et al. (2013)
	Fumonisins/moniliformin	Increased BBB permeability, fumonisin B1 level, sphingosine: sphingosine ratio, cerebral edema, and increased albumin and IgG in CSF	Kwon et al. (1997), Foreman et al. (2004), Osuchowski et al. (2005), Behrens et al. (2015), and Smith (2018)
	Ergot (ergotamine, ergocristinine, and others)	Ergot residue in CSF and reduced BBB integrity	Hovdal et al. (1982), Mulac et al. (2012), and Behrens et al. (2015)
	Pertussis toxin	Increased BBB permeability	Brückener et al. (2003)
	Endotoxin	Increased BBB permeability	Osuchowski et al. (2005) and De Vries et al. (1997)
Experimental chemicals/drugs	Lipopolysaccharide (LPS)	Increased BBB permeability, increased nitric oxide and prostaglandins, and increased α-synuclein expression	Wispelwey et al. (1988), Minami et al. (1998), Osuchowski et al. (2005), and Jangula and Murphy (2013)
	Monosodium glutamate	Increased BBB permeability	Škultétyová et al. (1998)
	N-Methyl D-aspartate (NMDA)	BBB breakdown	Zheng (2001)
	Kainic acid (Kainate)	Kainic acid and dihydrokainic acid residue in brain, BBB dysfunction, increased BBB permeability, and brain edema	Gynther et al. (2015) and Han et al. (2015)
	Pentylenetetrazole	BBB breakdown and increased BBB permeability	Sahin et al. (2003) and Oztas et al. (2007)
	Pilocarpine	Brain and plasma concentrations of pilocarpine and increased BBB permeability	Uva et al. (2008) and Römermann et al. (2015)

AFB1, aflatoxin B1; *HBMVEC*, human brain microvascular endothelial cells; *MAO*, monoamine oxidase; *MAPK*, mitogen-activated protein kinase; *MMP*, matrix metalloproteinase; *TEER*, transendothelial electrical resistance; *ZO*, zonula occludens.

and Cu homeostasis. Fe deficiency has a more profound effect on brain Cu levels than Fe overload. Fe deficiency increases Cu transport at the brain barriers and prompts Cu overload in the CNS. The BCSFB plays a key role in removing excess Cu from the CSF.

The potential biomarkers of Fe exposure and effects on brain, such a ferritin, transferrin, amyloid-β precursor protein, increased amyloid-β aggregates, ferroportin, ceruloplasmin, zyklopen, hephaestin, are listed in Table 56.1.

Lead

Lead (Pb), a well-known neurotoxicant, can enter the CNS either as free Pb^{2+} or via the exchange of $PbCO_3$ with an anion, or passively in the form of an inorganic complex, such as $PbOH^+$. Kinetic studies with ^{203}Pb continuously infused intravenously into adult rats revealed that ^{203}Pb uptake into different brain regions was linear with time up to 4 h after infusion (Bradbury and Deane, 1993). Pb accumulates in the choroid plexus of humans as well as animals, and the choroid plexus may be the primary target for Pb-induced neurotoxicity. This effect may alter the function of BCSFB. Behl et al. (2009) demonstrated that Pb significantly increased accumulation of intracellular amyloid-β ($Aβ_{1-40}$) in rat choroid plexus tissues in vivo and in immortalized choroidal epithelial Z310 cells in vitro. Mechanisms involved in lead-induced increase in Aβ level at the BCSFB may include (1) a diminished expulsion of Aβ molecules from the plexus cells to the extracellular milieu, (2) an increased uptake of Aβ from the CSF, blood, or both, (3) an increased synthesis of Aβ, and (4) a reduced metabolism or degradation of Aβ. Pb-induced inhibition of the production of LRP1 (a key intracellular Aβ transport protein in the choroid plexus) may be responsible for the accumulation of Aβ, a major risk factor for AD (reviewed in Gupta et al., 2015). Pb exposure has also been shown to cause abnormal protein kinase C (PKC-zeta) activity in both the BBB (Laterra et al., 1992) and the BCSFB (Zhao et al., 1998). In addition, Pb appears to accumulate in the same intramitochondrial compartment as Ca^{2+}, thereby disrupting intracellular Ca^{2+} metabolism as well as altering transepithelial transport processes (Zheng et al., 2003).

Manganese

Manganese (Mn) binds readily to transferrin without displacing iron (Fe) in plasma. Brain areas (pallidum, thalamic nuclei, and substantia nigra) having high Mn levels differ from those having high levels of transferrin receptors (nucleus accumbens and caudate putamen), suggesting that perhaps these sites may accumulate Mn through neuronal transport. Like Fe, Mn-loaded transferrin is taken up by receptor-mediated endocytosis at the luminal membrane of brain capillaries. Mn levels, transferrin in BBB and BCSFB, and neurodegeneration and dendritic damage may serve as biomarkers (Aschner and Gannon, 1994; Crossgove et al., 2003; Zheng et al., 2003; Milatovic et al., 2011).

Mercury

Mercury (Hg) exists in several forms, such as elemental (metallic), inorganic (e.g., mercurous chloride and mercuric chloride), and organic (e.g., methylmercury, ethylmercury, and phenylmercury). In the environment and mammalian systems, various forms of mercury are interchangeable. For example, inorganic mercury can be methylated to methylmercury (MeHg) and MeHg can change to inorganic or elemental Hg. Animals at the top of the food chain tend to bioaccumulated MeHg in their bodies. Therefore, poisoning by Hg is due to consumption of meat or grain contaminated with Hg. The US Food and Drug Administration estimates that, on an average, most people are exposed to about 50 ng Hg/kg body wt/day in the food they eat.

Metallic Hg can stay in the body for weeks to months. Because of its high lipophilicity, it can readily cross the BBB. When metallic Hg enters the brain, it is readily converted to an inorganic divalent Hg and is trapped there for an extended period. The inorganic divalent cation can, in turn, be reduced to metallic Hg. Inorganic Hg compounds do not readily cross the BBB. Yet, compounds such as mercuric chloride can act as direct BBB toxicants (Chang and Hartman, 1972; Zheng et al., 2003). The distribution of MeHg is similar to that of metallic Hg, i.e., a relatively large amount of Hg can accumulate in the brain because of its ability to penetrate the BBB either by diffusion or by utilizing a transport system. MeHg reacts with sulfhydryl groups of cysteine. The MeHg–cysteine complex acts as an amino acid analog, similar in structure to methionine, and is transported by the L system carrier across the BBB (Aschner and Clarkson, 1988). Developing fetuses and neonates are most sensitive to MeHg-induced neurotoxicity because MeHg is more readily transported across the immature BBB as well as because of its inhibitory effects on cell division. Microtubules are essential for cell division (main component of the mitotic spindle), and MeHg reacts with the SH groups of tubulin monomers and thereby disrupts the assembly process. The dissociation process continues, and this leads to depolymerization of the tubule.

It is noteworthy that Hg can be involved in the development of AD, multiple sclerosis (MS), and ALS (Mutter et al., 2007).

Zinc

Zn is a component of more than 300 different enzymes that function in many aspects of cellular metabolism, including metabolism of proteins, lipids, and carbohydrates. The BBB is important for zinc (Zn) homeostasis in the brain (Yorulmaz et al., 2013). Zn serves as a mediator of cell–cell signaling and functioning of channels and receptors in the CNS. The transport of Zn into the brain parenchyma occurs via the brain barrier system. Its deficiency or excess can contribute to alterations in behavior, abnormal CNS development, and neurological diseases. In a recent study, circulating Zn has been linked to pathophysiological changes occurring with aging rather than to its nutritional intake (Giacconi et al., 2017). Increased Zn and Na concentrations, inhibition of Na/K-ATPase, increased BBB permeability, and BBB breakdown may serve as biomarkers.

Pesticides

Organophosphates

Several organophosphate (OP) compounds have been reported to cross the BBB, but very little is known about alterations in the BBB structure or function, except that these compounds can increase its permeability. OP nerve agents, which are irreversible AChE inhibitors, have been studied to a greater extent than pesticides with regard to changes in the BBB. Drewes and Singh (1985) reported that the cerebral transendothelial carrier-mediated transport of glucose and amino acids was not affected in mongrel dogs poisoned by soman. Carpentier et al. (1990) investigated acute changes in BBB permeability to proteins, using EB-labeled serum albumin and plasmatic gamma-immunoglobulin G (IgG) as indicators in rats. An increased BBB permeability to the EB–albumin complex was macroscopically observed in two-thirds of the convulsive rats exposed to soman (85 μg/kg). Soman produced seizures and reversible BBB opening to the greatest extent after 30–60 min of paroxysmal electroencephalographic (EEG) discharges when signs of cerebral hyperactivity (epileptic EEG pattern, and hyperoxia) were also at their peak. Topographically, the protein leakage was bilateral and restricted to anatomically defined brain structures, some of which were thereafter sites of parenchymal edema and neuronal damage. Interestingly, the first signs of increased BBB permeability were shown to precede the onset of edema. OP nerve agents, being small lipophilic molecules, can easily penetrate the BBB by free diffusion and thereby inhibit AChE in the CNS (Mercey et al., 2012). Increased BBB permeability by OP nerve agents or other ChE inhibitors may lead to their enhanced entry into the brain resulting in greater AChE inhibition and possibly in subsequent maintenance of seizures and aggravation of their pathological consequences, such as edema and neuronal loss in certain brain structures. Carpentier et al. (1990) detected the BBB opening in the amygdaloid complex and in some cortical regions (cingulum, entorhinal, and piriform complex), and that the thalamus was the most frequently and intensely affected structure.

Observations from various studies suggest that soman-induced brain alterations are predominantly related to seizures or brain hyperactivity, or to a direct cytotoxic action of soman or acetylcholine (ACh) itself or due to ChE inhibition (McDonough et al., 1987). Domer at al. (1983) found increased permeability of the BBB by systemic administration of ACh. Of course brain hyperactivity alone appears inadequate to be responsible for the BBB opening. Obviously, the short duration of the transient protein leakage (Carpentier et al., 1990) contrasted with the well-known long-lasting brain AChE inhibition induced by soman (Petrali et al., 1985). Several other anti-ChE compounds, such as physostigmine and paraoxon, are also known to cause the BBB opening for macromolecules that were seizure-dependent and reversible and unrelated to brain ChE inhibition. Ashani and Catravas (1981) observed that in soman-intoxicated rats, induced damage to BBB integrity was significantly reduced, despite a high degree of AChE and BChE inhibition, and protected from seizures by Nembutal or atropine. In essence, endothelial AChE or BChE play no role in BBB opening, although they may function as an "enzymatic barrier" to ACh.

Additional factors, such as increased electrical activity, oxidative/nitrosative stress, decreased energy supply and store, deleterious action of excitatory amino acids, enhanced calcium intrusion, and brain edema, seemed to play significant roles in the brain damage (Misulis et al., 1987; Carpentier et al., 1990; Solberg and Belkin, 1997; Gupta et al., 2001a,b; Zaja-Milatovic et al., 2009; Prager et al., 2013). Other mechanisms in soman-induced damage to BBB integrity may be related to vasoactive substances (ACh, amines, amino acids, peptides, free radicals, and steroid hormones of the pituitary adrenal axis) and vasogenic events (acidosis, increased blood flow, and hypertension). It is noteworthy that increased BBB permeability may facilitate the entry of an antidote (oxime class) to the brain, which otherwise has limited access because of the BBB (Gupta et al., 2015).

Organochlorines and Pyrethrins/Pyrethroids

In a number of studies, increased BBB permeability has been reported for organochlorine (Sinha and Shukla, 2003; Malik et al., 2017) and pyrethrin/pyrethroid (Sinha et al., 2004; Amaraneni et al., 2016; Martin-Reina et al., 2017) insecticides.

Rotenone

Rotenone is one of the naturally occurring insecticides obtained from plants of the "*Derris*" species (Gupta, 2012). Rotenone causes inhibition of mitochondrial respiratory chain complex I, (NADH-ubiquinone oxidoreductase) and cell death by apoptosis due to excess generation of free radicals. Rotenone has received enormous attention from neuroscientists because it causes inhibition of mitochondrial respiratory chain complex I, followed by mitochondrial dysfunction, impaired proteostasis, degeneration of dopaminergic neurons, neuroinflammation, and finally Parkinsonian motor deficits (Higgins and Greenamyre, 1996; Heinz et al., 2017; Terron et al., 2018). Mitochondrial complex I inhibition, mitochondrial dysfunction, reduced ATP level, degeneration of dopaminergic neurons, and Parkinsonian motor deficits may serve as biomarkers of rotenone's neurotoxicity.

Herbicides

A bipyridyl herbicide paraquat resembles the structure of *N*-methyl-4-phenyl pyridinium (MPP^+). MPP^+ is an experimental agent, which causes PD. Paraquat reaches the brain by crossing the BBB possibly by the neutral amino acid transport system (Shimizu et al., 2001; Prasad et al., 2009). Further, paraquat is taken up into striatal neurons by a Na^+-dependent mechanism. Paraquat inhibits the activity of tyrosine hydroxylase and decreases striatal dopamine levels, thereby inducing Parkinson-like dopaminergic toxicity in the brain.

Mycotoxins

In general, the effect of mycotoxins on the BBB is to increase its permeability. Aflatoxin B_1 (AFB_1) has been reported to form AFB_1-DNA (AFB_1-N7-guanine) adducts and kill human BMVECs (Qureshi et al., 2015). One of the trichothecenes deoxynivalenol is reported to cause BBB and BCSFB impairment, reduced BBB integrity, increased BBB permeability, reduced TEER, increased MAPK, decreased claudin-3 and claudin-4 (proteins involved in tight junctions), and increased lactate dehydrogenase release (Prelusky et al., 1990; Maresca, 2013; Behrens et al., 2015). T-2 and HT-2 toxins are known to cause increased BBB permeability, reduced protein synthesis, enhanced matrix metalloproteinase-9 (MMP-9), and alterations in neurochemicals by inhibiting MAO activity (Wang et al., 1998; Ravindran et al., 2011; Weidner et al., 2013). Fumonisin B1 commonly affects the equine species causing equine leukoencephalomalacia, cerebral edema, and an elevation of the albumin and IgG levels in CSF (Foreman et al., 2004; Smith, 2018). Ergot alkaloids (such as ergotamine, ergocristinine, and others) are known to cross the BBB by BCRP/ABCG2 transporter system, accumulate in BBB cells, and impair its integrity (Hovdal et al., 1982; Mulac et al., 2012; Behrens et al., 2015).

Drugs of Abuse and Therapeutic Drugs

Methamphetamine

The BBB has been identified as a target of methamphetamine neurotoxicity (Northrop and Yamamoto, 2015). Methamphetamine causes damage to monoamine terminals and BBB structural proteins, the tight junction proteins, and BBB function. In experimental animals, methamphetamine has been shown to cause decreased expression of occludin, claudin-5, and zona occludens, as well as decreased TEER, indicative of increased paracellular permeability (Mahajan et al., 2008). These alterations in endothelial structures and function are associated with an increase in ROS. In addition, stress appears to exacerbate methamphetamine-induced increases in ROS, induced BBB disruption by potentiating methamphetamine-induced increases in ROS, inflammatory mediators, mitochondrial dysfunction, and extracellular glutamate. The consequences of methamphetamine-induced BBB disruption are significantly enhanced vulnerability to other disease states, such as stress, HIV infection, Hepatitis C, cognitive decline and depression, etc. (Northrop and Yamamoto, 2015).

Cocaine

Depending on the dose and route of administration, cocaine-induced profound hyperthermia and increased plasma and brain serotonin levels lead to BBB breakdown and brain edema (Sharma et al., 2009; Multani et al., 2014). Cocaine also induces cellular stress as seen by upregulation of heat shock protein (HSP 72 kD) expression and results in marked neuronal and glial cell damages at the time of the BBB dysfunction. Increased serotonin levels, hyperthermia, HSP 72 kD expression, and BBB dysfunction can be used as biomarkers of cocaine-induced damage to the BBB (Table 56.2).

TABLE 56.2 Drugs of Abuse That Can Cross the Barriers and May Cause Dysfunction/Damage to the Barriers

Drugs of Abuse	Biomarkers	References
Amphetamine and methamphetamine	Disruption of BBB function, increased BBB permeability, decreased BBB structural proteins (occludin, claudin-5, and zona occludens-1), increased matrix metalloproteinase-9 and nitric oxide, decreased TEER, and increased ammonia and glutamate levels	Mahajan et al. (2008), Silva et al. (2010), Martins et al. (2013, 2017), Multani et al. (2014), Northrop and Yamamoto (2015), and Jumnongprakhon et al. (2016)
MPTP/MPP	Mitochondrial complex I inhibition, mitochondrial dysfunction, reduced ATP level, nigrostriatal degeneration, and Parkinsonian motor deficits	Schildknecht et al. (2017) and Terron et al. (2018)
Cocaine	Increased plasma/serum and brain serotonin levels, increased HSP 72 kD, hyperthermia, increased BBB permeability, BBB dysfunction, and brain edema	Sharma et al. (2009) and Multani et al. (2014)
Alcohol/Ethanol	Myosin light chain kinase activation, decreased occludin, claudin-5, and zona occludens-1, and increased BBR permeability	Haorah et al. (2005), Multani et al. (2014), and Northrop and Yamamoto (2015)
Antipsychotics (chlorpromazine and clozapine)	Increased electrical resistance, apoptosis of BBB endothelia, and impaired BBB permeability	Elmorsy et al. (2014)

Alcohol/Ethanol

Ethanol produces activation of myosin light chain kinase and consequent phosphorylation and degradation of tight junction proteins, such as occludin, claudin-5, and ZO-1 (Northrop and Yamamoto, 2015). A decrease in BBB function evidenced by a decrease in TEER and an increase in monocyte migration are also observed in brain microvascular endothelial cells. It has been suggested that the effects of ethanol on tight junction proteins and barrier function are associated with increases in ROS, resulting from ethanol metabolism and increased calcium concentrations. Ethanol in combination with methamphetamine could potentiate methamphetamine-induced BBB disruption through its proinflammatory and prooxidant effects. Biomarkers of ethanol-induced effects on BBB are summarized in Table 56.2.

For details on biomarkers of drugs of abuse, readers are referred to Chapter 52 of this book.

Neurodegenerative Diseases and Other Conditions

Neurodegenerative Diseases

In a number of CNS diseases, including neurodegenerative diseases (AD, PD, MS, ALS, and Huntington), the structure and function of BBB and/or BCSFB are compromised. The BBB itself may play an active role in the mediation of the neuroimmune response either by production of inflammatory mediators or by the expression of adhesion molecules. In neurological diseases, various inflammatory mediators, such as cytokines (TNF, IL-1β, and IL-6), prostaglandins (PGD_2, PGE_2, $PGF_{1\alpha}$, and $PGF_{2\alpha}$), thromboxane A_2, and leukotrienes, can be used as biomarkers of inflammation. The common adhesion molecules include (1) intercellular adhesion molecule-1, (2) intercellular adhesion molecule-2, and (3) vascular cell adhesion molecule-1; alterations in these parameters can be used as biomarkers.

Brain barriers possibly contribute to the etiology of AD by three aspects: (1) the aging of cerebral vascular structure in the overall aging process of the brain, (2) as the site of transport of extracerebral Aβ into the brain, and (3) the ability to prevent aggregation by producing transthyretin (Zheng, 2001). The neuropathological hallmark of AD is the aggregation of Aβ peptide in senile plaques and neurofibrillary tangles and is often used as a biomarker of AD. Detailed biomarkers of AD are summarized in Table 56.3 and described in Chapter 50 of this book.

The quantification of α-synuclein in CSF has been proposed as a diagnostic biomarker for PD and other α-synuclein–related diseases, such as multiple system atrophy and dementia with Lewy bodies (Mollenhauer, 2014). Biomarkers of PD are described in detail in Chapter-51 of this book.

BBB disruption is one of the hallmarks of MS. Development of sensitive and specific biomarkers for MS has been a challenge as they could (1) improve the monitoring of disease activity, (2) improve the monitoring of response to MS therapies that target BBB disruption, and (3) advance our understanding of dynamic MS processes participating in BBB disruption (Waubant, 2006). The presence of gadolinium-enhancing lesions on serial brain MRI scans is frequently used to evaluate BBB disruption. In MS, cerebral infections, hypertension, or seizures, enhanced permeability of BBB and is considered to be the result of either opening of tight junctions or of enhanced pinocytotic activity and the formation of transendothelial channels (De Vries et al., 1997).

Several studies for ALS biomarkers have been conducted, while others are ongoing (Su et al., 2013). It has been demonstrated that protein biomarkers in both plasma and CSF may aid the diagnosis and prognosis of patients with ALS (Mitchell et al., 2009, 2010). In CSF, various classes of biomarkers have been investigated with special attention to neurofilament protein, *tau* protein, S100-β, and cystatin C. Cytoskeletal protein *tau* can be measured in CSF and blood and used as a prognostic biomarker of ALS. Su et al. (2013) measured plasma and CSF levels of 35 biomarkers using multiplex and immunoassay analysis for prognosis of ALS. In a recent study, Gendron et al. (2017) identified the use of phosphorylated neurofilament heavy chain as a prognostic biomarker for clinical trials that may likely be useful in developing a successful treatment for C9ALS. Currently, none of the biomarkers is specific to BBB damage or dysfunction induced by ALS.

Epilepsy

Biomarkers in epilepsy play multiple roles because they may (1) predict the development of an epilepsy condition, (2) identify the presence and severity of tissue capable of generating spontaneous seizures, (3) measure progression after the condition is established, (4) be used to create animal models for more cost-effective screening of potential antiepileptic and antiseizure drugs and devices, and (5) reduce the cost of clinical trials of potential antiepileptogenic interventions (Engel et al., 2013). The mechanisms involved in the development of epilepsies and the generation of spontaneous recurrent seizures are multifactorial. Biomarkers linked to a precipitating factor could be useful for seizure prediction and the possible development of interventions. In epilepsy, biomarkers can vary from imaging (MRI, PET, or SPECT) and electrophysiological measurements to changes in gene expression and metabolites in blood or tissues (Engel et al., 2013; Obenaus, 2013). Alterations

TABLE 56.3 Biomarkers of Brain Barriers in Neurodegenerative Diseases and Other Neurological Conditions

Disease	Biomarkers	References
Alzheimer's disease	β-amyloid (Aβ peptide) in senile plaques and neurofibrillary tangles, increased BBB permeability, aberrant protein kinase C (PKC-zeta), dysfunctional BCSFB, decreased transthyretin, haptoglobulin in CSF, IL-1, ICAM-1, increased albumin in CNS, CSF/Serum ratio of albumin, and altered occludin and claudins	Moore et al. (1998), De Vries et al. (1997), Zisper et al. (2007), Abdel-Rehman et al. (2004), Agyare et al. (2013), Erickson and Banks (2013), and Srinivasan et al. (2016)
Parkinson's disease	Intracellular α-synuclein deposition, CSF α-synuclein and β-amyloid, and α-synuclein/total protein ratio	Shaltiel-Karyo et al. (2013), Mollenhauer (2014), Stav et al. (2015), and Wang et al. (2015)
Multiple sclerosis (MS)	BBB disruption, increased CSF osteopontin, pathological legions, gliosis, gadolinium enhancement on MRI scans, and changes in MMPs	Kermode et al. (1990), De Vries et al. (1997), Minagar and Alexander (2003), Palmer (2010), Waubant (2006), and Szalardy et al. (2013)
Amyotrophic lateral sclerosis (ALS)	Increased phosphorylated neurofilament heavy chain in CSF, *tau* protein in CSF and blood, cytokines (IL-1β, IL-1RA, IL-5, IL-9, IL-8, IL-10, IL-12, IL-15), RANTES, high granzyme B, eotaxin, growth factors, plasma L-ferritin, plasma-CSF interferon-γ ratio, high Nogo-A, low cystatin C, genetic biomarkers, and many others	Mitchell et al. (2009, 2010), Su et al. (2013), Bakkar et al. (2015), and Gendron et al. (2017)
Huntington's disease	Mutant Huntingtin protein (mHtt) aggregates, increased transcytosis, increased blood vessel density, reduced blood vessel diameter, reduced occludin-5 and claudin, BBB disruption, and increased permeability	Drouin-Ouellet et al. (2015)
Schizophrenia	Impaired choroid plexus and BCSFB	Rudin (1979)
Allergic encephalomyelitis	BBB disruption, decrease in number of mitochondria, and increased permeability	Claudio et al. (1989) and Hawkins et al. (1991)
HIV-1 and AIDS dementia	Detection of HIV-1 in blood–CNS barriers, increased cytokines (TNF, IL-1β, and IL-6) in CSF, HIV-1 gp-120 proteins, fibrinogen, IgG, increased BBB permeability, and BBB breakdown	Harouse et al. (1989), Petito and Cash (1992), De Vries et al. (1997), Kanmogne et al. (2005), Banks et al. (2006), Ivey et al. (2009), and Northrop and Yamamoto (2015)
Gulf war illness	Increased BBB permeability	Gupta et al. (2015)
Traumatic brain injury (TBI)	Brain edema, BBB disruption, BBB breakdown, loss of tight junction proteins (occludin, claudin-5, ZO-1), reduced expression of PDGFR-β, increased BBB permeability, decreased TEER, increased glutamate, αII-spectrin, GFAP, and serum microRNA let-7i	Ling et al. (2009), Shlosberg et al. (2010), Chodobski et al. (2011), Balakathiresan et al. (2012), Hue et al. (2013), Tate et al. (2013), Tomkins et al. (2013), Shetty et al. (2014), and Dobbins and Pan (2015)
Chronic traumatic encephalopathy (CTE)	*Tau* protein aggregation, tauopathy, and myelinated axonopathy	Goldstein et al. (2014)
Convulsions, seizures, and epilepsy	Increased BBB permeability, BBB dysfunction, electrophysiologic and imaging biomarkers, TGF-β, and GLUT1 deficiency	Ashani and Catravas (1981), Oztas and Kaya (2003), Oby and Janigro (2006), Persidsky et al. (2006), Remy and Beck (2006), Cacheaux et al. (2009), Staba and Bragin (2011), Engel et al. (2013), Obenaus (2013), and Salar et al. (2014)
Reversible cerebral vasoconstriction syndrome (RCVS) (Thunderclap headache)	BBB breakdown, arterial vasoconstriction, and hyper intense vessels	Chen and Wang (2014) and Lee et al. (2017)
Brain edema	Increased BBB permeability	Jangula and Murphy (2013)
Stress (from heat, combat, chemicals, etc.)	Increased serotonin, and increased BBB permeability	Sharma and Dey (1986), Škultétyová et al. (1998), and Amourette et al. (2009)
Hypoxia	Claudin-5	Koto et al. (2007)

in RNA, protein, and metabolites are biomarkers of loss of neurons, gliosis, inflammation, changes in BBB, angiogenesis, neurogenesis, etc. A growing body of evidence has shown that inflammatory mechanisms may participate in the pathological changes observed in epileptic brain, with increasing awareness that blood-borne cells or signals may participate in epileptogenesis by virtue of a leaky BBB (Oby and Janigro, 2006).

In epilepsy, pharmacoresistance to antiepileptic drugs is a common therapeutic challenge, as it occurs in about 30%–70% patients (Salar et al., 2014). These authors and others demonstrated that a dysfunctional BBB with acute extravasation of serum albumin into the brain's interstitial space could contribute to pharmacoresistance (Remy and Beck, 2006). BBB dysfunction per se could serve as a biomarker to direct clinical decisions for alternative drugs with high efficiency and lower affinity or no affinity to albumin. Monitoring BBB integrity may therefore also be important in the management of epilepsy. In addition, modulation of drug transporters such as Pgp, MRP2, and BCRP at the BBB could help in easy entry of antiepileptic drugs (Löscher and Potschka, 2005).

HIV Infection

The BBB can be affected by HIV infection and is also involved in the progression of disease (Banks et al., 2006; Northrop and Yamamoto, 2015). HIV can increase the BBB permeability and infiltration of virus and immune cells. HIV proteins (Tat and gp-120) act on BMECs and result in increased neuroinflammation. HIV proteins cause BBB disruption via oxidative stress and inflammatory mechanisms, decreases in tight junction proteins (claudin-1, claudin-5, and ZO-2) in BMVECs, and cytotoxicity of BMVECs. These alterations are associated with increased expression of MMP-2 and MMP-9. Biomarkers of HIV-induced alterations in BBB are summarized in Table 56.3.

CONCLUDING REMARKS AND FUTURE DIRECTIONS

The CNS barriers, such as BBB and BCSFB, play significant roles in protecting the brain from endogenous and exogenous substances thereby maintaining brain homeostasis. These barriers, of course, allow required nutrients to reach the CNS. The structure of the BBB is very complex as it contains many molecules, enzymes, receptors, transporters, and proteins of known and unknown functions. The structure and functions of these barriers are known to be altered by a variety of toxicants (metals, pesticides, and mycotoxins), drugs of abuse and therapeutic drugs in higher doses, neurodegenerative diseases, and conditions of the CNS. These biomarkers may aid in diagnosis, prognosis, and disease interventions. The future of biomarkers of CNS barriers seems bright, and with newer technologies more specific and sensitive biomarkers will be developed that could detect chemical toxicosis and CNS disease very early and allow timely interventions.

Acknowledgments

The authors would like to thank Ms. Robin B. Doss for her technical assistance in preparation of this chapter.

References

Abbott, N.J., Ronnback, L., Hansson, E., 2006. Astrocyte-endothelial interactions at the blood-brain barrier. Nat. Rev. Neurosci. 7, 41–53.

Abbott, N.J., Patabendige, A.A., Dolman, D.E., et al., 2010. Structure and function of the blood-brain barrier. Neurobiol. Dis. 37, 13–25.

Abbott, N.J., Friedman, A., 2012. Overview and introduction: the blood-brain barrier in health and disease. Epilepsia 53 (Suppl. 6), 1–6.

Abdel-Rehman, A., Abou-Donia, S., El-Masry, E., et al., 2004. Stress and combined exposure to low doses of pyridostigmine bromide, DEET, and permethrin produce neurochemical and neuropathological alterations in cerebral cortex, hippocampus, and cerebellum. J. Toxicol. Environ. Health Part A 67, 163–192.

Agyare, E.K., Leonard, S.R., Curran, G.L., et al., 2013. Traffic jam at the blood-brain barrier promotes greater accumulation of Alzheimer's disease amyloid-β proteins in the cerebral vasculature. Mol. Pharm. 10, 1557–1565.

Amaraneni, M., Sharma, A., Pang, J., et al., 2016. Plasma protein binding limits the blood-brain barrier permeation of the pyrethroid insecticide, deltamethrin. Toxicol. Lett. 250–251, 21–28.

Amourette, C., Lamproglou, I., Barbier, L., et al., 2009. Gulf war illness: effects of repeated stress and pyridostigmine treatment on blood-brain barrier permeability and cholinesterase activity in rat brain. Behav. Brain Res. 203, 207–214.

Aschner, M., Clarkson, T.W., 1988. Uptake of methylmercury in the rat brain: effects of amino acids. Brain Res. 462, 31–39.

Aschner, M., Gannon, M., 1994. Manganese transport across the rat blood-brain barrier: saturable and transferrin-dependent transport mechanisms. Brain Res. Bull. 33, 345–349.

Ashani, Y., Catravas, G.N., 1981. Seizure-induced changes in the permeability of the blood-brain barrier following administration of anticholinesterase drugs to rats. Biochem. Pharmacol. 30, 2593–2601.

Bakkar, N., Boehringer, A., Bowser, R., 2015. Use of biomarkers in ALS drug development and clinical trials. Brain Res. 1607, 94–107.

Banks, W.A., Ercal, N., Price, T.O., 2006. The blood-brain barrier in neuroAIDS. Curr. HIV Res. 4, 259–266.

Balakathiresan, N., Bhomia, M., Chandran, R., et al., 2012. MicroRNA let-7i is a promising serum biomarker for blast-induced traumatic brain injury. J. Neurotrauma 29 (7), 1379–1387.

Behl, M., Zhang, Y., Monnot, A.D., et al., 2009. Increased β-amyloid levels in the choroid plexus following lead exposure and the involvement of low-density lipoprotein receptor protein-1. Toxicol. Appl. Pharmacol. 240, 245–254.

Behrens, M., Hüwel, S., Galla, H.-J., et al., 2015. Blood-brain barrier effects of the *Fusarium* mycotoxins deoxynivalenol, 3-acetyldeoxynivalenol, and moniliformin and their transfer to the brain. PLoS One 1–20. https://doi.org/10.1371/journal.pone.0143640.

Bharavi, K., Reddy, K.S., 2005. Effect of anticholinesterase compound phosalone on blood-brain barrier (BBB) permeability. Ind. J. Physiol. Pharmacol. 49 (3), 337–340.

Birngruber, T., Ghosh, A., Perez-Yarza, V., et al., 2013. Cerebral open flow microperfusion: a new *in vivo* technique for continuous measurement of substance transport across the intact blood-brain barrier. Clin. Exp. Pharmacol. Physiol. 40, 864–871.

Bondy, S.C., 2010. The neurotoxicity of environmental aluminum is still an issue. Neurotoxicity 31, 575–581.

Bowman, P.D., Ennis, S.R., Rarey, K.E., et al., 1983. Brain microvessel endothelial cells in tissue culture: a model for study of blood-brain barrier permeability. Ann. Neurol. 14, 396–402.

Bradbury, M.W.B., 1985. The blood-brain barrier: transport across the cerebral endothelium. Circ. Res. 57, 213–222.

Bradbury, M.W.B., Deane, R., 1993. Permeability of the blood-brain barrier to lead. Neurotoxicology 14, 131–136.

Brückener, K.E., el Bayâ, A., Galla, H.-J., et al., 2003. Permeabilization in a cerebral endothelial barrier model by pertussis toxin involves the PKC effector pathway and is abolished by elevated levels of cAMP. J. Cell Sci. 116, 1837–1846.

Cacheaux, L.P., Ivens, S., David, Y., et al., 2009. Transcriptome profiling reveals TGF-beta signaling involvement in epileptogenesis. J. Neurosci. 29, 8927–8935.

Carpentier, P., Delamanche, I.S., Bert, M.L., et al., 1990. Seizure-related opening of the blood-brain barrier induced by soman: possible correlation with the acute neuropathology observed in poisoned rats. Neurotoxicology 11, 493–508.

Chang, L., Hartman, H.A., 1972. Ultrastructural studies of the nervous system after mercury intoxication. Acta Neuropathol. 20, 122–138.

Chen, S.P., Wang, S.J., 2014. Hypertensive vessels: an early MRI marker of reversible cerebral vasoconstriction syndrome? Cephalagia 34, 1038–1039.

Chen, L., Liu, W., Wang, P., et al., 2015. Endophilin-1 regulates blood-brain barrier permeability via EGFR-JNK signaling pathway. Brain Res. 1606, 44–53.

Chodobski, A., Zink, B.J., Szmydynger-Chodobski, J., 2011. Blood-brain barrier pathophysiology in traumatic brain barrier. Transl. Stroke Res. 4, 492–516.

Choi, B.-S., Zheng, W., 2009. Copper transport to the brain by the blood-brain barrier and blood-CSF barrier. Brain Res. 1248, 14–21.

Claudio, L., Raine, C.S., Brosnan, C.F., 1989. Evidence of persistent blood-brain barrier abnormalities in chronic-progressive multiple sclerosis. Acta Neuropathol. 90, 228–238.

Crossgrove, J.S., Allen, D.D., Bukaveckas, B.L., et al., 2003. Manganese distribution across the blood-brain barrier. 1. Evidence for Carrier-mediated influx of manganese citrate as well as manganese and manganese citrate. Neurotoxicology 24, 3–13.

Daneman, R., Zhou, L., Agalliu, D., et al., 2010. The mouse blood-brain barrier transcriptome: a new resource for understanding the development and function of brain endothelial cells. PLoS One 5 (10), e13741.

De Vries, H.E., Kuiper, J., De Boer, A.G., et al., 1997. The blood-brain barrier in neuroinflammatory diseases. Pharmacol. Rev. 49 (2), 143–155.

Dehouck, M.-P., Fenart, L., Dehouck, B., et al., 1997. The in vitro blood-brain barrier: the assessment of drug transport to the brain. In: Van Zutphen, L.F.M., Balls, M. (Eds.), Animal Alternatives, Welfare and Ethics, pp. 869–872. Amsterdam.

Deli, M.A., Ábrahám, C.S., Kataoka, Y., et al., 2005. Permeability studies on in vitro blood-brain barrier models: physiology, pathology, and pharmacology. Cell. Mol. Neurobiol. 25, 59–127.

Delonche, R., Pages, N., 1997. Aluminum: on both sides of the blood-brain barrier. In: Yasui, M., Strong, M.J., Ota, K., Verity, A.M. (Eds.), Mineral and Metal Neurotoxicity. CRC Press, Boca Raton, FL, pp. 91–97.

Dobbins, D.L., Pan, X., 2015. Neurological effects and mechanisms of blast overpressure injury. In: Gupta, R.C. (Ed.), Handbook of Toxicology of Chemical Warfare Agents, second ed. Academic Press/ Elsevier, Amsterdam, pp. 159–166.

Domer, F.R., Boertje, S.B., Bing, E.G., et al., 1983. Histamine and acetylcholine-induced changes in the permeability of the blood-brain barrier of normotensive and spontaneously hypertensive rats. Neuropharmacology 22, 615–619.

Drees, A., Hollnack, E., Eisenblätter, T., et al., 2005. The multidrug resistance protein BMDP/ABCG2: a new and highly relevant efflux pump at the blood-brain barrier. Int. Congr. Ser. 1277, 154–168.

Drewes, L.R., Singh, A.K., 1985. Transport, metabolism and blood flow in brain during organophosphate induced seizures. Proc. West. Pharmacol. Soc. 28, 191–195.

Drouin-Ouellet, J., Sawiak, S.J., Cisbani, G., 2015. Cerebral and blood-brain barrier impairments in Huntington's disease: potential implications for its pathophysiology. Ann. Neurol. 78, 160–177.

Elmorsy, E., Elzalabany, L.M., Elsheikha, H.M., 2014. Adverse effects of antipsychotics on micro-vascular endothelial cells of the human blood-brain barrier. Brain Res. 1583, 255–268.

Engel Jr., J., Pitkänen, A., Loeb, J.A., et al., 2013. Epilepsy biomarkers. Epilepsia 54 (Suppl. 4), 61–69.

Erickson, M.A., Banks, W.A., 2013. Blood-brain barrier dysfunction as a cause and consequence of Alzheimer's disease. J. Cerebr. Blood Flow Metabol. 33, 1500–1513.

Eum, S.Y., András, I.E., Couraud, P.-O., et al., 2008. PCBs and tight junction expression. Environ. Toxicol. Pharmacol. 25 (2), 234–240.

Fenstermacher, J.D., 1989. Pharmacology of the blood-brain barrier. In: Neuwelt, E. (Ed.), Implications of the Blood-Brain Barrier and Its Manipulations. Plenum Publ. Corp., New York, pp. 137–155.

Foreman, J.H., Constable, P.D., Waggoner, A.L., et al., 2004. Neurologic abnormalities and cerebrospinal fluid changes in horses administered fumonisin B1 intravenously. J. Vet. Intern. Med. 18, 223–230.

Garberg, P., Ball, M., Borg, N., et al., 2005. *In vitro* models for the blood-brain barrier. Toxicol. Vitro 19, 299–334.

Garcia, T., Esparza, J.L., Nogues, M.R., et al., 2010. Oxidative stress status and RNA expression in hippocampus of an animal model of Alzheimer's disease after chronic exposure to aluminum. Hippocampus 20, 218–225.

Garrick, M.D., Dolan, K.G., Horbinski, C., et al., 2003. DMT1: a mammalian transporter for multiple metals. Biometals 16, 41–54.

Ge, X.T., Wang, F.F., Zhong, Y.Q., et al., 2018. Manganese in blood cells as an exposure biomarker in manganese-exposed workers healthy cohort. J. Trace Elem. Med. Biol. 45, 41–47.

Geier, E.G., Chen, E.C., Webb, A., et al., 2013. Profiling solute Carrier transporters in the human blood-brain barrier. Clin. Pharmacol. Ther. 94, 636–639.

Gendron, T.F., Daughrity, L.M., Heckman, M.G., et al., 2017. Phosphorylated neurofilament heavy chain: a biomarker of survival for *C9ORF72*-associated amyotrophic lateral sclerosis. Ann. Neurol. 82, 139–146.

Gerlach, C.V., Derzi, M., Ramaiah, S.K., et al., 2018. Industry perspective on biomarker development and qualification. Clin. Pharmacol. Ther. 103 (1), 27–31.

Ghersi-Egea, J.F., Stragielle, N., 2002. Choroid plexus transporters for drugs and other xenobiotics. J. Drug Taret. 10, 353–357.

Giacconi, R., Costarelli, L., Piacenza, F., et al., 2017. Main biomarkers associated with age-related plasma zinc decrease and copper/ zinc ratio in healthy elderly from zinc age study. Eur. J. Nutr. 56 (8), 2457–2466.

Goldstein, N., Goldstein, R., Terterov, D., et al., 2014. Blood-brain barrier unlocked. Biochemistry (Mosc.) 77, 419–424.

Gupta, R.C., Milatovic, D., Dettbarn, W.-D., 2001a. Nitric oxide (NO) modulates high-energy phosphates in rat brain regions with DFP or carbofuran: prevention by PBN or vitamin E. Arch. Toxicol. 75, 346–356.

Gupta, R.C., Milatovic, D., Dettbarn, W.-D., 2001b. Depletion of energy metabolites following acetylcholinesterase inhibitor-induced status epilepticus: protection by antioxidants. Neurotoxicology 22, 271–282.

Gupta, R.C., Milatovic, S., Dettbarn, W.-D., et al., 2007. Neuronal oxidative injury and dendritic damage induced by carbofuran: protection by memantine. Toxicol. Appl. Pharmacol. 219, 97–105.

Gupta, R.C., 2012. Rotenone. In: Gupta, R.C. (Ed.), Veterinary Toxicology: Basic and Clinical Principles, second ed. Academic Press/Elsevier, Amsterdam, pp. 620–623.

Gupta, R.C., Pitt, J., Zaja-Milatovic, S., 2015. Blood-brain barrier damage and dysfunction by chemical toxicity. In: Gupta, R.C. (Ed.), Handbook of Toxicology of Chemical Warfare Agents, second ed. Academic Press/Elsevier, Amsterdam, pp. 725–739.

Gynther, M., Petsalo, A., Hansen, S.H., et al., 2015. Blood-brain barrier permeability and brain uptake mechanism of kainic acid and dihydrokainic acid. Neurochem. Res. 40 (3), 542–549.

Han, J.-Y., Ahn, S.-Y., Yoo, J.H., et al., 2015. Alleviation of kainic acid-induced brain barrier dysfunction by 4-0-methylhonokiol in *in vitro* and *in vivo* models. BioMed Res. Int. 1–14. Article ID: 893163.

Haorah, J., Heilman, D., Knipe, B., et al., 2005. Ethanol-induced activation of myosin light chain kinase leads to dysfunction of tight junctions and blood-brain barrier compromise. Alcohol Clin. Exp. Res. 29, 999–1009.

Harouse, J.M., Wroblewska, Z., Laughlin, M.A., et al., 1989. Human choroid plexus cells can be latently infected with human immunodeficiency virus. Ann. Neurol. 25, 406–411.

Hawkins, C.P., MaKenzie, F., Tofts, P., et al., 1991. Expanding the definition of the blood-brain barrier to protein. Brain 114, 801–810.

Heinz, S., Freyberger, A., Lawrenz, B., et al., 2017. Mechanistic investigations of the mitochondrial complex I inhibitor rotenone in the context of pharmacological and safety evaluation. Sci. Rep. 7, 45465.

Herbert, J., Wilcox, J.N., Pham, K.C., et al., 1986. Transthyretin: a choroid plexus specific transport protein in human brain. Neurology 36, 900–911.

Higgins Jr., D.S., Greenamyre, J.T., 1996. [3H]dihydrorotenone binding to NADH: ubiquinone reductase (complex I) of the electron transport chain: an autoradiographic study. J. Neurosci. 16, 3807–3816.

Hovdal, H., Syversen, G.B., Rosenthaler, J., 1982. Ergotamine in plasma and CSF after i.m. and rectal administration to humans. Cephalalgia 2, 145–150.

Hue, C.D., Cao, S., Haider, S.F., et al., 2013. Blood-brain barrier dysfunction after primary blast injury *in vitro*. J. Neurotrauma 30, 1652–1663.

Ivey, N.S., MacLean, A.G., Lackner, A.A., 2009. Acquired immunodeficiency syndrome and the blood-brain barrier. J. Neurovirol. 15 (2), 111–122.

Jangula, A., Murphy, E.J., 2013. Lipopolysaccharide-induced blood-brain barrier permeability is enhanced by alpha-synuclein expression. Neurosci. Lett. 551, 23–27.

Jumnongprakhon, P., Govitrapong, P., Tocharus, C., et al., 2016. Melatonin promotes blood-brain barrier integrity in methamphetamine-induced inflammation in primary rat brain microvascular endothelial cells. Brain Res. 1646, 182–192.

Kanmogne, G.D., Primeaux, C., Grammas, P., 2005. HIV-1 gp 120 proteins alter tight junction protein expression and brain endothelial cell permeability: implications for the pathogenesis of HIV-associated dementia. J. Neuropathol. Exp. Neurol. 64, 498–505.

Kermode, A.G., Thompson, A.J., Tofts, P., et al., 1990. Breakdown of the blood-brain barrier precedes symptoms and other MRI signs of new lesions in multiple sclerosis. Pathogenetic and clinical implications. Brain 113 (Pt 5), 1477–1489.

Koto, T., Takubo, K., Ishida, S., et al., 2007. Hypoxia disrupts the barrier function of neural blood vessels through changes in the expression of claudin-5 in endothelial cells. Am. J. Pathol. 170 (4), 1389–1397.

Kwon, O.-S., Sandberg, J.A., Slikker Jr., W., 1997. Effects of fumonisin B1 treatment on blood-brain barrier transfer in developing rats. Neurotoxicol. Teratol. 19 (2), 151–155.

Laterra, J., Bressler, J.P., Induri, R.R., et al., 1992. Inhibition of astroglia-induced endothelial differentiation by inorganic lead: a role for protein kinase C. Proc. Natl. Acad. Sci. U.S.A. 89, 10748–10752.

Lee, M.J., Cha, J., Choi, H.A., et al., 2017. Blood-brain barrier breakdown in reversible cerebral vasoconstriction syndrome: implications for pathophysiology and diagnosis. Ann. Neurol. 81, 454–466.

Ling, G., Bandak, F., Armonda, R., et al., 2009. Explosive blast neurotrauma. J. Neurotrauma 26, 815–825.

Löscher, W., Potschka, H., 2005. Blood-brain barrier active efflux transporters: ATP-binding cassette gen family. NeroRx 2, 86–98.

Mahajan, S.D., Aalinkeel, R., Sykes, D.E., et al., 2008. Methamphetamine alters blood-brain barrier permeability via the modulation of tight junction expression: implication for HIV-1 neuropathogenesis in the context of drug abuse. Brain Res. 1203, 133–148.

Malik, J.K., Aggarwal, M., Kalpana, S., Gupta, R.C., 2017. Chlorinated hydrocarbons and pyrethrins/pyrethroids. In: Gupta, R.C. (Ed.), Reproductive and Developmental Toxicology. Academic Press/Elsevier, Amsterdam, pp. 633–655.

Maresca, M., 2013. From the gut to the brain: journey and pathophysiological effects of the food-associated trichothecene mycotoxin deoxynivalenol. Toxins 5 (4), 784–820.

Martins, T., Burgoyne, T., Kenny, B.-A., et al., 2013. Methamphetamine-induced nitric oxide promotes vesicular transport in blood-brain barrier endothelial cells. Neuropharmacology 65, 74–82.

Martins, T., Baptista, S., Gonçalves, J., et al., 2017. Methamphetamine transiently increases the blood-brain barrier permeability in the hippocampus: role of tight junction proteins and matrix metalloproteinase-9. Brain Res. 1411, 28–40.

Martin-Reina, J., Duarte, J.A., Cerrillos, L., et al., 2017. Insecticide reproductive toxicity profile: organophosphate, carbamate and pyrethroids. J. Toxins 4 (1), 1–7.

McCarthy, R.C., Kosman, D.J., 2015. Iron transport across the blood-brain barrier: development, neurovascular regulation and cerebral amyloid angiopathy. Cell. Mol. Life Sci. 72 (4), 709–727.

McDonough Jr., J.H., McLeod Jr., C.G., Nipwoda, M.T., 1987. Direct microinjection of soman or VX into the amygdala produces repetitive limbic convulsions and neuropathology. Brain Res. 435, 123–127.

Mercey, G., Verdelet, T., Renou, J., et al., 2012. Reactivation of acetylcholinesterase inhibited by organophosphorus nerve agents. Acc. Chem. Res. 45, 756–766.

Milatovic, D., Gupta, R.C., Yu, Y., et al., 2011. Protective effects of antioxidants and anti-inflammatory agents against manganese-induced oxidative damage and neuronal injury. Toxicol. Appl. Pharmacol. 256, 219–226.

Minagar, A., Alexander, J.S., 2003. Blood-brain barrier disruption in multiple sclerosis. Mult. Scler. 9, 540–549.

Minami, T., Okazaki, J., Kawabata, A., et al., 1998. Roles of nitric oxide and prostaglandins in the increased permeability of the blood-brain barrier caused by lipopolysaccharide. Environ. Toxicol. Pharmacol. 5, 35–41.

Misulis, K.E., Clinton, M.E., Dettbarn, W.-D., et al., 1987. Differences in central and peripheral neural actions between soman and diisopropylfluorophosphate, organophosphorus inhibitors of acetylcholinesterase. Toxicol. Appl. Pharmacol. 89, 391–398.

Mitchell, R.M., Freeman, W.M., Randazzo, W.T., et al., 2009. A CSF biomarker panel for identification of patients with amyotrophic lateral sclerosis. Neurology 72 (1), 14–19.

Mitchell, R.M., Simmons, Z., Beard, J.L., et al., 2010. Plasma biomarkers associated with ALS and their relationship to iron homeostasis. Muscle Nerve 42 (1), 95–103.

Mollenhauer, B., 2014. Quantification of α-synuclein in cerebrospinal fluid: how ideal is this biomarker for Parkinson's disease. Park. Relat. Disord. 20 (Suppl. 1), S76–S79.

Monnot, A.D., Behl, M., Ho, S., et al., 2011. Regulation of brain copper homeostasis by the brain barrier systems: effects of Fe-overload and Fe-deficiency. Toxicol. Appl. Pharmacol. 256, 249–257.

Moore, P., White, J., Christiansen, V., et al., 1998. Protein kinase C-zeta activity but not level is decreased in Alzheimer's disease microvessels. Neurosci. Lett. 254, 29–32.

Mulac, D., Hüwel, S., Galla, H.J., et al., 2012. Permeability of ergot alkaloids across the blood-brain barrier *in vitro* and influence on the barrier integrity. Mol. Nutr. Food Res. 56 (3), 475–485.

Multani, P.K., Saini, N., Kaur, R., Sharma, P., 2014. Biomarkers for drugs of abuse and neuropsychiatric disorders: models and mechanisms. In: Gupta, R.C. (Ed.), Biomarkers in Toxicology. Academic Press/Elsevier, Amsterdam, pp. 983–1001.

Mutter, J., Naumann, J., Guethlin, C., 2007. Comments of the article "the toxicology of mercury and its chemical compounds" by Clarkson and Magos (2006). Crit. Rev. Toxicol. 37 (6), 537–549.

Northrop, N.A., Yamamoto, B.K., 2015. Methamphetamine effects on blood-brain barrier structure and function. Front. Neurosci. 9, 1–11.

Obenaus, A., 2013. Neuroimaging biomarkers for epilepsy: advances and relevance to glial cells. Neurochem. Int. 63, 712–718.

Oby, E., Janigro, D., 2006. The blood-brain barrier and epilepsy. Epilepsia 47 (11), 1761–1774.

Osuchowski, M.F., He, Q., Sharma, R.P., 2005. Endotoxin exposure alters brain and liver effects of fumonisin B1 in BALB/c mice: implication of blood brain barrier. Food Chem. Toxicol. 43 (9), 1389–1397.

Oztas, B., Kaya, M., 2003. Blood-brain barrier permeability during acute and chronic electroconvulsive seizures. Pharmacol. Res. 48 (1), 69–73.

Öztaş, B., 1996. Asymmetrical changes in blood-brain barrier permeability during pentylenetetrazol-induced seizure in rats. In: Couraud, S. (Ed.), Biology and Physiology of the Blood-Brain Barrier. Plenum Publ. Corp., New York, pp. 335–337.

Oztas, B., Akgul, S., Seker, F.B., 2007. Gender difference in the influence of antioxidants on the blood-brain barrier permeability during pnetylenetetrazole-induced seizures in hyperthermic rat pups. Biol. Trace Elem. Res. 118, 77–83.

Pal, A., 2014. Copper toxicity induced hepatocerebral and neurodegenerative diseases: an urgent need for prognostic biomarkers. Neurotoxicology 40, 97–101.

Palmer, A.M., 2010. The role of blood-CNS barrier in CNS disorders and their treatment. Neurobiol. Dis. 37, 3–12.

Partridge, W.M., 1986. Blood-brain barrier transport of nutrients. Fed. Proc. 45, 2047–2049.

Partridge, W.M., 2002. Drug and gene delivery to the brain: the vascular route. Neuron 36, 555–558.

Persidsky, Y., Ramirez, S.H., Haorah, J., et al., 2006. Blood-brain barrier: structural components and function under physiologic and pathologic conditions. J. Neuroimmune Pharmacol. 1 (3), 223–236.

Petito, C.K., Cash, K.S., 1992. Blood-brain barrier abnormalities in the acquired immunodeficiency syndrome: immunohistochemical localization of serum proteins in postmortem brain. Ann. Neurol. 32 (5), 658–666.

Petrali, J.P., Maxwell, D.M., Lenz, D.E., et al., 1985. A study of the effects of soman on rat blood-brain barrier. Anat. Rec. 211, 351–352.

Prager, E.M., Aroniadou-Anderjaska, V., Almeida-Suhett, C.P., et al., 2013. Acetylcholinesterase inhibition in the basolateral amygdala plays a key role in the induction of status epilepticus after soman exposure. Neurotoxicology 38, 84–90.

Prasad, K., Tarasevicz, E., Mathew, J., et al., 2009. Toxicokinetics and toxicodynamics of paraquat accumulation in mouse brain. Exp. Neurol. 215 (2), 358–367.

Prelusky, D.B., Hartin, K.E., Trenholm, H.L., 1990. Distribution of deoxynivalenol in cerebral spinal fluid following administration to swine and sheep. J. Environ. Sci. Health Part B 25 (3), 395–413.

Qureshi, H., Hamid, S.S., Ali, S.S., et al., 2015. Cytotoxic effects of aflatoxin B1 on human brain microvascular endothelial cells of the blood-brain barrier. Med. Mycol. 53 (4), 409–416.

Ravindran, J., Agrawal, M., Gupta, N., et al., 2011. Alterations of blood brain barrier permeability by T-2 toxin: role of MMP-9 and inflammatory cytokines. Toxicology 280 (1–2), 44–52.

Remy, S., Beck, H., 2006. Molecular and cellular mechanisms of pharmacoresistance in epilepsy. Brain 129, 18–35.

Römermann, K., Bankstahl, J.P., 2015. Pilocarpine-induced convulsive activity is limited by multidrug transporters at the rodent blood-brain barrier. J. Pharmacol. Exp. Ther. 353, 351–359.

Ropp, S.A., Grunwald, W.C.J., Morris, M., et al., 2008. Pyridostigmine crosses the blood-brain barrier to induce cholinergic and non-cholinergic changes in mouse hypothalamus. J. Med. CBR Def. 6.

Rose, P.H., 2012. Nicotine and the neonicotinoids. In: Marrs, T.C. (Ed.), Mammalian Toxicology of Pesticides. RSC Pub., Cambridge, UK, pp. 184–220.

Rudin, D.O., 1979. Covert transport dysfunction in the choroid plexus as a possible cause of Schizophrenia. Schizophr. Bull. 5, 623–626.

Sahin, D., Ilbay, G., Ates, N., 2003. Changes in the blood-brain barrier permeability and in the brain tissue trace element concentrations after single and repeated pnetylenetetrazole-induced seizures in rats. Pharmacol. Res. 48, 69–73.

Salar, S., Maslarova, A., Lippmann, K., et al., 2014. Blood-brain barrier dysfunction can contribute to pharmacoresistance of seizures. Epilepsia 55 (8), 1255–1263.

Savory, J., Herman, M.M., Ghribi, O., 2006. Mechanism of aluminum-induced neurodegeneration in animals: implications for Alzheimer's disease. J. Alzheim. Dis. 10, 135–144.

Schildknecht, S., Di Monte, D.A., Rape, R., et al., 2017. Tipping points and endogenous determinants of nigrostriatal degeneration by MPTP. Trends Pharmacol. Sci. 38 (6), 541–555.

Shaltiel-Karyo, R., Frenkel-Pinter, M., Rockenstein, E., et al., 2013. A blood-brain barrier (BBB) disrupter is also a potent α-synuclein (α-syn) aggregation inhibitor. A novel dual mechanism of mannitol for the treatment of Parkinson disease (PD). J. Biol. Chem. 288, 17579–17588.

Sharma, H.S., Dey, P.K., 1986. Probable involvement of 5-hydroxytryptamine in increased permeability of blood-brain barrier under heat stress in young rats. Neuropharmacology 25, 161–167.

Sharma, H.S., Muresanu, D., Sharma, A., et al., 2009. Cocaine-induced breakdown of the blood-brain barrier and neurotoxicity. Int. Rev. Neurobiol. 88, 297–3334.

Shetty, A.K., Mishra, V., Kodali, M., et al., 2014. Blood brain barrier dysfunction and delayed neurological deficits in mild traumatic brain injury induced by blast shock waves. Front. Cell. Neurosci. 8, 232.

Shimizu, K., Ohtaki, K., Matsubara, K., et al., 2001. Carrier-mediated processes in blood-brain barrier penetration and neutral uptake of paraquat. Brain Res. 906 (1–2), 35–42.

Shlosberg, D., Benifla, M., Kaufer, D., et al., 2010. Blood-brain barrier breakdown as a therapeutic target in traumatic brain injury. Nat. Rev. Neurol. 6, 393–403.

Silva, A.P., Martins, T., Baptista, S., et al., 2010. Brain injury associated with widely abused amphetamines: neuroinflammation, neurogenesis and blood-brain barrier. Curr. Drug Abuse Rev. 3, 239–254.

Sinha, C., Shukla, G.S., 2003. Species variation in pesticide-induced blood-brain barrier dysfunction. Hum. Exp. Toxicol. 22 (12), 647–652.

Sinha, C., Agrawal, A.K., Islam, F., et al., 2004. Mosquito repellent (pyrethroid-based) induced dysfunction of blood-brain barrier permeability in developing brain. Int. J. Dev. Neurosci. 22 (1), 31–37.

Škultétyová, I., Tokarev, D., Ježová, D., 1998. Stress-induced increase in blood-brain barrier permeability in control and monosodium glutamate-treated rats. Brain Res. Bull. 45, 175–178.

Smith, G., 2018. Fumonisins. In: Gupta, R.C. (Ed.), Veterinary Toxicology: Basic and Clinical Principles. Academic Press/Elsevier, Amsterdam, pp. 1003–1018.

Solberg, Y., Belkin, M., 1997. The role of excitotoxicity in organophosphorus nerve agents central poisoning. Trends Pharmacol. Sci. 18, 183–185.

Song, H., Zheng, G., Shen, X.-F., et al., 2014. Reduction of brain barrier tight junctional proteins by lead exposure: role of activation of nonreceptor tyrosine kinase Src via chaperone GRP78. Toxicol. Sci. 138 (2), 393–402.

Srinivasan, V., Braidy, N., Chan, E.K.W., et al., 2016. Genetic and environmental factors in vascular dementia: an update of blood brain barrier dysfunction. Clin. Exp. Pharmacol. Physiol. 43, 515–521.

Staba, R.J., Bragin, A., 2011. High-frequency oscillations and other electrophysiological biomarkers of epilepsy: underlying mechanisms. Biomarkers Med. 5, 545–556.

Stav, A.L., Aarsland, D., Johansen, K.K., et al., 2015. Amyloid-β and α-synuclein cerebrospinal fluid biomarkers and cognition in early Parkinson's disease. Park. Relat. Disord. 21 (7), 758–764.

Su, X.W., Simmons, Z., Mitchell, R.M., et al., 2013. Biomarker-based predictive models for prognosis in Amyotrophic lateral sclerosis. JAMA Neurol. 70 (12), 1505–1511.

Szalardy, L., Zadori, D., Simu, M., et al., 2013. Evaluating biomarkers of neuronal degeneration and neuroinflammation in CSF of patients with multiple sclerosis-osteopontin as a potential marker of clinical severity. J. Neurol. Sci. 331, 38–42.

Tate, C.M., Wang, K.K.W., Eonta, S., et al., 2013. Serum brain biomarker level, neurocognitive performance, and self-reported symptoms changes in soldiers repeatedly exposed to low-level blast: a breacher pilot study. J. Neurotrauma 30, 1–11.

Terron, A., Bal-Price, A., Paini, A., et al., 2018. An adverse outcome pathway for parkinsonian motor deficits associated with mitochondrial complex I inhibition. Arch. Toxicol. 92, 41–82.

Tomkins, O., Feintuch, A., Benifla, M., et al., 2013. Blood-brain barrier breakdown following traumatic brain injury: a possible role in pasttraumatic epilepsy. Cardiovasc. Psychiatr. Neurol. 2011, 765923.

Uva, L., Librizzi, L., Marchi, N., et al., 2008. Acute induction of epileptiform discharges by pilocarpine in the *in vitro* isolated Guinea-pig brain requires enhancement of blood-brain barrier permeability. Neuroscience 151 (1), 303–312.

Vorbrodt, A.W., Dobrogowska, D.H., Lossinsky, A.S., 1994. Ultracytochemical studies of the effects of aluminum on the blood-brain barrier of mice. J. Histochem. Cytochem. 42, 203–212.

Wang, J., Fitzpatrick, D.W., Wilson, J.R., 1998. Effect of T-2 toxin on blood-brain barrier permeability monoamine oxidase activity and protein synthesis in rats. Food Chem. Toxicol. 36 (11), 955–961.

Wang, X., Yu, S., Li, F., et al., 2015. Detection of α-synuclein oligomers in red blood cells as a potential biomarker of Parkinson's disease. Neurosci. Lett. 599, 115–119.

Waubant, E., 2006. Biomarkers indicative of blood-brain barrier disruption in multiple sclerosis. Dis. Markers 22 (4), 235–244.

Weidner, M., Hüwel, S., Ebert, F., et al., 2013. Influence of T-2 and HT-2 toxin on the blood-brain barrier *in vitro*: new experimental hints for neurotoxic effects. PLoS One 8 (3), 1–10.

Wispelwey, B., Lesse, A.J., Hansen, E.J., et al., 1988. *Haemophilus influenzae* lipopolysaccharide-induced blood-brain barrier permeability during experimental meningitis in the rat. J. Clin. Invest. 82, 1339–1346.

Yokel, R.A., 2012. The pharmacokinetics and toxicology of aluminum in the brain. Curr. Inorg. Chem. 2, 54–63.

Yorulmaz, H., Seker, F.B., Demir, G., et al., 2013. The effect of zinc treatment on the blood-brain barrier permeability and brain element levels during convulsions. Biol. Trace Elem. Res. 151, 256–262.

Zaja-Milatovic, S., Gupta, R.C., Aschner, M., Milatovic, D., 2009. Protection of DFP-induced oxidative damage and neurodegeneration by antioxidants and NMDA receptor antagonists. Toxicol. Appl. Pharmacol. 240, 124–131.

Zhao, Q., Slavkovich, V., Zheng, W., 1998. Lead exposure promotes translocation of protein kinase C activity in rat choroid plexus *in vitro*, not *in vivo*. Toxicol. Appl. Pharmacol. 149, 99–106.

Zheng, W., 2001. Neurotoxicology of the brain barrier system: new implications. Clin. Toxicol. 39 (7), 711–719.

Zheng, W., Aschner, M., Ghersi-Egea, J.-F., 2003. Brain barrier systems: a new frontier in metal neurotoxicological research. Toxicol. Appl. Pharmacol. 192, 1–11.

Zisper, B.D., Johanson, C.E., Gonzalez, L., et al., 2007. Microvascular injury and blood-brain barrier leakage in Alzheimer's disease. Neurobiol. Aging 28, 977–986.

Zlocovic, B.V., 2008. The blood-brain barrier in health and chronic neurodegenerative disorders. Neuron 57, 178–201.

CHAPTER

57

Biomarkers of Oxidative/Nitrosative Stress and Neurotoxicity

Dejan Milatovic[1], *Snjezana Zaja-Milatovic*[2], *Ramesh C. Gupta*[3]

[1]Charlottesville, VA, United States; [2]PAREXEL International, Alexandria, VA, United States; [3]Toxicology Department, Breathitt Veterinary Center, Murray State University, Hopkinsville, Kentucky, United States

INTRODUCTION

Reactive oxygen species (ROS) and reactive nitrogen species (RNS) are molecules or molecular fragments containing one or more unpaired electrons in atomic or molecular orbitals, which characterize free radicals with high reactivity. Exogenous agents (such as pesticides, metals, xenobiotics, and ionizing radiation) and a variety of endogenous processes (cellular respiration, antibacterial defense, phagocytic oxidative bursts, and others) can generate significant amounts of ROS and RNS in the human body (Chakravarti and Chakravarti, 2007; Mangialasche et al., 2009; Il'yasova et al., 2012). Both species of free radicals are also products of normal cellular metabolism.

Mitochondrial oxidative phosphorylation generates the majority of free radicals in the cell (Federico et al., 2012; Morán et al., 2012; Quinlan et al., 2012; Trewin et al., 2018; Sas et al., 2018). ROS are produced mainly as superoxide, when electrons leak onto molecular oxygen in side reactions from prosthetic groups or coenzymes involved in these redox reactions. Superoxide further serves as a precursor to the formation of the hydroxyl radical (HO•), the most reactive radical. The hydroxyl radical is produced by the Fenton and Haber–Weiss reactions from hydrogen peroxide (H_2O_2) and metal species (iron, copper) (Dix and Aikens, 1993; Guéraud et al., 2010). Superoxide easily reacts with nitric oxide ($NO^{\bullet}$) and forms peroxynitrite ($ONOO^-$), an RNS with very high reactivity (Vatassery, 2004). Peroxynitrite is a powerful oxidant exhibiting a wide array of tissue-damaging effects ranging from lipid peroxidation and inactivation of enzymes and ion channels via protein oxidation and nitration to inhibition of mitochondrial respiration (Virag et al., 2003). Peroxynitrite, which dissipates during oxidation (Wang et al., 2003), has also been found to nitrate and oxidize adenine, guanine, and xanthine nucleosides (Sodum and Fiala, 2001). Low concentrations of peroxynitrite trigger apoptotic cell death, whereas higher concentrations induce necrosis with cellular energetics (ATP and NAD) serving as a switch between the models of cell death. Superoxide is also involved in the production of nitrogen dioxide radical and hydrogen peroxide, which can be further transformed into lipid peroxidation–initiating species, namely peroxyl and alkoxyl radicals (ROO• and RO•) (Dix and Aikens, 1993; Guéraud et al., 2010).

Furthermore, ROS and RNS are also normally generated by tightly regulated enzymes or enzyme systems located in or associated with cellular membranes or organelles, in both phagocytic and nonphagocytic cells. Nitric oxide synthases (NOSs) and nicotinamide adenine dinucleotide phosphate [NAD(P)H] oxidase isoform–produced ROS/RNS are also involved in cellular signaling, synaptic plasticity in the central nervous system (CNS), neuronal transmission, reactions to stress and various agents, and the induction of mitogenic and apoptotic responses (Beal, 2000; Valko et al., 2007; Mangialasche et al., 2009).

ROS and RNS, produced under various mechanisms, can react with lipids, sugars, proteins, and nucleic acids and inhibit normal cell functions (Deavall et al., 2012; Cheignon et al., 2018). This damage can compromise cell viability or induce cellular response leading to cell death by necrosis or apoptosis (Valko et al., 2007; Mangialasche et al., 2009; Du et al., 2018; Fang et al., 2017). Importantly, living systems have developed several mechanisms to control these harmful effects of ROS and RNS. These mechanisms are mainly based on the presence of antioxidants (enzymatic and

Biomarkers in Toxicology, Second Edition
https://doi.org/10.1016/B978-0-12-814655-2.00057-8

nonenzymatic) and the repair or removal of the injured molecules/systems. Antioxidants are natural or synthetic molecules preventing excessive formation of ROS or inhibiting their reaction with biological structures and/or molecules. Antioxidant defense involves a variety of strategies, both enzymatic, such as superoxide dismutase, catalase, thioredoxin reductase, and glutathione peroxidase (GPx), and nonenzymatic, involving tocopherols, carotenes, ascorbate, glutathione (GSH), ubiquinols, and flavonoids (Halliwell, 2007; Espinosa-Diez et al., 2015; Radomska-Leśniewska et al., 2016, 2017; Valko et al., 2007; Vallejo et al., 2017). Different systems/mechanisms cooperate in the regulation of ROS/RNS production and neutralization, which is essential for avoiding their detrimental effects and preserving equilibrium. This is termed *redox homeostasis* (Droge, 2002).

LIPID PEROXIDATION AND MARKERS OF OXIDATIVE STRESS

Because of the high concentration of substrate polyunsaturated fatty acids (PUFA) in cells, lipid peroxidation is a major outcome of free radical–mediated injury (Montine et al., 2002). Lipid peroxidation is the mechanism by which lipids are attacked by chemical species that have sufficient reactivity to abstract a hydrogen atom from a methylene carbon in their chain. Lipid peroxidation in vivo, through a free radical pathway, requires a PUFA and a reactant oxidant inducer that together form a free radical intermediate. The free radical intermediate subsequently reacts with oxygen to generate a peroxyl radical, which with unpaired electrons may additionally abstract a hydrogen atom from another PUFA. The greater number of double bonds in the molecule and the higher instability of the hydrogen atom adjacent to the double bond explain why unsaturated lipids are particularly susceptible to peroxidation (Pratico et al., 2004; Gao et al., 2006).

A critically important aspect of lipid peroxidation is that it will proceed until the oxidizable substrate is consumed or termination occurs, making this fundamentally different from many other forms of free radical injury in that the self-sustaining nature of the process may entail extensive tissue damage (Porter et al., 1995). Decreased membrane fluidity following lipid peroxidation makes it easier for phospholipids to exchange between the two halves of the bilayer, increase the leakiness of the membrane to substances that do not normally cross it other than through specific channels (e.g., K^+, Ca^{2+}), and damage membrane proteins, inactivating receptors, enzymes, and ion channels (Halliwell, 2007). Increases in Ca^{2+} induced by oxidative stress can activate phospholipase A_2, which then releases arachidonic acid (AA) from membrane phospholipids. The free AA can then both undergo lipid peroxidation (Farooqui et al., 2001) and act as a substrate for eicosanoid synthesis. Increased prostaglandin synthesis is immediately linked to lipid peroxidation because low levels of peroxides accelerate cyclooxygenase action on polyunsaturated fatty acids (Smith, 2005). Phospholipase A_2 can also cleave oxidized AA residues from membranes.

The use of reactive products of lipid peroxidation as in vivo biomarkers is limited because of their chemical instability and rapid and extensive metabolism (Gutteridge and Halliwell, 1990; Moore and Roberts, 1998). For these reasons, other more stable lipid products of oxidative damage have generated intense interest in recent years as in vivo markers of oxidative damage (De Zwart et al., 1999). These compounds include the F_2-isoprostanes (F_2-IsoPs), F_3-Isops, isofurans (IsoFs), and F_4-neuroprostanes (F_4-NeuroPs) (Morrow et al., 1990; Fessel et al., 2002; Yin et al., 2007; Janicka et al., 2010).

PROSTAGLANDIN-LIKE COMPOUNDS AS *IN VIVO* MARKERS OF OXIDATIVE STRESS

F_2-IsoPs are prostaglandin-like compounds that are produced by a noncyclooxygenase free radical–catalyzed mechanism involving the peroxidation of the PUFA, arachidonic acid (AA, C20:4, ω-6). Formation of these compounds initially involves the generation of four positional peroxyl radical isomers of arachidonate, which undergo endocyclization to PGG_2-like compounds. These intermediates are reduced to form four F_2-IsoP regioisomers, each of which can consist of eight racemic diastereomers (Morrow et al., 1990). In contrast to cyclooxygenase (COX)-derived prostaglandins (PGs), nonenzymatic generation of F_2-IsoPs favors the formation of compounds in which the stereochemistry of the side chains is *cis* oriented in relation to the prostane ring. A second important difference between F_2-IsoPs and PGs is that F_2-IsoPs are formed primarily in situ, esterified to phospholipids, and subsequently released by phospholipases (Famm and Morrow, 2003; Gao et al., 2006), whereas PGs are generated only from free arachidonic acid (Morrow et al., 1990). The basic formation process is depicted in Fig. 57.1.

F_2-IsoPs analog may be formed by peroxidation of docosahexaenoic acid (DHA, C22:6, ω-3), which generates F_4-IsoPs. These compounds are also termed neuroprostanes (F_4-NeuroPs), because of the high levels of their precursor in the brain (Roberts et al., 1998). In contrast to AA, which is evenly distributed in all cell types in all tissues, DHA is highly concentrated in neuronal membranes (Salem et al., 1986). DHA is obtained mainly through dietary means as the human body can only

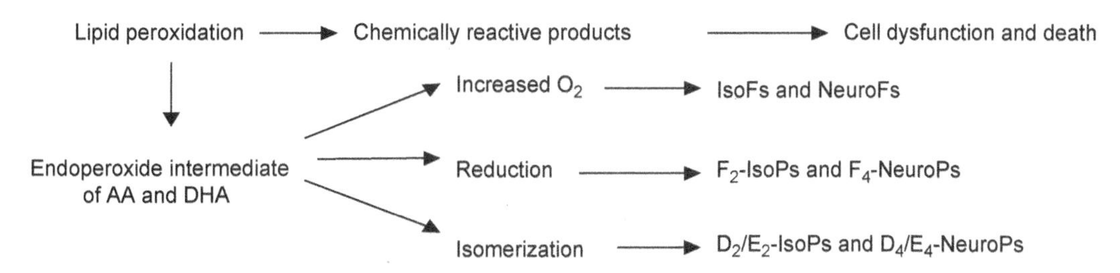

FIGURE 57.1 Schematic of the compound-forming process.

minimally synthesize this fatty acid. DHA deficiency has been linked to slow mental development (Connor et al., 1992), whereas DHA supplementation has been linked to a variety of health benefits including decreased rates of neurodegenerative diseases (Schaefer et al., 2006). DHA can undergo oxidation both in vitro and in vivo, and increased units of unsaturation suggest its higher susceptibility to lipid peroxidation than AA. Thus, although the measurement of F_2-IsoPs provides an index of global oxidative damage in the brain, integrating data from both glial and neuronal cells, determination of F_4-NeuroPs permits the specific quantification of oxidative damage to neuronal membranes in vivo. In fact, to our knowledge, F_4-NeuroPs are the only quantitative in vivo marker of oxidative damage that is selective for neurons. This is particularly important because of the implication of oxidative damage and lipid peroxidation being causative factors in numerous neurodegenerative diseases (Montine et al., 2004; Milatovic et al., 2005b).

Another F_2-IsoPs analog may be formed by peroxidation of eicosapentaenoic acid (EPA, C20:5, ω-3) that leads to the production of F_3-IsoPs. Levels of F_3-IsoPs can significantly exceed those of F_2-IsoPs generated from AA, perhaps because EPA contains more double bonds and is therefore more easily oxidizable (Gao et al., 2006). It has also been shown that in the presence of increased oxygen tension in the microenvironment in which lipid peroxidation occurs, an additional oxygen insertion step may take place (Fessel et al., 2002). This step diverts the IsoP pathway to form tetrahydrofuran ring–containing compounds termed isofurans (IsoFs), which are functional markers of lipid peroxidation under conditions of increased oxygen tension. Thus, measurements of IsoFs represent a much more robust indicator of hyperoxia-induced lung injury than measurements of F_2-IsoPs. Similar to IsoPs, the IsoFs are chemically and metabolically stable, so they are well suited to act as in vivo biomarkers of oxidative damage.

Stable isotope dilution, negative ion chemical ionization (NICI) gas chromatography/mass spectrometry (GC/MS) with select ion monitoring (SIM) was the first technique used in early discovery and quantification of isoprostanes by investigators at Vanderbilt University (Morrow et al., 1990). This methodology allows the lower limit of detection of the F_2-IsoPs to be in the low picogram range. Quantification of the F_2-IsoPs levels is achieved by comparing the height/areas of the peak containing derivatized F_2-IsoPs (m/z 569) with the height of the deuterated internal standard peak (m/z 573) (Fig. 57.2). These properties, along with the assay's high sensitivity and specificity, allow the F_2-IsoPs to be excellent biomarkers of and the most robust and sensitive measure of oxidative stress *in vivo*. However, mass spectrometry–based methods necessitate a skilled technical staff and are laborious to execute, because of extensive purification and appropriate derivatization procedures. In addition, over the past 20 years other methods such as enzyme immunoassays, radioimmunoassays, liquid chromatography (LC)–mass spectrometry, LC-MS-MS, and GC-MS-MS (Wang et al., 1995; Basu, 1998; Liang et al., 2003) have been developed and used to exploit the impact of this biomarker on human health and disease (Morrow et al., 1990; Wang et al., 1995; Pratico et al., 1998; Basu, 2008).

Immunoassays, although less specific and quantitative than GC-MS methods, have been found to be helpful tools for new discoveries in medical and pharmaceutical sciences. Immunoassays have a huge sample-analyzing capacity with fairly low expense. Thus, a well-validated technique could be a significant tool for evaluating free radical–mediated reactions in clinical research, where a large number of samples must be analyzed at an affordable cost. Specific antibodies against isoprostanes can also be used for in situ localization by immunostaining in oxidative stress–injured tissues. Immunostaining with specific antibodies opens possibilities for therapeutic application of various radical scavengers in disease-related damage (Basu, 2008).

ALDEHYDES AS LIPID PEROXIDATION PRODUCTS

The enzymatic and free radical peroxidation of PUFAs, which contain at least three double bonds, such as AA and DHA, could lead to malondialdehyde (MDA). This product can be generated by thromboxane synthase, but a report from the Biomarkers of Oxidative

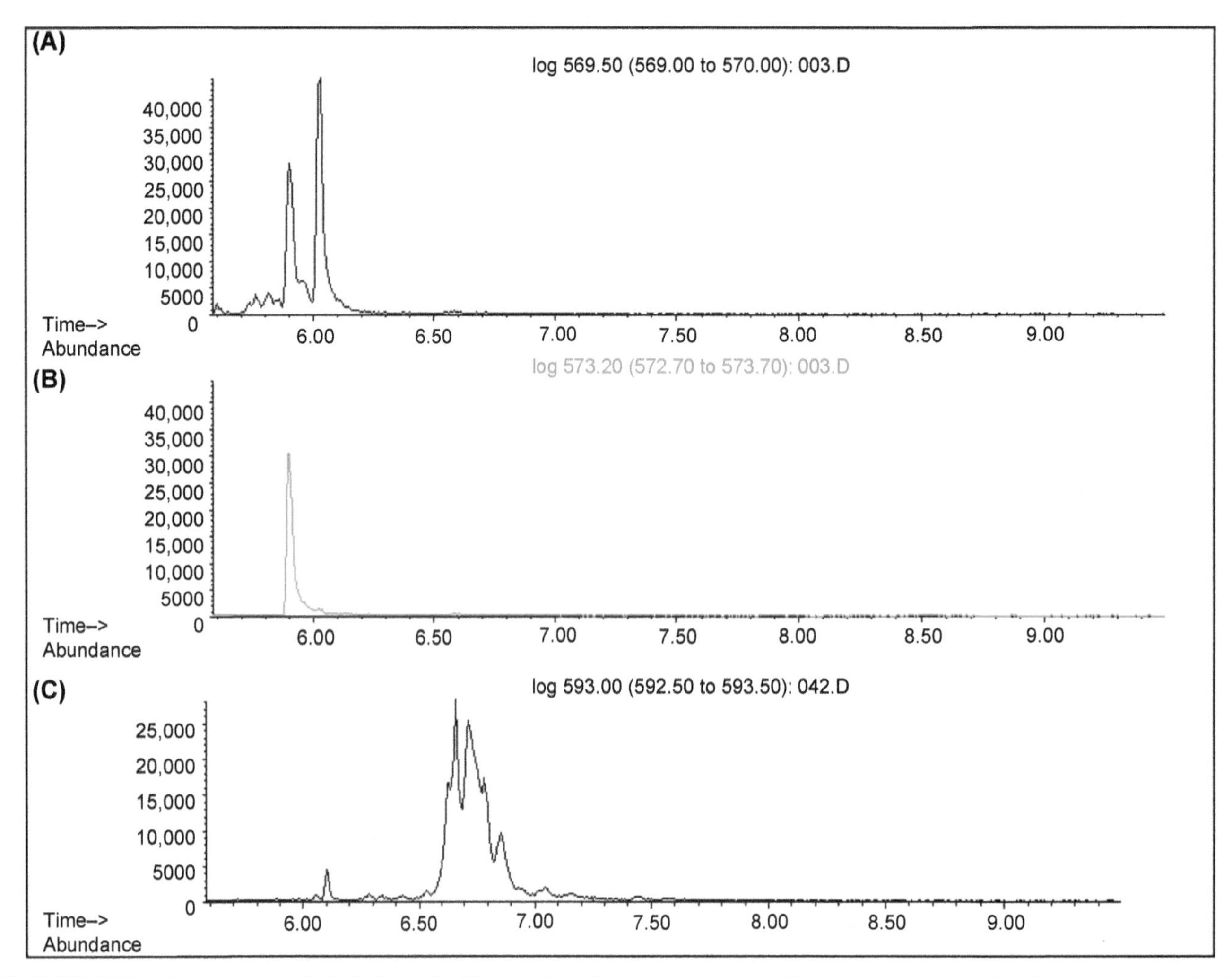

FIGURE 57.2 Chromatograms for F_2-IsoPs and F_4-NeuroPs from the mouse cerebrum. All chromatograms plot abundance versus time (min). (A) m/z 569 chromatogram showing F_2-IsoPs, (B) m/z 573 chromatogram showing internal standard, (C) m/z 593 chromatogram showing F_4-NeuroPs.

Stress Study showed that peripheral levels of MDA derive primarily from nonenzymatic peroxidative degradation of unsaturated lipids (Kadiiska et al., 2005). Because MDA reacts with thiobarbituric acid (TBA), MDA−TBA adducts are used to spectrophotometrically measure the levels of oxidative stress and consequent lipid peroxidation (Spickett et al., 2010; Fang et al., 2017). MDA and MDA−TBA complex assessment can be also measured by high-performance liquid chromatography (HPLC) or GC-MS (Maboudou et al., 2002). Although more specific, this approach does not deduce all the limitations of this biomarker (Devasagayam et al., 2003; Sultana et al., 2006).

4-Hydroxy-2-nonenal (HNE) is a reactive aldehyde arising from peroxidation of ω6 fatty acid (Uchida, 2003; Guéraud et al., 2010; Fang et al., 2017). HNE is formed under various conditions such as auto-oxidation and stimulated microsomal lipid peroxidation (Neely et al., 2005). HNE is found in food (Gasc et al., 2007) and detected in vivo (Neely et al., 2005). MDA and HNE are able to covalently modify proteins and alter their functions (Butterfield et al., 2006). In addition to protein modification, these lipid peroxidation products can also interfere with synthesis of DNA and RNA, alter cell metabolism and signaling, and mediate brain-induced oxidative damage. Several studies suggest MDA and HNE can promote the degeneration of cholinergic neurons, Aβ aggregation, and amyloidogenesis (Mark et al., 1995; Pedersen et al., 1999; Butterfield et al., 2006). In addition, increased levels of HNE and HNE-protein adducts (Montine et al., 1997; Markesbery and Lovell, 1998) have been described in the ventricular CSF and brain of AD patients compared with control subjects (Montine et al., 1997; Markesbery and Lovell, 1998).

Another highly reactive aldehydic product, acrolein, derived from the metal-catalyzed oxidation of polyunsaturated fatty acid, including AA and DHA, can also promote the formation of protein adducts and thus promote neuronal damage (LoPachin et al., 2008). Studies suggest that acrolein can modify DNA, promote lipid peroxidation, induce tau oligomerization, and promote neurofibrillary tangles formation (Uchida et al., 1998; Lovell et al., 2000; Kuhla et al., 2007).

REACTIVITY OF LIPID PEROXIDATION PRODUCTS

Metabolization of secondary lipid peroxidation products such as HNE in most cells and tissues is rapid and complete. GSH conjugation seems to be the primary and major step. GSH maintains protein sulfhydryl groups in the reduced form and has been implicated in the regulation of cytoskeletal function. SH groups of cystein constitute a main protein-associated target of HNE (Esterbauer et al., 1991). Indeed, cellular detoxification of HNE is mainly achieved through conjugation to the cystein GSH. This reaction is catalyzed by GSH S-transferase and leads to a transient decrease in cellular GSH (Johnson et al., 1993; Hayes and Pulford, 1995; Kinter and Roberts, 1996; Cheng et al., 1999; Radomska-Leśniewska et al., 2017). If GSH is depleted by buthionine sulfoximine, for instance, or by a concomitant oxidative insult, there is a reduction in GSH-HNE conjugate together with an increase in unmetabolized HNE and HNE toxicity (Picklo et al., 2002). On the other hand, when the oxidative insult occurs some time before treatment of cells by HNE, preconditioned cells acquire resistance to HNE-induced apoptosis by metabolizing and excluding HNE at a higher rate when compared with nonpreconditional cells (Rice et al., 1986; Rabinovitch et al., 1993). We previously demonstrated that another main cellular target to HNE is tubulin, the core protein of microtubules containing abundant cystein (Neely et al., 2005). Our studies demonstrated that the exposure of Neuro2A cells to HNE results in the inhibition of cytosolic taxol-induced tubulin polymerization and thus supported the hypothesis that HNE adduction to tubulin is the primary mechanism involved in the HNE-induced loss of the highly dynamic neuronal microtubule network (Neely et al., 2005).

ROS and RON can react with the DNA molecule and induce purine or pyrimidine base or sugar lesions, nitration and deaminations of purines, DNA–DNA, or DNA–protein cross-links (Dizdaroglu et al., 2002). These processes lead to mutations and impaired transcriptional and posttranscriptional processes and compromise protein synthesis (Colurso et al., 2003). In addition, DNA damage, oxidative phosphorylation, and altered cell metabolism may lead to apoptosis and promote neuronal death (Becker and Bonni, 2004; Fishel et al., 2007). The most investigated DNA adduct, 8-hydroxy-2′-deoxyguanosine (8-OHdG), can be evaluated by multiple techniques including immunoassay, HPLC, GC-MS, capillary electrophoresis, and LC-MS. However, LC-MS techniques for DNA damage evaluation permit the identification of a wide range of base adducts and offer a more complete picture of DNA oxidation (Lovell and Markesbery, 2007; Mangialasche et al., 2009; Fang et al., 2017).

Although ROS and RNS can attack any amino acid, sulfur-containing and aromatic amino acids are the most susceptible (Stadtman and Levine, 2003). The oxidation of amino acids leads to the formation of carbonyl derivatives, produced by fragmentation, due to an attack on several amino acid side chains, and by formation of adducts between some amino acids residues and products of lipid peroxidation. Protein carbonyls can be detected with 2,4-dinitrophenylhydrazine (DNPH) and used as biomarkers of oxidative stress (Dalle-Donne et al., 2003). Oxidation of protein also leads to protein fragmentation and protein cross-linking. In addition, peroxynitrite and a hydroxyl radical can react with tyrosine and form other indexes of protein oxidation, 3-nitrotyrosine and ortho-tyrosine, respectively. These products of oxidative/nitrosative modification of proteins are relatively stable with sensitive assays available for their detection (Chakravarti and Chakravarti, 2007).

EXCITOTOXICITY AND OXIDATIVE DAMAGE

The brain is especially susceptible to oxidative damage. The brain is relatively deficient in antioxidant systems, with a lower activity of GPx and catalase compared to other organs. It has high metabolic activity that requires large amounts of oxygen and contains redox-active metals (iron, copper) that can promote the production of free radicals. During normal physiological conditions, ROS are generated at a low rate in brain and are efficiently removed by scavenger and antioxidant systems. However, in pathophysiological conditions, such as seizures and acetylcholinesterase inhibitors (organophosphates and carbamates) toxicity, a high rate of ATP consumption is accompanied by increased generation of ROS. Previous studies have supported the role for oxidative stress and excessive generation of ROS and RNS in anticholinesterase-induced neurotoxicity (Dettbarn et al., 2001; Gupta et al., 2001a,b, 2007; Milatovic et al., 2005a). During sustained seizures, the flow of oxygen to the brain is greatly increased at a time when the use of ATP is greater than the rate of its generation. This metabolic stress results in a markedly increased ROS generation. A greatly increased rate of

ROS production, overwhelming the capacity of inherent cellular defense systems, results in an attack on the mitochondria and cell membranes, leading to peroxidation of lipids, cell lesions, and, in turn, cell death (Sjodin et al., 1990).

Processes such as excitotoxicity and inhibition of acetylcholinesterase (AChE) lead to unremitting stimulation of nervous tissue and muscle, which, in turn, causes depletion of high-energy phosphates (HEP), ATP, and phosphocreatine (PCr) and ROS generation (Dettbarn et al., 2001, 2006). If this stimulation is sufficiently low in intensity or brief in duration, cellular recovery will ensue without lasting consequences. If, however, intense cholinergic stimulation is allowed to persist, a self-reinforcing cycle of cellular damage is set in motion. ATP depletion for several hours to approximately 30%–40% of normal levels leads to a fall in the mitochondrial membrane potential that is associated with (1) reduced energy production (because of decrease in complex I and complex IV activities), (2) impaired cellular calcium sequestration, (3) activation of protease/caspases, (4) activation of phospholipases, (5) activation of nitric oxide synthase (NOS), and (6) excessive generation of ROS (Milatovic et al., 2006). Several of these steps are associated with exacerbation and propagation of the initial depletion of ATP; most notable are the decreases in complex I and IV activities, the impairment of mitochondrial calcium metabolism that regulates ATP production even in the face of a constant supply of substrates, and the generation of nitric oxide, which binds reversibly to cytochrome *c* oxidase (COx) in competition with oxygen, with subsequent sensitization to hypoxia. COx is the terminal complex in the mitochondrial respiratory chain, which generates ATP by oxidative phosphorylation, involving the reduction of O_2 to H_2O by the sequential addition of four electrons and four H^+. Electron leakage occurs from the electron transport chain, which produces the superoxide anion radical and H_2O_2. Under normal conditions, COx catalyzes more than 90% of the oxygen consumption in the cells. The chance of intermediate products, such as superoxide anion radical and H_2O_2 and hydroxyl radical, escaping is less under conditions where COx remains active. During the hyperactivity of brain or muscle, the activity of COx is reduced (Milatovic et al., 2001), leading to an increased electron flow within the electron transport chain, thereby increasing ROS generation, oxidative damage to mitochondrial membranes, and vulnerability to excitotoxic impairment (Soussi et al., 1989; Gollnick et al., 1990; Bose et al., 1992; Bondy and Lee, 1993; Yang and Dettbarn, 1998).

Excitotoxicity-induced cell damage is also thought to result from an intense transient influx of calcium leading to mitochondrial functional impairments characterized by activation of the permeability transition pores in the inner mitochondrial membrane, cytochrome *c* release, depletion of ATP, and simultaneous formation of ROS (Cadenas and Davies, 2000; Nicholls and Ward, 2000; Heinemann et al., 2002; Patel, 2002; Nicholls et al., 2003). In addition, increase in cytoplasmic calcium ions triggers intracellular cascades through stimulation of enzymes, including proteases, phospholipase A_2, and nitric oxide synthase, which also leads to increased levels of free radical species and oxidative stress (Lafon-Cazal et al., 1993; Farooqui et al., 2001). Because free radicals are direct inhibitors of the mitochondrial respiratory chain, ROS generation perpetuates a reinforcing cycle, leading to extensive lipid peroxidation and oxidative cell damage (Cadenas and Davies, 2000; Cock et al., 2002).

Our studies have shown that kainic acid (KA)–induced excitotoxicity and anticholinesterases, such as diisopropylphosphorofluoridate (DFP) and carbofuran (CF), cause neuronal injury by excessive formation of ROS (Yang and Dettbarn, 1998; Gupta et al., 2001a,b; Milatovic et al., 2000a,b, 2001, 2005a; Zaja-Milatovic et al., 2008). Exposure to DFP or carbofuran significantly suppressed AChE activity, induced severe seizure activity, and significantly increased biomarkers of global free radical damage (F_2-IsoPs) and the selective peroxidation biomarker of neuronal membranes (F_4-NeuroPs) (Gupta et al., 2007; Zaja-Milatovic et al., 2009). Although twofold elevations are seen in F_2-IsoPs levels, F_4-NeuroPs levels are fivefold higher compared to controls (Fig. 57.3). The selective increase in F_4-NeuroPs indicates that neurons are specifically targeted by this mechanism.

DFP or CF exposures also caused marked elevation in brain citrulline levels (an indicator of NO/NOS activity) (Gupta et al., 2001b, 2007). Many reports provide evidence that NO impairs mitochondrial/cellular respiration and other functions by inhibiting the activities of several key enzymes, particularly COx, and thereby causing ATP depletion (Yang and Dettbarn, 1998; Dettbarn et al., 2001; Gupta et al., 2001a; b; Milatovic et al., 2001). Results from our experiments also showed that 1 hour after DFP (1.5 mg/kg, s.c.) or CF (1.5 mg/kg, s.c.) treatment, the levels of ATP and PCr were significantly reduced in the cortex, hippocampus, and amygdala (Gupta et al., 2001a,b). The rapid decrease in energy metabolites at the onset of seizures indicates early onset of mitochondrial dysfunction, in turn further increasing ROS production and neuronal injury.

The most consistent pathological findings in acute experiments with anticholinesterases include degeneration and cell death in the pyriform cortex, amygdala, hippocampus (where the CA1 region is preferentially destroyed), dorsal thalamus, and cerebral cortex. The early morphological changes in AChEI-induced status epilepticus (SE) include dendritic swelling of pyramidal

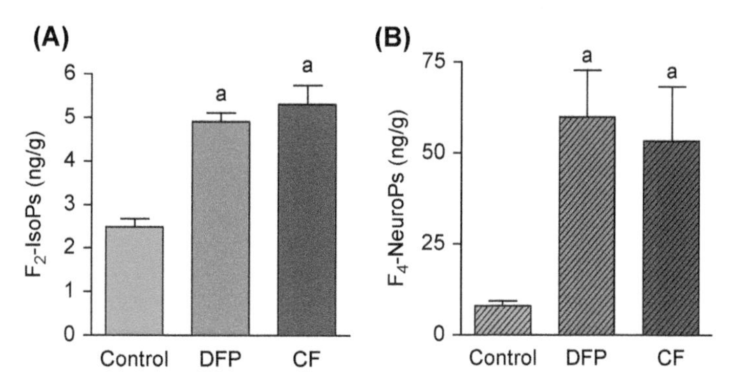

FIGURE 57.3 Effect of DFP (1.5 mg/kg, s.c.) and carbofuran (CF, 1.5 mg/kg, s.c.) on F_2-IsoPs (A) and F_4-NeuroPs (B) levels in rat brain. Rats were sacrificed 1 h after DFP or CF injection. Values are mean ± SEM (n = 4–6). [a]Significant difference between control and DFP- or CF-treated rats ($P < .05$).

neurons in the CA1 region of the hippocampus (Carpentier et al., 1991). Therefore, we have investigated whether seizure-induced cerebral oxidative damage in adult rats is accompanied by alterations in the integrity of the hippocampal CA1 dendritic system. Our results showed that anticholinesterase induced early increases in biomarkers of global free radical damage (F_2-IsoPs), and the selective peroxidation biomarker of neuronal membranes (F_4-NeuroPs) was accompanied by dendritic degeneration of pyramidal neurons in the CA1 hippocampal area. Anticholinesterase-induced brain hyperactivity targeted the dendritic system with profound degeneration of spines and regression of dendrites, as evaluated by Golgi impregnation and Neurolucida-assisted morphometry (Fig. 57.4).

Rats injected with DFP show a significant decrease in total dendritic length and spine density compared to pyramidal neuron from the hippocampal CA1 area of control rats (Fig. 57.4). Taken together, our results revealed that anticholinesterase exposure is associated with oxidative and nitrosative stresses, alteration in energy metabolism, and consequent degeneration of pyramidal neurons from the CA1 hippocampal region of rat brain. Ultimately, the additive or synergistic mechanisms of cellular disruption caused by anticholinesterase agents lead to cellular dysfunction and neurodegeneration.

Similar studies investigated the role of glutamatergic excitation, oxidative injury, and neurodegeneration in the model of KA excitotoxicity. We have used intracerebroventricular (icv) injection of KA and investigated whether F_2-IsoPs and F_4-NeuroPs formations correlated with the vulnerability of pyramidal neurons in the CA1 hippocampal area following KA-induced excitotoxicity.

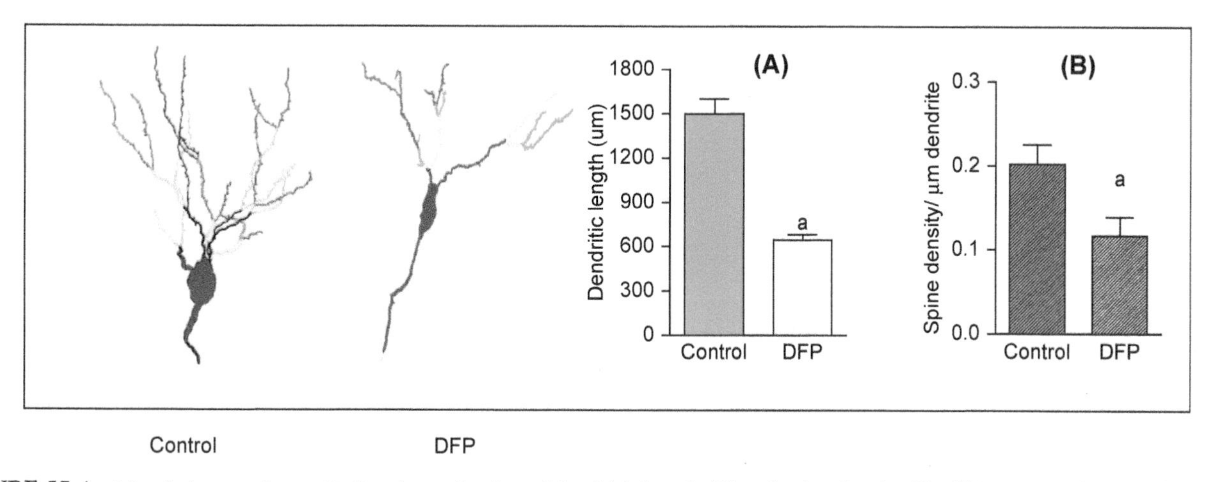

FIGURE 57.4 Morphology and quantitative determination of dendritic length (A) and spine density (B) of hippocampal pyramidal neurons from CA1 sector of rats treated with saline (control) or DFP (1.5 mg/kg, s.c.) and sacrificed 1 h after the treatment. Four to six Golgi-impregnated dorsal hippocampal CA1 neurons were selected and spines were counted using Neurolucida system. [a]Significant difference between control and DFP-treated rats ($P < .05$). Treatment with DFP induced degeneration of hippocampal dendritic system and a decrease in the total length of dendrite and spine density of hippocampal pyramidal neurons. Tracing and counting were done using a Neurolucida system at 100× under oil immersion (MicroBrightField, VT). Colors indicate degree of dendritic branching (blue = 1 degree, yellow = 2 degrees, green = 3 degrees, magenta = 4 degrees, orange = 5 degrees).

Our results showed that icv KA induced an early increase in biomarkers of oxidative damage, F_2-IsoPs and F_4-NeuroPs. Elevated levels of these in vivo markers of oxidative damage are in agreement with our previous findings (Montine et al., 2002; Milatovic et al., 2005b; Gupta et al., 2007), as well as those of others (Patel et al., 2001; Kiasalari et al., 2013, 2016) and indicate that KA injection leads to profound cerebral and neuronal oxidative damage in mice.

Our results also showed that the transient rise in F_2-IsoPs and F_4-NeuroPs is accompanied by rapid evolution of dendritic abnormalities, which is apparent in the significant decrease in dendritic length and spine density of pyramidal neurons as early as 30 min post KA injection. However, the recovery in oxidative damage biomarkers at 60 min following the injection was not paralleled by the rescue of damaged neurons from the CA1 hippocampal area. Extended seizure activity (60 min) induced the same level of dendritic length and spine density decrease when compared to 30 min following KA injection (Table 57.1). Together, these data suggest that both oxidative stress and neurodegeneration occur as an early response to seizures, but they do not establish whether oxidative stress is a cause or effect of seizure-induced CA1 cell damage. Neuronal damage processes triggered by sustained seizure activity may occur as a continuum, last longer than the formation of oxidative lipids and although not evident by the markers, and may already be in progress when the peak increases in F_2-IsoPs and F_4-NeuroPs occur. Thus, we investigated dynamic changes in lipid peroxidation and dendritic structures immediately after seizures, but future studies over the longer period should be able to determine the long-term time course of these spine and dendritic changes. It is very likely that the spine loss seen in our study is the initial phase of more chronic spine loss and progressive neurodegeneration reported in other studies (Isokawa and Levesque, 1991; Muller et al., 1993; Multani et al., 1994; Jiang et al., 1998; Zeng et al., 2007; Fang et al., 2017; Sabens Liedhegner et al., 2012).

NEUROINFLAMMATION AND OXIDATIVE INJURY

Neuroinflammation is a complex response to brain toxicity involving the activation of glia, release of inflammatory mediators, such as cytokines and chemokines, and the generation of ROS and RNS. The links among risk factors and the development of neuroinflammation are numerous and involve many complex interactions that contribute to vascular compromise, oxidative stress, and, ultimately, brain damage (Montine et al., 2002; Milatovic et al., 2003, 2004; Salinaro et al., 2018; Calabrese et al., 2016; Fang et al., 2017; Guangpin and Ping, 2016). Once this cascade of events is initiated, the process of neuroinflammation can become overactivated, resulting in further cellular damage and loss of neuronal functions.

Acute neuroinflammatory response resulting in phagocytic phenotype and the release of inflammatory mediators such as free radicals, cytokines, and chemokines (Tansey et al., 2007; Frank-Cannon et al., 2009) may be generally beneficial to the CNS because it tends to minimize further injury and contributes to repair of damaged tissue. In contrast, chronic neuroinflammation is a long-standing and often self-perpetuating neuroinflammatory response that persists long after an initial injury or insult. Sustained release of inflammatory mediators and increased oxidative and nitrosative stress activate additional microglia, promoting their proliferation and resulting in further release in inflammatory factors. Owing to this sustained nature of inflammation, the blood−brain barrier (BBB) may be compromised, thus increasing infiltration of peripheral macrophages into the brain parenchyma and perpetuating the inflammatory process further (Rivest, 2009). Rather than serving in a protective role, as does acute neuroinflammation, chronic neuroinflammation is most often detrimental and damaging to nervous tissue. Thus, whether neuroinflammation has beneficial or harmful outcomes in the brain may critically depend on the duration of the inflammatory response.

TABLE 57.1 Cerebral Concentrations of F_2-IsoPs and F_4-NeuroPs and Dendritic Degeneration of Hippocampal Pyramidal Neurons Following KA-Induced Seizures in Mice

	F_2-IsoPs (ng/g)	F_4-NeuroPs (ng/g)	Dendritic Length (μm)	Spine Density (Number of Spinal/100 μm Dendrite)
Control	3.07 ± 0.05	13.89 ± 0.58	1032.10 ± 61.41	16.45 ± 0.55
KA 30 min	4.81 ± 0.19*	34.27 ± 2.71*	363.44 ± 20.78*	8.81 ± 0.55*
KA 60 min	3.40 ± 0.18	18.55 ± 1.26	425.71 ± 23.04*	7.44 ± 0.56*

Data from KA-exposed mice were collected 30 min or 60 min postinjection. *One-way ANOVA showed $P < .0001$ for each end point. Bonferroni's multiple comparison test showed significant difference ($P < .001$) compared to vehicle-injected control.

Activation of innate immunity occurs simultaneously with several pathogenic processes and responses to stressors and injury, thereby greatly confounding any clear conclusion about cause-and-effect relationships. For these reasons, we have adopted a simple, but highly specific model of isolated innate immune activation: intracerebroventricular (ICV) injection of low dose lipopolysaccharide (LPS). LPS specifically activates innate immunity through a Toll-like receptor (TLR)–dependent signaling pathway (Imler and Hoffmann, 2001; Akira, 2003). Activation of proteins (CD14 and adaptor protein MyD88), signal transduction cascade, primarily via NF-κB activation but also through c-Fos/c-Jun-dependent pathways, culminate in the generation of effector molecules, including bacteriocidal molecules. Free radicals generated by NADPH oxidase and myeloperoxidase (MPO), as well as cytokines and chemokines, are known to attract an adaptive immune response (Milatovic et al., 2004).

We have employed an ICV model and identified the molecular and pharmacologic determinants of LPS-initiated cerebral neuronal damage in vivo (Montine et al., 2002; Milatovic et al., 2003, 2004). Interestingly, the degree of oxidative damage in this model was equivalent to what we observed in diseased regions of brain from patients with degenerative diseases (Reich et al., 2001). Results from our studies with mice showed that single ICV LPS injections induced delayed, transient elevation in both F_2-IsoPs and F_4-NeuroPs 24 h after exposure and then returned to baseline by 72 h postexposure (Table 57.2) (Milatovic et al., 2003). Although others have shown that altered gene transcription and increased cytokine secretion occur rapidly and peak within a few hours of LPS exposure, it is likely that the delay in neuronal oxidative damage observed in our experiments is related, at least in part, to the time required to deplete antioxidant defenses.

To address whether oxidative damage is related to neurodegeneration, we directly examined the dendritic compartment of neurons, which is largely transparent to the standard histological techniques used so far to investigate ICV LPS–induced damage. Using Golgi impregnation and Neurolucida-assisted morphometry of hippocampal CA1 pyramidal neurons (Leuner et al., 2003; Milatovic et al., 2010), we first determined the time course of dendritic structural changes following ICV LPS in mice. Our results show a time course similar to neuronal oxidative damage with maximal reduction in both dendrite length and dendritic spine density at 24 h post LPS and, remarkably, a return to baseline levels by 72 h (Table 57.2). Thus, these data strongly imply that neuronal oxidative damage is closely associated with dendritic degeneration following ICV LPS. We, along with others, have shown that primary neurons enriched in cell culture do not respond to LPS (Minghetti and Levi, 1995; Fiebich et al., 2001; Xie et al., 2002); therefore, our results also showed that LPS activated microglial-mediated paracrine oxidative damage to neurons.

It is becoming increasingly evident that neuroinflammation and associated oxidative damage plays a crucial role in the development and progression of brain diseases. Glia, and in particular microglia, are central to mediating the effects of neuroinflammation. Emerging evidence suggests that the number of activated microglia and the release of inflammatory mediators from these cells increase with age. This amplified or prolonged exposure to inflammatory molecules, including cytokines, chemokines, ROS, and PGs in the aged brain may impair neuronal plasticity and underlie a heightened neuroinflammatory response.

METAL TOXICITY AND OXIDATIVE INJURY

Essential metals are crucial for the maintenance of cell homeostasis. However, excessive exposure to some metals, including manganese (Mn) and mercury (Hg), present great health concerns and may lead to pathological conditions, including neurodegeneration.

TABLE 57.2 Cerebral Oxidative Damage and Dendritic Degeneration in Mice

	24 h	24 h	72 h	72 h
	ICV Saline	**ICV LPS**	**ICV Saline**	**ICV LPS**
F_2-IsoPs (ng/g tissue)	3.26 ± 0.19	4.77 ± 0.26*	3.13 ± 0.11	2.98 ± 0.17
F_4-NeuroPs (ng/g tissue)	13.91 ± 1.17	58.50 ± 5.98*	12.30 ± 1.18	16.80 ± 0.96
Dendritic length (μm)	1018 ± 113	324 ± 37*	848 ± 60	1030 ± 61
Spine density (spine no./100 μm dendrite)	16.89 ± 1.67	5.86 ± 0.57*	17.09 ± 1.13	16.77 ± 0.87

Effects of ICV saline (5 μL, control) and ICV LPS (5 μg/5 μL) treatment determined at 24 and 72 h following exposure. Each value represents mean ± SEM (n = 4–6). *One-way ANOVA showed $P < .001$ for each end point. Bonferroni's multiple comparison test showed significant difference ($P < .01$) compared to vehicle-injected control.

Neurodegenerative mechanisms and effects of Mn and Hg are associated with oxidative stress. Mn can oxidize dopamine (DA), generating reactive species, and also affect mitochondrial function, leading to accumulation of metabolites and culminating with oxidative damage. Cationic Hg forms have a strong affinity for nucleophiles, such as −SH, and target critical thiol and selenol molecules with antioxidant properties. Therefore, mediation of these processes and control of oxidative stress may provide a therapeutic strategy for the suppression of dysfunctional neuronal transmission and a slowing of the neurodegenerative process.

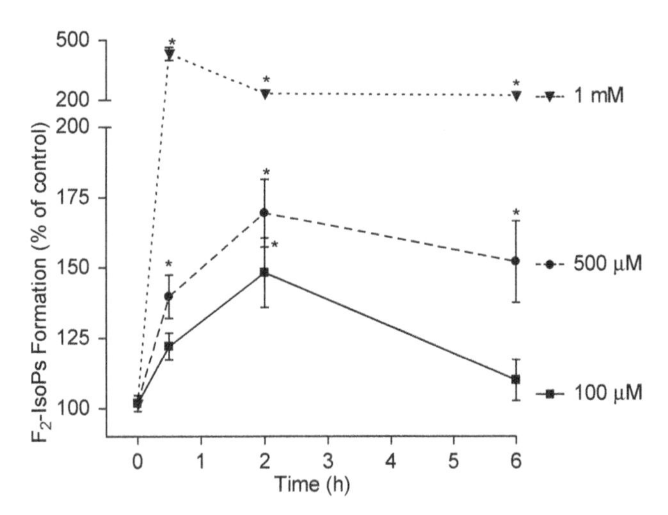

FIGURE 57.5 Effects of $MnCl_2$ on F_2-IsoPs formation in cultured astrocytes. Rat primary astrocyte cultures were incubated at 37°C in the absence or presence of $MnCl_2$ (100 μM, 500 μM, or 1 mM), and F_2-IsoPs levels were quantified at 30 min, 2 h, and 6 h. Data represent the mean ± SEM from three independent experiments. *Significant difference between values from control and Mn-treated astrocytes (*$P < .05$).

MANGANESE

Mn toxicity is primarily associated with neurological effects. Excessive accumulation of Mn in specific brain areas, such as the substantia nigra, the globus pallidus, and the striatum, produce neurotoxicity leading to a degenerative brain disorder. Although the mechanisms by which Mn induces neuronal damage are not well defined, its neurotoxicity appears to be regulated by a number of factors, including mitochondrial dysfunction, oxidative injury, and neuroinflammation.

Early studies on the cellular actions of Mn reported that mitochondria are the principal intracellular repository for the metal (Cotzias and Greenough, 1958). More recent data indicate that mitochondria actively sequester Mn, resulting in rapid inhibition of oxidative phosphorylation (Gavin et al., 1992). Manganese is bound to inner mitochondrial membrane or matrix proteins (Gavin et al., 1990) and thus directly interacts with proteins involved in oxidative phosphorylation. Mn directly inhibits complex II (Singh et al., 1974; Liu et al., 2013) and complexes I–IV (Zhang et al., 2003) in brain mitochondria and suppresses ATP-dependent calcium waves in astrocytes, suggesting that Mn promotes potentially disruptive mitochondrial sequestration of calcium (Tjalkens et al., 2006). Elevated matrix calcium increases the formation of ROS by the electron transport chain (Kowaltowski et al., 1995) and results in inhibition of aerobic respiration (Kruman and Mattson, 1999).

Our studies with primary astrocytes and neurons have shown that Mn exposure induces an increase in biomarkers of oxidative stress (Milatovic et al., 2007, 2009). We have measured F_2-IsoPs (Morrow and Roberts, 1999; Milatovic and Aschner, 2009) and showed that astrocytes exposed to Mn concentrations known to elicit neurotoxic effects (100 μM, 500 μM, or 1 mM) induced significant elevations in F_2-IsoPs levels at all investigated exposure times (Fig. 57.5). Thus, increases in ROS, potentially damaging mitochondria directly or through the effects of secondary oxidants such as superoxide, H_2O_2, or $ONOO^-$, mediate Mn-induced oxidative damage. Moreover, superoxide produced in the mitochondrial electron transport chain may catalyze the transition shift of Mn^{2+} to Mn^{3+} through a set of reactions similar to those mediated by superoxide dismutase and thus lead to the increased oxidant capacity of this metal (Gunter et al., 2006). Consequent oxidative damage produces an array of deleterious effects: it may cause structural and functional derangement of the phospholipid bilayer of membranes; disrupt energy metabolism, metabolite biosynthesis, and calcium and iron homeostasis; and initiate apoptosis (Attardi and Schatz, 1988; Uchida, 2003; Alaimo et al., 2011).

Consistently preceding the Mn-induced increase in biomarkers of oxidative damage (F_2-IsoPs) (Fig. 57.5), our study showed an early decrease in astrocytic ATP levels (Milatovic et al., 2007). As a consequence, ATP depletion or a perturbation in energy metabolism might diminish the ATP-requiring neuroprotective action of astrocytes, such as glutamate and glutamine uptake and free radical scavenging (Rao et al., 2001). In addition, depletion of high-energy phosphates may affect intracellular Ca^{2+} in astrocytes through mechanisms involving the disruption of mitochondrial Ca^{2+} signaling. This assertion is supported by data showing that Mn inhibits Na^+-dependent Ca^{2+} efflux (Gavin et al., 1990) and respiration in brain mitochondria (Zhang et al., 2004), both critical for maintaining normal ATP levels and ensuring adequate intermitochondrial signaling. A decrease in ATP following Mn exposure is also associated with excitotoxicity, suggesting a direct effect on astrocytes with subsequent impairment of neuronal function. Mn downregulates the glutamate transporter GLAST in astrocytes (Erikson and Aschner, 2002) and decreases levels of glutamine synthase in exposed primates (Erikson et al., 2008). Studies with a neonatal rat model indicated that both pinacidil, a potassium channel agonist, and nimodipine, a Ca^{2+} channel antagonist, reversed Mn neurotoxicity and loss of glutamine synthase activity, further indicating excitotoxicity in the mechanism of Mn-induced neurotoxicity. Excessive Mn may lead to excitotoxic neuronal injury by both decreased astrocytic glutamate uptake and loss of ATP-mediated inhibition of glutamatergic synapses.

Oxidative stress as an important mechanism in Mn-induced neurotoxicity is also confirmed in our in vivo model. Analyses of cerebral biomarkers of oxidative damage revealed that a one-time challenge of mice with Mn (100 mg/kg, s.c.) was sufficient to produce significant increases in F_2-IsoPs (Table 57.3) 24 h following the last injection, respectively.

Increased striatal concentrations of ascorbic acid and GSH, antioxidants that when increased signal the presence of an elevated burden from ROS, as well as other markers of oxidative stress have been previously reported (Desole et al., 1994; Dobson et al., 2003; Erikson et al., 2007). Mn-induced decrease in GSH and increased metallothionein were reported in rats (Dobson et al., 2003) and nonhuman primate studies (Erikson et al., 2007). ROS may act in concert with RNS derived from astroglia and microglia to facilitate the Mn-induced degeneration of DAergic neurons. DAergic neurons possess reduced antioxidant capacity, as evidenced by low intracellular GSH, which renders these neurons more vulnerable to oxidative stress and glial activation relative to other cell types (Sloot et al., 1994; Greenamyre et al., 1999; Filipov and Dodd, 2011). Therefore, the overactivation of glia and the release of additional neurotoxic factors may represent a crucial component associated with the degenerative process of DAergic neurons (Filipov and Dodd, 2011).

In addition to a decrease in mitochondrial membrane potential and the depletion of high-energy phosphates, Mn-induced ROS generation is also associated with inflammatory responses and release of inflammatory mediators, including prostaglandins. Our studies have confirmed that parallel to the increase in biomarkers of oxidative damage, Mn exposure induced an increase in the biomarker of inflammation, prostaglandin E_2 (PGE_2), in vitro and in vivo (Milatovic et al., 2007, 2009). Results from our in vivo study showed that Mn exposure induced a time-dependent increase in PGE_2 (Table 57.3). Previous studies have also shown an inflammatory response of glial cells following Mn exposure (Chen et al., 2006b; Zhang et al., 2009; Zhao et al., 2009). Mn potentiates lipopolysaccharide-induced increases in proinflammatory cytokines in glial cultures (Filipov et al., 2005) and an increase in nitric oxide production (Chang and Liu, 1999). An increase in proinflammatory genes, tumor necrosis factor-α, iNOS, and activated inflammatory proteins such as P-p38, P-ERK, and P-JNK have been measured in primary rat glial cells after Mn exposure (Chen et al., 2006b). However, data from our study indicated that release of proinflammatory mediators following Mn exposure is not only associated with glial response but also with neurons, suggesting that these two events are mechanistically related, with neuroinflammation either alone or in combination with activated glial response contributing to oxidative damage and consequent cell injury.

TABLE 57.3 Cerebral F_2-IsoPs and PGE_2 Levels in Saline (Control) or $MnCl_2$ (100 mg/kg, s.c.) Exposed Mice

Exposure	F_2-IsoPs (ng/g Tissue)	PGE_2 (ng/g Tissue)
Control (saline)	3.013 ± 0.03939	9.488 ± 0.3091
Single Mn	4.302 ± 0.3900*	12.030 ± 0.4987*
Multiple Mn	4.211 ± 0.4013*	14.220 ± 1.019*

Brains from mice exposed once or three times (days 1, 4, and 7) to $MnCl_2$ were collected 24 h post last injection. Each value represents mean ± SEM (n = 4–6). *One-way ANOVA showed $P < .001$ for each end point. Bonferroni's multiple comparison test showed significant difference ($P < .05$) compared to vehicle-injected control.

Features of Mn neurotoxicity reflect alterations in the integrity of DAergic striatal neurons and DA neurochemistry, including decreased DA transport function and/or striatal DA levels. The striatum is a major recipient structure of neuronal efferents in the basal ganglia. It receives excitatory input from the cortex and dopaminergic input from substantia nigra and projects to the internal segment of the globus pallidus (Dimova et al., 1993; Saka et al., 2002). Nigrostriatal dopamine neurons appear to be particularly sensitive to Mn-induced toxicity (Sloot and Gramsbergen, 1994; Sloot et al., 1994; Defazio et al., 1996). Intense or prolonged Mn exposure in adulthood causes long-term reductions in striatal DA levels and induces a loss of autoreceptor control over DA release (Autissier et al., 1982; Komura and Sakamoto, 1992). Nigrostriatal DA axons synapse onto striatal medium spiny neurons (MSNs), and these neurons have radially projecting dendrites that are densely studded with spines (Wilson and Groves, 1980).

Postmortem studies of PD patients have revealed a marked decrease in MSN spine density and dendritic length (Stephens et al., 2005; Zaja-Milatovic et al., 2005). Similar morphological changes in MSNs were seen in animal models of Parkinsonism (Arbuthnott et al., 2000; Day et al., 2006). Our study investigated the effects of Mn on degeneration of striatal neurons. Representative images of Golgi-impregnated striatal sections with their traced MSNs from control and Mn-exposed animals are presented in Fig. 57.6.

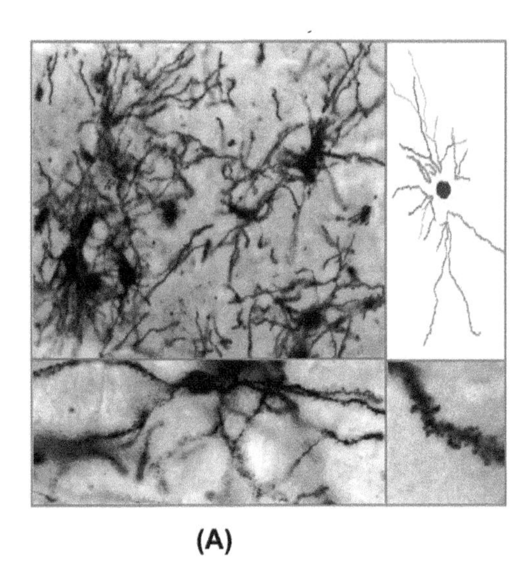

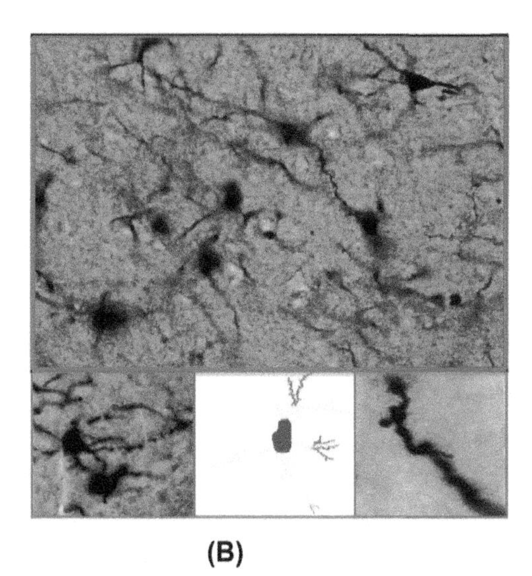

FIGURE 57.6 Photomicrographs of mouse striatal sections with representative tracings of medium spiny neurons (MSNs) from mice treated with saline (control) (A) or $MnCl_2$ (100 mg/kg, s.c.) (B). Brain from mouse exposed three times (days 1, 4, and 7) to $MnCl_2$ was collected 24 h post last injection. Treatment with Mn induced degeneration of the striatal dendritic system and a decrease in total number of spines and length of dendrites of MSNs. Tracing and counting were done using a Neurolucida system at 100× under oil immersion (MicroBrightField, VT). Colors indicate the degree of dendritic branching (yellow = 1 degree, red = 2 degrees, purple = 3 degrees, green = 4 degrees, turquoise = 5 degrees).

Images of neurons with Neurolucida-assisted morphometry show that Mn-induced oxidative damage and neuroinflammation targeted the dendritic system with profound dendrite regression of striatal MSNs. Although single Mn exposure altered the integrity of the dendritic system and induced a significant decrease in spine number (Fig. 57.7A) and total dendritic lengths (Fig. 57.7B) of MSNs, prolonged Mn exposure led to a further reduction in spine number and dendritic length. Our results indicate that MSNs neurodegeneration could result from loss of spines, removing the pharmacological target for DA-replacement therapy, without overt MSN death (Stephens et al., 2005; Zaja-Milatovic et al., 2005).

Together, several studies suggest that oxidative stress, mitochondrial dysfunction, and neuroinflammation are the underlying mechanisms in Mn-induced vulnerability of DAergic neurons. The mediation of any of these mechanisms and control of alterations in biomarkers of oxidative injury, neuroinflammation, and synaptodendritic degeneration may provide a therapeutic strategy for the suppression of dysfunctional DAergic transmission and slowing of the neurodegenerative process. In addition, multiple mechanisms of Mn action are not

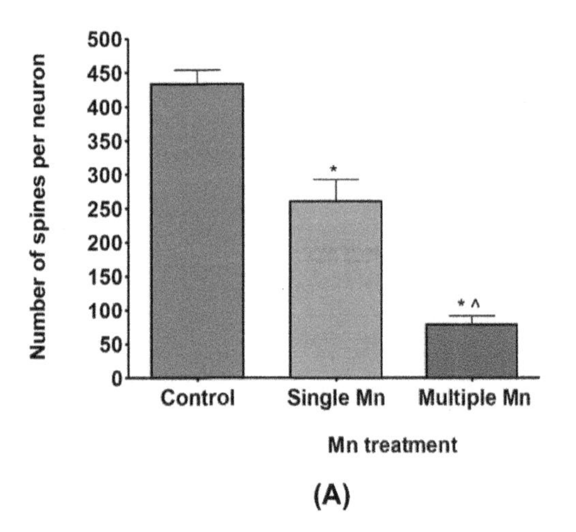

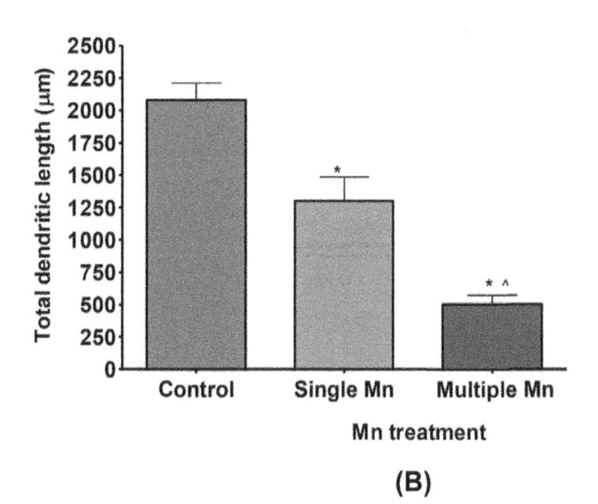

FIGURE 57.7 Total number of spines (A) and total dendritic lengths (B) of MSNs from striatal sections of mice exposed to saline (control), single Mn injection (100 mg/kg, s.c.), or multiple Mn injections (100 mg/kg, s.c.) on days 1, 4, and 7. Mice were sacrificed 24 h after the last injection. *Significant difference between values from control and Mn-treated mice (*$P < .01$). ^Significant difference between values from single Mn injection versus multiple (8 days) Mn treatment (^$P < .001$).

sufficiently known and may vary with environmental factors and susceptibilities, including single nucleotide polymorphisms that may alter Mn homeostasis, transport, and metabolism.

MERCURY

Mercury (Hg) is one of the most toxic elements in the periodic table. Although Hg is present in nature, it has also been released into the environment for centuries as a result of anthropogenic activities. It exists within the environment in three different chemical forms: elemental mercury vapor, inorganic mercury salts, and organic mercury. The distribution, toxicity, and metabolism of mercury are greatly dependent on its chemical form. Organic mercury compounds, such as methylmercury (MeHg), have been extensively studied because they are able to reach high levels in the central nervous system (CNS), leading to neurotoxic effects (Clarkson and Magos, 2006; Aschner et al., 2007; Dos Santos et al., 2018). Although the precise mechanisms of MeHg neurotoxicity are ill-defined, oxidative stress and altered mitochondrial and cell membrane permeabilities appear to be critical factors in its pathogenesis.

In vivo and in vitro biochemical studies employing glial and neuronal cultures have shown increased ROS formation with MeHg exposure (Yee and Choi, 1996; Sorg et al., 1998; Mundy and Freudenrich, 2000; Gasso et al., 2001; Shanker and Aschner, 2003; Dos Santos et al., 2018). Mitochondria are believed to be major targets of MeHg-induced toxicity (Limke and Atchison, 2002). Specifically, highly enriched Hg concentrations were found in mitochondrial fractions with the lowest Hg concentration found in the cytosol of livers from Hg-exposed animals (Chen et al., 2006a). Most of the bioenergetic experiments with Hg report the uncoupling of oxidative phosphorylation (Weinberg et al., 1982), inhibition of ATP synthesis (Atchison and Harem, 1994), impairment of the respiratory chain (Santos et al., 1997), and depletion of intracellular ATP and ADP (Palmeira et al., 1997). A study using a selective probe for mitochondrial reactive oxygen intermediates as well as other probes demonstrated a significant MeHg-induced increase in intracellular superoxide anion, hydrogen peroxide, and hydroxyl radicals, indicating that the mitochondrial electron transport chain is an early, primary site for ROS formation (Allen et al., 2001; Shanker et al., 2004, 2005). Additionally, MeHg exposure disrupts Ca^{2+} regulation in mitochondria derived from rat brains by decreasing Ca^{2+} uptake and inducing Ca^{2+} release (Levesque and Atchison, 1991). Results from our studies confirmed that MeHg induces increases in ROS and that the lipid peroxidation biomarkers of oxidative injury, F_2-IsoPs, are increased in MeHg-exposed astrocytes (Fig. 57.8).

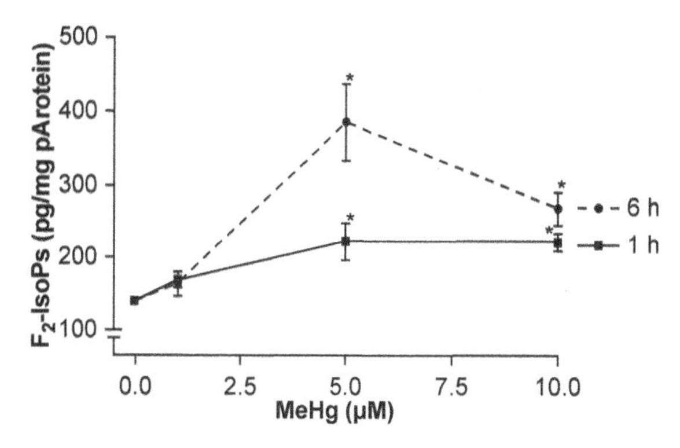

FIGURE 57.8 Effect of MeHg on F_2-IsoPs formation in cultured astrocytes. Rat primary astrocyte cultures were incubated at 37°C in the absence or presence of MeHg (1, 5, and 10 μM) and F_2-IsoPs levels quantified at 1 and 6 h, respectively. Data represent the mean ± S.E. from three independent experiments. $*P < .05$ versus control by one-way ANOVA followed by Bonferroni multiple comparison tests.

Another consequence of increased oxidative stress is the induction of the mitochondrial permeability transition (MPT), a Ca^{2+}-dependent process characterized by the opening of the permeability transition pore (PTP) in the inner mitochondrial membrane. This causes increased permeability to protons, ions, and other solutes ≤1500 Da (Zoratti and Szabo, 1995), leading to a collapse of the mitochondrial inner membrane potential ($\Delta\Psi_m$). Loss of the $\Delta\Psi_m$ results in colloid osmotic swelling of the mitochondrial matrix, movement of metabolites across the inner membrane, defective oxidative phosphorylation, cessation of ATP synthesis, and further generation of ROS. Our studies confirmed a concentration-dependent deleterious effect of MeHg on mitochondrial $\Delta\Psi_m$ in cultured astrocytes. It is generally believed that increased $[Ca^{2+}]_i$ triggers ROS formation and increased oxidative stress is a major factor for MPT induction (Castilho et al., 1995; Halestrap et al., 1997) and mitochondrial depolarization.

Of particular importance, in vivo studies with rats showed that MeHg combines covalently with sulfhydryl (thiol) groups from plasma cholinesterase, leading to the enzyme inhibition (Hastings et al., 1975). After this important observation, several in vitro and in vivo experimental studies showed that sulfhydryl-containing enzymes are inhibited by MeHg (Magour, 1986; Kung et al., 1987; Rocha et al., 1993). These observations led to the notion that the direct chemical interaction among MeHg and thiol groups from proteins and nonprotein molecules, such as GSH, plays a crucial role in MeHg-induced neurotoxicity (Aschner and Syversen, 2005).

Together, multiple studies demonstrated that MeHg exposure is associated with changes in membrane permeability and glutamine/glutamate cycling, increases in ROS formation, and consequent lipid peroxidation. Furthermore, lipid peroxidation increases mitochondrial and cellular permeability alterations involving GSH (Gstraunthaler et al., 1983; Strubelt et al., 1996) and calcium depletion (Strubelt et al., 1996). These outcomes work together to create a continuous cycle where acceleration of the mitochondrial chain induces oxidative stress, lipid peroxidation, and depletion of antioxidant defenses, which, in turn, diminish membrane permeability and accelerate the respiratory chain, thus generating more ROS. Ultimately, the additive or synergistic mechanisms of cellular disruption caused by MeHg lead to cellular dysfunction and cell death.

CONCLUDING REMARKS AND FUTURE DIRECTIONS

It is becoming increasingly evident that oxidative stress and associated neuronal damage plays a crucial role in the development and progression of brain diseases. However, measuring oxidative stress in biological systems is complex and requires accurate quantification of either free radicals or damaged biomolecules. One method to quantify oxidative injury is to measure nonenzymatic lipid peroxidation products, F_2-IsoPs. The quantification of F_2-IsoPs has provided a powerful approach to advance our understanding of the role of oxidative damage in a wide variety of research models and disease states. We have applied this methodology and explored cerebral oxidative damage in several models of neurodegeneration, including excitotoxicity generated by kainic acid, neurotoxicity associated with anticholinesterase agents and metals, and innate immune activation by lipopolysaccharide. Results from our studies supported an association between oxidative stress and neurotoxicity and suggested that oxidative stress, mitochondrial dysfunction, and neuroinflammation are underlying mechanisms in excitotoxicity- and metal-induced degeneration of dendritic systems in different brain areas. Future studies should be directed at deciphering the mechanisms of protection and addressing attenuation of neurotoxicity via radical scavenging mechanisms. Complementary studies should also guide the development of selective and efficacious antioxidant therapies that target neurotoxic mechanisms while maintaining neuroprotective actions.

References

Akira, S., 2003. Toll-like receptor signaling. J. Biol. Chem. 278, 38105–38108.

Alaimo, A., Gorojod, R.M., Kotler, M.L., 2011. The extrinsic and intrinsic apoptotic pathways are involved in manganese toxicity in rat astrocytoma C6 cells. Neurochem. Int. 59, 297–308.

Allen, J.W., El-Oqayli, H., Aschner, M., et al., 2001. Methylmercury has a selective effect on mitochondria in cultured astrocytes in the presence of [U-^{13}C] glutamate. Brain Res. 908, 149–154.

Arbuthnott, G.W., Ingham, C.A., Wickens, J.R., 2000. Dopamine and synaptic plasticity in the neostriatum. J. Anat. 196, 587–596.

Aschner, M., Syversen, T., 2005. Methylmercury: recent advances in the understanding of its neurotoxicity. Ther. Drug Monit. 27, 278–283.

Aschner, M., Syversen, T., Souza, D.O., et al., 2007. Involvement of glutamate and reactive oxygen species in methylmercury neurotoxicity. Braz. J. Med. Biol. Res. 40, 285–291.

Atchison, W.D., Harem, M.F., 1994. Mechanisms of methylmercury-induced neurotoxicity. FASEB J. 8, 622–629.

Attardi, G., Schatz, G., 1988. Biogenesis of mitochondria. Annu. Rev. Cell Biol. 4, 289–333.

Autissier, N., Rochette, L., Dumas, P., 1982. Dopamine and norepinephrine turnover in various regions of the rat brain after chronic manganese chloride administration. Toxicology 24, 175–182.

Basu, S., 1998. Radioimmunoassay of 8-iso-prostaglandin F2alpha: an index of oxidative injury via free radical catalysed lipid peroxidation. Prostaglandins Leukot. Essent. Fatty Acids 58, 319–325.

Basu, S., 2008. F2-isoprostanes in human health and diseases: from molecular mechanisms to clinical implications. Antioxidants Redox Signal. 10, 1405–1434.

Beal, M.F., 2000. Oxidative metabolism. Ann. N.Y. Acad. Sci. 924, 164–169.

Becker, E.B., Bonni, A., 2004. Cell cycle regulation of neuronal apoptosis in development and disease. Prog. Neurobiol. 1, 1–25.

Bondy, S.C., Lee, D.K., 1993. Oxidative stress induced by glutamate receptor agonists. Brain Res. 610, 229–233.

Bose, R., Schnell, C.P., Pinsky, C., Zitko, V., 1992. Effects of excitotoxin on free radical indices in mouse brain. Toxicol. Lett. 60, 211–219.

Butterfield, D.A., Reed, T., Perluigi, M., 2006. Elevated protein-bound levels of the lipid peroxidation product, 4-hydroxy-2-nonenal, in brain from persons with mild cognitive impairment. Neurosci. Lett. 3, 170–173.

Cadenas, E., Davies, K.J., 2000. Mitochondrial free radical generation, oxidative stress, and aging. Free Radic. Biol. Med. 29, 222–230.

Calabrese, V., Giordano, J., Signorile, A., et al., 2016. Major pathogenic mechanisms in vascular dementia: roles of cellular stress response and hormesis in neuroprotection. J. Neurosci. Res. 94, 1588–1603.

Carpentier, P., Lamrinis, M., Blanchet, G., 1991. Early dendritic changes in hippocampal pyramidal neurons (field CA1) of rats subjected to acute soman intoxication – a light microscopic study. Brain Res. 541, 293–299.

Castilho, R.F., Kowaltowski, A.J., Meinicke, A.R., et al., 1995. Permeabilization of the inner mitochondrial membrane by Ca^{2+} ions is stimulated by t-butyl hydroperoxide and mediated by reactive oxygen species generated by mitochondria. Free Radic. Biol. Med. 18, 479–486.

Chakravarti, B., Chakravarti, D.N., 2007. Oxidative modification of proteins: age-related changes. Gerontology 3, 128–139.

Chang, J.Y., Liu, L.Z., 1999. Manganese potentiates nitric oxide production by microglia. Brain Res. Mol. Brain Res. 68, 22–28.

Cheignon, C., Tomas, M., Bonnefont-Rousselot, D., et al., 2018. Oxidative stress and the amyloid beta peptide in Alzheimer's disease. Redox Biol. 14, 450–464.

Chen, C., Qu, L., Zhao, J., 2006a. Accumulation of mercury, selenium and their binding proteins in porcine kidney and liver from mercury-exposed areas with the investigation of their redox response. Sci. Total Environ. 366, 627–637.

Chen, M.K., Lee, J.S., McGlothan, J.L., 2006b. Acute manganese administration alters dopamine transporter levels in the non-human primate striatum. Neurotoxicology 27, 229–236.

Cheng, J.Z., Singhal, S.S., Saini, M., 1999. Effects of mGST A4 transfection on 4-hydroxynonenal-mediated apoptosis and differentiation of K562 human erythroleukemia cells. Arch. Biochem. Biophys. 372, 29–36.

Clarkson, T.W., Magos, L., 2006. The toxicology of mercury and its chemical compounds. Crit. Rev. Toxicol. 36, 609–662.

Cock, H.R., Tong, X., Hargreaves, I.P., 2002. Mitochondrial dysfunction associated with neuronal death following status epilepticus in rat. Epilepsy Res. 48, 157–168.

Colurso, G.J., Nilson, J.E., Vervoort, L.G., 2003. Quantitative assessment of DNA fragmentation and beta-amyloid deposition in insular cortex and midfrontal gyrus from patients with Alzheimer's disease. Life Sci. 14, 1795–1803.

Connor, W.E., Neuringer, M., Reisbick, S., 1992. Essential fatty acids: the importance of n-3 fatty acids in the retina and brain. Nutr. Rev. 50, 21–29.

Cotzias, G.C., Greenough, J.J., 1958. The high specificity of the manganese pathway through the body. J. Clin. Invest. 37, 1298–1305.

Dalle-Donne, I., Rossi, R., Giustarini, D., et al., 2003. Protein carbonyl groups as biomarkers of oxidative stress. Clin. Chim. Acta 329, 23–38.

Day, M., Wang, Z., Ding, J., 2006. Selective elimination of glutamatergic synapses on striatopallidal neurons in Parkinson disease models. Nat. Neurosci. 9, 251–259.

De Zwart, L.L., Meerman, J.H.N., Commandeur, J.N.M., Vermeulen, N.P.E., 1999. Biomarkers of free radical damage applications in experimental animals and in humans. Free Radic. Biol. Med. 26, 202–226.

Deavall, D.G., Martin, E.A., Horner, J.M., Roberts, R., 2012. Drug-induced oxidative stress and toxicity. J. Toxicol. 2012, 645460.

Defazio, G., Soleo, L., Zefferino, R., Livrea, P., 1996. Manganese toxicity in serumless dissociated mesencephalic and striatal primary culture. Brain Res. Bull. 40, 257–262.

Desole, M.S., Miele, M., Esposito, G., 1994. Dopaminergic system activity and cellular defense mechanisms in the striatum and striatal synaptosomes of the rat subchronically exposed to manganese. Arch. Toxicol. 68, 566–570.

Dettbarn, W.-D., Milatovic, D., Zivin, M., Gupta, R.C., 2001. Oxidative stress, acetylcholine and excitotoxicity. In: Marwah, J., Kanthasamy, A. (Eds.), Antioxidants and Free Radicals in Health and Diseases. Prominant Press, Scottsdale, AZ, pp. 183–211.

Dettbarn, W.-D., Milatovic, D., Gupta, R.C., 2006. Oxidative stress in anticholinesterase-induced excitoxicity. In: Gupta, R.C. (Ed.), Toxicology of Organophosphate and Carbamate Compounds. Academic Press/Elsevier, Amsterdam, pp. 511–532.

Devasagayam, T.P., Boloor, K.K., Ramasarma, T., 2003. Methods for estimating lipid peroxidation: an analysis of merits and demerits. Indian J. Biochem. Biophys. 40, 300–308.

Dimova, R., Vuillet, J., Nieoullon, A., et al., 1993. Ultrastructural features of the choline acetyltransferase-containing neurons and relationships with nigral dopaminergic and cortical afferent pathways in the rat striatum. Neuroscience 53, 1059–1071.

Dix, T.A., Aikens, J., 1993. Mechanisms and biological relevance of lipid peroxidation initiation. Chem. Res. Toxicol. 6, 2–18.

Dizdaroglu, M., Jaruga, P., Birinciogl, M., Rodriguez, H., 2002. Free radical-induced damage to DNA: mechanisms and measurement. Free Radic. Biol. Med. 11, 1102–1115.

Dobson, A.W., Weber, S., Dorman, D.C., 2003. Oxidative stress is induced in the rat brain following repeated inhalation exposure to manganese sulfate. Biol. Trace Elem. Res. 93, 113–126.

Dos Santos, A.A., López-Granero, C., Farina, M., et al., 2018. Oxidative stress, caspase-3 activation and cleavage of ROCK-1 play an essential role in MeHg-induced cell death in primary astroglial cells. Food Chem. Toxicol. 113, 328–336.

Droge, W., 2002. Aging-related changes in the thiol/disulfide redox state: implications for the use of thiol antioxidants. Exp. Gerontol. 37, 1333–1345.

Du, X., Wang, X., Geng, M., 2018. Alzheimer's disease hypothesis and related therapies. Transl. Neurodegener. 7 (2), 1–7.

Erikson, K.M., Aschner, M., 2002. Manganese causes differential regulation of glutamate transporter (GLAST) taurine transporter and metallothionein in cultured rat astrocytes. Neurotoxicology 23, 595–602.

Erikson, K.M., Dorman, D.C., Lash, L.H., et al., 2007. Manganese inhalation by rhesus monkeys is associated with brain regional changes in biomarkers of neurotoxicity. Toxicol. Sci. 97, 459–466.

Erikson, K.M., Dorman, D.C., Lash, L.H., Aschner, M., 2008. Duration of airborne-manganese exposure in rhesus monkeys is associated with brain regional changes in biomarkers of neurotoxicity. Neurotoxicology 29, 377–385.

Espinosa-Diez, C., Miguel, V., Mennerich, D., et al., 2015. Antioxidant responses and cellular adjustments to oxidative stress. Redox Biol. 6, 183–197.

Esterbauer, H., Schaur, R.J., Zollner, H., 1991. Chemistry and biochemistry of 4-hydroxynonenal, malonaldehyde and related aldehydes. Free Radical Biol. Med. 11, 81–128.

Famm, S.S., Morrow, J.D., 2003. The isoprostanes: unique products of arachidonic acid oxidation – a review. Curr. Med. Chem. 10, 1723–1740.

Fang, C., Gu, L., Smerin, D., et al., 2017. The interrelation between reactive oxygen species and autophagy in neurological disorders. Oxid. Med. Cell. Longev. 1–16.

Farooqui, A.A., Yi Ong, U., Lu, X.R., et al., 2001. Neurochemical consequences of kainate-induced toxicity in brain: involvement of arachidonic acid release and prevention of toxicity by phospholipase A(2) inhibitors. Brain Res. Rev. 38, 61–78.

Federico, A., Cardaioli, E., Pozzo, P.D., 2012. Mitochondria, oxidative stress and neurodegeneration. J. Neurol. Sci. 322, 254–262.

Fessel, J.P., Porter, N.A., Moore, K.P., et al., 2002. Discovery of lipid peroxidation products formed in vivo with a substituted tetrahydrofuran ring (isofurans) that are favored by increased oxygen tension. Proc. Natl. Acad. Sci. U.S.A. 99, 16713–16718.

Fiebich, B.L., Schleicher, S., Spleiss, O., et al., 2001. Mechanisms of prostaglandin E2-induced interleukin-6 release in astrocytes: possible involvement of EP4-like receptors, p38 mitogen-activated protein kinase and protein kinase C. J. Neurochem. 79, 950–958.

Filipov, N.M., Dodd, C.A., 2011. Role of glial cells in manganese neurotoxicity. J. Appl. Toxicol. 32, 310–317.

Filipov, N.M., Seegal, R.F., Lawrence, D.A., 2005. Manganese potentiates in vitro production of proinflammatory cytokines and nitric oxide by microglia through a nuclear factor kappa B-dependent mechanism. Toxicol. Sci. 84, 139–148.

Fishel, M.L., Vasko, M.R., Kelley, M.R., 2007. DNA repair in neurons: so if they don't divide what's to repair? Mutat. Res. (1–2), 24–36.

Frank-Cannon, T.C., Alto, L.T., McAlpine, F.E., Tansey, M.G., 2009. Does neuroinflammation fan the flame in neurodegenerative diseases? Mol. Neurodegener. 4, 1–13.

Gao, L., Yin, H., Milne, G.L., et al., 2006. Formation of F-ring isoprostane-like compounds (F3-isoprostanes) in vivo from eicosapentaenoic acid. J. Biol. Chem. 281, 14092–14099.

Gasc, N., Taché, S., Rathahao, E., 2007. 4-hydroxynonenal in foodstuffs: heme concentration, fatty acid composition and freeze-drying are determining factors. Redox Rep. 12, 40–44.

Gasso, S., Cristofol, R.M., Selema, G., 2001. Antioxidant compounds and Ca^{2+} pathway blockers differentially protect against methylmercury and mercuric chloride neurotoxicity. J. Neurosci. Res. 66, 135–145.

Gavin, C.E., Gunter, K.K., Gunter, T.E., 1990. Manganese and calcium efflux kinetics in brain mitochondria. Relevance to manganese toxicity. Biochem. J. 266, 329–334.

Gavin, C.E., Gunter, K.K., Gunter, T.E., 1992. Mn^{2+} sequestration by mitochondria and inhibition of oxidative phosphorylation. Toxicol. Appl. Pharmacol. 115, 1–5.

Gollnick, P.D., Bertocci, L.A., Kelso, T.B., 1990. The effect of high intensity exercise on the respiratory capacity of skeletal muscle. Pfluger's Arch Eur J Physiol 415, 407–413.

Greenamyre, J.T., McKenzie, G., Peng, T.I., Stephans, S.E., 1999. Mitochondrial dysfunction in Parkinson's disease. Biochem. Soc. Symp. 66, 85–97.

Gstraunthaler, G., Pfaller, W., Kotanko, P., 1983. Glutathione depletion and in vitro lipid peroxidation in mercury or maleate induced acute renal failure. Biochem. Pharmacol. 2, 2969–2972.

Guangpin, C., Ping, Q., 2016. ROS mediated inflammation and neurological diseases in central nervous system. Chinese Journal of Histochemistry and Cytochemistry 25 (3), 285–290.

Guéraud, F., Atalay, M., Bresgen, N., 2010. Advances in methods for the determination of biologically relevant lipid peroxidation products. Free Radic. Res. 44, 1098–1124.

Gunter, T.E., Gavin, C.E., Aschner, M., Gunter, K.K., 2006. Speciation of manganese in cells and mitochondria: a search for the proximal cause of manganese neurotoxicity. Neurotoxicology 27, 765–776.

Gupta, R.C., Milatovic, D., Dettbarn, W.-D., 2001a. Depletion of energy metabolites following acetylcholinesterase inhibitor-induced status epilepticus: protection by antioxidants. Neurotoxicology 22, 271–282.

Gupta, R.C., Milatovic, D., Dettbarn, W.-D., 2001b. Nitric oxide modulates high-energy phosphates in brain regions of rats intoxicated with diisopropylphosphorofluoridate or carbofuran: prevention by N-tert-butyl-α-phenylnitrone or vitamin E. Arch. Toxicol. 75, 346–356.

Gupta, R.C., Milatovic, S., Dettbarn, W.-D., Aschner, M., Milatovic, D., 2007. Neuronal oxidative injury and dendritic damage induced by carbofuran: protection by memantine. Toxicol. Appl. Pharmacol. 219, 97–105.

Gutteridge, J.M., Halliwell, B., 1990. The measurement and mechanism of lipid peroxidation in biological systems. Trends Biochem. Sci. 15, 129–135.

Halestrap, A.P., Woodfield, K.Y., Connern, C.P., 1997. Oxidative stress, thiol reagents, and membrane potential modulate the mitochondrial permeability transition by affecting nucleotide binding to the adenine nucleotide translocase. J. Biol. Chem. 272, 3346–3354.

Halliwell, B., 2007. Oxidative stress and neurodegeneration: where are we now? J. Neurochem. 97, 1634–1658.

Hastings, F.L., Lucier, G.W., Klein, R., 1975. Methylmercury-cholinesterase interactions in rats. Environ. Health Perspect. 12, 127–130.

Hayes, J.D., Pulford, D.J., 1995. The glutathione S-transferase supergene family: regulation of GST and the contribution of the isoenzymes to cancer chemoprotection and drug resistance. Crit. Rev. Biochem. Mol. Biol. 30, 445–600.

Heinemann, U., Buchheim, K., Gabriel, S., 2002. Cell death and metabolic activity during epileptiform discharges and status epilepticus in the hippocampus. Prog. Brain Res. 135, 197–210.

Il'yasova, D., Scarbrough, P., Spasojevic, I., 2012. Urinary biomarkers of oxidative stress. Clin. Chim. Acta 413, 1446–1453.

Imler, J.L., Hoffmann, J.A., 2001. Toll receptors in innate immunity. Trends Cell Biol. 11, 304–311.

Isokawa, M., Levesque, M.F., 1991. Increased NMDA responses and dendritic degeneration in human epileptic hippocampal neurons in slices. Neurosci. Lett. 132, 212–216.

Janicka, M., Kot-Wasik, A., Kot, J., Namiesnik, J., 2010. Isoprostanes-biomarkers of lipid peroxidation: their utility in evaluating oxidative stress and analysis. Int. J. Mol. Sci. 11, 4631–4659.

Jiang, M., Lee, C.L., Smith, K.L., Swann, J.W., 1998. Spine loss and other persistent alterations of hippocampal pyramidal cell dendrites in a model of early-onset epilepsy. J. Neurosci. 18, 8356–8368.

Johnson, J.A., Barbary, A., Kornguth, S.E., et al., 1993. Glutathione S-transferase isoenzymes in rat brain neurons and glia. J. Neurosci. 13, 2013–2023.

Kadiiska, M.B., Gladen, B.C., Baird, D.D., 2005. Biomarkers of oxidative stress study II: are oxidation products of lipids, proteins, and DNA markers of CCl4 poisoning? Free Radic. Biol. Med. 38, 698–710.

Kiasalari, Z., Roghani, M., Khalili, M., Rahmati, B., Baluchnejadmojarad, T., 2013. Antiepileptogenic effect of curcumin on kainate-induced model of temporal lobe epilepsy. Pharm. Biol. 51 (12), 1572–1578.

Kiasalari, Z., Khalili, M., Shafiee, S., Roghani, M., 2016. The effect of Vitamin E on learning and memory deficits in intrahippocampal kainate-induced temporal lobe epilepsy in rats. Indian J. Pharmacol. 48 (1), 11–14.

Kinter, M., Roberts, R.J., 1996. Glutathione consumption and glutathione peroxidase inactivation in fibroblast cell lines by 4-hydroxy-2-nonenal. Free Radical Biol. Med. 21, 457–462.

Komura, J., Sakamoto, M., 1992. Effects of manganese forms on biogenic amines in the brain and behavioral alterations in the mouse: long-term oral administration of several manganese compounds. Environ. Res. 57, 34–44.

Kowaltowski, A.J., Castilho, R.F., Vercesi, A.E., 1995. Ca^{2+}-induced mitochondrial membrane permeabilization: role of coenzyme Q redox state. Am. J. Physiol. 269, 141–147.

Kruman, I.I., Mattson, M.P., 1999. Pivotal role of mitochondrial calcium uptake in neural cell apoptosis and necrosis. J. Neurochem. 72, 529–540.

Kuhla, B., Haase, C., Flach, K., 2007. Effect of pseudophosphorylation and cross-linking by lipid peroxidation and advanced glycation end product precursors on tau aggregation and filament formation. J. Biol. Chem. 82, 6984–6991.

Kung, M.P., Kostyniak, P., Olson, J., Malone, M., Roth, J.A., 1987. Studies of the in vitro effect of methylmercury chloride on rat brain neurotransmitter enzymes. J. Appl. Toxicol. 7, 119–121.

Lafon-Cazal, M., Pietri, S., Culcasi, M., Bockaert, J., 1993. NMDA-dependent superoxide production and neurotoxicity. Nature 364, 535–537.

Leuner, B., Falduto, J., Shors, T.J., 2003. Associative memory formation increases the observation of dendritic spines in the hippocampus. J. Neurosci. 23, 659–665.

Levesque, P.C., Atchison, W.D., 1991. Disruption of brain mitochondrial calcium sequestration by methylmercury. J. Pharmacol. Exp. Therapeut. 256, 236–242.

Liang, Y., Wei, P., Duke, R.W., 2003. Quantification of 8-iso-prostaglandin-F(2alpha) and 2,3-dinor-8-iso-prostaglandin F(2alpha) in human urine using liquid chromatography-tandem mass spectrometry. Free Radic. Biol. Med. 34, 409–418.

Limke, T.L., Atchison, W.D., 2002. Acute exposure to methylmercury opens the mitochondrial permeability transition pore in rat cerebellar granule cells. Toxicol. Appl. Pharmacol. 178, 52–61.

Liu, Y., Barber, D.S., Zhang, P., Liu, B., 2013. Complex II of the mitochondrial respiratory chain is the key mediator of divalent manganese-induced hydrogen peroxide production in microglia. Toxicol. Sci. 132, 298–306.

LoPachin, R.M., Barber, D.S., Gavin, T., 2008. Molecular mechanisms of the conjugated alpha, beta unsaturated carbonyl derivatives: relevance to neurotoxicity and neurodegenerative diseases. Toxicol. Sci. 2, 235–249.

Lovell, M.A., Markesbery, W.R., 2007. Oxidative DNA damage in mild cognitive impairment and late-stage Alzheimer's disease. Nucleic Acids Res. 22, 7497–7504.

Lovell, M.A., Xie, C., Markesbery, W.R., 2000. Acrolein, a product of lipid peroxidation, inhibits glucose and glutamate uptake in primary neuronal cultures. Free Radic. Biol. Med. 9, 714–720.

Maboudou, P., Mathieu, D., Bachelet, H., et al., 2002. Detection of oxidative stress. Interest of GC-MS for malondialdehyde and formaldehyde monitoring. Biomed. Chromatogr. 16, 199–202.

Magour, S., 1986. Studies on the inhibition of brain synaptosomal Na^+/K^+-ATPase by mercury chloride and methyl mercury chloride. Arch. Toxicol. Suppl. 9, 393–396.
Mangialasche, F., Polidori, M.C., Monastero, R., 2009. Biomarkers of oxidative and nitrosative damage in Alzheimer's disease and mild cognitive impairment. Ageing Res. Rev. 8, 285–305.
Mark, R.J., Hensley, K., Butterfield, D.A., Mattson, M.P., 1995. Amyloid beta-peptide impairs ion motive ATPase activities: evidence for a role in loss of neuronal Ca^{2+} homeostasis and cell death. J. Neurosci. 9, 6239–6249.
Markesbery, W.R., Lovell, M.A., 1998. Four-hydroxynonenal, a product of lipid peroxidation, is increased in the brain in Alzheimer's disease. Neurobiol. Aging 1, 33–36.
Milatovic, D., Aschner, M., 2009. Measurement of isoprostanes as markers of oxidative stress in neuronal tissue. Curr Protocols Toxicol. 12, 1–12.
Milatovic, D., Radic, Z., Zivin, M., Dettbarn, W.-D., 2000a. Atypical effect of some spin trapping agents: reversible inhibition of acetylcholinesterase. Free Radic. Biol. Med. 28, 597–603.
Milatovic, D., Zivin, M., Dettbarn, W.-D., 2000b. The spin trapping agent phenyl-N-tert-butyl-nitrone (PBN) prevents excitotoxicity in skeletal muscle. Neurosci. Lett. 278, 25–28.
Milatovic, D., Zivin, M., Gupta, R.C., Dettbarn, W.-D., 2001. Alterations in cytochrome-c-oxidase and energy metabolites in response to kainic acid-induced status epilepticus. Brain Res. 912, 67–78.
Milatovic, D., Zaja-Milatovic, S., Montine, K.S., et al., 2003. Pharmacologic suppression of neuronal oxidative damage and dendritic degeneration following direct activation of glial innate immunity in mouse cerebrum. J. Neurochem. 87, 1518–1526.
Milatovic, D., Milatovic, S., Montine, K., et al., 2004. Neuronal oxidative damage and dendritic degeneration following activation of CD14-dependent innate immunity response in vivo. J. Neuroinflammation 1, 20.
Milatovic, D., Gupta, R.C., Dekundy, A., Montine, T.J., Dettbarn, W.-D., 2005a. Carbofuran-induced oxidative stress in slow and fast skeletal muscles: prevention by memantine. Toxicology 208, 13–24.
Milatovic, D., VanRollins, M., Li, K., et al., 2005b. Suppression of murine cerebral F2-isoprostanes and F4-neuroprostanes from excitotoxicity and innate immune response in vivo by alpha- or gamma-tocopherol. J. Chromatogr. B Analyt. Technol. Biomed. Life Sci. 827, 88–93.
Milatovic, D., Gupta, R.C., Aschner, M., 2006. Anticholinesterase toxicity and oxidative stress. Sci. World J. 6, 295–310.
Milatovic, D., Yin, Z., Gupta, R.C., 2007. Manganese induces oxidative impairment in cultured rat astrocytes. Toxicol. Sci. 98, 198–205.
Milatovic, D., Zaja-Milatovic, S., Gupta, R.C., et al., 2009. Oxidative damage and neurodegeneration in manganese-induced neurotoxicity. Toxicol. Appl. Pharmacol. 240, 219–225.
Milatovic, D., Montine, T.J., Zaja-Milatovic, S., 2010. Morphometric analysis in neurodegenerative disease. Curr. Protocols Toxicol. 12, 1–14.
Minghetti, L., Levi, G., 1995. Induction of prostanoid biosynthesis by bacterial lipopolysaccharide and isoproterenol in rat microglial cultures. J. Neurochem. 65, 2690–2698.
Montine, K.S., Olson, S.J., Amarnath, V., 1997. Immunohistochemical detection of 4-hydroxy-2-nonenal adducts in Alzheimer's disease is associated with inheritance of APOE4. Am. J. Pathol. 2, 437–443.
Montine, T.J., Milatovic, D., Gupta, R.C., 2002. Neuronal oxidative damage from activated innate immunity is EP2 receptor-dependent. J. Neurochem. 83, 463–470.
Montine, K.S., Quinn, J.F., Zhang, J., 2004. Isoprostanes and related products of lipid peroxidation in neurodegenerative diseases. Chem. Phys. Lipids 128, 117–124.
Moore, K., Roberts, L.J., 1998. Measurement of lipid peroxidation. Free Radic. Res. 28, 659–671.
Morán, M., Moreno-Lastres, D., Marin-Buera, L., 2012. Mitochondrial respiratory chain dysfunction: implications in neurodegeneration. Free Radic. Biol. Med. 53, 595–609.
Morrow, J.D., Roberts, L.J., 1999. Mass spectrometric quantification of F2-isoprostanes in biological fluids and tissues as measure of oxidant stress. Meth. Enzymol. 300, 3–12.
Morrow, J.D., Hill, K.E., Burk, R.F., 1990. A series of prostaglandin F2-like compounds are produced in vivo in humans by a non-cyclooxygenase, free radical-catalyzed mechanism. Proc. Natl. Acad. Sci. U.S.A. 87, 9383–9387.
Muller, M., Gahwiler, B.H., Rietschin, L., Thompson, S.M., 1993. Reversible loss of dendritic spines and altered excitability after chronic epilepsy in hippocampal slice cultures. Proc. Natl. Acad. Sci. U.S.A. 90, 257–261.
Multani, P., Myers, R.H., Blume, H.W., et al., 1994. Neocortical dendritic pathology in human partial epilepsy: a quantitative Golgi study. Epilepsia 35, 728–736.
Mundy, W.R., Freudenrich, T.M., 2000. Sensitivity of immature neurons in culture to metal-induced changes in reactive oxygen species and intracellular free calcium. Neurotoxicology 21, 1135–1144.
Neely, M.D., Boutte, A., Milatovic, D., Montine, T.J., 2005. Mechanisms of 4-hydroxynonenal-induced neuronal microtubule dysfunction. Brain Res. 1037, 90–98.
Nicholls, D.G., Ward, M.W., 2000. Mitochondrial membrane potential and neuronal glutamate excitotoxicity: mortality and millivolts. Trends Neurosci. 23, 166–174.
Nicholls, D.G., Vesce, S., Kirk, L., Chalmers, S., 2003. Interactions between mitochondrial bioenergetics and cytoplasmic calcium in cultured cerebellar granule cells. Cell Calcium 34, 407–424.
Palmeira, C.M., Serrano, J., Kuehl, D.W., Wallace, K.B., 1997. Preferential oxidation of cardiac mitochondrial DNA following acute intoxication with doxorubicin. Biochim. Biophys. Acta 1321, 101–106.
Patel, M.N., 2002. Oxidative stress, mitochondrial dysfunction, and epilepsy. Free Radic. Res. 36, 1139–1146.
Patel, M.N., Liang, L.P., Roberts, L.J., 2001. Enhanced hippocampal F2-isoprostane formation following kainate-induced seizures. J. Neurochem. 79, 1065–1069.
Pedersen, W.A., Cashman, N.R., Mattson, M.P., 1999. The lipid peroxidation product 4-hydroxynonenal impairs glutamate and glucose transport and choline acetyltransferase activity in NSC-19 motor neuron cells. Exp. Neurol. 1, 1–10.
Picklo, M.J., Montine, T.J., Amarnath, V., Neely, M.D., 2002. Carbonyl toxicology and Alzheimer's disease. Toxicol. Appl. Pharmacol. 184, 187–197.
Porter, N.A., Caldwell, S.E., Mills, K.A., 1995. Mechanisms of free radical oxidation of unsaturated lipids. Lipids 30, 277–290.
Pratico, D., Barry, O.P., Lawson, J.A., 1998. IPF2alpha-I: an index of lipid peroxidation in humans. Proc. Natl. Acad. Sci. U.S.A. 95, 3449–3454.
Pratico, D., Rokach, J., Lawson, J., Fitzgerald, G.A., 2004. F2-isoprostanes as indices of lipid peroxidation in inflammatory diseases. Chem. Phys. Lipids 128, 165–171.
Quinlan, C.L., Orr, A.L., Perevoshchikova, I.V., 2012. Mitochondrial complex II can generate reactive oxygen species at high rates in both the forward and reverse reactions. J. Biol. Chem. 287, 27255–27264.
Rabinovitch, P.S., June, C.H., Kavanagh, T.J., et al., 1993. Measurements of cell physiology: ionized calcium, pH, and glutathione. In: Bauer, U.D., Dunque, R.E., Shanley, T.V. (Eds.), Clinical Flow Cytometry: Principles and Application. Williams and Wilkins, Baltimore1993, pp. 505–534.
Radomska-Leśniewska, D.M., Hevelke, A., Skopiński, P., et al., 2016. Reactive oxygen species and synthetic antioxidants as angiogenesis modulators. Pharmacol. Rep. 68, 462–471.

Radomska-Leśniewska, D.M., Bałan, B.J., Skopiński, P., 2017. Angiogenesis modulation by exogenous antioxidants. Cent. Eur. J. Immunol. 42 (4), 370–376.

Rao, V.L., Dogan, A., Todd, K.G., 2001. Antisense knockdown of the glial glutamate transporter GLT-1, but not the neuronal glutamate transporter EAAC1, excerbates transient focal cerebral ischemia-induced neuronal damage in rat brain. J. Neurosci. 21, 1876–1883.

Reich, E., Markesbery, W., Roberts, I.L., 2001. Brain regional quantification of F-ring and D/E-ring isoprostanes and neuroprostanes in Alzheimer's disease. Am. J. Pathol. 158, 293–297.

Rice, G.C., Bump, E.A., Shrieve, D.C., et al., 1986. Quantitative analysis of cellular glutathione by flow cytometry utilizing monochlorobimane: some applications to radiation and drug resistance in vitro and in vivo. Cancer Res. 46, 6105–6110.

Rivest, S., 2009. Regulation of innate immune response in the brain. Nat. Rev. Immunol. 9, 429–439.

Roberts, L.J., Montine, T.J., Markesbery, W.R., 1998. Formation of isoprostane-like compounds (neuroprostanes) in vivo from docosahexaenoic acid. J. Biol. Chem. 273, 13605–13612.

Rocha, J.B., Freitas, A.J., Marques, M.B., 1993. Effects of methylmercury exposure during the second stage of rapid postnatal brain growth on negative geotaxis and on delta-aminolevulinate dehydratase of suckling rats. Braz. J. Med. Biol. Res. 26, 1077–1083.

Sabens Liedhegner, E.A., Gao, X.H., Mieyal, J.J., 2012. Mechanisms of altered redox regulation in neurodegenerative diseases—focus on S-glutathionylation. Antioxi. Redox Signal. 16 (6), 543–566.

Saka, E., Iadarola, M., Fitzgerald, D.J., Graybiel, A.M., 2002. Local circuit neurons in the striatum regulate neural and behavioral responses to dopaminergic stimulation. Proc. Natl. Acad. Sci. U.S.A. 99, 9004–9009.

Salem, N., Kim, H.Y., Lyergey, J.A., Martin, R.E., 1986. Docosahexaenoic acid: membrane function and metabolism. In: Martin, R.E. (Ed.), Health Effects of Polyunsaturated Acids in Seafoods. Academic Press, , New York, pp. 263–317.

Salinaro, A.T., Pennisi, M., Di Paola, R., Scuto, M., et al., 2018. Neuroinflammation and neurohormesis in the pathogenesis of Alzheimer's disease and Alzheimer-linked pathologies: modulation by nutritional mushrooms. Immun. Ageing 15 (8), 1–7.

Santos, A.C., Uyemura, S.A., Santos, N.A., et al., 1997. Hg(II)-induced renal cytotoxicity: in vitro and in vivo implications for the bioenergetic and oxidative status of mitochondria. Mol. Cell. Biochem. 177, 53–59.

Sas, K., Szabó, E., Vécsei, L., 2018. Mitochondria, oxidative stress and the kynurenine system, with a focus on ageing and neuroprotection. Molecules 23 (191), 1–28.

Schaefer, E.J., Bongard, V., Beiser, A.S., 2006. Plasma phosphatidylcholine docosahexaenoic acid content and risk of dementia and Alzheimer disease: the Framingham Heart Study. Arch. Neurol. 63, 1545–1550.

Shanker, G., Aschner, M., 2003. Methylmercury-induced reactive oxygen species formation in neonatal cerebral astrocytic cultures is attenuated by antioxidants. Brain Res Mol Brain Res. 110, 85–91.

Shanker, G., Aschner, J.L., Syversen, T., Aschner, M., 2004. Free radical formation in cerebral cortical astrocytes in culture induced by methylmercury. Mol. Brain Res. 128, 48–57.

Shanker, G., Syversen, T., Aschner, J.L., Aschner, M., 2005. Modulatory effect of glutathione status and antioxidants on methylmercury-induced free radical formation in primary cultures of cerebral astrocytes. Brain Res Mol Brain Res. 137, 11–22.

Singh, J., Husain, R., Tandon, S.K., et al., 1974. Biochemical and histopathological alterations in early manganese toxicity in rats. Environ. Physiol. Biochem. 4, 16–23.

Sjodin, B., Westing, Y.H., Apple, F.S., 1990. Biochemical mechanisms for oxygen free radical formation during exercise. Sports Med. 10, 236–254.

Sloot, W.N., Gramsbergen, J.B.P., 1994. Axonal transport of manganese and its relevance to selective neurotoxicity in the rat basal ganglia. Brain Res. 657, 124–132.

Sloot, W.N., Van der Sluijs-Gelling, A.J., Gramsbergen, J.B.P., 1994. Selective lesions by manganese and extensive damage by iron after injection into rat striatum or hippocampus. J. Neurochem. 62, 205–216.

Smith, W.L., 2005. Cyclooxygenases, peroxide tone and the allure of fish oil. Curr. Opin. Cell Biol. 17, 174–182.

Sodum, R.S., Fiala, E.S., 2001. Analysis of peroxynitrite reactions with guanine, xanthine, and adenine nucleosides by high-pressure liquid chromatography with electrochemical detection: C8-nitration and oxidation. Chem. Res. Toxicol. 14, 438–450.

Sorg, O., Schilter, B., Honegger, P., Monnet-Tschudi, F., 1998. Increased vulnerability of neurones and glial cells to low concentrations of methylmercury in a prooxidant situation. Acta Neuropathol. 96, 621–627.

Soussi, B., Idstrom, J.P., Schersten, T., et al., 1989. Kinetic parameters of cytochrome c oxidase in rat skeletal muscle: effect of endurance training. Acta Physiol. Scand. 135, 373–379.

Spickett, C.M., Wiswedel, I., Siems, W., et al., 2010. Advances in methods for the determination of biologically relevant lipid peroxidation products. Free Radic. Res. 44, 1172–1202.

Stadtman, E.R., Levine, R.L., 2003. Free radical-mediated oxidation of free amino acids and amino acids residue in proteins. Amino Acids (3–4), 207–218.

Stephens, B., Mueller, A.J., Shering, A.F., 2005. Evidence of a breakdown of corticostriatal connections in Parkinson's disease. Neuroscience 132, 741–754.

Strubelt, O., Kremer, J., Tilse, A., 1996. Comparative studies on the toxicity of mercury, cadmium, and copper toward the isolated perfused rat liver. J. Toxicol. Environ. Health 47, 267–283.

Sultana, R., Perluigi, M., Butterfield, D.A., 2006. Protein oxidation and lipid peroxidation in brain of subjects with Alzheimer's disease: insights into mechanism of neurodegeneration from redox proteomics. Antioxidants Redox Signal. 8, 2021–2037.

Tansey, M.G., McCoy, M.K., Frank-Cannon, T.C., 2007. Neuroinflammatory mechanisms in Parkinson's disease: potential environmental triggers, pathways, and targets from early therapeutic intervention. Exp. Neurol. 208, 1–25.

Tjalkens, R.B., Zoran, M.J., Mohl, B., Barhoumi, R., 2006. Manganese suppresses ATP-dependent intercellular calcium waves in astrocyte networks through alteration of mitochondrial and endoplasmic reticulum calcium dynamics. Brain Res. 1113, 210–219.

Trewin, A.J., Berry, B.J., Wojtovich, A.P., 2018. Exercise and mitochondrial dynamics: keeping in shape with ROS and AMPK. Antioxidants 7 (1), 1–8.

Uchida, K., 2003. 4-hydroxy-2-nonenal: a product and mediator of oxidative stress. Prog. Lipid Res. 42, 318–343.

Uchida, K., Kanematsu, M., Morimitsu, Y., 1998. Acrolein is a product of lipid peroxidation reaction. Formation of free acrolein and its conjugate with lysine residues in oxidized low density lipoproteins. J. Biol. Chem. 273, 16058–16066.

Valko, M., Leibfritz, D., Moncol, J., 2007. Free radicals and antioxidants in normal physiological functions and human disease. Int. J. Biochem. Cell Biol. 39, 44–84.

Vallejo, M.J., Salazar, L., Grijalva, M., 2017. Oxidative stress modulation and ROS-mediated toxicity in cancer: a review on in vitro models for plant-derived compounds. Oxid. Med. Cell. Longev. 1–9.

Vatassery, G.T., 2004. Impairment of brain mitochondrial oxidative phosphorylation accompanying vitamin E oxidation induced by iron or reactive nitrogen species: a selective review. Neurochem. Res. 29, 1951–1959.

Virag, L., Szabo, E., Gergely, P., Szabo, C., 2003. Peroxynitrite-induced citotoxicity: mechanism and opportunity for intervention. Toxicol. Lett. 140–141, 113–124.

Wang, Z., Ciabattoni, G., Creminon, C., et al., 1995. Immunological characterization of urinary 8-epi-prostaglandin F2 alpha excretion in man. J. Pharmacol. Exp. Therapeut. 275, 94–100.

Wang, J.Y., Shum, A.Y., Ho, Y.J., 2003. Oxidative neurotoxicity in rat cerebral cortex neurons: synergistic effects of H_2O_2 and NO on apoptosis involving activation of p38 mitogen-activated protein kinase and caspase-3. J. Neurosci. Res. 72, 508–519.

Weinberg, J.M., Harding, P.G., Humes, H.D., 1982. Mitochondrial bioenergetics during the initiation of mercuric chloride-induced renal injury. II. Functional alterations of renal cortical mitochondria isolated after mercuric chloride treatment. J. Biol. Chem. 257, 68–74.

Wilson, P., Groves, P.M., 1980. Fine structure and synaptic connections of the common spiny neuron of the rat neostriatum: a study employing intracellular inject of horseradish peroxidase. J. Comp. Neurol. 194, 599–615.

Xie, Z., Wei, M., Morgan, T.E., 2002. Peroxynitrite mediates neurotoxicity of amyloid beta-peptide1-42- and lipopolysaccharide-activated microglia. J. Neurosci. 22, 3484–3492.

Yang, Z.P., Dettbarn, W.-D., 1998. Lipid peroxidation and changes of cytochrome c oxidase and xanthine oxidase in organophosphorus anticholinesterase induced myopathy. Xth International Symposium on Cholinergic Mechanisms. J. Physiol. (Paris) 92, 157–162.

Yee, S., Choi, H., 1996. Oxidative stress in neurotoxic effects of methylmercury poisoning. Neurotoxicology 17, 17–26.

Yin, H., Gao, L., Tai, H.H., 2007. Urinary prostaglandin F2alpha is generated from the isoprostane pathway and not the cyclooxygenase in humans. J. Biol. Chem. 282, 329–336.

Zaja-Milatovic, S., Milatovic, D., Schantz, A., 2005. Dendritic degeneration in neostriatal medium spiny neurons in late-stage Parkinson disease. Neurology 64, 545–547.

Zaja-Milatovic, S., Gupta, R.C., Aschner, M., Montine, T.J., Milatovic, D., 2008. Pharmacologic suppression of oxidative damage and dendritic degeneration following kainic acid-induced excitotoxicity in mouse cerebrum. Neurotoxicology 29, 621–627.

Zaja-Milatovic, S., Gupta, R.C., Aschner, M., Milatovic, D., 2009. Protection of DFP-induced oxidative damage and neurodegeneration by antioxidants and NMDA receptor antagonist. Toxicol. Appl. Pharmacol. 240, 124–131.

Zeng, L.H., Xu, L., Rensing, N.R., 2007. Kainate seizures cause acute dendritic injury and actin depolymerization in vivo. J. Neurosci. 27, 11604–11613.

Zhang, S., Zhou, Z., Fu, J., 2003. Effect of manganese chloride exposure on liver and brain mitochondria function in rats. Environ. Res. 93, 149–157.

Zhang, S., Fu, J., Zhou, Z., 2004. In vitro effect of manganese chloride exposure on reactive oxygen species generation and respiratory chain complexes activities of mitochondria isolated from rat brain. Toxicol. Vitro 18, 71–77.

Zhang, P., Wong, T.A., Lokuta, K.M., 2009. Microglia enhance manganese chloride-induced dopaminergic neurodegeneration: role of free radical generation. Exp. Neurol. 217, 219–230.

Zhao, F., Cai, T., Liu, M., 2009. Manganese induces dopaminergic neurodegeneration via microglial activation in a rat model of manganism. Toxicol. Sci. 107, 156–164.

Zoratti, M., Szabo, I., 1995. The mitochondrial permeability transition. Biochim. Biophys. Acta 1241, 139–176.

CHAPTER

58

Cytoskeletal Disruption as a Biomarker of Developmental Neurotoxicity

Alan J. Hargreaves[1], *Magdalini Sachana*[2], *John Flaskos*[3]

[1]School of Science and Technology, Nottingham Trent University, Nottingham, United Kingdom; [2]Organization for Economic Cooperation and Development (OECD), Paris, France; [3]Laboratory of Biochemistry and Toxicology, Faculty of Veterinary Medicine, Aristotle University of Thessaloniki, Thessaloniki, Greece

INTRODUCTION

The eukaryotic cytoskeleton comprises a network of three interconnected protein filamentous arrays known as microtubules (MTs), microfilaments (MFs), and intermediate filaments (IFs). Many MT and MF arrays are dynamic structures that can undergo changes in organization, activity, and function at key stages in neural cell development. These phenomena are regulated by a variety of posttranslational modifications and interactions with a range of accessory proteins (Carlier, 1998; Joshi, 1998; Biernat et al., 2002; Ishikawa and Kohama, 2007; Akhmanova and Steinmetz, 2008; Janke and Kneussel, 2010; Svitkina, 2018). IFs, however, are relatively stable but may be modulated by cross-linking interactions with associated proteins or by their phosphorylation status (Herrmann and Aebi, 2000; Omary et al., 2006).

The cytoskeleton is involved in the control of key cellular processes in nervous system development and maintenance, such as cell division, cell migration, cell differentiation, intracellular transport, and structural support. Its disruption by the interaction of neurotoxins with core proteins or cytoskeletal regulatory systems can therefore be detrimental to a wide range of phenomena including neural development (Hargreaves, 1997; Flaskos, 2014).

There is a growing body of evidence for the induction of developmental neurotoxicity via disruption of cell signaling and cytoskeleton-dependent physiological processes by several groups of chemicals including organophosphorus esters (OPs), heavy metals, polybrominated diphenyl ethers (PBDEs), and solvents (Sachana et al., 2017; Pierozan et al., 2017). This chapter focuses on studies showing cytoskeletal disruption associated with impairment of neural development following exposure to such compounds, supporting the idea that cytoskeletal disruption can be a useful biomarker of developmental neurotoxicity.

MICROTUBULES

The main component of the MT network is the heterodimeric protein tubulin, which is composed of α and β subunits that form head-to-tail protofilaments, which in turn come together to make the tubular structure with an external diameter of 25 nm (Amos, 2004). MT assembly requires GTP binding to tubulin and MT dynamics are dependent on GTP hydrolysis, which occurs shortly after subunit addition to the growing MT end ("plus" end) (Carlier et al., 1984).

Both α- and β-tubulins have several isoforms encoded by different genes (Ludueña, 1998; Amos, 2004; Tischfield and Engle, 2010). Recent findings concerning congenital human neurological syndromes further emphasize the unique roles of specific α- and β-tubulin isoforms during nervous system development (Tischfield and Engle, 2010). For example, isotype III of β-tubulin (βIII-tubulin) appears almost exclusively in neuronal cells, plays an important role in neuritogenesis (Katsetos et al., 2003), and is among the earliest neuronal cytoskeletal proteins to be expressed during CNS development (Lee et al., 1990; Easter et al., 1993). Tubulin can be chemically modified by a variety of posttranslational adjustments, which may affect stability and location of MTs or act as a guidance cue for MT-binding proteins (Janke and Kneussel, 2010; Baas et al., 2016).

Biomarkers in Toxicology, Second Edition
https://doi.org/10.1016/B978-0-12-814655-2.00058-X

Many proteins can potentially interact with MTs, including the most studied microtubule associated proteins (MAPs). In neurons, these interactions influence MT dynamics, and it is believed that they are necessary for neuronal migration, differentiation, and axon guidance (Wade, 2009; Penazzi et al., 2016). Phosphorylation events have been found to regulate the association of MAPs with MTs, suggesting their potential involvement in cascade events relevant to neuronal development and degeneration (Biernat et al., 2002; Baas and Qiang, 2005).

An important member of the MAP family is the protein tau, which has tandem repeats of a tubulin-binding domain and contributes to tubulin assembly. Tau is abundant in neurons and is mainly located in axons, where it is closely associated with MTs. Changes in tau-protein levels and its phosphorylation state have been detected in numerous neurodegenerative diseases (Johnson and Stoothoff, 2004). Phosphorylation is also very important from a developmental point of view and is encountered extensively in fetal rather than adult tau (Watanabe et al., 1993). Indeed, increased phosphorylation of tau and dynamic MTs seem to coexist during brain development (Brion et al., 1994).

It has also been shown that MAP-1B is an essential protein for the development and function of nervous system both in vitro (Brugg et al., 1993; Di Tella et al., 1996) and in vivo (Meixner et al., 2000; Villarroel-Campos and Gonzalez-Billault, 2014). On the other hand, MAP1A dynamics are very closely associated to spine plasticity and any alterations in MAP1A may indicate changes in synaptic density (Jaworski et al., 2009). As a counter measure to MT-stabilizing MAPs, other groups of MAPs can bind to and destabilize MTs. An example is stathmin, which has been shown to regulate MT stability in the formation of dendritic branches in neuronal cells (Ohkawa et al., 2007). This low molecular weight protein can bind to and sequester tubulin heterodimers and hydrolyze GTP at the growing ends of MTs, leading to reduced MT stability (Howell et al., 1999).

The key role of MTs in intracellular transport (e.g., along developing axons) is regulated by another group of MAPs, which act as ATPase motor proteins. Such MAPs include kinesin and dynein, which direct anterograde and retrograde transport along axonal MTs respectively (Vale, 2003), a process which is essential for neurite growth and development. The interaction of such proteins with MTs or their ATPase activities could potentially be affected under conditions where ATP levels are depleted.

Several studies have dealt with the effects of established developmental neurotoxicants on MT assembly, organization, protein levels, posttranslational modifications, cell distribution, and gene expression. Altered status or intracellular distribution of MT proteins could reflect a range of adverse effects on the regulation of neural development. As discussed in the following text, most of the available experimental evidence from cell culture studies suggests that several well-established developmental neurotoxicants cause alterations in MTs of neuronal and glial cells under culture conditions, a common finding being a reduction in the levels of MAPs.

Effects of Organophosphorus Esters on Microtubules

The effects of several OP pesticides have been studied in the distant past, mainly in relation to posttranslational modifications of MT proteins and their role in organophosphate-induced delayed neuropathy (Abou-Donia, 1993, 1995). More recently, chlorpyrifos (CPF) and diazinon (DZ), the two OPs for which there are supporting data for developmental neurotoxicity, have been investigated regarding their potential effect on MTs both in vitro and in vivo. MAP-2 levels decreased following challenge with chlorpyrifos oxon (CPO) in organotypic slice cultures of immature rat hippocampus (Prendergast et al., 2007), whereas the parent compound CPF caused similar effects in the prefrontal cortex of Wistar rats (Ruiz-Muñoz et al., 2011) and in the mouse embryonic stem cells (mESCs) that were differentiated into neuronal cells (Visan et al., 2012). However, α-tubulin levels were not altered by 1–10 μM CPO exposure, suggesting that the general structure of MTs was not modified (Prendergast et al., 2007). Similar findings were reported in the case of N2a cells exposed to CPF at the time of the induction of cell differentiation, as well as 16 h after the induction of differentiation (Sachana et al., 2005). Interestingly, levels of total α-tubulin were also found unaltered in the case of DZ or diazoxon (DZO)-treated differentiating N2a cells (Flaskos et al., 2007; Harris et al., 2009a; Sidiropoulou et al., 2009a), whereas MAP-1B levels were reduced after 10 μM DZ exposure (Flaskos et al., 2007). No change in the levels of β-tubulin isotypes I and III were detected in differentiating N2a cells following exposure to DZ (Harris et al., 2009b). In contrast, in the same cell model, DZO induced a significant reduction in the levels of the βIII-tubulin isotype but had no effect on total β-tubulin levels, suggesting a neuron-specific effect (Sachana et al., 2014) In animal models, oral administration of CPF on postnatal days 1–6 did not affect the expression of the genes coding for the neuronal-specific marker βIII-tubulin (Betancourt et al., 2006).

The polymerization of bovine brain tubulin was inhibited by low concentrations of CPO (0.1–10 μM) (Prendergast et al., 2007). The ability of tubulin to polymerize is very important because, apart from contributing toward the maintenance of neuronal morphology, MTs also support axonal transport of mitochondria, components of ion channels, receptors, and scaffolding proteins. Perturbation of assembly and transport mechanisms in neurons due to CPF and CPO reaction with tubulin and its organophosphorylation has been suggested from several experimental works both in vivo (Terry et al., 2007; Jiang et al., 2010) and in vitro (Gearhart et al., 2007; Grigoryan et al., 2009b; Grigoryan and Lockridge, 2009).

In glial cells, 24 h exposure to CPF or CPO suppresses extension outgrowth in differentiating C6 cells (Sachana et al., 2008). CPO had a stronger morphological effect than CPF that has been associated with a significant decrease in the levels of tubulin and MAP-1B (Sachana et al., 2008). Similarly, only the in vivo metabolite of DZ, DZO, triggered inhibition of the development of C6 cell extensions, an effect linked also to the reduction in the levels of tubulin (Sidiropoulou et al., 2009b). Immunofluorescence staining revealed normal MT networks in control and CPF-treated cells and, although there was no evidence for a major collapse of the MT network, there were increased levels of localized patchy staining compared to the control, particularly in CPO-treated cells (Sachana et al., 2008).

Effects of Heavy Metals on Microtubules

MT proteins have also been investigated in relation to methylmercury–induced developmental abnormalities of the nervous system. Exposure of N2a neuroblastoma cells to methylmercury revealed significant disruption in MT organization after staining the cells with antibody that recognizes β-tubulin (Kromidas et al., 1990). The same research group further emphasized the predominant effect of methylmercury on MTs compared to IFs and MFs by using scanning electron microscopy (Trombetta and Kromidas, 1992). In the same cell line, cells demonstrated decreased reactivity against C-terminally tyrosinated α-tubulin following only 4 h exposure to a sublethal concentration of methylmercury chloride compared to controls, which was associated with inhibition of neurite outgrowth (Lawton et al., 2007).

The importance of the cytoskeleton in methylmercury neurotoxicity was further emphasized in a study by Castoldi et al. (2000) using primary cultures. In this study, rat cerebellar granule cells exhibited MT depolymerization within 1.5 h of exposure to 1 μM methylmercury and long before disturbance of neurite processes (Castoldi et al., 2000).

Methylmercury was found to suppress tubulin polymerization in vitro, to disrupt the MT network, and to reduce tubulin synthesis in mouse glioma cells (Miura et al., 1984; Miura and Imura, 1987). This corresponded to a decrease in tubulin mRNA levels but no effect on the transcription rates of β-tubulin genes were found, suggesting that exposure had disrupted the autoregulatory control of tubulin synthesis in a manner similar to that described for the antimitotic agent colchicine (Miura et al., 1998).

It is well documented that lead exposure can cause learning and memory impairment, particularly in developing organisms. However, the molecular mechanisms are not fully understood, although several studies investigated the potential role of MTs in disruption of memory formation. In human primary cultures, exposure to biologically relevant concentrations of lead (5, 10, 20, and 40 μg/dL) was associated with hyperphosphorylation of tau protein, as determined by Western blotting and immunocytochemistry, due to upregulation of protein phosphatases (Rahman et al., 2011). These findings were also confirmed in both Wistar rat and mouse pups, emphasizing the importance of tau hyperphosphorylation in cognitive impairment (Li et al., 2010; Rahman et al., 2012). Recent data further support this as the pre- and neonatal exposure to lead causes a significant increase in the phosphorylation of tau and upregulates tau protein level in the rat brain cortex and cerebellum (Gąssowska et al., 2016).

In a study by Scortegagna et al. (1998), the detectable levels of MAP-2b and MAP-2c were found to decrease 24 h after a 3 h exposure to 3 or 6 μM lead in serum-free medium maintained E14 mesencephalic rat primary cultures. However, in the same study, these protein levels remained similar to controls in serum-cultured cells, suggesting that a serum factor prevents cytoskeletal changes otherwise noted in this primary culture containing differentiating neurons and proliferating astrocytes (Scortegagna et al., 1998). On the other hand, lead (II) acetate reduced the number of MAP-2 stained cells and the mRNA levels of MAP-2 in a concentration-dependent manner, by applying default differentiation of mESCs (Beak et al., 2011; Visan et al., 2012). Inorganic lead had no effect on the in vitro assembly of MTs from porcine brain, whereas trimethyl lead blocked and completely inhibited MT assembly at 300 μM, as monitored by turbidity measurements and electron microscopy (Roderer and Doenges, 1983). In contrast, triethyl lead chloride inhibited MT assembly and depolymerized preformed MTs in porcine brain preparations (Zimmermann et al., 1988), whereas the same research group reported no change in MT network in mouse N2a neuroblastoma cells after exposure to the same organic lead compound (Zimmermann et al., 1987).

Regarding arsenic, it was shown that tau gene expression can be increased in ST-8814 schwannoma and SK-N-SH neuroblastoma cell lines by exposure to inorganic trivalent and monomethyl pentavalent arsenic metabolites, respectively (Vahidnia et al., 2007b). In addition, the phosphorylated state of MAP-tau has been found altered in both in vitro and in vivo studies (Vahidnia et al., 2007a). More specifically, Giasson et al. (2002) reported hyperphosphorylation of tau-protein in Chinese hamster ovary cells after treatment with inorganic trivalent arsenic. Similarly, subchronic exposure of rats to this arsenic metabolite increased the phosphorylation of tau in sciatic nerves (Vahidnia et al., 2008a). However, a more recent study shows that sodium arsenite causes neurite inhibition in N2a cells that was associated with alterations in cytoskeletal proteins. More specifically, sodium arsenite decreased the mRNA levels of tau and tubulin in a dose-dependent manner but had no significant effect on the mRNA levels of MAP-2 (Aung et al., 2013). The inhibitory effect of arsenic trioxide on the migration of primary neurons established from the brains of neonatal rats has also been recently investigated; the study revealed that it was associated with a decrease in the protein levels of doublecortin, which is a MT-associated protein expressed during the neuronal migration (Zhou et al., 2015).

Effects of Organic Solvents and Polybrominated Diphenyl Ethers on Microtubules

Ethanol is the only organic solvent studied in relation to its neurotoxic potential against MTs in *in vitro* models. Ethanol had no effect on the rate and extent of bovine tubulin polymerization in vitro, whereas the ethanol metabolite acetaldehyde inhibited MT formation (Jennett et al., 1980). Similarly, McKinnon et al. (1987) demonstrated that acetaldehyde had an inhibitory effect on the polymerization of MT protein derived from calf brains, further emphasizing the acetaldehyde-mediated alteration of cytoskeletal MTs. Continuous exposure of developing neural crest cells to ethanol has also been found to cause MT disruption (Hassler and Moran, 1986).

In PC12 cells, chronic exposure to ethanol led to alterations in the balance between free tubulin in the cytoplasm and tubulin polymerized into MTs, enhancing the content of the latter, possibly through phosphorylation (Reiter-Funk and Dohrman, 2005). In contrast, chronic alcohol exposure was found to decrease the levels of polymerized tubulin in cultured hippocampal neurons and simultaneously to reduce the amount of MTs and the levels of MAP-2 in dendrites (Romero et al., 2010). The same research group also described impairment of MT dynamics and reassembly in primary cultures of rat astrocytes treated with ethanol (Tomas et al., 2003). However, in a short-term exposure of rat C6 glioma cells to acute levels of ethanol (50, 100, and 200 mM), there was no detectable change in the MT network (Loureiro et al., 2011).

In whole cerebral hemisphere model, developing Layer 6 neurons exposed to ethanol exhibited diminished MAP-2 levels in dendritic processes (Powrozek and Olson, 2012). Similarly, MAP-2 immunostaining intensity was reduced in hippocampal slice cultures from neonatal Wistar rats subjected to ethanol treatment (200 mM) for up to 4 weeks (Noraberg and Zimmer, 1998). MAP-2 immunolabeling has also been used to determine the effect of ethanol (70 mM) on dendrites of rat embryonic hippocampal pyramidal neurons in culture, revealing decreases in both length and number of dendrites compared with controls (Lindsley et al., 2002; Lindsley and Clarke, 2004).

Stimulation of MAP-2 phosphorylation has been detected in MT preparations from rat brain exposed to low and biologically relevant doses of ethanol (6, 12, and 24 mM), whereas higher doses (48, 96, 384, and 768 mM) decreased phosphorylation (Ahluwalia et al., 2000). In the same study, MAP-1 was found to show increased phosphorylation with only 12 and 24 mM of ethanol and tubulin only from the lower dose tested (6 mM) (Ahluwalia et al., 2000).

More recently, the application of neural stem cell technology indicated disturbance of neuronal differentiation from noncytotoxic concentrations of alcohol (25–100 mM), as recorded by reduced immunostaining and levels of MAP-2 protein (Tateno et al., 2005). Immunolabeling of axons derived from rat hippocampal pyramidal neurons with a βIII-tubulin antibody was used to assess the effect of ethanol on the length of axons (VanDemark et al., 2009). Indeed, ethanol alone had no effect on axon length, whereas carbachol-treated cells in the presence of ethanol (50 and 75 mM) did cause shortening of axons compared to controls (VanDemark et al., 2009).

In the P7 rodent model, which is extensively used for mechanistic elucidation of ethanol-induced neurodevelopmental toxicity, it was found that ethanol elevated the phosphorylation state of tau, as demonstrated by two different phospho-specific antibodies (Saito et al., 2010). Similarly, in a human neuroblastoma cell line after tau induction, ethanol caused a dose-dependent increase in tau levels and cell mortality (Gendron et al., 2008). A more extensive overview on the effects of ethanol on the neuronal cytoskeleton, covering both in vivo and in vitro studies, can be found in Evrard and Brusco (2011).

MICROFILAMENTS

MFs are formed by the polymerization of monomeric G-actin into filaments with a diameter of 5 nm (Dominguez and Holmes, 2011). MF assembly is

ATP-dependent, and the dynamic properties of MFs are dependent on the hydrolysis of actin-bound ATP following incorporation at the filament plus end (Carlier, 1998). MF dynamics are also influenced by a wide variety of actin binding proteins (ABPs) that regulate its organization and function by acting as either nucleating factors (e.g., ARP 2/3, formin), cross-linking proteins (e.g., fascin), destabilizing factors (e.g., ADF/cofilin, gelsolin, fragmin), membrane cross-linkers (e.g., spectrin, GAP-43), or MF-associated motor proteins (e.g., myosin) (Ishikawa and Kohama, 2007; Dominguez, 2009; Lee and Dominguez, 2010; Dominguez and Holmes, 2011; Jansen et al., 2011; Svitkina, 2018). The interaction between actin and its ABPs is in turn regulated by cell signaling pathways (Endo et al., 2003; Ishikawa and Kohama, 2007). MFs play key roles in mitosis, neural cell differentiation, and the regulation of cell migration (Gungabissoon and Bamburg, 2003; Kunda and Baum, 2009). The inhibition of neurite outgrowth from explants of embryonic chick spinal cord cultured in the presence of the MT-stabilizing agent taxol and the MF-disrupting agent cytochalasin D clearly demonstrates the importance of both MT and MF integrity in the developmentally important process of neurite outgrowth (Rösner and Vacun, 1997). Given the important roles played by MFs in cytokinesis, receptor trafficking, and neurite outgrowth, their disruption by toxin exposure could have major effects on neural development.

Effects of Organophosphorus Esters on Microfilaments

In early work, Carlson and Ehrich (2001) tested the ability of several OPs including paraoxon, parathion, diisopropyl fluorophosphate (DFP), phenyl saligenin phosphate (PSP), tri*ortho*tolyl phosphate (TOTP), and triphenyl phosphite at 0.1–1 mM to disrupt the filamentous actin (F-actin) network in mitotic SH-SY5Y cells using a fluorescently labeled phalloidin probe. Significant decreases in the levels of F-actin were observed within 30 min exposure of cells treated with PSP and TOTP, whereas other OPs required longer exposure times and DFP had no observable effect. The data clearly support the idea that some OPs can disrupt the MF network, although the concentrations used were relatively high and cytotoxic within a few hours, as determined by protein assay (Carlson and Ehrich, 2001). However, proteomic studies on differentiating N2a neuroblastoma cells exposed to sublethal neurite inhibitory concentrations of the OP pesticide DZ showed that inhibition of neurite outgrowth by 10 μM DZ was associated with increased levels but decreased LIM kinase–mediated phosphorylation of the actin-destabilizing protein cofilin (Harris et al., 2009a). Although total actin levels were unaffected by DZ, the altered levels and phosphorylation status of cofilin, together with reduced staining intensity of neurites with antiactin antibody, suggest that DZ exposure leads to a reduction in the levels of F-actin in neurites due to disruption of MF dynamics caused by altered expression and phosphorylation of cofilin.

Another important modulator of MF organization in the axonal growth cone of developing neurons is growth associated protein 43 (GAP43), which has been shown to exhibit reduced mRNA and/or protein levels following exposure to sublethal neurite outgrowth inhibitory concentrations of OPs (Sachana et al., 2003; Flaskos et al., 2011; Ta et al., 2014). A transient reduction in GAP43 protein levels was also observed to be associated with the retraction of neurites in predifferentiated N2a cells exposed to sublethal concentrations of CPF and CPO (Sindi et al., 2016). It has been suggested that these phenomena may, at least in part, be the result of upstream events such as OP-induced autophagy or disruption of Ca^{2+} homeostasis, which could impact on the activity of Ca^{2+}-dependent ABPs and other MF-regulatory proteins (Chen et al., 2013; Fernandes et al., 2017).

Thus, OPs can disrupt MF organization and functions by interacting with ABPs and/or signaling pathways that modulate MF dynamics. However, in vitro studies by Grigoryan et al. (2009a) and Schopfer et al. (2010) demonstrated a covalent interaction of OPs with lysine and tyrosine residues on actin, raising the possibility that a direct interaction with actin might also be involved in the disruption of MFs by OPs. Further in vivo and in vitro studies to determine the role of OP-actin adduct formation would be worthwhile.

Effects of Heavy Metals on Microfilaments

Many cytoskeleton-related developmental neurotoxicity studies with heavy metals have focused on the MT network. However, micromolar concentrations of mercurial compounds can block SH groups on purified actin, thereby inhibiting its ability to interact with myosin and induce myosin ATPase activity in vitro (Perry and Cotterill, 1964; Martinez-Neira et al., 2005). The ability of mercury (as well as cadmium, copper, and zinc) ions to interfere with actin–ABP interactions has also been demonstrated by native gel electrophoresis (Kekic and dos Remedios, 1999), suggesting that direct binding of heavy metal ions to actin, ABPs, or other MF-regulatory proteins could induce toxic effects directed at the MF network.

Studies on the inhibition of glioma cell migration by arsenic suggested that this heavy metal compound may be able to disrupt F-actin by interference with

MF-regulatory cell signaling pathways such as protein kinase C (Lin et al., 2008). Sodium arsenite was also observed to inhibit neurite outgrowth in differentiating N2a neuroblastoma cells; although this was not found to involve alterations in mRNA levels for β-actin (Aung et al., 2013), the possibility that changes to MF dynamics may involve alterations to the levels and post-translational modification of ABPs or other regulatory pathways cannot be ruled out and warrants further investigation. Changes were observed in the levels of mRNA for NFL and NFM (both elevated) and for tubulin and tau (both reduced) in arsenic-treated cells; the possibility that these changes, if they are reflected at the protein level, may cause disruption of F-actin organization could also be further investigated.

In a study of the effects of methylmercury chloride on postnatal rat brain development, it was found that the degradation of the ABP α-spectrin by μ-calpain was significantly greater in cerebral cortex extracts from treated animals than in controls at postnatal day 16 (Zhang et al., 2003). Calpain activation could also target a range of other cytoskeletal proteins that act as substrates for calpain. In this respect it is interesting to note that exposure to other metal compounds has also been linked to the activation of calpain in brain (Zhang et al., 2012), neural cells (Vahidnia et al., 2008b; Rocha et al., 2011), or in other tissues (Lee et al., 2007).

Furthermore, studies in primary cultures of fetal mouse brain–derived cerebellar granule cells and human placental tissue showed that submicromolar levels of methylmercury chloride induced a significant decrease in the phosphorylation of the ABP cofilin and the translocation of actin and cofilin to mitochondria (Vendrell et al., 2010; Caballero et al., 2017). The fact that these changes were associated with elevated levels of protein carbonylation (detected by immunoassay) and could be blocked by cotreatment with the antioxidant probucol, suggest that they were triggered by elevated levels of protein oxidation (Caballero et al., 2017).

Moreover, evidence for the inhibition of both neurite outgrowth and neuronal cell migration via disruption of MFs through altered regulatory signaling pathways comes from a combination of cell culture and in vivo developing rodent brain studies following exposure to methylmercury chloride (Fujimura and Usuki, 2012; Usuki and Fujimura, 2012; Guo et al., 2013; Hernández et al., 2018). Thus, direct binding to SH groups, elevated protein oxidation, and disruption of MF-regulatory signaling pathways have a major impact on the regulation of MF dynamics, suggesting that exposure to heavy metal compounds can disrupt the regulation of MF dynamics in a variety of ways.

Effects of Organic Solvents and Polybrominated Diphenyl Ethers on Microfilaments

Chronic ethanol exposure has been shown to be associated with disruption of the MF network in cultured PC12 cells and in primary cultures of hippocampal neurons. In PC12 cells, chronic exposure was found to reduce dopamine release via protein kinase C–dependent pathways, an effect that was attenuated by cotreatment with the MF-disrupting agent cytochalasin, indicating the need for MFs in this exocytotic process (Funk and Dohrman, 2007). In ethanol-exposed hippocampal neurons, a reduction was observed in the levels of F-actin compared to untreated control cell cultures, as determined by FITC-labeled phalloidin staining, which corresponded to reduced protein levels of total Rac1, RhoA, and cdc42 (small GTPases known to be involved in the regulation of MF assembly and dynamics), as determined by G-LISA assays (Romero et al., 2010). However, although the levels of activated (GTP-bound) forms of Rac1 and cdc42 were reduced, ethanol had no significant effect on the levels of active RhoA, suggesting that only the inactive form of this GTPase was downregulated by ethanol treatment. Western blotting analysis indicated that the levels of total cofilin (which destabilizes MFs) were unchanged, although the levels of inactive (phospho-) cofilin were not assessed (Romero et al., 2010).

Chronic exposure to ethanol was also found to inhibit endocytotic uptake of serum albumin and transferrin by cultured fetal rat hippocampal neurons (Marín et al., 2010) This effect was associated not only with altered levels of proteins involved in vesicle formation and docking but also with proteins involved in the regulation of MF assembly and dynamics, including Arf6 (upregulated), cdc42, and RhoA (both downregulated). Furthermore, the ability of ethanol to inhibit neural cell differentiation in a human embryonic stem cell model of early brain development was also associated with disruption of the MF network (Taléns-Visconti et al., 2011).

In a study of chronic ethanol exposure on cultured rat astrocytes, the impairment of glucose uptake was associated with disassembly of actin stress fibers; the fact that lysophosphatidic acid attenuated these effects by stabilizing the MF network suggested that a major disruption of MF dynamics was caused by ethanol exposure (Tomas et al., 2003). Although there was no detectable change in the MT network following short-term exposure of rat C6 glioma cells to acute levels of ethanol, the resultant formation of reactive oxygen species was associated with major disruption of the MF network (Loureiro et al., 2011). Taken together, these data illustrate not only the fact that MFs are disrupted by this solvent but also that MF-dependent processes of importance in neural development are impaired when this occurs.

INTERMEDIATE FILAMENTS

The most abundant IF proteins found in differentiating and mature neurons are the three neurofilament (NF) triplet proteins (NF-L, NF-M, and NF-H), whereas in glial cells the most important IF protein is GFAP. Nestin is the main IF specifically expressed in immature neural cells. Indeed, most of the genes coding for IF proteins are expressed in a tissue- or cell type–specific manner, except for the nuclear lamins. Thus, the NF proteins are present only in neurons and GFAP only in glia (and more specifically in astrocytes). As a result, NF proteins and GFAP have been widely used in neurotoxicology as markers for specific effects on neurons and glia, respectively. In the context of developmental toxicology, changes in NF proteins and GFAP have been commonly employed as a measure of the capacity of a toxicant to interfere with neuronal and glial differentiation.

Although NFs are typically considered to be more stable than MFs and MTs, their dynamic capability is demonstrated by their reorganization that occurs during several neurodevelopmental stages including proliferation (mitosis), apoptosis, and axonogenesis (Omary et al., 2006). NFs can facilitate axonal elongation by stabilizing the axonal cytoskeleton and inhibiting the retraction of long axons (Lariviere and Julien, 2003). Apart from their use in developmental neurotoxicity studies as a neuronal differentiation marker, NF proteins may constitute a mechanistically relevant marker for assessing specific biochemical cytoskeletal effects in some cases. This is particularly true for the studies on OPs and arsenic discussed in the following paragraphs, where NFs have been proposed to be a direct target for these neurotoxicants.

NF parameters assessed in neurodevelopmental toxicity studies include NF protein levels, distribution, assembly, transport, phosphorylation, and expression of NF genes. A review of the available data on established developmental neurotoxicants (see the following) indicates that in most cases there is a decrease in the levels of at least one of the three NF proteins, whereas NF gene expression data are inconsistent, with both increases and decreases in NF mRNA levels caused by the toxicants. The distinct features of the three NF proteins in terms of structure, properties, and function and their differential expression during neuronal development imply that assessment of toxicant-induced changes in one NF protein cannot substitute for measurements in the other two. Finally, because NF protein phosphorylation is known to be the major factor in the regulation of the dynamics and function of NFs (Omary et al., 2006; Sihag et al., 2007), assessment of changes in NF phosphorylation-related parameters (determination of relevant kinases and upstream cell signaling molecules) may provide valuable markers for developmental neurotoxicity.

In developmental neurotoxicity studies both protein and mRNA levels of GFAP have been assessed. These studies have usually reported decreases in these parameters, indicating specific repression of glial cell differentiation, whereas any increases obtained have been attributed to reactive gliosis because of high dosing and primary damage to neurons.

Effects of Organophosphorus Esters on Intermediate Filaments

NF parameters have been assessed to a larger extent in toxicological studies involving OP pesticides than other developmental neurotoxicants. This may be partly because of the prior existence of data implicating NF (and other cytoskeletal) abnormalities as being etiologically important in delayed OP neurotoxicity (Abou-Donia, 1993; Jiang et al., 2010). In this context, in some OP neurodevelopmental studies NF parameters have not been employed as a common marker for neuronal cell-specific differentiation but as a mechanistically relevant marker for specific biochemical effects on the neuronal cytoskeleton.

Parameters assessed in OP neurodevelopmental studies include the levels and intracellular distribution and posttranslational modification (phosphorylation) of NF proteins and the expression of NF genes. Exposure of both mitotic and differentiating rat PC12 pheochromocytoma cells to a sub-lethal concentration of CPF (30 μM), led to the upregulation of nfl and nef3 genes (which encode NF-L and NF-M, respectively), whereas expression of the nefh gene (which encodes NF-H) was unaffected (Slotkin and Seidler, 2009). On the other hand, exposure of PC12 cells to 30 μM DZ under the same conditions caused upregulation of the nef3 but had no effect on the expression of nfl and nefh genes (Slotkin and Seidler, 2009). The expression of genes encoding NF-L and NF-H, used as a marker for neuronal differentiation, was studied in primary neuronal cultures of cerebellar granule cells prepared from 7-day-old rat pups treated with parathion (Bal-Price et al., 2010). Exposure to this OP, used at concentrations of 10–50 μM, for up to 12 days caused a concentration-dependent decrease in mRNA levels for both NF-L and NF-H.

A series of studies have assessed changes in several NF parameters following exposure of differentiating mouse N2a neuroblastoma cells to CPF and DZ, the two OPs for which there is the most evidence for developmental neurotoxicity. In these mechanistic studies of OP-induced developmental neurotoxicity, determination of NF (and other cytoskeletal) parameters were extended to include the influence of CPO and DZO, the two in vivo oxon metabolites of CPF and DZ, because there is now considerable evidence to suggest that these compounds can interfere with neuronal differentiation and development (Flaskos, 2012). These studies have demonstrated decreases in the levels of NF-H protein by both CPF and DZ. Thus, exposure of N2a cells to 3 μM CPF for 8 h (Sachana et al., 2001) and to 10 μM DZ for 24 h (Flaskos et al., 2007) leads to reduced NF-H protein levels. Indirect immunofluorescence studies also showed that, apart from an alteration in the expression of NF-H, there was a change in the intracellular distribution of this protein, with NF-H located mainly in cell body aggregates of DZ-treated N2a cells (Flaskos et al., 2007).

NF parameters were also affected by exposure to CPO and DZO. For example, at a concentration of 10 μM, CPO exposure for 24 h reduced the total levels of NF-H and disrupted NF-H intracellular distribution in differentiating N2a cells (Flaskos et al., 2011). On the other hand, CPO had no effect on the levels of phosphorylated NF-H under these experimental conditions, in which the OP was added at the point of induction of cell differentiation. However, in N2a cells induced to differentiate for 20 h prior to OP exposure, CPO induced a transient increase in ERK 1/2 activation and NF-H phosphorylation after 2 h, which preceded neurite retraction (Sindi et al., 2016), suggesting a mechanistic link between NF-H hyperphosphorylation and neurite destabilization. In contrast, the oxon metabolite DZO, applied at concentrations of 5 and 10 μM for 24 h, had no effect on total NF-H levels, but increased phosphorylated NF-H levels in differentiating N2a cells compared to the control (Sidiropoulou et al., 2009a). Under these conditions, DZO had no effect on the total levels of NF-L and NF-M (Sachana et al., 2014). Further work is required to determine the molecular basis of these effects in more detail.

Determination of GFAP in neurodevelopmental studies of OPs has involved measurement of both its protein and mRNA levels. Following developmental exposure to OPs, GFAP protein and mRNA levels exhibited both decreases and increases, the latter being attributed to the occurrence of reactive gliosis because of high dosing. Thus, in aggregating brain cells of fetal rat telencephalon, GFAP levels were found to be increased following parathion treatment, indicative of gliosis (Zurich et al., 2000). Postnatal exposure of rats to CPF for 4 days initially decreased GFAP levels, indicating specific repression of normal glial (astrocytic) development. At a later stage, however, increases in the levels of GFAP occurred, typical of reactive gliosis following neuronal cell damage (Garcia et al., 2002). Postnatal administration of CPF to rats for 6 days at doses high enough to cause significant cholinesterase inhibition also led to increased GFAP mRNA levels, reflecting increased astrocyte reactivity (Betancourt et al., 2006). On the other hand, administration of DZ to neonatal rats for 4 days resulted in a decrease in the expression of the gene coding for GFAP (Slotkin and Seidler, 2007). Similarly, the in vivo metabolite of DZ, DZO (at 1–10 μM) caused, after 24 h in N2a cells, a reduction in GFAP protein levels, indicating repression of specific glial cell/astrocytic differentiation (Sidiropoulou et al., 2009b).

Effects of Heavy Metals on Intermediate Filaments

Both NF-L and NF-M levels were found to be altered in a number of mammalian cell lines after exposure to mercuric oxide. Thus, exposure of differentiating human SK-N-SH neuroblastoma cells for 6 days to mercuric oxide decreased particularly NF-L levels (Abdulla et al., 1995). Because this effect correlated well with effects on neurite outgrowth, it was suggested that determination of NF-L might afford a rapid measure of effects on neuronal differentiation. More recently, methylmercury exposure was also found to decrease the mRNA levels for NF-L and NF-H in primary cultures of rat cerebellar granule cells, whereas GFAP mRNA expression was unaffected (Hogberg et al., 2010).

NF organization was disrupted following exposure of N2a cells to triethyl lead (Zimmermann et al., 1987). NF assembly was also disrupted. These effects led to suggestions that interaction of triethyl lead with NFs may be responsible for triethyl lead neurotoxicity in vivo. In addition, in rats exposed for 13 weeks to lead acetate, transport of NF proteins was retarded, indicating impairment of slow axonal transport (Yokoyama and Araki, 1992). Apart from NF organization, assembly, and transport, NF protein phosphorylation was also affected by lead; exposure of mice to lead acetate throughout gestation and postnatally led to increased phosphorylation of both NF-M and NF-H in auditory brainstem nuclei (Jones et al., 2008). Lead also affects GFAP, as suggested by decreased GFAP expression in four human and two rat glioma cell lines, indicating interference with glial cell differentiation (Stark et al., 1992). The mRNA levels for GFAP were also reduced after lead treatment of primary cultures of rat cerebellar granule cells for 12 days (Bal-Price et al., 2010).

In neurotoxicity studies involving arsenic, NF parameters constitute a mechanistically relevant marker because NFs (and other cytoskeletal elements) have been suggested to represent a possible target in arsenic neuropathy. Although these studies have been not always carried out in a developmental context, the known significance of NFs in neurodevelopment implies that any neurotoxic effects on NFs obtained in adult animals or differentiated neurons may be of potential relevance in neurodevelopment.

Both acute (Vahidnia et al., 2006) and subchronic (Vahidnia et al., 2008a) administration of arsenite to adult rats induced a reduction in NF-L levels in the sciatic nerve. In contrast, NF-M and NF-H expressions remained unchanged. In addition, NF-L was found to be hyperphosphorylated (Vahidnia et al., 2008a). These NF-L changes have been proposed to contribute to the disruption of the NF network, ultimately leading (in combination with other cytoskeletal changes) to the axonal degeneration seen in arsenic neuropathy (Vahidnia et al., 2007a, 2008a). However, in a study by the same group involving cell lines and assessment of NF gene expression, the results obtained were not compatible with the above data. In this case, exposure of SK-N-SH neuroblastoma and ST-8814 schwannoma cells to arsenite for up to 48 h had no effect on the expression of genes coding for NF-L and NF-M (Vahidnia et al., 2007b). However, the metabolites monomethyl- and dimethyl-arsenic induced alterations in the expression of NF genes in both cell lines and particularly in the expression of the gene coding for NF-H. Mechanistically important NF parameters assessed under the influence of arsenite also included axonal transport and phosphorylated NF distribution. For example, exposure of differentiated mouse NB2/dl neuroblastoma cells and dorsal root ganglion neurons cultured from embryonic day 12 chicks to arsenite decreased NF transport into axons and caused accumulation of phosphorylated NFs in the perikaryon leading to changes in NF dynamics that may contribute to arsenic neuropathy (De Furia and Shea, 2007).

Effects of Organic Solvents and Polybrominated Diphenyl Ethers on Intermediate Filaments

Ethanol exposure inhibited the expression of NF proteins in N2a cells, indicating disruption of neuronal differentiation (Chen et al., 2009). In contrast, at a concentration of 100 mM, ethanol induced an increase in GFAP levels in differentiating neural stem cells, which was thought to imply increased glial differentiation as a compensatory mechanism to repair the impaired neuronal differentiation (Tateno et al., 2005). However, exposure to environmentally relevant low levels of toluene (down to 0.2 ppb), decreased GFAP levels during differentiation of mouse embryo cells into an astrocytic lineage in serum-free medium (Yamaguchi et al., 2002, 2003).

In a study that adopted a proteomic approach, exposure of neonatal mice to 2,2′,4,4′,5-pentabromodiphenyl ether (PBDE 99) induced significant alterations in several cytoskeletal and other proteins in the cerebral cortex. However, one of the greatest changes noted was an increase in the levels of NF-L (Alm et al., 2008).

CONCLUDING REMARKS AND FUTURE DIRECTIONS

It is well established that the cytoskeleton plays a key role in a range of cellular processes involved in neural development. From the mechanistic studies of developmental neurotoxicity to date, there is now a significant body of evidence pointing to the disruption of one or more of the cytoskeletal networks following exposure to a range of developmental neurotoxicants, with changes at the protein level currently being more consistent than those at the level of gene expression. Taken together, the findings from studies discussed in this chapter strongly suggest that cytoskeletal disruption is a common feature of adverse outcome pathways associated with chronic exposure to many developmental neurotoxicants. Thus, despite the diversity of molecular initiating events associated with exposure to different developmental neurotoxins, subsequent molecular changes invariably converge on pathways that regulate the cytoskeleton, causing cytoskeletal disruption. In some cases, the molecular initiating event may be direct binding to cytoskeletal proteins themselves. A schematic view of cytoskeletal disruption as a convergence point in developmental neurotoxicity is summarized in Fig. 58.1.

Further work to characterize the ability of well-established developmental neurotoxins to disrupt cytoskeletal elements would be worthwhile, as this would help to identify the key events in each adverse outcome pathway in more detail. This could, for example, involve the study of upstream events such as kinase/phosphatase activities in cases where the phosphorylation status of cytoskeletal proteins is disrupted, or RT-PCR and/or proteolytic enzymes (calpain, proteasomes, etc.) cases where protein levels are significantly affected by exposure to toxin. The monitoring of these molecular events in high throughput screening platforms would help to establish a more comprehensive battery of rapid tests.

In summary, the monitoring of cytoskeletal disruption is an integral part of mechanistic studies of developmental neurotoxicity and is becoming an increasingly important component in a battery of in vitro tests to rapidly screen large numbers of compounds for their ability to induce developmental neurotoxicity.

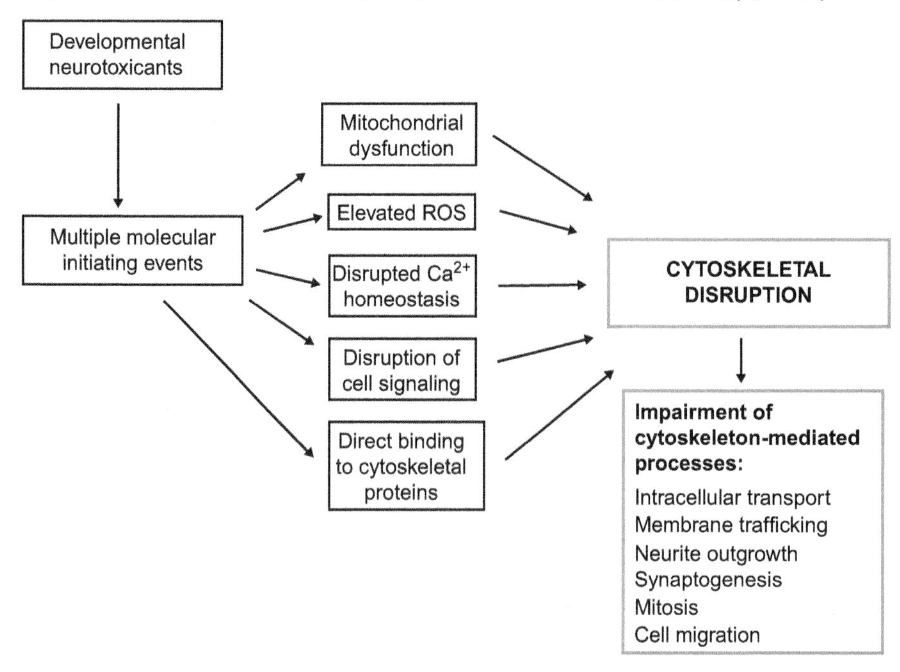

FIGURE 58.1 Simplified schematic representation of the involvement of cytoskeletal disruption in developmental neurotoxicity. Developmental neurotoxicants may act via many different molecular initiating events, such as direct binding to membrane or nuclear receptors and disruption of gene expression patterns. They may also induce the generation of increased ROS and/or cause disruption of mitochondrial function, both of which could interfere with a range of cellular activities, including intracellular transport, protein folding, and degradation. Similarly, the disruption of Ca^{2+} homeostasis could interfere with a host of Ca^{2+} dependent activities, including cell signaling pathways, proteolytic degradation (e.g., by calpain), and cytoskeletal dynamics. Finally, direct binding of toxins to cytoskeletal proteins could induce conformational changes that affect subunit assembly and/or the interaction of subunits with accessory proteins that regulate cytoskeletal dynamics.

References

Abdulla, E.M., Calaminici, M., Campbell, I.C., 1995. Comparison of neurite outgrowth with neurofilament protein subunit levels in neuroblastoma cells following mercuric oxide exposure. Clin. Exp. Pharmacol. Physiol. 22, 362–363.

Abou-Donia, M.B., 1993. The cytoskeleton as a target for organophosphorus ester-induced delayed neurotoxicity (OPIDN). Chem. Biol. Interact. 87, 383–393.

Abou-Donia, M.B., 1995. Involvement of cytoskeletal proteins in the mechanisms of organophosphorus ester-induced delayed neurotoxicity. Clin. Exp. Pharmacol. Physiol. 22, 358–359.

Ahluwalia, B., Ahmad, S., Adeyiga, O., et al., 2000. Low levels of ethanol stimulate and high levels decrease phosphorylation in microtubule-associated proteins in rat brain: an in vitro study. Alcohol Alcohol. 35, 452–457.

Akhmanova, A., Steinmetz, M.O., 2008. Tracking the end: a dynamic protein network controls the fate of microtubule tips. Nat. Rev. Mol. Cell Biol. 9, 309–322.

Alm, H., Kultima, K., Scholz, B., 2008. Exposure to brominated flame retardant PBDE-99 affects cytoskeletal protein expression in the neonatal mouse cerebral cortex. Neurotoxicology 29, 628–637.

Amos, L., 2004. Microtubule structure and its stabilization. Org. Biomol. Chem. 2, 2153–2160.

Aung, K.H., Kurihara, R., Nakashima, S., et al., 2013. Inhibition of neurite outgrowth and alteration of cytoskeletal gene expression by sodium arsenite. Neurotoxicology 34, 2226–2235.

Baas, P.W., Qiang, L., 2005. Neuronal microtubules: when the MAP is the roadblock. Trends Cell Biol. 15, 183–187.

Baas, P.W., Rao, A.N., Matamoros, A.J., Leo, L., 2016. Stability properties of neuronal microtubules. Cytoskeleton (Hoboken) 73, 442–460.

Bal-Price, A.K., Hogberg, H.T., Buzanska, L., 2010. In vitro developmental neurotoxicity (DNT) testing: relevant models and endpoints. Neurotoxicology 31, 545–554.

Beak, D.H., Park, S.H., Choi, Y., 2011. Embryotoxicity of lead (II) acetate and aroclor 1254 using a new end point of the embryonic stem cell test. Int. J. Toxicol. 30, 498–509.

Betancourt, A.M., Burgess, S.C., Carr, R.L., 2006. Effect of developmental exposure to chlorpyrifos on the expression of neurotrophin growth factors and cell-specific markers in neonatal rat brain. Toxicol. Sci. 92, 500–506.

Biernat, J., Wu, Y.-Z., Timm, T., 2002. Protein kinase MARK/PAR1 is required for neurite outgrowth and establishment of neuronal polarity. Mol. Biol. Cell 13, 4013–4028.

Brion, J.P., Octave, J.N., Couck, A.M., 1994. Distribution of the phosphorylated microtubule-associated protein tau in developing cortical neurons. Neuroscience 63, 895–909.

Brugg, B., Reddy, D., Matus, A., 1993. Attenuation of microtubule-associated protein 1B expression by antisense oligonucleotides inhibits initiation of neurite outgrowth. Neuroscience 52, 489–496.

Caballero, B., Olguin, N., Campos, F., et al., 2017. Methylmercury-induced developmental toxicity is associated with oxidative stress and cofilin phosphorylation. Cellular and human studies. Neurotoxicology 59, 197–209.

Carlier, M.F., 1998. Control of actin dynamics. Curr. Opin. Cell Biol. 10, 41–51.

Carlier, M.F., Hill, T.L., Chen, Y.D., 1984. Interference of GTP hydrolysis in the mechanism of microtubule assembly: an experimental study. Proc. Natl. Acad. Sci. U.S.A. 81, 771–775.

Carlson, K., Ehrich, M., 2001. Organophosphorous compounds alter F-actin content in human SH-SY5Y neuroblastoma cells. Neurotoxicology 22, 819–827.

Castoldi, A.F., Barni, S., Turin, I., et al., 2000. Early acute necrosis, delayed apoptosis and cytoskeletal breakdown in cultured cerebellar granule neurons exposed to methylmercury. J. Neurosci. Res. 15, 775–787.

Chen, G., Bower, K.A., Xu, M., 2009. Cyanidin-3-glucoside reverses ethanol-induced inhibition of neurite outgrowth: role of glycogen synthase kinase 3 Beta. Neurotox. Res. 15, 321–331.

Chen, J.X., Sun, X.J., Wang, P., et al., 2013. Induction of autophagy in differentiated human neuroblastoma cells lead to degradation of cytoskeletal components and inhibition of neurite outgrowth. Toxicology 310, 92–97.

De Furia, J., Shea, T.B., 2007. Arsenic inhibits neurofilament transport and induces perikaryal accumulation of phosphorylated neurofilaments: roles for JNK and GSK-3β. Brain Res. 1181, 74–82.

DiTella, M.C., Feiguin, F., Carri, N., et al., 1996. MAP-1B/TAU functional redundancy during laminin-enhanced axonal growth. J. Cell Sci. 109, 467–477.

Dominguez, R., 2009. Actin filament nucleation and elongation factors – structure-function relationships. Crit. Rev. Biochem. Mol. Biol. 44, 351–366.

Dominguez, R., Holmes, K.C., 2011. Actin structure and function. Annu. Rev. Biophys. 40, 169–186.

Easter, S.S., Ross, L.S., Frankfurter, A., 1993. Initial tract formation in the mouse brain. J. Neurosci. 13, 285–299.

Endo, M., Ohashi, K., Sasaki, Y., 2003. Control of growth cone motility and morphology by LIM kinase and slingshot by phosphorylation and dephosphorylation of cofilin. J. Neurosci. 23, 2527–2537.

Evrard, S.G., Brusco, A., Nixon, R.A., Yuan, A., 2011. Ethanol effects on cytoskeleton of nerve tissue cells. Cytoskeleton of the nervous system. In: Nixon, R.A., Yuan, A. (Eds.), Cytoskeleton of the Nervous System. Springer.

Fernandes, L.S., G. Dos Santos, N.A., Emerick, G.L., Cardozo Dos Santos, A., 2017. L- and T-type channel blockers protect against the inhibitory effects of mipafox on neurite outgrowth and plasticity-related proteins in SH-SY5Y cells. J. Toxicol. Env. Health Part B 80, 1086–1097.

Flaskos, J., 2012. The developmental neurotoxicity of organophosphorus insecticides: a direct role for the oxon metabolites. Toxicol. Lett. 209, 86–93.

Flaskos, J., 2014. The neuronal cytoskeleton as a potential target in the developmental neurotoxicity of organophosphorothionate pesticides. Basic Clin. Pharmacol. Toxicol. 115, 201–208.

Flaskos, J., Harris, W., Sachana, M., 2007. The effects of diazinon and cypermethrin on the differentiation of neuronal and glial cell lines. Toxicol. Appl. Pharmacol. 219, 172–180.

Flaskos, J., Nikolaidis, E., Harris, W., Sachana, M., Hargreaves, A.J., 2011. Effects of sub-lethal neurite outgrowth inhibitory concentrations of chlorpyrifos oxon on cytoskeletal proteins and acetylcholinesterase in differentiating N2a cells. Toxicol. Appl. Pharmacol. 256, 330–336.

Fujimura, M., Usuki, F., 2012. Differing effects of toxicants (methylmercury, inorganic mercury, lead, amyloid β and rotenone) on cultured rat cerebrocortical neurons: differential expression of Rho proteins associated with neurotoxicity. Toxicol. Sci. 126, 506–514.

Funk, C.K., Dohrman, D.P., 2007. Chronic ethanol exposure inhibits dopamine release via effects on the presynaptic actin cytoskeleton in PC12 cells. Brain Res. 1185, 86–94.

Garcia, S.J., Seidler, F.J., Qiao, D., Slotkin, T.A., 2002. Chlorpyrifos targets developing glia: effects on glial fibrillary acidic protein. Dev. Brain Res. 133, 151–161.

Gąssowska, M., Baranowska-Bosiacka, I., Moczydłowska, J., et al., 2016. Perinatal exposure to lead (Pb) promotes Tau phosphorylation in the rat brain in a GSK-3β and C DK5 dependent manner: relevance to neurological disorders. Toxicology 347–349, 17–28.

Gearhart, D.A., Sickles, D.W., Buccafusco, J.J., et al., 2007. Chlorpyrifos, chlorpyrifos-oxon, and diisopropylfluorophosphate inhibit kinesin-dependent microtubule motility. Toxicol. Appl. Pharmacol. 218, 20–29.

Gendron, T.F., McCartney, S., Causevic, E., et al., 2008. Ethanol enhances tau accumulation in neuroblastoma cells that inducibly express tau. Neurosci. Lett. 443, 67–71.

Giasson, B.I., Sampathu, D.M., Wilson, C.A., 2002. The environmental toxin arsenite induces tau hyperphosphorylation. Biochemistry 41, 15376–15387.

Grigoryan, H., Lockridge, O., 2009. Nanoimages show disruption of tubulin polymerization by chlorpyrifos oxon; implications for neurotoxicity. Toxicol. Appl. Pharmacol. 240, 143–148.

Grigoryan, H., Li, B., Xue, W., 2009a. Mass spectral characterization of organophosphate labeled lysines in peptides. Anal. Biochem. 394, 92–100.

Grigoryan, H., Schopfer, L.M., Peeples, E.S., 2009b. Mass spectrometry identifies multiple organophosphorylated sites on tubulin. Toxicol. Appl. Pharmacol. 240, 149–158.

Gungabissoon, R.A., Bamburg, J.R., 2003. Regulation of growth cone actin dynamics by ADF/cofilin. J. Histochem. Cytochem. 51, 411–420.

Guo, B.Q., Yan, C.H., Cai, S.Z., et al., 2013. Low level prenatal exposure to methylmercury disrupts neuronal migration in the developing rat cerebral cortex. Toxicology 304, 57–68.

Hargreaves, A.J., 1997. The cytoskeleton as a target in cell toxicity. Adv. Mol. Cell. Biol. 20, 119–144.

Harris, W., Sachana, M., Flaskos, J., Hargreaves, A.J., 2009a. Proteomic analysis of differentiating neuroblastoma cells treated with sub-lethal neurite inhibitory concentrations of diazinon: identification of novel biomarkers of effect. Toxicol. Appl. Pharmacol. 240, 159–165.

Harris, W., Sachana, M., Flaskos, J., Hargreaves, A.J., 2009b. Neuroprotection from diazinon-induced toxicity in differentiating murine N2a neuroblastoma cells. Neurotoxicology 30, 958–964.

Hassler, J.A., Moran, D.J., 1986. Effects of ethanol on the cytoskeleton and differentiating neural crest cells: possible role in teratogenesis. J. Craniofac. Genet. Dev. Biol. 2, 129–136.

Hernández, A.J.A., Reyes, V.L., Albores-García, D., et al., 2018. MeHg affects the activation of FAK, Src, Rac1 and Cdc42, critical proteins for cell movement in PDGF-stimulated SH-SY5Y neuroblastoma cells. Toxicology 394, 35–44.

Herrmann, H., Aebi, U., 2000. Intermediate filaments and their associates: multi-talented structural elements specifying cytoarchitecture and cytodynamics. Curr. Opin. Cell Biol. 12, 79–90.

Hogberg, H.T., Kinsner-Ovaskalainen, A., Coecke, S., et al., 2010. mRNA expression is a relevant tool to identify developmental neurotoxicants using an in vitro approach. Toxicol. Sci. 113, 95–115.

Howell, B., Larsson, N., Gullberg, M., Cassimeris, L., 1999. Dissociation of the tubulin-sequestering and microtubule catastrophe-promoting activities of oncoprotein 18/stathmin. Mol. Biol. Cell 10, 105–118.

Ishikawa, R., Kohama, K., 2007. Actin-binding proteins in nerve cell growth cones. J. Pharmacol. Sci. 105, 6–11.

Janke, C., Kneussel, M., 2010. Tubulin post-translational modifications: encoding functions on the neuronal microtubule cytoskeleton. Trends Neurosci. 33, 362–372.

Jansen, S., Collins, A., Yang, C., 2011. Mechanisms of actin bundling by fascin. J. Biol. Chem. 286, 30087–30096.

Jaworski, J., Kapitein, L.C., Gouveia, S.M., 2009. Dynamic microtubules regulate dendritic spine morphology and synaptic plasticity. Neuron 61, 85–100.

Jennett, R.B., Tuma, D.J., Sorrell, M.F., 1980. Effect of ethanol and its metabolites on microtubule formation. Pharmacology 21, 363–368.

Jiang, W., Duysen, E.G., Hansen, H., 2010. Mice treated with chlorpyrifos or chlorpyrifos oxon have organophosphorylated tubulin in the brain and disrupted microtubule structures, suggesting a role for tubulin in neurotoxicity associated with exposure to organophosphorus agents. Toxicol. Sci. 115, 183–193.

Johnson, G.V.W., Stoothoff, W.H., 2004. Tau phosphorylation in neuronal cell function and dysfunction. J. Cell Sci. 117, 5721–5729.

Jones, L.G., Prins, J., Park, S., 2008. Lead exposure during development results in increased neurofilament phosphorylation, neuritic beading, and temporal processing deficits within the murine auditory brainstem. J. Comp. Neurol. 506, 1003–1017.

Joshi, H.C., 1998. Microtubule dynamic in living cells. Curr. Opin. Cell Biol. 10, 35–44.

Katsetos, C.D., Legido, A., Perentes, E., Mork, S.J., 2003. Class III β-tubulin isotype: a key cytoskeletal protein at the crossroads of developmental neurobiology and tumor neuropathology. J. Child Neurol. 18, 851–866.

Kekic, M., dos Remedios, C.G., 1999. Electrophoretic monitoring of pollutants: effects of cations and organic compounds on protein interactions monitored by native gel electrophoresis. Electrophoresis 20, 2053–2058.

Kromidas, L., Trombetta, L.D., Jamall, I.S., 1990. The protective effects of glutathione against methylmercury cytotoxicity. Toxicol. Lett. 51, 67–80.

Kunda, P., Baum, B., 2009. The actin cytoskeleton in spindle assembly and positioning. Trends Cell Biol. 19, 174–179.

Lariviere, R.C., Julien, J.-P., 2003. Functions of intermediate filaments in neuronal development and disease. J. Neurobiol. 58, 131–148.

Lawton, M., Iqbal, M., Kontovraki, M., 2007. Reduced tubulin tyrosination as an early marker of mercury toxicity in differentiating N2a cells. Toxicol. Vitro 21, 1258–1261.

Lee, S.H., Dominguez, R., 2010. Regulation of actin cytoskeleton dynamics in cells. Mol. Cell. 29, 311–325.

Lee, M.K., Tuttle, J.B., Rebhun, L.I., et al., 1990. The expression and post-translational modification of a neuron specific β-tubulin isotype during chick embryogenesis. Cell Motil. Cytoskelet. 17, 118–132.

Lee, W.K., Torchalski, B., Thévenod, F., 2007. Cadmium induced ceramide formation triggers calpain-dependent apoptosis in kidney proximal tubule cells. Am. J. Physiol. Cell Physiol. 293, C839–C847.

Li, N., Yu, Z.L., Wang, L., 2010. Increased tau phosphorylation and beta amyloid in the hippocampus of mouse pups by early life lead exposure. Acta Biol. Hung. 61, 123–134.

Lin, T.-H., Kuo, H.-C., Chou, F.-P., et al., 2008. Berberine enhances inhibition of glioma tumor cell migration and invasiveness mediated by arsenic trioxide. BMC Canc. 8, 58.

Lindsley, T.A., Clarke, S., 2004. Ethanol withdrawal influences survival and morphology of developing rat hippocampal neurons in vitro. Alcohol Clin. Exp. Res. 28, 85–92.

Lindsley, T.A., Comstock, L.L., Rising, L.J., 2002. Morphologic and neurotoxic effects of ethanol vary with timing of exposure in vitro. Alcohol 28, 197–203.

Loureiro, S.O., Heimfarth, L., Reis, K., 2011. Acute ethanol exposure disrupts actin cytoskeleton and generates reactive oxygen species in C6 cells. Toxicol. Vitro 25, 28–36.

Ludueña, R.F., 1998. Multiple forms of tubulin: different gene products and covalent modifications. Int. Rev. Cytol. 178, 207–275.

Marín, M.P., Esteban-Pretel, G., Ponsoda, X., 2010. Endocytosis in cultured neurons is altered by chronic alcohol exposure. Toxicol. Sci. 115, 202–213.

Martinez-Neira, R., Kekic, M., Nicolau, D., dos Remedios, C.G., 2005. A novel biosensor for mercuric ions based on motor proteins. Biosens. Bioelectron. 20, 1428–1432.

McKinnon, G., Davidson, M., De Jersey, J., Shanley, B., Ward, L., 1987. Effects of acetaldehyde on polymerization of microtubule proteins. Brain Res. 416, 90–99.

Meixner, A., Haverkamp, S., Wassle, H., 2000. MAP1B is required for axon guidance and is involved in the development of the central and peripheral nervous system. J. Cell Biol. 151, 1169–1178.

Miura, K., Imura, N., 1987. Mechanism of cytotoxicity of methylmercury. With special reference to microtubule disruption. Biol. Trace Elem. Res. 21, 313–316.

Miura, K., Inokawa, M., Imura, N., 1984. Effect of methylmercury and some metal ions on microtubule networks in mouse glioma cells and in vitro tubulin polymerization. Toxicol. Appl. Pharmacol. 73, 218–231.

Miura, K., Kobayashi, Y., Toyoda, H., Imura, N., 1998. Methylmercury-induced microtubule depolymerization leads to inhibition of tubulin synthesis. J. Toxicol. Sci. 23, 379–388.

Noraberg, J., Zimmer, J., 1998. Ethanol induces MAP2 changes in organotypic hippocampal slice cultures. Neuroreport 9, 3177–3182.

Ohkawa, N., Fujitani, K., Tokunaga, E., et al., 2007. The microtubule destabilizer stathmin mediates the development of dendritic arbors in neuronal cells. J. Cell Sci. 120, 1447–1456.

Omary, M.B., Ku, N.-O., Tao, G.-Z., et al., 2006. 'Heads and tails' of intermediate filament phosphorylation: multiple sites and functional insights. Trends Biochem. Sci. 31, 383–394.

Penazzi, L., Bakota, L., Brandt, R., 2016. Microtubule dynamics in neuronal development, plasticity, and neurodegeneration. Int. Rev. Cell Mol. Biol. 321, 89–169.

Perry, S.V., Cotterill, J., 1964. The action of thiol inhibitors on the interaction of F-actin and heavy meromyosin. Biochem. J. 92, 603–608.

Pierozan, P., Biasibetti, H., Schmitz, F., et al., 2017. Neurotoxicity of methylmercury in isolated astrocytes and neurons: the cytoskeleton as a main target. Mol. Neurobiol. 54, 5752–5767.

Powrozek, T.A., Olson, E.C., 2012. Ethanol-induced disruption of Golgi apparatus morphology, primary neurite number and cellular orientation in developing cortical neurons. Alcohol 46, 619–627.

Prendergast, M.A., Self, R.L., Smith, K.J., 2007. Microtubule-associated target in chlorpyrifos oxon hippocampal neurotoxicity. Neuroscience 146, 330–339.

Rahman, A., Brew, B.J., Guillemin, G.J., 2011. Lead dysregulates serine/threonine protein phosphatases in human neurons. Neurochem. Res. 36, 195–204.

Rahman, A., Khan, K.M., Al-Khaledi, G., et al., 2012. Early postnatal lead exposure induces tau phosphorylation in the brain of young rats. Acta Biol. Hung. 63, 411–425.

Reiter-Funk, C.K., Dohrman, D.P., 2005. Chronic ethanol exposure increases microtubule content in PC12 cells. BMC Neurosci. 6, 16–23.

Rocha, R.A., Gimeno-Alcañiz, J.V., Martin-Abañez, R., 2011. Arsenic and fluoride induce neural progenitor cell apoptosis. Toxicol. Lett. 203, 237–244.

Roderer, G., Doenges, K.H., 1983. Influence of trimethyl lead and inorganic lead on assembly of microtubules from mammalian brain. Neurotoxicology 4, 171–180.

Romero, A.M., Esteban-Pretel, G., Marin, M.P., 2010. Chronic ethanol exposure alters the levels, assembly, and cellular organization of the actin cytoskeleton and microtubules in hippocampal neurons in primary culture. Toxicol. Sci. 118, 602–612.

Rösner, H., Vacun, G., 1997. Organotypic spinal cord culture in serum free fibrin gel: a new approach to study 3-dimensional neurite outgrowth and of neurotoxicity testing: effects of modulation the actin and tubulin dynamics and protein kinase activities. J. Neurosci. Meth. 78, 93–103.

Ruiz-Muñoz, A.M., Nieto-Escamez, F.A., Aznar, S., et al., 2011. Cognitive and histological disturbances after chlorpyrifos exposure and chronic Aβ (1-42) infusions in Wistar rats. Neurotoxicology 32, 836–844.

Sachana, M., Flaskos, J., Alexaki, E., Glynn, P., Hargreaves, A.J., 2001. The toxicity of chlorpyrifos towards differentiating mouse N2a neuroblastoma cells. Toxicol. Vitro 15, 369–372.

Sachana, M., Flaskos, J., Alexaki, E., Hargreaves, A.J., 2003. Inhibition of neurite outgrowth in N2a cells by leptophos and carbaryl: Effects on neurofilament heavy chain, GAP-43 and HSP-70. Toxicol In Vitro *17, 115–120.*

Sachana, M., Flaskos, J., Hargreaves, A.J., 2005. Effects of chlorpyrifos and chlorpyrifos-methyl on the outgrowth of axon-like processes, tubulin and GAP-43 in N2a cells. Toxicol. Mech. Meth. 15, 405–410.

Sachana, M., Flaskos, J., Hargreaves, A.J., Gupta, R.C., 2017. In vitro biomarkers of developmental neurotoxicity. In: Gupta, R.C. (Ed.), Reproductive and Developmental Toxicology, second ed. Academic Press/Elsevier, Amsterdam, pp. 255–288.

Sachana, M., Flaskos, J., Sidiropoulou, E., et al., 2008. Inhibition of extension outgrowth in differentiating rat C6 glioma cells by chlorpyrifos and chlorpyrifos oxon: effects on microtubule proteins. Toxicol. Vitro 22, 1387–1391.

Sachana, M., Sidiropoulou, E., Flaskos, J., et al., 2014. Diazoxon disrupts the expression and distribution of βIII-tubulin and MAP 1B in differentiating N2a cells. Basic Clin. Pharmacol. Toxicol. 114, 490–496.

Saito, M., Chakraborty, G., Mao, R.-F., 2010. Tau phosphorylation and cleavage in ethanol-induced neurodegeneration in the developing mouse brain. Neurochem. Res. 35, 651–659.

Schopfer, L.M., Grigoryan, H., Li, B., 2010. Mass spectral characterization of organophosphate-labeled, tyrosine-containing peptides characteristic mass fragments and a new binding motif for organophosphates. J. Chromatog. B Analyt. Technol. Biomed. Life Sci. 878, 1297–1311.

Scortegagna, M., Chikhale, E., Hanbauer, I., 1998. Effect of lead on cytoskeletal proteins expressed in E14 mesencephalic primary cultures. Neurochem. Int. 32, 353–359.

Sidiropoulou, E., Sachana, M., Flaskos, J., 2009a. Diazinon oxon affects the differentiation of mouse N2a neuroblastoma cells. Arch. Toxicol. 83, 373–380.

Sidiropoulou, E., Sachana, M., Flaskos, J., 2009b. Diazinon oxon interferes with differentiation of rat C6 glioma cells. Toxicol. Vitro 23, 1548–1552.

Sihag, R.K., Inagaki, M., Yamaguchi, T., Shea, T.B., Pant, H.C., 2007. Role of phosphorylation on the structural dynamics and function of types III and IV intermediate filaments. Environ. Cell Res. 313, 2098–2109.

Sindi, R.A., Harris, W., Arnott, G., et al., 2016. Chlorpyrifos- and chlorpyrifos oxon-induced neurite retraction in pre-differentiated N2a cells is associated with transient hyperphosphorylation of neurofilament heavy chain and ERK 1/2. Toxicol. Appl. Pharmacol. 308, 20–31.

Slotkin, T., Seidler, F.J., 2007. Comparative developmental neurotoxicity of organophosphates in vivo: transcriptional responses of brain cell development, cell signaling, cytotoxicity and neurotransmitter systems. Brain Res. Bull. 72, 232–274.

Slotkin, T., Seidler, F., 2009. Transcriptional profiles reveal similarities and differences in the effects of developmental neurotoxicants on differentiation into neurotransmitter phenotypes in PC12 cells. Brain Res. Bull. 78, 211–225.

Stark, M., Wolff, J.E., Korbmacher, A., 1992. Modulation of glial cell differentiation by exposure to lead and cadmium. Neurotoxicol. Teratol. 14, 247–252.

Svitkina, T., 2018. The actin cytoskeleton and actin-based motility. Cold Spring Harb. Perspect. Biol. 10 pii:a018267.

Ta, N., Li, C., Fang, Y., et al., 2014. Toxicity of TDCPP and TCEP on PC12 cell: changes in CAMKII, GAP43, tubulin and NFH gene and protein levels. Toxicol. Lett. 227, 164–171.

Taléns-Visconti, R., Sanchez-Vera, I., Kostic, J., 2011. Neural differentiation from human embryonic stem cells as a tool to study early brain development and the neuroteratogenic effects of ethanol. Stem Cell. Dev. 20, 327–339.

Tateno, M., Ukai, W., Yamamoto, M., 2005. The effect of ethanol on cell fate determination of neural stem cells. Alcohol Clin. Exp. Res. 29, 225S–229S.

Terry, A.V., Gearhart, D.A., Beck, W.D., 2007. Chronic, intermittent exposure to chlorpyrifos in rats: protracted effects on axonal transport, neurotrophin receptors, cholinergic markers, and information processing. J. Pharmacol. Exp. Therapeut. 322, 1117–1128.

Tischfield, M.A., Engle, E.C., 2010. Distinct α- and β-tubulin isotypes are required for the positioning, differentiation and survival of neurons: new support for the "multi-tubulin" hypothesis. Biosci. Rep. 30, 319–330.

Tomas, M., Lazaro-Diguez, F., Duran, J.M., 2003. Protective effects of lysophosphatidic acid (LPA) on chronic ethanol-induced injuries to the cytoskeleton and on glucose uptake in rat astrocytes. J. Neurochem. 87, 220–229.

Trombetta, L.D., Kromidas, L., 1992. A scanning electron-microscopic study of the effects of methylmercury on the neuronal cytoskeleton. Toxicol. Lett. 60, 329–341.

Usuki, F., Fujimura, M., 2012. Effects of methylmercury on cellular signal transduction systems. In: Ceccatelli, S., Aschner, M. (Eds.), Methylmercury and Neurotoxicity. Current Topics in Neurotoxicity, vol. 2. Springer, Boston, MA.

Vahidnia, A., Romijn, F., Tiller, M., et al., 2006. Arsenic – induced toxicity: effect on protein composition in sciatic nerve. Hum. Exp. Toxicol. 25, 667–674.

Vahidnia, A., van der Voet, G.B., de Wolff, F.A., 2007a. Arsenic neurotoxicity – a review. Hum. Exp. Toxicol. 26, 823–832.

Vahidnia, A., van der Straaten, R.J.H.M., Romijn, F., 2007b. Arsenic metabolites affect expression of the neurofilament and tau genes: an in-vitro study into the mechanism of arsenic neurotoxicity. Toxicol. Vitro 21, 1104–1112.

Vahidnia, A., Romijn, F., van der Voet, G.B., de Wolff, F.A., 2008a. Arsenic-induced neurotoxicity in relation to toxicokinetics: effects on sciatic nerve proteins. Chem. Biol. Interact. 176, 188–195.

Vahidnia, A., van der Straaten, R.J.H.M., Romijn, F., 2008b. Mechanism of arsenic-induced neurotoxicity may be explained through cleavage of p35 to p25 by calpain. Toxicol. Vitro 22, 682–687.

Vale, R.D., 2003. The molecular motor toolbox for intracellular transport. Cell 112, 467–480.

VanDemark, K.L., Guizzetti, M., Giordano, G., Costa, L.G., 2009. Ethanol inhibits muscarinic receptor-induced axonal growth in rat hippocampal neurons. Alcohol Clin. Exp. Res. 33, 1945–1955.

Vendrell, I., Carrascal, M., Campos, F., et al., 2010. Methylmercury disrupts the balance between phosphorylated and non-phosphorylated cofilin primary cultures of mouse cerebellar granule cells. Toxicol. Appl. Pharmacol. 242, 109–118.

Villarroel-Campos, D., Gonzalez-Billault, C., 2014. The MAP1B case: an old MAP that is new again. Dev. Neurobiol. 74, 953–971.

Visan, A., Hayess, K., Sittner, D., et al., 2012. Neural differentiation of mouse embryonic stem cells as a tool to assess developmental neurotoxicity in vitro. Neurotoxicology 33, 1135–1146.

Wade, R.H., 2009. On and around microtubules: an overview. Mol. Biotechnol. 43, 177–191.

Watanabe, A., Hasegawa, M., Suzuki, M., 1993. In vivo phosphorylation sites in fetal and adult rat tau. J. Biol. Chem. 268, 25712–25717.

Yamaguchi, H., Kidachi, Y., Ryoyama, K., 2002. Toluene at environmentally relevant low levels disrupts differentiation of astrocyte precursor cells. Arch. Environ. Health 57, 232–238.

Yamaguchi, H., Kidachi, Y., Ryoyama, K., 2003. Increased synthesis of GFAP by TCDD in differentiation-disrupted SFME cells. Environ. Toxicol. Pharmacol. 15, 1–8.

Yokoyama, K., Araki, S., 1992. Assessment of slow axonal transport in lead-exposed rats. Environ. Res. 59, 440–446.

Zhang, G.S., Ye, W.F., Tao, R.R., 2012. Expression profiling of Ca^{2+}/calmodulin dependent signaling molecules in the rat dorsal and ventral hippocampus after acute lead exposure. Exp. Toxicol. Pathol. 64, 619–624.

Zhang, J., Miyamoto, K., Hashioka, S., 2003. Activation of calpain in developing cortical neurons following methylmercury treatment. Dev. Brain Res. 142, 105–110.

Zhou, H., Liu, Y., Tan, X.J., et al., 2015. Inhibitory effect of arsenic trioxide on neuronal migration in vitro and its potential molecular mechanism. Environ. Toxicol. Pharmacol. 40, 671–677.

Zimmermann, H.-P., Plagens, U., Traub, P., 1987. Influence of triethyl lead on neurofilaments in vivo and in vitro. Neurotoxicology 8, 569–578.

Zimmermann, H.-P., Faulstich, H., Hansch, G.M., et al., 1988. The interaction of triethyl lead with tubulin and microtubules. Mutat. Res. 201, 293–302.

Zurich, M.G., Honegger, P., Schilter, B., et al., 2000. Use of aggregating brain cell cultures to study developmental effects of organophosphorus insecticides. Neurotoxicology 21, 599–605.

CHAPTER

59

MicroRNA Expression as an Indicator of Tissue Toxicity and a Biomarker in Disease and Drug-Induced Toxicological Evaluation

Gopala Krishna[1], *Saurabh Chatterjee*[2], *Priya A. Krishna*[1], *Ratanesh Kumar Seth*[2]

[1]Nonclinical Consultants, Ellicott City, MD, United States; [2]Department of Environmental Health Sciences, Arnold School of Public Health, University of South Carolina, Columbia, SC, United States

INTRODUCTION

Noncoding RNA sequences, termed microRNAs (or miRNAs or miRs), have been shown to posttranscriptionally regulate messenger RNA (mRNA). They contain a seed sequence (positions 2–7) that binds to the complementary sequence of mRNA causing an alteration in gene expression. miRNAs are short (~22 nucleotides) and relatively stable. Studies have shown their presence in plasma with a half-life of hours to days (Creemers et al., 2012). Because of their inherent stability and ease of sampling, the number of studies evaluating miRNAs as potential biomarkers both for disease- and drug-induced states has markedly increased. For a biomarker to be successfully utilized, it must satisfy a number of criteria such as ease of sampling, rapid detection, injury or disease sensitivity and specificity; discrimination between variations in the injury or disease; and stability within the biological matrix (Etheridge et al., 2011). miRNAs fulfill a number of these requirements.

Much effort has been spent in making correlations between various disease states and miRNA levels in both biological fluids and tissues (Mikaelian et al., 2013). These efforts have also been extended to include not just biomarker data in support of diagnosis but also more recently in the drug development process, where relationships between miRNA levels and toxicologic or pathologic signs (Wang et al., 2013) may be used to generate earlier safety data in advance of more expensive animal or clinical trials. Perhaps most importantly, a biomarker must be associated with the biological mechanisms of a disease or treatment, and it must be possible to make a statistical correlation between the biomarker and the clinical effect.

Although the bulk of data reported in the literature refers to the relationship between miRNAs and therapeutic effects, less has been published regarding the relationship between changes in miRNA expression following administration of xenobiotics and the utility these extracellular molecules may have a correlating drug effect with, for example, a toxicologic endpoint.

REGULATORY MECHANISMS OF MIRNA BIOGENESIS

miRNAs are an important regulatory molecule in many biological processes and pathways. Being said that, the biogenesis and expression profile of the miRNAs are changed considerably in various diseases (O'Reilly, 2016). The miRNA profiling has been extensively studied in several diseases especially in cancer and the results of these studies indicate that miRNA expression profiling is an important tool for diagnostic and therapeutic strategy. Therefore, it is important to illustrate the molecular mechanism that regulates the miRNA expression, and it would allow an explanation of variation in protein coding genes (Gulyaeva and Kushlinskiy, 2016). The two steps in such regulation involve (1) transcriptional regulation and (2) posttranscriptional regulation (Fig. 59.1). Endogenous and/or xenobiotic

Biomarkers in Toxicology, Second Edition
https://doi.org/10.1016/B978-0-12-814655-2.00059-1

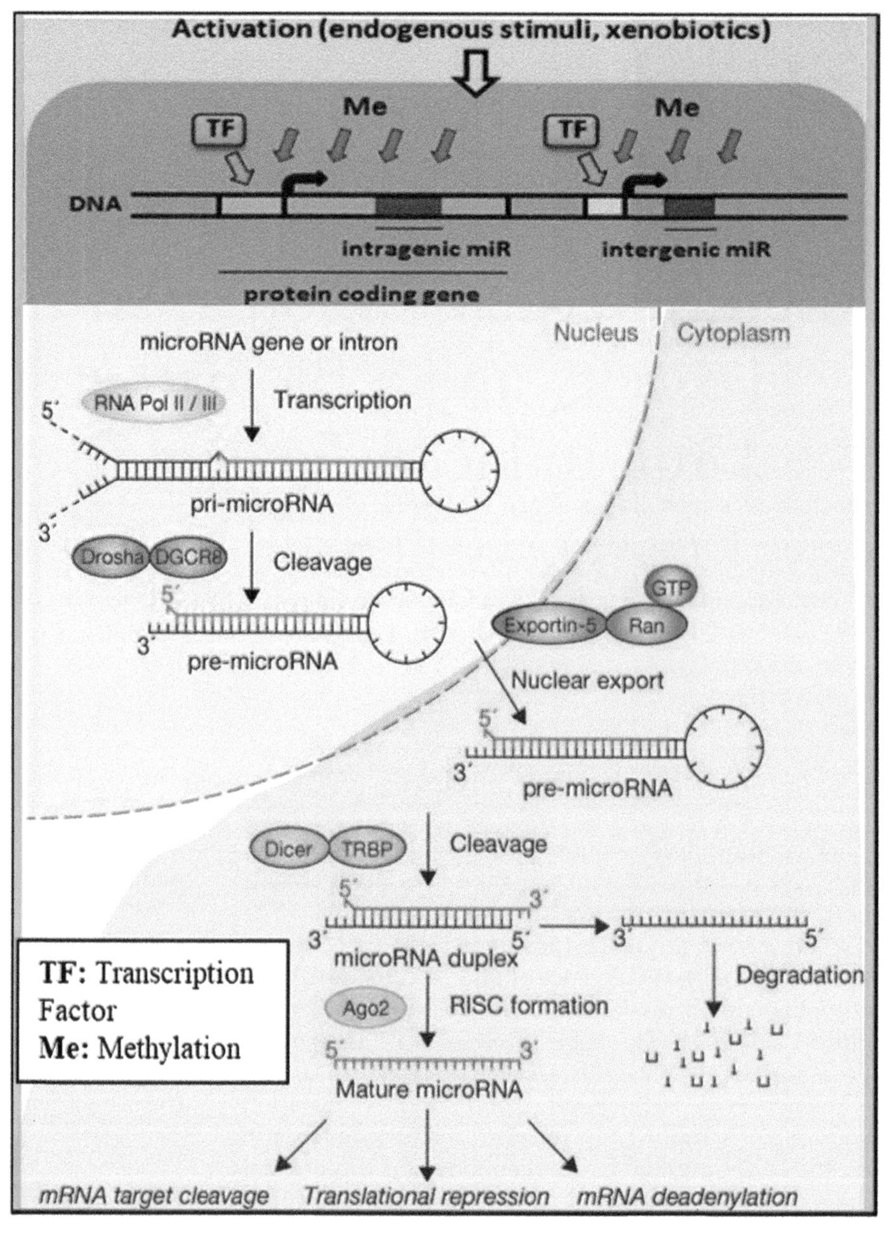

FIGURE 59.1 Schematic representation of miRNA biosynthesis and regulation: transcription, cleavage, transport to cytoplasm, and targeting of the target gene mRNA 3′UTR (Winter et al., 2009).

stimulation either activates transcription factor (TF) binding to the promotor region to facilitate miRNA transcription or methylation of promotor region of intergenic or intragenic miRNA genes to inhibit miRNA transcription (Gulyaeva and Kushlinskiy, 2016). The posttranscription regulation represents changes in miRNA processing and stability. The transcript of intronic region of host gene or independent miRNAs gene is known as mirtrons or primary miRNAs (pri-miRNAs). The Pri-miRNA further processed by Drosha and DGCR8 into precursor miRNA (pre-miRNA) inside the nucleus. DGCR8 contains RNA binding domain and it binds with pri-miRNA to stabilize it and facilitate the cleavage by Drosha (Yeom et al., 2006). The pre-miRNA was then exported to cytoplasm, where Dicer cleaved pre-miRNA to form mature miRNA (Merritt et al., 2010). Therefore, changes in miRNA profile at processing levels depend on availability and activity of these enzymes. The mature miRNA forms RNA-induced silencing complex (RISC), which binds with complementary sequence of mRNA and cleaves the mRNA using RNase or destabilize or inhibit translation of mRNA (Mendell and Olson, 2012; Vermeulen et al., 2005).

The Ingenuity Pathway Analysis tool by Qiagen Inc. can precisely predict the target mRNA (Bam et al., 2017). In summary, the endogenous molecules such as hormones and cytokines, and exogenous molecules such as xenobiotics, can alter miRNA expression. These comprehensive mechanisms of miRNA regulation solely depend on cell type, physiological conditions, and external or internal factors (Gulyaeva and Kushlinskiy, 2016). Thus, it is clear that the study of miRNAs expression profile enables us for the diagnosis, prognosis, and treatment of a wide variety of diseases.

STANDARD TOXICITY MEASUREMENTS (TISSUE AND CIRCULATING BIOMARKERS)

For many years, the standard continuum approach employed in the discovery and development of new chemical or molecular entities (NCEs or NMEs) has been to first screen large numbers of molecules using in vitro (biochemical and cell-based) assays followed by testing a smaller number of active compounds in preclinical efficacy, safety, and toxicity models. In general, it is only at the in vivo testing stage that biomarkers are given the attention they deserve. Within the drug development setting, it has long been standard practice to focus on easily accessible matrices, for example, blood, saliva, and urine, as potential reservoirs of information for drug safety and efficacy and as correlates for disease progression. It is perhaps interesting to note that although analytical technologies have advanced significantly in biomarker measurement (genomics, metabolomics, proteomics, etc.), the corresponding information technology tools required to process the large amounts of information generated have lagged (Starmans et al., 2012).

TOXICOLOGY

Toxicological evaluations typically include body weights, clinical observations, food consumption, ophthalmologic examination, clinical pathology, body temperatures, electrocardiographic evaluation (large animals only), neurobehavioral evaluation (usually only done in rodents), toxicokinetic evaluation, organ weights, and gross and microscopic pathology of tissues. Several of the items listed above are typical standard biomarkers utilized both preclinically, as part of a drug development program, and clinically, for both injury and disease diagnosis and monitoring. Further details are highlighted below.

CLINICAL PATHOLOGY

Overview

The standard clinical pathology analyses routinely performed include hematology and clinical chemistry evaluations and, less frequently, urine and coagulation analyses. The results of these noninvasive tests may suggest a pathology that impacts the function of various organs. Limitations exist in the sensitivity and specificity of these analyses and, as such, they are used primarily to screen for health problems. An advantage, however, is that information on general animal health may be gathered via a nonterminal bleed throughout the duration of a study.

HEMATOLOGY

Hematological tests are used to examine the components of blood, specifically erythrocytes, leukocytes, and platelets, as well as specific parameters related to each cell type. The most common test is the complete blood count/hemogram, which is used as a broad screening panel that provides information such as, but not limited to, the erythrocyte: red blood cell count, hematocrit, hemoglobin, mean cell volume, and reticulocyte percentage, and count; leukocyte: count and differential; platelet: count and mean platelet volume. This information is used to screen for processes such as anemia, infectious or noninfectious inflammation, and thrombocytopenia or thrombocytosis. Microscopic examination of the peripheral blood may also be performed. This provides information regarding erythrocyte, leukocyte, and platelet number and morphology. Certain insults may result directly or indirectly in alterations in cellular morphology (e.g., Heinz body or eccentrocyte formation in erythrocytes resulting from oxidative damage) that may provide additional information regarding the pathologic process.

CLINICAL CHEMISTRY

A variety of different serum clinical chemistries may be analyzed to provide an overview of the general health status of the individual including the electrolyte balance and the status of several organs. Interpretation of the quantitative levels of the individual analytes may be affected and complicated by factors such as the site of production (i.e., a single cell type/organ that is the origin vs. multiple potential cell types/organs) and physiological regulation critical for homeostasis, as well as potential sites of loss or processing by various organ systems. Generally, there is no single analyte that is specific in indicating the dysfunction of one organ system; as such, multiple different analytes are typically measured. For example, blood urea nitrogen (BUN) and creatinine classically increase with kidney failure (renal azotemia). In contrast, if there was a concern for hepatic toxicity, specific chemistry tests of interest would include analytes such as albumin, total protein, alanine aminotransferase (ALT), aspartate aminotransferase (AST), alkaline phosphatase (ALP), lactate dehydrogenase, and bilirubin. Therefore, if the target organ of interest is known, it is possible to rank the chemistry analytes based on the degree of importance.

URINALYSIS

Urine is produced by the kidney and is a means by which the body eliminates waste products and regulates fluid homeostasis. A complete urinalysis is composed of three types of examination (visual, chemical, and microscopic) during which the levels of various biochemical products and cellular components are determined. Therefore, a urinalysis may function as a screening or diagnostic procedure in the assessment of renal as well as various other metabolic functions. Currently, the chemical portion of urinalysis is semiquantitative and utilizes dry-chemistry methodology via a commercially manufactured reagent stick. Quantitative measurements of selected analytes may be performed using the Cobas c501 chemistry analyzer, a wet-chemistry based methodology.

COAGULATION TESTS

Prothrombin time (PT), activated partial thromboplastin time (aPTT), and fibrinogen levels are assays commonly performed in various species. These tests are ex vivo screening measures used to assess coagulation. Multiple factors are involved in the cascade of reactions that result in clot formation in the PT and aPTT test(s). A significant decrease ($\sim <30\%$) of a single or multiple factor(s) may result in the prolongation of clotting time in either PT or aPTT test(s). The PT and/or aPTT may increase in situations such as hepatic disease, vitamin K deficiency/antagonism, and consumption (e.g., disseminated intravascular coagulopathy). In contrast, fibrinogen is the precursor to fibrin, which is the end product of the coagulation cascade. Therefore, with hepatic dysfunction or secondary to increased consumption, fibrinogen levels may decrease. Additionally, fibrinogen is an acute phase protein, and its levels may increase with inflammation.

ANATOMIC PATHOLOGY

Pathological alterations by novel test agents need to be evaluated in laboratory animals before human testing. The alterations are monitored by macroscopic (gross) and microscopic (histologic) structural (morphologic) changes, as well as by conventional histopathological analysis, which evaluates tissue and cellular patterns. The ultimate outcome of this comprehensive evaluation is the integration of the information into meaningful biological conclusions. Commonly used tissues for a toxicology study and follow-up histopathological evaluation are listed in Table 59.1.

For morphological examination of both paraffin-embedded and frozen sections, hematoxylin-eosin staining (H&E) is commonly used, with hematoxylin for nuclear staining and eosin for cytoplasmic staining. This staining procedure often enables diagnostic determination because structural discriminations can be made as to normal or pathological changes in the tissue, though special staining techniques are used as required for further detailed subtyping.

TABLE 59.1 List of Organs and Tissues Typically Studied in a GLP Toxicology Study

Adrenal glands	Lungs
Aorta	Lymph nodes (bronchial, mandibular, mesenteric)
Bone, femoral head with articular surface	Mammary gland (when present in regular abdominal skin section)
Bone marrow (sternum)	Rectum
Bone marrow (rib, costochondral junction)	Pancreas
Bone marrow (sternum) slides for differential	Pituitary gland
Brain	Prostate gland
Cecum	Salivary gland, mandibular
Colon	Sciatic nerve
Duodenum	Skeletal muscle
Epididymides	Skin (ventral abdomen)
Esophagus	Spinal cord, thoracic, and spinal cord, lumbar
Eyes	Spleen
Gall bladder	Stomach (cardiac, fundic, and pyloric)
Gonads	Thymus
Gross lesions	Thyroid gland/Parathyroid gland
Heart	Tonsils
Ileum	Trachea
Jejunum	Urinary bladder
Kidneys	Uterus
Liver	

GLP, Good Laboratory Practice

In addition, these early preclinical studies may also provide valuable information on the kinetics and in vivo absorption, distribution, metabolism, and excretion properties of a new chemical entity (NCE). Typically, the use of a biochemical measurement as a surrogate indicator of efficacy is used to demonstrate the correlation between the pharmacokinetics (PK) of the NCE and the pharmacodynamic (PD) effect produced (PK/PD). As noted earlier, it is generally accepted that these biomarker measurements can be hugely predictive during the drug development process and thus are highly relevant for their potential contribution to understanding, for example, the mechanism of action of NCE.

There are currently no routinely accepted or standardized assays for miRNAs that are required in the regulatory submission, i.e., investigational new drug process. As described earlier, existing histological measurements are based on commonly used staining techniques, and clinical chemistry assays are well defined and documented and have a significant historical database to which to refer. For those of us involved in the development of new drugs, these techniques have been the backbone of understanding and recording toxicity as applied to various compartments—either tissues or circulating physiological markers. We believe that the correlation of single or multiple miRNAs with toxic effects studied temporally will aid significantly in the understanding of the pathophysiological nature of injury and disease and the effects of NCEs on specific test systems.

NOMENCLATURE

miRNAs were originally discovered in the model organism, *Caenorhabditis elegans* (*C. elegans*). *lin-4* was the first of the miRNAs to be implicated in gene regulation (Ambros, 2001). It was shown to bind to the 3′UTR of *lin-14*, which is involved in larval development in *C. elegans*. Ambros et al. (2003) further presented guidelines for new miRNA annotation for diverse organisms, thus preventing confusion when considering other forms of RNA (e.g., small interfering RNAs, or siRNAs) (Ambros et al., 2003). In a summary of annotation on the miRBase website (University of Manchester, 2013), it is noted that the numbering of newly assigned miRNA genes follows a sequential pattern (n+1) from the most recently assigned miRNA, whereas the 3 to 4 letter prefix indicates the species of miRNA. Differentiation of the mature miRNA and the miRNA gene are indicated by the capitalization of the r in miR. The capitalized miR references the mature miRNA, whereas the lowercase mir references the miRNA gene and refers to the pri-miRNA transcript's stem loop portion. miRNA with identical sequences that are located in different areas of the genome are indicated by numbering suffixes. miRNAs that are closely related are indicated by lettered suffixes. Interestingly, some exceptions to the rules created do exist; for example, different naming conventions are used for plants and viruses. Additionally, the historical miRNA such as *lin-4* will be maintained and homologues will continue with *lin-4* naming.

DATABASE

The primary online database for miRNA sequences and annotations is miRBase (Kozomara and Griffiths-Jones, 2011) This database coordinates the naming of new miRNA genes and is administered by the University of Manchester, United Kingdom, and can be utilized prior to publication of unique miRNA. miRBase provides users with the ability to search annotated and published miRNA sequences. For each miRNA, the database displays the predicted hairpin portion (mir) and the mature miRNA sequence (miR) and its location. For a full definition of hairpin and mature sequences, readers are referred to Ambros et al. (2003). There are many options for searching miRNA in miRBase, such as hairpin sequences, mature sequences, name, keyword, references, and annotation. These sequences and annotated data can be downloaded from the miRBase website (University of Manchester, 2013). Maintenance of a single repository for miRNA sequences enables consistency among the miRNA nomenclature. This facilitates the ability of researchers to search and download miRNA data, as well as providing access to data supporting the miRNA annotation. This also allows a central location for predicted miRNA targets and validations of these targets. miRNA expression studies generate large amounts of data, and this database provides a central source for computer-parsable annotation of miRNA sequences (for example, functional data, references, genome mappings). It is free and available online.

miRNEST (Laboratory of Functional Genomics, 2013) is a database generated by searching expressed sequence tags (ESTs), a short subsequence of a cDNA sequence (NCBI, 2013), to identify novel miRNA (Szczesniak et al., 2012). It is a database containing miRNA from diverse plants, animals, and viruses. This database was developed to address the need for a central location for a wide range of species, which is important because of the conservation of miRNA among species. For the analysis, 225 animal and 202 plant ESTs were analyzed for novel miRNAs. Szczesniak et al. (2012) identified over 10,004 potential miRNAs across 221 animal and 199 plant species. Interestingly, for 239 of the species used in this study, no miRNA had been previously identified and of the miRNAs identified, only 299 had been deposited in miRBase. Furthermore, to expand the analysis capability, miRNEST incorporates published data and 13 additional databases. The user-friendly website combined with the diverse species' data make miRNEST an integral miRNA evaluation tool.

The miR2Disease database, which is manually curated, links human diseases to miRNA alteration (Jiang et al., 2009). This database allows researchers to correlate miRNA expression changes with specific diseases. The database maintains information about the detection method used to identify the differentially expressed miRNA as well as gene targets that have been verified experimentally with references to the data.

There were over 20 different databases, many of which are curated in academic settings in countries far apart as China, Korea, Switzerland, Germany, Greece, Poland, and the United States. A list of many of these databases is provided by Mikaelian et al. (2013). Because of the rapid advancement of miRNA research and identification, it is difficult to maintain the current databases. For example, the number of submissions to miRBase has risen exponentially in the last 3 years alone, and the number of miRNA sequences has tripled.

A clear area of development for this field could be adding or linking in preclinical and clinical toxicity data with miRNA data based on studies performed in the development of new therapeutics in a variety of disease states.

SAMPLE PREPARATION

Numerous studies have shown that differences in sample collection, extraction methodology, normalization, and data analysis can lead to variation in expression profiles of miRNA, and these differences pose significant challenges with implementation of miRNA biomarkers in a clinical setting. Low RNA levels in the plasma and serum dictate that sampling protocols and normalization be carefully considered in the development of a biomarker assay.

Control of sample collection is critical because cross-contamination by different cell types can interfere with subsequent analysis and interpretation of results. In addition, the method of collection or storage can lead to up- or downregulation of miRNA, which in turn can lead to alterations in gene expression resulting from potential bias imposed on the experimental procedure. To ensure robust evaluation in the validation of fluid or tissue biomarkers, multiple site studies (i.e., at different laboratories) need to be undertaken, thus demonstrating assay reproducibility. Accordingly, this means that a consistent and well-developed protocol for handling samples is required and should be delineated before initiation of studies.

There are many extraction methods commercially available and optimized for RNA in general. These kits either extract total RNA, including small RNA molecules or specifically enrich for small RNA molecule extraction. Some kits offer protocols for both options. Commonly used kits include miRNeasy Mini Kit (Qiagen), MirVana miRNA Isolation Kit (Life Technologies), High Pure miRNA Isolation kit (Roche), PureLink miRNA Isolation Kit (Life Technologies), and miRCURY RNA Isolation Kit (Exiqon). Direct comparison of several commercially available kits showed that a majority of the differentially expressed miRNA were consistent between extractions (Ach et al., 2008). This study identified a small population of differentially expressed miRNA, which was found to be due to the specific extraction method. Therefore, in expression profiling, it is important to use the same extraction method throughout the entire experimental design.

Many methods are available for expression analysis of miRNA, and some include northern blotting, next-generation sequencing (NGS), and in situ hybridization; however, currently, the most commonly used methods are microarray analysis and real-time PCR analysis. Because the conversion of miRNA to cDNA is typically required with many of these methods, it is common practice to perform this in a sequence-specific manner, typically using stem-loop primers or with the addition of a common sequence to the end of the miRNA. However, microarray analysis can directly identify miRNAs without the generation of cDNA and analyze thousands of targets simultaneously. One of the drawbacks to microarray analysis is the inability to identify miRNA in low abundance, due to the amount of RNA required for the input—a fact that is important when analyzing serum or plasma that may yield low levels of RNA. Also, many microarrays have the inability to differentiate similar miRNA species, some of which will only differ by a single nucleotide change, as well as mature miRNAs and their precursor.

There are many commercially available microarray platforms, and multiple studies have demonstrated that similar results can be obtained across different microarray platforms; however, differences in input RNA concentrations should be factored in when choosing a platform (Git, 2012; Yauk et al., 2010). When specific miRNA targets are identified by microarray analysis, the results are typically confirmed with real-time PCR. The use of real-time PCR for miRNA expression profiling allows for the differentiation of isoforms of miRNA, as well as precursors of the mature miRNA using probe technology such as TaqMan MGB probes (Schmittgen et al., 2008). A drawback to real-time PCR analysis is the limited number (as compared to microarrays) of miRNAs that can be evaluated per reaction plate. Also, bias may be introduced during the cDNA amplification process, depending on the amplification efficiencies of the many targets. Currently, NGS is labor-intensive. Typically, before sequence analysis, the adaptor sequence needs to be removed; however, because this is not limited by probe design, identification of novel miRNA and differentiation of isoforms is an advantage. NGS is also able to detect miRNA in low abundance.

To aid in the identification of miRNA in the presence of the background RNA, many software tools have been developed, such as miRDeep, miRanalyzer, miRExpress, miRTRAP, DSAP, miRTools, MIReNA, MiRNAkey, and MIREAP. These tools have various advantages and disadvantages; therefore, depending on the organism and study needs, the software tools utilized may change. Comparison of these software tools using three organisms was performed by Li et al. (2012), which highlights the advantages of each program.

For expression profiling, many disease states or toxicity profiles are human specific and therefore require human tissue for analysis (Li et al., 2012). This can be problematic when the disease states contribute to the degradation of RNA or when postmortem degradation of the RNA is observed in the tissue. Therefore, tissue collected at tissue banks needs to provide high-quality RNA and be thoroughly analyzed by indicators such as RNA integrity and 260/280 absorbance ratio before use for miRNA expression profiling. Also, the clinical history should be properly documented for the tissue and biofluid sample.

Most analytical techniques compare the relative concentrations of the miRNA rather than the absolute quantities. This is due to the variation on binding efficiencies for the miRNA and variation in labeling efficiencies by ligase for microarray analysis (Bissels et al., 2009), as well as the difficulty identifying miRNA reference genes for normalization. For diagnostic purposes, an internal control for normalization of samples is imperative. However, differential regulation throughout the tissue, fluids, and changes due to disease state(s) makes identification of reference genes difficult. Currently, this is one of the main drawbacks of the use of miRNAs as biomarkers. When evaluating the use of miRNAs for drug toxicity assessment, additional confusion can occur because standard "reference" or "housekeeping" gene expression may also be affected by the drug. Typically, therefore, small RNAs RNU 48 and U6 are commonly used as reference genes for miRNA. Chen et al. (2011) confirmed that these were reliable reference genes during the treatment of tissue culture cells with various chemicals. However, levels of these RNAs are variable across different pathological conditions and, thus, they are not appropriate as reference genes for all studies (Chen et al., 2011). These authors also identified RNU44 and RNU47 as suitable reference genes and suggested that a combination of these reference genes would be ideal. Synthetic miRNA addition to reactions may be one way to circumvent the endogenous control issue similar in concept to the addition of an internal standard to a bioanalytical or pharmacokinetic analysis (Chen et al., 2011). These synthetic miRNAs could also be used as extraction controls and allow for normalization between samples. Because of the RNase activity in plasma, it has been noted that the synthetic miRNA should not be added to the sample until the RNase activity is inactivated (Mitchell et al., 2008). However, for plasma, it has also been suggested that normalization to the sample volume can be utilized like the standard for other biomarkers (Cheng et al., 2010). Obviously, more studies are needed to evaluate the most reliable method of normalization.

One of the greatest challenges is identifying miRNA target genes. This limits the ability to define the mechanism of action of miRNA deregulation leading to injury, toxicity, and/or disease. Typically, only 2–7 nucleotides (seed sequence) of the mature miRNA are found in the regulated target. This is further complicated by the fact that many miRNAs are cotranscribed as a single pri-miRNA and then cleaved into mature miRNA. The commonly used prediction algorithms include Targetscan, miRanda, DIANA-mircoT, PITA, Rna22, and PicTar. Many of these algorithms align the seed sequence with the 3′UTR. However, recent reports suggest that some miRNAs interact with 5′UTR and the open reading frame. Another challenge with target identification is that 100% alignment between the seed sequence and binding regions is not always required. However, Zhao et al. (2007) demonstrated that the secondary structure of mRNA is also important for regulation by miRNA. The miRNA requires accessibility to the binding site for regulation of the mRNA; therefore, a less structured 3′UTR is optimal. By evaluating the free energy required to unfold the surrounding nucleotides, binding sites may be eliminated because of inaccessibility of the binding site to the miRNA. Potential binding sites may also be evaluated based on conservation among closely related species, allowing the elimination of targets that are not conserved. Note that conservation is not always the case and needs to be applied with caution. Many of the prediction tools consider many of these factors, and thus it is important to use multiple prediction tools because of the variation in the algorithms; each algorithm will predict different sites within a single gene. Because of the complexity of identifying miRNA targets, it is optimal to evaluate targets predicted by multiple algorithms (Zhao et al., 2007).

Validation of the miRNA regulations can be done in many ways. Regulation can be assessed by reporter gene assays, such as cloning of the entire 3′UTR containing the binding site upstream of a reporter gene such as green fluorescent protein. Utilizing cells that do not express the miRNA of interest, the interaction can be evaluated by cotransfecting the reporter construct with a vector expressing the miRNA or a miRNA mimetic and measuring the reporter gene expression levels. This approach also can be expanded with the use of miRNA inhibitors. This method is limited in the number of targets that can be analyzed during an experiment. High-throughput analyses, including microarray and proteome analysis, have been utilized, but have issues with distinguishing direct and indirect regulation. Microarray analysis can be utilized by inhibiting the miRNA of interest and then evaluating the transcriptome, although not all interactions can be identified by this method. Proteome analysis employs the labeling of amino acids that are then quantified by mass spectrometry. However, this method can also identify proteins that are regulated in an indirect manner by the miRNA–mRNA complex. Immunoprecipitation has also been utilized with similar specificity issues.

The FDA defines changes in miRNA levels as a "genomic biomarker" (FDA Guidance for Industry E15, 2008). To date, the pertinent regulatory guidance documents (FDA Guidance for Industry E15, 2011, 2008) focus on definitions and submission processes without articulating specific requirements for testing or validation. It is our expectation that, as more potential miRNA biomarkers are identified by the scientific community and evaluated therein, specific requirements for validation of miRNA-based biomarkers will be defined.

KINETICS AND PHARMACOKINETICS/PHARMACODYNAMICS

It has been shown that reduced or elevated levels of certain circulating miRNAs can be related to certain disease states. It is also well known that treatment of a wide variety of disease states with xenobiotics can result in adverse effects related directly to the drug administered or to a biotransformation of the parent drug (metabolite). As noted earlier, biomarkers, whether applied as diagnostics or as indicators of toxicity, can be found in a wide variety of compartments (e.g., blood, urine, tissues). Furthermore, there are numerous examples of measurement of a biomarker following administration of a drug, resulting in correlation with the PK of the drug and a PD effect. It has been demonstrated that administration of a known cardiotoxic drug (isoproterenol) to rats results in perturbations in serum miRNA expression levels with a combination of up- and downregulation of over 60 individual miRNAs (Table 59.2). In a study of miRNA turnover in mouse fibroblasts following an inhibition in miRNA synthesis (Gantier et al., 2011), the half-life of miRNA was established as approximately 5 days or 10-fold greater than that of mRNA as modeled in a theoretically nondividing cell. They also noted significant variations in half-lives between individual miRNAs, suggesting that kinetic variations may be relevant if targeting an individual miRNA as, for example, a biomarker of toxicity, tissue injury, or disease state. Using saliva, Park et al. (2009) identified certain miRNAs as markers of oral cancer and further profiled their endogenous and exogenous degradation patterns over time. Krol et al. (2010) demonstrated that levels of specific miRNAs in mouse retina can change rapidly in response to light and dark exposure.

It has been a challenge to find many publications documenting the correlation of miRNA expression with toxic drug effects, although this may be a function of the somewhat guarded nature of pharmaceutical companies developing NCEs. Even less information is available in the public domain regarding measurement of tissue-specific miRNAs, either from clinical biopsy or preclinical terminal procedures.

MIRNA BIOMARKERS OF TOXICITY IN ORGAN SYSTEMS

In toxicologic pathology or disease etiology, biofluid-based biomarkers can pave the way for identifying and monitoring potential safety risks in nonclinical studies and human populations. In the recent years, many studies and reviews have been focused on whether miRNA profiling in biofluids more specific and sensitive biomarkers of tissue toxicity are as compared with conventional biomarkers. A summarization of candidate miRNA biomarkers and comparison to traditional clinical or research biomarkers is provided in Table 59.3. An overview of how toxin exposure regulates miRNA release in circulation has been given in Fig. 59.2.

ESOPHAGUS

Esophagus, commonly called as food pipe, connects pharynx to stomach. The inner lining of esophagus is consisting of three layers of stratified squamous epithelial cells unlike single layer of columnar cells of stomach. Prolonged esophagitis, especially gastric reflux, plays major role in replacing stratified squamous epithelial cells by columnar cells and develop Barrett's esophagus (BE), which is refered as metaplasia (abnormal changes) in the cells. The later stage outcome of Barrette's esophagus is known as esophageal adenocarcinoma (EAC).

TABLE 59.2 Summary of Up-(Positive) and Down-(Negative) Regulated miRNAs in Serum Following Administration of Isoproterenol in Rats

Assay	Fold-Change	Function	PubMed ID/Journal ID
Rn_miR-26a_1	75.6		
Rn_miR-30c_2	53.3	Regulates apoptosis	20062521
Rn_miR-301b_1	47.3		
Rn_miR-301a_1	35.8	NF-κB activator	PMC3020116
Rn_miR-26b_1	33.4		
Rn_miR-31_1	24.3	Factor-inhibiting hypoxia-inducible factor (FIH)	20145132
Rn_miR-338_1	24.2	Regulates cytochrome *c* oxidase IV	19020050
Rn_miR-142-3p_1	21.8	Restricts cAMP production by targeting AC9	19098714
Rn_miR-103_1	18.3		
Rn_miR-194_1	18.3		
Rn_miR-195_1	15	Overexpression can cause cardiomyopathy	PMC1838739
Rn_miR-151_2	14.5	Down regulates RhoGDIA	20305651
Rn_miR-101b_3	13.7	Regulates ATXN1	18758459
Rn_miR-101a_3	13.3	Regulates ATXN1	18758459
Rn_miR-16_1	13.1	Apoptosis	19232449
Rn_miR-363_1	12.9		
Rn_miR-142-5p_1	12.2		
Rn_miR-652_1	10.3		
Rn_miR-140_1	10.3	CD133; CD44 modification	19734943
Rn_miR-17-3p_1	9.8	Regulates OLIG2	21338882
Rn_miR-141_1	9.7	Target p38α and modulate the oxidative stress response	22101765
Rn_miR-451_1	9.4	Repress Myc expression	PMC3135352
Rn_let-7i_1	9.2	Regulates Toll-like receptor 4	jbc.M702633200
Rn_miR-192_1	9.2	Regulates Smad3 and TGF-beta	20488955
Rn_miR-148b_1	8.8	Regulates Mitf	20644734
Rn_miR-542-3p_2	8.8	Role as tumor suppressor	21756784
Rn_miR-30e_2	8.1	Regulates Ubc9	PMC2846614
Rn_miR-196b_1	8		
Rn_miR-126_1	7.9	Regulates angiogenic signaling	18694566
Rn_miR-30e*_1	7.8	Regulates Ubc9	PMC2846614
Rn_miR-424_1	7.8	Regulates HIF-a isoforms and promotes angiogensis	20972335
Rn_let-7d_2	7.5	Alpha-smooth muscle actin (ACTA2)	20395557
Rn_miR-27a_1	7.3	Fatty acid metabolism	21178770
Rn_miR-126*_1	7.2	Regulates angiogenic signaling	18694566
Rn_miR-21_2	6.8	Stimulates MAP kinase signaling	19043405
Rn_let-7e_2	6.5	Regulates caspase-3 during apoptosis	21827835
Rn_miR-29b_1	6.4	Targets p85 and CDC42	19079265

Continued

TABLE 59.2 Summary of Up-(Positive) and Down-(Negative) Regulated miRNAs in Serum Following Administration of Isoproterenol in Rats—cont'd

Assay	Fold-Change	Function	PubMed ID/Journal ID
Rn_let-7c_1	6.4	Suppresses androgen receptor expression	jbc.M111.278705
Rn_let-7b_1	6	Targets cell cycle molecules	18379589
Rn_miR-30a_1	5.9	Regulation of autophagy by beclin 1	19535919
Rn_miR-15b_2	5.8	Apoptosis	19232449
Rn_miR-32_1	5.2	Bim targeting and AraC-induced apoptosis	21816906
Rn_miR-742_1	−2238.8		
Rn_miR-708_1	−1127.5	Induces apoptosis and suppresses tumorgenicity	21852381
Rn_miR-494_1	−266	Regulation of survivin and apoptosis	20807887
Hs_RNU1A_1	−207		
Rn_miR-483_1	−57	Control angiogenesis and targets serum response factor	21893058
Rn_miR-501_2	−55		
Rn_miR-207_1	−37.7		
Rn_miR-92b_1	−16.2	Regulates oligopeptide transporter PepT1	21030610
Rn_miR-34c*_1	−14.9	Represses stress induced anxiety	21976504
Rn_miR-343_2	−14.9		
Rn_miR-434_1	−13.2		
Rn_miR-347_2	−11.9		
Rn_miR-323*_2	−9.9		
Rn_miR-21*_1	−8.2	Stimulates MAP kinase signaling	19043405
Rn_miR-291-5p_1	−6.8		
Rn_miR-124*_1	−5.6		
Rn_miR-671_1	−5.2		
Rn_miR-370_1	−4.9		
Rn_miR-877_1	−4.5		
Rn_miR-320_3	−4.3	Targets transferrin receptor 1 (CD71)	19135902

$P \leq .001$.

More than 90% EAC cases arise from delayed detection because the current method has failed to identify at BE stage (Corley et al., 2002; Shaheen and Hur, 2013). The urgent need of effective detection strategy for BE may pave the way to halt the significant increase in the rate of EAC. Because miRNAs are considered more specific than mRNA profiles for diagnosis in several cancers, targeting specific miRNA could be the best diagnostic biomarker for BE (Lu et al., 2005). A pilot study published by Bansal et al., in 2014 suggest that the "BE miRNA signature has shown significant promise and should be evaluated further as a specific marker for the presence of intestinal-type BE; this signature has the potential to improve our understanding of BE pathogenesis" (Bansal et al., 2013).

To discover miRNAs associated with BE pathogenesis, entire miRNA transcriptome has been sequenced and analyzed from chronic gastroesophageal reflux disease (GERD) and BE patients followed by verification of the sequencing results by quantitative real-time polymerase chain reaction in several independent patients (Bansal et al., 2013). After extensive analysis, the three miRNAs: miR-192-5p (240611 reads per million [RPM] in BE, fold change 7.9 vs. GERD), miR-215-5p (69250 RPM in BE, fold change 9.6 vs. GERD), and miR-194-5p (8209 RPM in BE, fold change 6.5 vs. GERD) showed promising result and excellent accuracy in BE diagnosis. In another study of metastatic transformation of esophagus, the circulating miRNA expression profile reveals the significant upregulation or downregulation of miRNAs at

TABLE 59.3 Comparison of Biofluid-Based miRNAs to Traditional Biomarkers of Tissue Toxicity (Harrill et al., 2016)

miRNA	Traditional Biomarker	Species	Toxicant/Condition	Matrix	miRNA Specific for Target Tissue?[a]	miRNA Better Correlated With Tissue Injury?[b]	miRNA Detected at Earlier Timepoint or Lower Dose?[c]	References
LIVER								
miRNA-122	ALT, AST	Rat	$CBrCl_3$, CCL_4	Blood	Yes	Yes	NA	Laterza et al. (2009)
miRNA-122	ALT, AST, α-GST, GLDH, SDH	Rat	Allylalcohol, APAP, ANIT, PB	Blood	NA	Yes or similar (biomarker and toxicant dependent, see ref)	Yes or similar (biomarker and toxicant dependent, see ref)	Starckx et al. (2013)
miRNA-122	ALT	Human, mouse	HBV infection, liver injury (human); D-GalN + LPS, ethanol (mouse)	Blood	Yes	Yes	Yes (D-GalN + LPS, ethanol)	Zhang et al. (2010)
miRNA-122	ALT	Human	APAP	Blood	NA	Yes	Yes	Antoine et al. (2013)
miRNA-122, miRNA-155	ALT, TNF-α	Mouse	Ethanol, APAP, LPS + CpG	Blood	NA	Similar	Similar	Bala et al. (2012)
miRNA-122, miRNA-192	ALT	Mouse	APAP	Blood	Yes	NA	Yes	Wang et al. (2009)
miRNA-122, miRNA-192, miRNA-182	ALT, AST, ALP, GGT, Tbil, total bile acids	Rat	ANIT, FP004BA	Blood	NA	Similar or no (biomarker and toxicant dependent, see ref)	Similar or no (biomarker and toxicant dependent, see ref)	Church et al. (2016)
miRNA panel	ALT	Human	Ischemic-hepatitis, APAP	Blood	NA	NA	Yes	Ward et al. (2014)
KIDNEY								
miRNA-34c-3p	Albumin	Rat	Doxorubicin	Urine	Yes	Similar	Yes	Church et al. (2014)
let-7g-5p, miRNA-93-5p, miRNA-191a-5p, miRNA-192-5p	BUN, sCr (blood), KIM-1, clusterin (urine)	Rat	Cisplatin	Urine	NA	Similar	Similar (clusterin) or no (bun, scr, kim-1)	Kanki et al. (2014)
HEART								
miRNA-1	cTnI	Human	AMI	Blood/urine	NA	Similar	Similar	Zhou et al. (2013)
miRNA-1	cTnI	Human	AMI	Blood	NA	No	NA	Li et al. (2014)
miRNA-208a	cTnI	Rat, human	AMI	Blood	Yes	Similar	Yes	Wang et al. (2010)
miR-208b	cTnI	Pig, human	AMI	Blood	NA	Similar	NA	Gidlof et al. (2011)

Continued

TABLE 59.3 Comparison of Biofluid-Based miRNAs to Traditional Biomarkers of Tissue Toxicity (Harrill et al., 2016)—cont'd

miRNA	Traditional Biomarker	Species	Toxicant/Condition	Matrix	miRNA Specific for Target Tissue?[a]	miRNA Better Correlated With Tissue Injury?[b]	miRNA Detected at Earlier Timepoint or Lower Dose?[c]	References
miR-208a	cTnI	Rat	Isoproterenol, allylamine	Blood	NA	Similar	Similar	Glineur et al. (2016)
miR-208a, miR-133a/b	cTnI	Rat	Isoproterenol, allylamine, metaproterenol	Blood	Yes	Yes	Similar	Calvano et al. (2016)
miR-208b, miR-1, miR-133a, miR-499	cTnI	Human	AMI	Blood	NA	Similar	NA	Li et al. (2013)
PANCREAS								
miR-216a	Amylase, lipase	Rat	L-Arginine, CLP	Blood	Yes	NA	Similar	Kong et al. (2010)
miR-216a, miR-217	Amylase, lipase	Mouse, rat	Cerulein, L-arginine, pancreatic duct ligation	Blood	NA	Yes/similar (L-arginine, mouse/rat), similar (cerulein, rat), or no/similar (ligation; mouse/rat)	Similar	Goodwin et al. (2014)
miR-216a, miR-375	Amylase, lipase	Rat	Cerulein, CHB	Blood	NA	Yes (CHB) or similar (cerulein)	Similar	Usborne et al. (2014)

For conciseness, only the miRNA(s) and traditional biomarker(s) of primary interest are compared based on referenced studies. Some evaluations were treatment or assay dependent, as specified in parenthesis.

[a] *"Yes" indicates that the miRNA was predominantly expressed in the target tissue.*

[b] *"Yes" indicates that miRNA alterations correlated with histopathological changes of target tissue more precisely than the protein biomarkers after injury. "Similar" indicates significant alterations of miRNA and protein biomarkers exhibited similar patterns. "No" indicates that the protein biomarkers correlated with histopathological changes of target tissue more precisely than the miRNA biomarker. When available, significant differences according to ROC curve analyses for biomarker performance were used to make assessments.*

[c] *"Yes" indicates that miRNA biomarker alterations occurred at a sooner time point or lower dose than protein biomarkers. "Similar" indicates these dynamics were similar. "No" indicates traditional biomarkers altered at an earlier time or lower dose than the miRNA biomarkers.*

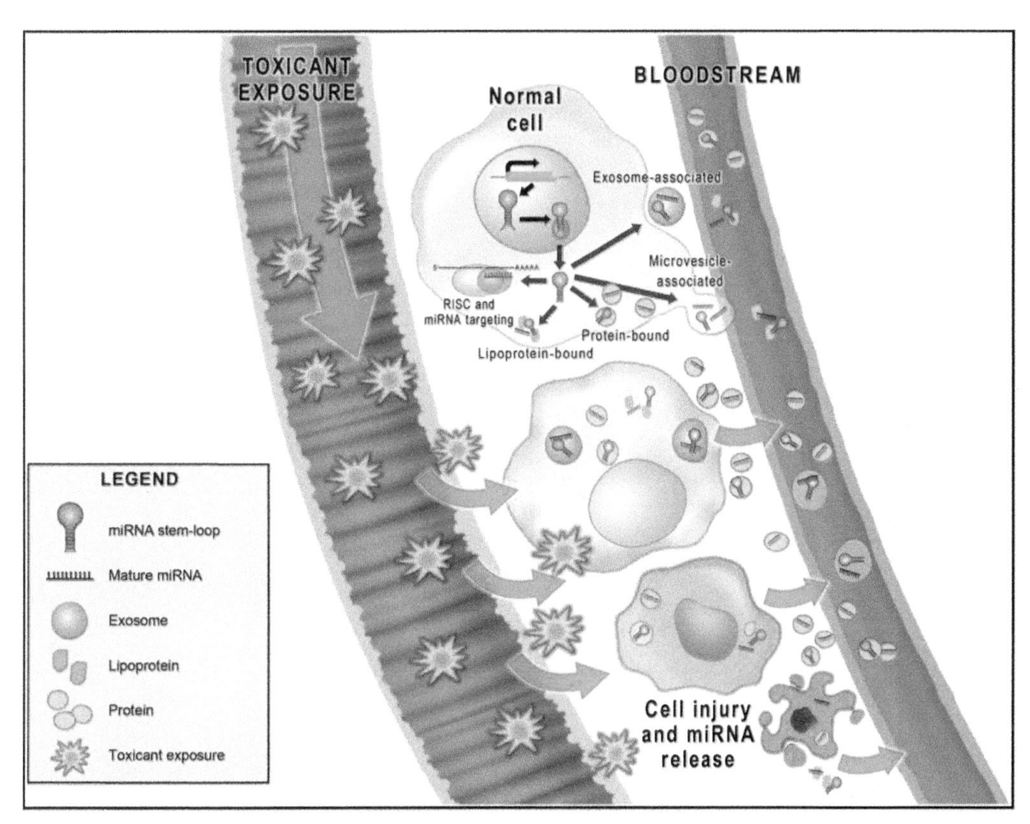

FIGURE 59.2 **Overview of miRNA release in blood circulation on toxin exposure.** Irrespective of toxicant exposure biological routes, the cells exposed to toxicant may experience stress and alteration in several biological pathways including miRNA transcription and processing. Further, processed and mature miRNA will incorporate into the RISC (expand what the abbreviations mean!) element and target mRNAs for degradation or protein translation inhibition. These mature miRNAs can be released from the producing cell after packaging them into membrane-bound vesicles such as exosomes, microvesicles by normal processes. The function of miRNA release is unknown, but it is speculated that this may serve as paracrine or endocrine signaling. Depending on type and levels of cellular stress, the amount and types of released miRNAs may alter. However, overt toxicity can result in cellular necrosis or apoptosis, leading to passive release of cellular miRNA contents or entrapment in apoptotic bodies, thereby increasing the presence of miRNAs in accessible biofluids. Released miRNAs encapsulated in vesicles or associated with lipoproteins or other proteins are protected from the highly RNase-rich environment and can be sampled and measured through several quantitative techniques. Those miRNAs that are predominantly expressed in certain tissues may therefore serve as specific biomarkers for tissue stress and toxicity (Harrill et al., 2016).

different stages of disease progression (Cabibi et al., 2016). The summary of expression profile in BE (n = 8), columnar-lined esophagus (CLO (n = 12)), and esophagitis (n = 10) has been provided in Table 59.4 and Fig. 59.3.

Changes in miRNAs could be important in the development of BE and its progression (Feber et al., 2008; Wijnhoven et al., 2010; Yang et al., 2009). Among them, miR-143, miR-145, miR-194, miR-203, miR-205, and

TABLE 59.4 Mean, Median, and Standard Deviation of miR-143, miR-145, miR-194, miR-203, miR-205, and miR-215 Relative Expression Values in Relation to the Histologic Phenotype (Barrett, CLO, Esophagitis) (log Scale) (Cabibi et al., 2016)

	Barrett (*n* = 8)			CLO (*n* = 12)			Esophagitis (*n* = 10)		
miRNA ID	**Mean**	**Median**	**SD**	**Mean**	**Median**	**SD**	**Mean**	**Median**	**SD**
miR-143	4.55	4.32	1.69	3.13	3.31	1.10	0.68	0.55	1.17
miR-145	4.79	4.59	1.20	3.27	3.34	1.20	0.72	0.57	0.75
miR-194	7.99	8.27	1.31	5.30	6.25	3.16	0.56	0.24	1.74
miR-203	−3.92	−4.32	1.41	−3.09	−3.13	2.59	−0.91	−0.49	1.28
miR-205	−5.62	−5.43	2.97	−4.60	−1.96	5.25	−0.39	−0.27	0.66
miR-215	8.23	8.56	2.33	5.06	5.56	2.53	0.14	−0.20	1.86

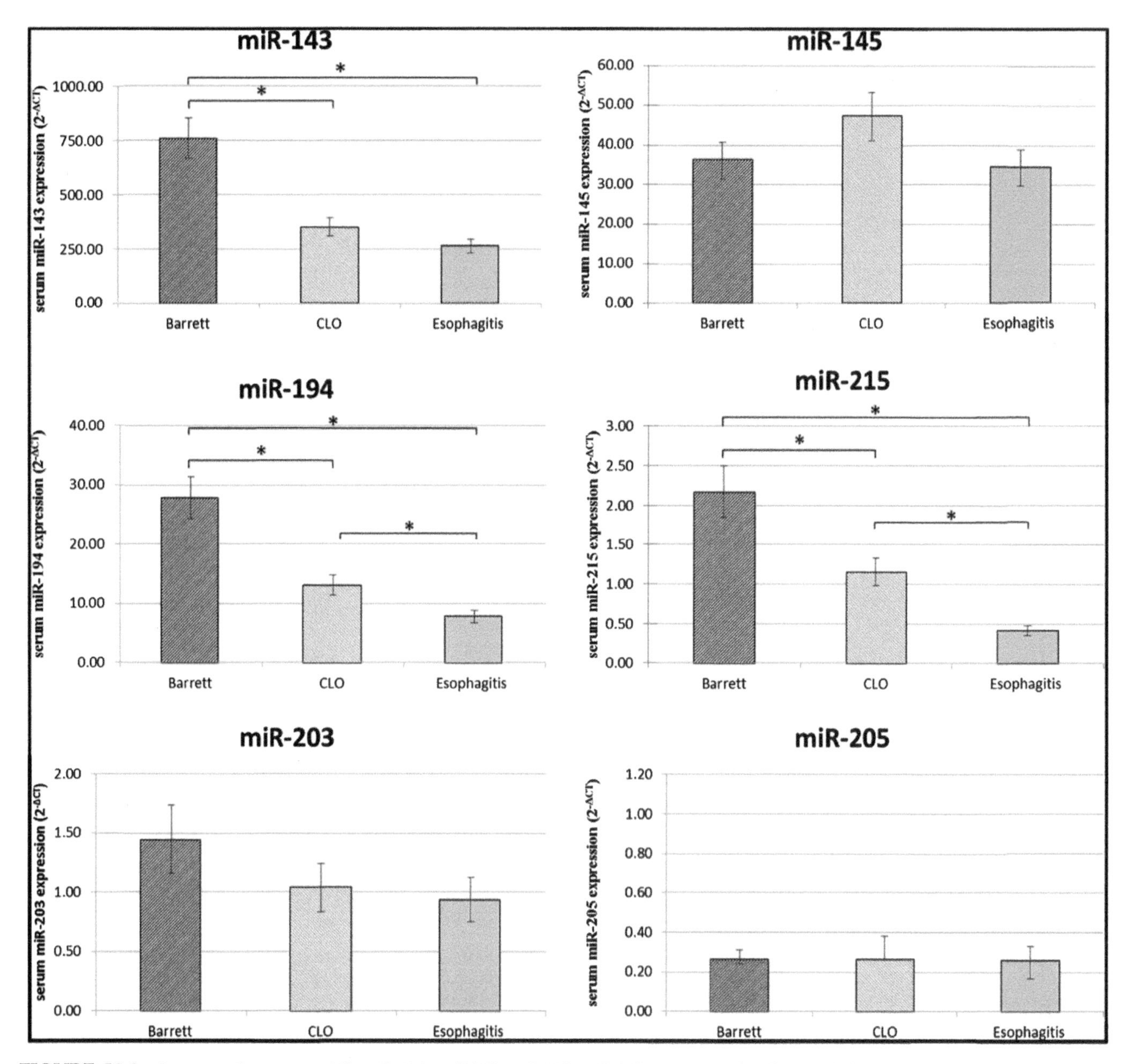

FIGURE 59.3 Serum miR-143, miR-145, miR-194, miR-203, miR-205, miR-215 expression levels (2−ΔCT) in Barrett, CLO, and Esophagitis groups. Differences in miRNA expression between the different groups were considered significant when the *P* value was under 0.05 (Cabibi et al., 2016).

miR-215 appear to have a key role in the metaplasia and in neoplastic progression. Numerous studies reported the downregulation of miR-143, miR-145, miR-203, miR-205, and miR-215 in EAC (Kan and Meltzer, 2009; Wijnhoven et al., 2010). However, few studies have been performed on lesions preceding the adenocarcinoma and on miRNA expression in different types of columnar mucosa and esophagitis. In conclusion, these findings suggest that miRNAs may be involved in the neoplastic/metaplastic progression and miRNA analysis might be useful for progression risk prediction as well as for monitoring of BE/CLO patients.

MIRNAS AS BIOMARKERS OF PANCREATIC TOXICITY

Pancreas is a large organ behind the stomach that produces digestive enzymes and several hormones. Pancreatic injury or pancreatic toxicity is generally characterized in two stages: (1) early or reversible stage with dysregulation of lipid metabolism and edema followed by (2) advance nonreversible stage with massive necrosis resulting in inflammation, with or without fibrosis (scarring of the tissue). The sequence of pancreatic injury follows a similar pattern for most of the

TABLE 59.5 Pancreas Enriched miRNAs Conserved Between Rat and Dog

Rat Pancreas	Sequence (5–3′)	Dog Pancreas	Sequence (5–3′)
miR-101c	TACAGTACTGTGATAACTGATC	miR-101c	ACAGTACTGTGATAACTGACC
miR-141-3p	TAACACTGTCTGGTAAAGATG	miR-141	TAACACTGTCTGGTAAAGATG
miR-193b-3p	AACTGGCCCACAAAGTCCCGCT	miR-193a-3p	AACTGGCCTACAAAGTCCCAGTT
miR-200c-3p	TAATACTGCCGGGTAATGATGGA	miR-200c	TAATACTGCCGGGTAATGATGGA
miR-216a-5p	TAATCTCAGCTGGCAACTGTG	miR-216a	TAATCTCAGCTGGCAACTGTGA
miR-216b-5p	AAATCTCTGCAGGCAAATGTGA	miR-216b	AAATCTCTGCAGGCAAATGTGA
miR-320-3p	AAAAGCTGGGTTGAGAGGGCGA	miR-320b; miR-320c	AAAAGCTGGGTTGAGAGGGAA
miR-4286	ACCCCACTCCTGGTACCA	miR-4286	ACCCCACTCCTGGTACCA
miR-5100	ATCCCAGCGGTGCCTCCA	miR-5100	GTTCGAATCCCAGCGGTG
miR-217-5p	TACTGCATCAGGAACTGACTGGA	miR-217	ATACTGCATCAGGAACTGATT
miR-375	TTGTTCGTTCGGCTCGCGTGA	miR-375	TTTGTTCGTTCGGCTCGCGTGA

Tissue enriched miRNAs conserved between rat and dog were identified and are shown above. These miRNAs were analyzed for changes in the serum in rat and dog caerulein toxicology studies. Bases underlined indicate bases that are not conserved between the rat and dog isomiRNAs with the highest sequencing counts.

etiologies and does not depend on the types of etiological agents. It is evident that a portion of patients with pancreatitis can also develop pancreatic cancer at the later stage (Kaphalia, 2014). Overall, chronic pancreatitis is a disastrous disease with significant psychological, social, and economic impact. The prognosis of pancreatitis and pancreatic cancer is very poor, and many patients die before reaching the clinical stage of the disease. Therefore, identification of biomarkers for both pre- and postpancreatitis stages could be important for clinical interventions. Early detection and prevention of pancreatic injury, progressing to inflammation and fibrosis and possibly pancreatic cancer, should be critical for the success of both clinical and surgical interventional therapies.

Although a significant literature available on biomarkers of pancreatic diseases is focused on various forms of pancreatic cancer, identification of biomarkers for early pancreatic injury and precancerous stages should be the key to early prevention and therapy. To better understand toxicities in preclinical and clinical settings and to speed the development of new therapies, body fluid–based biomarkers of organ damage have been sought to both identify and monitor toxicities. It is evident that identifying tissue specific, species translatable blood-based miRNA expression profiling could serve biomarkers of tissue injury. The potential pancreas biomarkers were tested in rat and dog toxicology studies using the pancreatic toxicant caerulein. In these proof-of-concept studies, miRNAs performed as well as or better than traditional markers of pancreatic injury, demonstrated a much larger dynamic range and increased tissue specificity over traditional biomarkers. The tissue specificity and/or enrichment of pancreas miRNA were identified and confirmed as: miRNAs – 216a-5p, 217-5p, 375-3p, 148a-3p, 141-3p and 200c-3p. (Bravo-Egana et al., 2008; Kong et al., 2010; Wang et al., 2010).

Eli Lilly investigated the isomiRNA expression of pancreas tissue specific and enriched miRNAs, and the analysis revealed that isomiRNAs generally mirror their parent miRNA expression, but some isomiRs are more tissue specific than others as shown for miRNA 217-5p in the pancreas. For example, the mature miRNA-217-5p sequence is 5′TACTGCATCAGGAACTGACTGG-3′ and the isomiRNA-217-5p sequence 5′-TACTGCATCAGGAACTGACTGGAC-3′ are both highly enriched in the pancreas, but the miR-217-5p isomiR lacks expression in the stomach, intestine, and ovary, whereas other isomiRNAs of miRNA-217-5p are expressed in the stomach, intestine, and ovary. Conserved pancreas tissue–enriched miRNAs were examined in the rat and dog models of caerulein pancreatic toxicity to begin to characterize miRNAs as species translatable serum-based biomarkers of pancreatic injury (Table 59.5).

LIVER

Liver injury, which is chronic in nature, increases in severity over time. Examples of chronic liver injury include steatosis (fatty liver), fibrosis, cirrhosis, and cancer. Acute liver injury occurs rapidly and may include drug-induced liver failure or cholestasis (Benichou, 1990). For the noninvasive diagnosis of liver injury, serum levels of ALT, AST, ALP, and bilirubin are commonly used. Although these parameters are typically part of a standard clinical pathology evaluation, they are not specific in nature and cannot be used to identify a specific type of liver injury. Additionally, Nathwani et al. (2005) showed that these factors may increase in cases of nonliver injury (e.g., muscle damage), whereas Shi et al. (2010) demonstrated that parameters such as ALT do not always correlate well with liver histopathology. Furthermore, serum aminotransferase levels may rise too late for therapeutic intervention (e.g., cases of acute acetaminophen [APAP] toxicity). Serum transaminase concentrations do not delineate between different types of liver injury; they are general rather than specific indicators. With these limitations, the identification of specific, reproducible, sensitive biomarkers of both acute and chronic liver injury is needed, and below a few examples are cited to show how miRNAs may serve as biomarkers for hepatic toxicity, injury, and disease states.

Of the many distinct miRNAs found in liver tissue, miR-122 is the most highly expressed, comprising 70% of the hepatic miRNA population (Lagos-Quintana et al., 2002). Girard et al. (2008) demonstrated the interplay between miR-122 and metabolism and stress–response pathway targets. Hepatocyte nuclear factors, among other liver-specific TFs, are thought to be involved in miR-122 transcriptional regulation (Coulouarn et al., 2009). Interestingly, Coulouarn et al. (2009) described miR-122 as defining the phenotype of a hepatocyte. Peters and Meister (2007) demonstrated that primarily protein complexes containing an Argonaute (Ago) protein carry serum miR-122, and Ago proteins and miRNAs interact in such a way as to ultimately target and repress mRNA translation in cells (Dueck et al., 2012). The cellular source of miR-122 in liver (e.g., sporadic diffuse, distinct, and pattern driven) is not yet known (Arroyo et al., 2011) nor is its release mechanism during a drug-induced liver injury (DILI) (Starkey Lewis et al., 2011). One hypothesis is that DILI-driven miR-122 release may occur in two phases, first export then leakage from necrotic cells (Starkey Lewis et al., 2011).

Plasma levels of miR-122 are significantly increased in chronic hepatitis B virus–infected patients and in mice treated with ethanol (8 g/kg body weight via gastric perfusion) and D-galactosamine (700 mg/kg body weight)/lipopolysaccharide (5 μg/kg body weight) treatment (Zhang et al., 2010). Furthermore, significant increases in plasma levels of miR-122 and another liver-enriched miRNA, miR-192, were identified in a mouse model of acetaminophen (APAP)–induced hepatotoxicity, wherein liver injury was induced following a single dose of 75, 150, or 300 mg/kg APAP with levels of miRNAs and ALT evaluated at 1 and 3 h posttreatment (Wang et al., 2009). The significantly increased levels of miR-122 and miR-192 preceded alterations in serum ALT both in time and by APAP dose level. Yamaura et al. (2012) compared changes in plasma miRNA levels in rat models of acute hepatocellular injury (APAP or methapyrilene treatment), cholestasis (α-naphthyl isothiocyanate treatment or through bile duct ligation), and chronic liver injury (steatosis [high fat diet], steatohepatitis [methionine choline-deficient diet], and fibrosis [carbon tetrachloride treatment]). Using miRNA array analysis, they identified plasma miRNA profiles that could potentially differentiate between the types of liver injury. These miRNAs included miR-122, which was induced more rapidly and dramatically than plasma ALT levels in the acute hepatic injury models. Clinically, Starkey Lewis et al. (2011) evaluated circulating miR-122 levels in healthy human controls, patients with acute liver injury resulting from APAP overdose, patients with APAP overdose without observable hepatic injury, and patients with acute liver injury without relation to APAP. To assess specificity, samples from chronic kidney disease patients also were evaluated in the study. Results showed that circulating levels of miR-122 were increased in the liver injury patients only and correlated with serum ALT levels, though miR-122 levels decreased prior to ALT levels. Studies by Dattaroy et al. (2015) and Pourhoseini et al. (2015) suggest that liver tissue–specific miRNA21 expression level increased in toxin-induced nonalcoholic steatohepatitis.

Elucidating the sequential steps resulting in miR-122 release, comparing this mechanism(s) to those of standard (e.g., ALT) and other potential liver injury biomarkers (e.g., haptoglobin, matrix metallopeptidase 8), and discerning differences between transitory damage and chronic hepatic injury, is critical for establishing whether circulating miR-122 levels can offer prognostic/diagnostic value.

HEART

Whether as part of normal physiological development, disease initiation/progression, or in response to cardiac injury, miRNAs have been recognized as important regulators of cardiovascular functions (Mayr et al., 2013; Wang et al., 2008b). Across many studies, increased plasma levels of various miRNAs have been reported with their concomitant reduction in infarcted myocardium. Examples of such miRNAs include miR-208a, miR-1, miR-133, and miR-499 (Devaux et al., 2012; Gidlof et al., 2011; Wang et al., 2010).

As reviewed by Oliveira-Carvalho et al. (2013), miR-208a levels are prominent in the heart and measurable in peripheral blood and may be valuable as a noninvasive biomarker of cardiovascular injury and/or disease, though it may not be the only circulating miRNA of utility in this regard. To illustrate, a heart attack (i.e., myocardial infarction, or MI) occurs when a disruption in blood flow to the heart is of sufficient duration to cause damage to or destruction of heart tissue. Of the aforementioned miRNAs, Wang et al. (2010) showed that plasma levels of miR-208a could successfully discriminate with 100% specificity and 90.9% sensitivity of those patients with acute MI. Importantly, in consideration of the timing of appearance in the circulation, cardiac troponin levels reportedly rise within 4–8 h of myocardial injury (Wu and Feng, 1998). In contrast, an increase in circulating levels of plasma miR-208a was observed in the aforementioned study in all MI patients within 1 h of coronary ligation, whereas cardiac troponin was measurable in only 85% of patients (Wang et al., 2010). Furthermore, Ji et al. (2009) showed in rats that miR-208 and cardiac troponin I levels were very comparable following isoproterenol-induced myocardial injury; wherein, miR-208 levels were not present at time zero (baseline), but significantly, increased over time following isoproterenol treatment (320 mg/kg body weight) through 12 h. Other studies in rat models of MI (Cheng et al., 2010; Wang et al., 2010) have shown that levels of other miRNAs, such as the muscle-enriched miR-1, miR-133a, and miR-499, have a similar temporal response with a quick rise in levels within 1 h after coronary artery ligation to reach highest levels between 3 and 12 h, decreasing thereafter, with a return to baseline levels by 3 days after MI induction. In an occlusion–reperfusion closed-chest pig model of MI, Gidlof et al. (2011) observed comparable time courses to those described above in that plasma levels of miR-208b, miR-1, miR-499-5p, and miR-133a were not detectable at baseline, were present at 1 h, and peaked at 2 (miR-208b, miR-1, miR-133a) or 2.5 (miR-499-5p) hours following initiation of the procedure.

Although levels of miR-499-5p persisted through the final time point evaluated (2.5 h), levels of the other miRNAs did not and, while still present in plasma, were at appreciably lower levels, illustrating a somewhat differential response between the circulating miRNAs. Gidlof et al. (2011) also evaluated the time course of miRNA release in patients with ST-Elevation Myocardial Infarction, a category of heart attack in which blood flow is impeded for a protracted period and found similarities to that observed in the preclinical animal models; miR-1 and miR-133a and -b levels were highest at 2.5 h after MI symptomatology appeared. Meanwhile, circulating levels of miR-499 and cardiac troponin I increased more slowly with highest levels observed at 12 and 6 h, respectively. Like the protraction observed in the previously noted pig model of MI, miR-499 levels were still present 48 h following MI, though levels of all miRNAs had returned to baseline after 3 days. Thus, these results indicate that plasma levels of certain miRNAs (e.g., miR-208, miR-1, miR-133a) may have greater sensitivity when compared to cardiac troponin levels for diagnosis of MI.

Coronary artery disease (CAD) results from a build-up of plaques in the coronary arteries. These plaques narrow the arterial lumen preventing blood flow to the heart. The first study evaluating levels of circulating miRNAs in CAD patients was conducted by Fichtlscherer et al. (2010) in which RNA was isolated from 36 CAD patients and 17 normal volunteers. In addition to perturbations in several different circulating miRNAs, they reported that plasma levels of miR-208a and miR-133 were elevated in CAD patients as compared to controls, which is reasonably consistent with results described above for MI. As reviewed by Tijsen et al. (2012), other studies investigating changes in miRNA levels in CAD patients have utilized sources alternative to plasma, including serum, whole blood, platelets, and peripheral blood mononuclear cells (Tijsen et al., 2012). Although miRNA levels were detected in all the aforementioned matrices, there were no consistent changes such that an overarching CAD signature was apparent, implying selectivity in miRNA changes dependent on the matrix assayed.

Failure of the ventricles to fill with or empty themselves of blood, because of any structural or functional abnormality, results in heart failure (HF). In two separate studies with relatively small cohorts of HF patients versus healthy controls, Tijsen et al. (2010) and Goren et al. (2012) identified circulating miR-423-5p levels as able to differentiate left ventricular HF patients from controls. In both studies, circulating miR-423-5p levels correlated with B-type natriuretic peptide, a commonly used clinical biomarker of HF (Tijsen et al., 2010; Goren et al., 2012). Notably, however, when circulating levels of miR-423-5p were evaluated in right ventricular HF patients versus healthy controls, it was not able to distinguish between the groups, implying perhaps a degree of selectivity of miR-423-5p for left ventricular HF (Tutarel et al., 2013). Other miRNAs were not evaluated in this study (Tutarel et al., 2013).

Thus, the possible utility of circulating miRNAs as biomarkers for cardiac injury and disease is quite promising, though significant additional work is required to assess levels prior to and following cardiovascular injury or disease initiation, as well as to further evaluate, confirm, and validate those miRNAs currently identified (larger cohorts of patients, inclusion of multiple disease states to assess specificity, etc.).

KIDNEY

Acute kidney injury (AKI) is a key risk factor in the disease process, potentially leading to the complete, or almost complete, failure of kidney function. Current biomarkers used for indications of kidney injury include BUN and serum creatinine levels; however, these parameters lack the sensitivity or specificity needed to appropriately detect renal damage at its earliest stages. Several other parameters have been proposed for AKI detection, including urinary measurements for interleukin (IL)-18, kidney injury molecule-1, liver-type fatty acid binding protein (L-FABP), and cystatin C and plasma and urinary measurements of neutrophil gelatinase-associated lipocalin. Although the aforementioned are promising, doubt remains as to their full utility in a clinical setting (for review see Schiffl and Lang (2012)), and whether they can be utilized reliably as biomarkers for either AKI or other types of kidney injuries. Thus, the need remains for the identification and validation of reliable biomarkers of renal damage to enable appropriate and timely treatment intervention, and efforts are focused on the potential utility of miRNAs in this regard, both in acute injury and chronic renal diseases.

Multiple miRNA studies in the kidney have dealt with tissue expression patterns in various renal diseases (e.g., renal allograft, renal cell carcinoma, polycystic kidney disease). Early work by Gottardo et al. (2007) identified miR-28, miR-185, miR-27, and let-7f-2 as being differentially expressed in renal cell carcinoma tissues as compared to normal kidney tissues, whereas Spector et al. (2013) reported the use of 24 miRNAs to differentiate/classify four types of renal cell carcinomas (i.e., clear cell, papillary, chromophobe, and oncocytoma) from one another. Additional renal diseases, such as lupus nephritis and IgA nephropathy, may benefit diagnostically using miRNA expression as well as acute renal allograft rejection. For example, levels of 35 of 132 miRNAs, identified by Dai et al. (2008) in IgA nephropathy renal biopsies versus control kidney biopsies, were increased. Preclinically, Bhatt et al. (2010) treated mice with cisplatin (single injection at 30 mg/kg body weight), a chemotherapeutic known to induce AKI, and found that miR-34a levels in the kidney were increased at 1, 2, and 3 days following treatment.

To date, few studies have focused on miRNAs in urine as potential biomarkers for detection of kidney injury and disease. One example, though, is a study conducted by Saikumar et al. (2012) in which miRNA levels of miR-21, miR-155, and miR-18a were evaluated in rat kidney tissue, blood, and urine following bilateral ischemia/reperfusion injury and after subcutaneous gentamicin treatment (200 or 300 mg/kg body weight). In kidney tissue, levels of miR-21, miR-155, and miR-18a were increased with levels of miR-21 and miR-155 decreased in blood and urine as compared to controls.

Given that the various cell types in kidneys are affected in different ways based on the type of renal injury or disease, it is important to understand which miRNAs are expressed in which cells and how the miRNA levels change during injury or disease and are thereafter released into circulation.

BRAIN

Brain biomarkers have not been as extensively studied as those for other tissues such as cardiac biomarkers. Although biomarkers for brain injury have been identified, many of them lack specificity for brain injury. These biomarkers include S100 calcium binding protein B (S100B), neuron-specific enolase (NSE), glial fibrillary associated protein (GFAP), and myelin basic protein (MBP). SB-100B increases in blood and cerebrospinal fluid because of multiple types of brain injury; however, it also increases with an injury to muscle, fat, and bone marrow (Laterza et al., 2006). Similarly, NSE levels increase following stroke but also increase in small-cell lung cancer, neuroblastoma, melanoma, and other cancers.

Biomarkers as early indicators of neurodegenerative diseases and in neuronal injury are important for many factors, but two of the most critical factors are diagnosis and appropriate treatment. Traditionally, clinical manifestations and imaging are widely used for diagnosis although, by the time these interventions are made, neuronal damage may have already occurred. There is a significant and unmet medical need for a rapid diagnostic that could assist in predictive assessment or in the identification of brain disease prior to neuronal damage.

It is known that miRNAs are important for the normal development of the brain and distinct profiles have been shown in brain tissue, which suggests their importance in neural development. This has been demonstrated using expression pattern changes in miRNAs during development versus maturation. Through these studies, miRNA-134 has been shown to contribute to synaptic plasticity (Schratt et al., 2006). In addition, miR-9, miR-124a, miR-125b, miR-127, miR-128, miR-131, miR-132, and the let-7 family were highly expressed in the mouse brain when compared to other tissue (Hohjoh and Fukushima, 2007; Bak et al., 2008). Studies have shown regional specificity for miRNAs in the brain, such as miR-195, miR-497, and miR-30b in the cerebellum; miR-34a, miR-451, miR-219, miR338, miR-10a, and miR-10b in the medulla oblongata; miR-7 and miR-7b in the hypothalamus; miR-218, miR-221, miR-222, miR-26a, miR-128a/b, miR-138, and let7c in the hippocampus; and miR-375, miR-141, and miR-200a in the pituitary (Bak et al., 2008; Landgraf et al., 2007).

Stroke is one of the leading causes of death in industrialized countries and many surviving cases result in severe disability due to tissue damage. The rapid development of apoptosis and neuronal death due to stroke has limited effective clinical therapies. Because of the complexity of ischemic strokes, identifying blood biomarkers for clinical applications has been difficult with current clinical practice including assessment by the physician in conjunction with imaging. Application of ischemic stroke biomarkers is important in facilities where imaging is not available, as well as in cases of ambiguous diagnosis. Current stroke biomarkers include S100B, NSE, MBP, and GFAP (Rothstein and Jickling, 2013). However, they are only released into systemic circulation on the breakdown of the blood–brain barrier; therefore, they do not directly correlate with the severity of the stroke. Inflammatory proteins, such as C-reactive protein, IL-6, matrix metallopeptidase 9, and vascular cell adhesion molecule also have shown potential as biomarkers of ischemic stroke. However, because they are inflammatory biomarkers, they may not differentiate stroke from other disease states that increase other biomarkers of inflammation, making clinical application difficult. Visinin-like protein 1 was shown to be present in both the blood and CSF of stroke patients, although it also has been shown to be increased in the CSF of Alzheimer's patients. Gene expression analysis with multiple gene panels shows promise for the diagnosis of stroke as well as an understanding of the cellular pathways involved, which may also lead to new treatments. Many multiple gene panels evaluating mRNA have been identified that diagnose stroke. However, the stability of the sample and variation due to sample processing make use of this as a diagnostic tool problematic. Because of the lack of specificity and stability of the previously identified biomarkers, identification and utilization of stroke miRNA biomarkers may dramatically enhance diagnosis and treatment.

In rats, miRNA-145 upregulation was observed in transient ischemia, which was shown to downregulate the expression of superoxide dismutase-2 (SOD2) (Dharap et al., 2009). SOD2 is important for neuroprotection. Inhibition of miRNA-145 led to increased SOD2 in neurons, further linking miRNA-145 to SOD2 expression. In mice, miRNA-497 was shown to be induced during transient ischemia (Yin et al., 2010b). This induction correlated with increased neuronal death, which was decreased with a miRNA-497 inhibitor. Furthermore, these studies showed that miRNA-497 repressed antiapoptotic proteins, Bcl-2 and Bcl-w. miRNA-15a was also shown to regulate *bcl-2* by directly binding to the 3′UTR and inhibiting translation (Cimmino et al., 2005). This study was further confirmed by Yin et al. (2010a), who found that peroxisome proliferator-activated receptor δ provided a protective role during stroke by inhibiting miRNA-15a. Not only do these miRNAs hold potential for rapid diagnosis of stroke, but they also show potential for therapeutic treatment.

miRNA-124 has been shown to be a brain-specific miRNA. Previous studies showed 260-fold or higher expression of miRNA-124 in the brain when compared to other tissues (Laterza et al., 2009). Using the middle cerebral artery occlusion model for brain injury due to stroke, miRNA-124 increased in the plasma 8 h after transient occlusions. Twenty-four (24) hours after both transient and permanent occlusions, miRNA-124 was increased 150-fold when compared to sham samples. However, miRNA-124 did not increase with liver or muscle injury, suggesting that miRNA-124 may be a specific brain biomarker for neuronal injury.

A study evaluating miRNA expression profiles in young stroke patients identified a panel of miRNAs present in the blood of these patients (Tan et al., 2009). This panel included the increased expression of miRNA-16, miRNA-19b, miRNA-23a, miRNA-103, miRNA-106b, miRNA-185, miRNA-191, miRNA-320, miRNA-451, and let7c and the decreased expression of let-7a. Therefore, considering the multitude of miRNAs that appear to be relevant, it may be the case that a panel of different miRNAs in blood ought to be utilized when searching for biomarkers of stroke.

Inflammation is an important factor in the neuronal death. Microglia cells become activated on brain injuries such as ischemia and traumatic brain injury (TBI) due to oxygen deprivation. This leads to inflammation of the neurons by the production of proinflammatory cytokine and neurotoxic products such as reactive oxygen species. Evaluation of these activated cells identified the downregulation of miRNA-181c, which leads to the upregulation of TNF-α (Zhang et al., 2012). TNF-α is an important proinflammatory cytokine that has been linked to neuronal damage in ischemia and in neurodegenerative and neurotoxic diseases.

TBI is damage due to the brain hitting the inside of the skull, or an object reaching the brain causing damage. Current diagnosis involves the use of head computed tomography scan, as well as monitoring of altered functions (e.g., cognitive, sensory, and motor). Magnetic resonance imaging (MRI) is also used, but because of the high cost, time involved, and limited availability of MRI machines, its use has been limited. Biomarkers would aid in monitoring the severity of the damage. S100B is one of the most extensively studied biomarkers of TBI. Very low levels are found in the CSF and serum under normal conditions. The levels have been shown to increase after an injury to astrocytes.

However, the correlation between levels in CSF and serum are inconclusive. As S100B does not cross the blood–brain barrier, identification in the serum may be an indicator of the loss of integrity of the blood–brain barrier. It also has issues with specificity as described earlier (Jeter et al., 2013). GFAP is released by astrocytes after TBI. It has been shown to be elevated in the serum within 1 h after injury. In adult rats, proinflammatory cytokines such as IL-α, IL-β, and TNF increased within 3 h after injury, likely due to microglia activation (Hernandez-Ontiveros et al., 2013).

Many patients with traumatic brain injuries have other bodily injuries, making differentiation difficult. Redell et al. (2010) compared TBI miRNA to those upregulated in orthopedic injuries. In this study, they found that repression of miRNA-16 and miRNA-92a and upregulation of miRNA-765 in plasma was specific for severe TBI when compared to healthy volunteers and orthopedic injury patients (Redell et al., 2010). In contrast, miRNA-16 and miRNA-92a were upregulated in mild TBI patients. Studies in rats by Balakathiresan et al. (2012) identified 47 differentially expressed miRNA in the serum when the rodents were exposed to multiple blasts over short (2 h) and long (24 h) intervals. It was found that miR-let-7i, miR-122, and miR-340-5p were upregulated in both groups. Therefore, there are several potential miRNAs for easier diagnosis of TBI (Balakathiresan et al., 2012).

Current biomarkers for the neurodegenerative disorder, Alzheimer's disease (AD), involve imaging tools that make use of radioactive tracers that bind to beta-amyloid plaques. Diagnosis is also enhanced by the presence of proteins total tau, phosphorylated tau, and beta-amyloid in the CSF and blood. Genetic testing can also be a powerful tool, given that the presence of the ApoE4 allele suggests a high risk for developing the disease. Because of the ability of the brain to adjust to neuronal death, clinical manifestations of AD occur at the late stages of the disease, limiting the success of the pharmacological intervention.

Differentially expressed miRNAs have been identified in neurodegenerative disorders, such as AD and Parkinson's disease (PD). Many studies are elucidating the role of miRNA in the sporadic AD and have identified BACE1 as a key enzyme regulated by miRNA (Saugstad, 2010). BACE1 is one of the key enzymes required for the aggregation of beta-amyloid peptides in the brain. miRNA-107 is predicted to bind to the 3′UTR of BACE1 and was shown to be decreased in AD patients. Wang et al. (2008) showed that expression of BACE1 and miRNA-107 is inversely correlated, such that as miRNA-107 decreases and BACE1 increases (Wang et al., 2008a). Alexandrov et al. (2012) showed miRNAs implicated in inflammation and AD were also upregulated in extracellular fluid surrounding brain cells and CSF of patients with the AD, which confirmed similar findings in brain tissue. All the patients evaluated had at least one ApoE4 allele. These miRNAs include miRNA-9, miRNA-125b, miRNA-146a, and miRNA-155 (Alexandrov et al., 2012). NF-κB inhibited the expression of these miRNAs. This suggests that these miRNAs may be suitable biomarkers for the AD and that inflammation plays an important role in neuronal death.

α-Synuclein (ASYN) is a major component of neuronal inclusions called Lewy bodies, which are a characteristic of PD. The presence of ASYN in CSF can differentiate α-synucleinopathies from other neurological disorders such as an AD. Similarly, to other complicated diseases, a panel of biomarkers accurately differentiated Parkinson's from other diseases using CSF; these markers include total tau, phosphorylated tau, beta-amyloid, Flt3 ligand, and fractalkine. Imaging has also been effectively used as a biomarker.

Minimal expression of miRNA-133b has been shown to be important for PD. Functional studies showed that it is important for the maturation of dopaminergic neurons (Kim et al., 2007). Let-7 and miRNA-184 are important in regulating LRRK2, an important protein involved in sporadic and familial PD (Gehrke et al., 2010). Furthermore, decreased levels of miRNA-34b/c in the brain tissue of PD patients have been shown to be important in multiple stages of disease development (Minones-Moyano et al., 2011). The aggregation of ASYN leads to decreased neuronal function and cell death. Junn et al. (2009) identified that the 3′UTR of ASYN is regulated by miRNA-7. The induction of miRNA-7 leads to decreased levels of ASYN. Because ASYN is a key protein in the development of PD, the therapeutic use of miRNA-7 has great potential (Junn et al., 2009). Using blood samples, Margis et al. (2011) identified that expression of miRNA-1, miRNA-22, and miRNA-29 can differentiate individuals with PD from healthy individuals. Furthermore, miRNA-16-2, miRNA-26a2, and miRNA-30a can differentiate treated and untreated individuals with PD. In a small set of patient samples, Sheinerman et al. (2012) identified biomarker pairs that were able to predict mild cognitive impairment before clinical presentation. They grouped these pairs into families. The mir-132 family includes miR-128/miR-491-5p, miR-132/miR-491-5p, and miR-874/miR-491-5p, and the miR-134 family includes miR-323-3p/miR-370, miR-382/miR-370, and miR-134/miR-370 (Sheinerman et al., 2012). Larger populations need to be analyzed for validation for AD or PD. Because of the extent of neuronal damage on diagnosis of AD and PD, earlier detection could dramatically aid in treatment options.

ENVIRONMENTAL EXPOSURE AND MIRNA

Our inbuilt environment is the major effector of our health or disease status. In combination with genetic factor, environmental exposure can lead to severe complication in most of the disease. In the last two decades, the noncoding gene product such as miRNA has gained huge interest in environment associated epigenetic effects. Consulting several review articles, experimental articles and database search, a strong relation has been observed in environmental pollution exposure and miRNA levels. The environmental pollution potentially regulates miRNA expression in disease pathogenesis, and therefore, miRNAs could potentially be novel biomarker of environmental associated disease (Vrijens et al., 2015). The review published by Vrijens et al (2015) clearly suggests the challenges of epidemiological studies. Thus, the collection of information on the tissue specific miRNAs is limited and depends on the levels of circulatory, free and exosomal miRNAs. The effects of environmental pollution and toxin and/or personal exposure on miRNA expression have been summarized in Table 59.6. It has been observed that the miRNAs frequently reported in response to personal or environmental exposure are directly associated with cancer etiology especially lung and breast cancer and leukemia or complication in cardiovascular system and also hypertension, HF, myocardial infarct, and atherosclerosis (Vrijens et al., 2015).

ROLE OF MIRNA IN ARSENIC-INDUCED CARCINOGENESIS

Skin cancer is the most common of all human cancers, and arsenic is second to UV in sunlight as a cause of skin cancer. In multiple epidemiologic studies, arsenical keratosis, hyperpigmentation, and multiple cutaneous malignancies were associated with arsenic exposure (Centeno et al., 2002; Maloney, 1996). The signs of arsenic toxicity appear clearly in the skin as arsenical keratosis in the palms, soles, and trunk (Centeno et al., 2002; Hunt et al., 2014). Chronic arsenic exposure causes both basal cell carcinoma and squamous cell carcinoma (SCC) of the skin. Several arsenic toxicity and carcinogenicity mechanisms have been suggested and proposed including abnormal signaling cascades, oxidative stress, and chromosomal aberrations as well as abnormal transcriptional activity and global gene expression (Isokpehi et al., 2012). Moreover, arsenic can induce environmentally driven epigenetic alterations that are known to influence disease development, including differential miRNA expression (Gonzalez et al., 2015; Su et al., 2011). An in vitro study by Al-Eryani et al. (2018) identified three of the six miRNAs with predicted targets that were differentially expressed after 3 and 7 weeks (miR-1228, miR-1254, miR-645). These miRNAs were associated with cancer in previous studies (Foss et al., 2011; Lin et al., 2015; Sun et al., 2015). MiR-1228 (suppressed at 3 weeks, induced at 7 weeks) was reported to be induced in breast cancer tissues and cell lines and regulates the levels of mRNA and protein of SCAI (suppressor of

TABLE 59.6 Expression Levels of miRNAs That Are Responsive to Personal and/or Environmental Exposure and Their Roles in Human Disease (Vrijens et al., 2015)

miRNA	Regulated	Exposure	Diseases
Let-7e	Down	TCDD	HCC, lung, pituitary, and breast cancer, GEP tumors
	Up	RDX	Heart failure, asthma
Let-7g	Down	BPA,	PM Lung carcinoma, GEP tumors, breast cancer
miR-9	Down	PM	Brain cancer, Huntingon's disease
	Up	Aluminum	Hodgkin lymphoma, breast cancer
miR-10b	Down	Formaldehyde, PM	Gastric cancer
miR-21	Down	Smoking	Diabetes type 2
	Up	DEP, metal-rich PM	Breast cancer, glioblastoma, neo-intimal lesions, cardiac hypertrophy, atherosclerosis
miR-26b	Down	DEP, BPA, PFOA	Schizophrenia, CRC, breast cancer
miR-31	Down	DEP, TCDD	Medulloblastoma, T-cell leukemia
miR-34b	Down	Smoking (2×)	CRC, pancreatic, mammary, ovarian, and renal cell carcinoma

Continued

TABLE 59.6 Expression Levels of miRNAs That Are Responsive to Personal and/or Environmental Exposure and Their Roles in Human Disease (Vrijens et al., 2015)—cont'd

miRNA	Regulated	Exposure	Diseases
miR-92b	Down	Smoking, DDT	Medulloblastoma
miR-122	Down	Smoking	HCC
	Up	TCDD	Hepatitis C, renal cell carcinoma, male infertility, sepsis, hyperlipidemia
miR-125b	Down	Smoking (2×)	Breast cancer, head and neck cancer
	Up	Aluminum sulfate (2×)	Endometriosis, cardiac hypertrophy, Alzheimer's disease
miR-135b	Down	DEP	Medulloblastoma
	Up	Smoking	CRC
miR-142	Down	Formaldehyde	Heart failure
	Up	Smoking	B-cell ALL
miR-143	Up	PM, ozone	Colon cancer
miR-146a	Down	Smoking	Postpartum psychosis, type 2 diabetes
	Up	BPA, aluminum sulfate (2×)	Alzheimer's disease, Creutzfeldt-Jakob disease, atherosclerosis, leukemia, protection against myocardial injury
miR-149	Up	BPA, DDT	Melanoma
miR-155	Down	PM	Hypertension
	Up	PM	Breast cancer, Hodgkin lymphoma, B-ALL
miR-181a	Down	Formaldehyde	Leukemia, glioma, NSCLC, breast cancer, metabolic syndrome, and CAD
	Up	TCDD	Severe preeclampsia, male infertility
miR-203	Down	Smoking, formaldehyde	Myeloma
miR-205	Up	Smoking (2×)	Heart failure, lung cancer
miR-206	Up	Smoking, RDX	Myocardial infarct, slows ALS progression, myotonic dystrophy
miR-222	Up	Metal-rich PM, BPA	Severe preeclampsia, thyroid carcinoma, prostate cancer, breast cancer
miR-223	Down	Smoking	AML
	Up	Smoking	Heart failure, atherosclerosis
miR-338-5p	Down	Formaldehyde	Melanoma
	Up	DEP	Oral carcinoma
miR-340	Down	Smoking	NA
	Up	Smoking	Heart failure, breast cancer
miR-638	Up	BPA, DDT, arsenic	Lupus nephritis
miR-663	Up	BPA, DDT, arsenic	CTCL, nasopharyngeal carcinoma, burns

ACC, acute lymphocytic leukemia; *ALS*, amyotrophic lateral sclerosis; *AML*, acute myeloid leukemia; *B-ALL*, B-cell acute lymphocytic leukemia; *BPA*, bisphenol A; *CAD*, coronary artery disease; *CRC*, colorectal carcinoma; *CTCL*, cutaneous T-cell lymphoma; *DDT*, dichlorodiphenyltrichloroethane; *DEP*, diesel exhaust particles; *GEP*, gastro-entero pancreatic; *HCC*, hepatocellular carcinoma; *NA*, not applicable; *NSCLC*, nonsmall cell lung carcinoma; *PFOA*, perfluorooctanoic acid; *PM*, particulate matter; *RDX*, hexahydro-1,3,5-trinitro-*s*-triazine; *TCDD*, 2,3,7,8-tetrachlorodibenzo-*p*-dioxin.

cancer cell invasion) (Lin et al., 2015). MiR-1254 (suppressed at 3 and 7 weeks) was reported to be induced in the sera of nonsmall cell lung cancer patients (Foss et al., 2011). MiR-645 (induced at 3 and 7 weeks) was induced in head and neck SCC, and its induction was found to promote cell invasion and metastasis (Sun et al., 2015). Arsenite dysregulated (mostly suppressed) several snoRNAs. SnoRNAs are 60–300 nucleotide small RNAs that are concentrated in the nucleolus (Williams and Farzaneh, 2012). SnoRNAs are involved in guiding premature ribosomal and other spliceosomal RNAs for nucleoside posttranscriptional modifications, crucial for accurate ribosomes. SnoRNAs have two main classes: C/D box and H/ACA box snoRNAs, both reported to be dysregulated in several diseases including cancer (Dong et al., 2009; Williams and Farzaneh, 2012). These results suggest that dysregulation of snoRNAs may play a role in arsenic-induced carcinogenesis.

CONCLUDING REMARKS AND FUTURE DIRECTIONS

There are several areas where it would seem obvious to continue to advance the study of miRNAs as potential biomarkers of injury, toxicity, and disease including compartment (e.g., retina, tears, saliva, amniotic fluid, and muscle). Clearly, the volume of data generated requires that there be systems available to handle such large amounts of data in a meaningful way. The work performed thus far in building databases to aid in correlations of miRNA levels with disease states would seem a logical place to start to extend this to biomarkers of injury and toxicity following administration of an NCE. Clearly, however, where commercial interests prevail, this may not happen to the extent that would be meaningful.

A second area that may be ready for further advancement is in the understanding of the relationships between the various miRNAs and the impact that these molecules themselves may have on a disease state.

A further area for research is in the design of novel sampling procedures that may allow small tissue samples to be taken that might permit the correlation of a pathologic value with a circulating biomarker such as miRNA.

Finally, these authors believe that the proliferation of human cell–based *ex vivo* systems to model efficacy and toxicity testing is ripe for exploitation in terms of validation of these approaches to replace the use of animals in preclinical drug development. Generation of robust biomarker data including miRNA information, such as may be generated from validated assays, could result in the faster acceptance and more routine use of these technologies, thus decreasing the time to bring a new drug to market.

The last two decades or so have seen a monumental rise in the amount of data generated because of continued efforts in miRNA biomarker development. Correspondingly, the value of these data in aiding our understanding of toxicology is not yet fully understood. It has been well documented that most biomarkers do not reach clinical use for many different reasons, such as failure to consider external factors, lack of understanding of temporal variations, or simply an analytical error (Starmans et al., 2012). However, the future of miRNAs as potential biomarkers to better facilitate the understanding of tissue injury and toxicology, and therefore aid in the drug development process, appears promising. In addition, miRNA technology is expected to play a significant role in environmental monitoring and in understanding the natural and/man-made insults to the biological systems over time.

References

Ach, R.A., Wang, H., Curry, B., 2008. Measuring microRNAs: comparisons of microarray and quantitative PCR measurements, and of different total RNA prep methods. BMC Biotechnol. 8, 69.

Al-Eryani, L., Waigel, S., Tyagi, A., et al., 2018. Differentially expressed mRNA targets of differentially expressed miRNAs predict changes in the TP53 axis and carcinogenesis-related pathways in human keratinocytes chronically exposed to arsenic. Toxicol. Sci. 162 (2), 645–654.

Alexandrov, P.N., Dua, P., Hill, J.M., et al., 2012. microRNA (miRNA) speciation in Alzheimer's disease (AD) cerebrospinal fluid (CSF) and extracellular fluid (ECF). Int. J. Biochem. Mol. Biol. 3, 365–373.

Ambros, V., 2001. microRNAs: tiny regulators with great potential. Cell 107, 823–826.

Ambros, V., Bartel, B., Bartel, D.P., et al., 2003. A uniform system for microRNA annotation. RNA (N.Y.) 9, 277–279.

Antoine, D.J., Dear, J.W., Lewis, P.S., et al., 2013. Mechanistic biomarkers provide early and sensitive detection of acetaminophen-induced acute liver injury at first presentation to hospital. Hepatology 58, 777–787.

Arroyo, J.D., Chevillet, J.R., Kroh, E.M., et al., 2011. Argonaute2 complexes carry a population of circulating microRNAs independent of vesicles in human plasma. Proc. Natl. Acad. Sci. U.S.A. 108, 5003–5008.

Bak, M., Silahtaroglu, A., Moller, M., et al., 2008. MicroRNA expression in the adult mouse central nervous system. RNA (N.Y.) 14, 432–444.

Bala, S., Petrasek, J., Mundkur, S., et al., 2012. Circulating microRNAs in exosomes indicate hepatocyte injury and inflammation in alcoholic, drug-induced, and inflammatory liver diseases. Hepatology 56, 1946–1957.

Balakathiresan, N., Bhomia, M., Chandran, R., et al., 2012. MicroRNA let-7i is a promising serum biomarker for blast-induced traumatic brain injury. J. Neurotrauma 29, 1379–1387.

Bam, M., Yang, X., Zumbrun, E.E., et al., 2017. Decreased AGO2 and DCR1 in PBMCs from War Veterans with PTSD leads to diminished miRNA resulting in elevated inflammation. Transl. Psychiatry 7, e1222.

Bansal, A., Lee, I.-H., Hong, X., et al., 2013. Discovery and validation of Barrett's esophagus microRNA transcriptome by next generation sequencing. PLoS One 8, e54240.

Benichou, C., 1990. Criteria of drug-induced liver disorders. Report of an international consensus meeting. J. Hepatol. 11, 272–276.

Bhatt, K., Zhou, L., Mi, Q.-S., et al., 2010. MicroRNA-34a is induced via p53 during cisplatin nephrotoxicity and contributes to cell survival. Mol. Med. 16, 409–416.

Bissels, U., Wild, S., Tomiuk, S., et al., 2009. Absolute quantification of microRNAs by using a universal reference. RNA (N.Y.) 15, 2375–2384.

Bravo-Egana, V., Rosero, S., Molano, R.D., et al., 2008. Quantitative differential expression analysis reveals miR-7 as major islet microRNA. Biochem. Biophys. Res. Commun. 366, 922–926.

Cabibi, D., Caruso, S., Bazan, V., et al., 2016. Analysis of tissue and circulating microRNA expression during metaplastic transformation of the esophagus. Oncotarget 7, 47821–47830.

Calvano, J., Achanzar, W., Murphy, B., et al., 2016. Evaluation of microRNAs-208 and 133a/b as differential biomarkers of acute cardiac and skeletal muscle toxicity in rats. Toxicol. Appl. Pharmacol. 312, 53–60.

Centeno, J.A., Mullick, F.G., Martinez, L., et al., 2002. Pathology related to chronic arsenic exposure. Environ. Health Perpect. 110 (Suppl.), 883–886.

Chen, D., Pan, X., Xiao, P., et al., 2011. Evaluation and identification of reliable reference genes for pharmacogenomics, toxicogenomics, and small RNA expression analysis. J. Cell. Physiol. 226, 2469–2477.

Cheng, Y., Tan, N., Yang, J., et al., 2010. A translational study of circulating cell-free microRNA-1 in acute myocardial infarction. Clin. Sci. 119, 87–95.

Church, R.J., McDuffie, J.E., Sonee, M., et al., 2014. MicroRNA-34c-3p is an early predictive biomarker for doxorubicin-induced glomerular injury progression in male Sprague-Dawley rats. Toxicol. Res. 3, 384–394.

Church, R.J., Otieno, M., McDuffie, J.E., et al., 2016. Beyond miR-122: identification of MicroRNA alterations in blood during a time course of hepatobiliary injury and biliary hyperplasia in rats. Toxicol. Sci. 150, 3–14.

Cimmino, A., Calin, G.A., Fabbri, M., et al., 2005. miR-15 and miR-16 induce apoptosis by targeting BCL2. Proc. Natl. Acad. Sci. U.S.A. 102, 13944–13949.

Corley, D.A., Levin, T.R., Habel, L.A., et al., 2002. Surveillance and survival in Barrett's adenocarcinomas: a population-based study. Gastroenterology 122, 633–640.

Coulouarn, C., Factor, V.M., Andersen, J.B., et al., 2009. Loss of miR-122 expression in liver cancer correlates with suppression of the hepatic phenotype and gain of metastatic properties. Oncogene 28, 3526–3536.

Creemers, E.E., Tijsen, A.J., Pinto, Y.M., 2012. Circulating microRNAs: novel biomarkers and extracellular communicators in cardiovascular disease? Circ. Res. 110, 483–495.

Dai, Y., Sui, W., Lan, H., et al., 2008. Microarray analysis of microribonucleic acid expression in primary immunoglobulin A nephropathy. Saudi Med. J. 29, 1388–1393.

Dattaroy, D., Pourhoseini, S., Das, S., et al., 2015. Micro-RNA 21 inhibition of SMAD7 enhances fibrogenesis via leptin-mediated NADPH oxidase in experimental and human nonalcoholic steatohepatitis. Am. J. Physiol. Gastrointest. Liver Physiol. 308, G298–G312.

Devaux, Y., Vausort, M., Goretti, E., et al., 2012. Use of circulating microRNAs to diagnose acute myocardial infarction. Clin. Chem. 58, 559–567.

Dharap, A., Bowen, K., Place, R., et al., 2009. Transient focal ischemia induces extensive temporal changes in rat cerebral microRNAome. J. Cerebr. Blood Flow Metabol. 29, 675–687.

Dong, X.-Y., Guo, P., Boyd, J., et al., 2009. Implication of snoRNA U50 in human breast cancer. J. Genet. Genomics 36, 447–454.

Dueck, A., Ziegler, C., Eichner, A., et al., 2012. microRNAs associated with the different human Argonaute proteins. Nucleic Acids Res. 40, 9850–9862.

Etheridge, A., Lee, I., Hood, L., et al., 2011. Extracellular microRNA: a new source of biomarkers. Mutat. Res. 717, 85–90.

FDA Guidance for Industry E16, 2008, Apr 8. Definitions for genomic biomarkers pharmacogenomics, pharmacogenetics, genomic data and sample coding categories. Fed. Regist. 73 (68), 19074–19076.

FDA Guidance for Industry E16, 2011, Aug 11. Biomarkers related to drug or biotechnology product development: context, structure, and format of qualification submissions. Fed. Regist. 76 (155), 49773–49774.

Feber, A., Xi, L., Luketich, J.D., et al., 2008. MicroRNA expression profiles of esophageal cancer. J. Thaorac. Cardiovasc. Surg. 135, 255–260.

Fichtlscherer, S., De Rosa, S., Fox, H., et al., 2010. Circulating microRNAs in patients with coronary artery disease. Circ. Res. 107, 677–684.

Foss, K.M., Sima, C., Ugolini, D., et al., 2011. miR-1254 and miR-574-5p: serum-based microRNA biomarkers for early-stage non-small cell lung cancer. J. Thorac. Oncol. 6, 482–488.

Gantier, M.P., McCoy, C.E., Rusinova, I., et al., 2011. Analysis of microRNA turnover in mammalian cells following Dicer1 ablation. Nucleic Acids Res. 39, 5692–5703.

Gehrke, S., Imai, Y., Sokol, N., et al., 2010. Pathogenic LRRK2 negatively regulates microRNA-mediated translational repression. Nature 466, 637–641.

Gidlof, O., Andersson, P., van der Pals, J., et al., 2011. Cardiospecific microRNA plasma levels correlate with troponin and cardiac function in patients with ST elevation myocardial infarction, are selectively dependent on renal elimination, and can be detected in urine samples. Cardiology 118, 217–226.

Girard, M., Jacquemin, E., Munnich, A., Lyonnet, S., Henrion-Caude, 2008 Apr. A miR-122, a paradigm for the role of microRNAs in the liver. J. Hepatol. 48 (4), 648–656. https://doi.org/10.1016/j.jhep.2008.01.019. Epub 2008 Feb 12.

Git, A., 2012. Research tools: a recipe for disaster. Nature 484, 439–440.

Glineur, S.F., De Ron, P., Hanon, E., et al., 2016. Paving the route to plasma miR-208a-3p as an acute cardiac injury biomarker: preclinical rat data supports its use in drug safety assessment. Toxicol. Sci. 149, 89–97.

Gonzalez, H., Lema, C., Kirken, R.A., et al., 2015. Arsenic-exposed keratinocytes exhibit differential microRNAs expression profile; Potential implication of miR-21, miR-200a and miR-141 in melanoma pathway. Clin. Cancer Drugs 2, 138–147.

Goodwin, D., Rosenzweig, B., Zhang, J., et al., 2014. Evaluation of miR-216a and miR-217 as potential biomarkers of acute pancreatic injury in rats and mice. Biomarkers 19, 517–529.

Goren, Y., Kushnir, M., Zafrir, B., et al., 2012. Serum levels of microRNAs in patients with heart failure. Eur. J. Heart Fail. 14, 147–154.

Gottardo, F., Liu, C.G., Ferracin, M., et al., 2007. Micro-RNA profiling in kidney and bladder cancers. Urol. Oncol. 25, 387–392.

Gulyaeva, L.F., Kushlinskiy, N.E., 2016. Regulatory mechanisms of microRNA expression. J. Transl. Med. 14, 143.

Harrill, A.H., McCullough, S.D., Wood, C.E., et al., 2016. MicroRNA biomarkers of toxicity in biological matrices. Toxicol. Sci. 152, 264–272.

Hernandez-Ontiveros, D.G., Tajiri, N., Acosta, S., et al., 2013. Microglia activation as a biomarker for traumatic brain injury. Front. Neurol. 4, 30.

Hohjoh, H., Fukushima, T., 2007. Expression profile analysis of microRNA (miRNA) in mouse central nervous system using a new miRNA detection system that examines hybridization signals at every step of washing. Gene 391, 39–44.

Hunt, K.M., Srivastava, R.K., Elmets, C.A., et al., 2014. The mechanistic basis of arsenicosis: pathogenesis of skin cancer. Canc. Lett. 354, 211–219.

Isokpehi, R.D., Udensi, U.K., Anyanwu, M.N., et al., 2012. Knowledge building insights on biomarkers of arsenic toxicity to keratinocytes and melanocytes. Biomark. Insights 7, 127–141.

Jeter, C.B., Hergenroeder, G.W., Hylin, M.J., et al., 2013. Biomarkers for the diagnosis and prognosis of mild traumatic brain injury/concussion. J. Neurotrauma 30, 657–670.

Ji, X., Takahashi, R., Hiura, Y., et al., 2009. Plasma miR-208 as a biomarker of myocardial injury. Clin. Chem. 55, 1944–1949.

Jiang, Q., Wang, Y., Hao, Y., et al., 2009. miR2Disease: a manually curated database for microRNA deregulation in human disease. Nucleic Acids Res. 37, D98–D104.

Junn, E., Lee, K.-W., Jeong, B.S., et al., 2009. Repression of alpha-synuclein expression and toxicity by microRNA-7. Proc. Natl. Acad. Sci. U.S.A. 106, 13052–13057.

Kan, T., Meltzer, S.J., 2009. MicroRNAs in Barrett's esophagus and esophageal adenocarcinoma. Curr. Opin. Pharmacol. 9, 727–732.

Kanki, M., Moriguchi, A., Sasaki, D., et al., 2014. Identification of urinary miRNA biomarkers for detecting cisplatin-induced proximal tubular injury in rats. Toxicology 324, 158–168.

Kaphalia, B.S., 2014. Biomarkers of acute and chronic pancreatitis. In: Gupta, R.C. (Ed.), Biomarkers in Toxicology. Academic Press/Elsevier, Amsterdam, pp. 279–289.

Kim, J., Inoue, K., Ishii, J., et al., 2007. A microRNA feedback circuit in midbrain dopamine neurons. Science 317, 1220–1224.

Kong, X.-Y., Du, Y.-Q., Li, L., et al., 2010. Plasma miR-216a as a potential marker of pancreatic injury in a rat model of acute pancreatitis. World J. Gastroenterol. 16, 4599–4604.

Kozomara, A., Griffiths-Jones, S., 2011. miRBase: integrating microRNA annotation and deep-sequencing data. Nucleic Acids Res. 39, D152–D157.

Krol, J., Busskamp, V., Markiewicz, I., et al., 2010. Characterizing light-regulated retinal microRNAs reveals rapid turnover as a common property of neuronal microRNAs. Cell 141, 618–631.

Laboratory of Functional Genomics, 2013. miRNEST 2.0 Website: An Integrative MicroRNA Resource.

Lagos-Quintana, M., Rauhut, R., Yalcin, A., et al., 2002. Identification of tissue-specific microRNAs from mouse. Curr. Biol. 12, 735–739.

Landgraf, P., Rusu, M., Sheridan, R., et al., 2007. A mammalian microRNA expression atlas based on small RNA library sequencing. Cell 129, 1401–1414.

Laterza, O.F., Lim, L., Garrett-Engele, P.W., et al., 2009. Plasma MicroRNAs as sensitive and specific biomarkers of tissue injury. Clin. Chem. 55, 1977–1983.

Laterza, O.F., Modur, V.R., Crimmins, D.L., et al., 2006. Identification of novel brain biomarkers. Clin. Chem. 52, 1713–1721.

Li, L.-M., Cai, W.-B., Ye, Q., et al., 2014. Comparison of plasma microRNA-1 and cardiac troponin T in early diagnosis of patients with acute myocardial infarction. World J. Emerg. Med. 5, 182–186.

Li, Y.-Q., Zhang, M.-F., Wen, H.-Y., et al., 2013. Comparing the diagnostic values of circulating microRNAs and cardiac troponin T in patients with acute myocardial infarction. Clinic 68, 75–80.

Li, Y., Zhang, Z., Liu, F., et al., 2012. Performance comparison and evaluation of software tools for microRNA deep-sequencing data analysis. Nucleic Acids Res. 40, 4298–4305.

Lin, L., Liu, D., Liang, H., et al., 2015. MiR-1228 promotes breast cancer cell growth and metastasis through targeting SCAI protein. Int. J. Clin. Exp. Pathol. 8, 6646–6655.

Lu, J., Getz, G., Miska, E.A., et al., 2005. MicroRNA expression profiles classify human cancers. Nature 435, 834–838.

Maloney, M.E., 1996. Arsenic in dermatology. Dermatol. Surg. 22, 301–304.

Margis, R., Rieder, C.R., 2011. Identification of blood microRNAs associated to Parkinsońs disease. J. Biotechnol. 152 (3), 96–101.

Mayr, M., Zampetaki, A., Willeit, P., et al., 2013. MicroRNAs within the continuum of postgenomics biomarker discovery. Arterioscler. Thromb. Vasc. Biol. 33, 206–214.

Mendell, J.T., Olson, E.N., 2012. MicroRNAs in stress signaling and human disease. Cell 148, 1172–1187.

Merritt, W.M., Bar-Eli, M., Sood, A.K., 2010. The dicey role of Dicer: implications for RNAi therapy. Canc. Res. 70, 2571–2574.

Mikaelian, I., Scicchitano, M., Mendes, O., et al., 2013. Frontiers in preclinical safety biomarkers: microRNAs and messenger RNAs. Toxicol. Pathol. 41, 18–31.

Minones-Moyano, E., Porta, S., Escaramis, G., et al., 2011. MicroRNA profiling of Parkinson's disease brains identifies early downregulation of miR-34b/c which modulate mitochondrial function. Hum. Mol. Genet. 20, 3067–3078.

Mitchell, P.S., Parkin, R.K., Kroh, E.M., et al., 2008. Circulating microRNAs as stable blood-based markers for cancer detection. Proc. Natl. Acad. Sci. U.S.A. 105, 10513–10518.

Nathwani, R.A., Pais, S., Reynolds, T.B., et al., 2005. Serum alanine aminotransferase in skeletal muscle diseases. Hepatology 41, 380–382.

NCBI, 2013. EST Database.

O'Reilly, S., 2016. MicroRNAs in fibrosis: opportunities and challenges. Arthr. Ther. Res. 18, 11.

Oliveira-Carvalho, V., Carvalho, V.O., Bocchi, E.A., 2013. The emerging role of miR-208a in the heart. DNA Cell Biol. 32, 8–12.

Park, N.J., Zhou, H., Elashoff, D., et al., 2009. Salivary microRNA: discovery, characterization, and clinical utility for oral cancer detection. Clin. Canc. Res. 15, 5473–5477.

Peters, L., Meister, G., 2007. Argonaute proteins: mediators of RNA silencing. Mol. Cell 26, 611–623.

Pourhoseini, S., Seth, R.K., Das, S., et al., 2015. Upregulation of miR21 and repression of Grhl3 by leptin mediates sinusoidal endothelial injury in experimental nonalcoholic steatohepatitis. PLoS One 10, e0116780.

Redell, J.B., Moore, A.N., Ward 3rd, N.H., et al., 2010. Human traumatic brain injury alters plasma microRNA levels. J. Neurotrauma 27, 2147–2156.

Rothstein, L., Jickling, G.C., 2013. Ischemic stroke biomarkers in blood. Biomarkers Med. 7, 37–47.

Saikumar, J., Hoffmann, D., Kim, T.-M., et al., 2012. Expression, circulation, and excretion profile of microRNA-21, -155, and -18a following acute kidney injury. Toxicol. Sci. 129, 256–267.

Saugstad, J.A., 2010. MicroRNAs as effectors of brain function with roles in ischemia and injury, neuroprotection, and neurodegeneration. J. Cerebr. Blood Flow Metabol. 30, 1564–1576.

Schiffl, H., Lang, S.M., 2012. Update on biomarkers of acute kidney injury: moving closer to clinical impact? Mol. Diagn. Ther. 16, 199–207.

Schmittgen, T.D., Lee, E.J., Jiang, J., et al., 2008. Real-time PCR quantification of precursor and mature microRNA. Methods 44, 31–38.

Schratt, G.M., Tuebing, F., Nigh, E.A., et al., 2006. A brain-specific microRNA regulates dendritic spine development. Nature 439, 283–289.

Shaheen, N.J., Hur, C., 2013. Garlic, silver bullets, and surveillance upper endoscopy for Barrett's esophagus. Gastroenterology 273–276.

Sheinerman, K.S., Tsivinsky, V.G., Crawford, F., et al., 2012. Plasma microRNA biomarkers for detection of mild cognitive impairment. Aging 4, 590–605.

Shi, Q., Hong, H., Senior, J., et al., 2010. Biomarkers for drug-induced liver injury. Expet Rev. Gastroenterol. Hepatol. 4, 225–234.

Spector, Y., Fridman, E., Rosenwald, S., et al., 2013. Development and validation of a microRNA-based diagnostic assay for classification of renal cell carcinomas. Mol. Oncol. 7, 732–738.

Starckx, S., Batheja, A., Verheyen, G.R., et al., 2013. Evaluation of miR-122 and other biomarkers in distinct acute liver injury in rats. Toxicol. Pathol. 41, 795–804.

Starkey Lewis, P.J., Dear, J., Platt, V., et al., 2011. Circulating microRNAs as potential markers of human drug-induced liver injury. Hepatology 54, 1767–1776.

Starmans, M.H., Pintilie, M., John, T., et al., 2012. Exploiting the noise: improving biomarkers with ensembles of data analysis methodologies. Genome Med. 4, 84.

Su, L.J., Mahabir, S., Ellison, G.L., et al., 2011. Epigenetic contributions to the relationship between cancer and dietary intake of nutrients, bioactive food components, and environmental toxicants. Front. Genet. 2, 91.

Sun, Q., Chen, S., Zhao, X., et al., 2015. Dysregulated miR-645 affects the proliferation and invasion of head and neck cancer cell. Canc. Cell Int. 15, 87.

Szczesniak, M.W., Deorowicz, S., Gapski, J., et al., 2012. miRNEST database: an integrative approach in microRNA search and annotation. Nucleic Acids Res. 40, D198–D204.

Tan, K.S., Armugam, A., Sepramaniam, S., et al., 2009. Expression profile of MicroRNAs in young stroke patients. PLoS One 4, e7689.

Tijsen, A.J., Creemers, E.E., Moerland, P.D., et al., 2010. MiR423-5p as a circulating biomarker for heart failure. Circ. Res. 106, 1035–1039.

Tijsen, A.J., Pinto, Y.M., Creemers, E.E., 2012. Circulating microRNAs as diagnostic biomarkers for cardiovascular diseases. Am. J. Physiol. Heart Circ. Physiol. 303, H1085–H1095.

Tutarel, O., Dangwal, S., Bretthauer, J., et al., 2013. Circulating miR-423_5p fails as a biomarker for systemic ventricular function in adults after atrial repair for transposition of the great arteries. Int. J. Cardiol. 167, 63–66.

University of Manchester, 2013. What Do the miRNA Names/Identifiers Mean? MiRBase Website.

Usborne, A.L., Smith, A.T., Engle, S.K., et al., 2014. Biomarkers of exocrine pancreatic injury in 2 rat acute pancreatitis models. Toxicol. Pathol. 42, 195–203.

Vermeulen, A., Behlen, L., Reynolds, A., et al., 2005. The contributions of dsRNA structure to Dicer specificity and efficiency. RNA (N.Y.) 11, 674–682.

Vrijens, K., Bollati, V., Nawrot, T.S., 2015. MicroRNAs as potential signatures of environmental exposure or effect: a systematic review. Environ. Health Perspect. 123, 399–411.

Wang, G.-K., Zhu, J.-Q., Zhang, J.-T., et al., 2010. Circulating microRNA: a novel potential biomarker for early diagnosis of acute myocardial infarction in humans. Eur. Heart J. 31, 659–666.

Wang, K., Yuan, Y., Li, H., et al., 2013. The spectrum of circulating RNA: a window into systems toxicology. Toxicol. Sci. 132, 478–492.

Wang, K., Zhang, S., Marzolf, B., et al., 2009. Circulating microRNAs, potential biomarkers for drug-induced liver injury. Proc. Natl. Acad. Sci. U.S.A. 106, 4402–4407.

Wang, W.-X., Rajeev, B.W., Stromberg, A.J., et al., 2008a. The expression of microRNA miR-107 decreases early in Alzheimer's disease and may accelerate disease progression through regulation of beta-site amyloid precursor protein-cleaving enzyme 1. J. Neurosci. 28, 1213–1223.

Wang, Z., Luo, X., Lu, Y., et al., 2008b. miRNAs at the heart of the matter. J. Mol. Med. 86, 771–783.

Ward, J., Kanchagar, C., Veksler-Lublinsky, I., et al., 2014. Circulating microRNA profiles in human patients with acetaminophen hepatotoxicity or ischemic hepatitis. Proc. Natl. Acad. Sci. U.S.A. 111, 12169–12174.

Wijnhoven, B.P.L., Hussey, D.J., Watson, D.I., et al., 2010. MicroRNA profiling of Barrett's oesophagus and oesophageal adenocarcinoma. Br. J. Surg. 97, 853–861.

Williams, G.T., Farzaneh, F., 2012. Are snoRNAs and snoRNA host genes new players in cancer? Nat. Rev. Canc. 84–88.

Winter, J., Jung, S., Keller, S., et al., 2009. Many roads to maturity: microRNA biogenesis pathways and their regulation. Nat. Cell Biol. 11, 228–234.

Wu, A.H., Feng, Y.J., 1998. Biochemical differences between cTnT and cTnI and their significance for diagnosis of acute coronary syndromes. Eur. Heart J. 19 (Suppl. N), N25–N29.

Yamaura, Y., Nakajima, M., Takagi, S., et al., 2012. Plasma microRNA profiles in rat models of hepatocellular injury, cholestasis, and steatosis. PLoS One 7, e30250.

Yang, H., Gu, J., Wang, K.K., et al., 2009. MicroRNA expression signatures in Barrett's esophagus and esophageal adenocarcinoma. Clin. Canc. Res. 15, 5744–5752.

Yauk, C.L., Rowan-Carroll, A., Stead, J.D., et al., 2010. Cross-platform analysis of global microRNA expression technologies. BMC Genom. 11, 330.

Yeom, K.-H., Lee, Y., Han, J., et al., 2006. Characterization of DGCR8/Pasha, the essential cofactor for Drosha in primary miRNA processing. Nucleic Acids Res. 34, 4622–4629.

Yin, K.-J., Deng, Z., Hamblin, M., et al., 2010a. Peroxisome proliferator-activated receptor delta regulation of miR-15a in ischemia-induced cerebral vascular endothelial injury. J. Neurosci. 30, 6398–6408.

Yin, K.-J., Deng, Z., Huang, H., et al., 2010b. miR-497 regulates neuronal death in mouse brain after transient focal cerebral ischemia. Neurobiol. Dis. 38, 17–26.

Zhang, L., Dong, L.-Y., Li, Y.-J., et al., 2012. The microRNA miR-181c controls microglia-mediated neuronal apoptosis by suppressing tumor necrosis factor. J. Neuroinflammation 9, 211.

Zhang, Y., Jia, Y., Zheng, R., et al., 2010. Plasma microRNA-122 as a biomarker for viral-, alcohol-, and chemical-related hepatic diseases. Clin. Chem. 56, 1830–1838.

Zhao, Y., Ransom, J.F., Li, A., et al., 2007. Dysregulation of cardiogenesis, cardiac conduction, and cell cycle in mice lacking miRNA-1-2. Cell 129, 303–317.

Zhou, X., Mao, A., Wang, X., et al., 2013. Urine and serum microRNA-1 as novel biomarkers for myocardial injury in open-heart surgeries with cardiopulmonary bypass. PLoS One 8, e62245.

CHAPTER

60

Citrulline: Pharmacological Perspectives and Role as a Biomarker in Diseases and Toxicities

Shilpa N. Kaore, Navinchandra M. Kaore
Raipur Institute of Medical Sciences, Raipur, India

List of Abbreviations

AA Amino acid
Ab Antibody
ACP Anti-cyclic Citrullinated peptide
ACPA Anticitrullinated protein antibodies
ACR American College of Rheumatology
ACR Acute cellular rejection
ADMA Asymmetric dimethylarginine
Ag Antigen
AKA Anti-keratin antibodies
AKI Acute kidney injury
ALT Alanine aminotransferase
AMCV Antibodies to modified citrullinized vimentin
Anti-CarP Anti-carbamylated proteins
Anti-CCP1 First generation of anti-CCP antibodies
Anti-CCP2 Second generation of anti-CCP antibodies
Anti-CCP3 Third generation of anti-CCP antibodies
APACHE Acute Physiology and Chronic Health Evaluation
APF Antiperinuclear factor
ARDS Acute respiratory distress syndrome
ARF Acute renal failure
Arg Arginine
ASL Argininosuccinate lyase
ASS Argininosuccinate synthase
AST Aspartate aminotransferase
ATP Adenine triphosphate
AUC Area under curve
BPD Bronchopulmonary dysplasia
BUN Blood urea nitrogen
C(max) Concentration maximum
C(min) Concentration minimum
CCP Cyclic citrullinated peptide
cGMP Cyclic guanylyl monophosphate
Cit Citrulline coupled with polyethylene glycol 5000
CPS Carbamoyl phosphate synthetase
CPS1D Carbamoylphosphate synthase 1 deficiency
CRP C-reactive protein
CTLN2- Citrullinemia type II
DBS Dried blood spot
DFP Diisopropylflourophosphate
ED Erectile dysfunction
ELISA Enzyme linked immunoassay
eNOS Endothelial Nitric oxide synthase
FTTDCD Failure to thrive and dyslipidemia caused by Citrin deficiency
GFD Gluten-free diet
GVHD Graft versus host disease
HCit Homocitrulline
HDL-C High density cholesterol
HU Hydroxyurea
HuArgI (Co)-PEG5000 Human recombinant arginase I cobalt [HuArgI (Co)]
IDDM Insulin-dependent diabetes mellitus
IgA Immunoglobin A
IgM Immunoglobin M
IL-10 Interleukin 10
IL-6 Interleukin 6
iNOS Inducible Nitric oxide synthase
IUGR Intra uterine growth retardation
KA Kainic acid
mTOR Mammalian Target of Rapamycin
NAGSD N-acetylglutamate synthase deficiency
NICCD Neonatal intrahepatic cholestasis caused by Citrin deficiency
nNOS Neuronal Nitric oxide synthase
NO Nitric oxide
NOS Nitric oxide synthase
NPV Negative predictive values
OAT Ornithine-δ-aminotransferase
OTC Ornithine transcarbamylase
OTCD Ornithine transcarbamylase deficiency
PAD Peptidylarginine deiminase
PAD2 Peptidylarginine deiminase 2
PCOS Polycystic ovary syndrome
PH Pulmonary hypertension
PI Protease inhibitors
PRISM Pediatric Risk of Mortality
RF Rheumatoid factor
RNS Reactive nitrogen species
ROS Reactive oxygen species
RVF Right ventricular function
SBS Short bowel syndrome
sGC Soluble guanylyl cyclase
SHR Spontaneously hypertensive rats
SIRS Systemic inflammatory response syndrome

Biomarkers in Toxicology, Second Edition
https://doi.org/10.1016/B978-0-12-814655-2.00060-8

SOFA Sequential Organ Failure Assessment score
T2DM Type 2 diabetes mellitus
TB Tuberculosis
TNF Tumor necrosis factor
TPN Total parenteral nutrition
UCD Urea cycle disorders

INTRODUCTION

L-Citrulline is a naturally occurring nonessential amino acid (AA), present in mammals and also in each living organism. It is produced by the body naturally and also found naturally in certain foods such as watermelons, cucumbers, pumpkins, muskmelons, bitter melons, squashes, and gourds. Historically, citrulline's (Cit) name was derived from the Latin word for watermelon *Citrullus vulgaris* (Wada, 1930), because it was first isolated from it (Collins et al., 2007). Watermelon is unusually rich in Cit (Kaore et al., 2013).

Cit, made from ornithine and carbamoyl phosphate, is a component of the urea cycle in the liver (Endo et al., 2004). Cit is synthesized from arginine (Arg) and glutamine in the enterocytes. Cit enters into the kidney, vascular endothelium, macrophages, and other tissues and can be readily converted to Arg and nitric oxide (NO). Arg is then made available for utilization by the peripheral tissues (Fig. 60.1).

All three isoforms of nitric oxide synthase (NOS)—neuronal NOS (nNOS), inducible NOS, and endothelial (eNOS)—convert L-arginine to NO and L-citrulline. The Arg/asymmetric dimethylarginine (ADMA) ratio is one of the determinants of NO production. NO activates soluble guanylylcyclase (sGC) in smooth muscles, leading to an increase in intracellular cyclic guanylyl monophosphate (cGMP) and causing vasodilation. This process is essential for endothelial function, hence disturbed NO production in the human endothelium attributes to endothelial dysfunction. Under ischemic

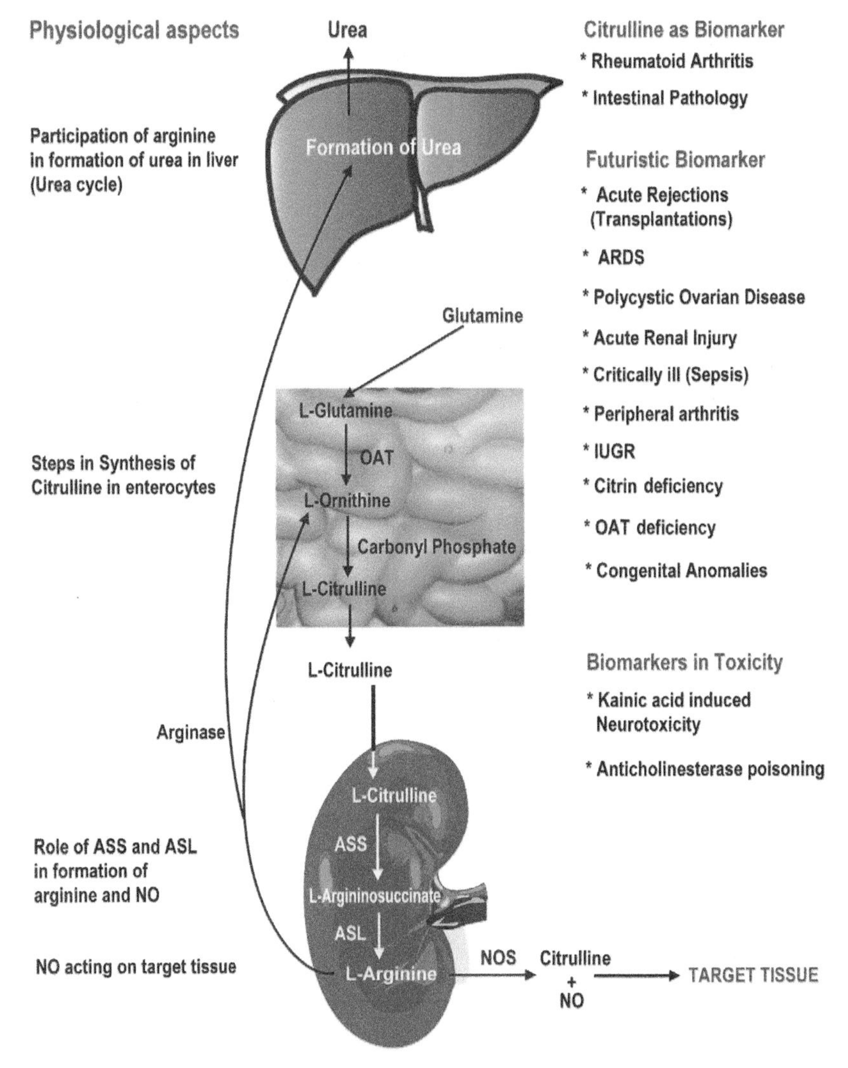

FIGURE 60.1 Schematic representation of synthesis of citrulline and its highlights as biomarker.

conditions, NO is synthesized from nitrite, in acidic conditions, via a nonenzymatic pathway mainly in tissues (Kaore et al., 2013).

Cit is formed exclusively in enterocytes with the liver having only a minimal effect on Cit production. A decrease in the functional mass of enterocytes decreases plasma Cit levels (Oliverius et al., 2010). About 80% of Cit is converted in the kidney to Arg (Luiking et al., 2009; Oliverius et al., 2010), finally converting/recycling Arg to Cit and NO and serving as a potent Arg precursor (Waugh et al., 2001; Flam et al., 2007; Moinard and Cynober, 2007; Luiking et al., 2009) (Fig. 60.1). This is carried out via argininosuccinate synthase (ASS) converting L-citrulline to L-argininosuccinate, which is subsequently converted to Arg by argininosuccinate lyase (ASL) (Fig. 60.1). The importance of the argininosuccinate pathway has been appreciated in the pig for the Cit to Arg conversion in the cerebral vascular endothelium is a constant neuronal source of NO for inducing cerebral vasodilatation (Lee and Yu, 2002). But it is now realized that lack of both, as well as excess NO production, can have very important implications in health and disease. Hence, inhibition of excessive NO production is also a promising future therapeutic target (Kaore et al., 2013) (Box 60.1).

CITRULLINE OR ARGININE SUPPLEMENTATION

To summarize, first, Arg is exposed to more presystemic/systemic degradation than is Cit, making oral delivery very ineffective for Arg. Secondly, Cit is superior to Arg in terms of increased bioavailability. Thirdly, the beneficial effects of Arg do not endure with chronic therapy. Lastly, diseased states such as heart failure, atherosclerosis, diabetic vascular disease, and ischemia-reperfusion injury (as well as Arg supplementation) enhance arginase expression and activity, reducing the effectiveness of Arg therapy. In contrast, Cit inhibits arginase activity. For these obvious reasons, Cit supplementation seems to be a better alternative for supplying Arg than Arg supplementation itself (Kaore et al., 2013).

Cit supplementation holds promise as a therapeutic adjunct in clinical conditions associated with Arg/NO deficiency and/or endothelial dysfunction. Supplementation is likely to benefit clinical conditions such as hypertension, heart failure, atherosclerosis, diabetic vascular disease, and ischemia-reperfusion injury, to name a few. This is likely to be exploited in the future as a therapeutic agent for other clinical conditions, such as sickle cell anemia, hypertension, hyperlipidemia, diabetes, erectile

BOX 60.1

SALIENT FEATURES OF CITRULLINE, MAJOR PHARMACOLOGICAL ACTIONS AND ITS CLINICAL APPLICATIONS

Citrulline

Natural sources
- Watermelon *(Citrullis vulgaris)*, cucumbers, pumpkins, muskmelons, bitter melons, squashes, gourds

Salient features
- A nonessential amino acid
- Donor of arginine & NO
- Conditionally essential amino acid in intestinal pathology and critically ill

Major pharmacological actions
- Antioxidant
- Vasodilatation
- Decreases leukocyte migration
- Restores nitrogen balance & increases muscle protein content as well as muscle protein synthesis
- Improves endothelial dysfunction
- Increases arginosuccinate synthase (ASS) expression in cancer cells
- Preserves anti-inflammatory mediator response (in sepsis)

Clinical applications
- Alzheimer's Disease
- Multi-Infarct Dementia
- Urea Cycle Disorder
- Diabetes
- Sickle Cell Disease
- Hypertension
- Atherosclerosis
- Intestinal Pathology
- Short Bowel Syndrome
- Villous Atrophy Disease
- Sepsis
- T-cell Dysfunction
- Cancer
- Erectile Dysfunction

dysfunction, sepsis (as an immunomodulator), arginase-associated T-cell dysfunction, short bowel syndrome (SBS) (to restore nitrogen balance), cancer chemotherapy, acute respiratory distress syndrome (ARDS), hyperoxic lung damage, urea cycle disorders (UCDs), Alzheimer's disease (AD), intrauterine growth retardation (IUGR) (amelioration), and multi-infarct dementia (MID). In addition, it has a role in athletic training and in parasitic diseases, such as severe *Falciparum* malaria. Many of these aspects, as well as the basis for therapeutic benefits, are discussed in the following sections.

SALIENT ASPECTS OF NITRIC OXIDE

NO is a free gaseous chemical messenger, which is thought to be involved in a number of physiological processes in the mammalian central nervous system, including memory, respiration, skeletal muscle activity, and cardiovascular homeostasis (Milatovic et al., 2005). Facts about NO as studied in brain reveal a half-life of less than a few seconds. The distance of NO diffusion is 300 μm, which is enough to reach neighboring neurons (Gupta, 2004). Estimates of the half-life of NO in blood range from 0.05 to 1.8 ms, and once it enters the circulation it is not only destroyed quickly but also transported to and interchanges with other compartments (Rassaf et al., 2002).

Growing evidence suggests the role of NO in regulation of insulin sensitivity and oxidation of energy substrates, a key role in the immune response, neurological functions, and hemodynamics in animals and humans. NO released from endothelium is an important regulator of vascular tone and an inhibitor of platelet and leukocyte aggregation and adhesion, as well as an inhibitor of cell proliferation (Kaore et al., 2013). NO undoubtedly plays a crucial role in the regulation of vascular disease. However, as discussed, Cit supplementation is more efficacious and reliable for increasing NO availability as compared to Arg, so Cit is a more promising and successful therapeutic agent. Thus, supplementation with Cit is a novel therapeutic approach in conditions of Arg and NO deficiency (Luiking et al., 2010). Thus, the accompanying rise in Cit levels (also NO/NO synthase and reactive nitrogen species [RNS]) can be used as markers in certain toxicities (Gupta, 2004).

This chapter describes various aspects of Cit: pharmacological perspectives, details about its exploration for therapeutic benefits, and its role as a biomarker, both now and in the future. Another area of interest is the exploration of Cit as a predictor of diseases by detection of the antibody (Ab) to citrullinated peptides. This is due to protein citrullination (i.e., posttranslational modification of Arg; Arg–Cit conversion, deaminated Arg) of certain body proteins that produce novel epitopes and give rise to autoantibodies, detection of which provides a novel way to diagnose a wide variety of diseases and serves as a marker for monitoring certain clinical conditions. Plasma Cit concentration has been extensively studied in research, and supportive results have established its concentration as a reliable quantitative biomarker of enterocyte absorptive capacity (Papadia et al., 2014), which is used by clinicians to assess functional enterocyte mass in various chronic and acute small bowel pathologies such as SBS (Oliverius et al., 2010). The plasma Cit values can aid in diagnosis and severity of intestinal failure, and possibly also in monitoring bowel function. Thus, decreased plasma Cit concentration reflects a decrease of functional mass of enterocytes provided renal function is normal (Moinard and Cynober, 2007). In addition to being pioneered for intestinal conditions, now it has been explored as a biomarker for rheumatoid arthritis (RA), transplant rejections, IUGR, peripheral arthritis, ornithine-δ-aminotransferase (OAT) deficiency, and congenital anomalies, as well as a marker of inflammation and ARDS.

BIOCHEMICAL ASPECTS

L-Citrulline has a molecular formula of $C_6H_{13}N_3O_3$ (chemical structure shown in Fig. 60.2) and molar mass of 175.19 g/mol. It is an α-AA. Cit exists in two forms: free Cit, which is a product of the NOS enzyme family, and Cit that results from the posttranslational modification of certain proteins at Arg residues, catalyzed by peptidylarginine deiminases (PADs) (Herrera-Esparza et al., 2013). It is considered to be a nonprotein AA in the context that Cit cannot be incorporated into proteins during the translation process (because Cit does not have its own trinucleotide codon). But it is found in some proteins due to posttranslational modification of these proteins and the proteins thus modified are called citrullinated proteins. Interestingly, there is now evidence to suggest that the posttranslational modifications may be responsible for the initial triggering of autoimmunity and the breaking of tolerance (Utz et al., 2002). Research suggests citrullination to be an inflammation-dependent process because it is present in many inflammatory tissues. Citrullination is found to occur with specific proteins, including filaggrin, vimentin, fibrin, fibrinogen, α-enolase, and

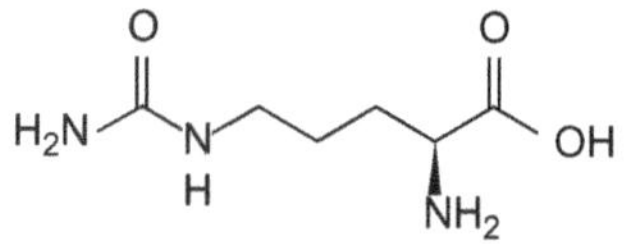

O O
H_2N N OH
H NH_2

FIGURE 60.2 Chemical structure of citrulline.

collagen II, producing novel epitopes that give rise to autoantibodies. Therefore, anticitrullinated protein antibodies (ACPA) are now recognized as specific biomarkers for RA, and assays that use cyclic citrullinated peptide (CCP) as an artificial antigen are showing high specificity and sensitivity (Kaore et al., 2013). Currently, the combination of anti-CCP Abs with rheumatic factor (RF) IgM and RF IgA is recommended either in parallel or for stepwise determination for RA (Conrad et al., 2010), and for erosive arthritis, suggesting that protein citrullination plays a role in the etiology of RA and that PAD inhibitors might be therapeutic (Duskin and Eisenberg, 2010). Its inclusion as a criterion in the 2010 American Rheumatoid Arthritis Classification emphasizes its importance and significance (CDC, 2013). Thus, these Abs could be used as predictors of disease.

Cit is also a posttranslationally modified Arg residue and, surprisingly, is an essential part of the B-cell epitope found in Abs in RA (For more details refer to the later section titled "Anticyclic citrullinated peptide (ACP) antibodies: a novel biomarker of rheumatoid arthritis"). Further exploration may establish Cit as a marker in a variety of disorders in the future.

PHARMACOKINETICS AND PHARMACODYNAMICS OF CITRULLINE

Pharmacodynamics of Citrulline

Citrulline in Sickle Cell Anemia

Cit (0.1 g/kg, p.o. twice daily) in sickle cell disease acts as an Arg precursor in palliative therapy of sickle cell disease, for which it has cleared Phase II trials, decreasing complications and increasing overall well-being of these patients. Arg serves for endogenous nitrovasodilation and vasoprotection (inhibition of adhesion and activation of leukocytes), which forms the basis of vasoprotection by Arg, partly mediated by NO-induced inhibition of endothelial damage and inhibition of adhesion and activation of leukocytes, decreasing the vasoocclusive crisis. With continued treatment with Cit, it causes symptomatic improvement and also maintains nearly normal total leukocyte and neutrophil counts (Waugh et al., 2001). Another study favored L-Arg (100 mg/kg *tid*) for significant and remarkable reduction in total parenteral opioid use by 54%. Lower pain scores at discharge were observed in the treatment arm compared to placebo. Hence Arg therapy represents a novel intervention for painful vasoocclusive episodes; being inexpensive and with narcotic-sparing effects, patients received L-arginine (100 mg/kg *tid*) or placebo for 5 days or until discharge. A significant reduction in total parenteral opioid use by 54% (1.9 ± 2.0 mg/kg *versus* 4.1 ± 4.1 mg/kg, $P = .02$) and lower pain scores at discharge (1.9 ± 2.4 vs. 3.9 ± 2.9, $P = .01$) were observed in the treatment arm compared with the placebo one. There was no significant difference in hospital length of stay (4.1 ± 01.8 vs. 4.8 ± 2.5 days, $P = .34$), although a trend favored the Arg arm, and total opioid use was strongly correlated with the duration of the admission ($r = 0.86$, $P < .0001$). No drug-related adverse events were observed. Arg therapy represents a novel intervention for painful vasoocclusive episodes. A reduction of narcotic use by >50% is remarkable. Arg is a safe and inexpensive intervention with narcotic-sparing effects that may be a beneficial adjunct to standard therapy for sickle cell–related pain in children. A large multicenter trial is warranted to confirm these observations (Morris et al., 2013). Further studies are needed to confirm these observations.

A recent systematic review showed that Arg (oral, topical or nasal), with or without hydroxyurea (HU), increased NO levels and improved patient clinical condition in sickle cell disease. The parameters shown to be improved were reduction in pain and opioid use [remarkably up to 50% reduction in hospitalizations and emergency visits (Morris et al., 2013)], improvement in heart rate indices, and healing of leg ulcers (topical use of butyrate Arg cured 30% of the patients at 12 weeks and 78% after 3 months). Because it is known that oral HU used for this disorder acts by multiple direct effects such as increasing the synthesis of fetal hemoglobin, reducing infraerythrocytes, and HbS polymerization in deoxygenation conditions, the number of neutrophils is decreased along with erythroid hydration and the expression of adhesion erythrocytes, and the synthesis of NO and its bioavailability is increased. However, being a chemotherapeutic agent HU is both cytotoxic and genotoxic, hence further strategies need to be explored to improve quality of life in these patients (Eleuterio et al., 2016).

Citrulline and Cardiovascular Effects

NO, synthesized by the endothelial cells using Arg as a substrate, contributes to regulation of vascular tone in humans that has been demonstrated by showing that inhibition of its synthesis results in significant vasoconstriction and impaired endothelial regulation of vascular tone in patients with systemic arterial hypertension (Cardillo and Panza, 1998). Thus, increasing Arg availability either by Arg therapy or arginase inhibition may provide therapeutic benefits in future patients with hypertension. NO activates sGC in smooth muscles, leading to an increase in intracellular cGMP, causing vasodilation. This process is essential for endothelial function, and disturbed NO production in the human endothelium is attributed to endothelial dysfunction.

Citrulline in Hypertension

It has been observed that hypertensive patients had a lower Arg–Cit ratio than normotensive patients associated with NO deficiency, oxidative stress, increased ROS, and a disturbed constrictor–dilator balance in the kidneys. Melatonin was found to exert an antihypertensive effect in young spontaneously hypertensive rats (SHRs) because of restoration of the NO pathway by reduction of plasma ADMA, preservation of renal Arg availability, and attenuation of oxidative stress (Kaore et al., 2013).

Studies with both Cit and nitrates consumption significantly decreased diastolic blood pressure, both decreased vascular compliance but had no effect on either pulse pressure or rate-pressure product, whereas Cit also decreased systolic pressure. Cit significantly decreased R–R interval 9% at rest and increased heart rate in addition to significantly decreasing pulse transit duration by 6%. QRS duration decreased by 5% with reduction in R–R interval (this seems to be a direct result of the modest tachycardia caused by Cit). HR augmentation in response to exercise was approximately reduced to half in Cit-treated subjects, without the desensitization observed for glyceryltrinitrate (Alsop and Hauton, 2016). Hence, Cit shows benefit in human subjects for the improvement of oxygen delivery in heart failure.

Another study for effective therapies for infants with forms of pulmonary hypertension that develop or persist beyond the first week of life needs to be worked out. Cit improves NO signaling and ameliorates pulmonary hypertension in newborn animal models. Hence, strategies that increase the supply and transport of Cit merit pursuit as novel approaches to managing infants with chronic, progressive pulmonary hypertension (Cit has proved better than Arg supplementation). More clinical trials are needed for safety and efficacy of Cit therapy in human infants with and at risk of developing chronic, progressive forms of pulmonary hypertension deserve consideration as a novel and potentially cost-effective approach to treat this life-threatening condition in term and preterm infants (Fike et al., 2014).

Therapeutic strategies that increase the transport of L-citrulline may represent a novel approach to the management of pulmonary hypertension in infants.

Another study evaluated Cit in preventing deterioration of pulmonary hypertension and explored combination therapy of tadalafil + Cit versus tadalafil + Arg in 4-week-old rats. Echocardiographical examination suggested that the ratios of RV to LV weight in a tadalafil group and a tadalafil + Cit group were significantly lower than other groups. The estimated pulmonary artery pressure in a Cit and tadalafil + Cit group seem to be lower than those in other groups because of Cit-induced vasodilation that reduces right ventricular function with improved survival rate attributed to tadalafil + Cit group due to low RV to LV ratio and decreased pulmonary artery pressure. The survival rate of tadalafil + Cit was higher than other groups with lower ratio of RV to LV weight (Ishikura et al., 2015).

Preliminary results from a case report clearly indicates that L-citrulline supplemetatation might be a potential therapy for chronic PH in infants with BPD. In this case report, for the first time a premature newborn (study in newborn animals was supporing) was given oral Cit along with sildenafil, hydrocortisone, inhaled nitric oxide. Oral Cit, from beginning of the sixth month of hospitalization, single dose of 150 mg/kg/day was introduced and continued for 70 days; patient was weaned from mechanical ventilation and never intubated again until discharge (Lauterbach et al., 2018).

A study explored and concluded that both NOS-dependent and -independent approaches in the prehypertensive stage toward augmentation of NO can prevent the development of hypertension in young SHRs. The study was done in male SHR and normotensive control rats where Cit was replaced with sodium nitrate. Rats were sacrificed at 12 weeks of age. It is noteworthy that Cit therapy also reduced levels of Arg and ADMA (endogenous inhibitor of NOS) and increased the ARG–ADMA ratio in SHR kidneys, whereas nitrate treatment reduced plasma levels of ARG and ADMA concurrently in SHRs (Chien et al., 2014). This further proves its beneficial effect in hypertension at the molecular level operating in this condition. Hence, Cit is a safer way of delivering Arg for endothelial and immune cells as well as preventing excessive uncontrolled NO production (Cynober et al., 2010) that exerts deleterious effects of its own and is an important coadjuvant in the treatment of stable systolic heart failure patients.

Citrulline in Diabetes

Cit is shown to benefit the underlying endothelial dysfunction (Hecker et al., 1990) of diabetes, in addition to producing a reduction in glucose levels. Watermelon juice exerts beneficial effects by increasing Arg availability, improving glycemic control and vascular dysfunction in type 2 diabetes mellitus through intake of Cit from watermelon (Wu et al., 2007). Cit (50 mg/kg/day, p.o.) and arginase inhibitors are found to be of benefit in endothelial dysfunction and improve impaired vasodilation of coronary arteries in diabetes (Romero et al., 2008).

Thus, studies suggest Cit may be of benefit in diabetes by improving glycemic control and in diabetes and metabolic syndrome by ameliorating underlying endothelial dysfunction. Large-scale clinical trials and mechanistic studies are required to evaluate the therapeutic strategies.

Citrulline as Immunomodulator

Cit production is significantly reduced in sepsis because of diminished de novo conversion of Arg and NO, reducing Cit and Arg availability in septic patients and transgenic mice, in part due to ADMA release. In sepsis, patients are found to have significantly decreased levels of Cit (by 56%), glutamate (by 48%), and Arg (by 47%) in comparison to healthy control subjects. A study in rats revealed that Cit supplementation reduced cell membrane lipid peroxidation and enhanced SOD activities in septic rats. It also decreased the release of TNF-α, IL-6, and IL-1β in early stages of sepsis (Cai et al., 2016). Thus, Cit may be a safe means of immunomodulation that preserves the antiinflammatory mediator response.

The concept of immunonutrition with formulation containing medium dose Arg and high dose Arg (20 g/d) is emerging, which significantly lessens the overall complications, and also improves local wound complications as well as reduces the length of hospital stay in postoperative cancer patients (De Luis et al., 2009; Rodera et al., 2012). This has been supported by another study (containing Arg, zinc, and antioxidants within a high-calorie, high-protein formula) that showed improvement in healing of pressure ulcers (PU) in malnourished patients resulting in greater reduction in PU area at 8 weeks (Cereda et al., 2015).

Citrulline Therapy May Decrease Severity of Sepsis

Sepsis is a major health problem due to its high morbidity and mortality rate and may be considered to be an Arg deficiency state (Luiking et al., 2004). Arg availability is increased in sepsis mainly by protein breakdown, but Arg catabolism is also increased due to its enhanced utilization for NO synthesis and higher arginase activity. Therefore, Arg availability is reduced in sepsis because of lower plasma Arg concentrations and a slower Arg production rate (Luiking et al., 2009). Because of this, Cit production is very low, but because data for Cit in intensive care units are not available, large clinical trials are warranted before any reliable recommendations can be made for Cit-enriched diets in this situation. At least reinforcement can be suggested for Cit supplementation. Its utility as a marker in sepsis or critically ill patients is doubtful because of associated vascular abnormalities, acute renal failure (ARF) and/or multiple organ failure, or comorbid conditions (Cynober, 2013). Further studies are warranted to either prove or disprove the use of Cit therapy as an Arg precursor in this condition.

Citrulline in Arginase-Associated T-Cell Dysfunction

It is interesting to note that T lymphocytes depend on Arg for their proliferation, zeta-chain peptide formation, T-cell receptor complex expression, and development of memory. This makes it obvious that T-cell abnormalities such as decreased proliferation and loss of zeta chain are observed in cancer and after trauma, and they may provide new insights into T-cell dysfunction (Kaore et al., 2013). Cit may serve as a substitute for arginase-associated T-cell dysfunction, the basis for which is the molecular capability of T-cells to increase Cit membrane transport and upregulate arginosuccinate synthase (ASS) expression and thus convert Cit to Arg in *in vitro* studies, escaping the ill effects of Arg depletion (Bansal et al., 2004). Cit supplementation can preserve T-cell proliferation and prevent the loss of CD3 zeta chain under conditions of low Arg. New drug targets for improving T-cell associated dysfunction may be explored in the future, and further research may open novel mechanistic insights in this regard.

Citrulline: Role in Dementia of Alzheimer's Disease and in Multi-Infarct Dementia

Cit levels in the cerebrospinal fluid were significantly higher in MID in comparison to a healthy control group (Mochizuki et al., 1996). Further, it was found that there is elevated NOS activity in microvessels of the brain in AD, raising vascular NO production, exerting neurotoxicity, and being responsible for susceptibility to neuronal injury and cell death in AD (Kaore et al., 2013). Some studies contradict this notion and suggest decreased NO production to play a role in neurodegenerative disorders.

Arg fortification with a lysine (Lys)-restricted diet is one of the emerging strategies for a rare autosomal recessive disorder, pyridoxine-dependent epilepsy (PDE), which is resistant to conventional antiepileptic drugs but responsive to pharmacological dosages of pyridoxine (van Karnebeek and Jaggumantri, 2015). Further research may prove or disprove the use of Cit modulators in brain lesions or in PDE.

Citrulline: Can It Improve Aerobic Function/Energy Production?

L-Citrulline malate can develop beneficial effects on the elimination of ammonia in the course of recovery from exhaustive muscular exercise (Sureda and Pons, 2012). Cit is also shown to be a more suitable substrate than Arg to restore NO production in microcirculation during endotoxemia (Wijnands et al., 2012). Cit is shown to enhance protein expression of the myofibrillar constituents (Faure et al., 2013), and protein synthesis occurs through a mechanism of action mediated by the mTOR signaling pathway (Bahri et al., 2013).

Citrulline malate (CM) ingestion significantly reduced fatigue sensation, improved aerobic energy production in muscles, and improved exercise tolerance (Sureda and Pons, 2012), thus greatly enhancing aerobic performance and prolongation of onset of muscular fatigue (Kaore et al., 2013). CM also limits muscle fatigue or skeletal muscle dysfunction induced by bacterial endotoxins (Kaore et al., 2013). Recent studies show ergogenic effects on Cit supplementation with improvement of muscular contraction efficiency and of athletic performance in mice (Giannesini et al., 2011; Sureda and Pons, 2012). Thus, Cit may be used as an agent to increase exercise capacity for various reasons. However, a recent study found conflicting results where CM supplementation (single 6 g dose preworkout) does not improve the muscle recovery process following a high-intensity resistance exercise session in untrained young adult men (Da Silva et al., 2017).

Citrulline for Urea Cycle Disorders

Cit may be used to treat the inborn errors of metabolism of the urea cycle responsible for removal of ammonia that, fortunately, are quite rare. The first step in the urea cycle is the reaction of ammonia and bicarbonate using ATP, catalyzed by the enzyme carbamoyl phosphate synthetase (CPS) to produce carbamoyl phosphate. The second step is the reaction of carbamoyl phosphate and L-ornithine to produce L-Cit, catalyzed by the enzyme ornithine transcarbamylase. It is interesting to note that deficiencies of either of these two enzymes lead to low serum levels of L-Cit. Treatment is with oral L-Cit. In long-term management, dietary treatment must be individualized (Kaore et al., 2013). Recently, a breakthrough in these disorders has emerged in the form of FDA approval of Cit for certain UCDs.

Citrulline for Acceleration of Wound Healing

NO is an integral part of the inflammatory phase, functioning as a regulatory mechanism to mediate epithelialization, angiogenesis, and collagen deposition crucial to the proliferative phase. NO-induced vasodilatation acted as a host-protective agent by killing pathogens in an in vitro study (Norris et al., 1995) and increasing blood flow to wounds (Stechmiller et al., 2005). This notion is supported by a clinical study in graft patients suggesting that Arg metabolism is involved in tissue repair, including angiogenesis, epithelialization, and collagen formation as increased levels of Cit and other metabolites were found in wound fluid versus plasma (Debats et al., 2009).

Citrulline in Intestinal Pathology

Studies suggest that Cit has a strong potential in total parenteral nutrition after massive intestinal resection (SBS), causing an increase in the Arg pool and acting to restore nitrogen balance. Long-term supplementation of Arg intraduodenally in rats failed to show beneficial results in intestinal ischemia and reperfusion injury (Lee et al., 2012). Cit pretreatment improves integrity of the gut barrier and helps to preserve ileum mucosa in experimental studies, thus reducing bacterial translocation. Similar effects are seen with Arg as well. This understanding of molecular mechanisms may help in developing new strategies that could be extended to malnourished patients with compromised intestinal functions.

Citrulline: Effects of Amino Acids on Hair Strength

Human studies report considerable amounts of Arg deposition on or in hair fibers from coloring agents, whereas a decreased amount of Arg and Cit was found in damaged hairs in users of "relaxers." Decreased Cit has been associated with inflammation. Study reports indicate that when coloring agents were partially replaced with Arg, it decreased the oxidative change in tensile strength of the hair by preventing the undesirable attack by hydrogen peroxide on hair proteins and hair surface lipids. Therefore, prospective studies need to be undertaken to understand whether or how "relaxers" induce inflammation and whether Arg or Cit can substantially reduce hair damage and fragility (Kaore et al., 2013).

Citrulline's Effect on Protein Synthesis

Cit is found to increase protein synthesis during refeeding in rodents with SBS, aging, and malnutrition, and it improves nitrogen balance in healthy humans (Thibault et al., 2011). Cit is a conditionally essential AA in stress, or when intestinal function is compromised. This notion is proved in SBS in rats, where Cit is able to restore nitrogen balance, increase mucosal protein content in the ileum, increase Arg levels, and increase muscle protein content, as well as muscle protein synthesis (+90%) in elderly malnourished rats. Moreover, it also has potential as a supplement for total parenteral nutrition in SBS patients because it additionally prevents muscle atrophy, not seen with Arg supplementation (Osowska et al., 2008; Kaore et al., 2013). Thus, Cit plays a crucial role in maintaining protein homeostasis and improving muscle mass related to malnutrition. Further understanding of molecular mechanisms may help in developing new strategies for malnourished patients with compromised intestinal functions (Box 60.2).

Citrulline as a Biomarker in Diseases

Cit is not involved in protein synthesis or nutrition products, but it may serve as a biomarker in various diseases such as RA, intestinal dysfunction, or sepsis. It is now an established biomarker of enterocyte functional metabolic mass in adults and children, which can be used in assessing the remnant length of small bowel in intestinal diseases such as short bowel, extensive enteropathies, intestinal toxicity of chemotherapy, and radiotherapy. Normal plasma Cit levels range

BOX 60.2

THERAPEUTIC APPLICATIONS OF CITRULLINE

Citrulline supplementation exploited for benefits in a variety of conditions

Cardiovascular conditions
- Hypertension
- Right ventricular dysfunction
- Hyperlipidemia

Hematological disorders
- Sickle cell anemia

Metabolic disorders
- Diabetes
- Urea cycle disorders

Neurodegenerative conditions
- Alzheimer's disease
- Multi-infarct dementia

Parasitic infections
- Malaria
- *Trypanosoma cruzi*
- *Toxoplasma gondii*

Others
- Wound healing
- As an antioxidant
- Erectile dysfunction
- Sepsis (as an immunomodulator)
- Arginase associated T-cell dysfunction
- Cancer chemotherapy (in certain tumors)

between 30 and 50 μmol/L independent of nutritional status. Levels less than 10 μmol/L may indicate an objective threshold for parenteral nutrition in case of intestinal failure that allows monitoring of intestinal function, except in cases of significant renal failure (Crenn et al., 2011). This reveals that decreased plasma Cit concentration reflects a decrease of functional mass of enterocytes provided renal function is normal (Kaore et al., 2013) (Box 60.3).

Citrulline as a Biomarker in Intestinal Disease

Serum Cit is emerging as an innovative biomarker candidate for assessment of intestinal function. It has been further assessed as a marker of intestinal failure and for monitoring of bowel function. Serial plasma Cit assay helps to monitor residual small bowel adaptation in children. Studies show serum Cit to be significantly lower in SBS patients, and thus it could be accurately taken as a simple, noninvasive biomarker for assessing the severity of intestinal failure. It may prove to be a candidate marker for the gut-trophic effects of bowel rehabilitation therapies. Studies show postabsorptive plasma Cit concentration to be a marker of functional absorptive bowel length that also allows distinction of transient to permanent intestinal failure in SBS patients, and also to be a marker of reduced enterocyte mass in patients with villous atrophy diseases (Kaore et al., 2013). Decreased Cit levels also correlate to the intensity of acute mesenteric ischemia and duration of intestinal damage, for which Cit has a potential future role (Cakmaz et al., 2013).

Plasma Cit is a reliable biomarker of enterocyte functional mass in HIV patients. It can discriminate between protease inhibitor (PI) toxic diarrhea and infectious enteropathy and quantify the functional consequences, which makes it an objective indicator of the need for parenteral nutrition. In HIV enteropathy, Cit concentration <10 μmol/L was associated with a clinical indication for parenteral nutrition. Observations show patients with PI-induced chronic toxic diarrhea had Cit levels in the range of 22–30 μmol/L, whereas in patients with HIV enteropathy Cit levels were <10 μmol/L. This helps to discriminate between PI-induced toxic diarrhea and infectious enteropathy. Further it was observed that 22 μmol/L was a cutoff point with 76% specificity, 95% sensitivity, and 93% negative predictive values (NPV). In addition, Cit levels <10 μmol/L could differentiate patients who needed parenteral nutrition from treatable patients without the need for parenteral nutrition, with 75% specificity, 87% sensitivity, and 87% NPV. Moreover, Cit levels normalized within 2–12 weeks with successful nutritional therapies and treatment for infectious diarrhea (Crenn et al., 2009).

In mice with schistosomiasis, Cit concentrations were reduced and found to be directly correlated with morphometry (height and area) of villi, supporting its role as a biomarker of intestinal function (Siqueira et al., 2010). Interestingly, schistosoma bind to Arg from surrounding tissue and thus escape the immune attack that requires Arg as an NO precursor (Haas et al., 2002).

BOX 60.3

STATUS OF CITRULLINE AS A BIOMARKER

Conditions detected by antibodies to citrullinated peptides as a biomarker	
1. Rheumatoid arthritis	ACP antibodies detection
2. Sjögren's syndrome	ACP antibodies detection
3. Brucellosis presenting with peripheral arthritis	ACP antibodies detection
Clinical conditions detected by citrulline as a biomarker	
1. Intestinal pathology	
• Short bowel syndrome	Low citrulline levels
• Celiac disease	Low citrulline levels $<10\ \mu mol/L$
• Total parenteral nutrition (indicator) in HIV patients on protease inhibitor	Low citrulline levels
2. Sepsis	Low citrulline levels $<13\ \mu mol/L$
3. Acute rejections (transplantations)	Low citrulline levels
4. Critically ill and correlated with severity of inflammation	Low citrulline levels $\approx 9\ \mu mol/L$
5. ARDS (acute respiratory distress syndrome)	Low citrulline levels
6. Polycystic ovary syndrome	Steady increase in citrulline levels preceding injury by about 36 hours (and of creatinine too)
7. Acute kidney injury/ failure	
8. IUGR	Low citrulline levels
9. Citrin deficiency	Low citrulline levels
10. Perinatal conditions	
• IUGR	Low citrulline levels
• Congenital anomalies (orofacial cleft)	Low citrulline levels ($<8\ \mu mol/L$),
• Citrin deficiency	NICCD – Increased citrulline levels
• OAT deficiency	FICCD – Increased citrulline levels
	CTLN2 – Increased citrulline levels
	Low citrulline levels Increased proline/ citrulline ratio
Toxicities for which citrulline acts as a biomarker *(as an indicator of nitrosoxidative stress)*	
1. Anticholinesterase poisoning	Increased citrulline levels
2. Kainic acid induced neurotoxicity	Increased citrulline levels

Thus, plasma Cit concentration has been extensively used in research studies with supportive results that establish it as a reliable quantitative biomarker of enterocyte absorptive capacity (Papadia et al., 2014) and also for monitoring of certain intestinal diseases.

Citrulline as a Biomarker in Celiac Disease

Plasma Cit levels are found to be lower in celiac disease, reflecting small bowel involvement. Cit levels increased rapidly after a short period of gluten-free diet (GFD), suggesting Cit to be a sensitive marker of the positive effect of GFD on intestinal repair (Ioannou et al., 2011). Plasma Cit levels were lower in untreated disease and after 1 month of GFD, the levels were significantly higher, and after 3 months of GFD, levels were similar to those of healthy controls. Hence, whole blood Cit concentration may be considered to be a simple biochemical marker for diagnosis and monitoring of enterocyte loss in celiac disease. In Canada self-administered home blood tests or blood-testing kits are available, and patients with positive results should have small intestine endoscopy to confirm the diagnosis before starting a GFD (Rashid et al., 2009).

Citrulline: Is It a Reliable Biomarker in Critically Ill Patients?

Questions have been raised on the acceptability of serum Cit as a biomarker in acute intestinal failure in critically ill patients. These questions need further evaluation because this can be a tricky situation to interpret because of sepsis, vascular abnormalities, and multiple organ failure (Cynober, 2013). On one hand, small bowel ischemia causes acute reduction of enterocyte mass and loss of barrier function. On the other hand, systemic inflammatory response syndrome and sepsis may be linked to an acute dysfunction of enterocytes without enterocyte reduction. As Cit is mainly synthesized by small bowel enterocytes, acute reduction of the enterocyte mass correlates with low plasma Cit concentration. This is exaggerated in shock or critical illness, and thus its utility in this area needs to be evaluated (Piton et al., 2011).

The two studies cited in the preceding paragraph are in accordance regarding a doubtful conclusion that Cit cannot be agreed on as a marker in the critically ill patient, but another recent study refutes this. In this study of critically ill patients, correlation between Cit levels and various parameters, such as Sequential Organ Failure Assessment (SOFA) score, Acute Physiology and Chronic Health Evaluation II (APACHE II) score, survival, inflammation (C-reactive protein [CRP]), inotrope use, serum levels of prealbumin and albumin, and renal failure in the critically ill patient, was analyzed. Findings revealed that only patients with intestinal dysfunction had significantly low plasma Cit levels <15 μmol/L, whereas there was no correlation between serum levels of CRP, albumin, or prealbumin; renal failure; inotrope use; SOFA score; or APACHE II score and plasma Cit levels (Noordally et al., 2012).

It is of immense importance to note that while both plasma Cit concentrations and glucose absorption were reduced in critical illness, fasting plasma Cit concentrations were not predictive of subsequent glucose absorption. Data suggest that fasting Cit concentration does not appear to be a marker of small intestinal absorptive function in patients who are critically ill (Poole et al., 2015). Hence, because of its metabolism and particular kinetics in the critically ill, plasma Cit concentration might not be as accurate as in stable patients for determining small bowel function. In the recent study, Poole and colleagues found that plasma Cit concentration and glucose absorption were reduced in 20 critically ill patients compared with 15 controls. However, the authors found no correlation between these two variables. Further studies should take into account the type of intestinal failure considered, the particular metabolism of Cit, the time of plasma Cit measurement, and the range of citrullinemia (Piton et al., 2015). Further studies are required to establish Cit as a marker in applicable cases.

Citrulline as a Biomarker to Identify Patients at High Risk of Developing a Catheter-Related Bloodstream Infection

Observations suggest that serum Cit concentration may serve as a useful biomarker to identify patients with intestinal failure who are at high risk of developing a catheter-related bloodstream infection (CRBSI) (Hull et al., 2011). All patients required parenteral nutrition support and had a venous catheter in situ >30 days and were comparable in terms of variables. The patients who developed CRBSIs had lower Cit concentrations (6.7 ± 4.6 μmol/L) than those who did not (11.3 ± 6.4 μmol/L). Cit serves the purpose of being used as a biomarker in this area, and lower Cit concentrations and longer central venous catheter placement are independent factors associated with CRBSI.

Citrulline as a Biomarker for Gastrointestinal Tolerance to Enteral Feeding

Another novel way of using Cit as a biomarker is measuring its urinary levels, which could serve as a noninvasive index of GI tolerance to enteral feeding and nutritional intake in neonates. Studies did not support its role as a biomarker in this condition. This also supports the notion that in neonatal gut, Cit is converted to Arg in situ rather than exported toward the kidneys as observed in adults (Bourdon et al., 2012). Researchers will have to further extrapolate this to measure urinary levels that may help in diagnosing diseases.

Anticyclic Citrullinated Peptide Antibodies: A Novel Biomarker of Rheumatoid Arthritis

Cit may be also considered as a nonprotein AA in that it cannot be incorporated into proteins during the translational process (Cit does not have its own trinucleotide codon), but it is found in some proteins because of posttranslational modification of these proteins. The proteins thus modified are called *citrullinated proteins*. Of interest is recent evidence that suggests the posttranslational modifications to be responsible for the initial triggering of autoimmunity and the breaking of tolerance. Citrullination appears to be an inflammation-dependent process because it is present in many inflammatory tissues (Kaore et al., 2013).

Cit is also a posttranslationally modified Arg residue and, surprisingly, an essential part of the B-cell epitope found in Abs in RA. Citrullination in RA is carried out by PAD, an enzyme that appears to be hormonally controlled. A genetic variation in PAD, PAD4, increases susceptibility to RA due to increased enzymatic activity. Further, the posttranslational protein modification unfolds the protein with a loss of positive charge in Arg residues that alters the antigenicity of self-proteins, leading to autoimmunity and chronic inflammation.

It is now understood that autoantibodies in RA containing the unusual AA Cit are specifically found in serum of RA patients. Antibodies against citrullinated proteins/peptides (filaggrin), i.e., detection of anticyclic citrullinated peptide (ACP) Abs, are highly specific biomarkers in RA. Other biomarkers for RA are RF, antiperinuclear factor, and anti-keratin antibodies (AKA). Moreover, the frequency of false positive results is lower with ACP Ab detection than with RF, so this modern biomarker of diagnostic value needs further evaluation, for early disease detection and for prevention of joint damage and deformity. It also has similar diagnostic value in patients with established or long-duration disease and is a reliable marker of severe erosive disease and in RF seronegative cases. Another related biomarker equally sensitive to ACP Abs that needs to be analyzed further is antibody to modified citrullinized vimentin, which is significantly correlated with early detection of RA and subclinical atherosclerosis in RA (Kaore et al., 2013). Recently, ACP or ACPA have been found prevalent within the atherosclerotic plaque, which may explain accelerated atherosclerosis in RA (Sokolove et al., 2013) and thus may be associated with increased cardiovascular risks (Cambridge et al., 2013). Results of a recent study also indicated that insulin resistance and anti-CCP may be helpful in the early detection of subclinical atherosclerosis in RA patients (Wahab et al., 2016).

These Cit-containing Abs are detected early, are specific, and are detected well before the appearance of other manifestations of the disease, which may also indicate its role in the autoimmune response and pathogenesis. Citrullinated Ags are expressed in inflamed joints and can be detected early before the manifestation of the disease. ACP-ELISA assays suggest diagnostic sensitivities between 69.6% and 77.5% and specificities between 87.8% and 96.4%. To conclude, an estimation of ACP Abs is highly specific but not absolutely specific in RA, can be exploited as a biomarker for RA, and can be detected early in the course of disease. This may improve the prognostic value in RA patients and new insights in etiology and pathogenesis of the disease may be revealed (Kaore et al., 2013).

ACP Abs provide a better prognostic significance than RF, but whether this will replace RF wholly still remains to be answered. A recent study has raised doubts, because it failed to improve the performance on the American College of Rheumatology (ACR) 1987 criteria in diagnosing early RA in combination with detection of ACP Ab. Future studies may provide an answer to this, or the strategy may be to use RF in combination with ACP Abs for definitive diagnosis of RA. At any rate, the concurrent detection of ACP Ab and RF will surely increase the chances of differentiating RA from other diseases. Further data collection may answer the question of whether the ACP titers might prove to be predictors of efficacy of antitumor necrosis factor (anti-TNF) therapy (Kaore et al., 2013) (Box 60.3).

Good biomarkers serve as an important aspect to guide decisions in the clinical management of RA. A recent study in RA patients documented Abs directed against carbamylated proteins that may predict joint damage and may serve as surrogate prognostic markers. Serum levels of anti-CarP (anti-carbamylated proteins) Abs were found to be much higher in RA patients with joint erosions/deformities than in those without any joint damage, suggesting anti-CarP Abs may have good prognostic value in RA patients with erosions. However, there was no significant correlation between titers of anti-CarP Abs and presence or absence of the rheumatoid factor (Kumar et al., 2017).

First generation of anti-CCP antibodies (anti-CCP1) analysis revealed a higher specificity for RA in comparison to the RF analysis. By the end of 2002, second generation anti-CCP Ab tests were developed with different cyclic peptides and improved performance characteristics, showing an even better specificity for RA. Following anti-CCP2, a third generation anti-CCP test (anti-CCP3) has been developed to increase sensitivity for the detection of patients with RA. Overall, it was observed that both sensitivity and specificity of the anti-CCP tests were significantly higher than those of the RF test (Vos et al., 2017).

Several scholars compared the diagnostic performance of anti-CCP2 and anti-CCP3, but conflicting evidence emerges from these studies. Several studies concluded that the anti-CCP3 test has no apparent diagnostic advantage compared with the anti-CCP2 test, whereas other studies showed a higher sensitivity with anti-CCP3 test. However, it has also been speculated that the reported higher sensitivity of CCP3 may only be found in cohorts with early RA, whereas the sensitivity may be similar in groups with established disease. It was further observed that discrimination between RA and non-RA patients was better using CCP3. The most pronounced difference between CCP2 and CCP3, however, was found in RF-negative patients with a disease duration of $\leq$5 years. In this cohort ($n = 31$), the sensitivity of CCP3 was 51.6% compared to 38.7% for CCP2 (Vos et al., 2017).

Logistic regressions were used to investigate whether the usage of both tests (anti-CCP2 and anti-CCP3) gives a better prediction of RA. At the manufacturer's cutoffs sensitivity and specificity were 79.4% and 61.0% for CCP3 and 80.9% and 69.5% for CCP2. No significant differences could be observed regarding the areas under the curve (AUC) of both ROC curves. The optimal cutoff point for CCP2 was 10.5 U/ml (sensitivity 75.0% and specificity 80.0%) and 5.6 U/ml for CCP3 (sensitivity 86.9% and specificity 61.0%). Binary logistic regressions indicated that the likelihood of having RA is significantly higher when testing positive on both CCP2 and CCP3 compared to CCP2 or CCP3 alone. In our cohort, comparable performance was found between the two CCP assays. Positive results for both CCP2 and CCP3 led to the most specific identification of RA patients. In patients with joint complaints suspected of having RA and with a weakly positive CCP2 ($\geq$7 and $\leq$16 U/mL), CCP3 testing could be of additional value for diagnosing RA (Vos et al., 2017).

Although the majority of studies comparing CCP2 and CCP3 detected no advantage of using CCP3 over CCP2, a few studies showed a higher sensitivity of the anti-CCP3 peptide assay compared to anti-CCP2.

Recently, the CCP3 test has been developed for analysis on an automated analyzer, the BIO-FLASH instrument, and this assay may have different properties as compared to ELISA-based assays (Vos et al., 2017).

Future studies may provide an answer to this, or the strategy may be to use RF in combination with ACP Abs for a definitive diagnosis of RA. At any rate, the concurrent detection of ACP Ab and RF will surely increase the chances of differentiating RA from other diseases. Further data collection may answer the question of whether the ACP titers might prove to be predictors of efficacy of anti-TNF therapy (Kaore et al., 2013) (Box 60.3).

Citrulline as a Biomarker in Periodontitis for Presymptomatic Rheumatoid Arthritis?

An interesting study correlated the finding that periodontitis may be a risk factor for RA because the autoantibody response in periodontitis can lead to epitope spreading to citrullinated epitopes, and this may evolve into presymptomatic RA (de Pablo et al., 2014).

Citrulline as a Biomarker in Rheumatoid Arthritis in Patients With Active Tuberculosis

A study validated the utility of anticyclic citrullinated peptide Abs, i.e., second-generation anti-CCP2, as a diagnostic marker for RA in patients with active tuberculosis (TB). The results suggested that the positive rate of anti-CCP2 in patients with newly diagnosed RA was 82.1% (87 of 106 cases), whereas the positive rate in healthy control subjects was 0.4% (one in 237 individuals). As far as the mean level of anti-CCP2 among the RA group, it was 159.3 U/mL versus 15.4 U/mL in the TB group and 0.7 U/mL in the healthy control groups. This supports anti-CCP2 detection being a reliable serological tool for identifying early RA in patients with active TB (Mori et al., 2009). Another recent study supports high sensitivity of anti-CCP and high specificity of both anti-CCP and anti-MCV Abs for RA, even in a population with high incidence of TB. This study concluded that the higher frequency of positive ACPA in TB observed in previous studies may be due to methodological factors (Lima et al., 2013). Thus, ACPA serves as a sensitive diagnostic marker for RA in patients with active TB with 78% sensitivity (confidence interval [CI], 63%–88%) and 97% specificity (CI, 89%–99%), whereas the sensitivity of anti-MCV was 50% (CI, 35%–64%) and specificity was 97% (CI, 89%–99%).

Conditions in Which Citrulline May Prove to Be a Biomarker in the Future

Prospective studies are indicated to develop rational, cost-effective, and risk-stratified guidelines for long-term follow-up of these patients. In addition, specificity and sensitivity needs to be explored in future so that the guidelines may be laid accordingly (Fig. 60.1 and Box 60.3).

Conditions Detected by Anticyclic Citrullinated Peptide

Citrulline as a Biomarker for Brucellosis Presenting With Peripheral Arthritis

The detection of anticyclic citrullinated peptide (anti-CCP) Abs is more specific for RA, but these may be positive in other autoimmune diseases and some infectious diseases. Clinical signs and symptoms of brucellosis (a zoonotic disease) are usually nonspecific and noncharacteristic, and further investigation will definitely aid in differential diagnosis. ACP Abs may serve as a positive biomarker for diagnosing brucellosis presenting with peripheral arthritis (BPA), which is common in children and young adults but which returns to normal values in the follow-up period after brucellosis treatment (Gokhan et al., 2014). It is also observed that the anti-CCP Abs are occasionally present in ankylosing spondylitis, and their presence may be helpful as a serum marker for predicting peripheral arthritis (Kim et al., 2013). Clinicians should be well versed in these findings and should evaluate for the differential diagnosis of BPA.

Citrulline as a Biomarker for Sjögren's Syndrome

The detection of anticyclic citrullinated peptide (anti-CCP) Abs can also be extrapolated to the diagnosis of another autoimmune disease, Sjögren's syndrome (SS). SS is a chronic, slowly progressive, inflammatory, autoimmune disease that is characterized by lymphocytic infiltration of the exocrine glands, characterized by generalized dryness, typically including xerostomia (dry mouth) and keratoconjunctivitis (dry eyes). Observations from a recent study revealed anti-CCP Abs, a weak presence of peptidylargininedeiminase 2 (PAD2) in normal salivary glands, and high expression of PAD2 in primary SS, which suggests the enzyme-dependent posttranslational modification of Cit. Moreover, the coexpression of Abs and PAD2 suggests autoantigen triggering. In a cohort of 134 patients, 7.5% of the serum samples were positive for anti-CCP Abs, whereas 5.2% were positive for AKAs without any radiographic evidence of erosion after a long follow-up. Therefore, ACP Abs may prove to be a diagnostic tool for detection of SS (Herrera-Esparza et al., 2013). Atzeni et al. (2008) documented that ACP Abs in SS may be associated with nonerosive synovitis, but only a minority of patients were ACP positive. Moreover, their coexistence may be a predictor of future progress to RA.

It is noted that ACPA are highly specific of RA, but are also detected in 5%–10% of primary Sjogren's syndrome. Another related recent study documented that for ACPA-positive patients, clinical and radiological reevaluation was systematically performed after at least 5 years of follow-up found that ACPA-positive patients had more frequent arthritis but not arthralgias and also had more frequent lung involvement. After a median follow-up of 8 (5–10) years, 43.8% patients developed RA, including 31.25% that developed typical RA erosions. Researchers concluded that median term follow-up of ACPA-positive patients with SS showed that almost half of them developed RA, particularly in the presence of elevated of acute phase reactants. These results support the usefulness of a close radiological monitoring of these patients for early detection of erosive change not to delay initiation of effective treatment. Indeed, a number of these patients with ACPA-positive SS may actually have RA and associated SS (Payet et al., 2015).

Another entity introduced, which might tentatively be called ACA-positive limited scleroderma/SS overlap syndrome, is characterized by a benign systemic sclerosis clinical course but poses a high risk of non-Hodgkin's lymphoma. It is noteworthy that the prevalence of non-Hodgkin's lymphoma in the "overlap patients" was higher than in SS (Baldini et al., 2013). There is much scope for further exploration.

Conditions Detected by Citrulline Levels

Citrulline as a Biomarker in Transplants

Intensive chemotherapy is needed to prepare a patient for hematopoietic stem cell therapy, often resulting in gastrointestinal mucositis. The concern therefore is to have a reliable biomarker to decrease the severity of intestinal damage. Findings suggest that Cit may be a more specific and better marker as compared to albumin for inflammation using CRP (van der Velden et al., 2013). This may be due to the supporting fact that Cit correlates better with inflammatory response, showing a strong but inverse relationship (van Waardenburg et al., 2007). After myeloablative regimens, low Cit levels were found to correlate with intestinal damage. Cit-based assessment of intestinal damage has been shown to be an objective and reliable marker in these patients. Cit needs to be evaluated as a biomarker that can have value in diagnosing as well as in clinical decision-making in critical situation. Clinicians should also be aware of findings of significantly lower levels of Cit in patients with intestinal graft versus host disease (GVHD) (Vokurka et al., 2013).

A 3-year study that began in 2004 observed that a cutoff level of Cit <13 μmol/L should alert the clinical team to a serious problem (rejection or infection), as it could be a warning sign of rejection in a previously stable intestinal transplant recipient, whereas levels ≥13 μmol/L practically rule out moderate or severe rejection. For this cutoff point, the sensitivity for detecting moderate or severe ACR and the negative predictive value were high (96.4% and >99%, respectively). Specificity was 54%–74% in children and 83%–88% in adults. Certain other characteristics found to be associated with lower Cit levels were the presence of mild/moderate/severe acute cellular rejection (ACR), the presence of bacteremia or respiratory infection, and pediatric age (David et al., 2007). Cit is thought to be a potent indicator and hence can serve as a danger signal or a marker of ACR in intestinal transplantation. A 16-year-long term study (1995–2011) investigated Cit samples from various transplant patients and found levels to be inversely proportional to the severity of ACR. NPV for any type of ACR (cutoff, 20 μmol/L) and moderate/severe ACR (cutoff, 10 μmol/L) were 95% and 99%, respectively. Moreover, when patients were divided according to graft size, diagnostic accuracy using the same cutoff was identical. Thus, Cit is a potent indicator for ACR, an exclusionary noninvasive biomarker with excellent NPV in the long term after pediatric intestinal/multivisceral transplant (Hibi et al., 2012). It may be emphasized further that time-to-normalization of Cit was delayed by the incidence of rejection, and in some cases with moderate-to-severe rejection, normalization of Cit levels never occurred (Pappas et al., 2004). Dried blood spot (DBS) was then used as a noninvasive method for monitoring graft function following transplantation. Further, an association of lower Cit levels within 30 days of acute rejection was found in patients who had received intestinal transplant (David et al., 2006) and can be taken as a predictive marker. The Cit levels in mild rejection were 15.0 ± 2.3 μmol/L prior to rejection and 18.8 ± 2.4 μmol/L during the rejection-free periods; whereas for moderate/severe rejection the level was 12.4 ± 1.1 μmol/L before rejection and 18.8 ± 2.0 μmol/L during the rejection-free periods (David et al., 2006). Thus, Cit serum levels can be utilized as a biomarker for assessing the grades of rejection. Because of high NPV, a moderate or severe ACR can be ruled out, based exclusively on knowledge of a high value for DBS Cit. It may prove to be a promising noninvasive biomarker for graft monitoring and increasing the success rates of transplants.

A study was done to evaluate correlation of plasma Cit and rejection episodes in intestinal transplantation. Results revealed significantly decreased average serum Cit levels when the patients presented a rejection episode. It is noteworthy that there was no rejection at Cit levels of 17.38 μm/L, mild rejection at 13.05 μm/L, moderate rejection at 7.98 μm/L, and severe rejection at 6.05 μm/L. Cit correlated significantly with the rejection status of the graft such that serial follow-up of the patients using this assay may indicate the possibility of increased alloreactivity and rejection episodes (Ruiz et al., 2010).

Another study to evaluate clinical and biochemical markers of gastrointestinal function in children undergoing hematopoietic cell transplantation (HCT) was done including weekly serum Cit concentrations obtained from 10 days earlier until 30 days after HCT. Mean Cit concentration was found to be statistically lower in patients with severe oral mucositis than in those without severe mucositis. Changes in Cit were not correlated with stool volume, CRP, tumor necrosis factor-α, leptin, or ghrelin in children undergoing HCT. Serum Cit correlates with measures of gastrointestinal function (oral mucositis severity, dietary intake, acute GVHD) and may reflect mucosal injury to the gastrointestinal tract (Gosselin et al., 2014).

Citrulline as a Biomarker in Acute Respiratory Distress Syndrome

It has been observed that in patients with ARDS and severe sepsis, the plasma Cit levels are low. In a study that included patients with severe sepsis, Cit levels were low (6.0 μM) in ARDS as compared with the levels in the group having no ARDS (Ware et al., 2013). The incidence of ARDS was 50% in the group with Cit in the lowest range, and it has been postulated that NOS deficiency may play a pathophysiological role. Thus, Cit can be used as a biomarker in this area, but whether its supplementation could prevent the development of ARDS needs to be determined.

Citrulline as a Biomarker in Polycystic Ovary Syndrome

Patients with polycystic ovary syndrome (PCOS) are found to have a unique metabolomic profile. The profile that helps to differentiate between control and PCOS were AAs (Arg, Lys, proline, glutamate, and histidine), organic acids (citrate), and significantly decreased levels of Cit, Arg, Lys, ornithine, proline, glutamate, acetone, citrate, and histidine in PCOS compared with controls. These findings suggest that Cit is one factor among the metabolomics spectrum that is altered in PCOS (Atiomo and Daykin, 2012), and this lends itself to further investigation in the future.

Citrulline as a Biomarker in Acute Kidney Injury

A need is realized for a biomarker to categorize acute kidney injury (AKI) that could help in the differential diagnosis of prerenal versus intrinsic renal AKI, and for one that might be a better indicator, especially in critically ill patients at risk of AKI (Schiffl and Lang, 2012), and a predictor of long-term kidney outcomes and mortality. Biomarkers for diagnosis of AKI are usually based on blood urea nitrogen (BUN) and serum creatinine and are not very sensitive or specific for the diagnosis of AKI because they are affected by renal and nonrenal factors that are independent of kidney injury or kidney function. Many new biomarkers are being tried in these patients such as interleukin-18, neutrophil gelatinase-associated lipocalin, and kidney injury molecule-1 (Edelstein, 2008). Decreased Arg levels correlate with renal ischemia-reperfusion injury seen after shock, trauma, and major vascular surgery leading to acute tubular necrosis (Prins et al., 2002).

New metabolomic biomarkers are emerging this area and are currently clinically governed by serum creatinine and BUN, markers with low sensitivity in the early stage of AKI. Cit is found to play the role of a novel indicator for acute kidney failure, shown by an overall direct correlation between Cit and creatinine. A steady increase in Cit levels is anticipated by about 36 h (associated with increased creatinine) along with decreased urine output (Chiarla and Giovannini, 2013). The study revealed that low serum Arg levels could facilitate the diagnosis and determine prognosis of AKI in hospitalized patients (Sun et al., 2012). This might support a role for Cit as an early marker of AKI, but needs further evaluation because some studies refute this (Noordally et al., 2012). Hence, future studies are required to explore this area of concern (Box 60.3).

Homocitrulline as a Biomarker for Renal Failure

A study was done with the aim to evaluate serum homocitrulline (HCit), which results from the carbamylation of ε-amino groups of Lys residues in ARF and to determine if it could be useful for differentiating acute from chronic renal failure (CRF). HCit concentrations increased in ARF correlated with urea concentrations, and serum HCit concentrations were statistically higher in the CRF group when compared to the ARF group. Regarding the predictive value of ARF, it was noted that sensitivity was 83% whereas specificity was 72%. Thus results demonstrate that HCit is a promising biomarker for distinguishing between ARF and CRF patients (Desmons et al., 2016).

Citrulline as a Biomarker in Critically Ill Patients and Correlation With Severity of Inflammation

Critical illness is often associated with a strong inflammatory response that is correlated with increased NO production, increased arginase activity, and decreased renal de novo Arg synthesis. Inflammation could therefore play a key role in Arg depletion and the anticipated need for Arg supplementation in these conditions (van Waardenburg et al., 2007; Luiking et al., 2009). This is due to utilization of Arg (another conditionally essential AA), which is catabolized primarily to Cit and ornithine to generate NO and other polyamines, and thus its nutritional requirements are increased during inflammation (Suh et al., 2012).

A cohort study was done in critically ill children (severity in children was assessed by the Pediatric Risk of Mortality [PRISM] score) having sepsis, trauma, and viral disease, and who were admitted to a pediatric intensive care unit. Findings suggested that plasma Cit and Arg were significantly lower (Arg: 33 ± 4, 37 ± 4, and 69 ± 8 μmol/L, respectively; Cit: 10 ± 1, 14 ± 1, and 23 ± 2 μmol/L, respectively) and correlated strongly and inversely with the severity of inflammation as indicated by raised plasma CRP concentrations (van Waardenburg et al., 2007). Thus, plasma concentrations of Arg and Cit are low during the acute phase of critical illness in children and normalize again during recovery. It is interesting to note that in severe inflammation (sepsis and trauma) and mild inflammation (viral disease), the plasma concentrations of Cit and glutamine (precursors of de novo Arg synthesis) were lower in the former than in the latter condition, whereas CRP was higher in the former than in the latter. Plasma Arg and Cit are strongly correlated to the severity of inflammation (van Waardenburg et al., 2007) and may be exploited as a biomarker in critically ill patients.

Citrulline as a Biomarker in Intrauterine Growth Retardation

Animal studies prove that plasma Cit concentrations are lower in IUGR. In addition, associated hyperammonemia is likely to be a major factor, contributing to high rates of fetal mortality in IUGR fetuses (Lin et al., 2012). Thus, it may gain importance as a promising candidate biomarker of abnormal embryogenesis (Hozyasz et al., 2010).

Citrulline as a Biomarker of Congenital Anomalies

Cleft lip with or without cleft palate is considered to be a very common congenital abnormality in humans. The Centers for Disease Control and Prevention estimates that each year 2651 babies in the United States are born with a cleft palate and 4437 babies are born with a cleft lip with or without a cleft palate. This imposes severe consequences on both physical and psychological development, as well as being a substantial economic and social burden (Parker et al., 2010). Observations suggest Cit as a biomarker in assessment of risk for congenital anomalies such as orofacial cleft, for which further studies are indicated. In addition, it is interesting to note that increased consumption of foods rich in Cit (or supplements) by women of childbearing age may bring down the incidence, indicating that the anomaly may possibly be associated with Cit status. Findings suggest that orofacial cleft incidence was higher in patients having lower concentrations (<8 μmol/L) of Cit, than in those having 8–16 μmol/L or still higher levels. Hence, lower Cit levels during pregnancy may serve as an indicator for congenital anomalies (Hozyasz et al., 2010) for which further studies need to be carried out.

Citrulline as a Biomarker for OAT Deficiency in Early Infancy

OAT deficiency is a rare congenital metabolic disorder in humans, and early diagnosis is delayed because of the nonspecific character of initial symptoms, into adulthood. Thus, earlier diagnosis will lead to a better prognosis because treatment could be started before the occurrence of irreversible damage. Observations from case reports show that reversed enzymatic flux in early infancy results in borderline low ornithine concentration—evoking urea cycle disturbances—and increased proline. Additionally, plasma Cit was low with an increased proline/Cit ratio in plasma versus controls. Proline concentrations (777 and 1381 μmol/L) were above the 99th percentile (776 μmol/L) of the general population, whereas Cit concentrations (4.5 and 4.9 μmol/L) were only just above the 1st percentile (4.37 μmol/L). Hence, early diagnosis of OAT deficiency can lead to earlier treatment and prevent visual impairment. Further studies are needed to evaluate whether newborns with OAT deficiency leads to metabolic disorders in humans (de Sain-van der Velden et al., 2012).

Further it can be mentioned that in UCDs, patients manifesting severe neonatal hyperammonemia are benefited very little from newborn screening because of poor prognosis. N-acetylglutamate synthase deficiency, carbamoylphosphate synthase 1 deficiency, and ornithine transcarbamylase deficiency are generally not screened for the given instability of glutamine and the low specificity and sensitivity for detection of decreases in Cit levels, while Cit supplementation is of great value in treating acute hyperammonemic decompensation (Cavicchi et al., 2009).

Citrulline as a Biomarker in Citrin Deficiency

Citrin is a mitochondrial aspartate–glutamate carrier primarily expressed in organs such as the liver, heart, and kidney. Citrin deficiency can manifest in newborns as neonatal intrahepatic cholestasis (NICCD), in older children as failure to thrive and dyslipidemia (FTTDCD), and in adults as recurrent hyperammonemia with neuropsychiatric symptoms in citrullinemia type II (CTLN2). Generally, citrin deficiency is characterized by a liking for protein-rich and/or lipid-rich foods and an aversion to carbohydrate-rich foods (Saheki et al., 2004).

In NICCD, the affected neonates have significantly higher blood levels of ornithine and Cit—AAs involved in the urea cycle—than healthy counterparts despite normal plasma ammonia levels. Elevated Cit levels can be considered as a biomarker and, more importantly, can also be a diagnostic marker even in the silent period

(Nagasaka et al., 2009; Treepongkaruna et al., 2012). Affected neonates also have associated hypercholesterolemia and augmented oxidative stress (Nagasaka et al., 2009). FTTDCD is also characterized by elevated Cit levels (Song et al., 2009). In adult-onset CTLN2, elevated plasma Cit and ammonia levels were observed, as shown by case reports (Funabe et al., 2009; Tazawa et al., 2013). Citrin deficiency, a novel cause of failure to thrive (FTTDCD), responds to a high-protein, low-carbohydrate diet. Its other manifestations, such as NICCD or CTLN2, are characterized by higher Cit levels. Thus, estimation of high Cit levels with or without increases in ammonia levels could serve as a biomarker for various manifestations of citrin deficiency.

Citrulline as a Biomarker in Toxicities

Citrulline as a Biomarker of Kainic Acid–Induced Neurotoxicity

It is well established that NO is implicated in various models of neurotoxicity, such as in kainic acid (KA)–induced excitotoxicity in rat brain. Cit has been used as a marker of NO and NOS. A rapid increase in NO occurred 30 min after KA injection, peaked at 2 h, and returned to normal within 24 h (except cortex), manifestations that were well correlated with neuronal damage and death. Vitamin E (an antioxidant) pretreatment prevented the rise in Cit levels and depletion of high-energy phosphates and attenuated the neurodegenerative changes without modifying the seizure activity of KA. This suggests the involvement of NO and peroxynitrite, responsible for mitochondrial dysfunction and neuronal death, for which Cit has been used as a biomarker in KA-induced neurotoxicity (Milatovic et al., 2002; Gupta and Dettbarn, 2003).

Involvement of NO is supported by findings of other studies that evaluated the anticonvulsant effect of 7-nitroindazole, a selective nNOS inhibitor (Gupta and Dettbarn, 2003). These data show that NO is involved in KA-induced neurotoxicity and Cit can be used as a biomarker of NO/NOS.

Citrulline as a Biomarker of Anticholinesterase Poisoning

Organophosphate and carbamate compounds exert their toxicity primarily by virtue of cholinesterase inhibition in the brain and skeletal muscles, commonly referred to as anticholinesterase agents (Gupta, 2006). Cit can serve as a biomarker of anticholinesterase poisoning, where Cit levels correlate well with signs of cholinergic toxicity (Liu et al., 2007). Cit has also been used as a marker of NO/NOS and RNS in assessment of myotoxicity induced by diisopropylphosphorofluoridate (Gupta et al., 2002). Similar findings were also noted with a carbamate compound, carbofuran, which induced skeletal muscle hyperactivity causing RNS and nitrosative stress, where Cit levels were significantly elevated. The increase in RNS was largely prevented by pretreatment with memantine due to prevention of nitrosative stress (Gupta et al., 2002; Milatovic et al., 2005).

Citrulline as a Biomarker of Ammonia Toxicity?

In an experimental study, mice exposed to acute ammonia toxicity showed elevated levels of NO in discrete brain regions, leading to NO-induced toxic effects and corresponding increases in ASS and ASL activities, suggesting effective recycling of Cit to Arg (Swamy et al., 2005). Thus, Cit may serve as a marker of ammonia toxicity.

CONCLUDING REMARKS AND FUTURE DIRECTIONS

Cit is a ubiquitous AA in mammals, is strongly related to Arg, and has a scope beyond that normally associated with an AA. Cit has generated tremendous interest and curiosity in the scientific community world due to its plethora of effects. Its role in the urea cycle was realized some time ago. In addition, it has a role in protein homeostasis, and further exploration is opening new horizons for Cit as a supplement as well as biomarker for various diseases and toxicities. To summarize regarding the metabolic pathways in tissues: (1) liver Cit is locally synthesized from OCT and metabolized by ASS to urea; (2) in most tissues, Cit is recycled into Arg and NO via ASS, increasing NO availability; and (3) Cit is synthesized in the gut from glutamine and released into the circulation with 80% being converted back into Arg in the kidneys by ASS made available for utilization by peripheral tissues. Circulating Cit is in fact a masked form of Arg to avoid degradation of Cit in the liver (Cit is readily taken up by liver and degraded). Thus, Cit is actually a better method of getting supplemental Arg into the blood than Arg itself, proving that Cit is a novel therapeutic agent in conditions of Arg and NO deficiencies. Cit supplementation offers an inexpensive, advantageous nutritional therapeutic objective that may prove beneficial in various diseases due to its multiple effects. There may be cardiovascular risk associated with Cit supplementation that needs further evaluation. Research is indicated to prove or disprove whether glutamine supplementation could be used as a precursor for Cit in the future (Lingthart-Melis and Deutz, 2011). Other AAs may also act as a Cit precursor such as glutamate or proline (Cynober, 2002).

Cit supplementation may be (1) highly exploited in therapeutic conditions and situations such as erectile dysfunction, sickle cell anemia, SBS, hyperlipidemia, cancer chemotherapy, and cardiovascular disease (sustained release preparation), (2) used to increase protein content in malnourished as well as elderly patients, (3) to increase endurance for athletics training, (4) used for the acceleration of wound healing, and (5) used for treatment of AD and MID, as well as many other conditions. Studies support use of Cit supplementation for athletics training due to the capability of oxidative stress reduction after cycling, so Cit may be utilized as an antioxidant in the future. Cit alters the systemic response of mediators and cytokines in sepsis and acts as an immunomodulator. It is also likely to be used as a therapeutic agent (treatment/prevention) in neonates or for childhood conditions such as citrin deficiency, congenital malformations, and IUGR. Moreover, multiple system atrophy is associated with elevated Cit levels, so the role of Cit modulators can be explored in certain brain lesions. Cit has also been tried as an adjunctive therapy in parasitic infections such as toxoplasma, trypanosomiasis, and malaria. Further studies are needed to explore the pleiotropic effects of Cit supplementation in different clinical conditions.

Another area of interest regarding Cit is as an Ab to citrullinated peptides that serve as predictors for diseases, thus serving as biomarkers in various diseases. Protein citrullination (from posttranslational modification of Arg) of certain body proteins produces novel epitopes that give rise to autoantibodies. Their detection provides a novel way to diagnose a variety of diseases and act as a marker for monitoring certain clinical conditions. Until now, Cit has been used on a large scale for diseases such as RA, intestinal pathology, and SBS. In the future, it is likely to emerge as a biomarker for other conditions such as parenteral nutrition in HIV patients, congenital anomalies, acute rejections in transplantations, IUGR, critically ill patients, and acute kidney failure. Its role in toxicities is realized as a biomarker of nitrosative stress (NO/NOS) and RNS in conditions such as anticholinesterase poisoning. In addition, its role as a biomarker is realized in kainic acid–induced neurotoxicity because oxidative stress increases RNS and ROS (when the rate of ATP demand exceeds production), which causes formation of peroxynitrite, a potent inducer of lipid peroxidation and DNA damage.

Much more exploration in this regard is indicated in all aspects mentioned and discussed here and large-scale studies are the need of the hour before Cit is used in therapy with confidence. Similarly, studies are required in the field of its use as a biomarker/predictor of diseases of concern and an effort should be made to determine clear-cut levels in each condition. This is essential in either proving or refuting the role of Cit as a marker and making accurate estimates of sensitivity and specificity. Future studies are indicated in parasitic diseases such as malaria as an adjuvant therapy and as a biomarker for the clinical conditions in question and others such as psoriasis or multiple sclerosis.

References

Alsop, P., Hauton, D., 2016. Oral nitrate and citrulline decrease blood pressure and increase vascular conductance in young adults: a potential therapy for heart failure. Eur. J. Appl. Physiol. 116 (9), 1651–1661.

Atiomo, W., Daykin, C.A., 2012. Metabolomic biomarkers in women with polycystic ovary syndrome: a pilot study. Mol. Hum. Reprod. 18 (11), 546–553.

Atzeni, F., Sarzi-Puttini, P., Lama, N., et al., 2008. Anti-cyclic citrullinated peptide antibodies in primary Sjögren syndrome may be associated with non-erosive synovitis. Arthritis Res. Ther. 10 (3), R51.

Bahri, S., Zerrouk, N., Aussel, C., et al., 2013. Citrulline: from metabolism to therapeutic use. Nutrition 29 (3), 479–484.

Baldini, C., Mosca, M., Della Rossa, A., et al., 2013. Overlap of ACA-positive systemic sclerosis and Sjogren's syndrome: a distinct clinical entity with mild organ involvement but at high risk of lymphoma. Clin. Exp. Rheumatol. 31 (2), 272–280.

Bansal, V., Rodriguez, P., Wu, G., et al., 2004. Citrulline can preserve proliferation and prevent the loss of CD3 zeta chain under conditions of low arginine. J. Parenter. Enter. Nutr. 28 (6), 423–430.

Bourdon, A., Rouge, C., Legrand, A., et al., 2012. Urinary citrulline in very low birth weight preterm infants receiving intravenous nutrition. Br. J. Nutr. 108 (7), 1150–1154.

Cai, B., Luo, Y., Wang, S., Wei, W., et al., 2016. Does Citrulline have protective effects on liver injury in septic rats? BioMed Res. Int. 2016, 1469590.

Cakmaz, R., Buyukasik, N., Kahramansoy, N., et al., 2013. A combination of plasma DAO and Cit levels as a potential marker for acute mesenteric ischemia. Libyan J. Med. 26 (8), 1–6.

Cambridge, G., Acharya, J., Cooper, J.A., et al., 2013. Antibodies to citrullinated peptides and risk of coronary heart disease. Atherosclerosis 228 (1), 243–246.

Cardillo, C., Panza, J.A., 1998. Impaired endothelial regulation of vascular tone in patients with systemic arterial hypertension. Vasc. Med. 3 (2), 139–144.

Cavicchi, C., Malvagia, S., la Marca, G., et al., 2009. Hypocitrullinemia in expanded newborn screening by LC-MS/MS is not a reliable marker for ornithine transcarbamylase deficiency. J. Pharmaceut. Biomed. Anal. 49 (5), 1292–1295.

Centers for Disease Control (CDC), 2013. Rheumatoid Arthritis, Centers for Disease Control and Prevention Website. Available at: http://www.cdc.gov/arthritis/basics/rheumatoid.htm.

Cereda, E., Klersy, C., Serioli, M., , et al.OligoElement Sore Trial Study Group, 2015. A nutritional formula enriched with arginine, zinc, and antioxidants for the healing of pressure ulcers: a randomized trial. Ann. Intern. Med. 162 (3), 167–174.

Chiarla, C., Giovannini, I., 2013. Citrulline and metabolomics in acute kidney injury. J. Chromatogr. B Analyt. Technol. Biomed. Life Sci. 137, 913–914.

Chien, S.J., Lin, K.M., Kuo, H.C., et al., 2014. Two different approaches to restore renal nitric oxide and prevent hypertension in young spontaneously hypertensive rats: l-citrulline and nitrate. Transl. Res. 163 (1), 43–52.

Collins, J.K., Wu, G., Perkins-Veazie, P., et al., 2007. Watermelon consumption increases plasma arginine concentrations in adults. Nutrition 23 (3), 261–266.
Conrad, K., Roggenbuck, D., Reinhold, D., et al., 2010. Profiling of rheumatoid arthritis associated autoantibodies. Autoimmun. Rev. 9 (6), 431–435.
Crenn, P., De Truchis, P., Neveux, N., et al., 2009. Plasma citrulline is a biomarker of enterocyte mass and an indicator of parenteral nutrition in HIV-infected patients. Am. J. Clin. Nutr. 90 (3), 587–594.
Crenn, M., Hanachi, N., Neveux, L., et al., 2011. Circulating citrulline levels: a biomarker for intestinal functionality assessment. Ann. Biol. Clin. 69 (5), 513–521.
Cynober, L., 2002. Plasma amino acid levels with a note on membrane transport: characteristics, regulation, and metabolic significance. Nutrition 18 (9), 761–766.
Cynober, L., 2013. Citrulline: just a biomarker or a conditionally essential amino acid and a pharmaconutrient in critically ill patients? Crit. Care 17 (2), 122.
Cynober, L., Moinard, C., De Bandt, J.P., 2010. The 2009 ESPEN Sir David Cuthbertson. Citrulline: a new major signaling molecule or just another player in the pharmaconutrition game? Clin. Nutr. 29 (5), 545–551.
David, A.I., Gaynor, J.J., Zis, P.P., et al., 2006. An association of lower serum citrulline levels within 30 days of acute rejection in patients following small intestine transplantation. Transplant. Proc. 38 (6), 1731–1732.
David, A.I., Selvaggi, G., Ruiz, P., et al., 2007. Blood citrulline level is an exclusionary marker for significant acute rejection after intestinal transplantation. Transplantation 84 (9), 1077–1081.
De Luis, D.A., Izaola, O., Cuellar, L., et al., 2009. High dose of arginine enhanced enteral nutrition in postsurgical head and neck cancer patients. A randomized clinical trial. Eur. Rev. Med. Pharmacol. Sci. 13 (4), 279–283.
de Pablo, P., Dietrich, T., Chapple, I.L., et al., 2014. The autoantibody repertoire in periodontitis: a role in the induction of autoimmunity to citrullinated proteins in rheumatoid arthritis? Ann. Rheum. Dis. 73 (3), 580–586. https://doi.org/10.1136/annrheumdis-2012-202701.
de Sain-van der Velden, M.G., Rinaldo, P., Elvers, B., et al., 2012. The proline/citrulline ratio as a biomarker for OAT deficiency in early infancy. JIMD Rep 6, 95–99.
Debats, I.B., Wolfs, T.G., Gotoh, T., et al., 2009. Role of arginine in superficial wound healing in man. Nitric Oxide 21 (3–4), 175–183.
Desmons, A., Jaisson, S., Pietrement, C., et al., 2016. Homocitrulline: a new marker for differentiating acute from chronic renal failure. Clin. Chem. Lab. Med. 54 (1), 73–79.
Duskin, A., Eisenberg, R.A., 2010. The role of antibodies in inflammatory arthritis. Immunol. Rev. 233 (1), 112–125.
Edelstein, C.L., 2008. Biomarkers of acute kidney injury. Adv. Chron. Kidney Dis. 15 (3), 222–234.
Eleuterio, R., Nogueira, N., Eleuterio, J., et al., 2016. Arginine in the treatment of patients with sickle cell disease: a systematic review. Adv. Biotech. Micro. 1 (4), 555566.
Endo, F., Matsuura, T., Yanagita, K.I., et al., 2004. Clinical manifestations of inborn errors of the urea cycle and related metabolic disorders during childhood. J. Nutr. 134 (6 Suppl.), 1605S–1609S.
Faure, C., Morio, B., Chafey, P., et al., 2013. Citrulline enhances myofibrillar constituents expression of skeletal muscle and induces a switch in muscle energy metabolism in malnourished aged rats. Proteomics 13 (14), 2191–2201.
Fike, C., Summar, M., Aschner, J.L., 2014. L-citrulline provides a novel strategy for treating chronic pulmonary hypertension in newborn infants. Acta Paediatr. 103 (10), 019–1026.
Flam, B.R., Eichler, D.C., Solomonson, L.P., 2007. Endothelial nitric oxide production is tightly coupled to the Citrulline–NO cycle. Nitric Oxide 17 (3–4), 115–121.
Funabe, S., Tanaka, R., Urabe, T., et al., 2009. A case of adult-onset type II citrullinemia with repeated nonconvulsive status epilepticus. Rinsho Shinkeigaku 49 (9), 571–575.
Giannesini, B., Le Fur, Y., Cozzone, P.J., et al., 2011. Citrulline malate supplementation increases muscle efficiency in rat skeletal muscle. Eur. J. Pharmacol. 667 (1–3), 100–104.
Gokhan, A., Turkeyler, I.H., Babacan, T., et al., 2014. The antibodies cyclic citrullinated peptides (anti-CCP) positivity could be a promising marker in brucellosis patients presented with peripheric arthritis. Mod. Rheumatol. 24 (1), 182–187.
Gosselin, K.B., Feldman, H.A., Sonis, A.L., et al., 2014. Serum citrulline as biomarker of gastrointestinal function during hematopoietic cell transplantation in children. J. Pediatr. Gastroenterol. Nutr. 58 (6), 709–714.
Gupta, R.C., 2004. Brain regional heterogeneity and toxicological mechanisms of organophosphates and carbamates. Toxicol. Mech. Meth. 14 (3), 103–143.
Gupta, R.C., 2006. Toxicology of Organophosphate and Carbamate Compounds. Academic Press/Elsevier, Amsterdam.
Gupta, R.C., Dettbarn, W.D., 2003. Prevention of kainic acid seizures-induced changes in levels of nitric oxide and high-energy phosphates by 7-nitroindazole in rat brain regions. Brain Res. 981 (1–2), 184–192.
Gupta, R.C., Milatovic, D., Dettbarn, W.D., 2002. Involvement of nitric oxide in myotoxicity produced by diisopropylphosphorofluoridate (DFP)-induced muscle hyperactivity. Arch. Toxicol. 76, 715–726.
Haas, W., Grabe, K., Geis, C., et al., 2002. Recognition and invasion of human skin by Schistosoma mansoni cercariae: the key-role of L-arginine. Parasitology 124 (Pt 2), 153–167.
Hecker, M., Sessa, W.C., Harris, H.J., et al., 1990. The metabolism of L-arginine and its significance for the biosynthesis of endothelium-derived relaxing factor: cultured endothelial cells recycle L-citrulline to L-arginine. Proc. Natl. Acad. Sci. U.S.A. 87 (21), 8612–8616.
Herrera-Esparza, R., Rodriguez-Rodriguez, M., Perez-Perez, M.E., et al., 2013. Posttranslational protein modification in the salivary glands of Sjögren's syndrome patients. Autoimmune Dis. 2013, 548064.
Hibi, T., Nishida, S., Garcia, J., et al., 2012. Citrulline level is a potent indicator of acute rejection in the long term following pediatric intestinal/multivisceral transplantation. Am. J. Transplant. 12 (4 Suppl.), S27–S32.
Hozyasz, K.K., Oltarzewski, M., Lugowska, I., 2010. Whole blood citrulline concentrations in newborns with non-syndromic oral clefts – a preliminary report. Asia Pac. J. Clin. Nutr. 19 (2), 217–222.
Hull, M.A., Jones, B.A., Zurakowski, D., et al., 2011. Low serum citrulline concentration correlates with catheter-related bloodstream infections in children with intestinal failure. J. Parenter. Enter. Nutr. 35 (2), 181–187.
Ioannou, H.P., Fotoulaki, M., Pavlitou, A., et al., 2011. Plasma citrulline levels in paediatric patients with celiac disease and the effect of a gluten-free diet. Eur. J. Gastroenterol. Hepatol. 23 (3), 245–249.
Ishikura, F., Egawa, M., Takano, Y., et al., 2015. Effects of citrulline combined with tadalafil on monocrotaline-induced pulmonary hypertension in rats compared with arginine. J. Nov. Physiother. 5, 269.
Kaore, S.N., Amane, H.S., Kaore, N.M., 2013. Citrulline: pharmacological perspectives and its role as an emerging biomarker in future. Fundam. Clin. Pharmacol. 27 (1), 35–50.
Kim, J.O., Lee, J.S., Choi, J.Y., et al., 2013. The relationship between peripheral arthritis and anti-cyclic citrullinated peptide antibodies in ankylosing spondylitis. Joint Bone Spine 18 (3), 399–401.
Kumar, S., Pangtey, G., Gupta, R., et al., 2017. Assessment of anti-CarP antibodies, disease activity and quality of life in rheumatoid arthritis patients on conventional and biological disease-modifying antirheumatic drugs. Reumatologia 55 (1), 4–9.
Lauterbach, R., Pawlik, D., Lauterbach, J.P., 2018. L-citrulline Supplementation in the Treatment of Pulmonary Hypertension Associated

With Bronchopulmonary Dysplasia in Preterm Infant: A Case Report. SAGE open Med case reports [Internet]. Available from: http://www.ncbi.nlm.nih.gov/pubmed/29854406.

Lee, T.J., Yu, J.G., 2002. L-Citrulline recycle for synthesis of NO in cerebral perivascular nerves and endothelial cells. Ann. N.Y. Acad. Sci. 962, 73–80.

Lee, C.H., Hsiao, C.C., Hung, C.Y., et al., 2012. Long-term enteral arginine supplementation in rats with intestinal ischemia and reperfusion. J. Surg. Res. 175 (1), 67–75.

Ligthart-Melis, G.C., Deutz, N.E., 2011. Is glutamine still an important precursor of citrulline? Am. J. Physiol. Endocrinol. Metab. 301 (2), E264–E266.

Lima, I., Oliveira, R.C., Atta, A., et al., 2013. Antibodies to citrullinated peptides in tuberculosis. Clin. Rheumatol. 32 (2), 685–687.

Lin, G., Liu, C., Feng, C., et al., 2012. Metabolomic analysis reveals differences in umbilical vein plasma metabolites between normal and growth-restricted fetal pigs during late gestation. J. Nutr. 142 (6), 990–998.

Liu, J., Gupta, R.C., Goad, J.T., et al., 2007. Modulation of parathion toxicity by glucose feeding: is nitric oxide involved? Toxicol. Appl. Pharmacol. 219, 106–113.

Luiking, Y.C., Poeze, M., Dejong, C.H., et al., 2004. Sepsis: an arginine deficiency state? Crit. Care Med. 32 (10), 2135–2145.

Luiking, Y.C., Poeze, M., Ramsay, G., et al., 2009. Reduced citrulline production in sepsis is related to diminished de novo arginine and nitric oxide production. Am. J. Clin. Nutr. 89 (1), 142–152. https://doi.org/10.3945/ajcn.2007.25765.

Luiking, Y.C., Engelen, M.P.K.J., Deutz, N.E.P., 2010. Regulation of nitric oxide production in health and disease. Curr. Opin. Clin. Nutr. Metab. Care 13 (1), 97–104.

Milatovic, D., Gupta, R.C., Dettbarn, W.D., 2002. Involvement of nitric oxide in kainic acid-induced excitotoxicity in rat brain. Brain Res. 957 (2), 330–337.

Milatovic, D., Gupta, R.C., Dekundy, A., et al., 2005. Carbofuran-induced oxidative stress in slow and fast skeletal muscles: prevention by memantine and atropine. Toxicology 208, 13–24.

Mochizuki, Y., Oishi, M., Hara, M., et al., 1996. Amino acid concentration in dementia of the Alzheimer type and multi-infarct dementia. Ann. Clin. Lab. Sci. 26, 275–278.

Moinard, C., Cynober, L., 2007. Citrulline: a new player in the control of nitrogen homeostasis. J. Nutr. 137 (6 Suppl. 2), 1621S–1625S.

Mori, S., Naito, H., Ohtani, S., et al., 2009. Diagnostic utility of anti-cyclic citrullinated peptide antibodies for rheumatoid arthritis in patients with active lung tuberculosis. Clin. Rheumatol. 28, 277–283.

Morris, C.R., Kuypers, F.A., Lavrisha, L., et al., 2013. A randomized, placebo-controlled trial of arginine therapy for the treatment of children with sickle cell disease hospitalized with vaso-occlusive pain episodes. Haematologica 98 (9), 1375–1382.

Nagasaka, H., Okano, Y., Tsukahara, H., et al., 2009. Sustaining hypercitrullinemia, hypercholesterolemia and augmented oxidative stress in Japanese children with aspartate/glutamate Carrier isoform 2-citrin-deficiency even during the silent period. Mol. Genet. Metabol. 97, 21–26.

Noordally, S.O., Sohawon, S., Semlali, H., et al., 2012. Is there a correlation between circulating levels of citrulline and intestinal dysfunction in the critically ill? Nutr. Clin. Pract. 27 (4), 527–532. https://doi.org/10.1177/0884533612449360.

Norris, K.A., Schrimpf, J.E., Flynn, J.L., et al., 1995. Enhancement of macrophage microbicidal activity: supplemental arginine and citrulline augment nitric oxide production in murine peritoneal macrophages and promote intracellular killing of Trypanosoma cruzi. Infect. Immun. 63 (7), 2793–2796.

Oliverius, M., Kudla, M., Balaz, P., et al., 2010. Plasma citrulline concentration – a reliable noninvasive marker of functional enterocyte mass. Cas. Lek. Cesk. 149 (4), 160–162.

Osowska, S., Neveux, N., Nakib, S., et al., 2008. Impairment of arginine metabolism in rats after massive intestinal resection: effect of parenteral nutrition supplemented with citrulline compared with arginine. Clin. Sci. (Lond.) 115 (5), 159–166.

Papadia, C., Sabatino, A.D., Corazza, G.R., et al., 2014. Diagnosing small bowel malabsorption: a review. Intern. Emerg. Med. 9 (1), 3–8.

Pappas, P.A., Tzakis, A.G., Gaynor, J.J., et al., 2004. An analysis of the association between serum citrulline and acute rejection among 26 recipients of intestinal transplant. Am. J. Transplant. 4 (7), 1124–1132.

Parker, S.E., Mai, C.T., Canfield, M.A., et al., 2010. Updated national birth prevalence estimates for selected birth defects in the United States, 2004–2006. Birth Defects Res. A Clin. Mol. Teratol. 88 (12), 1008–1016.

Payet, J., Belkhir, R., Gottenberg, J.E., et al., 2015. ACPA-positive primary Sjögren's syndrome: true primary or rheumatoid arthritis-associated Sjögren's syndrome? RMD Open 1 (1), e000066a.

Piton, G., Capellier, G., 2015. Plasma citrulline in the critically ill: intriguing biomarker, cautious interpretation. Crit. Care 19 (1), 204.

Piton, G., Manzon, C., Cypriani, B., et al., 2011. Acute intestinal failure in critically ill patients: is plasma citrulline the right marker? Intensive Care Med. 37 (6), 911–917.

Poole, A., Deane, A., Summers, M., et al., 2015. The relationship between fasting plasma citrulline concentration and small intestinal function in the critically ill. Crit. Care 19, 16.

Prins, H.A., Nijveldt, R.J., Gasselt, D.V., et al., 2002. .The flux of arginine after ischemia-reperfusion in the rat kidney. Kidney Int. 62 (1), 86–93.

Rashid, M., Butzner, J.D., Warren, R., et al., 2009. Home blood testing for celiac disease: recommendations for management. Can. Fam. Physician 55 (2), 151–153.

Rassaf, T., Preik, M., Kleinbongard, P., et al., 2002. Evidence for in vivo transport of bioactive nitric oxide in human plasma. J. Clin. Invest. 109 (9), 1241–1248.

Rodera, P.C., De Luis, D.A., Candela, C.G., et al., 2012. Immunoenhanced enteral nutrition formulas in head and neck surgery; a systematic review. Nur. Hosp 27 (3), 681–690.

Romero, M.J., Platt, D.H., Tawfik, H.E., et al., 2008. Diabetes induced coronary vascular dysfunction involves increased arginase activity. Circ. Res. 102 (1), 95–102.

Ruiz, P., Tryphonopoulos, P., Island, E., et al., 2010. Citrulline evaluation in bowel transplantation. Transplant. Proc. 42 (1), 54–56.

Saheki, T., Kobayashi, K., Iijima, M., et al., 2004. Adult-onset type II citrullinemia and idiopathic neonatal hepatitis caused by citrin deficiency: involvement of the aspartate glutamate carrier for urea synthesis and maintenance of the urea cycle. Mol. Genet. Metabol. 81 (1 Suppl.), S20–S26.

Schiffl, H., Lang, S.M., 2012. Update on biomarkers of acute kidney injury: moving closer to clinical impact? Mol. Diagn. Ther. 16 (4), 199–207.

Silva, D.K., Jacinto, J.L., de Andrade, W.B., et al., 2017. Citrulline malate does not improve muscle recovery after resistance exercise in untrained young adult men. Nutrients 9 (10), 1132.

Siqueira, L.T., Ferraz, A.A., Campos, J.M., 2010. Analysis of plasma citrulline and intestinal morphometry in mice with hepatosplenic schistosomiasis. Surg. Infect. 11, 419–426.

Sokolove, J., Brennan, M.J., Sharpe, O., et al., 2013. Citrullination within the atherosclerotic plaque: a potential target for the anti-citrullinated protein antibody response in rheumatoid arthritis. Arthritis Rheum. 65 (7), 1719–1724.

Song, Y.Z., Guo, L., Yang, Y.L., et al., 2009. Failure to thrive and dyslipidemia caused by citrin deficiency: a novel clinical phenotype. Zhongguo Dang Dai Er Ke Za Zhi 11, 328–332.

Stechmiller, J.K., Childress, B., Cowan, L., 2005. Arginine supplementation and wound healing. Nutr. Clin. Pract. 20 (1), 52–61.

Suh, J.H., Kim, R.Y., Lee, D.S., 2012. A new metabolomic assay to examine inflammation and redox pathways following LPS challenge. J. Inflamm. 9, 37.

Sun, J., Shannon, M., Ando, Y., et al., 2012. Serum metabolomic profiles from patients with acute kidney injury: a pilot study. J Chromatogr B Anal Technol Biomed Life Sci 893–894, 107–113.

Sureda, A., Pons, A., 2012. Arginine and citrulline supplementation in sports and exercise: ergogenic nutrients? Med. Sport Sci. 59, 18–28.

Swamy, M., Zakaria, A.Z., Govindasamy, C., et al., 2005. Effects of acute ammonia toxicity on nitric oxide (NO), citrulline-NO cycle enzymes, arginase and related metabolites in different regions of rat brain. Neurosci. Res. 53 (2), 116–122.

Tazawa, K., Yazaki, M., Fukushima, K., et al., 2013. Patient with adult-onset type II citrullinemia beginning 2 years after operation for duodenal malignant somatostatinoma: indication for liver transplantation. Hepatol. Res. 43 (5), 563–568.

Thibault, R., Flet, L., Vavasseur, F., et al., 2011. Oral citrulline does not affect whole body protein metabolism in healthy human volunteers: results of a prospective, randomized, double-blind, cross-over study. Clin. Nutr. 30 (6), 807–811.

Treepongkaruna, S., Jitraruch, S., Kodcharin, P., et al., 2012. Neonatal intrahepatic cholestasis caused by citrin deficiency: prevalence and SLC25A13 mutations among Thai infants. BMC Gastroenterol. 12, 141.

Utz, P.J., Gensler, T.J., Anderson, P., 2002. Death, autoantigen modifications, and tolerance. Arthritis Res. 2, 101–114.

van der Velden, W.J., Herbers, A.H., Brüggemann, R.J., et al., 2013. Citrulline and albumin as biomarkers for gastrointestinal mucositis in recipients of hematopoietic SCT. Bone Marrow Transplant. 48 (7), 977–981.

van Karnebeek, C.D., Jaggumantri, S., 2015. Current treatment and management of pyridoxine-dependent epilepsy. Curr. Treat. Options Neurol. 17 (2), 335.

van Waardenburg, D.A., de Betue, C.T., Luiking, Y.C., et al., 2007. Plasma arginine and citrulline concentrations in critically ill children: strong relation with inflammation. Am. J. Clin. Nutr. 86 (5), 1438–1444.

Vokurka, T., Svoboda, D., et al., 2013. Serum citrulline levels as a marker of enterocyte function in patients after allogeneic hematopoietic stem cells transplantation – a pilot study. Med. Sci. Mon. Int. Med. J. Exp. Clin. Res. 19, 81–85.

Vos, I., Mol, C.V., Trouw, L.A., et al., 2017. (2017) Anti-citrullinated protein antibodies in the diagnosis of rheumatoid arthritis (RA): diagnostic performance of automated anti-CCP-2 and anti-CCP-3 antibodies assays. Clin. Rheumatol. 36 (7), 1487–1492.

Wada, M., 1930. Über Citrullin, eine neue Aminosäure im presssaft der Wassermelone, Citrullus vulgaris schrad. Biochem. Z. 224, 420–429.

Wahab, M.A.K.A., Laban, A.E., Hasan, A.A., et al., 2016. The clinical value of anti-cyclic citrullinated peptide (anti-ccp) antibodies and insulin resistance (IR) in detection of early and subclinical atherosclerosis in rheumatoid arthritis (RA). Egy. Heart J. 68, 109–116.

Ware, L.B., Magarik, J.A., Wickersham, N., 2013. Low plasma citrulline levels are associated with acute respiratory distress syndrome in patients with severe sepsis. Crit. Care 17 (1), R10.

Waugh, W.H., Daeschner 3rd, C.W., Files, B.A., et al., 2001. Oral citrulline as arginine precursor may be beneficial in sickle cell disease: early phase two results. J. Natl. Med. Assoc. 93 (10), 363–371.

Wijnands, K.A., Vink, H., Briede, J.J., et al., 2012. Citrulline a more suitable substrate than arginine to restore NO production and the microcirculation during endotoxemia. PLoS One 7 (5), e37439.

Wu, G., Collins, J.K., Perkins-Veazie, P., 2007. Dietary supplementation with watermelon pomace juice enhances arginine availability and ameliorates the metabolic syndrome in Zucker diabetic fatty rats. J. Nutr. 137 (12), 2680–2685.

PART X

APPLICATIONS OF BIOMARKERS IN TOXICOLOGY

CHAPTER

61

Analysis of Toxin- and Toxicant-Induced Biomarker Signatures Using Microarrays

Sadikshya Bhandari, Michael A. Lynes

Molecular and Cell Biology, University of Connecticut, Storrs, CT, United States

BIOMARKERS AND BIOMARKER SIGNATURES IN DISEASE: A VARIETY OF MARKER CLASSES AND POSSIBLE INFERENCES

The National Institutes of Health Biomarkers Definitions Working Group has defined a biomarker as "a characteristic that is objectively measured and evaluated as an indicator of normal biological processes, pathogenic processes, or pharmacologic responses to a therapeutic intervention" (Biomarkers Definitions Working, 2001). The assessment of biomarkers can enable the diagnosis of disease, the assessment of disease prognosis, and may facilitate predictions of the response to therapeutic interventions in the disease process. Owing to the variability in baselines of the molecular concentrations of these markers, the wide range of concentrations that may be present in unexposed populations, the variability in both the exposure history of an individual, and the possibility that biomarkers may also be induced by a variety of different xenobiotics, it is often difficult to interpret with certainty the meaning of the appearance of an individual marker. Biomarker signatures are the aggregate of a set of individual biomarkers that together represent a more robust and potentially reliable indicator of exposures and the identity of the toxic agent.

Biomarkers can represent the unique changes in the chemistry of the initially exposed tissue or in tissues where the agent subsequently bioaccumulates. For example, inhaled tobacco smoke is associated with oxidative stress and that process can result in oxidation of deoxyguanosine. One product of DNA oxidation, 8-Oxo-2′-deoxyguanosine (8-Oxo-dG), has been used as a biomarker of other forms of oxidative stress (Pilger and Rudiger, 2006). However, 8-Oxo-dG is rapidly cleared (Hamilton et al., 2001) and has not been found to correlate well with the level of smoking exposures in humans (Azab et al., 2015). As a consequence, downstream biomarkers may be more informative for the characterization of the cellular response to the disease process that arises from xenobiotic exposure.

These downstream biomarkers may include different classes of molecules. Most commonly, a change in protein expression has been used as a hallmark of toxic effects of a xenobiotic or the onset of disease. Malignant and benign melanomas can be differentiated by characterizing the presence of insulin-like growth factor-II messenger RNA-binding protein-3 (IMP-3). The level of IMP-3 is much higher in metastatic melanoma when compared to benign lesions (Pryor et al., 2008). Cytokines, chemokines, and growth factors have been used as an important class of biomarkers in the field of toxicology. One of the biggest challenges in using cytokines as biomarkers is their short half-life in serum as well as a very low basal level expression (Tarrant, 2010). RNAs have been identified as another potential type of biomarkers, specifically plasma microRNAs (miRNAs) and messenger RNAs (mRNAs). Some of the advantages of using RNAs as biomarkers are the tissue-specificity, high level of expression at specific times, and ease of multiplexed assays for RNA measurements (Bartel, 2004).

Biomarkers can include specific changes in gene structure, in epigenetic marks, or to gene activity as indicated by RNA levels. For example, mutation of the fibroblast growth factor receptor gene (FGFR3) is used to predict the severity of urothelial cancer (Chow et al., 2000) and has a potential role as a therapeutic target for that disease process (Knowles, 2007). This kidney disease can result from exposure to agents such as aristolochic acid, melamine, and heavy metals (Chen et al., 2012;

Biomarkers in Toxicology, Second Edition
https://doi.org/10.1016/B978-0-12-814655-2.00061-X

Vervaet et al., 2017), as well as cyclophosphamide, acrolein, and phenacetin.

The immune response is a commonplace target that responds to changes evoked by many classes of xenobiotics. The environmental contaminant mercuric chloride has been shown to affect T cells and predispose to autoimmune phenotypes by shifting from a Th1 to Th2 response (Hemdan et al., 2007). This change to Th2 response includes the shift to production of specific cytokines such as interleukin 4 (IL-4) and IL-10 that can be used as a biomarker for development of autoimmune disease after mercury exposure. However, in the same report, it was found that Th1-polarized responses can be elicited when exposure to mercury occurs in the presence of *Salmonella* antigens. This observation underscores the challenges of interpretation of biomarkers obtained from patients who exist in complex environments that may be populated with multiple xenobiotics, as well as other environmental factors and contemporaneous health issues that can influence the presentation of a biomarker signature.

Rheumatoid arthritis (RAs) is an autoimmune disease associated with inflammation of joints. Meta analysis of a collection of 15 studies identified silica exposure as an important risk factor for RA (Khuder et al., 2002). In addition to the traditional RA biomarkers IL-6, IL-2, and oncostatin M, additional protein markers such as macrophage colony-stimulating factor (M-CSF), TNF receptor superfamily-9, CCL23, transforming growth factor α, and CXCL13 have been shown to identify patients with this disease and were found to be more accurate in defining RA when compared to the traditional set. Additionally, CXCL13 in particular proved to be an important predictor of successful clinical intervention in RA patients (Rioja et al., 2008).

The Predictive Safety Testing Consortium (PSTC) of the Critical Path Institute, US Food and Drug Administration (FDA), along with other federal and international agencies have characterized several biomarkers for detection and quantification of drug-related organ toxicities. For example, albumin, KIM-1 cystatin, β2-macroglobulin, clusterin, and trefoil factor 3 have been evaluated favorably as biomarkers of kidney injury. Markers of drug-induced liver injury include alanine aminotransferase and aspartate aminotransferase, but these markers do not always correlate with the damage observable by histologic evaluation. Other biomarkers such as glutamate dehydrogenase, malate dehydrogenase, paraoxonase/arylesterase 1, purine nucleoside phosphorylase, arginase, sorbitol dehydrogenase, and glutathione S-transferase are under investigation as potential biomarkers for liver injuries. Recent work done to identify more sensitive and specific biomarkers for drug-related skeletal muscle injury has identified skeletal troponins, creatinine kinase protein M, parvalbumin (Pvalb), myosin light chain 3 (Myl3), fatty acid-binding protein 3 (Fabp3), aldolase A (Aldoa), and myoglobin as additional biomarkers (Campion et al., 2013). Similarly, cardiac hormones (natriuretic peptides, NPs) that are synthesized and secreted in response to myofiber stretch are considered to be potential biomarkers for cardiac hypertrophy that can arise from exposure to some drug therapies.

SELECTING SAMPLE SOURCES FOR THE MEASUREMENT OF BIOMARKERS

The most commonly evaluated samples from which biomarkers are measured are blood, serum, saliva, and urine, owing to the minimal invasiveness of sample harvesting techniques. Other sources of samples from which to measure biomarkers include hair (Zhang et al., 2018) and fingernails (Ward et al., 2017), tissue biopsies, interstitial fluid (Samant and Prausnitz, 2018; Zhang et al., 2017), and exhaled breath condensate (Lopez-Sanchez et al., 2017), but these have greater challenges associated with their collection and/or interpretation. In the case of both hair and fingernails, there is some indication that temporal exposure profiles can be interpreted from the spatial location along the length of the hair shaft or fingernail.

Blood samples are one of the most commonly used sources for biomarker as it represents a source of the patient's genetic information (from circulating nucleated leukocytes) as well as physiologically relevant hormones and other regulatory molecules. Moreover, circulating blood passes through all tissues and thus can sample areas where a toxicant may naturally accumulate while being relatively inaccessible to direct sampling. Because collection and processing of blood samples typically requires trained professionals, it is not always cost-effective in large-scale studies. Saliva has proven to be another source of sample material that is readily accessible and contains proteins, lipids, peptides, and other small molecules, as well as sloughed cells and transient cell populations (Molony et al., 2012). Harvests of saliva samples are noninvasive and relatively inexpensive to collect, but the microbial community present in these samples, the nucleases, proteases, and other catabolic enzymes present in this fluid can produce changes in the composition of the sample as it is being collected and processed. Nevertheless, changes in salivary proteins have been shown to be associated with oral cancers as well as other systemic diseases. Higher levels of IgA and amylase are seen in saliva samples from diabetic patients (Aydin, 2007). Posttranslational modifications to proteins have also been identified as a useful class of biomarkers. For example, biomarkers in cystic fibrosis include a change in the glycosylation pattern of salivary mucin (Shori et al., 2001).

INDIVIDUAL BIOMARKERS AND AGGREGATED BIOMARKER SIGNATURES

Over the past few decades, extensive efforts have led to the identification of a host of biomarkers for different cancers as well as other disease conditions. However, the number of individual biomarkers approved by FDA to refine diagnoses and evaluate treatment efficacy remains small (Fuzery et al., 2013). One of the major reasons behind the failure to transition from biomarker identification to their use in diagnosis and treatment is the inability of biomarkers to go beyond differentiating patients from healthy individuals.

It is possible that biomarker signatures, which evaluate a larger set of molecular and cellular entities, will provide enhanced clinical information (Hanash et al., 2008). For example, detection of prostate cancer that is done by analyzing a set of genetic tumor markers in combination with evaluation of a set of plasma protein markers has proven to be more effective than measurements of prostate surface antigen levels alone (Gronberg et al., 2015). In breast cancer studies, a 21-gene signature was able to identify low-risk patients who had a good outcome even without adjuvant chemotherapy (Sparano et al., 2015). A study done by Pierce et al. identified systemic C-reactive protein and serum amyloid A, circulating markers of inflammation, as potential biomarker signatures to predict long-term survival of breast cancer patients (Pierce et al., 2009). A biomarker signature that includes the detection of multiple chromosomal rearrangements in cell-free circulating tumor DNA was found to be about 93% accurate for postsurgical discrimination of patients who would go on to develop metastatic disease (Olsson et al., 2015). A large study done to distinguish between pancreatic cancer and chronic pancreatitis was based on a metabolic biomarker signature panel with a much higher diagnostic accuracy than that of an individual biomarker (Mayerle et al., 2018). More accurate diagnosis of pancreatic cancer at an early stage can lead to increased patient survival rates of 30 to 40 percent, underscoring the importance of these signatures in health care (Mayerle et al., 2018). In the case of inflammatory bowel disease (IBD), several individual biomarkers have been identified as potential diagnostic tools without much success in the clinical setting. As a consequence, there have been attempts to identify biomarker signatures in the array of circulating proteins and from miRNA profiles isolated from colonic tissues to diagnose and differentiate between the different types and stages of IBD (Wu et al., 2008).

Biomarker signatures may provide information regarding the effectiveness and side effects of treatment at the point of toxicant exposure, or during the subsequent disease process that results from exposure. They may also enable physicians to minimize the use of harsh treatments that have severe side effects in specific individuals. Moreover, it may be possible to identify patient populations most responsive to therapeutic intervention. For example, a plasma inflammatory biomarker signature consisting of CXCL9, CXCL10, sIL-2R, and sCD14 has been identified as a possible marker to understand the effect of combination antiretroviral therapy (cART) for HIV patients (Kamat et al., 2012). In another example, RA patients are heterogeneous in the cytokine expression profile that they present with different individuals showing dramatic differences in the expression of TNF-α. Those RA patients that express little or no TNF-α are likely to be poor responders to treatment with TNF-α antagonists. Similarly, IBD patients can become refractive to treatment with TNF-α antagonists. Hence, ongoing monitoring of the expression profile of inflammatory cytokines may indicate when those changes are occurring and can be useful in selecting other potential biologics with which to treat the ongoing inflammation.

APPROACHES TO BIOMARKER SIGNATURE CHARACTERIZATION

Protocols for the detection and analysis of individual biomarkers are selected on the basis of size, molecular characteristics, and concentration of the biomarker, the influence the sample matrix will have on these measurements, and the specificity and sensitivity necessary to make meaningful measurements. There are advantages and disadvantages to the most commonly used assay systems which include factors such as cost, the requirements for skilled technicians, and an established record of reproducibility. These challenges are compounded for the analysis of biomarker signatures where combinations of different marker classes may multiply the assay requirements for both sample preparations and for successful measurement. The following sections describe some of the most common approaches to these biomarker signature measurements.

MULTIPLEXED HIGH CONTENT SOLID PHASE IMMUNOASSAYS

There are two major formats of multiplexed immunoassays: those done in a planar configuration on a sensor surface and those done in suspension. Planar formats employ capture ligands that have been immobilized on a solid surface such as a microplate. The enzyme-linked immunosorbent assay (ELISA) is one of the most frequently used immunoassays that enable qualitative as well as quantitative measurements of analytes from a heterogeneous sample mixture. In the most common assay design, ELISAs use a "sandwich" format that is organized as sequential additions of the immobilized capture antibody, analyte, and matched pair antibody–enzyme conjugate, followed by the addition of a chromogenic enzyme substrate. These assays are typically done in 96 well microplates but can be adapted to smaller well formats to economize on reagent use. These standard assays typically measure a single analyte in each well. Multiplexed ELISAs use a similar "sandwich" format but employ chemiluminescent or fluorescently labeled antibodies in the detection phase and can report on multiple analytes in each microplate well. These optical assays can be limited by spectral overlap of the tags used to label each detection antibody, but some multiplex ELISA formats evaluate analytes with capture antibodies that have been immobilized on a glass slide or membrane (e.g., Quantibody Cytokine Arrays) and are reported to be capable of quantifying 400 to 1000 analytes in a single sample volume.

There are other solid phase enzyme immunoassay formats that are also capable of multiplexed analysis. For example, Meso Scale Discovery (MSD) has developed a multiplex immunoassay platform that enables detection of biomarkers such as cytokines and intracellular signaling proteins in a biological sample. MSD assays incorporate electrochemiluminescent labels (SULFO-TAGs) conjugated to detection antibodies, which generate light when stimulated under the appropriate chemical conditions. The use of SULFO-TAG labeling increases the sensitivity and dynamic range as well as reduces the background signal compared to traditional immunoassay techniques like ELISA. In this format, up to 10 analytes can be interrogated in each microwell, and the sensitivity is reportedly two to three logs better than traditional ELISA analysis. The MSD assay has been used to assess proinflammatory cytokines in serum from HIV-infected and noninfected donors in comparison to cytometric microbead assays of the same samples (Dabitao et al., 2011). Studies have suggested that MSD technology's sensitivity in human cytokine profiling is similar to that of microbead assays described in the following section (Chowdhury et al., 2009). These cytometric bead assays represent a suspension immunoassay format that has become commonly used, for example, in the clinical measurement of circulating cytokine levels.

In the suspension format, capture ligands are immobilized on microbeads and are mixed with analytes in liquid samples. This microbead technology can analyze up to 500 analytes per sample. Each individual bead's population is initially labeled internally with a specific fluorescent dye at a unique concentration to create a distinct bead fluorescent profile that is used to identify the specific capture antibody that is subsequently immobilized on the bead surface. Other capture reagents such as aptamers, antigens, oligonucleotides, and receptors that have binding specificity for an individual analyte can be used in place of the immobilized antibody. A mixture of different beads representing different analyte specificities can then be incubated with a biological sample, and individual analytes present in the sample will be captured by the cognate beads in the mixture (Houser, 2012). Subsequent quantification of the analyte concentration is done by incubating the beads with a cocktail of matched-pair fluorescent antibodies with a fluorescent emission profile distinct from the internal fluor profile that will bind to a second epitope on each analyte. Once evaluated by flow cytometry, each individual bead identity is established by the level of internal fluorescence, and the amount of each analyte captured on that bead is determined by the amount of the matched-pair fluorescent antibody that is bound to the captured analyte on that particular bead. This technology is customizable, reproducible, and can be more accurate, sensitive, and economical with sample volume consumption when compared to traditional immunoassays such as ELISA. The most commonly used platforms for this bead-based analysis are the Luminex, Milliplex, and Bio-Rad Bio-Plex analyzer systems.

FLOW CYTOMETRY AND FLUORESCENCE-ACTIVATED CELL SORTING

In addition to its use in microbead analysis, flow cytometry is also used for single cell phenotypic analysis and is most often used to measure biomarkers in the context of intact cells. The level of specific surface markers can, for example, be measured with antibodies or recombinant major histocompatibility complex tetramer/peptide complexes that have been labeled with a specific fluorophore. Internal cytosolic and nuclear antigens can also be detected after fixation and permeabilization of the cell's plasma membrane. As cells bound by the detection reagents pass through the illuminating laser, fluorescent emission characteristics of the fluorophore(s) are detected and quantified. This technique has been used to study biomarkers such as cytokines (Pala et al., 2000) and stress response proteins (Wieten et al., 2009), as well as immunophenotyping of T-cell markers for autoimmune disease (Danke et al., 2005) and vaccine efficacy (He et al., 2008). One of the biggest limitations of flow cytometry in the measurement of multiple cellular characteristics is that the number of labels used to measure analytes associated with each cell is limited by the spectral overlap of available fluorophores that can limit their optical separation and individual quantification. Multicolor flow cytometry can evaluate as many as 18 fluorescent colors on and in cells passing through the illuminating excitation laser light at rates of 10,000 cells per second. A variant of this technology, mass cytometry, substitutes mass spectroscopy for the fluorescent detection system of a flow cytometer (Bendall et al., 2012; Leipold et al., 2015). In this system, the antibodies used to detect the target cellular analytes are coupled to heavy metal ions that can be readily distinguished from each other as the ions are analyzed in the mass spectrometer. Because the spectrometer can differentiate about three dozen heavy metal ions without spectral overlap, a greater ability to resolve unique cell populations according to their marker profile is possible by mass cytometry than by flow cytometry, but mass cytometry throughput is typically much slower than fluorescence-based cytometry.

Fluorescence-activated cell sorting (FACS) is a variant of flow cytometry that can be used to characterize and subsequently purify an individual subset of cells from a heterogeneous mixture. The cells of interest are identified based on the expressed surface or intracellular molecules that are detected by fluorescently labeled antibodies. Cells with particular characteristics can then be physically separated from the rest of the population. FACS can be used to both characterize the cells that are present in toxin- or toxicant-exposed individuals and subsequently isolate specific subpopulations for additional molecular analysis independent of other cells in the sample that otherwise might serve to dilute the biomarker signals that are associated with exposure.

IMMUNOHISTOCHEMISTRY AND TISSUE MICROARRAYS

Immunohistochemistry (IHC) utilizes antibodies to detect specific antigens on the surface of tissue sections. The tissue sections are processed using a microtome and can be then subsequently probed for different antigens using antibodies which are linked to fluorescent labels or to enzyme, radionuclide, or electron-dense labels for visualization.

The tissue microarray (TMA) is a histologic technique that combines tissue samples from different sources in an assembled array that enables simultaneous analysis of each tissue within the array (Camp et al., 2008). TMAs are made by embedding each tissue, and then producing small diameter cores of each tissue (the cores are as small as 0.6 mm diameter). Each of these cores is then embedded as a spatially encoded array and sectioned to produce slides that represent all of the multiple sections. These TMAs can be processed by IHC or in situ hybridization to study protein markers as well as DNA and RNA expression to determine analyte presence in each tissue section under the same experimental conditions. This technology has been used extensively in clinical studies to compare different types of tumors. The microarray can be used to study different stages of tumors or samples from multiple individuals in a single analysis. It can also be used to screen for novel markers from large groups of tumors (Schoenberg Fejzo and Slamon, 2001). Candidate biomarkers for pancreatic cancer have been identified using TMA (Winter et al., 2012), and TMA approaches were used to explore the association between estrogen and progesterone receptor expression (Hussein et al., 2008) and to create digitized images of kidney and liver from toxicological studies done on rats exposed to different drug candidates (Ryan et al., 2011).

MASS SPECTROMETRY ANALYSIS

In addition to its use in mass cytometry, mass spectrometry (MS) can be used as an analytical technique that generates multiple ions from sample analytes and separates them based on their intrinsic mass to charge ratio (m/z). MS is one of the most commonly used methods to detect, identify, and quantify a relative abundance of molecules within a complex biological sample mixture. Mass spectrometers all have three basic components: ion source, mass analyzer, and ion detector. However, there are differences in the nature of these components that gives rise to different type of mass spectrometers. The use of MS spans several fields such as proteomics, lipidomics, metabolomics, genomics, drug discovery, and structure validation.

MS is one of the most commonly used technologies for toxoproteomics. MS enables accurate measurement and identification of sample composition with sensitivity in the femtomolar (10^{-15}) to attomolar (10^{-18}) range (Mann and Kelleher, 2008). MS has been used to detect toxins from *Clostridium* and *Bacillus* species, for example. Although the most commonly used method to test for *Botulinum* toxins is the mouse lethality bioassay (Dressler et al., 2000), this assay takes several days and can only detect active toxins. MS enables rapid and more sensitive detection of not only the active toxin but also several different components of *Botulinum* neurotoxins (BoNTs) and has been used in clinical settings (Barr et al., 2005; Scarlatos et al., 2008). Additionally, MS-based analytics have been used to identify the enzymatic and biological activities of BoNT and anthrax lethal toxins produced by *Bacillus anthracis* (Kalb et al., 2015). Similarly,

MS techniques such as liquid chromatography–mass spectrometry (LC-MS) and matrix-assisted laser desorption/ionization time-of-flight mass spectrometry (MALDI-TOF) have been used to enhance the detection sensitivity for the plant toxin ricin that is produced by the castor bean plant, *Ricinus communis* (Darby et al., 2001).

MS can detect small molecule analytes such as the uremic toxins indoxyl sulfate and p-cresol sulfate that are involved in some forms of kidney disease (Niwa, 2009). A study done by Williamson et al. utilized MS to perform a comparative study of pertussis toxin (Ptx) and chemically modified toxoid (Ptxd) from licensed vaccines at the peptide sequence level, which could be used to determine how they differentially influence an immune response (Williamson et al., 2010). MS-based analysis has also been used to differentiate between Shiga and Shiga-like toxins that can be present at very low levels in human serum (Silva et al., 2015).

GRATING-COUPLED SURFACE PLASMON RESONANCE MICROARRAY ANALYSIS

Surface plasmons are electron density oscillations that are produced at a metal–dielectric interface (Homola et al., 1999). The generation of these plasmons in the metal (which is typically a thin film of gold) requires the interaction of illuminating light that excites the plasmon generation at the metal surface. To generate these plasmons, the illuminating light must meet three specific requirements (phase velocity, wavelength, and angle of incidence) to be in resonance with the metal electrons. In the most common configuration of an SPR instrument, the phase velocity and wavelength of the illuminating light are fixed at conditions which meet plasmon generation requirements, and the angle of incidence of the illuminating light is varied. Under those conditions, when the angle of the illuminating light does not meet resonance requirements, there is a high degree of reflected light. When in resonance, energy is passed into the metal to create the plasmon, and the reflected light diminishes. Typically, illuminating light is swept through a small series of angles that surround the critical angle at which resonance is maximal, and a U-shaped curve (an SPR curve) of reflected light intensity is produced. The plasmon is influenced by the refractive index very near the metal surface and the SPR phenomenon can be used to measure binding events caused by a molecular interactions that occur at the metal surface that increase the local refractive index (and hence the angle at which plasmon generation occurs). If the rate of change in the SPR curve is measured over time as binding and dissociation events are happening at the sensor chip surface, affinity constants can be calculated from the data produced.

Comparison of preanalyte and postanalyte resonance angles can be used to measure the amount of antigen capture. Typically, the phase velocity of the incident light is matched to the plasmon's requirements using a prism (i.e., a prism-coupled system). This limits the number of simultaneous analyses to the number of prisms and is typically small. In a different embodiment (called grating-coupled surface plasmon resonance, GCSPR), a holographic diffraction grating can be used to establish the required phase velocity, and as a consequence, the metal surface can be imaged and the image computationally divided into separate regions of interest (ROIs) where each SPR curve can be measured independently. If unique capture reagents are immobilized at different ROIs, the change in SPR can be used to indicate increases in refractive index after exposure of the sensor chip surface to the cognate analytes. Using this system, it is possible to measure both molecular and cellular analytes in the same microarray format, and the system is capable of several thousand assays on a single sensor chip (Jin et al., 2006; Unfricht et al., 2005). In a variation of this grating-coupled configuration, the captured analyte can be subsequently interrogated with a fluorescent-matched pair antibody, in a grating-coupled surface plasmon coupled emission (GCSPCE) assay system (Marusov et al., 2012; Molony et al., 2012; Reilly Sr. et al., 2006; Rice et al., 2012; Yuk et al., 2013). This version of the system retains the microarray characteristics, while increasing the limits of detection by almost two logs (to femtogram/ml levels) compared with standard fluorescent microarray immunoassays. Fig. 61.1 is an example of this GCSPCE analysis.

From the data, it is clear that while both patients were diagnosed with RA, the spectrum of cytokines present in each of their sera is substantially different. TNF-α levels, in particular, are different, suggesting that only patient 1 would be responsive to a therapeutic TNF-α antagonist. The technology is also capable of enumerating the capture of whole cells on the sensor chip surface as an indicator of population composition and can assess the functional phenotype (e.g., changes in surface antigen expression or secretion of proteins from the captured cells) over time in this microarray format.

Fig. 61.2 shows the sequence of steps to do functional phenotyping of protein secretion by analysis of immobilized cells in a microarray format. Fig. 61.3 is an example of data from a single set of ROIs in this microarray format. In this example, Jurkat T cells that had been stimulated with the T-cell mitogen phytohemagglutinin (PHA) in bulk culture were subsequently captured on the GCSPCE sensor chip using an immobilized anti-CD3. In some of these ROIs, there was an additional

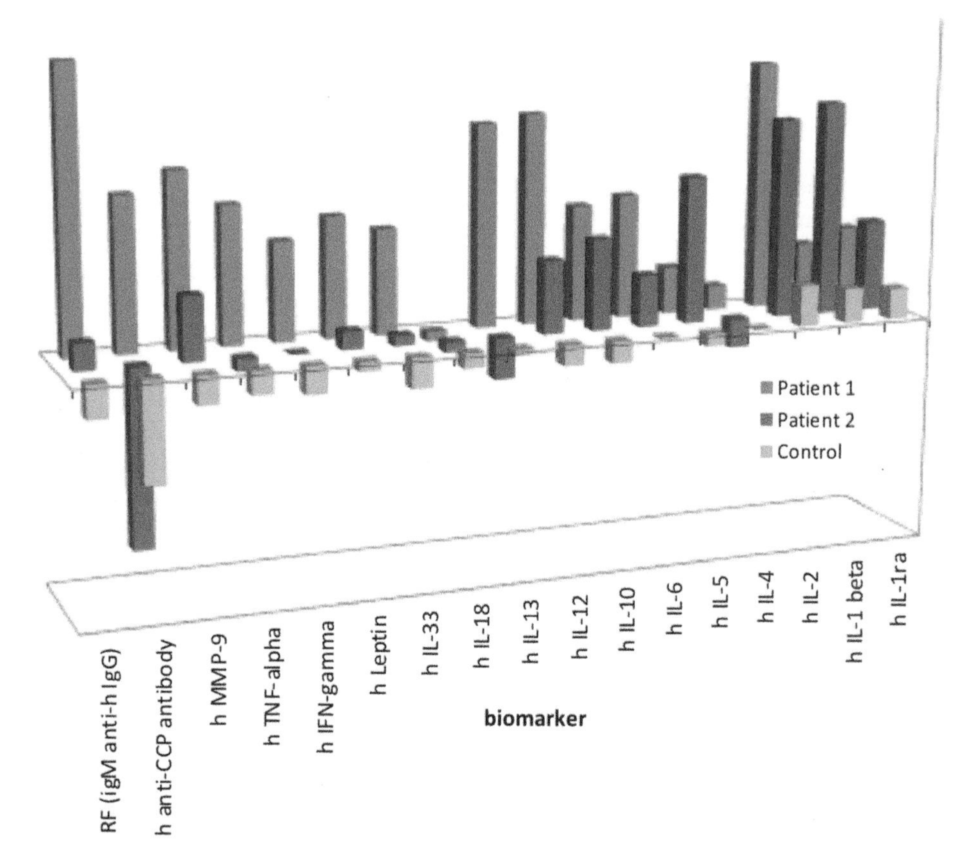

FIGURE 61.1 GCSPCE microarray analysis of rheumatoid arthritis (RA) patient and control sera. In this study, serum from two patients who meet the RA diagnostic criteria and one control were evaluated for the presence of 17 biomarkers by GCSPEF microarray. Bars indicate average fluorescent intensity values from five regions of interest on the sensor chip surface for each patient sample. Data provided by Dr. James M. Rice, personal communication.

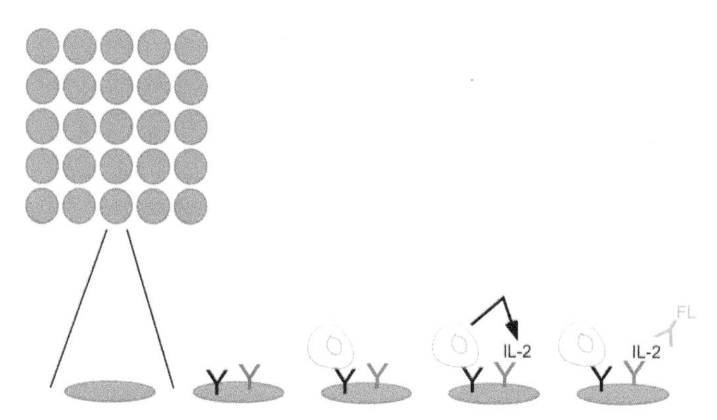

FIGURE 61.2 Capture of viable cells on a GCSPCE microarray and subsequent functional phenotyping. In this graphic representation, a single ROI from an array is shown in expanded detail. In the initial step, the ROI is populated with antibodies specific for a surface marker on the target cell population. After nonspecific binding is blocked, the ROIs in the array are incubated with target cells and are captured on these ROIs. Subsequent culture on the array (in the presence of agonists or antagonists of protein secretion) will produce secreted proteins (e.g., IL-2 cytokine shown in the figure) which will be captured by anti-cytokine antibodies that also populate the same ROI. This capture can subsequently be detected with a matched-pair anti-cytokine antibody that has been labeled with the appropriate fluor (FL).

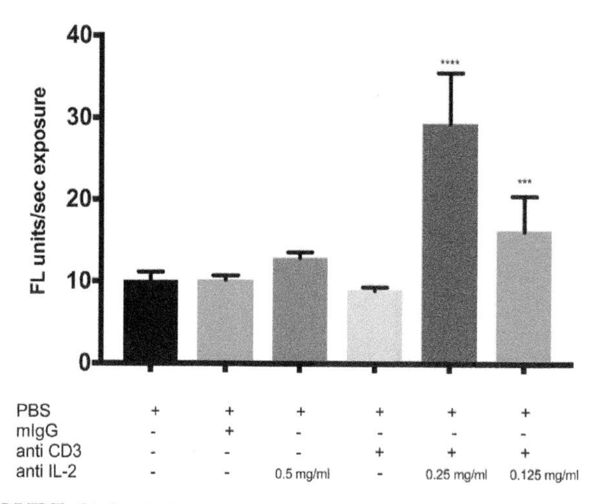

FIGURE 61.3 Jurkat T cells were prestimulated in culture for 3 h with 10 μg/mL phytohemagglutinin in complete RPMI 1640 at 37°C, 5% CO_2. After stimulation, cells were washed and resuspended in the stimulation media and flowed over the DSP-modified chip printed with capture antibody for 30 min. Unbound cells were washed off and the chip incubated overnight at 37°C, 5% CO_2. Cell capture was determined by SPR and IL-2 secretion was detected with biotinylated anti-IL2 antibody + streptavidin-Alexa fluor 647. One way ANOVA analysis, $^{****}P < .0001$, $^{***}P = .0003$.

immobilized anti-IL-2 antibody at one of two different concentrations. Once the cells were captured on the ROIs, they were incubated overnight before biotinylated matched pair anti-IL-2 was passed over the sensor chip followed by streptavidin-alexa fluor. Other ROIs (populated with no capture antibody or with whole mouse IgG) were used as negative controls to determine baseline values.

The data show that these conditions can evaluate cytokine secretion in a system that is versatile and high content capable. By simply employing different fluorescently tagged antibodies, it is also possible to interrogate the functional phenotype of captured cells by evaluating the presence of surface antigens on the cells and the shifts in the expression of those antigens over time as a different indicator of cellular differentiation. Finally, by exposing spatially encoded ROIs of captured cells in the microarray to a diffusing gradient of a toxin, toxicant or pharmaceutical, the different functional phenotype that can characterize cells that experience different exposure profiles according to their spatial location in the gradient will be indicative of how cells may respond to the xenobiotic. Fig. 61.4 is illustrative of this approach, where in a single biosensor chip the individual exposure conditions of multiple individual cell cultures each exposed to a single dose can be rapidly and efficiently modeled. Because a large number of ROIs are available, different kinds of cells captured from a heterogeneous cell suspension (e.g., different leukocytes from a patient's whole blood) at adjacent ROIs can be characterized simultaneously.

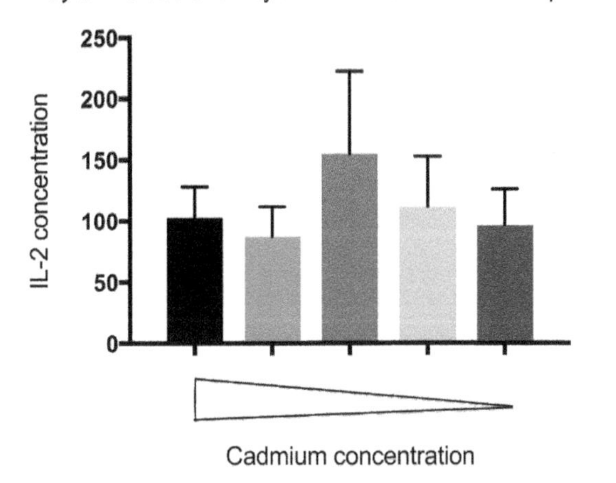

FIGURE 61.4 Jurkat T cells captured at spatially encoded ROIs were subsequently exposed to a diffusing gradient of cadmium chloride. Using GCSPCE, there was evidence of increased IL-2 expression by the immobilized cells at a midpoint in the diffusing gradient.

BIOMARKER SIGNATURES IN THE -OMICS ERA

Genomic and Transcriptomic Analyses

The accessibility of genomic sequences as a result of the human genome project has made whole genome data a commonly used approach to analyze complex phenotypes including those that arise from toxin and toxicant exposure. Additionally, genome-wide association studies (GWASs) have provided important information to identify genetic markers that can be used to predict the presence of the disease or the susceptibility to toxic effects of xenobiotic exposure. The use of genomics to identify and characterize biomarkers will enable better opportunities for prognosis, diagnosis, and therapeutic development. DNA microarrays, serial analysis of gene expression, and high-throughput sequencing are some of the most commonly used genomic high content multiplexed assays. DNA microarray analysis of gene expression profiles that are altered by chemicals provides a very effective way to identify biomarkers that can be used for toxicological studies. For example, the toxicity of low dose acetaminophen and subchronic exposures to three halogenated aromatic hydrocarbons was identified using microarray gene expression profiling (Heinloth et al., 2004; Kulkarni et al., 2008). This approach allowed for early detection of toxicity before appearance of a visible phenotype. In some cases, the study of genomics provides reproducible and homogenous data compared to histopathological readings (Heinloth et al., 2007). Studies done by Le Fevre et al. (2007) and Nie et al. (2006) identified marker genes for genotoxic and nongenotoxic carcinogens, which can be used as predictive models for toxicity studies. The two largest databases for toxicogenomics analyses are ArrayExpress and the Gene Expression Omnibus (GEOs).

Transcriptomics was first developed in the late 1990s using Sanger sequencing of concatenated random transcript fragments (Velculescu et al., 1995). The first human transcriptome was published in 1991 for the mRNA sequences of the human brain (Adams et al., 1991). In the past decade, two human transcriptomes consisting of over 16,000 genes have been published (Pan et al., 2008; Sultan et al., 2008). There are now transcriptomes of individuals with specific diseases, of single cells, and of tissues (Kolodziejczyk et al., 2015; Lappalainen et al., 2013; Mele et al., 2015; Sandberg, 2014).

The use of transcriptomics has played a very important role in the clinical field for disease diagnosis and subsequent treatment strategies. Disease association studies have identified regulatory elements as well as novel splice variants, which are associated with specific disease conditions and disease-associated single-nucleotide polymorphismss (Costa et al., 2013; Khurana et al., 2016). Similarly, RNA-Seq has enabled the transcriptional profiling of T cells and B cells from patients that can be used to study immune-related diseases (Proserpio and Mahata, 2016).

Several genomic studies have identified specific genes that are implicated in the cellular response to toxicant exposure. For example, microarray analysis was used to identify genes that were implicated in the hepatotoxic response to 15 different chemicals (Hamadeh et al., 2002). Transcriptomic data from cardiac biopsies has been used to identify biomarkers associated with cardiomyopathy and myocarditis. miRNA profiling has also been used as a biomarker signature for cardiovascular diseases such as myocardial infarction and hypertension (Li et al., 2009).

Proteomics

Proteomics, "the high-throughput separation, display, and identification of proteins" (Anderson and Anderson, 1998), is one of the most powerful and commonly used tools to study a biological system, as proteins are considered to be the end points of a biological process. Analysis of the proteome uses the many technologies already discussed. Some of the most commonly used tools for proteomic studies for the high-throughput studies are MS techniques such as SELDI (surface-enhanced laser desorption ionization), MALDI (matrix-assisted laser desorption ionization), and antibody microarrays. Numerous posttranslational modifications such as phosphorylation, methylation, glycosylation, S-nitrosylation, and N-acetylation add an additional level of complexity to proteomic analysis.

Proteomic biomarkers that have been used for clinical studies include proteins belonging to classes such as G protein-coupled receptors, enzymes, and components of cellular signaling pathways (Overington et al., 2006). The use of proteomics allows identification of a set of proteins that are associated with a disease or toxicological effect that can be used as potential biomarkers for subsequent studies. For example, a large-scale proteomic screening identified several S100 proteins that were overexpressed in breast cancer tissues, and which can be used as potential biomarkers for cancer progression (Hudler et al., 2014). Proteomics studies have been carried out to create prototype databases to study the effects of toxicological compounds on organs such as the liver (Anderson et al., 1996), kidney (Kennedy, 2001), as well as potential carcinogens in several cases of cancers. Proteomic analysis has also been used to identify biomarkers of acute nephrotoxicity induced by puromycin aminonucleoside and after exposure to benzene (Cutler et al., 1999; Joo et al., 2003).

Metabolomics

Metabolomics refers to the identification and quantification of small metabolites in a biological system. This branch of "omics" is considered to most closely represent the physiological state of an organism, as metabolites are the direct representation of underlying mechanisms (Gieger et al., 2008). The most commonly used techniques to analyze metabolites on a large scale are nuclear magnetic resonance (NMR) spectroscopy; gas chromatography mass spectrometry (GC-MS); and liquid chromatography MS (LC-MSs). Recent work done by Wishart et al. (2013) estimated over 19,000 metabolites that constitute the human metabolome.

The use of metabolomics in diseases and cancer provides a unique opportunity to identify both gene-derived endogenous biomarkers and environmentally associated exogenous metabolites (Wishart, 2016). Metabolites have been identified as possible biomarkers in diseases such as type 2 diabetes (Wang et al., 2011), pancreatic cancer (Mayers et al., 2014), and memory impairment (Mapstone et al., 2014). Similarly, metabolomic studies have shown that intestinal microbiota metabolism of L-carnitine (a component of red meat) is associated with increased risk of atherosclerosis (Koeth et al., 2013). Precision medicine is a developing area where metabolomics is being extensively used to refine the therapeutic treatments designed for individual patients with cancer (Wishart, 2016). Several studies have investigated the effect of cocaine on the neuronal metabolome of rats (Kaplan et al., 2013; Li et al., 2012).

Lipidomics

Lipidomics is the study of pathways and networks of cellular lipids and their interactions with other molecules within the biological system. The field of lipidomics has been rapidly growing since the early 2000s and this growth is mostly attributed to the development of analytical techniques such as MS and ultra-performance liquid chromatography. Lipidomics has been used to identify potential biomarkers for cardiovascular diseases and inflammatory diseases, as well as neurological disorders such as Alzheimer's and Huntington disease. Similarly, panels of lipids have also been implicated in different cases of breast, prostate, and lung cancers (Yang et al., 2016). For example, a correlation was seen between elevated levels of sphingomyelin and different cases of breast cancer (Nagahashi et al., 2016). Additionally, palmitic amide, oleamide, hexadecanoic acid, octadecanoic acid, eicosatrienoic acid, myristic acid, and LPC were identified as biomarkers for colorectal cancer and have been used for early diagnosis (Zhao et al., 2014). In the field of toxicology and drug discovery, lipidomic profiling has been used to investigate drug toxicity markers and to do risk assessment characterization. Lipidomic analysis has identified liver damage caused by Ximelagatran, an oral thrombin inhibitor (Sergent et al., 2009). There have been lipidomic profile changes associated with co-exposure to subtoxic doses of benzo [*a*]pyrene and cadmium (Jungnickel et al., 2014).

Genotype–Phenotype Associations

Phenotypes that are observable characteristics above the molecular level are used as the primary identifier of human diseases. Databases such as a Human Phenotype Ontology provide an avenue for annotating and analyzing the phenotypes associated with specific diseases (Robinson et al., 2008). GWASs have discovered over 8900 genetic variants associated with hundreds of human traits and diseases (Hindorff et al., 2009). The efficient cross-referencing of genotypic variants with different phenotypes can open new avenues for understanding biological processes as well as disease conditions. The genotype–phenotype associations enable a comprehensive study of genetic variation, its consequences, and the mechanism of action (Brookes and Robinson, 2015). There are several databases such as PhenoScanner, PhenomicDB, dbGap, and PheGenl that enable efficient cross-referencing of genotype–phenotype associations. These databases enable user-specified comparison and association of phenotypes with the existing GWAS catalogues such as NHGRI-EBI GWAS catalog and data repositories such as dbGaP along with existing genomic resources from the National Center for Biotechnology Information (NCBI). Additionally, PHenotypic Interpretation of Variants in Exomes (PHIVEs) is an algorithm that allows cross species in-depth comparison of human disease phenotypes with genetically modified mouse models (Robinson et al., 2014).

Data Analysis Considerations

High content analysis is capable of producing enormous datasets. It is very important to employ statistical tests that are compatible with high content data. For example, where there are large numbers of markers that are being evaluated in a biomarker signature analysis, there is a higher chance of random correlation between parameters. These correlations need to be adjusted for using statistical methods by using techniques such as "leave-one-out cross-validation". While analyzing multiple biomarkers, independently calculated *P* values cannot be used to define the biomarker combination analytical accuracy, as there may be synergistic effects. One method to get the most accurate assessment from multiple biomarkers is by identifying the biomarkers that have the maximum impact and by eliminating the least effective ones by methods such as "Kullback–Leibler divergence" (Borrebaeck, 2017). The Fisher Z test, which is a commonly used statistical method for multiparameter analysis, is another approach to biomarker signature studies. Another way to perform data analysis of biomarker signature datasets is the specific constraint-based local causal discovery algorithm proposed by Lagani et al. (2013). This algorithm incorporates Conditional Independence Tests (CITs) for continuous and multiple categories, thus known as the Multi-Class Conditional Independence Test (MC-CIT). MC-CIT is based on multiple multinomial logistic regressions, which are Generalized Linear Models (GLM) designed for multi-class outcomes. This method has been shown to have better predictive capability for biomarker signatures than the Fisher Z test.

CONCLUDING REMARKS AND FUTURE DIRECTIONS

The complexity of the biological system is reflected in the wide array of markers that can be used to infer the consequences of toxin or toxicant exposures. There are a number of different existing and developing technologies that can interrogate biological systems for the composition of individual classes of molecular markers of these exposures and the diseases that result from exposure, and new technologies on the horizon that will enable real-time assessment of multiple classes of markers simultaneously. The data that arise from these experiments will have invaluable benefit for the identification of specific instances of intoxication, for identification of the toxin or toxicant identity, and for our understanding of the molecular mechanisms that attend these exposures. The interpretation of these datasets will require continuing advances in the fields of information management and interpretation, as well as ongoing refinement of the technologies used to detect individual classes of signatures. The opportunities that arise from this work are compelling.

Acknowledgments

The authors would like to thank past members of the Lynes research group for their contributions to discussions of source material used for this chapter: Monalissa Doane, Gregory P. Marusov, Clare E. Melchiorre, Peter Reinhold, Kathryn M. Pietrosimone, and James R. Rice.

Conflict of Interest Declaration

M.A.L. has intellectual property related to SPR microarrays.

References

Adams, M.D., Kelley, J.M., Gocayne, J.D., et al., 1991. Complementary DNA sequencing: expressed sequence tags and human genome project. Science 252 (5013), 1651–1656.

Anderson, N.L., Anderson, N.G., 1998. Proteome and proteomics: new technologies, new concepts, and new words. Electrophoresis 19 (11), 1853–1861.

Anderson, N.L., Taylor, J., Hofmann, J.P., et al., 1996. Simultaneous measurement of hundreds of liver proteins: application in assessment of liver function. Toxicol. Pathol. 24 (1), 72–76.

Aydin, S., 2007. A comparison of ghrelin, glucose, alpha-amylase and protein levels in saliva from diabetics. J. Biochem. Mol. Biol. 40 (1), 29–35.

Azab, M., Khabour, O.F., Alzoubi, K.H., et al., 2015. Assessment of genotoxicity of waterpipe smoking using 8-OHdG biomarker. Genet. Mol. Res. 14 (3), 9555–9561.

Barr, J.R., Moura, H., Boyer, A.E., et al., 2005. Botulinum neurotoxin detection and differentiation by mass spectrometry. Emerg. Infect. Dis. 11 (10), 1578–1583.

Bartel, D.P., 2004. MicroRNAs: genomics, biogenesis, mechanism, and function. Cell 116 (2), 281–297.

Bendall, S.C., Nolan, G.P., Roederer, M., et al., 2012. A deep profiler's guide to cytometry. Trends Immunol. 33 (7), 323–332.

Biomarkers Definitions Working, G., 2001. Biomarkers and surrogate endpoints: preferred definitions and conceptual framework. Clin. Pharmacol. Ther. 69 (3), 89–95.

Borrebaeck, C.A., 2017. Precision diagnostics: moving towards protein biomarker signatures of clinical utility in cancer. Nat. Rev. Cancer. 17 (3), 199–204.

Brookes, A.J., Robinson, P.N., 2015. Human genotype-phenotype databases: aims, challenges and opportunities. Nat. Rev. Genet. 16 (12), 702–715.

Camp, R.L., Neumeister, V., Rimm, D.L., 2008. A decade of tissue microarrays: progress in the discovery and validation of cancer biomarkers. J. Clin. Oncol. 26 (34), 5630–5637.

Campion, S., Aubrecht, J., Boekelheide, K., et al., 2013. The current status of biomarkers for predicting toxicity. Expert. Opin. Drug. Metab. Toxicol. 9 (11), 1391–1408.

Chen, C.H., Dickman, K.G., Moriya, M., et al., 2012. Aristolochic acid-associated urothelial cancer in Taiwan. Proc. Natl. Acad. Sci. U. S. A. 109 (21), 8241–8246.

Chow, N.H., Cairns, P., Eisenberger, C.F., et al., 2000. Papillary urothelial hyperplasia is a clonal precursor to papillary transitional cell bladder cancer. Int. J. Cancer. 89 (6), 514–518.

Chowdhury, F., Williams, A., Johnson, P., 2009. Validation and comparison of two multiplex technologies, Luminex and Mesoscale Discovery, for human cytokine profiling. J. Immunol. Methods. 340 (1), 55–64.

Costa, V., Aprile, M., Esposito, R., et al., 2013. RNA-Seq and human complex diseases: recent accomplishments and future perspectives. Eur. J. Hum. Genet. 21 (2), 134–142.

Cutler, P., Birrell, H., Haran, M., et al., 1999. Proteomics in pharmaceutical research and development. Biochem. Soc. Trans. 27 (4), 555–559.

Dabitao, D., Margolick, J.B., Lopez, J., et al., 2011. Multiplex measurement of proinflammatory cytokines in human serum: comparison of the Meso Scale Discovery electrochemiluminescence assay and the Cytometric Bead Array. J. Immunol. Methods. 372 (1–2), 71–77.

Danke, N.A., Yang, J., Greenbaum, C., et al., 2005. Comparative study of GAD65-specific $CD4^+$ T cells in healthy and type 1 diabetic subjects. J. Autoimmun. 25 (4), 303–311.

Darby, S.M., Miller, M.L., Allen, R.O., 2001. Forensic determination of ricin and the alkaloid marker ricinine from castor bean extracts. J. Forensic Sci. 46 (5), 1033–1042.

Dressler, D., Dirnberger, G., Bhatia, K.P., et al., 2000. Botulinum toxin antibody testing: comparison between the mouse protection assay and the mouse lethality assay. Mov. Disord. 15 (5), 973–976.

Fuzery, A.K., Levin, J., Chan, M.M., et al., 2013. Translation of proteomic biomarkers into FDA approved cancer diagnostics: issues and challenges. Clin. Proteomics. 10 (1), 13.

Gieger, C., Geistlinger, L., Altmaier, E., et al., 2008. Genetics meets metabolomics: a genome-wide association study of metabolite profiles in human serum. PLoS Genet. 4 (11), e1000282.

Gronberg, H., Adolfsson, J., Aly, M., et al., 2015. Prostate cancer screening in men aged 50–69 years (STHLM3): a prospective population-based diagnostic study. Lancet Oncol. 16 (16), 1667–1676.

Hamadeh, H.K., Bushel, P.R., Jayadev, S., et al., 2002. Gene expression analysis reveals chemical-specific profiles. Toxicol. Sci. 67 (2), 219–231.

Hamilton, M.L., Guo, Z., Fuller, C.D., et al., 2001. A reliable assessment of 8-oxo-2-deoxyguanosine levels in nuclear and mitochondrial DNA using the sodium iodide method to isolate DNA. Nucleic Acids Res. 29 (10), 2117–2126.

Hanash, S.M., Pitteri, S.J., Faca, V.M., 2008. Mining the plasma proteome for cancer biomarkers. Nature 452 (7187), 571–579.

He, X.Ä., Holmes, T.H., Mahmood, K., et al., 2008. Phenotypic changes in influenza'ÄêSpecific CD8+ T cells after immunization of children and adults with influenza vaccines. J. Infect. Dis. 197 (6), 803–811.

Heinloth, A.N., Boorman, G.A., Foley, J.F., et al., 2007. Gene expression analysis offers unique advantages to histopathology in liver biopsy evaluations. Toxicol. Pathol. 35 (2), 276–283.

Heinloth, A.N., Irwin, R.D., Boorman, G.A., et al., 2004. Gene expression profiling of rat livers reveals indicators of potential adverse effects. Toxicol. Sci. 80 (1), 193–202.

Hemdan, N.Y., Lehmann, I., Wichmann, G., et al., 2007. Immunomodulation by mercuric chloride in vitro: application of different cell activation pathways. Clin. Exp. Immunol. 148 (2), 325–337.

Hindorff, L.A., Sethupathy, P., Junkins, H.A., et al., 2009. Potential etiologic and functional implications of genome-wide association loci for human diseases and traits. Proc. Natl. Acad. Sci. U. S. A. 106 (23), 9362–9367.

Homola, J.I., Ye, S., Gauglitz, G., 1999. Surface plasmon resonanace sensors: Review. Sensors. Actuators. B. Chem. 54 (1–2), 3–15.

Houser, B., 2012. Bio-Rad's Bio-Plex(R) suspension array system, xMAP technology overview. Arch. Physiol. Biochem. 118 (4), 192–196.

Hudler, P., Kocevar, N., Komel, R., 2014. Proteomic approaches in biomarker discovery: new perspectives in cancer diagnostics. Scient. World. J. 2014, 260348.

Hussein, M.R., Abd-Elwahed, S.R., Abdulwahed, A.R., 2008. Alterations of estrogen receptors, progesterone receptors and c-erbB2 oncogene protein expression in ductal carcinomas of the breast. Cell Biol. Int. 32 (6), 698–707.

Jin, G.B., Unfricht, D.W., Fernandez, S.M., et al., 2006. Cytometry on a chip: cellular phenotypic and functional analysis using grating-coupled surface plasmon resonance. Biosens. Bioelectron. 22 (2), 200–206.

Joo, W.A., Kang, M.J., Son, W.K., et al., 2003. Monitoring protein expression by proteomics: human plasma exposed to benzene. Proteomics 3 (12), 2402–2411.

Jungnickel, H., Potratz, S., Baumann, S., et al., 2014. Identification of lipidomic biomarkers for coexposure to subtoxic doses of benzo [a]pyrene and cadmium: the toxicological cascade biomarker approach. Environ. Sci. Technol. 48 (17), 10423–10431.

Kalb, S.R., Boyer, A.E., Barr, J.R., 2015. Mass spectrometric detection of bacterial protein toxins and their enzymatic activity. Toxins 7 (9), 3497–3511.

Kamat, A., Misra, V., Cassol, E., et al., 2012. A plasma biomarker signature of immune activation in HIV patients on antiretroviral therapy. PLoS One 7 (2), e30881.

Kaplan, K.A., Chiu, V.M., Lukus, P.A., et al., 2013. Neuronal metabolomics by ion mobility mass spectrometry: cocaine effects on glucose and selected biogenic amine metabolites in the frontal cortex, striatum, and thalamus of the rat. Anal. Bioanal. Chem. 405 (6), 1959–1968.

Kennedy, S., 2001. Proteomic profiling from human samples: the body fluid alternative. Toxicol. Lett. 120 (1–3), 379–384.

Khuder, S.A., Peshimam, A.Z., Agraharam, S., 2002. Environmental risk factors for rheumatoid arthritis. Rev. Environ. Health 17 (4), 307–315.

Khurana, E., Fu, Y., Chakravarty, D., et al., 2016. Role of non-coding sequence variants in cancer. Nat. Rev. Genet. 17 (2), 93–108.

Knowles, M.A., 2007. Role of FGFR3 in urothelial cell carcinoma: biomarker and potential therapeutic target. World J. Urol. 25 (6), 581–593.

Koeth, R.A., Wang, Z., Levison, B.S., et al., 2013. Intestinal microbiota metabolism of L-carnitine, a nutrient in red meat, promotes atherosclerosis. Nat. Med. 19 (5), 576–585.

Kolodziejczyk, A.A., Kim, J.K., Tsang, J.C., et al., 2015. Single cell RNA-sequencing of pluripotent states unlocks modular transcriptional variation. Cell. Stem. Cell. 17 (4), 471–485.

Kulkarni, K., Larsen, P., Linninger, A.A., 2008. Assessing chronic liver toxicity based on relative gene expression data. J. Theor. Biol. 254 (2), 308–318.

Lagani, V., Kortas, G., Tsamardinos, I., 2013. Biomarker signature identification in "omics" data with multi-class outcome. Comput. Struct. Biotechnol. J. 6, e201303004.

Lappalainen, T., Sammeth, M., Friedlander, M.R., et al., 2013. Transcriptome and genome sequencing uncovers functional variation in humans. Nature 501 (7468), 506–511.

Le Fevre, A.C., Boitier, E., Marchandeau, J.P., et al., 2007. Characterization of DNA reactive and non-DNA reactive anticancer drugs by gene expression profiling. Mutat. Res. 619 (1–2), 16–29.

Leipold, M.D., Newell, E.W., Maecker, H.T., 2015. Multiparameter phenotyping of human PBMCs using mass cytometry. Methods. Mol. Biol. 1343, 81–95.

Li, Y., Yan, G.Y., Zhou, J.Q., et al., 2012. (1)H NMR-based metabonomics in brain nucleus accumbens and striatum following repeated cocaine treatment in rats. Neuroscience 218, 196–205.

Li, Z., Wilson, K.D., Smith, B., et al., 2009. Functional and transcriptional characterization of human embryonic stem cell-derived endothelial cells for treatment of myocardial infarction. PLoS One 4 (12), e8443.

Lopez-Sanchez, L.M., Jurado-Gamez, B., Feu-Collado, N., et al., 2017. Exhaled breath condensate biomarkers for the early diagnosis of lung cancer using proteomics. Am. J. Physiol. Lung Cell Mol. Physiol. 313 (4), L664–L676.

Mann, M., Kelleher, N.L., 2008. Precision proteomics: the case for high resolution and high mass accuracy. Proc. Natl. Acad. Sci. U. S. A. 105 (47), 18132–18138.

Mapstone, M., Cheema, A.K., Fiandaca, M.S., et al., 2014. Plasma phospholipids identify antecedent memory impairment in older adults. Nat. Med. 20 (4), 415–418.

Marusov, G., Sweatt, A., Pietrosimone, K., et al., 2012. A microarray biosensor for multiplexed detection of microbes using grating-coupled surface plasmon resonance imaging. Environ. Sci. Technol. 46 (1), 348–359.

Mayerle, J., Kalthoff, H., Reszka, R., et al., 2018. Metabolic biomarker signature to differentiate pancreatic ductal adenocarcinoma from chronic pancreatitis. Gut 67 (1), 128–137.

Mayers, J.R., Wu, C., Clish, C.B., et al., 2014. Elevation of circulating branched-chain amino acids is an early event in human pancreatic adenocarcinoma development. Nat. Med. 20 (10), 1193–1198.

Mele, M., Ferreira, P.G., Reverter, F., et al., 2015. Human genomics. The human transcriptome across tissues and individuals. Science 348 (6235), 660–665.

Molony, R.D., Rice, J.M., Yuk, J.S., et al., 2012. Mining the salivary proteome with grating-coupled surface plasmon resonance imaging and surface plasmon coupled emission microarrays. Curr. Protoc. Toxicol. 11–19 (Chapter 18), Unit 18 16.

Nagahashi, M., Tsuchida, J., Moro, K., Hasegawa, M., Tatsuda, K., Woelfell, I.A., Takabe, K., Wakai, T., 2016. High levels of sphingolipids in human breast cancer. J. Surg. Res. 204 (2), 435–444. https://www.ncbi.nlm.nih.gov/pubmed/27565080.

Nie, A.Y., McMillian, M., Parker, J.B., et al., 2006. Predictive toxicogenomics approaches reveal underlying molecular mechanisms of nongenotoxic carcinogenicity. Mol. Carcinog. 45 (12), 914–933.

Niwa, T., 2009. Recent progress in the analysis of uremic toxins by mass spectrometry. J. Chromatogr. B. Analyt. Technol. Biomed. Life. Sci. 877 (25), 2600–2606.

Olsson, E., Winter, C., George, A., et al., 2015. Serial monitoring of circulating tumor DNA in patients with primary breast cancer for detection of occult metastatic disease. EMBO Mol. Med. 7 (8), 1034–1047.

Overington, J.P., Al-Lazikani, B., Hopkins, A.L., 2006. How many drug targets are there? Nat. Rev. Drug Discov. 5 (12), 993–996.

Pala, P., Hussell, T., Openshaw, P.J.M., 2000. Flow cytometric measurement of intracellular cytokines. J. Immunol. Methods. 243 (1–2), 107–124.

Pan, Q., Shai, O., Lee, L.J., et al., 2008. Deep surveying of alternative splicing complexity in the human transcriptome by high-throughput sequencing. Nat. Genet. 40 (12), 1413–1415.

Pierce, B.L., Ballard-Barbash, R., Bernstein, L., et al., 2009. Elevated biomarkers of inflammation are associated with reduced survival among breast cancer patients. J. Clin. Oncol. 27 (21), 3437–3444.

Pilger, A., Rudiger, H.W., 2006. 8-Hydroxy-2'-deoxyguanosine as a marker of oxidative DNA damage related to occupational and environmental exposures. Int. Arch. Occup. Environ. Health 80 (1), 1–15.

Proserpio, V., Mahata, B., 2016. Single-cell technologies to study the immune system. Immunology 147 (2), 133–140.

Pryor, J.G., Bourne, P.A., Yang, Q., et al., 2008. IMP-3 is a novel progression marker in malignant melanoma. Mod. Pathol. 21 (4), 431–437.

Reilly Sr., M.T., Nessing, P.A., Guignon, E.F., Lynes, M.A., et al., 2006. SPR surface enhanced fluorescence with a gold-coated corrugated sensor chip. Progr. Biomed. Optics Imag. Proce. SPIE 6099.

Rice, J.M., Stern, L.J., Guignon, E.F., et al., 2012. Antigen-specific T cell phenotyping microarrays using grating coupled surface plasmon resonance imaging and surface plasmon coupled emission. Biosens. Bioelectron. 31 (1), 264–269.

Rioja, I., Hughes, F.J., Sharp, C.H., Warnock, L.C., Montgomery, D.S., Akil, M., Wilson, A.G., Binks, M.H., Dickson, M.C., 2008. Potential novel biomarkers of disease activity in rheumatoid arthritis patients: CXCL13, CCL23, transforming growth factor alpha, tumor necrosis factor receptor superfamily member 9, and macrophage colony-stimulating factor. Arthritis Rheum. 58 (8), 2257–2267. https://www.ncbi.nlm.nih.gov/pubmed/18668547.

Robinson, P.N., Kohler, S., Bauer, S., et al., 2008. The Human Phenotype Ontology: a tool for annotating and analyzing human hereditary disease. Am. J. Hum. Genet. 83 (5), 610–615.

Robinson, P.N., Kohler, S., Oellrich, A., et al., 2014. Improved exome prioritization of disease genes through cross-species phenotype comparison. Genome. Res. 24 (2), 340–348.

Ryan, D., Mulrane, L., Rexhepaj, E., et al., 2011. Tissue microarrays and digital image analysis. Methods. Mol. Biol. 691, 97–112.
Samant, P.P., Prausnitz, M.R., 2018. Mechanisms of sampling interstitial fluid from skin using a microneedle patch. Proc. Natl. Acad. Sci. U. S. A. 115 (18), 4583–4588.
Sandberg, R., 2014. Entering the era of single-cell transcriptomics in biology and medicine. Nat. Methods 11 (1), 22–24.
Scarlatos, A., Cadotte, A.J., DeMarse, T.B., et al., 2008. Cortical networks grown on microelectrode arrays as a biosensor for botulinum toxin. J. Food Sci. 73 (3), E129–E136.
Schoenberg Fejzo, M., Slamon, D.J., 2001. Frozen tumor tissue microarray technology for analysis of tumor RNA, DNA, and proteins. Am. J. Pathol. 159 (5), 1645–1650.
Sergent, O., Ekroos, K., Lefeuvre-Orfila, L., et al., 2009. Ximelagatran increases membrane fluidity and changes membrane lipid composition in primary human hepatocytes. Toxicol. In. Vitro. 23 (7), 1305–1310.
Shori, D.K., Genter, T., Hansen, J., et al., 2001. Altered sialyl- and fucosyl-linkage on mucins in cystic fibrosis patients promotes formation of the sialyl-Lewis X determinant on salivary MUC-5B and MUC-7. Pflugers. Arch. 443 (Suppl. 1), S55–S61.
Silva, C.J., Erickson-Beltran, M.L., Skinner, C.B., et al., 2015. Mass pectrometry-based method of detecting and distinguishing type 1 and type 2 Shiga-like toxins in human serum. Toxins (Basel) 7 (12), 5236–5253.
Sparano, J.A., Gray, R.J., Makower, D.F., et al., 2015. Prospective validation of a 21-gene expression assay in breast cancer. N. Engl. J. Med. 373 (21), 2005–2014.
Sultan, M., Schulz, M.H., Richard, H., et al., 2008. A global view of gene activity and alternative splicing by deep sequencing of the human transcriptome. Science 321 (5891), 956–960.
Tarrant, J.M., 2010. Blood cytokines as biomarkers of in vivo toxicity in preclinical safety assessment: considerations for their use. Toxicol. Sci. 117 (1), 4–16. https://doi.org/10.1093/toxsci/kfq134.
Unfricht, D.W., Colpitts, S.L., Fernandez, S.M., et al., 2005. Grating-coupled surface plasmon resonance: a cell and protein microarray platform. Proteomics 5 (17), 4432–4442.
Velculescu, V.E., Zhang, L., Vogelstein, B., et al., 1995. Serial analysis of gene expression. Science 270 (5235), 484–487.
Vervaet, B.A., D'Haese, P.C., Verhulst, A., 2017. Environmental toxin-induced acute kidney injury. Clin. Kidney. J. 10 (6), 747–758.
Wang, T.J., Larson, M.G., Vasan, R.S., et al., 2011. Metabolite profiles and the risk of developing diabetes. Nat. Med. 17 (4), 448–453.
Ward, E.J., Edmondson, D.A., Nour, M.M., et al., 2017. Toenail manganese: a sensitive and specific biomarker of exposure to manganese in career welders. Ann. Work. Expo. Health. 62 (1), 101–111.
Wieten, L., van der Zee, R., Goedemans, R., et al., 2009. Hsp70 expression and induction as a readout for detection of immune modulatory components in food. Cell. Stress. Chaperones. https://doi.org/10.1007/s12192-009-0119-8.
Williamson, Y.M., Moura, H., Schieltz, D., et al., 2010. Mass spectrometric analysis of multiple pertussis toxins and toxoids. J. Biomed. Biotechnol. 2010, 942365.
Winter, J.M., Tang, L.H., Klimstra, D.S., et al., 2012. A novel survival-based tissue microarray of pancreatic cancer validates MUC1 and mesothelin as biomarkers. PLoS One 7 (7), e40157.
Wishart, D.S., 2016. Emerging applications of metabolomics in drug discovery and precision medicine. Nat. Rev. Drug Discov. 15 (7), 473–484.
Wishart, D.S., Jewison, T., Guo, A.C., et al., 2013. HMDB 3.0—The human metabolome database in 2013. Nucleic Acids Res. 41 (Database issue), D801–D807.
Wu, F., Zikusoka, M., Trindade, A., et al., 2008. MicroRNAs are differentially expressed in ulcerative colitis and alter expression of macrophage inflammatory peptide-2 alpha. Gastroenterology 135 (5), 1624–1635 e1624.
Yang, L., Li, M., Shan, Y., et al., 2016. Recent advances in lipidomics for disease research. J. Sep. Sci. 39 (1), 38–50.
Yuk, J.S., Guignon, E.F., Lynes, M.A., 2013. Highly sensitive grating coupler-based surface plasmon-coupled emission (SPCE) biosensor for immunoassay. Analyst 138 (9), 2576–2582.
Zhang, J., Hao, N., Liu, W., et al., 2017. In-depth proteomic analysis of tissue interstitial fluid for hepatocellular carcinoma serum biomarker discovery. Br. J. Cancer. 117 (11), 1676–1684.
Zhang, Q., Chen, Z., Chen, S., et al., 2018. Correlations of hair level with salivary level in cortisol and cortisone. Life Sci. 193, 57–63.
Zhao, Y.Y., Cheng, X.L., Lin, R.C., 2014. Lipidomics applications for discovering biomarkers of diseases in clinical chemistry. Int. Rev. Cell. Mol. Biol. 313, 1–26.

CHAPTER

62

Biomarkers Detection for Toxicity Testing Using Metabolomics

David J. Borts

Department of Veterinary Diagnostic and Production Animal Medicine, College of Veterinary Medicine, Iowa State University, Ames, IA, United States

INTRODUCTION TO METABOLOMICS

Metabolomics has been defined as the comprehensive and quantitative analysis of all metabolites of a biological system under study (Fiehn, 2001). Metabolomics emerged as a distinct science around the year 2000 and has grown rapidly up until today. While other similar terminologies, including "metabonomics" (Nicholson et al., 1999) and "metabolic profiling," have been used, metabolomics is now the most commonly used term (Robertson et al., 2011).

A typical metabolomics experiment involves measurement, using one or more analytical techniques, of a large number of small molecule metabolites in an extract from a biological fluid or tissue. Biological samples from a number of organisms in each of two or more groups are usually analyzed as part of a study. These groups commonly include at least one group of organisms that has a disease, has been exposed to a drug or toxin, or has been subjected to some other type of stimulus. A control group consisting of healthy organisms or those that have not been exposed to the drug, toxin, or other stimulus is also included. Once the raw analytical data are collected, differential analysis is performed using a variety of statistical and computational methods. The results of these differential analyses reveal metabolites whose concentrations are statistically significantly different between the groups of organisms in the study. These results can be interpreted and applied in a number of different ways, including elucidation of gene function, understanding of underlying mechanisms of disease, and monitoring therapeutic outcomes, among many others. These results can also lead to identification of biomarkers, which can be applied in a variety of ways, such as the diagnosis of disease, prediction of disease susceptibility, and in personalized medicine. Biomarkers identified using metabolomics approaches have also been widely applied in toxicology, including in preclinical and clinical drug development, in ecotoxicology, and to help understand pathways or signatures of toxicity, which can be correlated with specific modes of action (MOA). Several excellent general metabolomics reviews have appeared recently (Beger et al., 2016; Courant et al., 2014; Engskog et al., 2016; Goldansaz et al., 2017; Johnson et al., 2016; Patti et al., 2012; Wang et al., 2015; Wishart, 2016).

METABOLOMICS EXPERIMENTAL OVERVIEW

A typical metabolomics experimental workflow for identification of biomarkers involves the following steps: (1) study design and planning, (2) sample collection and quenching, (3) sample preparation, (4) instrumental analysis, (5) raw data processing, (6) statistical analysis of processed data, (7) identification of putative biomarkers, and (8) biomarker validation. Each of the steps in this workflow will be briefly reviewed.

Study Design and Planning

Study design is a crucial, and frequently overlooked or undervalued, part of a successful metabolomics study. There are myriad considerations necessary when designing metabolomics studies. A partial list of metabolomics study design considerations includes selection of study objective (hypothesis generating, hypothesis testing, biomarker identification, MOA

Biomarkers in Toxicology, Second Edition
https://doi.org/10.1016/B978-0-12-814655-2.00062-1

identification, etc.); subject selection; sample matrix or matrices selection; sample size/statistical power determination; number and timing of biological and/or analytical replicates; identification of appropriate matched control subjects and samples; randomization of sample collection, processing, and analysis; selection of all glass and plastic ware that will contact samples; choice of anticoagulant for blood samples; sample volume requirement; sample collection, quenching, handling, storage, and preparation protocols; selection of quality control samples; selection of analytical technology platform; selection of specific analytical instrumentation within a platform type; and selection of data processing tools and approaches, among many other necessary study design considerations (Dunn et al., 2012a,b; Hernandes et al., 2017).

Sample Collection and Quenching

A wide variety of sample types including biofluids (plasma, serum, urine, amniotic fluid, saliva, cerebrospinal fluid [CSF], etc.), tissues (brain, cardiac, liver, kidney, etc.), and cells (cultured mammalian, fungal, bacterial, etc.) have been used in metabolomics studies. Virtually any type of biological fluid or tissue can be amenable to metabolomics analysis.

Sample collection and quenching are critically important and frequently overlooked steps in a metabolomics study. There are a large number and variety of sample collection and preparation protocols for metabolomics studies of these various sample types (Bando et al., 2010; Chetwynd et al., 2017; Dudzik et al., 2018; Raterink et al., 2014; Vuckovic, 2012). Although these protocols can be quite different depending on sample type, they also have some common elements. One key element is the concept of quenching, defined as steps taken to ensure the cessation of enzymatic activity and metabolite turnover. If enzymatic activity is not stopped very quickly after sampling, continued metabolic turnover will lead to "metabolic pooling." Metabolic pooling will result in a "blurred," if not completely uninterpretable, picture of the system's metabolic state. The main techniques used to ensure enzymatic quenching are low temperature, adjustment of pH, or organic solvents. For example, a common protocol for quenching a blood sample involves rapidly spinning the blood to plasma in a refrigerated centrifuge, immediately freezing small (0.5–1.0 mL) aliquots of the plasma in liquid nitrogen, and quickly transferring the frozen plasma samples to a −80°C freezer for storage.

Examples of other important considerations for sample collection include collection time, anticoagulant selection for blood plasma, and appropriate steps to avoid possible bacterial contamination of samples. Sample collection time should take into account factors such as diurnal variation, fasting time, and time relative to administration of dietary supplements or pharmaceuticals. Common plasma anticoagulants include heparin, EDTA, and citrate. While there is some disagreement as to whether heparin or EDTA is most appropriate for metabolomics studies, citrate is inappropriate because it is also an endogenous metabolite. Bacterial contamination can negatively affect metabolomics results, as bacteria may metabolize and turn over metabolites that are endogenous to the study organism.

Sample Preparation

Sample preparation is another key step in metabolomics protocols and is typically very specific to a particular sample matrix. Sample preparation for urine is frequently a very simple "dilute and shoot" protocol. Sample preparation for plasma or serum typically involves protein precipitation using organic solvent. Sample preparation procedures for tissue samples and cultured cells tend to be more complex and specific for the particular tissue or cell type in question. At this point, sample preparation protocols for metabolomics are not standardized and an array of procedures is in use (Bando et al., 2010; Chetwynd et al., 2017; Dudzik et al., 2018; Raterink et al., 2014; Vuckovic, 2012).

Sample preparation protocols for metabolomics should be as unselective and reproducible as possible. However, the perfect sample preparation and extraction protocol does not exist and the range of metabolites extracted will depend on the protocol used. In some cases, multiple extraction protocols are employed in an attempt to enable the broadest possible metabolite coverage. For example, a common sample preparation protocol for plasma involves two separate extractions with one extract analyzed by liquid chromatography–mass spectrometry (LC-MS) using reversed phase chromatography and the other extract analyzed by LC-MS using HILIC (hydrophilic interaction chromatography).

Instrumental Analysis

A number of different analytical instrumentation platforms have been used for metabolomics analysis. Currently, the most widely used platforms include nuclear magnetic resonance (NMR) spectroscopy, gas chromatography–mass spectrometry (GC-MS), and LC-MS. Each platform possesses strengths and weaknesses relative to the others, and selection of the appropriate analytical instrumentation platform is dependent on several factors including sample matrix type and study objective. NMR's primary strength is that it is inherently quantitative, generally even without any sample preparation, whereas NMR's primary weaknesses are high cost and poor sensitivity, which result in detection of a limited number of metabolites. The primary strengths of GC-MS include its relatively low cost and the availability of large, well-curated electron ionization spectral library databases. The main weaknesses of GC-MS include limitations on the type and polarity of metabolites that can be made amenable to GC-MS analysis, the necessity to chemically derivatize sample components in many cases, and concerns regarding thermal lability of metabolites during analysis (Fang et al., 2015). The main strength of LC-MS platforms includes the wide metabolite coverage enabled by liquid chromatography separation coupled with electrospray ionization. The main weakness of LC-MS approaches at this point is challenges surrounding metabolite identification. Spectral libraries for LC-MS metabolite identification, while growing, are still significantly limited, and electrospray ionization generates a number of "features" resulting from isotopes, fragments, adducts, multimers, and combinations of these that can be difficult to deconvolve into clear metabolite identifications (a feature, in LC-MS and metabolomics terms, is a molecular entity with a unique m/z and retention time). Despite these challenges and because of the wide metabolite coverage enabled, LC-MS has emerged as the most frequently used analytical technology for metabolomics studies. Currently, the most widely used mass analyzer formats for LC-MS metabolomics work are quadrupole time-of-flight (Q-TOF) and quadrupole orbitrap instruments. These mass spectrometer configurations both provide high resolution and accurate mass capabilities that are beneficial in subsequent metabolomics data processing steps.

Raw Data Processing

There are two fundamentally different approaches to analytical data acquisition and raw data processing in metabolomics: targeted and untargeted metabolomics. Targeted metabolomics is defined as "the measurement of defined groups of chemically characterized and biochemically annotated metabolites," whereas untargeted metabolomics is defined as "the comprehensive analysis of all the measurable analytes in a sample, including chemical unknowns" (Roberts et al., 2012). Targeted and untargeted metabolomics studies require two very different raw data processing procedures. Targeted metabolomics uses a raw data processing workflow similar to typical multianalyte quantitative methods. For targeted metabolomics with GC-MS or LC-MS, analyte peaks are integrated and calibration curves are created using the ratios of peak areas of analytes to those of internal standard compounds. Untargeted metabolomics studies require more specialized raw data processing workflows.

Raw data processing workflows for untargeted metabolomics comprise of four essential steps: (1) identification of features, (2) grouping of the same features between samples, (3) integration of peak areas, and (4) generation of a "peak table," which consists of an array of features and peak areas for each sample. A significant challenge of untargeted metabolomics raw data processing is reduction of a large number of redundant features into an unambiguous entity corresponding to a single metabolite (Mahieu and Patti, 2017). A number of approaches and software tools for untargeted metabolomics raw data processing have been developed (Di Guida et al., 2016; Gowda et al., 2014; Spicer et al., 2017a,b; Tautenhahn et al., 2012; Uppal et al., 2016).

Statistical Analysis of Processed Data

Untargeted metabolomics studies generate large amounts of data. It is common for several thousand features to be measured in a single sample with LC-MS analysis. Multiplied by the dozens, or even hundreds, of samples that are included in a typical metabolomics experiment, data sets can easily contain hundreds of thousands, or even millions, of data points. Statistical analysis methods are necessary to help draw meaning and information from these very large data sets. The most commonly used statistical analysis methods for metabolomics data sets include a group of methods known as multivariate statistical analysis. Principal component analysis (PCA) is a frequently used unsupervised (meaning there is no identification of group membership of samples) statistical analysis tool. PCA allows viewing of groupings of data and can provide an overview of overall data quality. Partial least squares is a supervised (meaning group membership of samples is identified) tool that is frequently used to classify samples into groups and to identify specific metabolites responsible for the groupings (Barnes et al., 2016; Ebbels and De Iorio, 2011; Ren et al., 2015).

There are a number of software options available for statistical analysis of metabolomics data. Most analytical instrumentation vendors offer software packages that include statistical analysis modules. There are some excellent freely available software packages for statistical analysis (Spicer et al., 2017a,b). One freely available statistical analysis software package that is particularly widely used is MetaboAnalyst (Xia et al., 2015; Xia and Wishart, 2016). General commercial statistical software packages such as SIMCA-P, SAS, and Pirouette are also frequently used for metabolomics statistical analysis. Finally, there are many custom software packages for metabolomics statistical analysis, many based on the R programming language (Spicer et al., 2017a,b).

Identification of Putative Biomarkers

Statistical analysis of a processed metabolomics data set may reveal dozens of metabolites that show statistically significant differences in relative abundance between groups tested. However, it is not practical to include dozens of metabolites in a biomarker model. It has been proposed that small groups of 1–10 biomarkers lend themselves to generation of more mathematically robust and practical models (Xia et al., 2013). One common method to reduce the complexity of a biomarker model is to select highly ranked metabolites based on variable importance in projection scores or other selection methods. Multivariate models including selected metabolites are then typically evaluated using cross-validation and permutation testing techniques. Finally, receiver operator characteristic (ROC) curves, which plot one minus specificity (the true negative rate) against sensitivity (the true positive rate), are constructed to evaluate the predictive performance of a biomarker model. The area under the ROC (AUROC) curve represents the predictive performance of the model, with an AUROC value of 1 representing ideal model performance (Nagana Gowda and Raftery, 2013; Xia et al., 2013).

Although it is not necessary to know the identity of a metabolite in order for it to be used as part of a biomarker model, it is certainly desirable. An understanding of the identities of metabolites that make up a biomarker model can help rationalize the metabolites' roles in a particular disease or toxicity state. One of the biggest challenges in untargeted metabolomics has been metabolite identification. In many studies, only a small fraction of the features detected are confidently identified as corresponding to a specific molecular structure. A significant amount of work continues to be devoted to the development of metabolite databases and to approaches to automate and facilitate metabolite identification (Dunn et al., 2012a,b; Vinaixa et al., 2016). The Metabolomics Standards Initiative has developed guidance regarding evaluation and reporting of the level of confidence in metabolite identification that is to be used when publishing results (Sumner et al., 2007).

Biomarker Validation

There are currently no standardized protocols for biomarker validation. However, it is generally acknowledged that validated methods for biomarker measurement should be targeted quantitative methods that measure all of the metabolites in a biomarker set. These methods should make use of appropriate analytical standards, internal standards, and quality control checks as with any other validated quantitative bioanalytical method. These methods should be applied to larger, and more diverse, sample sets compared with those used for preliminary biomarker identification (Nagana Gowda and Raftery, 2013; Xia et al., 2013).

Metabolomics Experimental Overview Conclusions

Although information derived from metabolomics studies can be used in a wide variety of ways in a plethora of applications, the remainder of this chapter will focus on metabolomics studies where specific biomarkers, or sets of biomarkers, have been associated with specific toxins, or systems affected by, or diseases related to specific toxins.

APPLICATIONS OF METABOLOMICS BIOMARKERS IN TOXICITY TESTING

Toxicology was one of the primary application areas of metabolomics in the early 2000s. It has been calculated that 25%–35% of metabolomics publications in the years 2000–02 were related to toxicology (Robertson et al., 2011). Although the percentage of metabolomics publications related to toxicology has decreased over the years to about 10%, the total number of metabolomics publications related to toxicology has continued to grow along with the very rapid growth of metabolomics overall. Following the methodology of Robertson (Robertson et al., 2011), a Scopus search (www.elsevier.com/solutions/scopus) performed in early 2018 reveals that approximately 4000 publications have appeared since 1999, which include the words metabolomics, metabonomics, or metabolic profiling in the title, keywords, or abstract that also contain the root word "toxic." A large fraction of these studies have identified possible biomarkers of toxicity, MOA of toxicity, or pathways of toxicity. By now, metabolomics has been applied to identifying possible biomarkers in virtually every subfield of toxicology. An exhaustive review of these studies is obviously not possible here. A number of excellent reviews of metabolomics applications in toxicology have appeared in recent years (Bouhifd et al., 2013; Nassar et al., 2017; Ramirez et al., 2013; Robertson, 2005; Robertson et al., 2011; Rouquie et al., 2015; Roux et al., 2011). The remainder of this chapter will focus on some of the primary areas in which metabolomics has been applied in toxicology, including hepatotoxicity, renal toxicity, toxicity from chemical agents, neurotoxicity, biotoxins, nanomaterials, and drugs of abuse. Prior to examining each of these topics, a brief overview of a number of larger-scale cooperative efforts to apply metabolomics in toxicology is provided.

Metabolomics in Toxicology Consortia

The Consortium for Metabonomic Toxicology (COMET) was active from 2002 to 2005. This effort was led by Jeremy Nicholson from Imperial College London, UK, and involved five major pharmaceutical companies. The objective of this collaborative effort was to generate a metabolomics database using NMR spectroscopy of rodent urine and blood serum and to build a predictive expert system for target organ toxicity in the rat. This database was delivered to the sponsoring companies at the conclusion of the effort and has been reported as highly successful (Lindon et al., 2005).

BASF's metabolomics program at Metanomics Health, along with a consortium of 12 biopharmaceutical companies, has constructed a metabolomics database covering over 600 chemicals and identifying more than 100 metabolic fingerprints (groups of biomarkers) corresponding to different toxicological MOA. The database was constructed using MS and rat serum. This database is not open access and is available as a commercial product (van Ravenzwaay et al., 2012).

The InnoMed PredTox initiative involved a group of 19 partners, including 14 pharmaceutical companies, 3 universities, and 2 technology providers, and was funded in part by the European Union. The goal of this effort was to identify biomarkers to help predict the toxicity of drug candidates. The approach taken by this group involved the combination of conventional toxicology techniques, proteomics, transcriptomics, and metabolomics using NMR, LC-MS, and GC-MS. The full database resulting from this effort was made publicly available through the European Bioinformatics Institute (Suter et al., 2011).

The Human Toxome Project, funded through the NIH, involved a group comprised of members from government, university, and nonprofit research institutes. The goal of this initiative was to apply transcriptomics and metabolomics to map and annotate "pathways of toxicology" for a defined set of endocrine disruptor compounds. A number of publications have arisen from this effort (Bouhifd et al., 2015).

A recently announced initiative called Metabolomics Standards Initiative in Toxicology (MERIT) has been launched by the European Center for Ecotoxicology and Toxicology of Chemicals (ECETOC). This group includes partners from industry, government agencies, regulators, and academia from across Europe and the United States. Participating members include the US EPA, US FDA, the European Food Safety Authority (EFSA), the UK Health and Safety Laboratory (HSL), BASF, Syngenta, Unilever, Imperial College London, University of Birmingham, UK, VU University Amsterdam, and the Metabolomics Society Data Standards Task Group. The goal of this group is to accelerate the use of metabolomics technology to improve safety assessment of chemicals. The initial focus will be on developing best practice guidelines and minimal reporting standards for the acquisition, processing, and analysis of metabolomics data (Spicer et al., 2017a,b).

Metabolomics Biomarkers in Hepatotoxicity

Hepatotoxicity is frequently observed in the form of drug-induced liver injury (DILI). DILI is a major concern both in the pharmaceutical industry and for public health. DILI is a frequent cause for termination of drug development programs and withdrawal of approved drugs from the market. Worldwide, the estimated annual incidence of DILI is between 14 and 24 per 100,000 people. DILI is one of the leading causes of acute liver failure in the United States, accounting for approximately 13% of cases. A variety of biological mechanisms and a large number of pharmaceutical compounds have been associated with DILI (Araujo et al., 2017).

There are several conventional clinical biomarkers of hepatotoxicity, including total bilirubin, alanine aminotransferase (ALT), aspartate aminotransferase, and alkaline phosphatase, among others. All of these conventional biomarkers of DILI suffer from lack of specificity and sensitivity. These markers are only present at significant levels after extensive liver damage has occurred. Because of these limitations, there has been significant interest in identification of new and more sensitive and specific biomarkers of DILI using metabolomics.

Several authors have used in vitro approaches to assess and classify liver toxicity mechanisms and pathways associated with drug-induced hepatotoxicity (Garcia-Canaveras et al., 2016; Ramirez et al., 2018). For example, Ramirez et al. (2018) treated HepG2 human hepatocyte carcinoma cells with 35 different test substances and performed metabolomics instrumental analysis on cellular extracts using GC-MS and LC-MS. Data analysis revealed concentration-response effects and patterns of metabolome changes consistent with specific liver toxicity mechanisms, including liver enzyme induction/inhibition, liver toxicity, and peroxisome proliferation. Garcia-Canaveras et al. (2016) also used human-derived HepG2 cells exposed to nonhepatotoxic control compounds and to different hepatotoxic drug compounds with known toxicity mechanisms, including steatosis, phospholipidosis, and oxidative stress. Cellular extracts were analyzed using Q-TOF LC-MS in both reversed phase and HILIC chromatography modes. Several dozen metabolites were identified, which were shown to be correlated to each toxicity mechanism. Oxidative stress damage biomarkers were found in all three mechanisms including altered levels of metabolites associated with glutathione and γ-glutamyl cycle.

Phospholipidosis was characterized by a decreased lysophospholipids to phospholipids ratio, which suggested a phospholipid degradation inhibition. Steatosis was correlated with impaired fatty acids β-oxidation and a subsequent increase in triacylglycerides synthesis. These characteristic metabolite profiles were used to develop predictive models both to discriminate between nontoxic and hepatotoxic drugs and to propose general drug toxicity mechanisms.

Acetaminophen (APAP) is a common cause of acute liver failure, accounting for nearly 50% of cases in the United States. APAP has been considered a model toxicant for DILI, and a number of groups have performed metabolomics studies in an attempt to identify new biomarkers of overdose and acute liver failure. There have recently been several human clinical studies where APAP biomarkers have been identified or tested.

In the first study, Winnike et al. (2010) used ^{1}H NMR in an attempt to identify individuals who were susceptible to DILI caused by APAP. Thirty-five healthy adults received 4 g of APAP per day for 7 days. Urine metabolome profiles obtained 2 days prior to treatment and measured by NMR were not sufficient to predict which subjects were susceptible to development of liver injury. However, metabolomics profiles obtained shortly after APAP treatment, but before ALT elevation, were able to predict subjects that were susceptible ("responders") to liver injury. Levels of cysteine and mercapturate conjugates of APAP, glycine, alanine, and acetate were higher in the urine of responders than in nonresponders. This study confirms the use of biomarkers identified using metabolomics to identity subjects at risk for developing DILI soon after initiation of drug treatment and prior to liver injury.

In another study, Kim et al. (2013) used ^{1}H NMR to analyze urine samples from 20 healthy human subjects treated with 3 g of APAP per day for 7 days. Seven endogenous urinary metabolites (trimethylamine N-oxide [TMAO], citrate, 3-chlorotyrosine, phenylalanine, glycine, hippurate, and glutarate) were responsible for the PCA separation of pre- and postdosing urine samples. The authors concluded that NMR measurement of urinary biomarkers could be applied to predict or screen for susceptibility to hepatotoxicity in humans.

Schnackenberg et al. (2017) tested potential biomarkers of APAP DILI, identified in earlier nonclinical metabolomics studies, in children who had suffered an overdose of APAP. Urine samples from the overdose patients were analyzed using a Q-TOF–based LC-MS system. The results identified metabolites from metabolic pathways noted in the previous studies and pathway analysis indicated analogous pathways were significantly altered in both rats and humans after APAP overdose. Metabolites from arginine and proline metabolism, tricarboxylic acid (TCA) cycle, taurine and hypotaurine metabolism, glycine, serine and threonine metabolism, and glutathione metabolism pathways were significant in the APAP overdose group compared with the control group. The authors concluded that a metabolomics approach may enable the discovery of specific, translational biomarkers of drug-induced hepatotoxicity that may aid in the assessment of patients.

Many other studies have used animal models and metabolomics analysis to identify biomarkers of liver hepatotoxicity. Slopianka et al. (2017) used an approach based on LC-MS and a commercially available targeted metabolomics kit to quantitate 20 specific bile acids in plasma and liver tissue from rats that had been dosed the known hepatotoxic drug methapyrilene. These data indicated that cholic acid, chenodeoxycholic acid, glycocholic acid, taurocholic acid, and deoxycholic acid were sensitive biomarkers of hepatocellular/hepatobiliary damage in rats. The bile acid perturbations were observed at earlier time points, indicating better sensitivity than conventional clinical chemistry biomarkers.

In one example with a nondrug hepatotoxin, Bando et al. (2011) used GC-MS to profile plasma and urine from rats that had been dosed with hydrazine. Biomarkers identified included amino acid precursors of glutathione (cysteine, glutamate, and glycine) and 5-oxoproline, a product of glutathione metabolism. In addition, intermediates of the TCA cycle were decreased, while components of the urea cycle and other amino acids were increased. These results confirmed the role of oxidative stress in hydrazine-induced hepatotoxicity.

Metabolomics Biomarkers in Renal Toxicity

Nephrotoxicity (renal toxicity) occurs when kidney-specific detoxification and excretion do not function properly due to damage or destruction of kidney function by toxicants. Nephrotoxicity can be the result of hemodynamic changes, direct injury to cells and tissue, inflammatory tissue injury, and obstruction of renal excretion. Nephrotoxicity is induced by a wide variety of therapeutic drugs (including antibiotics, immunosuppressants, antineoplastic agents, nonsteroidal antiinflammatory drugs, drugs of abuse, and natural medicines) and environmental pollutants (including heavy metals, organic solvents, insecticides, and glycols).

There are a number of conventional clinical biomarkers of nephrotoxicity, including creatinine clearance, serum creatinine (SCr), and blood urea nitrogen

(BUN). These conventional biomarkers of nephrotoxicity suffer from lack of specificity and sensitivity. These markers are only present at significant levels after extensive kidney damage has occurred. Because of these limitations, there has been significant interest in identification of new, more sensitive and specific biomarkers of nephrotoxicity. There is a number of emerging protein biomarkers for nephrotoxicity and there is also significant interest in discovery of potential small molecule biomarkers of nephrotoxicity using metabolomics (Zhao and Lin, 2014).

A number of studies have been performed in attempts to identify biomarkers of nephrotoxicity of antibiotics using metabolomics approaches. For example, one study measured metabolome changes in newborn rat urine following dosing with the aminoglycoside antibiotic gentamicin using GC-MS and LC-MS. A number of putative biomarkers were identified, including tryptophan, kynurenic acid, xanthurenic acid, and hippuric acid. Tryptophan levels were increased following dosing with gentamicin, while kynurenic acid, xanthurenic acid, and hippuric acid levels were decreased. Statistically significant changes in these biomarkers were present 3 days after gentamicin dosing, while significant increases in BUN and SCr were not noted until 7 days following gentamicin dosing (Hanna et al., 2013).

Immunosuppressant drugs are also well-known causes of nephrotoxicity. The calcineurin inhibitor drugs cyclosporine and tacrolimus are the basis for most protocols for the prevention of organ rejection following transplantation. A summary of metabolomics studies to identify biomarkers of immunosuppressant nephrotoxicity has recently appeared (Bonneau et al., 2016). Numerous studies in rat and human urine using NMR have identified candidate biomarker molecules, including glucose, TMAO, citrate, lactate, 15-F_{2t}-isoprostane, hippurate, inositol, creatinine, succinate, α-ketoglutarate, creatine, trimethylamine (TMA), and taurine. A metabolomics study using rat plasma and NMR identified candidate biomarkers, including creatinine, TMAO, and glutathione. A study using human serum from transplant recipients and NMR identified putative biomarkers, including lipids, glucose, hypoxanthine, lactate, succinate, and taurine.

Cisplatin is an antineoplastic agent used to treat a variety of solid tumors. One of the major side effects of cisplatin therapy is nephrotoxicity. Cisplatin nephrotoxicity is related to triggering of reactive oxygen species among several other mechanisms. The conventional nephrotoxicity biomarkers SCr and BUN are used as markers of cisplatin toxicity but these only increase markedly after significant kidney injury. Several candidate biomarkers discovered using genomic and proteomic approaches have proven to be more sensitive than the conventional biomarkers. However, there remains significant interest in identification of metabolomics biomarkers of cisplatin toxicity. In an example study, GC-MS and LC-MS were used to find cisplatin nephrotoxicity biomarkers in rat plasma (Ezaki et al., 2017). GC-MS–derived biomarkers include cysteine and cystine and 3-hydroxybutyrate. LC-MS–derived biomarkers include three acylcarnitines (AC 14:0, AC 18:1, and AC 18:2) and a phosphatidylethanolamine (C18:2-C18:2). Plasma levels of all of the metabolomics identified biomarkers showed significant changes in plasma more quickly after cisplatin administration than the conventional clinical biomarkers SCr and BUN.

Toxicants contained in natural medicines, including traditional Chinese medicines, are also potential sources of nephrotoxicity. Aristolochic acids are a family of potent nephrotoxic acids found primarily in the plant genera *Aristolochia* and *Asarum*. These herbs have been used to relieve symptoms such as cough, arthritic pain, and gastrointestinal problems. In a recent study, the investigators used LC-MS to identify biomarkers of aristolochic acids nephrotoxicity in rat urine. A number of metabolites with significantly changed levels were identified including citrate, aconitate, fumarate, glucose, creatinine, p-cresyl sulfate, indoxyl sulfate, hippuric acid, phenylacetylglycine, kynurenic acid, indole-3-carboxylic acid, spermine, uric acid, allantoin, cholic acid, and taurine. The authors performed KEGG (Kyoto encyclopedia of genes and genomes) pathway mapping to conclude that aristolochic acids nephrotoxicity perturbed biochemical pathways, including the TCA cycle, gut microflora metabolism, amino acid metabolism, purine metabolism, and bile acid biosynthesis (Zhao et al., 2015).

Melamine is a small, nitrogen-rich molecule with a number of industrial uses. It also sometimes added to animal feed and food as an adulterant to increase the apparent protein level in the feed or food when nitrogen-based protein testing is conducted. Melamine is also severely nephrotoxic when it is coadministered with cyanuric acid, a related compound, and can cause acute renal failure. A 2007 pet food recall and a 2008 milk and infant formula scandal in China were both related to nephrotoxicity outbreaks caused by melamine and cyanuric acid contamination. Since these incidents, a number of studies have used metabolomics approaches to identify possible biomarkers of melamine and cyanuric acid nephrotoxicity. A study published in 2012 used NMR to measure alterations of endogenous metabolites in kidney tissue from rats dosed with melamine and cyanuric acid. Over 30 metabolites were observed to have dramatically altered levels compared with normal rat kidney tissue (Kim et al., 2012).

Metabolomics Biomarkers of Toxicity From Chemical Agents

Biomarkers of toxicity for a number of chemical agents have been thoroughly investigated using metabolomics approaches. These chemical agents include synthetic chemicals, such as pesticides and halogenated compounds, metals, solvents, and gases. Representative examples of metabolomics studies yielding candidate biomarkers of toxicity for each of these classes of chemical agents are reviewed in this section.

Organophosphorus and carbamate insecticides are acetylcholinesterase inhibitors with well-known toxic effects and biomarkers of toxicity. Wang et al. studied the effects of the organophosphorus pesticides dichlorvos and malathion, the carbamate pesticide pirimicarb, and their mixtures in subchronic low-level exposures to mice (Wang et al., 2014). Metabolomics analysis was based on NMR spectroscopy with comparison to conventional biochemical assays. Mouse serum analysis showed significant alterations of TMAO, lactate, acetone, very low-density and low-density lipoprotein, and 3-hydroxybutyrate. Mouse liver tissue analysis showed significant alterations of lactate, glucose, choline, glutathione, alanine, glutamine, and isoleucine. Perturbations in these metabolites, along with the conventional biochemical analyses, led the authors to conclude that glucose, fatty acid, and protein metabolism, as well as energy metabolism and oxidative balance, were disturbed by the pesticides tested.

Another study examined toxicity induced by the neonicotinoid pesticide imidacloprid on the central nervous system of the freshwater snail *Lymnaea stagnalis*. This study used a combination of analytical platforms including an LC-MS–targeted metabolomics analysis focused on neurotransmitters, an untargeted LC-MS analysis, and an untargeted GC-MS analysis. Many changes in the metabolome of *L. stagnalis* were observed, including changes in amino acid and nucleotide metabolites tryptophan, proline, phenylalanine, uridine, and guanosine. Many fatty acids levels were decreased and levels of the polyamines, spermidine and putrescine, were increased indicating neuronal cell injury (Tufi et al., 2015).

Toxicity of the herbicide glyphosate has been the subject of much debate and controversy. Mesnage et al. exposed female rats to environmentally relevant low doses of a glyphosate-based herbicide for a period of 2 years (Mesnage et al., 2017). At the end of the study period, rat liver tissues were extracted and metabolomics analysis was performed using GC-MS and LC-MS. Over 50 metabolites were found to be statistically significantly altered. This information, in conjunction with proteomic analysis, was interpreted to be consistent with development of nonalcoholic fatty liver disease and its progression to nonalcoholic steatohepatosis (NASH).

The fungicide vinclozolin has been associated with endocrine disruption and antiandrogenic effects. van Ravenzwaay et al. (2013) used vinclozolin as a model compound and explored metabolomics profiles of plasma from rats dosed with vinclozolin using GC-MS and LC-MS. Results of the metabolome profiling were matched against a proprietary database of metabolome profiles from compounds with known MOA. The best databases matches were with MOA that included liver toxicity and enzyme induction, inhibition of adrenal steroid synthesis, and antiandrogenic effects.

Potential toxicity of the halogenated flame retardants tetrabromobisphenol A (TBBPA) and tetrachlorobisphenol A (TCBPA) was investigated in a study in which embryos of the fish species *Oryzias melastigma* were exposed to environmentally relevant concentrations of the two compounds. Embryo tissue was extracted and analyzed using GC-MS. Lactate and dopa were identified as potential biomarkers of the developmental toxicity and related genetic effects of TBBPA and TCBPA. Moreover, analysis of offspring embryos indicated that disorders of the neural system and disruptions of a number of metabolic pathways induced by TBBPA and TCBPA exposure were heritable (Ye et al., 2016).

Wang et al. used NMR and LC-MS to investigate the effects of another brominated flame retardant, hexabromocyclododecane (HBCD), on the metabolome of mouse urine (Wang et al., 2016). Female mice were dosed with HBCD by oral gavage. Untargeted metabolomics was performed using NMR, whereas targeted metabolic profiling of 20 amino acids was performed using LC-MS. Analysis of NMR data showed perturbed levels of citrate, 2-ketoglutarate, alanine, acetate, formate, TMA, 3-hydroxybutyrate, and malonic acid. LC-MS data showed perturbed levels of alanine, lysine, and phenylalanine. Interpretation of these results led to the conclusion that HBCD exposure caused disturbances in metabolic pathways, including TCA cycle, lipid metabolism, gut microbial metabolism, and homeostasis of amino acids.

Metal toxicity is major concern worldwide. Correspondingly, significant effort has been devoted to metabolomics study and identification of new and sensitive biomarkers of toxic metal exposure. Three example studies involving arsenic, nickel, and thallium are described here.

Arsenic pollution is a serious environmental health risk in many locations, including areas in China where significant dietary and environmental exposure to arsenic is a concern. Zhang et al. (2014) sought to identify dose-dependent biomarkers of arsenic exposure in a Chinese male cohort. Urinary arsenic species were measured using high-performance liquid chromatography–inductively coupled plasma mass spectrometry. Urinary metabolomics data were acquired using LC-MS. Arsenic-related biomarkers were investigated by comparing results from the first and fifth quintiles of arsenic exposure. Five potential biomarkers related to arsenic exposure were identified: testosterone, guanine, hippurate, acetyl-*N*-formyl-5-methoxykynurenamine, and serine. Interpretation of these biomarkers suggested that oxidative stress and endocrine disruption were associated with increased urinary arsenic levels. Testosterone, guanine, and hippurate could discriminate the first and fifth quintiles of arsenic exposure with AUROC values of 0.89, 0.87, and 0.83, respectively.

Nickel exposure was investigated by measurement of metabolomics changes in rat urine using NMR. Male rats were dosed intraperitoneally with $NiCl_2$ at several levels. Metabolomics profiles showed significant changes in levels of citrate, dimethylamine, creatinine, choline, TMAO, phenylalanine, and hippurate. These biomarkers are consistent with the known mechanism of nickel-induced nephrotoxicity. The authors speculate that this combination of biomarkers could be useful in predicting nickel-induced nephrotoxicity (Tyagi et al., 2012).

The same research group also investigated thallium toxicity by NMR metabolomics measurement in mouse urine. Male mice were injected intraperitoneally with Tl_2SO_4 at three different doses. Metabolic profiles showed significant changes in levels of creatinine, taurine, hippurate, 3-hydroxybutyrate, TMA, and choline. These metabolic perturbations could be seen as soon as 3 h after thallium dosing, indicating that these compounds could potentially serve as rapid and sensitive biomarkers of thallium exposure (Tyagi et al., 2011).

Hydrogen sulfide is an extremely poisonous gas and is a leading cause of workplace gas inhalation deaths worldwide. There are currently no specific, direct tests that can be used to diagnose poisoned patients. In an attempt to identify biomarkers of hydrogen sulfide exposure, a GC-MS metabolomics approach has been used with serum from rats chronically poisoned with hydrogen sulfide (Deng et al., 2015). Adult male Sprague–Dawley rats were exposed to 20 ppm of hydrogen sulfide gas twice per day for 1 h for each exposure. Serum samples were measured after 40 days of twice daily exposures. Alterations in levels of a number of metabolites were observed, including citrate, galactose, lactate, mannose, inositol, urea, phosphate, alanine, valine, and hexadecanoic, linoleic, and arachidonic acids. These changes corresponded to changes in metabolic pathways, including lipid metabolism, energy metabolism, and amino acid metabolism. The authors postulate that the identified candidate biomarkers could provide a novel means for detection of chronic hydrogen sulfide poisoning.

Carbon tetrachloride is a well-known hepatotoxicant with a number of industrial uses, including as a solvent, cleaning agent, refrigerant, and fire suppression. A number of metabolomics studies with carbon tetrachloride have consistently shown alterations to lipids and bile acids in plasma. Sun et al. (2014) performed a comprehensive study with rats in which clinical chemistry, histopathological, and transcriptomic data were integrated with data from multiple analytical metabolomics platforms including untargeted LC-MS and targeted GC-MS and LC-MS. Metabolomics analysis was performed on plasma and liver tissue extracts. Metabolomics results showed that levels of total fatty acids increased in the liver but decreased in blood. Primary bile acids increased in both liver and blood, whereas secondary and conjugated bile acids decreased in the liver and increased in the blood, indicating that the bile acid conjugation pathway and bile acid uptake by the liver were inhibited by carbon tetrachloride.

Metabolomics Biomarkers in Neurotoxicity

Alzheimer's disease (AD) is the most common neurodegenerative disease worldwide. AD causes approximately 70% of dementia in elderly people and there are currently an estimated 50 million people worldwide with AD. This number is projected to increase to over 130 million by 2050. There are currently no treatments effective at slowing the progression of AD. A number of genetic mutations have been associated with AD; however, approximately 95% of cases are sporadic and without a definitive genetic link. Although there is general consensus around a number of signatures of AD, including deposition of β-amyloid plaques, hyperphosphorylation of tau protein, oxidative stress, inflammation, abnormal metal homeostasis, and others, the underlying molecular mechanisms of AD etiology and development are not well understood. There is tremendous interest in the potential of metabolomics to help identify biomarkers that may be useful for early detection, improved prognosis, and monitoring the progression of AD. There have been hundreds of metabolomics studies with AD performed to date. Recent reviews summarize the current state of this work (Gonzalez-Dominguez et al., 2017; Wilkins and Trushina, 2017).

In one noteworthy study, a targeted metabolomics LC-MS approach was used to analyze serum samples from AD patients and controls. Metabolomics data were correlated with a number of clinical measures of mild cognitive impairment and AD. Results showed that sphingomyelins and ether-containing phosphatidylcholines were perturbed in preclinical biomarker-defined

AD stages, whereas acylcarnitines and amines, including valine and α-aminoadipic acid, were perturbed in clinical AD stages (Toledo et al., 2017).

Parkinson's disease (PD) is the second most common neurodegenerative disease worldwide. An estimated 10 million people worldwide suffer from PD. PD is characterized by loss of dopaminergic neurons and by the time of clinical manifestation of disease, 60%−80% of the dopaminergic neurons are lost in the striatum. Currently, there is no way to definitively diagnose PD at its early stages. Extensive metabolomics studies have been undertaken with PD in hopes of identifying biomarkers which may lead to early diagnosis, improved prognosis, and monitoring the progression of PD. Recent reviews provide an overview of metabolomics studies of PD (Gill et al., 2018; Havelund et al., 2017).

In one example study, untargeted LC-MS was used to measure plasma and CSF metabolite profiles from early-stage PD patients. A total of 20 candidate biomarkers were identified in plasma and 14 were identified in CSF. These candidate biomarkers were used to successfully differentiate the test set from a control set with AUROC values of 0.8 for plasma and 0.9 for CSF (Stoessel et al., 2018).

Metabolomics Biomarkers of Biotoxins

Biotoxins are toxic substances that have a biological origin. Biotoxins include mycotoxins, made by fungi, and cyanotoxins, made by cyanobacteria (blue-green algae). Mycotoxins are typically associated with fungi that colonize crops and are capable of causing disease and death in both humans and animals. Cyanotoxins are produced in highest concentrations where cyanobacteria reproduce exponentially to form blooms. Cyanotoxins can cause poisoning and death in both humans and animals. Examples of applications of metabolomics to identify biomarkers of toxicity from mycotoxin and cyanotoxin exposure are described here.

Liu et al. used NMR to study the effects of zearalenone (ZEN) intragastric dosing on the urine and plasma metabolome of rats (Liu et al., 2013). In plasma, levels of glucose, lactate, *N*-acetyl glycoprotein, *O*-acetyl glycoprotein, and propionate were elevated, whereas levels of tyrosine, branched chain amino acids, and choline were reduced. In urine, levels of allantoin, choline, *N*-methylnicotinamide, 1-methylhistidine, acetoacetate, acetone, and indoxyl sulfate were elevated, whereas levels of butyrate, lactate, and nicotinate were reduced. The authors concluded that ZEN exposure caused oxidative stress and changed systemic metabolic processes, including cell membrane metabolism, protein biosynthesis, glycolysis, and gut microbiota metabolism.

Another study examined the combined toxic effects of ZEN and deoxynivalenol (DON) in liver and serum of mice. Mice were treated with ZEN only, DON only, or a combination of ZEN and DON. Histopathological and serum biochemical measurements were collected and metabolomics profiling was performed with GC-MS. A number of metabolomics biomarkers were identified for each treatment group. Interestingly, all data indicated that the combination ZEN + DON treatment had slightly weaker toxicity than the individual ZEN and DON treatments. The authors concluded that combined ZEN + DON treatment had an antagonistic toxicity effect in mice (Ji et al., 2017).

De Pascali et al. (2017) used NMR metabolomics to investigate the effects of a single bolus dose of a combination of five mycotoxins on piglet urine. Dosed mycotoxins included ZEN, DON, ochratoxin A, and fumonisin $B_1 + B_2$ ($FB_1 + FB_2$). Dosed mycotoxin levels were close to the established maximum guidance levels for animal feed. Perturbed urine levels of creatinine, *p*-cresol glucuronide, phenyl acetate glycine, betaine, and TMAO were detected. The postulated reason for these changes was alteration of the gut microbiome. Interestingly, the changes in metabolite levels persisted for at least a week after bolus dosing suggesting a sustained impact on gut microbiota.

The metabolomics of rats and mice exposed to microcystin-LR has been investigated using NMR profiling. Microcystin-LR is a cyclic heptapeptide that has been shown to have acute hepatotoxic effects. In a study published in 2013, rat urine metabolite levels of citrate, 2-oxoglutarate, succinate, hippurate, betaine, taurine, creatine, guanidinoacetate, dimethylglycine, urocanic acid, and bile acids were found to be altered. The extent of alteration of metabolites levels was correlated with the severity of histopathological findings (Cantor et al., 2013). A more recent study combined proteomics and NMR metabolomics data from liver and serum of mice exposed to chronic low doses of microcystin-LR. A large number of metabolite levels were found to be altered in both liver tissue and serum. Interpretation of these alterations, along with histopathological data, led to the assessment that chronic microcystin-LR dosing led to hepatic inflammation resulting in NASH disease (He et al., 2017).

Metabolomics Biomarkers of Nanomaterials

Nanoparticles are particles between 1 and 100 nanometers (nm) in size. Nanoparticles are of interest because of their small size and high surface to volume ratio. Nanoparticles frequently exhibit unique physical and chemical properties that are not observed with bulk material of the same composition. Nanoparticles are made from a variety of substances, including fullerenes, metals, ceramics, polymers, lipids, and others. Nanoparticles are used in applications including catalysis, imaging, energy research, environmental applications, and consumer and medical products. The unique properties of nanoparticles can also lead to toxicity that is unique compared with bulk material of the same composition. Because of their small size, nanoparticles can be readily inhaled, ingested, and absorbed through the skin (Hobson et al., 2016). As awareness of and interest in nanotoxicology has grown over the last several years, so has interest in applying systems biology approaches (including metabolomics) in an effort to develop a comprehensive understanding of the toxicological mechanisms of these materials. Costa and Fadeel (2016) have reviewed systems biology approaches in the context of nanotoxicology.

Titanium dioxide is a whitening agent that is used in many industrial and consumer products including toothpaste. While titanium dioxide nanoparticles are not typically used in toothpaste, they are used in dental cements to improve mechanical and antibacterial properties. Garcia-Contreras et al. (2015) examined the alteration of metabolomics profiles caused by titanium dioxide (TiO_2) nanoparticles in human gingival fibroblast (HGF) cells. HGF cells were treated with TiO_2, extracted, and metabolomics analysis performed using capillary electrophoresis–mass spectrometry. Results showed that the TiO_2 nanoparticles augmented metabolic changes in a number of pathways, including amino acid, urea cycle, polyamine, *S*-adenosylmethionine, and glutathione synthesis.

Hadrup et al. compared the effects of selenium nanoparticles and equimolar bulk sodium selenite administered by oral gavage on the rat urine metabolome using LC-MS (Hadrup et al., 2016) Evaluation of all significantly changed metabolites showed that selenium nanoparticles and sodium selenite induced similar dose-dependent changes in the metabolite pattern. Altered metabolites included decenedioic acid and hydroxydecanedioic acid for both selenium formulations, whereas dipeptides were altered only for sodium selenite. The authors concluded that these effects may reflect altered fatty acid and protein metabolism.

Silver nanoparticles are widely used in consumer and biomedical products. Due to their antimicrobial properties, they are used in medical products such as wound dressings, implants, and catheters. A study used NMR metabolomics, combined with conventional clinical chemistry and histological examination, to characterize organ and systemic response in mice (Jarak et al., 2017). Mice were intravenously dosed with silver nanoparticles and serum and tissues were analyzed by NMR. In serum, significant changes were observed in glucose, lactate, pyruvate, lysine, and phosphocreatine. These changes were interpreted as possibly related to oxidative stress caused by the silver nanoparticles.

Metabolomics Biomarkers of Drugs of Abuse

Metabolomics has also been applied to study acute and chronic toxic effects of drugs of abuse, including stimulants, opioids, and designer drugs. Metabolomics has not only assisted in understanding the molecular mechanisms behind drug dependence but can also be applied to elucidate the mechanism(s) of toxicity of new and emerging designer drugs. Examples of metabolomics applications to discern mechanisms of toxicity and identify related biomarkers for a stimulant (methamphetamine [MA]), an opioid (heroin), and a designer drug (MAM-2201) are described next (Zaitsu et al., 2016).

MA is an illegal stimulant drug of abuse with serious negative health consequences, including cardiovascular disease, psychosis, depression, and dental pathology. A study by McClay et al. examined the metabolomics of single and repeated exposures of MA in male mouse brain tissue extracts using GC-MS and LC-MS (McClay et al., 2013). Mice were dosed intraperitoneally with either a single dose of MA or with daily dosing of MA over a period of 5 days. After a single MA exposure, levels of compounds, including lactate, malate, succinate, fumarate, tryptophan, and 2-hydroxyglutarate, were altered significantly. After repeated, 5 days, MA exposure, additional specific compound alterations included phosphocholine and ergothioneine. The authors concluded that these data confirmed and extended existing models of MA action in the brain, in which an initial increase in energy metabolism gives way to disruption of mitochondria and phospholipid metabolism pathways and increased endogenous antioxidant response.

Heroin is the most widely used opioid in the world. Repeated heroin use changes the physical structure and physiology of the brain, creating long-term imbalances in neuronal and hormonal systems, which are not easily reversed. Li et al. (2017) investigated the metabolomics of repeated daily dosing and withdrawal of heroin in male mouse brain tissue extracts using LC-MS. Mice in the heroin-treated group were dosed intraperitoneally with heroin twice per day for 12 consecutive days. Mice in a heroin-withdrawal group were dosed twice per day for 10 consecutive days followed by a 2-day withdrawal period. Perturbed metabolites in the heroin-treated group included amino acids, TCA cycle intermediates, neurotransmitters, and nucleotides. In the heroin-withdrawal group, a marked reduction in histidine, increases in phenylalanine and tryptophan, and a decrease of catecholamines to baseline levels were measured. Melatonin was significantly decreased in the heroin-treated group, while N-acetylserotonin, a precursor of melatonin, was increased in the heroin-withdrawal group. The authors interpreted these results as an indication that heroin disrupts not only energy metabolism but also the biosynthesis of both catecholamines and melatonin in the mouse brain.

MAM-2201 is a synthetic cannabinoid that has recently been used as a designer drug and has been reported to show acute toxicity in humans. Zaitsu et al. (2015) studied the disruption of the metabolome of the male rat cerebrum using GC-MS. Mice were dosed intraperitoneally with a single dose of MAM-2201 at either a high- or low-dose level. The levels of 12 metabolites were significantly changed in a dose-dependent manner compared with the control group. Altered metabolites included creatinine, fructose, gluconic acid, glutamic acid, psicose, malic acid, phenylalanine, sorbitol, sorbose, succinic acid, tagatose, and valine. These results indicated that MAM-2201 disrupted energy metabolism in the rat cerebrum and also glutamatergic neurotransmission.

CONCLUDING REMARKS AND FUTURE DIRECTIONS

Metabolomics is a maturing science that is now nearly 20 years old. Metabolomics has been widely applied in toxicology with over 4000 publications in the field. Many of these publications have identified putative or candidate biomarkers correlated with susceptibility, exposure, and/or effects of toxicants. All of these studies demonstrate that metabolomics approaches for the identification of small molecule endogenous biomarkers of toxicity are extremely promising. However, while some areas, such as hepatotoxicity of drugs, have seen sufficient work that reproducible patterns of biomarkers have emerged, many of these studies have been very preliminary and their results are not validated. Many barriers and challenges exist to establishing validated metabolomics-derived biomarkers of toxicity. For example, there are currently no standardized study design, analytical instrumental analysis, or data analysis and reporting protocols for metabolomics. Indeed, the literature abounds with metabolomics studies, in which similar toxicants, species, and biological specimens are examined, but very different sets of putative biomarkers have been identified (Poste, 2011). In spite of the successes of metabolomics, there are currently no FDA-approved biomarkers for clinical use, which have been discovered with metabolomics approaches (Trivedi et al., 2017).

For biomarkers of toxicity discovered using metabolomics to become more practically useful, they must be validated. For these biomarkers to become validated, they must be quantitated more accurately and precisely and measured in larger groups of subjects. More accurate quantitation can be accomplished using targeted metabolomics approaches with appropriate internal standards and quality control samples, similar to conventional validated bioanalytical methods. An obvious barrier to performing larger-scale trials is the cost involved. To overcome this barrier, adequate funding for such trials must exist and, more than likely, government-industry partnerships will need to be formed. If this can be accomplished, and if these candidate biomarkers can become rigorously validated, then the great potential for application of small endogenous molecule biomarkers of toxicity identified using metabolomics may be realized.

References

Araujo, A.M., Carvalho, M., Carvalho, F., et al., 2017. Metabolomic approaches in the discovery of potential urinary biomarkers of drug-induced liver injury (DILI). Crit. Rev. Toxicol. 47 (8), 633–649. https://doi.org/10.1080/10408444.2017.1309638.

Bando, K., Kawahara, R., Kunimatsu, T., et al., 2010. Influences of biofluid sample collection and handling procedures on GC-MS based metabolomic studies. J. Biosci. Bioeng. 110 (4), 491–499. https://doi.org/10.1016/j.jbiosc.2010.04.010.

Bando, K., Kunimatsu, T., Sakai, J., et al., 2011. GC-MS-based metabolomics reveals mechanism of action for hydrazine induced hepatotoxicity in rats. J. Appl. Toxicol. 31 (6), 524–535. https://doi.org/10.1002/jat.1591.

Barnes, S., Benton, H.P., Casazza, K., et al., 2016. Training in metabolomics research. II. Processing and statistical analysis of metabolomics data, metabolite identification, pathway analysis, applications of metabolomics and its future. J. Mass Spectrom. 51 (8), 535–548. https://doi.org/10.1002/jms.3780.

Beger, R.D., Dunn, W., Schmidt, M.A., et al., 2016. Metabolomics enables precision medicine: "A White Paper, Community Perspective". Metabolomics 12 (10), 149.

Bonneau, E., Tetreault, N., Robitaille, R., et al., 2016. Metabolomics: perspectives on potential biomarkers in organ transplantation and immunosuppressant toxicity. Clin. Biochem. 49 (4–5), 377–384.

Bouhifd, M., Andersen, M.E., Baghdikian, C., et al., 2015. The human toxome project. ALTEX Altern. Anim. Exp. 32 (2), 112–124.

Bouhifd, M., Hartung, T., Hogberg, H.T., et al., 2013. Review: toxicometabolomics. J. Appl. Toxicol. 33 (12), 1365–1383.

Cantor, G.H., Beckonert, O., Bollard, M.E., et al., 2013. Integrated histopathological and urinary metabonomic investigation of the pathogenesis of microcystin-LR toxicosis. Vet. Pathol. 50 (1), 159–171.

Chetwynd, A.J., Dunn, W.B., Rodriguez-Blanco, G., 2017. Collection and preparation of clinical samples for metabolomics. Adv. Exp. Med. Biol. 965, 19–44.

Costa, P.M., Fadeel, B., 2016. Emerging systems biology approaches in nanotoxicology: towards a mechanism-based understanding of nanomaterial hazard and risk. Toxicol. Appl. Pharmacol. 299, 101–111.

Courant, F., Antignac, J.P., Dervilly-Pinel, G., et al., 2014. Basics of mass spectrometry based metabolomics. Proteomics 14 (21–22), 2369–2388.

De Pascali, S.A., Gambacorta, L., Oswald, I.P., et al., 2017. (1)H NMR and MVA metabolomic profiles of urines from piglets fed with boluses contaminated with a mixture of five mycotoxins. Biochem. Biophys. Rep. 11, 9–18.

Deng, M.J., Zhang, M.L., Huang, X.L., et al., 2015. A gas chromatography-mass spectrometry based study on serum metabolomics in rats chronically poisoned with hydrogen sulfide. J. Forensic Leg. Med. 32, 59–63.

Di Guida, R., Engel, J., Allwood, J.W., et al., 2016. Non-targeted UHPLC-MS metabolomic data processing methods: a comparative investigation of normalisation, missing value imputation, transformation and scaling. Metabolomics 12, 93.

Dudzik, D., Barbas-Bernardos, C., Garcia, A., et al., 2018. Quality assurance procedures for mass spectrometry untargeted metabolomics. A review. J. Pharm. Biomed. Anal. 147, 149–173.

Dunn, W.B., Erban, A., Weber, R.J.M., et al., 2012a. Mass appeal: metabolite identification in mass spectrometry-focused untargeted metabolomics. Metabolomics 9 (S1), 44–66.

Dunn, W.B., Wilson, I.D., Nicholls, A.W., et al., 2012b. The importance of experimental design and QC samples in large-scale and MS-driven untargeted metabolomic studies of humans. Bioanalysis 4 (18), 2249–2264.

Ebbels, T.M.D., De Iorio, M., 2011. Statistical Data Analysis in Metabolomics. Handbook of Statistical Systems Biology, pp. 163–180.

Engskog, M.K.R., Haglöf, J., Arvidsson, T., et al., 2016. LC–MS based global metabolite profiling: the necessity of high data quality. Metabolomics 12 (7), 114.

Ezaki, T., Nishiumi, S., Azuma, T., et al., 2017. Metabolomics for the early detection of cisplatin-induced nephrotoxicity. Toxicol. Res. 6 (6), 843–853.

Fang, M., Ivanisevic, J., Benton, H.P., et al., 2015. Thermal degradation of small molecules: a global metabolomic investigation. Anal. Chem. 87 (21), 10935–10941. https://doi.org/10.1021/acs.analchem.5b03003.

Fiehn, O., 2001. Combining genomics, metabolome analysis, and biochemical modelling to understand metabolic networks. Comp. Funct. Genom. 2 (3), 155–168.

Garcia-Canaveras, J.C., Castell, J.V., Donato, M.T., et al., 2016. A metabolomics cell-based approach for anticipating and investigating drug-induced liver injury. Sci. Rep. 6, 27239.

Garcia-Contreras, R., Sugimoto, M., Umemura, N., et al., 2015. Alteration of metabolomic profiles by titanium dioxide nanoparticles in human gingivitis model. Biomaterials 57, 33–40.

Gill, E.L., Koelmel, J.P., Yost, R.A., et al., 2018. Mass spectrometric methodologies for investigating the metabolic signatures of Parkinson's disease: current progress and future perspectives. Anal. Chem. https://doi.org/10.1021/acs.analchem.7b04084.

Goldansaz, S.A., Guo, A.C., Sajed, T., et al., 2017. Livestock metabolomics and the livestock metabolome: a systematic review. PLoS One 12 (5), e0177675.

Gonzalez-Dominguez, R., Sayago, A., Fernandez-Recamales, A., 2017. Metabolomics in Alzheimer's disease: the need of complementary analytical platforms for the identification of biomarkers to unravel the underlying pathology. J. Chromatogr. B Anal. Technol. Biomed. Life Sci. 1071, 75–92.

Gowda, H., Ivanisevic, J., Johnson, C.H., et al., 2014. Interactive XCMS online: simplifying advanced metabolomic data processing and subsequent statistical analyses. Anal. Chem. 86 (14), 6931–6939.

Hadrup, N., Loeschner, K., Skov, K., et al., 2016. Effects of 14-day oral low dose selenium nanoparticles and selenite in rat-as determined by metabolite pattern determination. PeerJ 4, e2601.

Hanna, M.H., Segar, J.L., Teesch, L.M., et al., 2013. Urinary metabolomic markers of aminoglycoside nephrotoxicity in newborn rats. Pediatr. Res. 73 (5), 585–591.

Havelund, J.F., Heegaard, N.H.H., Faergeman, N.J.K., et al., 2017. Biomarker research in Parkinson's disease using metabolite profiling. Metabolites 7 (3), 42.

He, J., Li, G.Y., Chen, J., et al., 2017. Prolonged exposure to low-dose microcystin induces nonalcoholic steatohepatitis in mice: a systems toxicology study. Arch. Toxicol. 91 (1), 465–480.

Hernandes, V.V., Barbas, C., Dudzik, D., 2017. A review of blood sample handling and pre-processing for metabolomics studies. Electrophoresis 38 (18), 2232–2241.

Hobson, D.W., Roberts, S.M., Shvedova, A.A., et al., 2016. Applied nanotoxicology. Int. J. Toxicol. 35 (1), 5–16.

Jarak, I., Carrola, J., Barros, A.S., et al., 2017. From the cover: metabolism modulation in different organs by silver nanoparticles: an NMR metabolomics study of a mouse model. Toxicol. Sci. 159 (2), 422–435.

Ji, J., Zhu, P., Cui, F., et al., 2017. The antagonistic effect of mycotoxins deoxynivalenol and zearalenone on metabolic profiling in serum and liver of mice. Toxins 9 (1), 28.

Johnson, C.H., Ivanisevic, J., Siuzdak, G., 2016. Metabolomics: beyond biomarkers and towards mechanisms. Nat. Rev. Mol. Cell Biol. 17 (7), 451–459.

Kim, J.W., Ryu, S.H., Kim, S., et al., 2013. Pattern recognition analysis for hepatotoxicity induced by acetaminophen using plasma and urinary H-1 NMR-based metabolomics in humans. Anal. Chem. 85 (23), 11326–11334.

Kim, T.H., Ahn, M.Y., Lim, H.J., et al., 2012. Evaluation of metabolomic profiling against renal toxicity in Sprague-Dawley rats treated with melamine and cyanuric acid. Arch. Toxicol. 86 (12), 1885–1897.

Li, R.S., Takeda, T., Ohshima, T., et al., 2017. Metabolomic profiling of brain tissues of mice chronically exposed to heroin. Drug Metab. Pharmacokinet. 32 (1), 108–111.

Lindon, J.C., Keun, H.C., Ebbels, T.M.D., et al., 2005. The Consortium for Metabonomic Toxicology (COMET): aims, activities and achievements. Pharmacogenomics 6 (7), 691–699.

Liu, G., Yan, T., Wang, J., et al., 2013. Biological system responses to zearalenone mycotoxin exposure by integrated metabolomic studies. J. Agric. Food Chem. 61 (46), 11212–11221.

Mahieu, N.G., Patti, G.J., 2017. Systems-level annotation of a metabolomics data set reduces 25 000 features to fewer than 1000 unique metabolites. Anal. Chem. 89 (19), 10397–10406.

McClay, J.L., Adkins, D.E., Vunck, S.A., et al., 2013. Large-scale neurochemical metabolomics analysis identifies multiple compounds associated with methamphetamine exposure. Metabolomics 9 (2), 392–402.

Mesnage, R., Renney, G., Seralini, G.E., et al., 2017. Multiomics reveal non-alcoholic fatty liver disease in rats following chronic exposure to an ultra-low dose of Roundup herbicide. Sci. Rep. 7, 39328.

Nagana Gowda, G.A., Raftery, D., 2013. Biomarker discovery and translation in metabolomics. Curr. Metab. 1 (3), 227–240.

Nassar, A.F., Wu, T., Nassar, S.F., et al., 2017. UPLC-MS for metabolomics: a giant step forward in support of pharmaceutical research. Drug Discov. Today 22 (2), 463–470.

Nicholson, J.K., Lindon, J.C., Holmes, E., 1999. 'Metabonomics': understanding the metabolic responses of living systems to pathophysiological stimuli via multivariate statistical analysis of biological NMR spectroscopic data. Xenobiotica 29 (11), 1181–1189.

Patti, G.J., Yanes, O., Siuzdak, G., 2012. Innovation: metabolomics: the apogee of the omics trilogy. Nat. Rev. Mol. Cell Biol. 13 (4), 263–269.

Poste, G., 2011. Bring on the biomarkers. Nature 469 (7329), 156–157.

Ramirez, T., Daneshian, M., Kamp, H., et al., 2013. Metabolomics in toxicology and preclinical research. ALTEX Altern. Anim. Exp. 30 (2), 209–225.

Ramirez, T., Strigun, A., Verlohner, A., et al., 2018. Prediction of liver toxicity and mode of action using metabolomics in vitro in HepG2 cells. Arch. Toxicol. 92 (2), 893–906.

Raterink, R.-J., Lindenburg, P.W., Vreeken, R.J., et al., 2014. Recent developments in sample-pretreatment techniques for mass spectrometry-based metabolomics. TrAC Trends Anal. Chem. 61, 157–167.

Ren, S., Hinzman, A.A., Kang, E.L., et al., 2015. Computational and statistical analysis of metabolomics data. Metabolomics 11 (6), 1492–1513.

Roberts, L.D., Souza, A.L., Gerszten, R.E., et al., 2012. Targeted metabolomics. Curr. Protoc. Mol. Biol. 98 (Unit 30.2), 1–24.

Robertson, D.G., 2005. Metabonomics in toxicology: a review. Toxicol. Sci. 85 (2), 809–822.

Robertson, D.G., Watkins, P.B., Reily, M.D., 2011. Metabolomics in toxicology: preclinical and clinical applications. Toxicol. Sci. 120, S146–S170.

Rouquie, D., Heneweer, M., Botham, J., et al., 2015. Contribution of new technologies to characterization and prediction of adverse effects. Crit. Rev. Toxicol. 45 (2), 172–183.

Roux, A., Lison, D., Junot, C., et al., 2011. Applications of liquid chromatography coupled to mass spectrometry-based metabolomics in clinical chemistry and toxicology: a review. Clin. Biochem. 44 (1), 119–135.

Schnackenberg, L.K., Sun, J., Bhattacharyya, S., et al., 2017. Metabolomics analysis of urine samples from children after acetaminophen overdose. Metabolites 7 (3), 46.

Slopianka, M., Herrmann, A., Pavkovic, M., et al., 2017. Quantitative targeted bile acid profiling as new markers for DILI in a model of methapyrilene-induced liver injury in rats. Toxicology 386, 1–10.

Spicer, R., Salek, R.M., Moreno, P., et al., 2017a. Navigating freely-available software tools for metabolomics analysis. Metabolomics 13 (9), 106.

Spicer, R.A., Salek, R., Steinbeck, C., 2017b. A decade after the metabolomics standards initiative it's time for a revision. Sci. Data 4, 170138.

Stoessel, D., Schulte, C., Teixeira dos Santos, M.C., et al., 2018. Promising metabolite profiles in the plasma and CSF of early clinical Parkinson's disease. Front. Aging Neurosci. 10 (51).

Sumner, L.W., Amberg, A., Barrett, D., et al., 2007. Proposed minimum reporting standards for chemical analysis. Metabolomics 3 (3), 211–221.

Sun, J., Schmitt, T., Schnackenberg, L.K., et al., 2014. Comprehensive analysis of alterations in lipid and bile acid metabolism by carbon tetrachloride using integrated transcriptomics and metabolomics. Metabolomics 10 (6), 1293–1304.

Suter, L., Schroeder, S., Meyer, K., et al., 2011. EU framework 6 project: predictive toxicology (PredTox)-overview and outcome. Toxicol. Appl. Pharmacol. 252 (2), 73–84.

Tautenhahn, R., Patti, G.J., Rinehart, D., et al., 2012. XCMS online: a web-based platform to process untargeted metabolomic data. Anal. Chem. 84 (11), 5035–5039.

Toledo, J.B., Arnold, M., Kastenmuller, G., et al., 2017. Metabolic network failures in Alzheimer's disease: a biochemical road map. Alzheimers Dement. 13 (9), 965–984.

Trivedi, D.K., Hollywood, K.A., Goodacre, R., 2017. Metabolomics for the masses: the future of metabolomics in a personalized world. New Horiz. Transl. Med. 3 (6), 294–305.

Tufi, S., Stel, J.M., de Boer, J., et al., 2015. Metabolomics to explore imidacloprid-induced toxicity in the central nervous system of the freshwater snail *Lymnaea stagnalis*. Environ. Sci. Technol. 49 (24), 14529–14536.

Tyagi, R., Rana, P., Gupta, M., et al., 2012. Urinary metabolomic phenotyping of nickel induced acute toxicity in rat: an NMR spectroscopy approach. Metabolomics 8 (5), 940–950.

Tyagi, R., Rana, P., Khan, A.R., et al., 2011. Study of acute biochemical effects of thallium toxicity in mouse urine by NMR spectroscopy. J. Appl. Toxicol. 31 (7), 663–670.

Uppal, K., Walker, D.I., Liu, K., et al., 2016. Computational metabolomics: a framework for the million metabolome. Chem. Res. Toxicol. 29 (12), 1956–1975.

van Ravenzwaay, B., Herold, M., Kamp, H., et al., 2012. Metabolomics: a tool for early detection of toxicological effects and an opportunity for biology based grouping of chemicals-from QSAR to QBAR. Mut. Res. Genet. Toxicol. Environ. Mut. 746 (2), 144–150.

van Ravenzwaay, B., Kolle, S.N., Ramirez, T., et al., 2013. Vinclozolin: a case study on the identification of endocrine active substances in the past and a future perspective. Toxicol. Lett. 223 (3), 271–279.

Vinaixa, M., Schymanski, E.L., Neumann, S., et al., 2016. Mass spectral databases for LC/MS- and GC/MS-based metabolomics: state of the field and future prospects. TrAC Trends Anal. Chem. 78, 23–35.

Vuckovic, D., 2012. Current trends and challenges in sample preparation for global metabolomics using liquid chromatography-mass spectrometry. Anal. Bioanal. Chem. 403 (6), 1523–1548.

Wang, D.Z., Zhang, P., Wang, X.R., et al., 2016. NMR- and LC-MS/MS-based urine metabolomic investigation of the subacute effects of hexabromocyclododecane in mice. Environ. Sci. Pollut. Res. 23 (9), 8500–8507.

Wang, P., Wang, H.P., Xu, M.Y., et al., 2014. Combined subchronic toxicity of dichlorvos with malathion or pirimicarb in mice liver and serum: a metabonomic study. Food Chem. Toxicol. 70, 222–230.

Wang, Y., Liu, S., Hu, Y., et al., 2015. Current state of the art of mass spectrometry-based metabolomics studies – a review focusing on wide coverage, high throughput and easy identification. RSC Adv. 5 (96), 78728–78737.

Wilkins, J.M., Trushina, E., 2017. Application of metabolomics in Alzheimer's disease. Front. Neurol. 8, 719.

Winnike, J.H., Li, Z., Wright, F.A., et al., 2010. Use of pharmaco-metabonomics for early prediction of acetaminophen-induced hepatotoxicity in humans. Clin. Pharmacol. Ther. 88 (1), 45–51.

Wishart, D.S., 2016. Emerging applications of metabolomics in drug discovery and precision medicine. Nat. Rev. Drug Discov. 15 (7), 473–484.

Xia, J., Broadhurst, D.I., Wilson, M., et al., 2013. Translational biomarker discovery in clinical metabolomics: an introductory tutorial. Metabolomics 9 (2), 280–299.

Xia, J., Sinelnikov, I.V., Han, B., et al., 2015. MetaboAnalyst 3.0–making metabolomics more meaningful. Nucleic Acids Res. 43 (W1), W251–W257.

Xia, J., Wishart, D.S., 2016. Using metaboAnalyst 3.0 for comprehensive metabolomics data analysis. Curr. Protoc. Bioinform. 55 (Unit 14.10), 1–91.

Ye, G.Z., Chen, Y.J., Wang, H.O., et al., 2016. Metabolomics approach reveals metabolic disorders and potential biomarkers associated with the developmental toxicity of tetrabromobisphenol A and tetrachlorobisphenol A. Sci. Rep. 6, 35257.

Zaitsu, K., Hayashi, Y., Kusano, M., et al., 2016. Application of metabolomics to toxicology of drugs of abuse: a mini review of metabolomics approach to acute and chronic toxicity studies. Drug Metab. Pharmacokinet. 31 (1), 21–26.

Zaitsu, K., Hayashi, Y., Suzuki, K., et al., 2015. Metabolome disruption of the rat cerebrum induced by the acute toxic effects of the synthetic cannabinoid MAM-2201. Life Sci. 137, 49–55.

Zhang, J., Shen, H.Q., Xu, W.P., et al., 2014. Urinary metabolomics revealed arsenic internal dose-related metabolic alterations: a proof-of-concept study in a Chinese male cohort. Environ. Sci. Technol. 48 (20), 12265–12274.

Zhao, Y.-Y., Lin, R.-C., 2014. Metabolomics in nephrotoxicity. Adv. Clin. Chem. 65, 69–89.

Zhao, Y.Y., Tang, D.D., Chen, H., et al., 2015. Urinary metabolomics and biomarkers of aristolochic acid nephrotoxicity by UPLC-QTOF/HDMS. Bioanalysis 7 (6), 685–700.

CHAPTER

63

Transcriptomic Biomarkers in Safety and Risk Assessment of Chemicals

David T. Szabo[1], *Amy A. Devlin*[2]

[1]PPG Industries Incorporated, Pittsburgh, PA, United States; [2]US Food and Drug Administration, Silver Spring, MD, United States

INTRODUCTION

Identification and use of transcriptomic biomarkers to distinguish physiological conditions or clinical stages is an emerging field that has advanced substantially this decade. This is partly due to several national and international agencies' recommendations to incorporate advanced technologies in the safety and risk assessment process. Transcriptomics is recognized as an unbiased, sensitive, and personalized approach with the potential to reveal new predictive biomarkers of disease and ultimately improve the decision-making process. When performed in a dose–response format, the observed transcriptional changes can provide both quantitative and qualitative information on the dose at which cellular processes are affected. Transcriptomic approaches have transformed the way in which physicians approach diagnosis, prognosis, and treatment and in which regulators approach risk assessment. This chapter provides fundamental insights into the promising technology of transcriptomics with multiple applications in both regulatory science and clinical research, where it can be used to discover novel biomarkers potentially useful for human safety and risk assessment (Fig. 63.1).

Transcriptomics

The transcriptome of an organism includes the total of all its RNA transcripts at one point in time. The term "transcriptome" was first used in the 1990s (Piétu et al., 1999; Velculescu et al., 1997). Although transcripts originate from less than 5% of the genome in humans and other mammals, each gene (locus of expressed DNA) may produce a variety of messenger RNA (mRNA) molecules using the process of alternative splicing. Therefore, the transcriptome has a level of complexity greater than the genome that encodes it. Modern transcriptomics is regarded as a high-throughput technology concerned with determining how the transcriptome changes with respect to various factors at a certain time point and at a given biological state. Regulation of gene expression is highly complex and underlies many fundamental biological processes such as growth, differentiation, and disease pathogenesis with the ability to adapt rapidly with tremendous variability in different tissues and in response to stimuli (Sandvik et al., 2006). Transcriptomics and global gene expression are powerful tools used in the field of toxicology and, as in most cases, toxicity is not expected to occur without alterations at the transcriptional level (Eun et al., 2008; Gibb et al., 2011). A common application of transcriptomics in toxicology is to compare gene expression in samples following chemical administration from exposed and nonexposed animals to provide a list of genes that demonstrate altered expression in the diseased group. Such findings not only advance our understanding of disease pathogenesis but they also reveal transcripts that can be qualitatively or quantitatively assessed as new biomarkers.

Within the National Institutes of Health, the term biomarker is defined as follows: a characteristic that is objectively measured and evaluated as an indicator of normal biologic processes, pathogenic processes, or pharmacologic responses to a therapeutic intervention (Ilyin et al., 2004). Although a single biomarker can be easily understood, transcriptomic studies have uncovered aggregate measures composed of multiple genes as informative biomarkers of complex disease. Efforts in transcriptomic biomarker development are not restricted to medical diagnostics but include environmental chemical risk assessment and determination of exposure to

Biomarkers in Toxicology, Second Edition
https://doi.org/10.1016/B978-0-12-814655-2.00063-3

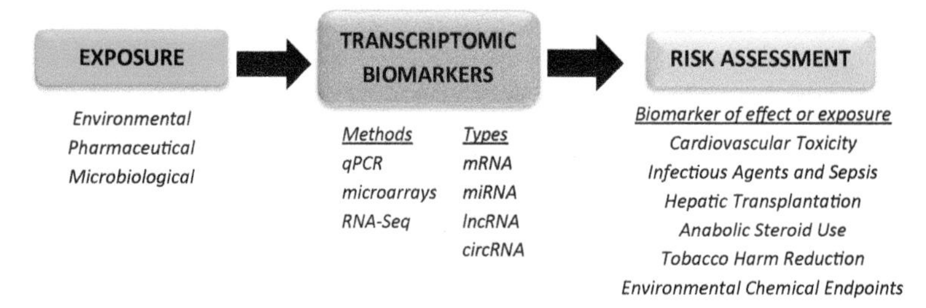

FIGURE 63.1 Chapter overview on the use of transcriptomic biomarkers for safety and risk assessment.

microbes or chemical residues in food (Riedmaier et al., 2009a,b,c; Pinel et al., 2010).

Transcriptomic Methods

There are three commonly used methods for assessing transcriptomes: quantitative real-time polymerase chain reaction (qPCR), microarrays, and RNA sequencing (RNA-Seq). Each method has advantages and disadvantages.

qPCR is a targeted method that involves the reverse transcription of mRNA to cDNA followed by the use of short DNA sequences called primers to anneal and amplify known genes from a biological sample. Rapid quantitation of hundreds of transcripts using qPCR is relatively inexpensive, but knowledge of sequences of the candidate transcripts is required.

Microarrays and RNA-Seq were developed more recently, within the last 30 years, with further advancements made in the 21st century (Wang et al., 2009; Nelson, 2001). Microarrays were the preferred method of transcriptional profiling until the late 2000s when the use of RNA-Seq increased (Medline trend, 2017; Nelson, 2001). Microarrays consist of probes, which are short nucleotide oligomers, arranged on a solid substrate, such as glass (Romanov et al., 2014). Relative transcriptional expression is determined by hybridization of fluorescently labeled transcripts to array probes. At each probe location, fluorescence intensity on the array signals transcript abundance for each probe sequence (Barbulovic-Nad et al., 2006). Unlike RNA-Seq, microarrays do require some knowledge of the organism of interest, such as its annotated genome sequence, that can be used to create probes for the array. Microarrays can be further subdivided into low-density spotted arrays or high-density short probe arrays (Heller, 2002). Single- or dual-channel detection of fluorescent tags can be used to record transcript presence. Low-density spotted arrays generally use a range of picoliter amounts of purified cDNAs arrayed on the surface of a glass slide (Auburn et al., 2005). Probes used are longer than those of high-density arrays, often lacking the transcript resolution of high-density arrays. Different fluorophores are used for exposed and nonexposed samples in spotted arrays; the ratio of fluorescence is used to determine a relative calculation of abundance (Shalon et al., 1996). High-density arrays, popularized by Affymetrix GeneChip array (Santa Clara, CA), quantify transcripts using several short 25-mer probes (Irizarry et al., 2003).

RNA-Seq combines high-throughput sequencing methodology with computational methods to assess and quantify transcripts present in an RNA extract (Ozsolak and Milos, 2011). As previously indicated, RNA-Seq usage exceeded the use of microarrays as the dominant transcriptomics technique in 2015 (Medline trend, 2017). Typical nucleotide sequences generated are approximately 100 base pairs (bp) in length, with a range of 30 bp to over 10,000 bp dependent on sequencing method used. RNA-Seq couples deep sampling of the transcriptome with many short fragments from a transcriptome to permit computational assembly of the original RNA transcript by aligning reads to a reference genome or to each other, the latter referred to as de novo assembly (Wang et al., 2009). RNA-Seq has a typical dynamic range of five orders of magnitude over microarray transcriptomes; additionally, input RNA quantities are much lower for RNA-Seq (nanogram quantity) compared with microarrays (microgram quantity). This allows more detailed examination of cellular structures down to single cells when used with linear amplification of cDNA (Hashimshony et al., 2012). Background signal is very low for 100 bp reads in nonrepetitive regions, and, in theory, there is no upper limit of quantification in RNA-Seq (Ozsolak and Milos, 2011). RNA-Seq can be used to identify genes within a genome or identify what genes are upregulated or downregulated at a specific point in time. Read counts and a simple quantitation approach allowing for the comparison across data sets can be used to correctly model relative gene expression.

These three methods of assessing the transcriptome all have their benefits and challenges when compared with one another, and each method's utility is dependent on the application and/or function of the study and outcome. Although RNA-Seq has already begun to replace microarrays in basic research, clinical studies will likely use both approaches depending on scientific goals, sample size, and cost. Looking forward, it is clear that advances in our fundamental understanding of gene transcription and rapidly advancing techniques for transcriptome assessment will have a continued impact on biomarker development. However, one hurdle in the field is to establish better platforms for the reproducibility and predictive accuracy between sites (Fielden et al., 2008).

Types of Transcriptomic Biomarkers

There are two frequently used types of transcriptomic biomarkers: mRNA and micro RNA (miRNA). Two other transcriptomic biomarkers less commonly used are noncoding RNA and circular RNA (circRNA).

mRNA biomarkers are already an established method in several scientific fields. Diseased tissue can be distinguished from nondiseased tissue by analyzing the expression of specific genes. Use of mRNA gene expression analysis is helpful when validating differentiation of types or stages of diseases. miRNAs are endogenous noncoding RNA of approximately 19–22 nt in length (Krol et al., 2010). Transcription of miRNAs occurs via RNA polymerase II, resulting in primary miRNA with 500−−3000 nucleotides (Benz et al., 2016). Primary miRNA is next cleaved into premature miRNA of 70–80 nucleotides in length by the "microprocessor complex," consisting of RNase III Drosha enzyme and the DiGeorge syndrome critical region 8 (DGCR8) protein (Lee et al., 2003). Premature miRNA is exported into the cytoplasm with aid of the nuclear export transporter, exportin 5, processing about 22 nucleotides by interacting with RNase III endonuclease Dicer protein and the cofounder double-stranded transactivation-responsive RNA-binding protein (Lund et al., 2004). This duplex is integrated into the "RNA-induced silencing complex" following binding to the argonaute protein and a glycine tryptophan repeat-containing protein, where they bind to partial or total complimentary sequences in the 3′ or 5′ untranslated region of the target mRNA (Ha and Kim, 2014; MacFarlane and Murphy, 2010). miRNAs are involved in posttranscriptional processing of mRNA. In this way, they are able to regulate physiological pathways and metabolic processes and therefore impact the entire cellular physiology, organ development, and tissue differentiation. The expression of miRNAs can be measured in cell culture samples. These miRNAs are also present in body fluids, such as urine, blood, and breast milk (Laterza et al., 2009; Kosaka et al., 2010; Kroh et al., 2010). Identification of single biomarkers on the mRNA or the miRNA level is not possible in most pathological disorders. In such cases, a set of multiple biomarkers must be present to distinguish between specific disease types, disease states, or applied treatments. Bioinformatics is one approach that can be used to integrate the data analysis of multiple biomarker levels of different types of RNA. This could help to generate an integrative gene expression pattern (Molloy et al., 2011). To achieve this goal, different multivariate analysis methods are available which are used for biomarker selection and validation, namely hierarchical cluster analysis and principal components analysis.

Noncoding RNAs can be subdivided into small noncoding RNAs, which are shorter in length than 200 nucleotides (nt), and long noncoding RNAs (lncRNA), which are greater than 200 nt in length (Gibb et al., 2011). LncRNA transcripts do not code for proteins; there are greater than 60,000 members of the lncRNA family that have been catalogued (Volders et al., 2012, 2014). While the primary sequence of lncRNAs is not well conserved, it can be partially compensated through a high level of structural conservation (Johnsson et al., 2014). LncRNAs can be transcribed from conserved genomic regions (Boon et al., 2016) and back-splicing of exons, also capable of forming circRNAs (Jeck et al., 2013; Memczak et al., 2013). LncRNAs play an important role in the pathogenesis of several diseases as they are involved in cell functioning, including roles in chromatin rearrangement, histone modification, modification of alternative splicing genes, and regulation of gene expression (Hu et al., 2017; Fan et al., 2017). CircRNAs have recently been demonstrated to be widespread and abundant within transcriptomes (Jeck et al., 2013; Memczak et al., 2013; Jeck and Sharpless, 2014). CircRNAs are not formed by the usual model of RNA splicing but primarily from the exons of protein-coding genes (Salzman et al., 2012). CircRNAs are characterized by covalently closed loop structures through joining the 3′ and 5′ ends together via exon or intron circularization (Jeck and Sharpless, 2014; Zheng et al., 2016). Formation of circRNAs occurs through two varied mechanisms of exon circularization: intron-pairing-driven circularization and lariat-driven circularization (Jeck et al., 2013). When introns between exons form circular structures, they are removed or retained to form exon-only circRNA or intron-retaining circRNA known as ElciRNA (Jeck et al., 2013; Li et al., 2015). CircRNAs can also be formed by the circularization of two flanking intronic sequences (Conn et al., 2015; Ashwal-Fluss et al., 2014). To date, several circRNAs have been identified in organs or tissues, with some linked to disease, indicating that circRNAs have a role beyond being by-products of mis-splicing or splicing errors (Wang et al., 2016; Zheng et al., 2016). Many circRNAs have been found to behave as "miRNA sponges" to regulate gene expression (Memczak et al., 2013; Zheng et al., 2016; Hansen et al., 2013).

TRANSCRIPTOMICS IN BIOMARKER DISCOVERY

Transcriptomic approaches have been used by various researchers to identify novel biomarkers, which are beginning to influence both clinical practice and environmental chemical risk assessment. This section focuses on biomarkers of cardiovascular disease and hepatotoxicity, exposure to infectious agents, tobacco products, anabolic steroids, beta-agonists, and use of biomarkers in pathway analysis for quantitative environmental chemical risk assessment.

Cardiovascular Biomarkers

Cardiovascular disorders are responsible for high morbidity and mortality and can lead to a substantial economic burden at the individual, institutional, and national levels. The use of the human cardiac transcriptome as a biomarker to improve clinical diagnosis is slow to develop, mostly due to risks and complications associated with endomyocardial biopsies (Zhang et al., 2003). However, recent transcriptomic experiments using human cardiac biopsies have been conducted and used to identify specific causes of cardiomyopathy (Farazi et al., 2011) including subtypes of myocarditis (Soga et al., 2002). The ST2 protein, encoded by the IL1RL1 gene, is markedly elevated in patients with heart failure and exists in a soluble form that can be measured in peripheral blood. ST2 was first identified as a potential biomarker upregulated in an in vitro transcriptomics study involving myocardial stretch. Follow-up studies in animals have suggested ST2 is part of a cardioprotective paracrine signaling axis between cardiac fibroblasts and myocytes (Sanada et al., 2007; Kakkar and Lee, 2008). ST2 signals the presence and severity of adverse cardiac remodeling and tissue fibrosis, which routinely occurs in response to myocardial infarction, acute coronary syndrome, or worsening heart failure. ST2 use as a biomarker has considerable prognostic value and is used as an aid for risk stratification in identifying patients who are either at high risk of mortality or need rehospitalization. As a result of these biomarker discovery studies, a commercial-grade soluble ST2 assay for use in assessing prognoses of chronic heart failure has been cleared by the United States Food and Drug Administration (US FDA) (Presage, Critical Diagnostics, San Diego, CA). Although the ST2 immunoassay measures protein levels in the blood, the use of transcriptomics was vital in its discovery. The initial findings of the transcriptomic screens have led to both a novel cardiac biomarker and therapeutic target for heart failure.

The use of miRNAs as a cardiovascular disease biomarker is promising across several endpoints. They are considered to be involved in fibroblasts proliferation, collagen synthesis, and connective tissue growth factor signaling (Angelini et al., 2015). The implication of miRNAs in heart failure was also demonstrated; its increased expression can depress cardiac function (Wahlquist et al., 2014). Other associations between miRNAs and cardiovascular endpoints include cardiac hypertrophy (Care et al., 2007; Bang et al., 2014), acute coronary syndrome (Widera et al., 2011), acute myocardial infarction (Ji et al., 2009; Wang et al., 2009; Corsten et al., 2010; D'Alessandra et al., 2010), hypertension (Li et al., 2011), and heart failure (Latronico et al., 2007; Tijsen et al., 2010). Above all, there are a large number of detectable miRNAs, but only a few that may currently provide valuable information. This area remains largely under investigation for cardiovascular endpoints.

Hepatic Biomarkers

Several studies have used transcriptomics to identify biomarkers of hepatotoxicity and to determine their mechanism of action (Kussmann et al., 2006; Mutlib et al., 2006; Nie et al., 2006; Tugendreich et al., 2006; Fielden et al., 2007; Eun et al., 2008). Liver gene expression profiling by microarray technology was used to develop biomarkers for nongenotoxic hepatocarcinogens (Fielden et al., 2007). Fielden et al. (2007) collected hepatic gene expression data from rats treated for 5 days with 47 test chemicals. They observed that this short-term in vivo rodent model was more sensitive, more accurate, and provided quicker results when compared with the traditional models for risk assessment of nongenotoxic carcinogens. qPCR has also been used for the identification of biomarkers of toxicity (Ellinger Ziegelbauer et al., 2009). These studies demonstrate that transcriptomic technologies are valuable sensitive tools for screening hepatotoxic and hepatocarcinogenic potential of suspected test agents, although their biomarkers and mechanisms of action may be different.

Liver transplantation is a surgical procedure in which a liver that no longer functions properly is removed and replaced with a healthy liver from a living or deceased donor. The unique characteristics of the liver transplant present opportunities to decipher the immunological mechanisms underlying allograft survival and to develop therapeutic targets aimed toward tolerance strategies. RNA-Seq is beginning to be capitalized on to provide insights into normal, pathological, and pharmacological processes (Mastoridis et al., 2016). The integration of clinical and molecular data becomes essential to the pursuit of advancing the field of transplantation and developing personalized therapy.

Biomarkers of Anabolic Agents

Screening for anabolic agents such as steroid hormones (testosterone or estrogen) and beta-agonists in meat-producing animals demonstrates an application of transcriptomics for use in safety and regulatory monitoring. In agricultural meat-producing animals, the myotropic, growth-promoting properties of steroid hormones and beta-agonists are beneficial because of increased productivity and reduced costs with weight gain and feed efficiency (Chwalibog et al., 1996). As a consequence, hormone and drug residues in meat are increased and adverse side effects to the consumer occur (Daxenburger et al., 2001; Lange et al., 2001; Maume et al., 2001; Swan et al., 2007). Questions and controversy over the impacts of these added hormones on human development and health have lingered for decades. The use of growth promoters is approved in the United States, Canada, Mexico, Australia, and South Africa, and although the US FDA does not currently regulate their use in cattle, these drugs are prohibited for over-the-counter use by humans. The European Union (EU) forbids the use of these substances, including the importation of meat from animals treated with these substances. To enforce the EU rules, the monitoring of anabolic agents in meat is necessary. Current methods to uncover the abuse of anabolic agents in animal residues are limited to immune assays or chromatographical methods (Mayer et al., 1991, 2011) which can only detect known substances. The use of transcriptomics is a promising approach to develop a biomarker pattern based on physiological changes that result after illegal application of anabolic agents. Discovery of a single transcription biomarker in cattle for a particular organ or tissue is challenging; therefore, gene expression profiles, the measurement of several genes at one time, is used (Riedmaier et al., 2009a,b,c). Promising candidate genes and gene profiles for the development of a biomarker screening method in cattle include insulin-like growth factor 1 in liver and muscle (Johnson et al., 1998; Pfaffl et al., 2002; Reiter et al., 2007), steroid hormone receptors in the liver, muscle, uterus, gastrointestinal tract, kidney, prostate and blood cells (Pfaffl et al., 2002; Toffolatti et al., 2006; Reiter et al., 2007; Riedmaier et al., 2009a,b,c), and various inflammatory, apoptotic and proliferative genes in blood cells (Chang et al., 2004; Cantiello et al., 2007; Riedmaier et al., 2009a,b,c). Agonists for the beta adrenergic receptor are known to effect mRNA expression of different muscle proteins, such as α-actin, myosin, or calpastatin in cattle. Overall, the use of transcriptomics and other omics technologies can be used to develop new screening methods for the detection of the misuse of anabolic steroids and beta-agonists for public health protection.

Infectious Agents and Sepsis

One of the most common causes of mortality among hospitalized patients in the intensive care unit (ICU) is sepsis. Sepsis can be challenging to diagnose among patients in the ICU due to multiple comorbidities and underlying conditions (Novosad et al., 2016; Vincent et al., 2009). In 2016, after 25 years of no change in definition, the Sepsis-3 conference redefined sepsis as a "life-threatening organ dysfunction caused by a deregulated host response to infection," and septic shock as a "subset of sepsis in which underlying circulatory and cellular/metabolic abnormalities are profound enough to substantially increase mortality" (Singer et al., 2016). Laboratory biochemical, microbiological, and hematological testing is fundamental in the diagnosis of sepsis. Because culture-based diagnosis is lengthy, considerable effort has been made to uncover biomarkers that allow early diagnosis of sepsis. Currently, biomarkers alone are not sufficient for diagnosis of sepsis and should accompany thoughtful clinical assessment and other supportive laboratory data (Faix, 2013; Vincent, 2016).

Most frequently in sepsis, serum protein inflammatory biomarkers are used. Sepsis can be separated into two successive phases, the initial hyperinflammatory phase, which may resolve or come before the secondary immunosuppressive phase, generally characterized by organ dysfunction. Markers are available for both phases but biomarkers of the hyperinflammatory phase predominate and include procalcitonin (PCT) and C reactive protein (CRP). The second immunosuppressive phase contains biomarkers such as human leukocyte antigen-D related (HLA-DR).

CRP is an acute-phase protein often manufactured in the liver and can rapidly rise in a few hours after exposure. Peripheral blood mononuclear cells have demonstrated upregulation of CRP mRNA in response to endotoxin, the most common cause of sepsis (Haider et al., 2006). CRP plasma concentration appears to reflect sepsis severity (Miglietta et al., 2015). Although CRP does not differentiate sepsis from other diseases, it is particularly useful to screen for early onset neonatal sepsis (Hofer et al., 2012) and to monitor patients after surgery (Watt et al., 2015). PCT is generally considered the most useful biomarker of sepsis (Riedel et al., 2011). Its production is stimulated similarly to that of CRP. PCT can discriminate between infectious and noninfectious systemic inflammation in critically ill patients (Harbarth et al., 2001) and may be able to differentiate between viral and bacterial infections (Ahn et al., 2011). PCT mRNA expression is elevated in septic tissues versus nonseptic tissues in the hamster model (Domenech et al., 2001).

HLA-DR expression on the cell surface is a key indicator of immune response due to its vital role in antigen presentation. Expression of HLA-DR is lower in septic patients than in nonseptic patients, but its expression may also be low due to other causes of a weakened immune system (Das, 2014). Reduced HLA-DR expression in septic patients has also been determined transcriptionally (Winkler et al., 2017). Biomarkers of organ dysfunction include the following: lactate, venous to arterial carbon dioxide pressure difference, proadrenomedullin (MR-proADM), and cell-free DNA (cf-DNA). Hyperlactatemia, high blood lactate level, is considered a severe sepsis marker of organ dysfunction, as it indicates poor tissue perfusion (Rello et al., 2017). Levels of CO_2 in the blood above 6 mmHg within the first 24 h in critically ill patients are correlated with higher mortality (He et al., 2016). Higher levels of MR-proADM are associated with greater mortality, and investigations demonstrate that it has also been observed transcriptionally (Becker et al., 2004). Cf-DNA levels are greater in sepsis patients than in healthy individuals, but cf-DNA is found after cell death.

Sepsis is a major public health concern due to its high mortality and morbidity. Protein biomarkers are commonly used in the diagnosis of sepsis, but several studies utilizing precursor biomarker mRNA are promising in the pursuit of more rapid diagnostics. Combinations of biomarkers are likely needed for sepsis diagnosis and treatment due to its high complexity in the patient. Combinatorial biomarker assessments can likely be performed more rapidly using transcriptional analyses in the future.

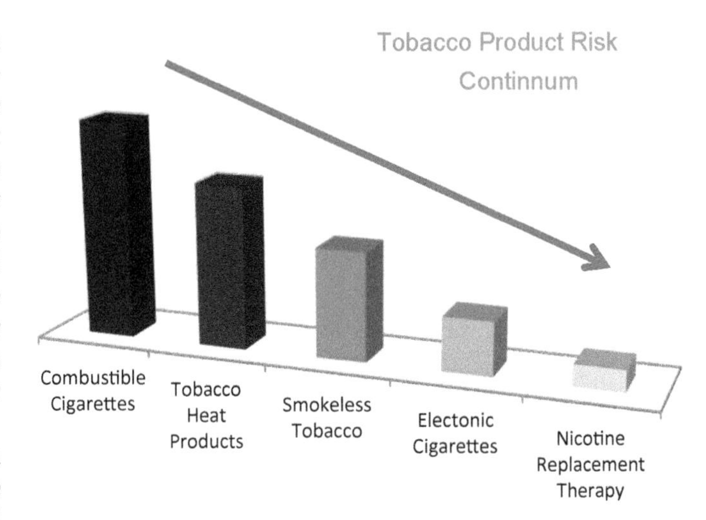

FIGURE 63.2 Hypothetical example of a tobacco product risk continuum.

Tobacco and Risk Continuum

Since 2009, the US FDA Center for Tobacco Products has had authority to regulate the manufacture, distribution, and marketing of tobacco products with the goal of reducing death/disease from consumer use. Biomarkers could play an important role, including transcriptomics, across a number of FDA regulatory activities, including assessing new and modified risk tobacco products and identifying and evaluating potential product standards. The concept that not all tobacco and nicotine products present the same risks to human health is not a novel concept as there is clearly a large difference in the health risks associated with cigarette smoking when compared to medicinal nicotine products. There have been attempts to describe a "risk continuum" placing various nicotine product categories within a paradigm (Fig. 63.2). On July 28, 2017, FDA noted, "A key piece of the FDA's approach is demonstrating a greater awareness that nicotine — while highly addictive — is delivered through products that represent a continuum of risk and is most harmful when delivered through smoke particles in combustible cigarettes" (FDA, 2017).

Smoking is a major risk factor for developing oral squamous cell carcinoma, but less attention has been paid to the effects of smokeless tobacco products. Using oral cell lines, potential biomarkers distinguishing biological effects of combustible tobacco products from those of noncombustible tobacco products have been identified (Woo et al., 2017). Microarrays revealed xenobiotic metabolism and steroid biosynthesis as two pathways upregulated by combustible but not by noncombustible products. Notably, aldo-keto reductase genes, *AKR1C1* and *AKR1C2*, were the genes in the top enriched pathways and were statistically upregulated more than eightfold by combustibles. *AKR1C1* was further supported by qPCR as a potential biomarker for differentiating the biological effects of combustible from noncombustible tobacco products.

Electronic Nicotine Delivery Systems (ENDS, electronic cigarettes or vapor products) are gaining in global popularity and generate a heated aerosol comprised of purified nicotine and generally recognized as safe ingredients. An expert review concludes that e-cigarettes are around 95% safer than smoked tobacco and they can help smokers to quit (Public Health England, 2015). RNA-Seq analysis has been used to examine the transcriptomes of differentiated human bronchial epithelial cells exposed to air, tobacco smoke, and ENDS aerosol in an in vitro air–liquid interface model for cellular exposure. Results indicate that ENDS do not elicit many of the cell toxicity responses observed from combustible cigarette smoke exposure. These transcriptomic studies establish a baseline for future analysis that can better inform the FDA of the risks and future regulation of these products while comparing them to the greater health-risk combustible cigarettes. Currently, there is a lack of systematic effort or FDA guidance by the Center for Tobacco Products to identify transcriptomic biomarkers that would have utility and validity for evaluating tobacco products. Omics research could further strengthen the applicability for tobacco regulatory science to better understand the risk continuum for overall public health protection.

Environmental Chemical Risk Assessment

The US EPA, the National Toxicology Program (NTP), and the National Research Council recommend the use of pathway analysis as a basis for toxicity testing and chemical risk assessment (Collins, 2008; Thomas et al., 2012). To date, transcriptomics have not yet translated to a toxicity value derivation for regulatory purposes. Transcriptomic data in risk assessment have focused on supportive roles, elucidating a chemical's mechanism or mode of action (Thomas et al., 2012) that underlies chemically induced noncancer and cancer endpoints. Female B6C3F1 mice were previously exposed to multiple concentrations of five chemicals (1,4-dichlorobenzene, propylene glycol mono-t-butyl ether, 1,2,3-trichloropropane, methylene chloride, and naphthalene) that had tested positive for lung or liver tumor formation in 2-year rodent cancer bioassays conducted by the NTP (Thomas et al., 2011). The dose–response changes in gene expression were analyzed using a standard benchmark dose modeling (BMD) approach and then grouped based on cellular/biological processes. The transcriptional points of departure for the five chemicals were compared with both cancer and noncancer points of departure based on apical endpoints. Overall, the transcriptomic responses corresponded closely with both noncancer and cancer apical responses as a function of dose, and their use as a point of departure is beginning to look promising.

As the field continues to evolve, additional studies have demonstrated that transcriptional points of departure correlate with those derived from apical endpoint changes, resulting in a suitable transcriptomic biomarker; however, there is currently no consensus on the process. Eleven approaches have been identified from apical data showing promise for chemical risk assessment (Farmahin et al., 2017). This was accomplished using published microarray data from dose–response studies across several chemicals in rats exposed orally for up to 90 days. It is promising for this field that transcriptional BMD values for all approaches aligned well with corresponding apical points of departure.

Cancer risk assessment from exposure to environmental chemicals has also benefited from transcriptomic biomarkers. Positive genotoxicity findings are a major challenge to industry and regulatory agencies when these tests produce irrelevant positive findings (i.e., positive results in vitro that are not relevant to human cancer hazard). In vitro transcriptomic biomarker-based approaches have also been developed that provide biological relevance to positive genotoxicity assay data, particularly for in vitro chromosome damage assays (Li et al., 2017).

Transcriptomic changes in canonical pathways are a promising approach to identify biomarkers to estimate the noncancer and cancer points of departure for use in quantitative risk assessments. The high correlation between these pathways and rodent toxicity indicates the need for further investigation of their role in these endpoints and examination of their relevance to human responses. If they are relevant, these pathways could be considered toxicity pathways and used as the basis for future high-throughput toxicity screen.

CONCLUDING REMARKS AND FUTURE DIRECTIONS

Transcriptomics as a discipline is still evolving, and although there are no clearly defined methods for the discovery of biomarkers using gene expression technologies, the potential of this new field seems promising. There is a growing demand for transcriptional biomarker development in a variety of applications, including the following: clinical diagnosis, pathological processes, biomarkers for environmental chemical exposure or response, and safety evaluation. The discovery of predictive markers that precede pathology should be able to lessen the adverse impact to health. Because gene transcription is a very dynamic process able to adapt rapidly to environmental, physiological, or pathological changes, the transcriptome is a preferred data set for the identification of sensitive and predictive biomarkers. It is conceivable that both a single biomarker and the concept of multimarker panels will become the standard in biomarker research (Ky et al., 2011). To validate an identified transcribed biomarker set with high significance, the application of bioinformatics is necessary. Considering that there are different levels at which biomarkers can be measured and evaluated, and as technologies advance, new categories of transcriptomic biomarkers will continue to emerge and offer opportunities to significantly improve risk assessment while reducing cost.

References

Ahn, S., Kim, W.Y., Kim, S.H., et al., 2011. Role of procalcitonin and C-reactive protein in differentiation of mixed bacterial infection from 2009 H1N1 viral pneumonia. Influenza Respir. Viruses. 5 (6), 398–403.

Angelini, A., Li, Z., Mericskay, M., et al., 2015. Regulation of connective tissue growth factor and cardiac fibrosis by an SRF/MicroRNA-133a axis. PLoS One 10, e0139858. https://doi.org/10.1371/journal.pone.0139858.

Ashwal-Fluss, R., Meyer, M., Pamudurti, N.R., et al., 2014. circRNA biogenesis competes with pre-mRNA splicing. Mol. Cell. 56 (1), 55–66.

Auburn, R.P., Kreil, D.P., Meadows, L.A., et al., 2005. Robotic spotting of cDNA and oligonucleotide microarrays. Trends Biotechnol. 23 (7), 374–379.

Bang, C., Batkai, S., Dangwal, S., et al., 2014. Cardiac fibroblast-derived microRNA passenger strand-enriched exosomes mediate cardiomyocyte hypertrophy. J. Clin. Invest. 124, 2136–2146. https://doi.org/10.1172/JCI70577.

Barbulovic-Nad, I., Lucente, M., Sun, Y., et al., 2006. Bio-microarray fabrication techniques—a review. Crit. Rev. Biotechnol. 26 (4), 237–259.

Becker, K.L., Nylen, E.S., White, J.C., et al., 2004. Procalcitonin and the calcitonin gene family of peptides in inflammation, infection, and sepsis: a journey from calcitonin back to its precursors. J. Clin. Endocrinol. Metab. 89 (4), 1512–1525.

Benz, F., Roy, S., Trautwein, C., et al., 2016. Circulating microRNAs as biomarkers for sepsis. Int. J. Mol. Sci. 17 (1), 78.

Boon, R.A., Jaé, N., Holdt, L., et al., 2016. Long noncoding RNAs: from clinical genetics to therapeutic targets? J. Am. Coll. Cardiol. 67 (10), 1214–1226.

Cantiello, M., Carletti, M., Cannizzo, F.T., et al., 2007. Effects of an illicit cocktail on serum immunoglobulins, lymphocyte proliferation and cytokine gene expression in the veal calf. Toxicology 242 (1–3), 3951.

Carè, A., Catalucci, D., Felicetti, F., 2007. MicroRNA-133 controls cardiac hypertrophy. Nat. Med. 13, 613–618.

Chang, L.C., Madsen, S.A., Toelboell, T., et al., 2004. Effects of glucocorticoids on Fas gene expression in bovine blood neutrophils. J. Endocrinol. 183 (3), 569–583.

Chwalibog, A., Jensen, K., Thorbek, G., 1996. Quantitative protein and fat metabolism in bull calves treated with beta-adrenergic agonist. Arch Tierernahr. 49 (2), 159167A.

Collins, F., 2008. Collins interview. Departing U.S. genome institute director takes stock of personalized medicine. Interview by Jocelyn Kaiser. Science 320 (5881), 1272.

Conn, S.J., Pillman, K.A., Toubia, J., et al., 2015. The RNA binding protein quaking regulates formation of circRNAs. Cell 160 (6), 1125–1134.

Corsten, M.F., Dennert, R., Jochems, S., et al., 2010. Circulating microRNA-208b and microRNA-499 reflect myocardial damage in cardiovascular disease/clinical perspective. Circ Cardiovasc Genet 3, 499–506.

Das, U., 2014. HLA-DR expression, cytokines and bioactive lipids in sepsis. Arch Med Sci. AMS 10 (2), 325.

Daxenberger, A., Ibarreta, D., Meyer, H.H., 2001. Possible health impact of animal oestrogens in food. Hum. Reprod. Update 7 (3), 340–355.

D'Alessandra, Y., Devanna, P., Limana, F., et al., 2010. Circulating micro-RNAs are new and sensitive biomarkers of myocardial infarction. Eur. Heart J. 31, 2765–2773.

Domenech, V.S., Nylen, E.S., White, J.C., et al., 2001. Calcitonin gene-related peptide expression in sepsis: postulation of microbial infection-specific response elements within the calcitonin I gene promoter. J. Invest. Med. 49 (6), 514–521.

Ellinger-Ziegelbauer, H., Gmuender, H., Bandenburg, A., et al., 2009. Prediction of a carcinogenic potential of rat hepatocarcinogens using toxicogenomics analysis of short-term in vivo studies. Mutat. Res. 637, 2339.

Eun, J.W., Ryu, S.Y., Noh, J.H., et al., 2008. Discriminating the molecular basis of hepatotoxicity using the large-scale characteristic molecular signatures of toxicants by expression profiling analysis. Toxicology 249 (2–3), 176183.

Faix, J.D., 2013. Biomarkers of sepsis. Crit. Rev. Clin. Lab Sci. 50 (1), 23–36.

Fan, R., Cao, C., Zhao, X., et al., 2017. Downregulated long noncoding RNA ALDBGALG0000005049 induces inflammation in chicken muscle suffered from selenium deficiency by regulating stearoyl-CoA desaturase. Oncotarget 8 (32), 52761.

Farazi, T.A., Spitzer, J.I., Morozov, P., et al., 2011. miRNAs in human cancer. J. Pathol. 223 (2), 102115.

Farmahin, R., Williams, A., Kuo, B., et al., 2017. Recommended approaches in the application of toxicogenomics to derive points of departure for chemical risk assessment. Arch. Toxicol. 91 (5), 2045–2065.

FDA, 2017. FDA News Release FDA Announces Comprehensive Regulatory Plan to Shift Trajectory of Tobacco-related Disease, Death. https://www.fda.gov/NewsEvents/Newsroom/PressAnnouncements/ucm568923.htm.

Fielden, M.R., Brennan, R., Gollub, J., 2007. A gene expression biomarker provides early prediction and mechanistic assessment of hepatic tumor induction by non-genotoxic chemicals. Toxicol. Sci. 99 (1), 90–100.

Fielden, M.R., Nie, A., McMillian, M., et al., 2008. Predictive safety testing consortium, carcinogenicity working group: interlaboratory evaluation of genomic signatures for predicting carcinogenicity in the rat. Toxicol. Sci. 103, 28–34.

Gibb, E.A., Brown, C.J., Lam, W.L., 2011. The functional role of long non-coding RNA in human carcinomas. Mol. Canc. 13, 1038.

Ha, M., Kim, V.N., 2014. Regulation of microRNA biogenesis. Nat. Rev. Mol. Cell Biol. 15 (8), 509.

Haider, D.G., Leuchten, N., Schaller, G., et al., 2006. C-reactive protein is expressed and secreted by peripheral blood mononuclear cells. Clin. Exp. Immunol. 146 (3), 533–539.

Hansen, T.B., Jensen, T.I., Clausen, B.H., et al., 2013. Natural RNA circles function as efficient microRNA sponges. Nature 495 (7441), 384.

Harbarth, S., Holeckova, K., Froidevaux, C., et al., 2001. Diagnostic value of procalcitonin, interleukin-6, and interleukin-8 in critically ill patients admitted with suspected sepsis. Am. J. Respir. Crit. Care Med. 164 (3), 396–402.

Hashimshony, T., Wagner, F., Sher, N., et al., 2012. CEL-Seq: single-cell RNA-Seq by multiplexed linear amplification. Cell Rep. 2 (3), 666–673.

He, H.W., Liu, D.W., Long, Y., et al., 2016. High central venous-to-arterial CO2 difference/arterial-central venous O2 difference ratio is associated with poor lactate clearance in septic patients after resuscitation. J. Crit. Care 31 (1), 76–81.

Heller, M.J., 2002. DNA microarray technology: devices, systems, and applications. Annu. Rev. Biomed. Eng. 4 (1), 129–153.

Hofer, N., Zacharias, E., Müller, W., et al., 2012. An update on the use of C-reactive protein in early-onset neonatal sepsis: current insights and new tasks. Neonatology 102 (1), 25–36.

Hu, G., Dong, B., Zhang, J., et al., 2017. The long noncoding RNA HOTAIR activates the Hippo pathway by directly binding to SAV1 in renal cell carcinoma. Oncotarget 8 (35), 58654–58667.

Ilyin, S.E., Belkowski, S.M., Plata-Salamán, C.R., 2004. Biomarker discovery and validation: technologies and integrative approaches. Trends Biotechnol. 22 (8), 411416.

Irizarry, R.A., Hobbs, B., Collin, F., et al., 2003. Exploration, normalization, and summaries of high density oligonucleotide array probe level data. Biostatistics 4 (2), 249–264.

Jeck, W.R., Sharpless, N.E., 2014. Detecting and characterizing circular RNAs. Nat. Biotechnol. 32 (5), 453.

Jeck, W.R., Sorrentino, J.A., Wang, K., et al., 2013. Circular RNAs are abundant, conserved, and associated with ALU repeats. RNA 19 (2), 141–157.

Ji, X., Takahashi, R., Hiura, Y., et al., 2009. Plasma mir-208 as a biomarker of myocardial injury. Clin. Chem. 55, 1944–1949.

Johnson, B.J., White, M.E., Hathaway, M.R., et al., 1998. Effect of a combined trenbolone acetate and estradiol implant on steady-state IGF-I mRNA concentrations in the liver of wethers and the longissimus muscle of steers. J. Anim. Sci. 76 (2), 491–497.

Johnsson, P., Lipovich, L., Grandér, D., et al., 2014. Evolutionary conservation of long non-coding RNAs; sequence, structure, function. Biochim. Biophys. Acta Gen. Subj. 1840 (3), 1063–1071.

Kakkar, R., Lee, R.T., 2008. The il-33/st2 pathway: therapeutic target and novel biomarker. Nat. Rev. Drug Discov. 7, 827–840.

Kosaka, N., Iguchi, H., Ochiya, T., 2010. Circulating microRNA in body fluid: a new potential biomarker for cancer diagnosis and prognosis. Cancer Sci. 101 (10), 2087–2092.

Kroh, E.M., Parkin, R.K., Mitchell, P.S., et al., 2010. Analysis of circulating microRNA biomarkers in plasma and serum using quantitative reverse transcription-PCR (qRT-PCR). Methods 50 (4), 298–301.

Krol, J., Loedige, I., Filipowicz, W., 2010. The widespread regulation of microRNA biogenesis, function and decay. Nat. Rev. Genet. 11 (9), 597.

Kussmann, M., Raymond, F., Affolter, M., 2006. OMICSdriven biomarker discovery in nutrition and health. J. Biotechnol. 124 (4), 758–787.

Ky, B., French, B., McCloskey, K., et al., 2011. High-sensitivity ST2 for prediction of adverse outcomes in chronic heart failure. Circ. Heart. Fail. 4, 180–187.

Lange, I.G., Daxenberger, A., Meyer, H.H.D., 2001. Hormone contents in peripheral tissues after correct and off-label use of growth promoting hormones in cattle: effect of the implant preparations Finaplix-H®, Ralgro®, Synovex-H® and Synovex Plus®. APMIS 109 (1), 53–65.

Laterza, O.F., Lim, L., Garrett-Engele, P.W., et al., 2009. Plasma microRNAs as sensitive and specific biomarkers of tissue injury. Clin. Chem. 55 (11), 1977–1983.

Latronico, M.V.G., Catalucci, D., Condorelli, G., 2007. Emerging role of microRNAs in cardiovascular biology. Circ. Res. 101, 1225–1236.

Lee, Y., Ahn, C., Han, J., et al., 2003. The nuclear RNase III Drosha initiates microRNA processing. Nature 425 (6956), 415.

Li, S., Zhu, J., Zhang, W., et al., 2011. Signature microRNA expression profile of essential hypertension and its novel link to human cytomegalovirus infection/clinical perspective. Circulation 124, 175–184.

Li, Z., Huang, C., Bao, C., et al., 2015. Exon-intron circular RNAs regulate transcription in the nucleus. Nat. Struct. Mol. Biol. 22 (3), 256.

Li, H., Chen, R., Hyduke, D.R., et al., 2017. Development and validation of a high-throughput transcriptomic biomarker to address 21st century genetic toxicology needs. Proc. Natl. Acad. Sci. USA 114 (51), E10881–E10889.

Lund, E., Güttinger, S., Calado, A., et al., 2004. Nuclear export of microRNA precursors. Science 303 (5654), 95–98.

MacFarlane, L.A., Murphy, P.R., 2010. MicroRNA: biogenesis, function and role in cancer. Curr Gemom 11 (7), 537–561.

Mastoridis, S., Martínez-Llordella, M., Sanchez-Fueyo, A., 2016. Biomarkers and immunopathology of tolerance. Curr. Opin. Organ Transplant. 21 (1), 81–87.

Maume, D., Deceuninck, Y., Pouponneau, K., et al., 2001. Assessment of estradiol and its metabolites in meat. APMIS 109 (1), 3238.

Mayer, F., Krämer, B.K., Ress, K.M., et al., 1991. Simplified, rapid and inexpensive extraction procedure for a high-performance liquid chromatographic method for determination of disopyramide and its main metabolite mono-N-dealkylated disopyramide in serum. J. Chromatogr. 572 (12), 339–345.

Mayer, G., Muller, J., Lunse, C.E., 2011. RNA diagnostics: real-time RT-PCR strategies and promising novel target RNAs. Wiley Interdisciplinary Rev RNA 2, 3241.

Medline Trend: Automated Yearly Statistics of PubMed Results for Any Query, 2017. http://dan.corlan.net/medline-trend.html.

Memczak, S., Jens, M., Elefsinioti, A., et al., 2013. Circular RNAs are a large class of animal RNAs with regulatory potency. Nature 495 (7441), 333.

Miglietta, F., Faneschi, M.L., Lobreglio, G., et al., 2015. Procalcitonin, C-reactive protein and serum lactate dehydrogenase in the diagnosis of bacterial sepsis, SIRS and systemic candidiasis. Infezioni Med. Le: Rivista Periodica di Eziologia, Epidemiologia, Diagnostica, Clinica e Terapia Delle Patologie Infettive 23 (3), 230–237.

Molloy, T.J., Devriese, L.A., Helgason, H.H., et al., 2011. A multimarker qPCR-based platform for the detection of circulating tumour cells in patients with early-stage breast cancer. Br. J. Canc. 104 (12), 1913–1919.

Mutlib, A., Jiang, P., Atherton, J., et al., 2006. Identification of potential genomic biomarkers of hepatotoxicity caused by reactive metabolites of N-methylformamide: application of stable isotope labeled compounds in toxicogenomic studies. Chem. Res. Toxicol. 19 (10), 1270–1283.

Nelson, N.J., 2001. Microarrays have arrived: gene expression tool matures. J. Natl. Cancer Inst. 93 (7), 492–494.

Nie, A.Y., McMillian, M., Parker, J.B., et al., 2006. Predictive toxicogenomics approaches reveal underlying molecular mechanisms of nongenotoxic carcinogenicity. Mol. Carcinog. 45, 914–933.

Novosad, S.A., 2016. Vital signs: epidemiology of sepsis: prevalence of health care factors and opportunities for prevention. Morb. Mortal. Wkly. Rep. 65.

Ozsolak, F., Milos, P.M., 2011. RNA sequencing: advances, challenges and opportunities. Nat. Rev. Genet. 12 (2), 87–98.

Pfaffl, M.W., Daxenberger, A., Hageleit, M., et al., 2002. Effects of synthetic progestagens on the mRNA expression of androgen receptor, progesterone receptor, oestrogen receptor alpha and beta, insulin-like growth factor-1 (IGF-1) and IGF-1 receptor in heifer tissues. J. Vet. Med. A Physiol. Pathol. Clin. Med. 49 (2), 57–64.

Piétu, G., Mariage-Samson, R., Fayein, N.A., et al., 1999. The Genexpress IMAGE knowledge base of the human brain transcriptome: a prototype integrated resource for functional and computational genomics. Genome Res. 9 (2), 195–209.

Pinel, G., Rambaud, L., Monteau, F., et al., 2010. Estranediols profiling in calves' urine after 17 beta-nandrolone laureate ester administration. J. Steroid Biochem. Mol. Biol. 121 (35), 626–632.

Public Health England, 2015. E-cigarettes: An Evidence Update a Report Commissioned by Public Health England, p. 260.

Reiter, M., Walf, V.M., Christians, A., et al., 2007. Modification of mRNA expression after treatment with anabolic agents and the usefulness for gene expression biomarkers. Anal. Chim. Acta 586 (1–2), 73–81.

Rello, J., Valenzuela-Sánchez, F., Ruiz-Rodriguez, M., et al., 2017. Sepsis: a review of advances in management. Adv. Ther. 34 (11), 2393–2411.

Riedel, S., Melendez, J.H., An, A.T., et al., 2011. Procalcitonin as a marker for the detection of bacteremia and sepsis in the emergency department. Am. J. Clin. Pathol. 135 (2), 182–189.

Riedmaier, I., Becker, C., Pfaffl, M.W., Meyer, H.H., 2009a. The use of omic technologies for biomarker development to trace functions of anabolic agents. J. Chromatogr. A 1216, 8192–8199.

Riedmaier, I., Tichopad, A., Reiter, M., Pfaffl, M.W., Meyer, H.H., 2009b. Identification of potential gene expression biomarkers for the surveillance of anabolic agents in bovine blood cells. Anal. Chim. Acta. 638, 106–113.

Riedmaier, I., Tichopad, A., Reiter, M., Pfaffl, M.W., Meyer, H.H., 2009c. Influence of testosterone and a novel SARM on gene expression in whole blood of Macaca fascicularis. J. Steroid. Biochem. Mol. Biol. 114, 167–173.

Romanov, V., Davidoff, S.N., Miles, A.R., et al., 2014. A critical comparison of protein microarray fabrication technologies. Analyst 139 (6), 1303–1326.

Salzman, J., Gawad, C., Wang, P.L., et al., 2012. Circular RNAs are the predominant transcript isoform from hundreds of human genes in diverse cell types. PLoS One 7 (2), e30733.

Sanada, S., Hakuno, D., Higgins, L.J., et al., 2007. Il-33 and st2 comprise a critical biomechanically induced and cardioprotective signaling system. J. Clin. Invest. 117, 1538–1549.

Sandvik, A.K., Alsberg, B.K., Nørsett, K.G., Yadetie, F., Waldum, H.L., Laegreid, A., 2006 Jan. Gene expression analysis and clinical diagnosis. Clin. Chim. Acta. 363 (1-2), 157–164. Epub 2005 Oct 5. Review.

Shalon, D., Smith, S.J., Brown, P.O., 1996. A DNA microarray system for analyzing complex DNA samples using two-color fluorescent probe hybridization. Genome Res. 6 (7), 639–645.

Singer, M., Deutschman, C.S., Seymour, C.W., et al., 2016. The third international consensus definitions for sepsis and septic shock (sepsis-3). J. Am. Med. Assoc. 315 (8), 801–810.

Soga, T., Sugimoto, M., Honma, M., et al., 2002. Expression and regulation of st2, an interleukin-1 receptor family member, in cardiomyocytes and myocardial infarction. Circulation 106, 2961–2966.

Swan, S.H., Liu, F., Overstreet, J.W., et al., 2007. Semen quality of fertile US males in relation to their mothers' beef consumption during pregnancy. Hum. Reprod. 22, 1497–1502.

Thomas, R.S., Clewell III, H.J., Allen, B.C., et al., 2011. Application of transcriptional benchmark dose values in quantitative cancer and noncancer risk assessment. Toxicol. Sci. 120, 194–205.

Thomas, R.S., Clewell III, H.J., Allen, B.C., et al., 2012. Integrating pathway-based transcriptomic data into quantitative chemical risk assessment: A five chemical case study. Mutat. Res. 746, 135–143.

Tijsen, A.J., Creemers, E.E., Moreland, P.D., et al., 2010. Mir423-5p as a circulating biomarker for heart failure. Circ. Res. 106, 1035–1039.

Toffolatti, L., Rosa Gastaldo, L., Patarnello, T., et al., 2006. Expression analysis of androgen-responsive genes in the prostrate of veal calves treated with anabolic hormones. Domest. Anim. Endocrinol. 30, 38–55.

Tugendreich, S., Pearson, C.I., Sagartz, J., et al., 2006. NSAID-induced acute phase response is due to increased intestinal permeability and characterized by early and consistent alterations in hepatic gene expression. Toxicol. Pathol. 34, 168–179.

Velculescu, V.E., Zhang, L., Zhou, W., et al., 1997. Characterization of the yeast transcriptome. Cell 88 (2), 243–251.

Volders, P.J., Helsens, K., Wang, X., et al., 2012. LNCipedia: a database for annotated human lncRNA transcript sequences and structures. Nucleic Acids Res. 41 (D1), D246–D251.

Volders, P.J., Verheggen, K., Menschaert, G., et al., 2014. An update on LNCipedia: a database for annotated human lncRNA sequences. Nucleic Acids Res. 43 (D1), D174–D180.

Vincent, J.L., 2016. The clinical challenge of sepsis identification and monitoring. PLoS Med. 13 (5), e1002022.

Vincent, J.L., Rello, J., Marshall, J., et al., 2009. International study of the prevalence and outcomes of infection in intensive care units. J. Am. Med. Assoc. 302 (21), 2323–2329.

Wahlquist, C., Jeong, D., Rojas-Muñoz, A., et al., 2014. Inhibition of miR-25 improves cardiac contractility in the failing heart. Nature 508, 531–535.

Wang, Z., Gerstein, M., Snyder, M., 2009. RNA-Seq: a revolutionary tool for transcriptomics. Nat. Rev. Genet. 10 (1), 57–63.

Wang, K., Long, B., Liu, F., et al., 2016. A circular RNA protects the heart from pathological hypertrophy and heart failure by targeting miR-223. Eur. Heart J. 37 (33), 2602–2611.

Watt, D.G., Horgan, P.G., McMillan, D.C., 2015. Routine clinical markers of the magnitude of the systemic inflammatory response after elective operation: a systematic review. Surgery 157 (2), 362–380.

Widera, C., Gupta, S.K., Lorenzen, J.M., et al., 2011. Diagnostic and prognostic impact of six circulating micro RNAs in acute coronary syndrome. J. Mol. Cell Cardiol. 51, 872–875.

Winkler, M.S., Rissiek, A., Priefler, M., et al., 2017. Human leucocyte antigen (HLA-DR) gene expression is reduced in sepsis and correlates with impaired TNFα response: a diagnostic tool for immunosuppression? PLoS One 12 (8), e0182427.

Woo, S., Gao, H., Henderson, D., et al., 2017. AKR1C1 as a biomarker for differentiating the biological effects of combustible from non-combustible tobacco products. Genes 8 (5) pii: E132.

Zhang, X., Zhou, Y., Mehta, K.R., et al., 2003. A pituitary-derived MEG3 isoform functions as a growth suppressor in tumor cells. J. Clin. Endocrinol. Metab. 88, 5119–5126.

Zheng, Q., Bao, C., Guo, W., et al., 2016. Circular RNA profiling reveals an abundant circHIPK3 that regulates cell growth by sponging multiple miRNAs. Nat. Commun. 7, 11215.

CHAPTER

64

Percellome Toxicogenomics Project as a Source of Biomarkers of Chemical Toxicity

Jun Kanno

Japan Bioassay Research Center, Japan Organization of Occupational Health and Safety, Hadano, Japan

INTRODUCTION

Toxicology is a field of science to study interaction of the living organisms and the chemicals or xenobiotics, and its aim is to ensure human safety. For this aim, chemicals to which we encounter daily are tested by experimental animals, and the toxicity data thereof are extrapolated to humans. This classical approach is based on an assumption that experimental animals and humans should respond similarly to an exogenous chemical. And this approach has been apparently robust for nonspecific toxicity targets such as respiratory chain disruption, DNA damage, and radical oxygen species generation. However, thalidomide showed us that this classical approach has a serious blind spot. Incorporating the recent achievements of the basic biology, new molecular toxicology deals with various kinds of toxicological phenomena via a specific receptor or transcription factor, or modifying epigenetic regulation systems. These targets are often related to delayed adverse effect when exposed to an organism in phases of development and maturation. New pharmaceuticals are also molecular targeted and drug evaluation authorities are in need of molecular biology-based toxicology.

Although "first in human" has been a trend in the field of drug development, the incidence of TGN1412 still reminds us that there are issues to be studied for species differences and similarities at molecular level. There are other factors that human is difficult to use for toxicity assessment. One example is the assessment of combined exposure of a chemical to drugs and quasi-drugs, supplements, and health foods. Another example is the embryo and child to which voluntary clinical trials are often difficult to conduct.

To modernize the experimental animal toxicity studies, we initiated a project to draw the comprehensive and dynamic changes in gene network of the liver and other organs induced by exogenous chemicals. By compiling the drawings into a database, we attempted to build a systems toxicology for the mechanism-based prediction of toxicity of both old and new chemicals. For this purpose, we developed a standardization method designated "Percellome method" (Kanno et al., 2006; Kanno, 2015). This method generates absolute copy number of mRNA in a per one cell bases as an average of a test organ or tissue. The standard animal study was designed to monitor time course and dose response at once, and to plot the gene expression data in a single 3-dimensional surface graph with dose levels in X-axis, time in Y-axis, and mRNA in copy number in Z-axis (Fig. 64.1). A series of homemade algorithms and software programs are prepared for handling and analyzing the 3D surface data. More than 140 chemicals (Table 64.1) chosen for various characteristics were tested for liver and other organs including lung, kidney, heart, hippocampus, etc., in more than 370 studies (one study consist of four time points and four dose levels with organ samples in triplicate). The variation in protocol includes single oral gavage with 2, 4, 8, and 24-h sampling, repeated dosing, and others. More than one organ is sampled in most studies up to six in special studies. The data represent the average of the sampled organs.

During the process of preparing the project, we found that the quality of gene expression data is highly dependent on the level of control over the animals' husbandry. The most important factor was the circadian change of the mRNA; some genes show 10-fold changes only by the time of day. After knowing the phenomenon, the light–dark cycle of the animal facility has been strictly controlled, and the treatment and sampling is conducted swiftly at the fixed clock time.

Biomarkers in Toxicology, Second Edition
https://doi.org/10.1016/B978-0-12-814655-2.00064-5

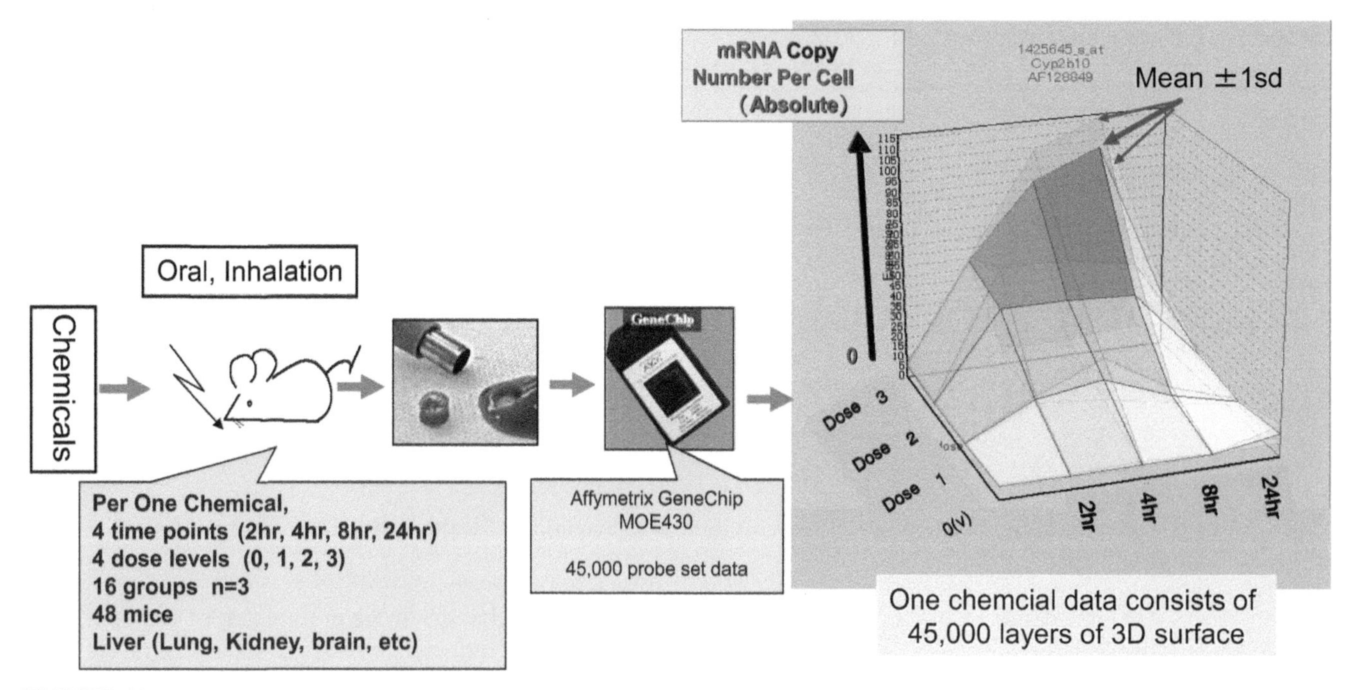

FIGURE 64.1 Standard experimental protocol of Percellome Project and 3-dimensional surface graph for visualization of the gene expression data.

We are still on the way to develop the fully operational systems toxicology. The high-resolution gene expression data are similar to the electron microscopy. When invented, the images first seen were new to anybody. Until the text book or the atlas is complied, many photographs are taken under a standardized sampling and staining method. The concept of data analysis is shown in Fig. 64.2. A chemical is considered as a stone thrown into a water surface of gene network, and Percellome project monitors the ripples by unsupervised clustering of the genes according to the shape of the 3D surface. The limitation is clear that the transcriptome monitors only the gene expression and misses the protein reaction. However, as shown below, we believe that at least some of the protein reactions are monitored by transcriptomics as a shadowgraph. The power of large database consisting of standardized data of similar quality would be demonstrated by cross-referring the individual data. As illustrated in Fig. 64.3, the gene network as a sphere of water is probed by multiple chemicals, and the data of each intersection are becoming available at high precision by our software programs.

Nevertheless, at this stage, we would like to show two examples of our analysis that new confirmative information is generated for chemicals near us. We show that proteins known to be controlled by phosphorylation, which is impossible to directly monitor by mRNA transcriptome, was monitored as upregulation of the relevant gene as a shadowgraph. At the end, joining the rat toxicogenomics data to mouse Percellome data and an ongoing analysis of the New repeated dosing experiments and Garuda Platform to facilitate data handling are briefly demonstrated.

By publicizing the Percellome database along with the homemade analysis tools for 3D surface data via the Garuda project, we hope the data analysis will be accelerated by the help of toxicologists and bioinformaticians worldwide.

MATERIALS AND METHODS FOR PERCELLOME DATA GENERATION AND EXAMPLE STUDIES

Experimental animals and dose setting

Male C57BL/6 Cr Slc (SLC, Hamamatsu, Japan) or male C57BL/6J (Charles River, Japan) mice (48 mice per study) maintained in a barrier system with a 12-h photoperiod (starts at 8:00) were given a single oral dosage (between 10:00 and 10:20) of test chemicals at a ratio of 0, 1, 3, and 10 and the liver was sampled at 2, 4, 8, and 24 h postgavage (three animals from each dosages groups). The top dose for each chemical is set, referring to the dose-setting preliminary study, as the maximum dose that is negative for treatment-related symptoms and for macroscopic and microscopic changes in target organs such as liver and lung in 24 h after single gavage. In general, the maximal doses in the Percellome studies are 10- to 100-fold lower than the reported maximum tolerated doses (MTDs). To minimize the effect of circadian oscillations in the transcriptome, 12 animals were processed at each time point

TABLE 64.1 List of Chemicals Tested in Percellome Project

No.	Chemical Name	CAS Number
1	Acephate	30560-19-1
2	Acetaldehyde	75-07-0
3	Acetaminophen	103-90-2
4	Agaritine	2757-90-6
5	Aloe arborescens extract	NA
6	Alpha lipoic acid	1077-28-7
7	Alprazolam	28981-97-7
8	Aluminum ammonium sulfate	7784-26-1
9	Aluminum lactate	18917-91-4
10	Aluminum sulfate	10043-01-3
11	2-Aminomethylpyridine	3731-51-9
12	3-Amino-1H-1,2,4-triazole	61-82-5
13	4-Amino-2,6-dichlorophenol	5930-28-9
14	AraC (cytosine arabinoside)	147-94-4
15	Aspirin	50-78-2
16	Azacytidine	320-67-2
17	2-(2H-Benzotriazol-2-yl)-4,6-bis(1-methyl-1-phenylethyl)phenol	70321-86-7
18	Benzene	71-43-2
19	2-(2-Hydroxy-5-methylphenyl) benzotriazole	2440-22-4
20	2-(2-Hydroxy-5-tert-octylphenyl) benzotriazole	3147-75-9
21	2-(3,5-Di-tert-butyl-2-hydroxyphenyl)-5-chlorobenzotriazole	3864-99-1
22	2-(5-Chloro-2-benzotriazolyl)-6-tert-butyl-p-cresol	3896-11-5
23	2-(5-tert-Butyl-2-hydroxyphenyl) benzotriazole	3147-76-0
24	2-Benzotriazole-2-yl-4,6-di-tert-butylphenol	3846-71-7
25	3-(2H-benzotriazolyl)-5-(1,1-dimethylethyl)-4-hydroxy-benzenepropanoic acid octyl esters	127519-17-9
26	Bisphenol A	80-05-7
27	Black No.401	1064-48-8
28	Bromobenzene	108-86-1
29	Caffeine	1958-8-2
30	Carbaryl	63-25-2
31	Carbon tetrachloride	56-23-5
32	2-Chloro-4,6-dimethylaniline	63133-82-4

TABLE 64.1 List of Chemicals Tested in Percellome Project—cont'd

No.	Chemical Name	CAS Number
33	Chlorpyrifos	2921-88-2
34	Cisplatin	15663-27-1
35	Citric acid-calcium salt	813-94-5
36	Clofibrate	637-07-0
37	Coenzyme Q10	303-98-0
38	Curcumin	458-37-7
39	Cyclopamine	4449-51-8
40	Daizein	486-66-8
41	Deet	134-62-3
42	DEHP	117-81-7
43	Dexamethasone	1950-2-2
44	1,2-Dichloro-3-nitrobenzene	3209-22-1
45	Diazinon	333-41-5
46	Dibutyltin dichloride	683-18-1
47	Diethylnitrosamine	55-18-5
48	Diethylstilbestrol	56-53-1
49	Digitoxin	71-63-6
50	2,4-Dinitrophenol	51-28-5
51	DMSO	67-68-5
52	Doxorubicin	23214-92-8
53	Estragole	140-67-0
54	Ethanol	64-17-5
55	2-Ethyl-1-hexanol	104-76-7
56	4-Ethylnitrobenzene	100-12-9
57	Ethinylestradiol	57-63-6
58	FD & C Blue No.1 (Food Blue No.1)	3844-45-9
59	Fenobucarb	3766-81-2
60	FK506	104987-11-3
61	5-fluorouracil	51-21-8
62	Food Red No.104	18472-87-2
63	Food Red No.40 (Allura Red AC)	25956-17-6
64	Food Yellow No.4 (FD & C Yellow No.5)	1934-21-0
65	Formaldehyde	50-00-0
66	Forskolin	66575-29-9
67	Genistein	446-72-0
68	Genistin	529-59-9
69	Glycyrrhizin 2K	1405-86-3

Continued

TABLE 64.1 List of Chemicals Tested in Percellome Project—cont'd

No.	Chemical Name	CAS Number
70	Green No.204 (Pyranine Conc)	6358-69-6
71	Houttuynia extract	NA
72	Hydroxycitric Acid	27750-10-3
73	Ibotenic acid	2552-55-8
74	Ibuprofen (dl-p-isobutylhydratropic acid)	15687-27-1
75	Imidacloprid	138261-41-3
76	Indigo	482-89-3
77	Indigo Carmine	860-22-0
78	Isoniazid	54-85-3
79	Kanamycin monosulfate	25389-94-0
80	Levothyroxine	51-48-9
81	Maltol	118-71-8
82	Mastic	NA
83	MEHP	4376-20-9
84	Menthyl Valerate	89-47-4
85	3-Methylcholanthrene	56-49-5
86	Methanol	67-56-1
87	Methoprene	40596-69-8
88	Methoprene acid	53092-52-7
89	Methyl dihydro jasmonate	24851-98-7
90	Monocrotaline	315-22-0
91	Monofluoroacetic Acid	144-49-0
92	Nerolidol	7212-44-4
93	N-ethyl-N-nitrosourea	759-73-9
94	N-Methylaniline	100-61-8
95	Omeprazole	73590-58-6
96	Orange No.403	2646-17-5
97	Paclitaxel (Taxol)	33069-62-4
98	Paraquat dichloride	1910-42-5
99	PCB153	35065-27-1
100	p-Dichlorobenzene	106-46-7
101	Pentachlorophenol	87-86-5
102	Permethrin	52645-53-1
103	Phenobarbital sodium	57-30-7
104	Phenytoin	57-41-0
105	Phytol	150-86-7
106	5-Pregnen-3beta-ol-20-one-16alpha-carbonitrile	1434-54-4

TABLE 64.1 List of Chemicals Tested in Percellome Project—cont'd

No.	Chemical Name	CAS Number
107	Pyriproxyfen	95737-68-1
108	Red No.102	2611-82-7
109	Red No.105	632-69-9
110	Red No.225	85-86-9
111	Red No.501	85-83-6
112	Red No.505	3118-97-6
113	9-cis retinoic acid	5300-3-8
114	All trans retinoic acid	302-79-4
115	Rifampicin	13292-46-1
116	Sesame seed oil unsaponified matter	NA
117	Sodium Arsenite	7784-46-5
118	Sodium dehydroacetate	4418-26-2
119	Sorafenib tosylate	475207-59-1
120	Styrene	100-42-5
121	Tamoxifen	10540-29-1
122	TCDD(2,3,7,8-Tetrachlorodibenzo-p-Dioxin)	1746-01-6
123	TCDF(2,3,7,8-Tetrachlorodibenzofuran)	51207-31-9
124	Tebufenozide	112410-23-8
125	Testosterone propionate	57-85-2
126	Tetradecane	629-59-4
127	Thalidomide	50-35-1
128	Toluene	108-88-3
129	1,2,3-Triazole	288-36-8
130	1,2,4-Triazole	288-88-0
131	Transplatin	14913-33-8
132	Tributyltin chloride/tributyltin	1461-22-9/56573-85-4
133	Troglitazone	97322-87-7
134	Valproic acid sodium salt	1069-66-5
135	Vat Red I(D&C Red No.30)	2379-74-0
136	Verbenone	18308-32-5
137	2-Vinylpyridine	100-69-6
138	Violet No.201	81-48-1
139	Violet No.401	4430-18-6
140	Warfarin	81-81-2
141	Xylene	1330-20-7

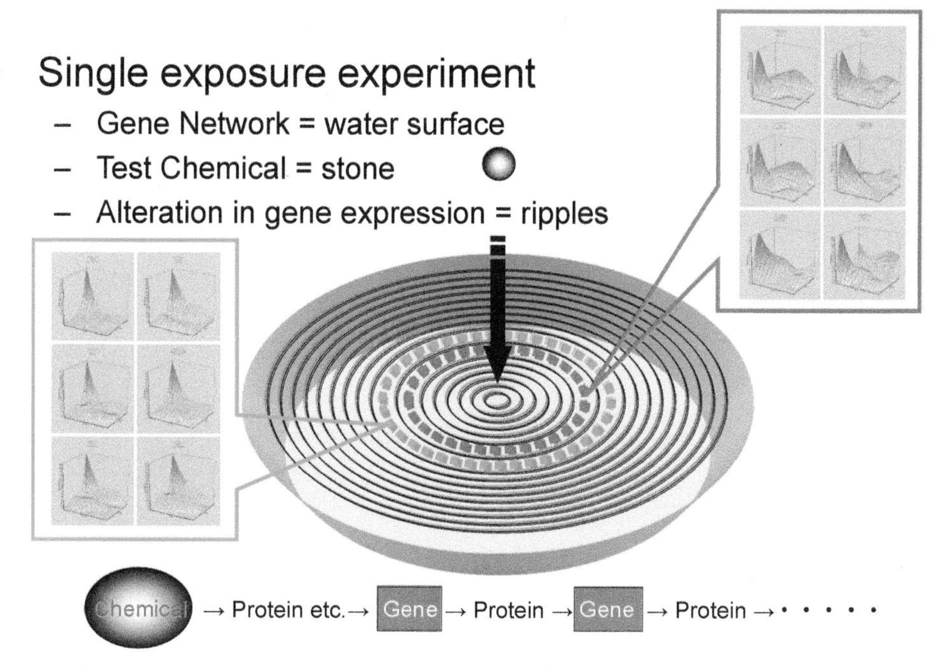

FIGURE 64.2 A conceptual illustration of Percellome project using 3D-surface data system as a fundamental for unsupervised and phenotype-independent clustering.

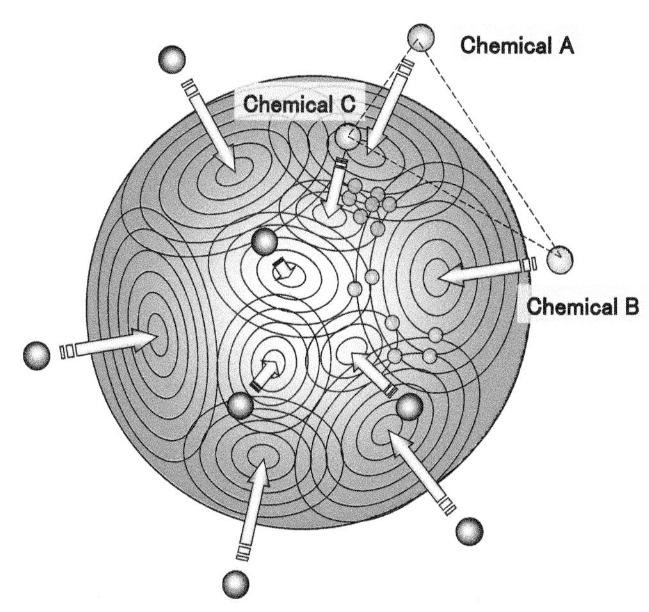

FIGURE 64.3 A conceptual illustration of the advanced state of Percellome project unsupervised network analysis.

at a speed of 2.5 min per mouse from initiation of anesthesia to immersion of the organ samples into RNAlater solution, keeping the total elapsed time within 30 min.

Microarray data generation

As reported previously (Kanno et al., 2006), tissue blocks soaked in RNAlater (Ambion Inc., TX) were kept overnight at 4°C until use. RNAlater was replaced with RLT buffer (Qiagen GmbH., Germany) and homogenized. A small aliquot of the homogenate was treated with DNAse-free RNase A and Proteinase K, and the DNA concentration was measured by PicoGreen fluorescent dye (Molecular Probes Inc., USA). In proportion to the DNA contents, each sample homogenates were spiked with the grade-dosed spike cocktail (GSC; cocktail of five *Bacillus subtilis* RNA sequences selected from the gene list of Affymetrix GeneChip arrays AFFX-ThrX-3_at, AFFX-LysX-3_at, AFFX-PheX-3_at, AFFX-DapX-3_at, and AFFX-TrpnX-3_at) and then processed by the Affymetrix Standard protocol to apply to Mouse 430 2.0 GeneChip. The raw readouts were converted to Percellome data. The mean value (m) with standard deviation (sd) was calculated for all of the probesets (PSs) for each dose time points. To better visualize the changes in 2 h, 2-h vehicle value was used for putative 0-h data and drawn a 5 × 4 surface graph with X-axis for dose, Y for time, and Z for expression, in which thick colored mean surface and two thin blue-colored ±1sd surfaces are overlaid (Fig. 64.1). The mean surface is rainbow-colored from blue, green, red, to yellow according to its peak values (cf. http://toxicomics.nihs.go.jp/db/).

COMPREHENSIVE SELECTION OF RESPONDING mRNAs

A homemade software program named RSort was used for automatic selection of treatment-responding mRNAs. This program firstly sorts the PSs based on

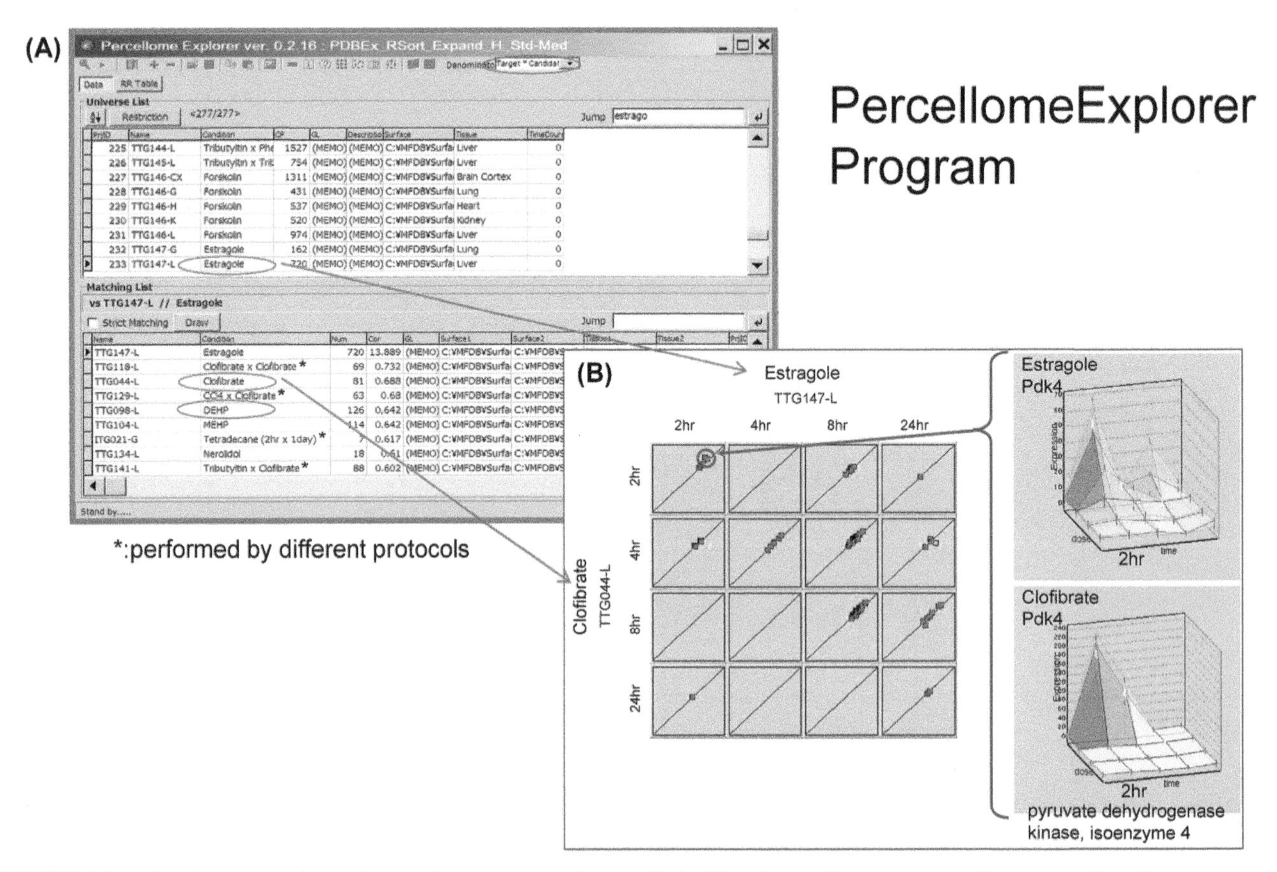

FIGURE 64.4 Finding chemicals sharing similar gene expression profile by PE software. Data automatically generated by RSort program are used in this program. (A) The lower window shows the result of search that the most similar chemical to estragole by the same protocol was clofibrate followed by DEHP. (B) A sub window generated by this software indicating the numbers of PSs with the peak time. One plot can be visually checked by several clicks as shown in the right-hand side for Pdk4 gene.

the roughness of the 3D surface; it calculates the numbers of peaks (upward and downward) in each surface and sort the surface by the number of the peaks (1–8 peaks in 4 × 4 grid). Secondly, it filters the PSs by the roughness (normally surface with more than three peaks are discarded) and several parameters such as maximum expression level (detection level of GeneChip, 1 copy per cell for liver samples), P values between vehicle and top dose groups ($P < .05$ or $P < .01$). In this study, RSort program selected surfaces with the number of peaks three or less, with first or second peaks in high doses (at any time) or with first or second peak in middle doses if its mean values are statistically insignificant to that of the neighboring high-dose peak at $P < .01$ by student's t-test, and the mean value of the selected highest or lowest peaks is significantly different from that of vehicle control at $P < .05$ by student's t-test. These auto-selected PSs were visually checked for its surface shape, eliminating noise data, and subdivide into those showed initial changes at 2, 4, 8, and 24 h. Another program named Percellome Explorer (PE) was used to select a list of chemicals that share PSs with a particular chemical of interest. The PE contains the gene lists automatically selected by the RSort of all data in our Percellome Project (341 datasets for liver samples, 200 for all samples), and automatically sorts out the chemicals sharing same PSs by its number (cf. Fig. 64.4). The automatically selected gene lists (product sets) were visually checked to remove noise surfaces.

ESTRAGOLE STUDY

The flavor compound estragole (1-allyl-4-methoxybenzene) (CAS No.: 140-67-0) is a major component of tarragon, as well as of oils obtained from basil and other plants. Estragole (98%) from Sigma–Aldrich (Catalog No. A29208-25G, lot No.: 05202AH) at a dose of 100 mg/kg caused no observable effects after 24 h in dose finding study. Percellome study was conducted at doses of 0, 10, 30, and 100 mg/kg dissolved in corn oil (Sigma–Aldrich, C8267) and delivered by gavage (Kanno et al., 2012).

PENTACHLOROPHENOL STUDY

The pesticide pentachlorophenol (PCP) (CAS No.: 87-86-5) at a dose of 100 mg/kg caused no observable effects after 24 h in dose finding study. Percellome study was conducted at doses of 0, 10, 30, and 100 mg/kg dissolved in 0.5% methyl cellulose (Shin-Etsu Chemical Co., Ltd. Tokyo, Japan) and delivered by oral gavage (Kanno et al., 2013).

MERGING OF TGP DATA TO PERCELLOME DATABASE

A government–private sector joint research project, Toxicogenomics Project, so-called TGP (2002–2011), was initiated by the National Institute of Health Sciences (NIHS) and an alliance of 17 Japanese pharmaceutical companies. Rat liver was chosen as a primary target because the member companies already had their own rat data. The project adopted the Percellome method and was set up by a group of scientists in the NIHS. Up to 150 chemicals were monitored for rat liver transcriptome on acute phase (3, 6, 9, and 24 h after single gavage), and repeated phase (3, 7, 14, and 28 days during repeated oral dosage) with four dosage levels (Affymetrix GeneChip, Rat Expression 230A, Rat Genome 230 2.0 Array and Human Genome U133 Plus 2.0 Array). In 2005, the whole project was moved from NIHS to the National Institute of Biomedical Innovation (NIBIO) in Osaka, Japan. The data (non-Percellome, 75 percentile global normalization) are available at TG-GATEs homepage (http://toxico.nibio.go.jp/english/datalist.html). As the raw data of TGP contain Percellome spike data, we started to transform TGP data into absolute values using Percellome method and developed a method to combine the rat TGP data and mouse Percellome data using the ortholog-based concatenated ID allowing many-to-many correspondence. Here, examples on clofibrate (CAS No.: 37-07-0) and phenobarbital (CAS No.: 57-30-7) are briefly shown. Doses of clofibrate were 0, 10, 30, and 100 mg/kg dissolved in MC 0.5% with DMSO 0.1% to mouse (http://percellome.nihs.go.jp/PercellomeWeb.dll/EXEC) and 0, 30, 100, 300 mg/kg dissolved in corn oil to rat, and phenobarbital was 0, 15, 30, 150 mg/kg in MC 0.5% to mouse and 0, 100, 150, 300 mg/kg in MC 0.5% to rat.

NEW REPEATED DOSING STUDY

All 48 mice were pretreated with 5.0 mg/kg of carbon tetrachloride (CCl_4) for 14 days and on the 15th day, 0, 0.7, 2.0, and 7.0 mg/kg of CCl_4 was given to 12 mice each and then sampled at 2, 4, 8, and 24 h ($n = 3$). Gene expression data were obtained as described above. The expression profile was compared to the single dose study performed at the same dose levels without any pretreatment, i.e., the study performed by the standard Percellome single exposure protocol (cf. Figs. 64.1 and Fig. 64.12).

RESULTS AND DISCUSSION

Estragole Study

As previously reported (Kanno et al., 2012), 1214 upregulated and 244 downregulated PSs were automatically selected and 167 of those were selected visually as biologically feasible. Among those upregulated PSs, Ingenuity Pathway Analysis (IPA) highlighted a certain number of genes encoding proteins associated with the PPAR-alpha. The metabolic enzymes upregulated were Cyp4a10, Cyp4a14, and Cyp4a31, reported to be regulated by PPAR-alpha (Bumpus and Johnson, 2011). The upregulated transporters Slc16a5, Slc25a20, and Slc27a1 are also considered to be regulated by PPAR-alpha. A limited number of genes were related to oxidative stress, apoptosis, and cell proliferation. The genes downregulated could not be assigned to any known pathway by IPA or by searching the literature.

To confirm the result, we scanned the Percellome database by the PE program, which automatically compares the PS lists with all the studies performed. As a result, PE identified clofibrate (81 PSs), followed by di(2-ethylhexyl)phthalate (DEHP) (126 PSs) as the two chemicals most similar to estragole under the same experimental protocol (Fig. 64.4). Clofibrate and DEHP (Lee et al., 1995; Yu et al., 2001) are known as PPAR alpha agonists. The PE program also showed that the time course in induction of PSs is slightly slower in estragole than clofibrate but considerably similar (Fig. 64.4B). Fig. 64.4B also shows that genes commonly induced at 2 h contained isozyme 4 of pyruvate dehydrogenase kinase (Pdk4). Such early-responding genes are considered to be located immediately downstream of the receptor. Dynein, axonemal and intermediate chain 1 (Dnaic1) is induced 8 and 4 h after administration of estragole and clofibrate, respectively, as well as by DEHP, and its metabolite mono-(2-ethylhexyl) phthalate (MEHP) at later time points (Fig. 64.5). Dnaic1 appears to play an important role in peroxisome biogenesis (Brocard et al., 2005; Kural et al., 2005). This delayed induction compared to that of early-response genes, such as Pdk4, might indicate that several mediators are located between PPAR-alpha and this gene. The nature of the pathway, together with a literature search indicate that HNF4A and PEX13 (Odom et al., 2004) may be involved, since the latter is induced slightly by estragole (not significantly) and clearly by clofibrate, DEHP, and MEHP (Fig. 64.5).

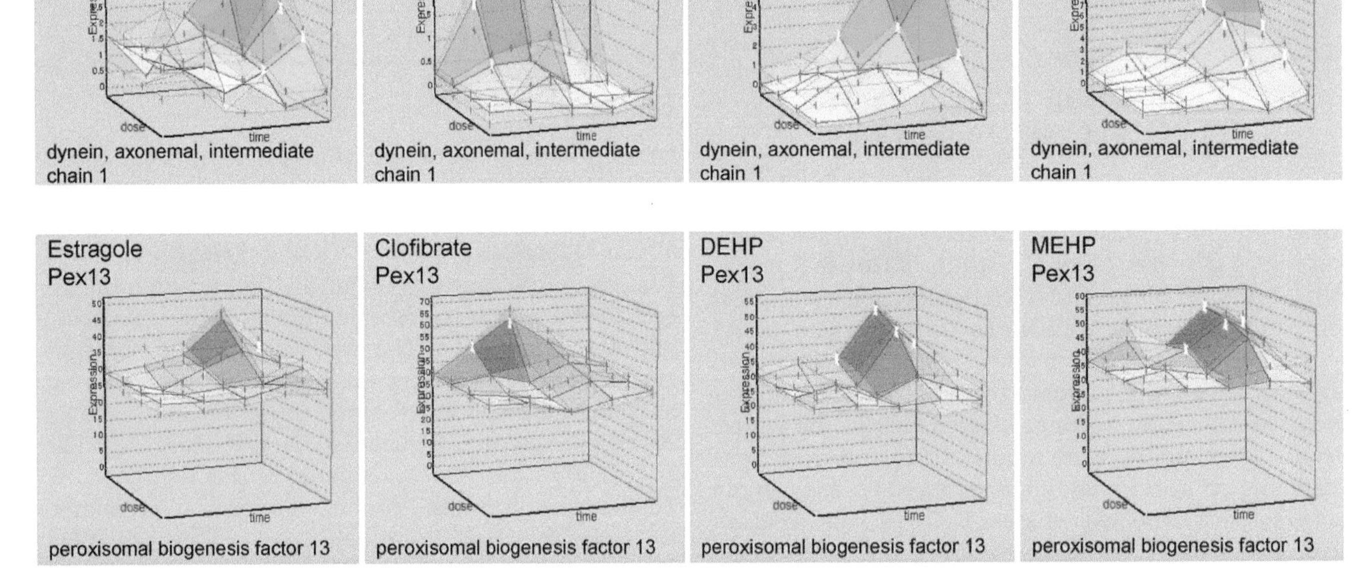

FIGURE 64.5 Dnaic1 and Pex13 induction by estragole, clofibrate, DEHP, and its metabolite MEHP. Dnaic1 and PEX13 appear to play an important role in peroxisome biogenesis.

Among the isozymes of cytochrome P-450, Cyp4a10, Cyp4a14, and Cyp4a31 are induced significantly by estragole, as well as by clofibrate and DEHP. On the other hand, Cyp1a2 which is induced slightly by Phenobarbital was not upregulated by estragole. All three of these Cyp4a's are considered to be regulated by PPAR alpha. Whereas DEHP also induces Cyp2b10 and Cyp51, estragole and clofibrate do not. Phenobarbital also induces Cyp2b10 and Cyp51, but not Cyp4a's. Thus, the present Percellome analysis indicates (Fig. 64.6) that DEHP activates two receptors, PPAR-alpha and constitutive androstane receptor (NR1I3 or CAR), which regulates Cyp2b (Ren et al., 2010) and probably Cyp51 (Dekeyser et al., 2011; Xu et al., 2005). In contrast, estragole and clofibrate activate PPAR-alpha, but not CAR.

Expression of the drug and/or metal transporters Slc12a7, Slc14a2, Slc16a1, Slc16a5, Slc22a5, Slc23a2, Slc25a20, Slc25a42, Slc27a1, and Slc29a3 appeared here to be altered by estragole. Again, PE analysis reveals that most of these Slc are also induced by DEHP and clofibrate (not shown). Among these, only Slc27a1 (Francis et al., 2003; Lemoine et al., 2009) has so far been reported to be regulated by PPAR-alpha. However, expression of most of these Slc's including Slc27a1 peaked after 8 h, indicating that these genes may be activated indirectly.

Our novel finding that estragole activates PPAR-alpha signaling may help elucidate the mechanism(s) of its hepatocarcinogenicity, i.e., peroxisome proliferation. Indeed, estragole increases liver weight at a dose lower than the carcinogenic dose (Sixty-Ninth Meeting of the Joint FAO/WHO, 2009). Although recent reports on estragole carcinogenicity suggest involvement of its metabolites (Ishii et al., 2011) or glucocorticoid pathways (Kaledin et al., 2009), our Percellome data did not support the involvement of such pathways or its genotoxicity which can be monitored indirectly as an enhancement in DNA repair and responses to oxidative stress. Interestingly, DEHP and Wyeth 14,643, well-characterized nongenotoxic rodent hepatocarcinogens that evoke tumors through peroxisome proliferation, gave mutation in Lac Z transgenic mice (Boerrigter, 2004).

An important advantage of determining the actual average number of mRNA molecule per cell is that the responses obtained in different studies can be compared directly. Estragole was comparable to clofibrate in copy numbers of mRNAs induced, and the studies were conducted at the same dose (i.e., 0, 10, 30, and 100 mg/kg). Therefore, these data strongly suggest that estragole appears to be as potent as clofibrate in activating PPAR-alpha signaling in per body weight basis.

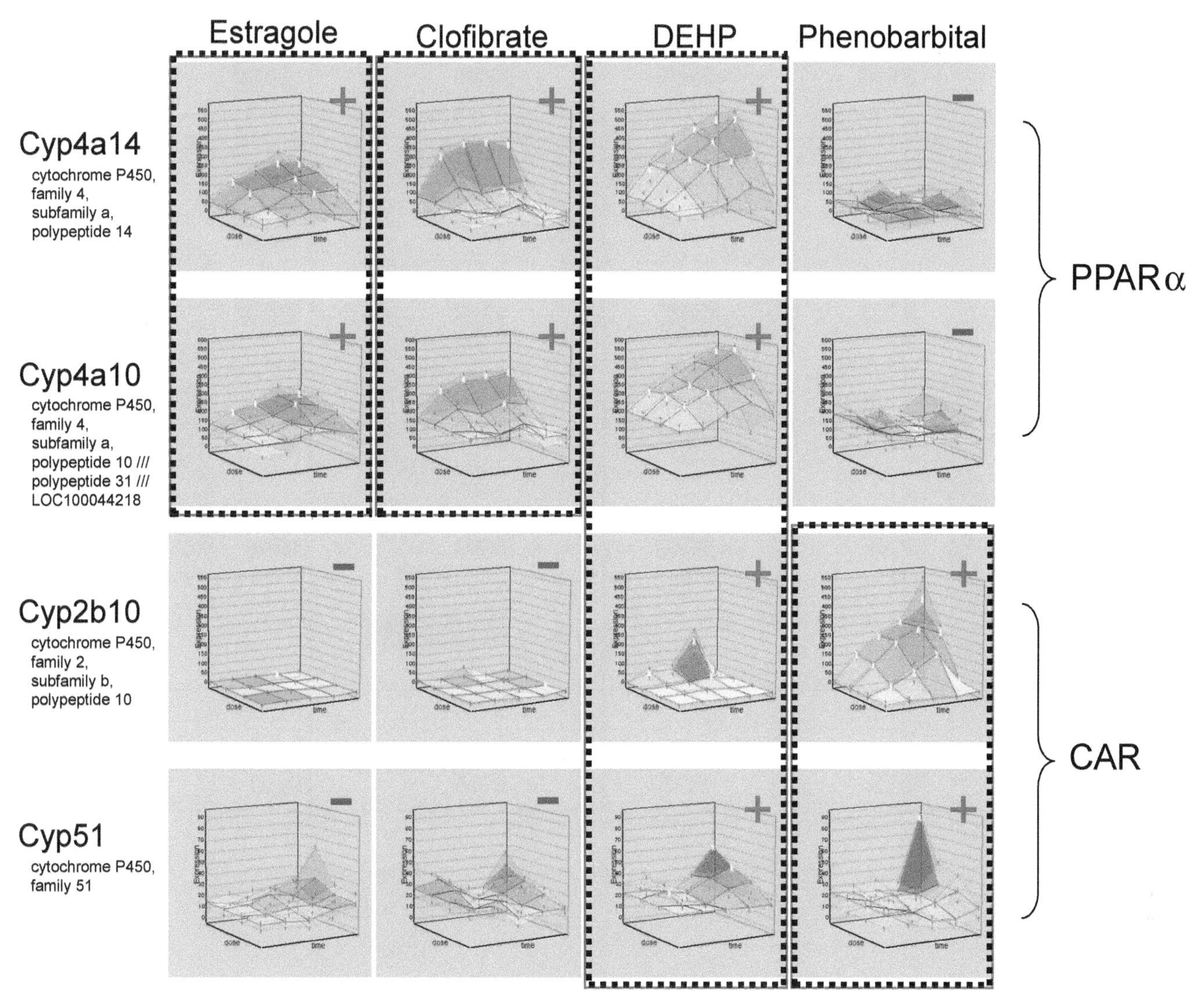

FIGURE 64.6 A summary of the pathways shared with estragole, clofibrate, DEHP, and MEHP illustrated by the responses of Cyp4a14, Cyp4a10, Cyp2b10, and Cyp51; the former two are PPAR-alpha-driven, and the latter two are CAR-driven.

PCP Study

The numbers of PSs that started to change their expression at 2, 4, 8, and 24 h were 98, 55, 127, and 1192, respectively (Kanno et al., 2013). Among 168 liver gene expression studies registered in Percellome database, chemicals that shared the PS list with PCP are listed in Table 64.2. The chemicals that shared the PSs with the 2-hr list of PCP was sodium dihydroacetate (TTG154-L); 51 PSs, followed by acephate (TTG109-L); 24 PSs, down to 5-fluorouracil (TTG160-L); 4 PSs. The sum set (union of the PS lists) from TTG154-L through TTG160-L was 75 (upregulated (Up) 59, downregulated (D) 16). Likewise, the sum set of the 4 h PS lists was 31 (Up 22, D 9), 8 h was 46 (Up 23, D 23), and 24 h 636 (all Up). The PS list unique to PCP (Unique list) at each time point was 23, 24, 81, and 556, respectively.

The PS list common to PCP and other related chemicals (Common list) at 2 h contained Gluconeogenesis Pathway of PGC-1A (PPARGC1A)/Foxo1/HNF4 (Fig. 64.7) (Puigserver et al., 2003). This finding is in concordance with the report in experimental animals that PCP acutely induces hyperglycemia (Deichman et al., 1942). PPARGC1A is reported to increase the expression of Lpin1 (Finck et al., 2006), which is also the case with the Common list.

A small set of genes of metabolic enzymes are induced during the first 8 h, including Cyp2a4, Cyp4f16, Cyp7a1, Cyp17a1, Cyp39a1, Fmo2, Fmo5, and Sult1d1. IPA indicated that they are likely to be induced by Nr1i3 (CAR), Nr1i2 (PXR/SXR), or Nr5a1. Although our RSort program with current parameter setting did not pick up those nuclear receptors, manual search showed, among the three receptors, PXR/SXR was

TABLE 64.2 List of Chemicals Sharing Probeset List With PCP at 2, 4, 8, and 24 h

2hr			4hr			8hr			24hr		
Percellome No.	Treatment	PS	Percellome No.	Treatment	PS	Percellome No.	Treatment	PS	Percellome No.	Treatment	PS
TTG016-L(C)	Pentachlorophenol	100	TTG016-L(C)	Pentachlorophenol	55	TTG016-L(C)	Pentachlorophenol	127	TTG016-L(C)	Pentachlorophenol	1192
TTG154-L	Sodium Dehydroacetate	54	TTG104-L	MEHP	21	TTG098-L	DEHP	15	TTG098-L	DEHP	258
TTG109-L	Acephate	26	TTG098-L	DEHP	16	TTG041-L	Valproic Acid	14	TTG032-L	3-Amino-1H-1,2,4-triazole	212
TTG059-L	Caffeine	21	TTG037-L	Phenobarbital	14	TTG104-L	MEHP	14	TTG104-L	MEHP	177
TTG062-L(C)	Dexamethasone	20	TTG032-L	3-Amino-1H-1,2,4-triazole	12	TTG109-L	Acephate	13	TTG037-L	Phenobarbital	160
TTG041-L	Valproic Acid	18	TTG144-L	Tributyltin x Phenobarbital	12	TTG160-L	5-fluorouracil	10	TTG041-L	Valproic Acid	109
TTG098-L	DEHP	17	TTG150-L	Valproic acid sodium salt x Thalidomide	8	TTG154-L	Sodium Dehydroacetate	9	TTG157-L	Valproic acid sodium salt	103
TTG019-L	2-Vinylpyridine	17	TTG141-L	Tributyltin x Clofibrate	8	TTG141-L	Tributyltin x Clofibrate	8	TTG031-L	2-Chloro-4,6-dimethylaniline	94
TTG104-L	MEHP	14	TTG074-L	Bromobenzene	8	TTG031-L	2-Chloro-4,6-dimethylaniline	8	TTG154-L	Sodium Dehydroacetate	77
TTG165-L	Chlorpyrifos	14	TTG151-L	Valproic acid sodium salt x Valproic acid sodium salt	7	TTG032-L	3-Amino-1H-1,2,4-triazole	8	TTG162-L	Sesame seed oil unsaponified matter	71
TTG034-L	4-Ethylnitrobenzene	14	TTG031-L	2-Chloro-4,6-dimethylaniline	7	TTG146-L	Forskolin	6	TTG044-L	Clofibrate	69
TTG166-L	Carbaryl	12	TTG044-L	Clofibrate	6	TTG062-L(C)	Dexamethasone	6	TTG074-L	Bromobenzene	47
TTG031-L	2-Chloro-4,6-dimethylaniline	10	TTG162-L	Sesame seed oil unsaponified matter	5	TTG054-L	Diethylnitrosamine (C57BL/6)	5	TTG109-L	Acephate	17
TTG141-L	Tributyltin x Clofibrate	9	TTG173-L	TCDD/AhRKO	0	TTG132-L	Curcumin	3	TTG160-L	5-fluorouracil	13
TTG032-L	3-Amino-1H-1,2,4-triazole	9				TTG136-L	Phytol	2			
TTG027-L	1,2,3-Triazole	9				TTG096-L	Omeprazole	0			
TTG160-L	5-fluorouracil	4									
	Sum Set (Total)	77		Sum Set (Total)	31		Sum Set (Total)	46		Sum Set (Total)	636
	Sum Set (Up)	61		Sum Set (Up)	22		Sum Set (Up)	23		Sum Set (Up)	636
	Sum Set (Dn)	16		Sum Set (Dn)	9		Sum Set (Dn)	23		Sum Set (Dn)	0
	PCP NOT Sum	23		PCP NOT Sum	24		PCP NOT Sum	81		PCP NOT Sum	556

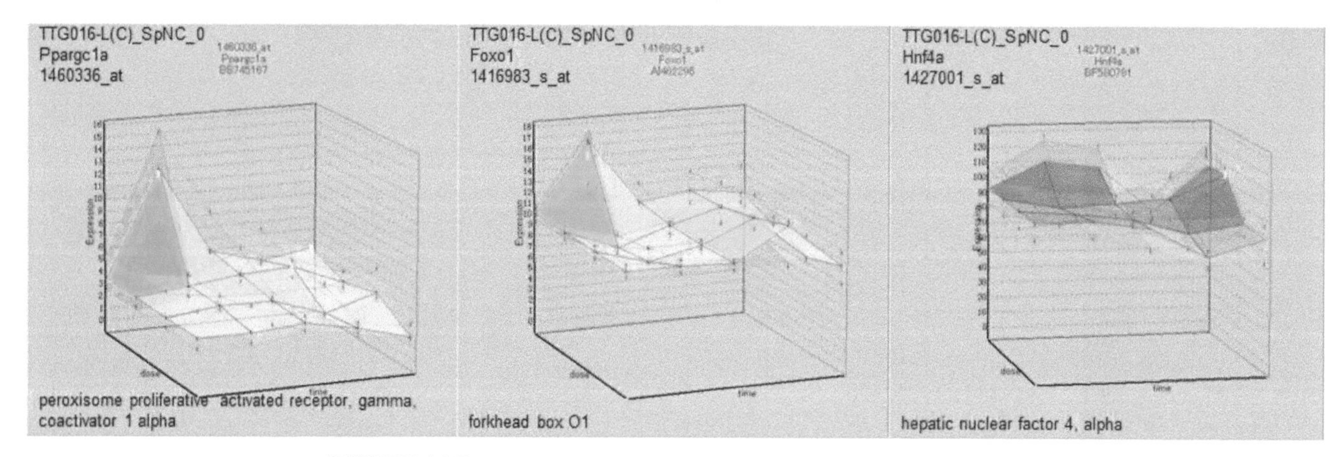

FIGURE 64.7 Early induction of PPARGC1A, Foxo1, and HNF4 by PCP.

induced by PCP. This series of change was not unique to PCP. Downregulation of Junb and Fos at 2, 4, and 8 h were uniquely found in PCP gene list.

The list of PSs induced at 24 h contained well-known networks. About half of the PSs were in the common list and assigned, by IPA, to NRF2-mediated Oxidative Stress Response accompanying clear induction of Keap1, and phase I and II metabolic pathways. The other half was unique to PCP and was assigned to interferon pathways (Figs. 64.8 and 64.9). Other networks were not effectively searched by the IPA.

Among the chemicals tested in the Percellome project, PCP is relatively a slow starter in induction of genes, i.e., less than 100 PSs are induced at 2, 4, 8 h and increases to 1200 at 24 h. It would be plausible to consider that metabolite(s) of PCP was generated during the first 8 h and the metabolite(s) induced the 24 h burst in genes of Nrf2 and interferon networks. Some metabolizing

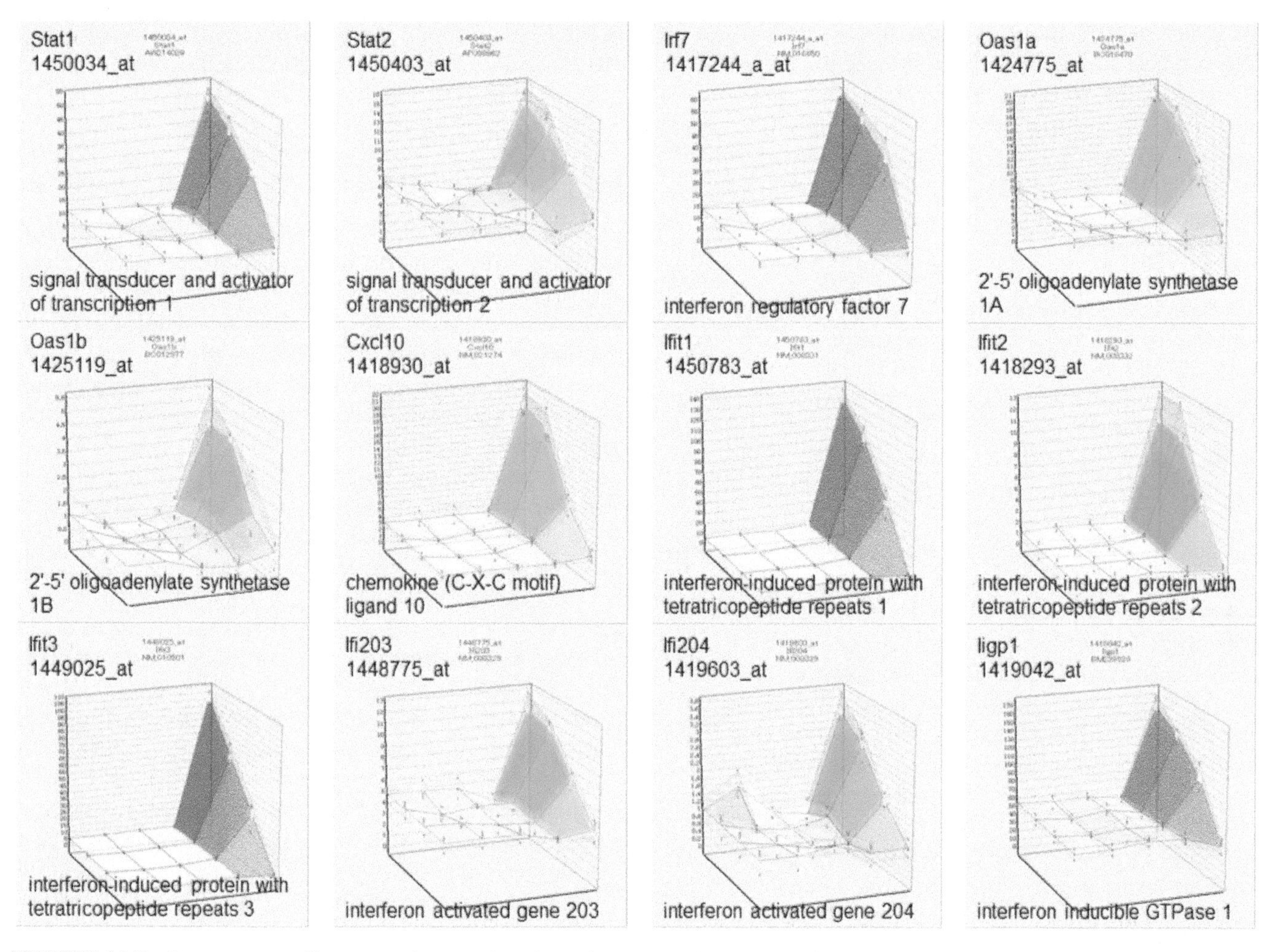

FIGURE 64.8 Representative illustration of PCP-induced interferon network genes. All of these genes are induced at 24 h in a dose-dependent manner. *From Kanno et al., 2013, with permission.*

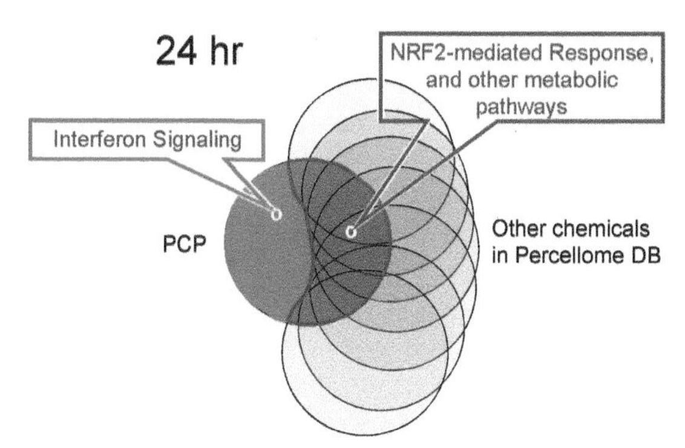

FIGURE 64.9 Venn diagram illustration of the uniqueness of PCP in terms of inducing interferon signaling shown by the PE results. Among 168 liver study data registered in the Percellome database, approximately 10 chemicals shared the list with PCP at 24 h. Intersection (purple) of the PCP and the Sum of the list (Common list in text) contained Nrf2-mediated response genes and those of other metabolic pathways. The rest of the list of PCP (red: PCP NOT Sum) are almost exclusively made of interferon signaling and related genes.

enzymes known to be located downstream of PXR/SXR are induced by 8 h. These enzymes might have metabolized PCP. IPA analysis indicated a possible activation of CAR and PXR/SXR. This finding may explain the engagement of DEHP to the top chemical list in Table 64.2. Weak signal to CAR and PXRSXR may have been transduced by PCP and cyp2a4 and other metabolic enzymes were turned on, which leads to PCP metabolism, and triggered Nrf2 network.

It is of an interest whether PCP or its metabolite(s) can be a ligand to PXR/SXR. Metabolites known are tetrachloro-p(o)-hydroquinone (TCpHQ and TCoHQ) and tetrachloro-p-benzoquinone (chloranil, TCpBQ). Furthermore, TCpHQ is reported to be metabolized, generating hydroxyl radicals with a help of H_2O_2 without Fenton reaction, to trichloro-hydroperoxyl-1,4-benzoquinone (TrCBQ-OOH) and trichloro-hydroxy-1,4-benzoquinone (TrCBQ-OH) (Zhu and Shan, 2009).

We have no Percellome data on those metabolites. To date there is no report on the interaction of PCP and its metabolites with the PXR/SXR. Since 8OH-dG, a type of DNA adduct was monitored in PCP-treated animal liver, mechanism for the hepatocarcinogenicity by PCP has been attributed to the hydroxyl radical formation during metabolism of PCP. The common list supports this mechanism.

It is reported that TLR4-mediated lipopolysaccharide-induced activation of the IFN-beta promoter was inhibited by PCP in a Myd88-independent way (Ohnishi et al., 2008). On the other hand, PCP was considered to trigger TLR4 via the induction of hydroxyl radicals (Lucas and Maes, 2013). Our data show that PCP had significantly upregulated Myd88, Irak1, Traf6, Tlr2, Tlr3, Tlr5, Tlr9, and although not significantly, Tlr4. Irf 3, Irf7, and Irf9 are also induced. This finding indicates that PCP might have triggered the TLR system via modifying the proteins in the cytoplasm either by itself, its metabolites, or hydroxyl radicals. TLRs or the pattern recognition receptors in the endosomes or on the cell surface may have sensed modified proteins. As Irf 3, 7, 9, and interferon alpha are increased, it would be possible that Irf mediated autocrine or paracrine of the interferon receptor ligand, i.e., interferon alpha is induced, and subsequent burst in interferon network had followed (Sato et al., 2000). It is noted that, although Myd88 is induced, NF-κB, TNF, IL12, and CD40 were not. This result might indicate that there may be a switch toward inflammatory cytokine production that was not triggered by PCP. There is a recent report that isopropanol impaired the AP-1 activation in TLR4-mediated lipopolysaccharide effect on TNF-alpha in monocytes (Carignan et al., 2011). The downregulation of Fos/Junb induced by PCP might be related to this switching; further analysis is needed for elucidating relation between suppression of Fos/Junb gene expression and segregation of the inflammatory cytokine pathway from the interferon pathway.

There is a relatively new low-molecular weight compound "imiquimod" sold as a TLR7 agonist for treatment of skin viral infection (Hemmi et al., 2002). A poly fluoromethylated compound, 8-(1, 3,4-oxadiazol-2-yl)-2, 4-bis (trifluoromethyl) imidazo [1, 2-a] [1, 8] naphthyridine (RO4948191) is reported as an orally available low-molecular weight interferon receptor agonist (Konishi et al., 2012) and shown to induce a set of genes similar to those induced by PCP, such as OAS1, Adar, Bst1, Stat1, Ifit3, Usp18, Osg15, Herc6, and CXCL10. These two reports suggested that synthetic (halogenated) hydrocarbons can be ligands to both TLR(s) and interferon receptor(s) and directly trigger the interferon responses. Further analyses on interaction of the PCP and related chemical(s) with those receptors are essential for the clarification of the mechanism of triggering interferon gene networks.

Merging the TGP Data to Percellome Database

A mouse Percellome data of clofibrate (TTG044-L) and a rat TGP data of clofibrate (TGP0098-L), both oral single gavage studies, are shown as example. The probe IDs were concatenated according to ortholog, and PE database was rebuilt for 323 projects consisting of 293 (liver 171) Percellome projects and 30 TGP project data (liver only) as a trial. The coverage of the PSs is about 55% for both species.

Clofibrate induced a large number of ortholog genes commonly found in both species. The PE graph indicated that rat is responding slightly slower; many genes induced at 4 and 8 h in mice are induced at 24 h in rat. For this chemical, which is known to hit a single target PPAR-alpha, it seems that the direct matching at ortholog level is effective in analyzing the response interspecies-wise (Fig. 64.10).

Phenobarbital induced less number of ortholog genes commonly found in both species. It is known that, for example, P450 genes are not the same, at least in its numbering. Therefore, for listing up the responding Cyp PSs needs a search at the level closer to Gene Ontology (Fig. 64.11).

In the Percellome database, using PE as a tool, there is no boundary between different organs and different study protocols in terms of searching common PS lists. However, between mouse and rat, because of the differences in annotation and differences in the numbers of PSs per one annotation makes the borderless search a challenge. The ortholog-based concatenated ID approach is effective in certain cases, but apparently not enough for more universal use. A tiered approach, i.e., ortholog level, gene ontology level, and probably more relevant binders between species based on the gene sequences will be considered in the near future for better performance in interspecies gene expression analysis.

New Repeated Dosing Study

This study design is a derivative of a study on gene knockout mouse. Instead of using knockout mouse, this new method starts with a process of creating a mouse in the "chemically induced transgenic state" by giving repeated dose of a chemical before applying the single dose Percellome standard protocol. The idea is that repeated daily dosing should gradually alter the

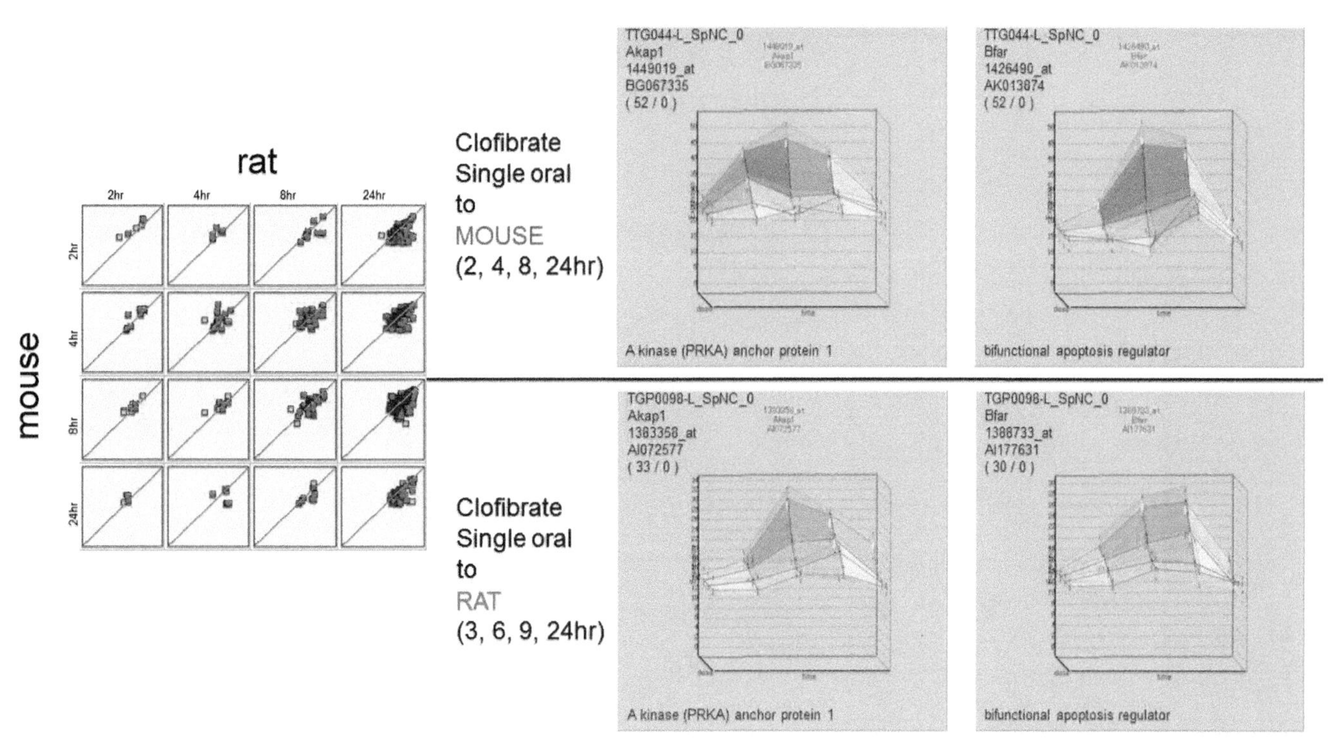

FIGURE 64.10 A few representative genes from the mouse Percellome data of clofibrate (TTG044-L) and a rat TGP data of clofibrate (TGP0098-L). Ortholog-based search was effective for clofibrate data.

gene network and eventually induce a new steady state where at least some part of the whole gene network is different from that of the untreated mice. Therefore, the gene expression profile generated by the single challenge on the next day of the last daily dosing contains the information of altered gene network of the chemically induced transgenic state caused by repeated dosing of a chemical (Fig. 64.12) just like we observe the effect of knocked out gene. Fourteen days of repeated dosing of CCl_4 at the dose of 5 mg/kg resulted in a suppression of responses to the final challenge of CCl_4 in almost all of the genes that showed clear response in the standard single dosing experiment. We found that there was a tendency in the suppressed genes that the baseline copy number of mRNA was also suppressed in such genes. A few genes that the response to the final challenge was enhanced showed a tendency that the baseline was elevated. Further analysis suggested that the effect of 14-day pretreatment can be assessed by the combination of two responses, i.e., the change in the "Transient Response (TR)" and the "Baseline Response (BR)." TR is a quick response that is monitored within the 24 h after the single dosing. BR is a gradual response that is monitored as a change in expression level of vehicle control groups caused by the repeated dosing (Fig. 64.12). In case of CCl_4, the result can be summarized as "there were many genes that showed diminished TR with lowering of the BR. A small number of genes showed enhanced TR with higher BR." The duration of pretreatment was shortened up to 4 days, and the effect of repeated dosing on both TR and BR was found to be similar to those by 14 day pretreatment. By the end of March of 2018, this New Repeated Dose Study protocol (4 days of repeated dosing and final challenge) was applied to CCl_4, valproic acid sodium salt, clofibrate, tributyltin, N,N-diethyl-3-methylbenzamide (Deet), acetaminophen, phenobarbital sodium, thalidomide, 5-fluorouracil, acephate, and pentachlorophenol. The effect of repeated dosing on the TR and BR was various among those chemicals. For example, CCl_4 was the most potent chemical to suppress the TR and BR of the genes. In contrast, valproic acid sodium salt was the extreme case that showed least effect on both TR and BR. Phenobarbital showed a tendency to boost the TR in many genes.

Comparison of the Percellome data between the single dose experiment data and the New Repeated Dose study data is generating a database on the relation between TR and BR in molecular basis. Again, Percellome analysis tools can list up a group of genes by the pattern of TR and BR and search for, for example, the upstream and downstream of the gene networks by checking the enhancer–promoter sequences for common transcription factor and the relevant binding sites. Detailed analysis is underway with a plan to test more

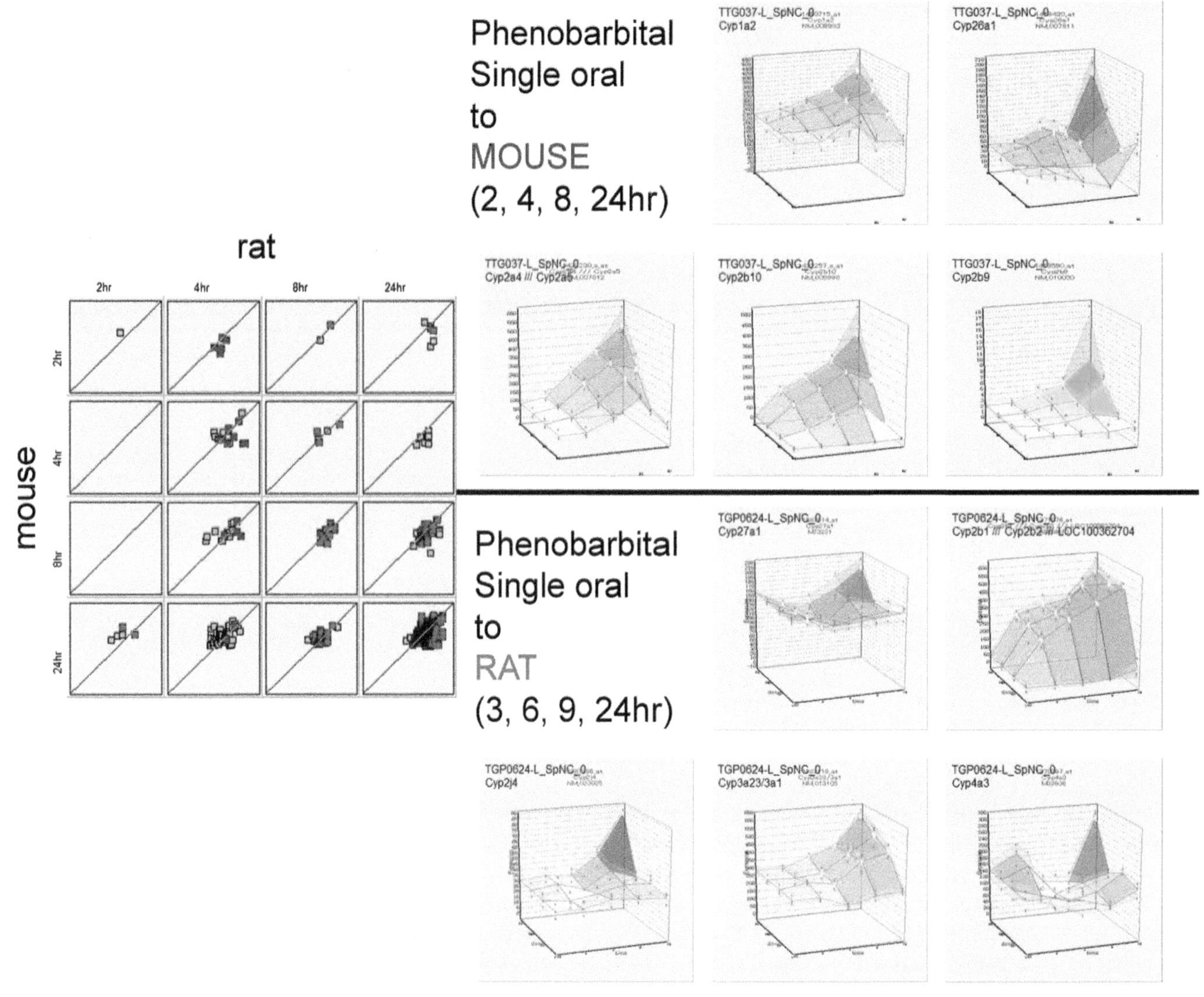

FIGURE 64.11 A few representative genes from the mouse Percellome data of phenobarbital (TTG037-L) and a rat TGP data of phenobarbital (TGP0624-L). Ortholog-base search was less effective for phenobarbital data and needed more broad level of matching list.

chemicals in different categories and to advance epigenetic search using whole genome bisulfite sequencing for DNA methylation analysis and ChIP-Seq study (using anti-H3K4me3, -H3K27Ac, -H3K27me3, and -H3K9me3) for histone modification analysis.

CONCLUDING REMARKS AND FUTURE DIRECTIONS

The Percellome Project for the mouse transcriptome had started from the fiscal year 2003, 1 year after the initiation of the TGP project. The TGP project was designed, launched, and maintained by us until the alliance members took over the project. The data structure of both projects is identical and therefore Percellome method can be applied. In the first 3 years, the Percellome project generated 90 chemicals database on mouse liver acute response, i.e., 2, 4, 8, and 24 h after single oral exposure of three dose levels (n = 3 for each point of data). From 2005, the second subproject, Inhalation Toxicogenomics project was launched for collecting lung and liver transcriptome data from mice exposed to a very low concentration of gases comparable to the so-called "sick building syndrome." From 2006, the second stage of Percellome Project was launched for repeated exposure studies and multiorgan relation studies. Fetus (developmental) Percellome and

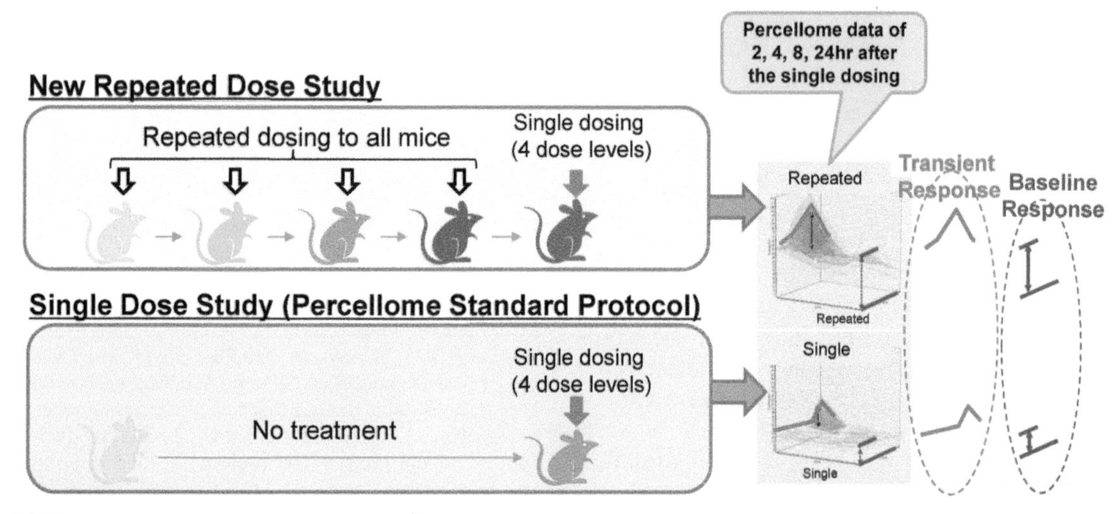

FIGURE 64.12 New Repeated Dose Study. The study is a modification of single dose study using gene knockout mice. In case of gene knockout study, shifts in gene networks caused by knocking out a gene can be evaluated by comparing the Percellome data of wild-type mice and the gene knockout mice challenged by a same chemical. The new repeated dose study is based on a hypothesis that pretreatment will create a "chemically induced transgenic state" in the repeatedly dosed mice. The response within 24 h triggered by the last single dosing is defined as "transient response (TR)". The shift in the expression level of vehicle control group, representing the slow response to the repeated dosing, is defined as the "baseline response (BR)."

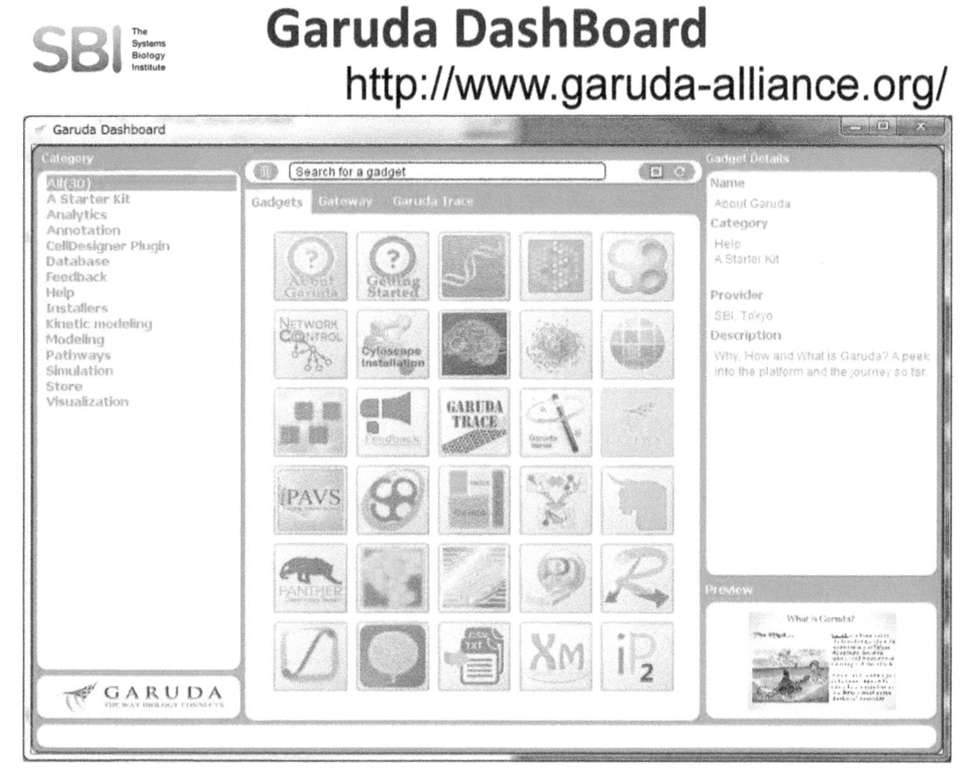

FIGURE 64.13 Garuda alliance. A project to publicize the Percellome data and analysis tools including our homemade software through the Garuda platform (http://ww.garuda-alliance.org/) is in progress.

Brain-behavior Percellome projects are also launched. Up to now (May 2018), the Percellome database contains a total of nearly 400 datasets (one dataset is 4 × 4, triplicate, one organ, total of 48 GeneChip data) tested on about 150 kinds of chemicals over multiple organs, such as liver, lung, kidney, thymus, testis, and brain (hippocampus, cortex).

The absolutized mRNA expression data were easy to handle and therefore greatly facilitated the data analysis including developing the softwares. And the 3-D graph (time x dose x mRNA in copies per cell) made the whole process biologist-friendly. At any time, the data quality and the biological plausibility of the response can be visually checked by the eyes of biologists.

The final goal of the project is to develop a simulation method of virtual mice and virtual human *in silico* to perform toxicity testing. It may take many years until the "Star Trek" era. The practical interim goal is to develop the method to instantly draw a comprehensive and dynamic gene network of each test chemical and construct a systems toxicology for molecular mechanism-based predictive toxicology. Combined effects of two or more chemicals, cross-interaction among organs in a body, and species differences are expected to be logically handled by the Percellome systems toxicology.

Examples shown here, especially the PCP data, indicated that, at least some of the regulation mediated by the protein phosphorylation can be considered as "monitored indirectly," like a shadowgraph, by the transcriptome, i.e., Stat1 and Stat2. Therefore, we believe that precise transcriptome can cover signaling network driven by protein-modification up to a certain degree.

The toxicology of repeated dosing is now gathered to the Percellome Project. New Repeated Dosing Protocol turns out to be very promising to characterize the molecular basis of the shift of response of an organism against repeated chemical exposure by comparing the response to the single dose exposure by dissecting the response to two factors, i.e., the transient response (TR) and the baseline response (BR) of each genes. CCl_4 New Repeated Dosing study instantly indicated the involvement of ER stress (eIF2 and Xbp1 pathway involved, data not shown).

The joining of two large databases, TGP (publicized as TG-GATEs) and Percellome database (now equal in data size to TGP), by Percellome method has been planned. The analysis tools including our homemade software through the Garuda platform (Fig. 64.13) will facilitate the process. We hope to have all data analyzed by many toxicologist and biologists worldwide for drawing the entire dynamic gene network activated in toxicological conditions of a living organism.

Acknowledgments

The author thanks the team Percellome method (2001 ~), Dr. Katsuhide Igarashi, Dr. Ken-ichi Aisaki, Dr. Atsushi Ono, Ms. Tomoko Ando, Ms. Noriko Moriyama, Ms. Yuko Kondo, Ms. Yuko Nakamura, and Ms. Maki Abe of Division of Cellular and Molecular Toxicology, NIHS, the startup team of The Toxicogenomics Project (2003–4), Dr. Akihiko Hirose (Risk Assess), Dr. Takayoshi Suzuki (Mutagen), Dr. Makoto Shibutani (Pathol), Dr. Katsuhide Igarashi (Tox), Atsushi Ono (Tox), Ken-ichi Aisaki (Tox), and Dr. Ken-ichi Aisaki for writing the homemade softwares and maintenance and enhancement of the Percellome Database. IT collaboration was performed with Drs. Shinya Matsumoto and Bun-ichi Tajima of Teradata, and with NTT Comware and NTT Data. The Systems Biology collaboration is conducted with Drs. Hiroaki Kitano, Natalia Polouliakh, Samik Ghosh, and other members of the Systems Biology Institute. The author thanks all the member of the division of Cellular and Molecular Toxicology, NIHS for strong support of the project. The project has been supported by MHLW Health Sciences Research Grants H30-Kagaku-Shitei-001, H27-Kagaku-Shitei-001, H24-Kagaku-Shitei-006, H21-Kagaku-Ippan-001, H19-Toxico-001, H18-Kagaku-Ippan-001, H15-Kagaku-002, H14-Toxico-001, and H13-Seikatsu-012.

References

Boerrigter, M.E., 2004. Mutagenicity of the peroxisome proliferators clofibrate, wyeth 14,643 and di-2-ethylhexyl phthalate in the lacZ plasmid-based transgenic mouse mutation assay. J. Carcinog. 3 (1), 7.

Brocard, C.B., Boucher, K.K., et al., 2005. Requirement for microtubules and dynein motors in the earliest stages of peroxisome biogenesis. Traffic 6 (5), 386–395.

Bumpus, N.N., Johnson, E.F., 2011. 5-Aminoimidazole-4-carboxyamide-ribonucleoside (AICAR)-stimulated hepatic expression of Cyp4a10, Cyp4a14, Cyp4a31, and other peroxisome proliferator-activated receptor alpha-responsive mouse genes is AICAR 5′-monophosphate-dependent and AMP-activated protein kinase-independent. J. Pharmacol. Exp. Ther. 339 (3), 886–895.

Carignan, D., Desy, O., et al., 2011. The dysregulation of the monocyte/macrophage effector function induced by isopropanol is mediated by the defective activation of distinct members of the AP-1 family of transcription factors. Toxicol. Sci. 125 (1), 144–156.

Deichman, W., Machle, W., et al., 1942. Acute and chronic effects of pentachlorophenol and sodium pentachlorophenate upon experimental animals. J. Pharmacol. Exp. Ther. 76, 104–117.

Dekeyser, J.G., Laurenzana, E.M., et al., 2011. Selective phthalate activation of naturally occurring human constitutive androstane receptor splice variants and the pregnane x receptor. Toxicol. Sci. 120 (2), 381–391.

Finck, B.N., Gropler, M.C., et al., 2006. Lipin 1 is an inducible amplifier of the hepatic PGC-1alpha/PPARalpha regulatory pathway. Cell Metab. 4 (3), 199–210.

Francis, G.A., Fayard, E., et al., 2003. Nuclear receptors and the control of metabolism. Annu. Rev. Physiol. 65, 261–311.

Hemmi, H., Kaisho, T., et al., 2002. Small anti-viral compounds activate immune cells via the TLR7 MyD88-dependent signaling pathway. Nat. Immunol. 3 (2), 196–200.

Ishii, Y., Suzuki, Y., et al., 2011. Detection and quantification of specific DNA adducts by liquid chromatography-tandem mass spectrometry in the livers of rats given estragole at the carcinogenic dose. Chem. Res. Toxicol. 24 (4), 532–541.

Kaledin, V.I., Pakharukova, M.Y., et al., 2009. Correlation between hepatocarcinogenic effect of estragole and its influence on glucocorticoid induction of liver-specific enzymes and activities of FOXA and HNF4 transcription factors in mouse and rat liver. Biochemistry (Mosc.) 74 (4), 377–384.

Kanno, J., 2015. Biomechanism-based innovation of toxicology by the fundamental concept of "signal toxicity", review in Japanese. Kokuritsu Iyakuhin Shokuhin Eisei Kenkyusho Hokoku 133, 21–28, 2015.

Kanno, J., Aisaki, K., et al., 2012. Application of percellome toxicogenomics to food safety. Issues in Toxicology No. 11. In: Pongratz, I., Bergander, L. (Eds.), Hormone-disruptive Chemical Contaminants in Food. Royal Society of Chemistry, London, pp. 184–198.

Kanno, J., Aisaki, K., et al., 2006. Per cell" normalization method for mRNA measurement by quantitative PCR and microarrays. BMC Genom. 7, 64.

Kanno, J., Aisaki, K., et al., 2013. Oral administration of pentachlorophenol induces interferon signaling mRNAs in C57BL/6 male mouse liver. J. Toxicol. Sci. 38 (4), 643–654.

Konishi, H., Okamoto, K., et al., 2012. An orally available, small-molecule interferon inhibits viral replication. Sci. Rep. 2, 259.

Kural, C., Kim, H., et al., 2005. Kinesin and dynein move a peroxisome in vivo: a tug-of-war or coordinated movement? Science 308 (5727), 1469–1472.

Lee, S.S., Pineau, T., et al., 1995. Targeted disruption of the alpha isoform of the peroxisome proliferator-activated receptor gene in mice results in abolishment of the pleiotropic effects of peroxisome proliferators. Mol. Cell Biol. 15 (6), 3012–3022.

Lemoine, M., Capeau, J., et al., 2009. PPAR and liver injury in HIV-infected patients. PPAR Res. 2009, 906167.

Lucas, K., Maes, M., 2013. Role of the toll like receptor (TLR) radical cycle in chronic inflammation: possible treatments targeting the TLR4 pathway. Mol. Neurobiol. 48 (1), 190–204.

Odom, D.T., Zizlsperger, N., et al., 2004. Control of pancreas and liver gene expression by HNF transcription factors. Science 303 (5662), 1378–1381.

Ohnishi, T., Yoshida, T., et al., 2008. Effects of possible endocrine disruptors on MyD88-independent TLR4 signaling. FEMS Immunol. Med. Microbiol. 52 (2), 293–295.

Puigserver, P., Rhee, J., et al., 2003. Insulin-regulated hepatic gluconeogenesis through FOXO1-PGC-1alpha interaction. Nature 423 (6939), 550–555.

Ren, H., Aleksunes, L.M., et al., 2010. Characterization of peroxisome proliferator-activated receptor alpha–independent effects of PPAR-alpha activators in the rodent liver: di-(2-ethylhexyl) phthalate also activates the constitutive-activated receptor. Toxicol. Sci. 113 (1), 45–59.

Sato, M., Suemori, H., et al., 2000. Distinct and essential roles of transcription factors IRF-3 and IRF-7 in response to viruses for IFN-alpha/beta gene induction. Immunity 13 (4), 539–548.

Sixty-Ninth Meeting of the Joint FAO/WHO, E. C. o. F. A. J, 2009. Safety Evaluation of Certain Food Additives. WHO Food Additives Series. International Programme on Chemical Safety, World Health Organization, Geneva.

Xu, C., Li, C.Y., et al., 2005. Induction of phase I, II and III drug metabolism/transport by xenobiotics. Arch Pharm. Res. (Seoul) 28 (3), 249–268.

Yu, S., Cao, W.Q., et al., 2001. Human peroxisome proliferator-activated receptor alpha (PPARalpha) supports the induction of peroxisome proliferation in PPARalpha-deficient mouse liver. J. Biol. Chem. 276 (45), 42485–42491.

Zhu, B.Z., Shan, G.Q., 2009. Potential mechanism for pentachlorophenol-induced carcinogenicity: a novel mechanism for metal-independent production of hydroxyl radicals. Chem. Res. Toxicol. 22 (6), 969–977.

CHAPTER

65

Proteomics in Biomarkers of Chemical Toxicity

Christina Wilson-Frank

Purdue University, College of Veterinary Medicine, Animal Disease Diagnostic Laboratory, Department of Comparative Pathobiology, West Lafayette, IN, United States

INTRODUCTION

Toxicoproteomics is emerging as an important tool in elucidating biochemical pathways of chemical toxicity and identifying potential biomarkers for establishing risk assessment, level of concern, and early detection of toxicant exposure. Proteomic analyses allow global simultaneous, high-throughput monitoring of proteins following exposure to chemical toxicants and provides an effective strategy to identifying useful biomarkers indicative of toxicity.

One of the major challenges in proteomics research, particularly when probing biological samples for biomarkers of chemical toxicity, is the inherent complexity of the proteome. However, advancements in sample preparation (to reduce sample complexity), analytical instrumentation, and automated sequence database searching have enabled for simplified and improved, high-throughput analysis and more accurate protein identification. When evaluating the proteome in complex biological samples, the proteomics platforms commonly employed include a technology that globally separates proteins, combined with the use of mass spectrometry for protein identification. To achieve global separation of proteins from complex samples, gel-based technologies are more commonly used in proteomics studies for chemical toxicity, such as 1-dimensional gel electrophoresis (1-DE), 2-dimensional gel electrophoresis (2-DE), and 2-dimensional-difference gel electrophoresis (2-DGE).

Recently, advances in analytical instrumentation have increased the separating capacity and resolution of chromatographic systems; therefore, the use of high-performance liquid chromatography (HPLC) and ultrahigh-performance liquid chromatography are emerging as popular platforms for high-throughput, complex protein separations. The most common analytical platforms used for protein sequencing and identification are matrix-assisted laser desorption/time-of-flight mass spectrometry (MALDI-TOF MS) and tandem mass spectrometry (MS/MS) which can include time-of-flight (TOF), quadrupole (Q), or ion trap mass spectrometry.

Using these technologies, researchers have been able to measure qualitative and quantitative changes in proteins in samples such as serum/plasma, blood, brain, liver, and kidney samples to distinguish unique biomarkers of chemical toxicity. Much of the proteomics research conducted to elucidate candidate biomarkers of chemical toxicity predominantly involves recognizing quantitative changes in proteins after biological systems or cells are subjected to acute or chronic exposure to toxicants. However, an alternative approach to defining potential, candidate biomarkers of toxicity involves using protein network building tools, such as MetaCore or Ingenuity Pathways Analysis software (Wang et al., 2013; Huang and Huang, 2011). Proteins that are differentially expressed in chemical toxicity studies can be subject to protein interaction network analysis, which predicts protein–protein interactions and highlights which molecular pathways in which those proteins play a key role. Not only does this highlight probable molecular mechanisms of toxicity but also may allow investigators to focus on a specific group of proteins or protein in that pathway to evaluate as prospective biomarkers.

Pesticides, herbicides, heavy metals, and most organic contaminants are well-studied chemical toxicants that are known to be potential occupational or environmental health hazards. This chapter describes proteomics investigations conducted in these biological samples that identify potential, predictive biomarkers of chemical toxicity and lead to an improved understanding of mechanisms of toxicity of these chemicals. The potential biomarkers identified in these investigations have been reported to be a result of either cause acute or chronic chemical toxicity and have been

Biomarkers in Toxicology, Second Edition
https://doi.org/10.1016/B978-0-12-814655-2.00065-7

recognized as environmental contaminants or have a high risk of occupational exposure.

HEAVY METALS

Arsenic

Arsenic is a carcinogenic metalloid and is recognized as one of the most toxic heavy metals (Tokar et al., 2013). Environmental exposure to arsenic primarily occurs via arsenic-contaminated drinking water, because trivalent and pentavalent forms of the heavy metal are widely distributed in the environment. Because arsenicals are used in the smelting industry and in the manufacturing of pesticides, herbicides, and products used in wood preservation, occupational exposure to arsenic is also a concern (Tokar et al., 2013). Individuals exposed to arsenic are at risk for developing cancer, hyperkeratosis, cardiovascular disease, and kidney and liver damage. For most of these cases, chronic exposure to arsenic is implicated. Due to the lack of adequate biomarkers to assess exposure to arsenic in these cases, various proteomics methodologies have been used to help reveal candidate biomarkers for early diagnosis of arsenic-induced diseases.

Arsenic-induced hyperkeratosis has been associated with skin cancer and other internal cancers and is endemic in parts of the world in which individuals are chronically exposed to arsenic in drinking water (Hong et al., 2017; Hsu et al., 2013). To identify useful biomarkers for early diagnosis of arsenic-induced, hyperkeratosis in exposed individuals, a proteomic study was conducted using skin samples from the palms and feet of individuals who had skin lesions who resided in an arsenic-contaminated area in Shanyin, China (Guo et al., 2016). One protein, DSG1 (desmoglein 1), was significantly suppressed in the skin from arsenic-exposed individuals. DSG1 is a cadherin-like protein that promotes anchoring of the keratin cytoskeleton to the cell membrane; therefore, suppression of this protein can result in abnormal epidermal cell differentiation. Proteins that were present in increased concentrations in affected individuals were FABP5 (fatty acid-binding protein 5) and KRT6C (keratin 6C). KRT6C and other keratins are responsible for maintaining epidermal cell-to-cell adhesion, and mutations in these proteins have been linked to other types of skin abnormalities (Guo et al., 2016). FABP5 plays an integral role in fatty acid metabolism and is known to be upregulated in various forms of human cancers, possibly promoting tumor development (Chen et al., 2011; Levi et al., 2013; Ogawa et al., 2008).

Cardiotoxicity can also occur with chronic arsenic exposure, as it causes cardiac arrhythmias and can alter myocardial depolarization (Chen and Karagas, 2013). To evaluate the mechanism of arsenic cardiotoxicity, rats were given sodium arsenate and changes in the proteomic profiles of heart tissue from treated rats were evaluated (Huang et al., 2017). Eighty-one proteins changed with arsenic exposure and 14 of the proteins were associated with cardiovascular function and development. Proteins that were upregulated included cardiac troponin T and creatinine kinase. Cardiac troponin T is present in cardiomyocytes and is responsible for myocardial contractility. It is a known, specific biomarker for acute cardiotoxicity and concentrations will increase after myocardial injury. Other proteins involved in cardiac contractility were also identified and were downregulated with arsenic exposure. These proteins include myosins, tropomyosin alpha-1 chain, and galectin-1. Other proteins identified that may serve as potential biomarkers include talin 1 and vinculin, both cytoskeletal proteins important for myocardium morphogenesis, and vimentin (filament protein important for enhancing protective effects on myocardial ischemic injury).

Cadmium

In addition to occurring naturally in the environment, cadmium is also a major environmental contaminant due to being widely used in the production of textiles, plastics, batteries, and fertilizers (Hulla, 2014). Because tobacco plants are known to accumulate cadmium from the soil, individuals smoking tobacco are known to have increased levels of cadmium in their blood and kidneys (Satarug and Moore, 2004). Prolonged exposure to cadmium can lead to accumulation in the proximal tubule epithelial cells and glomerulus, ultimately leading to nephrotoxicity (Pari et al., 2007). Measuring blood urea nitrogen and creatinine are the traditional biomarkers used to assess kidney damage; however, when high levels of these biomarkers are detected, significant kidney damage has already occurred. In an effort to identify early biomarkers of nephrotoxicity, researchers have used proteomics to investigate the effects of cadmium on renal cells. In one study using 2D-gel electrophoresis/MALDI-TOF MS and SILAC/LC-MS, HK-2 human kidney epithelial cells were treated with cadmium chloride (Kim et al., 2015). Of the several proteins identified in this study, HSPA8 (heat shock protein) and ENO1 (alpha-enolase) increased in the treated cells. The increase in ENO1 and HSPA8 was suggested to be due to cadmium-induced cell death through oxidative stress.

Another study investigating cadmium nephrotoxicity in human renal cells identified 27 proteins effected by cadmium exposure (Galano et al., 2014). Fifty percent of the proteins identified are involved in apoptosis, whereas others identified were found to be important for protein synthesis and cytoskeleton formation. Similar to the previous study, this group of investigators also identified an increase in ENO1. It is important to note that HSPA8 (and other heat shock proteins) and ENO1 are also increased when human kidney cells are damaged by other toxic agents (Kim et al., 2015). Therefore, these proteins may be candidates for biomarkers of early kidney damage as opposed to specific biomarkers for cadmium-induced nephrotoxicity.

Chromium (VI)

Hexavalent chromium (Cr (VI)) is a toxic, heavy metal commonly used as an oxidizing agent for stainless steel production, welding, chrome pigment production, and chromium plating (Ashley et al., 2003). Due to its widespread industrial use, concerns regarding environmental contamination and occupational exposure to Cr (VI) have been raised due to its proven toxicity to ecosystems and also because it is a known human carcinogen (Tokar et al., 2013). Recently, investigators have imposed the use of comparative proteomics and serum protein expression profiling to identify biomarkers of occupational exposure and also to gain a better understanding of the mechanism of carcinogenesis of Cr (VI).

In an attempt to identify novel serum proteins that change with occupational exposure to Cr (VI), blood samples were collected from 107 chromate workers and were analyzed using nano-flow HPLC/MS and critical proteins of interest that were identified were verified using ELISA (Hu et al., 2017). When compared with the control group, there were 44 differentially expressed serum proteins found in occupational Cr (VI) exposure, all involved in 16 important signaling pathways associated with the immune system, C-reactive protein (CRP), sonic hedgehog protein (SHH), and calcium. Although more studies need to be conducted to garner an understanding of the noted regulatory proteins and mechanisms involved in Cr (VI) exposure, the investigators concluded that CRP and SHH might be potential, novel biomarkers as higher levels of SHH and lower levels of CRP were noted in all of the workers exposed to Cr(VI).

Lead

Lead is a major, toxic heavy metal that has been historically used in variety of products including paints, batteries, pesticides, gasoline, and ammunition (Thompson, 2018; Tokar et al., 2013). Although workplace exposure to lead has been progressively reduced, environmental exposure to lead still remains a toxicological concern. Most exposures to lead are through contaminated food or water sources or lead-containing paint chips (Tokar et al., 2013). In mammals, the toxic effect of lead exposure can be widespread ranging from liver or kidney damage and neurotoxicity to hematologic effects (Tokar et al., 2013). Proteomic responses to lead neurotoxicity and hepatotoxicity have been conducted that identify possible biomarkers of exposure and also define metabolic pathways involved in lead-induced cell injury.

In addition to exposure via inhalation or orally, dermal exposure to lead (e.g., lead acetate) is also a concern (Cohen and Roe, 1991). One group evaluated the hepatotoxic risk caused by derma exposure to lead acetate using 2-DE and MALDI-TOF MS (Fang et al., 2014). In this study, dermal application of lead acetate was applied every 24 h to the back of nude mice and the harvested livers subjected to proteomic analysis and Western blots. The proteins that changed significantly with lead exposure were proteins associated with protein folding, ER stress, apoptosis, and oxidative stress. Specifically, the proteins GRP75 (mortalin) and GRP78 (glucose-regulated protein 78 kDa) were increased by 4.5-fold and 2.0-fold (respectively) with dermal lead exposure. These proteins are responsible for protein folding and targeting misfolded proteins for degradation and also are important indicators of oxidative stress. The investigators suggested that oxidative stress and proapoptotic signals from the ER as a result of lead toxicity were responsible for these proteins being elevated. Other proteins that increased in a dose-dependent manner with lead exposure were ATF6 (activated transcription factor 6), 1RE1α (inositol requiring enzyme), and PERK (protein kinase R-like endoplasmic reticulum kinase). Increased expression of these proteins would lead to cleavage of poly (ADP-ribose) polymerase ultimately resulting in apoptosis of hepatocytes. The investigators also noted that AST and ALT levels were elevated in this study, further confirming liver damage. This study concluded that dermal exposure to lead acetate leads to generation of reactive oxygen species culminating in hepatotoxicosis due to necrosis and apoptosis of hepatocytes.

In addition to lead hepatotoxicity, one group investigated the neurotoxic effects of lead exposure by identifying differentially expressed proteins in the hippocampus of juvenile mice given lead acetate in drinking water. Six proteins were upregulated and three proteins were downregulated in lead-exposed mice. Pdhb1 (pyruvate dehydrogenase E1β) and ATPase (proteins involved in cell energy metabolism), Hspd1 and Hspa8 (heat shock proteins involved in cell stress response), and Dpysl2 and Spna2 (proteins involved in protein binding and cytoskeleton development) all increased with oral lead exposure. There were three proteins that decreased with lead neurotoxicity, NADH dehydrogenase, Aars (alanyl tRNA synthetase), and Grb (growth factor receptor protein). The investigators surmised that the increase in ATPase in this study may be compensatory to the decrease in NADH dehydrogenase due to the increased demand for energy by the cells after lead toxicity. This group also conducted object recognition tests which revealed that lead administration reduced the memory ability and vertical activity of affected mice to show the biological responses to lead neurotoxicosis induced in these mice.

Mercury

Because of its ubiquitous presence in the environment and its propensity to bioaccumulate, mercury poses a significant health hazard to aquatic species and humans. Depending on the form of mercury, it can cause toxicity to the central nervous system, gastrointestinal tract, liver, and kidneys (Gupta et al., 2018a). There have been several studies conducted using proteomics technologies to help elucidate key pathways or biomarkers that may lead to an earlier assessment of exposure risk and to gain an understanding of the mechanism of toxicity of mercury in these organ systems.

Proteins involved in mercury neurotoxicity and hepatotoxicity have been studied in medaka fish (*Oryzias melastigma*). In both studies, chronic exposure to mercury chloride was investigated and potential biomarkers elucidated using 2D-gel electrophoresis/MALDI-TOF MS. Following mercury chloride exposure, there were 33 proteins identified in liver samples from medaka fish of which several proteins significantly changed in the treated fish versus the controls (Wang et al., 2013). Protein markers that increased significantly with mercury exposure included cathepsin D, peroxiredoxin-1 and 2, and natural killer enhancing factor. These proteins are involved in cellular responses to oxidative stress or mediate apoptosis of damaged hepatocytes, highlighting that chronic mercury exposure causes oxidative stress in the liver.

Keratin 15 and novel protein similar to vertebrate (PLEC), proteins involved in cytoskeleton assembly, decreased in the treatment group. Mercury has been previously shown to disrupt the cytoskeleton, particularly in cases of acute toxicosis (Wang et al., 2011). Studies investigating the neurotoxicity of mercury in medaka fish revealed similar results. Sixteen proteins were significantly different in the treatment group when compared to controls. As was observed in the chronic hepatotoxicity study, the proteins that changed were important in responses to oxidative stress, cytoskeletal assembly, and metabolic disorders (Wang et al., 2015).

The protein markers that increased were cathepsin D, peroxiredoxin-1 and 2, and natural killer enhancing factor. Whereas the proteins that decreased were keratin 15, formimidoyltransferase cyclodeaminase (FTCD), and novel protein similar to vertebrate (PLEC). Novel biomarkers for acute and chronic mercury nephrotoxicity were evaluated in vivo in Sprague–Dawley rats orally dosed with mercury chloride and also in vitro in rat kidney proximal tubular cells (Shin et al., 2017). Proteomic analyses revealed two key biomarkers to be considered for mercury-induced nephrotoxicity. Aldo-keto reductase (AKR7A1) and glutathione-S-transferase (GSTP1) were significantly elevated, in a dose-dependent manner, in the kidney and proximal tubular cells. Interestingly, these proteins were notably increased in the absence of detectable increases in BUN and creatinine. AKR7A1 is a protein that is integral in detoxifying metabolites, and GSTP1 is important in protecting cells from oxidative stress. To further corroborate these findings, the investigators were able to show that generation of reactive oxygen species in vitro was increased by mercury in a dose-dependent manner. The authors mention that it is not clear these proteins can serve as noninvasive biomarkers (measured in blood or urine); however, they may serve as biomarker candidates for early stages of nephrotoxicity. Collectively, in the aforementioned proteomic studies, the biomarkers identified for hepatotoxicity, neurotoxicity, and nephrotoxicity allude to mercury causing oxidative stress as a mechanism of toxicity in these organ systems.

In addition to tissue proteomics, there has been a study conducted to identify serum biomarkers for organic methylmercury toxicity. One group of researchers dosed mice with organic methylmercury, enriched for serum glycoproteins and analyzed the samples using nano-UPLC/MS/MS (Kim et al., 2013a). Although 21 proteins were differentially expressed in the organic methylmercury-treated mice, two serum protein biomarkers were identified and validated using Western blot analysis and immunohistochemistry. Amyloid P component (SAP) and inter-alpha-trypsin inhibitor heavy chain 4 (ITI-H4) were upregulated. SAP, a protein that exists in amyloid deposits and stabilized chromatin, was the protein that increased the most. ITI-H4 increased more than twofold and is responsible for regulating inflammatory conditions and acute phase response in cells. The authors note that previous proteomic investigations in serum from children with high levels of mercury in their blood (due to eating contaminated fish) also have detectable ITI-H4 (Gump et al., 2012).

PESTICIDES AND HERBICIDES

Methyl Parathion

Methyl parathion is an organophosphorus pesticide and is classified by the Environmental Protection Agency as a class I toxicant. Due to its widespread use in agriculture, occupational exposure in humans often occurs via inhalation or dermally. Environmental exposure occurs due to contamination of soil and water or due to runoff from sources where methyl parathion has been applied. The primary mechanism of toxicity of this organophosphate insecticide is through inhibition of acetylcholinesterase in the central nervous system and at the neuromuscular junction (Gupta and Kadel, 1990; Gupta et al., 2018b). Some studies have also shown that methyl parathion can cause DNA damage and induce oxidative stress in a variety of organ systems (Bartoli et al., 1991; Hai et al., 1997).

A group of researchers in China have taken an interest in studying changes in cell membrane proteomes to investigate the neurologic and oxidative effects of methyl parathion in brain and other organ systems to establish candidate biomarkers of methyl parathion in those tissues and gain a better understanding of the mechanisms of methyl parathion toxicity. Using 1-DE and 2-DE to separate the membrane proteins and MALDI-TOF MS/MS to sequence the peptides, candidate biomarkers for methyl parathion toxicity were evaluated in zebrafish (*Danio rerio*) brain and liver, pleural-pedal ganglia from sea slugs (*Aplysia juliana*), and kidney from scallops (*Mizuhopecten yessoensis*) (Huang and Huang, 2011, 2012a,b; Chen et al., 2014). The proteins that were identified to be upregulated or downregulated were further validated and confirmed using Western blots and real-time polymerase chain reaction (RT-PCR). Common in all of these proteomic studies, the cell membrane proteins that changed significantly with methyl parathion treatment play a role in oxidative stress, mitochondrial function, energy/cell metabolism, signal transduction, protein synthesis, and degradation and intracellular transport. A list of the membrane proteins that changed with methyl parathion treatment are listed in Table 65.1.

The proteins PDIA3 (protein disulfide isomerase-associated 3), ALDH5A1 (aldehyde dehydrogenase 5A1), and MDH (malate dehydrogenase) were found to have significantly changed in with methyl parathion treatment in zebrafish brain tissue (Huang and Huang, 2011). The investigators suggested that these three proteins are suitable candidate biomarkers as they have been previously reported to be associated with methyl parathion toxicity in other studies (Huang and Huang, 2011). Interestingly, SDH (succinate dehydrogenase) was shown to decrease in sea slug, pleural-pedal ganglia and increase with methyl parathion treatment in zebrafish liver cells (Huang and Huang, 2012a; Chen et al., 2014). Organophosphates have been reported in early studies in aquatic species to decrease SDH activity in tissues, suggesting that these pesticides effect aerobic oxidation in the TCA cycle (Samuel and Sastry, 1989). SDH was also shown to decrease in the pleural-pedal ganglia of sea slug with methyl parathion treatment. However, in zebrafish liver cells, SDH was shown to increase with methyl parathion treatment. The investigators surmised that this increase in SDH might be compensatory in light of aerobic oxidation being effected in the liver.

TABLE 65.1 Membrane Proteins Altered With Methyl Parathion Treatment

Organ System	Proteins Increased	Proteins Decreased
Brain[a]	GNB1L	GDI2
	GNB2	UCHL1
	ALDH5A1	Ependymin
	PDIA3	STXBP1
	DLST	
	IDH3	
	mMDH	
	EF-Tu	
Pleural-Pedal Ganglia[b]	NADH	SDH
	ALD	ICDH
	ANN11	MDP
	GPCR	TCR
	PHB	ILR
	β-tubulin	VDAC
		ABC transporters
Liver[c]	CD146	TPHR 1
	Annexin A2a	
	SDH	
	ACBD5A	
Kidney[d]	ATP synthase	GPx
	MAPRE1	MPP
	IOX	HPPD
		PEPCK

[a]*Huang and Huang (2011).*
[b]*Chen et al. (2014).*
[c]*Huang and Huang (2012a).*
[d]*Huang and Huang (2012b).*

Diazinon

In addition to acetylcholinesterase inhibition, noncholinergic targets of organophosphate pesticides are also of toxicological concern. Recent studies have suggested that diazinon, an organophosphate pesticide, may cause developmental defects in the absence of acetylcholinesterase inhibition (Harris et al., 2009). To investigate if sublethal doses of diazinon would affect cell development, a targeted proteomic study was conducted in mouse N2a cells to examine if diazinon would induce molecular changes in stress response and axonal cytoskeletal proteins (Harris et al., 2009). Using 2-DE and MALDI-TOF MS, 12 proteins involved in cell survival, differentiation, and metabolism changed significantly with diazinon treatment. Notably, the protein cofilin increased approximately fivefold. Cofilin is an actin-binding protein that dissembles actin filaments, suggesting that diazinon disrupts microfilament organization. The researchers concluded that the cytoskeleton may be a target for the neurite inhibitory levels of diazinon, and that proteins important for microfilaments, microtubules, and neurofilaments are affected by diazinon.

Paraquat

Paraquat is a potent, restricted, nonselective herbicide that has been associated with high mortality in humans and animals exposed to the product. Ingestion of paraquat can cause multisystem organ failure through production of reactive oxygen species as a result of inhibiting reduction of NADH to NADPH (Gupta, 2018). Paraquat is also selectively taken up into the lungs, causing pulmonary edema, alveolar hemorrhage, and lung fibrosis (Gupta, 2018; Kim et al., 2013c). The use of proteomics in identifying diagnostic biomarkers for pulmonary toxicity in bronchoalveolar lavage fluid has been somewhat successful (Govender et al., 2009). However, isolating biomarkers in noninvasive samples as a result of pulmonary toxicity caused by paraquat intoxication have been challenging. Serum uric acid and acute phase response gene pentraxin-3 have been suggested as prognostic biomarkers for paraquat toxicity in serum (Kim et al., 2011; Yeo et al., 2011); however, these biomarkers may not be ideal for early diagnosis. In an attempt to discover early, diagnostic biomarkers of acute paraquat poisoning, 2-DE and MALDI-TOF MS/MS were used to isolate and identify proteins that change with paraquat poisoning (Kim et al., 2012). In this study, male CD(SD)IGS rats were dosed orally with paraquat dichloride and serum samples were collected from treatment and control groups. Out of approximately 500 protein spots observed, eight proteins were differentially expressed with paraquat treatment. Proteins related to the inflammatory process which included ApoE (apolipoprotein E), Hp (haptoglobin), amd C3 (complement component 3) increased in the treatment group. Proteins that decreased with treatment were FGG (fibrinogen γ-chain) and Ac-158. Western blot and RT-PCR were used to further validate the usefulness of ApoE, Pphg, and FGG as diagnostic biomarkers of paraquat toxicity. These results concluded that ApoE, Hp, and FGG may be appropriate candidate biomarkers in serum samples for early diagnosis of acute paraquat toxicosis. The increased expression of the protein C3 is interesting in that C3 may play a role in acute inflammatory reactions caused by paraquat (Sun et al., 2011). Therefore, the investigators concluded that C3 expression in serum may serve as a diagnostic biomarker for paraquat-induced, acute pulmonary inflammation.

Proteomic profiling of lung tissue has also been conducted to evaluate changes in proteins expressed in the lung subsequent to paraquat exposure. In this study, protein biomarkers of acute pulmonary toxicosis were investigated using 2-DE and MALDI-TOF MS/MS in rats dosed intraperitoneally with paraquat. It was discovered that CaBP1 (calcium-binding protein 1 regulates calcium-dependent activity in cell cytoskeleton), FKBP4 (important in immunoregulation and protein folding), osteonectin (glycoprotein that binds calcium), and S100A6 (calcium-binding protein that regulates cell cycle progression and differentiation) all increased in rat lung tissue with paraquat treatment.

Glyphosate

Glyphosate is widely used as a broad spectrum herbicide and is present in approximately 750 commercial herbicide products (Landrigan and Belpoggi, 2018). The herbicidal mechanism of action is through inhibition of plant enolpyruvylshikimate-3-phosphate synthase, an enzyme involved in the synthesis of tyrosine, tryptophan, and phenylalanine. Without these amino acids, the plant cannot synthesize proteins needed for growth. Due to its widespread use to control broadleaf weeds and grasses, trace amounts of glyphosate can be found in soil, foodstuffs, and water (Landrigan and Belpoggi, 2018). The impact of glyphosate on the environment, particularly in aquatic organisms, is an emerging concern. Glyphosate-based herbicides have been shown to cause a variety of toxic effect in nontarget aquatic organisms, such as hemorrhagic anemia, oxidative stress, and genotoxic effects (Arnett et al., 2014; Rocha et al., 2015). To assess the early toxicological response of glyphosate in fish, guppies (*Poecilia reticulata*) were housed in aquarium water fortified with 1.82 mg glyphosate/liter and the gill proteome evaluated using 2-DE and MALDI-Q-TOF MS/MS (Rocha et al., 2015). After a 24-h exposure time, 14 proteins involved in energy metabolism, regulation, cytoskeleton maintenance, and stress were identified. Glyphosate exposure inhibited expression of α-enolase, a protein that plays a role in hypoxia tolerance in cells. This inhibition was thought to be due to glyphosate-dependent hypoxia-induced stress in the gills. Proteins important for cytoskeleton regulation were also suppressed in gills from glyphosate-treated fish. Arp4, cortactin-binding protein, myosin-VI-like isoform 4, and actin isoforms decreased with glyphosate exposure, suggesting that glyphosate may interfere with stability of actin filaments in the gills of guppies. The investigators concluded that glyphosate modifies expression of gill proteins and promotes changes in the cellular architecture of gill cells in response to hypoxia caused by glyphosate, and that the proteins identified could serve as potential biomarkers for biomonitoring water pollution by herbicides.

ORGANIC POLLUTANTS

Benzo(a)pyrene

Polycyclic aromatic hydrocarbons (PAHs) are known carcinogens, posing a significant health threat to humans and animals. Their ubiquitous presence in the environment is largely because they are produced by all types of combustion of organic materials such as the incomplete burning of fuels and other substances including tobacco (cigarette smoke) and other plant material (Kim et al., 2013b). One of the most studied PAHs is benzo(a)pyrene, which is metabolized by the cytochrome P450 monooxygenase system and epoxide hydrolase in mammals to form benzo(a)pyrene diol epoxide, a carcinogenic metabolite that forms an adduct with DNA (Di Giulio and Newman, 2013). The genotoxic effects of benzo(a)pyrene have been well studied; however, the effects of benzo(a)pyrene at the protein level are just beginning to be understood. Several researchers have taken advantage of proteomics techniques to begin to evaluate biomarker candidates of benzo(a)pyrene carcinogenicity in bladder cancer, prostate cancer, and lung cancer.

Tobacco smoking has been linked to bladder cancer cases in men (Castelao et al., 2001). One concern regarding tobacco is that cigarette smoke has been shown to contain 20–40 ng of benzo(a)pyrene per cigarette (Rodgman et al., 2000). Efforts to understand the possible role of benzo(a)pyrene in bladder cancer development have been attempted using proteomics. In one study, pig primary bladder epithelial cells were exposed to benzo(a)pyrene and the proteome isolated and analyzed using 2-DE and MALDI-TOF MS/MS (Verma et al., 2013). Twenty-five differentially expressed proteins were identified which are known to be important for either mitochondrial repair, DNA repair, or apoptosis. Proteins that were upregulated with benzo(a)pyrene exposure included RAD23, PMSD5, and PMSD4. These proteins play a crucial role in DNA repair and their upregulation suggests that benzo(a)pyrene caused DNA damage to the bladder epithelial cells (Verma et al., 2013). Many of the proteins identified that changed significantly were cathepsin D, VDAC 2 (voltage-dependent anion channel protein), HSP27, and HSP70 (heat shock proteins). These proteins are known to be involved in the mitochondrial death receptor pathway, ultimately inducing apoptosis. These proteins not only may serve as candidate biomarkers for benzo(a)pyrene-induced bladder cancer but also this study was able to show that at low concentrations and during short exposure time periods that benzo(a)pyrene can cause DNA damage in bladder epithelial cells leading to cell death due to apoptosis (Verma et al., 2013).

Benzo(a)pyrene in cigarette smoke has also been implicated in the development of lung cancer. Evidence has shown that benzo(a)pyrene diol epoxide-DNA adducts occur at the same codon positions that are known to be major, mutational hot spots in human lung cancers (Denissenko et al., 1996). The lack of biomarkers for early diagnosis of lung cancer has been a persistent, clinical challenge. Comparative proteomic analysis of the cellular response to (A549) human airway epithelial cells to benzo(a)pyrene was investigated in hopes of elucidating biomarkers and mechanisms of toxicity (Min et al., 2011). Using 2-DE and MALDI-TOF MS/MS, 23 proteins that play a role in signal transduction, antioxidation, energy and metabolism, and apoptosis were differentially expressed in cell lysates from the airway epithelial cells treated with benzo(a)pyrene. The proteins that increased were annexin A1, thioredoxin, cathepsin D, and heterogeneous nuclear ribonucleoprotein K. Peroxiredoxin I, vimentin, nucleoside diphosphate kinase A, poly(rC)-binding protein 1, and superoxide dismutase (Mn) all decreased in the benzo(a)pyrene-treated cells. Superoxide dismutase (Mn), which is a protein that is considered to be the cell's first line of defense against superoxides, was very significantly decreased with benzo(a)pyrene exposure in A549 cells (Min et al., 2011). The researchers suggested that accumulation of intracellular reactive oxygen species induced by benzo(a)pyrene exposure is why this protein likely decreased. This proteomic study helped clarify that benzo(a)pyrene can perturb the antioxidant status in airway epithelial cells and highlighted the proteins that change with benzo(a)pyrene toxicity in the lung.

In addition to being present in cigarette smoke, benzo(a)pyrene is also known to be present in grilled meat and that consumption of grilled meat has been linked to an increased risk of prostate cancer (Fatma and Kabadayi, 2005). Proteomic investigations conducted to assess benzo(a)pyrene-mediated prostate cancer have been conducted in vitro using PrEC normal prostate epithelial cells (Chaudhary et al., 2006). This investigation revealed 26 proteins that changed with benzo(a)pyrene exposure. Twenty-six proteins were differentially expressed in the prostate epithelial cells with benzo(a)pyrene, most having cellular function in metabolism, signal transduction, cytoskeleton protection, and oxidative stress. Peroxiredoxin I, which increased with benzo(a)pyrene, and peroxiredoxin II, which decreased with benzo(a)pyrene, were of particular interest to the researchers. These proteins are thiol-specific antioxidant enzymes and play a role in reducing the oxidative stress in cells (Kang et al., 2005). Peroxiredoxin I expression increase has been shown to increase in several types of cancers, and it is thought that elevation of this protein in cancer is suggestive of the cell's defense against tumorigenesis (Chaudhary et al., 2006). Peroxiredoxin II suppression can enhance prosurvival pathways through aberrant activation of growth factor receptors resulting in hyperproliferation; therefore, the investigators concluded that decreased peroxiredoxin II may be important for benzo(a)pyrene tumor promotion (Chaudhary et al., 2006). These two proteins may be involved in benzo(a)pyrene toxicity and could potentially serve as biomarkers of benzo(a)pyrene carcinogenicity.

Although much of the proteomics studies have been conducted on benzo(a)pyrene, there has been one study investigating plasma protein biomarkers in workers occupationally exposed to mixed PAHs (Kap-Soon et al., 2004). Monitoring occupational exposure to PAHs has historically required measuring for 1-hydroxypyrene, a urinary PAH metabolite (Jongeneelen, 2014). However, this marker is not suited for early detection of low-dose, PAH exposure (Kap-Soon et al., 2004). Kap-Soon et al. (2004) analyzed plasma samples from waste gas pollution measurers that worked in an automobile emission inspection center that are exposed to PAHs on a daily basis. To confirm exposure, the 1-hydroxypyrene was measured in the workers' urine and was detected at approximately 4 times higher in the PAH-exposed workers when compared to controls. Proteomic evaluation of the plasma samples from the exposed workers revealed six proteins that increased with PAH exposure which included serum albumin precursor, hemopexin precursor, fibrinogen γ-A chain precursor, TCR-β, and CCE channel protein. Serum albumin, hemopexin, and fibrinogen are plasma protein targets of oxidative stress. Their increased presence may be due to attempts to prevent oxidative stress in the cells caused by PAHs (Kap-Soon et al., 2004). TCR-β protein is important for recognizing foreign antigens and subsequently triggering a cascade of signals to activate cells to respond. The authors surmise that the overexpression of TCR-β may be due to this protein recognizing PAH as a foreign antigen and is consequently overexpressed TCR-β (Kap-Soon et al., 2004). CCE channel protein is a plasma membrane protein important for role in calcium hemostasis. The authors suggest that PAH metabolites may target immune cells altering calcium hemostasis leading to apoptosis of cells. In this case, the overexpression of CCE could be due to release from the cell membrane as a result of PAH exposure ultimately perturbing calcium hemostasis and causing apoptosis. As a result of this study, they concluded that these six proteins could be noninvasive, candidate biomarkers for PAH exposure.

2,3,7,8-Tetrachlorodibenzo-p-Dioxin

Polychlorinated dioxins, such as TCDD, are very toxic, environmental contaminants that can cause immune system modulation, teratogenesis, and tumor promotion (Sany et al., 2015).

TCDD is a by-product in the production of herbicides and can originate from industrial processes such as metal production, fossil fuels/wood combustion, and waste incineration. Similar to benzo(a)pyrene, occupational exposure to TCDD has been reported in individuals working in the waste incineration industry (Phark et al., 2016). Identification of novel biomarkers to use for biomonitoring TCDD exposure in these workers has been made possible through proteomics technologies. In one study, differentially expressed proteins were evaluated and compared in HepG2 cells exposed to TCDD, plasma from rats exposed to TCDD, and plasma from industrial incineration workers exposed to low and high doses of TCDD (Phark et al., 2016). The secreted proteome from the HepG2 cells had seven proteins that increased with TCDD exposure and one protein that decreased. Proteins that increased were GLO 1 (glyoxylase), HGD (homogentisate dioxygenase), peroxiredoxin I, PSMB5 and PSMB6 (proteasome subunit beta type protein), UDP-GlcDH (UDP-glucose-6-dehydrogenase), and HADH 9hydroxylacyl-coenzyme A dehydrogenase. STF (serotransferrin) decreased with TCDD treatment. Of these proteins, PSMB5 and peroxiredoxin I were identified in plasma from rats dosed with TCDD and increased in a time-dependent manner in the plasma proteome. GLO 1, HGD, peroxiredoxin I, and PSMB6 were present in the plasma from incineration workers, and protein expression of these proteins was greater in the high-dose exposure group versus the low-dose exposure group. Peroxiredoxin I was present in all three sample models and increased with exposure to TCDD. Interestingly, in proteomic studies exposing PrEC prostate epithelial cells to benzo(a)pyrene to model benzo(a)pyrene-induced prostate cancer, peroxiredoxin I was also overexpressed and increased expression of this protein has been implicated in several types of cancer (Chaudhary et al., 2006). The authors noted that although these proteins may serve as candidate biomarkers for TCDD exposure; more large-scale quantitative studies need to be conducted to further validate their use in biomonitoring occupational exposure to this toxicant.

CONCLUDING REMARKS AND FUTURE DIRECTIONS

Toxicoproteomics applies global protein separation and detection to profile complex biological samples in hopes of identifying candidate biomarkers and gain an understanding of the mechanism of toxicity of the acute and chronic effects of chemical toxicants. Proteomics is now in a unique position to provide meaningful data due to evolving advancements in protein separation and isolation, analytical instrumentation, and database sequence and pathway software programs. However, despite the advances in these technologies, there are still some limitations to these approaches. Future challenges for toxicoproteomics research will require further optimization and validation of candidate biomarkers for chemical toxicity. For example, biomarker peroxiredoxin I was shown to increase in plasma from incineration workers exposed to TCDD, in prostate epithelial cells exposed to benzo(a)pyrene, and in liver cell lysates from fish exposed to mercury (Chaudhary et al., 2006; Wang et al., 2011; Phark et al., 2016). As mentioned earlier, HSP8 (heat shock protein) and ENO1 (α-enolase) were shown to increase in human kidney proximal tubule epithelial cells (HK-2) treated with cadmium in a study evaluating biomarker candidates for cadmium-induced nephrotoxicity (Kim et al., 2015). HSP8 has also been shown to increase in HK-2 cells exposed to other nephrotoxic agents such as mercury chloride, cisplatin, cyclosporine, and sodium arsenite (Kim et al., 2015). Overall, some of the proteins identified to be candidate biomarkers for specific chemical toxicity may in fact be more appropriate biomarkers for organ-specific toxicity. Additionally, although the proteins recognized to change with chemical toxicity may differ, it is evident that these proteins in a few cases are involved in similar cellular responses. For example, across the chemicals reported in this chapter, several implicate differentially expressed proteins involved in oxidative stress response, cell metabolism, and cytoskeleton maintenance (cell structure). Although the proteins associated with these biological processes may differ, it highlights the general mechanisms by which most chemicals exert their toxicity on biological systems. Before differentially expressed proteins will be accepted as reliable biomarkers of chemical toxicity, validation studies on a larger scale will likely be required to ensure the results are repeatable and robust. These studies would also likely require standardization of sample preparation and analysis.

Historically, biomarkers for toxicity have included defining one or a few proteins that change with exposure. For example, BUN and creatinine increase with kidney injury; however, many chemicals can cause this increase. Further exploiting the global aspect of proteomics techniques, future definition of biomarkers of chemical toxicity may require identifying groups of proteins (e.g., protein profile) that may be related to a specific chemical exposure. Although there were similar proteins that were differentially expressed to comparable degrees in more than one chemical, the entire set of proteins that changed were different when comparing the chemicals evaluated. In addition, some of the researchers conducting the aforementioned proteomics investigations reached for histopathology, RT-PCR, and Western blots to confirm the biomarkers identified in their chemical toxicity studies, further adding creed to the data reported. Ultimately, toxicoproteomics is emerging to be an effective strategy in identifying useful, predictive biomarkers of chemical toxicity.

References

Arnett, R., Habibi, H.R., Hontela, A., 2014. Impact of glyphosate and glyphosate-based herbicides on the freshwater environment. J. Appl. Toxicol. 34, 458–479.

Ashley, K., Howe, A.M., Demange, M., et al., 2003. Sampling and analysis considerations for the determination of hexavalent chromium in workplace air. J. Environ. Monit. (5), 707–716.

Bartoli, S., Bonora, B., Colacci, A., et al., 1991. DNA damaging activity of methyl parathion. Res. Commun. Chem. Pathol. Pharmacol. 71, 209–218.

Castelao, J.E., Yuan, J.M., Skipper, P.L., et al., 2001. Gender and smoking-related bladder cancer risk. J. Natl. Cancer Inst. 93 (6), 538–545.

Chaudhary, A., Pechan, T., Willett, K.L., 2006. Differential protein expression of peroxiredoxin I and II by benzo(a)pyrene and quercetin treatment in 22Rv1 and PrEC prostate cell lines. Toxicol. Appl. Pharmacol. 220, 197–210.

Chen, R., Feng, C., Xu, Y., 2011. Cyclin-dependent kinase-associated protein Cks2 is associated with bladder cancer progression. J. Int. Med. Res. 39, 533–540.

Chen, Y., Karagas, M.R., 2013. Arsenic and cardiovascular disease: new evidence from the United States. Ann. Intern. Med. 159, 713–714.

Chen, Y., Huang, L., Zhang, Y., et al., 2014. Differential expression profile of membrane proteins in *Aplysia* pleural-pedal ganglia under the stress of methyl parathion. Environ. Sci. Pollut. Res. 21, 3371–3385.

Cohen, A.J., Roe, F.J., 1991. Review of lead toxicology relevant to safety assessment of lead acetate as a hair colouring. Food Chem. Toxicol. 29, 127–139.

Di Giulio, R.T., Newman, M.C., 2013. Ecotoxicology. In: Klaasen, C.D. (Ed.), Casarett & Doull's Toxicology, the Basic Science of Poisons, eighth ed. McGraw- Hill Education, China, pp. 1283–1284.

Denissenko, M.F., Pao, A., Pfeifer, G.P., 1996. Preferential formation of benzo(a)pyrene adducts at lung cancer mutational hotspots in P53. Science 274 (5286), 430–432.

Fang, J., Wang, P., Huang, C., et al., 2014. Evaluation of the hepatotoxic risk caused by lead acetate via skin exposure using a proteomic approach. Proteomics 14, 2588–2599.

Fatma, A.S., Kabadayi, F., 2005. Meat and meat mutagens and risk of prostate cancer in the agricultural health study. Cancer Epidemiol. Biomark. Prev. 17, 80–87.

Galano, E., Arciello, A., Piccoli, R., et al., 2014. A proteomic approach to investigate the effects of cadmium and lead on human primary renal cells. Metallomics 6, 587–597.

Govender, P., Dunn, M.J., Donnelly, S.C., 2009. Proteomics and the lung: analysis of bronchoalveolar lavage fluid. Proteonomics Clin. Appl. 3, 1044–1051.

Gump, B.B., MacKenzie, J.A., Dumas, A.K., et al., 2012. Fish consumption, low-level mercury, lipids, and inflammatory markers in children. Environ. Res. 112, 204–211.

Guo, Z., Hu., Q, Tian, J., et al., Proteomic profiling reveals candidate markers for arsenic-induced skin keratosis, Environmental Pollution. 218, 2016, 34-38.Gupta, R.C., Kadel, W.L., 1990. Methyl parathion acute toxicity: prophylaxis and therapy with memantine and atropine. Arch. Int. Pharmacodyn. Ther. 305, 208–221.

Gupta, P.K., 2018. Toxicity of herbicides. In: Gupta, R.C. (Ed.), Veterinary Toxicology; Basic and Clinical Principles, third ed. Academic Press/Elsevier, Amsterdam, pp. 495–508.

Gupta, R.C., Milatovic, D., Srivastava, A., et al., 2018a. Mercury. In: Gupta, R.C. (Ed.), Veterinary Toxicology; Basic and Clinical Principles, second ed., pp. 573–585.

Gupta, R.C., Sachana, M., Mukherjee, I.M., et al., 2018b. Organophosphates and carbamates. In: Gupta, R.C. (Ed.), Veterinary Toxicology; Basic and Clinical Principles, second ed., pp. 573–585.

Hai, D.Q., Varga, S.I., Matkovics, B., 1997. Organophosphate effects on antioxidant system in carp (*Cyprinus carpio*) and catfish (*Ictalurus nebulosus*). Comp. Biochem. Physiol. C Pharmacol. Toxicol. Endocrinol. 117, 83–88.

Harris, W., Sachana, M., Flaskos, J., et al., 2009. Proteomic analysis of differentiating neuroblastoma cells treated with sub-lethal neurite inhibitory concentrations of diazinon: identification of novel biomarkers of effect. Toxicol. Appl. Pharmacol. 240, 159–165.

Hong, Y., Ye, B., Kim, Y., et al., 2017. Investigation of health effects according to the exposure of low concentration of arsenic contaminated ground water. Int. J. Environ. Res. Publ. Health 14, 1461–1475.

Hsu, L., Chen, G., Lee, C., et al., 2013. Use of arsenic-induced palmoplantar hyperkeratosis and skin cancers to predict risk of subsequent internal malignancy. Am. J. Epidemiol. 177 (3), 202–212.

Hu, G., Wang, T., Liu, J., et al., 2017. Serum protein expression profiling and bioinformatics analysis in workers occupationally exposed to chromium (VI). Toxicol. Lett. (277), 76–83.

Huang, Q., Xi, G., Alamdar, A., et al., 2017. Comparative proteomic analysis reveals heart toxicity induced by chronic arsenic exposure in rats. Environ Pollut. 229, 210–218.

Huang, Q., Huang, H., 2011. Differential expression profile of membrane proteins in zebrafish (*Danio rerio*) brain exposed to methyl parathion. Proteomics 11, 3743–3756.

Huang, Q., Huang, H., 2012a. Alterations of protein profile in zebrafish liver cells exposed to methyl parathion: a membrane proteomics approach. Chemosphere 87, 68–76.

Huang, X., Huang, H., 2012b. Alteration of kidney membrane proteome of *Mizuhopecten yessoensis* induced low-level methyl parathion exposure. Aquat. Toxicol. 114–115, 189–199.

Hulla, J.E., 2014. Metals. In: Hayes, A.W., Kruger, C.L. (Eds.), Hayes' Principles and Methods of Toxicology. Taylor & Francis Group, Boca Raton, Florida, p. 849.

Jongeneelen, F.J., 2014. A guidance value of 1-hydroxypyrene in urine in view of acceptable occupational exposure to polycyclic aromatic hydrocarbons. Toxicol. Lett. 231 (2), 239–248.

Kang, S.W., Rhee, S.G., Chang, T.S., et al., 2005. 2-Cys peroxiredoxin function in intracellular signal transduction: therapeutic implications. Trends Mol. Med. 11, 571–578.

Kap-Soon, N., Do-Youn, L., Hak, C.J., et al., 2004. Protein biomarkers in the plasma of workers occupationally exposed to polycyclic aromatic hydrocarbons. Proteomics 4, 3505–3513.

Kim, B., Moon, P., Lee, J., et al., 2013a. Identification of potential serum biomarkers in mercury-treated mice using a glycoproteomic approach. Int. J. Toxicol. 32 (5), 368–375.

Kim, J.H., Gil, H.W., Yang, J.O., et al., 2011. Serum uric acid level as a marker for mortality and acute kidney injury in patients with acute paraquat intoxication. Nephrol. Dial. Transplant. 26, 1846–1852.

Kim, K., Jahan, S.A., Kabir, E., et al., 2013b. A review of airborn polycyclic aromatic hydrocarbons (PAHs) and their human health effects. Environ. Int. 60, 71–80.

Kim, S.Y., Lee, H.M., Kim, K.S., et al., 2015. Non-invasive biomarker candidates for cadmium-induced nephrotoxicity by 2DE/MALDI-TOF-MS and SILAC/LC-MS proteomic analyses. Toxicol. Sci. 148 (1), 167–182.

Kim, Y., Jung, H., Gil, H., et al., 2012. Proteomic analysis of changes in protein expression in serum from animals exposed to paraquat. Int. J. Mol. Med. 30, 1521–1527.

Kim, Y., Jung, H., Zerin, T., Song, H., 2013c. Protein profiling of paraquat-exposed rat lungs following treatment with Acai (*Euterpe oleracea* Mart.) berry extract. Mol. Med. Rep. 7, 881–886.

Landrigan, P.J., Belpoggi, F., 2018. The need for independent research on the health effects of glyphosate-based herbicides. Environ. Health 17, 51.

Levi, L., Lobo, G., Doud, M.K., et al., 2013. Genetic ablation of the fatty acid-binding protein FABP5 suppresses HER2-induced mammary tumorigenesis. Cancer Res. 73, 4770–4780.

Min, L., He, S., Chen, Q., et al., 2011. Comparative proteomic analysis of cellular response of human airway epithelial cells (A549) to benzo(a) pyrene. Toxicol. Mech. Methods 21 (5), 374–382.

Ogawa, R., Ishiguro, H., Kuwabara, Y., et al., 2008. Identification of candidate genes involved in the radiosensitivity of esophageal cancer cells by microarray analysis. Dis. Esophagus 21, 288–297.

Pari, L., Murugavel, P., Sitasawad, S.L., et al., 2007. Cytoprotective and antioxidant role of diallyltetrasulfide on cadmium induced renal injury: an *in vivo* and *in vitro* study. Life Sci. 80, 650–658.

Phark, S., Park, S., Chang, Y., et al., 2016. Evaluation of toxicological biomarkers in secreted proteins of HepG2 cells exposed to 2,3,7,8-tetrachlorodibenzo-p-dioxin and their expression in the plasma of rats and incineration workers. Biochim. Biophys. Acta 584–593.

Rocha, T.L., Rezende dos Santos, A.P., Yamada, A.T., et al., 2015. Proteomic and histopathological response in the gill of Poecilia reticulate exposed to glyphosate-based pesticide. Envrion. Toxicol. Pharmacol. 40, 175–186.

Rodgman, A., Smith, C.J., Perfetti, T.A., 2000. The composition of cigarette smoke: a retrospective with emphasis on polycyclic components. Hum. Exp. Toxicol. 19, 573–595.

Samuel, M., Sastry, K.V., 1989. In vivo effect of monocrotophos on the carbohydrate metabolism of freshwater shake head fish (*Channa punctatus*). Pestic. Biochem. Physiol. 34, 1–8.

Sany, S.B.T., Hashim, R., Salleh, A., et al., 2015. Dioxin risk assessment: mechanisms of action and possible toxicity in human health. Environ. Sci. Pollut. Res. 22, 19434–19450.

Satarug, S., Moore, M.R., 2004. Adverse health effects of chronic exposure to low-level cadmium in foodstuffs and cigarette smoke. Environ. Health Perspect. 112 (10), 1099–1103.

Shin, Y., Kim, K., Kim, E., et al., 2017. Identification of aldo-keto reductase (AKR7A1) and glutathione-S-transferase (GSTP1) as novel renal damage biomarkers following exposure to mercury. Hum. Exp. Toxicol. https://doi.org/10.1177/0960327117751234.

Sun, S., Wang, H., Zhao, G., et al., 2011. Complement inhibition alleviates paraquat-induced acute lung injury. Am. J. Respir. Cell Mol. Biol. 45, 834–842.

Thompson, L.J., 2018. Lead. In: Gupta, R.C. (Ed.), Veterinary Toxicology; Basic and Clinical Principles, third ed. Academic Press/Elsevier, Amsterdam, pp. 439–443.

Tokar, E.J., Boyd, W.A., Freedman, J.H., et al., 2013. Toxic effects of metals. In: Klaasen, C.D. (Ed.), Casarett & Doull's Toxicology, the Basic Science of Poisons, eighth ed. McGraw-Hill Education, China, pp. 986–999.

Verma, N., Pink, M., Rettenmeier, A.W., et al., 2013. Benzo(a)pyrene-mediated toxicity in primary pig bladder epithelial cells: a proteomic approach. J. Proteom. 85, 53–64.

Wang, M., Wang, Y., Zhang, L., et al., 2013. Quantitative proteomic analysis reveals the mode of action for chronic mercury hepatotoxicity to marine medaka *Oryzias melastigma*. Aquat. Toxicol. 130–131, 123–131.

Wang, M.H., Wang, Y.Y., Wang, J., et al., 2011. Proteome profiles in medaka (*Oryzias melastigma*) liver and brain experimentally exposed to acute inorganic mercury. Aquat. Toxicol. 103 (3–4), 129–139.

Wang, Y., Wang, D., Lin, L., et al., 2015. Quantitative proteomic analysis reveals proteins involved in the neurotoxicity of marine medaka *Oryzias melastigma* chronically exposed to inorganic mercury. Chemosphere 119, 1126–1133.

Yeo, C.D., Kim, J.W., Kim, Y.O., et al., 2011. The role of pentraxin-3 as a prognostic biomarker in paraquat poisoning. Toxicol. Lett. 212, 157–160.

CHAPTER

66

Biomarkers for Testing Toxicity and Monitoring Exposure to Xenobiotics

Jorge Estévez, Eugenio Vilanova, Miguel A. Sogorb
Instituto de Bioingeniería, Universidad Miguel Hernández de Elche, Spain

INTRODUCTION

A biomarker (or a biological marker) is an indicator of a biological state that offers information about the exposure of an organism to a xenobiotic, the effect of this exposure on the organism, or the susceptibility of the organism to the xenobiotic. According to this definition, biomarkers can be classified into three different categories: (1) biomarkers of exposure, which determine whether an organism has been exposed to a certain xenobiotic; (2) biomarkers of effect (response), which determine the effect of the exposed organism to the xenobiotics; and (3) biomarkers of susceptibility, which predict the susceptibility (resistance) of the organism to the deleterious effect of a specific xenobiotic.

Fig. 66.1 summarizes the main uses of biomarkers in toxicology. Sometimes there are doubts about whether or not an organism has been exposed to xenobiotics, especially if that exposure was not strong enough to cause adverse clinical effects. Biomarkers of exposure reveal if such organisms have been exposed because the presence of xenobiotics or their metabolites in biological samples (both free or bound to endogenous molecules) from the monitored individual is proof of exposure. Thus, biomarkers of exposure are suitable for assessing exposure to xenobiotics, especially when the organism is not in direct contact with the source of xenobiotics in the moment of the sampling and also when an exposure assessment based on chemical determinations in environmental samples representative of the media to which the individual has been exposed is not possible. Another advantage is that biomarkers of exposure always refer to internal doses (that the xenobiotic has been absorbed). Therefore, biomonitoring is a better indicator of how many xenobiotics might reach endogenous targets, thus making the expected intensity of the adverse effect more predictable. Aldridge (1996) divided toxic phenomena into five different stages, according to which, biomarkers of exposure might be used at any time after stage 1 (absorption) (Fig. 66.2).

Exposure is a necessary cause, but it does not suffice to alter exposed organisms. In this way, biomarkers of effect can help to determine if the intensity of exposure is strong enough to cause an effect or a response in exposed organisms. According to Aldridge's scheme (Aldridge, 1996), immediately before clinical consequences appear in an organism,

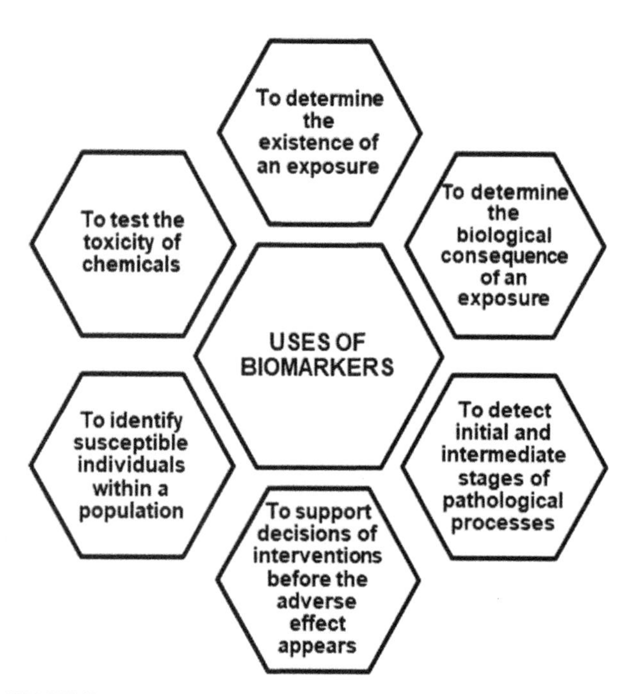

FIGURE 66.1 Uses and applications of biomarkers in toxicology.

Biomarkers in Toxicology, Second Edition
https://doi.org/10.1016/B978-0-12-814655-2.00066-9

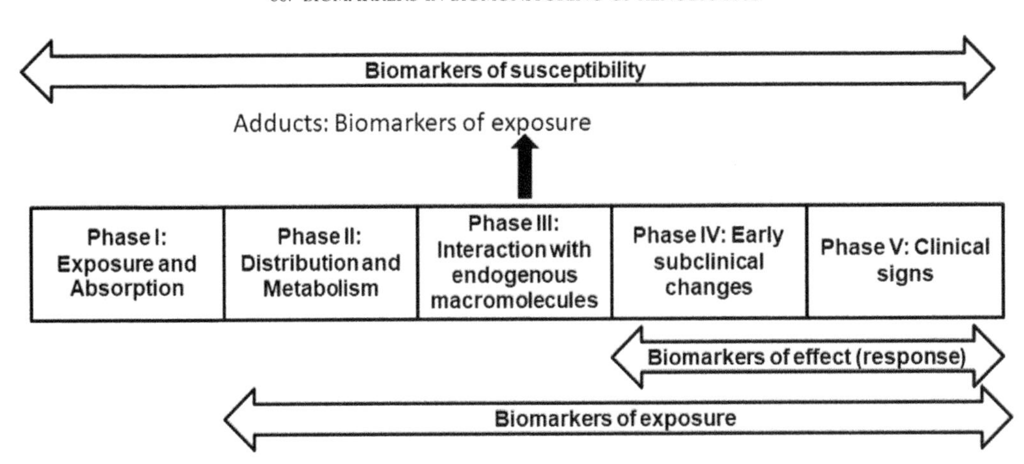

FIGURE 66.2 **Different stages of toxic phenomena in relation to different types of biomarkers.** The scheme is based on Aldridge's proposal (1996). Biomarkers of exposure can be determined at any time after stage 1 (exposure and absorption). Usually biomarkers of effect ideally address stage 4 (early subclinical effects) because they can help prevent clinical signs from appearing in stage 5 (clinical consequences of exposure). Biomarkers of susceptibility are related with bioavailability and/or endogenous targets; therefore, they can address the three first stages of the process, although measures in further stages also can yield relevant information.

there is an asymptomatic stage in which certain changes can be detected at the biochemical, cellular, tissue, or functional levels (Fig. 66.2). Biomarkers of effect are very useful for detecting these preclinical stages, which might lead to decision-making (basically decisions on acceptable levels of exposure and about whether allow the exposure or not) before adverse effects are declared in exposed organisms (Fig. 66.1). Biomarkers of effect are very important for in vivo and in vitro toxicology testing because in vivo assessments biomarkers are able to detect preclinical stages, whereas in vitro scenarios biomarkers offer molecular end points to measure the response of the exposed system to the assessed xenobiotics.

Finally, biomarkers of susceptibility allow us to identify, among all the people in a given population, those individuals who are particularly susceptible to xenobiotics. This would also allow us to provide greater protection to these individuals. Because biomarkers of susceptibility are related to bioavailability processes, these biomarkers can be determined for prevention purposes preferably at any point of the three first toxic phenomena stages (Fig. 66.2). Measures in the fourth and fifth stages can also be relevant for explaining the severity of the effects caused by the exposure. Biomarkers of susceptibility are also important for toxicology testing, despite the fact that they are not directly involved in exposure or response themselves. This can be highly relevant for explaining the intensity of the toxic responses.

The main limitations in employing biomarkers in toxicity toxicology testing and biomonitoring of xenobiotics exposure are related to the necessity of prior knowledge about the biomarker and the lack of utility for xenobiotics with either low absorption or immediate toxic effects. Previous knowledge is needed because information about toxicokinetics (i.e., route and rate of excretion, the tissue where the xenobiotic is potentially bioaccumulated, etc.) is very relevant to determine which biological sample is appropriate for biomonitoring and even which chemical has to be analyzed (the parental xenobiotics or some of its metabolites). Knowledge about mechanisms of action is also relevant because the closer its relationship with the basic toxicity mechanism, the more relevant the biomarker is.

Based on the information mentioned above, biomarkers are not usable if the xenobiotic offers a low absorption rate or if its toxic effects are immediate (i.e., irritation or corrosion) because, in these cases, there is no preclinical phase (stage 4 in Fig. 66.2). Therefore, it makes little sense because toxic effects become unavoidable after exposure.

REQUIREMENTS EXPECTED IN A BIOMARKER FOR TOXICOLOGY TESTING AND BIOMONITORING XENOBIOTICS EXPOSURE

The ideal biomarker should (1) be ethically acceptable, (2) imply simplicity for sampling, (3) be dependent on easy chemical analysis, (4) reflect a reversible change, (5) be relevant, and (6) be valid (Fig. 66.3).

The relevancy of a biomarker is also an important point when having to make a choice. The relevance of a biomarker will only be as good as its sensitivity, specificity, and relationship with the toxic process. Sensitivity refers to the biomarker's capability to display major changes after mild exposures. Specificity is the biomarker's capability to react only after a single

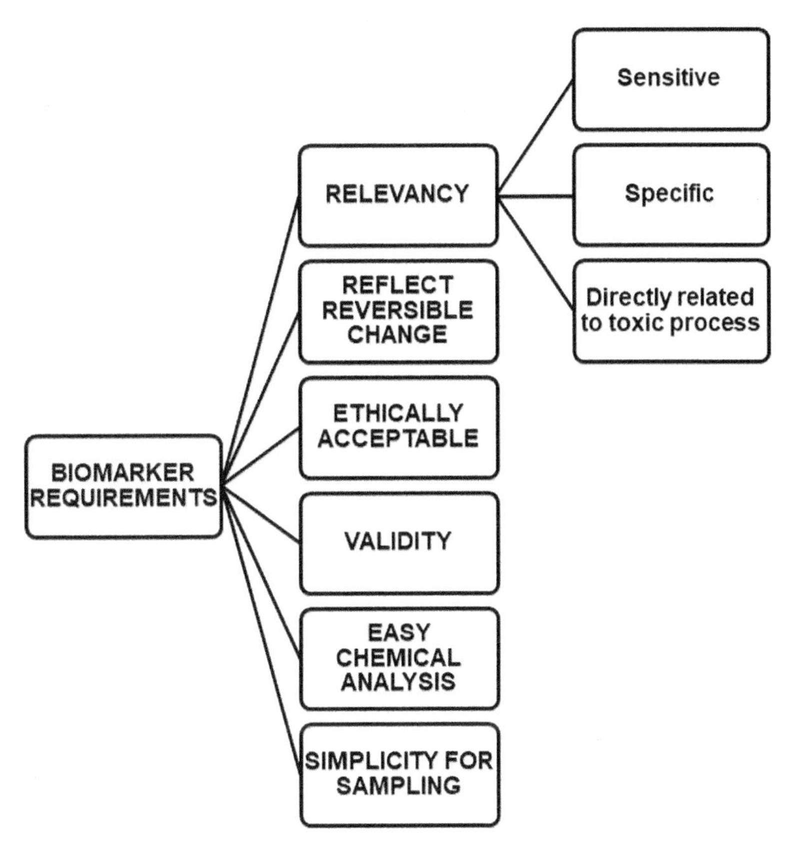

FIGURE 66.3 Requirements expected in a good biomarker for toxicology testing and biomonitoring xenobiotics exposure.

xenobiotic challenge; i.e., the level of metallothioneins is considered a nonspecific biomarker of response to heavy metals because the level of these proteins increases after exposure to most heavy metals. This does not allow us to determine the heavy metals to which organisms have been exposed. Conversely, the level of a specific heavy metal in an organism is a specific biomarker of exposure to this given heavy metal.

A good biomarker also has to address the basic toxicity mechanisms as closely as possible. According to Aldridge (1996), with clinical signs of poisonings, the xenobiotic (or its metabolite) has to interact with an endogenous biomolecule (stage 3 in Fig. 66.2) to cause a primary alteration or response. This response may cause secondary alterations. Evidently, the biomarker reflects primary alterations or responses instead of secondary or tertiary ones.

The validity of a biomarker is regulated by several factors (Fig. 66.4). The capability to distinguish among different levels of exposure is important. This directly addresses the existence of reliable dose–response relationships to support decisions relating to interventions before adverse effects appear, which is one of the most relevant applications of biomarkers (Fig. 66.1).

Certain biomarkers, especially those of enzymatic activities, present an endogenous expression in the absence of exposure. The existence of previous knowledge on levels of expression in unexposed subjects becomes necessary to validate the biomarker (i.e., serum cholinesterases display a wide range of activity in people without detected exposure to their inhibitors),

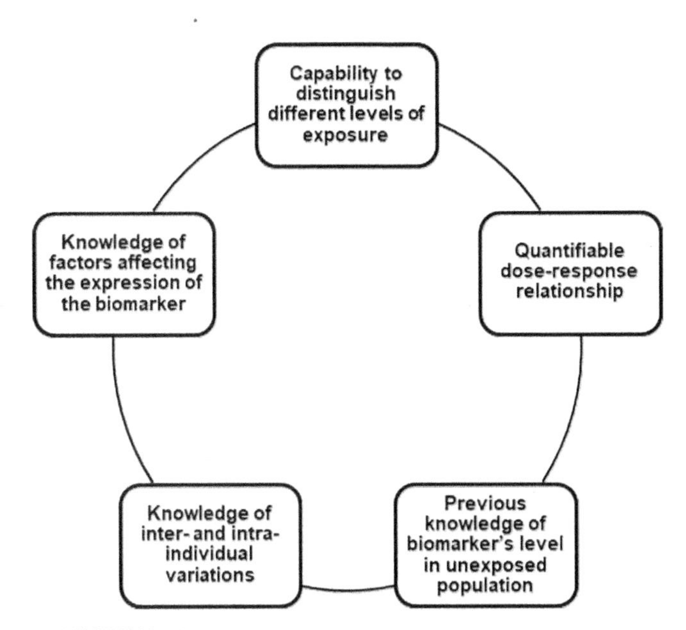

FIGURE 66.4 Factors affecting the validity of a biomarker.

which may then act as controls for comparisons with subjects who have been exposed. Moreover, variations in the expression of the biomarker in the absence of exposure and the factors affecting the expression of the biomarker are also required because protein expression can be affected by several temporary situations, such as pregnancy, age, nutritional state, exposure to other environmental contaminants, etc.

Biomarkers are based on determinations of chemicals or processes in biological samples. This implies, especially in humans, that not all biological samples are acceptable for ethical reasons. Hence, the most used biological samples are hair, blood, and urine. Other biological samples from biopsies are acceptable, only if the patient risks are lower than the expected benefit. The necessity of the biomarker to be analyzed in easily accessible samples is also closely linked to ethical reasons. The process to determine the biomarker in a biological sample should, ideally, be as easy as possible. It is also advisable that biomarkers of exposure and effect address the reversible toxic effect because this would allow taking preventive measures to avoid irreversible adverse effects.

BIOLOGICAL SAMPLES USED FOR BIOMONITORING EXPOSURE THROUGH BIOMARKERS

The main requisites for a biological matrix to be used for biomonitoring exposure are that it is available in sufficient amounts and does not pose a risk to the donor (Esteban and Castaño, 2009). The biological samples typically used for biomonitoring biomarkers of exposure and response in humans are urine and blood. Both samples are ethically acceptable, but their main disadvantage is their temporary validity. Urine only reflects exposure that took place in the few hours before sampling, whereas blood can be valid in certain situations, but for longer periods (maximum 3 or 4 months). Conversely, hair might prove to be an ideal biomarker of exposure because it is bloodless and represents an individual's cumulative exposures, the time of the exposure to the time the sample is taken. Expired air may also prove suitable for volatile xenobiotics.

Other biological samples are also suitable but are less frequently used given either the small amount available (saliva, sweat), the temporary availability (breast milk, meconium) or because the sample is not always ethically acceptable (i.e., biopsies that require major surgery, risk for fetus involved, etc.).

BIOMARKERS OF EXPOSURE FOR MONITORING XENOBIOTICS EXPOSURE

A biomarker of exposure is the presence of a xenobiotic, its metabolites, or the product of its interactions with endogenous targets. Table 66.1 provides a general overview of the most widely used biomarkers of exposure, whereas Table 66.2 displays specific biomarkers for several xenobiotics, together with the biological sample used for biomonitoring. It is remarkable that the chemical to be analyzed may differ for the same xenobiotic depending on the biological sample used; i.e., benzene is exhaled and circulates in blood as such (in an untransformed form), whereas it is found in urine as a free phenol (biotransformation product) or glucuronized phenol.

TABLE 66.1 The Most Popular Biomarkers of Exposure

Exposure to:	Biomarker
Xenobiotics	Parental xenobiotic or its metabolites
Alkylating xenobiotics	Adducts among xenobiotic and DNA and proteins
Oxidizing xenobiotics	Damaged macromolecules (DNA, lipid, or proteins) Increases in oxidized glutathione Induction of superoxide dismutases

TABLE 66.2 Some Biomarkers of Exposure

Exposure to:	Biological Sample	Chemical
Benzene	Exhaled air and blood	Benzene
Benzene	Urine	Phenol
Styrene	Exhaled air and urine	Styrene
Styrene	Urine	Mandelic acid
Dioxins	Fat	Dioxins
Lead	Blood	Lead
Aluminum	Bone	Aluminum
Toluene	Exhaled air	Toluene
Methylmercury	Hair	Methylmercury
Aniline	Urine	p-aminophenol
Hexane	Urine	2,5-hexanodione
Ethyleneglycol	Urine	Oxalic acid
Acetone	Urine and blood	Acetone

The biomonitoring of xenobiotic exposure can be performed using biomarkers in biological samples where the target of toxicity is not found. For example, the inhibition of butyrylcholinesterase in plasma is considered a biomarker of exposure to organophosphorus or carbamate insecticides and can be performed without necessity of performing a biopsy in the nervous system. This overcomes obvious bioethical considerations in the case of humans or even protected wild fauna that would not have to be sacrificed for environmental biomonitoring purposes. Moreover, unlike biomarkers of effect, a xenobiotics' exposure assessment can be performed using biomarkers with still unknown physiological roles.

A situation where both factors (the use of nontarget tissues and biomarkers with unknown roles) simultaneously appear is the use of the carboxylesterases found in the nervous system, which have to be discriminated from among the pool of esterases using inhibition kinetics with model organophosphorus compounds given the absence of a specific substrate. Three carboxylesterase enzymatic components were differentiated according to sensitivity to mipafox, paraoxon, and phenylmethanesulfonyl fluoride in chicken peripheral nerves (Estévez et al., 2004, 2011, 2012). Esterases with similar kinetic properties were found in serum, suggesting that they may be considered a "mirror" of the esterases of peripheral nerve for monitoring studies (García-Pérez et al., 2003; Estévez et al., 2011). Therefore, these esterases in the nervous system still cannot be biomarkers of effect because the physiological role is unknown and access to biopsies of the nervous systems is bioethically unacceptable in most cases. However, inhibition of their analogous esterases in plasma, a bioethically acceptable biomarker, warns about exposure to carboxylesterase inhibitors and may help prevent hazardous exposures because they are more sensitive to inhibition than classic biomarkers of exposure to carboxylesterase inhibitors with defined physiological roles as cholinesterases.

These same esterases have also been proposed for biomonitoring exposure to organophosphorus insecticides and carbamates in wild fauna, such as in birds (Sogorb et al., 2007), lizards (Basso et al., 2012; Sánchez-Hernández and Sánchez, 2002), earthworms (Sanchez-Hernandez et al., 2009, 2015), oysters (Ochoa et al., 2013), snails (Laguerre et al., 2009), crayfish (Vioque-Fernández et al., 2007), and other species like aquatic organisms such as fish and crustaceans (Solé and Sánchez-Hernández, 2015).

Adducts: A Relevant Case of Biomarkers of Exposure

Sometimes the xenobiotic or its metabolite binds to biological molecules (not necessarily a target of toxicity) to yield stable new structures called adducts, which can be used as proof that the organism has been exposed to the xenobiotic. The main requisite of the adducts to be used as a biomarker of exposure is that adduct stability is sound enough to allow the detection of exposures that occurred in the past and to resist handling of the samples with the subsequent treatments needed for chemical analysis.

Adducts of xenobiotics can be formed either with DNA or proteins. The usual biological samples taken for monitoring adducts of DNA are lymphocytes because they do not arouse ethical concerns. Nevertheless, any other cellular sample can also be useful; i.e., from biopsies. DNA adducts are especially suitable for the detection of exposures to carcinogenic chemicals. Table 66.3 depicts some examples of DNA adducts that are detectable in people exposed to several genotoxic carcinogenic xenobiotics.

Adducts formed between xenobiotics and proteins are usually monitored with hemoglobin, which is an abundant easily accessible protein in blood. Albumin is also an another abundant easily accessible protein in blood that has been proposed as a biomarker of exposure. Indeed, it has been reported that this protein is able to form stable adducts with lots of organic compounds, such as: (1) chemoprotective agents such as isothiocyanates (Sabbioni and Turesky, 2017); (2) therapeutic drugs such as β-lactam antibiotics and nonsteroidal antiinflammatory drugs, acetaminophen, or nevirapine (Sabbioni and Turesky, 2017); (3) carcinogens such as benzene, naphthalene, styrene, aromatic amines, and aflatoxin B1 (Sabbioni and Turesky, 2017); (4) organophosphorus insecticides such as azamethiphos(oxon), chlorfenvinphos (oxon), chlorpyrifos-oxon, diazinon-oxon, malaoxon, dichlorvos, tricresyl phosphate, cresyl saligenin phosphate, parathion, chlorpyrifos, chlorpyrifos-oxon (Tarhoni et al., 2008; Li et al., 2010, 2013; Schopfer et al., 2010; Liyasova et al., 2012; Noort et al., 2009; Jiang et al., 2012); (5) nerve agents with an organophosphorus structure such as soman, sarin, cyclosarin, and tabun (Williams et al., 2007; Li et al., 2008; Read et al., 2010); (6) drugs such as diazepam and ketoprofen (Watanabe et al., 2000); (7) cyanide (Fasco et al., 2011); and

TABLE 66.3 Xenobiotics Capable of Forming DNA Adducts

Xenobiotics	DNA Adduct
Ethylene oxide	N-2-hydroxyethylvaline
Mycotoxins	Aflatoxin B1, ochratoxin A
Heterocyclic amines	2-amino-3,8-dimethylimidazoquinoxaline
Aromatic amines	2-acetylaminofluorene
Polycyclic aromatic hydrocarbons	Benzopyrenes, anthracenes

(8) p-nitrophenyl esters (Means and Bender., 1975; Sakurai et al., 2004). Adducts formation also involves the deactivation of a molecule of organophosphorus insecticide that will never be available for further phosphorylation of acetylcholinesterase. Therefore in the case of albumin, an extremely abundant protein, this binding also becomes a mechanism of defense against organophosphorus insecticides, which, depending on the capability of phosphotriesterases to hydrolyze insecticides, might even prove critical, as demonstrated with paraoxon (Sogorb et al., 2008; Sogorb and Vilanova, 2010a; b). The relevance of the adducts of albumin as a biomarker of exposure is such that a transgenic mouse model expressing human serum albumin as biomarker of exposure to carcinogenic chemicals has been developed (Sheng et al., 2016). It was considered necessary because in vitro studies have reported differences between human and rodent albumins in their reactivity with certain xenobiotics that might potentially allow the formation of different adducts with subsequent shortcomings in the interspecies extrapolations. According to the authors, the use of this transgenic animal will facilitate the development and validation of adducts of albumin as biomarkers of exposure to carcinogenic chemicals.

BIOMARKERS OF EFFECT FOR TOXICOLOGY TESTING

Biomarkers of effect exhibit a biological response (biochemical, physiological, behavioral, etc.) to the exposed organisms. Table 66.4 provides examples of biomarkers of effect for different xenobiotics, whereas Table 66.5 shows some biomarkers of effect in several target tissues (Gupta, 2014).

Sometimes biomarkers of effect are not easy to differentiate from biomarkers of exposure. In general, however, a requirement for biomarkers of effect is that the state

TABLE 66.4 Some Examples of Biomarkers of Effect

Exposure to	Biomarker
Organophosphorus and carbamate insecticides	Inhibition of acetylcholinesterase in red blood cells
Lead	Inhibition of δ-aminolevulinic dehydratase and subsequent increases of δ-aminolevulinic acid in blood
Oxidizing compounds	Induction of antioxidant enzymes such as superoxide dismutase and others
Heavy metals	Induction of metallothioneins

TABLE 66.5 Examples of Some Biomarkers of Effect Typically Used for Biomonitoring Challenges to Different Targets Systems by Different Xenobiotics

Target System	Matrix	Biomarker
Central nervous system	Platelets	Monoamine oxidase b
	Cerebrospinal fluid	Glial fibrillary acidic protein, ubiquitin C-terminal hydrolase-1, α-synuclein, neuron specific enolase
	Plasma	miRNA-124
Respiratory system	Bronchoalveolar lavaged fluid and sputum	Protein, albumin, lactate dehydrogenase, N-acetyl glucosaminidase, γ-glutamyl transferase, lysozyme, alkaline phosphatase, neutrophils, lymphocytes, eosinophils, macrophages, monocytes
Liver	Blood, serum, plasma	Alanine aminotransferase, aspartate aminotransferase, alkaline phosphatase, total bilirubin, cholesterol/triglycerides, ammonia, albumin
Kidney	Urine	Proteinuria, enzymuria, cystatin C,
	Blood and urine	Neutrophil gelatinase–associated lipocalin
	Serum	Kidney injury molecule-1
Gastrointestinal system	Blood	Citrulline, diamine oxidase, gastrin, C-reactive protein, pepsinogen I and II
	Feces	Calprotectin, lactoferrin, polymorphonuclear neutrophil elastase, bile acids, fecal S100A12
Pancreas	Serum/plasma	Amylase, isoamylase, and total amylase, lipase, carbohydrate-deficient transferrin, trypsinogen activation products amylase, isoamylase, phospholipase A2
	Urine	Amylase, isoamylase, phospholipase A2
	Feces	Elastase 1
	Pancreatic juice	Cathepsins B, L, and S
Skin	Biopsy of stratum corneum	Keratin 1, 10, 11, desmoglein-1, desmocollin-1, corneodesmosin, total protein, adherins junction protein gene expression
Blood	Urine	δ-Aminolevulinic acid levels

TABLE 66.5 Examples of Some Biomarkers of Effect Typically Used for Biomonitoring Challenges to Different Targets Systems by Different Xenobiotics—cont'd

Target System	Matrix	Biomarker
	Blood	δ-Aminolevulinic acid dehydratase activity
	Blood	Mean cell hemoglobin concentration
	Urine	Corproporphyrin
	Blood	Zinc protoporphyrin
	Blood	Hematocrit/packed cell volume blood, hematocrit, hematocrit oxidants, mean cell volume, reticulocyte count, carboxyhemoglobin, methemoglobin, bleeding time, platelet count, leukocyte count
	Bone marrow	Altered myeloid/erythroid ratio, Morphologic alterations
Immune system	Blood	Leukocyte count
	Spleen, thymus, lymph node	Organ weight
	Serum, secretions, exudates	IgA, IgD, IgG, IgE, IgM
	Serum, cell supernatant, saliva, sputum	Cytokine profiling
	Urine	Leukotriene E(4)
	Serum, inflammatory exudate	Autoantibodies

of the exposed organism has to change as a result of the xenobiotic challenge (i.e., increasing the level of expression of a protein).

As in biomarkers of exposure, the effect can sometimes be monitored in tissues where the main toxicity target is not found. This can be said of the biomarkers of effect of organophosphorus insecticides. The main target of organophosphorus insecticides is acetylcholinesterase (EC 3.1.1.7) (World Health Organization, 1986), which is considered the main biomarker of effect to organophosphorus insecticides. The biological sample usually taken for biomonitoring acetylcholinesterase inhibition is red blood cells. Erythrocytes express an acetylcholinesterase that is genetically identical to acetylcholinesterase expressed in the central nervous system. Therefore, it is considered that the level of inhibition of acetylcholinesterase in red blood cells will be a reflection of the response of acetylcholinesterase in the nervous system. Organophosphorus insecticides also inhibit other esterases in the CNS that differ from acetylcholinesterase. This is the case with neuropathy target esterase (EC 3.1.1.5), an esterase target of a neurodegenerative syndrome caused by certain organophosphorus insecticides called organophosphorus inducing delayed polyneuropathy (Glynn, 2003). Neuropathy target esterase is also expressed in lymphocytes. Therefore the level of inhibition of the protein in these cells is considered a biomarker of response to the exposure to organophosphorus insecticides (Lotti et al., 1986; McConnell et al., 1999).

Molecular Biomarkers for In Vitro Testing Toxicity

Molecular biomarkers are a powerful tool for assessing the toxicity of chemicals, especially in vitro. In vitro and ex vivo systems offer an excellent opportunity for a fast and reliable assessment of the toxicity of chemicals through an appropriate selection of molecular biomarkers.

Omic methodologies have provided a great advance in toxicology during the 21st century (Choudhuri et al., 2018). Omic approaches allow designing sets of key molecular biomarkers that might play a notable role in toxicity testing. Within the omic methodologies, transcriptomics is probably the most promising tool for the discovery and validation of molecular biomarkers. Indeed, alterations in connectivity maps of the transcriptome after exposure to xenobiotics is able to play a double role in that it is able to help explain the mechanism of toxicity and build adverse outcome pathways (Leist et al., 2017). These alterations are also able to identify key genes in the development of the adverse outcome that will serve as biomarkers for testing toxicity.

Some examples of biomarker genes that have been proposed for toxicity testing (specifically developmental toxicity) are "van Dartel Heartdiff_24h" and "EST biomarker genes" that were able to correctly predict between 67% and 83% of embryotoxicants within a battery of model toxicants (van Dartel et al., 2011). The battery of 16 different genes proposed by Romero et al. (2015) that were able to predict the embryotoxicity potency of a set of five different embryotoxicants plus two negative controls with an in vivo—in vitro concordance of 100% is also very promising in this field of developmental toxicity.

It is quite obvious that a single in vitro test based on a single biomarker will never be able to replace an in vivo test. However, it is theoretically possible to develop *I*ntegrated *T*esting *S*trategies that efficiently combine

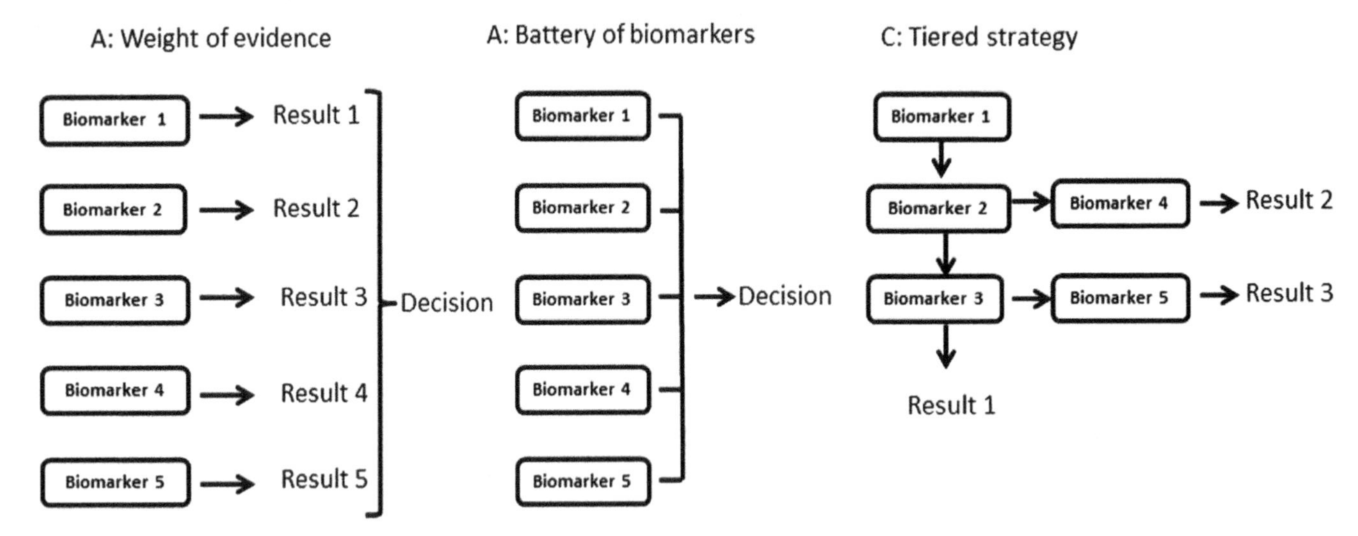

FIGURE 66.5 Schemes for the different approaches that are considered as form of Integrated Testing Strategy. *Based on Rovida C., Alépée N., Api A.M., et al., 2015. Integrated testing strategies (ITS) for safety assessment. ALTEX 32(1), 25–40.*

different information coming from different biomarkers and which might introduce notable enhancements in the process of safety assessment of chemicals (in terms of reduction of cost, time, and consumption of animals). This would theoretically allow (1) the preliminary assessment of the potential toxicity of the tested chemical, which might be especially relevant in screening processes; (2) the understanding of the mechanism of toxicity, which might allow for safer assessment of the hazards/risks associated with exposure because the adverse effect to be tested will be preliminarily well identified; (3) when negative results are obtained, the design of more focused in vivo experiments (e.g., using a single limit dose) will also be potentially possible; and (4) when results are positive, this will still allow a more rational approach for proposing which cellular, physiological, and biological end points will deserve further research in more complex in vitro or in vivo assays.

Three different approaches are currently being considered for developing *I*ntegrated *T*esting *S*trategies (Rovida et al., 2015): (1) the weight of evidence; (2) the battery of tests; and (3) the tiered strategy (Fig. 66.5). In the weight of evidence approach, independent assays providing the same number of independent results have to be analyzed to make a decision, as none of the available results are suitable for such a decision alone (Fig. 66.5A). In the battery of tests, the final decision is defined by the sum of the results from many analyses, all of which must be performed (Fig. 66.5B). However, the tiered strategy is an open system that allows one to decide the next test to perform based on the results of the first step (Fig. 66.5C). Each of the three strategies will be considered for defining a set of specific biomarkers that allow performing hazard identification and risk characterization of developmental toxicity in the fastest, safest, and more reliable way.

The development of biomarker signatures also allows the grouping of chemicals with common mechanisms of toxicity, which would notably cut the number of animals used in toxicity testing because only the most representative chemicals of a specific group can be assayed, and the further extrapolation of the results to the remaining members of the group is possible without animal experimentation.

The sets of in vitro biomarkers might also be part of *I*ntegrated *A*pproach for *T*esting and Assessment strategies with future consequences at the regulatory level, which would allow the assessment of chemicals exclusively on the basis of in vitro assays, as suggested specifically for neurodevelopmental toxicity testing by Fritsche et al. (2017).

CONCLUDING REMARKS AND FUTURE DIRECTIONS

Biomarkers are a powerful tool for risk assessments of aspects of exposure assessment and risk characterization. Biomarkers of exposure are indicated for exposure assessment, especially when the chemical species being analyzed is the adduct resulting from the interaction between the xenobiotic and an endogenous molecule (not necessarily the toxicity target). These adducts are usually eliminated from organisms more slowly than the free xenobiotic (or its free metabolites), which allows the detection of exposures, even when they have been totally eliminated from the exposed organism. Biomarkers of effect are useful for recording the responses of exposed organisms to the toxic insult and are used to detect exposures before clinical consequences appear. The development of integrated testing strategies based on molecular biomarkers is a promising and powerful approach for assessing the safety of chemicals on the basis of in vitro or ex vivo toxicity tests. Ideally, a risk assessment should be performed on the basis of the analysis of a complete battery of good biomarkers of exposure, effect, and susceptibility because a single biomarker does not usually yield enough information to allow a good risk assessment process.

References

Aldridge, N., 1996. Stages in the induction of toxicology. In: Mechanisms and Concepts in Toxicology. Taylor and Francis, London.

Basso, A., Attademo, A.M., Lajmanovich, R.C., et al., 2012. Plasma esterases in the tegu lizard *Tupinambis merianae* (Reptilia, Teiidae): impact of developmental stage, sex, and organophosphorus in vitro exposure. Environ. Sci. Pollut. Res. 19 (1), 214–225.

Choudhuri, S., Patton, G.W., Chanderbhan, R.F., et al., 2018. From classical toxicology to Tox21: some critical conceptual and technological advances in the molecular understanding of the toxic response beginning from the last quarter of the 20th century. Toxicol. Sci. 61 (1), 5–22.

Esteban, M., Castaño, A., 2009. Non-invasive matrices in human biomonitoring: a review. Environ. Int. 35 (2), 438–449.

Estévez, J., Barril, J., Vilanova, E., 2012. Kinetics of inhibition of soluble peripheral nerve esterases by PMSF: a non-stable compound that potentiates the organophosphorus-induced delayed neurotoxicity. Arch. Toxicol. 86 (5), 767–777.

Estévez, J., García-Pérez, A., Barril, J., Vilanova, E., 2011. Inhibition with spontaneous reactivation of carboxyl esterases by organophosphorus compounds: paraoxon as a model. Chem. Res. Toxicol. 24 (1), 135–143.

Estévez, J., García-Pérez, A., Barril, J., Pellín, M., Vilanova, E., 2004. The inhibition of the high sensitive peripheral nerve soluble esterases by mipafox A new mathematical processing for the kinetics of inhibition of esterases by organophosphorus compounds. Toxicol. Lett. 151 (1), 171–181.

Fasco, M.J., Stack, R.F., Lu, S., Hauer 3rd, C.R., et al., 2011. Unique cyanide adduct in human serum albumin: potential as a surrogate exposure marker. Chem. Res. Toxicol. 24 (4), 505–514.

Fritsche, E., Crofton, K.M., Hernandez, A.F., et al., 2017. OECD/EFSA workshop on developmental neurotoxicity (DNT): the use of non-animal test methods for regulatory purposes. ALTEX 34 (2), 311–315.

García-Pérez, A., Barril, J., Estévez, J., Vilanova, E., 2003. Properties of phenyl valerate esterase activities from chicken serum are comparable with soluble esterases of peripheral nerves in relation with organophosphorus compounds inhibition. Toxicol. Lett. 142 (1–2), 1–10.

Glynn, P., 2003. NTE: one target protein for different toxic syndromes with distinct mechanisms. Bioassays 25 (8), 742–745.

Gupta, R.G., 2014. In: Gupta, R.C. (Ed.), Biomarkers in Toxicology, first ed. Academic Press/Elsevier, Amsterdam.

Jiang, W., Duysen, E.G., Lockridge, O., 2012. Mice treated with a nontoxic dose of chlorpyrifos oxon have diethoxyphosphotyrosine labeled proteins in blood up to 4 days post exposure, detected by mass spectrometry. Toxicology 295 (1–3), 15–22.

Laguerre, C., Sanchez-Hernandez, J.C., Köhler, H.R., et al., 2009. B-type esterases in the snail Xeropicta derbentina: an enzymological analysis to evaluate their use as biomarkers of pesticide exposure. Environ. Pollut. 157 (1), 199–207.

Leist, M., Ghallab, A., Graepel, R., et al., 2017. Adverse outcome pathways: opportunities, limitations and open questions. Arch. Toxicol. 91 (11), 3477–3505.

Li, B., Eyer, P., Eddleston, M., et al., 2013. Protein tyrosine adduct in humans self-poisoned by chlorpyrifos. Toxicol. Appl. Pharmacol. 269 (3), 215–225.

Li, B., Ricordel, I., Schopfer, L.M., et al., 2010. Detection of adduct on tyrosine 411 of albumin in humans poisoned by dichlorvos. Toxicol. Sci. 116 (1), 23–31.

Li, B., Nachon, F., Froment, M.T., et al., 2008. Binding and hydrolysis of soman by human serum albumin. Chem. Res. Toxicol. 21 (2), 421–431.

Liyasova, M.S., Schopfer, L.M., Lockridge, O., 2012. Cresyl saligenin phosphate, an organophosphorus toxicant, makes covalent adducts with histidine, lysine, and tyrosine residues of human serum albumin. Chem. Res. Toxicol. 25 (8), 1752–1761.

Lotti, M., Moretto, A., Zoppellari, R., et al., 1986. Inhibition of lymphocytic neuropathy target esterase predicts the development of organophosphate-induced delayed polyneuropathy. Arch. Toxicol. 59 (3), 176–179.

McConnell, R., Delgado-Téllez, E., Cuadra, R., et al., 1999. Organophosphate neuropathy due to methamidophos: biochemical and neurophysiological markers. Arch. Toxicol. 73 (6), 296–300.

Means, G.E., Bender, M.L., 1975. Acetylation of human serum albumin by p-nitrophenyl acetate. Biochemistry 14 (22), 4989–4994.

Noort, D., Hulst, A.G., van Zuylen, A., et al., 2009. Covalent binding of organophosphorothioates to albumin: a new perspective for OP-pesticide biomonitoring? Arch. Toxicol. 83 (11), 1031–1036.

Ochoa, V., Riva, C., Faria, M., Barata, C., 2013. Responses of B-esterase enzymes in oysters (*Crassostrea gigas*) transplanted to pesticide contaminated bays form the Ebro Delta (NE, Spain). Mar. Pollut. Bull. 66 (1–2), 135–142.

Read, R.W., Riches, J.R., Stevens, J.A., et al., 2010. Biomarkers of organophosphorus nerve agent exposure: comparison of phosphylated butyrylcholinesterase and phosphylated albumin after oxime therapy. Arch. Toxicol. 84 (1), 25–36.

Romero, A.C., Del Río, E., Vilanova, E., Sogorb, M.A., 2015. RNA transcripts for the quantification of differentiation allow marked improvements in the performance of embryonic stem cell test (EST). Toxicol. Lett. 238 (3), 60–69.

Rovida, C., Alépée, N., Api, A.M., et al., 2015. Integrated testing strategies (ITS) for safety assessment. ALTEX 32 (1), 25–40.

Sabbioni, G., Turesky, R.J., 2017. Biomonitoring human albumin adducts: the past, the present, and the future. Chem. Res. Toxicol. 30 (1), 332–366.

Sakurai, Y., Ma, S.F., Watanabe, H., et al., 2004. Esterase-like activity of serum albumin: characterization of its structural chemistry using p-nitrophenyl esters as substrates. Pharm. Res. (N. Y.) 21 (2), 285–292.

Sánchez-Hernández, J.C., Sánchez, B.M., 2002. Lizard cholinesterases as biomarkers of pesticide exposure: enzymological characterization. Environ. Toxicol. Chem. 21 (11), 2319–2325.

Sánchez-Hernández, J.C., Mazzia, C., Capowiez, Y., Rault, M., 2009. Carboxylesterase activity in earthworm gut contents: potential (eco)toxicological implications. Comp. Biochem. Physiol. C Toxicol. Pharmacol. 150 (4), 503–511.

Sanchez-Hernandez, J.C., Notario del Pino, J., Domínguez, J., 2015. Earthworm-induced carboxylesterase activity in soil: assessing the potential for detoxification and monitoring organophosphorus pesticides. Ecotoxicol. Environ. Saf. 122, 303–312.

Schopfer, L.M., Furlong, C.E., Lockridge, O., 2010. Development of diagnostics in the search for an explanation of aerotoxic syndrome. Anal. Biochem. 404 (1), 64–74.

Sheng, J., Wang, Y., Turesky, R.J., et al., 2016. Novel transgenic mouse model for studying human serum albumin as a biomarker of carcinogenic exposure. Chem. Res. Toxicol. 29 (5), 797–809.

Sogorb, M.A., Vilanova, E., 2010a. Serum albumins and detoxication of anti-cholinesterase agents. Chem. Biol. Interact. 187 (1–3), 325–329.

Sogorb, M.A., Vilanova, E., 2010b. Detoxication of anticholinesterase pesticides. In: Satoh, T., Gupta, R.G. (Eds.), Anticholinesterase Pesticides: Metabolism, Neurotoxicity, and Epidemiology. John Willey & Sons, pp. 121–133.

Sogorb, M.A., García-Argüelles, S., Carrera, V., Vilanova, E., 2008. Serum albumin is as efficient as paraxonase in the detoxication of paraoxon at toxicologically relevant concentrations. Chem. Res. Toxicol. 21 (8), 1524–1529.

Sogorb, M.A., Ganga, R., Vilanova, E., Soler, F., 2007. Plasma phenylacetate and 1-naphthyl acetate hydrolyzing activities of wild birds as possible non-invasive biomarkers of exposure to organophosphorus and carbamate insecticides. Toxicol. Lett. 168 (3), 278–285.

Solé, M., Sanchez-Hernandez, J.C., 2015. An in vitro screening with emerging contaminants reveals inhibition of carboxylesterase activity in aquatic organisms. Aquat. Toxicol. 169, 215–222.

Tarhoni, M.H., Lister, T., Ray, D.E., Carter, W.G., 2008. Albumin binding as a potential biomarker of exposure to moderately low levels of organophosphorus pesticides. Biomarkers 13 (4), 343–363.

van Dartel, D.A., Pennings, J.L., Robinson, J.F., et al., 2011. Discriminating classes of developmental toxicants using gene expression profiling in the embryonic stem cell test. Toxicol. Lett. 201 (2), 143–151.

Vioque-Fernández, A., de Almeida, E.A., López-Barea, J., 2007. Esterases as pesticide biomarkers in crayfish (*Procambarus clarkii*, Crustacea): tissue distribution, sensitivity to model compounds and recovery from inactivation. Comp. Biochem. Physiol. C Toxicol. Pharmacol. 145 (3), 404–412.

Watanabe, H., Tanase, S., Nakajou, K., et al., 2000. Role of arg-410 and tyr-411 in human serum albumin for ligand binding and esterase-like activity. Biochem. J. 349 (Pt 3), 813–819.

Williams, N.H., Harrison, J.M., Read, R.W., Black, R.M., 2007. Phosphylated tyrosine in albumin as a biomarker of exposure to organophosphorus nerve agents. Arch. Toxicol. 81 (9), 627–639.

World Health Organization, 1986. Organophosphorous insecticides: a general introduction. In: World Health Organization (Ed.), Environmental Health Criteria 63. Available in: http://www.inchem.org/documents/ehc/ehc/ehc63.htm.

CHAPTER

67

Biomarkers in Epidemiology, Risk Assessment and Regulatory Toxicology

A.M. Fan[1], P. Cohn[2], S.H. You[3], P. Lin[4]

[1]California Environmental Protection Agency (retired), Oakland/Sacramento, CA, United States; [2]New Jersey Department of Health (retired), Trenton, NJ, United States; [3]Institute of Food Safety and Risk Management, National Taiwan Ocean University, Keelung, Taiwan; [4]National Institute of Environmental Health Sciences, National Health Research Institutes, Taiwan

INTRODUCTION

Biomarkers have been used in clinical medicine and molecular epidemiology, mainly as biologic indicators of exposure or risk of disease, and their use in toxicology, risk assessment, and environmental chemical regulations has been discussed by Fan (2014). The earlier discussion focused on the three types of biomarkers (namely biomarkers of exposure, effects, and susceptibility) related to the study of environmental chemicals and the associated human exposures and health effects, how biomarkers fit into the toxicological evaluation and risk assessment considerations, and how risk assessment with the potential use of biomarkers is integrated into the development of chemical guidance and regulations. It highlights some successes in advancing the understanding of some of the most critical elements of the risk assessment process, and in using biomarkers to address environmental and public health issues. Several examples were given for polycyclic aromatic hydrocarbons (PAHs), pesticides that are cholinesterase (ChE) inhibitors, triazines, lead, selenium (Se), cotinine, methylmercury (MeHg), and nitrate. It is encouraged that it should be read as a compendium to this chapter.

This chapter updates the discussion with additional relevant information on the definition and uses of biomarkers, and on MeHg from fish consumption, ChE-inhibiting pesticides (e.g., chlorpyrifos), Se, and formaldehyde. It focuses on environmental chemicals and the associated biomarkers (excluding clinical and diagnostic tools, surrogate endpoints in clinical research, and therapeutic intervention), and it includes recent reports from the National Research Council (NRC) that substantially advance the conceptual and experimental approaches in the fields of exposure science, biomonitoring, -omics technologies, and toxicology/toxicity testing and risk assessment, with the recognition that biomarkers are anticipated to play a role in these related advances.

A focus of this chapter is exposure science and its use in health risk assessments, in particular, risk from oral exposure. In brief, the usual risk assessment procedure involves developing health reference values such as an oral reference dose (RfD). The RfD is an estimate (with uncertainty spanning perhaps an order of magnitude) of a daily exposure to the human population (including sensitive subgroups) that is likely to be without an appreciable risk of deleterious effects during a lifetime (approximated as 70 years). It is expressed as mass per kilogram of body weight per day, e.g., microgram (μg)/kg-day. The first step is the selection of the best study or studies, followed by the determination of a point of departure (POD) such as a no-adverse-effect level (NOAEL), a lowest-observable-adverse-effect level (LOAEL), or a benchmark dose (BMD). The BMD is usually expressed as the dose at which there is a 10% or 5% benchmark response rate for the effect of interest, BMD_{10} or BMD_{05}. This process typically involves applied statistical software that also generates an upper and, in particular, lower 95th percentile limit, the $BMDL_{10}$ or $BMDL_{05}$. Application of part or all of a set of uncertainty factors (UFs) to the $BMDL_{10}$ or $BMDL_{05}$ establishes the RfD. The margin of exposure is the degree to which the actual exposure exceeds or is less than the RfD. This process is discussed below in examples that help to illustrate how exposure science informs risk assessment.

Biomarkers in Toxicology, Second Edition
https://doi.org/10.1016/B978-0-12-814655-2.00067-0

BIOMARKERS

Biomarkers are measures reflecting interactions between biological systems, as in the human body, and environmental chemicals to which the individual is exposed. A major factor is the need to identify critical toxicological endpoints with cause–effect and dose–response relationships for quantitative risk assessment. This can then provide a scientific health basis for regulatory and risk management decision-making, often achieved by setting guidance levels or regulatory standards.

Traditionally, biomarkers have been classified by various researchers and scientific bodies as markers of exposure, effect, or susceptibility, with some variations (Fan, 2014). These biomarkers, sometimes referred to as biologic indicators, generally include biochemical, molecular, genetic, immunologic, or physiologic signals of events in biologic systems. The events are presented as a continuum between an external exposure to a chemical and resulting biological effect in the body, which may or may not lead to clinical manifestations or diseases.

For environmental chemicals, biomarkers of exposure, effect, and susceptibility are defined as follows (WHO, 2001; NRC, 2006):

Biomarker of exposure. The chemical or its metabolite or the product of an interaction between a chemical and some target molecule or cell that is measured in a compartment in an organism.

Biomarker of effect. A measurable biochemical, physiologic, behavioral, or other alteration in an organism that, depending on the magnitude, can be recognized as associated with an established or possible health impairment or disease.

Biomarker of susceptibility. An indicator of an inherent or acquired ability of an organism to respond to the challenge of exposure to a specific chemical substance.

Various elements are involved before exposure biomonitoring can be used for risk assessment and management decisions (NRC, 2006). These include scoping, screening, exploratory investigations, source investigations, societal-hazard identifications, status and trends, exposure surveillance, population research, pathway research, decision validation and health surveillance, exposure and health research, epidemiologic research (ecologic and analytic), toxicologic research, pharmacokinetic and pharmacodynamic research, and community and occupational investigations. With the related relevant and adequate information, risk assessment, population risk characterization, individual risk characterization, and clinical applications can be considered. The interpretation and use of biomonitoring data will depend on the entire body of knowledge on these elements.

NRC (2006) described the properties for biomarkers of exposure for which all the relationships needed are clearly delineated and thus would probably be useful for risk assessment purposes. A framework to characterize biomarkers links the potential uses of biomarkers to their properties and is based on the simplified relationships between external dose, internal dose, and biological effects. In this framework, the seven qualitative properties (based on a weight-of-evidence approach) are described as follows:

1. reproducible sampling and analytical method;
2. known relationship of external dose to biomarker in animals;
3. known relationship of external dose to biomarker in humans;
4. known relationship of biomarker to biological effects in animals;
5. known relationship of biomarker to biological effects in humans;
6. known relationship of external dose to response in animals; and
7. known relationship of external dose to response in humans.

The advantages and limitations of biomarkers are presented by Shea (WHO, 2011). The advantages include the following:

1. confirming absorption into human body;
2. measuring integrated exposure from all routes and all sources—not dependent on models or assumptions;
3. detecting very low-level exposures because analytical techniques have become exquisitely sensitive over the past several decades;
4. helping to test and validate exposure models when the results of modeling predictions are compared to internal doses actually measured in exposed persons;
5. helping to follow exposure trends when individuals or representative samples of groups are followed with serial biomarker testing over time; and
6. helping to evaluate public health interventions.

The limitations of biomarkers include the following:

1. does not define sources or pathways of exposure;
2. cannot define toxic dose—unless toxicology and epidemiology studies have defined toxicity and the dose response curve;
3. susceptible to inferior or unscrupulous analytical laboratories;
4. lack of meaningful reference levels for most chemicals; and

5. lack of toxicological and epidemiological information about the vast majority of environmental contaminants.

The US Food and Drug Administration (FDA)-NIH Biomarker Working Group (USFDA, 2016a,b) recently developed the BEST (Biomarkers, EndpointS, and other Tools) resource with the first phase comprising a glossary that clarifies important definitions and describes some of the hierarchical relationships, connections, and dependencies among the terms it contains. It was created with the goal of harmonization of terms used in translational science and medical product development, with a focus on terms related to study endpoints and biomarkers. The BEST glossary aims to capture distinctions between biomarkers and clinical assessments and to describe their distinct roles in biomedical research, clinical practice, and medical product development fields. The glossary showed biomarkers to be a defined characteristic that is measured as an indicator of normal biological processes, pathogenic processes, or responses to an exposure or intervention, including therapeutic interventions. Thus molecular, histological, radiographic, or physiological characteristics are approaches to measuring biomarkers, and a biomarker is not an assessment of how an individual feels, functions, or survives.

New Advances in Biomonitoring, Exposure Science, Toxicology, Risk Assessment, and Decision-Making

NRC has released several reports that substantially advance the conceptual and experimental approaches in the fields of exposure science and biomonitoring, and the companion fields of toxicology and risk assessment (NRC, 2012). Biomarkers are anticipated to play a role in these related advances.

In the 2012 report Exposure Science in the 21st century (NRC, 2012), exposure science addressed the intensity and duration of contact of humans or other organisms with certain agents (defined as chemical, physical, or biological stressors) and their fate in living systems. Exposure assessment was described as being instrumental in helping to forecast, prevent, and mitigate exposures that led to adverse human health or ecological outcomes; to identify populations that had high exposures; to assess and manage human health and ecosystem risks; and to protect vulnerable and susceptible populations. For example, the ability to detect chemical contaminants in drinking water at low but biologically relevant concentrations can help to identify emerging health threats, and it is valuable for assessing the value of proposed public health actions in policy and regulatory decisions (or success following implementation).

Exposure science has evolved from other disciplines as industrial hygiene, radiation protection, and environmental toxicology into a theoretical and practical science that includes development of mathematical models and other tools for examining how individuals and populations come into contact with environmental stressors. It has played a fundamental role in the development and application of many fields related to environmental health, including toxicology, epidemiology, and risk assessment. In this regard, exposure information is critical in the design and interpretation of toxicology studies and is important in epidemiology studies to compare outcomes in populations that have different exposure levels. Increasing collection and evaluation of biomarker data through the Centers for Disease Control and Prevention National Health and Nutrition Examination Survey and other government efforts offers the potential for improving evaluation of source–exposure and exposure–disease relationships.

The concept of the exposome has been defined as the totality of exposures individuals experience over the course of their lives (including such factors as diet, stress, drug use, and infection) and how those exposures affect health (DeBord et al., 2016). The three domains of the exposome are internal, specific external, and general external. Internal factors are those that are unique to the individual, and specific external factors include occupational exposures and lifestyle factors. The general external domain includes sociodemographic factors such as education level and financial status. A variety of tools have been identified to measure the exposome. Biomarker measurements will be one of the major tools in exposomic studies. Biomonitoring and biomarker data from -omics or other techniques would be used in the internal domain. The exposomic data in epidemiologic studies can provide greater understanding of the relationships among a broad range of chemical and other risk factors and health conditions and ultimately lead to more effective and efficient disease prevention and control. Many of the techniques for assessing exposure and the exposome involve molecular epidemiologic studies which use biological markers (exposure, effect, susceptibility) in epidemiologic research.

The Toxicity Testing in the 21st century report (NRC, 2007) laid the foundation for a paradigm shift toward the use of new scientific tools to expand in vitro pathway-based toxicity testing (NRC, 2012). A key component of that report is the generation and use of population-based and individual human-exposure data for interpreting test results and using toxicity biomarker data with exposure data for biomonitoring, surveillance, and epidemiologic studies. The focus of the report on systems approaches to understanding human biology, coupled with information about systems-level perturbations resulting from human–environment interactions, is critical for understanding biologically relevant exposures. By emphasizing early perturbations of biological pathways that can lead to disease, the report moves the focus of risk assessment along the exposure–disease spectrum toward exposure, especially the role of prior and current exposures in altering vulnerability of individuals and communities to additional environmental exposures. The resulting toxicology focus is on early biomarkers of effects in the population. At the same time, such concepts as the exposome have moved the focus of exposure science along the exposure–disease spectrum toward the health-effects side, especially biologic perturbations that correlate with exposure and are predictive of disease. The interconnections between fields of exposure science and toxicology allows for better linkages between exposure and disease.

In the report Using 21st century Science to Improve Risk-Related Evaluations (NASEM, 2017), a committee discussion on the best ways to incorporate the emerging science into risk-based evaluations was presented. Several large-scale US and international programs have been initiated to incorporate advances in molecular and cellular biology, -omics technologies, analytical methods, bioinformatics, and computational tools and methods into the field of toxicology. In addition, the causal understanding for human health risk-based decision-making was discussed at a recent NAS panel meeting (NASEM, 2018.).

BIOMARKERS OF EXPOSURE

Measuring a biomarker of exposure starts with the choice of the biological media to study and whether the original chemical or its metabolite (or ionic state, species in the case of metals and metalloids) is more available and/or quantifiable. The choice of biological media for measuring biomarkers is linked to the pharmacokinetics of the contaminant of interest, though there are other factors, such as ease of collection and willingness to participate. Each chemical must be evaluated separately for validity of the media choice. However, there are some general approaches. While biomonitoring programs typically choose one biological media, many research studies look at more than one to better characterize exposure among what is typically a smaller group of participants. Blood and urine are the most common media, but in the case of metals (e.g., mercury) and metalloids (e.g., arsenic) that bind to sulfhydryl moieties in amino acids, exposure can also be measured in hair and nails, which is useful for estimating longer-term exposures and the overall body burden of the metal. Lipophilic chemicals, including persistent organic pollutants, are best measured in a lipid-rich environment, such as breast milk.

The choice of what to measure must also be well considered. For example, many volatile organic chemicals are quickly exhaled, so exhaled air could be the most appropriate method of exposure estimation if done shortly after exposure, such as at the end of a shift at a work site, whereas measurement of nonvolatile metabolites (e.g., after glycosylation) in blood or urine might be more appropriate at a later point.

Interpretation of biomonitoring data for use in exposure and risk assessment can be greatly assisted by application of pharmacokinetic models (Ruiz et al., 2011).

METHYLMERCURY

In this discussion, emphasis is placed on concentrations in blood and hair in fish-eating populations based on best available data, and how these internal exposures relate to external intakes from fish consumption, toxicology and dose–response assessment, health risk assessment, derivation of health-protective reference values (e.g., USEPA RfD), and implications in regulatory decision-making and public health risk management guidance. The update includes a recent study by the authors on estimates of MeHg intake through internal and external exposures, using measurement of total mercury (THg) in blood as a biomarker of exposure in young children and adolescents in Taiwan, plus intake estimates from food consumption data, based on two national biomonitoring databases (You et al., 2018). The current issues of interest relate to widespread MeHg contamination and exposure from fish consumption, the regulatory implications, and the need to provide awareness and guidance to the general public, especially the sensitive populations (women of childbearing age, infants, and children), along with risk–benefit considerations, and the increasing knowledge of dose–effect (response) relating to internal and external biomarkers. Emphasis is placed on quantitative data of biomarker concentrations and health concern levels.

In most epidemiologic and exposure studies, mercury exposure is assessed by analysis of hair, blood, or urine, most commonly in occupational contexts, and nail analysis has been used to assess body burdens of metals, often in relation to nutritional epidemiology (WHO, 2010a). The physiological and kinetic relationships between body burden and toenail levels are less understood, as compared to hair. Toenail mercury has not been used in studies of children, but it has been applied as a biomarker of exposure in studies of mercury exposures related to cardiovascular endpoints in adults. There have been relatively few direct comparisons of the informational value of results by compartment.

The major source of MeHg exposure is through the consumption of contaminated fish and seafood (WHO,

2010a,b,c). Human MeHg exposures are often assessed through the use of biomarkers, which serve as a surrogate for the biologically relevant internal dose of MeHg (NRC, 2000). The commonly used biomarkers are concentrations in blood, red blood cells (RBCs), hair, and urine (WHO, 1990; Groth, 2010; EFSA, 2012). Typically, blood and urine Hg levels are reported as THg, which comprises both inorganic and organic Hg. THg or organic Hg (total minus inorganic Hg) have been used to estimate MeHg in whole blood (WB) based on studies showing that 70%–95% of THg in blood is in the form of MeHg and is bound to hemoglobin (Mortensen et al., 2014), and the relationship of THg to MeHg has been assumed to be linear and constant across population demographics. MeHg is the toxic form responsible for adverse neurodevelopmental effects seen in fish-consuming human populations in contrast to effects due to inorganic mercury poisoning.

There is a direct relationship between Hg concentrations in human blood and consumption of fish contaminated with MeHg (WHO, 2008; Mahaffey et al., 2004; Grandjean et al., 1997; Karagas et al., 2012). In the general population, the total blood Hg concentration is due mostly to the dietary intake of organic forms, particularly MeHg (CDC, 2018), and it reflects short-term exposure, giving an estimate of exposure over the most recent 2–5 months, with an average half-life of MeHg in blood of 50 days (NRC, 2000). When measured at a specific point in time, it reflects contributions from both recent exposures, which may be increasing if intake is ongoing, and older exposures whose relative contribution decreases over time. Urinary Hg consists mostly of inorganic Hg.

Hg concentration in hair is often used to estimate MeHg exposure as the predominant form of Hg in hair is MeHg among persons exposed to MeHg (WHO, 2010a). The incorporation of Hg into the growing hair follicle is assumed to be directly proportional to the blood concentration, with a delay of 1–2 months between MeHg intake and measurement in the visible proximal hair shaft (Budtz-Jørgensen et al., 2004; WHO, 2010a; NRC, 2000). The relationship between location along the hair strand and timing of exposure can be ascertained by assuming a constant rate of hair growth, and 1.1 cm per month is commonly assumed.

Sheehan et al. (2014) reviewed and synthesized the evidence from published studies reporting total hair Hg and total blood Hg biomarkers to systematically compare global MeHg exposure among women and their infants from seafood-consuming populations. The authors compared the findings with the WHO provisional tolerable weekly intake (PTWI) reference level for fish (1.6 μg/kg bw/wk) and compared that intake to the approximate equivalent level in hair (2.2 μg/g). They found that pooled central THg biomarker distribution for various studied populations exceeded the WHO reference level based on the upper bound median (2.9–23.1 μg/g) or high-end biomarkers for women and infants in rural riverine communities, near tropical gold mining sites, Arctic traditional food consumers, industry and fishing categories, Pacific coastal subcategory, and Mediterranean and inland categories.

To relate the levels of MeHg biomarkers to the corresponding daily intakes of MeHg, an empirical one-compartment kinetic model describing the fate of MeHg in the human body has been used (WHO, 1990, 2007, 2010b; US EPA, 2001; You et al., 2018). The steady-state conditions (i.e., the intake equals the excretion) are assumed in establishing the relationship, resulting in a constant ratio between the MeHg daily intake and the total Hg measured in the biological matrices. In practice, the consumption pattern varies among individuals.

Various adverse health effects of MeHg have been shown in humans and in animal studies. The nervous system is the most sensitive target organ, particularly in developing organisms. More recent human health effects data were obtained from three studies that have been used for quantitative risk assessment. These are the three longitudinal developmental epidemiologic studies in Faroe Islands, New Zealand, and the Seychelles Islands that were conducted to study in utero MeHg exposure in the fish-eating populations (NRC, 2000; US EPA, 2001). The biomarkers of exposure are THg in maternal hair collected at delivery and Hg in infant cord blood and maternal blood, as internal dosimeters of MeHg exposure during pregnancy. The NRC conducted a comprehensive review of these studies on MeHg-related health effects and concluded that neurodevelopmental impacts from prenatal MeHg exposures are the most sensitive and best-documented endpoints (NRC, 2000).

For the purposes of risk assessment, biomarker concentrations of MeHg serve two functions (NRC, 2000). First, the concentration of a biomarker is used as a surrogate for the unknown biologically relevant dose of MeHg in the developing fetal brain, representing the dose in the development of a dose–response relationship. Second, in this dose–response relationship, the biomarker concentration identified as the critical (e.g., benchmark) concentration must be translated into an estimate of the ingested dose. The estimated ingested dose is used to guide public health interventions and regulatory measures. The latter step involves the use of toxicokinetic modeling to recapitulate the steps that precede the biomarker presence as measured.

In 1995, the USEPA derived an RfD for MeHg based on data from a poisoning episode in Iraq that reported clinical neurologic signs in 81 mother-and-child pairs. In its 2001 assessment, the US EPA used cord blood as the biomarker for a risk assessment of the Faroe Islands data (US EPA, 2001) and developed an RfD because a comparison of the analyses based on hair and cord blood in the study suggested that the cord-blood measure explained more of the variability in more of the outcomes than hair Hg. Maternal daily dietary intake levels were used as the dose surrogate for the observed developmental effects in the children exposed in utero. The daily dietary intake levels were calculated from blood concentrations measured in the mothers with supporting additional values based on their hair concentrations. The US EPA used the *K* power model for BMD analysis, with the constraint that $K \geq 1$. This model response was set at the lowest 5% (fifth percentile) of children. The BMDL values were converted into ingested daily amounts that would result in exposure to the developing fetus at the BMDL levels in terms of ppb Hg in blood. The one-compartment model for dose conversion is used. The concentration in blood (c) corresponds to the $BMDL_{05}$.

$BMDL_{05s}$ were developed from a number of endpoints, all indications of neuropsychological processes involved in a child's ability to learn and process information, in terms of cord-blood Hg. The BMDLs for these scores were all within a relatively close range. Hair mercury was converted into blood mercury using a 250:1 ratio and an assumption of equivalent maternal and cord levels. $BMDL_{05}$ values were developed in the range of 46–79 ppb in maternal blood for different neuropsychological effects in the offspring at 7 years of age, corresponding to a range of maternal daily intakes of 0.857–1.472 μg/kg-day. The RfDs were developed from $BMDL_{05s}$ with application of a UF of 10 and a modifying factor of 1 to account for pharmacokinetic variability and uncertainty in estimating an ingested mercury dose from cord-blood mercury concentration, and a factor of 3 for pharmacodynamic variability and uncertainty. The calculated RfD values of all three studies converge at the same point: 0.1 μg/kg/day. The PODs were the BMDL for a number of endpoints from the Faroe Islands study, with supporting analyses from the New Zealand study, and an integrative analysis of all three recent large epidemiological studies. This convergence provides reassurance that additional UFs, other than for intraspecies variability, are not needed.

WHO (2007) evaluated MeHg using data from studies included in the NRC (2000) evaluation plus studies published between 2004 and 2006, investigating correlations among a number of biomarkers. The summarized correlations included additional data on MeHg in maternal hair, hair at 7 and 14 yrs, blood at 7 and 14 yrs, maternal and cord blood, and placental tissues. The original PTWI for MeHg (3.3 μg/kg bw) was revised at the 61st meeting of the Joint FAO/WHO Expert Committee on Food Additives (JECFA) to 1.6 μg/kg bw, based on an assessment of results from various epidemiological studies involving fish-eating populations and developmental neurotoxicity (WHO, 2004) and confirmed in 2007 and 2010. Specifically children 5.5- to 7-years old were assessed for neurodevelopmental endpoints, and maternal hair Hg levels were measured. WHO (2007) stated: "An average BMDL/NOEL of 14 mg/kg (14 μg/g) was derived for concentrations of mercury in maternal hair in the studies of neurodevelopmental effects, which was calculated to arise from a daily Hg intake of 1.5 μg/kg bw. The PTWI was derived by dividing this intake by a total UF of 6.4 to give a value of 1.6 μg/kg bw. For adults, the Committee considered that intakes of up to two times higher than the existing PTWI would not pose any risk of neurotoxicity, although in the case of women of childbearing age, intake should not exceed the PTWI to protect the embryo and fetus. Concerning infants and children up to 17 years, no firm conclusions may be drawn regarding their sensitivity compared with that of adults. While they are clearly not more sensitive than the embryo or fetus, they may be more sensitive than adults due to continuing neurodevelopment in infancy and childhood. Therefore, the Committee could not identify a level of intake higher than the existing PTWI that would not pose a risk of developmental neurotoxicity."

In its scientific opinion on the risk for public health related to the presence of Hg and MeHg in food, the European Food Safety Authority (EFSA, 2012) Panel on Contaminants in the Food Chain reevaluated the WHO PTWI for MeHg of 1.6 μg/kg and set a tolerable weekly intake (TWI) of 1.3 μg/kg bw/week after considering the counteracting effects of omega-3 fatty acids in fish. The panel analyzed THg converted into MeHg inorganic Hg by applying conversion factors based on the MeHg/THg proportion derived from literature data, using a conservative approach. For fish meat, fish products, fish offal, and unspecified fish and seafood a conversion factor of 1.0 was used for MeHg and 0.2 for inorganic Hg. For crustaceans, molluscs, and amphibians the conversion factor was 0.8 for MeHg and 0.5 for inorganic Hg. For all other food categories apart from "fish and other seafood," THg was regarded as inorganic Hg.

More recent data from the Seychelles Child Development Study showed no consistent pattern of adverse associations present between prenatal MeHg exposure and detailed domain-specific neurocognitive and behavioral testing at 17 years of age (Davidson et al., 2011) and no impaired autonomic heart rate control in young Seychellois adults 19 years of age (Périard et al., 2015). Two prospective cohort studies provided quantitative data that showed neurodevelopment benefits of the omega-3 fatty acid, docosahexaenoic acid, in children following maternal fish consumption during gestation (WHO, 2010c). There is the Avon Longitudinal Study of Parents and Children (ALSPAC) (Hibbeln et al., 2007) which included 7223 mother–child pairs in England, and the Project Viva (Oken et al., 2008) which included 341 mother–child pairs in the United States, both of which showed dose–response relationships between maternal fish consumption and child verbal IQ gains.

Few data are available on estimates of internal (blood) versus external exposure (dietary) for MeHg from fish consumption, especially in children. You et al. (2018) evaluated the potential health risk of MeHg in Taiwanese children from fish and seafood consumption using estimations of internal exposures and dietary intakes. Seafood is a major part of the Taiwanese diet among all ages of the general population, and its consumption was estimated at approximately twice that of meat and twice that of eggs in 2014 (Taiwan CoA, 2017). Recent studies were reported in adults relating to measurements of MeHg in blood and hair and their possible association with fish and seafood consumption (Hsi et al., 2016; Lee et al., 2012); however, data on MeHg exposure assessment in children and adolescents are limited.

In the study of You et al. (2018), Hg was measured in WB and RBCs of children obtained from two biomonitoring programs. Internal exposures from blood measurements were assessed for preschool children aged 4–6 years, school children aged 7–12 years, and adolescents aged 13–18 years using data from the Nutrition and Health Survey in Taiwan (NAHSIT) 2005–08 (7- to 18-year-olds), and the Taiwan Maternal and Infant Cohort Study (TMICS) 2003 (4- to 6-year-olds). A total of 189 children were included in the study from the overall 815 subjects aged 3 to over 63 in the database. Consumption of fish and seafood was assessed using data from the Taiwan National Food Consumption Database (TNFCD). To assess external exposure from fish and seafood consumption, the daily intake rates of fish and seafood in children were reanalyzed individually for subjects 0–3, 4–6, 7–12, and 13–18 years old. Data on THg concentrations in edible fish and seafood species were collected from published studies in Taiwan on regional species. Three categories of fish and seafood were included as freshwater fish, saltwater fish, and shellfish, and cephalopods and crustaceans (SCC). The THg concentrations in each fish and seafood product were compiled and converted into MeHg concentrations by applying a conversion factor of 1.0 for freshwater and saltwater fish, and 0.8 for SCC (EFSA, 2012). A toxicokinetic model (WHO, 1990) was used to estimate the MeHg body burden in the study subjects. The potential risk for children from exposure to MeHg from fish and seafood consumption was estimated using the hazard index (HI) approach. The results indicated that the highest median daily MeHg intake was estimated in children consuming saltwater fish. A median HI of greater than one, based on WB-THg, was found in 28% of 4- to 6-year-old children. The internal exposure estimates based on WB-THg, though slightly higher, were comparable to intake estimates based on consumption data. The results support the use of dietary exposure estimates as surrogates for internal dose estimates using blood measurement of MeHg in children in Taiwan.

The fish consumption guidelines for women of childbearing age is particularly important as risks during pregnancy may exist from a single-meal exposure if the fish tissue concentration is high enough (Ginsberg and Toal, 2000). The Hg hair concentration associated with the US EPA RfD is 1.1 ppm. Ginsberg and Toal (2000) used model simulations of the single-meal scenario at different fish MeHg concentrations and found that concentrations of 2.0 ppm or higher can be associated with maternal hair concentrations elevated above the RfD level for days to weeks during gestation. A single meal of 2 ppm is projected to elevate hair concentrations to 1.1 ppm for a brief period (several days), whereas consumptions above 2 ppm are projected to cause elevated hair concentration for 1–2 weeks. This would be in addition to the body burden from intake of other fish sources.

The exposure–effect data on MeHg from human fish-eating populations are used in health risk assessments, and regulatory, risk management, and public health outreach and policy decisions. The US EPA RfD has been used to guide related activities providing development of fish advisories which are fish consumption guidelines that present restrictive and safe consumption advice to prevent excessive exposure to MeHg (US EPA, 2018b). US EPA (2018b) and US FDA (2018) have issued joint advice regarding eating fish. This advice is geared toward helping women who are pregnant or may become pregnant, as well as breastfeeding mothers and parents of young children, to make informed choices when it comes to fish that is healthy and safe to eat. Health advisories present recommendations on how much of the affected fish in the affected areas can be safely eaten. Women of childbearing age and children are encouraged to be especially careful about following this advice because of the greater sensitivity of fetuses and children to MeHg.

Currently, most states in the United States have issued advisories concerning consumption of certain fish in various water bodies due to findings of MeHg and its toxicity (US EPA, 2018c). On the other hand, it is well recognized that fish is a key source of dietary protein in much of the world, and important polyunsaturated fatty acids. MeHg contamination of fish has the potential to impact the health of geographically diverse populations, the market for fish and seafood and the dietary choices of Americans. The US FDA and American Heart association have developed dietary guidelines on safety of eating fish, relating to the type of fish with low or high mercury levels, frequency of consumption, and meal portions, with special precautionary guides to women of childbearing age and infants and children.

BIOMARKERS OF EFFECT

Biomarkers of effect are indicators of a change in biologic function in response to a chemical exposure and can more directly relate to insight into the potential for adverse health effects compared with biomarkers of exposure. These have been less universally used than exposure markers. One example of a biomarker of effect is blood ChE activity, which can become depressed following exposure to ChE-inhibiting pesticides such as organophosphate (OPs) and N-methyl carbamate pesticides.

Acetylcholinesterase (AChE) activity measured in RBCs is an example of a bioindicator with the adverse outcome/biological process being toxicity due to AChE inhibition (AChEI) (US EPA, 2018a). The effect may be reversible if the specific exposure is discontinued. As the biomarker events are in a continuum, in many cases, a clear distinction among some effective dose markers, early effect markers, and markers of adverse effects cannot be readily made. Various levels of RBC ChE activity inhibition represent different levels of severity of effects, which can eventually lead to death at very high effect levels.

CHOLINESTERASE INHIBITION

This discussion focuses on the recent status of cholinesterase inhibition (ChEI) as a critical endpoint for toxicology evaluation, risk assessment, and regulation. It relates to ChE-inhibiting pesticides as a group, and specifically recent developments on chlorpyrifos which has been getting intensive attention. The current issues of interest are as follows: (1) adequacy of using ChEI as an endpoint for risk assessment and health protection; (2) finding of neurodevelopmental effects at exposure levels lower than those producing ChEI for chlorpyrifos; (3) estimated risks following use of cord blood time-weighted average (TWA) as an internal biomarker with a first-time use of a physiologically based pharmacokinetic–pharmacodynamic (PBPK-PD) model; and (4) US EPA's regulatory actions on chlorpyrifos which have been viewed by some as not consistent with the scientific findings.

OPs, including chlorpyrifos, are known for their mode of action (MOA) being ChEI with toxicity manifested as central and peripheral cholinergic effects producing a steady-state AChEI (US EPA, 2018d; Fan, 2014). After repeated dosing at the same level, the degree of inhibition comes into equilibrium with the production of new, uninhibited enzyme. At this point, the amount of AChEI at a given dose remains relatively consistent across duration. In general, OPs reach steady state within 2–3 weeks. Therefore, for OPs it is appropriate to assess steady-state exposure durations (up to 21 days) instead of longer-term exposures. The steady-state POD is protective of longer exposure durations, including chronic exposures. In addition, blood ChEI by ChE-inhibiting pesticides can be also considered a biomarker of exposure in identifying the occurrence of exposure to these pesticides, stratifying individuals according to intensity of exposure, and identifying actions to prevent further exposure.

As discussed in Fan (2014), the US EPA derived an acute population oral adjusted dose of 0.0036 mg/kg using a $BMDL_{10}$ of 0.36 mg/kg as the POD associated with RBC ChEI in male and female rat pups exposed to chlorpyrifos. This includes the consideration of new epidemiological studies in mothers and children regarding the relation between gestational exposure to chlorpyrifos and adverse neurodevelopmental effects in infants and children, pending more evaluation. A chronic population oral adjusted dose of 0.0003 mg/kg was derived using a $BMDL_{10}$ of 0.03 mg/kg/d as the POD based on developmental neurotoxicity data and RBC ChEI in pregnant rats.

The question and concern has been raised for a long time as to whether human-exposure limits for ChE-inhibiting pesticides, based on the detection of ChEI, are sufficient to provide heath protection, especially in children. Recent findings on the effect of chlorpyrifos on brain development have intensified this concern and associated discussions (Rauh, 2018). Chlorpyrifos is one of the chemicals evaluated under US EPA's Cumulative Risk Assessment (US EPA, 2006). It is a broad spectrum, chlorinated OP insecticide, acaricide, and miticide with registered uses including a large variety of food crops, and nonfood use settings. General public and occupational exposures are two major areas of concern for potential adverse effects. A study of its MOA has identified brain ChEI as the most appropriate dose metric for risk assessments. The current issue is the new finding of neurodevelopmental effects in children below the exposure level that caused ChEI (USEPA, 2016, 2018d).

Since its first registration in 1965, chlorpyrifos has been reviewed by US EPA for tolerance reassessment, reregistration, and most recently, as part of its ongoing registration review (US EPA, 2018b). Some major regulatory and manufacturer volunteered activities were undertaken in 1996, 2000, 2001, 2002, and 2012 to address environmental, public, and worker exposure and the associated health risks. In 2014, as part of the registration review process, the US EPA completed a revised human health risk assessment for all chlorpyrifos uses, which updated the June 2011 preliminary human health risk assessment, and was further updated in 2016.

In March 2017, the USEPA denied a petition to revoke all pesticide tolerances (maximum residue levels in food) for chlorpyrifos and cancel all chlorpyrifos registrations (US EPA, 2018b). The agency concluded that despite several years of study, the science addressing neurodevelopmental effects remains unresolved and further evaluation of the science during the remaining time for completion of registration review is warranted. As a part of the ongoing registration review, the agency will continue to review the science addressing neurodevelopmental effects of chlorpyrifos and intends to complete the assessment by the statutory deadline of October 1, 2022, in the program that reevaluates all pesticides on a 15-year cycle.

For OPs, US EPA has used the guidelines for detecting AChEI to set safe exposure limits, including for chlorpyrifos. The recent concern focuses on data from studies in animals and more recent studies in humans show that OPs are developmental neurotoxicants even at exposure levels below the threshold for systemic toxicity due to AChEI (Rauh, 2018; US EPA, 2016). The results of these studies implicate other pathogenic mechanisms involving widespread disruption of neural cell replication and differentiation, axonogenesis and synaptogenesis, and synaptic function (Slotkin and Seidler, 2007). In light of such mechanisms, all related to processes required for later cognitive and behavioral tasks, there is good reason to be concerned about possible longer-term, potentially irreversible effects (Rauh, 2018). Rauh (2018) noted that among the most worrisome findings are the corroborative results from several prospective cohort studies of children, which show an inverse dose−response effect of prenatal exposure to chlorpyrifos on cognition at 7 years of age (Rauh et al., 2011). In a subgroup of inner-city children 5.9−11.2 years of age from one of the cohorts, investigators found structural anomalies in the brains of the most highly exposed children in the superior temporal, posterior middle temporal, and inferior postcentral gyri bilaterally, and in the superior frontal gyrus, gyrus rectus, cuneus, and precuneus along the mesial wall of the right hemisphere (Rauh et al., 2012). Cognitive and behavioral processes subserved by these cortical regions include attention, receptive language, social cognition, reward, emotion, and inhibitory control; complementary functional deficits were also observed in a number of auditory attention skills. Deformations were detected in the dorsal and mesial surfaces of the left superior frontal gyrus, which supports executive function—a critical set of mental skills that permit people to plan, organize, and complete tasks throughout life (Rauh et al., 2012, 2018). In addition, persistent motor deficits among children have been linked to high early exposure to chlorpyrifos (Silver et al., 2017).

The 2014 US EPA revised human health risk assessment used dose−response data on AChEI in laboratory animals and recent data in humans to derive a POD using a new approach. This assessment incorporated the following: (1) a PBPK-PD model for deriving toxicological PODs based on 10% RBC AChEI and (2) evidence on neurodevelopmental effects in fetuses and children resulting from chlorpyrifos exposure as reported in epidemiological studies, particularly the results from the Columbia Center for Children's Environmental Health (CCCEH) study on pregnant women which reported an association between fetal cord blood levels of chlorpyrifos and neurodevelopmental outcomes. The assessment retained the 10X Food Quality Protection Act (FQPA) Safety Factor (SF) for children because of the uncertainties that neurodevelopmental effects may be occurring at doses lower than those that cause 10% RBC AChEI and used for the POD.

For the 2016 assessment, the the use of US EPA vs. USEPA cord blood data quantitatively for deriving PODs (US EPA, 2016, 2018d). Following consultation with the FIFRA panel which concluded that epidemiology and toxicology studies suggest there is evidence for adverse health outcomes associated with chlorpyrifos exposures below levels that result in 10% RBC AChEI, which was used as the POD in the USEPA's 2014 assessment and for the 2015 proposed revocation rule, USEPA used the TWA blood level as the internal dose for determining separate PODs for infants, children, and adults exposed to chlorpyrifos. These separate PODs were calculated by PBPK modeling for dietary (food, drinking water), residential, and occupational exposures. The TWA blood concentration of chlorpyrifos for the CCCEH study cohort as the POD for risk assessment was recommended because the window(s) of susceptibility are currently not known for the observed neurodevelopmental effects, and the uncertainties associated with quantitatively interpreting the CCCEH cord blood data. With the exception of the acute (single day) exposure for non-occupational bystander post-application inhalation exposures, only steady state (repeat) exposure durations are considered in this assessment as assessing the steady-state exposure duration most closely matches the TWAs calculated for the PODs. The PODs derived from the TWA blood level are protective of any additional acute exposures to chlorpyrifos.

The TWA blood level resulting from chlorpyrifos exposure from the crack and crevice scenario is considered a LOAEL rather than a NOAEL, because this is the exposure level likely to be associated with neurodevelopmental effects reported in the CCCEH study (US EPA, 2016, 2018d). As a NOAEL has not been identified, the 10X factor for extra fetal/infant/young child protection (the FQPA factor) has been retained to account for the uncertainty in using a LOAEL in this assessment. The revised risk assessment also applies a 10X UF for intraspecies variability. With the use of the PBPK-PD model which accounts for the pharmacokinetic and pharmacodynamic differences between animals and humans to derive PODs, the interspecies factor is reduced to 1X from the typical 10X when animal data are used. Therefore, the total UF is 100X.

For dietary assessment, PODs are divided by the total UF (100) to derive a population adjusted dose (PAD). The chlorpyrifos exposure values resulting from dietary modeling are compared to the PAD. The steady-state PADs (ssPAD, μg/kg-d) are 0.002 for infants <1, 0.0017 for children 1–2 yrs, 0.0012 for children 6–12, and 0.0012 for females 13–49 yrs. Estimated exposures were found to be 6200 to 14,000% of the ssPAD.

The CCCEH study, with supporting results from two other US cohort studies and the seven additional epidemiological studies reviewed in 2015, provides sufficient evidence that there are neurodevelopmental effects occurring at chlorpyrifos exposure levels below that required for AChEI. The CCCEH study is primarily tested for the presence of chlorpyrifos in cord blood and is most relevant for the purposes of chlorpyrifos risk assessment. When comparing high to low exposure groups at 3 years of age in the CCCEH study, there were increased odds of mental delay, psychomotor delay, attention disorders, attention deficit hyperactivity disorder, and pervasive developmental disorders. In a follow-up study at age 11, CCCEH study authors observed increased odds of mild to moderate tremor when comparing high to low exposure groups. Evaluation of the relationship between prenatal chlorpyrifos exposure and neurodevelopment in 265 of the CCCEH cohort participants at age 7 years revealed that the log of Working Memory Index of children was linearly associated with concentration of chlorpyrifos in cord blood. For each standard deviation increase in exposure (4.61 pg/g), the authors observed a 1.4% reduction in Full-Scale IQ and a 2.8% reduction in Working Memory.

Presently, the US EPA will continue to review the science addressing neurodevelopmental effects of chlorpyrifos and plans to complete the assessment by October 1, 2022.

SELENIUM

For toxicity, the biomarker of effect for Se is selenosis (Fan, 2014). The present discussion updates recent toxicological findings and discusses the relationship of toxicity to the nutritional values of Se and biomarkers of Se status.

Selenium (Se) occurs naturally and is found ubiquitously in the environment, being released from both natural and anthropogenic sources (Fan and Vinceti, 2013; ATSDR, 2015; Combs, 2015). Se exists in various chemical forms and oxidation states that can affect its occurrence, properties, uses, and toxicity. Soils contain inorganic selenites and selenates that plants accumulate and convert to organic forms, mostly selenocysteine (SeCcys) and selenomethionine (SeMet) and their methylated derivatives. Ingestion of the organic form from food, including dietary supplements, is the main source of exposure for the general population. Dietary intake is mainly as the amino acid SeMet in grains, cereals, and forage crops; as SeCys in meats and dairy products to a lesser extent; and as the inorganic selenate and selenite, which are the main soluble forms in aqueous media. Se is an essential nutrient, and evaluation of human exposure and health effects or toxicity has to consider both essentiality and toxicity.

Selenosis, presented primarily as hair loss and nail changes, is the biomarker of Se toxicity which is used as the basis for risk assessment and regulation of the chemical in drinking water (Fan, 2014; US EPA, 2018f). Major findings on selenosis were documented in local populations in studies carried out in the endemic seleniferous areas of Enshi County, Hubei province in central China in the 1980s. Residents who consumed food grown in areas with unusually high environmental concentrations of Se in soil contaminated by stony coal showed a wide spectrum of hair, nail, and skin abnormalities related to Se exposure, plus other clinical signs and gastrointestinal symptoms (Yang et al., 1983, 1989a). Persistent clinical signs of selenosis were observed only in 5/349 adults, a potentially sensitive subpopulation. The dermatological abnormalities were ascribed to daily Se intake ranging from 1500 to 5000 μg/day, though nail brittleness was detected in a subsequent study following supplementation with 600 μg/day of Se.

The biomarker of effect was used for risk assessment to derive reference levels for regulations (e.g., 50 ppb in drinking water, 50 ppb in bottled water) and public health policies (e.g., health guidance on dietary intake and fish consumption advisories). Based on data obtained from the Se poisoning episode in China, the US FDA (Poirier, 1994) set a NOAEL for Se at a WB Se concentration of 1000 ng/mL, corresponding to a dietary intake of 853 μg/day in an adult male. The US EPA (1991, 2018f) established an RfD of 5 μg/kg/day based on clinical selenosis observed in China as described in the studies of Yang et al. (1983, 1989a, 1989b), with a NOEL of 15 μg/kg/day and an UF of 3. The US Institute of Medicine (IOM, 2000) set the LOAEL of Se intake at 900 μg/day, and NOAEL at 800 μg/day. A UF of 2 is used based on determination that the adverse health effects observed at the LOAEL as not severe, but likely not readily reversible, thus establishing a tolerable upper intake level (UL) of Se from food and supplements at 400 μg/day for adults 14 and older. The ULs for other age groups are 1–3 yrs, 90 μg/day; 4–8, 150 μg/day; 9–13, 280 μg/day.

The Norwegian Scientific Committee for Food Safety (Vitenskapskomiteen for mattrygghet, VKM, 2017) evaluated the intake of Se in the Norwegian population and reviewed four risk assessments undertaken by the Institute of Medicine (IOM, USA), Scientific Committee on Food (SCF, European Union), Expert Committee on Vitamins and Minerals (EVM, United Kingdom), and the Nordic Nutrition Recommendations (NNR, Nordic Project Group). It decided to use the tolerable Upper intake Levels (ULs) of 300 μg/day for adults set by the SCF, including pregnant and lactating women, later adopted by NNR. The UL was extrapolated to children and adolescents that range from 60 (1–3 years) to 250 μg Se/day (15–17 years).

Se was originally known mainly for its high toxicity, then recognized as an essential trace element and studied for its protective role in human health and diseases. More recently, interest has increased regarding its protective role and possible link to human diseases, including studies of the potential role of Se compounds in the prevention of specific cancers and antitumorigenic effects and its link to human diseases. Se supplementation in animal nutrition led to the consideration of possible inadequate dietary Se in human health.

The essential functions of Se for human and animal health are mediated through selenoproteins (NIH, 2018). Se is a constituent of more than two dozen selenoproteins that play critical roles in reproduction, thyroid hormone metabolism, DNA synthesis, and protection from oxidative damage and infection. Most Se is in the form of SeMet in animal and human tissues, where it can be incorporated nonspecifically with the amino acid methionine in body proteins. The most commonly used measures of Se status are plasma and serum Se concentrations. Concentrations in blood and urine reflect recent Se intake. Analyses of hair or nail Se content can be used to monitor longer-term intakes over months or years. Quantification of one or more selenoproteins (such as glutathione peroxidase and selenoprotein P) is also used as a functional measure of Se status. Plasma or serum Se concentrations of 8 μg/dL or higher in healthy people typically meet the needs for selenoprotein synthesis.

Biomarkers for Se status have recently been discussed by Combs (2015). In animal studies and accidental exposures of humans, clinical indicators were reported, but few biomarkers were available with predictive value. The default approach has been to use as risk indicators the highest Se tissue levels observed with no associated adverse effects. Most studies show no adverse effects in human subjects with plasma Se levels <1000 ng/mL. Some cases showed garlic odor of the breath due to the excretion of volatile Se metabolites across the lung, but measurement of such metabolites has not been used in cases of selenosis and diagnostic criteria have not been established.

Combs (2015) also noted that the multiple biological activities of Se call for biomarkers that can provide information about Se status relative to nutritional (functional) needs, antitumorigenic potential, and risk of adverse effects. He described two types of Se biomarkers. Se biomarkers such as GPX3, GPX1, and SEPP1 provide information about function directly. GPX3 is the extracellular glutathione peroxidase, and it reduces hydroperoxides using reducing equivalents from reduced glutathione. SEPP1 is the Se-transporter selenoprotein P produced and excreted by the liver, functioning as the primary transporter of Se to peripheral tissues, and comprising 40%–60% of total plasma Se. These biomarkers are useful under conditions of Se intake within the range of regulated selenoprotein expression, which for humans is less than about 55 μg/day (e.g., in parts of China, New Zealand, and Europe), and for animals is less than about 0.1 mg/kg diet. These biomarkers are also useful in identifying nutritional Se deficiency and tracking responses of deficient individuals to Se treatment. Selenocysteine occurs in only two plasma selenoproteins, SEPP1 and GPX3, which contain 10 and 4 SeCys residues, respectively. Other Se biomarkers provide information indirectly through inferences based on Se levels in foods, tissues, urine, or feces. They can indicate the likelihood of deficiency or adverse effects, but they do not provide direct evidence of either condition. For example, SeMet in plasma and tissues can indicate the amount of Se potentially available for functional use, the likelihood of Se deficiency, or the likelihood of adverse effects of Se, but it provides no direct evidence of either of these states. These biomarkers are useful in providing information about Se status over a wide range of Se intake, particularly from food forms, over which tissue retention of SeMet is unregulated.

Relatively low Se intakes determine the expression of selenoenzymes in which Se serves as an essential constituent, whereas higher intakes have been shown to have antitumorigenic potential; and very high Se intakes can produce adverse effects. Se status, as in nutrition, refers to the amount of biologically active or potentially active Se as a nutrient in the body. Nutritional status is a product of a nutrient's intake, retention, and metabolism. It includes the nutrient pool that is metabolically functional and, thus, the most nutritionally relevant, as well at that pool that can be readily mobilized to functional forms. Combs (2015) described Se status as having four components: Se intake, tissue Se, Se excretion, and Se function, and there are biomarkers associated with each status. Se status is assessed for several purposes such as to determine the risk of nutritional Se deficiency; to estimate the potential for reducing cancer risk; and to monitor the risk of adverse effects associated with excess Se. Such assessments are made in research and clinical care, and they may also provide useful information for public health programs. All these purposes can rely on different sets or interpretations of biomarkers.

In examining toxicity and beneficial effects and indicators of Se status, Vinceti et al. (2017) reviewed environmental studies conducted in populations characterized by abnormally high or low Se, and from high-quality and large randomized controlled trials (RCTs) with Se carried out in the United States and in other countries after the poisoning reports in China in the 1980s. The authors noted that these studies indicate that the minimal amount of environmental Se which is a source of risk to human health is much lower than anticipated on the basis of older studies, with toxic effects shown at levels of intake as low as around 260 μg/day for organic Se and around 16 μg/day for inorganic Se. On the other hand, populations with average Se intake of less than 13–19 μg/day appear to be at risk of Keshan disease, a severe cardiomyopathy. Regarding selenoproteins, the authors noted that the relations between Se exposure and selenoprotein activity are complex, and a clear relation between selenoprotein activity and health outcomes has not been established in epidemiologic studies. Although the amount and activity of antioxidant selenoproteins, particularly glutathione peroxidases and plasma selenoprotein P, have been used as indirect indicators of Se intake, and detection of low levels has been interpreted as a consequence and a biomarker of Se deficiency, a proteomic approach based on maximal upregulation of selenoprotein levels by ingested Se has not been adopted for regulatory purposes or used for setting the dietary reference values for this element.

A subject of intense interest is the potential antitumorigenic effect of Se. Vinceti et al. (2017) reported that recent trials have consistently shown that Se does not modify risk of overall cancer, prostate cancer, and other specific cancers, whereas it may even increase risk of cancers such as advanced or overall prostate cancer, nonmelanoma skin cancer, and possibly breast cancer in high-risk women. The authors showed that at Se exposure (baseline dietary intake plus supplementation) of around 250–300 μg/day, there is an increased risk of type-2 diabetes. Such excess diabetes risk linked to Se overexposure was first discovered in trial carried out in a population with a "low" baseline Se status and later confirmed in large trials. Finally, the largest of the Se RCTs, the Selenium and Vitamin E Cancer Prevention Trial (SELECT), whose overall Se intake in the supplemented group averaged 300 μg/day, has shown that such an amount of exposure induces "minor" adverse effects such as dermatitis and alopecia, a long-recognized sign of Se toxicity.

National drinking water regulations establish a Maximum Contaminant Level, MCL, of 50 ppb for Se in drinking water based on risk assessment of exposure to Se and findings on hair or fingernail loss, numbness in fingers or toes, and circulatory problems as observed in cases of selenosis (US EPA, 2018e). The public health goal in California, which is used as the health basis for California regulation, is established at 30 ppb based on the same toxicology data, but using drinking water consumption rate for infants (OEHHA, 2010). No special sensitivity to Se is found in infants or children (ATSDR, 2015).

BIOMARKERS OF SUSCEPTIBILITY

Kelly and Vineis (2014) defined biomarkers of susceptibility as indicators of an elevated sensitivity to the effects of an environmental agent that can be objectively measured in a biological system or a sample. These markers are used to identify "at-risk" individuals, where risk is either acquired (e.g., they indicate a disorder that renders people more susceptible to an environmental exposure) or, more usually, inherited (as indicated by genetic markers). Genetic markers in particular play a vital role in understanding disease risk. Most complex disorders, including cancers, arise not from single gene variants or single exposures, but rather from the interplay between an individual's genetic susceptibilities and their personal exposure histories. Underlying genetic differences may influence the uptake, metabolism, retention, and excretion of chemical carcinogens and therefore susceptibility to their effects. Variants of critical cell function proteins, including those underlying differentiation, cell division, and cell death, are also important and can represent important susceptibilities. Another susceptibility mechanism is the presence of higher or lower levels of antioxidants, such as glutathione, that reduce the level of oxidizing radicals in the tissue where damage occurs or in the liver, where a wide variety of metabolism of xenobiotics can be impacted. Cellular repair mechanisms, including DNA-adduct repair, also play a role in susceptibility. Consequently, variants in the genes of all of these processes are candidates for susceptibility biomarkers and gene–environment interactions.

Kelly and Vineis (2014) explored the issues involved in the study of gene–environment interactions which complicate the design and the interpretation of findings and consider how the results emerging from these studies can be utilized to identify novel biomarkers of susceptibility in cancer epidemiology, using non-

Hodgkin's lymphoma (NHL) as an example. The results suggest a modifying role for genetic susceptibility to a number of occupational and environmental exposures including organochlorines, chlorinated solvents, chlordanes, and benzene in the etiology of NHL. They noted that the potential importance of these gene–environment interactions in NHL may help to explain the lack of definitive carcinogens identified to date for this malignancy. Although a large number of genetic variants and gene–environment interactions have been explored for NHL, to date replication is lacking, and therefore the findings remain to be validated. The ability to select a subset of individuals at risk to development of disease caused by a specific chemical would improve the epidemiological ability to observe effects by reducing the "noise" from individuals not at risk or at much lower risk.

The -omics technologies that have been applied in epidemiological research have expanded beyond genomics to include epigenomics, proteomics, transcriptomics, and metabolomics (NASEM, 2017). -Omics technologies have advanced the paradigm of molecular epidemiology, which focuses on underlying biology (pathogenesis) rather than on empirical observations alone. Many studies have utilized -omics technologies in epidemiological research as exemplified by the incorporation of genomics. For example, the genetic basis of disease has been explored in genome-wide association studies in which the genomic markers in people who have and do not have a disease or condition of interest are compared. New studies are being designed with the intent of prospectively storing samples that can be used for existing and future -omics technologies. From this, obtaining data from human population studies that are parallel to data obtained from in vitro and in vivo assays or studies is already possible and potentially can help in harmonizing comparisons of exposure and dose. It was also noted that -omics technologies have the potential for providing a suite of new biomarkers for hazard identification and risk assessment.

The advantages and limitations of -omics technologies have been described by Vineis et al. (2009) and adapted by NAS (NASEM, 2017) as follows. The advantages include use in large, hypothesis-free investigations of the whole complement of relevant biological molecules; better understanding of phenotype–genotype relations; and potential for insights into the effects of interactions between environmental conditions and genotypes, and mechanistic insights into disease etiology. The limitations include cost of assays, quality of biological material available (such as instability of RNAs), and the amount of labor needed.

Many techniques are still in their discovery phase and need to be carefully investigated and compared with existing biological information from in vivo and in vitro tests. Moving from promising techniques to successful application of biomarkers in occupational and environmental epidemiology and medicine requires not only standardizing and validating techniques but also appropriate study designs and sophisticated statistical analyses for interpreting study results.

CANCER-RELATED BIOMARKERS

There are two general groups of cancer-related biomarkers. The focus in this chapter is on those used to predict and measure the impact of environmental chemicals. Readers are probably also familiar with clinical cancer biomarkers, used increasingly for cancer screening, monitoring the effect of treatment, and prognosis. Some of these may also become important in the study of gene–environment and susceptibility studies.

DNA adducts represent key events in mutagenesis and carcinogenesis (Poirer, 2012; Basu, 2018). They can serve as biomarkers of exposure that are by and large repairable, whereas mutations are biomarkers of unrepaired effect that are passed on to daughter cells. Chemicals forming adducts with DNA are chemically electrophilic, reacting with DNA purine or pyrimidine bases (which are nucleophilic) in a process called alkylation (Gates, 2009). Many of them are first enzymatically transformed to a reactive chemical before the adduct can be formed. There are similar reactions with nucleophilic targets in certain amino acids in proteins and with certain lipids. Bifunctional agents (with two or more reactive sites, such as dialdehydes) can crosslink DNA bases or crosslink DNA bases to amino acids in proteins. "Bulky" adducts, such as PAHs and the liver carcinogen, aflatoxin (produced by a fungus growing particularly in peanuts), were among the first to be studied (Poirer, 2012). Reactions with oxidizing radicals (reactive oxygen species) also occur and can result in adduct formation (Gates, 2009; Kasai, 2016).

There are also endogenously produced reactive chemicals that can cause adducts with DNA or proteins. Those endogenous adducts can result in mutations that are considered to be associated with the background level cancer in conjunction with other predisposing genetic and epigenetic factors. One such chemical is formaldehyde, discussed below. However, a critical issue is whether adduct formation alone adequately captures the risk from exogenous exposure. Systemic response at the portal of entry may provide other mechanisms that contribute to carcinogenesis.

FORMALDEHYDE

Formaldehyde is a very reactive compound, an electrophile, directly targeting nucleophilic chemicals like

DNA and protein. It can induce DNA adducts including N2-hydroxymethyl-deoxyguanosine (dG), N6-hydroxymethyldeoxyadenosine (dA), and N4-hydroxymethyl-deoxycytosine (dC) in vitro. These DNA adducts are considered to be promutagenic, and formaldehyde exposure has been shown to induce mutations in bacterial test systems and laboratory animals (NRC, 2014). DNA–protein crosslinks are also formed. Both have been used as biomarkers. Formaldehyde has been linked to sinonasal cancer in laboratory animal studies and nasopharyngeal cancer, Hodgkin's lymphoma and certain leukemias in occupational settings with activities like production/use of formaldehyde–melamine resin and embalming. Laboratory animal studies also found hematotoxicity, which would be consistent with leukemogenic effects.

Formaldehyde is also produced by endogenous metabolism. Adducts formed from exogenous exposure are chemically identical to endogenous DNA adducts created from reactive chemicals produced by the body during normal metabolism. Thus, exogenous and endogenous DNA adducts with formaldehyde have the same structure and likely the same mutational efficiency.

Starr and Swenberg (2013, 2016) and their coworkers (Lu et al., 2010; Yu et al., 2015) separately quantified endogenous and exogenous adducts induced by formaldehyde (by use of exogenous carbon-13 labeled formaldehyde) and derived the relationship between the number of exogenous DNA adducts derived from the inhaled formaldehyde exposure and the number of endogenous adducts present in bone marrow. Based on 28-day exposures of rats (up to 9 ppm) and 2-day exposures (up to 6 ppm) of macaques, they found that in bone marrow there was less than one exogenous DNA adduct present for approximately 14,000 endogenous formaldehyde adducts. This is evidence that formaldehyde is not transported systemically, due to its reactivity and its metabolism. They concluded that one additional identical DNA adduct would not likely drive the biology that leads to carcinogenesis, assuming that both endogenous and exogenous dG adducts give rise to proportional tissue-specific risks of cancer development (up to the point that local formaldehyde tissue concentrations become so high as to cause cytotoxicity, saturate DNA repair, or deplete detoxification). However, the authors noted that the validity of short-term formaldehyde exposures for this analysis is not yet proven.

In their model, the statistical 95% confidence limit upper bound dose–response of exogenous formaldehyde declined linearly toward the background endogenous level and the associated background cancer level. The nasopharyngeal cancer and leukemia/lymphoma risk estimates they developed were considerably smaller than the corresponding US EPA estimates, developed from epidemiologic data for exposed workers. (Linearity also assumes that exposures are below those that cause cytotoxicity and regenerative cell proliferation, deplete detoxification processes, e.g., glutathione depletion, or saturate DNA repair pathways.) In both epidemiology and risk assessment, enhanced understanding of endogenous and exogenous DNA adduct formation, especially at low doses, can potentially be used as a molecular dosimeter to place bounds on the plausibility and potency of exogenous exposures to induce disease.

Whether all of the attributable risk is dependent on the concentration of adducts as measured following short-term exposures is not clear. An NRC expert committee reviewing formaldehyde for the National Toxicology Program 12th Report on Carcinogens (NRC, 2014) found that

1. There is sufficient evidence of carcinogenicity from studies of humans based on consistent epidemiologic findings on nasopharyngeal cancer, sinonasal cancer, and myeloid leukemia for which chance, bias, and confounding factors can be ruled out with reasonable confidence.
2. There is sufficient evidence of carcinogenicity in animals based on malignant and benign tumors in multiple species, at multiple sites, by multiple routes of exposure, and to an unusual degree with regard to type of tumor, and.
3. There is convincing relevant information that formaldehyde induces mechanistic events associated with the development of cancer in humans, specifically genotoxicity and mutagenicity, hematologic effects, and effects on gene expression.

In addition, the NRC reviewers stated that, "while it would be desirable to have an accepted mechanism that fully explains the association between formaldehyde exposure and distal cancers, the lack of such mechanism should not detract from the strength of the epidemiological evidence that formaldehyde causes myeloid leukemia", given the uncertainties in the scientific understanding of the potential mechanisms of the systemic effects of formaldehyde.

Based on the NRC review (2014), areas that need further investigation include (1) whether inhalation exposure results in release of cytokines that might impact homeostasis in lymphohematopoiesis; (2) the extent of systemic oxidative stress due to formaldehyde exposure, (3) the nature and degree of immune responses to formaldehyde exposure; (4) the nature and degree of local effects on leukocytes at the portal of entry (also in relation to immune responses) that could lead to leukemia or lymphoma; and, (5) changes in adduct formation, gene expression, epigenetic effects during long-term versus short-term exposure. Variation of alcohol dehydrogenase (also known as formaldehyde dehydrogenase) enzyme activity in the population may also play a role in susceptibility by allowing more formaldehyde to move from the portal of entry to systemic circulation. An epidemiological study design incorporating this susceptibility marker would be useful to more accurately measure risk. Likewise, there may be different DNA repair enzyme differences that could be examined because there is a paucity of data on that topic (NRC, 2014).

Since that review, the validity and consistency of the epidemiology, including new studies, has come under criticism that has yet to be fully resolved. Overall, existing data indicate that for formaldehyde, the MOA and ultimately the biomarkers need more research for a better understanding of the related issues. The state of science in this area is an example of the difficulties that can occur in trying to apply data from biomarkers to risk assessment when the MOA is not yet clear.

CONCLUDING REMARKS AND FUTURE DIRECTIONS

Biomarkers for environmental chemicals are being used increasingly for assessing human exposure to toxic substances. It involves sampling and measuring environmental chemicals in human fluids and tissues, such as blood and urine, as humans are exposed to chemicals through air, water, food, soil, dust, and consumer products. Correlation of dose—effect relationship provides the basis for health risk assessments which, in turn, provide the basis for regulatory and risk management decision-making for health protection.

Examples are given for biomarkers of exposure such as MeHg in blood, and biomarkers of effect such as ChEI by ChE-inhibiting pesticides, and selenosis from selenium exposure. MeHg exposure from fish consumption is widespread and few data are available specifically for infants and children. Fish consumption guidance relating to MeHg has been developed with special attention to children and women of childbearing age. A concern for evaluating ChEI-inhibiting pesticides is whether neurodevelopmental effects occur at levels below those that cause ChEI, as in the case of chlorpyrifos. Consideration of selenium toxicity needs to include consideration of its nutritional status and new data on possible adverse health effects of low-dose overexposure. Study of biomarkers of susceptibility has involved the study of gene—environment interactions and more recently the -omics technologies that have been applied in epidemiological research have expanded beyond genomics to include epigenomics, proteomics, transcriptomics, and metabolomics, to name a few. The -omics technologies have the potential for providing a suite of new biomarkers for hazard identification and risk assessment. The case of formaldehyde is an example of the difficulties that can occur in trying to apply data from biomarkers to risk assessment when the MOA is not yet clear. More research for a better understanding of the related issues is needed.

New advancements in biomonitoring, exposure science, -omics technologies, toxicology, risk assessment, and decision-making have been made and are anticipated to be made, with the further development and use of biomarkers, increased biomonitoring efforts and research, improvement of methods and techniques, and development and implementation of the visions and strategies described in several recent NRC reports that advance the conceptual and experimental approaches in these fields. Biomarkers are anticipated to play an increasing role in these fields advancing the understanding of human health and diseases, and improving human health and welfare, partly through providing data to support regulatory and risk management actions.

References

ATSDR, 2015. Toxicological Profile for Selenium. Agency for Toxic Substances and Disease Registry. Centers for Disease Control and Prevention, Atlanta, GA. Updated January 21, 2015. https://www.atsdr.cdc.gov/toxprofiles/TP.asp?id=153&tid=28.

Basu, A.K., 2018. DNA damage, mutagenesis and cancer. Int. J. Mol. Sci. 19, 970.

Budtz-Jørgensen, E., Keiding, N., Grandjean, P., 2004. Effects of exposure imprecision on estimation of the benchmark dose. Risk Anal. 24, 1689—1696.

CDC, 2018. National Biomonitoring Program. Centers for Disease Control and Prevention. Updated December 23, 2016. https://www.cdc.gov/biomonitoring/Mercury_BiomonitoringSummary.html.

Combs Jr., G.F., 2015. Biomarkers of selenium status. Nutrients 7, 2209—2236.

Davidson, P.W., Cory-Slechta, D.A., Thurston, S.W., et al., 2011. Fish consumption and prenatal methylmercury exposure: cognitive and behavioral outcomes in the main cohort at 17 years from the Seychelles child development study. Neurotoxicology 32, 711—717.

DeBord, D.G., Carreón, T., Lentz, T.J., et al., 2016. Use of the "exposome" in the practice of epidemiology: a primer on -omic technologies. Am. J. Epidemiol. 184, 302—314.

EFSA, 2012. Scientific opinion on the risk for public health related to the presence of mercury and methylmercury in food. European Food Safety Authority (EFSA). EFSA. J. 10, 2985.

Fan, A.M., 2014. Biomarkers in toxicology, risk assessment and environmental chemical regulations. In: Gupta, R.C. (Ed.), Biomarkers in Toxicology, first ed. Elsevier Inc., Burlington, pp. 1057—1080.

Fan, A.M., Vinceti, M., 2013. Selenium and its compounds. In: Harbison, R.D., Bourgeois, M.M., Johnson, G.T. (Eds.), Hamilton & Hardy's Industrial Toxicology, sixth ed. John Wiley & Sons, New Jersey, pp. 203—226.

Gates, K.S., 2009. An overview of chemical processes that damage cellular DNA: spontaneous hydrolysis, alkylation, and reactions with radicals. Chem. Res. Toxicol. 22, 1747—1760.

Ginsberg, G.L., Toal, B.F., 2000. Development of a single-meal fish consumption advisory for methyl mercury. Risk Anal. 20, 41—47.

Grandjean, P., Weihe, P., White, R.F., et al., 1997. Cognitive deficit in 7-year-old children with prenatal exposure to methylmercury. Neurotoxicol. Teratol. 19, 417—428.

Groth, E., 2010. Ranking the contributions of commercial fish and shellfish varieties to mercury exposure in the United States: implications for risk communication. Environ. Res. 110, 226–236.

Hibbeln, J.R., Davis, J.M., Steer, C., et al., 2007. Maternal seafood consumption in pregnancy and neurodevelopmental outcomes in childhood (ALSPAC study): an observational cohort study. Lancet 369, 578–585.

Hsi, H.C., Hsu, Y.W., Chang, T.C., et al., 2016. Methylmercury concentration in fish and risk-benefit assessment of fish intake among pregnant versus infertile women in Taiwan. PLoS One 11, e0155704.

IOM, 2000. Dietary Reference Intakes for Vitamin C, Vitamin E, Selenium and Carotenoids. Institute of Medicine. National Academy Press, Washington, DC.

Karagas, M.R., Choi, A.L., Oken, E., et al., 2012. Evidence on the human health effects of low-level methylmercury exposure. Environ. Health Perspect. 120, 799–806.

Kasai, H., 2016. What causes human cancer? Approaches from the chemistry of DNA damage. Gene Environ. 38, 19.

Kelly, R.S., Vineis, K.P., 2014. Biomarkers of susceptibility to chemical carcinogens: the example of non-Hodgkin lymphomas. Br. Med. Bull. 111, 89–100.

Lee, C.C., Chang, J.W., Huang, H.Y., et al., 2012. Factors influencing blood mercury levels of inhabitants living near fishing areas. Sci. Total Environ. 424, 316–321.

Lu, K., Collins, L.B., Ru, J., et al., 2010. Distribution of DNA adducts caused by inhaled formaldehyde is consistent with induction of nasal carcinoma but not leukemia. Toxicol. Sci. 116 (2), 441–451.

Mahaffey, K.R., Clickner, R.P., Bodurow, C.C., 2004. Blood organic mercury and dietary mercury intake: National Health and Nutrition Examination Survey, 1999 and 2000. Environ. Health Perspect. 112, 562–570.

Mortensen, M.E., Caudill, S.P., Caldwell, K.L., et al., 2014. Total and methylmercury in whole blood measured for the first time in the U.S. population: NHANES 2011–2012. Environ. Res. 134, 257–264.

NASEM, 2017. Using 21st Century Science to Improve Risk-Related Evaluations. National Academy of Sciences, Engineering and Medicine. The National Academies Press, Washington, DC.

NASEM, 2018. Advances in Causal Understanding for Human Health Risk-Based Decision-Making: Proceedings of a Workshop—in Brief. National Academy of Sciences, Engineering and Medicine. The National Academies Press, Washington, DC.

NIH, 2018. Selenium. Fact Sheet for Health Professionals. https://ods.od.nih.gov/factsheets/Selenium-HealthProfessional/.

NRC, 2000. Toxicological Effects of Methylmercury. National Research Council, National Academies Press, Washington, DC.

NRC, 2006. Human Biomonitoring for Environmental Chemicals. National Research Council, National Academies Press, Washington, DC.

NRC, 2007. Toxicity Testing in the 21st Century: A Vision and a Strategy. National Research Council, National Academies Press, Washington, DC.

NRC, 2012. Exposure Science in the 21st Century: A Vision and a Strategy. National Research Council, National Academies Press, Washington, DC.

NRC, 2014. Review of the Formaldehyde Assessment in the National Toxicology Program 12th Report on Carcinogens. National Research Council, National Academies Press, Washington, DC.

OEHHA, 2010. Public Health Goal for Selenium in Drinking Water. Office of Environmental Health Hazard Assessment. California Environmental Protection Agency, CA.

Oken, E., Radesky, J.S., Wright, et al., 2008. Maternal fish intake during pregnancy, blood mercury levels, and child cognition at age 3 years in a US cohort. Am. J. Epidemiol. 167, 1171–1181.

Périard, D., Beqiraj, B., Hayoz, D., et al., 2015. Associations of baroreflex sensitivity, heart rate variability, and initial orthostatic hypotension with prenatal and recent postnatal methylmercury exposure in the Seychelles Child Development Study at age 19 years. Int. J. Environ. Res. Publ. Health. 12, 3395–3405.

Poirier, K.A., 1994. Summary of the derivation of the reference dose for selenium. In: Mertz, W., Abernathy, C.O., Olin, S.S. (Eds.), Risk Assess Ess Elem. Intl Life Sciences Inst, Washington, pp. 157–166.

Poirer, M.C., 2012. Linking DNA adduct formation and human cancer risk in chemical carcinogenesis. Environ. Mol. Mutagen. 57, 499–507.

Rauh, V.A., 2018. Polluting developing brains — EPA failure on chlorpyrifos. N. Engl. J. Med. 378, 1171–1174.

Rauh, V., Arunajadai, S., Horton, M., et al., 2011. Seven-year neurodevelopmental scores and prenatal exposure to chlorpyrifos, a common agricultural pesticide. Environ. Health Perspect. 119, 1196–1201.

Rauh, V.A., Perera, F.P., Horton, M.K., et al., 2012. Brain anomalies in children exposed prenatally to a common organophosphate pesticide. Proc. Natl. Acad. Sci. USA. 109, 7871–7876.

Ruiz, P., Ray, M., Fisher, J., et al., 2011. Development of a human physiologically based pharmacokinetic (PBPK) toolkit for environmental pollutants. Int. J. Mol. Sci. 12, 7469–7480.

Sheehan, M.C., Burke, T.A., Navas-Acien, A., et al., 2014. Global methylmercury exposure from seafood consumption and risk of developmental neurotoxicity: a systematic review. Bull. World Health Organ. 92, 254–269.

Silver, M.K., Shao, J., Zhu, B., et al., 2017. Prenatal naled and chlorpyrifos exposure is associated with deficits in infant motor function in a cohort of Chinese infants. Environ. Int. 106, 248–256.

Slotkin, T.A., Seidler, F.J., 2007. Comparative developmental neurotoxicity of organophosphates in vivo: transcriptional responses of pathways for brain cell development, cell signaling, cytotoxicity and neurotransmitter systems. Brain Res. Bull. 7, 232–274.

Starr, T.B., Swenberg, J.A., 2016. The bottom-up approach to bounding potential low-dose cancer risks from formaldehyde: an update. Regul. Toxicol. Pharmacol. 77, 167–174.

Starr, T.B., Swenberg, J.A., 2013. A novel bottom-up approach to bounding low-dose human cancer risks from chemical exposures. Regul. Toxicol. Pharmacol. 65, 311–315.

Taiwan CoA, 2017. Food Supply and Utilization Annual Report. Council of Agriculture, Taiwan. http://eng.coa.gov.tw/ws.php?id=2503642.

US EPA, 1991. Selenium and Compounds: Integrated Risk Information System (IRIS). US Environmental Protection Agency, Washington, DC. Update June 1, 1991. https://cfpub.epa.gov/ncea/iris/iris_documents/documents/subst/0472_summary.pdf.

US EPA, 2001. Methylmercury (MeHg): Integrated Risk Information System (IRIS). US Environmental Protection Agency, Washington, DC. Update July 27, 2001. https://cfpub.epa.gov/ncea/iris/iris_documents/documents/subst/00073_summary.pdf.

US EPA, 2006. Organophosphorus Cumulative Risk Assessment 2006 Update. US Environmental Protection Agency, Office of Pesticide Programs. http://www.epa.gov/pesticides/cumulative/2006-op/op_cra_main.pdf.

US EPA, December 29, 2014. Chlorpyrifos: Revised Human Health Risk Assessment for Registration Review. US Environmental Protection Agency, Washington, DC. Memorandum from Drew et al., 12/29/2014. D424485.

US EPA, 2016. Chlorpyrifos: Revised Human Health Risk Assessment for Registration Review. US Environmental Protection Agency, Washington, DC. Update November 3, 2016. Memorandum from Britton et al., 11/3/2016. D436317.

US EPA, 2018a. Defining Pesticide Biomarkers. US Environmental Protection Agency, Washington, DC. https://www.epa.gov/pesticide-science-and-assessing-pesticide-risks/defining-pesticide-biomarkers.

US EPA, 2018b. Fish and Shellfish Advisories and Safe Eating Guidelines. US Environmental Protection Agency, Washington, DC. https://www.epa.gov/choose-fish-and-shellfish-wisely/fish-and-shellfish-advisories-and-safe-eating-guidelines.

US EPA, 2018c. Advisories and Technical Resources for Fish and Shellfish Consumption. US Environmental Protection Agency, Washington, DC. Updated December 7, 2017. https://www.epa.gov/fish-tech.

US EPA, 2018d. Chlorpyrifos. US Environmental Protection Agency, Washington, DC. Updated February 16, 2018. https://www.epa.gov/ingredients-used-pesticide-products/chlorpyrifos.

US EPA, 2018e. National Primary Drinking Water Regulations. US Environmental Protection Agency. Update March 22, 2018. https://www.epa.gov/ground-water-and-drinking-water/national-primary-drinking-water-regulations.

US EPA, 2018f. Selenium and Selenium Compounds: Integrated Risk Information System (IRIS). Environmental Protection Agency. https://cfpub.epa.gov/ncea/iris/iris_documents/documents/subst/0472_summary.pdf.

US FDA, 2016a. USFDA-NIH Biomarker Working Group. (2016a) BEST (Biomarkers, EndpointS, and Other Tools) Resource. FDA-NIH Biomarker Working Group. Silver Spring (MD): Food and Drug Administration (US); Bethesda (MD): National Institutes of Health (US). https://www.ncbi.nlm.nih.gov/books/NBK338449/?report =reader.

US FDA, 2016b. The Best Resource: Harmonizing Biomarker Terminology. US Food and Drug Administration. In: https://www.fda.gov/downloads/Drugs/DevelopmentApprovalProcess/DrugDevelopmentToolsQualificationProgram/UCM510443.pdf.

US FDA, 2018. Eating Fish: What Pregnant Women and Parents Should Know. US Food and Drug Administration. Update November 29, 2017. https://www.fda.gov/Food/ResourcesForYou/Consumers/ucm 393070.htm.

Vineis, P., Khan, A.E., Vlaanderen, J., et al., 2009. The impact of new research technologies on our understanding of environmental causes of disease: the concept of clinical vulnerability. Environ. Health (Lond.) 8, 54 (Commentary, 10 pp.

Vinceti, M., Filippini, T., Cilloni, S., et al., 2017. Health risk assessment of environmental selenium: emerging evidence and challenges (review). Mol. Med. Rep. 15, 3323–3335.

VKM, 2017. Assessment of Selenium Intake in Relation to Tolerable Upper Intake Levels. Opinion of the Panel on Nutrition, Dietetic Products, Novel Food and Allergy of the Norwegian Scientific Committee for Food Safety. VKM Report 2017: 20, ISBN: 978-82-8259-277-2, Oslo, Norway. www.vkm.no.

WHO, 2008. Guidance for Identifying Populations at Risk from Mercury Exposure. World Health Organization, Geneva, Switzerland. http://www.who.int/foodsafety/publications/chem/mercuryexposure.pdf?ua=1.

WHO, 1990. Methylmercury. International Programme on Chemical Safety (IPCS), Environmental Health Criteria 101. WHO, Geneva, Switzerland. http://www.inchem.org/documents/ehc/ehc/ehc101.htm.

WHO, 2001. Biomarkers in Risk Assessment: Validity and Validation. International Programme on Chemical Safety (IPCS), Environmental Health Criteria 222. WHO, Geneva, Switzerland. http://www.inchem.org/documents/ehc/ehc/ehc222.htm.

WHO, 2004. WHO Technical Report Series 922. Evaluation of Certain Food Additives and Contaminants. Sixty-first report of the Joint FAO/WHO Expert Committee on Food Additives. World Health Organization, Geneva, Switzerland.

WHO, 2007. Safety Evaluation of Certain Food Additives and Contaminants. International Programme on Chemical Safety (IPCS), Prepared by the Sixty-seventh Meeting of the Joint FAO/WHO Expert Committee on Food Additives (JECFA). WHO, Geneva, Switzerland. http://www.inchem.org/documents/jecfa/jecmono/v58je01.pdf.

WHO, 2010a. Children's Exposure to Mercury Compounds. World Health Organization, Geneva, Switzerland. http://apps.who.int/iris/bitstream/handle/10665/44445/9789241500456_eng.pdf?sequence=1.

WHO, 2010b. Food and Agriculture Organization of the United Nations: Summary and Conclusions. Reported from Joint FAO/WHO Expert Committee on Food Additives Seventy-second meeting of February 16–25, 2010. http://www.fao.org/3/a-at868e.pdf.

WHO, 2010c. Report of the Joint FAO/WHO Expert Consultation on the Risks and Benefits of Fish Consumption. Rome, 25–29 January 2010. World Health Organization, Geneva, Switzerland.

WHO, 2011. Biomarkers & Human Biomonitoring. Children's Health and the Environment WHO Training Package for the Health Sector. World Health Organization. Prepared by Shea, K.M. http://www.who.int/ceh/capacity/biomarkers.pdf.

Yang, G.Q., Wang, S.Z., Zhou, R.H., et al., 1983. Endemic selenium intoxication of humans in China. Am. J. Clin. Nutr. 37, 872–881.

Yang, G., Yin, S., Zhou, R., et al., 1989a. Studies of safe maximal daily dietary Se-intake in a seleniferous area in China. Part II: relation between Se-intake and the manifestation of clinical signs and certain biochemical alterations in blood and urine. J. Trace Elem. Electrolytes Health & Dis. 3, 123–130.

Yang, G., Zhou, R., Yin, S., et al., 1989b. Studies of safe maximal daily dietary selenium intake in a seleniferous area in China. I. Selenium intake and tissue selenium levels of the inhabitants. J. Trace Elem. Electrolytes Health & Dis. 3, 77–87.

You, S.H., Wang, S.L., Pan, W.H., et al., 2018. Risk assessment of methylmercury based on internal exposure and fish and seafood consumption estimates in Taiwanese children. Int. J. Hyg Environ. Health 4, 697–703.

Yu, R., Lai, Y., Hartwell, H.J., Moeller, B.C., et al., 2015. Formation, accumulation, and hydrolysis of endogenous and exogenous formaldehyde-induced DNA damage. Toxicol. Sci. 146, 170–182.

Index

'*Note*: Page numbers followed by "f" indicate figures, "t" indicate tables.'

C

D

E

F

H

J

K

L

N

T

U

V

W

X